Textbook of Physiology and Biochemistry

The Authors

GEORGE H. BELL
B.Sc., M.D. Glasg., F.R.C.P. Glasg., F.R.S.E

Professor of Physiology in the University of Dundee; Honorary Fellow of the Accadèmia Anatomico-Chirurgica of Perugia

J. NORMAN DAVIDSON
C.B.E., M.D. Edin., D.Sc., F.R.C.P. Edin., F.R.C.P. Glasg., F.R.I.C., F.R.S.

Gardiner Professor of Biochemistry in the University of Glasgow; formerly Professor of Biochemistry in the University of London at St Thomas's Hospital Medical School

DONALD EMSLIE-SMITH
M.D. Aberd., F.R.C.P., F.R.C.P. Edin.

Reader in Medicine in the University of Dundee; Honorary Consultant Physician, the Dundee Royal Infirmary

Textbook of
Physiology and Biochemistry

GEORGE H. BELL
J. NORMAN DAVIDSON
DONALD EMSLIE-SMITH

Eighth Edition

CHURCHILL LIVINGSTONE · EDINBURGH AND LONDON · 1972

First edition	.	.	.	1950
Reprinted	.	.	.	1952
Second edition	.	.	.	1953
Reprinted	.	.	.	1954
Third edition	.	.	.	1956
Reprinted	.	.	.	1957
Fourth edition	.	.	.	1959
Spanish edition	.	.	.	1959
Italian edition	.	.	.	1959
Fifth edition	.	.	.	1961
Reprinted	.	.	.	1963
Sixth edition	.	.	.	1965
Seventh edition	.	.	.	1968
Eighth edition	.	.	.	1972

E.L.B.S edition first published 1965

ISBN 0 443 00984 8

Filmset by Keyspools Ltd, Golborne, Lancs.
Printed in Great Britain by C. Tinling & Co. Ltd, London and Prescot

Preface to the eighth edition

Every third year or thereby when we have despatched the last consignment of copy to the publishers we are tempted to hope that we have, at least momentarily, reached a stable state. But in fact this can never be so. This edition in particular has undergone very great alteration and revision.

First of all Professor Harold Scarborough had to give up his activities as one of the three authors in favour of his work as Dean of the Medical Faculty at Ahmadu Bello University. In his place we have been joined by Dr Donald Emslie-Smith of the Department of Medicine in the University of Dundee. The influence of his special interest in cardiology is to be seen in Chapter 26 and his interest in general medicine in other chapters with a clinical content. One effect of his arrival is that the short title (or acronym) of B.D.S. has become B.D.E-S.

A second change affects the publishers and readers more than the authors—namely the change to two-column pages and to lithographic methods, to both of which we are new recruits.

These changes will be noticed by teachers and others who have used the previous editions. We hope that our new readers will find our combination of physiology and biochemistry acceptable; it is in fact impossible to make a sharp division between the two. Certainly for the authors, the subjects become more and more exciting as knowledge advances, and more and more are we able to provide a description and explanation of events in man.

We have accepted the recommendations of the authoritative bodies on the various conventions used in physiology and biochemistry. For spelling we have followed the Oxford English Dictionary. For anatomical terminology we have used *Gray's Anatomy* (1967), and for physiological terminology Suggestions to Authors (1966), *Journal of Physiology* **182**, 1–33. For biochemical terminology we have adopted the system of nomenclature and abbreviations recommended by the Combined Commission of Biochemical Nomenclature (CBN) organized at the international level jointly by the International Union of Pure and Applied Chemistry (IUPAC) and the International Union of Biochemistry (IUB). The CBN collaborates closely with the IUB Commission of Editors of Biochemical Journals (CEBJ) and with the Office of the Biochemical Nomenclature (OBN) set up in the United States by the National Academy of Sciences and the National Research Council (W. E. Cohn (1967), *Journal of Chemical Documentation* **7**, 72–73). For enzyme nomenclature we have adopted the recommendations revised in 1964, of the Commission on Enzymes of the IUB. We have taken note of the SI nomenclature and abbreviations so far as possible.

G.H.B.
J.N.D.

Acknowledgements

We have been helped by information or advice from many of our colleagues and we wish especially to thank those whose names follow for help in this new eighth edition in the chapters indicated.

Dr B. L. Andrew (40), Professor E. L. Blair (16), Miss M. Browning (50), Professor W. Burns (38), Dr Sheila T. Callender (24), Dr C. Cameron (23, 53), Dr Ailsa M. Campbell (3, 6, 19), Dr Mary C. Coyle (51), Dr J. W. Crawford (51, 52), Dr B. Creamer (15), Dr R. R. Crichton (22), Professor J. Crooks (50), Dr G. S. Dawes (51), Dr W. W. Downie (50), Dr J. V. G. A. Durnin (13), Dr M. J. W. Faed (53), Dr W. Frain-Bell (35), Dr H. B. Goodall (24), Dr J. K. Grant (18, 50, 51), Mr J. P. Greaves (13), Dr D. M. Green (50), Professor A. Iggo (35), Dr R. H. Johnson (41), Professor A. C. Kennedy (32), Dr J. C. Kernohan (25, 31), Dr R. Keynes (39), Professor J. Knowelden (53), Dr G. Leaf (25), Dr J. A. R. Lenman (44, 45, 46, 47), Professor R. J. Linden (26), Dr N. E. Loveless (34, 49), Mr A. G. D. Maran (45), Dr P. B. Marshall (41, 50), Dr J. J. Misiewicz (17), Dr Margaret McKiddie (18), Mr H. R. Noltie (12, 40, 42), Dr Maureen Palmer (17), Dr W. W. Park (24), Dr C. R. Paterson (14, 50), Dr J. M. Patrick (30, 31), Dr J. D. Pitts (6, 20, 21), Dr J. W. G. Porter (13), Dr M. J. Purves (43), Dr P. M. H. Rack (40), Miss J. Robertson (13), Dr J. G. Robson (37), Professor I. C. Roddie (27, 28), Dr T. Semple (26), Professor R. M. S. Smellie (2), Dr D. A. Stansfield (43), Professor R. E. Steiner (26), Dr A. S. Todd (23, 24), Dr R. Y. Thomson (7, 8, 10, 11, 18), Dr W. S. T. Thomson (33), Dr G. R. Tudhope (13), Dr M. J. Turnbull (50), Dr I. C. Whitfield (38), Dr K. G. Wormsley (17), Professor O. M. Wrong (32, 33).

We have to thank also a number of friends who have gone to considerable trouble to provide new illustrations.

We should like to express our gratitude for the help we have received from the library staffs of our universities, and from Mr R. Callander and Miss M. Benstead who have prepared many of the illustrations. Mr G. Leslie has helped in proof reading; Mrs M. I. Glenday has earned our thanks for preparing the index, and we also thank her, with Mrs E. Hanton and Mrs C. Moir for secretarial assistance.

We have been met at all times by encouragement and frankness by Mr Henderson and his colleagues of Churchill Livingstone. For this we are duly grateful.

G.H.B.
J.N.D.
D.E-S.

1972

Contents

Preface   v

1   Introduction   1

2   The properties of water and solutions   9

3   The proteins   23

4   Carbohydrates   40

5   Lipids   50

6   Nucleotides and nucleic acids   63

7   Biochemical reactions   80

8   Enzymes   90

9   Intermediary metabolism: methods of study   118

10   Biological oxidation   125

11   Glycolysis and the citric acid cycle   139

12   Energy exchange   161

13   Food, nutrition and vitamins   176

14   Bone and minerals   208

15   Mouth, oesophagus and swallowing   222

16   The stomach   233

17   Digestion and absorption in the intestine   263

18   Organization and control of carbohydrate metabolism   311

19   Lipid metabolism   333

20   Nucleic acid metabolism   347

21   Enzymic control mechanisms   363

22   Protein metabolism   375

23   The blood   391

24   The formed elements of the blood   415

25   Blood pigments   444

26   The heart   456

27   The circulation   526

28   Vasomotor control   555

29 Circulation through lungs, liver
        and spleen   577
30 Respiration   588
31 Respiratory functions of the blood   612
32 The kidney   641
33 Water, electrolyte and acid–base
        balance   673
34 Special senses   705
35 The skin   709
36 The chemical senses   722
37 Structure of the eye   729
38 Speech and hearing   755
39 Neurone and synapse   776
40 Muscle   825
41 Autonomic nervous system   860
42 Temperature regulation   873
43 Nervous system   891
44 Spinal cord   901
45 The brain stem and postural
        reflexes   921
46 The cerebellum   941
47 The diencephalon   951
48 The telencephalon   957
49 Conditional reflexes   984
50 The endocrine glands   994
51 Reproduction   1027
52 The pituitary body   1080
53 Chromosomes and heredity   1098
54 Growth and senescence   1111
Units and measures   1116
Index   1123

1 Introduction

In the subjects of physiology and biochemistry we are concerned with living matter at two different levels. The biochemist is concerned mainly with molecular events in the cell, nowadays sometimes called molecular biology, while the physiologist is more concerned with the intact organ or the whole organism. There is no need to justify the treatment of the two subjects in one book; nearly every chapter in this book contains both chemical and physiological information. It could be argued that the difference between the two subjects lies in the different techniques used by the two groups of scientists who are investigating the same problem, namely how events occur in living material and how they are controlled. Physiologists and biochemists are concerned essentially with changes in the organism or its parts as it reacts to changes in the internal or external environment. They try to take a cinematographic view of living material; however, they frequently find it necessary to stop the cine-film from time to time to make a detailed examination of a single frame. For example, the chemical analysis of a tissue is a statement referring to one particular instant and a series of such analyses may indicate a progressive change from which the biochemist is able to build up a dynamic picture of the chemical activities of the tissue. Similarly the microscopic examination of a single section of a tissue is of value in showing its minute anatomy, but it may give little information as to its function. Such a section indeed represents only a momentary glimpse of what may be a continuously changing process. One must recognize too that there may be disturbances of function that seem to have no structural basis. For example, interference with an enzyme system may produce disease, and changes in the arrangement of molecular layers may significantly alter the properties of cells and tissues.

In this book the terms 'physiology' and

'biochemistry' are used in a relatively narrow sense. We are concerned primarily with human aspects of the two subjects but where necessary we shall refer to events in other mammals and even in the frog, so familiar to students of biology; especially in biochemistry it is often necessary to deal with the metabolism of bacteria because the investigation of their metabolism has produced much information relevant to the higher animals. However, any biologist actively engaged in research is soon made painfully aware of the great differences of function between different species and is reluctant to assume that what holds good for another animal applies without modification to man.

STRUCTURE AND FUNCTION

The single cell is the functional unit of the body and each tissue is made up of vast numbers of cells. We must not imagine, however, that a single cell is a simple uncomplicated structure. Although the ordinary light microscope may reveal little difference between one part of its cytoplasm and another, the electron-microscope and modern biochemical techniques have revealed an amazing complexity. It is, however, not yet possible to describe the subtle difference between a living cell and a mere aggregation of molecules.

In the case of the more complex animals such as birds and mammals, the first characteristics of life that come to mind are warmth and movement. If a non-hibernating animal is still and cold it is assumed to be dead. By taking in food and oxidizing it an animal obtains energy which is used to produce heat or movement. The energy is obtained by oxidation of body constituents by a process known as *catabolism*. The energy obtained by these catabolic processes may also be used for *anabolic* or synthetic processes such as are necessary for growth. Since both processes occur side by side it is convenient to use the word *metabolism* when referring to the total chemical changes occurring in the cell or in the body. So long as metabolic processes continue, however slowly, the cell is alive; their arrest is death. Since these chemical processes are under the control of enzymes the cells can move or grow only within certain limits of temperature. If the temperature is too high the enzymes are destroyed,

while at low temperatures enzymic reactions are retarded and finally cease.

Living material is *organized*, that is, it has a definite structure. Moreover, particular functions, such as movement or secretion or conduction, are carried out by cells or organs whose structure is peculiarly fitted to these purposes. For a cell to survive there must be *integration* of function within it. In multi-cellular organisms there must be *co-ordination* of the activities of various cells, either by chemical messengers (*hormones*) or by a system of *nerves*.

Growth is a characteristic feature of living material. Growth of a single cell, however, cannot go on indefinitely because, as the cell increases in volume, its surface through which oxygen and food materials are admitted becomes so far removed from its centre that the supply of essential material to the latter is endangered. Before this stage is reached the cell divides into two daughter cells, a process known as *reproduction*.

Because a cell does not live in isolation, and is dependent on obtaining its food from outside, it must be capable of *reacting* to changes in its environment. Such changes are called *stimuli*. If a stimulus increases the rate of chemical changes in the cell it is said to *excite*; if it decreases the metabolic rate it is said to *depress*, or to be an *inhibitory* stimulus.

Living organisms possess two properties which at first sight seem to conflict. These are best described under the headings of *adaptation* and *homeostasis*. Simple forms of life can survive over a wide range of temperature and can adapt themselves to changes in their environment and to the foodstuffs available. Indeed, if they could not do so they would soon die. The study of adaptation forms a large part of the subject of physiology, for the cells of the body can adjust themselves to a wide variety of changes. On the other hand, many physiological reactions are directed towards preserving the *status quo*. All the cells in the body except those on the surface are provided with a fluid environment of relatively constant temperature, hydrogen ion concentration and osmotic pressure. This permits many bodily activities, the functioning of nerve cells and the working of enzymes for example, to be carried out under optimum conditions. As Claude Bernard put it: 'La fixité du milieu intérieur est la condition de la vie libre, indépendante.' Small changes in the *milieu intérieur* produce reactions which quickly restore the internal environment to its original state. Many of the adjustments of the activity of the cells and organs of the body are examples of the working of this principle which Cannon called homeostasis.

It is generally appreciated that the body regulates its internal environment with some accuracy. For example we are apt to say it keeps its hydrogen-ion concentration remarkably constant. But to leave it at that is simply to admire it from a distance without going to the trouble of making enquiries about the working of the regulatory mechanisms. The first requirement of such a mechanism is a detector of deviation from the standard conditions; the appropriate regulator must then be 'instructed' to reduce the deviation. The new state of affairs is continuously assessed by the detector and the regulator is given fresh instructions. In other words the activity of the regulating device is constantly modified on the basis of information fed to it from the detector: such systems are termed 'feed-back' or 'control' mechanisms. Sensory receptors in muscles and joints send information to the central nervous system about length of muscles and angle of joints and movement and posture are regulated; cells sensitive to osmotic changes in the blood regulate the loss of water from the body; receptors in blood vessels detect changes in blood pressure and so the output of the heart and the calibre of the blood vessels are altered to regain the *status quo*. In many cases, however, the detecting mechanism is still to be discovered—for example we do not yet know how the level of blood volume, or the level of blood sugar, is detected. Both, however, are regulated within quite narrow limits.

Although the regulation of the internal environment reaches its highest development in man, he has gone still further by attempting to control his external environment. At first this involved the wearing of clothing, the building of shelters and houses, and the use of artificial heating, lighting and ventilation. These have not all proved to be unmixed blessings. Artificial lighting, for example, while it makes life more pleasant in the darkness of winter, introduces new problems of eyestrain, fatigue and extended hours of work.

With the advent of the industrial era, unfavourable environments were created in which men had to labour but, although many of these dangers have been mitigated, we are now, because of so-called advances in technology, struggling against the evils of pollution. Man now climbs high in the air and descends to great depths in the sea; a few have even ventured into space and experienced the low gravitational forces of the moon. A great many new biological problems have appeared, some of which may come under the heading of applied physiology. There is, however, no definite boundary between 'pure' and 'applied' in this or any other science.

Our knowledge of the properties and functions of living cells is incomplete, but we know that the laws of conservation of matter and energy apply to the animal body just as certainly as they apply to non-living material. Investigation of living matter is largely a matter of observation supplemented by the methods of physics and chemistry. Thus we may measure pressures and potentials, make chemical analyses, or trace the pathways of radioactive substances through the body. The results of these observations are correlated, interpreted in the light of previous knowledge, and used as evidence for or against a particular hypothesis. Thus physiology and biochemistry imply much more than a mere catalogue of the functions of the various parts of the body. However, neither has yet become a highly organized body of knowledge comparable to the exact sciences of physics or mathematics and many of the phenomena to be discussed can be given as yet only an incomplete explanation. Nevertheless their description must not be regarded as irrelevant since empirical findings may have immediate practical importance in the diagnosis and treatment of disease.

At one time the organic basis for a patient's symptoms could be established only at postmortem examination. The trend in modern medicine and surgery, however, is to study the living patient more and more intensely in order to understand the way in which normal physiological and biochemical processes have broken down, for disease is increasingly thought of as disordered physiological or biochemical processes which the homeostatic mechanisms have been unable to correct. The cure and prevention of numerous diseases are due to rapid advances in our knowledge of bacteriology and immunology: to many the methods of experimental physiology are being successfully applied. Biochemical methods have been used for many years for the examination of the body fluids, to provide important diagnostic information and to control treatment. Thus, for example, much of our recent knowledge of cardiac diseases, cardiac failure and the surgical treatment of cardiac abnormalities is based upon physiological studies made in the laboratory or in the clinic, while the treatment of diabetes mellitus depends almost entirely on material and methods originally developed in the laboratory.

The bodily activities are so closely dependent on one another that the workings of one part of the system cannot be comprehended without an understanding of the functioning of the whole. For example, in thinking of the activities of the heart we have to bear in mind the influence of the peripheral blood vessels, of the central nervous system, of respiration and of the chemical changes occurring in cardiac muscle. Our subject may, therefore, be likened to a circle: it is difficult to know where to enter it to begin our study, for it is only when we have completed the circle that we can fully understand our subject. For this reason it is necessary to consider briefly the subject as a whole before beginning a more detailed description of its various parts. Riley's *Introducing Biology* will be found helpful, especially to those whose training in biology is limited (see References, p. 8).

The source of all the energy required by the body for carrying out muscular activity, for respiration, for the beating of the heart and the working of the nervous system, is the food. This consists mainly of proteins, fats and carbohydrates which are oxidized (burned) in the tissues. In addition to sources of energy the food must contain inorganic substances which are necessary to make good the loss of salts in the excreta and to provide material for the formation of blood and bone. The food must also supply certain substances which the body cannot synthesize, such as vitamins and essential amino acids. Since fluid is lost continuously by way of the kidneys as well as by the skin and lungs, water must be drunk to make good this loss. When food is swallowed it reaches the stomach and small intestine, where it is broken

down by enzymes into substances of relatively simple chemical constitution which are absorbed through the lining of the small intestine into the blood stream and distributed throughout the body. In the case of fats, however, absorption may occur with less modification of the original chemical structure.

The oxygen required for combustion of the foodstuffs reaches the blood through the lungs. During breathing the chest expands and air flows into the lungs which are spongy organs richly supplied with blood vessels. Oxygen diffuses readily through the very thin walls of the capillary vessels of the lung tissue, becoming attached to the haemoglobin contained in the red cells in which it is distributed throughout the body by the circulation. The carbon dioxide produced in combustion in the tissues is taken up by the blood and carried to the lungs where it escapes from the blood and is exhaled. By-products of oxidation not needed by the body reach the kidneys in the blood and are excreted into the urine.

The heart is a two-sided muscular pump which drives the blood along the blood vessels. The left side pumps blood to the heart muscle itself, to the skeletal muscles, the brain and other organs. The blood from these parts returns to the right heart which sends the blood to the lungs where oxygen is taken up and carbon dioxide is eliminated. The oxygenated blood then returns to the left side of the heart and is pumped out to the tissues. The blood is conveyed away from the heart at a fairly high pressure in thick-walled tubes, the arteries. These vessels branch repeatedly and become smaller in diameter, with thinner and thinner walls. In the tissues the smallest blood vessels, the capillaries, are bounded by a single layer of cells through which gases, fluid, or chemical substances of small molecular size move easily. The blood is drained away from the tissues at low pressure in wide vessels (the veins) with relatively thin walls.

The skeletal muscles are the main effector tissues. By their contractions the position of the bones is altered and respiration and speech are made possible. The highly complex movements of the limbs in walking, and of the tongue in speech, are co-ordinated by the central nervous system, consisting of the brain and spinal cord. Nerves called *efferent* or *motor* nerves leave this system and pass to all the structures of the body and control muscular movement as well as the secretion of the glands, the heart beat and the calibre of the blood vessels. Central control is, however, of no value unless the centre has full information about events in the body and around it. This information is conveyed to the central nervous system by the *sensory* or *afferent* nerves which carry impulses from the eye, the ear, the skin, the muscles and joints, and the heart, the lungs and the intestines. The sensory nerves are actually much more numerous than the motor nerves. Although many of the activities occurring in the central nervous system are exceedingly complex, relatively few rise to consciousness. We are quite unaware, for example, of the muscular adjustments needed to maintain balance or to move our eyes so that images of the external world are kept fixed on the retinae. These adjustments are called *reflex* and the pathways involved, namely sensory nerves, central nervous system and motor nerves, are called *reflex arcs*. The lowest part of the brain, the medulla oblongata, is responsible for the muscular movements of respiration, for the control of the heart rate and the regulation of the blood vessels, and is concerned in the maintenance of posture. The cerebellum, which lies above the medulla, is concerned with co-ordination of muscular movements. The fore-part of the brain, the cerebrum, has a layer of grey matter (nerve cells) on its surface and also masses of grey matter within. The grey matter is interconnected by innumerable nerve fibres which together make up the white matter. The cerebrum is concerned in all the higher mental activities and with reading, writing and speaking, as well as in so-called voluntary movements, the perception of touch and in the special senses of vision and hearing.

In addition to the rapidly acting co-ordinating and integrating mechanism of the nervous system there is a chemical (*humoral*) system which operates more slowly. For example, during the digestion of food in the duodenum a chemical substance (*hormone*) called secretin is produced in the mucous membrane, absorbed into the blood and carried to the pancreas which responds by pouring out its digestive juices. The thyroid gland in the neck produces a chemical substance which is absorbed directly from the gland into the blood stream and influences the rate of metabolic activity of the cells of the body.

Under the heading of reproduction, we shall have to consider the processes necessary for the maintenance of the species. The male cells, the *spermatozoa*, are produced in the testis and when deposited in the female genital tract one of them may fertilize an *ovum* produced in the ovary. This sets off a series of complicated changes, mainly under hormonal control, to provide for the growth of the fertilized cell in the uterus. By repeated division the fertilized ovum develops into the embryo whose nutrition in the uterus is carried out by transfer of materials across the placenta. At the end of pregnancy the muscular wall of the uterus contracts, the fetus is delivered and then acquires oxygen directly, by breathing air into its lungs, instead of indirectly through the placenta. In the meantime the mother's mammary glands have enlarged in preparation for lactation and soon after parturition they produce milk for the nourishment of the infant. Many of the organs of the infant, especially the central nervous system and the kidneys, are somewhat undeveloped at birth, but as he grows these disabilities disappear and he is then able to lead an active independent existence. An important landmark in development is puberty. At this time the gonads complete their development and in so doing produce the physical and mental changes characteristic of the adult. The ovaries continue to produce ova till the menopause in the fifth decade, but the production of spermatozoa by the male persists much longer. With the approach of old age the arteries harden, the lens of the eye, the skin and probably many other tissues become less elastic, muscular power declines and there is some deterioration of mental activity. If the hazards of accident, infection, cardiac damage and malignant disease are avoided, death occurs in old age by a gradual process of degeneration.

THE COMPOSITION OF LIVING TISSUES

The living body contains, in addition to a large amount of water, protein which is the main nitrogenous constituent of all living material, a variable amount of fatty material known collectively as lipid, a small amount of carbohydrate, and mineral salts. In man the relative proportions of these constituents, especially fat and carbohydrate, vary greatly from one person to another, and in the one individual at different times in his life, according to his nutritional status. Nevertheless, the composition of the human body may be represented roughly as shown in Table 1.1.

Table 1.1 Approximate composition of a man weighing 70 kg (154 lb)

	Percentage	kg
Water	70	49
Fat	15	10·5
Protein	12	8·4
Carbohydrate	0·5	0·35
Minerals	2·5	1·75
	100	70

The chief constituent of living matter is water. Its importance in both the structure and functioning of the tissues is discussed in many chapters in this book, especially Chapters 2 and 33 but some preliminary considerations are given here.

The body of a healthy adult male consists of some 65 to 70 per cent of water and about 15 per cent of fat; the remainder is accounted for by the solid parts of cells and supporting structures. The considerable variations in total body water (TBW) as between one person and another are the result of differences in fat content since the amount of water in the fat-free parts of the body is remarkably constant. This accounts for the fact that the water content of the body of the female (50 to 55 per cent) is rather less than that of the male. In very fat people the body water may be no more than 40 per cent of the total body weight. In the fetus the relative proportion of water in the body is much higher, for example 94 per cent at the third month of fetal life. The reasons for these differences are not known.

An estimation of the water content of the various tissues of a 70 kg man is shown in Table 1.2. Most tissues contain more than 70 per cent water; a red blood cell contains 71·7 per cent water, and even the skeleton and adipose tissue contain quite large amounts. The latter, of course, does not consist entirely of fat; it contains connective tissue, for example, which has water in it and the spaces

Table 1.2 Water content of human tissues (analytical figures from one male human body, age 35)

Tissue	Per cent of total body wt.	Water content per cent
Skin	7·81	64·68
Skeleton	14·84	31·81
Teeth	0·06	5·00
Striated muscle	31·56	79·52
Brain, spinal cord, and nerve trunks	2·52	73·33
Liver	3·41	71·46
Heart	0·69	73·69
Lungs	4·15	83·74
Spleen	0·19	78·69
Kidneys	0·51	79·47
Pancreas	0·16	73·08
Alimentary tract	2·07	79·07
Adipose tissue	13·63	50·09
Remaining tissues—		
Liquid	3·79	93·33
Solid	13·63	70·40
Contents of alimentary tract	0·80	—
Bile	0·15	—
Hair	0·03	—
Total body, weighing 70·55 kg	100·00	67·85

After H. H. Mitchell, T. S. Hamilton, F. R. Steggerda & H. W. Bean (1945), *Journal of Biological Chemistry* **158**, 625.

between the fat cells also contain water. Table 1.3, shows that by far the greatest amount of water is to be found in muscle which accounts for the largest part of the body mass.

Table 1.3 Percentage of the total body water which is found in the various tissues and organs

Muscle	50·8	Brain	2·7
Skeleton	12·5	Lungs	2·4
Skin	6·6	Fatty tissue	2·3
Blood	4·7	Kidneys	0·6
Intestine	3·2	Spleen	0·4
Liver	2·8	Rest of body	11·0
			100·0

In its capacity as a solvent (Chap. 2) water plays a fundamental role in cellular reactions. A very large number of substances are soluble in water and many others, such as fats and fat-soluble compounds, can be carried in fine emulsions or be rendered water-soluble by combination with hydrophilic substances. Certain other properties of water are also of importance. Owing to the high heat capacity of water, large changes in heat production can take place in the body with very little alteration in body temperature. Since the latent heat of evaporation of water is high, the loss of a small amount of water in evaporated sweat means a relatively large loss of heat. Moreover the high latent heat of solidification is a protection against the freezing of the tissues.

In the normal course of events a large amount of water is lost from the body daily and a corresponding amount is taken in, so that water balance is maintained. As shown in Table 1.4, the amount of water gained and lost by an adult man engaged in a sedentary occupation in a temperate climate is about $2\frac{1}{2}$ litres per day.

Water is gained by the body from two main sources. Most of it is taken in by the mouth in the form of food and drink, but a small amount of water is normally formed in the tissues as the result of the oxidation of the hydrogen of foodstuffs. The amount of water ingested in the diet varies, of course, over very wide limits according to habit, climate and occupation. Studies in man with deuterium oxide (heavy water, Chap. 9) have shown that the absorption of water from the stomach is approximately 2·5 per cent of the dose per minute whereas from the small intestine it is much more rapid, about 26 per cent of the dose per minute. Patients with injuries of the limbs absorb water from the alimentary tract more slowly than normal, and in patients with abdominal injuries water absorption is grossly delayed.

The amount of metabolic water formed in the tissues as the result of the oxidation processes described in Chapter 10 is about 300 ml in man, that is about 14 per cent of his total daily fluid intake. This water is formed in the cells and is of great value to the organism, since its formation is not accompanied by the great osmotic changes associated with the intake of large amounts of fluid. It is of the utmost importance to the hibernating animal which lives for long periods on metabolic water, and to organisms such as the clothes moth which do not normally have ready access to water.

Table 1.4, gives no idea of the great turnover of fluid which takes place in the body in the

Table 1.4 Water balance of an adult man in a temperate climate

Daily intake		Daily output	
Drink	1300 ml	Urine	1500 ml
Food	850 ml	Expired air	400 ml
Formed in body		Skin	500 ml
by oxidation	350 ml	Faeces	100 ml
Total	2500 ml	Total	2500 ml

course of a day. In 24 hours a man secretes 1 to 1·5 l of saliva; 1 to 2 l of gastric juice; 0·5 to 1 l of bile; 0·6 to 0·8 l of pancreatic juice and 3 l of intestinal juice. All this fluid, with the exception of about 100 ml which escapes in the faeces, is reabsorbed.

The composition of the human body in terms of elements is shown in Table 1.5. The most abundant elements are carbon, oxygen and hydrogen, but minerals such as calcium and phosphorus are also plentiful. Other minerals such as iodine and iron are present only in small quantities. Nevertheless their presence in the diet is of great nutritional importance. The requirement of a mineral element varies according to age and activity, and is greater in the child, in relation to body weight, than in the adult. It is also increased during pregnancy to meet the needs of fetal growth, and is still further increased during lactation.

Table 1.5 Composition of the human body (after Sherman)

	Wet weight basis per cent		Wet weight basis per cent
Oxygen	65	Chlorine	0·15
Carbon	18	Magnesium	0·05
Hydrogen	10	Iron	0·004
Nitrogen	3	Iodine	0·00004
Calcium	1·5	Copper	
Phosphorus	1·0	Manganese	
Potassium	0·35	Zinc	traces
Sulphur	0·25	Fluorine	
Sodium	0·15	Molybdenum, etc.	

REFERENCES

Body Composition (1963). *Annals of the New York Academy of Sciences*, **110**, 1–1018.

BROOKS, C. McC. & CRANFIELD, P. S. (Eds.) (1959). *The Historical Development of Physiological Thought*. Symposium held at the State University of New York. New York: Haffner.

BROZEK, J. (Ed.) (1965). *Human Body Composition: Approaches and Applications*. London: Pergamon Press.

HORROBIN, D. F. (1970). *Principles of Biological Control*. Aylesbury: Medical & Technical Publishing Co.

MOORE, F. D., OLSEN, K. H., McMURREY, J. D., PARKER, H. V., BALL, MARGARET, R. & BOYDEN, C. M. (1963). *The Body Cell Mass and its Supporting Environment*. London: Saunders.

MOMMAERTS, W. F. H. M. (1966). The Maxwell demon in sheep's clothing. *Bulletin of the New York Academy of Medicine*, **42**, 991–995.

PASSMORE, R. & DRAPER, M. H. (1970). The chemical anatomy of the human body. In *Biochemical Disorders in Human Disease*, 3rd edn (Edited by R. H. S. Thompson & I. D. P. Wootton). London: Churchill.

RIGGS, D. S. (1971). *Control Theory and Physiological Feedback Mechanisms*. Edinburgh: Churchill Livingstone.

RILEY, J. F. (1967). *Introducing Biology*. Pelican books A886. London: Penguin Books.

STONE, K. (1966). *Evidence in Science*. Bristol: Wright.

WIDDOWSON, E. M. & DICKERSON, J. W. T. (1964). The chemical composition of the body in *Mineral Metabolism* (Edited by C. L. Comar and F. Bronner), Vol. II, Part A, pp. 2–248. New York: Academic Press.

YOUNG, J. Z. (1957). *The Life of Mammals*. Oxford: Clarendon Press.

2 The properties of water and solutions

A high content of water is a universal characteristic of living tissues; for example, 70 per cent of the mass of mammalian liver is water, while in muscle the proportion may be even higher. All biochemical reactions must be regarded as taking place in an aqueous medium and consequently a knowledge of the physical and chemical properties of water is fundamental in the study of biology. The present chapter is concerned with the more important of these properties.

Quantities and concentrations. It is convenient to express quantities and concentrations in units which allow direct comparisons between different compounds. Two basic units are useful for this purpose: the *mole* (or gram-molecular weight) and the *equivalent* (or gram-equivalent weight). For practical purposes these are often too large, and it is more convenient to use the *millimole* (mmol = 10^{-3} mole) and the *milli-equivalent* (m-equiv. = 10^{-3} equivalent). The units of concentration corresponding to the *mole* and *equivalent* are *molar* (M = 1 mol/litre) and *normal* (N = 1 equiv./litre) respectively. For practical purposes it is, however, more convenient to express concentrations in terms of *millimolar* (mM = 10^{-3}M) or *milli-equivalents per litre* (m-equiv./l). The SI system recommends that *all* concentrations be expressed as mol/l (or in mmol/l or μmol/l).

A few practical examples will illustrate the use of units based on the *mole* and the *equivalent*. Blood plasma contains the following cations in the concentrations shown:

Na^+	3300 mg/litre
K^+	200 mg/litre
Ca^{2+}	100 mg/litre
Mg^{2+}	30 mg/litre
Total	3630 mg/litre

Since a molar solution contains one gram molecular weight of solute in one litre, to convert these quantities into terms of *molarity*, each is divided by the corresponding atomic weight (Na = 23·0; K = 39·1; Ca = 40·0; Mg = 24·3) and the table then becomes:

$$Na^+ \quad 3300 \div 23\cdot0 = 143\cdot4 \text{ mM}$$
$$K^+ \quad 200 \div 39\cdot1 = 5\cdot1 \text{ mM}$$
$$Ca^{2+} \quad 100 \div 40\cdot0 = 2\cdot5 \text{ mM}$$
$$Mg^{2+} \quad 30 \div 24\cdot3 = 1\cdot25 \text{ mM}$$

$$\text{Total} \quad 152\cdot25 \text{ mM}$$

To convert these quantities into *m-equiv./l* we must multiply the figures by the valencies of the ions thus:

$$Na^+ \quad 143\cdot4 \times 1 = 143\cdot4 \text{ m-equiv./l}$$
$$K^+ \quad 5\cdot1 \times 1 = 5\cdot1 \text{ m-equiv./l}$$
$$Ca^{2+} \quad 2\cdot5 \times 2 = 5\cdot0 \text{ m-equiv./l}$$
$$Mg^{2+} \quad 1\cdot25 \times 2 = 2\cdot5 \text{ m-equiv./l}$$

$$\text{Total} \quad 156\cdot0 \text{ m-equiv./l}$$

The advantage of this system is that the concentrations of different compounds or ions can be compared directly either in terms of number of molecules or number of charges. We might, for example, want to compare blood plasma with the acid secreted by the parietal cells of the stomach. This secretion is virtually 0·6 'per cent' hydrochloric acid, that is 0·6 g HCl per 100 ml with relatively small amounts of Na^+, K^+, Mg^{2+} and Ca^{2+}. The figure of 0·6 'per cent' is not by itself very informative until it is translated into millimoles or milli-equivalents (HCl = 1 + 35·5 = 36·5):

$$0\cdot6 \text{ 'per cent'} = 6000 \text{ mg/l}$$
$$= 6000 \div 36\cdot5 = 160 \text{ mM}$$
$$= 160 \text{ m-equiv./l}$$

Only then is it apparent that the concentration of H^+ in this secretion is roughly equal (in terms of molecules or charge) to the total cation concentration (152·25 mM or 156 m-equiv./l) in blood plasma.

One advantage of this method of expression is that it enables us to equate anions with cations as shown in Table 2.1.

THE STRUCTURE OF THE WATER MOLECULE

The water molecule, H_2O, can be shown by physical measurements to have the triangular shape shown below.

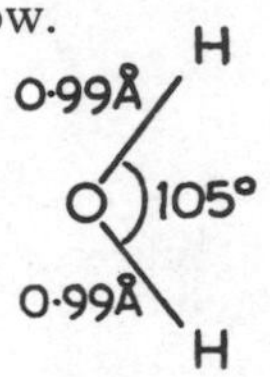

The O—H bonds are covalent, that is to say they are formed as the result of the oxygen atom and the hydrogen atom each sharing an electron with the other as shown in II below.

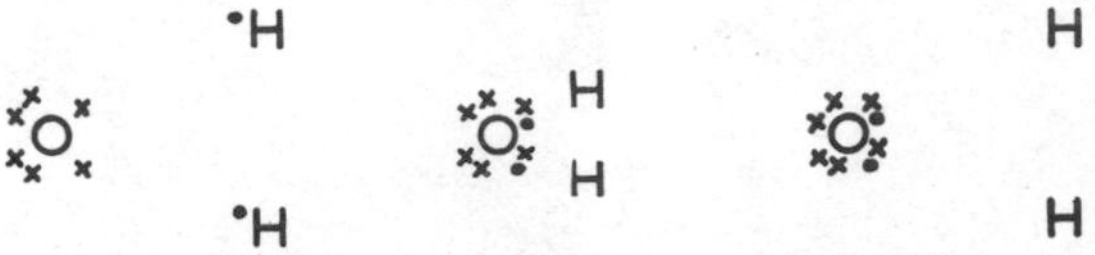

Table 2.1 Average composition of venous plasma in terms of anions and cations

Cations	mg/100 ml	m-equiv./l	Anions	mg/100 ml	m-equiv./l
Na^+	330	143·4	HCO_3^-	60 ml of CO_2	27·0
K^+	20	5·1	Cl^-	365	103·0
Ca^{2+}	10	5·0	HPO_4^{2-}	5	3·0
Mg^{2+}	3	2·5	SO_4^{2-}	1	1·0
			Proteins	5090	16·0
			Organic acids	—	6·0
Totals		156·0			156·0

This table makes it clear that the cations are exactly balanced by the anions and shows that sodium, though present in smaller weight than chloride, can balance all the chloride, bicarbonate and other acid radicals as well as some of the protein.

One millimole of a gas occupies 22·4 ml at s.t.p. Hence 60 ml CO_2/100 ml corresponds to 60/22·4 or 2·7 millimoles/100 ml or 27 m-equiv./l.

The oxygen atom thus attains its full complement of 8 valence electrons and each of the two hydrogen atoms attains its full complement of 2. But, as shown in III, the shared pairs of electrons are attracted more strongly toward the large nucleus of oxygen, with its 8 protons, than to the small nuclei of the hydrogen atoms, with only one proton each. For this reason the oxygen atom has a slight negative charge represented by $\delta-$ and the two hydrogen atoms have correspondingly slight positive charges which may be represented by the symbol $\delta+$ thus:

$$H\,\delta+$$
$$\delta - O$$
$$H\,\delta+$$

A molecule possessing this imbalance of charges is said to be *polar* and many of the special properties of water are attributable to the polarity of its molecules. For example the negatively-charged oxygen atom of one molecule exerts an electrostatic attraction on the positively-charged hydrogen atoms of neighbouring molecules. The full three-dimensional arrangement is difficult to show on paper but for present purposes the relationship between two neighbouring H_2O molecules can be shown thus:

$$H\,\delta+ \qquad\qquad\qquad H\,\delta+$$
$$O \overset{\delta+}{—} H \cdots\cdots \overset{\delta-}{O}$$
$$\delta- \qquad\qquad\qquad\qquad H\,\delta+$$

The type of electrostatic attraction indicated by the dotted line is the so-called *hydrogen bond*. It is much stronger than the intermolecular van der Waals forces which exist between uncharged molecules and is responsible for the very high boiling point of water compared with that of such compounds as ammonia ($-33°C$), hydrogen fluoride ($+19°C$) and methane ($-161°C$).

WATER AS A SOLVENT

For a solid to dissolve in a liquid, three things must happen:

1. The molecules (or ions) of the solid must separate from one another. In other words, the linkages holding them must be severed.

2. The molecules of the solvent must also undergo a separation to admit the molecules (or ions) of the solute. This necessitates the breaking of some of the linkages holding the solvent molecules together.

3. Some sort of linkage must be formed between the solvent molecules and the solute molecules or ions to balance the attractions which the solute molecules exert on one another and which the solvent molecules exert on one another.

VAN DER WAALS FORCES. Uncharged molecules like benzene or hexane or naphthalene are attracted to one another by these relatively weak non-specific intermolecular forces which operate between all molecules when they are close together. Consequently compounds such as the saturated hydrocarbons tend to have relatively low boiling points (methane $-161°C$; ethane $-89°C$; propane $-42°C$). Liquids, such as benzene and ether, in which the intermolecular forces are chiefly or exclusively of this type are readily miscible with one another. They can also dissolve large quantities of solids like natural fat, in which the intermolecular forces are of the same type. This sort of behaviour reflects the fact that the van der Waals attraction between two molecules of different structure is comparable in strength to that existing between two molecules of the same structure. On the other hand, benzene and similar liquids are only slightly miscible with water, because water molecules are attracted to one another by electrostatic forces. These are much stronger than the van der Waals forces which exist between a water molecule and a benzene molecule. Consequently water is only slightly soluble in benzene and benzene is only slightly soluble in water. Water and benzene are extreme examples of polar and non-polar solvents respectively. Many liquids possess intermediate properties. For example, ethanol, like all alcohols, can be regarded as a derivative of water in which one of the hydrogen atoms has been replaced by an alkyl group.

$$H\,\delta+ \qquad\qquad\qquad \begin{matrix} H & H \\ C—C—H \\ H & H \end{matrix}$$
$$\delta - O \qquad\qquad \delta - O$$
$$H\,\delta+ \qquad\qquad\qquad H\,\delta+$$

This structure gives ethanol polar properties but to a much smaller degree than water. Acetone also has polar properties since in the carbonyl group the oxygen atom has a slight negative charge and the carbon a slight positive charge.

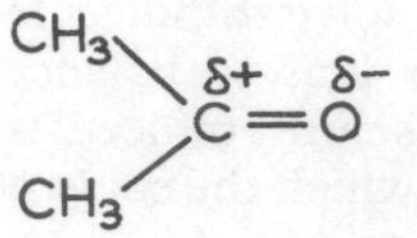

$$CH_3 \diagdown \atop CH_3 \diagup \overset{\delta+}{C} = \overset{\delta-}{O}$$

Such weakly polar molecules are electrostatically attracted to water molecules. For this reason they are, unlike benzene, readily miscible with water in all proportions. But because they are only weakly polar they are also, unlike water, miscible with non-polar liquids like benzene. The extent to which compounds like ketones and alcohols show polar characteristics depends on the number of polar groups in the molecule relative to its size. Thus while methanol, ethanol and *n*-propanol are miscible with water in all proportions the higher alcohols are not.

CH_3OH
Methanol

$CH_3CH_2CH_2CH_2OH$
n-Butanol

CH_3CH_2OH
Ethanol

$CH_2OHCHOHCHOHCH_2OH$
Erythritol

$CH_3CH_2CH_2OH$
n-Propanol

Conversely whereas the monohydric 4-carbon alcohol *n*-butanol is only moderately soluble in water, the corresponding tetrahydric alcohol, erythritol, is completely miscible with water.

Electrostatic Forces. Many salts, such as sodium chloride, are ionized even in the solid state. Thus a crystal of sodium chloride consists of a lattice in which Na^+ and Cl^- ions alternate, the lattice being held together by the electrostatic attraction between these oppositely charged ions. Because of the much larger charges involved, the attraction is very much stronger than that exerted by one water molecule on another. Consequently pure sodium chloride is a solid at room temperature and is exceedingly difficult to fuse or volatilize. In these respects it contrasts sharply with pure hydrogen chloride, in which the hydrogen is linked to the chloride by a covalent bond. This compound is gaseous at room temperature.

Since the electrostatic interionic forces in solid sodium chloride are so strong, it is not surprising that it is insoluble in such solvents as benzene. It is, however, soluble in water, because the Na^+ ions exert an electrostatic attraction on the negatively charged oxygen atoms of the water molecules, while the Cl^- ions exert a similar attraction on the positively charged hydrogen atoms. This is illustrated diagrammatically in Figs. 2.2 and 2.3.

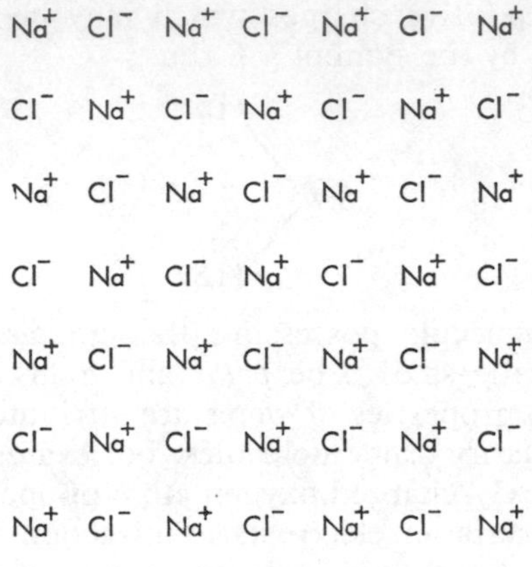

Na^+	Cl^-	Na^+	Cl^-	Na^+	Cl^-	Na^+
Cl^-	Na^+	Cl^-	Na^+	Cl^-	Na^+	Cl^-
Na^+	Cl^-	Na^+	Cl^-	Na^+	Cl^-	Na^+
Cl^-	Na^+	Cl^-	Na^+	Cl^-	Na^+	Cl^-
Na^+	Cl^-	Na^+	Cl^-	Na^+	Cl^-	Na^+
Cl^-	Na^+	Cl^-	Na^+	Cl^-	Na^+	Cl^-
Na^+	Cl^-	Na^+	Cl^-	Na^+	Cl^-	Na^+

Fig. 2.2 Arrangement of Na^+ and Cl^- ions in a crystal of sodium chloride.

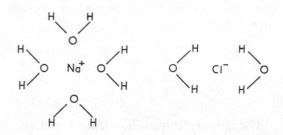

Fig. 2.3 Arrangement of water molecules around Na^+ and Cl^- ions in aqueous solution (purely schematic).

The situation can be summed up by saying that water, being a highly polar liquid, is an exceptionally good solvent for ions and for molecules which are themselves polar. But it does not readily dissolve molecules which are un-ionized and non-polar. It is, however, possible to obtain a stable emulsion of a non-polar liquid in water or of water in a non-polar liquid by using as emulsifying agent a compound in which one end of the molecule is polar and the other non-polar. A simple example is sodium stearate which in aqueous solution ionizes to give stearate ions. If benzene and water are shaken up together with a small quantity of sodium stearate the negatively-charged carboxyl end of the stearate ion attracts water molecules electrostatically. At the same time the uncharged hydrocarbon end of the ion attracts benzene molecules by van der Waals forces. In this way the stearate acts as a link between water and benzene. In practice the stearate molecules arrange themselves at the surface of the benzene droplets in such a manner that their carboxyl groups are orientated toward the surrounding water and their hydrocarbon chains toward the centre of the benzene droplet.

It is somewhat incorrect to represent the ionization of acids in the conventional way.

$$HCl \rightleftharpoons H^+ + Cl^-$$

$$CH_3COOH \rightleftharpoons H^+ + CH_3COO^-$$

Protons do not exist free in aqueous solution. They are always associated with one or perhaps more water molecules as H_3O^+ or $H_9O_4^+$ and so on. But for our present purposes we can ignore this complication.

Pure water itself ionizes thus:

$$H_2O \rightleftharpoons H^+ + OH^-$$

Though this ionization is of great importance, quantitatively it is minute; the proportion in

Hydrocarbon chain
attracts other hydrocarbons
by van der Waals forces

Carboxyl group
attracts water
electrostatically

IONIZATION IN AQUEOUS SOLUTIONS

It has been emphasized above that because of its polar properties water is an excellent solvent for ionized compounds like sodium chloride. For the same reason water tends to facilitate the ionization of compounds which can exist in either the ionized or the undissociated form. For example, there is good evidence that in the solid state mercury chloride, unlike sodium chloride, is not ionized. The mercury and chlorine atoms are held together apparently by covalent bonds. But when the mercuric chloride dissolves in water it does ionize, though to a limited extent.

$$HgCl_2 \rightleftharpoons Hg^{2+} + 2Cl^-$$

In the same way acetic acid, and similar organic acids, ionize slightly in water, but not at all in non-polar solvents like ether and benzene. A far more dramatic example is HCl. In the gaseous state the hydrogen and chlorine atoms are linked covalently, but in aqueous solution the interaction between ions and water molecules favours dissociation to such an extent that ionization is virtually complete.

pure water is only about one hydrogen ion and one hydroxyl for every 550 000 000 undissociated molecules of H_2O. As in other reversible reactions (see Chap. 7) equilibrium in this ionization is described by an equation of the form

$$\frac{[H^+][OH^-]}{[H_2O]} = K$$

where the symbols in square brackets represent the molar concentrations of the various reactants and K is the dissociation constant. Since the ionization is so very slight it does not produce a significant decrease in the number of undissociated H_2O molecules. The above equation can be written

$$[H^+][OH^-] = [H_2O]K$$

Since the term $[H_2O]$ can be regarded as constant and we can write

$$[H^+][OH^-] = K_w$$

Physical measurements show that at 20°C K_w is approximately 10^{-14}.

Thus $[H^+][OH^-] = K_w = 10^{-14}$

The Properties of Water and Solutions   13

This generalization holds for all dilute aqueous solutions. In effect it means that the concentrations of hydrogen and hydroxyl ions are rigidly linked by an inverse relationship. If the concentration of one is known, the concentration of the other can be calculated. And if the concentration of either is altered the concentration of the other is automatically altered in the opposite direction.

In pure water the only ions present are H^+ and OH^-. Since the net charge, as in any liquid, must be zero,

$$[H^+] = [OH^-] = \sqrt{K_w} = \sqrt{10^{-14}} = 10^{-7}$$

The concentrations of hydrogen and hydroxyl ions are equal. This is the situation which, in terms of acids and bases, we term neutrality. The concentrations are both very low. In terms of molarity they are 10^{-7} M = 0·0001 mM. This is minute compared with the concentrations of other ions with which we are concerned in the body (for example the cations in blood plasma). It is very much less even than the trace amounts of inorganic ions commonly found in public water supplies.

HYDROGEN ION CONCENTRATION. Although in pure water, the concentrations of hydrogen and hydroxyl ions are very small—about 10^{-7} M or 0·0001 mM—they can be greatly increased or decreased by the addition of quite small amounts of acid or alkali. The 'dilute hydrochloric acid' found in any laboratory commonly has a concentration of about 2 M. If 0·5 ml of this is added to a litre of pure water the final concentration of acid will be roughly

$$2 \text{ M} \times \frac{0·5}{1000} = 10^{-3} \text{ M} = 1·0 \text{ mM}$$

Since HCl is completely ionized, the hydrogen ion concentration will also be 1·0 mM. So the addition of this small amount of acid has increased the hydrogen ion concentration in the water 10 000-fold. Since

$$[H^+][OH^-] = 10^{-14}$$

the 10 000-fold increase in $[H^+]$ must be accompanied by a corresponding 10 000-fold decrease in $[OH^-]$. If we had added 2 M-sodium hydroxide instead of hydrochloric acid we would have produced exactly the opposite results. The $[OH^-]$ would have gone up from 10^{-7} M to 10^{-3} M and the $[H^+]$ would have gone down from 10^{-7} M to 10^{-11} M. A striking example of the large changes in $[H^+]$ and $[OH^-]$ brought about by relatively trivial causes is the fact that ordinary distilled water is preceptibly acid ($[H^+] = 10^{-5}$ M) and not, as one might expect, neutral. This is because distilled water is usually in contact with air which contains 0·04 per cent of CO_2, the partial pressure (P_{CO_2}) being about 0·03 mm Hg. Consequently the distilled water contains dissolved CO_2, some of which combines with water to give carbonic acid, which in turn ionizes liberating hydrogen ions.

$$CO_2 + H_2O \rightleftharpoons H_2CO_3 \rightleftharpoons H^+ + HCO_3^-$$

Because hydrogen ion concentration can vary so widely it is inconvenient to express it in terms of concentration units such as millimolar and it is usually, though not always, expressed in terms of $pH = -\log[H^+]$.

So that if $[H^+] = 10^{-3}$ M   pH = 3
if $[H^+] = 10^{-7}$ M   pH = 7
if $[H^+] = 10^{-9}$ M   pH = 9

This logarithmic notation allows us to describe the hydrogen ion concentration, that is, the acidity or alkalinity, of any aqueous solution on a convenient scale. However, since we may not be accustomed to thinking in logarithmic terms we may not always appreciate what a given change in pH means, especially in relation to changes in blood plasma. The hydrogen ion concentration of arterial blood plasma is normally about 0·04 µmol/l. It is maintained at this figure by quite elaborate and powerful mechanisms. An increase to, say, 0·10 µmol/l (acidaemia) or a fall to, say, 0·025 µmol/l (alkalaemia) would give cause for grave concern. Let us convert these values to the pH scale (see top of p. 15).

It should be noted that although acidaemia and alkalaemia represent quite substantial changes in hydrogen ion concentration, the corresponding changes in pH look quite small. This is a point which must be kept in mind in interpreting pH values. It can be illustrated in the reverse direction by considering the possible variation in the pH of urine. This can be as high as 7·5 or as low as 4·8. Translating these

Condition	[H$^+$]	pH
Normal	0·040 μmol/l = 4×10^{-8} mol/l	$-\log(4 \times 10^{-8})$
		$= -\log 4 - \log 10^{-8}$
		$= -0·6 - (-8)$
		$= 7·4$
Acidaemia	0·10 μmol/l = 10×10^{-8} mol/l	$-\log(10 \times 10^{-8})$
		$= -\log 10 - \log 10^{-8}$
		$= -1 - (-8)$
		$= 7·0$
Alkalaemia	0·025 μmol/l = $2·5 \times 10^{-8}$ mol/l	$-\log(25 \times 10^{-8})$
		$= -\log 2·5 - \log 10^{-8}$
		$= -0·4 - (-8)$
		$= 7·6$

values into hydrogen ion concentrations we have

pH [H$^+$]

$$4·8 = \text{antilog}\,(-4·8)$$
$$= \text{antilog}\,(0·2 - 5·0)$$
$$= 10^{0·2} \times 10^{-5}$$
$$= 1·6 \times 10^{-5}\ \text{mol/l}$$
$$= 16\ \text{μmol/l}$$

$$7·5 = \text{antilog}\,(-7·5)$$
$$= \text{antilog}\,(0·5 - 8·0)$$
$$= 10^{0·5} \times 10^{-8}$$
$$= 3·2 \times 10^{-8}\ \text{mol/l}$$
$$= 0·032\ \text{μmol/l}$$

Here again a modest difference in pH represents a much larger difference in hydrogen ion concentration.

The values of [H$^+$] encountered in ordinary chemical manipulations range from 0·1 mol/l in 0·1 M HCl to 10^{-13} mol/l in 0·1 M NaOH. This variation is so enormous that it is more convenient to indicate the value of [H$^+$] by the logarithmic expression pH.

For example in 0·01 M-HCl, [H$^+$] = 0·01 = 10^{-2}; but since pH = $-\log_{10}$[H$^+$], pH = $-\log_{10} 10^{-2} = 2$.

It follows that a neutral solution has pH 7, an acid solution has pH < 7, and an alkaline solution has pH > 7*. It should be noted that a small figure indicates a high acidity and conversely a larger figure indicates a low acidity; a solution of pH 2 is much more acid than one of pH 6.

Let us take the case of 0·1 M-HCl. If it is

* At 37° K_w = $2·45 \times 10^{-14}$ so that pH = 6·8 for pure water. Consequently at body temperature 'neutrality' is at pH 6·8.

completely dissociated the solution is 0·1 M with respect to hydrogen ions, that is, [H$^+$] = 10^{-1} and pH = $-\log 10^{-1} = 1$. Or if we have 0·1 M-NaOH then [OH$^-$] = 10^{-1} and since [H$^+$] $\times$ [OH$^-$] = 10^{-14} it follows that [H$^+$] = 10^{-13} and pH = 13.

On the other hand, if [H$^+$] = $2·0 \times 10^{-5}$, by definition

$$pH = -\log[H^+]$$
$$= -\log(2·0 \times 10^{-5})$$
$$= -\log 2 - \log 10^{-5}$$
$$= -0·3010 + 5 = 4·7.$$

The converse calculation is carried out as follows.

If we have a solution of pH 7·4 then, from the definition of pH,

$$-\log[H^+] = pH = 7·4 = 8·0 - 0·6$$
$$= \log 10^8 - \log 3·98$$
$$= -(\log 3·98 - \log 10^8)$$
$$= -(\log 3·98 + \log 10^{-8})$$
$$-\log[H^+] = -\log(3·98 \times 10^{-8})$$
$$[H^+] = 3·98 \times 10^{-8}\ \text{g ions per litre.}$$

Similarly it can be shown that at

pH 7·3 [H$^+$] = $5·01 \times 10^{-8}$ g ions per litre.

The pH of a solution is measured either by indicators or by electrical methods. For a description of these procedures a textbook of practical biochemistry should be consulted.

ACIDS AND BASES

The fluid in the cells and in the tissues of plants and animals is usually close to neutrality. Consequently the concentration of hydrogen ions in biological systems is exceedingly minute. Yet its precise value is of great importance to simple unicellar organisms and even more so to complex organisms like mammals. To understand why this should be so it is necessary to make a brief digression into elementary physical chemistry. At one time acids were defined as compounds which could yield H^+ ions in solution; and it was recognized that they varied in strength, that is in the extent to which they ionized. Hydrogen chloride (hydrochloric acid), which ionized completely, was a strong acid whereas lactic acid and acetic acid, which ionized only slightly, were weak. The terms base or alkali were correspondingly applied to compounds which released OH^- ions in solution. The shortcoming in these definitions was that they did not cover compounds like Na_2CO_3, which are undoubtedly basic but which cannot directly release OH^- ions. However, in 1923 this unsatisfactory situation was resolved by Brønsted and Lowry, who redefined acids as *compounds (or ions) which can donate protons* and bases as *compounds (or ions) which can accept them*. Since such transfers of protons are reversible, it follows that any acid which gives up its proton becomes a base; conversely any base which accepts a proton becomes an acid. We can therefore write a general equation:

$$\text{acid} \rightleftharpoons H^+ + \text{base}$$

An acid and a base related in this fashion are called conjugate. For example in the reactions

$$HCl \rightleftharpoons H^+ + Cl^-$$
$$CH_3COOH \rightleftharpoons H^+ + CH_3COO^-$$

HCl and CH_3COOH are acids and Cl^- and CH_3COO^- are their conjugate bases. Obviously when an acid is strong, that is it tends to ionize completely, its conjugate base is weak, that is it has little tendency to accept protons. In the two examples above HCl is a very strong acid and Cl^- a very weak base; conversely CH_3COOH is a weak acid and CH_3COO^- is a strong base.

Although the Brønsted-Lowry definitions may at first seem strange they have the great advantage that the single model equation

$$\text{acid} \rightleftharpoons H^+ + \text{base}$$

can be applied not only to obvious examples like the ionization of HCl, but to all acid-base reactions. For instance, the successive ioniza-

tions of carbonic acid can be represented thus

$$H_2CO_3 \rightleftharpoons H^+ + HCO_3^-$$
$$\text{and} \quad HCO_3^- \rightleftharpoons H^+ + CO_2^{2-}$$

In the first of these reactions HCO_3^- acts as base, in the second as acid. The basic character of Na_2CO_3 is easily fitted into the scheme since on solution it ionizes completely:

$$Na_2CO_3 \rightleftharpoons 2Na^+ + CO_3^{2-}$$

and the CO_3^{2-} ion takes up H^+ ions provided by dissociation of water (that is it acts as a base).

$$CO_3^{2-} + H^+ \rightleftharpoons HCO_3^-$$

The basic character of ammonia and of amines is also explained because they also take up H^+ ions

$$NH_4^+ \rightleftharpoons H^+ + NH_3$$
$$R-NH_3^+ \rightleftharpoons H^+ + R-NH_2$$

The term base is no longer applied to compounds like $NaOH$ and KOH but is reserved for the OH^- ion they yield on solution. Since the equilibrium of the reaction

$$H_2O \rightleftharpoons H^+ + OH^-$$

lies so far to the left OH^- must be regarded as a very strong base. Its conjugate acid is the undissociated water molecule, which may be regarded as a very weak acid.

Because the behaviour of any acid (or base) can be represented by the simple equation

$$\text{acid} \rightleftharpoons H^+ + \text{base}$$

it follows that the strength of the acid (or base) is indicated by the equilibrium of the reaction. If the acid is strong (as in HCl), at equilibrium $[H^+]$ and $[base]$ are high in relation to $[acid]$. If the base is strong (as in acetate) the reverse holds good. Such an equilibrium is described by an equation of the familiar form

$$\frac{[H^+][\text{base}]}{[\text{acid}]} = K_a$$

where K_a is a constant, the so-called *acidity constant*. The value of K_a varies enormously, as may be seen from the examples quoted in Table 2.4. For this reason it is often convenient to convert the equation to the negative logarithmic form

$$-\log_{10}K_a = -\log_{10}[H^+] - \log_{10}\frac{[\text{base}]}{[\text{acid}]}$$

Table 2.4 pK_a values for some acids and bases important in biochemistry

Dissociation	pK_a
Acetic acid $\rightleftharpoons$ acetate$^-$ + H$^+$	4·7
Lactic acid $\rightleftharpoons$ lactate$^-$ + H$^+$	3·8
Succinic acid $\rightleftharpoons$ succinate$^-$ + H$^+$	4·2
Succinate$^-$ $\rightleftharpoons$ succinate^{2-} + H$^+$	5·6
Citric acid $\rightleftharpoons$ citrate$^-$ + H$^+$	3·1
Citrate$^-$ $\rightleftharpoons$ citrate^{2-} + H$^+$	4·7
Citrate^{2-} $\rightleftharpoons$ citrate^{3-} + H$^+$	6·4
$H_3PO_4 \rightleftharpoons H_2PO_4^- + H^+$	2·0
$H_2PO_4^- \rightleftharpoons HPO_4^{2-} + H^+$	6·7
$HPO_4^{2-} \rightleftharpoons PO_4^{3-} + H^+$	12·4
$NH_4^+ \rightleftharpoons NH_3 + H^+$	9·2
$CH_3NH_3^+ \rightleftharpoons CH_3NH_2 + H^+$	10·6

and, by analogy with pH, to employ the symbol pK_a or $-\log_{10} K_a$. The equation then becomes

$$pK_a = pH - \log\frac{[base]}{[acid]}$$

or

$$pH = pK_a + \log\frac{[base]}{[acid]}$$

This is the important Henderson-Hasselbalch equation. Because pK_a (like K_a) is a constant for any given system, the equation in effect describes a relationship between two variables
  (1) The hydrogen ion concentration
  (2) The ratio of a particular base to its conjugate acid
Any agent which affects one of them necessarily affects the other.
  When [acid] = [base], pK_a = pH.

pH AND pK IN LIVING ORGANISMS

Cells and tissue fluids such as blood plasma are chemically very complex. They contain a wide variety of acids and bases (in the Brønsted-Lowry sense) with a wide range of pK_a values. But we can use the Henderson-Hasselbalch equation to reduce this situation to some sort of order. We can start with the fact that, at least in multicellular organisms, pH values are generally around 7, and fluctuations are not likely to be large. At such a pH, strong acids are almost completely dissociated. Many biologically important compounds contain carboxyl groups which ionize as follows

$$R - COOH \rightleftharpoons H^+ + R - COO^-$$

The pK_a of the carboxyl group of most of these compounds is about 4. In a cell at pH 7 therefore we can calculate as follows:

$$pH = pK_a + \log_{10}\frac{[base]}{[acid]}$$
$$7 = 4 + \log_{10}\frac{[R - COO^-]}{[R - COOH]}$$

hence

$$\log_{10}\frac{[R - COO^-]}{[R - COOH]} = 3$$

and

$$\frac{[R - COO^-]}{[R - COOH]} = 1000$$

In other words such carboxylic acid groups are almost completely ionized. This would still have been true if the pH had been one unit higher or lower. If the pH had been 6 the ratio would have been 100 to 1; if the pH had been 8 the ratio would have been 10 000 to 1. In either case the carboxylic acid group would still have been almost entirely ionized.
  Conversely many biologically important compounds contain amino groups which, as bases, can take up hydrogen ions

$$R - NH_3^+ \rightleftharpoons H^+ + R - NH_2$$

The pK_a for such compounds is generally, though not invariably, about 10. If we put this value in the equation as in the previous example

$$pH = pK_a + \log\frac{[base]}{[acid]}$$
$$7 = 10 + \log\frac{[R - NH_2]}{[R - NH_3^+]}$$
$$\therefore \quad \log\frac{[R - NH_2]}{[R - NH_3^+]} = -3$$
$$\therefore \quad \frac{[R - NH_2]}{[R - NH_3^+]} = \frac{1}{1000}$$

In this situation the acid predominates over the base, and once again a change in pH of say 1·0 unit in either direction still leaves the greater part of the compound in the acid form.
  The preceding paragraph can be generalized by saying that if $pK_a \ll pH$ (as in the case of many carboxylic acids) the compound will be almost entirely in the form of the conjugate base; that if $pK_a \gg pH$ as in the case of many amines the compound will be almost entirely in the form of the conjugate acid; and that in neither case will small changes in pH have any

great effect on the preponderance of acid over base or base over acid.

There remains, however, a third situation in which pH and pK_a are of nearly the same magnitude. For example all cells contain substantial amounts of inorganic phosphate which ionizes as follows

$$H_2PO_4^- \rightleftharpoons H^+ + HPO_4^{2-}$$

(Inorganic phosphate can exist in other ionic forms but at physiological pH values these can be neglected.) The pK_a of this reaction is about 7. If the pH inside the cell is 7 we can apply the Henderson-Hasselbalch equation

$$7 = 7 + \log \frac{[HPO_4^{2-}]}{[H_2PO_4]}$$

$$\frac{[HPO_4^{2-}]}{[HPO_4^-]} = \text{antilog } 0 = 1$$

In this case, therefore, in contrast to those we have considered before, *conjugate acid and base are present in roughly equivalent concentrations.* Now suppose that for some reason the pH falls from 7 to 6. Our calculation then becomes

$$6 = 7 + \log \frac{[HPO_4^{2-}]}{[H_2PO_4^-]}$$

$$\therefore \frac{[HPO_4^{2-}]}{[H_2PO_4^-]} = \text{antilog } (-1) = \frac{1}{10}$$

In other words, *a relatively small fall in pH has resulted in the conversion of a substantial proportion of the phosphate from the base to the acid form.* A similar calculation shows that a small rise in pH results conversely in a substantial conversion of phosphate from acid to base. These effects are in sharp contrast to the behaviour of strong acids and strong bases. It is important to notice that the hydrogen ion concentration of a solution is related to the ratio of acid to base and not to their absolute values.

In blood plasma the concentrations of $H_2PO_4^-$ and of HPO_4^{2-} are enormously greater than the hydrogen ion concentration (10^{-3} M compared to 10^{-7} M). Yet because H^+ takes part in the equilibrium

$$H_2PO_4^- \rightleftharpoons H^+ + HPO_4^{2-}$$

a knowledge of its concentration, although it is minute, allows us to calculate the ratio of $[H_2PO_4^-]$ to $[HPO_4^{2-}]$.

The system

$$H_2PO_4^- \rightleftharpoons H^+ + HPO_4^{2-}$$

is not by any means the only one in plasma or in cells with a pK_a close to the prevailing pH. Another important example with $pK_a = 6 \cdot 1$ is

$$H_2CO_3 \rightleftharpoons H^+ + HCO_3^-$$

It would be pointless to enumerate all the others but it is important to realize that all of them are necessarily in equilibrium with H^+. If we know $[H^+]$ or pH we can calculate the ratio of [base]/[acid] for any one of them. As will be shown in Chapter 8 there is good reason to think that the undoubted importance of pH for all cells is due, not to any direct action of the hydrogen ion as such, but to the fact that its concentration necessarily reflects the ratio of [acid] to [base] of systems with a pK_a around 7.

BUFFERS

Up to this point we have been trying to explain why the concentration of hydrogen ions should be so important for the normal life and function of cells and tissues. There is however another aspect to the matter. This is the fact that when acid or base is added to cells, or to tissue fluids such as blood plasma, the change in pH of the cells or fluid is much smaller than would have been the case if they had been added to distilled water. This so-called *buffering* is due to the presence of compounds, such as phosphate and bicarbonate, which have pK_a values of roughly the same magnitude as the prevailing pH. This property can best be explained by taking a specific example.

In an earlier section it was explained that addition of quite trivial amounts of acid or base to pure water produces very large changes in the pH because in pure water the concentrations of H^+ and OH^- are so low ($0 \cdot 1$ μmol/l in both cases). The addition of enough strong acid to give a final concentration of 1 μmol/l is sufficient to increase the hydrogen ion concentration tenfold, that is to reduce the pH from 7 to 6. Exactly the same is true if, instead of pure water, we use a dilute solution of, say, sodium chloride, because neither the sodium nor the chloride ion reacts with hydrogen or hydroxyl ions. But if we deal with a solution containing phosphate ions the result is quite different. Let us suppose we have a solution containing both $H_2PO_4^-$ and HPO_4^{2-} ions, both in a concentration of 10 mmol/l. If the pK_a of the dissociation

$$H_2PO_4 \rightleftharpoons H^+ + HPO_4^{2-}$$

is taken as 7, we can calculate the pH of the solution

$$pH = pK_a + \log \frac{[base]}{[acid]}$$
$$pH = 7 + \log \frac{[0 \cdot 010]}{[0 \cdot 010]}$$
$$= 7 + \log 1 = 7 + 0 = 7$$

It has therefore the same pH as pure water. But now the pH is linked rigidly to the base-to-acid ratio and we cannot alter one without altering the other. If we want to bring the pH of our hypothetical phosphate solution down from 7 to 6 we must supply enough hydrogen ions to alter the base to acid ratio accordingly. We can calculate the new ratio thus:

$$pH = pK_a + \log \frac{[base]}{[acid]}$$
$$6 = 7 + \log \frac{[HPO_4^{2-}]}{[H_2PO_4^-]}$$
$$\frac{[HPO_4^{2-}]}{[H_2PO_4^-]} = antilog - 1 = \frac{1}{10}$$

We started with 10 mmol/l HPO_4^{2-} and 10 mmol/l $H_2PO_4^-$. We now need approximately 1·8 mmol/l HPO_4^{2-} and 18·2 mmol/l $H_2PO_4^-$. To bring this about we need to convert 8·2 mmol/l HPO_4^{2-} to $H_2PO_4^-$. This obviously requires 8·2 mmol/l acid which is 8000 times as much as would be required to make the same pH change in pure water. Similarly if we had tried to raise the pH of the phosphate solution from 7 to 8 by addition of base we would have required many times more than is necessary to produce the corresponding change in pure water. Because of their capacity to take up acid or base, compounds with pK_a values around 7 are referred to as buffers. As we might expect, their effectiveness in this respect is greatest at pH values close to their individual pK_a's. It is because cells and organisms contain substantial amounts of compounds with pK_a values around 6 to 8 that they can accept quite significant amounts of acid or base without suffering much change in pH.

To sum up, cells and tissues contain a wide variety of compounds and groups with pK_a values around the prevailing pH. All of these are necessarily in equilibrium with the free hydrogen ions and therefore with one another. Consequently any change in pH must be accompanied by a corresponding change in the base-to-acid ratio for all these compounds; conversely a change in the ratio for any one compound must be accompanied by a corresponding change in all the ratios for all the other compounds and in pH. In such a system therefore pH is, as it were, the visible and measurable part of a much larger whole. So far as the cell is concerned what probably matters is the base-to-acid ratio in certain of the compounds taking part in the equilibrium and pH is useful to the biologist chiefly as an indication of this.

From a biological point of view the importance of buffers lies in the fact that the chemical processes of living cells are exceedingly sensitive to changes in pH. The buffer systems in cells and in body fluids tend to minimize these changes.

In the example quoted above it was shown that the Henderson-Hasselbach equation

$$pH = pK_a + \log \frac{[base]}{[acid]}$$

can be used to determine the pH of a solution if the pK_a and the ratio of base to acid are known. It can, of course, also be used to determine the ratio of base to acid if the pK_a and pH are both known, as in the following example. The pH of blood plasma is approximately 7·4. One of its principal buffer systems is represented by the equation

$$H_2PO_4^- \rightleftharpoons H^+ + HPO_4^{2-}$$

for which $pK_a = 6·7$ in the conditions found in blood. The ratio of $[HPO_4^{2-}]$ to $[H_2PO_4^-]$ can therefore be calculated as follows:

$$pH = pK + \log \frac{[HPO_4^{2-}]}{[H_2PO_4^-]}$$
$$7·4 = 6·7 + \log \frac{[HPO_4^{2-}]}{[H_2PO_4^-]}$$
$$\log \frac{[HPO_4^{2-}]}{[H_2PO_4^-]} = 7·4 - 6·7 = 0·7$$
$$\frac{[HPO_4^{2-}]}{[H_2PO_4^-]} = \frac{5}{1} \text{ approximately}$$

To sum up, the Henderson-Hasselbalch equation is a relationship between (a) pH (b) the pK_a of an acid and (c) the ratio of an acid to its conjugate base. If any two of these quantities are known it can be used to determine the third. The importance of these considerations in the control of the acid-base balance in the blood will be evident in Chapter 33.

COLLOIDAL SOLUTIONS

The foregoing discussion has been concerned with water as a solvent for small molecules, but one of the characteristics of living cells is that much of their chemical machinery is in the form of very large molecules. It is usual nowadays to apply the term 'the colloidal state' to particles varying in diameter between 1 and 500 nm (nanometres); they may be either single large molecules such as protein molecules or aggregates of smaller molecules.

Two aspects of colloid chemistry require some mention here.

Osmotic pressure. If a vessel A (Fig. 2.5) containing water is divided into two compartments, S and W, by a membrane P permeable to water, thermal agitation causes some water molecules to pass through the membrane from W to S and others from S to W. So long as the pressure in S and W are equal the flow in one

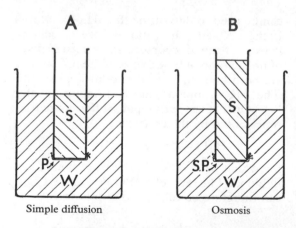

Fig. 2.5 S, solution. P, permeable membrane. W, water. SP, semi-permeable membrane.

direction equals the flow in the other and there is no net transfer. If now a small quantity of some simple solute (say sucrose) is dissolved in the water in S, thermal agitation causes some solute molecules to pass through the membrane

into W. As this process continues the concentration of solute in S falls and the passage of solute molecules from S to W gradually diminishes. At the same time, as the concentration of solute in W rises, an increasing number of solute molecules passes back into S. Eventually equilibrium is attained with equal concentrations of solute on both sides of the membrane and the passage of solute molecules from S to W is exactly balanced by a similar traffic in the reverse direction. Consequently there is no net transfer of solute. This is an example of simple diffusion.

A very different situation prevails if the membrane is permeable to the solvent but not to the solute as in B of Fig. 2.5. Suppose, for example, S contains dilute aqueous sucrose solution while W contains pure water. Once again thermal agitation causes both sides of the membrane to be bombarded with molecules. Those coming from W are all water molecules and pass through into S. The other side of the membrane is bombarded by sucrose and by water molecules. The water molecules pass through the membrane from S into W, but the sucrose molecules are turned back into S again. This means that the number of water molecules passing from S to W is smaller than the number passing from W to S. In other words there is a net transfer of water from W to S. Such a transfer can be prevented only if the pressure in S is increased to counterbalance this 'osmotic pressure'. The osmotic pressure of a solution depends on its molecular concentration. It can be calculated that the osmotic pressure of a solution containing 1 mole of solute in 22·4 litres of water is one atmosphere. Molar solutions of non-ionizable solutes all have the same osmotic pressure, although their concentrations expressed as g/litre are very different. Thus a molar solution of glucose (180 g/litre) has the same osmotic pressure as a molar solution of sucrose (342 g/litre), namely 22·4 atmospheres, but a 1 per cent solution of glucose has an osmotic pressure of 946 mm Hg, which is nearly twice the osmotic pressure of a 1 per cent solution of sucrose (498 mm Hg). In the same way a 1 per cent solution of a substance of very high molecular weight such as albumin (68 000) has a very low osmotic pressure, namely 2·5 mm Hg.

If a solute ionizes in water the osmotic pressure depends on the number of ions rather than on the number of molecules. Thus if we assume that sodium chloride and calcium chloride are completely dissociated in aqueous solution, then M-NaCl has twice the osmotic pressure, and M-CaCl$_2$ three times the osmotic pressure, of M-glucose. In other words, a litre of a molar solution of glucose contains one osmole, a litre of molar sodium chloride two osmoles, and a litre of molar calcium chloride three osmoles.

The expression milliosmole is used in connexion with osmotic pressure. For a non-electrolyte such as glucose 1 milliosmole = 1 millimole. The concentration of sodium (as Na$^+$ ion) in blood plasma is 330 mg/100 ml. The atomic weight of Na is 23 and, since the Na$^+$ ion is monovalent, one milliequivalent is 23 mg. Hence the concentration of Na$^+$ in plasma is 330/23 m-equiv./100 ml, i.e. 3300/23 m-equiv./l or 143·4 m-equiv/l which is the same as 143·4 millimoles/l or 143·4 milliosmoles/l.

Similarly the concentration of calcium in blood plasma is 10 mg/100 ml. The atomic weight of Ca is 40 and, since the Ca^{2+} ion is bivalent, one milliequivalent is 20 mg. Hence the concentration of Ca^{2+} in plasma is 10/20 m-equiv./100 ml, that is 100/20 m-equiv./l which is the same as 2·5 millimoles/l or 2·5 milliosmoles/l.

In the case of an electrolyte such as sodium chloride which dissociates completely in dilute solution into Na$^+$ ions and Cl$^-$ ions, 2 milliosmoles ≡ 1 millimole. For CaCl$_2$ 3 milliosmoles ≡ 1 millimole.

Dialysis. If a solution containing a simple salt, such as sodium chloride, and a protein is placed in the inner vessel (Fig. 2.6), and is separated from pure water in the outer vessel by a semi-permeable membrane (or microfilter) which allows the passage of small molecules, but not of large molecules, the osmotic pressure of the solution EP tends to cause water to pass from W to EP. At the same time salt passes out from EP to W, and if the water in the outer vessel is continually renewed the protein solution in the inner one eventually becomes salt-free. This is the process of dialysis.

Colloid osmotic pressure. Semi-permeable membranes are not uncommon in the body, and the bounding membrane of the red blood cell is an interesting example. If human red blood cells are suspended in 0·5 per cent sodium chloride solution which has a lower osmotic pressure than the cell contents (a *hypotonic* solution), water and salt pass into the cells until they swell and burst. On the other hand if red blood cells are suspended in 1·4 per cent sodium chloride solution which has a higher osmotic pressure (a *hypertonic* solution), water and salt pass out of the red cells so that they shrivel up and become *crenated*. At some intermediate concentration of sodium chloride (about 0·9 per cent) the cell neither swells nor shrinks, and at this concentration the sodium chloride solution is said to be *isotonic* with the cell contents.

Blood plasma contains both readily diffusible, small, non-colloidal molecules or ions and non-diffusible, large colloidal particles. The total osmotic pressure of normal human blood is equivalent to about 7 atmospheres, whereas the proportion due to the plasma proteins chiefly albumin (see Chap. 23) (the *colloid osmotic pressure*) is equivalent only to some 25 mm Hg. Nevertheless this proportion is of great physiological importance, since it determines the distribution of fluid between blood and tissue fluid. The difference between the osmotic pressure of blood and of lymph or tissue fluid is due entirely to plasma protein and is estimated to be about 22 mm Hg. This difference represents the effective osmotic pressure of plasma. As we shall see in Chapter 33, the osmotic pressure, together with the hydrostatic pressure in the capillaries, controls the flow of water between blood and tissue fluid.

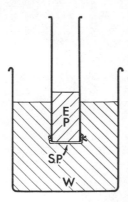

Fig. 2.6 E, electrolyte. W, water. P, protein. SP, semipermeable membrane or microfilter.

REFERENCES

BULL, H. B. (1964). *An Introduction to Physical Biochemistry*. Philadelphia: Davis.

DAWES, E. A. (1972). *Quantitative Problems in Biochemistry*, 5th edn. Edinburgh: Livingstone.

DORMANDY, T. L. (1967). Osmometry. *Lancet*, **i**, 267–271.

EDSALL, J. T. & WYMAN, J. (1958). *Biophysical Chemistry*. New York: Academic Press.

LIPSON, N. & VISOCHER, M. B. (1961). Osmosis in living systems. In *Medical Physics* (Edited by O. Glasser), pp. 869–892. Chicago: Year Book Publishers.

PAULING, L. (1970). *General Chemistry*, 3rd edn. San Francisco: Freeman.

VAN HOLDE, K. E. (1971). *Physical Biochemistry*. New Jersey: Prentice Hall.

WEST, E. S. (1956). *Textbook of Biophysical Chemistry*. New York: Macmillan.

WHIPPLE, H. E. (Ed.) (1965). Forms of water in biologic systems. *Annals of the New York Academy of Sciences*, **125**, 251–772.

3 The proteins

The proteins, all containing nitrogen, are among the most important constituents of cells. The complete hydrolysis of proteins yields some twenty different amino acids; since there are no other products of hydrolysis it is concluded that proteins are built up exclusively of amino acid units.

The simplest amino acid is amino-acetic acid or glycine.

$$
\begin{array}{ccc}
\text{H} & & \text{H} \\
| & & | \\
\text{H--C--COOH} & \quad\quad & \text{H--C--COOH} \\
| & & | \\
\text{H} & & \text{NH}_2
\end{array}
$$

acetic acid   amino-acetic acid
(glycine)

The next higher amino acid is amino-propionic acid. There are two amino derivatives of propionic acid depending on whether the amino group is attached to the α or the β carbon atom.

$$
\begin{array}{cc}
\text{H} & \text{H} \\
| & | \\
\text{H--C----C--COOH} \\
| & | \\
\text{H} & \text{H}
\end{array}
$$

propionic acid

$$
\begin{array}{cc}
\text{H} \quad \text{H} & \text{H} \quad \text{H} \\
| \quad | & | \quad | \\
\text{H--C----C--COOH} \quad\quad \text{H--C----C--COOH} \\
| \quad | & | \quad | \\
\text{H} \quad \text{NH}_2 & \text{NH}_2 \quad \text{H}
\end{array}
$$

α-amino-propionic acid β-amino-propionic acid
(alanine)   (β alanine)

In the same way butyric acid can give rise to three possible amino acids, and so on. The amino acids commonly found in proteins are all α-amino acids and thus they all have the general formula:

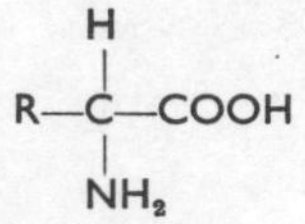

$$
\begin{array}{c}
\text{H} \\
| \\
\text{R—C—COOH} \\
| \\
\text{NH}_2
\end{array}
$$

where R is H in glycine and R is CH_3 in alanine.

The α-carbon atom in all amino acids except glycine is attached to four different atoms or groups and is therefore asymmetric. The amino acids are accordingly optically active and can exist in two stereoisomeric forms. All the amino acids which occur naturally in the tissue proteins belong to the L-series of configurationally related compounds—but some amino acids are dextrorotatory (+) and others are laevorotatory (−). Thus naturally occurring alanine is L(+)alanine, which is dextrorotatory although it belongs to the L-series, that is it is configurationally related to L-lactic acid. Amino acids of the D-series occur in some antibiotics and in the walls of bacteria.

The amino acids obtained by hydrolysis of proteins may be classified as shown below. The standard contractions are shown in brackets after the name of each amino acid:

ALIPHATIC AMINO ACIDS

1. *Monoamino–monocarboxylic acids*

Glycine (Gly)

(amino-acetic acid)

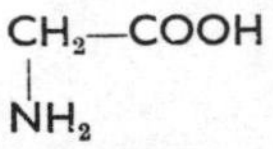

$$
\begin{array}{c}
\text{CH}_2\text{—COOH} \\
| \\
\text{NH}_2
\end{array}
$$

L(+)Alanine (Ala)

(α-amino-propionic acid)

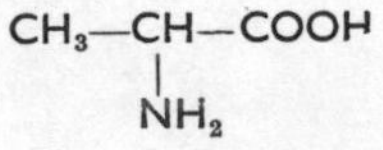

$$
\begin{array}{c}
\text{CH}_3\text{—CH—COOH} \\
| \\
\text{NH}_2
\end{array}
$$

L(−)Serine (Ser)

(α-amino-β-hydroxy-propionic acid)

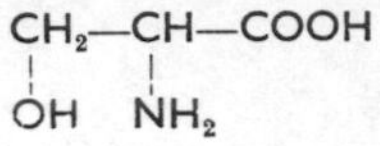

$$
\begin{array}{c}
\text{CH}_2\text{—CH—COOH} \\
| \quad\; | \\
\text{OH} \;\; \text{NH}_2
\end{array}
$$

L(+)Threonine (Thr)

(α-amino-β-hydroxy-butyric acid)

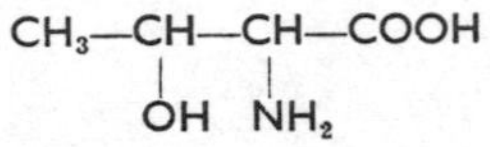

$$
\begin{array}{c}
\text{CH}_3\text{—CH—CH—COOH} \\
| \quad\; | \\
\text{OH} \;\; \text{NH}_2
\end{array}
$$

L(+)Valine (Val)
(α-amino-isovaleric acid)

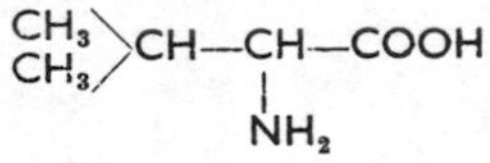

$$
\begin{array}{c}
\text{CH}_3 \\
\quad\;\; \diagdown \\
\qquad\text{CH—CH—COOH} \\
\quad\;\; \diagup \qquad\quad | \\
\text{CH}_3 \qquad\qquad \text{NH}_2
\end{array}
$$

L(−)Leucine (Leu)
(α-amino-isocaproic acid)

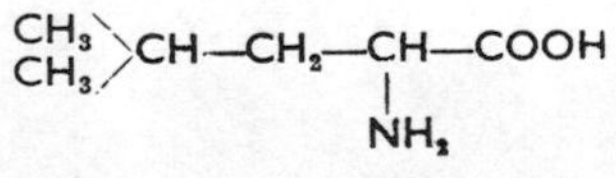

$$
\begin{array}{c}
\text{CH}_3 \\
\quad\;\; \diagdown \\
\qquad\text{CH—CH}_2\text{—CH—COOH} \\
\quad\;\; \diagup \qquad\qquad\quad | \\
\text{CH}_3 \qquad\qquad\qquad \text{NH}_2
\end{array}
$$

L(+)Isoleucine (Ile)
(α-amino-β-methylvaleric acid)

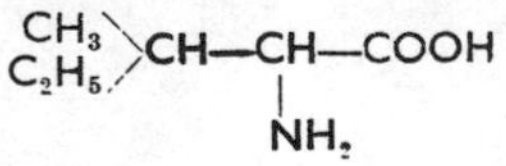

$$
\begin{array}{c}
\text{CH}_3 \\
\quad\;\; \diagdown \\
\qquad\text{CH—CH—COOH} \\
\quad\;\; \diagup \qquad\quad | \\
\text{C}_2\text{H}_5 \qquad\qquad \text{NH}_2
\end{array}
$$

2. Monoamino–dicarboxylic acids

L(−)Aspartic acid (Asp)

(α-amino-succinic acid)

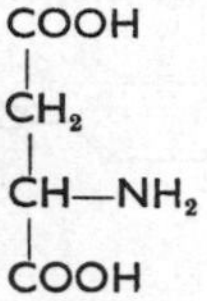

```
COOH
|
CH₂
|
CH—NH₂
|
COOH
```

and its amide asparagine (Asp(NH₂) or Asn)

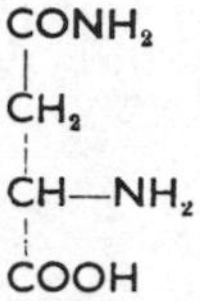

```
CONH₂
|
CH₂
|
CH—NH₂
|
COOH
```

L(+)Glutamic acid (Glu)

(α-amino-glutaric acid)

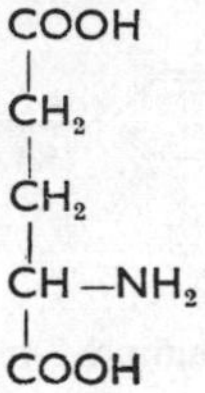

```
COOH
|
CH₂
|
CH₂
|
CH —NH₂
|
COOH
```

and its amide glutamine (Glu(NH₂) or Gln)

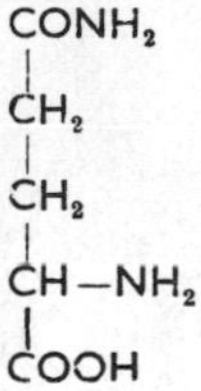

```
CONH₂
|
CH₂
|
CH₂
|
CH —NH₂
|
COOH
```

3. Diamino–monocarboxylic acids

L(+)Arginine (Arg)

(α-amino-δ-guanido-valeric acid)

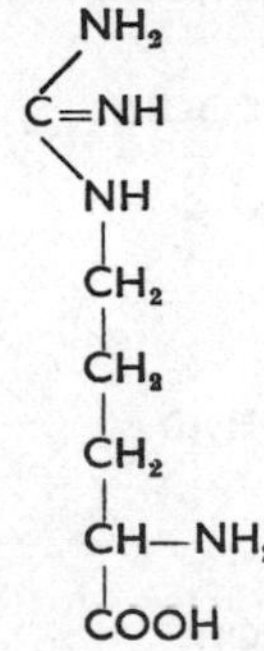

```
NH₂
|
C=NH
|
NH
|
CH₂
|
CH₂
|
CH₂
|
CH—NH₂
|
COOH
```

L(+)Lysine (Lys)

(α, ε-diamino-caproic acid)

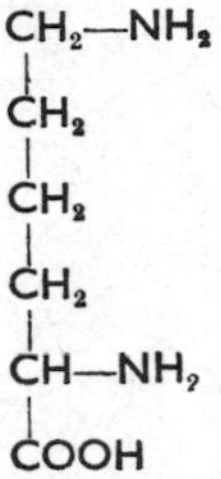

```
CH₂—NH₂
|
CH₂
|
CH₂
|
CH₂
|
CH—NH₂
|
COOH
```

4. Sulphur-containing amino acids

L(−)Cysteine (α-amino-β-thiol-propionic acid) (Cys) and L(−)Cystine (dicysteine) (CySSCy)

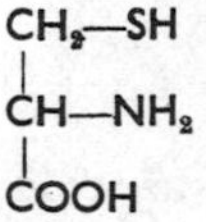

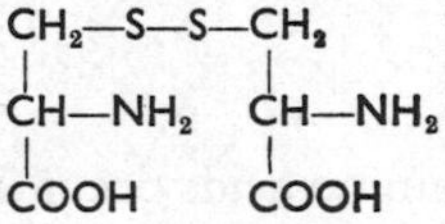

```
CH₂—SH
|
CH—NH₂
|
COOH
```

```
CH₂—S—S—CH₂
|              |
CH—NH₂   CH—NH₂
|              |
COOH       COOH
```

L(−)Methionine (Met)

(α-amino-γ-methylthiol-butyric acid)

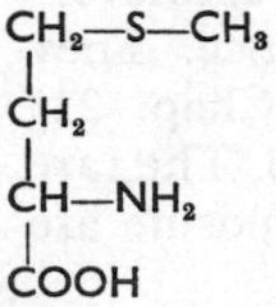

```
CH₂—S—CH₃
|
CH₂
|
CH—NH₂
|
COOH
```

Aromatic and heterocyclic amino acids

L(−)Tyrosine (α-amino-β-p-hydroxyphenyl-propionic acid) (Tyr)

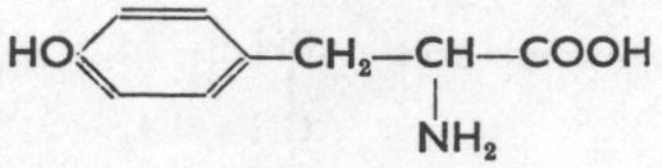

$$HO-\!\!\!\!\bigcirc\!\!\!\!-CH_2-CH-COOH$$
$$|$$
$$NH_2$$

L(−)Phenylalanine (α-amino-β-phenyl-propionic acid) (Phe)

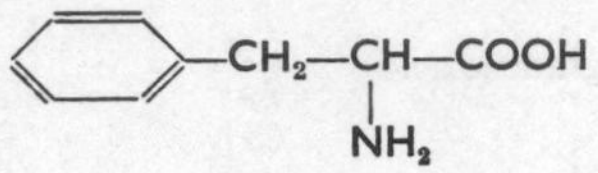

$$\bigcirc\!\!\!\!-CH_2-CH-COOH$$
$$|$$
$$NH_2$$

L(−)Histidine (α-amino-β-iminazolyl-propionic acid) (His)

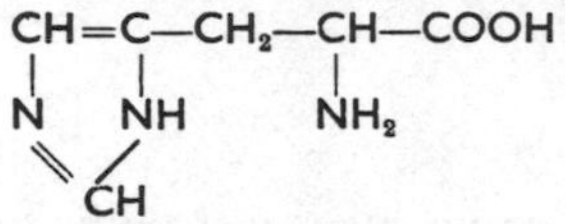

$$CH=C-CH_2-CH-COOH$$
$$|\quad\ |\qquad\quad |$$
$$N\quad NH\qquad NH_2$$
$$\diagdown\!\!=\!\!\diagup$$
$$CH$$

L(−)Tryptophan (α-amino-β-indole-propionic acid) (Trp)

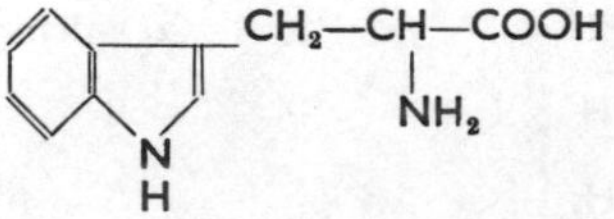

$$CH_2-CH-COOH$$
$$|$$
$$NH_2$$
$$N$$
$$|$$
$$H$$

L(−)Hydroxyproline (Hyp)

$$HO-CH-CH_2$$
$$|\qquad\quad |$$
$$CH_2\quad CH-COOH$$
$$\diagdown\!\!/$$
$$NH$$

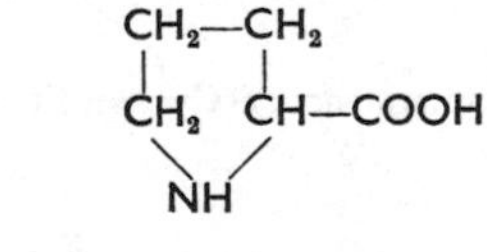

L(−)Proline (α-pyrrolidine-carboxylic acid) (Pro)

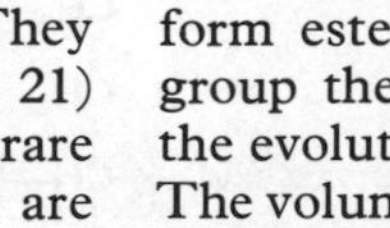

$$CH_2-CH_2$$
$$|\qquad\quad |$$
$$CH_2\quad CH-COOH$$
$$\diagdown\!\!/$$
$$NH$$

All the above amino acids occur in proteins. A few others, not found in proteins, are nevertheless of great biological importance. They include ornithine and citrulline (Chap. 21) and ε-amino caproic acid (Chap. 23). The rare amino acids desmosine and isodesmosine are found in elastin.

Amino acids are colourless, crystalline solids. Their carboxyl groups react with alcohols to form esters. Because they possess an amino group they can react with nitrous acid with the evolution of nitrogen.
The volume of nitrogen evolved is a measure of the amount of amino acid present.

$$\begin{array}{c} \text{R—CH—COOH} \\ | \\ \text{NH}_2 \end{array} + \text{ONOH}$$

$$\longrightarrow \begin{array}{c} \text{R—CH—COOH} \\ | \\ \text{OH} \end{array} + \text{H}_2\text{O} + \text{N}_2$$

Amino acids contain two reactive groups, the carboxyl group and the amino group. This may be shown by titrating a solution of an amino acid with HCl and with NaOH. Titration of a glycine solution (which has a pH of about 6·0) reveals two pK_a values (Fig. 3.1).

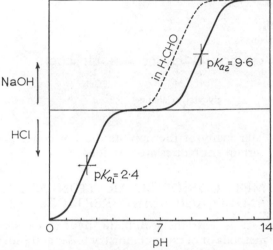

Fig. 3.1 Titration curve of glycine in absence (full line) and presence (dotted line) of formaldehyde. For explanation of pK see p. 17

If glycine solution is first treated with formaldehyde, which reacts with the amino group, and is then titrated with sodium hydroxide, the titration curve shows a fall in pK_{a2} as indicated in Fig. 3.1; however, the titration curve in acid remains unaltered. The group titrated with NaOH is therefore the amino group—not the carboxyl group. This apparent anomaly is explained by the fact that in neutral solution the amino and the carboxyl groups are both charged. This structure, described as a *zwitterion* or dipolar ion, may be regarded as the result of internal salt formation. The effects of HCl and NaOH are most appropriately represented by the scheme:

Addition of HCl gives the cationic form on the left as for instance in glycine hydrochloride. Addition of NaOH gives the anionic form on the right as in sodium glycinate.

Amino acids are therefore *amphoteric electrolytes* or *ampholytes*.

Titration of amino acids with alkali in presence of formaldehyde is the basis of a quantitative method of estimating amino acids. This is called *formol titration*.

$$\begin{array}{c} \text{R—CH—COOH} \\ | \\ \text{NH}_2 \end{array} + 2\,\text{CH}_2\text{O} \longrightarrow \begin{array}{c} \text{R—CH—COOH} \\ | \\ \text{HOCH}_2\text{—N—CH}_2\text{OH} \end{array}$$

Free amino acids react with ninhydrin to form blue derivatives. This reaction has wide applications in the detection of amino acids on paper chromatograms and in the estimation of amino acids in solution.

PEPTIDES

The carboxyl group of one amino acid can react with the amino group of another amino acid, with the elimination of water. Thus glycine and alanine can react together in one of two ways with the formation of the glycylalanine or alanylglycine.

$$\begin{array}{c} \text{NH}_2\text{—CH—COOH} \\ | \\ \text{H} \end{array} + \begin{array}{c} \text{HNH—CH—COOH} \\ | \\ \text{CH}_3 \end{array}$$

glycine   alanine

$$\longrightarrow \begin{array}{c} \text{NH}_2\text{—CH—CO—NH—CH—COOH} \\ | \qquad\qquad\qquad | \\ \text{H} \qquad\qquad\qquad \text{CH}_3 \end{array} + \text{H}_2\text{O}$$

glycylalanine

$$\begin{array}{c} \text{NH}_2\text{—CH—COOH} \\ | \\ \text{CH}_3 \end{array} + \begin{array}{c} \text{HNH—CH—COOH} \\ | \\ \text{H} \end{array}$$

alanine   glycine

$$\longrightarrow \begin{array}{c} \text{NH}_2\text{—CH—CO—NH—CH—COOH} \\ | \qquad\qquad\qquad | \\ \text{CH}_3 \qquad\qquad\qquad \text{H} \end{array} + \text{H}_2\text{O}$$

alanylglycine

$$\begin{array}{c} \text{NH}_3^+\text{—CH—COOH} \\ | \\ \text{R} \end{array} \underset{\pm\text{H}^+}{\rightleftharpoons} \begin{array}{c} \text{NH}_3^+\text{—CH—COO}^- \\ | \\ \text{R} \end{array} \underset{\pm\text{H}^+}{\rightleftharpoons} \begin{array}{c} \text{NH}_2\text{—CH—COO}^- \\ | \\ \text{R} \end{array}$$

zwitterion

Glycylalanine may then react either at its amino group or at its carboxyl group with another molecule of glycine to form glycylalanylglycine or glycylglycylalanine.

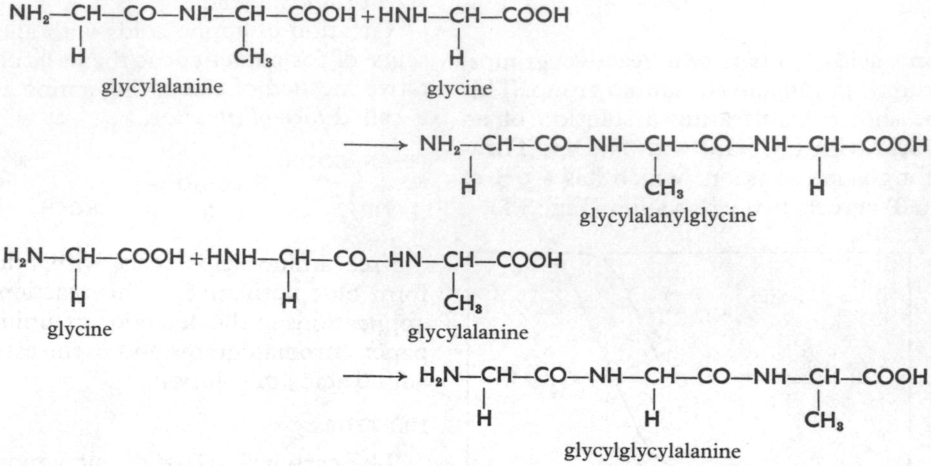

Some relatively simple peptides have important biological activities. For example glutathione (Chap. 11) is a tripeptide, glutamylcysteylglycine (Glu–Cys–Gly). As will be seen

$$NH_2-CH-CO-NH-CH-COOH + HNH-CH-COOH$$
$$|||$$
$$HCH_3H$$

glycylalanine · glycine

$$\longrightarrow NH_2-CH-CO-NH-CH-CO-NH-CH-COOH$$
$$|||$$
$$HCH_3H$$

glycylalanylglycine

$$H_2N-CH-COOH + HNH-CH-CO-HN-CH-COOH$$
$$|||$$
$$HHCH_3$$

glycine · glycylalanine

$$\longrightarrow H_2N-CH-CO-NH-CH-CO-NH-CH-COOH$$
$$|||$$
$$HHCH_3$$

glycylglycylalanine

The compounds formed by the condensation of amino acids in this way are known as *peptides* and the group —CO—NH— joining the amino acid residues is known as the *peptide bond*. When two amino acids are joined together in this way the product is known as a *dipeptide*, for example glycylalanine. Three amino acids form a *tripeptide*, for example glycylalanylglycine, and the product formed by the condensation of a large number of amino acids is called a *polypeptide*. A section of a polypeptide chain is shown below in which the sequence of amino acid *residues* is lysyl–glutamyl–tyrosyl–alanyl (Lys–Glu–Tyr–Ala).

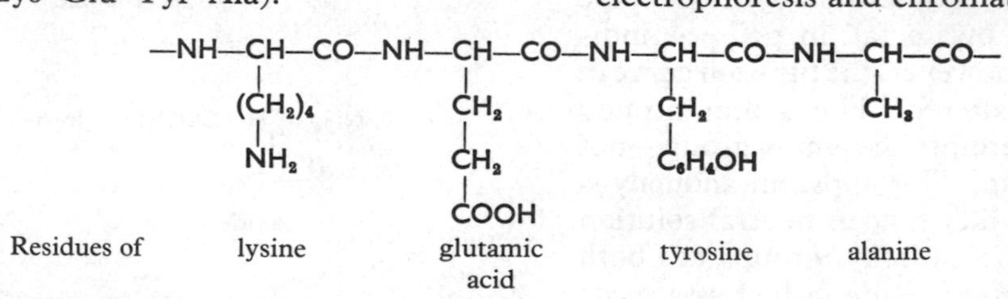

later many of the non-steroid hormones and certain antibiotics are peptides.

METHODS OF SEPARATION OF AMINO ACIDS AND PEPTIDES

It is extremely difficult by the classical methods of organic chemistry to separate such varied and complex molecules as amino acids and peptides, or to obtain and identify pure samples. For this reason modern amino acid and peptide chemistry depends to a large extent on two very valuable techniques known as electrophoresis and chromatography.

$$-NH-CH-CO-NH-CH-CO-NH-CH-CO-NH-CH-CO-$$
$$||||$$
$$(CH_2)_4CH_2CH_2CH_3$$
$$NH_2CH_2C_6H_4OH$$
$$COOH$$

Residues of · lysine · glutamic acid · tyrosine · alanine

Since some 20 amino acids, any one of which may occur more than once, are available for building up into peptides, and since the order of their occurrence in the peptide chains can vary, it is evident that the number of possible polypeptides is enormous. Thus, from 20 different amino acids it is theoretically possible to make 400 dipeptides and 8000 tripeptides.

Electrophoresis depends on the electrolytic properties of the amino acids. A mixture of amino acids is dissolved in a buffer solution at a pH at which all the carboxyl and amino groups are ionized and then electrodes connected to a battery are placed in the solution. If, as is usual, only small amounts of material are available, a solid medium such as filter paper or blocks of starch gel soaked in the solution may be used. Amino acids or peptides with a positive charge migrate through the buffer towards the cathode and those with a negative charge towards the anode. For example, in a mixture of leucine, lysine and glutamic acid, lysine migrates towards the cathode, glutamic acid towards the anode, while leucine tends to remain at or near the starting point (Fig. 3.2). The position of the amino acid on the solid medium can be established by appropriate staining, for example, with ninhydrin.

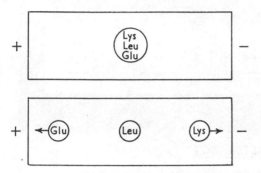

Fig. 3.2 Electrophoresis of a mixture of amino acids at neutral pH. The glutamic acid carrying a negative charge moves towards the anode, lysine carrying a positive charge moves towards the cathode, leucine which is electrically neutral moves only slightly.

The amino acids can be identified by running known 'marker' amino acids on the paper or starch gel alongside the unknown mixture. Since the distance moved by any one amino acid under the same conditions is always the same, the amino acid spot which migrates exactly the same distance as the 'marker' lysine can be identified as lysine. The same technique can be applied to peptides, for example glycylleucylarginine moves towards the cathode and alanylaspartylserine towards the anode. Any particular area can then be excised and the peptide eluted from the gel or paper and analysed further. Separation of large peptides from small ones can be achieved by the pore size of the gel; since the progress of the large peptides may be hindered, two peptides with exactly the same charge may migrate to different positions on the medium because of differences in size or shape.

Partition chromatography also relies on the fact that amino acids are electrolytes but this method depends on their different affinities for two solvents. Amino acids with many ionic groups tend to 'prefer' water as a solvent in exactly the same way that sodium chloride 'prefers' water to benzene as a solvent. Amino acids with large non-polar side groups such as leucine or phenylalanine, have slightly greater affinity for organic solvents. This can be shown by placing a complex mixture of amino acids at one end of a sheet of filter paper, and allowing an aqueous buffer to creep slowly down the paper from that end in a closed tank in which the atmosphere is saturated with vapour from an organic solvent. It is found that the ionic amino acids tend to move with the water at the 'solvent front' whereas those with large paraffinic or 'hydrophobic' side chains lag behind in the paper with their side chains protruding into the organic vapour. A mixture of phenylalanine, alanine and glutamic acid can readily be separated by this means (Fig. 3.3). For the separation of larger amounts of amino acids column chromatography is employed (Fig. 3.5).

The same protein, for example insulin or haemoglobin, can vary both within species and even within individuals of the same species. This difference can be in one or several amino acid residues but the vast majority of amino acid sequences will be the same for all species and individuals. Fortunately it is not necessary to determine the entire amino acid sequence for the same protein from each species since a valuable technique known as *fingerprinting* can be employed. The protein is digested with the enzyme trypsin to give large peptides which are separated by electrophoresis on paper in two dimensions to give a characteristic distribution of peptides. If two proteins differ only in one amino acid, the peptide in which this difference occurs behaves somewhat differently on electrophoresis. In the case of haemoglobin S (Chap. 25) which has a valine residue in place of glutamic acid, one of the peptides migrates more rapidly on electrophoresis than the corresponding peptide from normal haemoglobin (Fig. 3.4). It is then only necessary to determine the difference in amino acid sequence in this peptide.

The process of fingerprinting can also be carried out by using separation by electrophoresis in one direction and by chromatography in a direction at right angles to the first.

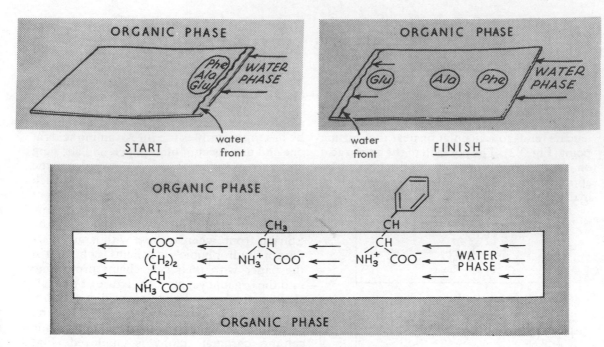

Fig. 3.3 Diagrammatic representation of the process of partition chromatography on paper. The mixture of amino acids is applied in the form of a spot on the end of a strip of paper down which a stream of water molecules (as indicated by the arrows) is allowed to flow in a tank in which the atmosphere is saturated with the vapour of the organic phase, which might, for example, be phenol. The amino acids separate out as shown in the lower diagram, those which are hydrophobic moving more slowly than those which are not.

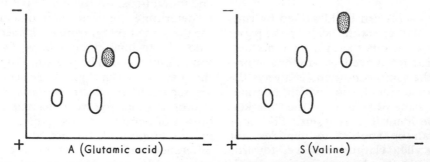

Fig. 3.4 The process of fingerprinting. A hypothetical case in which a protein A is broken down to a mixture of peptides, one of which contains glutamic acid, whereas the corresponding protein S, containing valine in place of glutamic acid, gives a similar mixture of peptides, one of which shows a difference in electrical mobility.

Fingerprinting is also very extensively employed in the study of the sequences of nucleotides in closely related samples of ribonucleic acid (Chap. 6).

PROTEINS

Proteins may be regarded as very large and elaborate polypeptides. There is, however, a great difference between proteins and other biologically important macromolecules such as the polysaccharides and lipids. The latter are composed of small units, often identical or at least very similar. Their functions in the living organism are largely structural or as reservoirs of essential nutrients. The units of which proteins are composed are, however,

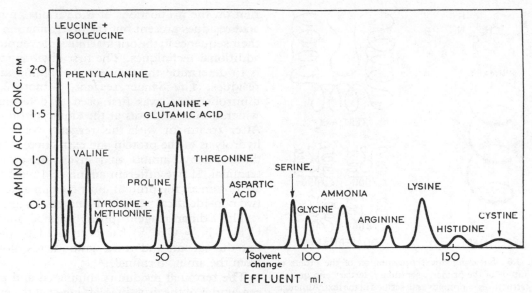

Fig. 3.5 Column chromatography of a hydrolysate of serium albumin. The eluate from the column was collected in a series of tubes and the amino acid content of each tube determined colorimetrically by developing a purple colour with ninhydrin.

very varied since all 20 different amino acids may be present in greater or smaller amounts in one protein. Consequently, there is a great variety and range of proteins, according to their composition, size and shape. They are of the utmost importance in biological processes since, although some have a purely structural function, the majority play not a static role in living tissues but a very dynamic one. Such proteins include the enzymes discussed in a later chapter.

The molecular weights of proteins range from the 6000 of insulin, through the 60 000 of bovine serum albumin to the 600 000 of myosin and even higher. The proportions of the various amino acid residues in proteins vary considerably; some such as serum albumin contain all the common amino acids, others such as insulin, which only lacks methionine, tryptophan and cysteine, have most of them. Collagen, the protein of connective tissue, contains 33 per cent glycine residues and 12 per cent each of alanine and proline with very low proportions of the other amino acids. The order of the amino acid residues along the chain is also variable so that proteins with similar gross amino acid compositions can have completely different biological functions. The shapes of proteins range from long ribbon-like structures such as fibrinogen,

to almost completely spherical structures such as the enzyme ribonuclease. They may be very acidic in character such as pepsin, or very basic, such as the histones which have a very high proportion of lysine residues or the protamines which are rich in arginine. Proteins may be highly insoluble, such as the keratins from hair and nails, or highly soluble such as the albumins of the blood plasma. The tremendous range in properties shows that any classification on the grounds of one characteristic is not likely to be useful. For example, proteins which have similar molecular weights may vary so greatly in shape or in solubility as to be totally dissimilar. The simplest approach to protein structure is therefore to define the level of organization, which may be described as primary, secondary, tertiary and quaternary. This distinction, though useful, is arbitrary and all the levels of organization are interdependent so all four must be determined to obtain a clear picture (Fig. 3.6).

If we could by some means of magnification see in detail the peptide linkages and individual amino acids along the protein chain, we should then be observing the primary structure. At a slightly lower magnification the chain would not be seen in such detail but we could observe the coils into which it is folded: this is the secondary structure. At still lower magnifica-

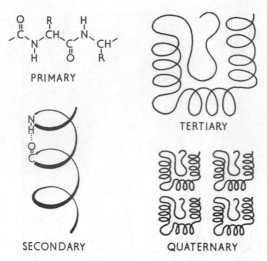

PRIMARY

SECONDARY

TERTIARY

QUATERNARY

Fig. 3.6 Diagrammatic representation of the orders of magnitude of the primary, secondary, tertiary and quaternary structures of proteins as described in the text. Although the subunits are shown above as identical in the quaternary structure, it is not uncommon for a protein to have several different kinds of subunits in its quaternary structure.

tion we could see these coils bend themselves round and entwine with each other (the tertiary structure). At an even lower magnification we should be able to see the number and nature of these entwined units in the entire molecule (the quaternary structure).

PRIMARY STRUCTURE

The primary structure of proteins refers to the number, nature and sequence of amino acid residues along the peptide chains. This is extremely difficult to determine since there may be several hundred residues in sequence. Indeed, the first successful attempt (on insulin) was concluded only in 1955 by Sanger and his colleagues in Cambridge. Total hydrolysis of a protein with separation of the products on starch or ion-exchange resins yields informa-

tion on the proportions of the various amino acid residues present but the determination of their sequence in the original molecule requires additional techniques. The first obvious step is to determine the two terminal amino acid residues. The Sanger reagent, 1-fluoro-2,4-dinitrobenzene, was first used to determine which amino acid was at the amino terminal. After treatment with this reagent, complete hydrolysis of the protein and extraction of the dinitrophenyl amino acid gives the amino terminal. If two different amino acid residues are obtained, the protein must contain at least two non-identical peptide chains (see Diagram I).

The Edman reagent (phenylisothiocyanate, $C_6H_5 - N = C = S$) has been used to pick out the second and third amino acids along from the amino terminal.

The terminal residue is eliminated and the remainder of the protein is left intact at the end of this reaction. The reagent can be used again on the fragment to give the second end amino acid, and so on (see Diagram II).

Two techniques are available to determine which amino acids are present at the carboxyl end of the protein. In one the protein is reacted with hydrazine. The protein is split and all the amino acids but the one at the carboxyl terminal are found to be present as hydrazides. In another method the enzyme carboxypeptidase is used to split off amino acids from the carboxyl terminal of the protein one by one. The rate of release of amino acids gives information about their sequence.

Once the basic information about the few amino acids at each end of the protein has been determined the protein is split into small peptides and the sequence of each of them determined. There are two main ways of doing this. The first is to use a specific enzyme which hydrolyses the protein only at certain points.

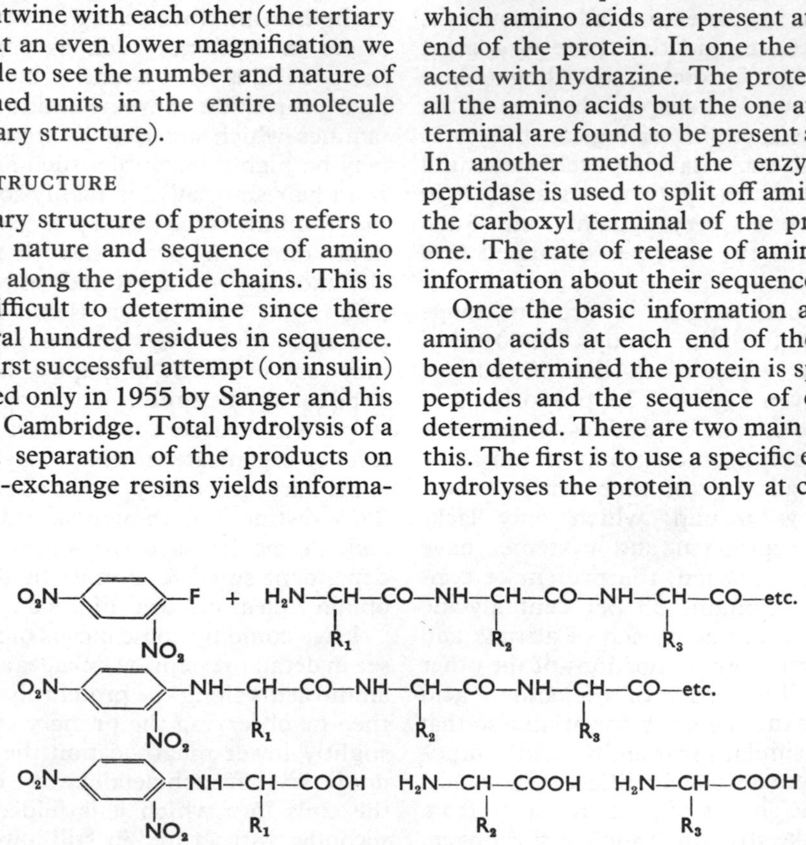

O_2N—⬡—F + H_2N—CH—CO—NH—CH—CO—NH—CH—CO—etc.
　　　　|　　　　　　|　　　　　　　|　　　　　　|
　　　　NO_2　　　　R_1　　　　R_2　　　　R_3

O_2N—⬡—NH—CH—CO—NH—CH—CO—NH—CH—CO—etc.
　　　　|　　　|　　　　　　|　　　　　　|
　　　　NO_2　R_1　　　　R_2　　　　R_3

O_2N—⬡—NH—CH—$COOH$　　H_2N—CH—$COOH$　　H_2N—CH—$COOH$
　　　　|　　　|　　　　　　　　　　|　　　　　　　　　|
　　　　NO_2　R_1　　　　　　　　R_2　　　　　　　R_3

DIAGRAM I

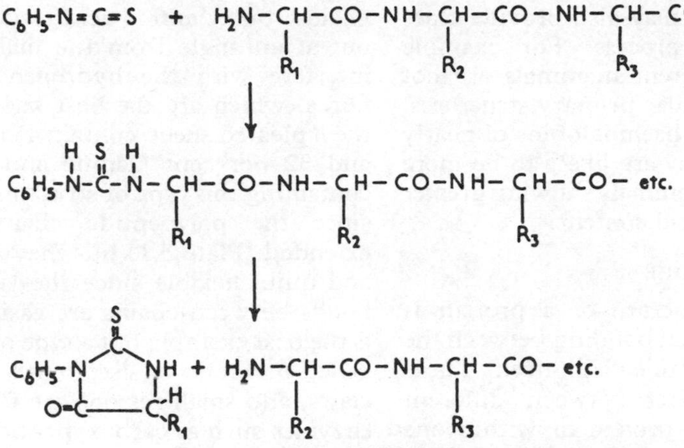

$$C_6H_5-N=C=S \quad + \quad H_2N-CH-CO-NH-CH-CO-NH-CH-CO- \text{ etc.}$$
$$\underset{R_1}{\quad} \quad \underset{R_2}{\quad} \quad \underset{R_3}{\quad}$$

$$\overset{H\;\;S\;\;H}{C_6H_5-N-C-N-CH-CO-NH-CH-CO-NH-CH-CO-}\text{ etc.}$$
$$\underset{R_1}{\quad} \quad \underset{R_2}{\quad} \quad \underset{R_3}{\quad}$$

$$C_6H_5-N\overset{S}{\underset{O=C}{\diagup\;\;\diagdown}}NH \quad + \quad H_2N-CH-CO-NH-CH-CO-\text{ etc.}$$
$$\underset{R_1}{\quad} \qquad \underset{R_2}{\quad} \quad \underset{R_3}{\quad}$$

DIAGRAM II

The best known of these is the enzyme trypsin which hydrolyses protein chains on the carboxyl side of a lysyl or arginyl residue. The second method is to use a specific chemical substance. The most successful is cyanogen bromide which splits protein chains only where a methionine residue is present. When the peptides produced by these techniques have been separated and their amino acid sequences determined, the results are pieced together to give the sequence in the whole protein. For example, if in three different hydrolyses the following peptides are obtained,

i. ABCDEA ii. DEAFGH iii. IJKA

and it is known that there are only two residues of A in the whole protcin, the sequence of this segment of the protein must be

IJKABCDEAFGH

It is in effect a very complex and fascinating biological jigsaw puzzle.

In this way Sanger and his colleagues worked out the entire primary structure of insulin shown in Fig. 3.7. The molecule contains 51 amino acids arranged in two polypeptide chains which are covalently linked by disulphide bridges. At the present time the structure of many other proteins has been solved, including myoglobin and haemoglobin (Chap. 25).

Studies of the variation in protein primary structure can yield valuable information. If the amino acid residue in one position of a specific protein differs between species and individuals, and yet the protein retains its normal biological properties, then it can be asserted that this residue is not necessary for the functional integrity of the protein and is not intimately involved in its action. Sometimes, however, such an apparently minor alteration may be detrimental to the function of the protein, as in the case of haemoglobin S found in persons with sickle-cell anaemia (Chap. 25).

Knowledge of the variations of primary

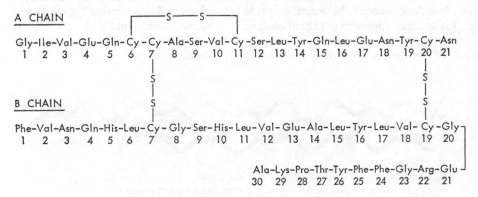

Fig. 3.7. Sequences of amino acids in the cattle insulin molecule (Sanger).

structures of proteins may also provide clues to the evolutionary process. For example haemoglobins of different mammals are not identical but have similar primary structures. If two mammals have haemoglobins of nearly the same structure they are likely to be more closely related than mammals showing greater differences in amino acid content.

SECONDARY STRUCTURE

The secondary structure of a protein is determined by hydrogen bonding between the components of the peptide chain itself. These bonds can occur either between different peptide chains of the protein or within one polypeptide chain. The great functional importance of secondary structure is shown by the fibrous or structural proteins. Three basic types of *helix* are used to define secondary structure.

1. THE α HELIX. In this highly stable structure the peptide chain assumes the form of a spiral staircase with three and a half amino acid residues per turn. The coils of the helix are held together by hydrogen bonds lying parallel to the main axis of the fibre (Plate 3.1). It is evident that such a structure should be both flexible and elastic since there are relatively few cross connexions between fibres. When the chain is extended the hydrogen bonds rupture but they reform when the chain is allowed to relax. The α helix is found in proteins such as α keratin in hair and nails, and in the muscle protein myosin, and also in short regions of many globular or non-fibrous proteins. For example in haemoglobin there are 8 such areas of α helix. The smallest is 7 residues long and the largest 26.

2. THE β PLEATED SHEET. In this structure hydrogen bonding occurs between peptide chains and is at right angles to the main chains. Fibres which have this structure are usually composed of amino acids with small side chains; bulky amino acid residues protruding out at an angle from the main chain would interfere with the hydrogen bonding. Silk fibres, which are the best known example of the β pleated sheet, contain 45 per cent glycine and 32 per cent alanine and serine. Fibres containing this type of structure are not elastic since the polypeptide chains are already extended (Plate 3.1) but they are very strong and quite flexible since the lateral hydrogen bonds between chains are easily broken. Silk is the best example but a type of beta structure called β keratin is also found in feathers and claws, and small regions are found in protein enzymes such as carboxypeptidase.

3. THE COLLAGEN HELIX. The structure of collagen is very complex. The basic unit is the tropocollagen molecule 280 nm long and 1·5 nm in diameter, consisting of three separate polypeptide chains each about 1000 amino acids long. Every fourth amino acid is glycine which is followed by either hydroxyproline or proline. No other protein has this abundance of hydroxyproline. The tropocollagen molecules become packed in a remarkably orderly array to form the collagen fibrils.

The triple-helix structure of tropocollagen is illustrated in Fig. 3.8. It consists of three separate strands each in itself a left handed helix, intertwined with each other to form a right handed super-helix. Each of the three chains is hydrogen bonded to the other two in a complex fashion which leads to a very strong and rigid structure which is not easily bent or extended. This type of structure is most easily formed by polypeptide chains containing many residues of glycine, proline and hydroxy-proline. Collagen fibrils are found in all tissues of the animal body particularly in bones, tendons and connective tissues (see Chap. 14).

None of the methods available at present for determining the secondary structure of a protein is entirely reliable. The only satis-

Fig. 3.8 Diagrammatic representation of the triple helix of the tropocollagen molecule showing the three left handed helices wound round each other to form a right handed super-helix.

factory method is an entire X-ray crystallographic analysis of the whole protein but this is too laborious for wide application. Polarized infra-red light can be applied to stretched-out fibres of protein to determine alterations caused by hydrogen bonding in the characteristic vibration frequencies of the components of the peptide bond. This method is limited in its applications; it cannot be used to study proteins in their normal aqueous environment because water absorbs at the same frequencies. Another method measures the rate at which deuterium can substitute for hydrogen on the nitrogen of the peptide bond. This substitution is slower if the hydrogen is also bonded to a carbonyl group. In the best method one measures the variation of the optical rotatory power of the protein with the wavelength of the light used. The parameters obtained from such experiments can be compared to those obtained from solutions of synthetic polypeptides known to exist in the alpha or beta forms and hence a measure of the degree and nature of the helix formation can be obtained.

TERTIARY STRUCTURE

The tertiary structure of a protein is defined as its total three-dimensional structure; that is the coiling of the long peptide chain with or without a helical region into the final compact structure in the case of the globular proteins. Although it is possible to tell by light scattering or by sedimentation and viscosity measurements the outline of the tertiary structure of a protein, the only way to determine the detailed tertiary folds is by the technique of X-ray crystallography. Thus hydrodynamic techniques give the information that ribonuclease is shaped more like an orange than a cigar, but X-ray crystallography can go into greater detail and, for example, can define the exact position of residue histidine number 119 in relation to the other amino acid residues in ribonuclease. The technique requires that the protein under investigation is pure and can form crystals. A narrow beam of X-rays is deflected from the crystals and, since the areas of the protein most dense in electrons (say tryptophan residue) scatter X-rays to the greatest extent, an electron density map of the protein in three dimensions can be built up. At good resolution every amino acid residue can be identified. So far most of the proteins

studied by this technique have been pigments such as myoglobin (Fig. 3.12) or small hydrolytic enzymes such as lysozyme, ribonuclease and chymotrypsin but work is now proceeding on the more complex enzymes.

Three main types of bond, hydrogen, ionic and hydrophobic, are responsible for the formation of the tertiary structure of a protein (Plate 3.2). The disulphide bond between two halves of a cystine residue could be said to form a fourth category. This bond is by definition included in primary structure since it is covalent, but it is more usually thought of as a strong bond maintaining the tertiary structure of a protein. But, since some proteins have no disulphide bonds at all, such links cannot be assigned a major general role.

Hydrogen bonds between side chains can be formed by a number of residues. Possible donors are the hydroxyl groups on serine, threonine or tyrosine or amido groups on histidine, tryptophan or the peptide backbone. Possible acceptors are the oxygen atoms on the aspartic and glutamic acid residues or carbonyl groups on the peptide chain.

Ionic bonds are formed between the basic and acidic amino acid residues as would be expected. They were originally thought to be one of the primary forces in the maintenance of tertiary structure but crystallographic studies show that the majority of these residues tend to remain on the exterior of the molecule so that, although they are strong bonds, comparatively few of them are involved in bonding within protein molecules. They are, however, of considerable importance in maintaining bonds between protein molecules. This can be readily seen from the fact that protein molecules behave as electrolytes in solution; the bonds between them can be broken or formed by variations in the salt concentration, and the solubility properties of proteins are largely dependent on the number and nature of the ionic residues.

The formation of tertiary structures of proteins is believed to be very dependent on the hydrophobic bonds. This dependence is largely due to the fact that the hydrophobic, or hydrocarbon-like side chains, tend to 'prefer' to be in the interior of the protein molecule where there is less water. The phenomenon can be illustrated by putting a few drops of benzene on the surface of water in a beaker.

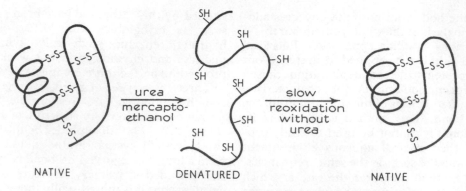

Fig. 3.9 In the controlled denaturation of a protein (ribonuclease) in the presence of urea and mercaptoethanol, the disulphide linkages break and the polypeptide chain uncoils. Slow re-oxidation without urea allows the original tertiary structure to be re-formed.

The drops tend to coalesce, not because they have great affinity for each other since the Van der Waals forces between them are relatively weak, but because they have in common a lack of affinity for water and are pushed together out of the network of water molecules strongly bound together by internal hydrogen bonds. This is believed to apply to protein molecules since both crystallographic and solubility evidence suggests that the apolar type of amino acid residues are in the centre of the protein. Hence most proteins are relatively insoluble in organic solvents.

None of the four types of bond concerned in the maintenance of protein structure can be said to be of greater importance than the others; their combined effect achieves the final structure of the protein. This final three-dimensional structure is probably thermodynamically the most stable, as shown by the ability of many proteins to refold to their 'native' or normal biological structure after they have been gently induced to unfold to a random chain by mild chemical treatment. The example of ribonuclease shown in Fig. 3.9 illustrates the process.

This unfolding process is called denaturation. The secondary and tertiary organization is completely lost without breaking the primary structure. Gentle treatment with urea leads to slow denaturation which is frequently reversible as shown above but violent treatment of proteins with acid or alkali, heat, organic solvents, or ultraviolet light, may produce totally irreversible changes. The protein alters completely in physical properties and becomes much less soluble at its isoelectric point. It also loses all its biological activity. A typical example of irreversible denaturation occurs in the boiling of an egg (Fig. 3.10).

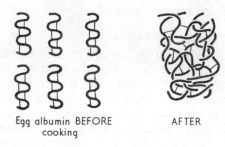

Egg albumin BEFORE cooking   AFTER

Fig. 3.10 In uncontrolled denaturation such as takes place during the boiling of an egg, the tertiary structure of the proteins is irreversibly destroyed with the formation of a disorganized mass of polypeptide chains.

QUATERNARY STRUCTURE

Quaternary structure involves the non-covalent association of one or more peptide chains. Quaternary structure can be divided into two groups: one contains proteins with several dissimilar peptide chains but one active site, and the other contains proteins with several similar or identical subunits each with its own active site. An example of the first class is aspartate carbamoyl transferase (see Chap. 20) which has two types of subunits, one involved in the actual activity of the enzyme and the other in modifying this activity according to conditions in the cell. A familiar protein in the second group is haemoglobin which has four similar subunits each able to carry an oxygen molecule. This association of similar subunits is the more common form of quaternary structure and it is believed that many proteins can exist in either monomeric or multimeric form according to conditions prevailing in the cell; the catalytic activities of the two forms are slightly different so that by controlling the interactions between monomers the cell can in fact control the catalytic activity of an enzyme.

The forces maintaining the quaternary structure are similar to those involved in tertiary structure but the association of the subunits is in general more flexible. Small molecules can influence quaternary structure by their association with one or more of the protein chains. Thus, when oxygen is bound to haemoglobin the chains move further apart from each other and when the oxygen is given up the chains move back again.

Variations in quaternary structure. Variations in the units involved in quaternary structure are important in relation to isoenzymes. These are enzymes which exist in different forms according to their tissue source within the same species (p. 116). Thus in the chicken lactate dehydrogenase from the heart exists in one form and in voluntary muscle it exists in a different form. Other tissues appear to have these two types in varying proportions and various other types intermediate between the two. It has been discovered that two different types of subunit (H and M) are associated in quaternary structure in varying amounts according to the tissues of origin. Thus in heart the enzyme is composed of four subunits HHHH, in muscle of four MMMM but the three intermediate structures HHHM, HHMM, HMMM are also found. Hence on electrophoresis of the mixture, 5 bands are obtained, one for each type. This difference has clear biological significance since the enzyme in heart muscle works fairly slowly so that the lactic acid does not rise to dangerous levels, whereas that from voluntary muscle is more suited to the relatively anaerobic conditions which can be tolerated by this tissue.

The proteins can be divided into three groups according to their function. The fibrous proteins have already been mentioned (p. 34) and the enzymes will be discussed in Chapter 8. The third group are the immunoglobulins responsible for antibody activity. Like enzymes these show great specificity in their action and can differentiate between proteins that cannot be distinguished by physical or chemical means. If a rabbit, for instance, is given repeated small injections of a 'foreign' protein, called an antigen, then an antibody of the foreign protein is produced by the rabbit and circulates in its blood. When the serum of the rabbit is added to a solution containing the antigen it reacts with it in an observable way. Antigen–antibody reactions are valuable in the diagnosis of infections; thus the serum of a patient suffering from typhoid fever contains specific antibodies which clump (agglutinate) the typhoid bacillus. Other antigen–antibody tests in which the antibody precipitates a soluble antigen from solution are useful in medico-legal work, for example, in tracing the species from which a sample of blood has come.

The proteins carrying the antibody activity are the immunoglobulins. There are five classes of immunoglobulins designated M, G, A, D, and E; M, G, A, and E show antibody activity. Immunoglobulin G (IgG), the major immunoglobulin found in mammals, has been studied more extensively than the others. It is a protein of molecular weight about 150 000 consisting of four polypeptide chains, two identical heavy chains (H-chains) each having a molecular weight of about 50 000 and two identical light chains (L-chains) each with a molecular weight of about 25 000. The chains are held together by disulphide bonds (Fig. 3.11). Each heavy chain has a carbohydrate component.

There are two binding sites for antigen at the amino terminal ends of the chains and each binding site is made up of the amino-terminal segment of one H- and one L-chain. The amino terminal segments made up of about 107 residues in each chain (shaded line in Fig. 3.11) are variable in composition whereas the remainder (solid line in Fig. 3.11) of both types of chain is constant in amino acid sequence. The structure of the IgG molecule has been investigated by biochemical methods and recently electron microscopy of the molecule has confirmed the shape shown in Fig. 3.11 overleaf.

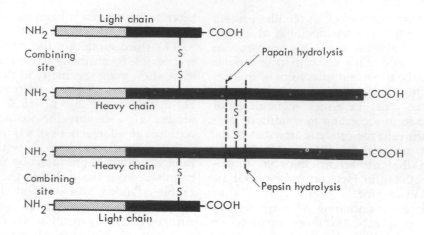

Fig. 3.11 Diagrammatic representation of an immunoglobulin molecule. For explanation see text. The position and number of both inter- and intra-chain disulphide bonds varies with the different species and classes of immunoglobulins.

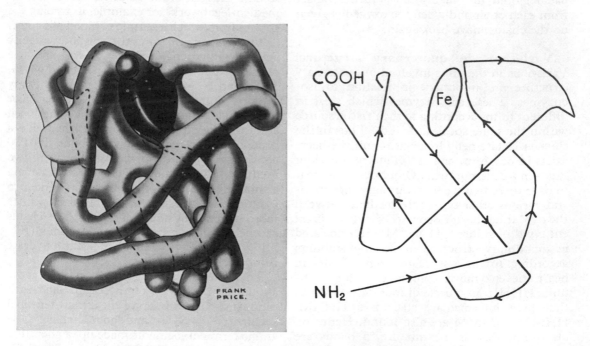

Fig. 3.12a   Fig. 3.12b

Fig. 3.12 (a) Drawing of the tertiary structure of myoglobin. (b) The course of the polypeptide chain.

J. C. Kendrew, R. E. Dickerson, B. E. Strandberg, R. G. Hart, D. R. Davies, D. C. Phillips, and V. C. Shore, 1960) Structure of Myoglobin, *Nature*, **185**, 422.

(By courtesy of J. C. Kendrew.)

REFERENCES

ANFINSON, C. B. (1965–66). The formation of the tertiary structure of proteins. *Harvey Lectures* **61**, 95–112.

ANFINSEN, C. B. (1960). *The Molecular Basis of Evolution.* New York: Wiley.

DAYHOFF, M. O. & ECK, R. V. (1968). *Atlas of protein sequence and structure.* National Biochemical Research Foundation, Maryland, U.S.A.

DICKERSON, R. E. & GEIS, T. (1969). *The Structure and Action of Proteins.* New York: Harper & Row.

EDELMAN, G. M. (1970). Structure and function of antibodies. *Scientific American* **223(2)**, 34–42.

HALL, D. A. (1961). *The Chemistry of Connective Tissue.* Springfield, Ill.: Thomas.

HARTLEY, B. S. (1970). Strategy and tactics in protein chemistry. *Biochemical Journal* **119**, 805–822.

HAUROWITZ, F. (1968). *Immunochemistry and the Biosynthesis of Antibodies.* New York: Interscience.

HUMPHREY, J. H. & WHITE, R. G. (1970). *Immunology In Students of Medicine,* 3rd edn. Oxford: Blackwell.

JUKES, T. H. (1966). *Molecules and Evolution.* New York: Columbia University Press.

KENDREW, J. C., DICKERSON, R. E., STRANDBERG, B. E., HART, R. G., DAVIES, D. R., PHILLIPS, D. C. & SHORE, V. C. (1960). The structure of myoglobin. *Nature (Lond.),* **185**, 422–427.

KOPPLE, K. D. (1966). *Peptides and Amino Acids.* New York: Benjamin.

MEISTER, A. H. (1965). *Biochemistry of the Amino Acids,* 2nd edn (2 Vols.). New York: Academic Press.

NEEDLEMAN, S. B. (Ed.) (1970). *Protein Sequence Determination.* London: Chapman & Hall.

NEURATH, H. (1963). *The Proteins: Composition, Structure and Function,* 2nd edn, Vol. I; (1964) Vol. II; (1965) Vol. III; (1966) Vol. IV. New York: Academic Press.

PORTER, R. R. (1967). The structure of immunoglobulins. In *Essays in Biochemistry* (Edited by P. N. Campbell and D. G. Greville) **3**, 1–24.

SANGER, F. (1964). The chemistry of insulin. In *Nobel Lectures. Chemistry.* 1942–1962, pp. 544–556. Amsterdam: Elsevier.

SCHROEDER, W. A. (1968). *The Primary Structure of Proteins.* New York: Harper & Row.

4 Carbohydrates

The carbohydrates are usually defined as substances containing carbon, hydrogen and oxygen, the hydrogen and oxygen being present in the same ratio as in water. Their general formula is, therefore, $C_m(H_2O)_n$. Although satisfactory for most purposes this definition is not strictly accurate, since in a few carbohydrates the proportion of oxygen is lower than that required by the general formula.

The easiest approach to the study of their structure is to consider all carbohydrates as either hydroxy-aldehydes or hydroxy-ketones. The simpler carbohydrates are known as sugars. Those containing an aldehyde group are known as aldoses, the simplest of these being hydroxy-acetaldehyde. The aldose sugar containing three carbon atoms is glyceraldehyde or glycerose.

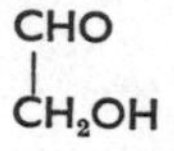

CHO
|
CH₂OH

hydroxy-
acetaldehyde

CHO
|
CHOH
|
CH₂OH

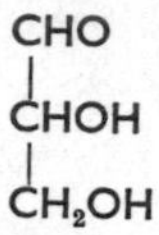

glycerose
(glyceraldehyde)

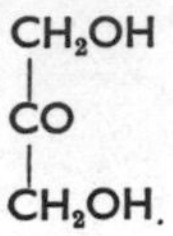

CH₂OH
|
CO
|
CH₂OH.

dihydroxy-
acetone

Ketoses are sugars containing a ketone grouping instead of the aldehyde grouping and the simplest of these is dihydroxy-acetone.

Glycerose and dihydroxy-acetone both contain three carbon atoms and are therefore termed trioses. Tetroses contain four carbon atoms, pentoses contain five, and hexoses contain six carbon atoms. Such simple sugars are known as *monosaccharides*.

When two monosaccharide units condense together with the elimination of water, a *disaccharide* is formed; when three condense together the product is a *trisaccharide*; while the product formed by the condensation of a large number of monosaccharide units is known as a *polysaccharide*.

MONOSACCHARIDES

The simplest hexose in the aldose series (aldohexose) might be expected to have the following structure in which the six carbon atoms form a chain with the aldehyde group at the end.

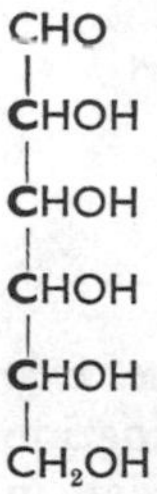

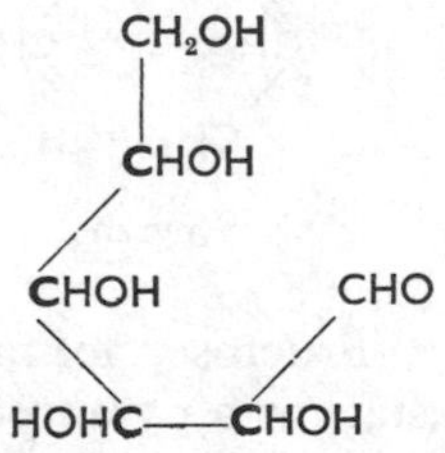

The reason for folding the chain as shown on the right will appear later.

Since there are four asymmetric carbon atoms in this structure (shown in bold type) the aldohexoses are optically active and occur in several sterioisomeric forms. Thus the aldohexoses include such sugars as glucose, mannose and galactose, each of which can exist in a dextro and a laevo form. Some of the properties of the hexoses, however, are inconsistent with this simple aldehyde structure which, for example, is insufficient to explain *mutarotation*.

If ordinary glucose is dissolved in water, the solution has a specific rotation of $(\alpha)_D +111°$ but, when the solution is allowed to stand, the rotation falls slowly to $+52·5°$. It is also possible to prepare glucose with an initial rotation of $+19°$ which rises to $+52·5°$ when the solution is allowed to stand. It seems, therefore, that glucose can exist in two forms. The change in rotation to a common value, which occurs when either of these two forms is allowed to stand in solution, is known as *mutarotation*. Each form can be prepared separately in solid state by special methods of crystallization.

To explain this phenomenon and to account for the unduly low reactivity of the aldehyde group in glucose, it is accepted that the carbon atom of the aldehyde group is linked by means of an oxygen atom to the fifth carbon atom in the chain, so that hexoses have a ring structure thus:

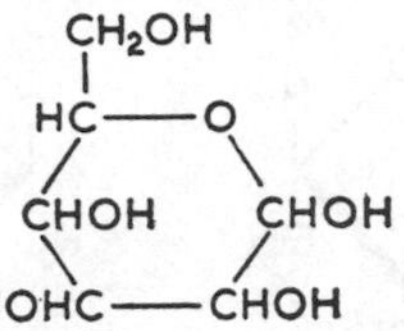

This representation of the hexose molecule in one plane is slightly misleading. If a model is made to show the relationship of the various atoms in space it is found that five of the carbon atoms and the oxygen atom of the ring form a hexagon, with the other atoms and groups arranged either above or below the plane of the hexagon. In the formulae shown below, this plane is supposed to be at right angles to the page with the forward edge indicated by the heavy line.

D-Glucose, for example, can be written in the following way:

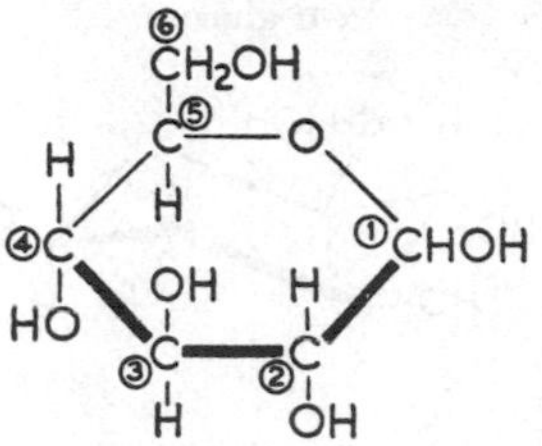

or more briefly

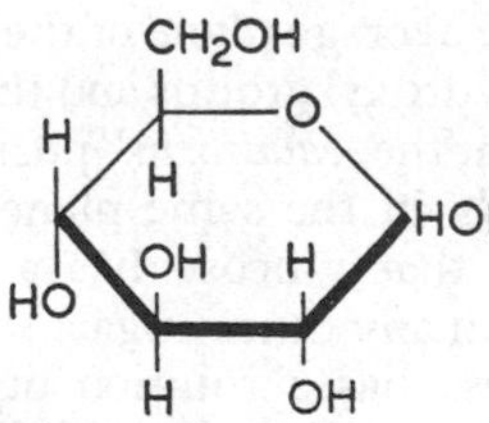

It is to be noted that carbon atom number one, which is the carbon atom associated with the aldehyde group, is now asymmetric. D-Glucose can therefore exist in two forms: one with the hydroxyl group on this carbon atom below the plane of the hexagon and the other with the hydroxyl group above the plane of the hexagon. These two forms, known as α- and β-glucose, have the following structures:

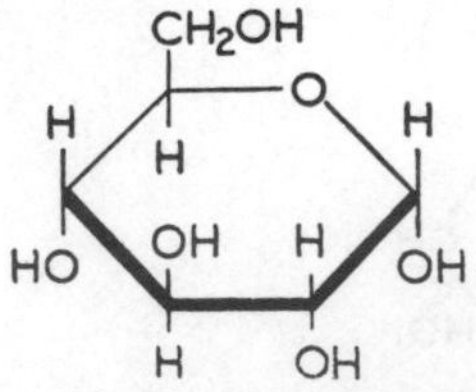

α-D-glucose 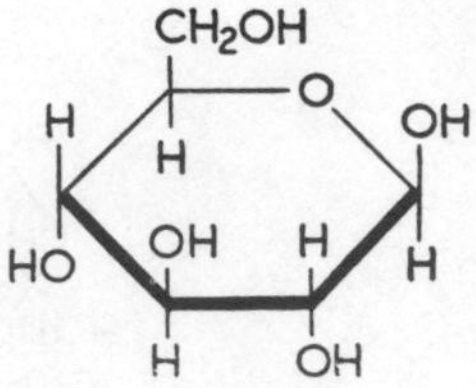   β-D-glucose

sugars containing a six-membered ring system are sometimes known as *pyranose* sugars. Sugars can also exist in the form of a five-membered ring and may therefore be formulated as derivatives of the parent substance furan. Such sugars are said to have the *furanose* structure.

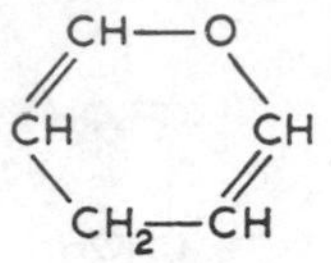

pyran   furan

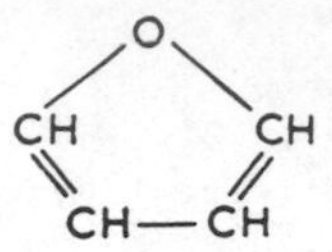

While this representation is commonly employed to illustrate the structure of the glucose molecule, it is not altogether accurate, for the pyranose ring is not flat with the five carbon atoms and the oxygen atom exactly in the same plane, but 'puckered' as shown in the diagram.

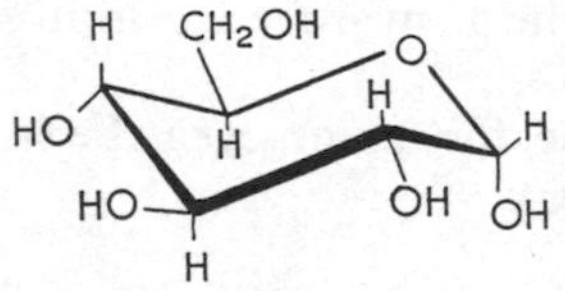

α-D-glucose

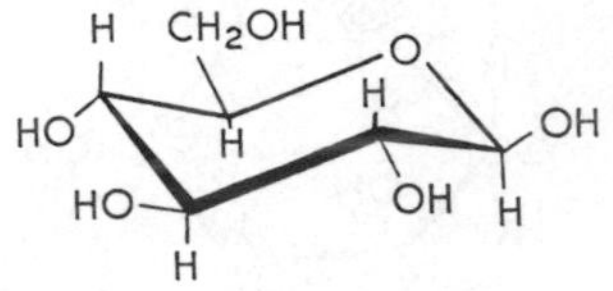

β-D-glucose

Fructose, for example, exists in the free state as a pyranose sugar, but in combination with glucose as cane sugar (sucrose) it exists in the furanose state (fructofuranose) (p. 44).

Glucose. Glucose (dextrose, grape sugar) is a white, crystalline solid easily soluble in water. Like all sugars it has a sweet taste. It occurs in the animal body and is present in human blood in a concentration of 1 g/l. Glucose also occurs in nature in the combined form as polysaccharides which on hydrolysis yield free glucose. For example glycogen, or animal starch, which is found in the liver and in the muscles, is built up from glucose units and so are the plant polysaccharides starch and cellulose.

Glucose belongs to the 'D' series of sugars; these are sugars in which the configuration at the carbon atom next to that carrying the primary alcohol group is the same as in D-glyceraldehyde:

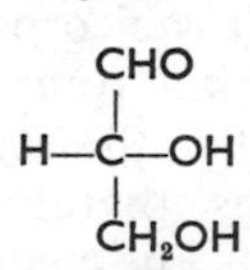

In glucose, but in no other sugar, all five hydrogen atoms attached to the carbon atoms of the ring, are in the *axial* position, perpendicular to the average plane of the ring, whereas the four hydroxyl groups and the —CH_2OH group are in the *equatorial* position, that is approximately in the same plane as the ring. This means that glucose has a more stable structure than any other sugar.

α-D-Glucose has a rotation of +111° and β-D-glucose a rotation of +19°. When a solution of either form is allowed to stand an equilibrium mixture of the two forms with a rotation of +52·5° eventually results. Because the equilibrium mixture contains traces of the open-chain aldehyde form (see p. 41), a solution of glucose gives certain of the reactions characteristic of aldehydes.

When the formula of glucose is written as a six-membered ring it is obvious that glucose is a derivative of the substance pyran; hence

D-glyceraldehyde 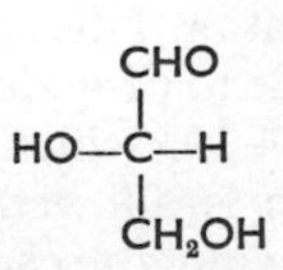   L-glyceraldehyde

Sugars of the 'L' series have the configuration at this carbon atom corresponding to that in L-glycerose. The correct name for glucose is therefore D (+) glucose, the letter 'D' indicating the dextro series, and the (+) sign indicating a dextro rotation. (It should be noted that all sugars of the 'D' series are not necessarily dextro-rotatory.)

Since glucose is a potential aldehyde it is a reducing agent, capable of reducing cupric compounds to the cuprous state, and potassium ferricyanide to ferrocyanide. These reducing properties are made use of in several common methods of detecting and estimating glucose in biological fluids.

It has been known for a very long time that glucose is fermented by yeast to yield alcohol and carbon dioxide. This reaction helps to characterize glucose, since some sugars, for example lactose, are not fermented by yeast.

Glucose, being an alcohol, forms esters with acids; for example phosphoric acid can react either with carbon atom number one or carbon atom number six in the glucose molecule to yield glucose 1-phosphate and glucose 6-phosphate. These phosphoric esters of glucose, as we shall see later, are of the utmost importance in carbohydrate metabolism.

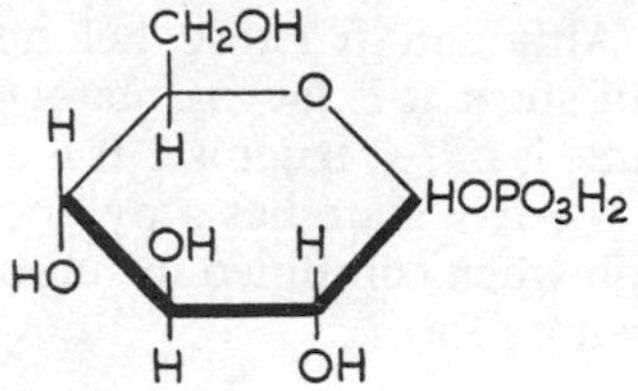

glucose 1-phosphate

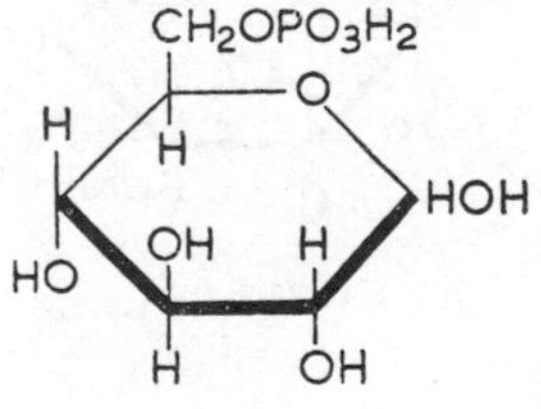

glucose 6-phosphate

Carbon atom number one in the glucose molecule is reactive, and the hydrogen atom of its hydroxyl group may be replaced by other radicals, with the formation of compounds known as *glucosides*. For example, it may be replaced by a methyl group with the formation of methyl-glucosides. There are of course two glucosides, α-methylglucoside and β-methylglucoside, corresponding to the α- and β-forms of glucose.

The general term for sugar derivatives of this sort is *glycoside*. Glycosides formed from glucose are *glucosides*, those from galactose are *galactosides*, and so on.

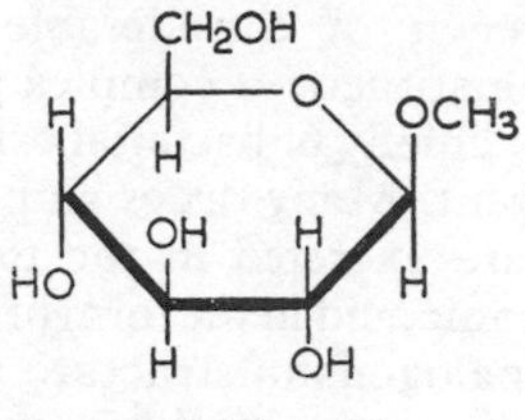

β-methylglucoside

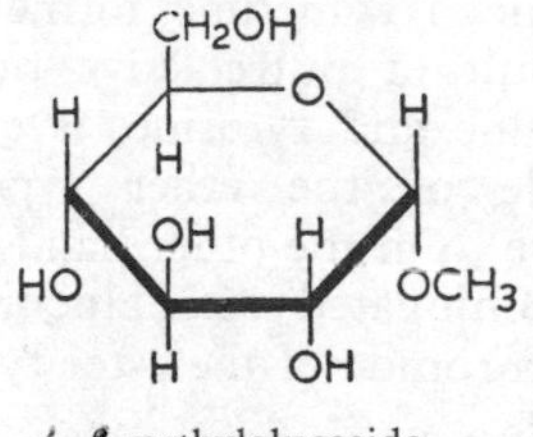

α-methylglucoside

The glycosides are important compounds. Many complex glycosides occur in nature, providing materials of great pharmacological value, such as digoxin which is extensively used in medicine for its action on the heart.

Carbon atom number one in the glucose molecule may be oxidized to a carboxyl group. When such an oxidation takes place in any hexose the resulting acid is known as *hexonic acid*, and the acid formed from glucose is known as *gluconic acid*. The glucose molecule may, however, also be oxidized at carbon atom six. When a carboxyl group is produced at this carbon atom in any hexose the resulting acid is known as a *uronic acid*, or *hexuronic acid*, and the uronic acid produced from glucose is *glucuronic acid*.

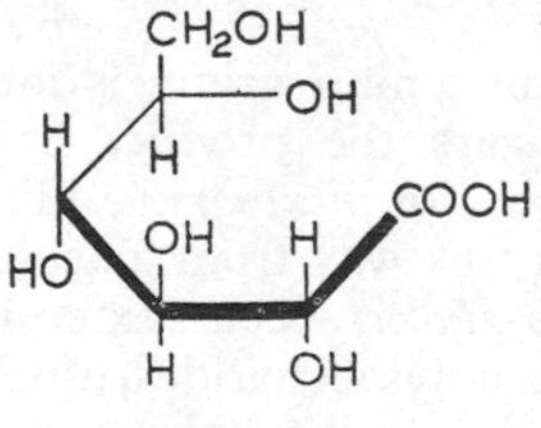

gluconic acid

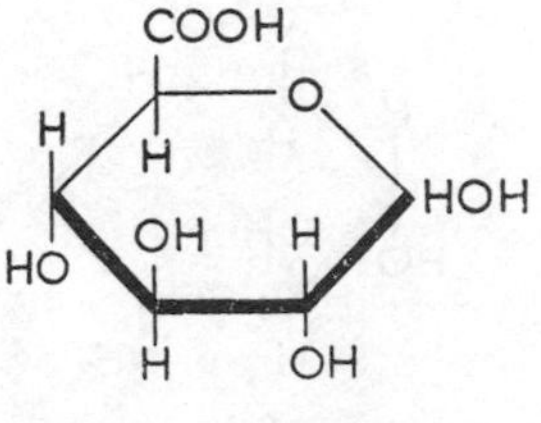

glucuronic acid

The latter is of considerable importance both as a constituent of complex polysaccharides (p. 48), chiefly of bacterial origin, and as a coupling agent. Many drugs and a number of hormones are excreted in the urine coupled with glucuronic acid in the form of *glucuronides*. They are analogous in structure to the glycosides and can exist in α and β forms. All urinary glucuronides are of the β type. For example phenol, which is sometimes formed in the body as the result of putrefactive breakdown of phenylalanine and tyrosine, is excreted as a glucuronide of the ether type (β-phenyl glucuronide). On the other hand benzoic acid becomes conjugated with glucuronic acid to give a glucuronide of the ester type (benzoylglucuronide).

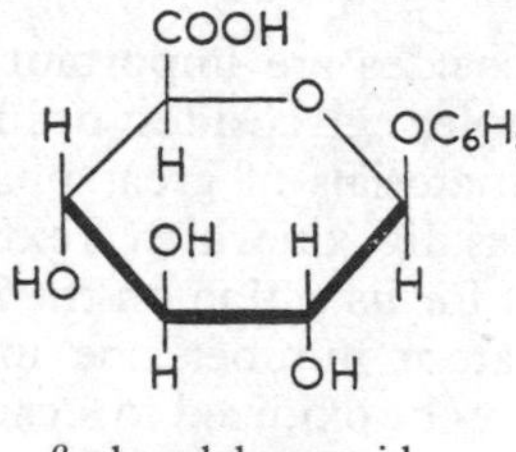

β-phenylglucuronide

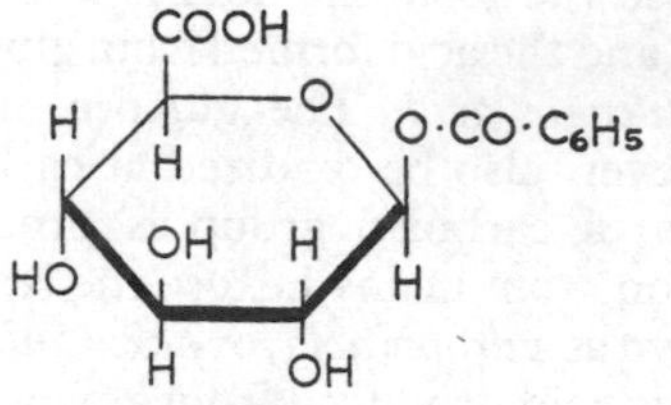

benzoylglucuronide

When an amino group is introduced into hexose sugars the product is known as an *amino sugar*, or *hexosamine*. The compound formed in this way from glucose, *glucosamine* or *2-amino-glucose*, occurs extensively in nature in complex polysaccharides, usually in the form of its acetyl derivative N-acetylglucosamine.

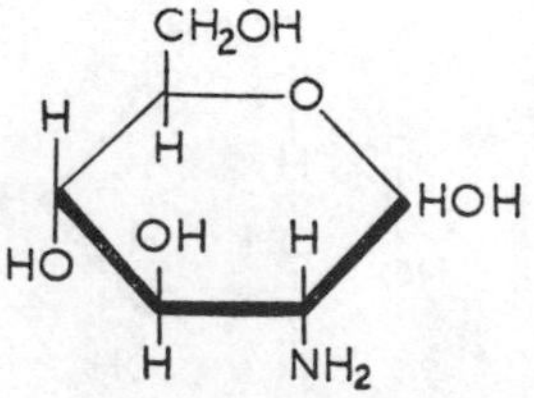

glucosamine

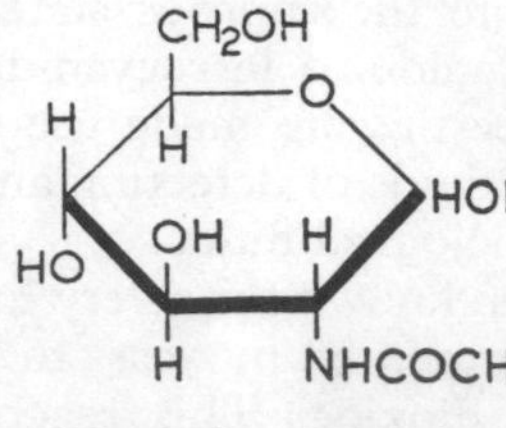

N-acetylgucosamine

Other monosaccharides. *Galactose* occurs in nature in the combined form as a constituent of lactose, in certain complex lipids and in some proteins. It is a dextro-rotatory stereoisomer of glucose. *Mannose* occurs in nature in the form of complex polysaccharides. *Fructose*, sometimes also called *laevulose* or *fruit sugar*, occurs naturally in fruits and honey. It is present in the free state in the seminal plasma and in the blood of the fetus of ruminants. Although fructose belongs to the 'D' series of sugar it is laevo-rotatory and its correct name is D (−) fructose. It is a ketose sugar and in the free state has a pyranose structure, although when combined in sucrose it exists in the furanose form.

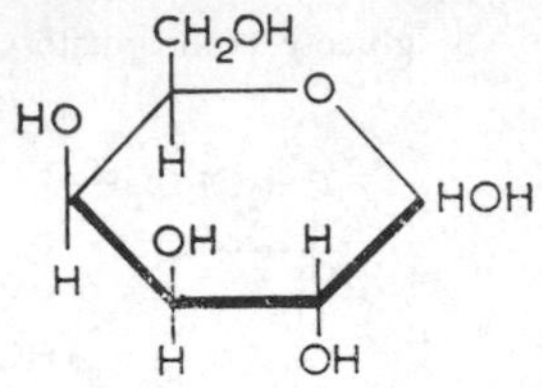

D-galactose

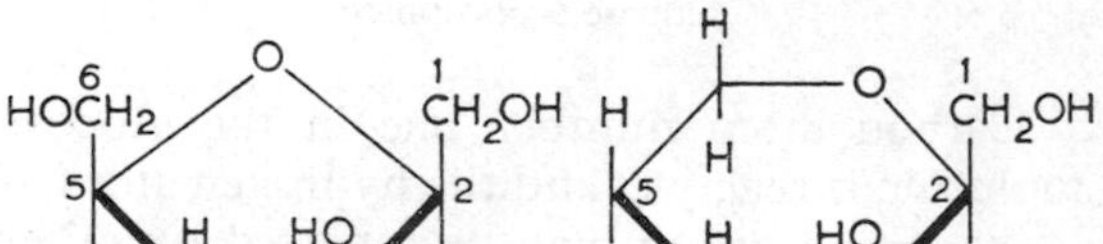

α-D-fructose

furanose form   pyranose form

The *pentoses* are monosaccharides with only five carbon atoms in the molecule. The most important pentose is *ribose*, found in the furanose form in the ribonucleotides and in RNA (Chap. 6).

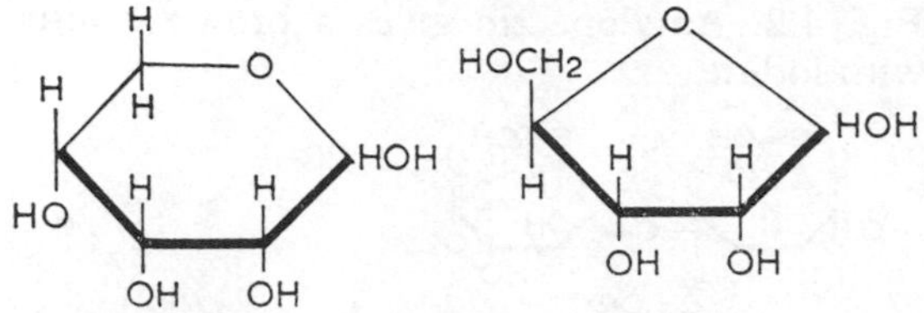

H
H
O
H
H
HO
H
H
H
OH OH

D-ribose

pyranose form

HOCH₂
O
H
H
HOH
OH OH

furanose form

CHOH
CHOH CHOH
CHOH CHOH
CHOH

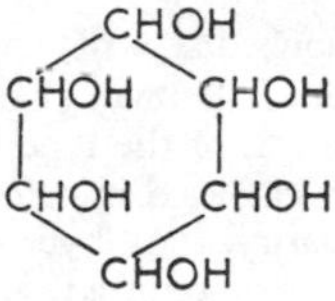

inositol

Inositol (hexahydroxycyclohexane), although not strictly a sugar, is worth mentioning here. It is found in liver, heart and brain; in the last it is a constituent of certain complex phospholipids (Chap. 5). In plants it occurs chiefly in the form of its hexaphosphate, *phytic acid*.

Disaccharides

The disaccharides consist of two monosaccharide molecules condensed together with the elimination of water. The three important disaccharides are *maltose, lactose,* and *sucrose.* Maltose is composed of two molecules of glucose, lactose of one molecule of glucose and one of galactose, and sucrose of one molecule of glucose and one of fructose.

Maltose does not occur free in the body but it is important as an intermediate stage in the breakdown of starch to glucose. It consists of two molecules of glucose condensed together by an α-linkage. The reducing group of one of the glucose molecules is involved in the linkage but, since the reducing group of the other is free, maltose is a reducing sugar. It forms a characteristic osazone.

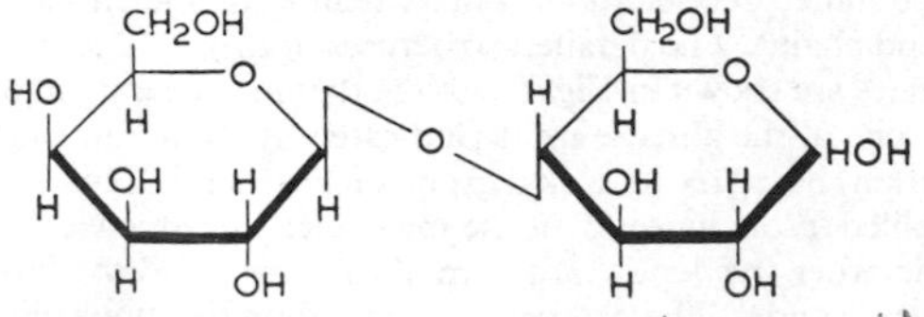

CH₂OH
HO
H
OH H
H
H OH

β-galactose residue   ~~maltose~~
~~glucose-α-glucoside~~

CH₂OH
H
H
OH H
HOH
H OH

glucose residue

lactose
(**β**-galactosido - glucose)

Lactose, or milk sugar, is found in milk and is synthesized in the mammary gland. On hydrolysis it yields one molecule of glucose and one of galactose. It is a β-galactoside and its structure is that of β-galactosido-glucose. Like maltose it has reducing properties.

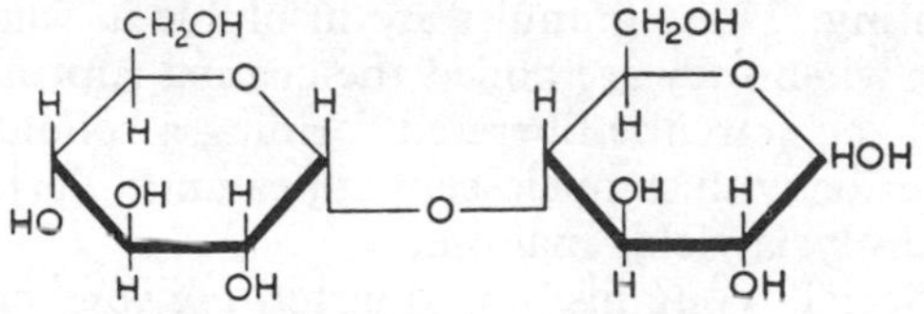

CH₂OH
H
H
HO OH H
H OH

glucose residue

CH₂OH
H
H
OH H
HOH
H OH

glucose
~~β-galactose~~ residue

~~lactose~~
maltose
(glucose - α - glucoside)

Sucrose (cane sugar or *beet sugar)* occurs naturally in certain plants and is one of the important carbohydrates of the diet. On hydrolysis it yields one molecule of glucose and one of fructose. Since the reducing groups of both the glucose and the fructose are involved in the linkage, sucrose has no reducing properties. When combined in sucrose, fructose exists in the furanose form. Sucrose is dextrorotatory but the mixture of glucose and fructose produced by hydrolysis is laevo-rotatory, since fructose is more laevo-rotatory than glucose is dextro-rotatory. Hence the rotation is changed from positive to negative during hydrolysis, a phenomenon sometimes referred to as 'inversion'.

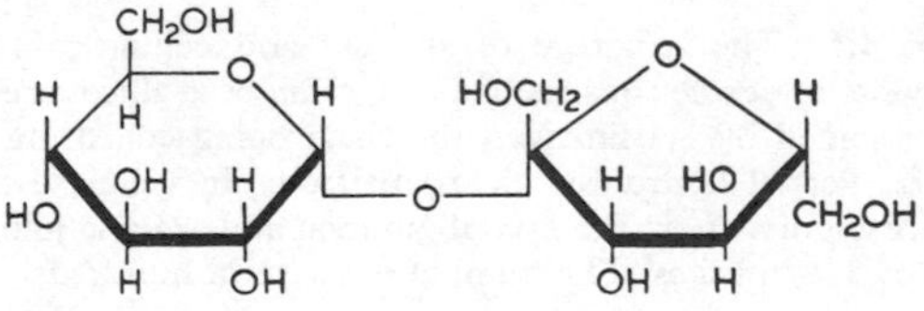

CH₂OH
H
H
HO OH H
H OH

α-glucose residue

HOCH₂
O
H
H HO
OH H
CH₂OH

fructofuranose residue

sucrose

Polysaccharides

Polysaccharides are formed by the condensation of large numbers of monosaccharide units which are joined together with the elimination of water in much the same way as the amino acids are joined together to form proteins (p. 27). Like the proteins, the polysaccharides have high molecular weights and are not usually

Carbohydrates 45

soluble in water though some may form colloidal solutions.

The three chief polysaccharides, starch, glycogen and cellulose, are built up solely from glucose units.

Starch is the form in which carbohydrate is stored in the plant. It occurs in the potato in the form of granules with a thin cellulose coating. These granules are insoluble in water but when they are boiled the coating ruptures and the starch is liberated forming a colloidal solution with an opalescent appearance. Partial hydrolysis yields maltose.

Starch contains one-quarter *amylose* and three-quarters *amylopectin*. Amylose consists of long unbranched chains (Fig. 4.1) of about

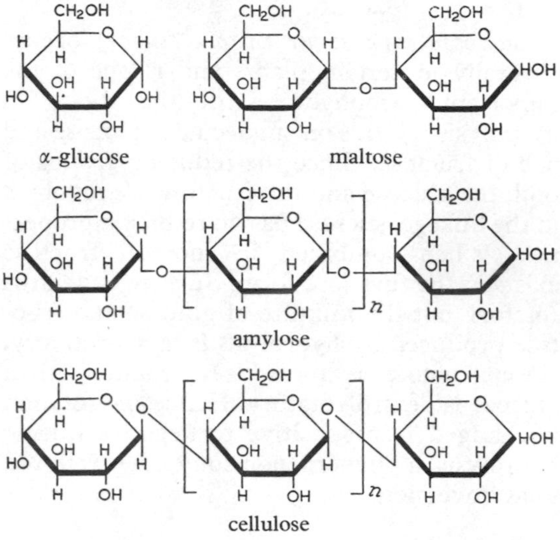

Fig. 4.2. Amylopectin gives a brown colour with iodine.

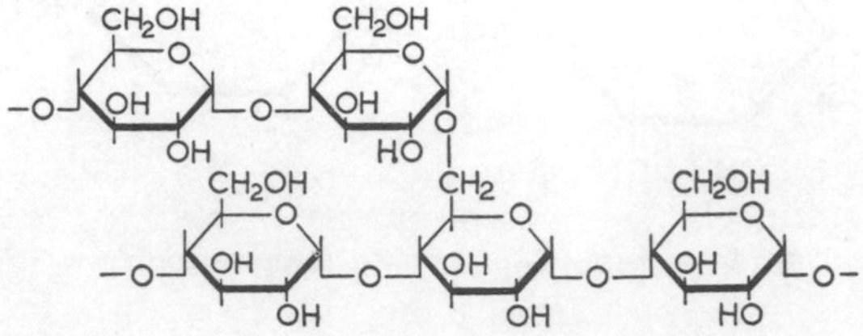

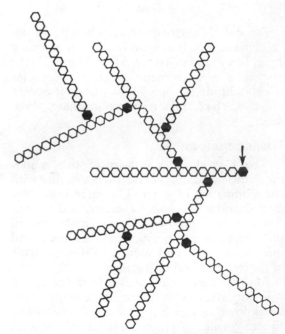

Fig. 4.2   The branched type of structure found in amylopectin and glycogen. The main chain, shown in the lower part of this diagram, is of the type found in amylose; the centre glucose unit carries a branch attached at carbon 6. For the sake of clarity, lone hydrogen atoms have been omitted.

Fig. 4.1   The structure of amylose and cellulose. The amylose molecule consists of a long chain of α-glucose residues joined by 1,4-linkages, the chain being coiled into a helix. Partial hydrolysis yields maltrose. In cellulose the glucose units are in the β-configuration and are also joined by 1,4-linkages. The number n is several hundreds.

300 to 400 α-glucose units forming a helical structure. It gives an intense blue colour with iodine. Amylopectin, on the other hand, contains more than 1000 glucose residues in a highly branched structure with about 24 to 30 units per branch. While most of the glucose units are joined by α-linkages between carbon 1 of one unit and carbon 4 of the next, the branches start as α-linkages between carbon 1 of one unit and carbon 6 of the unit in the chain from which the branch originates as shown in

Fig. 4.3   Diagrammatic representation of the structure of a glycogen molecule. Each hexagon represents one glucose unit. These form chains (on the average, 12 units long) in which each unit is attached through its own position 1 to position 4 of the unit in front of it. Branch points are formed where the unit at the 'head' of such a chain (indicated here by a solid black hexagon) is attached through its own position 1 to position 6 of a unit in the middle of another chain (to be more precise, about 4 units behind the 'head' of the second chain). The detailed structures of these 1, 4 and 1, 6 linkages are shown in Fig. 4.2. Note that in such a structure only one of the glucose units (indicated by the arrow in the diagram) has a free hydroxyl group on position 1. This is the so-called 'reducing-end' of the molecule. The glucose units at the other free 'ends' of the molecule are called the 'non-reducing ends'. The processes of extending the molecule or of breaking it down always start from the non-reducing ends (see Chap. 11).

Glycogen is the form in which carbohydrate is stored in the animal body. Glucose is readily converted into glycogen in the liver and muscles and kept there until required, when it may be broken down again into glucose. Glycogen dissolves fairly readily in water to yield an opalescent solution which gives a reddish colour with iodine. The glycogen molecule resembles amylopectin in containing many glucose units joined by α-linkages between carbon 1 of one unit and carbon 4 of the next with branches originating here and there involving 1,6-linkages (Fig. 4.2), but the ramifications are more extensive and each branch contains only about 12 glucose units (Fig. 4.3). The molecular weight of glycogen is of the order of several million.

Cellulose is a very stable polysaccharide which forms the supporting tissues of the plant. It is insoluble in water and gives no colour with iodine. The cellulose molecule consists of very long chains of glucose units in the β-configuration joined together by 1,4-linkages (Fig. 4.1).

If the glucose molecule is represented in the conformation shown on p. 42 with the four hydroxyl groups and the —CH_2OH group on C-6 in the equatorial position, the cellulose molecule (with the substituents omitted for clarity) would appear, side-on, thus:

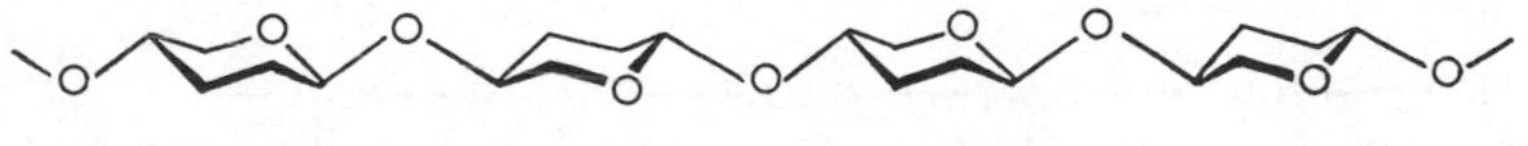

In plan (including substituents), it would look like this:

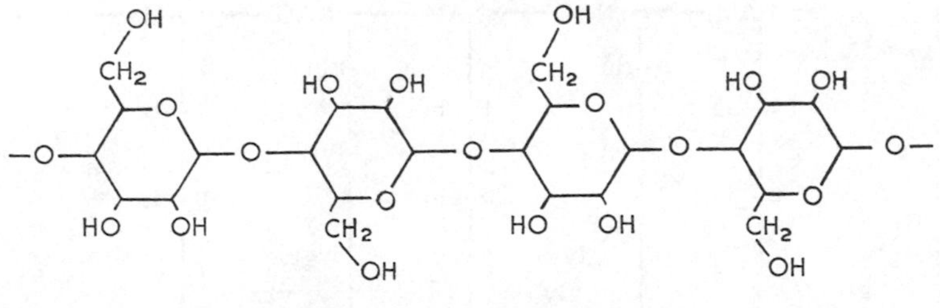

Although ingested in considerable quantities, it is not digested to any significant extent by man since his digestive tract does not contain the enzymes which attack β-linkages. Herbivorous animals, however, are able to make use of cellulose since the bacteria and protozoa in the rumen or colon convert it, not to glucose, but to smaller fragments such as short-chain fatty acids, carbon dioxide and methane.

Dextrins are intermediate between starch and maltose. They form an ill-defined group of substances with very feeble reducing properties. Dextrins of high molecular weight, such as amylodextrin, give a blue colour with iodine. Erythrodextrin gives a reddish-brown colour but achroodextrin which has a still smaller molecule, gives no colour with iodine. The dextrins form sticky solutions used as gums.

The polysaccharide *inulin* is composed entirely of fructose units. It is found in artichokes and in dahlia tubers and is used in a test of renal function (Chap. 31).

MUCOPOLYSACCHARIDES AND GLYCO-PROTEINS

The *mucopolysaccharides* are substances of complex macromolecular structure and great biological importance built up out of units which include amino sugars and uronic acids. They are acidic in nature and may be rich in sulphate ester groups. In nature they may occur in association with peptide groups (p. 28), the peptide and carbohydrate portions being held together by ionic linkages as well as by covalent bonds.

In the *glycoproteins*, on the other hand, the carbohydrate and peptide portions are linked by covalent bonds only. They contain no uronic acids and only a few contain sulphate esters. The glycoproteins include some of the important plasma proteins (Chap. 23) and the specific blood group substances (Chap. 24) as well as some of the proteins (sometimes termed mucoproteins) characteristic of the submaxillary gland, of the intestinal tract and of bone and connective tissue.

While the polysaccharide moiety of most mucopolysaccharides and glycoproteins may be very complex and contain more than one type of unit, a few cases are known in which a relatively simple pattern prevails. For example, the *polyuronides* are composed of uronic acid units and are mainly of plant and bacterial origin. They include many gums, such as gum arabic and gum tragacanth. The specific polysaccharides elaborated by certain type of bacteria also contain uronic acids; for example the capsular polysaccharide of the type-3 pneumococcus, which causes pneumonia, consists of alternating units of glucose and glucuronic acid.

The *polyhexosamines* include the substance chitin which forms the shells of crustaceans and the hard outer portions of insects. It is composed of glucosamine units joined together by β-linkages and is therefore very similar in structure to cellulose.

Of the polysaccharides containing both uronic acids and hexosamines, one of the most important is *hyaluronic acid* which consists of alternating residues of N-acetylglucosamine (NAG) and glucuronic acid. It forms aqueous solutions of high viscosity and is found in the skin, in the vitreous humour of the eye, in the umbilical cord and in certain bacteria. It exercises a cementing function in the tissues and probably also in the capillary walls, and it forms a coating gel round the ovum. Synovial fluid, which contains 0·02 to 0·05 per cent of hyaluronate, owes about 80 per cent of its viscosity to this substance. The enzyme *hyaluronidase* breaks down hyaluronic acid into a number of small molecules with a decrease in the viscosity of the solution and an increase in the concentration of reducing sugar. If fluid containing this enzyme is injected into a tissue it spreads rapidly from the site of injection. The enzyme is, therefore, sometimes referred to as the 'spreading factor'. It is found in relatively high concentration in the testis and seminal fluid, in the venoms of certain snakes and insects, and in some bacteria.

The *chondroitin sulphates* are the chief mucopolysaccharides of cartilage, bone, heart valves, tendons and the cornea. They are built up from units of N-acetylgalactosamine, glucuronic acid and sulphuric acid.

Another mucopolysaccharide containing sulphuric acid is *heparin*, a naturally occurring anticoagulant found in the liver, as its name suggests, and also in lung, spleen, kidney and intestinal mucosa. It can be highly purified and is used to prevent the clotting of blood (Chap. 23). In chemical structure it is a polymer of

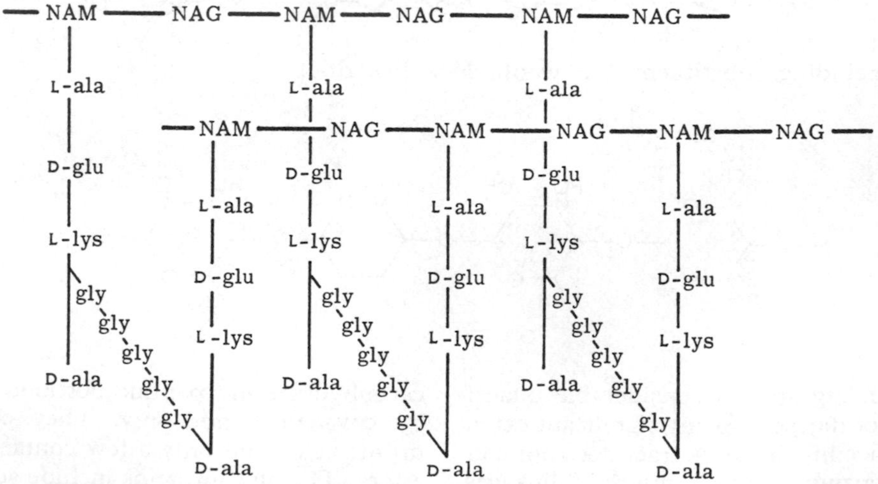

Fig. 4.4 The structure of part of the cell wall of the bacterium *Staphylococcus aureus* showing the chains of alternating residues of N-acetyl glucosamine (NAG) and N-acetyl muramic acid (NAM), the latter carrying short peptide chains in which D-amino acids are conspicuous. The individual chains are joined together by glycine residues to form an elaborate mesh-work.

glucuronic acid and N-acetylglucosamine and is strongly acid in character owing to the sulphate ester groupings which it contains.

The cell walls of many bacteria contain a polysaccharide consisting of alternating units of N-acetyl glucosamine (NAG) and another amino sugar N-acetyl muramic acid (NAM)

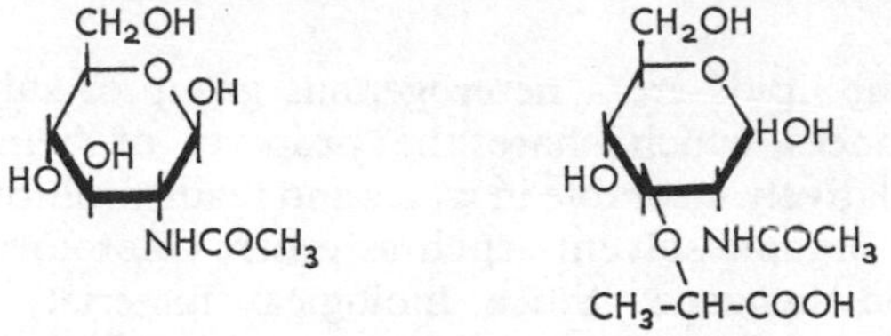

NAG

NAM

linked by its carboxyl group to short peptide chains (p. 28) of four amino acids. These units are connected into a complex pattern with cross-links forming a meshwork illustrated in Fig. 4.4. The antibiotic penicillin inhibits bacterial growth by preventing the formation of the cross-links.

The enzyme *lysozyme* (Chap. 8) brings about hydrolysis of the bond between a NAM unit and the oxygen atom attached to the next NAG unit so that the chain is broken into NAG-NAM disaccharides and the cell wall is disrupted.

Some mucoproteins contain members of an important group of compounds known as *sialic acids* which are acetyl derivatives of the substance *neuraminic acid*.

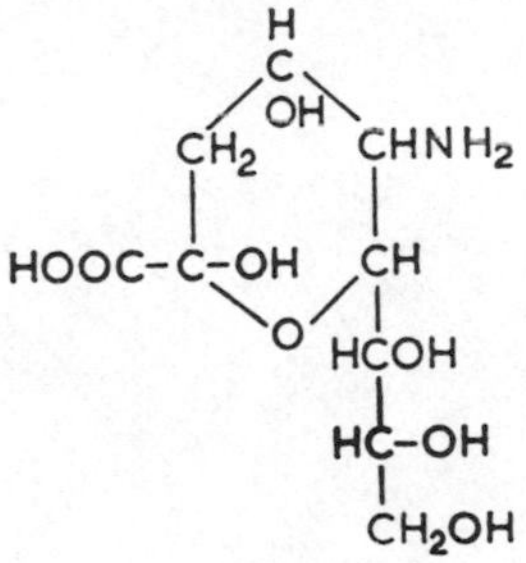

neuraminic acid

REFERENCES

BAILEY, R. W. (1964). *Oligosaccharides*. Oxford: Pergamon Press.
BELL, D. J. (1962). Natural monosaccharides and oligosaccharides. In *Comparative Biochemistry* (Edited by M. Florkin and H. S. Mason) **3**, Part A, 288–354. New York: Academic Press.
Biochemistry of the Mucopolysaccharides (1960). Biochemical Society Symposium, No. 20.
BRIMACOMBE, J. S. & WEBBER, J. M. (1964). *Mucopolysaccharides*. Amsterdam: Elsevier.
DAVIDSON, E. A. (1967). *Carbohydrate Chemistry*. New York: Holt, Rinehart & Winston.
DUTTON, G. J. (Ed.) (1966). *Glucoronic Acid*. New York: Academic Press.
GALE, E. F. (1967). How antibiotics work. *Science Journal* **3**, No. 1, 62–67.
GOTTSHALK, A. (1966). *Glycoproteins*. Amsterdam: Elsevier.
JEANLOZ, R. W. & BALAZS, E. A. (Eds.) (1965). *The Amino Sugars*. New York: Academic Press.
KENT, P. W. (1967). Structure and function of glycoproteins. In *Essays in Biochemistry* (Edited by P. N. Campbell and G. D. Greville) **3**, 105–152.
PERCIVAL, E. G. V. & PERCIVAL, E. (1962). *Structural Carbohydrate Chemistry*. New York: Academic Press.
PIGMAN, W. W. (1957). *The Carbohydrates*. New York: Academic Press.
STACEY, M. & BARKER, S. A. (1962). *Carbohydrates of Living Tissues*. London: Van Nostrand.

5 Lipids

The lipids are a heterogenous group of substances which share the property of being relatively insoluble in water and readily soluble in organic solvents such as ether, chloroform and benzene. Since biological material in general has a very high percentage of water (p. 6), the insolubility of the lipids contributes to their specialized roles in the body. They can be broadly classified according to their functions, for example:

(a) The storage of the body's reserve of energy.
(b) The maintenance of the structural integrity of the cell.
(c) Hormonal functions as in the case of the steroids.

ENERGY STORAGE

Energy is stored in the body mainly as saturated fatty acids of the general formula $CH_3(CH_2)_nCOOH$ where $n = 0$ in acetic acid, $n = 1$ in propionic acid, $n = 2$ in butyric acid and so on. The value of n lies usually between 10 and 16 but is usually an even number. Most of the common fatty acids have a long inert hydrophobic chain with a highly reactive hydrophilic acidic grouping at one end. In the neutral fats, the acidic grouping is joined by an ester linkage to the trihydric alcohol glycerol to give a completely inert fat molecule which has no charge and no reactive groupings.

$$
\begin{array}{ll}
CH_2OH & HOOCC_3H_7 \\
| & \\
CHOH \quad + & HOOCC_3H_7 \\
| & \\
CH_2OH & HOOCC_3H_7 \\
\end{array}
$$

Glycerol   Butyric acid

$$
\longrightarrow
\begin{array}{l}
CH_2.O.CO.C_3H_7 \\
| \\
CH.O.CO.C_3H_7 \quad + \quad 3H_2O \\
| \\
CH_2.O.CO.C_3H_7 \\
\end{array}
$$

Tributyrin

Glycerol reacts with one molecule of a fatty acid to form a monoglyceride, with two molecules to form a diglyceride and with three to form a triglyceride.

With stearic acid ($C_{17}H_{35}COOH$) glycerol forms glyceryl tristearate or tristearin, and with palmitic acid ($C_{15}H_{31}COOH$) it yields glyceryl tripalmitate or tripalmitin. Both tristearin and tripalmitin occur in large quantities in beef and mutton fat. With the unsaturated acid, oleic acid ($C_{17}H_{33}COOH$), glycerol yields glyceryl trioleate or triolein which is the main constituent of olive oil.

$$CH_2.O.CO.C_{17}H_{35}$$
$$CH.O.CO.C_{17}H_{35}$$
$$CH_2.O.CO.C_{17}H_{35}$$
Tristearin

$$CH_2.O.CO.C_{15}H_{31}$$
$$CH.O.CO.C_{15}H_{31}$$
$$CH_2.O.CO.C_{15}H_{31}$$
Tripalmitin

$$CH_2.O.CO.C_{17}H_{33}$$
$$CH.O.CO.C_{17}H_{33}$$
$$CH_2.O.CO.C_{17}H_{33}$$
Triolein

It can be seen, therefore, that the general formula for a fat is

$$CH_2.O.CO.R_1$$
$$CH.O.CO.R_2$$
$$CH_2.O.CO.R_3$$

Butyric acid

where R_1, R_2 and R_3 may be derived from the same or different fatty acids.

Some polyunsaturated fatty acids occur in small amounts in the body fats. They include linoleic acid

$$CH_3(CH_2)_4CH=CHCH_2CH=CH(CH_2)_7COOH$$

with 18 carbon atoms, linolenic acid

$$CH_3CH_2CH=CHCH_2CH=CHCH_2CH=CH(CH_2)_7COOH$$

with 18 carbon atoms, and arachidonic acid

$$CH_3(CH_2)_4CH=CHCH_2CH=CHCH_2CH=CHCH_2CH=CH(CH_2)_3COOH$$

with 20 carbon atoms.

These three are sometimes called *essential fatty acids* (Chap. 13) since they cannot be synthesized in the animal body and must be provided in the diet.

The *prostaglandins* (p. 62) are derivatives of polyunsaturated fatty acids containing 20 carbon atoms (Fig. 5.1).

Fat is particularly well suited to be the main form of energy reserve in man and other mammals because the fat stores are only slowly exhausted in fasting.

Fat storage provides economy in both weight and space. A gram of stearic acid combusted in a bomb calorimeter produces 40 kJ (9·5 kcal) whereas a gram of glycogen produces 16 kJ (3.8 kcal). In other words the amount of energy from a given weight of fat is very much greater than that from the same weight of carbohydrate. It is also economical in terms of bulk, since fatty acids with their flexible backbones can be stored in a much more compact form than the highly spatially oriented and rigid glycogen structure. Fat is an excellent form of energy store. It occupies relatively little space in the body; it has no affinity for water and therefore, once it has been carried to the fat depots by the specialized transport proteins in the plasma (Chap. 23), it is unlikely to break loose and travel away in the watery body fluids which bathe the adipose tissue. The fat remains as a stable and fixed reserve of energy until 'mobilized' by enzymes which break it into glycerol and fatty acids, that is until the reactive acidic groupings with their affinity for water and ability to form ionic bonds with proteins are set free. The enzymes are under the control of hormones which activate them to break down the triglycerides in the adipose tissue when energy expenditure is increased. Much of the adipose tissue of the body is sub-

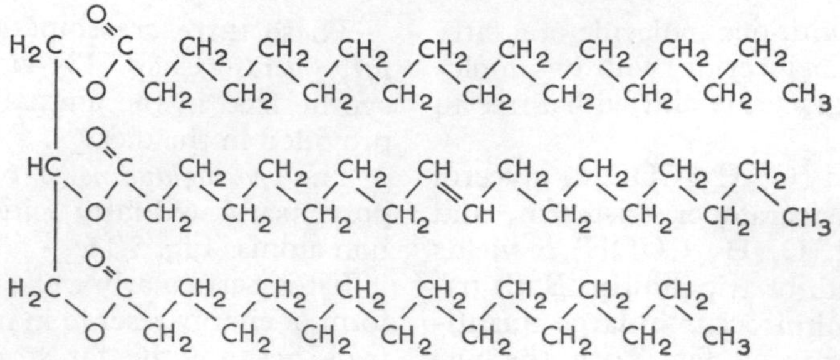

Fig. 5.1 Structure of a mixed triglyceride. The top fatty acid residue represents stearic acid, the lowest palmitic acid; in the middle an unsaturated fatty acid residue (oleic acid) is shown.

cutaneous and because fat is a bad conductor of heat it provides excellent insulation. This gives the fat store its own built-in economy, since in cold conditions, in which heat is lost to the environment, it provides both an insulating blanket and an energy source to replace the heat which does escape.

Most of the body fats contain saturated fatty acids, but appreciable amounts of mono-, di-, and tri-unsaturated fatty acids are always present. Plants often store large quantities of unsaturated fats, for example olive oil which is essentially triolein. Unsaturated fatty acids and also fats containing unsaturated fatty acids such as olive oil or corn oil are usually liquid at room temperature and are commonly referred to as oils. This term is, however, rather confusing since it is also applied to mineral oils which are not glyceryl esters but hydrocarbons. Some fats are solid at room temperature but are liquid at the temperature of the body; human fat is said to melt at about 17°C. The melting points of fats are always higher than their solidification points; for example, tristearin melts at 72°C but solidifies on cooling at 52°C. Fats may be hydrolysed by boiling with an alkali such as sodium hydroxide to yield glycerol and the sodium salts of fatty acids which are known as soaps. This process is termed *saponification*.

Because soap molecules have both a hydro-phobic and a hydrophilic end they have a dual affinity for water and for neutral fats or hydrocarbons (Fig. 5.2). The action of soap on a greasy frying pan is therefore to emulsify the insoluble fat by forming small globules with a hydrophobic interior but hydrophilic, or water soluble exterior as shown in Fig. 5.2. When soap is added to 'hard' water which contains a high proportion of calcium ions, the calcium salt of the fatty acid is precipitated and forms a scum on top of the water. More soap is required under these conditions to emulsify a given quantity of fat. The process of emulsification also occurs in the body; for example after a fatty meal the fat in the gut must be emulsified before it can be absorbed.

In the manufacture of hard margarine unsaturated plant fats are converted to more saturated and palatable solid fats by catalytic hydrogenation, usually over finely divided nickel. In this way tristearin can be obtained from triolein.

$$-\overset{\displaystyle \overset{H}{|}}{C}=\overset{\displaystyle \overset{H}{|}}{C}- \quad +2H \quad \xrightarrow{\text{Ni}} \quad -CH_2-CH_2-$$

$$\begin{array}{l} CH_2.O.CO.C_{17}H_{33} \\ CH.O.CO.C_{17}H_{33} \quad + \quad 6H \quad \longrightarrow \\ CH_2.O.CO.C_{17}H_{33} \end{array} \qquad \begin{array}{l} CH_2.O.CO.C_{17}H_{35} \\ CH.O.CO.C_{17}H_{35} \\ CH_2.O.CO.C_{17}H_{35} \end{array}$$

Triolein   Tristearin

$$\begin{array}{l} CH_2.O.CO.C_{17}H_{35} \\ CH.O.CO.C_{17}H_{35} \quad + \quad 3NaOH \quad \longrightarrow \\ CH_2.O.CO.C_{17}H_{35} \end{array} \qquad \begin{array}{l} CH_2OH \\ CHOH \quad + \quad 3C_{17}H_{35}COONa \\ CH_2OH \end{array}$$

Tristearin   Glycerol   Sodium stearate

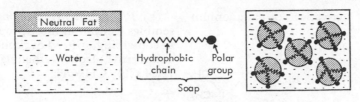

Fig. 5.2 The mechanism of emulsification of neutral fats by soaps.

The process of hydrogenation is not carried to completion because completely saturated fats are very hard, brittle and unpalatable.

When fats are exposed to the air for a month or so they undergo a certain amount of oxidation, especially at the double bonds, and they also suffer partial hydrolysis with liberation of free fatty acids. In this way fat becomes much less palatable and is said to have gone rancid. Oxidation can be prevented to a large extent by the addition of small amounts of organic substances termed *anti-oxidants*. Many phenols act in this way and are thus of considerable importance in the food industry. Highly unsaturated fats, such as linseed oil, undergo considerable oxidation on exposure to air forming tough waterproof films. Such oils, called drying oils, are used in the manufacture of paints and varnishes.

THE WAXES. The waxes are esters of fatty acids not with glycerol but with complex monohydric alcohols. Beeswax, for example, is an ester of palmitic acid with myricyl alcohol ($C_{37}H_{61}OH$), and spermaceti from the sperm whale is an ester of palmitic acid with cetyl alcohol ($C_{16}H_{32}OH$). Many animal waxes are esters of the steroid alcohol, cholesterol (p. 59).

THE MAINTENANCE OF THE STRUCTURAL INTEGRITY OF THE CELL

All animal cells are surrounded by a highly flexible and elastic membrane and many intracellular particles such as nuclei or mitochondria are also enclosed by a membrane. The endoplasmic reticulum is also a membrane-like structure (see p. 6). These membranes are very selective with respect to the ions and molecules which they allow to pass so that the composition of the intracellular fluids of the body is very different from that of the extracellular fluids (Chap. 33). This selectivity is dependent on the nature of the molecules rather than on their size; some quite large molecules such as small proteins can enter the cell whereas some relatively small ions such as sodium ions are to a large extent excluded. A membrane may be able to differentiate between molecules of the same size. Thus D-glucose can pass into some cells whereas its optical isomer L-glucose cannot.

This membrane specificity varies with the animal, the tissue, or even the intracellular organelle involved. Thus although the red cells of man and of primates are permeable to glucose, those of the pig and horse are not; and again, although human red cells are permeable to glucose, human muscle cells are relatively impermeable in the absence of the hormone insulin. The external membrane of the cell allows very little potassium to pass, whereas the membrane of the nucleus tends to permit the concentration of potassium in the nucleus and allows very little sodium to enter. It is apparent therefore that these membranes are not simple sieves but are highly selective and intricate permeability mechanisms.

It is obvious that the ideal barrier for preventing water soluble materials from passing freely between the intra- and extracellular fluids would be a lipid one, since it would have little affinity for these materials. This is, in fact, the main role of the *phospholipids*. These have, like the fatty acids, a polar region and a non-polar region, but the ionic functions are greatly increased by the presence of phosphoric acid and a nitrogenous organic base such as choline or ethanolamine.

THE PHOSPHOLIPIDS

(a) *Phosphatidylcholine (lecithin)*. The best known phospholipids are the lecithins which on hydrolysis yield glycerol, fatty acids, phosphoric acid and choline. Partial hydrolysis may yield phosphoglycerol in which the phosphate is attached to the α-carbon atom of the glycerol.

Choline is a derivative of ammonium hydroxide thus:

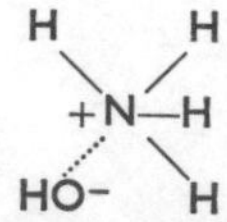

$$H_2N.CH_2.CH_2OH$$

Ammonium
hydroxide

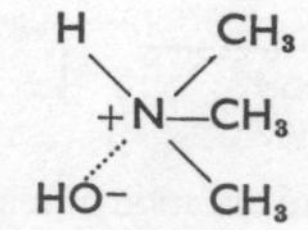

Trimethyl
ammonium
hydroxide

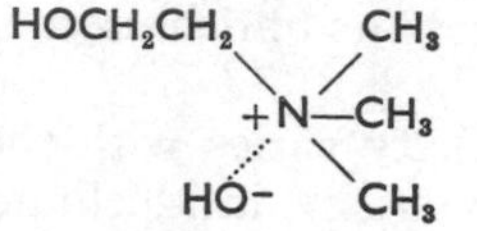

Choline
(trimethyl hydroxy ethyl
ammonium hydroxide)

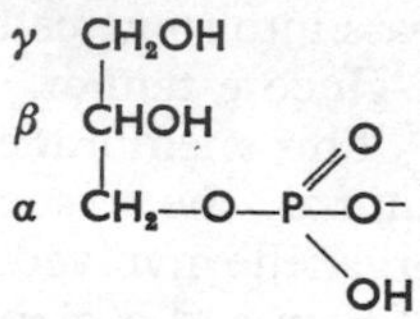

α-Phosphoglycerol

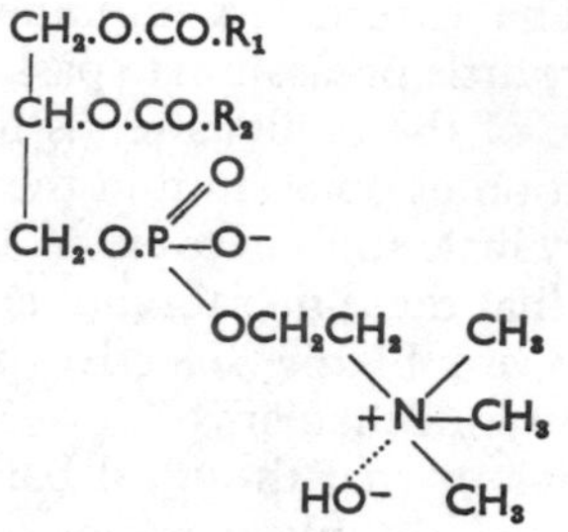

CH₂.O.CO.R₁
CH.O.CO.R₂
CH₂.O.P—O⁻
OCH₂CH₂ CH₃
+N—CH₃
HO⁻ CH₃

Lecithin

The various constituents combined in lecithin are shown above; R_1 and R_2 are fatty acid residues. Lecithins, like the other phospholipids, are yellowish, greasy solids, soluble in all the fat solvents except acetone. This property enables them to be easily distinguished from the fats. On exposure to air they rapidly darken in colour and absorb water, forming a dark, greasy mass.

The lecithins can be attacked by the enzyme, lecithinase A, which removes one of the fatty acid residues, leaving a product known as *lysolecithin*, a surface-active agent with the ability to cause haemolysis of red blood cells. Lecithinase A is present in the venoms of many snakes and poisonous insects.

(b) *Phosphatidylethanolamine (cephalin)* resembles lecithin in most properties but differs in containing, instead of choline, the substance aminoethyl alcohol (ethanolamine, colamine), $H_2N.CH_2.CH_2OH$:

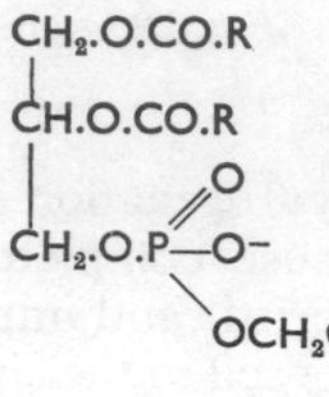

CH₂.O.CO.R
CH.O.CO.R
CH₂.O.P—O⁻
OCH₂CH₂NH₃⁺

Cephalin

(c) *Phosphatidylserine* contains the amino acid serine in place of ethanolamine.

(d) *Phosphatidylinositol* is found mainly in plants and in nervous tissue, and contains inositol in place of a nitrogenous base (Fig. 5.3).

(e) *Sphingomyelins* are more complicated phosphatides containing instead of glycerol the base sphingosine. On hydrolysis they yield fatty acids, phosphoric acid, choline and sphingosine.

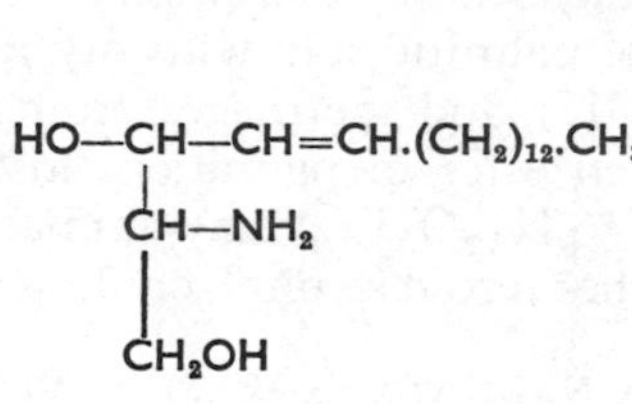

HO—CH—CH=CH.(CH₂)₁₂.CH₃
CH—NH₂
CH₂OH

Sphingosine

HO—CH—CH=CH.(CH₂)₁₂CH₃
CH—NH—CO—R
CH₂—O—P—O⁻
O.CH₂.CH₂ CH₃
+N—CH₃
HO⁻ CH₃

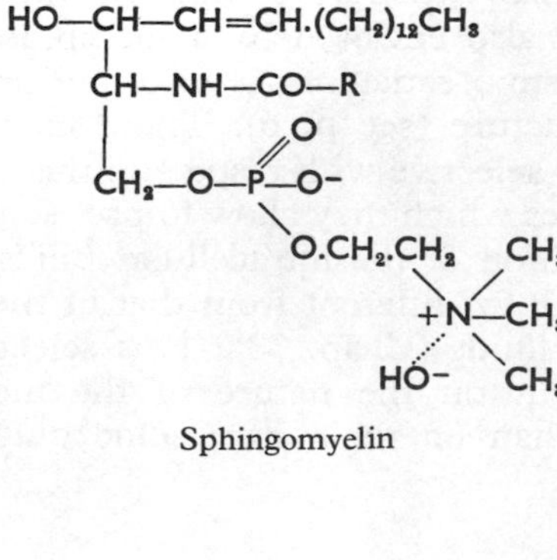

Sphingomyelin

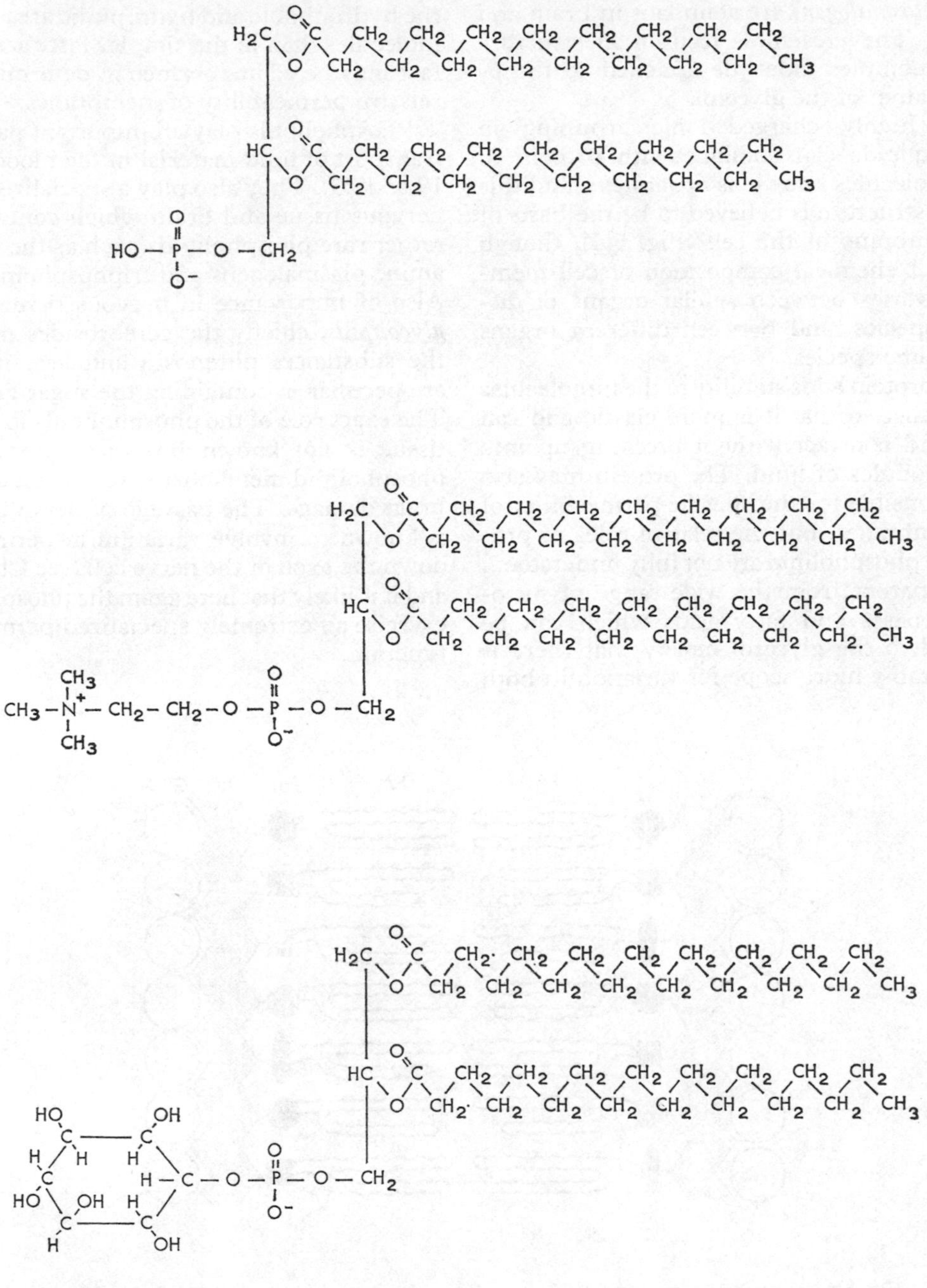

Fig. 5.3 Top, phosphatidic acid; middle, lecithin, showing the base choline; bottom, phosphatidyl inositol, showing the position of the inositol residue.
In all cases shown here the fatty acid residue is stearic acid, but it may be replaced in naturally occurring compounds by other fatty acids saturated or unsaturated.

(f) *Plasmalogens* are abundant in brain and muscle. They resemble lecithin in structure with a complex aldehyde attached to the β-carbon atom of the glycerol.

The highly charged ionic grouping in phospholipids can combine with protein to form molecules known as *lipoproteins* and this type of structure is believed to be the basis of the membrane of the cell (Fig. 5.4), though the exact chemical composition of cell membranes varies between similar organs of different species, and between different organs in the same species.

The protein adds stability to the bimolecular lipid leaflet so that it is more elastic and can swell and contract without breaking up into small globules of lipid. The protein may also be responsible for the specific permeability of the membranes, but the relative roles of protein and phospholipid are not fully understood. It is apparent from the wide range of nitrogenous bases and fatty acids which can be attached to the glycerol moiety that there is considerably more scope for variation in both the hydrophobic and hydrophilic areas of these molecules than in the simpler fatty acids. This fact may be of importance in determining the relative permeability of membranes.

Phospholipids play an important part in the transport of lipid material in the blood (Chap. 19 and 23). They also play a specialized role in nervous tissue and brain which contain some rather rare phospholipids such as the ethanolamine plasmalogens and triphosphoinositides. Also of importance in nervous tissue are the *glycolipids*, chiefly the cerebrosides including the substances phrenosin and kerasin which are peculiar in containing the sugar galactose. The exact role of the phospholipids in nervous tissue is not known but many disorders of phospholipid metabolism are accompanied by brain damage. The passage of nerve impulses is known to involve variation in permeability down the axon of the nerve cell (see Chap. 39), and it is likely that here again the phospholipids exercise an extremely specialized permeability function.

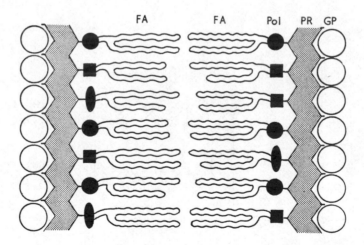

Fig. 5.4 Schematic representation of a membrane showing two layers of phospholipid with their hydrophobic fatty acid chains (FA) facing each other and the polar ends (Pol) in contact with the protein component.
Recent work suggests that the proteins may be only polypeptide chains in the extended β-configuration (PR) (Chap. 6) or may be only globular proteins (GP) or may be mixed. It is even possible that the membrane protein may be arranged within the lipid bilayer (see References at end of chapter).

STEROIDS

The steroids are conveniently considered along with the lipids, although they differ greatly from them in chemical structure. Like the lipids they are soluble in the fat solvents and, in general, insoluble in water. Because they are not hydrolysed by treatment with sodium hydroxide they form part of the *unsaponifiable fraction* of the lipid portion of tissues. The steroids include such substances as cholesterol and other sterols, the bile acids, the sex hormones and the hormones of the adrenal cortex. Although the steroids have a wide range of biological activities, they all contain the nucleus known as the perhydro*cyclo*pentenophenanthrene ring system. The four rings are labelled A, B, C and D as shown below.

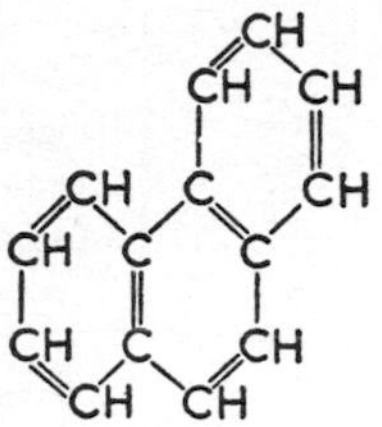

*cyclo*pentenophenanthrene ring system with a double bond between carbon atoms 5 and 6.

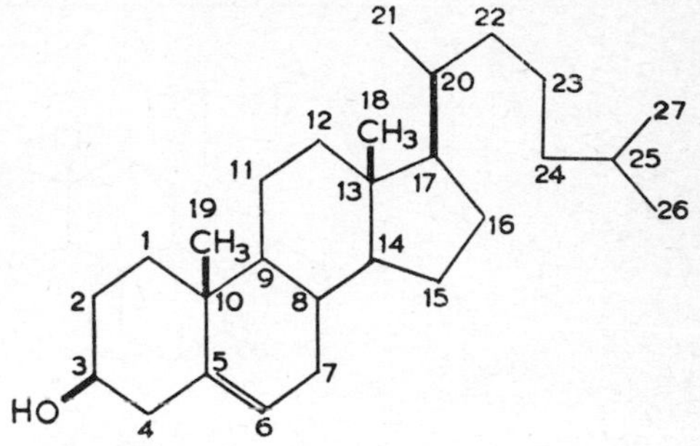

Cholesterol (cholest-5-en-3β-ol)
(skeleton structure)

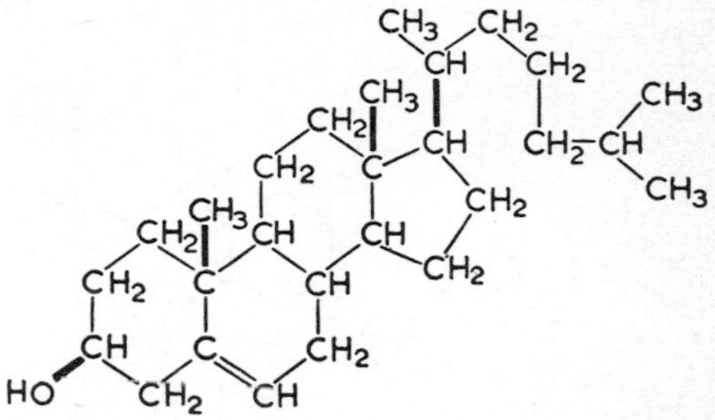

Phenanthrene

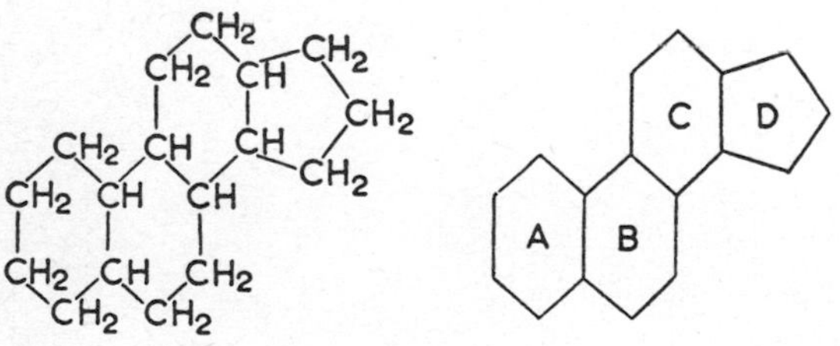

Cholesterol (full structure)

Perhydro*cyclo*pentenophenanthrene

(full formula)   (skeleton formula)

THE STEROLS. The *sterols* are steroid alcohols. The best known, *cholesterol*, is widely distributed in the tissues, being particularly abundant in brain and nervous tissue, in the adrenal glands and in the skin. It is found also in egg-yolk and gall stones. Cholesterol is a derivative of the parent hydrocarbon cholestane (Table 5.5) and contains the perhydro-

Like all sterols, cholesterol has 'angular' methyl groups attached to carbon atoms 10 and 13, a hydroxyl group at position 3 and an aliphatic side chain at position 17.

By virtue of its hydroxyl group cholesterol is an alcohol and can therefore form esters. In the tissues it exists partly in the free state and partly in the form of esters. Both forms unite with protein to give water-soluble complexes found in blood plasma.

Many other sterols are known but only a few are of recognized importance in the body. *7-Dehydrocholesterol* occurs in skin and is converted by ultraviolet light to one form of vitamin D (Chap. 13). *Ergosterol*, found in yeasts and moulds, is the precursor of another form of vitamin D. *Coprosterol* is found in the faeces.

NOMENCLATURE OF STEROIDS. Steroids are classified according to the number of carbon

Table 5.5 The steroids

No. of carbon atoms	Parent hydrocarbon		Biologically important compounds
18	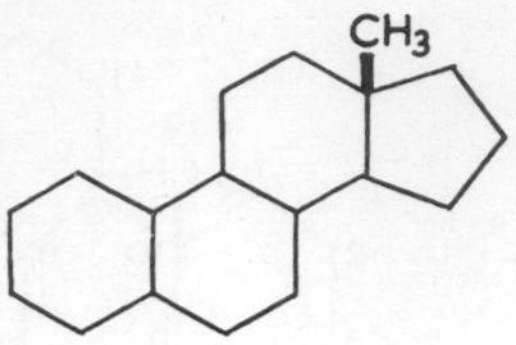	Oestrane	Oestrogens
19	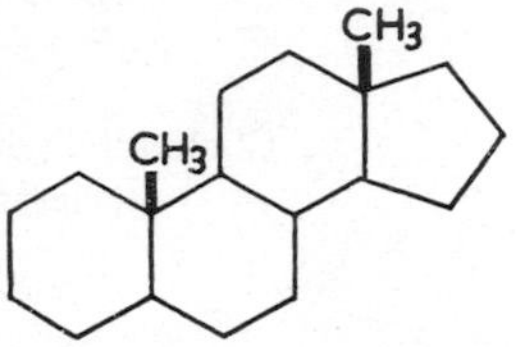	Androstane	Androgens
21	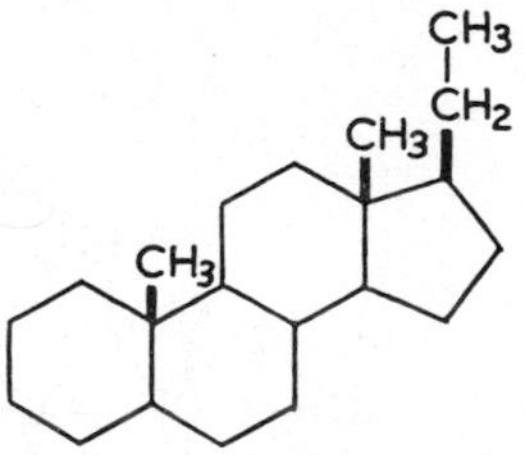	Pregnane	Progesterone Hormones of adrenal cortex
24	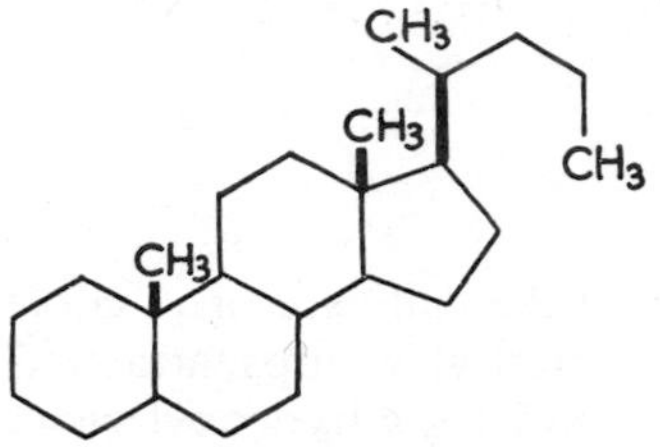	Cholane*	Bile acids
27	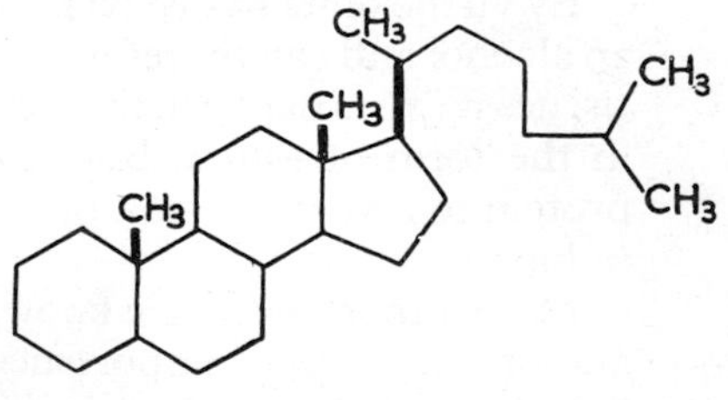	Cholestane*	Cholesterol

* Side chain methyl groups are shown as —CH_3 in this table, but in formulae elsewhere only angular methyl groups are shown in this way.

To avoid confusion the stereochemical configurations at all the asymmetric centres have not been included in the table. Variations at C-5 for example, are common but not shown here.

atoms present in the ring structures plus side chains. Parent saturated hydrocarbons are shown in Table 5.5. Unsaturated compounds

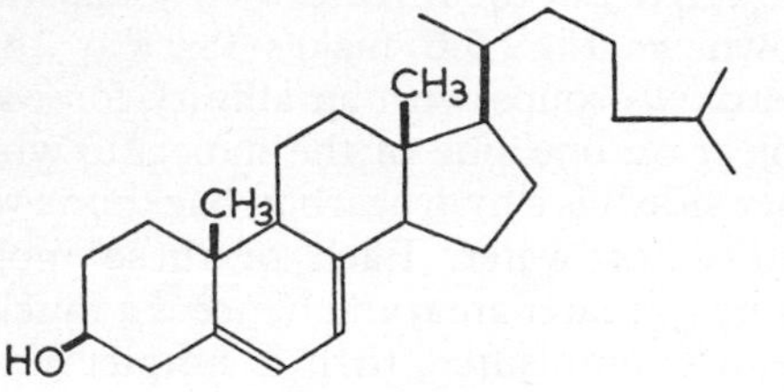

7-Dehydrocholesterol

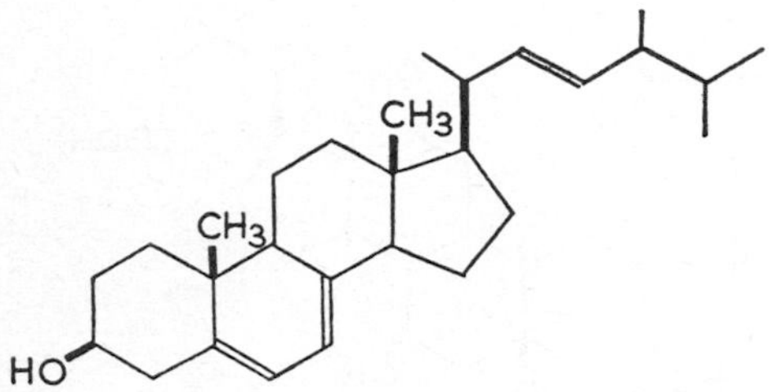

Ergosterol

are indicated by the suffix -ene or -en added to the root name of the hydrocarbon, and the position of the double bond is indicated by the number of the carbon atom at which it starts. The presence of an alcohol group is indicated by -ol and that of an oxo group by -one. These

groups are preceded by a number showing their position. Cholesterol can be described as cholest-5-en-3β-ol. The significance of the symbol β is explained below.

THE STEREOCHEMISTRY OF THE STEROIDS

The steroid ring system contains six asymmetric carbon atoms (C-5, 8, 9, 10, 13 and 14) and hence sixty-four stereoisomers are possible. If substituent groups are attached to carbon atoms 3, 11 and 17, then five hundred and twelve stereoisomers of the same steroid are possible. However, in all the naturally occurring steroids, with a few exceptions, the asymmetry at carbon atoms 8, 9, 10, 13 and 14 of the ring system is the same and stereoisomerism is restricted to carbon atoms 3, 5, 11 and 17.

X-ray diffraction studies show that the CH_3 group on C-10 projects above the plane of the rings (and of the paper). A hydrogen atom on C-5 may project either below or above the same plane. In the former case the two groups are *trans* to each other (as in cholestanol, see formulae). The configuration is referred to as the *allo, trans,* or α type, and the bond at C-5 is shown as a dotted line. If the two groups project above the plane they are in the *cis* relationship to each other and the configuration is referred to as the *normal, cis,* or β type, such as is found in coprostanol. The bond at C-5 is

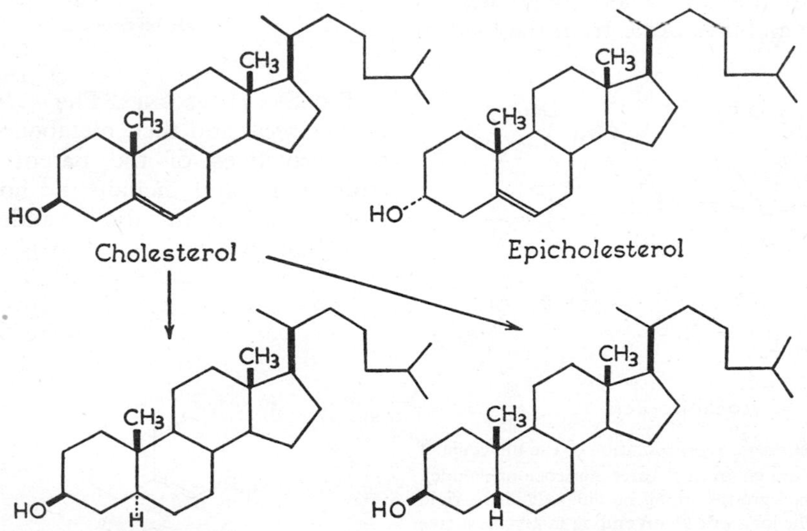

Cholesterol

Epicholesterol

Cholestanol

Coprostanol

shown in this case as a solid line. Both choles-tanol and coprostanol are formed by the reduction of cholesterol by addition of hydro-gen at the double bond between C-5 and C-6.

Isomerism may also occur at C-3. In cholesterol the hydroxyl group at C-3 is assumed to project above the plane of the rings, that is to be *cis* with respect to the methyl group at C-10. The bond is therefore shown as a solid line (p. 59), and the configuration is referred to as *β*. In an isomer of cholesterol known as epicholesterol, the hydroxyl group at C-3 projects below the plane of the rings and is therefore *trans* with respect to the methyl group at C-10. The bond is shown as a dotted line and the configuration is referred to as α.

Epicholesterol can of course give rise on reduction to epicholestanol and epicoprostanol.

THE BILE ACIDS. The bile contains a num-ber of steroid acids which are derivatives of the parent substance *cholanic acid*. They in-clude *cholic acid, deoxycholic acid* and *lithocholic acid*, all of which contain the perhydrocyclo-pentenophenanthrene ring system, with a five-membered side chain on position 17 and with hydroxyl groups attached at other positions on the nucleus. These acids which are found in the bile in combination with glycine and taurine are by far the most abundant of the steroids derived from cholesterol, and are the most effective biological emulsifiers (Chap. 17). They are excreted by the liver cells into the bile and pass into the gut where they help to solubilize the globules of fat from the food so that the water soluble enzymes can reach the fat molecules and split them to facilitate absorption. In a two-dimensional diagram it is impossible to see why these molecules are sur-face active but the three-dimensional structure shown in Fig. 5.6 makes it clear that the hydroxyl groups with an affinity for water all project on one side of the molecule while the other side has a hydrocarbon backbone with no affinity for water. Each of these molecules covers a greater area, and hence is a much more effective emulsifier, than a simpler molecule such as sodium stearate.

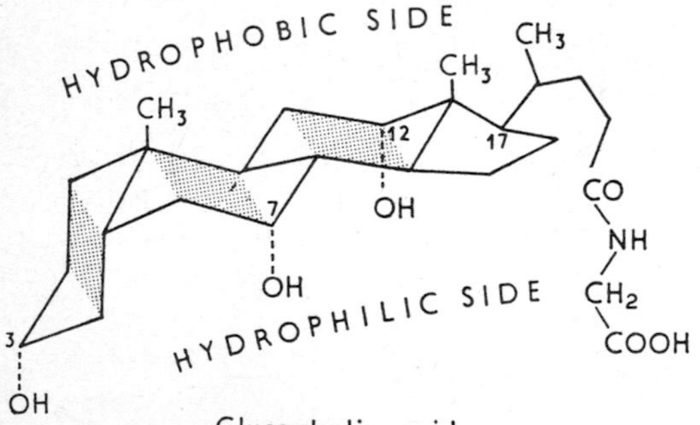

Cholic acid

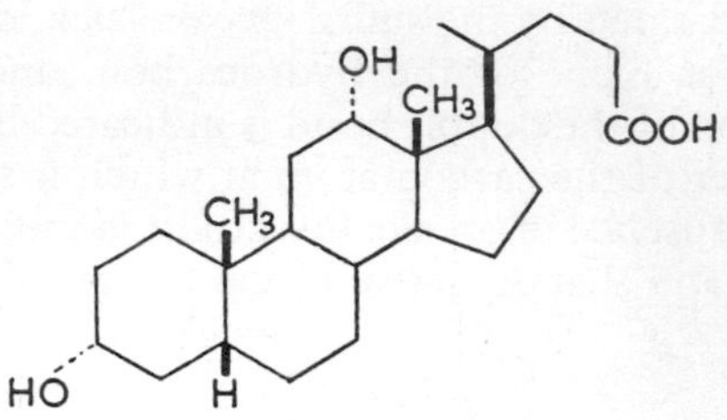

Deoxycholic acid

THE SEX HORMONES. The *male sex hormones* or *androgens* and their metabolic products are all derivatives of the parent hydrocarbon androstane, and include the hormone testo-sterone and its metabolite androsterone (see also Chap. 51).

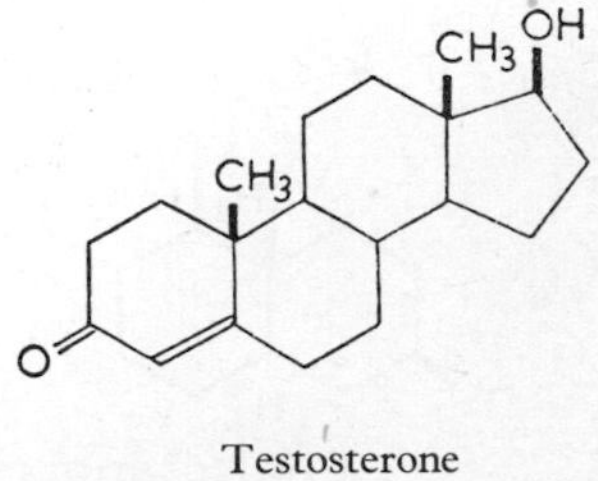

Glycocholic acid

Fig. 5.6 Diagrammatic representation of the molecule of glycocholic acid drawn from a three-dimensional model. The three hydroxyl groups are all on one side of the rigid ring structure. The long side chain ending in glycine is free to rotate at several points and may occupy many different positions.

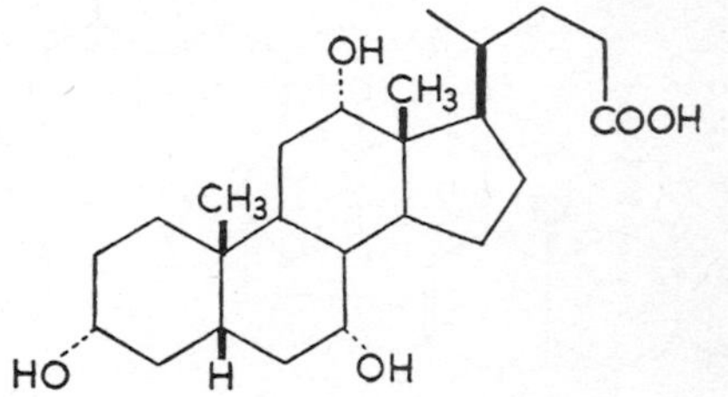

Testosterone
(17β-hydroxyandrost-4-en-3-one)

Androsterone
(3α-hydroxy-5α-androstan-17-one)

Pregnanediol
(5β-pregnane-3α,20α-diol)

These androgens have no side chain in position 17 but have oxo or hydroxyl groupings in positions 3 and 17. Those with oxo groups in position 17 are known as 17-oxosteroids (formerly 17-ketosteroids).

The *female sex hormones* are the *oestrogens* and *progesterone*. The *oestrogens* are derivatives of the parent hydrocarbon *oestrane*, and include *oestradiol* and *oestrone*. It should be noted that

The *hormones of the adrenal cortex* also contain a two-carbon side chain in position 17. They include such compounds as *cortisol* and *aldosterone*, discussed in Chapter 50.

Oestradiol-17β
(oestra-1,3,5-triene-3,17β-diol)

Oestrone
(3-hydroxyoestra-1,3,5-trien-17-one)

the A ring is of the benzene type. These steroids are phenolic and thus acidic in nature. There is no angular methyl group at C-10; a hydroxyl or an oxo group is found at C-17.

Progesterone, a derivative of the parent hydrocarbon *pregnane*, has a short side chain of two carbon atoms in position 17. An important metabolite is *pregnanediol*.

Aldosterone
(11β, 21-dihydroxy-3,20-dioxopregn-
4-en-18-al)

Cortisol
(11β, 17α, 21-trihydroxypregn-4-ene-3,
20-dione)

It has been established that, in mammals, cholesterol functions as a precursor of the bile acids, adrenal cortical hormones, progesterone, androgens, and oestrogens. Such transformations demand oxidative fission of the aliphatic side chain at C-17 at the appropriate position, together with oxidation at various sites in the steroid nucleus. In the case of the bile acids the configuration of the C-3 hydroxyl group is inverted (3β—3αOH).

PROSTAGLANDINS

The prostaglandins (PG) were first discovered in 1933 in extracts of human seminal

Progesterone
(pregn-4-ene-3,20-dione)

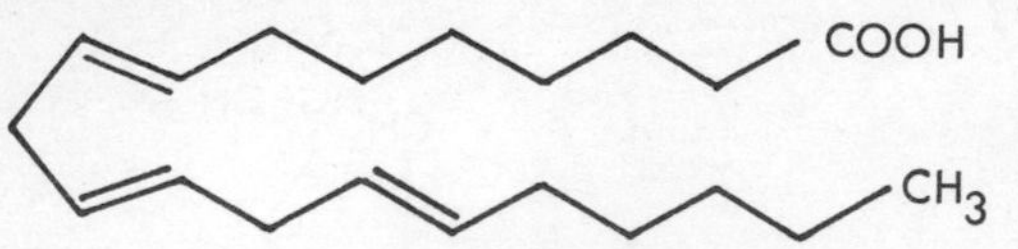

C-20 polyunsaturated fatty acid

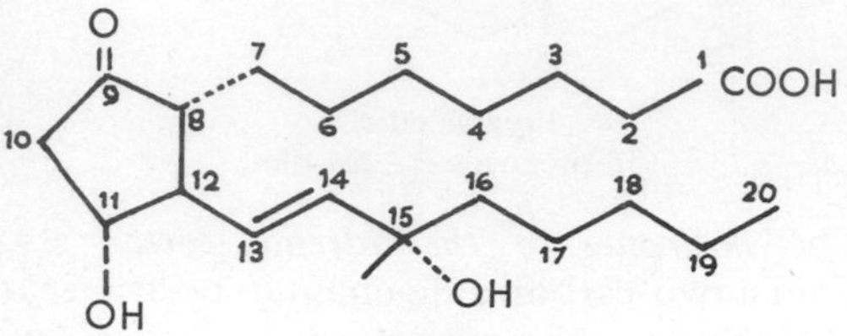

PGE$_1$

plasma and in the vesicular gland of sheep but they have since been found in a wide variety of tissues.

Many different prostaglandins are now known all with structures similar to that of PGE$_1$ which is a C$_{20}$ carboxylic acid with two hydroxyl groups, one *trans* double bond and one keto group in a five-membered ring.

Prostaglandins of the E and F series are formed in the seminal vesicles by cyclization and oxidation of polyunsaturated fatty acids. They have diverse pharmacological effects (see Chap. 50) including stimulation of myometrial contraction: PGE$_2$ has been used clinically to induce labour in pregnant women.

REFERENCES

ANSELL, G. B. & HAWTHORN, J. N. (1964). *Phospholipids: Chemistry Metabolism and Function*. Amsterdam: Elsevier.

BEAZLEY, J. M. (1971). Prostaglandins in human reproduction. *British Journal of Hospital Medicine* **5**, 535–540.

BELL, D. J. & GRANT, J. K. (1963). *The Structure and Function of the Membranes and Surfaces of Cells*. Biochemical Society Symposium No. 22. Cambridge: University Press.

BRONNER, F. & KLEINZELLER, A. (Eds.) (1970). *Current Topics in Membranes and Transport*. New York: Academic Press.

COOK, R. P. (1958). *Cholesterol: Chemistry, Biochemistry, Pathology*.

DANIELSON, H. (1963). Biosynthesis and metabolism and bile acids. *In Advances in Lipid Research* (Edited by D. Kritchevsky & R. Paoletti). New York: Academic Press.

DORFMAN, R. I. & UNGAR, F. (1965). *Metabolism of Steroid Hormones*. London: Academic Press.

FIESER, L. F. & FIESER, M. (1960). *Chemistry of the Steroids*, 4th edn. New York: Reinhold.

GRANT, J. K. (1967). The gas liquid chromatography of steroids. *Memoirs. Society for Endocrinology* **16**.

GURR, M. I. & JAMES, A. T. (1971). *Lipid Biochemistry*. London: Chapman & Hall.

HASELWOOD, G. A. D. (1967). *Bile Salts*. London: Methuen.

HEFTMANN, E. & MOSETTIG, E. (1960). *Biochemistry of Steroids*. New York: Reinhold.

HORTON, E. W. (1969). Hypotheses on physiological roles of prostaglandins. *Physiological Reviews* **49**, 122–161.

KARIM, S. M. M. (1971). Prostaglandins. *British Journal of Hospital Medicine* **5**, 555–560.

KLYNE, W. (1964). *The Chemistry of the Steroids*. London: Methuen.

KORN, E. D. (1970). Cell membranes: structure and synthesis. *Annual Review of Biochemistry* **38**, 263–288.

LOCKWOOD, A. P. M. (1971). *The Membrane of Animal Cells. Institute of Biology's Studies in Biology* No. 27. London: Arnold.

MASORO, E. J. (1968). *The Physiological Chemistry of Lipids in Mammals*. Philadelphia: Saunders.

Nomenclature of lipids (1967). *Biochemical Journal* **105**, 897–902.

NORTHCOTE, D. H. (1968). Structure and function of membranes. *British Medical Bulletin* **24**, 99–186.

ROTHFIELD, L. & FINKELSTEIN, A. (1968). Membrane biochemistry. *Annual Review of Biochemistry* **37**, 463–496.

SCHUMAKER, V. N. & ADAMS, G. H. (1970). Circulating lipoproteins. *Annual Review of Biochemistry* **38**, 113–136.

6 Nucleotides and nucleic acids

The nucleic acids are important cell constituents of high molecular weight. They are
constructed out of units known as *nucleotides*.
Each nucleotide consists of a purine or pyrimidine base linked to a pentose sugar which in
turn is esterified with phosphoric acid thus:

base—sugar—phosphate

Pentose sugars. Ribose and 2-deoxyri-
bose, found in nucleotides, have the structures
shown below. In the free state they are in the
pyranose form (see p. 45), but in the combined
state in nucleotides they exist in the furanose
form.

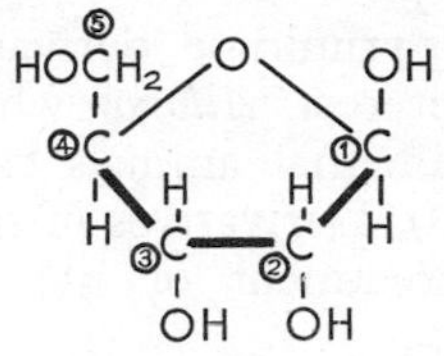

ribose

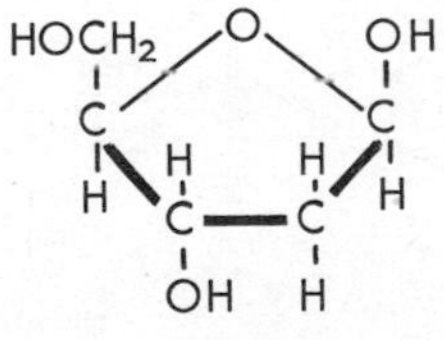

deoxyribose

Pyrimidine bases. The pyrimidine bases
are all derivatives of the parent compound
pyrimidine which is structurally a six-membered ring with two nitrogen and four carbon
atoms. They are numbered as follows:

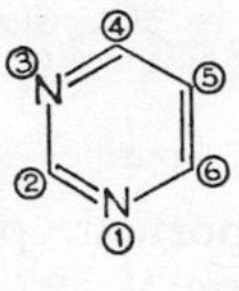

The pyrimidine derivatives commonly found in nucleotides and in nucleic acids are cytosine, uracil and thymine.

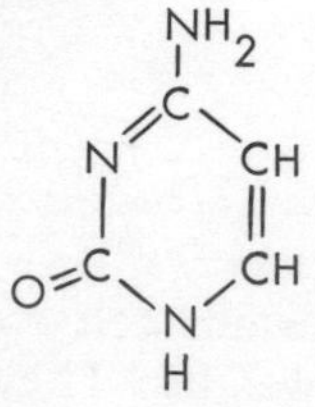

cytosine

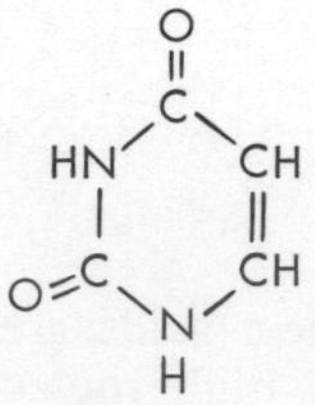

uracil

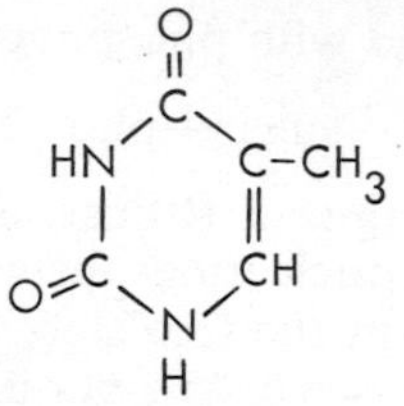

thymine
5-methyl-uracil

Two other pyrimidine derivatives are of physiological interest. *Alloxan* when administered to experimental animals causes glycosuria (Chap. 18). Derivatives of *thiouracil* are used in the treatment of hyperthyroidism (Chap. 50).

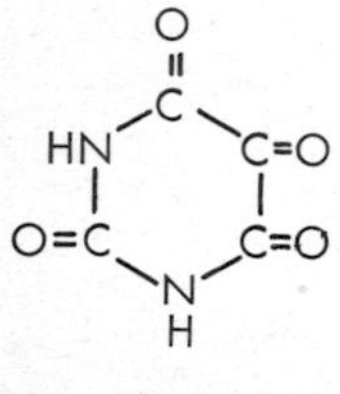

alloxan

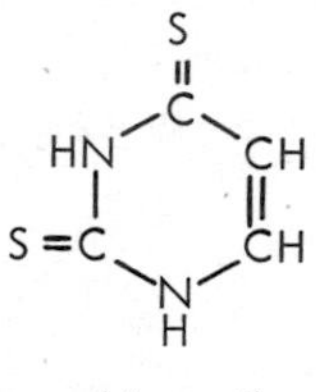

thiouracil

Purine bases. The purine bases are derivatives of the parent compound purine which contains a pyrimidine ring and an imidazole ring fused together.

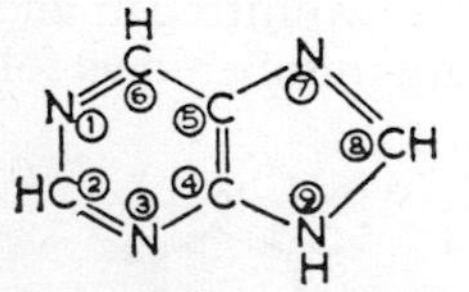

purine

The most important purine bases are adenine and guanine.

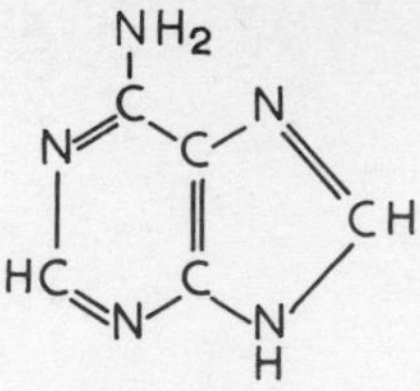

adenine

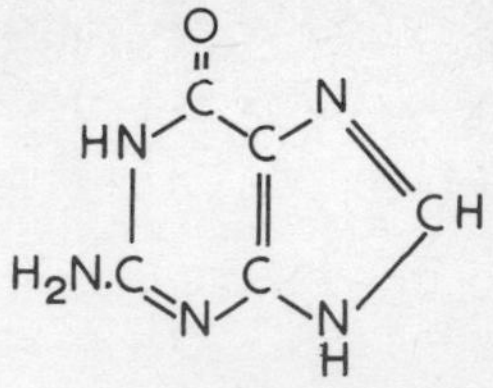

guanine

Other naturally occurring purine derivatives are hypoxanthine, xanthine and uric acid. The diuretic drugs *caffeine*, 1,3,7-trimethyl-xanthine, and *theobromine*, 3,7-dimethyl-xanthine are also purine derivatives.

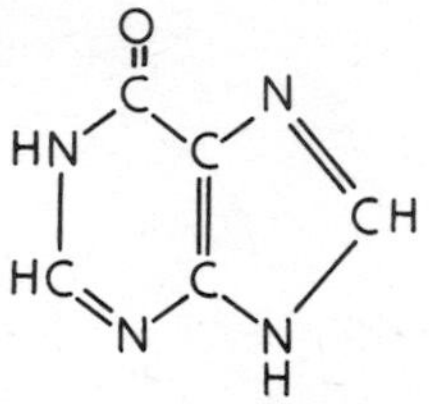

hypoxanthine

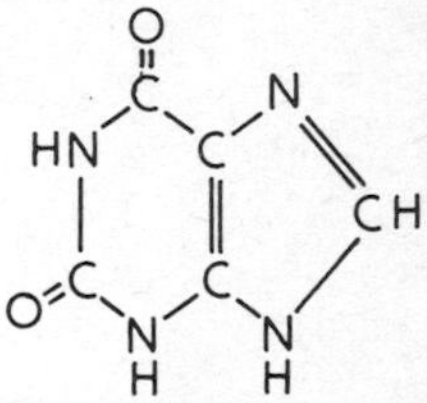

xanthine

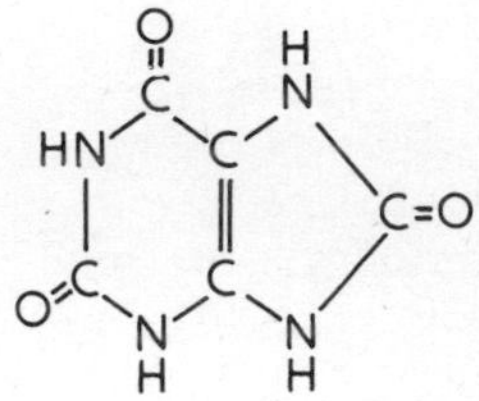

uric acid

Nucleosides. When a purine or a pyrimidine base condenses with a sugar a *nucleoside* is formed. Thus the condensation product of adenine with ribose is the ribonucleoside *adenosine*.

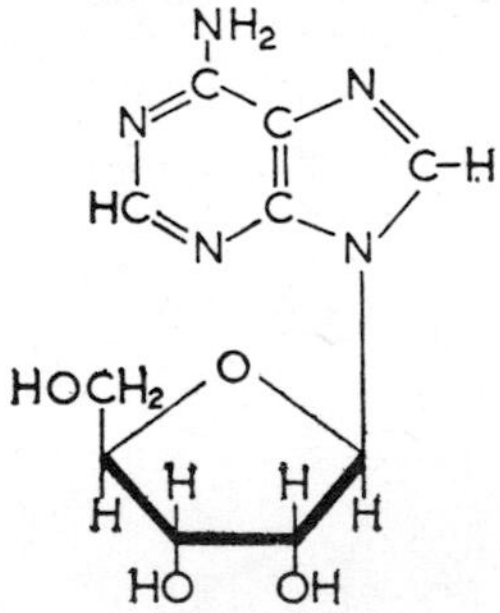

adenosine

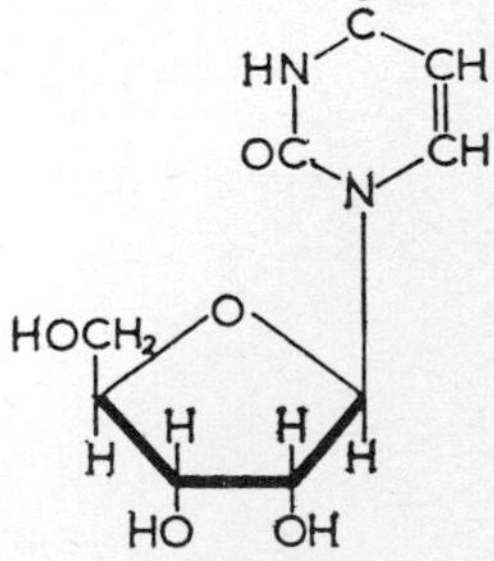

uridine

Nucleosides formed with ribose are termed ribonucleosides; while those formed with deoxyribose instead of ribose, are termed deoxyribonucleosides (see Table 6.1).

Table 6.1

Base	Ribonucleoside	Deoxyribonucleoside
Cytosine	Cytidine	Deoxycytidine
Uracil	Uridine	—
Thymine	—	Thymidine
Adenine	Adenosine	Deoxyadenosine
Guanine	Guanosine	Deoxyguanosine
Hypoxanthine	Inosine	—

In order to differentiate between carbon atoms in the sugar and the carbon and nitrogen atoms in the base, the carbon atoms in the sugar are numbered 1', 2', 3', and so on. In the pyrimidine nucleosides, the nitrogen atom at position 1 of the pyrimidine ring is linked to the carbon atom at position 1' of the sugar; in purine nucleosides the link is between nitrogen atom 9 of the purine and carbon atom 1' of the sugar.

Nucleotides. When the pentose residue of a nucleoside is esterified with phosphoric acid a *nucleotide* is formed. For example, guanosine can condense with phosphoric acid to give the ribonucleotide *guanosine monophosphate* (GMP), and similarly adenosine forms a ribonucleotide *adenosine monophosphate* (AMP),

adenine—ribose—phosphate.

Since ribose bound in nucleoside form possesses free hydroxyl groups on carbons 2', 3' and 5', the phosphate residue may be attached in any of these positions. All such ribonucleoside monophosphates have been isolated but those most commonly encountered are the 5'-phosphates and the 3'-phosphates. Cyclic 3':5'-adenosine monophosphate has an important function in hormonal control mechanisms (see Chap. 50).

Deoxynucleotides are formed from deoxyribose. Since this sugar has no hydroxyl group in position 2', only two deoxyribonucleotides can be formed, the 3' and 5' phosphate derivatives, e.g. deoxyguanosine 5'-monophosphate (dGMP).

The nucleoside 5'-monophosphates of ribose and deoxyribose respectively may be represented schematically thus:

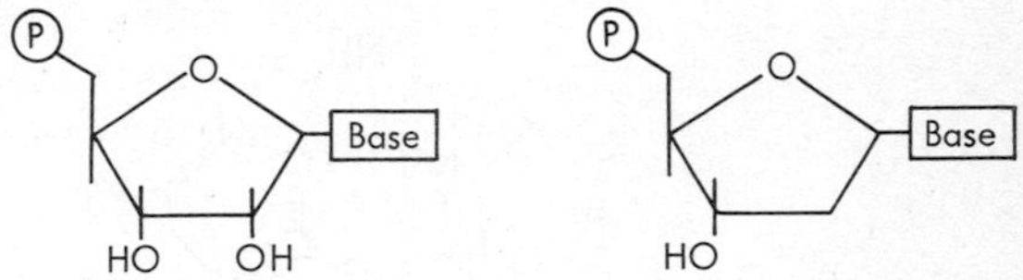

Nucleoside 5'-diphosphates and nucleoside 5'-triphosphates have one or two further molecules as phosphoric acid linked as acid anhydrides to the 5'-phosphate of the mononucleotide. For example, guanosine 5'-diphosphate (GDP) and guanosine 5'-triphosphate (GTP) are formed from GMP. The

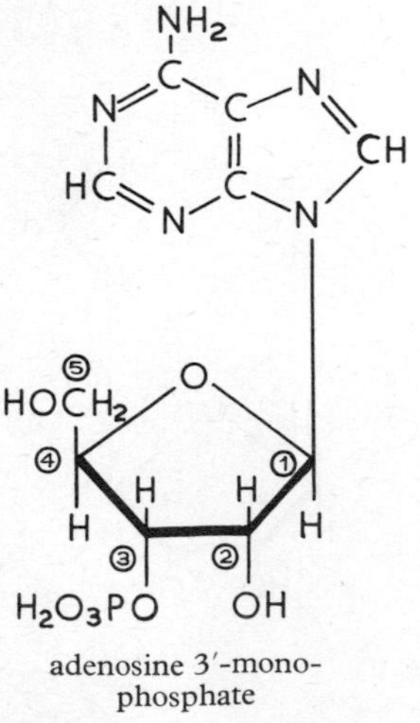

adenosine 3'-monophosphate

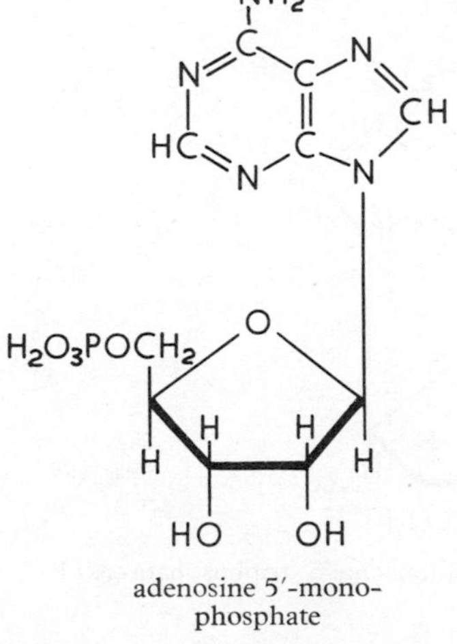

adenosine 5'-monophosphate

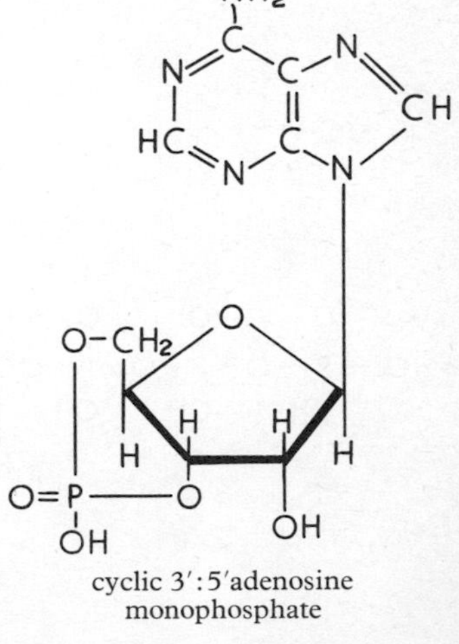

cyclic 3':5'adenosine monophosphate

standard abbreviations for nucleotides are shown in Table 6.2.

Table 6.2 Standard abbreviations for nucleotides

Adenosine 5′-monophosphate	AMP
Adenosine 5′-diphosphate	ADP
Adenosine 5′-triphosphate	ATP
Adenosine 3′:5′ cyclic monophosphate	cAMP
Guanosine 5′-monophosphate	GMP
Guanosine 5′-triphosphate	GTP
Inosine 5′-monophosphate	IMP
Cytidine 5′-monophosphate	CMP
Uridine 5′-monophosphate	UMP
Deoxyadenosine 5′-monophosphate	dAMP
Deoxyguanosine 5′-monophosphate	dGMP
Deoxycytidine 5′-monophosphate	dCMP
Thymidine 5′-monophosphate	dTMP

The nucleotide coenzymes. Many important coenzymes (see Chap. 8) are nucleotides and their structure may conveniently be discussed at this stage.

Adenosine di- and triphosphates. Adenosine diphosphate (ADP) and *adenosine triphosphate* (ATP) play a central role in metabolism (see p. 151). In ATP the phosphate nearest the ribose is termed the α-phosphate group while the other phosphates are labelled β and γ.

Nicotinamide nucleotides. We have seen that the nucleotides of nucleic acids contain purine and pyrimidine bases. In addition, certain dinucleotide coenzymes are found which contain the base *nicotinamide (nicotinic acid amide)*, and are called *nicotinamide nucleotides*. Nicotinamide is a vitamin of the B complex essential for the prevention of pellagra (Chap. 13).

adenosine 5′-diphosphate (ADP)

adenosine 5′-triphosphate (ATP)

One of these nucleotides, known as *nicotinamide-adenine dinucleotide* (NAD), formerly known as *diphosphopyridine nucleotide* (DPN) or *Coenzyme I*, contains the two bases adenine

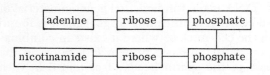

nicotinic acid   nicotinamide

and nicotinamide. It consists of AMP condensed through the phosphate grouping with a nucleotide containing nicotinamide to form a dinucleotide thus:

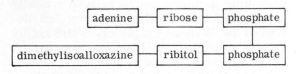

A second such nucleotide *nicotinamide-adenine dinucleotide phosphate* (NADP) differs from NAD in having a third phosphate in position 2' in the ribose residue which is attached to adenine. NADP was formerly known as *triphosphopyridine nucleotide* (TPN) or *Coenzyme II*.

Flavin nucleotides. A third important dinucleotide is *flavin-adenine dinucleotide* (FAD). which differs from NAD in containing *dimethylisoalloxazine* in place of nicotinamide thus:

The name *flavin* is used for the condensation product of dimethylisoalloxazine with *ribitol*, the alcohol corresponding to ribose. The phosphate ester of flavin is known as *flavin mononucleotide* (FMN).

Coenzyme A. One of the most important coenzymes in carbohydrate and fat metabolism (Chaps. 18 and 19) is coenzyme A, discovered by Lipmann. In structure it resembles the nicotinamide nucleotides, but in place of nicotinamide it contains the vitamin pantothenic acid bound to 2-mercaptoethylamine (thioethanolamine) ($HS—CH_2CH_2—NH_2$).

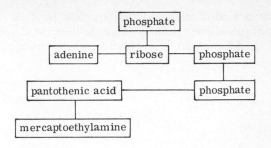

Since the reactive part of the molecule is the sulphydryl (—SH) group, coenzyme A is frequently denoted by the contraction HS—CoA.

Uridine coenzymes. The coenzyme nucleotides so far mentioned all contain adenine but others contain the pyrimidine base uracil. The most important of these is *uridine diphosphoglucose* (UDPGlucose) which is involved in the conversion of galactose to glucose. It is also known as uridine pyrophosphoglucose (UPPGlucose).

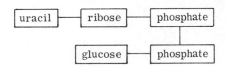

THE NUCLEIC ACIDS

Two types of nucleic acids occur in nature, deoxyribonucleic acid (DNA) and ribonucleic acid (RNA). They are large polymeric molecules the monomer units of which are mononucleotides; nucleic acids are also referred to as polynucleotides.

In DNA, the monomers are deoxyribonucleotides, and the chief bases are adenine (A), guanine (G), cytosine (C) and thymine (T). In RNA, the monomers are ribonucleotides and the chief bases are A, G, C and uracil (U). Small amounts of unusual bases (that is bases other than A, G, C, T and U), often bearing methyl groups or hydroxymethyl groups, are also found in both DNA and RNA.

The constituent nucleotides in the nucleic acids are joined together by ester linkages between the phosphate group on carbon-5' of the sugar of one nucleotide and 3'-hydroxyl of the sugar of the next nucleotide in the sequence. In other words, the polynucleotide chains are composed of monomer units linked by 3′, 5′-phosphodiester bonds. This is illustrated in Fig. 6.3 for part of a polynucleotide chain of

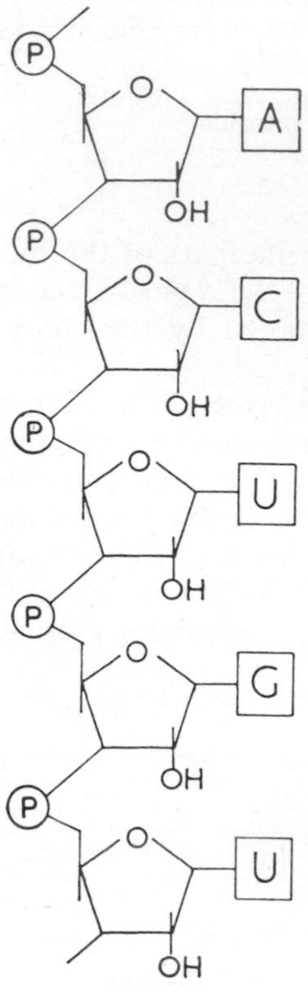

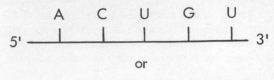

or

ACUGU

Fig. 6.4 Simplified method of representing the sequence of nucleotides in a stretch of polynucleotide. The 5′- and 3′ ends indicate the orientation of the chain in relation to carbon-5′ of the ribose residue at one end and carbon-3′ of the ribose at the other end, as shown in Fig. 6.3.

orientation apply to polydeoxyribonucleotides (DNA chains) (Plate 6.1).

DEOXYRIBONUCLEIC ACID (DNA)

DNA is the fundamental genetic material of all cells. In multicellular organisms it is found in the cell nucleus where it exists in combination with the basic protein *histone* or, in the case of the sperm heads of certain fish, with *protamine*. This complex of DNA and basic protein is termed *deoxyribonucleoprotein*. In bacteria DNA is associated with only a small amount of protein.

Each molecule of DNA is very large with a molecular weight of several millions, and consists of two single chains of polydeoxyribonucleotide held alongside each other by *hydrogen bonds*. (These are weak bonds each formed by the sharing of a hydrogen atom between two electronegative atoms, for example, N or O.) The configuration of the DNA molecule is such that the two chains are twisted around each other to form a double helix. A crucially important feature of this structure, first described by Watson and Crick, is the highly specific nature of the hydrogen bonding. A forms hydrogen bonds only with T, and G only with C (Plates 6.1, 6.2 and Fig. 6.7). Thus along the double helix, only AT, TA, GC and CG *base pairs* are found. Each strand of polydeoxyribonucleotide in the DNA molecule is said to be the complement of the other, and this complementarity is dictated by the specificity of formation of the hydrogen bonds. This has consequences of very great importance in the process of DNA replication.

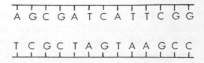

Fig. 6.3 Part of a polynucleotide chain in RNA showing the phosphate bonds connecting carbon atom 5′ of the sugar in one nucleotide to carbon atom 3′ of the sugar in the next nucleotide in sequence.

RNA. To simplify the diagrammatic representation of the nucleic acids, it is helpful to bear in mind that the monomer units all have in common the sugar (either ribose, in RNA or deoxyribose, in DNA) and phosphate radicals, and that differences can occur only in the base moieties. Therefore, the short stretch of polynucleotide shown in Fig. 6.4 may conveniently be represented as in Plate 6.2. It is important to note that the chain has a definite orientation, and that the convention of writing the sequence ACUGU indicates that A is situated at the 5′-end and U at the 3′-end. Although a polyribonucleotide (RNA chain) is illustrated in Fig. 6.3 the same considerations of 3′, 5′-phosphodiester bonds and chain

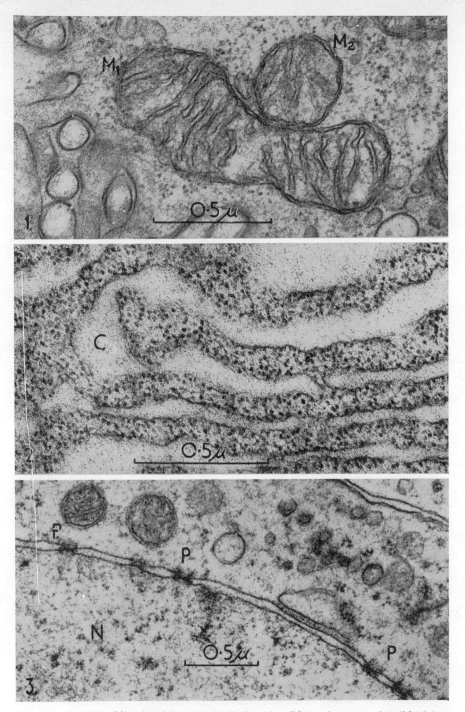

Fig. 6.5 *Top* (1).—Mitochondria in longitudinal section (M_1) and cross-section (M_2) lying close to the surface membrane of the cell.
Middle (2).—Cisternae of grandular endoplasmic reticulum showing cavities (C) bounded by membranes. Ribosomes are attached to the outer, or cytoplasmic surfaces of the cisternae, and also lie free in the cytoplasm.
Bottom (3).—A portion of the nucleus (N) is bounded by the inner and outer nuclear membrane. Several nuclear pores (P) are present, each of which is closed by a diaphragm. (*By courtesy of P. G. Toner and G. M. Wyburn.*)

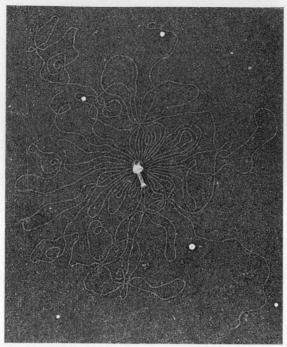

Fig. 6.6a Electron micrograph of a particle of T2 bacteriophage osmotically disrupted to release the thread of DNA which it contains. Note that there are only two 'ends' visible. The empty capsid comprising head and tail components is in the centre of the micrograph. Magnification ×72 000. A. K. Kleinschmidt, D. Lang, D. Facherts & R. K. Zahn (1962). *Biochimica biophysica Acta* **61**, 857–864. (*By courtesy of D. Lang.*)

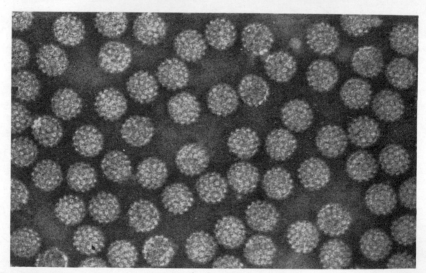

Fig. 6.6b Electron micrograph of human wart virus. Magnification ×150 000. Human wart virus can be readily extracted from warts and verrucas. The protein coat or capsid has an icosohedral symmetry and is composed of 72 subunits or capsomeres. The 12 capsomeres which lie at the points of the icosohedron are each made up of 5 protein subunits and the remaining 60 capsomeres which lie on the faces and edges of the icosohedron, are each made up of 6 protein subunits. Each subunit is a single protein molecule, that is the complete virus capsid is an assembly of 420 protein molecules. Each virus contains one molecule of double-stranded DNA molecular weight 4 000 000, in cyclic form. (*By courtesy of E. A. Follett.*)

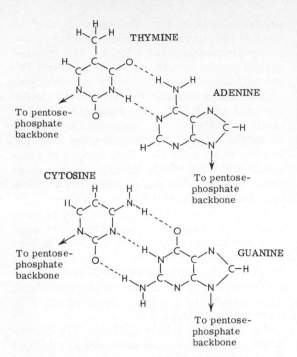

Fig. 6.7 Hydrogen bond formation between A and T, and between G and C. Note that the AT base pair is weaker than the GC pair because it can form two hydrogen bonds while the latter forms three. Since thymine is 5-methyluracil, it is implicit in this diagram that A could equally well form hydrogen bonds with U.

The two complementary polynucleotide strands representing a segment of DNA in Plate 6.1 lie parallel to each other. However, closer examination of the strands reveals that the orientation of the nucleotide units is 3′ to 5′ upwards in one strand and 3′ to 5′ downwards in the other. Because of this opposing orientation the strands are said to be anti-parallel.

In the double-helical condition, the molecule of DNA is said to be native. Individually,

each hydrogen bond is weak, but collectively the hydrogen bonds in the double helix give rise to a stable molecule. However, if the molecule is exposed to extremes of pH, or to high temperature (80° to 100°), the double helix collapses owing to disruption of the hydrogen bonds. This event is termed *denaturation* and occurs without scission of covalent bonds. Two separate, single chains of polydeoxyribonucleotide result from the denaturation; this single-stranded configuration is called the *random coil*. Under certain conditions and with certain species of DNA, it is possible to restore the random coils to the original double-helical state; in this process of renaturation the hydrogen bonds between the base pairs reform in such a way that the two complementary strands re-unite in precise register giving the original molecule of native DNA.

Variations in the relative proportions of bases in DNA's from different sources are shown in Table 6.8. Although the proportions vary widely from one DNA to another, they are constant for the DNA of any one species and show equimolar proportions of A and T, and of G and C (or a cytosine derivative such as 5′-methylcytosine). Such equivalence is, of course, essential to the DNA molecule with its double helical conformation comprising two complementary, anti-parallel strands.

The genetic information carried by DNA is encoded in the sequence of bases (that is nucleotides) along one strand of the DNA molecule (Chap. 20). On first consideration, it may seem highly improbable that molecules of DNA, located within a cell which is too small to be seen by the naked eye, could ever store the genetic information required for the elaboration of a complex living organism such as a mammal. Yet the DNA of one spermatozoon

Table 6.8 Molar proportions of bases in DNAs from several sources

Source of DNA	Adenine	Thymine	Guanine	Cytosine	5-Methyl cytosine
Bovine thymus	28·2	27·8	21·5	21·2	1·3
Rat bone marrow	28·6	28·4	21·4	20·4	1·1
Wheat germ	27·3	27·1	22·7	16·8	6·0
Yeast	31·3	32·9	18·7	17·1	—
Escherichia coli	26·0	23·9	24·9	25·2	—
Tubercle bacillus	15·1	14·6	34·9	35·4	—

and one ovum can indeed, after fusion of these two cells, initiate and control the production of an adult animal.

A gene, which represents the genetic information required to form one protein, is a stretch of DNA containing on an average some 1000 nucleotide pairs.

Further consideration reveals that DNA molecules are unquestionably large enough to be at least potentially capable of storing vast amounts of information. For example, given that there are 6×10^{-12} g of DNA per human diploid cell, that the molecular weight of a nucleotide pair is 600 and that Avogadro's number is 6×10^{23}, then number of nucleotide pairs per cell

$$= \frac{(6 \times 10^{-12}) \times (6 \times 10^{23})}{600}$$
$$= 6 \times 10^9$$

since the internucleotide distance in the double helix is 0·34 nm, therefore the total length of DNA per cell

$$= 6 \times 10^9 \times 0\cdot34 \text{ nm} = 2 \text{ metres}$$

that is the total length of DNA contained in the nucleus of one diploid cell is 2 metres.

The human body contains about 10^{13} cells, that is 10^{10} miles of DNA. The distance from the earth to the sun is a mere 9×10^7 miles.

The bacterium *Escherichia coli*, has one chromosome, a single molecule of DNA of mol. wt. about 2×10^9. The molecule is a cyclic double helix and its extended length is about 1 mm.

DNA of high molecular weight is found predominantly in cell nuclei but also, in small amounts, in mitochondria and chloroplasts. It seems probable that these cell organelles are self-replicating and this would be in keeping with each organelle having its own complement of DNA.

Ribonucleic Acid (RNA)

RNA exists in the nucleus of the animal cell and in the cytoplasm, but chiefly (in the quantitative sense) in the cytoplasm. It may be divided into several categories.

(a) RIBOSOMAL RNA (rRNA). This is found mainly in the cytoplasm in the minute ribonucleoprotein particles known as ribosomes (p. 76), and it accounts for up to 80 per cent of the total cell RNA. Two molecules of RNA occur in each ribosome, both of relatively high molecular weight, one approximately $0\cdot7 \times 10^6$ and the other $1\cdot6 \times 10^6$. Both species of ribosomal RNA are single-stranded, but possess some measure of *secondary* structure to allow tight packing within the intact ribosome. This is probably promoted by the existence of some double-helical stretches displaying equivalence of A and U, and of G and C, as shown in Fig. 6.9. These short stretches therefore resemble DNA in double-stranded and anti-parallel conformation; but it is emphasized that, unlike DNA, there is only one polynucleotide chain in the molecule.

(b) TRANSFER RNAs (tRNA). These contain 75 to 80 nucleotides and have a molecular weight of about 25 000. Several unusual bases are found in transfer RNAs (see Fig. 6.10). Transfer RNAs account for 15 to 20 per cent of the total RNA of the cell.

Transfer RNAs function as carriers of activated amino acids during protein synthesis (see Chap. 21). More than one transfer RNA molecule is specifically responsible for the transfer of each amino acid.

The entire *primary* structures (that is the sequences of the nucleotides) of several transfer RNAs are now known and one of these, alanyl-transfer RNA is given in Fig. 6.10. Like all other transfer RNA molecules alanyl-transfer RNA has the base sequence CCA in the three nucleotides at the 3′-end of the molecule, and has G in the nucleotide at the 5′-end. Other parts of the molecule are different in different transfer RNA species.

(c) MESSENGER RNA (mRNA). This class of RNA is of high molecular weight (perhaps up to several million) and accounts for only about 1 per cent of the total RNA of the cell. Its function is to convey genetic information from the cell nucleus to the protein-synthesizing centres in the cell, where, in collaboration with ribosomes and transfer RNA, it engages in the complex process of protein synthesis (Chap. 21).

The base compositions of ribosomal RNAs,

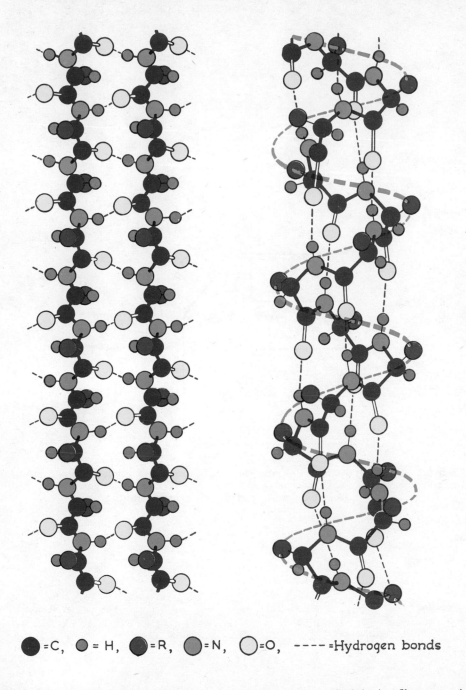

● =C, ● = H, ● =R, ● =N, ○ =O, ---- =Hydrogen bonds

Plate 3.1 *Left.* The three-dimensional arrangement of the polypeptide chains in a fibrous protein in the *β*-form showing the 'pleated sheet' structure. *Right.* The three-dimensional arrangement of the polypeptide chains coiled into the α-helix of Pauling. (*Modified from Doty.*)

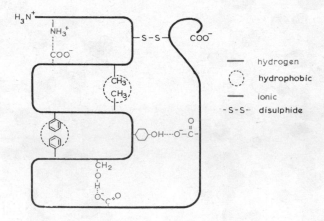

Plate 3.2 The bonds which hold together the tertiary structure of a protein are of three main types, hydrogen bonds, hydrophobic bonds and ionic bonds. The disulphide linkage which is also shown is essentially part of the primary structure.

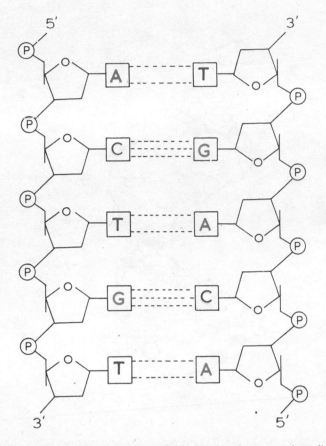

Plate 6.1 Part of a polynucleotide chain in DNA showing the two anti-parallel (and complementary) strands connected by hydrogen bonds (dotted lines), linking A (adenine) to T (thymine), or G (guanine) to C (cytosine).

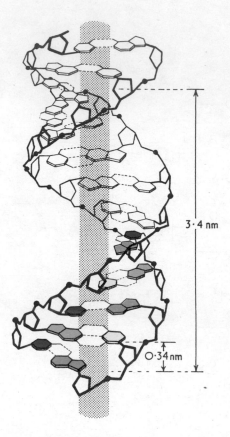

Plate 6.2   The double helix of DNA. The two polynucleotide strands are linked by hydrogen bonds (dotted lines) connecting adenine (green) to thymine (red), or guanine (blue) to cytosine (yellow). The deoxyribose residues are shown as pentagons and the phosphate groups as black discs. There are ten base-pairs for each complete turn (360°) of the helix.

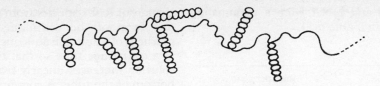

Fig. 6.9 Part of a strand of ribosomal RNA showing double-helical stretches within the single strand of the molecule. This intra-strand secondary structure arises from complementary regions occurring next to each other sequentially along the molecule. Thus, the equivalence of A with U, and G with C in these regions allows the molecule to fold back upon itself and to assume a double-helical configuration at these regions.

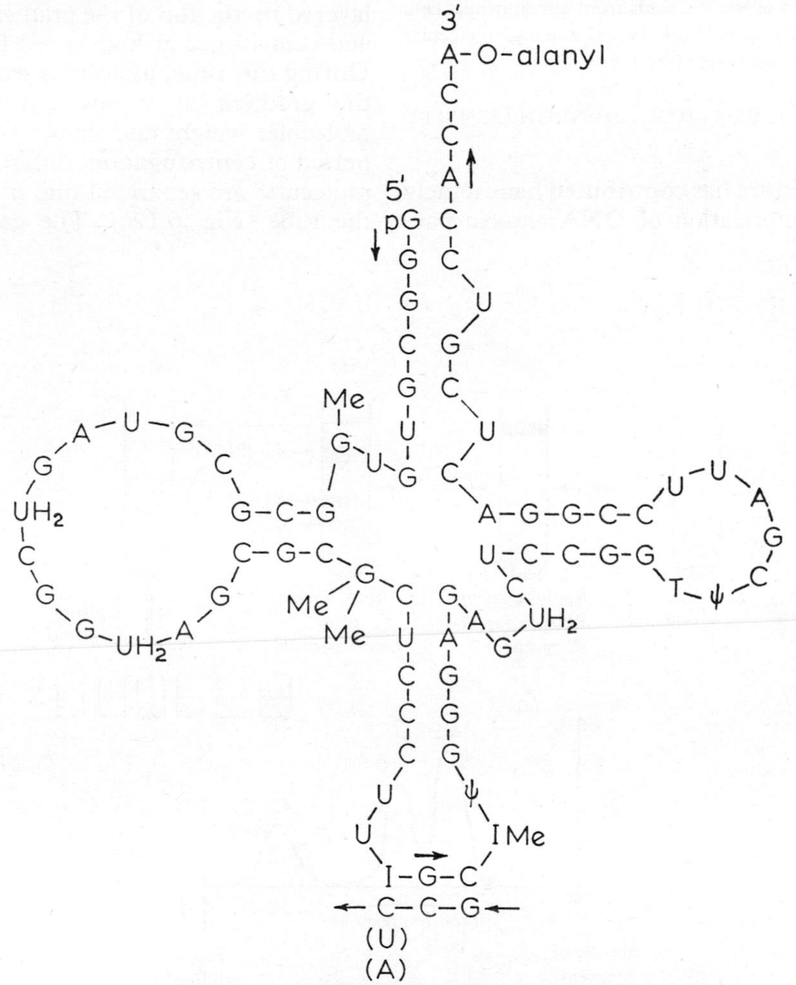

Fig. 6.10 The *primary* structure (nucleotide sequence) of alanyl-transfer RNA from yeast. The molecule is shown 'charged' or 'loaded' with alanine. In this diagram, the molecule has been 'looped out' in certain areas to allow complementary stretches to come into register (A with U and G with C). Note the presence of ten 'unusual' bases (I = hypoxanthine; UH$_2$ = dihydrouracil; ψ = pseudo-uridine; Me = methyl radical). The presence of the extraneous triplet of nucleotides at the bottom of the diagram is important and will be explained in Chapter 21.

transfer RNA and DNA from a mammalian cell are given in Table 6.11.

Table 6.11

	Moles per cent				
	A	U	G	C	T
Ribosomal RNA I	16·5	17·4	34·6	31·6	—
Ribosomal RNA II	20·3	22·3	32·2	25·5	—
Transfer RNA	21·1	24·3	29·5	25·1	—
DNA	29·8	—	20·1	20·0	30·1

Base composition of nucleic acids from a mammalian cell. Ribosomal RNA's I and II are the low and high molecular weight species, respectively (see Fig. 6.15).

ZONE CENTRIFUGATION THROUGH DENSITY GRADIENTS

This procedure has contributed handsomely to the characterization of DNA and, in par-ticular, of RNA species from many sources. In one form of the technique a concentration gradient of a suitable aqueous solution is set up in a centrifuge tube so that the density of the solution increases linearly from the top to the bottom of the tube. A common type of density gradient containing 5 per cent sucrose at the top of the tube gradually increasing to 25 per cent at the bottom of the gradient is shown in Fig. 6.12a. Gradients formed with solutions of a variety of salts, in particular caesium chloride, are also used.

The nucleic acid solution to be analysed is layered on the top of the gradient (Fig. 6.12b), and centrifuged at high speed for some hours. During this time, molecules sediment through the gradient at a rate governed by their molecular weight and shape; at the end of the period of centrifugation, different nucleic acid molecules are separated into discrete layers in the tube (Fig. 6.12c). The gradient may be

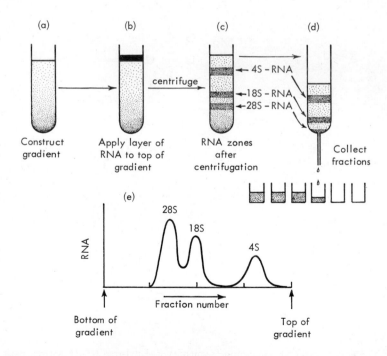

Fig. 6.12 The process of gradient centrifugation. A sucrose gradient is constructed in a centrifuge tube (a), the RNA solution is then applied carefully as a layer on top (b), and on centrifuging, the RNA separates out into its main components according to the molecular weight and shape of the molecules (c). When the tube is punctured (d), the gradient is collected a few drops at a time in tubes to give sequential fractions. The amount of RNA in each fraction is then determined by ultraviolet absorption measurements so as to give the pattern shown (e).

analysed by puncturing the bottom of the tube with a syringe needle, through which fractions are collected dropwise and analysed for their content of nucleic acid (Fig. 6.12d). The distribution of ribosomal and transfer RNA species is illustrated in Fig. 6.12c and d. The movement of each macromolecule through the gradient is related to its sedimentation coefficient which is expressed in Svedberg units (S). Thus the two ribosomal RNA species have sedimentation coefficients of 18S and 28S, while transfer RNA has a sedimentation coefficient of 4S (Fig. 6.12d).

Relatively large particles (for example virus particles and ribosomes) can be fractionated and characterized in this way.

CYTOCHEMISTRY

A schematic diagram of a 'typical' animal cell is shown in Fig. 6.13. The cell is surrounded (or limited) by a cytoplasmic membrane. In the centre is the *nucleus* containing a mesh-work of densely staining deoxyribonucleoprotein, the *chromatin* of the histologist. During preparation for cell division the chromatin condenses to form the *chromosomes* which contain all the DNA of the cell. The nucleus is surrounded by a double membrane pierced at intervals by pores, and contains one or more dense, spherical bodies termed *nucleoli* which are rich in RNA and are the sites of ribosome synthesis (see below).

Surrounding the nucleus is the *cytoplasm* in which are found various inclusions such as secretion granules, lysosomes etc. and the rod-like bodies known as *mitochondria* (Fig. 6.5) (dimensions 0·5 to 5 μm by 0·3 to 0·7 μm) bounded by a double membrane, the inner layer of which is folded to produce a number of partitions, the *cristae mitochondriales*, which divide the interior of each mitochondrion into compartments. Each membrane consists of a layer of protein molecules lined by a double layer of lipid molecules (Fig. 6.14). The mito-

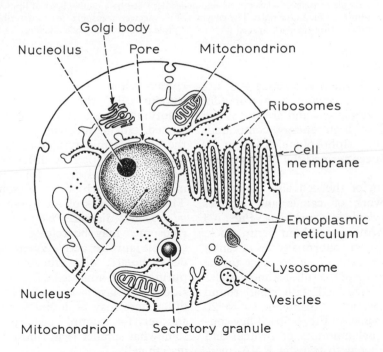

Fig. 6.13   Schematic diagram of a 'typical' animal cell (J. N. Davidson, 1972).

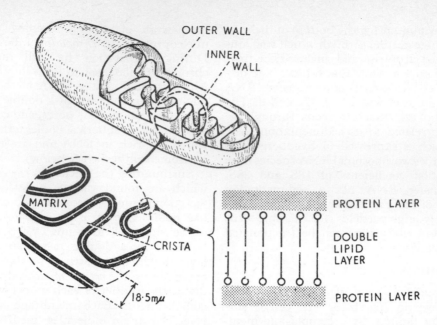

OUTER WALL

INNER WALL

MATRIX

CRISTA

18·5mμ

PROTEIN LAYER

DOUBLE LIPID LAYER

PROTEIN LAYER

Fig. 6.14 Mitochondrion (upper) showing double wall. The inner membrane is folded into a series of partitions, the cristae mitochondriales. The interior of the mitochondrion contains a fluid matrix. The lower diagram on the left shows an enlarged view of the double membrane. Each membrane consists of a layer of protein molecules lined by a double layer of lipid molecules as shown on the lower right. The enzymes of the electron transport system described in Chapter 10 are regularly spaced units in the protein monolayers. See also Fig. 5.4.

chondria (of which there are about 400 in a liver cell) contain the enzymes responsible for oxidative phosphorylation, and are the site of the production of high energy compounds (such as adenosine triphosphate, ATP) in the cell. They have been termed the 'power plants of the cell'.

The cytoplasm of the cell also contains a complex meshwork of canals and vesicles known as the *endoplasmic reticulum* (Figs. 6.5 and 13) with membranes about 5 nm thick. These tubules appear to form a series of canals leading from the exterior of the cell to the nucleus. The surface of the tubules and vesicles in some areas is studded with small round electron-dense particles (diameter 10 to 20 nm) known as *ribosomes* (Fig. 6.5b). They consist of about equal amounts of protein and RNA and, together with transfer RNA, messenger RNA, activated amino acids, enzymes and other factors, they engage in the complex process of protein synthesis (Chap. 21). Ribosomes also occur free in the cytoplasm, unattached to reticular membranes. Indeed, in

certain types of cell (intestinal epithelium, tumour cells) the endoplasmic reticulum is scanty and most of the ribosomes are unattached. During mechanical disruption of cells, the endoplasmic reticulum also is disrupted and, after isolation by differential centrifugation, is obtained in the form of fragments of membrane, each bearing some ribosomes. These fragments of membrane are known as *microsomes* (diameter 60 to 150 nm) and the ribosomes may be released from them by a suitable surface active agent such as sodium deoxycholate. The ribosomes may then be purified and analysed (Fig. 6.15). While all ribosomes have a basically similar structure, differences may be found from one source to another, for example bacterial ribosomes are somewhat smaller than ribosomes from mammalian cells.

When a ribosome engages in the process of protein synthesis (Chap. 21), it becomes attached to one end of a molecule of messenger RNA (mRNA). It then starts moving along the mRNA and, with the aid of tRNA molecules

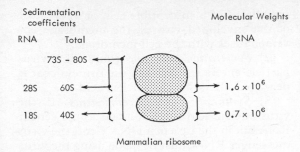

Mammalian ribosome

Fig. 6.15 The S values and molecular weights of a mammalian ribosome. The corresponding values for the appropriate RNA's are also shown.

and certain other factors, translates the nucleotide sequence of the mRNA into the amino acid sequence of a protein or proteins. At any one instant, several ribosomes may be found moving along a given molecule of mRNA, each ribosome being at a different stage of translation of the mRNA. This complex of mRNA with ribosomes is known as a polyribosome or *polysome*. A polysome may be likened to beads threaded on to a string.

The general cell matrix or *cell sap* in which the endoplasmic reticulum and mitochondria are embedded contains the low molecular weight transfer RNA.

The cell cytoplasm also contains, in most cells at least, small organelles known as *lysosomes* which are essentially little sacks of hydrolytic enzymes which can break down many different kinds of large molecules. The enzymes found in lysosomes include cathepsins which break down proteins, ribonuclease, deoxyribonuclease, β-glucuronidase and acid phosphatase. It is believed that they are responsible for the dissolution of damaged or dead cells during the process of autolysis (but

they appear also to play an important role in relation to immunology and disease).

Peroxisomes are particles somewhat smaller than mitochondria. They correspond to the micro-bodies seen in electron micrographs of liver and they contain the enzymes urate oxidase, D-amino acid oxidase and catalase. The first two enzymes produce hydrogen peroxide and the third destroys it.

Organized soluble macromolecules within the cytoplasm of the cell form the many enzyme groupings which are permanent and carry out organized chemical function. They are of the nature of large protein macromolecules which are soluble and are not normally considered as cell organelles. An example is the fatty acid synthetase (Chap. 19) which is an enzyme complex just visible in the electron microscope. It performs the whole function of fatty acid synthesis without liberating any intermediate products into the cell's metabolic pool.

It is now possible to isolate many of the subcellular components from disrupted cells, by centrifugation and other techniques. This allows a considerable degree of correlation to be made between morphology and biochemical function.

VIRUSES

Viruses are infective agents which multiply inside the cells of susceptible animals and are responsible for diseases such as smallpox, poliomyelitis, influenza etc. They are very small, ranging in size from 20 to 200 nm, and most consist only of one molecule of RNA or DNA surrounded by a protein coat (Table 6.16). The nucleic acid carries the genetic information necessary for virus multiplication;

Table 6.16 Properties of some viruses

Virus	Host	RNA (%)	DNA (%)	Mol. wt. of nucleic acid	Size
Poliomyelitis	Man	25	—	2×10^6	30 nm diam.
Influenza	Man	0·9	—	3×10^6	80 nm diam.
SV40	Vertebrates	—	13	3×10^6	40 nm diam.
Vaccinia	Vertebrates	—	3·2	180×10^6	200×300 nm
Tobacco mosaic virus	Tobacco plants	5	—	2×10^6	15×300 nm
Tomato bushy stunt virus	Tomato plants	15	—	$1·6 \times 10^6$	28 nm diam.
T2	*E. coli*	—	50	120×10^6	head 65×95 nm tail 20×95 nm
MS2	*E. coli*	32	—	$1·05 \times 10^6$	17 nm diam.

Nucleotides and Nucleic Acids   77

the protein coat protects the nucleic acid and also confers a host specificity on the virus particle.

All viruses depend absolutely on the host cell for energy (ATP) and a supply of precursors (for example, amino acids and nucleotides) necessary for the synthesis of new virus particles. Outside the cell, virus particles are quite inert.

Most living things, including animals (vertebrates and invertebrates), plants and microorganisms, are susceptible to viruses. Viruses which infect bacteria are known as bacteriophages.

STRUCTURE OF VIRUSES. Most viruses contain either a molecule of single-stranded RNA or a molecule of double-stranded DNA but some viruses contain unusual forms of nucleic acids. Reovirus contains double-stranded RNA which has a double helical structure similar to DNA. The minute virus of mice contains a single-stranded DNA molecule. Human wart virus (Fig. 6.6b) contains a cyclic DNA molecule.

The nucleic acid is sometimes complexed with basic proteins which may help it to adopt the compact conformation necessary to allow it to fit inside the protein coat. The coat (capsid) is usually a geometrical (often icosohedral) arrangement of subunits (capsomeres). Each capsomere is a group of coat protein molecules. The capsomeres are the morphological units of the capsid when viewed in the electron microscope (Fig. 6.6b). In some animal viruses the capsid is surrounded by a membrane envelope which is formed partly from the membrane of the host cell and partly from new viral components.

REPLICATION OF VIRUSES. The replication cycle of a virus can be conveniently divided into five stages: (1) absorption, (2) penetration and uncoating, (3) synthesis of virus components, (4) assembly, (5) release.

(1) *Absorption.* The virus capsid binds to specific receptor sites on the cytoplasmic membrane of a susceptible cell, or in the case of the enveloped viruses, the membrane envelope fuses with the cell membrane.

(2) *Penetration and uncoating.* The virus particle enters the cell and the protein coat is degraded liberating the nucleic acid.

(3) *Synthesis of virus components.* If the virus nucleic acid is RNA it acts as messenger RNA but in the case of a DNA virus, the virus messenger RNA must be made using the virus DNA as a template (see Chap. 21). Virus proteins are synthesized on the cellular ribosomes with the help of virus messenger RNA. These proteins are of two classes; enzymes for the replication of the viral nucleic acid and proteins for the structural components of the virus particle.

(4) *Assembly.* The nucleic acid interacts with the structural proteins to form new virus particles.

(5) *The cell* may die releasing the virus, or the virus particles may 'bud off' from the surface of the cell.

Depending on the type of virus, from 10^2 to 10^5 new virus particles are made in an infected cell.

TUMOUR VIRUSES. Some types of virus modify the properties of the infected cell to form tumour cells. The DNA of DNA tumour viruses such as simian virus 40 (SV40) becomes integrated into the cellular DNA, and thereafter replicates with the cellular DNA before each cell division. In this way, the constant production of virus messenger RNA and virus proteins leads to a modification of the cellular growth properties.

It is not known how the RNA of RNA tumour viruses, such as the feline leukaemia virus, permanently modifies the properties of the infected cell. Tumour virus give rise to a variety of solid tumours and leukaemias in experimental animals but as yet there is no definitive evidence that human cancers can be caused by viruses.

REFERENCES

ALLISON, A. C. (1965). Lysosomes in disease. *Science Journal* 1, 32–38.

ASHWELL, M. & WORK, T. S. (1970). The Biogenesis of mitochondria. *Annual Review of Biochemistry* 39, 251–290.

BIRNSTIEL, M. & CHIPCHASE, M. (1970). The nucleolus: pacemaker of the cell. *Science Journal* 6, 41–47.

BRACHET, J. & MIRSKY, A. E. (Eds.) (1959). *The Cell* (five volumes). New York: Academic Press.

CAMPBELL, P. N. (1966). *The Structure and Function of Animal Cell Components.* Oxford: Pergamon Press.

CHARGAFF, E. (1963). *Essays on Nucleic Acids.* Amsterdam: Elsevier.

CHARGAFF, E. & DAVIDSON, J. N. (Eds.) (1955). *The Nucleic Acids*, Vols. 1 and 2; (1960) Vol. 3. New York: Academic Press.

COX, R. (1970). The ribosome—decoder of genetic information. *Science Journal* 6, 56–63.

DAVIDSON, J. N. (1972). *The Biochemistry of the Nucleic Acids*, 7th edn. London: Chapman & Hall.

DAVIDSON, J. N. & COHN, W. E. (Eds.) (1963–1972). *Progress in Nucleic Acid Research*, Vols. 1–12. New York: Academic Press.

DAVIES, B. D., DULBECCO, R., EISEN, H. N., GINSBERG, H. S. & WOOD, W. B. (1971). *Microbiology.* New York: Harper & Row.

DE DUVE, C. & BAUDHUIM, P. (1966). Peroxisomes, microbodies and related particles. *Physiological Reviews* 46, 323–357.

DU PRAW, E. J. (1971). *DNA and Chromosomes.* New York: Holt, Rinehart & Winston.

FELL, H. & BRACHET, J. (Organizers) (1969). A discussion on cytoplasmic organelles. *Proceedings of the Royal Society* B, 173, 3–111.

GREEN, D. E. & GOLDBERGER, R. F. (1967). *Molecular Insights into the Living Process.* New York: Academic Press.

HAGGIS, G. H., MICHIE, D., MUIR, A. R., ROBERTS, K. B. & WALKER, P. M. B. (1964). *Introduction to Molecular Biology.* London: Longmans.

HEARST, J. E. & BOTCHAN, M. (1970). The eukaryotic chromosome. *Annual Review of Biochemistry* 39, 151–182.

HUTCHINSON, D. W. (1964). *Nucleotides and Coenzymes.* London: Methuen.

JUKES, T. H. (1966). *Molecules and Evolution.* New York: Columbia University Press.

KENDREW, J. (1966). *The Thread of Life.* London: Bell.

KURLAND, C. G. (1970). Ribosome structure and function emergent. *Science Journal* 169, 1171–1177.

LIMA-DE-FARIA, A. (1969). *Handbook of Molecular Cytology.* Amsterdam: North Holland Publishing Co.

LOEWY, A. G. & SIEKEVITZ, P. (1970). *Cell Structure and Function*, 2nd edn. New York: Holt, Rinehart & Winston.

NOMURA, M. (1969). Ribosomes. *Scientific American* 221, 28–35.

NORTHCOTE, D. H. (Ed.) (1968). Structure and function of membranes. *British Medical Bulletin* 24, 99–182.

ROSENBERG, E. (1971). *Cell and Molecular Biology.* New York: Holt, Rinehart & Winston.

SMELLIE, R. M. S. (1969). *A Matter of Life—DNA.* Edinburgh: Oliver & Boyd.

SMITH, E. L., DeLANGE, R. J. & BONNER, J. (1970). Chemistry and biology of the histones. *Physiological Reviews* 50, 159–170.

STEINER, R. F. (1965). *The Chemical Foundations of Molecular Biology.* Princeton: Van Nostrand.

STEINER, R. F. & EDELHOCH, H. (1965). *Molecules and Life.* Princeton: Van Nostrand.

TAYLOR, J. H. (Ed.) (1963). *Molecular Genetics, Part I*; (1967) *Part II.* New York: Academic Press.

TONER, P. G. & CARR, K. E. (1968). *Cell Structure.* Edinburgh: Livingstone.

ULBRICHT, T. L. V. (1964). *Purines, Pyrimidines and Nucleotides.* Oxford: Pergamon Press.

VOGEL, H. J., BRYSON, V. & LAMPEN, J. O. (Eds.) (1963). *Informational Macromolecules.* New York: Academic Press.

WATSON, J. D. (1968). *The Double Helix.* New York: Atheneum.

WATSON, J. D. (1970). *Molecular Biology of the Gene*, 2nd edn. New York: Benjamin.

7 Biochemical reactions

Previous chapters have provided a brief description of the main chemical components of living tissues. This chapter and those which follow are concerned with the mechanisms by which these components are formed, the roles they play in the life of the organism and the manner in which they are eventually broken down. One of the most significant conclusions which has emerged from biochemical research in the present century is that all these processes follow the same laws as the simple chemical reactions which occur in the test-tube. There is thus no need to postulate the existence of some special 'vital force' controlling the chemical activities of living organisms. A discussion of these activities can therefore start with a consideration of chemical reactions in general. The student of biochemistry or physiology does not at this stage need to know very much about thermodynamics or reaction mechanisms, but he will find that one or two quite simple ideas will make it a good deal easier to understand his own subject.

CHEMICAL EQUILIBRIUM

It is convenient to start with the observation, first made in the last century, that when reactions take place in solution—as almost all chemical reactions in cells and tissues do—the rate of the reaction is proportional to the concentrations of the reactants. There are exceptions to this generalization, or rather apparent exceptions, especially where catalysts are involved, but for present purposes we can ignore them. If we represent a chemical reaction in the general form

$$A + B + C + \ldots \longrightarrow products$$

we can write

$$Rate\ of\ reaction = k[A][B][C]\ldots$$

where

$[A]$ = concentration of A, usually expressed in molar terms

[B] = concentration of B
[C] = concentration of C

and k is a constant, the velocity constant of the reaction.

The classical example of a reaction which can be treated in this way is the esterification of ethanol and acetic acid

ethanol + acetic acid $\longrightarrow$
$\qquad\qquad\qquad$ ethyl acetate + water

If we mix ethanol and acetic acid the initial rate of the reaction is given by the expression

$$k_{+1}[\text{ethanol}][\text{acetic acid}]$$

As the reaction proceeds the ethanol and acetic acid are used up, their concentrations fall, and so consequently does the rate of the reaction. At the same time ethyl acetate and water, produced by the reaction, accumulate. Since the reaction is reversible, they can in turn react to reform the original reactants

ethyl acetate + water $\longrightarrow$
$\qquad\qquad\qquad$ ethanol + acetic acid

The rate of this reaction is given by the expression

$$k_{-1}[\text{ethyl acetate}][\text{water}]$$

As ethyl acetate and water continue to accumulate, the rate of this 'back reaction' necessarily increases until it equals the rate of the 'forward reaction'. When this happens, a dynamic equilibrium is attained in which the rate of conversion of ethanol and acetic acid into ethyl acetate and water is exactly balanced by the reverse process. There will, therefore, be no further change in the concentrations of any of the participants. We can write

$$k_{+1}[\text{ethanol}][\text{acetic acid}]$$
$$= k_{-1}[\text{ethyl acetate}][\text{water}]$$
$$\therefore \frac{[\text{ethyl acetate}][\text{water}]}{[\text{ethanol}][\text{acetic acid}]} = \frac{k_{+1}}{k_{-1}} = K$$

Since k_{+1} and k_{-1} are constants, their ratio must also be a constant; K is called the *equilibrium constant* of the reaction. The equilibrium described by the above equation could, of course, have been reached equally well if we had started with water and ethyl acetate instead of ethanol and acetic acid, and allowed the reaction to proceed in the reverse direction. Although it is customary to call the compounds on the left-hand side of an equation the reactants and those on the right-hand side the products, this is purely a matter of convention.

THE SIGNIFICANCE OF K.—It is important to be perfectly clear that K represents the ratio of products to reactants when equilibrium has been attained. It is also the *ratio* of the velocity constants of the forward and back reactions but it tells us nothing about the *magnitude* of either of these constants. In other words it tells in what proportions reactants and products will be in equilibrium, but it gives no indication whatever of the *rate* at which this equilibrium will be approached. The expression

$$K = \frac{[\text{ethyl acetate}][\text{water}]}{[\text{ethanol}][\text{acetic acid}]}$$

can be derived from a purely thermodynamic argument without introducing velocity constants at all. Consequently, K is not dependent on the route or mechanism by which reactants are converted into products or *vice versa*. In particular, it is not altered by the presence of catalysts, since these affect the rates of the forward and back reactions to an equal degree. It is important also to appreciate that there is no sharp distinction between reversible and irreversible reactions. In theory, all reactions are reversible, and what is commonly termed an irreversible reaction is simply one in which the back reaction is so slow in comparison with the forward reaction that it can be neglected. Such a reaction would none the less have a definite K value, and in some circumstances it might be important to know its magnitude.

From the biological point of view the equilibrium constant of a reaction is useful because it indicates the direction in which the reaction is most likely to proceed *in vivo*. This is illustrated by the following examples:

(a) One of the best-known of all biochemical reactions is the phosphorylation of glucose at the expense of ATP

glucose + ATP $\rightleftharpoons$ glucose 6-phosphate
$\qquad\qquad\qquad\qquad\qquad$ + ADP

Since $K = 1600$ equilibrium is reached when

$$\frac{[\text{glucose 6-phosphate}][\text{ADP}]}{[\text{glucose}][\text{ATP}]} = 1600$$

that is, when the reactants have been almost completely converted into products. Consequently, this reaction would be an effective method for converting glucose to glucose 6-phosphate or ATP to ADP. On the other hand, if we tried to operate the reaction in the reverse direction, equilibrium would be reached as soon as a very small portion of glucose 6-phosphate and ADP had been converted to glucose and ATP. If, therefore, we consider the reaction in isolation we would have to regard it as virtually irreversible and accordingly summarize it as follows:

glucose + ATP → glucose 6-phosphate + ADP

(b) Theoretically, pyruvate can be phosphorylated at the expense of ATP in the same way as glucose:

pyruvate + ATP ⇌ phosphopyruvate + ADP

But for this reaction $K = 0.0005$, so that equilibrium is attained when

$$\frac{[\text{phosphopyruvate}][\text{ADP}]}{[\text{pyruvate}][\text{ATP}]} = 0.0005$$

that is when only a minute fraction of the reactants has been converted into products. Obviously, therefore, this is not likely to be a practical method of converting pyruvate to phosphopyruvate or ATP to ADP. On the other hand, if we tried to operate the reaction in reverse, equilibrium would not be reached until almost all the phosphopyruvate and ADP had been converted to pyruvate and ATP. So this reaction, like the preceding one, can be regarded as virtually irreversible, but in the opposite direction.

pyruvate + ATP ← phosphopyruvate + ADP

(c) In the two previous examples the equilibrium was clearly in favour of either the forward or the back reaction. But there are many examples in which the balance is more evenly poised. For example, *in vivo* glucose 6-phosphate and fructose 6-phosphate are interconvertible

glucose 6-phosphate ⇌ fructose 6-phosphate

For this reaction $K = 0.45$. Consequently, at equilibrium

$$\frac{[\text{fructose 6-phosphate}]}{[\text{glucose 6-phosphate}]} = 0.45$$

Thus if we start with glucose 6-phosphate, equilibrium is not attained until a substantial proportion has been converted to fructose 6-phosphate. Conversely, if we start with fructose 6-phosphate, equilibrium is not reached until a substantial proportion has been converted into glucose 6-phosphate. In other words, the reaction can be regarded as freely reversible.

The equilibrium constant, then, can be regarded as a numerical expression of the reversibility of a reaction. A value near unity indicates a readily reversible reaction like (c) above; a value much greater than unity indicates that the forward reaction is favoured as in (a). The equilibrium constant is nearly always a reliable guide to the direction in which a reaction proceeds in a living organism. Thus, of the three examples cited above

(a) almost always proceeds in the forward direction,
(b) almost always in the reverse direction and
(c) in either direction.

There are, however, occasional exceptions to this general rule. A most interesting and important one is the reduction of pyruvate to lactate at the expense of the nucleotide coenzyme NADH:

pyruvate + NADH + H$^+$ ⇌ lactate + NAD$^+$

For this reaction $K = 20\,000$ at pH 7. Consequently, it would be expected to proceed *in vivo* only in the forward direction but in living cells, in spite of the unfavourable equilibrium constant, it has been found to operate also in the reverse direction. This could perhaps best be understood by considering what would happen if we started out with lactate and NAD$^+$. Obviously, a small proportion of the lactate would be reduced by the NAD$^+$ but very soon enough NADH and pyruvate would have been generated to give the ratio required for equilibrium:

$$\frac{[\text{lactate}][\text{NAD}^+]}{[\text{pyruvate}][\text{NADH}]} = \frac{20\,000}{1}$$

and the reaction would come to a stop. If, however, the pyruvate and NADH were not allowed to accumulate but were removed as soon as they were formed, more lactate and NAD$^+$ would be converted to pyruvate and NADH. Continuous removal of the pyruvate and NADH would allow the process to go on indefinitely. The pyruvate and NADH might be removed by being transported away from the site of the reaction or, alternatively, they might be removed chemically by being converted into other compounds. The nature of the removal mechanism does not matter. It must, however, be highly efficient since even a very small accumulation of pyruvate and NADH establishes an equilibrium and halts the reaction. Perhaps it is because the requirements for removal mechanisms are so stringent that it is rare for biochemical reactions to be operated in this way, namely in the opposite direction to that favoured by the equilibrium constant.

ENERGY CHANGES IN CHEMICAL REACTIONS

One of the obvious facts about chemical reactions is that they are associated with energy changes. Indeed, the greater part of the large energy requirements of modern industrial societies are met by the reaction, with oxygen, of fossil fuels like coal and mineral oil and natural gas. Unfortunately, the relationship between chemical reactions and energy changes is not a simple one; but it is so important in biochemistry that a short explanation is worthwhile. We can start from the well-known fact that energy can exist in a variety of forms, thermal, electrical, mechanical, and so on. Generally speaking, biochemists and physiologists are particularly interested in *the capacity of energy to do useful work*. We use this term in the common everyday sense of the work done in, say, raising a weight or generating electricity. It is a matter of common experience that some forms of energy are not available for performing useful work. For example, in a beaker of water at room temperature the individual molecules have a considerable kinetic energy in the form of thermal agitation, but this energy cannot be harnessed to do work. This is not because of technical difficulties which might one day be overcome; it is a corollary of the basic laws of thermodynamics. Energy which is capable of doing useful work is designated 'free energy'. In general this free energy can be obtained from any process which takes place spontaneously. For example, water spontaneously runs downhill and this process can be used to generate electricity. A compressed spring will return spontaneously to its natural shape; in doing so it can be made

to drive a clockwork toy. In a steam turbine heat flows spontaneously from the boiler, which is at a high temperature, to the condenser, which is at a low temperature; the turbine itself is a device for obtaining the free energy made available by this spontaneous heat transfer. Conversely, to bring about a process which is not itself spontaneous, free energy must be supplied in a suitable form. Water flows uphill only if it is supplied with the necessary energy by means of some sort of pump. A spring can be compressed only by doing mechanical work on it. Free energy must be supplied to transfer heat from the cold interior of a refrigerator to its relatively warm surroundings. In short, non-spontaneous processes take place only if they are supplied with free energy.

These generalizations apply equally to chemical reactions. In any reaction, unless the reactants and the products are initially in equilibrium, the reaction will continue, in one direction or the other, until equilibrium is attained. This may occur very slowly, perhaps so slowly as to be imperceptible. None the less, it is a spontaneous process in the sense that it takes place of its own accord, and not because it is being pushed or driven by any outside agency. Because it is spontaneous it can, like any other spontaneous process, provide free energy. Conversely, if we want the reaction to proceed in the direction which takes the reactants and products farther from equilibrium, free energy must be supplied. The amount of free energy supplied in the first case or required in the second is, as might be expected, greater the farther the system is from equilibrium.

These theoretical ideas can be illustrated from everyday experience. The engine of a motor car is started by means of an electric motor (the starter) which is driven by an accumulator in which the following reaction takes place

$$Pb + PbO_2 + 2H_2SO_4 \rightleftharpoons 2PbSO_4 + 2H_2O$$

The equilibrium of this reaction is very far over to the right, and it is never, in fact, reached in normal use. Consequently, as the reaction proceeds from left to right the system is approaching equilibrium and free energy is made available. The accumulator is essentially a device for converting this free energy into an electrical form from which it can be converted into mechanical form to operate the starter. Here then we have an example of free energy being obtained from a reaction which is proceeding to equilibrium. When the car is running the sequence of events is reversed. A dynamo driven by the engine produces electrical power which is fed into the accumulator where it provides the energy necessary to drive the reaction from right to left (that is away from equilibrium).

The free energy change associated with a chemical reaction is not identical with the amount of heat liberated. This is most strikingly illustrated by the fact that when many salts dissolve in water the temperature of the system falls. In other words, although the process is spontaneous, and therefore capable of yielding free energy, it involves uptake of heat and not its liberation. The relation between heat and free energy is too complex to be explored here. What matters is that if the free energy made available by a reaction (or any other spontaneous process) is not utilized in some way, it appears as heat. In practice, it is not easy to devise mechanisms which make it possible to utilize directly the free energy made available by chemical reactions. Electric batteries are, so far, the major examples of such mechanisms. In relation to the amount of power they provide, they are heavy and clumsy, and they require expensive materials. The main source of chemical energy available to meet the large requirements of modern industrial societies is the oxidation, by molecular oxygen, of coal, oil and natural gas. Although the free energy of these processes is very large, no means has yet been devised by which it can be converted directly into mechanical or electric form. Instead it has to be liberated as heat, and the heat energy used to drive, say, a steam turbine or a diesel engine. Unfortunately the efficient conversion of heat energy into other forms of energy depends on the creation of a large temperature difference, as in the steam turbine or the internal combustion engine. Unless this temperature difference is made very large, only a small proportion of the available heat energy is converted into the useful form of 'free energy'. One of the most remarkable properties of living cells is that they can convert the free energy of chemical reactions directly into other forms of energy without using the clumsy and wasteful expedient of converting it first into heat. The most obvious example is muscle. A muscle fibre alternately contracting and relaxing does mechanical work just as a steam engine does, but it derives the necessary energy directly from chemical reactions. It is, therefore, nearly analogous to a car battery. Other examples of such direct utilization of chemical energy *in vivo* could be cited. One of the most important, though least conspicuous, is the ability of many cells to bring about the transfer of solutes. For instance, the parietal cells of the gastric mucosa can be regarded as pumping H^+ ions into the lumen of the gastric glands; the cells of the kidney tubule pump Na^+ and other ions. These transfers require energy which is obtained directly from the free energy of chemical reactions. Perhaps rather surprisingly the energy-requiring activities of cells which have so far been investigated have all been shown to be driven by the same reaction, namely, the hydrolysis of ATP:

$$ATP + H_2O \rightleftharpoons ADP + P_i \quad K = 250\,000$$
(P_i stands for inorganic phosphate.)

CHEMICAL REACTIONS INSIDE THE CELL

In considering in a little more detail the chemical reactions going on inside the cell it is convenient to begin with two simple generalizations:

1. Most of these reactions go on more or less continuously throughout the life of the cell.
2. At least in the short term, the concentrations of reactants and products remain fairly constant.

As an example, we might take the hydrolysis of ATP

$$ATP + H_2O \rightleftharpoons ADP + P_i \quad K = 250\,000$$

as it occurs in the brain. In subsequent chapters it will be shown that the breakdown of ATP by this and other processes is continuous, and that it is balanced by other processes by which ATP is formed. Similarly the reactions in which ADP and P_i are liberated are balanced by those in which they are removed. In rat brain these balances must be exactly maintained, because the concentrations of all three remain remarkably constant at the following values:

$$\text{ATP } 0.002 \text{ M}$$
$$\text{ADP } 0.0005 \text{ M}$$
$$\text{P}_i \text{ } 0.005 \text{ M}$$

The ratio of products to reactants is therefore

$$\frac{[\text{ADP}][\text{P}_i]}{[\text{ATP}]} = \frac{[0.0005][0.005]}{[0.002]} = 0.00125$$

This is very different from the ratio (250 000) required for equilibrium; the ATP concentration is too high relative to the ADP and P_i. Consequently, under these conditions, hydrolysis of ATP makes available free energy. As already mentioned on p. 83 the amount of free energy depends on how far the ratio of reactants to products differs from that which obtains at equilibrium. To be precise it is given by the relationship

Free energy change
$$(\Delta G)$$
$$= 5.9 \log_{10} \frac{\text{actual ratio}}{\text{equilibrium ratio}}$$
$$\text{kilojoules (kJ)/mole}$$

In the present instance this is

$$5.9 \log_{10} \frac{0.00125}{250\,000} = 5.9 \log 5 \times 10^{-9}$$
$$= -49.0 \text{ kilojoules (kJ)/mole}$$
$$(-13\,000 \text{ calories/mole})$$

It should be noted that although energy is available, ΔG is, by convention, negative.

Unfortunately, it is seldom that the exact concentrations of products and reactants are known. In these circumstances it is convenient to calculate the free energy change when the concentrations of products and reactants are all 1M. The equation then becomes:

$$\Delta G = 5.9 \log \frac{1}{K} = -5.9 \log K$$

This value of ΔG called the standard free energy change (ΔG°) represents a highly artificial situation. Concentrations as high as 1M are not normally encountered inside the cell. None the less, ΔG° does represent a very useful approximation. For the hydrolysis of ATP it has the value:

$$\Delta G^\circ = -5.9 \log 250\,000$$
$$= -31.8 \text{ kJ/mole } (-7600 \text{ calories/mole})$$

THE BIOCHEMICAL SIGNIFICANCE OF ΔG°

In the preceding section the standard free energy change (ΔG°) was introduced as a convenient approximation of the amount of free energy theoretically available from a reaction *in vivo*. Indeed, the number of reactions from which cells can directly or indirectly utilize the free energy theoretically available is very limited. The biochemist's main interest in ΔG° arises from its relationship with K:

$$\Delta G^\circ = -5.9 \log K$$

If one is known, the other can easily be calculated. For example:

$$\text{glucose} + \text{ATP} \rightleftharpoons \text{glucose 6-phosphate} + \text{ADP}$$
$$K = 1600$$
$$\therefore \ \Delta G^\circ = -5.9 \log 1600$$
$$= -18.9 \text{ kJ/mole}$$
$$(-4400 \text{ calories/mole})$$

$$\text{pyruvate} + \text{ATP} \rightleftharpoons \text{phosphopyruvate} + \text{ADP}$$
$$K = 0.00016$$
$$\therefore \ \Delta G^\circ = -5.9 \log 0.00016$$
$$= +22.4 \text{ kJ/mole}$$
$$(+5400 \text{ calories/mole})$$

$$\text{glucose 6-phosphate} \rightleftharpoons \text{fructose 6-phosphate}$$
$$K = 0.45$$
$$\therefore \ \Delta G^\circ = -5.9 \log 0.45$$
$$= +2.0 \text{ kJ/mole}$$
$$(+500 \text{ calories/mole})$$

$$\text{pyruvate} + \text{NADH} \rightleftharpoons \text{lactate} + \text{NAD}$$
$$K = 20\,000$$
$$\therefore \ \Delta G^\circ = -5.9 \log 20\,000$$
$$= -25.4 \text{ kJ/mole}$$
$$(-6000 \text{ calories/mole})$$

$$\text{sucrose} + \text{H}_2\text{O} \rightleftharpoons \text{glucose} + \text{fructose}$$
$$K = 10\,000$$
$$\therefore \ \Delta G^\circ = -5.9 \log 10\,000$$
$$= -23.6 \text{ kJ/mole}$$
$$(-5500 \text{ calories/mole})$$

(The experimental determination of K or of

ΔG° is liable to substantial errors so the figures quoted are approximate.)

These examples show that readily reversible reactions have a small ΔG°; reactions in which the equilibrium favours the forward direction have a negative ΔG° and are termed *exergonic*; reactions in which the equilibrium favours the reverse direction have a positive ΔG° and are termed *endergonic*.

Thus the value of ΔG° indicates the reversibility (or otherwise) of a reaction in the same way as K. Its great advantage over K is that it is additive. Thus phosphopyruvate can react with ADP to give pyruvate and ATP. The ATP produced in this reaction can then react with glucose to give glucose 6-phosphate and ADP.

$$\text{phosphopyruvate} + \text{ADP} \rightleftharpoons \text{pyruvate} + \text{ATP}$$
$$\Delta G^{\circ} = -22 \cdot 6 \text{ kJ/mole } (-5400 \text{ calories/mole})$$

$$\text{glucose} + \text{ATP} \rightleftharpoons \text{glucose 6-phosphate} + \text{ADP}$$
$$\Delta G^{\circ} = -18 \cdot 4 \text{ kJ/mole } (-4400 \text{ calories/mole})$$

When these equations are added the ADP's and ATP's cancel out and the overall result is:

$$\text{phosphopyruvate} + \text{glucose} \rightleftharpoons \text{glucose 6-phosphate} + \text{pyruvate}$$

Similarly the ΔG° for this overall process is the sum of the ΔG°'s for the two stages viz.

$$(-22 \cdot 6) + (-18 \cdot 4) = -41 \cdot 0 \text{ kJ/mole}$$
$$(-9800 \text{ calories/mole})$$

This sort of approach is particularly useful in elucidating the mechanism by which complex molecules are built up *in vivo*. A good example is provided by the synthesis of glycogen from glucose (p. 154). In theory this could be achieved by coupling one glucose molecule to another with elimination of water. Repetition of this process would give a chain of any desired length. However, the equilibrium of such a reaction is very unfavourable. Extension of a glycogen chain by a single glucose unit:

$$\text{glucose} + (\text{glucose})_n = (\text{glucose})_{n+1} + H_2O$$

has a ΔG° of $+12 \cdot 6$ kJ/mole. Clearly, therefore, some other mechanism must be employed. The complex process can be summarized thus (the ΔG° values are all approximate):

$$\text{glucose} + \text{ATP} \rightleftharpoons \text{glucose 6-phosphate} + \text{ADP}$$
$$\Delta G^{\circ} = -18 \cdot 4 \text{ kJ/mole}$$

$$\text{glucose 6-phosphate} \rightleftharpoons \text{glucose 1-phosphate}$$
$$\Delta G^{\circ} = +6 \cdot 7 \text{ kJ/mole}$$

$$\text{glucose 1-phosphate} + \text{UTP} \rightleftharpoons \text{UDP} - \text{glucose} + \text{PP}_i$$
$$\Delta G^{\circ} = 0 \text{ kJ/mole}$$

$$\text{PP}_i + H_2O \rightleftharpoons 2P_i$$
$$\Delta G^{\circ} = -31 \cdot 4 \text{ kJ/mole}$$

$$\text{UDP glucose} + (\text{glucose})_n \rightleftharpoons (\text{glucose})_{n+1} + \text{UDP}$$
$$\Delta G^{\circ} = -16 \cdot 7 \text{ kJ/mole}$$

These five reactions added together give:

$$\text{glucose} + (\text{glucose})_n + \text{ATP} + \text{UTP} + H_2O \longrightarrow (\text{glucose})_{n+1} + \text{ADP} + \text{UDP} + 2P_i$$

$$\Delta G^{\circ} = -18 \cdot 4 + 6 \cdot 7 + 0 - 31 \cdot 4 - 16 \cdot 7$$
$$= -59 \cdot 8 \text{ kJ/mole}$$

What is important in this complex sequence is that the overall equilibrium is overwhelmingly in favour of glycogen synthesis. The use of ΔG° reveals why this should be. In effect, the reaction sequence brings about two opposite results. On the one hand, two glucose residues are linked together, and this, as already seen, has a ΔG° of $+12 \cdot 6$ kJ; on the other hand, the terminal phosphoryl groups of ATP and UTP are split off, and this makes energy available ($31 \cdot 8$ kJ for each molecule, $63 \cdot 6$ kJ in total). The second process can be regarded as providing the energy required for the first. Because it provides far more energy than is necessary, the reaction sequence as a whole is irreversible.

In subsequent chapters it will be shown that the biosynthesis not only of polysaccharides, but of other complex molecules like lipids and proteins and nucleic acids, is driven in this way by the breakdown of ATP and other nucleoside triphosphates. For these reactions to continue there must be a continuous supply of nucleoside triphosphates and especially ATP. The source of this supply will be described in later chapters. For the moment it suffices to say that a good deal of modern biochemistry is concerned with the way in which endergonic reactions are coupled to exergonic ones. Compounds like ATP are particularly important in this respect. The reason for their importance lies in the unusually large amount of energy made available by the hydrolysis of the pyrophosphate bonds of ATP. This exceeds the ΔG° required to form most other biochemical linkages and because of this the conversion of ATP to ADP and P_i or to AMP and $2P_i$ makes available sufficient energy to drive most synthetic reactions. For this reason ATP is classed as a *high energy or energy-rich compound*; it may be compared with glucose 1-phosphate which has a ΔG° of hydrolysis of only -12.6 kJ/mole. Examples of similar energy-rich compounds will be given in subsequent chapters.

CHEMICAL REACTIONS AND FREE ENERGY CHANGES

What factors determine the ΔG° of a reaction? The largest of these, by a substantial margin, is the nature of the existing bonds broken and the new bonds formed. A simple example, of considerable biological importance, is the oxidation of fat by molecular oxygen. Animal fats consist predominantly of long, saturated, paraffin hydrocarbons. Their oxidation can be represented as follows:

$$C_{17}H_{35}COOH + 26O_2 \longrightarrow 18CO_2 + 18H_2O$$

The very large favourable free energy change of this reaction is almost entirely attributable to the fact that the bonds which are broken (C—C, C—H, and O=O) are less strong and less stable than those which are formed. This, in turn, is an example of the general phenomenon that bonds such as C—H, C—C, and O=O, in which the bonding electrons are equally or almost equally attracted by the atoms they link, are less strong and stable than bonds such as H—O and C=O in which the bonding electrons are attracted to one atom (in this case oxygen) rather than the other.

A second major factor determining the K or ΔG° of biochemical reactions is the degree to which the reaction in question involves the breakdown of a complex, ordered structure or conversely the formation of such a structure from simpler units. If other things are equal, the equilibrium of such a reaction will tend to favour the breakdown of the complex structure rather than its formation. While this question of order is of great importance in the case of

such highly ordered structures as proteins, we shall not encounter it very frequently in most of the reactions with which we shall be concerned in subsequent chapters. Its relative importance can perhaps be gauged from an example. The oxidation of glucose

$$C_6H_{12}O_6 + 6O_2 \longrightarrow 6CO_2 + 6H_2O$$

has a ΔG° of 2870 kJ/mole. Of this 2820 kJ can be attributed to the substitution of C=O and H—O bonds for C—H, C—C and O=O. Only 50 kJ can be attributed to the break-up of the relatively elaborate glucose structure to give the smaller, simpler structures of CO_2 and H_2O.

Most of the individual biochemical reactions with which we shall be concerned are far less drastic and extensive than the complete oxidation of fat and carbohydrate by molecular oxygen and their ΔG°s are of the order of only 40 kJ/mole. Frequently the new bonds formed are identical with the old ones which are broken as, for example, in the hydrolysis of an ester:

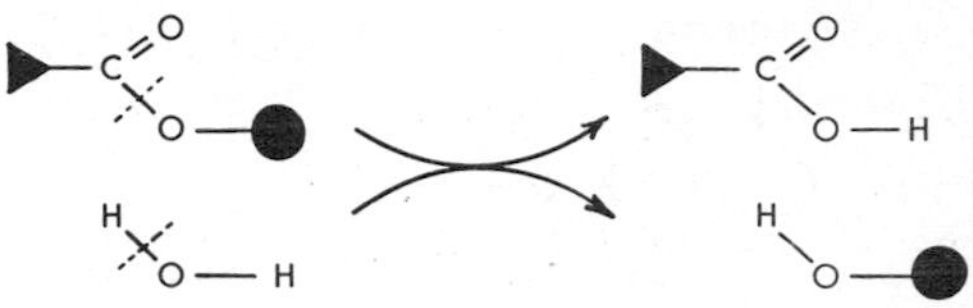

The ΔG° values of such reactions can be influenced by a variety of relatively small effects. One of the most important of these is resonance.

It has long been accepted that the conventional structural formula with single and double bonds is in many cases an incomplete and sometimes misleading representation of the structure and properties of a compound. For example, the acetate ion is not accurately represented by either I or II but rather by a hybrid between them represented by III, in which the negative charge is shared equally by

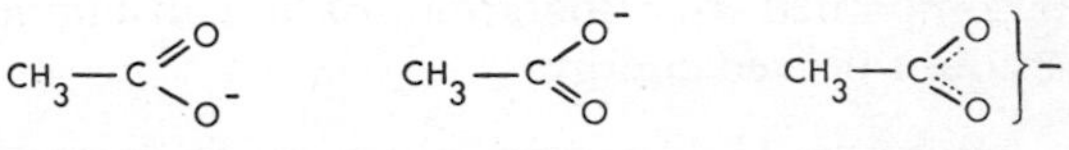

the two oxygen atoms, and the two carbon-to-oxygen bonds are, to an equal degree, hybrids between single bonds and double bonds. Similarly we can regard acetic acid itself and its esters as analogous resonance hybrids though

$$CH_3-C\overset{O}{\underset{OH}{\diagdown}} \qquad CH_3-C\overset{O^-}{\underset{\overset{OH}{+}}{\diagdown}}$$

$$CH_3-C\overset{O}{\underset{O-Et}{\diagdown}} \qquad CH_3-C\overset{O^-}{\underset{\overset{O-Et}{+}}{\diagdown}}$$

in these cases the uncharged form on the left predominates. It has also been accepted for many years that such a hybrid is more stable than a similar compound which can be accurately represented by a single formula. A striking example of the energy which can be made available when a molecule is rearranged in such a way as to make resonance possible is provided by the process of glycolysis which is described in detail in Chapter 11. This is a sequence of reactions in which the constituent atoms of glucose are rearranged to form two molecules of lactic acid thus:

$$\text{glucose} \longrightarrow 2 \begin{array}{c} CH_3 \\ | \\ C=O \\ | \\ C\overset{O}{\underset{OH}{\diagdown}} \end{array}$$

The equilibrium lies very heavily to the right ($\Delta G° = -200$ kJ) and this is almost entirely attributable to the fact that lactic acid and, to an even greater extent, the lactate ion are stabilized by resonance whereas glucose is not.

ACTIVATION ENERGY

Up to this point we have been considering at what state a reaction or a system of reactions reaches equilibrium and how much energy can be obtained from a system which is not at equilibrium. We have not so far considered the rate at which equilibrium is approached. As already stated the hydrolysis of sucrose is accompanied by liberation of a substantial amount of free energy:

$$\text{sucrose} + H_2O \rightleftharpoons \text{glucose} + \text{fructose}$$
$$\Delta G° = -23\cdot0 \text{ kJ/mole } (-5500 \text{ calories/mole})$$

Equilibrium, therefore, is not reached until the concentrations of glucose and fructose far exceed that of sucrose. Yet it is a matter of common experience that solutions of sucrose in water are quite stable and can be kept for very long periods without appreciable hydrolysis, or more precisely equilibrium is being approached with imperceptible slowness. On the other hand some reactions with a relatively small $\Delta G°$ attain equilibrium quite rapidly. For example α-glucose dissolved in water is converted to an equilibrium mixture of the α and β forms although the $\Delta G°$ of the reaction is very small.

$$\alpha\text{-glucose} \rightleftharpoons \beta\text{-glucose}$$
$$\Delta G° = -1\cdot7 \text{ kJ/mole } (-400 \text{ calories/mole})$$

Clearly, therefore, the rate of a chemical reaction is not determined by its $\Delta G°$. Physical chemists have elaborated more than one theory to explain what does determine the rate of a

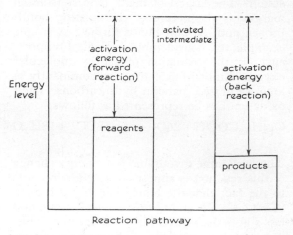

Fig. 7.1 Schematic representation of the activation energy concept.

chemical reaction. It is, however, universally agreed that, before a reactant molecule can undergo any reaction, it must possess an amount of energy in excess of a level characteristic of the reaction in question—the *activation energy* (Fig. 7.1). Most biochemical reactions take place in aqueous solution in which the molecules of the reactant are being continuously jostled by collision with solvent molecules. The amount of energy which any individual molecule possesses varies from moment to moment depending on the number and angle of collisions it has recently experienced. Very occasionally a molecule reaches an energy content equal to or greater than the activation energy for a particular reaction. Then, and only then, is it capable of undergoing that reaction. The rate of the reaction, other things being equal, depends on how frequently molecules do reach this very high energy content. If the activation energy of a reaction is low, molecules will reach it relatively often and the reaction will be rapid. If it is high, molecules will attain it only rarely and the reaction will be slow. In general a reaction can be accelerated in two ways. The first is by raising the temperature of the solution and so increasing the kinetic energy of all the molecules. An increase of 10°, say from 20°C to 30°C, brings about only a small (3 per cent) increase in the average energy of the molecules with very high energy content and it often doubles the frequency with which they reach the level of activation energy for the average sort of reaction. It is partly for this reason that the Bunsen burner is such an essential piece of laboratory equipment. It is important to recognize that increasing the temperature is a quite undiscriminating method of accelerating reactions. A wide range of reactions is equally affected.

The other means of accelerating a reaction is in effect to lower its activation energy. This can be done by using a catalyst. For example the activation energy of the decomposition of hydrogen peroxide is about 75 kJ/mole (18 000 calories/mole).

$$2H_2O_2 \longrightarrow 2H_2O + O_2$$

In the presence of a platinum catalyst this is reduced to about 50 kJ/mole (12 000 calories/mole). Notice that in a reversible reaction the velocity of the back reaction is likewise determined by the height of the activation energy barrier. Consequently, when the height of the barrier is diminished by the presence of a catalyst, both forward and back reactions are accelerated to an equal degree. In other words, the presence of a catalyst does not alter the equilibrium of a reaction: it merely accelerates the rate at which the equilibrium is approached. In the living cell, the catalysts are *enzymes*. They are discussed in the next chapter.

REFERENCES

CLARK, W. M. (1952). *Topics in Physical Chemistry*, 2nd edn. Baltimore: Williams & Wilkins.
DAWES, E. A. (1972). *Quantitative Problems in Biochemistry*, 5th edn. Edinburgh: Livingstone.
EDSALL, J. T. & WYMAN, J. (1958). *Biophysical Chemistry*. New York: Academic Press.
GREEN, D. E. & BAUM, H. (1970). *Energy and the Mitochondrion*. New York: Academic Press.
INGRAM, L. L. (1962). *Biochemical Mechanisms*. New York: Wiley.
KAPLAN, N. O. & KENNEDY, E. P. (1966). *Current Aspects of Biochemical Energetics*. New York: Academic Press.
LEHNINGER, A. L. (1965). *Bioenergetics*. New York: Benjamin.
RACKER, E. (1965). *Mechanism in Bioenergetics*. New York: Academic Press.
VAN HOLDE, K. E. (1971). *Physical Biochemistry*. New Jersey: Prentice Hall.
WEST, E. S. (1956). *Textbook of Biophysical Chemistry*. New York: Macmillan.

8 Enzymes

The difference between a plant or an animal, on the one hand, and a stone on the other, is obvious to a small child, but it is only in the last century that we have begun to understand the causes underlying this difference. It was discovered in prehistoric times that if grape juice is allowed to stand at room temperature it ferments: that is to say, the sugar it contains is converted to carbon dioxide and alcohol.

$$C_6H_{12}O_6 \longrightarrow 2C_2H_5OH + 2CO_2$$

This process of fermentation is accompanied by the appearance of a grey sediment called yeast. If some of the yeast is added to a further quantity of fresh grape juice, or to a sugar solution, these too will undergo fermentation and from the fermentation more yeast can be obtained. To the naked eye, yeast looks like some sort of chemical and the early chemists regarded it as such. They classified it as a catalyst, since it brought about reactions without itself being used up or destroyed, though they had difficulty in explaining how it was formed, apparently spontaneously, in grape juice. After the invention of the compound microscope it was found that yeast is really a unicellular plant commonly found on the surface of grapes (as well as in other places) and is particularly well adapted to growth in media which, like grape juice, have a high concentration of sugar. Few other micro-organisms tolerate the high osmotic pressure of such solutions. When the grapes are pressed some yeast cells pass into the juice, where they immediately start to grow and to convert the sugar to alcohol and carbon dioxide. Indeed we now know that this conversion is the source from which the yeast obtains the energy necessary for growth.

The identification of yeast as a living organism explained why fermentation differed from other chemical reactions but it left open the question of how yeast was able to bring it about. The early chemists, having accepted that yeast was not itself simply a catalyst, wondered whether its activity was due to the catalysts within it. Soluble catalysts had already been obtained from animals and plants. For example gastric juice and pancreatic juice had been shown to catalyse the hydrolysis of proteins and an extract of malt had been shown to catalyse the hydrolysis of starch. Biologists, on the other hand, tended to regard fermentation as a process peculiar to living cells. For many years the biologists' view could be justified on the ground that no one had succeeded in extracting the supposed catalysts from yeast cells and demonstrating their action *in vitro*. But in 1897 the brothers Büchner prepared an extract of yeast for medicinal purposes, and added sugar in order to preserve it. To their surprise the cell-free yeast extract fermented the sugar. This observation heralded an enormous advance in the progress of biology; it meant that the process, hitherto thought inseparable from living material, could now be studied like any other catalysis by already known chemical methods. Because of the role which yeast had played in their discovery the catalysts of living cells were called 'enzymes'.

Initially enzyme isolation proved exceedingly difficult and the early studies of enzyme action had to be carried out with very crude preparations. This led to a prolonged debate on the chemical nature of enzymes. It was relatively easy to demonstrate that they behaved like colloids and that they were very easily inactivated; but efforts to decide whether or not they were proteins were repeatedly frustrated because their catalytic activity could easily be demonstrated at concentrations so low that the standard protein tests gave negative results. It was not until 1926 that the first enzyme was obtained in reasonably pure crystalline state and shown to be a protein. This was *urease* which Sumner isolated from jack bean meal. Since then several hundred enzymes have been obtained in pure crystalline form and many more have been purified to less exacting standards. All so far have proved to be proteins. Enzymes, therefore, might be defined as protein catalysts formed by living cells.

ENZYMES AS CATALYSTS

Although enzymes fall quite clearly within the chemist's definition of a catalyst they differ from all other catalysts in a number of important respects. They are, in the first place, far more powerful. For example, disaccharides and polysaccharides can readily be hydrolysed by hydrochloric acid or any other mineral acid as catalyst but the acid concentration has to be quite high, about 1 M, and the reaction is slow unless the temperature is near 100°C. The same reaction can be brought about at room temperature by the appropriate enzyme in a concentration of 0·00001 M.

Even more striking than the very high catalytic activity of enzymes is their specificity. Metallic and other catalysts often act on a wide variety of compounds. Acids, for instance, catalyse the hydrolysis not only of di- and polysaccharides but also of proteins and triglycerides, and a great number of other reactions. In contrast to this, each enzyme acts only on a single compound or a small group of closely related compounds. For instance, the enzyme urease catalyses the hydrolysis of urea. It has no effect on any of the hundreds of other compounds with which it has been tested, even those closely related to urea. Invertase, the enzyme which catalyses the hydrolysis of sucrose, also attacks several related compounds in which the glucose moiety of the sucrose has been replaced by, for example, a methyl group. But it does not attack any compound in which the fructose moiety of the molecule has been altered, even though the alteration may be merely a small stereochemical change. A similar degree of specificity is shown by the enzymes which break the polypeptide chains in the protein molecule. Thus trypsin attacks the bond between the carboxyl group of lysine or arginine and the amino group of the next amino acid in sequence; while chymotrypsin preferentially acts on the bond between the carboxyl group of an aromatic amino acid and the amino group of its neighbour. The action of these enzymes is illustrated in Fig. 8.1.

Enzymes are specific also in the sense that almost invariably they catalyse only one of the reactions which their substrates can undergo. For example, oxaloacetic acid, an important intermediate in metabolism, can undergo reduction to give malic acid, or decarboxylation to give pyruvic acid, or it can accept an amino group to give aspartic acid, or an acetyl residue to give citric acid and so on (Fig. 8.2). Each of these reactions is promoted by its own particular enzyme which catalyses only that re-action and none of the others. This again is in contrast to the behaviour of non-enzymic catalysts which often convert the reagents on

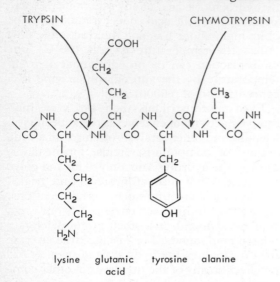

Fig. 8.1 Specificity of proteolytic enzymes.

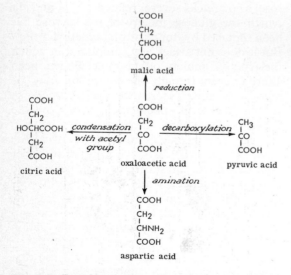

Fig. 8.2 Enzymic reactions of oxaloacetic acid. Each reaction is catalysed by a separate enzyme. If only one enzyme is present only one product will be formed.

which they act into a mixture of several different products (Fig. 8.3). Partly because of this specificity and partly because they take place at relatively low temperatures, enzyme-cata-lysed reactions are much more likely to give a

nearly theoretical yield of end-product than reactions brought about by other agents. In

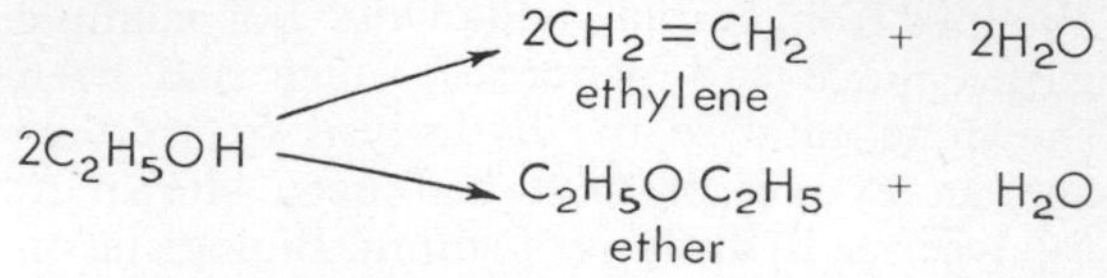

$$2C_2H_5OH \longrightarrow \begin{cases} 2CH_2 = CH_2 \quad + \quad 2H_2O \\ \text{ethylene} \\ C_2H_5O\,C_2H_5 \quad + \quad H_2O \\ \text{ether} \end{cases}$$

Fig. 8.3 Reactions of ethanol catalysed by sulphuric acid. Both reactions are catalysed to an extent which depends on the conditions of the reaction.

most organic reactions carried out without the help of enzymes, only a proportion of the start-ing material is converted to the desired pro-duct; the rest is lost in a variety of side re-actions. An organic synthesis usually involves a sequence of such reactions, each incurring some loss. Obviously the cumulative effect of such losses may be so great that the final pro-duct represents only a half or a third of the starting material. By contrast, the enzymes present in muscle can convert glucose to lactic acid by a sequence of no less than eleven con-secutive reactions with an overall efficiency of 100 per cent. The extraordinary power and efficiency of enzymes allow very minute quantities to catalyse the rapid and quantita-tive conversion of large quantities of substrate to a wide variety of products through an elaborate network of reactions. Examples of such networks will be found in subsequent chapters.

FACTORS AFFECTING ENZYME ACTIVITY

Enzymes are not only more powerful and specific than other catalysts; they are also far more sensitive to their environment. They are particularly affected by temperature and pH and by the presence or absence of small cations and anions.

The Effect of Temperature. With certain exceptions, the rates of chemical reactions are increased as the temperature is raised. A crude, but useful, rule is that an increase in tempera-ture of 10°C roughly doubles the rate of most reactions. Between about 0°C and 50°C, this rule applies also to enzyme-catalysed re-actions. This explains why, for example, bacterial cultures generally grow more rapidly at 37°C than at 20°C, why food can be preserved in a refrigerator and why it is possible to reduce the rate of a patient's metabolism by lowering his temperature (artificial hypothermia, Chap. 42). The rates of enzyme-catalysed reactions do not, however, increase indefinitely as the temperature is increased. At some point, usually in the range 40° to 60°C, the enzyme itself, being a protein, undergoes denaturation with concomitant loss of activity. As the enzyme is inactivated the reaction which it catalyses slows down and ultimately stops. The relationship of enzyme activity to temperature can therefore usually be represented by a curve of the form shown in Fig. 8.4.

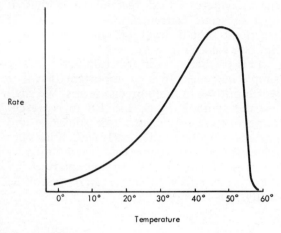

Fig. 8.4 Schematic representation of the effect of temperature on the rate of an enzyme-catalysed reaction.

The precise shape and position of the curve depends on the enzyme, the substrate, and the conditions of the experiment, but its general shape is determined by the fact that protein denaturation is much more temperature-dependent than the great majority of reactions. A 10°C rise in temperature increases the rate of denaturation not twofold but a hundredfold or even a thousandfold. Consequently, an enzyme which may be to all intents and purposes stable at 30°C may be almost instantaneously inactivated at 60°C or 70°C. For this reason few organisms can survive temperatures of 80°C or above.

The Effect of pH. Like other proteins, enzymes are denatured and therefore inactivated at extremes of pH. But quite apart from this irreversible inactivation, the activity of intact enzymes varies markedly with quite small changes in pH. A graph of activity against pH usually takes the characteristic bell-shaped form shown in Fig. 8.5 with an

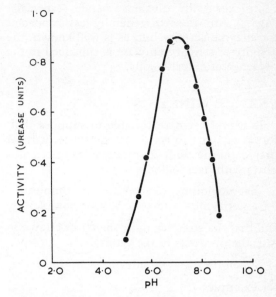

Fig. 8.5 Influence of pH on the activity of urease in phosphate buffer. (Modified from S. F. Howell & J. B. Sumner (1934), *Journal of Biological Chemistry* **104**, 619.)

optimum which is characteristic of the particular enzyme. Generally it lies in the range 4 to 9, but occasionally more extreme values are found.

The explanation of this pH effect seems to lie, at least partly, in the fact that, like all proteins, enzyme molecules possess a multiplicity of ionizable groups. The state of these ionizable groups obviously depends on pH. It would seem reasonable to assume that for an enzyme molecule to be active as a catalyst certain of these groups have to be ionized while certain others are not. This state of affairs would obviously prevail only within a limited pH range which would depend on the pK_a values of the groups concerned.

Effects of other ions. H^+ is the only cation whose concentration is important for all enzymes, but many enzymes are inactive unless a cation such as Mg^{2+}, Ca^{2+}, Mn^{2+}, Zn^{2+}, Na^+ or K^+ is present. Some enzymes apparently require a loosely-bound cation (often a divalent one) as part of their structure. Thus, phosphopyruvate hydratase (p. 143) contains a loosely bound Mg^{2+} without which it is inactive. In other cases the ion may combine with substrate. Thus nucleoside di- and triphosphates participating in enzyme-catalysed reactions do so not in the free form, but as complexes with divalent metals (especially Mg^{2+}). Anions seem generally less important for enzyme activity but, as is well known, the activity of salivary amylase is enhanced in the presence of chloride ions.

ENZYME KINETICS

It would seem reasonable to suppose that, for an enzyme to act on its substrate, it must first combine with it, much as oxygen combines with haemoglobin:

$$\text{haemoglobin} + \text{oxygen} \rightleftharpoons \text{complex}$$
$$\text{enzyme} + \text{substrate} \rightleftharpoons \text{complex}$$

and that the complex must then break down to give the products of the catalysis:

$$\text{complex} \longrightarrow \text{enzyme} + \text{products}$$

for example:

$$\text{invertase} + \text{sucrose} \rightleftharpoons \text{complex}$$
$$\text{complex} \xrightarrow[\text{H}_2\text{O}]{} \text{enzyme} + \text{glucose} + \text{fructose}$$

If enzymes do operate in this way it should be possible to predict how they will be affected by variations in substrate concentration. In broad terms the argument runs as follows:

1. Since the enzyme can act on the substrate only after combining with it, the rate of the enzyme-catalysed reaction, other conditions being constant, must depend on the frequency with which enzyme molecules encounter substrate molecules.

2. When the concentration of substrate is low, enzyme molecules encounter substrate molecules only rarely. Consequently the rate of the reaction is low. At any given instant only a small proportion of enzyme molecules are acting on substrate molecules. The remainder are unoccupied and waiting for the arrival of substrate molecules.

3. At higher substrate concentrations enzyme molecules encounter substrate molecules more frequently and the rate of the reaction is proportionately higher. At any given instant a larger proportion of enzyme molecules are being used.

4. As the substrate concentration is increased still further, enzyme molecules encounter substrate molecules so frequently that as soon as an enzyme molecule has dealt with one it immediately finds another on which to act. At any given instant virtually all the enzyme molecules are acting on substrate molecules. In this situation the maximum rate of operation of the enzyme molecules is almost attained. A further increase in substrate concentration has virtually no effect because there are no unoccupied enzyme molecules available.

Qualitatively, therefore, if this is a valid model it can be predicted that as substrate concentration increases so also should the rate of the enzyme-catalysed reaction; but that at high substrate concentrations the rate of the reactions should level off and approach a limiting value (Fig. 8.6).

This prediction is in fact fulfilled by many simple enzymes. It is so important that it is worth putting in more precise terms. We shall use the symbols E, S, and ES to represent respectively free enzyme, free substrate and enzyme–substrate complex. If we confine ourselves to the very simple case, exemplified by invertase, in which the reaction catalysed by the enzyme is irreversible, we can represent the catalysis as follows:

$$E + S \underset{k_{-1}}{\overset{k_{+1}}{\rightleftharpoons}} ES \xrightarrow{k_{+2}} E + \text{products}$$

where k_{+1}, k_{-1} and k_{+2} are the velocity constants of the individual steps.

Let

 e = total concentration of enzyme (both free and in the form of ES)

 p = concentration of ES

 s = concentration of free substrate

then

e − p = concentration of enzyme in the free form E.

Usually the concentration of substrate is enormously greater than that of the enzyme. Consequently s is correspondingly very large compared to e or p and virtually identical with the *total* concentration of substrate.

It is a matter of experimental observation that if the experimental conditions are properly controlled the rate of the overall reaction is initially constant, though it may fall off

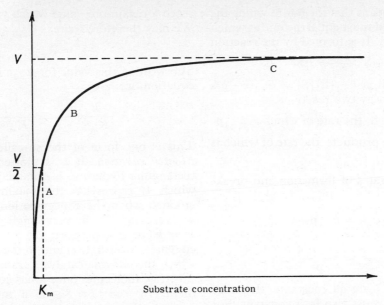

Fig. 8.6 The relationship between substrate concentration and the rate of an enzyme-catalysed reaction. At low substrate concentrations (A) enzyme molecules encounter substrate molecules only occasionally, so the reaction rate is low. Increasing the substrate concentration increases the frequency with which enzyme molecules encounter substrate molecules and therefore increases the rate of the reaction (B). If the substrate concentration is increased still further, each enzyme molecule encounters substrate molecules so frequently that as soon as it has dealt with one it finds another on which to act. At this stage the enzyme is working to full capacity and the reaction rate can only be infinitesimally raised by increasing substrate concentration (C).

subsequently due, for example, to substrate being used up (Fig. 8.7). This clearly implies that in this initial phase the three component

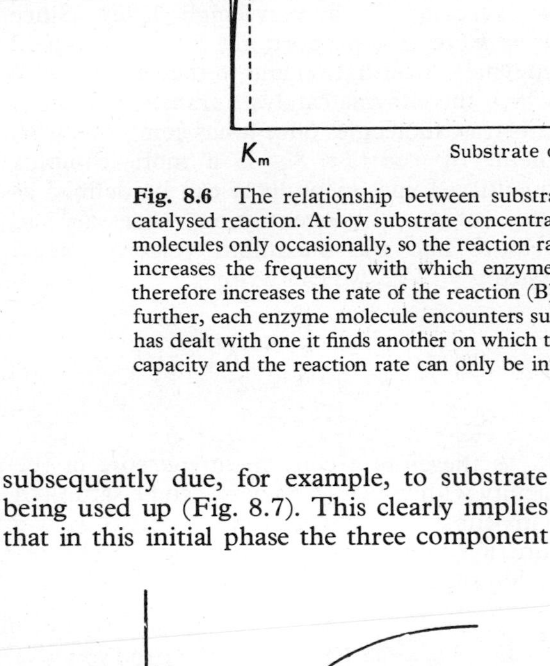

Amount of substrate converted to products

Time

Fig. 8.7 The course of an enzyme reaction expressed in terms of the amount of substrate disappearing per unit of time.

reactions must likewise be proceeding at constant rates and that the concentrations of E, S and ES must be constant. We assume that the period in which we make our observations is so short that the change in S is negligible. This sort of situation is often described as a *steady state*. So long as it prevails the rate (v) of the overall reaction

$$S \longrightarrow products$$

must be identical with that of the second stage

$$ES \xrightarrow{\ k_{+2}\ } E + products$$

that is

$$v = k_{+2}[ES] = k_{+2}p$$

The rate of conversion of substrate to products is therefore proportional to the concentration of the enzyme–substrate complex. In order to evaluate the latter we make use of the fact that it is necessarily constant in the steady-state situation we are considering.

This in turn means that the rate at which the complex is formed must equal the rate at which it is broken down. It is formed by the reaction

$$E + S \longrightarrow ES$$

the rate of which is $k_{+1}[E][S] = k_{+1}(e-p)s$ and broken down by two reactions

$$ES \longrightarrow E + S, \text{ the rate of which is } k_{-1}p$$

and $ES \longrightarrow E + \text{products}$, the rate of which is $k_{+2}p$.

Equating the rates of formation and breakdown, we have

$$k_{+1}(e-p)s = k_{-1}p + k_{+2}p$$
$$k_{+1}(e-p)s = (k_{-1} + k_{+2})p$$
$$\frac{k_{-1} + k_{+2}}{k_{+1}} = \frac{(e-p)s}{p}$$

k_{-1}, k_{+2} and k_{+1} are all constants, therefore $(k_{-1} + k_{+2})/k_{+1}$ must also be a constant. For convenience we will denote it by K_m, the *Michaelis constant*.

$$K_m = \frac{k_{-2} + k_{+2}}{k_{+1}} = \frac{(e-p)s}{p}$$

or $\quad K_m/s = \dfrac{(e-p)}{p} = \dfrac{(E)}{(ES)}$

The ratio K_m/s therefore represents the ratio at any given moment of free enzyme to enzyme combined with substrate. We can rearrange this equation as follows to give p in terms of e, s and K_m.

$$pK_m = es - ps$$
$$p = \frac{es}{K_m + s} = \frac{e}{K_m/s + 1}$$

We can substitute this value for p in the equation derived above

$$v = k_{+2}p = \frac{k_{+2}e}{K_m/s + 1}$$

In many cases we cannot determine the values of either k_{+2} or e, but this difficulty can easily be circumvented. At very high concentrations of substrate the reversible reaction

$$E + S \rightleftharpoons ES$$

will be pushed to the right and virtually all the enzyme will be combined with substrate. Under these conditions p = e and v will

reach a maximum value which we will call V. We may therefore write

$$V = k_{+2}e$$

Substituting this value for $k_{+2}e$ in our former equation, we have

$$v = \frac{k_{+2}e}{(K_m/s) + 1} = \frac{V}{(K_m/s) + 1}$$

This is one form of the so-called *Michaelis-Menten equation*. If v is plotted against s a rectangular hyperbola is obtained (Fig. 8.6) in which V represents the maximum velocity attained when the concentration of substrate is increased to a very high level. Since $V = k_{+2}e$, it is proportional both to the total enzyme concentration and to the velocity with which the enzyme catalyses transformation of substrate molecules once it has combined with them. By contrast K_m is a more complex quantity. Experimentally it can be defined as the substrate concentration necessary to achieve half the maximum velocity, since, when $v = V/2$, we can write

$$\frac{V}{2} = \frac{V}{(K_m/s) + 1}$$
$$\therefore \quad K_m/s + 1 = 2$$
$$K_m = s$$

K_m is therefore a convenient measure of the facility with which an enzyme can be saturated with substrate. Theoretically whereas V is a function only of the second stage in the reaction sequence

$$E + S \underset{k_{-1}}{\overset{k_{+1}}{\rightleftharpoons}} ES \xrightarrow{k_{+2}} E + \text{products}$$

K_m is a function of the rate constants of all three separate steps

$$K_m = \frac{k_{-1} + k_{+2}}{k_{+1}}$$

In many cases, however, it has been observed experimentally that variations in, for example, pH can produce changes in V (and therefore in k_{+2}) without affecting K_m. In such cases, presumably, k_{+2} is small compared to k_{+1} and k_{-1} so that the second stage

$$ES \xrightarrow{k_{+2}} E + \text{products}$$

is slow compared to the rapid establishment of equilibrium in the first stage

$$E + S \underset{k_{-1}}{\overset{k_{+1}}{\rightleftharpoons}} ES$$

In such cases $K_m = k_{-1}/k_{+1}$ and can be regarded as a function solely of the first stage as V is of the second.

As we shall see later, many enzymes catalyse reactions which are much more complicated than the very simple case we have considered above and in many cases their kinetics can only be investigated thoroughly by computer techniques. Nevertheless in very many cases the factors influencing the rates of enzyme-catalysed reactions can profitably be analysed in terms of K_m and V and; at least in some of these, K_m can be regarded as a measure of an enzyme's affinity for its substrate and V as a measure of its ability to transform the substrate once it has combined with it.

For practical purposes it is often convenient to invert the Michaelis-Menten equation in the form first suggested by Lineweaver and Burke:

$$v = \frac{V}{(K_m/s) + 1}$$

$$\therefore \frac{1}{v} = \frac{K_m}{V}\frac{1}{s} + \frac{1}{V}$$

this gives a straight-line relationship between $1/v$ and $1/s$ which is easier to plot than the hyperbolic relationship (Fig. 8.8). The inter-

cepts of the line on the $1/s$ and $1/v$ axes are equal to $-1/K_m$ and $1/V$ respectively.

INHIBITION, COMPETITIVE AND NON-COMPETITIVE

Considerable support for the simple model of enzyme action set out above was obtained from the study of enzyme inhibitors. The activity of enzymes can be diminished by a great variety of chemical agents. Such inhibition is quite distinct from the inactivation which accompanies denaturation of the enzyme protein since there is no large-scale disruption of the enzyme molecule. For example, many enzymes are inhibited by reagents such as mercury and trivalent arsenic which combine with sulphydryl (HS—) groups.

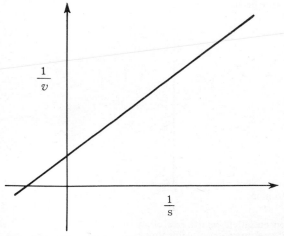

This is why mercury and arsenic are so poisonous to all organisms and why they have a limited use as antibacterial agents. Similarly, the cyanide ion (CN^-) inhibits many copper- or iron-containing enzymes because it can combine with these metals and prevent them from playing their normal role in the operation of the enzyme. The poisonous properties of cyanide are due to its inhibition of the cytochrome oxidase which is essential for virtually all mammalian cells.

Much useful information has been obtained from the type of inhibitor described as competitive. Such inhibitors usually show a marked structural resemblance to the substrates of the enzymes they inhibit. It therefore seems reasonable to assume that they combine with the enzyme in the same way as the substrate does, but are unable to undergo the reaction which the enzyme catalyses. On the lines of the previous model, substrate and inhibitor may be envisaged as competing for the enzyme thus

enzyme + substrate $\longrightarrow$
    enzyme–substrate complex
enzyme + inhibitor $\longrightarrow$
    enzyme–inhibitor complex.

The inhibition of the enzyme-catalysed reaction can then be explained by supposing that,

Fig. 8.8 The relationship between $1/v$ and $1/s$ for an enzyme which exhibits Michaelis-Menten kinetics. The intercept on the $1/s$ axis is numerically equal to $1/K_m$; the intercept on the $1/v$ axis is numerically equal to $1/v$.

since some of the enzyme molecules are occupied by inhibitor molecules, they are not free to combine with substrate molecules. If this is so, the inhibition should be overcome simply by increasing the concentration of substrate to enable it to compete more effectively with the inhibitor. If the substrate concentration is sufficiently increased, the inhibitor should be almost entirely displaced from the enzyme, all the enzyme molecules should be occupied by substrate, and the same maximum velocity should be attained as in the absence of inhibitor. In other words the presence of the inhibitor should not alter V, but it should increase K_m (Fig. 8.9). These predictions are in fact borne out experimentally and the correctness of the model is thus confirmed. The classical example is the case of succinate dehydrogenase, which catalyses the reaction

```
COOH              COOH
 |                 |
CH_2              CH
 |        ⟶        ‖
CH_2              CH
 |       2H        |
COOH              COOH
succinic          fumaric
 acid              acid
```

Malonic acid, malic acid and oxaloacetic acid all act as competitive inhibitors of the enzyme presumably by virtue of their structural resemblance to the substrate.

```
COOH        COOH        COOH        COOH
 |           |           |           |
CH_2        CH_2        CH_2        CH_2
 |           |           |           |
CH_2        COOH        CHOH        CO
 |                       |           |
COOH                    COOH        COOH
succinic    malonic     malic       oxaloacetic
 acid        acid        acid         acid
```

Since a competitive inhibitor must have a very close resemblance to the substrate, it is inevitably highly selective in its action. In theory it should be possible to inhibit a single enzyme in a complex organism by administering the appropriate competitive inhibitor. It will be shown in a subsequent section that the action of several important drugs can be explained in this way.

A different, but rather striking, example of a practical application of competitive inhibition is the treatment of methyl alcohol or ethylene glycol poisoning with ethyl alcohol. Methyl alcohol is a widely-used industrial

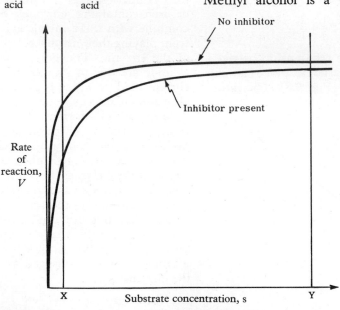

Fig. 8.9 Competitive inhibition. The effect of substrate concentration on the rate of an enzyme-catalyzed reaction in presence and absence of a fixed concentration of competitive inhibitor. At low substrate concentrations (X) the reaction rate is substantially reduced in presence of the inhibitor. But as the concentration of substrate is increased (Y) it displaces inhibitor from the enzyme molecules and the effect of the inhibitor on the reaction rate is correspondingly diminished.

solvent; ethylene glycol is the main constituent of antifreeze solutions used in automobile engines. These substances are occasionally drunk accidentally, suicidally, or as substitutes for ethyl alcohol. Both are oxidized by enzymes in the liver in essentially the same way as ethyl alcohol. The oxidation of ethyl alcohol gives rise to acetaldehyde, which can be oxidized further (by the same pathway as fat and carbohydrate) to carbon dioxide and water.

$$CH_3CH_2OH \longrightarrow CH_3CHO \longrightarrow$$
$$2CO_2 + 2H_2O.$$

But the oxidation of methanol stops at the highly toxic substance formaldehyde:

$$CH_3OH \longrightarrow HCHO.$$

By a similar process ethylene glycol gives rise to oxalic acid which crystallizes out in the kidney tubules and causes renal failure.

$$\begin{array}{ccc} CH_2OH & & COOH \\ | & \longrightarrow & | \\ CH_2OH & & COOH \end{array}$$

Methanol and ethylene glycol, therefore, are poisonous, not in themselves, but because of the metabolic changes they undergo. If their oxidation could be stopped or slowed down so that they were excreted rather than metabolized, they would be relatively harmless. The theory of competitive inhibition suggests that if the enzymes in the liver responsible for the oxidation of alcohols are kept supplied with ethyl alcohol it should compete with the methanol or ethylene glycol and slow down their oxidation. In practice cases of poisoning with methanol and ethylene glycol have been effectively treated by the intravenous infusion of ethyl alcohol.

THE MECHANISM OF ENZYME ACTION

It will become apparent in subsequent chapters that the extraordinarily high catalytic activity of enzymes and their equally extraordinary specificity go far to explain how cells and organisms work. Consequently it is important to try to find out why enzymes possess these properties. The complexity of enzyme structure made progress in this field slow and difficult until very recently. Even now our knowledge extends only to a small group of enzymes which have been intensively studied because they are relatively small and readily isolated and purified. Nonetheless the results obtained have been of importance.

If the simple model of enzyme action discussed above is correct we are faced by two questions.

(1) How does the enzyme combine with its substrate?

(2) Combination having taken place, how is the catalysis brought about?

Answers to these questions require knowledge of the exact structure of the enzyme and, if possible, of the position occupied by the substrate. The combination of X-ray crystallography and amino acid sequence determination is beginning to provide information of this sort for a few of the smallest and simplest enzyme molecules.

 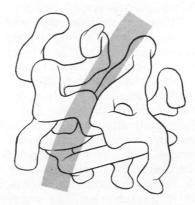

Fig. 8.10 Schematic representation of lysozyme molecule (left) and of the manner in which the polysaccharide substrate (shaded) is accommodated (right).

The first enzyme for which such information became available was *lysozyme*, which was discovered by Alexander Fleming, better known perhaps for his discovery of penicillin. He noticed that nasal mucus was able to dissolve certain bacteria. In the hope that this might lead to the discovery of a new antibacterial agent he isolated the active principle and showed it to be an enzyme which he called lysozyme because it dissolved or lysed bacteria. Lysozyme has not proved useful as an antibiotic because the major pathogens are not dissolved by its action but it has been a useful tool in microbiological research for investigating the structure of the bacterial cell. More recently it has attracted attention from biochemists because, with a molecular weight of only 14 600, it is one of the smallest enzymes known. It is, moreover, readily available from egg white. X-ray crystallography has shown that, in spite of its small size, its secondary and tertiary structure are quite complex. For present purposes it is sufficient to say that the lysozyme molecule has a compact but irregular shape and that one of its most striking features is a deep cleft or crevice (Fig. 8.10). The component of the cell wall hydrolysed by lysozyme could be described as an infinite array of very long polysaccharide chains cross-linked by short peptide chains (p. 48). Lysozyme attacks the polysaccharide chains which are made up of amino-sugar units (NAM and NAG) (p. 48) derived from glucose and linked $\beta(1-4)$. In other words the polysaccharide chain has a structure rather like that of cellulose (p. 47). The crosslinking peptide chains are attached to every second saccharide unit (Fig. 8.11). X-ray crystallography suggests that when the substrate combines with the enzyme it fits into the cleft in the enzyme molecule as shown schematically in Figure 8.10. From the crystallographic data it is possible to calculate the exact position of the substrate relative to the complicated folds of the enzyme's peptide chain and to deduce which bonds are likely to bind substrate to enzyme. Figure 8.11 represents diagrammatically some of the main features of this very complicated relationship. The cleft in the enzyme molecule is long enough to accommodate six monosaccharide units of the substrate. The upper three of these are held in position by a total of six hydrogen bonds, shown schematically in

Figure 8.11. The bond broken is that between the fourth and the fifth of the six units inside the cleft.

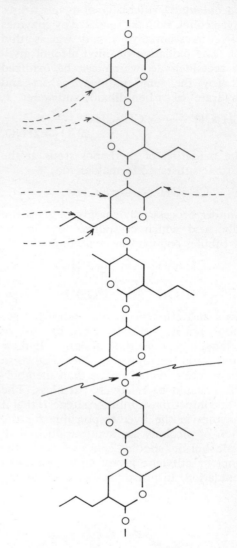

Fig. 8.11 Schematic representation of the attack of lysozyme on its substrate. The dashed lines in the upper part of the diagram indicate points at which the substrate is attached to the enzyme by hydrogen bonds. The two arrows below indicate the bond which the enzyme attacks.

To explain how the bond is broken we must digress briefly. Glycoside bonds, including those attacked by lysozyme, are susceptible to hydrolysis even in the absence of enzymes. This non-enzymic hydrolysis is greatly accelerated in the presence of acid. In Chapter 2 (p. 13) brief mention was made of the fact that

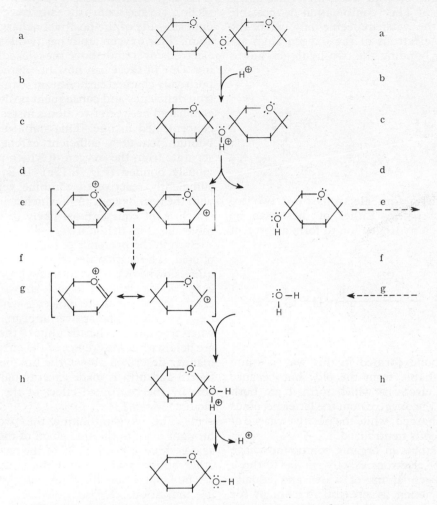

Fig. 8.12 Mechanism of acid-catalysed hydrolysis of a glycoside linkage in a polysaccharide.
Note:

(a) Lone pairs of electrons on the oxygen atoms.

(b) Attachment of a hydrogen ion to the oxygen atom of the linkage by means of a lone pair.

(c) The consequent positive charge on the oxygen atom.

(d) Breakage of the glycoside bond, the bonding electrons being taken over by the oxygen atom which thus loses its positive charge.

(e) (right) The electrons of the broken bond are taken over by the oxygen atom, which thus loses its positive charge.

(e) (left) The carbon atom at the other end of the bond broken in (d) is left electron-deficient and positively charged but is stabilized because both the electron deficiency and the positive charge are shared with the ring oxygen (as shown within the square brackets).

(f) Because of this stabilization the two parts of the original molecule do not necessarily re-combine but may move apart.

(g) The positively charged fragment is, however, sufficiently reactive to combine rapidly with a water molecule.

(h) The process of combination with water is the exact reverse of the dissociation of the original molecule.

H^+ ions do not exist free in aqueous solution, but are combined with water molecules in the form H_3O^+. This combination is possible because in the water molecule only two of the six valence electrons of the oxygen atom are involved in bonding the two hydrogen atoms thus

The remaining four electrons form two so-called 'lone pairs'. Either of these can be shared with a hydrogen ion to form a covalent bond thus

The new bond formed in this way is quite indistinguishable from the old; but because the shared electrons which form it are both donated by the oxygen atom the latter becomes positively charged, while the positive charge of the hydrogen is neutralized.

Oxygen atoms in organic compounds have lone pairs of electrons exactly similar to those on the oxygen atoms of water and equally capable of acting as sites of attachment for hydrogen ions. Consequently under acid conditions the oxygen atoms of an organic compound can acquire a hydrogen ion and thus become positively charged (Fig. 8.12a, b, c on previous page).

Such a positively charged oxygen atom exerts a strong attraction for the electrons bonding it to the carbon atoms on either side. If a pair of these bonding electrons were entirely taken over by the oxygen the bond would be broken (Fig. 8.12d, e). This would, however, leave the carbon atom concerned short of two electrons and positively charged. A carbon atom in this state (a so-called carbonium ion) is normally so reactive and unstable that one would expect it to recombine instantly with the oxygen, thus re-establishing

the original bond. In the case of glycosides this is made less likely because of the presence of a second oxygen atom (the ring oxygen) on the opposite side of the positively charged carbon atom. This oxygen atom has two lone pairs of electrons over and above those used for bonding. One of these can now be shared with the positively charged carbon atom so that the electron deficiency and consequent positive charge are shared between two atoms instead of being concentrated in one. This stabilizes the carbonium ion to a sufficient extent for it to separate from the oxygen to which it was previously bonded (Fig. 8.12e). It is, however, sufficiently reactive to combine very rapidly with the next oxygen atom which it encounters, which is, of course, most likely to be a water molecule (Fig. 8.12g, h).

Even in the presence of high concentrations of acid, the reaction described above proceeds quite slowly. We want to know how lysozyme is able to accelerate it to such an extraordinary extent in the substrate molecule with which it has combined. The answer appears to be that when six monosaccharide units of the substrate are held within the crevice of the enzyme in the manner described above, the link between the fourth and fifth is made exceptionally vulnerable by the combined effect of the following factors (Fig. 8.13).

(1) The oxygen linking the two units is brought close to the side chain of the glutamic acid residue at position 35 of the enzyme. The other amino acids around this glutamic acid are non-polar and therefore tend to suppress its ionization. It is, however, favourably situated to donate its hydrogen ion to the substrate oxygen. It therefore does for this specific linkage what high concentrations of acid would do for all such linkages.

(2) The shape of the crevice is such that the ring of the fourth substrate unit is distorted from its normal puckered configuration to a planar form. This is an unnatural shape for the intact ring, but is natural for the carbonium ion which forms when the bond between the fourth and fifth units is severed. The distortion imposed by the shape of the crevice favours therefore the otherwise rather unlikely transition from a normal sugar ring to the carbonium ion.

(3) This transition is favoured also because the ring of the fourth substrate unit is held

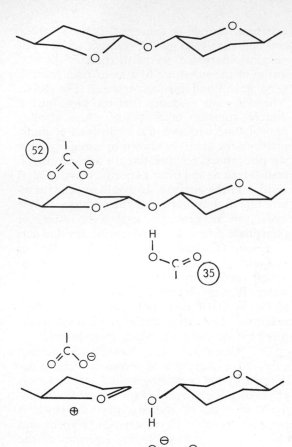

Fig. 8.13 Schematic representation of the mechanism by which lysozyme catalyses hydrolysis of the linkage between two of the subunits in its substrate. The *upper diagram* represents the two subunits intact and in their natural puckered configuration. For the sake of simplicity all the side chains have been omitted. The *middle diagram* represents the subunits in the crevice of the enzyme, showing (i) the un-ionized side chain of the glutamate residue at position 35 in the polypeptide chain of the enzyme, which is thought to provide a hydrogen ion for the oxygen of the bond, (ii) the flattening of the subunit on the left which is necessary to fit it into the very narrow crevice in the enzyme, (iii) the ionized side chain of the aspartate residue at position 52 of the enzyme. The *lower diagram* shows the situation immediately after the bond has broken. The hydrogen ion of glutamate-35 of the enzyme has been transferred to the oxygen of the bond which has broken. Breakage of the bond has left the carbon at the other end positively charged. This highly reactive structure is, however, stabilized because the charge is shared with the oxygen atom of the monosaccharide ring. The enzyme helps to stabilize it further by providing the adjacent negative charge of aspartate 52 and by forcing the ring into a flattened shape which is more natural for the charged structure than for the intact monosaccharide unit.

close to the side chain of the aspatic acid residue at position 52 of the enzyme chain. The other amino acids in the vicinity are all polar in character. The aspartic acid residue is therefore ionized and negatively charged. Consequently it tends to stabilize the carbonium ion formed when the link between the fourth and fifth units of the substrate is severed.

We can sum all this up by saying that, when lysozyme binds a substrate molecule, it does so by a large number of individually weak attractive forces. These reflect an elaborate and specific correspondence between the structures of enzyme and substrate and they serve to align the substrate in precisely the correct position for a number of factors to render a particular bond quite exceptionally vulnerable. Lysozyme is far more efficient than acid as a catalyst because it acts not at random but at exactly the right place, and because it facilitates the reaction not by one mechanism but by three.

The other enzymes for which we have comparable information from X-ray crystallography belong to the so-called serine esterases and proteases. It has long been known that a wide range of enzymes which catalyse hydrolysis of esters can be inactivated by the so-called organophosphorus inhibitors.

The essential features of the structure of these are exemplified by diisopropylphosphofluoridate (DFP), one of the best known.

$$
\underset{\text{phosphoric acid}}{\overset{\displaystyle HO}{\underset{\displaystyle HO}{\diagdown}}\overset{\displaystyle O}{\underset{\displaystyle OH}{P}}}
\qquad
\underset{\text{DFP}}{\overset{\displaystyle CH_3}{\underset{\displaystyle CH_3CHO}{\overset{\displaystyle CH_3CHO}{\diagdown}}}\overset{\displaystyle O}{\underset{\displaystyle F}{P}}}
$$

The organophosphorus inhibitors are extremely poisonous both to insects and to mammals. They produce their toxic effects by inhibiting the enzyme *cholinesterase* which catalyses the hydrolysis of *acetylcholine* to acetic acid and choline. Acetylcholine thus accumulates in muscle and nerve.

$$
\begin{array}{c}
CH_3\text{-}C\overset{O}{\diagup}\\
\underset{H_2O}{\diagdown}OCH_2CH_2\overset{\oplus}{N}\text{-}CH_3\\
CH_3
\end{array}
\longrightarrow
\begin{array}{c}
CH_3\text{-}C\overset{O}{\diagup}\quad\text{acetic}\\
\diagdown OH\quad\text{acid}\\
\\
H\diagdown OCH_2CH_2N\text{-}CH_3\\
\text{choline}\quad CH_3
\end{array}
$$

acetyl choline

The organophosphorus compounds are equally powerful inhibitors of the digestive enzymes trypsin and chymotrypsin which catalyse the hydrolysis of peptides *in vivo*. These enzymes have the capacity, apparently not used *in vivo*, to catalyse the hydrolysis of esters, presumably at the same active site and by the same mechanism. Analysis of enzymes inhibited with organophosphorus compounds shows that only one inhibitor residue is attached to each enzyme molecule. Since the presence of the substrate gives the enzyme some protection against the inhibitor it seems safe to infer that the inhibitor residue is attached at the active site of the enzyme. If the inhibited enzyme is hydrolysed to its constituent amino acids, the inhibitor residue is recovered attached to one of the serine residues in the hydrolysate (Fig. 8.14). By methods of

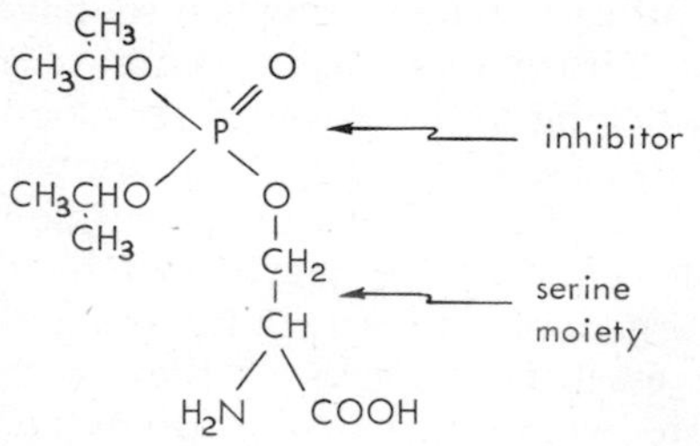

Fig. 8.14 DFP residue attached to serine.

partial hydrolysis (p. 33), it is possible to determine the serine residues to which the inhibitor has become attached. Thus in chymotrypsin, which has about 30 serine residues in its structure, the inhibitor is bound to that at position 195. In trypsin it is bound at position 183.

It would appear that in these two enzymes and in others like them the phosphoryl group of the inhibitor is attached to the reactive seryl residue of the enzyme with simultaneous release of HF (Fig. 8.15). It seems reasonable to assume that when the enzyme reacts with its substrate·the acyl group of the substrate is transferred to the reactive seryl residue, in the same way as the phosphoryl group of the inhibitor, with simultaneous liberation of the alcohol group. But whereas the inhibitor, once

attached, cannot be removed, the acyl group can be hydrolysed off the reactive serine. The enzyme, therefore, seems to transfer the acyl group of the substrate first to its own reactive seryl group and thence to water (Fig. 8.16). There is some evidence that the same sort of double transfer takes place when alkaline phosphatase catalyses the hydrolysis of phosphate esters. It can be shown by using ^{32}P, that the phosphoryl residue, like the acyl residue, is transferred to and from a specific reactive seryl residue in the enzyme. It would appear therefore that in some of the enzymes which catalyse hydrolysis of peptides, carboxylic esters or phosphate esters, an active serine residue acts as a point of attachment to which the acyl or phosphoryl group of the substrate is transferred, and from which it is transferred to water. It must, however, be emphasized that serine by itself does not catalyse hydrolytic reactions. The rest of the enzyme must therefore provide some activating mechanism.

The nature of this mechanism is still speculative. Amino acid sequence determination and X-ray crystallography have, however, given us a fairly detailed picture of the structures of some of the enzymes in which it operates, notably chymotrypsin, trypsin and elastase, protein-hydrolysing enzymes of the digestive tract. They have a similar primary structure, folded in a broadly similar manner; they all hydrolyse both peptide and ester linkages, but they differ in their specificities (Fig. 8.1). Trypsin attacks ester and peptide bonds in which the carboxyl residue is carried by the basic amino acids lysine and arginine; chymotrypsin attacks similar bonds in which the carboxyl residue is carried by an aromatic amino acid (phenylalanine, tyrosine or tryptophan); elastase is relatively non-specific. X-ray crystallography has shown that all three enzymes are roughly spherical and that the active serine residue lies on the surface. In chymotypsin this serine is adjacent to a deep recess lined by hydrophobic amino-acid side chains. Presumably, when the enzyme and substrate combine this accommodates the aromatic side-chain of the substrate (Fig. 8.17). The trypsin molecule has a similar recess but in this case there is a negatively charged aspartate residue at the carboxyl end which presumably can form an ionic bond with the positively charged lysine or arginine side-chain of the

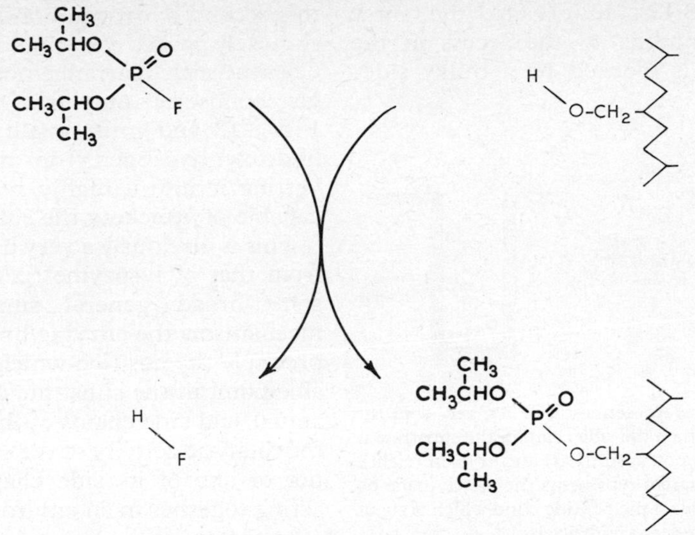

Fig. 8.15 Reaction of DFP with reactive seryl residue in trypsin, chymotrypsin and similar enzymes which hydrolyse esters. The zig-zag line represents the protein backbone.

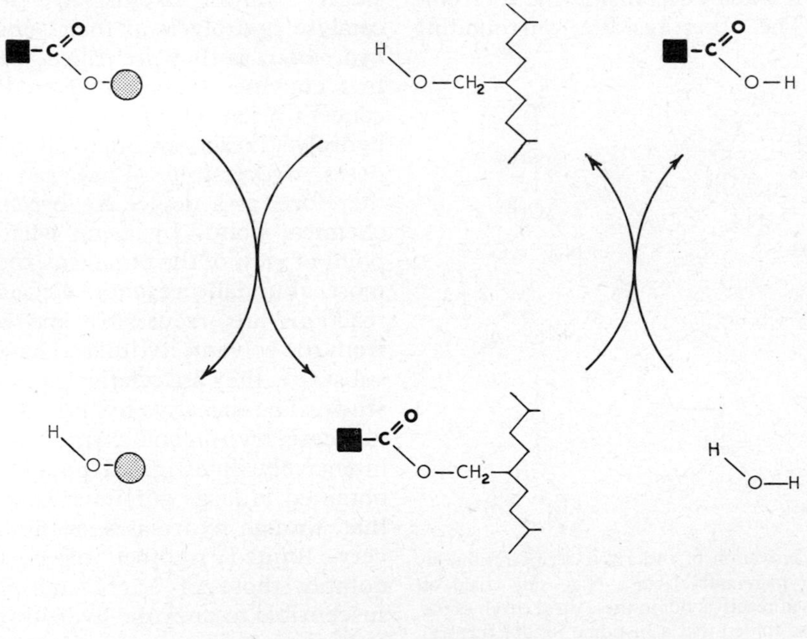

Fig. 8.16 Reaction of an ester with the reactive seryl residue of trypsin or chymotrypsin (*cf*. Fig. 8.15). The zig-zag line represents the protein backbone.

substrate (Fig. 8.17). In elastase the entry position corresponding to the recess in the other molecules is blocked by a bulky side-chain.

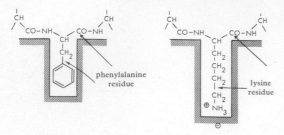

Fig. 8.17 Schematic representation of the recess in the enzyme molecule in which the side chain of the substrate is accommodated in chymotrypsin (left) and trypsin (right). The arrow in each diagram represents the attack made by the active serine residue on the peptide bond which is about to undergo hydrolysis.

In the case of chymotrypsin it appears that a histidine residue lies behind the active serine in such a position that a hydrogen bond can form between the hydroxyl of the serine side chain and one of the nitrogen atoms of the histidine side chain. The other nitrogen of the serine side chain is thought in turn to be hydrogen-bonded to the ionized side chain of an aspartate residue deep inside the molecule (Fig. 8.18). The other residues surrounding

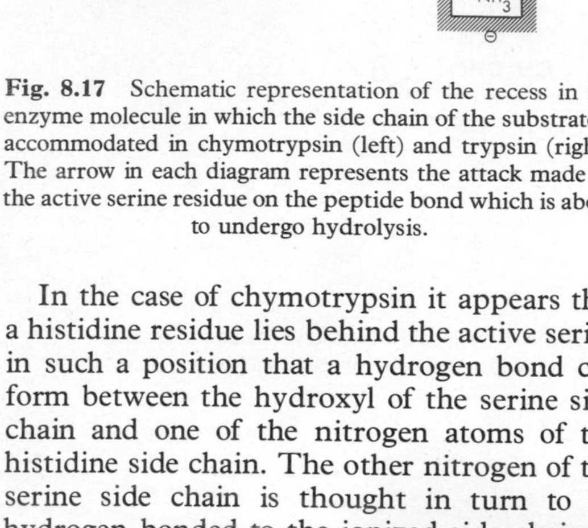

Fig. 8.18 The activation of the active serine residue in chymotrypsin, trypsin and elastase. The side chain of serine is not by itself reactive, but in these three enzymes the peptide chain is so folded that a histidine residue (centre) lies behind the serine residue (right) on the catalytic site of the molecule and in turn an aspartate residue (left) lies behind the histidine residue.

this second hydrogen bond are non-polar and so closely packed as to shield it from the solvent. Consequently rearrangement of the two hydrogen bonds can only take the form shown in Fig. 8.18 and must result in removal of the hydroxyl hydrogen from the serine thus converting it into a highly reactive *alkoxide* ion capable of attacking the substrate.

This is obviously a very different mechanism from that of lysozyme (p. 100) but there are some broad general similarities. In both mechanisms the enzyme binds the substrate in precisely the position which brings the vulnerable point in the substrate close to the reactive amino acid side chains of the enzyme. In both, the catalytic activity of the enzyme is a function not of one of its side chains, but of several acting together in an environment provided by other parts of the molecule.

TYPES OF REACTIONS CATALYSED BY ENZYMES

Because enzymes are most readily identified by the reactions they catalyse and by the substrates on which they act, they are customarily classified on this basis.

Hydrolases. Up to this point we have considered almost exclusively enzymes which catalyse hydrolysis of their substrates. These *hydrolases*, as they are called, were among the first enzymes to be discovered. Because the concentration of water is so high *in vivo*, hydrolytic reactions are to all intents and purposes irreversible. They can be regarded, therefore, as a device for breaking particular chemical bonds by using what is, from the point of view of the organism, the cheapest and most abundant reagent. Because hydrolytic reactions are irreversible and because, apart from the solvent, hydrolases have only a single substrate, they are relatively easy to detect and study. The digestive hydrolases, notably ribonuclease, trypsin and chymotrypsin, have been intensively investigated because they can be obtained in large amounts. It is worth noting that, though hydrolases are numerous, only a very limited number of chemical bonds, notably those of esters and glycosides, are susceptible to enzymic hydrolysis (Fig. 8.19).

Lyases. A second major group of enzymes, the lyases, catalyse quite a different type of reaction in which two parts of a molecule are

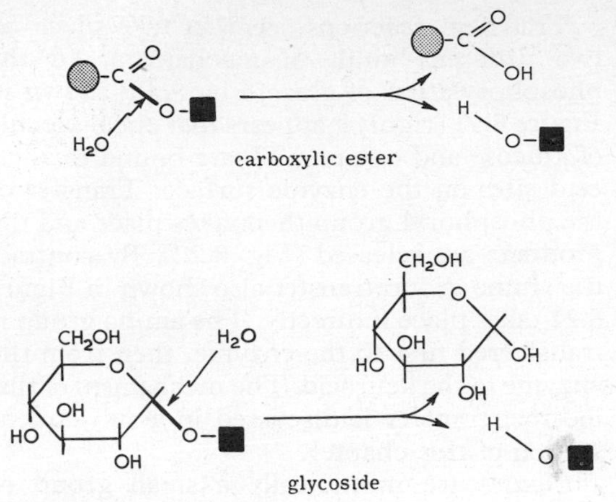

carboxylic ester

glycoside

Fig. 8.19 Two examples of enzyme-catalysed hydrolysis.

The bond broken may lie between carbon and oxygen, carbon and nitrogen, or carbon and sulphur (just as in the case of hydrolases); but it may also be a carbon to carbon bond. The double bond formed is between carbon and oxygen or between two carbon atoms. An example of each type is given in Figure 8.20.

Transferases catalyse the transfer of a functional group from one substrate to another. The group transferred may be a small structure like $-NH_2$, or an entire sugar residue, or even a polysaccharide chain. Two examples are given in Figure 8.21. The reversibility or otherwise of transferase reactions depends on the nature of the participants. When the donor and recipient have similar chemical structures, the reaction is likely to be reversible, as in the amino group transfer shown in Figure 8.21. Where the participants are chemically very different there is more likelihood of a substantial free energy change (see

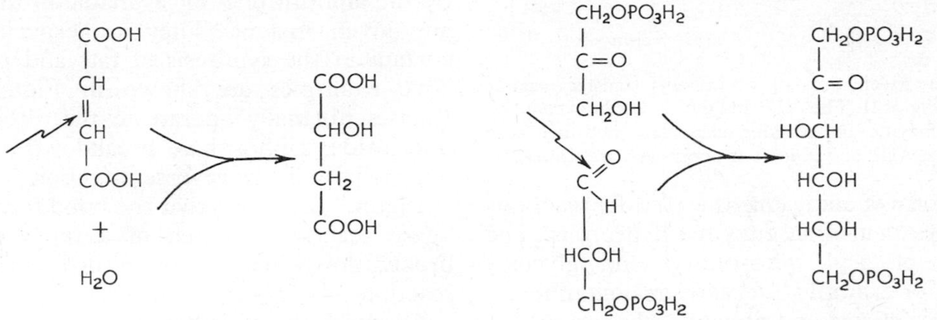

separated with the formation of a double bond in one of them. Unlike enzymic hydrolysis, lyase reactions are quite commonly reversible.

Fig. 8.20 Two examples of lyase reactions. Fumarate hydratase (left) and aldolase (right) are both important in carbohydrate metabolism (see Chap. 11).

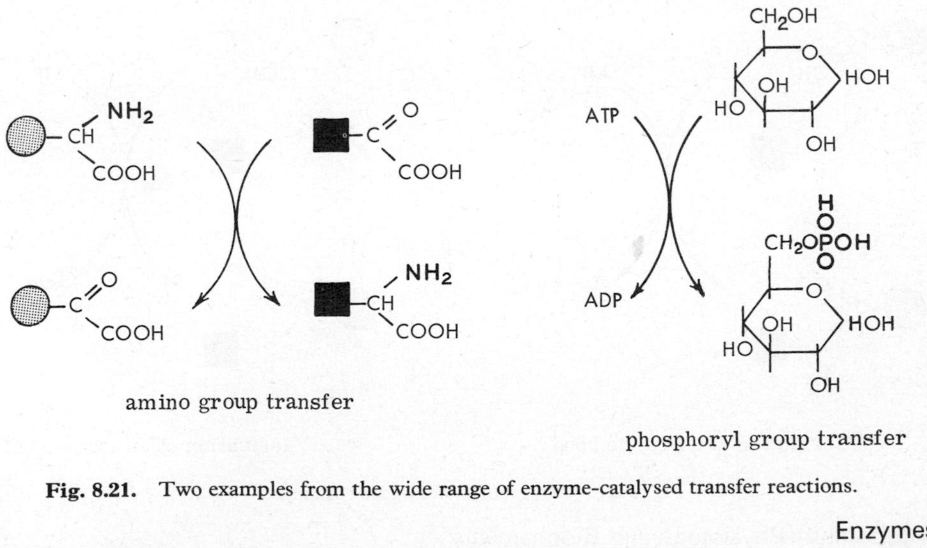

amino group transfer

phosphoryl group transfer

Fig. 8.21. Two examples from the wide range of enzyme-catalysed transfer reactions.

Enzymes   107

p. 85). For example, in many phosphoryl transfer reactions there is a large negative free energy change. In other words, the transfer is virtually irreversible. This is sometimes expressed by saying that the terminal phosphate of ATP is attached to the rest of the molecule by a 'high-energy' bond.

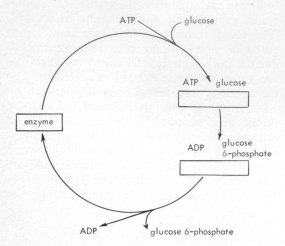

Fig. 8.22 Mechanism of phosphoryl transfer reaction shown in Fig. 8.21. The ATP and the glucose both combine with the enzyme, the transfer takes place, and then both products, ADP and glucose 6-phosphate, are released.

The form of many enzymic transfer reactions resembles that of an enzymic hydrolysis. The transfer of acyl, phosphoryl and glycosyl groups, for example, is exactly analogous to the hydrolysis of acyl and phosphoryl esters and of glycosides. A few other transfer reactions have mechanisms resembling that of the lyases.

Transfers reactions seem to take place by two different kinds of mechanism. In the phosphorylation of glucose by ATP shown in Figure 8.21 (right) it appears that one molecule of glucose and one of ATP are bound to adjacent sites on the enzyme surface. Transfer of the phosphoryl group then takes place and the products are released (Fig. 8.22). By contrast the amino group transfer also shown in Figure 8.21 takes place indirectly. The amino group is transferred first to the enzyme, then from the enzyme to the keto acid. The mechanism of this indirect transfer is discussed in a subsequent section of this chapter.

Ligases are numerically a small group of enzymes. They usually link two reactants together at the expense of the breakdown of ATP or some other nucleoside triphosphate. Since, as will be shown later, the supply of ATP in the cell is continually replenished by a variety of mechanisms, synthesis by ligases is not limited by the amount of ATP available in the cell at any given instant. They are especially important in the synthesis of fats and proteins. Two examples are shown in Figure 8.23. Ligases normally operate in the direction of nucleoside triphosphate breakdown and only occasionally in the reverse direction.

Figure 8.23 shows that the bond formed in a ligase reaction is often of a type which is broken down in a hydrolase (or lyase) type of reaction.

Oxidoreductases differ from all the other enzymes so far described in that they are concerned, as their name implies, with oxidations

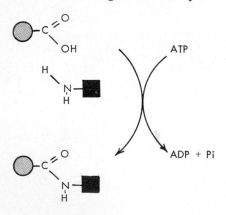

formation of a peptide bond

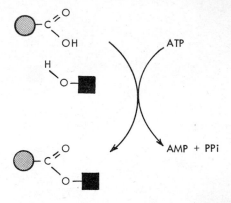

formation of an ester bond

Fig. 8.23 Two examples of ligase reactions.

and reductions rather than with addition or removal of groups of atoms. Much more detail will be given in Chapter 10 but for present purposes it is sufficient to say that a majority of enzymes of this class catalyse reactions which can be represented as transfers of a hydrogen atom and an electron from an alcohol or an amine to one or other of the nucleotide co-enzymes NAD and NADP (see Chap. 6). A typical reaction can be represented as shown in Figure 8.24.

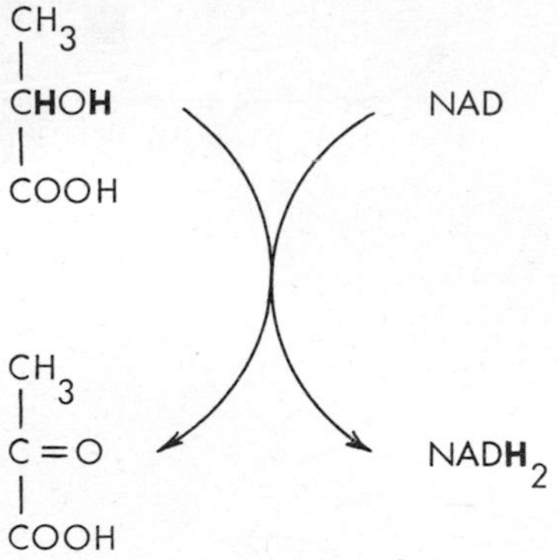

Fig. 8.24 An example of an oxidoreductase reaction involving NAD.

The mechanism revealed by the use of deuterium labelled alcohol is shown in detail in Figure 8.25. Before reduction the pyridine

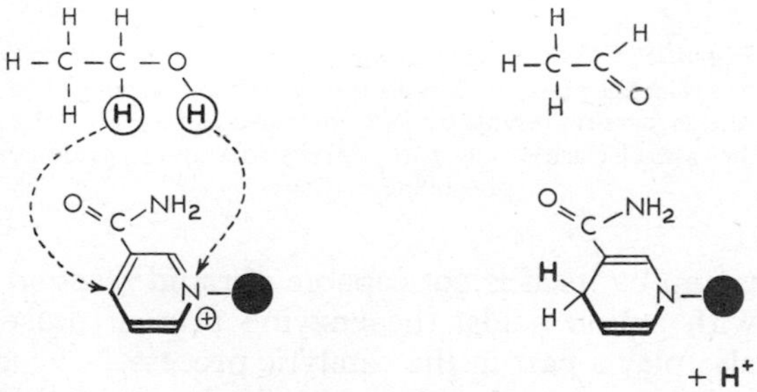

Fig. 8.25 Enzyme catalysed oxidation of alcohol by NAD^+.

ring is planar and the hydrogen atoms attached at position 4 lie one above and one below the plane of the ring. It should be noted that the hydrogen bound directly to the carbon of the ethanol is always directly transferred to the coenzyme and always to the upper face (as represented here) of the pyridine ring. The hydrogen of the hydroxyl group is always liberated as a hydrogen ion. It seems clear, therefore, that the enzyme must always hold the alcohol and the coenzyme in the same geometrical relationship to one another in order that the reaction may take place. Investigation of other reactions involving NAD or NADP has shown that some enzymes transfer the hydrogen and acceptors in the same way that ADP and ATP can react with phosphoryl donors and acceptors. It is worth noting that alcohols, amines and aldehydes are the commonest types of electron donor in such reactions.

Other oxidoreductases have not yet been so well characterized but two main divisions can be distinguished: the *flavoproteins* and the *cuproproteins*. The flavoproteins are so called because they contain as prosthetic group either flavin mononucleotide (FMN) or flavin adenine dinucleotide (FAD) (p. 67). Unlike the nicotinamide nucleotides, which accept two electrons simultaneously, the flavins can accept one electron at a time. Perhaps for this reason, the flavoproteins catalyse a much wider range of oxidations.

The cuproprotein oxidoreductases are almost equally varied. Since copper ions can exist in two oxidation states:

$$Cu^{2+} + e \rightleftharpoons Cu^+$$

it is tempting to suppose that in cuproprotein oxidoreductases they undergo alternate oxidation and reduction, but there is no conclusive evidence for this. The other distinguishing feature of cuproproteins is that the electron acceptor is always molecular oxygen. An important example is tyrosinase which catalyses the first step in the conversion of the amino acid tyrosine to the pigment melanin (see Chap. 21).

Isomerases catalyse a variety of internal rearrangements in their substrates, especially changes in configuration and internal oxidation-reductions (Fig. 8.26 overleaf).

COENZYMES AND PROSTHETIC GROUPS

The enzymes, whose presumed mechanisms of action were described above, lysozyme, chymotrypsin trypsin and elastase, are all simple proteins composed exlusively of amino

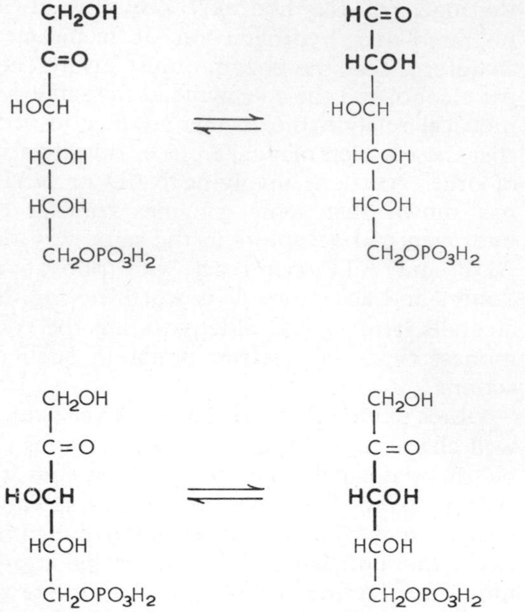

CH_2OH
|
C=O
|
HOCH
|
HCOH
|
HCOH
|
CH_2OPO_3H_2

⇌

HC=O
|
HCOH
|
HOCH
|
HCOH
|
HCOH
|
CH_2OPO_3H_2

CH_2OH
|
C = O
|
HOCH
|
HCOH
|
CH_2OPO_3H_2

⇌

CH_2OH
|
C = O
|
HCOH
|
HCOH
|
CH_2OPO_3H_2

Fig. 8.26 Two examples of isomerase reactions. Glucose phosphate isomerase (above) and ribulose phosphate epimerase (below).

acids. Many enzymes, however, include in their structure a prosthetic group analogous to the haem of haemoglobin. In most cases this appears to play an important part in the mechanism of the enzyme-catalysed reaction. For example, in a previous section of this chapter (p. 107) it was remarked that the enzymic transfer of an amino group from an amino acid to a keto acid (Fig. 8.21) is not direct. As shown in Figure 8.27 the amino

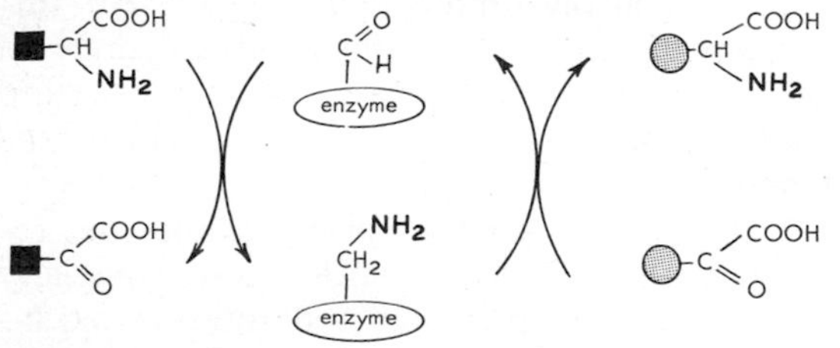

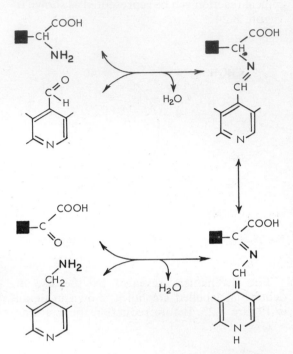

Fig. 8.27 Mechanism of transamination. The amino group is transferred from donor to enzyme and then from enzyme to recipient.

group is transferred first to the prosthetic group of the enzyme (pyridoxal phosphate) and from there to the keto acid.

The reaction is complex and depends on the ability of pyridoxal phosphate (Chap. 13 and Chap. 22) to take up reversibly an amino group from an amino acid and is related to complex system of alternating single and double bonds in the pyridoxal molecule and to the well-known capacity of such systems to undergo rearrangement (Fig. 8.28). The prosthetic group therefore endows the enzyme molecule with a property which it would not otherwise possess. On the other hand, since the prosthetic

Fig. 8.28 Transfer of an amino group from an amino acid to pyridoxal phosphate, as in transaminase reactions. The black square represents the side chain of an amino acid. For the sake of clarity, only part of the structure of pyridoxal phosphate is shown.

group by itself is not capable of rapid reaction with amino acids, the enzyme protein must also play a part in the catalytic process.

It is not yet possible to explain the role of the prosthetic groups in all the transferases that possess them but prosthetic groups in general seem to serve as points of attachment for the group to be transferred. In other words, they have the same role as the active serine residue in chymotrypsin. Their special importance seems to lie in the fact that they enable the enzyme to perform chemical manipulations which cannot be brought about solely by the agency of amino acid side chains.

Coenzymes. The term coenzyme was coined early in the history of enzymology to denote a small molecule whose presence was necessary to allow a particular enzyme to act. The concept was never precisely defined, but a coenzyme was assumed to be in the nature of a detachable prosthetic group. Although this early view has proved to be rather misleading, the term is still used to describe a rather miscellaneous group of important compounds. Their essential features can be shown by a simple illustration. The vast majority of phosphotransferases catalyse reactions in which ATP acts as phosphoryl donor or ADP as phosphoryl recipient. For example, creatine phosphokinase catalyses the transfer of a phosphoryl group from phosphocreatine to ADP:

phosphocreatine + ADP $\longrightarrow$ creatine + ATP

Similarly hexokinase catalyses the transfer of a phosphoryl group from ATP to glucose:

ATP + glucose $\longrightarrow$ ADP + glucose 6-phosphate

There is no known mechanism by which a phosphoryl group can be transferred direct from phosphocreatine to glucose, but this can obviously be achieved by creatine kinase and hexokinase working in sequence. The process can be represented thus:

phosphocreatine ⟶ ADP ⟶ glucose 6-phosphate
creatine ⟵ ATP ⟵ glucose

Only a very small, indeed catalytic, amount of the nucleotide is required to link the two reactions together. If they are considered separately, ADP and ATP must be regarded as substrates on exactly the same footing as phosphocreatine and glucose. It is only when the two reactions are linked together that the nucleotide can be regarded as part of the catalytic mechanism and then it fully merits the name coenzyme. Because so many phosphotransferase reactions involve ADP and ATP in the same way as hexokinase and creatinekinase do, these nucleotides provide a common centre through which phosphate can be transferred from any one of a wide variety of phosphate donors to any one of an equally wide variety of recipients (Fig. 8.29).

Like prosthetic groups, coenzymes do not merely provide a point of attachment for the group which is being transformed; they also influence the chemical properties of the group. For example, acyl groups in process of transfer and generally attached to coenzyme A (p. 67). Some of these transfer reactions are shown in

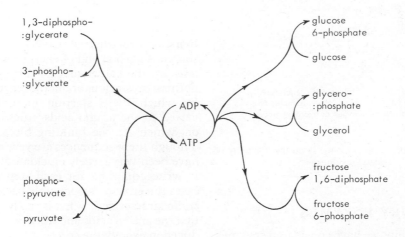

Fig. 8.29 A number of enzymes catalyse the transfer of phosphoryl residues to ADP (left) or from ATP (right). In this way ADP and ATP provide a common centre for the exchange of phosphoryl residues.

Figure 8.30. Attachment to coenzyme A endows acyl groups with a much greater reactivity than they possess in the form of the free acid or carboxylate ion or as esters or peptides. Thus the first step in the metabolism

substrates, by pH, and sometimes by the availability of metal and other ions. The effects of these factors can be shown *in vitro* but in the living organism the rates of chemical reactions are subject to additional controls. To take a

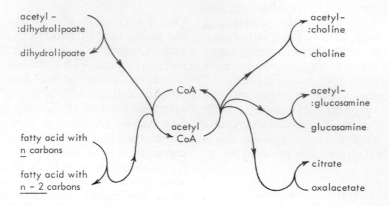

Fig. 8.30 A variety of enzymes transfer acetyl residues to coenzyme A or from acetyl coenzyme A. These two compounds, therefore, provide a common centre for the transfer of acetyl residues.

of fatty acids is generally their conversion to the coenzyme A derivative (Fig. 8.31).

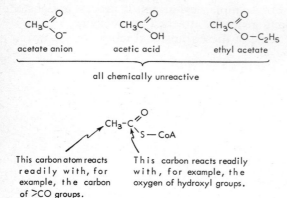

Fig. 8.31 The effect of coenzyme A on the reactivity of acetyl residues.

REGULATORY ENZYMES

Up to this point we have considered enzymes simply as catalysts, albeit extraordinarily powerful and selective. The chemical reactions they catalyse proceed at rates determined by temperature, by the concentrations of their

simple example, after a meal glucose absorbed from intestine is stored in the liver in the form of glycogen. Subsequently the glycogen is broken down again as required to yield glucose. Why should the glucose–glycogen conversion proceed in one direction at one time and in the opposite direction at another?

A more fundamental question is posed by growing cells. The bacterium *Escherichia coli*, which is abundant in the gut, is capable of living on a medium in which glucose is the sole source of carbon and of energy and ammonium ions are the sole source of nitrogen. There is nothing mysterious about the reaction sequence by which these starting materials are converted to the amino acids, nucleotides and so on, which are the building blocks of the cell. Though these sequences are very complex they have been very largely elucidated; it is possible to write equations for each step and in many cases the enzyme responsible has been purified and characterized. What is really remarkable is that the amino acids and nucleotides are all produced in exactly the correct proportions for the formation of new cells. Furthermore if the cells are supplied with a preformed amino acid (say valine), in addition to glucose and ammonium ions, they will stop synthesizing that particular amino acid while continuing to synthesize all the others. Quite clearly therefore the cells have some mechanism by which they can control the synthesis of each amino acid (and of every other cell constituent) to exactly the level required. How this is achieved is not yet fully understood; our knowledge is most extensive in the case of bacterial cells, but there is no doubt that similar controls operate in plant and animal cells.

In general it can be said that many metabolic products (in particular amino acids and nucleotides) inhibit one of the enzymes in the pathway by which they are synthesized. Thus in *Escherichia coli* isoleucine is formed from threonine by the sequential action of six enzymes. If more isoleucine is formed than can be used, its accumulation inhibits the first enzyme in the sequence (threonine deaminase) so that further isoleucine production is switched off.

Similarly the nucleotide cytidine triphosphate inhibits the enzyme aspartate carbamoyl transferase which catalyses one of the first steps in its synthesis (Chap. 20). This end-product inhibition of a synthetic pathway is known as *feedback inhibition* and the enzyme concerned is called a *regulatory enzyme*. The mechanism of the phenomenon is not yet understood. Since the end-product need bear no structural resemblance to the normal substrate of the enzyme it inhibits, it is termed an *allosteric* inhibitor (Fig. 8.32). Presumably it combines

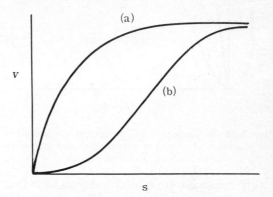

Fig. 8.32 The structures of threonine (left), and an allosteric inhibitor of threonine deaminase (right).

with the enzyme molecule at a site quite separate from that occupied by the substrate but somehow it modifies the relationship between enzyme and substrate. Regulatory enzymes have molecular weights ten to a hundred times greater than those of lysozyme and

chymotrypsin and many of them are apparently built up from subunits. In spite of their complexity we have some clues to their mode of action.

Most regulatory enzymes show a sigmoid relationship in contrast to the simple hyperbolic Michaelis-Menten relationship between substrate concentration and reaction rate (Fig. 8.9). The sigmoid curve (Fig. 8.33) could be

Fig. 8.33 The relationship of v and s for (a) an ordinary enzyme, and (b) a regulatory enzyme.

explained by assuming that each enzyme molecule combines, not with a single substrate molecule, but with several, say n, so the reaction could be represented as follows:

$$E + nS \rightleftharpoons ES_n \longrightarrow E + n \text{ products}$$

This could happen if, after the enzyme had combined with a single substrate molecule, its affinity for additional substrate molecules is greatly increased. We have an illuminating model of such a situation in two well-known non-enzymic proteins, haemoglobin and myoglobin (see p. 35). Both combine rapidly and reversibly with molecular oxygen. Myoglobin consists of a single polypeptide chain and has a single oxygen-binding site. Its combination with oxygen appears to be exactly analogous in speed and reversibility to the combination of enzyme and substrate:

$$My + O_2 \rightleftharpoons MyO_2$$
$$E + S \rightleftharpoons ES$$

If the percentage saturation of myoglobin with oxygen is plotted against partial pressure of oxygen a rectangular hyperbola is obtained (Chap. 31), exactly analogous to the relationship of v to s in an enzyme catalysed reaction (Fig. 8.34).

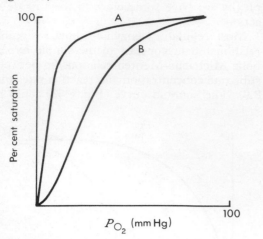

Fig. 8.34 Relationships of percentage saturation to partial pressure of oxygen in (a) myoglobin and (b) haemoglobin. Compare with Fig. 8.33.

The haemoglobin molecule differs from that of myoglobin in being formed from four polypeptide subunits (Chap. 25), each with its own oxygen-binding site. If the percentage saturation of haemoglobin with oxygen is plotted against the partial pressure of oxygen, the relationship is not hyperbolic, as with myoglobin, but sigmoid (Fig. 8.34 and Chap. 25). It has long been accepted that this is because attachment of an O_2 molecule at one of the four binding sites of a haemoglobin molecule enormously facilitates the binding of other O_2 molecules at the other three sites (Chap. 31). Exactly how binding of O_2 at one site affects the properties of the other sites is by no means clear. The four sites are quite separate from one another. When O_2 is bound there is a slight, but probably significant, change in the relative positions of the four polypeptide chains and this may be responsible for the change in properties of the binding sites.

The anomalous kinetics of regulatory enzymes have important practical consequences.

For an ordinary enzyme for which

$$\cdot\, v = \frac{V}{K_m/s + 1}$$

it can easily be shown that an increase in s can never produce *more* than a proportional increase in v. If, for example, s is increased by say

50 per cent, v will be increased by less than 50 per cent. This generalization does not, however, apply to enzymes with sigmoid kinetics; in these a moderate increase in substrate concentration can produce a quite disproportionate increase in enzyme activity (Fig. 8.35). Thus

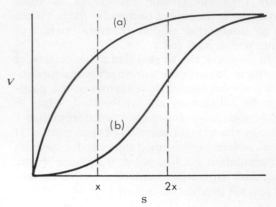

Fig. 8.35 Sensitivity of enzymes to changes in substrate concentration. Ordinary enzymes (a) are highly active even at very low substrate concentrations and the effect of increasing substrate concentration has less and less effect as V is approached. At substrate concentration x the rate is already close to the maximum and doubling the concentration produces only 30 per cent increase in this example. Regulatory enzymes with sigmoid kinetics (b) are relatively inactive at low substrate concentrations but increases in substrate concentration produce a disproportionate increase in enzyme activity. In the example shown, increasing the substrate concentration from x to $2x$ increases the regulatory enzyme activity sixfold.

regulatory enzymes can be in effect switched on and off according to the concentration of substrate available.

An allosteric inhibitor can affect the behaviour of a regulatory enzyme either by diminishing its affinity for its substrate or by depressing the rate at which it acts on the substrate after formation of the enzyme–substrate complex. In the first case, which is the more common, the sigmoid curve relating v to s is extended to the right. In the second, it is compressed downwards (Fig. 8.36).

Allosteric inhibition can be regarded in most instances as a mechanism by which a particular metabolic pathway is regulated in accordance with the demand, or lack of demand, for its product. There is an exactly analogous phenomenon of allosteric activation in which the activity of an enzyme is enhanced by accumulation of a metabolite. For example, when the

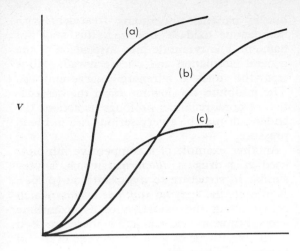

Fig. 8.36 Changes in the relationship of v to s in a regulatory enzyme brought about by allosteric inhibitors. (a) No inhibitor present. (b) Inhibitor affecting affinity of enzyme for substrate. (c) Inhibitor affecting rate of breakdown of enzyme–substrate complex.

requirement of an organism for energy outruns the available supply, certain metabolites, notably AMP and ADP, tend to accumulate. In turn this accumulation activates certain key enzymes in the energy-yielding metabolic pathways and so increases the supply of energy until the demands are met.

The best known and best understood regulatory enzyme at present is the aspartate carbamoyl transferase of *E. coli* which catalyses the first step in pyrimidine biosyntheses (Chap. 20). Its action is inhibited by cytidine triphosphate (CTP), one of the ultimate products of its action. Treatment with *p*-hydroxymecuribenzoate (which reacts with thiol groups) causes the native enzyme (mol. wt. 310 000) to dissociate into two *catalytic subunits* (mol. wt. 100 000) and three *regulatory subunits* (mol. wt. 30 000). The catalytic subunits combine with substrate and catalyse the reaction. They are quite insensitive to CTP, and follow normal Michaelis-Menten kinetics, unlike the intact enzyme. The regulatory subunits have no catalytic activity by themselves, but can combine reversibly with CTP.

THE SIGNIFICANCE OF ENZYMES

Although they are complex in structure, enzymes obey the well-known laws of chemistry. A rational explanation can now be provided for the action of certain drugs which are enzyme inhibitors. The first notable examples were the sulphonamides. Once the principles were clearly recognized the empirical testing of drugs began to be supplemented by the deliberate synthesis of compounds which were likely to be competitive inhibitors of enzymes present in pathogenic organisms or in cancer cells.

The mechanism of the action of sulphonamides was not immediately apparent, but it was noticed that pus and other natural materials contained something eventually identified as *p*-aminobenzoic acid which protected sensitive micro-organisms from sulphonamides.

p-aminobenzoic acid

p-aminobenzene sulphonamide (sulphanilamide)

Many micro-organisms require this compound as a precursor of the folic acid coenzymes (Chap. 13). Sulphonamides competitively inhibit its utilization; consequently folic acid production is prevented and growth of the micro-organisms stops. The host, however, is unaffected because he obtains his folic acid preformed in his diet and does not need to synthesize it. Rather surprisingly the micro-organisms which are sensitive to sulphonamides cannot get round their difficulties by making use of the host's preformed folic acid. Hence sulphonamides can be toxic to micro-organisms but virtually harmless to their host and so are effective because they inhibit a reaction which is essential to the invading bacteria but not to the host. Unfortunately, biochemical differences of this nature are rare. In spite of enormous differences in shape and size normal cells, cancer cells and even bacteria share a common set of biochemical mechanisms and competitive inhibitors are likely to disturb the function of all three.

The action of many drugs, not merely those

that act against infection, can be explained in terms of competitive inhibition of enzymes. For example, *monoamine oxidase* (MAO), catalyses the oxidative deamination of monoamines such as adrenaline, noradrenaline and 5-hydroxytryptamine (Chap. 50), to the corresponding carboxylic acids.

$$R-CH_2-CH_2-NH_2 \xrightarrow[\text{deamination}]{\text{oxidative}} R-CH_2-COOH$$

Ephedrine and amphetamine prolong the effects of these hormones by acting as competitive inhibitors of monoamine oxidase. (Though they are themselves amines they are very resistant to attack by the enzyme.) A similar explanation has been put forward for the action of cocaine and of antidepressive drugs such as iproniazid and tranylcypromine.

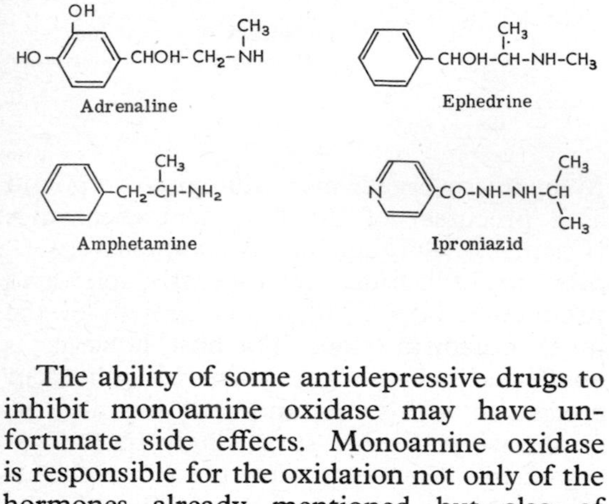

Adrenaline

Ephedrine

Amphetamine

Iproniazid

The ability of some antidepressive drugs to inhibit monoamine oxidase may have unfortunate side effects. Monoamine oxidase is responsible for the oxidation not only of the hormones already mentioned but also of amines such as tyramine present in the diet.

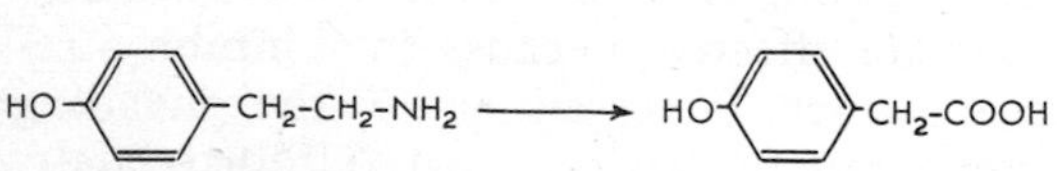

tyramine

p-hydroxyphenyl acetic acid

An average portion (say 30 g) of cheddar cheese contains as much as 20 mg tyramine. Normally it is oxidized to *p*-hydroxyphenyl acetic acid by the monoamine oxidase of the gut and the liver but in patients undergoing treatment with monoamine oxidase inhibitors this may not happen. The tyramine may instead enter the general circulation and release noradrenaline from the stores at adrenergic nerve endings. The inhibitor, by slowing down the destruction of noradrenaline, prolongs its action; the end-result may be a very serious rise in blood pressure.

Another example of a competitive inhibitor used as a drug is *allopurinol* which is very similar in structure to hypoxanthine (p. 64). It inhibits the enzyme *xanthine oxidase* which brings about the oxidation of hypoxanthine and xanthine to uric acid and is therefore used in the treatment of gout to prevent uric acid formation (Chap. 20).

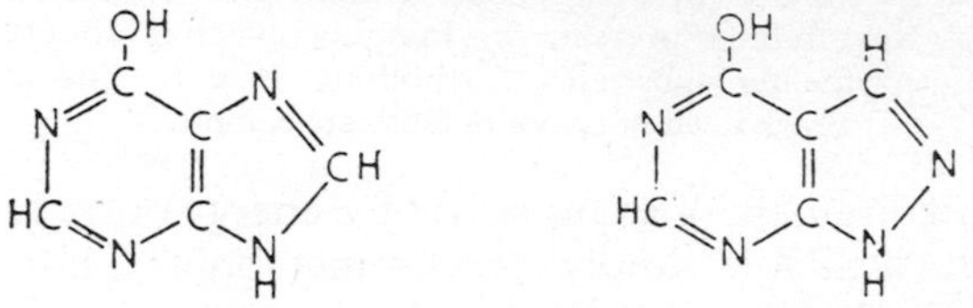

Hypoxanthine

Allopurinol

DIAGNOSTIC USE OF ENZYMES

Human serum contains a remarkable range of enzymes which serve no obvious function there; they are presumably derived either from leakage from living cells or from the debris of dead or dying cells. Several diseases are accompanied by marked changes in the levels of certain enzymes in the serum.

For many years alkaline phosphatase activity in the serum has been measured in the investigation of liver and bone diseases (Chap. 14). Other serum enzymes now estimated for diagnostic purposes include: aminotransferases (transaminases) in liver disease and myocardial infarction; amylase in pancreatic disease; lactate dehydrogenase in myocardial infarction; and creatine phosphokinase in conditions involving damage to skeletal or cardiac muscle.

Isoenzymes. Certain enzymes may exist in tissues in two or more varieties which have the same catalytic activity but are chemically, immunologically and electrophoretically distinct. These are known as *isoenzymes* or *isozymes*.

A well-known example is lactate dehydrogenase which catalyses the oxidation of lactate to pyruvate with the aid of NAD as coenzyme (p. 144). It can readily be separated by electrophoresis on starch gels or polyacrylamide gels (p. 37) into five isoenzymes usually referred to as LD_1, LD_2, LD_3, LD_4 and LD_5, all of the same molecular weight (about 135 000). The distribution of these isoenzymes varies according to the tissue of origin, LD_1 and LD_2 being particularly abundant in heart muscle and LD_4 and LD_5 predominating in liver.

The existence of these five isoenzymes can be explained on the basis of the quaternary structure (p. 37) of lactate dehydrogenase. Each isoenzyme is a tetramer, combining with 4 molecules of coenzyme, which may be split into four monomers, each of molecular weight about 35 000, on treatment with solutions of 12 M-urea or 5 M-guanidine hydrochloride. The monomer subunits, themselves catalytically inactive, are of two types, H and M, differing slightly in amino acid composition.

Each of the five isoenzymes is a hybrid composed of mixed subunits as follows:

LD_1	LD_2	LD_3	LD_4	LD_5
H	H	H	H	M
H	H	H	M	M
H	H	M	M	M
H	M	M	M	M

The estimation of lactate dehydrogenase activity in blood serum and the separation of the isoenzymes are of considerable diagnostic importance in medicine. For example, after the heart muscle has been damaged as the result of coronary thrombosis there is a rise in the activity of LD_1 and LD_2 in the blood serum. In liver disease, on the other hand, there is a rise in LD_5 and to a lesser extent in LD_4 in the plasma.

Other enzymes existing in multiple molecular forms include malate dehydrogenase, isocitrate dehydrogenase, glucose-6-phosphate dehydrogenase, aspartate amino transferase (glutamate-oxaloacetate transaminase, GOT), creatine kinase, esterases and phosphatases.

REFERENCES

BERGMEYER, H. U. (1963). *Methods of Enzymatic Analysis.* New York: Academic Press.

BERNHARD, S. (1968). *The Structure and Function of Enzymes.* New York: Benjamin.

BOYER, P. D., LARDY, H. & MYRBACK, K. (1959). *The Enzymes,* 2nd edn. New York: Academic Press.

DAWES, E. A. (1972). *Quantitative Problems in Biochemistry,* 5th edn. Edinburgh: Livingstone.

DICKENSON, R. E. & GEIS, I. (1969). *The Structure and Action of Proteins.* New York: Harper & Row.

DIXON, M. & WEBB, E. C. (1964). *Enzymes,* 2nd edn. London: Longman.

Enzyme Nomenclature: Recommendations (1964) of the International Union of Biochemistry on the nomenclature and classification of enzymes, together with their units and the symbols of enzyme kinetics in *Comprehensive Biochemistry* (Edited by M. Florkin and E. H. Stotz) 1965, **13** (2nd edn), 1–219.

FLORKIN, F. & STOTZ, E. H. (Eds.) (1964–65). *Comprehensive Biochemistry.* Vols. 12, 14, 15 & 16. Amsterdam: Elsevier.

GRAY, C. J. (1971). *Enzyme Catalysed Reactions.* London: Van Nostrand.

GUTFREUND, H. (1965). *An Introduction to the Study of Enzymes.* Oxford: Blackwell.

GUTFREUND, H. & KNOWLES, J. R. (1967). The foundations of enzyme action, *Essays in Biochemistry* (Edited by P. N. Campbell and G. D. Greville) **3**, 25–72. London: Academic Press.

KING, J. (1965). *Practical Clinical Enzymology.* Princeton: Van Nostrand.

NEURATH, H. (1964). Protein digesting enzymes. *Scientific American,* **211** (12), 68–79.

NORTH, A. C. T. (1966). The structure of lysozyme. *Science Journal,* **2** (11), 55–60.

PHILLIPS, D. C., BLOW, D. M., HARTLEY, D. S. & LOWE, G. (Organizers) (1970). A discussion on the structure and function of proteolytic enzymes. *Philosophical Transactions of the Royal Society* (B) **257**, 63–266.

THEORELL, H. (1967). Function and structure of liver alcohol dehydrogenase. In *The Harvey Lectures* 1965–1966, *Series* 61, p. 17–42. London: Academic Press.

WILKINSON, J. H. (1965). *Isoenzymes.* London: Spon.

WILLIAMS, A. (1969). *Introduction to the Chemistry of Enzyme Action.* London: McGraw Hill.

9 Intermediary metabolism: methods of study

The study of intermediary metabolism aims at an understanding of the changes undergone by substances in the process of their utilization by the body and of the reactions in which they take part after their absorption. It is not, perhaps, surprising that biochemistry is largely concerned with the investigation of the way in which particular compounds are broken down in the living cell and of the steps by which others are built up. Moreover, if the sequence of the chemical reactions which normally take place in a particular tissue is known, disordered metabolism in that tissue may be treated in a logical and systematic manner, perhaps by administration of substances which may be lacking, or of drugs which slow down the production of harmful substances.

The investigation of metabolic pathways by simple chemical analysis of organs and tissues can yield only limited information. It may tell us where particular compounds are stored in the body, and it may even disclose the existence of intermediate compounds formed during breakdown or synthesis. However, in many tissues metabolic processes go on actively without accumulation of intermediate products since these are quickly used up in subsequent reactions. The detection of minute quantities of intermediate products by conventional chemical methods is rather difficult.

Analysis of urine may give information about intermediary metabolism. For example, when the amount of protein in the diet is increased the amount of urea in the urine is also increased, and conversely the urea excretion is decreased when the protein intake is decreased. It is therefore fairly safe to assume that urea is one of the end-products of protein metabolism. We must, however, keep an open mind on two questions, first whether protein gives rise to other end-products in addition to urea, and secondly whether part of the urea arises from substances other than protein.

If a substance A is broken down through two

intermediates B and C to a final product D (Fig. 9.1), it is a reasonable assumption that if excessively large amounts of A are fed so much B and/or C may be produced that one or other or both of these compounds may accumulate and appear in the urine. Hence excess feeding has sometimes been used in an attempt to identify intermediates in metabolic reactions. The main objection to the validity of conclusions based on this method is that, when the body is overloaded with substance A, emergency mechanisms may be called in to deal with it (see also Chap. 22). Thus the compounds found in the urine may not be the normal intermediates but those produced by the emergency mechanism.

The results of feeding possible intermediates may provide useful information. It is reasonable to assume that if compound B is fed to an animal which normally metabolizes A to D via B and C, B will enter the normal metabolic route and will disappear. If, on the other hand, the animal is given a substance which is not a normal intermediate, we may expect it to be excreted unchanged, or else to be metabolized at a different rate. For example, since butyric acid is oxidized more rapidly than is β-hydroxybutyric acid, presumably butyric acid is not oxidized by way of the hydroxy derivative.

If we know the enzymes which catalyse the intermediate steps in a sequence of reactions and also if we know substances which inhibit these enzymes specifically, we are in a good position to study the course of the reaction. Let us consider for example the scheme shown in Fig. 9.1 in which enzymes E_2 and E_3 catalyse the steps B $\rightarrow$ C and C $\rightarrow$ D respectively, and in which these enzymes are inhibited respectively by inhibitors I_2 and I_3. If I_2 is administered the metabolism of A is effectively stopped at B since the conversion of B to C is inhibited, and B tends to accumulate and can be identified. In the same way administration of I_3 results in the accumulation of intermediate C. The mechanisms of feedback inhibition and repression are discussed in Chapters 8 and 21.

So far we have been considering investigations carried out on the intact animal, but experiments can also be carried out with organs or portions of organs. If an organ is removed from the body and maintained at body temperature while being kept moist with a suitable solution, it remains alive and can carry out the chemical reactions which it normally accomplishes in the intact body. If a solution of the substance under investigation is perfused through the blood vessels of such an isolated organ, the perfusate can be examined for intermediate breakdown products.

In the case of tissues it is usual to cut thin slices only a few cells thick and to incubate them with the material under test. Alternatively, the test substance may be incubated with finely minced tissue, known as a *homogenate*, and the breakdown products identified and estimated in the incubation mixture. The slice method has the advantage that the normal spatial arrangement of the enzymes and other constituents in the cell is preserved, whereas in a homogenate the cells are much more broken up and this arrangement is destroyed. Indeed, a homogenate may not be able to carry out certain reactions demonstrable in the intact organ or in tissue slices.

Isolated organs are difficult to keep alive for any length of time while tissue slices and homogenates represent dying, damaged and broken cells. Many types of metabolic experiments may be carried out by the valuable technique of tissue culture by which cells can be maintained almost indefinitely. Under suitable conditions they grow and multiply.

A number of surgical methods have been used to investigate intermediary metabolism. By the removal of an organ, or by interference with its blood supply we may learn something about the part played by that organ in metabolism. London and his pupils in Leningrad devised the technique of angiostomy; by inserting cannulae leading to blood vessels blood may be withdrawn or material injected at any time. Dogs with cannulae in the hepatic,

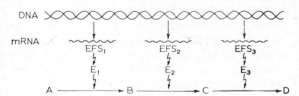

Fig. 9.1 The sequence of reactions by which metabolite A is converted to B, C and D is controlled by the three enzymes E_1, E_2 and E_3. Each enzyme is produced by an enzyme-forming system (EFS) associated with the messenger RNA (mRNA) of the cell (Chap. 21) and controlled in turn by the DNA of the appropriate gene in the nucleus.

renal, and portal veins have lived apparently normally for many months. In the Eck fistula operation a communication is made between the portal vein and the inferior vena cava to short-circuit the liver so that the portal blood from the small intestine flows directly into the inferior vena cava. It is thus possible to investigate the part played by the liver in the metabolism of many substances. A more drastic measure employed by Mann and his colleagues is the removal of the whole liver (*hepatectomy*). During the short time that animals remain alive urea formation ceases. The liver thus appears to be the site of urea formation in the intact animal.

Certain drugs interfere with metabolic processes. Carbon tetrachloride, chloroform, or phosphorus damage liver cells. The glycoside phlorizin (Chap. 18) which poisons the renal tubules and prevents reabsorption of glucose has yielded much information about the intermediary metabolism of carbohydrates. The excretion of glucose in the urine of a phlorizinized dog after administration of an amino acid shows that it is metabolized by way of glucose after deamination.

Feeding an animal on a diet which is deficient in the substance under investigation may, especially if the material in question is one of the so-called essential nutrients, yield useful information about the role of such a substance in metabolism. Thus when the diet is deficient in nicotinic acid (Chap. 13) the amount of nicotinamide nucleotides (p. 67) in the blood and tissues falls.

Some diseases may provide information about intermediary metabolism. For example, the investigation of diabetes mellitus has yielded much information about carbohydrate metabolism.

Certain rare hereditary metabolic disorders, called by Garrod 'inborn errors of metabolism', are of great fundamental interest since they illustrate how a genetic defect may affect a metabolic process. Each enzyme in the cell is controlled by a gene, a segment of nuclear DNA. If, for example, the gene controlling the enzyme E_2 in Figure 9.1 is defective or absent owing to a hereditary abnormality, the enzyme E_2 will be deficient or absent. Products C and D are not formed and substances A and B accumulate in the tissues and may be excreted in the urine. Examples of such disorders are alkaptonuria in which tyrosine metabolism is abnormal (Chap. 22), phenylketonuria in which the metabolism of phenyalanine is upset (Chap. 22), and galactosaemia (Chap. 18).

After the constituents of food—proteins, fats, and carbohydrates—have been digested and absorbed, the breakdown products merge with similar compounds in the body in what has come to be known as a metabolic 'pool'. None of the methods so far described enables us to follow the metabolic fate of a particular batch of a substance fed to an animal or introduced into a biological system. Some sort of label must be attached to molecules to allow of the identification of substances derived from them. In early attempts to do this a 'signpost' or marker group was introduced into the molecule of the administered substance. Knoop, for instance, put a phenyl group into fatty acid molecules. Since the body finds difficulty in breaking down the benzene ring he was able to obtain information about fatty acid oxidation (Chap. 19). Other substances have been 'labelled' by the introduction of bromine or iodine atoms. All such labels are somewhat limited in their use, and it can always be objected that the introduction of the labelling group alters the physical, chemical, and physiological properties of the substance. What is required is a label which does not alter these properties and nowadays isotopes provide tools which are very near to ideal. It is difficult to exaggerate the importance of the use of isotopes as labels or the advances which have been made by this technique. Isotope methods can, of course, be applied to the intact animal, to tissue slices or to homogenates.

ISOTOPES

Elements of the same atomic number, and hence the same nuclear charge, may occur as *isotopes* with different atomic weights. The important point is that, since the nuclear charges are the same, the number of planetary electrons must also be identical, and it is on the planetary electrons that the chemical properties of an element depend. Isotopic forms of the same element are therefore *chemically* identical, though they may be distinguished by physical means.

Isotopes can be divided into two groups. Some are classed as *stable isotopes*, typified in the case of carbon by ^{12}C and ^{13}C shown in Figure 9.2. In these isotopes the atomic arrangement is stable and the atoms show no tendency to disintegrate. *Radioactive isotopes* are exemplified by ^{11}C and ^{14}C, seen in Figure 9.3. In these the atom is unstable and tends to disintegrate, emitting in the process α-particles, β-particles (electrons) or γ-radiation. Those emitting α-particles are not important in biological laboratory work. Pure β-emitters such as ^{32}P, ^{14}C and ^{3}H are commonly employed in biology and medicine as also are β- and γ-emitters, such as ^{131}I. Radioactive isotopes differ in the rates at which they disintegrate or decay but the rate of decay is in all cases exponential. It is convenient, therefore, to refer to the rate at which a radioactive isotope disintegrates in terms of its *half-life*, that is the time taken for half of the isotope originally present to disappear. Thus the half-

life of radioactive phosphorus, ^{32}P, is 14·3 days; this means that after 14·3 days half of the amount of ^{32}P originally present is still present, after 28·6 days only one quarter remains, after 57·2 days one-eighth, and so on. The half-lives of radioisotopes vary enormously; some decay so rapidly that they are useless for biochemical studies. The useful range is covered by ^{11}C at one end with a half-life of about 20 minutes and ^{14}C at the other extreme with a half-life of about 5600 years.

Although the techniques for measuring stable isotopes and radioactive isotopes are quite different, the principle of their use as biochemical labels or tracers is the same. It is necessary to have a sample of the substance under investigation which has a high proportion of one of its elements in an isotopic form. For example, glucose might be used in which the ordinary carbon atoms of atomic weight 12 are replaced by atoms of radioactive carbon of atomic weight 14. Such a sample behaves chemically, and therefore also metabolically, exactly like ordinary glucose. If this radioactive glucose is fed to an animal, or used in a tissue slice experiment, it undergoes the same chemical reactions as ordinary glucose. Hence when the compounds in the tissue are subsequently isolated and examined for the presence of ^{14}C, we can say quite conclusively that those which contain ^{14}C must have been derived from glucose, while those which do not cannot have been produced in this manner. Moreover, by determining precisely the amount of radioactive carbon in the different compounds, much useful information can be gained regarding the rates at which different substances are formed from glucose, and even, in some cases, the order in which they are synthesized.

The radioactive isotopes used as tracers are forms of the element which do not occur naturally—they therefore appear in a biological system only in so far as they have been introduced in an experiment. On the other hand stable isotopes occur universally in all compounds. Thus carbon as it occurs in nature is a mixture of its two stable isotopes ^{12}C and ^{13}C with the former preponderating. In fact, approximately 99 atoms of ^{12}C are found for every atom of ^{13}C. The mean atomic weight is therefore nearer to 12 than to 13, and is, in fact 12·01. It is important to note that the ratio of ^{12}C to ^{13}C is constant for all carbon com-

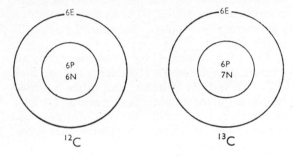

Fig. 9.2 Stable isotopes of carbon. ^{12}C has a nucleus consisting of 6 protons (P) and 6 neutrons (N) and an outer shell of 6 electrons (E). ^{13}C has 6 protons and 7 neutrons in the nucleus and 6 planetary electrons. Both have a nuclear charge, and therefore an atomic number, of 6, but their atomic weights are $6+6 = 12$ and $6+7 = 13$ respectively.

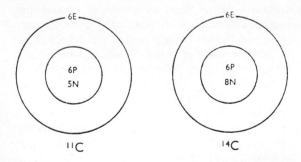

Fig. 9.3 Radioactive isotopes of carbon. ^{11}C has a nucleus containing 6 protons and 5 neutrons, that is an atomic weight of 11. ^{14}C has a nucleus containing 6 protons and 8 neutrons, that is an atomic weight of 14. Both ^{11}C and ^{14}C have a nuclear charge, and therefore an atomic number, of 6, and both have 6 planetary electrons.

pounds occurring naturally. In the same way all naturally occurring nitrogen, whether free or combined in chemical compounds, consists of 99·62 per cent of ^{14}N and 0·38 per cent of ^{15}N; the latter is often referred to as *heavy nitrogen*. Hydrogen consists of 99·98 per cent of ^1H and 0·02 per cent of ^2H; ^2H is known as *heavy hydrogen* or *deuterium*, and is sometimes given the symbol D.

To use a stable isotope as a tracer it is necessary first of all to prepare a starting material containing more than the normal abundance of the isotope. For example, an amino acid might be used in which 60 per cent instead of the usual 0·38 per cent of the nitrogen atoms are atoms of heavy nitrogen. If such a compound is administered to an animal and the tissues subsequently fractionated, the indication that a substance has been formed from the administered compound is the presence in it of a higher than normal concentration of heavy nitrogen. Thus any compound which contains more than 0·38 per cent of ^{15}N must have derived its additional heavy nitrogen from the administered amino acid.

In experiments with deuterium as a label it is possible to combust the compounds and estimate the concentration of deuterium by measuring the density or refractive index of the water formed; but the measurement of all other stable isotopes requires the use of a *mass spectrometer*. The principle of this is that particles of different mass but carrying the same electrical charge are deflected to differing degrees when they are made to pass through a magnetic field. Stable isotopes are, however, difficult to estimate and are now seldom used.

Radioactive isotopes are constantly emitting radiation and they are measured by the various effects which this radiation can produce. Perhaps the best known detector of radioactivity is the *Geiger counter*. Radiation entering this counter ionizes molecules of gas contained in it and the negative ions are attracted towards the positively charged anode. If the voltage on the anode is high enough (of the order of 1000 V) the negative ions are accelerated so rapidly towards the anode that they collide with other molecules of gas on the way and in this way produce still more ions. The final result is that a very large number of negative particles discharge themselves on the anode and produce a temporary drop in its potential

which is passed on to the recording instrument as a pulse. By counting the number of pulses produced in a given time by a known amount of the compound being assayed, the *specific activity* of the compound can be worked out. This is really a measure of the proportion of the element present in the radioactive form. Thus, in the case of a compound labelled with ^{32}P, the specific activity is given by

$$\frac{\text{Amount of } ^{32}P}{\text{Amount of } ^{31}P + \text{Amount of } ^{32}P}$$

The numerator in this fraction is usually given as the number of pulses recorded in unit time, usually described as disintegrations per minute, while the denominator is obtained by a standard chemical analytical determination of phosphorus and is expressed in μg, mg, μmoles, or m-moles. Thus the specific activity of our compound might be stated as so many disintegrations per minute per 100 μg phosphorus (d.p.m./100 μg P) or as disintegrations per minute per μmole phosphorus (d.p.m./μmole P).

The conventional Geiger counter is not an efficient detector of the low energy radiation of ^{14}C, ^{35}S and ^3H (tritium) and other isotopes frequently used as biological tracers. To deal with such isotopes a different type of apparatus, the scintillation counter, or scintillation spectrometer, is commonly used.

Certain substances, known as phosphors, when bombarded by radiation, undergo a scintillation reaction and emit light. There are two main types of *scintillation counting*. In one type, which is applicable to isotopes emitting γ-radiation, the radiation is allowed to strike a suitable crystal such as a crystal of sodium iodide activated with thallium. This crystal produces flashes of light which are detected by a photomultiplier tube. The second type, liquid scintillation counting, is commonly used to measure ^{14}C and the radioactive isotope of hydrogen, ^3H (tritium). In this method the compound to be assayed is dissolved in a suitable solvent along with a soluble organic phosphor which undergoes scintillation reactions.

The detection of radioisotopes by the technique of *autoradiography* depends on the ability of the emitted radiation to blacken a photographic emulsion. For example, after a separation of a mixture of substances by paper

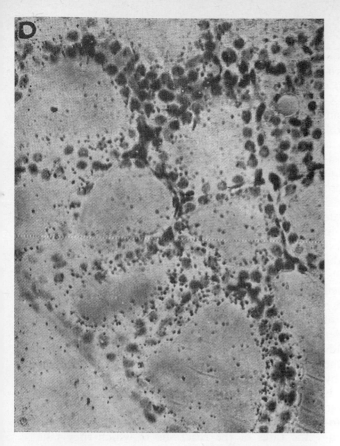

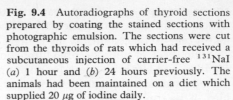

Fig. 9.4 Autoradiographs of thyroid sections prepared by coating the stained sections with photographic emulsion. The sections were cut from the thyroids of rats which had received a subcutaneous injection of carrier-free ^{131}NaI (a) 1 hour and (b) 24 hours previously. The animals had been maintained on a diet which supplied 20 μg of iodine daily.

The radioactivity in these sections is due to the presence of newly-formed radioactive thyroglobulin, and is indicated by the black silver granules in the emulsion. Note that soon after the injection (a) the highest concentration of radioactivity occurs in a ring overlying the innermost part of the cells and the outer margin of the colloid. At the later time (b), the radioactivity has moved uniformly through the colloid. This indicates that thyroglobulin is formed in the apex of the cell, and is secreted into the lumen of the follicle where it mixes with the pre-existing protein; this process is much more rapid in more active glands (C. P. Leblond & J. Gross (1948) in J. Gross & R. Pitt-Rivers (1952). *British Medical Bulletin* **8**, 136).

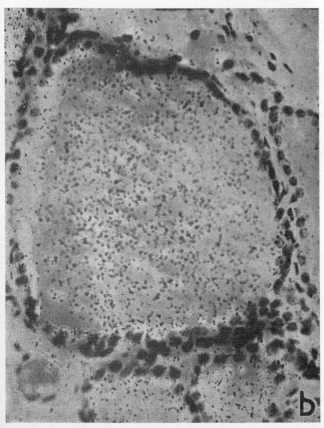

chromatography (p. 29) we may wish to know which of the spots are radioactive. The chromatogram is placed firmly in contact with a suitable piece of photographic film, such as X-ray film, and left in the dark for an appropriate period. On subsequent development of the film, a picture known as a autoradiogram is obtained. The blackened areas show the position of the radioactive spots and the degree of blackening gives an indication of the amount of radioactivity present.

This technique can be applied to tissue sections. If a histological section is placed in contact with a photographic film, blackening is produced in the positions corresponding to the locations of radioactive materials in the cells. If subsequently the autoradiogram is examined under the microscope and compared with a conventionally stained histological section of the same tissue, it is often possible to discover the cells which contain the labelled compounds (Fig. 9.4). It may even be possible to decide the location of certain compounds *within the cell* by this method, but the resolution attainable is limited by the fact that the radiation tends to penetrate through the emulsion on the photographic film and so produce, at high magnification, somewhat diffuse spots.

In this discussion of the use of isotopic tracers we have supposed that only one element in the compound was labelled in this way but more than one label can be incorporated into the same molecule. It is possible to use double labelling with stable isotopes, for example, deuterium and ^{15}N, labelling with stable and radioactive isotopes, for example, ^{15}N and ^{14}C, and also labelling with different radio-isotopes, for example, ^{14}C and ^{3}H. The modern liquid scintillation counter is an excellent tool for estimating the activities of two radioactive isotopes in the one chemical substance.

The usefulness and advantages of the isotopic tracer method are based, as we have said, on the assumption that the labelled compound behaves chemically and physiologically in a manner identical with that of the normal, unlabelled compound. Such an assumption is true, except in the case of low molecular weight compounds, particularly water, labelled with deuterium (D) or tritium (T). Such labelling increases the molecular weight from 18 for ordinary water to 20 for D_2O or 22 for T_2O; the consequent change in specific gravity alters the rate of transport across cell membranes and the rates of the reactions in which they take part. These are, however, exceptional cases.

REFERENCES

ARNSTEIN, H. R. V. & GRANT, P. T. (1957). The use of isotopes in the study of intermediary metabolism. *Progress in Biophysics and Biophysical Chemistry* **7**, 165–224.
BELL, C. G. & HAYES, F. N. (1962). *Liquid Scintillation Counting.* New York: Pergamon.
DAWES, E. A. (1972). *Quantitative Problems in Biochemistry*, 4th edn. Edinburgh: Livingstone.
FAIRES, R. A. & PARKS, B. H. (1958). *Radioisotope Laboratory Techniques.* London: Newnes.
FRANCIS, G. E., MULLIGAN, W. & WORMALL, A. (1959). *Isotope Tracers*, 2nd edn. London: Athlone Press.
KAMEN, M. D. (1957). *Isotopic Tracers in Biology*, 3rd edn. New York: Academic Press.
QUIMBY, E. H. & FEITELBERG, S. (1963). *Radioactive Isotopes in Medicine and Biology.* Philadelphia: Lea & Fibiger.
WANG, C. H. & WILLIS, D. L. (1965). *Radiotracer Methodology in Biological Science.* New Jersey: Prentice Hall.
WILLIAMS, R. T. (1959). *Detoxication Mechanisms*, 2nd edn. London: Chapman & Hall.
WILSON, B. J. (Ed.) (1966). *The Radiochemical Manual*, 2nd edn. Amersham: The Radiochemical Centre.

10 Biological oxidation

The classical researches of Priestley, Lavoisier and others at the end of the 18th century showed that in respiration animals take up oxygen from the air and give off carbon dioxide in exchange. The obvious analogy with the chemical process of combustion suggested that animals used oxygen to oxidize some of their food to carbon dioxide, and that this oxidation might be employed as a source of energy more or less as the oxidation of coal is used in a steam engine. At this point the analogy broke down. In the steam engine, as in the internal combustion engine and gas turbine, chemical energy is converted to heat energy and some of this heat energy is in turn converted to mechanical energy. This second conversion, however, can only be brought about if one part of the engine is at a different temperature from the rest. It can be shown by a simple exercise in thermodynamics that the extent of the conversion depends on the magnitude of this temperature difference. But the temperature differences which can exist inside a living animal are relatively small. Clearly, therefore, animals must be endowed with some other mechanism by which the energy liberated in oxidation of food can be turned to useful account, a mechanism which does not involve heat as an intermediate.

The first step toward the elucidation of this mechanism was the recognition that respiration—the uptake of oxygen and the release in exchange of carbon dioxide—is a function of the individual cell. Lavoisier had thought that 'biological oxidation' was a special function of the lungs, but this view was soon discarded when it was realized that the lungs serve only as a mechanism for the exchange of gases between the blood and the atmosphere. By about 1880 it was generally recognized that most tissues take up oxygen from the blood and release carbon dioxide into it. Moreover, the uptake of oxygen and the release of carbon dioxide in exchange was shown to take place also in plant cells and in many micro-organisms. Since most nutrients and metabolites are not spontaneously oxidized by molecular oxygen, some catalytic mechanism had to be postulated and, as the concept of enzymes gradually emerged, the mechanism came to be thought of as enzymic in character. The earliest hypotheses envisaged either an enzyme which activated molecular oxygen or alternatively an enzyme which activated the metabolite which was being oxidized. In the event the enzymic mechanism has proved a good deal more complicated than this, but before it is described it will be convenient to discuss the nature of oxidation and the amount of energy which can be obtained from it.

It is not easy to produce a short general definition of oxidation, but in general a compound is said to be oxidized (a) when it loses one or more electrons (b) when it loses some atom or atoms such as hydrogen which have a relatively weak attraction for electrons or (c) when it combines with one or more atoms such as chlorine or oxygen which have a relatively strong attraction for electrons. Reduction is the reverse of this process. Fortunately the oxidations and reductions which are important to biochemistry are restricted to three simple classes:

(a) loss of one or more electrons, for example

$$Fe^{2+} \xrightarrow{\;-e\;} Fe^{3+}$$

or (b) loss of one or more atoms of hydrogen, for example

$$CH_3CH_2OH \xrightarrow{\;-2H\;} CH_3CHO$$

or (c) addition of one or more atoms of oxygen, for example

$$CH_3CHO \xrightarrow{\;+O\;} CH_3COOH$$

The fundamental identity of these can be most easily appreciated by reformulating (b) and (c) as a loss of electrons

$$CH_3CH_2OH \xrightarrow{\;-2H^+ -2e\;} CH_3CHO$$

$$CH_3CHO \xrightarrow{\;+H_2O\; -2H^+\; -2e\;} CH_3COOH$$

Biological oxidations can accordingly be represented as follows:

$$A_{red} \xrightarrow{\;-ne\;} A_{ox}$$

where A_{red} represents the reduced form of the compound, A_{ox} represents the oxidized form of the compound and n is the number of electrons involved.

Reduction can be represented as the reverse of oxidation:

$$A_{ox} \xrightarrow{\;+ne\;} A_{red}$$

Chemical reactions do not generally involve free electrons or atoms. Consequently a compound can only be oxidized if an equivalent amount of another compound is simultaneously reduced. An obvious example is the reaction which takes place when hydrogen is passed over heated copper oxide. The hydrogen is oxidized to water and the copper oxide is reduced to copper.

Another example is the reaction which takes place when metallic zinc is added to an aqueous solution of copper sulphate. The zinc is oxidized to zinc ions at the expense of the copper ions which are reduced to metallic copper.

Oxidations and reductions can therefore be represented in the following way, where A and B are used to denote the two compounds taking part (subscripts indicate their oxidized and reduced forms).

This reaction can be regarded as a competition for electrons between two 'oxidation–reduction systems', namely, A_{red}/A_{ox} and B_{red}/B_{ox}. The equilibrium constant of the reaction, and the standard free energy change associated with it, depend on the relative affinity of the two systems for electrons. If one system has a very much greater affinity for electrons than the other, equilibrium is attained only when the former is almost all in the reduced form and the latter almost all in the oxidized form. Consequently, the standard free energy change associated with such a reaction is very large. This is the case in the copper/zinc example quoted above since the system Cu/Cu^{2+} has a much greater affinity for electrons than the system Zn/Zn^{2+}. The equilibrium of the reaction is so much in favour of zinc ions and metallic copper that for practical purposes it can be regarded as irreversible.

The affinity of an oxidation–reduction system for electrons can be represented on an arbitrary scale as an *oxidation–reduction potential* ($E°$) (p. 137). In some cases (such as the reaction between Zn and Cu^{2+} discussed above) $E°$ can be measured electrically and it is therefore expressed in terms of volts. A system with a strong tendency to take up electrons (that is to oxidize other systems) has a large positive oxidiation-reduction potential. When two oxidation–reduction systems react with one another

$$A_{red} \diagdown \diagup B_{ox}$$
$$A_{ox} \diagup \diagdown B_{red}$$

the standard free energy change of the reaction is related to the difference in potential between the two systems ($\Delta E_0'$) by the equation

$$\Delta G° = -nF\,\Delta E_0'$$

where n is the number of electrons involved in the transfer, and F is a constant (the Faraday, the electrical charge carried by 1 gram equivalent). If ΔG is expressed in kilojoules (kJ), F has the numerical value 96·5

$$\therefore \Delta G° = -96{\cdot}5n\,\Delta E_0'$$

In the example quoted above the values of E_0' for the Zn/Zn^{2+} and Cu/Cu^{2+} system are $-0{\cdot}76$ and $+0{\cdot}34$ volts respectively and $n = 2$.

Therefore, for the reaction between Zn and Cu^{2+}.

$$\Delta G° = -2F[+0{\cdot}34(-0{\cdot}76)]$$
$$= -212\,kJ/mole.$$

BIOLOGICAL OXIDATIONS

The overall reaction by which mammalian organisms derive energy from the oxidation of food can be represented by equations such as

$$C_6H_{12}O_6 + 6O_2 \longrightarrow 6CO_2 + 6H_2O$$

for oxidation of carbohydrate and

$$C_{16}H_{34}O_2 + 23\tfrac{1}{2}O_2 \longrightarrow 16CO_2 + 17H_2O$$

for oxidation of fat. These oxidations, as will be shown in Chapters 11 and 20, are not brought about in a single step but by long sequences of reactions. For the present it suffices to say that the oxidative steps in these sequences all take the form of simple dehydrogenations which can be shown thus

$$AH_2 \xrightarrow{\;-2H\;} A$$

The nature of the compounds represented by AH_2 will be explained in Chapters 11 and 18. A simple example is the dehydrogenation of malic acid to oxaloacetic acid:

$$\begin{array}{ccc}
COOH & & COOH \\
| & & | \\
CHOH & \xrightarrow{\;-2H\;} & CO \\
| & & | \\
CH_2 & & CH_2 \\
| & & | \\
COOH & & COOH
\end{array}$$

In many reactions of this type the two hydrogen atoms removed from the substrate are transferred to either of the coenzymes NAD or NADP (p. 67) thus:

$$AH_2 \diagdown \diagup NAD \qquad AH_2 \diagdown \diagup NADP$$
$$\qquad\qquad or$$
$$A \diagup \diagdown NADH_2 \qquad A \diagup \diagdown NADPH_2$$

When these coenzymes act as oxidizing agents or 'hydrogen acceptors' in reactions of the type shown above, the pyridine ring of the molecule undergoes reduction which we can represent as follows:

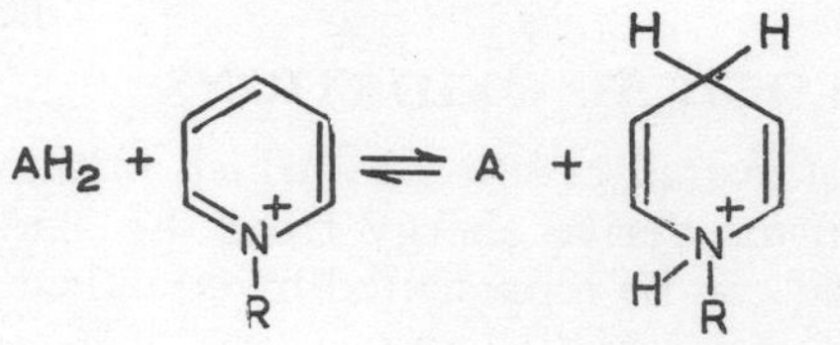

$$AH_2 + \text{(pyridine ring)} \rightleftharpoons A + \text{(reduced pyridine ring)}$$

The reduced form of the coenzyme at physiological pH is, however, in the ionized state.

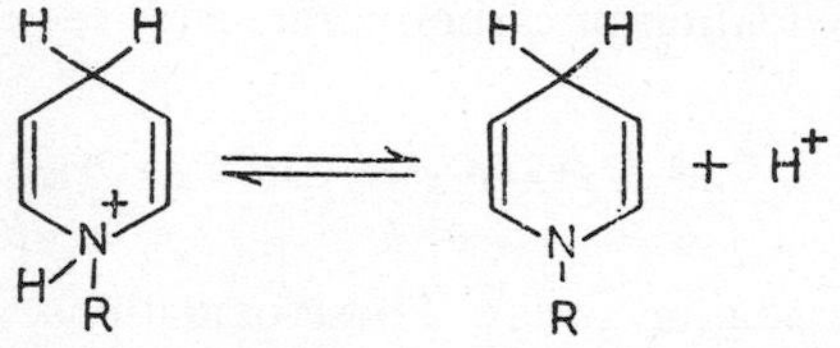

$$\text{(reduced form)} \rightleftharpoons \text{(ionized form)} + H^+$$

The reaction may be more accurately represented thus:

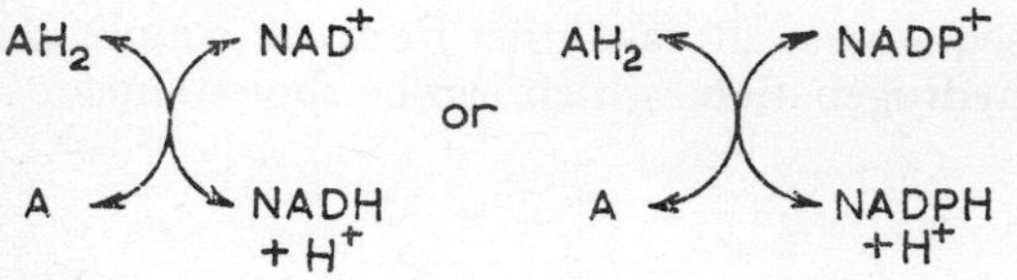

$$AH_2 \text{ / } NAD^+ \qquad \text{or} \qquad AH_2 \text{ / } NADP^+$$
$$A \text{ / } NADH + H^+ \qquad \qquad A \text{ / } NADPH + H^+$$

Such transfers of two hydrogen atoms from substrate to NAD^+ or $NADP^+$ are catalysed by *dehydrogenases* which are generally highly specific both with regard to substrates and to coenzyme. *Malate dehydrogenase*, for example, catalyses only the reaction

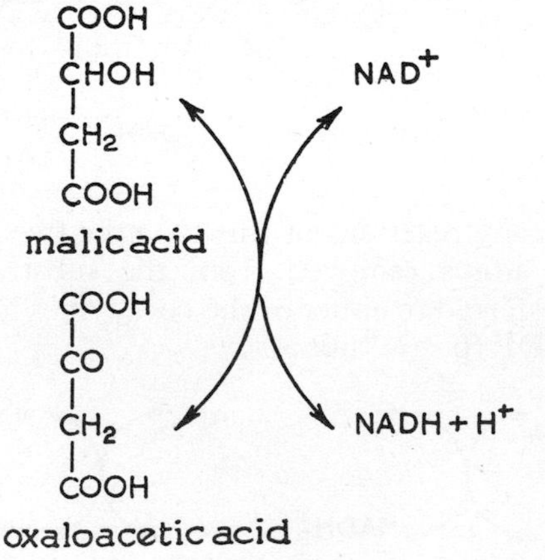

COOH
|
CHOH
|
CH$_2$ NAD$^+$
|
COOH
malic acid

COOH
|
CO
|
CH$_2$ NADH + H$^+$
|
COOH
oxaloacetic acid

The Flavoproteins

The extent to which a substrate can be dehydrogenated by the reactions described above necessarily depends on the availability of NAD^+ or $NADP^+$. The total amount of these coenzymes in the oxidized forms in the tissues is very small and the supply would soon be exhausted if there were no mechanism by which NADH and NADPH could be reoxidized back to NAD^+ and $NADP^+$. So far as NADH is concerned, such a mechanism is supplied by certain members of the *flavoprotein* group of enzymes. The enzymes are so called because of their yellow colour which is due to the presence as prosthetic groups of either *flavin mononucleotide* (FMN) or *flavin adenine dinucleotide* (FAD) (p. 67). They catalyse oxidation–reduction reactions, in which the dimethylisoalloxazine moiety of their prosthetic group undergoes reversible reduction and oxidation as follows:

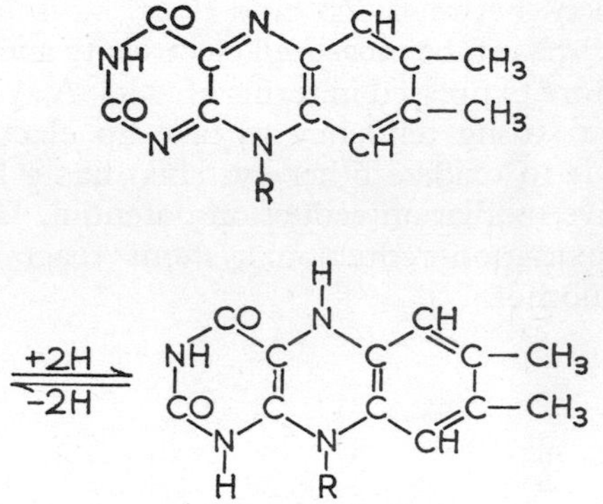

$$\xrightarrow[-2H]{+2H}$$

which can briefly be represented:

$$FP \xrightarrow[-2H]{+2H} FPH_2$$

An enzyme of this group, *NADH dehydrogenase*, appears to reoxidize NADH.

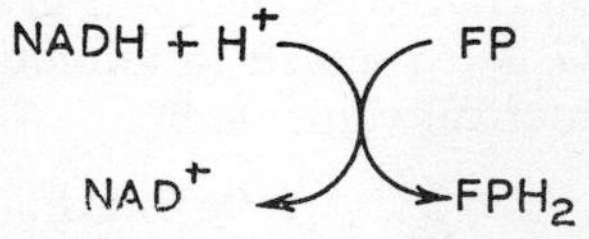

$$NADH + H^+ \text{ / } FP$$
$$NAD^+ \text{ / } FPH_2$$

The overall reaction for the transfer of 2H from substrate to nicotinamide nucleotide and thence to flavoprotein can be represented thus:

$$AH_2 \quad NAD^+ \quad FPH_2$$
$$A \quad NADH+H^+ \quad FP$$

But, just as in the case of the nicotinamide nucleotides, the cell contains very little flavoprotein and the oxidation outlined above can take place on a large scale only if some mechanism is available by which the reduced flavoprotein can be oxidized. This is achieved by the cytochrome system.

The Cytochrome System

The cytochromes are a group of iron-containing protein pigments, the iron atom being held in a porphyrin ring system (Chap. 25). They were discovered by MacMunn in 1886 in muscle but little attention was paid to his observation until it was repeated in 1925 by Keilin who was the first to use the term cytochrome. The special feature of the cytochromes is their ability to undergo a reversible oxidation which probably involves a change in the valency of the iron:

$$Fe^{3+} \underset{-e}{\overset{+e}{\rightleftharpoons}} Fe^{2+}$$

They are easily identifiable by their absorption spectra which alter as they are oxidized and reduced. A considerable number has now been isolated from various plants, animals and micro-organisms. In mammalian cells five of them, cytochromes b, c_1, c, a and a_3 are believed to have important functions in oxidation reactions. Cytochrome c is relatively easy to isolate in pure form and has been extensively studied. It has a molecular weight of 13 000 and contains one atom of iron per molecule in the

form of haematin (Chap. 25). Cytochromes a and a_3 have proved very difficult to separate and may be a single protein with two prosthetic groups. Cytochrome a_3 is also called *cytochrome oxidase*. The cytochromes appear to reoxidize the flavoproteins in the following way, two molecules of cytochrome being required for each molecule of flavoprotein:

$$FPH_2 \quad 2 \text{ cyt. } Fe^{3+}$$
$$FP \quad 2 \text{ cyt. } Fe^{2+} + 2H^+$$

The cytochromes are then reoxidized by molecular oxygen:

$$2 \text{ cyt. } Fe^{3+} \quad H_2O$$
$$2 \text{ cyt. } Fe^{2+} + 2H^+ \quad \tfrac{1}{2}O_2$$

They thus form a link in the process by which the reduced flavoprotein is oxidized by molecular oxygen:

$$FPH_2 \quad 2 \text{ cyt. } Fe^{3+} \quad H_2O$$
$$FP \quad 2 \text{ cyt. } Fe^{2+} + 2H^+ \quad \tfrac{1}{2}O_2$$

The details of the reaction are obscure but it is generally thought that the flavoprotein is first oxidized by cytochrome b, cytochrome b in turn is oxidized by cytochrome c and cytochrome c by cytochrome a, thus

$$FPH_2 \quad 2 \text{ cyt. } b\ Fe^{3+} \quad 2 \text{ cyt. } c\ Fe^{2+} \quad 2 \text{ cyt. } a\ Fe^{3+}$$
$$FP \quad 2H^+ \quad 2 \text{ cyt. } b\ Fe^{2+} \quad 2 \text{ cyt. } c\ Fe^{3+} \quad 2 \text{ cyt. } a\ Fe^{2+}$$

Finally cytochrome a is oxidized by cytochrome a_3 which in turn is oxidized by molecular oxygen.

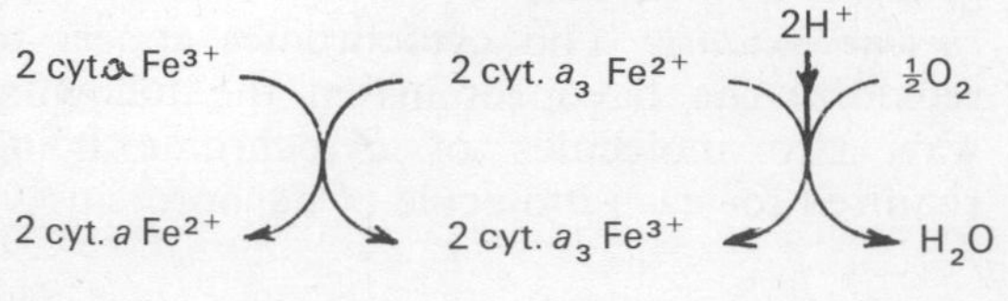

cytochromes b and c or between flavoprotein and cytochrome b:

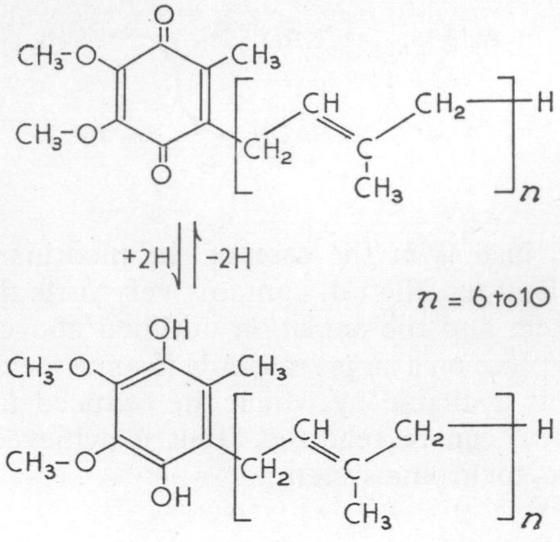

$$n = 6 \text{ to } 10$$

Cytochrome a_3 is the only member of the series which can react with molecular oxygen in this way.

We can now represent the mechanism by which substrates such as malate are oxidized by the following diagram:

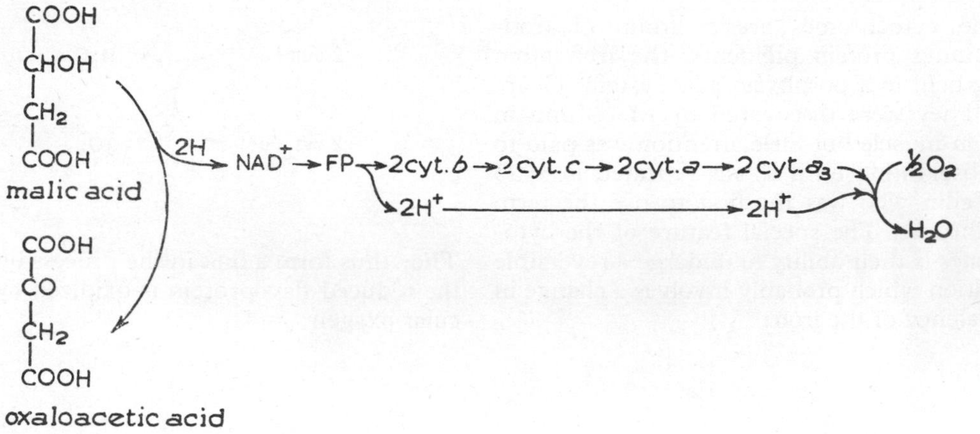

malic acid

oxaloacetic acid

The overall reaction is the oxidation of substrate by molecular oxygen:

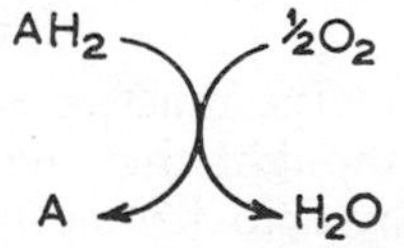

but it is brought about by transfer of electrons or hydrogen atoms via the mechanism shown above. It must be emphasized that the details of the mechanism are not yet clear but they are under continuing investigation. There is reason to think that the compounds generically known as *ubiquinones* (*coenzyme Q*) also take part in electron transfer probably between

There is also some doubt about the position of cytochrome b in the mechanism, and there is

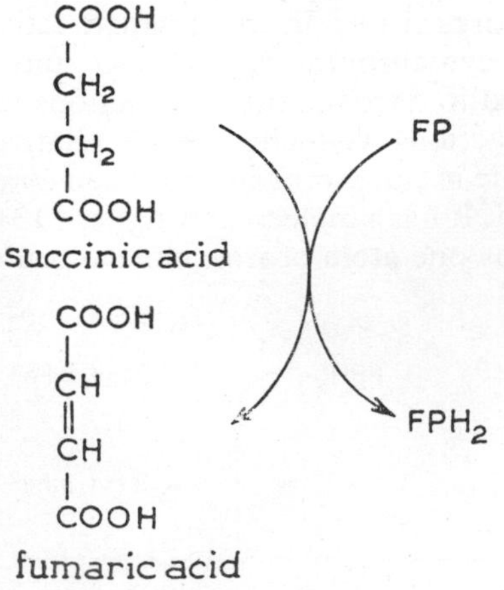

succinic acid

fumaric acid

probably an additional pigment, cytochrome c_1 between cytochromes b and c.

Various inhibitors act at different stages in the chain. For example, the drug amylobarbitone acts between NAD and FP; other narcotic drugs and the antibiotic actinomycin A act between cytochromes b and c; cyanides inhibit cytochrome a_3.

Not all oxidations of this general type involve NAD^+ or $NADP^+$. Some substrates are directly oxidized by flavoproteins. For example, succinic acid is oxidized by the flavoprotein *succinate dehydrogenase.*

The succinate dehydrogenase is then re-oxidized by the cytochrome system in the same way as its fellow flavoprotein, NADH dehydrogenase. The relationship of the two enzymes may be illustrated thus

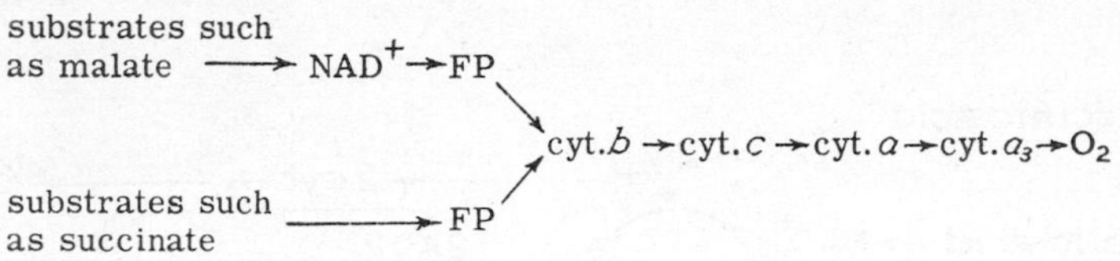

substrates such
as malate $\longrightarrow$ $NAD^+ \rightarrow FP$

cyt.$b \rightarrow$ cyt.$c \rightarrow$ cyt.$a \rightarrow$ cyt.$a_3 \rightarrow O_2$

substrates such
as succinate $\longrightarrow FP$

OXIDATIVE PHOSPHORYLATION

The foregoing section has shown that even an apparently simple reaction such as the dehydrogenation of succinic acid or malic acid at the expense of molecular oxygen

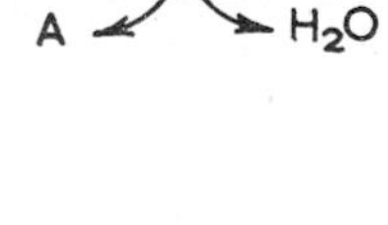

AH_2 $\quad$ $\frac{1}{2}O_2$

A $\quad$ H_2O

requires *in vivo* an exceedingly complex mechanism involving enzymes, co-enzymes, cytochromes and so on, and one might well be

inclined to wonder what advantage the organism derives from all this complexity and how the energy of the oxidation is turned to useful account.

If a suitable tissue preparation (a homogenate or a preparation of mitochondria) is allowed to respire *in vitro* it can be shown that inorganic phosphate disappears, being used to phosphorylate a variety of metabolites, for example sugars and other carbohydrate derivatives. If respiration is stopped by cutting off the oxygen supply, or by inhibiting cytochrome a_3 with cyanide, the phosphorylation stops also. Closer examination shows that the process is a two-stage one. The first stage is the phosphorylation of ADP to ATP:

$$ADP + Pi \longrightarrow ATP$$

which can occur only when respiration is pro-

ceeding. The second stage is the transfer of phosphate from ATP to other molecules, for example glucose:

glucose $+ ATP \rightarrow$ glucose-phosphate $+ ADP$

The first reaction is the important one in this sequence because it shows that respiration, the transfer of two electrons from the substrate to oxygen by way of the scheme outlined above, is somehow coupled to the formation of ATP from ADP and Pi.

If a respiring tissue preparation is studied under carefully controlled conditions it is found that for every 2 H atoms which pass from NAD to oxygen (that is for every O atom utilized) about 3 P atoms appear in esterified form. This is sometimes expressed by saying that the P/O ratio is about 3. The situation can be represented thus:

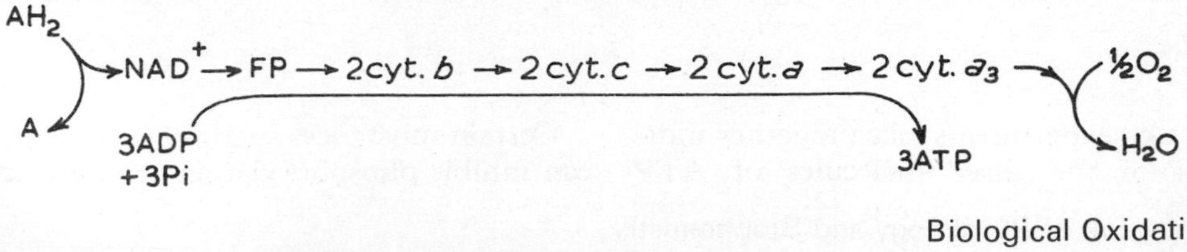

AH_2

$\rightarrow NAD^+ \rightarrow FP \rightarrow 2$cyt.$b \rightarrow 2$cyt.$c \rightarrow 2$cyt.$a \rightarrow 2$cyt.$a_3 \rightarrow \frac{1}{2}O_2$

A

3ADP
+3Pi $\qquad$ 3ATP $\qquad H_2O$

It has been possible to identify the steps in the electron transfer mechanism which are associated with ATP synthesis. The enzymes responsible for the first step, the transfer of hydrogen from substrate to NAD^+ can be isolated in reasonably pure form and it can be shown that their action is not associated with ATP synthesis. The subsequent components of the sequence have proved much more difficult to isolate and consequently they have been investigated mainly by indirect methods. For example, when succinic acid is oxidized by a respiring tissue preparation the P/O ratio obtained is approximately 2 instead of 3. This difference is presumably related to the fact that the mechanism for oxidizing succinic acid does not include NAD^+. If this is so, it follows that one of the three ATP molecules generated in the electron transfer system must be formed in the reaction between NAD^+ and flavoprotein.

generated when one molecule of nicotinamide nucleotide is reoxidized by molecular oxygen, one is produced in the reaction between nicotinamide nucleotide and flavoprotein and another in the sequence of steps between cytochrome c and molecular oxygen. By a process of elimination it follows that the third must be produced in the reactions connecting the flavoproteins to cytochrome c (see Diagram I on opposite page).

Confirmation of this conclusion can be found in the fact that when intramitochondrial NADH is oxidized by a tissue preparation in which cytochrome a_3 is inhibited by cyanide and in which cytochrome c acts as the ultimate oxidizing agent instead of molecular oxygen, two molecules of ATP are formed for every molecule of NADH oxidized, as would be predicted from the scheme outlined (in Diagram II on opposite page).

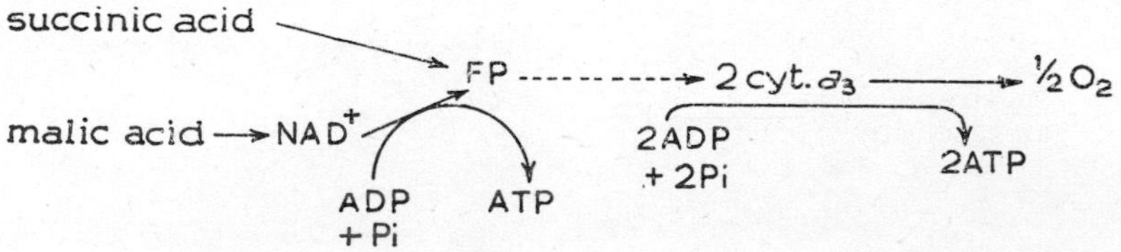

The other ATP molecules must then be formed somewhere in the series of reactions by which flavoproteins are oxidized at the expense of molecular oxygen. Some light is thrown on this question by the observation that ascorbic acid and p-phenylene diamine can be oxidized directly by cytochrome c without the intervention of nicotinamide nucleotides or flavoproteins. The P/O ratio for this oxidation is approximately 1. It could therefore be represented as follows:

In suitable tissue preparations and under properly controlled conditions it is possible to show that the relationship or 'coupling' between electron transport and ATP synthesis works in both directions. The formation of ATP takes place only during electron transport; and conversely the electron transport mechanism only functions as long as it is supplied with ADP (and Pi) which it can convert to ATP. The cellular supply of ADP may thus control the rate of respiration *in vivo*.

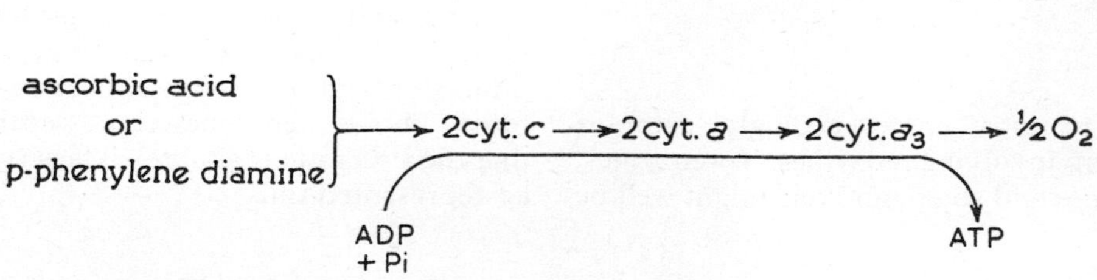

These two experiments taken together indicate that of the three molecules of ATP

Certain substances such as 2,4-dinitrophenol can inhibit phosphorylation without decreas-

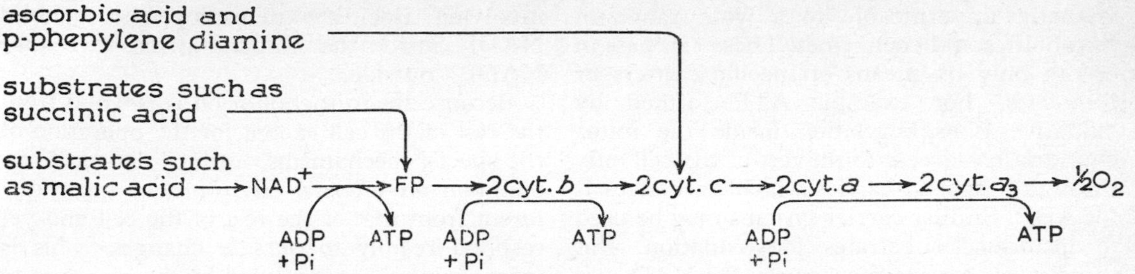

ascorbic acid and
p-phenylene diamine

substrates such as
succinic acid

substrates such
as malic acid $\longrightarrow$ NAD$^+$ $\longrightarrow$ FP $\longrightarrow$ 2cyt.b $\longrightarrow$ 2cyt.c $\longrightarrow$ 2cyt.a $\longrightarrow$ 2cyt.a_3 $\longrightarrow$ ½O$_2$

ADP + Pi   ATP   ADP + Pi   ATP   ADP + Pi   ATP

DIAGRAM I.

NADH + H$^+$ $\longrightarrow$ FP $\longrightarrow$ 2cyt.b $\longrightarrow$ 2cyt.c

ADP + Pi   ATP   ADP + Pi   ATP

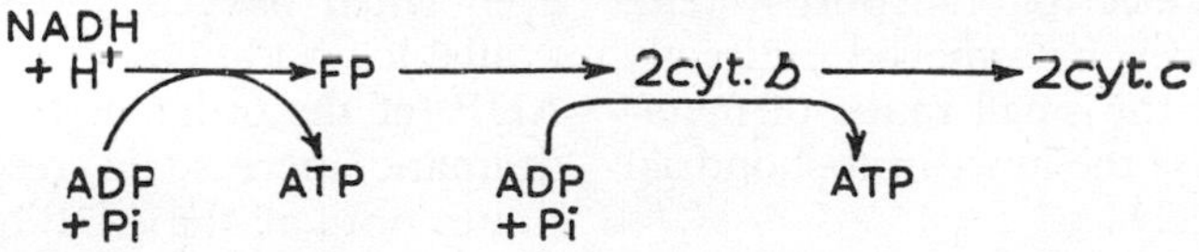

DIAGRAM II.

ing oxygen utilization. Often, in fact, they *increase* the rate of oxygen consumption. This indicates that phosphorylation of ATP is a controlling process. Such substances which disconnect phosphorylation from oxidation are known as uncoupling agents, since they uncouple the energy-yielding mechanism (oxidation) from the energy-harnessing system (phosphorylation).

MITOCHONDRIA

Up to this point we have considered biological oxidation and oxidative phosphorylation simply as chemical reactions. They are, however, distinguished from most other chemical reactions in living cells by the fact that the necessary enzymes are localized within the mitochondria from which they can be extracted only with difficulty. Attempts to obtain them in pure solution have given preparations capable of catalysing oxidation and reduction but incapable of generating ATP. It would appear that oxidative phosphorylation is in some way a property of the manner in which the enzymes are organized in the mitochondria. Consequently attempts to investigate oxidative phosphorylation generally start with the investigation of intact mitochondria. The appearance of mitochondria in the electron microscope has already been described (p. 75). It varies in detail from one cell type to another but a common pattern is always discernible: within a smooth outer membrane

there is an inner membrane which is usually folded inward to form *cristae* and inside the inner membrane there is a protein gel, the *matrix*. Mitochondria can readily be isolated in bulk from tissue homogenates by differential centrifugation (p. 77). Treatment of the isolated mitochondria with detergent removes the outer membranes which can be separated from the inner membranes by further differential centrifugation. The inner membranes can then be broken ultrasonically to release the matrix. Analysis of mitochondrial fractions obtained by these and similar means has shown that while many of the dehydrogenases are located in the matrix, the rest of the mechanism of biological oxidation and oxidative phosphorylation is in the inner membrane (where indeed it accounts for about a quarter of the total protein). The individual protein components (total flavoprotein, cytochrome *b*, cytochrome *c* and cytochrome *a*) are present in roughly equimolar amounts, and this has led to the belief that they are not scattered at random throughout the membrane but are organized in assemblies each containing a single molecule of each component. It is moreover possible to fractionate inner mitochondrial membrane fragments into four complexes, each containing several components of the electron transport system.

Most small molecules can diffuse readily across the outer membrane of the mitochondrion. The inner membrane, however, is virtually impermeable to a wide range of metabolites and coenzymes. These can pass in or out only by means of specific carriers or *permeases*. For example ATP formed by oxidative phosphorylation inside the mitochondria has access to the rest of the cell only through a special carrier which exchanges it for ADP. Similar carrier systems may be used to introduce substrates for oxidation. But there is no carrier mechanism for NAD^+ or NADH. Consequently extramitochondrial NADH cannot transfer its electrons directly to the mitochondrial electron transport system. Transfer can, however, be achieved indirectly by means of one of the small range of metabolites which can pass the inner mitochondrial membrane fairly freely.

For example, the extramitochondrial NADH can transfer its electrons to dihydroxyacetone phosphate thus converting it to glycerol phosphate, which can penetrate the inner mitochondrial membrane.

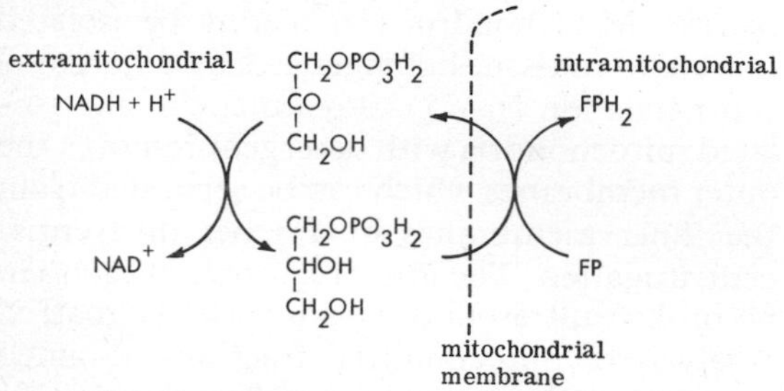

Inside the mitochondrion the glycerol phosphate transfers its electrons, by way of a flavoprotein, to the electron transport chain. The resulting dihydroxyacetone phosphate can then pass back through the membrane again. This indirect transfer of electrons from NADH to flavoprotein, the glycerol phosphate 'shuttle', is not associated with a phosphorylation. Consequently the transfer of electrons from extramitochondrial NADH to molecular oxygen yields only two molecules of ATP instead of three. Similarly while NADH inside the mitochondrion cannot transfer its electrons directly to metabolites outside, it can transfer them to oxalacetate which is thus converted to malate. The malate can readily escape from the mitochondrion and in turn reduce extramitochondrial NAD^+. The oxalacetate formed in this reaction can then return to the mitochondrion. A similar sort of mechanism

involving isocitrate dehydrogenase enables NADH inside the mitochondrion to reduce $NADP^+$ outside.

Because the mitochondrion is isolated from the rest of the cell except for the operation of the special mechanisms outlined above, it can maintain an internal environment quite different from that of the rest of the cell and yet respond readily to outside changes. This is most strikingly exemplified by the response of isolated mitochondria to added ADP. Such mitochondria normally respire quite slowly even when oxygen and substrate are readily available. Addition of small concentrations of ADP (of the order of 10^{-5} M) will produce a dramatic increase in respiration which continues until all the ADP has been converted to ATP. Respiration then falls back to its previous low level.

FREE ENERGY CHANGES AND OXIDATIVE PHOSPHORYLATION

The theoretical maximum amount of useful energy available from the overall reaction in which a substrate is oxidized by molecular oxygen

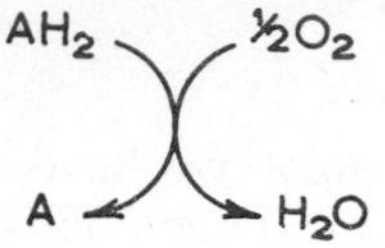

can readily be calculated from the difference in oxidation–reduction potential between the two systems AH_2/A and $H_2O/\frac{1}{2}O_2$. Where the substrate is malic acid, this difference (see Fig. 10.1) is $+0.82 - (-0.19) = +1.01$ volts, which corresponds to a free energy change ($\Delta G°$) of -197 kJ/mole. This is an enormous amount of energy compared to the $\Delta G°$ values of most biochemical reactions which seldom exceed ± 50 kJ/mole.

It has, however, been shown above that the oxidation of substrates like malate or succinate by molecular oxygen is not achieved directly but is brought about by the transfer of hydrogen atoms or electrons along a sequence of intermediates. Each individual step in the sequence is associated with its own $\Delta G°$ which can be calculated from the difference in oxidation–reduction potentials involved. The potentials for some of the substrates and some of the principal components of the electron transfer sequence are shown diagrammatically in Figure 10.1. From this it is clear that although the overall reaction is always strongly

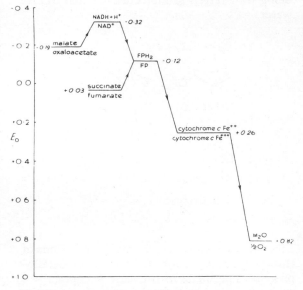

Fig. 10.1 Approximate values of E'_0 for some of the components of the electron transfer system. Transfer of electrons from systems at the upper (that is negative) end of the scale to systems at the lower (that is positive) end liberates free energy.

exergonic, the initial dehydrogenation reaction is in some cases actually endergonic. The succeeding stages, however, are all strongly exergonic. If we confine our attention to those components whose place in the system is definitely established we can divide these stages as shown in Table 10.2.

In other words, each of the three reactions (or reaction sequences) coupled to the synthesis of an ATP molecule has a standard free energy change sufficient to supply the 29 kJ

Table 10.2

	ΔE_0 (V)	ΔG° kJ
Malate to nicotinamide nucleotide	−0·13	+25
Nicotinamide nucleotide to flavoprotein	+0·20	−38
Flavoprotein to cytochrome c	+0·38	−75
Cytochrome c to molecular oxygen	+0·56	−109
Total	+1·01	−197

required for the synthesis of one mole of ATP.

The importance of oxidation in the life of the organism lies chiefly in the very large amount of useful energy which it can yield. The elaborate arrangements described above ensure that this energy is not released in a single reaction but is divided up among a series of reactions. Three of these reactions, each yielding 36 kJ or more (Table 10.2) are coupled to the synthesis of ATP molecules, each synthesis requiring 29 kJ. Of the theoretical yield of 197 kJ (Table 10.2), therefore, $3 \times 29 = 87$ kJ (about 40 per cent) are conserved in chemical form instead of being lost as heat. Moreover, the energy is conserved in a very serviceable form. Though the ΔG° of 29 kJ of the reaction

$$ADP + Pi \longrightarrow ATP + H_2O$$

is small compared with the energy made available in the oxidation, it is large in comparison with the ΔG° of most reactions *in vivo*. It can, therefore, be coupled to a great variety of endergonic reactions and so used to drive them.

In subsequent chapters it will be shown that ATP serves as the immediate source of the energy required not only for the chemical activities of the cell but also for the contraction of muscle and the transmission of nerve impulses.

This account of biological oxidation has been concerned almost entirely with its role as a mechanism for the production of ATP. Most of the oxidation which takes place in mammalian cells does indeed seem to have this function. But it should be emphasized that some of the steps in the major metabolic pathways in all cells take the form of oxidations or reductions. For example, the conversion of glucose to lactic acid and *vice versa*

$$C_6H_{12}O_6 \; \rightleftharpoons \; 2C_3H_6O_3$$

involves both an oxidation and a reduction (see Chap. 11). The conversion of carbohydrate to fat involves two oxidations and two reductions. The oxidations are brought about by the usual electron transport mechanism. The reductions are catalysed by dehydrogenases working so to speak 'in reverse' and using NADH and NADPH as hydrogen donors. (The NADH and NADPH are of course formed as a result of dehydrogenation of other metabolites.) A simple example of such a reduction is the conversion of pyruvic acid to lactic acid:

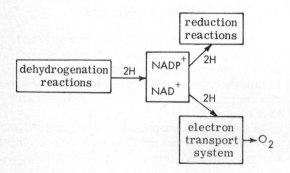

$$\begin{array}{ccc}
CH_3 & & CH_3 \\
| & \xrightarrow{\;NADH + H^+ \quad NAD^+\;} & | \\
CO & & CHOH \\
| & & | \\
COOH & & COOH
\end{array}$$

The role of NAD^+ and $NADP^+$ in relation to such reactions and to electron transport can be represented as follows:

```
                        ┌──────────┐
                     2H │reduction │
                   ↗────│reactions │
dehydrogenation  ┌─────┐└──────────┘
  reactions  ──2H│NADP+│
             │   │     │
             └───│NAD+ │
                 └─────┘2H
                   ↘────┐
                        │electron │
                        │transport│──O₂
                        │ system  │
                        └─────────┘
```

In other words NAD^+ and $NADP^+$ are not merely agents for transferring hydrogen from certain metabolites to the electron transport system. They can also act as agents for transferring hydrogen from one metabolite to another. It follows, therefore, that they can play an important role even in cells in which, for lack of oxygen, the electron transport system is not functioning. In general NAD^+ is responsible chiefly for transferring hydrogen from metabolites to the electron transport system, whereas $NADP^+$ is responsible chiefly for transferring hydrogen from one metabolite to another.

OTHER OXIDATIVE REACTIONS

Although the electron transport system is the main mechanism by which oxidations are catalysed in the living organism, it is by no means the only one. Quite early in the investigation of biological oxidation a variety of

enzymes was isolated which catalysed relatively simple oxidations. They include the aerobic oxidases; for example, *polyphenol oxidase*—an enzyme found in potatoes—catalyses the oxidation of polyphenols such as catechol, the two hydrogen atoms removed from the polyphenol being transferred to molecular oxygen to give water:

catechol

½O₂

o-quinone

H₂O

Like other enzymes which catalyse reactions of this type polyphenol oxidase is a metalloprotein. The metal in this case is copper (0·2 per cent) which is thought to undergo cyclic oxidation and reduction in the course of the reaction thus:

2Cu⁺⁺   H₂O

2Cu⁺ + 2H⁺   ½O₂

A similar enzyme, *tyrosinase*, containing 0·3 per cent of copper, is responsible in animal tissues for the oxidation of tyrosine to dihydroxyphenylalanine, which is the first step in the series of reactions which leads to the production of *melanin*, the dark pigment of hair and skin (Chap. 21). *Ascorbic acid oxidase* is a similar copper-containing enzyme. But otherwise such metalloprotein oxidases do not appear to be of much importance in the economy of the cell.

The term *oxidase* is also applied to certain

flavoproteins because they are capable of catalysing oxidations in which molecular oxygen acts as a hydrogen acceptor. Such flavoproteins are sometimes called aerobic dehydrogenases. A well-known example is D-*amino acid oxidase* which catalyses the oxidation of D-amino acids as follows:

$$R-CHNH_2-COOH \underset{}{\overset{}{\rightrightarrows}} O_2$$
$$\underset{R-\overset{NH}{\underset{\|}{C}}-COOH}{} \longleftarrow H_2O_2$$

The other product of the reaction is not water but hydrogen peroxide. The significance of such reactions in the life of the cell is problematical, except in the case of *xanthine oxidase* (Chap. 20), a flavoprotein containing molybdenum which catalyses the reaction:

$$\text{xanthine} + H_2O + O_2 \longrightarrow \text{uric acid} + H_2O_2$$

One flavoprotein oxidase is, however, of considerable practical importance. This is glucose oxidase (from the mould *Penicillium notatum*) which because of its very high specificity is widely used as a means of detecting and estimating glucose:

$$\text{glucose} + O_2 + H_2O \longrightarrow \text{gluconic acid} + H_2O_2$$

Finally, the *hydroperoxidases* catalyse oxidations in which hydrogen peroxide acts as hydrogen acceptor and in the process is reduced to water:

$$AH_2 \underset{}{\overset{}{\rightrightarrows}} H_2O_2$$
$$A \longleftarrow 2H_2O$$

The best-known examples of this class are the *peroxidase* of horse-radish and the *catalase* found in mammalian tissues. The latter catalyses the reaction:

$$H_2O_2 \underset{}{\overset{}{\rightrightarrows}} H_2O_2$$
$$O_2 \longleftarrow 2H_2O$$

The place of such enzymes in the metabolism of the cell is uncertain—they may serve to get rid of the hydrogen peroxide produced by the flavoprotein oxidases.

Catalase is an iron protein of molecular weight 248 000 containing 4 iron atoms per molecule. The iron is incorporated in the protein as a ferriprotoporphyrin (Chap. 25) prosthetic group. Catalase from liver tissue and erythrocytes has been highly purified and obtained in the crystalline state. It is one of the most powerful enzymes known. One molecule of catalase can decompose 2 640 000 molecules of hydrogen peroxide per minute at 0°C.

OXIDATION–REDUCTION POTENTIALS

Oxidation–reduction potentials have already been mentioned on page 127 and 135 but some further information may be of value at this stage.

We have already seen that the process of oxidation involves loss of electrons whereas reduction involves gain of electrons. The reaction in the case of iron may be represented thus:

$$\underset{\text{ferrous iron}}{Fe^{2+}} \underset{\text{reduction}}{\overset{\text{oxidation}}{\rightleftharpoons}} \underset{\text{ferric iron}}{Fe^{3+} + e}$$

At equilibrium when the solution contains both ferrous and ferric ions we have in accordance with the law of mass action,

$$\frac{[Fe^{3+}][e]}{[Fe^{2+}]} = K.$$

where the brackets denote the concentration of each reactant and K is a constant indicating the oxidation–reduction balance.

If an inert electrode, say of platinum, is immersed in such a solution electrons tending to escape from the ferrous ions give a negative charge to the platinum. If such a half cell is connected through a salt bridge to a normal hydrogen electrode (that is an electrode of hydrogen gas at pressure of one atmosphere in equilibrium with a solution of pH = 0 and whose potential is arbitrarily taken to be zero) then a potential difference between the two electrodes is set up. The charge on the inert electrode increases as the tendency of electrons to escape to it becomes greater and is therefore a function of the reducing tendency of the system. The potential in volts set up between the inert electrode and a normal hydrogen electrode is known as the *oxidation–reduction potential*, or *redox-potential*, E_h. It is given by the following equation,

$$E_h = E° + \frac{RT}{nF} \ln \frac{(\text{oxidant})}{(\text{reductant})}$$

where $E°$ is the standard e.m.f. of the system.

R is the gas constant (in joules per mole per degree).

T is the absolute temperature.

F is the faraday (96 500 coulombs).

n is the number of electron equivalents.

ln is the logarithm to the base e.

When the concentrations of oxidant and reductant are equal $\ln[\text{oxidant}]/[\text{reductant}]$ becomes 0 and $E_h = E°$. When the value of $E°$ is known it is possible to calculate the potential E_h for any degree of oxidation or reduction of the system.

These relationships apply only when the hydrogen-ion concentration is normal, that is at pH 0. When the pH is at some other value $E°$ becomes E'_0 and the pH value must be quoted.

E'_0 is a measure of the oxidation or reduction *intensity* of a system just as pH is a measure of acid or alkali *intensity* but not of buffering *capacity*. Thus a system of $E'_0 + 0·1$ volts is potentially able to oxidize a system of $E'_0 - 0·1$ volt but is itself oxidized by a system of $E'_0 + 0·2$ volt.

The E_h value of a system may be measured either by electrical methods, as already mentioned, or by means of indicator dyes (redox indicators). Thus methylene blue, which has an E'_0 value at pH 7·0 of $+0·011$ volt, is reduced to leucomethylene blue by systems more negative. The dye 2:6-dichlorophenolindophenol ($E'_0 = +0·217$ at pH 7·0) is used in the quantitative determination of ascorbic acid ($E'_0 = +0·06$) which is a powerful reducing agent.

REFERENCES

BALDWIN, E. (1967). *Dynamic Aspects of Biochemistry*, 5th edn. Cambridge University Press.

DAWES, E. A. (1972). *Quantitative Problems of Biochemistry*, 5th edn. Edinburgh: Livingstone.

GREEN, D. E. and BAUM, H. (1970). *Energy and the Mitochondrion*. New York: Academic Press.

HEWITT, L. F. (1950). *Oxidation–reduction Potentials in Bacteriology and Biochemistry*, 6th edn. Edinburgh: Livingstone.

LEHNINGER, A. L. (1964). *The Mitochondrion*. New York: Benjamin.

MAHLER, H. R. and CORDES, E. H. (1971). *Biological Chemistry*, 2nd edn. New York: Harper & Row.

MORTON, R. A. (1961). Ubiquinone. *Vitamins and Hormones* **19**, 1–42.

SINGER, T. P. (Ed.) (1968). *Biological Oxidations*. New York: Interscience.

11 Glycolysis and the citric acid cycle

In previous chapters it has been emphasized that one of the main functions of food is to supply the energy which the body requires for the mechanical, chemical, osmotic and electrical activities of its various tissues. It has also been shown (Chap. 10) that this energy is released in useful form mainly (though not exclusively) as a result of oxidation of constituents of the food by molecular oxygen to yield carbon dioxide and water. Closer examination of this oxidative process has shown that the energy is obtained, in the form of ATP, as a result of transfer of hydrogen atoms from a substrate, via the electron transport system, to molecular oxygen (see Diagram A, p. 140).

The present chapter and Chapters 18, 19 and 21 are concerned chiefly with the mechanisms by which carbohydrate, fat and protein can be used to provide substrates for this energy-yielding hydrogen transfer. It is convenient to start with a consideration of carbohydrate, since it is by far the most important source of energy. This is hardly surprising since ultimately the energy which animals and man require is supplied by the sun through photosynthesis (p. 159) in green plants (Fig. 11.1). The equation by which photosynthesis is commonly represented

$$6CO_2 + 6H_2O \longrightarrow C_6H_{12}O_6 + 6O_2$$

is in some respects misleading, but it is at least correct in indicating that the product of the reaction is carbohydrate. The process is discussed further on p. 159. Fat and protein can be regarded, so far as plants are concerned, as derivatives of carbohydrate which are elaborated in limited amounts and for special purposes (see Fig. 11.1 overleaf).

Although the metabolism of carbohydrates includes a great network of interconnected

$$AH_2 \searrow \nearrow NAD^+ \searrow \nearrow FPH_2 \searrow \nearrow 2\ cyt.\ c\ Fe^{3+} \searrow \nearrow 2\ cyt.\ a_3\ Fe^{2+} \searrow \nearrow {\scriptstyle 2H^+} \searrow \nearrow {\scriptstyle \frac{1}{2}O_2}$$

$$A \nwarrow \quad NADH \quad FP \quad 2\ cyt.\ c\ Fe^{2+} \quad 2\ cyt.\ a_3\ Fe^{3+} \quad H_2O$$

$$+H^+ \qquad\qquad 2H^+$$

DIAGRAM A

reactions there are only three major pathways (Fig. 11.2).

(a) Glycolysis or the Embden-Meyerhof pathway.

(b) The Krebs tricarboxylic acid cycle, known also as the citric acid cycle.

(c) The pentose phosphate pathway.

These will be dealt with in turn. In each, glucose is taken as the starting point since it is the principal monosaccharide involved in carbohydrate metabolism in mammals.

GLYCOLYSIS

In most tissues the process of obtaining energy from glucose begins with the sequence of reactions known as *glycolysis*. In this

mechanism useful energy can be obtained by splitting the glucose molecule into identical halves. It takes place in four stages and is brought about by enzymes in the cell sap (p. 77).

Stage I is preparatory. Before it can be split, the rather asymmetric glucose molecule is converted to the almost symmetrical fructose 1,6-diphosphate by donation of two phosphate

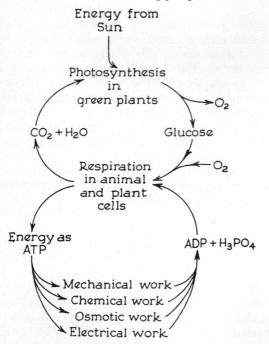

Fig. 11.1 The sun provides energy in the form of light for the photosynthetic reaction by which carbon dioxide and water are converted to glucose and oxygen. The recombination of glucose and oxygen in plant or animal cells provides energy in the form of ATP which can be used for the activities of the organism.

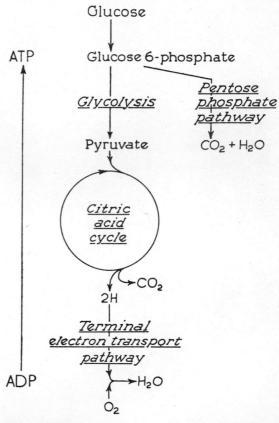

Fig. 11.2 The breakdown of glucose in the cell takes place by stages (1) the conversion of glucose to pyruvate in glycolysis (2) the oxidation of pyruvate in the citric acid cycle with production of carbon dioxide and (3) the final conversion of hydrogen to water by molecular oxygen in the electron transport pathway. In each of these stages energy is provided for the phosphorylation of ADP to ATP. Alternatively glucose may be oxidized by the pentose phosphate pathway to carbon dioxide and water.

groups from ATP. The conversion is brought
about in three steps:

(a) glucose is phosphorylated to glucose
6-phosphate by the enzyme *glucokinase*.

glucose   glucose 6-phosphate

(b) glucose 6-phosphate undergoes an internal rearrangement to give fructose 6-
phosphate. The enzyme responsible is *glucose-
phosphate isomerase*.

(c) A second phosphorylation, this time by
phosphofructokinase, gives fructose 1,6-dipho-
sphate

fructose   fructose
6-phosphate   1,6-diphosphate

The overall effect of stage I can therefore be
represented thus:

It is to be noted that, far from yielding useful energy, this stage involves the expenditure of two ATP molecules: these are, however, recovered later at stage IV.

Stage II is the splitting of the fructose 1,6-diphosphate molecule by the enzyme *aldolase*.

$$H_2O_3POCH_2 \quad CH_2OPO_3H_2$$

aldolase

CH₂OPO₃H₂		CH₂OPO₃H₂
CO		CHOH
CH₂OH		CHO
dihydroxy acetone phosphate		**glyceraldehyde 3-phosphate**

The two triosephosphates produced in this reaction, glyceraldehyde 3-phosphate (right) and dihydroxyacetone phosphate (left), are interconvertible in the presence of the enzyme *triosephosphate isomerase*. Since only the aldehyde is used in the next step of the reaction sequence, the overall effect of this stage is to split the fructose 1,6-diphosphate into identical halves

$$H_2O_3POCH_2 \quad CH_2OPO_3H_2 \;\longleftrightarrow\; 2 \begin{array}{l} CH_2OPO_3H_2 \\ CHOH \\ CHO \end{array}$$

This stage involves neither expenditure nor production of ATP.

Stage III is the energy-yielding reaction. Glyceraldehyde 3-phosphate is oxidized to the corresponding carboxylic acid by *glyceraldehyde 3-phosphate dehydrogenase*. Reactions of this type in which an aldehyde group is oxidized to an acid are accompanied by the liberation of large amounts of potentially useful energy. In the present case some of the energy instead of being wasted as heat is used to form ATP from ADP + Pi. (The symbol Pi is used here and subsequently to indicate one molecule

of inorganic phosphate). In this reaction a sulphydryl group ($-SH$), firmly bound to the enzyme, is attached to the aldehyde group by a dehydrogenation reaction in which NAD is reduced to NADH.

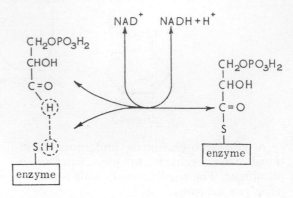

In the second step the complex so formed is split by addition of inorganic phosphate to give 1,3-diphosphoglyceric acid thus:

1,3-diphospho-glyceric acid

This then loses its newly acquired phosphate to ADP under the influence of the enzyme *phosphoglycerate kinase* to yield 3-phospho-glyceric acid.

phosphoglycerate kinase

The overall reaction is therefore:

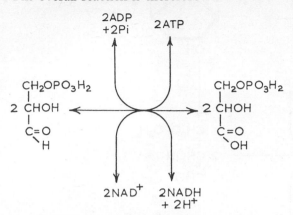

It is to be noted that the production of ATP at this stage in no way involves the phosphate at C-3.

The $NADH + H^+$ produced is extramitochondrial and is not interchangeable with that produced in the mitochondrion.

Stage IV is the recovery of the phosphate groups from 3-phosphoglyceric acid. The two molecules of 3-phosphoglyceric acid, the end-product of the previous stage, still contain the phosphate groups originally derived from ATP in stage I. These phosphate groups are now transferred back to ADP as follows:

(*a*) The phosphate residue moves from C-3 to C-2. The enzyme responsible is *phosphoglyceromutase*.

$$
\begin{array}{ccc}
CH_2OPO_3H_2 & & CH_2OH \\
| & & | \\
CHOH & \xrightarrow{\hspace{1cm}} & CHOPO_3H_2 \\
| & phospho & | \\
COOH & glyceromutase & COOH \\
\text{3-phosphoglyceric} & & \text{2-phosphoglyceric} \\
\text{acid} & & \text{acid}
\end{array}
$$

(*b*) The 2-phosphoglyceric acid under the influence of the enzyme *phosphopyruvate hydratase (enolase)* loses the elements of water to form phospho(enol)pyruvic acid.

$$
\begin{array}{ccc}
CH_2\lceil OH \rceil & & CH_2 \\
| & phosphopyruvate & \| \\
C\lceil H \rceil OPO_3H_2 & \xrightleftharpoons{\hspace{0.5cm}} & C-OPO_3H_2 \\
| & hydratase & | \\
COOH & & COOH \\
& & \text{phosphoenol-} \\
& \searrow & \text{pyruvic acid} \\
& H_2O &
\end{array}
$$

(*c*) The phospho(enol)pyruvate then loses its phosphate group to ATP. This reaction is catalysed by *pyruvate kinase*.

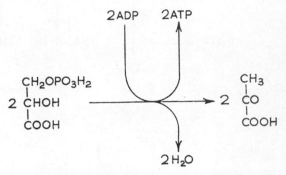

pyruvic acid

The overall result of stage IV can be represented thus:

$$
\begin{array}{ccc}
& 2ADP \quad 2ATP & \\
CH_2OPO_3H_2 & & CH_3 \\
| & & | \\
2\ CHOH & \xrightarrow{\hspace{1.5cm}} & 2\ CO \\
| & & | \\
COOH & & COOH \\
& \downarrow & \\
& 2H_2O &
\end{array}
$$

This stage can be regarded as the recovery of the two molecules of ATP used up in stage I. (See Diagram B.)

If the four stages are now combined as in Diagram D it can be seen that breakdown of one glucose molecule is accompanied by the net formation of two ATP molecules in stage III as shown here in simplified form (Diagram C). In other words, the glucose molecule has been split and partially oxidized to yield two pyruvic acid molecules. Part of the energy made available by this conversion has been used to synthesize two new ATP molecules. Clearly the reaction can continue only so long as NAD is available to take part in stage III. Since the amount present in any tissue is very small, some mechanism must be available to convert NADH back to NAD.

If glycolysis takes place in aerobic conditions the two NADH molecules can transfer their hydrogen atoms to molecular oxygen via the electron transport system with the consequent production of 2×2, that is 4, additional ATP molecules. In this case only 2 ATP molecules are produced for each $NADH + H^+$ oxidized

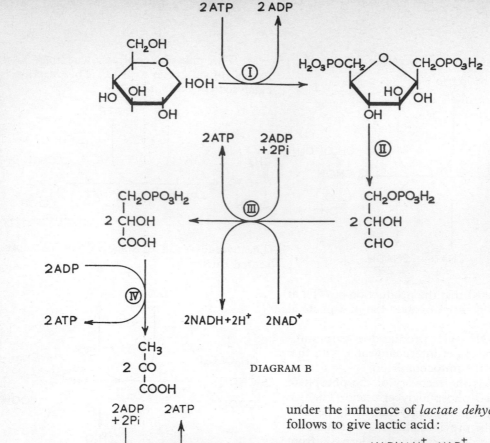

DIAGRAM B

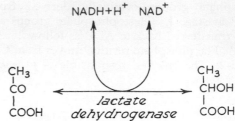

DIAGRAM C

because the electrons have to pass into the mitochondrion by way of the glycerol phosphate shuttle (see p. 134), and are passed straight to the flavoprotein enzyme inside the mitochondrion. The first of the 3 ATP molecules normally formed on oxidation of NADH $+H^+$ produced in the mitochondrion is therefore not formed in the case of glycolysis. The total ATP yield obtained on aerobic glycolysis is thus $4+2$, that is 6 molecules. In anaerobic conditions this mechanism for conversion of NADH back to NAD is, of course, not available. Instead in mammalian tissues the NADH reacts with the pyruvic acid

under the influence of *lactate dehydrogenase* as follows to give lactic acid:

This reaction is essentially the conversion of a ketone to an alcohol. Unlike the conversion of an aldehyde to an acid, as in reaction III above, it involves no great liberation or uptake of energy and there is therefore no uptake or production of ATP. The overall reaction is shown in Diagram D.

The whole series of reactions can be summarized as follows:

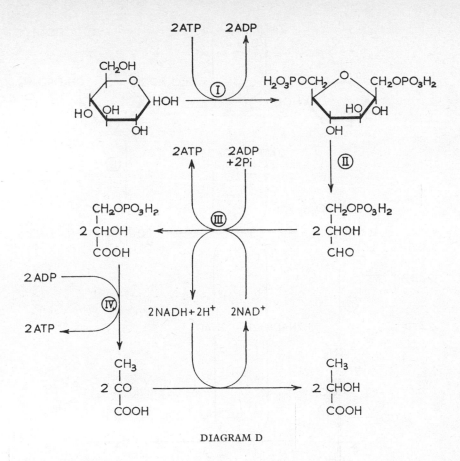

2ATP   2ADP

CH$_2$OH

HO OH
HOH ①   →   H$_2$O$_3$POCH$_2$ CH$_2$OPO$_3$H$_2$

OH   HO   OH

                                        OH

②

2ATP   2ADP
        +2Pi

CH$_2$OPO$_3$H$_2$ CH$_2$OPO$_3$H$_2$
2 CHOH   ③   ←   2 CHOH
COOH   CHO

2ADP

④   2NADH+2H$^+$ 2NAD$^+$

2ATP

CH$_3$ CH$_3$
2 CO   →   2 CHOH
COOH   COOH

DIAGRAM D

Thus, even under anaerobic conditions glyco-lysis can still yield a small but useful quantity of ATP.

Aerobic or anaerobic breakdown of glucose by the glycolytic pathway occurs also in plants and micro-organisms. In these instances the end-product is not necessarily pyruvic acid or lactic acid. For example, in anaerobic condi-tions yeast breaks down glucose by a series of reactions similar to that described above, but the re-oxidation of NADH to NAD$^+$ is brought about by a different mechanism. The

pyruvic acid is not reduced directly to lactic acid; instead, it is decarboxylated to yield acetaldehyde and it is the acetaldehyde which reoxidizes the NADH to NAD. In the process the acetaldehyde is itself converted to ethyl alcohol under the influence of *alcohol de-hydrogenase* (Diagram E).

By this mechanism yeast produces ethyl alcohol from carbohydrate in the process of alcoholic fermentation. In summary it is represented thus:

2ADP   2ATP
+2Pi

CH$_2$OH

HO OH
HOH → 2 CH$_3$CH$_2$OH+ 2CO$_2$

OH

Glycolysis and the Citric Acid Cycle   145

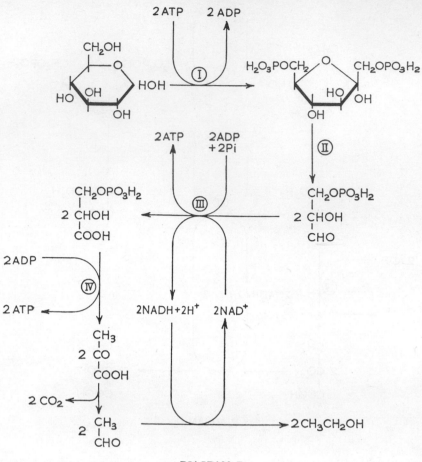

DIAGRAM E

Synthesis of Glucose by Reversal of Glycolysis

The reactions of the glycolytic pathway can be regarded as reversible with two exceptions:

(a) the phosphorylation of glucose;
(b) the phosphorylation of fructose 6-phosphate.

This means that, provided these reactions can be by-passed, the glycolytic pathway can be used to form glucose from lactic acid or pyruvic acid. A suitable by-pass is provided by the hydrolytic enzymes *glucose 6-phosphatase* and *hexosediphosphatase* (*fructose 1,6-diphosphatase*) which respectively catalyse the following reactions:

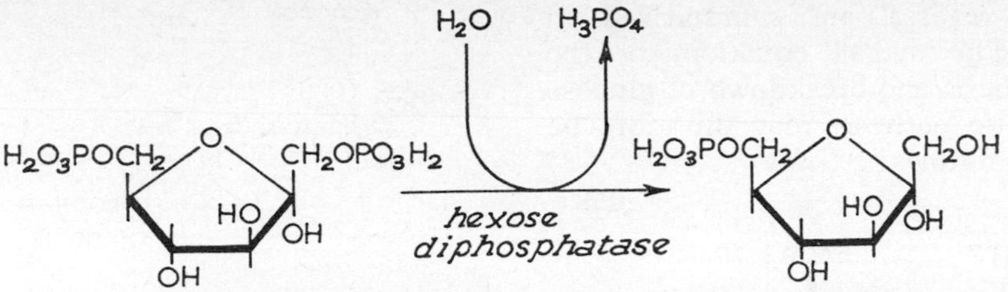

A third reaction, the conversion of phospho-(enol)pyruvate to pyruvate, is usually also by-passed when the glycolytic pathway operates in reverse. The by-pass is a rather complex one but it can be represented as follows:

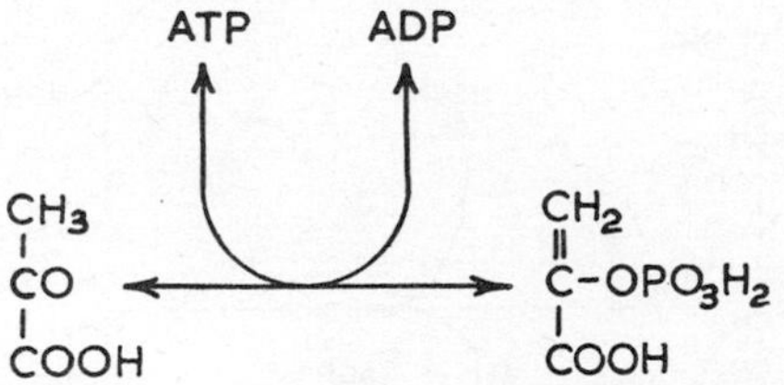

In essence the pyruvic acid is converted by uptake of CO_2 to oxaloacetic acid. This reaction is driven by the breakdown of a molecule of ATP. The oxaloacetic acid then loses CO_2 and takes up a phosphate group from GTP (p. 66). The GTP may then be regenerated by transfer of a phosphoryl residue from ATP.

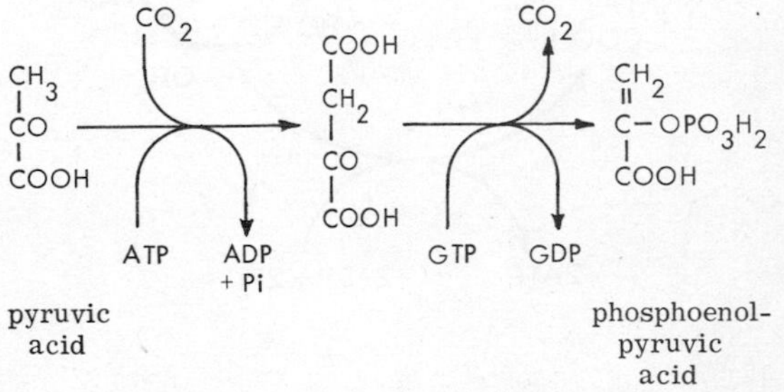

The essential steps in the process of glyco-

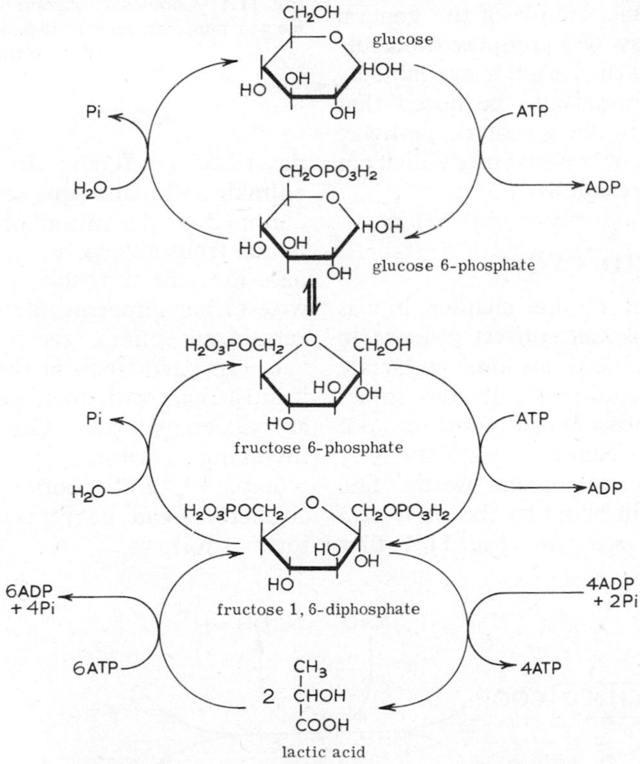

Fig. 11.3 Summary of the forward and reverse steps in glycolysis.

lysis and its reversal are summarized in Figure 11.3. The overall equation for the anaerobic synthesis and breakdown of glucose by the glycolytic pathway may therefore be summarized as follows:

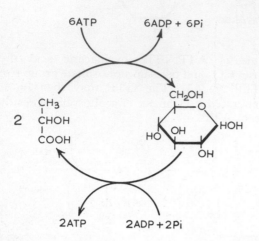

While the breakdown of a glucose molecule yields only 2 ATP molecules its resynthesis requires 6. This is an example of the general rule that the synthesis of a complex molecule requires more energy than is made available by its breakdown. It should also be noted that synthesis of glucose by the glycolytic pathway requires the presence of two enzymes which are not involved in its breakdown.

THE CITRIC ACID CYCLE

In the earlier part of this chapter it was shown that glycolysis can convert glucose to pyruvate (under aerobic conditions) or lactate (under anaerobic conditions). In the intact animal, however, glucose metabolism does not stop at this point but continues until the only products are carbon dioxide and water. The final steps are brought about by the *citric acid cycle* or *tricarboxylic acid cycle* (Fig. 11.4) first

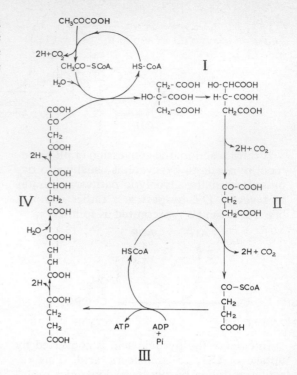

Fig. 11.4 Composite diagram of the citric acid cycle. The Roman numerals refer to the four stages of the cycle described in the text.

described by Krebs. In the cells of higher animals and plants this sequence of reaction is confined to the mitochondria. Pyruvate produced from glucose in the cell sap (p. 77) can pass into the mitochondria which are otherwise rather impermeable to metabolites (with certain exceptions, see p. 134). Before pyruvate can participate in this series of reactions it must react with coenzyme A (p. 67) to form acetylcoenzyme A. This complex reaction, involving thiamine pyrophosphate (TPP) (Chap. 13) and lipotic acid (Chap. 13) as co-factors, can be represented in simplified form as follows:

The acetyl moiety of the pyruvic acid is transferred to coenzyme A while the carbon of the carboxyl group is liberated as carbon dioxide. (This transfer like others of the same type, occurs via a complex with thiamine pyrophosphate.) The remaining two hydrogen atoms (one from the —COOH of pyruvate and one from the HS— of coenzyme A) are transferred to NAD by way of a mechanism in which lipoic acid participates. Since the overall reaction involves both oxidation and loss of carbon dioxide it is termed an oxidative decarboxylation.

The tricarboxylic cycle proper begins by acetyl coenzyme A transferring its acetyl group to oxaloacetic acid to give citric acid. By stepwise dehydrogenations and loss of two molecules of CO_2 accompanied by internal rearrangements, the citric acid is reconverted to oxaloacetic acid which can then take up another acetyl group from acetyl coenzyme A to form another citric acid molecule which goes through the whole process again (Fig. 11.5). In this way a very small catalytic amount of oxaloacetic acid can bring about the complete oxidation of a much larger amount of pyruvic acid. The hydrogen atoms removed in the successive dehydrogenations which the citric acid undergoes are passed to the electron transport system and thence to oxygen to give water. As described in Chapter 10, this results in the production of ATP.

The reactions of the cycle can be divided into four main stages (Fig. 11.4):

In **Stage I** the acetyl group of acetyl-coenzyme A is transferred to oxaloacetate to form citric acid. The reaction is catalysed by *citrate synthase (condensing enzyme)*. No oxidation or decarboxylation is involved but a molecule of water is required to hydrolyse the linkage between the acetyl group and the coenzyme A.

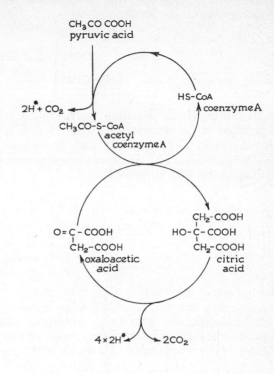

Fig. 11.5 Simplified outline of the citric acid cycle. The pairs of hydrogen atoms marked * are transferred by the usual oxidative pathways (Chap. 10) to form water.

In this way the coenzyme A, of which only a minute amount is present in the tissues, is liberated and can react with more pyruvate. In the meantime, the citric acid undergoes an internal rearrangement in which the hydroxyl group migrates to an adjacent carbon atom to give isocitric acid by way of aconitic acid as an

CH₃CO-S-CoA   *citrate synthase*   HS-CoA

O=C-COOH
|
CH₂-COOH

oxaloacetic
acid

CH₂-COOH
|
HO-C-COOH
|
CH₂-COOH

citric acid

intermediary. The enzyme responsible is *aconitate hydratase (aconitase)*.

$$
\begin{array}{c}
\text{CH}_2-\text{COOH} \\
| \\
\text{HO-C - COOH} \\
| \\
\text{CH}_2-\text{COOH} \\
\text{citric acid}
\end{array}
\quad
\xleftrightarrow[\textit{hydratase}]{\textit{aconitate}}
\quad
\begin{array}{c}
\text{HO-CH-COOH} \\
| \\
\text{CH-COOH} \\
| \\
\text{CH}_2-\text{COOH} \\
\text{isocitric acid}
\end{array}
$$

In **Stage II** the 6-carbon isocitric acid is converted to a derivative of the 4-carbon succinic acid. The isocitric acid undergoes oxidation followed by decarboxylation to give α-oxoglutaric acid as follows:

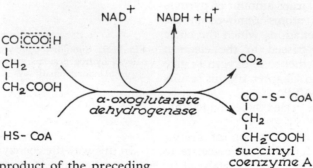

Both steps of the reaction seem to be catalysed by the same enzyme, *isocitrate dehydrogenase*. The α-oxoglutaric acid then undergoes an oxidative decarboxylation exactly analogous to oxidative decarboxylation of pyruvic acid.

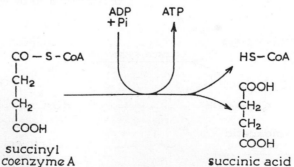

In **Stage III** the product of the preceding stage, succinyl coenzyme A, is split to succinic acid and free coenzyme A. The energy of this hydrolysis is utilized indirectly to form ATP from ADP and Pi. GTP acts as an intermediary in this reaction.

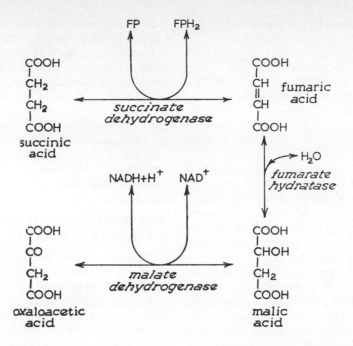

Stage IV is a succession of three reactions in which succinic acid is oxidized to oxaloacetic acid. Succinic acid is dehydrogenated by the flavoprotein *succinate dehydrogenase* to give the corresponding unsaturated fumaric acid. Fumaric acid, in presence of the enzyme *fumarate hydratase (fumarase)*, takes up water to give malic acid which is dehydrogenated by *malate dehydrogenase* and NAD to give oxalo-acetic acid. The oxaloacetic acid then combines with a fresh molecule of acetyl coenzyme A and the whole process is repeated.

The sequence of reactions is shown in Figure 11.4, which gives in outline the order in which carbon dioxide and pairs of hydrogen atoms are removed in the course of the cycle. The hydrogen atoms are transferred by the electron transport system to combine eventually with molecular oxygen. The reaction can be summarized.

If it is assumed that the transport of 2H from NADH or NADPH to combine with $\frac{1}{2}O_2$ and give H_2O yields 3 ATP molecules, (p. 131), and that transport from flavoprotein in the same way yields 2 ATP molecules it can be calculated that the oxidative steps involved in the breakdown of one pyruvate molecule yield the following:

oxidation of pyruvate via NAD to acetyl CoA	3 ATP
oxidation of isocitrate via NAD	3 ATP
oxidation of α-oxoglutarate via NAD	3 ATP
oxidation of succinate via flavoprotein	2 ATP
oxidation of malate via NAD	3 ATP
	Total 14 ATP

To this sum must be added a further ATP molecule produced in the conversion of succinyl coenzyme A to free succinic acid. The oxidation of one pyruvic acid molecule is thus accompanied by production of $14+1$ or 15 ATP molecules. The total yield of ATP obtained when one molecule of glucose is converted to pyruvic acid by glycolysis (under aerobic conditions) and the pyruvic acid is subsequently oxidized by the citric acid cycle is therefore:

conversion of glucose to two pyruvic
  acid molecules   6 ATP
oxidation of two pyruvic acid molecules by the cycle   30 ATP
                                        ─────────
                              Total 36 ATP

Under anaerobic conditions the glucose is converted only to lactic acid with a yield of 2 ATP molecules. Clearly, therefore, much more useful energy can be obtained by complete oxidation of a glucose molecule than by simply splitting it into two 3-carbon molecules.

As already mentioned (p. 131) the oxidation of a substrate with concomitant phosphorylation of ADP to ATP is known as *oxidative phosphorylation*. This process takes place in the mitochondria (p. 133), the enzymes of the citric acid cycle being located in the spaces between the cristae and those involved in electron transport being arranged in appropriate patterns in the cristae themselves.

The complete pathway of glycosis and oxidation is summarized in Plate 11.1.

It should be noted that two of the reactions of the cycle are irreversible, namely the oxidative decarboxylations of pyruvic acid and α-oxoglutaric acid. Since these reactions cannot be by-passed the cycle is irreversible. In other words, it can be used to degrade pyruvic acid and acetyl coenzyme A, but not to synthesize them.

FORMATION OF OXALOACETIC ACID

An important feature of the tricarboxylic acid cycle is that acetyl coenzyme A can be oxidized only if oxaloacetic acid is available to combine with it. The tissues must, therefore, have some means of maintaining and, if necessary, increasing the amount of oxaloacetic acid available for this purpose. This can be achieved by the addition of CO_2 to pyruvate.

It is important to note that the citric acid cycle is involved in the oxidation of fats and of amino acids, as well as of carbohydrate. Fatty acids are broken down stepwise (Chap. 19) to yield acetyl coenzyme A which enters the citric acid cycle and is dealt with in precisely the same way as acetyl coenzyme A derived from pyruvate. Several amino acids can undergo transamination (Chap. 21) to yield components of the citric acid cycle. For example, aspartate yields oxaloacetate and glutamate yields α-oxoglutarate.

THE PENTOSE PHOSPHATE PATHWAY

The mechanism just described is the principal but not the only pathway by which animal tissues oxidize glucose to carbon dioxide and water with the production of useful energy in the form of ATP. One of the most important alternative routes is the *pentose phosphate pathway*, a cyclic mechanism analogous to the tricarboxylic acid cycle, which can be best understood by considering the individual reactions step by step. (For the sake of simplicity all sugars will be shown in the straight chain form.)

The cycle can be divided into two phases:

1. CONVERSION OF HEXOSE TO PENTOSE

(*a*) Glucose 6-phosphate is oxidized to 6-phosphogluconic acid by *glucose-6-phosphate dehydrogenase* and NADP.

(*b*) In a second dehydrogenation by NADP, 6-phosphogluconic acid is further oxidized to an unstable intermediate which loses CO_2 to give the pentose, ribulose 5-phosphate. The enzyme is 6-*phosphogluconate dehydrogenase*.

COOH
|
HCOH
|
HOCH
|
HCOH   6-*phosphogluconate*
| *dehydrogenase*
HCOH
|
CH₂OPO₃H₂
6-phospho-
gluconic acid

NADP⁺ NADPH+H⁺

COOH
|
HCOH
|
C=O
|
HCOH
|
HCOH
|
CH₂OPO₃H₂

CO₂

CH₂OH
|
C=O
|
HCOH
|
HCOH
|
CH₂OPO₃H₂
ribulose
5-phosphate

The sum of these reactions is:

H₂O   CO₂

CHO
|
HCOH
|
HOCH
|
HCOH
|
HCOH
|
CH₂OPO₃H₂

CH₂OH
|
C=O
|
HCOH
|
HCOH
|
CH₂OPO₃H₂

2NADP⁺ 2NADPH+2H⁺

Ribulose 5-phosphate is readily converted into a variety of other pentoses such as xylulose 5-phosphate by means of epimerase enzymes.

2. CONVERSION OF PENTOSE TO HEXOSE

Theoretically pentose could be converted to hexose by reversing the above reactions. Normally, however, this is achieved by a complex series of rearrangements among several pentose molecules. The following is a brief description of the three main types of reaction:

(*a*) *Transketolation* is a reaction between two sugar phosphate molecules in which a —CO—CH₂OH group is transferred from one to another. In this way, for example, 2 pentose phosphate molecules can react together to give a triose phosphate and a heptose phosphate thus: $C_5 + C_5 \rightleftharpoons C_3 + C_7$.

(*b*) *Transaldolation*, a similar reaction in which the unit transferred is

$$-CHOH-CO-CH_2OH,$$

allows a triose phosphate and a heptose phosphate to react together to give a hexose phosphate and a tetrose phosphate:

$$C_3 + C_7 \rightleftharpoons C_6 + C_4.$$

(*c*) Finally the aldolase reaction (p. 142) can be used to combine two triose phosphate molecules to yield one hexose diphosphate:

$$C_3 + C_3 \rightleftharpoons C_6.$$

These three reactions can be employed to convert pentose quantitatively to hexose in the following manner. It is convenient to take, as starting point, six pentose phosphate molecules arranged in pairs as in Figure 11.6. Two

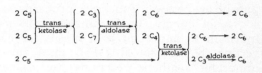

Fig. 11.6   Summary of the transketolase and transaldolase reactions.

of these pairs undergo transketolation followed by transaldolation to yield two hexose phosphates and two tetrose phosphates. The latter then undergo transketolation with the remaining pair of pentose phosphates to give two more hexose phosphates and two triose phosphates. Under the influence of aldolase the two triose phosphates combine to give fructose 1,6-diphosphate, which is hydrolysed by *fructose 1,6-diphosphatase* to give another hexose monophosphate. Thus the overall reaction is the conversion of 6 molecules of pentose phosphate to 5 of hexose phosphate: $6\ C_5 \rightarrow 5\ C_6$.

In the pentose phosphate pathway the two reaction sequences described above are combined. Six hexose monophosphates are oxidized to give six pentose monophosphates and $6CO_2$. The six pentose monophosphates are then reconverted to five hexose monophosphates. The latter can then, in association with a fresh molecule of hexose monophosphate, go through the cycle again thus:

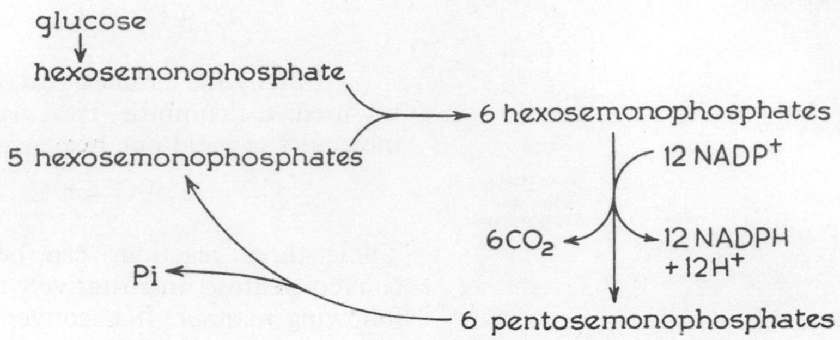

The overall effect of the reaction is the complete oxidation of one glucose molecule. It is, however, very doubtful whether the pentose phosphate pathway functions primarily as a source of energy. In mammalian cells it is much more important as a means for converting NADP to NADPH since several important synthetic reactions (for example biosynthesis of fatty acids (Chap. 19)) require large amounts of NADPH.

It is difficult to be certain what proportions of the glucose oxidized in a tissue follow the glycolytic and pentose phosphate pathways respectively. It seems probable, however, that the pentose phosphate pathways may be important in the liver and the adrenal cortex. Muscle, on the other hand, seems to use only the glycolytic pathway.

STORAGE OF CARBOHYDRATE— GLYCOGEN SYNTHESIS AND BREAKDOWN

The reaction mechanisms described above are all concerned with the breakdown of glucose to yield energy. There remains the problem of the storage of glucose. Obviously some storage system is essential. If none existed, the tissues would be flooded with excess glucose immediately after a meal and starved of it at all other times. But the direct accumulation of large quantities of such a small molecule is not possible. For instance the carbohydrate reserve in the adult human liver (1·8 kg) is about 100 g. If it were all in the form of glucose the glucose concentration inside the liver cells would be of

the order of 0·3 M. This would approximately double the osmotic pressure inside the cells with disastrous results. It is, therefore a great advantage to the organism to store its glucose in the form of the polymer glycogen which has a very high molecular weight and correspondingly low osmotic pressure.

Glucose is converted to glucose 6-phosphate by *glucokinase*. Under the action of *phosphoglucomutase* the phosphate residue is transferred from position 6 in the glucose molecule to position 1.

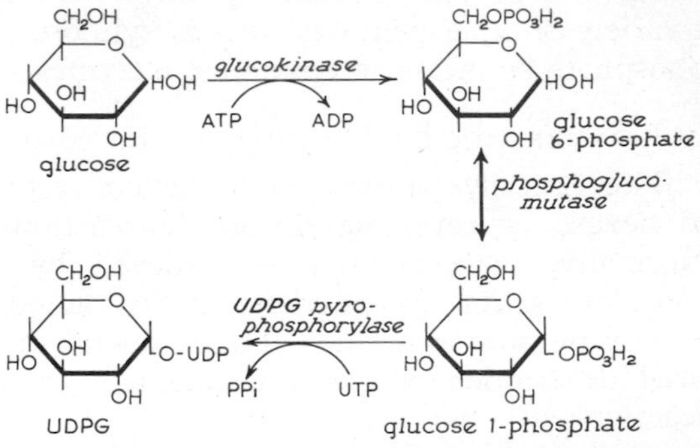

The glucose 1-phosphate then combines with uridine triphosphate (UTP) to give uridine diphosphoglucose (UDPG) under the influence of *UDPG pyrophosphorylase* with the elimination of pyrophosphate (PP_i). The uridine diphosphoglucose can then transfer the glucose molecule to the end of an existing glycogen chain under the influence of the *glycogen-UDP glucosyl transferase* thus:

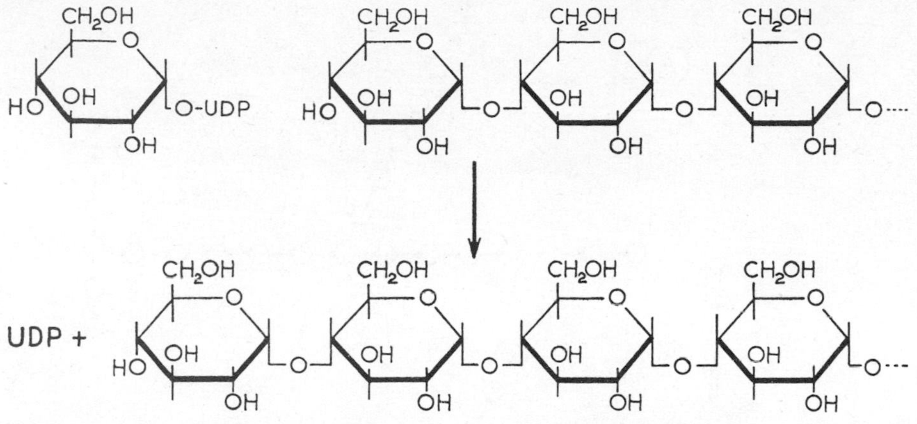

which breaks off the terminal segment, containing several glucose residues, of the free end

UDP +

Finally UDP can be reconverted to UTP by transfer of phosphate from ATP

$$UDP + ATP \rightleftharpoons UTP + ADP$$

In this way an existing glycogen chain can be repeatedly extended by one glucose unit at a time. In each extension 2 ATP molecules are expended—one in the formation of glucose 6-phosphate and another in the regeneration of UTP.

The glycogen molecule has a highly branched structure (Figs. 4.2 and 4.3), the branching being achieved by the presence of occasional α 1,6 linkages in addition to the more common α 1,4. Glycogen synthesis must therefore involve the creation of new branches as well as the elongation of existing ones. This is achieved by the so-called 'branching enzyme'

of an existing chain and re-attaches it by means of an α 1,6 linkage to a glucose residue in the middle of another chain. The existing branches of the molecule are extended by the addition of new glucose molecules. As they grow longer they become more open to attack by the branching enzyme which transplants their terminal segments to start new branches. In this way the molecule is extended without losing its characteristic structure (Fig. 11.7).

The breakdown of glycogen is catalysed by the enzyme *phosphorylase*. The reaction takes the form of a phosphorolysis, the linkage between the terminal glucose and its neighbour being split by the introduction of inorganic phosphate. In this way a glycogen chain can be shortened by one glucose unit at a time. The reaction is readily reversible *in vitro*, but *in*

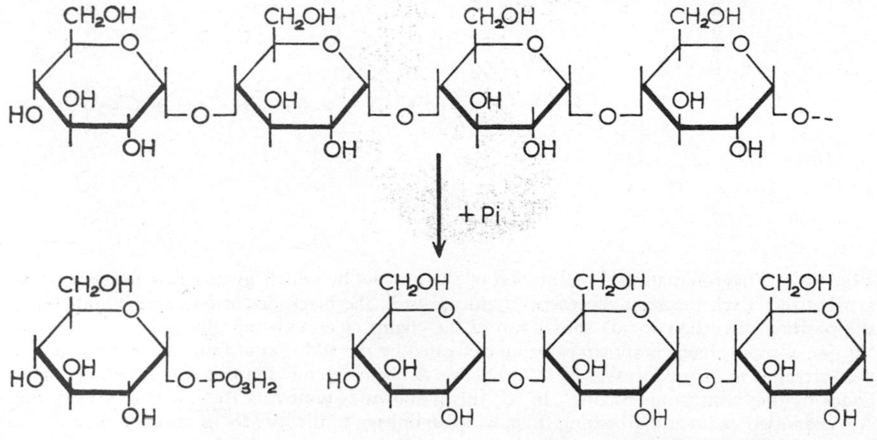

+ Pi

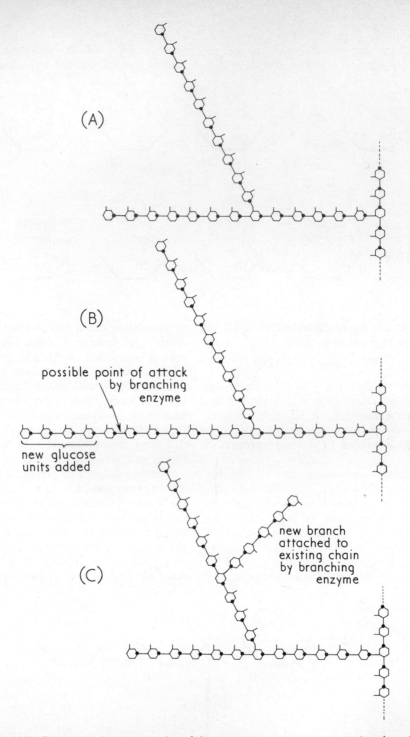

(A)

(B)

possible point of attack
by branching
enzyme

new glucose
units added

(C)

new branch
attached to
existing chain
by branching
enzyme

Fig. 11.7 Diagrammatic representation of the process by which glycogen is thought to be synthesized. Each hexagon represents a glucose unit, the black dot at one corner indicating the position of carbon 1. (*A*) shows two of the chains of an existing glycogen molecule. In (*B*) one of these chains is shown as being elongated by the addition of four new glucose units, each attached to its predecessor by a 1,4 linkage. A chain extended in this way is vulnerable to attack by the 'branching enzyme'. In (*C*) this is shown as removing the last five units of the extended chain and attaching them by a 1,6 linkage to the middle of another chain.

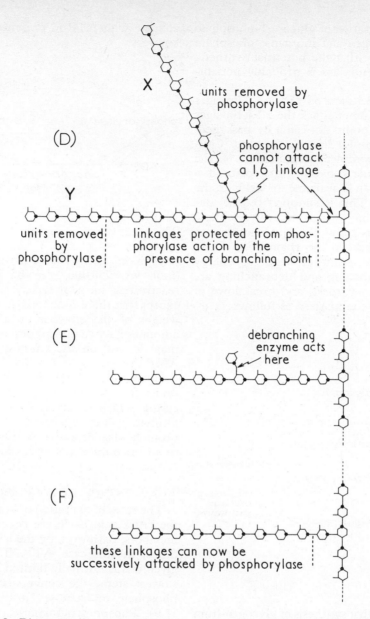

Fig. 11.8 Diagrammatic representation of the process by which glycogen is thought to be broken down. Each hexagon represents a glucose unit, the black dot at one corner indicating the position of carbon 1. (D) shows two chains X and Y in a glycogen molecule in which X can be regarded as a 'daughter' chain attached to its 'parent' Y by means of a 1,6 linkage. The free ends of both chains can be attacked by phosphorylase as shown, but the action of the enzyme is impeded in the region around the branching point. It cannot attack the 1,6 linkage itself nor can it by-pass it and attack the 1,4 linkages beyond it on the parent chain. Moreover, its attack on the free end of the parent chain Y is inhibited as soon as it approaches within 5 glucose units of the branching point. By itself, therefore phosphorylase cannot bring about the complete breakdown of a glycogen molecule; it can only 'prune' the ends of the chains in the manner shown in (E). A chain 'pruned' in this way can, however, be attacked by the 'debranching enzyme' which hydrolyses the 1,6 linkage at the branching point. This leaves an unbranched chain, no longer protected by the presence of a branching point and therefore open to attack by phosphorylase (F). Thus a glycogen molecule can be completely broken down by the alternating action of phosphorylase and 'debranching enzyme'.

vivo the concentration of glucose 1-phosphate relative to glycogen and inorganic phosphate is usually so low that the potential synthetic action of phosphorylase is probably not important.

Phosphorylase attacks only the α 1,4 linkages in glycogen. It cannot attack the α 1,6 linkages at the branching points nor can it by-pass them and attack the α, 1,4 linkages beyond. Consequently the complete degradation of the glycogen molecule can only take place if a second mechanism is available to deal with the α, 1,6 linkages. This is supplied by the 'debranching enzyme' which hydrolyses these linkages and thus allows the phosphorylase to continue its attack on the chain (Fig. 11.8). Since, from the point of view of energy expenditure 'branching' and 'debranching' are unimportant the synthesis and breakdown of glycogen can be summarized as follows:

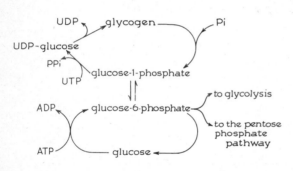

It is to be noted that synthesis of glycogen from glucose or glucose phosphate involves expenditure of ATP, but no ATP is synthesized by the reverse process.

Phosphorylase can exist in the tissues in both an active '*a*' form and an inactive '*b*' form. Muscle phosphorylase '*a*' has a molecular weight of 500 000. Under the influence of a 'phosphate rupturing' enzyme *phosphorylase phosphatase (PR enzyme)* it is converted to 2 molecules of phosphorylase '*b*' with a mole-molecular weight of 250 000. This inactivation can be reversed by rephosphorylating the enzyme with ATP in presence of a phospho-kinase.

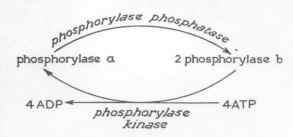

The reactivation is greatly accelerated by the presence of cyclic 3′,5′-adenosine monophosphate (3′:5′-AMP). Liver phosphorylase (mol. wt. 250 000) can be inactivated and reactivated in a precisely similar manner, except that there is no change in the molecular weight of the enzyme in the process. The amount of active phosphorylase in both tissues may depend on the amount of cyclic 3′:5′-AMP present.

Cyclic 3′:5′-AMP is formed from ATP under the influence of the enzyme *adenyl cyclase* which is activated by adrenaline. The ultimate effect of adrenaline is therefore to promote the formation of phosphorylase and so to encourage the breakdown of glycogen (Chap. 18).

THE CONTROL OF ATP PRODUCTION

The rates of operation of both glycolysis and the citric acid cycle are regulated, like other metabolic pathways, by the levels of their products, in this case ATP. Thus the rate of glycolysis is normally limited by the rate of its slowest steps, the conversion of fructose 6-phosphate to fructose 1,6-diphosphate (Fig. 11.9). Phosphofructokinase, which catalyses this reaction, is a regulatory enzyme whose activity is inhibited by ATP and increased by ADP and inorganic phosphate. Consequently an increased demand for energy with a drain of ATP and accumulation of ADP and Pi has the effect of 'switching on' glycolysis and thus making more ATP available. Conversely, if the cell's demand for energy falls so that ATP accumulates, glycolysis is slowed down.

While this seems to be the main mechanism by which the rate of ATP production is controlled, the operation of glycolysis and of the

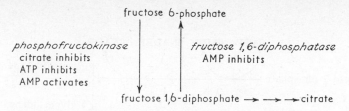

Fig. 11.9 The control by AMP of the interconversion of fructose 6-phosphate and fructose 1,6-diphosphate, a key reaction in the breakdown and formation of carbohydrate.

citric acid cycle is influenced by a number of other controls. Thus, for example, accumulation of citrate, which might be the result of glycolysis proceeding more rapidly than the citric acid cycle, inhibits phosphofructokinase and so retards glycolysis.

As one might expect the biosynthesis of glucose is controlled in such a way that it is inhibited in conditions which favour glycolysis. Thus, while AMP accelerates the conversion of fructose 6-phosphate to fructose 1,6-diphosphate in glycolysis (catalysed by phosphofructokinase), it inhibits the reverse conversion (catalysed by fructose diphosphatase) (Fig. 11.9).

The interconversion of glycogen and glucose 1-phosphate is similarly controlled (Fig. 11.10). Phosphorylase, which catalyses the breakdown of glycogen, is activated by AMP (which will be abundant when the supply of energy is more than equal to the demand). Glycogen synthetase, which converts UDP-glucose to glycogen, on the other hand, is activated by an accumulation of glucose 6-phosphate, which is the product both of the phosphorylation of glucose and of the formation of monosaccharide from lactate.

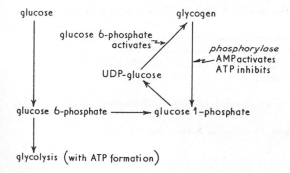

Fig. 11.10 Factors controlling the activity of enzymes involved in the interconversion of glycogen and glucose.

DIRECT OXIDATION OF GLUCOSE

Glucose cannot be oxidized to gluconic acid in animal cells but certain micro-organisms, such as *Penicillium notatum,* contain an enzyme *glucose oxidase* (*notatin*) which brings about this oxidation with the formation of hydrogen peroxide thus:

$$C_6H_{12}O_6 + H_2O + O_2 \longrightarrow C_6H_{12}O_7 + H_2O_2.$$

This enzyme is now commonly used for the detection and estimation of glucose in biological materials such as urine.

PHOTOSYNTHESIS

The process of photosynthesis in green plants is exceedingly complicated and only the briefest outline can be given here. It takes place in two stages—the *light reaction* and the *dark reaction.*

The *light reaction* occurs in the *chloroplasts,* cell organelles which are slightly larger than the mitochondria (p. 75) which they resemble in being surrounded by a double membrane, in containing cytochrome pigments, flavoproteins, and a small amount of DNA, and in being the site of formation of ATP. They differ from the mitochondria, however, in producing instead of consuming oxygen, and in containing *chlorophyll* (Chap. 25), *carotenoids* (Chap. 13) and the iron-containing pigment *ferredoxin.*

Chloroplasts consist of parallel lamellae of lipoprotein in which are found areas of increased density called *grana* containing carotenoid, chlorophyll and lipid molecules in orderly array surrounded, above and below, by a double layer of protein molecules. These proteins include the enzyme complexes for photosynthetic electron transport and for carbon dioxide reduction.

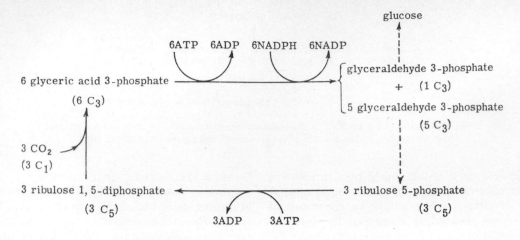

Fig. 11.11 The dark reaction in photosynthesis.

The excitation of chlorophyll by light starts up a complex series of reactions resulting in the splitting of water thus:

$$H_2O \longrightarrow 2H^+ + \tfrac{1}{2}O_2 + 2e^-$$

with the liberation of molecular oxygen. At the same time the hydrogen ions and electrons are used in the formation of ATP from ADP and in the reduction of NADP to NADPH.

In the *dark reaction* carbon dioxide is fixed; the reaction is illustrated in Figure 11.11. One molecule of CO_2 reacts with a molecule of ribulose 1,5-diphosphate to give two molecules of glyceric acid 3-phosphate. These are then reduced to two molecules of glyceraldehyde 3-phosphate with the aid of two molecules each of ATP and NADPH which have been formed in the light reaction. Of each six molecules of glyceraldehyde 3-phosphate formed in this way one is used for conversion to glucose by what is essentially the reversal of glycosis, while the remaining five rise to three molecules of ribulose 1,5-diphosphate which can react with a new molecule of CO_2.

The final result of the dark and light reactions, therefore, is the fixation of carbon dioxide with the production of molecular oxygen and glucose.

REFERENCES

BALDWIN, E. (1967). *Dynamic Aspects of Biochemistry*, 5th edn. Cambridge: University Press.
GREEN, D. E. & BAUM, H. (1970). *Energy and the Mitochondrion*. New York: Academic Press.
GREENBERG, D. M. (1968–70). *Metabolic Pathways*, 3rd edn, 3 vols. New York: Academic Press.
GRIFFITHS, E. E. (1965). Oxidative phosphorylation. *Essays in Biochemistry* (Edited by P. N. Campbell & G. D. Greville) **1**, 91–120. London: Academic Press.
HILL, R. (1965). The biochemist's green mansions: the photosynthetic electron transport chain in plants. *Essays in Biochemistry* (Edited by P. N. Campbell & G. D. Greville) **1**, 121–152. London: Academic Press.
KREBS, H. A. & KORNBERG, H. L. (1957). *Energy Transformations in Living Matter*. Berlin: Springer.
LEHNINGER, A. L. (1964). *The Mitochondrion*. New York: Benjamin.
LEHNINGER, A. L. (1965). *Bioenergetics*. New York: Benjamin.
SLATER, E. C. (1966). Oxidative phosphorylation. *Comprehensive Biochemistry* (Edited by M. Florkin & E. H. Stotz) **14**, 327–396. Amsterdam: Elsevier.
STETTEN, DEW. & STETTEN, M. R. (1960). Glycogen metabolism. *Physiological Reviews* **40**, 505–534.

12 Energy exchange

The energy required for the maintenance of body temperature, for the muscular movements of the heart and of respiration, for the performance of external work, and for the various synthetic processes which occur in the tissues, for example, the building up of proteins and hormones, is derived ultimately from the oxidation of organic foodstuffs. The potential energy of a foodstuff can be determined by measuring the amount of heat produced when the foodstuff is burned in a *bomb calorimeter* (Fig. 12.1). The 'bomb' is a strong cylindrical steel vessel fitted with a screw top which can resist very high internal pressures. A weighed amount of the foodstuff is placed in the bomb which is then filled with oxygen under pressure. The bomb is placed in a can of water, surrounded in turn by a vessel insulated with cork to reduce heat loss. The foodstuff is ignited electrically and the amount of heat produced by oxidation is calculated from the rise in temperature of the known volume of water surrounding the bomb and the water equivalent of the calorimeter.

The unit of heat used in nutrition was until recently the calorie or kcal. However, this long tradition is coming to an end with the introduction of absolute units (S.I.) and it is now usual to use joules (J) as a measure of heat. One kilocalorie is equal to 4184 J, or 4·184 kJ; 1 MJ (megajoule) $= 10^6$ J. One joule per second is one watt (W).

The energy derived from the complete combustion of the usual mixture of carbohydrates in human foodstuffs has been found to be 17 kJ or 4·1 kcal per g. The value for fat is 39 kJ or 9·3 kcal per g. In the bomb carbohydrates and fats are oxidized to carbon dioxide and water just as they are in the body. Protein is oxidized to carbon dioxide, water, and oxides of nitrogen and other elements, whereas in the body protein is incompletely oxidized, part being excreted as urea which has an appreciable energy value. When the poten-

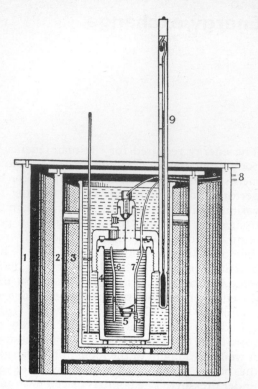

Fig. 12.1 The bomb calorimeter consists of three vessels—(1), (2) and (3). The innermost vessel (3) contains water in which is immersed the bomb (4) containing the crucible (5) with the foodstuff to be burned. The foodstuff is ignited by means of the leads (6) and (7) which pass out of the casing at (8). Combustion of the material in the bomb produces a rise in temperature of the water surrounding it, which is measured by the thermometer (9).

tial energy of the urea excreted is subtracted from the figure for the complete oxidation (23 kJ or 5·4 kcal per g) a value of 17 kJ or 4·1 kcal per g is obtained for protein.

Of more direct use to the physiologist, however, is the oxycalorimeter (Benedict) which operates on the same principle as the bomb calorimeter but it allows measurement of both heat evolved and volume of oxygen used. This gives the calorific value of a litre of oxygen used in oxidizing the particular foodstuff combusted.

METHODS FOR ESTIMATING ENERGY OUTPUT

1. Direct calorimetry. To determine energy expenditure by direct calorimetry the subject is placed in a calorimeter, a small room with heavily insulated walls. The heat generated by the subject is taken up by water pumped through a series of finned pipes which pass through the calorimeter. By multiplying the difference in temperature between the ingoing and outgoing water by the volume of water flowing his heat output can be obtained; the oxygen consumption can be read on a meter. Very few laboratories possess this complicated and very costly apparatus. By its use Atwater showed that energy expenditure balances energy intake in experiments lasting a few days within 1 per cent or so. He also showed that measurements by direct calorimetry agreed well with measurements by indirect calorimetry which is much more convenient and relatively cheap, and very nearly as accurate.

More recently the gradient calorimeter has come into use, particularly for measuring directly the heat output of domestic animals. A system of many thermocouples in series is used to record the temperature difference on a galvanometer and hence to calculate the heat flow across the walls of an insulated box large enough to house the animal. The heat flow Q is given by $Q = k(t_1 - t_2)$ where t_1 and t_2 are the temperatures inside and outside the walls of the calorimeter. The constant k may be found by burning a known quantity of some suitable substance of known heat of combustion, for example ethanol, or by running a low voltage lamp in it for some time.

The shell containing the calorimeter must of course be kept at a constant temperature and arrangements made for the respiratory exchange as well as for the absorption of water vapour given out by the subject or animal.

2. Indirect calorimetry. In indirect methods the heat output of the subject is calculated from his oxygen consumption. Because the relationship between oxygen consumption and heat production depends on the type of food being oxidized, we must briefly consider the *respiratory quotient* (R.Q.) which is by definition the *volume* of carbon dioxide produced divided by the *volume* of oxygen used in the same time.

The oxidation of carbohydrate can be represented by the equation

$$C_6H_{12}O_6 + 6O_2 = 6CO_2 + 6H_2O + \text{heat}$$

The respiratory quotient is 6/6, that is 1.

Oxidation of the fat triolein represented by the equation

$$C_{57}H_{10}O_6 + 80O_2 = 57CO_2 + 52H_2O + heat$$

has an R.Q. of 57/80 or 0·71; the R.Q. of human fat is 0·718. The R.Q. for an average protein is 0·802. A subject on an ordinary mixed diet containing carbohydrate, fat and protein has an R.Q. of the order of 0·85. In other words the R.Q. gives us a clue to the type of food which is being oxidized by the body at any given time if we assume that the carbon dioxide collected and measured is derived solely from the oxygen taken into the body during the same period.

When carbohydrate is oxidized in the body to carbon dioxide and water, 21·1 kJ or 5·047 kcal are produced for each litre of oxygen used; thus the calorific value of oxygen used in the combustion of starch is 21·1 kJ or 5·047 kcal per litre and for protein the value is 18·7 kJ or 4·463 kcal per litre at s.t.p.d. It is therefore clear that for each litre of oxygen used by the body between 18·7 and 21·1 kJ are produced, the actual value depending on the types and proportions of the foodstuffs being metabolized. This information is given by the R.Q. The calorific values of oxygen used by the body in the combustion of mixtures of fat and carbohydrate for different levels of the R.Q. between 0·7 and 1·0 are shown in Fig. 12.2. It is usually

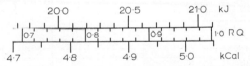

Fig. 12.2 A figure (nomogram) from which may be read off the energy value of a litre of oxygen (at s.t.p.) in kJ and kcal at non-protein R.Q.S. from 0·7 to 1·0.

sufficient to assume that the body is using a mixture of fat and carbohydrate only and to derive the energy value of a litre of oxygen from the non-protein R.Q. An error of less than 1·5 per cent is made by neglecting protein metabolism but the correction for this can easily be made.

The R.Q. can lie outside the limits 0·71–1·0. When carbohydrate is transformed into fat the non-protein R.Q. is greater than 1, since carbon dioxide is produced and oxygen is not used.

$$13C_6H_{12}O_6 \rightarrow$$
$$C_{55}H_{104}O_6 + 23CO_2 + 26H_2O.$$

In one experiment overfeeding of thin men with carbohydrate gave a respiratory quotient as high as 1·3. Conversely when fat is being transformed to carbohydrate in excess of that being oxidized, the non-protein R.Q. is less than 0·71 because oxygen is used up without a corresponding production of carbon dioxide.

(a) CLOSED CIRCUIT METHODS. In the more elaborate type of closed circuit method the subject is kept in a respiration chamber through which a continuous circulation of air is maintained by a rotary blower. Oxygen is added through a meter to make good the oxygen consumed in respiration; the water in the expired air is caught in a sulphuric acid trap in the ventilation circuit; the carbon dioxide is caught in a soda-lime trap and is estimated by weighing or by a chemical method. The chamber is airtight but because it is not insulated the heat production cannot be obtained directly. The subject remains within the chamber for a period of several hours; since his oxygen consumption and car-

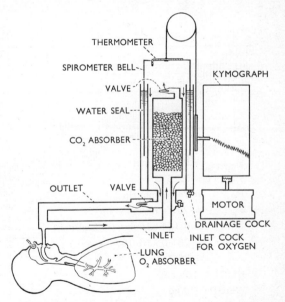

Fig. 12.3 Diagram of recording spirometer. The subject wears a nose-clip and breathes through a mouthpiece which is connected to the apparatus by two tubes. He breathes in oxygen through the inspiratory valve and breathes out into the carbon dioxide absorber and then, through the expiratory valve, into the spirometer bell. A pump may be introduced in place of the valves shown in the diagram to ease the circulation of air through the carbon dioxide absorber. The amount of oxygen used is recorded on the revolving drum by the pen attached to the counterweight. (Diagram by J. B. de V. Weir.)

bon dioxide output are measured his R.Q. can be calculated. Protein metabolism can be taken into account by measuring the nitrogen content of the urine excreted during the time of the experiment.

Much simpler methods have been evolved for clinical work. The Benedict-Roth spirometer (Fig. 12.3) consists of a gasholder (spirometer bell) which is filled with oxygen. As the oxygen is used up by the subject the level of the gasholder falls and from the record of its movements on the rotating drum the amount of oxygen consumed in a given period can be measured. Because the carbon dioxide is not measured, the R.Q. cannot be calculated but an arbitrary value of 0·8 is usually assumed. At this R.Q. the calorific value of oxygen is 20·2 kJ (4·825 kcal) per litre, and by multiplying this value by the number of litres of oxygen consumed the energy output during the experimental period is obtained with an error of no clinical importance. A typical record with the accompanying calculations is shown in Fig. 12.4.

(b) OPEN CIRCUIT METHOD. In the open circuit method the subject's nose is closed by a clip and he breathes through a mouthpiece fitted with valves so that he inspires atmospheric air and expires through a wide tube into a rubberized canvas or polythene bag (Douglas bag) of capacity 100 or 200 litres (Fig. 12.5). The subject breathes into the bag for a measured time and the total volume of the air in the bag is subsequently measured with a gas-meter. A sample is taken for analysis in a gas-analysis apparatus in which its carbon dioxide and oxygen content are determined by absorption in turn by potassium hydroxide and alkaline pyrogallol. The volume of oxygen used, the volume of carbon dioxide produced and the respiratory quotient are calculated. The energy value of oxygen, obtained from Fig. 12.2, multiplied by the volume of oxygen used gives the heat produced during the period of the experiment.

The calculation of the metabolic rate in the Douglas bag method becomes very simple and equally accurate if, instead of making the cal-

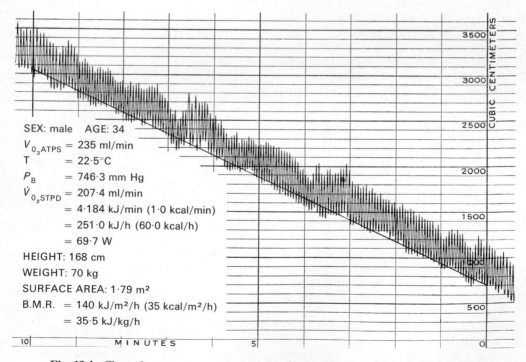

SEX: male   AGE: 34

$V_{O_2 ATPS}$ = 235 ml/min

T   = 22·5°C

P_B = 746·3 mm Hg

$\dot{V}_{O_2 STPD}$ = 207·4 ml/min

    = 4·184 kJ/min (1·0 kcal/min)

    = 251·0 kJ/h (60·0 kcal/h)

    = 69·7 W

HEIGHT: 168 cm

WEIGHT: 70 kg

SURFACE AREA: 1·79 m²

B.M.R. = 140 kJ/m²/h (35 kcal/m²/h)

    = 35·5 kJ/kg/h

Fig. 12.4 Chart of oxygen consumption and data for the estimation of B.M.R. V_{O_2} = rate of oxygen consumption; ATPS = atmospheric temperature and pressure saturated with water vapour; STPD = standard temperature and pressure, dry; P_B = barometric pressure. The calculations are explained in the text. Note that as the spirometer bell falls the writing lever rises; the zero time on this chart is on the right and the graph must be read from right to left.

Fig. 12.5 The subject is wearing a 100 l capacity Douglas bag which is partly filled with expired air. The cylindrical box attached to the mouth-piece contains valves so arranged that the subject breathes atmospheric air in and breathes out through the corrugated tubing into the bag.

culations described above, the volume of expired air is multiplied by the calorific value corresponding to 1 litre of expired air. It is necessary to find only (1) the volume of the expired air and (2) the percentage of oxygen in the expired air. For the usual range of protein intake, that is where protein contributes 10 to 15 per cent of the total energy, Weir has shown that the energy value of a litre of expired air is $0·209(O_i - O_e)$ kJ where O_i is the percentage of oxygen in the inspired air (normally 20·93) and O_e is the percentage of oxygen in the expired air. The following example illustrates the method.

The subject of an experiment breathed into a Douglas bag for 10 min and the volume of expired air measured by a gas-meter was 80·25 litres at 16·8°C and 750 mm Hg atmospheric pressure, equivalent to 73·19 litres at s.t.p.d. A sample of this air contained 16·86 per cent oxygen. From Weir's formula we find the energy value of a litre of this expired air was 0·209 (20·93 − 16·86) or 0·851 kJ. The energy output was therefore 73·19 × 0·851 = 62·4 kJ in 10 min or 373·4 kJ per h. The subject weighed 50 kg and his height was 163 cm; his surface area obtained from the formula on

p. 166 was 1·52 m² and his metabolic rate was therefore 373·4 ÷ 1·52 = 246·4 kJ per m² per h = 68·4 W/m².

Because of its limited capacity the Douglas bag can be used for only 2 or 3 min if the subject is carrying out strenuous activity. It is sometimes more convenient to use the respirometer devised in 1940 at the Max-Plank Institute at Dortmund for the study of the physiology of work. This apparatus, weighing less than 4 kg, is really a small accurately-built gas meter. It measures directly the volume of expired air and at the same time diverts a small fraction (0·3 or 0·6 per cent) of the air into a rubber bag for subsequent analysis. The energy expenditure can be calculated from Weir's formula. An alternative to this apparatus is the integrating motor pneumotachograph, usually known as IMP, devised by Wolff, which also meters the expired air and extracts a predetermined proportion for later analysis.

In one individual a fairly close correlation exists between simultaneous measurements of oxygen consumption and heart rate at different exercise levels. Thus a small heart-rate counting apparatus (SAMI, socially acceptable monitoring instrument) inconspicuously attached to the subject may be used to give sufficiently accurate estimates of his energy expenditure throughout the day including normal working hours. Each electrocardiographic deflexion (Chap. 26) is converted to a standard sized pulse which is applied to an electrochemical integrator.

So many figures for the energy expenditure during different kinds of activity are now tabulated in the literature (see references at the end of this chapter) that reasonably accurate estimates (as distinct from laborious measurements) of the energy expended in different jobs can be made by an observer equipped with stop watch, paper and pencil, recording the duration of each kind of activity—this is the diary method discussed in more detail on pages 169 and 170.

BASAL METABOLIC RATE

The output of energy and the rate of oxygen consumption depend on many factors including body size, muscular activity, the nature and amount of food eaten, changes in environmental and body temperature, thyroid activity and emotional excitement. To reduce the influence of these factors to a minimum and allow valid comparison of the metabolism of one individual with that of another, it is highly desirable to measure the metabolism under basal conditions. It is well known, however, that it is difficult, especially for an untrained subject, to achieve the desired mental and physical relaxation. The oxygen consumption is measured in the morning while the subject is lying warm and comfortable in bed completely relaxed, 12 to 15 h after the last meal. The expenditure of energy under these conditions is called the *basal metabolism*. A certain amount of energy must continually be produced to maintain essential processes such as the beating of the heart and breathing and to maintain the body temperature at about 37°C.

The heat production of animals of various sizes (Table 12.6) is closely correlated with the surface area of the body and not with body weight. Smaller animals have a greater surface area per unit of weight than larger ones and therefore a relatively greater surface from which heat can be lost. These findings led to the convention established about 60 years ago by which the basal metabolism of large and small individuals were 'standardized' by referring them to the surface area as *basal metabolic rate* (B.M.R.). It is impossible to justify this convention (especially in the case of a clothed man) since the metabolism goes on about the same rate even when little heat is required to keep the body temperature at 37°C. It is simpler and quite adequate to refer basal metabolism to body weight (see Fig. 12.6).

The daily fasting heat production of adult mammals, including animals of such diverse size as mice and elephants, is on the average 293 (body wt. in kg)$^{\frac{3}{4}}$ kJ or 70 (body wt. in kg)$^{\frac{3}{4}}$ kcal. No satisfactory explanation of this relationship has yet been found; it is particularly puzzling to find that it applies equally well to homeotherms and poikilotherms.

The surface area of the body was determined directly by Du Bois by clothing his subjects in thin, wax-impregnated garments which were subsequently removed, cut up and measured. The formula giving the best fit to his data is:

$$S = W^{2.425} H^{0.725} \times 0.007184,$$

where S is the surface area in m^2, W is the nude weight in kg, H is the height in cm. The value of S can be obtained from a nomogram (see references, p. 175, B. L. Andrew (1972)). Because of the scatter of the data there is considerable likelihood that the value obtained from the formula in any one person deviates seriously from the true value and that the error in estimation of surface area is greater than that in estimating oxygen consumption. The surface area of an adult man is of the order of $1.8 \, m^2$.

The *basal metabolic rate* (B.M.R.) is usually expressed as a percentage of the standard value of a subject of the same sex and age. Variations within the range from 85 to 115 per cent or, as it is usually expressed, minus 15 to plus 15 per cent are accepted as being within normal limits. Occasionally healthy persons deviate more than 20 per cent above or below the standard. The test has therefore little diagnostic value and is seldom used clinically. Figure 12.7 shows the variations in B.M.R. and in total energy requirements with age. For a man the average B.M.R. is about 167 kJ/m^2/h (46W/m^2 or 40 kcal/m^2/h) and for a woman about 150 kJ/m^2/h (42 W/m^2 or 36 kcal/m^2/h).

Table 12.6 Energy expenditure by different species

	Weight (kg)	kcal/day		kJ/day	
		Per kg	Per m^2 surface	Per kg	Per m^2 surface
Pig	128·0	19·1	1078	80	4510
Man	64·3	32·1	1042	134	4360
Dog	15·2	51·5	1039	216	4347
Goose	3·5	66·7	967	279	4046
Fowl	2·0	71·0	943	297	3946
Mouse	0·018	654·0	1188	2736	4971

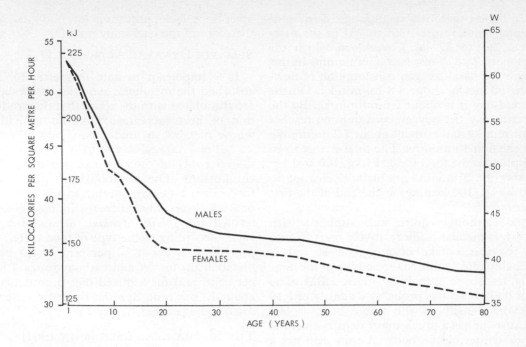

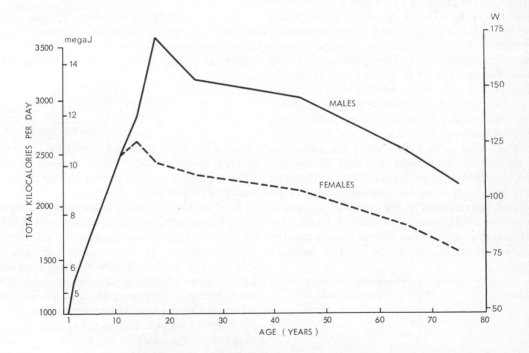

Fig. 12.7 The upper graph shows the average basal metabolic rate in kJ or kcal/m²/h from one year onwards. The lower figure shows the total daily requirement in MJ and kcal from 1 year onwards. The lower values for women are almost certainly due to the greater fat content of the female body (see p. 000). The data for construction of these graphs are contained in A. Fleisch (1951). *Helvetica medica Acta* **18**, 23–44 and S. Davidson & R. Passmore (1966). *Human Nutrition and Dietetics,* 3rd edn. Edinburgh: Livingstone.

In babies there is a small range of environmental temperature—36° to 38° in the newborn, 32° to 37° at 1 week—in which the metabolic rate at rest is at a minimum. In this range the basal oxygen consumption of newborn infants is about 4·8 ml/min/kg. On the second day it is about 6·6 ml/min/kg. By the seventh day the oxygen consumption reaches 7 ml/min/kg and remains about 7·2 ml/min/kg for the first 18 months. The total energy to be supplied in the food is 500 kJ/kg (120 kcal/kg) in the first 3 months gradually decreasing to 420 kJ/kg (100 kcal/kg) at the end of the first year.

Fat, although quite well supplied with blood, is metabolically relatively inert. Hence the oxygen uptake of a person at rest is probably mostly due to the activity of the non-fatty tissues or lean body mass, LBM. The lean body mass has probably a density of 1·10 but the density of the whole body is less because fat has a much lower density than the other tissues of the body. A thin man has a density of the order of 1·075 and a plump man about 1·046. Density is measured by weighing the subject in air and then, after maximum expiration, by weighing him in water during complete submersion, due allowance being made for the residual volume of air in the lungs (Chap. 30). From the figure for density the LBM can be calculated. Passmore measured the resting oxygen consumption of 49 medical students (Table 12.8) and found that although there was a significant sex difference in oxygen consumption referred to surface area, the difference in oxygen consumption per kg of lean body mass was not significant. This method is unsuitable for routine measurements but subcutaneous tissue is largely fat and many workers have shown that measurement of skinfold thickness by suitable callipers

Table 12.8

Students	Resting oxygen consumption	
	ml per m² per min	ml per kg lean body mass per min
24 men	141	4·22
25 women	115	4·31

(*From* Margaret G. MacMillan, Catriona M. Reid, D. Stirling & R. Passmore (1965). *Lancet* **i**, 728–729.)

gives a reliable indication of body fat and therefore of the lean body mass.

SPECIFIC DYNAMIC ACTION (S.D.A.)

It is important to note that after food is absorbed the metabolic activities of the body are stimulated with the result that the production of heat increases. Thus an individual whose metabolism under basal conditions is 7·1 MJ or 1700 kcal/day actually requires more than 7·1 MJ of food to maintain himself in equilibrium. The metabolism of 7·1 MJ of food raises his energy output to about 7·9 MJ or 1900 kcal. This increase in metabolism, termed the *specific dynamic action* of food, varies according to the type of food taken; for proteins it is about 30 per cent of the basal metabolism, for fat and carbohydrates 4 to 5 per cent, and for a mixed diet it amounts to about 10 per cent. Specific dynamic action is discussed again in Chapter 13.

THE MECHANICAL EFFICIENCY OF THE HUMAN BODY

The efficiency of a steam engine, or an internal combustion engine, is given by the ratio of the useful work done to the total potential energy (P.E.) of the fuel used. Thus

$$\text{efficiency} = \frac{\text{useful work}}{\text{P.E. of fuel}}$$
$$= \frac{\text{useful work}}{\text{useful work} + \text{heat}}.$$

The difference between the total potential energy and the useful work appears as heat. The efficiency of the petrol engine is of the order of 25 per cent.

In man the situation is somewhat different since his metabolism must continue even when he is at complete rest and when no external work is done. In these circumstances the efficiency is obviously zero. Values approaching the true efficiency can be obtained if the work brings a very large mass of muscles into action. In such experiments the subject pedals a stationary bicycle connected to an ergometer to measure the external work; the oxygen consumption is measured at the same time to get a value for potential energy of the fuel. The gross efficiency in such experiments is of the order of 10 to 20 per cent when care is taken to avoid, or correct for, an oxygen debt (Chap. 40).

The Energy Requirement of Man

Food provides energy for three main purposes. First it is required to maintain basal bodily activities such as heart beat, breathing, muscle tone and body temperature. For a man with a basal metabolic rate of 167 kJ/m^2/h or 40 kcal/m^2/h, and a surface area of 1·8 m^2, the energy output for basal purposes is 318 kJ/h or 7·7 MJ per day (or 1728 kcal per day). If 10 per cent is allowed for specific dynamic action his daily requirements to keep him in equilibrium are 8 MJ.

Secondly, additional food is needed to cover the expenditure of energy in the simple activities of everyday life, such as sitting, standing, walking, dressing and undressing, and for this purpose an allowance of 1·5 MJ may be made for the 8 hours of the day that are occupied by these simple activities.

Thirdly, an allowance, depending on the nature of the occupation, has to be made for work carried out during the day. People engaged in sedentary occupations expend 84 to 188 kJ/h at their work, whereas men engaged in hard muscular work may expend as much as 1·6 MJ/h over their basal requirements.

The results of such simple calculations have often been shown by detailed investigations to be grossly in error for three main reasons. First, the energy expenditure varies considerably from minute to minute in any particular job—few people work steadily at a given rate for an hour. Secondly, even within one sort of job, say domestic work or light engineering, the work varies from sedentary up to heavy. Thirdly, very high expenditure cannot be kept up for very long. The average man doing physical work for 8 h a day is not likely to show evidence of fatigue if the intensity of the work and the length of the rest pauses are adjusted to give an average total metabolism of less than 21 kJ/min. This is equivalent to walking on the level at 3·8 miles per hour (6 km/h). If he works twice as hard, he must have rest pauses equal to the actual working time. If we allow 2·1 MJ for sleep and 5·9 MJ as energy expenditure off duty, the total energy expenditure (including 10 MJ for 8 h of work) for this upper limit activity is of the order of 18 MJ/day (4300 kcal/day or 208 W). This is much less than was at one time given as the daily requirements for blacksmiths or lumbermen but even these strong men must have rest pauses which bring down their average requirements in spite of their high peak performance. Detailed observation of officer cadets in strenuous training with a great deal of drill and games has shown that they spend much of their time actually sitting down.

The two tables, Tables 12.9 and 12.10, show the results of investigations of energy expenditure in a sedentary worker and in a heavy manual worker. To assess the requirements accurately, detailed minute to minute diaries of the activities of men were kept for a complete week because omission of week-end activities gives an entirely false impression of the average needs. About one-third of the day is spent sleeping. The amount of muscular movement in sleep varies from person to person, but it is considerable and no one sleeps like the proverbial log. On the average the energy requirement is at the basal level—for an adult man 5·4 kJ/min. The performance of personal necessities, such as dressing and undressing, requires in different people 10 to 17 kJ/min. Walking on the level for an adult requires at 2 miles per hour (3 km/h) 12 kJ or 2·9 kcal/min, at 4 miles per hour (6·5 km/h) 24 kJ or 5·7 kcal/min on the average. Walking up an incline of 15 per cent at 2 miles per hour (3 km/h) requires 25 kJ or 6 kcal/min. Mental 'work', for example adding or multiplying figures, has no significant effect on metabolism. Light indoor recreations may require up to 17 kJ or 4 kcal/min while hard exercise may involve 84 kJ or 20 kcal/min. Domestic work for women varies between 8 and 33 kJ/min (2 to 8 kcal/min). Typing by women requires about 5·9 kJ or 1·4 kcal/min. The jobs in any one industry, as already stated, are quite variable and are best investigated by individual determinations of metabolic rate. Christensen's gradings of severity for different forms of work are generally accepted:

Unduly heavy	over 2·5 litres oxygen =	850 W
Very heavy	over 2·0 litres oxygen =	680 W
Heavy	over 1·5 litres oxygen =	510 W
Moderate	over 1·0 litres oxygen =	340 W
Light	over 0·5 litres oxygen =	170 W

His observations on Swedish ironworkers led him to point out that in a hot environment grading by oxygen uptake does not measure the real stress on the subject. Pulse rate then

Table 12.9 Energy output and intake of a clerk over 1 week (Ian C., age 29, ht., 165 cm, wt., 66 kg, occupation, clerk)

Activity	Total time spent hr min		kcal/min	Total kcal	kJ/min	Total MJ
In bed	54	4	1·13	3670	4·72	15·4
Daytime dozing	1	43	1·37	140	5·73	0·6
Recreational and off work :						
Light sedentary activities	31	14	1·48	2810	6·19	11·8
Washing, shaving, dressing	3	18	3·0	590	12·5	2·5
Light domestic work	7	14	3·0	1300	12·5	5·4
Walking	8	35	6·6	3400	27·6	14·2
Gardening	2	48	4·8	810	20·1	3·4
Standing activities	6	45	1·56	630	6·5	2·6
Watching football	2	10	2·0	260	8·4	1·1
Total recreational and off work	62	32	..	9800		41·0
Working :						
Sitting activities	22	22	1·65	2210	6·90	9·2
Standing activities	25	27	1·90	2960	7·9	12·4
Walking	1	22	6·6	540	27·6	2·3
Total working	49	41	..	5710		23·9
Grand total	168		..	19320		80·8
Daily average	24		..	2760		11·5
Food intake (daily av. determined by diet survey)	..	..	..	2620		11·0

(*From* R. C. Garry, R. Passmore, G. M. Warnock & J. V. A. Durnin (1955). Studies on expenditure of energy and consumption of food by miners and clerks. *Special Report Series, Medical Research Council, London* No. 289. London: H.M.S.O.)

Table 12.10 Energy output and intake of a coal miner over 1 week (John H., age 32, ht., 175 cm, wt., 67 kg, occupation, stripper)

Activity	Total time spent hr min		kcal/min	Total kcal	kJ/min	Total MJ
In bed	58	30	1·05	3690	4·39	15·4
Recreational and off work :						
Light sedentary activities	38	37	1·59	3680	6·65	15·4
Washing, shaving, dressing	5	3	3·3	1000	13·8	4·2
Walking	15		4·9	4410	20·5	18·5
Standing	2	16	1·8	250	7·5	1·0
Cycling	2	25	6·6	960	27·6	4·0
Gardening	2		5·0	600	20·9	2·5
Total recreational and off work	65	21	..	10900		45·6
Working :						
Loading	12	6	6·3	4570	26·4	19·1
Hewing	1	14	6·7	500	28·0	2·1
Timbering	6	51	5·7	2340	23·8	9·8
Walking	6	43	6·7	2700	28·0	11·3
Standing	2	6	1·8	230	7·5	1·0
Sitting	15	9	1·68	1530	7·0	6·4
Total working	44	9	..	11870		50·0
Grand total	168		..	26460		110·7
Daily average	24		..	3780		15·8
Food intake (daily av. determined by diet survey)	..	..	..	3990		16·7

(*From* R. C. Garry, R. Passmore, G. M. Warnock & J. V. A. Durnin (1955). Studies of expenditure of energy and consumption of food by miners and clerks. *Special Report Series, Medical Research Council, London* No. 289. London: H.M.S.O.)

becomes a better indicator and the above gradings would correspond to pulse rates of: over 175, 150 to 175, 120 to 150, 100 to 125 and 75 to 100 respectively.

For children the energy requirements relative to the body weight are high, since children expend much energy in everyday activities, and in addition require food for growth. Thus a girl of 14 requires as much food as does a woman engaged in active physical work and a boy of 16 requires more food than does his father, unless the latter is engaged in severe physical work.

The daily energy requirements could be apportioned in a wide variety of ways between carbohydrates, fats and proteins. Although it is theoretically possible to replace a given amount of one foodstuff by an isocaloric amount of another, such a process is limited by the considerations mentioned in Chapter 13. They might for example be distributed as follows: 400 g carbohydrate, 100 g fat and 100 g protein. This would provide 12 MJ or 2900 kcal. Most diets contain a large proportion of carbohydrate because of the relative abundance and cheapness of this component.

Tables 12.9 and 12.10 show that the potential energy of the food matches quite closely the energy requirements measured by adding together the various amounts of work done in the week. Since body weight in adult life remains relatively constant from year to year we are forced to conclude that the body possesses some mechanism which normally adjusts the intake of food so that it balances the energy output (see also Chap. 13). We must therefore enquire into the nature of appetite and its control.

HUNGER AND APPETITE

Hunger is usually taken to mean the sensation of emptiness resulting from abstinence from food, an unconditional phenomenon (Chap. 49). On the other hand appetite is thought of as the desire for food or the pleasure in taking food which is conditional, or learned. Thus hunger may be thought of as a physical or physiological phenomenon and appetite as a psychological one.

Cannon in 1911 found that the sensation of hunger was often associated with gastric contractions and later Carlson and others tried to show that the sensation arose from the stomach.

However, after complete vagotomy the sensation of hunger is modified but does not disappear and even after surgical removal of large portions of the stomach patients feel hungry much as other people do. Insulin in sufficient dose to produce a marked hypoglycaemia gives rise to a sensation of hunger even after double vagotomy or removal of the sympathetic supply of the stomach.

It has often been claimed that hunger decreases and even disappears during starvation but a recent experiment suggests that this is not true. A group of obese subjects was put on a 4·2 MJ (1000 kcal) diet for a few days and then starved for 14 days. The sensation of hunger increased during the first day or two of starvation and then returned to the level experienced on the initial low calorie diet. If ketosis occurs during starvation the appetite is depressed.

Experiments on rats, cats and monkeys have shown that food intake is controlled by the hypothalamus; bilateral damage to the ventromedial nuclei (Fig. 47.1) of the tuber cinereum of the hypothalamus causes obesity but bilateral lesions placed more laterally abolish hunger and the animals die of starvation. The medial nuclei, or satiety centres may act as brakes on the more laterally placed 'feeding centres'. The satiety centre may be activated by impulses arising from chewing or from distension of the stomach; experiments show, however, that these have only a small influence on food intake. The increased metabolism due to S.D.A. (Chap. 13) occurring during a meal may cause a small rise in body temperature but not sufficient to stimulate the hypothalamic centres. Passmore and Ritchie carried out experiments to see if S.D.A. itself could be connected with satiety. Their subjects were given milk powder enriched with protein (2·8 MJ and 46 g protein) which they consumed in about 3 min. The expired air was collected in 5-min periods and the energy expenditure calculated (Fig. 12.11). The expenditure increased immediately after taking the milk powder and reached a maximum in about half an hour. The feeling of satiety also arose immediately after eating the enriched milk provided that the subjects were previously stuffed (that is grossly overfed). Furthermore the S.D.A. was greater after 'stuffing' than after starvation. How S.D.A. could affect the

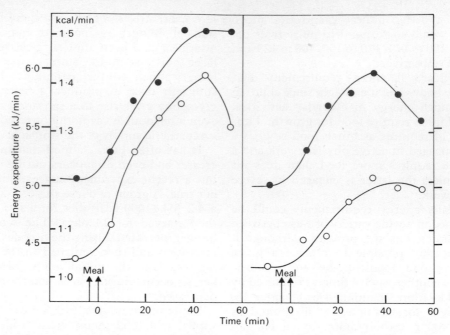

Fig. 12.11 Changes in energy expenditure in two subjects after a standard meal. The rise in the curves shows the specific dynamic effect of the meal. ○, when the subjects were previously starved and the meal did not satisfy appetite; ●, when the subjects had previously overfed and were satiated. (R. Passmore & F. J. Ritchie (1957). *British Journal of Nutrition* **11**, 79–85.)

hypothalamic centres is not known. It is worth noting that the S.D.A. is increased by physical activity; in other words the waste of calories is greater when the individual is active than when he is at rest. The value of exercise in the treatment of obesity may be partly explained in this way.

Hyperphagia occurs in man and experimental animals when the cerebrospinal fluid pressure in the third ventricle is increased; this effect is probably due to compression of the ventromedial nuclei. Pentobarbitone infused into the third ventricle also causes overeating.

Since the body's store of carbohydrate is small (Chap. 18) it is reasonable to ask if the cause of hunger is to be found in carbohydrate metabolism and in particular in the level of blood sugar. Certainly a sensation of hunger is present when the blood sugar is markedly lowered by the administration of insulin but in hungry, healthy, fasting persons the blood sugar is always in the normal range of 70 to 90 mg/100 ml blood. Hunger does not seem, therefore, to be related to the absolute level of blood glucose. Jean Mayer thinks that hunger

is due to diminished utilization of glucose in peripheral tissues as shown by a low arteriovenous (A-V) glucose difference (the glucostatic theory). In untreated diabetes blood glucose is not readily transferred across cell membranes and therefore, although the absolute levels in venous and arterial blood are very high, the A-V difference is small and the patient is hungry; this is in marked contrast to the high A-V glucose difference observed immediately after a meal which dispels the hunger sensation (Chap. 18). Support for the theory is given by the effects of insulin. An injection of insulin produces a large A-V difference at first while the cells take up glucose but after the uptake stage the A-V difference is small and now the subject feels hungry. The glucostatic theory is applicable only to short term regulation of food intake. A clue to the problem of the cause of the sensation of hunger may be provided by the observation that in fasting men the sight of food causes a fall in the free fatty acids of the blood. The composition of the diet seems to affect the level of satiety. According to Yudkin, an obese person who is made to reduce his carbohydrate intake with-

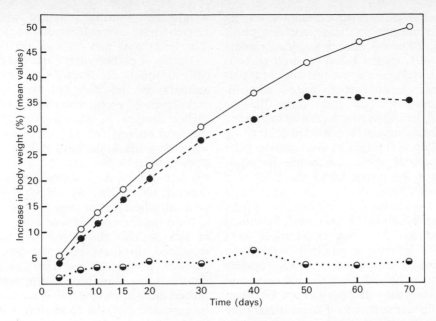

Fig. 12.12 Growth curve ○ — ○ of adult rats injected with growth hormone and given unlimited access to food. Growth curve ○ . . . ○ of control rats and growth curve ● - - - - ● of hormone-treated rats given the same quantity of food as the control group. (A. L. Greenbaum (1953). *Biochemical Journal* **54**.

out any restriction on protein and fat consumption may greatly reduce his total energy intake without feeling hungry. If glucose is added to a 'reducing' diet, the total energy intake may be disproportionately increased as though glucose stimulated the appetite.

The balance between body needs and hunger has long been suspected to be under hormonal influence since, to quote one example, eunuchs tend to become rather fat. A clue to another hormonal influence may be furnished by the experiments of Greenbaum (Fig. 12.12). Adult rats injected with growth hormone (Chap. 52) while obtaining access to unlimited supplies of food increased their weight by 50 per cent in 70 days. A second group of rats also injected with growth hormone, but restricted to the dietary intake of a third control uninjected group, increased their weight by 28 per cent in 30 days much as did the first group; after 30 days the gain in weight of the second group slowed up. In other words, the animals injected with growth hormone made a different use of the same quantity of food; they converted a certain amount of it to body tissue whereas the uninjected controls (the third group) oxidized

more and did not gain in weight. Rats treated with progesterone also gain in weight.

It is often said that exercise increases appetite and this must be true since the body weight of active people and animals usually remains constant. Various surveys of healthy persons with normal body weight in the range 45 to 80 kg show that heavier individuals have only a slightly greater energy intake than lighter ones. This is a rather surprising finding since undoubtedly extra body weight requires greater energy expenditure at rest or during exercise. It seems that heavy people are less active than lighter ones; it also points to the importance of exercise in keeping body weight within normal limits.

Some experiments on animals seem to show that food intake is increased only when exercise exceeds a certain amount. It is difficult to prove that this applies to man. Mayer has shown that when adult men reduce their activities below a certain level the food intake is not reduced and body weight rises. Obese persons often claim that they eat very little and this has been verified by observation of obese adolescents; at the same time it has been found that such adolescents expend much less energy during exercise periods than their non-obese colleagues. Glucose disappears less rapidly than normal from the blood of such persons after an intravenous glucose test and it may be that the utilization of carbohydrate is delayed.

To account for the long term balance between intake and output Kennedy suggested that the cells in the hypothalamic feeding centres (Chap. 47) might be able to detect the concentration of some metabolite in equilibrium with the stored fat of the body and Hervey (1959) supported this idea as the best explanation of the results of his experiments on pairs of rats which had been surgically joined side by side so that their circulations were anastomosed (parabiotic pairs). If one of the partners had lesions placed in the appropriate position in its hypothalamus it over-ate and became obese; the other partner ate less than before and became thin as if some metabolite had passed over in the circulation from the obese partner to affect the intact hypothalamus of the thin partner. If later this rat had similar lesions placed in its hypothalamus it too became obese.

Information obtained by life insurance companies shows that body mass in many adults remains constant (within a kilogram or so) for many years in spite of variations in energy intake and energy output. This long term balance between energy intake and body weight must be capable of explanation without violating the law of conservation of energy which states that the potential energy of the food ingested can be completely accounted for by work done, heat lost and the potential energy of materials lost in urine, faeces and skin. Kekwick has pointed out, however, that this does not mean that the metabolic fate of ingested food is always the same in all individuals or in any one individual at all times. For example, a group of rats placed on a high-fat diet deficient in calories lost more body weight than a similar group of rats on a high-carbohydrate diet equally deficient in calories. The latter obtained most of their energy by oxidation of carbohydrate and did not need to call so much on their body stores as the animals on the high-fat diet. The 'high-carbohydrate' group excreted more carbon as carbon dioxide; in other words the fraction of the food broken down to carbon dioxide and water depends on the kind of food eaten. It has already been shown (pp. 162 and 163) that the R.Q. depends on the composition of the diet; it depends also on the way oxygen is used. It may be used, when the carbohydrate intake is high, to form water by combining with the hydrogen of fats so that the depot fat becomes desaturated. In these circumstances a certain amount (perhaps 10 per cent) of the oxygen absorbed is not combined with carbon to form carbon dioxide and the R.Q. is lower than would be expected if all the foodstuffs were metabolized to carbon dioxide and water. The use made of ingested food depends also on the hormonal state. Increase of circulating thyroid hormone decreases body weight; increase of cortisone is liable to increase body weight by increasing the body fat. The effect of growth hormone has already been mentioned. The intact animal seems to be able to alter its metabolic pathways to balance energy needs and body mass.

A few fortunate healthy persons remain thin in spite of a relatively high food intake. In one experiment thin young men who consumed low-protein diets in excess put on much less weight than obese persons. Perhaps the thin men retained less water than those inclined to be obese or it may be that their metabolic processes were quantitatively or qualitatively different as suggested above. The fundamental explanation of the way in which the normal individual regulates his calorie intake to balance his output still eludes us.

REFERENCES

ANDREW, B. L. (1972). *Experimental Physiology*, 9th edn. Edinburgh: Livingstone.

BEATON, G. H. (1964–66). *Nutrition*, Vols. 1–3. London: Academic Press.

BENEDICT, F. G. & CATHCART, E. P. (1913). *Muscular Work*. Carnegie Institute Publications No. 187.

BENZINGER, T. H., HUEBSCHER, R. G., MINARD, D. & KITZINGER, C. (1958). *Journal of Applied Physiology*, Suppl. No. 1 in **12**, Pt. 2.

CHRISTENSEN, E. H. (1953). *Ergonomics Research Society Symposium 'Fatigue'* (Edited by Floyd and Welford), Chap. 10, p. 93. London: Lewis.

CONSOLAZIO, C. F., JOHNSON, R. E. & PECORA, L. J. (1963). *Physiological Measurements of Metabolic Function in Man*. London: McGraw-Hill.

DAVIDSON, S. & PASSMORE, R. (1972). *Human Nutrition and Dietetics*, 7th edn. Edinburgh: Livingstone.

VON DOBELN, W. (1956). Human standard and maximal metabolic rate in relation to fat-free body mass. *Acta physiologica scandinavica* **37** Suppl. 126, 1–79.

DU BOIS, E. F. (1927). *Basal Metabolism in Health and Disease*. London: Bailliere, Tindal & Cox.

DURNIN, J. V. G. A. & PASSMORE, R. (1967). *Energy, Work and Leisure*. London: Heinemann.

DURNIN, J. V. G. A. & RAHAMAN, M. H. (1967). The assessment of the amount of fat in the human body from measurements of skinfold thickness. *British Journal of Nutrition* **21**, 681–689.

HERVEY, G. R. (1959). The effects of lesions in the hypothalamus in parabiotic rats. *Journal of Physiology* **145**, 336–352.

HILL, JUNE R. & RAHIMTULLA, KULSUM A. (1965). Heat balance and the metabolic rate of newborn babies in relation to environmental temperature: and the effect of age and of weight on basal metabolic rate. *Journal of Physiology* **180**, 239–265.

HOLLANDER, F. (Ed.) (1955). The regulation of hunger and appetite. *Annals of the New York Academy of Sciences* **63**, 1–144.

KEKWICK, A. & PAWAN, G. L. S. (1969). Body-weight, food and energy. *Lancet* i, 822–825.

KENNEDY, G. C. (1966). Food intake, energy balance and growth. *British Medical Bulletin* **22**, 216–220.

KEYS, A. & BROZEK, J. (1953). Body fat in adult man. *Physiological Reviews* **33**, 245–325.

KLEIBER, M. (1947). Body size and metabolic rate. *Physiological Reviews* **27**, 511–541.

LEHMANN, G., MULLER, E. A. & SPITZER, H. (1949–50). Der Calorienbedarf bei gewerblicher Arbeit. *Arbeitsphysiologie* **14**, 166–235.

McHARDY, G. J. R., SHIRLING, D. & PASSMORE, R. (1967). *Basic Techniques in Human Metabolism and Respiration*. Oxford: Blackwell Scientific Publications.

MACWORTH, N. H. (1950). Researches on the measurement of human performance. *Special Report Series, Medical Research Council, London*, No. 268. London: H.M.S.O.

PASSMORE, R. & DURNIN, J. V. G. A. (1966). *Human Energy Expenditure*. London: Heinemann.

ROBERTSON, J. D. & REID, D. D. (1952). Standards for the basal metabolism of normal people in Britain. *Lancet*, i, 940–943.

MAYER, JEAN (1968). *Overweight, Causes, Cost and Control*. New Jersey: Pentice-Hall.

SLOAN, A. W. & WEIR, J. B. DE V. (1970). Nomograms for prediction of body density and total body fat from skinfold measurements. *Journal of Applied Physiology* **28**, 221–222.

Symposium of nutrition balance techniques and their limitations (1967). *Proceedings of the Nutrition Society* **26**, 86–124.

Symposium on energy expenditure in man (1956). *Proceedings of the Nutrition Society* **15**, 72–99.

Symposium on obesity (1957). *Metabolism* **6**, 404–465.

TATA, J. R. (1964). Basal metabolic rate and thyroid hormones. *Advances in Metabolic Disorders* **1**, 153–189.

WHIPPLE, H. E. (Ed.) (1965). Adipose tissue: metabolism and obesity. *Annals of the New York Academy of Sciences* **131**, 1–683.

WIDDOWSON, E. M. (1965). Assessment of the energy value of human foods. *Proceedings of the Nutrition Society* **14**, 142–154.

YUDKIN, J. (1959). The causes and cure of obesity. *Lancet* ii, 1135–1138.

13 Food, nutrition and vitamins

The food we eat is composed of animal and plant tissues or products derived with them. This chapter is concerned with the properties of the principal foodstuffs, then with vitamins and lastly with the composition of a normal diet.

THE COMPOSITION OF FOODS

Milk. The human infant starts life on a diet consisting solely of milk (Tables 13.1 and 13.2). The chief proteins of milk are casein and the whey proteins, α-lactalbumin and β-lacto-globulin. Casein is a phosphoprotein present in milk as calcium caseinate. When acid is

Table 13.1 Comparison of human milk with cow's milk

	Human g per 100 g	Cow's g per 100 g
Water	88	88
Protein	1·2	3·3
Fat	3·7	3·5
Lactose	6·8	4·6
Ash	0·2	0·75
Casein:whey protein ratio	1:3	4:1
Vitamin C	4 mg	2 mg

The composition of cow's milk varies with the breed of cow, with the time of lactation and with the nutritional state of the animal.

Table 13.2 The composition of cow's milk,* content per 100 g (1 pint is approximately 570 g)

Energy value	280 kJ (66 kcal)	Carbohydrate	4·6 g
Total solids	12·4 g	Calcium	0·12 g
Water	87·6 g	Phosphorus	0·10 g
Protein	3·3 g	Magnesium	0·01 g
Fat	3·5 g	Iron	0·03 mg
Vitamin A Carotene }	150 i.u.	Vitamin B_6	40 µg
		Biotin	2·5 µg
Thiamin	45 µg	Folacin	6·5 µg
Riboflavin	180 µg	Inositol	18 mg
Nicotinic acid	80 µg	Vitamin C	2 mg
Pantothenic acid	320 µg	Vitamin D	2 i.u.
		Vitamin E	0·08 µg

* S. K. Kon (1947). *British Medical Bulletin* **5**, 170.

added to milk casein is rendered insoluble (its isoelectric point is pH 4·6) and the milk curdles. Milk goes 'sour' and curdles because bacteria in it produce lactic acid. On the other hand, when milk comes into contact with the enzymes rennin or pepsin *coagulation* or *clotting* occurs and a coagulum or gel of calcium caseinate separates. On standing the clot contracts, expressing a clear fluid known as whey. The clot retains all the milk fat which contains a high proportion of lower fatty acids such as butyric acid.

Milk is an excellent source of protein and is a good source of calcium and phosphorus but it is deficient in iron. However, many young mammals, including the human infant, are born with a store of iron sufficient to tide them over the period of suckling.

Milk is an important dietary source of vitamin A and of certain vitamins of the B complex, particularly riboflavin; it is a poor source of vitamin C.

Raw milk is a highly perishable commodity since its composition makes it an ideal substrate for the growth of micro-organisms. Freshly-drawn milk may contain spoilage organisms, some of them pathogens, and it may be further infected during handling before reaching the dairy or factory. Thus to ensure that a bacteriologically safe product of good keeping quality is delivered to the consumer it is necessary to destroy the pathogens and to reduce the population of the other bacteria. This is done effectively by appropriate heat treatment.

Most milk destined for the liquid market is pasteurized by the high temperature short time (HTST) process, in which milk is heated to 71° to 73° for not less than 15 seconds. Pasteurized milk should contain no pathogenic micro-organisms and the content of other micro-organisms should be so low that the milk will remain fresh for several days if kept cool. It is well-established that pasteurization has no noteworthy effect on the nutritional value of milk and that the only significant change is a loss of up to about one-quarter of the content of vitamin C.

For longer term storage it is necessary to sterilize milk completely. In the traditional in-bottle sterilization procedure homogenized milk is bottled and the bottles heated in steam at 110° to 115° C for 20 to 40 min. This relatively drastic treatment imparts to milk a marked cooked flavour and a richer colour and texture than does pasteurization. In-bottle sterilization is associated with some slight overall reduction in nutritive value and it causes a measurable loss of protein quality and some destruction of vitamins.

More recently continuous sterilization processes have been introduced in which milk is heated to 130 to 150°C for 1 to 2 seconds, and aseptically filled into aluminium foil-lined waxed cartons. The resulting product, ultra heat treated (UHT) milk, has suffered little loss of nutritive value and keeps for up to six months, even when stored without refrigeration.

Dried milk. In the preparation of dried milk, water is removed either by spray-drying or by roller-drying. The resulting product is of high nutritive value but cannot be stored for very long periods since the fat tends to be oxidized and become rancid. Dried skim milk is a valuable source of protein and calcium but is, of course, almost free from fat and the fat-soluble vitamins A and D.

Condensed milk is prepared by removing part of the water *in vacuo*. The concentrate is then sterilized at a high temperature in tins (*evaporated milk*), or it is treated with sufficient sugar (about 40 per cent) to prevent the growth of bacteria (*sweetened condensed milk*).

In the preparation of *cream* the fat is separated from milk either by gentle centrifugation or by allowing it to rise to the surface. Cream contains a high proportion (40 to 50 per cent) of fat. Milk from which most of the fat has been removed is known as *skim milk*.

When cream is submitted to prolonged shaking (churning) the fat globules coalesce to form a solid mass of *butter* which is, like cream, an important source of the fat-soluble vitamins, especially vitamin A. The vitamin A and D content of butter is higher in summer, when the cows are fed on grass, than in winter. The residual fluid, known as *buttermilk*, has a composition similar to that of skim milk and is a rich source of protein, lactose and the inorganic salts of milk.

Cheese is made by coagulating the proteins of milk with rennet (an enzyme preparation from calf stomach) usually at 30°C. Most of the fat is included in the coagulated mass which is pressed out and allowed to 'ripen' under the influence of bacteria. The characteristic texture and taste of the finished cheese depends on the particular bacteria and moulds involved in the ripening process. In some cheeses during the ripening process tyramine may be formed from tyrosine; in some varieties more than 0·5 mg of tyramine per g may be present. Cheese is an important source of protein, fat, and mineral elements such as calcium, and has a very high nutritive value.

Margarine has a fat content and consistency similar to that of butter. It is prepared from blends of vegetable and marine oils and animal fats some of which are highly unsaturated and require hardening by hydrogenation. Developments in the application of hydrogenation procedures to fats have led to the manufacture of some margarines containing 30 to 50 per cent of *cis-cis* polyunsaturated fatty acids. In other margarines the unsaturated fats will have been modified by the hardening process. Vitamins A and D are added to all margarines sold by retail, in the proportions of 760 to 940 i.u.

vitamin A per oz and 80 to 100 i.u. vitamin D per oz. This is equal on average to about 960 µg retinol equivalents and 8 µg vitamin D per 100 g (see also Table 13.3).

Eggs are a rich source of protein, vitamins and minerals. One egg supplies 6·8 g protein, 5·5 g fat, 30 mg Ca, 110 mg P, 1·6 mg Fe (all of which is available), 200 to 800 i.u. vitamin A, 0·1 mg thiamin, 100 to 350 mg riboflavin, 10 to 50 i.u. vitamin D and 335 kJ (80 kcal).

Meat. Meat is essentially skeletal muscle, Table 13.4. When fresh meat is allowed to hang carbohydrate (glycogen) disappears and acids such as lactic acid are produced which tend to soften the muscle fibres and make the meat more tender. When meat is cooked the proteins are coagulated, some water is lost and some collagen is converted to gelatin. If meat is boiled the soluble constituents such as inorganic salts, gelatin and extractives are lost, but these substances tend to be retained when meat is roasted or fried.

The *extractives* in meat are soluble organic

Table 13.3 Composition of dairy produce

	Per 100 g						
	Protein g	Fat g	Carbo-hydrate g	kJ	Calcium mg	Iron mg	kcal
Cow's milk, whole	3·4	3·7	4·8	276	120	0·1	66
Cow's milk, skimmed	3·5	0·2	5·1	146	124	0·1	35
Butter	0·4	85·1	0·0	3318	15	0·2	793
Margarine, vitaminized	0·2	85·3	0·0	3326	4	0·3	795
Cheese, Cheddar	25·4	34·5	0·0	1778	810	0·6	425
Egg, whole, fresh	11·9	12·3	0·0	682	56	3·0	163

The figures in this and subsequent tables are taken from The Composition of Foods by R. A. McCance and E. M. Widdowson (1960). *Special Report Series, Medical Research Council, London* No. 297. London: H.M.S.O.

Table 13.4 Composition of meat and fish (edible portion)

	Per 100 g						
	Protein g	Fat g	Carbo-hydrate g	kJ	Calcium mg	Iron mg	kcal
Beef, sirloin, roast, lean only	26·8	12·3	0·0	937	7	5·3	224
Mutton, leg, boiled	25·8	16·6	0·0	1088	4	5·1	260
Bacon, back, fried	24·6	53·4	0·0	2498	12	2·8	597
Liver, ox, fried	29·5	15·9	4·0	1188	9	20·7	284
Cod, grilled	27·0	5·3	0·0	669	31	1·0	160
Sole, steamed	17·6	1·3	0·0	351	113	1·0	84
Herring, fried	21·8	15·1	0·0	983	38·6	1·9	235
Salmon, fresh, steamed	19·1	13·0	0·0	833	29	1·0	199
Salmon, canned	19·7	6·0	0·0	537	66·4	1·3	137
Haddock, smoked, steamed	22·3	0·9	0·0	418	58	1·0	100
Kipper, baked	23·2	11·4	0·0	841	64·8	1·4	201

substances, mostly of unknown composition. They are of little nutritive value but since they are largely responsible for the flavour of meat they may stimulate the flow of digestive juices.

Other parts of animals usually described as offal are used for food. Liver and kidney are both rich in nucleoprotein and contain less fat than most meat. Liver is rich in vitamin A and in iron. Sweetbreads (pancreas and thymus) are rich in nucleoprotein and tripe (boiled ox stomach and intestine) in gelatin.

Fish is an important source of animal protein (Table 13.4). White fish such as cod, haddock and sole contain only a small proportion (less than 2 per cent) of fat, but fat fish, herring, mackerel and salmon, contain 5 to 18 per cent of fat and can provide important amounts of the fat-soluble vitamins A and D.

Cereals. Cereals contain 11 per cent of protein and approximately 70 per cent of carbohydrate in the form of starch granules with a thin indigestible cellulose coat. The amount of fat varies widely (0·5 to 8 per cent) from one cereal to another, being particularly high in oatmeal. The inorganic matter, about 2 per cent consists chiefly of calcium, phosphorus and iron. The various nutrients contained in the wheat grain are not evenly distributed (Table 13.5). The endosperm in the centre is mainly starch with some protein, but the outer layer of the endosperm contains the important proteins glutelin and gliadin as well as minerals and nicotinic acid. The germ is particularly rich in vitamins of the B group.

In the process of milling the degree of extraction (the percentage of the whole grain retained in the flour) can be varied widely. 'Wholemeal' (92 per cent extraction) or flour of 100 per cent extraction contains a large proportion of indigestible fibrous matter. Roller-milling usually gives only 70 to 80 per cent extraction. When the figure falls below 80 per cent the loss in minerals and vitamins is considerable. White flour (72 per cent extraction) has the advantage of good keeping qualities but much of the minerals, vitamins and proteins have been removed with the bran and embryo. When white bread was first introduced, in the second half of the nineteenth century, the intake of minerals and vitamins of the B group by the population of Great Britain was much reduced.

'Brown bread' is usually made from white flour to which a proportion of bran has been added. It has a higher vitamin content than has white bread, has more flavour and contains more indigestible fibrous matter or 'roughage'.

The Bread and Flour Regulations (1963), allow millers to produce flour at any extraction rate provided it contains at least 0·24 mg thiamin per 100 g, 1·60 mg nicotinic acid per 100 g and 1·65 mg iron per 100 g. Since these levels are not usually reached in white flour of extraction rate below 80 per cent, such white flour has to be 'fortified' by the addition of appropriate supplements. The Regulations also require that calcium carbonate should be added at the rate of 14 oz per 280 lb flour (31·4 g per 100 kg), to all except wholemeal flour and self raising flour which has a calcium compound as a raising ingredient.

At one time it was thought that the phytic acid content of cereals, especially oatmeal, interfered with the absorption of calcium. This is not now believed to be of importance in human nutrition (see p. 45).

Bread. When wheat flour is kneaded with water a sticky mass of dough is formed. To

Table 13.5 Composition of different parts of wheat grain

	Per cent by weight (approx.)	Thiamin µg per g	Riboflavin µg per g	Nicotinic acid µg per g	Iron mg per 100 g
Whole wheat	100	3·6	1·6	50	3·5
Clear bran, including aleurone layer	12	4·8	5	250	12
Scutellum	1·5	165	15	60	9
Embryo	1	9	15	60	9
Outer endosperm	3	4·5	1·8	150	10
Bulk endosperm, including outer endosperm	85·5	0·6	0·7	22	2·1

(*From* T. Moran & Sir Jack Drummond (1945). *Lancet* **ii**, 698.)

make bread the dough is puffed up (allowed to rise) by carbon dioxide produced either by the action of yeast on the starch or by the addition of 'activators'. During baking the carbon dioxide trapped in the mass makes it spongy. The starch grains are ruptured by the heat and some of the starch is converted into soluble starch and dextrins; caramelized products are formed in the crust. When bread is toasted it loses water, further caramelization occurs and the starch on the surface is largely converted into degradation products which give toast its brown colour. The composition of bread is given in Table 13.6.

Gluten is a mixture of proteins, mainly gliadins and glutelins, present in wheat, barley and rye and the flours made from these cereals produce a dough on the addition of water. Cornflour (wheat starch), oatmeal and flour made from maize do not contain gluten and therefore can be made into bread only with considerable difficulty. Some people suffer from steatorrhoea (Chap. 17) if they eat foodstuffs containing gluten.

Fruits and vegetables. Most green vegetables contain a large proportion of indigestible fibrous material but they are important sources of minerals and of ascorbic acid and carotene, the precursor of vitamin A (Table 13.10).

Potatoes contain large amounts of starch in the form of granules which swell up and burst on cooking. In the United Kingdom they provide only 6 per cent of the total energy intake of the population but as much as one-third of

Table 13.6 Composition of cereal products

	Per 100 g						
	Protein g	Fat g	Carbo-hydrate g	kJ	Calcium mg	Iron mg	kcal
Bread, brown	8·7	2·1	49·9	1013	95	2·4	242
Bread, white, batch loaves	8·0	1·4	51·7	1004	91	1·8	240
Flour, Manitoba, 85 per cent extraction	13·6	1·7	74·0	1464	18·5	2·7	350
Flour, Manitoba, 70 per cent extraction unfortified	12·8	1·2	6·9	1473	13	2·2	352
Oatmeal, raw	12·1	8·7	72·8	1690	55	4·1	404
Pearl barley, raw	7·7	1·7	83·6	1506	10	0·7	360
Biscuits, water	10·7	12·5	75·9	1858	121	1·6	444

Table 13.7 Composition of fruits and vegetables (edible portion)

	Per 100 g						
	Protein g	Fat g	Carbo-hydrate g	kJ	Calcium mg	Iron mg	kcal
Apples	0·3	0·0	12·2	197	4	0·3	47
Bananas	1·1	0·0	19·2	332	7	0·4	7
Black currants	0·9	0·0	6·6	121	60	1·3	29
Oranges	0·8	0·0	8·5	146	41	0·3	35
Tomatoes	0·9	0·0	2·8	59	13	0·4	14
Prunes, dried	2·4	0·0	40·3	674	38	2·9	161
Raisins	1·1	0·0	64·4	1033	61	1·6	247
Almonds	20·5	53·5	4·3	2502	247	4·2	598
Beans, broad, boiled	4·1	0·0	7·1	180	21	1·0	43
Cabbage, boiled	1·1	0·0	0·8	33	30	0·5	8
Carrots, boiled	0·9	0·0	4·5	88	29	0·4	21
Lettuce	1·1	0·0	1·8	46	26	0·7	11
Parsley	5·0	0·0	0·0	88	325	8·0	21
Peas, fresh, boiled	5·8	0·0	7·7	205	13	1·2	49
Potatoes, new, boiled	1·6	0·0	18·3	314	5	0·5	75
Spinach	5·1	0·0	1·4	109	595	4·0	26

the ascorbic acid intake. They are also an important source of iron. The pulses (beans, peas, lentils) are important sources of protein (20 to 25 per cent in the dried state). In the fresh form (but not when dried) they supply ascorbic acid and carotene. Most fruits are of little nutritive value except as sources of ascorbic acid. When ripe they contain varying amounts of sugar and the banana is unique in supplying starch as well.

Chocolate and cocoa have a relatively high protein, fat and carbohydrate content and have thus a high energy value, but the energy value of tea and coffee depends entirely on the added sugar and milk. Tea and coffee contain the stimulant caffeine. Cocoa is a valuable source of iron (Table 13.8).

Alcohol has a potential energy of 29 kJ or 7 kcal/g but its value as a food is restricted by its depressant action on the central nervous system and toxicity towards the liver. Since alcohol can be mixed with water in all concentrations the concentration in the brain is related directly to its concentration in the blood. Its solubility in fat allows it to diffuse into the central nervous system. Deep coma occurs when the blood alcohol is 400 mg/100 ml. Ethyl alcohol is absorbed very quickly from the gastrointestinal tract and is oxidized to acetaldehyde in the liver by the enzyme alcohol dehydrogenase (p. 99) operating in conjunction with NAD (p. 127) which is reduced to NADH. The acetaldehyde is subsequently rapidly oxidized to acetic acid which enters the general metabolic pool as acetyl-CoA (Chap. 19) prior to oxidation to carbon dioxide and water in the citric acid cycle. Some may be converted to fat (Chap. 19) with the aid of the NADH formed by the action of the alcohol dehydrogenase. Since the enzyme is saturated at low concentrations of alcohol—corresponding to a blood concentration of 80 mg ethanol per 100 ml—alcohol disappears quite slowly from the blood; only about 7 g of alcohol can be metabolized per hour. This corresponds to a fall in blood alcohol concentration of 15 mg per 100 ml per hour. No way has yet been found of accelerating this process but it is well known that persons who habitually drink large amounts of alcohol develop an increased tolerance. This is due in part to a change in the response of the central nervous system and in part to an adaptive increase in the amount of the enzyme in the liver. Alcohol by inhibiting hepatic glucogenesis (Chap. 18) is a powerful hypoglycaemic agent; hypoglycaemia appears 4 to 12 hours after cessation of drinking.

Small amounts of alcohol (say 3 per cent of the intake) are excreted in the urine and breath. This fact is made use of in the breathalyser test for motorists. Beer contains, in addition to 4 per cent of alcohol, a little carbohydrate and significant amounts of calcium and of members of the vitamin B group derived chiefly from the yeast used in its fermentation. Spirits contain about 40 per cent alcohol, fortified wines 20 per cent and ordinary wines 15 per cent alcohol.

Most foods of plant origin contain indigestible matter, chiefly cellulose, which is usually referred to as *roughage*. Although small quantities may be digested by bacterial action in the human large intestine, the chief value of cellulose in the diet is to increase the bulk of the intestinal contents and so stimulate peristalsis (Chap. 17).

VITAMINS

At the beginning of the present century it

Table 13.8 Composition of miscellaneous foodstuffs

| | Per 100 g | | | | | | |
	Protein g	Fat g	Carbo-hydrate g	kJ	Calcium mg	Iron mg	kcal
Honey	0·4	0·0	76·4	1205	5	0·4	288
Chocolate, milk	8·7	37·6	54·5	2460	246	1·7	588
Chocolate, plain	5·6	35·2	52·5	2276	63	2·9	544
Cocoa	20·4	25·6	35·0	1891	51	14·3	452
Jams	0·4	0·0	69·3	1092	12	1·0	261
Beer, 4 per cent alcohol	0·3	0·0	2·0	160	9	0·02	38

was beginning to be realized that disease in laboratory animals and in man could be produced by a diet deficient in certain respects. The idea of accessory food factors (now called vitamins) is strikingly illustrated by a series of experiments carried out by Hopkins between 1906 and 1912 (Fig. 13.9). He showed that young rats fed on diets consisting of purified casein, starch, cane-sugar, lard and inorganic salts invariably ceased to grow in a very short time and then became ill and died. If the purified diet was supplemented with a small daily allowance of milk, no more than 4 per cent of the total food eaten, the health of the animals was maintained and they began to grow again.

The vitamins, organic substances of known structure and low molecular weight, are essential components in metabolic processes. In their absence a 'biochemical lesion' develops which may or may not be accompanied by

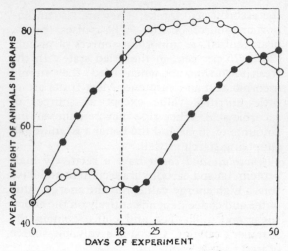

Fig. 13.9 An experiment of Hopkins illustrating the conception of accessory food factors. 'Lower curve (up to 18th day) eight male rats upon pure dietary; upper curve eight similar rats taking 3 ml of milk each day. On the 18th day, marked by vertical dotted line, the milk was transferred from one set to the other.' (F. G. Hopkins (1912). *Journal of Physiology* **44**, 425.)

Table 13.10 The vitamins

Vitamin	Name	Active derivative	Biochemical role	Result of deficiency
A	Retinol	Retinal	Vision	Night blindness
	Dehydroretinol		Control of keratin production	Hyperkeratosis
B complex				
B_1	Thiamin	TPP	Decarboxylation	Beri-beri
—	Lipoic acid	—	Decarboxylation	—
B_2	Riboflavin	FMN FAD	Hydrogen transport	—
—	Niacin (Nicotinic acid Nicotinamide)	NAD NADP	Hydrogen transport	Pellagra
B_6	Pyridoxine Pyridoxal Pyridoxamine	Pyridoxal phosphate	Amino acid metabolism	—
—	Pantothenic acid	Coenzyme A	Transport of acetyl and succinyl units	—
—	Biotin	—	Carboxylation	—
—	Folacin	Tetrahydro folic acid	Transport of one-carbon units	Anaemia
B_{12}	Cobalamin	5-deoxyadenosine derivative	Nucleotide reduction Methyl malonyl CoA metabolism	Pernicious anaemia
C	Ascorbic acid	—	Connective tissue formation	Scurvy
D	Calciferol	Ergocalciferol Cholecalciferol	Calcium absorption and deposition in bone	Rickets
E	Tocopherol	—	Antioxidants of unsaturated fatty acids	Sterility (?)
K	Menaphthone derivatives	—	Blood clotting	Impaired blood coagulability

See Tentative rules for generic descriptions and trivial names for vitamins and related compounds recommended by the IUNS Committee on Nomenclature. *Nutrition Abstracts and Reviews* **40**, 395 (1970).

structural changes in tissues. The amount that has to be ingested to maintain health is different for each vitamin but is usually quite small. The *daily requirement* for any one vitamin depends on a number of factors and may be increased during growth, pregnancy and lactation.

So far as possible, the biochemical actions of the vitamins as coenzymes are described in appropriate places in this book but in this chapter a brief account of their chemical properties and of the effects of dietary deficiencies is given (Table 13.11).

VITAMIN A (RETINOL)

Two forms of vitamin A, retinol and dehydroretinol (formerly called vitamin A_1 and vitamin A_2) are both unsaturated primary alcohols containing a β-ionone ring, dehydroretinol having one more double bond than retinol. Vitamin A exhibits cis-trans isomerism and several isomerides are known, all biologically active.

Table 13.11 Vitamin A potency of foods* (i.u. per 100 g)

Halibut-liver oil	2 000 000 to 36 000 000
Cod-liver oil	40 000 to 400 000
Fresh herring	90 to 120
Ox liver	12 000 to 40 000
Beef and mutton (lean)	30 to 50
Cow's milk† (raw	
summer	140
winter	70
Human milk	200 to 500
Butter	2000 to 5000
Cheese (cheddar type)	1300
Margarine (vitaminized)	2300 to 3000
Carrots	10 000 to 16 000‡
Spinach	10 000‡

* The values in this and subsequent tables are mainly from 'Nutritive Values of Wartime Foods' (1945). *M.R.C. War Memorandum No.* 14. London: H.M.S.O. The vitamin content of any individual sample may differ widely from the value found in the table.
† Large variations in content are found according to breed of cow, yield of milk, pasture, season, etc.
‡ Potency is due to provitamin.

Vitamin A has a characteristic absorption spectrum which is commonly used for its estimation, and its presence in animal tissues

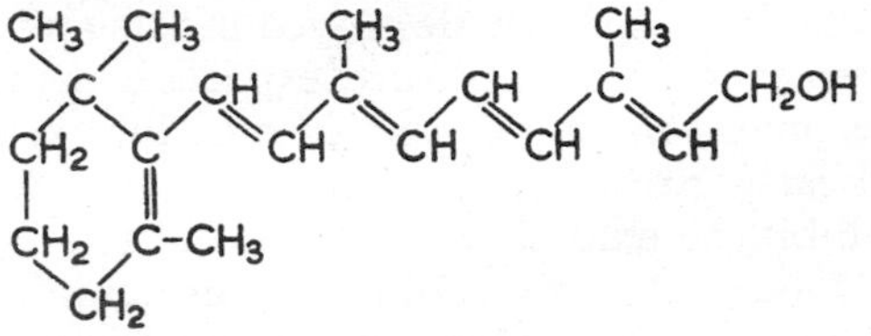

retinol

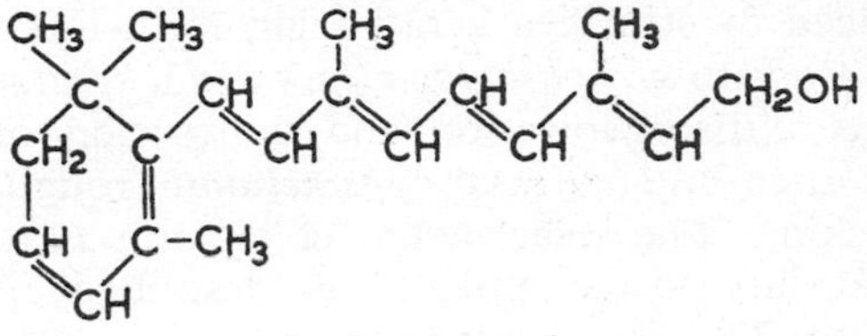

3-dehydroretinol

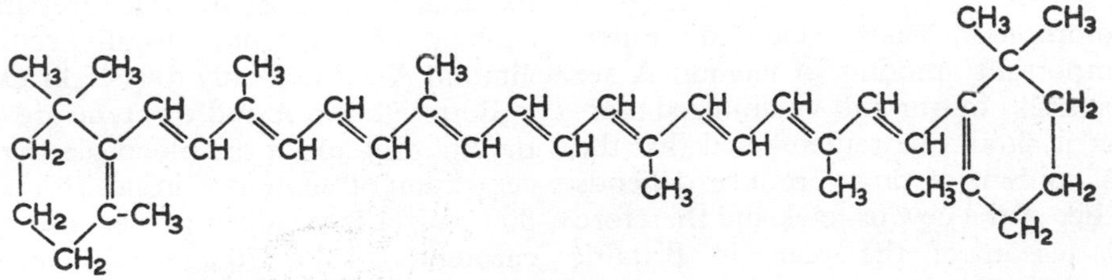

β-carotene

Plants do not contain vitamin A but they synthesize precursors of the vitamin which are found particularly in the yellow and green parts. Animals eat plant material and in their intestinal epithelium these precursors are converted to vitamin A which is carried in the intestinal lymphatics to the tissues mainly in the form of esters. Vitamin A precursors belong the class of carotenoids, unsaturated pigments the molecules of which contain forty carbon atoms. These substances are soluble in fat and fat oils but not in water.

has been shown by its fluorescence in ultraviolet light. Vitamin A gives a blue colour with antimony trichloride in chloroform solution (Carr-Price reaction) and this reaction is used for its quantitative estimation.

Vitamin A deficiency may result from inadequate intake or from defective absorption of the vitamin or provitamins, or from interference with the conversion of provitamins into vitamin A. With the exception of that in the retina the nature of the biochemical lesion of vitamin A deficiency is unknown, but in the

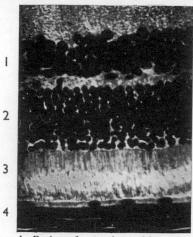

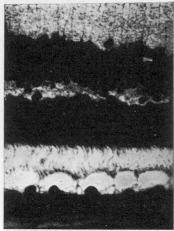

A. Retina of control rat. Masson's trichome stain. × 600.

B. Similar but from vitamin A deficient rat, showing atrophy of the rods and cones and swelling of the pigment cells.

Fig. 13.12 V. Ramalingaswami, E. H. Leach & S. Sriramachari (1955), Ocular structure in vitamin A deficiency in the monkey. *Quarterly Journal of Experimental Physiology* **40**, 337–347. 1, inner nuclear layer; 2, outer nuclear layer; 3, layer of rods and cones; 4, pigment epithelium.

absence of the vitamin there are widespread changes in epithelium which tends to be replaced by stratified keratinizing epithelium. This leads to *xerophthalmia* ('dry eye'), *keratomalacia* (disorganization and destruction of the cornea) and *follicular hyperkeratosis* (rough, dry skin). The relationship of vitamin A to rhodopsin (visual purple) is described in Chapter 37. (See also Fig. 13.12.)

The commonest foods containing nutritionally important amounts of vitamin A are liver, eggs, milk, butter and vitaminized margarine, but it should be remembered that the vitamin A content of farm produce depends upon the diet of the cow or fowl, and therefore upon the season of the year: in Britain, vitamin A is added to margarine to the amount of 760 to 940 i.u. (international units) per ounce (28 g). Fish-liver oils contain variable but relatively large amounts of vitamin A. The provitamins are widely distributed in vegetables and fruits where they are found in association with chlorophyll and other pigments so that in general the greener or more highly coloured parts are the best sources of the provitamin (Table 13.11).

Vitamin A is easily oxidized in the presence of molecular oxygen to biologically inactive compounds but in an inert atmosphere it is quite stable to heat. Most foodstuffs retain their vitamin A content if they are stored under conditions in which oxygen is excluded. Vitamin A is slowly destroyed in fats and oils, even at room temperature, especially if the fat becomes rancid, but certain oils, for example wheat-germ oil, contain 'anti-oxidants' which inhibit the oxidation of vitamin A. The frying of food in deep fat results in considerable loss of vitamin A, not only in the fat but also in material cooked in it. The preparation and cooking of vegetables usually reduces their vitamin A content only to a slight extent.

Both vitamin A and carotene are present in the lipoproteins of the blood plasma; the concentration of vitamin A in health is from 50 to 300 i.u. (17 to 100 mg) per 100 ml and of carotenoids 15 to 370 μg per 100 ml. In xerosis in children (Plate 13.1) the plasma vitamin A is only 2 or 3 μg/100 ml. One international unit is defined as the activity of 0·344 μg of retinol acetate (equal to 0·300 μg of retinol). The international unit of provitamin is 0·6 μg of β-carotene. The estimated daily requirements are given in Table 13.27.

The absorption of vitamin A from the alimentary tract is favoured by the simultaneous absorption of fat and the presence of bile salts. After absorption vitamin A is stored in the liver and to a small extent in the body fat generally. Excessive intake of vitamin A leads to bone resorption and hypercalcaemia.

THE VITAMIN B COMPLEX

Since 1926 vitamin B, as it was then called, has been shown to consist of about a dozen different water-soluble substances.

Thiamin (vitamin B$_2$). Thiamin consists of a pyrimidine and a thiazole ring system joined by a methylene bridge. In acid solution it withstands prolonged heating at 120°C but in alkaline solution it is rapidly destroyed. Under ordinary conditions thiamin is not

pyrimidine ring   thiazole ring
thiamin chloride hydrochloride

oxidized by atmospheric oxygen but is converted by mild oxidizing agents to the pigment thiochrome which has no vitamin activity.

The physiological actions of thiamin are due to its pyrophosphoric ester, co-carboxylase (thiaminpyrophosphate, TPP), which is formed by the phosphorylation of the vitamin under the influence of ATP and magnesium ions. Thiaminpyrophosphate, often in associa-

thiamin pyrophosphate, TPP, or co-carboxylase

tion with lipoic acid (p. 186), is a coenzyme in the enzyme systems needed for the decarboxylation of α-keto acids such as pyruvic acid (p. 148). When thiamin is lacking, pyruvic acid cannot be metabolized and so accumulates in the fluids and tissues of the body. The estimation of the pyruvate level in the blood may be helpful in detecting thiamin deficiency but it is not specific since the pyruvate level may be raised in other conditions.

Prolonged deficiency of vitamin B in man results in the disorder known for a long time as *beri-beri* (Fig. 13.13). This condition, still common in many parts of the world, and occasionally seen in this country in chronic alcoholics living on a very poor diet, is characterized by: (a) polyneuritis, that is to say muscle weakness and atrophy, incoordination of movements, and disturbances of sensation; (b) enlargement of the heart, generalized caso-

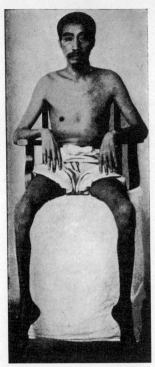

Fig. 13.13 A patient with 'dry' beri-beri suffering from wrist-drop, foot-drop and marked wasting of the lower extremities. (*Courtesy of B. S. Platt.*)

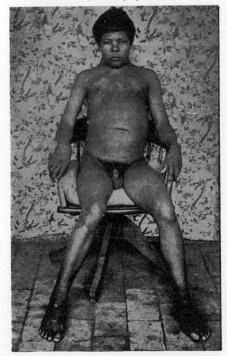

Fig. 13.14 A case of beri-beri with extensive oedema. (*Courtesy of Wellcome Museum of Medical Science.*)

Food, Nutrition and Vitamins   185

dilatation, and cardiac failure; (c) oedema (Fig. 13.14).

Thiamin is present in many plants, in particularly high concentrations in the seeds and the outer coats of grains (bran); the pulses and nuts are among the richest natural sources. White flour contains little thiamin because most of it is removed in the milling process. On the other hand wholemeal flour contains 350 to 400 μg per 100 g and is therefore a relatively good source. The vitamin is widely distributed in animal tissues and the lean parts of meat, especially pork, may be important dietary sources. Yeast and yeast extracts contain very large amounts of thiamin and many bacteria

Table 13.15   Thiamin in foods (mg per 100 g)

Wheat (germ)	2·5 to 5·6
Flour, wholemeal	0·35 to 0·4
85 per cent extraction	0·30 to 0·4
70 per cent extraction (white)	0·05
Oatmeal	0·5
Yeast	1·8 to 3·0
Milk	0·04 to 0·06
Eggs	0·1 to 0·15
Beef and mutton	0·05 to 0·1
Pulses	0·5 to 0·9
Rice, polished	0·02 to 0·04
Rice, whole	0·1

are capable of synthesizing it. Although high temperatures and an alkaline medium favour destruction of the vitamin, losses during the preparation and cooking of food are not large. Freezing, canning and dehydration likewise result in small losses (Table 13.15).

The blood contains some 5 to 11 μg of thiamin per 100 ml almost entirely in the form of TPP. The total amount of the vitamin in the body is of the order of 25 mg; the heart has the highest concentration (230 μg per 100 g), brain, liver and kidney contain about 100 μg per 100 g and skeletal muscle about 50 μg per 100 g. Since the body cannot store large quantities of thiamin the period of survival on a deficient diet is not significantly prolonged by prior administration of large amounts of the vitamin.

The recommended thiamin intake is of the order of 1 mg (333 i.u.) per day but it increases as the intake of carbohydrate increases. Diets consisting mainly of polished rice supply much less and, because the body stores only about 25 mg thiamin, continued subsistence on such a diet produces beri-beri.

Lipoic acid (thioctic acid). α-Lipoic acid, a growth factor for certain protozoa and microorganisms, is usually classified among the vitamins of the B group. It is a fat-soluble material containing eight carbon and two sulphur atoms and hence is also known as thioctic acid.

Lipoic acid is a cofactor in the oxidative decarboxylation of α-keto acids. For example in the decarboxylation of pyruvic acid, $CH_3COCOOH$ (p. 148), the first step is the removal of carbon dioxide with the formation of the TPP derivative $CH_3CO\text{-}TPP$. This reacts with lipoic acid to form an acetyl derivative in which the acetyl group is attached to one of the sulphur atoms thus: $CH_3CO\text{-}S\text{-}lip\text{-}SH$. This product now reacts with coenzyme A (p. 67) to form acetyl coenzyme A thus:

This in turn enters the tricarboxylic acid cycle as described in Chapter 11.

Riboflavin. In 1932 Warburg and Christian isolated from yeast a 'yellow enzyme' which they later showed could be separated into a protein component and a coloured substance neither of which was active without the other. The yellow compound proved to be riboflavin phosphate. Riboflavin can be estimated quantitatively by the greenish-yellow fluorescence

riboflavin (6:7-dimethyl-9-D-ribityl-isoalloxazine)

shown in ultraviolet light. A microbiological assay depends on the fact that the growth of a strain of *Lactobacillus casei* and the resulting production of acid is proportional to the concentration of riboflavin in the culture medium.

In the living cell riboflavin is converted into riboflavin phosphate or into flavin-adenine dinucleotide (FAD) (p. 67), both of which combine with proteins to form the *flavoproteins* that act as important hydrogen carriers in biological oxidation systems (p. 128).

Riboflavin is readily absorbed from the upper alimentary tract. It is present in blood mainly in combination with plasma globulin to the extent of some 0·5 µg per 100 ml. High concentrations of the vitamin are found in the liver, heart and kidneys, which retain relatively large amounts even when the body is seriously depleted. Riboflavin is eliminated in the urine in the form of a pigment, uroflavin, the excretion of which fluctuates with the intake of the vitamin.

Deficiency of riboflavin (*ariboflavinosis*) has been produced experimentally in man. The skin becomes rough and scaly; the pink parts of the lips are bright red, swollen and cracked (*cheilosis*); and the tissues at the corners of the lips are swollen and fissured (*angular stomatitis*). The tongue is enlarged, tender and magenta in colour. In minor degrees of riboflavin deficiency, microscopic examination of the patient's eyes with a slit-lamp shows that the transparent cornea is invaded by minute blood vessels.

Riboflavin is synthesized by plants and occurs particularly in germinating seeds and young growing green plants. Peas, beans, grains and yeast are the most important plant sources. Animal sources of value are milk, egg-yolk, liver, kidney and heart muscle: the riboflavin found in most plant and animal tissues, as well as in yeast, is present mainly in the combined form but in milk the greater part is in the free state.

The vitamin is relatively stable in acid solution in the dark. Ordinary cooking processes cause little destruction of riboflavin although a certain amount may be lost by extraction in the cooking water. Considerable loss of riboflavin may occur in milk exposed to bright sunlight.

Table 13.16 Riboflavin in foods (mg per 100 g)

Yeast	5·0	Cheese	0·50
Wheat	0·18 to 0·25	Egg	0·4
Pulses	0·15 to 0·3	Meat	0·25
Milk	0·15 to 0·20	Liver	3·0

The recommended daily allowance is 1·5 to 1·8 mg per day, but a more generous intake is advised in pregnancy.

Niacin. The term 'niacin is used as the generic name to cover both nicotinic acid and nicotinamide. Nicotinic acid has been known to chemists for a hundred years but the discovery that it is the factor preventing pellagra was made only in 1937. The beneficial effects produced by the administration of nicotinic acid to human patients or to dogs or pigs suffering from deficiency are due to its conversion into the amide which is then converted into nicotinamide adenine dinucleotide (NAD) and nicotinamide adenine dinucleotide phosphate (NADP) (p. 67). The intake of nicotinic acid is supplemented by its synthesis from tryptophan by the intestinal flora and probably by the tissues. In man about 60 mg of tryptophan can replace 1 mg of nicotinic acid in the food.

nicotinic acid
(β-pyridine-carboxylic acid)

nicotinic acid amide
(nicotinamide)

Nicotinamide is one of the most stable vitamins since it is not destroyed by heat, light, oxidation or alkali. It is an essential growth factor for many micro-organisms so microbiological, as well as chemical, methods are available for measuring its concentration in body fluids.

The condition known as *pellagra* (Fig. 13.17 (a) and (b)) is rapidly relieved by the administration of nicotinic acid or of nicotinamide and the disease is usually considered to be due to a deficiency of this vitamin although it may also be caused by the presence in the diet of an abnormally high content of leucine which destroys the nicotinamide pattern of red blood cells. Pellagra affected millions of people in the Southern United States between 1900 and 1940. Extreme poverty made these people dependent on maize and molasses. Improvement in the diet during the 1930s led to a great decline in the incidence of the disease and to its virtual disappearance there in 1940.

After a prolonged period of general ill-health, characterized by irritability, depres-

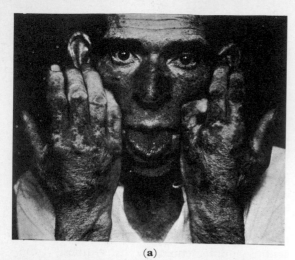

(a)

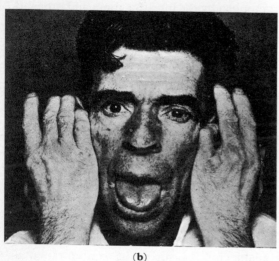

(b)

Fig. 13.17(a) and (b) Showing the dermatitis of face and hands in a case of pellagra subsequently cured by administration of nicotinic acid amide. Note the difference in the appearance of the tongue, which before treatment was red and tender (R. W. Vines & A. M. Olsen (1943). *Proceedings of Staff Meetings of the Mayo Clinic* **18**, 389).

sion, loss of weight and strength and lack of appetite, the patient often develops diarrhoea which accelerates the loss of weight and strength. About the same time the skin becomes reddened and later roughened and scaly, especially on the exposed parts of the body normally subjected to the action of light and heat (Fig. 13.17 (a) and (b)). Extreme redness and soreness of the mouth, with swelling and redness of the tongue, further increase the patient's difficulties and a vicious

circle occurs, in which the more severe the deficiency the more difficult it becomes for the patient to eat a curative diet and the more rapidly therefore the disorder progresses. Mental symptoms such as irritability, anxiety, depression and confusion appear and the patient may become demented. Most of these symptoms can be rapidly corrected by the administration of nicotinic acid or its amide. Nearly all patients with pellagra, however, have signs of other vitamin deficiencies.

Maize contains structural analogues of tryptophan which act as competitive inhibitors in its metabolism and may prevent its conversion to nicotinic acid.

Table 13.18 Nicotinic acid in foods (mg per 100 g)

Barley, whole	5·5
Maize, whole	2·0
Rice, brown	4·6
milled (polished)	1·6
Oatmeal, rolled oats	0·9
Wheat, whole	2·8 to 5·0
Flour, white	0·6 to 0·8
85 per cent extraction	0·15 to 0·20
Bran	30 to 35
Yeast, dry	10 to 50
Liver, fresh	10 to 27
Kidney	6 to 10
Fresh meat	3 to 10
Cow's milk	0·1 to 0·5
Fish	3 to 8

In its combined form as NAD and NADP, nicotinic acid is widely distributed in animal tissues. Liver, kidney and muscle are the best animal sources. About 0·5 mg per 100 ml is found in blood. Yeast and the outer parts of grain are good natural sources of the vitamin, but the valuable part of the grain is almost completely removed as bran in the milling process. Like riboflavin, nicotinic acid is synthesized by intestinal bacteria. Since nicotinic acid is very stable losses in the preparation and cooking of food are small. Significant losses may, however, occur by extraction into the cooking water.

The daily requirement of nicotinic acid is probably of the order of 12 mg per day but usually a higher intake, up to 20 mg per day, is recommended. In spite of its chemical relationship to nicotine nicotinic acid is non-toxic even in large doses.

Vitamin B$_6$ (pyridoxine). The term

vitamin B_6 is the generic name officially given to a group of naturally occurring derivatives of pyridine. The principal members of the group are pyridoxol, pyridoxal and pyridoxamine. It is probable that pyridoxol and pyridoxamine are converted to pyridoxal in

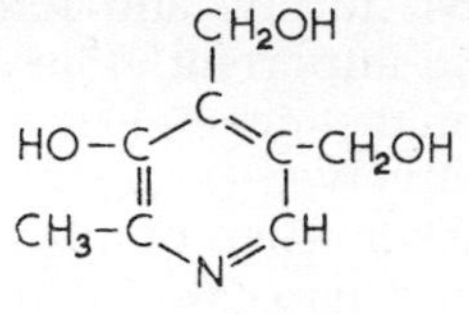

pyridoxol

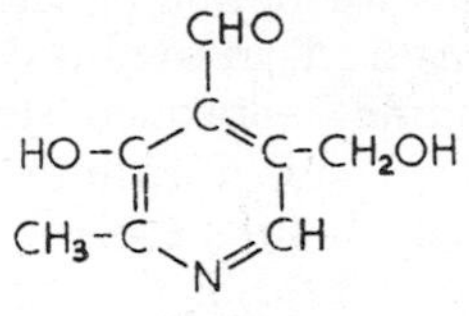

pyridoxal

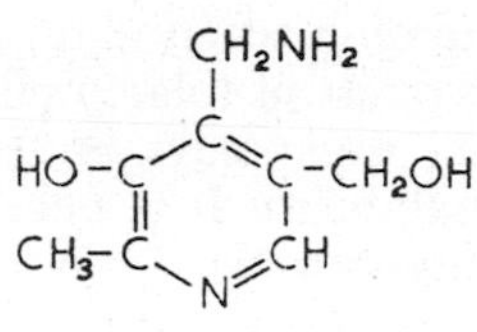

pyridoxamine

the tissues. Pyridoxal phosphate acts as co-enzyme for some amino-acid decarboxylases (Chap. 22) and for transaminase (Chap. 22). Pyridoxal phosphate catalyses the synthesis of tryptophan from indole and serine and is necessary for the conversion of tryptophan to nicotinic acid. It is also required in the haem synthetic pathway (Chap. 25), acting specifically as a coenzyme at the stage of formation of δ-aminolaevulinic acid.

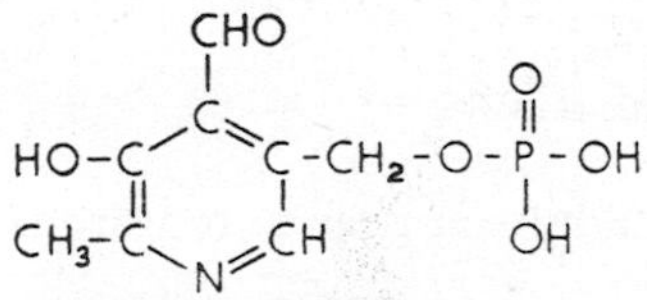

pyridoxal phosphate

Lack of pyridoxine in the diet of some animals produces dermatitis, anaemia or convulsions but deficiency very rarely if ever occurs in man since he is supplied with pyridoxine by the activity of his intestinal bacteria.

Deficiency symptoms can be produced by administering an anti-vitamin, deoxypyridoxine (competitive inhibition, p. 97). Isoniazid (isonicotinic acid hydrazide) frequently used in the treatment of tuberculosis has occasionally produced peripheral neuritis which has been ascribed to lack of pyridoxine.

Antituberculosis therapy with isoniazid, particularly if combined with pyrazinamide or cycloserine, may also occasionally be complicated by anaemia which responds to pyridoxine. A similar type of anaemia may be found in alcoholics and this may require folic acid as well as pyridoxine for adequate therapy. (The morphological characteristics of this type of anaemia will be discussed later, Chap. 24.)

The minimum daily requirement for pyridoxine is probably 0·2 to 1·0 mg. The vitamin is widely distributed in both animal and vegetable foods.

Pantothenic acid. Pantothenic acid, so called because of its wide distribution, was isolated in 1939 and identified as α, γ-dihydroxy-β, β-dimethylbutyryl-β-alanine:

pantothenic acid

Since it is universally available it is not astonishing that human deficiency symptoms have never been described.

In the living cell pantothenic acid combines with mercaptoethylamine to form pantetheine which in turn is built up into the important substance, coenzyme A (p. 67) to play an essential role in the metabolism of both fat and carbohydrate (Fig. 11.6).

Biotin. Biotin, a relatively stable cyclic ureide, has the structure:

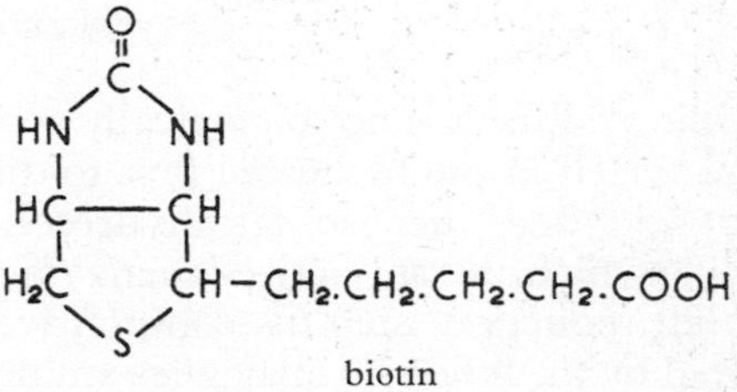

biotin

It acts as a coenzyme in the fixation of carbon dioxide in carboxylation reactions such as the conversion of pyruvic acid to oxaloacetic acid (p.152):

$$CO_2 + CH_3COCOOH \rightleftharpoons$$
$$HOOCCH_2COCOOH$$

or the carboxylation of acetyl-coenzyme A to form malonylcoenzyme A in fatty acid biosynthesis (Chap. 19).

Biotin deficiency has been produced in man by a diet consisting mainly of uncooked egg-white. The protein avidin of egg-white forms with biotin a stable compound which, since it is not hydrolysed in the intestine, cannot be absorbed. Biotin may therefore be described as an essential food factor but biotin deficiency has been found only in the quite exceptional dietary regimen just described.

Biotin is found in yeasts, bacteria of the human intestine, in liver and kidney and to a small extent in cereals.

Folacin. The term *folacin* is used to describe folic acid and related compounds. Folic acid itself, monopteroylglutamic acid, has a pterin nucleus linked to *p*-aminobenzoic acid to form pteroic acid which, in turn, is linked to a glutamic acid residue. Pteroyl triglutamate and pteroyl heptaglutamate also occur in living material.

Folic acid, first obtained from spinach, is synthesized by bacteria in the large intestine but this is of little significance since folic acid is absorbed in the jejunum. The best food sources are fresh green vegetables and liver.

such as leukaemia since they inhibit the formation of thymine and so of DNA (Chap. 20).

FH_4 acts as a coenzyme capable of carrying one-carbon units. For example it can accept the β-carbon atom of serine (which yields glycine, Chap. 22) to form N-5, N-10 methylene tetrahydrofolic acid, but methyl, hydroxymethyl, formyl and formimino derivatives are also important. These derivatives are implicated in the metabolism of the one-carbon fragments involved (a) in the biosynthesis of purines (carbon atom numbers 2 and 8 in the purine ring) (Chap. 20), (b) in the methylation of the pyrimidine ring to give thymine (Chap. 20), (c) in the formation of serine from glycine (Chap. 22) and (d) in the biosynthesis of histidine. Histidine is broken down by way of urocanic acid to formiminoglutamic acid (FIGLU) which in the presence of folic acid yields glutamic acid. If the supply of folic acid in the tissues is inadequate FIGLU accumulates and is excreted in the urine. A large oral dose of histamine accentuates this and is used clinically as a test of folate deficiency.

Folic acid thus plays an essential role in cellular metabolism. It is necessary especially for production of red cells in the bone marrow (haemopoiesis, Chap. 24) and severe deficiency results in the megaloblastic anaemia. In the United Kingdom this kind of anaemia used to be a common complication of pregnancy, but daily prophylaxis with 300 µg folic acid has been largely successful in eliminating it. Non-

pterin residue   *p*-aminobenzoic   glutamic acid residue
acid residue

folic acid (pteroyl-monoglutamic acid)

Folic acid itself is not biologically active but is converted in the living cell first to dihydrofolic acid and then to tetrahydrofolic acid (FH_4) which is an active coenzyme. The second reduction step is competitively inhibited by the folic acid analogues aminopterin and amethopterin which are therefore termed anti-folic acid agents. They are used in the treatment of certain forms of malignant disease

pregnant adults appear to require 50 to 100 µg daily.

Cobalamin (vitamin B$_{12}$). Vitamin B_{12} belongs to a group of compounds known collectively as *corrinoids*, all of which contain the *corrin* nucleus consisting of four linked pyrrole rings numbered in the same way as is the closely related porphyrin nucleus (Chap. 25).

folinic acid
(5-formyl-5, 6, 7, 8-tetrahydrofolic acid)

When the corrin nucleus with an appropriate series of substituent groupings on the periphery and a cobalt atom in the centre is linked to a ribofuranose (p. 45) phosphate residue, *cobamide* is formed. In vitamin B_{12} dimethylbenzimidazole is attached to cobamide to form *cobalamin*. When the cobalt is linked to —CN the compound is cyanocobalamin and when to —OH, hydroxocobalamin.

Vitamin B_{12} like other vitamins of the B group, is converted in the living cell into a coenzyme known in this instance as coenzyme B_{12}, the best known member of the group of cobamide or corrin coenzymes. In the conversion of cyanocobalamin to coenzyme B_{12}, the cyanide group is replaced by a 5-deoxyadenosine group. The coenzyme is necessary for the conversion of methylmalonyl-coenzyme A to succinyl-coenzyme A by the enzyme methylmalonyl mutase (Chap. 19) both in animal tissues and in bacteria. The excretion of methyl malonic acid (MMA) is greatly increased in the urine of patients who have vitamin B_{12} deficiency with megaloblastic anaemia. Coenzyme B_{12} is also involved in the conversion of glutamate to β-methyl asparate in bacteria, in the regulation of the conversion of homocysteine to methionine (Chap. 22), and in the reduction of nucleoside diphosphates to the deoxyribonucleoside derivatives (Chap. 20).

Cobalamin is a growth-promoting factor for a number of micro-organisms and algae. Two of these, *Euglena gracilis* and *Lactobacillus leichmannii* are commonly used to measure the concentration of cobalamin in blood or urine, the amount of growth under appropriate conditions being proportional to the concentration of cobalamin in the medium.

After absorption cobalamins (especially hydroxocobalamin) become bound to β-globulin although small amounts remain free in the plasma. In health blood serum has a concentration of 200 to 1000 pg/ml. If not utilized in cell metabolism cobalamins are stored bound to β-globulin in the liver which may contain as much as 1 to 2 mg. The influence of vitamin B_{12} on haemopoiesis is described in Chapter 24.

Vitamin B_{12} (cyanocobalamin or α-(5,6-dimethylbenzimidazolyl) cobamide

No more than about 1 µg per day is required to maintain health provided there is no interference with absorption. Normal (not vegetarian) diets contain very much more than this. Bovine kidney and liver are the best sources of vitamin B_{12} (15 to 20 µg/100 g); fresh meat and dairy produce contain from 1 to 3 µg/100 g. Plant tissues contain no vitamin B_{12}.

Choline (p. 54) is used in the synthesis of acetycholine and phospholipids. The methyl groups of choline play an important role in intermediary metabolism since they are concerned in the process of transmethylation (Chap. 21). Provided some other source of utilizable methyl groups, such as methionine, is available choline can be synthesized in the body. In any case, choline deficiency is extremely unlikely to occur because lecithin, which contains choline, is widely distributed in all animal and vegetable foods.

Ascorbic Acid (Vitamin C)

Scurvy was at one time a common disorder in seamen on long voyages and many died until Lind proved in 1747 that it could be prevented and cured by eating fresh fruit or vegetables. The antiscorbutic vitamin ascorbic acid was isolated in 1928 and synthesized in 1933.

L-ascorbic acid   L-dehydroascorbic acid

Ascorbic acid possesses an asymmetrical carbon **C** atom and is therefore optically active. L-Ascorbic acid and L-dehydroascorbic acid are the only known naturally occurring biologically active substances, the D-compounds being inert.

Pure ascorbic acid is a white crystalline solid freely soluble in water. It is a powerful reducing agent, giving up two hydrogen atoms to become dehydroascorbic acid. This oxidation is reversible in the body. Ascorbic acid may, however, be irreversibly oxidized beyond the stage of dehydroascorbic acid to diketogulonic acid and oxalic acid. Ascorbic acid is estimated in biological material by its reduction of the blue dye 2, 6-dichlorophenolindophenol to a colourless compound. The reaction is not, however, specific for ascorbic acid.

moulds, fungi, all the higher plants, and by animals except the guinea-pig, primates and man. The biosynthetic pathway is as follows: D-glucose → D-glucuronolactone → L-gulonolactone → L-ascorbic acid. In the rat the process takes place in the liver, but in man, primates and the guinea-pig the liver lacks the enzyme necessary for the final step. These species therefore develop deficiency symptoms (scurvy) if the diet is lacking in the vitamin.

Ascorbic acid belongs to a group of compounds which form reversible oxidation-reduction systems and the vitamin has therefore been supposed to act as a hydrogen carrier or respiratory catalyst. There is evidence that it may act in this capacity along with glutathione.

Ascorbic acid *in vitro* inhibits the formation of pigment from adrenaline and from 3, 4-dihydroxyphenylalanine (DOPA). It prevents, too, the darkening of cut fruit and potatoes which is the result of the oxidation of phenols to catechols and quinones. There is evidence that in ascorbic acid deficiency the end-product of tyrosine is *p*-hydroxyphenyl pyruvic acid, and ascorbic acid has been found necessary for the oxidation of this substance by liver tissue. Premature human infants fed on cow's milk and full-term infants fed tyrosine or phenylalanine as supplements excrete *p*-hydroxyphenyl pyruvic acid and *p*-hydroxyphenyl lactic acid unless they are also given ascorbic acid. These facts suggest that ascorbic acid has a role in the metabolism of tyrosine and phenylalanine.

Ascorbic acid is present in all body fluids

ascorbic acid   2: 6-dichlorophenol-indophenol (blue)   dehydroascorbic acid   reduced dye (colourless)

In the presence of light, oxygen and traces of copper, ascorbic acid is rapidly oxidized at pH values above 4, especially if the solution is warmed.

Ascorbic acid is synthesized by certain

and tissues. The adrenal glands contain 100 to 200 mg ascorbic acid per 100 g aggregated near the Golgi apparatus of the cells. The amount is greatly reduced after injection of ACTH (see Chap. 52). The pituitary, corpus luteum and thymus are all rich sources; liver, lung and heart muscle possess successively lower concentrations and skeletal muscle only a trace. The total amount of ascorbic acid in the body

Table 13.19 Ascorbic acid in foods, mg per 100 g (approximately)

Rose hip syrup	150 to 200
Blackcurrant syrup	60
Leafy vegetables (fresh)	50
Eggs, meat, milk	Trace
Citrus fruits	25 to 60
Tomato	20
Peas (dried)	0
Peas (sprouting)	20 to 50
Potatoes (uncooked)	10 to 30
Potatoes (boiled)	5 to 15

of a man replete with the vitamin is about 5 g. Normally some 15 to 30 mg of ascorbic acid is contained in each 100 g of the buffy layer of the blood which is composed of leucocytes and platelets; half of the ascorbic acid is in the leucocytes and half in the platelets. However, in subjects depleted of ascorbic acid the ascorbic acid content of the leucocytes and platelets falls progressively over several months to reach very low values 3 to 6 weeks before the signs of scurvy appear. Normally blood plasma contains from 0·5 to 2 mg/100 ml, but values below 0·5 mg/100 ml may be found over a long time in an individual without any sign of ill-health. In man ascorbic acid appears in the urine when the concentration in the plasma reaches approximately 1 mg/100 ml.

Deficiency of ascorbic acid leads to defective formation of the collagen fibres of connective tissue. Since the laying-down of new connective tissue is essential to the healing of wounds, this process is retarded in man and in guinea-pigs made deficient in ascorbic acid (Plate 13.2). Formation of bone is also abnormal partly because of the abnormal collagen formation and partly because without ascorbic acid the specialized cells that lay down bone (osteoblasts) become functionless. The teeth of guinea-pigs made deficient in ascorbic acid become soft and spongy but they are quickly restored to normal by giving the animals ascorbic acid. There is no good evidence, however, that ascorbic acid is essential for healthy teeth in man.

The disorder that occurs in guinea-pigs, monkeys and man as the result of prolonged deficiency of ascorbic acid is called *scurvy*. Its main features are bleeding into the skin and deeper tissues and swollen bleeding gums (Plate 13.3). In young children there is bleeding under the periosteum of the bones and into the joints. The reason for the haemorrhage is not known but it has been attributed to deficiencies of the intercellular cement substance between the endothelium cells of capillaries (Chap. 27). Scurvy is rapidly and completely cured by the giving of ascorbic acid.

Ascorbic acid is found in nearly all fruits and vegetables; especially rich sources are rose hips, blackcurrants, green vegetables and citrus fruits (Table 13.19). Although potatoes contain smaller amounts they are an important dietary source especially when they are new. The ascorbic acid content of potatoes and vegetables diminishes gradually if they are stored and much more rapidly when they are cooked.

A daily intake of about 10 mg/day is sufficient to prevent scurvy. Most people in this country have more than 20 mg/day. To allow for individual variations in requirements and to provide a margin of safety, three times this dose, that is 30 mg/day, has been suggested as the daily requirement for a healthy adult. This advice is sound in spite of the fact that with this dose the body is by no means saturated.

Vitamin D (Calciferol)

Rickets, a disorder of children characterized by stunted growth and bowing of the limbs (Fig. 13.20) has been recognized for hundreds of years, but it has been known to be a deficiency disease only since 1918. Adults may have a corresponding deficiency disease, known as osteomalacia. Both disorders can be prevented or cured either by administration of vitamin D or by exposure to ultraviolet light.

The name vitamin D has been applied to a number of chemically similar heat-stable compounds, but the most important are vitamin D_2 (ergocalciferol) and vitamin D_3 (cholecalciferol). Ergocalciferol does not occur naturally but is produced artificially by the ultraviolet irradiation of a plant sterol, ergosterol. Cholecalciferol is the naturally occurring antirachitic compound of animals and, at least in mammals, is produced by the irradiation of a provitamin, 7-dehydrocholesterol.

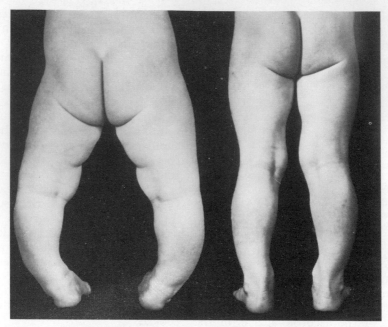

Fig. 13.20 Effect of vitamin D on infantile rickets before treatment and 2 years later after vitamin D therapy. (*By courtesy of* I. D. G. Richards, E. M. Sweet & G. C. Arneil (1968). *Lancet*, April 13, 803.)

Both provitamins, ergosterol and 7-dehydrocholesterol, are converted into the corresponding vitamins by cleavage of the bond between C-9 and C-10. In birds and mammals the vitamin D_3 is formed on the feathers or fur and licked off in 'preening'. In man the vitamin is formed in the skin and in dark-skinned races the melanin serves to limit the penetration of excessive ultraviolet light.

Vitamin D is found in a limited number of

ergosterol

ergocalciferol (vitamin D_2)

7-dehydrocholesterol

cholecalciferol (vitamin D_3)

194   Textbook of Physiology and Biochemistry

foods (Table 13.21) but fish livers are by far

Table 13.21   Vitamin D in foods (i.u. per 100 g)

Tunny liver oil	85 000 to 25 000 000
Halibut liver oil	20 000 to 400 000
Cod liver oil	2000 to 30 000
Fat fish	200 to 1800
Ox liver	40 to 50
Egg yolk	150 to 500
Butter	8 to 100
Cow's milk (raw)	1 to 4
Margarine	80 to 350
Meat, white fish	Trace

the richest source. The vitamin D content of eggs and butter is very variable and depends on the diet and exposure to sunlight of the hens and cows. Ergocalciferol is added to margarine during manufacture. Milk for infants is usually fortified by the addition of the vitamin.

Vitamin D is absorbed in the upper half of the small intestine (at least in rats) and bile salts are required. Some vitamin D is stored in the tissues. Both vitamin D_2 and D_3 are hydroxylated in the liver at the 25 position to give active metabolites: in the case of cholecalciferol, 25-hydroxy-cholecalciferol (25 HCC). Recent evidence suggests that 25 HCC is not itself the active metabolite but is further hydroxylated in the kidney to give 1,25 dihydroxy-cholecalciferol (1,25 DHCC). This substance appears to act as a hormone as its output by the kidney is governed by the serum calcium and it operates by controlling the absorption of calcium in the intestine. It operates at the cell nucleus by influencing the transcription of messenger RNA and the formation of calcium transport systems including those controlled by parathormone (Chap. 14). 1,25 DHCC probably acts on bone and kidney as well as on the intestine. The ill-effects of deficiency of vitamin D seem to be due to the limitation of the output of 1,25 DHCC in much the same way as severe iodine deficiency leads to diminished thyroxine production.

Vitamin D deficiency occurs whenever there is *both* dietary deficiency and lack of exposure to sunlight. Thus as a result of the reduction of the vitamin D content of dried milk since 1956 rickets has become more common in large cities such as Glasgow and London where atmospheric pollution filters off the ultraviolet light. Nutritional osteomalacia is recognized among house-bound elderly women in Britain and among Moslem women in purdah. Osteomalacia is also recognized as a consequence of disorders which impair the absorption of vitamin D: in coeliac disease, in obstructive jaundice, and after partial gastrectomy.

In both rickets and osteomalacia there is inadequate deposition of calcium salts in newly-formed bone matrix. In babies with rickets ossification is retarded or ceases but the cartilage cells continue to grow. This produces swellings adjacent to joints and of the coastal cartilages ('rickety rosary'). Since cartilage is transparent to X-rays, radiographs of rickets are easily recognized (Fig. 13.22). The process of bone remodelling (Chap. 14), which normally maintains the shape of the bones, is abnormal and the bones tend to bend under the child's weight (Fig. 13.20).

Supplements of calciferol to the diets of children may diminish the spread of dental decay already present and reduce the amount of new caries forming subsequently. However, since calciferol does not completely prevent dental caries it cannot be the only factor involved in this disease.

Dairy produce is the chief source of calciferol in the diet (Table 13.21); its vitamin D content depends on the diet of the animals and on the length of their exposure to sunlight. Ergocalciferol is added to margarine during manufacture. Milk may be fortified or enriched by irradiation of the milk or of the cow or by adding the vitamin directly to the milk. Cod liver oil, which consists mainly of polyunsaturated fats, contains both vitamin A and vitamin D. Fish livers all contain large amounts of vitamin D_3. Plant tissues contain no vitamin D. Calciferol is a relatively stable substance so that losses occurring during the storage preparation, canning or dehydration of food are probably small.

Infants and young children require some 400 to 700 i.u. (10 to 18 μg) of calciferol per day. It is important to note that toxic effects may be produced by higher doses—as little as 2000 i.u. per day in some infants. Adults probably require no dietary supplement except in pregnancy.

Because of the low vitamin D potency of most foods it is important that man should benefit by the natural method of obtaining the

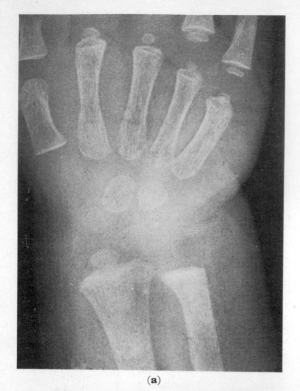

(a)

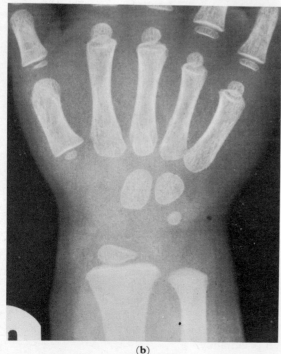

(b)

Fig. 13.22(a) and (b) Radiological appearances (a) before and (b) 3 months after administration of calciferol to a child with active rickets. (*By courtesy of G. C. Arneil.*)

vitamin, namely by the action of the sun's ultraviolet radiation upon the sterols in the skin. The enormous reduction in the incidence of rickets during the last 30 years is probably due as much to changed ways of bringing up children as to improvements in their diets.

VITAMIN E

Vitamin E is the generic name given to a number of chemical compounds, the most active of which is α-tocopherol. Several other tocopherols, with one or two CH_3 groups attached to the aromatic ring, instead of three as in α-tocopherol, have been isolated from natural sources, but they have lower biological activity. All tocopherols are methyl derivatives of the parent compound tocol. α-Tocopherol is a yellow oil, insoluble in water.

The richest sources of vitamin E are vegetable oils and wheat germ; the vitamin is present in small amounts in most vegetables, in dairy produce, meat and ox liver. It is not found in fish liver oils. Naturally occurring vitamin E is relatively stable, and is not destroyed by cooking. Vitamin E was isolated from wheat-germ oil in 1936, and has since been synthesized.

Vitamin E behaves as a biological antioxidant; this effect is particularly evident in the protection from oxidation of unsaturated fatty acids, vitamin A and carotene. The full biochemical role of vitamin E is not yet clear, and it is possible that in addition to its antioxidant (p. 53) property, it may play a part in protecting certain intracellular enzymes necessary for normal respiration; it also appears to exert some controlling influence on the rate of DNA biosynthesis.

Various pathological manifestations have been produced in different animal species by feeding diets deficient in vitamin E; these deficiency states have included testicular degeneration in male rats, death and resorption of the fetus in female rats, muscular dystrophy in rabbits and guinea-pigs, encephalopathy in chickens and macrocytic anaemia in monkeys. In vitamin E deficiency, lipid peroxidation of fatty acids leads to the formation of brown pigment ('ceroid') in various tissues. As vitamin E is itself oxidized and inactivated in the process of inhibiting the peroxidation of lipids, increased amounts of polyunsaturated fatty acids in the diet accelerate the depletion and increase the requirement of vitamin E. Most of the natural sources of unsaturated fatty acids contain tocopherol in proportion to the amount of linoleic acid present; however, storage or cooking may upset this relationship.

Despite the multiplicity of deficiency states in different animal species, the evidence for the occurrence of overt deficiency disease in man is less impressive. Abnormally low plasma levels of tocopherol occur in newborn, particularly premature, infants, and in children and adults who suffer from defects of fat absorption. The red blood cells of such subjects show abnormal susceptibility to the haemolytic effect of hydrogen peroxide *in vitro*, and *in vivo* there is shortening of the red-cell life-span, which is restored to normal by the administration of tocopherol. Overt haemolytic anaemia has been described in infants with severe deficiency of vitamin E.

From experimental dietary studies in man, the daily requirement of vitamin E to prevent any evidence of deficiency is probably in the range 5 to 15 mg; larger amounts may be required when the polyunsaturated fat content of the diet is high.

In vitamin K deficiency the process of blood coagulation (Chap. 23) is disturbed as the result of a reduction in prothrombin and factors VII, IX and X in the blood. A useful clinical index of this disturbance is the 'prothrombin time' (Chap. 23) which, normally 12 to 15 sec, may increase to over 30 sec. This impaired coagulability of the blood may actually lead to haemorrhage. Synthesis of these factors takes place mainly in the liver so that in severe diseases of the liver their concentration in the blood may be low no matter how much vitamin K is given.

Vitamin K is fat-soluble and its absorption from the alimentary tract therefore depends upon bile salts. In obstructive jaundice (Chap. 25), in which the amount of bile salt reaching the intestine may be much reduced, vitamin K, whether taken in the food or synthesized by the intestinal flora, is not absorbed in adequate amounts so that bleeding may occur. When the absorption of fat from the intestine is deficient the absorption of vitamin K is impaired and a tendency to haemorrhage may develop. Bleeding due to a combined deficiency of prothrombin and factors VII, IX and X is occasionally seen in new-born infants. It may

α-tocopherol, R_1, R_2 and R_3 = CH_3; β-tocopherol, R_1, R_3 = CH_3, R_2 = H; γ-tocopherol R_2, R_3 = CH_3, R_1 = H

VITAMIN K

At least two forms of the vitamin are found in nature. Phylloquinone (vitamin K_1) is synthesized in the green parts of the plants, and the menaquinones such as farnoquinone (vitamin K_2) by certain micro-organisms. Numerous active substances containing the 2-methyl-1,4-naphthoquinone structure (menadione, menaphthone) have now been synthesized; several are more potent biologically than either of the naturally occurring vitamins. The bacterial flora in the human intestine can synthesize farnoquinone and probably for this reason deficiency symptoms have never been produced in man or other mammals solely as the result of diets lacking in the vitamin.

be caused by reduced permeability of the placenta to vitamin K, inadequate maternal intake of the vitamin, or by the absence from the infant intestine of the bacterial flora which synthesize vitamin K.

The best natural sources of vitamin K are the dark green parts of plants, especially those of alfalfa, spinach, kale and cabbage. The dietary requirements of man are unknown; it is doubtful if deficiency ever occurs in adult man.

Cattle eating 'spoiled' sweet clover develop a haemorrhagic condition owing to the presence in the clover of the substance dicoumarol which acts as a vitamin K antagonist and interferes with prothrombin formation. Related synthetic coumarin deriva-

$$In\ 2\text{-methyl-}1:4\text{-naphthoquinone}:\ R\ is\ H$$

$$In\ phylloquinone:\ R\ is\ -CH_2.CH=C.CH_2.[CH_2.CH_2.\overset{CH_3}{\underset{|}{CH}}.CH_2]_2.CH_2CH_2.\overset{CH_3}{\underset{|}{CH}}.CH_3$$

$$In\ farnoquinone:\ R\ is\ -CH_2.[\overset{CH_3}{\underset{|}{CH}}=C.CH_2.CH_2]_5.CH=\overset{CH_3}{\underset{|}{C}}.CH_3$$

tives such as warfarin are used clinically as 'anticoagulants' to prevent the clotting of blood in cases of coronary artery occlusion or venous thrombosis. Massive doses cause internal bleeding so warfarin is employed also as a rat poison.

EFFECTS OF COOKING

The chemical composition of food may be considerably altered in the process of cooking which consists essentially in the application of heat, either direct (grilling, baking, roasting) or by means of hot water (boiling and steaming) or of hot fat (frying). Such treatment, apart from rendering most foodstuffs more palatable and more digestible, has the effect of destroying bacteria.

The most obvious effect of cooking on protein is coagulation such as occurs during the boiling of an egg. When meat is cooked water is lost, soluble protein is coagulated and collagen is converted to gelatin. At the same time the fibres shrink and become softer and looser so that the meat is easier to chew. Moderate cooking thus makes meat more readily digested; overheating has the opposite effect.

During cooking the starch granules of cereals and potatoes swell up and burst and in this form they are more easily digested; the cellulose framework of vegetables is softened and loosened. Soluble materials such as vitamins and salts are dissolved out of food during the process of boiling and may be lost unless the water is used to make soup. Loss of soluble materials is much less during roasting and baking, and less still during grilling and frying, in which the heat is usually applied for a short time and the surface rapidly sealed by the heat.

Nutritionally important amounts of calcium may be found in the diet if hard water is used in cooking or for drinking.

Cooking has little or no effect on any of the vitamins except thiamin and ascorbic acid. Thiamin is destroyed by heating, especially in the presence of alkali, and it does not survive the temperature used for example in the baking of biscuits. It is, however, not destroyed in the baking of bread unless baking soda is used. As much as 60 per cent of the ascorbic acid content of foods may be lost during cooking because (a) it is destroyed by the enzyme ascorbic acid oxidase (p. 135) liberated from damaged cells during the preparation of vegetables, (b) it is very soluble and thus is easily dissolved out by the water in which vegetables are cooked, and (c) part of it is destroyed by heat, especially in an alkaline medium. Vegetables should therefore be prepared immediately before they are to be used and cooked for as short a time as possible in the minimum of water.

WASTE. A certain part, sometimes a large proportion, of the food is discarded as inedible (bones, bacon-rind, peelings, outer leaves of vegetables). To allow for this *food-waste* the figures given in the tables refer to the edible portions of the foods. There is usually also some *plate-waste*, that is edible residue left on plates and in cooking utensils. In assessing the value of any diet, therefore, it is important to measure as exactly as possible the food eaten by the individuals concerned rather than the food supplied to the kitchen. If this cannot be done, an allowance of 10 per cent may be made for cooking and plate wastage. This is probably a liberal estimate but of course the fraction lost varies widely according to circumstances.

THE 'NORMAL' DIET

The energy requirements for various ages and occupations have been dealt with in Chapter 12 and we must now discuss the quantities of the proximate principles, protein, fat and carbohydrate that are needed.

PROTEIN REQUIREMENTS. The growing child requires an abundant supply of protein to provide the amino acids out of which new tissue protein is constructed; the adult requires protein to maintain the proper level of body protein to make good tissue lost by wear and tear, to build up new tissue protein after a wasting illness, and to supply amino acids essential for the synthesis of enzymes and certain hormones.

The nutritional value of any particular protein depends upon the nature of its constituent amino acids and their relative proportions. Some amino acids are readily synthesized from ammonia and simple carbon compounds; their presence in the diet is therefore not essential. Other amino acids required by the body cannot be synthesized by the tissues and must be supplied in the diet. These are known as the *essential* or *indispensable amino acids* and have been defined by Rose as 'not synthesized by the animal organism out of the materials ordinarily available at a speed commensurate with normal growth'. For example lysine and threonine must be supplied in the diet. In some cases a particular grouping, or nucleus, may be all that is necessary to allow the body to form a complete amino acid. For example, histidine is an essential amino acid but the body can complete its synthesis if the iminazole ring is supplied in the form of iminazole pyruvic acid, the keto acid corresponding to histidine. In such a case the indispensable unit appears to be the carbon chain, the amino group being introduced by the body as required. Indeed, all the indispensable amino acids except lysine and threonine take part in the reversible transfer of amino groups from amino acids to keto acids.

The division between the essential and the non-essential amino acids is not precise. The amino acid glycine is readily synthesized by the tissues of the growing chick but its glycine requirements are so great that the rate of utilization exceeds the rate of synthesis and glycine becomes one of the essential amino acids in its diet. It should also be noted that if certain essential amino acids are present in suboptimal amounts the deficiency in the diet may be made good by others not normally essential. Thus a portion of the essential amino acid methionine may be converted to cystine in the body (Chap. 22) and used in reactions involving cystine. Although cystine cannot be transformed to methionine its presence in the diet releases dietary methionine for other purposes. To take another example, the tissues can synthesize tyrosine from phenylalanine but not phenylalanine from tyrosine. Phenylalanine is therefore an essential amino acid while tyrosine is not, but the addition of tyrosine to a diet containing suboptimal amounts of phenylalanine may prevent the appearance of deficiency symptoms.

The amino acid requirements of the young animal or child differ in some respects from those of the adult. In the absence of certain amino acids from the diet growth may be retarded. For example when young rats were fed an otherwise adequate diet in which the sole protein is zein, from maize, which is deficient in tryptophan and lysine, they not only ceased to grow but lost weight. When tryptophan was added the fall in weight was arrested and when both tryptophan and lysine were given normal growth was resumed. Tryptophan and lysine are therefore essential amino acids for the growing rat (Fig. 13.23). Rose has used mixtures of the known amino acids in place of whole proteins; he withdrew amino acids one by one and observed the effect on growth and on nitrogen balance. Those whose withdrawal interfered with growth or caused negative nitrogen balances were classified as essential amino acids. Conversely, an amino acid which could be removed without affecting growth was non-essential.

The amino acid requirements of the adult are not easily determined, but it is found that, when certain of the amino acids are missing from an otherwise adequate diet, the output of nitrogen in urine and faeces exceeds the intake. This negative nitrogen balance (p. 201) can be made positive by the addition of the appropriate amino acid to the diet.

Dietary studies in adult man have shown that the amino acids essential for the maintenance of nitrogen balance are lysine, tryptophan, phenylalanine, leucine, isoleucine, threonine,

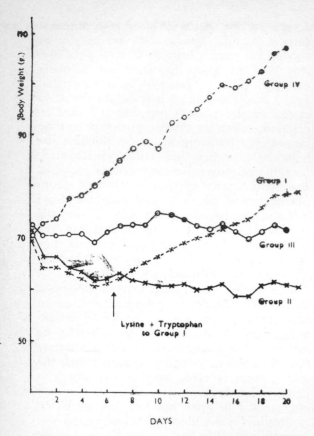

acids (alanine, aspartic acid, citrulline, cystine, glutamic acid, glycine, hydroxyproline, serine, tyrosine) are certainly not required either by man or by the growing rat.

The dietary value of any protein thus depends upon its content of essential amino acids. A protein such as lactalbumin containing all the essential amino acids is clearly nutritionally superior to one, such as zein of maize, which is deficient in tryptophan and lysine, or gelatin, which is deficient in both tryptophan and tyrosine. In other words those proteins have different *biological values*. Proteins of high biological value supply all the essential amino acids approximately in optimal proportions and are capable of supporting life and maintaining nitrogen balance if they are the sole source of nitrogen in the diet. Most animal proteins with the exception of gelatin fall into this category. Most plant proteins, on the other hand, are of low biological value because they are deficient in one or more of the essential amino acids. The distinction between these two types of protein, however, is seldom of practical importance for man does not normally eat a diet containing a single protein. In a mixed diet the lack of particular amino acids in one protein is made up by their presence in others. It is usually recommended that at least 50 per cent, and preferably 60 per cent, of the protein in a mixed diet should be of animal origin.

Attempts have been made to assess the minimum protein requirement by nitrogen balance requirements. On any given diet, there are three possible results: (i) nitrogen output may exceed intake; the subject is then in negative balance; (ii) output may be less than intake; in which case he is storing nitrogenous compounds and is said to be in positive balance; (iii) nitrogen intake may be just adequate to offset excretion and the subject is then in nitrogen equilibrium. Nitrogenous equilibrium is reached at a wide range of protein intake but below a certain level it cannot occur for the following reasons. If a human subject is given a diet of adequate energy value but containing no protein, the output of nitrogenous compounds in the urine is reduced to the equivalent of 2 to 3 g of nitrogen per day (Fig. 13.24). This is so called endogenous output of nitrogen is of course derived from the breakdown of tissue protein. The obli-

Fig. 13.23 An experiment carried out by H. N. Munro to show the effect of adding lysine and tryptophan to a zein diet fed to young rats. The rats in Group II were fed on a diet containing 15 per cent zein as sole source of protein. They did not grow and tended to lose weight. The rats in Group I were similarly fed but lysine and tryptophan were added at the point indicated by the arrow. Good growth then occurred. The rats in Group III were pair-fed with those in Group II on a diet containing 15 per cent casein in place of zein, that is they were given only that amount of food which the rats in Group II ate spontaneously. They showed poor growth, but they did not lose weight. The rats in Group IV received the casein diet in the same amounts as those in Group III but they were allowed to eat a non-protein diet *ad libitum* as well. They showed excellent growth. It is clear that, while addition of the missing amino acids improves the deficient diet, the animals on the deficient diet show poor appetite and this tends to aggravate their condition.

methionine and valine. About 1 g of each is the minimum daily requirement. Histidine and arginine are apparently not essential for nitrogen equilibrium in man but their absence may have specific effects. Thus the absence of arginine from the diet is followed by a reduction in the number of spermatozoa in the seminal plasma. The remaining nine amino

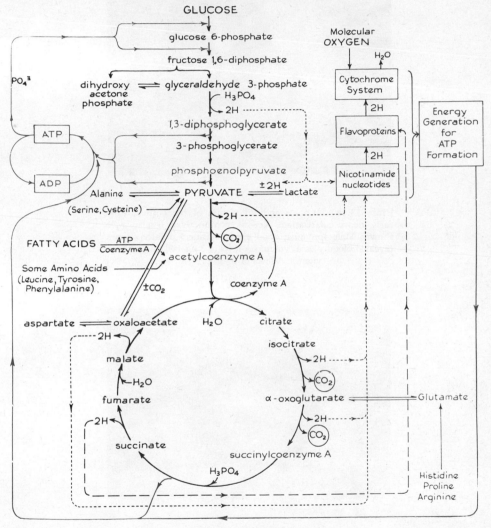

Plate 11.1 The overall pattern of glycolysis and the citric acid cycle. High energy compounds and pathways leading to them are shown in red.

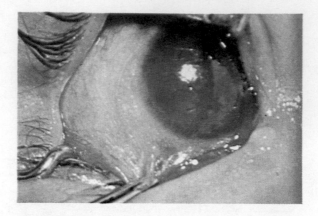

Plate 13.1a Haziness and infiltration of the cornea, with early neovascularization. Xerosis conjunctivae is also well shown. Male Jordanian child aged 2 years 3 months. Tests revealed blood-plasma vitamin A level of 3 μg/100 ml and liver vitamin A level of 0 μg/g liver.

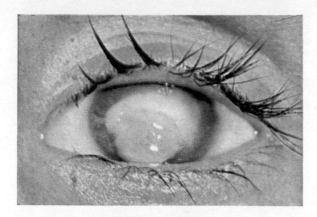

Plate 13.1b Perforation of the cornea and extrusion of the lens in an 18-month-old Jordanian male child. Tests revealed blood-plasma vitamin A level of 2 μg/100 ml. Comparative values for healthy children: plasma vitamin A 20–50 μg/100 ml; liver vitamin A 50 μg/g. (McLaren, Oomen & Escapini (1966). *Bulletin of the World Health Organization* **34**, 360.)

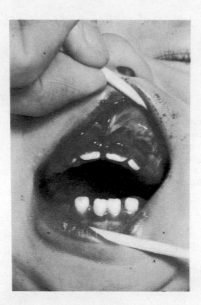

Plate 13.2   A 10-month old baby with scurvy. The gums were bleeding and there was a faint petechial rash on the trunk and extremities. The child was very sensitive to touch. X-ray examinations showed no skeletal abnormalities. The condition responded to administration of ascorbic acid. (*By courtesy of Dr H. Moll and Boehringer Ingelheim Ltd.*)

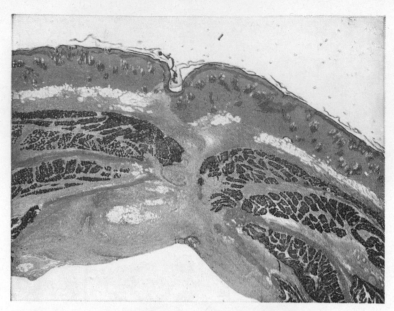

Plate 13.3a Section through a wound experimentally produced six days previously in a normal guinea-pig. This and the section below (×20) have been stained with the azan method which stains collagen blue; muscle fibres and skin are coloured red. Note the collagen formation throughout the wound, the absence of exudate and the compact nature of the healing process. The tensile strength of this wound was 674 g per cm.

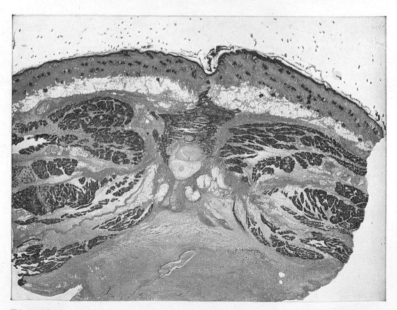

Plate 13.3b A similar wound in a guinea-pig deficient in ascorbic acid. There is still much exudate (red) lying between the muscle fibres; collagen formation is defective and such as is present in the deeper parts of the wound is loose and bulky. Tensile strength, 291 g per cm. (J. B. Hartzell & W. E. Stone (1942). *Surgery, Gynecology and Obstetrics* **75**, 1.)

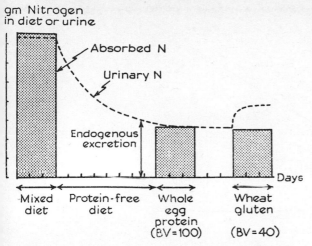

gm Nitrogen in diet or urine

Absorbed N

Urinary N

Endogenous excretion

Days

Mixed diet | Protein-free diet | Whole egg protein (BV=100) | Wheat gluten (BV=40)

Fig. 13.24 The urinary nitrogen output of a human subject on a protein-free diet gradually falls to a fairly constant level, the so-called 'endogenous output', representing obligatory loss of body protein. If the subject is now fed with whole egg protein, equivalent in amount to the loss of body protein, the urinary nitrogen output does not increase above the endogenous level; this indicates that the egg protein has been fully utilized by the body. Whole egg protein thus has a Biological Value of 100. However, if wheat gluten is fed in similar quantity, the urinary nitrogen output rises because this protein is poorly utilized, having a biological value of 40. (H. N. Munro (1964) in *Mammalian Protein Metabolism* (edited by H. N. Munro & J. B. Allison), Vol. 1, Introduction. New York: Academic Press.)

gatory loss in the faeces is about 1 g nitrogen per day. Loss of skin cells, hair, nails and sweat in adult man causes a further loss of over 1 g of nitrogen per day. In women the skin loss is smaller but 2 or 3 g nitrogen are lost at each menstrual bleeding. Even minor infections cause large and prolonged loss of urinary nitrogen. When the intake of protein is zero, the subject is in negative nitrogen balance to the extent of the nitrogen lost. A protein containing an amount of nitrogen equal to this negative balance may now be added to the diet. If this protein contains amino acids in the optimal proportions for utilization by the body, it will completely eliminate the negative nitrogen balance. This is found to apply to the mixed proteins of whole egg when fed to the rat. On the other hand, the protein gluten is not of the same quality as whole egg protein and rather more is needed to restore nitrogen equilibrium of the essential amino acids (Fig. 13.24). The biological values of some common proteins are given in Table 13.25.

Table 13.25 The biological value of some proteins for growing rats, mature rats and adult man (after H. H. Mitchell in *Protein and Amino Acid Requirements of Mammals* (Edited by A. A. Albanese). New York: Academic Press, 1950)

Protein	Biological value		
	Growing rat	Mature rat	Adult man
Egg albumin	97	94	91
Beef muscle	76	69	67
Wheat gluten	40	65	42
Caseinogen	69	51	56

Block and Mitchell express the nutritional value of a protein by calculating a *chemical score* in terms of whole egg protein which is almost completely utilized by the young rat. Fairly reliable figures are available for the essential amino acid content of most common foods and the chemical score is obtained by calculating the concentration of each essential amino acid in the protein of a foodstuff as a percentage of the concentration of the same amino acid in whole egg protein. The lowest percentage obtained is used as the chemical score since the limiting amino acid is held to determine the value of the whole protein. Figure 13.26 shows the chemical score of various proteins plotted against their biological value for rats as determined by nitrogen balance methods already described. The two methods of evaluation show a high correlation and it is clear that animal materials are superior to those of plant origin.

Several committees have from time to time discussed the protein requirements of man but, since no easy criterion is available, it is not surprising that their reports vary quite considerably.

The most reliable figures are those published by the Department of Health and Social Security in 1969 and reproduced in Table 13.27.

If the diet contains carbohydrate and fat for the provision of energy much less protein is required to produce nitrogen equilibrium than when protein alone is fed. Carbohydrate and fat are therefore said to exert a *protein-sparing action*. If his energy requirements are met, man can adjust himself to wide variations in the level of protein intake. In temperate climates he usually elects to take between 10

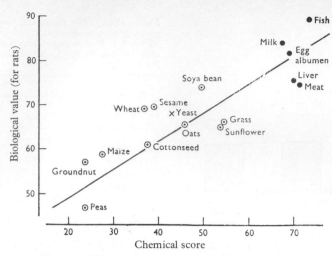

Fig. 13.26 Correlation diagram of the biological value and chemical score of five animal (⊙) and ten vegetable (●) materials and yeast (×). (*From* K. J. Carpenter (1951). *British Journal of Nutrition* **5**, 243.)

and 12 per cent of his total energy intake in the form of protein, of which about 60 per cent is animal protein. Some races take more protein than this without any apparent deleterious or beneficial effect. Protein is, however, the most expensive article of diet and in many countries where the family income is small the intake may be dangerously low. In Britain the cheapest sources of protein are milk, cheese and herring.

The specific dynamic action (p. 168) of protein is high and the consumption of protein increases the heat production of the body. Thus people exposed to cold usually have a particularly high protein intake; on a low protein diet they tend to feel cold. The increased energy expanded on a high protein diet is obtained from oxidation of fat and carbohydrate.

The popular impression that men engaged in hard physical work require more protein and consequently need to eat more meat than those engaged in sedentary occupations has no scientific support. The fuel used in muscular work is carbohydrate, not protein, and the increased energy needs of the manual worker can generally be met out of an increased consumption of carbohydrate and fat. But the man who is habitually engaged in hard physical work is generally larger, better developed and heavier than his sedentary colleague, and to that extent his protein requirements are

greater. It is of interest that the National Research Council recommends the same protein intake irrespective of physical activity (Table 13.27). The British Medical Association Committee of Nutrition on the other hand in an earlier report related the protein intake to the energy intake, so that a man engaged in very hard work is estimated to need 150 g protein.

MANIFESTATIONS OF INADEQUATE PROTEIN INTAKE. An animal given a diet low in protein loses protein from different organs at a rate depending on the carbohydrate content of the diet. The liver loses protein rapidly and prolonged protein deficiency can cause the death of large numbers of liver cells (liver necrosis). The kidney is less rapidly depleted of protein than the liver, and the loss of protein from skeletal muscle is much slower and less severe. Some organs, for example the eye, may actually continue to grow during protein depletion. Furthermore, a skin wound heals quite rapidly in a protein-deficient animal, though the process is less rapid than in the case of a well-fed animal. Malignant tumours also continue to grow in animals given protein-free diets. Since enzymes are protein in nature, it would be expected that they would be affected by severe protein deficiency. The effect varies from enzyme to enzyme; thus liver phosphatase is hardly altered, whereas xanthine oxidase practically vanishes from the

Table 13.27 Recommended daily intakes of energy and nutrients for the U.K.

(a) Age range	Occupational category	(c) Body weight kg	(d) Energy kcal	Energy MJ (e)	(f) Protein g	(g) Thiamin mg	Riboflavine mg	Nicotinic acid (h) mg equivalents	Ascorbic acid mg	Vitamin A (i) μg retinol equivalents	Vitamin D (j) μg cholecalciferol	Calcium mg	Iron mg
BOYS AND GIRLS													
0 up to 1 year (b)		7·3	800	3·3	20	0·3	0·4	5	15	450	10	600 (l)	6 (l)
1 up to 2 years		11·4	1200	5·0	30	0·5	0·6	7	20	300	10	500	7
2 up to 3 years		13·5	1400	5·9	35	0·6	0·7	8	20	300	10	500	7
3 up to 5 years		16·5	1600	6·7	40	0·6	0·8	9	20	300	10	500	8
5 up to 7 years		20·5	1800	7·5	45	0·7	0·9	10	20	300	2·5	500	8
7 up to 9 years		25·1	2100	8·8	53	0·8	1·0	11	20	400	2·5	500	10
BOYS													
9 up to 12 years		31·9	2500	10·5	63	1·0	1·2	14	25	575	2·5	700	13
12 up to 15 years		45·5	2800	11·7	70	1·1	1·4	16	25	725	2·5	700	14
15 up to 18 years		61·0	3000	12·6	75	1·2	1·7	19	30	750	2·5	600	15
GIRLS													
9 up to 12 years		33·0	2300	9·6	58	0·9	1·2	13	25	575	2·5	700	13
12 up to 15 years		48·6	2300	9·6	58	0·9	1·4	16	25	725	2·5	700	14
15 up to 18 years		56·1	2300	9·6	58	0·9	1·4	16	30	750	2·5	600	15
MEN													
18 up to 35 years	Sedentary	65	2700	11·3	68	1·1	1·7	18	30	750	2·5	500	10
	Moderately active		3000	12·6	75	1·2	1·7	18	30	750	2·5	500	10
	Very active		3600	15·1	90	1·4	1·7	18	30	750	2·5	500	10
35 up to 65 years	Sedentary	65	2600	10·9	65	1·0	1·7	18	30	750	2·5	500	10
	Moderately active		2900	12·1	73	1·2	1·7	18	30	750	2·5	500	10
	Very active		3600	15·1	90	1·4	1·7	18	30	750	2·5	500	10
65 up to 75 years	Assuming a sedentary life	63	2350	9·8	59	0·9	1·7	18	30	750	2·5	500	10
75 and over	Assuming a sedentary life	63	2100	8·8	53	0·8	1·7	18	30	750	2·5	500	10
WOMEN													
18 up to 55 years	Most occupations	55	2200	9·2	55	0·9	1·3	15	30	750	2·5	500	12
	Very active		2500	10·5	63	1·0	1·3	15	30	750	2·5	500	12
55 up to 75 years	Assuming a sedentary life	53	2050	8·6	51	0·8	1·3	15	30	750	2·5	500	10
75 and over	Assuming a sedentary life	53	1900	8·0	48	0·7	1·3	15	30	750	2·5	500	10
Pregnancy, 2nd and 3rd trimester			2400	10·0	60	1·0	1·6	18	60	750	10 (k)	1200 (m)	15
Lactation			2700	11·3	68	1·1	1·8	21	60	1200	10	1200	15

Footnotes to table:

(a) The ages are from one birthday to another: e.g. 9 up to 12 is from the 9th up to, but not including, the 12th birthday. The figures in the table in general refer to the mid-point of the ranges, though those for the range 18 up to 35 refer to the age 25 years, and for the range 18 up to 55, to 35 years of age.

(b) Average figures relating to the first year of life.

(c) The body weights of children and adolescents are averages and relate to London in 1965 (taken from Tanner, Whitehouse & Takaishi, 1966; Tables IV A and IV B, 50th centile). The body weights of adults do not represent average values; they are those of the FAO (1957) reference man and woman, with a nominal reduction for the elderly.

(d) Average requirements relating to groups of individuals.

(e) Megajoules (10^6 j). Calculated from the relation 1 kcal = 4·186 kJ, and rounded to 1 decimal place.

(f) Recommended intakes calculated as providing 10 per cent of energy requirements.

(g) The figures, calculated from energy requirements and the recommended intake of thiamin of 0·4 mg/1000 kcal, relate to groups of individuals.

(h) 1 nicotinic acid equivalent = 1 mg available nicotinic acid or 60 mg tryptophan.

(i) 1 retinol equivalent = 1 μg other biologically active carotenoids.

(j) No dietary source may be necessary for those adequately exposed to sunlight, but the requirement for the housebound may be greater than that recommended.

(k) For all three trimesters.

(l) These figures apply to infants who are not breast fed. Infants who are entirely breast fed receive smaller quantities; these are adequate since absorption from breast milk is higher.

(m) For the third trimester only.

liver. Protein insufficiency causes a considerable reduction in the plasma albumin level which, unlike the other manifestations of protein deficiency, can easily be measured. The colloid osmotic pressure of the plasma falls and consequently the subject may develop oedema (Chap. 33). Plasma globulin is not reduced.

Severe protein malnutrition, called *kwashiorkor*, affects millions of young children in the tropics, notably in Asia and Africa. It occurs in infants after weaning when instead of their mothers' milk they take a mainly carbohydrate diet. Although the energy (calorie) intake may be adequate the children cease to grow and they lose weight. They become listless, apathetic and peevish. The skin and hair become depigmented; the body becomes oedematous. The fundamental defect is lack of amino acids to synthesize proteins. The serum albumin concentration falls and pathological changes are found mainly in the liver and pancreas, organs with a high protein turn-over. The children can be cured by giving them milk or other protein-containing foods. The condition is, however, very often fatal and is one of the main causes of the high mortality among children in the tropics.

Immediately after physical injuries, for example a fracture of a bone or a surgical operation, patients go into negative nitrogen balance, often to the extent of 15 g per day for several days. It seems to be due to increased secretion of ACTH and adrenocortical hormones. Cuthbertson named this phenomenon the *catabolic response to injury*. Excessive protein catabolism may also occur in some chronic diseases. All such patients therefore should receive a high-protein, high-energy diet to make good this loss.

FAT REQUIREMENTS. The fat of the diet is important not only on account of its high energy value but because it is the vehicle for the fat-soluble vitamins A, D, E and K, and contains the essential fatty acids, *linoleic acid* and *linolenic acid*. Both are unsaturated acids which cannot be synthesized by the tissues. However, in contrast to rats and domestic animals, man has never been proved, even on very poor diets, to suffer from a deficiency of essential fatty acids (EFA).

Provided that the supply of fat-soluble vitamins and essential fatty acids is adequate, fat itself is not an absolutely essential constituent of the diet. The fat intake in human diets varies over a wide range according to climate, race and dietary custom. In Britain the fat intake is about 80 to 150 g per head per day, and in Europe generally, and in U.S.A. the fat content of the diet usually accounts for 35 to 40 per cent of the total energy. In tropical countries and in the East much smaller quantities of fat are normally eaten, say 15 per cent or even less of total energy, while in circumpolar regions the fat intake may be as high as 300 g per day. Although hard-and-fast rules cannot be laid down it may be said that when the food intake is less than 12·6 MJ or 3000 kcal/day fat should account for at least 25 per cent of the total energy; if the intake is greater than 12·6 MJ/day fat should account for at least 30 per cent of the total energy. A diet lacking in fat tends to be bulky if it supplies adequate energy. Fat adds to the palatability of foods and for this reason is used extensively for baking, cooking, frying and for spreading on bread.

CARBOHYDRATE REQUIREMENTS. When the requirements for protein and fat have been met, the remaining energy needs are derived from carbohydrate. The proportion of carbohydrate in the diet is subject to wide variations but in Europe and America less than 50 per cent of the total energy intake may come from carbohydrate; it is usually recommended that between 55 to 65 per cent should be obtained from this source. Since carbohydrate foods are generally cheap they tend to form a large proportion of the diet of the poorer sections of the community. In the East and in much of Africa carbohydrate, derived mainly from cereals, may account for as much as 90 per cent of the total intake.

It is of course desirable to eat sufficient carbohydrate to prevent ketosis (Chap. 19), but the amount required for this purpose is very low and is in practice always exceeded unless deliberate measures are taken to construct a ketogenic diet.

THE EFFECTS OF STARVATION. When the organism is deprived of food it calls first on its stores of carbohydrate and fat for the provision of energy. Stores of carbohydrate—glycogen in liver and muscle (Chap. 18)—are never great and are soon exhausted but, in the well nourished subject, fat stores may be sufficient, if activity is reduced, to supply the needs of the body for several weeks. When fat is being metabolized in large amount, ketosis (Chap. 19) results and the urine may contain large amounts of ketone bodies and may have a high titratable acidity and ammonia content. Even before the reserves of non-nitrogenous fuel are exhausted, the tissue proteins begin to be broken down to provide energy. The skeletal, cardiac and visceral muscle all atrophy but the brain does not. When the fat stores are exhausted protein is the sole source of energy. Thus, while nitrogen excretion is low in the early stages of starvation, it rises sharply after the exhaustion of carbohydrate and fat stores, when the tissue proteins are being drawn upon. This rise in nitrogen excretion occurs shortly before death.

MEALS

Having considered the requirements of the various nutrients in the diet, we may examine the composition of two sample meals (Table 13.28) to find how requirements are met. Meal 1 is a light lunch which might be consumed during the course of a morning's work; meal 2

Table 13.28 Analysis of two meals (*Manual of Nutrition* (1951). Ministry of Food)

Meal 1	Weight oz	g	kJ	kcal	Protein g	Fat g	Calcium mg	Iron mg
Roll	3·5	98	1054	252	8·8	1·1	109	1·8
Butter	0·3	8	271	65	0·0	7·3	0	0·0
Cheese	2·0	56	979	234	14·2	19·6	460	0·4
Tea (with milk and sugar)	10·0	280	84	20	1·0	1·0	30	0·0
Total			2389	571	24·0	29·0	599	2·2

Meal 2	Weight oz	g	kJ	kcal	Protein g	Fat g	Calcium mg	Iron mg
Mutton	2·4	67	941	225	8·9	21·1	7	1·4
Cabbage	3·0	84	88	21	1·2	0·0	54	0·9
Potato	4·0	112	351	84	2·4	0·0	8	0·8
Apples	3·0	84	151	36	0·3	0·0	3	0·3
Custard	2·0	56	259	62	1·8	2·2	70	0·0
Total			1791	428	14·6	23·3	142	3·4

Table 13.29 Daily diet of women doing light engineering at a factory in Birmingham

Meal	kJ	kcal	Protein g	Fat g	Calcium mg	Iron mg	Vitamin A i.u.	Thiamin mg	Ribo-flavin mg	Nicotinic acid mg	Ascorbic acid mg
Breakfast,* 7 a.m.	2971	710	21	26	240	4	630	0·51	0·30	2·9	5
Snack† 10 a.m.	879	210	9	8	140	1	230	0·12	0·12	0·6	0
Dinner,‡ 1 p.m.	3473	830	30	27	350	4	1100	0·87	1·90	5·2	30
Snack,§ 4 p.m.	1004	240	5	9	50	1	50	0·06	0·05	0·3	0
Tea,‖ 7 p.m.	2385	570	19	21	100	4	560	0·51	1·12	3·9	31
Total day's nutrients	10711	2560	84	91	880	14	2570	2·07	3·49	12·9	66

* Substantial breakfast.   † Piece of cake.   ‡ Meat, vegetables and a substantial pudding.   § Scone.
‖ A good meal, including chipped potatoes.

is of the type provided by a factory canteen as a midday dinner. A comparison of the two meals shows some striking differences. Meal 1, which appears superficially to be much less substantial than meal 2, is in fact much richer in energy, in protein, in fat and in calcium. It is also much richer in vitamin A on account of the high vitamin A content of butter and cheese. The only feature in which meal 2 is superior is in ascorbic acid content.

Table 13.29 shows the figures obtained for a whole day's diet by adding together the figures for individual meals of a group of women engaged in factory work. The total intake of 10·7 MJ (2560 kcal) compares favourably with the allowance of 10·5 MJ (2500 kcal) proposed in Table 13.27 for women engaged in medium work. For such women the proposed protein allowance is 60 g and the amount provided (84 g) is therefore liberal. It accounts for 13 per cent of the total energy. Fat provides 32 per cent of the total energy, which is over the minimum figure recommended on p. 204. Iron and calcium are both above the recommended allowances. The amount of vitamin A (2570 i.u.) is less than the recommended allowance of 5000 i.u. but is slightly greater than the minimum safe figure of 2500 i.u. Thiamin (2·07 mg) and riboflavin (3·49 mg) are both present in amounts exceeding the recommended allowances of 1·0 and 1·5 mg respectively. Nicotinic acid (12·9 mg) is above the allowance of 10 mg and the amount of ascorbic acid (66 mg) is ample. This diet is therefore satisfactory.

REFERENCES

Valuable reviews of various problems of nutritional importance are to be found from time to time in the *Proceedings of the Nutrition Society*. The following publications will also be found useful:

ALBANESE, A. A. (Ed.) (1970). *Newer Methods of Nutritional Biochemistry*, Vol. 4. New York: Academic Press.

ARNSTEIN, H. R. V. & WRIGHTON, R. J. (1971). *The Cobalamins. A Glaxo Symposium*. Edinburgh: Churchill Livingstone.

BARKER, H. A. (1967). Biochemical functions of corrinoid compounds. *Biochemical Journal* **105**, 1–10.

BEATON, G. H. & McHENRY, E. W. (Eds.) (1964). *Nutrition, A Comprehensive Treatise in Three Volumes*. New York: Academic Press.

BROCK, J. F. (1961). *Recent Advances in Human Nutrition*. London: Churchill.

CAMPS, F. E. (1968). Alcohol. *Journal of the Royal College of Physicians, London* **2**, 311–326.

DAVIDSON, S. & PASSMORE, R. (1972). *Human Nutrition and Dietetics*, 5th edn. Edinburgh: Livingstone.

Department of Education and Science (1964). *Requirements of Man for Protein*. London: H.M.S.O.

Domestic food consumption and expenditure: 1950 to 1964. *Annual Reports of the National Food Survey Committee*. London: H.M.S.O.

DRUMMOND, J. C. & WILBRAHAM, A. (1958). *The Englishman's Food*. Revised edition. London: Cape.

F.A.O. (1954). Food composition tables—minerals and vitamins for international use. *Nutritional Studies* No. 11. Rome: F.A.O.

Feeding the World. (1968). *Science Journal* **4**, No. 5, 3–106.

Food Composition Tables for International Use (1953). 2nd edn. Food and Agricultural Organization of the United Nations. Rome: F.A.O. *Nutritional Studies* No. 3.

FOURMAN, P. & ROGER, P. (1968). *Calcium Metabolism and the Bone*. Oxford: Blackwell.

GOLDSMITH, G. A. & DUNCAN, G. G. (Eds.) (1964). *Diseases of Metabolism*. London: Saunders.

GOODWIN, T. W. (1963). *The Biosynthesis of Vitamins and Related Compounds*. New York: Academic Press.

HALPERN, S. L. (Ed.) (1964). Symposium on recent advances in applied nutrition. *Medical Clinics of North America* **48**, 1111–1281.

KEYS, A., BROZEK, J., HENSCHEL, A., MICHELSON, O. & TAYLOR, H. L. (1950). *The Biology of Human Starvation*. Oxford: University Press.

KITCHIN, A. H. & PASSMORE, R. (1949). *The Scotsman's Food*. Edinburgh: Livingstone.

KNOX, W. E. & GOSWAMI, W. N. D. (1961). Ascorbic acid in man and animals. *Advances in Clinical Chemistry* **4**, 121–205.

KON, S. K. & COWIE, A. T. (1961). *Milk*. New York: Academic Press.

LATHAM, M. (1965). *Human Nutrition in Tropical Africa*. Rome: F.A.O.

LOOMIS, W. F. (1970). Rickets. *Scientific American* **223**, 76–91.

McCANCE, R. A. & WIDDOWSON, E. M. (1967). The Composition of Foods. *Special Report Series, Medical Research Council, London*, No. 297. London: H.M.S.O.

McCANCE, R. A. & WIDDOWSON, E. M. (1956). *Breads White and Brown: Their Place in Thought and Social History*. London: Pitman.

McCance, R. A. & Widdowson, E. M. (Eds.) (1968). *Calorie Deficiencies and Protein Deficiencies*. London: Churchill.

McGillivray, W. A. & Porter, J. W. G. (1958). Nutritive value of milk and milk products. *Journal Dairy Science* **25**, 344–363.

Marks, J. (1968). *The Vitamins in Health and Disease*. London: Churchill.

Ministry of Health (1964). *Ready Reckoner of Food Values*. London: H.M.S.O.

Moore, T. (1957). *Vitamin A*. London: Cleaver House Press.

Moran, T. (1959). Nutritional significance of recent work on wheat, flour and bread. *Nutrition Abstracts and Reviews* **29**, 1–16.

Munro, H. N. & Allison, J. B. (1963–1970). *Mammalian Protein Metabolism*, Vols. I to IV. New York: Academic Press.

Platt, R. S. (1956). Protein malnutrition. *Lectures on the Scientific Basis of Medicine* **4**, 145–166.

Platt, R. S. (1962). Tables of representative values of foods commonly used in tropical countries. *Special Report Series, Medical Research Council, London* No. 302. London: H.M.S.O.

Protein requirements (1965). Report of a joint F.A.O.–W.H.O. expert group. *World Health Organization Technical Report Series* No. 301. Geneva: W.H.O.

Pyke, M. (1964). *Food Science and Technology*. London: John Murray.

Rasmussen, E. (Ed.) (1970). *International Encyclopedia of Pharmacology and Therapeutics*, Section 51, Vol. 1, Chaps. 4 and 5.

Requirements of Man for Protein (1964). *Reports on Public Health and Medical Subjects*, No. 111. London: H.M.S.O.

Rose, W. C. (1957). The amino acid requirements of adult man. *Nutrition Abstracts and Reviews* **27**, 631–647.

Sebrell, W. H. & Harris, R. S. (1967) *The Vitamins*, 2nd edn., Vol. 1; (1969) *The Vitamins*, 2nd edn., Vol. 2; (1970) *The Vitamins*, 2nd edn., Vol. 3. New York: Academic Press.

Sinclair, A. M. & Hollingsworth, Dorothy F. (Eds.) (1969). *Hutchison's Food and the Principles of Nutrition*, 12th edn. London: Arnold.

Smith, E. Lester (1960). *Vitamin B_{12}*. London: Methuen.

Symposium, Advances in the detection of nutrition deficiencies in man. (1967). *American Journal of Clinical Nutrition* **20**, 513–658.

Symposium on the place of food science and technology in the campaign against malnutrition (1961). *Proceedings of the Nutrition Society* **20**, 91–137.

Symposium on the values of animal and vegetable fats in nutrition (1961). *Proceedings of the Nutrition Society* **20**, 138–173.

Thompson, R. H. S. & Wootton, I. D. P. (1971). *Biochemical Disorders in Human Disease*. London: Churchill.

Thomson, A. M. (1971). Nutrition in pregnancy. *British Journal of Hospital Medicine* **5**, 600–612.

Underwood, E. J. (1956). *Trace Elements in Human and Animal Nutrition*. New York: Academic Press.

Vitamins and Hormones. New York: Annual publication of Academic Press Inc.

Von Wartburg, J. P. (1966). Metabolism of alcohol. *Science Journal* (June) 2–7.

Wadsworth, G. R. & McKenzie, J. C. (1963). The potato. *Nutrition Abstracts and Reviews* **33**, 327–344.

Weissbach, H. & Dickerman, H. (1965). Biochemical role of vitamin B_{12}. *Physiological Reviews* **45**, 80–97.

Whipple, H. E. (Ed.) (1964). Vitamin B_{12} coenzymes. *Annals of the New York Academy of Sciences* **112**, 547–921.

14 Bone and minerals

The term connective tissue is used to denote the material which joins up the other three primary tissues, the epithelial, muscle and nervous tissues. Some kinds of connective tissue cells secrete solid intercellular substances—cartilage and bone—which provide protection and support. Other kinds form loose or dense fibrous tissue, tendons, blood vessels, adipose tissues and possibly also the blood cells. Apart from cells most connective tissues also contain extracellular proteins such as collagen and elastin, mucopolysaccharides and, in the case of bone and teeth, inorganic salts.

ORGANIC CONSTITUENTS OF CONNECTIVE TISSUE

Collagen derives its name from the fact that when boiled in water it yields gelatin or glue. It is a protein of unusual amino acid composition. One third is glycine and another third is proline and hydroxyproline (p. 26). It contains alanine but no tryptophan or cysteine. Hydroxyproline and hydroxylysine are found only in collagen and elastin. Vitamin C is required for their formation and in its absence collagen formation is defective. Since collagen contains 14 per cent hydroxyproline, the collagen content of a tissue can be deduced from the estimation of its hydroxyproline content. Similarly the urinary excretion of hydroxyproline provides a measure of the rate of collagen turnover in the body; high values are found in growing children and in some patients with bone disease. Collagen molecules consist of three chains of amino acids wound round one another to form a triple helix (Fig. 3.8). The molecules are released from the fibroblasts in which they have been formed and aggregate extracellularly to give long fibrils which may be recognized under the electron microscope by the cross bands at regular intervals of 64 nm (Fig. 14.1).

Elastin. Very much less is known about

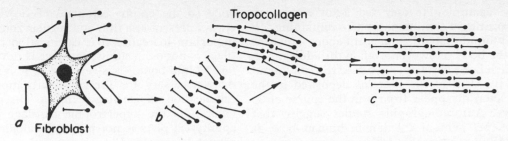

Fig. 14.1 Tropocollagen molecules are extruded from fibroblast (or osteoblast) and aggregate extracellularly to give collagen fibres with the regular banding which may be detected by electron-microscopy. (*After* J. Gross (1961). *Scientific American* **204**, 120.)

elastin, the protein of the elastic, or yellow, fibres of connective tissue. The cell which forms elastin is not known but some fibroblasts may produce both elastin and collagen. Its amino acid composition is unusual in that it contains two unique amino acids desmosine and isodesmosine (p. 26) which probably provide the cross-linking between polypeptide chains. The cross-linkages are essential for the characteristic elasticity of elastin which may be extended 100 per cent without breaking. Elastin is found particularly in ligaments, in the skin, and in the wall of the aorta and large blood vessels where its extensibility and brisk rebound are important (see Chap. 27).

Mucopolysaccharides. Acid mucopolysaccharides (p. 47) are found in varying quantity in all connective tissues; tendon contains 0·5 per cent and hyaline cartilage 20 per cent. The types of mucopolysaccharides extractable from a tissue depend on its site and on its age. The principal mucopolysaccharides of cartilage are the chondroitin sulphates; their combination with collagen gives cartilage its rigidity. In several rare inherited disorders, excess quantities of mucopolysaccharides are laid down in the tissues.

CARTILAGE

Hyaline cartilage contains cells known as chondrocytes which produce an amorphous intercellular material consisting mainly of sulphated mucopolysaccharides and collagen. If the chondrocytes are active they form more and more intercellular substance, and cartilage grows in volume interstitially. If the cells surrounding the cartilage, collectively termed the perichondrium, lay down intercellular substance cartilage grows by apposition. Since cartilage has no blood vessels, nutrient materials can arrive at the chondrocytes only by diffusion through the matrix.

BONE

Dried and fat-free bone contains 30 to 35 per cent organic material and 65 to 70 per cent inorganic material. Collagen accounts for 90 per cent of the organic matter. Much of the remainder consists of mucopolysaccharides, protein complexes, glycoproteins (protein-oligosaccharide complexes) and lipids.

The mineral of bone to which its hardness and rigidity are due is a crystalline material which can be shown by X-ray crystallography to possess a lattice structure similar to that of the mineral apatite. The elementary composition is denoted approximately by $Ca_{10}(PO_4)_6(OH)_2$. The Ca/P ratio in bone is 1·5 which is lower than that of apatite, namely 1·67, but this slight difference can be accounted for by various substitutions in the lattice which has small amounts of Na, Mg, Sr, K, Cl and F together with bicarbonate, citrate and water. About 60 per cent of the apatite is crystalline in form and the surface area presented by the small crystals must be very large. About 40 per cent of adult bone mineral is not crystalline but is an amorphous tricalcium phosphate which has a distinct appearance (doughnut) on electron microscopy. The amorphous material seems to be converted *in vivo* to crystalline apatite; the proportion of crystalline material increases with age.

Experiments with radioactive labels show that there is a very large continuous exchange of calcium and phosphate ions between the inorganic material and the interstitial fluid.

For example Hevesey fed adult rats small quantities of sodium phosphate containing ^{32}P and found that a P atom spent about 2 months in the body. However, even in the adult rat, in which bone has virtually ceased to grow, about 30 per cent of the P atoms deposited in the skeleton disappear from it in the course of 20 days. Autoradiographic studies suggest that 0·65 per cent of calcium in human bone is readily exchangeable whereas *in vitro* experiments show that 27 per cent is exchangeable. This probably means that only the linings of Haversian canals or resorption cavities produced by osteoclasts can exchange ions freely. In spite of these great exchanges of ions the shape and size of the bones remain fixed.

BONE GROWTH AND DEVELOPMENT. Many of the bones appear in the embryo as cartilaginous models to be replaced sooner or later by bone. This process is seen particularly easily at the epiphyseal regions of long bones where the chondrocytes are actively producing cartilage. A long bone grows in length by the interstitial growth of the epiphyseal cartilage, in which the cells are arranged in regular longitudinal columns produced by repeated cell division (Fig. 14.2). Each row of cells is enclosed in a tunnel of cartilage with thin partitions between the cells. As the cells approach the diaphyseal (shaft) side of the epiphyseal plate they enlarge and the cartilage around them becomes calcified. Most of the cells next to the marrow cavity are dead and the thin transverse cartilage partitions disappear leaving tunnels which are invaded by capillaries and by osteoblasts. The osteoblasts put down a layer of bone on the inner walls of the tunnels, and this process is repeated. In this way only a narrow channel containing a blood vessel and some cells remains. This concentric arrangement is called a Haversian system. The cells contained in tiny spaces in the Haversian system—descendants of the osteoblasts—are called osteocytes. These early Haversian systems are remodelled by osteoclasts, which remove bone, and by osteoblasts, which lay down fresh bone. The process of remodelling goes on throughout life. Narrow canaliculi pass radially and circumferentially in the Haversian systems to convey nutrients to the enclosed osteocytes. During the growing period the interstitial growth of epiphyseal cartilage keeps pace with its replacement by

bone so the epiphyseal plate of cartilage is always present as a thin and regular zone. It is important to realize that calcification of cartilage is not ossification; cartilage is not *transformed* into bone—it is *replaced* by it. Vitamin A is necessary for growth, maturation and calcification of cartilage. In the absence of vitamin D the hypertrophic cartilage in the epiphyseal plate is not properly calcified and cannot be removed. As a result the cartilage layer becomes much thickened and disorganized (Fig. 13.27).

The shafts of long bones are made of a hollow cylinder of hard compact bone containing marrow. In the adult the marrow is mostly yellow and fatty with a small amount of red (haemopoietic) marrow at the ends. The flat bones, for example the sternum and ilium, are the usual sites for puncture to obtain a specimen of haemopoietic tissue. The ends of long bones and also the vertebrae and flat bones are supported internally by spongy bone (spongiosa) arranged in a pattern of trabeculae which is apparently determined by the load carried by the bone. Compact bone is nearly as strong as cast iron but much more flexible.

Because bone is rigid, an increase in size by interstitial growth is impossible—it can grow only by the laying down of new bone on the surface of already existing bone, that is by apposition. The osteoid matrix laid down by the osteoblasts is calcified to form bone as soon as it is formed unless there is a deficiency of vitamin D. In contrast to cartilage, bone is a vascular tissue and no bone cell is more than 0·1 mm from a capillary. Adult bone receives about 10 ml of blood per 100 g per min; the whole skeleton must therefore receive about 7 per cent of the resting cardiac output.

As the child grows so do his bones and they have to be remodelled by the removal of bone and the laying down of new bone. At the sites of resorption there are large cells called osteoclasts containing six or more nuclei. They are probably formed by fusion of many separate cells; no mitotic figures have been seen in them. The electron microscope shows that the osteoclast has a complex system of folds (brush border) in the cytoplasm lying nearest to the bone. Bone crystals and fragments of collagen have been seen extracellularly between the folds. It is not clear how osteoclasts remove bone but they possess many enzymes in-

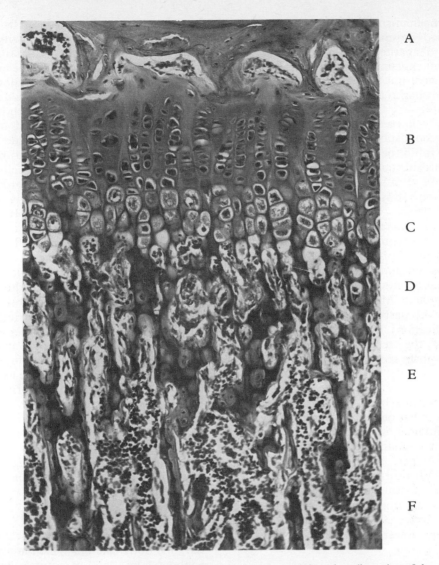

Fig. 14.2 (× 140). A longitudinal section through the upper epiphyseal cartilage plate of the tibia in a normal young rat, showing the normal growth processes (see p. 210). (*By courtesy of H. A. Sissons.*) A, Epiphyseal bone. B, Epiphyseal cartilage with regular columns of dividing cells. C, Enlarged dying or dead cells surrounded by calcified cartilage. D, Transverse cartilage partitions have disappeared. The tunnels are invaded by capillaries and osteoblasts. E, Bone is being laid down on cartilage and removed and bony trabeculae are laid down. F, Marrow cavity containing blood vessels and bone cells.

cluding collagenase. Osteoclastic activity is stimulated by parathyroid hormone. It is generally accepted that in resorption of bone the inorganic and the organic materials are removed at the same time. In vitamin D deficiency, which gives rise to rickets in children and osteomalacia in the adult, osteoid tissue is laid down on bone surfaces but is not ossified. The osteoid seams are *not* caused by removal of salts from the osteoid matrix.

MECHANISM OF CALCIFICATION. At one time it was thought that calcification resulted from a local increase in the concentration of calcium and phosphate. It was suggested that alkaline phosphatase, by breaking down phosphate esters into inorganic phosphate, provided this booster mechanism. Later when it was realized that blood plasma was supersaturated with respect to apatite and bone mineral it was difficult to explain why large amounts of bone mineral were not laid down all over the body and why collagen in other parts of the body did not calcify.

Recently it has been suggested that inorganic pyrophosphate, $(O_3P—O—PO_3)^{4-}$ is the local inhibitor of calcification at bone surfaces. This substance, found in both blood and urine, is *in vitro* a very powerful inhibitor of calcification. It is split by *pyrophosphatases*, some of which have been identified with alkaline phosphatase, and whose local concentration could regulate bone formation. In the urine pyrophosphate also probably inhibits crystallization and some patients with recurrent renal stones seem to be deficient in pyrophosphate. Many theories have been proposed in the past to account for calcification and ossification; few have stood the test of time but the recent evidence implicating pyrophosphate seems more promising.

TEETH

Dentine, like bone, contains collagen and apatite with various substitutions in the crystal lattice. Enamel is almost entirely inorganic apatite; it contains 3 per cent of water and about 0·4 per cent of protein, probably keratin. Substitution of a little fluorine in the lattice of apatite of dentine and enamel appears to be important; in areas where the drinking water is deficient in fluoride, the teeth have a lowered resistance to caries.

JOINTS

Movements between adjacent long bones take place at their apposed cartilagenous ends. The scanning electron microscope has shown that the apparently smooth surface of articular cartilage is in fact gently undulating and is many times 'rougher' than engineering bearings. The peak to peak distance of the undulations is about 25 μm and the valleys are about 2·5μm deep. The superficial zone of the cartilage matrix contains tightly packed small collagen fibres lying parallel to the surface forming a 'skin'; the deeper fibres are longer and form an open meshwork; the fibres next to the bone are disposed radially towards the joint surface. A fibrous capsule lined internally by synovial membrane passes from one bone to the other forming a closed cavity which contains a small quantity of lubricating fluid; this synovial fluid in the human knee joint amounts to only 0·2 ml. It is a clear yellow fluid containing only a few cells; its viscous, elastic and thixotropic properties depend on its content of hyaluronic acid (p. 48) linked to a protein. Fluorescent antibody technique shows that both the protein and the hyaluronate are produced by the synovial lining cells. The large molecules of hyaluronate are probably coiled up when the joint movement is slow but are uncoiled in rapid movement. It seems likely that, when a load is applied to a joint, fluid is squeezed out of the spongy, flexible, almost gelatinous, articular cartilage.

By sliding the femoral condyles of an amputated human knee joint against the corresponding articular surface of the tibia it has been found that the coefficient of friction is about 0·01 which is three times better than the coefficient of ice sliding on ice and more than ten times better than the value found in a plain lubricated bearing. A complete explanation of this very successful form of lubrication has not yet been found. Since the viscosity of the lubricating film is increased as the articulating surfaces approach under load it is likely that the liquid phase of the synovial fluid, trapped in the undulations of the cartilage, is squeezed away and a thickened gel remains.

CALCIUM METABOLISM

The skeleton contains at least 99 per cent of the total body calcium which in a young adult is about 1 kg (2600 m-equiv.). As mentioned earlier skeletal calcium is constantly exchanging with the calcium of the extracellular fluid. In an adult in balance, the rate of bone mineral deposition is about 400 mg/day or 10 to 15 per cent of the total skeletal calcium per year. The rate of bone resorption is similar. In older subjects, especially in women, bone resorption exceeds bone accretion and there is a progressive loss of calcium from the skeleton. This process of 'osteoporosis' appears to be universal, at least in Western communities, but the cause is not known.

DIETARY CALCIUM. In Europe and U.S.A. the average daily calcium intake in adults is 800 to 1000 mg. In developing countries the intake is often considerably lower (200 to 400 mg/day). The main sources of calcium are milk and cheese, green vegetable and (in Britain and the United States) artificially enriched bread. Drinking water is seldom a significant source of calcium but very hard water may provide up to 200 mg per day.

At one time it was thought that dietary lack of calcium was a factor contributing to poor development of bone and teeth. This may be true in some very underprivileged communities but the body has a remarkable facility for adapting to a low calcium intake (see below) and it is unlikely that calcium intakes of as little as 200 mg/day have any harmful effects in otherwise normal subjects. More calcium than usual is needed during periods of rapid growth and during pregnancy and lactation. While it is reasonable to ensure an adequate diet at these times, the healthy subject readily adapts by increasing the proportion of calcium absorbed from the diet.

INTESTINAL ABSORPTION. In an adult in calcium balance, and on a diet containing 1000 mg/day, the faeces contain about 900 mg/day and the urine about 100 mg/day. Thus the absorption from the diet appears to be 100 mg/day or one tenth of the dietary calcium. In fact the true intestinal absorption is about 350 mg/day but about 250 mg/day is secreted into the gut with the intestinal secretions.

Calcium can be absorbed from all parts of the small intestine by an active transport mechanism. In normal subjects the proportion of dietary calcium absorbed is greatly increased after a period on a low calcium diet. This 'adaptation' which becomes evident within a week of the change in diet is not seen in the absence of vitamin D. Nicolaysen in 1943 suggested that there was an 'endogenous factor', a hormone, which was responsible for regulating calcium absorption, in relation to calcium needs. This factor is probably 1,25-dihydroxycholecalciferol, a recently-discovered metabolite of vitamin D (Chap. 13). Parathyroid hormone in excess also stimulates calcium absorption, but not in the absence of vitamin D.

THE PLASMA CALCIUM. The total plasma calcium in man is normally between 8·8 and 10·3 mg/100 ml (4·4 to 5·2 m-equiv./litre). Just under half of this amount is bound to plasma proteins, particularly albumen, a small propor-

tion (less than 0·5 mg/100 ml) is complexed with citrate, and the remainder (about 5 mg/100 ml) circulates as ionized calcium. The complexed and the ionized calcium are together known as the 'diffusible' calcium. A constant concentration of ionized calcium is necessary for the normal function of muscles and nerves, and a later section will deal with the control of the level of ionized calcium.

URINARY EXCRETION OF CALCIUM. About 15 g calcium daily passes into the glomerular filtrate from the diffusible fractions of the plasma calcium. All but 50 to 200 mg is reabsorbed in the tubules but the exact site is not yet known. Calcium reabsorption is diminished in vitamin D deficiency and in the presence of a sodium diuresis.

The calcium lost in the sweat is very variable, between 20 and 350 mg/day. This limits the accuracy of calcium balance experiments.

FACTORS AFFECTING THE PLASMA CALCIUM

A constant and normal concentration of ionized calcium in the extracellular fluid is of great importance in, among other things, muscular contraction, neural and neuromuscular transmission and the activity of several enzymes. If the ionized calcium is low tetanic spasms may occur and may be fatal. If the plasma calcium is high cardiac function is disturbed and calcium may be deposited in the kidney or other tissues.

PHYSICO-CHEMICAL FACTORS. If EDTA (ethylene-diamine tetra-acetate, a calcium-complexing agent) is injected intravenously into man or the dog the plasma calcium falls rapidly but returns to a normal value (10 mg/100 ml) within a few hours. If EDTA is given after removal of the parathyroid glands the plasma calcium again returns to the pre-infusion value (7 to 8 mg/100 ml) but not so rapidly. Thus chemical equilibrium between the labile part of the bone mineral and the interstitial fluid, quite independent of the parathyroid glands, keeps the plasma calcium up to 7 to 8 mg/100 ml. In intact animals parathyroid hormone is responsible for maintaining the normal serum calcium level of around 10 mg/100 ml.

PARATHYROID HORMONE (PTH). The parathyroid glands develop from the third and fourth pharyngeal pouches of the embryo.

They lie in the neck immediately adjacent to the posterior surface of the thyroid gland with which, however, they have no physiological relationship. As a rule they consist of four oval bodies about 6 mm long each weighing 20 to 50 mg, but they are variable in number, size and position; accessory parathyroid tissue is not uncommon lower in the neck or even in the thorax. The secretory cells are arranged in cords, separated into imperfect lobules by thin septa of connective tissue. They are of two types, the *chief cells* with large nuclei nearly filling the cell and *oxyntic cells* with small nuclei and acidophil granules in the cytoplasm.

The vascular supply is rich but the nervous connexions are scanty. Secretory nerve fibres have never been demonstrated and transplanted glands can function in the absence of all nervous attachments.

Bovine PTH is a single-chain polypeptide, containing 84 amino acids (Fig. 14.4); it has a molecular weight of 8500. Assay of PTH by radioimmunoassay has not yet been perfected but has demonstrated that the secretion of PTH increases linearly with the fall in the plasma calcium (Fig. 14.3). PTH secretion

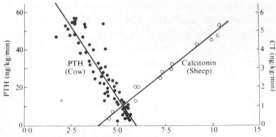

Fig. 14.3 Effect of plasma calcium on secretion rate of parathormone and calcitonin. (From D. H. Copp (1969), *Journal of Endocrinology* **43**, 137–161.)

apparently ceases when the plasma calcium exceeds 12 mg/100 ml. Plasma calcium appears to be the sole stimulus for PTH secretion. Since the half-life of the hormone in the blood is about twenty minutes, changes in hormone secretion may play an important part in the minute to minute regulation of the plasma calcium.

Quite independent of its action on blood calcium PTH has effects on bone and kidney, and probably also on the gastrointestinal tract and on the mammary gland in lactation. The effects on the bone and intestine, but not on the kidney, require the presence of vitamin D.

PTH exerts its action on bone by stimulating osteoclasts to mobilize bone. PTH decreases the reabsorption of phosphate by the kidney tubules, and so increases phosphate excretion. It also increases the urinary excretion of sodium, potassium and bicarbonate, and decreases the excretion of hydrogen ions. In the intestine, PTH enhances calcium absorption. The exact mechanism of action of PTH is not known but recent evidence suggests that cyclic 3′:5′-AMP may be involved.

Clinical disorders of both excess and deficiency of parathyroid activity are recognized. Actively secreting tumours of the parathyroid glands produce excessive amounts of PTH, and the resulting disturbance of phosphorus and calcium metabolism leads to withdrawal of large amounts of these elements from the bones which therefore become weak and deformed and liable to fracture (Fig. 14.5). The absorption of calcium from the gut is increased. The serum calcium may be 16 mg per 100 ml or more and the urinary excretion may be greatly raised so that stones may form in the kidneys.

Parathyroid deficiency sometimes occurs after accidental removal of parathyroid glands during thyroidectomy. Neuromuscular excitability and muscular spasm (tetany) are the main symptoms (Fig. 14.6). The condition improves with administration of PTH but large doses of vitamin D are given in long-term management.

CALCITONIN. In 1962 Copp and his colleagues produced evidence for the existence of a hormone which lowered the plasma calcium. He called it calcitonin. Subsequent work has amply confirmed this: the hormone has been isolated and its chemical structure has been determined. Recently human calcitonin has been synthesized. The hormone is a lipophilic single chain polypeptide of 32 amino acids with the sequence shown in Fig. 14.7.

In man calcitonin is secreted by the C cells of the thyroid, parathyroids and thymus. The parafollicular (C cells) of the thyroid have been known since 1932. They contain granules which increase in number during a period of prolonged hypocalcaemia. Calcitonin has been identified within these cells by the immunofluorescent antibody technique.

Calcitonin secretion is stimulated by hypercalcaemia (Fig. 14.3) and its best documented

H$_2$N—Ala—Val—Ser—Glu—Ile—Gln—Phe—Met—His—Asn—Leu—Gly—Lys—

His—Leu—Ser—Ser—Met—Glu—Arg—Val—Glu—Trp—Arg—Lys—Lys—Leu—

Gln—Asp—Val—His—Asn—Phe—Val—Ala—Leu—Gly—Ala—Ser—Ile—Ala—

Tyr—Arg—Asp—Gly—Ser—Ser—Gln—Arg—Pro—Arg—Lys—Lys—Glu—Asp—Asn—

Val—Leu—Val—Glu—Ser—His—Gln—Lys—Ser—Leu—Gly—Glu—Ala—Asp—Lys—

Ala—Asp—Val—Asp—Val—Leu—Ile—Lys—Ala—Lys—Pro—Gln—COOH

Fig. 14.4 Amino acid sequence of bovine PTH. The fragment 1 to 44 (heavy type) is biologically active, *in vivo* and *in vitro*, on both bone and kidney receptors. (J. T. Potts, J. M. Murray, M. Peacock, H. D. Niall, G. W. Tregear, H. T. Kaufman, D. Powel & L. J. Deftos (1971). *American Journal of Medicine* **50**, 639–649.)

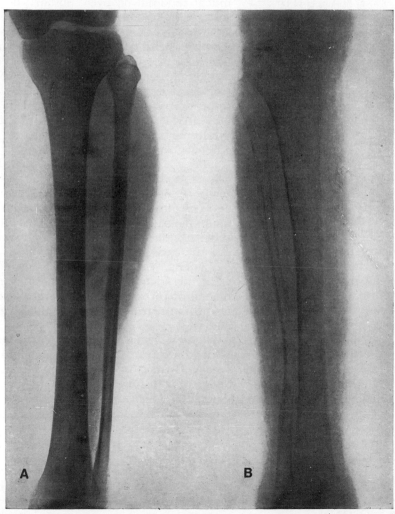

Fig. 14.5 X-ray photographs of the right lower legs of two women, age 25. A is from a normal subject, B from a patient with a parathyroid tumour. In B the tibula is deformed and both bones cast a poor shadow because of extensive resorption of calcium salts. (*By courtesy of W. T. Cooke.*)

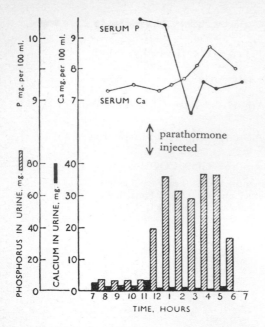

Fig. 14.6 Effect of parathyroid extract on serum levels of calcium and phosphorus and urinary calcium and phosphorus excretion in a patient with idiopathic hypoparathyroidism; 75 units of parathormone were injected at the arrow. (F. Albright & E. C. Reifenstein (1948). *Parathyroid Glands and Metabolic Bone Disease*. Baltimore: Williams & Wilkins.)

action is the inhibition of parathormone-induced bone resorption (Fig. 14.8). Calcitonin also reduces bone resorption even in the absence of parathormone, and promotes phosphate and sodium excretion in the kidney. Recently calcitonin release has also been shown to be stimulated by gastrin, pancreozymin and glucagon. Thus as calcium is absorbed from a meal, there is little or no rise in serum calcium.

Calcitonin excess has been described in patients with medullary-cell carcinomas of the thyroid. Surprisingly these patients seldom have hypocalcaemia. There is as yet no convincing evidence of a syndrome due to calcitonin deficiency.

1,25-DIHYDROXYCHOLECALCIFEROL. This hormone, a metabolite of vitamin D (Chap. 13), is produced by the kidney. It promotes the absorption of calcium in the intestine and probably has some actions on the kidney and bone. It seems mainly to be concerned with the long-term control of calcium balance and the adaptation to chronic dietary lack of calcium.

OTHER HORMONES AFFECTING CALCIUM METABOLISM. An adequate supply of growth hormone (Chap. 52) from the anterior pituitary is necessary for proliferation of the cells of the epiphyseal cartilage and, therefore, for the growth in length of a long bone. In hypophysectomized animals epiphyseal activity is much reduced or even absent; it can be restored by administration of growth hormone. There is no evidence that this hormone influences the time of closure of the epiphyses. Excessive activity of the thyroid gland is associated with loss of bone; the rate of bone formation is increased but the rate of bone resorption is increased to a greater extent. Increased loss of bone is a feature of Cushing's syndrome due to excessive activity of the adrenal cortex.

THE PLASMA CALCIUM IN DISEASE. Hypercalcaemia of hyperparathyroidism, has already been mentioned. Hypercalcaemia may also be caused by tumours which cause rapid bone destruction, by tumours secreting a PTH-like substance, by vitamin D poisoning and the excessive ingestion of milk and alkali by patients with peptic ulcer. Signs of hypercalcaemia include thirst, tiredness, weakness, mental disturbances and, if severe, coma and death. Untreated hypercalcaemia causes renal damage.

Hypocalcaemia is found in hypoparathyroidism, osteomalacia, rickets, and renal failure. Tetany is a prominent feature. The outstanding feature of tetany is neuromuscular irritability which manifests itself first as hypertonicity of muscles and then as fibrillary

$$
\begin{array}{c}
\text{S}\!-\!\!-\!\!-\!\!-\!\!-\!\!-\!\!-\!\!-\!\!-\!\!-\!\!-\!\!-\!\!-\!\!-\!\!-\!\!-\!\!-\!\!-\!\!-\!\!-\text{S} \\
| \qquad\qquad\qquad\qquad\qquad\qquad | \\
\end{array}
$$

NH$_2$—Cys—Gly—Asn—Leu—Ser—Thr—Cys—Met—Leu—Gly—Thr—Tyr—Thr—
Gln—Asp—Phe—Asn—Lys—Phe—His—Thr—Phe—Pro—Gln—Thr—Ala—Leu—Gly—
Val—Gly—Ala—Pro—CONH$_2$

Fig. 14.7 Human calcitonin.

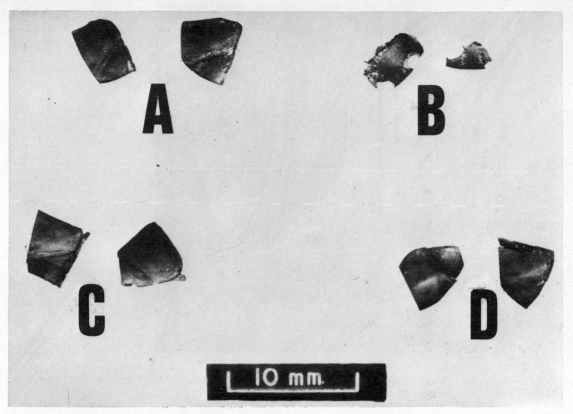

Fig. 14.8 Mouse calvaria in tissue culture to show effect of calcitonin on the bone resorption induced by parathormone. B, calvaria incubated with parathormone; D, incubation with parathormone and calcitonin; A, untreated control half-calvaria for B; C, untreated control half-calvaria for D. (*By courtesy of Jennifer Nisbet.*)

twitchings of muscle fasciculi, leading finally, especially in children, to generalized clonic movements. The muscle hypertonia produces the characteristic attitude of the hand in tetany, the *main d'accoucheur*. Simultaneously the feet are held firmly flexed at the ankles with all the toes plantar-flexed. The condition, called *carpopedal spasm* (Fig. 14.9) may be accompanied in infants by spasm of the glottis (*laryngismus stridulus*), which can be severe enough to cause alarming cyanosis. In the early stages of tetany sensory phenomena are present, such as widespread tingling feelings and sensations of heat and flushing (paraesthesiae). The neuromuscular hyperexcitability can be demonstrated before tetany occurs (*latent tetany*). Tapping over the facial nerve in front of the ear produces twitching of the facial muscles (Chvostek's sign), and the motor nerves are unduly excitable to electrical stimu-

lation. Carpal spasm can be induced by inflating a blood pressure cuff round the upper arm to a pressure exceeding the systolic blood pressure and maintaining the occlusion for 3 minutes (Trousseau's test, Fig. 14.9A); ischaemia of nerve trunks itself increases their excitability and thus reinforces the effect of a low concentration of calcium in the serum.

The factor most closely related to the onset of tetany is a reduction in the plasma concentration of ionized calcium. This may be caused by a reduced total plasma calcium in the disorders mentioned above or by a reduced proportion of ionized calcium due to alkalosis caused by vomiting or by over-ventilation. Tetany has more rarely been attributed to a low plasma magnesium, a high plasma potassium or other metabolic changes. Tetany is abolished by curare and probably depends on the integrity of the spinal reflex arcs.

Bone and Minerals 217

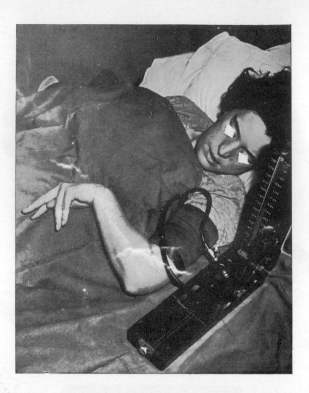

Fig. 14.9A Main d'accoucheur induced by arrest of the circulation in the forearm in a patient with parathyroid tetany due to the accidental extirpation of the parathyroid glands during thyroidectomy for hyperthyroidism.

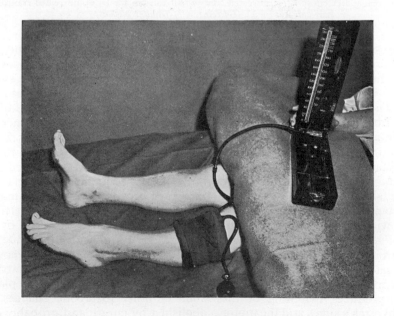

Fig. 14.9B Pedal spasm induced in the same patient.

218   Textbook of Physiology and Biochemistry

Phosphorus

The mean phosphorus content of an adult man is 800 g; four-fifths of this is in the bones, the remainder being in the cells as phosphates or nucleic acids. The inorganic phosphorus concentration in the plasma in fasting adults is between 2·5 and 4·5 mg/100 ml. Higher values are found in infants. Almost all the inorganic phosphate is dialysable, only 12 per cent being protein-bound. At normal blood pH (7·4) 85 per cent of the ionized inorganic phosphate is present as HPO_4^{2-} and 15 per cent as $H_2PO_4^-$. Small amounts of the phosphate may be bound to calcium and magnesium. It has been stated that the amount of non-ionized phosphate may be 50 per cent of the dialysable phosphate. The uncertainty about the true plasma concentration of inorganic phosphate ions leads to difficulties in determining the true ionic product of calcium and phosphate. Since phosphorus is present in all animal and vegetable cells dietary deficiency never occurs in man. Phosphate depletion may occur as a result of renal tubular disorders and, very rarely, in patients who consume excessive quantities of aluminium hydroxide, an antacid which binds phosphorus in the gut.

Phosphorus is excreted by the kidney, and in the steady state the amount excreted is equal to that which the gut absorbs. Ninety per cent of the phosphorus filtered at the glomerulus is reabsorbed in the tubule and there is no good evidence for tubular secretion of phosphorus. Phosphate depletion may occur with tubular defects and a high plasma phosphate may be a consequence of renal (glomerular) failure.

Other Minerals

Magnesium. The adult human body contains only 25 g (2000 m-equiv.) of magnesium, about half being in the bones and half in the cells. The cells contain 30 to 40 m-equiv per litre where it is an essential part of many enzymes, including phosphoglucomutase in the mitochondria. The plasma and the gastro-intestinal secretions contain only 1·7 to 2·3 m-equiv per litre; about 55 per cent of the plasma magnesium is ionized. The average urinary secretion is usually over 100 mg or 10 m-equiv. per day; the amount depends very much on the intake, which may lie between 17 and 34 m-equiv. per day. Since green vegetables and cereals contain much magnesium a dietary deficiency is unlikely unless there is intestinal loss of magnesium because of diarrhoea or some form of malabsorption. If the serum level falls epileptic fits, tetany and muscular weakness are seen in man; these disturbances may in fact be due to an accompanying hypocalcaemia.

Copper. The daily intake of copper is about 2 mg. If copper is absorbed it is rapidly removed from the blood by the liver and excreted in the bile. About 15 μg of copper are excreted per day in the urine. When absorbed from the upper small intestine copper is at first loosely attached to albumin but within 12 to 18 hours it becomes firmly bound to a non-dialysable $α_2$-globulin (caeruloplasmin) which has weak oxidase activity. Human liver contains about 40 μg Cu per g dry matter; in cirrhosis or Wilson's disease there may be up to ten times as much.

Copper is an essential trace element. Its most important role is as a component of cytochrome oxidase, which is involved in the final step for the reduction of molecular oxygen (p. 129). It is concerned also in phospholipid synthesis and in the copper protein enzymes, tyrosinase, ascorbic acid oxidase and monoamine oxidase. Deficiency in domestic animals produces anaemia and damage to the central nervous system.

Zinc. The daily intake of zinc is between 10 and 15 mg, almost all of which is excreted in the faeces. Urinary excretion is 0·4 mg daily. Blood contains about 0·75 mg per 100 ml, nearly all of which is present in the red cells in the enzyme carbonic anhydrase; blood plasma contains about 90 μg of zinc per 100 ml. Zinc is also present in carboxypeptidase, alcohol dehydrogenase, liver glutamic dehydrogenase and alkaline phosphatase. Zinc deficiency is common in domestic animals and may retard growth and development of the gonads. The plasma zinc is lowered in man in many conditions including disorders of the skin and liver; in some patients administration of zinc has produced improvement. Zinc deficiency may not be uncommon in man.

Manganese. The intake of manganese is about 5 mg daily, mainly derived from cereals and tea. Plasma contains 2·5 μg per litre carried by $β_1$-globulin; whole blood contains about 12 μg per litre. Manganese is accumulated in both mitochondria and microsomes. It is excreted mainly in the faeces; the urine contains negligible amounts. The effects, if any, of deficiency in man are not known but overdoses of manganese produce effects in man resembling Parkinsonism (Chap. 48).

Cobalt. The body contains about 80 μg of cobalt in cyanocobalamin (Chap. 13). A deficiency syndrome has not been encountered in man but lack of cobalt in ruminants causes wasting.

Strontium. The body contains about 350 mg of strontium in the skeleton where it behaves much like calcium. Strontium has no known physiological function, but is of possible clinical importance because ^{90}Sr is radioactive and is produced as a result of atomic fission. This isotope has a long half-life (25 years) and accumulates in the skeleton. Fortunately the amount of ^{90}Sr in human bone has declined since a peak figure was reached in 1964.

Fluorine. Fluorine is present in animal soft tissues in amounts ranging from 0·2 to 1·5 mg per 100 g of the dry material. Bones and teeth contain larger amounts (20 to 30 mg per 100 g ash) and dentine may contain more than the enamel of teeth. The incidence of caries tends to be lower in geographical areas where the drinking water contains one part per million of fluorine than in places where the fluorine content is much less. In districts where the water contains three parts per million or over of fluorine the enamel of the teeth may show bands of brown pigmentation between chalkish white patches but they are still resistant to caries; the bones may be thickened but without any functional disability. In districts where the water supply is not fluoridated the intake of fluorine lies between 0·6 and 1·8 mg per day. Deliberate fluoridation of the water supply may double these figures but this dose is well below the level known to cause toxic effects. Tea and sea-fish are the only other significant sources of fluorine in the diet.

Iron metabolism is discussed in Chapter 24 and iodine metabolism in Chapter 50.

REFERENCES

Annotation. Calcitonin and metabolic bone disease (1971). *Lancet* **i**, 1168–1169.

BELL, G. H. (1966). Rheology of bone. *Laboratory Practice* **15**, 71–76.

BELL, G. H. (1969–70). Living bone as an engineering material. *Advances in Science* **26**, 75–85.

BOURNE, G. H. (Ed.) (1972). *The Biochemistry and Physiology of Bone*, 2nd edn. New York: Academic Press.

BOWEN, H. J. M. (1966). *Trace Elements in Biochemistry*. London: Academic Press.

BRONNER, F. (Ed.) (1969). Symposium on calcium absorption. *American Journal of Clinical Nutrition* **22**, 376–446.

BROOKES, M. (1971). *The Blood Supply of Bone*. London: Butterworths.

Calcium requirements (1962). Report of an FAO/WHO Experimental Group. *World Health Organization Technical Report Series* No. 230.

CAMPBELL, J. (1958). Fluorine and dental decay. *Advances in Science* **15**, 221–224.

COMAR, C. L. & BRONNER, F. (Eds.) (1960–1969). *Mineral Metabolism*. New York: Academic Press.

COPP, D. H. (1969). Control of calcium homeostasis. *Journal of Endocrinology* **43**, 137–161.

DEPARTMENT OF HEALTH AND SOCIAL SECURITY (1969). The fluoridation studies in the United Kingdom and the results achieved after eleven years. *Reports in Public Health and Medical Studies* No. 122. London: H.M.S.O.

DUCKWORTH, R. (1966). Fluoridation of water supplies. *British Medical Journal* **ii**, 283–286.

EASTOE, J. E. (1968). Review of chemical aspects of the matrix concept in calcified tissue organization. *Calcified Tissue Research* **2**, 1–19.

FOURMAN, P. & ROYER, P. (1968). *Calcium Metabolism and the Bone*, 2nd edn. Oxford: Blackwell.

HALL, D. A. (1961). *The Chemistry of Connective Tissue*. Springfield, Ill.: Thomas.

HALL, D. A. (Ed.) (1963–1970). *International Review of Connective Tissue Research*, Vols. I to V. London: Academic Press.

HARKNESS, R. D. (1966). Rheological problems of collagenous tissues. *Laboratory Practice* **15**, 166–170, 183.

JACKSON, W. P. U. (1967). *Calcium Metabolism and Bone Disease*. London: Edward Arnold.

JENKINS, G. N. (1966). *The Physiology of the Mouth*, 3rd edn. Oxford: Blackwell.

McLEAN, F. C. & BUDY, ANN M. (1964). *Radiation, Isotopes and Bone*. New York: Academic Press.

MALM, O. J. (1958). *Calcium Requirement and Adaptation in Adult Man*. Oslo: University Press.

PIEZ, R. A., MILLER, E. J. & MARTIN, G. R. (1965). The chemistry of elastin and its relationship to structure. In *Advances in Biology of the Skin* (Edited by W. Montagna), pp. 245–253. Oxford: Pergamon.

POSNER, A. S. (1969). Crystal chemistry of bone material. *Physiological Reviews* **49**, 760–792.

RASMUSSEN, H. (Ed.) (1970). Parathyroid hormone, thyrocalcitonin and related drugs. *International Encyclopaedia of Pharmacology and Therapeutics*, Section 51, Vol. I. Oxford: Pergamon.

RAY, R. D. (Ed.) (1968). Circulation and the skeletal system. Instructional course lectures. *Journal of Bone and Joint Surgery* **50A**, 764–824.

SIRI, W. (1956). The gross composition of the body. *Advances in Biological and Medical Physics* **4**, 239–261.

SMITH, J. G., SAMS, W. M. & FINLAYSON, G. R. (1966). Biochemistry and pathology of cutaneous elastic tissue. In *Modern Trends in Dermatology*, 3rd edn (Edited by R. MacKenna), pp. 119–142. London: Butterworth.

Symposium on detection of nutritional deficiencies in man. (1967). *American Journal of Clinical Nutrition* **20**, 513–612.

TAYLOR, S. (Ed.) (1971). *Symposium on Endocrinology, 1971*. London: Heinemann.

VAUGHAN, JANET M. (1970). *The Physiology of Bone*. Oxford: Clarendon Press.

WILSON, J. C. (Ed.) (1968). Introductory course on circulation and the skeletal system. *Journal of Bone and Joint Surgery* **50A**, 764–824.

WRIGHT, V. (Ed.) (1969). *Lubrication and Wear in Joints. A Symposium*. London: Sector Publishing.

WYKE, B. (1967). The neurology of joints. *Annals of the Royal College of Surgeons* **41**, 25–50.

15 Mouth, oesophagus and swallowing

The oral cavity and tongue are covered by a stratified squamous epithelium richly supplied with pain, tactile and temperature nerve endings; in addition the tongue has taste buds distributed over its surface. If, when food enters the mouth, the impulses from these nerve endings and those in the olfactory area indicate that the food is acceptable, mastication prepares it for swallowing. Chewing breaks up the solid parts of the food and at the same time increases the salivary secretion and mixes up the food with it to make a bolus suitable for swallowing. Mastication may increase slightly the digestibility of some foodstuffs, but no advantage seems to accrue from prolonged mastication. Vigorous mastication is said to be essential for the health of the teeth and other oral structures.

SALIVARY DIGESTION

THE SALIVARY GLANDS. In man and in many mammals the saliva is secreted mainly by three pairs of glands: the *parotid*, the *submandibular* and the *sublingual*. The contribution of the numerous small salivary glands scattered over the oral mucosa is much smaller. The duct of the parotid gland enters the mouth opposite the second upper molar tooth. Each submandibular gland has a single duct which reaches the mouth at the side of the frenulum of the tongue, but the secretion of the sublingual glands passes by way of some 10 to 20 small ducts which open into the floor of the mouth. In man saliva may be collected from the parotid or submandibular glands by placing a soft plastic cannula in the duct; saliva can be got from the parotid gland with less discomfort by using a cannula with a wide end placed over the opening of the duct and kept in place by evacuating an outer concentric tube. In dogs Pavlov (Chap. 49) freed the terminal portion of the duct, with a small area of mucous membrane around it, and transferred it to the external skin surface to form a salivary fistula.

THE HISTOLOGICAL STRUCTURE OF THE SALIVARY GLANDS. The salivary glands are typical compound tubular glands in which the glandular cells lie within a supporting framework of connective tissue carrying the blood vessels, lymphatics and nerves. The glandular cells are arranged in a single layer around a central cavity which receives the secretion from the surrounding cells (Fig. 15.1 and 2). Such a unit, called an *acinus* or an *alveolus*, may be serous, mucous or mixed. Small ducts, intercalated ducts, usually surrounded by myoepithelial cells, from a number of adjacent acini join together to form larger striated ducts which are lined by tall cells whose basal membrane has numerous infoldings giving the appearance of striation. All the duct cells possess numerous mitochondria. The secretion from the gland finally reaches the mouth along a large excretory duct. This arrangement, in which the acini are related to one another like the grapes in a bunch, is called *racemose*. Because the saliva reaches the surface by a duct it is said to be an external secretion and a salivary gland is a gland of *external secretion* or an *exocrine gland*.

The parotid gland consists almost entirely of serous cells which are small and granular with well-stained nuclei; saliva from this gland is clear and watery, being much less viscous but richer in amylase than that from the other salivary glands. Most of the cells in the sublingual gland are mucous cells which are larger than serous cells; in the usual histological preparations they are clear and transparent. The sublingual gland secretes a thick, sticky, opalescent material rich in mucin. Submandibular gland has both serous and mucous acini in roughly equal proportions.

THE NERVE SUPPLY OF THE SALIVARY GLANDS. The salivary glands are supplied by both sympathetic and parasympathetic fibres (Fig. 15.3). The central origin of the latter, the superior and inferior salivary nuclei, is in the medulla, between the VIIth nucleus and the lateral vestibular (Deiter's) nucleus. Probably every secreting cell receives both sympathetic and parasympathetic fibres. Stimulation of the

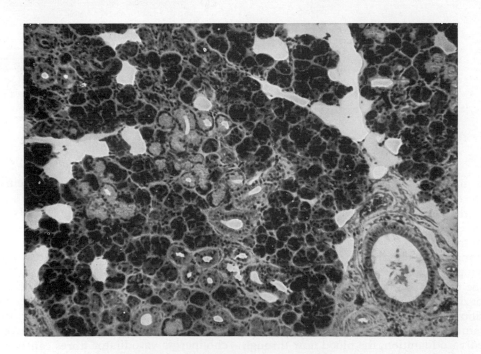

Fig. 15.1 Human salivary gland (submandibular). The mucous acini are lightly stained. The serous acini are darkly stained. Magnification × 110. Stain: haemalum, phloxin and tartrazine. (*By courtesy of A. C. Lendrum.*)

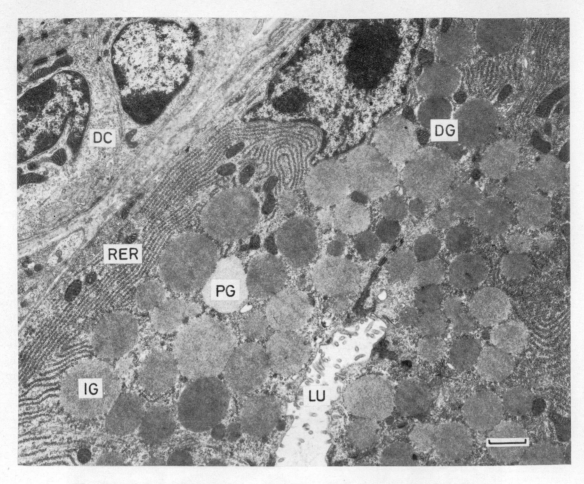

Fig. 15.2 Guinea pig submaxillary gland. Type: serous. Acinar cells lying adjacent to striated duct cells (DC). The acinar cells contain granules which appear to belong to three different types; large pale granules (PG) smaller intermediate granules (IG) and small dense granules (DG). They tend to congregate near the apical end of the cell near the acinar lumen (LU). The basal part of the cell is filled with rough endoplasmic reticulum (RER). × 9500. The scale (bottom right) represents 1 μm. (*By courtesy of Gundula Dorey.*)

nerves to the salivary glands produces extremely divergent results in different laboratory animals. In the absence of direct information we can only assume that, since parasympathomimetic drugs produce a copious watery secretion, serous cells in man are innervated by parasympathetic nerve fibres.

If the chorda tympani (parasympathetic) of a cat or dog is stimulated electrically a copious secretion of saliva occurs at once and may be maintained for a long period. Owing to the marked vasodilatation, the blood flow through the submandibular gland is increased up to fivefold and the gland swells and reddens. The vasodilatation may be so great that the arterial

pulsations are transmitted to the veins leaving the gland.

If atropine is given intravenously, stimulation of the chorda produces no secretion but still causes an increase in the metabolic activity of the gland and a marked but smaller vasodilatation. When botulinus toxin (Chap. 39) is injected into the submandibular gland stimulation of the chorda produces no secretion and no vasodilatation. Since this toxin blocks cholinergic fibres the presence of noncholinergic vasodilator fibres can be excluded but atropine-resistant vasodilatation is left unexplained. Hilton and Lewis believe that the vasodilatation occurring in the submandibular

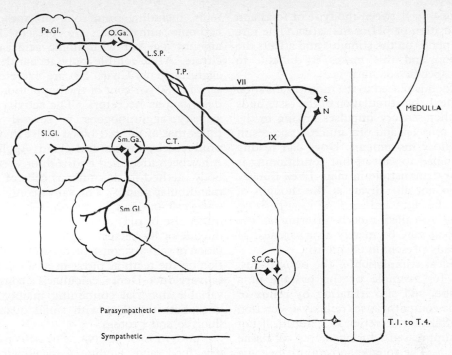

Fig. 15.3 Diagram to illustrate the nerve supply to the salivary glands. Pa.Gl., parotid gland. Sl.Gl., sublingual gland. Sm.Gl., submandibular gland. O.Ga., otic ganglion. S.C.Ga., superior cervical ganglion. Sm.Ga., submandibular ganglion. L.S.P., lesser superficial petrosal nerve. T.P., tympanic plexus. C.T., chorda tympani. VII, facial nerve. IX, glossopharyngeal nerve. S N, salivary nuclei. T.1. to T.4., first to fourth thoracic segments of the spinal cord.

gland on chorda stimulation is due not to the action of vasodilator nerves or to the direct action of acetylcholine but to the escape from active gland cells into the interstitial spaces of an enzyme (kallikrein) which acts upon proteins in the interstitial fluid to form a vasodilator polypeptide. Schachter cannot accept this theory since in a gland deprived of kallikrein or perfused without a substrate for kallikrein stimulation of the chorda still produces maximum vasodilatation. He thinks there may be true vasodilator nerves in the chorda and attributes the differential effect of atropine to variations in the sensitivity of different cholinergic receptors to atropine.

Sympathetic stimulation, or injection of adrenaline or noradrenaline, produces a flow of viscous saliva containing a large proportion (6 per cent) of solids from the submandibular gland of the cat. During stimulation there is vasoconstriction with a reduction in blood flow, but when the stimulation is ended there is a prolonged vasodilatation which is paralleled by an increase in the oxygen consump-

tion of the gland. This may be an adrenergic effect.

Although many drugs influence the amount of saliva produced, atropine is the only one which is much used in medicine. Atropine, because it blocks the action of acetylcholine, the humoral substance of the parasympathetic nerves (Chap. 41), inhibits secretion. Acetylcholine, pilocarpine and eserine all produce a thin watery juice, while adrenaline and ephedrine are said to cause the secretion of saliva containing mucus.

During sleep the human salivary glands produce only a few ml of saliva per hour but during the waking state the secretion is continuous. The parotid gland, weighing about 25 g, can produce 25 g saliva in fifteen minutes. When paraffin wax is chewed about 250 ml/hour may be secreted but it is usually much less. Pavlov in experiments on dogs found that when dry food was chewed the saliva was thin and watery; meat (which was swallowed whole) induced a flow of thick saliva containing much mucin. In man, however, little is known about the relationship between the type of food and the composition or pH of the saliva. The rate of flow depends on the stimulus and affects the composition and this makes it difficult to recognize specific adaptations.

When food is placed in the mouth salivation occurs reflexly by stimulation of the taste buds and by other sensory impulses arising in the teeth and muscles and oral mucosa as a result of masticatory movements. Dogs can readily be conditioned to salivate but conditioning is difficult to demonstrate in man. Even hungry persons do not all salivate at the thought of food or at the sight and smell of an appetizing meal. The so-called mouth-watering at the sight of food may be merely an awareness of saliva already present in the mouth.

Mechanical stimulation of the oesophagus, produced for example by the passage of a rubber tube, and also irritation by reflux of acid gastric contents causes reflexly a secretion of saliva. Salivary secretion is reduced, if not actually suppressed, during states of tissue dehydration. The tongue and mouth become dry and there may be a sensation of thirst. The secretion of saliva is also inhibited by exercise or by emotional stress.

COMPOSITION OF SALIVA. The mixed saliva from all the glands is viscous, colourless and opalescent. The specific gravity is 1002 to 1010 (usually about 1003). The pH varies with increasing rate of flow from 6·2 to 7·4 but it may be more acid if contact with air is prevented. Under the microscope the fluid shows nucleated squamous cells from the buccal lining together with disintegrating leucocytes and gland cells, as well as numerous microorganisms of many varieties. About 750 ml are produced per day by man; the submandibular glands are said to contribute 70 per cent, the parotids 25 per cent and the sublinguals 5 per cent of the resting volume.

The solid matter of this complex and very variable fluid is about 0·5 per cent; about 0·3 per cent is protein. The osmotic pressure is half to three-quarters that of blood plasma. The main organic constituents are the glycoprotein mucin, which gives saliva its viscosity and lubricating properties, and the enzyme ptyalin, an α-amylase probably identical with pancreatic amylase. Ptyalin (optimum pH 6·8) which is activated by chlorides catalyses the breakdown of starch to maltose. Other pro-teins, including numerous enzymes, such as carbonic anhydrase, are present in small amount together with amino acids, urea and citrate. ABO soluble polysaccharides (blood agglutinogens, Chap. 23) are excreted in the saliva of 80 per cent of the population who are described as 'secretors'. The activity of these salivary agglutinogens is several hundred times that of the red blood cells; they are not produced by the parotid gland. Agglutinogen A has been identified by the fluorescence antibody method in the mucous cells of the submandibular glands. The saliva contains about 6 mg calcium and 17 mg phosphorus per 100 ml. A rise of pH, produced by loss of carbon dioxide or by bacterial action, causes precipitation of salivary constituents and their deposition on the teeth as tartar or as a calculus in a salivary duct. Dental calculus is a complex and variable material containing mainly calcium phosphate together with small quantities of fluoride and protein.

FUNCTIONS OF SALIVA. The saliva moistens dry food and facilitates swallowing by a lubricating action. Since water evaporates slowly from saliva it prevents desiccation of the oral mucosa. It provides an enzyme for the digestion of starch. Since this action is short and not important it may be that the main effect of ptyalin is the removal of food debris lodged between the teeth. Saliva keeps the mouth and teeth clean; the bactericidal effect of the enzyme lysozyme may be partly responsible for this. During fever, when salivary secretion is suppressed, the lips, teeth and mouth become coated with a mixture of food particles, dried mucus and dead epithelium which, if it is not removed mechanically, occasionally becomes the site of bacterial infection. Certain substances, for example lead, mercury and iodides are excreted in the saliva but this excretion cannot be important since these substances are reabsorbed when the saliva is swallowed. By facilitating movements of the tongue and lips saliva makes rapid articulation possible. Drying of the mouth produces reflexly a flow of saliva. Extreme nervousness, or an injection of atropine, may inhibit secretion to such a point that rapid speech becomes impossible. Saliva subserves the sense of taste by acting as a solvent. The taste buds can be stimulated only when the sapid substance is actually in solution (Chap.

36). Saliva contains three buffering systems, bicarbonate, phosphate and mucin of which the first is the most important. The concentration of bicarbonate, and the buffering power, rises when salivary flow increases as during eating. Only if food accumulates after salivary flow has diminished does the pH of the tooth surface fall by bacterial action to levels which allow the calcium of the teeth to go into solution. At pH 7 the saliva is saturated with calcium so that the teeth do not lose calcium unless the pH falls to about 5·5.

SALIVARY DIGESTION. Chewing mixes the food with saliva and brings ptyalin into intimate contact with the starch of the food. The food normally remains for too short a time in the mouth to allow the digestion of starch to proceed very far but, even after the food is swallowed, the pH remains for some time favourable to the continued action of amylase and the starch molecules are hydrolysed to dextrins and then to maltose. Hydrolysis is finally arrested by a fall in pH and destruction of amylase as the acid gastric juice gradually penetrates through the food mass in the body of the stomach.

MECHANISM OF SECRETION. The fact that a very large and continuous secretion of saliva can be obtained by stimulation of the parasympathetic nerves shows that there must be a considerable transference of fluid from the capillaries around the secreting acini into the tissue fluid and then into the secreting cells.

During stimulation of the chorda tympani a manometer placed in the duct of the submandibular gland may show a secretion pressure greater than the systolic pressure in the carotid artery (Fig. 15.4). Since the pressure in the capillaries around the acini must be much lower than this, the secretion of saliva cannot be explained as a simple process of filtration from the blood plasma. The submaxillary secretion in the rat has been sampled by glass microcapillaries inserted into various parts of the gland. These experiments show that the primary acinar secretion resembles an ultrafiltrate of plasma and that the secretion of the acinar cells is considerably modified as it passes along the ducts. The duct cells absorb Na and Cl from the fluid and secrete K into it; the original isotonic fluid becomes markedly hypotonic, the extent of this change being related to the time spent in passing through the ducts; the reduction in osmolality is greatest when the gland is actively secreting. The output of K is greatest in the resting saliva. The ducts contribute more urea than the acini to the saliva.

The salivary glands take up almost as much iodide from the blood plasma as the thyroid gland. Autoradiography after injection of ^{131}I shows that the cells of the striated ducts are responsible. In man experiments with ^{132}I have given a salivary inorganic iodide/plasma inorganic iodide ratio of about 100 in resting conditions, falling to 20 or less at high flow

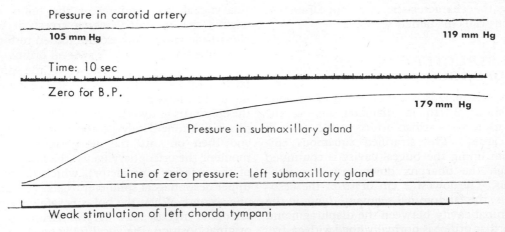

Pressure in carotid artery

105 mm Hg   119 mm Hg

Time: 10 sec

Zero for B.P.

179 mm Hg

Pressure in submaxillary gland

Line of zero pressure: left submaxillary gland

Weak stimulation of left chorda tympani

Fig. 15.4 A record of the rise of pressure in the submandibular gland during prolonged stimulation of the chorda tympani nerve. The submandibular duct was connected directly to the manometer so that no saliva could escape. (*By courtesy of E. W. H. Cruickshank.*)

rates. The salivary iodide concentrating mechanism is quite independent of thyroid function. The iodide is not lost from the body; it is absorbed when the swallowed saliva reaches the small intestine.

If the volume of the submandibular gland is recorded during stimulation of the chorda tympani it is found that there is an initial shrinkage of the gland in spite of the increased volume of its blood vessels. Thus the first effect produced by stimulation is a discharge of the fluid contained in the acinar cells (so accounting for the very short latent period); this fluid is then made good from the tissue fluid. This is in marked contrast to the findings after atropinization, when stimulation of the chorda tympani causes an increase in volume of the gland due to vasodilatation and probably to an increase in tissue fluid.

HISTOLOGICAL BASIS OF SECRETION. The enzymes and other proteins are formed by the endoplasmic reticulum of the acinar cells and released as microsomes which are converted to the secretion granules which are easily seen under the light microscope when living cells from a salivary gland are teased out in blood serum.

The granules in the serous cells of the salivary glands contain a precursor of ptyalin which is discharged into the acinus as zymogen and later converted to the active enzyme. The granules in the mucous cells contain a precursor of mucin, called *mucinogen*, and not mucin itself. Mucin is coagulated by acetic acid, or any of the usual histological fixatives, thus giving the characteristic clear appearance of mucous cells in stained sections.

THE STRUCTURE OF THE OESOPHAGUS

The oesophagus is a tube about 25 cm long and about 2 cm in diameter joining the pharynx to the cardiac orifice of the stomach (the cardia). The stratified squamous epithelium lining the buccal cavity is continued through the pharynx down into the oesophagus. The lowest 2 cm or so of the oesophagus, the 'abdominal segment', lying in the abdominal cavity between the diaphragm and the cardiac orifice is normally lined with gastric mucosa and is covered by peritonetum. In the undistended state the mucosa of the oesophagus is thrown into several fine longitudinal folds which disappear when it is distended. The submucous coat contains mucous glands. Outside this there is a considerable amount of muscular tissue (circular fibres internally and longitudinal externally) chiefly striated in the upper third, but consisting almost entirely of smooth muscle fibres in the lower third; both striated and smooth muscle fibres are found in the middle third.

SWALLOWING

This complex process which occupies only a few seconds may be divided into three stages; the first buccal, the second pharyngeal and the third oesophageal. Only the first is under voluntary control. The mouth is closed and the bolus, collected on the upper surface of the tongue, is forced past the anterior pillars of the fauces (oropharyngeal isthmus) by an upward and backward movement of the tongue caused mainly by the contraction of the mylohyoid and styloglossus muscles. At the same time the soft palate rises and approaches the posterior pharyngeal wall which is brought forward to close off the nasopharynx. Respiration is reflexly inhibited. The larynx begins to rise as the bolus passes over the back of the tongue. From this moment the process of deglutition is involuntary. Since the pharynx communicates with both oesophagus and trachea, the movements occurring in the next stage are designed to allow the bolus to enter the oesophagus while avoiding the air passages. The tongue moves back like a piston towards the posterior pharyngeal wall and forces the bolus back against the epiglottis, which arrests it for a short time and then becomes folded to form a cowl-like hood over the laryngeal orifice. The entrance to the larynx is closed by the sphincteric action of the girdle of muscles surrounding it. The food passes over the lateral edges of the epiglottis in two streams into the part of the pharynx immediately posterior to the larynx and then on into the oesophagus. At this moment the cricopharyngeus (the lower fibres of the inferior constrictor), which forms the upper oesophageal sphincter, relaxes for about one second. When the bolus is safely past the cricopharyngeus, the larynx drops to its original position, the vocal folds open and the epiglottis quickly resumes its initial position. The cricopharyngeal sphincter then contracts.

This is the end of the second stage. If a number of swallowing movements are made in rapid succession the soft palate and epiglottis do not return to their resting positions until the series is completed. The epiglottis can be removed without any harmful effect on swallowing.

During the first stage of swallowing the lips and jaws are normally closed. If the lips are not closed swallowing becomes difficult; but if the jaws cannot be closed swallowing is nearly impossible. During dental operations, when the jaws must be kept open, saliva accumulates in the mouth and has to be removed by suction. Breathing is reflexly arrested, no matter the actual phase of respiration, during the act of swallowing. During this respiratory inhibition the larynx is drawn upwards—without this movement swallowing is very difficult or even impossible. Swallowing, however, can be carried out after the whole of the larynx has been removed to eradicate malignant disease.

The oesophagus at rest is relaxed but closed off at the top and bottom by sphincters. The upper sphincter is well defined anatomically, the cricopharyngeal sphincter being a thick muscular band about 2 to 3 cm long. The lower sphincter is difficult to define anatomically but it can be demonstrated physiologically as the lower 3 cm of the oesophageal tube. It normally lies so that about half is in the thorax and half in the abdomen. Oesophageal activity can be observed radiologically by a barium swallow but in more detail by pressure recording. This is normally performed with open-tip, water-filled polythene tubes connected to pressure sensitive manometers placed beside the subject. With this equipment a detailed picture of oesophageal function has been obtained.

The cricopharyngeal sphincter is easy to recognize physiologically because it produces at rest a band of high pressure, about 20 to 40 cm of water above atmospheric pressure. On swallowing it opens and the pressure falls briefly to atmospheric levels. After about 1 second the sphincter contracts with a rise in pressure above the resting levels (Fig. 15.5). This contraction, in peristaltic sequence with the rise of pressure in the pharynx, progresses down the oesophagus as the primary peristaltic wave (Fig. 15.5 and 6).

At rest the intra-oesophageal pressure is a true index of intrathoracic pressure; it shows a swing with respiration, a fall with inspiration

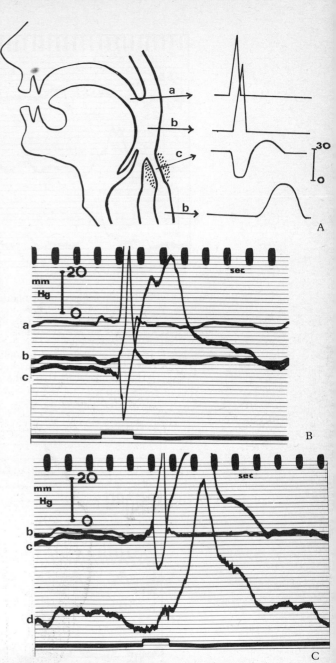

Fig. 15.5 Diagram A shows the pressure changes during swallowing a, in the mouth; b, in the pharynx; c, in the cricopharyngeal sphincter; and d, in the upper oesophagus. In the pressure records, B and C, the ends of the pressure recording tubes are 5 cm apart. In B the tubes end at positions a, b, c, whereas in C the tubes end at positions b, c, d. Note that the record of the pressure at c stands at 30 mm Hg between swallows, it falls to 0 (atmospheric pressure) when the sphincter opens during swallowing. The rise in the signal line at the foot of each trace denotes the moment at which the subject was instructed to swallow.
(*By courtesy of B. Creamer.*)

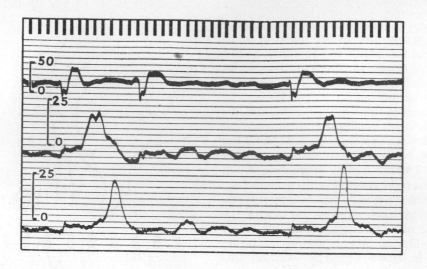

Fig. 15.6 Pharyngeal swallows and oesophageal propulsive waves. Upper trace: pressure in pharynx. Middle and lower traces: pressures 9 and 18 cm down oesophagus. Note that a pharyngeal swallow repeated too soon fails to initiate a propulsive wave. The waves in the lowest trace are about 2 sec later than those in the middle trace. The speed of the propulsive wave is therefore about 4 cm/sec. The small simultaneous rises of pressure to 0 (that is atmospheric pressure) preceding each oesophageal pressure wave are due to the opening of the cricopharyngeal sphincter. Time is marked at the top in seconds and pressures in mm Hg.
(A. C. Dornhorst, K. Harrison & J. W. Pierce (1954). *Lancet* **i**, 696.)

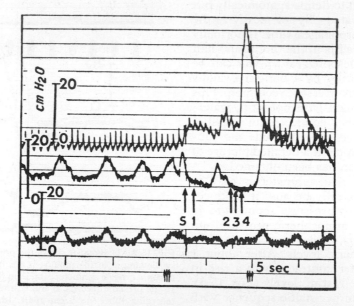

Fig. 15.7 Pressure during swallowing of a suspension of barium. Upper trace is from the lower oesophagus; middle trace is from the gastro-oesophageal junction below the point of reversal of inspiratory pressure swing (in this position there is a rise of pressure at inspiration); lower trace is from the stomach. S, Act of swallowing. 1. Arrival of barium at junction. 2. Opening of junction. 3. Junction maximally open. 4. All barium has left oesophagus. (These events were recorded by cineradiography.) Note the fall in pressure in the middle trace after the swallow. Time in 5 seconds. (B. Creamer & J. W. Pierce (1957). *Lancet* **ii**, 1310.)

and a rise with expiration. It is frequently used by respiratory physiologists as a convenient measure of intrathoracic pressure. After a swallow a single peristaltic wave of contraction moves down the oesophagus and can be recorded as a rise of pressure as it passes the open tip of a tube. The wave passes down at a rate of 2 to 4 cm/sec, being faster in the lower half of the oesophagus. The peak of the wave is usually above 50 cm of water pressure but there is considerable individual variation. In the erect position swallowed liquids pass quickly down the oesophagus at a fast rate under the influence of gravity and may enter the stomach well in advance of the peristaltic wave. More solid food forms a bolus which is actively pushed down by peristalsis. Secondary peristaltic waves arise from the stimulation of the oesophagus by distension and serve to move sticky lumps of food or refluxed material back into the stomach.

The lower oesophageal (gastro-oesophageal) sphincter is weaker than the upper sphincter; its pressure at rest is about 10 cm of water above intra-gastric pressure (Fig. 15.7). On swallowing this sphincter relaxes well ahead of the peristaltic wave; the pressure drops about one second after the act of swallowing and stays low until the peristaltic wave enters this area. The sphincter is then closed by a slow contraction that may take 7 to 10 sec. Barium emulsion, gently introduced by a tube into the lower oesophagus, rests above the sphincter and on swallowing passes through into the stomach when the sphincter relaxes. With repeated swallowing the oesophageal peristalsis is inhibited until the final swallow. During this time the gastro-oesophageal sphincter remains relaxed.

It is clear that this lower sphincter is only part of the gastro-oesophageal mechanism because the junction can withstand a backward pressure gradient far in excess of its resting tone without allowing reflux. The retrograde passage (reflux) of gastric contents is undesirable as a smart chemical inflammation can result from continued reflux. There is good evidence that a valvular function is involved and the intra-abdominal segment of oesophagus could well perform this. The walls of the segment exposed to the higher intra-abdominal pressure collapse and the higher the pressure about it the firmer is the segment compressed to form an effective valve. During the act of vomiting the abdominal segment is raised above the diaphragm and the valve is abolished.

Oesophageal motor activity is under the control of the vagus nerve. If this is sectioned in the neck peristalsis is lost. The sympathetic system has no appreciable control over the oesophagus. Motor function can become deranged, a well known example being *achalasia*. In this condition degeneration of the myenteric (Auerbach's) plexus in the oesophagus effectively causes denervation which results in a loss of peristalsis over most of the oesophagus and a failure of relaxation of the lower sphincter. The oesophagus dilates and food accumulates in it, trickling only slowly through into the stomach. In the commoner hiatus hernia the upper part of the stomach slides up through the diaphragmatic hiatus and the whole gastro-oesophageal junction lies in the thorax. The tube-valve function is lost and free reflux occurs with resulting chemical inflammation of the oesophagus.

REFERENCES

ARDRAN, G. M. & KEMP, F. H. (1952). The protection of the laryngeal airway during swallowing. *British Journal of Radiology* **25**, 406–416.

ATKINSON, M., EDWARDS, D. A. W., HONOUR, A. J. & ROWLANDS, E. N. (1957). Comparison of cardiac and pyloric sphincters. *Lancet* **ii**, 918–922.

BOSMA, J. F. (1957). Deglutition: pharyngeal stage. *Physiological Reviews* **37**, 275–300.

BOSMA, J. F. (1967). *Symposium on Oral Sensation and Perception.* Springfield, Ill.: Thomas.

BURGEN, A. S. V. & EMMELIN, N. G. (1961). *Physiology of the Salivary Glands.* London: Arnold.

CODE, C. F. (Ed.) (1968). *Handbook of Physiology. Section 6: Alimentary Canal.* Volume IV, Motility. Washington: American Physiological Society.

CREAMER, B., HARRISON, G. K. & PIERCE, J. W. (1959). Further observation on the gastro-oesophageal junction. *Thorax* **14**, 132–137.

DAVENPORT, H. W. (1961). *Physiology of the Digestive Tract. An Introductory Text.* Chicago: Year Book Publishers.

HILTON, S. M. & LEWIS, G. P. (1957). Functional vasodilatation in the submandibular salivary gland. *British Medical Bulletin* **13**, 189–196.

INGLEFINGER, F. J. (1958). Esophageal motility. *Physiological Reviews* **38**, 533–584.

JAKOWSKA, S. (1963). Mucous secretions. *Annals of the New York Academy of Sciences* **106**, 157–809.

JENKINS, G. N. (1966). *The Physiology of the Mouth,* 3rd edn. Oxford: Blackwell.

KERR, A. C. (1961). *The Physiological Regulation of Salivary Secretions in Man. A Study of the Response of Human Salivary Glands to Reflex Stimulation.* London: Pergamon.

MILSTEIN, B. B., EDWARDS, D. A. W. & BERRIDGE, F. R. (1961). The mechanism at the cardia. *British Journal of Radiology* **34**, 471–498.

MORGANE, P. J. (Ed.) (1969). Neural regulation of food and water intake. *Annals of the New York Academy of Sciences* **157**, 531–1216.

SCHNEYER, L. H. & SCHNEYER, CHARLOTTE A. (Eds.) (1967). *Secretory Mechanism of Salivary Glands.* London: Academic Press.

SREEBNY, L. M. & MEYER, J. (1964). *Salivary Glands and their Secretions.* Vol. 3 of *Monographs on Oral Biology.* Oxford: Pergamon.

Symposium on Nutrition and Teeth (1959). *Proceedings of the Nutritional Society* **18**, 54–95.

WHITELOCK, O. V. ST. (Ed.) (1960). The metabolism of oral tissues. *Annals of the New York Academy of Sciences* **85**, 1–499.

16 The stomach

GASTRIC FUNCTION

The major functions of the stomach in man can be summarized under four heads (1) By storing food temporarily and controlling its entry into the small intestine it spreads out the time during which food is presented to the digestive and absorptive processes of the upper small intestine. In this way the time spent in the actual ingestion of food is conveniently reduced. (2) It liquefies the ingested food and passes it on to the small intestine in a quantity and form suitable to small intestinal function. (3) It reduces the risk of swallowed noxious agents entering the small intestine. (4) The gastric mucosa produces hydrochloric acid and pepsin, the hormone gastrin and the intrinsic factor.

The final effector tissues which endow the stomach with its special properties are its musculature and its mucosa which control respectively its movements and its secretions.

GASTRIC MUSCLE AND GASTRIC MOTILITY

The musculature is responsible for the stomach's action as a reservoir and for the mixing and onward propulsion of the chyme. The word chyme is used to describe food that has undergone partial digestion so that it is semifluid.

Muscular structure and innervation. The muscular coat covers the whole of the stomach and is composed of four layers:

1. an external longitudinal layer (immediately under the serosa) part of which is continuous with the longitudinal fibres of the oesophagus,
2. a circular layer (continuous with the circular layers of the oesophageal muscle) which increases in thickness in the antrum and pylorus,
3. in the proximal portion only, a well-

developed oblique layer lying inside the circular layer,

4. a thin layer of muscle (muscularis mucosae) in the mucous coat.

The tissue between the inner layers of the circular or oblique muscle and the mucosa is referred to as the submucosa (Fig. 16.1).

The stomach is innervated by the terminal branches of the right and left vagi (parasympathetic) and also by sympathetic fibres from the coeliac plexus. It is generally accepted that the parasympathetic nerves relay in the two nerve plexuses within the stomach, the myenteric plexus lying between the two main muscle coats and the submucous plexus lying in the submucosa (Fig. 16.1). The sympathetic nerves which supply the stomach are already postganglionic.

Reservoir function of the stomach. The shape of a full stomach is illustrated in Fig. 16.2A. The outline of the human stomach can be observed by X-rays after the administration of a meal containing barium sulphate which is

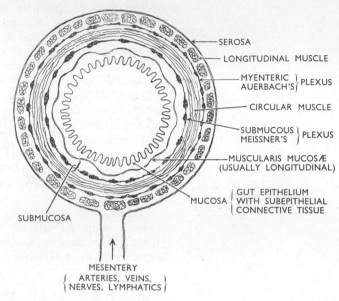

Fig. 16.1   A simplified plan of the main features of the wall of the intestinal tract.

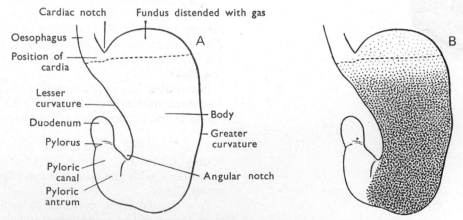

Fig. 16.2A and B   An outline of the full stomach to show the names of the various parts. The dotted line represents the level to which a large barium meal fills the stomach. The shaded area of the diagram on the right indicates the density of distribution of oxyntic cells according to Berger. There are almost none in the antrum and relatively few in the fundus.

234   Textbook of Physiology and Biochemistry

opaque to X-rays (Fig. 16.3B). The shape and position of the normal stomach vary considerably according to the posture of the subject, the skeletal build, the tone of the abdominal wall and the state of adjacent viscera. (Fig. 16.3A and B.)

As food enters the stomach and fills it, the smooth muscle fibres increase in length so that between peristaltic contractions the intragastric pressure remains nearly constant at low values not very different from those found in the empty stomach (<5 mm Hg or <68 mm H_2O). The intragastric pressure remains low unless the stomach becomes very large. When the stomach of an animal has been outlined with silver-wire sutures which can be seen on an X-ray photograph, it has been found that the lesser curvature remains nearly constant in position and contour whereas the greater cur-

vature of the body and fundus increase markedly in length as the organ distends.

The low pressure at which the stomach holds its contents can in part be explained on physical grounds. The Law of Laplace states that the transmural pressure P, across the wall of a cylinder is directly proportional to the tension in its walls, T, and is inversely proportional to its radius, R, so that $P = T/R$. Therefore although the tension in the wall of the stomach increases as it fills, the transmural pressure alters relatively little because its radius is also increasing. The stomach is often said to undergo *receptive relaxation* as it fills but this explanation of the low pressure in the full stomach would not be justified if the Law of Laplace alone explained these low pressures. When, however, smooth muscle is held in a constant state of stretch for some time its ten-

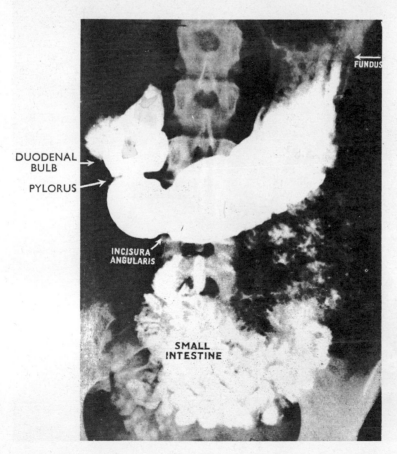

Fig. 16.3A Radiograph of abdomen twenty minutes after a barium meal. Patient erect.
(*By courtesy of R. C. Garry.*)

The Stomach 235

sion decreases and furthermore impulses in the vagus nerve probably cause muscular relaxation in the filling stomach. Relaxation as well as contraction of gastric muscle can be induced by appropriate stimulation of efferent fibres in the vagus and splanchnic nerves. This has been illustrated in animal experiments in which the vagus nerve is stimulated in the lower thorax while intragastric pressure is maintained constant and the volume of the gastric contents is recorded (Fig. 16.4). In these circumstances it is assumed that the external pressure on the stomach is constant and therefore the transmural pressure is constant so that a decrease in volume of the gastric contents indicates that contraction is the predominant response in the gastric muscle whereas an increase in volume indicates that the predominant effect is relaxation. Stimulation of the splanchnic nerves also causes gastric relaxation. Some of the tone of the gastric musculature undoubtedly originates

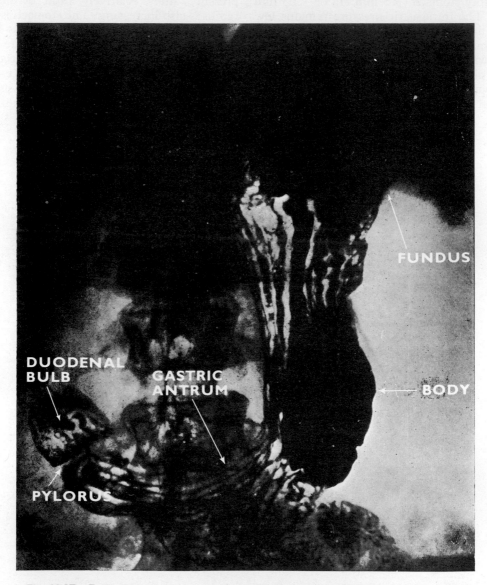

Fig. 16.3B Radiograph of stomach taken twenty minutes after a barium meal to show the mucosal pattern of stomach and duodenal cap. Patient supine. (*By courtesy of C. Pickard.*)

from intrinsic cholinergic mechanisms, probably the nerve plexuses in the stomach wall. In experimental animals with the vagus and splanchnic nerves to the stomach divided and the transmural pressure held constant, an increase of intragastric volume follows the

administration of atropine (Fig. 16.5a).

The relaxation of gastric muscle which accompanies vagal and splanchnic stimulation can be shown to be confined to the proximal part of the stomach and not to occur in the antrum. Relaxation caused by splanchnic

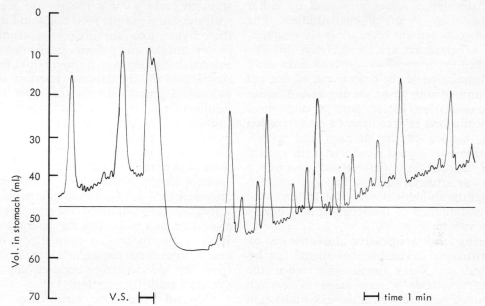

Fig. 16.4 Illustrating changes of intragastric volume at constant intragastric pressure (50 mm H_2O) in the anaesthetized cat with its pylorus occluded. After 60 seconds vagal stimulation (VS) there is an initial decrease in 'background' volume lasting 1 to 2 minutes, followed by a more prolonged period during which the 'background' volume is increased. Superimposed on these 'background' changes are additional short-lived alterations in volume.

(a)　　　　　　(b)

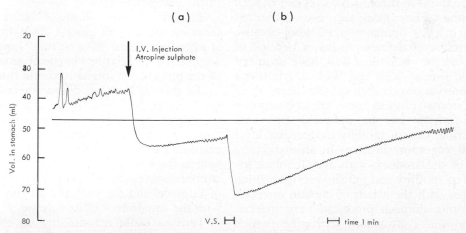

Fig. 16.5 (a) Illustrating the increased gastric volume which occurs at constant intragastric pressure (50 mm H_2O) after the intravenous administration of atropine sulphate. (The experimental conditions were the same as those described in Fig. 16.4.) Atropine also abolishes 'spontaneous' motility changes. (b) Illustrating that the relaxation of gastric musculature which follows vagal stimulation (VS) occurs in the atropinized animal. The stimulus was the same as that used in the experiment illustrated in Fig. 16.4. Note the absence of contraction in response to vagal stimulation after atropine and the immediate onset of relaxation.

stimulation is probably due to the release of noradrenaline and is abolished by sympathetic blocking agents such as guanethidine. The chemical transmitter released from the vagal nerve endings to cause gastric relaxation has not been identified and relaxation following vagal stimulation is not prevented by either atropine (Fig. 16.5b) or guanethidine. The physiological stimuli responsible for originating this relaxation are incompletely investigated but experiments on cats and dogs show that stimulation of the central end of one cut vagus immediately above the diaphragm causes gastric relaxation. Experiments on dogs show that swallowing is accompanied by relaxation in the fundus of the stomach and gastric relaxation after a meal is reduced if the meal is passed directly into the stomach through a fistula (an artificial opening into the stomach through the abdominal wall). Relaxation of the body and fundus of the stomach also occurs during vomiting (p. 261).

Mixing and propulsive movements of the stomach. Peristaltic movements can be observed on X-ray examination when the stomach contains a suspension of barium sulphate. These are seen as waves which begin in the body of the stomach and travel towards the pylorus as shallow indentations becoming deeper in the pyloric antrum where they may be so marked as to appear to bisect the organ. As a rule from one to three such waves can be seen at a time. They have also been observed directly in a few human subjects who, because of a stricture of the oesophagus or because of damage to the abdominal wall, have acquired a gastric fistula. Wolf and Wolff made many such observations on their subject Tom.

The stomach has an autonomous peristaltic rhythm which can be modified by vagal stimulation. Gastric motility is decreased after bilateral vagotomy in man. In animals, perfusion of the stomach with acetylcholine increases its motility and putting eserine (which interferes with the action of choline esterase) into Tom's stomach produced very marked and vigorous contractions. Atropine (which blocks the action of acetylcholine) stops gastric contractions for a long time both in animals and man.

The peristaltic movements in the upper two-thirds of the stomach are relatively weak but are much more intense over the more muscular pyloric antrum. Many of the peristaltic waves which originate near the cardia do not proceed along the antrum but when they do they may be sufficiently strong to produce a localized obliteration of its lumen. The peristaltic movements in the upper two-thirds of the stomach serve to mix the gastric secretions with the surface of the food bolus and to 'milk' the chyme into the pyloric antrum. The pyloric antrum and pylorus can be regarded as a peristaltic pump which moves fluid from the stomach to the duodenum. The fact that the pylorus is narrow means that only suitably liquified chyme is able to pass through. In contrast to the gastro-oesophageal junction, the pylorus is usually relaxed. It closes only as the peristaltic wave passes over it. While each wave of peristalsis in the antrum moves a small spurt of chyme forward into the duodenum it also squeezes a considerable amount backwards into the stomach so that antral peristalsis also mixes and macerates the gastric contents. Because the pylorus is normally open duodenal contents can regurgitate into the stomach from the duodenum whenever appropriate pressure gradients develop.

Control of gastric emptying. The intensity of peristalsis of the gastroduodenal pump affects the rate at which the stomach empties and the action of the pump is in its turn modified by the volume and composition of the gastric contents.

The effect of the volume of the gastric contents on the rate of gastric emptying can be studied by means of a special liquid meal. Despite wide variations in the original volume of the liquid meal introduced into the stomach under standardized conditions there is a linear relationship between the square root of the volume of the meal remaining in the stomach and the passage of time (Fig. 16.6). This relationship may depend on alterations in tension in the wall of the stomach. The tension is proportional to the radius of the stomach (Law of Laplace) and the radius of a cylinder varies with the square root of its volume.

Tension could act directly on the gastric muscle or could alter reflexly the number of impulses travelling in the efferent vagus and splanchnic nerves to the stomach. The finding that distension of the stomach causes an increase in the frequency of impulses travelling in afferent fibres of the vagus suggests that

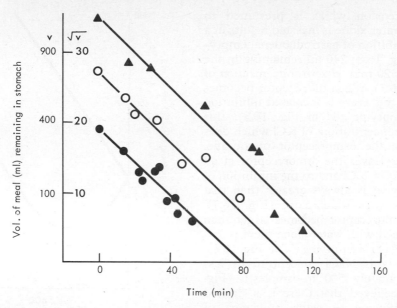

Fig. 16.6 Illustrating the relationship between the volume of a standard liquid meal remaining in the stomach and the passage of time, in a single human subject. The initial volumes of the meals introduced into the stomach were ▲ 1250 ml, ○ 750 ml and ● 330 ml. (Redrawn from A. Hopkins (1966). *Journal of Physiology* **182**, 144–149.)

there are tension receptors in the stomach wall.

Distension of the stomach causes it to contract and inhibits electrical and motor activity in the duodenum through an efferent nervous pathway which is probably adrenergic. Such reciprocal effects of gastric distension on the stomach and duodenum have the advantage of aiding the onward propulsion of chyme; unless the tension in the wall of the narrower duodenum is less than that in the stomach, with its larger diameter, intraduodenal pressure will be greater than intragastric pressure (Law of Laplace). The force of the gastroduodenal pump is decreased and gastric emptying time is greatly prolonged after bilateral vagotomy which is often performed to reduce gastric acid secretion in man. To avoid gastric stasis the operation is usually combined with measures designed to decrease the normal resistance at the pylorus. The muscle in the region of the pylorus may be incised, *pyloroplasty*, or a connexion may be made between the pre-pyloric region of the stomach and the upper small intestine, a *gastroenterostomy*, to short circuit the pylorus.

The composition of the gastric contents affects the rate of gastric emptying. A number of feedback mechanisms from the duodenum influence the rate of gastric emptying according to the nature of the chyme delivered from the stomach.

(a) The effect of the osmolarity of the gastric contents entering the duodenum. The effects on the rate of gastric emptying when water and two different solutes are introduced into the stomach are illustrated in Fig. 16.7. Hunt has suggested that these and similar observations could be explained by the presence of osmoreceptors in the walls of the duodenum and jejunum which exert a control over those factors which modify the strength of the gastroduodenal pump. It is assumed that the membrane of the osmoreceptor has permeability characteristics like those of the red blood cell but unlike the red cell the osmoreceptor is not disrupted by increased intracellular volume. It is presumed that the osmoreceptor alters in size according to the tonicity of the duodenal contents. Hunt believes that when relatively large volumes of solute are placed in the stomach and their rate of emptying is studied over short periods the composition of the instilled fluid is a reasonable indication of the nature of the solute reaching the duodenum despite dilution by the gastric and intestinal secretions. The moderate degree of osmoreceptor distension which is presumed to occur when water alone is ingested results in a moderate inhibition of gastroduodenal emptying ('b' in Fig. 16.8; 240 ml remaining in the stomach after 20 min, at zero concentration of solute, Fig. 16.7). When the receptor becomes smaller than this there is increased inhibition of gastric emptying ('c' in Fig. 16.8): this occurs at all concentrations of KCl which does not penetrate the osmoreceptor membrane. That is water leaves the osmoreceptor at all concentrations of KCl and so the inhibition of gastric emptying is always greater than that which occurs with water alone (Fig. 16.7). When the osmoreceptor becomes larger than the size attained with water alone there is less inhibition of gastric emptying ('a' in Fig. 16.8). This occurs with NaCl concentrations less than approximately 500 m-osmoles/l. (Fig. 16.7). It is believed that there is a limited facilitated diffusion or active transport of NaCl into the osmoreceptor: as the NaCl concentration increases from zero to approximately 300 m-osmoles/l increasing amounts of NaCl enter the cell and these are accompanied by water so that more water enters the cell than in the presence of water alone. At concentrations of NaCl above 300 m-osmoles/l however the membrane behaves towards NaCl as it does towards KCl and so the facilitated

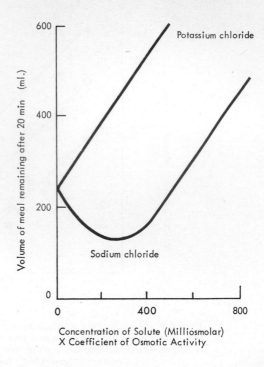

Fig. 16.7 Diagram showing the relationship between the volume of 750 ml test 'meals' of two different solutes and of water remaining in the stomach after 20 minutes and the effect of varying concentrations of the solutes in the meals. The volume remaining at zero concentration on the abscissa represents the effect of water alone. (From J. N. Hunt (1961). *Gastroenterology* **41**, 49–51.)

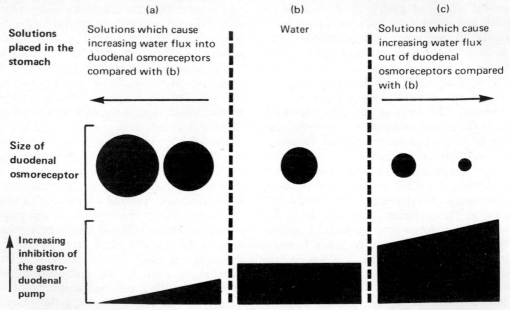

Fig. 16.8 A diagrammatic representation of the duodenal osmoreceptor hypothesis in the control of gastric emptying.

transfer of NaCl becomes less and less effective in causing distension of the receptor because the entry of water into the receptor is opposed by the increasing external concentration of NaCl. If the external NaCl concentration continues to increase a point is reached (approximately 500 m-osmoles/l, Fig. 16.7) when the osmotic effect of the internal and external concentrations of NaCl balance one another and the net flux of water across the cell membrane is the same as in the presence of water alone. If the NaCl concentration is increased still further then relatively more water will leave the cell than in the presence of water alone and so there is increased inhibition of gastric emptying (Fig. 16.7).

When the stomach is filled with fluids known not to reduce the rate of emptying by the osmoreceptor mechanism, gravity has, at least during the first 20 min, a significant effect on the rate of gastric emptying. In the head-down position the rate of emptying of such solutions is 80 per cent of that which occurs in the head-up position. With meals of similar concentration which do not penetrate the osmoreceptor and which therefore markedly reduce the rate of emptying by their osmotic effect, the rate is the same in the head-up position and in the head-down position.

It is believed that these receptors are in the duodenum rather than in the stomach. Starch and glucose meals of equal carbohydrate concentration, placed in the stomach through a tube, empty at similar rates. It is therefore assumed that starch can exert its effect on emptying only after it has been hydrolysed in the upper small intestine.

(b) The effect of acid entering the duodenum. The presence of acid in the duodenum reduces gastric motility and the rate of gastric emptying. On a molar basis acid is much more effective than increased osmolarity in slowing gastric emptying. When citrate solutions of similar osmolarity but different pH are placed in the stomach in man, more acid solutions leave less rapidly (Fig. 16.9). In these experiments a constant amount of acid titratable to pH 6·0 is delivered to the duodenum in unit time and the duodenum seems to have receptors sensitive to pH which inhibit gastric emptying when they are at pH 6·0 or less. Many of the features of inhibition of gastric emptying by acid are, however, as yet unexplained. Thus acetic acid, although a weaker acid than citric acid, is effective at lower concentrations. Also, although acetic acid is two thousand times less fat-soluble than hexanoic acid and would therefore seem less likely to penetrate a receptor cell, it is effective at lower concentrations. Furthermore, salts of myristic acid (14C atoms) are on a molar basis four times more effective in slowing gastric emptying than HCl and the salts of long chain fatty acids (with 12 to 18C atoms) are considerably more effective than the salts of fatty acids with 2 to 10C atoms. The rate of gastric emptying

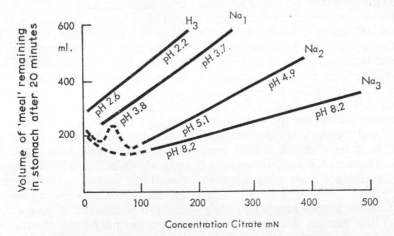

Fig. 16.9 The influence of salts of citric acid on gastric emptying. H_3, citric acid; Na_1, dihydrogen monosodium citrate; Na_2, monohydrogen disodium citrate; Na_3, trisodium citrate. The volume of the 'meal' instilled was 750 ml in each instance. (From J. N. Hunt & M. T. Knox (1961). *Journal of Physiology* **163**, 34–45.)

must depend not only on the rate of delivery of acid into the duodenum but also on the neutralizing capacity of the duodenal contents. The presence of acid in the duodenum increases the buffering capacity of the duodenal contents because it increases the secretions of the pancreas (p. 266) and of Brunner's glands (p. 290), both of which contain bicarbonate.

(c) The effect of the products of digestion of fat and protein in the duodenum. The presence of emulsified fat in the duodenum decreases gastric motility and the rate of gastric emptying in animals. It is unlikely that even in its most finely divided state fat could exert its effect through the osmoreceptor mechanism.

The inhibition of gastric motility by amino acids and polypeptides also is not entirely explicable as an osmotic effect.

(d) The effect of duodenal distension on gastric emptying. There is evidence that duodenal receptors respond to distension and that distension causes the inhibition of electrical and motor activity in the pyloric antrum.

Probably duodenal stimulation affects gastric motility by both nervous and humoral mechanisms which may act separately or in combination. In some instances vago-vagal reflexes have been identified; in others local enterogastric reflexes are involved whose pathway may pass through the coeliac ganglion. A polypeptide which stimulates gastric motor activity in dogs has been extracted from the duodenal mucosa of hogs. It remains to be determined whether this polypeptide which has been named motilin is concerned in the physiological control of gastric emptying. Since inhibitory effects can be demonstrated in denervated and transplanted gastric pouches (p. 248) they can in part at least be explained by a humoral mechanism. A humoral agent 'enterogastrone' has been postulated but it has not been identified with certainty. Both secretin (p. 269) and cholecystokinin-pancreozymin (p. 269) have been shown to have some enterogastrone-like activity and the existence of a polypeptide distinct from these has been suggested.

Emotion affects the rate of gastric emptying. Observation of the human stomach through gastric fistulae has shown that the emotional state of the subject has a great influence on emptying time and motility. In a state of fear, food may remain in the stomach for 12 hours or more whereas excitement may reduce emptying time.

GASTRIC MUCOSA AND GASTRIC SECRETION

Mucosal structure and growth. The gastric mucosa consists of simple branched tubular glands packed tightly together and arranged perpendicularly to the surface. Groups of glands, usually four, open into gastric pits which in turn open on to the mucosal surface. In the mucosa of the distal part of the stomach in man (the pyloric area) the glands are lined by mucous secreting cells only. In the mucosa of the remainder of the stomach the glands are lined not only with mucous cells but also with parietal (oxyntic) cells (Fig. 16.2B) and peptic (chief or zymogen) cells. These different cell types are arranged as illustrated in Fig. 16.10. The glands are described as having three parts, base, neck and isthmus. The neck region has irregularly-shaped mucous cells with basal nuclei and most of the parietal cells. A few parietal cells are also found in the isthmus and basal regions. Peptic cells (zymogen cells with granular cytoplasm) are present in the base of the glands. Special staining techniques reveal a few scattered argentaffin cells in the basal region. The gastric pits and the mucosal surface throughout the stomach are lined with characteristic mucus-laden, columnar epithelial cells with oval basal nuclei.

The mucosal structure is the outcome of a dynamic equilibrium between mucosal cell production and destruction. These cells are produced by cellular division probably from a common 'mother' cell in the mucous neck cell region. From this region gradually maturing cells pass up to the surface or down into the glands. The process of cell renewal has been studied by the use of tritiated thymidine (Chap. 20). The cells take 2 to 3 days to migrate from the neck region to the surface; the oldest surface cells are extruded into the gastric lumen and are digested. Estimation of the DNA content of gastric washings shows that in normal people about half a million cells are lost per minute. The cells which pass down into the glands give rise to the parietal and peptic cells; they are probably destroyed eventually in the base of the glands.

The mechanisms controlling this cell proliferation are not fully understood but they are known to be influenced by hormonal changes and dietary factors affecting somatic growth. The anterior pituitary probably plays an important part through its growth hormone. Mucosal growth may also be influenced by nervous and hormonal stimuli responsible for controlling gastric secretion (p. 249).

The surface cells of the stomach (Fig. 16.10) together with the slimy (visible) mucus which they secrete protects the gastric mucosa against physical and chemical damage. Without this protection the mucosa is susceptible to

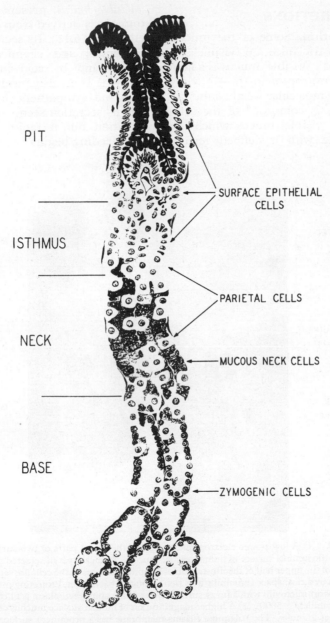

PIT

ISTHMUS

NECK

BASE

SURFACE EPITHELIAL
CELLS

PARIETAL CELLS

MUCOUS NECK CELLS

ZYMOGENIC CELLS

Fig. 16.10 A drawing of a section of the body of the stomach of a monkey. Mucus and nuclei are shown in black. (From A. W. Ham. *Textbook of Histology*, 5th edn. Preparation by C. P. Leblond. Figs. 14–28, p. 683.)

digestion by the gastric juice. The mucus adheres to the stomach wall and provides a slippery surface over which the chyme can move freely; furthermore, while electrolytes can diffuse through it, it contains bicarbonate and in effect holds a buffering and diluting layer close to the mucosa.

GASTRIC SECRETIONS

Mucous secretion. Some of the mucous constituents of gastric juice are 'visible' and some are 'dissolved'. 'Visible' mucus is a large molecular weight mucoprotein (p. 47) which is a polymer of low molecular weight subunits. 'Dissolved' mucus is composed of the same chemical units as 'visible' mucus which is a hydrated molecule with a defined tertiary

structure that forms a water-insoluble gel. Both polymerization and hydration are essential for the formation of the gel. Proteolysis or chemicals with reducing groups may reduce the viscosity of mucus by splitting the mucoprotein into subunits. The tertiary structure of the mucoprotein in the hydrated gel can be collapsed by high salt concentrations.

Visible mucus, present even in the resting stomach, is derived from the surface epithelial cells (Fig. 16.11). Its secretion is increased by mechanical and chemical irritation of the mucosa and by acid in the lumen of the stomach. It is also produced in response to vagal and sympathetic nerve stimulation. In dogs its secretion seems to be reduced during starvation but it rapidly returns to normal when feeding begins again.

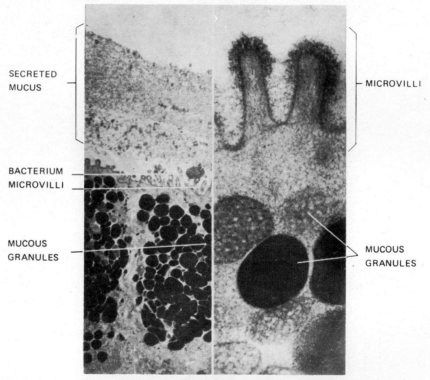

SECRETED MUCUS

MICROVILLI

BACTERIUM MICROVILLI

MUCOUS GRANULES

MUCOUS GRANULES

Fig. 16.11 (1) A low-power electron micrograph of the apical parts of two surface mucous cells from the luminal surface of human gastric mucosa. A thick layer of secreted mucus occupies most of the upper half of the illustration. The apical surfaces of the cells are separated from the mucus by a clear space containing a section through a bacterium. Projecting into the lumen are a few short microvilli with a fuzzy surface. Most of the apical cytoplasm is filled with dense mucous granules ($\times$5000). (2) A higher magnification of human surface mucous cell microvilli and mucous granules. The trilaminar plasma membrane has a prominent surface coat of fine filamentous material. Mucous granules of varying density are found in the apical cytoplasm ($\times$43 000). (From S. Ito (1967). *Gastric Secretion, Mechanisms and Control*, pp. 3–24. Oxford: Pergamon.)

Intrinsic factor (Chap. 24) originates in the parietal cell region of the glandular mucosa of the stomach. When mucosal sections are incubated with labelled vitamin B_{12} ($^{57}CoB_{12}$), autoradiographs show radioactivity localized to this region of the gastric mucosa; the uptake of the vitamin can be prevented if the tissue is first incubated with intrinsic factor antibody. Gastric intrinsic factor is probably not normally absorbed from the gut but in abnormal circumstances cellular damage may result in its entry into the circulation with the subsequent formation of antibodies. Intrinsic factor antibodies are found in the circulation of persons suffering from pernicious anaemia (Chap. 24). The injection of a variety of stimulants of gastric acid secretion increases the rate of secretion of intrinsic factor and the quantity of intrinsic factor normally secreted is con-siderably in excess of that required for the complete absorption of dietary vitamin B_{12}.

Blood group substances A, B, H (Chap. 24) are found in gastric mucous secretions in high concentration in about three-quarters of the population who are described as 'secretors'. These people are differentiated from those who secrete only minute quantities of these substances ('non-secretors'). Secretor status is genetically determined with 'secretor' dominant and 'non-secretor' recessive.

Acid secretion. The gastric juice contains hydrochloric acid and there is strong circumstantial evidence that it is secreted by the parietal cells (Fig. 16.10). In mammals, acid secretion occurs only from the glandular region of the body and fundus of the stomach which contains parietal cells (Fig. 16.2B). The dye neutral red is converted to its acid form in

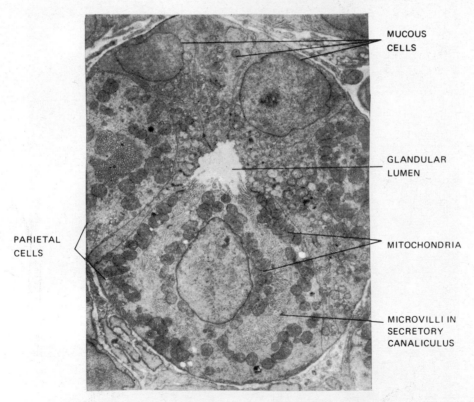

Fig. 16.12A A low-power electron micrograph of a transverse section through the neck region of a bat gastric gland. The specimen was obtained when the tissue was secreting acid. The section passes through a large parietal cell and parts of two others: note the numerous microvilli in the secretory canaliculi and the numerous large mitochondria around the nucleus and in the peripheral cytoplasm. Several mucous cells containing a few mucous granules surround the glandular lumen ($\times 4300$). (From S. Ito (1967). *Gastric Secretion, Mechanisms and Control*, pp. 3–24. Oxford: Pergamon.)

the canaliculi of the parietal cell. The electron-microscope shows that these cells have an elaborate system of canaliculi lined with microvilli; this suggests that they are well adapted for the secretion of large volumes of fluid (Fig. 16.12A and B).

Histochemical studies of biopsy specimens of gastric mucosa reveal intense concentrations of the flavoprotein enzymes and of cytochrome oxidase in the parietal cells (Fig. 16.13); the presence of these enzymes is very appropriate for the production of the large amounts of energy necessary to elaborate a concentrated acid solution.

For technical reasons uncontaminated parietal cell secretion has not yet been obtained. Gastric juice is, of course, a mixture of secretions from a variety of cell types. However, maximal stimulation of the mucosa might be expected to yield a concentration of acid in the juice closely approximating to that secreted by the parietal cell. Such secretion can be collected in intact man by continuous suction through a tube passed from the nose or mouth through the oesophagus into the stomach; contamination by swallowed saliva is prevented by simultaneous aspiration of saliva through another tube placed in the mouth. In

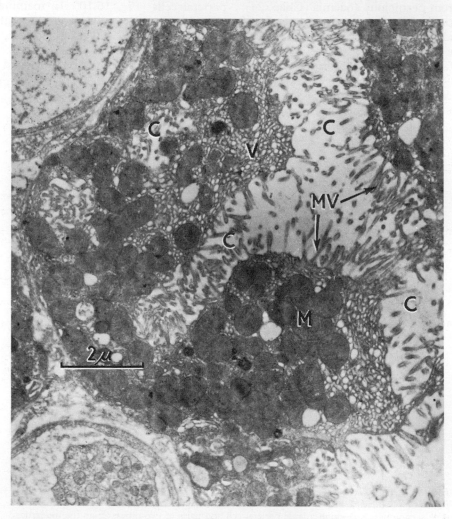

Fig. 16.12B A high-power electron micrograph of a portion of a parietal cell after a histamine injection. The intracellular canaliculi (C) are distended and therefore the microvilli (MV) are readily recognized. Numerous small cytoplasmic vacuoles (V) of the smooth endoplasmic reticulum lie adjacent to the canaliculi. ($\times 11\,000$). (*By courtesy of A. D. Hally.*)

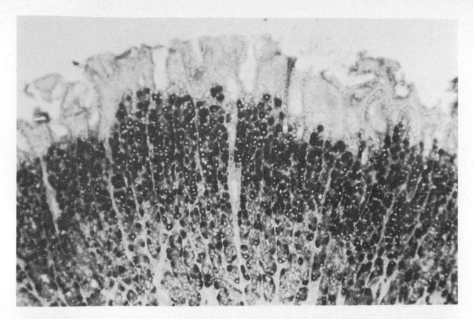

Fig. 16.13 Low-power photomicrograph of monkey stomach. Cytochrome oxidase reaction (*black granules*). Parietal cells are stained strongly and peripheral epithelium and chief cells give a trace reaction. This is a black colour stain and the appearance is similar for all oxidative enzymes, succinate dehydrogenase, reduced nicotinamide adenine dinucleotide, and nicotinamide adenine dinucleotide phosphate dehydrogenases (× 100). (From M. H. Floch, S. V. Noorden & H. M. Spiro (1967). *Gastroenterology* **52**, 230–238.)

animals it is possible to obtain samples of juice secreted by the parietal cell glandular region of the stomach without contamination by secretion from the mucosa of the pyloric gland area by isolating a pouch of mucosa of the body of the stomach as illustrated in Fig. 16.14. There are probably small variations in the concentration of chloride secreted by the parietal cells in different species but in man the value appears to be about 160 m-equiv./l. Of this chloride secretion, about 140 m-equiv./l is HCl and about 20 m-equiv./l is KCl.

The hydrogen ion concentration of parietal cell secretion is therefore a million times greater than that found in plasma. The energy required for the secretion of hydrogen ions against this concentration gradient is derived mainly from aerobic oxidation. The energy-rich phosphate compounds produced by these mechanisms are of great importance. If they are rendered unavailable by dinitrophenol, cell respiration continues but acid secretion is inhibited. Since both at rest and during acid secretion the mucosal surface of the body of the stomach is electronegative to the serosal surface, secretion of chloride must take place against an electrical gradient as well as against a concentration gradient (from 107 m-equiv./l in the intracellular fluid to 160 m-equiv./l in the canaliculus). Movement of chloride ions across the canalicular membrane, like the movement of hydrogen ions, requires an active transport mechanism (Fig. 16.15).

The existence of the hydrogen ion pump can be demonstrated if the chloride potential is first abolished by replacing the chloride on the surface of the mucosa with sulphate, to which the mucosal cells are impermeable. If the mucosa is now stimulated to secrete acid, the mucosal surface becomes electropositive with respect to the serosa as the action of the hydrogen ion pump is unmasked. During resting conditions in the dog, the mucosal surface of the mucosa of the body of the stomach is approximately − 70 mV to the serosal surface; this value is reduced by almost half when the preparation is stimulated to secrete acid and the hydrogen ion pump as well as the chloride pump is operating strongly.

An unlimited supply of H^+ is available from the dissociation of intracellular water. Uncontrolled accumulation within the parietal cell of

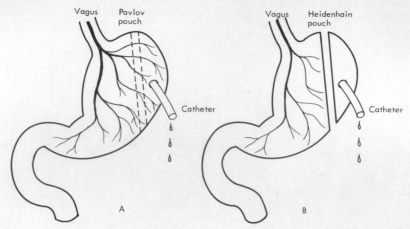

Fig. 16.14 A diagram of two commonly used types of gastric pouch. A. *A Pavlov (vagally innervated) pouch.* This is constructed by sewing the opposing mucosal surfaces together beneath the muscular layer in the region indicated by the two dotted lines. The nerve and blood supply to the pouch which reaches the mucosa through the muscular coat, is preserved intact. Note the vagal innervation of the pouch. B. *A Heidenhain (vagally denervated) pouch.* This is constructed by completely severing the mucosa and muscle of the stomach in the region indicated by the two solid lines. Separate whole stomach and body pouches are created. The sympathetic nerves and blood supply continue to reach the Heidenhain pouch along the greater curvature but the vagal innervation which normally spreads to this region from the lesser curvature is severed.

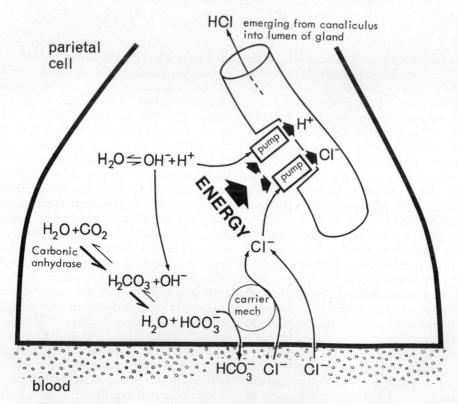

Fig. 16.15 Simplified scheme to illustrate current views on the method of production of hydrochloric acid by the parietal cell of the stomach.

the OH^- produced during the formation and secretion of hydrogen ions would, however, result in a rise of intracellular pH sufficient to interfere with cell metabolism and with acid secretion. The accumulation of OH^- is prevented by its combination with H_2CO_3, which is also in abundant supply, to form HCO_3^-; the supply of H_2CO_3 is facilitated by the enzyme carbonic anhydrase which is present in high concentration in the parietal cell. The bicarbonate ions are transported into the blood (Fig. 16.15) so that the blood leaving the stomach during acid secretion shows an increase in pH. This is presumably responsible for the postprandial 'alkaline tide', that is the slight increase in plasma and urine pH which occurs after meals. The chloride required for the formation of HCl enters the parietal cell from the extracellular fluid. This chloride diffuses down a concentration gradient but its entry into the cell is in part facilitated by a carrier mechanism and coupled to the transport of bicarbonate out of the cell (Fig. 16.15).

Enzyme secretion. The only gastric enzymes of any consequence in digestion are the pepsins. The gastric pepsins are a complex of several enzymes each with its own distinct chemical characteristics; at least seven have been detected electrophoretically. These enzymes are formed by the action of acid and of pepsin itself on corresponding inactive precursors, pepsinogens, which are secreted by the gastric mucosa. The major pepsinogen, sometimes known as pepsinogen I or pepsinogen A, occurs throughout the stomach, as do some of the other pepsinogens. The remainder come only from the mucosa of the body and fundus in the stomach but are known also to occur in the duodenum. In the mucosa of the body of the stomach the peptic or chief cells in the base of the gastric glands (Fig. 16.10) are believed to be the source of pepsinogen. Microchemical analysis of tiny blocks of gastric mucosa reveals that pepsinogen is abundant in the deeper parts of the glands. After prolonged stimulation of pepsin secretion the number of granules in the peptic cells is decreased.

The pepsins are unusual enzymes in that their pH optima are in the range pH 1·5 to 3·5. Pepsin breaks the polypeptide chain at many sites, the most susceptible link involving tyrosine or phenylalanine. The pepsins thus act on proteins to produce short peptide chains without liberating a significant quantity of amino

acids. The products of pepsin digestion are later broken down to amino acids by the proteolytic enzymes of the pancreatic juice and intestinal cells. After complete removal of the human stomach, digestion of proteins to amino acids in the intestine proceeds quite well in spite of the absence of preliminary digestion. Pepsin also has the property of clotting milk.

The physiological significance of the different forms of pepsin is at present unknown. Their proportions in gastric juice vary from person to person and certain patterns of their occurrence appear to be associated with predisposition to ulceration of the stomach and duodenum. Normal human plasma contains all the pepsinogens in small quantities but urine appears to contain only those which in the stomach are limited to the body and fundus (uropepsinogens).

CONTROL OF GASTRIC SECRETION

The secretion of gastric juice is controlled by a variety of mechanisms, nervous and chemical, excitatory and inhibitory, which are integrated during the process of digestion in a complex manner not yet completely understood.

Nervous excitation. Stimulation of the peripheral end of a cut vagus nerve is followed by the secretion of acid, mucus and pepsin. Acetylcholine is liberated at the vagal nerve endings; the effect of vagal stimulation can be simulated by the injection of stable cholinergic drugs and can be markedly reduced by anticholinergic drugs like atropine.

The importance of the vagus nerves for gastric secretion was demonstrated by Pavlov's experiments on dogs in which the oesophagus was cut in the neck and the two ends made to open separately on the surface of the skin (oesophagostomy). The sight and smell of food alone was found to initiate a flow of gastric juice (psychic or appetite juice) when the vagi were intact but not when they were cut. This phase of gastric secretion does not appear to be as important in man as it is in the dog. In dogs with oesophagostomy, food which was chewed and swallowed was diverted to the exterior and did not enter the stomach (sham feeding) but gastric juice was secreted provided that the vagi were intact. These initial stages of stimulation are described as the *cephalic phase* of gastric secretion.

Reflex stimulation of gastric secretion is not, however, limited to these reflexes alone and the presence of food in the stomach and upper small intestine as well as distension in these regions stimulates gastric secretion by nervous reflexes whose afferent and efferent nerve fibres are in the vagus (see vagovagal reflexes already referred to on pp. 238 and 242 in relation to the control of gastric motility).

The excitatory effect of vagal stimulation occurs partly by a direct stimulation of the secretory cells and partly by an indirect stimulation of these cells consequent upon the liberation of a hormone *gastrin* from the mucosa of the pyloric antrum and possibly also from the upper small intestine (see p. 252). The release of gastrin in response to vagal stimulation has been demonstrated by the finding that vagally denervated pouches (Heidenhain pouches) of gastric body mucosa show an increase of acid secretion in response to sham feeding (Fig. 16.16A). This acid secretion occurs even when complete denervation of the 'body' pouch is ensured by transplanting it into the body wall where it acquires an entirely new blood supply. Vagal denervation of the antrum abolishes this response as does the intravenous administration of a drug like hexamethonium which blocks both sympathetic and parasympathetic transmission at ganglionic synapses in the autonomic nervous system.

Chemical excitation. The presence of certain food materials, especially meat extracts,

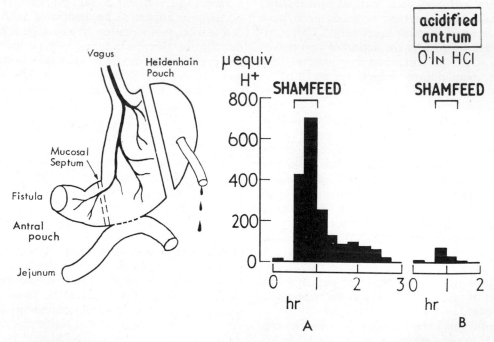

Fig. 16.16A and B (A) Response of Heidenhain pouch to sham feeding in a dog with an isolated vagally innervated pouch of the pyloric antrum. (B) Response of Heidenhain pouch to the same sham feeding stimulus as in A but with the antral pouch irrigated with 0·1N HCl. (Adapted from Maung Pe Thein and B. Schofield (1959). *Journal of Physiology* **148**, 291–305.)

within the pyloric antrum provokes acid secretion even in the absence of vagal innervation of the antrum; this is part of the so-called *gastric phase* of gastric secretion. These materials probably produce their effects by liberating the hormone *gastrin* from the mucosa of the pyloric antrum into the blood which carries it to the glands of the mucosa of the body of the stomach and possibly also in some instances by absorption into the blood stream of materials which themselves directly stimulate acid secretion (*secretagogues*).

Unequivocal proof of the existence of an antral hormone which stimulates acid secretion was first provided by experiments in dogs which showed that distension of a denervated pouch of the pyloric antrum in the absence of food caused secretion of acid from a denervated or transplanted pouch of the body of the stomach.

If extracts are made of thin slices of antral mucosa cut parallel with the surface, it is found that gastrin activity is located in the deeper portions of the pyloric glands. Cytological and cytochemical studies have demonstrated a gastrin containing cell (G cell) in the surface epithelium of the mid zone region of the pyloric glands of a number of species including man. These cells are thought by some workers to bind antibodies to gastrin, they contain numerous basal granules and on their apical surface microvilli protrude into the lumen of the pyloric gland (Fig. 16.17). The G cells belong to a group of endocrine polypeptide cells which possess cytochemical characteristics described as APUD (Amine Precursor Uptake and Decarboxylation) and Pearse and his colleagues believe they migrate to the gastrointestinal tract from the neural crest during embryonic development.

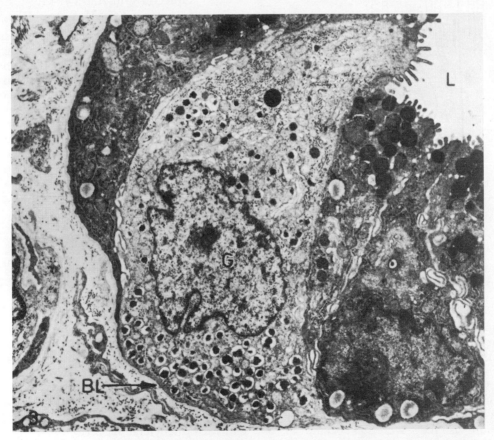

Fig. 16.17 A low-power electron micrograph of a G cell in the pyloric gland epithelium of the cat. Gastrin (G cell) (G). Basal lamina (BL). Lumen (L). (G. Vassallo, E. Solcia & C. Capella (1969). *Zeitschrift für Zellforschung und mikroskopische Anatomie* **98**, 337.)

The simulation of gastric acid secretion by meat extracts placed in pouches of the pyloric antrum can be prevented if the antral mucosa is first exposed to local anaesthetics such as cocaine. These anaesthetics do not, however, prevent the stimulation of acid secretion in response to irrigation of the antral mucosa with acetylcholine. For these reasons some workers believe that meat extracts act on surface receptors which communicate by nerve fibres with the gastrin cells. These workers believe there is a synapse in the pathway because, in the dog, administration of hexamethonium prevents the release of gastrin by meat extracts but not by acetylcholine which it is assumed stimulates the G cell directly (Fig. 16.18). The evidence for this nerve pathway is

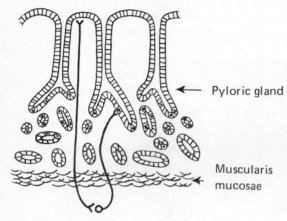

Fig. 16.18 Diagram of the antral mucosa to show the postulated excitatory nervous pathway for gastrin release from the G cell in the gastric gland by distension and by food in the stomach. (Redrawn from B. Schofield, Mary Redford, A. H. Grabham & K. Nuaimi (1967). *Gastric Secretion Mechanisms and Control*. Oxford: Pergamon.)

not, however, conclusive. It has not been demonstrated histologically; it seems unecessary in view of the superficial position of the G cell in the surface epithelium; and there is some evidence that meat extracts in the pyloric antrum stimulate gastrin release even after atropinization.

The gastric glands can also be excited by the products of digestion and by distension of the small intestine. When a dog with its oesophagus anastomosed to the second part of the duodenum is fed, secretion occurs in an isolated vagotomized stomach or in a transplanted pouch of body mucosa. This represents part of the so-called *intestinal phase* of

gastric secretion and is probably due to the release of gastrin from the mucosa of the upper small intestine.

Gregory and Tracy have isolated from the antral mucosa of a number of species including man, a polypeptide capable of stimulating gastric acid secretion. This polypeptide consists of seventeen amino acid residues (a heptadecapeptide) and in each species it occurs in two slightly different forms described as gastrins I and II. The physiological actions of the endogenously liberated hormone are very similar to the responses which follow the injection of these peptides. The structures of these peptides have been elucidated by Kenner and his colleagues. Human gastrin I (mol. wt. 2114) isolated from the antral mucosa in man has the following structure:

*Glu-Gly-Pro-Trp-Leu-Glu-Glu-
Glu-Glu-Glu-Ala-Tyr-Gly-
Trp-Met-Asp-Phe.NH$_2$

Human gastrin II differs in that tyrosine is replaced by tyrosine-*O*-sulphate. Hog gastrins differ from human gastrin in having methionine instead of leucine in position 5; in sheep gastrin valine replaces leucine and one of the glutamic acid residues in positions 6–10 is replaced by alanine; dog gastrin is similar to human gastrin except that one of the glutamic residues in position 6–10 is replaced by alanine. Both termini of the polypeptide molecule are blocked; the N-terminal residue (indicated by the asterisk) is pyroglutamyl (pyrrolidone carboxyl)

$$H_2C \text{———} CH_2$$
$$O=C \qquad CH—C—$$
$$N$$
$$H$$

and as in a number of other biologically active polypeptides there is no free C-terminal carboxyl group, since the chain ends with an amide (see secretin, pancreozymin and vasopressin). While the C-terminal residue of secretin is valine.NH$_2$ and of vasopressin is glycine.NH$_2$, both gastrin and pancreozymin end with phenylalanine.NH$_2$.

These gastrins have been synthesized by Kenner and his co-workers and it has been found that all of the physiological actions of the total molecule are possessed by the C-terminal tetrapeptide (indicated above in heavy type) which appears to be the 'active centre' of the molecule. Pentagastrin, a synthetic analogue containing five amino acids is used in clinical medicine (p. 260).

Pure gastrin injected intravenously has a surprising number of actions on the gastro-intestinal tract. In addition to stimulating gastric acid secretion it stimulates the secretion of pepsin and intrinsic factor by the stomach, the secretion of water, bicarbonate and en-zymes by the pancreas and the secretion of hepatic bile and bicarbonate. Gastrin also increases the tone of the musculature of the stomach and small intestine. The other gastro-intestinal hormones, secretin and pancreo-zymin (p. 269) when injected intravenously also have more than one action on gastro-intestinal function. The physiological sig-nificance of these multiple effects is uncertain but many hormones which are similar in chemical structure show an overlap in their biological effects. This overlap is shown not only after injection of gastrin but also by endo-genously released gastrin. Irrigation of the pyloric antrum of the dog with acetylcholine results in a significant increase in the secretion of pancreatic enzymes as well as of gastric acid. The old idea that each gastrointestinal hor-mone has a single action on one target tissue has to be reconsidered.

In most physiological conditions the con-centrations of gastrin in the circulation, which are measured in picograms (10^{-12} g) per ml, are too small to be detected by bioassay. They can, however, be measured by the technique of radioimmunoassay which, because of its sensi-tivity and precision, has revolutionized the study of hormones in plasma. In this technique the plasma of which the gastrin content is to be determined is incubated *in vitro* with anti-bodies to gastrin in the presence of radio-labelled gastrin. The gastrin in the plasma competes with the added radio-labelled gastrin for binding with the antibody and less label is bound to the antibody in proportion to the amount of unlabelled hormone present. A sig-

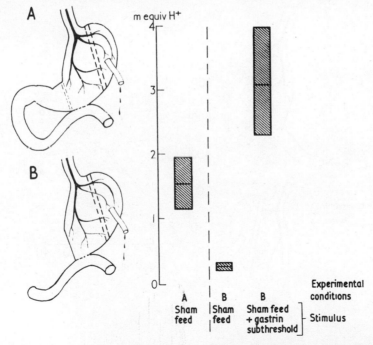

Fig. 16.19 Comparison of the acid secretory responses from Pavlov pouches in 5 dogs in which the pyloric antrum and duodenum were excluded from gastrointestinal continuity (A) and from the same dogs after the pyloric antrum and first part of the duodenum had been excised (B). The mean acid output is indicated by the mid-line of the hatched area which include ±1 S.E. of these mean values. (Based on data from L. Olbe (1964). *Acta physiologica scandinavica* **62**, 169–175.)

nificant increase in the immunoreactive gastrin concentration in plasma has been detected after feeding protein meals.

Biochemical studies of the size and electrophoretic properties of immunoreactive gastrin in plasma and in the mucosa of the pyloric antrum and upper small intestine reveal that it exists in two forms; one has the characteristics of the Gregory and Tracy heptadecapeptide gastrin and the other, a larger component, is more basically charged. This larger component is estimated to have a molecular weight of about 7000 and is believed to be a complex of heptadecapeptide gastrin covalently linked with a more basic peptide just as proinsulin is a peptide linked to chains of insulin (p. 319). Whether this complex is a precursor of heptadecapeptide gastrin and whether its secretion is independently regulated remains uncertain.

During digestion there is undoubted interdependence of the nervous and chemical factors which control gastric acid secretion; these are not simply additive in their effects, they potentiate one another. The acid secretory response to sham-feeding a dog with a Pavlov pouch can be markedly reduced if the gastrin bearing areas of the stomach and small intestine are removed. This response can be restored by giving, simultaneously with the sham-feed, subthreshold doses of gastrin (doses of gastrin which by themselves are ineffective in stimulating gastric acid secretion) (Fig. 16.19). Furthermore, the response to injected gastrin can be shown to be considerably greater if the stomach is innervated by the vagus nerve even under conditions of basal vagal activity (Fig. 16.20).

Additional nervous mechanisms in the wall of the stomach. Local cholinergic mechanisms within the wall of the body of the stomach, independently of vagal stimulation, also potentiate responses to gastrin stimulation. The rate of acid production in a vagally denervated secreting pouch in response to a given dose of gastrin is considerably increased by distension of the pouch (Fig. 16.21). This effect can be prevented by atropine.

The excitatory mechanisms which result in

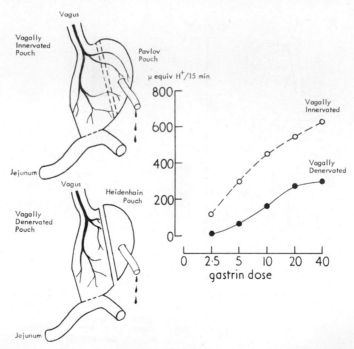

Fig. 16.20 Comparison of acid secretory responses to graded doses of gastrin from vagally innervated (Pavlov pouches) and vagally denervated (Heidenhain pouches) pouches of gastric body mucosa in the same 4 dogs after their pyloric antra had been excised. The mean results only are illustrated but the responses from the innervated pouches were significantly greater than those from the denervated pouches at all dose levels of gastrin. (Adapted from A. Andersson & M. I. Grossman (1965). *Gastroenterology* **48**, 449–462.)

gastric secretion are summarized in Fig. 16.22). Stress must be included as a cephalic influence on gastric acid secretion because there is no doubt that emotional factors modify gastric secretion. In the subject Tom, who had a gastric fistula, emotions of aggression and hostility were associated with increased acid secretion. In the absence of recognizable stimuli the human stomach secretes acid at low rates (basal secretion) and this secretory rate decreases during the sleeping hours and is minimal about 2 a.m. Stimulation of the anterior hypothalamus in monkeys increases gastric secretion provided that the vagus nerves are intact. It is not surprising that anticholinergic drugs markedly reduce the normal acid secretory responses in digestion and that they reduce the responses in both nervous and chemical phases of stimulation. It must now be clear why vagotomy is effective in reducing gastric acid secretion. It is for this reason that vagotomy is frequently carried out in patients

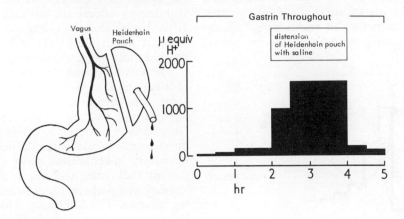

Fig. 16.21 Illustrating the effect during gastrin stimulation in the dog of distending a vagally denervated pouch of the body of the stomach with saline. (Adapted from M. I. Grossman (1961). *Gastroenterology* **41**, 385–390.)

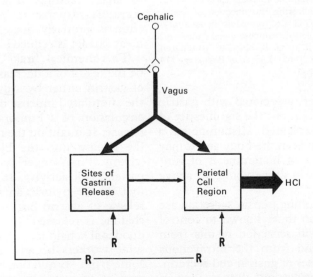

Fig. 16.22 A diagrammatic representation of the physiological excitatory mechanisms which cause gastric acid secretion. Receptor mechanisms, R, within the walls of the stomach and upper small intestine.

who suffer from ulcers of the stomach or duodenum.

For almost fifty years it has been known that histamine has a powerful stimulatory effect on gastric acid secretion and, until pure preparation of gastrin became available, it was the most powerful known stimulant of acid secretion. Gastrin II is 250 times more potent (on a molar basis) as a secretagogue than histamine. Inevitably, attempts were made to implicate histamine in the physiology of gastric acid secretion. There is, however, no convincing evidence that histamine is the final common mediator in parietal cell stimulation. Alterations in histamine metabolism in the gastric

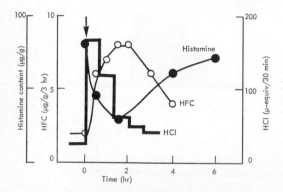

Fig. 16.23 Illustrating the relationship in the rat between alterations in gastric acid secretion and alterations in the histamine content and histamine forming capacity (HFC) of the gastric mucosa following the injection of gastrin (injected at the arrow). HFC is expressed as the amount of ^{14}C-histamine formed when the mucosa is incubated with ^{14}C-histidine. (Adapted from G. Kahlson, E. Rosengren, D. Svahn & R. Thimberg (1964). *Journal of Physiology* **174**, 400–416.)

mucosa are, however, associated with gastrin stimulation (Fig. 16.23) but the significance of this remains to be explained. Histamine has a ubiquitous distribution in the body and it may be that the turnover of histamine is of fundamental importance in all metabolically active tissues.

The factors controlling pepsin secretion are less well defined than those known to control acid secretion. Pepsin secretion results from vagal stimulation and from the intravenous injection of large doses of gastrin and secretin. In man, although vagotomy reduces pepsin production, the spontaneous output of pepsin remains substantial.

Inhibition of gastric secretion. Gastric acid secretion is controlled not only by the stimulatory mechanisms just described but also by inhibitory mechanisms, both nervous and chemical. The details of these inhibitory controls are still uncertain.

In the subject Tom, who had a gastric fistula, emotions of sadness and fear and sensations of nausea were associated with decreased acid secretion. There is no experimental evidence that efferent vagal stimulation can inhibit gastric acid secretion even when the stimulatory effects are blocked with atropine. It is likely that this kind of inhibition is brought about simply by a reduction of the normal tonic vagal stimulation.

There is no doubt that mechanisms capable of 'braking' acid secretion can be activated by large concentrations of H^+ in the pyloric antrum and by large concentrations of H^+, fatty acids or osmotically active particles in the duodenum. While some if not all of these inhibitory mechanisms have a neural component they may also act in the absence of vagal or sympathetic innervation of the secreting mucosa. The inhibitory effects of high H^+ concentrations are the most thoroughly investigated. Acid at low pH bathing the pyloric antrum or duodenum suppresses the acid secretion of vagally denervated or transplanted pouches of gastric mucosa secreting in response to sham feeding (Fig. 16.16B). The acid secretory response to meat extract in the antrum is similarly decreased if the pH of the meat extract is reduced below 2·5 (Fig. 16.24).

This chemical 'brake' to acid secretion could be the result of a decrease in the rate of release of gastrin either by a direct inhibition within the stimulated mucosa or by the entry into the circulation of a humoral inhibitor of gastrin release. It might on the other hand result from the release into the blood of an inhibitory 'chemical messenger' (chalone) which inhibits parietal cell activity. It is generally accepted that in the pyloric antrum acid inhibits the release of gastrin but that in the duodenum a chalone is released (Figs. 16.25 and 26). The duodenal chalone, called bulbogastrone by some workers because it seems especially to come from stimulation of the duodenal bulb, may or may not be the same as that which slows gastric emptying (see enterogastrone, p. 242). Within 5 min of a single intravenous

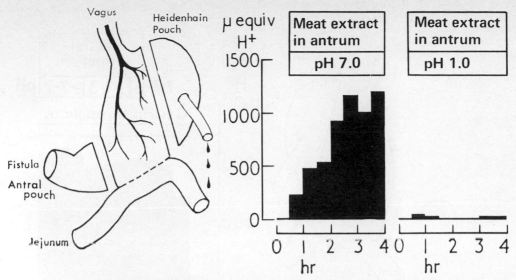

Fig. 16.24 Illustrating the effect on gastric acid secretion from a Heidenhain pouch (dog) of placing meat extract at pH 7·0 and pH 1·0 in a denervated pyloric antrum. (Adapted from E. H. Longhi, H. B. Greenlee, J. L. Bravo, J. D. Guerro & L. R. Dragstedt (1957). *American Journal of Physiology* **191**, 64–70.)

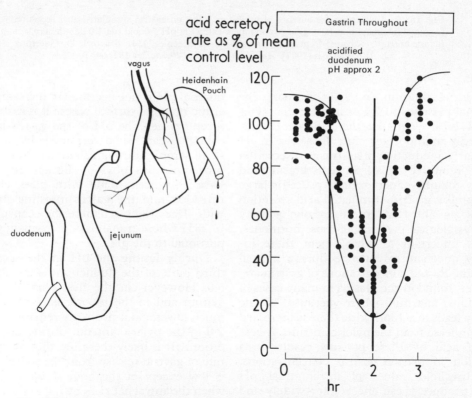

Fig. 16.25 Illustrating the inhibitory effect of the installation of HCl into the duodenum on gastrin-stimulated acid secretion from Heidenhain pouches (three dogs). The dogs each had a pylorojejunostomy and a cannula was placed in the duodenum. During the 1-h period of acid instillation the duodenal contents were at approximately pH 2. The control intraduodenal pH was between 7 and 8. The mean H⁺ secretory output from the pouches/15 min during the hour immediately preceding duodenal acid infusion is taken as 100 per cent and all the 15-min rates of H⁺ output are expressed as a percentage of this mean value. (Adapted from S. Andersson (1960). *Acta physiologica scandinavica* **50**, 105–111.)

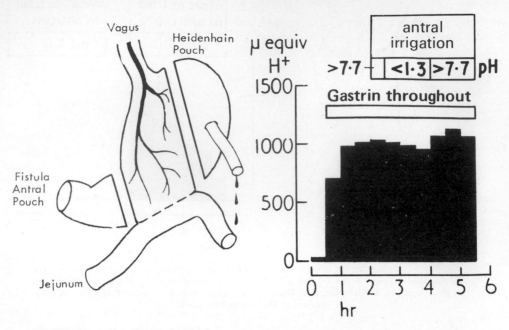

Fig. 16.26 Illustrating the absence of inhibitory effect on gastrin stimulated acid secretion from a Heidenhain pouch (dog) of placing meat extract at pH 7·0 and pH 1·0 in a denervated pyloric antrum. (Adapted from E. H. Longhi, H. B. Greenlee, J. L. Bravo, J. D. Guerro & L. R. Dragstedt (1957). *American Journal of Physiology* **191**, 64–70.)

injection of secretin (p. 272) in man, the duodenal pH rises to 7·4 or higher: the secretion of bicarbonate by the pancreas is undoubtedly an important cause of the rise in pH but clearly a reduction of gastric acid secretion would reinforce this. Since secretin and cholecystokinin–pancreozymin (p. 269) in large doses inhibit gastrin-stimulated acid secretion it is just possible that 'enterogastrone' activity is a physiological property of these hormones.

It is uncertain to what extent these inhibitory mechanisms operate during normal digestion. Raised concentrations of gastrin are, however, found in the plasma in many cases of pernicious anaemia where because gastric atrophy leads to achlorhydria there is increased gastrin release from the pyloric antrum. Ingestion of acid by these patients results in a reduction in plasma gastrin concentrations. The threshold values of intraluminal pH which produce these effects vary widely and there are technical difficulties in following pH changes within the gastro-intestinal tract during digestion. It is generally accepted that these inhibitory mechanisms operate at intraluminal pH values below 2·5; however, intra-

luminal pH is not necessarily the same as that at the mucosal surface where it is assumed the receptors lie. The pH of the gastro-intestinal contents can be determined by aspirating small samples through a stomach tube but this procedure has obvious limitations. Studies have also been made with glass electrodes threaded into the gastro-intestinal tract and with free radio-telemetering capsules (see p. 285) whose transmission frequency is proportional to the pH.

During fasting the pH in the second and third parts of the duodenum is usually above 6·0. However, in the first part of the duodenum and in the pyloric antrum the pH is much lower and it probably remains below pH 2·0 in the pyloric antrum during fasting conditions. It is likely therefore that normally the rate of gastrin release from the pyloric antrum is low except in the hour or so after a meal when the antral pH rises owing to the buffering effect of the meal. If acid chyme is unable to reach the pyloric antrum the secretion of acid by the stomach increases considerably; this situation may be brought about in man if the stomach is joined to the jejunum (gastrojejunostomy) so that the pyloric antrum and duodenum are excluded from the main stream of acid chyme; the resulting increase of acid secretion may lead to ulceration of the intestinal mucosa.

Control of the blood flow through the gastric mucosa. It is likely that mucosal blood flow and secretion in the stomach are closely related and that the blood flow is disturbed in disease processes in the stomach. However, little is known about the control of the gastric mucosal circulation. There are extensive arterial anastomoses within the mucosa; consequently interruption of flow even in a major gastric blood vessel is unlikely to affect the mucosal flow significantly. The flow of blood through the mucosal capillaries must depend to an important extent on the state of the numerous submucosal arteriovenous anastomoses as well as on the arterial perfusion pressure and flows. Sympathetic stimulation and vasoconstrictor drugs may interfere with mucosal blood flow sufficiently to depress secretion.

Changes in mucosal blood flow may be assessed very roughly from the colour of the mucosa if it can be seen through a gastric fistula. The basic lipid-soluble drugs amidopyrine and ^{14}C aniline can be used in the quantitative measurement of mucosal blood flow. Acidic lipid-soluble drugs remain in the undissociated state in the acid of the stomach lumen and are rapidly absorbed whereas basic drugs like amidopyrine and aniline are completely ionized and are absorbed very slowly. In blood, however, these drugs are unionized at the pH of blood so that as they circulate through the gastric mucosa they diffuse rapidly and completely into the acid gastric lumen where they are trapped and ionized. If, therefore, amidopyrine or ^{14}C aniline is injected to maintain a constant plasma concentration and if at the same time the rate of gastric excretion of these drugs is measured the rate of flow in the gastric mucosa can be calculated. Mucosal blood flow probably accounts for more than 50 per cent of the total gastric blood flow. Extremely rapid rates of blood flow may occur in the gastric mucosa as in other secretory tissues and values thirty times the secretory rate may occur during maximal secretion produced by histamine.

ABSORPTION FROM THE STOMACH

Reference has already been made to the absorption of lipid-soluble substances. Their rapid absorption in the undissociated form is the result of the mucosal cell membranes behaving as though they are complexes of lipid and protein. There is a continuous *flux* of water and water soluble substances across the water permeable pores of the gastric mucosa but normally their *net* rate of absorption from the lumen is small.

TESTS OF GASTRIC FUNCTION

Increased gastrin production, increased vagal activity, reduction of the normal inhibitory controls or increased reactivity of the mucosal cells may all cause hypersecretion. The balance of these factors seems to have an effect on the rate of growth of parietal cells (p. 242) so that in diseases with acid hypersecretion (hyperchlorhydria) the number of parietal cells is usually increased. The maximum output of acid produced by a given stimulant is remarkably constant in any one subject and this output appears to be correlated with the number of parietal cells (Fig. 16.27). The maximum output of acid, therefore, provides a useful guide to the parietal cell mass (PCM or total number of parietal cells).

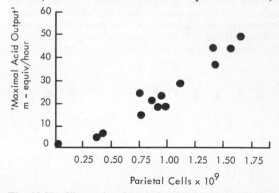

Fig. 16.27 Illustrating the relationship between the number of parietal cells in human gastric mucosa and the maximal acid output obtained from the same mucosa in response to histamine stimulation. The parietal cells were counted in the mucosa removed at operation from patients with gastroduodenal diseases. The maximal output of this excised mucosa was assumed to be the difference between the maximal output of acid secreted by the patients before and after operation. (Adapted from W. I. Card & I. N. Marks (1960). *Clinical Science* **19**, 147–163.)

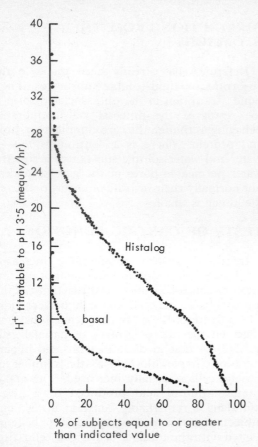

Fig. 16.28 Illustrating the wide range of values of acid secretion by 387 human subjects aged 20 to 49 years whose gastric secretory function seemed to be normal. 'Basal' secretion was obtained during a period of 1 hour after an overnight fast. Histalog-stimulated secretion was collected during a period of 1 hour after the subcutaneous injection of 0·5 mg Histalog/kg of body weight. (Adapted from M. I. Grossman, J. B. Kirisner & I. E. Gillespie (1963). *Gastroenterology* **45**, 114–26.)

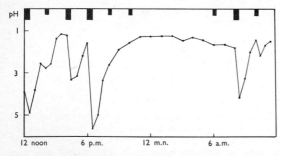

Fig. 16.29 The pattern of gastric acid as shown by a healthy male subject, aged 36. The pH of samples of gastric juice obtained by a Ryle's tube lying in the stomach was measured by a glass electrode. The large black rectangles indicate meals, the small rectangles drinks of milk (A. H. James & G. W. Pickering (1949). *Clinical Science* **8**, 181).

The term maximal acid output (MAO) is now conventionally applied to the maximum amount of acid which can be obtained by continuous aspiration of the stomach in the hour immediately following the injection of a stimulant such as histamine (after the administration of a protective dose of antihistamine), histalog (a pyrazole isomer of histamine), and pentagastrin, an analogue of the active C-terminal tetrapeptide of the gastrin molecule (p. 252) with β-alanine as the fifth amino-acid added at the N-terminus. Pentagastrin is now widely used in clinical practice to test the maximum ability of the parietal cells to secrete acid and so give an indication of the PCM. In pernicious anaemia no hydrochloric acid is produced (achlorhydria) because of atrophy of the mucosal cells, and antibodies to parietal cells may be detected in the blood by immunoassay techniques. Men tend to have bigger MAO than women and the values decrease after the age of 50. However, the values for acid secretion even in subjects without known disorders of gastric secretory function vary widely (Fig. 16.28).

During gastric acid secretion an immediate acid deficit occurs in the body, equivalent to the amount of acid secreted into the lumen of the stomach. On the assumption that relatively little acid is reabsorbed during the first hour after a meal the base excess (Chap. 33) which develops in the body during this time (estimated from arterial blood) can be used to determine the rate of gastric acid secretion. Such studies suggest that the amount of acid secreted during a meal is similar to the maximum acid output obtained in response to histamine but, of course, the food itself produces some neutralization and buffering of the secreted acid (Fig. 16.29) within the lumen of the gut.

Insulin is used to test the integrity of the vagal innervation of the stomach. When a small dose of soluble insulin (for example, 20 units) is given intravenously to man, there is, after a latent period of 40 minutes, a copious secretion of acid gastric juice during the period in which the blood sugar falls to about 50 mg/ 100 ml. This effect can be prevented by maintaining the blood sugar at its normal level with glucose, or by bilateral vagotomy or by giving large doses of atropine. The hypoglycaemia is assumed to act centrally, probably on the hypothalamus, and the administration of insulin can be regarded as a method of stimulating the parietal cells via the vagus nerves both directly and by causing release of gastrin from the pyloric antrum. The concentration of plasma gastrin increases after insulin hypoglycaemia. The insulin test is used to study the completeness of vagotomy in man.

THE PROTECTIVE FUNCTION OF THE STOMACH

The stomach stands guard over the entrance to the small intestine, controlling access by swallowed materials to that part of the gut from which nutrients actually gain entry to the body. In the sense that man can survive without his stomach, it is not an essential organ and many persons have had large portions of their stomachs removed for therapeutic reasons. Most of these patients remain well but some develop disturbances of digestive function or nutritional disorders.

The strong acid and pepsin secreted by the stomach destroys or limits the growth of ingested micro-organisms. Thus, after gastrectomy, development of abnormal bacterial flora in the lumen of the small intestine, together with the removal of the source of intrinsic factor (Chap. 24) may lead to deficient absorption of vitamin B_{12} and so cause anaemia. Similar changes may occur in disease states when the acid, pepsin and intrinsic factor-secreting region of the gastric mucosa becomes atrophied.

Vomiting. Vomiting, by emptying the contents of the stomach and upper small intestine, limits the possibility of damage from ingested noxious agents by preventing their access to the major absorptive region of the alimentary tract.

Vomiting may be initiated by mucosal irritation or abnormal distension of the gastrointestinal tract or gall-bladder or it may result from a great variety of causes such as feelings of disgust or pain, or from abnormal stimulation of the labyrinths or their central nervous pathways, as in sea-sickness. Vomiting may be initiated as a first-aid measure by irritating the back of the patient's throat or by making him drink warm salt-and-water or mustard-and-water. Such preparations are called *emetics*. The afferent impulses from the gastric mucosa

are probably carried by the parasympathetic and also the sympathetic nerves to an *emetic centre* situated in the lateral reticular formation of the central nervous system (Chap. 45). The threshold dose of the emetic substance copper sulphate given by mouth to cats is increased two-fold if the vagi are cut and eight-fold if the stomach is completely denervated. After complete denervation of the stomach, copper sulphate produces vomiting only after a long latent period which is the time taken for the absorption of copper into the blood from the gastrointestinal tract; once in the blood stream it acts on the brain stem. The *emetic centre* is stimulated also by impulses from higher centres and from near-by brain-stem nuclei and vomiting excited by visual and labyrinthine stimuli can be explained in this way.

In the decerebrate cat, Wang and Borison have shown that there is a bilateral sensory *trigger zone* in the reticular formation of the medulla near the fasciculus solitarius (Fig. 45.5) lying just below the vagal triangle near the inferior angle of the fourth ventricle (Fig. 45.5). Electrical stimulation of this area produces vomiting; ablation abolishes the emetic response to apomorphine and raises the threshold to a number of other drugs and metabolic agents to which it is sensitive. Ablation of the trigger zone does not abolish the emetic response to irritating stimuli from the gut and it is therefore supposed that the trigger zone sends impulses directly to the *emetic centre*.

Drugs such as phenothiazines inhibit only the *trigger zone* and do not suppress vomiting due to direct stimulation of the *emetic centre*. Drugs such as atropine and its derivatives on the other hand inhibit the *emetic centre* and can ameliorate many forms of vomiting.

Vomiting is usually accompanied by nausea and preceded by retching, in which the stomach is periodically compressed by a series of more or less violent contractions of the diaphragm and abdominal muscles. At the same time there are autonomic disturbances such as pallor, slowing of the heart rate, sweating and excessive salivation. The act of vomiting is brought about by a complex series of movements of the respiratory and abdominal muscles. Vomiting can be produced after complete motor and sensory denervation of the stomach and intestine in animals and even after the replacement of the stomach by a rubber bag.

X-ray observations in man show that the stomach contracts at the angular notch and that the contents of the pyloric antrum are forced into the relaxed body and fundus. A deep inspiration is taken and the glottis closes; the soft palate rises and prevents vomitus entering the nose. A sharp increase in intra-abdominal pressure, brought about by a violent contraction of the abdominal muscles, forces the gastric contents through the relaxed cardia and along the oesophagus into the mouth. In an unconscious subject the glottis may not close completely and vomited material may be inhaled into the lungs. Prolonged vomiting may endanger life because of the loss of fluid, hydrogen ion, sodium chloride and potassium.

REFERENCES

BADENOCH, J. & BROOKE, B. N. (Eds.) (1965). *Recent Advances in Gastroenterology*. London: Churchill.

BERSON, S. A. & YALOW, R. S. (1971). Nature of immunoreactive gastrin extracted from tissues of gastrointestinal tract. *Gastroenterology* **60**, 215–222.

CODE, C. F. (Ed.) (1967). *Handbook of Physiology*. Section 6, Alimentary Canal; Volume II, Secretion. American Physiological Society.

CREAN, G. P. (1963). The endocrine system and the stomach. *Vitamine und Hormone* **21**, 215–280.

DAVENPORT, H. W. (1966). *Physiology of the Digestive Tract*, 2nd edn. Chicago: Year Book Medical Publishers.

Discussion on polypeptide hormones (1968). Gastrin. *Proceedings of the Royal Society* B, **170**, 81–111.

GREGORY, R. A. (1962). Secretory mechanisms of the gastrointestinal tract. *Monograph of the Physiological Society*. London: Arnold.

GREGORY, R. A. (1968). Recent advances in the physiology of gastrin. *Proceedings of the Royal Society* B, **170**, 81–88.

GROSSMAN, M. I. (Ed.) (1966). *Gastrin*. UCLA Forum in Medical Sciences, No. 5, University of California Press.

HOGBEN, D. A. M. (Ed.) (1965). Gastric secretion of hydrochloric acid. Symposium. *Federation Proceedings* **24**, 1353–1395.

HUNT, J. N. & KNOX, M. T. (1968). Regulation of gastric emptying. *Handbook of Physiology*. Section 6, Alimentary Canal; Volume IV, Motility. American Physiological Society.

KENNER, G. W. & SHEPPARD, R. C. (1968). Chemical studies on some mammalian gastrins. *Proceedings of the Royal Society* B, **170**, 89–96.

MARTINSON, J. (1965). Studies on the efferent vagal control of the stomach. *Acta physiologica scandinavica* **65**, Suppl. 255.

PEARSE, A. G. E. & POLAK, JULIA M. (1971). Neural crest origin of the endocrine polypeptide cells (APUD) cells of the gastrointestinal tract and pancreas. *Gut* **12**, 783–788.

REHM, W. S. (1967). Ion permeability and electrical resistance of the frog's gastric mucosa. *Federation Proceedings* **26**, 1303–1313.

SCHNITKA, T. K., GILBERT, J. A. L. & HARRISON, R. C. (1967). *Gastric Secretion, Mechanisms and Control*. Oxford: Pergamon Press.

SIRCUS, W. (1966). Homeostasis in gastric acid secretion. *Scottish Medical Journal* **11**, 411–422.

THOMPSON, C., BERKOWITZ, D., POLISH, E. & MOYER, J. H. (Eds.) (1967). *The Stomach*. London: Grune & Stratton.

THOMPSON, T. J. & GILLESPIE, I. E. (1966). *Postgraduate Gastroenterology*. London: Bailliere, Tindall & Cassell.

WANG, S. G. & BORISON, H. L. (1950). The vomiting centre. *Archives of Neurology and Psychiatry* **63**, 928–941.

WOLF, S. & WOLFF, H. G. (1943). *Human Gastric Function*. Oxford University Press.

YALOW, R. S. & BERSON, S. A. (1971). Further studies on the nature of immunoreactive gastrin in human plasma. *Gastroenterology* **60**, 203–214.

17 Digestion and absorption in the intestine

In this chapter the processes of digestion and absorption are discussed. Since the pancreas is probably the most important source of digestive juices it is considered first.

THE PANCREAS

The functions of the pancreas depend on the capacity of the different types of pancreatic cells to secrete specific proteins either into the blood stream (internal secretions) or into the lumen of the alimentary tract (external secretions).

The internal (endocrine) secretions of the pancreas, which are derived mainly from the cells of the pancreatic islets, regulate the disposal of materials absorbed from the intestine (Chap. 18). The external (exocrine) secretions of the pancreas contain the most important digestive enzymes which catalyse the hydrolysis of food materials too complex to be absorbed unchanged. Pancreatic juice also contains much sodium bicarbonate which provides a buffered solution of optimal pH for activity of the pancreatic enzymes.

Functional anatomy of the pancreas. Throughout embryonic development, the endocrine and exocrine elements of the pancreas are closely intermingled. However, differentiation of the different cell types occurs at a very early stage of development; for example the islet cells may be derived from the neural crest. Ultimately, the cells with endocrine function are found mainly in clumps (interalveolar cell islets or islets of Langerhans) interspersed throughout the much greater mass of the cells which produce the exocrine secretions (Fig. 17.1).

Light- and electron-microscopy have demonstrated a number of different types of cell with presumed endocrine function (Fig. 17.2). The morphological differentiation has been based especially on the characteristics of the intracytoplasmic granules which are

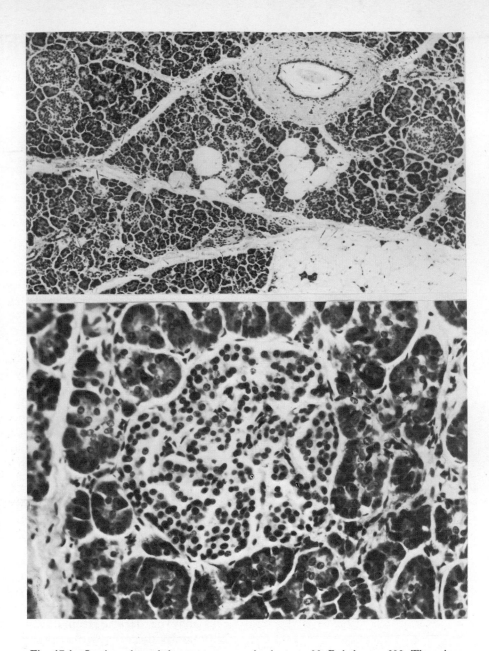

Fig. 17.1 Sections through human pancreas. A, above, ×90. B, below, ×390. The acinar tissue is deeply stained whereas the islet tissue is pale with dark nuclei. A shows (upper right) a duct, and also pale strands of connective tissue. Three islets are seen and also clear spaces which contained fat. B shows more details of the islet and acinar tissue. It should be compared with Plate 18.1A. (*By courtesy of A. C. Lendrum.*)

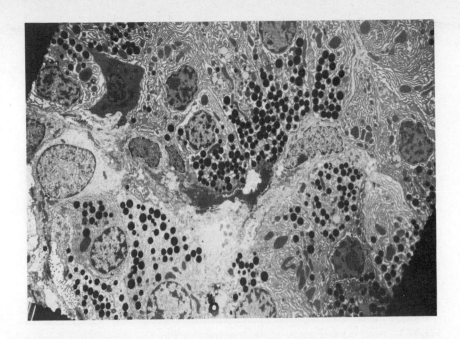

Fig. 17.2A Human exocrine pancreas with characteristic endoplasmic reticulum and zymogen granules in acinar cells. ×2200. From a case of Zollinger-Ellison syndrome. (*By courtesy of A. G. Everson Pearce.*)

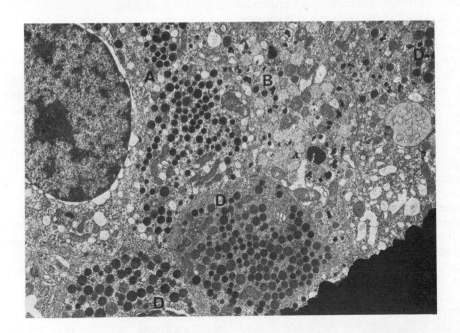

Fig. 17.2B Human endocrine pancreas from a case of Zollinger-Ellison syndrome with hyperplasia of the gastrin-secreting D cells. In normal pancreas it is difficult to find these cells. (A) glucagon-secreting cell. (B) insulin-secreting cell. ×7500. (*By courtesy of A. G. Everson Pearce.*)

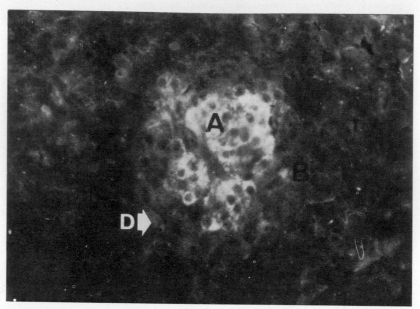

Fig. 17.3 α-cells of pancreatic islets (which secrete glucagon) have reacted with a fluorescein-labelled anti-glucagon antibody. A, glucagon-secreting cell; B, insulin-secreting cell; D, gastrin-secreting cell. (*By courtesy of A. G. Everson Pearce.*)

thought to represent the storage form of the hormone specific to each type of cell. Immuno-fluorescent antibody staining has confirmed the cellular origin of the pancreatic hormones —insulin, glucagon and gastrin (Fig. 17.3).

The cells which secrete the pancreatic enzymes are arranged in acini around small ducts. The acinar cells contain much rough-surfaced endoplasmic reticulum and large zymogen granules and have microvilli on their apical (ductular) surfaces. A different type of cell, with few microvilli and organelles (centro-acinar cells), abuts on the lumen of the ductules and is structurally similar to the low columnar cells lining the large (intralobular) ducts. The intralobular ducts drain into the interlobular ducts which, in turn, join to become the main pancreatic duct, which opens into the duodenum.

The pancreatic enzymes are formed on the ribosomes of the rough-surfaced endoplasmic reticulum of the acinar cells and pass into the cisternal spaces, from which vesicles containing the enzymes bud off to reach the Golgi region of the cells. The vesicles coalesce and grow to become mature zymogen granules, containing the concentrated secretory proteins. During the process of secretion, the membranous sacs of the zymogen granules fuse with the apical cell membrane and the contents of the granules are extruded from the cell (emiocytosis).

PANCREATIC SECRETION

A. **Electrolytes.** Most of the information about the electrolytes of pancreatic juice has been obtained from animal experiments. It is very difficult to obtain uncontaminated human pancreatic juice because introduction of drainage tubes into the human pancreatic duct may result in severe pancreatic inflammation. Human duodenal juice is a mixture of the secretions of the pancreas, hepatobiliary system and duodenal mucosa.

The electrolytes of pancreatic juice are mainly sodium, bicarbonate and chloride, with much smaller concentrations of potassium and calcium. The concentrations of sodium and potassium in pancreatic juice are usually fairly constant while small, but regular, increases in the concentration of calcium occur with increases in the concentration of enzymes. The concentrations of the two main anions—bicarbonate and chloride—vary widely in a reciprocal fashion, so that the sum of the concentrations of the two anions remains constant. In human duodenal juice the concentration of bicarbonate decreases, and chloride increases, with increasing rate of secretion (Fig. 17.4). In most animals, the converse relationship is found.

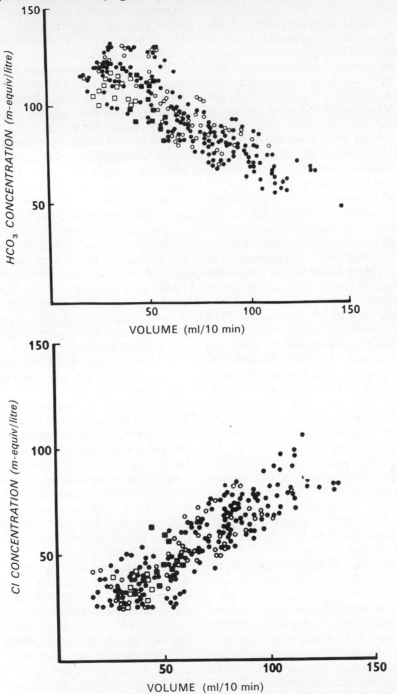

Fig. 17.4 Decrease in bicarbonate and increase in chloride concentration of duodenal aspirate with increase in secretory rate during maximal stimulation of human pancreas by secretin. Squares represent the response to intravenous injection and circles the response to intravenous infusion of secretin. (K. G. Wormsley (1968). *Gastroenterology* **54**, 197–209.)

Three models have been proposed to explain the variations in the anionic composition of the pancreatic juice.

1. Ductal modification of the primary acinar secretion. It has been suggested that the primary secretion of the electrolyte-secreting cells consists of sodium bicarbonate and that, at low rates of flow, bicarbonate exchanges with chloride across the pancreatic ductal epithelium, the bicarbonate diffusing out of the lumen and the chloride ions diffusing in. Alternatively, it has been postulated that the primary secretion consists of sodium chloride and that bicarbonate ions are secreted into the lumen by the ductal epithelial cells.

2. Admixture of two secretions. It has been proposed that different cell types within the pancreas (acinar and centro-acinar) secrete juice containing predominantly sodium chloride and sodium bicarbonate, respectively. The overall anionic composition of the pancreatic juice would depend on the relative activity of the two cell types.

3. Unicellular origin of the anions. The reciprocal relationship between bicarbonate and chloride has been explained by assuming that the cells secrete both bicarbonate and chloride, the relative proportions of which change with the degree of stimulation of the cells.

The experimental evidence is still inconclusive, but at present favours the ductal modification hypothesis.

Despite the uncertainties surrounding the source of the electrolytes of pancreatic juice and even more complex origin of the contents aspirated from the human duodenum during clinical tests, demonstrable reduction of bicarbonate secretion is the best evidence of pancreatic disease in man.

B. **Enzymes.** The proteases secreted by the pancreas comprise trypsinogens, chymotrypsinogens, procarboxypeptidases and proelastase. All have related amino acid compositions, with a serine residue at the enzymically active centre (see Chap. 8) and all are secreted as inactive zymogens. Trypsinogen with 249 amino acids in known sequence is converted into the active proteolytic enzyme, trypsin, by the loss of a small N-terminal peptide chain of 6 amino acids under the influence of the enzyme enteropeptidase (enterokinase), which originates in the microvillous brush border

membrane of the duodenal mucosal cells and is released into the duodenal lumen during digestion. Trypsin subsequently activates chymotrypsinogen and proelastase, by a more elaborate multistage process involving a peptide bond cleavage.

Proteases hydrolyse the peptide bonds of proteins to produce shorter peptide chains and amino acids. The individual proteases tend to be selective in the type of peptide bond which is hydrolysed (see p. 92).

Starch and related polysaccharides in food are hydrolysed at α 1,4-glucosidic bonds by pancreatic amylase. The resultant disaccharide, maltose, is further split to glucose by maltases situated in the microvillous membrane of the epithelial cells of the small intestine.

The neutral, long chain fats in food are hydrolysed by pancreatic lipase to fatty acids, di- and mono-glycerides and glycerol (Fig. 17.5). Pancreatic lipase, in the presence of a co-lipase, acts especially at the lipid-water interphase which characterizes the surface of fat droplets in emulsions of the type formed under the influence of bile salts in the lumen of the upper small intestine.

The ribo- and deoxyribonucleases (Chap. 20) secreted by the pancreas break down the nucleic acids in the food.

The pancreatic enzymes all act best in neutral or mildly alkaline solution. Exposure to acid gastric secretion rapidly and irreversibly inactivates amylase and lipase, while trypsin is rapidly destroyed by pepsin in acid solution. The bicarbonate secreted by the pancreas is therefore very important for the satisfactory functioning of the pancreatic enzymes, both by buffering the acid and by inactivating the pepsin of the gastric juice which has passed into the duodenum.

All the pancreatic enzymes are concentrated in the zymogen granules of the pancreatic acinar cells and are secreted together. However, the relative proportions of the individual enzymes in pancreatic juice is not constant but can be influenced by the composition of the diet, the rate of secretion of pancreatic juice and by disease processes affecting the pancreas. After discharge into the duodenum, a proportion of the pancreatic enzymes is adsorbed on to the glycocalyx of the cells lining the small intestine. In spite of cellular adsorption a marked loss of enzyme activity occurs during

Course of hydrolysis of triglyceride by pancreatic lipase

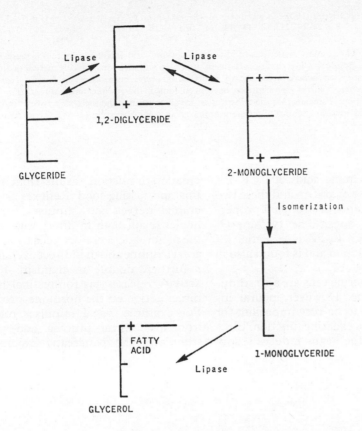

Fig. 17.5 Schematic representation of the hydrolysis of long chain triglycerides by pancreatic lipase. (A. M. Dawson (1967). *British Medical Bulletin* **23**, 247.)

transit through the small intestine. Amylase and lipase are destroyed by the proteases. Trypsin is more labile than chymotrypsin in intestinal contents and although the pancreas secretes more trypsin than chymotrypsin, much more chymotrypsin is excreted in the faeces. Indeed, the faecal content of chymotrypsin is used as an index of the ability of the pancreas to secrete enzymes.

Stimulants of pancreatic exocrine secretion. The most powerful stimulants of the pancreatic exocrine cells are the gastrointestinal hormones secretin and cholecystokinin-pancreozymin.

Secretin is a polypeptide composed of 27 amino acids (Fig. 17.6) with a cyclic secondary structure. The whole molecule is required for the pancreatic stimulant action. The composition of secretin is closely related to that of glucagon (Fig. 17.6 and p. 323). Secretin stimulates principally the secretion of electrolytes and water from the pancreas and is only a weak stimulant of the secretion of enzymes.

Pancreozymin is a polypeptide containing 33 amino acids which stimulates the secretion of pancreatic enzymes (Fig. 17.7). Pancreozymin has been shown to be identical with the small intestinal polypeptide cholecystokinin which elicits contraction of the gall-bladder. The hormone is therefore usually referred to as cholecystokinin-pancreozymin (CCK-PZ).

Glucagon

His-Ser-GLN-Gly-Thr-Phe-Thr-Ser-*Asp*-TYR-Ser-*Lys*-TYR-*Leu*-Asp-Ser-ARG-Arg-ALA-Gln-*Asp*-*Phe*-*Val*-Gln-*Trp*-Leu-*Met*-Asn-Thr
 1   2   3   4   5   6   7   8   9   10   11   12   13   14   15   16   17   18   19   20   21   22   23   24   25   26   27   28   29

Secretin

His-Ser-ASP-Gly-Thr-Phe-Thr-Ser-*Glu*-LEU-Ser-*Arg*-LEU-*Arg*-Asp-Ser-ALA-Arg-LEU-Gln-ARG-*Leu*-*Leu*-Gln-*Gly*-Leu-*Val*-NH$_2$
 1   2   3   4   5   6   7   8   9   10   11   12   13   14   15   16   17   18   19   20   21   22   23   24   25   26   27

Fig. 17.6 The primary structures of glucagon (porcine) and secretin (porcine) arranged for direct correspondence. For those positions in which no substitutions have occurred, ordinary characters are used; where 'conservative' substitutions have been observed, the residues are *italicized*; where 'radical' substitutions have been found, the residues are CAPITALIZED. The term 'conservative' means that the substitution involves similar chemical structures: 'radical' substitutions involve chemically different amino acids. (B. Weinstein (1968). *Experientia* **24**, 406.)

The five C-terminal amino acids of CCK-PZ are identical with those of gastrin and these two hormones share a number of stimulant properties, particularly in dogs. The tyrosine-*O*-sulphate residue in CCK-PZ is essential for its cholecystokinetic action and is found also in human gastrin II (Fig. 17.7).

The pancreatic exocrine cells are also stimulated by acetylcholine. However, neural influences do not seem to be very important for satisfactory pancreatic exocrine function. Electrical stimulation of the vagus induces scanty creatic stimulation results from seeing, smelling and tasting food. Reflexes arise in various cranial nerves and impulses from the vagal nuclei result both in direct vagal excitation of the pancreatic secretory cells by local release of acetylcholine and in indirect excitation through a further, double mechanism. Firstly, vagal activity releases gastrointestinal hormones by direct action on the hormone-producing cells. For example, vagal impulses release gastrin from the antral mucosa and gastrin then stimulates the pancreatic exocrine cells (in

$$SO_3H$$
Human gastrin II *Glu-Gly-Pro-Trp-Leu-Glu-Glu-Glu-Glu-Glu-Ala-Tyr-*Gly*-*Trp*-*Met*-*Asp*-*Phe*-*NH$_2$*

Porcine cholecystokinin-pancreozymin

$$SO_3H$$
Lys-(Ala,Gly,Pro,Ser)-Arg-Val-(Ile,Met,Ser)-Lys-Asn-(Asn,Gln,His,Leu$_2$,Pro,Ser$_2$)-Arg-Ile-(Asp,Ser)-Arg-Asp-Tyr-Met-*Gly*-*Trp*-*Met*-*Asp*-*Phe*-*NH$_2$*

Fig. 17.7 The end of chain, Phe-NH$_2$, carries a free —COOH group. Phe-NH$_2$ stands for the amide of phenyl alanine. The first amino acid residue, marked *Glu, is pyroglutamyl.

pancreatic secretion, particularly of enzymes, after total removal of the stomach and small intestine. Probably more important, vagal stimulation has been shown to potentiate the pancreatic secretory response to secretin. On the other hand, vagotomy only slightly reduces the pancreatic responsiveness to direct stimulation with hormones.

Control of pancreatic secretion. The so-called 'phases' of pancreatic secretion denote where the stimulus to pancreatic secretory activity is operating. Thus, the cephalic phase of pan- dogs, but not in man). Secondly, interaction between direct vagal excitation and vagally released gastrin (p. 252) on the gastric parietal cells promotes the secretion of acid which acts as stimulant of the release of secretin and CCK-PZ from the small intestinal mucosa. The net result of 'cephalic' stimuli is the secretion of both electrolytes and enzymes by the pancreas.

The gastric phase of pancreatic stimulation has been well documented in dogs, but not yet in man. Distension of the stomach elicits both long (vago-vagal) and local neural reflexes which activate the pancreas both directly and indirectly as in the cephalic phase. Similarly, the chemical stimulus of food materials in the antrum elicits neural reflexes, particularly through local nerve nets, resulting both in local release of gastrin and in more distant neural activity.

The intestinal phase of pancreatic stimulation is elicited by distension and, chemically, by acid and food materials in the lumen of the small intestine. This phase depends especially on the release of the hormones secretin and CCK-PZ from the mucosa of the small intestine. The magnitude of the pancreatic response to acid is related to the total amount of acid which enters the small intestine and depends on the length of small intestine (duodenum and jejunum) exposed to contents which are more acid than pH 4·5. Amino acids and fatty acids release mainly CCK-PZ from the small intestine. Only the L-forms of the essential amino acids (especially phenylalanine, valine, methionine and tryptophan in man) are capable of evoking a pancreatic response from the small intestine.

The release of the small intestinal hormones in response to intraluminal stimuli appears to be mediated by neural reflexes, since vagotomy, atropine and local anaesthetics markedly decrease the pancreatic response to the intraluminal stimuli, while not greatly affecting the direct response of the pancreas to the hormones of the small intestine.

Neurohormonal and interhormonal interactions have also been shown to be largely responsible for the secretory stimulation of the pancreatic exocrine cells. In addition to the potentiation of the stimulant effect of secretin by vagal impulses, it has been shown that a small dose of secretin greatly increases the stimulant effect of CCK-PZ on both pancreatic secretion of electrolytes and enzymes (Fig. 17.8). The potentiation occurs under physiological circumstances, since the introduction of amino acids into the small intestine greatly increases the pancreatic response to small doses of secretin (Fig. 17.9). Similarly, the pancreatic response to secretin liberated by acid in the small intestine is potentiated by endogenous CCK-PZ liberated in response to L-amino acids in the small intestine.

Tests of pancreatic function. Tests of pan-

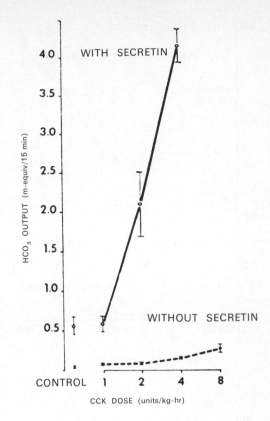

Fig. 17.8 Pancreatic bicarbonate responses to exogenous porcine CCK alone or CCK plus 1·0 unit/kg per hour of secretin (four observations in two dogs). In this figure responses to 8 units/kg per hour of CCK alone were obtained in separate experiments (four observations in two dogs). Note that secretin greatly augments the bicarbonate stimulating effect of CCK (cholecystokinin-pancreozymin) and that CCK increase the stimulating effect of secretin. (J. H. Meyer, L. J. Spingola & M. I. Grossman (1971). *American Journal of Physiology* **221**, 742.)

creatic function are used to confirm the clinical diagnosis of pancreatic disease, to define prognosis—that is, how the patient and his disease are likely to fare over a period of time—and to guide treatment. Pancreatic disease presents clinically as pain which indicates inflammation of the gland and by failure of normal endocrine and exocrine function.

The principal tests of pancreatic function indicate either what the pancreas can do (tests of exocrine secretory capacity) or what the pancreas does do (in response to small intestinal stimulation). Both types of test require intubation of the duodenum and aspiration of the duodenal contents. In tests of exocrine secretory capacity, the pancreas is stimulated

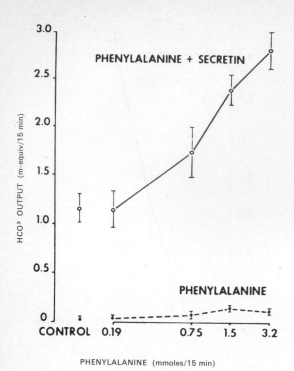

directly by intravenously administered secretin and CCK-PZ, in quantities sufficient to elicit maximal output of bicarbonate and pancreatic enzymes. A low output suggests that the exocrine secretory cells of the pancreas are diseased or have been destroyed (Fig. 17.10). Tests involving indirect stimulation of the pancreas from the small intestine depend on measuring the response to meals, or to acid or amino acid solutions, introduced into the small intestine. The magnitude of the indirect response is normally as great as that to direct stimulation of the pancreas, but is impaired either if the pancreas is diseased or if the small intestine does not release normal amounts of hormones as in coeliac disease (Fig. 17.11).

Fig. 17.9 Pancreatic bicarbonate outputs to phenylalanine alone or phenylalanine plus 1·0 unit/kg per hour of secretin (four observations in two dogs). Note that phenylalanine increases the bicarbonate-stimulating effect of secretin and vice versa. (J. H. Meyer, L. J. Spingola & M. I. Grossman (1971). *American Journal of Physiology* **221**, 742.)

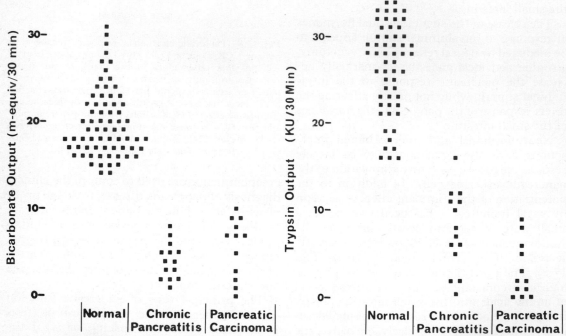

Fig. 17.10 A tube was passed down the oesophagus and into the duodenum to aspirate the duodenal contents for estimation of bicarbonate and trypsin. Each point represents the response of one individual to stimulation with intravenous infusion of secretin plus cholecysto-kinin–pancreozymin. The pancreas has been severely damaged in the patients with chronic pancreatitis and pancreatic carcinoma. (K. G. Wormsley (1970). *British Journal of Clinical Investigation* **24**, 271.)

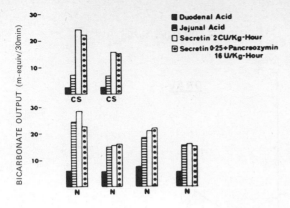

Fig. 17.11 Comparison of bicarbonate outputs (in duodenal contents) in response to duodenal and jejunal acidification and to intravenous infusion of secretin (2·0 units/kg per hour) and a combination of secretin (0·25 units/kg per hour) and pancreozymin (16 units/kg per hour). Each set of four bars represents results from one subject. CS = coeliac syndrome; N = normal. Note that the pancreatic response of normal subjects to intravenous secretion and to jejunal acidification is similar. In patients with coeliac syndrome, the response to acid in the jejunum is significantly less than the pancreatic response to direct stimulation with secretin. (K. G. Wormsley (1970). *Scandinavian Journal of Gastroenterology* **5**, 353.)

Effects of gastrointestinal hormones on the endocrine pancreas. In addition to their stimulant effect on the exocrine pancreas, the gastrointestinal hormones stimulate the islet cells of the pancreas. Glucose taken orally causes a smaller rise of blood sugar, but a greater rise of blood insulin, than the same amount given intravenously (Fig. 17.12; see also Chap. 18). Similarly, after absorption of both acid and amino acids in the small intestine, the serum levels of insulin rise. The 'entero-insular' axis involves the gastrointestinal hormones gastrin, secretin, CCK-PZ and glucagon, all of which promote the release of insulin from the pancreas (Fig. 17.13) and potentiate insulin release in response to the direct, islet-cell-stimulant effect of glucose or arginine.

BILIARY SYSTEM

Functional anatomy. Bile is secreted by the parenchymal cells of the liver into the canaliculi which are spaces bounded by two adjacent liver cells (Fig. 17.14). The canaliculi join to form intralobular biliary ductules which, in turn, empty into the interlobular biliary canals in the portal spaces. The biliary canals drain into the right and left hepatic ducts, which join in the hilum of the liver to form the common hepatic duct. The total length of the biliary tract in man is about 2 km.

The gall-bladder has a thin wall composed of mainly longitudinal muscle and elastic tissue, bounded externally by peritoneum and lined by a layer of columnar epithelial cells. The capacity varies, but is on average a little less than 50 ml. The cystic duct, which arises from the neck of the gall-bladder, joins the common hepatic duct to form the bile duct. At its lower end, the bile duct is considerably thickened, over a length of about 1·5 cm, by muscle fibres forming a sphincter (of Oddi), which usually also embraces the termination of the pancreatic duct (Fig. 17.15).

COMPOSITION OF BILE

A. *Electrolytes.* The concentration of sodium, the principal cation of bile, is related to the total concentration of anions (chloride, bicarbonate and bile salts). Although bile is always isosmotic with extracellular fluid, the concentration of sodium may be as great as 300 m-equiv./l, owing to the large amounts of bile salts in the micellar complexes. The concentrations of chloride and bicarbonate increase as the concentration of bile salts falls and also increase with greater rates of secretion of bile.

B. *Bile salts.* The bile acids (p. 60)—cholic and chenodeoxycholic—are synthesized from cholesterol in the liver cells (Fig. 17.16), where they are conjugated with taurine $(NH_2CH_2CH_2SO_3H)$ and glycine, in a ratio of 1 (taurine) to 3 (glycine) in man, and secreted into the biliary canaliculi as sodium salts. The bile salts, in a concentration up to 40 mM, comprise about two-thirds of the solid matter in hepatic bile.

After transit through the small intestine, 90 per cent or more of the conjugated bile salts are actively reabsorbed in the ileum and carried back to the liver by the portal vein, for re-excretion by the liver cells. (Fig. 17.17). The process, which permits continual re-utilization of the bile salts, is termed 'the entero-hepatic

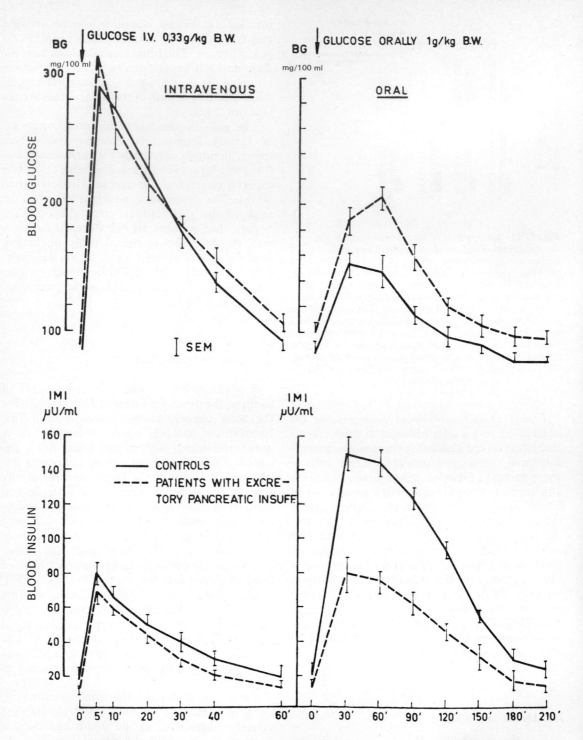

Fig. 17.12 Blood glucose (BG) and blood insulin (IMI) concentrations in patients with excretory pancreatic insufficiency ($n = 11$) and in normal controls ($n = 30$) (mean $\pm$ S.E.M.) after intravenous and oral administration of glucose. (S. Raptis, R. M. Rau, K. E. Schröder, W. Hartman, J. D. Faulhaber, P. H. Clodi & E. F. Pfeiffer (1971). *Diabetologie* **7**, 160.)

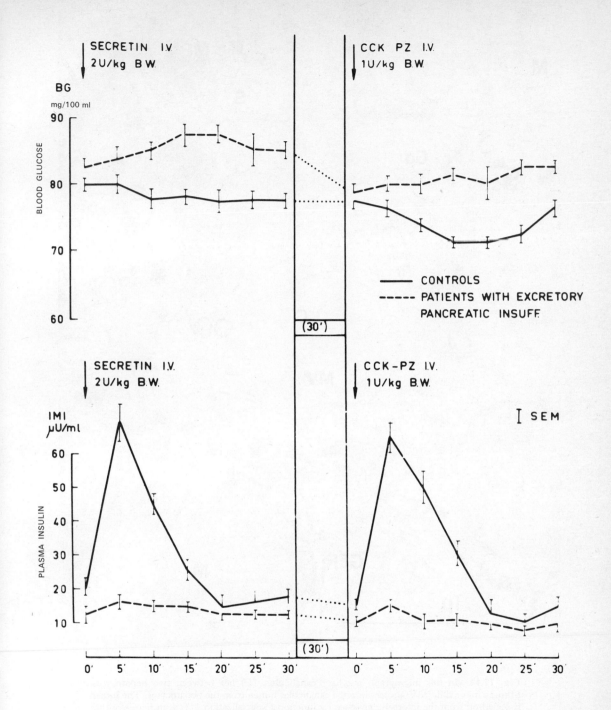

Fig. 17.13 Blood glucose (BG) and immunologically measurable insulin (IMI) after intravenous injection of secretin and pancreozymin (CCK-PZ) in plasma of patients suffering from excretory pancreatic insufficiency ($n = 11$) and in normal controls ($n = 30$) (mean $\pm$ S.E.M.) respectively. (S. Raptis (1971). *Diabetologia* **7**, 160.)

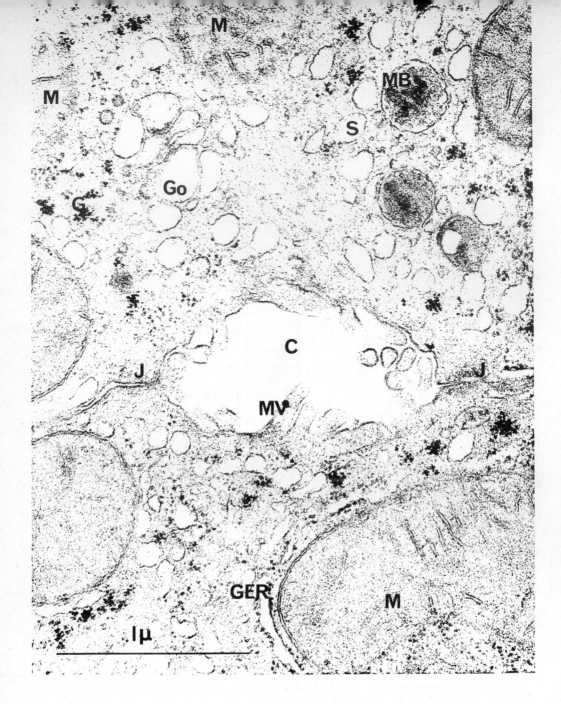

Fig. 17.14 In this micrograph a biliary canaliculus, (C) lies between two hepatocytes. Stumpy microvilli, (MV) project into the canalicular lumen from the hepatocytes. The lumen is sealed off from the intercellular space by junctional specializations, (J) seen here at either side of the canaliculus. Within the hepatocytes can be identified mitochondria, (M) some of which are tangentially sectioned, microbodies, (MB) containing amorphous material with a central dense nucleoid and glycogen, (G) the individual particles of which are aggregated into rosettes. Cisternae of the granular endoplasmic reticulum, (GER) with associated ribosomes can be seen: other membrane limited spaces without attached ribosomes are classified as smooth, (S) or agranular endoplasmic reticulum. One small group of membranes can be identified from their layout as part of the Golgi apparatus, (Go). (*By courtesy of K. E. Carr and P. G. Toner.*)

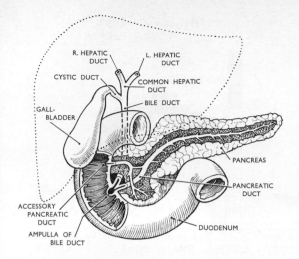

Fig. 17.15 Connexions of the liver, gall-bladder, and pancreas, with the duodenum. The anatomical arrangement of the entry of the pancreatic duct and the bile duct is not always as shown here; they enter the duodenum separately in 29 per cent of persons.

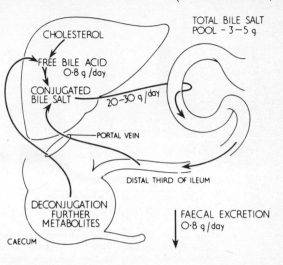

Fig. 17.17 Formation and circulation of human bile salts (after Bergström *et al.*, 1960). From A. M. Dawson (1968). In *Fourth Symposium on Advanced Medicine* (Edited by O. Wrong). London: Pitman.

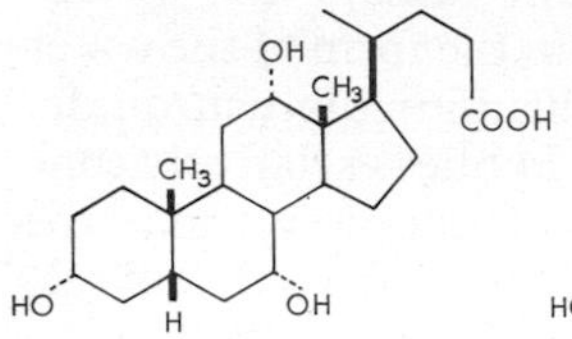

cholic acid

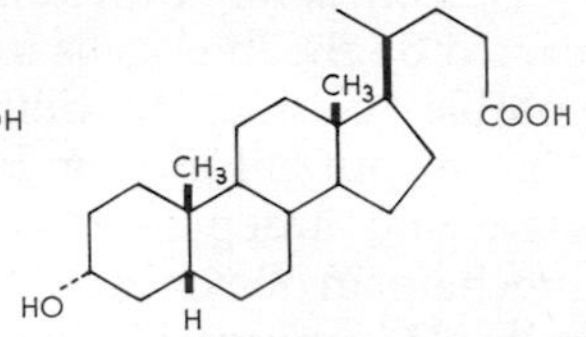

chenodeoxycholic acid

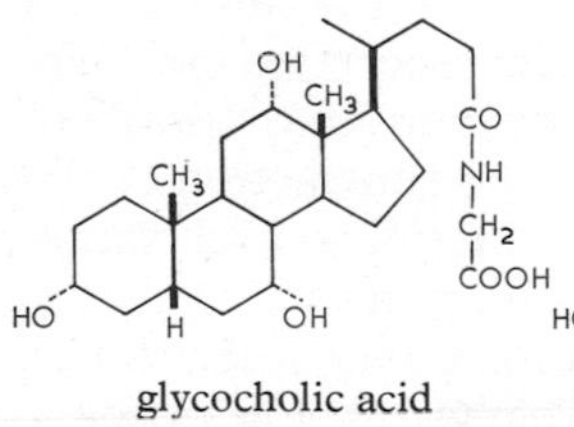

glycocholic acid

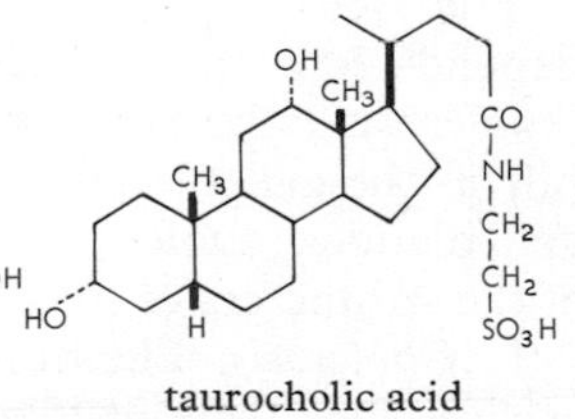

taurocholic acid

Fig. 17.16 Structures of the two chief bile acids, cholic acid and chenodeoxycholic acid. Both of them can be conjugated with glycine or taurine to give, for example, glycocholic acid or taurocholic acid.

circulation' of bile salts. Less than 10 per cent of the bile salts escape into the colon where bacterial action results in deconjugation and dehydroxylation to deoxycholate which is passively reabsorbed or excreted in the faeces.

The extent of faecal excretion determines the rate of synthesis of bile salts, usually about 200 to 400 mg per day in a normal bile salt pool of 2 to 4 g. Interruption of the enterohepatic circulation of bile salts, by disease or surgical removal of the ileum, markedly increases the conversion of cholesterol to bile salts in the liver cells. Bile salts affect not only the self-regulation of hepatic synthesis, but also control the synthesis of cholesterol by the small intestinal mucosa. When the concentration of bile salts in the small intestine is abnormally low, the formation of cholesterol by the small intestine increases markedly, so that more cholesterol becomes available to the liver for conversion to bile salts (Fig. 17.18).

C. *Phospholipids.* Lecithin, the predominant phospholipid in bile, forms approximately one-sixth of the total solids of bile (approximately 7 g/l).

D. *Cholesterol.* Cholesterol occurs in hepatic bile in concentrations of just over 1 g/l. Cholesterol is insoluble in water but does not precipitate in bile because in association with lecithin and bile salts it forms a micellar solution. A change in the relative proportions of the three major organic components of bile (particularly a decrease in one or all of the bile salts)

Intestinal Digestion and Absorption   277

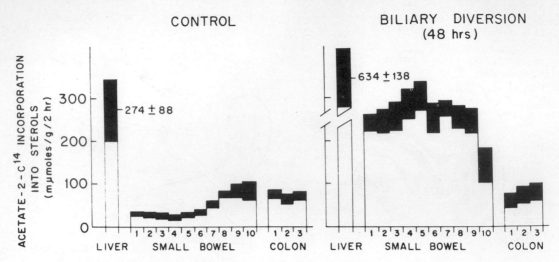

CONTROL

BILIARY DIVERSION
(48 hrs)

ACETATE-2-C^{14} INCORPORATION INTO STEROLS (mμmoles/g/2 hr)

300

200

100

0

274 ± 88

634 ± 138

1 2 3 4 5 6 7 8 9 10   1 2 3
LIVER   SMALL BOWEL   COLON

1 2 3 4 5 6 7 8 9 10   1 2 3
LIVER   SMALL BOWEL   COLON

Fig. 17.18 Effect of biliary diversion on sterol synthesis in the gastrointestinal tract of the New World monkey. Bile was diverted from the gastrointestinal tract of four monkeys by means of indwelling bile duct catheters; four control monkeys had laparotomies but no diversions. After 48 h slices were prepared from the liver, all levels of the small bowel, and three levels of the colon. These were assayed for their ability to incorporate acetate-2-^{14}C into cholesterol. (J. M. Dietschy (1967). *Federation Proceedings* **26**, 1594.)

may result in precipitation of the cholesterol in the form of gall-stones (Fig. 17.19).

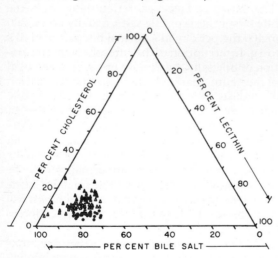

100   0

80   20

60   40

40   60

20   80

0   100

PER CENT CHOLESTEROL

PER CENT LECITHIN

100   80   60   40   20   0
PER CENT BILE SALT

Fig. 17.19 The composition of gall-bladder bile from normal subjects and patients with gall-stones. The composition of gall-bladder bile, in terms of bile salts, phospholipids, and cholesterol from each of 25 normal subjects and 66 patients with cholesterol or mixed gall-stones are plotted on triangular coordinates. The closed circles represent biles from normal subjects. The triangles represent biles from gall-stone patients. The closed triangles indicate the presence of cholesterol microcrystals, whereas the open triangles represent biles without microcrystals of cholesterol. (W. H. Admirand & D. M. Small (1968). *Journal of Clinical Investigation* **47**, 1046.)

E. *Bilirubin*. Bilirubin (Chap. 25) is excreted by the liver cells in the form of the water soluble conjugate, bilirubin diglucuronide. The output of bilirubin in bile is relatively constant and independent of bile flow, since the mechanism of excretion differs from that of bile salts. Deconjugation by bacterial action in the gall-bladder or the biliary tract may precipitate bilirubin as a calcium salt.

DISCHARGE OF BILE

The delivery of bile into the duodenum depends on the hepatic secretion of bile (choleresis), the concentration and storage of bile in the gall-bladder, the contraction of the gall-bladder (cholecystokinesis) and relaxation of the sphincters of the biliary tract.

1. *Choleresis*. The human liver secretes bile at a pressure of about 25 cm water. The primary driving force for the formation of bile is the active secretion of organic anions, especially bile salts, by the liver cells into the biliary canaliculi. The high canalicular concentration of bile salts (more than 2000 times greater than that in blood) promotes a secondary flow of water and electrolytes by osmosis into the canaliculi. Ninety to 95 per cent of the rate of secretion of bile salts is determined by the rate of clearance of reabsorbed bile salts from the portal venous blood in the hepatic sinusoids;

the rate of hepatocellular synthesis of new bile salts accounts for the remainder. Interruption of the enterohepatic circulation of bile salts therefore markedly reduces bile flow. In man, bile is secreted at rates of 13 to 65 ml per hour, with a mean daily output of about 600 ml.

The composition of canalicular bile is modified, during passage through the bile ducts, by the secretion of electrolytes and water quite independently of the secretion of bile salts. The biliary ductal component of chloresis is under neurohormonal control. Thus vagal stimulation promotes choleresis both by a weak direct effect and by a stronger indirect action, secondary to the secretion of gastric acid and subsequent release of gastrointestinal hormones. Secretin and glucagon, as well as gastrin and CCK-PZ (in dogs) have been shown to increase the secretion of water and electrolytes, particularly bicarbonate, without change in the output of bile salts (Fig. 17.20a and b).

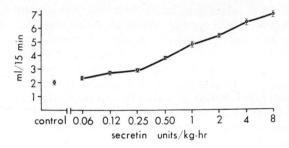

Fig. 17.20a Increase in flow rate of bile in response to secretin.

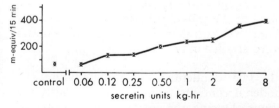

Fig. 17.20b Increase in biliary bicarbonate output in response to secretin (R. S. Jones & M. I. Grossman (1969). *American Journal of Physiology* **217**, 533.)

2. *Control of gall-bladder contraction.* Bile, secreted by the liver and bile ducts, is concentrated about ten-fold by the gall-bladder, by active absorption of sodium, chloride and bicarbonate, with associated passive parallel movement of water from the gall-bladder lumen. The concentration of total solids increases, but the osmotic pressure of the contents of the gall-bladder remains similar to that of

extracellular fluid, because the micellar particles increase in size.

Filling of the gall-bladder depends on the positive pressure gradient brought about by the active secretion of bile together with the contraction of the sphincter at the choledochoduodenal junction, and on the adaptive relaxation of the gall-bladder wall. The resting intraluminal pressure of the human gall-bladder ranges from 0 to 16 cm water.

Food in the upper small intestine induces contraction of the gall-bladder and relaxation of the choledochal sphincter by causing the release of CCK-PZ and secretin from the small intestinal mucosa. The CCK-PZ exerts a direct stimulant action on the muscle of the gall-bladder, so that the pressure in the gall-bladder increases up to 30 cm water. The cholecystokinetic effect of CCK-PZ is greatly augmented by small amounts of secretin and by vagal stimulation (Fig. 17.21).

3. *Control of the choledocho-duodenal sphincter.* The pressure gradient between the lumen of the bile duct and the duodenal lumen is determined by contraction of the sphincteric muscle at the choledocho-duodenal junction. The sphincteric contraction is mediated by cholinergic mechanisms, so that the flow of bile into the duodenum decreases when the antrum contracts or when hydrochloric acid reaches the duodenum in high concentration. On the other hand, sphincteric resistance is reduced by vagotomy and atropine, and by CCK-PZ.

FUNCTIONS OF BILE SALTS

A. *Stimulation of pancreatic solution.* The introduction of bile salts into the human duodenum elicits a dose-dependent stimulation of pancreatic exocrine secretion, so that the amount of lipase and other pancreatic enzymes in the duodenum increases.

B. *Activation of pancreatic lipase.* Bile salts are adsorbed on to the surface of lipid droplets and there increase the lipolytic activity of pancreatic lipase.

C. *Formation of micelles.* Bile salts act as wetting agents and disperse polar lipids such as phospholipids and monoglycerides into micellar form. Bile salts, together with lecithin from bile, convert the emulsion of fat leaving the stomach into a finer emulsion which is hydrolysed by pancreatic lipase. The resultant mixture of bile salts, monoglycerides and fatty

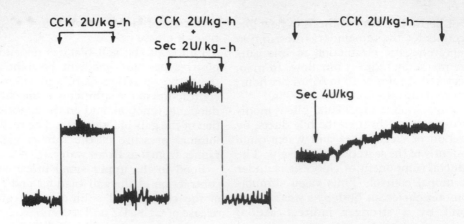

CCK 2U/kg–h   CCK 2U/kg–h
                        +
                   Sec 2U/kg–h

Sec 4U/kg

CCK 2U/kg–h

Fig. 17.21 Pressure responses of gall-bladder of dog to CCK alone and to CCK plus secretin. Left panel shows response to infusion of CCK alone and then CCK plus secretin. Right panel shows effect of adding secretin to a constant infusion of CCK. Secretin alone did not cause contraction of the gall-bladder but it augmented contraction when given with CCK. From M. I. Grossman (1970). *Origin, Chemistry, Physiology and Pathophysiology of the Gastrointestinal Hormones*, p. 137 (Edited by W. Creutzfeld). Stuttgart: Schattauer.

acids rapidly forms a micellar solution if the concentration of the bile salts is greater than the 'critical micellar concentration' (about 2 mM).

D. *Release of enterokinase.* Since bile salts increase the release of enterokinase from the brush border of the duodenal mucosal cells, the intraluminal concentration of enterokinase is greatly increased during the period of secretion of pancreatic enzymes.

TESTS OF BILIARY FUNCTION. In the absence of bile salts from the small intestine some fat can be absorbed but fat absorption is incomplete and the faeces contain large amounts of fat. The concentration of bile salts, at all levels of the small intestine, can be measured by aspiration and analysis of the contents. In patients with biliary obstruction, no bile salts may reach the small intestine, while in patients with interrupted enterohepatic circulation of bile salts, the concentration of bile salts may be insufficient for the formation of micelles in the lumen of the small intestine. The low concentration of bile salts may become manifest only with the second meal of the day, since the first meal may utilize all the bile salts which have been synthesized overnight by the liver.

The size of the bile salt pool (that is the total bile salts in the body) and rate of synthesis of bile acids can be measured by intravenous injection of radioactively labelled bile salts. The ratio of labelled to unlabelled bile salts is then followed during the course of a few days by aspirating the contents of the duodenum after stimulation of the gall-bladder with CCK-PZ. In patients with interruption of the enterohepatic circulation, the turn-over of the radioactively labelled bile salts is very markedly increased and the size of the bile salt pool is usually reduced by up to 80 per cent (Fig. 22a and b) because the hepatic synthesis of bile salts has been unable to compensate for the loss of bile salts in the faeces.

TESTS OF GALL-BLADDER FUNCTION. The ability of the gall-bladder to contract and discharge concentrated bile into the duodenum can be studied in two ways.

1. The duodenum is intubated and the duodenal contents aspirated continuously for 1 hour. During the intravenous infusion of secretin, bile pigment disappears from the duodenal contents of normal subjects so that persistence of bile pigment in the aspirate denotes functional abnormality of the gall-bladder and biliary ductal system (Fig. 17.23). Subsequent intravenous administration of CCK-PZ in normal subjects induces contraction of the gall-bladder with the appearance of large amounts of bile pigment in the duodenal contents. The bile pigment response to CCK-PZ is impaired in patients with disease of the

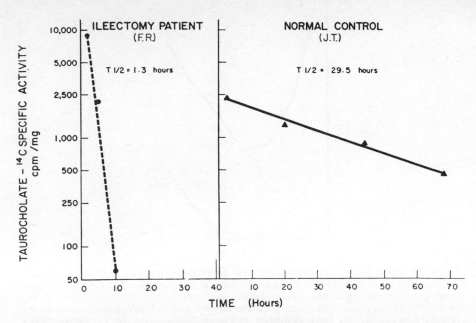

Fig. 17.22a Decreased half-life of bile salts due to failure of reabsorption with rapid faecal excretion in a patient with interruption of enterohepatic recirculation of bile salts after ileectomy. (B. W. Van Deest, J. S. Fordtran, S. G. Morawski & J. D. Wilson (1968). *Journal of Clinical Investigation* **47**, 1321.)

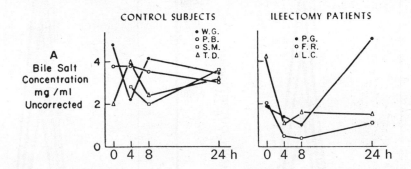

Fig. 17.22b Decreased bile acid due to interruption of enterohepatic recirculation of bile salts after ileectomy, presenting as low concentration of bile salts in the intestine after meals. (B. W. Van Deest, J. S. Fordtran, S. G. Morawski & J. D. Wilson (1968). *Journal of Clinical Investigations* **47**, 1320.)

gall-bladder and is absent in patients with obstruction of the bile duct (Fig. 17.24).

2. The absorptive capacity of the gall-bladder can be estimated by giving a radio-opaque iodine-containing compound which is excreted by the liver cells and subsequently concentrated by the gall-bladder. Failure to concentrate is shown radiologically by failure

of the gall-bladder to opacify (Fig. 17.25A and B).

MOVEMENTS OF THE SMALL INTESTINE

Movements of the small intestine mix and spread the contents over the absorptive surface. The contractions produce small segmental

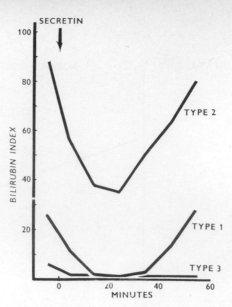

Fig. 17.23 Type 1 represents normal duodenal aspirate after injection of secretin. Type 2 represents duodenal aspirate after injection of secretin in patients with a diseased gall-bladder or after cholecystectomy. Type 3 represents duodenal aspirate after injection of secretin in patients with obstruction of the biliary tract. The bilirubin index (Meulengracht) is a measure of the yellow colour of the duodenal aspirate after injection of secretin. (H. T. Howat (1965). In *The Biliary System* (Edited by W. Taylor), p. 252. Oxford: Blackwell.

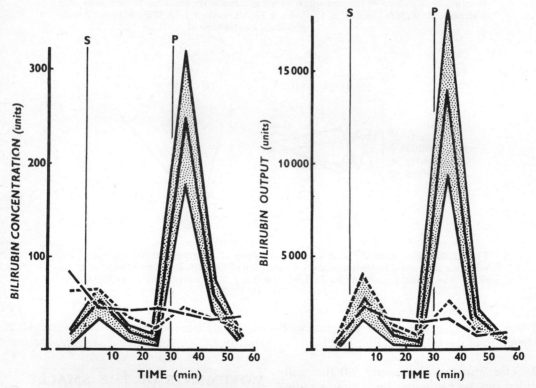

Fig. 17.24 Cholecystokinesis in response to CCK-PZ in normal subjects and patients with disease of the gall-bladder. At S, intravenous infusion of secretin; at P, intravenous infusion of CCK-PZ. ———— Normal (29 subjects); ——— post cholecystectomy (14); ————— gall-bladder disease (27). (H. T. Howat (1965). In *The Biliary System* (Edited by W. Taylor), p. 258. Oxford: Blackwell.)

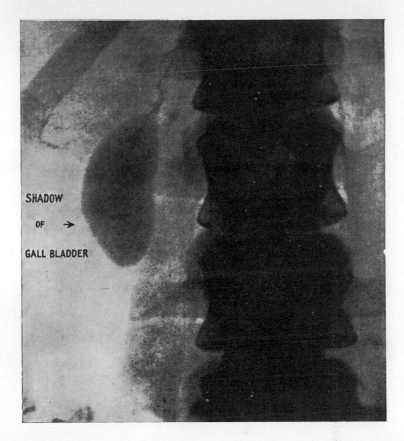

SHADOW

OF →

GALL BLADDER

Fig. 17.25A X-ray photograph showing the gall-bladder filled with radio-opaque iodine-containing contrast medium. The contrast medium has been secreted by the liver cells, has passed to the gall-bladder in the hepatic bile and has become concentrated in the gall-bladder lumen (and therefore become visible) owing to the absorption of sodium chloride and water by the mucosa of the gall-bladder. If the gall-bladder is diseased absorption of electrolytes and water does not occur and the dye does not become sufficiently concentrated to produce an X-ray shadow. (*By courtesy of R. C. Garry.*)

narrowings from which complex patterns are constructed. Peristalsis, as seen in the oesophagus, occurs in the human small intestine but peristaltic activity is intermittent and interspersed with non-propulsive contractions. The length of the intestine (mouth to anus) after death is usually given as 7 m but it may be as little as 4 or as much as 8 m. The length in life is about 4·5 m but a narrow rubber tube 2·5 m long can be passed from mouth to anus in living man as if the intestine has gathers on it like an accordion pleating. The great increase in length of the intestine which occurs after death is due to loss of tone.

The basic electrical rhythm arises and is conducted in intestinal muscle similarly to the electrical activity of the heart. In the stomach the basic electrical rhythm (BER) is constant at three waves per minute and spreads into the first few centimetres of the duodenum in man. The basic electrical rhythm of the small intestine arises at a point (pacemaker) in the duodenum just above the entry of the bile and pancreatic ducts. Recordings in animals show a small intestinal gradient of electrical activity with a slow wave propagated down the intestine from the duodenum, the duodenal rate of 12 per minute decreasing to 8 per minute in the ileum; a similar gradient of mechanical activity has been shown in man. If the small intestine is cut the electrical rhythm below the level of the transection falls to 8 per minute. A muscular contraction is accompanied by spike potentials superimposed on the slow wave.

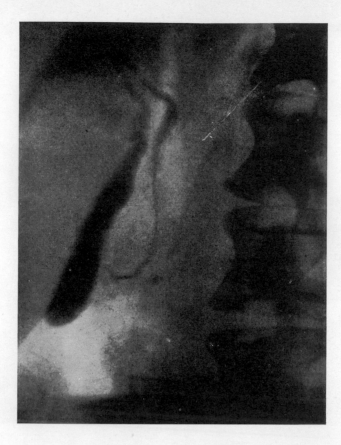

Fig. 17.25B X-ray photograph of right subcostal region taken 1 hour after a fatty meal showing decrease in size of the gall-bladder together with filling of the cystic and bile ducts. This normal response to cholecystokinin, released from the small intestine by fat, does not occur if the gall-bladder is diseased or if the small intestine is abnormal. (*By courtesy of C. Pickard.*)

The simplest form of a small intestinal contraction is of a short segment 1 to 2 cm long, variously called pendular movement, rhythmic segmentation, or to-and-fro movement. For much of the time, particularly in the fasting state, there is little contractile activity. Contractions may occur singly, with a duration of about 8 seconds and an amplitude between 3 and 75 mm Hg. Other contractions produce monophasic pressure waves superimposed on a baseline rise which may last up to several minutes. These sustained complexes may progress in a caudal direction (the peristaltic rush). Transit of small intestinal contents depends upon the gradient of the basic rhythm and the occurrence of propulsive contractions. Examples of some of the rhythms are shown in Fig. 17.26.

Neural control and co-ordination. The extrinsic nerves of the small intestine, the vagus and the sympathetic, exert a modifying influence on motor activity but after division of these in man there may be no apparent disturbance. Experiments in animals suggest that the vagus increases motor activity and sympathetic activity inhibits it.

The rich intrinsic nerve supply in the myenteric plexuses co-ordinates local activity. Distension of the intestine results in contraction orad and relaxation caudad. During muscular activity 5-hydroxytryptamine is released and this sensitizes the intestine to stimulation by distension. If a loop of intestine is cut in two places, then reversed and sewn up in continuity, the intestinal contents are held up although the lumen remains patent. This

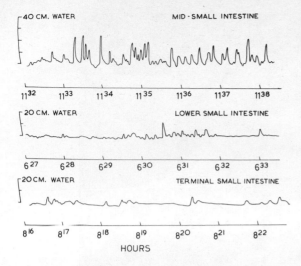

Fig. 17.26 Typical recordings of the pressure activity of the small intestine in man obtained by 'radio pill'. The radio pill is a small capsule containing a miniature radio transmitter whose frequency varies with the external pressure. The records are from three different patients. The time scale gives hours and minutes after swallowing the pill. Since the time the pill stays in the stomach varies greatly from patient to patient the time of arrival at a given part of the small intestine is also extremely variable. (A. N. Smith & M. Ridgway (1962). *Gut* **3**, 373.)

shows that the correct orientation of the small intestine is necessary for the propulsion of the intestinal contents.

In man intestinal transit is usually studied by observing the passage of a radio-opaque barium meal through the gut. The barium suspension is propelled into the duodenum by gastric contractions and is then shuttled to and fro by segmenting activity so that the meal is rapidly spread out over a large area of mucosa. The time of transit from stomach to ileocaecal valve is normally about 3 hours but there is great individual variation. The rate of transit slows down distally in the small intestine. The transit time is influenced by the rate of gastric emptying so that fast emptying produces fast transit. Intestinal activity increases after eating and speeds the transit of barium already present in the small intestine.

The ileocolic junction. A short segment, about 4 cm long in man, acts as a sphincter and a band of raised intraluminal pressure has been found in this position but not as constantly as in other sphincters. Inspection through the opened caecum suggests that the lips of the

ileocaecal valve remain closed unless the ileum is discharging its contents. *In vitro* observations show that muscle from this region contracts under the effect of acetylcholine and α-adrenergic stimulation but relaxes to β-adrenergic stimulation. The ileocaecal sphincter also responds to polypeptide hormones; it is contracted by secretin but relaxed by gastrin. It is interesting to note that these hormones have diametrically opposite actions on the lower oesophageal sphincter. However, when the colon is filled with a barium enema, reflux into the ileum is common.

SMALL INTESTINE

The intestinal villi. The surface of the mucous membrane of the small intestine shows an enormous number of villi, estimated at five million (Figs. 17.27 and 17.28). Every mm^2 of mucosa carries 20 to 40 of these finger-like processes, each about 1 mm long. Each villus (Fig. 17.29) contains an arteriole and a venule with their communicating capillary plexus and also a blind-ending lymphatic vessel or lacteal. There are apparently no arterio-venous anastomoses in the small intestine although they are present in the stomach. The veins ultimately open into the portal vein going to the liver, about 1·4 litres of blood flowing through this vein per minute; during the digestion of a meal this amount increases by one third. The lacteals lead into the thoracic duct which discharges into the large veins in the neck at the junction of the left internal jugular and subclavian veins. The thoracic duct flow is only one or two ml per min between meals but during a meal it may increase five or ten times.

Blood passing through the minute vessels of the small intestine is brought into close proximity with the fluid in the intestine over an area estimated to be 10 m^2. The electron microscope has shown that the striated free border (the brush border) of each columnar epithelial cell in man has about 1000 minute processes or microvilli from 0·9 to 1·3 μm in height, with a maximal width of 0·12 μm, which project into the intestinal lumen (Fig. 17.30). The microvilli increase the absorptive area at least thirty times.

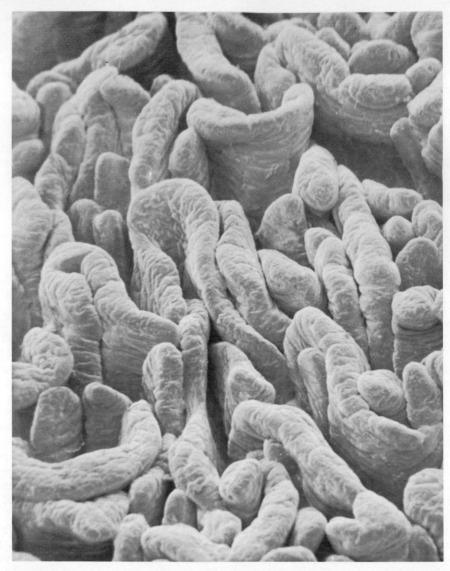

Fig. 17.27 Scanning electron micrograph of human small intestinal villi (× 72). (*By courtesy of K. G. Carr and P. G. Toner.*)

Fig. 17.28 Movements of the intestinal villi in a dog. Every eighth frame of a cine-film is reproduced. Note the changes in the villus marked with an arrow (E. von Kokas and G. von Ludány (1930). *Pflügers Archiv für die gesamte Physiologie des Menschen und der Tiere* **225**, 421).

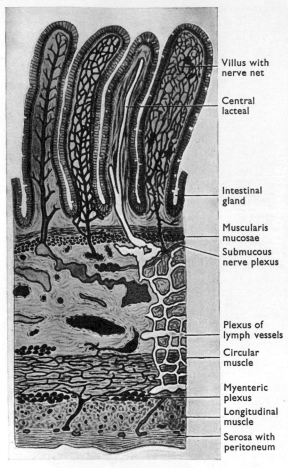

Villus with
nerve net

Central
lacteal

Intestinal
gland

Muscularis
mucosae

Submucous
nerve plexus

Plexus of
lymph vessels

Circular
muscle

Myenteric
plexus

Longitudinal
muscle

Serosa with
peritoneum

Fig. 17.29 Diagram of part of a cross-section through human small intestine. The villus on the extreme left shows the venous drainage and the adjacent villus show the arterial supply. The villi are covered by simple columnar cells among which are mucus-secreting goblet cells. The intestinal glands are lined by similar cells and they possess, usually in the deepest part, argentaffin (Kulchistsky) cells whose granules contain 5-hydroxytryptamine and also granular Paneth cells whose function is not known. (*After Ramon y Cajal.*)

During digestion and absorption the villi contract fairly quickly at irregular intervals and relax slowly. These contractions are brought about by smooth muscle fibres derived from the muscularis mucosae which pass into the villi to be attached to the lacteals. Contraction of these fibres probably helps to pump lymph into the lacteals of the submucosa (Fig. 17.28).

In the upper part of the small intestine (jejunum) the villi are long and closely packed and the area of the mucosa is further increased by circular folds. The stroma of the mucosal layer contains large number of lymphocytes which may be grouped together as solitary lymph follicles. These are particularly well developed in the ileum as *aggregated lymphatic follicles* or Peyer's patches (Fig. 17.31).

Microscopic examination of a section through the small intestine shows many *intestinal glands* (crypts of Lieberkühn), simple tubular glands which open into spaces between neighbouring villi. Many of the cells show mitotic figures. Mucosal replacement has been studied by the use of tritiated thymidine, a specific precursor of DNA (Chap 20). It releases a low energy β-particle which, on account of its restricted range, provides high radio-autographic resolution (p. 122). Tritiated thymidine given to an animal is incorporated into the DNA of cells which happen to be at the pre-mitotic stage. The nuclei of these cells, and of their descendants, are marked permanently. In the mouse the cells of the bottom of the intestinal glands (crypt cells) are labelled within an hour; they move up to the surface epithelium in an escalator-like fashion to the tips of the villi in two or three days and are then shed from the extrusion zone which is a definite break in the epithelial sheet. The functional significance of this method of mucosal renewal is obvious; the tips of the villi undergo constant abrasion and were it not for the cell renewal they would become gradually reduced in length. From observations of human biopsy material, and from measurements of the DNA content of the intestinal lumen it has been estimated that 50 to 200 g of the human gastro-intestinal mucosa are renewed daily. The fastest rate of renewal in man occurs in the ileum; total cell renewal takes 5 to 7 days. With such a rapid rate of renewal the intestinal epithelium is particularly susceptible to diverse factors affecting mitosis. In normal subjects the mitotic rate is approximately equal to the extrusion rate and thus the length of the villi remains constant. Ionizing radiation and radio-mimetic cytotoxic drugs, for example colchicine, inhibit mitosis and as a result the mucosa is flattened as cell extrusion continues. Starvation and protein depletion markedly slow the mitotic cycle and cause thinning of the mucosa and sometimes flattening. In coeliac disease (p. 300) the rate of mitosis is increased because the cells are damaged by the leakage of acid hydrolytic enzymes from the lysosomes

Intestinal Digestion and Absorption 287

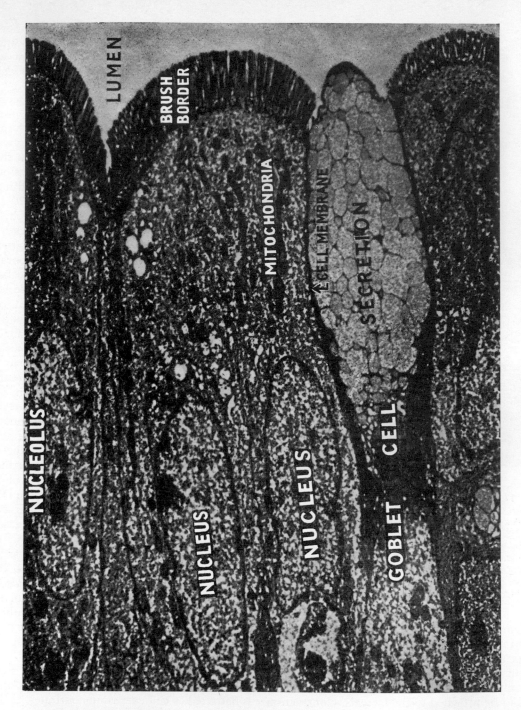

Fig. 17.30 Electronmicrograph of the epithelium of the small intestine of a rat. Magnification 8000 times. Fixed in osmium tetroxide. Note the brush border on the free edge of the cells. *(By courtesy of M. S. C. Birbeck.)*

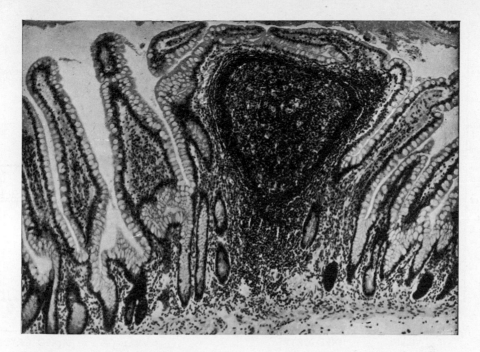

Fig. 17.31 Section of human intestine (ileum) to show solitary lymph follicle with germinal centre. A few villi, pointing up into the lumen of the intestine, are seen on each side of the lymph follicle. (*By courtesy of A. C. Lendrum.*)

The villi become stunted and the crypts elongated so that the ratio villus: crypt length is 1:2 instead of the 4:1 of the normal intestine (Table 17.32).

Three types of cells are present in the intestinal glands: (a) columnar cells which are undifferentiated whilst in the crypts, but which become capable of absorption during their upward migration, (b) goblet cells, which produce mucus; mucus production occurs while the cells are in the crypts but as the cells migrate upwards mucus is extruded (Fig. 17.30), (c) Paneth cells, which are confined to the crypts where they appear to secrete glycoprotein; their function is still unknown.

As the columnar cells migrate from the crypt on to the villus their enzymes become differentiated and thus their function (Figs. 17.33 and 17.34). In the crypts the cells have a high concentration of RNA and a predominantly basophilic cytoplasm. The RNA is probably associated with the production of the enzymes (Chap. 21) which appear during the migration of the cell. In the proximal third of

Table 17.32 Dimensions of microvilli in human jejunal epithelial cells

Cell location	Height (μm)	Diameter (μm)	Density, i.e. microvilli/μm^2 cell surface	Microvilli/cell*
Villous crest	1·36 ± 0·24l	0·08 ± 0·01	10·7 ± 3·4	1717
Intervillous space i.e. sides of villus	1·01 ± 0·23	0·10 ± 0·02	4·7 ± 1·7	331
Intestinal glands (Crypts of Lieberkühn)	0·67 ± 0·24	0·15 ± 0·12	3·9 ± 2·1	225

* Area of free surface of cell taken as 15 μm^2 (A. L. Brown (1962). *Journal of Cell Biology* **12**, 624).

Intestinal Digestion and Absorption   289

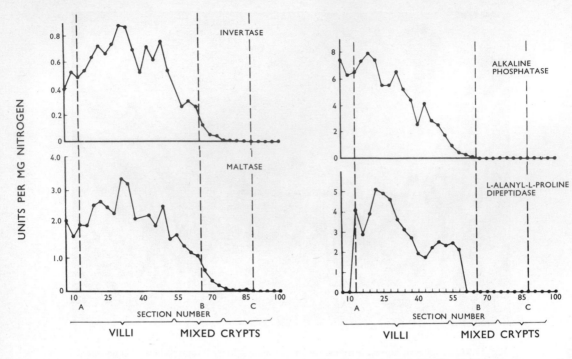

Fig. 17.33 Location of various enzymes on the villus of the jejunum. (C. Nordström, A. Dahlquist & L. Josefson (1968). *Journal of Histochemistry and Cytochemistry* **15**, 718.)

the villus much of the RNA has disappeared from the cell. The activities of all the cellular enzymes reach a maximal concentration in the distal part of the villus (Figs. 17.33 and 17.34).

The enzymes can be shown by histochemical methods to be located in the brush border of the cell, a highly differentiated and specialized structure. Recent electron microscopy has revealed that the microvilli are surrounded by a glycocalyx or 'fuzz', a polysaccharide–protein complex. Within the glycocalyx attached to the membrane are large numbers of particles, approximately spherical in shape, which appear to be 'packets' of enzymes. The membrane also has binding sites for the mobile carrier systems. On the inner side of the membrane is an ATP-ase which is of the type thought to be associated with ion movement.

The intestinal juice (succus entericus). Two distinct types of glands are found in the small intestine: (1) the duodenal glands (glands of Brunner), found only in the submucosa of the duodenum (Fig. 17.35), produce an alkaline secretion containing mucus, (2) the intestinal

glands or crypts of Lieberkühn (Fig. 17.29), present throughout the small intestine, are simple tubular glands which open into the spaces between neighbouring villi. The glands possess a number of different types of cells which secrete mucus and a few enzymes. The intestinal juice or succus entericus, produced in the intestinal glands, has a pH of about 7·6 with an electrolyte composition similar to that of extracellular fluid (Table 17.36). The amount of enzymes produced by the intestinal glands is almost negligible and the small quantities of disaccharidases and dipeptidases, at one time thought to be part of the secreted juice, are now known to be of cellular origin. The most important enzyme present in succus entericus is enteropeptidase (enterokinase) which is derived from the microvillus membrane. Enteropeptidase converts trypsinogen to trypsin, an essential step in the activation of the proteolytic enzymes secreted by the pancreas, by removing the terminal Val-$(Asp)_4$-Lys hexapeptide, the resulting polypeptide being the activated digestive enzyme, trypsin. A rare inborn error of metabolism causing deficiency of enterokinase results in very severe impairment of protein digestion.

The digestive changes in the small intestine. The chyme as it enters the duodenum from the stomach consists of a mixture of coarsely emulsified fat, protein (together with protein derivatives produced by the action of pepsin) and carbohydrates, including starch, a large proportion of which escape hydrolysis by salivary amylase. The acid chyme is buffered by the bicarbonate in pancreatic juice and in bile so that the reaction of the intestinal contents varies from pH 6·5 to 7·6. While digestion normally begins in the mouth and stomach, the enzymes secreted into the duodenum by the pancreas can, together with the bile, initiate and carry through the whole digestive process.

The proteins in the chyme are attacked first by the pancreatic endopeptidases, trypsin and chymotrypsin which split the protein molecules into polypeptides. These in turn are broken down by the exopeptidases and carboxypeptidases (p. 268) which split off the terminal amino acids. The final breakdown of tri- and dipeptidases to amino acids takes place within the epithelial cells, the dipeptidases

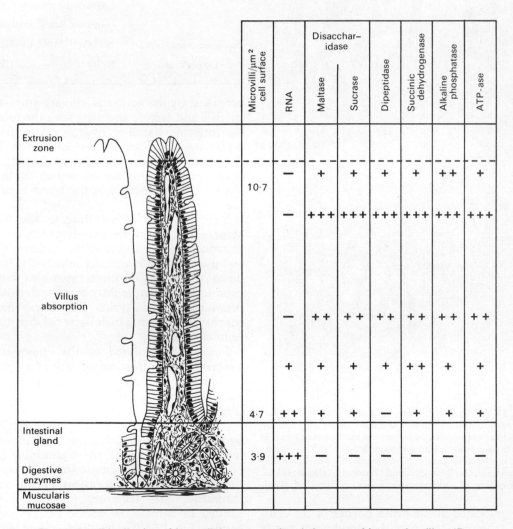

| | Microvilli/μm^2 cell surface | RNA | Disaccharidase | | Dipeptidase | Succinic dehydrogenase | Alkaline phosphatase | ATP-ase |
			Maltase	Sucrase				
Extrusion zone	10·7	−	+	+	+	+	++	+
		−	+++	+++	+++	+++	+++	+++
Villus absorption		−	++	++	++	++	++	++
		+	+	+	+	++	+	+
	4·7	++	+	+	−	+	+	+
Intestinal gland / Digestive enzymes	3·9	+++	−	−	−	−	−	−
Muscularis mucosae								

Fig. 17.34 Distribution of intracellular enzymes in relation to position on the villus. (*By courtesy of Maureen Palmer.*)

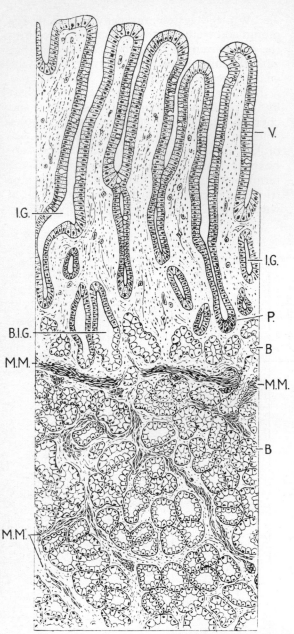

Fig. 17.35 Section through the duodenum to show villi and the duodenal (Brunner's) glands. Most of these are deep to or surrounded by the muscularis mucosae which is split up into many irregularly running strands. V. = villus; I.G. = intestinal glands; P. = Paneth cell; M.M. = muscularis mucosae; B. = duodenal (Brunner's) glands; B.I.G. = emptying of a Brunner gland into an intestinal gland.

being located in the brush border of the cells.

Carbohydrates enter the duodenum partly in the form of disaccharides such as maltose,

Table 17.36 Composition of intestinal juice (succus entericus) (pH 6·5 to 7·6)

Organic constituents	Enteropeptidase (enterokinase)
0·6 per cent	Amylase
	Mucin
	Nucleases
	Traces of disaccharidases, e.g. maltase, lactase, sucrase
	Traces of dipeptidases and aminopeptidases
	Traces of lipase, esterases, phosphatases
	Intracellular enzymes released by shedding of mucosal cells
Inorganic constituents	Na^+ Ca^{2+} Mg^{2+} Cl^- HCO_3^-
1·0 per cent	K^+ HPO_4^{2-}

produced by the action of salivary amylase, or sucrose and lactose ingested with the food. In the intestine starch is hydrolysed to maltose under the influence of amylase derived from the pancreas. The conversion of disaccharides to monosaccharides is mediated by the appropriate enzymes located in the brush border of the cells.

The digestion of cellulose in the human alimentary canal is not important but it occurs to a much greater extent in the rumen of the ox and goat. In these animals cellulose is broken down by a variety of micro-organisms with the production of simple fatty acids such as acetic, propionic and butyric. The horse has an enormous colon in which bacterial digestion of cellulose occurs.

Fats are hydrolysed by the lipases of the pancreatic and intestinal secretions and under the influence of the bile salts the ingested fat is finely dispersed as small particles or 'micelles' (Fig. 17.37).

Nucleic acids are believed to be broken down by nucleases, nucleotidases and nucleosidases to a mixture of purine and pyrimidine bases, phosphoric acid and pentose sugars (Chap. 20).

Control of the secretion of the small intestine. The secretory processes can be studied in animals by means of Thiry-Vella fistulae (Fig. 17.38) or by blind-ending pouches of small lengths of intestine.

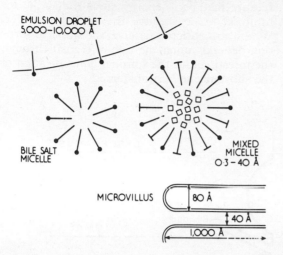

SURFACE EFFECTS OF BILE SALTS

EMULSION DROPLET
5,000–10,000 Å

BILE SALT
MICELLE

MIXED
MICELLE
0·3–40 Å

MICROVILLUS   80 Å

40 Å

1,000 Å

Fig. 17.37 Relative size of emulsion, micelle, and small gut brush border. Note that when a bile salt micelle has been converted into a mixed micelle with, for example, an unsaturated monoglyceride, then non-polar sterols and saturated fatty acids can be more freely solubilized. (A. M. Dawson (1968). *Fourth Symposium on Advanced Medicine* (Edited by O. Wrong), p. 258. London: Pitman. 1A = 10^{-1} nm.

(a) *Duodenal glands (Brunner's glands).* There is a small basal secretion from these glands in most animals and feeding, particularly of fatty foods such as cream or egg yolk, causes a very large secretion both in innervated and denervated pouches. Injection of secretin causes secretion but the existence of a further stimulating hormone, duocrinin, has been postulated. In addition the increased

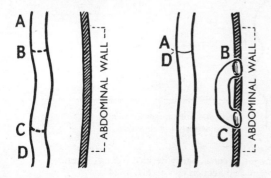

Fig. 17.38 Thiry-Vella fistula. An isolated loop of intestine is made by cutting the gut at two places, B and C, and then restoring the continuity at AD. The loop BC is then made to open on the abdominal wall.

intestinal motility produced by secretin stimulates the secretion of the duodenal glands. Stimulation of the vagus nerves in the neck of decerebrate cat causes secretion but since the response to feeding is identical whether the pouches used are innervated or denervated the vagus probably plays only a minor role.

(b) *Secretion from the small intestine.* Electrolytes and water pass continuously across the mucosa of the small intestine. When there is a net accumulation of fluid in the intestinal lumen, secretion is said to have occurred. Secretion occurs most strikingly under the influence of the toxin produced by the cholera vibrio and, to a lesser extent, under the influence of the hormones glucagon and gastrin, and after mechanical stimulation of the mucosa. The splanchnic nerves may exert an inhibitory effect on the secretion of intestinal juice, since a denervated loop at first secretes more juice than an innervated loop, perhaps due to the vasodilatation following on splanchnic section. Stimulation of the vagus gives inconsistent results but injection of eserine or pilocarpine causes secretion of succus entericus. Nausea induced by injection of apomorphine causes a very large secretion from a Thiry-Vella loop which is abolished by atropine or by vagal section.

Absorption in the small intestine. In the mouth and oesophagus no appreciable absorption of foodstuffs occurs although some drugs, for example trinitrin, morphine and steroid hormones, are absorbed through the oral mucous membrane. Absorption through the gastric mucosa is very limited but small amounts of water, undissociated organic acids such as acetyl salicylic acid, and alcohol, may be absorbed.

The site of absorption in the small intestine depends upon the relationship between the rate of transit and the rate of absorption. Despite the fact that transit through the jejunum is much faster than that through the ileum a rapidly absorbed substance can be cleared completely from the jejunal lumen. More slowly absorbed materials are absorbed more distally in the ileum (Fig. 17.39).

The site of absorption also depends on whether the substance is transferred by an active transport mechanism or by diffusion. Active transport implies that substances are absorbed through the intestinal mucosa against a concentration gradient. The rate of active transfer is reduced by metabolic inhibitors because the processes require energy. The absorption due to active transport is usually rapid and therefore occurs in the jejunum. Passive absorption is due to diffusion through the intestinal mucosa in the same direction as the concentration gradient. Diffusion does not require metabolic energy since metabolic inhibitors do not reduce the rate of transfer. Where absorption is passive, the site of absorption depends upon the rate of transit through the intestine and the luminal concentration of the substance to be transferred.

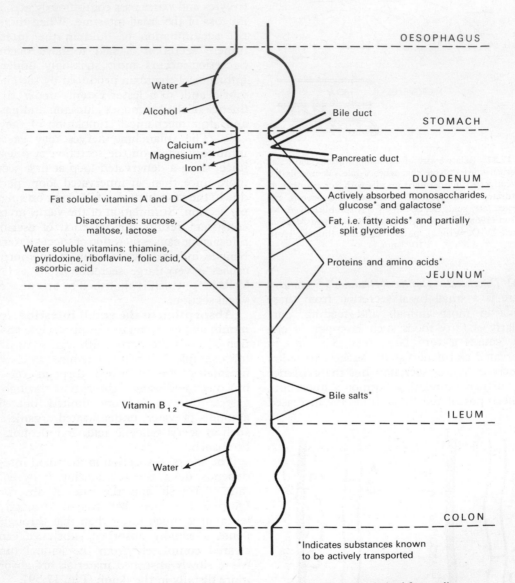

Fig. 17.39 Known sites of absorption in the alimentary canal, partly derived from a diagram by C. C. Booth (1968). *Handbook of Physiology* Section 6, Vol. III, 1524.
* Indicates substances known to be actively transported. Passively absorbed substances begin to be transferred in the duodenum but their movement relies on the maintenance of a concentration gradient. Bile salts are absorbed passively along the whole of the intestine by ionic and non-ionic diffusion.

Substances which are actively transported require a specialized carrier mechanism and therefore the site of absorption is limited to the site of the carrier. For this reason, vitamin B_{12} and bile salts are absorbed specifically in the terminal ileum.

Absorption of water and food materials occurs most actively in the upper part of the small intestines where the structure of the mucous membrane is specially adapted for the purpose. Substances absorbed across the villous mucosa pass through (1) the outer or free border of the epithelial cell, (2) the body of the cell itself, and (3) the lateral border of the cell before they reach the venule or lacteal. Some sugars (for example glucose and disaccharides) and essential amino acids are transported into the cell from the lumen of the gut by an inward-acting carrier at the microvillous membrane, and then move passively down the concentration gradient into the blood. Sodium ions are implicated in the uptake of amino acids and glucose and the sodium is then actively pumped out of the cell. Small particles may be taken up by the intestinal cells by pinocytosis.

In health the absorption of foodstuffs is virtually complete during passage through the small intestine. After a mixed meal all of the carbohydrate, about 95 per cent of the fat and 90 per cent of the protein are absorbed.

In the colon absorption is confined to water and substances of low molecular weight such as glucose, inorganic salts and perhaps short chain fatty acids. It is impossible to maintain metabolic equilibrium by introducing into the colon fluids containing protein and fat since these cannot be absorbed by the colonic mucous membrane.

At least one third of the intestine can be excised and, provided that the continuity is restored, no permanent harm ensues; but removal of more than 50 per cent reduces absorption. A few patients have survived with less than 25 per cent of their intestine. Loss of the ileum is more serious than loss of the jejunum because active reabsorption of bile salts occurs only in the ileum. After a massive resection of the small intestine the absorption of fat is grossly reduced, whereas that of amino acids and carbohydrates is only slightly diminished. The fat soluble vitamins and calcium are poorly absorbed.

When a large portion of the small intestine of the rat is removed the remainder increases in weight per unit length owing to the increased thickness of the mucosa and the muscle coats. The number of epithelial cells per villus increases because of an increase in the mitotic rate and the villi therefore increase in length (Table 17.40) since the rate of cell extrusion remains constant. The increased cell renewal reduces the age of the villous cells and the enzyme activity, particularly of disaccharidases, is reduced. Nevertheless, glucose absorption does improve probably because of the increase in the number of absorbing cells, although each cell may have a reduced capacity for glucose transport. In man there is no evidence that the rate of mitosis is altered, although during recovery from massive resection the capacity to absorb glucose is con-

Table 17.40 Changes in villus height and glucose absorption after resection of part of the small intestine

	Part resected	Villus height	Glucose absorption (mg/cm intestine/hour)
A. *Rat*	Ileum	Jejunum 369 μm	Jejunal remnant 2·55
	Control	Jejunum 323 μm	Jejunal remnant 2·15
	Jejunum	Ileum 354 μm	Ileal remnant 1·74
	Control	Ileum 232 μm	Ileal remnant 0·89
B. *Man*			Mean glucose absorption (mg/25 cm intestine/min)
		After resection	60·9
		Control	48·3

Figures from various papers of R. H. Dowling & C. C. Booth.

siderably increased, and fat absorption eventually improves (Table 17.40).

The nervous system appears to have little influence on absorption. After vagotomy in man intestinal absorption of monosaccharides and of electrolytes soon returns to pre-operative values. The faecal excretion of fat is often increased after vagotomy but the increase may be due to defective digestion rather than to changes in absorption.

The absorption of carbohydrate. After digestion by pancreatic amylase, dietary carbohydrates are found in the small intestine mainly as the disaccharide maltose, together with sucrose and lactose which are present as such in the food. The disaccharides are further digested to monosaccharides during transfer across the microvillous membrane.

Monosaccharides are seldom encountered in food but their transport has been intensively investigated. In the small intestine glucose and galactose are transported much more rapidly than fructose, mannose and pentoses although all have the same molecular weight. Metabolic inhibitors reduce the rate of transport of glucose and galactose across the small intestinal mucosa, whereas the transfer of the more slowly absorbed sugars is unaffected, presumably because an active transport mechanism operates for glucose and galactose but not for the slowly absorbed sugars.

Wilson and Crane have shown that sugars which are actively transported have several chemical features in common. First, they all have a six-membered ring. Second, they all have one or more carbon atoms attached to carbon 5. Third they have a hydroxyl group at carbon 2 with the same stereoconfiguration as occurs in D-glucose (Fig. 17.41).

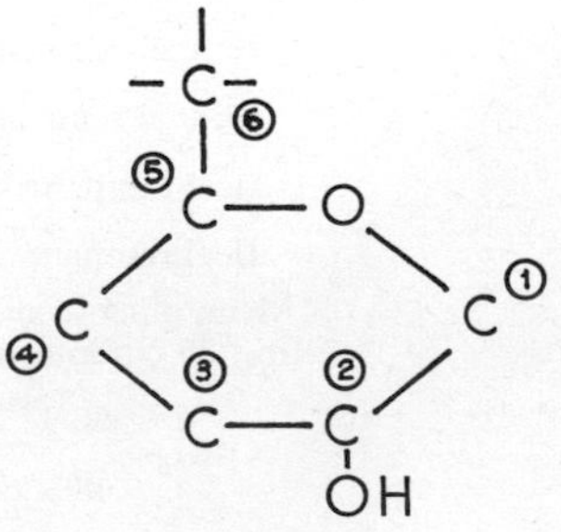

It was at one time thought that active transport through the epithelial cell depended on reduction of the concentration of the sugar by phosphorylation at the free border of the cell but this hypothesis has had to be abandoned.

Glucose is absorbed very rapidly and completely even from a very dilute solution, that is, against a concentration gradient. Crane and his collaborators explain active transport by supposing that in the brush border there is a sodium-dependent carrier, specific for the configuration in Fig. 17.41 which transports sugars across the membrane in either direction in the presence of sodium ions. The carrier is not energy-dependent; by the interaction of the sodium-dependent sugar carrier and the sodium pump, actively transported sugars are concentrated within the cell without any back-leakage of the sugar into the lumen. The dependence of the active transport of glucose upon the presence of sodium ions has been demonstrated in isolated loops of rat intestine by replacing the sodium of the bathing medium by lithium. In these circumstances

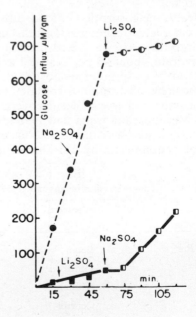

Fig. 17.42 Effect of Na^+ on Absorption *in vivo*. Glucose influxes μmole/g dry tissue, from small intestine of rat perfused *in situ* with isosmotic solution of either Na_2SO_4 (broken line) or Li_2SO_4 (solid line). Composition of the perfusate was changed at 60 min in each case. (T. Z. Csáky & L. Zollicoffer (1960). *American Journal of Physiology* **198**, 1056–1058.)

Fig. 17.41.

the rate of glucose transfer is markedly reduced (Fig. 17.42). Drugs such as strophanthin and ouabain which inhibit active ion transport (sodium pump) also inhibit active transport of sugars. Substances preventing the liberation of metabolic energy, such as dinitrophenol, also inhibit active transport of sugars. Phloridzin, a glycoside, inhibits glucose transport by competing for the carrier site at the brush border (Fig. 17.43). Glucose is transported at a greater rate than galactose because although they share a carrier site, glucose has a greater affinity for the carrier than galactose.

Fructose absorption is passive, that is, it cannot be absorbed against a concentration gradient and the presence of inhibitors does not affect the rate of transfer. However, the absorption of fructose is rapid because fructose is converted to glucose or lactate within the mucosal cells so that a steep concentration gradient is maintained. Fructose has been infused into the jejunum of human subjects undergoing laparotomy and subsequent analysis of mesenteric blood for glucose and fructose has shown that between 20 and 50 per cent of fructose is converted to glucose. The conversion perhaps involves glucose-6-phosphatase and alkaline phosphatase (p. 146 and 312) which are found in the brush border of the intestine. This conversion is described in detail in Chapter 18. The conversion to glucose is not essential since subjects with hereditary malabsorption of glucose and galactose who lack the active hexose carrier mechanism absorb fructose at a normal rate although the rate of absorption of glucose is very much reduced and the conversion of fructose to glucose does not take place.

The hydrolytic activity of the intestinal juice is too low to account for the rate of absorption of sucrose and lactose. Hydrolysis of disaccharides to the constituent monosaccharides does not occur within the lumen of the small intestine but occurs during transport into the absorptive cell across the 'digestive-absorptive' villous membrane. The disaccharides are absorbed more rapidly in the brush border, the monosaccharides are actively transported into the cell. It has been postulated recently that one of the functions of the glycocalyx is to prevent a rapid diffusion of monosaccharides away from the outer membrane of the microvillus where final hydrolysis has taken place.

The enzymes sucrase, maltase and lactase have all been shown by histochemical techniques and by analysis of microquantities of various fractions of the cell to be located within the brush border area. The enzymes isolated are highly specific, there being three types of maltase, one of which attacks sucrose (sucrase), a specific trehalase and two forms of lactase. All the disaccharidases are developed at the time of birth but in most species after weaning there appears to be regression of lactase activity which cannot be prevented by maintaining the animal on a high lactose diet.

Deficiency, either inherited or acquired, of disaccharidase can lead to malabsorption of the

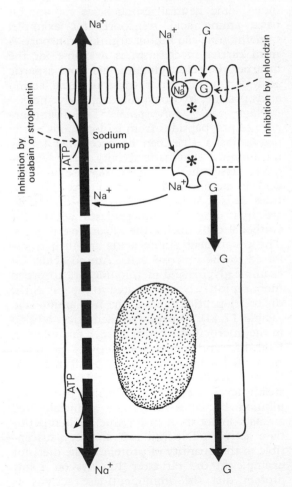

Fig. 17.43 Diagram to show the active transport of glucose based on the hypothesis of Crane. (*By courtesy of Maureen Palmer.*)

corresponding sugar and malabsorption may cause diarrhoea owing to bacterial breakdown of the unabsorbed sugar to lactic and other acids. Lactase deficiency is not uncommon and exclusion of milk from the diet may completely relieve diarrhoea. Deficiency of a specific disaccharidase is tested clinically by giving a test dose (generally about 50 g) of the specific disaccharide to a patient and observing whether diarrhoea develops. When a disaccharidase is present, the blood glucose rises after the test dose but remains unchanged in the absence of the disaccharidase.

Carbohydrates are absorbed in the proximal part of the small intestine and both monosaccharides and disaccharides have completely disappeared from the lumen by the time the meal reaches the ileum. However, since resection of the jejunum does not impair the absorption of glucose and galactose, the specific carrier systems for the actively absorbed sugars must be present throughout the small intestine. Disaccharidase activity is greatest in the jejunum and diminishes throughout the ileum.

The absorption of amino acids. The protein of food is digested in the stomach and intestine to peptides and amino acids. In addition much nitrogenous material passed into the intestine, partly as proteins in the gastrointestinal secretions (approximately 140 g) and partly as desquamated epithelial cells (approximately 25 g) undergoes digestion. In health approximately 90 per cent of the ingested and endogenous protein is absorbed. In man radioactive human serum albumin and milk proteins are 80 to 90 per cent absorbed in the upper 100 cm of the jejunum. Both digestion and absorption are very rapid in man since $^{15}NH_3$ appears rapidly in the urine with a peak at one hour after the ingestion of either ^{15}N-labelled protein or ^{15}N-labelled hydrolysate.

Most of the naturally occurring amino acids are L-isomers. L-Amino acids are transported against concentration gradients by carrier mechanisms while D-amino acids are absorbed passively much more slowly. Active transport by the known carriers (for neutral, basic, acidic and cyclic amino acids) requires correct stereoisomerism.

Active transport of amino acids is sodium-dependent. Experiments with everted sacs of golden hamster intestine have shown that replacement of the sodium of the bathing fluid by lithium or potassium reduces the rate of transfer of L-tryptophan. However, if the chloride is replaced by sulphate the rate of transfer is not altered. Transport of L-amino acids is oxygen dependent and in the absence of oxygen the rate of absorption is slowed to that of the D-isomers.

All the neutral L-amino acids, including L-histidine, are actively transported. Experiments with mixtures of neutral amino acids have shown that they compete with each other for transport because they share a common carrier system. Transport of neutral amino acids is unaffected by basic or acidic amino acids. A second carrier mechanism is shared by the basic amino acids, lysine, ornithine, arginine. Some neutral amino acids can inhibit basic amino acid transport; for example, L-methionine can inhibit arginine transport. A third carrier mechanism is available for the dicarboxylic acids, glutamic acid and aspartic acid. A fourth carrier transports the amino acids proline, hydroxyproline and glycine.

Absorption of polypeptides. The disappearance of polypeptides from jejunal loops of the dog is faster than can be accounted for by intraluminal enzymatic activity. After feeding dipeptidases there is a great increase of amino acids in the portal blood but no trace of dipeptides. It is thought, therefore, that dipeptides are hydrolysed by dipeptidases within the epithelial cells during the absorptive process. The constituent amino acids are then transferred to the mucosal cells. Amino acids, for example glycine and methionine, are absorbed more rapidly when presented to the gut as the di- or tri-peptide than as the free amino acid (Table 17.44). Dipeptidase activities develop in the microvillous brush border fairly late in fetal life and reach full activity at birth. Breast-fed animals then show a fall in dipeptidase activity followed by a rise to a new steady level. The fall is probably due to inhibition by colostrum since artificially fed animals do not show such reduction in dipeptidase activity. Dipeptidases are very susceptible to the quantity of protein in the diet; for example in the rat after 10 days on a low protein diet the aminopeptidase activity is greatly depleted.

All undigested peptides with a molecular weight of over two or three hundred are poorly

Table 17.44 Rate of absorption of glycine, methionine and their peptides

Substance absorbed	Concentration	Absorption µmoles/cm/10 min
Methionine	100	2·42
Dimethionine	100	3·42
		µmoles/cm/5 min
Glycine	267	4·24
Diglycine	267	6·88
Triglycine	267	6·63

Methionine data I. L. Craft, R. F. Crampton, M. T. Lis & D. M. Matthews (1969). *Journal of Physiology* **200**, 111P.

Glycine data, D. M. Matthews *et al.* (1968). *Clinical Science* **35**, 415.

absorbed. The small amounts of undigested proteins shown to be absorbed, perhaps by pinocytosis, are found in vacuoles in the cytoplasm and in tubular spaces in the mucosal cells. Some new-born animals acquire passive immunity by absorbing antibodies (proteins) (p. 37) from the mother's colostrum, but the human baby obtains very little in this way. In the young animal unchanged dietary protein is absorbed less well than maternal antibodies. Adult man absorbs extremely small amounts of unchanged protein, quite insignificant from the nutritional point of view, but such protein may act as an antigen with formation of antibody. When the concentration of the antibody has been raised sufficiently by the repeated absorption of small amounts of the antigen, the reaction between the antibody and a further amount of antigen may produce asthma or skin rashes. The patient is then said to be sensitive, or hypersensitive, to the protein.

In many gastrointestinal diseases, and in obstruction of the intestinal lymphatics, loss of plasma protein into the gut may lead to hypoproteinaemia. Loss of plasma protein can be demonstrated by intravenous administration of plasma proteins labelled with ^{51}Cr which cannot be reabsorbed and so is excreted in the faeces. In health less than 2 per cent of intravenously administered ^{51}Cr albumin is found in the faeces. Increased intestinal loss of plasma protein is called exudative enteropathy and as much as 10 to 20 per cent of an intra-venous dose of labelled protein may be excreted in the faeces.

The absorption of fat. In the stomach the triglycerides, which make up most of the dietary fat, form a coarse suspension. During digestion the triglycerides and other fats are rendered water-soluble so that they can pass into the intestinal mucosal cells.

In the duodenum triglycerides, which have been emulsified by bile salts, undergo lipolysis under the influence of pancreatic lipase in the presence of co-lipase and bile salts. Since this process is very pH sensitive, the finely adjusted secretion of bicarbonate in pancreatic juice is of extreme importance. Lipase acts specifically at the interface between the fat globule and water and hydrolyses the triglyceride (p. 268) at the two primary positions, producing fatty acids and monoglycerides. Under the influence of the salts and monoglycerides the emulsion of fat is broken down into water soluble micelles which are aggregates of molecules containing both water-soluble polar groups (hydroxyl and carboxyl) and fat-soluble hydrocarbon chains (long chain fatty acid monoglycerides). The molecules orientate themselves so that the water soluble hydrophilic groups face outwards and the fat-soluble (hydrophobic) groups form the interior of the aggregate. The micelles are stable water-soluble structures which can dissolve other water insoluble (hydrophobic) compounds in their interior (Fig. 17.37). Substances made soluble in this manner include fatty acids, cholesterol and fat soluble vitamins.

After a meal the concentration of bile salts in the upper small intestine is normally sufficient for the formation of micelles (p. 279). In the upper small intestine the micelles pass across the brush border by a mechanism which has not yet been defined but the bile salts remain in the intestinal lumen because they are poorly absorbed in the jejunum. The bile salts are actively absorbed in the terminal ileum, from which they return to the liver by the portal vein to be re-excreted in the bile (p. 273). The entero-hepatic circulation is extremely efficient and the whole bile salt pool is recirculated about six times in 24 hours.

Within the intestinal epithelial cell triglycerides are resynthesized and rendered water-soluble as chylomicrons with a covering of lipoprotein (Fig. 17.45). The chylomicrons

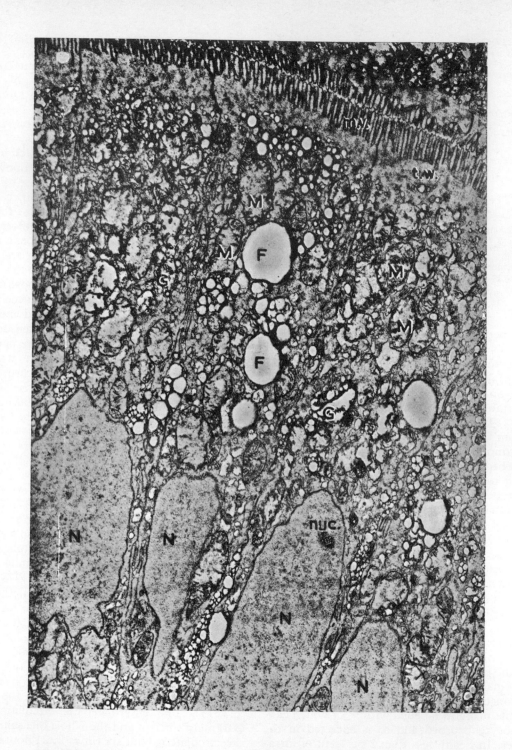

Fig. 17.45 General appearance of the principal cell of rat small intestine during fat absorption. Fixed: $KMnO_4$. Mag. $\times 10\,000$. m.v., microvilli. M, mitochondrion. F. fat droplet. G, Golgi complex. N, nucleus. nuc., nucleolus. t.w., terminal web. (Electronmicrograph by P. F. Millington, Department of Medical Biochemistry and Pharmacology, University of Birmingham.)

pass out of the cells into the lacteals, from which they pass into the thoracic duct and the systemic circulation.

Fatty acids with chains longer than 14 carbon atoms are transported as triglycerides in the chylomicrons while fatty acids with shorter chains, being water soluble, are absorbed directly into the portal blood stream.

Disorders of fat absorption. Fat digestion and absorption is assessed in man by collecting the faeces over a period of 3 or 5 days and measuring the amount of fat excreted. Normal subjects excrete less than 5 g of fat per day, not all of which is residue from the dietary fat since a considerable amount of endogenous fat, about 25 g, has been added from bile and desquamated cells during intestinal transit. Excessive excretion of fat in the faeces (steatorrhoea) indicates either maldigestion due to deficient secretion of pancreatic lipase or bile salts (p. 280) or to defective absorption resulting from disease of the small intestinal mucosa, such as the coeliac syndrome, or obstruction of the intestinal lymphatics. In the coeliac syndrome sensitivity to gluten in cereals, particularly wheat, causes mucosal damage and the villi disappear particularly in the upper small intestine. The condition can be greatly improved and the villi can regrow if gluten is withheld.

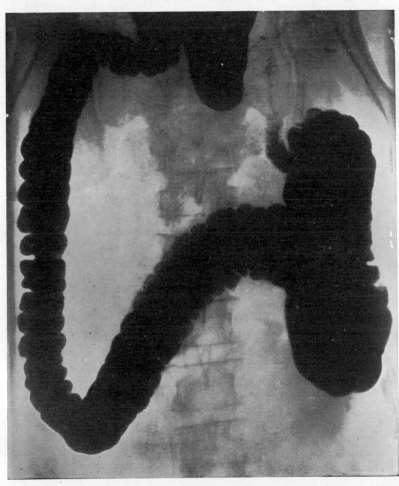

Fig. 17.46A Radiograph of the human colon outlined by barium sulphate introduced in the form of an enema. A small amount of barium has entered the ileum. The indentations of the column of barium, visible particularly in the descending colon, are known as *haustrations*. They are the results of muscular contraction and vary greatly both in number and form from time to time.

THE LARGE INTESTINE

The calibre of the large intestine is greater than that of the small, the transverse measurement being greatest at the caecum and becoming narrower towards the rectum. The longitudinal muscle is gathered into three longitudinal strands (taeniae coli) which, since they are shorter than the other coats, produce puckering (Fig. 17.46). The mucous coat is smooth without villi. The mucosal glands are 0·07 mm long and closely packed and are composed largely of mucus-secreting goblet cells (Fig. 17.47). As in the small intestine, the epithelium is in a state of continuous renewal, migration and loss, the turnover time being 1 to 2 days. Lymph nodules are present in the proximal part of the colon and especially in the vermiform appendix. The transition from the columnar epithelium of the rectum to the stratified squamous epithelium of the anal canal is not sharply defined but it usually occurs just above the anal valves. The rectum about 13 cm long, by no means straight, is sensitive to stretch but not to tactile or thermal stimuli. The anal canal (as just defined) is sensitive to tactile, thermal and painful stimuli but localization is poor.

The size of the large intestine relative to the rest of the gut varies considerably in different species and is apparently related to the habitual diet of the animal. In the horse

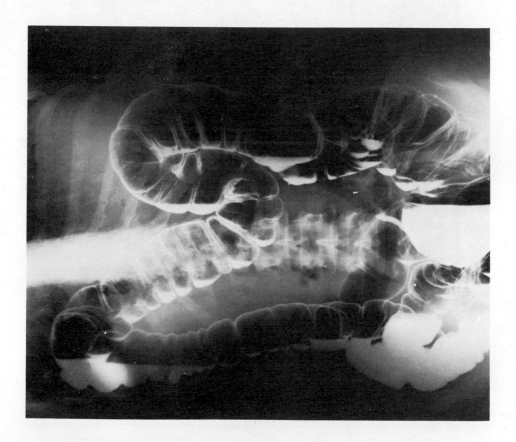

Fig. 17.46B The normal colon demonstrated by the double contrast radiological technique. The colon is first filled with a barium emulsion. When most of the emulsion has passed out carbon dioxide is introduced to distend the colon. In this picture the patient was lying on his left side. Hence the level surfaces of pools of barium emulsion run horizontally in this picture.
(*By courtesy of G. F. A. Howie.*)

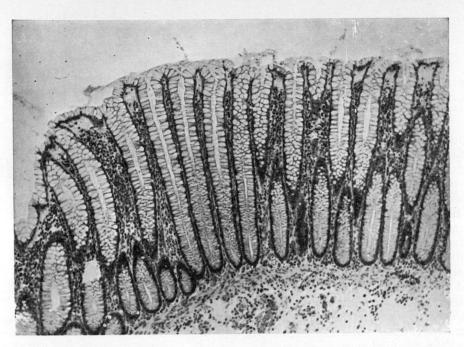

Fig. 17.47 Transverse section of human large intestine showing large numbers of mucus-secreting cells. Haemalum and eosin; ×115. (*By courtesy of A. C. Lendrum.*)

digestion of cellulose occurs in the enormous colon and caecum (p. 292). In carnivorous animals the large intestine is relatively small. The caecum and colon of man, being about 150 cm long, are intermediate in size, as might be expected since his normal diet contains both flesh and vegetables. The capacity of the large bowel in man, as measured by a contrast enema, ranges from 0·9 to 1·8 litres.

Movements of the large intestine. The caecum and colon usually fill with chyme from the small intestine when food is eaten as the result of a peristalsis in the small intestine accompanied by a relaxation of the ileocolic sphincter. However, even when the subject remains without food a barium meal usually enters the caecum about 4 hours after it has been taken and soon after this the meal quickly fills the ascending and transverse colon. On the average a barium meal reaches the hepatic flexure of the colon in 6 hours, the splenic flexure in 9 hours and the pelvic colon in 12 hours; three-quarters of a barium meal are expelled in 72 hours, the remainder escaping in the succeeding 4 or 5 days. Food residues probably pass more slowly than the X-ray

would indicate. There are undoubtedly quite large individual and day to day variations in the rate of passage of residues through the intestinal tract.

In man, there are two modes of colonic motor activity: propulsive and non-propulsive. The non-propulsive mode predominates and consists of segmental, non-peristaltic contractions which produce localized elevations of colonic intraluminal pressure. These pressure waves occur at the rate of 2 to 3 per min, their average amplitude being approximately 30 cm H_2O in the proximal and 50 cm H_2O in the distal colon. Segmental pressure activity is increased by ingestion of food and by cholinergic stimuli but is unaffected by the antral hormone gastrin; it diminishes during sleep. Although normal segmental motor activity occurs for about 50 per cent of recording time, it is unperceived by the subject. On the other hand excessive segmentation may cause abdominal pain. Segmental contractions probably delay rather than accelerate the transit of colonic contents. Segmental activity tends to be high in constipated and low in diarrhoeal patients: there is, however, an overlap between the two groups. Gross thickening of colonic muscle in the sigmoid region is a common finding in the older population in the Western countries. This abnormality is accompanied by colonic pressure waves of high amplitude and by the formation of numerous mucosal diverticula, which may become inflamed (diverticulitis).

Propulsion of faeces is brought about by a different type of colonic muscle activity, namely mass action or mass peristalsis. This consists of a ring of muscular contraction which passes for a variable distance along the colon. The haustral folds disappear momentarily ahead of the peristaltic contraction, only to reform behind it. The frequency of mass movements increases after meals and during somatic activity. If haustration of the colon is destroyed by disease (for example ulcerative colitis), distal resistance to the flow of faeces is diminished with resultant fast transit time, diminished water absorption from the colonic lumen and the production of diarrhoea.

Certain laxatives (senna, bisacodyl) act by stimulating propulsive activity of the colon. Studies on patients with colostomies suggest that these substances activate mucosal receptors. The propulsive activity is probably mediated through the myenteric nerve plexus, and the plexus can be damaged by prolonged usage of high doses of these drugs. The habitual use of certain purgatives may cause a considerable loss of electrolytes, especially potassium, and cause hypokalaemia.

Nerve supply of colon and rectum. Fibres from the vagus nerves reach the proximal colon since section of the vagi leads to degeneration of nerve fibres in the proximal, but not in the distal colon. The outflow from the spinal cord to the distal colon in man is shown in Fig. 17.48. The thoracolumbar sympathetic outflow leaves the lumbar sympathetic chains, forming plexuses on the aorta, and passes eventually in the mesentery to the colon as the inferior mesenteric nerves. The presacral nerve is in fact a nerve plexus, the superior hypogastric plexus, formed by an extension of the aortic plexus together with branches from the lumbar sympathetic trunks. Below it is continued as two narrow plexiform strands, the right and left hypogastric nerves, which in turn become continuous with the corresponding inferior hypogastric or pelvic

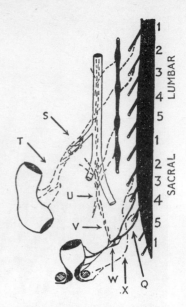

Fig. 17.48 The nervous outflow to the distal colon in man (R. C. Garry (1934). *Physiological Reviews* **14**, 107). Q = pelvic splanchnic nerves; S = inferior mesenteric plexus; T = inferior mesenteric nerves; U = superior hypogastric; V = hypogastric nerve; W = inferior hypogastric (pelvic) plexus; X = pudendal nerve.

plexus. The main sacral (parasympathetic) outflow in man (pelvic splanchnic nerves) comes from the second, third and fourth sacral nerves and joins the hypogastric plexus. The parasympathetic outflow joins the hypogastric plexus before arriving at the rectum and internal sphincter. The striated muscle of the external sphincter is supplied by the pudendal nerve (Figs. 17.48 and 17.49). This muscle is bilaterally innervated since section of one pudendal nerve does not affect its reflex behaviour.

Stimulation of the lumbar outflow (sympathetic) in the cat inhibits the colon and contracts the anal sphincter. When it is cut, or when the cord is anaesthetized, the colon is more active and the internal anal sphincter relaxes. Transection of the spinal cord in man by trauma or disease is also followed by hyperactivity of the pelvic colon. On the other hand bilateral removal of the lumbar chains in man seems to have no long-term effects on the motility of the colon. Stimulation of the pelvic nerves in the cat causes contraction of the colon and inhibition of the smooth muscle of the anal sphincter. After section of this parasympathetic outflow the movements of the distal colon become weak and the internal anal sphincter contracts.

Spinal cord and defaecation. Throughout the greater part of its length the intestinal tract shows a large degree of autonomy. Co-ordinated movements continue after complete destruction of the spinal cord. As its caudal extremity, however, the gut comes directly under control of the nervous system. If the lumbo-sacral region of the cord is destroyed the anal sphincters, the internal sphincter of smooth muscle and the external sphincter of striped muscle, are patulous and the caudal portion of the large intestine is paralysed. In time there may be some return of activity and the internal sphincter may recover tone. The caudal region of the gut may thus eventually recover its autonomy. Under those conditions faecal matter may be expelled at intervals but the patient is unaware of it and can exert no voluntary control.

If the spinal cord is transected craniad to the sacral region of the cord the condition of the gut, so long as spinal shock persists, resembles that seen after destruction of the sacral region of the cord. Subsequently, when the spinal shock has passed off, reflex defaecation is established. Expulsion of faecal matter may be complete but the patient is unaware of the act and is unable to exert any voluntary control over it.

The reflex pathways subserving defaecation are shown in Fig. 17.49. Stimulation of the distal colon in the cat causes contraction of the colon through the pelvic splanchnic nerve and relaxation, through the pelvic splanchnic and pudendal nerves, of the internal and external sphincters respectively. When the faeces pass through the sphincter these contractions and relaxations are reinforced by reflexes involving afferents and efferents in both pelvic splanchnic and pudendal nerves. A continuous inflow of impulses passes along the pelvic nerves in the cat; if these nerves are cut the external sphincter closes more firmly and distension of the colon no longer causes it to relax. The anal sphincter is normally closed and remains so in man even if he is deeply anaesthetized. The afferent impulses which reflexly maintain the tone of the external sphincter probably arise in the muscle spindles in the muscle and enter the cord mainly by the dorsal root of the second sacral segment (Fig. 17.49).

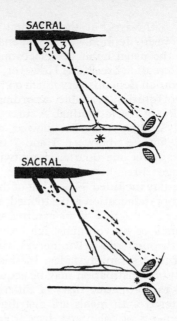

Fig. 17.49 Diagram of the supposed course of the nerve impulses in defaecation. The sacral cord of the cat: continuous line the pelvic nerve; interrupted line the pudendal nerve; * the point of stimulation. (*After* R. C. Garry (1933). *Transactions of the Royal Medico-Chirurgical Society of Glasgow* in *Glasgow Medical Journal* **28**, 9, Fig. 6.)

Defaecation is a complex act involving not only contraction of the rectum and caudal portion of the colon with inhibition of both internal and external sphincters, but also contraction of the diaphragm and of the muscles of the abdominal wall which are, of course, under voluntary control. The levator ani muscles in the pelvic floor also aid defaecation by pulling up the anal canal over the faecal mass. During defaecation the pressure in the rectum may rise as high as 200 mm Hg with violent expulsive efforts. Such high intra-abdominal pressures have important effects on the circulation, especially on the venous return to the heart.

Massive peristalsis in the transverse and descending colon drives the contents onwards and the subject may experience a sensation of fullness in the rectum and a desire to defaecate. Such a peristaltic movement frequently follows the eating of a meal and was thought to be due to the filling of the stomach (gastrocolic reflex) but gastric emptying, filling of the small intestine, somatic movement and colonic activity all play a part. The gastrocolic response may be only partly dependent on

Intestinal Digestion and Absorption   305

nervous pathways but the humoral pathway has not been defined. This response is the basis of the usual habit of defaecating after a meal, often after breakfast. However, the time of defaecation depends very much on habit and the opportunities available according to the occupation of the individual. A survey in 1965 of 1500 healthy persons showed that two thirds defaecated between 5 and 7 times per week and about one-quarter about twice a day. The range three bowel actions per week to three per day included 99 per cent of the group. If the act of defaecation is postponed, either by voluntary contraction of the external sphincter of the anus or by voluntary inhibition of the rectum through the pelvic nerves, the sensation of fullness rapidly fades. In this fashion the habit of constipation may be acquired.

Small radio-opaque discs of different sizes administered with meals are distributed over long segments of colon and discs from several meals are often out of chronological order. A great deal of mixing must occur in the colon. No particular distribution or change of distribution of the discs can be correlated with the desire to defaecate; quite often there is no urge to defaecate even when the rectum contains faeces. The length of colon emptied at defaecation varies considerably; in some subjects the whole of the left side is evacuated while in others even the rectum is not completely emptied. Purgatives in full doses may cause the large bowel to empty completely and two or three days may elapse before the distal colon becomes filled with faeces and the defaecation reflex is elicited once more.

Distension of the lower colon in man by means of a balloon produces a sensation of colic in the abdomen. The nerve impulses pass into the spinal cord along the sympathetic pathways. Distension of the rectum on the other produces a desire to pass flatus or faeces and a reflex contraction of the external sphincter, the afferent pathway being in the pelvic nerves (Fig. 17.49). The rectum is able to distinguish between flatus and faeces probably because flatus produces less distension. When these sensations rise to consciousness a voluntary decision is made as to whether the contraction of the external sphincter, initially reflex, should be maintained or relaxed. The external sphincter shows continuous electrical activity. When the rectum is distended by a balloon, at first the electrical activity increases along with the sensation of distension; then the electrical activity declines as the sphincter relaxes (see also Chap. 40). All the factors maintaining rectal continence are not yet known but it is generally agreed that the external sphincter allows a voluntary emergency control effective up to a minute; the internal sphincter apposes the sides of the anal canal which allows some mechanism (perhaps a flutter valve, p. 231) to maintain continence.

ABSORPTION AND SECRETION IN THE LARGE INTESTINE

About 1000 ml of chyme passes from the small intestine through the ileocolic valve into the caecum each day. In healthy subjects with well-established ileostomies about 60 m-equiv. of sodium and about 4 m-equiv. of potassium are lost per day together with 1·5 g of nitrogen and 2 or 3 g of fat. (An ileostomy is a surgically created opening through the abdominal wall into the ileum.) The volume of the chyme is much reduced during its stay in the colon by absorption of about 850 ml of water mainly in the caecum and ascending colon. The retention of this water is of importance for the maintenance of fluid equilibrium but it has to be remembered that the absorption of water in the small intestine, about 7 l per day, is very much greater. Persistent diarrhoea, that is the passage of frequent watery stools, leads ultimately to loss of water and electrolytes, especially potassium (Chap. 33) which may, especially in infants, be quickly fatal. Sodium chloride is readily absorbed from the colon but chloride ions are more rapidly absorbed than sodium ions—a fact of some importance if the ureters are transplanted into the colon. Amino acids, certain anaesthetic drugs, aspirin and some steroids are also well absorbed but protein, fat, sugar and calcium are not. Large amounts of electrolytes, especially potassium, can be removed from the body by introducing ion exchange resins into the distal colon. Especially in patients with megacolon or chronic constipation there is a serious danger of water intoxication after tap-water or soap-and-water enemas. A large absorption of water with lowering of the plasma sodium leads to drowsiness, convulsions and coma. Some of the absorptive power attributed to the colon may, in fact, be due to the passage of the substance to be absorbed through the ileocolic valve into the lower ileum.

An isolated loop of colon secretes a small amount of mucus which is produced by the goblet cells; mechanical stimulation arouses a small quantity of watery secretion with practically no enzymic activity. The mucus lubricates the faeces and neutralizes any acids formed in them. The pH of the contents of the colon is about 8. If one pelvic splanchnic nerve is cut, stimulation of either the peripheral or the central end produces secretion of mucus in the large intestine. The former is a direct effect, the second a reflex effect. Secretion can also be obtained by injection of vagomimetic drugs such as pilocarpine and can be prevented by atropine.

The bacteria of the intestine. The intestine of the new-born child is bacteriologically sterile and the material excreted from it is a semi-fluid greenish mass known as *meconium*. Sterility persists for a few days only; thereafter the intestine is invaded and colonized by ingested bacteria. It is well known that large intestine has a high bacterial content but only in the last decade has it been appreciated that the small intestine also has a substantial flora. This is now recognized as a useful symbiosis and not merely a contamination.

In healthy subjects the acid in the stomach normally acts as a barrier against a massive invasion by organisms derived from food and water or from the mouth. However, bacteria are constantly invading the small intestine from above. The organisms that flourish in the small intestine are mutually selected; their biochemical performance matches the local conditions. Some live in the lumen but it is thought that the major site is in the intestinal mucus. They hold a position against the flow of fluid and the propulsive action of the gut; they probably ebb and flow in numbers with meals. The fluid of the small intestine contains about 10^3 to 10^5 organisms per ml. Many different kinds of bacteria have been isolated but the difficulty in mimicking the conditions in which they exist in the intestine probably limits recovery. In the upper small intestine the flora is mainly streptococci, staphylococci, lactobacilli and fungi; in the lower small intestine rather more organisms are found and they are similar to those in the colon, particularly the anaerobes, bacteroides and bifidobacteria.

If an intestinal cul-de-sac (blind loop) is created in a patient, for example in the course of an operation to relieve obstruction, he is liable some time later to develop steatorrhoea (see p. 301) and anaemia. Instead of being relatively free of organisms his small intestine harbours numerous bacteria. These hydrolyse conjugated bile acids and the free acids so liberated have a toxic effect on the mucosa and do not promote the normal formation of micelles, necessary for the absorption of fat (p. 279). The steatorrhoea is more pronounced the greater the number of bacteria. The anaemia is explained by the competition between the host and the large numbers of bacteria for the available vitamin B_{12}.

Many intestinal bacteria can synthesize substances that are nutritionally useful but it is uncertain whether the amounts formed are significant. It is generally held that in man vitamin K is principally derived from intestinal bacteria. The role of intestinal bacteria has been helped by the study of germ-free animals (gnotobiotics) which develop in much the same way as conventionally reared ones except for a protuberant abdomen due to an enormous caecum which may be ten times the normal size. In germ-free animals rather more protein than usual appears in the faeces. It may be that some protein is utilized by the intestinal bacteria of the normal animal or that its bacteria enhance the absorption of protein. Clearly some aspects of small intestinal structure and function are determined by the normal flora. In the germ-free animal the villi are slender and the total surface area is reduced and cell turnover is slowed so that cells take 4 days to reach the top of the villus instead of 2. The number of lymphocytes and plasma cells in the intestinal wall is also reduced. Furthermore mucus accumulates so that it seems likely that bacteria normally degrade it.

In spite of its small molecular size urea passes only slowly from the circulation into the lumen of the colon but when it enters the lumen it is immediately hydrolysed to ammonia by urease produced by non-sporing aerobes lying on the mucosa and quickly reabsorbed. About 7 g urea is degraded per day and the process is so efficient that normally no urea is found in the faeces. Most of the ammonia in the body is produced in the colon. After reabsorption it is partly changed to urea in the liver.

Certain products of bacterial metabolism are normally absorbed and excreted in the urine; phenol, cresol, indole, skatole and pyrocatechol (p. 386). The level of indole excretion in the urine is used as an index of small intestinal bacterial activity in man.

Colonic bacteria are predominantly bacteroides and bifidobacteria but lactobacilli and streptococci are also present. They bring about fermentation of carbohydrates with the production of carbon dioxide and hydrogen sulphide and also combustible gases such as hydrogen and methane. More than half of the gas found in the colon is, however, derived from the atmospheric air which can pass surprisingly quickly down the alimentary canal; air which was swallowed has been seen by X-rays to reach the caecum in 14 min and to be passed as flatus in 30 min. About 50 per cent of the flatus is nitrogen derived mainly from air; a small amount of nitrogen diffuses out of the intestine into the lumen. Even the gas which accumulates behind an intestinal obstruction is derived mainly from the air. Normally part of the gaseous content of the colon is absorbed and is excreted in the lungs and the remainder, about 500 ml per day, is passed per rectum as flatus. The composition of flatus is extremely variable depending on the microorganisms and the food eaten. Since the volume of the gas in the stomach and intestines doubles at 5 km (15 000 ft) it is easy to see that pain due to distension may occur in a sudden rise to high altitude.

Many animals, notably the ruminants, have evolved elaborate symbiotic systems for obtaining nourishment from the plant material of their diet which cannot be digested by their own alimentary enzymes. Ruminants possess a complex stomach with an extensive flora that can break down cellulose and similar plant polymers. Passage through the stomach is slowed by rumination–regurgitation and chewing the cud. However, other animals, especially rodents and rabbits, have a relatively large caecum and colon which have a flora capable of breaking down cellulose. This might be thought to be of no value to the animal but coprophagy is now recognized to be usual in these animals and indeed there are two kinds of faeces, soft faeces which are eaten and hard

which are not. Furthermore urea passes into the caecum and is converted into protein by the bacteria; if coprophagy is prevented nitrogen balance falls by 50 per cent.

Immunity and the gastrointestinal tract. The presence of bacteria and of numerous foreign substances in the diet results in a large and highly organized immune system. As with all epithelial surfaces the major humoral line of defence is IgA (p. 37) which is formed by plasma cells which are found in thin sheets around the crypts of the intestinal glands. IgA passes through the epithelial cells where two molecules are bound together by a secretory piece, a glycoprotein, which probably facilitates the transport of IgA molecules through the epithelial cell. This complex then passes into the lumen where it acts as an effective antibody, at one time called a 'coproantibody'. This IgA is specific and people can be immunized against certain diseases, such as poliomyelitis, by an oral vaccine which gives rise to IgA production in the gut.

Cellular immunity is also important and the gut has a large population of lymphocytes and plasma cells. The lymphocytes often occur in follicles and large collections in the small intestine are called aggregated lymph follicles or Peyer's patches. Germ-free animals have very few lymphocytes in these sites; and these are presumably a response to antigens in their food. The plasma immunoglobulins of germ-free animals are only one-third to one-tenth those of conventional animals. Flora and host combine to form the fully immunologically competent animal.

THE FAECES

At the end of about thirty-six hours the contents of the large intestine have become solid or semisolid and are known as faeces. The amount produced in twenty-four hours varies widely, according to the diet, from about 80 to 200 g, but is usually about 100 g. The reaction is usually slightly alkaline on the surface (pH 7·0 to 7·5). The colour is due to the presence of the pigment *stercobilin* derived from bile pigment (Chap. 25), but in disease the colour may be greatly changed, for example to black (melaena) in the presence of altered blood or to clay-coloured in the absence of bile. In steatorrhoea the stool is pale because the pigment is decolorized by the action of the intestinal bacteria. The faecal odour is due mainly to the presence of indole and skatole (p. 386) and to gases produced by fermentation of carbohydrate.

Since chromic oxide or sodium chromate taken by mouth is not absorbed, it can if labelled with ^{51}Cr be used to time the passage of material through the gastrointestinal tract. In normal persons the time to the initial appearance of ^{51}Cr in the faeces is 10 to 28 hours; time to maximal appearance 34 to 117 hours; time to final appearance 68 to 165 hours. The mean amount excreted in 24 hours is 22 per cent, in 48 hours 54 per cent and in 96 hours 76 per cent. Non-radioactive chromium sesquioxide is also used as a marker. By measuring the recovery of chromium in the stools, faecal losses (on toilet paper and in manipulation) can be allowed for and the accuracy of balance studies improved.

The water content of the faeces varies from 60 to 80 per cent according to the length of time they are retained in the colon. In health, and on a relatively normal diet, the proportion of the daily faecal mass attributable to undigested or unabsorbed food is small. The faeces consist chiefly of the residues of bile and other intestinal secretions, mucus, leucocytes and desquamated epithelium and enormous numbers of bacteria. The chemical composition of faeces is as follows:

1. Inorganic material. This accounts for 10 to 20 per cent of the total solid and consists mainly of calcium and phosphates.

2. Nitrogenous material containing between 1 and 2 g N per day. On a protein-free diet the faecal N is about 1 g and even in starvation it lies between 0·1 and 0·8 g per day. A large part of the nitrogen in the faeces must therefore be derived from non-dietary sources, namely digestive enzymes, desquamated cells (see p. 287) and bacteria. Normal persons given protein or protein hydrolysate labelled with ^{15}N excrete in their faeces, in the course of the following three days, nitrogenous material with a much lower ^{15}N content than that of the food. The quantitative results can only be explained by assuming that non-dietary sources such as gastric juice, pancreatic juice, succus

entericus, contribute more than half of the faecal N. Gross disease of the pancreas may be accompanied by defective secretion of trypsin and the faeces therefore often contain more than the normal amount of nitrogen.

3. Fatty material extractable by ether. In normal subjects most of this arises from sources other than dietary fat. The amount of fat in the faeces can be increased by taking a diet containing large amounts of fat but with an ordinary intake of some 50 to 100 g per day the faecal fat content is less than 5 g and some 10 to 20 per cent of the total solids consists of fatty acids (so-called 'split' fat), neutral fats, phospholipids and sterols. The neutral fat in faeces consists largely of fat derived from cellular debris and from bacteria. The sterols in faeces consist of cholesterol, coprosterol, produced by the bacterial reduction of cholesterol, derivations of bile acids and of plant sterols (phytosterols), which are not normally absorbed.

4. Undigested material. A variable proportion of the faecal solids consists of cellulose and other undigested food residues such as the seeds and skins of fruit.

The composition of the faeces depends very little on the constitution of the diet, except that when the food contains much insoluble indigestible material (roughage), such as cellulose, the bulk of the intestinal contents is greater; intestinal peristalsis is stimulated so that the faeces pass more quickly through the large intestine and less time is allowed for the absorption of water. It may be that one of the important advantages of a cellulose-containing vegetable diet is its stimulating action and people with infrequent bowel action may benefit by taking a diet which contains a certain amount of roughage. The wet weight of the faeces is quite unaffected by the drinking of large additional amounts of water. In health and in the absence of diarrhoea the energy value of the faeces is from 3 to 8 per cent of the total energy intake.

Faeces continue to be formed during starvation; though somewhat smaller in quantity their composition remains essentially unaltered. This shows clearly that the faeces are produced chiefly within the intestinal tract and are not simply unabsorbed residues of foodstuffs.

REFERENCES

ADMIRAND, W. H. & SMALL, D. M. (1968). The physiochemical basis of cholesterol gallstone formation in man. *Journal of Clinical Investigation* **47**, 1043–1052.

BECK, I. & SINCLAIR, D. (Eds.) (1971). *The Exocrine Pancreas*. London: Churchill.

BERK, J. E. (Ed.) (1968). Gastrointestinal gas. *Annals of the New York Academy of Sciences* **150**, 1–190.

BISHOP, B., GARRY, R. C., ROBERTS, T. D. M. & TODD, J. K. (1956). Control of the external sphincter of the anus in the cat. *Journal of Physiology* **134**, 229–240.

BOOTH, C. C. (1970). Enterocyte in coeliac disease. *British Medical Journal* **iii**, 725–731; and **iv**, 14–17.

BURLAND, W. L. & SAMUEL, PAMELA D. (Eds.) (1971). *Transport across the Intestine. A Glaxo Symposium*. Edinburgh: Churchill Livingstone.

CARD, W. I. & CREAMER, B. (1970). *Modern Trends in Gastroenterology*, Vol. 4. London: Butterworth.

CREUTZFELD, W., FEURLE, G. & KETTERER, H. (1970). Effect of gastrointestinal hormones on insulin and glucagon secretion. *New England Journal of Medicine* **282**, 1139–1141.

DAVENPORT, H. W. (1966). *Physiology of the Digestive Tract*, 2nd edn. Chicago: Year Book Medical Publishers.

DAWSON, A. M. (Ed.) (1971). Intestinal absorption and its derangements. Symposium Proceedings. *Supplement, Journal of Clinical Pathology*. London: Royal College of Pathologists.

DIETSCHY, J. M. (1967). Effects of bile salts on intermediate metabolism of the intestinal mucosa. *Federation Proceedings* **26**, 1589–1598.

DIETSCHY, J. M. (1968). Mechanisms for the intestinal absorption of bile acids. *Journal of Lipid Research* **9**, 297–309.

DOUGLAS, A. P. (1968). Small intestinal biopsy in diagnosis and research. *Hospital Medicine* **2**, 1400–1410.

FERGUSON, ANNE (1972). Immunological roles of the gastrointestinal tract. *Scottish Medical Journal* **17**, 111–118.

FRAZER, A. C. (1962). The fate of dietary fat in the body. Seventh Leverhulme Memorial Lecture. *Chemistry and Industry* July–Sept. 1962, 1438–1446.

GARRY, R. C. (1957). Innervation of abdominal viscera. *British Medical Bulletin* **13**, 202–206.

GRACE, W. J., WOLF, S. & WOLFF, H. G. (1951). *The Human Colon*. London: Heinemann.

GURR, M. I. & JAMES, A. T. (1971). *Lipid Biochemistry*. London: Chapman & Hall.

HARPER, A. A. (1968). Hormonal control of pancreatic secretion. *Handbook of Physiology*, Section 6, Alimentary Canal, Vol. II, pp. 969–995.

HOFMANN, A. F. (1967). Functions of bile in the alimentary canal. *American Handbook of Physiology*, Chapter 117, pp. 2507–2533.

JORPES, J. E. (1968). The isolation and chemistry of secretin and cholecystokinin. *Gastroenterology* **55**, 157–164.

Lipids in Health and Disease (1960). Proceedings of the Fifth International Congress on Nutrition, Washington.

MATTHEWS, D. M. (1968). Intestinal absorption. *Hospital Medicine* **2**, 1382–1398.

MUNRO, H. N. (Ed.) (1967). *The Role of the Gastrointestinal Tract in Protein Metabolism*. Oxford: Blackwell.

NEALE, G. (1968). Absorption and malabsorption of carbohydrates. *Hospital Medicine* **2**, 1372–1381.

READ, A. E. (Ed.) (1967). *The Liver*. London: Butterworth.

ROUILLER, C. (Ed.) (1967). *The Liver*, Vol. 1. New York: Academic Press.

SMITH, A. N. (Ed.) (1962). *Surgical Physiology of the Gastrointestinal Tract*. Edinburgh: Royal College of Physicians.

SMITH, R. & SHERLOCK, SHEILA (Eds.) (1964). *Surgery of the Gall Bladder and Bile Ducts*. London: Butterworth.

SMYTH, D. H. (Ed.) (1967). Intestinal absorption. *British Medical Bulletin* **23**, 205–296.

Symposium on absorption of nutrients from the intestine (1967). *Proceedings of the Nutrition Society* **26**, 1–72.

TAYLOR, W. (Ed.) (1965). *The Biliary System. A Symposium of the NATO Advanced Study Institute*. Oxford: Blackwell.

THOMAS, J. E. (1951). *The External Secretions of the Pancreas*. Oxford: Blackwell.

THUREBORN, E. (1962). Human hepatic bile. *Acta Chirugia Scandinavica* Suppl. 303.

TORSOLI, A. (1971). The function of biliary 'sphincters'. *Journal of the Royal College of Surgeons of Edinburgh* **16**, 270–273.

TRUELOVE, S. C. (1966). Movements of the large intestine. *Physiological Reviews* **46**, 457–512.

UNGER, R. H. (1971). Glucagon: physiology and pathology. *New England Journal of Medicine* **285**, 443–448.

WASSERMAN, R. H. (1968). Calcium transport by the intestine. *Calcium Tissue Research* **2**, 301–313.

WISEMAN, G. (1964). *Absorption from the Intestine*. New York: Academic Press.

WOLSTENHOLME, G. E. W. & CAMERON, M. P. (Eds.) (1962). *Intestinal Biopsy. Ciba Foundation Study Group*, No. 14. London: Churchill.

WORMSLEY, K. G. (1968). The source of duodenal aspirate in man. *Gut* **9**, 398–404.

WORMSLEY, K. G. (1971). Reactions to acid in the intestine in health and disease. *Gut* **12**, 67–84.

WORMSLEY, K. G. (1972). Tests of pancreatic function. *Clinics in Gastroenterology* **1**, 27–51.

18 Carbohydrate metabolism, organization and control

CARBOHYDRATE METABOLISM IN THE BIOCHEMICAL ECONOMY

For most of the world's population carbohydrate is much the most important source of energy in the diet. There are, however, exceptions to this generalization. In societies such as the Eskimos or the Masai, based on hunting rather than agriculture, the diet may contain very little carbohydrate. Among the wealthy nations of Europe and North America, where food can be selected on the basis of individual preference, carbohydrate may supply only half the energy requirement. But for the vast majority of mankind now and throughout recorded history, eating habits have been and are largely determined by the fact that cereals (which are chiefly starch) yield up to six times more energy per acre than cattle. Consequently, in poor countries, particularly in the tropics, carbohydrates provide up to 90 per cent of the energy intake.

Starch is the main carbohydrate in the diet of adults and lactose in that of unweaned infants. Both may be accompanied by varying but substantial quantities of sucrose. All three are hydrolysed to monosaccharides in digestion and are absorbed and carried by the portal vein to the liver (p. 296). A normal European or North American diet yields approximately 100 g glucose, 20 g galactose and 80 g fructose over a 24 h period. Fructose and galactose are converted to glucose or glycogen. So far as fructose is concerned there are several processes available for this conversion, and these vary from one tissue to another. In the liver the main pathway seems to be that shown in Fig. 18.1. The first step is a phosphorylation by a specific phosphofructokinase to give fructose 1-phosphate. This is a substrate for aldolase which splits it to an equimolar mixture of dihydroxyacetone phosphate and glyceralde-

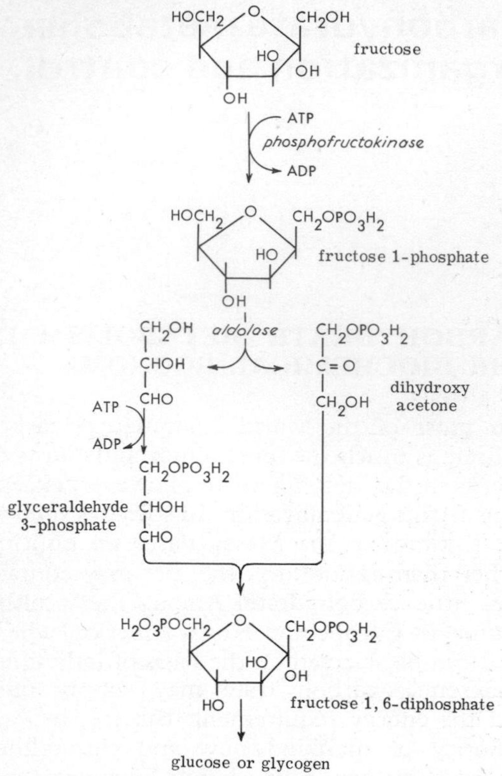

Fig. 18.1 The conversion of fructose to triose phosphates and to fructose 1,6-diphosphate. The pathway shown above is believed to operate in the liver.

hyde. The latter can be phosphorylated by a specific enzyme at the expense of ATP to give glyceraldehyde 3-phosphate. This, like dihydroxyacetone phosphate, is an intermediate in the glycolytic pathway and both can be converted by that pathway to glucose and glycogen. In the liver, galactose is also phosphorylated in position 1 by a specific galactokinase. The galactose 1-phosphate so formed participates in an exchange reaction with UDP glucose under the influence of the enzyme *phosphogalactose uridyl transferase* to give UDP-galactose and glucose 1-phosphate (Fig. 18.2). The glucose 1-phosphate produced in this reaction can be converted to either glucose or glycogen. In the meantime the UDP-galactose undergoes epimerization to UDP-glucose which can then combine with a further molecule of galactose 1-phosphate. Children with a congenital lack of phosphogalactose uridyl transferase cannot metabolize

galactose and so suffer from the hereditary disease galactosaemia in which galactose and galactose 1-phosphate accumulate in the blood and are excreted in the urine.

The supply of carbohydrate, and other nutrients, is not continuous but is obtained at three or four meals a day. The varying need for energy bears no necessary relation to the timing of these meals. There is therefore a physiological requirement for a mechanism by which carbohydrate provided by digestion in the gut can be stored until it is needed. This is done in two ways. A limited amount of carbohydrate can be stored in the form of glycogen either in the liver or in the muscles. The processes by which glycogen is formed and broken down are described in detail in Chapter 11. Liver glycogen can be broken down again to free glucose which can be made available to all the tissues by way of the blood stream. Muscle glycogen, by contrast, is available only to the cell in which it is located. Alternatively, circulating glucose can be taken up by the adipose tissue and converted to triglycerides. The fuel reserves contained in the adipose tissue are very much larger than those stored in the form of glycogen in liver and muscle. An average man might have the equivalent of 800 kJ in glycogen in his liver and another 1700 kJ in his muscles, whereas the triglycerides of his adipose tissue might well amount to 400 000 kJ.

The manner in which glucose and glycogen are metabolized varies from tissue to tissue but the major metabolic patterns are as follows:

1. Complete oxidation to CO_2 and water by way of the glycolytic pathway and the citric acid cycle. This yields approximately 36 molecules of ATP per molecule of glucose or 6 molecules of ATP per molecule of oxygen. This is the main pathway used in brain and peripheral nerve.

2. Anaerobic conversion to lactate, for example in tissues which lack either mitochondria or an adequate oxygen supply. This yields only 2 molecules of ATP (3 if the glucose is in the form of glycogen) but the lactate can be oxidized to provide energy in other tissues.

3. Complete oxidation to carbon dioxide and water by the pentose phosphate pathway. This is sufficient to reduce 12 molecules of NADP per molecule of glucose. It is important in adipose tissue, for example, where

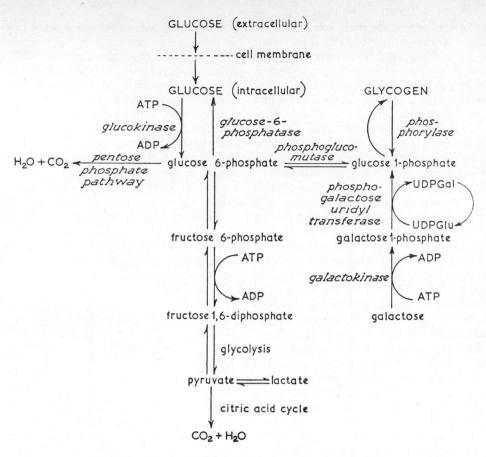

Fig. 18.2 The interconversion of glucose, glycogen and galactose.

large quantities of $NADPH_2$ are required for the synthesis of fatty acids.

UTILIZATION OF CARBOHYDRATES BY THE TISSUES

All cells and tissues require energy and all or almost all utilize carbohydrate in some degree. A few organs are of outstanding quantitative importance: brain, skeletal muscle, heart muscle, liver, kidney and the red and white blood cells. Each uses carbohydrate in a different and characteristic manner. The simplest pattern is that found in nervous tissue. The brain has a constant requirement for energy at the rate of approximately 1700 kJ per day, or about 25 per cent of the energy requirement of a normal resting adult. It is met almost entirely by the conversion to carbon dioxide and water of approximately 120 g glucose per day (=5 g/h or 83 mg/min). This requirement is extraordinarily constant. It is

unaffected by intellectual exertion or by sleep. It is maintained even when the level of glucose available in the blood is moderately depressed. On the other hand interruption of the supply of oxygen or of glucose, for even a brief period, results in irreversible brain damage. Though the brain does contain some glycogen the amount (about 1 g) is small in comparison to the requirement.

The erythrocytes and leucocytes provide an interesting comparison with nervous tissue. Although their total mass (2·5 kg) far exceeds that of the brain, their energy requirements, though constant like those of brain, are quite small. They are met, since erythrocytes have no mitochondria, by the conversion to lactate of about 36 g glucose per day.

Skeletal muscle, unlike nervous tissue and erythrocytes, has an extraordinarily variable requirement for energy. In the resting subject it is metabolically fairly inert. Its circulation is sluggish. Its metabolic rate is so slow that it is difficult to be quite certain from what sources its modest energy requirements are met. The available evidence suggests that it depends largely on the oxidation of fat rather than carbohydrate. Though its total mass in a 75 kg man may well amount to about 25 or 30 kg, its glucose requirement in the resting state is probably less than 30 g per day. The mild exertion involved in standing or sitting rather than lying, or in sedentary activities or in walking, does not notably change this pattern; the energy output of muscles as a whole increases but it is still satisfied mainly by the oxidation of fatty acids and their derivatives. Running or other vigorous exercise, brings about a spectacular change. The mechanical output of the muscles can increase by as much as 64-fold. This is accompanied by a greatly increased blood supply which provides the numerous muscle mitochondria with a correspondingly increased supply of oxygen. At the same time there is a dramatic shift in the metabolic pattern. The enormously increased energy output is met not by accelerating the existing process of fatty acid oxidation but by switching to the oxidation of carbohydrate. Some of this carbohydrate is derived from the blood. Muscular exertion increases the passage of glucose from the blood into muscle cells: indeed in the days before insulin vigorous muscular exercise was advocated as a palliative measure for patients with diabetes mellitus. The major reserve of carbohydrate in the body is, however, the muscles' own store of glycogen. The extent of this store is difficult to measure since different muscles have different concentrations of glycogen but it appears to support 60 to 90 min of very hard physical exertion, for example, long distance running. When this reserve is used up the blood sugar concentration declines steeply. This decline is accompanied by a sense of fatigue, although the muscles continue to maintain a high energy output at the expense once again of fatty acids rather than glucose. Marked relief of fatigue can indeed be obtained by administration of a quite small amount of glucose which has the effect of raising the blood sugar; the muscles continue to depend on their supply of fatty acids. Muscle thus possesses the capacity for prolonged exertion whether it is supplied with carbohydrate or with fat. The yield of ATP

per molecule of oxygen is slightly higher for carbohydrate than for fat. An important advantage of carbohydrate as a fuel is that anaerobic glycolysis can be an important source of energy especially in brief periods of very vigorous exertion when the supply of oxygen is inadequate to meet the demand made on the muscles. The upper limit of sustained exertion is set by the ability of the muscles to maintain the oxidation of fat or carbohydrate and by the ability of the heart and lungs to maintain the supply of oxygen. These limits can be exceeded for a short time by extensive glycolytic breakdown of the muscles' own stores of glycogen. The extent of this extraordinary activity is limited because the production of ATP by glycolysis is accompanied by the liberation of large amounts of lactic acid (Fig. 18.3).

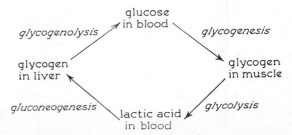

Fig. 18.3 The Cori cycle. In sudden vigorous muscular exercise muscle glycogen is broken down anaerobically to lactic acid which diffuses out into the blood. The liver can convert this back to glycogen by reversal of the glycolytic pathway. In turn liver glycogen can break down to free glucose which is carried by the blood to the muscles and can be used to replenish muscle glycogen.

While a highly trained athlete can tolerate levels of up to 200 mg lactate/100 ml blood, equivalent to the conversion of 100 g muscle glycogen to lactate, most individuals cannot tolerate more than half this amount. Brief intense exertion of this sort is presumably a function mainly of the white muscle fibres which have relatively few mitochondria whereas prolonged exertion is presumably a function mainly of the red muscle fibres which, as will be seen in Chapter 40, have a very large number of mitochondria and considerable reserves of myoglobin.

Unlike all the tissues considered so far, the liver is important in carbohydrate metabolism chiefly because of functions which it performs for other organs. First among these is the storage in the form of glycogen of carbohydrate absorbed in the intestine after meals. The glycogen reserve in the liver of a well-nourished man amounts to about 90 g or half a day's supply of carbohydrate for the entire organism. The glycogen can be broken down at need and liberated in the blood in the form of free glucose which can be used by all the tissues, particularly the brain. If this store of glycogen were not available hypoglycaemia would probably develop in the intervals between meals, particularly if they were prolonged, with serious impairment of brain function. It has long been known that after total hepatectomy, the dog dies of hypoglycaemia. In a rare genetic deficiency (von Gierke's disease) the liver lacks the enzyme glucose-6-phosphatase so that although glycogen is formed in the liver it cannot be converted to glucose to be made available to other tissues. Children suffering from this condition require frequent feeding of carbohydrate throughout the day and night if they are not to die of hypoglycaemia. Liver glycogen can therefore be regarded as a fuel tank for the brain.

Another important role of the liver is in gluconeogenesis. We have seen that the formed elements of the blood convert approximately 36 g glucose a day to lactate and that in vigorous exertion the muscles can produce 50 or even 100 g lactate in a few minutes. Some of this lactate is taken up by other tissues such as the heart, and perhaps by resting skeletal muscles, and oxidized to carbon dioxide and water; about half, or perhaps more, is converted to glucose or glycogen in the liver by reversal of the glycolytic pathway in what is called the Cori cycle (Fig. 18.3). The rate of this process depends on the lactate concentration in the blood. An athlete at maximum exertion can in $2\frac{1}{2}$ min perform mechanical work equivalent to 300 kJ with the liberation of about 100 g lactate, most of which is removed from the blood in about an hour. The liver is capable also of forming glucose or glycogen from amino acids (Fig. 18.4) and, as will be seen below, this glucose is important for the brain in brief periods of fasting. The liver itself does not appear to make extensive use of carbohydrate as an energy source. In a normal well-fed individual the liver appears to subsist mainly by the oxidation of amino acids and in

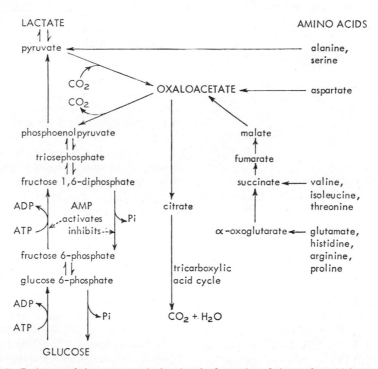

Fig. 18.4 Pathways of gluconeogenesis showing the formation of glucose from (a) lactate and (b) the glucogenic amino acids. In both pathways oxaloacetate is a key intermediate.

starvation, when the amino acids are required for glucose formation, it can maintain itself by the oxidation of fatty acids.

The kidneys, like the liver, are capable of extensive formation of glucose from amino acids. This capacity is, as we shall see, particularly important in prolonged starvation.

We can now consider the carbohydrate metabolism in the whole organism. It is convenient to consider first a fasting subject at rest. The brain and other nervous tissue in such a subject requires about 144 g glucose per day, virtually all of which is completely oxidized to carbon dioxide and water. A further 36 g glucose is required by the red and white blood cells and platelets, which convert it to lactate. The lactate is to a large extent converted back to glucose in the liver so that the blood cells and platelets do not constitute a drain on the body's carbohydrate resources. The energy requirements of all other tissues (heart, muscle, kidney and the liver itself) are met almost exclusively by the metabolism of fatty acids liberated from adipose tissue. It is therefore only brain and nervous tissue which deplete the body of glucose. Their requirements appear to be met, when the liver's reserve of glycogen is exhausted, from two sources:

1. The breakdown of triglycerides in the adipose tissue, which provides the fatty acids needed by muscle, liver and kidney, incidentally provides the equivalent of 16 g carbohydrate per day in the form of glycerol.

2. The slow breakdown of muscle protein to amino acids, most of which are convertible to glucose.

If starvation is sufficiently prolonged, and initially obese individuals have been known to fast for several months without ill-effect, the brain and nervous tissue adapt their metabolic pathways to use increasing amounts of fatty acid metabolites; thus the body's carbohydrate resources are spared.

A subject fasting and at rest is in a sense in a steady state. Though his muscle protein and his adipose tissue are being depleted to maintain his metabolism, these processes continue for a long time at a steady rate. Eating a meal on the other hand produces a transient sequence of biochemical events. If we consider a well nourished 75 kg man in the 6 h period after a really substantial dinner made up of, say, 200 g carbohydrate, 75 g protein and 75 g fat we could expect to find that the carbohydrate had been disposed of as follows.

1. The fructose and galactose would all have been converted to glucose or glycogen.

2. The nervous system and blood cells and platelets would have metabolized a proportionate quantity of their constant daily requirements of carbohydrate. The quantity involved would be about 42 g.

3. As carbohydrate began to be absorbed from the gut, the liver would stop releasing glucose and would instead start taking up carbohydrate from the portal blood and converting it to glycogen. About 20 g glucose would be added to the liver's reserve of glycogen, though more would be laid down in this form if the liver glycogen reserve were depleted as a result of either fasting or exertion. The amount of glucose laid down as glycogen in muscle would similarly depend on the preexisting muscle glycogen reserve. It would also be increased on a high carbohydrate diet and diminished on a diet high in protein and fat.

4. The remaining glucose would largely be taken up by the adipose tissue and converted to fat.

Vigorous exertion alters carbohydrate metabolism even more profoundly. The circulation in resting skeletal muscle is sluggish. Although it can increase enormously in muscular exercise, this adaptation is not instantaneous, and at least for a brief period the muscles' output of energy exceeds that which can be derived from their oxygen supply. The deficit is met by glycolysis of the muscles' own reserve of glycogen and within perhaps ten minutes of moderate exercise the blood lactate level can rise from a resting level of 10 mg/100 ml to 80 to 90 mg/100 ml. Since lactate can probably diffuse through half the body mass, this rise represents liberation of perhaps $(80/1000) \times (40/0 \cdot 1) = 8$ g lactate from the muscles, equivalent to about 2 per cent of the entire muscular reserve of glycogen. If the exertion is not too severe (for example walking uphill) the lactate level of the blood reaches a maximum within about 10 min and then starts to decline as it is removed from the blood, by the liver, and to a smaller extent by the heart and kidneys, and by those muscles not actively employed. This amounts to 50 to 100 mg/min. About half the lactate of the blood entering the liver is removed in transit and presumably converted to glucose or glycogen. Though the blood flow through the liver may fall to $\frac{1}{3}$ or $\frac{1}{4}$ of its normal value more oxygen is extracted from the blood as it passes through and the liver's oxygen supply remains constant. In vigorous exercise the liver may produce as much as 0·5 or 1·0 g glucose a minute. About half the lactate initially liberated at the start of moderate exercise has been removed from the blood half an hour later and about 50 per cent of this removal is accounted for by the liver.

If the exertion is sufficiently vigorous the liver and other tissues may be unable to remove lactate from the blood as rapidly as the muscles produce it, and lactate accumulates until it reaches a level, 100 to 200 mg/100 ml, at which the subject can no longer tolerate exertion. Lactate production is, as we have seen, a very inefficient means of obtaining useful energy from glucose but it allows the muscles a certain independence of their blood supply. It enables them to develop full power almost instantaneously and to exceed, for short spells, the limits normally placed on their output of power. In summary, vigorous exertion lasting less than about 10 seconds is supported by the breakdown of the muscles' existing ATP and creatine phosphate. If it lasts for about 2 min it is supported mainly by breakdown of muscle glycogen to lactate. Thereafter it must be maintained by oxidation in the first instance of glycogen and, when this is exhausted, of fatty acids.

Sustained muscular exertion, in so far as it requires an increased blood supply to the skeletal muscles, requires a substantial increase in the work of the heart. The energy necessary for this is provided by the oxidation chiefly of fatty acids but also to some extent of glucose and lactate. Indeed the heart can oxidize about 10 g of the 42 g of lactate produced daily by blood cells and platelets. In vigorous exercise the output may increase five-fold. This is accompanied by a four-fold increase in oxygen uptake. Fat is still the main contributor to the increased energy requirement but the uptake of lactate is also increased.

CONTROL MECHANISMS—AN INTRODUCTION

We can distinguish several levels of control of carbohydrate metabolism. At the level of the individual cell, metabolic pathways are largely controlled by the allosteric mechanisms described in Chapter 8. Thus the rate of the overall glycolytic process is determined principally by the activity of the enzyme phosphofructokinase. This activity is enhanced by ADP and AMP and inhibited by ATP. Similarly the generation of ATP in mitochondria is largely determined by the availability of ADP. Thus when, for example, in muscle there is a sudden increase in energy output and therefore increased ATP breakdown the consequent increase in ADP in itself stimulates the ATP generating reaction systems. High levels of ATP have the opposite effect, that is, they tend to inhibit the breakdown of carbohydrate, and by activating the enzyme fructose diphosphatase in the liver they enhance the reconversion of lactate to glucose.

Storage of glucose as glycogen and the subsequent breakdown of glycogen are under more elaborate control.

Glycogen is broken down by *phosphorylase* and other enzymes to glucose 1-phosphate, which is converted by *phosphoglucomutase* to glucose 6-phosphate which can then be metabolized by either the glycolytic or pentose phosphate pathway. In liver, but not in muscle, glucose 6-phosphate can be converted to free glucose by the enzyme glucose 6-phosphatase. Muscle phosphorylase exists in two forms *a* and *b* which appear to be respectively a dimer and a tetramer of the same subunit. The *b* form requires AMP as an activator and is inhibited by ATP and glucose 6-phosphate. Its activity is therefore increased when the demand for ATP or glucose 6-phosphate is increased. Phosphorylase *a*, on the other hand, is *not* dependent on AMP for activity and its function is therefore independent of the demand for either glucose 6-phosphate or energy. It can be formed from phosphorylase *b* by the action of an enzyme *phosphorylase kinase* which catalyses the transfer of a phosphoryl group from ATP to a serine residue on each of the phosphorylase subunits. This conversion can be reversed by hydrolysis.

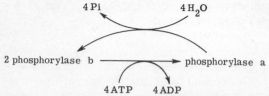

$$2 \text{ phosphorylase } b \longrightarrow \text{phosphorylase } a$$

Conversion of *b* to *a* is accelerated, apparently indirectly, by the presence of the cyclic nucleotide cAMP. The direct effect of cAMP is to activate an enzyme *kinase kinase* (Fig. 18.5). A minute quantity of cAMP can by this 'cascade system' rapidly transform a large quantity of muscle phosphorylase from the *b* to the *a* form and therefore quickly free the

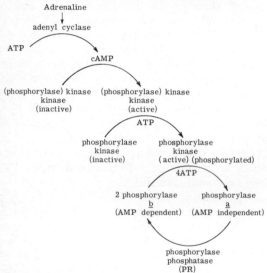

Fig. 18.5 Mechanism of the activation of phosphorylase by cAMP cascade.

rate of glycogen breakdown from its normal dependence on AMP (Fig. 18.5). In other words cAMP allows glycogen to be broken down before it is actually required. cAMP is formed from ATP by the enzyme adenyl cyclase which is activated by adrenaline.

The phosphorylase in liver also exists in two forms *a* and *b* which have properties analogous to the muscle enzymes and are interconvertible by similar mechanisms. In this case however the *b* form is a dimer not a tetramer as in muscle.

The synthesis of glycogen is active when glycogen breakdown is inhibited and less active when glycogen breakdown is stimulated. The details of the control mechanism are not yet clear but glycogen synthetase appears to

exist in a phosphorylated 'D' form which is *dependent* on the presence of glucose 6-phosphate for activation and in a non-phosphorylated 'I' form which is *independent* of glucose 6-phosphate. Conversion of the independent 'I' to the dependent 'D' form is brought about by a kinase (activated by cAMP) whose presence inhibits glycogen synthesis as well as favours its breakdown.

At this level metabolic pathways are largely self-regulating. They are activated by high concentrations of their starting materials or low concentrations of their products and inactivated under the reverse conditions. A second level of control which regulates the activity of whole tissues can be illustrated by the *glucose tolerance test* (see also p. 327), in which 50 g glucose is given orally to a fasting subject. This amount of glucose, about one fifth of the normal dietary carbohydrate intake, is normally absorbed within half an hour. If it were evenly distributed throughout half the body mass, as lactate appears to be, the blood glucose concentration would increase by about 140 mg/100 ml, and return to the fasting level only after 4 hours or so. In fact the increase in blood sugar level is normally much less (of the order of 50 mg per 100 ml over the fasting level of about 80 mg per 100 ml) and the return to the fasting level is normally complete in about 90 min (Fig. 18.6). In diabetes mellitus

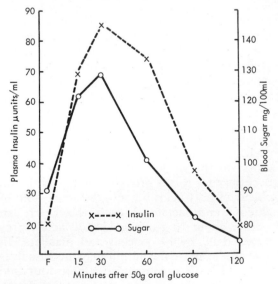

Fig. 18.6 Blood sugar and plasma insulin response in thirty-four control subjects. (K. D. Buchanan & Margaret T. McKiddie (1967). *Diabetologia* **3**, 461.)

(see below), the increase in blood sugar is much greater. The explanation of these differences is that although glucose is continually taken up by the brain and can readily diffuse into the liver, it cannot by itself enter muscle cells. In a healthy subject, a small increase in blood glucose level causes the β-cell of the islets of Langerhans in the pancreas to liberate the hormone, *insulin*, which facilitates the entry of glucose into muscle cells and adipose tissue and also appears to promote its phosphorylation in the liver. In diabetes mellitus insulin formation is either diminished or entirely lost and consequently glucose absorbed from the gut is only slowly removed from the circulation.

The secretion of insulin by the β-cells can therefore be regarded as a mechanism by which an influx of carbohydrate activates the mechanisms by which it is stored as glycogen in either liver or muscle or converted to fat in the adipose tissue. Consequently the blood sugar level does not normally rise above say 150 mg per 100 ml. If the blood sugar level falls below 90 mg/100 ml a second mechanism comes into play.

Glucagon, formed in the α-cells of the islets of Langerhans in the pancreas, is secreted when the blood glucose falls. Glucagon appears to act on carbohydrate metabolism chiefly by increasing formation of cAMP in the liver. This, as we have seen, increases breakdown of glycogen to glucose and inhibits the reverse process.

Insulin and glucagon together regulate the storage of incoming glucose and the release of glucose from liver glycogen to meet the demands of brain and other tissues. In an emergency, particularly one requiring violent muscular exertion, there seems to be a need for some means of overriding this normal homoeostatic control. This is provided by *adrenaline*, released in such situations from the adrenal medulla. Like glucagon, adrenaline stimulates cAMP formation, but unlike glucagon it acts on muscle as well as liver, so that it increases breakdown of muscle glycogen as well as liver glycogen.

HORMONAL CONTROL OF CARBOHYDRATE HOMOEOSTASIS

The hormonal control of carbohydrate metabolism is of such importance in physiology and medicine as to warrant a considerable amplification of the brief account just given.

1. Insulin. The structure of the pancreas has been described on p. 263. It contains about two million interalveolar cell islets (islands of Langerhans) which together weigh about 1 g. Each of these small, highly vascular bodies (Plate 18.1) contains many β cells which secrete insulin, a smaller number of α cells which secrete glucagon, and a few D cells which secrete gastrin.

A normal human pancreas contains about 8 mg insulin and secretes about 2 mg or 50 units per day. Insulin is a small protein (mol. wt. 6000) containing an A chain of 21 amino acid residues, and a B chain of 30 residues. The chains are linked by disulphide bonds (Fig. 3.6). Insulin is a derivative of a much larger single polypeptide chain of 84 amino acids known as pro-insulin (Fig. 18.7) which was first found in study of islet cell tumour tissue. A trypsin-like enzyme in the pancreas splits off the connecting chain of 33 amino acids to give the active hormone. Pro-insulin has a low biological activity but it reacts like insulin in

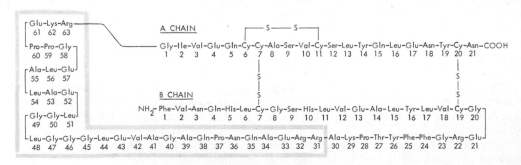

Fig. 18.7 Amino acid sequence in the proinsulin molecule. During conversion to insulin the shaded portion is discarded (cf. Fig. 3.6).

radioimmunoassays. Only traces are found in plasma. Although the amino acid sequences differ in the insulins of different species, fortunately, pig and ox insulins are very similar to the human hormone, and although patients treated with these insulins develop measurable amounts of antibodies, they show little or no adverse antigenic reactions. Insulin has been synthesized but it cannot yet be produced in the large quantities needed to replace preparations from animals. Isolated natural insulins may contain zinc, and zinc is often deliberately added because it increases the insolubility of the hormone, probably by causing aggregation of the molecules. Insulin is rapidly metabolized in the liver and has a half-life of about 5 min. The S—S bonds are cleaved by an enzyme known as glutathione–insulin transhydrogenase (GIT) found in the liver and probably also in renal tubules. It is possible, but by no means proved, that there is genetic control over pro-insulin synthesis, conversion of pro-insulin to insulin and the degradation of insulin, all of which may have a bearing on the aetiology of diabetes mellitus.

Insulin is normally stored in a relatively insoluble granular form, possibly as a complex with zinc and proteins, in the islet cells. The granules are thought to migrate to the cell surface and fuse with the plasma membrane at the time of insulin release.

Insulin is destroyed by enzymes of the alimentary tract, and when it is given therapeutically it must therefore be administered by injection, usually subcutaneously. It diffuses freely from the site of injection and its duration of action is only 6 to 12 hours, so two or three injections are needed per day. This practical drawback has been overcome to some extent by combining the insulin, which is an acidic protein, with the basic fish sperm protein protamine (p. 68) and by adding zinc. Alternatively zinc can be added under carefully controlled conditions to form a slowly soluble insulin–zinc complex. In this way the action of insulin may be sufficiently delayed so that a single injection may be able to keep the blood sugar within satisfactory limits throughout the entire day.

Plasma insulin concentrations have been measured by bio-assays based on the fact that insulin increases the uptake of glucose by isolated rat tissues, for example diaphragm or epididymal fat pad. Values of 30 to 100 micro-units per ml (25 units = 1 mg) have been reported for normal fasting individuals. Shortly after a carbohydrate meal these values may be increased tenfold. Bio-assays measure the total 'insulin-like activity' rather than insulin alone and thus give falsely high values. A more satisfactory technique is that of *immuno-assay* which is specific for insulin and is readily applicable to the measurement of large batches of samples. To fluid containing either an unknown concentration or a standard amount of insulin, a known amount of insulin labelled with radioactive iodine (^{125}I or ^{131}I) is added and the mixture is treated with an anti-insulin serum prepared by injecting guinea pigs with human or pig insulin, purified by repeated recrystallization. The insulin–antibody complex so formed is soluble and various techniques have been devised to separate it from the surplus free radioactive insulin. These include chromato-electrophoresis and precipitation by a second antibody, an anti-guinea-pig γ-globulin. The radioactivity of the insulin–antibody complex is then compared with standards to determine the concentration of insulin in the original sample.

Under the best conditions, less than 0·1 micro-units of insulin may be measured by immunoassay. Immunoassay methods, however, have the disadvantage that they measure immunological rather than biological activity and the two are not necessarily the same. Thus they may measure circulating protein fragments and also pro-insulin which have little or no biological activity.

With the radio-immunoassay method normal fasting subjects have a fasting plasma insulin level of about 20 micro-units per ml. After oral administration of 50 g glucose, values of 85, 74, 37 and 19 micro-units have been found at 30, 60, 90 and 120 min (Fig. 18.6), the shape of the plasma insulin response closely resembling the blood sugar curve. The intravenous administration of glucose, though producing higher circulating blood glucose levels, has a less potent effect on insulin secretion. The greater effect of oral glucose is thought to be due to a 'local hormone' which is released immediately on entry of glucose into the gut, and stimulates insulin secretion. This 'local hormone' is not yet identified but may be allied to secretin, pancreozymin or enteroglucagon, all of which stimulate insulin release.

It seems likely that the secretion of insulin in response to rising blood sugar concentration takes place in two phases: an initial rapid rising phase, reaching a maximum in 1 or 2 min, then a fall, followed by a slow rise to a higher level, which is maintained for longer periods. *In vitro* experiments show that the second rise may be abolished by inhibition of protein synthesis by puromycin. This suggests that during this time new hormone is being synthesized. The insulin response to glucose is enhanced when growth hormone is administered or when it is naturally present in excess as in acromegaly (Chap. 52). It has been shown *in vitro* that the β-cell of the pancreatic islets responds directly to increasing concentrations of glucose to which it is fully permeable. High glucose concentrations appear to increase the concentration of ATP which is probably converted by adenyl cyclase to cyclic $3':5'$-AMP. The hormones, glucagon, corticotrophin (ACTH) and growth hormone, and β-adrenergic stimuli (Chap. 41) probably stimulate insulin secretion by their action on the adenyl cyclase. Cyclic AMP is normally destroyed by conversion to $5'$-AMP by a phosphodiesterase. This may be inhibited, for example, by caffeine. If the breakdown of cAMP is inhibited in this way, the glucose-induced secretion of insulin is increased.

Some species differences in the stimulation of insulin secretion are noteworthy. In dogs, though not in man, acetoacetic acid and β-hydroxybutyric acid (acetone bodies, p. 338). are powerful stimulants. Their action may serve to limit ketosis in animals which have a low carbohydrate diet. Short chain fatty acids such as butyric acid stimulate insulin secretion in ruminants, and may be even more effective than glucose. In man, oral administration or intravenous infusion of amino acids also increases insulin secretion, and the simultaneous ingestion of starch and protein is disproportionately much more effective in raising plasma insulin than is either nutrient alone.

Adrenaline inhibits the first stage of the secretion of insulin stimulated by rising blood glucose levels, possibly by decreasing cAMP formation. Since the effect of orally administered glucose is unaffected, it appears that the 'local hormone' stimulatory action referred to above, overrides the inhibitory effect of the adrenaline. Pancreatic glucagon is also a powerful stimulant of insulin secretion, acting no doubt through cAMP. Its action is not affected by adrenaline and it may thus be of importance only when the more powerful stimulus of rising blood glucose is inhibited by adrenaline.

Insulin secretion may be stimulated by sulphonylureas such as tolbutamide which are extensively used as oral agents in the treatment of mild diabetes in older obese patients. They are of little or no value in young diabetics who are deficient in insulin or show frank β-cell failure.

Experiments with insulin labelled with ^{131}I have shown that insulin is to some extent bound to the globulin fraction of the plasma protein and also that it is taken up from the blood mainly by the kidney, liver and muscle tissue in which it is concentrated in the microsome material of the cytoplasm. Insulin is destroyed in the tissues by a heat labile proteolytic enzyme, which is particularly abundant in liver and kidney. This enzyme, insulinase, or glutathione insulin transhydrogenase, cleaves insulin into its A and B chains.

When soluble insulin is given by injection to a healthy man the blood-sugar falls rapidly (Fig. 18.8). If it should fall below about 40 mg per 100 ml there are symptoms and signs of *hypoglycaemia*. The subject becomes apprehensive, irritable and tremulous; he finds it difficult to concentrate and often has a headache; he feels hungry, sweats and may become alternately hot and cold. Since the metabolism of the brain is, as we have seen, exclusively dependent on a supply of sugar, hypoglycaemia produces confusion, convulsions, loss of consciousness and death. The changes in the electroencephalogram are described in Chapter 48.

Hypoglycaemia can be quickly remedied by giving glucose or cane sugar by mouth or by injecting glucose intravenously or, less certainly, by injecting adrenaline subcutaneously. Adrenaline is effective only if there is sufficient glycogen in the body from which glucose may be produced. This condition is, however, not always fulfilled. Many of the symptoms of hypoglycaemia are due to the adrenaline liberated from the adrenal glands in response to the low blood-sugar level rather than to the latter itself. In the treatment of schizophrenia

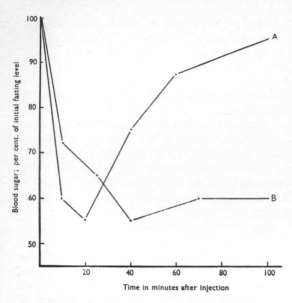

Fig. 18.8 The alteration in concentration of blood sugar after *intravenous* injection of insulin (0·1 unit/kg) in two patients. A had no disturbance of carbohydrate metabolism; B had a disorder of the adenohypophysis (Simmonds' disease) as the result of which gluconeogenesis was impaired and the blood sugar remained low.

insulin may be injected in large enough amounts to produce prolonged coma but how this produces a beneficial effect is not clear.

Very occasionally an actively secreting tumour (adenoma) of the islet tissue occurs or the cells of an otherwise normal gland may become overactive. The excessive secretion of insulin (hyperinsulinism) produces hypoglycaemia with the symptoms already described.

The lowering of blood sugar by insulin was for long regarded as its fundamental action. Insulin was supposed to facilitate the passage of glucose through cell membranes but this theory does not explain all the actions of the hormone. The liver cell is freely permeable to glucose, yet insulin influences its ability to take up or secrete glucose by inhibiting or stimulating various enzymes involved in carbohydrate metabolism in the liver cell. Several metabolic activities of insulin apparently do not depend on glucose metabolism. These include the suppression of the release of fatty acids from adipose tissues and augmentation of protein synthesis. It is difficult to find a single hypothesis to explain all these diverse actions. Some effects may be explained by reduction of the concentration of cAMP in insulin-sensitive

tissues such as the liver. Cyclic AMP is required for the phosphorylation and consequent activation of glycogen phosphorylase. Thus, if the concentration of cAMP is decreased, glycogenolysis is decreased and the amount of glycogen in the cell increases. Adrenaline and glucagon raise the concentration of cyclic AMP in tissues. Insulin antagonizes this action, perhaps by interfering with the stimulation of adenyl cyclase by adrenaline and glucagon, or by activating the phosphodiesterase which destroys cAMP. The former theory is supported by the observation that insulin linked to polymers can lower blood sugar although in this form it cannot enter the cell. The adenyl cyclase is located in the cell membrane and substances which interfere with its activity may be able to do so without actually entering the cell.

Insulin antagonists. Changes in the concentration of plasma insulin are closely related to changes in concentrations of certain other hormones (see Fig. 18.9), and the action of insulin is antagonized by several hormones including cortisol, adrenaline and, in some circumstances, thyroxine. An explanation at the molecular level cannot yet be given, but in general these insulin antagonists favour catabolic processes while insulin increases anabolism. Growth hormone, the most important hormonal antagonist of insulin, is diabetogenic in adults, causing impaired glucose tolerance (p. 327) and lack of response to insulin. Placental lactogen, found in plasma during human pregnancy, acts immunologically like growth hormone and is also an insulin antagonist. Its function might reasonably be thought to prevent the protein sparing action of insulin. A decreased response to insulin and the need for increased protein are characteristics of normal pregnancy.

Adrenalectomy (removal of the adrenal glands) relieves the symptoms of experimental insulin lack (diabetes mellitus) in animals, and the administration of cortisol exacerbates diabetes. Administration of cortisol to human subjects decreases glucose tolerance (p. 327) despite the fact that it also stimulates the secretion of insulin. Thus, cortisol appears to antagonize the action of insulin. It may do so by inhibiting the phosphorylation of glucose and thus decreasing its utilization in muscle and adipose tissue. Cortisol also influences

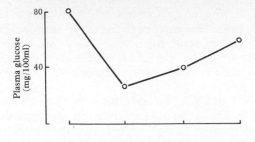

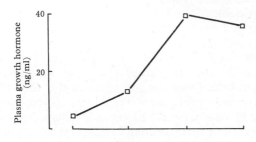

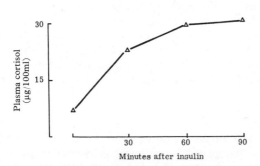

Fig. 18.9 Typical responses of blood glucose, plasma growth hormone and cortisol after intravenous injection of insulin in a normal subject. (M. Friedman & F. C. Greenwood (1968). In *Memoirs of the Society of Endocrinology* No. 17, p. 256.)

carbohydrate metabolism in many ways not directly related to insulin (p. 324).

The anti-insulin effect of the thyroid hormones is only evident when they are in excess.

Adrenaline blocks insulin release and at the same time promotes liver glycogenolysis, thus further antagonizing the action of insulin.

2. Glucagon. Glucagon was discovered as an impurity in insulin preparations. It is a polypeptide of 29 amino acid residues (Fig. 17.6), and has been synthesized. In response to

hypoglycaemia, the α-cells of the pancreatic islets secrete glucagon which produces rapid liver glycogenolysis. Glucagon may thus be regarded as part of a first line defence against hypoglycaemia. An intravenous injection in man of 1 mg of purified glucagon preparation doubles the blood glucose concentration in a few minutes. It returns to normal values in about 90 min. The rise may be higher and more prolonged in diabetic patients. As might be expected, both the rise and duration of effect are less pronounced in circumstances such as starvation in which liver glycogen stores are low. These observations all support the view that glucagon opposes hypoglycaemia. It is, however, notable that, in man, administration of glucose by mouth causes a small increase in enteroglucagon secretion (p. 273) but is unlikely to have a hyperglycaemic effect. Glucagon probably acts by enhancing the production of the ubiquitous cAMP. This in turn stimulates liver phosphorylase which catalyses the first step in glycogenolysis. Glucagon, acting in conjunction with cAMP, causes a large increase in the phosphorylation of certain histones in the liver cell nucleus. This reaction inhibits the repressive effect normally exerted by histones on DNA (Chap. 21) and allows the initiation of a sequence of events leading to the synthesis of new enzyme protein involved in gluconeogenesis. Such a sequence of events might well be slower and possibly secondary to the more rapid stimulation of glucose production by glycogenolysis. Another, and better established effect of glucagon is its stimulation of insulin production. The situation is somewhat complicated by the fact that the gut produces a glucagon, enteroglucagon, which by stimulating insulin production, prepares the tissues for an influx of glucose from the intestines. Enteroglucagon and pancreatic glucagon cross react in the radioimmunoassay; thus it is difficult to interpret the results of hormone measurements in blood plasma.

3. Adrenaline. Adrenaline, like glucagon, stimulates the release of glucose from storage by activation of glycogenolytic enzymes, and it might be regarded as another hormone in the first line of defence against hypoglycaemia. Gluconeogenesis is also stimulated by the influence of adrenaline on the enzymes involved and on the availability of substrate. Insulin secretion is inhibited. Profound hypoglycaemia markedly increases adrenalin secretion. Some doubts have been expressed about the physiological significance of these actions of adrenaline in the control of blood sugar levels, in view of the normally low concentrations of adrenaline in blood. Nevertheless, it has been suggested that secretion of adrenaline from adrenergic nerve endings to neighbouring receptors (Chap. 41) may provide enough hormone to play a significant role in blood sugar homeostasis.

4. Growth hormone. Observations by Houssay in the Argentine, in 1924 and by F. G. Young in London, led to the discovery that growth hormone is the substance in pituitary extracts which, on prolonged administration to dogs, causes permanent diabetes with degeneration of islet cells. Human growth hormone (Chap. 52) injected into patients with diabetes mellitus makes their condition worse. Indeed, a few such patients may be found to have raised plasma concentrations of growth hormone (Chap. 52). An overproduction of growth hormone may thus be a factor in at least some diabetics. Dogs, made diabetic by a long course of injections of growth hormone, require larger amounts of insulin to restore their blood sugar to normal levels than animals made diabetic by pancreatectomy. The reason for this is not known. Young believes that growth hormone promotes insulin secretion and at the same time prevents or diminishes the normal action of insulin in promoting utilization of glucose. According to this view, growth hormone causes diabetes or sustains growth according to the extent to which additional insulin is available from the pancreas. As might be expected, extreme insulin sensitivity occurs after hypophysectomy in experimental animals. This is remedied by administration of growth hormone. Large amounts of insulin have been observed in the plasma of patients who have an excessive production of growth hormone (Chap. 52). Even mild hypoglycaemia is sufficient to promote growth hormone secretion. At least one role of growth hormone seems to be to spare glucose when it is in short supply. During exercise, muscles readily use glucose, even when there is a lack of insulin. Growth hormone tends to prevent this by providing an alternative source of energy in the form of free fatty acid from the triglycerides of adipose tissue. Ultimately,

this action spares amino acids which would otherwise become precursors of glucose in gluconeogenesis. Such amino acids are essential units to be spared for processes of growth and repair. Thus, growth hormone favours protein anabolism.

5. Cortisol. In early investigations of the hormones of the *adrenal cortex*, it was usual to class them as glucocorticoids or mineralocorticoids depending on whether they acted primarily on carbohydrate or on salt and water metabolism. Since it is now well established that quantitatively the most important steroid secreted by the human adrenal cortex is cortisol, and since it is the most potent steroid influencing carbohydrate metabolism, we should refer to cortisol rather than glucocorticoid when discussing the influence of the adrenal cortex on carbohydrate metabolism. Cortisone, the 11-oxoderivative of cortisol (Chap. 50), is secreted in small amounts and is active only on reduction to cortisol. Aldosterone has a slight influence on carbohydrate metabolism but, since the amount secreted is one hundred times less than cortisol, its influence is almost entirely on mineral metabolism. Various synthetic analogues of cortisol, such as prednisolone or dexamethasone (Chap. 50) have many clinical applications, and have potent effects on carbohydrate metabolism. The action of cortisol is in general catabolic. When cortisol concentrations rise above the mean normal adult concentration of about 10 μg per 100 ml plasma, protein synthesis is reduced and an increased amount of amino acids becomes available to the liver for gluconeogenesis. In addition, in experimental animals after 2 or 3 hours, phosphoenolpyruvate carboxykinase activity (Chap. 11) increases, probably by the synthesis of new enzyme protein. Glycogen synthetase also seems to be activated, probably because of the increasing amounts of insulin secreted in response to the rising blood sugar. The various hormones which play a role in glucose homeostasis do not function independently of one another but in a concerted manner.

It may be concluded that the effect of cortisol on carbohydrate metabolism is to maintain gluconeogenesis by provision of amino acids and by induction of enzymes as required. It may thus be regarded as a somewhat slow, second line defence against hypoglycaemia adrenaline. Cortisol may also be regarded as a hormone that facilitates the actions of glucagon, adrenaline and growth hormone by inhibiting peripheral glucose utilization, and by its synergistic effect on lipolysis in adipose tissue.

6. Thyroxine and triiodothyronine. The thyroid hormones increase gluconeogenesis, glycogenolysis and amino acid production from protein and so tend to bring about a rise in the blood sugar level. However, they also stimulate insulin secretion with resultant glucose uptake by peripheral tissues and promotion of lipid synthesis. Thus, overall, they have little effect on blood sugar concentration. Marked disturbances of carbohydrate metabolism are, however, observed in thyroidectomized animals and in patients with excess thyroid hormone. Hyperthyroidism exacerbates diabetes mellitus.

GLYCOSURIA

When urine gives positive tests for reducing sugar with Benedict's reagent the sugar is almost certainly glucose and the condition *glycosuria*; if a method involving glucose oxidase, such as the popular 'Clinistix' procedure, gives a positive result, the sugar is certainly glucose. However, in certain circumstances reducing sugars other than glucose may be present. Lactose, for example, is frequently found in the urine of pregnant or lactating women and galactosuria has been reported in breast-fed infants. Galactose and fructose (laevulose) may appear in the urine during the liver function tests in which they are used, and either sugar may be found persistently in the urine of persons with rare congenital anomalies of carbohydrate metabolism. Pentoses may appear in the urine after ingestion of large amounts of certain fruits; pentosuria is seen very occasionally as an inborn error of metabolism peculiar to the Jewish race.

Although the concentration of sugar in the blood of healthy people varies in the course of the day between 60 and 150 mg/100 ml only small amounts of reducing substances are found in the urine and the usual clinical tests for glycosuria are negative. The explanation is that although sugar is freely filtered from the blood as it passes through the glomeruli in the kidney (Chap. 32) it is subsequently almost completely reabsorbed by the renal tubules. The capacity of the tubules to reabsorb glucose is, however, limited and, when the concentration of glucose in the blood rises above about 160 mg per 100 ml, the maximum tubular reabsorptive capacity is exceeded and glycosuria occurs.

The reabsorptive capacity of the renal tubules for glucose varies somewhat from one person to another, tending for example to be higher in patients with uncontrolled diabetes mellitus, in whom the blood sugar concentration is habitually raised. In some otherwise healthy individuals (0.3 per cent of army recruits) the reabsorptive capacity is less than normal so that glycosuria is present even when the blood sugar concentration is within normal limits ('renal glycosuria') (Figs. 18.10 and 18.11). Glycosuria may occur in pregnancy because of a temporary reduction in the maximum tubular reabsorptive capacity for glucose.

We have already seen that the concentration of sugar in the blood is greatly influenced by the rate at which sugar enters the blood from the alimentary tract and the rate at which it is removed from the blood by the cells of the liver. Sometimes, after surgical removal of a large part of the stomach, glucose may be absorbed from the small intestine into the blood at a rate greater than that at which, initially, the liver and muscles can convert it into glycogen; a temporary hyperglycaemia with glycosuria therefore occurs (Fig. 18.11). A similar phenomenon is occasionally observed in persons whose daily intake of carbohydrate food has been exceptionally low for some time, for example, as the result of taking a 'reducing' diet.

Transient hyperglycaemia, and therefore glycosuria, may follow emotional excitement. It is observed, for example, in students sitting examinations, at medical examinations for insurance purposes or for the services, and in similar conditions of mental stress. This temporary disturbance of carbohydrate metabolism is the result of stimulation through the sympathetic nervous system of the adrenal medulla with consequent liberation of adrenaline. As we have already seen adrenaline increases the blood sugar concentration by promoting glycogenolysis.

The French physiologist, Claude Bernard,

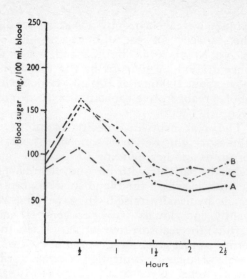

Fig. 18.10 Blood-sugar curves (venous) after 50 g glucose in three subjects with renal glycosuria. Periods of glycosuria are indicated by interrupted lines.

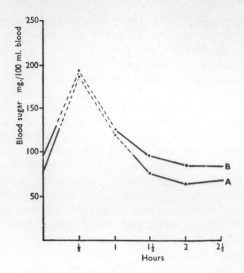

Fig. 18.11 Blood-sugar curves (venous blood) in two post-gastrectomy subjects after 50 g glucose. Note the unusually high peak in the ½-hour specimen and the subsequent rapid fall. Periods of glycosuria are indicated by interrupted lines.

found that puncture of the floor of the fourth ventricle in the brain of the rabbit was followed by glycosuria, provided that the animal had adequate stores of glycogen in the liver (diabetes piqûre). In this experiment also hyperglycaemia is brought about by stimulation of the adrenal glands through the sym-

pathetic nervous system, for it is not observed if the splanchnic nerves are cut or the adrenal glands removed. The glycosuria that sometimes occurs in patients with intracranial disease may be produced in a similar way.

Experimental glycosuria. Glycosuria can be produced in animals by alloxan, by strepto-

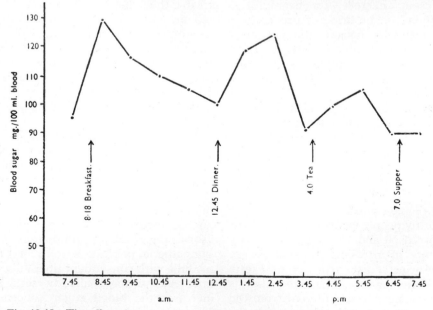

Fig. 18.12 The effect of meals on the capillary blood-sugar concentration of a healthy man during a period of 12 hours.

326 Textbook of Physiology and Biochemistry

zotocin and by phlorizin and the investigation of their actions has thrown much light on carbohydrate metabolism.

Alloxan, a ureide derivative (p. 64), when given by injection to animals causes necrotic changes in the pancreatic islets and a condition resembling diabetes mellitus ensues. Dehydro-ascorbic acid has been shown to produce diabetes mellitus in a similar fashion (Plate 18.1).

Phlorizin causes glycosuria when taken by mouth or injected into animals or man. Its does not produce hyperglycaemia but since it inhibits phosphorylase it may act by reducing the ability of the renal tubule cells to reabsorb glucose from the glomerular filtrate. After the administration of phlorizin the blood sugar level falls as the result of the continuous excretion of glucose and the carbohydrate stores eventually become exhausted. A fasting phlorizinized animal attempts to maintain its blood sugar level by converting the antiketogenic amino acids (p. 381) of its own tissue proteins into glucose (gluconeogenesis) with consequent serious loss of tissue. An index of the severity of the condition can be obtained by measuring the G/N or D/N ratio, that is the ratio of glucose (or dextrose) to nitrogen excreted in the urine. While carbohydrate stores are still available the amount of glucose relative to nitrogen is high; but as the carbohydrate stores diminish, and as tissue proteins are utilized, the amount of glucose tends to fall, while the amount of nitrogen rises. The ratio falls until it reaches the value of 3·65 which indicates that tissue protein only is being utilized.

When protein is fed to such an animal antiketogenic amino acids are converted to glucose which is at once excreted in the urine. Thus information can be obtained about the gluconeogenic properties of any particular protein or amino acid (p. 381) by studying its effect upon the D/N ratio in a phlorizinized animal. An amino acid which causes an increase in the D/N ratio is obviously antiketogenic.

THE GLUCOSE TOLERANCE TEST: TECHNIQUE AND INTERPRETATION

Glucose was formerly measured in whole blood or plasma by its ability to reduce cupric or ferric salts. Since blood contains other reducing substances in addition to glucose, newer more specific procedures employ glucose oxidase which catalyses the reaction:

$$glucose + H_2O + O_2 \rightarrow gluconic\ acid + H_2O_2$$

A peroxidase is used to decompose the H_2O_2 in the presence of an oxygen acceptor which becomes coloured on oxidation. While older methods gave values for fasting 'blood sugar', or rather total reducing substances, ranging from 70 to 120 mg per 100 ml blood, the range of fasting blood sugar with the more specific oxidase method is 60 to 100 mg per 100 ml. Significantly lower concentrations of glucose are found in whole blood than in plasma when modern methods arc used. If blood or plasma is to be used for blood sugar estimations, sodium fluoride is commonly added to stop glycolysis.

After a meal containing carbohydrate, the glucose concentration in the blood rises (Fig. 18.13). The *glucose tolerance*, that is the

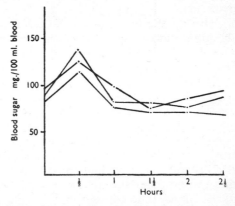

Fig. 18.13 Blood-sugar curves from three normal human subjects after the ingestion of 50 g of glucose (glucose tolerance test).

capacity of the body to deal with ingested glucose, is shown by the *glucose tolerance test*. After a fast of at least 8 hours, a blood sample is taken from a vein or finger stab for determination of the fasting blood or plasma glucose concentration. A sample of urine is obtained and a solution of 50 g glucose in 200 ml water (or preferably 0·75 g glucose per kg body weight) is drunk. Samples of blood and urine are collected at 30 min intervals over a period of 2 hours and their glucose concentration determined. The results obtained in three normal subjects are shown in Fig. 18.13.

The following points should be noted:

(1) The fasting blood sugar at the start of the test is about 80 mg/100 ml.

(2) Ingestion of glucose is followed by a rise in blood-sugar level, since the rate of absorption of glucose is greater than the rate at which it is removed from blood by oxidation or conversion to tissue glycogen.

(3) Maximum values about 140 to 150 mg per 100 ml are reached around 30 min after ingesting the glucose.

(4) In the later part of the test, blood-sugar levels fall as a result of uptake and utilization of glucose by the tissues. After 2 hours, normal levels have been restored. In the course of the return to normal, the blood-sugar frequently falls below the initial fasting level. A similar rise and fall of blood-sugar follows the ingestion of a meal containing carbohydrate; the height of the curve varies with the amount of glucose produced as a result of carbohydrate digestion, with the rate of digestion and rate of removal of glucose from the blood by tissues (Fig. 18.12).

(5) During the test, glucose does not normally appear in the urine, since the ability of the renal tubular cells to reabsorb glucose is not exceeded (Chap. 32).

(6) When there is an effective lack of insulin (diabetes mellitus), there is an exaggerated rise and unusually slow fall in the concentration of blood glucose.

In fasting subjects the arterial blood glucose is only 2 or 3 mg per 100 ml higher in the arterial than in the venous blood, because during fasting the tissues take very little sugar from the blood. The concentration of glucose in free flowing capillary blood, obtained by stabbing a warmed finger-tip, is virtually the same as found in arterial blood. The arterio-venous difference is increased to 30 mg/100 ml or more after a carbohydrate meal, on account of the rapid uptake of sugar by the tissues from the blood (Fig. 18.14).

The shape of the glucose tolerance test curve is influenced by the type of food taken in the days before the test. If it contained little carbohydrate the curve rises more sharply and to a higher level, indicating a relatively poor glucose tolerance. On the other hand, if the meals taken before the test have been largely carbohydrate, the rise in blood sugar is smaller indicating a relatively better glucose tolerance (Fig. 18.15).

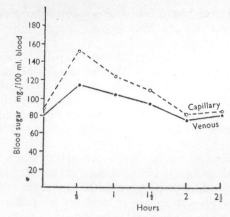

Fig. 18.14 Capillary and venous blood-sugar levels in a healthy human subject after ingestion of 0·75 g glucose per kg body weight.

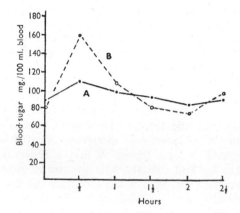

Fig. 18.15 Blood-sugar curves from the same healthy human subject on (A) a high-carbohydrate low-fat diet and (B) a high-fat low-carbohydrate diet.

Normally the dose of glucose given in the glucose tolerance test causes a rapid rise in plasma insulin concentration. Obese people have high basal plasma insulin levels and secrete very much more than normal persons (Fig. 18.16) when confronted with the rise in blood sugar that results from ingesting 50 g glucose. Resistance to insulin and a compensatory hypersecretion of insulin are characteristics of obesity (Fig. 18.16 overleaf).

DIABETES MELLITUS

Although, as we have seen, sugar may appear in the urine for a variety of reasons, if it is secondary to hyperglycaemia, the patient is usually suffering from diabetes mellitus. This is a disorder of carbohydrate metabolism,

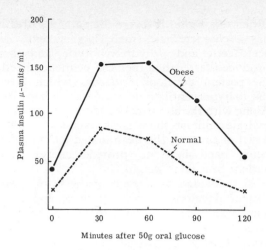

Fig. 18.16 Comparison of plasma insulin response in obese subjects with the normal response. The fasting level is elevated and the response to glucose excessive. (*By courtesy of Margaret McKiddie*)

(Fig. 18.17). Most patients with this type of diabetes are not absolutely deficient in insulin.

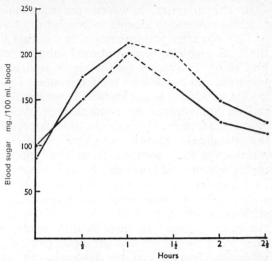

Fig. 18.17 Blood-sugar curves from two mildly diabetic patients. The dotted lines indicate glycosuria. Although the fasting blood sugar is nearly normal the response to 50 g glucose is abnormal. Compare with the normal curves in Fig. 18.13.

characterized by hyperglycaemia and glycosuria, accompanied by alterations, probably secondary, of fat and protein metabolism. It is the result of lack of effective insulin action, which may be either absolute or relative. Absolute insulin deficiency occurs after total pancreatectomy which is very occasionally performed for the removal of a malignant tumour. In the majority of patients, however, the disease develops apparently unprovoked, probably as a result of a hereditary predisposition. There are two main clinical groups though the distinction between the two is not absolute.

The '*juvenile-onset*' type occurs characteristically with an abrupt onset in patients under 25 years of age who are usually underweight. These patients are deficient in insulin: there is little, if any, circulating insulin in their plasma and no insulin response to a glucose load. They are liable to ketosis (see below) and require insulin therapy for the control of their diabetes.

The '*maturity-onset*' type develops insidiously in middle-aged, usually obese, patients; symptoms, if present, have often lasted several months before medical advice is sought. Recent population surveys have shown that many apparently healthy people have very mild diabetes. Glucose tolerance tests in these people show the characteristic abnormality of diabetes in spite of the absence of symptoms

The total amount of islet tissue in the pancreas may be greater than normal and the plasma

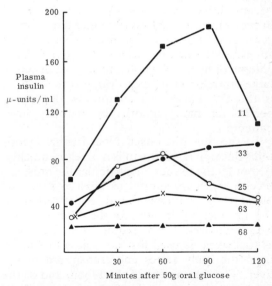

Fig. 18.18 Type of plasma insulin responses in 200 diabetic subjects. The diabetics were assigned to groups depending on whether their response was flat, low, normal, or high. The normal group were subdivided depending on whether or not the response was delayed. The means of each group and number of patients in each group are shown. (Margaret T. McKiddie & K. D. Buchanan (1969). *Quarterly Journal of Medicine* **38**, 445–465.)

insulin levels may be normal or elevated (Fig. 18.18).

In the majority, however, the maximum insulin response to glucose is delayed until 90 or 120 min after administration in contrast to the brisk response seen in normal subjects (cf. Fig. 18.6). The reason for this delay in insulin secretion and the relative ineffectiveness of the insulin produced is not known. Insulin action can be antagonized by a number of hyper-glycaemic factors: for example, glucagon from the pancreas, growth hormone from the pituitary gland, steroids from the adrenal cortex and thyroid hormone. Except, however, for those rare cases in which diabetes mellitus is secondary to some other endocrine disease such as acromegaly (Chap. 52) or Cushing's syndrome (Chap. 50) there is little evidence that these hormones are of aetiological significance. Insulin can be destroyed by insulinases from the liver or its action interfered with by insulin antagonists as well as by defects in the storage or release of glycogen by the liver. Again, there is no clear evidence which, if any, of these mechanisms are important in the majority of cases.

The initial symptoms of diabetes develop because large amounts of glucose are excreted in the urine (as much as 100 g per day in some instances). The loss of so much solute causes an osmotic diuresis (Chap. 32) and a large volume of urine (*polyuria*). In spite of drinking large amounts of fluid (*polydipsia*) the patient is thirsty. These two symptoms alone may persist for many months in maturity-onset patients.

In the 'juvenile-onset' type, further symptoms develop if treatment is not begun quickly. Although the tissues, in particular the muscles, receive a liberal supply of glucose from the blood, they are unable to utilize it efficiently in the absence of insulin so the diabetic feels weak and tired. Carbohydrate cannot be used as fuel so fat is used instead. Fat is mobilized from the body stores and transferred to the liver. The fat content of the blood, and of the liver, rises and a sample of plasma taken from an untreated severe diabetic is often opaque and fatty (*lipaemia*). The disproportionate metabolism of fat in diabetes results in the overproduction of ketone bodies (acetone, acetoacetic acid and β-hydroxybutyric acid) so there is ketonaemia and ketonuria (p. 338).

When this *ketosis* is severe the patient's breath has a characteristic smell, like acetone. As acetoacetic and β-hydroxybutyric acids are produced faster than they can be metabolized, the patient develops *acidaemia* which gives rise to hyperventilation (*air hunger*) (Chap. 33). Along with the abnormal metabolism of carbohydrate and fat there is an excessive breakdown of protein, the deaminated amino acids being catabolized to provide energy: the patient thus loses weight.

At this stage, as a result of the ketosis, the patient develops *anorexia* (lack of appetite), nausea and vomiting, and therefore the continued loss of water and electrolytes in the urine leads to increasing 'dehydration'. The 'keto-acidosis' is associated with increasing drowsiness and, if untreated, the patient may become unconscious (diabetic coma) and die from a combination of hyperglycaemia, ketosis, acidaemia and 'dehydration'. Most patients, however, are now treated with insulin before this stage is reached. Diabetics of the 'maturity-onset' type usually do not need insulin therapy. Those who are overweight generally respond to dietary restriction and weight reduction. Most of the remainder can be controlled by oral hypoglycaemic drugs, either sulphonylureas, such as tolbutamide and chlorpropamide, which act partly by increasing the

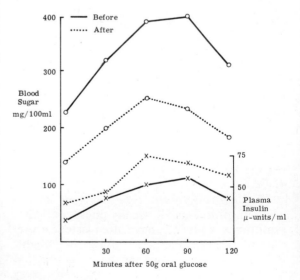

Fig. 18.19 The improved glucose tolerance after treatment is associated with a rise in plasma insulin levels. (Margaret T. McKiddie & K. D. Buchanan (1969). *Quarterly Journal of Medicine* **38**, 459.)

release of insulin from the pancreas (Fig. 18.19) or diguanides, such as phenformin or metformin, which are thought to act mainly by increasing the peripheral utilization of glucose.

Virtually all juvenile-onset diabetics and a number of older, thin diabetics require life-long insulin therapy. Soluble insulin is always used in the emergency situation of diabetic ketosis. For long-term therapy it has the disadvantage of a rather short action so that two or three injections are required every day. For patients requiring less than 60 units of insulin per day a single daily injection of one of the modified insulins such as protamine zinc insulin or insulin zinc suspension may give satisfactory control.

CARBOHYDRATE METABOLISM IN THE RUMINANT

Carbohydrate metabolism in ruminant animals is very different from that in other mammals, including man. In the rumen of the cow or goat complex carbohydrates including cellulose are broken down by bacteria and protozoa to yield large quantities of short-chain fatty acids, chiefly acetic acid, which are absorbed into the blood. The concentration of glucose in the blood of the cow is only half that found in man but there is a remarkably high concentration of acetate which is used with great efficiency in the cow's tissues by conversion to acetyl-coenzyme A which enters the citric acid cycle (Chap. 11). It has been calculated that 90 per cent of the cow's energy requirements are supplied by acetate produced in the rumen.

Any sugar formed in the rumen is fermented before it reaches the small intestine. The ruminant must therefore obtain its glucose not from its food but by synthesis from acetate and other short-chain fatty acids which are also used in the formation of milk fat.

REFERENCES

BERSON, S. A. & YALOW, R. S. (1966). Insulin in blood and insulin antibodies. *American Journal of Medicine* **40**, 676–690.

BESSMAN, S. P. (1966). A molecular basis for the mechanism of insulin action. *American Journal of Medicine* **40**, 740–749.

CLARKE, B. F. (1970). The pathogenesis of diabetes mellitus. *Scottish Medical Journal* **15**, 339–349.

DICKENS, F., RANDLE, P. J. & WHELAN, W. J. (Eds.) (1967). *Carbohydrate Metabolism and its Disorders.* New York: Academic Press.

GREENGARD, P. & COSTA, E. (1970). *Role of Cyclic AMP in Cell Function.* New York: Raven Press.

HALES, C. N. (1967). Some actions of hormones in the regulation of glucose metabolism. *Essays in Biochemistry* (Edited by P. N. Campbell & G. D. Greville), Vol. **3**, pp. 73–104. London: Academic Press.

HALES, C. N. & RANDLE, P. J. (1963). Immunoassay of insulin with insulin-antibody precipitate. *Biochemical Journal* **88**, 137–146.

HARRIS, H. (1970). Genetical theory and the inborn errors of metabolism. *British Medical Journal* **i**, 321–327.

JOSLIN, E. P., ROOT, H. F., WHITE, P. & MARBLE, A. (1959). *The Treatment of Diabetes Mellitus,* 10th edn. London: Kimpton.

KATSOYANNIS, P. G. (1966). The chemical synthesis of human and sheep insulin. *American Journal of Medicine* **40**, 652–661.

KREBS, H. A. (1964). Gluconeogenesis. *Proceedings of the Royal Society* B **159**, 545–564.

KREBS, H. A. (1965). Some aspects of gluconeogenesis. *Energy Metabolism* (Edited by K. L. Blaxter), pp. 1–9. New York: Academic Press.

LARNER, J. (1966). Hormonal and nonhormonal control of glycogen metabolism. *Transactions of the New York Academy of Sciences* Ser. II, Vol. 29, pp. 192–209.

LEVINE, R. (1966). The action of insulin at the cell membrane. *American Journal of Medicine* **40**, 691–694.

LEVINE, R. & HAFT, D. E. (1970). Carbohydrate homeostasis. *New England Journal of Medicine* **283**, 175–183, 237–246.

POPE, C. G. (1966). Immunology of insulin and radioimmunoassay. *Advances in Immunology* **5**, 231–238.

PYKE, D. A. (Ed.) (1962). *Disorders of Carbohydrate Metabolism.* London: Pitman.

RANDLE, P. J. (1965). *Insulin Mechanisms of Hormone Action* (Edited by P. Karlson), pp. 94–103. New York: Academic Press.

RENOLD, A. E. (1970). Insulin biosynthesis and secretion—a still unsettled topic. *New England Journal of Medicine* **282**, 173.

ROBISON, G. A., BUTCHER, R. W. & SUTHERLAND, E. W. (1971). *Cyclic AMP.* New York: Academic Press.

SANGER, F. (1959). The chemistry of insulin: Nobel Lecture. *Chemistry and Industry* **104**, 109.

SMITH, L. F. (1966). Species variation in the amino acid sequence of insulin. *American Journal of Medicine* **40**, 662–666.

SUTHERLAND, E. S., OYE, I. & BUTCHER, R. W. (1965). The action of epinephrine and the role of the adenyl cyclase system in hormone action. *Recent Progress in Hormone Research* (Edited by G. Pincus), Vol. 40, pp. 651–772.

Symposium on Insulin (1966). *American Journal of Medicine* **40**, 651–772.

THOMPSON, R. H. S. & WOOTTON, I. D. P. (1970). *Biochemical Disorders in Human Disease*, 3rd edn. London: Churchill.

TRAYNER, IRIS M., WELBORN, T. A., RUBENSTEIN, A. H. & RUSSELL FRASER, T. (1967). Serum and urine insulin in late pregnancy and in a few pregnant latent diabetics. *Journal of Endocrinology* **37**, 443–450.

VALLANCE-OWEN, J. (1964). Insulin inhibitors and antagonists. *Advances in Metabolic Disorders*, Vol. 1, pp. 191–217.

YOUNG, F. G. (1962). On insulin and its action. *Proceedings of the Royal Society* B **157**, 21–26.

19 Lipid metabolism

The plasma lipids are a complex mixture of neutral fats, phospholipids and cholesterol. In health the total concentration varies between 400 and 1240 mg per 100 ml plasma. The wide variation (Table 19.1) among the different constituents cannot yet be fully explained.

The fraction which is smallest in amount, the non-esterified fatty acids (NEFA), is one of the most important for it is metabolically exceedingly active with a half life of less than 3 min in man. This is the form in which fat is liberated from the depots and conveyed to the liver and other tissues for oxidation. It is sometimes referred to as 'unesterified fatty acids' (UFA) or 'free fatty acids' (FFA) but this latter term is misleading for the fatty acids are not in the free state in plasma but are attached to the plasma proteins, about two thirds to plasma albumin and one third to the lipoproteins (see Plasma proteins, Chap. 23).

The concentration of NEFA in the plasma is increased very rapidly by administration of adrenaline or noradrenaline and more slowly by growth hormone. It is decreased by insulin.

After ingestion of fat, the fat content of the blood increases and may remain above the fasting level for up to 5 hours. Much of this increase is due to the presence of chylomicrons (see p. 300). Blood fat is also increased after

Table 19.1 The lipid content of human blood plasma

	mg/100 ml
Total lipids	385 to 675
Neutral fat	80 to 120
Total fatty acids	110 to 500
Non-esterified fatty acids (NEFA)	8 to 50
Total phospholipids	150 to 250
Total cholesterol	130 to 260
Cholesterol esters	90 to 200
Free cholesterol	40 to 70

exercise, in starvation, pregnancy and lactation. In certain pathological states, for example nephrosis, the lipoproteins of the blood are often much increased and are qualitatively altered.

Although the lipids are in general insoluble in water, plasma and serum are limpid solutions. The reason for this is that lipids occur in blood plasma as lipid-protein complexes (lipoproteins) in which the lipid is bound to the protein fraction, chiefly the α- and β-globulins (Chap. 23). Such lipids cannot be extracted from plasma with ether at ordinary temperatures, but are readily extracted after denaturation of the protein.

Lipid stores. The digestion and absorption of fat from the intestine has already been discussed in Chapter 17. After absorption fat is either oxidized immediately in the liver and the muscles, or is stored in the body until required.

The chief fat stores are the subcutaneous and retroperitoneal depots which may in obese individuals contain enormous amounts of fat (Chap. 33). As a rule, fat is stored in the form which is characteristic of the species, and ingested fat is converted into this characteristic form; thus, mutton fat is different from beef fat, both differ from human fat in melting point, degree of unsaturation and so on, depending on the nature of the fatty acids in the triglycerides. In certain cases, however, an animal may deposit in its depots the fat of another species, as in the classical example of the starved dog which was fed such large amounts of mutton fat that it deposited in its subcutaneous tissues fat characteristic of the sheep.

Not all fat in the body comes from dietary lipid. Fat is readily synthesized from carbohydrate and any excess carbohydrate in the diet which is not immediately oxidized or stored as glycogen is converted to fat by the process discussed on p. 341. The fattening properties of a diet rich in sugar and starch are well known. Some people store but little fat on overfeeding with fat or carbohydrate, while others again lay down fat readily when their dietary intake exceeds their energy requirements only slightly Chap. 13).

Herbivorous animals synthesize most of their fat from cellulose which is broken down by micro-organisms in the alimentary tract to yield short chain fatty acids which are the ultimate sources of the longer chain fatty acids of the stored fat.

The fat in the animal body serves as an insulating blanket to retain body heat and also as a storehouse of food. Contrary to common belief, adipose tissue is by no means merely inert storage material but has an active metabolism. Indeed, it has been described as not an insulating blanket, but rather an electric blanket. The metabolism of adipose tissue is under hormonal control and slices of such tissue as the epididymal fat pad in the rat may be used in the assay of certain hormones, for example insulin (p. 320).

Slices of adipose tissue have a metabolic activity about the same as that of kidney tissue and about half that of liver. About half the adipose tissue of rats and mice is metabolized and reconstituted in a week or so. The turnover has not been measured in man and while it is unlikely to be as rapid as that of the rat it must be very high. An indirect confirmation of this is to be found in the high rate of turnover of non-esterified fatty acids (NEFA) in the plasma.

Adipose tissue of two types. White adipose tissue, found all over the body, is by far the commoner; it is made up of large cells each containing a single large fat droplet within a thin rim of cytoplasm. The electron microscope shows capillaries closely applied to these cells but only a few nerve fibres. Brown adipose tissue, which has a rich blood and nerve supply, consists of cells with round nuclei, granular cytoplasm and numerous fat droplets. The granularity of the cytoplasm is due to the high concentration of iron-containing cytochrome pigments. Most newborn animals, including the human baby, possess masses of brown fat around the neck and between the shoulder blades; in most adult mammals very little brown fat can be found, but in hibernating animals this tissue is abundant and is of great physiological importance (see Chap. 42).

The brown adipose tissue has a large capacity for generating energy in the form of heat. In the newborn rabbit this tissue may comprise as much as 6 per cent of the total body weight and is responsible for the newborn animal's increase in heat production when it is suddenly exposed to cold. This response is dependent on oxygen, for if the animal is deprived of oxygen the brown adipose tissue promptly cools to the same temperature as the rest of the body. If the brown adipose tissue is excised, the animal can no longer increase its heat production in response to exposure to cold. Cold causes a release of noradrenaline which activates the enzyme that splits triglyceride fat into glycerol and free fatty acids. Since adipose cells do not contain the enzymes required for the metabolism of glycerol, it is discharged into the blood stream together with a small proportion of the free fatty acids which are then metabolized in other tissues, chiefly liver and muscle. However, more than 90 per cent of the fatty acid molecules remain in the adipose cells where they combine with coenzyme A under the influence of ATP to form the corresponding acyl coenzyme A compounds (p. 337). Some of these are then oxidized (p. 148) to provide energy for the regeneration of ATP, but most are converted to the original triglyceride fat by combination with α-glycerol phosphate derived by way of glucose 6-phosphate from the glucose of the blood (see Fig. 19.2). This cycle of changes is a device for turning the chemical bond energy of fatty acids into heat and is responsible for more than 80 per cent of the increased body heat produced when a newborn

rabbit is exposed to the cold. During such exposure, the blood flow through the brown adipose tissue is increased several fold and as much as a third of the total cardiac output of blood may be diverted through it.

Fat is present in most tissues as well as in the depots. While depot fat is mainly neutral fat (triglycerides), the lipids of the tissues consist of both neutral fat and phospholipids. After prolonged fasting the neutral fat content of tissues becomes greatly depleted; the neutral fat can, therefore, be regarded as stored fat. The phospholipid content of the brain and other tissues, however, remains high during starvation; the cerebral phospholipids do not represent stored material but are essential for the proper functioning of the tissue. It is desirable, nevertheless, that not too sharp a distinction should be made between the fats, mainly phospholipids, which are essential components of cell and membrane structure (p. 56) and which do not disappear on fasting (the 'constant element'), and the storage fat, mainly neutral fat, which can be utilized by the fasting animal (the 'variable element').

In poisoning with arsenic, phosphorus, chloroform, carbon tetrachloride and some other drugs, and in many infective processes the fat content of the liver is greatly increased.

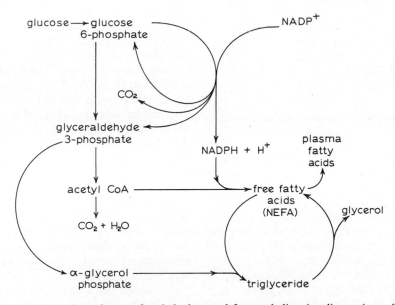

Fig. 19.2 The main pathways of carbohydrate and fat metabolism in adipose tissue. The fatty acids liberated into the blood are non-esterified. In liver the processes are similar but the fatty acids liberated into the blood are usually esterified (after F. G. Young, 1962).

Poisons such as carbon tetrachloride act on the mitochondria in the liver cell, partly by altering the permeability of the mitochondrial membrane so that nicotinamide nucleotides leak out and are lost, and partly by disorganizing the chain of respiratory enzymes associated with the tricarboxylic acid cycle. The oxidation of fat then becomes defective and fat accumulates in the liver cells.

FAT IN IN THE LIVER

The liver plays a special role in the metabolism of fat. The normal liver contains about 4 per cent of lipid of which only one-quarter is neutral fat while three-quarters are phospholipid. The amount of neutral fat rises markedly

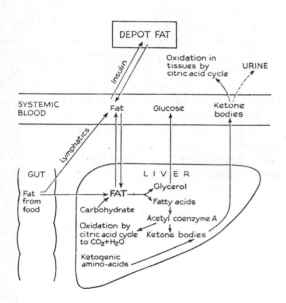

Fig. 19.3 Diagrammatic representation of fat metabolism.

during the early stages of starvation as fat is conveyed from the depots to the liver to be oxidized. As the stores become exhausted the fat content of the liver falls. The fat content of the liver is also increased in diabetes mellitus, in which carbohydrate metabolism is impaired (p. 328) and fat metabolism is increased (p. 330). The fat in the liver is very active metabolically. In the rat its half-life is about 2 days whereas the half-life of triglyceride fat in the brain and in the subcutaneous and abdominal depots is about 12 and 7 days respectively.

THE OXIDATION OF FATS

When fat is to be oxidized in the animal body for the provision of energy, it is mobilized in the fat stores and conveyed to the blood in the form of non-esterified fatty acids (NEFA) to those tissues such as liver and muscle in which fatty acid oxidation readily takes place.

The liver normally contains a store of fat and phospholipid; it also receives from the food or from the fat depots a continuous supply of fat with which it deals by the process of oxidation. In the liver, whether triglycerides or phospholipids are to be metabolized, the essential part of the process consists in the oxidation of the long-chain fatty acids in the mitochondria. The glycerol portion of the fat molecule reacts with ATP to form glycerol phosphate which is oxidized to glyceraldehyde 3-phosphate. This in its turn may either be converted to glycogen by reversal of part of the glycolysis cycle or be converted to pyruvate (p. 142).

Fatty acid oxidation also occurs in muscle; indeed in cardiac muscle fatty acids are an important fuel of respiration. Diaphragmatic muscle also oxidizes fatty acids but glucose, if it is available, is preferentially oxidized.

Several theories have been proposed to explain the mechanism of the oxidation of the fatty acid chains. The classical theory of *β-oxidation* was the outcome of the work of Knoop. According to this mechanism, fatty acid chains are oxidized by the removal of two carbon atoms at a time. The carbon atom in the β-position to the carboxyl group is assumed to be attacked with the formation of the corresponding β-keto acid; then the two terminal carbon atoms are split off as acetic acid. A new carboxyl group (—COOH) is formed at the site of the keto (=CO) grouping so that a fatty acid remains with two carbon atoms fewer than the original. Again the new β-carbon atom is attacked and two more carbon atoms are split off. In this way the fatty acid is broken down by the removal of two carbon atoms at a time, until finally the stage of aceto-acetic acid (β-ketobutyric acid) is reached.

Although this theory has had to be modified in the light of recent discoveries, the essential concept of the stepwise removal of 2-carbon units remains. According to the modern view of Lynen and others the oxidation of fatty acids requires their preliminary activation by the formation of acyl coenzyme A derivatives.

The stages in the breakdown of a fatty acid are illustrated in Fig. 19.4, for the acid $R—CH_2—CH_2—CH_2—COOH$ in which R may have any even number of carbon atoms. The first step consists of the combination of the fatty acid with coenzyme A under the influence of a *thiokinase* in the presence of ATP and magnesium ions. The compound so formed, $RCH_2CH_2CH_2CO—S—CoA$, is then oxidized by removal of two atoms of hydrogen under the influence of an *acyl-CoA dehydrogenase* with flavin-adenine dinucleotide (FAD) as coenzyme.

The resulting compound takes up water under the influence of *enoyl CoA hydratase* to form a hydroxy acid in which the hydroxyl group is attached to the carbon atom in the β position to the original carboxyl carbon. This hydroxy acid is then oxidized by a *β-hydroxy acyl dehydrogenase* acting along with NAD (p. 67) to yield a keto acid with the keto group in the β-position to the original carboxyl carbon (Fig. 19.4). The keto acid finally reacts with another molecule of coenzyme A forming acetyl coenzyme A and the coenzyme A derivative ($RCH_2CO—S—CoA$) of the fatty acid containing two carbon atoms fewer than the original. This derivative then enters the cycle again as shown in Fig. 19.4, and loses a further two carbon atoms as acetyl coenzyme A. In this way any fatty acid containing an even number of carbon atoms is degraded two carbon atoms at a time to yield acetyl coenzyme A.

The oxidation of fatty acids takes place in the mitochondria. The esterification of the fatty acid ($R—COOH$) with coenzyme A takes place at the outer mitochondrial membrane, but in order to cross the inner mitochondrial membrane the fatty acyl group is transferred to the carrier molecule *carnitine*.

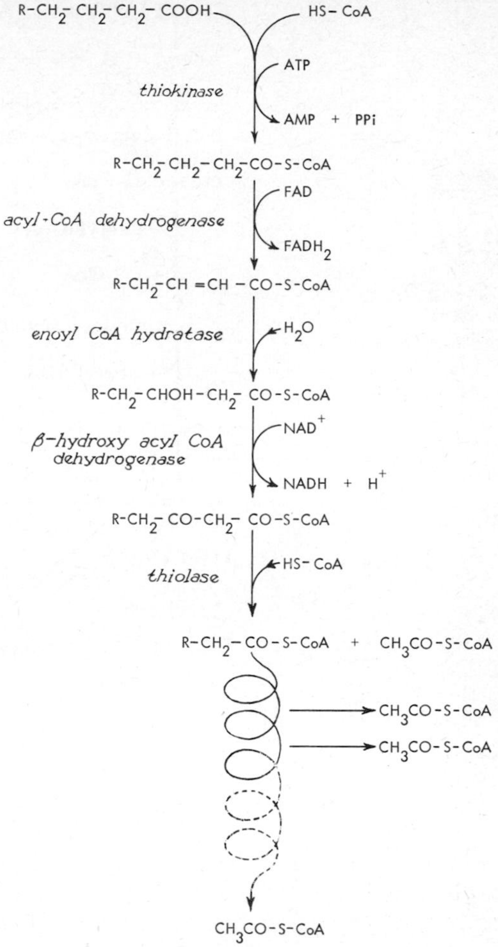

Fig. 19.4 Scheme of fatty acid oxidation.

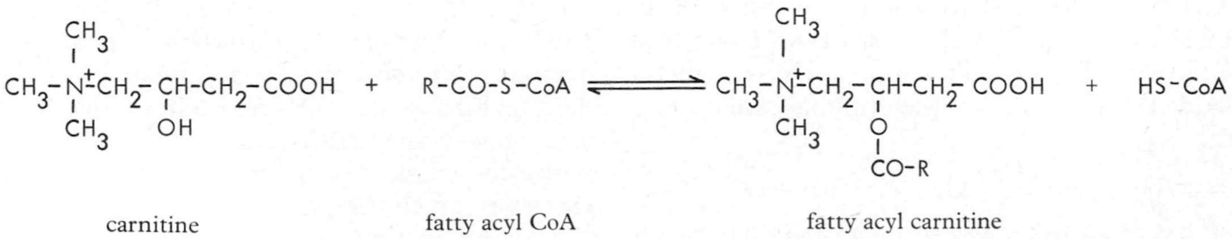

The fatty acyl group is then transferred from carnitine to coenzyme A within the mitochondrion by reversal of this reaction.

The Fate of the Acetyl Coenzyme A (Fig. 19.5). The acetyl coenzyme A produced in this way is of course identical with acetyl coenzyme A formed from carbohydrate by way of pyruvate as described in Chapter 11. Most of it combines with oxaloacetate to form citrate and

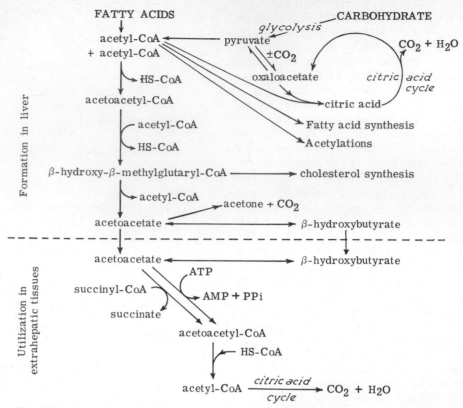

Fig. 19.5 Formation and utilization of acetyl coenzyme A.

is oxidized via the citric acid cycle (tricarboxylic acid cycle) (Fig. 11.6). The final oxidation pathway for fat and carbohydrate is therefore the same and the end result is the formation of carbon dioxide and water with the production of ATP. The coenzyme A released on citric acid formation enters the cycle again with another molecule of fatty acid while the reduced FAD and NAD are reoxidized by the usual hydrogen transport systems.

THE KETONE BODIES

Acetoacetyl coenzyme A is formed in the mammalian liver partly from the last four carbon atoms of long chain fatty acids which have been oxidized by successive removal of acetyl coenzyme A and mainly by the condensation of two molecules of acetyl coenzyme A under the influence of the enzyme *thiolase*:

$$CH_3CO{-}S{-}CoA + CH_3CO{-}S{-}CoA$$
$$\Longleftrightarrow CH_3COCH_2CO{-}S{-}CoA + HS{-}CoA$$

The acetoacetyl coenzyme A can then react with a further molecule of acetyl coenzyme

A to form β-hydroxy-β-methylglutaryl coenzyme A which is also an intermediate in teroid biosynthesis (p. 344). The reactions concerned are shown at the top of the next page.

The β-hydroxy-β-methylglutaryl coenzyme A breaks down to yield acetoacetic acid and acetyl coenzyme A.

The free acetoacetate so formed may be reduced to D-β-hydroxybutyrate thus:

$$CH_3COCH_2COOH + NADH + H^+ \Longrightarrow$$
$$CH_3CHOHCH_2COOH + NAD^+$$

or it may be decarboxylated to yield acetone

$$CH_3COCH_2COOH \rightarrow CH_3COCH_3 + CO_2$$

These three substances, which are formed in the liver, acetoacetate, β-hydroxybutyrate and acetone, are sometimes referred to as the 'ketone bodies' or 'acetone bodies' (Fig. 19.5), and their concentration in the blood is normally less than 1 mg per 100 ml. The blood carries them to the peripheral tissues in which β-hydroxybutyrate is oxidized to acetoacetate which in turn is converted into acetoacetyl

$$CH_3COCH_2CO-S-CoA \;+\; CH_3CO-S-CoA \longrightarrow HOOC\,CH_2\underset{OH}{\overset{\overset{\displaystyle CH_3}{|}}{C}}CH_2CO-S-CoA \;+\; HS-CoA$$

acetoacetyl-CoA   acetyl-CoA β-hydroxy-β-methylglutaryl-CoA

$$HOOC\,CH_2\underset{OH}{\overset{\overset{\displaystyle CH_3}{|}}{C}}CH_2CO-S-CoA \longrightarrow CH_3COCH_2COOH \;+\; CH_3CO-S-CoA$$

β-hydroxy-β-methylglutaryl-CoA   acetoacetic acid

coenzyme A either by reaction with succinyl coenzyme A thus:

acetoacetate + succinyl-CoA $\rightarrow$
         acetoacetyl-CoA + succinate

or by activation with ATP thus:

acetoacetate + ATP + HS-CoA $\rightarrow$
         acetoacetyl-CoA + AMP + PPi

The acetoacetyl coenzyme A is then converted into two molecules of acetyl coenzyme A under the influence of the enzyme *thiolase*.

acetoacetyl-CoA + HS—CoA $\rightleftharpoons$
         2 acetyl-CoA

The acetyl coenzyme A so formed may then enter the tricarboxylic acid cycle.

The utilization of ketone bodies by the extrahepatic tissues is considerable. When carbohydrate is not being utilized, fat alone cannot supply the fuel needs of the tissues, and such needs are met in part by ketone body utilization in muscle, kidney, heart, brain and adrenal gland all of which possess the necessary enzymes (unlike liver which cannot convert acetoacetate to acetoacetyl coenzyme A). Human brain can utilize ketone bodies to the extent of 20 per cent of the total energy requirement after an overnight fast, 60 per cent after an 8-day fast and about 80 per cent after a 40-day fast.

Although the concentration of ketone bodies in the blood is normally quite low, under certain circumstances it may be greatly increased. During starvation or even during heavy exertion, glycogen stores are rapidly depleted with the result that the use of fatty acids as a source of energy is greatly increased. Similarly in diabetes mellitus, the proper utilization of carbohydrate is impaired owing to absence of insulin, again with resultant increase in fat utilization. This increase in fatty acid oxidation results in the formation of ketone bodies in the liver in excess of the capacity of the peripheral tissues to utilize them. Their concentration in the blood increases (*ketosis*) and they appear in the urine (*ketonuria*).

Ketosis occurs most commonly in starvation and clinical or experimental diabetes mellitus, when the stores of liver glycogen are low. The ketosis of diabetes mellitus is accompanied by profound metabolic disturbances (p. 330) which lead to gradually deepening coma and finally to death. The odour of acetone can readily be detected in the patient's breath and urine, and more than 200 mg of β-hydroxybutyric acid may be excreted in the urine in 24 hours instead of the normal 5 to 10 mg.

Severe forms of ketosis occur when gluconeogenesis is very rapid, for example in the terminal stage of diabetes mellitus. Gluconeogenesis, mainly from amino acids, is greatly increased in this condition to provide glucose for the use of the tissues and, by making good the loss by excretion in the urine, to maintain the blood level at a high value. In cattle, ketosis is associated with the onset of lactation, the increased carbohydrate requirement arising from the synthesis of lactose being met by the conversion of amino acids to glucose.

A key substance in these considerations is oxaloacetate which is an intermediate metabolite associated both with gluconeogenesis and ketogenesis (Fig. 19.6). As has already been mentioned on p. 315, oxaloacetate is an obligatory intermediate in the synthesis of carbohydrate from most precursors including lactate and the glucogenic amino acids.

Lipid Metabolism   339

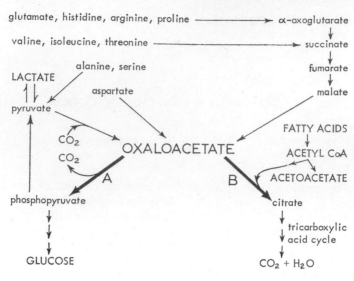

AMINO ACIDS

glutamate, histidine, arginine, proline ⟶ α-oxoglutarate

valine, isoleucine, threonine ⟶ succinate

fumarate

malate

LACTATE
alanine, serine
aspartate

pyruvate

CO_2

CO_2

OXALOACETATE

FATTY ACIDS

ACETYL CoA

ACETOACETATE

A

B

phosphopyruvate

citrate

tricarboxylic
acid cycle

GLUCOSE

$CO_2 + H_2O$

Fig. 19.6 The relationship between ketosis and gluconeogenesis from lactate and from the glucogenic amino acids. This figure should be compared with Fig. 18.4.

Oxaloacetate also plays an important role in metabolism by combining with acetyl coenzyme A to form citrate in the initial stages of the citric acid cycle (p. 149). If oxaloacetate is not present in adequate amounts, acetyl coenzyme A is diverted to the formation of acetoacetate. Oxaloacetate can prevent ketosis by taking up acetyl coenzyme A to form citrate; in its absence ketone bodies accumulate.

In those tissues in which gluconeogenesis occurs, mainly liver and the cortex of the kidney, oxaloacetate undergoes, apart from transamination, two major possible reactions. One is the formation of phosphopyruvate by reaction A in Fig. 19.6 (see also p. 147) and the other is combination with acetyl coenzyme A to form citrate in accordance with reaction B in Fig. 19.6 (see also p. 147). The citrate then breaks down to carbon dioxide and water by means of the tricarboxylic acid cycle. When the demands for carbohydrate are great, as in diabetes, starvation or lactation, reaction A predominates and oxaloacetate is preferentially used up in the formation of glucose, either from lactate or from the glucogenic amino acids. The result is that acetyl coenzyme A formed by the oxidation of fatty acids cannot be oxidized by way of the citric acid cycle and is diverted to the formation of acetoacetate in unusually large amounts.

When there is no great demand for gluconeogenesis, for example, when the carbohydrate intake is high, reaction B predominates and oxaloacetate is preferentially metabolized by combination with acetyl coenzyme A to form citrate which is then oxidized to carbon dioxide and water. Under these conditions the formation of acetoacetate is not great.

When the demand for glucose is high, as in diabetes, starvation or lactation, the liver cannot satisfactorily meet all physiological needs. It may make the maximum possible amount of glucose by glycogenolysis or by gluconeogenesis and as a result, the normal operation of the tricarboxylic acid cycle is interfered with so that ketone bodies appear as by-products of liver respiration. Alternatively, if some of the oxaloacetate is used in cell respiration, the rate of gluconeogenesis is lowered.

The suppression of glucose oxidation during periods of carbohydrate deprivation leading to the release of fatty acids (NEFA) from the fat depots into the blood and their subsequent oxidation, has been termed by Randle the 'glucose-fatty acid cycle'.

340   Textbook of Physiology and Biochemistry

THE BIOSYNTHESIS OF LIPIDS

Apart from dietary fat, the major source of fatty acids in the animal body is carbohydrate which is broken down to pyruvate by the mechanisms already described in Chapter 11. Pyruvate yields acetyl CoA (p. 148) by means of oxidative decarboxylation under the in-influence of thiamine pyrophosphate (TPP) (p. 185), lipoate (p. 186), NAD, coenzyme A and Mg^{2+} ions. Acetyl CoA (p. 148), 'active acetate', is the starting material for the synthesis of fatty acids.

The mechanism by which units of coenzyme A are built up into fatty acid chains by a multi-enzyme complex or 'fatty acid synthetase' has been elucidated by the work of Lynen and his colleagues. A preliminary step is the formation of *malonylcoenzyme A*.

$$\underset{\text{malonic acid}}{\overset{\displaystyle COOH}{\underset{\displaystyle}{\overset{|}{CH_2-COOH}}}} \qquad \underset{\text{malonyl-CoA}}{\overset{\displaystyle COOH}{\underset{\displaystyle}{\overset{|}{CH_2-CO-SCoA}}}}$$

Malonylcoenzyme A is formed by carboxylation of acetyl coenzyme A in the presence of ATP and manganese ions under the influence of the biotin-containing enzyme *acetyl-CoA carboxylase*, thus:

$$CH_3CO-S-CoA + CO_2 + ATP \xrightarrow[\text{biotin}]{Mn^{++}} \overset{\displaystyle COOH}{\underset{\displaystyle}{\overset{|}{CH_2-CO-S-CoA}}} + ADP + Pi$$

$$\text{serine} \atop \text{residue} \qquad \overset{\displaystyle NH}{\underset{\displaystyle CO}{CH-CH_2-O-\text{phosphate-pantothenic acid-}}} \atop \qquad\qquad\qquad \text{mercaptoethylamine (-SH)}$$

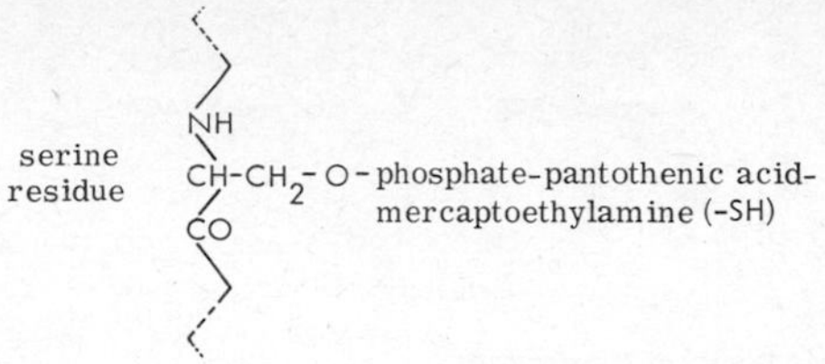

The acyl groups of acetyl CoA and of malonyl CoA are transferred to the —SH group of the ACP thus:

$$CH_3CO-S-CoA \; + \; ACP-SH \rightarrow \underset{\text{acetyl-S-ACP}}{CH_3CO-S-ACP} \; + \; CoA-SH$$

$$\underset{\text{malonyl-S-ACP}}{\overset{\displaystyle COOH}{\overset{|}{CH_2CO-S-CoA}}} \; + \; ACP\;SH \rightarrow \underset{\text{malonyl-S-ACP}}{\overset{\displaystyle COOH}{\overset{|}{CH_2CO-S-ACP}}} \; + \; CoA-SH$$

Acetyl-S-ACP and malonyl-S-ACP now react together with the release of CO_2 and coenzyme A to form acetoacetyl-S-ACP which is reduced by NADPH (p. 128) to the β-hydroxybutyryl derivative under the influence of *β-ketoacyl-ACP-reductase* (Fig. 19.7). The removal of water by *enoyl ACP dehydratase* yields crotonyl-S-ACP which undergoes further reduction by NADPH under the influence of *crotonyl-ACP reductase* to yield butyryl-S-ACP. This in turn reacts with malonyl-S-ACP with a repetition of the same

The synthesis of fatty acids such as palmitic acid occurs in the soluble fraction of the cytoplasm under the influence of the *fatty acid synthetase complex* consisting of six enzymes associated with a small protein the *acyl carrier protein* or *ACP* of molecular weight about 10 000. Such a complex has been studied chiefly in pigeon liver, *E. coli*, and yeast, in the last of which all seven proteins are tightly bound in a cluster of molecular weight $2 \cdot 3 \times 10^6$.

The polypeptide chain of the ACP contains a serine residue which is linked through a phosphate residue to pantetheine (p. 67 and p. 189) (pantothenic acid joined to mercaptoethylamine) thus:

sequence of events to yield a fatty acyl-S-ACP derivative with two more carbon atoms in the chain, and so the process is repeated until the palmityl derivative with 16 carbon atoms is formed. ACP then splits off and palmitic acid is released. It is important to note that the whole process of chain elongation has taken place while the fatty acyl residue remains attached to the ACP.

It is very important to note that NADPH, the reduced form of nicotinamide adenine dinucleotide phosphate, is essential for fatty acid biosynthesis. It is produced by the operation of the pentose phosphate cycle (p. 152) which is particularly active in liver and adipose tissue (Fig. 19.2). The supply of

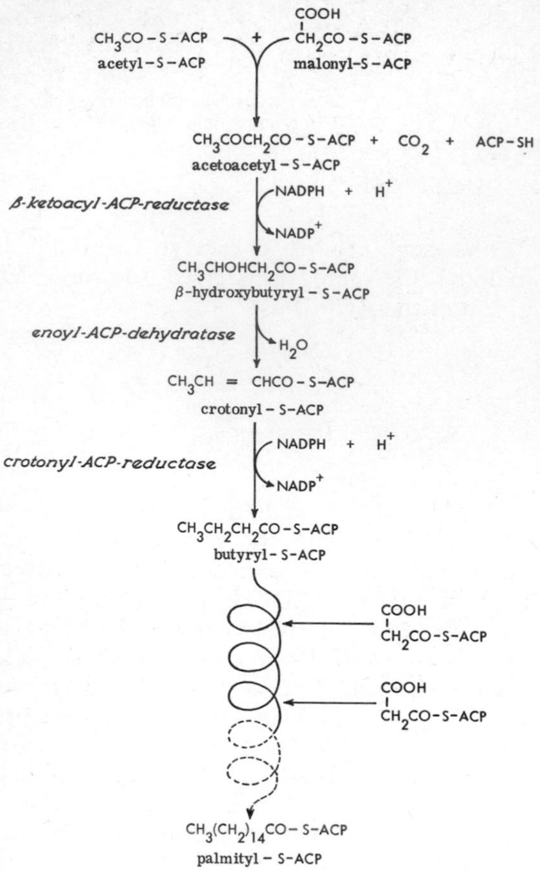

Fig. 19.7.

$$3 \text{ glucose 6-phosphate} + 6\,NAD^+ \rightarrow$$
$$\text{glyceraldehyde 3-phosphate} + 3CO_2$$
$$+ 2\text{-glucose 6-phosphate} + 6\,NADPH + 6H^+$$

Glycerol is formed in the tissues as α-glycerol phosphate by the reduction of glyceraldehyde 3-phosphate produced in the pentose phosphate pathway (Fig. 19.2) or in glycolysis or by the phosphorylation of glycerol by a phosphoglycerokinase and ATP. The α-glycerophosphate reacts with two molecules of the coenzyme A derivatives of a fatty acid to yield a *phosphatidic acid* which in turn is dephosphorylated by a phosphatase to form a diglyceride. This in turn can react with a third molecule of an acyl CoA derivative to yield a triglyceride.

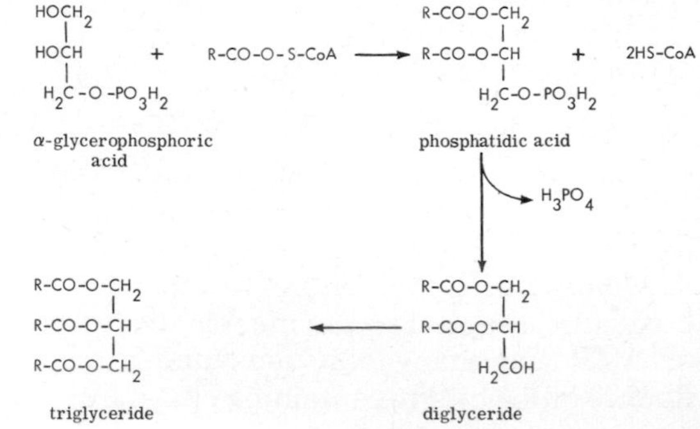

In the biosynthesis of the phospholipids, phosphatidic acid reacts with cytidine triphosphate (CTP) (p. 66) to form a CDP-diglyceride derivative which then reacts with serine to form phosphatidyl serine while CMP

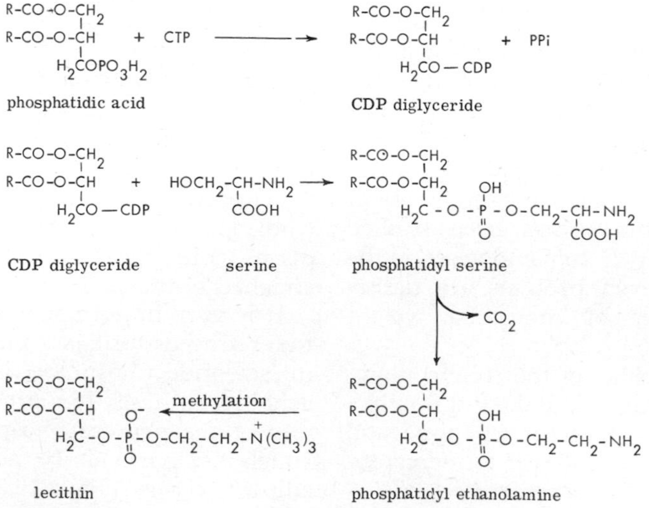

NADPH in adipose tissue is increased when glucose utilization is stimulated by insulin.

is released. The phosphatidyl serine may be decarboxylated to yield phosphatidyl ethanolamine which in turn can be methylated under the influence of S-adenosylmethionine (p. 388) as methyl donor to yield lecithin.

In ruminant animals cellulose is fermented by micro-organisms in the rumen to yield not glucose but short chain fatty acids such as acetate and butyrate, which can be used for the biosynthesis of long chain fatty acids, and propionate which acts as a major source of glucose.

After conversion to the coenzyme A derivative by a thiokinase reaction, propionate is carboxylated to yield methylmalonyl-CoA:

Growth hormone has a similar effect but acts much more slowly. ACTH (Chap. 52), TSH (Chap. 52) and glucagon (p. 323) also stimulate fat mobilization. Prostaglandin E counteracts these effects.

The glucocorticoids (Chap. 50) exercise an indirect action on fat metabolism through their effect on carbohydrate metabolism.

THE METABOLISM OF CHOLESTEROL

The total amount of cholesterol in the body of a man weighing 70 kg is about 140 g and the intake of cholesterol on the average diet in Western Europe is between 0·4 and 0·8 g per

$$CH_3CH_2CO\text{-}S\text{-}CoA \ + \ CO_2 \ + \ ATP \rightleftharpoons \underset{\underset{CH_3CH\text{-}CO\text{-}S\text{-}CoA}{\overset{COOH}{|}}}{} \ + \ ADP \ + \ Pi$$

propionyl-CoA   methylmalonyl-CoA

The methylmalonyl-CoA is then converted by *methylmalonyl mutase* to succinyl-CoA in the presence of the 5′-deoxyadenosyl derivative of vitamin B_{12} Chap. 13).

$$\underset{CH_3CHCO\text{-}S\text{-}CoA}{\overset{COOH}{|}} \quad \overset{\textit{methylmalonyl}}{\underset{\textit{mutase}}{\rightleftharpoons}} \quad \underset{CH_2CH_2CO\text{-}S\text{-}CoA}{\overset{COOH}{|}}$$

methylmalonyl-CoA   succinyl-CoA

The succinyl-CoA then enters the citric acid cycle and can yield glucose by glycolysis reversal (p. 146). Ruminants have a very active methylmalonyl mutase and a pronounced need for vitamin B_{12} or cobalt.

HORMONAL CONTROL OF FAT METABOLISM

The various phases of fat metabolism are coordinated and controlled by the endocrine system in a manner which may be summarized thus:

Insulin decreases the concentration of NEFA in the blood plasma by diminishing the rate of release of free fatty acids from adipose tissue. It stimulates the utilization of glucose 6-phosphate by the pentose phosphate pathway, so increasing the supply of NADPH and thus promoting fatty acid synthesis.

Adrenaline stimulates the mobilization of fat from the fat depots and so causes a rise in the concentration of NEFA in the plasma.

day. Ingested cholesterol is absorbed along with other lipids. Many other sterols, however, including most plant sterols are not absorbed from the gut. There appears to be some regulatory mechanism controlling the absorption of cholesterol; only a small fraction of the ingested cholesterol is absorbed and even on a high cholesterol diet man absorbs only 15 mg per kg of body weight per day. Such cholesterol as is absorbed inhibits the endogenous production which in man accounts for about 14 mg per kg of body weight per day. Since the total cholesterol content of the body is 2000 mg per kg body weight, the turnover as a percentage of the total content is 0·7.

In the rat, on the other hand, the endogenous production is much higher and may reach 70 mg per kg body weight per day, while the total cholesterol content is also 2000 mg per kg body weight. The turnover in the rat is therefore much higher.

In man, cholesterol is normally present in blood to the extent of 150 to 250 mg per 100 ml, being equally distributed between the cells and the plasma. In the cells, cholesterol occurs mainly in the free form, whereas in the plasma about 70 per cent is in the form of cholesterol esters. In the plasma, cholesterol, like other lipids, is transported in combination with proteins as water soluble lipoproteins, which are separable by electrophoresis into alpha and beta fractions. Some 70 per cent of the plasma cholesterol is found in the β-lipoprotein fraction and the remaining 30 per cent in the α-lipoprotein fraction. The total amount of cholesterol in the blood plasma of a 70 kg man is therefore between 4 and 6 g and it is reckoned to have a half-time of 8 to 12 days.

Cholesterol and its esters in blood plasma are essential for the welfare of the tissues, since they are needed for the repair of membranes, for the production of hormones and for other purposes. It is therefore undesirable to lower the blood cholesterol artificially below values of 150 mg per 100 ml.

A plasma cholesterol concentration greater than 300 mg per 100 ml is undesirable in man, but it may go up to 700 mg per 100 ml in certain diseases including diabetes mellitus (p. 330) and myxoedema (Chap. 50). Patients suffering from these disorders are frequently found to have a high incidence of atheromatous plaques containing cholesterol in the intima of their arteries and there is believed to be a relationship between arterial disease and the metabolism of cholesterol and fat.

In man the level of plasma cholesterol normally bears little relationship to the amount consumed in the diet but, by adding cholesterol to the diet of the chicken and the rabbit, the blood cholesterol can be greatly increased with the production of atheroma.

Patients with a tendency to high blood cholesterol levels should be advised to avoid the consumption *in excess* of such foodstuffs as butter and eggs, but in general the ordinary European diet is not likely to induce high blood cholesterol values in normal human subjects. Abnormally high values are due more to excessive endogenous production than to excessive intake.

It is probable that cholesterol is the source from which the body synthesizes other steroids such as sex hormones and bile acids. Cholesterol is synthesized in the body from two-carbon units in the form of acetyl coenzyme A formed either from fatty acids or from the metabolism of carbohydrate through pyruvate. Two molecules of acetyl coenzyme A condense to form acetoacetyl coenzyme A which reacts with a third molecule of acetyl coenzyme A to form β-hydroxy-β-methylglutaryl coenzyme A. This in turn, gives rise to the important intermediate, mevalonic acid which is activated

by two molecules of ATP to yield 5-diphosphomevalonic acid (mevalonic acid 5-pyrophosphate).

5-Diphosphomevalonic acid in the presence of ATP now loses carbon dioxide and water to form isopentenyl pyrophosphate which can exist also in an isomeric form 3, 3-dimethylallyl pyrophosphate. These compounds are the precursors of many important biological compounds including rubber and the carotenoid pigments as well as cholesterol.

One molecule of 3,3-dimethylallyl pyrophosphate now reacts with one of isopentenyl pyrophosphate to yield geranyl pyrophosphate which in turn reacts with another molecule of isopentenyl pyrophosphate to form farnesyl pyrophosphate, with the elimination of inorganic pyrophosphate at each stage.

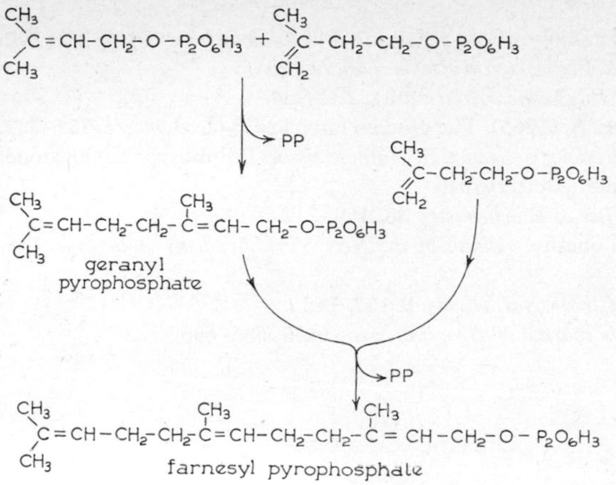

$$\begin{array}{c}CH_3\\ \diagdown\\ C=CH-CH_2-O-P_2O_6H_3\\ \diagup\\ CH_3\end{array} + \begin{array}{c}CH_3\\ \diagdown\\ C-CH_2-CH_2-O-P_2O_6H_3\\ \diagup\\ CH_2\end{array}$$

blood cholesterol is a well-known feature of this disease.

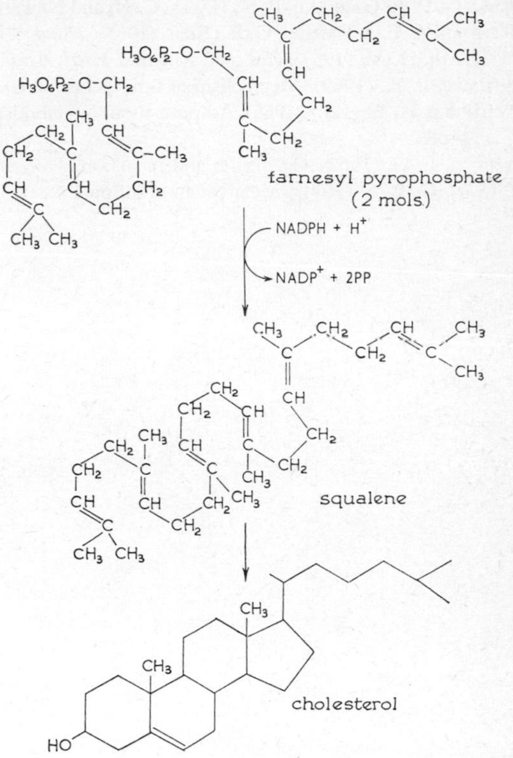

$$\begin{array}{c}CH_3 \qquad\qquad CH_3\\ \diagdown\qquad\qquad\qquad\\ C=CH-CH_2-CH_2-C=CH-CH_2-O-P_2O_6H_3\\ \diagup\\ CH_3\end{array}$$

geranyl
pyrophosphate

$$\begin{array}{c}CH_3\\ \diagdown\\ C-CH_2-CH_2-O-P_2O_6H_3\\ \diagup\\ CH_2\end{array}$$

PP

$$\begin{array}{c}CH_3 \qquad\qquad CH_3 \qquad\qquad CH_3\\ \diagdown\qquad\qquad\qquad\qquad\qquad\\ C=CH-CH_2-CH_2-C=CH-CH_2-CH_2-C=CH-CH_2-O-P_2O_6H_3\\ \diagup\\ CH_3\end{array}$$

farnesyl pyrophosphate

farnesyl pyrophosphate
(2 mols)

NADPH + H$^+$

NADP$^+$ + 2PP

squalene

Two molecules of farnesyl pyrophosphate finally condense to form the hydrocarbon squalene which by ring closure and loss of methyl groups is readily converted into cholesterol by enzymes in the cells of the liver.

The defective metabolism of carbohydrate in diabetes mellitus results in excessive oxidation of fatty acids (p. 330). Some of the acetyl coenzyme A so produced is diverted to cholesterol production (Fig. 19.5) and a high

cholesterol

REFERENCES

BALL, E. G. & JUNGAS, R. L. (1964). Some effects of hormones on the metabolism of adipose tissue. *Recent Progress in Hormone Research* (Edited by G. Pincus) **20**, 183–197.

BLOCH, K. (1960). *Lipid Metabolism*. New York: Wiley.

COOK, R. P. (1958). *Cholesterol: Chemistry, Biochemistry, Pathology*. New York: Academic Press.

DAWKINS, M. J. R. & HULL, D. (1965). The production of heat by fat. *Scientific American* **213** (2), 62–69.

DAWSON, R. M. C. & RHODES, D. N. (1964) (Eds.). *Metabolism and Physiological Significance of Lipids*. London: Wiley.

DEUEL, H. J. (1955). *The Lipids*, Vol. 2. London: Interscience.

DEUEL, H. J. (1957). *The Lipids*, Vol. 3. London: Interscience.

GOODMAN, DE W. S. (1965). Cholesterol ester metabolism. *Physiological Reviews* **45**, 747–839.

GRANT, J. K. (Ed.) (1964). The Control of Lipid Metabolism. *Biochemical Society Symposia* No. 24. London: Academic Press.

GREVILLE, G. D. & TUBBS, P. K. (1968). The catabolism of long-chain fatty acids in mammalian tissues. *Essays in Biochemistry*, Vol. 4, pp. 155–212.

GURR, M. I. & JAMES, A. T. (1971). *Lipid Biochemistry*. London: Chapman & Hall.

HULL, D. (1966). The Structure and Function of Brown Adipose Tissue. *British Medical Bulletin* **22**, 92–96.

KING, H. K. (1960). *The Chemistry of Lipids in Health and Disease*. Springfield: Thomas.

KINSELL, L. W. (1962). *Adipose Tissue as an Organ*. Springfield: Thomas.

KREBS, H. A. (1961). The physiological role of the ketone bodies. *Biochemical Journal* **80**, 225–233.

KREBS, H. A. (1964). Gluconeogenesis. *Proceedings of the Royal Society* B **159**, 545–564.

KREBS, H. A. (1965). Some aspects of gluconeogenesis. In *Energy Metabolism* (Edited by K. L. Blaxter), pp. 1–9. New York: Academic Press.

LYNEN, F. (1961). Biosynthesis of saturated fatty acids. *Federation Proceedings* **20**, 941–945.

LYNEN, F. (1967). The role of biotin dependent carboxylations in biosynthetic reactions. *Biochemical Journal* **102**, 381–390.

MASORO, E. J. (1966). Effect of cold on metabolic use of lipids. *Physiological Reviews* **46**, 67–101.

Nomenclature of Enzymes of Fatty Acid Metabolism (1956). *Biochemical Journal* **64**, 782–784.

RANDLE, P. J., GARLAND, P. B., HALES, C. N. and NEWSHOLM, E. A. (1963). The glucose fatty acid cycle. *Lancet* **i**, 785–789.

RENOLD, A. E. & CAHILL, G. F. (Eds.) (1965). *Handbook of Physiology. Section 5. Adipose tissue.* Edinburgh: Livingstone.

SINCLAIR, H. M. (1957). (Editor). *Essential Fatty Acids.* London: Butterworth.

STUMPF, P. K. (1969). Metabolism of fatty acids. *Annual Review of Biochemistry* **38**, 159–212.

WHIPPLE, H. E. (Ed.) (1965). Adipose tissue metabolism and obesity. *Annals of the New York Academy of Sciences* **131**, 1–683.

YOUNG, F. G. (1962). On insulin and its action. *Proceedings of the Royal Society* B **157**, 1–26.

PAWAN, G. L. S. (1971). Metabolism of adipose tissue. *British Journal of Hospital Medicine* **5**, 686–695.

The biosynthesis of the large, complex molecules (Chap. 6) of the nucleic acids can conveniently be divided into synthesis of the monomers (mononucleotides) and synthesis of the polymers (DNA and RNA). Since the synthesis of DNA and that of RNA are intimately concerned in it, the synthesis of protein is described in the next chapter.

BIOSYNTHESIS OF THE MONONUCLEOTIDES

Pyrimidine mononucleotides. The complete series of enzymic reactions giving rise to the parent pyrimidine mononucleotide (uridine 5′-monophosphate, UMP) is shown in Fig. 20.1. The starting compounds are aspartic acid and carbamoyl-phosphate (p. 380) which combine to form carbamoyl-aspartate. Formation of the pyrimidine ring is then effected by the action of *dihydro-orotase* giving dihydro-orotic acid, dehydrogenation of which produces the important pyrimidine intermediate, orotic acid. A pyrophosphorylase reaction then follows in which orotic acid accepts a ribose 5-phosphate group from 5-phosphoribosyl-1-pyrophosphate (PRPP):

5-phosphoribosyl–1–pyrophosphate where (P) represents a phosphate residue.

The resulting product is orotidine 5′-monophosphate; inorganic pyrophosphate is elimi-

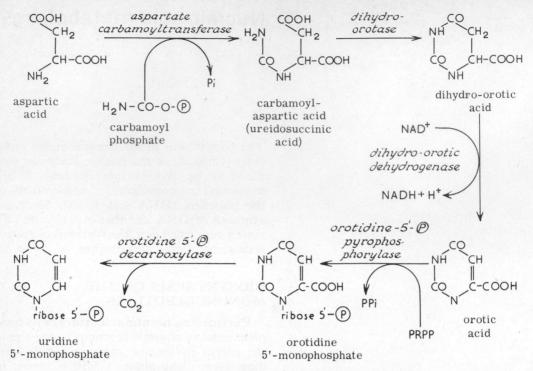

Fig. 20.1 Biosynthesis of uridine 5′-monophosphate (UMP), the parent pyrimidine nucleotide. In this diagram, Ⓟ represents a phosphate residue. PRPP = Phosphoribosyl pyrophosphate. PPi = Inorganic pyrophosphate.

nated. Decarboxylation of orotidine 5′-monophosphate gives uridine 5′-monophosphate, UMP, the parent pyrimidine nucleotide.

UMP is then converted (Fig. 20.2) by *kinases* through uridine 5′-diphosphate (UDP) to uridine 5′-triphosphate (UTP). The uracil moiety of UTP may then be aminated to give cytidine 5′-triphosphate (CTP). UTP and

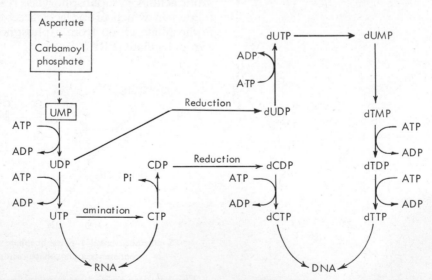

Fig. 20.2 Biosynthetic pathways leading to the immediate pyrimidine precursors of DNA and RNA. For explanation of contractions see text and Chapter 6.

CTP are the immediate pyrimidine precursors that are incorporated into RNA under the influence of *RNA polymerase*.

CTP may be dephosphorylated to CDP which can then undergo reduction to the deoxy form, dCDP. Phosphorylation of dCDP in a kinase reaction gives dCTP.

UDP can also undergo reduction to the deoxy form, dUDP, which is phosphorylated to the triphosphate dUTP. This in turn is dephosphorylated to dUMP. The action of a *synthetase* on dUMP in the presence of a derivative of tetrahydrofolic acid produces deoxythymidine 5′-monophosphate (dTMP) by the addition and reduction of a one-carbon unit to the dUMP at carbon-5 of the pyrimidine ring. Kinases then raise the level of phosphorylation of the dTMP through dTDP to dTTP (Fig. 20.2).

dCTP and dTTP are the immediate pyrimidine precursors that are incorporated into DNA by the action of *DNA polymerase* (p. 353).

The formation of dTMP from dUMP is inhibited by folic acid antagonists (p. 000) such as aminopterin and amethopterin which can thus prevent the formation of DNA. They are used in the treatment of certain forms of malignant disease.

Purine mononucleotides. It is known from experiments with isotopes that the sources of the atoms in the purine ring are as shown in Fig. 20.3.

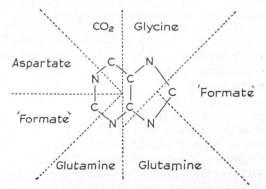

Fig. 20.3 Origin of the separate atoms in the purine ring. 'Formate' stands for the formyl derivative of tetrahydrofolic acid.

The starting material in purine biosynthesis is PRPP which accepts the γ-amino group of glutamine to give 5-phosphoribo-

sylamine, 5-PRA (Fig. 20.4). Glycine then reacts with 5-PRA to give glycinamide ribonucleotide, GAR, a nucleotide-like compound in which the amide of glycine takes the place of the usual purine or pyrimidine base. The reaction sequence continues by formylation from N-formyltetrahydrofolic acid to give formylglycinamide ribonucleotide (FGAR), then amination from glutamine to give formylglycinamidine ribonucleotide (FGAM). Ring closure ensues, producing the imidazole ring compound 5-aminoimidazole ribonucleotide (AIR). Carboxylation of this compound gives 5-aminoimidazole-4-carboxylic acid ribonucleotide (carboxy-AIR). The corresponding amide, 5-amino-imidazole-4-carboxamide ribonucleotide (AICAR) is produced in the subsequent two reactions via an intermediate compound, 5-aminoimidazole-4-succinocarboxamide ribonucleotide (SAICAR). This reaction may be compared with synthesis of arginine via argininosuccinic acid (p. 380). The purine ring system is completed when N-formyl-tetrahydrofolic acid donates its formyl group to the 5-amino group of the imidazole carboxamide ribonucleotide. The complete parent ribonucleotide is inosinic acid (inosine 5′-monophosphate, IMP).

Amination of IMP to AMP proceeds in two stages with the intermediate formation of adenylosuccinic acid (Fig. 20.5). This reaction, in which the amino group of aspartate is transferred to carbon-6 of IMP to give AMP, resembles the reaction above in which 5-aminoimidazole-4-carboxamide ribonucleotide is formed from 5-aminoimidazole carboxylic acid ribonucleotide (Fig. 20.4). One difference, however, is the requirement for GTP as coenzyme in the reaction forming adenylosuccinic acid from IMP.

The formation of GMP from IMP is also a two-stage reaction in which xanthosine 5′-monophosphate (XMP) is initially formed and then aminated to give GMP (Fig. 20.5).

The two purine mononucleotides, AMP and GMP, are phosphorylated by kinases through the diphosphate stage to give ATP and GTP. These triphosphates together with CTP and UTP are used as monomers in the RNA polymerase reaction (p. 355).

ADP and GDP are reduced in a ribonucleotide reductase reaction (compare reduction of CDP and UDP to dCDP and

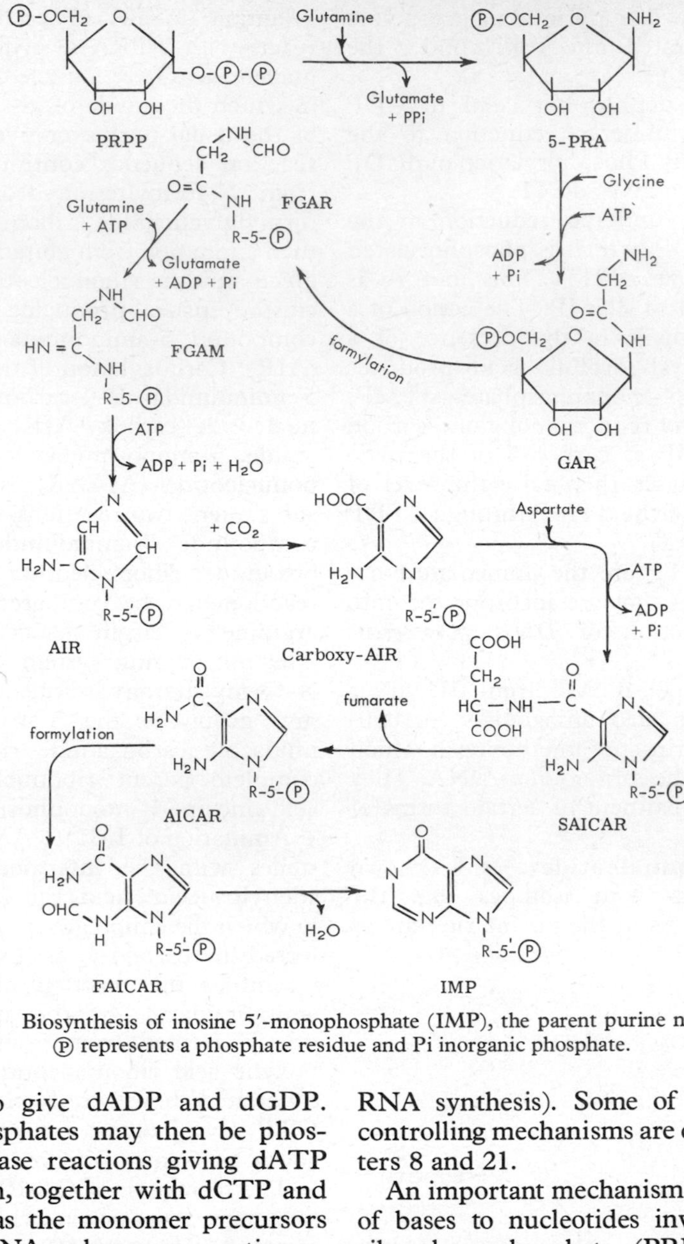

Fig. 20.4 Biosynthesis of inosine 5'-monophosphate (IMP), the parent purine nucleotide. ℗ represents a phosphate residue and Pi inorganic phosphate.

dUDP, p. 66) to give dADP and dGDP. These two diphosphates may then be phosphorylated in kinase reactions giving dATP and dGTP, which, together with dCTP and dTTP, are used as the monomer precursors of DNA in the DNA polymerase reaction.

Since the four deoxyribonucleoside triphosphates are required in approximately equal amounts for the synthesis of DNA, their different synthetic pathways must be strictly controlled in an integrated manner. The synthesis of the ribonucleoside triphosphates is also under strict metabolic control and is somewhat more complicated as they are required in different amounts (for example ATP is used in many reactions apart from

RNA synthesis). Some of the very efficient controlling mechanisms are described in Chapters 8 and 21.

An important mechanism for the conversion of bases to nucleotides involves 5-phosphoribosylpyrophosphate (PRPP) and the enzymes termed *nucleotide pyrophosphorylases*. Under their influence bases can react with PRPP to form nucleotides and pyrophosphate.

The reactions catalysed by *nucleotide pyrophosphorylases* are illustrated in Fig. 20.6. One such enzyme, *adenine phosphoribosyltransferase* (AMP pyrophosphorylase) converts adenine to AMP in the presence of PRPP. A second enzyme, *hypoxanthine-guanine phosphoribosyltransferase* (IMP-GMP pyrophosphorylase)

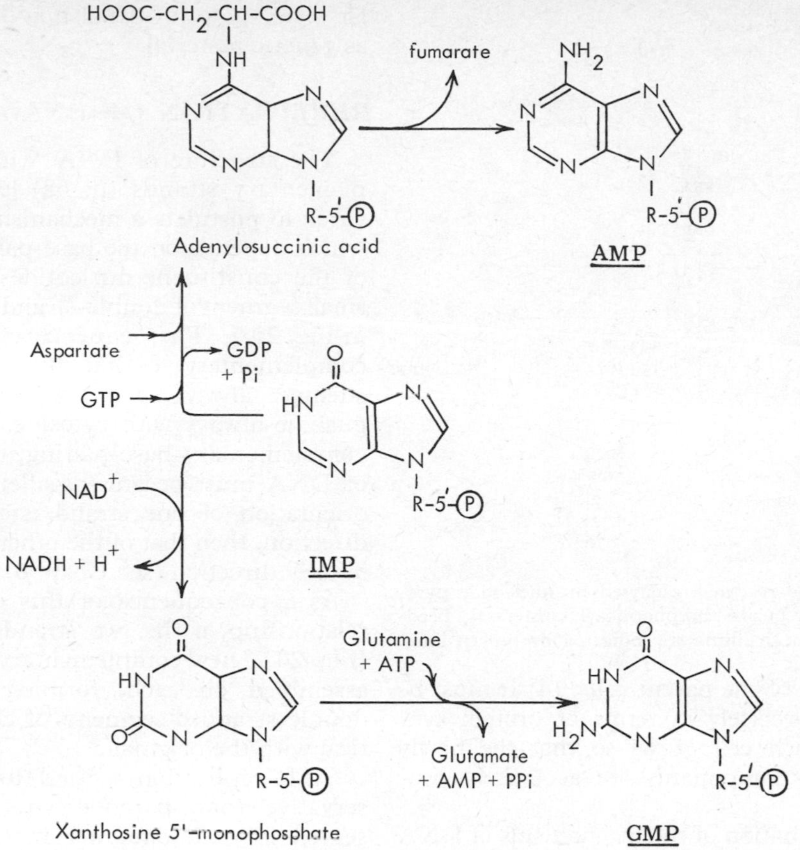

Fig. 20.5 Biosynthesis of adenosine 5'-monophosphate (AMP) and guanosine 5'-monophosphate (GMP). ℗ represents a phosphate residue.

converts hypoxanthine and guanine to IMP and GMP respectively in the presence of PRPP and also converts the drugs 6-mercaptopurine and azathioprine (imuran) into the corresponding nucleotides which in turn inhibit phosphoribosyl pyrophosphate amidotransferase and so prevent biosynthesis of purines. Since such drugs inhibit purine (and therefore nucleic acid) biosynthesis they are sometimes used as immunosuppressive or cancerostatic agents. Azathioprine is also of value in the treatment of gout by inhibiting purine formation.

In the rare condition in children known as the *Lesch-Nyhan syndrome* there is a deficiency of the enzyme hypoxanthine-guanine phosphoribosyltransferase. Consequently the condition is associated with excessive uric acid synthesis but is resistant to the action of azathioprine.

BIOSYNTHESIS OF THE NUCLEIC ACIDS

It is now clearly established that DNA is the carrier of genetic information in living cells. To fulfil this vitally important function DNA must have a number of properties. (1) It must be stable so that successive generations of species maintain their individual characteristics, but not so stable that evolutionary changes cannot take place. (2) It must be able to store a very large amount of information. An animal cell contains genetic information for the synthesis of over a million proteins. Furthermore DNA is the genetic material of all the 10^9 or so different living species, each of which is different from the others by at least one characteristic. (3) It must be duplicated exactly before each cell division, so that both daughter cells contain an accurate copy of the genetic

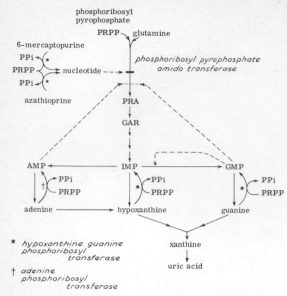

Fig. 20.6 The reactions catalysed by nucleotide pyro-phosphorylases (purine phosphoribosyl transferases). Feedback control mechanisms are indicated by broken lines.

information of the parent cells. (4) It must be expressed precisely in terms of protein synthesis as each cell grows so that the newly synthesized components are accurately produced.

An examination of the mechanisms of DNA replication and protein synthesis at the molecu-

lar level shows in detail how DNA functions as genetic material.

REPLICATION OF DNA

The structure of DNA, with its two complementary strands (p. 68) led Watson and Crick to postulate a mechanism of replication which depends on the base-pairing properties of the constituent nucleotides. Consider the small segment of double-stranded DNA shown in Fig. 20.7. The sequence of one strand is complementary to that in the other, that is adenine always pairs with thymine, and guanine always with cytosine. To allow such complementary base-pairing, the two strands of DNA must be anti-parallel, that is, if the orientation of one strand is in the 3′ to 5′ direction, then that of the other must be in the 5′ to 3′ direction (see Chap. 6).

As a consequence of this complementary relationship, if the two strands are separated (Fig. 20.7) new complementary strands can be assembled on each, forming two identical double-stranded segments of DNA each identical with the original.

Such replication is said to be semi-conservative (one parental strand being conserved in each daughter molecule) and has been shown to be sequential, the two double

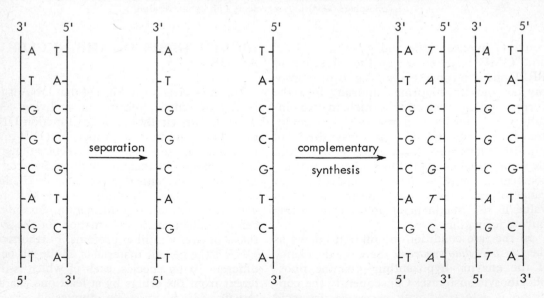

Fig. 20.7 Schematic representation of DNA synthesis. The two strands separate and on each is built a new complementary strand (*italics*). The result is two identical daughter DNA molecules each of which is identical with the parental DNA molecule.

stranded daughter molecules being synthesized progressively by a polymerizing enzyme which moves along the parental DNA molecule from one end to the other (Plates 20.1 and 20.2).

A more accurate representation of the process illustrated in Plates 20.1 and 20.2 is given in Fig. 20.8 which takes into account the fact that native DNA is a double helix.

It is now known that this model of DNA replication is essentially correct but the nature of the polymerizing enzyme, probably part of a larger replicating complex, is unknown. However, an enzyme discovered in 1957 by Arthur Kornberg and his colleagues, DNA polymerase (DNA nucleotidyltransferase) is obviously closely related to the polymerizing enzyme of DNA replication. DNA polymerase catalyses the sequential polymerization of the deoxyribonucleotide 5′-triphosphates (dATP, dCTP, dGTP and dTTP) in the presence of a template strand of DNA, to form a new complementary DNA strand (Plate 20.1). The 3′-hydroxyl of the growing strand reacts with the 5′-triphosphate group of the incoming deoxyribonucleotide. The incoming deoxyribonucleotides are selected by complementarity with the bases of the template strand.

The energy required during the formation of each 3′, 5′-phosphodiester bond of the polydeoxyribonucleotide is provided by the

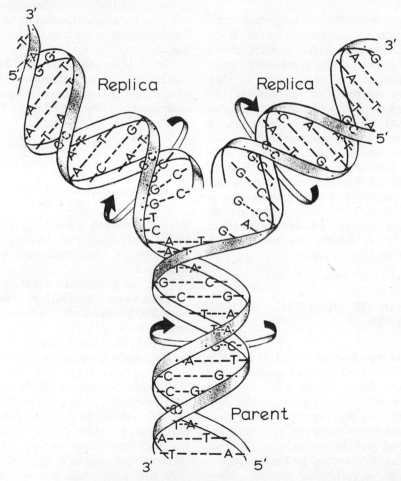

Fig. 20.8 A portion of native, double-helical DNA shown in the process of replication. The arrows indicate the directions of rotation that are necessary to allow unwinding of the parent helix with concomitant formation of the daughter helices. The two newly synthesized strands are growing in the downwards direction. (From *Molecular Biology of Bacterial Viruses* by Gunther S. Stent. W. H. Freeman and Company. Copyright © 1963.)

high energy bonds of the triphosphates (compare ATP, p. 66); as each monomer is incorporated into the new chain it loses its terminal pyrophosphate unit (PPi). The DNA template dictates the sequence of nucleotides in the newly synthesized strand, which is assembled in the 5′ to 3′ direction along the template, the template having an anti-parallel 3′ to 5′ orientation.

Replication (as depicted in Fig. 20.8) requires the coincident synthesis of the two daughter strands, one in the 5′ to 3′ direction and the other in the 3′ to 5′ direction. Kornberg's DNA polymerase corresponds to the former, but no enzyme has been discovered which will synthesize a DNA strand in the 3′ to 5′ direction.

DNA is very stable and is not synthesized in a non-dividing cell (that is there is no turnover). Furthermore, DNA replication is a very accurate process and consequently spontaneous mutations are very rare. If they occurred with anything more than an extremely low frequency, the daughter molecules of DNA would, after several generations, be quite different from the original parent molecule. In other words, there would be little resemblance between parent and offspring. The frequency of mutation can be markedly increased by means of certain agents such as ionizing radiation, ultraviolet radiation, alkylating agents and certain dyestuffs, all of which are known to alter the structure of DNA or interfere with normal replication.

EXPRESSION OF GENETIC INFORMATION

The genetic information stored in DNA is known to reside in the linear sequence of bases in the molecule. Since genetic information in an organism is manifested mainly in the proteins synthesized by that organism, there must be some relationship between the base sequence in DNA and the amino acid sequence in proteins. In other words, the base sequence must constitute a code—*the genetic code*—that can be translated into protein. Since there are twenty different amino acids in proteins, but only four different bases in DNA, the genetic code obviously cannot consist of a one-to-one relation between the amino acids and the bases, that is one base signifying one amino acid. Nor can there be a two-to-one relation, because permutation of the four bases in twos along a DNA strand allows only $4^2 = 16$ different pairs, four short of the minimum requirement of twenty. In fact, the genetic code is translated linearly in groups of three bases along the DNA; each group of three bases is known as a *codon*, for example:

$$-C—T—A-|-A—C—T-|-A—G—C-$$

Permutation of the four bases in threes allows $4^3 = 64$ possible different codons, 44 in excess of the minimum requirement. The code has been described as *degenerate*, because almost all amino acids are coded for more than one codon.

The code in DNA may be likened to a tape which contains information written linearly in a language of three-letter words drawn from an alphabet of four letters A, C, G and T. Each

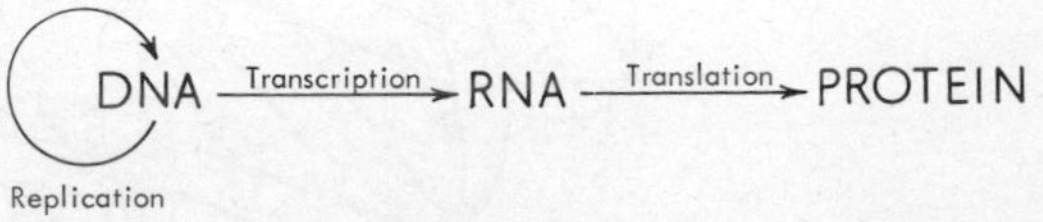

Fig. 20.9 The processes of replication, transcription and translation.

three-letter word is the code for one of the 20 amino acids so that as the tape is read the information is translated into a sequence of amino acids.

Thus the genetic information is encoded in the base sequence of DNA. This information is used to programme the biosynthesis of protein in two stages. In the first, the information is *transcribed* from DNA on to a large intermediate molecule of messenger RNA (mRNA). In the second stage, the messenger RNA-intermediary transports the information to the protein-synthesizing centres of the cell where the information is *translated* from the linear sequence of codons in the RNA into a linear sequence of amino acids which are concurrently polymerized into protein.

At this point, it is convenient to integrate these concepts with the replication of DNA in the form of Fig. 20.9.

This representation of the overall process is sometimes termed the *Central Dogma* of modern biology. The following points are now clearly established.

(i) DNA is a self-replicating molecule, that is it acts as its own template (p. 352). This is indicated by the cyclic arrow in Fig. 20.9.

(ii) All RNA molecules of the cell (p. 72) acquire their individual base sequences from DNA by being synthesized on DNA as template.

(iii) All protein molecules acquire their individual amino acid sequences from the codon sequences in messenger RNA molecules.

(iv) DNA itself cannot serve as the direct template for synthesis of protein.

The process of *replication* has been dealt with on p. 352; the molecular mechanisms of *transcription* will now be described. Protein synthesis will be dealt with in Chap. 21.

RNA SYNTHESIS

Transcription, like replication, is an enzyme-catalysed reaction; the enzyme concerned is known as *RNA polymerase* (RNA nucleotidyl polymerase). The mechanism of action of RNA polymerase is closely similar to that of DNA polymerase; in both, a DNA template and four nucleoside triphosphates are required. Transcription differs from DNA replication in that (*a*) the triphosphates are ribonucleoside 5'-triphosphates, (*b*) the product of the reaction is a single strand of RNA, and (*c*) the double stranded DNA template remains unaltered.

3',5'-Phosphodiester bonds are synthesized and inorganic pyrophosphate is eliminated during the incorporation of the monomer units into the RNA product. An important feature of transcription by RNA polymerase is that only one of the two strands of the DNA template is transcribed; this strand is termed the *informational strand* because its nucleotide sequence determines the nucleotide sequence in the transcribed RNA. The DNA strand complementary to the coding strand has merely a supporting function which ensures that DNA is replicated accurately before cell division (Chap. 53). The mechanism of transcription is illustrated in Plate 20.3 which shows the enzyme *RNA polymerase* engaged in transcribing the informational strand of a segment of DNA. It is clear that the RNA product

has a nucleotide sequence complementary to that of the informational strand of the DNA template. The complementarity arises because the process follows the usual rules of hydrogen-bonding and antiparallelism. The RNA contains uracil (U) in place of thymine (T), but each of these pyrimidine bases forms the same specific hydrogen-bonding arrangement with adenine (A) (see Fig. 6.7).

The detailed molecular mechanism of transcription (Fig. 20.10) shows that the RNA polymerase starts transcribing at a specific point on the DNA template, and that the linear 5'-triphosphate unit of the first nucleotide incorporated remains on the newly-synthesized RNA product. The direction of transcription is therefore 5' → 3', that is, the 5'-end is synthesized first and the 3'-end last.

All of the RNA of the cell, the rRNA, the tRNA and the mRNA (p. 72), is transcribed from DNA. Although the bulk of the RNA of any one cell consists of ribosomal RNA (rRNA) and transfer RNA (tRNA) only a very small proportion of the DNA of the cell serves as template for transcription of these two categories of RNA. The bulk of the DNA serves as template for synthesis of mRNA which comprises only about 1 per cent of the total cell RNA. This apparent paradox is explained by the fact that the turnover rate of mRNA is very much greater than that of rRNA or that of tRNA. In other words, once rRNA and tRNA have been synthesized, they tend to survive for long periods, whereas mRNA exists only long enough to fulfil its purpose as a template for protein synthesis (Chap. 21).

The RNA molecules synthesized during transcription are smaller than the DNA template (p. 356) and usually carry only enough information for one or a few proteins. The system must therefore have built-in signals to indicate where the RNA polymerase has to start transcribing and where it has to stop. The exact nature of these signals is not yet clear, but it is fairly certain that each stretch of mRNA transcribed from DNA proceeds to programme protein synthesis. If an mRNA molecule programmes the synthesis of one protein, it is known as a *monocistronic messenger*, having been transcribed from one *cistron* or *gene* (protein-coding unit), on the DNA template. If an mRNA molecule programmes

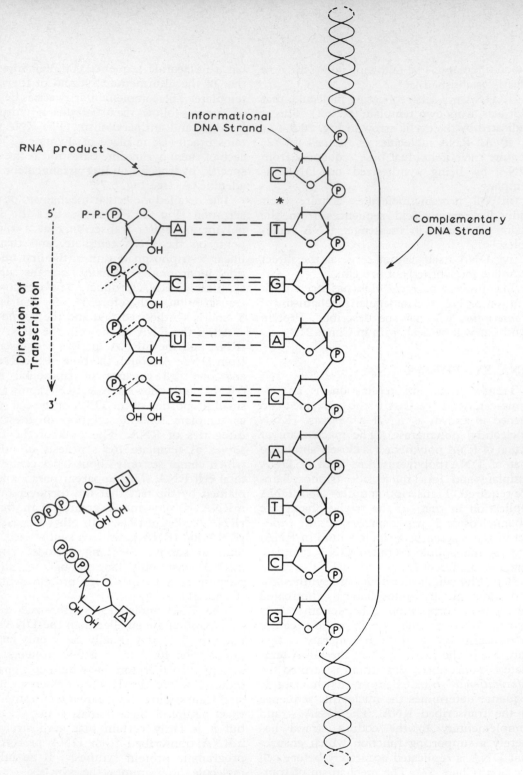

Fig. 20.10 Mechanism of biosynthesis of RNA on a DNA template. For the sake of clarity only one of the two strands (the coding strand) of the DNA is drawn out in detail in the region of transcription. This diagram should be compared with Plate 20.3.

The initiation point for transcription is indicated by the asterisk. Note that the biosynthesis is proceeding in the 5′ → 3′ direction and in accordance with the rules of complementarity and antiparallelism.

Ⓟ represents phosphate.

the synthesis of two or more proteins, it is known as a *polycistronic messenger*, that is, it has been transcribed from a consecutive sequence of cistrons or genes on the DNA template.

The synthesis of an enzyme can be controlled in the cell at the level of transcription. Protein regulator molecules attach themselves to the DNA between the point where transcription is initiated and the DNA sequence which codes for the enzyme (the gene for the enzyme). The regulator thus blocks the progression of the RNA polymerase along the DNA and prevents the synthesis of mRNA from the gene. Transcriptional control will be described in more detail in Chapter 21.

In the cells of higher organisms the DNA is associated with histones and other proteins in a deoxyribonucleoprotein complex called *chromatin* (p. 75). It is not easy to visualize how transcription can proceed unimpeded when the DNA template is so linked; however, not all of the DNA in chromatin is transcribed. Moreover, different stretches of the DNA are transcribed in the cells of different tissues of the same animal. This implies that different species of mRNA are elaborated and go on to programme the synthesis of different proteins in the various tissues, and in conformity with this it is found that different cell types do indeed synthesize different sets of protein. For example, muscle cells (but no other cells) produce vast amounts of actin and myosin, and cells of the pancreas produce certain digestive enzymes not synthesized by the cells of other tissues. It is possible that the proteins in the chromatin mask certain regions of the DNA (different regions of the DNA in different tissues), thereby preventing transcription from these regions.

Since all cellular synthesis of RNA occurs on a DNA template, the process has been described as *DNA-dependent synthesis of RNA* and the enzyme has been termed *DNA-dependent RNA polymerase*. Events are quite different when an RNA-virus (p. 77) invades a cell. In these circumstances, the viral RNA serves as mRNA for synthesis of virus-specific proteins and also as template for its own replication. Among the virus-specific proteins synthesized is an RNA polymerase which uses the parental viral RNA as template in a replicative reaction which synthesizes many daughter copies of the parent molecule. This process is termed *RNA-dependent synthesis of RNA* and the virus-induced polymerase is called an *RNA-dependent RNA polymerase*.

NUCLEASES

Nucleases catalyse the hydrolysis of internucleotide bonds. Some are specific for DNA, and others for RNA, while some are capable of hydrolysing both types of nucleic acid. Nucleases specific for DNA are termed *deoxyribonucleases*, and those specific for RNA are termed *ribonucleases*.

Some nucleases hydrolyse only the internucleotide bonds located at the ends of the nucleic acids, and thus release mononucleotides, one at a time, from the end. These enzymes are known as *exonucleases*. Some of these attack the internucleotide bonds consecutively from the 5'-end of the nucleic acid, others from the 3'-end. In contrast, other enzymes hydrolyse internucleotide bonds located at points throughout the length of the nucleic acid chains. These enzymes are termed *endonucleases*.

A well-known ribonuclease from bovine pancreas (pancreatic RNase) hydrolyses internucleotide bonds within RNA chains to give mono- and oligonucleotide products bearing 3'-phosphoryl groups. The specificity of this enzyme is such that bonds between purine nucleotides *are not* hydrolysed, bonds between adjacent pyrimidine nucleotides *are* hydrolysed; bonds between purine (Pu) nucleotides adjacent to pyrimidine (Py) nucleotides are hydrolysed only if the sequence in the RNA chain is 5'—Py—Pu—3' and *not* 5'—Pu—Py—3'.

Endonucleases may hydrolyse internal internucleotide bonds in nucleic acids to produce oligonucleotides bearing either 3'-phosphoryl-terminal or 5'-phosphoryl-terminal groups. For example, a DNA-endonuclease from bovine pancreas (pancreatic DNase) has a pH optimum of 7 to 8, and hydrolyses internucleotide bonds in double-helical DNA (single-chain scission) to yield oligodeoxyribonucleotides terminated by 5'-phosphoryl and 3'-hydroxyl groups. This enzyme has been termed *deoxyribonuclease I*. A DNA-endonuclease from bovine spleen has a pH optimum of 4·5 and hydrolyses both chains in double-helical DNA at the same point (double chain scission) to yield oligodeoxyribonucleotides terminated by 3'-phosphoryl and 5'-hydroxyl groups. This enzyme has been termed *deoxyribonuclease II*.

Exonucleases may hydrolyse the terminal internucleotide bonds in nucleic acids to produce either nucleoside 3'-monophosphates or nucleoside 5'-monophosphates. Well-known examples of exonucleases are the DNA-exonucleases, (*a*) from bovine spleen, which starts at the 5'-end of DNA chains and consecutively releases deoxyribonucleoside 3'-monophosphates, and (*b*) from snake venom, which starts at the 3'-end of DNA chains and consecutively releases deoxyribonucleoside 5'-monophosphates. (These two enzymes have also been referred to as *phosphodiesterases*.)

Some deoxyribonucleases show a preference for, or in some cases an absolute specificity for, double-helical DNA; others show the reverse type of specificity, that is they prefer single-stranded, or denatured DNA as substrate.

In summary, the nucleases show specificity in action that relates to one or more of the following:

DNA and/or RNA as substrate;
exo- or endo-nucleolytic action;
3'- *or* 5'-phosphoryl-terminal groups produced;
single-stranded and/or double-helical condition of the substrate.

Nucleases are involved in many of the important reactions of DNA. They are required for the repair of damaged DNA (p. 359), for genetic recombination, where sections of the double-stranded DNA of homologous chromosomes are interchanged, and also probably for the complex process of DNA replication.

Endo- and exonucleases with lower specificity also catalyse the hydrolysis of nucleic acids to nucleotides, in the living cell (for example, degradation of mRNA) and in the course of degradation of dead cells. Foodstuff derived from cellular material also contains nucleic acids which are degraded by pancreatic nucleases in the duodenum.

REPAIR OF DAMAGED DNA

Single-strand scissions in double-stranded DNA which have a 3'-hydroxyl group on one side and a 5'-phosphate group on the other (like those formed by pancreatic DNase, p. 357), can be repaired by the enzyme polynucleotide ligase (Fig. 20.11).

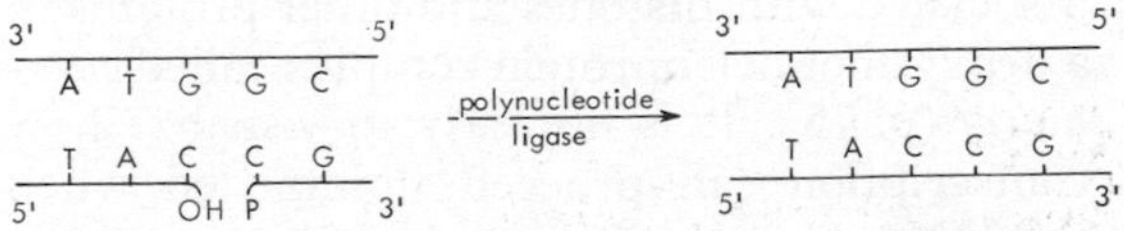

Fig. 20.11 The mode of action of polynucleotide ligase.

Similar breaks are produced in DNA *in vivo* by X-irradiation and many of them are repaired by a polynucleotide ligase.

Ultraviolet radiation, on the other hand, gives rise to thymine dimers in DNA. Absorption of energy at 260 nm causes the 5,6 double bonds of adjacent thymine bases in one strand of the DNA to rearrange and form dimers (Fig. 20.12).

Thymine dimers locally distort the regular form of the DNA double helix and block replication. The reaction is enzymically reversible in the presence of light, but in the dark such thymine dimers are removed by a complicated series of reactions (Fig. 20.13).

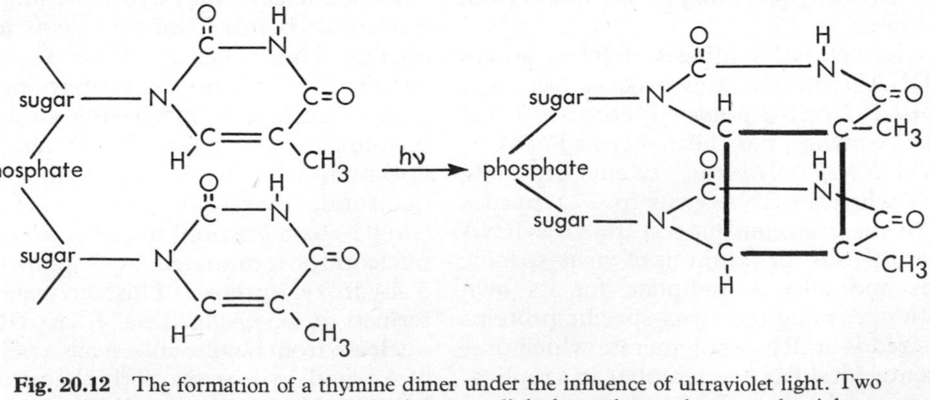

Fig. 20.12 The formation of a thymine dimer under the influence of ultraviolet light. Two adjacent thymine residues (left) in DNA become linked-together as shown on the right.

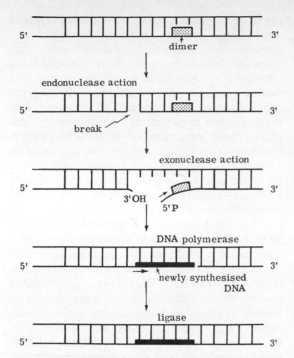

5' [diagram] 3'
dimer

endonuclease action

5' [diagram] 3'

break

exonuclease action

5' [diagram] 3'
3'OH   5'P

DNA polymerase

5' [diagram] 3'
newly synthesised DNA

ligase

5' [diagram] 3'

Fig. 20.13 The repair of damaged DNA. An endonuclease cuts the affected strand on the 5' side of the dimer. An exonuclease then excises a length of the affected strand including the dimer leaving a gap which is filled in by the polymerase. The final join between the newly synthesized DNA filling the gap, and the remainder of the original strand is brought about by the ligase.

An endonuclease recognizes the distortion in the double helix and forms a single strand scission (3'-hydroxyl, 5'-phosphate) on the 5' side of the thymine dimer. An exonuclease, working from the scission in the 5' to 3' direction, removes nucleotides and the thymine dimer. Concurrently, a DNA polymerase (with the properties of Kornberg's DNA polymerase, p. 352) adds nucleotides sequentially (using the deoxyribonucleoside triphosphates, dATP, dCTP, dGTP and dTTP) starting at the free 3'-hydroxyl of the original scission, and using the other strand of the double helix as template. Eventually, polynucleotide ligase joins the gap left by the joint action of the exonuclease and the DNA polymerase. The product has the same structure as the DNA prior to ultraviolet irradiation. This process, which is probably carried out by a complex of enzymes, is summarized in Fig. 20.13.

Repair can be envisaged as the continual monitoring of the DNA by the repair complex, which cuts out and replaces irregularities in the double helix. A heritable disease in man, xeroderma pigmentosum, which in homozygous individuals is characterized by an extreme sensitivity to sunlight and the eventual development of multiple carcinomas on the exposed parts of the body, is caused by a defect in DNA repair. The lesion has been identified with the absence of the specific endonuclease which forms single-strand scissions as the first step in the repair process.

ENZYMIC MODIFICATION OF NUCLEIC ACIDS

Some enzymes can modify the nucleic acids at the polynucleotide level. Thus, after synthesis of nucleic acids, methylation, glucosylation, phosphorylation, acylation, thiolation and other alterations of the macromolecule may take place. Of these reactions methylation is quantitatively the most important.

Methylation of DNA and RNA is effected by specific enzymes (*methyltransferases*) which recognize not only the ribo- or deoxyribo-type of nucleid acid but also specific nucleotides at specific sites within the nucleic acid chain. Only a small proportion of the bases is methylated. The general reaction is:

nucleic acid + S-adenosylmethionine →
CH_3-nucleic acid + S-adenosylhomocysteine.

The sites of methylation are predominantly on the amino groups of adenine and cytosine residues giving 6-methylaminopurine and 5-methylcytosine respectively. Any given DNA has a definite, that is characteristic, amount of methylated bases.

tRNA contains a number of methylated bases (p. 72) and also contains methyl groups at the 2'-hydroxyl position of the ribose moiety at certain specific sites in the polynucleotide chain.

CATABOLISM OF NUCLEOTIDES

Nucleotidases and nucleosidases. Some of these enzymes have very low specificities and hydrolyse both 5'-monophosphates and 3'-monophosphates, while others display a much higher degree of specificity, degrading only the 5'- or the 3'-monophosphates. Nucleotidases are known which hydrolyse the phosphate from one specific nucleotide only.

The collective action of these enzymes gives

rise to pentose, pentose phosphate, and to the free purine and pyrimidine bases.

Catabolism of pyrimidines. The pyrimidine bases are catabolized in mammalian tissues by preliminary reduction of uracil and thymine to the corresponding dihydro-derivatives, followed by ring opening to give the appropriate ureido-acid, and removal of ammonia and carbon dioxide to give β-alanine or its methylated derivative β-aminoisobutyric acid (Fig. 20.14).

Catabolism of purines. The purine bases arising from degradation of the nucleic acids are conveyed in the blood to the liver where they are catabolized still further through a well-known series of changes resulting in the formation of uric acid (Fig. 20.15). Adenine is deaminated by *adenine deaminase* (adenase) to yield hypoxanthine. Guanine is deaminated by *guanine deaminase* (guanase) to give xanthine. Further oxidation is brought about by *xanthine oxidase* in the liver. This enzyme oxidizes hypoxanthine to xanthine, and xanthine to uric acid which, in primates, is excreted by the kidney. The amount of uric acid excreted per day in human urine varies widely from 0·1 to 2·0 g.

Uric acid is excreted in large amounts by birds and reptiles, but in them it is the end-product not only of purine catabolism but also of protein catabolism. Instead of excreting urea, they excrete uric acid as a semi-solid paste. They can thus conserve water by avoiding the necessity of producing urine in the form of a dilute aqueous solution.

Human blood normally contains 3·0 to 6·0 mg of uric acid per 100 ml. This value tends to rise whenever cells are being destroyed and nucleoprotein liberated. Raised levels are therefore encountered in such conditions as leukaemia, polycythaemia, and in pneumonia during the process of resolution. In chronic renal disease, the excretion of uric acid, like that of other nitrogenous compounds, is impaired and retention of uric acid occurs. Probably the highest values for uric acid in plasma are found in gout immediately before an acute attack. In this disease, salts of uric acid are actually deposited in the tissues, causing the typical chalky swellings (tophi) but it is not clear why such large amounts of uric acid should be formed in gouty persons; there may be a defect in the feed-back control (p. 113) of the enzyme that forms 5′-phosphoribosylamine (PRA) (Fig. 20.4), a precursor of purine nucleotides. The disorder is found mainly in males and appears to be genetically determined. Gout may be controlled effectively by the substance allopurinol which inhibits xanthine oxidase and so prevents the formation of uric acid. The end products of purine metabolism excreted in the urine of patients treated with allopurinol are the much less insoluble compounds hypoxanthine and xanthine (p. 64).

Most mammals, apart from man and other primates, oxidize uric acid further by the action of *uricase* to give allantoin. Hydrolysis of allantoin by the enzyme *allantoinase* yields allantoic acid, the end-product of purine catabolism in certain teleost fishes. In most fishes and in amphibia, however, allantoic acid

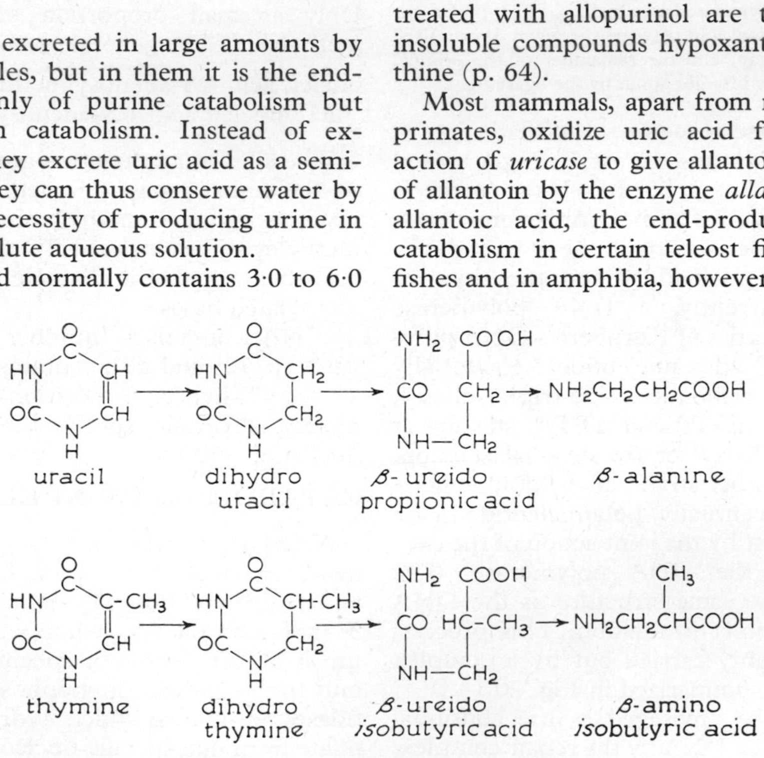

Fig. 20.14 The catabolism of pyrimidine bases.

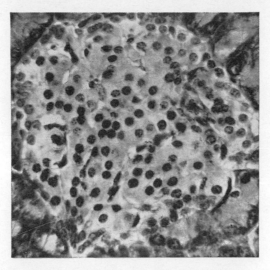

Plate 18.1A From a control (healthy) rat showing relatively few α-cells (with red cytoplasm) mainly at the edge of the islet which contains a large number of β-cells. (M. K. MacDonald & S. K. Bhattacharya (1956). *Quarterly Journal of Experimental Physiology* **41**, 153–161.) × 500.

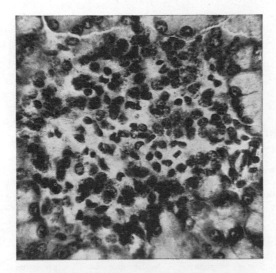

Plate 18.1B From a rat made hyperglycaemic by repeated intravenous injections of dehydroascorbic acid. There are relatively few β-cells remaining towards the centre of the islet but numerous α-cells surrounding them. (M. K. MacDonald & S. K. Bhattacharya (1956). *Quarterly Journal of Experimental Physiology* **41**, 153–161.) × 500. Stain in both A and B: Gomori's chrome alum haematoxylin and phloxin.

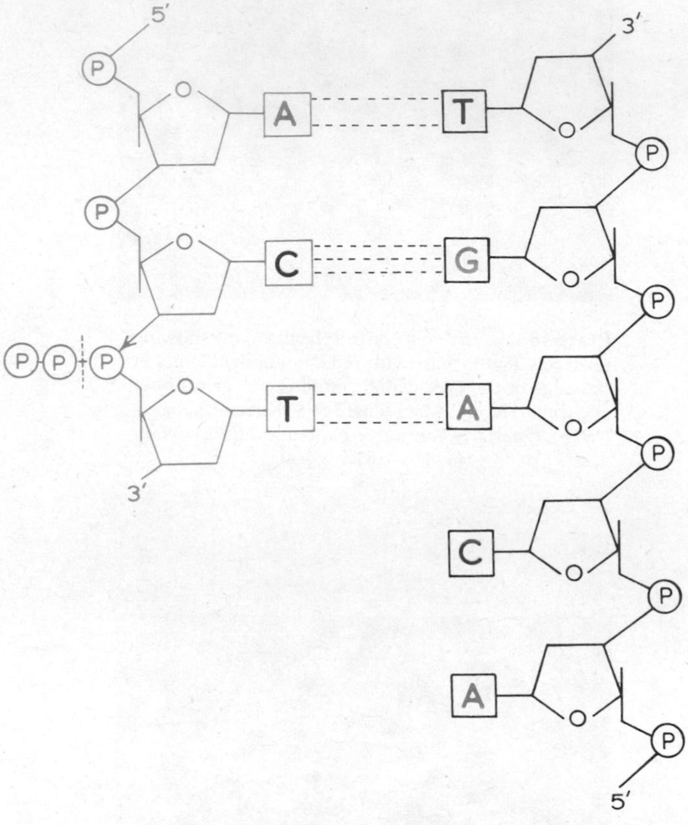

Plate 20.1 Biosynthesis of DNA. A single-stranded template is shown on the right. Mononucleotides in the form of deoxyribonucleoside 5′-triphosphates align themselves in appropriate sequence, and in keeping with the antiparallel orientation, along the template DNA. In the diagram, the nucleotide containing thymine is about to be linked to carbon-3′ of the preceding nucleotide (containing cytosine), and inorganic pyrophosphate is about to be eliminated. In the same way, the next mononucleotide in the sequence to be synthesized will align itself opposite the cytosine nucleotide in the template; it will therefore be dGTP and the new chain is thus extended by one further unit. Dotted lines between the bases represent hydrogen bonds. A = adenine, C = cytosine, G = guanine, T = thymine and P = phosphate.

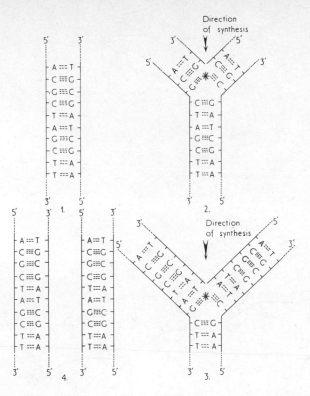

Plate 20.2 Replication of double-stranded DNA. 1. A small segment of double-stranded DNA showing the bases joined by hydrogen bonds. 2. An early phase of replication of the segment; newly synthesized DNA is shown in RED and the actual growing point is indicated by the RED asterisk. 3. A later stage of the replication process. 4. The two complete daughter segments are shown with the newly synthesized strands and appropriate complementary base-pairing. These double-stranded segments are identical with each other and with the parent segment shown in 1.

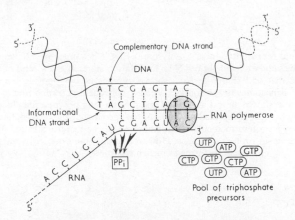

Plate 20.3 Biosynthesis of RNA by transcription from a DNA template. The double-helical configuration of the DNA has become 'loosened' at the point of active transcription catalysed by the RNA polymerase. Transient base-pairing (A with U, C with G, G with C and T with A) is thus permitted between the coding strand of the DNA and the incoming ribonucleoside 5′-triphosphates. Polymerization of the monomers ensues by formation of 3′,5′-phosphodiester bonds. The polymerase advances to the right along the double-helix, opening new transcription sites. Concurrently, the section of 'loosened' DNA at the left resumes its original 'tight', double-helical configuration, and the product RNA peels off the template. Note that the nucleotide sequence of the RNA product is complementary to that of the coding strand. PP_i = Inorganic pyrophosphate.

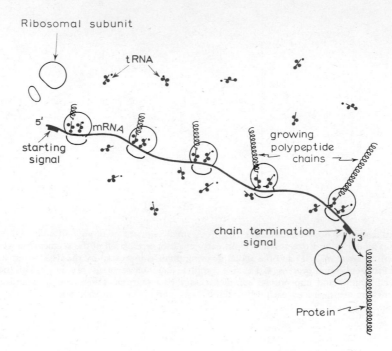

Ribosomal subunit

tRNA

5'

starting
signal

mRNA

growing
polypeptide
chains

chain termination
signal

3'

Protein

Plate 21.1 Diagrammatic representation of a polysome. Five ribosomes are shown at different stages of translation on the same molecule of mRNA. Translation starts at the 5'-end of the mRNA. The ribosomes nearer to that end therefore have short stretches of polypeptide bound to them through tRNA molecules. Ribosomes nearer to the 3'-end are approaching completion of the translation process; one ribosome is shown almost at the end and it has a long polypeptide chain associated with it. Another ribosome has just completed translation, has recognized the termination signal and has discharged its completed protein molecule. That ribosome is free to return to the 3'-end of the mRNA molecule to re-engage in protein synthesis. However, at any one time, only one polypeptide chain can be formed on each ribosome.

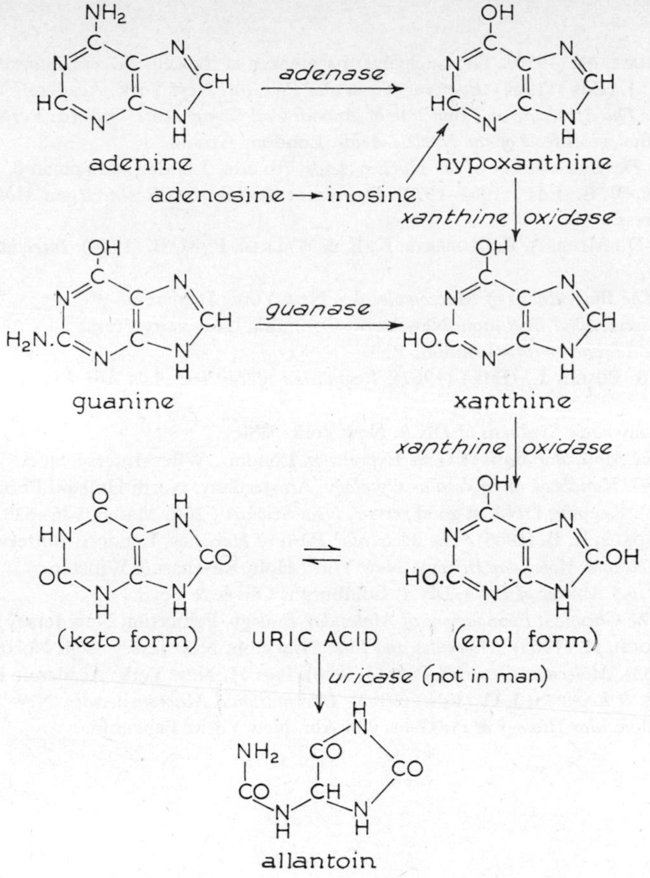

Fig. 20.15 The catabolism of purine bases.

is hydrolysed to glyoxylic and urea and in these species purine catabolism does not proceed further. Some marine invertebrates degrade urea to ammonia and carbon dioxide and in these species, the original purine-nitrogen is excreted as ammonia. The relationship of allantoin to these other metabolites is shown below.

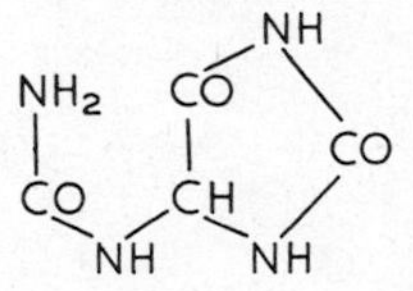

Allantoin

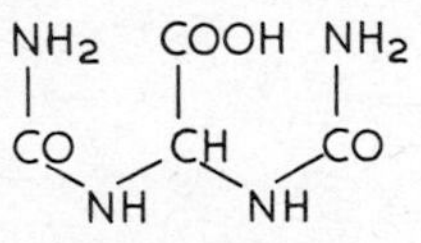

Allantoic acid

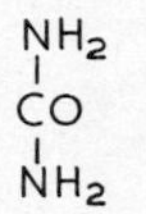

Urea

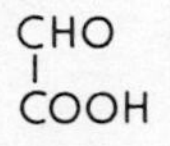

Glyoxylic acid

REFERENCES

BIRNSTIEL, M. & CHIPCHASE, M. (1970). The nucleolus: pacemaker of the cell. *Science Journal* **6** (7), 41–47.

BRYSON, V. & VOGEL, H. J. (Eds.) (1965). *Evolving Genes and Proteins*. New York: Academic Press.

CAMPBELL, P. N. (1966). *The Structure and Function of Animal Cell Components*. Oxford: Pergamon Press.

COHEN, D. (1965). *The Biological Role of the Nucleic Acids*. London: Arnold.

DAVIDSON, J. N. (1972). *The Biochemistry of the Nucleic Acids*, 7th edn. London: Chapman & Hall.

DAVIDSON, J. N. & COHN, W. E. (Eds.) (1963–1972). *Progress in Nucleic Acid Research and Molecular Biology*, Vols. 1–12. New York: Academic Press.

HAGGIS, G. H., MICHIE, D., MUIR, A. R., ROBERTS, K. B. & WALKER, P. M. B. (1964). *Introduction to Molecular Biology*. London: Longmans.

INGRAM, V. M. (1965). *The Biosynthesis of Macromolecules*. New York: Benjamin.

JUKES, T. H. (1966). *Molecules and Evolution*. New York: Columbia University Press.

KENDREW, J. (1966). *The Thread of Life*. London: Bell.

KONINGSBERGER, V. V. & BOSCH, L. (Eds.) (1967). *Regulation of Nucleic Acid and Protein Biosynthesis*. Amsterdam: Elsevier.

KORNBERG, A. (1961). *Enzymatic Synthesis of DNA*. New York: Wiley.

LEWIN, B. M. (1970). *The Molecular Basis of Gene Expression*. London: Wiley-Interscience.

LIMA-DE-FARIA, A. (1969). *Handbook of Molecular Cytology*. Amsterdam: North Holland Publishing Co.

MOSELEY, B. E. B. (1969). Keeping DNA in good repair. *New Scientist*, 20th March, 626–628.

PEACOCKE, A. R. & DRYSDALE, R. B. (1965). *The Molecular Basis of Heredity*. London: Butterworth.

ROSENBERG, E. (1971). *Cell and Molecular Biology*. New York: Holt, Rinehart & Winston.

SMELLIE, R. M. S. (1969). *A Matter of Life—DNA*. Edinburgh: Oliver & Boyd.

STEINER, R. F. (1965). *The Chemical Foundations of Molecular Biology*. Princeton, New Jersey: Van Nostrand.

STEINER, R. F. & EDELHOCH, H. (1965). *Molecules and Life*. Princeton, New Jersey: Van Nostrand.

TAYLOR, J. H. (Ed.) (1963). *Molecular Genetics*, Part I, (1968). Part II. New York: Academic Press.

VOGEL, H. J., BRYSON, V. & LAMPEN, J. O. (Eds.) (1963). *Informational Macromolecules*. New York: Academic Press.

WATSON, J. D. (1970). *Molecular Biology of the Gene*, 2nd edn. New York: Benjamin.

21 Protein synthesis

THE GENETIC CODE

Genetic information resides in the linear sequence of nucleotides in DNA, and is transcribed from the DNA into mRNA before protein is synthesized. The genetic code consists of three-letter code-words drawn from an alphabet of four letters A, C, G and U (Chap. 21) so that there are 64 (4^3) possible code-words or codons. Sixty-one of the codons have been assigned to amino acids (Table 21.1).

Table 21.1 The genetic code

5'-OH terminal base	Middle base				3'-OH terminal base
	U	C	A	G	
U	Phe	Ser	Tyr	Cys	U
	Phe	Ser	Tyr	Cys	C
	Leu	Ser	CTS	CTS	A
	Leu	Ser	CTS	Trp	G
C	Leu	Pro	His	Arg	U
	Leu	Pro	His	Arg	C
	Leu	Pro	Gln	Arg	A
	Leu	Pro	Gln	Arg	G
A	Ile	Thr	Asn	Ser	U
	Ile	Thr	Asn	Ser	C
	Ile	Thr	Lys	Arg	A
	Met	Thr	Lys	Arg	G
G	Val	Ala	Asp	Gly	U
	Val	Ala	Asp	Gly	C
	Val	Ala	Glu	Gly	A
	Val	Ala	Glu	Gly	G

CTS = Chain termination signal.

In a stretch of mRNA grouping of the nucleotides in threes gives the codon sequence in the mRNA, for example:

—CAC|CUG|AAG|UCA|GUU|GAU|GAA—

The codon assignments presented in Table 21.1 show that the above segment of mRNA corresponds to, or can be translated into, the amino acid sequence:

—His—Leu—Lys—Ser—Val—Asp—Glu—

The sequence of codons cannot be recognized directly by the corresponding free amino acids. However, base pairing (p. 71) allows specific association of two complementary polynucleotide strands, and it is by such complementarity, and therefore indirectly, that the amino acids 'recognize' their correct codons on the mRNA. This recognition process is achieved through the mediation of the tRNA molecules.

TRANSFER RNA

Each tRNA molecule contains in its nucleotide sequence a stretch of three nucleotides (the anti-codon) *complementary* to the three nucleotides of an amino acid codon, and each amino acid can be specifically attached to a tRNA molecule containing the corresponding anti-codon.

Before amino acids can be attached to the correct tRNA molecules, they must be *activated* through their carboxyl groups. The activation is catalysed by enzymes known as *aminoacyl-tRNA synthetases*, there being at least one synthetase specific for the activation of each amino acid. The first step consists of the formation of a high-energy acyl bond between the carboxyl group of the amino acid and the α-phosphate of ATP, the β- and γ-phosphates of the ATP being removed as inorganic pyrophosphate. The energy for the reaction is derived from the high-energy pyrophosphate bond in the ATP, and the aminoacyl-AMP so formed is held as a complex on the surface of the synthetase (Fig. 21.2). In the second step, the activated amino acid is passed on to its specific tRNA, a high-energy ester bond being formed between its carboxyl group and the 3'-hydroxy-group on the terminal adenosine moiety of the tRNA. The tRNA is then said to be *loaded* or *charged* with the amino acid, and is therefore primed to pass the amino acid into the protein-synthesizing machinery.

It is clear that the *aminoacyl-tRNA synthetase* must possess two binding sites. The first site recognizes the specific amino acid, and the second site selects the specific tRNA molecule to which that amino acid is to be covalently bonded. In other words each aminoacyl-tRNA synthetase is a highly specific enzyme capable of selecting one amino acid, and only one, out of 20, and then of selecting a species of tRNA that corresponds to the amino acid but no other. In this way, many molecules of tRNA become *loaded* with their specific amino acids, and the system is set to provide amino acids for the protein-synthesizing machinery.

Each tRNA molecule likewise has two selection sites in its structure. The first recognizes its specific aminoacyl-tRNA synthetase and therefore ensures that the tRNA species concerned becomes loaded with the correct amino acid (Fig. 21.2). The second site (the anti-codon) recognizes the corresponding codon on mRNA molecules by base pairing in an anti-parallel way, and thereby ensures that the correct amino acid is presented at the appropriate instant for incorporation into the growing polypeptide chain during protein synthesis.

In other words, the tRNA molecules, each loaded with its specific amino acid, serve as adaptors for assembling amino acids in the correct sequence along mRNA molecules. It is therefore through the agency of the tRNA molecules that translation of a nucleotide sequence into an amino acid sequence is accomplished.

RIBOSOMES

As described in Chapter 6, ribosomes contain two subunits, each a complex of RNA and protein. One subunit is about twice the size of the other. mRNA binds to a site on the small subunit and then the mRNA-small subunit complex binds to the large subunit, that is ribosomes are assembled on mRNA. Two codons (six nucleotides) of mRNA are available in the ribosome at any one time. The large subunit has two binding sites for tRNA molecules, one, the amino-acyl-tRNA site, accepts loaded tRNA molecules corresponding to the next codon on the mRNA, and the other, the peptidyl-tRNA site, binds the tRNA carrying the growing polypeptide chain (see p. 366).

TRANSLATION

Translation is the process which culminates in the synthesis of protein molecules. In it all species of RNA, namely rRNA (as it exists in inact ribosomes), tRNA and mRNA, participate.

The overall molecular mechanism of translation is quite complex. The fundamental principle is presented diagrammatically in Fig. 21.3 which illustrates the sequence of events in the translation of a segment of mRNA into the corresponding segment of polypeptide. The chronological sequence may be summarized thus:

(i) All species of tRNA are loaded with their specific amino acids in readiness for the synthesis of protein.

(ii) A molecule of mRNA is attached to a ribosome as described above: at one end of the molecule a sequence of nucleotides (the starting signal) indicates where translation must begin.

(iii) The sequence of codons that follows the starting signal, always beginning with AUG (the initiation codon, which codes for N-formyl methionine), is translated, one codon at a time, into a sequence of amino acids. Thus,

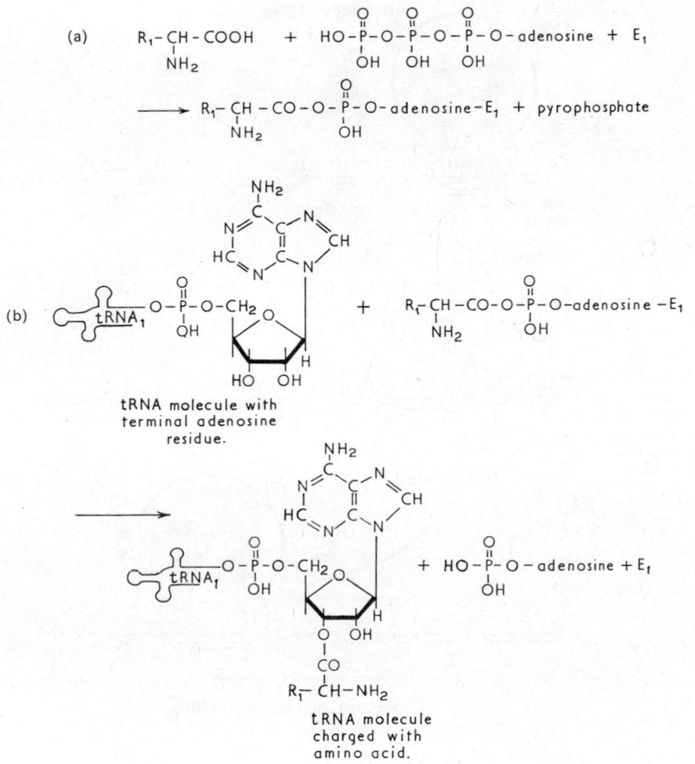

Fig. 21.2 Activation of an amino acid in preparation for protein synthesis. In the first step, (a), the aminoacyl-tRNA-synthetase (E_1) catalyses synthesis of a high-energy aminoacyl bond between the α-phosphate of ATP and the carboxyl group of amino acid$_1$; pyrophosphate is eliminated in the process. The aminoacyl complex remains bound on the surface of the synthetase at this stage. In the second step (b), the aminoacyl complex selects its specific tRNA molecule (tRNA$_1$) and transfers the activated amino acid residue to the 3′-hydroxyl group located on the terminal adenosine moiety of the tRNA. The ester bond so formed is a high-energy bond. AMP and the free synthetase are liberated from the reaction. Note that the aminoacyl-tRNA-synthetase catalyses both reactions, the initial formation of the aminoacyl-AMP complex (a), and then the transfer of the activated amino acid to tRNA (b). The overall process is highly specific, amino acid$_1$ being activated only by aminoacyl-tRNA-synthetase$_1$ (E_1); the aminoacyl complex selects only a tRNA specific for amino acid$_1$, that is, it selects tRNA$_1$. The energy for the activation comes from the ATP present initially in reaction (a).

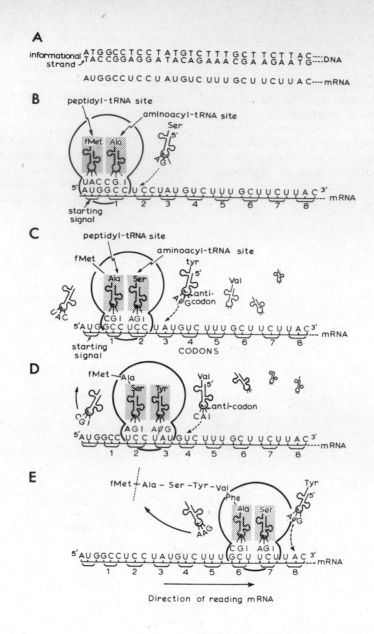

Fig. 21.3 Synthesis of a polypeptide directed by a segment of a molecule of mRNA transcribed from a segment of DNA (A). In B the starting signal AUG of the mRNA has become associated with formylmethionyl-tRNA (fMet-tRNA) on the peptidyl-tRNA site on the ribosome while the aminoacyl tRNA site is occupied by Ala-tRNA corresponding to codon 1(GCC). In C the ribosome has moved one stop to the right and codon 1 now carries the tRNA bearing the dipeptide fMet-Ala while the aminoacyl-tRNA site bears Ser-tRNA associated with codon 2(UCC). In D the process has gone further and the peptidyl-rRNA site is occupied by the tRNA bearing the peptide fMet-Ala-Ser-Tyr-Val-Phe-Ala and associated with codon 6 (GCU for alanine) while the aminoacyl-tRNA site is occupied by Ser-tRNA associated with codon 7 (UCU for serine). The symbol ψ represents pseudouridine, a modified form of uridine.

AUG is the first codon in Fig. 21.3, and a molecule of loaded N-formyl methionyl tRNA whose anti-codon is CAU, complementary to AUG, enters the amino acyl-tRNA site on the ribosome (Fig. 21.3A). The ribosome then moves along the mRNA by one codon (three nucleotides) and the N-formyl methionyl tRNA moves to the peptidyl-tRNA site. The anti-codon of N-formyl methionyl-tRNA appears in Fig. 21.3 as UAC but it is important to recall that the convention for writing nucleotide sequences in the text (Fig. 6.2) is to name the sequence from left to right and from the 5′-end to the 3′-end. The examples quoted above for N-formyl methionine, AUG for the codon and CAU for the anti-codon therefore meet this requirement,

5′–AUG–3′; 5′–CAU–3′.

However, the diagram in Fig. 21.3, makes it clear that the rules of complementary and anti-parallel configuration are obeyed. Other suitable diagrammatic representations are therefore

$\overleftarrow{\text{UAC}}$ 3′–UAC–5′

or

$\overrightarrow{\text{AUG}}$ 5′–AUG–3′.

The same considerations apply to all codon–anticodon relationships.

(iv) The second codon (GCC) corresponds to alanine (Table 21.1) and therefore a tRNA molecule loaded with alanine enters the amino acyl site on the ribosome. Its anti-codon is IGC (I (p. 66) pairs with C).

(v) A peptide bond is now formed between N-formyl methionine and alanine. The alanine remains attached through its carboxyl group to its tRNA molecule but, since the carboxyl group of the N-formyl methionine is now involved in peptide bond formation (—CO—NH—) with the amino group of alanine, the unloaded tRNA for N-formyl methionine is cast off from the system (Fig. 21.3B). The formylated amino group of the N-formyl methionine remains free. It can therefore be seen that polymerization of the amino acids is proceeding in the $H_2N— \rightarrow$ —COOH direction.

(vi) At this point, the ribosome again moves one codon to the right along the mRNA molecule, and the tRNA molecule bearing the dipeptide (N-formyl methionylalanine) now occupies the peptidyl-tRNA site just vacated by the unloaded tRNA molecule that originally carried N-formyl methionine. At the same time, seryl-tRNA enters the vacant aminoacyl-tRNA site on the ribosome, and this becomes associated through its anticodon (IGA) with the third codon (UCC) on the mRNA molecule.

(vii) The next step of polymerization consists in the formation of a peptide bond between the amino group of serine and the carboxyl group of alanine. The tRNA molecule for alanine is cast off, now unloaded, while the tRNA for serine, carrying the tripeptide, N-formyl methionyl alanylserine, bonded to its 3′-end, moves to the peptidyl-tRNA site as the ribosome again moves on by one codon. Concurrently, tyrosyl-tRNA takes up position in the aminoacyl-tRNA site opposite the next codon (UAU).

(viii) The process continues with stepwise lengthening of the peptide, and in Fig. 21.3E synthesis has come to the heptapeptide stage.

(ix) The polymerization continues until a point on the mRNA is reached which gives the stop signal or polypeptide chain termination signal. At present, three chain termination signals are known (Table 21.1), in bacterial systems. It is not certain whether these or other chain termination signals exist in the cells of higher organisms.

(x) Finally, when the chain termination signal is reached, the completed polypeptide chain or protein molecule is released together with the free ribosomal subunits which carried it along the mRNA molecule during its growth. The subunits are of course available to return to the starting signal to reassemble, and to conduct the synthesis of another molecule of the protein.

A molecule of mRNA is very long relative to the diameter of a ribosome, and it is therefore possible for several ribosomes to be bound to a molecule of mRNA at any one instant. Threads of mRNA bearing several ribosomes have been isolated from certain systems (Fig. 21.4). Such complexes are known as polyribosomes or polysomes. Each component ribosome of the polysome is engaged in protein synthesis but is at a different stage of translation of the molecule of mRNA. Thus, ribosomes which have newly become bound to a molecule of

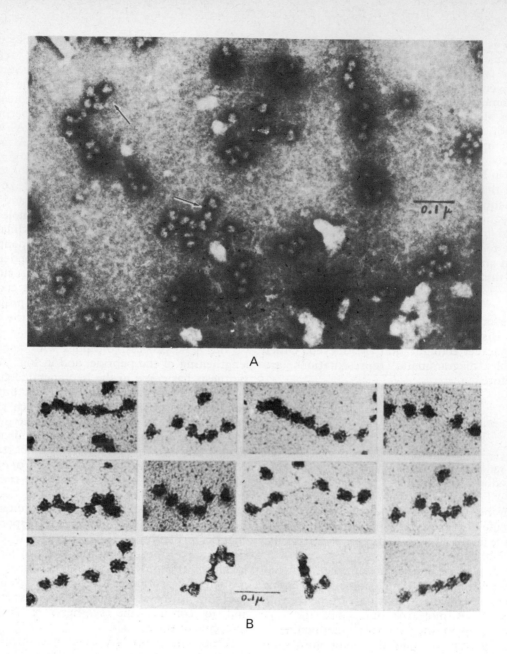

Fig. 21.4 Electron micrograph of rabbit reticulocyte polysomes. A. Negatively-stained polysomes. Occasionally, light strands (shown by the arrows) can be seen connecting adjacent ribosomes. B. Positively-stained polysomes. Examples of the extended configuration showing ribosomes threaded on to mRNA strands. The average picture shows five or six ribosomes associated with one molecule of mRNA at any one instant. This pentamer or hexamer structure is characteristic of reticulocyte polysomes engaged in synthesis of polypeptide chains for assembly into haemoglobin. Since there are 150 amino acids per polypeptide chain (= 450 nucleotides in mRNA) the inter-ribosome distance is some 80 nucleotides in the mRNA strand. (A. Rich, J. R. Warner & H. M. Goodman (1963). *Cold Spring Harbor Symposia on Quantum Biology* **28**, 269, Fig. 4.)

mRNA are at an early stage of the translation process. Ribosomes which have moved along the mRNA molecule almost to the end have virtually completed the translation; they must therefore have a long chain of polypeptide bound to them through a tRNA molecule bound in the peptidyl-tRNA site. This concept is illustrated in Plate 21.1.

Molecules of mRNA vary in length according to the size of the protein molecules to be synthesized under their direction. Further, some of the larger mRNA molecules may have several initiation and termination codons, and may programme the synthesis of several protein molecules. These mRNAs are known as polycistronic messengers.

If it is assumed that a typical protein contains 500 amino acids then, since each amino acid is represented by one codon of three nucleotides in the mRNA, the mRNA corresponding to that typical protein must contain some 1500 nucleotides. Moreover, it must have been transcribed from a stretch of DNA corresponding in length to 1500 nucleotide pairs and with a molecular weight of 1500×660, that is 10^6. The typical gene, or cistron, therefore, which is the smallest piece of a DNA helix which can carry sufficient information to determine the composition of one protein, corresponds to a segment of DNA of molecular weight about 10^6. Since four different types of base are available, the total possible number of different genes of molecular weight 10^6 is 4^{1500}, an astronomical figure.

Herpes virus causes cold sores in man and its DNA (molecular weight, 8×10^7) contains 120 000 nucleotide pairs, enough information for the formation of some 120 000/1500, that is 80 proteins. The bacterium *Escherichia coli* contains, in the resting state, a single chromosome consisting of one cyclic DNA molecule about 1 mm long. Its molecular weight is about $2 \cdot 5 \times 10^9$ and it contains $(2 \cdot 5 \times 10^9)/660$, that is 4×10^6 nucleotide pairs $= (4 \times 10^6)/1500 = 2000$ to 3000 genes. Thus it carries information for the synthesis of 2000 to 3000 proteins.

The component polypeptide chains of haemoglobin each contain about 150 amino acids (Chap. 25). This corresponds to 450 nucleotides in mRNA for each polypeptide chain. The polysomes concerned with the biosynthesis of haemoglobin are believed each to contain five or six ribosomes (Fig. 21.4). Polysomes involved in the synthesis of larger polypeptide chains (for example, 500 amino acids long, corresponding to 1500 nucleotides in mRNA) may carry up to 20 ribosomes, each of which is actively engaged in translating the mRNA of molecular weight about 500 000.

CONTROL OF PROTEIN SYNTHESIS

Some of the most important recent developments in our knowledge of the physiology of the cell relate to the mechanisms by which enzyme reactions are controlled. Most of our information about these control mechanisms has been derived from the study of bacterial systems, but there is reason to believe that similar systems operate in the mammalian cell. One form of control mechanism, known as feedback inhibition, has already been described (Chap. 8). Feedback inhibitions means that the distal product, or the end product of a series of reactions, blocks the first step in its own production by binding itself to the enzyme and inhibiting its function through allosteric modification of the enzyme structure.

Another, completely different type of control mechanism controls the rate of synthesis of the enzyme. The synthesis of an enzyme can be *repressed* or *induced*: an increase in concentration of the first substrate of a series of reactions induces the enzymes required for its metabolism, whereas an increase in concentration of an end-product of a series of reactions represses the enzymes responsible for its synthesis. Together, induction and repression of enzyme synthesis allows a cell to respond to the fluctuations of supply and demand of substrates and end-products with considerable economy. These different forms of control are illustrated in Fig. 21.5.

ENZYME INDUCTION

The classical example of enzyme induction is the utilization of lactose by yeast. Yeast cells cannot normally use lactose because they lack the enzyme β-galactosidase necessary for the conversion of lactose to glucose and galactose. However, after yeast cells have been exposed to lactose for a few hours they develop the appropriate enzyme. The presence of lactose has included the cells to produce the enzyme β-galactosidase necessary for its utilization.

As a result of their studies of this type of

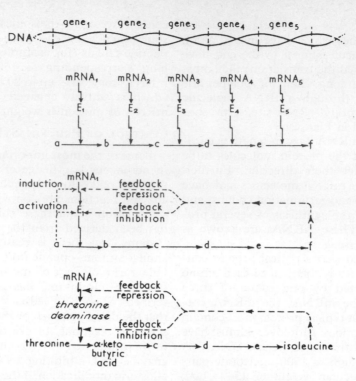

Fig. 21.5 *Top.* A series of five enzyme-controlled reactions. Each enzyme E is formed by the appropriate mRNA derived from the corresponding gene. *Middle.* Feedback inhibition and repression by the final product of the reaction. *Bottom.* A specific example from the conversion of threonine to isoleucine (see text).

reaction in the micro-organism *Escherichia coli*, Monod and Jacob in Paris developed their now famous concept of the way in which enzyme synthesis is controlled. They showed that two separate genes are involved. One of them, regulating the production of the enzyme, is known as the *regulator gene*, while the second, governing the amino acid sequence of the enzyme, is known as the *structural gene*. Genetic analysis shows that these two genes may be contiguous or may be quite widely separated on the bacterial chromosome. Mutations in the structural gene result in the production either of no enzyme at all or of an altered, and consequently inactive, enzyme. On the other hand, mutations of the regulator gene usually lead to the production of uncontrolled amounts of enzyme quite unrelated to the needs of the cell.

Each set of structural genes contained in a length of DNA is found closely adjacent to two other sections of DNA which do not code for proteins. These are the *promotor site* and

the *operator site*. The promotor site, the operator site and the associated structural genes form a complex which Monod and Jacob called an *operon*. The operon in turn is under the control of the regulator gene, which is the structural gene for the regulator protein (Fig. 21.6).

When the regulator protein is bound to the operator site, the RNA polymerase, which initially binds to the promotor site, cannot transcribe the structural genes.

When the regulator protein is not bound to the operator site, the RNA polymerase can move along the DNA and thus is able to transcribe the structural genes. The mRNA so formed is translated to form the polypeptide chains of the enzymes corresponding to the genetic information in the operon. A gene may be controlled independently and produce its own mRNA molecule or a group of genes may be controlled together and produce one long mRNA molecule (a polycistronic messenger) corresponding to all the enzymes represented in the operon.

In repression, the control is negative, in the sense that the regulator protein in its *active* *form inhibits* mRNA synthesis and therefore the subsequent synthesis of the corresponding enzyme.

The activity of a regulator protein is

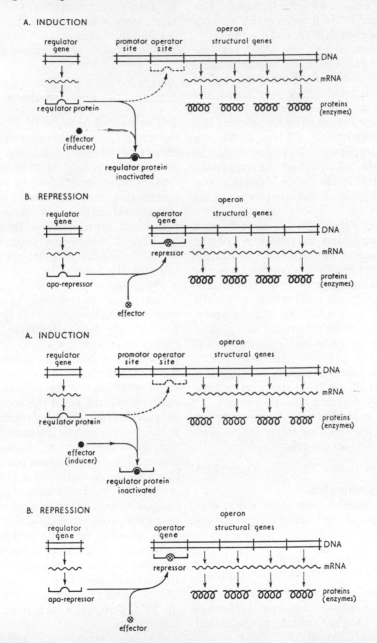

Fig. 21.6 The operon. A. The regulator gene produces a repressor which blocks the operator gene and prevents the structural genes from producing mRNA. In the presence of an effector (inducer) the repressor is inactivated and the operator gene allows the structural genes to come into action. B. The repressor does not become fully operative until it is combined with the effector which may be a product of one of the enzymes produced by the structural genes (feed-back repression).

governed by a specific metabolite known as an effector. In the formation of inducible enzymes the *inducer* acts as an effector and inactivates the regulator protein so that the operator gene ceases to be repressed. The genes in the operon are therefore able to produce the appropriate mRNA so that the polypeptides represented in the DNA of the operon can be synthesized. In mutant strains of yeast, which have a defective regulator gene no inducer is needed; they synthesize the enzymes in an uncontrolled manner.

In a repressible system the end-product of a sequence of reactions acts as a type of effector known as a corepressor. It reacts with the regulator protein (often called the repressor) and the complex binds to the operator and blocks the synthesis of the structural genes (Fig. 21.6).

An example of repression has already been given in the inhibition by isoleucine of the formation of the enzyme threonine deaminase (Fig. 21.5). Another example comes from the study of the histidine operon in certain microorganisms. Histidine is synthesized by a well-established pathway involving ten enzymes, the structural genes for which are in a cluster on the bacterial chromosome. The mRNA for this system is polycistronic and controls the formation of all ten enzymes in the operon. The production of all these enzymes is prevented by histidine acting as a corepressor.

A few other examples may be mentioned. Formation of the enzyme *tryptophan synthetase* which catalyses the reaction:

$$\text{indole-3-glycerol phosphate} \xrightarrow{+\text{serine}} \text{tryptophan}$$

is repressed by tryptophan.

Arginine represses the formation of *ornithine carbamoyl transferase* (p. 368) which catalyses the reaction:

$$\text{Ornithine} + \text{carbamoyl phosphate} \rightarrow \text{citrulline} \rightarrow \text{arginine.}$$

Cytidine triphosphate (CTP) represses the formation of the enzyme *aspartate carbamoyl transferase*, an enzyme which is discussed further elsewhere (Chap. 8).

Although the operon concept has been established by the study of bacterial systems, there are good reasons to suppose that it can be extended, perhaps with certain modifications, to mammalian systems. Indeed enzyme induc-

tion can readily be demonstrated in mammalian cells. One of the best known examples is *tryptophan pyrrolase* which converts tryptophan to N-formylkynurenine and is produced in large amounts in the livers of rats fed on a high tryptophan diet. Other enzymes inducible in mammalian systems are thymidine kinase, alkaline phosphatase and arginase.

ANTIBIOTICS AND BIOSYNTHESIS OF NUCLEIC ACIDS AND PROTEINS

The overall process of biosynthesis of the nucleic acids and proteins is a complex affair clearly dependent upon unerring co-operation among a relatively large number of factors. This complexity is matched by the number of agents, that interfere with the system, including certain antibiotics and related drugs. By the use of the appropriate antibiotic it is possible to inhibit selectively some aspects of the overall biosynthetic process without interfering with the others.

Actinomycin D and *mitomycin C* are drugs which bind themselves to DNA. Actinomycin associates by forming specific hydrogen-bonds with deoxyguanosine residues in the polymer; it does not associate with guanosine residues and therefore does not bind itself to RNA. Its main action is therefore to inhibit replication and transcription. The effect is easily explained since the movement of DNA polymerase and RNA polymerase along the DNA template must be seriously impaired by the molecules of drug bound to the DNA chains. The transcription process is at least 10 times more sensitive to actinomycin than is replication and, therefore, with the appropriate concentration of drug, it is possible to eliminate transcription selectively.

In contrast, mitomycin inhibits replication by forming covalent cross-links *in vivo* between the two strands of double-helical DNA. The strands are therefore not separable and the DNA cannot replicate semi-conservatively (p. 368). Transcription is much less affected by mitomycin.

Although actinomycin and mitomycin have proved to be useful experimentally, they have little or no medical application since bacterial invaders and mammalian cells are affected with equal severity.

Certain synthetic drugs such as the acridine

and ethidium derivatives bind themselves to double-helical DNA and so interfere with replication and transcription. Like actinomycin and mitomycin, their action is not selective because they cannot differentiate among different species of DNA.

The antibiotics *chloramphenicol, streptomycin* and *tetracycline* inhibit synthesis of protein by attaching themselves to ribosomes thereby interfering with the proper binding and orientation of mRNA and tRNA to the ribosomes (p. 366). Chloramphenicol binds itself to the larger ribosomal subunit, while streptomycin and tetracycline associate (at different sites) with the smaller subunit. While the precise actions of chloramphenicol and tetracycline are not clear, the inhibitory action exerted by streptomycin arises from errors of 'reading' of the codons of the mRNA by the anticodons of the tRNA during binding of the drug to the ribosomal surface. Thus, mistakes are made in the insertion of amino acids into newly synthesized proteins with the result that the proteins are non-functional.

Chloramphenicol and streptomycin selectively inhibit bacterial growth by their ability to attach themselves specifically to *bacterial* ribosomes. These antibiotics are not bound by mammalian ribosomes and therefore mammalian protein synthesis proceeds unimpaired in their presence. This is the basis of the therapeutic application of these drugs. In contrast, tetracyclines, which attach themselves equally strongly to bacterial and mammalian ribosomes, are selectively active against bacteria because they can enter intact bacterial cells much more easily than mammalian cells.

Cycloheximide is a drug which also affects protein synthesis. It appears to associate with the ribosomes of cells of organisms higher than the bacteria (including mammalian ribosomes) but not with ribosomes of bacterial cells.

The drug *puromycin* is a nucleoside derivative which closely resembles the 3'-terminal nucleoside residue of tRNA esterified with an amino acid. It competes with aminoacyl-tRNA molecules in its capacity to serve as an acceptor for the peptidyl group of peptidyl-tRNA (p. 367) during protein synthesis on the ribosome. The consequence is that synthesis of complete proteins is prevented and, instead, peptides are produced which bear a puromycin residue covalently bonded to the carboxyterminal group. These peptides are of course non-functional.

Other antibiotic substances act at sites different from those described above. For example, the *penicillins*, and certain other groups of antibiotics, inhibit the formation of components of the bacterial cell wall (p. 48) so that the cells are lysed and die. Since mammalian cells do not possess the same cell-wall structure, they are unharmed by these drugs. Thus in the last few years detailed explanations of the selective toxicity of antibiotic substances have emerged. These explanations, although they add to our understanding, do not exclude the necessity for empirical investigations of new plant or animal extracts for antibiotic activity.

Bacteria may acquire resistance to antibiotic substances by altering the permeability of the cell wall so that the drugs are excluded, or alternatively by elaborating enzymes which act on the drugs to produce derivatives devoid of antibiotic activity.

REFERENCES

ANFINSEN, C. B. (Ed.) (1970). *Aspects of Protein Biosynthesis*. New York: Academic Press.

ATKINSON, D. E. (1966). Regulation of enzyme activity. *Annual Review of Biochemistry* **35**, 85–124.

BIRNSTIEL, M. & CHIPCHASE, M. (1970). The nucleolus: pacemaker of the cell. *Science Journal* **6** (7), 41–47.

COX, R. (1970). The ribosome—decoder of genetic information. *Science Journal* **6** (11), 56–60.

CRICK, F. H. C. (1967). The genetic code. *Proceedings of the Royal Society* B **167**, 331–339.

DAVIDSON, J. N. (1972). *The Biochemistry of the Nucleic Acids*, 7th edn. London: Chapman & Hall.

DAVIDSON, J. N. & COHN, W. E. (Eds.) (1963–1972). *Progress in Nucleic Acid Research and Molecular Biology*, Vols. 1–12. New York: Academic Press.

HAGGIS, G. H., MICHIE, D., MUIR, A. R., ROBERTS, K. B. & WALKER, P. M. B. (1964). *Introduction to Molecular Biology*. London: Longmans.

INGRAM, V. M. (1965). *The Biosynthesis of Macromolecules*. New York: Benjamin.

JUKES, T. H. (1966). *Molecules and Evolution*. New York: Columbia University Press.

KENDREW, J. (1966). *The Thread of Life*. London: Bell.

KONINGSBERGER, V. V. & BOSCH, L. (Eds.) (1967). *Regulation of Nucleic Acid and Protein Biosynthesis*. Amsterdam: Elsevier.

KURLAND, C. G. (1970). Ribosome structure and function emergent. *Science Journal* **169**, 1171–1177.

LEWIN, B. M. (1970). *The Molecular Basis of Gene Expression*. London: Wiley-Interscience.

MUNRO, H. N. & ALLISON, J. B. (Eds.) (1964). *Mammalian Protein Metabolism*. New York: Academic Press.

NOMURA, M. (1969). Ribosomes. *Scientific American* **221** (4), 28–35.

PERUTZ, M. F. (1967). Some molecular controls in biology. *Endeavour* **26**, 3–8.

ROSENBERG, E. (1971). *Cell and Molecular Biology*. New York: Holt, Rinehart & Winston.

SMITH, E. L., DELANGE, R. J. & BONNER, J. (1970). Chemistry and biology of the histones. *Physiological Reviews* **50**, 159–170.

SPIRIN, A. S. & GAVRILOVA, L. P. (1969). *The Ribosome*. Berlin: Springer.

STEINER, R. F. (1965). *The Chemical Foundations of Molecular Biology*. Princeton: Van Nostrand.

STEINER, R. F. & EDELHOCH, H. (1965). *Molecules and Life*. Princeton, New Jersey: Van Nostrand.

TAYLOR, J. H. (Ed.) (1963). *Molecular Genetics*, Part I, (1968). Part II. New York: Academic Press.

VOGEL, H. J., BRYSON, V. & LAMPEN, J. O. (Eds.) (1963). *Informational Macromolecules*. New York: Academic Press.

WATSON, J. D. (1970). *Molecular Biology of the Gene*, 2nd edn. New York: Benjamin.

WISEMAN, A. (1965). *Organization of Protein Synthesis*. Oxford: Blackwell.

WOESE, C. R. (1967). *The Genetic Code*. New York: Harper & Row.

YANOFSKY, C. (1967). Gene structure and protein structure. *Scientific American* **216** (5), 80–86.

22 Protein metabolism

The problem of protein biosynthesis has been discussed in the previous chapter. Here we are concerned with the question of protein breakdown and utilization.

The ultimate products of protein digestion are the amino acids, formed in the intestine by the action of endopeptidases and exopeptidases as outlined previously (Chap. 17). Most of these are absorbed from the intestine into the portal blood and thus conveyed to the liver; a small proportion, however, may be deaminated (see below) by the intestine. Almost all of the absorbed amino acids are removed from the blood by the liver and the muscles so that the level of amino acid nitrogen in the systemic circulation rises but little over the normal value of about 4 mg per 100 ml after a protein meal, although the blood urea nitrogen is somewhat increased.

Amino acids in solution injected intravenously into the systemic circulation disappear very rapidly from the blood. They are very quickly taken up by the kidneys, by the muscles, and especially by the liver. As the amino acids are removed from the blood by the liver an amount of urea corresponding to their nitrogen is liberated into the blood by the mechanism considered in detail on p. 381.

In contrast to the well-known ability of the body to store large amounts of both carbohydrate (Chap. 18) and fat (Chap. 19) the tissues have relatively little capacity for storing either amino acids or protein. However, after a rich protein meal at the end of a fast, temporary storage of amino acids takes place in the liver and in other tissues and is followed by a small increase in protein in the tissues which may be regarded as restoration of protein lost during fasting. Such amino acids as are not required for the synthesis of tissue proteins or of other nitrogenous substances of special importance are broken down by deamination or transamination as described later. During metabolism of amino acids there is an increase in the production of heat by the body which Rubner referred to as *specific dynamic action* (p. 168). It is apparently due to reactions which the amino acids undergo in the liver, or to the stimulation of other reactions in the cells by products formed in the liver. The phenomenon of specific dynamic action does not occur after removal of the liver. It is elicited by proteins and by individual amino acids, and to a much smaller extent by urea and glucose.

Proteins are continuously being synthesized in the body. In the adult it is necessary to replace protein lost in the process of normal wear and tear, but in growing children, and in adults recovering from wasting illnesses, protein synthesis must obviously be much greater. In the liver the protein synthesized from amino acids (after a protein meal) is deposited in the cytoplasm of the cells in the form of a labile store of protein which is used up rapidly in the early stages of starvation or in other conditions of deficient protein intake, or in helping to make good the loss of plasma protein after a severe haemorrhage or loss of blood at a surgical operation. In starvation, or when the intake of protein is inadequate (as after certain surgical operations), the liver loses protein to a greater extent than does any other tissue and in experimental animals on a protein-free diet the loss of liver protein is accompanied by a parallel loss of liver enzymes. This loss of actual liver substance suggests that there is no protein reserve or store in the usual sense of the term. The liver, when depleted of protein in these ways, seems to be more susceptible to injury by such agents as phosphorus, arsenic or alcohol which may reach it in the blood; in the treatment of liver disease it is therefore important to supply a diet containing a relatively high proportion of protein.

With the exception of gamma globulin (Chap. 23) the plasma protein are synthesized in the liver, whence they are liberated into the blood. In liver disease the concentration of plasma protein tends to fall and alterations in the relative concentrations of the different proteins can be detected by electrophoresis on paper.

Dynamic equilibrium of body proteins. Amino acids are constantly being liberated into the blood from tissue proteins which are broken down during normal metabolism and, to a much greater extent, in starvation, in chronic wasting diseases, in fever and in other states of increased metabolism. Such amino acids together with those derived from the food form a common stock or 'metabolic pool' in the blood and tissues. In this metabolic pool the body cannot distinguish between amino acids derived from dietary sources and those derived from protein catabolism. The pool serves as a source of amino acids some of which may be used, whatever their origin, to build nitrogenous substances such as proteins or certain hormones, while the vast majority are degraded by a process to be described later, their nitrogen being excreted as urea (Fig. 22.1).

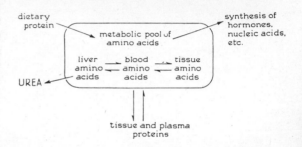

Fig. 22.1.

It was at one time supposed that there was a clear-cut distinction between exogenous and endogenous protein metabolism, that is between catabolism of amino acids derived from dietary protein on the one hand, and catabolism of amino acids derived from breakdown of tissue proteins on the other. This idea is no longer tenable, largely because of the pioneering work on the metabolism of amino acids and proteins carried out between 1939 and 1946 by Rudolf Schoenheimer and his colleagues at Columbia University, New York, using amino acids labelled with the nitrogen isotope ^{15}N. He fed such labelled amino acids to rats in nitrogen equilibrium (Chap. 13) on a diet containing adequate amounts of protein; the excreta and the body tissues were subsequently analysed. According to classical theory most of the ^{15}N should have appeared directly in the urine as labelled urea; in fact much of the ^{15}N was found in the tissue proteins of all organs examined, including even the skin, the metabolism of which is known to be relatively slow. When labelled leucine was fed, ^{15}N was subsequently found not only in the leucine of the tissue proteins but also in the glycine, tyrosine and arginine, and in particularly large amounts in the dicarboxylic amino acids, glutamic acid and aspartic acid. Lysine and threonine, however, never became labelled. It appears, therefore, that the nitrogen of one amino acid can be transferred readily to other amino acids (except threonine and lysine).

These experiments led Schoenheimer to conclude that 'all constituents of living matter, whether functional or structural, of simple or of complex constitution, are in a steady state of rapid flux'.

The process of transfer of nitrogen from one amino acid to another is known as *transamination* and is discussed later (p. 379).

With the advent of radioisotopes such as ^{14}C and ^{35}S these studies were extended and the earlier results obtained with ^{15}N confirmed. The labile nature of body proteins is one of the most striking discoveries to result from the use of isotopes in the investigation of intermediary metabolism. The life span of proteins can be determined by feeding an isotopically labelled amino acid, and then following the loss of isotope from a given protein as a function of time (Fig. 22.2). Such experi-

ments usually yield results like those shown for protein A in the figure. The isotopic content rises rapidly and after reaching a maximum, usually after a few hours or days, decreases at an exponential rate, from which the half-life (that is, the time for half of the protein to be broken down and replaced) can be calculated mathematically. However, there are some proteins with fixed life spans which give curves like B in Fig. 22.2. Thus when Shemin and Rittenberg in 1946 followed the turnover of haemoglobin, that is the rate at which the protein is renewed, in an adult human subject after labelling the porphyrin of the molecule with ^{15}N-glycine (Chap. 25), they found that the isotope content rose to a maximum at 25 days and then remained constant for about 70 days before decreasing rapidly over the next 30 days or so. The constant part of the graph is due to the fact that haemoglobin is metabolically inert in the red blood cell and its catabolism reflects the catabolism of the cell of which it is a major constituent—once the red cell is destroyed, so is the haemoglobin.

The half-life values of the major tissue proteins have been determined by Sprinson and Rittenberg using glycine labelled with ^{15}N. In

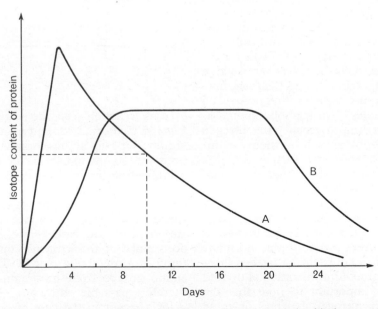

Fig. 22.2 Protein life span. Curve A represents the typical relationship between isotope content and time for a protein that is randomly destroyed. The half-life in this case is about 10 days. Curve B shows the relationship for a protein with a fixed life span. (*After* Neuberger & Richards, 1964.)

man the half-life of the total body protein is 80 days, that of the liver and serum proteins being about 10 days, whilst that of proteins of other tissues, mainly muscle, is 160 days. In general structural proteins gave a low turnover rate whereas metabolic proteins have a rapid rate of turnover.

Dogs on a protein-free diet can be kept in good health in nitrogen equilibrium for many months if whole dog plasma is given intravenously at regular intervals. Patients also may be kept in nitrogen equilibrium for limited periods by intravenous infusions of human plasma or solutions containing suitable amino acids so long as a high energy intake is also provided.

REGULATION OF PROTEIN METABOLISM

The chemical changes occurring in protein metabolism can be shown by a series of diagrams but it is not easy to describe the dynamic aspects of these changes. The rate at which amino acids are taken up by the cells of different tissues varies, and so does the rate of tissue protein breakdown by which amino acids are released. The currency of protein metabolism is thus the pool of free amino acids, and all the tissues draw from and discharge into this pool. Both synthesis and breakdown are affected by a variety of factors, particularly by hormones secreted into the blood by the endocrine glands. The effects of the hormones on protein metabolism can be classed according to whether the overall action on the body is to retain nitrogen (anabolic hormones) or to cause loss of body nitrogen (catabolic hormones). An anabolic action, as revealed by an improvement in nitrogen balance, is shown by growth hormone, by insulin given with sufficient carbohydrate to prevent hypoglycaemia, by testosterone, and by a number of synthetic anabolic steroids such as methandienone (Chap. 51). A catabolic effect (negative nitrogen balance) occurs after treatment with adrenocortical hormones or with sufficient adrenocorticotrophic hormone (ACTH) to cause an outpouring of these hormones, and also after administration of thyroid hormones. It is important to note that the overall effect on nitrogen balance of these anabolic and catabolic hormones is a summation of their effects on different tissues since all tissues are not equally affected by each hor-

mone named; thus, although adrenocortical hormones cause an overall loss of nitrogen from the body and breakdown of body protein, some organs such as the liver may gain protein during treatment. In addition, it has to be remembered that all these hormones have actions on the metabolism of nutrients other than amino acids. Thus growth hormone and testosterone cause a loss of body fat through oxidation at the same time as they promote nitrogen retention; it is well known that insulin induces desposition of glycogen in the body. Adrenocortical hormones cause a net loss of body protein, but reduce the oxidation of fat. These multiple effects on intermediary metabolism indicate that the action of hormones on protein metabolism has to be viewed as part of a larger effect on metabolism generally. At present too little is known of the mode of action of hormones on cell metabolism to provide a reasoned picture for the effects observed on metabolism as a whole.

AMINO ACID CATABOLISM

The first stage in the catabolism of amino acids is the formation of the corresponding keto acid, the nitrogen ultimately appearing in the form of urea.

This reaction can be catalysed by the L-*amino acid oxidase* found mainly in liver and kidney, but even in these organs the concentrations are too low to have a significant effect. It is interesting that the liver and kidney possess a very active and readily soluble amino acid oxidase which is virtually exclusive in its specificity towards D-amino acids. Since D-amino acids are rather rare in nature, the occurrence of this enzyme constitutes something of a metabolic puzzle. The removal of amino groups from most amino acids must therefore be achieved by another method (see below) except in the case of one amino acid, glutamic acid, for which a *glutamate dehydrogenase* of high activity is found abundantly in many tissues. This important enzyme requires

NAD as coenzyme and brings about the oxidation of glutamic acid to α-oxoglutaric acid (α-ketoglutaric acid) thus:

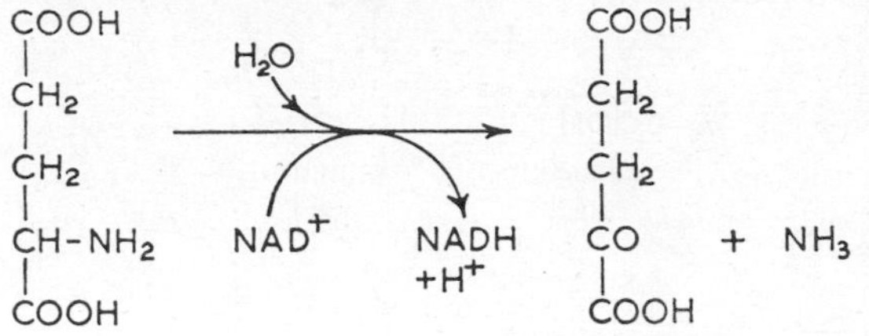

$$
\begin{array}{ccc}
\text{COOH} & & \text{COOH} \\
| & \text{H}_2\text{O} & | \\
\text{CH}_2 & & \text{CH}_2 \\
| & & | \\
\text{CH}_2 & & \text{CH}_2 \\
| & & | \\
\text{CH}-\text{NH}_2 \quad \text{NAD}^+ \quad \text{NADH} & \text{CO} & + \text{NH}_3 \\
| & +\text{H}^+ & | \\
\text{COOH} & & \text{COOH}
\end{array}
$$

glutamic acid   α-oxoglutaric acid

Transamination. The most important mechanism for the conversion of an amino acid to a keto acid is transamination in which an amino group is transferred from a donor amino acid to a recipient keto acid under the influence of a *transaminase* or *aminotransferase* thus:

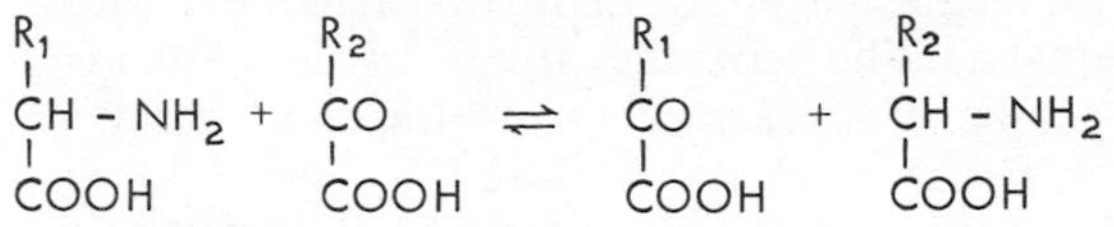

$$
\begin{array}{cccc}
R_1 & R_2 & R_1 & R_2 \\
| & | & | & | \\
\text{CH}-\text{NH}_2 + \text{CO} & \rightleftharpoons & \text{CO} & + \text{CH}-\text{NH}_2 \\
| & | & | & | \\
\text{COOH} & \text{COOH} & \text{COOH} & \text{COOH}
\end{array}
$$

$$ \text{I} \qquad \text{II} \qquad \text{III} \qquad \text{IV} $$

The donor amino acid thus becomes a keto acid and the recipient keto acid becomes an amino acid. The coenzyme required for the reaction is pyridoxal phosphate (cf. Chap. 13). In the process of transamination the amino acid reacts with pyridoxal phosphate to form a complex of the Schiff base type which then yields the keto acid and pyridoxamine phosphate. The latter then reacts with a second keto acid to produce a similar Schiff base complex which decomposes forming a second amino acid and regenerating the pyridoxal phosphate.

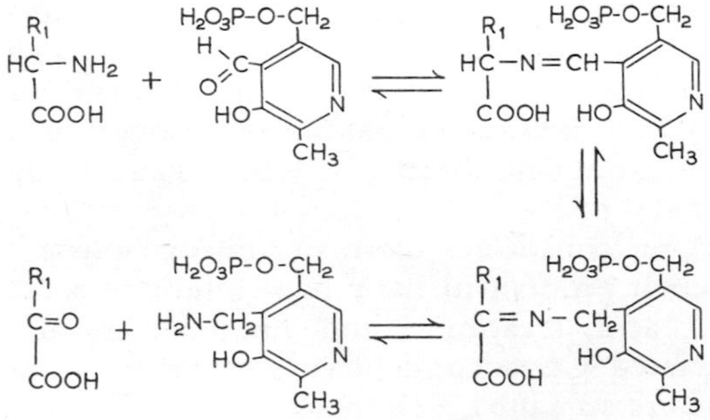

There are, however, certain limitations to the reaction. While most amino acids may act

as donor (I) the recipient (II) must in general be either α-oxoglutaric acid, or oxaloacetic acid or pyruvic acid. It is important to note that all of these keto acids are components of the tricarboxylic acid cycle and are, therefore, common metabolites of the cell. The amino acids (IV) formed from them are respectively glutamic acid, aspartic acid and alanine. To date most transaminases studied have been shown to contain bound pyridoxal phosphate, which plays a key role in the reaction as shown above. The amino acids threonine and lysine do not appear to participate in transamination reactions. A few transaminases do not utilize the three keto acid recipients (II) mentioned above.

While there are thus three main types of transaminase the most important reaction is that involving glutamic acid and its corresponding keto acid, α-oxoglutaric acid. For example, the reaction between aspartic acid and α-oxoglutaric acid is catalysed by the *aspartate aminotransferase* sometimes referred to as *glutamate-oxaloacetate transaminase* (*GOT*). See Diagram I.

The activity of this transaminase in the blood serum (SGOT) in man is increased in the first few days after a coronary thrombosis because the enzyme is released from damaged cells in the myocardium. The normal value is 30 to 40 units per 100 ml serum but there may be as much as 530 units per 100 ml when there is much tissue damage.

The widespread occurrence and broad spectrum of transamination reactions is clearly consistent with a significant metabolic role. Braunstein and Bychkov in 1939 suggested that the coupled action of an amino acid-α-oxoglutaric acid transaminase and glutamate dehydrogenase might explain the oxidative deamination of L-amino acids. Though unequivocal proof of this has not been forthcoming, it seems clear that this mechanism (summarized below) forms the major pathway for removal of amino groups from amino acids. The process takes place chiefly in the liver but occurs also in the kidneys. In the liver the ammonia is converted into urea; the keto acid is oxidized to yield energy. These processes must be considered separately. See Diagram II.

UREA FORMATION

Ammonia is highly toxic to the mammalian

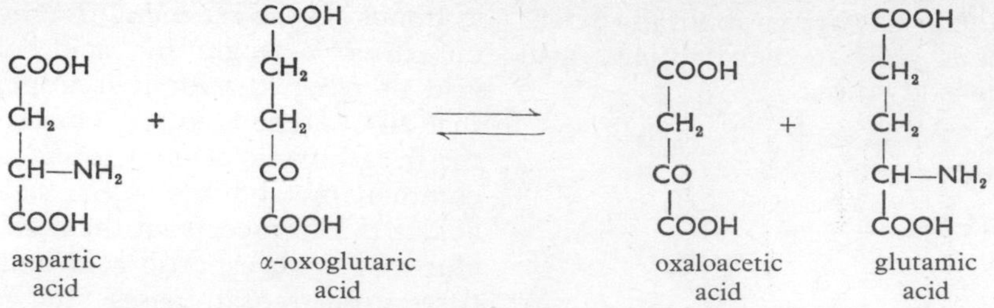

COOH   COOH   COOH   COOH

$$CH_2 + CH_2 \rightleftharpoons CH_2 + CH_2$$

Reactants: aspartic acid (CH_2, $CH-NH_2$, $COOH$) + α-oxoglutaric acid (CH_2, CH_2, CO, $COOH$) $\rightleftharpoons$ oxaloacetic acid (CH_2, CO, $COOH$) + glutamic acid (CH_2, CH_2, $CH-NH_2$, $COOH$)

DIAGRAM I

$$R-CH(NH_2)-COOH \xrightarrow{\text{aminotransferase}} \text{α-oxoglutarate} \xrightarrow{\text{glutamate dehydrogenase}} NH_3$$

$$R-CO-COOH \longleftarrow \text{glutamate}$$

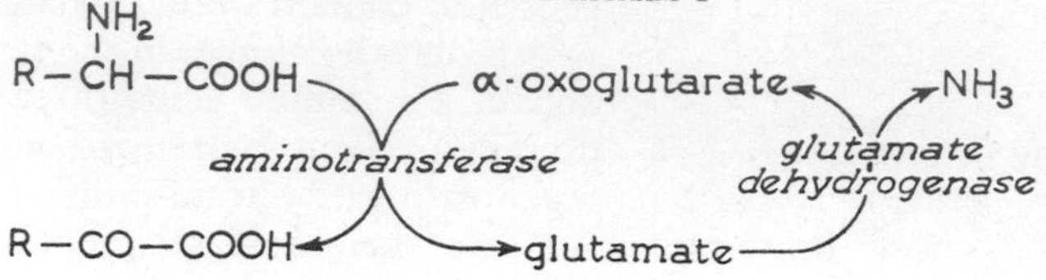

DIAGRAM II

organism because it reacts with α-oxoglutarate in liver mitochondria to form glutamate. The loss of α-oxoglutarate inhibits respiration and leads to excessive formation of ketone bodies from acetyl CoA.

While the portal blood of mammals contains a relatively high concentration of ammonia, most of which appears to be absorbed from the caecum, the concentration of ammonia in the peripheral blood, other body fluids and the tissues of mammals is very low. In man the free ammonium ion concentration of fresh plasma is less than 20 μg/100 ml. Such low concentrations suggest that the mechanism for removal of this highly toxic substance is extremely efficient. The removal of excess ammonia derived from amino acid catabolism in the tissues or from bacterial action in the gut is accomplished by the production of urea and its excretion. Early studies (1932) by Krebs and Henseleit established that urea formation takes place by means of a cyclic series of reactions (Fig. 22.3) involving arginine, ornithine and citrulline and studies with isotopically labelled compounds on intact animals have confirmed this general mechanism. The process is confined in mammals to the liver and all of the enzymes involved have been isolated from liver tissue.

The reactions of the ornithine arginine cycle, shown in Fig. 22.3, are brought about by a complex co-ordinated system of enzymes in the mitochondria and in the cytoplasm of the liver cell.

Ammonium ions, from the oxidative de-amination of glutamate, and bicarbonate ions react together with the hydrolysis of two molecules of ATP to form carbamoyl phosphate. The enzyme involved is *carbamoyl phosphate synthetase*. Ornithine combines with this in a reaction catalysed by *ornithine carbamoyltransferase* to form citrulline. The formation of arginine from ornithine specifically requires aspartate and ATP, and occurs in two steps. First, *argininosuccinate synthetase* catalyses the formation of argininosuccinate with concomitant hydrolysis of ATP; then the argininosuccinate is hydrolysed by *argininosuccinase* to yield arginine and fumaric acid. The enzyme *arginase*, which has been known since 1904, cleaves the arginine to give urea and ornithine; the latter can enter the cycle again. The fumarate formed may be converted back to oxaloacetic acid by a series of reactions (as in the tricarboxylic acid cycle) and this compound may undergo transamination with glutamic acid to regenerate the aspartic acid. Thus in each turn of the cycle one molecule of ammonia and the amino group of one glutamic acid molecule are converted to urea. The concentration of urea in normal blood plasma from a fasting human subject is between 24 and 36 mg per 100 ml.

Urea constitutes the major nitrogenous excretion product in mammals, although some uric acid, creatinine, and ammonia are also excreted. Certain aquatic vertebrates also form and retain urea. Tadpoles excrete mainly ammonia; during development this mode of excretion is replaced by urea formation. Meta-

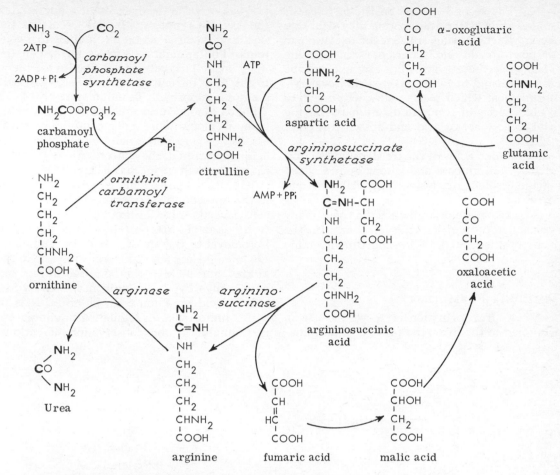

Fig. 22.3 The arginine ornithine cycle for urea synthesis.

morphosis is accompanied by the appearance of arginase in the liver. Birds and reptiles excrete nitrogen mainly in the form of uric acid: ammonia is formed in the early development of the chick embryo; later urea is formed and eventually, in the adult, uric acid.

THE FATE OF KETO ACIDS

We must now consider the fate of the keto acids yielded by removal of the amino groups of the amino acids.

If amino acids are fed one at a time to a starving phlorizinized dog (Chap. 18) it is found that some give rise to glucose in the urine, while others give rise to acetoacetic acid (Chap. 19). A few give rise to neither. In such an animal about 60 g of glucose are formed and excreted in the urine for each 100 g of protein metabolized. In other words, 60 per cent of protein is potentially glucogenic. When the liver is removed, however, no such glucose formation takes place. Amino acids can be divided therefore into three groups:

(1) Those of the first group, termed *glucogenic* or *antiketogenic* amino acids, are deaminated to form keto acids from which glycogen may be produced when necessary (Chap. 18). They are oxidized to carbon dioxide and water by the tricarboxylic acid cycle as are carbohydrates. For example, alanine is deaminated to give the keto acid, pyruvic acid, an intermediary product of carbohydrate breakdown. The antiketogenic amino acids are glycine, alanine, serine, aspartic acid, glutamic acid, valine, histidine, arginine, proline, hydroxyproline, threonine, tryptophan, methionine and cystine.

(2) The second group consists of *ketogenic* amino acids. After deamination they yield keto acids which, during subsequent oxidation to carbon dioxide and water, pass through the stage of acetoacetic acid. For example, leucine is deaminated to yield the keto acid isovaleryl-formic acid which reacts with coenzyme A to yield a series of compounds ultimately giving rise to acetoacetic acid and acetyl CoA (Fig. 22.4).

(3) Amino acids of the third group can give rise to both glucose and ketone bodies. They include isoleucine, lysine, phenylalanine and tyrosine.

The ketogenic or antiketogenic value of any particular protein depends of course on the sum of the effects of its constituent amino acids.

AMMONIA FORMATION

While urea formation is, as we have seen, confined to the liver, the process of deamina-tion can occur in the liver, in the kidney and in the intestinal mucosa. It also occurs as the result of bacterial action in the intestinal contents. The ammonia which is formed in the intestine and in the kidney passes in the blood to the liver for conversion to urea; very little ammonia is normally found in systemic blood. Unlike blood, urine contains appreciable amounts of ammonia and in conditions such as acidosis, where abnormally large amounts of acid are excreted, the concentration of ammonium salts in the urine may be greatly increased (Chap. 32). The enzyme *glutaminase*, located in the mitochondrial fraction of kidney tubule cells, is responsible for producing the ammonia (to act as an H^+ acceptor) by deamidation (that is removal of the amide group) of glutamine. *Glutaminase* is present in large amounts in the kidney, and its level can be raised in rats and guinea pigs by the administration of hydro-chloric acid or ammonium chloride. It is of major importance in regulating ammonia excretion and in the conservation of cations.

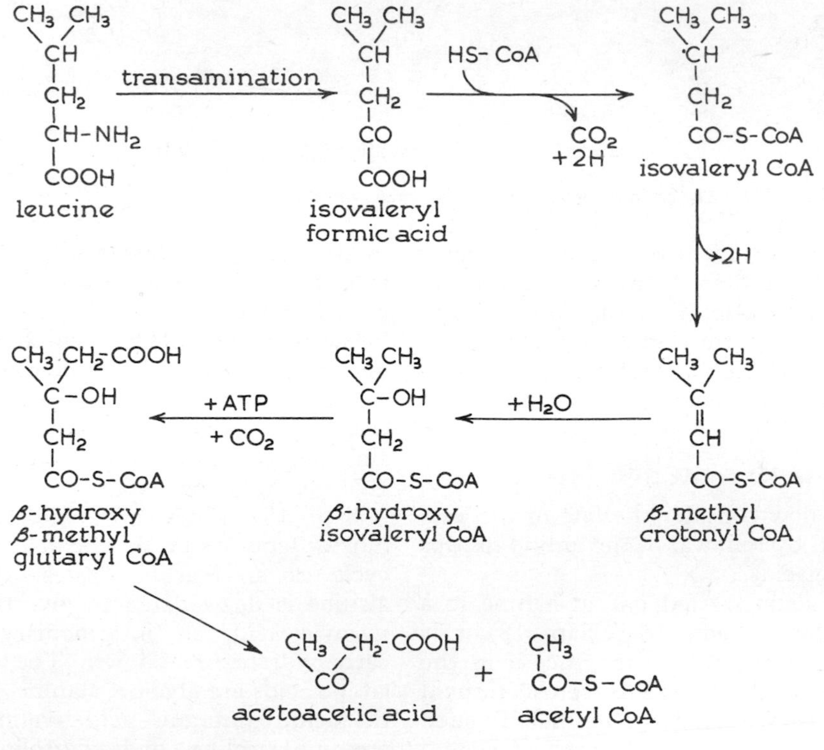

Fig. 22.4 Metabolic pathway in the breakdown of leucine.

382 Textbook of Physiology and Biochemistry

Some of the ammonia formed in the kidney comes also from the amino group of glutamine.

$$
\begin{array}{c}
\text{CONH}_2 \\
| \\
\text{CH}_2 \\
| \\
\text{CH}_2 \quad + \text{H}_2\text{O} \rightleftharpoons \\
| \\
\text{CH}-\text{NH}_2 \\
| \\
\text{COOH}
\end{array}
\qquad
\begin{array}{c}
\text{COOH} \\
| \\
\text{CH}_2 \\
| \\
\text{CH}_2 \quad + \text{NH}_3 \\
| \\
\text{CH}-\text{NH}_2 \\
| \\
\text{COOH}
\end{array}
$$

The reactions involved in protein metabolism which have already been discussed are summarized in Fig. 22.5.

DECARBOXYLATION

Certain physiologically important amines are formed by the removal of carbon dioxide dioxide from the carboxyl groups of amino acids by amino acid decarboxylases all of which seem to require pyridoxal phosphate as coenzyme. Thus histidine gives rise to histamine, 3,4-dihydroxyphenylalanine gives dopamine, a precursor of adrenaline, and 5-hydroxytryptophan gives 5-hydroxytryptamine (5-HT). The chemistry and the physiological effects of these three hormones are considered in Chap. 50.

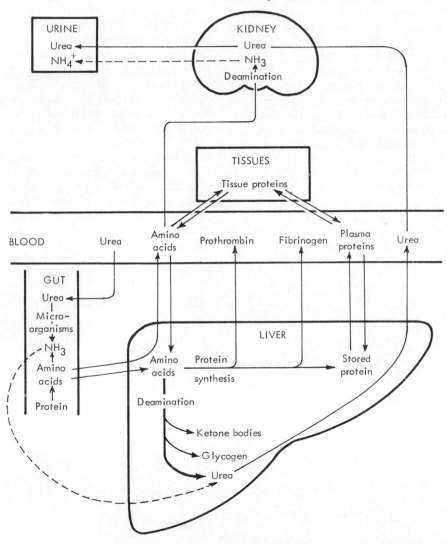

Fig. 22.5 Diagrammatic representation of protein metabolism. The balance of evidence suggests that plasma protein and tissue protein are not directly interconvertible but that each is built up from, and broken down into, the common metabolic pool of amino acids.

Protein Metabolism 383

Reactions Peculiar to Individual Amino Acids

Glycine, the simplest amino acid, is a precursor of the ring system in both purines and porphyrins. It is glucogenic in character and in its conversion to glycogen it reacts with hydroxymethyl tetrahydrofolic acid (Chap. 13) to form serine. The pathways for the interconversion of glycine, serine, alanine and glucose are shown in Fig. 22.6.

Glycine also acts as a *conjugating agent*; by combining with potentially harmful substances in the body it renders them harmless. The process is sometimes called *detoxication* and the compounds so formed are excreted in the urine. In the process of conjugation water is eliminated and the peptide bond

$$-CO-NH-$$

is formed. Thus, for example, benzoic acid and other aromatic acids absorbed from the alimentary tract or arising during the processes of metabolism are conjugated with glycine before being excreted in the urine.

$$C_6H_5COOH + NH_2CH_2COOH$$
benzoic acid · · · · · · · glycine

$$\longrightarrow C_6H_5CO \cdot NH \cdot CH_2COOH + H_2O$$
hippuric acid

In a similar way cholic acid is conjugated with glycine to form glycocholic acid which is excreted in the bile.

Phenylalanine and tyrosine. The amino acids phenylalanine and tyrosine, both of which are precursors of the hormones of the thyroid gland and of adrenaline and noradrenaline, contain an aromatic nucleus and are therefore metabolized by routes that are quite different from those followed by the simple aliphatic amino acids. There is little doubt that tyrosine can be formed in the body from phenylalanine but the reverse process does not occur to any extent. On deamination or transamination these acids yield the corresponding keto acids, *p*-hydroxy-phenylpyruvic acid and phenylpyruvic acid respectively (Fig. 22.7). The former can be further oxidized to homogenistic acid which, after the ring is opened, yields acetoacetic acid and fumaric acid. The latter is oxidized to carbon dioxide and water; acetoacetic acid reacts with coenzyme A and so enters the citric acid cycle (Fig. 11.6).

In his Croonian lectures of 1908 Sir Archibald Garrod clearly recognized the existence of *inborn errors of metabolism*. Four such disorders, all of them associated with the absence of, or with a marked reduction of, a particular enzyme, have been observed in the metabolism of phenylalanine and tyrosine (see Fig. 22.7 overleaf).

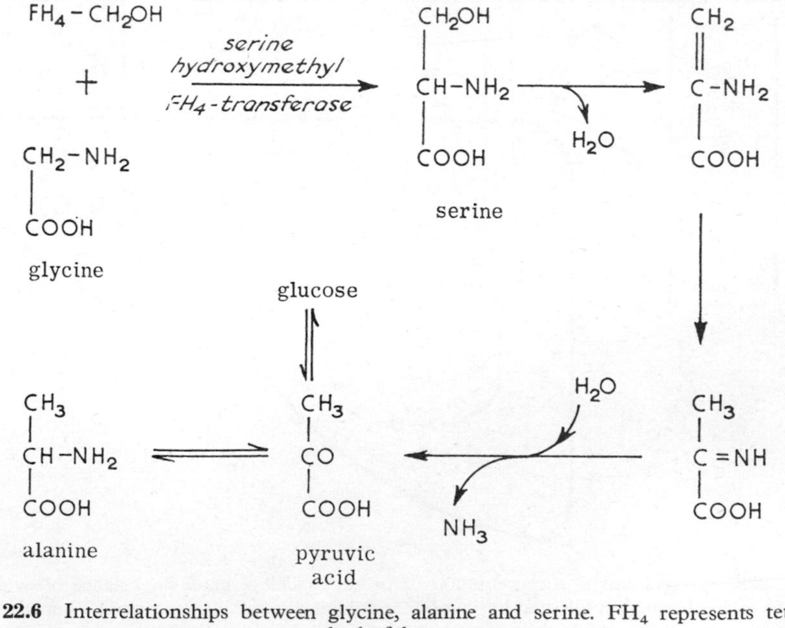

Fig. 22.6 Interrelationships between glycine, alanine and serine. FH_4 represents tetrahydrofolate.

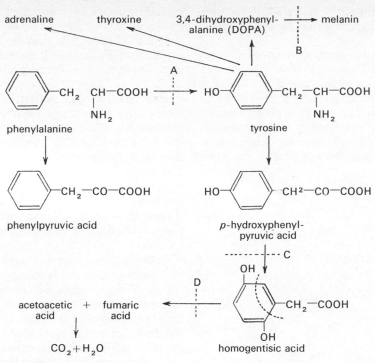

Fig. 22.7 Metabolic pathways for phenylalanine and tyrosine. Blockage at A, due to absence of the appropriate gene and appropriate enzyme, results in phenylketonuria, at B in albinism, at C in tyrosinosis and at D in alkaptonuria. Blockage at C is found in ascorbic acid deficiency.

In persons with the apparently harmless abnormality called *alkaptonuria*, a rare inborn error or hereditary defect of metabolism of these amino acids, the enzyme *homogentisate oxygenase* is missing from the cells. Homogenistic acid therefore accumulates in the tissues and appears in the urine. Since the homogenistic acid, like many derivatives of tyrosine, is readily oxidized to black pigments of the melanin group, the urine of alkaptonuric persons slowly turns black when it is exposed to the air.

Tyrosine can also be oxidized by a complex series of changes to the black pigment *melanin* which occurs in the hair, in the choroid coat of the eye, in the dark skin of the negro, and in smaller quantity in the skin of white people in which it increases after exposure to strong sunlight. The first step in this process is the oxidation of tyrosine by the enzyme tyrosinase to dihydroxyphenylalanine (DOPA). This compound undergoes further changes to form indole derivatives which polymerize to melanin. In certain tumours the production of melanin by such a series of reactions may be very active and *melanuria*, the excretion of a black pigment in the urine, may actually occur. Melanin formation is defective in albinos.

In patients suffering from a type of mental defect usually referred to as *phenylketonuria*, the enzyme involved in the introduction of a hydroxyl group into the ring in phenylalanine is missing and therefore phenylalanine accumulates in the blood. It undergoes deamination in kidney to phenylpyruvic acid which is immediately excreted in the urine. The amount of phenylpyruvic acid excreted is increased by administration of phenylalanine but not of tyrosine. A diet low in phenylalanine, if started soon enough after birth, may prevent a child with this condition from becoming mentally defective.

Further evidence of the metabolic pathway has been provided by the study of a patient with *tyrosinosis* in whom p-hydroxyphenylpyruvic acid could not be oxidized and appeared in the urine.

Under the influence of putrefactive bacteria in the intestine, tyrosine may be broken down to phenol.

Tryptophan is the precursor of 5-hydroxytryptamine (5-HT) discussed in Chapter 50. Under certain conditions such as are found in the colon, tryptophan may be broken down to yield the evil-smelling compounds *skatole* and *indole* which are responsible for the characteristic smell of faeces, and the related substance *indoxyl* which may appear in the urine as indican.

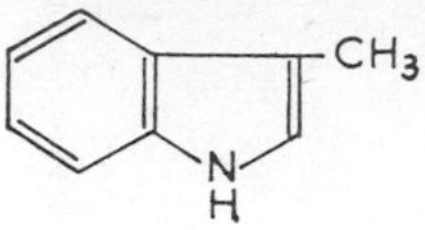

skatole

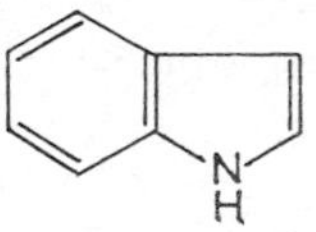

indole

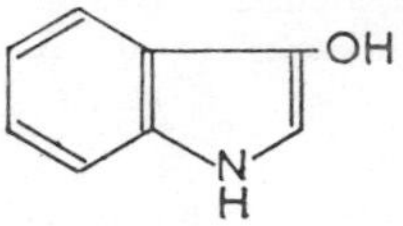

indoxyl

In certain species tryptophan gives rise to nicotinic acid by way of the intermediate *kynurenine*.

It has been shown that *arginine, ornithine, proline* and *glutamic acid* can be converted one into another, the carbon chain remaining intact (Fig. 22.8).

Essential amino acids. From the point of view of nutrition amino acids are of two types, dispensable and indispensable (Chap. 13).

The indispensable amino acids, both for man and for the rat, are valine, methionine, threonine, leucine, isoleucine, phenylalanine, tryptophan, lysine, arginine and histidine. The last two are not essential in the adult organism. Tyrosine, as we have seen, can be formed from phenylalanine and cysteine from methionine (p. 388).

Some of these amino acids, for example histidine, can be replaced in the diet by the corresponding keto acid. In such a case the indispensable unit appears to be the carbon chain; the amino group can be introduced by the body as required. Indeed, all the indispensable amino acids except lysine and threonine take part in the reversible transfer of amino groups from amino acids to keto acids. These two amino acids form 'a single indispensable chemical unit which has to be supplied as such to the animal' (Schoenheimer).

CREATINE AND CREATININE

Two closely related compounds which must be discussed in relation to protein metabolism

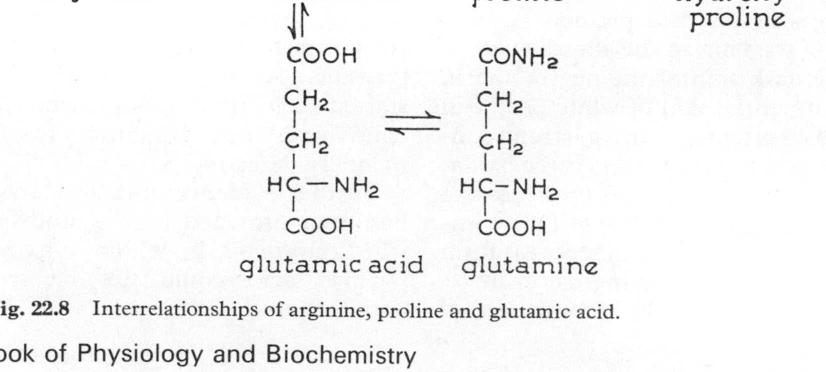

Fig. 22.8 Interrelationships of arginine, proline and glutamic acid.

are *creatine* and *creatinine*. Creatine is methyl guanidoacetic acid; creatinine is its anhydride.

$$
\begin{array}{llll}
NH_2 & NH_2 & NH_2 & NH\,\text{---} \\
| & | & | & | \\
C=NH & C=NH & C=NH & C=NH \\
| & | & | & | \\
NH_2 & NH & N\text{---}CH_3 & N\text{---}CH_3 \\
& | & | & | \\
CH_2 & CH_2 & CH_2 & CH_2 \\
| & | & | & | \\
COOH & COOH & COOH & CO\,\text{---} \\
\text{guanidine} & & & \\
CH_2 & & & \\
| & & & \\
COOH & & & \\
\text{acetic acid} & \text{guanido-acetic} & \text{creatine} & \text{creatinine} \\
& \text{acid} & & \\
& \text{(glycocyamine)} & &
\end{array}
$$

Creatine is found abundantly in muscle but not in other tissues. Skeletal muscle contains about 0·5 per cent creatine, mainly in the form of creatine phosphate or phosphagen, which is discussed in Chapter 40 in connexion with the chemical processes involved in muscular contraction.

Creatine is synthesized in the body and is not a necessary constituent of the diet. It is present in the muscles of herbivorous animals although none is contained in the vegetable tissues they eat. It has been shown that creatine is formed in the animal body from glycine, arginine and methionine (Fig. 22.9). The glycine supplies the —NH—CH_2—COOH portion of the molecule, arginine supplies the amidine group

$$
\begin{array}{c}
\quad\quad NH \\
\quad\quad \parallel \\
\text{---C} \\
\quad\quad \diagdown \\
\quad\quad NH_2
\end{array}
$$

and methionine supplies the methyl group (CH_3—) by the process of transmethylation. When synthetic glycine containing the nitrogen isotope ^{15}N is fed to rats, the labelled nitrogen appears in the appropriate position in the molecule of the muscle creatine. In a similar way it has been shown that arginine is the source of the other two nitrogen atoms in the creatine molecule. It is concluded that arginine and glycine react to form guanido-acetic acid, the transfer of the amidine group being known as *transamidination*.

Guanido-acetic acid is known to occur in small amounts in both tissues and urine. The feeding of labelled guanido-acetic acid results in an immediate formation of labelled creatine and creatinine.

Creatine is rapidly formed when liver slices are incubated with guanido-acetic acid and methionine which, in its active form, S-adenosyl methionine, acts as methyl donor in the process of *transmethylation*.

When rats are fed methionine containing heavy hydrogen (deuterium) in the methyl group, the heavy hydrogen appears subsequently in the methyl group of creatine isolated from the muscles.

Creatine formation therefore occurs in two stages, the formation of guanido-acetic acid and then its methylation, and there is evidence that the first stage occurs in the kidney and the second in the liver.

Some of the creatine phosphate in muscle loses water to yield phosphoric acid and the anhydride *creatinine* which appears in the urine. The use of labelled nitrogen has revealed that the urinary creatinine is formed from creatine (or its phosphate) and from no other source. Creatine, however, cannot be obtained from creatinine; the needs of the body for creatine are met by synthesis from methionine, arginine and glycine.

When creatinine is ingested most of it is rapidly eliminated in the urine, but when creatine is taken some is retained in the body, its fate being unknown. Some 20 to 30 per cent of the ingested creatine is, however, excreted unchanged, only about 2 per cent appearing in the urine as creatinine.

The urine of the normal healthy adult male contains creatinine but no creatine. The amount of creatinine excreted is about 1·0 to 1·5 grams per day, and is independent of the amount of protein in his diet. It is greater in muscular persons and appears to be related to the muscular development of the individual rather than to his muscular activity. After severe exercise creatinine excretion may be increased so that the total output remains remarkably constant from day to day.

Creatine occurs normally in the urine of children along with creatinine. Creatine appears also in the urine of adults in whom breakdown of muscle tissue is occurring, that is in fevers, in prolonged starvation, and in hyperthyroidism. In women, creatine is excreted during menstruation and during pregnancy and puerperium.

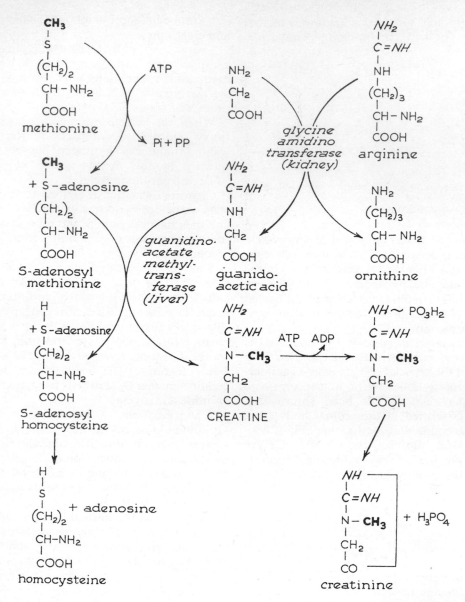

Fig. 22.9 The pathway of creatine formation. Note the transference of the amidine group (italics), and the methyl group (bold type).

Creatinuria may be found in almost all types of muscle disease, including poliomyelitis and muscular dystrophies.

SULPHUR METABOLISM

The metabolism of sulphur compounds is conveniently considered along with that of the proteins. Sulphur is ingested chiefly in the form of the sulphur-containing amino acids *cysteine, cystine* and *methionine*. Methionine may lose its methyl group in the process of *transmethylation*, leaving *homocysteine* (Fig. 22.10). This in its turn may be broken down to cystine by way of the intermediate compound cystathionine.

The fates of these acids are as follows:

(1) Some of the ingested sulphur-containing amino acids are used to repair or to build up the tissue proteins. Hair, for example, contains large amounts of cystine.

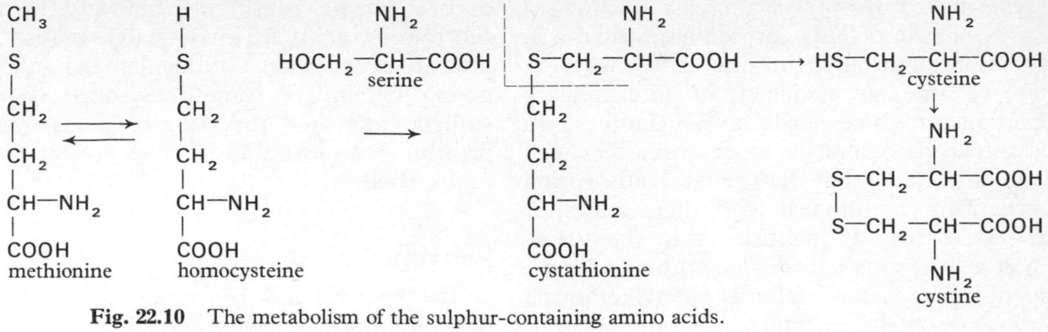

$$
\begin{array}{lll}
\underset{\text{methionine}}{\begin{array}{l}CH_3\\|\\S\\|\\CH_2\\|\\CH_2\\|\\CH-NH_2\\|\\COOH\end{array}}
\rightleftarrows
\underset{\text{homocysteine}}{\begin{array}{l}H\\|\\S\\|\\CH_2\\|\\CH_2\\|\\CH-NH_2\\|\\COOH\end{array}}
\end{array}
$$

$$HOCH_2-\underset{\text{serine}}{CH}-COOH$$

$$\underset{\text{cystathionine}}{\begin{array}{l}S-CH_2-CH-COOH\\|\\CH_2\\|\\CH_2\\|\\CH-NH_2\\|\\COOH\end{array}} \longrightarrow HS-CH_2-\underset{\text{cysteine}}{CH}-COOH$$

$$\underset{\text{cystine}}{\begin{array}{l}NH_2\\|\\S-CH_2-CH-COOH\\|\\S-CH_2-CH-COOH\\|\\NH_2\end{array}}$$

Fig. 22.10 The metabolism of the sulphur-containing amino acids.

(2) Small amounts of the sulphur-containing amino acids are used in the synthesis of compounds of special significance, such as the hormone insulin, the tripeptide glutathione (glutamylcysteylglycine) (p. 28), and β-mercaptoethylamine which is a constituent of coenzyme A. Cysteine is utilized in the formation of taurine, which combines with cholic acid (p. 60) to form the bile acid, taurocholic acid.

(3) The majority of the sulphur-containing amino acids, however, are catabolized in the liver like other amino acids. The nitrogen goes to form urea and the sulphur is oxidized to sulphuric acid which appears in the urine as sulphate. A diet high in protein therefore increases the output of sulphur in the urine. In these circumstances, the sulphuric acid is neutralized by ammonia derived from glutamine (p. 382), and therefore the output of ammonia is also increased (cf. Table 22.11).

EXCRETION OF NITROGEN AND SULPHUR

The preceding discussion makes it clear that the output of nitrogen and sulphur in the urine is determined by the protein intake. Table 22.11 shows the effect on a human subject of a change from a diet containing a large amount of protein to one low in protein. The result is a sharp reduction in the excretion of nitrogen involving urea, ammonia and to a smaller extent uric acid (derived from purines as explained in the preceding chapter). The excretion of sulphur is similarly affected. It is important to note that the excretion of creatinine is unaltered.

The effects of starvation on the excretion of nitrogen have been discussed in Chap. 13. The effect of changing from a high protein diet to a protein-free diet is also discussed in Chap. 13.

A normal man excretes about 1 g of free amino acids daily and about 2 g of amino acids

Table 22.11 Excretion of nitrogen and sulphur in the urine of a normal male student (Glasgow University, October 1955) on a high and a low protein diet

	Day	Estimated intake in diet in grams		Total daily output in urine in grams					Nitrogen balance assuming faecal N = 10 per cent of intake
		Protein	N	Urea N	Ammonia N	Creatinine N	Total N	Sulphate S	
High	2	176	28·1	18·0	0·991	0·65	20·7	1·66	+4·6
protein	3	198	31·6	23·7	1·15	0·63	26·0	1·44	+2·4
intake	4	193	30·9	22·7	1·18	0·64	25·9	1·52	+1·9
Low	2	71	11·3	11·1	0·79	0·64	13·4	1·42	−3·2
protein	3	59	9·4	8·5	0·53	0·63	10·2	0·60	−1·7
intake	4	39	6·2	6·0	0·38	0·65	7·7	0·66	−2·1

in conjugated form. An increase in the excretion of amino acids, termed aminoaciduria, may be brought about in one of two ways. It may be due to a defect in intermediary metabolism which leads to an increase in plasma levels of one or more amino acids to such an extent that the renal reabsorption mechanism cannot deal with them and they pass in increased quantities into the urine. These are the so called *overflow* aminoacidurias. Examples are, hepatic failure; phenylketonuria (A in Fig. 22.7), in which not only phenylpyruvic acid, but also phenylalanine, appear in the urine; and 'maple syrup urine', in which the branched-chain amino acids leucine, isoleucine and valine are found in high concentrations in the plasma because of a deficiency of the decarboxylase for conversion of branched chain α-keto acids to the corresponding acyl coenzyme A derivative (Fig. 22.4).

The second type, *renal* aminoaciduria, is due to defective renal tubular reabsorption so that even at normal plasma concentrations one or more amino acids are inadequately reabsorbed. Examples are renal tubular damage and congenital cystinuria (sometimes called cystinylsinuria) in which the diamino acids cystine, ornithine, arginine and lysine are all poorly reabsorbed.

PROTEIN METABOLISM IN THE RUMINANT

The bacteria and protozoa which abound in the rumen of the cow, sheep and goat can synthesize amino acids from simple sources such as short-chain fatty acids and ammonia (or urea). These amino acids are subsequently available to the animal for the synthesis of milk or muscle protein. These domestic animals are therefore exceedingly valuable producers of human food since they can convert plant materials with proteins of low biological value into the first-class proteins of meat and milk.

REFERENCES

ALBANESE, A. A. (1967). *Newer Methods of Nutritional Biochemistry*, Vol. 3. New York: Academic Press.

COHEN, G. N. (1969). *The Regulation of Cell Metabolism*. London: Holt, Rinehart & Winston.

CUTHBERTSON, D. P. (1958). Digestion in the ruminant. *Advances in Science* 15, 140–144.

GREENBERG, D. M. (1969). *Metabolic Pathways*, Vol. III, 3rd edn. *Amino Acids and Tetrapyrroles*. New York: Academic Press.

HARRIS, H. (1970a). *The Principles of Human Biochemical Genetics*. Amsterdam: North Holland Publishing Co.

HARRIS, H. (1970b). Genetical theory and the inborn error of metabolism. *British Medical Journal* (i), 321–327.

KREBS, H. A. (1964a). Glucogenesis. *Proceedings of the Royal Society* B 159, 545–564.

KREBS, H. A. (1964b). The metabolic fate of amino acids. *Mammalian Protein Metabolism* (Edited by H. N. Munro & J. B. Allison), 1, 125–177. New York: Academic Press.

MEISTER, A. H. (1965). *Biochemistry of the Amino Acids*, 2nd edn. New York: Academic Press.

MILNE, M. D. (1970). Some abnormalities of amino acid metabolism. *Biochemical Disorders in Human Disease*, 3rd edn (Edited by R. H. S. Thompson & I. D. P. Wootton). London: Churchhill.

MUNRO, H. N. (1970). *Mammalian Protein Metabolism*, Vol. IV. New York: Academic Press.

MUNRO, H. N. & ALLISON, J. B. (Eds.) (1964). *Mammalian Protein Metabolism*, Vols. I and II. New York: Academic Press.

NEUBERGER, A. & RICHARDS, F. F. (1964). Protein biosynthesis in mammalian tissues—II. Studies on turnover in the whole animal. *Mammalian Protein Metabolism* (Edited by H. N. Munro & J. B. Allison), 1, 243–297. New York: Academic Press.

ROTHSCHILD, M. A. & WALDMAN, T., Eds. (1970). *Plasma Protein Metabolism*. New York: Academic Press.

SCHOENHEIMER, R. (1946). *The Dynamic State of Body Constituents*. Harvard University Press.

SCHULTZE, H. E. & HEREMANS, J. F. (1966). *Molecular Biology of Human Proteins*. Amsterdam: Elsevier.

SCRIVER, C. R. (1969). Inborn errors of amino-acid metabolism. *British Medical Bulletin* 25, 35–41.

TATA, J. R. (1969). Hormonal control of protein synthesis. *Scientific Basis of Medicine Annual Review* 112–132.

WILLIAMS, R. T. (1959). *Detoxication Mechanisms*, 2nd edn. London: Chapman & Hall.

YOUNG, L. & MAW, G. A. (1958). *The Metabolism of Sulphur Compounds*. London: Methuen.

23 The blood

The chief functions of the blood are the transport of oxygen and food materials to all the cells of the body and the removal of carbon dioxide and waste materials to the organs of excretion. The blood in the vessels, although red to the naked eye, actually consists of a pale yellow fluid called *plasma* in which *formed elements*, the red corpuscles, the white (that is transparent) corpuscles and the platelets are suspended.

Since the coagulation of blood, which occurs a short time after it leaves the blood vessels, interferes with the examination of blood and its constituents, it is best to consider this aspect of the physiology of blood first.

BLOOD COAGULATION

When blood is withdrawn from a vein in the arm, and emptied into a test-tube, clotting takes place in about five minutes at room temperature, complete coagulation being indicated by the failure of the blood to move when the tube is tilted or inverted. The time of coagulation depends to some extent on the precise conditions under which the estimation is made. The main factors determining the rapidity of clotting are: the amount of damage to the tissues during venepuncture; the nature of the surface of the container used in the test, and the temperature. The reactions leading to blood coagulation are accelerated by factors from injured tissue, by surfaces such as glass, and by increases of temperature up to 40°C (Fig. 23.1). When blood clots in a test-tube a solid red mass is formed but, if the tube is left for some time, the clot contracts ('retracts') under the influence of a contractile protein in blood platelets and the force of gravity acting on the cells, to leave a supernatant yellow fluid called *serum*. Microscopic examination of the clot shows it to be composed of irregularly arranged fibrils of a protein material called *fibrin*, in the interstices of which are found the formed elements; the red

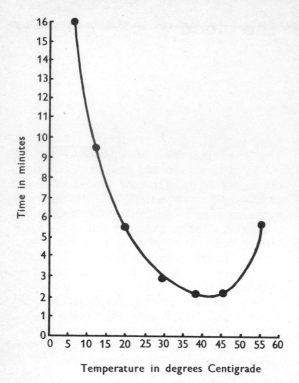

Fig. 23.1 Curve showing the effect of temperature on the coagulation time of normal whole blood. Above 47°C clotting is poor since fibrinogen (the protein of the clot), which coagulates at 54°C, is partly precipitated before clotting can occur. Below 1°C the action of thromboplastin is much retarded and clotting is greatly delayed, so that the cells sink to the bottom of the container and the plasma can be removed. (R. G. Macfarlane (1948). *Journal of Clinical Pathology* **1**, 126.)

blood corpuscles entangled in the fibrin give the clot its red colour. The formed elements of blood as such are, however, not necessary for coagulation since plasma free from erythrocytes, leucocytes and platelets can still be made to clot within 3 to 5 min. The platelets do, however, release factors which accelerate coagulation since for example in siliconed tubes the clotting time of platelet-rich plasma is shorter than that of plasma from which the platelets have been carefully removed.

Fibrin cannot be present as such in any quantity in the circulating blood; it is produced by polymerization of a soluble precursor *fibrinogen* normally present in the blood plasma. This protein material, fibrinogen, can be precipitated from plasma by salting out or by alcohol precipitation. By repeated precipitation fairly pure fibrinogen with a molecular weight of 340 000 can be obtained. Work with rat tissues maintained by perfusion has shown that fibrinogen is formed in the liver, the stimulus to increased production being a fall in the fibrinogen level of the perfusate, and the presence of fibrin degradation products in the circulating fluid. The fibrinogen molecule (Fig. 23.2) is a dimer, each sub-unit being composed of three peptide chains, united near the free ends of the dimer by a complex of disulphide linkages known as the 'disulphide knot'. The peptide chains are designated α, β, γ and those parts distal to the knot play an important part in the formation of fibrin. A solution of fibrinogen clots rapidly when fresh

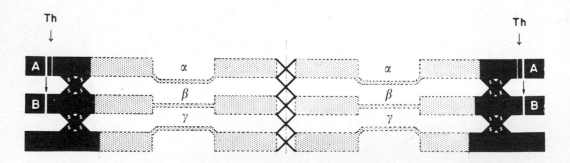

Fig. 23.2 Schematic model of the fibrinogen molecule. As here the usual form is a symmetrical dimer (molecular weight 680 000). The heavy shaded portions at each end show the region of the disulphide knot. The action of thrombin is to release fibrinopeptides (A and B) from their corresponding peptide chains α and β (the lightly shaded portions). The latticework at the centre indicates the end to end linkages between the two halves of the dimer. (Modified from B. Blombäck, M. Blombäck, A. Henschen, B. Hessel, S. Iwanaga & K. R. Woods (1968). *Nature, London* **218**, 131.)

serum is added to it. This reaction is due to the enzyme, *thrombin*, which is capable of coagulating many hundred times its own weight of fibrinogen. Thrombin splits arginyl–glycine bonds to liberate fibrinopeptides A and B from the chains distal to the disulphide knots (Fig. 23.2). This 'unmasks' charged sites which allow weak electrostatic linkages to form between adjacent fibrinogen molecules. Polymers of fibrin formed in this way are relatively unstable and may be dissolved in strong urea solution; they repolymerize if the concentration of urea is reduced. On the other hand, fibrin formed in whole blood or from plasma containing calcium ions is insoluble in urea; the fibrin polymer has been 'stabilized' by the formation of peptide linkages between adjacent monomers. The cross-links are formed by another enzyme, variously called *fibrin stabilizing factor*, *fibrinase* or *fibrinoligase*. This enzyme, an aminotransferase, forms (glutamyl)-lysine linkages at three or four specific sites on the fibrin monomer. The resulting 'stable fibrin' is mechanically stronger, more compact structurally and more resistant to attack by proteolytic enzymes. The diminution or absence of stable fibrin may explain the impairment of wound healing seen in persons with an inborn deficiency of fibrin stabilizing factor. The precursor of the aminotransferase is activated by thrombin in the presence of calcium ions. Undoubtedly these linkages account for the characteristic periodicity of fibrin seen in electron micrographs (Fig. 23.3).

Thrombin also has a powerful accelerating effect on blood coagulation because of its ability to increase the rate of formation of other enzymes of the coagulation system.

The conversion of fibrinogen to stable fibrin by thrombin, fibrinoligase and calcium ions is but the end-stage of a complex chain of enzyme reactions involving at least ten other factors. Knowledge of these was in the past gained by studying patients with inborn, or acquired, deficiency of single factors and it was customary to describe the coagulation system by presenting the history of these discoveries. However, modern biochemical methods are furnishing so much new information that the historical description has to be abandoned and accordingly the stages of the coagulation system will be described in the form of a chain of reactions. Traces of the history of blood coagulation research remain in the numerical nomenclature of the clotting factors which relates to the order of their discovery rather than to their position in the train of reactions (Table 23.4). It should be noted: (1) that the substrate for each enzyme or enzyme complex is a proenzyme which becomes the active enzyme for the next stage of the reaction; (2) that this arrangement can function as an 'enzymic amplifier', so that a small change at the beginning of the system may result in the formation of large amounts of thrombin at the end; (3) that the amplifier can be modified by negative or positive feed-back. Positive feed-back is provided by the ability of thrombin to

Fig. 23.3 Electron micrograph of fibrin showing characteristic periodicity. Negative staining, i.e. the uranyl acetate has been precipitated around the bands rather than within them. ×233 000. (*By courtesy of A. S. Todd.*)

Table 23.4 International nomenclature of blood coagulation factors with synonyms

Factor	Synonyms
I	Fibrinogen
II*	Prothrombin
III	Thromboplastin
IV	Calcium
V	Labile factor, proaccelerin, Ac-globulin
VII*	Stable factor, proconvertin, serum prothrombin conversion accelerator (SPCA)
VIII	Antihaemophilic globulin (AHG), antihaemophilic factor A
IX*	Christmas factor, plasma thromboplastin component (PTC), antihaemophilic factor B
X*	Stuart-Prower factor
XI	Plasma thromboplastin antecedent (PTA), antihaemophilic factor C
XII	Hageman factor
XIII	Fibrin stabilizing factor

The first four factors are well known by their names which are likely to remain in use, but the international nomenclature (Roman figures) for the other factors is now becoming general.

* These factors are affected by the anticoagulant drugs (coumarins and indanediones) much used in the treatment of thrombotic diseases.

accelerate the rate of formation of intrinsic factor X activator which directly increases the rate of prothrombin conversion. Negative feedback is provided by the neutralization of thrombin by fibrin; (4) that the system has two 'inputs' (the intrinsic and extrinsic systems of thromboplastin generation, see below) (5) that calcium ions are required at several stages of the reaction.

The coagulation reaction may be initiated in two different ways (Fig. 23.5). The first is by exposure of the plasma to a foreign surface, that is a surface bearing a negative electrical charge, which in some way causes alteration and activation of factors XI and XII. The product of this reaction is activated factor XI — usually designated as XIa which then acts on factor IX to give IXa. Factor IXa then combines with factor VIII and with phospholipid from platelets to activate factor X. Factor X similarly complexes with factor V, calcium and

phospholipid to form thrombokinase (sometimes called thromboplastin). Thrombokinase activates factor II (prothrombin) to form thrombin. It will be noted that apart from the 'foreign surface' all the components of this reaction chain are contained in the blood, hence the name *intrinsic coagulation system*. In the second coagulation reaction, the *extrinsic system*, the surface-activated reactions of the intrinsic system are by-passed; phospholipid and protein from injured tissue (also sometimes called thromboplastin) combine with calcium to activate factor X; from this point onwards the reactions are as in the intrinsic system. In both systems the function of the phospholipid seems to be physical rather than chemical, the lipid-protein micelles forming a suitable surface on which the reactions may take place.

Some of the methods used to study these reactions in the blood of patients will now be described. It is first necessary to obtain blood in the unclotted state usually by removing calcium ions from the system. Oxalate salts precipitate the calcium as insoluble oxalate; citrate and salts of ethylene-diamine-tetra-acetic acid (EDTA, 'sequestrene') sequester or chelate calcium ions; passage of the blood over a suitable ion exchange resin removes calcium ions. When blood treated in this way is allowed to stand or is centrifuged the supernatant yellowish fluid is called *plasma*. In some cases *heparin* (see below) may be used to prevent clotting both *in vitro* and *in vivo*. Substances which prevent blood clotting are collectively known as anticoagulants. It should be noted that this term is also applied to drugs which inhibit the synthesis of clotting factors by the liver (see Fig. 23.5).

Apart from measurement of the clotting time of whole blood, the oldest routine test of coagulation function is the *one-stage prothrombin time* in which tissue extract and calcium are added to plasma and the time to complete coagulation of the mixture measured. The coagulation reaction scheme (Fig. 23.5) shows that this test can also detect deficiency of fibrinogen, and factors V, VII and X. In fact the test is rather insensitive to variations in prothrombin itself but despite this it is useful in detecting deficiencies in the later part of the reaction sequence. The prothrombin time test is commonly used to assess the effect of anti-

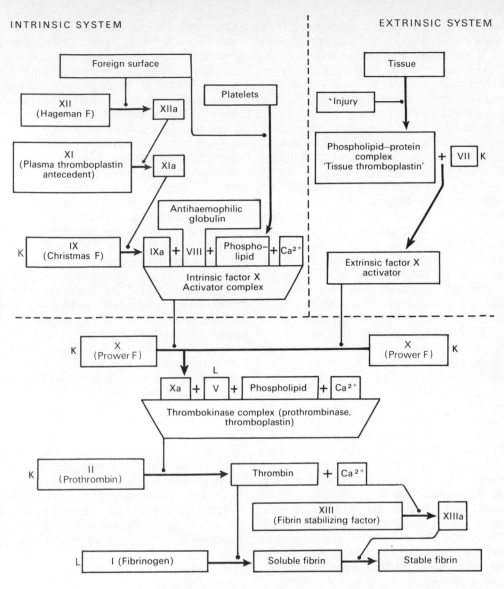

INTRINSIC SYSTEM

EXTRINSIC SYSTEM

Fig. 23.5 Scheme of the coagulation system: a, after the number of a factor indicates the activated form of that factor, for example IXa is activated factor IX; K, factors requiring vitamin K for their synthesis in the liver; L, factors synthesized in the liver, not needing vitamin K. (*Courtesy of A. S. Todd.*)

coagulant drugs such as dicoumarin. By adding purified clotting factors to the system the test can be modified to make it sensitive only to the factors likely to be affected by the therapy. Prothrombin in plasma is more accurately assayed by the *two-stage prothrombin time test* in which the prothrombin is measured by the amount of thrombin produced in the presence of adequate added thrombokinase complex.

The thrombin is assessed by adding samples of the thrombokinase–plasma mixture to purified fibrinogen solution. The clotting time of this fibrinogen substrate gives a measure of the amount of thrombin generated.

The *thromboplastin (thrombokinase) generation test* illustrates the use of two stage systems, and also the technique of partial fractionation of the blood under examination, in identifying coagulation deficiencies. Three fractions are used: (1) fresh plasma containing factor V and factor VIII, other factors having been removed from the plasma by adsorption on to aluminium hydroxide; (2) serum containing factor VII and factors IX and X, other factors having been destroyed by incubating the serum at 37°C; (3) platelet extract providing phospholipid. Phospholipid extracted from tissue may be used instead of that from platelets. When these three fractions are incubated at 37°C together with calcium ions thrombokinase is formed. The amount of thrombokinase formed after any given time may be measured by adding a sample of the incubated mixture to normal plasma and measuring the rate of coagulation. By substituting each normal fraction in turn by the corresponding fraction from the patient's plasma, the probable position of the defect in the reaction scheme can be detected. This type of test is frequently used in the diagnosis of haemophilia (factor VIII deficiency) and Christmas disease (factor IX deficiency), but the identification of other defects in the clotting system and assay of individual clotting factors are also possible by using variations of this technique. Another method often used to identify a clotting disorder is to add the plasma under investigation to a plasma from a patient with a known single deficiency, to see whether the plasmas are mutually corrective. Correction of the clotting function implies a disorder different from that in the 'known' sample, failure to correct implies identity of deficiency.

Disorders of the coagulation system may be inborn, may be acquired through disease processes or may be deliberately caused for therapeutic reasons.

The best known of the genetically determined disorders is haemophilia (factor VIII deficiency). The mode of inheritance of this disease is discussed in Chapter 53. The haemophiliac patient suffers from crippling haemorrhage into tissues and joints after minor injury and bleeds excessively (often for a week or more) if wounded. Factor IX deficiency (Christmas disease) is indistinguishable clinically from haemophilia and is inherited in the same way (as a sex-linked recessive character).

These two conditions may be distinguished by substitution tests, since haemophilia and Christmas disease plasma are mutually corrective. The only effective treatment for haemophilia is the administration of factor VIII from either human or animal sources. Whole blood is inefficient for this purpose because the volume needed to supply sufficient factor VIII to correct the deficiency is very large and unless the blood is very fresh the content of factor VIII is negligible. Animal concentrates of factor VIII are expensive and may be antigenic since they are 'foreign' proteins. A very useful material is *cryoprecipitate* from human plasma. If plasma is frozen, then allowed to thaw at 4°C the fibrinogen is precipitated, carrying with it most of the factor VIII from the plasma. The precipitate can be redissolved by warming to 37°C and may be administered intravenously in approximately one tenth of the original plasma volume. Major difficulties in the treatment of haemophilia by replacement are the lability of factor VIII *in vitro* and its short half-life in the circulation (12 to 14 hours). It is estimated that there are 2000 to 3000 haemophiliacs in Great Britain who may each require factor VIII from 90 to 100 blood donations each year. Christmas disease is only one tenth as common.

Of the acquired coagulation disorders, those associated with liver disease have provided useful information about the coagulation system. Fibrinogen, prothrombin, factors V, VII, IX and X are all synthesized in the liver and their production may be severely depressed in liver disease. Of these factors all but factor V and fibrinogen seem to require vitamin K for their synthesis. Thus clotting abnormalities closely resembling those due to liver disease may be seen in patients with vitamin K deficiency due to biliary obstruction and failure of fat absorption from the gut.

The vitamin K-dependent factors are also found to be deficient after administration of some types of anticoagulant drugs. (Fig. 23.6). The discovery of these agents followed the observation that cattle fed on badly cured hay developed a haemorrhagic disease—*sweet clover disease*. Analysis of the clover in the hay showed the active agent to be dicoumarin—a substance resembling vitamin K in its chemical structure (p. 197). Dicoumarin and its derivatives are now widely used to reduce the coagulant power of the blood in cases of thrombosis. The therapeutic effects can be reversed

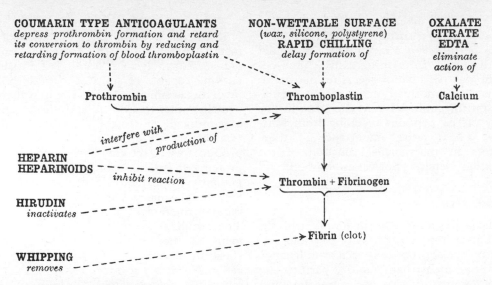

Fig. 23.6 Simplified scheme of the factors involved in the coagulation of the blood with an indication of the principal mode of action of some anticoagulants.

by administration of vitamin K_1 in fairly large doses. The fact that dicoumarin has no effect when added to an *in vitro* coagulation system indicates that its effect is on the synthesis of coagulation factors in the liver.

Other anticoagulants are: *heparin*, an acid (sulphated) mucopolysaccharide (mol. wt. 16 000), first obtained from the liver (hence the name), now prepared from beef lung or intestinal mucosa. It is active *in vitro* and *in vivo* but must be given parenterally. The anticoagulant activity of heparin is attributed to its sulphate groups which react with free basic groups on the proteins concerned in blood coagulation; it interferes with the formation of thrombokinase and also with the thrombin–fibrinogen reaction. The action of heparin in the circulation may be neutralized by administration of *protamine*—a low molecular weight basic protein from fish sperm heads (p. 68). Other substances capable of neutralizing heparin are hexadimethrene bromide (polybrene) and the dye toluidine blue which, like protamine, neutralize the acidic groups of the heparin molecule. Although heparin is a highly effective anticoagulant drug and is widely distributed in the mast cells, there is little evidence that, in man, endogenous heparin functions as a natural anticoagulant. Heparin-like activity previously described in the blood is more likely to be due to peptides released during the proteolytic degradation of fibrin and fibrinogen. The anticoagulant activity of these peptides is, like that of heparin, neutralized by protamine. The cervical glands of the leech contain hirudin, a powerful antithrombin. An extract from the venom of the Malayan pit viper has been used to produce a different kind of anticoagulant effect, *defibrination*. This extract (arvin) splits fibrin peptide A (Fig. 23.2) from the fibrinogen molecule to produce a weakly aggregated unstable fibrin which is rapidly removed from the circulation by the reticulo-endothelial system. In this way all the circulating fibrinogen is consumed. However, the antigenicity of arvin in man is delaying its widespread use as an anticoagulant.

In normal circumstances clotting does not occur in the blood vessels partly because the tendency to clot is prevented by inhibiting reactions described below and partly because any fibrin formed is removed by fibrinolysis to be described later (p. 398). Inhibitory control of the coagulation reaction occurs mainly at the stage of fibrinogen conversion, in the form of *antithrombin* effects. Six antithrombin effects have been described—not all of them well characterized or generally recognized. The best established are: *antithrombin I* effect, that is the neutralization of thrombin by adsorption on to fibrin; antithrombin II, a protein cofactor in plasma necessary for the anticoagulant action of heparin, sometimes called *heparin co-factor*; antithrombin III or *progressive antithrombin,* a factor which combines with thrombin in a progressive, time-dependent way (this factor seems to increase in patients with abnormally active blood coagulation); antithrombin VI refers to heparin-like activity of the peptides resulting from fibrinolysis. The relative importance of these inhibitors in the normal function of the coagulation system is not yet established.

FIBRINOLYSIS

In the living animal fibrin formation is an intermediate step in a variety of physiological and pathological processes; such as in haemostasis during menstruation and at parturition; in the inflammatory response to infection and trauma; and in reactions to cancer cells in the circulation and tissues. In many cases fibrin formed within blood vessels is rapidly dissolved to restore the fluidity of the blood, in others the fibrin becomes hyalinized or is removed by phagocytes and replaced by connective tissue. The process of liquefaction of fibrin is known as fibrinolysis; it may occur as rapidly as blood coagulation itself. Fibrinolysis also resembles blood clotting in being the end result of a chain of enzyme reactions involving components from both plasma and tissues (Fig. 23.7).

The main plasma component of the fibrinolytic system is *plasminogen* (profibrinolysin), inactive precursor of the proteolytic enzyme *plasmin* (fibrinolysin). Plasmin is a trypsin-like enzyme capable of splitting peptide linkages involving arginine or lysine. Plasmin also splits proteins other than fibrin and fibrinogen such as gamma globulin, casein, gelatin, blood clotting factors V, VIII and IX, kininogens (to produce vasoactive kinins) and chondromucoproteins. Plasmin is rapidly destroyed by *antiplasmins*, from the α_1- and α_2-globulin fractions of plasma. α_2-Antiplasmin acts by the rapid formation of a complex with plasmin. The complex is readily dissociated by attachment of the plasmin to insoluble protein substrates, especially fibrin. α_1-Antiplasmin acts more slowly to form a more stable, non-reversible complex with plasmin. There is sufficient of these two inhibitors in the plasma to neutralize the plasmin resulting from the total transformation of all of the blood plasminogen. For this reason plasmin is not detectable in blood unless fibrin is added to dissociate the α_2-globulin–plasmin complex. These unique inhibitory arrangements render the action of this non-specific protease highly selective for fibrin. Many drugs have been used as inhibitors of fibrinolysis, the most effective are ε-aminocaproic acid (εACA), tranexamic acid (*p*-aminomethyl-cyclohexane carboxylic acid, AMCHA) and trasysol (a polypeptide

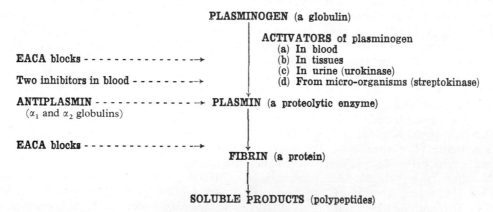

PLASMINOGEN (a globulin)

ACTIVATORS of plasminogen
   (a) In blood
   (b) In tissues
   (c) In urine (urokinase)
   (d) From micro-organisms (streptokinase)

EACA blocks - - - - - - - - - - - - - - →

Two inhibitors in blood - - - - - - - - →

ANTIPLASMIN - - - - - - - - - - - - → PLASMIN (a proteolytic enzyme)
  (α_1 and α_2 globulins)

EACA blocks - - - - - - - - - - - - - - →

FIBRIN (a protein)

SOLUBLE PRODUCTS (polypeptides)

Fig. 23.7 The fibrinolytic system of plasma. EACA is ε-amino-n-caproic acid which interferes with the system at two points. Its action in blocking the activation of plasminogen is an example of competitive inhibition, that is it competes with plasminogen for activator. It inhibits the action of plasmin on fibrin in a non-competitive way. Two inhibitors oppose plasminogen activator or the process of activation; one, stable at 37°C appears in circulating plasma and a second labile at 37°C is formed during coagulation or thrombus formation in the presence of calcium.

extract from bovine salivary glands, which also inhibits trypsin and kallikrein).

The transformation of the enzyme precursor, plasminogen, to the active form, plasmin, requires the presence of an activator simply referred to as *plasminogen activator*. Substances having an activating effect on plasminogen are found in small amounts in the blood and in high concentration in the tissues, especially in the endothelium of blood vessels. Urine contains an activator, urokinase; the activity of urine is more than that of blood but less than that of endothelial cells. Small amounts of proteolytic enzymes, such as trypsin and leucocyte protease, are also able to activate plasminogen, as can the lysosomal enzymes of many types of cell. The activating effect of these substances is, however, weaker and less specific than that of the plasminogen activators found in blood, urine and endothelium. In addition to endogenous agents, activators of plasminogen have been obtained from various bacteria, moulds and plants. Of these the most important is *streptokinase*, a protein from cultures of haemolytic streptococci. In purified form this material is widely used to promote fibrinolytic activity in the circulating blood of patients with thrombosis. The mechanism of action of streptokinase on human plasminogen is still debated but appears to be more complex than that of endogenous activators, and may involve an auto-catalytic mechanism.

Fibrinolysis has been studied by innumerable techniques. Fibrinolytic activity in whole blood is commonly measured by the *dilute whole blood clot lysis time technique*. Blood is diluted to diminish the effect of inhibitors, clotted by thrombin and then incubated. The time taken for the clots to dissolve is a measure of spontaneous fibrinolytic activity. The other widely used method is the *euglobulin lysis time technique*. The plasma under test is diluted and acidified to precipitate the euglobulin fraction containing fibrinogen, plasminogen and plasminogen activator; much of the inhibitory activity remains in the supernatant. The precipitate is redissolved, clotted by the addition of thrombin, and the time taken for the clot to dissolve at 37°C is measured.

For testing the fibrinolytic activity of fluids other than plasma, such as urine and extracts of tissue the *fibrin plate* method is used. The substrate is fibrin prepared as a uniform layer in a petri dish. The material to be tested is placed on the fibrin and the dish is incubated at 37°C to allow digestion of the fibrin substrate to take place. The diameter of the zone of digestion in the fibrin layer gives a measure of the fibrinolytic activity. A modification of this technique has been used to reveal the distribution of plasminogen activator in tissues. A thin section of fresh tissue is incubated on a thin layer of fibrin. Wherever there is fibrinolytic activity in the section the fibrin is digested. The areas of digestion show as clear spots when the preparation is stained and can be related to structures in the overlying section (see Fig. 23.8).

Fibrinolytic activity is detectable in the blood of normal subjects; is increased after exercise, after injection of adrenaline and in emotional stress; it shows a circadian rhythm being increased by day and diminished at night. Arterial blood contains less activator than venous and the activator content of the endothelium in arteries is generally less than that in the venous intima. The endothelium of arteries supplying the retina, myocardium and kidneys is an exception to this rule and has a high content of activator.

Some fibrinolysis seems to go on continuously in the circulating blood since fibrin degradation products (FDP) can be detected in the blood of healthy subjects. In cases of thrombosis and of excessive fibrinolysis the level of FDP may be greatly increased from the normal level of 2 to 10 µg/ml of plasma to as much as 760 µg/ml. At these high levels FDP have a heparin-like action on the thrombin-fibrinogen reaction and impair the stabilization of fibrin by factor XIII.

The physiological effects of fibrinolysis are still uncertain. There is no doubt about the existence of a thrombolytic effect but the influence of fibrinolysis on kinin systems and on the control of blood viscosity and flow have yet to be fully assessed.

PLATELETS

Platelets (p. 434) are essential components of the coagulation system but thus have so far been mentioned only in connexion with their ability to release phospholipid for the forma-

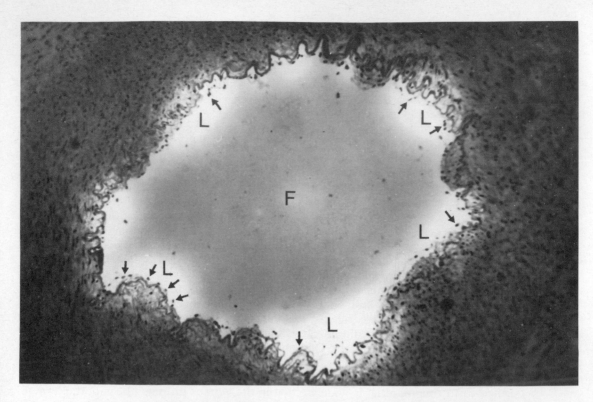

Fig. 23.8 Histological demonstration of plasminogen activator. The tissue is a small branch of a coronary artery. The section has been mounted on a thin layer of plasminogen rich fibrin, incubated then fixed and stained. The fibrin shows as a uniform dark background (F). The light areas (L) show where the fibrin has been digested and are related to endothelial cells (indicated by arrows). The activator in the endothelium has been converted from plasminogen to plasmin and fibrinolysis has followed. ×170. (*By courtesy of A. S. Todd.*)

tion of intrinsic factor X activator. Platelets have other actions important in haemostasis: (1) *Adhesion reactions.* Platelets have the ability to adhere to foreign surfaces. The most important of these are collagen (*in vivo*) and glass (in laboratory experiments). (2) *Reversible aggregation.* Platelets adhere to each other in a reversible fashion under the influence of certain substances especially adenosine diphosphate, noradrenaline and 5-hydroxytryptamine. This reaction is not accompanied by changes in fine structure. (3) *Irreversible aggregation.* This reaction is triggered by thrombin and is accompanied by morphological changes such as rupture of the cytoplasmic membrane and degranulation of the platelets. The reaction is also known as *viscous metamorphosis* and, unlike reversible aggregation, it is recognizable by light microscopy. *In vivo* masses of irreversibly aggregated

platelets form small haemostatic plugs in damaged blood vessels. (4) *The release phenomenon.* During the aggregation reactions substances important in coagulation and in the control of vascular tone are released from the platelets. The release phenomenon is most complete during irreversible aggregation by thrombin and least during reversible aggregation by ADP. The released components include ATP and ADP; 5HT; K^+; platelet factor 3 (phospholipid); platelet factor 4 (a basic protein, like protamine, with an antiheparin action); and lysosomal enzymes. (5) *Contractile function.* Platelets contain an actomyosin-like protein, thrombosthenin, which contracts in the presence of magnesium ions, using ATP as its energy source. This protein is responsible for the phenomenon of *clot retraction* (contraction) which expresses serum from blood clots. The contractile protein may have important functions in altering the configuration of the platelet membrane and in discharging cytoplasmic contents during the release reaction. Clot retraction is an exceedingly weak effect and may be little more than a vestigial function, of use only as a laboratory indicator of platelet activity.

Methods of studying platelets and their functions. In the *clot retraction* test the proportion of serum expressed from whole blood by clot retraction under standard conditions is measured. Clot retraction is impaired if either the number, or the contractile function, of the platelets is diminished. The *proportion of adhesive platelets* is assessed when whole blood is passed over a large surface conducive to platelet adhesion (for example glass beads) and the consequent reduction in the platelet count is measured. This test can be modified by addition of aggregating agents such as ADP which increase adhesion. *Platelet aggregation* can be measured by the fall in optical density in platelet suspensions resulting from the reduction in the number of suspended particles. This technique is widely used in assessing the effect of aggregating reagents such as ADP, collagen, noradrenaline and thrombin. A *blood film* made from a fresh capillary sample normally shows platelets aggregated in clumps but if their function is impaired the platelets are distributed singly throughout the film.

important platelet disorder is a reduction in number—*thrombocytopenia*. This may be due to excessive utilization in abnormal coagulation processes; excessive destruction by the reticuloendothelial system; impaired production from megakaryocytes damaged by poisons, immune reactions or the presence of neoplasms. Major disorders of platelet function are rare and are usually hereditary; they are referred to as *thrombasthenia*. Minor impairment of functions such as aggregation are usually transient and are often associated with drugs, for example aspirin and other analgesics. Platelet disturbances lead to haemorrhage from small vessels, a prolonged bleeding time and spontaneous capillary haemorrhages into the skin, known as purpura.

HAEMOSTASIS

The action of platelets and the coagulation of the blood modified by fibrinolysis are directed towards the achievement of haemostasis (Fig. 23.9). A fourth element is also concerned, namely the response of the blood vessels not only to injury but also to some products of the coagulation and fibrinolytic reactions.

Thrombosis. Thrombosis can be regarded as due either to over-activity of the coagulation and haemostatic mechanisms or to under-activity of the fibrinolytic process. Its starting point is often, or perhaps always, an area of damage to the vascular endothelium to which platelets adhere. A thrombus forming at one place in a vein or an artery has a different structure from a clot produced in a glass tube, since it consists of a 'head' of adherent and fused platelets and a 'tail' of fibrin containing trapped red and white cells. It occurs more often in veins than in arteries, for in the former, of course, the rate of flow is slower. Venous thrombosis is much commoner in the veins of the legs than elsewhere, particularly in the deep veins of the calves. Prolonged rest in bed and varicosity of the veins, which are both accompanied by further slowing of the blood flow, favour certain changes in the blood maximal some 8 to 12 days after surgical operations which also favour thrombosis. These are an elevated platelet count, increased stickiness of the platelets and a raised concentration of fibrinogen. The tendency to venous thrombosis is reduced by the administration of anticoagulants.

THE PLASMA PROTEINS

Blood plasma contains about 9 per cent of solids, some 7 per cent being protein in nature. According to the classical view, the plasma proteins are divided into three main types: *fibrinogen*, *albumin* and *globulin*. Each of the two latter, as shown in Table 23.10, is actually a group of proteins with similar electrical properties (isoelectric point and surface charge, for example). Thus 'plasma albumin' consists of at least two proteins while 'plasma globulin' has been further subdivided into α_1, α_2, β and γ fractions. It is commonly stated that the globulins are precipitated at half saturation of plasma with ammonium sulphate and albumins at full saturation, but the method of fractional precipitation is in fact much less crude than this. Fibrinogen precipitates at 0·25

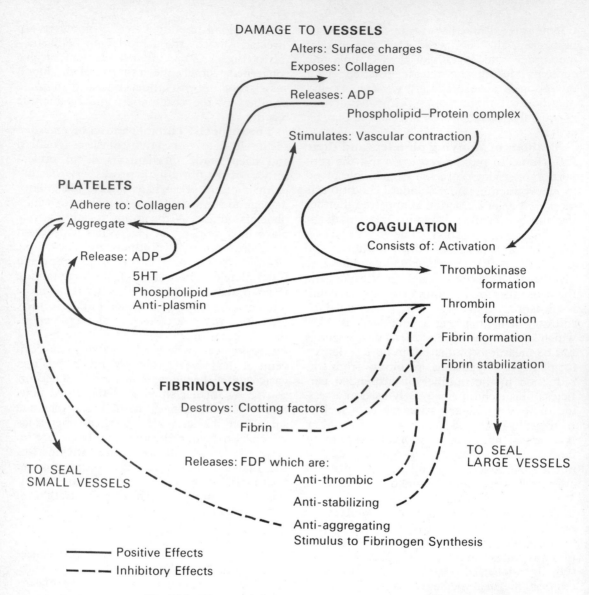

DAMAGE TO **VESSELS**
Alters: Surface charges
Exposes: Collagen
Releases: ADP
Phospholipid–Protein complex
Stimulates: Vascular contraction

PLATELETS
Adhere to: Collagen
Aggregate

Release: ADP
5HT
Phospholipid
Anti-plasmin

COAGULATION
Consists of: Activation
Thrombokinase formation
Thrombin formation
Fibrin formation
Fibrin stabilization

FIBRINOLYSIS
Destroys: Clotting factors
Fibrin
Releases: FDP which are:
Anti-thrombic
Anti-stabilizing
Anti-aggregating
Stimulus to Fibrinogen Synthesis

TO SEAL
SMALL VESSELS

TO SEAL
LARGE VESSELS

———— Positive Effects
– – – – Inhibitory Effects

Fig. 23.9 Haemostatic interactions. (*By courtesy of A. S. Todd.*)

saturation, γ-globulin at 0·34 saturation, α- and β-globulin at 0·5 saturation and albumin at 0·68 saturation.

The plasma proteins can be separated by precipitation with various concentrations of alcohol at temperatures below 0°. At such low temperatures denaturation does not occur and it is possible to obtain plasma proteins in bulk in a relatively pure state.

The plasma proteins may also be separated from one another by electrophoresis on paper or cellulose acetate (Fig. 23.11) as described already in Chapter 3 is now commonly employed in clinical work. A very sensitive method for the detection of proteins in plasma and other biological fluids is provided by the technique of immuno-electrophoresis (Fig. 23.12). The plasma proteins give serum a viscosity between 1·4 and 1·8 times that of water. Occasionally abnormal proteins appear in the blood and if the viscosity is much increased bleeding from various sites may occur

Table 23.10   Human plasma proteins

Protein	Concentration g/100 ml	Mol. wt.	Special functions
Prealbumin	0·3	61 000	Transport: osmotic regulation
Albumin	2·8–4·5	69 000	
α_1-Globulins	0·3–0·6		
α_1-Globulin (orosmucoid)	0·075	41 000	40% carbohydrate
α_1-Globulin (glycoprotein)	0·030	54 000	14% carbohydrate
α_1-Lipoproteins			
Density 1·093	0·05–0·13	435 000	67% lipid: transport
Density 1·149	0·3–0·4	195 000	43% lipid: transport
Haptoglobin	0·1	85 000	Binds haemoglobin
α_2-Globulins	0·4–0·9		
α_2-Globulin (glycoprotein)			16% carbohydrate
α_2-Globulin (macroglobulin)	0·2	820 000	10% carbohydrate
Caeruloplasmin	0·03	160 000	Copper transport
Plasminogen		143 000	Forms plasmin
Prothrombin (bovine)	0·01	68 000	Forms thrombin
β-Globulins	0·6–1·1		
β_1-Lipoproteins			
Density 0·98–1·002	0·13–0·20	5×10^6	90% lipid: transport
Density 1·03	0·20–0·25	3×10^6	79% lipid: transport
Transferrin	0·4	90 000	Iron transport
Fibrinogen	0·3	341 000	Blood clotting
γ-Globulins	0·7–1·5	150 000	Antibodies

Data from F. W. Putnam (Ed.) (1960). *The Plasma Proteins*, Vols. 1 and 2. New York: Academic Press.

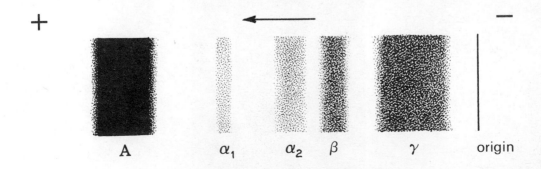

Fig. 23.11   Record of paper electrophoresis of the serum from a normal human subject showing the separation of albumin, A, which moves most rapidly, from globulins α_1, α_2, β and γ. The direction of migration is shown by the arrow. The run was made in veronal buffer at pH 8·6 for 18 h at 4V/cm.

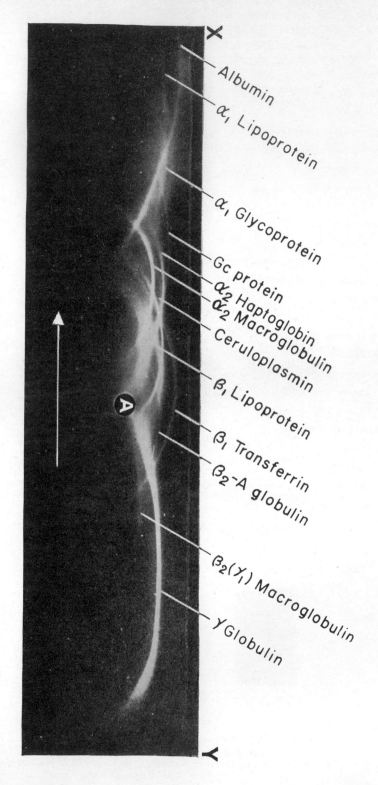

Albumin

α₁ Lipoprotein

α₁ Glycoprotein

Gc protein
α'₂ Haptoglobin
α'₂ Macroglobulin
Ceruloplasmin

β₁ Lipoprotein

β₁ Transferrin

β₂-A globulin

β₂(γ₁) Macroglobulin

γ Globulin

Fig. 23.12 The protein pattern of normal human serum identified by immunoelectrophoresis. The sample of serum is put into a well in the centre (A) of a strip of agar-gel; the albumin migrates towards the left. A shallow trough XY at the top of the figure holds horse serum containing immune bodies to the various proteins which it is intended to identify (see p. 402). These specific immune bodies have been produced in the horse by injecting an animal with the various components of human serum. The names and positions of some of the proteins shown in the figure are provisional.

with interruption of the retinal circulation and damage to the central nervous system.

Fibrinogen (molecular weight 330 000) is essentially a globulin which is unusually easy to precipitate. It is readily coagulated by heat and can be precipitated by 0·25 saturation with ammonium sulphate. Fibrinogen is the precursor of the fibrin of the blood clot and there is good evidence that fibrinogen is formed almost exclusively in the liver. Solutions of fibrinogen are some six times as viscous as albumin solutions of the same concentration and the viscosity of plasma is largely due to fibrinogen.

The *globulins*, designated α_1-, α_2, β- and γ-globulins, cover a wide range of molecular weights from 90 000 to 1 300 000 or more and can readily be separated by electrophoresis (Fig. 3.2). They are produced mainly in the liver but γ-globulins are formed in reticuloendothelial tissue and in plasma cells and lymphocytes.

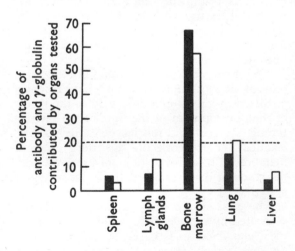

Fig. 23.13 The contribution of various organs to the production of antibodies after intravenous immunization of a rabbit. Black columns, anti-ovalbumin; white columns other γ-globulins (B. A. Askonas & J. H. Humphrey (1958). *Biochemical Journal* **68**, 252).

α_1-Globulins include a fraction which combines with bilirubin and another fraction, the α_1-lipoprotein of the plasma, responsible in part for the carriage of lipids and steroids. The α_2-globulin fraction includes the mucoproteins and also the haptoglobins.

Among the β-globulins are β-lipoprotein, a very high molecular weight protein responsible for the transport of lipids in the blood, and also globulins which bind iron (*transferrin*, p. 425) and copper (*caeruloplasmin*) for transport. Prothrombin (p. 394) is a β-globulin. All known antibodies are included in the γ-globulin fraction. Antibodies are produced mainly in the bone marrow (from the plasma cells), and in the spleen, liver and lymph nodes (Fig. 23.13).

The new-born child possesses few plasma cells in lymph nodes or bone marrow and accordingly he produces very little γ-globulin but he is protected by γ-globulin transferred across the placenta from his mother. However, by the time he is nine months old he is able to produce adequate amounts of γ-globulin. In a few male children a sex-linked recessive trait expresses itself as an inability to form plasma cells and γ-globulin. The plasma γ-globulin may be less than 50 mg per 100 ml (hypogammaglobulinaemia). Since these children cannot form antibodies in sufficient amount they are apt to suffer from recurring infections, especially those produced by pyogenic organisms.

Solutions of γ-globulin are frequently used to confer passive immunity. For example, the γ-globulin fraction of the plasma protein of a person who has recently had German measles contains a high concentration of antibody to the virus of the disease. Given in the form of 'convalescent serum' or of a preparation of γ-globulins it may prevent, or cut short, an attack of the disease in a patient. Such preparations have now been used in a number of virus diseases in man, in poliomyelitis, for example, and in distemper in dogs.

The *albumins* (molecular weight about 68 000) are the most abundant of the plasma proteins and have the highest electrophoretic mobility. The albumin fraction is formed in the liver. Its half-life in the plasma is about 18 days and some 10 to 12 g are produced each day. Serum albumins have the important property of binding other substances such as lipids, hormones, bilirubin and many drugs. Plasma albumin, partly because it is the most abundant of the plasma proteins and partly because of its low molecular weight, is responsible for about 80 per cent of the total osmotic pressure of the plasma. It is, therefore, very important in preserving the fluid balance between blood and tissues (Chap. 33). Concentrated solutions of human albumin have been used instead of blood plasma for transfusion to increase the plasma protein concentration and therefore the osmotic pressure with the purpose of attracting oedema fluid into the vascular compartment of the extracellular space and so facilitating its excretion. Such measures are not usually very successful because the transfused albumin rapidly leaves the circulation. A few persons have been found who were completely lacking in serum albumin without disturbance of water or electrolyte balance. Their low plasma osmotic pressure was compensated for by a marked drop in the capillary hydrostatic pressure (see Chap. 33).

Haptoglobins. The haptoglobins are a group of α_2-globulins having the distinctive property of binding haemoglobin. By starch electrophoresis three α_2-globulin bands can be identified which depend on two alternative (allelic) genes which produce three phenotypes Hp 1–1, Hp 2–2, and Hp 2–1. These are inherited in the same way as the MN blood groups (see Chap. 53). When red cells break down haemoglobin is released and up to 150 mg haemoglobin per 100 ml plasma can combine firmly with haptoglobin to be quickly picked up by the reticulo-endothelial system. When the rate of destruction of red blood cells exceeds the rate of haptoglobin production haemoglobin may appear in the urine.

Normal capillary filtrates and normal lymph from the limbs contain small amounts of albumin (mol. wt. 68 000) and even of globulin (mol. wt. 90 000 to 175 000) and fibrinogen (mol. wt. 330 000). It was difficult to understand how these proteins, especially those of very high molecular weight, could pass through the capillary wall until Cohn showed that the molecules of plasma proteins are sausage-shaped bodies, almost all of which have nearly the same diameter, namely 3·3 to 3·8 nm. They vary in length, however, according to their molecular weight (Fig. 23.14). It appears that, although the majority of pores in the capillary walls measure from 0·7 to 2·0 nm, there are some at least 3·8 nm in diameter. All these pores allow simple salts to pass quite freely while the largest of them allow the passage of albumin and even of globulin and fibrinogen provided that these large molecules strike the capillary wall in the correct orientation, that is 'end-on' to the pores. In this way we can account for the small amounts of protein which are normally found in capillary filtrates and in urine. The albumin which passes through the capillary walls returns to the circulation via the lymphatics; about half of the total plasma albumin circulates in this way every 24 hours.

Hypoproteinaemia may be said to exist when the total plasma protein concentration is less than 5·5 g per 100 ml. It may be caused by excessive protein catabolism or by loss of protein into the urine, into the alimentary tract, or from the skin after burns or scalds. When this loss is not fully balanced by a simultaneous production of new plasma protein either because of a much reduced protein intake in the food, or as the result of liver disease, the concentration of protein in the plasma gradually falls. The colloid osmotic pressure of the plasma is diminished and the tendency for water to be attracted from the tissues into the blood is reduced. Water may therefore accumulate in the tissues; if this is excessive it may become obvious as *oedema*.

The plasma lipoproteins. The lipoproteins of blood plasma (see p. 333) have recently attracted much attention, largely as the result of the development of the lipoprotein phenotyping system of Fredrickson and his colleagues. Their work on disorders of lipid metabolism has proved to be of great value in the study of diseases of the heart and vascular system.

The plasma lipoproteins may readily be separated by ultracentrifugation, but the simplest classification is based on paper electrophoresis which reveals four main classes of lipoproteins.

1. The *α-lipoprotein* fraction shows greatest electrophoretic mobility and consists of about 50 per cent protein, 30 per cent phospholipid and 20 per cent cholesterol (and its esters).

2. The *β-lipoproteins* have a rather lower electrophoretic mobility, and are of such low density that they float to the top of the tube upon ultracentrifugation. Their composition is about 25 per cent protein, 20 per cent phospholipid, 10 per cent triglycerides and 45

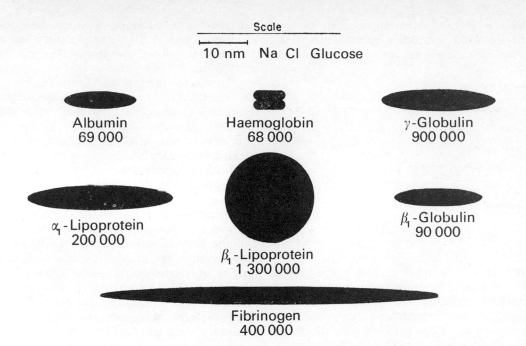

Scale

10 nm   Na Cl   Glucose

Albumin
69 000

Haemoglobin
68 000

γ-Globulin
900 000

α_1-Lipoprotein
200 000

β_1-Globulin
90 000

β_1-Lipoprotein
1 300 000

Fibrinogen
400 000

Fig. 23.14 Diagram showing relative dimensions and molecular weights of blood proteins. The molecules of albumin, globulin and fibrinogen are elongated bodies of about the same diameter but of very different lengths according to their molecular weights (E. J. Cohn (1945). *Science in Progress* **4**, 319).

per cent cholesterol and its esters. They are, in fact, the main carriers of cholesterol in the plasma.

3. The *pre-β-lipoprotein* fraction migrates slightly more rapidly than the β-fraction; it is even less dense than the β-fraction. It contains some cholesterol but its main function is as a carrier of triglyceride of endogenous origin which may account for between 50 and 80 per cent of its composition.

4. The fourth class of 'lipoproteins' are the chylomicrons which are important in relation to fat absorption from the gut (p. 299). They are large particles which do not migrate on paper and show up as a band at the origin. They consist from dietary fat together with a small amount of cholesterol.

In various disorders of lipid metabolism one or more of the lipoprotein fractions may be increased, giving rise to the conditions known as *hyperlipoproteinaemias*.

The formation of plasma proteins. Experiments on dogs by Whipple have shown that the plasma proteins are formed from amino acids derived from the food. By a technique called *plasmapheresis* he has investigated some of the factors involved in the generation of plasma protein. Plasmapheresis consists in withdrawing a quantity of blood from which the red blood corpuscles, after separation by centrifugation, are washed with isotonic saline and then returned to the dog by intravenous injection. In this way the animal is deprived of plasma proteins without loss of haemoglobin. If the animal is kept on a standard diet the plasma proteins can be gradually reduced by daily plasmapheresis from the normal level of 6 per cent to 4 per cent and can be kept at this lower level.

If now a protein is fed as a supplement to the standard diet without altering the amount of plasmapheresis, the percentage of protein in the plasma rises and the total amount of extra plasma protein manufactured can be estimated if the blood volume is known. Some experimental results are given in Table 23.15, which is extracted from Whipple's data.

Such experiments indicate that the plasma

protein is formed from amino acids derived from the food and other experiments along similar lines show further that cystine is a key

Table 23.15

Amount of protein fed	Amount of plasma protein formed
100 g serum protein (ox)	38 g
100 g meat (skeletal muscle)	18 g
100 g casein	12 g
100 g gelatin	5 g
100 g zein	0 g

amino acid in plasma protein regeneration. Both animals and human patients can be kept in nitrogen equilibrium by plasma given intravenously. Such experiments, and others in which protein tagged with isotopic tracers has been used, suggest that there is a continuous rapid exchange of protein molecules between the tissues and the blood plasma (p. 376). It is unlikely, however, that plasma protein is directly converted into tissue protein or vice versa, except through the amino acid pool. Plasma protein is not stored as such, but there is a considerable reserve of haemoglobin- and plasma-protein-building material, so that even after a severe haemorrhage about half the amount lost can be re-formed in a day. The existence of this reserve store is clearly shown by the fact that after severe depletion by plasmapheresis, followed by a period of starvation, plasma protein is regenerated at the expense of tissue protein (see also p. 376). An adult man synthesizes about 15 g of plasma protein per day.

Blood Volume

The blood volume of an animal can be assessed by collecting all the blood which flows from a cut artery. Bleeding-out by this method yields only part of the total blood in the body but the blood remaining in the tissues can be washed out by perfusing Locke's solution through the vessels. The fluid that emerges is blood diluted by Locke's solution but the true blood volume may be obtained by comparing the depth of colour of the animal's undiluted blood with that of the diluted blood

and making an allowance for the extent of the dilution. The blood volume of a dog measured in this way is about one-thirteenth, or 7·7 per cent, of the body weight.

The volume of plasma and blood in man can be measured by the principle of dye dilution. A known volume of a substance, such as the non-toxic azo-dye Evans Blue, is injected into the circulation. After a time during which mixing occurs the extent of its dilution in the circulating plasma is found by measuring the concentration of the dye in the plasma. If the relative proportion of plasma and corpuscles (the haematocrit) is known then the total blood volume can be calculated from the plasma volume. Dyes usually begin to leave the circulation within a few minutes, but this difficulty can be overcome by the use of human serum albumin labelled with radioactive iodine. ^{131}I, the half-life of which is 8 to 14 days, combines with tyrosine molecules of human serum albumin, thus labelling the protein without disturbing any of its properties. The amount of dilution is found by comparing the counts of radioactive emission of the tracer dose and of a specimen of blood withdrawn 10 min after the injection.

These dilution methods make the assumption that the ratio of plasma to corpuscles is the same throughout the body. However, the haematocrit value of the total body is about 10 per cent lower than the haematocrit in a large vessel. This may be because of the preferential accumulation of red cells in the more rapidly moving axial stream of smaller vessels or because some capillaries contain plasma but no red cells. The volume of the red cells can, however, be obtained independently by incubating a sample of the subject's blood with tracers such as ^{32}P, ^{51}Cr, ^{42}K or CO which combine with the red cells. A known amount of the sample is injected intravenously and the amount of dilution is measured.

In a normal adult the haematocrit rises about 3·5 per cent on changing from the lying to the standing position, presumably because fluid leaves the capillaries in which the hydrostatic pressure is increased. At the same time the plasma protein level rises about 0·7 g per 100 ml. The plasma Ca level does not change on standing, presumably because it diffuses easily out of the capillaries.

In adults the blood volume may be pre-

dicted from the height and weight thus:
$$BV = 0.3669 H^3 + 0.03219 W + 0.6041 \text{ (men)}$$
$$BV = 0.3561 H^3 + 0.03308 W + 0.1833 \text{ (women)}$$
where BV is the total blood volume in litres; H the height in metres and W the weight in kg (Nadler, Hiladgo and Block (1962). *Surgery* **51**, 224–232).

In children there is a gradual increase in blood volume with age (Table 23.16). The blood volume increases in pregnancy (Chap. 51).

The 5 litres of blood are, of course, contained in the heart, arteries, capillaries and veins (the volume in the heart is discussed in Chap. 26). Of the 1300 ml in the pulmonary circulation about 400 ml are in the arteries, 60 ml in the capillaries and 840 ml in the venules and veins. Of the 3000 ml in the systemic circulation about 550 ml are in the arteries, about 300 ml in the capillaries and 2150 ml in the venules and veins. The veins are to be regarded as blood reservoirs, partly because they contain a large proportion of the blood and partly because they can accommodate an increased amount with little rise in pressure (p. 530). The veins are often termed capacity vessels.

The effects of haemorrhage. The effects of loss of blood depend, of course, on the amount of blood lost and the rapidity of bleeding. The loss of small amounts of blood, for example from bleeding haemorrhoids, intermittently or continuously over many months does not disturb the circulation but it may produce anaemia.

In healthy normal subjects quite a large amount of blood can be lost before the arterial pressure falls. In one experiment 11 normal human subjects were bled about 17 per cent of their blood volume in 35 minutes. The changes in heart rate, blood pressure and cardiac output were insignificant. However, the splanchnic blood volume was reduced by 30 per cent so that the splanchnic viscera contributed about half the volume of blood removed. These findings show the importance of the splanchnic

circulation as a blood reservoir in man; they can best be explained by supposing that there is active venous vasoconstriction without arterial vasoconstriction.

For a short time after the haemorrhage the concentrations of haemoglobin and of plasma protein may remain almost normal but within two or three hours both fall because of the entry into the circulation of protein-free fluid as a result of the fall in capillary filtration pressure produced by arteriolar constriction. In this way the circulating fluid volume is restored within about 24 hr and during this period the concentrations of plasma protein and haemoglobin fall progressively. Between 2 and 4 days after the loss of blood the plasma protein concentration is restored and thereafter the concentration of haemoglobin returns to normal more slowly by increased production of red cells by the bone marrow. The period needed for complete restoration of haemoglobin depends on the amount of blood lost and upon the availability of iron in the body stores; thus about 20 days are required after the loss of 200 ml of blood and some 50 days after the loss of 500 ml.

When quantities of blood of the order of 2 litres (equal to 30 to 40 per cent of the blood volume) are rapidly lost from the body a state of shock is produced (Chap. 28). The circulation is little upset by alterations in blood volume of ± 10 per cent but alterations of ± 30 per cent have serious repercussions. Increases of blood volume of 1 to 2 litres commonly occur in heart failure. There is apparently no need for a precise regulation of blood volume. The mechanism by which it is regulated is not known.

Antigens and antibodies. Before going on to discuss blood groups it is necessary to describe the body's reaction to the introduction of foreign protein. This topic has already been mentioned briefly on p. 35. The phenomenon was first studied in connexion with the immunity which develops, for example, after infection with the bacilli of diphtheria. If

Table 23.16 The increase of blood volume with age

Age in years	0	$\frac{1}{2}$	1	$2\frac{1}{2}$	4	7	12	19	Adult male	Adult female
Blood volume (litres)	0·3	0·5	0·6	1	1·3	1·7	2·5	3·2	5	3·5

the patient survives the disease his serum is found to contain *antitoxins* which can neutralize the *toxins* produced by the bacilli. Nearly all proteins act as *antigens* which call forth *antibodies* provided that they are injected into an animal which does not already possess the protein in its tissues, that is, the antigens must be foreign to the animal (Fig. 23.12). The antibodies which are formed are highly specific. By injecting human serum into one rabbit, ox serum into another, and so on, we can obtain rabbit sera which can be used in a precipitation test to identify the animal from which a serum comes even if only a minute quantity of protein, such as might be obtained from a blood stain, is available for test. Anti-human rabbit serum mixed with human serum causes a precipitate to appear but no other antiserum has this effect. If in much the same way red blood cells are injected into an animal of another species, specific *haemolysins* are formed. Thus if monkey red cells are injected into a rabbit, the rabbit's serum is found after some days to possess haemolytic properties which are revealed by mixing monkey corpuscles and the rabbit's serum *in vitro*. The cells are first agglutinated and then lysed so that haemoglobin is released and the serum becomes a transparent red colour. Corpuscles from any other species mixed with this rabbit serum are unaffected and simply sink to the bottom of the test tube.

Blood Groups

The *blood groups* refer to the presence on or in the red cell envelope of certain antigens, the blood group substances. Up to the present over 100 of these antigens have been demonstrated and no doubt others remain to be discovered. They are inherited characteristics and the majority of them have been shown to belong to genetically independent blood group systems, of which there are at least 14. The best known are ABO, Rh, MNSs, P and Lewis. One, discovered in 1962, Xg, has been shown to be sex linked, the gene responsible for the antigen, Xg^a, being carried on the X chromosome. All other known red-cell antigens are controlled by autosomal genes (Chap. 53).

In addition to being present on the surface of the red blood cells, some blood group substances may be detected in other tissues, for example in certain epidermal and epithelial cells. Furthermore, A, B, H and Lewis blood-group substances can occur in soluble form in certain secretions such as saliva, gastric mucin, ovarian cyst fluid. Chemical studies of these soluble blood-group substances have shown them to be glycoproteins but it is thought that the blood-group antigens on the red cell surface are in the form of glycolipids. In both types of molecule the antigenic specificity is associated with differences in the carbohydrate moiety, for example Group A with N-acetylgalactosamine and Group B with D-galactose. In general a blood-group antigen (for example D) possessed by an individual is detected only when a suspension of his red blood cells is mixed with a serum containing the equivalent antibody (in this case, anti-D). Agglutination of the red cells then occurs and the individual concerned is classified as 'D-positive'. The red cells of another person lacking the D antigen are not agglutinated by this antibody and in this case the individual is called 'D-negative'.

The agglutination which occurs when red cells containing a blood-group antigen are mixed with a serum containing the equivalent antibody is followed in some cases by lysis of the cells (haemolysis), if complement is present. Intravascular lysis of incompatible red cells may also occur *in vivo*. More commonly, however, the interaction between blood group antibody and antigen, after the transfusion of incompatible red cells, results in these cells being removed intact from the circulation by the reticulo-endothelial system before being destroyed. In man, the chief sites of destruction are the liver and the spleen.

Blood-group antibodies. These are found in the globulin fraction of the serum, and may be divided into two broad classes: immune and naturally occurring.

I. *Immune antibodies*. If red cells containing a blood-group antigen (for example the Rh antigen, D) are introduced into the circulation of a person who lacks that antigen (in this case a D-negative person) that person may be *immunized* and the equivalent antibody (anti-D) will appear in his serum.

This type of immunization can be caused by blood transfusion, and particularly by repeated transfusions, but even the intramuscular injection of a small quantity of blood may stimulate the formation of antibodies. Moreover, red cells from a fetus carrying an antigen, inherited from its father but absent in the mother, may enter the maternal circulation and immunize the mother through the placenta.

Immune blood group antibodies are usually γG globulins (synonyms IgG, 7Sγ). They have a molecular weight of about 150 000 and can cross the placenta easily. *In vitro* they are more active at 37°C than at lower temperatures and may fail to agglutinate red cells suspended in saline, but may agglutinate them when they are suspended in colloid media (for example albumin or plasma).

II. *Naturally-occurring antibodies.* A blood-group antibody may be present in the serum of an individual who, as far as can be determined, has never been immunized by the equivalent antigen. It was the regular occurrence of anti-A and anti-B which led to the discovery by Landsteiner in 1900 of the A and B substances —the first antigens to be recognized in human red cells.

Naturally-occurring antibodies are usually γM globulins (synonyms IgM or 19Sγ). They have a molecular weight of about 900 000 and do not readily cross the placenta. They agglutinate red cells suspended in saline and are more active at lower temperatures (for example 4°C) than at 37°C. Some may be completely inactive at 37°C and therefore do not in the great majority of cases destroy red cells *in vivo*. However, the distinction between immune and naturally-occurring antibodies is not always clear cut.

ABO system. The ABO group of a person depends on whether his red cells contain one, both, or neither of the two blood group antigens A and B. There are therefore four main ABO groups, AB, A, B and O.

If the A antigen is *not* present in the person's cells (that is, he is group B or group O) his serum contains naturally-occurring anti-A (α). Similarly, if the cells lack the B antigen (that is, he is group A or group O) the serum contains anti-B (β). The serum of a group AB person contains neither of these antibodies.

Subgroups of A. As far back as 1911 it was discovered that group A individuals could be divided into two categories, A_1 and A_2. Similarly group AB can be subdivided into A_1B and A_2B. There are two points of practical importance with regard to these groups. Firstly, cells of subgroup A_2, and more par-

Group	Antigens present in red cells	Antibodies present in serum
AB	A and B	—
A	A	anti-B (β)
B	B	anti-A (α)
O	—	anti-A + anti-B ($\alpha + \beta$)

ticularly, of subgroup A_2B, react weakly with anti-A serum and great care must be taken to avoid grouping these wrongly as O or B respectively. Secondly, the serum of the subgroups A_2 and A_2B may contain a naturally-occurring antibody, anti-A_1 (α_1) which reacts with A_1 and A_1B cells.

In recent years even weaker subgroups of A have been recognized, but they are of rare occurrence. Similarly, very occasional weak subgroups of group B have been discovered.

Group O. The red cells of a group O person lack both the A and B antigens, but they are not devoid of all blood-group substance. (They, of course, contain blood-group antigens belonging to the other systems—Rh, MNSs, etc.) Up to the present, no unequivocal evidence has been discovered that a true 'O antigen' or specific anti-O antibody occurs. With rare exceptions all human red cells contain a blood-group substance, called H substance and the corresponding antibody, anti-H, has been found in the sera of certain human beings and animals.

When tested with anti-H, red cells of the various ABO groups generally arrange themselves in the following order according to the strength of their reactions:

$$0 \geqslant A_2 \geqslant A_2B \geqslant B \geqslant A_1 \geqslant A_1B$$

Group O cells therefore contain most H substance. There is good evidence that H may be the basic substance from which the A and B substances are made under the influence of the A and B genes.

Rhesus system. In 1940 Landsteiner and Wiener discovered that an antibody, produced in rabbits by injecting them with the blood of the rhesus monkey, agglutinated not only the red cells of the monkey but also the cells of 85 per cent of a large series of human blood samples. The antibody was called anti-rhesus (anti-Rh) and the cells agglutinated by it were called rhesus-positive (Rh-positive).

Soon after this discovery, immune anti-Rh was demonstrated in the serum of certain Rh-negative human beings. These were Rh-negative persons who had been transfused with Rh-positive blood, or Rh-negative women who had borne Rh-positive children.

Fisher postulated in 1943 that the inheritance of Rh antigens was controlled by three pairs of closely linked allelic genes—*Cc, Dd, Ee*, each gene determining an antigen denoted by the same letter. The antigen D corresponded to the original Rh antigen and an Rh-positive individual was one whose red cells contained D. A person inherited three Rh genes (one of each pair) from each of his parents, for example *CDe* from one parent and *cde* from the other, giving a genotype which could be expressed *CDe/cde*. This theory not only satisfied all the facts observed up to that time but predicted other reactions which were duly observed during the next few years, for example an antibody with the specificity anti-e, and red cells expressing the gene combination *CdE*, both of which were unknown in 1943.

Fisher's original theory no longer satisfies all the accumulated observations of many workers over the years· since 1943. For example, it postulates the existence of an antigen 'd' controlled by the gene '*d*' allelic to '*D*', but no antibody with the specificity anti-d has ever been discovered and so there is no direct evidence for the existence of the '*d*' gene. While Fisher's theory of separable Rh genes is now accepted with reserve, his notation still remains the most convenient and the least confusing in use.

Recently it has been shown that anti-rhesus antibody produced in animals such as rabbits or guinea-pigs by injection of rhesus monkey red cells is not of the same specificity as the anti-Rh (anti-D) found in Rh-negative humans who have been immunized by human Rh-positive red cells. Anti-rhesus antibody can in fact be produced by injecting animals not only with human Rh-positive cells but also with Rh-negative cells. This antibody, while it does not agglutinate human Rh-negative cells to the same degree as it does Rh-positive cells, is adsorbed by Rh-negative cells and can be eluted from them. It now appears that the antigen in human red cells which this antibody detects is quite distinct from the human Rh antigen D and that the genes responsible for these two antigens segregate independently. It has been suggested that the animal anti-rhesus antibody should be named anti-LW and the equivalent antigen LW—the symbol LW referring to the original findings of Landsteiner and Wiener.

Irrespective of their Rh group nearly all human red cells contain the LW antigen but a few individuals, both Rh-positive and Rh-negative, have been found whose cells lack this antigen and anti-LW is present in the serum of some of these persons.

Haemolytic disease of the newborn. Blood-group antibodies of the immune type present in the blood of a pregnant woman can cross the placenta and gain access to the fetal circulation. Then, if the fetal red cells contain the equivalent antigen, haemolytic disease may ensue. It should be noted however that, whereas most children possess blood-group antigens, inherited from their father, which the mother lacks, it is comparatively uncommon for a woman to develop immune antibodies, even after repeated 'incompatible' pregnancies. Moreover, when immunization of the mother does occur and the fetus is affected, the severity of the condition varies very greatly. Destruction of the fetal red cells may be so rapid and widespread that death *in utero* occurs. Conversely no evidence of destruction may be seen and the child may be born, and remain, normal. Most cases are of a severity intermediate between these extremes.

Many blood-group antigens and their equivalent antibodies may cause haemolytic disease of the newborn but some are much more frequently implicated than others. The most important is the Rh antigen, D. Until recently the incidence of this disease in the United Kingdom, due to this antigen, was approximately 1 in 150 of all pregnancies. Cases of haemolytic disease of the new-born due to ABO incompatibility (in particular where the mother is group O and the child group A) are probably as common as those due to Rh. Only rarely, however, are such cases severe.

Prevention of Rh immunization by passively administered Rh antibody. Rh(D) positive fetal red cells, in numbers sufficient to cause immunization of an Rh(D) negative mother, may cross the placenta at any time during pregnancy but more commonly when the placenta is separating from the uterus during delivery or abortion. In the absence of any treatment between 10 and 20 per cent of Rh(D) negative women delivered of Rh(D) positive babies were actively immunized and anti-D could be detected in their sera.

It has been shown that D-positive fetal red cells can be cleared rapidly from the circulation of a D-negative mother by injecting her with a concentrated immunoglobulin (IgG) fraction prepared from plasma containing powerful anti-D antibody. In many countries it is now routine practice to give all D-negative women an injection of such an anti-D immunoglobulin preparation as soon as possible after delivery of a D-positive infant. This has resulted in a dramatic reduction in the number of women who, in a second or subsequent pregnancy, show evidence of active Rh(D) immunization and, therefore, in the number of babies being afflicted with haemolytic disease of the newborn. In a recent survey of over 10 000 women given such injections less than 1 per cent developed Rh antibody.

Blood transfusion. Before a transfusion is given precautions must be taken to ensure that the donor's and patient's bloods are compatible. The greatest danger is that the patient's plasma may contain an antibody, or antibodies, active against the equivalent antigen or antigens present in the donor's red cells. The effect on the patient of such an incompatible transfusion varies, but a severe and even fatal reaction may be caused.

ABO incompatibility is particularly dangerous since the anti-A and/or anti-B normally present in the plasma of all but group AB people are usually capable of causing rapid destruction of transfused red cells containing the equivalent A and/or B antigens. Next in importance is the Rh antigen, D. Although D-negative persons rarely possess naturally-occurring anti-D they may have been immunized by a transfusion of D-positive blood or, in the case of a woman, during pregnancy. Any subsequent transfusions of D-positive blood would then be liable to cause a severe and perhaps fatal reaction. Serious reactions due to blood-group antigens other than A, B or D are uncommon, but may occur. Antibodies present in the *donor's* plasma, active against antigens in the recipient's red cells do not usually cause reactions, mainly because they are rapidly diluted in the patient's circulation. Serious reactions due to this cause have occurred, however, particularly when group O blood containing powerful anti-A has been given to a group A recipient.

For these reasons it is advisable for transfusion to select blood of the same ABO and D groups as those of the patient and in addition to *cross-match* a sample of the patient's serum against a suspension of the donor's red cells. The cross-matching tests should reveal the presence of any antibody in the patient active against a blood-group antigen in the donor's cells.

A statistical association has occasionally been found between the ABO blood groups and certain diseases. For example, a person of group O is more likely to develop peptic ulcer than an individual of any other group. There is some association, too, between group A and cancer of the stomach.

REFERENCES

ALBERT, S. N. (1963). *Blood Volume*. Springfield: Thomas.

ANTONINI, E. (1965). Interrelationship between structure and function in hemoglobin and myoglobin. *Physiological Reviews* **45**, 123–170.

BIGGS, ROSEMARY (Ed.) (1971). *Human Blood Coagulation, Haemostasis and Thrombosis*. Oxford: Blackwell.

DIMOPOULLOS, G. T. (Ed.) (1961). Plasma proteins in health and disease. *Annals of the New York Academy of Sciences* **94**, 1–336.

FEARNLEY, G. R. (1965). *Fibrinolysis*. London: Arnold.

GREGERSEN, M. I. & RAWSON, R. A. (1959). Blood volume. *Physiological Reviews* **39**, 307–342.

JOHNSON, SHIRLEY A. (Ed.) (1971). *The Circulating Platelet*. New York: Academic Press.

KEKWICK, R. A. (1956). The plasma proteins. *Lectures on the Scientific Basis of Medicine* **4**, 112–122.

KOCHWA, S. & KUNKEL, H. G. (Eds.) (1971). Immunoglobulins. *Annals of the New York Academy of Sciences* **190**, 1–584.

KONTTINEN, Y. P. (1968). *Fibrinolysis, Chemistry, Physiology, Pathology and Clinics*. Tampere, Finland: Oy Star Ab.

MACFARLANE, R. G. (1966). The clotting of blood. *Science Journal* **2**, 58–63.

MACFARLANE, R. G. (Ed.) (1970). The haemostatic mechanism in man and other animals. *Symposium of the Zoological Society of London* No. 27. London: Academic Press.

MACFARLANE, R. G. & ROBB-SMITH, A. H. (1961). *Functions of the Blood*. Oxford: Blackwell.

MAYERSON, H. S. (1965). Blood volume and its regulation. *Annual Review of Physiology* **27**, 307–322.

MOLLISON, P. L. (1967). *Blood Transfusion in Clinical Medicine*, 4th edn. Oxford: Blackwell.

MOURANT, A. E. (1954). *The Distribution of Human Blood Groups*. Oxford: Blackwell.

O'BRIEN, J. R. (1966). Platelet stickiness. *Annual Review of Medicine* **17**, 275–290.

POLLER, L. (Ed.) (1969). *Recent Advances in Blood Coagulation*. London: Churchill.

PUTNAM, F. W. (1960). *The Plasma Proteins* (2 Vols.). New York: Academic Press.

RACE, R. R. & SANGER, R. (1968). *Blood Groups in Man*, 5th edn. Oxford: Blackwell.

RIFKIND, B. M. (1970). The familial hyperlipoproteinaemias. *Scottish Medical Journal* **15**, 223–230.

SCHULTZE, H. & HEREMANS, J. F. (1966). *The Molecular Biology of Human Proteins with Special Reference to Plasma Proteins*. Amsterdam: Elsevier.

SHERRY, S., FLETCHER, A. P. & ALLKJAERSIG, N. (1959). Fibrinolysis and fibrinolytic activity in man. *Physiological Reviews* **39**, 343–382.

SMELLIE, R. M. S. (Ed.) (1972). Plasma lipoproteins. *Biochemical Society Symposia* No. 33. London: Academic Press.

STAFFORD, J. L. (Ed.) (1964). Fibrinolysis. *British Medical Bulletin* **20**, 171–250.

Symposium on blood clotting (1969). *Proceedings of the Royal Society* B **173**, 257–445.

Symposium on blood volume (1963). *Annals of the Royal College of Surgeons of England* **33**, 137–197.

Symposium on plasma lipoproteins (1969). *Proceedings of the National Academy of Sciences of the U.S.A.* **64**, 1107.

THOMSON, W. A. R. (Ed.) (1965). Symposium in blood transfusion. *Practitioner* **195**, 147–206.

TRIA, E. & SCANU, A. M. (1969). *Structural and Functional Aspects of Lipoproteins in Living Systems*. New York: Academic Press.

TURNER, M. W. & HULME, B. (1971). *The Plasma Proteins: An Introduction*. London: Pitman Medical.

WHIPPLE, G. H. (1956). *Haemoglobin, Plasma Proteins, Organ and Tissue Proteins*. Oxford: Blackwell.

WHIPPLE, H. E. (Ed.) (1964a). Bleeding in the surgical patient. *Annals of the New York Academy of Sciences* **115**, 1–542.

WHIPPLE, H. E. (Ed.) (1964b). *In vivo* and *in vitro* behaviour of clotting factors in blood and tissues. *Annals of the New York Academy of Sciences* **105**, 983–1004.

24 The formed elements of the blood

Normal human erythrocytes or red blood corpuscles are circular biconcave discs which have lost their nuclei during the process of maturation. Although their nuclei have disappeared these cells have been endowed with highly complex enzyme systems. Their main constituent, however, is haemoglobin, and from it they derive their principal function, namely the transport of oxygen from the lungs, where it enters the body, to the other tissues where it is used. The haemoglobin in the erythrocyte is in a higher concentration (approximately 34 g per 100 ml of red cell mass) than could be achieved in simple solution. The enclosure of haemoglobin in the red corpuscles has several important consequences. So much haemoglobin free in the plasma would cause such an increase in viscosity that the circulation would be impaired; the rate of uptake of oxygen would be reduced; interaction of the free protein with other plasma proteins would occur, particularly the specific combination with haptoglobin after which the haemoglobin would be rapidly removed; and free haemoglobin would pass out in the renal glomerular filtrate.

When a specimen of blood, in which coagulation has been prevented by some method which does not alter the volume of the red cells, is centrifuged (at 3000 rev/min for 30 min) it separates into two main layers. The clear, normally straw-coloured upper layer of plasma accounts for some 55 per cent of the volume and the lower layer of packed erythrocytes with a small amount of trapped plasma between them for some 45 per cent. Between these two layers is a narrow band (buffy coat) comprising about 1 per cent of the volume and consisting of an upper creamy-white layer of platelets or thrombocytes (p. 434) and a lower grey-white layer of white cells or leucocytes (p. 429). The percentage of blood volume occupied by the packed red corpuscles, the packed cell volume (P.C.V.) or haematocrit, is used in clinical medicine as an index of anaemia, since the more anaemic a patient is the lower is the haematocrit; or, in a subject who is not anaemic, the P.C.V. is an index of the relative amount of fluid in the plasma (dehydration, Chap. 33).

There are normally about 5 million (5×10^6) red cells per mm^3 or 5×10^{12} per litre of circulating blood. Since the blood volume is approximately 5 litres, there are about 25×10^{12} red cells (or 2·5 kg) in the circulation. In the healthy adult the red cell count varies from 4·2 to $6·4 \times 10^6$ per mm^3, the average count in the male being $5·5 \times 10^6$ and in the female $4·8 \times 10^6$. In healthy subjects there are circadian variations equivalent to 4 per cent of the mean haemoglobin for the day, although these are said to be absent if the subject is in bed.

In health the erythrocytes are biconcave discs (Fig. 24.1) with an average diameter of 7·2 μm (measured on a fixed and stained film of blood) and some 1·9 μm thick. In disease these proportions may be altered so that in pernicious anaemia for example the mean corpuscular diameter is greater than normal (macrocytes) and there is a larger than normal variation between the cells (Fig. 24.2).

From the haematocrit and the red cell count it is possible to calculate the mean corpuscular volume (M.C.V.) which is normally from 78 to 94 μm^3.

Erythrocyte sedimentation rate (E.S.R.). When blood to which an anticoagulant has been added is allowed to stand in a narrow tube the red blood corpuscles form aggregates or rouleaux in which the cells are arranged face to face like a pile of coins and the aggregates gradually settle down (sediment) to leave a clear zone of plasma above. The sedimentation rate is measured as the length of the column of clear plasma above the sedimenting erythrocytes at the end of one hour. The E.S.R.

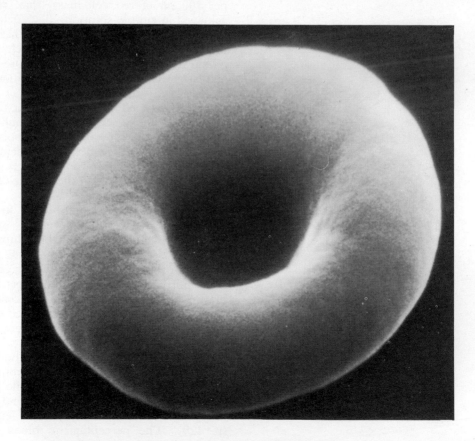

Fig. 24.1 Electron scanning micrograph of a normal red blood corpuscle. × 14 000. (*By courtesy of P. G. Toner.*)

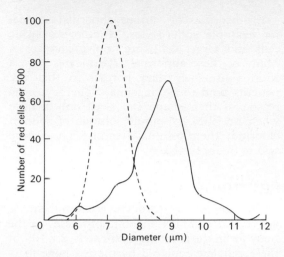

Fig. 24.2 Frequency distribution curves of diameters of healthy red cells and of cells from a case of pernicious anaemia.

depends almost entirely on the relative concentrations in the plasma of fibrinogen, and the α_2- and γ-globulins. The asymmetry of these macromolecules promotes the aggregation of red cells into rouleaux and clumps. Almost all infections, whether acute or chronic, are accompanied by a rise in α_2- and γ-globulins and most wasting and destructive diseases by an increase in fibrinogen. The E.S.R. is thus a simple, rapid procedure for revealing such changes. It is useful in detecting the presence and activity of disease but it is quite unspecific. The sedimentation rate also varies according to the red cell count, the lower the red cell count the greater the sedimentation rate.

The structure and metabolism of erythrocytes. Since the red cell can be bisected without its contents escaping it seems likely that it consists of a close network or of a jelly which is none the less flexible enough to allow the cells

to assume the roughly paraboloid form in which they pass through capillaries as a succession of tiny plugs separated by plasma (Fig. 27.26). The usual biconcave disc shape is regained immediately when the cells pass into a larger vessel. The elongated paraboloid form of the erythrocyte when passing through a capillary presents a much larger surface area next to the capillary wall for the transference of oxygen.

The stroma or framework of the erythrocyte consists of some 50 per cent protein and 10 per cent lipid, mainly phospholipid arranged in three layers. The outer layer of sialic acid determines the cell's electrical charge. It is traversed by pores about 0·4 nm wide which allow the passage of ions and molecules not exceeding a molecular weight of about 100.

Although erythrocytes have no nuclei and normally contain no glycogen they are metabolically quite active utilizing some 1·5 to 2·2 mmoles of glucose per litre per hour equivalent to the liberation of 100 J (25 cal). This energy is required to pump out sodium and pump in potassium against electrochemical gradients and to reduce methaemoglobin to haemoglobin.

The metabolism of glucose in the red blood cell, which is essential for its survival, takes place along the two common pathways, glycolysis (p. 139) and the pentose phosphate pathway (p. 153), 80 per cent of the glucose being broken down by the former and 20 per cent by the latter. The red cell has no mitochondria and therefore no citric acid cycle.

In the pentose phosphate pathway, the first enzyme involved is the *glucose-6-phosphate dehydrogenase (G-6-PD)* (p. 152) in the operation of which large amounts of reduced NADP are formed. This reduced NADP is utilized by the red blood cell to maintain glutathione (p. 28) in the reduced state

$$\text{glucose 6-phosphate} + \text{NADP}^+ + \text{H}_2\text{O} \xrightarrow{G\text{-}6\text{-}PD} \text{6-phosphogluconic acid} + \text{NADPH} + \text{H}^+$$

$$\text{GSSG} + \text{NADPH} + \text{H}^+ \xrightarrow[\text{reductase}]{\text{glutathione}} 2\ \text{GSH} + \text{NADP}^+$$

Formed Elements of the Blood 417

The function of the reduced glutathione within the erythrocyte is not clearly understood, but it probably prevents the oxidation of haemoglobin to methaemoglobin and its sulphydryl groups play a part in maintaining the structure of the red cell and also of the globin part of the haemoglobin molecule.

In the population of many parts of the world, there is a deficiency of the (G-6-PD) in the red blood cells. This is an inherited condition which affects more than 100 000 000 people in the world at the present time. This deficiency is a common cause of jaundice in babies, it also is responsible for haemolysis after the administration of certain drugs, particularly the antimalarials such as primaquine, and certain antibacterial drugs. Haemolysis also occurs in G-6-PD-deficient persons after the consumption of the bean of *Vicia faba* which is a common foodstuff in many countries. Many enzyme deficiencies have been discovered but only a few have serious effects.

Hydrogen peroxide is formed in small amounts in normal red cells during metabolic processes and also is generated within the cells as a result of the action of certain drugs, including those which cause haemolysis in G-6-PD deficiency. If not destroyed as rapidly as it is produced, hydrogen peroxide would oxidize haemoglobin to methaemoglobin and then, irreversibly, to oxidized degradation products; it is probable that other substances essential for the integrity of the red cell would also be oxidized, and haemolysis would follow.

There are two protective mechanisms against hydrogen peroxide within the red cell. Glutathione peroxidase uses hydrogen peroxide to oxidize reduced glutathione. Normal activity of glutathione peroxidase depends on the presence of adequate amounts of reduced glutathione which requires normal functioning of the pentose phosphate pathway, including G-6-PD. The other protective agent is the enzyme catalase which decomposes hydrogen peroxide to water and oxygen. Chemically, catalase resembles haemoglobin in being an iron-containing haemoprotein. The relative importance of glutathione peroxidase and of catalase is not fully understood.

At least one hundred enzyme systems have been identified in the red cell. But because it has no nucleus it cannot synthesize proteins (including enzymes).

Erythrocytes of various abnormal shapes, for example spherocytes, elliptocytes, sickle cells and target cells, are known in disease. Many of these abnormal erythrocytes have a shorter life-span than normal so that the patients tend to be anaemic. Spherocytes by reason of their shape can swell only a little when put into hypotonic solution of sodium chloride; their osmotic fragility is therefore said to be increased.

ERYTHROPOIESIS

The supply of erythrocytes is one of the most active and complex anabolic processes in the body for in the adult approximately 2×10^{11} of these cells are replaced by the red bone marrow every 24 hours. Specimens of red bone marrow can be made by puncturing the sternum or the ilium with a stout needle. For many years studies on erythropoiesis were mainly morphological, but more recently much information has been gained from isotopic labelling techniques, especially labelling with radioactive iron. In the classical methods of studying haemopoiesis the cells are stained with various mixtures of methylene blue (a basic dye) and eosin (an acid dye). Hence nuclei, because of their DNA content, stain blue (basophilic) and early red cell precursors have blue (basophilic) cytoplasm from RNA. The concentration of haemoglobin is shown by the degree of eosinophilic (acidophilic) staining. The term polychromatic is applied when the cytoplasm stains with both basic and acid dyes owing to the presence of both RNA and haemoglobin.

The erythroid elements are constantly replenished from a pool of undifferentiated stem-cells the nature of which is unknown, under the action of a number of specific inductors, the principal of which is the hormone *erythropoietin*, along an irreversible course of division and maturation. As the cells undergo maturation and successive division, they receive a variety of names, pro-erythroblast (E_1), large basophilic (early) normoblast (E_2), small basophilic normoblast (E_3), polychromatophilic (intermediate) normoblast (E_4), late non-dividing normoblast (E_5), reticulocyte (E_6) and erythrocyte (E_7) (Fig. 24.3). Cells in the early stages of their development (marrow proliferating pool) contain a relatively large

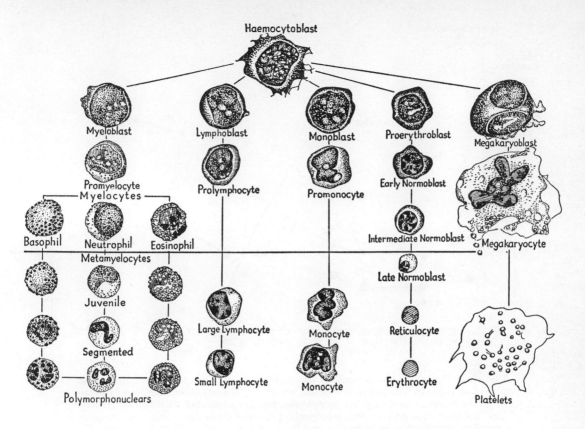

Fig. 24.3 All cells below the horizontal line may be found normally in the blood stream with the exception of the late normoblast. All cells above the horizontal line and the late normoblast, may be found in normal bone marrow. Stages E_2 and E_3 (p. 418) are represented here by one early normoblast. This is the key to the colour picture of Plate 24.1. This figure and the colour picture are from C. J. C. Britton (1963). *Disorders of the Blood*, 9th edn. London: Churchill.

amount of RNA (more than 5 per cent) and are basophilic; at the E_4 stage they have reached a degree of differentiation at which further divisions cease as the RNA content falls to 0·5 per cent. Subsequently (marrow non-proliferating pool) the cells lose their nuclear material so that the mature erythrocyte contains neither RNA nor DNA.

The term reticulocyte is used to denote an erythrocyte which shows a blue staining network when exposed to certain basic dyes, for example 'brilliant cresyl blue' or 'new methylene blue'. The reticulum, an aggregation of traces of RNA, appears in either nucleated or non-nucleated red cells and indicates immaturity. When for any reason the bone marrow produces red cells at a greater rate than normal, immature red cells tend to appear in the blood and the percentage of reticulocytes may rise (Fig. 24.4).

It seems likely that cells do not proceed along their maturation pathway (Fig. 24.3 and Plate 24.1) as a cohort but that the number of divisions as well as the maturation times may vary. Furthermore, a small number of cells normally die during the process (ineffective erythropoiesis), but the death rate is much greater in a number of diseases, for example pernicious anaemia. Normally some 60 E_5 cells are produced per hour per 1000 precursors but in pernicious anaemia only 5 to 10. It is possible that mitosis produces two identical cells so that differentiation occurs between mitoses; the interval between mitoses at the pro-erythroblast stage (E_1) is 12 to 15 hours and at the E_4 stage approximately 12 hours.

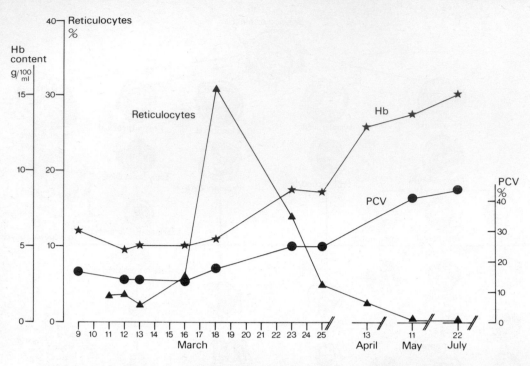

Fig. 24.4 This chart shows response to 1000 μg of vitamin B_{12} in a patient with Addisonian pernicious anaemia. The patient is unusual, being only 25 years old whereas most patients with such anaemias are elderly, but the following findings established the diagnosis—macrocytic normochromic anaemia with leucopenia, hypersegmented neutrophils and thrombocytopenia; megaloblasts in buffy coat and bone marrow smears; low serum vitamin B_{12} (60 pg per ml); normal serum folate (6·4 ng per ml); pentagastrin fast achlorhydria; poor absorption of oral radioactive vitamin B_{12} (Schilling test) corrected by 'intrinsic' factor given by mouth; antibodies to gastric cells and 'intrinsic factor'; normal faecal fat excretion. Treatment began on March 12. (*By courtesy of H. B. Goodall and D. G. Adamson.*)

The average number of divisions occurring during maturation is estimated to be 3 or 4 and the time for the whole process 120 hours.

THE LIFE-SPAN OF THE ERYTHROCYTE

The life-span of red cells in the circulating blood would appear to be about 120 days. Early transfusion experiments using compatible, but antigenically different erythrocytes have been confirmed by radioactive techniques (Fig. 24.5). Thus when the haem (iron-containing prosthetic group, p. 449) of the haemoglobin molecule is labelled by the incorporation of glycine containing ^{15}N the life-span can be shown to be 127 days (p. 377). The most frequently used isotope in clinical work is [51]Cr. A sample of blood is incubated with radioactive sodium chromate, $Na_2{}^{51}CrO_4$. The red cells take up ^{51}Cr rapidly and, after

they have been washed with saline to remove the radioactive plasma, the labelled red cells are returned to the subject by intravenous injection. The rate at which radioactivity leaves the blood is a measure of the survival time of the erythrocytes in circulation. By this method the apparent half life is 28 to 38 days in normal persons; this is less than the true value (about 60 days) because the decline in radioactivity is due partly to loss of old cells and partly to loss of ^{51}Cr from the cells. Nevertheless the test is very useful in the diagnosis of haemolytic disease and also, from the radioactivity of the stools, in the detection of bleeding into the alimentary tract.

Since the mean life span of erythrocytes is normally 120 days, one hundred-and-twentieth of the mass of circulating red cells (about 25 g) must be removed each day from the circulation by the reticulo-endothelial

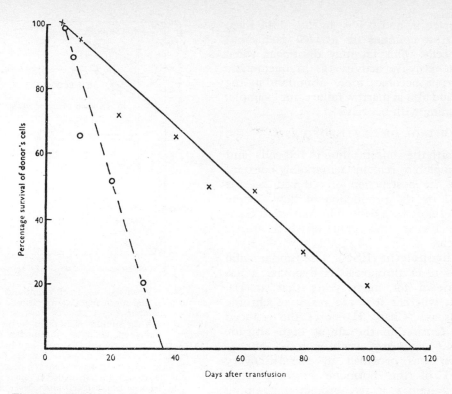

Fig. 24.5 The survival of transfused erythrocytes (group O) in two recipients (group A). The continuous line represents normal survival; the patient was a female aged 50. The interrupted line was obtained from a male aged 62 with a haemolytic type of anaemia. Forty days after transfusion all the donor's cells had disappeared from the circulation of the patient, but there were still 65 per cent of the donor's cells in the blood of the control in which some survived for about 120 days.

system. The ^{51}Cr of labelled red cells eventually leaves the blood and is concentrated in the liver and spleen. Since ^{51}Cr salts in the plasma are not taken up preferentially by the spleen it is possible, but not proved, that this organ is an important site of red cell destruction. Some cells, too, are destroyed by lytic agents to which they are exposed during the circulation through capillaries.

The daily destruction of so many erythrocytes produces about 250 mg of bilirubin which is released into the plasma to give a fairly constant concentration of 0·5 mg per 100 ml and an output in the faeces and urine of 250 mg per day of urobilinogen (p. 454).

When the life-span of the erythrocytes is reduced a state of *haemolytic disease* exists (Fig. 24.5). This is not necessarily followed by anaemia since the haemopoietic tissue in the bone marrow (the erythron) may produce new cells at a rate sufficient to compensate even if

the survival time of circulating erythrocytes is reduced to 20 or even 15 days.

The destruction of erythrocytes is to some extent random, but is mainly progressive with senescence. Normally the ageing of these cells is not accompanied by any morphological change but, in subjects from whom the spleen has been removed surgically or in whom it is congenitally absent or destroyed by disease, the red cells show a variety of shapes, for example target forms, burr forms and irregular contracted and fragmented forms. Ageing is accompanied by a progressive loss of lipid and diminution in glycolytic enzyme activity. By differential centrifugation it can be shown that red cells increase in density with age.

The sites of erythropoiesis gradually change with development: in blood vessels in the yolk sac membrane in the embryo; in the liver and spleen in the fetus; throughout the skeleton in the infant; and in membrane bones (thoracic cage, vertebrae and pelvis) in the adult. Even red marrow contains fat as well as haemopoietic cells. This fat may disappear when there is excessive activity, as in haemolytic anaemia; it becomes more abundant in the elderly and also in marrow failure, for example, in poisoning with benzene.

THE CONTROL OF ERYTHROPOIESIS

In health the concentration of red cells (and of haemoglobin) remains remarkably constant because the destruction of old red cells is balanced by the production of new. There must therefore be a delicately controlled feedback mechanism, the details of which are far from clear.

Erythropoietin (ESF). The fundamental stimulus to erythropoiesis is hypoxia; it has been known for many years that arterial hypoxia, whether it is the result of chronic lung disease (Chap. 31) or of the reduced oxygen tension in the air at high altitude (Fig. 24.6) increases erythropoiesis. An explanation was provided by the discovery in 1953 of the hormone erythropoietin. Erythropoietin is a glycoprotein (containing sialic acid) with a molecular weight of about 28 000 which moves electrophoretically as an α_2-globulin. It has not yet been crystallized but highly potent concentrates have been prepared. Several methods for the bio-assay of erythropoietin have been devised.

Erythropoietin is produced mainly in the kidney possibly by the juxtaglomerular cells (Chap. 32). It cannot normally be detected in blood but it is present within a few hours of the onset of hypoxia in experimental animals and it disappears again when the hypoxia is relieved. Erythropoietin is also present in the blood of animals made anaemic by repeated bleedings. It is presumably responsible for the reticulocytosis and blood regeneration in such animals because serum taken from them can induce a reticulocyte response in a normal animal.

High levels of erythropoietin have been demonstrated in the plasma or urine in most types of anaemia in man, the only notable, but not unexpected, exception being the anaemia of renal failure. Erythropoietin may be increased in patients with renal tumours. In the bone-marrow the chief action of erythro-

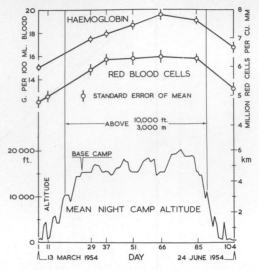

Fig. 24.6 Mean red blood cell count and blood haemoglobin level and mean night camp altitude of the ten expedition members (aged 24 to 38) during the attempt in 1954 to reach the summit of Makalu on the Nepal–Tibet border. They spent 62 days continuously above 4·6 km (15 000 ft). The increase in red cells and haemoglobin did not reach its maximum until 6 or 7 weeks had been spent at high altitude. On the other hand the descent to sea level brought the red cell and haemoglobin values in 17 days very nearly to the original sea-level values. Even if it is assumed that on descent erythropoiesis ceased entirely the rate of disappearance of the red cells is greater than can be accounted for on the basis of normal red cell destruction. Evidence for increased erythrolysis is also given by the reduced number of reticulocytes in the blood and increased serum bilirubin level (N. Pace, L. B. Meyer & B. E. Vaughan (1956). *Journal of Applied Physiology* **9**, 141–144). It is of interest to note that in the Andes and Himalayas people live and work in permanently occupied settlements at 4 to 4·6 km (13 000 to 15 000 ft) above sea level.

poietin is to increase the rate of formation of pro-erythroblasts from primitive stem-cells (Fig. 24.3 and Plate 24.1) after which proliferation and maturation proceed in the normal way.

It may well be that there are other humoral factors (erythrocytosis-stimulating factors, ESF) governing erythropoiesis. There is evidence that there is a lipid which increases the number of cellular divisions undergone by the erythrocyte precursors during their maturation. An unknown mechanism, not erythropoietin, operates in the fetus.

Cobalamin and folic acid (pp. 190 and 191). In man two other important factors are necessary for the normal development of red cells. In the absence of cobalamin (vitamin B_{12}) or of folic acid the pro-erythroblasts of the bone-marrow, instead of maturing into normoblasts give rise to immature cells called megaloblasts which later lose their nuclei and appear as large red cells, called macrocytes. These cells enter the peripheral circulation which therefore contains red cells on the average bigger than normal (Fig. 24.2). This type of anaemia is thus called *macrocytic* or better, *megaloblastic*. When it is due to deficiency of vitamin B_{12} and is associated with atrophy of the gastric mucosa it is called pernicious or Addisonian anaemia. Although most patients with vitamin B_{12} deficiency have this specific lesion, others become deficient because of other gastric or ileal lesions or from radical surgical removal of these parts of the alimentary canal.

As early as 6 hours after the injection of a liver extract or of vitamin B_{12} into a patient with pernicious anaemia cytological changes are observed in the bone-marrow, the numerous primitive megaloblasts being replaced by more mature cells. Three to four days later the bone-marrow contains only a few megaloblasts, the predominant erythroblastic cell being the normoblast. In the circulating blood the number of reticulocytes usually begins to rise on the fourth or fifth day and reaches a maximum at 7 days or a little later, thereafter falling and eventually reaching the normal level (less than 2 per cent) (Fig. 24.4).

Normally, the body gets it cobalamin from the diet, the average daily intake being of the order of 5 μg (p. 191). Doses of 0·5 μg of cyanocobalamin given by mouth labelled with ^{60}Co (or better with ^{56}Co or ^{58}Co which have a much shorter half-life) can be traced to the faeces by their radioactivity. In healthy people the radioactive cobalt is absorbed mainly in the distal ileum and can be detected in the blood in a few hours; only one-third of the test dose is found in the faeces. In patients with pernicious anaemia, however, 90 per cent or more of the test dose is recovered in the faeces. The most practical method of assessing cobalamin absorption is by testing the radioactivity of the urine of the patient after saturation with parenteral cobalamin and ingestion of a small dose of radioactive cobalamin (Schilling test). The absorption of cobalamin can be restored to normal by the simultaneous administration of normal human gastric juice or of watery extracts of gastric mucosa which contain *intrinsic factor*. The mechanism of this im-

proved absorption is unknown. Several compounds with intrinsic factor activity have been isolated from hog's stomach, namely a mucopolypeptide (mol. wt. 5000) and three mucoproteins. In man intrinsic factor is produced in the fundus and cardia of the stomach possibly by the cells that secrete pepsinogen; it is found in gastric but not in duodenal juice. The failure of secretion of intrinsic factor is associated with gastric atrophy and both may be related to the presence of auto-antibody formation; antibodies to intrinsic factor and gastric parietal cells can be found in the serum in the majority of patients with pernicious anaemia.

After absorption cobalamin accumulates in the body which may contain as much as 1 mg. The largest single store is the liver, depletion of which may take 18 months. The biological half-life of radioactive vitamin B_{12} in the liver is about 1 year. In pernicious anaemia the amounts of cobalamin are reduced proportionally in all organs and tissues.

In addition to pernicious anaemia deficiency of cobalamin causes demyelinization in the nervous system—in the brain (dementia), in the spinal cord (subacute combined degeneration of the cord) and in the peripheral nerves (peripheral neuropathy). The treatment of all three disorders requires the regular injection of cobalamin for the duration of life. Cobalamin is in fact required for the synthesis of nucleoprotein throughout the body and giant epithelial cells resembling megaloblasts may be found in many organs in pernicious anaemia, and indeed in other megaloblastic anaemias.

Deficiency of folic acid also produces megaloblastic anaemia. This can arise from lack of folic acid in the diet or of its impaired absorption in the upper part of the small intestine in patients with steatorrhoea (p. 300); the body's stores of folic and folinic acids are small. As mentioned earlier pregnant women require an increased intake of folic acid.

In cobalamin deficiency and folic acid deficiency the nuclei of the megaloblasts are large and finely stippled compared with those of normoblasts. Chromosomes derived from them are longer, more slender and less tightly coiled than normal. The time required for DNA synthesis is prolonged but, since other synthetic processes are not slowed down, more haemoglobin than usual accumulates in each cell and leads to macrocytosis. Cobalamin and folic acid appear to be required as coenzymes in the early stages of DNA synthesis but these relationships are highly complex and as yet incompletely understood. The complexity is reflected in a recent review of megaloblastic anaemia which quoted more than 3000 references.

Other erythropoietic factors. The value of copper in the regeneration of blood has been demonstrated in animals and also in children. Pyridoxine plays an essential role in the synthesis of haem, but it is only rarely that deficiency of this vitamin is the cause of anaemia in man. Anaemia is a feature of scurvy (p. 192) and is cured by giving ascorbic acid. Apart from its effects in promoting the absorption of iron it is doubtful if ascorbic acid is useful in the treatment of other types of anaemia. Anaemia is sometimes found in diseases due to deficiencies of hormones (Chap. 50), for example in myxoedema and in Addison's disease, but the causes are complex and the anaemia cannot necessarily be attributed directly to the hormonal deficiency.

Whipple produced anaemia in dogs by repeated bleeding and he then studied the effects of various substances on the rate of regeneration. He found that the intake of protein was an important factor but in adult man protein deficiency does not seem to lead to anaemia.

IRON METABOLISM

Where iron is not available in sufficient quantity, red cells can be formed but they lack the normal content of the iron-containing pigment haemoglobin, and anaemia characterized by a low mean corpuscular haemoglobin concentration (M.C.H.C.) results.

The total amount of iron in the body of a 70 kg man is of the order of 3 to 4 g. The bulk of this (about 70 per cent) is in the haemoglobin and myoglobin; the remainder consists of storage iron, transport iron and enzyme iron.

Iron released from the breakdown of red cells is not excreted but returns to the iron pool and is re-utilized. In the absence of external haemorrhage less than 1 mg of iron is lost from the body each day. This obligatory iron loss occurs from desquamated cells from the mucosae and skin and from traces in the bile and urine. In the adult male subject and

post-menopausal female this is the total amount of iron which needs to be replaced from the diet, but the physiological requirements for growth, menstruation and pregnancy make additional demands (Table 24.7). The

Table 24.7 Estimated physiological iron requirements

	mg/day
Minimal obligatory loss	0·5–1·0
Growth: 0–1 year	0·7–0·8
1–11 y-ars	0·3
Adolescence	0·5
Normal menstruation	0·5–1·6
Median loss 30 ml	
Upper limit of normal 80 ml	
Lactation	0·5

Pregnancy	Average total *iron requirement
Obligatory loss	180 mg
Fetus	250 mg
Placenta	70 mg
Increase in maternal red cells	400 mg
	900 mg

* This is accumulated mainly in the second half of pregnancy when the average daily iron requirement could be as high as 6 mg per day.

median loss of blood at menstruation is 30 ml but many women lose far more than this. The upper limit of normality is 80 ml; above this signs of iron deficiency are likely to become apparent. In a normal pregnancy about 300 mg of iron are needed for the fetus and placenta, and the very considerable increase in maternal red cell mass requires a further 400 to 500 mg. Most of this is required during the latter half of pregnancy and is achieved by an increase in the absorption of iron by the gut and mobilization of storage iron. Some of the iron from the increased red cell mass may go back into storage after delivery but some is lost in the bleeding occurring after parturition. Breast feeding accounts for an iron loss of about 0·5 mg per day. The physiological iron requirements may be, therefore, as much as 2 to 3 mg

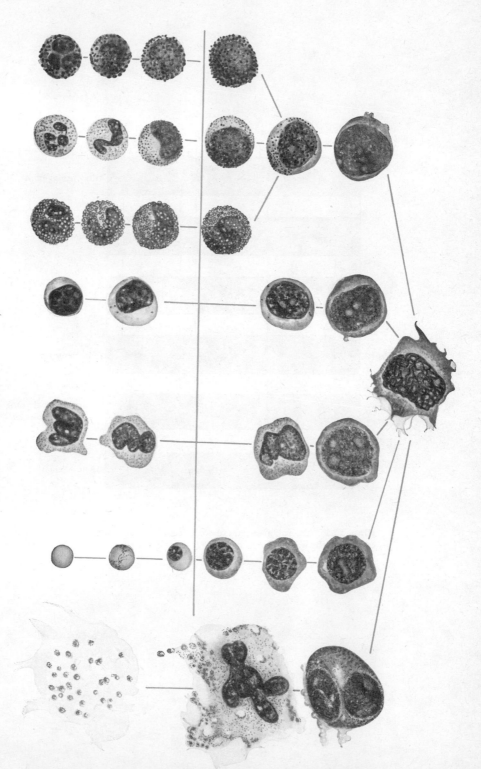

Plate 24.1   Origin of blood cells (Leishman's Stain).

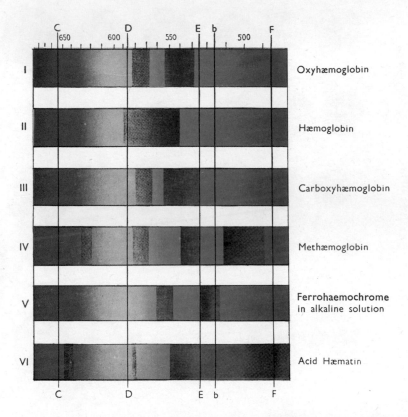

Plate 25.1 Absorption spectra of some haemoglobin derivatives. The scale at the top shows the wavelength in nm. The letters denote the position of the Fraunhofer lines of the solar spectrum.

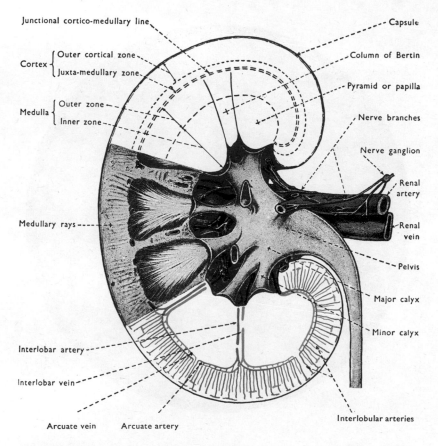

Junctional cortico-medullary line

Cortex { Outer cortical zone
          Juxta-medullary zone

Medulla { Outer zone
          Inner zone

Medullary rays

Interlobar artery

Interlobar vein

Arcuate vein   Arcuate artery

Capsule

Column of Bertin

Pyramid or papilla

Nerve branches

Nerve ganglion

Renal artery

Renal vein

Pelvis

Major calyx

Minor calyx

Interlobular arteries

Plate 32.1 Human kidney in L.S. Upper third to show areas and zones; middle third to show detailed appearance after removal of fat as seen with naked eye; and lower third to illustrate arrangement of blood vessels. A kidney measures 13 cm by 6 cm, the left being somewhat larger than the right. The two kidneys together weigh in the male about 300 g and in the female 250 g. (From H. S. D. Garven (1957). *A Student's Histology*. Edinburgh: Livingstone.)

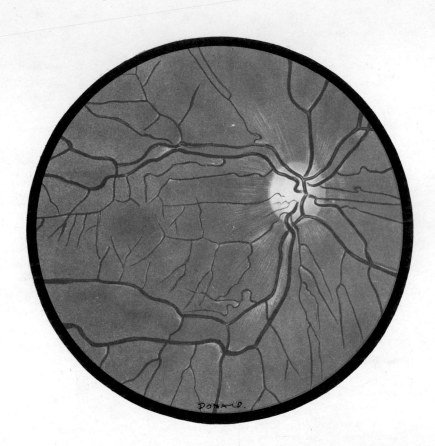

Plate 37.1 The fundus of the normal human right eye as seen through an ophthal-moscope. This view of the interior of the eye shows on the right the yellowish optic disc with the retinal vessels emerging from its centre. The macula is seen as a slightly darker area on the left of the figure. Note that minute vessels converge upon it but do not cross it. The thin dark vessels are veins and the narrower lighter vessels are arteries. The pink background is due to blood in the choroid seen through the transparent retina.

a day in a non-pregnant woman or up to 6 mg per day towards the end of a pregnancy (Table 24.7). Pathological losses from haemorrhage, notably from menorrhagia, ulcerative lesions in the gastrointestinal tract, or haemorrhage induced by drugs such as aspirin may make the requirement even higher.

In the United Kingdom the average daily iron intake is of the order of 14 mg and this might be thought to provide a generous margin, even for pathological iron losses. However, normally only about 10 per cent or less of the iron ingested is absorbed (Table 24.8). This amount is usually sufficient for adult men and post-menopausal females but other sections of the population may be in precarious balance.

Table 24.8 Iron absorption in normal and iron deficient subjects measured with radioactive iron

	Mean percentage absorption of iron ingested	
	Normal	Iron deficiency
Ferrous ascorbate		
(5 mg Fe)	9·2	40·1
White bread	2·2	7·3
Oat cakes and porridge	4·0	10·0
Eggs	2·2	5·6
Chicken muscle	6·9	17·0
Haemoglobin	10·0	22·0

The data for white bread were obtained by adding radioactive iron to the flour. The remaining foods were labelled biologically with radioactive iron. Iron absorption was measured with a Total Body Counter. (*Courtesy of Sheila Callender.*)

The iron in food is found either in the form of *ferric iron complexes* or as *haem iron*. The latter is absorbed as haem and the iron is split off within the intestinal mucosal cell. Non-haem iron needs to be brought to the absorbing surface of the duodenum or upper part of the jejunum as a soluble chelate of small molecular weight. Reducing agents and substances such as ascorbic acid which produce soluble chelates enhance the absorption of non-haem iron while phytates and phosphates reduce absorption. None of these factors affects the absorption of haem iron. The *gastric secretion* may also be important in relation to the absorption of non-haem iron; acid aids solubility and chelating substances in normal gastric juice

help to keep the iron in solution as the pH is raised. *Duodenal secretions* are of importance in the absorption of haem iron because they prevent the formation of large polymers which are not well absorbed. Conversely the *pancreatic secretion* by virtue of its bicarbonate content may reduce iron absorption by encouraging the formation of precipitates of non-haem iron and large polymers of haem iron.

The absorption of iron is adjusted to some extent in response to increased demand for iron. Thus absorption of a simple ferrous iron salt may be increased four-fold in iron deficiency. Absorption of iron from food is usually not more than doubled (Table 24.8). The change in absorption is probably determined by the amount of iron incorporated into ferritin of the mucosal cells as they are formed in the intestinal glands. In the normal situation some of the iron entering the intestinal mucosal cell is taken by an unidentified carrier through the transport pool to be absorbed into the plasma; the rest is trapped in the ferritin apparatus and is ultimately lost when the cell is sloughed into the lumen of the bowel (Fig. 24.9).

The amount of iron incorporated into the mucosal cells as they are formed is in turn determined by the rate of erythropoiesis and the state of the body iron stores. Increased erythropoiesis or diminished iron stores both increase iron absorption probably by diverting iron from the mucosal cell. Conversely reduced erythropoiesis or increased iron stores result in diminished iron absorption.

When the iron enters the plasma it is attached to the specific iron binding protein *transferrin*, a β_1-globulin, which transports iron to and from the various compartments, for example from gut to bone marrow and from destroyed red cells back to the bone marrow. Transferrin is normally about one third saturated with iron and the iron binding capacity is about 300 µg/100 ml. The total transport iron is of the order of 3 to 4 mg.

Iron which is not utilized for erythropoiesis is stored, mainly in the liver and the reticuloendothelial cells. The amount of storage iron depends upon the state of iron balance. It constitutes a reserve which may be called upon where iron requirements are increased, for example after blood loss. Storage iron consists of *ferritin* and *haemosiderin*. Ferritin is a water

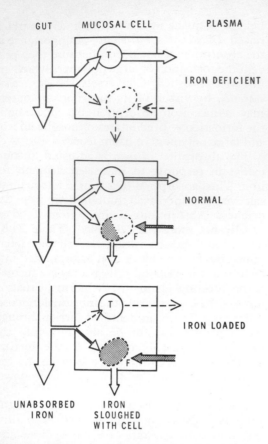

GUT   MUCOSAL CELL   PLASMA

IRON DEFICIENT

NORMAL

IRON LOADED

UNABSORBED   IRON
IRON   SLOUGHED
WITH CELL

Fig. 24.9 Diagrammatic representation of the control of iron absorption by the mucosal cell. Absorption is determined primarily by the amount of iron incorporated into the ferritin apparatus (F) of the mucosal cell. In iron deficiency mucosal cells are deficient in ferritin and virtually all the iron entering the mucosal cell enters the transport pool (T) and passes on into the plasma. Little iron is shed into the lumen of the gut. In normal mucosa some ferritin is present hence less iron enters the cells. Of this smaller amount some enters the transport pool (T), the remainder enters the ferritin pool (F) and is shed as the cells slough off. In iron loaded states the increased quantity of ferritin in the cell allows little iron to enter from the gut. Of that which does most goes to the ferritin and is lost with the death of the cell. Little, if any, enters the transport pool. (Modified from M. E. Conrad (1970). In *Iron Deficiency* Edited by L. Hallberg, H. G. Harwerth & A. Vannotti. London and New York: Academic Press.)

soluble crystalline protein. It consists of the protein apoferritin, which is synthesized in response to the presence of iron in cells, combined with micelles of a colloidal iron hydroxide-phosphate complex. The amount of iron is variable but when fully saturated the ferritin molecule contains 26 per cent of iron. Ferritin can be precipitated by ammonium sulphate and crystallized by cadmium sul-

phate. It can be visualized with the electron microscope and has a characteristic 'tetrad' appearance. Haemosiderin consists of iron-rich water insoluble granules which are visible with the light microscope. They stain blue with potassium ferrocyanide and hydrochloric acid (Perl's reaction).

Iron is also intimately concerned with the *respiratory enzymes*, for example the catalases and cytochromes. Reduction in the amount of these enzymes may be responsible for the degenerative tissue changes seen in association with iron deficiency.

Anaemia due to iron deficiency is particularly likely to occur in children and adolescents and in women during reproductive life. Premature infants are born with inadequate stores of iron and become anaemic if not given iron supplements. Such tiny infants may also develop folate deficiency. Surveys made in Western Europe show a prevalence of iron deficiency in women of between 15 and 25 per cent. In the tropics iron deficiency is very common, often from infestation with worms, particularly hookworms in the duodenum. In general iron in vegetables and cereals is less well absorbed than that from animal sources (Fig. 24.10), thus vegetarian diets are likely to contribute towards iron deficiency by making the balance more precarious. Before iron deficiency results in anaemia the tissue stores are depleted and the serum iron falls below 60 µg/100 ml. This is accompanied by an increase in the total iron binding capacity of the serum. With further iron depletion the haemoglobin falls and anaemia with low M.C.H.C. results. Treatment with iron corrects the abnormalities but in order to replenish the iron stores, treatment has to be continued for some time after the haemoglobin has reached normal levels.

Failure of a hypochromic anaemia to respond to iron may be due to the patient's failure to take iron, to blood loss, to the presence of an organic lesion, for example a malignant neoplasm, or to malabsorption of iron. Parenteral iron may be required, but before such therapy is undertaken it is essential to estimate the serum iron and iron-combining power in order to be sure that the hypochromia is not due to causes other than sideropenia, because anaemia with poorly haemoglobinized erythrocytes (low M.C.H.C.) may occur in individuals with

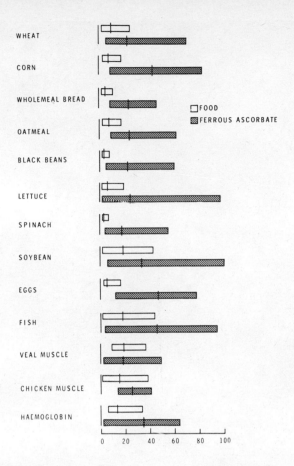

Fig. 24.10 Percentage absorption of iron from foods labelled biologically with radioactive iron compared with that from a similar dose of iron as a ferrous iron salt (ferrous ascorbate) in the same subjects. The vertical lines indicate the mean values and the horizontal bars indicate the limits. The vegetable sources were labelled with radioactive iron by growing them in hydroponic tanks with radioactive iron added to the nutrient solution. In the case of the animal sources, the radioactive iron was given by injection and thus incorporated into the tissues. (Data derived from M. Layrisse, J. D. Cook, C. Martinez, M. Roche, I. N. Kuhn, R. B. Walker & C. A. Finch (1969). *Blood* **33**, 430–443; and S. T. Callender (1971). *Geronto-logical Clinica* **13**, 44–51; and also unpublished data by S. T. Callender.)

adequate iron stores, but with disorders of synthesis of either the haem or the globin part of the haemoglobin molecule. The former is associated with massive accumulation of ferritin in the mitochondria round the nucleus of the erythroblast—the so-called ring sideroblast. This relatively rare type of anaemia, sideroblastic anaemia, is of complex aetiology but may sometimes respond to pyridoxine (vitamin B_6) (see p. 188). Disorders of globin synthesis (the so-called haemoglobinopathies) are much more common and are usually inherited. They may be of qualitative type, as in

sickle cell disease, or of quantitative type, as in thalassaemia (Mediterranean anaemia).

Sickling of the erythrocytes is due to the presence of an abnormal haemoglobin (sickle haemoglobin—HbS, see p. 447). The sickling phenomenon is precipitated by exposure to low oxygen tension (Fig. 24.11). Only slight hypoxia is required if the condition is inherited from both parents (homozygous) but more severe hypoxia is required to cause sickling in heterozygotes. In the homozygote severe hypochromic, haemolytic anaemia occurs from childhood, though not in the neonate who is

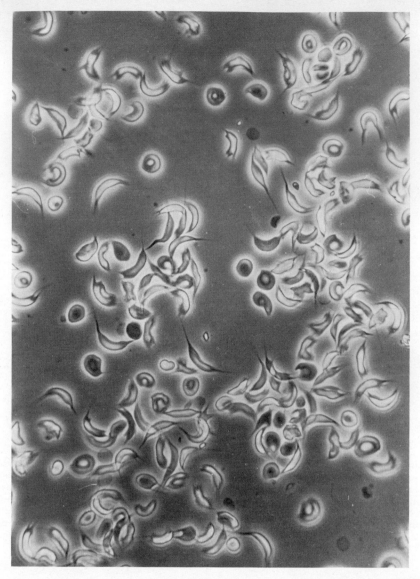

Fig. 24.11 Positive sickle test. This preparation was made by incubating a drop of blood in 2 per cent sodium metabisulphite for a few hours. The sample was from a Nigerian man with the sickle cell trait. Phase contrast. ×280. (*By courtesy of A. S. Todd.*)

protected by persistence of fetal haemoglobin; crises with thrombosis from sickling occur frequently, especially in relation to infection, particularly malaria. Yet curiously the heterozygote appears to succumb to malignant (falciparum) malaria less readily than normal individuals; this may help to explain the high concentration of sickling in tropical Africa and Greece. Figure 24.11 shows sickling in a West African.

Thalassaemia (Mediterranean anaemia) is due to quantitative failure of synthesis of either the α- or β-globin chains of the haemoglobin molecule. The classical type is β-thalassaemia which is mild in the heterozygote yet severe in the homozygote. In α-thalassaemia the condition is insignificant in the heterozygote yet lethal in the homozygote from severe haemolytic anaemic and oedema of the fetus.

LEUCOCYTES

The circulating blood contains colourless cells called *white cells* or *leucocytes* which, unlike red cells, possess nuclei. There are three varieties of white cells—the *granular* or *granulocyte* series (so called because of their numerous cytoplasmic granules), the *lymphocytes* and the *monocytes*. From the time of birth onwards all the granular cells are formed in the bone-marrow and they are, therefore, often termed the myeloid series. Although a few lymphocytes are formed in the marrow, the majority are produced in the lymphatic tissue of the body. The site of production of the monocytes is not known with certainty.

Myeloid (Granular) Leucocytes

The granular leucocytes are formed in the bone-marrow, outside the blood vessels, from non-granular myeloblasts, themselves derived from a more primitive 'stem-cell'. The myeloblasts give rise to three varieties of myelocyte, neutrophil, eosinophil and basophil, so-called from the behaviour of the granules toward Romanowsky stains which contain both eosin and (basic) methylene blue. When the myelo-

cytes mature they become the neutrophil, eosinophil and basophil polymorphonuclear leucocytes of the blood (Fig. 24.3 and Plate 24.1).

The *polymorphonuclear leucocytes* have lobed nuclei which vary greatly in shape, the number of lobes increasing with the age of the cell up to four or more. The neutrophil polymorphs, about 10 to 12 μm in diameter, are amoeboid and phagocytic when examined fresh on a warm stage. (It is to be noted that the diameter of these, and all other leucocytes, as seen in a blood film depends on the thickness of the film.) The small cytoplasmic granules stain faintly pink with Romanowsky stains and are, therefore, not strictly neutrophil. The electron microscope shows that the granules are 0·02 to 0·5 μm in diameter, and that some contain much lipid. The metabolism of polymorphonuclear leucocytes is much more active than that of erythrocytes (p. 417) and glucose is metabolized mainly aerobically. They contain glycogen, glutathione, histamine and a number of enzymes including amylase, lipase, protease, catalase, nucleotidases, β-glucuronidase and phosphatases. In all, some thirty enzymes have

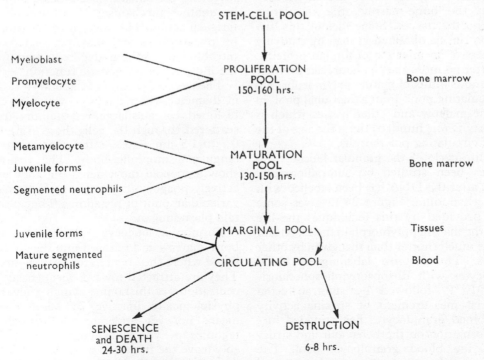

Fig. 24.12 Diagram of the life-cycle of myeloid leucocytes. (Modified from E. P. Cronkite & T. M. Fliedner (1964). *New England Journal of Medicine* **270**, 1347–1352.)

been isolated from human neutrophil granules and almost all chemical bonds known in nature can be broken by them. The glycogen content of human neutrophil polymorphonuclear leucocytes appears to be required for the energization of lymphocytes transforming to macrophages in experimental inflammatory lesions. This function would appear to be impaired in diabetes mellitus. Neutrophil alkaline phosphatase (NAP) in circulating leucocytes is limited normally to a variable proportion of mature neutrophil polymorphs. NAP seems to be controlled by a gene (or genes) on chromosome G-21, for NAP is high in mongols with trisomy-21 and virtually absent in chronic myeloid (granulocytic) leukaemia in which one of the chromosomes of the G-group is tiny and considerably deleted—the Philadelphia chromosome. Infections may lead to an increase in NAP, possibly from some direct effect, but possibly indirectly from increased secretion of corticosteroids or other steroids by the adrenals. Medication with such hormones has a similar effect and also causes an increase in the neutrophil count in the peripheral blood.

The mature granulocytes are distributed between the bone-marrow, the circulating blood and the tissues. Some idea of this distribution can be obtained in man by studying the effects of an injection of foreign proteins. This, after no more than an hour, can produce an enormous influx of mature polymorphs into the 'circulating pool' from a 'marginal pool' in the bone-marrow and other tissues which is from forty to one hundred times the size of the mass of circulating polymorphs.

The life-cycle of the granular leucocyte in man has been studied by autoradiography (p. 122) after the DNA has been labelled with titrated thymidine. Figure 24.12 gives some figures provided by this technique; the life span of the mature polymorph in the blood and tissues is much shorter than that given by other methods. The *in-vitro* labelling of blood granulocytes with diisopropylphosphofluoridate ($DF^{32}P$) followed by autotransfusion and serial measurement of specific activity in the blood granulocytes has provided further information on the behaviour and structure of the blood granulocyte pool. The level of granulocytes in the peripheral blood depends on a complex balance between pro-

liferation and mobilization in the marrow and specific utilization and random loss from the blood.

The main function of the polymorphonuclear leucocytes is the protection of the body from disease, particularly infection, because where the circulating and marginal pools are small (*leucopenia*) infection, general or local, readily occurs. Phagocytosis is the main way by which these leucocytes exert this function. The phagocytic process involves the surrounding of a particle, for example a bacterium, by pseudopodia, the formation of a phagocytic vacuole by fusion of membranes and the release of hydrolytic enzymes from lysosomal granules into the phagocytic vacuole, where they may act on the ingested particle without destroying the rest of the cell. Not all microorganisms are ingested and destroyed. Viruses are probably not even ingested. Some bacteria are ingested, but not destroyed. The ingestion of foreign particles or even some endogenous materials, for example damaged erythrocytes, is enhanced by a coating of specific antibodies or non-specific opsonins. The production of hydrogen peroxide by a complex process involving the oxidation of glucose appears to be another mechanism of intracellular bactericidal action. This mechanism is stimulated by phagocytosis and may be related to the nitroblue tetrazolium dye reduction test for the integrity of phagocytic function.

The *eosinophil leucocyte*, about 10 to 12 μm in diameter, the nucleus of which is usually bi-lobed, contains large, red-staining granules scattered through the cell; these granules are 0·7 to 1·3 μm across, often containing a rectangular osmophilic body. The eosinophils show amoeboid movements but they are not actively phagocytic. They contain a small but variable amount of histamine. They also contain plasminogen.

Eosinophil leucocytes are formed in the bone-marrow and released into the circulating blood where they remain for a few hours only. They are attracted towards local (tissue) concentrations of histamine which they render physiologically inactive by some chemical means not yet understood. They also antagonize 5-hydroxytryptamine. They probably leave the tissues to which they have been attracted through the lymphatics.

The *basophil leucocyte* is somewhat smaller,

8 to 10 μm, with large cytoplasmic granules about 1 μm in diameter which stain a deep blue and contain heparin, 5-hydroxytryptamine and relatively large amounts of histamine. These cells, and similar cells found in the tissues, are often called 'mast' cells.

LEUCOCYTOSIS AND LEUCOPENIA

The number of leucocytes in the circulating blood is very variable but ranges from 4000 to 11 000 per mm^3, 60 to 70 per cent of which are neutrophil polymorphs (Table 24.13). Somewhat higher polymorph counts are found in

Table 24.13 A list of the white cells found in normal adult blood. (The numbers occurring in the circulating blood usually fall within the ranges shown)

	Number per mm^3	Percentage of total white cells
All leucocytes	5000–9000	100
Neutrophil polymorphonuclears	3000–6000	60–70
Lymphocytes	1500–2700	20–30
Monocytes	100–700	2–8
Eosinophil polymorphonuclears	100–400	2–4
Basophil polymorphonuclears	25–200	0·5–2

pregnancy and in the newborn. An increase in the number of leucocytes above 11 000 per mm^3 is called a *leucocytosis* and a fall below 4000 a *leucopenia*. As a rule abnormal fluctuations in the total leucocyte count are produced largely by changes in the number of neutrophil polymorphs but the factors responsible for such changes are not known in any detail (see Fig. 24.14). Violent exercise, fever and haemorrhage are followed by a leucocytosis and some increase occurs a few hours after a protein meal. Damage to, or destruction of, tissue as in accidents, fractures, surgical operations as well as infections lead within a few hours to a considerable and often prolonged increase in circulating polymorphonuclear cells. It seems likely that leucocytosis is brought about by humoral factors which affect both release of leucocytes from the marginal pool as well as granulocytopoiesis—the production of more leucocytes by the bone-marrow. It is difficult at present to separate these two effects. A number of extracts prepared from various tissues, some of them containing a high proportion of nucleic acids, have been claimed to contain LPF (leucocyte-promoting factor) and a water-soluble, heat-stable, dialysable substance, leucopoietin G has been found in plasma. Bacterial pyrogen (Chap. 42) also causes leucocytosis. Adrenocortical steroids (Chap. 50) produce an increase in the polymorph count and a decrease in the number of lymphocytes and eosinophils.

The lungs may influence the number of circulating leucocytes. Thus in man after intravenous injection of 0·2 mg of adrenaline higher concentrations of white cells were found in blood taken from the femoral artery than in

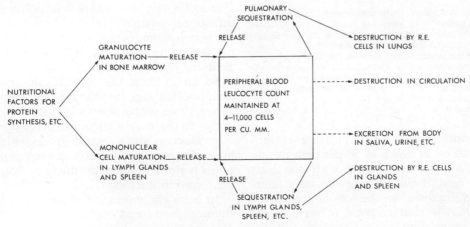

Fig. 24.14 Diagram to show some of the factors governing the numbers of leucocytes in the blood. (L. Whitby (1954). *British Medical Journal* **i**, 1279.)

samples simultaneously withdrawn (by catheter) from the pulmonary artery. On the other hand, there is evidence both in man and in the rabbit that large numbers of leucocytes can be lost from the circulating blood as it passes through the lungs.

Comparatively little is known about the ultimate fate of granular leucocytes and it is not known whether there is a feed-back mechanism governing granulocytopoiesis. Considerable numbers pass into the alimentary canal and are destroyed in the reticulo-endothelial system (p. 434 and Fig. 24.14). In certain circumstances leucocytes disappear very rapidly from the circulating blood (Fig. 24.15).

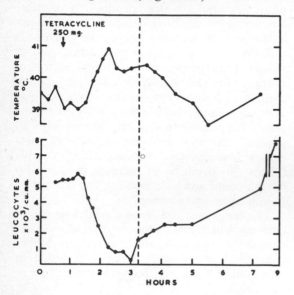

Fig. 24.15 Record of body temperature and leucocyte count in a patient with relapsing fever. With 3 hours of treatment by the antibiotic tetracycline leucocytes had almost disappeared from the patient's blood. The disappearance of leucocytes coincided with the disappearance of the infecting organism (a spirochaete) indicated by the vertical dotted line. It has been suggested that the raised temperature is due to the liberation of 'bacterial' pyrogen (see Chap. 42). (T. P. C. Schofield, J. M. Talbot, A. D. M. Bryceson & E. H. O. Parry (1968). *Lancet* 1, 58–62.)

The concentration of circulating eosinophils varies by as much as 50 per cent during the day, the lowest counts with the least variation being found during the early afternoon. An increase in the eosinophil count (*eosinophilia*) is characteristic of certain skin diseases, infestation with parasites and the allergic group of disorders which includes asthma and hay-fever.

Leucopenia occurs in starvation or debilitating conditions and occasionally after administration of certain drugs, for example sulphonamides. Malnutrition may also lead to leucopenia. This is commonly seen in deficiency of vitamin B_{12} or folic acid and may also occasionally be due to pyridoxine deficiency. Infection in a patient with megaloblastic anaemia may rapidly cause a further lowering of the leucocyte count to dangerously low levels. The leucocytes (particularly the neutrophils) may be severely reduced also in patients with enlargement of the spleen—so-called hypersplenism. Prolonged, severe reduction in the number of leucocytes is a serious threat to life because of the danger of infection.

THE LYMPHOCYTES

There are several different cell types to which the name 'lymphocyte' has been given, the most characteristic being a small round cell some 7 μm in diameter with a relatively large densely-staining nucleus and very little cytoplasm—the so-called small lymphocyte. Two main types of large lymphocyte are recognized microscopically: the so-called irritation lymphocyte (activated lymphocyte or Türk cell) has a characteristically deeply basophilic cytoplasm due to the presence of RNA; the so-called stress lymphocyte has cytoplasm which contains numerous azurophil granules but shows virtually no general basophilia. The irritation lymphocyte is an activated lymphocyte and is seen sometimes in large numbers in infections, particularly viral infections. The stress lymphocytes may well be effete; they are not seen in imprints of lymph nodes nor of the germinal centres of the spleen.

So far as their life-cycle is concerned the lymphocytes can be divided into two categories, those with a short life-span of the order of two to three days and a relatively rapid rate of production and a second group with a long life-span (perhaps 200 days or even more) and a much slower rate of production. Both types are found in the blood.

In both the young child and the adult the thymus gland (see Chap. 50) is the main source of lymphocytes. In the early stages of fetal development lymphoid cells are 'seeded' from the thymus to other organs of the body and it is likely that in adult life the thymus exercises a humoral control over other lymphoid tissues.

Lymphocytes are present in all tissues except the central nervous system. The lymphatic tissue contains 100 g of lymphocytes, the bone marrow 70 g and the blood 3 g but the total amount in the body is said to be 1300 g. Stress or administration of adrenal corticoids inhibits lymphopoiesis.

The metabolism of lymphocytes is similar to that of the granular leucocytes in that they use the oxidative pathway and are also capable of aerobic or anaerobic glycolysis.

The fate of lymphocytes was unknown until relatively recently. The unexplained rapid disappearance of lymphocytes from the circulation was shown by Gowans of Oxford to be due to the passage from the blood stream into the lymph node and thence into the lymph vessels and back to the blood. This recirculation of lymphocytes is shown in Fig. 24.16.

Lymphocytes appear to have two main functions in relation to immunity (see also p. 405). Those of thymic origin are concerned with cellular immunity and migrate to sites of foreign antigens as in tuberculous lesions, malignant neoplasms and homograft rejections. The lymphocytes which have been 'processed' in the para-alimentary lymphoid masses, tonsils, aggregated lymph nodules (Peyer's patches) and appendix are concerned with humoral immunity. When exposed to foreign antigens they synthesize RNA actively. Some of these activated cells (immunoblasts) differentiate into plasma cells (immunocytes) and secrete immune globulins (see p. 405). One cannot recognize which particular lymphocyte from the blood or lymph nodes will undergo such changes, but lymphocytes from both sources are capable of such metamorphoses. Adrenal steroid hormones and the synthetic agent azothioprine may be used to suppress such functions in patients with autoimmune disorders and in tissue grafting.

The monocyte series. The monocyte, 16 to 22 μm in diameter, the largest cell in the blood, is recognized by its oval or horseshoe-shaped nucleus. The cytoplasm of the monocyte is

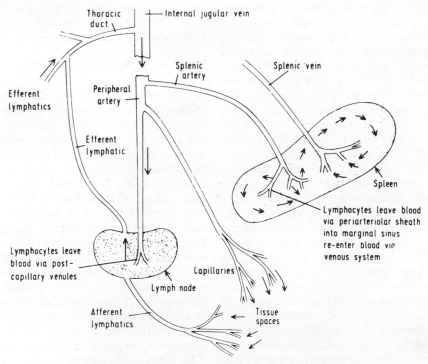

Fig. 24.16 Schematic view of main pathways of lymphocyte circulation. (D. M. Weir (1971). *Immunology for Undergraduates*, 2nd edn. Edinburgh: Churchill Livingstone.)

Formed Elements of the Blood 433

pale-staining and frequently contains vacuoles and a few small reddish granules. Its staining reactions suggest that the monocyte is more closely related to the myeloid series than to the lymphocytes. The monocytes probably originate mainly in the bone marrow and spleen. Some haematologists regard these cells as histiocytes which have migrated into the blood, implying that they are 'free' reticulo-endothelial cells but there is still controversy about their origin. Monocytes show slow amoeboid movements and are phagocytic.

THE BLOOD PLATELETS

The platelets (thrombocytes) are the smallest of the formed elements of the blood, measuring 2 to 3 μm in diameter. They are formed by division of the cytoplasm of *megakaryocytes*, large cells 40–150 μm in diameter found in bone marrow, lungs and to a less extent in other tissues.

The megakaryocytes are polyploid cells containing, on average, four times the diploid number of chromosomes in a large, usually lobulated nucleus. The separation of individual platelets is an orderly process involving the formation and fusion of small cytoplasmic vesicles to form the platelet membranes. The rate of platelet production is partly controlled by the number of circulating platelets (normally 100 000 to 500 000 per mm^3) and it now seems likely that a humoral substance *thrombopoietin* mediates this influence in much the same way as does erythropoietin in haemopoiesis. Since a megakaryocyte has a volume 3000 to 4000 times that of a single thrombocyte it seems likely that each may produce more than a thousand platelets.

The mature platelet is a biconvex disc with a complex structure, a living cell which, like the erythrocyte, consumes oxygen in spite of the absence of a nucleus. It shows the following elements (Fig. 24.17A and B): an exterior coat rich in glycoprotein and acid mucopolysaccharide; a unit membrane (cell membrane) with the usual trilaminar structure; a system of microfilaments (probably of actin-like protein) dispersed through the cytoplasm; an equatorial ring of 8 to 24 *microtubules* close to the cell membrane, each tubule consisting of 12 to 15 helical subfilaments; granules containing lysosome-like enzymes; glycogen particles; *electron dense granules* (probably serotonin

storing bodies); a few mitochondria; a Golgi apparatus; a *dense tubular system* of membranes.

In man platelets have a life span of 7 to 14 days. Some 10 to 15 per cent are lost by consumption in haemostasis and the maintenance of vascular integrity. The remainder are destroyed by the reticulo-endothelial (macrophage) system, especially in the spleen; thus the platelet count rises, sometimes permanently, if the spleen is removed. Information about platelet life span has come from experiments using diisopropylphosphofluoridate (DFP) labelled with either ^{32}P or ^{3}H. When injected DFP labels the platelets and enables their rate of destruction to be followed. Similar information has been obtained by labelling the platelets *in vitro* with ^{32}P or ^{51}Cr and transfusing them. The main functions of platelets seem to be in haemostasis (see p. 401) but it has now been shown that labelled platelets can contribute part of their cytoplasm to endothelial cells and may therefore have a role in supporting the metabolism of injured intima. Platelets also contain, and can release, a protein able to increase vascular permeability and thus contribute to the humoral control of inflammatory reactions.

THE MACROPHAGE SYSTEM

Certain cells throughout the body are highly phagocytic of particulate material including micro-organisms, effete erythrocytes, disintegrating fibrin, tissue debris and any finely divided or colloidal substance that may be introduced parenterally into the body whether therapeutically, experimentally or accidentally. Thus, if an animal is killed some days after the intravenous or intraperitoneal injection of, say, Indian ink or the dye Trypan Blue, it is found that the particles of carbon or dyestuff are not distributed generally throughout the body but are concentrated in cells in certain viscera depending on their content of phagocytic cells. These cells comprise the reticulo-endothelial system, named as such in 1924 by Ludwig Aschoff; others since have suggested the designation *macrophage system*. The term macrophage system is preferable because it clearly names the main function of the cells. The term reticulo-endothelial is confusing for it suggests not only that all such cells

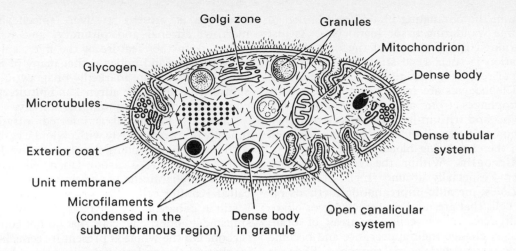

Golgi zone   Granules

Glycogen   Mitochondrion

Dense body

Microtubules

Exterior coat   Dense tubular
system

Unit membrane

Microfilaments   Dense body   Open canalicular
(condensed in the   in granule   system
submembranous region)

Fig. 24.17A A schematic cross-section of a platelet showing the fine structure. (Redrawn after J. G. White (1971). Platelet morphology. In *The Circulating Platelet* (Edited by S. Johnson), p. 76. Washington: Academic Press.

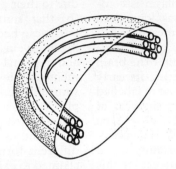

Fig. 24.17B Scheme of half-platelet showing the equatorial arrangement of microtubules.

are either supported by or actually form reticulin fibres but also that all endothelial tissues are phagocytic. Neither suggestion is wholly correct.

There are differences in the distribution of macrophages between species and also, within one species, in the behaviour of its macrophages towards different kinds of particulate substance. These differences demand great caution in arguing from circumstances in the experimental animal to those in man and from one type of substance or particle to another.

Tissue histiocytes. Cells of macrophage type occur in most connective tissues. Many of the cells appear to be long-term residents and are sometimes referred to as 'fixed' tissue histiocytes or macrophages. In most sites they

are scanty but they are relatively abundant in the omentum of the fetus and new-born where cells beneath the peritoneum appear as pale milky aggregates against the yellowish adipose tissue and have thus acquired the name of 'tâches laiteuses'. When the need arises for histiocytes in greater number, as in certain forms of inflammation, those of 'fixed' origin are rapidly supplemented by others, the 'wandering' histiocytes, that appear to be constantly on patrol in most tissues. These in turn are supplemented by yet others, formerly circulating blood monocytes but now, in response to the stimulus, extravascular. It may be that most if not all 'wandering' histiocytes are in fact of intravascular origin.

Monocytes. Monocytes (see p. 433) arise mostly in bone marrow and spleen. Many of these in the circulating blood are destined to become wandering tissue histiocytes. As just mentioned, in any area where the form of tissue irritation is such as to stimulate reaction by local or relatively fixed tissue histiocytes, these local leucocytes are rapidly supplemented by macrophages from afar. Observations with carbon- and tritium-labelled cells have shown that most if not all of these cells are derived from the circulating blood monocytes.

Microglia. Within the central nervous system, especially around small arteries and arterioles, metallic-impregnation methods reveal cells that are neither astrocytes nor oligodendrocytes. Such cells, in areas of brain damage, change their appearance and become clearly recognizable as macrophages. They are known collectively as the microglia or central nervous system histiocytes; another commonly used name derived from the frequent foamy appearance of their cytoplasm is compound granular corpuscle. It has long been believed that these cells are of mesenchymal origin, derived from the tissues of the perivascular sheaths, and that they enter the brain along with the blood vessels during its early development. More recently the use of labelled blood monocytes has shown that in areas of injury cells of apparently microglial type rapidly become supplemented by macrophages that were at one time certainly blood monocytes. The extent and frequency of this process is uncertain. The pericyte (p. 544) has also been regarded as possibly capable of becoming a macrophage in these circumstances. The exact histogenesis of the cells presently designated as the microglia is thus still debatable. It should be noted that leucocytes do not enter the substance of the brain from the blood stream during health, and also that even the microglial cells fail to engulf material from dyestuffs or colloidal suspensions injected intravenously into a healthy animal. These observations emphasize the existence of a so-called blood–brain barrier to which reference will be made later (Chap. 33) in connection with transfer between blood and cerebro-spinal fluid of various drugs or other substances.

Cells lining sinusoids. Sinusoids are relatively wide channels, both haemic and lymphatic, through which the flow of blood and lymph is correspondingly slow. This type of structure is present in liver, spleen, bone marrow, adrenal and pituitary, and lymph nodes. A further feature of the liver, spleen, bone marrow and nodes is that many of their endothelial cells are markedly phagocytic; the sinusoidal cells of the adrenal and pituitary by comparison show only minimal activity.

Liver. If an animal is injected intraperitoneally or intravenously with colloidal carbon, sections of liver obtained only minutes later will show numerous carbon-laden cells within the endothelial lining of the sinusoids, the so-called Kupffer cells. The origin of these cells is still in debate as is their relation to the other sinusoidal endothelial cells that do not contain carbon. On the whole at present it seems likely (a) that all the endothelial cells have a highly phagocytic ability, and are therefore potentially 'Kupffer' cells, but that some react more promptly or more avidly than others, perhaps due to their age or metabolic status at the time, (b) that some such cells, when laden with particles, become detached and pass in the hepatic veins or the lymphatics to the lungs and regional lymph nodes, (c) that replacement is achieved mainly by division of adjacent sinusoidal endothelial cells but also by lymphocytes that pass from the spleen and become transformed and incorporated into the sinusoidal lining. It seems likely also that the different forms of endothelial phagocytic cell claimed to exist are but morphological variants of the one endothelial cell type with occasional transforming immigrant lymphocytes.

Spleen. The structure of the spleen is described in Chapter 29 when it is remarked that there is probably a double form of circulation of blood, one rapid and one slow. The slow type of circulation leads in effect to a sequestration of blood cells and affords ample opportunity for phagocytosis whether of aged erythrocytes or abnormal materials and substances of the kinds mentioned earlier. The exact mode of final disappearance of an erythrocyte is principally fragmentation and ingestion of the particles by the phagocytic endothelial cell; it is unlikely to be engulfed whole by these cells or to disappear by relatively sudden rupture. The erythrocyte certainly becomes increasingly brittle as it ages, an alteration to be correlated probably with a decline in the effectiveness of its enzyme systems, and is thus less able to withstand the relatively severe mechanical stresses involved in traversing capillaries. The relative hypoxia suffered by cells sequestered in the splenic sinusoids is possibly the ultimately fatal factor for the ageing erythrocytes and thus the reason why the spleen is so predominantly their graveyard.

Bone marrow. The experimental intravenous injection of appropriate suspensions has shown that the disposition of phagocytes in the bone marrow is similar to that in the spleen. The endothelium of the marrow is discontinuous, and so particulate material (and ageing erythrocytes) has ready access to the extravascular or parenchymatous compartment of the marrow. Experimental observations of this kind have also shown that phagocytic cells are present not only in the endothelium but also in the non-endothelial parenchyma.

Lymph nodes. For present purposes the lymph node may be considered as having three types of cell (a) lymphoblasts/lymphocytes, (b) reticulum cells; that is, the cells which, with their reticulin fibres, comprise the permanent population and skeleton of the node, and (c) the endothelial cells of the peripheral and penetrating sinuses. Of these, the lymphocyte series possesses virtually no phagocytic ability; the reticulum cells possess some but of low degree only. Lymph nodes are major sites of phagocytosis due almost wholly to the activity of the endothelial cells (endothelium of the immediately related afferent and efferent lymphatics has no such activity, a sharp, interesting and ill-understood functional difference). In some states of inflammation, macrophages containing lipid material may accumulate in large number in the substance of a lymph node as well as in the endothelium of the sinuses. These cells might be reticulum cells, stimulated to greater-than-normal phagocytosis by unusually great demand, but seem more likely to be endothelial cells that have passed, fully laden, from the wall of the sinus into the pulp of the node until such time as they can digest and dispose of the engulfed substance.

Other situations. A few endothelial cells in the sinusoids of the adrenals and pituitary show phagocytic activity; they have little if any physiological significance. In the lung, phagocytes are abundant. They do not comprise a formal 'system' as do the Kupffer cells in the liver but are present as a relatively large and widespread patrol of wandering histiocytes. If no particulate material or micro-organisms ever entered the lungs, the macrophages might never be seen. As it is, macrophages with engulfed carbon or other material are virtually always present in the alveoli and airways and can be easily seen in bronchial secretion or expectoration. The origin of these cells is still uncertain: it is still unsure whether and to what extent they are derived from circulating blood monocytes (which become wandering tissue histiocytes) or alveolar epithelial cells or the fibroblastic mesenchymal cells that support the alveolar epithelium. Most inhaled particulate material is trapped directly in the mucus of the respiratory tract and then expectorated or swallowed; much is engulfed by the macrophages which are likewise removed in the mucus; some is carried by the macrophages into lymphatics from which it eventually becomes deposited in the small foci of lymphoid tissue in the lung or in the regional (mediastinal) lymph nodes. In persons heavily exposed to dusty conditions, the material may travel even further; thus, in miners, carbon is commonly found in the lymphatics and lymph nodes of the diaphragm, posterior abdominal wall and neck.

The extent to which one form of mesenchymal cell may change or 'transform' into another is an interesting problem. At one time or another, each of the lymphocyte, monocyte, endothelial cell, fibroblast and macrophage has been regarded as capable of transforming into any of the others; and this may be true. However, since most of the evidence so far has been mainly morphological, by which 'transitional' forms are easy to find but almost impossible to identify in a definitive way, some doubt remains. Continuing studies of the use of isotope-labelled cells, in the manner described earlier, give some hope that the uncertainty will gradually be dispelled.

Functions of the macrophage system. The earliest recognizable precursor of the macrophage is the primitive mesenchymal cell already mentioned as part of the population of a lymph node, a *reticulum cell*. This type of cell occurs also in the bone marrow and is the precursor of the haemocytoblast. It is thus, in turn, the precursor of all the formed elements of the blood and lymphoid tissue including the lymphocyte and its derivative, the plasma cell. Because of this common ancestry, it has been usual hitherto to regard these cells as all comprising one 'system', the reticulo-endothelial system, whose function was thus a combination of phagocytosis (macrophages and polymorphonuclear leucocytes, formerly microphages), antibody manufacture and transport with maintenance of immunity (lymphocytes and plasma cells), and haemopoiesis (the haemocytoblast). With increasing knowledge of these three processes, this association of functions has become increasingly artificial, hence the desirability, mentioned in the opening paragraph, of considering as a separate macrophage system those cells of which a major function is the phagocytosis of particulate matter (Fig. 24.18).

surgeon's catgut can be gradually destroyed by enzyme action and lysis. Most forms of inorganic material, whether exogenous or, like sterol crystals, endogenous, cannot be destroyed by lysis: these, therefore, remain undisturbed either throughout life or (possibly because of some tissue reaction) until surgical removal, or they may be transported to lymph nodes via the lymphatics by the macrophages. Bile pigment, haemosiderin, melanin, lipid substances and carbon are all materials that may be seen in different circumstances within macrophages; so also, in association with tissue damage, are dead and dying micro-organisms and polymorphonuclears, tissue debris and fibrin. Besides phagocytosis, cells of the macrophage system are intimately concerned with the catabolism of haemoglobin, that is,

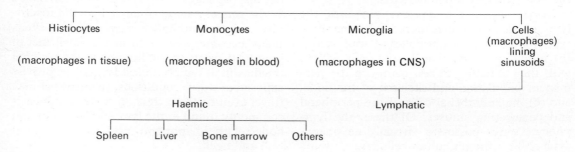

Fig. 24.18   The macrophage 'system'.

Phagocytosis. Macrophages differ from polymorphonuclear leucocytes in their much greater avidity for or ability to engulf inorganic particulate material such as carbon and dust particles in general. Examination of bronchial secretion shows dust in abundance within macrophages but not in other leucocytes (see Fig. 24.19). Macrophages have also the unusual capacity, when confronted by particularly insoluble or foreign material, to fuse together with the formation of multinucleated giant cells (see Fig. 24.20). The presence within a tissue of some foreign material such as a surgical suture or a thorn is not, of course, a physiological situation; but neither also, perhaps, is the presence of dust in the bronchial tree. Macrophages can destroy many types of micro-organisms, mostly types that cannot be adequately or efficiently destroyed by the polymorphonuclears. Most forms of organic foreign material such as a spicule of fish-bone or the

the further breakdown of the erythrocyte and its main constituent after phagocytosis has occurred (see p. 452).

Formation of bile pigment (bilirubin, haematoidin). Bile pigment is formed by and within the cells of the macrophage system, especially those of liver, spleen and bone marrow as a result of enzyme action upon the haemoglobin derived from effete or otherwise imperfect or damaged erythrocytes. Iron derived from the haemoglobin is retained within the macrophage as haemosiderin and ferritin; the bile pigment is released into the plasma. It was at one time believed that bile pigment was formed only by the cells of the liver and this belief appeared to derive sound support from an experiment (Minkowski & Naunyn, 1886) showing that haemolytic anaemia produced by the administration of arsine (AsH_3) to a goose produced jaundice in intact animals but not in those previously hepatectomized. It appeared

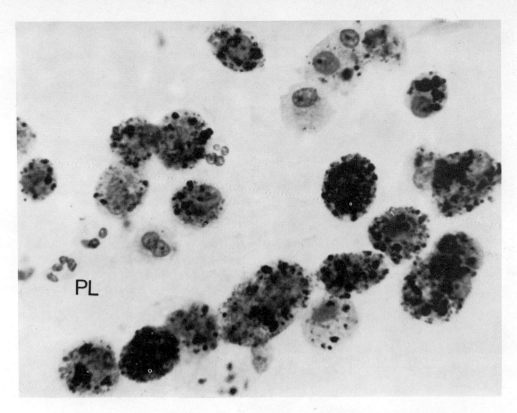

Fig. 24.19 Macrophages of bronchial origin in sputum. The cells are relatively large and contain different amounts of particulate material (mostly soot and dust). The amount is so great in some cells as to obscure the nucleus completely. At left-centre may be seen the multilobed nuclei of two polymorphonuclear leucocytes (PL); their cytoplasm is devoid of particles. Papanicolaou. ×900. (*By courtesy of W. W. Park.*)

that without a liver the goose could not produce bile pigment, and this was true. As later experiments with other species showed, the reason was indeed the absence of the liver, not, however, as a mass of functioning hepatic cells but because virtually the whole of the macrophage system in the goose is located in the liver. Animals of other species when hepatectomized could, because of the abundance of macrophages elsewhere than in the liver, easily be made jaundiced by the lysis of erythrocytes with arsine. Such experiments illustrate clearly the importance of species difference in experimental observations and results.

Erythrocyte destruction in man takes place predominantly in the spleen, liver and bone marrow but mostly within the spleen by reason of its action in sequestering cells and thus acting as a filter. Erythrocyte breakdown continues at a normal rate after splenectomy but a peripheral blood film will then usually show a greater than usual number of erythrocytes with minor morphological abnormalities such as undue thinness (leptocytosis) or retention of nuclear remnants (Howell-Jolly bodies, Fig. 24.21); such cells are presumably filtered out by the spleen in the intact individual.

When blood is extravasated or escapes into the tissues from within the vascular compartment, as in bruising, the released haemoglobin is disposed of in essentially the same way as when it is released inside the vascular compartment; that is, by absorption into macrophages and conversion to bile pigment and iron. In these circumstances bile pigment may accumulate locally in crystalline form both within macrophages and (presumably after extrusion from the cell) without. This form of the pigment has long been known to pathologists as haematoidin. The range of colours in the

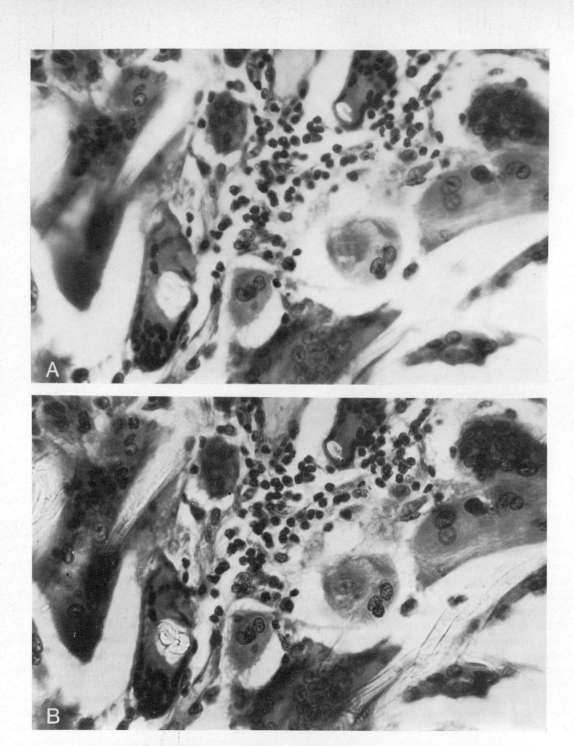

Figs. 24.20A and B The multinucleated masses of cytoplasm present in the depth of a healing wound, are 'foreign body' giant cells. They have either engulfed a foreign substance or they are applied to the surface of a foreign substance, probably in-driven clothing material. The stranded nature of the material and its relation to the giant cells is emphasized in B; reduced illumination has made the material mildly refractile. Haemalum and eosin. ×540.
(By courtesy of W. W. Park.)

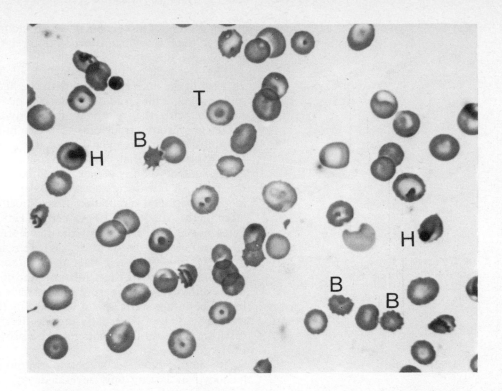

Fig. 24.21 A blood film made 1 year after splenectomy. Note the many morphologically abnormal erythrocytes. There is general aniso- and poikilo-cytosis, while some erythrocytes show undue thinness (the 'target' (T) or 'Mexican hat' forms), a crenellated outline ('burr' cells) (B) or, retention of pyknotic nuclear remnants (Howell-Jolly bodies) (H). Leishman × 900. (*By courtesy of W. W. Park.*)

familiar 'black eye', from purple through dark and light green to brownish, is produced by the extravasated blood, its progressive conversion to bile pigment and gradual absorption. A massive escape of blood into the tissues or one of the body cavities may result in jaundice of varying intensity depending on the amount of bile pigment formed and transported in the blood (p. 452).

Relation to antigens and antibodies. Almost as soon as the macrophage system was discovered and defined by the ability of its cells to take up and concentrate inert particulate matter and various dyestuffs, attempts were made to find out whether it behaved towards other substances in the same way. In many instances it does, and amongst the materials ingested and stored are certain types of lipid and foreign protein. We thus know that macrophages can ingest and store antigens, and from this the question immediately arises whether they also manufacture antibodies. In short, the answer is, almost certainly they cannot; recognition of antigen and manufacture of antibody remain the prime responsibility of the lymphocyte/plasma cell system. It is possible, however, that the macrophage may in some way 'process' antigenic protein and by either excreting the modified protein or altering its own surface RNA facilitate in some way the immunogenetic activity of the lymphocytes. The matter is one of intensive current research.

PHYSIOLOGICAL CHANGES IN BLOOD IN INFANCY AND CHILDHOOD

The haemoglobin content of the blood (Fig. 24.22) at birth is about 21 g per 100 ml, but it declines from this high value during the first two or three months of infancy to about 12 g, rising again to about 13 g in the succeeding three months. The iron released by breakdown

Formed Elements of the Blood 441

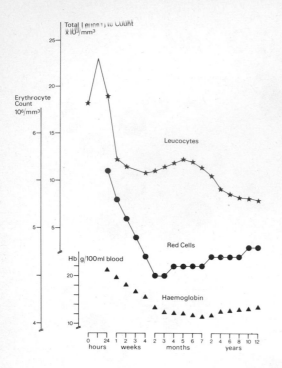

Fig. 24.22 A graphic representation of mean values only for erythrocyte count (★), total leucocyte count (●) and haemoglobin (Hb) content of whole blood (▲) from birth to twelve years. Bar representations of ±S.D. have been omitted for clarity but were of the order ±0·45 × 10⁶, ±2·5 × 10³ and ±1·3 for erythrocyte count, total leucocyte count and haemoglobin content respectively. (Data obtained from *Documenta Geigy: Scientific Tables*, 7th edn. Basle: Geigy.)

haemoglobin. This property is no doubt of value to the fetus which is exposed to relatively low oxygen tensions.

The red cell count is high at birth (Fig. 24.22) but declines in parallel with the haemoglobin level. This rapid destruction of red cells together with the relatively low activity of UDP glucuronyl transferase in the infants' 'immature' liver is responsible for the 'physiological jaundice' that is seen in some 30 per cent of all new-born babies, maximal on the third day. A few nucleated red cells are found in the peripheral circulation for a short time after birth. About 5 to 10 per cent of reticulocytes occur in the blood of the newborn but adult numbers (1 per cent or less) are reached in a few days. The red cells at birth contain only one quarter of the adult amount of carbonic anhydrase; premature infants have even less. The infant starts off with 300 to 500 mg of iron provided, of course, by the mother: after birth the iron requirements must be provided in the food.

At birth there is a polymorphonuclear leucocytosis lasting only several days; this is followed by a lymphocytosis which persists for a year or so (p. 431). The former is probably the result of deficient supply of oxygen before birth and its disappearance follows improved oxygen supply after birth.

The platelets, unlike the other cellular elements, are present at birth in reduced numbers, normal adult values being attained within the first 10 days of life.

The changes in blood volume in infancy are described on p. 409.

The haematological changes in premature infants are of particular importance. At birth many such infants have a very high haemoglobin level, often of the order of 20 or 21 g/100 ml. This appears to be due to fetal anoxia, particularly in the so-called dysmature infant suffering from placental failure. The tiny premature infant is born with poor iron stores and if his mother has not received prophylactic or therapeutic folic acid he often develops megaloblastic anaemia. Tocopherol may be important in protecting the erythrocyte of the premature infant from haemolysis. Soluble vitamin K analogues may cause haemolysis, and are thus of potential danger to the central nervous system from damage to the basal nuclei by unconjugated bilirubin.

of the haemoglobin of the destroyed red cells remains within the body and is re-utilized for the synthesis of haemoglobin, but the initial fall in the red cell count is not prevented by the administration of iron even in large amounts. A proportion of the haemoglobin of the foetus differs both chemically and spectroscopically from the haemoglobin of adults, the difference lying in the globin fraction (p. 444); this fetal haemoglobin (HbF) forms 55 to 98 per cent of the total circulating haemoglobin in the normal full-term infant. Small amounts (1 to 2 per cent) of a similar material are found in the blood of many adults. HbF can be detected in larger quantities in the blood of patients with certain blood disorders. Fetal haemoglobin takes up oxygen more readily and gives up carbon dioxide more easily than does adult

REFERENCES

ARNSTEIN, H. R. V. & WRIGHTON, R. J. (Eds.) (1971). *The Cobalamins. A Glaxo Symposium.* London: Churchill Livingstone.

BISHOP, C. & SURGENOR, D. M. (Eds.) (1964). *The Red Blood Cell. A Comprehensive Treatise.* London: Academic Press.

BRITTON, C. J. C. (1969). *Disorders of the Blood*, 10th edn. London: Churchill.

CHANARIN, I. (1969). *The Megaloblastic Anaemias.* Oxford: Blackwell.

CLARKE, W. J., HOWARD, E. B. & HACKETT, P. L. (Eds.) (1970). *Myeloproliferative Disorders of Animals and Man.* A.E.C. Symposium No. 19.

CLINE, M. J. (1965). Metabolism of the circulating leukocyte. *Physiological Reviews* **45**, 674–720.

ELVES, M. W. (1966). *The Lymphocytes.* London: Lloyd- Luke.

FISHER, J. W. (Ed.) (1968). Erythropoietin. *Annals of the New York Academy of Sciences* **149**, 1–583.

GLASS, G. B. J. (1963). Gastric intrinsic factor and its function in the metabolism of vitamin B_{12}. *Physiological Reviews* **43**, 529–849.

GOWANS, J. L. (1970). Lymphocytes. *Harvey Lectures* **64**, 87–119.

HALLBERG, L., HARWORTH, H. G. & VANNOTTI, A. (Eds.) (1970). *Iron Deficiency, Pathogenesis, Clinical Aspects and Therapy.* London and New York: Academic Press.

HARRIS, J. W. & KELLERMEYER, R. W. (1970). *The Red Cell.* Cambridge: Harvard University Press.

HUMPHREY, J. H. & WHITE, R. G. (1970). *Immunology for Students of Medicine*, 3rd edn. Oxford: Blackwell.

JOHNSON, SHIRLEY A. (Ed.) (1971). *The Circulating Platelet.* London: Academic Press.

KOWALSKI, E. & NIEWIAROWSKI, S. (1967). *The Biochemistry of Platelets.* London: Academic Press.

MARCUS, A. J. & ZUCKER, M. B. (1965). *The Physiology of Blood Platelets.* New York: Grune & Stratton.

PUGH, L. G. C. E. (1964). Man at high altitude. *Scientific Basis of Medicine Annual Review* 32–54.

RILEY, J. F. (1959). *The Mast Cells.* Edinburgh: Livingstone.

SELYE, H. (1965). *The Mast Cells.* London: Butterworth.

SHINTON, N. K. (1972). Red cell production. *British Medical Journal* i, 433–436.

STUART, A. E. (1970). *The Reticuloendothelial System.* Edinburgh: Livingstone.

Symposium on Disorders of the Red Cell. (1966). *American Journal of Medicine* **41**, 657–830.

WEIR, D. M. (1971). *Immunology for Undergraduates*, 2nd edn. Edinburgh: Churchill Livingstone.

WHIPPLE, H. E. (Ed.) (1964). Vitamin B_{12} coenzymes. *Annals of the New York Academy of Sciences* **112**, 547–921.

WHIPPLE, H. E. (Ed.) (1964). Leukopoiesis in health and disease. *Annals of the New York Academy of Sciences* **113**, 511–1092.

WHITTAM, R. (1964). *Transport and Diffusion in Red Blood Cells:* Monograph of Physiological Society. London: Arnold.

25 Blood pigments

Haemoglobin is a conjugated protein consisting of the protein globin united to the iron-containing substance haem. Haemoglobin contains 0·338 per cent of iron so that, if it is assumed that each molecule contains one atom of iron, the molecular weight must be about 16 750. However, when the molecular weight is determined by the ultracentrifuge, a value of 67 000 is found and it therefore appears that the haemoglobin molecule contains four atoms of iron, being composed of four units, each of molecular weight 16 750 and each containing one atom of iron.

Perutz and his colleagues using X-ray crystallography have shown that a haemoglobin molecule consists of four polypeptide chains each carrying a haem molecule (Fig. 25.1).

At least three physiological haemoglobins are known in man. Haemoglobin A comprises about 98 per cent of the total adult haemoglobin in the average human being and haemoglobin A_2 the remaining 2 per cent, but the proportion of A_2 is either raised or lowered in certain abnormalities of haemoglobin production. Haemoglobin F forms more than half of the haemoglobin in the human fetus and the newborn child but disappears almost completely during infancy, its place being taken by haemoglobin A.

The various forms of human haemoglobin all contain two identical α polypeptide chains, but differ in the other polypeptide chains. Thus adult haemoglobin contains two β chains, the A_2 form two δ chains, haemoglobin F two γ chains. Consequently adult human haemoglobin A is frequently designated $\alpha_2{}^A\beta_2{}^A$, the A_2 form as $\alpha_2{}^A\delta_2{}^{A_2}$ and haemoglobin F as $\alpha_2{}^A\gamma_2{}^F$.

These different forms can be separated by electrophoresis on paper and the differences in amino acid content are revealed by partial hydrolysis to peptides which are separated by chromatography on paper (see 'fingerprinting', p. 30). For example fetal haemoglobin (HbF)

(a)

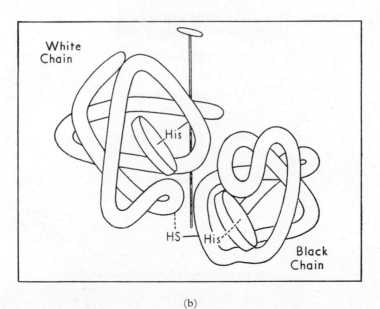

(b)

Fig. 25.1 (a) A model of the haemoglobin molecule. The haem groups, of which two are visible, are indicated by grey disks. (b) Diagrammatic representation of the general configuration in the two sub-units facing the observer. The haem groups are linked to histidine residues (His). HS—represents the sulphydryl group of cysteine. (M. F. Perutz, M. G. Rossman, A. F. Cullis, H. Muirhead, G. Will & A. C. T. North (1960). *Nature, London* **185**, 416–422.)

can be distinguished from ordinary adult haemoglobin (HbA) both by its electrophoretic mobility and by its amino acid composition.

The α chain of haemoglobin contains 141 amino acid residues while the β, γ and δ chains all contain 146 amino acid residues. A molecule of haem is attached to each chain. The complete amino acid sequence in the α and β chains of haemoglobin A has now been established (Fig. 25.2a and b). The amino acids are

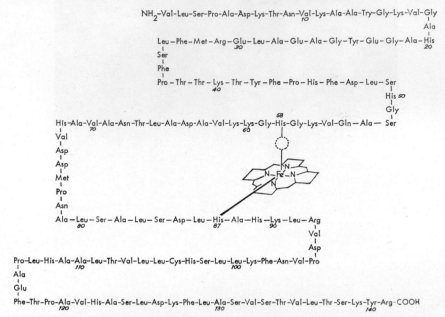

Fig. 25.2a   The α chain of human haemoglobin showing the haem moiety suspended between histidine residues numbers 58 and 87.

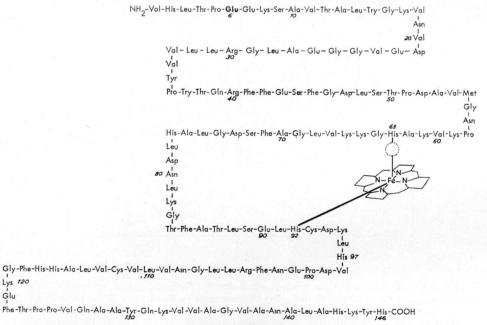

Fig. 25.2b   The β chain of human haemoglobin showing the haem moiety suspended between histidine residues numbers 63 and 92.

arranged in the form of an α-helix interrupted by regions which are non-helical. Proline residues, which do not easily fit into the α-helix structure, are commonly found at the ends of each helical region.

The haem molecule is suspended between two residues of histidine with its iron atom directly linked to one of the histidine residues (number 87 in the α chain and number 92 in the β chain). The other histidine residue involved (number 58 in the α chain and number 63 in the β chain) is associated also with the iron atom, but there is a gap between them into which the oxygen molecules appear to be introduced when haemoglobin becomes oxygenated.

Some of the haemoglobin variants are of considerable medical importance. One of the best known, sickle cell haemoglobin (HbS), is common in certain parts of the world. In West Africa for instance it is found in as many as 23 per cent of the population. The life span of HbS and other unusual haemoglobins is shorter than normal (p. 420) and consequently some of these people may be anaemic. The general term for such conditions is haemoglobinopathy.

In HbA position number 6 of the β chain is occupied by glutamic acid; in HbS by valine; in HbC, another haemoglobin variant, by lysine. The molecules are otherwise identical. The amino terminal portions of these β chains are as follows:

oxygen to form an easily dissociated compound known as oxyhaemoglobin (HbO_2). When haemoglobin is exposed to oxygen or air, it takes up two atoms of oxygen for each atom of iron. This combination with oxygen is not an oxidation; the oxidation product of haemoglobin, ferrihaemoglobin (methaemoglobin), is quite distinctly different from oxyhaemoglobin, which is simply a loose complex between haemoglobin and oxygen without any change in valency; the iron is considered to be in the ferrous state in both haemoglobin and oxyhaemoglobin. Oxyhaemoglobin gives up its oxygen very readily when blood is exposed to low oxygen tension:

$$HbO_2 \rightleftharpoons Hb + O_2$$

The ease with which haemoglobin unites with oxygen and with which the oxyhaemoglobin gives up its oxygen again is, as we shall see later, of the utmost importance in the carriage of oxygen by the blood (Chap. 31).

When the four polypeptide chains of haemoglobin A are put together to form the complete molecule, a large area of contact exists between the α and β chains, but there is little contact between one α chain and the other α chain, and between one β chain and the other β chain. When the haemoglobin becomes oxygenated, however, the amino end of one polypeptide chain comes into contact with the carboxyl end of its identical twin and when the haemoglobin is reduced again the complete molecule ex-

HbA ($\alpha_2{}^A\beta_2{}^A$) NH$_2$-Val-His-Leu-Thr-Pro-Glu-Glu-Lys . . .
HbS ($\alpha_2{}^A\beta_2{}^S$) NH$_2$-Val-His-Leu-Thr-Pro-Val-Glu-Lys . . .
HbC ($\alpha_2{}^A\beta_2{}^C$) NH$_2$-Val-His-Leu-Thr-Pro-Lys-Glu-Lys . . .

Since glutamic acid is a dicarboxylic acid and lysine is a diamino acid while valine has no additional acidic or basic group, the three amino acids carry different charges and the polypeptides containing them differ greatly in electrophoretic mobility.

Many other slight variations in the amino acid content (usually in the β chain) occur and give rise to many distinct varieties of haemoglobin. All of these variations—and indeed HbA itself—are genetically determined. Such variations are due to small differences in the composition of the DNA in the nucleus.

The most important property of haemoglobin (Hb) is its power of combining with

pands and the contact is released. The shape of the molecule therefore varies according to the state of oxygenation.

The quaternary structure of haemoglobin is consequently of great functional importance and is responsible for the sigmoid shape of the oxygen dissociation curve (Chap. 31). Haemoglobin is, in fact, a protein which exhibits allosteric changes (p. 114), as has been shown by Perutz with the aid of molecular models. When the haemoglobin molecule expands on losing oxygen, certain imidazole groups on histidine residues and also the terminal amino groups on the β-chains become more exposed. Conversely when the haemoglobin contracts on oxygenation these groups again become masked. It has been suggested that the degree of ionization of certain of these groups depends on whether they are exposed or covered, that is on whether the haemoglobin is reduced or oxygenated. Such a mechanism could account for the oxygen-linked buffering of haemoglobin (Chap. 31) and may also be responsible for the greater ability of reduced haemoglobin to carry carbon dioxide bound to the terminal amino groups as carbamino compounds (Chap. 31).

Haemoglobin can combine with gases other values obtained for the concentration of haemoglobin in the blood of normal healthy adults. In Britain, the normal range for adult males may be taken as 14 to 17 g per 100 ml of blood with a mean at 15·6 g. In the female the range is from 12 to 15·5 g per 100 ml with a mean at 13·7 g.

The mean corpuscular haemoglobin concentration, that is the mean concentration of haemoglobin per 100 ml of cells, is an index that is independent of the size of the red cells and is therefore a true expression of their haemoglobin load:

$$\text{M.C.H.C.} = \frac{\text{Haemoglobin concentration (g per 100 ml)}}{\text{Vol. of red cells in 100 ml of blood (haematocrit)}} \times 100$$

than oxygen to form addition compounds. For example, it combines reversibly with carbon monoxide forming carboxyhaemoglobin (COHb):

$$Hb + CO \rightleftharpoons COHb$$

Not only is this compound more stable than oxyhaemoglobin but the affinity of haemoglobin for carbon monoxide is some 250 times greater than the affinity of haemoglobin for oxygen.

The breathing of carbon monoxide (from coal gas) is therefore dangerous because the carboxyhaemoglobin deprives the blood of its ability to transport oxygen and the patient may die from anoxia. The carbon monoxide may be displaced from the haemoglobin by making the patient breathe oxygen and carbon dioxide (to stimulate respiration, Chap. 30). Since the rate at which oxygen displaces carbon monoxide is proportional to the pressure of oxygen, the patient may recover more quickly if he is put into a high pressure oxygen chamber (hyperbaric oxygen).

Up to 5 per cent of the haemoglobin in the blood of heavy tobacco smokers may be in the form of carboxyhaemoglobin.

METHODS OF ESTIMATION OF HAEMOGLOBIN

The concentration of haemoglobin in a sample of blood can be estimated by measuring its iron content or its oxygen capacity (Chap. 31) but much simpler colorimetric methods have been invented for clinical use.

There is considerable variation between the

Normal values lie between 32 and 36 g per 100 ml. Figures below this range indicate that the stroma of the red blood cells is not carrying its full quota of haemoglobin and strongly suggest that the patient will benefit from the administration of iron.

PIGMENTS RELATED TO HAEMOGLOBIN

METHAEMOGLOBIN. When haemoglobin is treated with mild oxidizing agents such as potassium chlorate or potassium ferricyanide it becomes oxidized to a stable oxidation product known as methaemoglobin (or ferrihaemoglobin). In this process a negative ion such as hydroxide is taken up to balance the extra negative charge on the iron atom:

$$HbO_2 + K_3Fe(CN)_6 + KOH \rightarrow$$
$$(Fe^{2+})$$
$$\qquad\qquad HbOH + K_4Fe(CN)_6 + O_2$$
$$\qquad\qquad (Fe^{3+})$$

Since the loosely combined molecular oxygen in HbO_2 is driven off, this reaction with ferricyanide is used to determine the oxygen capacity of blood (Chap. 31). Whereas oxyhaemoglobin is merely a loose association of haemoglobin with molecular oxygen, methaemoglobin is a true oxidation product. In both haemoglobin and oxyhaemoglobin iron is present in the ferrous state, while in methaemoglobin iron is in the ferric state. Methaemoglobin is occasionally found in the blood after the administration of preparations containing phenacetin and certain other drugs, and rarely as a familial anomaly. The condition is known as *methaemoglobinaemia*. From 2 to 12 per cent

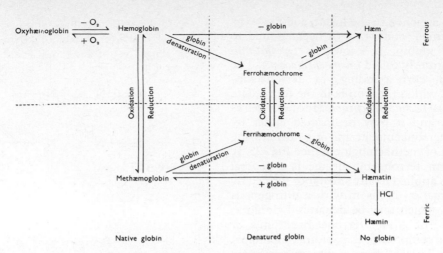

Fig. 25.3 Relationships of the blood pigments.

of the haemoglobin in the blood of healthy persons may be present in an 'inactive' form, probably as methaemoglobin. Methaemoglobin may be reduced *in vitro* to haemoglobin by strong reducing agents such as sodium hydrosulphite ($Na_2S_2O_4$) (Fig. 25.3). Since methaemoglobin can be reduced to haemoglobin *in vivo* by ascorbic acid this substance is useful in the treatment of methaemoglobinaemia.

Another abnormal pigment occasionally found in the blood is *methaemalbumin* in which the haem is combined with plasma albumin. Its presence in the plasma may be an important indication of a haemolytic process.

Sulphaemoglobin may be found in the blood after administration of certain drugs. It is a stable compound which, unlike methaemoglobin, cannot be reduced to haemoglobin. Sulphaemoglobin, once formed, remains as such in the red cells during their life span.

Haemoglobin and its derivatives may readily be distinguished from one another by means of their absorption spectra (Plate 25.1).

HAEM AND HAEMATIN. Haemoglobin, as already mentioned, consists of the pigment haem bound to the protein globin. Haem, like haemoglobin, contains iron in the ferrous state and it is easily oxidized to the corresponding compound *haematin* which contains ferric iron. Haematin reacts readily with hydrochloric acid to form a hydrochloride which is known as *haemin*. Haematin can be obtained not only from haemoglobin but also from other bio-

logical materials such as myoglobin (muscle haemoglobin) (p. 114), the enzymes catalase and peroxidase (p. 137), and from cytochromes (p. 129).

The haemin molecule is composed of four pyrrole rings, united by four CH groups to form a flat closed ring system. The iron atom occupies the centre of this large ring and is united to the nitrogen atoms of the pyrrole rings (Fig. 25.4).

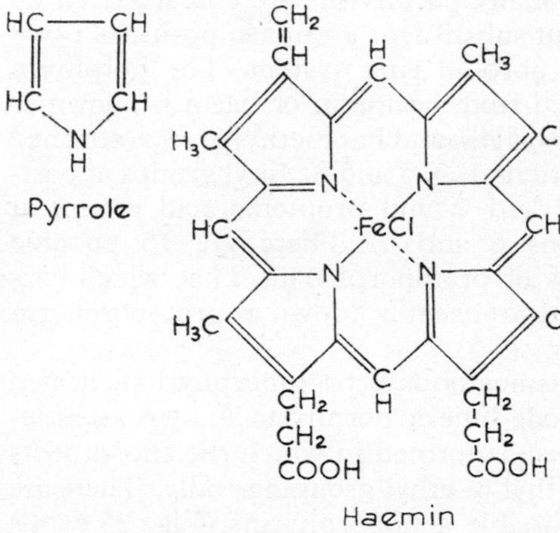

Fig. 25.4 The carbon atoms of the pyrrole rings carry substituent groups which are either methyl (CH_3—), vinyl (CH_2=CH—), or propionic acid ($HOOC.CH_2$. CH_2—) groupings.

FERROHAEMOCHROME. When haemoglobin is treated with alkali the protein is denatured but

Blood Pigments 449

it remains attached to the haem, and the compound formed, consisting of haem united to denatured globin, is known as *ferrohaemochrome* (formerly called *haemochromogen*). The compound corresponding to ferrohaemochrome in which the iron is in the ferric state is known as *ferrihaemochrome* (formerly *parahaematin*), and it may be formed either by oxidizing ferrohaemochrome or by denaturing the globin of methaemoglobin (see Fig. 25.3).

The term ferrohaemochrome is a general one, and is applied to any substance formed by the combination of haem with a nitrogenous substance, which may be denatured globin as in the case of globin ferrohaemochrome, or some such compound as pyridine, ammonia, or glycine. They all have a characteristic type of absorption spectrum. The cytochromes (p. 129) have this type of absorption spectrum and are considered to have structural features in common with these artificial ferrohaemochromes.

PORPHYRINS. When iron is removed from either haem or haematin a ring system is left composed of four pyrrole nuclei joined together by the =CH—bridges. A compound of this sort is known as a *porphyrin* and all porphyrins are derivatives of the parent substance *porphin*, the full formula of which is shown in Figure 25.5.

The many porphyrins are characterized by different substituent groups in positions 1 to 8 in the porphin ring system. The porphyrin obtained from haematin or haem is known as *protoporphyrin* and has methyl groups attached in positions 1, 3, 5 and 8, vinyl groups at positions 2 and 4, and propionic acid groups at positions 6 and 7. There are 15 possible isomers of protoporphyrin. That which corresponds to haem is known as protoporphyrin IX (Fig. 25.5).

If haemoglobin or protoporphyrin is heated with soda-lime a porphyrin known as *aetioporphyrin* is formed in which the substituents are methyl or ethyl groupings only. There are four possible aetioporphyrins (Fig. 25.6). In aetioporphyrin of type I the methyl groups are in positions 1, 3, 5 and 7; in type II the methyl groups are in positions 1, 4, 5 and 8; in type III they are in positions 1, 3, 5 and 8; and in type IV they are in positions 1, 4, 6 and 7. The aetioporphyrin derived from protoporphyrin belongs to type III. The naturally occurring

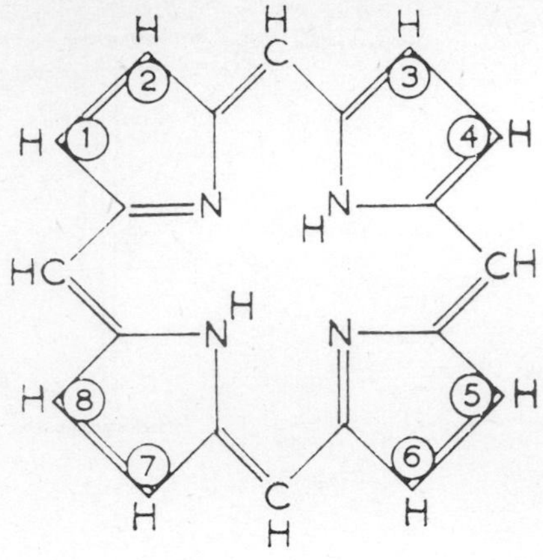

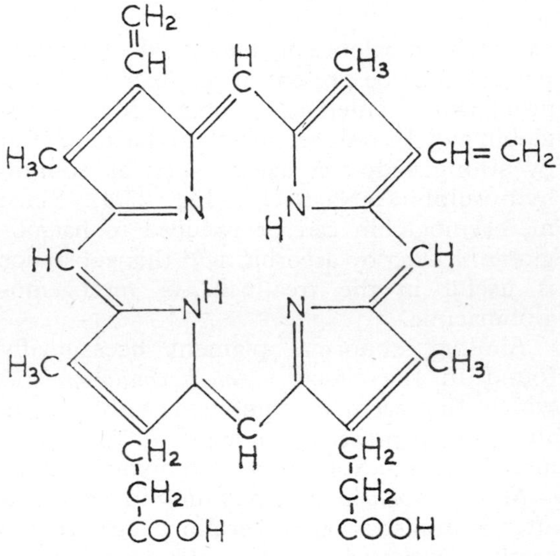

Porphin

Protoporphyrin IX

Fig. 25.5.

porphyrins all belong to type I or to type III.

Chlorophyll is a magnesium porphyrin in which complex substituent groups are attached in positions 1 to 8, but on distillation with soda-lime chlorophyll, like haematin, yields aetioporphyrin III.

THE BIOSYNTHESIS OF HAEM. The porphyrin ring is synthesized in the body from glycine which reacts with succinyl coenzyme A, produced in the tricarboxylic acid cycle, to form α-amino-β-oxo-adipic acid which loses carbon dioxide under the influence of the enzyme

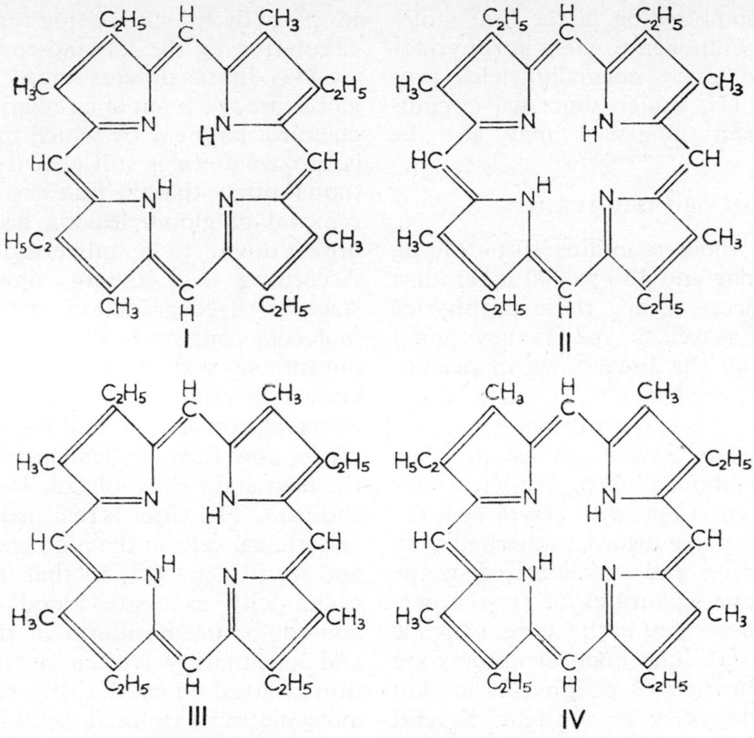

Fig. 25.6 The four possible aetioporphyrins.

δ-*aminolaevulinic acid synthetase* (*ALA synthetase*) to form δ-aminolaevulinic acid (ALA). Pyridoxal phosphate is required as coenzyme.

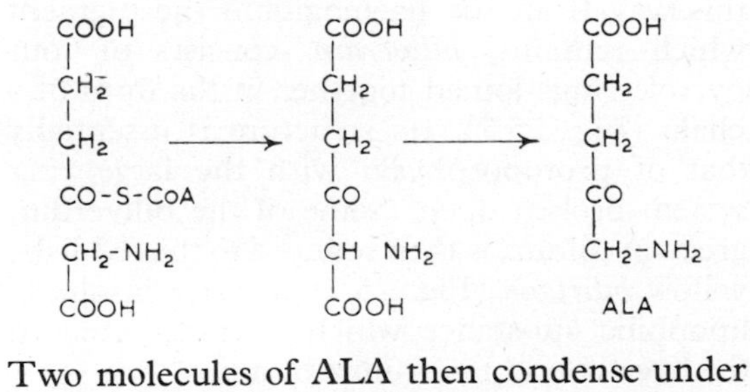

```
COOH            COOH            COOH
 |               |               |
CH₂              CH₂             CH₂
 |               |               |
CH₂        →     CH₂      →      CH₂
 |               |               |
CO-S-CoA         CO              CO
 |               |               |
CH₂-NH₂          CH  NH₂         CH₂-NH₂
 |               |
COOH             COOH           ALA
```

Two molecules of ALA then condense under the influence of δ-*aminolaevulinate dehydratase* (*porphobilinogen synthase*) to yield the important pyrrole intermediate *porphobilinogen*:

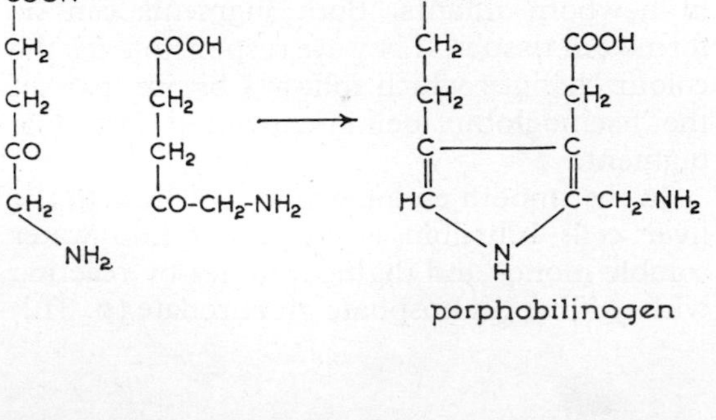

```
COOH                                COOH
 |                                   |
CH₂           COOH                  CH₂        COOH
 |             |                     |          |
CH₂           CH₂                   CH₂        CH₂
 |             |        →            |          |
CO            CH₂                    C          C
 |             |                    / \\       / \\
CH₂           CO-CH₂-NH₂         HC     C-CH₂-NH₂
  \                                      
   NH₂                              N
                                    |
                                    H
               porphobilinogen
```

Four molecules of porphobilinogen under the influence of two enzymes, porphobilinogen deaminase and an isomerase, now combine to yield the cyclic derivative *uroporphyrinogen* III which then undergoes decarboxylation to *coproporphyrinogen* III. This in turn is decarboxylated and dehydrogenated to *protoporphyrin* IX (Fig. 25.5). Under the influence of the enzyme *ferrochelatase* an atom of iron is introduced into this protoporphyrin to yield haem.

By side reactions the porphyrinogens may be dehydrogenated to yield the corresponding porphyrins. The process may be summarized as follows:

4 porphobilinogen

↓

uroporphyrinogen III → uroporphyrin III

↓

coproporphyrinogen III → coproporphyrin III

↓

protoporphyrin IX

↓ + Fe²⁺

haem

Although the combination of the four molecules of porphobilinogen to yield porphyrinogen referred to above normally yields uroporphyrinogen III, under abnormal circumstances uroporphyrinogen I may also be formed.

THE EXCRETION OF PORPHYRINS

The urine normally contains 10 to 200 μg porphyrin per day and 150 to 300 μg are lost daily in the faeces. Since these porphyrins belong to type I as well as type III they cannot all be formed by the breakdown of haemoglobin; those of type I may represent products of side reactions in haem synthesis.

The porphyrias. Several clinical disorders affecting the metabolic pathways of haem synthesis are known. *Congenital (erythropoietic) porphyria* is a very rare disorder inherited as an autosomal recessive and characterized by the excretion of large quantities of type I porphyrins in the faeces and in the urine, which is often coloured red. The main symptoms are due to the deposition of porphyrins in skin which is very sensitive to sunlight. Several other disorders are also associated with skin photosensitivity: the commonest is *porphyria cutanea tarda* in which there appears to be a defect in haem synthesis which becomes apparent only in patients who also have a defect in the hepatic excretion of porphyrins.

Another not uncommon inherited disorder is *acute intermittent porphyria* in which skin disease is not found but patients develop disorders of the nervous system such as peripheral neuropathy or episodic psychiatric disturbances, or may have recurrent abdominal pain. Acute attacks can be precipitated by barbiturates, sulphonamides and other drugs. This disorder is inherited as an autosomal dominant and is characterized by excessive urinary excretion of δ-aminolaevulinic acid and porphobilinogen. It is associated with an increase in the activity of the hepatic ALA-synthetase, the rate-controlling enzyme early in the biosynthetic chain.

Porphyrin metabolism may also be affected in liver disease, in haematological disorders and in lead poisoning, which affects, among other enzymes, aminolaevulinate dehydratase.

THE BREAKDOWN OF HAEMOGLOBIN

The red blood cells have a limited existence in the body (p. 420), being removed from the circulation by the reticulo-endothelial system (p. 434). In this process some 7 to 8 g of haemoglobin are taken out of circulation per day. The chemical pathway by which this haemoglobin is broken down is still obscure. It was at one time thought that the first step consisted in the removal of globin leaving haem which then broke down to a biliverdin-iron complex. According to Lemberg, however, the first stage in the degradation of the haemoglobin molecule consists in the opening of the porphyrin ring system. The resulting compound, known as *choleglobin (verdohaemoglobin)*, still contains iron and protein but the four pyrrole nuclei now form a chain instead of a ring. In the next stage the molecule loses both protein and iron. The latter is retained in the reticulo-endothelial cells in the pigments haemosiderin and ferritin (p. 425) so that in diseases associated with excessive blood destruction the amount of haemosiderin in the spleen, liver and bone marrow is often much increased. The iron derived from the original haemoglobin molecule must be used again in the synthesis of new haemoglobin (p. 451) since a daily turnover of 8 g Hb corresponds to 27 mg Fe whereas the daily loss from the body is estimated to be 1 mg (p. 424) in the adult male.

After iron and protein have been removed in this way from the haemoglobin the pigment which remains, *biliverdin*, consists of four pyrrole rings joined together in the form of a chain (Fig. 25.7). Its structure is essentially that of protoporphyrin with the large ring system broken open. Some of the biliverdin, green in colour, is then reduced to the reddish-yellow *bilirubin* (Fig. 25.7), a water insoluble lipophilic substance which is transported to the liver bound to serum albumin.

Some 300 mg of the two pigments biliverdin and bilirubin are excreted each day in bile. The latter predominates in human bile but the former is more abundant in the herbivora and in newborn infants. Both pigments can be formed in tissues. They are responsible for the colour changes which follow a bruise, part of the haemoglobin being converted into bile pigment.

In the smooth endoplasmic reticulum of the liver cells bilirubin is converted into water soluble mono- and diglucuronides by reacting with uridine diphosphate glucuronate (p. 312)

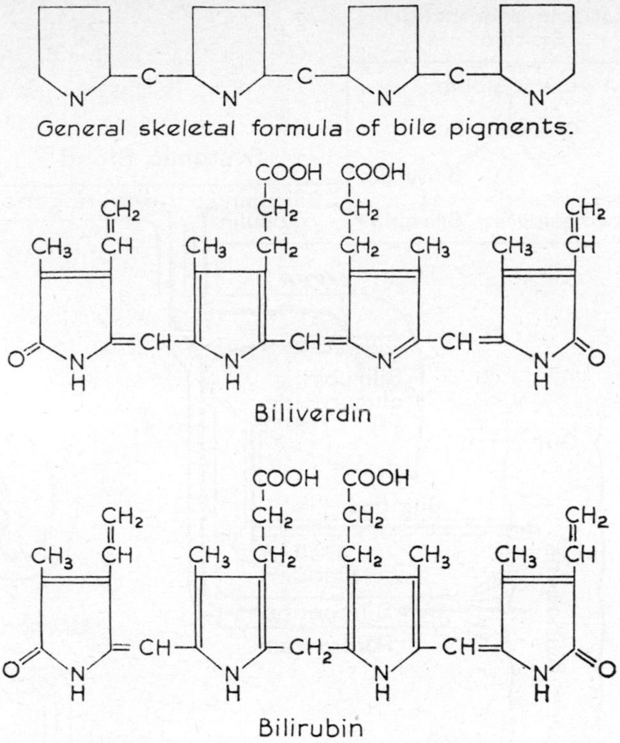

General skeletal formula of bile pigments.

Biliverdin

Bilirubin

Fig. 25.7 Structures of the bile pigments.

under the influence of a UDP *glucuronyl transferase*:

Bilirubin + UDP glucuronate →
    bilirubinglucuronide + UDP

Bilirubinglucuronide is excreted into the bile and passes into the intestine where it is hydrolysed and reduced by bacterial enzymes to a series of colourless compounds of which the most important are *mesobilirubinogen* and *stercobilinogen*. These, in turn, undergo autooxidation to a class of brown pigments of which the most important is *stercobilin*, the chief pigment of the faeces (Fig. 25.8). The faeces contain both stercobilinogen and stercobilin (100 to 200 mg per day) and the darkening of faeces exposed to air is due to the oxidation of stercobilinogen to stercobilin.

Experiments in which glycine labelled with ^{15}N has been ingested by human subjects have shown that some bilirubin is formed from sources other than worn out red blood corpuscles and that only about 70 per cent of the stercobilin of the faeces is formed by the breakdown of haemoglobin. Most of the remainder is probably produced by a rapid synthetic mechanism allied to that involved in haemoglobin biosynthesis.

Part of the stercobilinogen in the gut is reabsorbed and passes by the portal circulation to the liver where most of it is re-excreted either in the same or in a modified form, but a small amount reaches the systemic circulation and is excreted by the kidney as urobilinogens. When urine is exposed to air this bilinogen is oxidized to a bilin known as *urobilin* (Fig. 25.8). Normally the amount of urobilinogen excreted in the urine (0·5 to 2 mg per day) is too small to give strongly positive chemical tests but the quantity excreted is greatly increased in certain pathological conditions such as haemolytic jaundice (see below).

Blood serum normally contains up to 1 mg per 100 ml of bilirubin but not more than 0·2 mg per 100 ml is conjugated. Bilirubin is responsible for the faintly yellow colour of blood plasma and serum. The conjugated, water soluble, pigment gives an immediate red colour with diazotized sulphanilic acid whereas the unconjugated, lipid-soluble bilirubin does

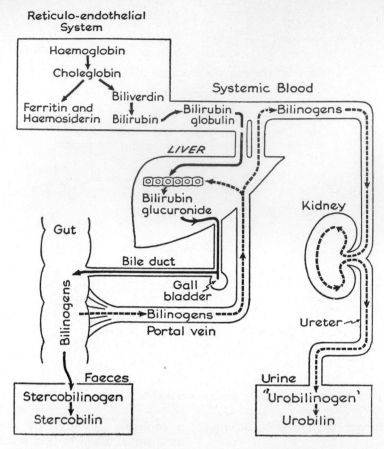

Fig. 25.8 A scheme of the changes involved in the formation of bile pigments.

so only after alcohol has been added (Van den Bergh reaction). This reaction thus serves to distinguish bile pigment which has, or has not, passed through the liver cells.

Unconjugated bilirubin, being bound to serum albumin, does not readily appear in the urine but the conjugated water soluble glucuronides are readily excreted through the kidneys. Simple tests for bile pigment in the urine of a jaundiced patient can thus distinguish *haemolytic* (sometimes called *pre-hepatic*) *jaundice*, due usually to excessive destruction of red cells, from other types of jaundice, namely *hepatic* in which the liver cells are diseased or *obstructive* (sometimes called *post-hepatic*) when the flow of bile from the liver is obstructed, for example by a gall-stone in the bile duct.

Jaundice is the result of an increased concentration of circulating bilirubin—usually more than 2 mg per 100 ml. The pigment,

some of which is bound to tissue protein, can be seen in the skin and, most easily in the

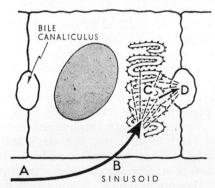

Fig. 25.9 Diagrammatic representation of the mechanisms which may result in jaundice. (A)—Following an increased load of bilirubin as the result of haemolysis. (B)—Defective uptake into the liver cell. (C)—Defective conjugation in the endoplasmic reticulum of the liver cell. (D)—Defective excretion into the bile canaliculi, or the result of a mechanical block in the bile duct system (after Sherlock).

454 Textbook of Physiology and Biochemistry

sclera of the eye which becomes tinged yellow.

In Fig. 25.9 are summarized the various ways in which jaundice can be produced. Although instances of each variety have been described in clinical practice some are very rare. The mechanism of one of the commonest causes of jaundice, namely a virus infection of the liver, is not exactly known. Another common form of jaundice is due to obstruction of the bile duct outside the liver by, for example, inflammation or gall-stones.

In haemolytic disease of the new-born (p. 412) the serum bilirubin may reach 30 mg per 100 ml or more and bile pigment is deposited in the basal nuclei of the brain which suffer permanent damage (kernicterus).

REFERENCES

BOUCHIER, I. A. D. & BILLING, B. H. (1967). *Bilirubin Metabolism.* Oxford: Blackwell.

DEAN, G. (1971). *The Porphyrias*, 2nd edn. Bristol: Pitman.

GERALD, P. S. & INGRAM, V. M. (1961). Recommendations for the nomenclature of haemoglobins. *Journal of Biological Chemistry* **236**, 2155–2156.

GOLDBERG, A. (1971). *Recent Advances in Haematology* (Edited by A. Goldberg & M. L. Brain), p. 302. Edinburgh: Livingstone.

GOLDBERG, A. & RIMINGTON, C. (1962). *Diseases of Porphyrin Metabolism.* Springfield, Ill.: Thomas.

GOODWIN, T. W. (Ed.) (1968). *Porphyrins and Related Compounds Biochemical Society Symposia, No. N28*, London: Academic Press.

GRAY, C. H. (1961). *Bile Pigments in Health and Disease.* Springfield, Ill.: Thomas.

INGRAM, V. M. (1959). Chemistry of the abnormal human haemoglobins. *British Medical Bulletin* **15**, 27–32.

INGRAM, V. M. (1960). *Hemoglobin and its Abnormalities.* Springfield, Ill.: Thomas.

INGRAM, V. M. (1963). *The Haemoglobins in Genetics and Evolution.* New York: Columbia University Press.

INGRAM, V. M. (1965–66). On the Biosynthesis of Haemoglobin. *Harvey Lectures*, Series 61, 43–64.

LEHMANN, H. (1959). Variations in human haemoglobin synthesis and factors governing their inheritance. *British Medical Bulletin* **15**, 40–46.

LEHMANN, H. & HUNTSMAN, R. G. (1966). *Man's Haemoglobins.* Amsterdam: North Holland Publishing Company.

LEMBERG, R. & LEGGE, J. W. (1949). *Haematin Compounds and Bile Pigments: Their Constitution, Metabolism, and Function.* New York: Interscience.

MACALPINE, I., HUNTER, R. & RIMINGTON, C. (1968). Porphyria in the Royal Houses of Stuart, Hanover and Prussia. *British Medical Journal* **i**, 7–18.

PERUTZ, M. F. (1969). The haemoglobin molecule. *Proceedings of the Royal Society* B, **173**, 113–140.

PERUTZ, M. F., ROSSMANN, M. G., CULLIS, A. F., MUIRHEAD, H., WILL, G. & NORTH, A. C. T. (1960). The structure of haemoglobin. *Nature (London)* **185**, 416–422.

RIMINGTON, C. (1959). The biosynthesis of haemoglobin. *British Medical Bulletin* **15**, 19–26.

SHERLOCK, S. (1968). *Diseases of the Liver and Biliary System.* Oxford: Blackwell.

STEWART, C. P. (Ed.) (1961). *The Formation and Breakdown of Haemoglobin.* Amsterdam: Elsevier.

THOMSON, R. H. S. & WOOTTON, I. D. P. (1970). *Biochemical Disorders in Human Disease*, 3rd edn. London: Churchill.

WHITE, J. C. & BEAVEN, G. H. (1959). Foetal haemoglobin. *British Medical Bulletin* **15**, 33–39.

WITH, T. K. (1968). *Bile Pigments.* New York: Academic Press.

26 The Heart

William Harvey's great discovery of the circulation of the blood is the basis of modern physiology. The circulation is the main transport system of the body, supplying oxygen, heat, metabolic fuels, hormones and vitamins to every cell and removing their metabolic end-products, all in accordance with the needs of the individual cells.

Apart from introducing the revolutionary idea of the circulation of the blood, Harvey's work was important in another and wider sense: the methods he used were essentially those of modern science—verifiable physiological experiment yielding qualitative and quantitative data and the subsequent use of these data, by inductive reasoning and suitable mathematical treatment, to develop a hypothesis.

The circulation of the blood through the body has so long been accepted as a fact of everyday life that Harvey's original observations are often forgotten. His experiments were first recorded in his book *De motu cordis*, published in 1628. His main observations are given here in modern terminology.

When the heart is held in the hand it becomes harder and smaller as it contracts just as does a muscle of the arm. Because the contraction of the heart occurs a very short time before the expansion of the arteries it is likely that the contraction of the heart is responsible for the arterial pulse. When an artery is cut the blood gushes out from the end nearer the heart; the force of the gush is increased at each contraction of the ventricles so that the blood appears to leave the artery in spurts. Once the blood has passed into either of the great arteries it cannot flow backward into the heart because of the semilunar valves, as Leonardo da Vinci had demonstrated. The flow into the arteries must, therefore, always be from the heart to the periphery.

A simple calculation based on this drives us inevitably to the idea of a circulation. Suppose

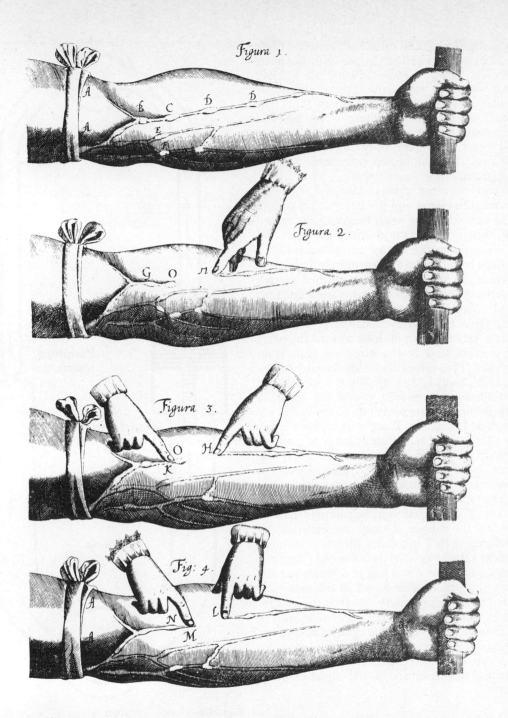

Fig. 26.1 Illustrations from Harvey's *De motu cordis* (Frankfurt: 1628). A, tight bandage above the elbow. B, C, D, E, F, valves in the veins. If the vein is milked down with the finger (Fig. 2, O to H) no blood is drawn through the valve O. Now if the vein is milked from H to O (Fig. 3) the vein becomes swollen above the valve O but O to H remains empty. If one finger is put at L (Fig. 4) another finger M can milk the vein towards and beyond the valve N; the part N to L remains empty, the blood being unable to pass back through this valve. When the finger L is removed the vein fills up from below. This emptying of the vein and filling from L towards N can be repeated any number of times.

the ventricle ejects 60 ml at each beat and there are 72 beats per minute. In one hour the heart will eject $60 \times 72 \times 60 = 259\,200$ ml $= 259 \cdot 2$ litres which is more than three times the volume of a man and very many times the volume of his blood. There is no possible mechanism for supplying this large volume of fluid every hour—the only way such a large flow could be maintained is by some form of circulation.

When the vena cava is clamped, the vein between the clamp and the heart collapses and at the same time the heart becomes empty and pale. When the aorta is clamped the heart and the aorta up to the clamp are distended. If a bandage is put around the arm tight enough to obstruct the superficial veins but not enough to obstruct the arteries the veins swell up. This blood cannot come from the veins; it must come from the arteries. Moreover the veins contain valves which all look toward the heart and prevent flow in the direction away from the heart. This can easily be shown in the fore-arm (Fig. 26.1; 1, 2, 3, 4). When a moderately tight bandage is placed round the arm the veins fill up and the positions of the valves are recognized by slight dilatations at short intervals along the veins. If an attempt is made to massage the blood away from the heart the valves stand out more prominently but the blood cannot be impelled backward unless the valves are forcibly ruptured. Further, if a portion of distended vein is compressed distally by the finger it may be emptied by massaging it in a proximal direction towards the next valve (Fig. 26.1; 4). The vein does not fill again until the distal finger is removed; this shows that the flow of blood in the vein proceeds from the periphery towards the centre. Harvey also noticed the difference in the thickness of the walls of the arteries and veins; the greater thickness of the arteries was, he supposed, to sustain the shock of the impelling heart and to accommodate the blood expelled from it.

THE HEART AND CIRCULATION

The circulation includes a muscular organ, the heart, that pumps the blood through the arterial system to perfuse all the tissues of the body and thence back through the veins to the heart. The arterial side of this system has a

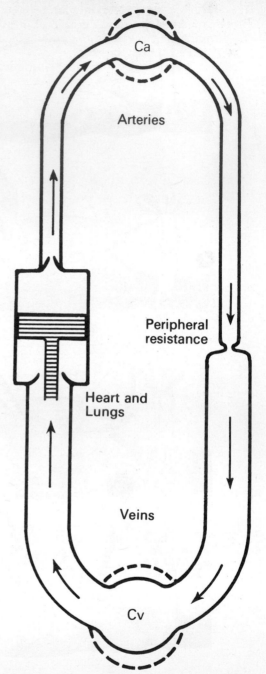

Fig. 26.2 Greatly simplified schema of the circulation of the blood indicating some important features. The pump represents both heart and lungs, as in the 'pump-oxygenators' used during surgical operations on the empty heart ('cardio-pulmonary bypass'). Ca indicates capitance of the arteries, Cv that of the veins. The system contains about 5 litres of blood. If the pump were to stop, the pressures in the system would theoretically equalize at about $+7$ mm Hg (see p. 506).

certain variable capacitance, but on the venous side the capacitance is much greater. During its flow round the circulation the blood meets a peripheral resistance (Fig. 26.2).

This schema provides a greatly simplified description, there being in fact two pumps and two main circulations arranged in series. The right ventricle pumps venous blood through the 'lesser' or pulmonary circulation to the lungs where gas exchange takes place at the alveolo-capillary surface, and whence the re-oxygenated blood returns to the left atrium. The left ventricle pumps the oxygenated blood by way of the many divisions of the 'greater' or systemic circulation (Fig. 26.3) to all parts of the body, whence it returns to the right atrium through the veins.

Harvey had only a vague notion of the manner of the flow of the blood between arteries and veins through the 'pores of the flesh'. Malpighi of Bologna about 1661 was the first to see the circulation through the capillaries in the lung of the frog. He described how the blood is forced from the arteries into a network of capillaries and then passes into the receiving branches of the veins.

The circulation of the blood is particularly easily observed in the vessels of the mesentery, where it may be seen that the blood in a small artery is flowing in one direction, whereas in the companion vein it flows in the opposite direction.

THE STRUCTURE OF THE HEART

The size of the human heart is quite variable (Table 26.4). It depends on the size of the body and on the sex of the subject, the heart in the female weighing as a rule about 75 per cent of that in the male.

The mammalian heart possesses two *atria* separated by a thin *interatrial septum*, and two *ventricles* separated by the thicker *interventricular septum*. All four chambers lie at approximately the same level in the human thorax (Fig. 26.5). Blood flows from each atrium into the immediately adjacent part of the ventricle, the *inflow tract*, and passes out via the *outflow tract* into the aorta or pulmonary artery: the inflow tracts have raised muscular ridges on their walls whereas the outflow tracts tend to be smooth-walled. The whole muscular structure, together with a certain amount of *epi-*

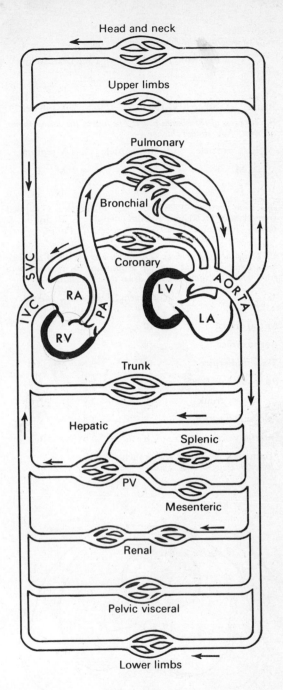

Fig. 26.3 Diagram to show the heart and circulation displayed as two pumps ('left' and 'right' hearts) and two circulations (systemic and pulmonary) arranged in series. Various important divisions of the systemic circulation are also indicated. Blood flows from arteries through capillary beds to veins. The renal circulation has two capillary beds, glomerular and tubular. PV = portal vein.

Table 26.4

A. *The weight of the human heart in grams*

	Muscle mass		Total heart weight	
	M.	F.	M.	F.
Mean	254	186	328	244
Range	207 to 319	153 to 217	256 to 390	198 to 279

B. *Thickness of the human heart. The figures are means in mm at point of maximum thickness*

Left ventricle		Right ventricle		I-V septum	
M.	F.	M.	F.	M.	F.
15	14	4	4	16	14

Based on a study of 26 males and 19 females. Compiled from data of L. Reiner, A. Mazzoleni, F. L. Rodriquez & R. R. Freudenthal (1959). *Archives of Pathology and Laboratory Medicine* **68**, 58–73.

C. *Dimensions of orifices. The heart and vessels of a young male adult were laid out flat to measure the circumference from which the diameters and areas were calculated.*

	Circumference (cm)	Diameter (cm)	Area (cm^2)
Mitral orifice	10	3·2	8
Aortic orifice	8	2·5	6
Tricuspid orifice	12	3·8	11·4
Pulmonary orifice	9	2·9	6·5
Pulmonary trunk	9	2·9	6·5
Abdominal aorta*	8	2·5	6
Inferior vena cava*	8	2·5	6

*Immediately above renal branches.
(*By courtesy of W. W. Park.*)

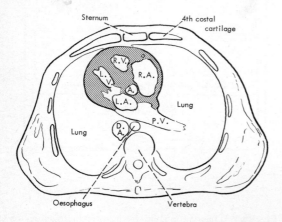

Fig. 26.5 Cross-section through the thorax at the level of the fourth costal cartilage. The atria lie to the right of the ventricles and behind (not above) the ventricles. The interventricular septum lies at an angle of about 45° from the autero-posterior axis of the thorax. (*After* R. Walmsley (1958). *British Heart Journal* **20**, 441.)

cardial fat, is enclosed with a serous membrane, the *serous pericardium* (p. 462).

The walls of the atria are thin, corresponding to the small amount of work required to force their contents into the ventricles. The ventricles do most of the work of the heart, expelling the blood into a system of branching tubes (arteries) which offer considerable resistance to the onflow of the blood, and accordingly the walls of the ventricles are much thicker than those of the atria. The wall of the left ventricle 2·5 cm below the atrio-ventricular ring is about 1·5 cm thick. This is three or four times thicker than the wall of the right ventricle at the corresponding position (Table 26.4B) and is related to the greater amount of work required of the left ventricle.

The volume of the human heart during life is some 700 ml at the end of diastole whereas the actual volume of muscle is about 300 ml. Thus the cavities may contain roughly 400 ml of blood, an amount that is much greater than the quantity expelled by both ventricles each time they contract (about 140 ml). In other words the cavities are not completely empty even at the end of systole; the ratio end-systolic volume/end-diastolic volume is about 0.55. (The rounded off figures used above give a ratio of 0·65.) The left ventricular residual volume at the end of systole in untrained persons is estimated to be about 50 ml and in endurance athletes about 120 ml. These values have been obtained by measurement of X-ray cinephotographs taken of living persons while the heart contained a concentrated solution of an iodine-containing compound and also by calculation from the dye dilution method of measuring cardiac output (p. 503).

Each ventricle has the same capacity and each expels the same amount of blood at systole or contraction. If this were not so the circulation would eventually cease. Let us consider what would happen if the right ventricle were to deliver, say, 0·1 ml more blood per contraction than the left. If the left expels 60 ml per beat and the right 60·1 ml per beat then in one hour, if the heart beats 80 times a minute, the left expels 288 000 ml and the right 288 480 ml. The difference, 480 ml, would have to be accommodated in the pulmonary vessels and the lungs, with grave effects upon the pulmonary circulation and respiration. This does not mean, however, that for *short* periods of time the amounts of blood expelled by the two ventricles must be exactly the same; at certain phases of the respiratory cycle, for example, there are small, temporary differences between the stroke volumes of the right and left ventricles. In health, then, the average outputs must be equal. In heart failure a discrepancy between the two outputs has serious consequences. In some types of congenital heart deformities in which blood is shunted from one side of the circulation to the other there may be quite large differences between the ventricular outputs.

The atria and the ventricles are completely separated by a *fibrotendinous ring* which gives attachment to most of the muscular fibres and to the atrio-ventricular valves. The only muscular connexion between atria and ventricles is the atrio-ventricular (junctional) bundle of His. Closely connected with the main fibrous ring are other fibrous rings that surround the arterial orifices and serve for the attachment of the great vessels and semilunar valves and some of the ventricular muscle fibres. The arrangement of the cardiac muscle (the *myocardium*) in both atria and ventricles is complicated and includes the *papillary muscles* that project into the cavities of the ventricles, and to which the chordae tendineae are attached.

Heart valves. The heart valves (Figs 26.6 and 7) are formed by cusps of fibrous tissue attached to the fibrous rings and covered on both sides by endocardium; they have no blood vessels in health except perhaps at the bases of the A.V. valves. All the valves are so constructed that they permit the blood to flow only from the atria to the ventricles and onward into the arteries. The points of entry of the veins into the atria are not provided with functional valves. Back-flow of blood is known to take place, however, in the coronary sinus in man.

The *atrioventricular valves* are large, sail-like structures that arise from the atrioventricular rings and pass down into the cavity of the ventricles (Fig. 26.6). Their margins are connected by thin cords (*chordae tendineae*) to the papillary muscles that form part of the muscular mass of the ventricles. As the ventricles contract the papillary muscles pull on the chord tendineae so that the flaps (cusps) of the valves cannot be pushed back into the atria. On the left side, between the left atrium and left ventricle, two cusps form the mitral valve, closure being brought about mainly by the more mobile anterior cusp. On the right

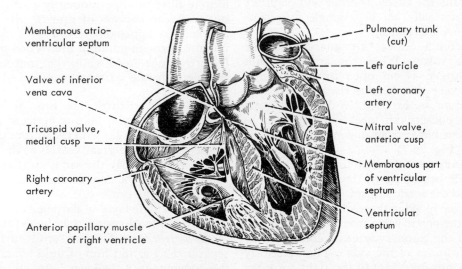

Membranous atrio-ventricular septum

Valve of inferior vena cava

Tricuspid valve, medial cusp

Right coronary artery

Anterior papillary muscle of right ventricle

Pulmonary trunk (cut)

Left auricle

Left coronary artery

Mitral valve, anterior cusp

Membranous part of ventricular septum

Ventricular septum

Fig. 26.6 Section through the human heart to show the valves. On the right side the atrioventricular orifice has been cut across. On the left side the section passes through the aortic orifice and in front of the atrio-ventricular orifice. The AV valve systems include cusps, chordae and papillary muscles. The cusps make the open valves funnel-shaped (After *Gray's Anatomy*, 31st ed. (1954), p. 694).

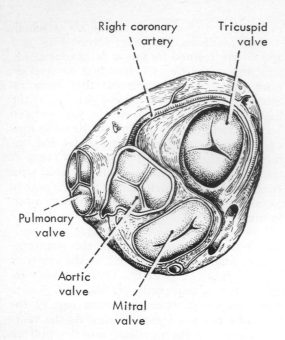

Right coronary artery

Tricuspid valve

Pulmonary valve

Aortic valve

Mitral valve

Fig. 26.7 The bases of the ventricles have been exposed by removal of the atria to show the heart valves (After *Gray's Anatomy*). The diagram has been oriented so that the antero-posterior direction is in the long axis of the page. The right coronary artery arises from the anteriorly placed aortic sinus (R. Walmsley.)

side the tricuspid valve has three cusps. Functionally the cusps, chordae and papillary muscles normally act as a single atrioventricular valve mechanism.

The arterial orifices are guarded by the semilunar valves. When the aorta, or pulmonary artery, is cut open three semilunar cusps of very thin fibrous tissue covered by endocardium are exposed. In the intact heart the three pockets can meet along three lines and prevent blood from flowing back into the ventricle (Fig. 26.7).

Coronary vessels. The three bulges of its wall at the beginning of the ascending aorta are termed the aortic sinuses. It is thought that eddies in the blood in these sinuses prevent the cusps of the aortic valve from obstructing the orifices of the coronary arteries, and may play a part in closure of the valve. The anteriorly situated sinus gives rise to the right coronary artery whereas the left coronary artery arises from the left posterior sinus. The right coronary artery most frequently supplies the right atrium, right ventricle and diaphrag-

matic surface of the left ventricle, while the left coronary artery supplies the greater part of the left heart. The coronary arteries and their larger branches lie in the sub-epicardium but from them smaller vessels pass into the substance of the myocardium where their branches form rich anastomosing intramyocardial plexuses of arterioles, capillaries and veins. From these arterioles, capillaries and venules arise luminal vessels which may open into any of the four chambers of the heart. From 30 up to 60 per cent of the coronary artery blood flow may be returned to the heart by these luminal vessels rather than by the veins leading to the coronary sinus. The luminal vessels may either drain or nourish the myocardium, according to the difference between the pressure in them and the pressure in the chambers of the heart. The coronary arterial tree is illustrated in Fig. 26.52.

The coronary arteries are not end-arteries; numerous arterial anastomoses of the order of 100 to 300 μm in diameter occur in all areas of the heart, especially in the deeper layers of the left ventricle and intraventricular septum. The capillary supply of heart muscle is about six times as rich as that of skeletal muscle, and greater in the inner than in the superficial parts of the myocardium.

In the human heart at birth the capillary to muscle-fibre ratio is frequently as high as 9:5 but with the process of growth this ratio is reduced to approximately 1:1 at which it remains throughout adult life. When the heart hypertrophies in disease the muscle fibres enlarge but the existing capillaries do not proliferate.

The pericardium. The heart is enclosed in a conical sac, the fibrous pericardium, the neck of which merges with the fibrous covering of the walls of the great vessels at the base of the heart; inferiorly the sac is attached to the central tendon of the diaphragm. Its inner surface and the outer surface of the heart are lined by a single layer of flattened cells resting on a layer of loose connective tissue. This is the *serous pericardium*. The two serous surfaces are separated by a thin film of fluid which allows free movement of the layers over one another.

THE SIZE OF THE HEART

In the adult human male the heart is about 0·43 per cent of the body weight and in the

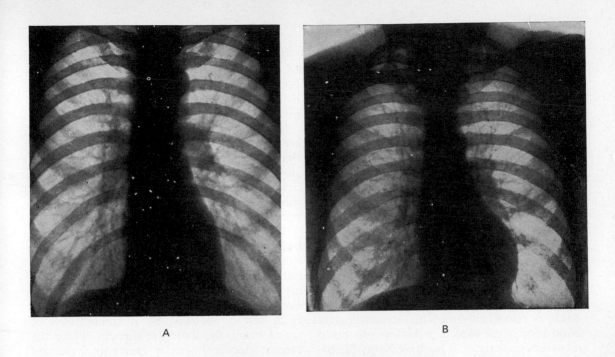

A   B

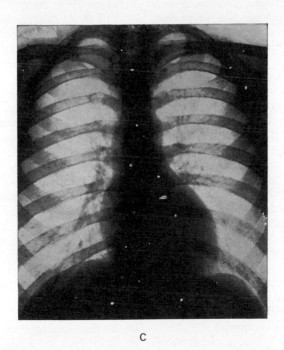

C

Fig. 26.8 X-rays of the chests of three healthy men of varying body build to show the extent of variation of the normal cardiac shadow. The subject of A was tall and slender, while the subject of C was short and stocky. All the photographs were taken as described on p. 464 with the subjects standing.

female about 0·4 per cent but in active animals like the deer or greyhound the heart is about 1 per cent of the body weight. Although at the bedside attempts may be made to estimate the size of the heart by palpation and percussion, exact knowledge of cardiac size can be derived only by X-rays. Since the X-ray picture is really the shadow of the heart is has to be remembered that the size of a shadow depends on the distance of the light source (in this case the X-ray tube) from the object, and also upon the distance from the object, that is the heart, to the surface of the photographic film on which the shadow is cast. Photographs taken with a tube-to-film distance of two metres, at the end of a quiet expiration with the patient as close as possible to the film, give negligible distortion with a magnification of the cardiac area of some 2 to 10 per cent.

The shadow of the normal heart and blood vessels is remarkably variable, depending on the position of the subject and on weight, build and muscular activity. The three X-ray photographs of Fig. 26.8, illustrate some variations in the shape of the heart shadow in normal persons. The shape of the heart as judged by X-rays alters between deep inspiration and deep expiration. There is usually little alteration with exercise, but long-distance runners tend to have large hearts. Elevation of the diaphragm by a pregnant uterus or increased intra-abdominal pressure, however, modifies the shape of the cardiac shadow and alters the position occupied by the heart in the chest. The transverse diameter of the cardiac shadow averages 12·2 cm in the male and 11 cm in the female but there is a wide range of normality— from 8 to 14·5 cm according to Roesler. The antero-posterior diameter of the cardiac shadow can be measured on a lateral film of the chest. The mean of many such measurements is 9 cm with a range of 7 to 11 cm.

A useful index of cardiac size is the *cardio-thoracic ratio*, that is the transverse diameter of the heart shadow divided by the maximal transverse diameter of the shadow of the chest; this should not exceed 0·5, but in children up to four years of age this ratio is relatively greater than it is in adults. Information about enlargements and movements of the individual chambers of the heart can be obtained by watching the shadow of the heart on a fluorescent X-ray screen.

THE MYOCARDIAL CELLS

Heart muscle is not a syncytium, as used to be thought. It consists, like most other tissues, of discrete cells of different types. The two main varieties may be regarded as 'working' myocardial cells and specialized 'conducting' cells. Within each of these two groups there are differences: for example, working cells in the atrium are anatomically and physiologically different from working cells in the ventricle, and there are similar differences between the cells of different parts of the specialized conducting tissue (p. 469).

Working myocardial cells. The main bulk of the atria and ventricles consists of working myocardial cells arranged in columns. Each cell contains one central nucleus, many myofibrils aligned along the cell's axis and an exceptionally large number of mitochondria. Each cell is enclosed by its membrane, or sarcolemma, the structure through which the cardiac electrical activity exerts its important function (Fig. 26.9).

The myofibrils, as in skeletal muscle (Chap. 40), are both structurally and functionally the fundamental contractile units of the myocardium. The microstructure of the cardiac sarcomere resembles that of skeletal muscle, with sets of actin and myosin myofilaments hexagonally arranged. The modulating proteins tropomyosin and troponin are also present in cardiac muscle though there is some evidence that all four proteins may not be chemically identical with those in skeletal muscle. It is probable, however, that they act in essentially the same way as described for skeletal muscle. (Chap. 40).

In the large and numerous mitochondria lying close to the myofibrils the energy of the metabolic substrates, glucose, fatty acids, lactate and pyruvate, is converted by oxidative phosphorylation to the terminal-bond energy of creatine phosphate (CP) and adenosine triphosphate (ATP), (p. 66). The utilization of the high-energy bonds of the ATP involves its hydrolysis by a Mg^{2+}-activated myofibrillar enzyme, *myosin ATPase,* which is inactive in the absence of Ca^{2+}. The breakdown of ATP at the myofibrillar bridges releases energy for the contraction of the myofibrils. A transphorylation equilibrium between ATP and CP ensures that ATP is rapidly replaced

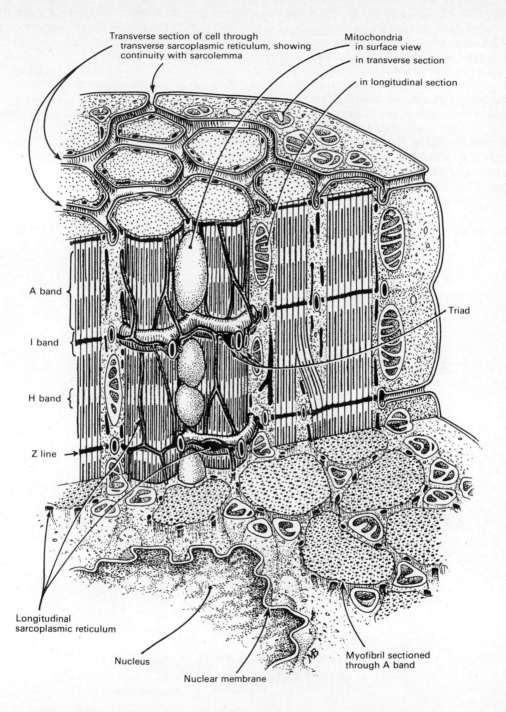

Transverse section of cell through
transverse sarcoplasmic reticulum, showing
continuity with sarcolemma

Mitochondria
in surface view
in transverse section

in longitudinal section

A band

I band

H band

Z line

Triad

Longitudinal
sarcoplasmic reticulum

Nucleus

Nuclear membrane

Myofibril sectioned
through A band

Fig. 26.9 Three-dimensional reconstruction of part of a myocardial cell based on electron microscopy showing the relationship between the transverse tubular system of the sarcoplasmic reticulum, the terminal cisternae of the longitudinal sarcoplasmic reticulum (triads) and the Z lines of sarcomeres.

through the hydrolysis of CP, which acts as a reserve of high-energy bonds.

The sarcolemma. The myocardial cell membrane probably has the usual tripartite structure of proteins and lipoproteins (Fig. 5.4). Adjacent cells are held together by a complicated system of interdigitating projections, which appear on ordinary light microscopy as the 'intercalated discs' (Fig. 26.10) and which consist of the two surface membranes of adjacent cells within the same column and the intercellular space between them. They are similar in structure to the rest of the cell membranes and continuous with them. The thin actin myofilaments seem to be attached to the inside surfaces of these areas,

and here and there desmosomes are present.

As in skeletal muscle invaginations of the sarcolemma extend into the substance of the cell to form the '*transverse tubules*' of the sarcoplasmic reticulum, lying in relation to the Z-lines of the sarcomeres as shown in Fig. 26.9. The *longitudinal sarcoplasmic reticulum*, a true intracellular organelle, is finer and scantier in cardiac muscle than in skeletal muscle but in the neighbourhood of the Z-lines it forms flattened *terminal cisternae* which are separate from, but closely apposed to, the transverse tubules. The terminal cisternae and transverse tubules appear in electron microscope sections as '*triads*' or '*diads*' (Fig. 26.9).

Another important modification of the sar-

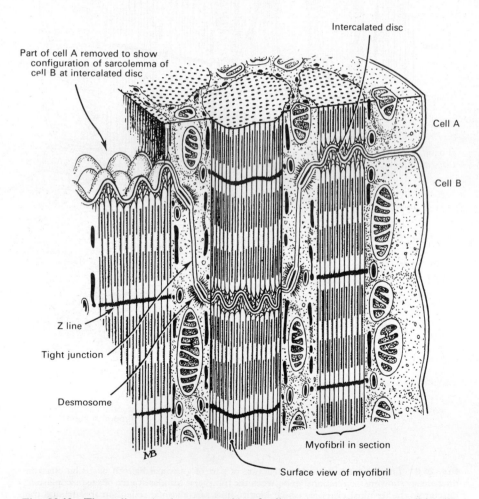

Fig. 26.10 Three dimensional reconstruction of adjacent parts of two myocardial cells separated by part of the intercalated disc, showing the tight junctions (nexuses) where the surface membranes of the two cells seem, on electron microscopy, to be fused.

colemma is found especially in areas along the longitudinal surfaces of the cells where the contact between two adjacent cells is closer than elsewhere and the surface membranes seem to be fused with obliteration of the intercellular space. These regions are called 'nexuses' or 'tight junctions' and may contain a system of intercellular pores (Fig. 26.10). The intercalated discs and the nexuses probably account for the relatively low electrical resistance between adjacent cardiac muscle cells.

The physiological differences between cardiac and skeletal muscle are probably related to three ways in which the membrane of the two kinds of striated muscle cells differ. First, unlike skeletal muscle, which contracts in response to a stimulus conveyed by a motoneurone (Chap. 40), cardiac muscle of all kinds has, at least in certain circumstances, the fundamental property of contracting with *innate rhythmicity*. This property is particularly well developed in the modified myocardial cells forming the '*specialized conducting tissue*', and especially in cells in certain regions, called '*pace-makers*', that generate electrical current, spontaneously and fairly regularly, at inherent rates characteristic for different sites in the system. Secondly, the myocardium is very sensitive to the *direct action of neurohumoral transmitters* such as adrenaline and acetylcholine. Thirdly, the *duration of the action potential* of myocardial cells is very much greater than that of skeletal muscle: the upstroke is no slower, but the return to the resting potential takes much longer, especially in ventricular muscle (Fig. 26.11).

These three features have an important bearing on the functioning of the heart. The innate rhythmicity of a hierarchy of potential pace-makers ensures the continued beating of the heart; the direct action of neurohumoral transmitters powerfully affects the contractility of the muscle (p. 484); the plateau of the action potential prolongs the *refractory period* of the cell (Fig. 26.12), making it incapable in ordinary physiological conditions of producing summated contractions. Tetanic contraction of the myocardium would prevent the coordinated rhythmic contraction and relaxation that constitute the normal beating of the heart (see p. 490).

THE ELECTRICAL EXCITATION OF THE HEART

The special conducting system. The innate rhythmicity of cardiac muscle contraction is normally controlled by the electrical activity of the heart initiated by *pace-making cells* in the specialized myocardial cells of the *conducting system*. Although experimentally, in unphysiological conditions, working myocardial cells are capable of spontaneous depolarization they are normally excited by an electrical current passing along the surface of the cell membrane and from cell to cell through the intercalated discs and tight junctions.

Cells of the special conducting tissue vary greatly in size and character, some being smaller and some larger than working myocardial cells, but they have a higher glycogen content and appear paler on staining than working cells; the electron microscope shows

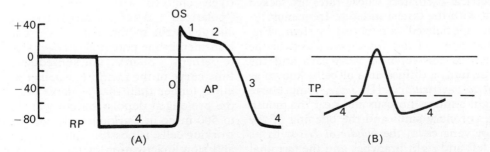

Fig. 26.11 Resting and action membrane potentials from two types of myocardial cell, working ventricular muscle (A) and nodal pacemaker (B). RP, resting potential. AP, action potential. TP, threshold potential. OS, overshoot. Phase 0 of the action potential represents the depolarization of the cell. Phases 1, 2 and 3 represent stages of repolarization. Phase 4 = RP in A, the pacemaker potential in B.

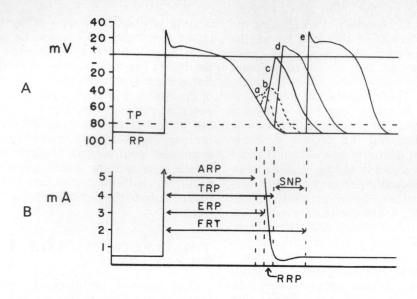

Fig. 26.12 Diagram to show the effect of the refractory states of a myocardial cell on attempts to elicit an action potential. A shows a normal membrane action potential (MAP). a–e, MAPs evoked by stimuli applied at times as indicated. B shows the excitability of the cell to cathodal stimulation. ARP, absolute refractory period. ERP, effective refractory period. TRP, total refractory period. RRP, relative refractory period. FRT, full recovery time. TRP–ERP = RRP. FRT–TRP = supranormal phase (SNP). The first potential to be *propagated* at all (c) arises at the end of ERP. The response, d, is not propagated normally. The first normal response (e) occurs at the end of FRT. (B. F. Hoffman & D. Singer (1964). *Progress in Cardiovascular Disease* **7**, 226–260.)

that they contain scantier myofibrils and mitochondria and a poorly developed sarcoplasmic reticulum. With the exception of a group of cells at the centre of the atrio-ventricular node, which seem to exert only a delaying effect on conduction, it is likely that all the cells in the special conducting tissue are potential pacemaker cells, and together they act like a series of electrical oscillators whose rates are locked to that with the fastest inherent frequency.

The specialized conducting system (Fig. 26.13) consists of the *sinu-atrial node* near the junction of the superior vena cava and the right atrium; a diffuse area of cells, known as the *atrio-ventricular (AV) node,* lying above the right annulus fibrosus and near the mouth of the coronary sinus and the opening of the inferior vena cava; the *bundle of His* with its main left and right branches and the terminal ramifications of the *Purkinje network* in the subendocardial muscle.

Transmembrane resting and action potentials can be obtained from myocardial cells by the

same techniques that are applied to skeletal muscle and nerve and are thought to be the result of differences of ionic concentrations and ionic fluxes across the cell membranes similar to those described in Chap. 40.

The transmembrane *resting potential* recorded from inside cardiac cells is about −90 mV, evidence that the inside is negatively charged with respect to the outside ·('polarized'). After excitation the polarity of charge on the membrane is reversed, and the transmembrane potential changes very rapidly to perhaps +30 mV. The subsequent voltage-time curve of the membrane *action potential* is much longer than that of nerve (Fig. 26.11); the prolonged depolarization, which lasts 400 to 500 msec in Purkinje cells and ventricular working cells, is possibly due to an incomplete and slow inactivation of the sodium permeability and perhaps also to an absence of the rise of potassium permeability seen in nerve at the crest of the action potential. For the greater part of this time the cardiac cell is

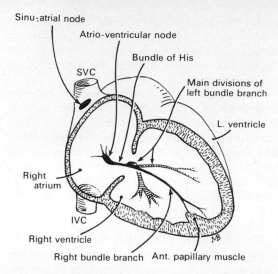

Fig. 26.13 Diagram to show the main parts of the cardiac specialized conducting system. The heart is viewed from the front and right, into the right atrium and ventricle, and has been cut in a plane parallel with the septa, which, in this display, are assumed to be transparent. The AV nodal tissue lies in the right atrium, the bundle of His runs below the membranous part of the interventricular septum, the two main divisions of the left bundle-branch lie beyond the septum. The Purkinje network, which ramifies in the subendocardium of both ventricles, is not shown.

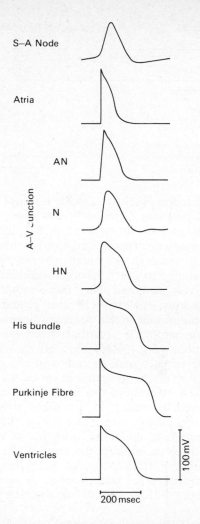

Fig. 26.14 Characteristic shapes of the action potential from various types of myocardial cell. The duration is prolonged progressively from the atria to the peripheral Purkinje fibre. N denotes cells in the central part of the AV node; AN, cells nearer the atrium; HN, cells nearer the bundle of His. (Y. Watanabe & L. S. Dreifus (1968). *American Heart Journal* **76**, 114–135.)

refractory to further stimuli (Fig. 26.12). The duration of the action potential in cells of the special conducting tissue increases progressively from the atrio-ventricular node to the Purkinje fibres (Fig. 26.14). Thus, as the electrical impulse is conducted down the system, the distal fibres with their longer effective refractory period may fail to respond to the higher rate of stimulation of proximal fibres. The abnormal condition called 'heart block' (p. 522) may result.

Another important feature of the special conducting tissue is that the inherent rate of discharge of the pace-maker cells declines progressively from the sinu-atrial node to the cells of the Purkinje network. Thus under ordinary conditions the sinu-atrial node, which has the fastest intrinsic rate, controls the rate of excitation of the whole heart because of the extremely rapid conduction of the excitatory current along the special conducting tissue. Table 26.15 shows conduction velocities through various parts of the conduction system and the working myocardium.

The action potential of pace-making cells differs from that of working myocardium; for example the rate of rise of the upstroke is less, the peak is more rounded and there is no great overshoot. The most important physiological difference, however, lies in phase 4 (Fig. 26.11). In working muscle cells and non-pace-making special conducting tissue phase 4 is the stable resting action potential of the cell. In active pace-making cells, however, the

Table 26.15 Speeds of conduction in the mammalian heart

Tissue	Approx. diameter cell (μm)	Velocity (m per sec)
Atrial fibres	10	0·3 to 0·5
AV node	5	0·05
Purkinje fibres	30	2·0 to 5
Ventricular cells	9 to 16	0·4 to 1

resting action potential steadily lessens ('pace-maker potential'), probably because of increasing permeability to Na^+ or K^+. The cell is progressively depolarized until a threshold level is reached at which the membrane permeability to Na^+ becomes suddenly and explosively increased; Na^+ move rapidly into the cell down its electro-chemical gradient and a new membrane action potential begins.

Transmembrane potentials from the sinu-atrial node (Fig. 26.16) have all the characteristics of pace-making cells and the rate of automatic discharge can be varied in three ways. The threshold can be altered, as with changes in external $[Ca^{2+}]$ concentration; the slope of the diastolic depolarization (Phase 4) can change, as under the influence of neurotransmitter substances; the level of the initial membrane resting potential can change, as with an altered K^+ concentration or under the influence of drugs such as the digitalis glycosides.

Normally electrical current spreads from the pace-maker cells in the sinu-atrial node to excite the neighbouring atrial cells and then spreads from cell to cell along the working myocardial fibres of the atrium, 'depolarizing' them. The excitation leads to contraction of the sarcomeres.

The excitatory local membrane currents run from cell to cell through the atrial muscle towards the atrio-ventricular (AV) ring. This structure forms the fibrous skeleton of the heart and would act as a complete barrier to the spread of electrical activity from the atria to the ventricles were it not for a gap through which the specialized *atrio-ventricular junctional system* runs in the form of the bundle of His.

The AV node is a rather complicated sub-endocardial structure on the atrial side of the right annulus fibrosus and consists of successive layers of cells. Those near the atrial working muscle (AN in Fig. 26.14) propagate the impulse increasingly slowly, a property that seems to be associated with the slow up-stroke and blunt peak of the action potential (Fig. 26.14). Propagation through the middle zone of the node (N in Fig. 26.14) is remarkably slow and accounts for about 30 msec of the total delay in atrio-ventricular conduction. After the impulse crosses this central zone of the AV node its conduction velocity increases progressively to the main branches of the bundle of His. The first fibres to leave the bundle of His are those that form the postero-inferior division of the left bundle-branch. The next to leave are those that form the antero-superior division of the left bundle-branch. Finally the remaining fibres become the right

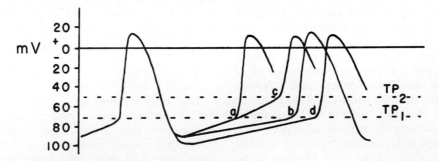

Fig. 26.16 Diagram to show how the rate of firing of transmembrane action potentials recorded from a cell in the SA node may be slowed by three different mechanisms. When the pacemaker potential (phase 4) reaches the normal threshold potential (TP_1) at a the cell fires. If the rate of diastolic depolarization is slower (slope of pacemaker potential less) it fires at b. If the threshold potential is raised to TP_2 the cell fires at c. If the resting potential is increased it fires at d. (B. F. Hoffman & D. H. Singer (1964). *Progress in Cardiovascular Disease* **7,** 226.)

bundle-branch. The two main divisions of the left bundle-branch fan out under the endocardium of the left side of the interventricular septum and adjacent ventricle and end in the Purkinje network in the septum and near the two left-ventricular papillary muscles. The right bundle-branch runs to the Purkinje network near the anterior papillary muscle of the right ventricle.

Ventricular excitation. The first parts of the ventricles to be excited are small subendocardial areas on the left side of the interventricular septum. A few milliseconds later the subendocardial muscle of the right ventricle is excited near the anterior papillary muscle. During the next 60 msec or so the electrical activity spreads through the ventricular muscle as shown in Fig. 26.17.

The summated unbalanced electrical activity of the whole heart is transmitted through the surrounding thoracic contents to the skin of the thorax and limbs, where by appropriate computer-aided techniques it can be mapped at different times in terms of potential differ-ence on the body surface (Fig. 26.18). Such 'isopotential maps' form the basis of the electrocardiogram (ECG) which is extremely useful in clinical medicine (see p. 484).

Excitation-contraction coupling. The electrical excitation of a myocardial cell membrane is the result of the generation of a local circuit current by the membrane action potential, similar to that in unmyelinated nerve. This current, by positive feedback, is responsible for the propagation of the explosively self-perpetuating action potential. In addition, the tight junctions and intercalated discs provide a low electrical resistance between cells probably because of their high permeability to current-carrying K^+. The membrane current is transmitted toward the interior of the cell along the surfaces of the transverse tubular system. In the region of the 'triads' the electrical impulse in some way allows Ca^{2+}, sequestered in the nearby cisternae of the longitudinal system, to move out into the myofibrils. Ca^{2+} activates the myofilaments, probably by binding ATP into the reactive

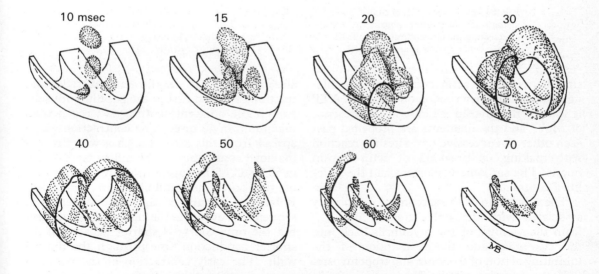

Fig. 26.17 Diagram to show the wavefronts of electrical excitation spreading through the human ventricles at successive instants after the start of ventricular depolarization. When a wavefront reaches the epicardium it is no longer moving and is therefore not represented in this diagram. The first part of the epicardium to be excited is the apical part of the right ventricle at about 20 msec. The septum is excited from both sides, and excitation in both ventricles moves from endocardium to epicardium, in the left fairly concentrically, in the right more tangentially.

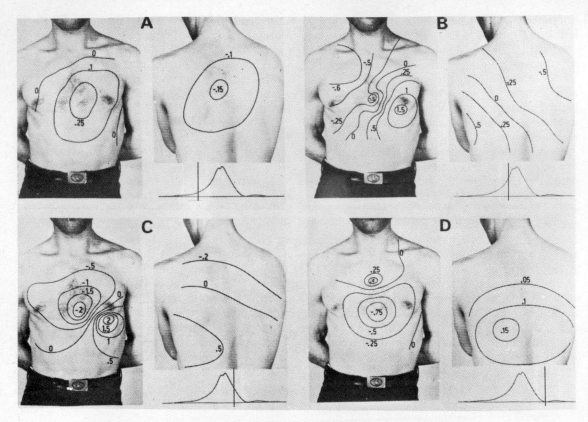

Fig. 26.18 Distribution of cardiac potentials on the chest surface of a normal subject, at four instants of time during ventricular activation ('isopotential field map'). The vertical lines intersecting the electrocardiograms at the bottom of the figures indicate the instant of time to which the maps are related (B. Taccardi (1963). *Circulation Research* **12**, 341.)

sites between the actin and the myosin filaments, possibly by its action on troponin. ATP is split by the myosin ATPase in the presence of Mg^{2+} and the filaments are propelled past each other to successive new sites for reaction with making or breaking of actin–myosin bonds. The sarcomere shortens and the myofibril contracts.

The relaxation of the sarcomere is probably associated with the active transfer of Ca^{2+} from the vicinity of the myofibrils back into the cisternae, and the resumption of the inhibiting action of troponin and tropomyosin.

If the function of the heart and lungs is taken over by a pump and oxygenator (cardiopulmonary bypass) it is possible to open the ventricles of the dog and study the mechanical movements of the myocardium by high-speed cinephotography. The order of contraction agrees with the order of electrical excitation already described. In the left ventricle contractions are seen first in the papillary muscles and in the adjacent septum and reach their maximum in 40 msec. The contraction waves spread from this area in such a way that the maximal contraction of the inflow tract occurs in 60 msec, of the outflow tract in 100 msec and of the aortic infundibulum in 120 msec. Relaxation occurs in the same order. The sequence of events in the right side is the same but the time course is shorter because of the shorter conduction time through the thinner walls. The early contraction of the papillary muscles tightens the chordae and so prevents the cusps of the AV valves from bulging much into the atrium during systole. This sequential contraction of the muscular wall of the ventricles is reminiscent of the peristaltic wave in the coiled tube of the embryonic heart and it is likely to aid the pumping mechanism.

MYOCARDIAL METABOLISM

The muscular tissue of the heart resembles skeletal muscle cells in containing glycogen, ATP and phosphocreatine, but its metabolic behaviour is somewhat different from that of skeletal muscle.

By means of cardiac catheterization (p. 493) it has been found that the human heart uses about 11 g glucose and 10 g lactate per day. Even if it is assumed that these substances are completely oxidized, only one-third of the total oxygen consumption of the heart can be accounted for. Fatty acids, both esterified and free, are the most important fuel of the heart. The presence of lipase in the myocardium suggests that triglycerides are metabolized by the heart. During severe exercise, when lactate is produced in large amounts in skeletal muscle and diffuses into the blood, the heart abstracts the lactate from circulating blood and uses it as a source of energy. Normally the heart functions aerobically and unlike skeletal muscle it cannot incur an oxygen debt, but metabolism may be partly anaerobic in the subendocardial myocardium in conditions of stress.

The myocardium contains a higher concentration than any other tissue in the body of the enzyme aspartate aminotransferase or 'glutamate–oxaloacetate transaminase (GOT)' (Chap. 22) (about 300 000 units per gram of homogenized tissue in the dog). When part of the muscle dies as the result of interruption of its blood supply (coronary thrombosis) this enzyme is liberated into the blood and the serum aminotransferase (SGOT) activity rises for 2 to 3 days from the normal of 10 to 40 units to values of 100 to 200 units. The increase in the serum of this and other cardiac intracellular enzymes is often useful in the diagnosis of myocardial infarction (coronary thrombosis).

Because the heart muscle is constantly active the myocardial oxygen extraction is higher than that of any other tissue, ranging between 70 and 75 per cent. The oxygen consumption of the myocardium is about 10 per cent of the total body O_2 consumption. Because cardiac metabolism is normally aerobic it relies on the delivery of oxygen by the coronary arteries. When the demand for oxygen increases, the myocardium cannot incur an oxygen debt and further extraction of O_2 from the blood is limited, the venous blood in the coronary sinus containing only about 6 ml of O_2 per 100 ml blood over a wide range of heart rates. The demand is met by an increase in the coronary blood flow (p. 518).

When ventricular contraction is increased to overcome a raised arterial impedance (pressure load) the oxygen consumption is greater than that involved in a corresponding contraction to expel an increased end-diastolic volume of blood (volume load). (See also Chap. 18, p. 317, and Chap. 19, p. 336.)

MYOCARDIAL CONTRACTION

The length-tension relation. Spiral and transverse muscle bundles have been described in the myocardium but since these bundles were seen after crude methods of preparation such as boiling they are likely to be artifacts. However, since in a papillary muscle all the fibres are aligned in much the same direction, these thin cardiac muscles provide simpler samples than the more complex ventricular wall. Sonnenblick and others have studied isolated papillary muscles with a myograph (Fig. 26.19). With such a preparation, the unstimulated muscle is progressively stretched and the 'resting (passive) tension' rises. When the muscle is then held at a fixed length and electrically stimulated, an isometric force can be measured at one end. If the muscle is stimulated at various lengths active contraction is superimposed on the resting *length–tension relationship* for each length, giving a curve of 'total tension'. The 'actively developed' tension can be derived from the total tension by subtracting the resting tension. Figure 26.20 shows a *'length–tension curve'* for papillary muscle. The range of muscle lengths at which the active tension is almost maximum (L_{max}) is relatively wide, and corresponds to the rounded peak of the active tension curve. As the muscle is shortened the actively developed tension and the resting tension both fall.

The length–tension curves of cardiac and skeletal muscle are similar in shape. Figure 26.20 also shows the curve for the sartorius muscle of the frog, corrected for the different cross-sectional areas and for the proportions of the contractile material in the two types of striated muscle cell. The peak active tension at L_{max} is generally the same for cardiac and

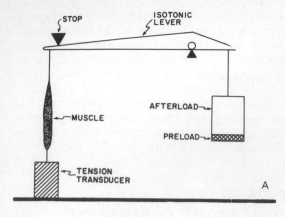

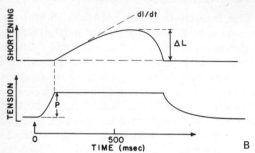

Fig. 26.19 Diagram showing (A) experimental arrangement for studying the loaded contraction of papillary muscle with a myograph. The initial length is set by a 'preload' and held constant by the stop. When the muscle then shortens it lifts the afterload. B illustrates tension and shortening of a typical after-loaded contraction. Starting at 0 the isometric tension increases until, when it reaches P, the muscle begins to shorten and the afterload is lifted. The initial velocity of shortening of the subsequent isotonic contraction is given by dl/dt. ΔL indicates the maximum shortening during the contraction. (E. H. Sonnenblick (1966). The mechanics of myocardial contraction. In *The Myocardial Cell* (Edited by S. A. Briller & H. J. Conn), p. 173. Philadelphia: University of Pennsylvania Press.)

skeletal muscle. By contrast the resting tension at L_{max} is considerable in cardiac muscle and exists during the major part of the ascending limb of the length–active-tension curve, while in skeletal muscle resting tension is minimal at L_{max}, rising only appreciably during the descending limb of the length–active-tension curve. Cardiac muscle therefore operates along the ascending limb of the length–active-tension curve while skeletal muscle operates mainly near the apex. It is not certain why cardiac muscle has this relatively high resting tension.

The length–tension curves in Figure 26.20 are related to sarcomere length and Figure

26.21 indicates average sarcomere lengths determined during systole and diastole in an

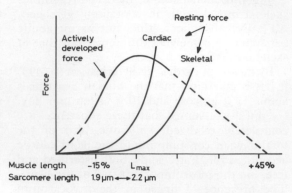

Fig. 26.20 The relationship between tension and initial muscle length for cat papillary muscle and the sartorius muscle of the frog. Muscle length has been normalized relative to that length where actively developed tension is maximum (L_{max}). Active tension is normalized relative to the tension developed at L_{max}. Resting tension is the tension produced by the passively stretched muscle. Active tension is the tension the striated muscle generates when stimulated. (R. A. Leyton & E. H. Sonnenblick (1971). The sarcomere as the basis of Starling's law of the heart in the left and right ventricles. In *Functional Morphology of the Heart: Methods and Achievements in Experimental Pathology*. (Edited by E. Bajusz & G. Jasmin) Vol. 5, p. 22. Basel: Karger.)

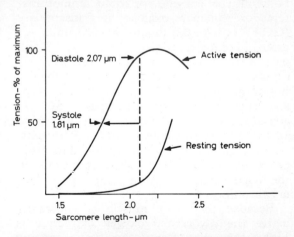

Fig. 26.21 Average sarcomere lengths determined in the left ventricle of the intact heart in diastole and systole have been plotted on the length tension curve derived from isolated feline papillary muscle. (R. A. Leyton & E. H. Sonnenblick (1971). The sarcomere as the basis of Starling's law of the heart in the left and right ventricles. In *Functional Morphology of the Heart: Methods and Achievements in Experimental Pathology* (Edited by E. Bajusz & G. Jasmin, Vol. 5, p. 22. Basel: Karger).

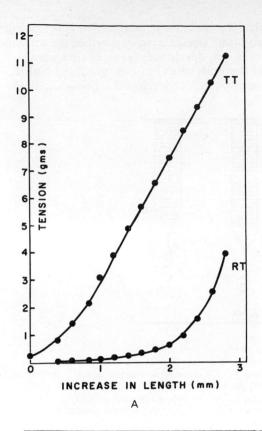

A

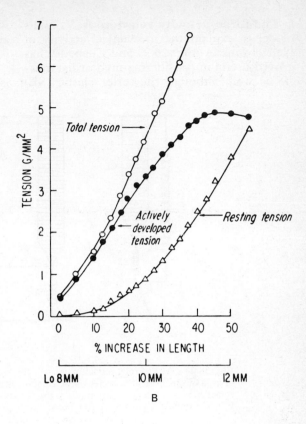

B

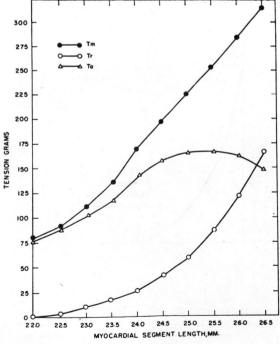

C

Fig. 26.22 Length–tension curves for cardiac muscle. (A) Cat papillary muscle, isometric curves; RT, resting tension; TT, total developed tension (S. E. Downing & E. H. Sonnenblick (1964). *American Journal of Physiology* **207**, 705). (B) Human papillary muscle removed at cardiac operation (E. H. Sonnenblick, E. Braunwald & A. G. Morrow (1965). *Journal of Clinical Investigation* **44**, 966). (C) Human ventricular myocardium *in situ* at operation; Tr, resting tension; Ta, active tension; Tm, maximal tension (M. M. Aygen & E. Braunwald (1962). *Circulation* **26**, 516).

intact heart. These lengths correlated closely with the theoretical limits to the shortening of sarcomeres in the ventricles. These relationships, and the fact that the length of the sarcomere is directly proportional to muscle length along the ascending limb of the length–active-tension curve, form the ultrastructural basis of Starling's 'Law of the Heart' which is considered later.

Length–tension curves of this general form have been derived experimentally from mammalian and human papillary muscle, and from the human heart *in situ* (Fig. 26.22, A, B, C).

The force-velocity relation. A. V. Hill's model to explain the mechanical activity of skeletal muscle (Fig. 26.23 and Chap. 40) has proved useful in considering myocardial muscle as well, although the series elastic (SE) which the contractile mechanism is operating at any time irrespective of whether the muscle is, or is not, developing force, or shortening.

If the ends of the muscle are fixed and the isometric tension developed is maximal, all the

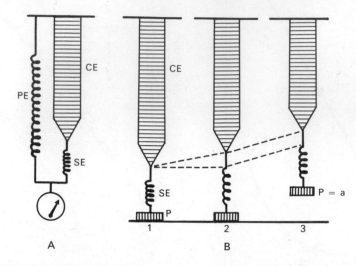

Fig. 26.23 A. V. Hill's model of muscle contraction. (A) CE, contractile element (actin–myosin mechanism); SE, series elastic component; PE, parallel elastic component. (B) during contraction—1, initial resting state; 2, the CE shortens, the SE is stretched but the preload, P. is not moved (isometric contraction); 3, further shortening of CE lifts the preload, which now becomes an afterload (a) (isotonic contraction).

component of myocardium is more compliant than that of skeletal muscle, and the parallel elastic component (PE) is stiffer. Hill studied the relation between the load on a muscle and the velocity of its shortening, the *force–velocity relation*, which is the most fundamental property of the contractile element (CE), that is the actin–myosin mechanism (Fig. 26.24). When the force, or load, (P) approaches zero the velocity of initial shortening (V) is maximal and is termed V_{max}. As the load is increased the amount and the velocity of contraction both decrease progressively until no shortening can occur and the maximum isometric force (P_0) is reached. A muscle acts most intensely when it operates on this *force–velocity curve*. The activation of heart muscle is relatively slow and the force–velocity relations shown on the curve are only approached gradually. Similarly, when the active state declines, the force–velocity relationship is no longer as depicted on the curve. The 'active state' is thus an expression of the degree at

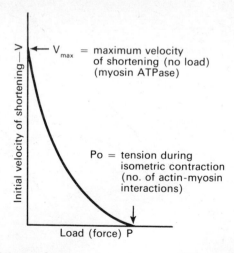

Fig. 26.24 Diagram of a force–velocity curve for muscle. A hyperbolic relationship exists between initial contraction velocity (V) and the muscle tension (P). P_0, maximum tension at zero shortening (isometric); V_{max}, maximum velocity of shortening at zero load.

active sites that generate tension and bring about shortening would tend to be in the 'combined' state: thus the maximum tension (P_0) should be determined by the total number of actin–myosin interactions, which in turn may be related to the amount of Ca^{2+} present, to the action of troponin and to the sarcomere length. With progressively lighter loads fewer and fewer actin–myosin interactions would occur until eventually both the rate of release of energy and the speed of shortening of the muscle are maximal, and the muscle then shortens freely with no load. Because the number of actin–myosin interactions is now minimal, V_{max} will be determined by the maximal intrinsic rate at which energy can be released to allow any one actin–myosin interaction to produce the cyclic changes in position that lead to shortening. This is related to myosin ATPase, and the rate of activity of this enzyme has been found to provide an excellent index to the V_{max} of muscle. Force–velocity curves for cardiac muscle have been obtained experimentally from animal papillary muscle, human papillary muscle, canine left ventricle and intact human ventricle (Fig. 26.25).

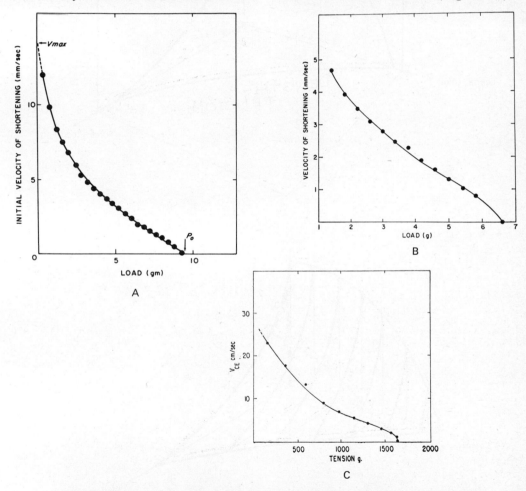

Fig. 26.25 Force–velocity curves for cardiac muscle. (A) Cat papillary muscle. V_{max} and P_0 as in Fig. 26.24. (E. H. Sonnenblick (1966). in *The Myocardial Cell* (edited by S. A. Briller & H. L. Conn), p. 187. Philadelphia: University of Pennsylvania Press. (B) Human papillary muscle removed at cardiac operation. Initial velocity of isotonic shortening on ordinate, afterload on abscissa. (E. H. Sonnenblick, E. Braunwald & A. G. Morrow (1965). *Journal of Clinical Investigation* **44**, 966.) (C) Force–velocity curve of isovolumetric contraction (p. 491) of canine left ventricle. V_{CE}, velocity of contractile element. (J. W. Covell, J. Ross, E. H. Sonnenblick & E. Braunwald (1966). *Circulation Research* **19**, 364.)

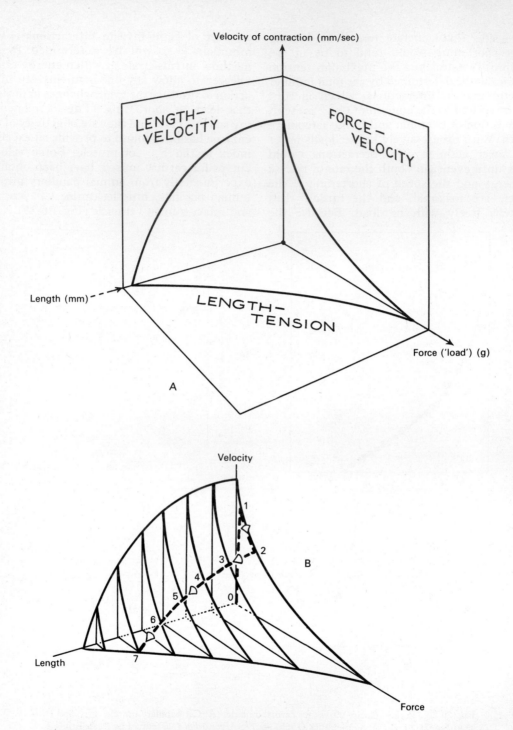

Fig. 26.26 (A) Length-tension and force-velocity curves combined in a three-dimensional construction to illustrate the contraction of a papillary muscle. Note that the usual convention regarding magnitude has been reversed in the axis representing length. (B) A family of force–velocity curves appropriate to different muscle lengths occupies the framework shown in A. During contraction, which starts at 0, the changing relationship between initial velocity, length and force is indicated by the dashed line and solid arrows.

Changes in force–velocity relations characterize the two most important general properties of cardiac muscle: the change in *initial muscle length*, the basis of the 'Starling mechanism' to be considered later, and changes in the *contractile state* of the myocardium. An increase in the initial length of the muscle produces a rise in P_0 without a change in $V_{\text{mäx}}$. A change in the contractile state of the myocardium alters $V_{\text{mäx}}$ with or without a change in P_0.

It is interesting to contrast this behaviour of the myocardium with that of skeletal muscle. Under physiological conditions a skeletal muscle cannot alter its force–velocity curve, its force is altered mainly by its innervation, which controls the number of contracting fibres, each of which contracts almost maximally. The nerve supply of the heart, however, cannot change the total number of contracting fibres; the contractile force of myocardium depends on the point on the ascending part of the length–tension curve at which the cardiac muscle cells are operating.

Length, force and velocity. Let us consider a papillary muscle preparation at a given contractile state. Both the force–velocity and the length–tension relations apply to it: that is to say, the greater the load, the slower is the contraction, and, within limits, the longer the fibre the greater is the force developed. It has been shown that during myocardial contraction the velocity of shortening at any instant is a function of the length of the muscle and also of its afterload at that moment. Three variables are therefore concerned: the force or 'load', the velocity of contraction and the fibre length. A three-dimensional plot is therefore needed to demonstrate adequately the course of events during a single contraction. The framework for such a three-dimensional representation can be constructed by displaying *a family of force–velocity curves* for different lengths of the muscle (Fig. 26.26).

Bearing in mind A. V. Hill's model, let us now consider the moment-to-moment state of the papillary muscle during a single contraction. In Figure 26.26 the muscle is activated at 0. When the active state has developed maximally the force–velocity relationship is as shown at 1. Here the contractile element starts to shorten, stretching the SE and building up isometric force. As the force increases the velocity of shortening of the CE drops towards 2. Here the force developed by the CE equals the load, which is therefore lifted, the contraction now becoming isotonic. The speed of shortening then drops according to the length of the CE, and the pathway of the three-dimensional relationship moves from one force–velocity curve to another. The plane on which this pathway runs, encompassing force, velocity and length, may be regarded as defining the mechanical correlates of the contractile state of the muscle.

The contraction of an intact left ventricle is, of course, more complicated. Loading conditions are different: the preload, or circumferential fibre length at the onset of contraction, is a function of the end-diastolic ventricular volume, and the afterload, or aortic impedance, also varies. Despite these complications the three-dimensional framework can in theory be suitably modified to form a useful basis for understanding several important general principles.

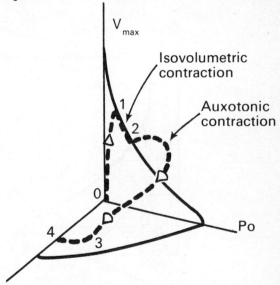

Fig. 26.27 Diagram to illustrate theoretical relations between force (load), velocity of contraction and circumferential fibre length during the contraction of an intact left ventricle. 0 denotes the end-diastolic length of the fibre. 1–2 represents isovolumetric contraction (p. 491). At 2 the aortic valve opens and, since the aortic impedance increases during the early ejection phase (p. 493), contraction is not isotonic, but auxotonic, that is against an increasing load. 3–4, isovolumetric relaxation. 4–0, diastolic increase in fibre length. (*After* E. Braunwald, J. Ross & E. H. Sonnenblick (1967). *Mechanisms of the Contraction of the Normal and Failing Heart*. London: Churchill.)

In Fig. 26.27 a left ventricular contraction starts at 0, where the circumferential fibre is at its end-diastolic length. Ventricular contraction is isovolumetric (p. 491) at the start, so that the velocity of contraction rapidly increases and then follows the force–velocity relation for isometric contraction until the aortic valve opens (2). The subsequent ejection phase (see p. 493) is characterized by contraction against an increasing resistance ('auxotonic'), the afterload or aortic impedance; at each instant during ejection the length of the muscle fibres and the velocity of contraction are correspondingly related. At the end of

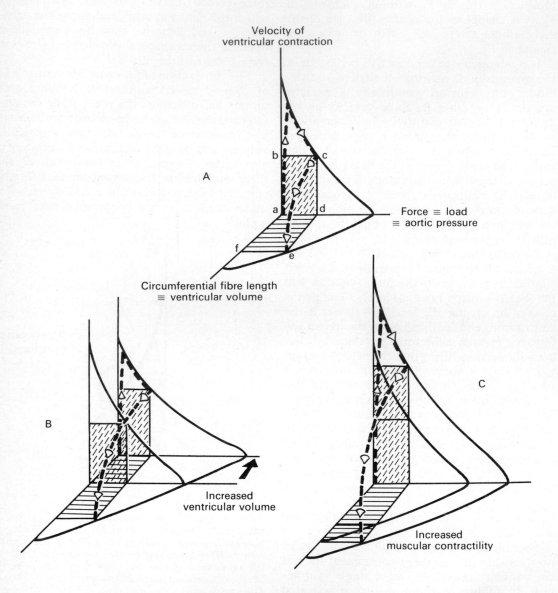

Fig. 26.28 Simplified diagrams to illustrate some important aspects of ventricular contraction. (A) af represents stroke volume. Power = afterload (aortic pressure) × velocity (abcd). Work = stroke volume × pressure (adef). (B) The effect of increased ventricular volume, the Starling mechanism. (C) The effect of increased muscular contractility. (*After* E. H. Sonnenblick (1966). The mechanics of myocardial contraction. In *The Myocardial Cell*. (Edited by S. A. Briller & H. J. Conn), p. 173. Philadelphia: University of Pennsylvania Press.)

ejection, when muscular shortening is maximal, the pathway that has characterized the ventricular contraction has reached point 3 (Fig. 26.27). Isovolumetric relaxation then occurs (3 to 4) and during diastole (4 to 0) the ventricle fills and the fibre length increases.

In this three-dimensional framework the axis representing the muscular tension or load can also represent the aortic pressure (or impedance), and the axis representing the circumferential length can also represent the ventricular volume. In Fig. 26.28A af can therefore represent the *stroke volume,* or the volume of blood ejected during one complete ventricular contraction (p. 490). There is an almost linear relationship between ventricular systolic pressure and the length of end-diastolic muscle segments, or the ventricular volume, in the physiological range. Because of this, and because power = aortic pressure × velocity, and work = stroke volume × pressure, the 'power' and 'work' of the ventricle during the contraction can be indicated in the three-dimensional diagram as shown in Figure 26.28. However, the 'power', 'work' and 'mechanical efficiency' of the heart are not nearly such useful concepts as the Starling mechanism and the contractile state of the myocardium.

To illustrate these two important mechanisms Figure 26.28 also provides three-dimensional diagrams showing (B) the effects of an increased ventricular end-diastolic volume (Starling mechanism) and (C) the effect of increased muscle contractility. Similar analyses can be made of contraction against an increased aortic impedance (afterload) and during an increased heart rate (tachycardia).

Starling's law of the heart. In E. H. Starling's famous experiments with the isolated and denervated *'heart–lung preparation'* (Fig. 26.29) the venous return to the heart or the arterial resistance could each be increased (Figs. 26.30 and 31). As a result of either change the heart responded by an increase in both the end-diastolic volume and the contractile activity. As a result of his experiments Starling enunciated his 'Law of the Heart': *'that the energy of contraction, however measured, is a function of the length of the muscle fibre'.* McMichael and Sharpey-Schafer first produced experimental evidence to suggest that in the intact human heart the end-diastolic length of ventricular muscle is a major deter-

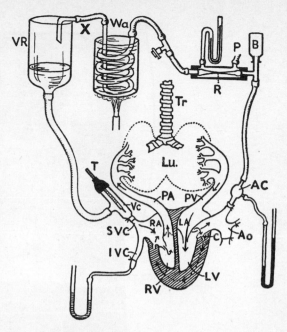

Fig. 26.29 Diagram of circulation in heart-lung preparation. AC = arterial cannula. B = pressure bottle. R. = arterial resistance. P = pump to control arterial resistance. Wa = warming coil. VR = venous reservoir. Vc = venous cannula. SVC and IVC = superior and inferior venae cavae. RA = right atrium. PA = pulmonary artery. Lu = lungs. PV = pulmonary vein. LA = left atrium. RV and LV = right and left ventricles. C = coronary blood supply. Ao = aorta. At **X** in the artificial circulation is inserted a recorder to measure the rate of flow, that is the cardiac output. (*After* S. W. Patterson & E. H. Starling (1914). *Journal of Physiology* **48**, 363, and Y. Bogue in C. Lovatt Evans (1945). *Principles of Human Physiology*, 9th edn, p. 545. London: Churchill.)

minant of the strength of contraction. In general this concept is now widely accepted (Fig. 26.32).

Starling believed that the importance of his law lay in the fact that it described the way in which the cardiac output of the healthy heart was automatically adjusted to the varying needs of the body, but in fact that adjustment is mainly achieved by the nervous and hormonal control of myocardial contractility. The function of the Starling mechanism is to ensure that over any period of time the outputs of the left and the right ventricle are approximately equal, so that the total blood volume is distributed in proper proportions between the systemic and pulmonary circulations. If a system of two pumps in series had not an automatic feed-back mechanism such as de-

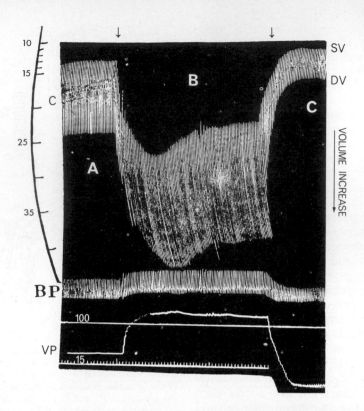

Fig. 26.30 Effect of alteration in venous supply on volume of heart. Dog, 5·15 kg. Heart, 67 g. The venous pressure was varied by altering the height of the reservoir (VR in Fig. 26.29) at the arrows.

Period	Venous inflow ml per min	Arterial pressure mm Hg	Venous pressure mm H$_2$O	Output of heart ml per min	Rate per minute
A	516	124	95	516	132
B	840	130	145	840	132
C	198	122	55	198	132

Time (sec)

C = cardiometer trace with calibration curve in ml. on the left. BP = arterial blood pressure. VP = pressure in the inferior vena cava. SV = systolic volume. DV = diastolic volume. Note: Since an increase in volume is indicated by a downward line in the trace the volume of the ventricles at the end of systole is recorded by the upper limit of the trace and the volume at the end of diastole by the lower limit. The distance between these limits thus represents the systolic output per beat (stroke volume) (S. W. Patterson, H. Piper & E. H. Starling (1914). *Journal of Physiology* **48**, 498).

scribed by Starling's Law even a very small difference in stroke volume, such as regularly occurs during continued respiration, would over a period of time inevitably lead to the overloading of one vascular bed at the expense of the other.

Non-contractile components in heart muscle. Any consideration of the measurement of the force of contraction of cardiac muscle is complicated by the fact that some of the resting tension is maintained by noncontracting elastic structures, Hill's SE and PE (Fig. 26.23). Anatomically they cannot yet be identified, and, indeed, need not even be structurally separate. Further, during a cardiac contraction more and more of the force exerted in the resting state may be transferred from the elastic elements to the myofilaments, and

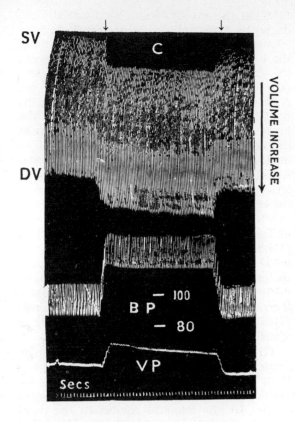

SV

C

DV

VOLUME INCREASE

— 100
B P
— 80

V P

Secs

Fig. 26.31 Tracing to show the effect of increased arterial pressure on the volume changes of the isolated heart with a steady inflow of 924 ml of blood per minute.

Dog, 5·15 kg. Heart, 67 g. Output per beat 10 ml throughout. The arterial pressure was altered by changing the peripheral resistance (R in Fig. 26.29) at the arrows. C = cardiometer trace. BP = arterial blood pressure with scale in mm Hg. VP = pressure in inferior vena cava. SV = systolic volume. DV = diastolic volume. (S. W. Patterson, H. Piper & E. H. Starling (1914). *Journal of Physiology* **48**, 483.)

during ventricular systole a significant amount of the tension may be borne by the PE, though the SE probably plays no part in alterations to the contractile state of the myocardium. The PE, moreover, determines the resting (diastolic) pressure–volume relationship of the ventricle.

VENTRICULAR PERFORMANCE

Starling's work showed that the stroke volume of a heart not in failure is a function of the diastolic fibre length, or ventricular dia-

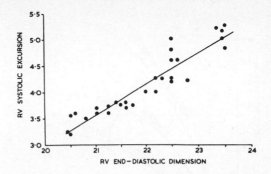

Fig. 26.32 Beat-to-beat relations between right ventricular (RV) end-diastolic dimensions and systolic shortening during the subsequent cycle in a patient with sinus rhythm during and after release of a Valsalva manoeuvre. Both variables were measured from cineradiographs of patients whose ventricular epicardium had been previously marked by radio-opaque clips at the time of a cardiac operation. (E. Braunwald (1965). *British Heart Journal* **27**, 1–16.)

stolic volume, which in turn is related to the diastolic filling pressure. This last relationship is, however, complicated by the law developed by Laplace for arteries, modified to apply to a ventricle: it implies that the tangential force in the myocardial wall needed to produce a given intraventricular pressure is a direct function of the ventricular radius ($T = (P \times R)/2h$, where T = tangential intramyocardial tension, P = intraventricular pressure, R = internal radius of ventricular cavity, h = thickness of ventricular wall). If the diastolic volume of the ventricle is increased a greater intramyocardial tension is needed to keep the intraventricular pressure constant.

Ventricular function curves. Sarnoff and his colleagues have correlated the ventricular stroke volume with mean atrial, or ventricular end-diastolic, pressures, using these as a measure of the diastolic volume. Ventricular performance, measured in this way, can be expressed in the form of *'ventricular function curves'* (Fig. 26.33). Alterations in the ventricular performance are represented in this graph as a family of ventricular function curves, the position of each curve corresponding to a different state of ventricular contractility. Rushmer's work on conscious, fully active animals, to whose hearts various measuring instruments had previously been attached at a surgical operation, has suggested that the situation is even more complicated. In

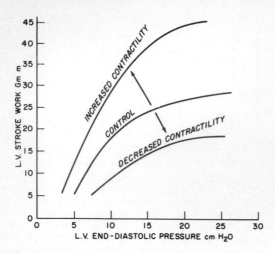

Fig. 26.33 Diagrammatic representation of ventricular function curves obtained under control conditions, during the administration of a positive inotropic agent (increased contractility) and during a negative inotropic state (decreased contractility). LV, left ventricle. (E. Braunwald, J. Ross & E. H. Sonnenblick (1967). *Mechanisms of Contractions of the Normal and Failing Heart.* London Churchill.)

such animals Starling's length–tension relationships appeared to hold during some simple activities: for example, when a dog lies down from the sitting position the diameter of the left ventricle increases for a few beats and during this increase the rate of flow through the aortic valve and the ventricular stroke volume also increase; minor alterations in heart rate such as sinus arrythmia are accompanied by similar changes. On the other hand the cardiac response to exercise in these animals, and probably in man, does not seem to conform to any simple intrinsic mechanism. It seems clear that changes in the contractility of the muscle and the heart rate, brought about by reflex and humoral mechanisms, and modified by posture, often over-rule the effects of the simple Starling mechanism.

Despite these complexities it is still useful to think of cardiac performance in terms of ventricular function curves, if we remember that many factors play a part in both the main variables—the ventricular end-diastolic volume and the myocardial contractility. Factors that affect the ventricular end-diastolic volume are shown in Figure 26.34a, those that affect the contractile state of the myocardium in Figure 26.34b.

When the blood volume is drastically reduced by a sudden severe haemorrhage ventricular performance is impaired. On standing, the blood tends to pool in the lower parts of the body so the intrathoracic volume and ventricular end-diastolic blood volume are reduced; lying with the legs raised, however, tends to raise the cardiac output unless the heart is failing. The normal negative intrathoracic pressure is an important factor in maintaining the diastolic filling of the heart. Artificial positive–pressure ventilation can interfere with this mechanism and, by impairing the venous return to the heart, may reduce the cardiac output. During exercise the smooth muscle in the walls of the veins is constricted, probably by sympathetic stimulation, and tends to increase the filling of the heart, as does the pumping action of skeletal muscles on the veins (p. 505). The so-called '*atrial transport function*' can increase the ventricular end-diastolic volume (p. 491) by acting like a 'booster-pump'.

The contractile state of the myocardium, or 'inotropy', can be increased ('positive inotropy') or decreased ('negative inotropy'), by physiological, pharmacological or pathological causes. In physiological conditions the main positive inotropic factor is the noradrenaline released by the sympathetic nerve endings in the heart and acting upon the β-adrenergic receptors in the myocardium. The amount of this noradrenaline is of course dependent on the frequency of sympathetic nervous impulses arriving at the heart, but circulating catecholamines also have a direct effect on the myocardium and increase its contractility. Many drugs have a positive or negative inotropic action: for example, the glycosides of digitalis are frequently used to increase the power of ventricular contraction in heart failure, and drugs that block the β-adrenergic receptors in the myocardium may impair ventricular contractility. In pathological conditions hypoxia, hypercapnia and acidosis of the myocardium and actual damage to the muscle itself all reduce the contractile power of the ventricle.

THE ELECTROCARDIOGRAM

We have already briefly considered the generation of the electrical currents and potential differences in cardiac pacemaker cells and

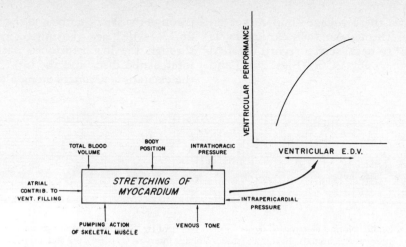

Fig. 26.34a Diagram of a Frank-Starling curve, relating ventricular end-diastolic volume (EDV) to ventricular performance (top right) and the major influences that determine the degree of stretching of the myocardium, that is, the magnitude of the EDV (bottom left). (E. Braunwald, J. Ross & E. H. Sonnenblick (1967). *Mechanisms of Contraction of the Normal and Failing Heart.* London: Churchill.)

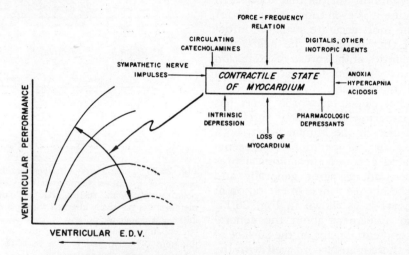

Fig. 26.34b Diagram showing the major influences that elevate or depress the contractile state of the myocardium (top right), and the manner in which alterations in the contractile state of the myocardium affect the level of ventricular performance at any given level of ventricular end-diastolic volume (bottom left). (E. Braunwald, J. Ross & E. H. Sonnenblick (1967). *Mechanisms of the Normal and Failing Heart.* London: Churchill.)

their conduction from cell to cell during the electrical excitation leading to mechanical contraction. This intrinsic cardiac electrical activity is to the heart what the impulse in the motoneurone is to skeletal muscle, an essential prerequisite for normal contraction. The electrical activity can be recorded as a potential difference on the surface of the body by electrodes applied to the skin and linked to a system of amplifiers and galvanometer called an '*electrocardiograph*'. The isopotential map on the skin of the thorax varies throughout each cardiac cycle (Fig. 26.18), and accordingly an electrode at any one site on the body records potential differences that vary as a function of time. The electrocardiograph usually records these voltage–time curves on moving paper throughout each cardiac cycle and from beat to beat: such a record is called an '*electrocardiogram*' (ECG) (Figs. 26.35 and 41).

through the sarcolemma and the currents (Fig. 26.36a) which are transmitted from cell to cell through the low resistance pathways of the intercalated discs and the tight junctions. As the excitation advances along a cell the distri-

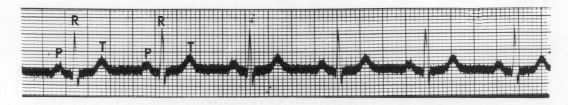

Fig. 26.35 Normal electrocardiogram (ECG) recorded by 'Standard lead II' (p. 490). Time intervals (horizontal), 40 and 200 msec. Voltage intervals (vertical), 0·2 and 1·0 mV.

The electrocardiogram is useful to the physiologist because its various 'deflections' or 'waves' provide useful referents for the timing of various events in the cardiac cycle. In clinical medicine it is even more valuable, for it provides the most accurate available information about the rhythm of the heart, and about abnormalities of the spread of excitation through the myocardium, for example, when it is damaged by disease.

The electrical excitation of the heart has been studied experimentally in dogs and lower primates and also in the human heart, both exposed during surgical operations, and recently as an isolated perfused physiological preparation. The findings agree generally and suggest that the normal course of excitation in the human heart is as shown in Fig. 26.17. There is also agreement about the general form of the isopotential maps on the surface of the body and these maps accord well with the specific shapes of the human ECG recognized as characteristic for certain electrode positions. There is therefore no uncertainty about the observed facts of cardiac electrophysiology. However, the precise manner in which the cellular membrane action potentials, the flow of current and potential (excitation) through the myocardium and the arrival of the potential fields on the body surface are related is still uncertain.

The electrical excitation of a cardiac muscle cell spreads from the point of stimulation, accompanied by the regenerating membrane action potential set up by the ion fluxes

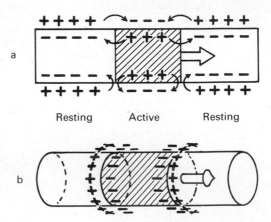

Fig. 26.36 Diagram of a myocardial fibre during the passage of excitation (depolarization) and recovery (repolarization). In the resting state the outside of the cell membrane is positively charged with respect to the inside; in the active state the polarity is reversed. Local membrane currents flow as indicated by the small arrows in (a) at the junctions of active and resting parts of the cell. These junctions form charged wavefronts of excitation and recovery (depolarization and repolarization). These wavefronts have direction, polarity and magnitude and so each can be represented as a vector: the solid arrow in (b) represents the advancing wavefront of excitation.

bution of charges at the junction between the resting and active regions is as shown in Fig. 26.36b. The excitation is conducted so rapidly from cell to cell that this distribution of charges may be regarded as an advancing 'wavefront' of excitation. Such a wavefront carries opposite electrical charges on its two surfaces and therefore produces appropriate fields of potential and current.

Electrical current seems to flow in the heart much as it does between two terminals immersed in a *'volume conductor'*, such as a large volume of electrolyte solution, through which the current can flow in three dimensions. The

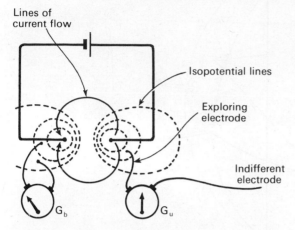

Lines of current flow

Isopotential lines

Exploring electrode

Indifferent electrode

G_b

G_u

Fig. 26.37 Diagram to show the lines of current flow and the lines of equal potential (isopotential field) between two terminals immersed in a large volume of electrolyte solution ('volume conductor'). G_b represents the galvanometer of a 'bipolar' lead system recording the potential difference between two points of relatively high potential that are close together. G_u represents the galvanometer of a 'unipolar' lead system where the electrodes are at points far apart, and one ('indifferent electrode') is at a point of relatively very low potential.

lines of current-flow in a volume conductor are accompanied in the usual manner by appropriate isopotential lines, also distributed in three dimensions (Fig. 26.37). Although the thorax contains organs with electrical resistances different from that of the heart, for example the lungs, it is probable that the intrathoracic contents also act as a volume conductor, extending that of the heart and providing a more or less homogeneous conducting medium for the three-dimensional fields of current and potential. The potentials are greatest near the positive and negative terminals of the 'cardiac electrical generator' and diminish as the square of the distance from them. When the isopotential lines meet a limiting surface such as the skin they are slightly distorted, but the small increases in magnitude thus produced seems to be balanced by the internal shunting effect of the blood inside the heart.

During the excitation of the heart charged wavefronts move through the muscle of the atria and the ventricles (Fig. 26.17). At any instant the apparent cardiac generator is the vectorial resultant of several complex and differently directed wavefronts. Figure 26.38 shows the *resultant vectors* at moments 10, 40, and 70 msec after the onset of ventricular excitation. Fig. 26.39 shows how these early,

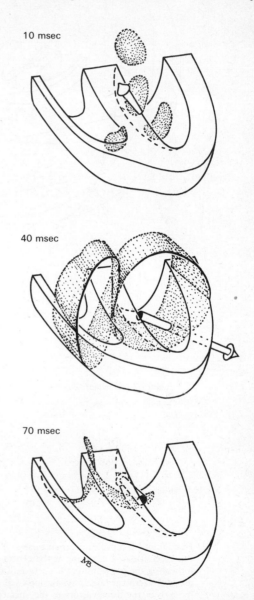

10 msec

40 msec

70 msec

Fig. 26.38 Diagrammatic representation of the wavefronts of excitation of the ventricles at 10, 40 and 70 msec after the onset of ventricular depolarization (see Fig. 26.17). The solid arrows indicate vectorial representations of the total unbalanced electrical activity at each instant.

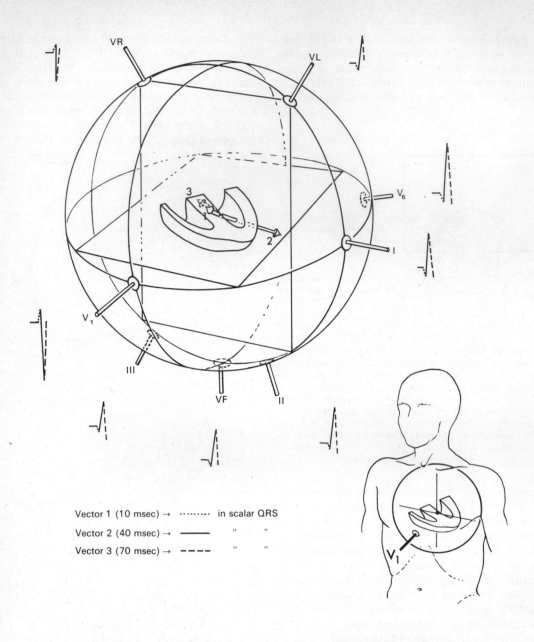

Vector 1 (10 msec) → ⋯⋯⋯ in scalar QRS

Vector 2 (40 msec) → ——— " "

Vector 3 (70 msec) → ----- " "

Fig. 26.39 Simplified diagram to show how the three vectorial forces shown in Figure 26.38 can be considered as arising at the centre of a spherical volume conductor in the thorax, and indicating the positions on the surface of this sphere of some conventional electrode positions used in electrocardiography. Beside each electrode is an indication of the contribution made by each of the three vectors to the surface (scalar) QRS complex recorded at each site. The unipolar exploring electrodes VR, VL and VF are actually applied to the right and left arms and the left leg respectively; V_1 is sited on the precordium as shown in the smaller diagram, V_6 is sited in the left axillary region. The standard bipolar leads (Roman numerals) record complexes similar to those recorded by unipolar exploring electrodes sited as shown, though the actual lead arrangements are more complex (p. 490).

middle and late vectorial forces contribute to the formation of characteristic deflections in the electrocardiogram recorded from the conventional electrode sites on the surface of the body during ventricular depolarization. Similar principles can be applied to the consideration of atrial depolarization.

The same general principles also apply to the repolarization of both the atria and the ventricles. Since the polarity of the charged wavefront that accompanies the restoration of the resting state is opposite to that of the depolarizing wavefront (Fig. 26.36), one might expect that the deflections in the ECG representing repolarization would always be opposite in direction to those that represent depolarization. However, repolarization does not proceed through the intact heart in the same order as depolarization.

During the first rapid phase (0) of the membrane action potential (Fig. 26.11) the depolarization of the cell is associated with the sudden increase in the permeability of the sarcolemma that allows the passive inward Na^+ flux. Repolarization, on the other hand, involves the active extrusion of Na^+ by the much slower mechanism of the Na^+-K^+ pump, which involves metabolic work and the action of ATPase, and is much more readily disturbed by changes such as altered electrolyte concentrations and oxygen tension in the myocardium. It has been suggested that repolarization may be delayed in the subendocardial myocardium because the high blood pressure in the ventricles affects the inner part of the myocardium more than the outer. There is,

however, no direct experimental evidence about the course of repolarization through the heart under normal conditions, but it is certainly true that the segment of the ECG that reflects repolarization is much more readily altered by comparatively small changes in the state of the myocardium than that which reflects depolarization. (Fig. 26.40).

Atrial repolarization is not usually recognized in the normal ECG because it occurs during the inscription of the QRS deflections that accompany ventricular depolarization. When AV conduction is delayed atrial repolarization can sometimes be recognized as a low-voltage negative deflection following that of atrial depolarization.

'Lead systems' and terminology. To record the ECG two electrodes and a galvanometer in the electrocardiograph apparatus are used to detect potential differences between the sites of the electrodes. Since in a volume conductor the potential far away from the source of current is negligible, an electrode sited far from the source can be regarded as providing a reference point of virtually zero potential and is then called the 'indifferent' electrode. The other, 'exploring', electrode can then be used to measure potentials in any part of the volume conductor or on its surface (Fig. 26.37). This is the basis of the 'unipolar lead' system devised by Wilson for experimental work and for clinical electrocardiography, sometimes called 'V leads', because they record values that approach meaningful voltages.

Where both electrodes are comparatively

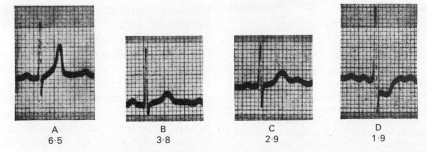

A	B	C	D
6·5	3·8	2·9	1·9

Fig. 26.40 Electrocardiograms (lead II) to show alterations in the segments (ST and T) reflecting repolarization with variations in concentration of potassium in the plasma (indicated as m-equiv/l. below the tracings). The normal value is 3·5 to 5 m-equiv/l. (A) In hyperkalaemia the T wave becomes high and spiked. (B) Normal pattern for comparison. (C) Moderate hypokalaemia may produce a prominent U wave after the T. (D) In severe hypokalaemia the ST segment becomes depressed. A change similar to that shown here is more often due to myocardial hypoxia. See Fig. 26.41 for terminology.

near the source of current the galvanometer records the difference between two potentials of considerable size, each of which is constantly changing throughout the cardiac cycle. In this case the potentials recorded by the electrocardiograph represent the differences between two variables, and are therefore less meaningful. This is, however, the basis of the original 'bipolar lead' systems, such as were used by Waller and Einthoven in their early work on the ECG. Bipolar leads are still used, in addition to unipolar recording, in clinical practice, for example the three 'standard limb leads', for which the electrodes are connected to the right arm, left arm and left leg as follows: Lead I, right arm and left arm; Lead II, right arm and left leg; Lead III, left arm and left leg. Because the dry skin surface has a considerable electrical resistance it is usual to reduce the resistance between the body and the metal electrode by briskly rubbing the skin with a jelly containing an electrolyte.

A normal electrocardiogram (lead I) is represented diagrammatically in Fig. 26.41, which is enlarged for ease in labelling. The standard record is made with the apparatus adjusted so that a vertical movement of 1 cm is produced by a potential change of 1 millivolt (mV).

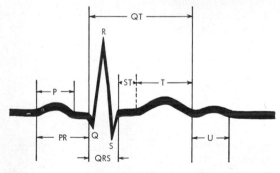

Fig. 26.41 A diagram showing the beginning and end of the various parts of the electrocardiogram, namely the P wave, QRS complex, T and U waves, the PR and QT intervals, and the PQ and ST segments. The TP segment from the end of T to the beginning of P is not labelled. By an agreed convention an upward deflection in the finished record indicates electrical negativity of RA in leads I and II and of LA in lead III.

The P wave begins as the excitatory process passes from the SA node to the atria, its duration indicating the time taken for excitation to spread throughout the atrial muscle.

The QRS complex indicates the depolarization of the ventricles. The T wave signals the last part of recovery (repolarization) in the ventricles; the U wave, not always seen, is probably related to after potentials (See Chap. 39). The PR interval, measured from the beginning of P to the beginning of Q, gives the time taken by the excitatory process to travel over the atria and through the AV junctional tissue to the ventricular muscle. During the ST segment all parts of the ventricles have been depolarized; the record is, however, not quite isoelectric (horizontal) but inclined slightly in the same direction as the T wave, because repolarization begins very slowly during this period. Ventricular repolarization is completed during the T wave. In the TP or UP interval, that is between two cycles of cardiac activity, the whole heart is in the resting polarized state and the record is isoelectric.

The P wave lasts from 60 to 110 msec. The PR interval usually lasts from 120 to 200 msec and, like QT, varies inversely with the heart rate. The QRS complex varies considerably in form, lasting from 60 to 110 msec. Values greater than this indicate either that the electrical events are prolonged in one ventricle or that the ventricles are not being excited simultaneously. The amplitude of the QRS complex is usually from 0·5 to 1·5 mV in leads I, II and III, but is considerably higher in unipolar leads from electrodes sited on the precordium ('chest leads').

THE CARDIAC CYCLE

In considering the 'cardiac cycle' we are concerned with the working of the pump, not simply in terms of the contraction of one ventricle, but with the cyclical contraction (systole) and relaxation (diastole) of the two atria and the two ventricles (Fig. 26.42). During diastole each chamber fills with blood, during systole the blood is expelled. Although the blood moves through the two cardiac pumps in series, the right and left hearts are anatomically arranged in parallel and both atria and both ventricles contract almost simultaneously. Although the thin walled atria act as central venous reservoirs their contraction produces comparatively small, but important, increases in atrial blood pressure (Fig. 26.43). Atrial systole begins at about the peak

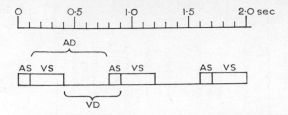

Fig. 26.42 Diagram drawn to scale to illustrate the events in the human cardiac cycle when the heart is beating 75 times per minute. AS = 0·1 sec (atrial systole). AD = 0·7 sec (atrial diastole). VS = 0·3 sec (ventricular systole). VD = 0·5 sec (ventricular diastole). Atrial cycle = 0·8 sec. Ventricular cycle = 0·8 sec. All parts of the heart are in diastole during the 0·4 sec between the end of VS and the beginning of AS. The contractions of the two atria and the two ventricles are not in fact exactly synchronous.

of the P wave of the ECG, contraction of the right atrium usually very slightly preceding that of the left. The muscular contraction forces blood from the atria through the funnels of the AV valves into the ventricles during the last phase of passive ventricular filling; it causes small increases in the pressures in both the atria (atrial 'a wave', Figs. 26.49 and 50) and ventricles, because at this stage the AV valves are still open. Since there are no valves between the right atrium and the venae cavae, some blood is expelled backwards during atrial systole into the superior vena cava where it causes a rise both in the pressure and in the volume. These changes are transmitted upward to the internal jugular vein and cause the 'a wave' in the jugular phlebogram (p. 553 and Fig. 27.31).

In normal hearts, beating with a normal rhythm, the atrial transport function varies with the heart rate. When the heart is beating slowly diastole is long and passive ventricular filling is almost complete (ventricular diastasis); atrial transport is small. When the heart rate is faster ventricular diastasis may never be attained and the atria may contract soon after the end of the rapid filling phase of the ventricles when there is a relatively high pressure gradient between the atria and the ventricles; the atrial transport function is then considerable. At even faster rates, when the atria contract during the phase of rapid ventricular filling, it may be the transport function of the atria that maintains the adequate filling of the ventricle and so prevents a fall in cardiac output. In the common cardiac dysrhythmia

called *atrial fibrillation* (p. 521) atrial contraction is not coordinated and there is therefore no transport function. The patient is rarely handicapped while the ventricular rate remains low, but when it is fast the patient's cardiac output may be reduced and can only be restored to normal when the ventricular rate is reduced by a drug, such as digitalis, that slows the discharge rate of the AV nodal pacemakers. Clinical studies of patients with atrial fibrillation and complete heart block (p. 522) indicate that the atrial transport function can be responsible for increases in the cardiac output varying from about 10 to about 40 per cent.

Atrial function may also normally play a part in closing the AV valves. At the end of atrial systole the blood continues to move through the valve because of its inertia, leaving for a moment behind it a negative pressure which tends to 'suck' the valve cusps towards each other.

After the electrical excitation of the atria is complete the excitation of the ventricles begins after a delay caused mainly by the slow conduction velocity through the central part of the AV node. This delay is responsible for most of the PR interval in the ECG and normally allows the atria to contract completely before ventricular contraction starts.

Ventricular contraction begins soon after the start of ventricular excitation denoted by the beginning of the QRS complex of the ECG. The pressure of blood in the ventricles begins to rise while that in the relaxing atria is falling. The cusps of the atrio-ventricular valves close and then bulge momentarily backward into the atria. In ordinary circumstances the valves do not leak, the surfaces of their cusps being held in apposition by the pull of the papillary muscles on the chordae tendineae attached to the cusp edges. This momentary backward bulging of the AV valve cusps produces slight transient increases of pressure in the atria ('c wave', Figs. 26.49, 50). These intra-atrial c waves are not the same as the c waves in the jugular phlebogram, which are caused by the transmitted pulsation of the underlying carotid artery (Fig. 27.31).

After the closure of the AV valves the blood pressure rises in both ventricles but because both the AV and the semilunar valves are closed the volume of intraventricular blood remains constant. During this *'isovolumetric'*

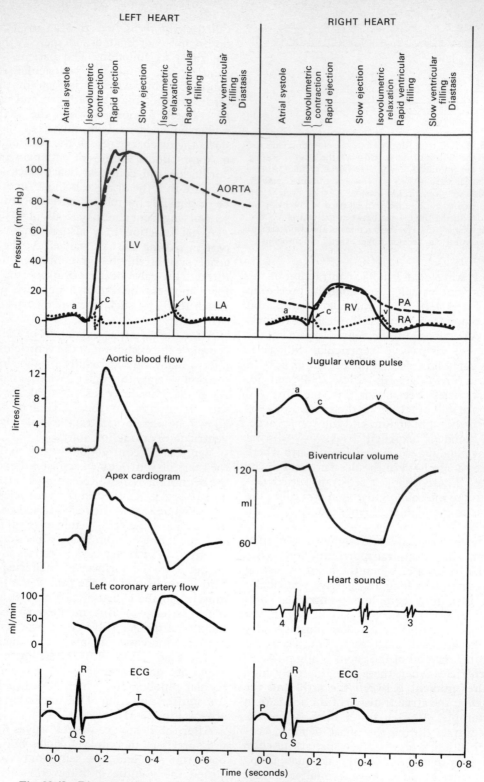

Fig. 26.43 Diagram to illustrate the main haemodynamic events associated with the cardiac cycle in man. See also Figs. 26.47–50 (pressure pulses), 53 (angiocardiograms), 54 and 55 (phonocardiograms), 27.31 (jugular phlebogram) and Table 26.46.

(isometric) phase of ventricular contraction the ventricles alter their shape, becoming plumper, and the apex and AV ring move nearer to one another. This 'descent of the base' stretches the right atrium and the consequent reduction in intra-atrial pressure contributes to the descending limb of the atrial *a wave*.

When the rising ventricular pressure exceeds the pressure in the aorta and pulmonary artery the semilunar valves open, the isometric phase ends and the '*ejection*' (isotonic) phase of contraction begins.

The ejection phase of ventricular contraction is strongly influenced by changing impedances in the great arteries. The early, shorter, rapid ejection phase is one of auxotonic contraction and the maximum velocity of blood flow is in fact attained earlier than the peak of the ventricular pressure curve. The volume of blood ejected (stroke volume) from each ventricle is partly accommodated in the root of the great vessels by a transient expansion which is the origin of the arterial pulse wave. It also displaces the blood expelled by the previous systole (p. 531). The later, longer phase ('reduced ejection') characterized by a fall in both the ventricular and the aortic pressures takes place during the inscription of the T wave of the ECG. During this time the flow of blood from the aorta to the periphery exceeds that from the ventricle to the aorta, the shortened muscle fibres generating progressively less tension. At the end of this phase the ventricular muscle relaxes, and when the ventricular pressure falls below that in the aorta and pulmonary artery the semilunar valves shut. The closure of the aortic valve may be assisted by the effects of vortices in the blood in the aortic sinuses.

Throughout ventricular systole the ventricular volume, as measured by a bell cardiometer (Fig. 26.44), falls steeply. At the same time blood has been returning to the atria, assisted by the negative intrathoracic pressure; because the AV valves are closed the intra-atrial pressure gradually rises to form the ascending limb of the intra-atrial '*v wave*' (Figs. 26.49, 50).

After the closure of the semilunar valves isovolumetric ventricular relaxation continues rapidly, the pressure in the ventricle soon falling below that in the atria. At this point,

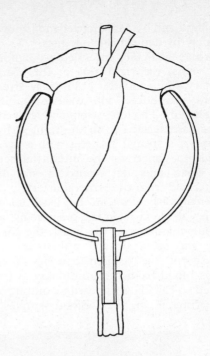

Fig. 26.44   The bell cardiometer for measuring changes in volume of the exposed heart.

the summit of the v wave in the atrial pressure curve, the atrio-ventricular valves open and the blood flows passively from the atria to the ventricles, at first very fast, later more slowly.

In diastole, during the first rapid phase of ventricular filling, the pressures in both atria and ventricles continue to fall but the ventricular volume increases sharply. At normal heart rates the first rapid phase of ventricular filling is succeeded by a slower phase in which the slope of the curve of the ventricular volume is flatter. If the ventricular rate is slow enough the volume curve may eventually become almost level because the ventricles are full (ventricular diastasis).

CARDIAC CATHETERIZATION

In 1929 a young German surgeon, Werner Forssman, passed a long, thin, flexible ureteric catheter from a vein at the bend of the elbow into his own heart, watching its progress by the help of an X-ray screen and a mirror held by a nurse (Fig. 26.45). This brave experiment opened a new era in human cardiovascular physiology based on the technique of cardiac catheterization, which was later developed by

Cournand and by McMichael and is now used routinely by clinical cardiologists.

Cardiac catheters intended to explore the great veins, the right side of the heart, the coronary sinus and the pulmonary artery are usually introduced via the antecubital or femoral vein. The left atrium and left ventricle can be catheterized from the right atrium by the use of a special catheter with a tip sharp enough to penetrate the interatrial septum. The aorta and left ventricle are usually catheterized from the femoral artery.

Special catheters can be used to obtain samples of blood from the great veins, any of the four chambers of the heart, the coronary sinus, the aorta or the pulmonary artery. Analysis of the oxygen content of mixed venous blood from the pulmonary artery can be used to calculate the cardiac output by the Fick principle (p. 502). Blood samples from

other sites allow the calculation of blood flows between cardiac chambers. The catheters can be attached to manometers to record intravascular pressures and to demonstrate pressure

Table 26.46 Normal average values of intracardiac pressures in man (from various authors). LV peak systolic pressure is approximately the level of the systolic blood pressure recorded in the brachial artery, that is about 110 mm Hg (See also Figs. 26.47–50)

Site	Mean	Range
RA	3 mm Hg	1–5 mm Hg
RV (peak systolic)	26 mm Hg	19–31 mm Hg
(end-diastolic)	4 mm Hg	2–6 mm Hg
PA (mean)	14 mm Hg	10–18 mm Hg
(peak-systolic)	23 mm Hg	16–29 mm Hg
(end-diastolic)	9 mm Hg	5–13 mm Hg
LA (mean)	8 mm Hg	2–12 mm Hg
LV (end-diastolic)	9 mm Hg	5–12 mm Hg

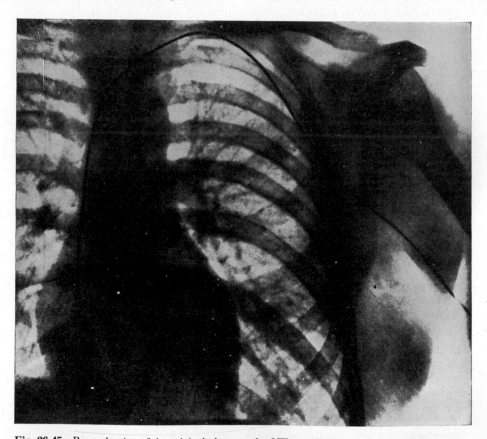

Fig. 26.45 Reproduction of the original photograph of Werner Forssmann's self-catheterization. The radio-opaque catheter can be seen in the left axillary vein, the innominate vein and the superior vena cava. Its tip lies in the right atrium. (W. Forssmann (1929). *Klinische Wochenschrift* **8**, 2085).

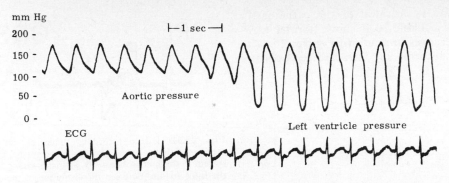

Fig. 26.47 Pressure records from the aorta and left ventricle in a female patient aged 39 with a normal aortic valve. As the catheter is gradually withdrawn from the aorta into the ventricle there is a fairly sudden change in the record which is normal in this patient. (H. A. Fleming, F. W. Hancock, B. B. Milstein & D. N. Ross (1958). *Thorax* **13**, 97–102.)

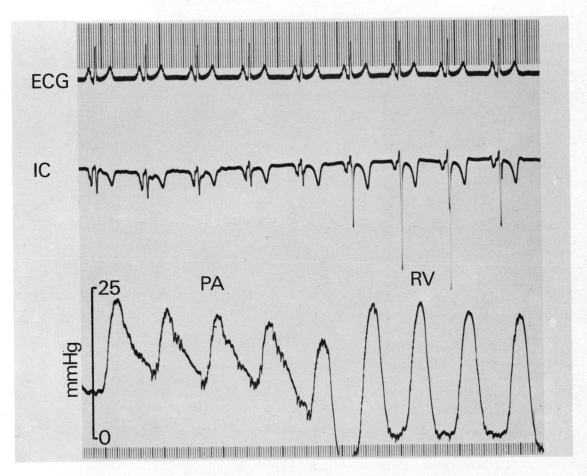

Fig. 26.48 Pressure pulse recorded as a cardiac catheter is withdrawn from the pulmonary artery (PA) to the right ventricle (RV). An electrode at the tip of the catheter records the intracardiac electrocardiogram (IC). As the catheter crosses the tricuspid valve the baseline of the pressure trace is momentarily displaced. The pressure pulses are normal and there is no significant gradient across the valve. ECG, Lead II. Time intervals, 40 msec, 200 msec and 1·0 sec. (*By courtesy of H. Watson.*)

gradients by withdrawal of the catheter tip across valves. Table 26.46 shows normal intracardiac pressures in man and Figures 26.47 to 50 show records of intracardiac pressure

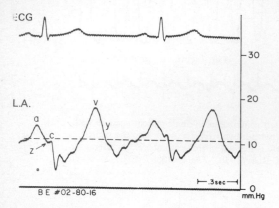

Fig. 26.49 Pressure pulse in the left atrium of a normal subject obtained by transeptal catheterization via the right atrium. (E. Braunwald, E. C. Brockenbrough, C. J. Frahm & J. Ross (1961). *Circulation* **24**, 267.)

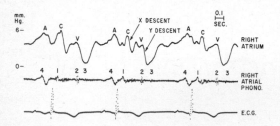

Fig. 26.50 Normal pressures and heart sounds simultaneously recorded from inside the right atrium. Above down: right atrial pressure pulse, intra-atrial phonocardiogram, surface ECG for reference. (H. L. Moscovitz, E. Donoso, I. J. Gelb & R. J. Wilder (1963). *An Atlas of Haemodynamics of the Cardiovascular System*, p. 31. New York: Grune & Stratton.)

curves obtained by cardiac catheterization. Special catheters with electrodes at the tip are used to record the intracardiac electrocardiogram and direct recordings can be made of the electrical impulse travelling through the AV nodal tissue, the bundle of His and the right and left bundle–branches of the special conducting system (Fig. 26.51). Tiny microphones at the tips of cardiac catheters have been used to identify the sources of normal and abnormal cardiac sounds (p. 502) (Fig. 26.50). Other cardiac catheters have at the tip a special platinum electrode capable of producing an electric potential in the presence of hydrogen:

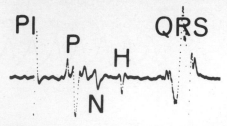

Fig. 26.51 Direct recording made by an electrode catheter of the electrical impulse in the special conducting system in the human heart. PI, artificial pacing impulse in the right atrium. Between the excitation of the atrium (P) and that of the ventricles (QRS) the electrode records the activity of the AV node (N) and the bundle of His (H) as smaller polyphasic deflections. The recording speed is very fast. (A. N. Damato, S. H. Lau, W. D. Berkowitz, K. M. Rosen & K. R. Lisi (1969). *Circulation* **39**, 435.)

with such a 'hydrogen electrode' sited in a cardiac chamber a lung-to-electrode circulation time can easily be determined by making the subject inhale a breath of pure hydrogen gas.

A radio-opaque fluid ('contrast medium') can be injected into any chamber of the heart or into the great vessels (*selective angiocardiography*). The passage of this opaque medium through the heart can be recorded by rapid still, or cine, X-ray photography, and gives useful information about the anatomy of the chambers and the flow of blood between them (Fig. 26.53). The right and left coronary arteries can be selectively catheterized, and the '*coronary arteriograms*' thus obtained demonstrate the coronary arterial circulation (Fig. 26.52 (a) and (b)).

Not only has cardiac catheterization yielded a great deal of knowledge about the physiology of the normal heart but it is an invaluable part of the investigation of many patients with heart disease, providing information obtainable by no other means.

HEART SOUNDS

The heart and great vessels may be considered as an elastic fluid-filled system that responds to a change in equilibrium of any part by oscillating as a whole. Rushmer has used the analogy of a fluid-filled balloon, tapped sharply: although the tap is only applied at a single point, the elastic walls of the balloon and the fluid it contains vibrate as a whole.

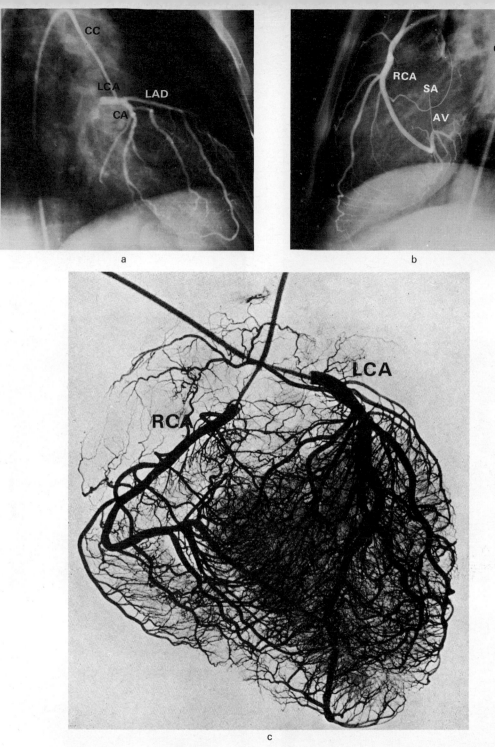

Fig. 26.52 Radiographs of the human coronary arterial system. In (a) and (b) contrast medium has been injected into the left (a) and right (b) coronary arteries of a beating heart through a cardiac catheter (CC) sited accurately in the ostium (*selective coronary arteriography*). In (a) the 'trunk' of the tree is the left coronary artery (LCA), the uppermost main branch is the left anterior descending artery (LAD) and its smaller branches, the lower main branch is the circumflex artery (CA) and its branches. In b the right coronary artery (RCA) is shown with its branches including those to the sinu-atrial node (SA) and AV node (AV). (By courtesy of the Cardiac Department, Victoria Infirmary, Glasgow). (c) Shows the normal coronary arterial tree in a slightly enlarged but otherwise normal heart removed at autopsy. The picture was obtained by X-raying the heart after both right and left coronary arteries had been injected with an opaque medium (bismuth–gelatin) under controlled pressure. The rich anastomosis of small vessels in the interventricular septum is well shown. (W. F. M. Fulton (1956). *British Heart Journal* **18**, 341–354.)

POSTERO-ANTERIOR　　　　　　　　LEFT LATERAL

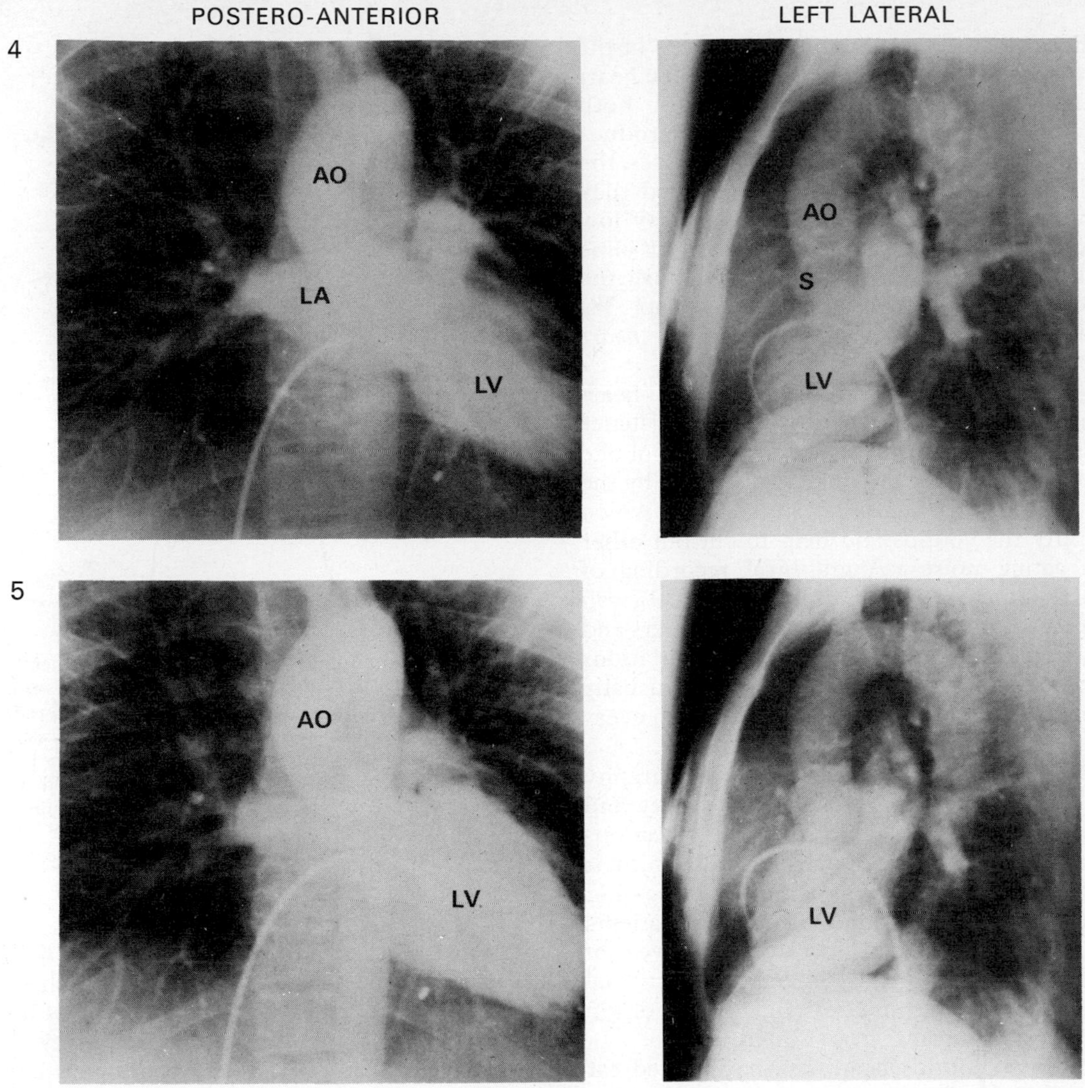

POSTERO-ANTERIOR   LEFT LATERAL

Fig. 26.53 Frames from simultaneous postero-anterior and left lateral rapid still radiographs during angiocardiography of a normal human heart. The contrast medium was injected through a cardiac catheter (CC) sited in the right ventricle (RV) during a previous cardiac cycle, and fills various cardiac chambers and vessels at subsequent stages of the cardiac cycle: (1) right ventricular diastole, with slight filling of the right atrium (RA) because of tricuspid incompetence caused by the catheter; (2) RV systole expels contrast into the pulmonary arteries (PA); (3) the contrast has passed through the pulmonary capillaries and is returning through the pulmonary veins (PV) to fill the left atrium (LA); (4) systole expels blood from the left ventricle (LV) into the aorta (AO) and in the lateral view an aortic sinus (S) is evident; (5) by the time the LV has relaxed in diastole the contrast has moved farther down the descending aorta. (*By courtesy of the Royal Postgraduate Medical School, London.*)

Because of the haemodynamic events of the cardiac cycle the blood flows through the heart with sharply varying velocities and both accelerations and decelerations of flow produce vibrations of the cardiac muscle and valves, the walls of the great vessels and the blood they contain. The vibrations have a relativity low frequency and are rapidly damped, but some of their frequency components lie above the threshold of human hearing (see Chap. 38) and can therefore be appreciated as 'heart sounds'.

The two main groups of normal heart sounds can usually be heard if the listener presses his ear firmly against the skin of the precordium. They are more easily heard by the use of various types of stethoscope which amplify the sounds and help to cut out other distracting noises. A graphical recording of heart sounds, a *phonocardiogram* (PCG), can be made by using microphones, amplifiers and filters to overcome some of the limitations imposed by the human ear. The PCG usually shows four main groups of vibrations in every cardiac cycle (Figs. 26.54 and 43).

Studies by intracardiac phonocardiography strongly suggest that changes of velocity and turbulence in the blood produce vibrations that are most intense 'downstream' from their site of production. This technique, together with the study of the heart sounds of patients who have had diseased valves removed at surgical operation and replaced by artificial ones, has thrown some fresh light on the origin of these four normal heart sounds.

The two sounds heard by the unaided ear are called the 'first' and 'second' heart sounds. The *first heart sound* is the longest and loudest of the normal heart sounds. The PCG shows a group of vibrations of mixed frequency occurring at the end of the QRS complex. It is associated with the beginning of ventricular systole and is mainly caused by vibrations set up by the abrupt rise in ventricular pressure and the closure of the AV valves. Often there are two main groups of vibrations and on auscultation with the stethoscope the first sound is heard to be 'split'. This splitting may be the result of the closure of, first, the mitral, and then the tricuspid, valve, but although this idea is a useful one in clinical practice it is likely that it is an oversimplification. Because the vibrations are most intense

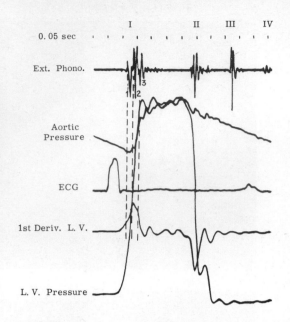

Fig. 26.54 Relationship of the three rapid components, marked 1, 2, 3, of the first sound to left-sided pressure pulses and first derivative of left ventricular pressure. From above: external phonocardiogram (filter, 200 cycle/sec); aortic pressure; ECG; first derivative LV pressure; LV pressure. The first derivative of pressure is the rate of change of pressure, that is dp/dt. The four heart sounds are marked I, II, III and IV. (P. M. Shah, M. Mori, P. M. McCanon & A. A. Luisada (1963). *Circulation Research* **12**, 388.)

'downstream' from the AV valves the first heart sound is best heard at an area of the chest wall near the apex beat (p 502). Because of its length and the comparatively low frequency of its components it is traditionally represented by the syllable, '*lub*'.

The *second heart sound* shows on the PCG as two shorter groups of vibrations of higher frequency occurring about the end of the T wave in the ECG. These sounds are caused by the sudden closing of, first, the aortic, and then the pulmonary valve. They are best heard 'downstream' from these valves in an area on the chest wall about the second intercostal space just to the right and left of the sternum. The untrained ear may not appreciate the split between the aortic and pulmonary components. The second heart sound, being shorter and of higher pitch than the first heart sound, is traditionally represented by the syllable '*dupp*'. Careful auscultation, however, during slow continued respiration, or phono-

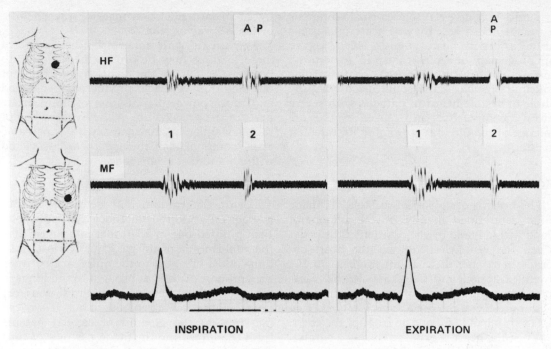

Fig. 26.55 Phonocardiogram of a normal young adult. One microphone at the left sternal edge provides a high-frequency (HF) record, the other, at the region of the cardiac apex provides a medium frequency (MF) record. 1 and 2, first and second heart sounds. A and P, vibrations associated with closure of the aortic and pulmonary valves. Lead II ECG at bottom. On inspiration the increased negative intrathoracic pressure increases the filling of the right ventricle in diastole. The increased right ventricular volume delays closure of the pulmonary valve: the pulmonary valve closure sound (P) is delayed, causing audible splitting of the second heart sound (0·04 sec. in this record). On expiration, pulmonary valve closure occurs so soon after aortic valve closure that the second heart sound sounds single on auscultation.

cardiography, confirms that the second heart sound is only single during the phase of full expiration. During the inspiratory phase the increase in the negative intrathoracic pressure increases the volume of blood in the right ventricle and this significantly delays the closure of the pulmonary valve, and hence the pulmonary component of the second heart sound. The second heart sound then sounds more like '*trupp*'. At the height of inspiration the delay in the closure of the pulmonary valve is maximal, the split between the aortic and pulmonary components is widest and the sound may be represented as '*tu-rupp*'. During expiration the split narrows again. This 'movement' of the splitting of the second heart sound is most easily heard in young people in the erect position (Fig. 26.55).

The *third heart sound* is heard in children and young adults. In the PCG it is shown as a short group of vibrations of very low frequency in early diastole, occurring towards the end of the phase of rapid ventricular filling. These vibrations are probably set up in the blood at a time when the left ventricular wall is changing from a state of active relaxation to one of passive distension and when the mitral valve system, consisting of the papillary muscles, chordae and cusps, reaches a certain tension (Fig. 26.43).

The fourth heart sound, or '*atrial sound*' is inaudible in normal people and can usually only be demonstrated by phonocardiography. It consists of a few low frequency vibrations during the PR interval of the ECG and is caused by the contraction of the atria and the flow of blood through the AV valves. In abnormal conditions, such as delayed AV conduction when the PR interval is long, the atrial sound may be heard on auscultation as a faint low thud.

In abnormal hearts altered cardiac haemodynamics may give rise to abnormal vibrations or turbulence. These may be heard on auscultation as extra heart sounds, such as the 'opening snap' of a pliant cusp in a narrowed mitral valve (mitral stenosis), or as cardiac 'murmurs', such as the sound occupying the whole of systole heard in patients with severe mitral incompetence, and caused by the turbulence of the blood regurgitated into the left atrium.

MOVEMENTS OF THE HEART

The heart is quite soft in diastole, its shape being determined mainly by the amount of blood within it and by the position of the body. When the ventricles contract the heart becomes hard and there is an increase in the antero-posterior, and a decrease in the transverse, diameter. The movement of the heart within the chest, the *cardiac impulse,* can often be felt by a hand laid on the front of the chest.

When the heart is exposed during a surgical operation it can be seen that the apex rotates to the right at each systole. This movement is largely prevented in the intact animal by the attachment of the pericardium to the central tendon of the diaphragm. Leonardo da Vinci passed needles through an animal's thoracic wall into the apical, middle and basal areas of the heart and found that at each ventricular systole the needle at the base moved most and the needle in the apex hardly moved at all. This very old experiment shows that at ventricular systole the base of the heart moves downwards and forwards while the apex moves relatively little. The movement of the dog's atrio-ventricular ring towards the apex in systole has been confirmed by making X-ray photographs after marking the AV groove with silver wire. These movements of the heart impart a localized pulsation (*apex beat*) to the chest wall in the fifth left intercostal space about the mid-clavicular line. X-rays show that the actual apex is about an inch lower down.

The pulsations of the apex beat can be more critically analysed if they are graphically recorded by appropriate techniques in the form of the '*apex cardiogram*'. The main outward pulsation is caused by ventricular systole, but this is preceded by the smaller pulsation associated with atrial contraction and followed by an inward pulsation marking the opening of the AV valves (Fig. 26.43).

In patients with abnormal hearts some of these features may be pronounced enough to be seen, or felt by the hand: hypertrophy of one or other ventricle produces a prolonged and more powerful pulsation of appropriate parts of the precordium, while hypertrophy of the atria may be associated with a palpable atrial component of the apex beat, associated with an audible atrial sound.

CARDIAC OUTPUT IN MAN

For the determination of the cardiac output in man only indirect methods are applicable. Fick pointed out in 1870 that in a given time the total amount of any gas gained or lost in the lungs must be equal to the difference between the amounts of the gas brought to the lungs in the arterial blood and the amount leaving the lungs in the venous blood. Thus if we can measure (a) the amount of oxygen passing from the lungs into the blood per unit of time (b) the oxygen content of the arterial blood and (c) the oxygen content of mixed venous blood, then the output of the heart is easily calculated. An average value for (a) is 250 ml per minute, for (b) is 19 ml per 100 ml blood, and for (c) is 14 ml per 100 ml blood. Each 100 ml of blood in passing through the lungs gains 5 ml of oxygen, that is the arteriovenous difference. Since 250 ml of oxygen actually passes into the blood per minute (see Fig. 26.56)

$$\frac{250}{5} \times 100 \text{ ml} = 5000 \text{ ml or 5 litres}$$

of blood are needed to carry that amount of oxygen, and this is the cardiac output per minute. If the heart contracts 72 times per minute then the stroke volume of each ventricle is $5000 \div 72 = 70$ ml.

The oxygen consumption can be determined by a spirometer or by the Douglas bag technique (p. 164). The oxygen content of arterial blood can be measured directly in a specimen obtained by puncturing an artery with a fine needle attached to a syringe. The main technical difficulty is to get a specimen of mixed venous blood. This can only be obtained from the pulmonary artery since there are large regional differences in the oxygen content of blood from superficial veins, and even between streams in the right ventricle.

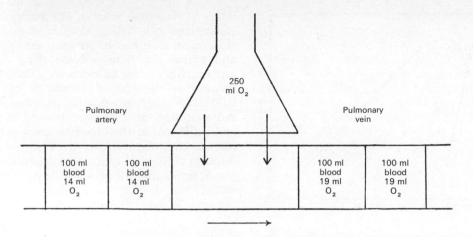

Fig. 26.56 Scheme to illustrate the determination of cardiac output by the Fick principle. Each rectangle represents 100 ml of blood which gains 5 ml of oxygen in passing through the lungs. 250 ml of oxygen disappears from the lungs per minute. This volume would be accounted for by 250/5 or 50 rectangles. The flow is therefore 250/5 × 100 ml, or 5 litres per minute.

The results obtained in this way represent the effective output of each ventricle but do not include any run-back of blood into the ventricles such as may occur in valvular disease. If all the oxygen taken up from the inspired air is carried in the blood and is not utilized in the lungs the results should be correct for normal persons. The average arteriovenous oxygen difference at rest is 4·5 ml O_2 per 100 ml blood. Since the average oxygen consumption is 240 ml per minute the average cardiac output is 240/4·5 × 100 = 5·3 litres per minute. The range in normal adults under basal conditions is from 4 to 7 litres per minute, with an average of 5·3 litres. This corresponds to a stroke output from each ventricle of 70 to 80 ml.

In severe anaemia it has been found that the arteriovenous oxygen difference may be only 2 ml per 100 ml of blood. If the oxygen usage is 240 ml per minute then the cardiac output is 240/2 × 100 = 12 litres per minute. This considerable increase in cardiac output is associated with a high pressure in the right atrium. The cardiac output in exercise may be increased to as much as 40 litres per minute, partly by an augmented stroke volume and partly by an increase in cardiac rate. The rise in cardiac output with exercise is, however, not accompanied by a parallel rise in venous pressure (see also Chap. 40). Apprehension and anxiety may be accompanied by an increase of some 10 to 20 per cent over the basal value. The output is some 10 per cent lower during sleep. In healthy persons there is a substantial increase in cardiac output (up to about 25 per cent) when the erect is changed to the supine position, presumably the result of an increased venous return. It seems reasonable to expect that the cardiac output would be closely correlated with body size and for many years it has been customary to express the *cardiac index* as the minute volume per m² of body surface (about 3·8 1). The cardiac output, however, is more closely correlated with height and weight. Smulyan finds that the best prediction is given by the formula: cardiac output in 1/min = 2·125 + 0·337 × weight in kg + 0·0397 × the height in cm with a standard error of 1·46 1/min.

The cardiac output in man may also be measured by the indicator dilution method. A known amount (X mg) of a non-diffusible dye is injected into a central vein and the rapid rise and fall of its concentration in arterial blood is followed by needle puncture of, say, the brachial artery. The falling concentration, before recirculation of the dye begins, can be plotted as a graph which, if extrapolated to zero concentration or 0·1 mg/l in Fig. 26.57, gives the time after injection at which all the dye must have been ejected from the heart if there

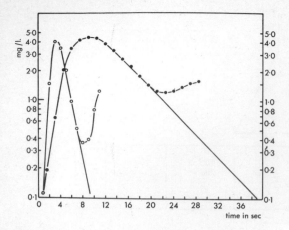

Fig. 26.57 Dye concentration curves (Evan's Blue) plotted on semi-logarithmic paper. ●——● Rest experiment. Cardiac output 4·73 litres per minute. ○——○ Work experiment. 1260 kg m per minute. Cardiac output 21·9 litres per minute. The second upward trend of the curves is due to recirculation of dye. (E. Asmussen and M. Neilsen (1952–53). *Acta physiologica scandinavica* **27**, 217.)

had been no recirculation (Y seconds). Since the concentration falls off exponentially it can be plotted as a straight line on semilog paper (Fig. 26.57). If the average concentration of the dye in the blood is calculated for this period (Z mg/litre) then X/Z litres is the amount of blood required for the ejection of X mg of dye. Since this took place in Y seconds the cardiac output in litres per min must have been $X/Z \times 60/Y$. Since the indicator does not enter the red cells but is carried in the plasma this formula gives the plasma flow from which the blood flow can be calculated (see Chap. 23). Brachial artery puncture can be avoided by measuring the rise and fall of the concentration of the dye by shining a light through the vasodilated lobe of the ear on to a photocell (oximetry).

If the dye is injected rapidly into the left ventricle the end-diastolic volume EDV (as well as the cardiac output) can be calculated from the exponential curve provided that the heart beat is recorded simultaneously. If the concentration of the dye in the blood ejected at one ventricular contraction is C_N and the concentration at the previous beat is C_{N-1} then

$$EDV = \frac{SV}{1 - \dfrac{C_N}{C_{N-1}}}$$

when SV is stroke volume. For example if the concentration is halved at each beat the ventricle must eject half its contents into the aorta and admit the same volume from the left atrium; EDV must be twice SV. The residual volume at the end of systole is of course EDV − SV. Serum albumin, labelled with a radio-isotope, may be used instead of a dye, and estimated in the blood by an appropriate isotope-counting technique. The indicator–dilution method of estimating the cardiac output is unreliable for giving a single 'absolute' value, but useful to compare short-term changes in one subject as shown in Figure 26.57.

Table 26.58 Distribution of the cardiac output at rest and during the steady state of supine leg exercise in a normal subject and a patient with mitral stenosis and cardiac failure (oxygen uptake during exercise 500 ml/min/m²)

Blood flow (ml/min)	Normal subject		Patient with mitral stenosis	
	Rest	Exercise	Rest	Exercise
Splanchnic	1400	1300	800	400
Renal	1100	900	650	300
Cerebral	750	750	600	600
Coronary*	250	450	300	500
Resting skeletal muscle	700	700	700	350
Leg—muscle	500 ⎫	5000	500 ⎫	3400
skin	250 ⎭		50 ⎭	
Skin (other than leg)	250	1000	100	50
Other organs	600	400	300	200
Cardiac output	5800	10 500	4000	5800

*It is likely that when the rise in cardiac output due to exercise reaches high levels the proportion passing through the coronary circulation falls below 4 per cent.

O. L. Wade & J. M. Bishop (1962). *Cardiac Output and Regional Blood Flow*. Oxford: Blackwell.

RETURN OF BLOOD TO THE HEART

Blood flows from the veins to the atria because the pressure in them is higher than it is in the atria and it is usually held that the heart, unlike mechanical pumps, exerts no suction on the entering fluid. However, ventricular filling is very fast in early diastole, blood rushing into the ventricles faster than it is ejected in systole. Furthermore, immediately after the end of systole the ventricles are not just flaccid bags distended by the pressure within them; there is an elastic recoil of the contracted muscle which rapidly lowers the pressure in the chambers promoting an increased flow of blood from atria to ventricles. Negative pressures up to 11 cm H_2O have been recorded in the left ventricle of the dog with the mitral orifice temporarily clamped.

An important factor in returning blood to the heart is, of course, the action of the heart itself in forcing blood into the blood vessels, a system of tubes with a limited capacity. If the capacity of the vascular system exceeds the ability of the heart to fill it as the result of excessive dilatation of the vessels the venous return is greatly diminished, the output of the heart falls and the circulation may fail.

The pressure within the chest, but outside the lung, that is the *intrathoracic pressure*, is normally negative, that is a little below atmospheric pressure, and this reduced pressure is exerted on the large veins and atria. Accordingly blood flows from the abdomen and other parts where the pressure is above atmospheric into the thoracic veins where the pressure is below atmospheric. The effective filling pressure of the right side of the heart is therefore the difference between the extrathoracic venous pressure and the diastolic pressure in the right atrium (the pressure in the relaxed atrium is the same as the intrathoracic pressure). The intrathoracic pressure becomes more negative during inspiration and the filling pressure therefore higher. During inspiration the descent of the diaphragm causes an increase of intra-abdominal pressure which not only aids the return of blood to the thorax but also augments the flow of portal blood through the liver into the inferior vena cava.

The importance of this *respiratory pump* is shown in the performance of static effort, such as pulling on a tug-of-war rope or blowing a trumpet (Fig. 26.59). Both require considerably expiratory effort so the pressure in the thorax becomes positive (see Valsalva manoeuvre, p. 564). The return of blood to the heart is much reduced or ceases entirely, the superficial veins of the head and neck become engorged and the mean arterial blood pressure falls. The blood flow through the brain may be so much reduced that the subject first feels giddy and then loses consciousness (*syncope*, p. 573). A similar state of affairs can occur during a severe spasm of coughing (*cough*

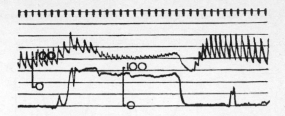

Fig. 26.59 Professional trumpet player blowing Concert A for 17 seconds. Upper curve, arterial pressure. Lower curve, mouth pressure. Calibration in mm Hg and time marker in seconds. (M. Faulkner & E. P. Sharpey-Schafer (1959). *Lancet* **i,** 685–686.)

syncope) in which the intrathoracic pressure may be over 200 mm Hg.

The return of the blood to the heart is also assisted by the contraction of muscles all over the body, especially in the legs. When the muscles contract blood is squeezed out of the capillaries and smaller veins within the muscles into the larger veins between the muscles. This blood must flow towards the heart because the valves in the veins prevent back flow. The alternate contractions and relaxations of the leg muscles that occur in walking serve to drive the blood back to the heart. The large external veins of the legs drain into the deep veins within the muscles. The intramuscular pressure in the calf muscles in man can be measured by passing a fine catheter into them; at rest it is about 10 mm Hg, on standing 30 mm Hg, while a maximal contraction may raise the pressure to 200 mm Hg. Thus the *muscle pump*, as Fig. 26.60 shows, is quite powerful. A person standing rigidly at attention for a prolonged period may eventually faint because the reduced venous return to the heart caused by inactivity of the muscle pump results in a diminished output of the ventricles and a decreased blood supply to the brain.

The size and distensibility of the pericardial sac obviously set a limit to the filling of the heart in diastole. If the sac is occupied with fluid or blood neither atria nor ventricles can dilate fully, and so the heart contains less blood when it contracts. The output per beat is therefore diminished and a form of cardiac failure (cardiac tamponade) may follow. If the pericardial pressure rises above the venous pressure, as may happen after a stab wound of a ventricle, blood cannot enter the heart and the circulation ceases.

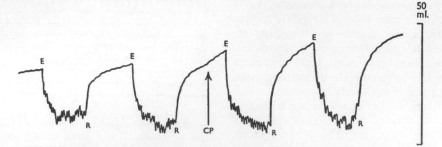

Fig. 26.60 A record of the changes in volume of the calf of the leg measured with a plethysmograph (p. 541) to show the effects of the 'muscle pump'. A fall in volume is indicated by a downward movement of the trace. At E a pedal was pressed down once per second for 10 sec; R indicates the beginning of a rest period of 10 sec. At CP a cuff just above the knee was inflated to 90 mm Hg until the end of the trace. During each period of exercise blood was forced out of the calf and its volume diminished. Even when the venous pressure was raised to 90 mm Hg blood was forced out of the calf by the exercise. (H. Barcroft & A. C. Dornhorst (1949). *Journal of Physiology* **109**, 402.)

CARDIAC OUTPUT AND VENOUS RETURN

The heart and vessels contain about 5 litres of circulating blood. If the heart were suddenly stopped and all the pressures throughout the circulation suddenly became equal the mean pressure in the system would be solely a function of the blood volume and the vascular capacitance, and would be about +7 mm Hg. It has been calculated that this pressure would fall to the atmospheric level if 1 pint of blood were removed, and that if the same volume were added the pressure would be raised to about +14 mm Hg. (Fig. 26.2). If the heart is beating and the blood circulating, the addition or removal of blood changes the venous return and influences the cardiac output.

The transport function of the atria has been described earlier: we must now consider the central role played by the *right atrial pressure* in controlling both the venous return and the cardiac output. At first its action may seem paradoxical, for a rise in right atrial pressure both increases the cardiac output and reduces the venous return. The cardiac output is increased, by the Starling mechanism, in response to the increased filling of the ventricle caused by the raised right atrial pressure. The venous return tends to be reduced because the raised right atrial pressure slows down the inflow of blood from the periphery. Except for periods of time lasting not more than a few seconds, the cardiac output is equal to the venous return, equilibrium being maintained by the overall velocity of blood flow through the heart–lung system.

The most important cause of the venous return is the pumping action of the heart but peripheral factors, such as are mentioned above and further considered in the following chapters, are also important. Guyton has proposed an elegant graphical model to display the relationships between the venous return, the cardiac output and the right atrial pressure. Figure 26.61a shows the relationship between the right atrial pressure and the venous return, a *'venous pressure curve'*. The venous return decreases when the right atrial pressure increases. Fig. 26.61b shows a *'cardiac output curve'*, where the right atrial pressure is plotted against the cardiac output, the cardiac output increasing with right atrial pressure. Because the right atrial pressure or *'central venous pressure'* is plotted along the abscissa in each graph and the units are the same (mm Hg for pressure and litres per minute for blood flow) the venous pressure curve may be superposed on the cardiac output curve as in Fig. 26.61c. The point at which the curves intersect represents the 'equilibrium point' of the system where the cardiac output equals the venous return. The regulation of the circulatory system, when considered in this way, involves the recovery of the equilibrium point after transient disturbances of the system, or the establishment of new equilibrium points when homeostatic mechanisms are ineffective:

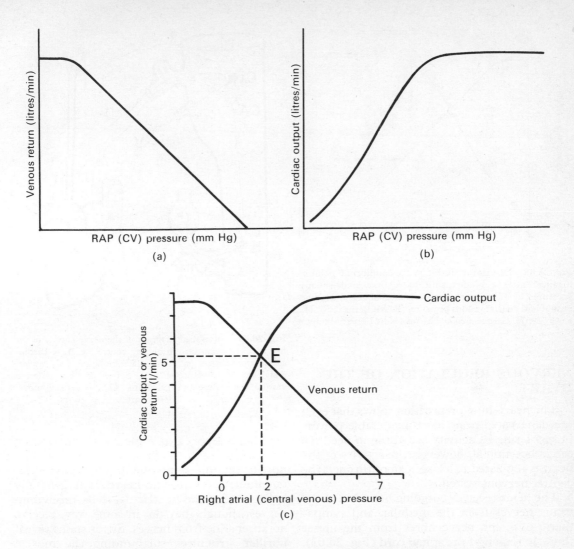

Fig. 26.61 Diagrams to show the relationship between cardiac output and venous return to the heart. (a) Shows the relation between venous return and right atrial (central venous) pressure. Contrary to the usual convention the central venous pressure is plotted on the abscissa. (b) Shows the relation between cardiac output and right atrial pressure. (c) Shows the curves in (a) and (b) superposed. The axes remain the same. E, equilibrium point, where cardiac output equals venous return. (*After* A. C. Guyton (1963). *Circulatory Physiology: Cardiac Output and its Regulation.* London: Saunders.)

for example, in Fig. 26.62 if the venous pressure rises from the equilibrium point to point A, the total blood volume remaining constant, the cardiac output of the next ventricular contraction would be increased to B because of the Starling mechanism. This increased stroke volume would reduce the venous pressure to point C, and so the adjust-

ment would go on during the next few beats until the original equilibrium point is reached again. If we consider again the family of ventricular function curves shown in Fig. 26.33 we can see that a change in the contractile state of the muscle will alter the cardiac output in Figures 26.61 and 62 and the equilibrium point will be a different one.

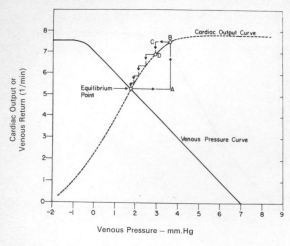

Fig. 26.62 Diagram to show how the equilibrium point is regained after a sudden shift to A (increased venous pressure) by subsequent beat-to-beat changes in cardiac output and central venous pressure. (R. M. Berne & M. N. Levy (1967). *Cardiovascular Physiology*. St Louis: Mosby.)

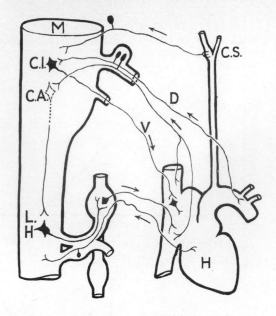

Fig. 26.63 Simplified diagram of the nerve supply of the heart. C.I. = cardio-inhibitory centre. C.A. = cardio-accelerator centre. M = medulla oblongata. L.H. = lateral horn of spinal cord. D = depressor fibre (afferent vagal). V = efferent vagal fibre. C.S. = carotid sinus. H = heart.

NERVOUS REGULATION OF THE HEART

The heart–lung preparation shows that even the denervated heart has considerable powers of regulating its activity according to need. In the intact animal, however, the activity of the heart is regulated to a large extent through the central nervous system.

The heart receives outgoing branches of the vagus nerves from the medulla, and sympathetic outgoing nerve fibres from the upper thoracic region of the spinal cord (Fig. 26.63). The former come from the dorsal nucleus of the vagus, the autonomic vagal nucleus, and travel down the vagus trunk in the neck, passing by way of the superficial and deep cardiac plexuses to the heart. The fibres from the right vagus end mainly round the SA node and those from the left chiefly round the AV node. In these positions they arborize with ganglion cells and new post-ganglionic fibres pass in the coronary plexuses to the atria and to the AV bundle. The sympathetic fibres arise chiefly from the second and third thoracic segments of the spinal cord and pass in the anterior roots as myelinated nerve fibres to the cervical sympathetic ganglia, where they form synapses. From here post-ganglionic non-myelinated nerve fibres pass to the super-

ficial and deep cardiac plexuses, where they intermingle with the fibres of the vagus. The post-ganglionic sympathetic fibres pass without interruption to the heart. It is generally believed that the ventricular cells receive no nerves, though they do, in some way, receive noradrenaline from nerves. Silver stains reveal fibrillar structures surrounding the muscle cells which diminish after denervation of the heart, but silver stains are notoriously unspecific and the fibres may be merely collagenous.

Afferents conveying pain sense run in the sympathetic nerves and enter the sympathetic chain and then pass by way of the white rami to the first four thoracic segments of the spinal cord. Bilateral removal of the upper four thoracic segments of the sympathetic chain relieves cardiac pain and improves effort tolerance partly by removal of pain afferents, and partly by removal of sympathetic efferents.

The vagus nerve. If the vagus nerve is divided in a dog or a cat, and the peripheral end is stimulated it is found that after a latent period of less than a second the heart beats more slowly and the contractility of the atria (but not that of the ventricles) is diminished (negative inotropic response). Thus the contribution by the atria to ventricular filling is reduced with a fall in cardiac output and blood pressure. The atrial pressure rises, the diastolic interval is prolonged, ventricular filling is greater and the stroke volume is increased

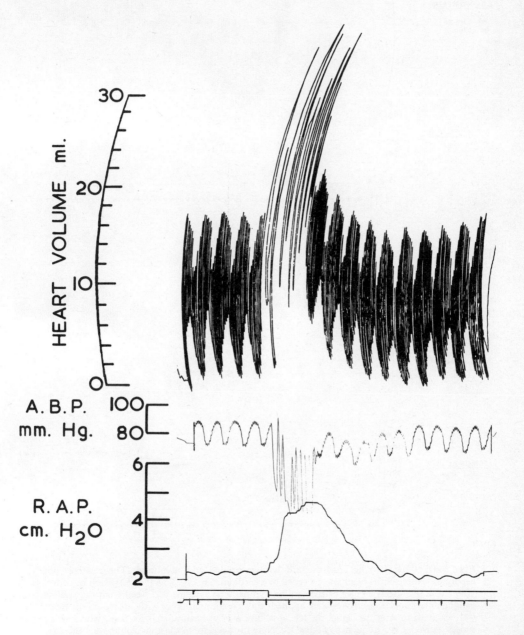

Fig. 26.64 Dog anaethetized with chloralose. Records from above downwards: heart volume; arterial blood pressure; right atrial pressure recorded by a saline manometer attached to 2 mm polythene tubing introduced into right atrium through the external jugular vein; signal, stimulation of peripheral end of right vagus (other buffer nerves intact); time 10-sec intervals. (*By courtesy of A. Hemingway.*)

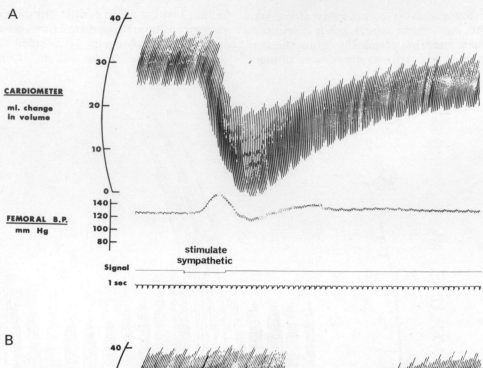

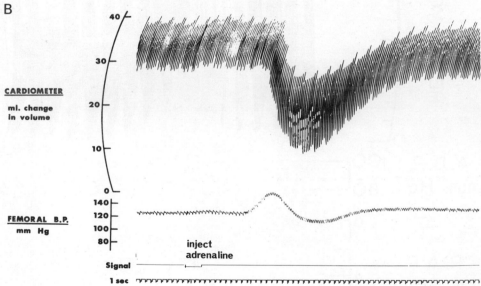

Fig. 26.65 Dog 10·4 kg anaesthetized with morphine and nembutal. The relative (not absolute) volume of the heart was recorded by a cardiometer (Fig. 26.44) and a piston recorder. The left ansa subclavia (part of the cervical sympathetic system) was freed from the stellate ganglion and bipolar electrodes were applied; thus on stimulation only the nerves to the heart were excited. In A at the signal the left ansa subclavia was stimulated for 8 sec with 2 msec pulses delivered at a frequency of 9 per sec. Note the decrease in diastolic volume and increase in stroke output without any alteration in the rate of beating and therefore an increased cardiac output. In this experiment the heart rate was constant at 175 beats per min (stimulation of the right and left sympathetic nerves to the heart usually produces acceleration also; a relatively greater effect is usually observed on the right). In B 4 μg of adrenaline was injected into a femoral vein. The effect closely resembles that shown in A. (*By courtesy of R. J. Linden.*)

(Fig. 26.64). When the stimulation is stronger the heart may stop in diastole but it beats again after the stimulation ceases. Even if stimulation is continued the ventricles (but not the atria) resume contractions. This is called *vagus escape,* and indicates that the ventricles, when deprived of impulses from the atria, can beat independently at their own rhythm. This escape also suggests that the vagus has little direct effect on ventricular activity, and that other factors operate when the heart is slowed. For example, the fall of aortic pressure may bring the aortic and carotid sinus reflexes (p. 512) into play. The main effect of stimulation of the right vagus, distributed chiefly to the SA node, is a reduction of the atrial and therefore the ventricular rate, and stimulation of the left vagus depresses mainly the AV conducting tissue. If the atrial frequency of an anaesthetized preparation is kept constant by applying electrical impulses directly to the atrium (atrial 'pacing'), stimulation of the peripheral end of the vagus causes a decrease in cardiac output due solely to depressed contractility of the atrial muscle. The heart receives fibres from both vagus nerves and stimulation of either the right or the left produces similar effects, namely slowing, diminutation of strength and duration of systole and an increase in the time taken by the excitatory process to pass from the SA to the AV node and along the AV bundle.

Impulses are continually passing down the vagus nerves to exert a retarding action on the heart and it is chiefly by variations in this *vagal tone,* as it is called, that alterations in heart rate are produced. If the vagus nerves to the heart of an animal are cut, or if in man an injection of atropine is given to inhibit the effect of the vagal impulses on the heart, the heart rate rises. The effect of vagal section is most marked in animals capable of periods of sustained exertion (such as the horse and the hare) which normally have a slow resting pulse. In athletes the heart rate is often slow (35 to 40 beats per min) and it may be that the tonic effect of the vagus is more marked in them than in non-athletic persons.

The sympathetic nerves. Stimulation of the sympathetic nerves affects mainly the SA node. After a latent period of six to ten seconds the rate of the heart increases at the expense of a shortening of the diastolic interval. The force of contraction of both atria and ventricles is increased (positive inotropic response) (Fig. 26.65); the cardiac output is increased and the atrial pressure is lowered. The sympathetic nerves appear (like the vagus nerves) to be constantly transmitting impulses to the heart because after they are cut the heart rate decreases slightly.

When adrenaline or noradrenaline is added to the blood of a denervated heart-lung preparation the ventricles beat very forcibly, both systole and diastole are shortened, the heart volume diminishes, the venous pressure falls, the arterial pressure rises, the oxygen usage increases and the coronary flow increases. In the intact animal, however, the effects of these two substances are different because of the regulatory power of the cardiac reflexes

Sarnoff has found that in the dog the cardiac sympathetic nerves, particularly those from the left stellate ganglion, increase cardiac contractility. This can be shown by relating, at constant heart rate, length of fibre to work done with and without increase in sympathetic activity produced by electrical stimulation of the stellate ganglion (Fig. 26.66). The cardiac output at any venous filling pressure is then dependent on the degree of sympathetic tonus. It is, of course, well known that fear, anger and fright can produce sudden striking increases in cardiac output although as a rule the rate is also increased. Adrenaline, however, can produce in man an increase in output without any change in rate (Fig. 26.67).

The importance of the cardiac nerves can be shown if the heart is completely denervated *in vivo* by section of the vagi and ansae subclaviae, or by giving the animal both atropine, which blocks the vagus, and propranolol, which blocks the sympathetic β-receptors (Chap. 41). The heart rate is steady at rest and the cardiac output varies according to Starling's law. With exercise the rate increases only slightly but the stroke output can increase considerably according to Starling's law; the maximum cardiac output is less than in the intact animal and the extraction of oxygen from the blood is relatively greater. If propranolol is given alone, so that the vagus nerves are intact, the cardiac output and heart rate are both reduced.

The effect of the autonomic innervation of a cardiac pace-maker can be appreciated by observing the membrane action potentials

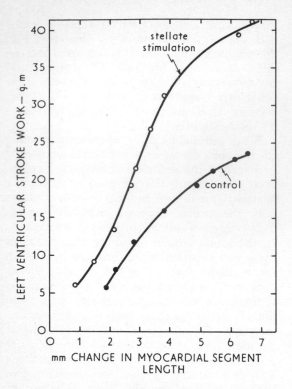

Fig. 26.66 The relation between changes in end diastolic myocardial segment length and left ventricular stroke work before (solid circles) and during (open circles) stellate ganglion stimulation. Heart rate held steady at 171 per minute by atrial pacemaker. Dog; bilateral cervical vagotomy; right stellate ganglion intact. (S. J. Sarnoff & J. H. Mitchell (1961). *American Journal of Medicine* **30**, 747–771.)

recorded by intracellular microelectrodes from spontaneously beating cells in the sinus venosus of the frog's heart. The slow diastolic depolarization characteristic of pace-making cells (Fig. 26.11) causes the resting potential to slope upward towards the threshold (pace-maker potential) (Fig. 26.68). When the vagus nerve is stimulated weakly, or a small dose of acetylcholine is given, the pace-maker potential rises more slowly so the critical value for the sudden rise in depolarization is reached more slowly and the heart beats more slowly. Stronger vagal stimulation (Fig. 26.69) suppresses the pace-maker potentials and stabilizes the membrane potential at an even greater value and so arrests the heart; this effect is possibly due to an increased permeability to potassium ions.

When the sympathetic nerve is stimulated

(Fig. 26.70) the results are in obvious contrast to those obtained on vagal stimulation; the pace-maker potential rises more quickly, the threshold for impulse propagation is reached more quickly and the heart beats follow one another more closely. A full analysis of this effect in terms of ionic permeability has not yet been achieved.

Afferent nerves and cardiac reflexes. Since the efferent vagal fibres arise in the dorsal nucleus of the vagus, and since afferent fibres of the vagus end in the same region, it is convenient to speak of this area of the medulla oblongata as the cardio-inhibitory centre (Fig. 26.63). The vasomotor centre controlling the calibre of the blood vessels is also situated in this part of the brain and the two centres have numerous interconnecting fibres.

It has not been possible to define exactly the position of the cardio-accelerator centre which may lie near the cardio-inhibitory centre in the medulla or perhaps in the hypothalamus. Nerve fibres pass down from the centre to the cells in the intermediolateral tract of the grey matter of the spinal cord from which arise the sympathetic cardiac nerves.

The two chief afferent nerves concerned in the regulation of cardiac activity are (1) those which arise in the heart and aorta and pass up the vagus, and (2) those which arise in the carotid sinus and pass by the glossopharyngeal nerve to the medulla oblongata (Fig. 26.63). In addition, myelinated nerve fibres pass from the heart by various routes through the inferior cervical and the upper four thoracic ganglia and by the upper four thoracic posterior nerve roots into the spinal cord. These afferents probably subserve pain sensation.

Cardio-aortic nerve. The main ingoing or afferent nerve, the cardio-aortic nerve, arises from endings in the left ventricle and aortic arch. In the rabbit the fibres form a separate nerve bundle called the depressor nerve. If this nerve is cut and the peripheral end stimulated there is no effect, but stimulation of the central end causes a reflex slowing of the heart and a fall of blood pressure. In most animals the cardio-aortic fibres are carried in the vagus trunk and stimulation of the central end of a cut vagus produces cardiac slowing and a fall of blood pressure. The afferent impulses travel in the vagus to the cardio-

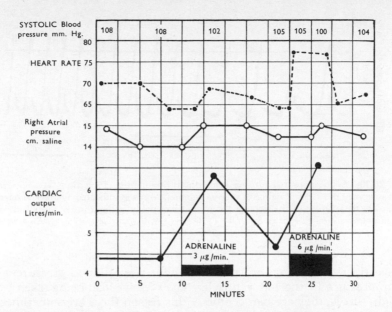

Fig. 26.67 Effect of adrenaline infusion at 3μg a minute in a healthy volunteer. Cardiac output (Fick principle) increases without change in heart rate or right atrial pressure or rise in blood pressure. When the dose of adrenaline was doubled, the heart rate increased. (J. McMichael & E. P. Sharpey-Schafer (1944). *British Heart Journal* **6**, 38.)

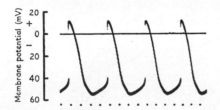

Fig. 26.68 Action potentials of a pace-maker fibre. Record obtained with an intracellular electrode from a frog sinus venosus beating spontaneously. The electrical events during four cardiac cycles are shown. Each cycle starts with a slow depolarization (upward deflection) during diastole. At a membrane potential of about −40 mV the rate of depolarization increases suddenly and the oscilloscope beam flies steeply upwards. The horizontal line indicates zero membrane potential. Time scale: 5 dots represent 1 sec. (O. F. Hutter & W. T. Trautwein (1956). *Journal of General Physiology* **39**, 715–733.)

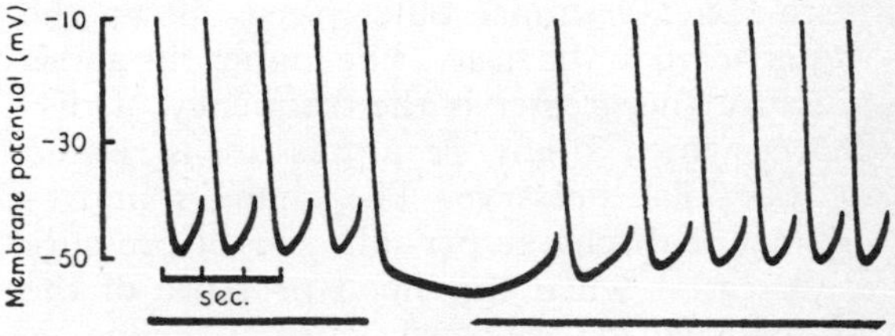

Fig. 26.69 Inhibitory polarization of frog's sinus venosus during vagal stimulation. The record is taken at an amplification that shows only the lower part of the action potential. During the fifth cycle, vagus stimulation at 20/sec starts and continues for about 3 sec as indicated by the break in the reference line. This causes a suppression of pace-maker potentials and a gradual increase in the membrane potential (downward deflection) to a value 9 mV higher than that usually reached during diastole (O. F. Hutter & W. T. Trautwein (1956). *Journal of General Physiology* **39**, 715–733.)

inhibitory centre and efferent impulses pass down the other intact vagus nerve and produce cardiac slowing. If both vagus nerves are cut the heart is not slowed when the central end of one vagus is stimulated but there is usually a fall of blood pressure due mainly to a dilatation of the abdominal vessels supplied by the splanchnic nerves. It might be expected that cutting of both vagus nerves would cause a rise of blood pressure but in an experiment in a conscious man blocking of both vagi did not do so; stimulation of vagal afferents in man does not always produce a fall in arterial pressure.

The Heart 513

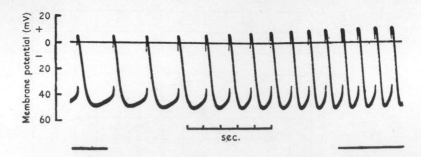

Fig. 26.70 Action potentials during sympathetic stimulation. Frog sinus venosus treated with atropine 1 in 10⁶. Stimulation of vago-sympathetic trunk as indicated by break in reference line. Note the increase in the slope of the pace-maker potentials and in the amplitude of the action potentials (O. F. Hutter & W. T. Trautwein (1956). *Journal of General Physiology* **39**, 715–733.)

The cardio-aortic nerve endings are stimulated when the pressure in the left ventricle and aorta is high, that is, they are baroceptors (Fig. 28.10). In the dog these endings are stimulated so long as the pressure is above 60 mm Hg. If a record of the nerve impulses is made at the same time as a record of the blood pressure it is found that a burst of impulses passes up the cardio-aortic nerve at each systole and a similar burst passes down the vagus nerve to the heart. The higher the aortic pressure the greater is the frequency of discharge which at any *mean* pressure is related to the pulse pressure. The changes in frequency of discharge per unit rise of pressure are greatest when the blood pressure of the animal is within the usual ('normal') range.

Carotid sinus nerve. The dilatation at the beginning of the internal carotid artery, the carotid sinus (Fig. 26.71), has a very rich sensory innervation; the receptors, which lie among the collagen fibres of the tunica adventitia, are sensitive to stretching like those of the cardio-aortic nerve. If the carotid sinus is encased in a rigid cement that does not allow it to expand no reflex effects are obtained. The fibres pass to the medulla in a branch of the glossopharyngeal nerve. Electrical stimulation of the carotid sinus or its nerve, or distension of the sinus by increase of the pressure within it, produces a reflex slowing of the heart and a fall of blood pressure exactly as described in the case of the cardio-aortic nerve. In fact, the cardio-aortic and the carotid sinus nerves should be regarded as two afferents of one regulating mechanism for preventing any

undue rise of blood pressure and for avoiding an excessive load being taken by the heart. For this reason these are sometimes called 'buffer' nerves. The effect of stimulation of the endings in man can be shown by digital pressure over the bifurcation of the carotid artery lying behind the medial edge of the sternomastoid muscle at the level of the hyoid bone (Fig. 26.71). Such pressure will often produce some

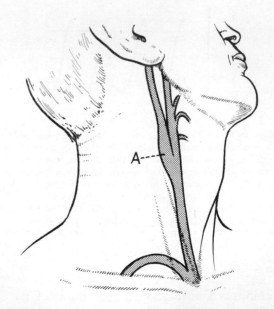

Fig. 26.71 Diagram showing the position of the carotid sinus (A) in the neck.

slowing of the heart beat even in normal individuals but the effect is more striking if the beat is rapid (Fig. 26.72). After denervation of

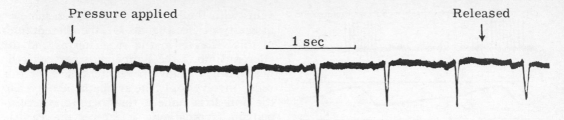

Fig. 26.72 Effect of left carotid pressure on heart rate. The rate at the beginning of the trace is about 150/min and at the end 75/min. The nerve endings in the adventitia of the carotid sinus normally discharge when the pressure in the sinus rises but these baroceptor ending also discharge if the sinus is distorted by traction on the carotid artery or, as in this case, by direct pressure.

the carotid sinus in man baroceptor activity disappears at once and recovers very little; the systemic blood pressure is usually raised about 25 mm Hg.

The reflex effects upon the heart of alterations of pressure within the carotid sinus have been demonstrated by Heymans. He isolated the carotid sinus of a dog from the rest of the circulation, leaving the carotid sinus nerve intact and perfusing the sinus by a pump. When the pressure in the isolated carotid sinus was raised the systemic pressure fell and the heart slowed; a further rise of carotid sinus pressure was followed by a further fall of systemic blood pressure. When the carotid sinus pressure was reduced to its original value the heart accelerated and the systemic blood pressure rose.

The impulses in the carotid sinus nerve are shown beautifully in the tracing of Bronk and Stella from a single nerve ending (Fig. 26.73). It can be seen that the number of afferent impulses passing along the nerve is proportional to the carotid sinus pressure and that during diastole, when the pressure is falling, the frequency of the impulses is less. (See also Fig. 26.74).

The carotid body. The carotid body situated at the bifurcation of the common carotid artery contains numerous sinusoidal capillaries which bring the blood into close contact with glomus cells richly supplied with afferent nerve fibres that join the carotid sinus nerve. The nerve endings are sensitive to alterations in the composition of the blood in the sinusoids and are, therefore, called *chemoceptors*. Similar chemoceptor tissue is found at various positions on or near the arch of the aorta.

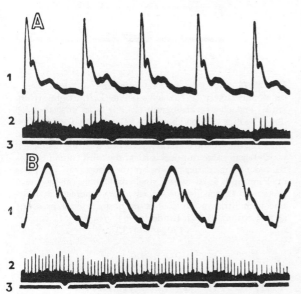

Fig. 26.73 In each tracing curve 1 shows the arterial pressure registered by a membrane manometer; curve 2 shows the discharge from a single end-organ in the carotid sinus of a rabbit; curve 3 gives the time in fifths of a second. In A, B.P. = 55 mm Hg. In B, B.P. = 135 mm Hg. (D. W. Bronk & G. J. Stella (1932). *Journal of Cellular and Comparative Physiology* **1**, 113.)

The chemoreflexes can be studied by isolating the carotid bifurcation, care being taken to keep the nerve supply and the venous drainage of the carotid body intact. In animals, perfusion of the bifurcation with blood of low Po_2 (low oxygen tension) causes a rise of blood pressure by increasing the sympathetic outflow. In man hypoxia produces tachycardia before and after removal of the carotid bodies; the increased heart rate could be due to stimulation of aortic bodies or due to liberation of adrenaline or a direct effect on the SA node.

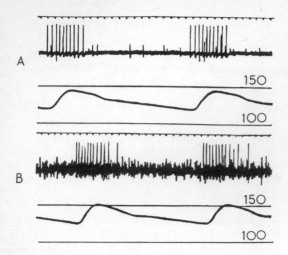

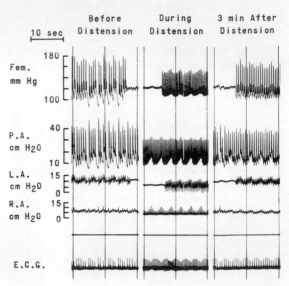

veins with the left atrium, causes an increase in heart rate (see Fig. 26.75); the afferent limb of this reflex is in the vagal nerves and the efferent limb in the sympathetic nerves to the heart. Surprisingly the efferent limb of this reflex involves only the sympathetic nerves to the sinu-atrial node so that there is no accompanying sympathetic effect on the cardiac muscle, only an effect on heart rate.

Fig. 26.74 Cat. Thiopentone anaesthesia. Right sinus nerve cut centrally and a thin slip laid on saline wick electrodes. Action potentials recorded on oscillograph via resistance-capacity coupled amplifier. Blood pressure recorded from femoral artery by condenser manometer. Records from above downwards: time trace 20 mse, electroneurogram of this slip of sinus nerve, blood pressure. Calibration lines for pressures of 150 and 100 mm Hg are shown. A = Cat breathing air spontaneously. B = Cat breathing 10 per cent. O_2 in N_2 spontaneously. The large spikes, about 7 mm in the diagram, are baroceptor discharges; the smaller (2 mm) spikes are chemoreceptor discharges. Note the increase in chemoreceptor activity during hypoxia. (C. Heymans & E. Neil (1958). *Reflexogenic Areas of the Cardiovascular System.* London: Churchill.)

Fig. 26.75 Records of pressure in femoral artery (Fem), pulmonary artery (P.A.), left atrium (L.A.), right atrium (R.A.) and electrocardiogram (EGG) in an anaesthetized dog. The first column shows the values immediately before distending the balloons in the pulmonary-vein-atrial junction; the second column was recorded during distension which began three minutes earlier; the third column shows the record 3 min after removal of the distension. The parts of Fem and L.A. with small oscillations show mean pressures. Distension produced a large increase in heart rate with little or no change in mean arterial pressure. (J. R. Ledsome & R. J. Linden (1964). *Journal of Physiology* **170,** 456–473.)

It is possible that in man, as in the dog, the carotid bodies are mainly concerned with respiratory reflexes while the aortic bodies take part mainly in circulatory reflexes. Reduction of Po_2 causes a great increase in the number of impulses in the nerves from the chemoceptors (Fig. 26.74). A fall of blood pressure, as after severe haemorrhage, reduces the carotid blood flow so severely that the glomus cells are stimulated by hypoxia (see above). Chemoceptor impulses produced by hypoxia usually cause the heart to slow if ventilation is controlled but if the ventilation is allowed to increase secondary reflexes from the lungs may cause an acceleration of the heart.

Atrial reflexes. Linden and his colleagues have shown that stimulation of atrial receptors situated on the right side of the heart at the junctions of the superior and inferior venae cavae with the right atrium and on the left side of the heart at the junction of the pulmonary

THE RATE OF THE HEART

The heart rate counted under basal conditions in young healthy adult males lies between 60 and 65 beats per minute, the figures for women being somewhat higher. However, under ordinary conditions the heart rates of men average about 78, and of women

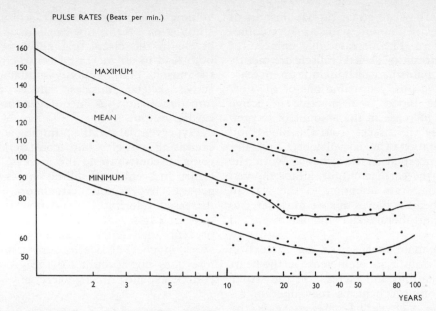

PULSE RATES (Beats per min.)

Fig. 26.76 Diagram to show variation of resting pulse rate with age constructed from data of Tigerstedt, *Lehrbuch der Physiologie des Kreislaufes.*

84 beats per min and the American Heart Association accepts between 50 and 100 heart beats per minute as the normal range. The heart rate tends to fall as age advances (Fig. 26.76). The cardiac rate increases progressively with increasing exercise up to some 200 beats per min but the rate of increase is less in trained subjects. The rate at which the cardiac frequency returns after exercise to the pre-exercise level is used as an index of physical fitness, the initial rate ideally being restored within 2 minutes of the end of exercise.

The cardiac rate is increased by all kinds of emotional disturbances (Fig. 26.77) and, conversely, it falls progressively during sleep. The difference between the waking and sleeping pulse rates may therefore be of some significance as an index of emotional tension. The heart rate is affected by body temperature and observations on a large number of individuals have shown that the heart rate can be predicted with reasonable accuracy by the formulae:

$$11 \times °F - 1018 = \text{heart rate,}$$
$$\text{or} \quad 20 \times °C - 674 = \text{heart rate.}$$

In other words, the heart rate increases on the average by 11 beats per minute for each °F rise in temperature (20 beats per min per °C).

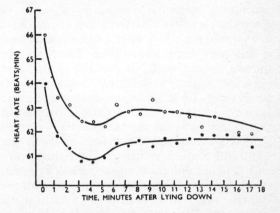

Fig. 26.77 Mean heart rate of forty-six healthy young men for 18 min after lying down. ○——○ , on first visit to laboratory; ●——● , on second visit, approximately 8 days later. (J. M. Tanner (1951). *Journal of Physiology* **115**, 391–409.)

THE CARDIAC RESPONSE TO EXERCISE

Muscular exercise is the main challenge to the transport function of the circulation. During exertion a man's oxygen consumption can increase twenty-fold and the cardiac output during exercise is directly related to the oxygen consumption. Studies on athletes have

The Heart 517

shown that they can attain cardiac outputs of over 40 lites per minute with stroke volumes of over 200 ml. The increased oxygen needs of actively contracting skeletal muscle are met by several mechanisms which include an increase in cardiac output, redistributions of the flow of available blood from inactive to active tissues, an increase in the amount of oxygen extracted by the tissues from the blood and the accumulation of an oxygen debt (Chap. 40). We are concerned in this chapter with the increase in the cardiac output, although even while limiting our attention to the way in which the heart achieves this we must refer to important peripheral factors that are discussed elsewhere.

The two main mechanisms for increasing the cardiac output are an increase in the heart rate (tachycardia) and increase in the stroke volume; both mechanisms may operate together. The tachycardia of exercise is the result of adrenergic influences on the cardiac pace-maker. The heart rate of a resting subject can be increased, without the action of the autonomic nervous system, by the use of electrical artificial pace-makers. Braunwald has shown that when the heart rate is increased in this way from about 80 to 120 beats per minute the cardiac index does not change, and even falls slightly at the faster heart rates, with an inverse linear relationship between the stroke volume and the heart rate. However, in these circumstances the tachycardia is accompanied by reduced dimensions of both the right and the left ventricles, the reduction in end-diastolic volume being about half a normal stroke volume. These findings accord with the fact that tachycardia is achieved mainly at the expense of diastole when the ventricles are filling. Moreover, during this experimental tachycardia the force–velocity curve is displaced upward, V_{max} being increased with no change in P_0. This indicates an increased contractility of the ventricular muscle attributable purely to the heart rate and independent of any direct effect of neurohumoral stimuli on the myocardium.

During exercise, however, only part of the displacement of the force–velocity curve is produced by an increase in rate alone, and during severe exertion the tissue requirements can only be met by a cardiac output that is further augmented by an increased stroke volume. In addition, the increased sympathetic stimulation of the myocardium and the increase in the circulating adrenergic amines both lead to an increased contractile state of the muscle. Peripherally, vasodilatation in the active skeletal muscles reduces the aortic impedance and thus encourages the increased cardiac output.

Hyperventilation, the pumping action of the exercising muscles and constriction of the veins all contribute to the increase of venous return and ventricular filling but, as part of the general 'hyperkinetic circulatory state', the increased velocity of blood flow through the heart and lesser circulation is great enough to prevent the central venous pressure from rising much. The Starling mechanism therefore does not account for the greater stroke volume expelled during exercise.

THE CORONARY CIRCULATION

About 4 per cent of the output of the left ventricle passes into the coronary vessels so the blood flow in the aorta is not the total cardiac output; that can only be found by measuring the flow in the pulmonary artery.

Because of the peculiarities of the venous drainage (p. 462) it is difficult to make an accurate estimate of the coronary flow in man. Samples of blood can be obtained from the coronary sinus through a cardiac catheter (p. 494) introduced into the mouth of the sinus through the right atrium. If a sample of aterial blood is obtained at the same time, while the patient is breathing nitrous oxide or radioactive krypton, then the arterio–venous gas difference can be obtained. This method measures only the blood flow through the left ventricle but it is unlikely that the flow through the right ventricle is substantially different. The coronary blood flow at rest is about 200 ml per min (65 ml per 100 g per min), about 4 per cent of the cardiac output.

During systole the coronary vessels lying within the heart muscle are compressed. This compression increases the total resistance of the coronary vascular bed and actually momentarily reverses the flow in early systole (Fig. 26.43). Since the blood pressure in the left ventricle during systole is slightly higher than it is in the aorta the intramyocardial tissue pressure in the inner half of the ventricular wall is greater than the arterial blood pressure and the total blood flow, systolic and diastolic, in subendocardial muscle is about half that in the superficial part of the myocardium. This flow gradient is partly offset by the greater density of capillaries in the deeper myocardium. However, ischaemic necrosis is most widespread in the subendocardial muscle although the diffusion capacity for ^{86}Rb of the coronary capillary bed is ten times greater than that in skeletal muscle and myocardial metabolism is anaerobic in the deepest layers of the myocardium of dogs severely exercised in a treadmill.

The pressure in the right ventricle is, of course, at all times much lower than the pressure in the aorta so perfusion of the coronary vessels in the right ventricle is much more efficient. It is probably for these reasons that ischaemic necrosis affects mainly parts of the left ventricle rather than the right. About 70 per cent of the total coronary blood flow takes place during diastole.

When the heart beats faster and more strongly its oxygen consumption is increased, so during exercise the myocardial blood flow is increased, like the blood flow to the skeletal muscle and the skin. The coronary blood flow during severe exercise with a cardiac output of 25 litres per minute would amount to 1 litre per minute (but see Table 26.58). The coronary vessels are dilated or constricted by an autoregulatory system in which the myocardial oxygen tension determines the blood flow by altering the coronary vascular resistance.

Hypoxia can cause vasodilatation and an increase of flow of as much as 500 per cent. Since neither carbon dioxide nor lactic acid can increase the flow by more than 50 per cent it is assumed that the hypoxic muscle fibres liberate a vasodilator substance. A vasodilator metabolite has not been identified but adenosine (p. 64) is a likely candidate. If the coronary circulation is occluded the Po_2 of the myocardium falls and when the occlusion is released the blood flow increases (reactive hyperaemia, Chap. 28) and the Po_2 rises above normal. This suggests that the capillaries, not arterio–venous shunts, carry the increased flow. Large changes of pressure in the aorta can also influence coronary perfusion. When the arterial pressure falls the consequent release of adrenergic substances from the adrenal medulla produces coronary vasodilatation and increased coronary blood flow while the flow through the skin, kidneys and splanchnic vessels is severely restricted. The circulation through the heart, like that through the brain, is therefore maintained.

The coronary blood vessels are supplied with both sympathetic and para-sympathetic nerves. In animals sympathetic stimulation and noraderenaline and adrenaline produce an increase in blood flow but whether this is due to vasodilatation, to alteration in heart rate and aortic pressures, or to metabolic effects has not yet been clarified. Drugs that block either the α or β-adrenergic receptors have been used to study the effect of sympathetic stimulation on the coronary vascular bed of the dog. Sympathetic vasoconstriction in cardiac muscle is only slight, in contrast to that in skeletal muscle, and is largely overcome by local chemical vasodilatation. Unlike skeletal muscle vessels, the coronary vascular bed is not supplied by sympathetic vasodilator fibres. During exercise, although diastole is shortened because of tachycardia a normal heart does not suffer a reduced blood supply. If the coronary arteries are diseased, however, the myocardium can become ischaemic during exercise: this is the underlying mechanism of cardiac pain on exertion (angina pectoris).

DISORDERED CARDIAC RHYTHMS

The clinical aspects of cardiac arrhythmias lie outside the scope of this textbook but their mechanisms are of interest to students of physiology because they illustrate some important general principles.

The mechanism of many of the commonest arrhythmias was first suggested by Mackenzie after careful observation of the jugular venous pulse and later by recording the venous pulse at the same time as the arterial pulse with his famous 'polygraph'. It was, however, Lewis's use of the electrocardiograph that confirmed their nature. The precise mechanisms of many of the more complicated arrhythmias are now being elucidated by the technique of intracardiac electrocardiography, excitation being directly recorded from various parts of the special conducting system (Fig. 26.51).

The premature discharge of a pace-maker

cell lower than the sinus node (ectopic focus) produces premature excitation and contraction of the chamber where it is sited (extrasystole). When the ectopic focus lies above the AV node there is a premature excitation of the whole heart (*supraventricular ectopic beat*) (Fig. 26.78).

ventricles (*ventricular ectopic beat*). An abnormal QRST occurs with no preceding P wave. If the excitation spreads backward up the AV conducting system there will be retrograde excitation of the atria, and a P wave may follow the QRS complex (Fig. 26.79a). During

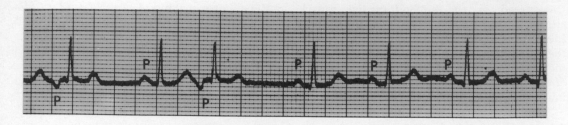

Fig. 26.78 Electrocardiogram, lead II, showing two *supraventricular extrasystoles*. The ectopic focus is low in the atrium or in the AV junctional tissue. Atrial excitation occurs in a direction the reverse of normal. The first and third P waves are therefore inverted. Ventricular conduction is normal so the abnormal premature P waves are followed by normal premature QRST complexes.

Because the spread of atrial excitation is abnormal it may cause an abnormal P wave, but ventricular excitation occurs normally through the AV node and bundles; the QRST complex is therefore normal. When the ectopic focus lies below the division of the bundle of His into its main right and left bundles, or in the potential pace-making cells of the Purkinje network in the ventricles, there is a premature and abnormal spread of excitation in the

ventricular ectopic beats the abnormal ventricular contraction is inefficient, and the intraventricular pressure and stroke volume are low. The distension of the aortic root is therefore small and a pulse wave may not be propagated to the wrist.

The premature excitation of the myocardium after the discharge of an ectopic focus may leave the muscle refractory to the next pace-making stimulus arising in the sinu-atrial

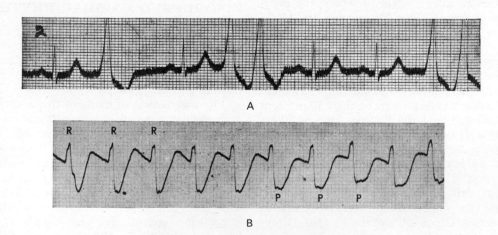

A

B

Fig. 26.79 (A) Electrocardiogram, lead II, showing *ventricular extrasystoles*. The bizarre ventricular complexes are not preceded by P waves. (B) *Ventricular tachycardia*. Bizarre rapid ventricular complexes, not quite regular. P waves occur at a slower rate than QRS complexes. An ectopic ventricular focus is discharging spontaneously at a fast rate.

node. Sometimes retrograde atrial excitation prematurely discharges the sinus node. Ectopic beats are then followed by a longer diastole than usual ('compensatory pause'). This may be felt by the patient as a 'missed beat'. A series of ectopic beats arising from a focus in the atria or ventricles can produce a paroxysm of tachycardia (atrial or ventricular) that may impede filling of the heart and so lower the cardiac output (Fig. 26.79b).

During atrial systole in a normally functioning heart all the atrial muscle fibres contract practically simultaneously and, in the ventricular systole that follows, all the ventricular muscle fibres contract in an orderly sequence. In certain diseases of the heart, or after electrical or chemical stimulation this orderly contraction may be replaced by very irregular activity, in which the individual muscle fibres contract in a disorderly sequence all out of step with one another so a fine rapid tremulous movement (*fibrillation*) is seen over the surface of the atria or ventricles or both. The affected chambers of the heart do not contract as a whole and their pumping action is lost. If the ventricles fibrillate (Fig. 26.80) the blood is not pumped out into the systemic

or pulmonary circulations and death rapidly results from anoxia of the vital centres. If, however, fibrillation is confined to the atria the ventricles still force the blood round the circulation and a patient with atrial fibrillation may have a reasonably efficient circulation (p. 491).

If the atrium of the exposed heart of the dog is subjected to a brief period of rapid electrical stimulation (300 to 600 per minute) the atria may change their rhythm and beat about 300 times per minute (*atrial flutter*). This abnormal rhythm also occurs in diseased, or even normal human hearts (Fig. 26.81).

The mechanism of atrial flutter and fibrillation was studied by Prinzmetal and his colleagues with high-speed cinephotography (2000 frames per second) and a cathode-ray oscillograph capable of recording electrical changes occurring as rapidly as 20 000 per minute. In atrial fibrillation minute irregular contractions (M) and also larger rhythmic contractions (L) are continuously present. The M contractions occur throughout the atrial musculature, each one involving an area of the wall from 0·03 to 3 mm in diameter. Superimposed on these are the L contractions which sweep across the atrium fairly regularly at

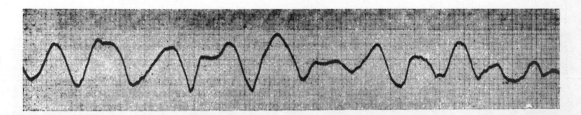

Fig. 26.80 *Ventricular fibrillation.* There are no recognizable P waves or QRS complexes. The chaotic cardiac electrical activity gives rise to random oscillations.

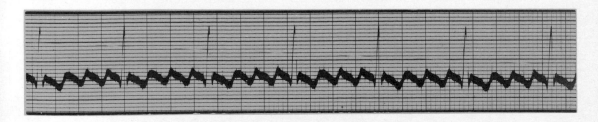

Fig. 26.81 Electrocardiogram, lead II, showing *atrial flutter.* A regular succession of abnormal P (flutter) waves at a frequency of 324 per minute indicates that the atrium is beating regularly at this rate but the ventricle responds only to each fourth beat. This is 4:1 heart block. Time intervals, 40 and 200 msec.

rates of 400 to 600 per minute giving an appearance of fibrillation to the unaided eye. Prinzmetal claimed that all the atrial arrhythmias arose from the discharge of a single ectopic focus. At rates of impulse formation in the ectopic focus below the normal sinus rate premature atrial systoles are produced; when the rate of impulse formation is in the range 100 to 250 per minute atrial tachycardia is the result. Flutter and fibrillation are found at rates of impulse formation of approximately 300 to 400 per minute respectively. Not all workers agree with Prinzmetal. For example Rytand has marshalled much evidence in support of the earlier idea that atrial flutter, when it occurs in patients and is not induced experimentally, is caused by 'circus movement'; that is to say the wave of excitation is trapped in a circular pathway in the atria because of the refractory period and so becomes perpetuated.

In both atrial flutter and fibrillation the rate of stimulation is so rapid that the ventricles are incapable of responding to every impulse reaching them from the atria. When flutter is present the ventricles usually respond in a regular manner to every third or fourth atrial contraction (Fig. 26.81) while in atrial fibrillation the response of the ventricles is irregular

(Fig. 26.82) and often rapid, giving an irregular rapid pulse, and a systolic blood pressure that varies from beat to beat because of the varying length of diastole.

Any disorder of cardiac rhythm that causes very rapid ventricular contractions tends to reduce the cardiac output and may lead to heart failure (see below). Many fast abnormal rhythms, however, can be converted to normal sinus rhythm by a single direct-current electric shock applied across the patient's thorax. The high voltage shock, which lasts only a few milliseconds, is usually triggered by the R wave of the ECG and has an energy value of between 50 and 400 joules.

Since the conduction of excitation in the AV junctional tissue does not normally occupy more than 210 msec a PR interval above this value denotes impaired AV conduction (*incomplete heart block*) (Fig. 26.83). *In complete heart block* (Fig. 26.84) the contractions of the atria and ventricles are completely dissociated from one another, the atria being activated from the sinus node and the ventricles from the slower rate of impulse formation in the AV node. In such a case the ECG shows regularly recurring P waves of normal form and, superimposed upon this, a regular succession of ventricular

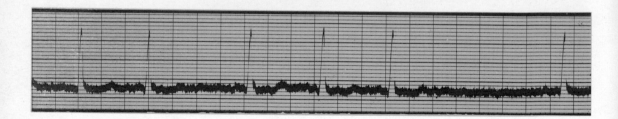

Fig. 26.82 Electrocardiogram, lead II, showing *atrial fibrillation*. The base line is very irregular (see text). No true P waves are present. The interval between QRST complexes is very variable. Time intervals, 200 msec.

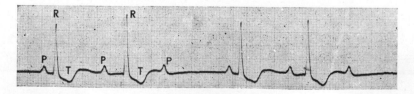

Fig. 26.83 *Incomplete heart block*. Gradual increase of AV block, with progressive increase of PR from complex to complex until the impulse is not conducted to the ventricles. The cycle then begins again. This form of partial AV block is called Wenckebach Periods.

522 Textbook of Physiology and Biochemistry

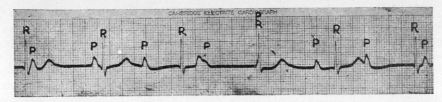

Fig. 26.84 *Complete heart block.* Atrial rate 70 per min. Ventricular rate 50 per min. The atrial and ventricular activities are completely dissociated. The ventricular complexes are normal and this is almost certainly congenital AV block. In such cases the ventricular rate is faster than in cases of acquired block in adults in whom the usual rate is 30 to 35 per min.

complexes at a slower rate (idio-ventricular rhythm).

If large parts of the heart muscle are so damaged that they cannot be excited, the remaining muscle may exhibit spontaneous electrical activity, as shown in Figures 26.85 and 26.86.

When the ventricles either do not contract at all, or do not do so efficiently enough to maintain an adequate cerebral circulation, consciousness is lost and artificial ventilation and cardiac massage may temporarily maintain a circulation.

By rhythmic pressure at a rate of 70 per min on the lower third of the sternum in man the heart can be compressed between the sternum and the vertebral bodies, lateral movements being restricted by the pericardium. Relaxation of the pressure allows the heart to fill again. By repetition of this *external cardiac massage* 60 to 100 times per minute it

is usually possible to produce a radial pulse and a systolic blood pressure of 60 to 100 mm Hg (mean 10 to 20 mm Hg). If the heart stops when the patient is on the operating table one hand can be passed through an incision into the patient's chest and his heart compressed rhythmically between the surgeon's palm and the patient's sternum (*internal cardiac massage*).

The appearance of the ECG gives no information about the force of cardiac contraction: indeed, evidence of electrical activity can be present in the ECG when the heart has lost all function as a pump.

CARDIAC FAILURE

In his experiments with the isolated dog's heart Starling related the cardiac output to the venous pressure. He also emphasized the

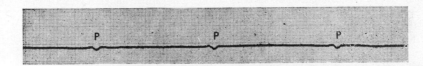

Fig. 26.85 *Ventricular standstill* (or asystole). There are no QRS complexes but atrial activity causes P waves to occur slowly and regularly.

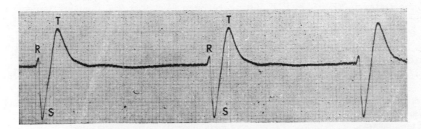

Fig. 26.86 *Ventricular automatism* or idioventricular rhythm with atrial standstill. There are no P waves. The ventricular complexes are bizarre and occur regularly at a slow rate.

difference between a good, that is to say healthy, heart and a fatigued or failing heart. For a given output the healthy heart was working at a low venous pressure and a low diastolic size while the failing heart worked at a high venous pressure and a large diastolic size. Although intact man differs considerably from a heart–lung preparation this state of affairs is still the essence of cardiac failure, which has been defined by Wood as 'a state in which the heart fails to maintain an adequate circulation for the needs of the body despite a satisfactory venous filling pressure'. Starling also pointed out that the right and left sides of the heart might fail separately according to the load imposed upon them and this also to some extent has analogies in clinical practice.

Cardiac failure is a complicated and somewhat controversial subject. Briefly however, whatever the actual cardiac output at the time, failure involves a reduction of output from previous levels and an increase in the venous pressure, apparent at first only during exercise, but later also at rest. The fall in output is accompanied by a redistribution of blood throughout the body (Table 26.58) and an alteration in the function of many organs and tissues. Retention of sodium and water occurs early in cardiac failure and leads to an increased blood volume that further overloads the heart. In some patients the secretion of aldosterone (Chap. 50) is increased and in many there is an increased release of antidiuretic hormone (Chap. 52). Both may contribute to the fluid retention, the mechanism of which is not yet fully understood. The increase of systemic venous pressure must also be a factor in producing oedema and may interfere with hepatic function. Increased pulmonary venous pressure and volume is associated with pulmonary oedema, which reduces the compliance of the lungs and may thus cause the breathlessness that is such a prominent feature of cardiac failure.

REFERENCES

Aviado, D. M. & Schmidt, C. F. (1955). Reflexes from stretch receptors in blood vessels, heart and lungs. *Physiological Reviews* **35,** 247–300.

Berne, R. M. (1964). Regulation of coronary blood flow. *Physiological Reviews* **44,** 1–29.

Berne, R. M. & Levy, M. N. (1964). Heart. *Annual Review of Physiology* **26,** 153–186.

Berne, R. M. & Levy, M. N. (1967). *Cardiovascular Physiology.* St Louis: Mosby.

Bing, R. J. (1965). Cardiac metabolism. *Physiological Reviews* **45,** 171–213.

Brady, A. J. (1964). Excitation and excitation–contraction coupling in cardiac muscle. *Annual Review of Physiology* **26,** 341–356.

Braunwald, E. (1965). The control of ventricular function in man. *British Heart Journal* **27,** 1–16.

Braunwald, E., Ross, J. & Sonnenblick, E. H. (1967. *Mechanisms of Contraction of the Normal and Failing Heart.* London: Churchill.

Brooks, D. K. (1967). *Resuscitation.* London: Spottiswoode, Ballantyne & Co.

Burch, G. E. & de Pasquale, N. P. (1964). *A History of Electrocardiography.* Chicago Year Book Publishers.

Burton, A. C. (1965. *Physiology and Biophysics of the Circulation.* Chicago: Year Book Publishers.

Cowan, C. M., Akbari-Fard, M. & Bing, R. J. (1969). The metabolism of heart muscle in health and disease, In *The Biological Basis of Medicine* (Edited by E. E. Bittar & N. Bittar), Vol. 6, p. 257. London: Academic Press.

Dawes, G. S. (1951). Reflexes from the heart and lungs. *British Medical Bulletin* **8,** 324–330.

Deuchar, D. C. (1964). *Clinical Phonocardiography.* London: English Universities Press.

Durrer, D., Van Dam, R. T., Freud, G. E., Janse, M. J., Meijler, F. L. & Arzbaecher, R. C. (1970). Total excitation of the isolated human heart. *Circulation* **41,** 899–912.

Emslie-Smith, D. (1955). The intracardiac electrogram as an aid in cardiac catheterization. *British Heart Journal* **17,** 219–224.

Fishman, A. P. & Richards, D. W. (Eds.) (1964). *Circulation of the Blood.* New York: Oxford University Press.

Folkow, B. & Neil, E. (1971). *Circulation.* London: Oxford.

Green, D. E. & Goldberger, R. F. (1961). Pathways of metabolism in heart muscle. *American Journal of Medicine* **30,** 666–678.

Guyton, A. C. (1963). *Cardiac Output and its Regulation.* London: Saunders.

HAMILTON, W. F. (Ed.) (1962). *Handbook of Physiology, Section 2 : Circulation*, Vol. I. London: Bailliere, Tindall & Cox.

HARLEY, H. R. S. (1966). Reflections on cardiopulmonary resuscitation. *Lancet* **ii,** 1–4.

HARVEY, W. (MDCXXVII). *Exercitatio Anatomica de Motu Cordis et Sanguinis*. Frankfurt: Fitzer. Translated into English by K. J. FRANKLIN (1957). Oxford: Blackwell.

HAWTHORNE, E. W. (1965). Heart. *Annual Review of Physiology* **27**, 351–394.

HENRY, J. P. & MEEHAN, J. P. (1971). *The Circulation*. Chicago: Year Book Publishers.

HEYMANS, C. & NEIL, E. (1958). *Reflexogenic Areas of the Cardiovascular System*. London: Churchill.

HOLLAND, W. C. & KLEIN, R. L. (1960). *Chemistry of Heart Failure*. Springfield, Ill.: Thomas.

HUTTER, O. F. (1957). Mode of action of autonomic transmitters on the heart. *British Medical Bulletin* **13**, 176–180.

JACKSON, J. R. (1972). Mechanism of ventricular ejection. *British Medical Journal* **iv**, 166–169.

JAMES, T. N. & SHERFF, L. (1968). Ultrastructure of myocardial cells. *American Journal of Cardiology* **22**, 389–416.

KATZ, A. M. & BRADY, A. J. (1971). Mechanical and biochemical correlates of cardiac contraction. *Modern Concepts of Cardiovascular Disease* **40**, 39–45.

KELMAN, G. R. (1971). *Applied Cardiovascular Physiology*. London: Butterworths.

LANGER, G. A. (1966). The ionic basis for control of myocardial contractility. *Progress in Cardiovascular Disease* **9**, 194–203.

LEYTON, R. A. & SONNENBLICK, E. H. (1971). The sarcomere as the basis of Starling's law of the heart in the left and right ventricles. In *Methods and Achievements in Experimental Pathology*, Vol. 5: *Functional Morphology of the Heart* (Edited by E. Bajusz & G. Jasmin), p. 22. Basel: Karger.

LINDEN, R. J. (1965). The regulation of the output of the mammalian heart. *Scientific Basis of Medicine Annual Review*, 164–185.

LUISADA, A. A. (1965). *From Auscultation to Phonocardiography*. St Louis: Mosby.

McMICHAEL, J. (Ed.) (1968). *Circulation*. Harvey Tercentenary Congress. Oxford: Blackwell.

MARCHETTI, G. & TACCARDI, B. (1967). *Coronary Circulation and Energetics of the Myocardium*. Basel: Karger.

MARSHALL, R. J. & SHEPHERD, J. T. (1968). *Cardiac Function in Health and Disease*. Philadelphia: Saunders.

MEESSEN, H. (1968). Ultrastructure of the myocardium. *American Journal of Cardiology* **22**, 319–327.

OLSEN, R. E. (1961). Contractile proteins of heart muscle. *American Journal of Medicine* **30**, 692–707.

OPIE, L. H. (1968–69). Metabolism of the heart in health and disease. Parts 1–3. *American Heart Journal* **76**, 685–698; and **77**, 100–122, 383–410.

PETERSON, L. H. (1960). Symposium on regulation of the cardiovascular system in health and disease. *Circulation* **21**, 739–748.

PICKERING, G. (1960). Starling and the concept of heart failure. *Circulation* **21**, 323–331.

RANDALL, W. C. (Ed.) (1965). *Nervous Control of the Heart*. Baltimore: Williams & Wilkins.

RUSHMER, R. F. (1970). *Cardiovascular Dynamics*. London: Saunders.

RYTAND, D. A. (1966). The circus movement (entrapped circuit wave) hypothesis and atrial flutter. *Annals of Internal Medicine* **65**, 125–159.

SARNOFF, S. J. & MITCHELL, J. H. (1961). The regulation of the performance of the heart. *American Journal of Medicine* **30**, 747–771.

SONNENBLICK, E. H. (1966). The mechanics of myocardial contraction. In *The Myocardial Cell* (Edited by S. A. Briller & H. J. Conn), p. 173. Philadelphia: University of Pennsylvania Press.

STENGER, R. J. & SPIRO, D. (1961). Structure of the cardiac muscle cell. *American Journal of Medicine* **30**, 653–665.

STEPHENSON, H. E. (1964). *Cardiac Arrest and Resuscitation*, 2nd edn. St Louis: Mosby.

Symposium on current problems in cardiac muscle physiology (1965). *Federation Proceedings* **24**, 1396–1437.

Symposium on the physiology of cardiac muscle (1961). *American Journal of Medicine* 649–771.

TACCARDI, B. & MARCHETTI, G. (1965). *Electrophysiology of the Heart*. Oxford: Blackwell.

TIEDT, N., LITWIN, J & SKOLASINSKA, K. (1966). The dynamics of the coronary venous pressure in the dog. *Pflügers Archiv für die gestamte Physiologie des Menschen und der Tiere* **288**, 27–42.

WADE, O. L. & BISHOP, J. M. (1962). *Cardiac Output and Regional Blood Flow*. Oxford: Blackwell.

WATSON, H., EMSLIE-SMITH, D. & LOWE, K. G. (1967). The intracardiac electrocardiogram of human atrio-ventricular conducting tissue. *American Heart Journal* **74**, 66–70.

WEIDMANN, S. (1967). Cardiac electrophysiology in the light of recent morphological findings. *Harvey Lectures* Series **61**, 1–16.

WHIPPLE, H. E. (Ed.) (1964). Cardiac pace-makers. *Annals of the New York Academy of Sciences* **111**, 813–1122.

WHIPPLE, H. E. (Ed.) (1965). Comparative cardiology. *Annals of the New York Academy of Sciences* **127**, 1–875.

27 The circulation

The object of the circulatory system is to bring blood close enough to individual cells so that the local cellular environment can be kept constant by diffusion between blood plasma and extracellular fluid. Diffusion takes place across the thin-walled vessels and the rest of the circulatory apparatus is arranged to keep these vessels adequately supplied with blood. At first sight the blood vessels of the circulation seem very complex in structure and arrangement and, therefore, it is often helpful to think of the circulation in functional rather than in anatomical terms. One way of doing this is shown in Fig. 27.1. We can think of the blood vessels as arranged in a large number of circuits in parallel. Thus there will be one circuit for the renal circulation, one for the skin circulation, one for the muscle circulation and so on. Each of these individual circuits consists of a number of sections arranged in series. The chief function of the first section of each circuit is to damp the huge pressure fluctuations generated by the heart to give a fairly steady driving pressure. Anatomically, these *damping* (or *windkessel*—see below) vessels correspond fairly closely with the arteries. The second section is composed of vessels which offer a high and variable resistance to the flow of blood around the circuit. They act as taps which regulate the flow in any particular circuit. They may be referred to as *resistance vessels* and correspond anatomically with the arterioles. The third section in each circuit is composed of the vessels which permit exchange of material by diffusion across their walls. These may be referred to as the *exchange vessels* and correspond anatomically with the capillaries and some of the thin-walled venules. The final section in each circuit consists of the vessels which contain the bulk of the blood volume. These are referred to as the *capacity vessels* and correspond anatomically with the veins and venules. By contraction and relaxation of the muscle in their walls, they adjust the capacity

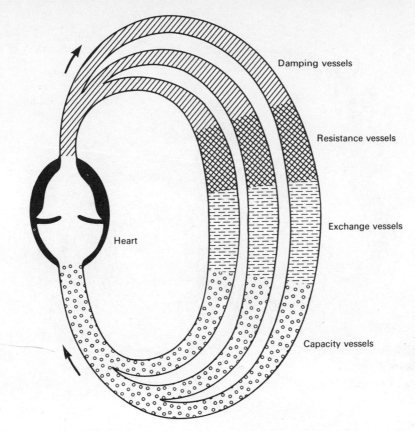

Fig. 27.1 Functional components of the circulation.

of the circulation to meet variations in the blood volume.

Although it is useful to think about each section of the circulation in terms of the chief feature it presents, that is damping, resistance, exchange or capacity, it should be stressed that all vessels present all these features to some extent. For example, all vessels offer some resistance to flow but the greater part of the resistance lies in the arterioles. Similarly, all vessels have the capacity to hold blood but a very large fraction of the blood volume is contained in the venules and veins. Each functional variety of blood vessel and its features will now be considered in turn.

THE ARTERIAL OR DAMPING VESSELS

These vessels have three main features. They convert the intermittent pressure generated by the heart into a fairly steady pressure

head. They distribute blood to the tissue at very low energy cost and the pressure head within them is maintained at a fairly constant value by a complex control system.

STRUCTURE OF ARTERIES

All the blood vessels of whatever size have a smooth inner lining of flattened *endothelial cells* joined edge to edge. This inner lining is continuous from the arteries through the capillaries to the veins and to the internal lining of the heart (the *endocardium*).

The arteries have three main coats and the following description refers mainly to a medium-sized artery (Fig. 27.2). The endothelial layer is surrounded by an elastic layer, the two together forming the *tunica intima*, or inner coat. After death, when the internal pressure falls, the elastic tissue contracts and forms the wavy inner border of an artery seen in histological preparations. The tunica intima is surrounded by smooth muscle fibres which

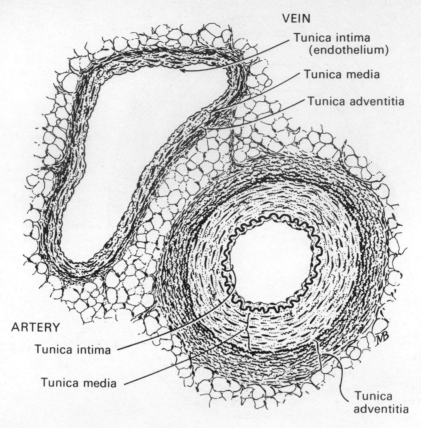

VEIN

Tunica intima
(endothelium)

Tunica media

Tunica adventitia

ARTERY

Tunica intima

Tunica media

Tunica
adventitia

Fig. 27.2 Cross-section of a medium sized artery and its accompanying vein (× 80). The cross-sectional area of a major vein is more than twice that of the corresponding artery; small veins draining a tissue may have six or seven times the cross-section of the little arteries supplying it. Recent measurements suggest that the post-capillary venules may have a larger total cross-sectional area than the capillaries (M. P. Wiedman (1963). *Circulation Research* **12**, 375–378).

pass circularly round the vessel, forming the *tunica media*, or middle coat. Finally there is an external coat, or *tunica adventitia*, of connective (fibrous) tissue which blends with the general fibrous tissue of the body. In the larger arteries the elastic layer takes the form of a fenestrated membrane and there is also a certain amount of subendothelial connective or fibrous tissue. The muscle fibres of the middle coat are reinforced by a network of elastic fibres and the outer coat contains elastic fibres in addition to white fibrous tissue. The large arteries, and particularly the aorta, thus contain a relatively greater proportion of elastic tissue in their walls than do the smaller arteries. The walls of the smallest arteries of all, the arterioles, are almost entirely muscular.

At the point where an artery gives off a branch, or divides into two, the combined cross-sectional area of the two divisions is somewhat greater than that of the original artery. The total cross-sectional area of the vascular system, therefore, gets gradually larger as the vessels themselves become smaller (Fig. 27.15) so that at the capillaries it may be one thousand times that at the aorta. It is probable, however, that under normal conditions a large number of the smaller vessels are closed, so that the total effective cross-section is reduced.

The walls of the veins are very much thinner than those of the corresponding arteries and their internal diameter is generally greater (Fig. 27.2). The elastic tissue is usually quite

Textbook of Physiology and Biochemistry

inconspicuous and some veins, for example those of the brain, bones and placenta, are said to have little or no muscular tissue.

The veins are easily distended by a small rise of pressure (Fig. 27.3b); the arteries, on the other hand, withstand a considerable

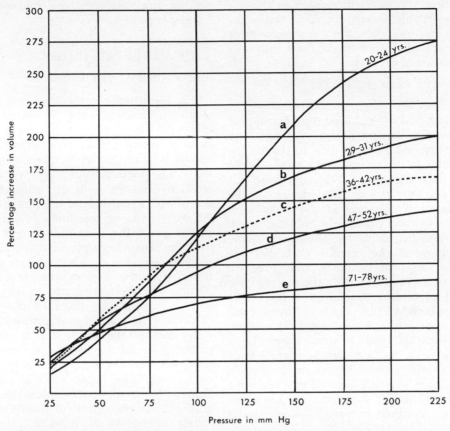

Fig. 27.3a Mean volume–elasticity curves of thoracic aortas in various age groups. The volume at various pressures is in all cases expressed as a percentage of the volume at 7·3 mm Hg (P. Hallock & I. C. Benson (1937). Elastic properties of human aorta, *Journal of Clinical Investigation* **16**, 597).

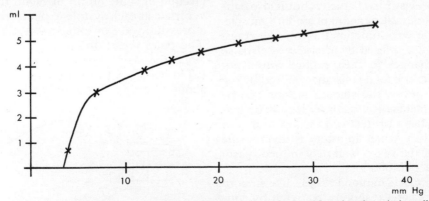

Fig. 27.3b The volume-pressure curve of the vena cava of a dog. Note that the vein is easily distended by a relatively small pressure compared to that required to distend an artery, but once distended it becomes resistant to further stretch. A distended vein can be shown to be stiffer than an artery. (*After* H. Blömer (1954–55). *Zeitschrift für Biologie* **107**, 468–480.)

increase of internal pressure with a relatively small increase in volume. This is indicated in Fig. 27.3a, which also shows that the elasticity of the aorta gradually diminishes as the age increases. In other words, the veins, unlike the arteries, can accommodate relatively large volumes of added (transfused) fluid with only small changes in pressure.

ROLE OF THE ELASTIC TISSUE

When the left ventricle contracts, all the energy imparted to the blood does not appear immediately as kinetic energy moving along the large arteries. Some is used to stretch the elastic fibres in the arterial walls and is stored there as potential energy. During systole this 'give' in the arterial wall limits the rise in the systolic arterial pressure. If the walls were rigid, the systolic pressure would be much higher. With increasing age the elastic tissue degenerates and the arterial walls become stiffer. As this happens the systolic pressure tends to rise. During diastole the recoil of the elastic tissue returns the energy stored in the wall to the blood and this tends to limit the fall in arterial pressure. If the walls were rigid, diastolic pressure would be much lower. Thus the elastic tissue tends to damp pressure fluctuations in the arterial system to provide a fairly constant pressure head. The properties of the large arteries described in this paragraph are sometimes referred to as *windkessel* effects.

ROLE OF THE ARTERIAL SMOOTH MUSCLE

The media in the major arteries is predominantly elastic in character, but it also contains a considerable amount of smooth muscle. It is not clear what function this muscle serves. The resistance offered by arterial vessels is so low that changes in their calibre would not greatly affect the total resistance offered by the circuit. However, the smooth muscle can be excited by mechanical damage. It may be that the muscle has a protective function so that in a severe injury when an artery is severed the irritation of the vessel wall may induce spasm which reduces blood loss.

Smooth muscle in blood vessels, like smooth muscle elsewhere, responds to stretch by contracting. When a person stands up the arterial pressure in his feet rises by about 100 mm Hg, the hydrostatic equivalent of the column of blood from the heart to the feet. This would tend to distend the arteries. However, the smooth muscle contraction induced by stretch tends to prevent excessive distension. (But see Syncope, p. 573.)

THE ARTERIAL PULSE

THE CAUSE OF THE PULSE

The pulse wave in the arteries is dependent on three factors: (1) the *intermittent inflow* of blood from the heart, (2) the *resistance* to outflow of blood from the arterioles into the capillaries and (3) the *elasticity* of the arterial walls. The phenomenon may be readily understood by analogy with the action of a water pump with a non-return valve (stirrup pump). At each stroke of the pump (systole) a pulse passes down the hose-pipe and at the same time the elastic hose-pipe is distended because the fine nozzle at the end prevents free exit of water, that is to say, it offers a resistance to outflow. Between the strokes of the pump (that is in diastole) the hose-pipe contracts (because it is elastic) and continues to force water through the nozzle so that, if the rate of pumping is not too slow, a steady stream of water flows from the nozzle. And so it is in the circulation. The pulse wave is almost non-existent in the capillaries unless the arterioles, analogous to the nozzle of the hose-pipe, are widely dilated. In the veins the rate of flow is uniform at rest except in the large veins near the heart.

The rapid rise of pressure following the ejection of some 60 ml of blood from the left ventricle at each systole expands the aorta and a pressure wave or pulse passes along its wall. The pulse is *not* due to the passage of blood along the arteries—the blood travels at only 0·5 m per s—but to a pressure wave travelling about 7 m per s. If the piston P in Fig. 27.4 is

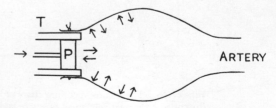

Fig. 27.4 Diagram to illustrate the initiation of a pulse wave in the aorta.

driven towards the right in the tube T tied into an artery the vessel is distended locally, because the blood is incompressible. This produces a local increase of pressure that has little effect on the advancing piston, which is in fact a mass of incompressible blood; instead the next section of the artery is stretched so that a wave of pressure travels along the vessel wall without involving actual transmission of blood. Indeed the pulse wave can travel down such an elastic tube even if the distant end is closed; there it may be reflected back in an attenuated form. Reflexion of pressure waves occurs at every branching point in the arterial tree and may make records of the arterial pulse very complex.

Pulse wave velocity. A pressure wave is transmitted through a pipe with a velocity c given by

$$c = \sqrt{\left(\frac{Eh}{2R\rho}\right)}$$

where h is wall thickness, R is radius, ρ is density of blood and E is the modulus of elasticity which increases with increasing stiffness. If the pulse is recorded simultaneously from the carotid and the radial arteries the velocity of the pulse wave is found to be about 5 m per sec at 5 years and 8 m per sec at 60 years. This shows that these arteries become stiffer with increasing age.

It is useful to know that the pulse wave arrives at almost the same moment at the beginning of the femoral artery and at the radial artery at the wrist. If the femoral pulse is later than the radial as judged by palpation it is probable that the wave is reaching the femoral artery along collateral channels developed as the result of a congenital narrowing of the aorta (coarctation).

The shape of the pulse wave. The shape of the pulse wave can be studied by applying a pressure sensitive capsule to the artery. This records only the general shape of the pulse wave, without in any way measuring the pressures involved. More exact information can be obtained by using an electronic manometer communicating through a needle with the inside of the artery. The record (Fig. 27.5) of the pulse wave in the brachial artery shows a steeply ascending limb, due to ventricular systole, followed by a more gradual fall in pressure corresponding to the elastic recoil of the arterial walls. However, the curve is not smooth as would be expected if it were merely a slow leak away of pressure, but contains several minor oscillations. The most constant of these is the dicrotic wave, the largest wave on the descending line of the trace. It has been ascribed to rebound from the closed aortic valves but it is more probable that this wave, together with other smaller subsidiary waves often observed, represents oscillations of the aorta at its own natural frequency. The pulse wave in the aorta has already been described, but the form of the pulse wave is changed as it passes to the periphery by the frictional resistance of the arterial walls and by reflected waves, so that even with an ideal recording device interpretation of the wave form is difficult. For clinical purposes, however, useful information about the shape of the pulse wave can be obtained by feeling the carotid artery. A wave that rises slowly and dies away slowly (plateau pulse) is a good indication that the aortic valve is narrowed (stenosed); on the

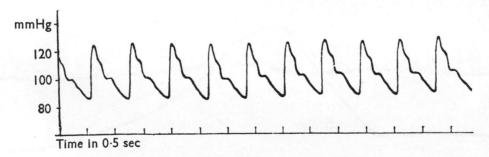

Fig. 27.5 Continuous trace of pulse waves in the brachial artery of a healthy young man recorded with a transducer manometer in communication with the lumen of the artery. (*By courtesy of G. Cumming.*)

other hand, a pulse wave that rises and falls quickly (collapsing pulse) may mean that the aortic valve is incompetent, allowing a proportion of the systolic discharge to run back (regurgitate) into the left ventricle.

The difference between the systolic and diastolic pressures (approximately 125 − 90 = 35 mm Hg in Fig. 27.5) is called the *pulse pressure*.

With the surge of blood into a limb at each heart beat there is a momentary increase in volume of the muscles. This can readily be observed in a person sitting with his legs crossed; the raised foot gives a little kick forward with each pulse beat as the calf muscles expand. This volume pulse occurs a little later than the corresponding arterial pulse. It can be recorded by putting the limb in a rigid box, or plethysmograph, connected with a tambour the movements of which are recorded by a mirror or a lever (Fig. 27.17).

The pulse rate. The pulse rate varies considerably from individual to individual and in the same subject at different times, depending on a large number of factors which are considered in Chapter 26. The resting pulse rate tends to fall as age advances, as shown in Fig. 26.76.

In heart disease when the cardiac rhythm is irregular (atrial fibrillation, premature beats, p. 522) the rate of the heart may exceed the pulse rate because the volume of some of the systolic discharges is insufficient to promote a pulse wave in the arterial walls great enough to reach the wrist. This is called a *pulse deficit*.

Sinus arrhythmia. The pulse, judged by the fingers, is not quite regular. Accurate records show that there is a slight acceleration during inspiration and a slight deceleration during expiration (Fig. 27.6) provided that the respiratory rate is low. This is termed *sinus arrhythmia*. In children this fluctuation in rhythm is much more marked than in adults. During inspiration the thoracic pressure falls while the abdominal pressure rises and, therefore, blood flows more readily into the thoracic veins, producing a rise of atrial and central venous pressure and an acceleration of the heart. This is only a partial explanation of sinus arrhythmia, since alterations of heart rate related to respiratory activity can occur independently of any alteration in venous pressure.

ARTERIAL PRESSURE

It is well known that blood escapes from a cut artery under considerable pressure, and the first attempt to measure the pressure was made by Stephen Hales in 1732. After occluding the artery with a temporary ligature he tied a brass pipe into the femoral artery of a mare and connected to it a glass tube 9 ft long. When he loosened the ligature on the artery the blood rose 8 ft 3 in (about 250 cm) above the level of the heart. He noted that alterations in pressure occurred with each beat of the heart and with the movements of respiration. These results are of great historical interest but the method is not very useful because the blood clots in the manometer in a few minutes.

Nowadays accurate and continuous records

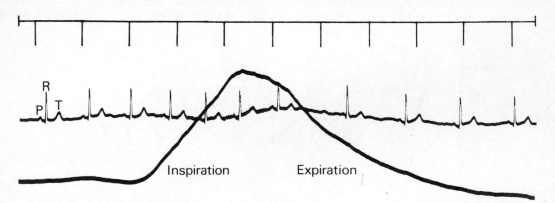

Fig. 27.6 Sinus arrhythmia. ECG (Lead II) from a healthy male aged 24 showing cardiac acceleration during inspiration and slowing during expiration. Timing marks, 0·04, 0·2 and 1·0 second.

of arterial blood pressure in man can be obtained with strain-gauge or condenser manometers communicating directly with an artery through a small needle. Such records show large oscillations between the maximum or systolic pressure and the minimum or diastolic pressure (Figs. 27.5 and 29.6). Because of the rapid rise and the relatively slower fall of the pressure wave, a point half-way between the maximum and minimum pressures does not represent the mean arterial pressure which is, in fact, closer to the diastolic than to the systolic pressure. For convenience it may be taken as the diastolic pressure plus one-third of the pulse-pressure (p. 533) but more strictly it is the level of the line halving the area between the pulse wave contour and the diastolic pressure level. This value is usually given automatically by the recording machine.

The direct method just referred to is obviously unsuitable for routine clinical use but the blood pressure can be estimated in man indirectly, although less accurately, by a development of the method originally invented by Riva-Rocci in 1896. The sphygmomanometer in general use nowadays consists of a mercury manometer with one wide and one narrow limb. The pressure to be measured is applied to the wide limb; the mercury in it moves down so little that the calibrated divisions on the narrower limb are nearly twice as far apart as in the original type of manometer which had limbs of equal cross-section. A rubber bag of standard size (18 cm × 12 cm for adults) covered with cloth is wrapped round the upper arm, leaving the cubital fossa exposed, and connected to the manometer and pump with which it is rapidly inflated to a point above the systolic blood pressure. The pressure in the bag is allowed to fall gradually. The arterial pressure can then be assessed in one of three ways. The first appearance of pulsation in the arteries distal to the cuff may be detected with the fingers, the pressure in the cuff at that time being some 10 to 15 mm Hg below the true systolic blood pressure. Secondly the pulsations of the air in the cuff may be recorded by some kind of oscillometer; then the first appearance of oscillations indicates the systolic blood pressure and the point of maximum oscillation the diastolic blood pressure.

More usually the blood pressure is measured indirectly by the *auscultatory method* (Fig. 27.7). The bell of a stethoscope is placed lightly over the brachial artery at the bend of the elbow. The pressure in the cuff is raised rapidly until the radial pulse disappears. When the pressure is allowed to fall a series of sounds is heard, first described by Korotkov in 1905 (Fig. 27.8). If the pressure in the bag is above the systolic pressure, S.B.P., the artery is occluded throughout the pulse cycle. When the pressure falls a little below S.B.P. (line A) the artery opens momentarily when the internal pressure is greater than the external (cuff) pressure and a blowing sound is heard at each pulse. As the pressure is further reduced the blowing sound is gradually succeeded by a distinct tapping which becomes louder as the pressure is reduced. As the pressure in the cuff is allowed to fall, preferably about 2 or 3 mm Hg per sec, the character of the sound suddenly changes from a sharp tapping to a softer blowing or muffled noise (Fig. 27.8, line C). This is usually taken as diastolic blood pressure, D.B.P. Finally the sounds disappear,

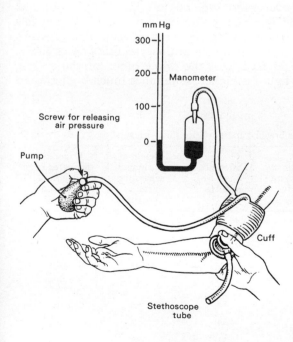

mm Hg

300—

200— Manometer

100—

0—

Screw for releasing air pressure

Pump

Cuff

Stethoscope tube

Fig. 27.7 Auscultatory method of measuring blood pressure in man. The cuff should be at heart level.

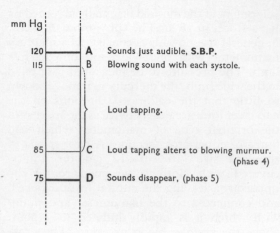

mm Hg

120 — A Sounds just audible, **S.B.P.**
115 — B Blowing sound with each systole.

        Loud tapping.

85 — C Loud tapping alters to blowing murmur.
                                    (phase 4)
75 — D Sounds disappear, (phase 5)

Fig. 27.8 The auscultatory method of determining arterial blood pressure. Diagram explained in the text.

usually quite suddenly, when the pressure in the bag falls below a certain pressure (line D) because below this pressure the internal pressure in the brachial artery is always greater than the pressure applied externally by the bag; the flow of blood along the artery is not interrupted and usually no sound is heard. Several investigators have compared the pressure readings obtained from a cannula inserted into the brachial artery in one arm with the readings obtained in the other by the auscultatory method. Considerable discrepancies have been observed. The indirect method usually gives readings of S.B.P. about 25 mm

Hg lower than the 'true' S.B.P. given by the direct method. The D.B.P. reading by the indirect method is usually only a few mm Hg lower at line C but about 13 mm Hg lower at line D. Although the indirect (cuff) method does not give an accurate (absolute) measurement of either S.B.P. or D.B.P. yet, since the direct and indirect readings are highly correlated, the indirect method is of great practical value in clinical medicine.

When the subject is supine the arterial pressures in the brachial and femoral arteries are approximately the same. With the subject standing however, the femoral pressure is higher than that in the brachial artery, the difference being due to the pressure of the column of blood equal in height to the vertical distance between one artery and the other. The effect of posture on the blood pressure is shown in Table 27.9. Slight variations, say 5 mm Hg, in the readings obtained on the two arms are not uncommon; in 10 per cent of persons the systolic reading on the right arm is 20 to 30 mm Hg higher than on the left.

In any individual arterial pressure is not constant but subject to appreciable variations over short intervals of time (Fig. 27.10 and Table 27.11). During sleep the S.B.P. may fall below 80 mm Hg (Fig. 27.10). The pressure recorded under ordinary conditions of life is referred to as the *casual* blood pressure. The figure is higher, sometimes much higher, than

Table 27.9 Comparison of the systolic blood pressure in the brachial and posterior tibial artery

Subject	Posture	Brachial artery pressure (in mm Hg)	Posterior tibial artery pressure (in mm Hg)	Difference in mm Hg	Height of column separating armlets (in cm)	Difference calculated from height of column (in mm Hg)
H.H.R.	Horizontal	140	138	2	—	—
	Standing	136	204	68	89	68·5
	L posture, legs up	122	76	46	60	46·1
	Vertical head down	148	70	78	101	77·7
P.H.R.	Horizontal	126	126	—	—	—
	Standing	140	204	64	86	66·1
	L posture, legs up	132	78	54	65	50·0
	Vertical head down	116	42	74	91	70·0

(M. Flack & L. Hill (1909), Measurement of systolic pressure in man. *British Medical Journal* **i**, 273.)

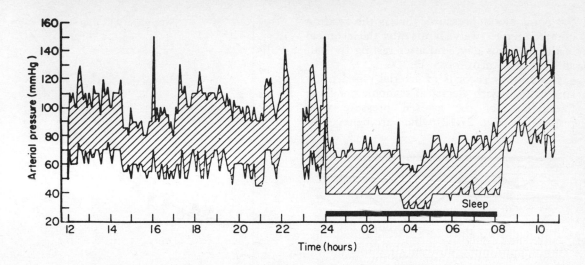

Fig. 27.10 Continuous record of arterial blood pressure of a healthy man during 24 hours made by means of an indwelling arterial cannula. Note the fall of S.B.P. and D.B.P. during sleep. The high pressure shown at 16.00 is due to a painful stimulus. (A. T. Bevan, A. J. Honour & F. H. Stott (1969). *Clinical Science* **36**, 329–344.)

Table 27.11 Variations in arterial pressure in an apparently healthy female aged 32. This table illustrates the considerable alterations that can occur under ordinary circumstances and emphasizes the care that may be required to obtain a steady reading.

Time (min)	Arterial pressure (mm Hg)		Circumstances	Time (min)	Arterial pressure (mm Hg)		Circumstances
	Systolic	Diastolic			Systolic	Diastolic	
0	140	90	Patient sitting, immediately after walking into the room for the first time	41¼	150	90	The response to the discomfort produced by cold
				41½	165	95	
				42	140	90	
7	130	80	Sitting talking	45	130	85	
10	135	90	Patient lies on couch	55	120	75	The basal blood pressure
25	120	80	After lying quietly on couch for 15 min	55½	140	85	Immediately after an unexpected loud noise is made behind the patient's head
30	125	80	After lying quietly on couch for 20 min				
35	120	75	After lying quietly on couch for 25 min	65	140	90	The blood pressure remains elevated
40	120	75	After lying quietly on couch for 30 min	85	125	80	20 min later the pressure is still not basal
41	—	—	Right hand suddenly immersed in ice-cold water for 30 sec	85½	125	85	Patient suddenly stands erect

the *basal* blood pressure, that is the reading obtained ten to twelve hours after the last meal of the previous day, and after resting for half an hour in a warm room.

Continuous records of arterial blood pressure in man with electrical manometers show fluctuations of the arterial pressure with respiration (Fig. 27.12) which are believed to

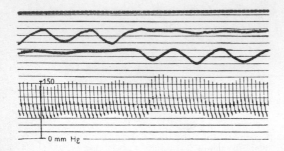

Fig. 27.12 Tracings in a normal man of, from above downwards, time in seconds, thoracic and abdominal breathing and arterial blood pressure. The respiratory records were made with spirometers; a downward movement indicates inspiration. The blood pressure was measured with an electrical (capacitance) manometer communicating with the inside of an artery. The arterial pressure shows variations corresponding to respiration which do not alter in phase when the breathing changes from thoracic to abdominal (A. C. Dornhorst, P. Howard & G. L. Leathart (1952). *Circulation* **6,** 554).

be due to variations in both stroke volume and peripheral resistance. The precise relationship between arterial blood pressure and respiration depends on the respiratory rate; at normal rates of breathing the blood pressure falls during most of inspiration but when breathing is slow inspiration is accompanied by a small rise in blood pressure.

Many attempts have been made to find values between which a subject's blood pressure could be regarded as normal, but all such efforts have been, for various reasons, unsatisfactory. The pressure obtained by the sphygmomanometer is affected by the thickness of the arm, the thicker the arm the higher the value obtained. Arterial pressure increases with age (Fig. 27.13), more in some subjects than in others, and no dividing line can properly be set at any age below which the pressure is normal and above which it is abnormal. Indeed, it is considered by some that arterial pressure is inherited as a graded

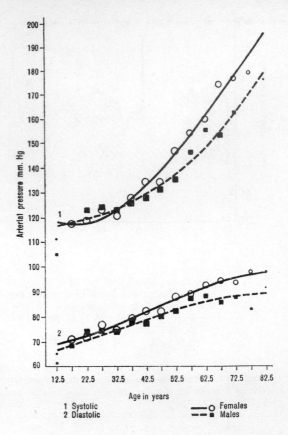

Fig. 27.13 Systolic and diastolic pressures for females (open circles) and males (black squares) for each five-year age group of the population sample, together with the fitted curves. The area of each circle or square is proportional to the number of subjects in that age group (M. Hamilton, G. W. Pickering, J. A. F. Roberts & G. S. C. Sowry (1954). *Clinical Science* **13,** 11–35.)

characteristic (like height) and that the range of variation includes values hitherto regarded as abnormal. It seems likely that at least three factors are concerned in the production of high blood pressure, namely age, heredity and environment.

FACTORS DETERMINING ARTERIAL PRESSURE

Blood enters the arterial system from the left ventricle and leaves through the arterioles. The amount entering is determined by the cardiac output and the amount leaving is determined by the resistance offered by the arterioles (peripheral resistance). If more blood enters, that is if the cardiac output increases or if less blood leaves, that is if the peripheral resistance rises, the pressure in the arterial system rises. Conversely if cardiac output or peripheral resistance falls, the arterial pressure decreases. Blood pressure is therefore directly proportional to cardiac output and peripheral resistance. Changes in the elasticity of the arteries affect pulse pressure more than mean pressure in the arterial system.

FACTORS REGULATING ARTERIAL PRESSURE

Control systems within the body tend to maintain mean arterial pressure within fairly strict limits. In this way the perfusion pressure head is maintained relatively constant despite the changes that frequently occur in both cardiac output and peripheral resistance. Stretch receptors are found in the wall of the proximal arterial tree, especially in the region of the aortic arch and carotid sinuses. When the arterial pressure rises there is increased stimulation of these nerve endings. The increased traffic of impulses up the vagus and glossopharyngeal nerves leads to reflex vagal slowing of the heart and reflex release of vasoconstrictor tone in the peripheral blood vessels. The resulting fall in cardiac output and reduction of peripheral resistance tend to restore the blood pressure to the normal value. Similarly a fall in arterial pressure decreases the stimulation of the arterial stretch receptors. The reflex tachycardia and vasoconstriction that ensue tend to raise blood pressure towards its normal value.

Even when the arterial baroreceptors are denervated a rise of blood pressure can still be regulated to some extent. The exact mechanism of the control system responsible is not understood. However, it is known that a fall in renal blood flow, such as would occur if arterial pressure fell, results in the release of the hormone renin from the kidney. Renin converts the plasma protein angiotensinogen to the angiotensin polypeptides which can constrict blood vessels and cause the release of aldosterone from the adrenal cortex. The salt and water retention brought about by aldosterone together with the vasoconstriction produced by angiotensin tend to raise arterial pressure to the set value.

ARTERIES AS TRANSPORT VESSELS

Arteries transport blood at high velocity and low energy cost from the heart to the periphery. The high velocity of flow (about 0·5 m per sec) is due to the relatively small cross-sectional area of the arterial system. The total cross-sectional area of the arteries is smaller than any other part of the vascular system. If the mean pressure is compared in the aorta and in a peripheral artery such as the radial artery, it is found that the value in the radial is only 5 mm Hg or so lower than in the aorta (Table 27.14).

Table 27.14 Table of mean blood pressure at various points in the vascular system in man (supine) (referred to the pressure in the right atrium)

	mm Hg
Large artery (for example carotid)	90
Medium artery (for example radial)	85
Capillary	10 to 30
Small veins of arm	9
Intestinal capillaries	over 10
Portal vein	10
Inferior vena cava	3
Large veins of neck	0 to −2
Pulmonary artery	15

This means that blood is transported from the heart to the wrist for the loss of only 5 mm Hg pressure. This is explained by the extremely low resistance offered by the arteries because of their wide calibre.

THE ARTERIOLES OR RESISTANCE VESSELS

The changes in systolic and diastolic pressures at various points of the vascular system are indicated diagrammatically in Fig. 27.15, and the mean intravascular pressures in various vessels are shown in Table 27.14. The biggest fall in arterial pressure occurs in the arterioles which offer the largest resistance to flow. Regional blood flow is regulated by variations in the resistance they offer.

PERIPHERAL RESISTANCE

The mean arterial pressure depends on the amount of blood which enters the arterial

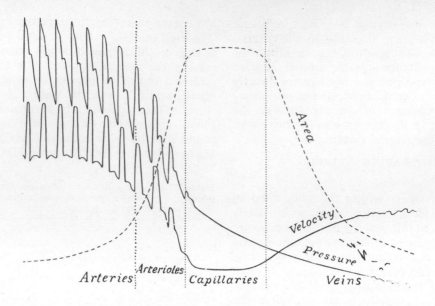

Fig. 27.15 Diagram showing the relations between pressure, velocity and total cross-sectional area in arteries, capillaries and veins. (*From* L. Fredericq & J. P. Nuel (1920), *Elements de Physiologie humaine* 7th edn., p. 136. Liege: Vaillant-Carmanne. Partly *after* J. Gad.)

system from the heart (cardiac output) and the rate at which blood leaves the arterioles to pass into the capillaries and veins, that is, on the resistance to the outflow of blood from the arteries, or *peripheral resistance*, perhaps better called the pressure-flow ratio. The relation between these three factors can be expressed thus:

$$\text{Peripheral resistance} \propto \frac{\text{Mean arterial pressure}}{\text{Cardiac output}}$$

Both mean arterial pressure and cardiac output can now be measured in man with considerable accuracy. The pressure can be expressed in dynes per cm^2 by multiplying the reading in mm Hg by 1332; the output is expressed in ml per sec. Then,

average resting cardiac output as 5 l per minute or 83 ml per second, then,

$$\begin{aligned} \text{Peripheral resistance} &= \frac{90 \times 1332}{83} \\ &= 1445 \text{ dynes sec cm}^{-5}. \end{aligned}$$

The total peripheral resistance in man is normally about 1300 dynes sec cm^{-5}.

The equation for peripheral resistance suggests that the flow-pressure graphs should be straight lines passing through the origin. In fact, however, this is not so; flow is not appreciable until the critical closing pressure is exceeded (p. 558).

The layer of fluid immediately in contact with the wall of the artery can be regarded as being stationary. The layer of fluid just inside

$$\text{Peripheral resistance} = \frac{\text{Mean arterial pressure in mm Hg} \times 1332 \text{ dynes cm}^{-2}}{\text{Cardiac output cm}^3 \text{ sec}^{-1}}$$

For example, if we take the mean arterial pressure at rest as equal to the diastolic blood pressure plus one-third of the pulse pressure (a reasonable approximation to the figure actually found by an intra-arterial manometer), say $80 + \frac{1}{3}(110-80)$ or 90 mm Hg, and the

this stationary layer slides over the latter, and so on in a series of concentric layers, each moving faster than the more peripheral one (laminar flow). The resistance to flow due to this moving of one layer of fluid past another is present even when the arteries are wide and straight. It varies with the viscosity of the blood and the size of the vessel. In 1842 the French physician Poiseuille studied the factors governing the laminar flow of viscous fluids through capillary tubes with rigid walls and Poiseuille's law is an expression of the relationship between these factors:

$$\text{Rate of flow} = \frac{P_1 - P_2}{8l} \times \frac{\pi r^4}{\eta}$$

where P_1 and P_2 are the pressures at the ends of the vessel of length l and radius r and where η is the viscosity of the blood. In other words, if a blood vessel constricts so that its initial diameter is reduced by one-half, the same pressure difference across its ends produces only one-sixteenth of the flow. However, if the diameter of a blood vessel is less than 1 mm η decreases quite markedly. The viscosity of blood increases greatly as the number of cells per mm^3 increases. For example if the haematocrit (p. 416) increases from 45 to 60 per cent η is increased one-and-a-half times. Polycythaemia, that is an increase in the number of red cells, is associated with an increase in arterial pressure.

As the rate of flow of a fluid in a tube is gradually increased there comes a point at which laminar flow is replaced by turbulence and the conditions mentioned above no longer apply. By injecting a stream of dye into a stream of water flowing in a long straight tube Reynolds (1883) found that turbulence developed if the expression rVd/η, in which r is the radius of the tube, V the mean velocity of flow, d the density and η the viscosity, exceeded a certain value. Reynolds' number is about 1000 for water and slightly less for blood. It is obvious that turbulent flow occurs more readily when the fluid is of low viscosity and the velocity relatively great and that it tends to develop first in large vessels. These conditions are found in the heart and aorta of severely anaemic patients and turbulence possibly accounts for the 'haemic' murmurs readily audible with the stethoscope over the heart and large arteries in such people. The relatively high velocity of the flow of blood through narrowed valves in the heart, for example, may be sufficient in itself to cause turbulence and the development of audible murmurs (organic murmurs). Turbulent motion in the blood greatly increases resistance to flow.

It can be imagined that in a liquid in motion there are tubes of flow, the boundary surfaces of which are formed by stream lines. In such a 'stream tube' the volumetric rate of flow is constant across any normal section (if there is no change in fluid density) or:

$$av = \text{constant}$$

where a is the cross-sectional area of the stream tube at any point and v is the mean velocity of the fluid at that point.

Further (if losses are neglected) the total energy of the fluid in the tube remains constant, the total energy being given by the sum of the pressure, kinetic, and potential energies of the fluid at a point. This is expressed in Bernoulli's theorem:

$$p + \tfrac{1}{2}\rho v^2 + \rho gh = \text{constant}$$

Where p is the pressure, ρ the density, and v the velocity of the fluid, g is the acceleration due to gravity and h the vertical height of the point considered above some given datum. The above equation may be simplified to read

$$p + \tfrac{1}{2}\rho v^2 = \text{constant}$$

since the effect of variation of height above the datum is simply to increase or decrease the hydrostatic pressure.

It follows from the first equation that an increase in cross-sectional area of the tube causes a decrease in velocity and from the last equation that a decrease in velocity causes an increase in pressure. Therefore if the cross-sectional area of the tube increases the pressure increases and if it decreases the pressure decreases. Consequently, the pressure inside the localized dilatation of an artery (aneurysm) is higher than it is in the remainder of the vessel. Conversely the pressure in a localized narrowing of an artery is reduced.

The experiments of Poiseuille, Reynolds and Bernoulli were carried out with homogeneous fluids and rigid tubes. Blood with its cellular content is certainly not homogeneous and the arteries are elastic. Although their results may not be exactly applicable to the circulation they must be a good approximation.

REGULATION OF PERIPHERAL RESISTANCE

Since most of the resistance offered by the circulation lies in the arterioles, changes in the calibre of these vessels cause much greater changes in peripheral resistance than comparable changes in other vessels. If perfusion pressure is held constant, flow is inversely proportional to peripheral resistance. This is the situation which obtains in man because control systems hold the arterial pressure steady. Thus changes in the calibre of arterioles, more than changes in the calibre of any other vessels, determine the amount of blood flowing around the circulation. Through regulation of the calibre of arterioles local peripheral resistance and hence local blood flow is controlled. It should be noted that arterioles control blood flow not blood velocity. Flow depends on resistance whereas velocity depends on the total cross-sectional area of the vascular bed.

The arterioles are particularly well designed for the precise regulation of their internal diameters. Their walls are very muscular. In the transverse section of an arteriole, the ratio of the area occupied by muscle to the area occupied by the lumen is about one to one. This ratio, known as the *wall to lumen ratio*, is higher in arterioles than in other types of blood vessel. The muscle is very reactive and its tone is influenced by nervous and other factors. Although changes in arteriolar calibre are usually brought about to alter the rate of blood flow, they are sometimes made to serve blood pressure regulation. Peripheral resistance influences blood pressure and a reflex increase in resistance may be induced when the baroreceptors detect a fall in arterial pressure.

The reflex vasoconstriction of the arterioles, which occurs when blood pressure falls, reduces the amount of blood entering, and thus the blood pressure in, the capillaries. This causes a reduction in the amount of fluid leaving the capillaries. Since the colloid osmotic pressure of the plasma proteins is not affected, this results in a net gain of fluid by the capillaries.

MEASUREMENT OF BLOOD FLOW

The simplest method of determining the blood flow through an organ is to collect and measure over a given period of time all the blood emerging from its veins. Various methods of measuring blood flow without loss of blood have been invented. In Rein's flow

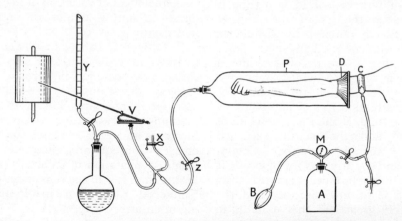

Fig. 27.16 Diagram of apparatus for determining the rate of blood flow through the arm. The arm is placed in the plethysmograph (P) the opening of which is closed by a piece of rubber dam (D) and the connexion with the skin made tight with soapsuds. The narrow pressure cuff (C) is placed round the arm about 3 cm above the opening into the plethysmograph. The pressure cuff is inflated by opening the stopcock connecting it with the large bottle (A) in which the pressure has previously been raised by the rubber bulb (B). Pressures are read by the spring manometer (M). The plethysmograph is connected with the volume recorder (V) which writes upon a moving drum. Air can be let out of the system by the stopcock X, and water can be introduced from the burette (Y) so that the writing point of the volume recorder can be adjusted. The stopcock (Z) serves to disconnect the plethysmograph from the recording apparatus during adjustments of the former. The recording apparatus is graduated by allowing 5 ml of fluid at a time to flow in from the burette, and marking the elevation of the volume recorder thus produced. (A. W. Hewlett & J. G. van Zwaluwenburg (1909–10). *Heart* **10**, 88.)

meter heat is applied to an artery in the living animal by a coil situated midway between two thermocouples placed upstream and downstream on the artery. If the flow is fast the temperature difference between the two couples is small, if slow the difference is large. The apparatus is calibrated by noting the galvanometer deflexions at various known rates of flow through an excised artery of the same size. The electromagnetic flow meter depends on the fact that an electromotive force is induced in a conductor (in this case the blood) moving through a magnetic field. The e.m.f. is picked up by two electrodes placed on opposite sides of the unopened artery at right angles to the direction of flow and to the magnetic field.

The changes of volume of a limb or part of a limb can be recorded in man or animals by means of a *plethysmograph* (Fig. 27.16). This is a rigid box which can be slipped over the part so as to make a closed system in which small changes of volume can be recorded. If the veins are obstructed for a *short* time the record shows an increase in volume which indicates the flow of blood into the part during the period of obstruction (Fig. 27.17).

In animal experiments this plethysmographic method can be applied to the measurement of the blood flow through an organ such as the kidney. If the venous outflow is temporarily obstructed the kidney swells and displaces air from the rigid box enclosing it into a volume recorder which indicates the flow into the organ.

Since the heat lost from a part is directly related to its blood flow it is possible to assess the latter in a hand or a finger in which the heat loss is high by measuring the heat gained by a water-calorimeter in which the part is immersed. Since the temperature of the skin is determined by the rate of blood flow through it, measurements of surface temperature may be used as an index of blood flow in superficial tissues. The Fick principle (p. 502) can be applied to the determination of the blood flow through an organ, for example the heart (cardiac output), the kidney (Chap. 32) or the brain.

The blood flow through various organs is given in Table 27.18. These values are

Table 27.18 Approximate values for resting blood flow through different organs in man in ml per 100 g per min obtained by various methods

	Blood flow (ml/100 g/min)
Carotid bodies	2000
Thyroid	560
Kidney	150
Liver	150
Heart (coronary circulation)	100
Intestine	70
Brain	65
Spleen	40
Stomach	25
Hand	7 to 12
Finger	15 to 40
Forearm (mainly muscle)	1 to 3
Leg (mainly muscle)	1 to 2

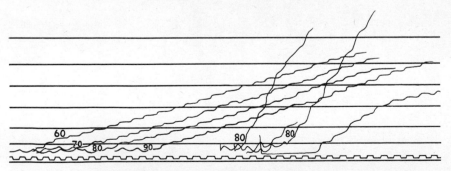

Fig. 27.17 Plethysmograph records. The first records were made during rest; the following ones at short intervals after exercising the hand within the plethysmograph. The figures on the graph give the pressure in mm Hg in cuff of Fig. 27.16. Each space between horizontal lines represents 5 ml increase in arm volume. Time mark in seconds. (A. W. Hewlett & J. G. van Zwaluwenburg (1909–10). *Heart* **10**, 92.)

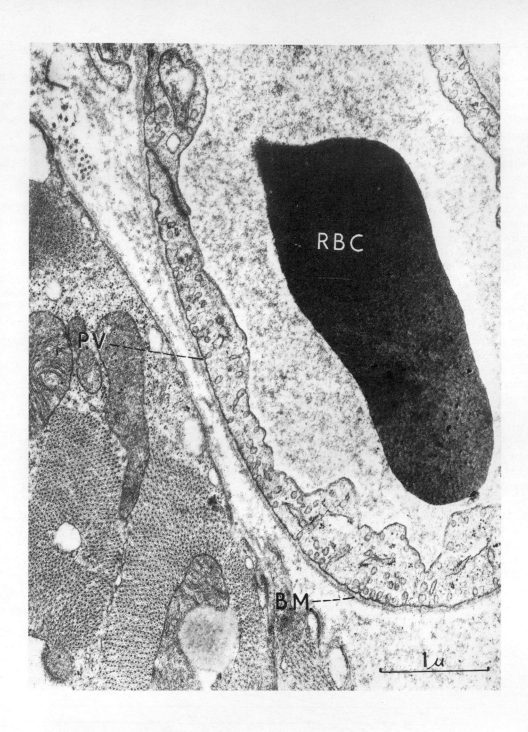

Fig. 27.19   Capillary in cardiac muscle. A red blood corpuscle (RBC) lies in the lumen. The capillary basement membrane (B.M.) is seen. Flask-shaped pinocytotic vesicles or caveolae (PV) are present at various points on the inner and outer surfaces and the endothelial cytoplasm contains many small vesicles. (*By courtesy of G. M. Wyburn.*)

approximate only, since they depend upon the state of activity of the part.

THE CAPILLARIES OR EXCHANGE VESSELS

The circulatory system conveys substances in solution to and from the capillaries where the exchanges between blood and tissue cells take place. The capillaries are minute vessels of the order of 10 μm in diameter connecting arterioles with venules but differing from them in having no muscular coat. Single capillaries differ in diameter, length, and in the nature of their connexions with the smallest arteries and veins. Each capillary consists of a tube of endothelium composed of a single layer of flat cells, about 30 μm by 10 μm. The edge of one cell is fitted closely into the edge of its neighbour by slight overlapping or by interdigitations which leave only a very narrow slit between adjacent cells. Each endothelial cell possesses mitochondria and other organelles like those in other cells; its thickness varies from 0·2 μm at the periphery to 3 μm at the bulge containing the nucleus. Even with the electron microscope no structures suggesting pores or interstices through the continuous endothelial surface have been seen (Fig. 27.19). A basement membrane about 0·025 μm (25 nm) thick completely surrounds the endothelial tube and outside it is a pericapillary sheath supporting the vessels and continuous with the surrounding connective tissue matrix. Embedded in the pericapillary sheath, or possibly attached to the surface of the endothelium, are at intervals large branched cells called *pericytes* or *Rouget cells* (Fig. 27.20).

Zweifach, by microscopic observation of the circulation in the capillary beds of the dog's omentum and the meso-appendix of the rat, has shown that the capillaries are arranged in functional units (Figs. 27.21 and 27.22). A metarteriole, which contains muscular elements, leads into a *thoroughfare channel*, or *AV-bridge*, larger than a true capillary but resembling it in having no muscle tissue in its walls. The thoroughfare channels which may

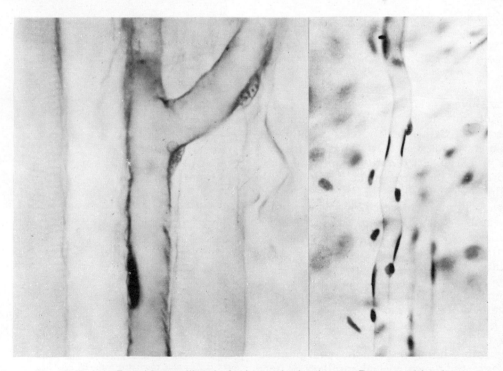

Fig. 27.20 (A) Branching capillary in frog's muscle showing two Rouget nuclei and one endothelial nucleus. (B) Mesenteric capillary of the rat showing Rouget nuclei sessile on the wall. Endothelial nuclei are also shown. × 1000. (*By courtesy of H. P. Gilding.*)

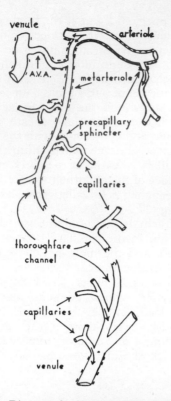

Fig. 27.21 Diagram of a functional unit of the capillary bed, together with a metarteriolar-venous anastomosis (A.V.A.) and a pre-capillary branching directly from an arteriole. (R. Chambers & B. W. Zweifach (1944). *American Journal of Anatomy* **75**, 173–200.)

be as long as 1·5 mm lead directly from arterioles to venules while the true capillaries, shorter in length, open from them and from the metarterioles and arterioles themselves. The true capillaries may be open or closed according to the needs of the tissue for at the origin of each capillary there is a *precapillary sphincter* by means of which blood flow through the capillary may be slowed or arrested. The capillaries themselves are probably not contractile. In tissues such as muscle, which require a very variable blood flow, the proportion of true capillaries to thoroughfare channels is high (of the order of four to one) while in others, in which the flow is relatively much more constant, there is a higher proportion of thoroughfare channels. Certain organs, for example the nail bed and the renal glomerulus have their own special capillary pattern. Whereas the arterioles are mainly under nervous control the more peripheral metarterioles

Fig. 27.22 Topographical view (approx. ×100) of portion of capillary bed in meso-appendix of a rat photographed during moderate hyperaemia to render visible most of the blood vessels in the field. Prominently displayed is the venule which drains this region. Two metarterioles, I and II, are shown—I starting at upper left and II at middle of right margin. Metarteriole I, along its course, gives off three precapillaries with sphincters, then passes down as the a–v capillary which gives off one precapillary at its right, and further down receives, in succession, two post-capillaries. At lower left corner it joins with another a–v capillary to form the non-muscular venule. The venule courses up as a prominent vessel and passes out of figure at the right after receiving several post-capillaries. The true capillaries carry the flow between pre- and post-capillaries. Metarteriole II, shortly after appearing at the right of figure, is connected with the neighbouring venule by a short anastomosing vessel (A.V.A.) which may serve to short-circuit the blood of this region. (R. Chambers & B. W. Zweifach (1944). *American Journal of Anatomy* **75**, 173–200.)

are mainly, but not entirely, under hormonal control; the precapillary sphincters are almost entirely under humoral control. It is thought that at rest the metabolic needs of a tissue can be met by exchange through thoroughfare channels and that when the tissue becomes active the precapillary sphincters open according to the metabolic needs.

Capillary pressure. Since the circulation of the blood is dependent on a pressure gradient the pressure in the capillaries must be lower than that in the smallest arteries and greater than that in the smaller veins (Fig. 27.15). The first satisfactory method of measuring intracapillary pressure is that devised by Landis in 1930. He inserted a micropipette of pyrex glass with an aperture of about 10 µm into various parts of the capillary loops of the skin of the nail bed. The pipette, moved by a micromanipulator and observed through a binocular microscope, was connected by a system of tubes containing citrate solution to a mercury manometer. When the pipette was inserted into a capillary the pressure was adjusted until blood neither entered nor left the micropipette. This pressure was taken as the internal capillary pressure. The average pressure (Fig. 27.23) in the arterial limb was

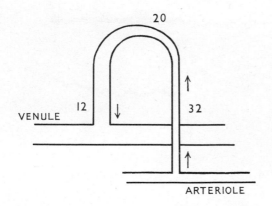

20

VENULE 12 ↓ 32

ARTERIOLE

Fig. 27.23 Average pressures in mm Hg in a skin capillary in man. (*After* E. M. Landis (1930). *Heart* **15**, 213.)

32 mm Hg with a range of 21 to 48 mm Hg: the summit of the loop gave a mean pressure of 20 (range 15 to 32) and the venous end a mean of 12 mm Hg (range 6 to 18). There is thus a considerable gradient of pressure along the capillaries near the root of the nail. The pulse pressure at the arteriolar end of the loop was at least 5 to 10 mm Hg and at the venous end

usually zero. When a pneumatic cuff was put round the arm to occlude the venous return the pressure in the arterial end of the loop rose at once to a level about 10 mm Hg above that in the cuff. When the hand was held above the head the venous pressure fell below atmospheric. The veins then collapsed and offered a considerable resistance to flow, thus keeping the capillary pressure gradient normal. When hand was lowered again the capillary pressure rose almost as much as would be calculated from hydrostatic considerations. Warming the hand produced arterial dilatation and the pressure in the capillary loop rose to 20 to 30 mm Hg, whereas cooling the skin resulted in a fall of about 10 mm Hg. In other words the capillary pressure varies considerably according to the state of contraction of the precapillary sphincters which, as we have seen, may alter from moment to moment. It is therefore incorrect to speak of a mean capillary pressure except in a statistical sense.

Because of the inertia of the mercury column, Landis's method cannot detect rapid fluctuations in capillary pressure. Recently a microelectrode with a tip diameter of about 0·5 µm has been used to measure capillary pressure. The microelectrode is filled with twice M-NaCl solution, which is a good electrical conductor. The microelectrode is introduced into the capillary. If the capillary pressure rises, plasma is forced up the tip of the microelectrode. This increases electrode resistance and the change in resistance can be easily detected. This system has a good frequency response and rapid fluctuations in capillary pressure can be recorded (Fig. 27.24). The pressure fluctuations observed in the minute vessels of the frog's mesentery are shown in (Fig. 27.25).

Blood flow in capillaries. When the frog's tongue, lung, or web of the foot is observed under a microscope a network of vessels in seen. In the arterioles the red cells move forward in jerks corresponding to the beats of the heart but the pulsatile nature of the flow is lost in the capillaries unless there is peripheral vasodilatation. If the capillaries are narrow the red cells follow one another in single file and may become deformed (Fig. 27.26), but they recover their shape on reaching a wider vessel. The capillaries show great variations in calibre, with corresponding alterations in flow. In the venules the red blood

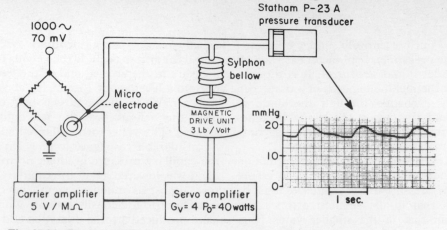

Fig. 27.24 Block diagram of micropressure transducer. The carrier amplifier has a sensitivity of 5 v/megohm. The servo amplifier has a voltage gain (G_v) of 4 and a maximal power output (P_0) of 40 w. The magnetic drive unit delivers a force of 3 lb/v. (C. A. Wiederhielm, J. W. Woodbury, S. Kirk & R. F. Rushmer (1964). *American Journal of Physiology* **207**, 173.)

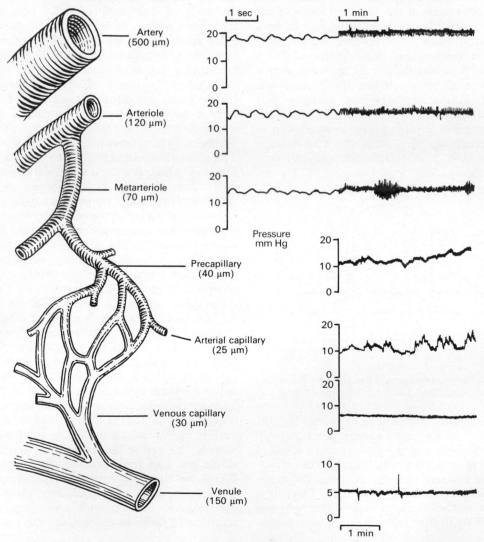

Fig. 27.25 Pressure records obtained at various points in the microcirculation by the method shown in Fig. 27.24.

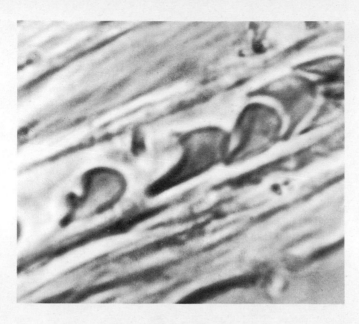

Fig. 27.26 Red cells in a mesenteric capillary of the dog: direction of flow left to right: rate of flow 1 mm per sec: magnification ×3000: exposure time 5 μsec. The red cells which are biconcave discs in the large vessels become paraboloid with a bell-shaped hollow in the capillaries. The leading part becomes convex and the trailing part becomes concave so that the shape resembles a thimble or parachute. Further details are to be found in M. M. Guest, T. P. Bond, R. G. Cooper & J. R. Derrick (1963). *Science, New York* **142**, 1319–1321. (*By courtesy of M. Mason Guest.*)

cells pass along in a steady axial stream, leaving a clear peripheral stream of colourless plasma. The pattern of blood flow in small vessels is beautifully shown in Fig. 27.27.

The capillaries in the nail bed of the human finger are easily observed under the microscope if the skin is illuminated by an oblique beam of light. A drop of oil makes the skin more translucent by replacing the air in the surface layers of the epidermis. The capillaries are of the order of 10 μm in diameter but the capillary loops vary greatly in size, in shape and in lumen, the venous side being usually much wider than the arterial side. Spontaneous alterations in the capillary diameter and the rate of flow of corpuscles occur often at the bases of the nail, and frequently in the back of the hand, capillaries opening and closing in the absence of any apparent stimulus. Thus although the pulsatile flow characteristic of the arterial system is absent in the capillaries the flow through these vessels is anything but continuous.

The velocity of blood flow in capillaries, about 0·5 mm/sec, is about 1000 times lower than that in the aorta. This is because the total cross-sectional area of the capillary bed is so enormous. It is not because the capillaries offer very much resistance to flow. The low rate of flow in capillaries provides a relatively long time for exchange to take place between the plasma and tissue fluid.

Capillary permeability. The rate at which fluid passes through the capillary membrane is determined not only by its permeability under different conditions but also by the pressure and flow within it. Hence in defining capillary permeability it is necessary to know (a) the volume or mass of substances passing through (b) unit area in (c) unit time, under the influence of (d) unit hydrostatic or osmotic pressure (e) per unit thickness of membrane. Finally, the rate of fluid transfer is affected by the hydrostatic and osmotic pressures of the fluid outside the capillary. There are therefore considerable difficulties in the way of our understanding the factors that govern the passage of materials across capillary

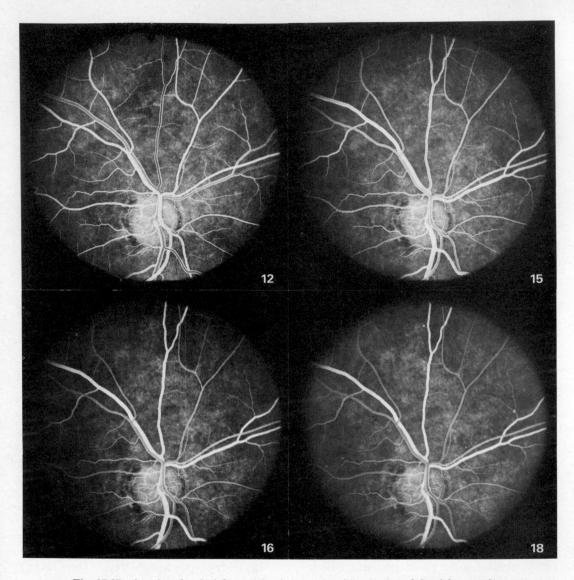

Fig. 27.27 A series of retinal fluorescein photographs of the fundus of the right eye taken after a single intracarotid injection of 4 ml 5 per cent fluorescein in a healthy male aged 30 years. Fluorescein appears white in the photographs. 12. Twelve seconds after the injection of fluorescein. The arteriolar system is filled with fluorescent blood. The main veins are seen filling from the smaller branches and the fluorescein maintains its position at the edge of the vein, demonstrating laminar flow. Just inferior to the disc, two veins showing laminar flow unite. A thin central stream has formed in the main vein from the merging of the streams at the walls of the tributary veins. 15, 16, 18. The remaining three photographs were taken at 15, 16 and 18 seconds. Dyed blood coming from the periphery fills the centre of the veins. A centre column of undyed blood enters the arterioles, again demonstrating laminar flow. (*By courtesy of D. F. J. Archer.*)

walls, and it is seldom that all these factors are known.

Three pathways through capillary endothelium have been proposed, (1) for fat-soluble molecules, (2) for relatively small numbers of large molecules and (3) high permeability path for water ions and small molecules. The resistance to the passage of molecules out of the capillaries and lymphatics is provided by the endothelial cells and their basement membranes (Fig. 27.19). Proteins may be able to pass in small amounts through endothelial intercellular junctions even where they are apparently adherent. Large molecules have been seen in vesicles in endothelial cells which probably carry them slowly through the cells; this may be an alternative pathway to the intercellular junction route. The number of vesicles is too small to account for the volume of water and the number of small molecules known to pass through the capillary endothelium. Since some substrates pass rapidly through the endothelium without entering the cytoplasm it must be supposed that they go through intercellular junctions. Pappenheimer calculated that transference could take place if capillary endothelium had pores about 300 nm long and under 9 nm in width. This looks like a description of the intercellular slits. Florey invented the technique of placing one ion inside a capillary and a second, which precipitated the first, outside the capillary; when the tissue is examined in the electron microscope the site of precipitation shows the pathway of the ions (Fig. 27.28). In capillaries and in lymphatics ions pass through intercellular clefts and also in vesicles. The passage through junctions is quantitatively more important. Molecules up to 1000 molecular weight, or about 3 nm, can pass through junctions but molecules greater than 2000 cannot. Lipid-soluble substances probably dissolve in plasma membranes and pass through or around cells. If substances known to damage capillary endothelium are added to the blood perfusing capillaries water and proteins pass rapidly out of the capillaries. There is no doubt, therefore, that the living cells have a great influence on permeability. When capillaries are perfused with artificial solutions containing no protein, water leaves the capillaries and accumulates in the surrounding tissue. This accumulation of water

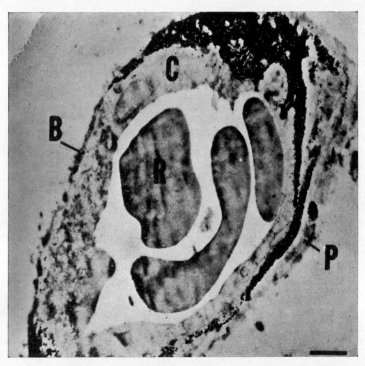

Fig. 27.28 Electronmicrograph (× 13 000) of a capillary from an area of mouse skin treated with histamine and fixed 20 min after injection of colloidal HgS. Note the considerable quantity of HgS which has crossed the endothelial cell during the 20 min interval. Many particles have accumulated between the basement membrane (B), or peri-endothelial cell (P) and the external surface of the endothelial cell (C). A few particles of HgS are seen in the endothelial cytoplasm, most of them within definite membrane-lined vesicles. R, red cell. (J. F. Alksne (1959). *Quarterly Journal of Experimental Physiology* **44**, 51–66.)

(oedema) can be reduced greatly by adding serum protein or dextran (a mixture of polysaccharides of high molecular weight) to the perfusion fluid.

Leucocytes are able to leave a capillary by inserting a pseudopodium into the cleft between endothelial cells and may do so without allowing leakage of plasma protein. Small lymphocytes may be carried through the cytoplasm of the venules of lymph nodes.

Marks and Shuster (1966) have measured 'capillary' permeability by measuring the plasma volume (A) by dilution of ^{131}I human serum albumin (HSA) and (B) by ^{51}Cr-labelled red cells and whole body packed cell volume. Since HSA leaks out of the vessels the concentration of HSA falls with time and the apparent plasma volume increases. The labelled red cells cannot escape from the circulation in the 10 minutes of the experiment and therefore the value of (B) remains constant. The leakage of plasma is accordingly (A)–(B). In collagen vascular diseases or inflammatory conditions the permeability is greatly increased and oedema may result. These experiments give no clue as to the region of leakage of the protein but it probably escapes at the venous end of the capillaries or at the venules. The dye (fluorochrome) brilliant sulfoflavin has a high affinity for the β-globulins and albumins. After injection of dyed plasma fluorescence indicating the passage of protein is seen only at the venous end of long capillaries (Fig. 27.29). This is not a contradiction of the Starling hypothesis of 1894, described in Chapter 33, which is concerned with the passage of small molecules. It is likely that there is an increase in the size of the pores, that is a 'gradient of permeability', from small arteries to small veins as described over forty years ago by Gilding and his colleagues.

Experiments with ^{131}I-labelled protein in man show that there is an overall exchange rate of 40 per cent of the total plasma albumin per day. In this way proteins concerned with immunity or carrying hormones can reach all the tissues.

Capillary filtration. Normally there is a dynamic balance between loss of fluid from the capillaries and reabsorption into the capillaries and removal of excess by the lymphatics so that the fluid does not accumulate. If, however, the venous pressure is increased the balance is

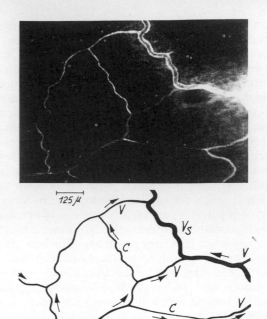

Fig. 27.29 Part of the capillary bed of the mesentery of the cat 10 min after injection of homologous plasma saturated with brilliant sulfoflavin. The escape of labelled protein takes place almost entirely from the venules (V) and small collecting veins (V_s). The arteriole (A) supplying the capillary bed (C) shows no perivascular fluorescence. (G. Hauck & H. Schroer (1969). *Pflugers Archiv fur die gesamte Physiologie des Menschen und der Tiere* **312**, 32–44.)

upset and fluid accumulates. The filtration in the forearm can be measured by blowing up a sphygmomanometer cuff around the upper arm and recording the increase in volume of the forearm by a plethysmograph. The initial rapid increase in volume measures the filling of the veins (Fig. 27.17) but the later slow increase depends on filtration. In healthy young persons the coefficient of capillary filtration is about 0.0028 ml/min mm Hg/100 ml tissue. It is increased slightly in older people and in persons with rheumatoid arthritis especially if oedema is present. In these experiments neither the surface area of the capillary membrane nor the tissue pressure outside the capillaries is known and thus the effective pressure cannot be calculated (see also Chapter 33).

Capillary fragility. It is important to understand that capillary permeability and capillary fragility (or resistance) refer to different, and probably distinct, properties of the capillary wall. In clinical parlance capillary fragility refers to the production of minute haemorrhages (*petechiae*) in the skin when the capillary pressure has been raised in the forearm by the application of a venous occluding pressure to the upper arm through a sphygmomanometer cuff; or to the production of petechiae by the application of suction to the skin.

Landis has pointed out that frequently there is a striking lack of association between increased capillary permeability and increased fragility. Thus, the oedema fluid of inflammation contains few red cells but a great deal of protein representing, therefore, a relatively pure increase in permeability. On the other hand, the petechiae that occur in certain blood diseases consist chiefly of red cells with little or no oedema fluid or protein.

Despite the fact that the walls of capillaries are so thin, they can withstand very large distending pressures. For example, when a person is standing without making rhythmical contractions of his leg muscles, the pressure in the capillaries in his feet rises to about 100 mm Hg. This is because of the hydrostatic effect of the column of blood above the feet. The reason why this pressure does not burst the capillaries can be seen by considering the law of Laplace. This states that the tension in the wall of a cylindrical tube (T) depends on the pressure difference across the wall of the tube (P) and the radius of the tube (r), that is $T = Pr$. In a capillary high internal pressures do not increase the wall tension very much because the radius is so small.

THE VEINS OR CAPACITY VESSELS

Veins have thinner walls, less elastic tissue and a larger internal diameter than the corresponding arteries and they are more easily distended than arteries (Fig. 27.3b). The valves made of folds of internal endothelium have already been described (p. 456). They have two main functions. They transport blood at high velocity and low energy cost from the tissues back to the heart. Secondly, by altering the tone in the smooth muscle in their walls they adjust the capacity of the circulation to meet the blood volume and so regulate the filling pressure of the heart.

VENOUS PRESSURE

The venous pressure in man is best measured directly by connecting a long thin, flexible cannula, introduced into a vein, to a manometer filled with a 3 per cent solution of sodium citrate. Measurement of the actual hydrostatic pressure within a vein, say at the level of the elbow, is of little value since the figure obtained depends on the position of the vein relative to the heart. Thus if the arm is raised the pressure falls and if the arm is lowered it rises. The upper level of fluid in the manometer tube does, however, occupy a constant relation to the level of the heart, or to the sternal angle which is a convenient reference point whose relation to the heart is reasonably constant. Thus if there is no obstruction or constriction between the manometer tube and the right atrium the difference in level between the height of the meniscus and the sternal angle represents the pressure in the right atrium. As a rule this lies between zero and -2 cm of water but there is a great deal of individual variation in *central venous pressure*, values as high as 9 and 12 cm being found not infrequently in apparently healthy young men; values over 20 cm H_2O may be found in congestive heart failure. The effective venous pressure for filling the heart is of course greater than the venous pressure in the neck because of the sub-atmospheric pressure within the thorax. The latter amounts during inspiration to -5 mm Hg ($\equiv 67$ mm water). Thus the *effective filling pressure* is $6·7 - (-2)$, or $8·7$ cm of water.

The internal jugular vein may be made to act as its own manometer since it usually has no functioning valves. The vessel collapses at the point at which the external pressure (that is the atmospheric pressure) is greater than the internal pressure (the venous blood pressure). The upper level of the column of blood in the internal jugular vein is conventionally measured from a horizontal plane passing through the sternal angle (Fig. 27.30) which is taken as the reference point. With the patient reclining and the head supported on a pillow the level is from 0 to 2 cm below this line and the venous pressure is said to be, therefore, 0 to -2 cm (of blood). Higher values are found when the heart is failing or when the blood volume is much increased. The use in this way of the external

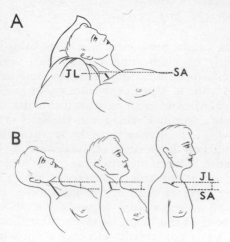

Fig. 27.30 The upper diagram, A, represents the level of blood in the internal jugular vein when the subject is lying with the head supported on pillows. The upper level of the column of blood in the jugular vein (JL) lies a little below the level of the sternal angle (SA). In the lower diagram, B, the upper level of the column of blood in the internal jugular vein is above the sternal angle, indicating an increase of venous pressure. Note that as the position of the patient alters the upper level of the column of blood occupies different positions in the neck although its relation to the sternal angle remains the same. (T. Lewis (1949). *Diseases of the Heart*. London: Macmillan.)

jugular vein as a manometer is unreliable.

The pressure in the atria and in the great veins is not constant but varies with the cardiac cycle and with respiration. It also varies with the position of the patient, being somewhat higher if the patient is recumbent.

The venous pressure in the superficial veins of the dorsum of the foot with the subject erect and motionless is from 70 to 100 mm Hg, which is the same as the calculated hydrostatic pressure from the right atrium. On exercise, owing to the action of the muscle pump (p. 505), the reading falls to less than 30 mm Hg. If, however, the leg veins are varicose with incompetent valves, there may be little difference between the readings during exercise and at rest.

VENOUS PULSE

The flow of blood in the larger veins at rest is continuous except near the heart. The variations of atrial pressure usually pass back a short way into the great veins near the heart since they are not equipped with valves. A venous pulsation at the root of the neck in the

internal jugular vein is therefore often seen in normal people. The pulsations of the internal jugular vein, which lies deep to the sterno-mastoid muscle, are seen as lateral, not as vertical movements. The venous pulse can be recorded by placing a metal cup on the right side of the neck above the sternal end of the clavicle. A rubber tube leads from the cup to a tambour attached to a lever the movements of which can be recorded on a moving paper or a photographic plate. Such a record is shown in Fig. 27.31; it cannot be analysed properly unless either an electrocardiogram or a record of the arterial pulse is made simultaneously.

In the venous pulse there are three positive waves during each cardiac cycle. The *a* wave is produced by atrial systole. It is replaced by a rapid fluttering movement when the atria flutter and disappears when they fibrillate (p. 521). The *a* wave becomes more prominent when the pressure in the right side of the heart is raised because of valvular disease or chronic lung disease, and very prominent should the right atrium contract against a closed tricuspid valve as sometimes happens with complete heart block. The *a* wave is followed shortly by the *c* wave. In the jugular pulse this is a disturbance imparted to the recording apparatus by the pulse in the underlying carotid artery. Some time later the *v* wave appears. This is due to the gradual increase of venous pressure during ventricular systole and the summit of the *v* wave is coincident with the opening of the mitral and tricuspid valves. The first negative wave, between *a* and *c*, is the result of atrial diastole. The trough between *c* and *v* is produced by descent of the heart during ventricular systole and the third negative wave (after *v*) is due to the sudden release of pressure in the veins as the tricuspid valve opens.

The tracing of the jugular pulse in Fig. 27.31 should be compared with the record of the changes in intra-atrial pressure obtained from a membrane manometer communicating directly with the interior of the right atrium.

VEINS AS TRANSPORT VESSELS

After passing slowly through the capillaries the blood speeds up as it passes through the venules and veins. This is because the total cross-sectional area of the venous bed is so much smaller than that of the capillaries. The blood is transported from the periphery to the

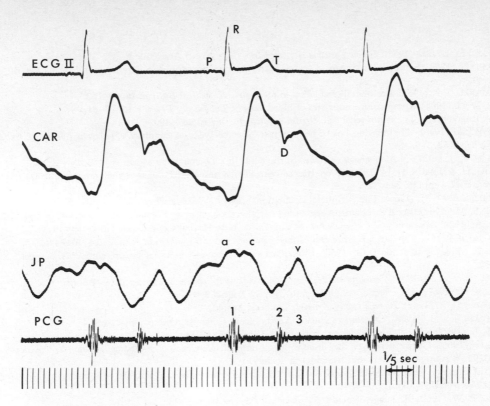

Fig. 27.31 Simultaneous records from a normal male aged 22 years. From above downwards electrocardiogram (standard lead II) (ECG II), external carotid arteriogram (CAR), jugular phlebogram (JP) and apical phonocardiogram (PCG). D marks the dicrotic notch, *a, c, v* the three main peaks of the phlebogram. 1, 2, 3. First, second and third heart sounds.

heart with very little expenditure of energy. The pressure gradient of about 10 mm Hg is enough to return blood from a small vein in the arm to the right atrium. This is because the veins offer very little resistance to blood flow.

Gravity can affect the transport function of the veins. Above heart level, gravity aids the return of blood to the heart. Below heart level the hydrostatic pressure of the dependent blood tends to distend the venous system. This would impair the venous return to the heart but this tendency is counteracted by the muscle pump. When muscles in the leg contract they compress the veins lying in the fascial planes between them. Because of the system of valves the blood is normally driven towards the heart.

VEINS AS CAPACITY VESSELS

Although the veins are easily distensible and have little muscle in their walls they should not be thought of as simple inert tubes. They have a resting tone and are capable of constriction and dilatation, in part governed by baroceptor reflexes from the aorta and right atrium and in part, as during exercise, by chemical substances in the blood. Venoconstriction is produced by a rise in venous pressure in the right side of the heart seen in cardiac failure or induced by the Valsalva manoeuvre (p. 505) or by a fall in arterial blood pressure. Adrenaline and noradrenaline both constrict veins. When the effective blood volume is reduced by approximately 500 ml either by removing this quantity of blood or by inflating cuffs round the thighs the hand volume decreases and the distensibility of the hand vessels, mainly veins, decreases. If the changes occur generally in the body they would be useful since they tend to reduce the amount of blood in the vascular dead space of the veins and so maintain the central venous pressure in spite of the reduction in blood volume.

REFERENCES

ABRAMSON, D. I. (Ed.) (1962). *Blood Vessels and Lymphatics.* New York: Academic Press.

BARCROFT, H. (1960). Sympathetic control of vessels in the hand and forearm skin. *Physiological Reviews Rev.* **40**, suppl. **4**, 81–91.

BARCROFT, H. (Ed.) (1963). Peripheral circulation in man. *British Medical Bulletin* **19**, 97–162.

BERGEL, D. H. (1966). Stress strain properties of blood vessels. *Laboratory Practice* **15**, 77–81.

FLOREY, LORD (1966). The endothelial cell. *British Medical Journal* **ii**, 487–490.

FLOREY, LORD (1968). The missing link. The structure of some types of capillary. *Quarterly Journal of Experimental Physiology* **53**, 1–5.

FRANKLIN, K. J. (1937). *A Monograph on Veins.* Springfield, Ill.: Thomas.

GAUER, O. H. & THRON, H. L. (1962). Properties of veins *in vivo*: integrated effects of their smooth muscle. *Physiological Reviews* **42**, suppl.j5, 283–327.

GREENFIELD, A. D. M. (1965). The peripheral circulation. *Annual Review of Physiology* **27**, 323–350.

HILTON, S. M. (1962). Local mechanisms regulating peripheral blood flow. *Physiological Reviews* **42**, suppl. 5, 265–282.

KROGH, A. (1929). *Anatomy and Physiology of the Capillaries.* New Haven, Conn.: Yale University Press.

LANDIS, E. M. (1934). Capillary pressure and capillary permeability. *Physiological Reviews* **14**, 404–426.

LANDIS, E. M. & HORTESTINE, J. C. (1950). Functional significance of venous blood pressure. *Physiological Reviews* **30**, 1–32.

LEWIS, T. (1927). *The Blood Vessels of the Human Skin and their Responses.* London: Shaw & Sons.

McCALLY, M. (1967). The effect of weightlessness. *Science Journal* **3**, 39–43.

McDONALD, D. A. (1960). *Blood Flow in Arteries.* London: Arnold.

McMICHAEL, J. (Ed.) (1958). *Circulation.* Harvey Tercentenary Congress. Oxford: Blackwell.

MAGGIO, E. (1965). *Microhaemocirculation.* Springfield, Ill.: Thomas.

MAJNO, G. (1965). Ultrastructure of the vascular membrane. In *Handbook of Physiology.Section 2, Circulation*, vol. 3, p. 2293. Washington: American Physiological Society.

MARKS, J., BIRKETT, D. A. & SHUSTER, S. (1972). 'Capillary permeability' in patients with collagen vascular diseases. *British Medical Journal* **i**, 782–784.

MERRILL, E. W. (1969). Rheology of blood. *Physiological Reviews* **49**, 862–888.

PICKERING, G. W. (1965). Hyperpiesis: high blood pressure without evident cause: essential hypertension. *British Medical Journal* **ii**, 959–968, 1021–1026.

RHODIN, J. A. G. (1962). Fine structure of vascular walls in mammals. *Physiological Reviews* **42**, suppl. 5, 48–57.

SPECTOR, W. G. (Ed.) (1964). The acute inflammatory response. *Annals of the New York Academy of Sciences* **116**, 747–1084.

STEWART, W. K. (1952–53). The physiological effects of gravity. In *Lectures on the Scientific Basis of Medicine*, vol. 2. London: Athlone Press.

WOOD, J. E. (1965). *The Veins.* London: Churchill.

WIEDERHIELM, C. A., WOODBURY, J. W., KIRK, S. & RUSHMER, R. F. (1964). Pulsatile pressures in the microcirculation of the frog's mesentery. *American Journal of Physiology* **207**, 173–176. ?

ZWEIFACH, B. W. (1961). *Functional Behaviour of the Microcirculation.* Springfield, Ill.: Thomas.

ZWEIFACH, B. W. (Ed.) (1965). Symposium: Blood flow and exchange at the microcirculatory level. *Federation Proceedings. Federation of American Societies for Experimental Biology* **24**, 1074–1111.

28 Vasomotor control

Just as the behaviour of the heart is modified by local, humoral and nervous factors, so is that of the blood vessels. The behaviour of blood vessels is usually modified by contraction or relaxation of the smooth muscle in their walls. Sometimes other mechanical factors may influence behaviour, but active control of the smooth muscle is thought to be the most important mechanism. However, just as vessels differ in structure and function, the regulatory mechanisms controlling all types of blood vessel are not identical. It is therefore best to think in terms of the regulatory systems for each functional variety of blood vessel, the damping, resistance, exchange and capacity vessels.

CONTROL OF THE RESISTANCE VESSELS

Although all blood vessels offer some resistance to the flow of blood, the main resistance lies in the small arteries and arterioles (Chap. 27). These vessels are controlled by the local action of chemical or physical factors, hormones circulating in the blood or by autonomic nerves which supply the vessels. Although changes in the calibre of resistance vessels tend to change the pressure levels and gradients in the circulation, the main effect is on blood flow. Blood flow to any tissue is normally regulated to serve the needs of that tissue by local factors but it may be subordinated to serve the needs of the entire body by hormonal or nervous factors.

LOCAL CONTROL

1. EFFECT OF METABOLITES

Figure 28.1 shows schematically the effect of temporary occlusion of blood flow in a limb on resistance blood vessels. On release of the occlusion the blood flow is raised well above

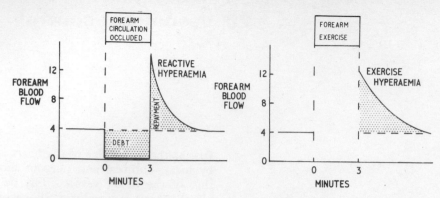

Fig. 28.1 Change in forearm blood flow (ml/100 ml/min) after occlusion of the circulation (left) and exercise of the forearm muscles (right). (*By courtesy of I. C. Roddie.*)

the resting level and then gradually returns towards the resting level. The increase in blood flow after occlusion is known as *reactive hyperaemia*. The excess blood flow after release is sometimes referred to as the *blood flow repayment* for the *blood flow debt* incurred during occlusion. Though there is not a precise relationship, the bigger the debt incurred, the bigger the repayment made. Increasing the period of occlusion tends to increase both the height and duration of the reactive hyperaemia. This phenomenon is confined to those tissues whose circulation has been reduced. It is not mediated by vasomotor nerves since it is unaffected if the autonomic nerve supply to the limb is severed.

A similar increase in blood flow is seen after a period of exercise in a limb. When the exercise stops the blood flow is increased greatly above the resting level and then falls in an exponential fashion towards the resting level. This phenomenon is called *exercise hyperaemia*. Like reactive hyperaemia it is confined to the exercising tissues and still occurs after section of the autonomic nerves to the tissue. The size of the exercise hyperaemia is related to the severity and duration of the exercise.

In both exercise and reactive hyperaemia, the increase in blood flow is thought to be due to an increased concentration of products of metabolism (metabolites) in the tissues which can cause blood vessels to dilate. In the case of circulatory arrest, metabolism continues to produce metabolites even when the blood flow is stopped. So the concentration of metabolites, and hence the vasodilatation in the tissue, increases with an increase in the period of cir-

culatory arrest. When the circulation is reestablished, the increased blood flow through the dilated vessels washes away the metabolites in an exponential fashion until their concentration in the tissues, and hence the blood flow, has returned to normal. The increase in blood flow that occurs in muscles during exercise is not adequate to supply the needs of the increased metabolism. Metabolites therefore accumulate in the tissues and the resulting vasodilatation is responsible for the hyperaemia which follows exercise. When the raised blood flow has cleared the excess of metabolites the blood flow returns to normal.

It is the local control of resistance blood vessels through metabolites which ensures that the blood supply to tissues is precisely regulated to meet their metabolic needs. If heart muscle has to do more work, the increased local metabolism that this entails increases the local production of metabolites and hence local coronary blood flow. The parts of the heart muscle which experience the greatest increases in work get the greatest increases in blood flow. The occurrence of metabolic hyperaemia is most evident in tissues such as muscle and liver where blood flow is mainly determined by metabolic needs. In tissues such as skin and kidney, where blood flow is related to functions other than metabolism, metabolic hyperaemia is less evident.

There is still no agreement as to the metabolites responsible for dilating the blood vessels. Among those suggested are carbon dioxide and hydrogen ion excess, oxygen lack, potassium excess, an increase in the osmolarity of the local tissue fluids, adenosine triphosphate, adenosine, phosphate and bradykinin. All can dilate blood vessels; some, such as adenosine triphosphate and bradykinin, do so strongly and others, such as potassium and phosphate, weakly. It is likely that most of the above substances contribute to metabolic hyperaemia, the contribution of any particular substance varying in different tissues and under different conditions. Oxygen lack seems to be particularly important in regulating myocardial blood flow whereas carbon dioxide tension seems to have an important influence in the regulation of cerebral blood flow (Chap. 43).

2. EFFECT OF LOCAL TEMPERATURE

If an extremity such as the hand or foot is put into warm water at 45°C the blood flow through the part increases several fold. The increase is local, being restricted to the part that is immersed. It is independent of the autonomic nerve supply since it can occur even if the part is denervated. This vasodilatation serves to protect the extremity from the damaging effect of heat. If the extremity is immersed in water at 45°C with the circulation occluded, it becomes painful as the tissue temperature rises. The vasodilatation that normally occurs tends to keep the tissues relatively cool by increasing their perfusion with blood at central body temperature. Since heat causes vasodilatation in denervated tissues it is likely that its effect is a direct local one on the smooth muscle in the wall of the resistance vessel. In intact tissues the stimulating effect of heat on local tissue metabolism may contribute to the dilatation.

The immersion of an extremity in moder-ately cold water normally causes vasoconstriction which reduces the loss of heat from the blood to the environment. In people whose limbs are exposed to cold water for prolonged periods, say after shipwreck, the prolonged decrease in blood flow may lead to local tissue death (necrosis) in the toes and feet (immersion foot syndrome).

An extremity exposed to near freezing temperatures, for example 0 to 4°C, shows a characteristic pattern of events (Fig. 28.2). The local blood flow falls to about zero initially and the part becomes painful. Blood flow then starts to rise rapidly to a value well above the resting level. At this time pain disappears. After another short interval flow falls again and pain reappears. This cyclic pattern of blood flow persists while the immersion in ice continues. It is referred to as the *hunting reaction*.

This vasodilatation in response to cold may be elicited locally in the fingers, toes, ears and the tip of the nose, parts which are particularly rich in A–V anastomoses. It does not depend on an intact nerve supply, but the exact mechanism underlying it is not understood. Its effect is clearly protective since it tends to prevent tissue damage due to freezing of the extremities.

3. EFFECT OF TRANSMURAL PRESSURE

The pressure difference across the wall of a blood vessel is measured by subtracting the external tissue pressure from the intravascular pressure. It is referred to as the transmural pressure. Changes in transmural pressure cause local changes in the resistance vessels.

As the transmural pressure in the vessels

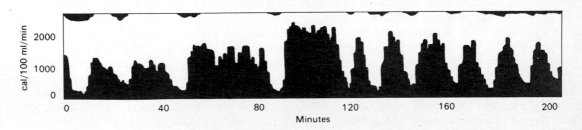

Fig. 28.2 The heat loss in cal/100 ml tissue/min from the right index finger to water at 0 to 6°C. The clear area represents the heat derived from the tissues of the finger, the black area the heat from the circulating blood. Pain is represented on a roughly quantitative scale at the top of the figure. Pain was felt when heat loss was at a minimum. (A. D. M. Greenfield, J. T. Shepherd & R. F. Whelan (1951). *Irish Journal of Medical Science* **309**, 415.)

rises, one would expect the resistance vessels to dilate due to distension and the blood flow to increase. However, as Figure 28.3 shows vascular beds but is a marked feature in kidney, brain and muscle. It is a useful phenomenon since it tends to keep blood flow to an organ

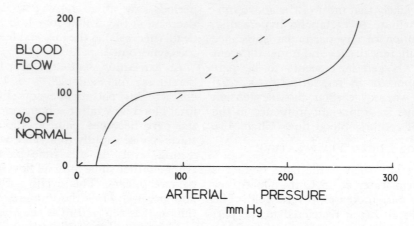

Fig. 28.3 The effect of arterial pressure on blood flow. The broken line shows the relationship which would be expected if the blood vessels behaved as rigid tubes. (*By courtesy of I. C. Roddie.*)

schematically, this expectation is not fulfilled. If the arterial pressure is raised above the normal value of about 100 mm Hg, blood flow rises only very slowly until, at a high critical pressure, it rises sharply. The fact that flow does not change very much when the pressure is raised suggests that the increase in transmural pressure elicits an increase in vascular resistance. The large increase in flow at very high pressures is thought to be due to the distending forces overcoming the vasoconstrictor response.

If the pressure is lowered below normal, the blood flow at first does not fall proportionately; this indicates that a fall in transmural pressure decreases the vascular resistance. However, at very low perfusion pressures the flow stops even while there is still a positive perfusion pressure head. This closure of the vessels below a certain critical transmural pressure has been referred to as *critical closure* by Burton. He predicted such instability at low pressure values from the law of Laplace.

The diagram also shows that over a very wide range of perfusion pressures the blood flow is relatively independent of the pressure head. This phenomenon is referred to as *autoregulation*. It is seen to some extent in most

fairly constant despite fluctuations in blood pressure provided its metabolic requirements do not change.

A number of mechanisms have been suggested to account for it. According to the *myogenic theory* a rise in transmural pressure tends to stretch the walls of the resistance vessels. Smooth muscle is excited by stretching and responds by contracting. The resulting vasoconstriction increases vascular resistance and limits the tendency for flow to increase. The converse reaction would explain the decrease in resistance when transmural pressure is lowered. A second possible mechanism is contained in the *tissue pressure theory*. In an encapsulated organ such as the kidney a rise in transmural pressure would result in a rise in tissue fluid formation. The accumulation of tissue fluid would increase tissue pressure and result in a mechanical compression of the resistance vessels. This would also tend to limit the increase in blood flow when transmural pressure rises. A third possibility is contained in the *tissue metabolite theory* of autoregulation. Here the increase in blood flow resulting from a rise in transmural pressure would tend to wash away local metabolites and reduce the metabolite concentration in the tissues. This in turn would cause vasoconstriction and so limit the increase in blood flow.

It is likely that all these mechanisms contribute to different degrees to the autoregulation that is seen in different tissues.

4. Effect of Oxygen Partial Pressure in Pulmonary Alveoli

As mentioned earlier oxygen lack may cause vasodilatation. In the myocardium oxygen seems to be the most important factor in the coronary vasodilator response to an increase in myocardial work. However, in the lungs, when the pressure of oxygen in the alveoli of one part of the lungs falls, a vasoconstriction occurs in the blood vessels perfusing that part. In this situation hypoxia acts as a vasoconstrictor agent. This response ensures that blood is not sent to poorly ventilated alveoli and so helps to maintain the normal ventilation to perfusion ratio in the lungs. It may also contribute to the pulmonary hypertension seen in people with chronic lung disease.

NERVOUS CONTROL

1. Vasoconstrictor Nerves

These are nerves which, when stimulated, cause blood vessels to constrict. Impulses reaching their nerve terminals are thought to cause release of noradrenaline which excites the smooth muscle in the walls of the blood vessel.

The existence of a nervous control of the blood vessels was first shown in 1852 by the French physiologist Claude Bernard, who cut the cervical sympathetic nerve in the neck of a rabbit and found that the blood vessels in the ear of the same side became dilated and the ear became warmer. Since electrical stimulation of the peripheral end of this nerve, that is the end going to the ear, caused a constriction of the vessels and a lowering of the temperature of the ear the blood vessels must normally have been kept constricted by impulses continually passing along the nerve.

The central origin of the vasoconstrictor nerve impulses can be traced in animals by making sections through the brain stem at various levels, starting at the cranial end and gradually passing caudad. The blood pressure is not altered if the cut passes above (cranial to) the fourth ventricle, but if the lower (caudal) part of this area is damaged there is a widespread vasodilatation and a fall of blood pressure. Destruction of a small region near the facial colliculi, half-way down the floor of the fourth ventricle (see Fig. 45.1), causes a large and immediate fall of blood pressure. Stimulation of the facial colliculus, or of the grey matter near it, gives a rise of blood pressure. Stimulation of the grey matter at the inferior angle of the fourth ventricle causes a fall of pressure but this is not satisfactory proof that the vasodilator centre is situated there. A large number of cells and fibres in the medulla are concerned with the transmission of impulses to blood vessels and this area is, therefore, called for convenience the *vasomotor centre*. Afferent impulses are constantly reaching the vasomotor centre from all parts of the body, especially from the pressor receptors in the carotid sinus and aortic arch (Fig. 28.4) and from the cardiac and respiratory centres in the medulla itself; efferent impulses are thus initiated or modified and appropriate adjustments made in the circulatory system. When there is disease of the afferent nervous system, as for example in the neuropathy of diabetes mellitus, reflex alterations of vascular tone may be interfered with.

The activity of this important vasomotor centre is modified by cortical and diencephalic autonomic centres. These may in fact have some direct connexions with preganglionic sympathetic fibres. The higher centres may initiate vasomotor changes accompanying, for example, alterations of muscular activity, while the vasomotor centre, since it receives information from the peripheral vasculature, may be concerned in fine control.

When the spinal cord is cut through in the cervical region the outflow of vasomotor impulses is stopped and the blood pressure falls steeply, but if the animal can be kept alive for several days the pressure rises again. It is usually assumed that this rise is produced by increased activity of the spinal vasomotor centres in the lateral columns of grey matter. These spinal centres are normally subordinate to the main vasomotor centre, being controlled by fibres which pass down from the medulla in the lateral columns of the cord. Section through the lower end of the cord, although it cuts off part of the vasomotor outflow, has little

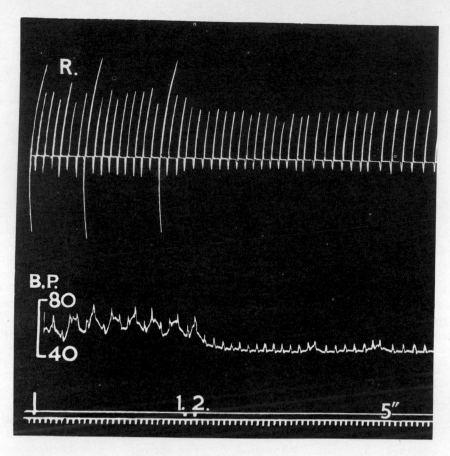

Fig. 28.4 Cat 2·9 kg. Bilateral vagotomy. Previously bled 65 ml. Records from above downwards: Respiration; arterial blood pressure; signal line; time in 5 sec intervals. The first part of the blood pressure trace shows Mayer waves each about 20 sec; the interval between respirations on the upper trace is 7·5 sec. At 1 and 2 the carotid sinus nerves were sectioned. Note fall in blood pressure and the disappearance of the Mayer waves following this procedure.
(B. Andersson, R. A. Kenney & E. Neil (1950). *Acta physiological scandinavica* **20**, 206.)

effect on blood pressure because of reflex compensatory mechanisms. If, however, the section passes above the sixth thoracic segment only partial compensation is possible and the blood pressure falls.

Electrical stimulation of the lateral and posterior parts of the hypothalamus in animals, or of the premotor cerebral cortex, produces vasoconstriction and a rise of blood pressure. In man, mental work as well as all kinds of emotional disturbances are accompanied by alterations of the circulation—the pallor of anger, for example, and the blush of shame—that seem to indicate connexions between the higher and lower centres.

The vasoconstrictor impulses pass down from the medulla into the cord (Chap. 45) and leave by the thoracic and upper lumbar nerves since electrical stimulation of the anterior spinal roots of the thoracic and upper lumbar part of the cord, or of the corresponding white rami communicantes, or of any of the sympathetic nerves arising from them, causes a vasoconstriction in the area supplied.

The position of the synapses on the vasoconstrictor pathway was investigated by Langley, who painted the sympathetic ganglia with nicotine which blocks the transmission of the nerve impulse across the synapses. The fibres leave the lateral horn of the grey matter in the anterior (ventral) root, passing by way of the white ramus communicans to the sympathetic ganglion (*preganglionic fibres*) (Fig. 28.5). Here

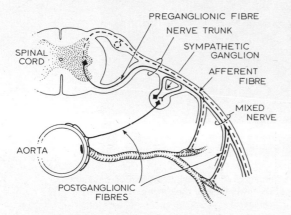

Fig. 28.5 Sympathetic nerve supply to the blood vessels.

new fibres (*postganglionic fibres*) arise, some going straight to the main blood vessels, others returning to the mixed spinal nerves, to be carried in them to the blood vessels in the periphery Although the blood vessels anastomose across the mid-line these vasoconstrictor fibres are confined strictly to their own side of the body. In the case of the forearm blood vessels, and perhaps other vessels, the vasoconstrictor impulses act on α receptors only whereas noradrenaline given intravascularly acts on α and β receptors (see also Chap. 41).

The abdominal organs are richly supplied with sympathetic fibres through the splanchnic nerves (Chap. 41). The fibres arise from the lower thoracic part of the cord, passing through the sympathetic chain to the coeliac ganglion and to the inferior mesenteric ganglia. From here postganglionic fibres reach the abdominal organs along the blood vessels. Denervation of a single organ in the body does not affect the general mean arterial pressure because the resulting vasodilatation produces only a small alteration of the circulation, for which reflex adjustments can easily compensate. Section of the splanchnic nerves, however, is followed by an immediate and profound fall of pressure. This operation was at one time employed for the treatment of high blood pressure in man; however, the fall of blood pressure that follows splanchnicectomy in either animals or man is not usually permanent. Electrical stimulation of the peripheral ends of the splanchnic nerves (Fig. 28.6) causes a marked vasoconstriction, which is shown by an increase in arterial pressure and a fall in volume in many of the abdominal viscera. These observations indicate the importance of the splanchnic area in the maintenance of arterial pressure.

If the sympathetic nerves to the extremities of man are cut, the blood flow through the extremities increases. This increase is most marked in the skin of the hands and feet, but is also seen in the skeletal muscles of the forearm and calf. It can be concluded that the blood vessels of these parts are normally subjected to considerable vasoconstrictor tone. It also follows that blood flow to these parts may be increased by release of vasoconstrictor tone.

The frequency of impulses in vasoconstrictor nerves is usually quite low compared with that in somatic motor fibres. It has been shown that almost complete vasoconstriction is produced when vasoconstrictor nerves are stimulated at rates of about 8 impulses per second. The resting level of vasoconstrictor tone could be mimicked by about 1 impulse per second. In somatic fibres, maximum contractions of skeletal muscle are achieved with impulse frequencies of about 50 per sec.

2. VASODILATOR NERVES

These are nerves which, when stimulated, cause vasodilatation in the blood vessels they innervate. They are thought to do this by releasing acetylcholine at their nerve endings. Vasodilator nerve fibres occur in both sympathetic and parasympathetic systems.

In certain animals and probably in man cholinergic vasodilator nerve fibres are distributed in the sympathetic nerve supply to voluntary muscles and they can be activated by stimulation of the frontal region of the anterolateral cerebral cortex. These fibres synapse in the anterior hypothalamus and mid-brain and the nerve impulses then pass through the medulla and the intermediate grey column of the spinal cord where the preganglionic fibres of the sympathetic systems, both vasodilator and vasoconstrictor, begin (Fig. 28.5). It seems unlikely that such impulses have an important role in man in the muscle vasodilatation of exercise because this is unaffected by sympathectomy but they may initiate the vasodilatation that occurs at the beginning of

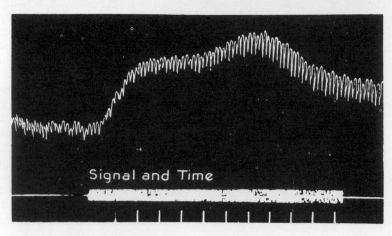

Fig. 28.6 The effect on blood pressure of stimulation of the peripheral end of the right splanchnic nerve. The time is marked in 10 sec at the lower edge of the tracing. (*By courtesy of E. W. H. Cruickshank.*)

voluntary effort, or even before it has started. It is more likely, however, that the hyperaemia produced by emotion, fright and by mental arithmetic is produced in this way.

Stimulation of the following parasympathetic nerves produces a vasodilatation in the part supplied: chorda tympani to the submandibular gland, small petrosal nerve to the parotid gland, lingual nerve to the tongue and the pelvic splanchnic nerves to the external genitalia. The nerves to the salivary glands might not be regarded as true vasodilators since vasodilatation may be brought about by bradykinin (Chap. 50).

After a muscle has been treated with atropine to make its blood vessels insensitive to acetylcholine, stimulation of the vasodilator nerves to that muscle is without effect. However, atropinization of the muscles in a limb of an intact animal does not reduce the limb blood flow. This shows that vasodilator fibres, unlike vasoconstrictor fibres, are not tonically active under normal circumstances. They appear to be brought into action whenever the need for them arises.

3. ANTIDROMIC IMPULSES AND VASODILATATION

Stimulation, either electrical or mechanical, of the peripheral end of the divided posterior root of a spinal nerve in an animal or in man produces dilatation of the vessels in the corresponding areas. The nerve impulses do not travel in efferent fibres (that is fibres originating in cells lying in the cord itself), for the phenomenon still occurs even when the posterior nerve root has been divided between the ganglion and the spinal cord and after time has been allowed for efferent fibres to degenerate. When the posterior root is divided distal to the ganglion and the fibres are allowed to degenerate, stimulation of the peripheral end of the cut nerve no longer produces dilatation at the periphery. In other words the cell stations for the fibres responsible for vasodilatation are in the posterior root ganglia and these fibres are indistinguishable from ordinary afferent fibres. Since the impulses that produce dilatation are evidently passing in afferent nerves in a direction opposite to that taken by sensory impulses they are said to be *antidromic*. It is very doubtful if antidromic impulses arising in spinal ganglia play any part in regulating circulatory changes in peripheral tissues, but the cutaneous vasodilatation that follows, for example, the application of an irritant like mustard to the skin is probably brought about by an *axon reflex* occurring in afferent (sensory) nerve fibres as indicated in Figure 28.7. When, for example, a pain nerve ending is stimulated by a painful stimulus, an impulse is conveyed by the sensory neurone not only back to the spinal cord but also by a collateral branch to adjacent blood vessels. Such dilatation no longer occurs if the afferent fibres have degenerated after section either of the peripheral

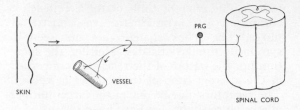

Fig. 28.7 Diagram to illustrate the probable pathway (shown by the arrows) of impulses set up when an irritant is placed on the skin. PRG, posterior root ganglion.

nerves or of the posterior roots distal to the spinal ganglia.

Antidromic impulses afford an explanation of the vasodilatation in herpes zoster (shingles), which is associated with an inflammatory disturbance of one or more posterior root ganglia. In this condition both the reddening of the skin with formation of vesicles and the pain are confined to the distribution of one or more sensory nerves.

4. VASOMOTOR REFLEXES

These permit the general circulatory requirements of the body to override the local requirements of the tissues. Though in some reflexes the effects may be fairly well localized, in most cases their effects are seen throughout the body. Vasomotor reflexes are most evident in those tissues such as skin, muscle and gut where the blood supply can be restricted without an immediate severe threat to the tissue. The flow through some tissues, such as the heart and the brain, which cannot tolerate prolonged restriction of flow, is only partly controlled by vasomotor nerves (see p. 518 and Chap. 43). Hundreds of vasomotor reflexes have been described but only some of the more important ones that can be elicited in man are described below.

(a) THERMOREGULATORY REFLEXES. When a subject is heated by immersing his legs in warm water, the blood flow to his hands increases. This is reflexly mediated by release of sympathetic vasoconstrictor tone since it does not occur if the sympathetic nerves to the hand are cut or if the vasoconstrictor fibres are selectively blocked (Fig. 28.8). A similar release of vasoconstrictor tone occurs in the other extremities, such as the ears, nose and lips. Vasodilatation also occurs in the skin of other parts of the body but this is the result of

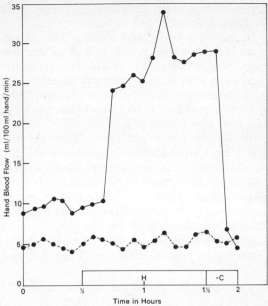

Fig. 28.8 Blood flow in a normal and a sympathectomized hand during body heating. H represents time during which feet were in hot water (45°C); C represents time during which feet were in cold water (12 to 16°C). (G. T. C. Hamilton (1947).*Ulster Medical Journal* **16**, 18–26.)

activity in the sympathetic cholinergic nerves which supply the sweat glands in these parts. The dilatation may be secondary to the release of bradykinin-forming enzyme by these active glands. Reflex vasodilatation in response to body heating is not seen in vascular beds such as muscle which lie deep to the skin. The skin vasodilatation, by raising skin temperature, normally leads to increased heat loss from the body.

The co-ordinating centres for these reflexes are thought to lie in the hypothalamus. The centres are controlled not only by changes in the temperature of the blood impinging upon them but also by afferent impulses travelling from thermoreceptors in the heated parts. The introduction of warm saline into the internal carotid artery causes reflex vasodilatation in the skin. However, if, in the experiment described in the previous paragraph, blood warmed in the legs is prevented from passing to the brain by inflating cuffs on the thighs the reflex vasodilatation in the hand is reduced but not abolished. This suggests that the residual vasodilatation in the hand results from impulses arising from temperature receptors in the heated skin.

If the body is cooled a reflex increase in vasoconstrictor tone occurs in most of the skin areas of the body. The vasoconstriction, by decreasing skin temperature and therefore the temperature gradient between the body and the environment, helps the body conserve heat (see also Chap. 42).

(b) BLOOD SHIFT REFLEXES. Many stimuli, which have in common the effect of shifting blood towards or away from the chest, can produce reflex alterations in vasoconstrictor tone. These reflex changes are most evident in the blood vessels of the muscles. Tilting a person into the foot-down position, the application of negative pressure to the lower part of the body, breathing at positive pressure (Fig. 28.9) and

tation. The efferent limb of these reflexes is composed of sympathetic vasoconstrictor fibres. The reflex changes are not seen in sympathectomized tissues. Neither the location nor the nature of the receptors concerned in the afferent limb of the reflexes is known. The receptors may be one of the large number of stretch receptor that have been identified in the walls of the low pressure vessels in the thorax. They are not thought to be the baroreceptors in the arterial system, since alteration of the activity in these does not have much effect on vasoconstrictor tone in man. When the blood pressure at the carotid sinus is altered in man the changes in blood flow in the limbs are small and can be explained by the changes that occur in arterial pressure (Fig. 28.10). However, in animals such as the dog

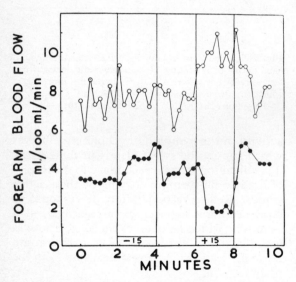

Fig. 28.9 Blood flow in normal (●) and nerve-blocked (○) forearms during pressure breathing. During the two periods indicated, the subject breathed air at pressures 15 mm Hg below and 15 mm Hg above atmospheric pressure, respectively. (D. A. Blair, W. E. Glover & B. S. L. Kidd (1959). *Clinical Science* **18**, 9.)

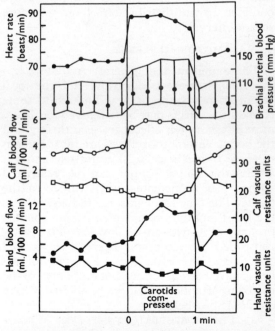

Fig. 28.10 The effect of bilateral carotid artery compression on heart rate, brachial arterial pressure, and blood flow, and resistance to flow in the calf and the hand. (I. C. Roddie & J. T. Shepherd (1957). *Journal of Physiology* **139**, 381.

the Valsalva manoeuvre all tend to shift blood away from the chest towards the feet. These stimuli are associated with a reflex increase in vasoconstrictor tone in muscle blood vessels. Tilting a person into the foot-up position, the application of positive pressure to the lower limbs and squatting, all tend to move blood towards the chest and result in reflex vasodila-

arterial baroreceptor stimulation does lead to reflex alterations in vasoconstrictor tone. This species difference may be an expression of the very different gravitational stresses experienced by the circulations of man and dogs.

(c) CHEMORECEPTOR REFLEXES. As mentioned earlier the direct effect of carbon dioxide on the blood vessels is to dilate them. However, when carbon dioxide is breathed by man in high concentration a reflex increase in peripheral vascular resistance occurs in the muscles. This is mediated by way of sympathetic vasoconstrictor fibres. If these fibres are blocked, peripheral vasodilatation occurs when high concentrations of carbon dioxide are breathed. Carbon dioxide is thought to act directly on the vasomotor centre in the medulla but it probably acts on the peripheral chemoreceptors also. Severe oxygen lack also produces some reflex increase in peripheral resistance but its effect is small relative to that of carbon dioxide. It is thought to act by way of the peripheral chemoreceptors.

(d) EXERCISE REFLEXES. When a subject exercises his muscles, strong local vasodilatation is brought about by metabolites in the active muscles. However, in other tissues, such as the muscles which are not taking part in the exercise, there is a reflex vasoconstriction. Figure 28.11 shows the changes which occurred when a subject exercised his legs on a bicycle ergometer. Blood pressure and heart rate rose. Blood flow rose in the forearm, whose sympathetic nerves had been blocked, probably because the increase in the arterial blood pressure drove more blood through the tissue. However, in the normally innervated forearm, the blood flow fell slightly in spite of the rise in mean arterial pressure. This showed that peripheral vascular resistance in the forearm rose reflexly during leg exercise. This reflex helps to redistribute blood from the non-active to the active muscles during exercise. The efferent limb of the reflex consists of sympathetic vasoconstrictor fibres but the nature and location of the receptors on the afferent limb and the reflex centre are not known.

If all the sympathetic vasoconstrictor fibres are blocked pharmacologically, exercise cannot be sustained for very long. The peripheral vasodilatation produced by the exercise results in such a fall in total peripheral resistance that the arterial pressure falls dramatically and unconsciousness ensues.

(e) EMOTIONAL STRESS REFLEXES. When a subject is frightened or given severe mental arithmetic to do a dramatic vasodilatation occurs in the muscle tissue (Fig. 28.12). This

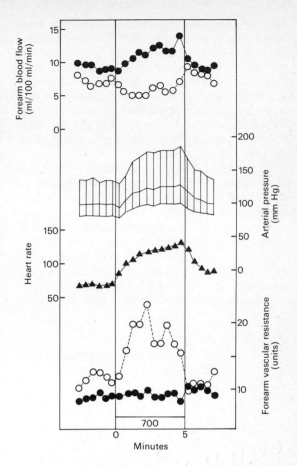

Fig. 28.11 The effect of deep nerve block on forearm blood flow and forearm vascular resistance during 5 min leg exercise. Nerve-blocked forearm (●); normal forearm (○); heart rate (▲). The rectangle represents the period of leg exercise. The increase in oxygen consumption above the resting level during exercise was 700 ml. (D. A. Blair, W. E. Glover & I. C. Roddie (1961). *Circulation Research* **9**, 264–274.)

response is reduced or abolished by blocking the sympathetic nerves supplying the tissue and by atropine. These results indicate that sympathetic cholinergic vasodilator nerves are involved in the response. It is likely also that adrenaline released from the adrenal medulla contributes to the response. Cholinergic vasodilator fibres to skeletal muscle blood vessels have been described in animals but the adequate stimulus for their reflex excitation has not been identified. They are not involved in the reflex responses to baroreceptor stimulation. However, stimulation through electrodes implanted in the hypothalamus in cats in areas which excite 'flight or fight' responses also excite the sympathetic cholinergic vasodilator pathway. This suggests that, as in man, these fibres are involved in the muscle vasodilatation accompanying stressful situations. It is interesting in this respect that vasodilator fibres to muscle seem to be involved in the vasodilatation in muscle seen during fainting (Fig. 28.13).

(f) LUNG INFLATION REFLEXES. Taking a deep breath causes a reflex vasoconstriction in the skin of the peripheral parts in man. It does not occur if the sympathetic fibres have been severed (Fig. 28.14). In fact, a large number of relatively trivial stimuli, such as a sudden noise, a pinch or the inflation of a cuff on an arm, can cause well marked reflex vasoconstrictions in the hand. The muscle blood vessels are not involved in these responses.

Some of these responses form part of the alerting reaction in man but their physiological significance is obscure.

HORMONAL CONTROL

Compared with the local and nervous control of resistance vessels, hormonal control is relatively unimportant. Some of the more important humoral agents which affect resistance vessels are mentioned below.

Adrenaline and noradrenaline. When the splanchnic nerves are stimulated the initial rise of blood pressure is probably due to vasoconstriction in the abdominal organs. This is followed about half a minute later by a second rise due to the liberation of adrenaline and noradrenaline from the adrenal medulla. A similar effect has already been described in asphyxia (p. 626). Again, when the pressure in

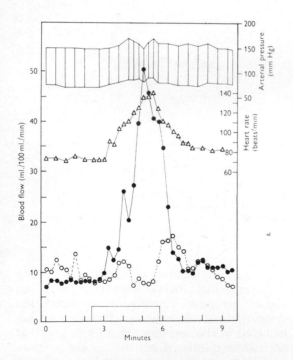

Fig. 28.12 Results showing that active cholinergic vasodilator nerves to human muscle contribute to the vasodilatation in the forearm muscles during stress. ○, hand blood flow; ●, forearm blood flow; △, heart rate. During the time represented by the rectangle it was suggested to the subject that he was suffering from severe blood loss. (D. A. Blair, W. E. Glover, A. D. M. Greenfield & I. C. Roddie (1959). *Journal of Physiology* **148**, 633.)

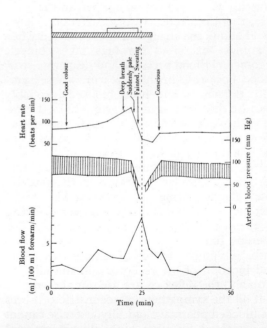

Fig. 28.13 Post-haemorrhagic fainting. Typical symptoms and typical changes in the heart rate, in the arterial blood pressure and in the blood flow in the *normal forearm*. During the venesection (open rectangle at top of diagram) about 700 ml of blood were removed. The subject fainted and the venesection was stopped. Note the great rise in blood flow in spite of the marked fall in blood pressure due to vasodilatation in the forearm muscles. Shaded rectangle: venous tourniquets on both thighs. (H. Barcroft & O. Edholm (1945). *Journal of Physiology* **104**, 161.)

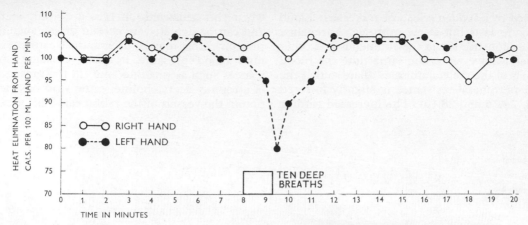

Fig. 28.14 Calorimetric records of the heat elimination from the right and left hands separately of a healthy man aged 38. Before the record began the median and ulnar nerves on the right side had been infiltrated with procaine. Beginning at the eighth minute as indicated by the open rectangle the subject took ten deep breaths, following which the heat elimination from the left hand fell, indicating vasoconstriction. The absence of vasoconstriction in the right (blocked) hand shows that this is mediated through efferent nerves.

the carotid sinus is reduced the rise of arterial blood pressure which follows is due partly to the reflex activity of the vasoconstrictor nerves and partly to liberation of adrenaline and noradrenaline. Liberation of these hormones

also occurs in flight or fight reactions or when the subject is exposed to emotional stress.

When it is injected intravenously into an animal adrenaline produces an immediate and transient fall of arterial pressure at once suc-

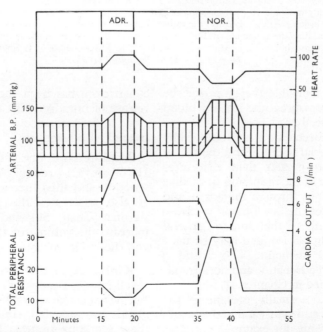

Fig. 28.15 Diagram showing the effects of intravenous infusions of adrenaline and noradrenaline on heart rate, arterial blood pressure, cardiac output and total peripheral resistance. (The initial transient drop in blood pressure that occurs during infusions of adrenaline is not shown.) H. Barcroft & H. J. C. Swan (1953). *Sympathetic Control of Human Blood Vessels*, Monographs of the Physiological Society, No. 1. London: Arnold.

ceeded by a sudden rise. The increase in blood pressure lasts but a few minutes. Adrenaline causes constriction in the cutaneous and other vessels; however, at the same time the blood vessels of the skeletal muscles dilate so that the total peripheral resistance is slightly lowered (Fig. 28.15 and 28.16). The increased cardiac

when the constrictor nerves are stimulated, and can be depleted by certain drugs, such as reserpine, that may be used in the treatment of high blood pressure in man. Certain substances such as nicotine, and (in the presence of atropine) acetylcholine, cause vasoconstriction in the vessels of the rabbit ear. This vaso-

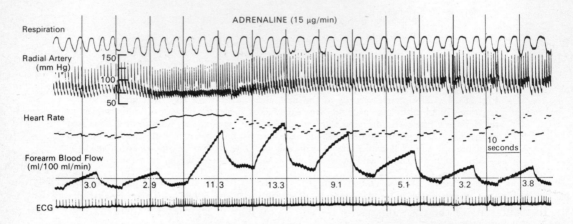

Fig. 28.16 Intravenous infusion of adrenaline bitartrate at 15 μg/min. Respiration was recorded by a nasal thermocouple. The pressure in the radial artery was measured by a strain-gauge manometer. Note the calibration 50, 100, 150 mm Hg. The heart rate was obtained electronically from the electrocardiogram (ECG). The forearm blood flow was obtained by venous occlusion plethysmography; the values in ml/100 ml/min are given below, 3·0, 2·9, etc. The vertical lines are 10 sec apart. The infusion began 30 sec before the first vertical line and continued throughout. (M. J. Allwood, E. W. Keck, R. J. Marshall & J. T. Shepherd (1962). *Journal of Applied Physiology* **17**, 71.)

output must be responsible therefore for the rise in systolic pressure; the mean blood pressure may not be altered.

Injection of noradrenaline is not followed by an initial fall in arterial pressure and the rise in systolic pressure is greater than that after adrenaline. In spite of the reduction in cardiac output brought about through the aortic and carotid sinus reflexes the mean blood pressure rises. Noradrenaline is therefore a general vasoconstrictor (Fig. 28.15), and may be used as such in clinical medicine. The α and β receptors, on which adrenaline and noradrenaline act, are described in Chapter 41.

Noradrenaline is normally present in the walls of the arteries and veins, and in the heart; the aorta of the rabbit, for example, contains an average amount of 0·5 μg per gram. The noradrenaline in the walls of vessels is thought to lie in stores associated with vasoconstrictor nerve endings. These stores are mobilized

constriction has been shown to be due to the release of noradrenaline from the artery wall, since it does not occur in the ears of rabbits treated with reserpine.

Angiotensin. The octapeptide angiotensin II (Chap. 32) produces generalized vasoconstriction and thus increases arterial pressure. It also stimulates the production of aldosterone (Chap. 50) and therefore tends to increase the volume of the extracellular fluid and the blood volume.

Kinins. A number of peptides are powerful vasolidators. Bradykinin (Chap. 50) formed during the active secretion of sweat and salivary glands enormously increases the blood flow in these secreting tissues. The intra-arterial injection of as little as 0·01 μg may, and 0·1 μg always does, increase the blood flow in the forearm (muscle) blood flow or in the hand (blood flow through the skin); intravenous infusion of bradykinin produces large increases in blood flow in both forearm and hand and a visible flushing of the face. On a molar basis, bradykinin (mol. wt. 1934) is the most active substance known to produce vasodilatation in man. The kinins may take part in the vasodilatation associated with the activity of all tissues, not only glandular tissues.

Histamine has very pronounced effects on the circulation. When injected intra-arterially or intravenously it causes vasodilatation of the resistance vessels, flushing of the skin and a fall in arterial pressure. It also increases the permeability of the capillaries which then permit loss of protein and fluid from the circulation (see also Chap. 50).

CONTROL OF THE DAMPING VESSELS

The large elastic arteries convert the large pressure fluctuations generated by the heart into a fairly steady perfusion pressure. They also contain the sensory nerve endings which detect stretching of the arterial wall for the control system which stabilizes arterial pressure.

The arterial wall contains smooth muscle and there is good evidence that it is innervated. However the precise role of this muscle in arterial function is not clear. The large elastic arteries offer so little resistance to blood flow that changes in their calibre make practically no difference in the total resistance offered by the circulation. Changes in muscle tone in the large arteries have, therefore, no appreciable effect on blood flow. Arterial muscle responds to circulating hormones such as adrenaline by constricting and to local metabolites such as carbon dioxide by relaxation. The muscle is attached to the elastic fibres in the artery wall and by contracting it increases the elastic tension in the wall and reduces the distensibility of the arteries. This reduces the amount of blood contained in the arteries at any arterial pressure and also changes the characteristics of the arterial pulse wave. The smooth muscle might also regulate the sensitivity of the stretch receptors in the aortic arch and carotid sinus by altering the distensibility of their walls. Arteries respond vigorously to local mechanical irritation by contraction. This may limit blood loss when the arterial wall is cut or damaged.

CONTROL OF THE EXCHANGE VESSELS

The true capillaries do not seem to have either nerve fibres or muscle cells but the precapillary sphincters and arterio-venous anastomoses are supplied by non-medullated nerve filaments. Stimulation of a sympathetic vasoconstrictor nerve probably constricts the metarterioles and the precapillary sphincters and prevents blood flowing into and through the capillaries (Fig. 27.21). The blood cells and plasma gradually drain away into the veins and the capillaries close but they open immediately if the precapillary sphincters relax. It seems likely that nervous control provides a background tone but it is much less important than the hormonal control by hormones such as adrenaline and by tissue metabolites such as histamine, 5-hydroxytryptamine, acetylcholine and carbon dioxide and the various kinins. The small vessels without nerves are much more sensitive to hormones than are the innervated arterioles. The reflex responses arising from the baroceptors depend on changes in the calibre of the arterioles, the microvessels do not show any active change in calibre.

CHEMICAL CONTROL OF THE CAPILLARIES

Vasomotion is the term applied to the spontaneously occurring periodic relaxation and constriction of the thoroughfare channel and of its precapillaries (p. 544). In uninjured resting tissue the blood-flow through the thoroughfare channel is fairly constant and its hydrostatic pressure relatively high. Vasomotion of the precapillary offshoots, manifested by an opening and closing of their sphincters, governs flow through the true capillaries and produces areas of varying hydrostatic pressure which are maintained for periods of time which differ greatly under different conditions. The fluid exchange in the capillary bed is thus greatly influenced by vasomotion. This delicately balanced activity depends on vasoconstrictor nervous influences and on humoral factors, both vasodilator and vasoconstrictor. The origin in the tissue itself of some of the humoral factors provides a local mechanism for the distribution of blood and, therefore, the extent and duration of fluid exchange.

The number of capillaries per unit area of tissue can be approximately determined by examining microscopically sections of the tissues of animals injected before death with India ink. In active muscles the capillaries are not only much more numerous than in resting muscles but also much wider. Thus in the resting muscle of a guinea-pig Krogh found 200 capillaries per mm^2 with an average diameter of 3·5 μm, against 2500 per mm^2 in an active muscle with an average diameter of 5·0 μm. A transverse section of an active skeletal muscle may show three or four times as many capillaries as muscle fibres. Activity is thus accompanied by capillary dilatation and in addition the size of the capillary bed is enormously increased by the opening up of vessels which were closed during the resting state. This opening up of new capillaries is believed to be initiated by the local release of chemical substances.

The modern view, as stated above, is that the capillaries, at least in certain parts of the body, are controlled by precapillary sphincters. However, the following classical observations are most easily explained by supposing that the skin capillaries are able to contract effectively on their own. If the circulation through the arm is completely cut off by inflation of a cuff round the arm the injection of vasopressin (Chap. 52) or of adrenaline into the skin of the forearm produces a white area which must be due to capillary contraction. Again, when a sphygmomanometer cuff is placed round the arm and inflated to, say, 40 mm Hg the venous pressure increases to this value within 60 sec and the capillary pressure rises about 10 mm Hg higher (see p. 546). Since an intradermal injection of adrenaline can still produce a white area the capillaries must be capable of contracting against an internal pressure of 50 mm Hg. Experiment shows that they cannot contract against higher pressures. Since normally the pressure at the arteriolar end of a capillary loop rarely exceeds 50 mm Hg the capillary should be able to act independently of the arteriole in spite of the high pressure at the arteriolar end. If, however, the skin capillaries are first constricted by an injection of adrenaline the venous pressure must be raised to nearly 100 mm Hg before the capillaries give way under the pressure and allow blood to flow back into them from the veins. In other words, once constricted, the capillaries can offer considerable resistance to distension from the venous end, and presumably also from their arteriolar ends.

NERVOUS CONTROL OF CAPILLARY FUNCTION

Though the capillaries are not innervated, the vessels upstream and downstream from them are. Nervous influences, by controlling the inflow and outflow from the capillary beds, can control the hydrostatic pressure in the capillaries and thus the net movement of fluid across the capillary wall. Stimulation of the vasoconstrictor fibres to a muscle vascular bed (Fig. 28.17) causes three main effects, (1) a decrease in blood flow due to constriction of the resistance vessels, (2) a decrease in the volume of blood contained in the muscle due to constriction of the capacity vessels and (3) mobilization of fluid from the tissue spaces into the blood. The third effect is thought to be due to the fall in capillary pressure which follows constriction of the pre-capillary resistance vessels.

In certain animals it has been found that stimulation of the carotid sinus nerve causes not only a reflex vasodilatation but also an increase in the rate of tissue fluid formation (Fig. 28.18). Such a change may also result from a shift of blood to the chest. However, all situations in which blood flow is increased by vasodilatation are not associated with increased tissue fluid formation. Unless the pre-capillary sphincters are relaxed blood flow will not flow through the true capillary bed.

CONTROL OF THE CAPACITY VESSELS

Contraction of the smooth muscle in the walls of the veins has a number of effects. It reduces the capacity of the venous system to hold blood at any particular venous pressure. By raising the mean pressure in the venous system it tends to raise the filling pressure of the heart. The resulting increase in filling leads to stronger contractions and an increase in cardiac output. The rise in mean venous pressure also tends to cause capillary pressure to rise with a consequential increase in the formation of tissue fluid.

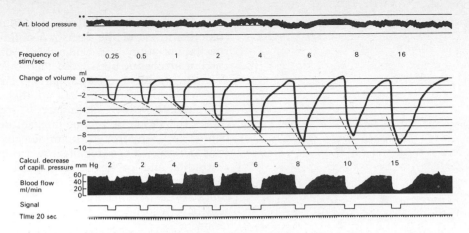

Fig. 28.17 Effects on resistance and capacitance vessels and net transcapillary fluid shift produced in the hindquarters of a cat by maximal lumbar vasoconstrictor fibre stimulation at different frequencies. Changes in blood flow reflect effects on resistance vessels (inflow and outflow pressures kept constant). The initial and rapid decreases in volume reflect effects on capacitance vessels and the subsequent slower and continuous decreases in volume (slopes indicated by dashed lines), transcapillary influx of extravascular fluid. Reductions in mean hydrostatic capillary pressure calculated in approximate figures. (S. Mellander (1960). *Acta physiologica scandinavica* **50**, Suppl. 176, 35.)

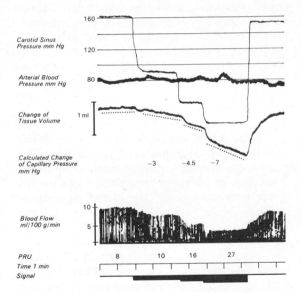

Fig. 28.18 Effects of stepwise reductions of pressure in the isolated, bilaterally perfused carotid sinus regions upon tissue volume and blood flow in the cat's hindquarters, while arterial inflow pressure and venous outflow pressure are kept constant in this region. Note how there occurs a reflex decrease of hindquarters blood flow at every stepwise decrease of carotid sinus pressure. Also, tissue volume decreases in a stepwise fashion, where each step is characterized by a first rapid phase and a second, slow, but continuous phase. The slope of this second phase is well graded to the reflex decrease in blood flow. (B. Öberg (1964). *Acta physiologica scandinavica* **62**, Suppl. 229, 1.)

MEASUREMENT OF VENOUS TONE. There is no completely satisfactory way of doing this but a number of methods may be employed. One is to compress both ends of a superficial segment of a vein under the skin which does not show any side branches opening from, or draining into it. If the pressure in the segment is measured through an indwelling needle or catheter, venoconstriction is signalled by a rise, and venodilatation by a fall of pressure in the segment. Another method is to expose the venous system to gradually increasing or decreasing steps of pressure and construct pressure–volume curves. Flattening of the pressure–volume curves indicates venoconstriction (Fig. 28.19). A third method is to fill the veins to a constant high pressure by venous congestion. A decrease in the volume of the tissue at constant venous pressure indicates venoconstriction. It is important to realize that distension by itself is not evidence that the smooth muscle in the venous wall has relaxed. If the resistance vessels dilate, the great inflow of blood into the capacity vessels raises the pressure in these vessels and distends them.

VENOMOTOR RESPONSES

Since the venous system is mainly concerned

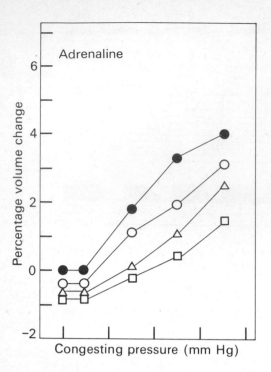

Fig. 28.19 Pressure–volume curves illustrating the effect of intra-arterial infusions of adrenaline on the high- and low-pressure capacity vessels of the forearm. Infusion of saline, ●; adrenaline infusions 0·1 μg/min, ○; 0·4 μg/min, ▲; 1·0 μg/min, □. Increasing doses of adrenaline progressively decrease the resting volume of the forearm and the distension of the forearm caused by raised venous pressure. This suggests that adrenaline constricts capacity blood vessels. (W. E. Glover, A. D. M. Greenfield, B. S. L. Kidd & R. F. Whelan (1958). *Journal of Physiology* **140**, 113.)

in integrated responses of the entire circulatory system, the responses of the venous system are usually integrated through vasomotor nerves or hormones.

LOCAL FACTORS. The local application of cold causes venoconstriction which may be intense and prolonged. This constriction may be proportionately greater than the constriction of the resistance vessels and account for the oedema seen in hands being rewarmed after exposure to cold. Local hypoxia and carbon dioxide excess relax venous smooth muscle, but these responses may be outweighed by the reflex effects of these stimuli acting centrally.

HUMORAL FACTORS. The pharmacological properties of the circular and longitudinal smooth muscle of the veins are different and species variation is considerable. Adrenaline and noradrenaline both cause venoconstriction (Fig. 28.19) due to the stimulation of α-receptors which are predominant in this muscle. Isoprenaline which stimulates β-receptors causes relaxation of venous muscle. Both histamine and 5-hydroxytryptamine cause venoconstriction. The female sex hormones oestradiol and progesterone tend to reduce spontaneous activity in veins and make their contractions weaker. Nitrites and adenylic acid have also been found to cause venodilatation.

NERVOUS FACTORS. Most of the stimuli that cause constriction of the resistance vessels seem also to cause constriction of the capacity vessels. As mentioned in the section on exchange vessels, stimulation of the sympathetic nerves to a tissue causes a fall in blood flow and blood capacity in that tissue. Stimuli which are thought to cause reflex venoconstriction include systemic hypoxia and hypercapnia, carotid sinus hypotension, body cooling, emotional stress, the shifting of blood away from the chest, exercise and the taking of deep breaths. Reflex venodilatation has been described in carotid sinus hypertension, fainting and sleep. Some of these responses are considered in more detail below.

1. *Exercise*. When a subject exercises his legs on a bicycle ergometer the capacity of the forearm veins is reduced (Fig. 28.20). The response is graded to the severity of the work. This does not occur if the sympathetic nerves to the arm have been cut. It can be demonstrated that the venoconstriction is generalized and occurs also in the muscles taking part in the exercise. Such an increase in venous tone is likely to play an important part in the circulatory adaptation to exercise. When the resistance vessels in the exercising muscles dilate the capillary and post-capillary pressures rise. If this were not accompanied by an adjustment of tone in the capacity vessels, blood would tend to pool in them. This tendency is antagonized by the pumping effect of the contracting muscles. The increase in venous tone acts in the same direction.

2. *Blood shifts*. Stimuli which shift blood away from the chest cause reflex reductions in venous capacity. Thus venoconstriction has

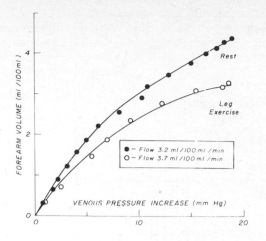

Fig. 28.20 Typical relationship between increase in forearm volume and pressure in a large forearm vein at rest (●) and during supine leg exercise (○). Note that forearm blood flow was similar in both circumstances. (B. S. Bevegard & J. T. Shepherd (1968). *Journal of Applied Physiology* **20**, 1.)

been reported after the Valsalva manoeuvre and the application of suction to the lower body. Figure 28.21 shows the decrease in the volume of a congested forearm when −30 mm Hg suction was applied to the body below the xiphisternum. This response was abolished by drugs which blocked the sympathetic nerves to the arm. This type of response would help to maintain the filling pressure of the

heart when blood tended to pool in the lower extremities.

3. *Emotional stress.* Figure 28.22 shows the result of an experiment in which pressure was measured in a segment of superficial forearm vein. With the subject at rest, the pressure in the segment was about zero mm Hg. When he became anxious about the nature of the forthcoming test, the pressure rose to about 40 mm Hg and then fell again. Immersing one hand in ice water had a similar effect. The increase in vasomotor tone during emotional stress would help to prevent pooling of blood in the extremities which might result from the concomitant dilatation of the resistance vessels.

Syncope. Syncope, usually called fainting, means a transient loss of consciousness due to reduction in cerebral blood flow. At the same time skeletal muscular tone is diminished and the patient slumps or falls to the ground if unsupported; the faint is thus usually self-limiting.

Cerebral blood flow is remarkably constant over a wide range of blood pressure (Chap. 43), thus fainting only occurs when the fall of blood pressure is severe. The reduction in blood pressure is usually the result of either a reflex vasomotor depression which produces marked vasodilatation or the loss of the baroceptor reflexes that normally maintain cerebral perfusion in the upright posture. Fainting therefore occurs most frequently when the person is

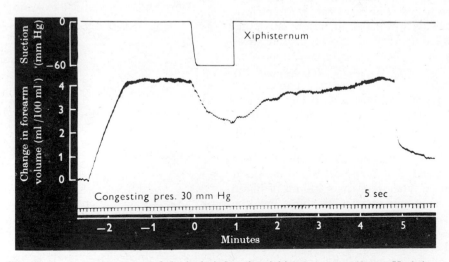

Fig. 28.21 Suction on parts of the body below the xiphisternum at −60 mm Hg below atmospheric: its effect on the volume of a congested limb. (B. L. Ardill, V. M. Bhatnagar & P. H. Fentem (1968). *Journal of Physiology* **194**, 627.)

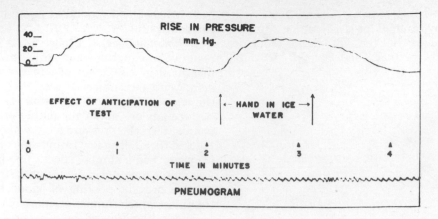

Fig. 28.22 Pressure changes recorded from a segment of a superficial vein of the forearm that had been isolated between wedges and kept at constant volume. (J. J. Duggan, V. Logan Love & R. H. Lyons (1953). *Circulation* **7**, 869.)

upright and consciousness usually returns quickly when he lies flat, unless the syncope is caused by a failure of cardiac output as for example in ventricular fibrillation or arrest or extreme bradycardia as in heart block (p. 522).

The commonest cause of fainting is probably strong emotions, particularly conflicting ones, as those aroused by sudden confrontation with an unpleasant sight such as a road traffic accident or surgical operation. Such emotions cause vasodilatation of skeletal muscle, perhaps because of an increase in circulating adrenaline and the activation of cholinergic vasodilator fibres. The heart rate is often slow, and this bradycardia is abolished by atropine, so the syndrome of bradycardia, muscle vasodilatation and systemic hypotension with fainting was named *vaso-vagal syncope* by Lewis. Pallor, 'cold' sweating, nausea and yawning may be associated with fainting, and other precipitating factors may include fatigue, hunger or standing in a crowded room. Syncope can also be induced by depletion of the circulating blood volume. A sudden loss of blood may lead to syncope: in blood donors it is uncommon when less than 400 ml of blood are withdrawn, but occurs in about half the subjects if 1 litre is removed (Fig. 28.23).

When a subject stands up, blood is normally prevented from pooling in the legs under the influence of gravity by reflex adrenergic vasoconstriction. If for some reason the vasoconstriction fails to oppose the tendency of the

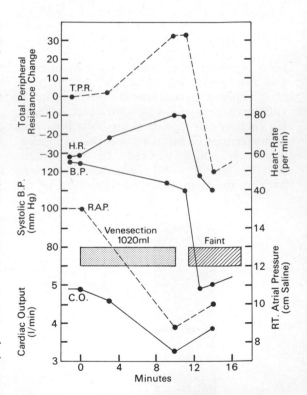

Fig. 28.23 Faint induced by venesection. Up to the end of the venesection, arterial pressure was maintained by peripheral vasoconstriction (increased total peripheral resistance) in spite of a decreasing cardiac output. During the faint, the cardiac output increased slightly, and the fall in blood pressure was, therefore, due to decrease in peripheral resistance. (H. Barcroft, O. T. Edholm, J. McMichael & E. P. Sharpey-Schafer (1944). *Lancet* **i**, 189.)

blood to pool, *orthostatic hypotension* with syncope may result. This is a common complication induced by the adrenergic-blocking drugs used to treat high blood pressure (hypertension), and sometimes after the reduction of blood volume and sodium produced by powerful diuretic drugs.

Rarely syncope is caused by cardiac inhibition due to reflex vagal effects triggered by pressure on the carotid sinus, by micturition or by prolonged coughing that virtually amounts to a Valsalva manoeuvre.

THE PERIPHERAL CIRCULATION IN CIRCULATORY FAILURE

The 'failure' of an organ or tissue is the state of affairs that arises when homeostatic mechanisms have ceased to ensure that function is maintained adequate to demand. The inability of the heart to provide an adequate output, *heart failure*, has been dealt with on p. 523. But *circulatory failure* may occur, even when the heart is working normally, because of maladjustments in various parts of the vascular system. This is sometimes referred to as *peripheral circulatory failure* of which a faint (syncope) is a special variety. However, the conventional distinction between cardiac failure and peripheral circulatory failure is a crude over-simplification. The whole circulatory system, heart, arteries, arterioles, capillaries, veins, functions as an integrated whole; a break-down in part of the complex, interconnected pattern of homeostatic processes can hardly fail to influence events in other parts of the system.

Shock is the term given to the clinical condition in which the patient is pale and greyish in colour; his skin is cold and shows drops of sweat. The pulse at the wrist is rapid and of low volume and the arterial pressure is low, being less than 80 mm Hg. The cardiac output is reduced. The patient has little pain or interest in his surroundings. Bodily and mental activities are depressed, the metabolic rate reduced and the temperature low. No satisfactory explanation of these phenomena has yet been discovered.

The most important factor in this complex condition is a reduction in the effective circulating blood volume. This may occur from sudden external loss of blood (as in acute haemorrhage, see p. 409); from pooling of blood in capillary beds; as a result of plasma loss from extensive burning of the surface of the body; or because of severe salt depletion (Chap. 33). The so-called traumatic shock that occurs after road accidents or under battle conditions is probably the result of a combination of several of these factors. It has to be pointed out, however, that a similar clinical state can occur in severe heart disease, when there is no reduction in the circulating blood volume, but when the cardiac output is suddenly reduced, for example as the result of a coronary thrombosis (infarction). Shock is increased in severity in the presence of infections; some bacterial toxins have a vasoconstrictor effect. However, germ-free animals suffer from shock after haemorrhage just as conventional animals.

Low blood pressure with a shock-like state is a feature of Addison's disease (Chap. 50), the result of failure of the adrenal cortex to produce corticosteroids. Patients with Addison's disease are liable to develop shock from relatively minor injuries or infections.

If reduction in blood volume were the only feature of shock after haemorrhage restoration of blood volume would be sufficient to save life. It is found, however, that if transfusion is delayed beyond a certain time shock may become 'irreversible'. This situation has been studied by subjecting dogs to severe haemorrhage. Irreversibility has been attributed to anoxia, caused by prolonged vasoconstriction, affecting the liver and kidneys and especially the intestine. Haemorrhage causes increased sympathetic activity and secretion of catecholamines which are responsible for the vasoconstriction; the cerebral and coronary circulations are relatively unaffected. Arteriolar vasoconstriction reduces the perfusion of the tissues, reduces their supply of oxygen, reduces the urinary output and causes local metabolic acidosis. Because the venules are in spasm, blood stagnates in the capillaries and the increased pressure in them causes fluid to move into the interstitial spaces and as a result the blood volume is reduced. Transfusion at this stage may merely increase the engorgement of the capillaries.

If the dog is transfused without delay after bleeding the cardiac output is increased, the blood pressure is raised and the vasoconstriction passes off. After a prolonged period of shock, say two or three hours, the animal passes into a refractory state attributed to irreversible changes in tissues. The survival period after haemorrhage can be prolonged by giving blood to increase the blood volume and an adrenolytic vasodilator drug. Vasodilatation increases the capacity of the circulation and further transfusion may be necessary to fill it up.

Shock in man is not identical with that in the dog; for example, the intestinal changes are not so marked. Thus the cause of irreversibility in man is not certain. Transfusion is undoubtedly a necessary part of treatment, but obviously if vasoconstriction is an important feature in man as well as in the dog it would seem to be unwise to combat hypotension by vasoconstrictor drugs such as noradrenaline or angiotensin. It seems likely that anti-adrenergic drugs combined with adequate transfusion may be most useful but this has yet to be proved clinically.

REFERENCES

AVIADO, D. (1965). *The Lung Circulation*. Vol. 1, *Physiology and Pharmacology*. Vol. 2, *Pathological Physiology*. Oxford: Pergamon Press.

BARCROFT, H. & SWAN, H. J. C. (1953). Sympathetic control of human blood vessels, *Monographs of Physiological Society* No. 1. London: Arnold.

BARCROFT, H. (Ed.) (1963). Peripheral circulation in man. *British Medical Bulletin* **19**, 97–162.

BOCK, K. D. (Ed.) (1961). Shock: Pathogenesis and therapy. International symposium. New York: Academic Press.

BYRNE, J. J. (Ed.) (1968). Symposium on the management of shock and unconsciousness. *Surgical Clinics of North America* **48**, 245–466.

Central Nervous System Control of Circulation (1960). A symposium. *Physiological Reviews* **40**, Suppl. 4, 1–311.

CHIEN, S. (1967). Role of sympathetic nervous system in haemorrhage. *Physiological Reviews* **47**, 214–288.

FOLKOW, B. & NEIL, E. (1971). *Circulation*. New York: Oxford University Press.

GREEN, H. D. & KEPCHAR, J. H. (1959). Control of peripheral resistance in major systemic vascular beds. *Physiological Reviews* **39**, 617–686.

GREENFIELD, A. D. M. (1963). In *Handbook of Physiology*, Section 2, Vol. 2, 1325–1351. Washington, D.C.: American Physiological Society.

HERSHEY, S. G. (Ed.) (1964). *Shock*. London: Churchill.

HEYMANS, C. & NEIL, E. (1958). *Reflexogenic Areas in the Cardiovascular System*. London: Churchill.

HERTZMAN, A. B. (1959). Vasomotor regulation of cutaneous circulation. *Physiological Reviews* **39**, 280–306.

HILTON, S. M. (1962). Local mechanisms regulating peripheral blood flow. *Physiological Reviews* **42**, Suppl. 5, 265–282.

JOHNSON, P. C. (Ed.) (1964). *Autoregulation of Blood Flow*. New York: American Heart Association.

LINDEN, R. J. (1963). Efferent nerves of the peripheral circulation. *Recent Advances in Physiology* (edited by R. Creese), Chap. 11. London: Churchill.

MAGGIS, E. (1965). Microhemocirculation. Observable variables and their biologic control. Springfield, Ill.: Thomas.

MELLANDER, S. & JOHANNSON, B. (1968). *Pharmacological Reviews* **20**, 117–196.

MOORE, F. D. & SHIRES, G. T. (1967). Editorial. *Annals of Surgery* **166**, 300–301.

RAPELA, C. E. & GREEN, H. D. (1964). Autoregulation of canine cerebral blood flow. *Circulation Research* **15**, Suppl., 205–212.

RODDIE, I. C. (1966). Nervous control of limb blood flow. *Scientific Basis of Medicine Annual Reviews*, 260–278.

SHARPEY-SCHAFER, E. P. (1952). Peripheral circulation in circulatory failure. *British Medical Bulletin* **8**, 331–332.

SHEPHERD, J. T. (1963). *Physiology of the Circulation in Health and Disease*. London: Saunders.

SJÖSTRAND, T. (1953). Volume and distribution of blood and their significance in regulating circulation. *Physiological Reviews* **33**, 202–228.

UVNAS, B. (1954). Sympathetic vasodilator outflow. *Physiological Reviews* **34**, 608–618.

WHELAN, R. F. (1967). *Control of the Peripheral Circulation in Man*. Springfield, Ill.: Thomas.

ZWEIFACH, B. W. (1961). *Functional Behaviour of the Microcirculation*. Oxford: Blackwell.

29 Circulation through lungs, liver and spleen

PULMONARY CIRCULATION

The lungs receive blood from two sources. The pulmonary arteries convey much the larger supply of blood (this is, of course, only partly oxygenated) from the right ventricle, while the bronchial arteries contribute less than one per cent of the supply as arterial blood, arising in the aorta. Blood leaves the lungs mainly by the large pulmonary veins which take oxygenated blood to the left atrium; a small amount leaves by the bronchial veins which drain venous blood mainly from the larger bronchi and interstitial tissues of the lungs near the hila. Communications exist between the two circulations; quantitatively these are quite unimportant in health but they may become much enlarged in disease (Fig. 30.4). Although there are precapillary bronchopulmonary anastomoses in the newborn it is rather doubtful if they are present in the healthy adult human lung. The walls of the large branches of the pulmonary artery are thin relative to those of the aorta and its main branches, but they are highly elastic. The elastic wall is continued down to branches of about 1000 μm in diameter but at this point there is a definite muscular coat which continues until the calibre is about 100 μm. Smaller branches of the pulmonary artery have thin elastic walls closely resembling those of the smaller pulmonary veins with which they are connected by an enormous mesh of capillaries spread out through the whole of both lungs (Fig. 29.1). This structure is consistent with experimental findings, namely that the pulmonary vascular bed can accommodate large changes in volume with small changes in pressure and that vasomotor activity plays a minor role in the control of pulmonary blood flow.

The pressure in the pulmonary artery in man is measured directly by catheterization through the right side of the heart (Fig. 29.2). The systolic pressure is, of course, the same as

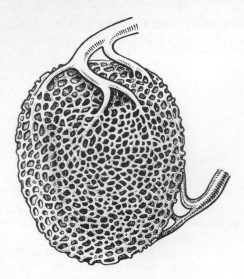

Fig. 29.1 A concept of the pulmonary capillary bed as an almost continuous meshwork of capillaries surrounding alveoli, represented here by the dark spaces containing air. (*After* W. S. Miller (1893). *Journal of Morphology* **8**, 186.) It is estimated that if the blood in the lung were spread out as a thin sheet of capillaries it would cover an area of 30 to 50m².

the systolic pressure in the right ventricle, that is 19 to 26 mm Hg, equal to 26 to 35 cm of blood. The diastolic pulmonary artery pressure is from 6 to 12 mm Hg or 8 to 16 cm of blood. The pressure in the pulmonary artery is thus barely sufficient to perfuse the apices of the lungs which are 15 cm above the pulmonary

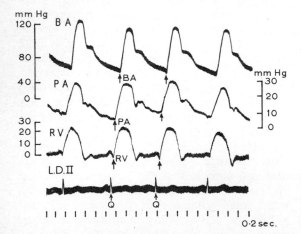

Fig. 29.2 Simultaneous records of pressure in brachial artery (BA), pulmonary artery (PA) and right ventricle (RV) obtained by optical manometers attached to cardiac catheters. An electrocardiographic record, lead II, is also shown (ECG). (*By courtesy of A. Cournand.*)

artery when the body is erect. The difference in blood flow throughout the lung can be investigated quantitatively. The subject is seated between a pair of scintillation counters which are sensitive to the radiation of a certain antero-posterior volume of lung. If he takes a single breath of carbon dioxide containing (radioactive) $^{15}CO_2$ the counting rate rises quickly indicating the amount of gas coming into the part of the lung being investigated. If he holds his breath the $^{15}CO_2$ is carried off in the pulmonary circulation and so the count falls, the rate of fall indicating the rate of blood flow. The results of many such experiments are shown in Figure 29.3. The blood flow is maximal at the bottom of the lung and is minimal (nearly zero) at the top of the lung when the subject is in the upright position. When the subject is lying on his back the flow in the bottom (base) of the lung is almost unchanged but the apical flow is much increased. In exercise the flow increases in all parts of the lungs but especially in the apical regions.

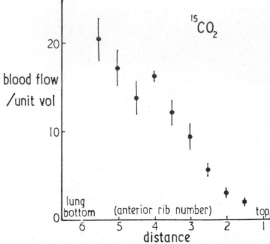

Fig. 29.3 Distribution of blood flow in the normal upright lung as measured with radioactive carbon dioxide. Data from 16 normal subjects; means and standard errors of clearance rates. Note that blood flow decreases steadily from the bottom to the top of the lung, there being virtually no flow at the apex. (J. B. West (1970). *Ventilation Blood Flow and Gas Exchange*, 2nd edn, p. 19. Oxford: Blackwell.)

Since the mean pulmonary pressure is about 15 mm Hg the total resistance in the pulmonary circuit (see formula, p. 538) is less than one-sixth of that in the systemic circuit and thus the right ventricle has less work to do than the left. When the catheter is advanced to the point at which it can pass no further, oxygenated blood can be withdrawn from it. Its tip must block a small branch of the pulmonary artery and the pressure recorded (the *wedge pressure*), 6 to 12 mm Hg, was originally assumed to be the pressure in the pulmonary capillaries; however, the contour of the pulse wave and the mean pressure usually correspond closely with those of the left atrium.

The pressure in the lung capillaries must lie between the pulmonary arterial pressure and the left atrial pressure, that is in the range 8 to 10 mm Hg, which is well below the plasma protein osmotic pressure of 25 mm Hg. If the capillary pressure rises above 25 mm Hg fluid escapes into the alveoli (pulmonary oedema). When a deep breath is taken the intrathoracic pressure falls (p. 596), perhaps as much as 30 mm Hg below atmospheric. The capillary transmural pressure must then for a short time exceed the plasma protein pressure; presumably the lymphatics drain away any fluid exuded into the alveoli.

By elegant studies of subjects rebreathing a gas mixture containing nitrous oxide in a whole-body plethysmograph Lee has shown that blood flow through the pulmonary capillaries is pulsatile.

The low resistance in the pulmonary circuit is accounted for by peculiarities of structure of the blood vessels of the lungs; the arterioles contain relatively little smooth muscle and the capillaries and veins are large, short and readily distensible. The low peripheral resistance and the large blood flow mean that the time taken for blood to pass through the pulmonary vascular bed (pulmonary circulation time) is very short, about 5 seconds. This time has been obtained in man by subtracting the time taken by a dye injected into the left atrium to reach a femoral artery from the time taken by the dye injected into the pulmonary artery to reach the same femoral artery.

Figure 29.4 shows the variations in pressure in the pulmonary system that occur with respiration. With the fall of intrathoracic pressure during inspiration there is increased filling of the right ventricle and therefore a larger stroke volume. The pulse pressure in the right ventricle is increased and the mean pressure in the pulmonary artery rises. On inspiration either the resistance to flow through

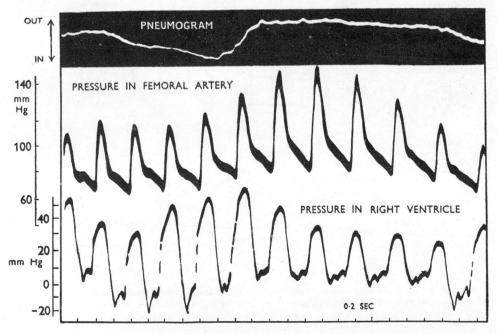

Fig. 29.4 Simultaneous records of respiration (top trace), blood pressure in femoral artery (middle trace) and pressure in right ventricle (lowest trace) during deep breathing. (*After* A. Cournand (1947). *Bulletin of the New York Academy of Medicine* **23**, 34.)

the lung capillaries increases or the pressure gradient through the lungs is reduced so that a smaller quantity of blood leaves by the pulmonary veins than is received through the pulmonary artery. The inflow to the left side of the heart is reduced and therefore the output from the left ventricle is temporarily reduced. The pulse pressure in the femoral artery therefore falls and the mean arterial pressure is reduced. In expiration, relaxation of the lungs improves the venous return through the pulmonary veins, the output of the left ventricle is increased and the pulse pressure and mean arterial pressure in the femoral artery are raised. Simultaneously the higher intrathoracic pressure reduces the return to the right ventricle and there is a fall in pulse pressure and mean pressure in the pulmonary circuit. This point was referred to on p. 533.

Acute hypoxia produces a rise in pulmonary arterial pressure due to constriction of the small pulmonary vessels, probably both arterioles and venules (Fig. 29.5). This seems to be a direct effect of hypoxic alveolar air or of unsaturated blood on these vessels, but the precise location of the reactive vessels is not known. The constriction of the pulmonary arteries may occur reflexly. In man chronic hypoxia, occurring at high altitudes, causes a rise of pulmonary artery pressure (Table 29.6);

Table 29.6 Pressure in pulmonary artery in relation to altitude and arterial hypoxia

Altitude (ft)	(m)	Mean pulmonary arterial pressure (mm Hg)	Arterial oxygen saturation (%)
0	0	13	97
5 280	1 610	16	95
10 150	3 094	25	92
14 900	4 542	28	78

After J. H. K. Vogel, W. F. Weaver, R. L. Rose, S. G. Blount & R. F. Grover (1962). Quoted from *Lancet* (1962), **ii**, 233–234.

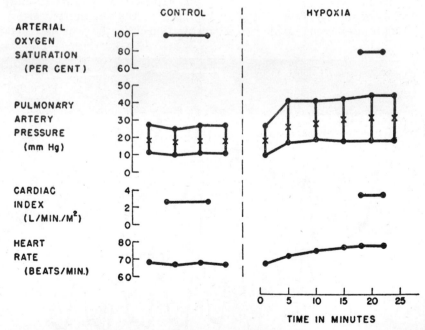

Fig. 29.5 Effect of acute induced hypoxia on the arterial oxygen saturation, pulmonary artery pressure, cardiac index, and heart rate. The study was performed on a man aged 35 believed to have a normal pulmonary circulation breathing a mixture of 12 per cent oxygen in nitrogen for 20 min. The 35 per cent increase in blood flow (cardiac index) that occurred during this study was not considered to be sufficient to produce the increase in pressure in the pulmonary artery, which is therefore taken to be evidence of vasoconstriction. (A. Cournand (1959). In *Pulmonary Circulation: A Symposium*. New York: Grune & Stratton.)

the pressure returns to normal if the subject is brought down to sea level. At high altitudes breathing is much deeper than at sea level and this (p. 632) combined with increased capillary pressure may account for pulmonary oedema in mountain climbers. However, pulmonary hypertension at high altitudes is not always incompatible with athletic prowess. Chronic disease of the lungs with hypoxia leads to generalized pulmonary vasoconstriction, persistent increase in the mean pulmonary artery pressure and, ultimately, to failure of the right ventricle. Localized hypoxia plays a role in the distribution of the total pulmonary blood flow. In hypoxic (diseased) areas of lung the blood vessels are constricted so that part of the blood that flows through them normally now flows through better aerated (healthy) parts of the lung.

Mitral stenosis (narrowing of the left atrioventricular valve) is apt to be associated with pulmonary arterial hypertension. The rise in left atrial pressure is accompanied by a rise in capillary pressure which leads to exudation of fluid (p. 579) and to alveolar hypoxia. This, as has been explained, may be a cause of pulmonary hypertension.

The facts mentioned above allow the amount of blood in the pulmonary circuit to be calculated. If 4·8 litres are pumped through the lungs per minute then in 5 seconds $5/60 \times 4·8 = 0·4$ litre passes completely through the circuit. Under basal conditions the pulmonary circuit must therefore contain at least 400 ml or approximately 10 per cent of the total blood volume. Some 75 to 100 ml of this is accommodated in the capillary bed which is so readily expansible that as much as 30 litres can flow through it per min during severe exercise with only a slight increase in pulmonary arterial pressure which is not nearly sufficient to increase the capillary pressure to a value exceeding the colloid osmotic pressure of the plasma proteins. In exercise the pulmonary capillaries may contain 170 ml of blood. At rest each red cell passes through the capillaries in

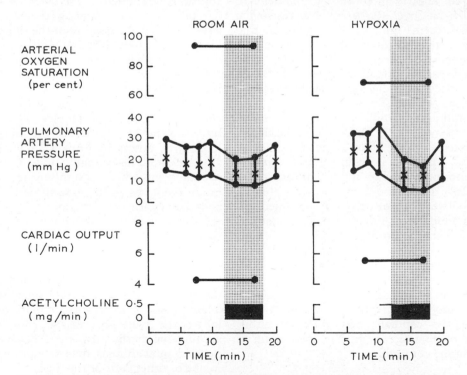

Fig. 29.7 Effect of acetylcholine and acute induced hypoxia on the arterial oxygen saturation, pulmonary artery pressure, and cardiac output. The study was performed on a man believed to have a normal pulmonary circulation. The shaded areas indicate the infusion of 0·5 mg per min of acetylcholine into the main pulmonary artery. (A. Cournand (1959). In *Pulmonary Circulation: A Symposium*. New York: Grune & Stratton.)

about 0·75 sec; during exercise the transit time is reduced to about 0·3 sec.

The blood vessels of the lungs have a sympathetic supply (both afferent and efferent) and a parasympathetic supply from the vagus but the results of stimulation in animal experiments are variable. In cats or dogs when either the pulmonary artery pressure or the pulmonary venous pressure is raised there is often a small fall in systemic arterial pressure and in heart rate which can be abolished by cutting the vagi (*pulmonary depressor reflex*). The afferent impulses probably arise in both pressure and stretch receptors in the atria and great veins. Daly and Daly found that stimulation of the carotid body chemoreceptors of the anaesthetized dog causes a decrease in pulmonary vascular resistance which is abolished by atropine or by cutting the carotid sinus nerves or the cervical vago-sympathetic trunks. This provides direct evidence for the reflex control of the pulmonary circulation in the experimental animal. The role of the nerves in man is not known: in contrast to the delicate control it has over the peripheral resistance (p. 539) it seems unlikely that the autonomic nervous system has much influence on the pulmonary circuit. Acetylcholine causes pulmonary vasodilatation in man especially when vasoconstriction has been previously produced by hypoxia (Fig. 29.7).

The pulmonary capillaries form a continuous mesh of tissue (Fig. 29.1). The air in the alveoli is separated from the blood in the capillaries by 0·3 to 0·7 μm which is occupied by two membranes normally closely applied to each other (Fig. 31.2). In disease, however, the potential space between them may be occupied by oedema fluid or an inflammatory exudate so that the distance through which oxygen and carbon dioxide have to diffuse is greatly increased. Normally about 20 ml oxygen per minute diffuses across the 'membrane' for each 1 mm Hg pressure gradient of oxygen but the amount increases greatly if the blood flow through the capillary bed increases either because of dilatation or by the opening up of new channels or both.

CIRCULATION THROUGH THE LIVER

The liver parenchyma in man is composed of cribriform sheets or plates of cells one cell thick. The sheets are continuous with one another and in the spaces between are the sinusoids. The whole arrangement is pervaded by two systems of tunnels, the portal tracts and the central hepatic veins.

Blood is brought to the liver by the portal vein and by the hepatic artery. The portal vein is formed by the union of the superior mesenteric vein, bringing blood from a large part of the alimentary tract, and the splenic vein conveying blood (about 45 per cent of the total) from the spleen, pancreas, gall-bladder and stomach. The portal vein and its main tributaries can be visualized in man by taking serial X-ray photographs after a concentrated solution of an iodine-containing compound has been injected into the spleen (Fig. 29.8).

The portal vein divides into branches lying in the portal tracts. These have smaller branches which leave the portal tracts and pass through holes in the sheets of hepatic cells and then become sinusoids. The walls of the sinusoids are lined by a layer of flattened (endothelial) cells. Among them are found large stellate cells floating in the blood stream but attached by delicate processes to the wall of the sinusoid. These stellate Kupffer cells are more actively phagocytic than the flat endothelial cells and pick up dyestuffs and particulate matter. After blood destruction they contain haemosiderin granules. The sinusoids lead into the central vein of the lobule which joins with others to form a sublobular vein (Fig. 29.9). These finally unite to form the hepatic vein which communicates with the inferior vena cava. In some species but not in man the hepatic veins possess sphincters made up of muscle fibres running longitudinally. All the vascular connexions to the lobule are capable of contraction or dilatation.

Injection studies in animals show that the blood from the superior mesenteric vein enters mainly the right lobe of the liver, while the splenic blood is distributed mainly to the left lobe of the liver. This absence of mixing may account for the fact that disease, especially in experimental animals, may be more prominent in one or other lobe of the liver.

The hepatic artery supplies the structures in the portal tracts including the bile ducts and anastomoses in a complicated fashion both with the sinusoids and with branches of the

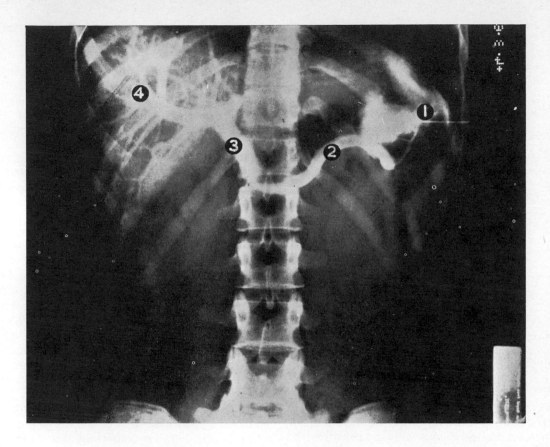

Fig. 29.8 Radiograph obtained by injecting radio-opaque material through the skin into the spleen. The normal appearances are shown: 1. needle in splenic pulp; 2. splenic vein; 3. portal vein; 4. branches of the portal vein within the liver. (*By courtesy of Sheila Sherlock.*)

portal vein (Fig. 29.9). About one-third of the hepatic arterial blood passes into the portal venous channels before the arterial blood reaches the sinusoids and increases the rate of flow in them. The flow in the hepatic artery and in the portal vein of the dog has been measured by electromagnetic flow-meters applied externally to the exposed but uncannulated vessels. If the superior mesenteric artery is tied, the flow in the portal vein is reduced but that in the hepatic artery is increased. If the hepatic artery is tied the portal venous flow is decreased. These findings are easily explained mechanically as the superimposition of slow-flowing and fast-flowing streams in shared channels. Partial occlusion of the portal vein causes pooling of blood in the intestines and a decrease of systemic blood pressure. Ligature of the hepatic artery in man

and in some experimental animals causes serious damage to the liver. In long-lasting hypoxia the liver is more easily damaged by poisons like chloroform.

The portal venous pressure can be measured in man during an abdominal operation by putting a needle connected to a manometer directly into the portal vein. It can also be measured in the conscious subject by inserting a fine needle through the abdominal wall into the splenic pulp, for the intrasplenic pressure is only slightly higher than the portal venous pressure (p. 585). Alternatively, a cardiac catheter can be passed from the cubital vein through the right atrium until its tip blocks a small radicle of the hepatic vein within the liver. The pressure then recorded approximates to the intrasinusoidal pressure and hence the pressure in the portal vein. The

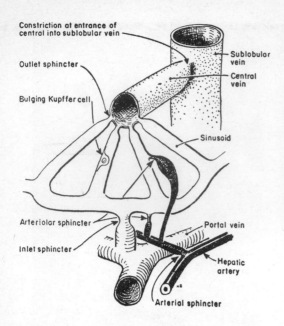

Constriction at entrance of
central into sublobular vein

Outlet sphincter

Bulging Kupffer cell

Sublobular
vein

Central
vein

Sinusoid

Arteriolar sphincter

Inlet sphincter

Portal vein

Hepatic
artery

Arterial sphincter

Fig. 29.9 Diagram of the arrangement of blood channels in the liver. (H. Elias & J. E. Pauly (1960). *Human Microanatomy*. Chicago: Da Vinci.)

portal venous pressure in man is normally some 5 to 15 mm Hg. Since the hepatic blood flow is large and the portal blood pressure low, the resistance to flow through the liver must be very small. Portal pressure varies little with posture unless the diaphragm and abdominal muscles contract when it may rise to 40 mm Hg. When the outflow of blood from the portal vein is obstructed, the pressure may increase to 30 or 40 mm Hg (*portal hypertension*).

The hydrostatic pressure in the hepatic veins is about 6 mm Hg and the blood is only about 67 per cent saturated with oxygen.

The blood flow through the liver in man can be estimated by the Fick principle. The dye bromsulphalein is infused intravenously at a rate such that its concentration in the arterial blood remains constant. Blood is then withdrawn for estimation of dye concentration from the hepatic vein through a catheter introduced into it from an antecubital vein. Catheterization of the hepatic vein can be avoided by injecting intravenously a colloid of uniform particles of human serum albumin aggregated with heat and labelled with [131]I. Below a certain dose this colloid is picked up quickly by the Kupffer cells as it passes through

the liver and almost none reaches the vena cava. Blood samples taken from another vein at several minutes interval show the rate of disappearance of ^{131}I and from this the hepatic blood flow can be calculated. These methods show that in an adult man, whose liver weight is 1·5 kg, the hepatic blood flow is about 1·5 litres per minute, that is about half the flow in the inferior vena cava.

In man, if the hepatic artery is occluded for a short time at a surgical operation and the hepatic blood flow measured, the hepatic artery is usually found to have been providing 35 per cent of the total flow and about 50 per cent of the oxygen supply but the contributions to the oxygen requirements of the liver made by the hepatic artery and portal vein vary considerably. After a meal the intestinal oxygen consumption is increased and the oxygen saturation of the blood in the portal vein falls; the flow in the hepatic artery increases three- or fourfold and thus supplies the higher proportion of the oxygen requirements. It seems that the liver by the accumulation or production of vasodilator metabolites secures its oxygen requirements. The vessels of the liver are supplied by vasoconstrictor fibres (but not by vasodilator) but the nervous control is probably not so important as the control by vasodilator metabolites.

The capillaries and venous tributaries of the portal vein and the liver (that is the entire splanchnic area) contain about 1 litre of blood in man; the liver itself contains about 400 ml. This 'circulating' volume can be drawn upon in emergencies. When the splanchnic sympathetic nerves are stimulated the portal pressure falls at first, due to diminished flow through the constricted mesenteric arteries, and then rises to nearly double the basal level. This is due partly to the damming up of blood by the contraction of the small portal venules within the liver and partly to a rise of hepatic artery pressure following the rise in systemic pressure resulting from the splanchnic vasoconstriction.

The liver produces about 0·5 ml of lymph per kg per min. This is about ten times the amount produced by the same mass of skeletal muscle. The protein content is over 90 per cent of that of the blood plasma. The 'membrane' separating plasma and lymph exerts only a slight filtering action.

THE CIRCULATION THROUGH THE SPLEEN

The framework of the spleen consists of trabeculae of white and yellow elastic tissue together with smooth muscle fibres which, though a prominent feature in the dog, pig and cat, are rather scanty in the human spleen. The spaces between the trabeculae are filled with the splenic pulp which is a reticulum of delicate fibrous tissue containing large numbers of cells. These are partly blood cells—red cells, lymphocytes and a few granulocytes—and partly cells belonging to the spleen itself, namely splenic or pulp cells, branched reticulum cells and giant cells. The splenic cells, which are phagocytic and belong to the reticulo-endothelial system, have a single large nucleus and often contain particulate matter and red cells in various stages of disintegration. In the fetus the spleen is an important haemopoietic organ.

The splenic artery enters at the hilum and breaks up into numerous branches with few anastomoses, that is, they are functionally endarteries. Near their terminations the arterioles are surrounded by a sleeve of lymphoid tissue (Malpighian bodies). It is difficult to trace the pathway of the blood through the spleen from this point but probably the arterioles, after a terminal dilatation, link up in a loose way with sinuses; the flow by this direct route is probably fairly rapid. The walls of these sinuses, which are incomplete, are made up of rod-like cells with narrow chinks between them lying parallel with the long axis of the sinus; occasional circularly placed bands of tissue keep the structure together like the hoops of a barrel (Fig. 29.10). Blood seems to pass from the sinus through the chinks in the wall into the pulp and back again into the sinus but circulation by this indirect route must be very slow indeed. Most of the plasma and some of the red cells go by the direct path to the sinuses where they are joined by red cells which have taken the indirect path. The sinuses either link up with one another or join a venule which has a definite complete wall. The veins unite in the hilum to form the splenic vein which is the largest radicle of the portal vein. There is considerable divergence of opinion among experts as to the detailed histology of the spleen. Certainly the structure and behaviour of the organ differ considerably from one species to another. In man injections of ^{51}Cr labelled red cells causes a rise in radioactivity over the healthy spleen which rises sharply to a maximum in one or two minutes; this indicates complete mixing of the cells within the spleen. It is only in enlarged spleens that slower mixing and slower release of red cells are observed. In man the intrasplenic pressure obtained by connecting a manometer to a needle inserted through the skin into the spleen varies between 3 and 17 mm Hg. This agrees well with other methods of measuring portal pressure and suggests that the splenic pulp in man is in free communication with the splenic vein. The circulatory arrangements described above allow of the removal of old or damaged erythrocytes; normal undamaged cells are allowed to pass through the spleen into the splenic vein. After splenectomy various somewhat abnormal red cells have been seen in the peripheral circulation (Chap. 24). After splenectomy in man iron granules and other inclusions are found in 1 or 2 per cent of the circulating red cells. The spleen stores platelets and destroys senescent ones. However, after splenectomy the life of red cells and platelets is unchanged.

The spleen of the dog can be brought through the abdominal wall without disturbing its vascular and nervous connexions and made to lie between the skin and superficial muscles so that its movements can be observed in the unanaesthetized animal. It contracts rhythmically at a much slower rate than the respiratory rate. Stimulation of the splanchnic nerves, exercise, asphyxia and haemorrhage all cause a contraction of the organ which adds a considerable number of red cells to the circulation. In the cat and dog a reflex contraction of the spleen occurs when the pressure in the carotid sinus is lowered by occlusion of the carotid artery.

In man the spleen can be visualized by radiographic methods. It is often stated that the human spleen contracts and that the red cell count rises after the injection of 1 mg adrenaline but recent investigations do not confirm earlier findings. By the use of red cells labelled with ^{11}CO it has been found that the healthy human spleen contains between 20 and 45 ml of red cells. The spleen undoubtedly acts as a reservoir of red blood cells in the dog, cat and horse but the human spleen, weighing

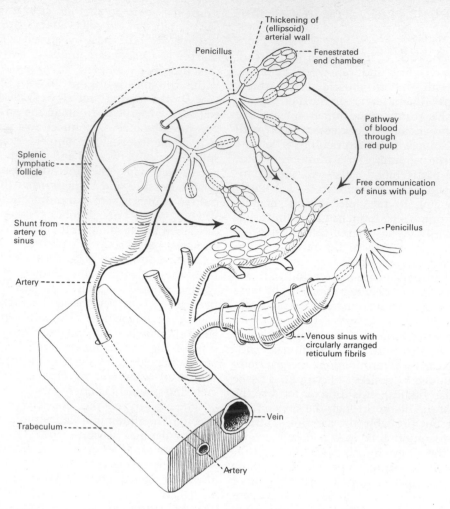

Fig. 29.10 A schematic representation of the blood vessels of the spleen. The arrows indicate the probable routes for passage of blood from splenic artery to vein. (E. B. Krumbhaar (1926). Functions of the spleen. *Physiological Reviews* **6**, 166.)

about 150 g in the adult and containing red cells corresponding to less than 100 ml of whole blood, could hardly be an important blood reservoir in man with a blood volume of 5·5 litres. Of course, contraction of an enlarged spleen might make a significant contribution.

REFERENCES

AVIADO, D. M. (1965). *The Lung Circulation*, Vol. I & II. Oxford: Pergamon Press.

BAIN, W. H. & HARPER, A. M. (1968). *Blood Flow Through Organs and Tissues*. Edinburgh: Livingstone.

BLAUSTEIN, A. (Ed.) (1963). *The Spleen*. New York: McGraw-Hill.

BRAUER, R. W. (1963). Liver circulation and function. *Physiological Reviews* **43**, 115–213.

DALY, I. DE BURGH & HEBB, C. O. (1966). *Pulmonary and Bronchial Vascular Systems*. London: Arnold.

DALY, R., GOODWIN, J. F. & STEINER, R. E. (1960). *Clinical Disorders of the Pulmonary Circulation*. London: Churchill.

DE REUCK, A. V. S. & O'CONNOR, M. (Eds.) (1961). *Problems of Pulmonary Circulation*. London: Ciba Foundation.

DONALD, K. W. (1960). Pulmonary haemodynamics. *British Journal of Diseases of the Chest* **54**, 175–185.

FISHMAN, A. F. (1961). Respiratory gases in the regulation of the pulmonary circulation. *Physiological Reviews* **41**, 214–280.

GORESKY, C. A. (1965). The nature of transcapillary exchange in the liver. *Canad. med. Ass. J.* **92**, 517–522.

GRAYSON, J. & MENDEL, D. (1965). *Physiology of the Splanchnic Circulation*. London: Arnold.

HARRIS, P. & HEATH, D. (1962). *The Human Pulmonary Circulation: Its Form and Function in Health and Disease*. Edinburgh: Livingstone.

READ, A. E. (Ed.) (1967). The liver. *Colston Papers*, **1a**.

ROUILLER, C. (Ed.) (1963). *The Liver*, Vol. 1. New York: Academic Press.

SHERLOCK, S. (1968). *Diseases of the Liver and Biliary System*, 4th edn. Oxford: Blackwell.

WEST, J. B. (1970). *Ventilation Blood Flow and Gas Exchange*, 2nd edn. Oxford: Blackwell.

30 Respiration

In unicellular organisms oxygen can diffuse from the watery environment to all parts of the cell. With increasing size and complexity of the animal, however, the distance between the surface and the tissues becomes so great that special means are required to carry oxygen to the individual cells. In mammals the oxygen is taken up in the capillaries of the lungs and is conveyed by the blood to the tissues. At the same time the carbon dioxide produced in the tissues is carried to the lungs and there leaves the body. A scheme of the processes involved in respiration is given in Fig. 30.1.

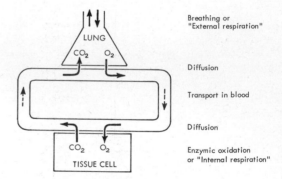

Fig. 30.1 Respiratory processes.

Air is moved in and out of the lungs by their 'bellows action' (*ventilation*). It is then distributed within each lung so that alveoli are adequately ventilated and also perfused by blood (*distribution*). Oxygen and carbon dioxide are then exchanged between the gas mixture in the alveoli and the blood in the pulmonary capillaries (*gas transfer*). The gases are then *transported* in the blood to the tissues. This chapter is mainly concerned with ventilation. The other stages of respiration are described in Chapter 31. The part played by respiration in maintaining the acid–base equilibrium of the body is outlined in Chapter 33.

RESPIRATORY TRACT

The respiratory tract may conveniently be thought of as consisting of two parts, the upper respiratory tract extending from the anterior nares to the vocal folds and the lower respiratory tract below the vocal folds.

The nose. The nasal passages are lined by a very vascular mucous membrane characterized, except at the entrance, by ciliated columnar epithelium which warms and moistens the entering air and removes a certain amount of dust. The warming is very effective; air inspired at 6°C, for example, has been heated to 30°C by the time it reaches the back of the nose. As it passes down the trachea its temperature rises to body temperature and it is completely humidified by the time it arrives at the bifurcation. The temperature of the expired air is always below 37°C because the air loses heat as it passes out through the nose.

Electrical, chemical or mechanical irritation of the nasal mucous membrane causes apnoea, closure of the larynx and constriction of the bronchi, and also slowing of the heart rate and variable changes in blood pressure. When the mucous membrane is congested or inflamed there is considerable resistance to inflow and breathing through the nose is correspondingly difficult. Exposure of the skin of the face to infra-red rays, from an electric radiator for example, causes a swelling of the mucosa. Short-wave radiation or cooling of the skin produces a shrinkage. Discomfort in artificially heated rooms (stuffiness) may be at least partly explained by the absence of short-wave radiation, and heating systems should ideally be designed to avoid swelling of the nasal mucosa and the discomfort so produced.

The air passages. The *trachea* is a nearly circular tube about 16 mm in diameter and 12 cm long. It is kept patent by incomplete rings of cartilage, the deficiency on the posterior surface being filled in with fibrous and muscular tissue. This structure allows free movement of the head and neck without the danger of kinking and obstructing the airway. During deep inspiration the trachea increases in diameter by about one-tenth and in length by about one-fifth. The trachea divides into two main *bronchi* and these in turn subdivide repeatedly, the smallest air passages being called *bronchioles*. The total cross-sectional area of the airways increases at each branching. All these air passages have a fibrous outer layer; in addition the larger bronchi are supported by irregular pieces of cartilage. Internal to this is a layer of smooth muscle, the *bronchial musculature*, placed both transversely and obliquely. Between the muscle fibres there is a considerable amount of elastic tissue. The bronchi are easily distended; an increase of 10 cm H_2O in the internal pressure can increase their volume by 50 per cent. The bronchi are lined by a ciliated columnar epithelium. The turnover time of tracheal and bronchial epithelium is about seven days. In the trachea and larger bronchi numerous muco-serous glands innervated by the vagus are found below the epithelium. The mucus of the bronchi is produced by large numbers of goblet cells: the bronchioles have neither glands nor goblet cells but club-like apocrine cells have been found between the ciliated cells. The terminal bronchioles, the muscle of which is particularly well developed, open into *respiratory bronchioles* of equal diameter (Fig. 30.2). A

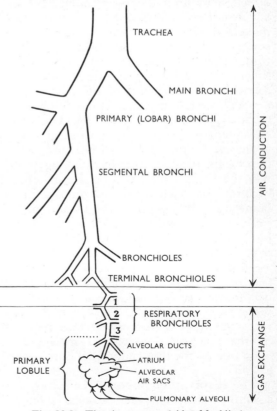

Fig. 30.2 The air passages. (*After* Macklin.)

small number of these open directly into *alveoli* but for the most part alveolar ducts form the final portion of the respiratory tree, connecting the respiratory bronchioles with the pulmonary alveoli. In man about 14×10^6 alveolar ducts open into 300×10^6 alveoli from 75 to 300 μm in diameter; the alveoli in the more dependent parts of the lungs have a somewhat smaller resting volume than those in the apical parts. An adult lung removed shortly after death weighs about 500 g.

The respiratory tract from the nose to the bronchioles is lined by ciliated columnar epithelium. The free border of each cell has several hundred hair-like processes or cilia 1 to 5 μm long and 0·3 μm in diameter. The electron microscope shows that each cilium has nine peripheral filaments and two central ones. The ciliated cells occur in little groups which beat synchronously. No variations in membrane potential of these cells has been detected and synchronous activity may depend simply on mechanical interference between cilia. The cilia impel the fluid on the surface in one direction by a rapid forward movement (the effective beat) followed by a slower return movement. This movement can be seen on the surface of a nasal polypus recently removed from the human nose. The cilia beat about 20 times per minute and particles of powder are carried along the trachea *in situ* about 0·4 mm per sec; the rate depends on the depth of the mucus layer lying over the cilia. The ciliary movement propels mucus and foreign bodies to the main bronchi and trachea from which they are finally cleared out by coughing; in this way the healthy lung is kept free of inhaled bacteria. The cilia of excised mucous membrane are readily depressed by noxious agents but *in vivo* they are hardier and are unaffected by cigarette smoking. The ciliary activity depends on the oxygen in the trachea; if this is less than 0·5 per cent ciliary activity is reduced. Ciliary activity continues independently of all nervous connexions and indeed continues for some hours after death. Small particles that reach the lower respiratory tract are removed by the phagocytic activity of macrophages lying free on the thin endothelial lining of the alveoli.

The respiratory parts of the lungs. The structures distal to each alveolar duct are spoken of as a *primary lobule*, which is the physiological unit of the lung (Fig. 30.2). The respiratory bronchioles are lined by cubical, non-ciliated epithelium, which becomes progressively flatter as it is traced towards the pulmonary alveoli. Minute vents connect adjacent alveoli within each primary lobule. The existence of these vents in man has been proved by passing a catheter with a wide terminal flange into a bronchus until the bronchus is blocked. The other end of the catheter is connected to a volume recorder. If the catheter is closed during inspiration and opened during expiration the volume recorder rises; this shows that air must have entered the blocked bronchi from the distal end. Epithelium-lined tubules connect terminal or respiratory bronchioles with adjacent alveoli; a communication is thus provided between the ultimate alveoli and their parent bronchioles.

The exchange of gases between the air and blood takes place in the alveoli where, as the electron microscope shows (Fig. 31.2a), the air and the blood are separated by two very thin cellular membranes, the endothelium of the pulmonary capillary and the flattened epithelium of the alveolus. The thickness of the tissue between the alveolar air and the blood is about 0·5 μm overall. The total area of the alveolar membrane of both lungs is estimated to be 70 to 80 m² in adult man; the area depends on the state of expansion of the lungs.

The passage of fluid across a tissue capillary is discussed on p. 682 but in a pulmonary capillary the situation is much more complicated (p. 579). The force tending to keep the alveoli dry is the osmotic pressure of the plasma proteins (say 25 mm Hg) less the capillary blood pressure (say 10 mm Hg), that is a total of about 15 mm Hg. This readily explains the occurrence of pulmonary oedema when the blood pressure in the left atrium rises above 20 mm Hg but it takes no account of the surface tension.

The surface tension of the fluid in a Langmuir trough tends to reduce its surface area but when a liquid surface is curved as in an alveolus there is a resultant force towards its centre. The pressure P is given by Laplace's equation $P = 2\gamma/r$ where γ is the surface tension and r is the radius of the sphere. A small alveolus with a radius of 50 μm, or 50×10^{-4} cm, if lined with plasma with a surface tension of 55 dyn/cm would have a pressure of 110 dyn/cm $\div 50 \times 10^{-4}$ cm $= 2.2 \times 10^4$ dyn/cm^2 $= 16.5$ mm Hg. (For conversion factors see end pages.) This pressure, acting in the same direction as the pulmonary capillary pressure, would overcome the osmotic pressure of the plasma proteins and would not only draw fluid out of the capillaries but would cause the alveolus to collapse. But in fact the fluid lining the alveoli contains a 'surfactant' (Fig. 31.2b), a lipoprotein containing much fully saturated di-palmitoyl lecithin which reduces γ to nearly zero (0.5 to 1.0 dyn/cm). The pressure towards the centre of the alveolus would, in the example given, be only 0.15 to 0.3 mm Hg. When the oedema foam (that is lining fluid) from the lungs is spread on a Langmuir trough the surface tension rises somewhat as the area is increased; this effect is probably present in the lungs as the alveoli expand on inspiration but the increase of surface tension may be partly balanced by the increase in radius so that P is not much altered. On expiration the alveoli become smaller and since their surface area is reduced γ falls; in this way collapse during expiration is prevented. The surfactant is thought to arise from the mitochondria of the giant alveolar cells which lie in the septa between the alveoli. It appears first at the twenty-eighth week of intrauterine life just at the time the fetus becomes viable. If excised gas-free lungs are filled with saline to abolish the air/fluid interface in the alveoli only half the pressure is required to expand them to a given volume. Since this treatment cannot affect the elastic tissue it shows clearly that the surface tension at the air/fluid interface is as important as the elastic tissue in recoil of the lungs.

The capillaries in the alveolar walls are derived from the pulmonary artery (p. 592). The bronchial arteries, branches of the thoracic aorta, supply the larger bronchi, the bronchioles and their muscles, and the pleura. In the region of the respiratory bronchioles there is an anastomosis between the two systems so that some of the blood conveyed in the bronchial arteries returns to the heart by the pulmonary veins (Fig. 30.4).

From an anatomical point of view the fundamental unit of lung structure could be considered to be the *secondary lobule* (Fig. 30.3) from many hundreds of which the whole lung is made up. The secondary lobules are divided from one another by interlobular septa in which run the small pulmonary veins and the majority of the lymphatics. In the centre of each lobule is a system of bronchi together with ramifications of the bronchial artery and branches of the pulmonary artery. Each secondary lobule is thus a lung in miniature with a bronchial tree and a circulation of its own. Arterial blood enters the lobule in the bronchial arteries which ramify through the lobule in close relation to the bronchi. Venous blood in the branches of the pulmonary artery is likewise distributed. After oxygenation the blood passes into tributaries of the pulmonary vein running in the interlobular septa but some blood leaves the lobule in bronchial veins running alongside the branches of the artery (Fig. 30.4). The main lymphatic drainage is in lymphatics in the interlobular septa though there are some lymphatics along the bronchi.

Like every other tissue the lungs consume

Fig. 30.3 Photograph (actual size) of part of a section of the lung of a man, aged 60, in which the secondary lobules are outlined by deposits of coal dust in the lymphatics of the interlobular septa. (*By courtesy of J. Gough.*)

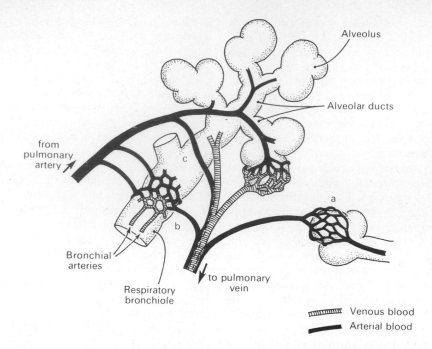

Labels on figure:
- Alveolus
- Alveolar ducts
- from pulmonary artery
- c
- a
- b
- Bronchial arteries
- Respiratory bronchiole
- to pulmonary vein
- Venous blood
- Arterial blood

Fig. 30.4 General scheme of a primary lobule (*after* W. S. Miller) showing vascular communications between the pulmonary and bronchial vascular systems in the region of the respiratory bronchiole and some intrapulmonary causes of venous admixture with arterial blood. The latter include (1) uneven alveolar ventilation of different parts of the lungs. Under-ventilation of alveoli with a normal pulmonary capillary blood flow (indicated by adjacent lobule (a)) leads to anoxaemia. (2) Venous blood from the bronchial circulation draining into the pulmonary veins (b). (3) Pulmonary artery-pulmonary vein shunts (c). The bronchial artery blood flow is exceedingly difficult to measure but is probably about 1 per cent of the pulmonary flow. The bronchial circulation has few veins of its own, the drainage is mainly as at (b) into the pulmonary veins which go to the left atrium. The bronchial arteries also supply a small amount of blood to mediastinal structures which drain into the azygos vein and left atrium. (*By courtesy of M. de Burgh Daly.*)

oxygen and the adult lung can account for as much as 4 per cent of the total oxygen uptake of the body.

The pleura. The outer surface of the lung is covered by a delicate serous membrane, the *visceral pleura*, which is reflected from the roots of the main bronchi on to the inside of the thoracic walls and upper surface of the diaphragm. This part is called the *parietal pleura*. The surface of each layer of pleura consists of a single layer of flattened cells, and normally the two layers are separated by a thin film of serous fluid, sufficient only to lubricate the surfaces so that they move over one another easily during breathing. So long as this film is intact the two layers cannot be separated except by considerable force. The potential space between the two layers is known as the pleural cavity. When the two layers of

pleura are separated by an accumulation of fluid, or the entry of air, the pleural cavity becomes apparent.

MECHANICS OF VENTILATION

The movement of air into and out of the lungs (*ventilation*) is brought about by alterations in the size of the thoracic cavity, the lungs following these variations passively. When the thoracic cavity enlarges during inspiration the lungs must also expand because the visceral pleura cannot part from the parietal pleura on the thoracic wall. Air flows through the nose into the lungs because at that instant the air in the atmosphere is at a higher pressure than the air in the lungs. On expiration the changes are in the opposite direction.

The most important muscle of inspiration is

the *diaphragm*, a dome-shaped sheet separating the thoracic and abdominal cavities; it is made up of a central tendon with muscle fibres radiating from it to the circumference of the lower parts of the thoracic wall. When it contracts the central tendon is pulled down and the intra-abdominal pressure is raised; at the same time the costal margins move up and out (bucket-handle movement), because of the vertical position of the fibres attached to them which expand the base of the thorax. In quiet inspiration the diaphragm may be the only respiratory muscle in action but some activity has been recorded in the internal intercostals (opposite). However, the diaphragm can be paralysed in man by cutting both phrenic nerves without producing serious respiratory difficulty. The liver is so much denser than the lungs that its upper edge can be clearly seen on the fluorescent X-ray screen. Since the liver is pressed against the diaphragm by the abdominal muscles, the upper edge of its shadow can be used to mark the level of the diaphragm. On quiet inspiration the downward movement is approximately 12 mm but on violent inspiration it may be 70 to 100 mm. The height of the diaphragm, and thus the capacity of the chest, varies with the position of the body—the diaphragm is highest in the supine position, a little lower in the erect position, and still lower when the subject is sitting because the abdominal muscles are relaxed in this position. Since the cross-sectional area of the thorax increases from above downwards, any given air intake can be accomplished by a smaller diaphragmatic movement the lower the resting position of the diaphragm. The patient with respiratory disease is, therefore, often most comfortable when sitting up. The diaphragm is the last muscle to go out of action during progressively deepening anaesthesia.

The movements of the ribs on inspiration are of two kinds: (1) 'pump-handle' in which the anterior end of the downward sloping rib is raised and (2) the 'bucket-handle' in which the middle of the rib is moved upwards and outwards about an anteroposterior axis. If the intercostal nerves supplying the intercostal muscles are cut in man the amplitude of these rib movements is reduced and the tissues of the intercostal spaces bulge or recede according to the variations in intrathoracic pressure.

Action potentials have been recorded in the intercostal muscles by inserting into them very fine bipolar needles which picked up impulses from a small volume of either external or internal intercostals. The intercostal muscles seem to contribute very little to the effort of quiet respiration. No activity could be detected in the external intercostals in these circumstances. The internal intercostals showed some activity in two restricted areas—in the parasternal region (upper four interchondral spaces) on inspiration and in the lower lateral part of the chest on expiration. In vigorous respiration, however, the two layers of muscle showed reciprocal activity all over the chest wall, the external intercostals contracting on inspiration and the internal intercostals on expiration.

When the subject is lying completely relaxed no electrical activity can be recorded during inspiration from the muscles of the anterior abdominal wall. The elastic properties of the abdominal wall keep up the viscera and the diaphragm. Although the abdominal muscles are true muscles of expiration they are little used in quiet breathing and the variations in intragastric, that is intra-abdominal, pressure are correspondingly small. They seem to go into vigorous action only when the ventilation is high or when the subject is coughing or straining; even then they are much less active than they are during postural movements. In maximum voluntary breathing efforts, such as occur in the vital capacity test, the lower ribs are drawn downwards and medially by the abdominal muscles with a large rise of intragastric pressure. A limit may be put to the possible depth of inspiration by the abdominal muscles which contract at the end of a maximal inspiration.

The only accessory muscles of any importance in inspiration are the scaleni and the sternomastoids. The former may be in action in some persons in quiet inspiration; they elevate and fix the first two ribs. The sternomastoids are used only in deep breathing and in dyspnoea; they raise the sternum and increase the anteroposterior diameter of the thorax. In deep breathing the extensors of the vertebral column may help inspiration. In the deepest inspiration that can be made the circumference of the chest is increased by 5 to 11 cm; in quiet inspiration the increase is only 1 cm. The increase in the size of the chest obtained with a tape measure is not a reliable indication of the volume of air entering the lungs.

In expiration the inspiratory muscles relax and the movements of the thorax and the lungs just described are reversed. During this largely passive phase the ribs fall and untwist partly by their own elasticity and partly by the contractions of the muscles attached to them. Even when there is obstruction to respiration this is true because the usual effect of obstruction is to increase inflation: the obstruction is thus overcome by increased passive recoil rather than by contraction of the abdominal muscles, or indeed of the other, accessory muscles of respiration.

Radiographic and other indirect methods of measuring the relative contributions of the descent of the diaphragm and of the movements of the rib cage have led to the conclusion that the diaphragm is responsible for three-quarters of the tidal volume.

Expansion of the lungs. During inspiration the vocal folds move apart, the bronchi dilate and, as the lungs expand, the bronchial tree is elongated, and the alveoli become wider. At the same time the intrathoracic pressure falls, the venous return increases and the right ventricular output increases and the amount of blood in the pulmonary circuit increases. Alterations in calibre of the bronchi during a respiratory act may be entirely passive but they are probably influenced by constrictor and dilator impulses passing respectively down the vagus and sympathetic nerves (see p. 598).

Since the lungs expand quite passively, the parts next to the most mobile portions of the thoracic cage expand most and the mediastinal portions least. As the lungs expand they occupy the potential space lined by pleura between the diaphragm and the lower ribs (*costo-dia-phragmatic recess*) which is obliterated at expiration by the coming together of the two layers of the parietal pleura.

Recording of respiratory movements. A simple way to obtain qualitative records of chest movements is to use a *stethograph* made of corrugated rubber hose-pipe (respirator tubing) which is tied round the chest and connected by a tube to a tambour writing on a smoked drum. Quantitative records can be obtained by using a recording *spirometer* (Fig. 30.5) or a pneumotachograph. The respiratory

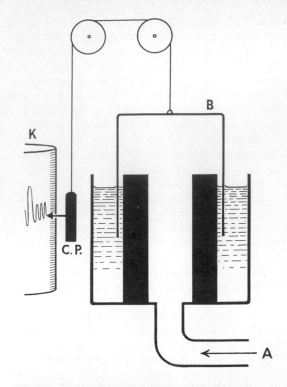

Fig. 30.5 Spirometer designed for high respiratory rates. The subject breathes into A and raises the bell B and the index attached to the counterpoise C.P. moves down the kymograph K. (L. Bernstein, J. L. D'Silva & D. Mendel (1952). *Thorax* **7**, 255.)

movements may also be registered by recording the changes in intrapleural pressure (Fig. 30.8).

Respiratory rate. The respiratory rate at rest varies considerably from subject to subject but in the adult it is 10 to 18 per minute. Faster rates are normally found in young children (Table 30.6). The respiratory rate is increased along with tidal volume by almost any respiratory stimulus such as exercise. It is under voluntary control and for this reason a true resting rate is not likely to be obtained if the subject is aware that his breathing is being observed. The relationship between ventilation, tidal volume and respiratory rate is shown in Figure 30.7.

Intrapleural pressure. The pleural space contains only a very thin film of fluid which because of surface tension makes it very difficult to separate the two pleural layers. The situation is comparable to that in which a wet rubber 'sucker' is stuck to a piece of glass.

Table 30.6 Respiratory data at different ages

Age	Respiratory rate (breaths per min)	Tidal volume (V_T) (ml)	Basal oxygen requirement ($\dot{V}_{O_2}$) (ml per min)
At birth	14–60	10–20	23
First year	25–35	—	78
2–4 years	20–30	—	87–100
5–14 years	20–25	200–350	100–175
Adult man	10–18	350–900 av. 500	240
Adult woman	10–18	200–650 av. 340	200

The figures in the last column have been calculated from the data in Table 12.7. The range of respiratory data is very wide indeed (see for example *The Handbook of Respiration* (1958). London: Saunders). The figures given in this table can only be regarded as representative.

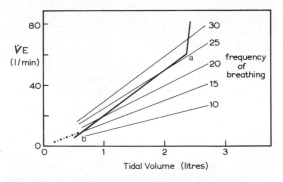

Fig. 30.7 The relation between ventilation and tidal volume in human subjects when breathing is stimulated by any factor except hyperthermia. Note that the observed relation appears to be linear over a wide range of tidal volumes and crosses the iso-frequency lines, and that there is usually an inflection a, beyond which the tidal volume rises little above half the vital capacity. There may be a further inflection b at low ventilations where the frequency is relatively constant (dotted line). (*After* E. N. Hey, B. B. Lloyd, D. J. C. Cunningham, M. G. M. Jukes & D. P. G. Bolton (1966). *Respiration Physiology* **1**, 193–205.)

When the sucker is applied firmly to the glass and released the atmospheric pressure holds it in position. If the film of water between the rubber and glass evaporates, the sucker is easily dislodged. Similarly, in the case of the lungs, the parietal and visceral layers of the pleura are kept pressed together by the atmospheric pressure and remain in firm contact. Alterations in the volume of the lungs

during breathing occur because of the movements of the thoracic cage. In the process of expansion the elastic tissue of the lung is stretched and provides a force tending to collapse the lungs—this is the so-called *negative intrapleural pressure*. It is greater, the greater the expansion of the lungs and the stretch of the elastic tissue. The intra-pleural pressure can be measured by injecting a *very small* quantity of air between the two layers of pleura and connecting this air pocket to a water manometer. Such a record is shown in the upper portion of Fig. 30.8. In quiet respiration the intrapleural pressure varies between plus 1 to minus 4 cm water; in forced inspiration it may decrease to minus 20 cm or even minus 40 cm water and in forced expiration it may increase to plus 30 cm. It is more convenient to measure the intrapleural pressure with a fine polythene tube attached to a thin-walled balloon lying in the lower third of the oesophagus. In the upright position this has been shown to give pressure readings that agree closely with those recorded in the intrapleural space.

The abdominal pressure, measured through a tube passed into the stomach, varies in the opposite direction to that in the thorax; on inspiration, when the intrapleural pressure becomes more negative, the diaphragm descends, so increasing the intra-abdominal pressure.

The elasticity of the lungs can be demonstrated by passing a hollow needle through the chest wall to allow air to enter and break the 'seal' between the two layers of the pleura. Air continues to pass in until the intrapleural pressure rises to atmospheric pressure and the lung on that side collapses by the retraction of its elastic tissue (see also p. 592). This condition is called a *pneumothorax*.

The volume of air contained in the chest at the resting respiratory level is called the functional residual capacity (F.R.C., see Fig. 30.10). If this volume is to be increased or decreased, work must be done (1) against the tendency of the lungs and chest wall to recoil to their resting position, (2) to overcome the resistance to air flow through the airways, and (3) to overcome the resistance to movement of the tissues themselves. These three forces, all tending to oppose the change of volume of the thorax, are called elastic resistance, airway

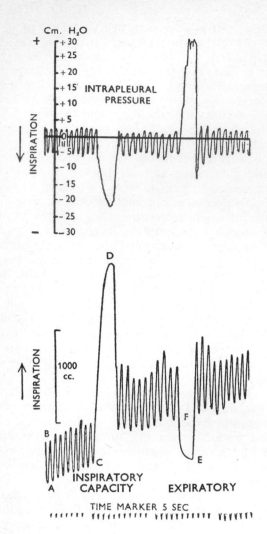

INTRAPLEURAL
PRESSURE

Cm. H₂O

INSPIRATION

INSPIRATION

1000
cc.

INSPIRATORY
CAPACITY   EXPIRATORY

TIME MARKER 5 SEC

Fig. 30.8 Simultaneous traces of intrapleural pressure recorded by a water manometer (upper trace) and respiratory record made with a recording spirometer (lower trace). Note on the respiratory trace that AB = tidal volume; CD = inspiratory capacity; EF = expiratory reserve volume; CD + EF = vital capacity (p. 599) (R. V. Christie & C. A. McIntosh (1934). *Journal of Clinical Investigation* **13**, 292).

resistance and tissue ('viscous' or 'non-elastic') resistance. The first depends only on the change of volume itself, while the latter two are proportional to the rate of change of volume, that is, on air flow. During normal quiet breathing, the active contraction of muscles provides the energy to overcome these resistances during inspiration during which potential energy is stored in the stretched elastic tissues. This energy is then

available to overcome the airway and tissue resistance during expiration. This combination of elastic and viscous resistance makes the relationship between intrapleural pressure and tidal volume somewhat complex. If the lungs were perfectly elastic and the resistance to air flow through the bronchi negligible, the two curves would be in phase so that the maximum negative pressure coincided with the maximum volume at the end of inspiration and the least negative pressure with the minimum volume at the end of expiration. But, as Fig. 30.9 shows, the peaks of the pressure curve occur a little before the peaks of the volume curve; the curves are not in phase. In other words, because of the viscous resistance of the lungs and the airway resistance, the volume curve lags behind the pressure curve. When simultaneous values of pressure and volume for a single cycle are read off from these curves and plotted in a graph with pressure as ordinate and volume as the abscissa an 'ellipse' OIAE is obtained. The area OIAN is the total work of inspiration, the area OAN representing the work done against elastic forces and the area OIA the work done against viscous resistance during inspiration. The energy OAN stored in the lungs at the end of inspiration is available to do the work OAE necessary to overcome the viscous resistance to expiration.

Such experiments show that a relatively small amount of work is expended in respiration at rest; it is of the order of 0·4 kg.m (4 J) per min, of which 60 per cent is elastic work and 40 per cent is used in overcoming viscous resistance including resistance to the air flow. Even in severe exercise the work of breathing is only about 3 per cent of the total energy output.

In the disorder *bronchial asthma* the airway resistance is much increased by reason of narrowing of the bronchi and the work of breathing is much increased. The shape of the pressure/volume curve is markedly altered in emphysema (see insert in Fig. 30.9).

Conscious subjects can easily recognize quite small increases, of the order of 25 per cent, in elastic resistance produced experimentally by breathing into and out of rigid drums or of viscous resistance produced by breathing through a narrow airway. In both cases increased respiratory effort is needed.

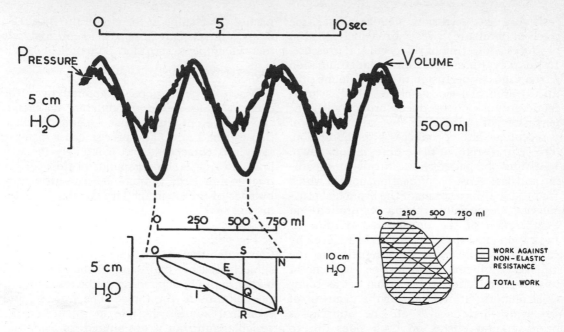

Fig. 30.9 A simultaneous tracing of intraoesophageal pressure and tidal volume in a normal subject at rest. The pressure-volume diagram derived from one breath is shown below. The pressure exerted against non-elastic resistance is obtained by subtracting the elastic pressure from the total pressure exerted at any point of the inspiratory cycle. At the point S this is SR − SQ = QR. Inserted on the lower right is a pressure-volume diagram obtained from a patient with emphysema (M. B. McIlroy, R. Marshall & R. V. Christie (1954). *Clinical Science* **13**, 128 and 148).

Vagal block does not relieve the feeling of distress associated with breathing against resistance. The sensation may arise in the muscles of the thoracic cage or more probably in the joint receptors.

Compliance as used by pulmonary physiologists is taken as $\Delta V/\Delta P$ or, in words, the change in volume produced by unit pressure. A high compliance means that a given change of pressure moves a large volume of air; the compliance is lowered if the lungs become stiff or if the surface tension of the fluid lining the alveoli is raised. The pressure required is the transpleural pressure, usually obtained by measuring the pressure difference between the intrathoracic oesophagus and the mouth. Measurements of compliance must be made under static conditions if the elastic properties only of the lungs are to be measured since in breathing part of the air pressure is used to overcome the viscous resistance of the tissues and the resistance to flow of air and part to expand the lungs against elastic resistance. If the measurements are made at the end of

inspiration (A in Fig. 30.9) there is no air flow and the non-elastic tissue and airway resistances are zero. In Fig. 30.9, about 750 ml of air move into the lungs for a change of pressure of about 3·5 cm of water. This gives a compliance of 0·22 litres/cm water. The range of compliance in the adult healthy male is from 0·09 to 0·26. The value in the newborn child is only 0·005; as in other small mammals the lungs are less easily inflated. At age 10 years the value is about 0·06. In old age the compliance decreases.

Bronchial muscle. The activity of the bronchial muscle has been studied in the intact animal under artificial respiration provided by a pump with a constant stroke volume. If the bronchi constrict, a manometer connected to the trachea shows an increase of pressure; bronchodilatation is indicated by a fall of tracheal pressure. Excised lungs can be perfused through the trachea with warm Locke's solution (a solution resembling mammalian plasma) which escapes through scratches made on the lung surface. If the bronchi constrict the resistance to perfusion increases and the tracheal pressure rises; bronchial dilation produces a lowering of the perfusion pressure. In this way it has been found that stimulation of the vagus or perfusion of acetylcholine produces bronchial constriction while sympathetic stimulation produces bronchodilation; perfusion of adrenaline by stimulation of β-receptors also produces bronchodilation. The central origin of these fibres is not known. In asthma the patient has difficulty in breathing and has a sense of suffocation, both of which are produced by a spasm of the bronchial musculature, especially during expiration—an exaggeration of the normal constriction on expiration. The spasm can often be relieved by subcutaneous injection of adrenaline or the inhalation of fine particles or droplets of related chemical substances that act as bronchial dilators. Stimulation of the vagus or of the sympathetic nerves on one side causes bronchomotor effects on both sides due to crossing over of the fibres in the chest.

The injection of histamine in the guinea-pig causes a powerful bronchoconstriction which is probably the cause of death in anaphylactic shock in this animal. Histamine has usually very little effect on the bronchi in man but in asthmatics histamine inhaled as an aerosol can produce bronchoconstriction and an asthmatic attack.

If a conscious man is given atropine the dead space (p. 600) increases and the airway resistance decreases; vagotomy in man increases the bronchial calibre. These facts show that the muscle of the trachea and bronchi is normally in tonic contraction produced by cholinergic vagal nerves. Asphyxia causes bronchoconstriction in several ways. The raised blood P_{CO_2} acts on the centres in the brain and the increased P_{CO_2} in the airways may act directly on the bronchial muscle; also the accompanying hypoxaemia may add to the bronchoconstriction by impulses set up in the chemoreceptors of the carotid body. However when bronchial tone is high CO_2 may cause bronchodilation. Inflation of the lungs or a deep inspiration relaxes the bronchial muscle whereas deflation contracts it. These reflex effects may account for the changes in calibre of the bronchi in breathing (p. 594) but the changes in calibre in ordinary quiet respiration are probably merely passive, that is they are

part of the change in the size of the lungs. So many interacting reflexes affect bronchial tone that no simple description is possible.

BREATH SOUNDS

When a stethoscope is applied over the chest, various sounds of different character, and of different temporal and spatial distribution can be heard. The precise cause of these sounds is conjectural but it is likely that the loud blowing *bronchial* sounds heard over the trachea and large bronchi are caused by turbulence of the air in the larynx, while the soft hissing *vesicular* sound is due to the opening of air sacs.

The velocity of sound in hydrogen is nearly four times that in air, and consequently if a subject breathes hydrogen all sounds become higher in pitch, depending on the resonance of gas columns in the lungs. The tracheal and bronchial sounds do rise in pitch but the vesicular murmur is not altered, so its causation must be quite different from that of the other sounds. In specific lung disorders there are sometimes characteristic local variations in these sounds that may be helpful in diagnosis.

VOLUME OF AIR BREATHED

The volume of air breathed in and out in a single quiet respiration is about 500 ml. This *resting tidal volume* $(\dot{V}_T)$ is measured by causing the subject to breathe from a recording spirometer (Fig. 30.5). A violent inspiratory effort can take into the lungs about 2500 ml measured from the resting respiratory level which by convention is taken as the end of a normal expiration (Fig. 30.10, and A, C, or F in Fig. 30.8). This is called the *inspiratory capacity*; it includes the resting tidal volume and the *inspiratory reserve volume* as shown in Figure 30.10. After a quiet expiration (500 ml) it is possible by a violent expiratory effort to blow out approximately 1300 ml of air; this is the *expiratory reserve volume*. Even after the deepest possible expiration the lungs and respiratory passages still contain about 1600 ml of air; this is the *residual volume*. At the end of a quiet expiration the lungs contain the expiratory reserve volume and the residual volume which together are called the *functional residual capacity* (2900 ml) (see Fig. 30.10). This large volume prevents rapid

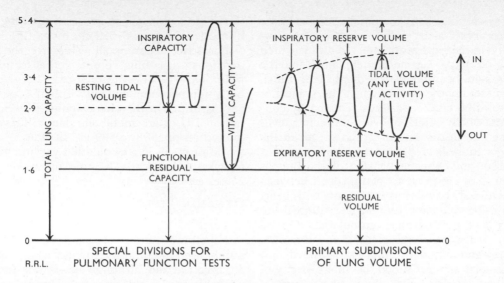

Fig. 30.10 A representation of a recording obtained by a spirometer (Fig. 30.5) to illustrate the subdivisions of the lung volume. The figures on the left are the average values in litres for an adult man. (*Federation Proceedings* (1950). **9**, 602.) R.R.L. resting respiratory level. Note that the word 'volume' is used for the simple subdivisions; where two or more of these subdivisions are involved the word 'capacity' is used.

changes in the composition of the alveolar air and collapse of the alveoli. Normally (see p. 600) only 360 ml of fresh air (tidal volume minus dead space) is added to the functional residual capacity on inspiration.

The amount of air breathed in per minute is called the respiratory minute volume or pulmonary ventilation ($\dot{V}_E$). Since the respiratory rate at rest is about 12 per minute and the tidal volume 500 ml, the minute volume is about 6 litres. In exercise it may go up to 100 litres.

Functional residual capacity. This can be measured by a dilution technique. The subject breathes at the end of expiration into a spirometer filled with oxygen. After a few minutes of quiet breathing the nitrogen originally in his lungs is distributed evenly between his lungs and the spirometer. The functional residual capacity is calculated from the reduction in nitrogen content of the spirometer.

In another method the subject breathes a mixture of hydrogen (or helium) and oxygen from a spirometer, into which oxygen is run to keep the trace level.

Vital capacity. If, after the deepest possible inspiration, a subject makes as forcible an expiration as possible into a spirometer we obtain a reading of the *vital capacity* (Figs. 30.10 and 11). Using the figures already quoted we find that this is composed of

Inspiratory capacity	2500
Expiratory reserve volume	1300
	3800 ml

The vital capacity depends on the size, age, sex and ethnic origin of the subject but because of its variability accurate prediction of the vital capacity from height, weight or surface area is not possible. However, although it has little diagnostic value, the vital capacity in any one individual is sufficiently reproducible to be a useful guide to the progress of certain diseases and is used in certain clinical tests of ventilatory performance (Fig. 30.11). Chest expansion as given by a tape-measure during a deep inspiration is not a reliable indication of the vital capacity.

The estimates of the subdivisions of the lung volume depend somewhat on the position of the subject. The slight reduction in vital capacity in the supine position as compared with the sitting position may be due to the increase in the amount of blood in the pulmonary vascular bed. The small increase in the inspiratory capacity and the considerable reduction in the expiratory reserve volume which occur on lying supine are explained by the changes in position of the diaphragm which lies higher in the thorax when the subject is supine. The vital capacity is reduced by conditions which interfere with thoracic movements, and in diseases in which the air in the spongy lung is replaced by fluid or solid material, and when expansion of the lung is impaired by the presence of fluid or air between the layers of the pleura. In patients with heart disease the vital capacity may be reduced 10 to 30 per cent, partly by an accumulation of blood in the engorged vessels of the pulmonary system, and partly by fluid extravasated into the alveolar spaces. When the vital capacity of a patient is reduced to about 40 per cent of normal, he can no longer perform even the simplest movements without becoming breathless. Such patients are much more breathless lying down than sitting up (*orthopnoea*), partly because of the decline in vital capacity in the recumbent position (see also p. 599) and partly because of the decrease in compliance which increases the work of respiration in the horizontal position.

The vital capacity (3800 ml) and the residual volume (1600 ml) added together give the *total lung capacity* (5400 ml).

Dead space. Of the 500 ml of air taken in at a quiet inspiration only a part reaches the respiratory parts of the lungs; the remainder occupies the mouth, pharynx, trachea (25 to 30 ml) and air passages up to the terminal bronchioles (that is the 'anatomical' dead space) and is unchanged in composition because no respiratory exchange takes place. On expiration the air in the dead space (unchanged air) is breathed out first, to be followed by air from the alveoli; the mixed expired air contains both dead space air and alveolar air. If 500 ml of expired air contains 4 per cent carbon dioxide, the total amount of carbon dioxide breathed out is $4/100 \times 500 = 20$ ml. But this 20 ml carbon dioxide came only from X ml of alveolar air which contained 5·5 per cent carbon dioxide (for method see p. 602); therefore $X \times 5·5 \div 100 = 20$ and $X = 364$ ml. The dead space in this simplified example is 136 ml which is approximately the value in young men. The value varies with the position of the subject and it increases if the lungs are distended (see also p. 589).

For strict accuracy all values for the subdivisions of the lung volume and all results of pulmonary function tests should be given at body temperature and atmospheric pressure and saturated with water vapour, BTPS, and not at $0°C$ as are metabolic data. The values quoted on pp. 598 to 599 are the mean values for adult males in a population varying widely in age and size. The values in women are about 25 per cent lower.

DYNAMIC TESTS OF VENTILATION

Tests of ventilation, designed to measure the 'bellows action' of the lung, are commonly used in clinical medicine. Many factors are involved and, although airway resistance is probably the most important, lung size and compliance, muscular co-ordination and power are also included. Because airway resistance is so often raised in the common lung diseases, such as asthma and chronic bronchitis, tests of ventilatory efficiency are the most widely used of all the tests of lung function. The maximum breathing capacity, M.B.C., measures the greatest volume of air that can be breathed by the subject per minute. It is usually measured over a 15 sec period with the subject breathing as deeply and as fast as possible; the result is expressed in litres per min. The value in a young adult is of the order of 150 l/min. This exhausting procedure is not suitable for disabled patients and is not much used now since similar information can be obtained from the analysis of a single forced expiration after a maximal inspiration. The volume expelled in the first second of expiration (FEV_1) has been found to correlate well with the subject's disability. This is an easier test and gives reproducible measurements. As with the M.B.C., normal values for FEV_1 vary with age, sex and to a smaller extent with body size but healthy persons can expel 75 per cent or more of the vital capacity within the first second. When the resistance of the airways is greatly increased the proportion may fall to 30 per cent or less (Fig. 30.11).

Normally, of course, the volume of air breathed per min (about 6 l), falls far short of the maximum breathing capacity, and the difference may be regarded as the *breathing reserve*. It has been shown that when the

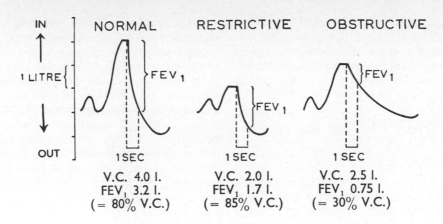

Fig. 30.11 *Forced expiratory spirograms.* The three records are taken from spirometer tracings. Inspiration is up and expiration down, and the records read from left to right. After a deep inspiration the subjects breathed out as rapidly and forcibly as they could. The total volume expired is the *(forced) vital capacity* and the volume expired in the first second is the *forced expired volume* in 1 second (FEV_1). The 'restrictive' record was obtained from a patient with kypho-scoliosis; the 'obstructive' record from a patient with airways obstruction. Kypho-scoliosis is a deformity of the bony structure of the chest that interferes with movement; although the vital capacity is more severely reduced in the patient with kypho-scoliosis than in the patient with airways obstruction the proportion of the vital capacity expired in the first second is normal. (E. J. M. Campbell, C. J. Dickinson & J. H. D. Slater (1968). *Clinical Physiology*, 3rd edn, p. 124. Oxford: Blackwell.)

breathing reserve is reduced to some 60 to 70 per cent of the maximum breathing capacity dyspnoea is usually present.

FETAL AND NEONATAL RESPIRATION

The fetus even from the early age of 12 weeks makes occasional respiratory movements. If radio-opaque material is injected into the amniotic fluid before Caesarean section, X-ray photographs of the fetus immediately after birth show shadows in its lungs and intestinal tract. Since the radio-opaque material is concentrated in the lungs, the water of the amniotic fluid must be absorbed through the lungs into the fetal circulation. During intra-uterine life respiratory movements are intermittent and irregular. At birth, air is breathed instead of fluid and the respirations become deeper and more regular. The extent of the expansion of the lungs is indicated by the change in specific gravity; lungs removed from the body before respiration has begun sink in water (sp. gr. 1·06) whereas aerated lungs float (sp. gr. 0·34).

When air enters the lungs at the first breath an air–liquid interphase is formed, the surface tension of which resists ingress of air. Since the resistance is inversely proportional to the radius of the tube it is not surprising to find that an air pressure of the order of 30 cm H_2O is needed to inflate fully an excised lung of a full-term baby. This resistance does not seem to present undue difficulty to the baby since intraoesophageal swings of 90 cm H_2O have been recorded during the first breath. The first inspiratory effort must therefore be violent but subsequent breaths need not be so vigorous as the residual air builds up. The low surface tension fluid lining of the alveoli (p. 591) must form at the first breath; if it does not the infant experiences respiratory distress due to its efforts to expand the lungs. The amniotic fluid in the lungs is carried off by the lymphatics in the first three or four hours after birth.

The respiratory rate at birth is usually about 30 per minute (range 14 to 60); since the average respiratory minute volume is 500 ml the average tidal volume is about 17 ml. Breathing may be quite irregular in the first few days of extra-uterine life but later becomes regular. The Hering-Breuer reflex (p. 606) is active during the first week of life only. The respiration of the newborn infant is stimulated by administration of carbon dioxide just as is the respiration of the adult. Other respiratory data in the newborn are: dead space is 4·4 ml, functional residual capacity is 75 ml, crying vital capacity is 56 to 110 ml.

COMPOSITION OF THE RESPIRED AIR

The effect of respiration, so far as oxygen is concerned, is that an adult man at rest acquires about 250 ml of oxygen per min. At the same time about the same quantity of carbon dioxide is breathed out.

Expired air can be collected by making the subject breathe through valves into a rubberized canvas bag (Douglas bag) (p. 164). A sample of *alveolar air* can be obtained by the method devised by Haldane and Priestley. A tube 120 cm long and 2·5 cm in diameter is fastened to a mouthpiece provided with an evacuated gas sampling tube as in Fig. 30.12.

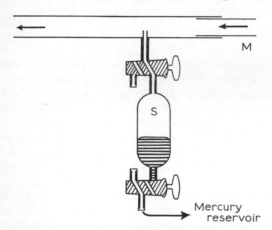

Fig. 30.12 Apparatus for collecting a sample of alveolar air. M = mouthpiece; S = sampling tube. (*After* J. S. Haldane & J. G. Priestley (1905). *Journal of Physiology* **32**, 227.)

At the end of a normal expiration the subject closes his nose and breathes out as deeply and as quickly as possible through the tube, finally closing the mouthpiece by putting his tongue against it. Immediately, the upper tap of the sampling tube is opened momentarily to draw in a sample of the air last expelled from the lungs. In this method the dead space air coming first into the long tube is washed out by the alveolar air coming later, so that alveolar

air only is taken into the sampling tube. This technique has been criticized since, by interrupting the rhythm of respiration, it may alter the composition of the alveolar air. This difficulty may be overcome by alternative methods that sample a small quantity of the gas at the end of each normal expiration over a period of time ('end-tidal sampling'). These methods have the advantage of providing continuous and sensitive monitoring of the alveolar composition in the undisturbed subject. Neither method gives accurate results when there is abnormal distribution of gases in the lungs such as occurs in many patients with chronic disease of the lungs or even in normal patients being ventilated by artificial means (p. 608).

Air samples can be analysed by a variety of methods, but the most precise is Lloyd's modification of Haldane's volumetric apparatus. The volume of the sample is first accurately measured and it is then passed over caustic potash solution to absorb carbon dioxide and the volume is measured once more. Finally the oxygen is absorbed by alkaline pyrogallol solution and the oxygen content is shown by the further reduction in the volume of the sample. Rapid methods, less precise but more convenient, have been introduced recently. They depend for example on the paramagnetic properties of oxygen, the infra-red absorption of carbon dioxide or on the thermal properties of gases (catharometry).

The following table of experimental results shows how the percentage composition of typical expired and alveolar air samples from a resting subject differs from that of inspired (atmospheric) air which is remarkably constant the world over.

Percentage composition of dry gas	Inspired air	Expired air	Alveolar air
Oxygen	20·93	16·89	14·08
Nitrogen	79·04	79·61	80·37
Carbon dioxide	0·03	3·50	5·55

Nitrogen is not given out from the body as the increase in nitrogen content of the expired air might at first sight suggest. Since the respiratory quotient (R.Q.),

$$\frac{\text{Volume of CO}_2 \text{ produced}}{\text{Volume of O}_2 \text{ used}},$$

is usually less than 1 (see p. 162), the volume of carbon dioxide escaping from the blood is less than the volume of oxygen taken into the blood. The R.Q. calculated from the expired air composition in the example given above is 0·83. Thus the volume of air expired is less than the volume of air inspired although both contain the same absolute amount of nitrogen.

The alveolar air is not so variable in composition as the tidal air. During quiet breathing the small quantity of air drawn into the alveoli with each breath, about 360 ml, is not sufficient to disturb very greatly the composition of the large volume, about 2900 ml, of air in the functional residual capacity but in deep breathing the composition of the expired air approaches more closely that of the alveolar air. It might be thought that such a system of ventilation of the alveoli is inefficient since the alveoli are not swept through with fresh air at each breath. However, the known rates of diffusion of oxygen and carbon dioxide are more than sufficient to allow the rapid transfer of carbon dioxide and oxygen through the space occupied by the functional residual air.

CENTRAL CONTROL OF RESPIRATION

The complexity of the control of breathing is indicated by Dornhorst's description of the respiratory centre as a computer which receives information from the brain, muscles and tissues via nerves and from the blood by chemical stimuli, from which it calculates the demand for ventilation (Fig. 30.13). The important chemical control of respiration is described in Chapter 31 (p. 626) but a rather crude idea of the organization of this computer, gained from surgical experiments on the central nervous system of animals, is outlined here, and the nervous control of respiration is described on page 605.

If the upper parts of the brain down to the mid-brain are cut away respiration continues but section through the posterior part of the medulla brings respiration to a stop. When the spinal cord is cut across in the thoracic region the intercostal muscles innervated from levels below the section are paralysed but the diaphragm, supplied by the phrenic nerves continues to move. Division above the roots of the phrenic nerves (3rd, 4th and 5th cervical) arrests diaphragmatic movements and the animal dies of asphyxia, although for a time movements of the alae nasi may be seen. The respiratory centre must therefore lie somewhere in the hind-brain and must send impulses out along the phrenic and intercostal nerves.

Lumsden found that, after cutting the vagus nerves in the cat (Fig. 30.14B, a–b), respiration became slower and deeper. A subsequent cut through 1 (Fig. 30.14A) produced no change in respiration. Section between 2 and 3 produced prolonged inspiration as at (c). Section at 4 produced gasping respiration (d). Section at 6 abolished all respiratory movements. These results suggest that there is a controlling or *pneumotaxic centre* in the upper (cranial) part of the pons which controls the activities of lower (caudal) groups of neurones, that is, those causing prolonged inspiration and gasping. If the vagal fibres which pass mainly to the

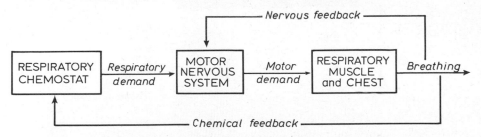

Fig. 30.13 Organization of breathing. In this simple model the organization of breathing is represented as two control systems, one within the other. The outer one (Chap. 31) maintains the long-term stability of the blood gases and in fact depends for its operation on information about them, notably the partial pressure of carbon dioxide in arterial blood ('chemical feedback'). The inner system attends to the performance of each breath and depends upon proprioceptive information ('nervous feedback'). (E. J. M. Campbell (1965). *British Medical Journal* i, 1452.)

expiratory centre are left intact after cutting between 2 and 3 (Fig. 30.14A) breathing is still regular. Thus to maintain rhythmic breathing the respiratory centres in the lower part of the medulla must be influenced either by the pneumotaxic centre or by afferent impulses through the vagus nerves (Fig. 30.15). Experiments on the dog have amplified Lumsden's findings. The apneustic state after section at level 2 (Fig. 30.14A) is greatly reduced by denervation of the carotid bodies. Dogs transected at level 4 with the vagus nerves cut can breathe regularly for many hours.

Pitts has explored the medulla by passing electrodes into its substance. Electrical stimulation of an area (Fig. 30.16) dorsal to the pyramids and deep to the posterior (caudal)

end of the fourth ventricle produced a prolonged inspiration: this area he called the 'inspiratory centre'. The 'expiratory centre' was found immediately dorsal to the 'inspiratory centre' and only a short distance below (that is, ventral to) the floor of the posterior part of the fourth ventricle. Stimulation of this area gave a prolonged contraction of the expiratory muscles.

Delisle Burns has explored the medulla with microelectrodes connected to an amplifier and has detected discharges arising in cell bodies on inspiration and on expiration in the area indicated in Fig. 30.16. He found no segregation of inspiratory from expiratory neurones. Burns thinks it likely that Pitts stimulated a dense track of efferent fibres lying medially to

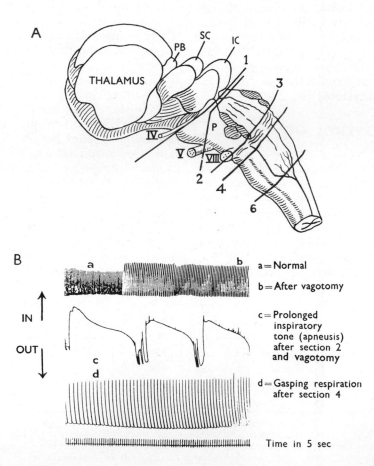

Fig. 30.14 Diagram (not to scale) of the brain stem of a cat to show the position of sections which produced the alterations in respiratory rhythm shown in the lower part of the figure. PB, pineal body; SC and IC superior and inferior colliculi; P, pons; IV, V, VIII, cranial nerves (T. Lumsden (1923). *Journal of Physiology* **57**, 153, 354; **58**, 81, 111).

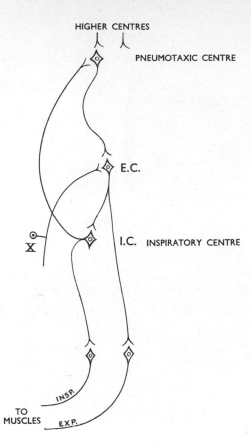

HIGHER CENTRES

PNEUMOTAXIC CENTRE

E.C.

X

I.C. INSPIRATORY CENTRE

INSP.

TO
MUSCLES EXP.

Fig. 30.15 A much simplified scheme of the various parts of the central respiratory connexions. X, vagus nerve. (*After* R. F. Pitts, H. W. Magoun & S. W. Ranson (1939). *American Journal of Physiology* **126**, 702.)

analogous intrinsic rhythm in mammals after complete denervation is abolished by artificial respiration with oxygen and increased if carbon dioxide is added to the inspired air.

Stimulation of certain parts of the premotor cortex increases respiratory movements, whereas stimulation of other cortical areas inhibits them. Since lesions in the lateral hypothalamus diminish respiration it is likely that this region exerts a facilitatory action on the respiratory centre. Cooling the medulla slows respiration. If the body temperature of an animal is raised, or if the blood going to the medulla is warmed, respiration becomes faster but not deeper (*tachypnoea*). During sleep the respiratory centre is less active so the alveolar carbon dioxide may rise above the waking value.

Motor fibres from the respiratory centre nuclei cross over in the medulla and pass down through the anterior and antero-lateral columns of the cord to the nuclei of the phrenic and intercostal nerves in the anterior horns of the grey matter. Commissural fibres must join the nuclei across the spinal cord, and thus hemisection of the cervical cord does not always cause paralysis of the diaphragm on the same side.

NERVOUS REGULATION OF RESPIRATION

The most important factor in the regulation of respiration is undoubtedly the carbon dioxide tension of the blood supplying the central chemoreceptors. The activity of the respiratory centre, however, is modified by afferent impulses which are continuously arriving at the medulla in the vagus nerves and in the nerves from the carotid body and carotid sinus and from the cerebral cortex; the centre is also influenced, especially in man, by impulses arising in receptors in the chest-wall.

On inspiration the lungs are stretched and impulses travel up the vagus to the respiratory centre. The frequency of the impulses increases with increasing stretch of the lungs (Fig. 30.17) until the peak of inspiration is reached when the frequency declines as expiration occurs. The receptors giving rise to these impulses lie in the bronchioles close to the bronchial muscles since drugs which increase bronchial tone increase the discharge and

the true respiratory neurones. Over-ventilation decreased or even stopped activity in the respiratory neurones, while administration of carbon dioxide to the animals increased their activity. There are, according to Burns, two independent networks of neurones, inspiratory and expiratory, with reciprocal innervation between each network since activity rises in one as it declines in the other.

While both the rate and the depth of respiration can to some extent be controlled voluntarily, and even arrested altogether for a short time, the respiratory centre seems to possess a rhythmic activity of its own. This was first clearly demonstrated in the goldfish since after removal of the brain from the body the medulla shows rhythmic electrical variations with the same frequency as the gill movements. An

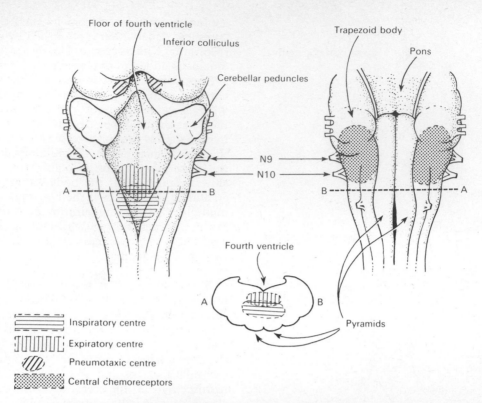

Floor of fourth ventricle

Inferior colliculus

Cerebellar peduncles

Trapezoid body

Pons

N9

N10

Fourth ventricle

Pyramids

⎯⎯⎯ Inspiratory centre

⫿⫿⫿⫿⫿ Expiratory centre

Pneumotaxic centre

Central chemoreceptors

Fig. 30.16   Diagrams of the medulla and pons of the cat to show the areas involved in control of respiration. (Compiled from diagrams by various authors.)

drugs which cause bronchodilatation decrease the discharge. The following experiments show that these impulses inhibit the respiratory centre. Inflation of the lungs or stimulation of the central end of a cut vagus nerve in an anaesthetized animal arrests inspiration. This inhibition of inspiration by lung inflation was first described in 1868 by Hering and Breuer. This reflex is an important mechanism for the control of breathing in many laboratory animals and also for a few days in the newborn baby but in the conscious man it is seen only if the inflation is large provided that the vagi are intact. After section of both vagus nerves in animals (see Fig. 30.14A and B) inspiration is prolonged and the rate of breathing is reduced. In man, however, bilateral vagus block does not alter the pattern of quiet breathing at rest in spite of the discharge of stretch receptors just as in the cat (Fig. 30.17). It does, however, substantially affect the degree and the pattern of the response to respiratory stimuli like CO_2, virtually abolishing the usual increase in

frequency of breathing and thus reducing the total ventilation achieved. Under these conditions, the relation between ventilation and tidal volume lies along one of the frequency iso-pleths of Fig. 30.7 (p. 595). Vagal block also diminishes the unpleasant sensation associated with rebreathing and breath-holding.

Breathing can be made more difficult by making a subject breathe in and out of a closed tank to simulate increase of elastic forces or by making him breathe through narrow orifices to simulate increase of viscous forces. An increase in the expiratory load causes an immediate increase in the volume of the lungs without any change in the rate of breathing (Fig. 30.18). The adjustment to inspiratory loading also occurs immediately—too quickly to be a response to alteration in blood P_{CO_2}. Since vagal block in man does not affect the ability to detect additional loads it seems likely that adjustments to loading (Fig. 30.18) are dependent on receptors in the muscles and joints of the chest wall.

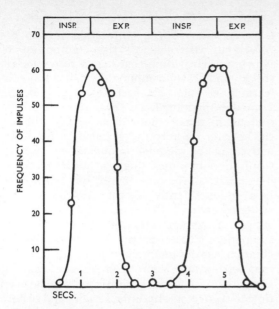

Fig. 30.17 Frequency of afferent impulses in a single fibre of the vagus of a decerebrate cat during normal breathing. The frequency of nerve impulses at the end of expiration need not necessarily be zero. For example another fibre discharged at the rate of 50 impulses per sec at the end of expiration and at 80 impulses per sec at the height of a normal expiration. The response of these stretch receptors is also affected by the physical state of the lung, for example by changes in pressure in lung vessels, size of bronchi and speed of inflation. (*After* E. D. Adrian (1933). *Journal of Physiology* **79**, 337.)

Many other reflex alterations of breathing have been described; they may arise from receptors in the lungs or heart or large vessels. If the lungs of an anaesthetized animal are deflated the rate and force of inspiration are increased if the vagus nerves are intact. In man deflation or collapse of the lungs as in the formation of a pneumothorax is followed by increase of ventilation. In animals increased pressure in the pulmonary capillaries, brought about by temporary occlusion of the mitral valve orifice, is associated with rapid shallow breathing which is abolished by section of the vagus nerves. Small pieces of blood clot (emboli) lodged in the small blood vessels of the lungs, have variable reflex effects, usually apnoea and then rapid shallow breathing and pulmonary vasoconstriction and oedema. Part of this complicated response may be due to stimulation of receptors in the lungs which send impulses up the vagus nerves.

A rise of venous pressure in the great veins near the heart, produced for example by rapid injection of fluid, or occurring in cardiac failure, stimulates respiration. Swallowing, which usually occurs during expiration, is attended by a reflex inhibition of respiration (p. 228) via the glosso-pharyngeal nerve. Application of cold water to the skin causes gasping followed by hyperventilation.

A *cough* is a sudden inspiratory effort with a wide open glottis which is followed by an expiratory effort against a closed glottis which suddenly opens to let the air rush out under pressure. During the expiratory effort the intrathoracic pressure rises to 100 or even 300 mm Hg and narrows the trachea and bronchi as in any forced expiration. Coughing is useful in bringing foreign bodies up the bronchi and trachea. The velocity of airflow in the trachea of a normal subject during a vigorous cough may reach 600 miles/hour (Mach 1). *Sneezing*

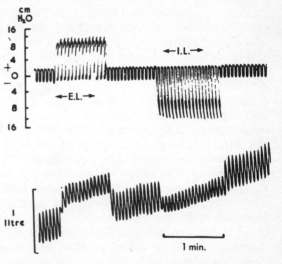

Fig. 30.18 The effect of loading on the breathing of an anaesthetized man. The upper record shows the pressure at the mouth relative to atmospheric pressure. The lower record shows the movements of a spirometer bell (inspiration upwards); the general upward slope is due to the gradual absorption of oxygen. Expiration was loaded (10 cm water) at E.L. and inspiration was loaded at I.L. The expiratory load immediately increased both end-expiratory and end-inspiratory lung volumes (lower record); when the load was removed they both decreased to the previous value; the respiratory rate was unaffected. Inspiratory loading reduced end-inspiratory lung volume without affecting the respiratory frequency. The effect on conscious subjects is similar. (E. J. M. Campbell, C. J. Dickinson, O. P. Dinnick & J. B. L. Howell (1961). *Clinical Science* **21**, 311.)

is a somewhat similar reflex arising from irritation of the nasal mucosa through the nasal branch of the fifth nerve; the expiratory blast passing through the nose may propel droplets up to 6 m. Stimulation of the nasal mucosa in dogs by water or tobacco smoke depresses or abolishes breathing; it also causes bradycardia, reduction of cardiac output, vasoconstriction in most areas of the body but not in the head. *Hiccup* is a repetitive spasmodic contraction of the diaphragm and external (inspiratory) intercostal muscles in the course of which the glottis closes suddenly and further entrance of air into the chest is prevented. The sudden arrest of the column of entering air produces the characteristic sound and sensation. The peculiar involuntary, 'infectious' respiratory act, the *yawn* is accompanied by facial and limb-stretching gestures and is remarkable for a marked dilatation of the pharynx. Apart from the temporary vasoconstriction associated with a large inspiration (p. 567), there is no accompanying circulatory alteration in the brain or elsewhere. The cause and function of yawning are quite obscure.

The cough reflexes in man are difficult to analyse. The results of experiments with cats can only be assumed to apply to man; these reflexes vary greatly between species. In the cat coughing can be brought about by mechanical irritation of the larynx or trachea but not of the main bronchi. Such irritation also produces broncho-constriction, bradycardia and hypertension. The reflex is not obtained if the mucosa is first sprayed with a local anaesthetic. The receptors are rapidly adapting and are probably knob-like subepithelial endings derived from large afferent fibres. Inhalation of sulphur dioxide or other noxious gases also causes coughing but in this case the receptors must be situated more deeply in the lungs since the reflex is not abolished by anaesthetizing the trachea. The afferent pathways for both reflexes are mainly in the vagus but the sympathetic pathway may also be involved to a small extent.

ARTIFICIAL VENTILATION

Artificial ventilation may be required during the administration of an anaesthetic. When a muscle relaxant similar to curare is administered all skeletal muscles including the diaphragm are paralysed, so artificial ventilation is essential. Artificial ventilation is also required in emergencies, the commonest of which are drowning, cardiac arrest, electric shock and carbon-monoxide poisoning; at least 5000 such emergencies occur in the United Kingdom every year. It is obviously difficult in these emergencies to say whether the heart or the respiration fails first but if the activity of the medulla oblongata is depressed respiration becomes sluggish and may cease. The heart, on the other hand, can continue to beat independently of the central nervous system, provided that it receives an adequate supply of oxygen. Thus, if oxygen can reach the patient's blood promptly enough, the circulation can be maintained until the cause of the respiratory depression is removed or alleviated so that the respiratory centre is given time to recover its rhythmic activity and restore natural respiration.

An equally important reason for prompt action is that asphyxia of only a few minutes' duration does irreparable damage to the cortical cells of the brain. The oxygen content of the blood falls from 20 to 15 ml per 100 ml blood during the first few minutes, to 2 or 3 ml after 6 min, and to zero in 8 min. Although it may be possible to restore breathing after periods of apnoea of 8 to 10 min there is rarely complete recovery of cerebral function after 2 or 3 min. *Not a single second* should therefore be lost in applying artificial ventilation. No time should be wasted in loosening clothing or in prolonged attempts to drain water out of the lungs or in moving the victim to a more convenient situation. These matters can be attended to later.

In fresh water drowning (see p. 610) death is usually due primarily to cardiac arrest. In electric shock death is due primarily to ventricular fibrillation (occasionally asystole). In these two emergencies and, of course, in cardiac arrest from any cause, both external cardiac massage (p. 523) and artificial ventilation will be required.

The essentials of a good first-aid method of artificial ventilation are: (1) it should produce adequate ventilation of the lungs; (2) if any apparatus is needed it must be of the simplest kind which should be readily available anywhere; (3) no damage should be done to the patient by violent handling; (4) the strain on the operator should be minimal so that he can apply the method for long periods; (5) the position of the patient should be such that the airway is kept open.

Expired air resuscitation. This ancient method, performed by Elisha (2 Kings 4, 34), was 'rediscovered' about 1957; it is easy to learn and is in every way superior to manual methods. The patient is placed on his back on a firm surface, if possible with his head at a slightly lower level than his stomach. The operator should quickly clear out mucus, food or other materials from the mouth and throat. It is most important to extend the patient's head as far as possible (Fig. 30.19) so that the

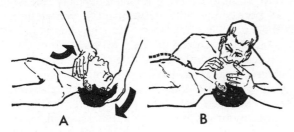

Fig. 30.19 Intermittent positive-pressure breathing. (a) Head hyperextended and chin raised. (b) Mouth-to-mouth resuscitation. Look for chest rise with each inflation. (M. H. Brook, J. Brook & G. M. Wyant (1962). *British Medical Journal* ii, 1564–1566.)

airway is opened up. The operator pinches the patient's nostrils to prevent leakage, takes a deep breath and blows into the patient's mouth until the chest is seen to rise; the operator then takes his mouth quickly away to allow passive exhalation. The process is repeated about 10 times a minute in adults. Mouth-to-nose ventilation is superior to mouth-to-mouth because full extension of the neck can be combined with complete closure of the mouth and lifting forward of the lower jaw; also inflation of the stomach is less likely. In resuscitating an infant the operator can place his mouth over the subject's nose and mouth and the lungs can be inflated with little puffs at say 30 per min.

There is no doubt that this is an effective method; investigation has shown that a tidal volume of 1 litre can readily be maintained. The expired air of the operator though containing 16 per cent of oxygen (instead of 21 per cent as in air) is capable of maintaining adequate oxygenation of the subject's blood with adequate carbon dioxide elimination. Aesthetic objections should never stand in the way of saving a life but may be overcome by the use of an airway; if this is not available a handkerchief with a hole in it may be placed over the patient's mouth.

If breathing does not start again within two or three minutes after recognition of the emergency external cardiac massage (p. 523) should be applied; but artificial ventilation must be continued since external cardiac massage alone does *not* ventilate the lungs. Artificial ventilation and external cardiac massage must be continued until either the patient recovers, or a doctor pronounces him dead.

It is of some importance to decide whether pure oxygen or oxygen with carbon dioxide should be administered in asphyxia. Inhalation of a mixture of oxygen and carbon dioxide by a healthy person increases the respiration so that there is practically no change in body carbon dioxide. If, however, the respiratory centre is damaged or depressed, the appropriate increase in respiration may not take place and there may be a dangerous rise in the carbon dioxide in the body. High concentrations of carbon dioxide have an anaesthetic effect. In carbon monoxide poisoning, as in all other forms of anoxia and asphyxia, pure oxygen is always to be preferred to oxygen-carbon dioxide mixtures. Normal man at rest breathing air eliminates half of the carbon monoxide in his blood in 250 min; if he breathes pure oxygen, half of his carbon monoxide is eliminated in 40 min. Even more efficient is the oxygen pressure chamber (hyperbaric oxygen) in which pure oxygen is supplied at a pressure of two atmospheres; not only does the oxygen drive off the carbon monoxide but the plasma itself is able to carry twice as much oxygen at two atmospheres.

Although carboxy-haemoglobin is cherry coloured the skin of a patient poisoned with carbon monoxide is cyanosed and pale. It is only at necropsy that the cherry colour is seen. Unless the patient is moribund hyperventilation is usual; it reduces the P_{a,CO_2} and because of hypoxia the arterial pH is low. Administration of oxygen with CO_2 is therefore contraindicated since it would lower the pH still further.

Prolonged artificial ventilation is sometimes

required when spontaneous respiration is depressed. It is necessary when the chest is unable to expand the lungs, for example during and after thoracic surgery; or when the muscles of the thorax are paralysed, for example after high injuries to the cervical spine, in poliomyelitis, and after administration of muscle relaxants during general anaesthesia for surgical operations. Mechanical ventilators are then employed. Almost all such machines now operate on the principle of 'intermittent positive pressure', the lungs being inflated at a predetermined pressure and rate through a tube inserted into the trachea either through the mouth or nose or directly through a small incision below the cricoid cartilage (tracheostomy). This endotracheal tube has an inflatable cuff that blocks the rest of the tracheal lumen. Old methods by which the chest was expanded by negative pressure applied to the outside of the thorax are now rarely used.

DROWNING

Over 1000 cases of drowning are reported every year in the United Kingdom. If lives are to be saved artificial ventilation must be used frequently so it is worth while to look at the mechanism of drowning especially as it provides good illustrations of several important physiological principles. Most results have been obtained from experiments on dogs and may not be directly applicable to man. The immediate effect is an increase in cardiac rate and output, and an increase in coronary flow due to arterial desaturation.

Fresh water drowning. After a period of breath-holding and struggling an involuntary inspiratory effort draws large volumes of water into the lungs. The fresh water passes rapidly into the pulmonary capillaries so that more than the equivalent of the blood volume may enter the circulation in a few minutes with a corresponding reduction of the concentration of electrolytes which causes haemolysis with consequent liberation of haemoglobin and of potassium. These electrolyte disturbances, plus the overloading of the circulation, plus anoxia, lead to fatal ventricular asystole. No animals survive when given artificial respiration after more than two minutes' submersion. At post-mortem examination very little fluid is found in the lungs.

Salt water drowning. In this situation the inhaled fluid is hypertonic and water moves from the pulmonary alveolar capillaries to the fluid in the air spaces of the lungs. The blood volume is reduced with a rise in electrolyte concentration but there is no haemolysis or release of potassium and consequently ventricular fibrillation is not seen but the heart fails because of anoxia in five to eight minutes. Protein exuding from the blood into the alveoli makes a tenacious froth in the air passages and a considerable quantity of water may be seen in them post mortem. These experiments suggest that submersion in fresh water is more dangerous than submersion in sea water.

Dry drowning. In about one-fifth of all cases of drowning the patient does not inhale water but becomes unconscious simply on account of anoxia. Provided that artificial ventilation is begun without delay the prognosis is very good and probably most persons who recover from alleged drowning are of this type. Artificial ventilation is the only treatment necessary.

Late drowning. This term is sometimes used for the pulmonary complications which result from inhalation of sand, oil or chemicals. These complications which include collapse, pneumonitis and irritation usually arise several hours after the accident and for this reason persons who have been resuscitated after near drowning should be under observation in hospital for some hours and preferably for one or two days after the accident.

REFERENCES

AVERY, M. E. & NORMAND, C. (1965). Respiratory physiology in the new-born infant. *Anaesthesiology* **26**, 510–521.

BATES, D. V. & CHRISTIE, R. V. (1964). *Respiratory Lung Function in Disease. An Introduction to the Integrated Study of the Lung.* London: Saunders.

BROWN, J. H. U. (Ed.) (1963). *Physiology of Man in Space.* London: Academic Press.

CAMPBELL, E. J. M., AGOSTINI, E. & DAVIS, J. N. (1971). *The Respiratory Muscles: Mechanics and Neural Control*, 2nd edn. London: Lloyd-Luke.

CARO, C. G. (Ed.) (1966). *Advances in Respiratory Physiology.* Oxford: Pergamon.

COMROE, J. H. (1965). *Physiology of Respiration.* Chicago: Year Book Medical Publishers.

COTES, J. E. (1968). *Lung Function—Assessment and Application in Medicine*, 2nd edn. Oxford: Blackwell.

CUNNINGHAM, D. J. C. & LLOYD, B. B. (Eds.) (1962). *The Regulation of Human Respiration.* Oxford: Blackwell.

DAFT, F. S. (Ed.) (1962). Science in the space age. *Federation Proceedings* **21**, 679–691.

DALY, I. DE BURGH (1966). *Pulmonary and Bronchial Vascular Systems.* London: Arnold.

DICKENS, F. & NEIL, E. (Eds.) (1964). *Oxygen in the Animal Organism.* Oxford: Pergamon.

EDHOLM, O. G. & BACHARACH, A. C. (Eds.) (1965). *Exploration Medicine, being a Practical Guide for Those Going on Expeditions.* Bristol: Wright.

FARHI, L. E. (1965). Respiration. *Annual Review of Physiology* **27**, 233–256.

FENN, W. O. & RAHN, H. (Eds.). *Handbook of Physiology*, Section 3: *Respiration.* Washington, D.C.: American Physiological Society.

FORGACS, P. (1969). Lung sounds. *British Journal of Disorders of the Chest* **63**, 1–12.

HALDANE, J. S. & PRIESTLEY, J. G. (1935). *Respiration*, 2nd edn. Oxford: Clarendon Press.

HEINEMANN, H. O. & FISHMAN, A. P. (1969). Non-respiratory functions of mammalian lung. *Physiological Reviews* **49**, 1–47.

HOWELL, J. B. L. & CAMPBELL, E. J. M. (Eds.) (1966). *Breathlessness.* Oxford: Blackwell.

HUGH-JONES, P. & CAMPBELL, E. J. M. (Eds.) (1962). Respiratory physiology. *British Medical Bulletin* **19**, 1–89.

KINOSITA, H. & MURAKAMI, A. (1967). Control of ciliary motion. *Physiological Reviews* **47**, 53–82.

NAHAS, C. G. (Ed.) (1963). Regulation of respiration. *Annals of the New York Academy of Sciences* **109**, 411–948.

PACE, W. R. (1965). *Pulmonary Physiology in Clinical Practice.* Oxford: Blackwell.

PATTLE, R. E. (1965). Surface lining of lung alveoli. *Physiological Reviews* **45**, 48–79.

PORTER, RUTH (Ed.) (1970). *Breathing: Hering-Breuer Centenary Symposium.* London: Churchill.

RANDEL, H. W. (1971). *Aerospace Medicine.* Edinburgh: Churchill Livingstone.

RIVERS, J. F., ORR, G. & LEE, H. A. (1970). Drowning. Its clinical sequelae and management. *British Medical Journal* **i**, 157–161.

SCHAEFER, K. E. (1962). *Environmental Effects on Consciousness.* New York: Macmillan.

SPALDING, J. M. K. & SMITH, A. C. (1963). *Clinical Practice and Physiology of Artificial Respiration.* Oxford: Blackwell.

STRANG, L. B. (1965). Changes in the lungs at birth. *Annual Review of the Scientific Basis of Medicine* 202–216.

WEYER, E. M. (Ed.) (1966). Interdisciplinary investigation of mucus production and transport. *Annals of the New York Academy of Science* **130**, 869–973.

WIDDICOMBE, J. G. (1963). Regulation of tracheobronchial smooth muscle. *Physiological Reviews* **43**, 1–37.

31 Respiratory functions of the blood

The fact that carbon dioxide is in greater concentration in the expired air than in the inspired (atmospheric) air was at one time taken to mean that combustion took place in the lungs. This idea had to be given up when it was shown by H. G. Magnus, about a hundred years ago, that the blood leaving the lungs contained a lower concentration of carbon dioxide and a higher concentration of oxygen than the blood reaching them. Thus it was clear that gas exchange took place in the lungs, carbon dioxide leaving the blood and oxygen entering it. A second exchange of gases takes place in the tissues which take up oxygen from the blood and give up carbon dioxide to it (Fig. 30.1).

SOLUTION OF GASES IN FLUIDS

Before discussing the gases carried in blood the conditions governing their solution in water must be mentioned. If the pressure of a gas above a volume of water is kept constant, the amount of the gas in solution diminishes as the temperature is increased. However, as the body temperature is remarkably constant, we need for most physiological purposes to consider only the factors which modify the solubility at constant temperature. If the temperature is held constant, the amount of a gas dissolved in water depends directly on its pressure, or, if the gas under consideration is mixed with other gases, on its partial pressure. When equilibrium is established between the gas and the water as many molecules of the gas enter the water as leave it. The tendency of a gas to leave the liquid is called its *tension*—it is well demonstrated by the effervescence which follows the opening of a bottle of aerated (carbonated) water. The tension of a gas in water is measured by finding the pressure (or partial pressure) of the gas in the atmosphere with which it is in equilibrium.

The partial pressure of a gas, P, in a mixture of gases is obtained by multiplying the pressure of the mixture by the percentage of the gas in it. For example, the partial pressure of oxygen, P_{O_2}, in dry atmospheric air (21 per cent oxygen) at 760 mm Hg is $21/100 \times 760 = 160$ mm Hg. The amount of a gas dissolved in water at 37°C exposed to a gas is given by the partial pressure multiplied by the solubility coefficient; for oxygen 0·00003, for carbon dioxide 0·0007, for nitrogen 0·000016 ml per ml water per mm Hg. Since arterial blood is nearly in equilibrium with alveolar air the tensions of the gases in arterial blood (p. 617) are approximately the same as their partial pressures in the alveolar air. The table shows the amounts of gases held in solution by 100 ml of water at 38°C exposed to alveolar air.

	Oxygen (ml)	Carbon dioxide (ml)	Nitrogen (ml)
100 ml water in equilibrium with alveolar air	0·3	2·6	0·9

When blood is exposed to a vacuum the whole of its contained gas is removed and the average volumes obtained from 100 ml of blood are as follows:

	Oxygen (ml)	Carbon dioxide (ml)	Nitrogen (ml)
Arterial blood	19	54	0·9
Venous blood	14	58	0·9

Thus the amount of oxygen in the blood is greater than can be accounted for by simple solution; most must be in some combination. Similarly, the greater part of the carbon dioxide must be present in combination. Nitrogen, on the other hand, is in simple solution.

DISTRIBUTION AND THE VENTILATION–PERFUSION RELATIONSHIP

Since about 4 litres of air enter the alveoli per minute and about 5 litres of blood pass through the lungs per minute the overall ventilation–perfusion ratio $(\dot{V}_A/\dot{Q})$ is about 0·8 but this average or ideal value does not apply to all alveoli even in a healthy man. Since the quantity of oxygen taken up by the blood is the same as the quantity lost by the alveolar air, and the quantity of carbon dioxide added to the alveolar air is the same as that lost by the body, the overall R.Q. of the blood and ideal alveolar air are the same, and this would be true for every alveolus if they all had the same $\dot{V}/\dot{Q}$ ratio. However, even in normal lungs there is a wide range of ratios from one extreme where no blood perfuses a well-ventilated alveolus $(\dot{V}/\dot{Q} = \infty)$ to the other where no air ventilates a well-perfused alveolus $(\dot{V}/\dot{Q} = 0)$. Because of the different shapes of the dissociation curves of carbon dioxide and oxygen, these extremes do not in general balance each other out, and even in normal lungs there is a difference between the P_{O_2}s, the P_{N_2}s and to a lesser extent the P_{CO_2}s of mixed alveolar gas and mixed pulmonary venous blood. This also results in a difference between the R.Q.s of individual alveoli (Fig. 31.1).

All parts of the lungs are not ventilated uniformly even in normal persons; the distribution of ventilation in disease may be extremely uneven. The nitrogen content of an alveolar air sample taken after breathing pure oxygen for 7 min should be less than 2·5 per cent; in emphysema, in which the ventilation is very uneven, it may be 10 per cent because the oxygen has not succeeded in washing the nitrogen out of the less well ventilated parts of the lungs. Uneven ventilation may also be detected by recording continuously the nitrogen content of the air expired immediately after a single maximal inspiration of oxygen. After the oxygen in the dead space is blown out the nitrogen level in normal persons climbs rapidly to a nearly flat plateau: a slow increase of nitrogen content up to a high level indicates defective distribution or mixing of the large breath of oxygen. This is called the 'nitrogen washout' test.

Measurement of the physiological dead space (p. 600) gives a clue to the ventilation–perfusion ratio since the space may be considered to include the anatomical dead space (where no gas exchange takes place) together with a contribution from alveoli with a high ventilation–perfusion ratio (where less than average exchange occurs). If the dead space is more than 30 per cent of the tidal volume a significant proportion of the alveoli is better ventilated than perfused.

By taking samples of alveolar air through catheters introduced into different parts of the lung it is possible to obtain direct information about regional ventilation–perfusion relationships. By using oxygen or carbon dioxide labelled with $^{15}O_2$ (half-life 2 min) ventilation and perfusion can be assessed without the disturbance and discomfort of catheterization. The gamma radiation produced by decay of the isotope is detected by scintillation counters, one on the front and one on the back of the chest. The subject takes a single breath of air

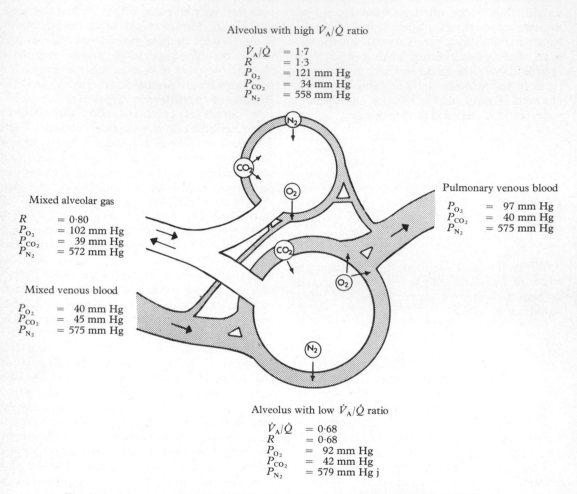

Alveolus with high $\dot{V}_A/\dot{Q}$ ratio

$$\begin{aligned}
\dot{V}_A/\dot{Q} &= 1{\cdot}7 \\
R &= 1{\cdot}3 \\
P_{O_2} &= 121 \text{ mm Hg} \\
P_{CO_2} &= 34 \text{ mm Hg} \\
P_{N_2} &= 558 \text{ mm Hg}
\end{aligned}$$

Mixed alveolar gas

$$\begin{aligned}
R &= 0{\cdot}80 \\
P_{O_2} &= 102 \text{ mm Hg} \\
P_{CO_2} &= 39 \text{ mm Hg} \\
P_{N_2} &= 572 \text{ mm Hg}
\end{aligned}$$

Mixed venous blood

$$\begin{aligned}
P_{O_2} &= 40 \text{ mm Hg} \\
P_{CO_2} &= 45 \text{ mm Hg} \\
P_{N_2} &= 575 \text{ mm Hg}
\end{aligned}$$

Pulmonary venous blood

$$\begin{aligned}
P_{O_2} &= 97 \text{ mm Hg} \\
P_{CO_2} &= 40 \text{ mm Hg} \\
P_{N_2} &= 575 \text{ mm Hg}
\end{aligned}$$

Alveolus with low $\dot{V}_A/\dot{Q}$ ratio

$$\begin{aligned}
\dot{V}_A/\dot{Q} &= 0{\cdot}68 \\
R &= 0{\cdot}68 \\
P_{O_2} &= 92 \text{ mm Hg} \\
P_{CO_2} &= 42 \text{ mm Hg} \\
P_{N_2} &= 579 \text{ mm Hg j}
\end{aligned}$$

Fig. 31.1 Diagram to illustrate possible variations in the ventilation–perfusion ratio, $\dot{V}_A/\dot{Q}$. R, respiratory quotient. The stippled pathways represent the pulmonary capillaries; the two open circles represent two alveoli, one with a high $\dot{V}_A/\dot{Q}$ and the other with a low $\dot{V}_A/\dot{Q}$. The respiratory quotients of the two alveoli vary considerably from each other and from that of an alveolus which is normally ventilated and perfused ($R = 0{\cdot}8$). In the alveolus with high $\dot{V}_A/\dot{Q}$ ratio the P_{A,CO_2} is relatively low while the P_{A,O_2} is high. On account of the slopes of the dissociation curves the volume exchange of CO_2 is large relative to that of O_2 (that is R is high). The N_2 in this alveolus is diluted by evolution of CO_2 to a greater extent than it is concentrated by absorption of O_2; the final P_{A,N_2} is therefore relatively low. In the alveolus with low $\dot{V}_A/\dot{Q}$ ratio P_{A,CO_2} is relatively high while P_{A,O_2} is relatively low. On account of the slope of the dissociation curves the volume exchange of CO_2 is low relative to that of O_2, that is R is low. The N_2 in the alveolus is concentrated to a greater extent than it is diluted by the evolution of CO_2; the final P_{A,N_2} is therefore high. (J. E. Cotes (1965). *Lung Function—Assessment and Application to Medicine*. Oxford: Blackwell.)

containing a trace of the labelled gas and holds his breath for 12 sec. The counting-rate rises in about 5 sec to a peak value, falls slowly during breath-holding and more rapidly when breathing of room air is resumed. The peak value indicates the amount of air entering the volume of lung between the counters, that is the ventilation. The slow fall during breath-holding measures the uptake of gas by the blood, that is the perfusion.

Such experiments show, when the subject is standing erect, a progressive fall in ventilation–perfusion ratio from 3·3 to 0·63 from apex to base of the lung. West has calculated that the alveolar P_{O_2} falls by 40 mm Hg from apex to base but because of the shape of the oxygen dissociation curve of haemoglobin the change in oxygen saturation of blood leaving the alveolar capillaries in the lower parts of the lungs is reduced by only 4 per cent. The P_{CO_2} and P_{N_2} increase on passing from the apex to the base of the lungs. The ventilation–perfusion inequality produced on changing from the horizontal to the vertical position has however little effect on the overall gas exchange.

In disease ventilation of an area of lung is reduced if the airways are narrowed or if the lung tissue becomes stiffer, that is its compliance is reduced. Neither condition allows the alveoli to fill up within the time of a single inspiration. Impaired alveolar ventilation with normal capillary perfusion produces a *physiological right-to-left shunt* of under-oxygenated blood ('*venous admixture*') causing arterial hypoxia which may be severe enough to produce 'central' cyanosis (p. 636). Many disease processes affect the pulmonary blood vessels and restrict blood flow locally.

ALVEOLO-CAPILLARY GAS TRANSFER

Although respiratory gas exchange is commonly referred to as occurring across the 'alveolar' or 'alveolo-capillary' membrane the gases actually pass through five separate media (Fig. 31.2a): the alveolar epithelium, the interstitial space, the capillary endothelium, the blood plasma and the membrane of the red cell. Surfactant (p. 589) is also present on the epithelial surface of the alveoli (Fig. 31.2b). It is the first three of these media that are regarded as the 'alveolo-capillary membrane'. The alveolar membrane acts in a purely passive manner and the transference of oxygen or carbon dioxide takes place by diffusion. To calculate if diffusion can account for the passage of the gases it is necessary to know (1) the partial pressures of the gases in the alveolar air and in the blood of the pulmonary artery and pulmonary veins, and (2) the rate of diffusion of gases through the alveolar membrane and the distances involved.

The methods for obtaining samples of alveolar air and for calculating the partial pressure of the gases are given on pp. 613 and 627. The partial pressures of gases in blood or plasma are measured by electrode methods (Fig. 31.5 and 6).

Krogh showed that the P_{a,CO_2} (carbon dioxide tension of the arterial blood) of the rabbit was nearly the same as the P_{A,CO_2} (partial pressure of carbon dioxide in the alveolar air); usually the former was slightly the greater. Although the free diffusion coefficient of carbon dioxide in water is a little less than that of oxygen, the solubility coefficient of carbon dioxide in water is twenty-four times greater than that of oxygen. This explains how it is that carbon dioxide is transferred about thirty times more rapidly than oxygen across the alveolar membrane and that a pressure difference of only 0·03 mm Hg is needed to transfer across the membrane all the carbon dioxide lost from the lungs.

When mixed venous blood with an oxygen tension of about 40 mm Hg enters the alveolar capillary it is exposed to alveolar air with an oxygen tension of 105 mm Hg, that is an alveolar-capillary difference of 65. As the blood flows through the capillary its oxygen tension rises and within 0·3 sec at rest approaches that in the alveolar air up to a final value of almost 105 mm Hg. The alveolo-capillary gradient at the venous end of the capillary is therefore less than 1 mm Hg (that is, it is not measurable) although the mean gradient over the whole length of the capillary is about 15 mm Hg. The membrane gradient at the venous end of the capillary is little altered in exercise, but is greatly increased at low barometric pressures at which the oxygen content of the blood is reduced. The value of this gradient at an altitude of 5000 metres (16 500 feet) at rest is about 8 mm Hg although the average gradient along the whole of the capillary is still 15 mm Hg. This pressure is

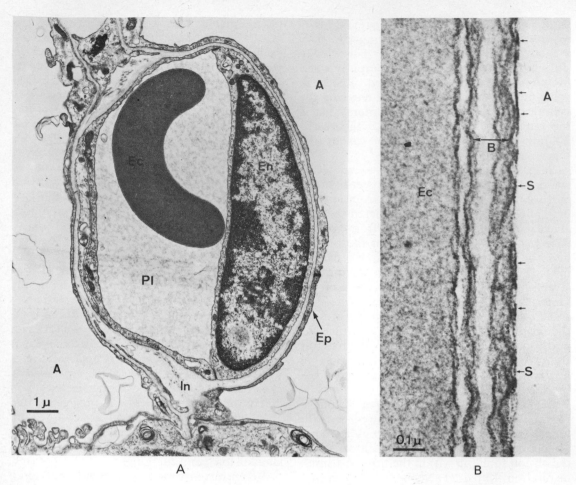

Fig. 31.2 Electron micrographs of mammalian lung. a Shows two alveoli (A) in monkey lung with the capillary between them in cross section. Ep, epithelium; In, interstitial space; En, endothelial cell; Pl, plasma; Ec, erythrocyte. (E. R. Weibel (1969). In *The Pulmonary Circulation and Interstitial Space* (Edited by A. P. Fishman & H. H. Hecht, p. 11. Chicago: University of Chicago Press.) b Shows the alveolar endothelium of rat lung at higher magnification. B, tissue barrier (endothelium, interstitial space and epithelium). S, surfactant (arrows point to osmophilic surface layer). (*By courtesy of Joan Gil.*)

needed to transfer the required volume of oxygen across the alveolar membrane.

However, at sea-level the P_{O_2} of the arterial blood in young adults breathing atmospheric air is about 5 mm Hg lower than that of the alveolar air. In old age the difference may be about 15 mm Hg. The gradient across the alveolar membrane, as just mentioned, is virtually nil. A small part of the difference in P_{O_2} between alveolar air and arterial blood is due to right to left anatomical shunts in the lungs, and a larger part is due to bronchial and Thebesian drainage to the left side of the heart.

About half of the total difference is due to unevenness in the ventilation/perfusion ratio in the different parts of the lungs (venous admixture, p. 615).

In exercise the oxygen requirements are greatly increased and large quantities of oxygen must pass across the alveolar membrane. The pulmonary capillary bed enlarges considerably but, although the overall rate of blood flow through the pulmonary circuit is increased many times, the time spent by each red corpuscle in passing through a pulmonary capillary is reduced by only a half (p. 581). The oxygen saturation of arterial blood in exercise is as high as it is in the resting state unless the work is severe. The increased demand of the muscles is met partly by the opening up of capillaries closed in the resting state, by the greatly augmented blood flow through the muscle as the result of arteriolar dilatation. The increased uptake of oxygen by the muscles reduces the oxygen content of the blood returning to the lungs. For example in mild exercise requiring 800 ml oxygen per min the arteriovenous oxygen difference may be 10 ml per 100 ml of blood instead of the resting value of about 4 ml. The hyperventilation is more or less balanced by the increased oxygen consumption, so P_{A,O_2} and P_{a,O_2} remain about constant.

The characteristics of the alveolar membrane of whole lung can be stated in terms of its transfer factor (T_L) which is the number of ml of a gas which can traverse it per minute per mm Hg gradient of pressure across the membrane. To measure T_L it is necessary to use a gas which is much more soluble in blood than in tissue fluids and plasma, otherwise the rate of uptake of the gas in the lungs would depend more on the amount of blood flowing past the alveoli than on the diffusion characteristics of the membrane. Two such gases exist, namely oxygen and carbon monoxide, and their increased solubility in blood is due to their chemical combination with haemoglobin. Of these two gases carbon monoxide uptake is easier to measure since the capacity of haemoglobin for it is such that even if the pulmonary blood flow were stopped during the period of measurement (less than one minute) at the concentrations used, the haemoglobin would still remain unsaturated at the end of the period. Furthermore in calculating the pressure difference between the gas in the alveolus and the plasma, it can be assumed (in non-smokers) that the plasma value is zero since this gas is not normally present in blood. (In smokers it is often necessary to measure the 'back-tension' of CO.) When oxygen is the measuring gas this of course is not the case and the main difficulty rests in calculating the mean alveolar to capillary gradient. Both gases measure the same function and the resting adult value of T_{CO} is 25 ml per min per mm Hg. According to the laws of diffusion and solubility T_{O_2} must be $1\cdot23 \times 25 = 31$ ml/min/mm Hg. Individual values vary with the size of the subject. The T_{CO_2} on the other hand is of the order of 500 ml/min/mm Hg.

The speed with which these gases can pass from alveolar air to the haemoglobin of the red cells is obviously affected by the permeability of the structures crossed, the area over which the haemoglobin is spread and the rate of uptake by haemoglobin. It is now possible to measure separately two of the factors limiting

Table 31.3 Table giving the approximate differences between arterial and mixed venous blood in a healthy subject at rest

	Arterial blood	Mixed venous blood
Oxygen		
content (ml per 100 ml whole blood)	19	14
partial pressure (mm Hg) (P_{O_2})	95	40
saturation of haemoglobin (percentage)	95	70
ml dissolved (in plasma of 100 ml blood)	0·3	0·12
Carbon dioxide partial pressure (mm Hg) (P_{CO_2})	40	46
Carbon dioxide as ml		
total in 100 ml blood	48	52
content in plasma of 100 ml blood (60 ml plasma)	36	38
as bicarbonate in plasma of 100 ml blood	34	36
dissolved in plasma of 100 ml blood	1·6	1·8
total in ml carried by red cells in 100 ml blood	12·5	13·8
carbamino content of red cells of 100 ml blood	2·2	3·2
as bicarbonate in red cells of 100 ml blood	9·1	9·9
dissolved in red cells of 100 ml blood	0·8	0·9
pH	7·4	7·38
Chloride ions (mmoles per litre of plasma)	59·6	58·7

For some purposes it is convenient to use moles. The conversion is easily made since 1 mmole $CO_2 = 22\cdot3$ ml CO_2 or 1 ml $CO_2 = 0\cdot045$ mmole CO_2.

the diffusion of gases: that due to the alveolo-capillary membrane itself, and that due to the red blood cell and the haemoglobin. It can be calculated that the membrane T_{O_2} is of the order of 70 ml/min/mm Hg.

The transfer factor is increased up to three times in healthy persons in exercise probably because many capillaries open up as a result of an increase in cardiac output and this increases the available surface area for diffusion. The resting value is reduced if the lung parenchyma is reduced in diseases which particularly affect the interstitial tissue of the lungs (alveolar-capillary block). Where there is a reduction in the normal pulmonary capillary reserve, as in emphysema, low values for T_L in exercise have been reported.

ESTIMATION OF OXYGEN AND CARBON DIOXIDE IN BLOOD

The oxygen and carbon dioxide content of blood can be estimated by Haldane's blood gas apparatus (Fig. 31.4). The oxalated blood sample must be introduced into the apparatus without exposure to the atmospheric air to avoid changes due to loss of carbon dioxide or gain of oxygen; the blood is therefore collected in a syringe without contact with air. After the red cells have been haemolysed by saponin, potassium ferricyanide is added and all the oxygen is liberated (p. 448). This volume of oxygen is its *oxygen content*; to find the amount of oxygen carried by the haemoglobin it is necessary to subtract the oxygen dissolved in the plasma (0·003 ml per 100 ml per mm Hg). If the estimation is repeated on another sample of the same blood which has first been shaken with air to saturate it with oxygen, the figure obtained is higher than the oxygen content of the first sample and represents the *oxygen capacity*. The *percentage saturation* of the first sample is given by (Oxygen content/Oxygen capacity) × 100. The carbon dioxide content of a blood sample can be measured in the same apparatus by adding tartaric acid which drives off all the carbon dioxide.

The Van Slyke apparatus has been used for many years for blood gas analysis. The blood is put into an acidified air-free ferricyanide solution and subjected to a vacuum. The ferricyanide drives off the oxygen, the acid drives off the carbon dioxide and the vacuum the nitrogen. The gas evolved is analysed by absorbing first the carbon dioxide and then the

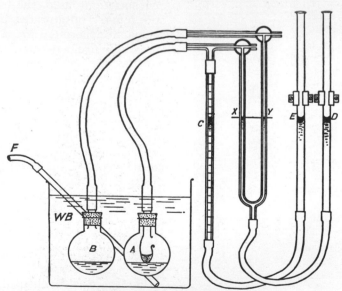

Fig. 31.4 Haldane's blood gas apparatus. The reaction vessel A contains blood in a borate buffer. The inner vessel of A contains ferricyanide. The reaction vessel B, containing water, compensates for changes in temperature and pressure. If vessel A is tilted the ferricyanide mixes with the blood, oxygen is evolved and depresses the level in the burette C. X, Y are gauge tubes, E, D levelling tubes. The water in the water-bath, WB, is stirred by air blown in through F (C. G. Douglas & J. G. Priestly (1949). *Human Physiology*, 3rd edn, p. 150. Oxford: Clarendon Press).

oxygen. Newer, more rapid, methods are illustrated in Figures 31.5 and 6.

Since the light transmissions, as shown in Plate 25.1 of haemoglobin and oxyhaemoglobin are different, the oxygen saturation of blood can be approximately determined by shining a light through the lobe of the ear on to two photocells one of which is covered with a red filter and the other with a blue filter (the *oximeter*). The ratio of the photo-cell currents is related to the saturation of the blood with oxygen. This method which is very rapid and convenient has been successfully used in the investigation of hypoxia during flying at high altitudes and has found increasing application in clinical medicine.

The partial pressures of oxygen and carbon monoxide in the tissues may be measured directly by suitably modified electrodes similar to those shown in Figures 31.5 and 6.

TRANSPORT OF OXYGEN

The amount of oxygen which a given amount of haemoglobin takes up, that is its percentage saturation, is a function of the partial pressure of the oxygen P_{O_2} in the atmo-

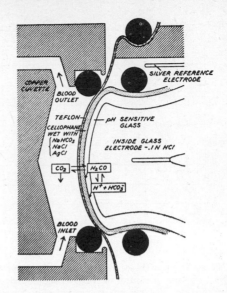

Fig. 31.5 Schematic design of a CO_2 electrode. The pH sensitive glass membrane is covered with cellophane soaked in bicarbonate solution saturated with silver chloride. This in turn is covered with teflon which is permeable to CO_2 gas but not to hydrogen ion. The blood sample passes through the apparatus as shown and CO_2 passes through the teflon from the blood into the bicarbonate solution and equilibrium is reached in a few minutes. The voltage measured between the two electrodes is linearly related to P_{CO_2}.

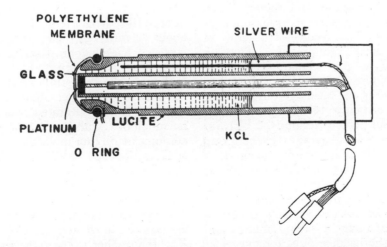

Fig. 31.6 The polarographic oxygen electrode. Blood is made to flow past a platinum electrode (charged to -0.5 V) covered with polyethylene membrane which is permeable to oxygen molecules. At the platinum surface the dissolved molecules receive electrons, becoming either OH^- or H_2O_2. The current passing from the platinum electrode to the silver-silver chloride electrode in saturated KCl is linearly related to the dissolved oxygen molecules at its surface. If there is no oxygen there is practically no current. (J. W. Severinghaus (1959). *Symposium on pH and Blood Gas Measurement* (Edited by R. F. Woolmer), Chap. 10. London: Churchill.)

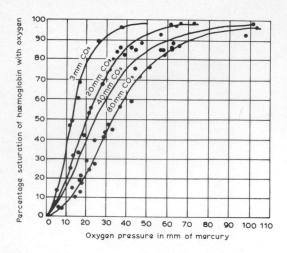

Fig. 31.7 Oxygen dissociation curves of human blood exposed to partial pressures of 3, 20, 40 and 80 mm Hg of carbon dioxide. The percentage saturation was measured by Van Slyke's apparatus (A. V. Bock, H. F. Field & G. S. Adair (1924). *Journal of Biological Chemistry* **59**, 366).

sphere in contact with it. The graph (Fig. 31.7) showing the relation between the two, the *dissociation curve*, is obtained in the following way. A small quantity of blood, say 5 ml, is placed in a 250 ml bottle (tonometer) filled with gas of known oxygen concentration. The bottle is turned on its side and rotated for 20 minutes in a water-bath at 37·5°C, so that a large area of blood is exposed to the gas. Then the percentage saturation is measured as described above. The results of a series of determinations at different pressures of oxygen can be plotted to give a sigmoid curve as illustrated in Figure 31.7.

The structure of the haemoglobin molecule has been described on p. 446. Each molecule is composed of four protein subunits, 2 α-chains and 2 β-chains, each having a haem group associated with it. Each haem group is a binding site for oxygen so the haemoglobin molecule can combine with up to four molecules of oxygen. When some of the binding sites combine with oxygen a change occurs in the arrangement of the protein subunits and the affinity of the haem groups for oxygen is increased. Because the sites of oxygen binding are at some distance from one another, this co-operation of the four sites in binding oxygen is an example of an allosteric effect (p. 113).

By having four binding sites which interact,

haemoglobin has a sigmoid dissociation curve (Chap. 25). The oxygen dissociation of myoglobin (Fig. 31.8) which has only one haem group is a simple hyperbola like the curve for enzyme action (Fig. 8.34). This sigmoid relationship makes oxygen transport by the blood a much more efficient process than it would be if blood contained a simple oxygen carrier like myoglobin. To dissociate half of the bound oxygen from arterial blood, the oxygen partial pressure need only be reduced to about 24 mm Hg at 38°C and pH 7·4. Under these conditions myoglobin is more than 80 per cent saturated (Fig. 31.8) and to dissociate half of its bound

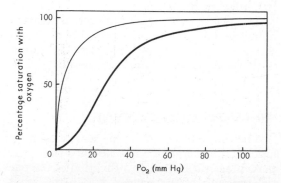

Fig. 31.8 Oxygen dissociation curves of myoglobin (thin line) and haemoglobin (thick line) at $P_{CO_2} = 40$ mm Hg.

oxygen the partial pressure must be reduced to about 6 mm Hg. The oxygen dissociation curve of haemoglobin thus allows the tissues to function at a relatively high oxygen tension without at the same time limiting too severely the amount of oxygen which is delivered.

When fully saturated one molecule of haemoglobin with molecular weight 66 800 combines with four molecules of oxygen which at s.t.p. occupy $4 \times 22·4$ l. Each gram of haemoglobin can therefore carry 1·34 ml oxygen.

Bohr found in 1910 that the position of the dissociation curve depended on the P_{CO_2} in the gas mixture in the tonometer—the greater the P_{CO_2} the further to the right was the dissociation curve (Fig. 31.7). Barcroft shortly afterwards found that this displacement could be produced by any acid; a fall in pH moves the curve to the right. A rise in temperature also shifts the dissociation curve to the right (Fig. 31.9); blood parts with its oxygen more readily

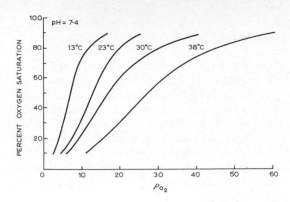

Fig. 31.9 The effect of temperature on the oxygen dissociation curve of oxyhaemoglobin of human blood at pH 7·4. (From data in P. Astrup, K. Engel, J. N. Severinghaus & E. Munson (1965). *Scandinavian Journal of Clinical and Laboratory Investigation* **17**, 520.)

at higher temperatures. Conversely a fall of temperature shifts the dissociation curve to the left.

The modification of the dissociation curve produced by the carbon dioxide tension of the blood (the Bohr effect) has biological advantages. Since the P_{CO_2} is lower in the lungs than in the tissues, haemoglobin has a higher affinity for oxygen where oxygen is being taken up than it has where oxygen is being set free. The reduction of P_{CO_2} of the blood as it passes through the pulmonary capillaries enables the haemoglobin to take up only a small additional amount of oxygen. (The dissociation curves (Fig. 31.7) are very close to one another at P_{O_2} of 95 mm Hg.) However, in the tissues where the P_{O_2} is between 10 and 40 mm Hg the effect of P_{CO_2} on the oxygen carrying capacity of the blood is quite large as shown by the wider separation of the oxygen dissociation curves (Fig. 31.7). An active tissue has a relatively high P_{CO_2}, low pH and raised temperature and all of these changes set free more oxygen for its respiratory processes.

Another factor with an important influence on the oxygen dissociation curve is the concentration of 2,3-diphosphoglycerate (DPG) in the red cell. The reduced haemoglobin molecule can bind one molecule of DPG very tightly in a site between its two β-chains. Oxyhaemoglobin with its altered arrangement of subunits binds DPG much less tightly if at all. Hence in the presence of DPG the reduced form of haemoglobin is stabilized and the affinity of haemoglobin for oxygen is greatly diminished, in other words the oxygen dissociation curve is shifted to the right.

The DPG content of the red cell is not constant. If blood is stored *in vitro* at 37° for 24 hours most of the red cell DPG is lost. DPG is synthesized in the red cell from a metabolite of the normal glycolytic pathway, 1,3-diphosphoglycerate, by the enzyme diphosphoglyceromutase. The synthesis is somehow stimulated by hypoxia such as is found in anaemia or at high altitudes. The increased DPG levels achieved in such conditions are important in facilitating the unloading of oxygen in the tissue capillaries. Factors which influence DPG levels in the red cell are being actively studied at this time.

In man the oxygen tension in the systemic arteries (P_{a,O_2}) is about 95 mm Hg and the carbon dioxide tension 40 mm Hg. Figure 31.7 shows that haemoglobin is 95 per cent saturated in these circumstances. In mixed venous blood the oxygen and carbon dioxide tensions are about 40 and 46 mm Hg respectively. The haemoglobin in mixed venous blood is, therefore, about 70 per cent saturated.

By means of an oxygen electrode similar to that shown in Figure 31.6 through which blood from the brachial artery is passed continuous records of arterial P_{O_2} can be obtained (Fig. 31.10).

The speed of the reaction $Hb + O_2 \rightarrow HbO_2$ has been measured by driving a dilute haemoglobin solution through one jet into a mixing chamber and water containing oxygen in solution through another jet into the same mixing chamber from which the mixed fluids travel steadily down a glass observation tube. While the mixed fluids pass down the tube the haemoglobin becomes oxyhaemoglobin. The point at which 50 per cent saturation occurs can be detected spectroscopically (Plate 25.1). The rate of flow can be calculated from the diameter of the tube and the volume flowing out of it. The distance from the mixing chamber to the point of half-saturation divided by the rate of flow gives the time to reach half-saturation. This is about 0·003 seconds but if a suspension of red cells is used the half-time is about 0·07 sec because the oxygen has to penetrate the membrane and diffuse through the substance of the cell. Nevertheless this time is so much less than the half-second or so

Respiratory Functions of the Blood   621

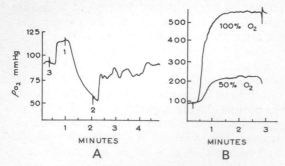

Fig. 31.10 A and B A. Polarographic electrode records. Effect of breath holding and hyperventilation on arterial oxygen tension while subject breathes room air. (1) Voluntary breath holding for 72 seconds during which the arterial P_{O_2} dropped to 51 mm Hg. (2) Recovery followed by normal respiration. (3) Preparatory hyperventilation lasting 36 seconds during which the P_{O_2} increased 21 mm Hg. The time lag after (1), (2) and (3) to the change of P_{O_2} is due to the dead space of the electrode and the circulation time. B. Continuous measurement of arterial oxygen tension of two individuals during administration of 50 and 100 per cent oxygen. The downward deflexion in the beginning indicates the time of application of the face mask. The interval between this deflexion and the upswing of the arterial oxygen tension (10 seconds) represents the dead space of the electrode plus the circulation time of the left brachial artery. (S. T. Koeff, M. U. Tsao, A. Vadney, T. O. Wilson & J. L. Wilson (1962). *Journal of Clinical Investigation* **41**, 1125.)

spent by a corpuscle in the alveolar capillary that the speed of this reaction cannot normally be a factor limiting the uptake of oxygen. In severe anaemia, however, when the oxygen-carrying capacity of the blood is reduced and when the rate of blood flow through the pulmonary capillary increases, the P_{O_2} of the plasma in the capillaries increases, and the P_{O_2} gradient across the alveolo-capillary membrane is diminished with a consequent reduction in oxygen diffusion. In health the rate of diffusion across the alveolar membrane is not a limiting factor but when the characteristics of this membrane are altered by disease the prolonged diffusion rate results in arterial unsaturation (p. 613). The time for the reverse reaction, $HbO_2 \rightarrow Hb + O_2$, to reach half completion is 0·004 sec in solution and 0·038 sec in the red cell, so that release of oxygen to the tissues from the blood can be very rapid.

THE CARRIAGE OF CARBON DIOXIDE

As indicated already, the carbon dioxide of the blood cannot be in simple physical solu-

tion. At 38°C and 40 mm Hg 100 ml of water can dissolve about 2·6 ml of carbon dioxide, whereas blood may contain 50 or 60 ml carbon dioxide per 100 ml. Since all the carbon dioxide can be obtained from blood by addition of an acid the carbon dioxide must be carried partly as dissolved gas but mainly as bicarbonate. Further, since the total anions (chloride, phosphate, and others) are insufficient to combine with the sodium ions found in plasma, the combined carbon dioxide must be present as sodium bicarbonate. Both corpuscles and plasma transport carbon dioxide; the former carries about one-quarter of the total and the latter about three-quarters (Table 31.3).

By exposing blood to various partial pressures of CO_2 we can obtain data to draw a dissociation curve showing the relationship between the pressure of carbon dioxide and the total amount of carbon dioxide contained in the blood sample. Figure 31.11 shows dissociation curves for both oxygenated and reduced blood. When an aqueous solution of sodium bicarbonate is exposed to a vacuum (zero P_{CO_2}) half the total carbon dioxide content is retained, the other half being removed according to the equation:

$$2NaHCO_3 \rightarrow Na_2CO_3 + CO_2 + H_2O.$$

If, however, the same solution is treated with an acid all the carbon dioxide is liberated thus:

$$2NaHCO_3 + 2HCl \rightarrow 2NaCl + 2CO_2 + 2H_2O.$$

The dissociation curves for blood (Fig. 31.11) show that at zero P_{CO_2} no carbon dioxide is left. Obviously there must be some factor which aids the liberation of carbon dioxide. This can be shown by exposing blood to a vacuum and then centrifuging it to separate plasma and red cells. If red cells are added to this *separated plasma* a further amount of carbon dioxide is liberated as if the corpuscles contained an acid.

The metabolism of the tissues produces carbon dioxide and when the P_{CO_2} in the tissues is greater than the P_{CO_2} in the arterial end of the capillaries carbon dioxide diffuses into the capillaries. The greater part of the carbon dioxide goes directly into the red cell as shown in Fig. 31.14 but a small additional quantity, due to the increase in P_{CO_2}, is dissolved in the plasma where it is *slowly* hydrated to form bicarbonate and hydrogen ions.

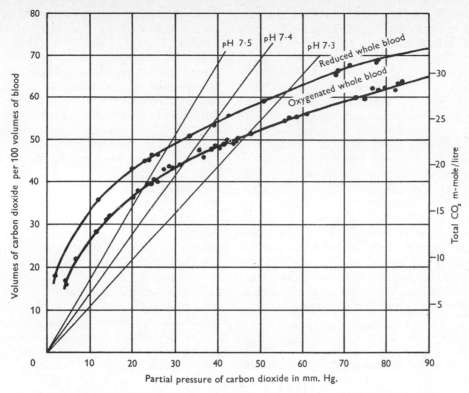

Fig. 31.11 Carbon dioxide dissociation curves for reduced and oxygenated blood. (*After* A. V. Bock, H. F. Field & G. S. Adair (1924). *Journal of Biological Chemistry* **59**, 371.) The pH values are calculated from the Henderson-Hasselbach equation (p. 17). Note that oxygenated blood can hold less CO_2 than reduced blood. This is called the Haldane effect.

$$CO_2 + H_2O \rightleftharpoons HCO_3^- + H^+$$

Some carbon dioxide is taken up by the plasma proteins to form carbamino compounds.

$$CO_2 + PrNH_2 \rightleftharpoons PrNHCOOH \rightleftharpoons$$
$$PrNHCOO^- + H^+$$

The H^+ ions produced in these reactions are buffered by the proteins and phosphate of the plasma. At pH 7·4, which is on the alkaline side of their iso-electric point the proteins are weak acids.

$$H^+ + Na\ proteinate \rightarrow Na^+ + H\ proteinate$$

When carbon dioxide enters the red cell bicarbonate and hydrogen ions are formed *very rapidly* by the enzyme *carbonic anhydrase* discovered by Roughton. The speed of the reaction $CO_2 + H_2O \rightleftharpoons H^+ + HCO_3^-$ is about 13 000 times greater in the red cells than in the plasma. In this way, and because of the high solubility of CO_2 in aqueous media, we can account for the rapid removal of carbon dioxide from the blood in the short transit time in the pulmonary capillaries and also for the rapid uptake when the blood is traversing the capillaries in the tissues. The H^+ ions formed in the red cells by the reaction

$$CO_2 + H_2O \rightleftharpoons H^+ + HCO_3^-$$

are buffered by the protein part of the haemoglobin molecule

$$H^+ + KHb \rightarrow K^+ + HHb.$$

Because haemoglobin is present in higher concentration in the red cells and because it is a more powerful buffer than the proteins of the plasma and because bicarbonate is formed more quickly in the red cells, the concentration of HCO_3^- ions is greater in the red cells than in the plasma; accordingly HCO_3^- ions pass into the plasma and, since cations such as

Na$^+$ and K$^+$ cannot pass easily through the red-cell membrane, anions (mainly Cl$^-$) pass into the cell to maintain electrical neutrality. At the same time a small amount of water moves into the red cells to maintain osmotic equilibrium. The plasma of venous blood contains about 1·5 per cent less chloride than the plasma of arterial blood (Table 31.3). This *chloride shift* was discovered by Hamburger in 1918. The result of this shift of HCO$_3^-$ ions from the red cells back into the plasma is that the plasma of the capillary blood carries the greater part, about two-thirds, of the additional carbon dioxide and the red cells only one-third although the major part of the buffering occurs in the red cells. When the venous blood is oxygenated in the lungs the chloride shift is reversed.

Some of the carbon dioxide entering the red cells from the tissues is taken up by the haemoglobin to form carbaminohaemoglobin.

$$CO_2 + HbNH_2 \rightleftharpoons HbNHCOOH \rightleftharpoons HbNHCOO^- + H^+$$

This rapid reaction does not require a catalyst.

Although the carbaminohaemoglobin accounts for only 5 to 10 per cent of the total carbon dioxide of blood (Table 31.3 and Fig. 31.12) it is mobile and easily exchanged and hence of considerable physiological importance. About one-quarter of the change in carbon dioxide content of the whole blood (the arterio-venous difference) in ordinary respiration can be accounted for by the change in carbon dioxide bound as carbaminohaemoglobin.

The H$^+$ ions produced by the hydration of CO$_2$ and by the ionization of carbaminohaemoglobin are buffered by haemoglobin and by the phosphate esters which the red cells contain. Haemoglobin is an effective buffer in the pH range 7·0 to 7·6 because it contains many histidine residues whose imidazole groups buffer in this range. However, the greater part of the buffering needed during respiratory exchange is carried out by a few of the ionizing groups on haemoglobin which are influenced by the oxygenation and deoxygenation of the haem groups. These oxygen-linked groups have now been tentatively identified; the α-amino groups of valines 1α and the

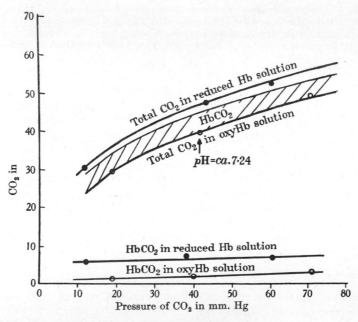

Fig. 31.12 The upper two curves are carbon dioxide dissociation curves of reduced and oxygenated human haemoglobin solution respectively. In the change from the oxygenated to the reduced state the carbon dioxide represented by the shaded area is taken up in the carbamino form—this is quite a considerable part of the change in the carbon dioxide content. The absolute amounts of carbamino compounds are shown in the lower two curves (J. K. W. Ferguson (1937). *Journal of Physiology* **88**, 49).

imidazole side chains of histidines at 122α and 146β all contribute to the effect (for positions see Fig. 25.2). On oxygenation of the haemoglobin these groups become less basic and release hydrogen ions. When the oxygen dissociates they become more basic again and hydrogen ions are taken up. To keep the pH constant when 1 gram molecule of oxygen dissociates, 0·7 gram molecule of hydrogen ions must be added. These hydrogen ions could be produced by the hydration of 0·7 gram molecule of CO_2. Thus when the R.Q. is 0·7 the red cell can bring about respiratory exchange without changing its internal pH. At higher R.Q.s the remainder of the buffering is carried out by the other buffers in the red cell. When haemoglobin takes up oxygen in the lungs, H^+ ions are given to the solution since oxyhaemoglobin is a stronger acid, the reaction catalysed by carbonic anhydrase, $H^+ + HCO_3^- \rightarrow CO_2 + H_2O$, occurs and the CO_2 so formed is got rid of through the lungs.

The amount of carbon dioxide which can be carried by Hb is affected by the oxygen tension (Haldane effect). Just as reduced whole blood (Fig. 31.11) can contain more carbon dioxide than oxygenated blood so reduced haemoglobin (Fig. 31.12) can carry more carbamino-bound carbon dioxide than oxyhaemoglobin. If in dogs the carbonic anhydrase is inhibited by giving them acetazolamide intravenously carbon dioxide transport and excretion are severely disturbed. Since the blood is moving slowly through the tissue capillaries there may be time for the hydration of carbon dioxide but since the blood (p. 581) passes through the lung capillaries in only 0·75 sec the reaction $H^+ + HCO_3^- \rightarrow H_2O + CO_2$ is too slow to allow transfer of the normal amount of carbon dioxide to the alveolar air. Therefore carbon dioxide is dammed back and the P_{CO_2} of the blood and the tissues rises until a new steady state is reached.

The factors involved in transport of carbon dioxide have been summarized by Roughton in a diagram (Figs. 31.13 and 14) which illustrates both its uptake in the tissues and its release in the lungs.

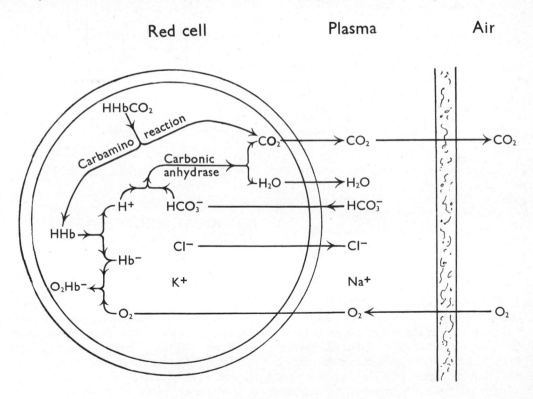

Fig. 31.13 A schematic summary of the chemical processes which occur when haemoglobin takes up oxygen in the lungs. See also Fig. 31.14.

Respiratory Functions of the Blood 625

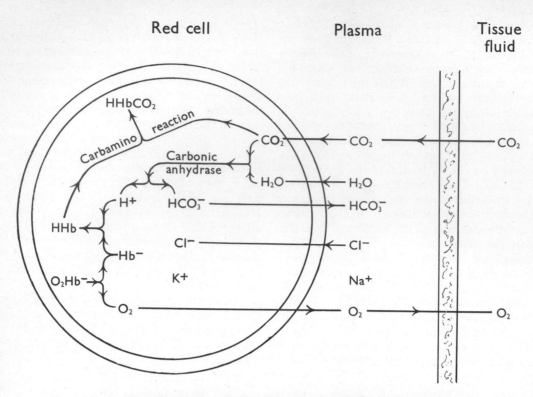

Fig. 31.14 A schematic summary of the chemical processes which occur when haemoglobin parts with its oxygen in the tissues. The chloride shift is represented as a movement of ions. The shaded part represents the capillary wall. Most of the CO_2 goes into the red blood cell as shown; a small quantity is dissolved in the plasma and some links up with H_2O slowly since there is no carbonic anhydrase in the plasma. Some is taken up by the plasma proteins to form carbamino compounds. These two reactions produce H^+ ions (see pp. 623 and 624) which are taken up by the plasma buffers. Inside the red cell HCO_3^- is formed very quickly in the presence of carbonic anhydrase. The H^+ ions resulting from the hydration of CO_2 and the formation of carbaminohaemoglobin are buffered by haemoglobin.

The importance of haemoglobin in the carriage of carbon dioxide can be illustrated by considering the effect of breathing oxygen under pressure. At three atmospheres $(3 \times P_B)$ of pure oxygen the P_{O_2} of the alveolar air is

$$(3 \times P_B) - P_{CO_2} - P_{H_2O} \text{ or } (3 \times 760) - 40 - 47.$$

The amount of oxygen carried in 100 ml plasma is therefore $2193 \times 0.003 = 6.6$ ml. Since this is sufficient (p. 168) to supply the metabolic needs of the body at rest if the cardiac output is in the usual range of about 5 l/min there is little or no need for oxyhaemoglobin to part with its oxygen. Since the carbon dioxide carrying capacity of HbO_2 is less than that of Hb the tissue P_{CO_2} rises and stimulates the respiratory centre. This paradoxical stimulation of ventilation, with its consequent

lowering of alveolar and arterial P_{CO_2}, is also seen when pure oxygen is breathed at normal atmospheric pressure.

CHEMICAL REGULATION OF RESPIRATION

The chemical regulation of respiration is maintained by central (p. 628) and peripheral chemoreceptors (p. 630). Its importance is illustrated by *asphyxia*, in which there is a simultaneous lack of oxygen and an excess of carbon dioxide in the alveolar gas mixture. Asphyxia has long been studied in experimental animals and in human subjects who are allowed to rebreathe their own expired air from a spirometer without the removal of the accumulating carbon dioxide by soda lime.

The respiratory events occur in three stages.

(1) The respiratory movements are first increased in depth and rate. The breathing becomes very violent and the expiratory efforts more and more powerful. Consciousness is lost at the end of this stage which is followed by (2) the stage of convulsions involving practically all the muscles of the body, with stimulation of many central neurones, resulting in vasoconstriction, salivation, contraction of pupils and contraction of muscles of alimentary canal and bladder. This is followed quite suddenly by (3) a stage of slow deep inspirations in which most bodily activities are in abeyance and reflexes are absent. Finally the respirations become infrequent and ineffective and the animal dies in a few minutes of acute oxygen lack.

EFFECT OF CARBON DIOXIDE

The effects of asphyxia on the circulation are produced partly by excess of carbon dioxide and partly by lack of oxygen in the blood and in the tissues, and the same is true of the respiratory phenomena of asphyxia. The effect of carbon dioxide was first demonstrated in the experiments of Haldane and Smith in which the subject re-breathed air containing rising concentrations of carbon dioxide up to 6 per cent. The depth and rate of respiration both increase. The subject may not notice the effect of low concentrations of carbon dioxide in the inspired air, but at concentrations above 3 per cent the subject is well aware of the increased respiratory effort. If air containing between 3 and 5 per cent carbon dioxide is breathed respiration becomes very deep and very rapid; at concentrations over 6 per cent cardiac and respiratory functions are depressed, confusion and headache occur .

No satisfactory explanation has yet been offered for the sensation of breathlessness and distress, usually called dyspnoea, which occurs if the lungs are stiff or if the airways are obstructed. Impulses going up from the lungs in the vagus nerves or from the joint receptors in the thoracic cage to the central nervous system have been implicated, as has stimulation of chemoreceptors by altered blood-gas tensions. It has been suggested that this dyspnoea arises when the actual expansion of the chest achieved is not appropriate to the ventilation which the central nervous system demands.

The total ventilation of the lungs per minute ('minute volume', $\dot{V}_A$) and the alveolar $P_{CO_2}(P_{A,CO_2})$ are interdependent. When air containing added CO_2 is inspired the P_{A,CO_2} rises, and so does the P_{CO_2} in the pulmonary venous blood and therefore in the arterial blood. This rise stimulates the central chemoreceptors and the ventilation is increased. There is a remarkably linear relation between $\dot{V}_A$ and the P_{A,CO_2} under these conditions. However, at a given inspired $P_{CO_2}(P_{I,CO_2})$, the greater the ventilation the lower the P_{A,CO_2}, because the extra volume of air breathed effectively dilutes the CO_2, produced by metabolism which is diffusing into the alveoli at a constant rate.

Algebraically this can be expressed as

$$\dot{V}_A.(P_{A,CO_2} - P_{I,CO_2}) = \dot{V}_{CO_2}.P_B$$

or

$$P_{A,CO_2} = \frac{\dot{V}_{CO_2}.P_B}{\dot{V}_A} + P_{I,CO_2}$$

(where P_B = barometric pressure and $\dot{V}_{CO_2}$ = volume of metabolic CO_2 per minute). Graphically, this relation between $\dot{V}_A$ and P_{A,CO_2} is a hyperbola that is shifted to the right by progressive increments of inspired CO_2 (Fig. 31.15A). In Figure 31.15B the straight CO_2-response line has been superimposed on these hyperbolae, and the air-breathing point is marked X. The points of intersection are the $\dot{V}_A$, P_{A,CO_2} co-ordinates in the steady-state at each different P_{I,CO_2}, and the double-ended arrows (marked 40, 21 and 6) indicate that although the P_{A,CO_2} rises with the P_{I,CO_2} when CO_2 is added to the inspired air, the two come closer together. The overall effect of this increased ventilation is to minimize the change in the alveolar and the arterial P_{CO_2}. The relation between ventilation and inspired P_{CO_2} is distinctly non-linear: in this example, the first increment of 20 mm Hg produces only a slight change in ventilation while a further identical increment in inspired P_{CO_2} gives rise to a substantial increase.

The carbon dioxide tension in the alveoli (P_{A,CO_2}) is calculated as follows. In alveolar air at sea level the carbon dioxide content is about 5·5 per cent and the pressure of water-vapour 47 mm Hg. Since the atmospheric pressure is 760 mm Hg the P_{CO_2} is equal to 5·5 per cent of (760 minus 47) mm Hg, that is 39 mm Hg. At

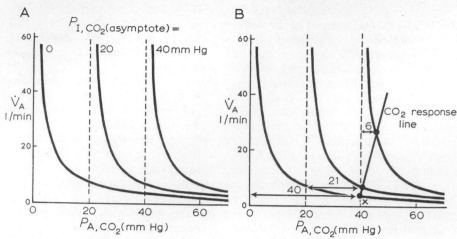

Fig. 31.15 A shows the hyperbolic relations between alveolar ventilation ($\dot{V}_A$) and alveolar CO_2 (P_{A,CO_2}) at three different partial pressures (0, 20 and 40 mm Hg) of inspired CO_2. B also shows the straight CO_2 response line; the horizontal bars represent the difference between inspired and alveolar P_{CO_2}. X shows the position when the subject is breathing atmospheric air. The dotted vertical lines are the asymptotes representing P_{I,CO_2}.

high altitudes or in a pressure chamber where the barometric pressure is much less or much greater than normal, the carbon dioxide content of alveolar air is correspondingly different but the alveolar P_{CO_2} remains the same at a given alveolar ventilation. For most purposes the really important measurement is the arterial P_{CO_2} and good electro-chemical methods are available for direct measurement of blood P_{CO_2} (p. 619). Where these methods are not available or desirable, however, indirect methods may be used. Carbon dioxide diffuses so rapidly through the lung tissues that the arterial and alveolar partial pressures are equal, and therefore measurement of P_{A,CO_2} gives a theoretically good indirect estimate of P_{a,CO_2}. However, the difficulties in obtaining the true alveolar air have already been mentioned (p. 602). An alternative indirect approach to the arterial P_{CO_2} has been via the mixed venous $P_{CO_2}(P_{\bar{V},CO_2})$. Direct studies with cardiac catheters and arterial needles have shown that $P_{\bar{V},CO_2}$ is about 6 mm Hg greater than P_{a,CO_2}. Many bloodless methods of measuring $P_{\bar{V},CO_2}$ indirectly have been described but none is able to give a continuous estimate. One way is to use the lungs as an equilibrating device. The subject is asked to rebreathe from a bag containing 1 or $1\frac{1}{2}$ l oxygen for a minute and a half or until an increase in depth or rate of respira-

tion is noticed. The bag then contains a mixture of gases with a P_{CO_2} slightly higher than that in mixed venous blood. After a rest (at least 2 minutes) the subject rebreathes the mixture for 20 seconds only to make a fine adjustment between the bag P_{CO_2} and mixed venous P_{CO_2} before recirculation of the blood can take place. For greater accuracy this 20-second rebreathing may be repeated. The P_{CO_2} of the gas in the bag and of the mixed venous blood are now very nearly the same.

STIMULATION OF CENTRAL CHEMORECEPTORS

When excess carbon dioxide is breathed the P_{CO_2} of all the tissues rises and may alter, for example, the state of the blood vessels in them. However, only at two sites, the carotid bodies and the medulla, is there an influence on breathing. The carotid body chemoreceptors, if the rise of P_{CO_2} is high enough, cause reflexly an increase in breathing, but the changes represented by the straight line in Fig. 31.15B can almost certainly be accounted for by the chemoreceptors of the medulla. The peripheral chemoreceptors in man may normally have only an emergency function or a supporting role to the more important medullary centre. With long residence at high altitude the medullary carbon dioxide chemoreceptors become more sensitive because of a fall in the bicarbonate concentration of the cerebro-spinal fluid (c.s.f.) while the carotid and aortic body chemoreceptors retain their sensitivity to hypoxia except in chronic mountain sickness. Since a rise of P_{CO_2} in the blood and an increase of acids in the c.s.f. both stimulate the medullary chemoreceptors, while both cause a fall of pH, it is pertinent to ask whether the effect of carbon dioxide is a specific property or whether it is due simply to the alteration of pH. Pappenheimer has shown by perfusing the cerebral ventricles of goats with mock-c.s.f. at different P_{CO_2} and bicarbonate concentrations that the ventilation is best related to the hydrogen ion concentration close to the surface of the medulla. In intact man, however, the commonest cause of alterations in the pH c.s.f. is a change in the arterial P_{CO_2}. The apparent specificity of carbon dioxide as a respiratory stimulant is due to the fact that carbon dioxide passes through cell membranes very rapidly whereas H^+ ions penetrate very slowly. This slow penetration could account for the slow effect of lactic and other acids. But since H^+ ions are formed when the carbon dioxide reaches the interior of a cell the final stimulus may in fact be an increase in $[H^+]$.

There is good evidence that the chemo-receptor cells sensitive to carbon dioxide are quite distinct from the cells of the respiratory centre responsible for the reflex adjustment of breathing. For example, chloralose and other anaesthetics reduce the stimulating effect of carbon dioxide without reducing the stimulating effect of afferent impulses. Perfusion of the fourth ventricle with cerebrospinal fluid of high P_{CO_2} stimulates breathing and local application of such solutions or of acetylcholine or nicotine to the anterior lateral medullary area (see Fig. 30.16) increases respiratory movements within a few seconds. Similar experiments show that such chemoreceptors are situated at various positions in the brain stem, that is at varying distances from the cells of the respiratory centre itself. When air rich in carbon dioxide is breathed the P_{CO_2} and the $[H^+]$ of the cerebrospinal fluid rise a little more slowly than the P_{CO_2} of the arterial blood; but since it has little protein to act as a buffer, the cerebrospinal fluid shows eventually a much greater rise in $[H^+]$ than does the well-buffered blood. The slow rise of $[H^+]$ in the cerebrospinal fluid fits in very well with the slow rise in depth and rate of respiration on breathing air rich in carbon dioxide; the slow return to the normal level of respiration when the subject breathes room air once more is explained by the relatively slow fall in P_{CO_2} and in $[H^+]$ in the cerebrospinal fluid.

Not all variations in ventilation are due to variations in the CO_2 stimulus. For example breathing is deeper in exercise although the P_{CO_2} of the alveolar air is at its usual level (Chap. 40). The excessive breathing of fever may be accompanied by a low arterial P_{CO_2}. If the respiratory centre is severely depressed by morphine or barbiturate, adding CO_2 to the inspired air may not stimulate respiration. The effect of a given rise of alveolar P_{CO_2} is greater if the P_{O_2} of the arterial blood is low, that is, the CO_2-response line (Fig. 31.15B) is made steeper by hypoxia. Lack of oxygen has a depressant action on the cells of the respiratory centre but in the intact animal inhalation of a gas with a low oxygen content results in increased ventilation. The stimulant effect of hypoxia on respiration is a reflex effect through the chemoreceptors of the carotid body (p. 630) (Fig. 31.16).

Breath holding. At the end of an inspiration the breath can be held for 30 to 50 sec. This is called the period of *voluntary apnoea* or *breath-holding time*. The subject is forced to breathe again when the alveolar P_{CO_2} is 47 to 50 mm Hg. When breathing is resumed there is an almost immediate relief of respiratory distress as the alveolar air composition returns to normal. If instead of breathing air at the end of a period of voluntary apnoea the subject is allowed a few breaths of a gas mixture containing, say, 8·2 per cent oxygen and 7·5 per cent carbon dioxide (P_{CO_2} = 54 mm Hg) relief of respiratory distress is obtained which is sufficient to allow another short period of apnoea in spite of the continuing decrease in oxygen saturation of the blood and increased alveolar P_{CO_2}. The period of voluntary apnoea is prolonged if the vagus nerves are blocked. If a subject is completely paralysed with tubocurarine except for one hand used for signalling it is found that the time during which he has no desire to breathe is prolonged. Afferent impulses from the respiratory muscles must therefore contribute to the desire to breathe. Therefore although the breaking point of apnoea depends mainly upon raised

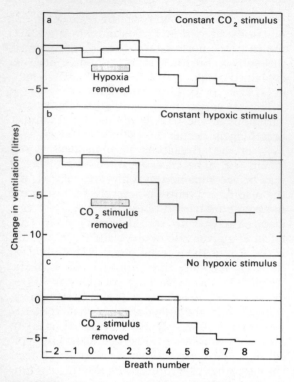

Fig. 31.16 The time course of the fall in ventilation in man when respiratory stimuli are removed for three breaths during the period marked by the bars. The subjects were previously in a steady-state of hyperventilation caused by hypercapnia (inspiring 6 per cent CO_2) with or without hypoxia (10 per cent or 94 per cent O_2). (a) Hypoxia is removed, leaving a constant background CO_2 stimulus. Ventilation is falling by the third breath. (b) CO_2 stimulation removed, leaving a constant background hypoxic stimulus. Time-course of fall in ventilation is the same as in (a), and is therefore probably under the control of the peripheral chemoreceptors. (c) As (b) but in hyperoxia. The fact that the ventilatory depression is delayed for about 6 seconds longer suggests that the peripheral chemoreceptors are not responding to CO_2 in the absence of hypoxia. (*After* D. J. C. Cunningham, B. B. Lloyd, J. P. Miller & J. M. Young (1965). *Journal of Physiology* **179**, 68P.)

P_{a,CO_2} (hypercapnia) other factors are involved.

After a short period of forced breathing (*voluntary hyperpnoea*) the period of voluntary apnoea is much longer. This cannot be due to an increased oxygen content of the blood, since it is normally very nearly (95 per cent) saturated with oxygen. The hyperpnoea flushes out the pulmonary alveoli with air and reduces their carbon dioxide content. Since this increases the gradient between the blood and the alveolar air more carbon dioxide than normal escapes from the blood. When this blood of reduced P_{CO_2} reaches the central chemoreceptors it does not possess its normal stimulating effect and the breath is easily held. The P_{CO_2} of the blood gradually builds up again as carbon dioxide from tissue metabolism enters it and finally, at the 'breaking-point', the subject is compelled to breathe once more. The breath-holding time immediately after exercise is relatively short because the pH of the blood is lowered by its increased content of carbon dioxide and by the appearance in it of organic acids produced by muscle metabolism. If oxygen is breathed during the preparatory hyperpnoea the apnoea is not greatly prolonged. There is thus no doubt that the carbon dioxide tension of the blood is the main factor determining the length of the breath-holding time although when there is hypoxia the breaking-point is reached somewhat sooner.

If hyperpnoea is carried to excess so much carbon dioxide is blown off that the pH of the blood rises and a state of alkalaemia occurs, with signs of tetany, including spasm of the muscles and hyper-excitability of the peripheral nerves (see Chap. 39). Unanaesthetized men, whether hyperventilation is carried out voluntarily or by a pump, rarely become apnoeic even when the arterial P_{CO_2} is as low as 16 mm Hg; hyperventilated anaesthetized men may become apnoeic when the pump is stopped.

Drowning accidents have been reported in which the victim hyperventilated before diving in an attempt to prolong the duration of underwater swimming. Such exercise after hyperventilation may reduce the blood P_{O_2} to less than 34 mm Hg, a degree of hypoxia which may result in unconsciousness before the P_{CO_2} rises to the level that would force him to surface.

STIMULATION OF PERIPHERAL CHEMORECEPTORS

Chemoreceptors sensitive to changes in the P_{CO_2} and P_{O_2} of the blood are found in the carotid bodies, ellipsoidal structures usually about 6 mm long at the bifurcation of the common carotid artery. Their weight is very variable but is on the average about 15 mg. They are heavier in people living at high altitudes. Similar aortic bodies lie between the ascending aorta and the pulmonary artery. Histological preparations of the carotid body show large glomus cells, presumably chemoreceptors, containing catecholamines. One pole of these cells lies close to the sinusoidal capillaries, at the opposite pole are the afferent endings of the glossopharyngeal nerve. Efferent sympathetic fibres come from the superior cervical ganglion. The glossopharyngeal nerve also contains efferent fibres. Vagal fibres also end in the carotid body.

Oxygen lack induced in man by breathing, for example, 10 per cent oxygen in nitrogen produces hyperventilation but after removal of the carotid bodies the same degree of hypoxia produces very little if any change in breathing. It can be inferred that the human aortic bodies, like those in the dog, have very little effect on respiration; they are more concerned in cardiovascular reflexes (p. 630). The respiratory centre is itself depressed by lack of oxygen but the impulses sent to the respiratory centre by the chemoreceptors of the carotid bodies are able to overcome the central depression. Inhalation of 3 per cent carbon dioxide in air increases ventilation in man before and after removal of the carotid bodies, presumably by stimulation of the central medullary chemoreceptors. If a person is given pure oxygen to breathe at sea level breathing is depressed slightly for a few minutes. This suggests that even at normal arterial P_{O_2} there is some discharge from the carotid and aortic bodies.

The carotid body contains acetylcholine and catecholamines which could act as transmitters to stimulate the nerve endings but the nature of the transmitter is still uncertain. Eyzaguirre believes that in hypoxia or hypercapnia an acetylcholine-like substance is responsible for initiating the nerve impulses but in more intense chemoreceptor activity, induced by exposure to nitrogen or sodium cyanide, acetylcholine and other substances are released. The blood flow through a 2 mg carotid body of the cat is about 40 µl/min; this is much greater than the supply to any other tissue in the body (Table 27.18). It is at first sight difficult to explain how a tissue with such a vast blood supply could be affected by mild hypoxia but it is likely that only a small fraction of the total blood flow traverses the capillaries supplying the glomus cells and even this may be reduced by opening up of arteriovenous anastomoses when the sympathetic is stimulated. Hypoxia and hypercapnia cause a rise in blood flow but a fall in oxygen consumption provided that the sympathetic nerves are intact. After section of the sympathetic nerves the total blood flow may increase by one-third. Stimulation of sympathetic nerve to the carotid body reduces the blood flow and increases the chemoreceptor discharge and an increase in the tidal volume. Limb movements cause increase in activity in cervical sympathetic and increase in chemoreceptor activity; this may account for the increase in respiration occurring a few seconds after the beginning of exercise. The rate of discharge from the carotid body chemoreceptors is related inversely to its oxygen usage and not to its total blood flow. Oxygen consumption is related mainly to sympathetic activity. It is likely that in hypoxia anaerobic metabolites accumulate and stimulate the nerve endings. This would account for the chemoreceptor discharge which continues for 30 min or so after death.

Carotid sinus. A rise in blood pressure tends to depress, and a fall to stimulate, respiration. These effects are initiated by the pressure receptors (baroreceptors) in the aorta and carotid sinuses. When a large dose of adrenaline is injected into an intact animal there is, along with the rise of blood pressure, a cessation of respiration, usually termed 'adrenaline apnoea'. This is a sinus and an aortic reflex and not a direct effect of adrenaline on the respiratory centre, because it is abolished after the vagus nerves and sinus nerves are cut, or if a compensator is included in the circulation to prevent the rise of blood pressure after adrenaline is injected. If the two common carotid arteries are clamped, the pressure within the carotid sinuses falls temporarily to zero; respiration is stimulated and the blood pressure rises (Fig. 31.17A). After denervation of both carotid sinuses this effect is absent (Fig. 31.17B). Clamping the carotid arteries reduces the number of impulses from the baroreceptors and, since it makes the carotid bodies anoxic, it increases the discharge from the chemoreceptors.

Both adrenaline and noradrenaline given intravenously in man at the rate of 10 to 20 µg per min have a stimulating effect on respiration. Also in the hypertensive crises produced

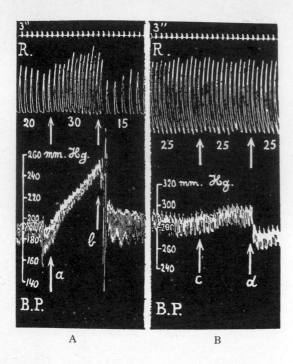

A   B

Fig. 31.17 Dog 10·5 kg anaesthetized with chloralosane. Depressor nerves cut. R, respiratory movements. B.P. femoral blood pressure recorded with a mercury manometer. ↑ a, common carotid arteries clamped. This was followed by increased rate and depth of respiration with acceleration of the heart and rise of B.P. ↑ b, common carotid arteries opened. This was followed by a return to original conditions. The two carotid sinus nerves were cut between A and B. This was followed by a rise of B.P. with acceleration of the heart and increase in rate and depth of respiration. ↑ c, common carotid arteries closed. ↑ d, common carotid arteries opened. Only very slight mechanical effects on the B.P. (C. Heymans & J. J. Bouckaert (1930). *Journal of Physiology* **69**, 261.)

by sudden discharge of these two substances from a tumour of the adrenal medulla the breathing is deep and rapid. These amines increase the sensitivity to hypoxia of the respiratory apparatus by stimulating the carotid bodies; they do not appear to act directly on the respiratory centre since injection into the vertebral or carotid arteries of a man at a rate of 1 to 2 μg per min has no such effect.

EFFECTS OF LACK OF OXYGEN

The alveolar oxygen tension is not at all rigidly regulated. It depends very largely on the percentage of oxygen in the atmosphere, upon the atmospheric pressure (Fig. 31.18)

and on the depth of ventilation. At sea-level the percentage of oxygen can be reduced from 21 to 13 without producing any marked alteration of breathing. This is because the arterial chemoreceptors (carotid bodies) are less active over this range than when the hypoxia is more intense, and because any reflex hyperventilation is immediately offset by the simultaneous reduction in P_{CO_2} as the carbon dioxide is blown off. The effects of lowering the oxygen tension still further can be seen by making a subject breathe in and out of a spirometer through a soda-lime container to absorb carbon dioxide. He becomes blue (cyanosis, p. 636) as the oxygen is used up but he may not show any respiratory distress. When the oxygen percentage falls below 12 breathing is stimulated by the intense hypoxia and more oxygen is taken into the alveoli. If CO_2 is allowed to build up in the spirometer by removing the soda-lime, the ventilation is substantially stimulated by the interaction of hypoxia and hypercapnia.

Effect of high altitude. Although the percentage of oxygen in the air remains the same at all levels, the atmospheric pressure and therefore the partial pressure of oxygen, P_{O_2}, vary according to the height above sea-level.

The newcomer who has ascended rapidly to, say 4 km (15 000 ft) is dyspnoeic on exertion and may become unconscious if he attempts to go up a stair quickly. He may also suffer from mountain sickness, the symptoms of which, namely weakness, nausea, loss of appetite, headache and sleeplessness, are due to cerebral hypoxia. At first respiration is stimulated by the chemoreceptors so that the alveolar P_{O_2} is raised and the alveolar P_{CO_2} is reduced; in consequence the blood and c.s.f. pH rise. In a few hours active transport of HCO_3^- out of the c.s.f. brings the c.s.f. pH back to its normal value, respiration is then once more under the control of both central and peripheral chemoreceptors.

The kidneys, in a slower compensatory adaptation, excrete more HCO_3^- and bring back the arterial pH to its normal value. Alkalosis by itself cannot explain the symptoms of mountain sickness; there may be an accompanying potassium deficiency. It has been recommended that men exposed to high altitudes should have a high intake of K but a limited intake of Na. Continued residence at a

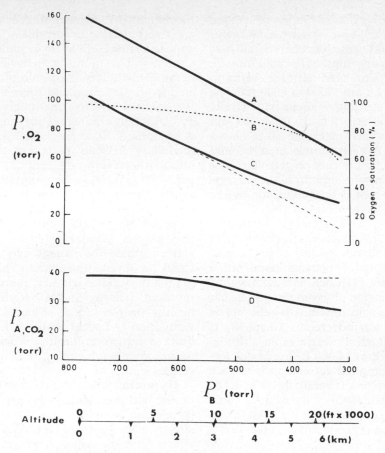

Fig. 31.18 The effect of acute exposure to simulated altitudes while breathing air. Curves show (A) partial pressure of oxygen in the inspired air, (B) percentage oxygen saturation of haemoglobin in arterial blood, (C) alveolar partial pressure of oxygen and (D) alveolar partial pressure of carbon dioxide. The broken lines on curves C and D depict the alveolar gas pressures which would occur in the absence of any ventilatory response to hypoxia. *N.B.* The barometer scale is linear but the altitude scales are not. (*By courtesy of the Royal Air Force, Institute of Aviation Medicine.*)

high altitude produces acclimatization and when this is complete the blood volume may have increased as much as 25 per cent, mainly because of the increase of red cells (Fig. 24.6). The resulting increase in viscosity of the blood is offset by a general vasodilatation so that the resting blood pressure may fall a little. The right ventricle hypertrophies probably because of the pulmonary arterial hypertension. These adaptive changes have enabled fit men to carry out for a short period moderately hard physical work on mountains at 6 km (20 000 ft). The changes in P_{O_2} and P_{CO_2} of the alveolar air and in the oxygen saturation of the blood at various heights are shown graphically in Figure 31.18.

When the change from sea-level to high altitude is very rapid, as in modern jet fighter aircraft, there is no time for acclimatization and the pilot must breathe oxygen from sea-level. Even in the slow fighter aircraft used at the beginning of the 1939–1945 war, pilots, if they did not breathe oxygen, were apt to behave abnormally above 3 km (10 000 ft) without their being aware of it. Above this altitude increasing cerebral hypoxia produces defects in judgment and reasoning with a spurious feeling of confidence, reduction in visual acuity and visual field, weakness of arms and legs with muscular inco-ordination and twitching and hilarity or pugnacity, and eventually unconsciousness.

Respiratory Functions of the Blood   633

Jet passenger aircraft fly at 9 to 12 km (30 000 to 40 000 ft). Since sudden exposure at this altitude would cause loss of consciousness in approximately one minute the cabin must be pressurized, the simulated 'altitude' being in the range 1·5 to 2·5 km (5000 to 8000 ft). The usual operating 'altitude' is about 1·5 km, only very rarely is it necessary to 'go up' to 2·5 km. Inspection of curve B of Figure 30.18 shows that the arterial oxygen saturation at 1·5 and 2·5 km is 92 and 90 per cent respectively. The small reduction in saturation from the sea level value of 97 per cent is accounted for by the 'flatness' of the oxygen dissociation curve of haemoglobin (Fig. 31.7) from 100 to 60 mm Hg, in other words from sea level to 3 km (10 000 ft). The small amount of hypoxia at these 'altitudes' is of no importance to a healthy person but even before these 'heights' are reached expansion of gas in the middle ear may give rise to pain unless it is relieved by repeated swallowing movements (Chap. 38). If a defect occurs which allows the cabin 'altitude' to rise above 4·3 km (14 000 ft) masks delivering oxygen are supplied automatically to each passenger. Supersonic aircraft fly at 18 to 21 km where the atmospheric pressure is only 33 mm Hg. Ionizing radiation is a small risk to passengers but to regularly flying crew it is a potential hazard.

Cheyne-Stokes respiration. The commonest type of *periodic breathing* is the occurrence of a series of respirations which gradually increase in depth from a barely perceptible movement to a maximum and then progressively decline to terminate in a period of apnoea lasting from a few seconds to as long as a minute. This 'hunting' behaviour is possible because medullary chemoreceptors are separated from the lungs, and there may be a time lag between effective medullary and alveolar P_{CO_2}. Usually this lag is small and the hunting is smoothed. If there is damage to supramedullary nervous pathways that normally inhibit the respiratory centres on the motor pathways to respiratory muscle there is an excess ventilatory response to CO_2 that overcomes the normal smoothing effect. A prolonged lung-to-brain circulation time may also play a part (Fig. 31.19). During the apnoeic phase the patient may become confused or even unconscious but breathing gradually begins again and the whole cycle is repeated. Cheyne-Stokes breathing occurs in healthy people at high altitudes, after hyperventilation and occasionally in deep sleep; it is quite common in healthy infants. When it occurs in an adult during waking hours it is a serious sign.

Hypoxia. When the oxygen supply to the tissues becomes inadequate, hypoxia is said to be present. Some aspects of this subject have already been discussed on p. 632. Although the effects upon the subject depend to some extent on the way in which the hypoxia is produced, and on the rapidity with which it develops, the commonest effects are cerebral, namely, confusion, excitement, hallucinations, restlessness

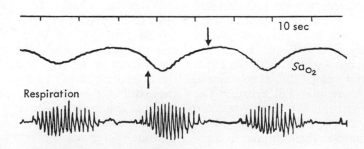

Fig. 31.19 Periodic 'Cheyne-Stokes' respiration. When the oxygen saturation of the blood (S_{a,O_2}) and therefore the P_{O_2} of the peripheral chemoreceptors falls to a certain level (↑) the breathing is stimulated, the S_{a,O_2} rises and breathing diminishes or ceases (↓). The oxygen saturation was determined in the ear, see oximeter, p. 619. (R. Hegglin (1961). *Triangle* **5**, 130–137.)

and unconsciousness. Hypoxia may be classi-fied as four types.

1. *Hypoxic hypoxia*. This type differs from others to be mentioned subsequently in that the oxygen saturation of the arterial blood is reduced below its normal level of about 19 ml O_2 per 100 ml because the arterial P_{O_2} is reduced. Obstruction to the air passages or paralysis of the respiratory muscles by re-ducing ventilation may interfere with the oxygenation of the blood. Breathing air at high altitudes where the oxygen tension is low or mixtures of anaesthetic gases with too low a content of oxygen also leads to this type of hypoxia. It may occur also when parts of a lung are thrown out of action by collapse or com-pression or pneumonia or when the mixing of gases in the lungs is imperfect as in emphysema. Hypoxia may also result from impaired dif-fusion of oxygen across the boundary zone between blood and alveolar air when the pul-monary tissues are infiltrated by disease: the hypoxia is then not accompanied by retention of carbon dioxide because this gas is trans-ferred about thirty times faster than oxygen across the alveolar membrane (p. 615). Finally when relatively large amounts of blood are shunted from the right to the left side of the heart as sometimes happens when there are congenital anomalies of the heart or great vessels the oxygenated blood in the aorta becomes mixed with poorly oxygenated blood that has not traversed the lungs (anatomical right-to-left shunt).

In all these circumstances oxygen is sup-plied to the tissues of the body at a tension lower than normal and may be inadequate for their needs, especially during activity. Acute hypoxic hypoxia is accompanied by an increase in cardiac output of 10 to 20 per cent and the hypoxia stimulates respiration through the carotid and aortic bodies; hyperventilation is produced and the elimination of carbon dioxide is increased. However, these compen-sating effects often fail to keep the P_{a,O_2} normal. Chronic hypoxia of this type stimu-lates the production of erythropoietin (p. 418), with a consequent increase in the numbers of red cells in the circulating blood ('secondary polycythaemia'). It may also cause constric-tion of the pulmonary arterioles and pul-monary arterial hypertension which can cause right ventricular failure. Anoxic pulmonary heart disease often afflicts patients with chronic airways obstruction, such as chronic bron-chitis. Cattle moved from low to high altitudes can develop a similar syndrome, named 'brisket disease' because of the site of the oedema.

2. *Anaemic hypoxia*. This results from interference with oxygen transport either because the amount of circulating haemo-globin is reduced (anaemia), or because the haemoglobin is rendered unavailable by being combined with a substance with which it forms a stable compound. In this type of hypoxia the oxygen load, that is the oxygen carried per 100 ml of blood, is reduced but, since the oxygen tension in the alveolar air and in the arterial blood is normal, the tissues receive their oxygen at a normal tension. The carotid bodies are little affected by this type of hypoxia, so there is no compensating hyper-ventilation. As the blood flows through the capillaries in the tissues the oxygen tension falls very rapidly so that, if no compensation occurred, the greater part of the oxygen would be supplied to the tissues at a much reduced tension. In anaemia this disadvantage is com-pensated for by an increase in the cardiac output; the circulation rate is more rapid, and each 100 ml of blood is required to part with less of its oxygen load to the tissues than would otherwise be the case. Oxygen is thus supplied to the tissues at tensions higher than could occur with the normal circulation rate. In patients with slowly developing severe anaemia the oxyhaemoglobin dissociation curve (cf. Fig. 31.7) may be displaced to the right probably by an increased concentration of 2·3 DPG (p. 621) so that a greater proportion of the oxygen is released in the tissue capil-laries. As a result of such compensatory mechanisms anaemic patients are rarely breath-less at rest, although they may rapidly become so with exercise.

Haemoglobin may be rendered unavailable for oxygen transport by drugs such as chlorates and nitrates which convert it to methaemo-globin, or by carbon monoxide which has an affinity for haemoglobin about 250 times greater than that of oxygen. For this reason concentrations of carbon monoxide as low as 0·1 per cent in the atmosphere produce dan-gerous hypoxia; where there is prolonged exposure, 0·01 per cent is the maximum concentration which can be tolerated. According to J. B. S. Haldane 10 per cent of COHb in the blood causes breathlessness on exertion; 30 per cent COHb causes headache, irritability and impaired judgement; 60 per cent unconsciousness and convulsion; 80 per cent death by asphyxia. The blood of a healthy non-smoker may contain 0·4 per cent COHb derived from destruction of red cells. The blood of cigarette smokers may contain 15 per cent COHb. The treatment of carbon monoxide poisoning is discussed on p. 000.

3. *Stagnant hypoxia*. This type of hypoxia is the result of slowing of the circulation, such as may occur locally as the result of vasoconstriction by cold or obstruction to the venous outflow from a part, or because of cardiac failure or peripheral circulatory failure. In this condition arterial oxygen saturation, the partial pressure of oxygen and the oxygen load are all normal, but the circulation rate is slowed and the blood dwells longer in the capillaries. The result is that the arteriovenous oxygen difference is increased and much of the oxygen is delivered to the tissues at a reduced tension. A small degree of compensation can, however, occur since the retarded flow leads to increase in the capillary blood carbon dioxide tension which encourages the dissociation of oxyhaemoglobin (p. 620) and thus increases the supply of oxygen to the tissues.

4. *Histotoxic hypoxia*. Cyanide poisoning is the only important example of histotoxic hypoxia. Cyanide interferes with tissue oxidation by paralysing the cytochrome system and other iron-containing enzyme systems. The cells are, therefore, unable to utilize the oxygen conveyed to them and the arteriovenous oxygen difference is much reduced. Narcotics may have a similar effect by interfering with dehydrogenase systems.

Cyanosis. As we have already seen the colour of the skin is determined by the amount and condition of the blood in the minute vessels and, especially, in the veins of the subpapillary plexus (Chap. 35). *Cyanosis* is a violet, bluish or sometimes greyish colour of the skin produced by abnormally large amounts of reduced haemoglobin, or, less frequently, by abnormal pigments such as methaemoglobin and sulphaemoglobin.

Arterial hypoxia, that is an abnormally low saturation of the arterial blood with oxygen, arises usually from chronic disease in the lungs associated with abnormal $\dot{V}/\dot{Q}$ ratios that allow an overall venous-admixture effect, or, less commonly where the transfer factor is reduced to the extent that blood reaches the pulmonary veins still insufficiently oxygenated. When cyanosis occurs in this way the patient's extremities are warm and pulsatile, the peripheral blood flow is usually relatively rapid, the heart rate rapid and the pulse pressure increased. There is, in short, vasodilatation. The cyanosis is said to be *central* and indicates hypoxic hypoxia (p. 635). The oxygen saturation may fall to 80 per cent before cyanosis is readily apparent; the P_{O_2} is then about 45 mm Hg.

In contradistinction some cyanosed patients may be cold and blue, with peripheral pulses difficult to feel. Blood flow through the tissues is reduced as the result of vasoconstriction and cyanosis occurs because of the longer time available in the tissues for gaseous exchange. This is *peripheral* cyanosis which is an index of the stagnant hypoxia that may occur in cardiac and peripheral circulatory failure. The oxygen saturation of the arterial blood is normal.

The cyanosis sometimes seen in persons with polycythaemia in whom the concentration of red cells is over eight million per mm^3 and of haemoglobin over 23 g per 100 ml is mainly the result of stagnant hypoxia produced by the increased viscosity imparted to the blood by its much greater content of cells. In anaemic hypoxia although the venous P_{O_2} is much reduced, the concentration of reduced haemoglobin may yet be insufficient to produce cyanosis.

Carbon dioxide retention. An abnormally high partial pressure of CO_2 in the arterial blood (P_{a,CO_2}) is referred to as hypercapnia. Patients with chronic lung disease and airway obstruction, such as chronic bronchitis, often have an impairment of ventilation severe enough to cause hypercapnia. When the P_{a,CO_2} rises to a certain level, which varies in individual patients, the clinical features of 'CO$_2$ narcosis' may be evident: change of personality, drowsiness and coarse tremor. The skin, though cyanosed, may be warm, and the pulse bounding, because of a high pulse pressure. The drowsiness may proceed to stupor, convulsions, coma and death.

Respiratory failure. Respiratory failure may be defined as a state in which respiration is unable to keep the P_{a,O_2} and P_{a,CO_2} within normal limits. Many patients with chronic respiratory disease have chronic mild respiratory failure according to this definition, which is widely accepted in clinical practice. If the causal condition worsens, of course, severe hypoxia or hypercapnia, and often both, may lead to death.

OXYGEN THERAPY

The administration of oxygen often benefits strikingly certain kinds of hypoxia and relieves cyanosis. When the arterial oxygen saturation is normal, as in the stagnant, anaemic and histotoxic forms, less is to be expected of oxygen therapy than in the hypoxic type, in which the arterial oxygen saturation is reduced. Although efficient administration of oxygen can raise the partial pressure of this gas in the alveolar air from 100 to 650 mm Hg it does not necessarily follow that the arterial oxygen tension is correspondingly increased. However, in the hypoxia due to paralysis of the muscles of respiration, to rapid shallow breathing and to many diseases of the lung in which a considerable proportion of the pulmonary blood flow perfuses areas that are not well aerated, the administration of oxygen relieves cyanosis rapidly and the effects of hypoxia more slowly. Such a dramatic relief is not to be expected when there is a large shunt of venous blood from the right to the left side of the heart, or when the diseased pulmonary tissue has been largely deprived of its blood, because in these conditions the blood leaving the functioning alveoli is already 95 per cent saturated.

Efficient oxygen administration at atmospheric pressure is capable of increasing the concentration of oxygen in solution in the plasma from the normal level of about 0·3 vol. per 100 ml to values of the order of 1·8 vol. per 100 ml, which is quite a significant increase. Administration of oxygen may, therefore, by increasing the plasma concentration, make an important contribution to the oxygen supply of the tissues, equivalent perhaps to an increase of 30 per cent in cardiac output. For this reason the administration of oxygen is useful in all kinds of hypoxia, however produced; its value can be judged in practice by observing its effect in dispelling cyanosis.

When it is intended that the concentration of oxygen in the inspired air should be as high as possible it should be administered through a mask that covers both nose and mouth and has a non-return valve. The mask is used either with a reservoir bag, or a demand-valve on the supply of oxygen.

A lower concentration, about 60 per cent, of oxygen in the inspired air is often adequate and can be provided by cheap, disposable plastic masks which fit the patient's face fairly well and into which oxygen is delivered at not less than 6 litres/min.

Children and very restless patients may sometimes need to be nursed in an 'oxygen tent' of flexible transparent plastic within which they can move fairly freely. The temperature of the enclosed atmosphere must be kept at about 18·5°C. The flow rate of O_2 is about 8 litres/min and there is a risk of fire. The oxygen must be humidified when any of these methods is used.

In certain chronic respiratory diseases such as chronic bronchitis the inadequate ventilation causes a raised P_{aCO_2} as well as a reduced P_{aO_2}. This chronic hypercapnia leads to a reduced central chemo-sensitivity to carbon dioxide, and the ventilation is largely driven by the hypoxic stimulus acting through the carotid and aortic body reflexes. If the hypoxia is suddenly corrected by administration of oxygen respiration may become dangerously depressed, causing a higher P_{a,CO_2}, carbon dioxide narcosis and even death. In these circumstances the percentage of oxygen in the inspired air should be easily controlled and closely supervised to provide the highest concentration of oxygen that will not cause a rise of P_{a,CO_2}: that concentration may vary between about 23 and 30 per cent. Ideally the control should be absolutely accurate, but in clinical practice it cannot be. Nasal cannulae can deliver 25 to 30 per cent O_2 into the nostrils with a flow-rate of 1 litre O_2/min; control is not very accurate but the method avoids masks, and may be convenient. More accurate methods involve the use of loosely-fitting masks designed on the Venturi principle to deliver a more finely-controlled oxygen concentration. To avoid hypoxia and hypercapnia the P_{a,CO_2} and P_{aO_2} should ideally be measured at regular intervals; where facilities for the direct estimation of P_{CO_2} in arterial blood are not available, a rebreathing method may be used (p. 628).

The value of oxygen inhalation in flying at high altitudes was demonstrated in the 1939–1945 War; the signs of oxygen lack have already been described on pp. 632, 633 and 634. It is essential to breathe through an oxygen mask when flying in non-pressurized aircraft at altitudes above 15 000 feet (5000 metres). At great heights when the barometer reads less than 150 mm Hg even pure oxygen must be administered under positive pressure; it is the partial pressure of oxygen rather than its concentration that determines the arterial oxygen saturation.

Oxygen in a chamber at a pressure of two atmospheres (hyperbaric oxygen) is sometimes used in the treatment of circulatory disorders. It increases the concentration of oxygen dissolved in the plasma and may sometimes allow the survival of tissues which have had their blood supply seriously impaired. Hyperbaric oxygen is probably the best treatment for carbon monoxide poisoning.

Effect of increased oxygen supply. It has been suspected for a long time that it is dangerous to breathe pure oxygen at hyperbaric pressures, but the first experiments demonstrating this were made during the last war to investigate casualties in divers carrying self-contained oxygen apparatus. Fit men were exposed in a 'dry' pressure tank to oxygen at 3·7 atmospheres (corresponding to a depth of 27 m). After exposure times from 6 to 96 min nausea, vertigo, bradycardia, severe twitching of muscles or even convulsions occurred for which the experiment had to be stopped. For some unknown reason the oxygen tolerance at the same pressure under water was greatly reduced. Thus in the 'dry' experiments 50 per cent of the subjects were eliminated by such symptoms in 20 min whereas in the 'wet' experiments 50 per cent were eliminated in 10 min. There is a wide range in individual tolerance to oxygen and in any one individual tolerance varies greatly and unpredictably from day to day. Oxygen tolerance is markedly reduced by work but the reason for this is quite unknown. Diving on pure oxygen to 7·6 m for 2 hours is quite safe but at greater depths diving on pure oxygen is a hazardous gamble.

The so-called toxicity of oxygen at high pressure may be explained by the finding that tissue enzymes, just as they have optimum pH requirements (p. 93), have also optimum P_{O_2} requirements. For example the enzymes involved in the oxidation of pyruvate (p. 148) work within a narrow P_{O_2} range and are inhibited if oxygen is lacking or if the P_{O_2} is too high. At sea-level breathing of pure oxygen raises the arterial oxygen tension from below 100 to about 600 mm Hg (Fig. 31.10B). The effects of exposure to a high P_{O_2}, unlike those of exposure to cyanide, are usually reversible. Rats given succinate intraperitoneally before exposure to high P_{O_2}, survive longer than controls and have higher ATP concentration in their tissues. Studies on beef heart suggest that mitochondria operate at a maximal rate at a P_{O_2} of 1 to 10 mm Hg.

New-born infants are particularly susceptible to oxygen poisoning; prolonged administration of oxygen for resuscitation causes a proliferation of the retinal vessels into the vitreous with excess formation of fibrous tissue (*retrolental fibroplasia*) and blindness. Oxygen administered in high concentrations (over 70 per cent) to kittens produces first a constriction and obliteration and later a proliferation of the retinal vessels. In man inhalation of pure oxygen causes constriction of the retinal arteries and veins; at two atmospheres pressure the diameter of these vessels may be reduced by one third. Ozone in a concentration as low as one or two parts per million damages the lungs; concentrations of up to 0·12 parts per million have been found in high flying aircraft; ozone damage is conceivably a risk of supersonic flight.

Decompression sickness. This condition was first noted in divers and caisson workers who had been brought from a pressure of several atmospheres too quickly to the atmospheric pressure at the surface of the water. Either immediately or after a latent period which may be several hours the victim experiences pain in the limbs (the 'bends') or a choking sensation and may in more severe cases develop various forms of paralysis. The symptoms are due to the presence of gas bubbles either in the blood vessels or in the nervous system.

The tissues of a man working under increased pressure gradually come into equilibrium with the increased partial pressure of nitrogen and oxygen. If he is suddenly decompressed by rising too quickly to the surface these gases come out of solution and form bubbles which, as they expand, tear through tissues or block blood vessels. The oxygen is quickly used up and disappears but the bubbles of nitrogen remain. Since nitrogen is five times more soluble in lipid material than in water, fatty tissues take up relatively more nitrogen during compression and give up correspondingly more during decompression. For this reason particularly large bubbles are formed in the central nervous system and give rise to randomly distributed neurological disturbances.

Decompression sickness can be avoided to a large extent by slow decompression. If a diver suffers from bends on arrival at the surface he should be sent down again and slowly decompressed. A very sudden rise in altitude as may occur in fighter aircraft can also produce similar disturbances but not so severe as those seen in divers.

REFERENCES

ADDIS, G. J. (1965). Effect of '100 per cent.' oxygen breathing on mixed venous oxygen pressures. *Scottish Medical Journal* **10**, 215–217.

Advisory group for aeronautical research and development (1955). *Collected Papers on Aviation Medicine*. London: Butterworth.

ARMSTRONG, H. G. (Ed.) (1962). *Aerospace Medicine*. London: Baillière, Tindall & Cox.

BARCROFT, J. (1928). *Respiratory Function of the Blood*. Cambridge: University Press.

BARTELS, H. (1964). Comparative physiology of oxygen transport in mammals. *Lancet* **ii**, 499–604.

BATES, D. V. & CHRISTIE, R. V. (1964). *Respiratory Function in Disease*. London: Saunders.

BHATTACHARJYA, B. (1964). *Mountain Sickness*. Bristol: Wright.

BROOKS, C. M., KAO, F. F. & LLOYD, B. B. (1965). *Cerebrospinal Fluid and the Regulation of Ventilation*. Oxford: Blackwell.

CAMPBELL, E. J. M. (1965). Respiratory failure. *British Medical Journal* **i**, 1451–1460.

COTES, J. E. (1968). *Lung Function: Assessment and Application in Medicine*, 2nd edn. Oxford: Blackwell.

DAVENPORT, H. W. (1961). *The ABC of Acid–Base Chemistry. The elements of physiological blood-gas chemistry for medical students and physicians*. Chicago: University Press.

FISHMAN, A. P. & HECHT, H. H. (Eds.) (1959). *The Pulmonary Circulation and Interstitial Space*. Chicago: University Press.

FORSTER, R. E. (1957). Exchange of gases between alveolar air and pulmonary capillary blood: pulmonary diffusing capacity. *Physiological Reviews* **37**, 391–452.

FRYER, D. I. (1969). *Subatmospheric Decompression Sickness in Man*. Slough: Technivision Services.

FULTON, J. F. (1951). *Decompression Sickness, Caisson Sickness, Diver's and Flier's Bends and Related Syndromes*. London: Saunders.

GILLIES, J. A. (Ed.) (1965). *A Textbook of Aviation Physiology*. Oxford: Pergamon.

HAUGAARD, N. (1968). Cellular mechanisms of oxygen toxicity. *Physiological Reviews* **48**, 311–373.

LEDINGHAM, I. M. (Ed.) (1965). *Hyperbaric Oxygenation. Proceedings 2nd International Congress*. Edinburgh: Livingstone.

LEDINGHAM, I. M. (1967). Hyperbaric oxygenation. *Scientific Basis of Medicine Annual Review* 90–108.

VAN LIERE, E. J. & STICKNEY, J. C. (1963). *Hypoxia*. University of Chicago Press.

MAREN, T. H. (1967). Carbonic anhydrase: chemistry, physiology and inhibition. *Physiological Reviews* **47**, 595–781.

MARGARIA, R. (Ed.) (1967). *Exercise at Altitude*. Amsterdam: Excerpta Medica Foundation.

MILES, S. (1969). *Underwater Medicine*, 3rd edn. London: Staples Press.

PAPPENHEIMER, J. R. (1967). The ionic composition of cerebral extracellular fluid and its relation to control of breathing. *The Harvey Lectures* 1965–1966, Series 61, pp. 71–94.

PETERS, J. P. & VAN SLYKE, D. D. (1946). *Quantitative Clinical Chemistry*, 2nd edn., Vol. 1. London: Baillière, Tindall & Cox.

PERUTZ, M. F. (1970). Stereochemistry of cooperative effects in haemoglobin. *Nature, London* **228**, 726–739.

RAHN, H. (Ed.) (1965). *Physiology of Breath-hold Diving and the Ama of Japan*. Washington: NAS-NRC.

ROUGHTON, F. J. W. (1935). Recent work on carbon dioxide transport by the blood. *Physiological Review* **15**, 241–296.

ROUGHTON, F. J. W. (1963). Kinetics of gas transport in blood. *British Medical Bulletin* **19**, 80–89.

SEMPLE, S. J. G. (1967). Respiration and the cerebrospinal fluid. *Scientific Basis of Medicine Annual Review* 109–127.

Scottish Home and Health Department (1969). *Uses and Dangers of Oxygen Therapy*. Edinburgh: H.M.S.O.

TORRANCE, R. W. (Ed.) (1968). *The Arterial Chemoreceptors*. Oxford: Pergamon.

WEIHE, W. H. (Ed.) (1964). *The Physiological Effects of High Altitudes*: Proceedings at Interlaken, Sept. 1962. Oxford: Pergamon.

WEST, J. B. (1970). *Ventilation/Blood Flow and Gas Exchange*, 2nd edn. Oxford: Blackwell.

WHIPPLE, H. E. (Ed.) (1965a). Hyperbaric oxygenation. *Annals of the New York Academy of Science* **117**, 647–890.

WHIPPLE, H. E. (Ed.) (1965b). Respiratory failure. *Annals of the New York Academy of Science* **121**, 651–958.

32 The kidney

The kidneys, each of which weighs between 120 and 170 g, lie behind the peritoneum on either side of the vertebral column extending from the 12th thoracic to the 3rd lumbar vertebrae. Their position in relation to the vertebral column, however, varies with posture and respiration. The size of the kidneys in life can be gauged by measuring the length of their shadows on an X-ray film after the subject has had an injection of a radio-opaque substance which is concentrated in the kidneys (intra-venous pyelogram (Fig. 32.1)). In health the kidneys measure from 11 to 13 cm in length, the left being as a rule the larger.

The *hilum* of the kidney is an indentation on its medial border (Plate 32.1) through which pass the renal arteries and veins and nerves and the pelvis of the ureter, the funnel-shaped upper end of the ureter formed by the joining together of about ten small branches or *calyces*.

When cut longitudinally the renal substance is seen to consist of an outer *cortex* and an inner, paler, *medulla* made up of pyramids, the apices or papillae of which project into the calyces. The medullary pyramids are separated by the renal columns (of Bertin) continuous with the cortex (Plate 32.1).

The minute anatomy of the kidney is difficult to understand from an ordinary histological section because it consists of an enormous number of intertwined tubules. Each kidney unit, called a *nephron* (Fig. 32.2 and 3), is a tubule about 5 cm long which begins as a blind dilated end (*glomerular capsule*) and then pursues a very tortuous course through the cortical and medullary substances, finally opening into a collecting tubule. The total length of all the tubules of both kidneys is estimated to be about 110 km (70 miles). The glomerular capsule can be thought of as a sphere, lined internally by epithelium lying on a basement membrane and invaginated at one place by a tuft or *glomerulus* of 4 to 6 capillary loops springing

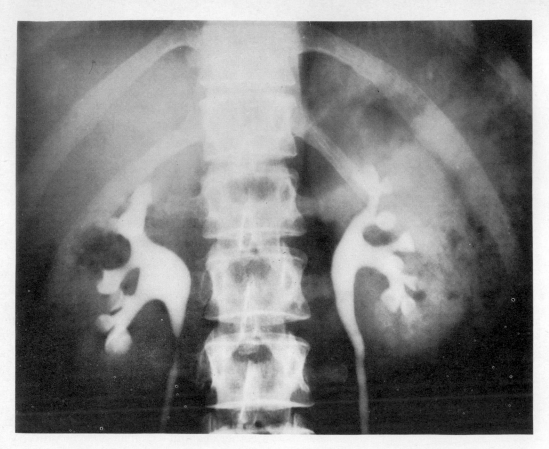

Fig. 32.1 A normal intravenous pyelogram obtained by injecting an organic iodine compound intravenously and taking an X-ray photograph at a suitable time (10 min) after the injection. From such a picture the size, shape and position of the two kidneys can be judged as well as the gross anatomy of the calyces and upper urinary tract.

from an *afferent arteriole* and ending in a smaller *efferent arteriole*. A glomerulus surrounded by its glomerular capsule is called a *Malphighian* or *renal corpuscle* about 0·2 mm in diameter. There are about a million of these bodies in each human kidney; it has been calculated that the total surface area of all the glomerular capillaries of both kidneys exceeds 1 m². The wall of a glomerular capsule is made of flattened cells but these change fairly abruptly to cuboidal cells at a point where the neck of the capsule becomes the first segment of the proximal tubule (Fig. 32.3).

The first or proximal convoluted tubule, about 14 mm long, is continuous with the relatively straight descending limb of the loop of Henle which passes into the medulla. The tube then forms a U-bend and returns to the

cortex as the ascending limb of the loop which eventually lies close to the renal corpuscle in the space between the afferent and efferent arterioles (Fig. 32.3 and 6). Because this part of the tubule has many prominent nuclei it has been named the *macula densa*. The tubule continues on to form the second or distal convoluted tubule which terminates as the junctional tubule by joining a collecting tubule which opens into one of the calyces of the ureter (Plate 32.1).

The filtering membrane of the renal corpuscle consists of three layers: (1) the endothelial layer, (2) a basement membrane and (3) an epithelial layer. The endothelial cells forming the walls of the capillaries are flat and their cytoplasm contains numerous pores some 20 and 90 nm in size each closed by a fine

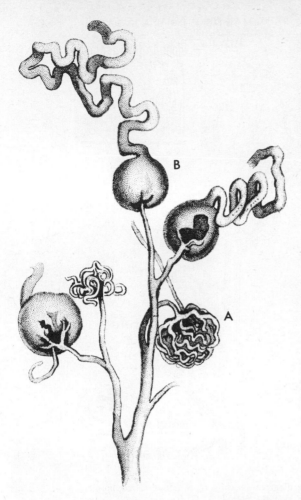

Fig. 32.2 Saturated solutions of potassium dichromate and of lead acetate were injected in succession into the renal artery and a deposit of lead chromate was produced in the glomeruli (A). At B the capillaries have given way and the lead chromate has passed into the glomerular capsule and into the first part of the renal tubule. (W. Bowman (1842). On the structure and use of Malpighian bodies of the kidney. *Philosophical Transactions* **132**, 79.)

membrane (Fig. 32.4). The basement membrane is an almost structureless mucopolysaccharide layer 80 nm thick. The epithelial cells or podocytes of the glomerular capsule have long tentacle-like processes from each of which spring small processes, the *pedicels*, which surround the capillary and are attached to its basement membrane, frequently interdigitating with each other. The spaces between the pedicels are known as slit-pores, the minimum distance between adjacent pedicels being 7nm.

The cells of the proximal tubules are large columnar-type cells with a well-marked brush border (Fig. 32.3). Each cell has about 150 thin microvilli (about 1 µm long) per square micron; the surface area of the brush border of the proximal tubules of each human kidney is of the order of 30 m². The cells of the distal convoluted tubule also have microvilli and contain large numbers of microvesicles.

Renal blood vessels. The kidneys are supplied with blood by two short renal arteries arising directly from the abdominal aorta. The main divisions of the renal artery within the kidney, called *arcuate arteries*, run between cortex and medulla parallel to the surface of the organ. At frequent intervals branches (*interlobular arteries*) arise from them and run in the cortex at right angles to the surface (Fig. 32.5). From each of these vessels numerous afferent glomerular arterioles arise which almost at once break up into the capillaries of the glomerular tuft. Blood leaves the glomerulus by the efferent glomerular arteriole which immediately breaks up into a second capillary network surrounding the convoluted tubules in the cortex from which the blood eventually passes into the renal veins (Fig. 32.5). The afferent arteriole is, because of the greater thickness of muscle in the tunica media, about twice the external diameter of the efferent but the lumen of the former is not much greater. The tunica media of the afferent arteriole, in the immediate neighbourhood of the renal corpuscle and close to the macula densa, is made up of highly granular cells, the *juxtaglomerular cells* (Fig. 32.6). The granules increase in them after adrenalectomy or sodium deprivation because in these circumstances renin production is increased.

The glomeruli situated in the deeper layers of the cortex (juxta-medullary glomeruli) are somewhat larger in size and have a different plan. Arising in many cases directly from a large artery the afferent glomerular arteriole breaks up into the glomerular capillaries which are then drained by an efferent vessel at least as large in calibre as the afferent arteriole. After a short straight course towards the medulla this vessel breaks up into a number of long vessels (the straight arterioles or *vasa recta*) which follow the course of the loops of Henle bending back upon themselves to drain into the arcuate veins (Fig. 32.5), an arrangement

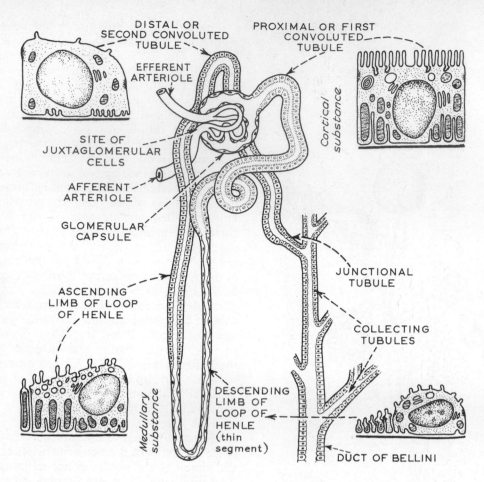

Fig. 32.3 The nephron and collecting tubule diagrammatically represented. (*After* Homer Smith.) The details of the cells given on either side are from J. A. G. Rhodin (1958). *International Review of Cytology* **7**, 485.

which subserves the function of a counter current exchanger (p. 651). The kidney has a very large lymphatic drainage.

There is normally no anastomosis between branches of the renal artery but numerous communications exist between the intertubular capillaries, and arteriovenous anastomoses analogous to those in the skin (p. 569) are also found in the kidney. Blood issuing from the renal veins has perfused two capillary beds, namely the glomerular tuft and the peritubular capillaries, since the direct supply of blood to tubules from the renal artery is insignificant. The blood flow in the cortex is much greater than that in the medulla.

The renal blood flow amounts to some 1200

ml per min, about one-quarter of the cardiac output. The blood reaches the kidneys through the relatively short renal arteries arising directly from the aorta. The kidneys are thus perfused with arterial blood at the highest available pressure. The flow per gram of kidney greatly exceeds the metabolic requirements of the tissue and oxygen is available in abundance to the vascular smooth muscle cells. Some factor other than oxygen lack must initiate the autoregulatory process. In the dog, the renal blood flow remains relatively constant even when the blood pressure is varied between 120 and 70 mm Hg but the autoregulation mechanism has not yet been completely elucidated. The following is a tentative ex-

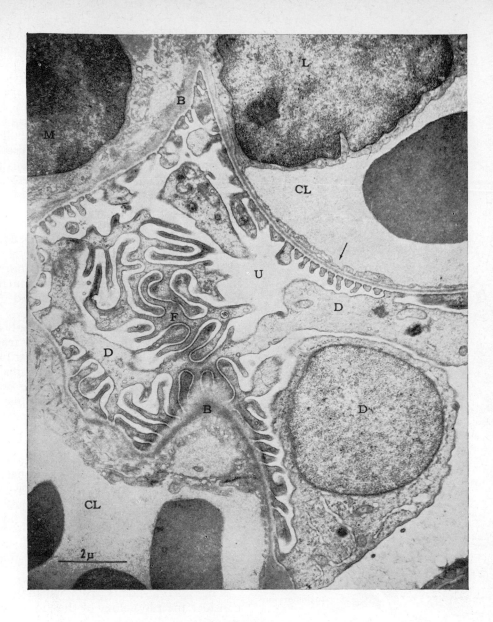

Fig. 32.4 An electron micrograph of part of a glomerulus (× 9000). CL is the lumen of a capillary containing an erythrocyte and L one of its endothelial cells. D is a glomerular epithelial cell; and U the space receiving filtrate (urinary space); B is its basement membrane and F its foot processes. At the arrow the section has passed exactly perpendicular to the division between epithelial cell and capillary. It shows the cytoplasm of the epithelial cell spreading along one side of the basement membrane and on the other side the foot processes of the epithelial cell attached to the basement membrane. The plasma in the capillary is separated from filtrate in the urinary space by basement membrane and a thin layer of endothelial cytoplasm. Some workers consider that there are gaps or 'dimples' even in the endothelial cytoplasm. The remaining foot processes of the epithelial cell have been cut tangentially (at F). M is a mesenchymal cell. (J. A. G. Rhodin from D. A. K. Black (1967). *Renal Disease,* 2nd edn, p. 38. Oxford: Blackwell).

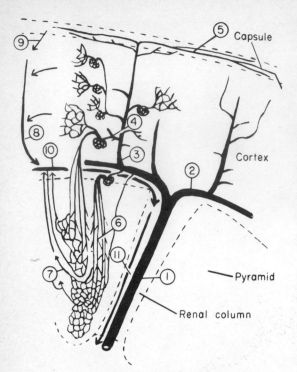

Fig. 32.5 Diagram showing blood supply of renal parenchyma, based on work of Fourman and Moffat, Morison, and Trueta *et al.* Veins indicated by arrows. (1) Interlobar artery, (2) Arcuate artery, (3) Interlobular artery, (4) Intralobular artery, (5) Capsular plexus, (6) Arteriolae rectae spuriae, (7) Venae rectae, (8) Interlobular vein, (9) Stellate vein, (10) Arcuate vein, (11) Interlobar vein. (From R. E. Coupland in D. A. K. Black (1967). *Renal Disease*, 2nd edn. Oxford: Blackwell.)

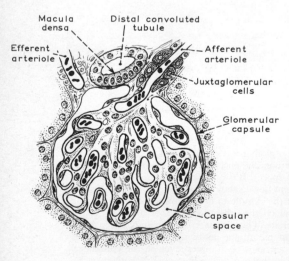

Fig. 32.6 Diagram showing relation of macula densa and juxtaglomerular cells.

planation. An increase of blood pressure (that is filtration pressure) increases the glomerular filtration rate (GFR). The more rapid flow in the proximal tubule and ascending loop of Henle provides less time for reabsorption of sodium. The increased concentration of sodium is detected by the cells of the macula densa (Fig. 32.6) in the distal convoluted tubule lying next to the juxtaglomerular cells of the afferent arteriole. Renin moves from these cells to the lumen of the afferent arteriole where angiotensin (p. 659) is formed; the afferent artery constricts and the GFR returns to normal. A fall in filtration pressure would, of course, be expected to produce the opposite effect. In man a profound and sustained fall of blood pressure, such as follows a severe haemorrhage, may reduce renal blood-flow so much that urine volume is markedly reduced (acute circulatory renal failure).

The renal nerves. The nerve supply of the kidney is mainly sympathetic, chiefly from the eleventh and twelfth thoracic and first lumbar segments of the spinal cord, but a small and variable number of parasympathetic fibres is derived from the vagus and pelvic splanchnic nerves. The renal nerves run in a plexus along the renal arteries to supply the blood vessels, but it is doubtful if the tubular cells have a nerve supply. Experimental stimulation of the nerves to the kidney produces vasoconstriction and a fall in the output of urine. However, in man section of the renal nerves has only slight and inconstant effects on renal blood flow.

RENAL FUNCTION

The starting-points from which we must consider the physiology of the kidney are the structure of the nephron and a comparison of the composition of the fluid reaching the kidney (blood plasma) and the fluid leaving it, that is the urine (Table 32.7). We have, for example, to account for the virtual absence of protein and glucose in urine, for the very high concentration in urine of ammonia and creatinine when other substances such as sodium and calcium may appear in nearly the same concentration in urine and plasma.

GLOMERULAR FUNCTION

Both Ludwig and Bowman in the middle of

Table 32.7 Comparison of representative values for the concentration of certain substances in plasma and urine of man
(The values given in this table are approximate. The range of variation in healthy persons is considerable)

	Plasma (g per 100 ml)	Urine (g per 100 ml)	Ratio $\dfrac{\text{Urinary conc.}}{\text{Plasma conc.}}$
Water	90 to 93	95	—
Proteins and other colloids	7 to 8	0*	—
Urea	0·03	2	60
Uric acid	0·003	2	15
Glucose	0·1	0·0*	—
Creatinine	0·001	0·1	100
Ammonia	0·0001	0·05	500
Sodium	0·32	0·6	2
Potassium	0·02	0·15	7
Calcium	0·01	0·015	1·5
Magnesium	0·0025	0·01	4
Chloride, as Cl	0·37	0·6	2
Phosphate, inorganic as P	0·003	0·12	40
Sulphate, as H_2SO_4	0·003	0·18	60

*The usual tests are negative but more delicate tests show that traces of glucose are present in normal urine. About 100 mg protein are excreted per day.

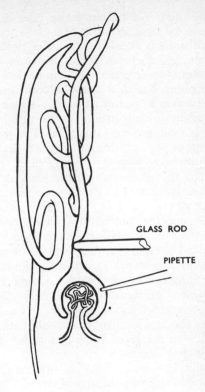

Fig. 32.8 Method of obtaining glomerular fluid by a pipette (10 μm in diameter) inserted into the glomerular capsule. The tubule is shut off by pressure with a blunt-ended glass rod.

the last century guessed that the function of the glomerular capsule was to produce a filtrate of blood plasma, but they did not agree about the role of the tubules. The renal corpuscle seems designed for filtration and receives its blood almost directly from the aorta. Thus the hydrostatic pressure in the glomerulus, being greater than the osmotic pressure of the plasma proteins (p. 406) is sufficient to cause water and simple solutes of low molecular weight to pass through the three layers of the filtering membrane.

By the use of a micromanipulator, a tiny pipette can be inserted into the glomerular capsule and a specimen of glomerular filtrate obtained for analysis (Fig. 32.8). It consists essentially of plasma without protein; the concentrations of sodium, chloride, potassium, glucose and urea are almost identical with those in blood plasma. The polysaccharide inulin, after injection into the frog's circulation, is also found in the same concentration in the plasma and in the glomerular filtrate. This relatively large molecule (mol. wt. 5200)

is thus transferred across the glomerular membrane in the same way as urea and chloride. Since large molecules pass as quickly across the membrane as small molecules the transference cannot be a process of diffusion but must be one of filtration in which all molecules smaller than the pores of the filter pass through; the only difference between the fluid to be filtered (plasma) and the filtrate is the absence in the latter of substances of large molecular size. Experiments with dextran, another polysaccharide of which solutions can be prepared containing molecules of different sizes, have shown that the pores in the glomerular membrane are not of equal sizes. All are permeable to dextran molecules of 3 nm diameter (mol. wt. 10 000) which pass just as freely as water molecules across the glomerular membrane. Larger molecules have a progressively lower clearance; molecules 14 nm in diameter (mol. wt. 50 000) are almost entirely absent from the glomerular filtrate.

The Kidney 647

The glomerular membrane does not normally allow large amounts of plasma protein to pass, although small quantities of protein are present in glomerular filtrate. Since only traces of protein (100 mg per day) can be detected in the urine in health it must be concluded that such protein as passes through the glomerular membrane is reabsorbed by the tubules, and there is evidence that this is so. The presence of protein in the urine, a common finding in renal disease, implies an increased leakage through the glomeruli rather than defective tubular reabsorption. Proteinuria may also be found where there is no renal disease. Examples are the proteinuria which occurs in congestive cardiac failure as a consequence of increased pressure in the renal veins and the orthostatic proteinuria which is found in a small proportion of healthy young people. Orthostatic proteinuria occurs only when the subject is erect and disappears when he lies down; the mechanism is probably increased pressure in the renal veins of a rather greater degree than normal.

About 125 ml of fluid are filtered through the glomerular membranes of the two kidneys per minute (GFR p. 653). Since the total surface area of the glomerular capillaries is about 1 m^2 the amount filtered per cm^2 per minute is only $125 \div 10^4 = 0.0125$ ml. Since there are about 2×10^6 glomeruli about 60×10^{-6} ml must be filtered through each per minute. This is an average value and does not mean that all glomeruli filter the same amount. Micropipettes placed in a superficial nephron of the rat kidney show that about 20×10^{-6} ml is filtered per minute whereas in a juxtamedullary glomeruli about 40×10^{-6} is filtered per min. The GFR in the rat depends on the sodium content of the diet; a high sodium diet increases the GFR in the superficial nephrons but lowers the GFR in the juxtamedullary glomeruli. Since the blood leaving the glomeruli has lost one-fifth of its water but no plasma proteins the osmotic pressure due to plasma proteins is raised.

Glomerular pressure. The pressure in the afferent arteriole of the glomerulus of the frog has been measured directly by Hayman. A micropipette connected to a manometer was first inserted into the fluid space in a glomerular capsule and, after blocking the neck of the tubule by pressing on it with a glass rod (Fig.

32.8), the pressure in the capsule was raised first until (1) the flow of blood was arrested in some of the capillaries of the glomerulus and secondly until (2) no red cells entered the glomerulus from the afferent arteriole. Measurement (1) is taken as the mean pressure in the glomerulus, usually about 50 per cent of aortic pressure, and (2) as the afferent arteriolar pressure, usually 85 per cent of aortic pressure (which is 21 to 61 cm H_2O in the frog). The pressure in the glomerulus is thus adequate to allow the production of a protein-free filtrate of plasma. If the fluid in the glomerular capsule is really an ultrafiltrate of plasma then a rise in arterial pressure might be expected to augment the volume of filtrate and an increase in the ureteric pressure, a back pressure, to diminish it. Both expectations have been shown to be correct.

Alterations in renal blood flow in man have little effect on glomerular filtration rate (see p. 656). It can be taken as a general principle that, no matter the volume of the glomerular filtrate, the volume of the urine is governed almost entirely by the amount of water reabsorbed by the tubules.

Effective filtration pressure. The hydrostatic pressure of the blood in the glomerular capillaries forces water and solutes through the glomerular membrane against the osmotic pressure of the plasma proteins. Pressure is also necessary to force the filtrate from the glomerular capsule along the lumen of the tubule. Thus the effective filtration pressure P_{ef} is given by

$$P_{ef} = P_g - P_{pp} - P_c$$

where P_g = blood pressure in glomerular capillaries,

P_{pp} = osmotic pressure of plasma proteins (about 25 mm Hg),

and P_c = intrarenal pressure, that is the smallest pressure which must be applied to the ureter to reduce urine flow (about 20 mm Hg).

From this it can be seen that P_{ef} can be reduced to zero, when filtration ceases, by lowering the arterial pressure (and therefore P_g) or by raising ureteric pressure, and thus increasing P_c. Since P_{ef} is then zero, P_g must equal $P_{pp} + P_c$. In other words, P_g, the pressure in the glomerular capillaries, must always exceed 45 mm Hg. The level to which the arterial pressure must be lowered to suppress urinary flow varies according to the conditions of the experiment, but even the lowest values, about 40 mm Hg, are in excess of the osmotic pressure of the plasma proteins. By comparing the effects of raising venous and ureteric pressures Winton calculated that the pressure in the glomerular capillaries in the excised mammalian kidney is about 70 per cent of the aortic pressure. This is more than enough to overcome the osmotic pressure of the plasma proteins. If, however, the blood pressure falls considerably the output of urine may be markedly reduced.

TUBULAR FUNCTION

Glomerular filtrate obtained by a micropipette is, as we have seen, a filtrate of blood plasma but, since the concentrations of substances in the filtrate are very different from their concentrations in urine (Table 32.7), the composition of the tubular fluid must be altered considerably during its passage to the collecting tubules. These changes have been studied in great detail in both frog and small mammals by microperfusion techniques. A single nephron can be observed by injection of lissamine green into a proximal tubule near the glomerulus, the dye identifying first the proximal tubule, the distal tubule and then the collecting tubule. Fluid can then be removed by micropipettes from various parts of the same nephron.

In the proximal tubules about two thirds of the water and sodium are reabsorbed from the glomerular filtrate in such a way that the fluid remaining in the lumen of the tubule is isotonic to plasma. There is thus no concentration gradient for sodium across the tubular cells. The absorption must be an active process because the oxygen consumption increases with increasing sodium reabsorption and also because it occurs against the potential gradient referred to below. It has been suggested that sodium and water are reabsorbed by passing through intercellular channels in the proximal tubule which are lined by adenosine triphosphatase, an enzyme known to be involved in sodium transport.

In spite of large changes in GFR the urinary excretion of sodium is kept relatively constant mainly by compensatory changes in the reabsorption of sodium in the proximal tubule. By injection of hypertonic dextran into the peritubular venous capillaries with a micropipette and by taking samples of the tubular fluid it was found that sodium and water reabsorption were inversely related to the plasma protein osmotic pressure. This effect occurs when the filtration fraction rises and causes a rise in plasma protein concentration in the blood leaving the glomerulus to pass to the peritubular capillaries.

The other solids are reabsorbed in unequal proportions and at different situations; thus the first part of the proximal tubule reabsorbs amino acids, glucose and almost completely the small amounts of protein that pass through the glomerulus; to a large extent reabsorption of phosphate also occurs in this area. Sodium, chloride and bicarbonate are probably reabsorbed uniformly along the length of the proximal tubule, as well as in the distal tubule, while potassium, and also uric acid, are reabsorbed in the proximal and secreted into the distal tubule. The fluid in the tubular lumen and the collecting duct lumen is 10 to 60 mV negative to the peritubular fluid but in the loop of Henle the difference is only 1 to 3 mV. This gradient favours the passage of Cl^- ions into the peritubular fluid but not Na^+ and K^+. These two must therefore be carried by an active transport mechanism.

Disorder of the reabsorption mechanisms in disease, often congenital, is frequently accompanied by glycosuria, phosphaturia and the loss of amino acids in the urine.

The loop of Henle reabsorbs only 5 per cent of the water. The distal tubules, where the fluid is approximately isotonic, remove about 20 per cent of the water and rather less sodium (Fig. 32.9). The cortex is therefore the major site of reabsorptive activity. Wirz found in the unipapillary kidney of the golden hamster that fluid withdrawn by micro-puncture from the proximal tubule was isotonic with plasma. Fluid from the first part of the distal tubule was, however, invariably hypotonic. It is therefor deduced that the function of the loop of Henle is to absorb sodium chloride in preference to water. Even when small amounts of concentrated (hypertonic) urine are being produced (*antidiuresis*) Wirz finds that fluid from the distal part of the distal convoluted tubule is

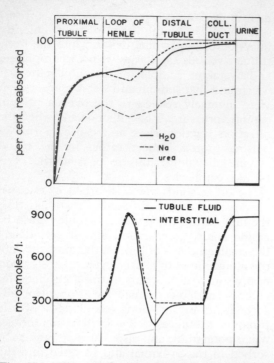

Fig. 32.9 This diagram shows above the amount of reabsorption of the various constituents of the glomerular filtrate in the different parts of the nephron. The lower diagram shows the osmolarity of the tubular fluid in the different parts of the nephron. (J. N. Barker (1963). *Medical Clinics of North America* **47**, 883.)

still only isotonic with plasma so that the final concentration to produce a markedly hypertonic urine must take place in the collecting tubules or ducts. Moreover, in such kidneys there is progressive concentration of fluid in the loops of Henle and in the collecting ducts from cortex to the tip of the papilla. The *counter-current theory* proposes a mechanism of urinary concentration which takes this evidence into account. In Fig. 32.10, the density of shading represents the osmotic pressure in each limb of a simplified loop of Henle; D, the descending limb, and A, the ascending limb, are shown juxtaposed so that S represents two layers of cells, some or all of which are supposed to produce a small osmotic gradient such that the fluid in D is rendered more concentrated than the fluid in A (stage 1). The process is, of course, continuous but for ease of understanding it can be considered in stages: in Figure 32.10 the flow is supposed to be interrupted at stages 1, 3 and 5. We can

suppose that the loop of Henle is first filled with isotonic fluid from the proximal convoluted tubule and then the flow is stopped (stage 1) while Na^+ ions are actively transferred from A to D. The fluid in D becomes hypertonic, that in A hypotonic. If a small amount of fluid is allowed to flow (stage 2) the osmotic gradient at the lower end of the loop disappears but if the flow is stopped again (stage 3) further transfer of Na^+ ions from A to D makes the fluid at the lower end of D even more hypertonic and repetition of this multiplier process (stages 4 and 5) increases the hypertonicity at the lower end of the loop of Henle. This effect is achieved without producing a large osmotic difference at any given level between the fluid in the two limbs of a loop (see stage 5 in Fig. 32.10). Since Na^+, Cl^- and also urea pass out actively or passively into the tissue around the loop of Henle this countercurrent process produces a great difference in osmotic pressure between cortex and medulla. The countercurrent mechanism depends on the fact that the ascending limb of the loop of Henle, which is thicker-walled than the descending limb, is impermeable to water so that when sodium is actively transferred from it, water does not move with the sodium as occurs for example in the proximal tubule. It is the movement of sodium without water which is responsible for the essential alterations in tonicity of the fluid within the loop.

The maximum extent to which different animals are able to concentrate the urine increases with the length of the loop of Henle. For example the desert rat, which depends for its existence on a remarkable ability to conserve water, has a particularly long loop. The efficiency of the concentrating system requires a nice balance between the fluid flow along the loop and the blood flow in the vasa recta. The maximum concentration of solute (mainly sodium) in the urine, and in the medulla, depends finally on the active transport of sodium in the loop. This is increased by aldosterone (p. 663) and diminished by adrenalectomy and cold. It also depends upon an adequate amount of sodium in the fluid reaching the loop. Similarly if the supply of urea is diminished, as for example will occur if a low protein diet (20 to 30 g per day) is taken the concentrating ability of the kidney is impaired.

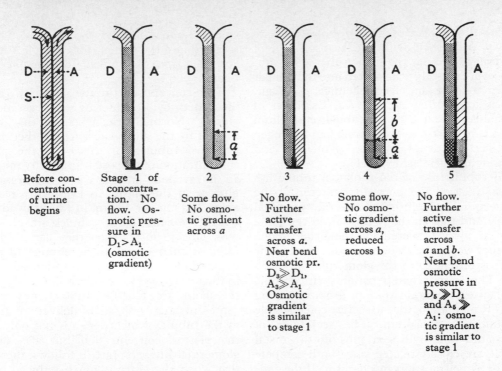

D‥	‥A	D	A	D	A	D	A	D	A	D	A

S‥

| Before con-centration of urine begins | Stage 1 of concentra-tion. No flow. Os-motic pres-sure in $D_1 > A_1$ (osmotic gradient) | 2 Some flow. No osmo-tic gradient across a | 3 No flow. Further active transfer across a. Near bend osmotic pr. $D_3 \gg D_1$, $A_3 \gg A_1$ Osmotic gradient is similar to stage 1 | 4 Some flow. No osmo-tic gradient across a, reduced across b | 5 No flow. Further active transfer across a and b. Near bend osmotic pressure in $D_5 \gg D_1$ and $A_5 \gg A_1$: osmo-tic gradient is similar to stage 1 |

Fig. 32.10 A simplified scheme of the countercurrent process described in the text. The depth of shading indicates the relative tonicity of the fluid in the loop of Henle. At stages 1, 3 and 5 the flow is supposed to be stopped (by a tap at the end of the loop) while ions are actively transferred from the ascending limb A to the descending limb D of the loop of Henle. Note that the maximum tonicity occurs at the tip of the loop. (*After* P. Wirz.) (*By courtesy of* P. Antonis.)

Vasa recta. These blood vessels (Fig. 32.5) run alongside the loops of Henle so that the blood flows in a countercurrent fashion towards the tip of the papillae and back again to the arcuate vessels running between the cortex and medulla. The blood flow in those vessels is comparatively slow and their walls are of capillary thickness. The fluid within them is therefore in equilibrium with the interstitial fluid of the medulla and thus with the fluid in Henle's loop. This arrangement further promotes a high concentration in the medulla of material (solute or gas) passing from the tubular lumen into the medullary interstitial tissue; in other words, it acts as an exchange multiplier for Na^+ and urea. Furthermore there is experimental evidence suggesting that antidiuretic hormone aids the reabsorption of water by reducing the blood flow through the vasa recta as well as by increasing the permeability of the cells lining the collecting ducts (p. 661). The ascending vasa recta and the medullary capillaries are unusually permeable to protein. The significance of this rapid protein leak is not yet known.

Distal and collecting tubules. The fluid reaching the distal tubule is hypotonic and the function of the distal convoluted and collecting tubule is to produce a concentration of the urine appropriate to the needs of the body, a process which is dependent on antidiuretic hormone (ADH) (p. 661).

Water and salt is reabsorbed in the distal convoluted tubules to render the fluid in the collecting tubules approximately isotonic. As the collecting ducts pass through the increasingly hypertonic environment of the medulla the fluid within them loses water passively until it reaches the renal pelvis as the finally concentrated urine (Table 32.7). In water diuresis ADH is lacking and the collecting ducts are relatively impermeable to water; the fluid within them remains dilute in spite of the high osmolality of the medulla.

Tubular secretion. The tubular cells are largely concerned with the reabsorption of material filtered at the glomeruli. Some substances are actively transported into the tubular lumen against an electrochemical gradient, that is to say they are *secreted*. Examples of substances secreted by the tubular epithelium in man are creatinine, potassium and hydrogen ion; certain drugs such as penicillin, and *p*-aminohippuric acid (PAH).

Micropuncture of the proximal tubule has shown that all the potassium filtered at the glomerulus is reabsorbed in the proximal tubule and that all the potassium in the urine has been actively secreted in the distal tubule. In other words the amount of potassium in the urine is unrelated to the glomerular filtration rate. This is important in patients with chronic renal disease who, in spite of gross reduction in GFR may still be able to secrete considerable amounts of potassium. The secretions of potassium and hydrogen ions in the distal tubule are closely related since both compete for the same transport mechanism. If the concentration of H ions is high in the tubular cells, H-ion tends to be transported; if the intracellular concentration of potassium ion is high and that of H-ion low, potassium occupies the transport mechanism to the exclusion of H-ion.

Apart from active reabsorption and secretion large movements of water and solute occur by passive diffusion through the tubular cells both inwards from the lumen to the blood and in the reverse direction. The active reabsorption of glucose and sodium in the proximal tubule creates osmotic and electrical gradients which account for the simultaneous reabsorption of large amounts of water and cation, mainly chloride with bicarbonate. Urea is reabsorbed throughout the length of the nephron. Many weak acids and bases (such as aspirin and phenobarbitone) move passively along electrochemical gradients.

Tubular capacity. The capacity of the tubular epithelium to reabsorb certain substances from the glomerular filtrate, or to secrete them into the tubular lumen, is limited because the transport mechanisms cannot deal with more than a limited amount of solute per minute. The maximum rate of tubular transport or maximum tubular capacity is referred to as T_m. This term can be applied to many substances handled by the tubules, whether secreted into the lumen (for example PAH) or absorbed from the lumen (for example phosphate).

The concept of tubular capacity can be illustrated by considering the excretion of glucose. Normally the whole of the glucose filtered by the glomeruli is reabsorbed by the proximal tubules; if, however, the plasma concentration of glucose is gradually raised by an intravenous infusion a concentration of glucose in the plasma is eventually reached, about 180 mg per 100 ml in a healthy adult, at which the cells of the tubules are presented with glucose faster than they can absorb it. Glucose then begins to be eliminated in the urine. This limiting concentration of glucose in the blood used to be referred to as the 'renal threshold' for glucose. In fact, however, it represents a critical rate of delivery of glucose to the tubules which depends not only upon the plasma concentration but also on the glomerular filtration rate. It follows, therefore, that when the latter is much reduced, as for example in a severely dehydrated diabetic patient, the tubular cells may be able to reabsorb all the glucose even though the concentration of glucose in the blood is much above the normal; this is the explanation of the apparent paradox of hyperglycaemia without glycosuria. A method for measuring the maximum rate of tubular reabsorption of glucose (glucose-T_m) is given below.

These findings are best explained by a carrier mechanism. The carrier combines with glucose in the inner (tubule) side of the cell and transports it to the outer, where the glucose separates from the carrier which is then free to transport more glucose. The transport mechanism, like any transport system, can carry only a limited number of passengers in a given time. At this value it is said to be saturated.

In man glucose-T_m is measured by raising the plasma glucose by intravenous injection to a high level and then measuring the inulin clearance (p. 653) and the glucose excretion. If the inulin clearance is 125 ml per minute when the plasma sugar is 400 mg per 100 ml then $125 \times 400/100 = 500$ mg of glucose is filtered per min at the glomeruli but 150 mg of glucose is found in the urine per minute. The maximum tubular reabsorption of glucose (glucose-

T_m) is, therefore, $500 - 150 = 350$ mg per minute. It serves as a measure of the total mass of functioning tubular cells.

Just as there is a maximum reabsorptive capacity for each substance reabsorbed from the glomerular filtrate, so there is a maximum secretory capacity, that is to say, the rate at which materials can be transferred from the blood across the tubular epithelium into the lumen is limited. Because it is easy to measure and has a high rate of secretion para-amino-hippuric acid (PAH) is the substance normally employed to measure secretory T_m. The procedure is similar to that for measuring glucose-T_m.

The reabsorption of inorganic phosphate resembles that of glucose except that the amounts reabsorbed depend upon the concentration of phosphate in the plasma, upon pH (acidosis and alkalosis, p. 699) and on the activity of the parathyroid gland (p. 213). When the plasma concentration of phosphate is low, the phosphate in the glomerular filtrate is almost completely reabsorbed. Large doses of parathyroid hormone or an excess of hormone as in hyperparathyroidism (p. 216) cause a great loss of phosphate in the urine.

RENAL CLEARANCE

The concept of clearance has proved useful in estimating kidney function. The clearance rate of any substance is the volume of plasma which contains the amount of the substance which is excreted in the urine in 1 minute. The term clearance can be applied to any substance in the plasma which appears in the urine by whatever process it reaches the urine, filtration, secretion or excretion or a combination of these. By choice of suitable substances clearance rates reveal the glomerular filtration rate (see below) and the renal plasma flow (p. 655).

Glomerular filtration rate (GFR). This can be measured by any non-toxic substance which is small enough to pass through the glomerular membrane but yet is neither excreted into the tubules nor absorbed from the tubular urine. Inulin, a polysaccharide of mol. wt. 5200 (p. 47), containing some 30 fructose residues, meets these requirements. A large dose of inulin is given intravenously and then a continuous infusion to keep the plasma concentration approximately constant.

The data required for the calculation of *inulin clearance* are the volume of urine V passed in ml per minute, the concentration P of inulin in the plasma in g per 100 ml and its concentration U in the urine in g per 100 ml. The amount of inulin excreted per minute is obviously $UV/100$ g. Since the glomerular fluid is an ultrafiltrate of blood plasma each ml of glomerular filtrate contains $P/100$ g inulin. If x is the volume of glomerular filtrate per minute then x ml of glomerular filtrate contain $xP/100$ g inulin. But $UV/100$ g of inulin have been excreted per minute therefore $xP/100 = UV/100$ or $x = UV/P$ ml per min.

An arithmetical example may make the idea of clearance easier to understand. Suppose the plasma to contain 50 mg (that is 0·05 g) of inulin per 100 ml and the urine 6·25 g inulin per 100 ml, while the urine is excreted at the rate of 1 ml per min. The inulin clearance or glomerular filtration rate (GFR) is then

$$\frac{6\cdot25 \times 1}{0\cdot05} = 125 \text{ ml/min (180 l/day)}.$$

This value is referred to as the *plasma clearance* for inulin, since it can be thought of as that volume of plasma from which all the inulin has been removed (cleared) in one minute. The evidence that the inulin clearance actually measures the GFR is as follows. The plasma clearance of inulin is not affected by the concentration of inulin in the plasma. The concentration of inulin in the glomerular filtrate obtained by direct puncture is the same as its concentration in the plasma, and the large molecule of inulin is not likely to diffuse out of the renal tubules. Aglomerular kidneys of certain fish do not secrete inulin. The plasma clearance of inulin is the same as that for sorbitol and mannitol, which have much smaller molecules (mol. wt. 182 in both cases). Furthermore, the inulin clearance in the dog is equal to the clearance of creatinine, ferro-cyanide and thiosulphate. These facts are difficult to explain unless all these substances are unabsorbed on their way down the tubules. If an animal or a man is injected with the glycoside phlorizin, glucose appears in the urine, since the tubules are then unable to reabsorb glucose (cf. p. 327). The glucose and inulin clearances are then for practical purposes identical; this suggests that in these circumstances both are filtered and no tubular

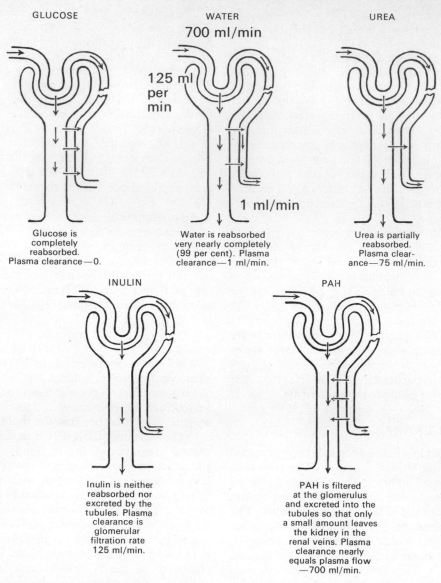

GLUCOSE

WATER
700 ml/min

125 ml
per
min

1 ml/min

UREA

Glucose is
completely
reabsorbed.
Plasma clearance—0.

Water is reabsorbed
very nearly completely
(99 per cent). Plasma
clearance—1 ml/min.

Urea is partially
reabsorbed.
Plasma clear-
ance—75 ml/min.

INULIN

PAH

Inulin is neither
reabsorbed nor
excreted by the
tubules. Plasma
clearance is
glomerular
filtration rate
125 ml/min.

PAH is filtered
at the glomerulus
and excreted into the
tubules so that only
a small amount leaves
the kidney in the
renal veins. Plasma
clearance nearly
equals plasma flow
—700 ml/min.

Fig. 32.11 Diagram of a nephron with afferent and efferent glomerular vessels and a tubular capillary. The arrows show the direction of movement of the substance under discussion and their length indicates the amount which is moving. The middle diagram above shows the amount of fluid passing into the kidney (700 ml per min), the amount filtered at the glomerulus (125 ml per min) and the amount eliminated as urine (1 ml per min). Since the 700 ml of plasma flowing through the glomerulus per min carries the red cells the volume of blood flowing through the kidneys is 1200 ml/min.

reabsorption occurs. When the blood pressure is too low for glomerular filtration inulin does not appear in the tubular fluid, but phenol red which is secreted by the tubules does appear.

Creatinine is completely filtered at the glomerulus and is not reabsorbed in the tubules and the endogenous creatinine clear-

ance is very widely used in clinical practice to give an estimate of GFR. The clearance values are as a rule somewhat higher than those obtained by inulin because, in man, some creatinine is secreted into the tubules (increasing the value for U). The error is lessened, however, by the fact that the value obtained for the plasma level of creatinine (P) is usually rather greater than the true value because some plasma chromogens give a similar reaction. In practice an accurate 24 hour urine specimen is collected and a blood sample is taken any time during the period of collection. No restrictions on diet, fluid intake or activity are necessary.

The clearance of urea, about 75 ml per minute, is less than the inulin clearance because much of the urea in the glomerular filtrate diffuses back into the blood as the filtrate passes along the tubules. This passive process depends on the fact that the concentration of urea in the tubular fluid exceeds its concentration in the plasma. The passage of urea from the tubules into the peritubular capillaries is greatest (and the elimination of urea from the body is least) when the urine is most concentrated and the flow along the tubules the slowest. When the urine flow is about 2 ml per minute the urea clearance is about 60 per cent of the GFR (60 per cent of 125 ml per minute is 75 ml per minute). As the urine flow falls below this level progressively more urea diffuses back into the blood and the urea clearance is therefore lower. This critical dependence on urine flow has rendered the test of limited value in clinical practice.

Because of their ease of measurement radioactive materials, such as ^{57}Co-cyanocobalamine or ^{51}Cr-edetic acid, have been used to measure GFR but the accuracy of the results is likely to be prejudiced by the binding of such substances to plasma protein to an unknown extent, thus making them incompletely filterable.

RENAL BLOOD FLOW

p-Aminohippuric acid, PAH, is used to estimate renal plasma flow. A large dose is given intravenously and then a constant infusion to make up for the loss of PAH in the urine. If the plasma concentration of p-aminohippurate (P_{PAH}) is relatively low (between 1 and 8 mg/100 ml) about 90 per cent of it is cleared, by proximal tubular secretion, in one passage through the kidney. PAH clearance can thus be used to give a reasonable approximation of renal plasma flow (RPF).

$$RPF = \frac{U_{PAH} \times V}{P_{PAH}}$$

where U_{PAH} is the concentration of PAH in the urine, V is the volume per minute, and P_{PAH} is the concentration in mg per 100 ml in the peripheral venous blood. In normal man the RPF thus calculated is 660 ml per minute per 1·73 m^2 surface area. The true RPF is about 10 per cent higher than this value, say 700 ml/min. From the RPF and the haematocrit value (p. 416) (normal 45 per cent) the renal blood flow (RBF) can be calculated as follows

$$RBF = RPF \left(\frac{1}{1-\text{haematocrit}}\right)$$

$$RBF = 660 \left(\frac{1}{1-0·45}\right) = 1200 \text{ ml/min}$$

The PAH clearance is an acceptable approximation to RPF in the normal kidney but in renal disease it is necessary to measure the reduced extraction of PAH which occurs and thereafter to make the appropriate correction.

The substance diodone (diodrast), 3,5-diiodo-4-pyridon-N-acetic acid, is cleared in a very similar way to PAH and was widely used as a measurement of RPF in earlier studies (for example Fig. 32.12) but it has the disadvantage that its estimation, in contrast to that of PAH, is difficult.

The concept of plasma clearance is illustrated diagrammatically in Figure 32.11. The range of plasma clearance is from zero for glucose, at normal plasma levels completely absorbed, up to 700 ml/min for PAH which is almost entirely removed from the plasma. The ratio inulin clearance/PAH clearance, that is, glomerular filtrate/plasma flow through kidney, is called the *filtration fraction*. It amounts to some 20 per cent of the renal plasma flow and represents the proportion of the latter filtered.

The following are approximate figures for a normal man (ml/min).

Blood flow through the kidneys	1200
Plasma flow through the kidneys (PAH clearance)	660
Glomerular filtrate (inulin clearance)	125
Urine	1
Filtration fraction ($125 \div 660 \times 100$)	19 per cent

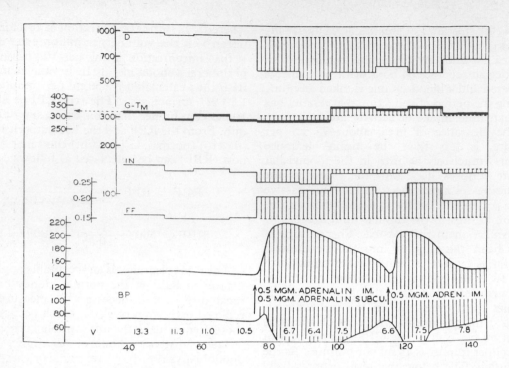

Fig. 32.12 Action of adrenaline on effective renal plasma flow, etc. D = diodone clearance in ml per minute. IN = inulin clearance in ml per minute. Filtration fraction, FF = inulin/diodone clearance ratio. Blood pressure BP in mm Hg and urine flow V in ml per minute. G-Tм = maximum rate of tubular reabsorption of glucose under conditions of tubular saturation, in mg per minute. G-Tм is indirect index of the number of active glomeruli in the kidney (Homer W. Smith (1943). *Lectures on the Kidney.* Lawrence, Kansas: University of Kansas).

The clearance tests devised by Homer Smith and his group have made it possible to investigate renal physiology in unanaesthetized normal man. Figure 32.12 shows an experiment in which a large amount of adrenaline was given. The renal plasma flow (diodone clearance D) was reduced from 926 ml per min to a minimum of 510 ml per min, although the mean blood pressure (BP) was probably unaffected. Obviously vasoconstriction had taken place and the site can be inferred thus. Narrowing of the afferent arterioles would have produced a fall of glomerular pressure with little change in the filtration fraction. In fact the filtration fraction (FF) rose and the filtration (IN) was almost unaffected. This can be explained only by a constriction of the efferent glomerular arteriole. Experiments of this kind show that glomerular filtration rate in man is remarkably constant under basal conditions. Variations in urinary output are largely dependent on variations in tubular reabsorption of water. The glucose-T_m (G-T_m) was measured during the experiment illustrated in Figure 32.12 and was almost unaffected (306 mg per min before adrenaline and 325 mg per min after). This shows that no glomeruli were shut down by this large dose of adrenaline. Other experiments in man show that exercise, by producing a constriction of the renal arterioles, may cause a reduction of 20 per cent or so in the glomerular filtration rate. Dehydration at rest in a hot environment may reduce it by 10 per cent. If exercise is carried out by a dehydrated subject in a hot environment the glomerular filtration may be reduced to one-half.

Constriction of the efferent arterioles with a marked reduction in urinary output has been found to occur when the subject of the experiment becomes alarmed. This reaction may be due to the outpouring of adrenaline from the adrenal medulla or to impulses passing along the sympathetic nerves to the kidneys (see also p. 646).

Renal vasoconstriction also occurs in syncope produced by standing still in the upright position, but here the constriction affects the afferent arterioles. In traumatic shock there is also renal vasoconstriction; since the kidneys normally receive a very large part (about one-quarter) of the cardiac output this vasoconstriction may be beneficial by diverting blood to other organs, but sometimes the resulting renal ischaemia leads to tubular necrosis and severe impairment of renal function which may be fatal.

Renal blood flow in man can be greatly increased by infusing acetylcholine or bradykinin into the renal artery. Dopamine given intravenously produces an increase in renal blood flow. In artificial pyrexia produced by intravenous injection of a foreign protein there is increased renal blood flow with a reduction in the filtration fraction but no change in GFR.

THE VOLUME AND COMPOSITION OF THE URINE

Even in health the volume and composition of the urine varies widely from day to day, such variations being governed by, among other things, the type of food consumed, the volume of fluid taken and the amount of fluid loss by other channels—a factor which itself depends on environmental temperature and humidity, exercise and sweating (Chap. 42).

Usually the daily output of urine is between 1000 and 1600 ml. The urine is not produced at a constant rate for the volume of urine during the day (8 a.m. to 8 p.m.) is two to four times greater than that secreted during the night (8 p.m. to 8 a.m.). In chronic renal disease there is a continuous osmotic diuresis (p. 666); the greater amount of urine which is produced (*polyuria*) is often brought to the patient's notice because of the need to pass urine during the night (*nocturia*) (Fig. 32.13). An osmotic diuresis similarly occurs in patients with untreated diabetes mellitus (p. 330).

The specific gravity of the urine in health can vary between 1·001 and 1·040 (50 to 1300 m-osmoles per kg), but it usually lies between 1·010 and 1·025 representing the elimination of some 40 to 60 g of dissolved solids in 24 hours. In renal disease the kidney becomes unable to alter the concentration of the urine and the specific gravity remains fixed about 1·010 (*isosthenuria*) equivalent to an osmolality of 300 m-osmoles per kg, the same as that of plasma.

The yellow colour of normal urine is due mainly to the pigment urochrome. The chemical nature and source of the substance are unknown but it is formed probably endogenously, since its output increases with tissue protein destruction or increase of metabolism. Traces of other pigments, such as urobilin and uroerythrin, are also normally present. The colour of the urine in general varies according to its concentration. Freshly passed urine has hardly any smell, but when it is allowed to stand it develops an ammoniacal odour owing to the conversion by bacteria of urea to ammonia.

The average concentration of the principal constituents of normal urine is given in Table 32.7. It is, however, generally more useful to know the amount of any constituent excreted in 24 hours (Tables 32.14 and 15). Variations from day to day are such that it may sometimes be difficult to make any absolute distinction between normal and abnormal; a substance usually present in small amounts may, when the kidneys or other organs are diseased be excreted in large quantities.

Urine from normal persons contains a few leucocytes and non-squamous epithelial cells, the combined output being between 20 000 and 200 000 per hour. The urine normally contains only very small amounts of protein of the order of 100 mg per day, too little to be detected by the usual clinical tests. After severe exercise the proteinuria is transiently increased with occasional casts (cell débris) and red cells.

The Renin–Angiotensin–Aldosterone System

In 1898 Tigerstedt and Bergman observed a prolonged rise of blood pressure in the unanaesthetized rabbit after intravenous injection of extracts of normal renal cortex. They named the active principle in these extracts *renin*. However, renin is not itself a pressor substance; it is a proteolytic enzyme capable of hydrolysing a substrate *angiotensinogen*, an α_2-globulin in the plasma protein produced in the liver, to give a vasoconstrictor substance

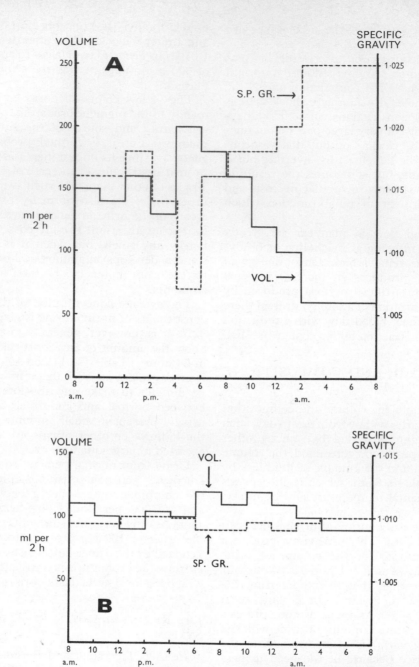

Fig. 32.13 Diurnal variations in volume of urine and specific gravity in a healthy person (A) and in a patient with advanced renal disease (B). The volume of urine produced during the night is relatively small in A; there is no such change in B. The rhythm shown in A is not present in the newborn infant. The rhythm is disorganized in persons who have lived in prolonged darkness or who are blind. An alternating light and dark environment seems necessary for the establishment of a urinary rhythm but some blind persons show a normal rhythm and persons in continuous daylight may establish normal excretory rhythms. (Recently the words *circadian* or *nychthemeral* have come into use instead of diurnal.)

called *angiotensin*. The steps in its production appear to be as follows:

$$\alpha_2\text{-globulin} \xrightarrow{\ renin\ } \text{angiotensin I (decapeptide)}$$

$$\text{angiotensin I} \xrightarrow[enzyme]{hydrolysing} \text{angiotensin II (octapeptide)} + \text{histidyl-leucine.}$$

The second step occurs mainly in the lungs. Angiotensin is destroyed by the enzyme, angiotensinase. In the presence of a synthetic pentapeptide called bradykinin-potentiating peptide which inhibits the conversion of angiotensin I to angiotensin II, the hypertensive effect of angiotensin I is much reduced. It seems that the hydrolysis must be complete to obtain the maximum vasoconstrictor effect.

The structure of angiotensin II is Asp-Arg-Val-Tyr-Ile-His-Pro-Phe if human material is used. When infused intravenously in man

Table 32.14 Principal constituents of urine (grams per 24 hours)

Urea	20 to 30	≡ 9 to 14 g nitrogen
Uric acid	0·6	≡ 0·2 g nitrogen
Creatinine	1·2	≡ 0·4 g nitrogen
Ammonia	0·5 to 0·9	≡ 0·4 to 0·7 g nitrogen
Sodium	6	≡ 260 m-equiv
Potassium	2	≡ 52 m-equiv
Calcium	0·2	
Chlorate, as Cl	7	≡ 197 m-equiv
Phosphate, as P	1·7	
Sulphate, as H_2SO_4	1·8	
Free amino acids	1	
Conjugated amino acids	2	

These figures are approximate only. The range of variation in healthy persons is considerable.

Table 32.15 To show the difference in constitution between a day's urine and the glomerular filtrate from which it was derived

Substance	Amount in glomerular filtrate (g)	Amount eliminated in the urine (g)
Sodium	600	6
Potassium	60	2
Calcium	5	0·2
Glucose	200	trace
Urea	60	35
Water	180 litres	1·5 litres

J. R. Robinson (1954). *Reflections on Renal Function.* Oxford: Blackwell.

synthetic angiotensin octapeptide is, molecule for molecule, 40 to 50 times as potent as L-noradrenaline in raising arterial pressure; it is in fact by far the most powerful pressor substance known. The rise in pressure begins immediately the infusion (2 to 3 µg per min) is started and continues for about 20 minutes after infusion is stopped (Fig. 32.16). However, prolonged infusion of angiotensin in man does not produce high blood pressure, although continuous infusion of a very small dose has done so in the rabbit.

It is now believed that the juxtaglomerular cells (Fig. 32.6) are the sole source of renin. The large highly refractile granules from individual juxtaglomerular cells of mice have been obtained by micromanipulation (Fig. 32.17A) and assayed for pressor activity after incubation with renin-substrate (Fig. 32.17B).

There is now good evidence that the physiological role of angiotensin is not to raise the blood pressure but rather to control the output of aldosterone from the adrenal cortex. This specific effect of angiotensin entitles it to be regarded as an intermediate or 'trophic' hormone. The renin–angiotensin–aldosterone system consists therefore of release of renin from the juxtaglomerular apparatus in the kidney with the production of angiotensin in the plasma and the subsequent secretion of aldosterone from the adrenal cortex. The actions of aldosterone are described on pp. 650 and 663 and in Chap. 50.

On a normal sodium intake (180 m-equiv./day) the plasma renin activity (PRA) of a recumbent subject on waking is about 100 ng/100 ml. Four hours later after walking quietly the level rises to 270 ng/100 ml. Both values are higher, and the aldosterone excretion is greater, if the subject is on a low sodium diet. The explanation seems to be that in the upright position blood pools in the legs and the reduction of effective blood volume causes a fall in renal blood flow with increased renin and aldosterone production leading to retention of sodium. Retention of sodium re-expands the

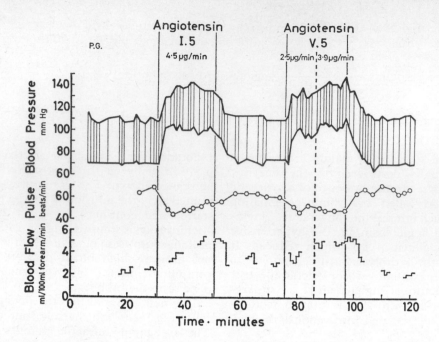

Fig. 32.16 Arterial pressure, pulse rate and forearm blood flow before, during and after infusions of two different preparations of angiotensin (labelled I.5 and V.5) into a foot vein of a healthy man of 24. (E. de Bono, G. de J. Lee, F. R. Mottram, G. W. Pickering, J. J. Brown, H. Keen, W. S. Peart & P. H. Sanderson (1963). *Clinical Science* **25**, 123–157.)

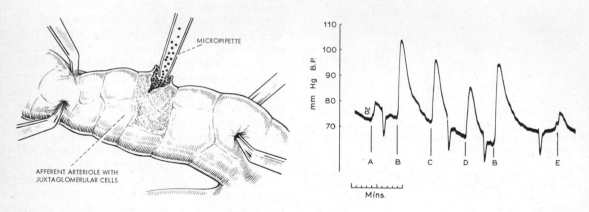

Fig. 32.17A Sampling granules in individual juxtaglomerular cells. The afferent arteriole of an isolated mouse glomerulus held with microforceps. Some of the granules in the juxtaglomerular cells can be taken up by micropipette when the cell membrane is broken.

Fig. 32.17B The detection of renin in juxtaglomerular cells. Blood pressure record from an anaesthetized rat treated with the autonomic ganglion-blocking drug pentolinium. A, 0·2 ml saline (control). B, 5 ng (nanogram) angiotensin. C, 0·2 ml dose of fluid obtained by incubating with renin substrate for 18 h at 37°C. D, 0·2 ml dose obtained by incubating the remainder of the glomerulus with renin substrate for 18 h at 37°C. E, 0·2 ml dose obtained by incubating surrounding medium similarly with renin substrate. (The depressor responses are due to spontaneous gasps.) (*By courtesy of W. F. Cook.*)

blood volume and so counteracts the effect of upright posture. If the lower limbs and the abdomen are firmly bound before the subject stands upright no rise in plasma renin is observed.

The precise relationship of the renin–angiotensin–aldosterone system to the problem of hypertension in man is still obscure. Part of the problem relates to the complexities of accurate measurement of all the components of the system in clinical circumstances. The following points appear to have been reasonably established. In the great majority of patients with hypertension (essential hypertension) there is no demonstrable abnormality of the renin–angiotensin–aldosterone system. In some patients with very severe forms of essential hypertension (malignant hypertension) there is an excess of aldosterone but it is believed that this is secondary to the hypertension and not the cause of it (secondary hyperaldosteronism). In a very small percentage of patients with hypertension the cause is a primary excess of aldosterone due to a tumour or hyperplasia affecting the adrenal glands (primary aldosteronism). In a proportion of severely hypertensive patients with disorders associated with renal ischaemia (renovascular hypertension) there is a demonstrable excess of renin, an increase in aldosterone secretion rate, and an inverse relationship between plasma renin and plasma sodium. Correction of the renovascular abnormality either by reconstructive surgery or by nephrectomy, if the condition is unilateral, may be followed by cure of the hypertension and a return of renin and aldosterone values to normal.

The kidney also produces, especially in hypoxic conditions, the hormone erythropoietin which is necessary for the formation of red cells (p. 418).

TUBULAR CONTROL OF WATER AND SODIUM LOSS

125 ml of glomerular filtrate are produced each minute. Since the total volume of extracellular fluid (p. 674) in the body is about 15 litres, the equivalent of the whole of this fluid passes through the kidney every 2 hours. Of the 125 ml of glomerular filtrate only about 1 ml appears as urine; in other words 99 per cent of the fluid that has passed out of the body into the renal tubules is reabsorbed on its way through the kidney. These enormous movements of fluid provide the means whereby the kidney can control the amount, composition and pH of the extracellular fluid.

When a healthy person drinks a large volume of water, say a litre, fairly rapidly his urine flow begins to increase almost at once. Within about 30 minutes the urine flow has risen to say 20 ml per minute and the urine osmolality is minimal (50 m-osmoles per kg; specific gravity 1001); the water load is all eliminated within 2 to 4 hours by water *diuresis* (Fig. 32.18). On the other hand if a healthy subject is deprived of fluid for over 12 hours, his urine output falls below 1 ml per minute and the osmolality reaches 1300 m-osmoles per kg; specific gravity 1040.

The finer details of all the homeostatic processes are not known but it is clear that the behaviour of the kidney is governed mainly from two sources, namely the posterior pituitary and the adrenal cortex.

The posterior lobe of the pituitary and water loss. The posterior pituitary releases into the blood an antidiuretic hormone (ADH), the chief action of which is to increase water reabsorption in the distal tubules and collecting ducts (Fig. 32.3). This has been shown very elegantly by perfusion of isolated fragments of collecting tubules of the rabbit. Water permeability, measured by THO, was increased when ADH, or cyclic 3′:5′-AMP, was added to the solution bathing the outside of the tubule but was quite unaltered when ADH was added to the fluid perfusing the lumen. In water diuresis ADH is lacking and the collecting ducts are relatively impermeable to water; the fluid in the ducts remains hypotonic in spite of the high osmolality of the medulla. A clue to the actual amount of ADH secreted under normal circumstances is given by the finding that a dog (10 to 15 kg) deprived of its posterior pituitary lobe but given an intravenous infusion of 1 to 5 milli-units of pituitary extract per hour reacts to ingestion of water like a normal animal.

Much information about the secretion of ADH was obtained by Verney by injection of solutions of different osmolality into the internal carotid artery which supplies the supraoptic nucleus and adjacent areas of the hypothalamus. His experiments in dogs (Fig. 32.19) show that there are receptors sensitive to alterations is osmotic pressure (*osmoreceptors*) in the anterior part of the hypothala-

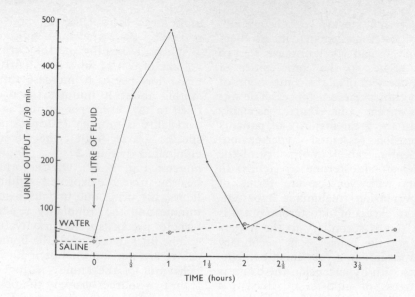

Fig. 32.18 Response to the drinking of 1 litre of water and, on the following day, 1 litre of 0·9 per cent sodium chloride solution by a healthy man. At the end of 2 hours all the water, but little of the saline, had been eliminated. It has been estimated that in man 1 litre of water drunk at 37°C is absorbed from the intestine into the blood in 25 minutes.

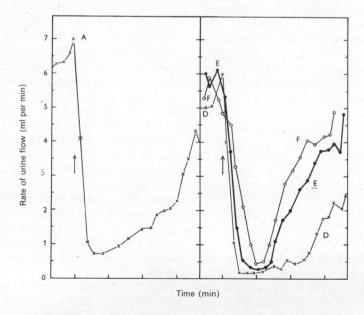

Fig. 32.19 The inhibition of water diuresis by a rise of osmotic pressure in the carotid artery. At A 10 ml of 2·5 per cent sodium chloride was injected into the right carotid artery in 12 seconds. Injection of 10 ml of 15·4 per cent glucose in 11 seconds had almost the same effect. On the right an attempt is made to imitate the antidiuretic effect of the hypertonic solution by injecting at the arrow, D, 3 m-u.; E, 2 m-u, and F, 1 m-u. of posterior lobe extract into the malleolar vein (1 m-u. = 1 milli-unit). (E. B. Verney (1947). *Proceedings of the Royal Society* B, **135**, 67.)

mus and the preoptic area of the brain (Chap. 52), probably in the supra-optic nucleus, which control the secretion of antidiuretic hormone (ADH) by the posterior lobe of the pituitary through the supraoptico-hypophysial tract. The supraoptic neurones are cholinergic as shown by the antidiuretic effects of nicotine.

The behaviour of the osmoreceptors depends on the nature of the substance responsible for the increase in the osmotic pressure of the blood. If a hypertonic solution of urea is injected rapidly (10 sec) into the carotid artery the flow of urine does not alter. Although a rapid injection of hypertonic glucose (Fig. 32.19, A) is effective in reducing urine flow, a long-continued infusion of glucose sufficient to produce a rise of 80 mg/100 ml in the carotid blood sugar has no effect on water elimination. The slow injection of sucrose produces the same secretion of antidiuretic hormone as a prolonged injection of sodium chloride of the same hypertonicity. To explain these findings Verney suggested that the osmoreceptors are freely permeable to urea, less freely permeable to glucose and relatively impermeable to sodium and sucrose. It has been calculated that an increase of osmotic pressure in the plasma of only 2 per cent causes the pituitary to secrete about 1 micro-unit of ADH per second. Such an increase of osmotic pressure is equivalent to a rise in the concentration of NaCl of 3 m-equiv. per litre.

Although the half-life of ADH in the blood is only about a minute, an injection of ADH may reduce the urinary output for an hour or so. It is therefore supposed that the hormone is very quickly bound at critical sites on the cells of the distal and collecting tubules. ADH is excreted in the urine. The ADH activity of the urine depends both on the water intake and the sodium chloride intake. On a modest salt intake (say 5 g per day) and average water intake (1·5 l) the ADH activity in the urine is about 30 milli-units per day. On a high fluid intake it is reduced to say 10 milli-units; on a low fluid intake it may be 100 milli-units. On higher salt intakes the daily output of ADH is much increased.

The ability of the renal tubules to respond to ADH depends on the integrity of the loop of Henle and the collecting ducts. The response may be impaired because of a congenital defect, long-standing potassium deficiency or hypercalcaemia. Also in chronic infections of the urinary tract involving the medulla of the kidney (chronic pyelonephritis) a dilute (less than 300 m-osmoles per kg) urine is produced even when the patient is dehydrated.

Adrenal cortex and water loss. The loss of electrolytes and the loss of water from the body are closely interconnected. It is not surprising therefore that the adrenal cortex (Chap. 50), which produces a number of substances affecting the elimination of sodium and potassium plays an important role in governing the reabsorption of water by the renal tubules. The excretion by the kidney of sodium and potassium is controlled by the steroid hormone *aldosterone* secreted by the zona glomerulosa of the adrenal cortex. In man, aldosterone first increases the elimination of potassium and hydrogen ion and then decreases the excretion of sodium without any change in GFR. However, when adrenal steroids (Chap. 50) are given to man for periods longer than 10 to 14 days the excretion of sodium rises again to normal levels although the increased excretion of potassium continues. The mechanism of this 'escape' phenomenon is not understood; it shows that aldosterone does not act simply by promoting the exchange of sodium for potassium but by increasing independently the reabsorption of sodium and the secretion of potassium (and H ion). Aldosterone acts mainly at the distal tubule but its effect on sodium reabsorption may be partly at the proximal tubule. Apart from its action on the renal tubule, aldosterone increases the reabsorption of sodium from the secretions of the intestinal mucosa and of the salivary and sweat glands. Thus, the body content of sodium rises and that of potassium falls.

Unlike other hormones of the adrenal cortex, the secretion of aldosterone is largely independent of adrenocorticotrophic hormone (ACTH, Chap. 50). The control of aldosterone secretion seems to depend on two main factors. In the first place it is increased by sodium deprivation; thus it may vary from 50 µg per day with a high intake of sodium to 1000 µg per day when the intake is low. Secondly, aldosterone secretion is increased by reduction in the fullness of blood vessels (*hypovolaemia*) such as may occur, for example, after a sudden haemorrhage. How this is brought about is uncertain but stimulation of pressure-sensitive nerve endings (baroceptors) in the common carotid artery produces a rise in aldosterone secretion. Stimulation of stretch-receptors in the right atrium produces a fall in aldosterone secretion. There is also evidence for a humoral mechanism governing the secretion of aldosterone since the adrenal glands can produce aldosterone in response to a fall in blood volume even when they are transplanted to another part of the body and are therefore without any nervous connexion. These responses are difficult to obtain in nephrectomized animals. Two critical experiments give the clue to the situation. Angiotensin activates aldosterone synthesis in tissue slices of adrenal cortex and the intravenous infusion of angiotensin II in man produces a consistent increase in aldosterone production. The system therefore consists of release of renin by the kidney which produces angiotensin in the plasma. This then stimulates the adrenal cortex to secrete aldosterone. It seems likely that the initial stimulus to renin production is the pulse pressure in the afferent arteriole of the juxta-glomerular apparatus, since conditions under which the arterial wall is not stretched lead to increased renin production.

Aldosterone is not the only cortical hormone affecting water balance. The diuretic response to a water load is impaired in patients whose adrenal glands have been destroyed by disease (Addison's disease) or removed at operation. The ability to deal normally with water is, however, restored by the administration of cortisone or hydrocortisone (Chap. 50).

When the extracellular fluid volume of the body is increased (expanded) by intravenous infusions even of isotonic saline the reabsorption of sodium by the renal tubules is decreased even although there is no change in the two factors known to influence sodium reabsorption namely glomerular filtration rate (as it falls percentage reabsorption of sodium increases) and aldosterone. There is some evidence that the kidney itself contains a pressure-sensing mechanism, sometimes termed the third factor, which responds to a rise of pressure by releasing an intrarenal hormone which is described as 'natriuretic' because it increases sodium excretion. The precise nature of the third factor and its site and mode of action remain uncertain.

Effect of emotion on urinary output. In his dogs Verney found that the emotional stress produced, for example, by exercise, loud noise, or electrical stimulation of the skin inhibited the production of urine but only if the posterior pituitary was intact.

In man emotional stress also reduces the output of urine. This could be attributed to an outpouring of adrenaline from the adrenal medulla or to the discharge of antidiuretic hormone. Adrenaline produces a marked but transient reduction in urinary output by constricting the efferent glomerular arterioles (p. 656). These effects on diuresis must not be confused with emotional alterations of the micturition reflexes; emotion often causes frequency of micturition without actual increase in the volume of urine passed.

THE KIDNEY AND OEDEMA

The retention of salt by the kidney is an important factor is the production of oedema and a low-salt diet is commonly employed as a therapeutic measure. Many powerful diuretic drugs are available and most of them act by reducing sodium reabsorption from the renal tubule with consequent natriuresis and diuresis. Retention of sodium is followed by retention of water, so that the osmotic pressure remains normal at the expense of an increase in the extracellular fluid and an increased blood volume (*hypervolaemia*). The retention of fluid finally becomes evident as oedema. Because similar changes can be produced experimentally by large doses of sodium bicarbonate or sodium salicylate, but not by potassium or ammonium chloride, it is concluded that sodium is the important factor.

The evidence is against variations in sodium excretion being due to changes in glomerular filtration rate. Since the amount of sodium in the urine is only some 1 to 2 per cent of that filtered at the glomerulus a change in glomerular filtration rate of as little as 1 per cent could halve or double the rate of sodium excretion provided that the reabsorption of sodium by the tubules remained constant and unrelated to the sodium load. However, reabsorption of sodium does not remain constant, it seems to be controlled by the volume and distribution of blood or extracellular fluid in the upper part of the body and, in part, by the concentration of sodium in the extracellular fluid. These factors affect the output of aldosterone from the adrenal cortex (Chap. 50), and so alter the reabsorption of sodium by the renal tubules indirectly through a humoral mechanism which is possibly angiotensin. It is not yet certain that an increased production of aldosterone always plays a part in producing sodium retention and oedema.

TUBULAR REGULATION OF ACID–BASE BALANCE

The hydrogen ion concentration of the plasma is maintained remarkably constant at pH 7·4 in the face of a constant daily production of some 40 to 60 m-equiv of H-ion, mainly by the incomplete oxidation of fats and carbohydrates and by the oxidation of sulphur-containing amino acids. In abnormal metabolic states, for example diabetic ketosis (p. 330), as much as 400 m-equiv. of H-ion may be produced with a comparatively slight fall in plasma pH. This subject is dealt with more fully in Chap. 19. The kidney is able to excrete an acid urine in acidaemia and a more alkaline urine in alkalaemia. The kidneys' ability to vary the pH of the urine depends upon the secretion of H-ion in the proximal tubule and the production of ammonia in the distal and collecting tubules.

As the glomerular filtrate passes down the nephrons its pH falls by some 0·3 to 0·5 pH units in the proximal and distal tubule but by 1 to 2 pH units in the collecting ducts. Although the secretion of H-ion occurs mainly in the proximal tubules, the volume of fluid in the collecting ducts is so small relatively that considerable changes in pH can be effected by small additions of H-ion. The secretion of free H-ion is accompanied by the reabsorption from the tubular fluid of sodium bicarbonate under the influence of the enzyme carbonic anhydrase which increases the concentration of H-ion in the tubule cell in much the same way as it does in the gastric mucosa (p. 248). The free H-ions pass into the tubular fluid in exchange for Na-ions which enter the tubule cell to be combined with the intracellular bicarbonate ion as sodium bicarbonate. This is then removed in the blood (Fig. 32.20). In the tubular lumen most of the H-ion combines with bicarbonate-ion to form carbonic acid,

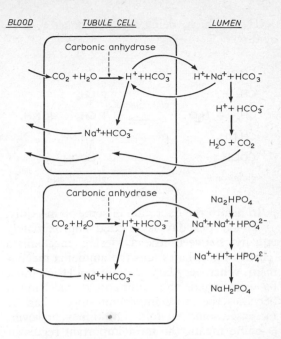

Fig. 32.20 Excretion of hydrogen ion and reabsorption of bicarbonate in the distal tubule.

which is then dehydrated to H_2O and CO_2. The CO_2 diffuses into the tubule cell and is either re-utilized to make more H-ion or enters the blood. The net effect of these exchanges is that nearly all the bicarbonate of the glomerular filtrate is reabsorbed and that H-ion is available for excretion. This H-ion is buffered by phosphate or combined with ammonia.

The unabsorbed phosphate of the glomerular filtrate takes part in the following reaction.

$$HPO_4^{2-} + H_2CO_3 = H_2PO_4^- + HCO_3^-$$

The dihydrogen phosphate is then excreted and the bicarbonate reabsorbed to provide a further supply of base to the plasma. In plasma the ratio of $H_2PO_4^-$ to HPO_4^{2-} is about 1 to 5 but in urine it is usually about 9 to 1. When large amounts of H-ion have to be eliminated the ratio may rise as high as 50 to 1.

Part of the H-ion in the tubular fluid is combined with ammonia formed from glutamine in the cells of distal and collecting tubules by the enzyme glutaminase and from glutamic acid by glutamate dehydrogenase (p. 378). The ammonia passing into the tubular lumen combines with free H-ion⁺ to form ammonium ion

(NH_4^+) which, being a proton donator, is an acid.

$$
\begin{array}{ccc}
\begin{array}{l}
\text{CO.NH}_2 \\
| \\
\text{CH}_2 \\
| \\
\text{CH}_2 \quad + \text{H}_2\text{O} \\
| \\
\text{CH.NH}_2 \\
| \\
\text{COOH} \\
\text{glutamine}
\end{array}
&
\xrightleftharpoons{\text{\textit{glutaminase}}}
&
\begin{array}{l}
\text{COOH} \\
| \\
\text{CH}_2 \\
| \\
\text{CH}_2 \quad + \text{NH}_3 \\
| \\
\text{CH.NH}_2 \\
| \\
\text{COOH} \\
\text{glutamic acid}
\end{array}
\end{array}
$$

In health the excretion of some 50 m-equiv. of H-ion per day is divided approximately equally between the buffering mechanism (chiefly phosphate) and the ammonia mechanism. When very large amounts of H-ion must be eliminated the ammonium mechanism becomes the more important since it can be increased some ten-fold. The kidney, however, is by no means the most important regulator of acid–base balance. A normal person at rest excretes 1500 m-equiv. of CO_2 through the lungs.

The excretion of H-ion by the kidney can be gauged by measuring the ammonium excretion plus the *titratable acidity,* that is to say the amount of alkali that is required to raise the pH of the urine to that of plasma.

For the successful maintenance of acid–base equilibrium by the kidney an adequate flow of urine is essential and the importance of giving large amounts of fluid to patients with acidaemia or alkalaemia cannot be over-emphasized, provided of course that the kidneys are capable of response.

RENAL FAILURE

Acute renal failure may be caused by rapidly progressive irreversible disease processes but more commonly it is the result of some acute disturbance to renal blood flow (as in shock) and from this the patient may recover. The production of urine falls below 400 ml per day (oliguria) and the quality of the urine is also impaired. The urine volume may be as low as 20 to 30 ml per day and, rarely, there may be no urine at all (anuria). The concentration of nitrogenous products such as urea and creatinine in the blood rapidly increases, the plasma potassium rises and acidaemia develops. Death

is likely in a week or two if these biochemical abnormalities are not corrected.

Chronic renal failure develops more gradually over a number of months or years as a result of progressive destruction of nephrons. Here again acidaemia and a gradually increasing blood urea and creatinine are the result. The increase in blood urea and of other nitrogenous substances in the blood is proportional to the loss of nephrons, usually measured in terms of creatinine clearance (p. 654) but it should be appreciated that the kidney has a very considerable reserve of function and the blood urea, for example, seldom shows any increase until the creatinine clearance is less than half the normal value. Thereafter the rise is progressively steeper but life is still possible even with a creatinine clearance as low as 2 ml per minute. Once the creatinine clearance has fallen significantly the surviving nephrons, subjected to a greater load of material to be excreted (filtered load), are in a state of persistent osmotic diuresis so that the patient passes relatively large amounts of dilute urine (polyuria). The fractional excretion of electrolytes as well as of water is increased and at a GFR of 3·0 ml per minute some 80 per cent of the fluid filtered and 50 to 60 per cent of the filtered load of sodium and chloride may be excreted. Finally, of course, as the number of functioning nephrons falls still further the volume of urine falls and death occurs from 'uraemia'. The main features of uraemia are nausea and vomiting, pigmentation of the skin (urochrome), drowsiness, stupor and coma. The biochemical disturbance in uraemia is complex involving acidaemia, retention of many nitrogenous end-products of which urea is probably the least harmful, and alterations in calcium, magnesium, potassium and many other substances.

Patients with severe renal failure may now be treated by *dialysis,* a procedure which seeks to replace temporarily some of the functions of the patients' kidneys. This may be done outside the body by connecting one of the patient's arteries to a cellophane tube lying in a bath of dialysing solution, isotonic with blood, containing physiological concentrations of electrolytes but no urea or other nitrogenous substance. The other end of the cellophane tube is connected to a vein. During this process of *haemodialysis* urea and other nitrogenous bodies diffuse from the blood into the dialysing fluid and electrolyte concentrations on either side of the membrane are equalized. A similar principle can be applied inside the body by circulating dialysing fluid through the abdominal cavity. In this method the peritoneal membrane is used as a dialysing surface (peritoneal dialysis). Both forms of dialysis were originally used as a means of tiding a patient over a severe episode of potentially reversible acute renal failure. In recent years haemodialysis has been employed twice or thrice weekly as a definitive form of treatment for patients with permanent renal failure and some have now been kept alive for many years. An alternative form of treatment for some patients with end-stage chronic renal failure is to replace the damaged kidneys with a kidney transplanted from a healthy relative or from a cadaver.

RENAL FUNCTION IN THE NEWBORN

The steady rise of urea and uric acid in the liquor amnii during pregnancy suggests that the kidneys normally begin to function before birth, although only to a small extent. The urine of babies examined immediately after birth is generally acid with a small amount of urea and a low osmotic pressure; albuminuria is not uncommon in the first five days of extra-uterine life.

Immediately after birth the body becomes dependent on the kidney for elimination of nitrogenous end-products of metabolism and to a large extent also for the regulation of its internal environment. The plasma protein level is low in infancy and, therefore, the colloid osmotic pressure of the plasma is lower than in the adult, a fact which favours the filtration of fluid through the glomerular capillaries as well as through capillaries in all parts of the body. The resulting tendency to accumulation of extracellular fluid and development of oedema is, however, offset by the low fluid intake during the first few days of extra uterine life, and the volume of the extracellular fluid actually tends to fall; at the same time the blood urea concentration rises. In a few days, as the baby's fluid intake increases, the extra-cellular fluid volume rises and the blood urea falls. Another factor accounting for the rela-tively large extracellular fluid volume of the human infant, especially if it is premature, is the low renal clearance of sodium and chloride.

In babies a few days old glomerular filtration rates measured by inulin clearance are about 30 ml/m^2/min compared with about 75 ml/m^2/min in the adult. Tubular function is also much less efficient than in adults for the renal clearance of p-aminohippurate is only about one-tenth of adult value when allowance is made for the differences in surface area. Adult levels are attained towards the end of the first year.

In newborn children the volume of urine passed per day is of the order of 20 ml. Water given to babies up to three months of age is excreted very slowly; the specific gravity and osmotic pressure of the urine are low; and the administration of posterior pituitary extracts containing antidiuretic hormone has little effect in making the urine more concentrated.

Infants easily become dehydrated and the administration of fluids readily produces oedema. The low phosphate clearances and low phosphate excretion make compensation for acidaemia inadequate. The limited powers of renal function explain how the chemical composition of the plasma in the newborn may be quickly upset by disease and why the condition so easily becomes irreversible.

MICTURITION

The urine passes into the pelvis of the kidney and is carried by peristaltic waves down the ureter so that the pressure in the pelvis is kept low. These waves occur every 10 seconds or so and travel down at 2 or 3 cm per sec, each wave sending a little spurt of urine into the bladder. This can readily be observed in man by injecting intravenously a dye which is excreted by the kidney and simultaneously observing the ureteric orifices through a cytoscope inserted along the urethra into the bladder.

The waves originate in a pace-maker at the renal end of the ureter. The ureteric muscle is fairly powerful since if the ureter is obstructed the ureteric pressure may rise to 50 mm Hg. The ureters pass obliquely through the thick muscular wall of the bladder forming a valvular opening so that urine is prevented from regurgitating into the ureter when the intra-vesical pressure is raised during micturition.

The bladder is lined by transitional epithelium, the cells of which under the electron microscope show scalloped margins with complex interdigitations between adjacent cells. These presumably allow easy alteration of the shape of the cells when the bladder is distended. The epithelium is firmly adherent to the wall of smooth muscle fibres, the *detrusor* muscle, which is supplied by the hypogastric nerves via the inferior hypogastric plexuses and by the pelvic splanchnic nerves through their ganglia on or near the bladder wall. At the base of the bladder the smooth muscle fibres pass around the urethral opening loopwise to form the *internal sphincter* (sphincter vesicae) supplied by hypogastric nerves (lumbar or sympathetic outflow) and the pelvic splanchnic nerves (sacral or parasympathetic outflow) (Fig. 32.21). Dissection of this region does not,

however, show a definite anatomical sphincter in either male or female and it may be that the sphincteric region is merely an extension of the bladder muscle which is stretched and therefore closed as the bladder fills; according to this interpretation when the detrusor muscle contracts at micturition the sphincteric region also contracts and the urethra is shortened and opened up. The *external sphincter* (sphincter urethrae) surrounding the membranous urethra is composed of two striated muscles, the compressor urethrae and the bulbocavernosus; it is under voluntary control through the pudendal nerves which are somatic nerves. This anatomical arrangement should be compared with that for the rectum and colon (p. 304). In males the external sphincter is able to maintain continence even when the internal sphincter is damaged. The mucosa of the

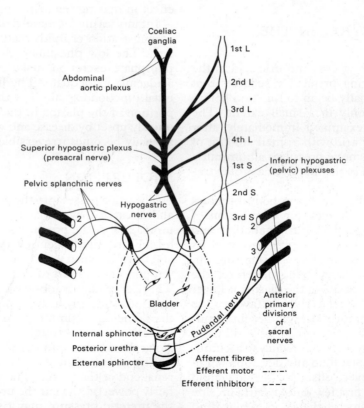

Fig. 32.21 Diagram of innervation of the bladder and sphincters in man (*after J. R. Learmonth*). The pelvic nerves (nervi erigentes) are parasympathetic. The hypogastric nerves are sympathetic. The pudendal nerves are somatic nerves. Some hypogastric fibres synapse with short adrenergic neurones originating in intramuscular ganglia in the base of the bladder, others synapse with short adrenergic neurones in the prostrate and vas deferens and seiminal vesicles. (See also G. Genser, C. Owman and L. Wehlin (1969), *Lancet* **i**, 154.)

human bladder is insensitive to light touch but vigorous pressure causes pain; it is sensitive to large changes in temperature. Nerve fibres from the muscle wall give information about bladder fullness. Most of these sensory fibres must travel in the parasympathetic nerves since sympathectomy does not affect these sensations. Sensory fibres subserving pain run in the sympathetic and the effects of cord injury or partial anaesthetization in man indicate that these fibres run up as high as the ninth thoracic segment before entering the cord. Since both sympathetic and parasympathetic fibres pass through the inferior hypogastric plexuses (Fig. 32.22), section of all nerves passing from the bladder to the plexuses produces complete anaesthesia of the bladder. The mucosa of the urethra is sensitive to painful and thermal stimuli.

In the empty bladder at rest the sphincters are tonically contracted so that the urine from the ureters gradually expands the bladder without, however, causing the intravesical pressure to rise very much (Fig. 32.21), This may be explained by supposing that the detrusor muscle relaxes as the bladder expands

but, while this may be true, the small rise in pressure in the bladder is partly the result of the elastic properties of the bladder wall, since it can be shown that in any elastic bag the pressure, except in the early stages of filling is almost independent of the volume of the contents (see Laplace equation p. 235). The tension in the walls is, however, proportional to the contained volume. If the pressure is increased suddenly by introducing, say, 200 ml of fluid through a catheter into the bladder, rhythmic contractions are set up, though they quickly die out. Bladder distension insufficient to produce a sensation of fullness produces vasoconstriction of the skin; if the pressure is sufficient to cause discomfort the blood pressure may increase by 20 or 30 mm Hg. In normal circumstances the subject becomes aware of bladder fullness when say 150 ml of urine has accumulated. The bladder is usually allowed to fill until it contains nearly 300 ml. If more urine is allowed to accumulate a sense of urgency arises and when the bladder contains over 600 ml the contractions become powerful enough to raise the pressure to 100 cm of water and cause pain. If voluntary efforts at restraint are made the pressure may fall for a short time, but when the pressure exceeds 100 cm of water again discomfort becomes acute and micturition inevitably occurs. The level of pressure at which the desire to micturate occurs depends partly on the rate of filling—if it is slow the bladder can accommodate more fluid—and partly on the irritability of the nerve endings in the bladder and their central connexions. For example when the bladder lining is inflamed (cystitis) the volume at which micturition occurs is much reduced. Afferent impulses travel in the pelvic splanchnic nerves (and probably in the hypogastric nerves) and emptying occurs by impulses passing down the pelvic splanchnic nerves which cause the detrusor muscle to contract and open the 'internal sphincter'. The older idea that a reduction in sympathetic (hypogastric) tone causes at the same time a relaxation of the internal sphincter is difficult to substantiate since division of the hypogastric nerves in man causes little or no interference with the evacuation of urine. Ejaculation is, however, impaired. It is of interest that adrenergic-blocking drugs used to control hypertension cause retrograde ejaculation

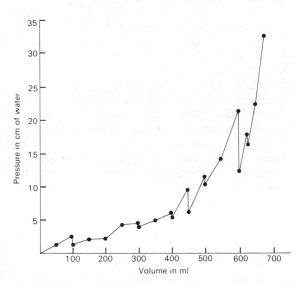

Fig. 32.22 Graph (cystometrogram) of the relationship of the volume of the vesical contents to their pressure in the course of distension of the human bladder. Each vertical fall of pressure indicates the degree of adaptation at a constant volume. (D. Denny-Brown & E. Graeme Robertson (1933). *Brain* **56**, 158.) If for any reason the bladder is overdistended the micturition reflex is absent and the cystometrogram shows no terminal rise.

The Kidney   669

probably because of a disturbance of the function of the internal sphincter which normally shuts to prevent it. The importance of the afferent impulses in the process of micturition is shown by the fact that, after section of the dorsal sacral roots in cats, the bladder may enlarge up to ten times its normal size. The pelvic splanchnic nerves are certainly the most important in emptying because section of both pelvic splanchnic nerves in the cat causes immediate paralysis of micturition. Although partial recovery may eventually take place the detrusor muscle shows only feeble contractions so that true micturition can scarcely be said to occur. This statement is probably true also for man. Section of both pudendal nerves (supplying the external sphincter) does not disturb the act of micturition, nor does it cause incontinence. In pregnancy the bladder shows less tone and may increase in capacity up to 1500 ml with considerable residual urine.

The urinary bladder is apt to be invaded by organisms passing backwards from the urethra; this invasion is easier through the short female urethra. Serious infection is prevented partly by the washing out of organisms at micturition, partly by the bactericidal activity of normal bladder mucosa and partly, in the case of the male, by the antibacterial properties of prostatic secretion.

The effect of spinal anaesthesia in man is to relax slightly the internal sphincter and the ureteric orifices; the external sphincter remains closed; the subject cannot initiate micturition. In transverse lesions of the spinal cord in man which result in paraplegia (Chap. 44) reflex micturition is at first abolished, the detrusor is relaxed and the bladder distends greatly as urine accumulates, but eventually the pressure rises high enough to force some urine through the sphincter. After small amounts of urine have been voided the pressure is somewhat reduced and urine accumulates again in the bladder until the process repeats itself. This is *overflow incontinence*. In treatment of this condition it is important to catheterize the bladder to prevent overdistension. After some months micturition starts to occur automatically, usually at hourly intervals, by contraction of the detrusor muscle. The capacity of the bladder is then smaller than normal and a small increase in its contents causes a considerable rise of pressure. Provided the sacral cord and its

nerve connexions with the bladder are intact the bladder reflexes in man with a transverse lesion of the spinal cord eventually occur in the usual way, except that they are not under restraint from the higher centres which normally exercise a constant inhibition on the detrusor muscle. Destruction of the cauda equina in man leaves the external sphincter flaccid. The internal sphincter gives way at much lower pressure than normal when spontaneous contractions of the bladder occur. A rudimentary micturition mechanism is supplied by voluntary contractions of the abdominal muscles which reinforce the weak myogenic contractions of the detrusor.

The external sphincter, although it is a striated muscle, cannot be relaxed voluntarily although it may be consciously contracted. Efforts to micturate do not produce a relaxation of the external sphincter unless the intravesical pressure is above a threshold value of from 20 to 40 cm of water. Inhibition of the sphincters occurs only when the detrusor muscle is contracting, the internal sphincter relaxing a little before the external. The stream of urine can, however, be stopped by a voluntary contraction of the external sphincter.

De Groat and Ryall (1969) have given a summary of their views on the activities of the bladder. At all stages of bladder filling the vesical musculature shows small contractions which may be a trigger mechanism upon which all the micturition reflexes depend. At low bladder pressures each small contraction sets up an afferent discharge but this is insufficient to elicit a reflex discharge of the parasympathetic neurones and may even actively inhibit them. At a certain critical pressure the afferent discharge is sufficiently great to overcome the inhibitory mechanisms and the autonomic neurones discharge, causing a contraction of the bladder which leads in turn to the opening of the proximal urethra. Simultaneously the external urethra relaxes by a spinal visceral somatic inhibitory reflex. The passage of urine through the external urethra causes a discharge of afferent fibres in the pudendal nerve which, by a supraspinal mechanism, reinforces the excitation of the parasympathetic neurones. The increase in bladder pressure brought about by the reflex autonomic discharge leads to an increase in the discharge from tension receptors and a continued discharge of the parasympathetic neurones. There is thus a progressive and self-reinforcing cycle which increases and maintains the contraction of the vesical musculature during urination.

Although, as described above, the micturition reflexes are present in man after transection of the spinal cord, the bladder and its sphincters are normally influenced by centres in the forebrain. A voluntary effort to micturate evokes powerful contractions of the bladder after a very short latent period and voluntary restraint of micturition results in an inhibition of these contraction waves. Normal micturition depends entirely on active contraction of the bladder; voluntary control, that is the ability to delay or start micturition when desired, should be thought of as inhibition or removal of inhibition respectively of the normal reflex

response to distension which is dependent on the lower parts of the spinal cord. According to this view the infant develops voluntary control of micturition by acquiring the power to inhibit the spinal reflexes.

Certain associated movements occur in micturition which are not regarded as essential parts of the process. An attempt to micturate may be accompanied by relaxation of the perineal muscles, but not of the sphincters, contraction of the abdominal muscles and occasionally movements of the lower colon and rectum. Voluntary restraint is associated with a contraction of the perineal muscles and closure of the external sphincter. After surgical removal of an enlarged prostate (prostatectomy) development by exercises of these associated muscles is important.

REFERENCES

ADLER, S., ROY, ARLENE & RELMAN, A. S. (1965). Intracellular acid–base regulation. *Journal of Clinical Investigation* **44**, 8–30.

ALLISON, MARJORIE E. M. & KENNEDY, A. C. (1971). Diuretics in chronic renal disease: a study of high dosage frusemide. *Clinical Science* **41**, 171–187.

BERLINER, R. W. & BENNETT, C. M. (1967). Concentration of urine in mammalian kidney. *American Journal of Medicine* **42**, 777–789.

BLACK, D. A. K. (1965). Renal rete mirabile. *Lancet* **ii**, 1141–1152.

BLACK, D. A. K. (Ed.) (1967). *Renal Disease*, 2nd edn. Oxford: Blackwell.

BRITTON, K. E. (1968). Hypothesis: renin and renal autoregulation. *Lancet* **ii**, 329–333.

BROWN, J. J., DAVIES, D. L., LEVER, A. F., ROBERTSON, J. I. S. & VERNIORY, A. (1966). The effect of acute haemorrhage in the dog and man on plasma–renin concentration. *Journal of Physiology* **182**, 649–663.

BÜNNING, E. (1967). *The Physiological Clock*, 2nd edn. Berlin: Springer.

CHRISTENSEN, H. N. (1964). *Body Fuids and the Acid–Base Balance: A Learning Program for Students of the Biological and Medical Sciences*. Philadelphia: Saunders.

DICKER, S. E. (1969). *Mechanisms of Urine Concentration and Dilution in Mammals*. London: Arnold.

ELKINGTON, J. R. (1962). Hydrogen ion turnover in health and disease. *Annals of Internal Medicine* **57**, 660–684.

FISHER, J. W. (Ed.) (1971). *Kidney Hormones*. New York: Academic Press.

FRASER, R., BROWN, J. J., CHIN, R., LEVER, A. F. & ROBERTSON, J. I. S. (1969). The control of aldosterone secretion and its relationship to the diagnoses of hyperaldosteronism. *Scottish Medical Journal* **14**, 420–440.

DE GROAT, W. C. & RYALL, R. W. (1969). Reflexes to sacral parasympathetic neurones concerned with micturition in the cat. *Journal of Physiology* **200**, 87–108.

GRUBER, C. M. (1933). The automatic innervation of the genito–urinary system. *Physiological Reviews* **13**, 497–609.

KURU, M. (1965). Nervous control of micturition. *Physiological Reviews* **45**, 425–494.

MILLS, J. N. (1963). Mechanisms of renal homeostasis (Chap. 8); and Other aspects of renal function (Chap. 9) from *Recent Advances in Physiology* (Edited by R. Creese). London: Churchill.

PALTON, A. J. & HAGNENAU, FRANCOISE (1967). *Ultrastructure of the Kidney*. London: Academic Press.

PEART, W. S. (1965). The renin–angiotensin system. *Pharmacological Reviews* **17**, 143–182.

PITTS, R. F. (1968). *Physiology of the Kidney and Body Fluids*, 2nd edn. Chicago: Year Book Medical Publishers.

PRATHER, G. C. (1949). *Urological Aspects of Spinal Cord Injuries*. Oxford: Blackwell.

SCHWARTZ, I. L. & SCHWARTZ, W. B. (Eds.) (1967). Symposium on antidiuretic hormones. *American Journal of Medicine* **42**, Suppl. No. 1, 651–858.

SIGNY, A. G. (Ed.) (1965). Symposium on renal function. *Journal of Clinical Pathology* **18**, 491–578.

SMITH, H. W. (1951). *The Kidney: Structure and Function in Health and Disease.* Oxford: University Press.

Symposium on hypertension (1966). *Postgraduate Medical Journal* **42**, 148–206.

Symposium on the role of hormones in heart failure (1962). *Circulation* **25**, 1001–1023.

VANDER, A. J. (1967). Control of renin release. *Physiological Reviews* **47**, 359–382.

DE WARDENER, H. (1967). *The Kidney: An Outline of Normal and Abnormal Structure and Function,* 3rd edn. London: Churchill.

DE WARDENER, H. (1969). Control of sodium reabsorption. *British Medical Journal* **iii**, 611–616, 676–683.

WEYER, E. M. (Ed.) (1966). The physiology of diuretic agents. *Annals of the New York Academy of Sciences* **139**, 273–539.

WHIPPLE, H. E. (Ed.) (1965). Biochemical aspects of the renin–angiotensin system. *Transactions of the New York Academy of Sciences* **27**, 445–449.

WINDHAGER, E. E. (1968). *Micropuncture Techniques and Nephron Function.* London: Butterworths.

WINDHAGER, E. E. & GIEBISCH, G. (1965). Electrophysiology of the nephron. *Physiological Reviews* **45**, 214–244.

WOOTON, I. D. P. (1963). Retention of aromatic compounds in acute renal failure. *Scientific Basis of Medicine Annual Review* 235–248.

33 Water, electrolyte and acid—base metabolism

WATER

Water is so familiar and so large a constituent of the body that its fundamental importance in both the structure and functioning of all the tissues tends to be overlooked.

Deprivation of water brings about death much more rapidly than deprivation of food. When water is completely withheld death occurs in a few days after the body has lost 10 to 20 per cent of its water content. If, however, water is given but no food, life may continue for several weeks in spite of the loss of most of the body fat and 50 per cent of the tissue protein.

Most tissues contain more than 70 per cent water (p. 6); even bone is nearly one-third water and adipose tissue (fat) contains quite large amounts of water in the connective tissue and in the spaces between the fat cells. About half of the body water is to be found in muscle which accounts for about one-third of the body mass.

In its capacity as a solvent water plays a fundamental role in cellular reactions (see Chap. 2). Water can dissolve a very large number of substances and many others, such as fats and fat-soluble compounds, can be carried by water in fine emulsions or be rendered water-soluble by combination with hydrophilic substances. Because of the high heat capacity of water large changes in heat production can take place in the body with very little alteration in body temperature. Since the latent heat of evaporation of water is high, the loss of a small amount of water in evaporated sweat means a relatively large loss of heat. Moreover the high latent heat of solidification is a protection against freezing of the tissues.

The cells of the body consist largely of water and they are embedded in a gel or matrix of protein and water. The water (fluid) of the body may therefore be thought of as being in two compartments or spaces—the water within the cells, *intracellular fluid*, and the water outside them, *extracellular fluid*. That part of the extra-cellular fluid circulating in the blood vessels as plasma water is known as *intravascular fluid* which is separated by the capillary endo-thelium from the remainder, *interstitial fluid* (Fig. 33.1 and Table 33.2).

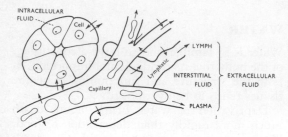

Fig. 33.1 Diagram to show the main subdivisions of the body fluid. The term extracellular fluid includes inter-stitial fluid, lymph and the protein-free fraction of plasma. (*After* Cowdry (1938). *Textbook of Histology*, p. 36. London: Kimpton.)

Table 33.2 Distribution of body water in a man of average build

	Volume (litres)	Percentage of body weight
Total body water (T.B.W.)	42	60
Intracellular fluid (I.C.F.)	26·5	38
Extracellular fluid (E.C.F.)	15·5	22
Interstitial fluid	12	17
Intravascular fluid	3·5	5

Based on data in *Metabolic Homeostasis* by N. B. Talbot, R. H. Richie & J. D. Crawford (1959). Harvard University Press.

Table 33.3 The relationship between fatness and body water at different relative body weights

	Wasted	Not wasted	Fat
Body Weight (kg)	33	54	(80)
Fat (kg)	3·5	16·5	(36)
Lean Tissue Solids (LTS)	6·5	9·5	(12)
Total Body Water (l) (TBW)	23	28	(32)
Body Water as Percentage of Body Weight	70	52	(40)

The figures in the first two columns refer to two patients; the figures in the third column are estimated. The relationships are expressed graphically on the right of the table. (*Courtesy of Brian McConkey.*)

Tissues containing very little fat vary little among themselves in their water content. Adipose tissue, however, is exceptional in having a lower water content that other tissues (except compact bone) because a large part of each cell is occupied by fat instead of by water. Thus the proportion of water in a tissue, or in the whole body, depends on the proportion in it of fat. It is therefore misleading to express the amount of water in the body (*total body water*) as a percentage of body weight unless at the same time the composition of the body in terms of fat is also stated (Fig. 33.3). In men of average build about 14 per cent of the body weight is fat; in women about 18 per cent. If the weight of fat is subtracted from the total body weight, the weight of lean tissue (*lean body mass*) is derived. The proportion of water in the lean body mass of man and animals is remarkably constant, about 73 per cent. In men of average build, containing some fat, 55 to 65 per cent of the body weight is water; in women of average build 45 to 55 per cent. The proportion of water in the body decreases pro-gressively as age advances (Table 33.4).

The total amount of water in the body of an animal can be found by drying the carcass completely and measuring the weight lost. In a similar way the water content of a piece of tissue, or of a whole organ such as the liver, can be found directly. In man only indirect methods of measurement based on the prin-ciple of dilution are applicable. In a few instances the results of indirect methods have been checked against direct measurements and found to be reasonably accurate (Table 33.5).

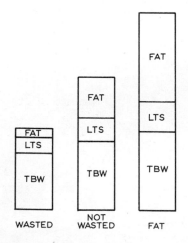

Table 33.4 Body water (as per cent of body weight) at various ages

Fetus	0– 3 months	94·3	by desiccation[1]
	3– 6 months	89·1	
	6– 9 months	84·5	
	0– 6 months	72·2	by dilution[2]
	½– 2 years	69·5	
	2– 7 years	63·1	
	7–16 years	58·4	
	22–58 years	51·7 by dilution[3]	
	71–84 years	50·8 by dilution[4]	

The progressive fall in body water when calculated as a percentage of body weight is not accounted for by increasing degrees of fatness. It represents real differences in the composition of the body.

[1] Moulton, C. R. (1923). *Journal of Biological Chemistry* **57**, 79.
[2] Friis-Hansen, B. (1957). *Acta paediatrica, Stockholm* **46**, Suppl. 110.
[3] McConkey, B. (1959). *Chemical Science* **18**, 95.
[4] Parker, H. V., Olesen, K. H., McMurrey, J. & Friss-Hansen, B. (1958). In *Water and Electrolyte Metabolism in relation to Age and Sex*, Vol. 4, p. 102. Ciba Foundation Colloquia on Ageing.

Table 33.5 Comparison of measurements of the body water of a number of animals measured by a dilution technique (antipyrine) and by desiccation of carcasses

	Total body water per cent	
	Dilution technique	Desiccation
5 Monkeys[1]	68·5	69·4
4 Dogs[1]	70·8	71·1
4 Rabbits[1]	74·3	71·3
3 Men[2]	65·7	—

[1] Soberman, R. J. (1950). *Proceedings of the Society of Experimental Biology, New York* **74**, 789.
[2] Widdowson, E. M., McCance, R. A. & Spray, C. M. (1951). *Clinical Science* **10**, 113.

A known amount of a suitable substance is injected into a fluid system and after diffusion throughout the system its concentration is measured. The extent to which it has been diluted is a measure of the volume of the system. Substances suitable for measuring total body water are antipyrine, deuterium oxide (heavy water) and tritium oxide (p. 124) since they readily diffuse throughout all the water in the body. Substances suitable for measuring the volume of extracellular fluid are inulin, stable or radioactive bromide, thiosulphate and thiocyanate, since they do not enter cells. The volume of intracellular fluid is found by subtracting the volume of extracellular fluid from the total body water. The values given in Table 33.9 have been obtained by such methods. Intravascular volume is discussed on p. 408.

ELECTROLYTES

The water in the body contains, of course, electrolytes in solution. These ions exert an effective osmotic pressure of approximately 285 m-osmoles per kg (see p. 21). The total osmolality (see Units) of both intravascular and intracellular fluid is very slightly greater than that of interstitial fluid. In the case of intravascular fluid this is due to the plasma proteins (about 1 m-osmole per kg) and in the case of intracellular fluid to the presence within the cells of ions which possess more than one charge per molecule, for example magnesium, phosphate and protein (see Chap. 2).

To maintain electrical equilibrium the number of equivalents (Chap. 2) of anions and cations in the body fluids must balance. The chief cation of the extracellular fluid is sodium and the chief anions are chloride and bicarbonate (Table 33.6). The composition of

Table 33.6 Composition of the body. Concentrations of anions and cations in intracellular fluid and extracellular fluid (in m-equiv. per litre). See also Fig. 33.27.

Extracellular fluid		Intracellular fluid	
Cations	Anions	Cations	Anions
Na^+ 145	Cl^- 100	Na 10	Cl 10
K^+ 5	HCO_3^- 27	K^+ 150	HCO_3^- 10
Ca^{2+} 2	PO_4^{3-} 2	Ca^{2+} 2	PO_4^{3-} 90
Mg^{2+} 2	SO_4^{2-} 1	Mg^{2+} 15	SO_4^{2-} 15
	Organic 5		Organic —
	Protein 19		Protein 52
Total 154	154	177	177

interstitial and intravascular fluids is almost the same except that interstitial fluid contains only a trace of protein, whereas the concentration of protein in intravascular fluid is about 7 g per 100 ml. Because of this similarity the chief ions in extracellular fluid can be determined with sufficient accuracy by measuring the concentration of sodium, potassium, chloride and bicarbonate in blood plasma (Table 33.7).

Table 33.7 Concentration of certain constituents of the blood of normal persons (the figures obtained in 80 per cent of normal healthy fasting individuals fall between the limits given in this table when the determinations are made by the methods described by King)

	mg per 100 ml		m-equiv. per litre	
	Lower limit	Upper limit	Lower limit	Upper limit
Whole blood				
Cholesterol	140	215	—	—
Chloride (as Cl⁻)	275	311	77	87
Creatinine	0·1	1·2	—	—
Glucose	68	96	—	—
Non-protein N	29	43	—	—
Urea	16	35	—	—
Uric Acid	1·6	3·9	—	—
Serum				
Calcium	9·6	10·9	4·8	5·4
Potassium	15·1	19·6	3·9	5·0
Sodium	318	342	137	148
Chloride (as Cl⁻)	360	378	101	106

(*After* E. J. King & I. D. P. Wootton (1956). *Microanalysis in Medical Biochemistry*, 3rd edn. London: Churchill.)

It is difficult to offer a complete explanation of the different ionic compositions of intra- and extracellular fluid. The differences are, however, preserved by expenditure of energy (for example the sodium pump Chap. 39). Thus variation in metabolic processes such as carbohydrate metabolism or oxidation may alter the relative concentration of ions inside and outside the cells so that potassium leaves the cells and sodium enters them. Two examples of this are the leakage of potassium from the red cells into the plasma of blood that has been stored in a blood bank and the high concentration of potassium in the plasma of patients with diabetic ketosis (p. 339). The changes in the concentrations of sodium and potassium in nerve cells during the conduction of nerve impulses and in the muscle cells during muscle contraction are both accompanied by depolarization of the cell membranes (Chaps. 39 and 40).

Exchangeable sodium and potassium. The dilution technique (p. 408) can be used to measure the mass of a substance in the body, provided of course that the substance introduced into the body is freely diffusible throughout that mass. Radioactive isotopes are useful for this purpose since they mix or ex-

change freely with the stable isotopes already in the body, and their concentration is easy to measure (Table 33.8). When ^{22}Na is used to measure total body sodium, the results are about 17 per cent lower than obtained by chemical analysis of cadavers. The reason is that only one-third of the large amount of

Table 33.8 Amounts of sodium and potassium present in the human body

Element	Total amount in body	Exchangeable
Na⁺	3900 m-equiv. (90 g)	3000 m-equiv. or 80 per cent of total
K⁺	3400 m-equiv. (135 g)	3200 m-equiv. or 95 per cent of total
Approximately		30 per cent of the sodium in the body is in bone 35 per cent of the sodium in bone is readily exchangeable

(*After* A. W. Wilkinson (1969). *The Body Fluids in Surgery*, 3rd edn. Edinburgh: Livingstone.)

sodium in bone is 'exchangeable' with the injected isotope, the remainder being incorporated in the apatite crystals of bone. The term 'exchangeable sodium' refers to that part of the body sodium not fixed in bone. Body potassium is almost entirely intracellular and almost all of it is exchangeable. Only a small amount is fixed in bone. It can be measured by muscle biopsy or from the γ emission of ^{40}K.

The composition of a healthy young man as measured by the methods outlined is given in Table 33.9.

Daily turnover of water and electrolytes. The water in the body is being continually lost and replenished. The amounts entering and leaving the body are shown in Table 1.4 which represents a daily balance, or equilibrium between intake and output. In metabolic studies the term positive balance means that the intake is greater than the loss, that is the body is gaining fluid; the subject is in negative balance when the total losses exceed the intake, so that he is gradually losing fluid.

A healthy man in a temperate climate can maintain his water balance by drinking about 650 ml a day; he has in addition the water in his food and the water produced by the oxidative metabolism of carbohydrate, fat and protein. Many foodstuffs contain considerable

Table 33.9 Body composition as measured by dilution in a normal man*

Body constituent and method used	Dilution volume or mass (litres or m-equiv.)	Percentage body weight	By surface area (per m²)
Direct measurements			
Plasma volume (blue dye)	3·67 litres	5·1	1·92 litres
Extracellular volume (thiocyanate)	16·8 litres	23·5	8·80 litres
Extracellular volume (thiosulphate)	13·4 litres	18·8	7·03
Total body water (deuterium)	40·6	57·0	21·3 litres
Total exchangeable sodium (^{24}Na)	3075 m-equiv.	43·2 m-equiv./kg	1610 m-equiv.
Total exchangeable potassium (^{42}K)	3440	48·2 m-equiv./kg	1801 m-equiv.
Derived data			
'Intracellular' water (by difference)†	27·2 litres	38·1	14·2 litres
Extracellular potassium (4·3 m-equiv./l thiosulphate space)	57·6 m-equiv.	0·81 m-equiv./kg	—
'Intracellular' potassium (by difference)†	3382 m-equiv.	47·4 m-equiv./kg	—
'Av intracellular' potassium concentration†	124 m-equiv./litre		
Extracellular sodium (142 m-equiv./l thiosulphate space)	1903 m-equiv.	26·7 m-equiv./kg	—
'Intracellular' sodium (by difference)†	1172 m-equiv.	16·4 m-equiv./kg	—
'Av intracellular' sodium concentration‡	43·2 m-equiv./litre		

* Subject: D. C.; male; age, 32; weight, 71·4 kg; height, 181·5 cm; surface area, 1·91 m².

† Intracellular water and intracellular potassium concentration are dependent on the characteristics of the methods used for estimating extracellular fluid volumes and do not represent exact values in any single tissue. It is particularly important to emphasize that the true intracellular water is not obtained merely by subtracting an extracellular dilution volume from total body water. The transcellular water—at present unknown—must not be subtracted.

‡ This figure for 'average intracellular' sodium includes some bone matrix sodium and is an average of all the sodium not in solution in extracellular fluid. It should not be construed as indicating the exact sodium concentration in, for example, muscle or liver cells.

(I. S. Edelman, J. M. Olney, A. H. James, L. Brooks & F. D. Moore (1952). *Science* **115**, 447.)

amounts of water—bread 40 per cent, potatoes 80 per cent, cooked meat 50 per cent—and a normal diet may provide 350 ml of water. A hundred grams of carbohydrate if completely oxidized provide 60 g of water, and 100 g of fat produces 109 g of water. Protein produces relatively less water (100 g produces 80 g) but urea, the end-product of its metabolism, requires water for its excretion in the urine. The water intake in health varies over a wide range and is more than sufficient to cover losses in the urine and sweat or deficiencies in the water obtained from food.

A minimum amount of urine must be passed each day to allow the kidneys to eliminate the end-products of protein metabolism. This volume (500 to 700 ml) is called the obligatory urine output but this quantity varies enormously depending on protein catabolism and electrolyte to be excreted. Approximately 100–200 ml of water are lost daily in faeces. Water is continually lost through the skin by evaporation (about 500 ml per day in a temperate climate). In a hot environment sweating can increase the loss of water and electrolytes considerably. Since the expired air is saturated with water vapour, breathing is inevitably accompanied by a loss of water (300 ml or more per day) but not of electrolyte. This loss is increased if the breathing becomes rapid as in exercise or in fever. In summary, about 900 ml is lost per day through the lungs, skin and faeces, and about 700 ml must be lost in the urine, so that the total minimum loss from the body is about 1600 ml of water (equivalent to about 4 per cent of the total water in the body). In theory, therefore, the minimum intake should be this amount less the metabolic water. Usually, however, the water intake is much larger, because of human drinking habits.

The average daily loss of electrolytes in a temperate climate is indicated in Table 33.10. Apart from the very small loss in the faeces and an obligatory loss of sodium and chloride in sweat (25 to 50 m-equiv. per day) electrolyte is lost mainly in the urine, but in hot dry con-

Table 33.10 Ionic composition of body excretion (m-equiv./litre) from a healthy man of 70 kg in a temperate climate. Approximate total external losses per day are also given.

	Na$^+$ (m-equiv./ day)	K$^+$ (m-equiv./ day)	Cl$^-$ (m-equiv./ day)
Urine	80 to 100	30 to 100	92 to 200
Sweat	4	5	50 to 100
Faeces	1·5	4	0·5
Approx. total	100 ≡ 2·2 g	60 ≡ 2·4 g	250 ≡ 8·8 g

(Figures based on H. F. Weisberg (1953). *Water, Electrolyte and Acid-base Balance*. Baltimore: Williams & Wilkins.)

ditions a man can lose in a day 10 litres of sweat containing 900 m-equiv. of sodium chloride and some 50 m-equiv. of potassium. Table 33.10 in the last line gives the approximate daily requirements in ordinary circumstances. These quantities would be provided by taking 8·3 g NaCl and 4·4 g KCl.

Simultaneously with the 'external' daily turnover of water and electrolyte there is a considerable 'internal' daily turnover of water and electrolyte within the alimentary tract. ·Large amounts of fluid and electrolyte are secreted in the digestive juices into the upper part of the alimentary canal (Table 33.11). Since the loss of fluid and electrolytes in the faeces is quite small, nearly all this material must be reabsorbed.

The volume of water involved, about 8 litres,

is rather more than twice the total circulating plasma volume. In Table 33.12 the quantities

Table 33.12 Daily intake, alimentary turnover, and body content of water and electrolytes

	Water (l)	Sodium (m-equiv.)	Potassium (m-equiv.)
Usual intake	2	150	100
Alimentary turnover (reabsorbed)	8	600	50
Body content	50	3000	4000
Daily 'turnover' as per cent of body content	4	5	2·5
Usual intake + daily turnover as per cent of body content	20	25	4

The last line in the table gives an approximate assessment of the possible effect in a single day of gastro-intestinal upset on water, sodium and potassium balance, if both loss of secretions and deprivation of the usual intake are taken into account. (R. H. S. Thompson & E. J. King (Eds.) (1964). *Biochemical Disorders in Human Disease*, 2nd edn. London: Churchill.)

of water, sodium and potassium exchanged through the alimentary tract are compared with the total amounts present in the body. The amounts of water and Na$^+$ are relatively great—some 20 per cent of the total body content.

It is obvious from these figures that a patient who is vomiting copiously and has diarrhoea can incur a serious loss of water and electrolyte within a few days or even hours.

Table 33.11 Average daily volume and composition of main alimentary secretions

	Mixed saliva	Gastric juice Parietal	Gastric juice Non-parietal	Pancreatic juice	Bile Hepatic	Bile Gall bladder
Volume (ml)	1500	2500		700	1000	400
pH	6·5	0·9	7·6	7·5–8·8	7·7	7·4
Osmolality (m-osmoles/kg)	90–180	285	285	285	285	285
Na$^+$ (m-equiv./l)	30	6	155	130	150	150
K$^+$ (m-equiv./l)	20	10	15	10	10	12
Cl$^-$ (m-equiv./l)	30	165	125	70	95	17
HCO$_3^-$ (m-equiv./l)	20	0	45	85	22	10
Remarks	Na$^+$ and Cl$^-$ concentrations rise steeply with increasing flow	H$^+$ is main cation (1555 m-equiv./l) in parietal secretion		Composition depends largely on whether, and how, stimulated	Bile anions 20–30 m-moles/l	Bile anions 150–210 m-moles/l

From O. M. Wrong (1970). *Biochemical Disorders in Human Disease* (Edited by R. H. S. Thompson & I. D. Wootton), 3rd edn. Churchill: London.

THIRST

Shortage of water causes more immediate and more intolerable distress than shortage of food. A person completely deprived of water soon feels his mouth dry, he complains of thirst and the craving for fluid rapidly becomes compelling. As time goes on the dryness of the mouth increases, the production of saliva ceases and swallowing of food becomes impossible. Finally delirium is followed by death within a day or two in a dry climate or a little longer in a moist environment.

Normally the intake of water in food and drink is so regulated that it balances the loss of water in the urine, faeces, sweat and breath. The body weight and the amount of water in the body thus remain constant. In exceptional circumstances this may not be quite true since men, actively working during the day in the desert and sweating profusely, drink only a half of the water they are losing. The consequent dehydration of the body reduces their physical efficiency. The body weight is regained by an increased fluid intake in the evening.

The sensation of thirst is produced by a general bodily need for water. It is not the result merely of a dry mouth because water deprivation produces thirst long before the mouth becomes dry. Furthermore patients who have *xerostomia* as a result of disease or of absence of the salivary glands do not complain of thirst. An injection of atropine, sufficient to stop the secretion of saliva, though producing a dry mouth, does not make a person thirsty.

On the other hand, the intravenous injection of hypertonic solutions, for example 5 per cent sodium chloride, regularly produces thirst. Such solutions increase the osmotic pressure of the extracellular fluid and bring about a movement of fluid out of the cells, the osmolality of which is relatively increased. The thirst following the ingestion of alcohol may have a similar explanation. Alcohol is believed to inhibit the release of antidiuretic hormone from the posterior lobe of the pituitary gland (Chap. 52). The consequent water diuresis leads to water depletion which results in thirst.

It is therefore likely that the sensation of thirst is produced by an increased osmotic pressure of the fluid inside cells and calculations based on experiments in man with hypertonic saline show that thirst is produced when about 1 per cent of the intracellular water of the body has passed into the extracellular space. The osmoreceptor mechanism for ADH release and the central thirst mechanism may not be identical. The drinking effect may be due to a stimulation of nervous elements specifically sensitive to an elevated sodium chloride concentration of the internal environment. The thirst mechanism is probably situated more posteriorly than the osmoreceptors.

The experiments with hypertonic saline just described give a threshold which is much the same as that found by Verney (p. 661) for the osmoreceptors in the hypothalamus. This resemblance led to a search for central receptors for thirst. From 1952 onwards, Andersson has described his experiments on unanaesthetized goats. Injection of tiny quantities of 2 or 3 per cent sodium chloride into the mid-hypothalamic region near the paraventricular nucleus (caudal to the osmoreceptors) induced drinking within a minute; drinking continued for several minutes. Electrical stimulation of this area produced drinking in 10 to 30 seconds; drinking stopped two or three seconds after the end of electrical stimulation. Destruction of this area caused complete loss of thirst. Hypertonic saline (0·85M NaCl) injected into the third ventricle of goats induced drinking of water within two minutes; injection into a lateral ventricle had no effect. Hypertonic solutions of NH_4Cl or glucose were also without effect.

Normally, then, the thirst centres are stimulated by a rise in the tonicity of the extracellular fluid but they may be stimulated in hypercalcaemia or in hypokalaemia by a direct effect. Reduction of extracellular volume, for example by haemorrhage or by salt depletion, produces thirst probably through the renin-angiotensin mechanism (p. 659).

WATER AND ELECTROLYTE HOMEOSTASIS

Because water can diffuse freely throughout the body, gains (or losses) of water alter the amount of water in the different compartments roughly in proportion to their volumes. An increase (or decrease) in the volumes of water results in corresponding alterations in the concentrations of solute; a gain in body water is followed by a decrease of osmotic pressure, and a loss of body water results in an increased osmotic pressure. Such changes in osmolality cannot easily be measured directly but an indirect assessment (osmometry) can be made by measuring the freezing point of a solution (cryoscopy). A convenient index is the serum Na^+ concentration, an increase above normal (135 to 145 m-equiv. per litre) indicating hyperosmolality and a decrease hypo-osmolality.

An increased concentration of solutes in the blood results in increased secretion of the antidiuretic hormone (p. 661) which increases reabsorption of water by the renal tubules. The limiting factor in this process is the osmotic gradient at the distal renal tubule. When the urine has a specific gravity of 1·036 the limiting osmotic gradient of 1100 m-osmoles per kg is attained, and the kidney cannot reabsorb any higher proportion of water.

The loss of body secretions from the alimentary tract is equivalent to a loss of fluid which is approximately isotonic with extracellular fluid. Although such a loss does not alter the osmotic pressure of the body fluids, the decline in volume of the extracellular fluid if sufficiently severe leads to an increased secretion of the water-retaining antidiuretic hormone. The reduction of extracellular fluid volume induces, by a similar mechanism, the secretion of the salt-retaining hormone aldosterone, which in turn increases the re-absorption of sodium and water by the renal tubules. The volume and osmolality of the extracellular fluid is controlled by the sensation of thirst, by the antidiuretic hormone and aldosterone and by other mechanisms not yet understood. There are also specific regulating mechanisms for calcium (parathyroid glands, Chap. 14) and for magnesium and potassium. The regulation of hydrogen ion concentration is discussed later in this chapter.

The normal balance of water and electrolytes can be upset in various ways by disease. Such disturbances usually involve both water and electrolytes.

In clinical practice states of combined salt and water loss (*saline depletion*) and excess (*saline overload*) are commoner than those involving water alone.

Saline depletion is caused by loss of salt and water, commonly from the alimentary tract by vomiting or diarrhoea, less often in the urine,

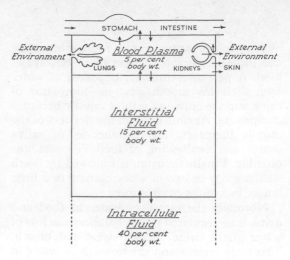

Fig. 33.13 Interrelationship of body fluids. (From J. L. Gamble (1951). *Companionship of Water and Electrolytes and the Organization of Body Fluids.* Lane Medical Lectures. Stanford University Press.)

either because of the long-continued administration of diuretic drugs or because of renal disease; occasionally the saline is lost in excess sweat or the exudate from severe burns. Only rarely is the dietary intake of sodium insufficient. Saline depletion causes cramps, weakness and faintness, and clinical features of 'dehydration', diminished tissue turgor, hypotension and peripheral circulatory failure.

Saline overloading is most often the result of a primary disturbance of the function of the heart, kidneys or liver, leading in each case to complex causes for the saline overload. Occasionally it is caused by excess intravenous infusion of saline. Oedema and raised central venous pressure commonly result.

Water depletion is usually the result of an inadequate water intake and often affects patients who are mentally confused or in prolonged coma, unless their need for fluids is recognized. Rarely it is caused by abnormal loss, as when the kidney fails to conserve water because of the impaired production of vasopressin (diabetes insipidus) or in some forms of primary renal disease. There are no signs of 'dehydration', but often fever and a raised concentration of Na^+ in the plasma (*hypernatraemia*). Conscious patients are intensely thirsty, and, except in diabetes insipidus, the urine is concentrated.

Water overload ('water intoxication') occurs

when the intake of water exceeds the ability of the kidneys to excrete it. This may be the case in renal failure or, more dramatically, in a patient in whom the release of antidiuretic hormone has been stimulated by the stress of a surgical operation and who has been given intravenous infusions of a solution, such as 5 per cent glucose in water, that becomes hypotonic in the body because the solute is used up in metabolism. In renal failure water overload may occur unless the intake is controlled. Water intoxication causes cerebral swelling which gives rise to nausea and vomiting, headache, confusion, convulsions and coma. The concentration of Na^+ in the plasma is low (*hyponatraemia*).

Potassium depletion. Since potassium resides mainly in the cells the extent to which this important ion may be lacking is not easily measured. Deficiency is often shown by a low plasma K but the plasma deficit is not an accurate guide to the deficit in the whole body. The electrocardiographic changes (p. 489) are also poorly related to the body's deficiency. Hypokalaemia causes tiredness, muscle weakness and polyuria. It is commonly seen after the prolonged administration of powerful diuretic drugs. These drugs provoke an increased flow of urine by inhibiting the activity of enzymes in the renal tubule that normally facilitate the absorption of Na^+ and Cl^-. When, however, there is a low dietary intake of Na^+ and a high level of circulating aldosterone these diuretics promote the exchange of Na^+ for K^+ in the distal tubule and lead to an increased excretion of K^+. Chronic diarrhoea (p. 306) and repeated vomiting also cause hypokalaemia.

The amount of metabolic water produced by the oxidation of foodstuffs can be readily calculated but it must not be forgotten that water is needed to dissipate the heat and, in the case of protein, to carry off the end-products, mainly urea, in the urine. The water deficit in protein metabolism is seven or eight times greater per kcal than that of carbohydrate or fat metabolism (Table 33.14). In young children a change from a mainly carbohydrate to a mainly protein diet can cause dehydration.

The case of shipwreck survivors adrift in a lifeboat with restricted water and food supplies is of practical importance and serves to illustrate the principles just discussed. The

Table 33.14 Grams of water needed for complete metabolism of 418 kJ (100 kcal) of some food substances

Food material	Water gained by oxidation	Lost in dissipating heat	Lost in excreting end-products	Water deficit
Protein	10·3	60	300	350
Starch	13·9	60	0	46
Fat	11·9	60	0	48

(From A. A. Albanese (1959). *Protein and Amino Acid Nutrition*, p. 301. New York: Academic Press.)

recommendations of a Medical Research Council Committee (1943) may be summarized as follows. Water is much more important than food but, unless there has been much loss of water by sweating, none should be issued in the first twenty-four hours; thereafter 18 oz (0·5 litre) are issued per man per day. This is not sufficient to meet the water requirements but a reasonable state of fitness can be maintained on this ration; smaller intakes lead to rapid deterioration. Although rinsing out the mouth with sea-water may be helpful, sea-water should never be drunk; a large intake of hypertonic sea-water is fatal and small amounts are harmful. The lifeboat rations should consist mainly of fats, starches and sugars. Since salt and the end-products of nitrogen metabolism require water for their excretion the salt and protein content of the diet should be low. A fat diet is valuable because it reduces hunger for a longer time than any other foodstuff (p. 204). Defaecation occurs infrequently and the volume of the urine is also reduced; in these ways body water is conserved. The drinking of urine must be prohibited; this fluid is hypertonic and contains substances which may be toxic.

FLUID EXCHANGE IN THE TISSUES

We must now consider the exchange of water and dissolved substances through the capillary endothelium, that is to say between the blood and the interstitial (intercellular) fluid. Quantitatively this exchange is very large; it has been calculated that as much as three-quarters of the plasma water is exchanged with interstitial fluid every minute.

When a solution containing protein and salt is separated from an electrolyte solution of the same concentration by a semi-permeable membrane through which sodium, chloride and water may diffuse, but protein may not, then the protein produces an osmotic pressure effect known as its colloid osmotic pressure (see p. 21). Starling pointed out that a similar situation exists in the capillaries of the vascular system where blood plasma is separated from interstitial fluid by a capillary membrane. In the capillaries, however, the additional factor of the blood pressure must be considered.

At the arteriolar end of capillaries in the human subject the blood pressure is about 32 mm Hg and at the venous end about 10 mm Hg. This pressure, of course, acts outwards, tending to produce an ultrafiltrate of plasma (compare p. 647). The colloid osmotic pressure acts in opposition to this, inwards, and is constant (about 25 mm Hg) throughout the length of the capillary; the net pressure is the difference between the two (Fig. 33.15). At the arteriolar end, the net hydrostatic pressure is 7 mm Hg outwards, and water and electrolytes therefore pass out of the capillary into the

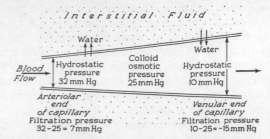

Fig. 33.15 Capillary vessel showing osmotic and hydrostatic pressures with resultant flow of water.

interstitial fluid. At the venous end water and electrolytes pass back into the capillary from the interstitial fluid.

This view (Starling's hypothesis) that, owing to the balance of osmotic and hydrostatic pressure, water must leave the capillaries at the arteriolar region and return at the venous region, and that under normal conditions of blood flow the interstitial fluid is being continuously exchanged has been clearly demonstrated in the frog (Fig. 33.16). At a hydrostatic

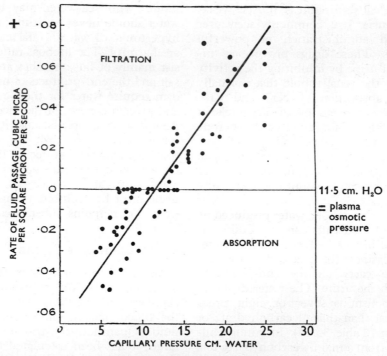

Fig. 33.16 The effect of pressure on the passage of fluid through the capillary wall. By means of a micro-manipulator a mesenteric vessel of the frog was blocked by pressure with a blunt-ended glass rod. The red cells moved away from the block if absorption took place. The pressure in the capillary was measured by a fine pipette attached to a water manometer. Positive values of the rate of fluid passage indicate filtration of fluid; negative values indicate absorption (E. M. Landis (1927). *American Journal of Physiology* **82**, 217).

pressure equal to the colloid osmotic pressure no movement of fluid takes place. When the blood pressure is greater than the colloid osmotic pressure filtration occurs and when the blood pressure is less than the colloid pressure reabsorption takes place.

In disease the concentration of plasma proteins, in particular albumin, may fall to a very low level. If the concentration of plasma albumin falls so low that the colloid osmotic pressure of the plasma does not exceed the blood pressure at the venous end of the capillary, the net hydrostatic pressure throughout the capillary is then positive and fluid passes into the interstitial space from all regions of the capillary. The volume of the extracellular fluid is increased and the tissues are swollen and waterlogged: the patient has *oedema* or dropsy. The oedema fluid is an ultra-filtrate of plasma and has a very low protein content.

The capillary membrane is not, as Starling supposed, completely impermeable to proteins since plasma proteins labelled with ^{131}I, injected intravenously, have been shown to pass into the interstitial fluid, the overall exchange rate being about 140 per cent of the total plasma albumin per day. The total amount of albumin in the extravascular pool is greater than the total amount in the blood plasma. This large reserve is called on when protein is lost from the blood. This 'leakage' of protein from the circulation may be useful in carrying antibodies and hormone-carrying proteins to the tissues. The rate of leakage depends on the size of the molecule (albumin leaks faster than globulin) and on the size of the 'pores'. The capillary 'pores' in the skeletal muscles are very small but so large in the liver that even red cells and lymphocytes may pass from the hepatic vessels to the lymphatics. Attempts have been made to find the size of the 'pores' through which proteins escape by introducing proteins or dextrans of various sizes labelled with a fluorescent dye or with ^{131}I. The electron microscope shows fenestrations either in the endothelial cells or between them in the capillaries of liver and spleen but other capillaries are not fenestrated. But the basement membrane 20 to 50 nm thick may be the real filtering membrane. The problem of capillary permeability at present is still a long way from being solved. See also p. 550.

The lymphatics. The distal lymphatics form a closed system of tubes (Fig. 33.1) consisting of an endothelial lining supported by fibrous tissue. The larger vessels have muscle fibres in their walls. The lymphatic vessels possess numerous valves and the flow of lymph from the periphery to the thoracic duct is brought about by muscular and respiratory movements in the same way as the flow of blood in the veins. The lacteals of the intestine show rhythmic contractions, ranging from 1 per min to 30 per min, which, because of the numerous valves in these vessels, propel the lymph on to the thoracic duct. In the sheep the lymph pressure may be as high as 25 mm Hg; it is increased by stimulation of the lumbar sympathetic nerves. Lymph has the same concentration of salts as interstitial fluid and plasma, a lower concentration of protein than plasma and a slightly higher concentration of protein than interstitial fluid (Table 33.17). Complete

Table 33.17 Chemical composition of plasma and lymph (after Drinker & Yoffey)

	Plasma	Interstitial fluid	Lymph
Protein (g/100 ml)	6·9	0·1	2·6
Chloride (m-equiv./litre)	100	104	116
Calcium (m-equiv./litre)	5·2	5·0	4·6
Urea (mg/100 ml)	22	22	23·5

Table adapted from H. Davson (1959). *Textbook of General Physiology*, 2nd edn., p. 261. London: Churchill.

obstruction of the lymphatic vessels draining a part of the body leads to oedema of the area. The oedema fluid has a protein concentration similar to that of plasma.

When albumin or globulin labelled with ^{131}I is injected intravenously one-thousandth of the amount injected leaks out per minute; the labelled protein reaches the lymphatics within a few minutes and is then returned to the circulation. It is calculated that in man half of the plasma protein escapes per day from the capillaries to be returned to the blood stream by the lymphatics. Lipids, chylomicrons and other large molecules, after passing through the capillary wall are picked up by the lymphatics. Hormones and enzymes (for example renin, amylase and lipase) formed by tissue cells enter the lymphatics to be carried to the blood.

Lymph nodes. Before reaching the blood the lymph passes through one or more lymph nodes (Fig. 33.18). The lymph enters a node through the capillary endothelium into the blood by which it is rapidly carried into the general circulation to be excreted in the urine.

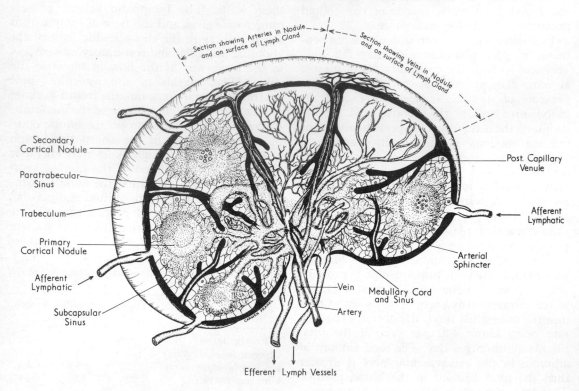

Fig. 33.18 Diagram of a typical lymph node. The lymph enters the peripheral sinus by many afferent lymphatics guarded by valves. It percolates in the medullary sinuses between the masses of lymphoid tissue and leaves at the hilum by a large efferent vessel. (*By courtesy of T. Gillman.* For further details see T. Gillman, R. M. Binns & J. J. Ruskin (1970). *Journal of Physiology* **210**, 117–118 P.)

by several afferent lymphatic vessels and flows into the *subcapsular lymph space* (peripheral sinus) which is lined by macrophages. Back flow is prevented by a valve at the point of entry of each afferent vessel. The lymph spreads along the subcapsular space and is guided by the trabecular framework through the cortex of the node, which is densely packed with lymphocytes, to the more loosely packed *medulla* from which it escapes by a single efferent lymphatic vessel. In the dense cortical area are paler germ centres containing developing lymphocytes which do not stain deeply with basic dyes. The majority of the lymphocytes are formed in these centres.

When a dye dissolved in isotonic saline is injected under the skin part of it passes

Some of the dye solution, however, enters the lymphatics along which it travels very slowly. Particulate matter in suspension, such as particles of India ink or bacteria, is carried in the lymphatics to the lymph nodes which act as filters preventing the entry of such material into the blood. Mild injury to a limb increases the lymph flow but severe injury such as burning causes a great increase in the flow of lymph which contains enzymes which have leaked out of the damaged cells.

In the fasting state the lymph from the thoracic duct is a pale yellow transparent fluid but after a fatty meal it becomes milky (p. 300) and this milky appearance is even more pronounced in the lymphatics of the small intestine which are often called *lacteals*.

The lymph from all over the body is collected into larger and larger lymphatic vessels formed by the union of smaller vessels. From most of the body and the whole of the intestine, the lymph passes to the *thoracic duct* which enters the large veins at the root of the neck on the left side. The lymph from the right upper half of the body enters the great veins via the *right lymphatic duct*. Back flow of venous blood into the lymphatic system is prevented by valves placed at the orifices of the ducts. In man the lymph flow from a cannulated thoracic duct is from 4 to 10 ml per min (or 600 to 1400 ml per day). If the thoracic duct lymph is drained away and so prevented from reaching the blood the concentration of plasma protein falls and the blood volume

decreases. The lymph in the thoracic duct is rich in lymphocytes the majority being T-lymphocytes.

CEREBROSPINAL FLUID

The central nervous system is covered by three meninges (Fig. 33.19). Beneath the tough outer coat, the *dura mater*, is the narrow *subdural space*, containing a very small amount of tissue fluid. This space is bounded internally by the leptomeninges, namely the *arachnoid mater*, separated from the *pia mater* by the *subarachnoid space* containing cerebrospinal fluid. The subarachnoid space does not communicate with the subdural space, but is continuous with the ventricular system of the

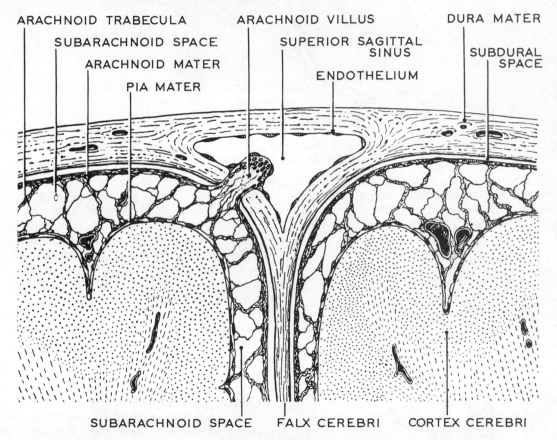

ARACHNOID TRABECULA   ARACHNOID VILLUS   DURA MATER

SUBARACHNOID SPACE   SUPERIOR SAGITTAL SINUS   SUBDURAL SPACE

ARACHNOID MATER   ENDOTHELIUM

PIA MATER

SUBARACHNOID SPACE   FALX CEREBRI   CORTEX CEREBRI

Fig. 33.19 A drawing of a coronal section through the vertex to show the arrangement of the leptomeninges. (L. H. Weed (1923). *American Journal of Anatomy* **21**, 191.) According to Welch and Friedman the arachnoid villi contain a labyrinth of tubes 4 to 12 μm in diameter which act as valves. When the pressure in the subarachnoid space is greater than the pressure in the superior sagittal sinus the tubules of the villus open up and cerebrospinal fluid passes into the sinus. When the subarachnoid pressure is lower than that in the sinus reflux of blood is prevented by collapse of the tubules.

brain through medial and lateral apertures in the roof of the fourth ventricle. The arachnoid and pia covering the brain and spinal cord are connected together by innumerable strands of arachnoid tissue; the pia closely invests the surface of the brain and dips down into the fissures and sulci (Fig. 33.20). At the base of

The cerebral blood vessels passing through the subarachnoid space are invested with two layers of arachnoid. The branches that penetrate into the brain are accompanied by prolongations of the subarachnoid space so that cerebrospinal fluid is carried some distance into the substance of the brain. There are also

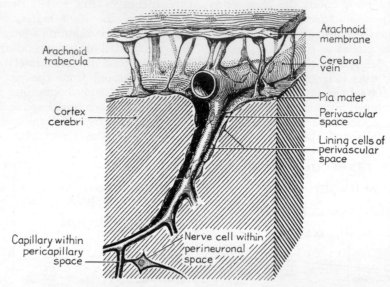

Fig. 33.20 A drawing to show the arrangement of blood vessels in the subarachnoid space. (L. H. Weed (1923). *American Journal of Anatomy* **21**, 191.)

the brain the two membranes are widely separated and form large spaces known as cisterns, the largest being the cisterna cerebello-medullaris or *cisterna magna* (Fig. 33.21).

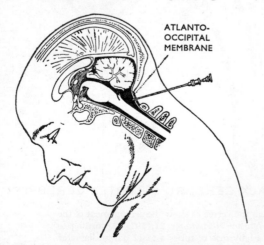

Fig. 33.21 Cisternal puncture. The diagram shows the direction of the needle introduced into the cisterna magna (Hamilton Bailey (1956). *Pye's Surgical Handicraft*, 17th edn. Bristol: Wright).

short prolongations of the subarachnoid space along the cranial and spinal nerve roots.

A sample of cerebrospinal fluid can be obtained in man by passing a long needle between the occipital bone and the atlas into the cisterna magna at the base of the brain (*cisternal puncture*) (Fig. 33.21) or between the third and fourth lumbar spinous processes into the spinal subarachnoid space (*lumbar puncture*) (Fig. 33.22). The latter is the safer procedure because, although the dura and the arachnoid meninges end at the second sacral vertebra, the cord comes to an end at the level of the first lumbar vertebra and cannot be damaged by a needle entering the subarachnoid space an inch or so lower.

The cerebrospinal fluid is formed by the choroid plexuses. These rich networks of blood vessels projecting into the cavities of the ventricles are covered only by the pia mater and a single layer of the cells lining the ventricular system of the brain (*ependyma*). The choroid plexuses are supplied with blood by anterior and posterior choroidal arteries from

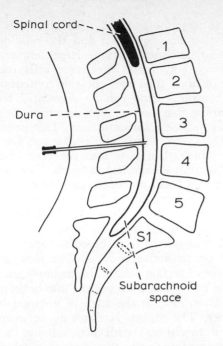

Fig. 33.22 Route of needle in performing lumbar puncture. The cord is out of danger. The cauda equina is not shown. (A. Lee McGregor (1957). *Synopsis of Surgical Anatomy*, 8th edn. Bristol: Wright).

the internal carotid and posterior cerebral arteries. There is no lymphatic system in the brain and spinal cord.

The choroid plexuses in the ventricles of the brain produce cerebrospinal fluid but since a considerable amount of fluid is produced after their removal they cannot be the only source. In the dog the lateral ventricles provide about 35 per cent, the fourth ventricle about 23 per cent and the brain tissue adjacent to the subarachnoid space 42 per cent of the total production, and similar values have been obtained in human infants. The cerebrospinal fluid formed in the lateral ventricles passes into the third ventricle by the inter-ventricular foramina and then by the aqueduct of the midbrain to the fourth ventricle, from which it escapes by three foramina into the subarachnoid space round the brain and spinal cord. That this is the manner of flow is shown by the fact that blockage of any of the foramina causes the ventricles upstream to dilate. Blockage of the cerebral aqueduct, for example, causes distension of the third ventricle and the two lateral ventricles (*internal hydrocephalus*) unless the choroid plexuses have previously been

removed. These experiments also show that the fluid is continuously produced at a rate of 0·5 ml per min (720 ml per 24 h). Since the total volume of cerebrospinal fluid is 120 ml the whole of the cerebrospinal fluid must be exchanged every four hours or so. Furthermore since the pressure of the cerebrospinal fluid remains relatively constant the production must be balanced by an equal absorption of fluid. The movement of cerebrospinal fluid from the lateral ventricles into the ventricular system and subarachnoid space is no doubt helped by the pulsations of the choroid plexuses and is maintained by its continuous absorption as it passes over the cortex and down the spinal cord.

It was at one time thought that the cerebrospinal fluid was absorbed into the venous system mainly through the arachnoid villi, which are invaginations of the subarachnoid space, through gaps in the dura mater into the large venous sinuses of the cranium. Welch has described valve-like structures in the villi which open when the cerebrospinal fluid pressure is 25 mm H_2O above that in the superior longitudinal sinuses (see Fig. 33.19). This difference of pressure may exist at each arterial pulse. The venous sinuses remain patent in spite of the low hydrostatic pressure within them because they are enclosed between layers of tough dural membrane. In chronic obstructive hydrocephalus fluid must be absorbed by the walls of the ventricles since the increase in brain size per day is only 2 per cent of the amount of fluid formed per day. It seems likely, therefore, that the major proportion of cerebrospinal fluid is absorbed into the venous system from the extensions of the subarachnoid space along the roots of the spinal nerves, into the spinal veins, through the ependymal linings of the ventricles. Reabsorption of cerebrospinal fluid into the blood is favoured by the colloid osmotic pressure of the plasma protein, since cerebrospinal fluid normally contains very little protein.

Normal cerebrospinal fluid (c.s.f.) is a clear colourless fluid, sp. gr. 1005, containing not more than 5 lymphocytes per cu. mm. The pH, about 7·33, remains nearly constant in the presence of large changes of plasma pH. Its composition (Table 33.23) is similar to that of plasma except that it contains very much less protein (only 0·02 g per 100 ml). The ven-

Table 33.23 Comparison of chemical composition of blood plasma and cerebrospinal fluid

	Blood plasma (mg/100 ml)	Cerebrospinal fluid (mg/100 ml)
Protein	7000–8000	20–30
Urea	16–35	10–40
Glucose	68–96	50–80
Sodium	318–342	350
Potassium	15–20	8
Calcium	10–11	5
Chloride (as Cl^-)	360–378	437–455
Bicarbonate (vol. CO_2 per 100 ml)	50–75	50–75
Inorganic phosphate	2–5	1–2

The calcium content of the cerebrospinal fluid is relatively low but only half of the calcium of the blood plasma is ionized, so that the two fluids can be considered to be in equilibrium in respect of calcium. Note the higher chloride and lower glucose content of cerebrospinal fluid as compared with blood plasma.

tricular fluid has chloride added to it and K^+ and bicarbonate subtracted from it as it passes to the cisterna magna. The cerebrospinal fluid is about 5 mV positive to blood; this suggests that the cells of the brain provide energy to maintain the relative acidity of the c.s.f. The protein content in the lumbar region (0·04 g per 100 ml) is greater than in the ventricles (0·01 g per 100 ml). If a vertical tube is attached to the lumbar puncture needle the fluid rises about 10 cm when the patient is lying on his side and about 30 cm when he is sitting. A rise of intracranial venous pressure produced by coughing or by digital compression of the internal jugular veins causes the cerebrospinal-fluid pressure to rise rapidly above these values and the pressure falls rapidly when the compression is removed.

The cells of the choroid plexuses seem to exert a selective action on the substances passing through them because the composition of plasma and cerebrospinal fluid is not quite identical and the distribution of ions between them is not that expected from the Donnan equilibrium (Table 33.24). Isotype studies suggest that cerebrospinal fluid is formed from blood plasma both by secretion and by ultrafiltration. It is secreted by the cells of the choroid plexuses but it also enters the ventricles through the ependymal lining as an ultrafiltrate. A relatively large volume is exchanged as an ultrafiltrate through the blood vessels of

the subarachnoid space. The glial cells surrounding the capillaries and the cells of the choroid plexuses contain large amounts of carbonic anhydrase; the nerve cells contain little or none. The high chloride content of the cerebrospinal fluid can be accounted for by assuming that carbon dioxide enters the glial cells producing carbonic acid. This produces HCO_3^- ions which are exchanged for Cl^- ions which then pass into the cerebrospinal fluid against a chemical gradient. The greater content of NaCl makes the cerebrospinal fluid slightly hypertonic to blood plasma so that water passes out of the choroid plexuses into the cerebrospinal fluid. Both Na and Mg seem to be transported from blood to cerebrospinal fluid by an active process. The slowness of passage of certain solutes, for example sucrose, inulin and ^{24}Na, from plasma into brain tissue has given rise to the idea of a blood–brain barrier. The plasma K can vary between 2·5 and 8 m-equiv./l with little change in the K concentration in the cerebrospinal fluid. Radioisotope investigations show that oxygen, carbon dioxide, alcohol, barbiturates, glucose and lipophilic substances pass quickly into the brain from the blood. Inorganic ions and most highly dissociated compounds, amino acids and penicillin pass very slowly into the brain. Raising the blood glucose to high levels increases the brain glucose only slightly.

The cerebrospinal fluid acts as a cushion between the soft and delicate brain substance and the rigid cranium. It supports the weight of the brain and distributes the force of blows on the head. The volume of the brain and its blood vessels changes from time to time and such changes take place at the expense of the volume of cerebrospinal fluid; when the volume of blood in the brain is increased the volume of cerebrospinal fluid is diminished and when there is degeneration or atrophy of the brain there is an increase in the volume of cerebrospinal fluid. The cerebrospinal fluid plays little part in cerebral metabolism.

A cerebral tumour which increases the volume of the contents of the cranium may cause a rise of intracranial pressure by interfering with the circulation of the cerebrospinal fluid. Increased intracranial pressure may be temporarily relieved by the intravenous injection of hypertonic solutions, such as 20 per cent mannitol. The resultant increase in the osmotic pressure of the blood drains fluid from the brain into the cerebral capillaries, and produces a shrinkage of the brain and a reduction of intracranial pressure which lasts for several hours. The increased osmotic pressure in the vessels of the choroid plexuses may be great enough to reverse the direction of flow so that cerebrospinal fluid may be absorbed through the plexuses into the blood stream.

The shape, size and position of the cerebral ventricles often given valuable clues to the position of tumours in the cranial cavity. To make a ventriculogram a needle passed through a trephine hole in the skull is pushed into the lateral ventricle; the cerebrospinal fluid drains off and is replaced by air. An X-ray photograph (Figs. 33.25, 33.26) shows lighter areas (in the

Table 33.24 Comparison of arterial blood and cerebrospinal fluid obtained simultaneously from thirteen subjects without metabolic or respiratory changes

	Arterial blood	Cisternal cerebrospinal fluid	Lumbar cerebrospinal fluid
pH	7·397	7·346	7·325
P_{CO_2}	40·5	46·5	49·1

(Data taken from A. N. P. van Heijst, H. J. Maas & B. F. Visser (1966). *Pflügers Archiv für die gesampte Physiologie des Menschen und der Tiere* **287**, 242–246.)

tion of the cations; its concentration in extracellular fluid is very small $(4 \times 10^{-5}$ m-equiv./l), the pH being 7·40. In the intracellular fluid, the concentration of hydrogen ions is somewhat greater, giving a pH of 7·00.

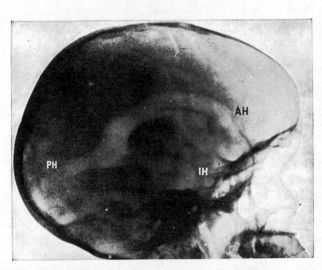

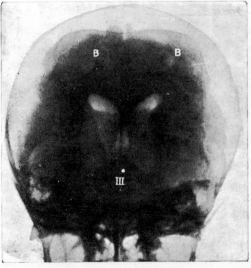

Fig. 33.25 Ventriculogram (lateral view). The ventricular cavities, being filled with air, show as lighter areas. AH, anterior horn. PH, posterior horn. IH, inferior horn. (*By courtesy of J. Sloan Robertson.*)

Fig. 33.26 Ventriculogram (antero-posterior view). The anterior horns can be recognized above the inferior horns. The third ventricle is above the point marked III. B, burr hole. (*By courtesy of J. Sloan Robertson.*)

positive) corresponding to the cerebral ventricles.

HYDROGEN-ION CONCENTRATION OF THE EXTRACELLULAR FLUID

The sum of cation concentrations of the extracellular fluid (154 m-equiv./l) must be balanced by the total concentration of anions (Fig. 33.27) to maintain electrical neutrality. Hydrogen ion accounts for a minute propor-

Comparatively little is known about the homeostatic processes controlling the pH of the fluid in the cells but it is clear that the cells participate in buffering changes in pH of the fluid outside them. The remainder of this chapter will be concerned exclusively with the extracellular fluid and erythrocytes.

The hydrogen ion concentration of the extracellular fluid is maintained within narrow limits by the buffering action of blood and interstitial fluid, by adjusting the pulmonary

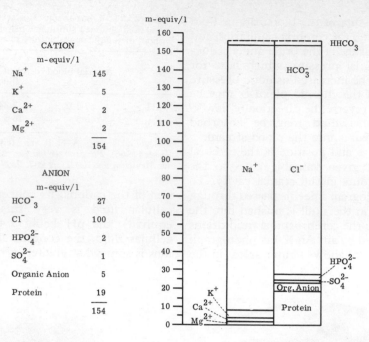

Fig. 33.27 The ionic composition of normal arterial plasma (after Gamble).

excretion of carbon dioxide and by regulating the renal secretion of hydrogen ions.

It is impossible to differentiate between the hydrogen ions produced in the body but it does simplify the understanding of acid base chemistry if arbitrarily they are divided into two groups, respiratory and metabolic hydrogen ions.

Respiratory hydrogen ion. This is derived from the ionization of carbonic acid formed by the hydration of CO_2, the source of which is the oxidation of carbon in glucose, fat and protein.

ions must be excreted by the kidney. Their major source is the oxidation of sulphur in the amino-acids methionine and cysteine to the non-oxidizable acid, sulphuric acid.

In addition, if oxygen is not freely available, glucose metabolism terminates at the stage of lactic acid. Starvation or insulin lack leads to the production of acetoacetic acid from fat. Under normal metabolic conditions these organic acids are further oxidized to carbon dioxide and water.

Some 50 m-equiv. of hydrogen ions are excreted daily in the urine. This covers the

$$\text{Alveolar air} \quad CO_2 \text{ (gas)}$$
$$\Updownarrow$$
$$\text{Blood} \quad CO_2 \text{ (dissolved)} + H_2O \rightleftharpoons H_2CO_3 \rightleftharpoons H^+ + HCO_3^-$$

If alveolar ventilation is increased, alveolar P_{CO_2} falls, the above reactions move from right to left and the hydrogen ion concentration decreases. On the other hand if alveolar ventilation is reduced, alveolar P_{CO_2} rises, the reactions move from left to right and the hydrogen ion concentration increases.

About 15 000 millimoles of CO_2 are produced and excreted daily.

Metabolic hydrogen ion. These hydrogen

production of both sulphuric acid and some of the organic acids which escape oxidation and require renal excretion.

In disease, excesses and deficits of both respiratory and metabolic hydrogen ions are encountered. These abnormal clinical states are termed respiratory or metabolic acidosis and alkalosis. Changes then occur in the various components of the plasma and whole blood buffer systems and the blood pH may

690 Textbook of Physiology and Biochemistry

alter. Suitable buffer systems must be chosen for chemical analysis so that changes due to metabolic hydrogen ion can be clearly and easily differentiated from those due to respiratory hydrogen ion.

PLASMA BUFFERS

The plasma buffer systems are as follows:

$$\text{BUFFER ACID} \rightleftharpoons H^+ + \text{BUFFER BASE}$$
$$H_2CO_3 \rightleftharpoons H^+ + HCO_3^-$$
$$H_2PO_4^- \rightleftharpoons H^+ + HPO_4^{2-}$$
$$HPr \rightleftharpoons H^+ + Pr^-$$
(Protein)

In metabolic acidosis, the excess hydrogen ions cause these reactions to move from right to left and the concentration of bicarbonate and the other buffer bases decreases. The converse occurs in metabolic alkalosis.

Although these systems are of necessity in equilibrium with each other, the carbonic acid bicarbonate buffer system is quantitatively the most important, and moreover is unique in that its conjugate acid, carbonic acid, is in equilibrium with the dissolved carbon dioxide of the body fluids and is thus under respiratory control. Any excess formed when strong acid is added can therefore be readily eliminated.

The pH of a buffer system, as has been shown on p. 19, depends on the relationship

$$pH = pK_a + \log_{10} \frac{[\text{base}]}{[\text{acid}]}$$

which in the case of the carbonic acid/bicarbonate system becomes

$$pH = pK_1 + \log_{10} \frac{[HCO_3^-]}{[H_2CO_3]}$$

(the Henderson-Hasselbalch equation, p. 17).

Carbonic acid is formed by the hydration of CO_2.

$$CO_2 \text{ (dissolved)} + H_2O \rightarrow H_2CO_3$$

However, the concentration of carbonic acid is negligible in comparison with the concentration of dissolved CO_2 gas which is proportional to the partial pressure of CO_2 and so the equation can be written:

$$pH = pK_1 + \log_{10} \frac{[HCO_3^-]}{[s\,P_{CO_2}]} \qquad (1)$$

where pK_1 is a constant and has a value of 6·1 and s is the solubility factor of CO_2 in plasma at 37°C, namely 0·03 m-equiv./l per mm Hg.

It is important to appreciate that calculation of one of the three variables, pH, P_{CO_2} or bicarbonate from the other two is valid only for a single phase such as plasma. The Henderson–Hasselbalch equation cannot be applied to whole blood.

In equation (1), bicarbonate is termed the metabolic component since it reflects changes due to excess or deficit of metabolic hydrogen ion and P_{CO_2} is termed the respiratory component as it reflects changes due to excess or deficit of respiratory hydrogen ion.

Thus the following general relation can be derived:

$$\text{Blood pH} \propto \frac{[\text{Metabolic component}]}{[\text{Respiratory component}]}$$

As will be seen in the next section on blood buffers, plasma bicarbonate, in addition to varying with metabolic hydrogen ion, also varies with the P_{CO_2}, and therefore in evaluating the degree of abnormality in the metabolic component, due allowance must be made for this, for example in blood with a normal metabolic component and a P_{CO_2} of 80 mm Hg the plasma bicarbonate is 31 m-equiv./l and in blood with a normal metabolic component and a P_{CO_2} of 20 mm Hg the plasma bicarbonate is 19 m-equiv./l.

CHARACTERIZATION OF BLOOD ACID BASE STATUS. It is most important that values for the appropriate indices be determined in arterial blood because venous blood only reflects local changes. However, capillary blood, provided that it has been obtained without congestion in a warm extremity yields values identical with arterial blood and can be used if suitable microtechniques are available. A very sensitive pH meter is also required as quite small changes in blood pH may be of great importance to the patient, for example the apparently small fall in pH from 7·42 to 7·38 represents an increase in hydrogen ion concentration from 38 to 42 nmoles/l (see p. 15).

Two techniques are in general use:
1. The pH of a portion of the sample is measured, then the total CO_2 content of the remainder is estimated by a Van Slyke gasometric apparatus. As

bicarbonate concentration =

$$\text{total } CO_2 - sP_{CO_2}, \quad (2)$$

the Henderson–Hasselbalch equation can be rearranged to allow P_{CO_2} to be calculated thus:

$$P_{CO_2} = \frac{[\text{total } CO_2]}{s\{[\text{antilog}(pH - pK_1)] + 1\}}$$

Bicarbonate concentration can now be calculated from equation (2).

2. It is now common practice to measure P_{CO_2} by the recently developed P_{CO_2} electrode (p. 619) whilst simultaneously measuring pH by the usual glass electrode method. Bicarbonate can then be calculated by a further rearrangement of the Henderson–Hasselbalch equation.

$$\text{bicarbonate} = sP_{CO_2}[\text{antilog}(pH - pK_1)]$$

Thus by these two techniques, values are obtained for pH, P_{CO_2} and plasma bicarbonate. Their interpretation is simplified when the results are plotted in a graphical form, for example the pH bicarbonate plot of Davenport. Examples are shown in Fig. 33.28.

BLOOD BUFFERS

Blood contains, for practical purposes, six different buffer systems unevenly distributed between the red cells and the plasma. For simplicity they can be separated into two main classes: (a) the bicarbonate system located principally in the plasma, and (b) the non-bicarbonate system consisting principally of haemoglobin located in the red cells.

$$\text{BUFFER ACID} \rightleftharpoons H^+ + \text{BUFFER BASE}$$

$$H_2CO_3 \rightleftharpoons H^+ + HCO_3^- \quad \text{bicarbonate system}$$

$$HBuf \rightleftharpoons H^+ + Buf^- \quad \text{non-bicarbonate system}$$

Buffer base. This is the sum of buffer anions in the blood and consists of bicarbonate plus Buf^-. At pH 7·40 and P_{CO_2} of 40 mm Hg it totals 48 m-equiv./l.

Characteristic changes occur in metabolic and respiratory disturbances.

Metabolic. (a) In metabolic acidosis, where acids such as acetoacetic acid may have accumulated to the extent of 16 m-equiv./l, the buffer base as it takes up hydrogen ions, will fall from 48 m-equiv./l to 32 m-equiv./l. The difference between the observed buffer base and the normal buffer base is termed the base

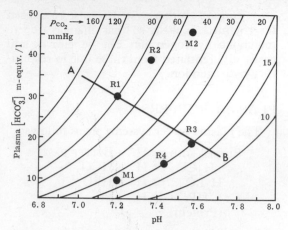

Fig. 33.28 AB is termed the normal buffer line and is the CO_2 dissociation curve of blood with a normal metabolic component. The P_{CO_2} isobars are calculated from the Henderson–Hasselbalch equation. If a point is above the normal buffer line, the metabolic component is increased. If a point is below the normal buffer line, the metabolic component is decreased. If a point is to the left of the isobar representing P_{CO_2} of 40 mm Hg the respiratory component is increased. If a point is to the right of the isobar representing P_{CO_2} of 40 mm Hg the respiratory component is decreased.

excess, namely $32 - 48 = -16$ m-equiv./l. The negative base excess value signifies a deficit of base or excess of acid.

Respiratory acidosis		
uncompensated R1	Plasma pH	7·20
	Plasma P_{CO_2}	80 mm Hg
	Plasma bicarbonate	30 m-equiv./l
compensated R2	Plasma pH	7·36
	Plasma P_{CO_2}	71 mm Hg
	Plasma bicarbonate	39 m-equiv./l
Respiratory alkalosis		
uncompensated R3	Plasma pH	7·55
	Plasma P_{CO_2}	20 mm Hg
	Plasma bicarbonate	17 m-equiv./l
compensated R4	Plasma pH	7·44
	Plasma P_{CO_2}	20 mm Hg
	Plasma bicarbonate	13 m-equiv./l
Metabolic acidosis		
partially compensated M1	Plasma pH	7·20
	Plasma P_{CO_2}	25 mm Hg
	Plasma bicarbonate	9 m-equiv./l
Metabolic alkalosis		
partially compensated M2	Plasma pH	7·54
	Plasma P_{CO_2}	55 mm Hg
	Plasma bicarbonate	47 m-equiv./l
Normal ranges	Plasma pH	7·36–7·44
	Plasma P_{CO_2}	35–45 mm Hg
	Plasma bicarbonate	Normal range depends on P_{CO_2}

(b) Conversely, in a metabolic alkalosis, the buffer base may rise from 48 to 66 m-equiv./l, resulting in a base excess value of $+18$ m-equiv./l. The positive base excess signifies an excess of base or deficit of acid.

Respiratory

(a) CO_2 (gaseous)
$$\text{\Updownarrow}$$
CO_2 (dissolved) $+ H_2O$
$$\text{\Updownarrow}$$
$H_2CO_3 + Buf^- \rightleftharpoons HBuf + HCO_3^-$
Buffer base $= Buf^- + HCO_3^-$

In respiratory acidosis, the excess carbonic acid causes a fall in Buf^- with an equivalent rise in bicarbonate.

(b) Conversely, in respiratory alkalosis the reduction in carbonic acid results in a fall in bicarbonate with an equivalent rise in Buf^-.

As Buffer base $= Buf^- + HCO_3^-$

the total buffer base value does not change in respiratory disorders although it is obvious that the bicarbonate ion is P_{CO_2} dependent.

Base excess is the change which occurs in total buffer base in states of metabolic imbalance and indicates directly in m-equiv./l the surplus amount of acid or base present. Since, as shown above, it is independent of the P_{CO_2} level it is an excellent index of the degree of abnormality in the metabolic component.

The Henderson–Hasselbalch equation cannot be applied to whole blood because of the complex nature of its buffering systems. However, the general relationship

$$pH \propto \frac{[\text{Metabolic component}]}{[\text{Respiratory component}]}$$

still applies. Base excess now becomes the metabolic component and P_{CO_2} remains as the respiratory component.

MEASUREMENT OF BLOOD ACID–BASE STATUS. The whole blood buffer system can be used for this purpose.

The technique developed by Astrup and his colleagues has now gained general acceptance because it is rapid and accurate, and it permits the relevant acid base data to be computed with only 0·2 to 0·4 ml blood. It depends on the observation that the CO_2 equilibration line for whole blood, when plotted on a log P_{CO_2}/pH co-ordinate system is a straight line over the physiological range (Fig. 33.29A). Since the

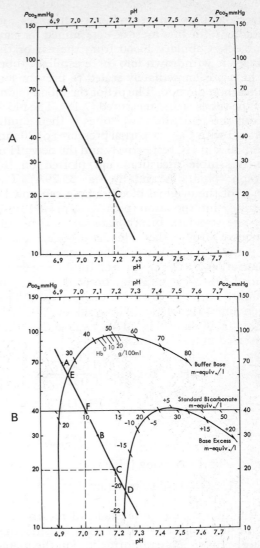

Fig. 33.29 (A) pH/log P_{CO_2} line for a blood sample equilibrated at a P_{CO_2} of 70 mm Hg and P_{CO_2} 30 mm Hg. The measured pH values were 6·91 and 7·09 respectively. The pH of the patient's blood was 7·18 and therefore its P_{CO_2} must be 20 mm Hg. (B) The same graph with curves added for obtaining values of buffer base and base excess. The point of intersection, E, shows observed buffer base to be 26 m-equiv./l. The point D shows a base deficit of 20 m-equiv./l (or a base excess of -20 m-equiv./l).
Normal buffer base $=$ observed buffer base $-$ base excess
$$= 26 + 20$$
$$= 46$$
This indicates a haemoglobin value of 10 g/100 ml. At P_{CO_2} 40 mm Hg the pH would be 7·03 (see line projected down from F) and the standard bicarbonate is read off as 10 m-equiv./l. (P. Astrup, K. Jurgensen, O. Siggard-Andersen & K. Engel (1960). *Lancet*, **i**, 1035–1039 as modified by O. Siggard-Andersen (1962). *Scandinavian Journal of Clinical and Laboratory Investigation*, **14**, 598–604.)

haemoglobin concentration influences its gradient, the line must be determined for each subject. Capillary blood from the ear or the finger is withdrawn into three capillary tubes which are immediately sealed to prevent loss of carbon dioxide. The pH of the blood in one of the tubes is first measured (7·18). The blood from the remaining two tubes is then equilibrated with CO_2 at partial pressures of 70 mm and 30 mm Hg respectively and the new pH of each sample measured and plotted on the graph (points A and B, Fig. 33.29A). The P_{CO_2} of the original blood sample can now be read by interpolation (point C at pH 7·18).

The standard bicarbonate can also be calculated from the Henderson–Hasselbalch equation. It is defined as the bicarbonate concentration in the plasma of fully oxygenated blood which has been equilibrated at a P_{CO_2} of 40 mm Hg at 37°C. From the CO_2 equilibration line, the blood pH would be 7·03 at a P_{CO_2} of 40 mm Hg. These values are then inserted in the equation:

$$pH = 6·1 + \log \frac{[HCO_3^-]}{[0·03 P_{CO_2}]}$$

$$7·03 = 6·1 + \log \frac{[HCO_3^-]}{1·2}$$

or $\qquad 0·93 = \log [HCO_3^-] - \log 1·2$

or $\qquad 1·01 = \log [HCO_3^-]$

or $\qquad [HCO_3^-] = 10$ m-equiv./l

The normal range of standard bicarbonate is 22 to 26 m-equiv./l. Values outside this range must denote abnormalities in the metabolic component because the contribution of the respiratory component has been eliminated by fixing the P_{CO_2} at 40 mm Hg.

So far, the only additional information furnished by this technique over that obtained by the pH-bicarbonate plot is a value for the standard bicarbonate. However, it becomes possible to identify base excess and buffer base values directly if curves for these variables are superimposed on the log P_{CO_2}/pH graph. The result is termed the Siggaard–Andersen curve nomogram and is constructed as follows:

(a) The base excess curve is constructed experimentally by plotting the points of intersection of the CO_2 equilibration lines for plasma and whole blood of the same base

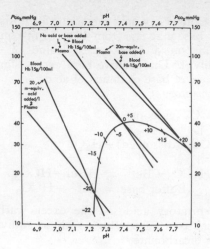

Fig. 33.30 The construction of a base excess curve. Twenty m-equiv. of strong acid are added to plasma and to blood with a haemoglobin concentration of 15 g/100 ml. The intersection of their respective CO_2 equilibration lines allows a point to be plotted at −20 m-equiv./l. The addition of 20 m-equiv. of strong base yields another point at +20 m-equiv./l. Repetition of this process with different additions of acid and base allows a base excess curve to be drawn through the points of intersection of their CO_2 equilibration lines. This figure also demonstrates that base excess unlike total buffer base is independent of haemoglobin concentration.

excess value (Fig. 33.30). This figure also demonstrates that addition of strong acid displaces the CO_2 equilibration line to the left and addition of strong base displaces it to the right.

(b) The buffer base line is constructed in a similar manner by plotting the points of intersection of the CO_2 equilibration lines for plasma and whole blood of the same buffer base value.

(c) A standard bicarbonate line calculated from the Henderson–Hasselbalch equation can also be added.

The point of intersection of the CO_2 titration line with these scales indicates the appropriate values for these indices (Fig. 33.29B).

If a blood sample has a pH of 7·40 and a P_{CO_2} of 40 mm Hg, the normal buffer base is approximately 48 m-equiv./l, the precise value depending on the haemoglobin concentration. The relationship between the buffer base and haemoglobin concentration can be expressed as follows:

Normal buffer base =
(41·7 + 0·42 × haemogloben in g/100 ml), m-equiv./l

This variation with haemoglobin is indicated by a haemoglobin subscale on part of the buffer base curve and this allows results obtained by the equilibration method to be checked as follows:

Base excess = observed buffer base
$-$ normal buffer base

Therefore, normal buffer base = observed buffer base $-$ base excess. Substituting values obtained from the example shown in Fig. 33.29B we have:

Normal buffer base = $26+20 = 46$ m-equiv./l

This indicates a haemoglobin value of 10 g/ 100 ml. A difference of more than ±3 g haemoglobin/100 ml between this predicted value and the haemoglobin concentration found by measurement suggests an error in the acid–base determination.

It will be noted that with this system there is a choice of two indices to characterize the metabolic component, the base excess and the standard bicarbonate. The base excess is the most widely used because it gives a quantitative measure of the surplus amount of acid or base present. Moreover, standard bicarbonate can be confused with actual bicarbonate.

With this Astrup technique accurate values can rapidly be obtained for blood pH, blood P_{CO_2} and blood base excess. Blood base excess, unlike plasma bicarbonate, is independent of P_{CO_2}. Thus abnormalities in this component, which must be of metabolic origin, can be detected immediately from perusal of its actual numerical value whereas the correct interpretation of plasma bicarbonate values requires graphical procedures. Duplicate examples of the acid–base disorders displayed in Fig. 33.28 are shown in Fig. 33.31.

To summarize, two systems are in current use for characterizing the acid–base status of the blood. In both, P_{CO_2} is used as a measure of the respiratory component but one method uses plasma bicarbonate to represent the metabolic component whereas the other uses whole blood base excess.

While the buffer mechanisms of the plasma and red cells can prevent excessive change of pH within the limits of their capacity, continuing control of pH in the presence of excess or deficit of metabolic or respiratory hydrogen ion can be achieved only by altering the pul-

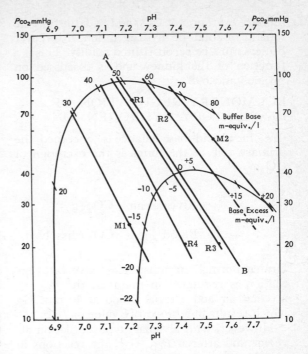

Fig. 33.31 AB = CO_2 equilibration line for blood with a normal metabolic component. If a point lies to the right of line AB the metabolic component is increased. If a point lies to the left of line AB the metabolic component is decreased. If a point lies above a line representing a P_{CO_2} of 45 mm Hg the respiratory component is increased. If a point lies below a line representing a P_{CO_2} of 35 mm Hg the respiratory component is decreased.

Respiratory acidosis		
uncompensated R1	Blood pH	7·20
	Blood P_{CO_2}	80 mm Hg
	Blood base excess	Zero
compensated R2	Blood pH	7·36
	Blood P_{CO_2}	71 mm Hg
	Blood base excess	+10 m-equiv./l
Respiratory alkalosis		
uncompensated R3	Blood pH	7·55
	Blood P_{CO_2}	20 mm Hg
	Blood base excess	−2 m-equiv./l
compensated R4	Blood pH	7·44
	Blood P_{CO_2}	20 mm Hg
	Blood base excess	−8 m-equiv./l
Metabolic acidosis		
partially	Blood pH	7·20
compensated M1	Blood P_{CO_2}	25 mm Hg
	Blood base excess	−17 m-equiv./l
Metabolic alkalosis		
partially	Blood pH	7·54
compensated M2	Blood P_{CO_2}	55 mm Hg
	Blood base excess	+20 m-equiv./l
Normal ranges	Blood pH	7·36–7·44
	Blood P_{CO_2}	35–45 mm Hg
	Blood base excess	±2·5 m-equiv./l

monary excretion of CO_2 and by adjusting the renal excretion of hydrogen ion. Moreover buffers must be reconstituted readily. This is a function of the kidney and is discussed on pp. 665 and 698.

PULMONARY REGULATION OF RESPIRATORY HYDROGEN ION

As far as acid–base balance is concerned, the primary role of the lungs is the excretion of CO_2.

blood pH therefore begins to rise but does not return completely to normal, the compensatory process stopping at a point where stimulation of ventilation by the low blood pH is balanced by depression caused by the decreased P_{CO_2}. The outcome is a partially compensated metabolic acidosis.

(b) *Metabolic alkalosis*. This is precisely the converse of the above. Compensation takes the form of an increase in arterial P_{CO_2} and the blood pH falls but does not return completely

$$
\begin{array}{ll}
\text{Alveolar air} & CO_2 \ (\text{gas}) \\
& \quad \updownarrow \\
\text{Blood} & CO_2 \ (\text{dissolved}) + H_2O \rightleftharpoons H_2CO_3 \rightleftharpoons H^+ + HCO_3 \qquad (3)
\end{array}
$$

Under normal circumstances, alveolar ventilation is regulated to maintain the P_{CO_2} in alveolar air and arterial blood at 40 mm Hg which results in a blood pH value of 7:40.

However, with alveolar hypoventilation alveolar and arterial P_{CO_2} rise, the reactions in equation (3) move to the right and the hydro-

to normal—a partially compensated metabolic alkalosis exists.

An important relationship exists between the rate of CO_2 production, the rate of alveolar ventilation and the partial pressure of CO_2 in alveolar air and arterial blood. This can be derived as follows:

$$CO_2 \text{ excretion rate} = \text{percentage of } CO_2 \text{ in alveolar air} \times \text{rate of alveolar ventilation}$$

Arterial P_{CO_2} = alveolar P_{CO_2} and, in a steady state of respiratory exchange, the rate of CO_2 production = the rate of CO_2 excretion

$$\text{Therefore, } CO_2 \text{ production rate} = \frac{\text{Arterial } P_{CO_2}}{\text{Barometric pressure}} \times \text{Rate of alveolar ventilation}$$

$$\text{Therefore arterial } P_{CO_2} = \frac{CO_2 \text{ production rate}}{\text{Rate of alveolar ventilation}} \times \text{bar pressure}$$

gen ion concentration increases, that is the blood pH falls.

The converse occurs in alveolar hyperventilation; alveolar and arterial P_{CO_2} fall and blood pH rises.

These changes can take place very rapidly and are potentially very large since a normal subject excretes 15 000 m-moles CO_2 per day. They may occur as primary disturbances (respiratory acidosis and alkalosis) which are described on page 699 or as compensatory responses to primary metabolic disorders.

COMPENSATORY PROCESSES: (a) *Metabolic acidosis*. The decrease in blood pH in metabolic acidosis stimulates ventilation and results in a compensatory fall in the arterial P_{CO_2}. The That is the arterial P_{CO_2} level is quantitatively

and inversely related to the rate of alveolar ventilation.

RENAL REGULATION OF METABOLIC HYDROGEN ION

About 50 to 80 m-equiv. of metabolic hydrogen ion are produced daily, mainly in the form of sulphuric acid from the oxidation of sulphur in proteins. The hydrogen ions are buffered by both the bicarbonate and non-bicarbonate buffer systems.

EXCESS METABOLIC HYDROGEN ION
The hydrogen ions in carbonic acid are absorbed into water as carbon dioxide is

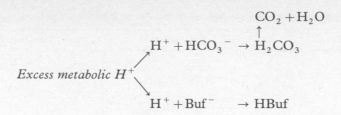

$$CO_2 + H_2O$$
$$\uparrow$$
$$H^+ + HCO_3^- \rightarrow H_2CO_3$$

Excess metabolic H^+

$$H^+ + Buf^- \rightarrow HBuf$$

excreted by the lungs but those in HBuf (which is mainly haemoglobin) cannot be filtered by the glomerulus because of the large size of the red blood cells.

Nevertheless urine can be acidified to around pH 4·5 at which point the hydrogen ion concentration is about 800 times that of plasma at pH 7·40. The urinary hydrogen ions are thought to arise within the renal tubular cells from the dissociation of carbonic acid formed by the hydration of CO_2 under the influence of carbonic anhydrase. This reaction is P_{CO_2}-dependent, a high arterial P_{CO_2} level accelerating and a low level inhibiting the renal tubular generation of hygrogen ions. The hydrogen ions are then secreted into the tubular lumen in exchange for sodium ions.

It is convenient to consider tubular hydrogen ion secretion as divided into three phases, though actually these phases overlap to some degree (Table 33.32).

1. *The reabsorption of filtered bicarbonate.* All the bicarbonate that appears in the glomerular filtrate is returned to the plasma.

2. *Excretion of titratable acid (acidification of urinary buffer).* Hydrogen ions added to the tubular fluid after all the bicarbonate has been reabsorbed are taken up by the other buffer bases, principally Na_2HPO_4 and creatinine which are converted to their respective buffer acids.

	BUFFER BASE $+ H^+ \rightarrow$ BUFFER ACID
Phosphate	$HPO_4^{2-} + H^+ \rightarrow H_2PO_4^-$
Creatinine	$Cr^- + H^+ \rightarrow HCr$

Titratable acid is the sum of the buffer acids in urine and is determined by titrating the urine back to the pH of blood with NaOH. It is this phase of hydrogen ion secretion that renders the urine acid. Normally titratable acid amounts to 20 to 30 m-equiv./day.

3. *Excretion of ammonium ion.* Ammonia is formed in the tubular cell from glutamine and amino acids. Ammonia diffuses out of the cell and unites with hydrogen ion in the tubular lumen to form ammonium ion. The significance of this process is that it permits the secretion of hydrogen ion to continue without further reduction in the urinary pH. Urinary ammonium excretion is inversely related to urinary pH, so the more acid the urine, the greater the ammonium content. The normal daily output is 30 to 50 m-equiv. Much higher rates can be attained in acidotic states but several days must elapse before this is achieved.

In phases 2 and 3, for every hydrogen ion secreted in the form of titratable acidity or ammonium, 1 m-equiv. of bicarbonate is added to the plasma.

RESTORATION OF BLOOD BUFFER BASE (RENAL CORRECTION) IN (a) *Metabolic acidosis.* In a severe metabolic acidosis such as diabetic ketosis, titratable acidity may rise to 100 or 150 m-equiv./day to which acetoacetic and β-hydroxybutyric acids make a significant contribution and ammonium production may amount to 500 m-equiv./day. If effective treatment is started, the kidney continues to secrete hydrogen ions and the bicarbonate ions which are added to the plasma restore the normal buffering capacity of the body fluids by (a) replacing the bicarbonate which has been destroyed by metabolic hydrogen ions and (b) reversing the reactions by which hydrogen ion has been buffered by the non-bicarbonate buffer system.

$$CO_2 + H_2O$$
$$\nearrow$$
$$Buf^- + H_2CO_3$$

Thus the blood buffer base returns to a normal level.

(b) *Metabolic alkalosis.* In metabolic alkalo-

Table 33.32 Tubular hydrogen ion secretion

Interstitial fluid	Tubular cell	Tubular lumen
(1) Na^+ HCO_3^-	$C.A.\star$ $CO_2 + H_2O \rightleftharpoons H_2CO_3$ $\Updownarrow$ HCO_3^- $+$ H^+	Na^+ HCO_3^- $H_2CO_3 \longrightarrow CO_2 + H_2O$
(2) Na^+ HCO_3^-	$C.A.\star$ $CO_2 + H_2O \rightleftharpoons H_2CO_3$ $\Updownarrow$ HCO_3^- $+$ H^+	Na^+ HPO_4^{2-} Na^+ Na^+ $H_2PO_4^-$
(3) Na^+ HCO_3^-	$C.A.\star$ $CO_2 + H_2O \rightleftharpoons H_2CO_3$ $\Updownarrow$ HCO_3^- $+$ H^+ Glutamine $\Big\}$ Amino acids $\Big\} \longrightarrow NH_3$	Na^+ X^- NH_4^+ X^-

$\star$ C.A. — carbonic anhydrase.

sis the plasma bicarbonate is high and the filtration of bicarbonate by the glomerulus is increased more than the tubular secretion of hydrogen ion. All the filtered bicarbonate is therefore not reabsorbed and an alkaline urine results. Titratable acidity and ammonium production amount to nothing. If the primary disturbance is treated effectively, the kidney continues to secrete bicarbonate, the bicarbonate level in the plasma falls; this in turn leads to a reduction in the level of the other buffer bases. Thus the normal buffering capacity of the body fluids is restored.

RENAL COMPENSATION. This term is applied to the changes in blood buffer base induced by the kidney in response to primary respiratory disturbances and is discussed on pp. 699, 701.

CLINICAL DISTURBANCES OF ACID-BASE EQUILIBRIUM

In order to eliminate as far as possible the difficulties in terminology commonly associated with this topic, for example in relation to the use of the ambiguous expressions acidosis and alkalosis, the system adopted here is that recommended in 1964 in a report made by a special committee set up under the auspices of the New York Academy of Sciences. These recommendations have been widely accepted. They include the suggestion that the terms acidosis and alkalosis should not be used to describe deviations in the blood pH or in any of the other acid base indices but should rather refer to the presence of abnormal processes or conditions, such as pulmonary or renal disease, diabetes, vomiting or diarrhoea that lead to body excesses or deficits of metabolic or respiratory hydrogen ion with resulting alterations in the acid base state of the blood. These abnormal processes are termed metabolic or respiratory acidosis and alkalosis. In the case of metabolic disorders, more specific adjectives such as diabetic, renal, lactic or diarrhoeal are often used.

In describing the blood acid–base variables, the parameters used should be stated along with the values obtained and the normal range. It is also permissible to use the adjectives, high, low or normal to qualify these variables. If descriptive terms are required the following are suggested, acidaemia and alkalaemia to indicate pH deviation, hypercapnia and hypocapnia to indicate P_{CO_2} deviation and hyperbasaemia and hypobasaemia to indicate deviations of the metabolic component.

The secondary or compensatory response to a primary disturbance that serves to maintain the blood pH within near normal limits should never be referred to as acidosis or alkalosis but should be related either to a change in composition in the blood, for example a secondary fall in the P_{CO_2}, or the base excess, or to a process such as an alteration in ventilation or the renal secretion of an acid or alkaline urine. Three degrees of compensation are recognized.

(a) Uncompensated—where no change has occurred in the compensatory component.

(b) Partially compensated—where a change has occurred in the compensatory component but the blood pH has not returned to normal.

(c) Fully compensated—where the change in the compensatory component has been of sufficient degree to restore the blood pH to normal.

RESPIRATORY DISORDERS

RESPIRATORY ACIDOSIS

1. *Aetiology.* The primary defect is alveolar hypoventilation due to pulmonary disease such as chronic bronchitis, interference with the normal action of the respiratory muscles or to depression of the respiratory centre due for example to overdose of morphine or barbiturates.

The primary disturbance leads to a rise in the alveolar and arterial P_{CO_2} and the resulting increase in carbonic acid brings the blood buffers immediately into action.

2. *Blood buffers.*

$$CO_2 \text{ (gas)}$$
$$\downarrow$$
$$CO_2 \text{ (dissolved)} + H_2O \qquad\qquad CO_2 \text{ gain}$$
$$\downarrow$$
$$H_2CO_3 + Buf^- \rightarrow HBuf + HCO_3^-$$

Note that this results in a rise in the plasma bicarbonate but the buffer-base and hence the base excess remains unaltered. Typical figures for the blood indices at this stage would be, pH 7·20, P_{CO_2} 80 mm Hg, plasma bicarbonate 30 m-equiv./l and base excess 0 m-equiv./l (R1 in Fig. 33.28; Fig. 33.31). As no compensatory change has yet occurred in the metabolic component, the blood pH remains very low—an uncompensated respiratory acidosis is present.

3. *Renal compensation.* The buffering mechanisms have led to a raised plasma bicarbonate and therefore filtration of bicarbonate by the glomerulus is increased. However, the elevated P_{CO_2} increases the tubular secretion of hydrogen ion to a greater extent so that all the filtered bicarbonate is reabsorbed and sufficient hydrogen ions are left over to increase urinary titratable acidity and ammonium production. As every m-equiv. of hydrogen ion excreted in the form of titratable acidity and ammonium results in the addition of 1 m-equiv. bicarbonate to the extracellular fluid, the plasma bicarbonate rises still further and the base excess begins to increase. This process may take 5 to 7 days to reach a maximum, when

representative blood findings may be as follows, pH 7.36, P_{CO_2} 71 mm Hg, plasma bicarbonate 39 m-equiv./l and base excess +10 m-equiv./l. Due to the secretion of an acid urine a compensatory change in the metabolic component has now occurred sufficient to restore the blood pH to normal—a compensated respiratory acidosis is now present (R2 in Fig. 33.28; Fig. 33.31).

RESPIRATORY ALKALOSIS

1. *Aetiology.* The primary defect is alveolar hyperventilation. This is encountered in hysterical hyperventilation, in anoxia, in various types of cerebral disease, in hepatic failure (generally attributed to excess ammonium ion) and salicylate poisoning. It is also a feature of certain types of pulmonary disease.

The alveolar and arterial P_{CO_2} fall and the blood buffers immediately respond to the reduced carbonic acid concentration as follows:
2. *Blood buffers.*

CO_2 (gas

↑

CO_2 (dissolved) + H_2O *CO$_2$ loss*

↑

$H_2CO_3 + Buf^- \leftarrow HBuf + HCO_3^-$

The plasma bicarbonate concentration falls but the buffer base and base excess remain unaltered. At this stage the following blood findings would be typical, pH 7.55, P_{CO_2} 20 mm Hg, plasma bicarbonate 17 m-equiv./l and base excess −2 m-equiv./l (R3 in Fig. 33.28; Fig. 33.31). No compensatory change has occurred in the metabolic component and the pH remains raised—an uncompensated respiratory alkalosis is present.

3. *Renal compensation.* Blood buffering results in a lowering of the bicarbonate concentration in the plasma so that the amount filtered by the glomerulus is reduced. However, the low P_{CO_2} reduces the tubular secretion of

hydrogen ions to a greater extent so that all the filtered bicarbonate is not reabsorbed and an alkaline urine results. This bicarbonate loss in the urine leads to a further reduction in the plasma bicarbonate and to a decrease in the base excess. This process may take several days to reach completion. The blood acid base status may now be pH 7.44, P_{CO_2} 20 mm Hg, plasma bicarbonate 13 m-equiv./l and base excess −8 m-equiv./l. A compensatory change has now occurred in the metabolic component due to the secretion of an alkaline urine and this has restored the blood pH to normal—a compensated respiratory alkalosis is now present (R4 in Fig. 33.28; Fig. 33.31).

METABOLIC DISORDERS

METABOLIC ACIDOSIS

1. *Aetiology.* The primary disturbance is an excess in the extracellular fluid of any acid apart from carbonic acid (that is excess metabolic hydrogen ion) or the loss of bicarbonate.

Thus it may be due to excess acetoacetic acid in diabetes or starvation, excess lactic acid in hypoxic states such as cardiac arrest or ventilatory failure, retention of sulphuric acid in renal failure, or excess hydrochloric acid after administration of ammonium chloride.

A decrease in bicarbonate can also occur when alkaline alimentary secretions are lost, as in diarrhoea or from intestinal, biliary or pancreatic fistulae. Bicarbonate loss can also occur in the alkaline urine of patients with renal tubular acidosis.

2. *Blood buffers.* Depending on the cause of the metabolic acidosis, two types of buffer system operate immediately.

(a) Excess metabolic hydrogen ions are buffered by both the bicarbonate and non-bicarbonate buffer systems.

$$CO_2 + H_2O$$
$$\uparrow$$
$$H^+ + HCO_3^- \rightarrow H_2CO_3$$

Excess metabolic H$^+$

$$H^+ + Buf^- \rightarrow HBuf$$

As the excess hydrogen ions are buffered, a decrease occurs in the concentration of both HCO_3^- and Buf^- (that is there is a reduction in both plasma bicarbonate and blood base excess) and the blood pH falls. A small amount of CO_2 is generated and this results in a slight increase in alveolar ventilation which serves to maintain the alveolar and arterial P_{CO_2} at 40 mm Hg.

(b) The buffer systems respond to HCO_3^- loss as follows,

$$CO_2 + H_2O$$
$$\downarrow$$
$$H_2CO_3 + Buf^- \rightarrow HBuf + HCO_3^-$$
$$\downarrow$$
$$\textit{loss of } HCO_3^-$$

Once again there is a reduction in both plasma bicarbonate and base excess with concomitant fall in blood pH but as carbon dioxide is absorbed in this reaction, alveolar ventilation decreases slightly to retain CO_2 and maintain alveolar and arterial P_{CO_2} at 40 mm Hg.

It is clear from the above reactions that in all cases of metabolic acidosis a decrease occurs in the metabolic component which in turn leads to a reduction in the blood pH

$$\left(pH \propto \frac{Metabolic\ component}{Respiratory\ component} \right)$$

3. *Respiratory compensation.* The low blood pH stimulates the respiratory centre and the resulting alveolar hyperventilation lowers the arterial P_{CO_2} which reduces the fall in blood pH. Respiratory compensation is apparent almost immediately and is maximal at 24 hours. Though very quick, respiratory compensation, unlike renal compensation, is rarely complete and typical blood acid–base indices would be blood pH 7·20, P_{CO_2} 25 mm Hg, plasma bicarbonate 9 m-equiv./l and base excess −17 m-equiv./l (M1 in Fig. 33.28; Fig. 33.31).

The base excess figure of −17 m-equiv./l is due to the primary disturbance and indicates that there is an excess 17 m-equiv. of metabolic hydrogen ion in every litre of blood. There has been an attempt at respiratory compensation, the P_{CO_2} falling to 25 mm Hg but this has not restored the blood pH to normal levels—a partially compensated metabolic acidosis is present.

4. *Renal correction.* The plasma bicarbonate level is low and filtration of bicarbonate by the glomerulus is reduced to a greater extent than the renal tubular secretion of hydrogen ions which is diminished by the compensatory decrease in the arterial P_{CO_2}. All the filtered bicarbonate is therefore reabsorbed and sufficient hydrogen ions are left over to increase titratable acidity and ammonium production. An appreciable time lag occurs before maximal hydrogen ion excretion is achieved; titratable acidity responds quickly but has a relatively low capacity whereas the ammonium mechanism which makes a much larger contribution requires several days to reach its peak.

If effective measures are taken to control or eliminate the primary disturbance, for example by the administration of insulin to a diabetic patient, the excessive hepatic production of acetoacetic acid is curtailed. The kidney meanwhile continues to excrete hydrogen ions and bicarbonate is added to the plasma. This restores the normal reaction and buffering capacity of the body fluids by (a) replacing the bicarbonate which has been destroyed by metabolic hydrogen ions and (b) reversing the reactions by which hydrogen ions have been buffered by the non-bicarbonate buffer system,

$$HBuf + HCO_3^- \rightarrow Buf^- + H_2CO_3$$

Thus both the concentration of bicarbonate and Buf^- increase and the base excess gradually returns to normal.

Some further points must be mentioned.

(a) In renal disease, the impairment of hydrogen ion excretion is due to (1) a decrease in titratable acidity because the low glomerular filtration rate reduces the filtered load of phosphate and creatinine, or (2) a failure of the tubular cells to synthesize ammonia which is generally attributed to lack of substrate, the result of poor renal blood flow. This results in a decrease in the ammonium ion content of the urine.

(b) The values for the acid base indices quoted above are fairly typical but they vary in degree depending on the severity of the primary disturbance, for example in diabetic coma it is common to encounter blood pH values of less than 7·00 and base excess values of −30 m-equiv./l or greater. However, the disturbance can always be characterized as a partially compensated metabolic acidosis.

(c) Excesses of the non-oxidizable acids,

hydrochloric acid and sulphuric acid can only be eliminated by renal excretion. In contrast, the organic acids, acetoacetic and lactic acid, are potentially oxidizable to CO_2 and H_2O if the primary disturbance can be treated effectively.

Metabolic Alkalosis

1. *Aetiology.* The primary disturbance is the loss of any acid except carbonic acid from the extracellular fluid (that is a deficit of metabolic hydrogen ion) or the gain of bicarbonate. Thus metabolic alkalosis may be due to the loss of hydrochloric acid as occurs with repeated vomiting, a frequent consequence of pyloric stenosis.

A gain of bicarbonate occurs when sodium bicarbonate or sodium salts of organic acids are administered (sodium lactate and citrate are oxidized in the body to sodium bicarbonate). Because stored blood contains sodium citrate to prevent coagulation, large transfusions can give rise to a metabolic alkalosis.

2. *Blood buffers.* Depending on the aetiology of metabolic alkalosis two types of buffer system operate immediately:

at 40 mm Hg. It is clear from the above reactions that in all cases of metabolic alkalosis a rise occurs in the metabolic component which of course leads to a rise in the blood pH.

3. *Respiratory compensation.* The increase in the blood pH depresses ventilation and the consequent rise in the arterial P_{CO_2} reduces the rise in pH. Respiratory compensation is apparent almost immediately and is maximal at 24 hours. Although it is very quick, respiratory compensation, unlike renal compensation, is rarely complete and typical blood acid–base figures would be: blood pH 7·54, P_{CO_2} 55 mm Hg, plasma bicarbonate 47 m-equiv./l, blood base excess +20 m-equiv./l (M2 in Fig. 33.28; Fig. 33.31).

The base excess figure of +20 m-equiv./l is due to the primary disturbance. In the example respiratory compensation has begun, the P_{CO_2} rising to 55 mm Hg, but this has not restored the blood pH to normal—a partially compensated metabolic alkalosis is present.

4. *Renal correction.* The plasma bicarbonate level is high and filtration of bicarbonate by the glomerulus is increased more than the tubular secretion of hydrogen ions. Not all the filtered bicarbonate is reabsorbed and an alkaline

$$CO_2 + H_2O$$
$$\downarrow$$
$$H_2CO_3 \rightarrow HCO_3^- + H^+ \searrow$$

loss of metabolic hydrogen ion

$$HBuf \rightarrow Buf^- + H^+ \nearrow$$

Both plasma bicarbonate and blood base excess rise but, as a small amount of CO_2 is consumed in this reaction, alveolar ventilation decreases slightly to retain CO_2 and maintain alveolar and arterial P_{CO_2} at 40 mm Hg.

(b) *Gain of HCO_3^-*
$$\downarrow \qquad\qquad\qquad CO_2 + H_2O$$
$$\qquad\qquad\qquad\qquad\qquad \uparrow$$
$$HCO_3^- + HBuf \rightarrow Buf^- + H_2CO_3$$

Once again, plasma bicarbonate and blood base excess rise, but in this case a small amount of CO_2 is generated and this results in a slight increase in alveolar ventilation which serves to maintain alveolar and arterial P_{CO_2}

urine is excreted. If the primary disturbance is treated effectively, the kidney continues to excrete bicarbonate and the bicarbonate level in the plasma, and the levels of the other buffer bases, fall.

It is important to note that respiratory compensation for metabolic disorders delays renal correction but this is of little consequence because the changes in P_{CO_2} maintain the blood pH within reasonable limits.

On the other hand, in disturbances of respiratory origin, the effect of P_{CO_2} on tubular hydrogen ion secretion provides an automatic compensation.

REFERENCES

ANDERSON, J., OSBORN, S. B., TOMLINSON, R. W. S., NEWTON, D., RUNDO, J., SALMON, L. & SMITH, J. W. (1964). Neutron-activation analysis in man *in vivo*. *Lancet* ii, 1201–1205.

ASSCHER, A. W. & JONES, J. H. (1965). Capillary permeability to plasma proteins. *Postgraduate Medical Journal* 41, 425–434.

ASTRUP, P., JORGENSEN, K., ANDERSEN, O. S. & ENGEL, K. (1960). The acid–base metabolism: a new approach. *Lancet* i, 1035–1039.

BITTAR, E. E. (1964). *Cell pH*. London: Butterworths.

BLACK, D. A. K. (1967). *Essentials of Fluid Balance*, 4th edn. Oxford: Blackwell.

BOWSHER, D. (1960). *Cerebrospinal Fluid Dynamics in Health and Disease*. Springfield, Illinois: Thomas.

BROOKS, C. M., KAO, F. F. & LLOYD, B. B. (1965). *Cerebrospinal Fluid and the Regulation of Ventilation*. Oxford: Blackwell.

BROZEK, J. (1965). *Human Body Composition. Symposium of the Society for Study of Human Biology*, Vol. 7. Oxford: Pergamon.

BUNTON, G. L. (1965). *Fluid Balance without Tears*. London: Lloyd-Luke.

CAMIEN, M. N., SIMMONS, D. H., GONICK, H. C. (1969). A critical reappraisal of 'acid–base' balance. *American Journal of Clinical Nutrition* 22, 786.

CAMPBELL, E. J. M. (1968). *Hydrogen Ion (Acid : Base) Regulation* (Edited by E. J. M. Campbell, C. J. Dickinson & J. D. H. Slater). In *Clinical Physiology*, 3rd edn., pp. 198–221. Oxford: Blackwell.

CASLEY-SMITH, J. R. & FLOREY, H. W. (1961). The structure of normal small lymphatics. *Quartery Journal of Experimental Physiology* 46, 101–106.

CHRISTENSEN, H. N. (1964). *Body Fluids and the Acid–base Balance. A Learning Program for Students of Biological and Medical Sciences*. Philadelphia: Saunders.

CREESE, R., NEIL, M. W., LEDINGHAM, J. M. & VERE, D. W. (1962). The terminology of acid–base regulation. *Lancet* i, 419–424.

CRONE, C. & LASSEN, N. A. (Eds.) (1970). *Capillary Permeability*. Copenhagen: Munksgaard.

DAVENPORT, H. W. (1969). *The ABC of Acid–Base Chemistry*, 5th edn. (revised). Chicago University Press.

DAVSON, H. (1967). *Physiology of the Cerebrospinal Fluid*. London: Churchill.

DORMANDY, T. L. (1967). Osmometry. *Lancet* i, 267–271.

FITZSIMMONS, J. T. (1966). The hypothalamus and drinking. *British Medical Bulletin* 22, 232–237.

FOURMAN, P. & LEESON, P. M. (1959). Thirst and polyuria. *Lancet* i, 268–270.

HOLMES, J. H. & MONTGOMERY, A. V. (1953). Thirst as a symptom. *American Journal of Medical Science* 225, 281–286.

KERNAN, R. P. (1965). *Cell K*. London: Butterworths.

LEAF, A. (1962). The clinical and physiologic significance of the serum sodium concentration. *New England Journal of Medicine* 267, 24–30, 77–83.

MOORE, F. D., OLSEN, K. H., McMURREY, J. D., PARKER, H. V., BALL, MARGARET R. & BOYDEN, C. M. (1963). *The Body Cell Mass and its Supporting Environment*. London: Saunders.

MORGAN, H. G. (1969). Acid–base balance in blood. *British Journal of Anaesthiology* 41, 196–212.

MORGAN, H. G., OGILVIE, R. R. & WALKER, W. F. (1963). Acid–base monitoring of open-heart surgery. *Journal of Clinical Pathology* 16, 545.

NAHAS, G. G. (Ed.) (1966). Current concepts of acid–base measurement. *Annals of the New York Academy of Sciences* 133, 1–274.

PITTS, R. F. (1963). *Physiology of the Kidney and Body Fluids*. London: Lloyd-Luke.

ROBINSON, J. R. (1967). *Fundamentals of Acid–Base Balance*, 3rd edn. Oxford: Blackwell.

ROOTH, G. (1966). *Introduction of Acid–Base and Electrolyte Balance*. Lund, Student Litteratur (Via Radiometer, London).

RUSZNYAK, I., FOLDI, M. & SZABO, G. (1960). *Lymphatics and Lymph Circulation: Physiology and Pathology*. Oxford: Pergamon.

SEVERINGHAUS, J. W. (1966). Blood Gas Calculator. *Journal of Applied Physiology* 21, 1108.

SIGGARD-ANDERSEN, O. (1969). *The Acid–Base Status of the Blood*. Copenhagen: Munksgaard.

STARLING, E. H. (1909). *The Fluids of the Body*. London: Constable.

STEWART, C. P. & DUNLOP, D. M. (1962). *Clinical Chemistry in Practical Medicine*, 6th edn. Edinburgh: Livingstone.

TSCHIRGI, R. D. (1962). Blood–brain barrier: fact or fancy? *Federation Proceedings* 21, 655–671.

VAN HARREVELD, A. (1966). *Brain Tissue Electrolytes*. London: Butterworths.

WADDELL, W. J. & BATES, R. G. (1969). Intracellular pH. *Physiological Reviews* 49, 285–329.

WAYNER, M. J. (Ed.) (1964). *Thirst*. Oxford: Pergamon.

WEYER, E. M. (Ed.) (1966). Current concepts of acid–base measurement. *Annals of the New York Academy of Sciences* 133, 1–274.

WHITEHEAD, T. P. (1967). Blood hydrogen ion: terminology, physiology and clinical applications. In *Advances in Clinical Chemistry* (Edited by H. Sobotka & C. P. Stewart), Vol. 9, pp. 195–226.

WILKINSON, A. W. (1960). *Body Fluids in Surgery*, 2nd edn. Edinburgh: Livingstone.

WINTERS, R. W., ENGEL, K. & DELL, R. B. (1967). *Acid Base Physiology in Medicine*. Westlake, Ohio: The London Company.

WOOTTON, I. D. P. (1964). *Microanalysis in Medical Biochemistry*. London: Churchill.

WOLF, A. V. (1958). *Physiology of the Urge to Drink and the Problems of Water Lack*. Oxford: Blackwell.

WOOLMER, R. F. & PARKINSON, J. (Eds.) (1959). *Symposium on pH and Blood Gas Measurement*. London: Churchill.

WRONG, O. M. (1970). Water and electrolyte metabolism. In *Biochemical Disorders in Human Disease* (Edited by R. H. S. Thompson & I. D. P. Wootton), 3rd edn., p. 662. Churchill: London.

YOFFEY, J. M. & COURTICE, F. C. (1970). *Lymphatics, Lymph and Lymphoid Tissue. Lymphomyeloid Complex*. New York: Academic Press.

34 Special senses

Vision, hearing, taste, olfaction, and the cutaneous senses are included in the general title 'special senses'.

The sensory processes of an animal may be investigated by training it to respond differentially to two stimuli, and then testing the limits of this discrimination. One method of doing this uses the technique of conditioning (see Chap. 49). Similar methods may be used with human subjects, but it is usually more convenient to rely upon previous learning by obtaining verbal reports, provided that suitable precautions are taken to avoid mistaken inferences. Thus a colour-blind person may have learned to call a leaf 'green' and blood 'red', but may fail to distinguish these hues when appropriately tested (see p. 746). When a subject reports that light of a certain wavelength is of one hue rather than another, he is not communicating the essence of his unique experience, and there is no way of determining whether his experience is the same as that of another person. The subject's words, like the animal's responses, merely indicate the ability to discriminate between two stimuli, and thus provide objective evidence from which it may be inferred that the sensory processes aroused by these stimuli differ in some respect.

There is no direct relationship between a stimulus and the sensory effect it produces although in ordinary usage the two are often taken to be the same. For instance, a sound is often described as loud, but the word 'loud' properly refers to the psychological effect and not to the stimulus, which is more correctly described as intense or powerful. Loudness indeed, although it is determined mainly by the intensity of a sound, is also a function of its frequency. There is not a precise correspondence between physical and psychological dimensions.

Although end-organs are specialized to be sensitive to different forms of stimulus energy, the quality of a sensation does not depend upon the type of energy which arouses it. A vibrating tuning-fork placed on the lower end of the radius gives rise to a sensation of vibration, whereas the same fork placed on the head or near the ear arouses a sensation of sound. If a variety of stimuli is applied to any one end-organ the same kind of sensation is produced. In the case of the eye, an electric current, pressure and light all evoke sensations of light. Experiments with single nerve-fibres show that all nerve impulses, however they are initiated, are the same. In other words, the same stimulus applied to different end-organs evokes different sensations: different kinds of stimuli applied to one end-organ all call forth the same sensation. Müller called this the law of specific irritability or the law of specific nerve energy. A reasonable conclusion is that the quality of the sensation aroused by the stimulation of any end-organ or nerve depends on the route taken by the nerve impulses and on their destination in the central nervous system.

It may be asked whether there is any lower limit to the sensitivity of a sensory system. The classical view has been that the subject either experiences a sensation or does not, and that there is some minimal stimulus energy (the *absolute threshold*) below which a sensation cannot be evoked. The concept of a psychological threshold is clearly allied to that of a physiological threshold, such as the minimal energy required to excite a neurone. However, it is clear that such a threshold cannot have a constant value, since if we repeatedly present a sound of a given low intensity, and ask the subject whether he can hear it, he will sometimes report that he can, and sometimes that he cannot. This variability could reasonably be attributed to physiological fluctuations in the sensory system; but it means that the threshold can be defined only statistically, by stating the energy required to evoke a positive response with some arbitrary probability (usually 50 per cent).

Considerable doubt has been thrown on the concept of the absolute threshold as a result of methodological studies directed at two related defects in the traditional procedure. First, the threshold estimate is affected by another source of variation, namely the attitude of the subject, who may adopt a more or less cautious criterion for reporting a stimulus. Secondly,

the traditional procedure provides no means of controlling or estimating the effect of the subject's criterion. To do this, it is necessary to obtain an estimate of the probability that the subject will report a stimulus when none is presented. This method makes it possible to eliminate the effect of the criterion and obtain a relatively pure measure of sensitivity. Results obtained in this way suggest that although, under certain circumstances, sensitivity decreases at low stimulus intensities, it may not reach an absolute limit. The apparent limit obtained by the traditional procedure results from restricting the subject to two categories of response. If methods are used which allow the subject to convey more information, for example by expressing his degree of confidence that a stimulus has been presented, he is always capable of doing so. Thus even if an absolute threshold exists, it is difficult to measure.

A related problem is the measurement of differential sensitivity, for example the subject's ability to decide whether one sound is louder than another. The traditional procedure has been to determine a *difference threshold*, estimated as the difference in energy required to produce some arbitrary proportion (usually 75 per cent) of judgements that a variable stimulus is more intense than a standard. Many measures of sensory acuity take this form. Clearly the same objections which apply in the study of absolute sensitivity apply in the study of differential sensitivity also, but less research has been done on this problem largely because attention has been diverted from it to the problem of deriving scales of sensation.

When differential sensitivity was first studied by the threshold method, it was noted by Weber that as the intensity of the standard stimulus increases, the size of the difference threshold increases also. Weber's law states that the size of the difference threshold is a constant proportion of the intensity level at which it is measured. This law is not precisely valid for any sensory modality. It does not hold at very low intensities where it is difficult to detect the stimuli, and has sometimes been found not to hold at very high intensities either. Thus although we may be able to distinguish between 20 and 21 g by the sense of touch, we cannot distinguish between 2·0 and 2·1 mg, nor between 200 and 210 kg. Over the major portion of the intensity range the Weber ratio is, however, approximately constant, typical values being for touch about $\frac{1}{10}$, for hearing about $\frac{1}{5}$, for olfaction about $\frac{1}{3}$, and for vision about $\frac{1}{100}$ of the magnitude of the stimuli which are being compared.

Fechner, accepting the Weber law as true, derived from it a scale of sensation and related it to the physical scale of stimulus intensity by assuming that all difference thresholds, since they rest upon an equal probability that two stimuli can be discriminated, are subjectively equal. Then by integration he obtained the formula:

$$\text{sensation intensity} = \text{constant} \times \text{logarithm of the stimulus intensity}.$$

A constant ratio of stimulus energies is thus required to produce a constant difference in sensory magnitude.

Such a relationship has been found useful especially in the measurement of sound intensity, where a logarithmic unit, the decibel (p. 761) is in standard use. However, the decibel scale, though satisfactory for the measurement of sound transmission, is of more doubtful value in psychoacoustics, since it has been noted that a tone of 100 decibels, for example, sounds much more than twice as loud as a tone of 50 decibels. Because of such practical difficulties, and because of doubt about the theoretical basis of Fechner's scale, which rests on the summation of differential thresholds which are assumed to be subjectively equal, a number of methods have been devised which attempt to construct sensory scales more directly. For example, the *category scaling method* presents the subject with about twenty stimuli, which he is asked to assign to a smaller number of categories, for example the first seven integers, in such a way that the intervals between categories are subjectively equal. The results of this procedure provide good support for Fechner's logarithmic scale. Unfortunately, other procedures do not. The *method of magnitude estimation* presents the subject with a similar set of stimuli, but requires him to estimate their magnitude on a scale ranging, for example, from 0 to 100. The subject is able to perform this operation quite consistently, but the resulting sensory scale is not a logarithmic function of stimulus intensity, but a power function of the form:

$$\text{sensation intensity} = \text{constant} \times (\text{stimulus intensity})^{\text{constant}}.$$

Theoretically, this function implies the subjective equality of ratios rather than of differences.

Thus different scaling methods, between which we have no *a priori* reason to choose, produce conflicting results. Moreover, each method, though producing consistent results for a standard procedure, is affected by procedural variations. In magnitude estimation, the exponent of the power function varies to some extent with the range and spacing of the stimuli, and with the set of numbers available as responses. Evidently, the subject's response depends not only on his perception of the stimulus, but also on his means of expressing this perception. No wholly satisfactory way of eliminating response bias has so far been devised for scaling procedures, and therefore the functions they produce cannot be regarded as reflecting merely the characteristics of the sensory system involved.

If we are concerned solely with the characteristics of a sense organ, we may correlate stimulus intensity with an electrophysiological measure. In general, while sensory quality is related to the place where afferent fibres terminate, intensity is coded in terms of their discharge frequency. Thus it is known that the rate of firing of receptors in the eye of *Limulus* (king crab) is logarithmically related to the intensity of light; and in the case of a muscle-spindle the frequency of impulses in the afferent nerve is roughly a logarithmic function of the load. However, in the cutaneous system the frequency of impulses in a single afferent nerve fibre is a power function of the degree of indentation of the skin; and it may be noted that the value of the exponent ranges from 0·27 to 1·17 in different nerve fibres, a range appreciably greater than that caused by variations in psychological scaling procedures. Furthermore, the function may not be identical at all levels of the sensory system.

Stimulus intensity has also been correlated with more central responses, such as the amplitude of the cortical evoked potential; and it has sometimes been claimed that this technique offers a more objective approach to the measurement of sensory magnitude in man. For example, the amplitude of the visual evoked potential recorded in response to a grating is linearly related to the logarithm of contrast when the mean light flux to the eye is controlled; and it has been suggested that this result supports the Fechner law. On the other hand, it has been shown that the amplitudes of the main components of the auditory evoked response to a tone are power functions of sound intensity. Moreover, the value of the exponent varies considerably for different components. It is salutary to remember that when we have related even so central a physiological response as this to the physical stimulus, it is still necessary to determine the relevance of its various parameters for the behaviour of the organism.

REFERENCES

CAMPBELL, F. W. & MAFFEI, L. (1970). Electrophysiological evidence for the existence of orientation and size detectors in the human visual system. *Journal of Physiology* **207**, 635–652.

GALANTER, E. (1962). Contemporary psychophysics. In *New Directions in Psychology* (R. Brown, E. Galanter, E. H. Hess & G. Mandler). New York: Holt, Rinehart & Winston.

GREEN, D. M. & SWETS, J. A. (1966). *Signal Detection Theory and Psychophysics.* London: Wiley.

HARTLINE, H. K. (1941). The neural mechanisms of vision. *The Harvey Lectures,* Series 37. New York: The Harvey Society.

KEIDEL, W. D. & SPRENG, M. (1965). Neurophysiological evidence for the Stevens power function in man. *Journal of the Acoustical Society of America* **38**, 191–195.

VAN LEEUWEN, S. (1949). The response of a frog's muscle spindle. *Journal of Physiology* **109**, 142–145.

POULTON, E. C. (1968). The new psychophysics. *Psychological Bulletin* **69**, 1–19.

STEVENS, S. S. (1961). The psychophysics of sensory function. In *Sensory Communication* (Edited by W. A. Rosenblith). London: Wiley.

STEVENS, S. S. (1970). Neural events and the psychophysical law. *Science, New York* **170**, 1043–1050.

SWETS, J. A. (1961). Is there a sensory threshold? *Science, New York* **134**, 168–177.

SWETS, J. A., TANNER, W. P. & BIRDSALL, T. G. (1961). Decision processes in perception. *Psychological Reviews* **68**, 301–340.

THOMPSON, R. F. (1967). *Foundations of Physiological Psychology.* London: Harper & Row.

WERNER, G. & MOUNTCASTLE, V. B. (1965). Neural activity in mechanoreceptive cutaneous afferents: stimulus–response relations, Weber functions, and information transmission. *Journal of Neurophysiology* **28**, 359–397.

35 The skin

In lower animals the skin is concerned with processes such as respiration, alimentation, and secretion, but in mammals these functions are taken over by specialized organs. In man the chief functions of the skin are to provide an effective barrier to a wide variety of substances, to regulate heat loss, and to mediate sensation. The skin is divided into three distinct tissue layers—the epidermis, the dermis (corium) and the subcutaneous fatty layer (Fig. 35.1). The epidermis in the early embryo

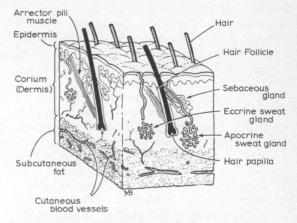

Fig. 35.1 Low power view of the structures found in hairy skin.

consists of a single layer of cells; it soon thickens and two main layers—the stratum corneum on the surface and below it the stratum germinativum—can be distinguished. The stratum germinativum shows mitotic activity which is greatest in the adult between midnight and 4 a.m. The cells so produced migrate or are displaced outwards to end in the most superficial layer of the stratum corneum as dead horny cells which have lost their nuclei (Fig. 35.2). Nearly 1 g of these dead cells is shed per day as surface scales (dandruff); the thickness of this horny layer is, however, maintained by new cells produced by the basal layer (stratum germinativum) (Fig. 35.2). The stratum corneum can be stained

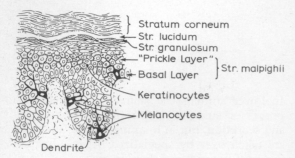

Fig. 35.2 Microscopic appearances of skin.

with a fluorescent dye which binds only to cornified cells. In the forearm the fluorescence disappears after two weeks; this must be the transit time of cornified cells. Immediately below the stratum corneum is a layer of granular cells (stratum granulosum) which possess a high energy system responsible for the synthesis of the insoluble protein keratin; the granules are precursors of keratin. Sulphydryl groups, phospholipids and glycogen are in high concentration in this layer but are not found in the immediately overlying stratum corneum. Presumably at this site the polypeptide chains in the cytoplasm are unfolded and broken and built up into keratin molecules (p. 34). When the granules are converted from keratohyalin into keratin the cells assume a homogeneous appearance and are recognizable, especially if the skin is thick, as the stratum lucidum, a layer of cells lying between the stratum granulosum and the more superficial stratum corneum.

The stratum corneum is an effective barrier against external noxious agents and also against water loss from within the body. If this main barrier layer is stripped off the water loss from the underlying tissues can increase 10 to 20 times. Although it has long been accepted that the junction between the stratum corneum and the stratum granulosum is the most probable site of the barrier function, new evidence suggests that the whole stratum corneum has this function. Within the stratum corneum hygroscopic water-soluble substances are responsible for much of the water-binding capacity of the horny layer and thus for its extensibility. Their removal by excessive exposure to water, or by increased permeability as a result of extraction of lipids by surfactant detergent solutions, results in drying and cracking of the protective horny layer and exposure of the highly sensitive epidermal cells. Free amino acids in the keratinous layer act as buffers and protect the skin from the action of acids and alkalis.

Projections from the basal layer cells extending down into the dermis are thought to have anchoring and nutritional functions. The cells of the epidermis (keratinocytes, Fig. 35.2) have been described as being joined together by 'prickles' or intercellular bridges linking the cytoplasm of one cell with that of its neighbours; however, the electron microscope has shown that the cells are separate but closely apposed, being firmly adherent at a few thickened nodes called desmosomes. An interlocking system of fine filaments (tonofibrils, Fig. 35.3) attached at the desmosomes crisscrosses the cells and gives the epidermis

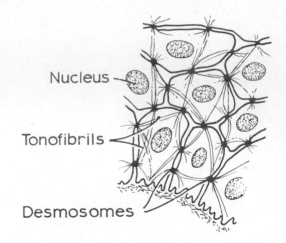

Fig. 35.3 Fine structure of epidermis.

mechanical stability. The function of the basement membrane lying between the basal layer and the dermis is uncertain but there is no doubt that the dermis influences the cells of the epidermis; separated epidermis when grown on dermis at a different site takes on the features of the skin at the receptor site. The thickness of the epidermis varies in different areas of the body; it is thin in the flexures and thick and compact on the palms, soles, and at sites of friction.

The colour of the skin depends on four pigments—oxyhaemoglobin, reduced haemoglobin, carotene (precursor of vitamin A), melanin—and on light scatter. Melanin is responsible for the dark pigmentation of the skin of the nipple, mammary areola, the margins of the anus, and sites exposed to sunshine and friction. It is formed by special cells, melanocytes, derived probably from the neural crest of the embryo, found mainly in the basal layer of the epidermis (Fig. 35.2). These cells inject melanin granules into the keratinocytes of the stratum germinativum through their dendrites and in this way pigment granules migrate towards the surface. The dark colour of certain races is due not to a greater number of melanocytes but to a greater amount of melanin in the epidermis than is found in pale-skinned races. All races have some melanin in the skin. Further details of skin coloration are given on p. 385. Exposure to ultraviolet rays increases the amount of melanin in the skin and also increases the thickness of the stratum corneum. In these ways the skin is protected from damage and the increase in melanin acts by absorbing wavelengths up to 300 nm whereas the thickened stratum corneum absorbs most of the carcinogenic wavelengths which are less than 320 nm. The mechanism of sunburn, that is the effect of solar radiation between 290 and 320 nm, is only partly understood. When ultraviolet radiation impinges on the skin, the skin becomes red after 7 to 12 hours because of the dilatation of the papillary venous complexes. This erythema may persist from 24 hours to 7 days depending on the dose; with higher doses it may be accompanied by pain, oedema, and even blistering, with systemic effects in the form of chills, nausea and fainting. Mild reactions are followed by increased pigmentation and scaling of the horny layer.

Cellular damage in the epidermis is due to photochemical formation of thymine dimers, to reparable lesions in nuclear deoxyribonucleic acid, and to membrane effects, particularly the labilization of lysosomes. Although there is considerable absorption of radiation in the horny layer and in the living cells of epidermis by protein, by urocanic acid (a derivative of histidine) and by melanin, sufficient radiation penetrates to the papillary dermis to have a direct photochemical effect at the level of the blood vessels; damage similar to that occurring in the epidermis can also occur at this level. Collagen may also undergo chemical change by direct action of radiation. It is likely that a vasoactive agent is released or activated in the epidermis and diffuses from there to the underlying vessels; this agent may well be a prostaglandin rather than a vasoactive amine or kinin.

Increased pigmentation after exposure to ultraviolet radiation is something more than a post-inflammatory reaction. During exposure melanin present in a bleached form is oxidized and pigment in the melanocytes is transferred to keratinocytes. A sulphydryl inhibitor of tyrosinase is inactivated by exposure and melanin synthesis is increased after a latent period of a day or so, the increase in skin pigmentation being apparent after transfer of the melanin to keratinocytes. There is no doubt that the increased pigmentation after exposure affords protection against further insult but thickening of the horny layer and increased amounts of urocanic acid within this layer also give protection.

DERMIS (CORIUM)

The electron microscope shows that the corium is made up of a network of robust collagen fibres of fairly uniform thickness with regularly spaced cross-striations. The network may, however, be an artefact since examination of unfixed corium by fluorescence microscopy suggests that it is a gel containing oriented tropocollagen (polypeptide) macromolecules. Histological fixatives change the tropocollagen into the collagen fibres seen in the ordinary histological preparation. The network or gel is responsible for the elastic properties of the skin since comparatively few elastic fibres are to be found. The ability of the skin to hold and to lose water rapidly is primarily attributable to the hydrophilic properties of dermal collagen and mucopolysaccharides. The collagenase enzymes in the skin are of importance in wound healing and in the remodelling of injured dermis. The surface of the dermis is formed into ridges or papillae projecting into the epidermis. These papillae contain blood vessels, lymphatics and nerve endings. Although the terminal capillaries of the blood vessels do not enter the epidermis they do in fact nourish the epidermal cells. Beneath the corium the fibrous tissue opens out and merges with the fat-containing subcutaneous tissue. The subcutaneous fatty layer beneath the dermis insulates the underlying structures from excessive environmental heat and reduces body heat loss in cold climates. Collagen bundles are distributed around the fat cells and in this way a flexible linkage is produced between the superficial skin layers and the underlying structures. The subcutaneous fat acts as a cushion for the dermis and epidermis and allows lateral displacement.

In man there are two different types of sweat glands—eccrine and apocrine (Fig. 35.1). The ordinary or eccrine sweat glands are simple tubular glands found all over the skin, being especially numerous in thick skin. The secretory part of the tubule, about 0·1 mm in diameter, lying deep in the corium, is coiled and twisted on itself. The initial coiled portion of the tubule is composed of a single layer of epithelial cells with externally placed myo-epithelial cells. The myoepithelial cells probably function as a support to the secretory cells rather than as a propelling mechanism. The epithelial cells are classified as 'dark' (baso-philic) cells and 'clear' cells in about equal numbers, the latter being responsible for the secretion of sweat. The clear cells have many mitochondria and are presumably capable of intense metabolic activity; intercellular canaliculi empty into the lumen of the tube. The function of the smaller dark cells is not clearly understood; they contain mucoid substances and nucleic acids; they have relatively few mitochondria and no canaliculi. Further on in the coil ductal cells appear mixed up with secretory cells. The duct itself has two layers of cuboidal epithelium, the outer layer possessing numerous mitochondria. The nerves ending on the sweat glands contain cholinesterase. The sweat glands have a rich capillary blood supply. Micropuncture studies in man have shown that the fluid in the secretory coil is isotonic and that sodium is abstracted from it by the cells of the ducts as the sweat passes to the exterior. The duct pursues a spiral course through the corium and epidermis to open on the surface of a dermal ridge at which point its cells closely resemble keratinized epidermal cells. Additional information about sweat glands is given in Chapter 42.

Apocrine glands, about 1 mm in diameter, are found in the axilla, the pubic region, in the areola of the breast and as modified glands in the eyelid and external meatus; the mammary gland itself is a modified apocrine gland. In abnormal circumstances apocrine glands can appear elsewhere in the skin. An apocrine gland consists of a coiled tube lying in the deepest part of the corium or in the subcutaneous tissue with a duct opening into the hair follicle above the sebaceous gland (Fig. 35.1). These glands do not become active until puberty and their secretion diminishes in old age. The glands are innervated by adrenergic sympathetic fibres which supply the myo-epithelial cells but may not reach the secretory cells. For this reason their secretion, which is continuous, is apparently independent of any nervous control but its delivery on to the skin surface, which is intermittent, may be provoked either by the sympathetic nerve supply or by circulating adrenaline and noradrenaline. The gland has a single layer of cuboidal or columnar cells resting on a basement membrane surrounded by myoepithelial cells similar to those seen in eccrine glands. Recent electron microscopic evidence supports the long held view that apocrine secretion results from pinching off of the cytoplasm of the cells but this process is probably responsible for only a part of the secretion. It is frequently stated, but it is by no means certain, that the axillary apocrine glands in women enlarge premenstrually. The axillary apocrine glands produce a viscid milky fluid, with a pH of 6, which is odourless when fresh but decomposition by the bacteria of the skin produces the characteristic unpleasant axillary odour of the human adult. The eccrine sweat contributes very little to body odour.

Hair, nails, sweat glands and sebaceous glands are all formed by downward growth of the epidermis in the embryo. Unlike hair, the nails grow continuously throughout life and are not normally shed. Growth is greatest in childhood and decreases slowly with age, the normal rate of growth of finger nails being between 0·5 and 1·2 mm per week. Due to pressure from the posterior nail fold the nail grows forward instead of upwards. A temporary slowing of the growth may follow general body disorders. Nail growth is also affected by local disturbances in the nail fold or by abnormal keratinization of the nail plate. General or local factors may result in the development in the nail of thickening, ridging, pitting, discoloration, brittleness, splitting, and even separation of the nail from its bed (onycholysis). A transverse groove may result from severe illness, for example coronary thrombosis, or an infection such as measles, mumps or pneumonia. The nail may change in colour for a variety of reasons; white spots in the nail plate, which are seen in 62 per cent of normal people, are due to imperfect keratinization with retention of nuclear material.

In many mammals hair acts essentially to protect the skin against physical trauma and cold and as part of sexual display and as part of the tactile apparatus; however, in man this is no longer the case. The fine soft 'lanugo' hair of the human fetus is replaced before birth by (a) vellus hair which is soft, occasionally pigmented and usually not more than 2 cm long—this is the characteristic body hair of the child and the facial hair of the adult woman; and (b) terminal hair which is longer, coarser and often pigmented. Genetic constitution, age and the level of androgens all play a part in determining hair pattern. At puberty terminal hair progressively replaces vellus hair and continues to do so up to the fourth decade and sometimes later. It appears in the pubis, upper lip, axilla, beard, trunk and limbs. The difference between male and female patterns is only one of degree, and some terminal hair is found on the face in at least 40 per cent of caucasoid women over the age of 50. At the end of the active growing period (anagen) there is a short transition phase (catagen) during which there is a gradual involution of the follicles which lasts approximately two weeks in the human scalp. The hair is then retained for a resting period (telogen) without further growth and is then shed. The rate of hair growth varies between 1 and 3 mm per week according to the region of the skin and also with sex and age; with increasing age the number of hair follicles gradually declines. In man each follicle acts independently of the others. Temporary hair loss can be precipitated by physical and mental stresses, so-called 'telogen defluvium'. This occurs most commonly after febrile illness or after childbirth. In certain genetically predisposed males (see Chapter 51) baldness develops. Anagen is reduced and an increased percentage of follicles is found in the telogen phase; terminal hairs are gradually replaced by vellus hairs and ultimately the vellus follicles disappear. By the age of 50, 60 per cent of caucasoid males have some degree of baldness of the temples and vertex.

Each sebaceous gland consists of a series of lobes or acini which open into a hair follicle by a short wide duct. Not infrequently, especially on the face, sebaceous glands are found independently of hair follicles. There are about 900 sebaceous glands per cm^2 on the face and scalp but only about 100 per cm^2 elsewhere; they are not found on the palms, soles and lower lip. In the eyelid sebaceous glands (tarsal glands) open directly on to the skin surface; some also open directly, free from hair follicles, on the female genitalia and the areola of the nipples and occasionally on the tongue and cervix uteri. The secretion of the sebaceous gland, sebum, is holocrine, being formed by the complete disintegration of the glandular cells. Sebum can be obtained by immersing a limb in acetone and recovering the extracted material. It is a mixture of lipid substances derived both from sebum and from epidermal lipids. Sebum contains triglycerides, wax esters and squalene; sterol and sterol esters are contributed mainly by keratinizing epidermis. It seems likely that free acids are absent from pure sebum. The water-holding power of the stratum corneum depends in part on this lipid film; it may also help to protect the skin from fungal and bacterial infection. The sebaceous glands are under control of the sex hormones and show increased activity after puberty. Their secretion is increased by androgens and inhibited by oestrogens, mainly by affecting the rate of cell division. The production of sebum is reduced for some months after the end of pregnancy. The sebaceous glands have recently been shown to have fine nerve fibres which end in the sebum.

A significant portion of the total daily nitrogen loss—1 g in men and 0·6 g in women —occurs by loss of cutaneous cells, hair and nails. In minimal sweating about 0·4 g nitrogen is lost but the loss is usually about 1·5 g daily in the average male; in severe sweating up to 4 g nitrogen may be lost per day.

Many substances applied to the skin are unable to penetrate further than the stratum corneum; some, however, pass through the skin without causing a reaction whereas others produce systemic effects once they have reached the blood stream. The rate of penetration varies for different substances and different sites; it can be altered by changes in temperature or by abrasion or inflammation of the skin. The physicochemical characteristics of the penetrant, for example solubility, molecular size and particle size, are most important; small molecules penetrate faster than large ones. If the stratum corneum is made moist by covering it with a waterproof bandage penetration is increased and corticosteroids applied to it may accumulate in the stratum corneum with obvious advantages in the treatment of certain skin diseases.

Substances enter the skin either by diffusion directly through the mass of the horny layer or by way of the appendages and in particular the pilosebaceous unit (Fig. 35.4). The cells of

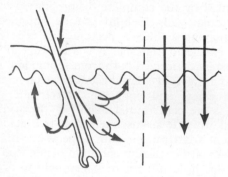

Fig. 35.4 Routes of penetration of the skin.

the sebaceous glands are more permeable than the keratinocytes of the epidermis and thus substances can reach the dermis by passing through the sebaceous glands from the hair follicles and also through the epithelium of the follicular root sheath. Penetration through the pilosebaceous unit is facilitated by massage. Substances arriving in the dermis by this route may also pass upwards to reach the epidermal cells. However, this pilosebaceous route is much less important than that through the stratum corneum.

CUTANEOUS SENSES

Stimulation of the skin gives rise to a sensation of touch, of warmth, of cold, or of pain,

and these are described as the four 'modalities' of cutaneous sensation. On the basis of these sensations we attribute properties to external objects.

TEMPERATURE SENSE

Although we are apt to think of the skin as being uniformly sensitive all over its surface, each modality has in fact a punctate distribution. In the case of the temperature sense this can be shown by passing a warm metal rod over the skin. At some places no sensation of warmth is aroused, while at other places, called 'warm spots', a sensation of warmth is quite distinct. The 'cold spots' can be mapped out by passing a cold metal rod over the skin. When the two maps for any one area are examined it is found that the cold spots are the more numerous and that they do not coincide with the warm spots. Furthermore, tests made from day to day show that the distribution of these spots is continually altering (Fig. 35.5). The

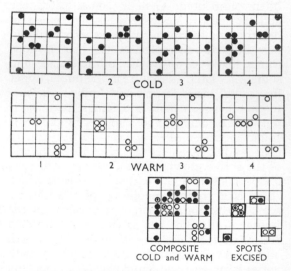

Fig. 35.5 Maps of cold and warm spots on an area of 1 cm^2 on the upper arm. The successive maps were made at intervals of two days. All the spots are combined in the diagram marked 'Composite Cold and Warm'. The spots in the last diagram were excised but no specialized nerve-endings were found. (K. M. Dallenbach (1927). *American Journal of Psychology* **39**, 416.)

punctate distribution is lost if the skin becomes red, as in sunburn, when all parts give rise to a sensation of warmth on being touched with a warm rod. A further observation suggesting that the two receptors are distinct is that a warm rod passed over a cold spot may give rise paradoxically to a sensation of cold. This may be explained by Zotterman's electrophysiological experiments described below. It has been suggested that the variability of the cold and warm spots is due to variations in the calibre of the skin vessels, and even that the vessels themselves are responsible for temperature sensations.

A piece of cloth and a piece of metal at the same temperature, either high or low, arouse quite different sensations. When an object is touched by the finger the sensation aroused depends on the rate of withdrawal or addition of heat. A finger of one hand is placed in hot water and a finger of the other in cold; after a little both are placed in water at an intermediate temperature. Since this intermediate stimulus feels warm to the latter and cold to the former it is reasonable to conclude that the cause of the sensation, in this experiment, is a rise or a fall in the temperature of the nerve-endings. This cannot, however, account for the sensation of cold in the following experiment. If a cold object such as a coin is pressed on the forehead and then removed a sensation of cold persists for a little while, although at this time the temperature of the end-organs must be rising.

The experiment with the hot and cold water shows a feature common to all sensations—adaptation. The finger kept in cold water for some time does not feel cold; it becomes adjusted to the new situation which, since it arouses no sensation, becomes a new zero. The finger in the hot water also shows adaptation but in the opposite direction. This experiment also shows that the temperature sense is not to be regarded as a thermometer; with its shifting zero it cannot measure absolute temperature. It can, however, detect quite small temperature differences. The sensitivity of the exposed areas of the skin (face, hands and scalp) is less than that of areas normally covered, such as the arms and forearms. At ordinary skin temperatures differences of more than $0.2°C$ can be appreciated by the arms and differences of 0.5 to $1.0°C$ by the fingers, the sensitivity of which is greater on the dorsal than on the palmar surface. The mucous membrane of the mouth is relatively insensitive to heat; tea can be drunk at a temperature which is painful to the finger.

Zotterman and his colleagues in Stockholm have recorded impulses in single nerve fibres coming from the tongue of the cat and have found two sets of thermoreceptors (see Fig. 35.6). If the temperature of the surface of the

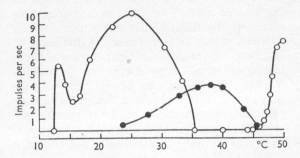

Fig. 35.6 Graph showing the frequency of the steady discharge of a single cold fibre (open circles) and of a single warm fibre (solid circles) when the receptors on the surface of the cat's tongue were exposed to constant temperatures within the range of 10° to 50°C. (Y. Zotterman (1953). Special senses: thermal receptors. *Annual Review of Physiology* **15**, 363.)

tongue is held steady the 'cold' receptors discharge continuously but the actual rate of discharge depends on the temperature. The discharge begins about 10°C and reaches a maximum of about 10 impulses per second between 20 and 30°C; above 30°C there are very few impulses until the tongue temperature reaches 45 to 50°C when the discharge rate increases again. If cold receptors in man behave similarly it is easy to account for the paradoxical sensation of cold mentioned above. Zotterman found that the 'warm' receptors behave quite differently, discharging when the tongue temperature is in the range 25 to 45°C. The actual rate of discharge, usually less than 3 impulses per sec, depends on the temperature. Both receptors show marked adaptation. A sudden fall in temperature produces a sudden rise in the frequency of discharge up to 100 or more impulses per sec from the cold receptors but this quickly falls to the steady value of 10 or less. Similarly a sudden rise in tongue temperature produces an immediate increase in the discharge from the warm receptors which falls quickly to a new resting value. In the cat distinctive cold receptor terminals have been seen in electron microscopic pictures. The nerve fibres supplying them are unmyelinated.

Iggo has found that the 'cold' receptors of primates have a very restricted receptor field on both hairy and glabrous skin. The speed of conduction of nerve impulses to the nerve trunk remote from the receptors ranged from 0·6 to 15·3 m/sec; this speed corresponds to myelinated axons 1 to 3 μm in diameter. The receptors discharged in bursts which showed adaptation but they continued to discharge at constant thermoneutral temperatures. Electrophysiological experiments on conscious human subjects are now providing confirmation of the results obtained from non-human primates that establish modality-specific receptors and mechanoreceptors. The anatomical nature of the receptors in man is not known.

THE SENSE OF TOUCH

This sense allows us to distinguish between hard and soft bodies and to judge their shape. In practice, however, muscle-joint sense may also be used to estimate hardness and, although we judge the roughness of a surface by passing the tips of the fingers gently along it, our conclusion may really depend on the auditory stimuli produced at the same time. The threshold of the tactile sense is measured by von Frey's bristles which consist of a series of fibres of different thicknesses mounted on match-sticks by which they can be held between the fingers. Each is calibrated by pressing one end of the bristle on a balance till it just bends. This weight divided by the cross-sectional area of the bristle gives the pressure exerted on the skin.

If the skin is explored with a bristle it is found that the tactile sense is, like the other skin senses, distributed in a punctate fashion. The map of the distribution of the touch spots in any given area varies from day to day, but if the observations are made with a careful technique on trained subjects the pattern of the touch spots becomes more nearly constant than in a poorly managed experiment. This variability in position of the touch spots makes it difficult to believe that there is a fixed anatomical basis for the touch sense. By means of a large series of bristles von Frey found that the smallest pressure in g/mm^2 required to elicit a sensation of touch varied in different

parts of the skin, as shown in the middle column of Table 35.7. If these values are converted into lb/in^2 (third column in Table 35.7)

Table 35.7 Threshold pressure in g/mm^2 required to elicit a sensation of touch on various areas of the skin surface as measured by von Frey. In the third column the values have been converted to lb/in^2 to show that there is a fallacy in expressing the threshold in terms of pressure (see discussion in text)

Area	g/mm^2	lb/in^2
Tongue and nose	2	2·8
Lips	2·5	3·5
Finger-tip and forehead	3	4·3
Back of finger	5	7·1
Palm, arm, thigh	7	10
Forearm	8	11
Back of hand	12	17
Calf, shoulder	16	23
Abdomen	26	37
Outside of thigh	26	37
Shin and sole	28	40
Back of forearm	33	47
Loins	48	68
Thick parts of sole	250	356

quite ridiculous values are obtained; the lips (second entry in the table) are certainly sensitive to much less than 3·5 lb on an area of 1 in^2. Thus the adequate stimulus for eliciting a sensation of touch cannot be mere pressure, but probably involves skin deformation, as is shown by placing a finger in a bowl of mercury. Contact is felt only at the air/mercury junction where the skin is deformed by the abrupt change of pressure, and not deeper in the mercury at the tip of the finger where, although the pressure is actually much greater, there is no deformation. Pressure in the skin causes movement of nerve terminals and depolarization which sets off nerve impulses.

Touch spots are thickly distributed around hair follicles and the root of each growing hair is surrounded by a network of nerve fibres; the nerve elements become very scanty when the hair is shed. Each hair pivoted at the skin surface acts as a lever to transmit pressure to the nerve-endings round the shaft near the root. When the hairs are removed the sensitivity to touch is greatly reduced.

Afferent fibres which arise from receptors in hair follicles have, characteristically, a large receptive field (perhaps several centimetres in diameter) but some organized tactile receptors with a field not more than 0·5 mm in diameter may be supplied by a single nerve fibre.

The tactile sensation is unique in lasting only a very little longer than the stimulus. If a finger is held lightly against a revolving cog wheel each stimulus gives rise to a separate sensation until the contacts follow one another at more than 500 or 600 per second. Above this frequency of stimulation the rotating cog wheel is described as smooth, the separate sensations having fused into a continuous sensation. If, nevertheless, the amplitude of vibration of an object applied to the skin is very high, a sensation of vibration may persist up to 8000 Hz and, with practice, amplified speech vibrations can be felt by the fingers and recognized. Although this may be of practical value to a deaf person it is possible that recognition depends on the intensity of the vibrations rather than on their frequency.

The tactile sense shows adaptation which is partly peripheral and partly central. We are aware of the contact of our clothes with the skin when they are first put on but this sensation quickly disappears. It can, however, be brought to consciousness by an act of the will or by a slight movement of the body. In the same way a new denture is very obvious at first but after a time the wearer becomes unconscious of its presence.

Localization. When a blindfolded person is touched, not only does he have an impression of the strength of the stimulus but he can also indicate the point touched with a considerable degree of accuracy. The power of localization depends partly on the position at which the nerve fibres from the tactile end-organs enter the spinal cord and on their higher connexions, and partly on experience. Aristotle's experiment shows the influence of experience. A pencil laid between two crossed fingers of a blindfolded observer gives the sensation of contact by two pencils. Experience also plays a part in visual localization. After removal of a congenital cataract in persons blind from birth the patient is not able at first to localize his visual impressions; the power to do so grows as experience of his new sense develops.

Discrimination. If two points on the skin are touched simultaneously and with approximately the same degree of pressure by the points of a pair of dividers (aesthesiometer) the

subject will report that two separate points have been touched, provided that the distance exceeds a value depending on the area of skin under test (Table 35.8). If the skin is carefully

Table 35.8   The two-point threshold*

Area	Separation in mm
Tip of tongue	1
Anterior surface of finger-tip	2
Posterior surface of third phalanx	6
Palm of hand	11
Back of hand	32
Back of neck	54
Middle of back	67

* The distance at which two points on the skin touched simultaneously with approximately the same pressure must lie apart to be recognized as separate points of contact. A recent reinvestigation of the two-point threshold is to be found in S. Weinstein (1968) in Chap. 10 of *The Skin Senses* (Ed. D. R. Kenshalo). Springfield: Thomas.

explored for touch spots it is found that in the area showing the lowest two-point threshold they are less than 1 mm apart but at no part of the body is the distance between these spots more than 6 mm. Scarcity of touch spots is not the explanation of the poor discrimination of the less mobile parts of the body surface.

This two-point threshold is usually three or four times the error of localization in the same region. This apparent anomaly is probably due to the diffuse nature of the stimulus. A blunt point on the skin stretches the surrounding skin and, according to Woodworth, localization is probably determined by the centre of strongest stimulation; in the case of two-point stimulation the whole stretched area is effective in giving a broad impression and obscuring the presence of the two points. In other words when the two points of contact are fairly close to one another the stimulus is not very different from that produced by one contact.

Projection. This is most highly developed in vision but it occurs also with the tactile sensation. If a pudding is stirred with a spoon, lumps may be detected and the bottom of the pan feels hard. The stimuli which originate these sensations are applied to the skin in contact with the upper end of the spoon but the sensations are projected to the lower end of the spoon.

THE PAIN SENSE

Pain is often referred or projected to the surface of the body when internal organs are diseased (Chap. 41) but here we are concerned only with pain actually arising in the skin. Noxious stimuli call forth withdrawal reflexes which occur even when the brain is removed from an animal and only the spinal cord remains intact. In the intact animal the same reflexes occur and it is reasonable to suppose that a painful sensation is evoked as in man.

Pain elicited by stimulation of the skin has a pricking, burning or itching quality and is well localized. The pain threshold may be raised by one-third by distracting the subject's attention but in sunburnt skin it may be lowered by half. If the finger just below the nail-bed is pressed against a hot electric bulb a double pain is produced. The first sensation arises abruptly and is pricking in quality; it is probably carried in moderately large fibres at about 10 metres per second. The second sensation follows a second or more later, has a burning quality but is less abrupt in onset and disappearance; it is probably carried in non-myelinated fibres. If the circulation through the finger is obstructed by a bandage to produce asphyxia of the skin the first pain but not the second is abolished; the underlying mechanisms must therefore be different. Some observers have produced evidence to show that pain is carried in A and C fibres (Chap. 39).

Quite gentle exploration of the skin surface with a needle produces a painful sensation at certain places. Once again the pain end-organs seem to be distributed in a punctate fashion, independently of the end-organs of touch or temperature. Any given area of skin may show ten times as many pain spots as touch spots. These pain endings have different characteristics from the tactile end-organs; the threshold stimulus (Table 35.9) is quite different and the order of increasing sensitivity is completely different from that given for touch in Table 35.7. The pain sense has a relatively slow rate of adaptation. Asphyxia reduces the response of the other senses but may increase the response to painful stimuli. In the disease of the spinal cord called syringomyelia pain may disappear while the other cutaneous senses remain. There is no doubt therefore that

Table 35.9 The minimum stimulus required to elicit a sensation of pain at various parts of the body. Note the want of parallelism with the figures for the sense of touch in Table 35.7 (von Frey)

Area	Threshold (g/mm^2)
Cornea	0·2
Conjunctiva	2
Abdomen	15
Front of forearm	20
Back of forearm	30
Calf of leg	30
Back of hand	100
Sole	200
Finger-tip	300

pain is a separate modality and that it is not simply due to excessive stimulation of nerve-endings.

Pain sensation is frequently tested by using thermal radiation. The threshold temperature of the skin which gives rise to pain lies between 43° and 47°C with slow or no adaptation. Application of water at 37°C may cause a sensation of pain which disappears in a few seconds. It is difficult to say how a stimulus gives rise to pain. Presumably it depolarizes the end-organ or suddenly increases its permeability. Mechanical or chemical stimuli could produce either effect directly or perhaps indirectly by releasing a chemical substance from damaged cells. Keele and his colleagues have investigated this possibility by the simple expedient of raising a skin blister with cantharidin and then applying various substances to be tested to the exposed base of the blister. Blood, plasma or inflammatory exudate withdrawn into a siliconed syringe have very little pain-producing activity; on transfer to a glass vessel, pain-producing activity appears in a few minutes but declines and disappears in about an hour. The pain-producing substance (P.P.S.) seems to be a polypeptide with pharmacological properties resembling those of bradykinin (Chap. 50) and in fact pure bradykinin in a concentration of 0·1 µg/ml produces pain when applied to a blister base. Skin possesses many proteolytic enzymes which are known to be released when the skin is damaged. It is therefore reasonable to speculate that tissue injury sets free an enzyme capable of acting on some component of the plasma proteins to give P.P.S. which then produces pain.

Haemolysis of red cells (and indeed damage to any cell) releases potassium ions and application of the haemolysate to the base of a blister produces pain which is, however, somewhat greater than that produced by a solution of potassium chloride made isotonic with NaCl containing the same amount of potassium as the haemolysate. Since 5-hydroxytryptamine (5-HT, Chap. 50) causes pain when it is placed on the base of a blister it may be that damage to blood platelets by releasing 5-HT also gives rise to pain. 5-HT potentiates the algogenic effect of bradykinin and of potassium. Acetylcholine even in high dilution gives rise to pain immediately. It is not likely that histamine is of any importance in this respect because pain is caused only when high concentrations, unlikely to be produced by injury, are applied.

Pruritus or itching may be defined as 'an unpleasant cutaneous sensation which provokes the desire to scratch'. The sensation may be mild, in which case it is almost pleasurable, and this is quite common in elderly people; sometimes it may be so severe as to cause the sufferer to attempt suicide. Pruritus may be generalized, as in jaundice, or it may be confined to quite a small area of skin or to a particular part of the body such as the perineum or the nostrils. The sensation arises in nerve endings in the epidermis since pruritus cannot be elicited in areas denuded of epidermis. It can be produced in a sensitive area by chemical, mechanical, thermal or electrical stimuli, and mild forms of such stimuli may elicit itching in a localized area of skin long after the initial pruritus has subsided. Thus, light stroking of an insect bite may produce itching weeks after the local reaction has subsided. In other words, the skin remains in a state of increased excitability. If the itching so produced leads to vigorous scratching, the area of excitable skin enlarges and the threshold is further reduced.

By inserting stinging hairs of a tropical plant into the skin Shelley and Arthur showed that the most effective site for itch production was the dermo-epidermal junction; they believe that itch results from the liberation of endopeptidases by damage to the skin. Contact with a variety of stimuli liberates chemomediators which act on peripheral nerve filaments and produce the sensation of itch. This sensation is carried in the sensory spinal nerves to the spinothalamic tract and then to the thalamus and sensory cortex. Because no itching can be elicited in analgesic skin itching has been thought to be a kind of pain sensation. In syphilitic disease of the posterior roots of the spinal cord (tabes dorsalis) the sense of touch may be lost without loss of the sense of pain; itching can be elicited in the affected areas of skin. On the other hand the fact that removal of the epidermis and the subepidermal nerve network abolishes itch but not pain suggests that itch is a distinct sensory experience. If this is accepted it is easy to understand how it is that itch and pain can be experienced simultaneously and that immersion of the skin in water at 40° to 41°C quickly abolishes itch but intensifies pain. The itch threshold may be affected by psychological factors. It is not clear why scratching relieves itching but it may do so by disturbing the rhythm of afferent impulses travelling towards the spinal cord or by local depletion of kinins. Tickle can be demonstrated all over the body by light touch with a wisp of cottonwool. It has been shown that after a tickle stimulus adjacent areas of skin are more sensitive with the result that a single light touch stimulus applied within them will arouse tickle. The stimulus for tickle is so slight that involvement of an end organ deeper than the most superficial layer of the skin seems very unlikely.

ANATOMICAL BASIS OF CUTANEOUS SENSATIONS

The experiments described in this chapter provide some evidence for the hypothesis that there are four 'modalities' of cutaneous sensation, and indeed the punctate distribution of the sensitive areas makes it logical to search for special receptors. The older histologists found many organized (that is elaborate) endings in the skin and, as it appears now, rather rashly labelled them 'pressure' endings, 'cold' endings and so on, simply on the basis of their morphology. Since then many workers, having found a spot on the skin consistently sensitive to a particular sensation, have marked it, excised it and examined it with a variety of histological techniques. Almost invariably no organized endings were found. A group of workers at Oxford have compared several areas of skin both experimentally and histologically.

They found no important difference in sensitivity to each of the four modalities when they tested the skin of the auricle (external ear) or the front of the forearm or the palmar aspect of the fingers. A complicated nerve plexus was found in the excised skin of the auricle ending either in a basket-like network around hair follicles or by repeated breaking-up into beaded branches which ended freely in the superficial layer of the corium and deeper layers of the epidermis; no organized endings were seen. The same techniques applied to the skin of the fingers revealed many organized endings as well as free terminations. Thus at least in the auricle organized endings are not necessary for the perception of touch, pain, cold and warm stimuli, and this finding applies to all hairy skin which has been investigated. This does not mean that the organized endings, where they occur, are functionless. Opinions differ considerably as to whether adult human epidermis possesses nerve endings which penetrate between epidermal cells. The Merkel cell (a tactile receptor complex) lies at the base of the epidermis.

The validity of the idea that there are four modalities of skin sensation each subserved by specific receptors and fibres is now questioned. The simplest model for such specificity could be a set of receptors for any given modality which had a much lower threshold than all other receptors for a given stimulus and that they sent a particular pattern of impulses (Chap. 39) as a sort of code to the central nervous system; and furthermore that each of the four kinds of receptors would require its own individual pathway to its own receiving area of the brain. The idea of an individual pathway is however suspect. Some time after an anterolateral cordotomy has successfully abolished pain, pain sensation may return. In the case of the peripheral nerves, experimenters have found some single fibre units whose discharge varies according to the stimulus applied to the receptor; thus the firing frequency may be higher with tactile than with thermal stimuli. As mentioned above, biopsies have not revealed specific receptors in human subjects but animal experiments have now shown convincingly that there are several kinds of afferent units with morphologically specific receptors. Furthermore, the pattern of impulses in a peripheral nerve is much modified in passing across synapses so that the oscillograph record obtained in the central nervous system appears to the experimenter to have lost all resemblance to the pattern originated in the skin receptor. Again, an individual cell of the posterior horn of the grey matter can receive impulses arriving by both myelinated and unmyelinated fibres, in other words these cells in the spinal cord are not 'modality specific' in terms of the simple model described above. One might speculate that some centrally placed cells in the brain are able to 'recognize' particular patterns arriving from many afferent paths.

The discussion in the previous paragraph suggests that we must abandon the idea of 'modality-specific' pathways but there is now evidence that in animals individual 'modality-specific' afferent fibres enter both 'relatively specific' as well as 'non-specific' centripetal pathways. The recently discovered divergence and richness of pathways for afferent units excited by mechanical stimuli (that is within the touch/pressure modality) could provide for two kinds of pathway. Also certain cells in the trigeminal nucleus in monkeys are excited by impulses from thermoreceptors but not by impulses from mechanoreceptor units. Therefore some cells at several levels in the sensory pathway display an input specificity. Whether or not there is some 'temporal pattern specificity' or 'decoding' will be more difficult to establish.

REFERENCES

CATTON, W. T. (1970). Mechanoreceptor function. *Physiological Reviews* **50**, 297–318.

CHAMPION, R. H., GILLMAN, T., ROOK, A. J. & SIMS, R. T. (1970). *An Introduction to the Biology of the Skin.* Oxford: Blackwell.

CHRENKO, F. A. (1964). Threshold intensities of thermal radiation evoking sensations of warmth. *Journal of Physiology* **173**, 1–12.

DANIELS, F. JR., VAN DER LEUN, J. C. & JOHNSON, B. E. (1968). Sunburn. *Scientific American* **219**, 38–46.

DE REUCK, A. V. S. & KNIGHT, J. (Eds.) (1966). *Touch, Heat and Pain.* Ciba Foundation Symposium. London: Churchill.

FRAIN-BELL, W. (1969). Anhydrotics. *Practitioner* **202**, 79–87.

HURLEY, H. J. & SHELLEY, W. B. (1960). *The Human Apocrine Sweat Gland in Health and Disease.* Springfield, Ill.: Thomas.

IGGO, A. (1969a). Cutaneous thermoreceptors in primates and subprimates. *Journal of Physiology* **200**, 403–409.

IGGO, A. (1969b). The structure and function of a slowly adapting touch corpuscle in hairy skin. *Journal of Physiology* **200**, 763–796.

JARRETT, A. (1964). *Science and the Skin.* London: English Universities Press.

JOHNSON, B. E., DANIELS, F. JR & MAGNUS, I. A. (1968). Response of human skin to ultraviolet light. In *Photophysiology. Current Topics in Photobiology and Photochemistry* (Edited by A. C. Giese), Vol. 4. New York: Academic Press.

KEELE, C. A. & ARMSTRONG, D. (1964). *Substances Producing Pains and Itch.* London: Arnold.

KEELE, C. A. & SMITH, R. (Eds.) (1962). *The Assessment of Pain in Man and Animals.* Edinburgh: Livingstone.

KENSHALO, D. R. (Ed.) (1968). *The Skin Senses.* Springfield, Ill.: Thomas.

MELZACK, K. & WALL, P. D. (1965). Pain mechanisms: A new theory. *Science* **150**, 971–979.

MONTAGNA, W. (1962). *The Structure and Function of Skin,* 2nd edn. New York: Academic Press.

MONTAGNA, W. & ELLIS, R. A. (1958). *The Biology of Hair Growth.* New York: Academic Press.

MONTAGNA, W. & LOBITZ, W. C. (Eds.) (1964). *The Epidermis.* London: Academic Press.

MONTAGNA, W. *et al.* (Eds.) (1960–66). *Advances in Biology of the Skin.* Vols. 1–7. Oxford: Pergamon.

NOORDENBOS, W. (1959). *The Problem of Pain.* London: Elsevier.

ROOK, A. (1965). Endocrine influences on hair growth. *British Medical Journal* i, 609–614.

ROOK, A. & CHAMPION, R. H. (Eds.) (1964). *Progress in the Biological Sciences in Relation to Dermatology.* Cambridge University Press.

ROOK, A. J. & WALTON, G. S. (Eds.) (1965). *Comparative Physiology and Pathology of the Skin.* Oxford: Blackwell.

ROOK, A., WILKINSON, D. S. & EBLING, F. J. G. (Eds.) (1968). *Textbook of Dermatology.* Oxford: Blackwell.

ROTHMAN, S. (Ed.) (1959). *The Human Integument, Normal and Abnormal.* Washington: American Association for the Advancement of Science.

SAMMAN, P. D. (1965). *The Nail in Disease.* London: Heinemann.

SINCLAIR, D. (1967). *Cutaneous Sensation.* Oxford University Press.

Symposium on dermatological drugs and dressings (1969). *Practitioner* **202**, 5–151.

THEOBALD, G. W. (1965). Cortical pain image or pain-sensitivity panel. *British Medical Journal* ii, 330–333.

TREGEAR, R. T. (1966). *Physical Functions of Skin.* London: Academic Press.

WINKELMANN, R. K. (1960). *Nerve Endings in Normal and Pathologic Skin. Contributions to the Anatomy of Sensation.* Oxford: Blackwell.

WOODWORTH, R. S. & SCHLOSBERG, H. (1955). *Experimental Psychology,* 3rd edn. London: Methuen.

36 The chemical senses

Taste and smell are referred to as the chemical senses because the receptor organs are remarkably sensitive to a wide variety of chemical substances. However, the nose, mouth and genital apertures, and many mucous membranes, are sensitive to irritants, such as ammonia. This third form of chemical sense is referred to as 'common chemical sensibility'. A sensation described as a taste may in fact arise from the simultaneous stimulation of the taste and smell receptors; this is the case with onions and with wine. The close association between the two senses is shown by the lack of flavour of many food substances when the sense of smell is lost during a cold in the head. It is convenient to use the word flavour for the sensation aroused by the simultaneous stimulation of these receptors.

The senses of taste and smell do not seem to be as well developed in man as they are in some other mammals. Man can detect quite minute amounts of certain substances but he has come to depend so much on his visual equipment that it is not surprising that the olfactory part of his brain is relatively small. The sense of smell is, like the sense of vision, normally projected on to the environment but the sense of taste is not so projected. Although an odorous substance stimulates end-organs in the nose the sensation is referred to or projected on to the source of the odour. Thus we say that the rose has a sweet perfume, not 'there is a sweet perfume in the nose'.

SENSE OF SMELL (OLFACTION)

The greater part (or respiratory region) of the nasal cavities is lined by columnar ciliated epithelium interspersed with goblet cells which are capable of producing large quantities of mucus. In man the olfactory area, about 3 cm^2 in area on each side, is situated in a narrow cleft in the highest part of each nasal cavity above the superior concha. Each area possesses fifty million olfactory cells which lie among supporting cells (Fig. 36.1). Non-medulated nerve fibres arise from the olfactory cells and form a plexus in the deeper parts of the mucosa and then pass through the cribriform plate of the ethmoid bone to the olfactory bulb where contact is made with second order neurones in complicated synapses (glomeruli). These second order neurones are mainly mitral cells going to the prepiriform cortex and tufted cells sending axons to the opposite bulb. To

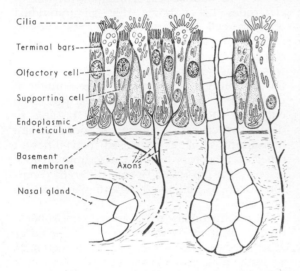

Cilia
Terminal bars
Olfactory cell
Supporting cell
Endoplasmic reticulum
Basement membrane
Axons
Nasal gland

Fig. 36.1 Olfactory epithelium and component cells. The nasal glands keep the epithelium moist. Each olfactory cell has about a dozen cilia each up to 200 μm long on its free surface. The microvilli of the supporting cells project into the layer of mucus which covers the olfactory epithelium. (*After* H. S. D. Garven and A. J. D. de Lorenzo.)

establish conditional reflexes to odours, the olfactory receptors, the olfactory bulbs and tracts and the piriform cortex are necessary but no other parts of the rhinencephalon.

The route taken by the stream of air through the nose has been demonstrated in several ways. For example the nasal mucosa of a cadaver was covered with red litmus and air containing ammonia was blown through the nose in imitation of inspiration or expiration. The air pathway was shown by a blue colour. Again, direct inspection of the nasal cavities in the living subject can be carried out after inhalation of light magnesia powder. These methods show that the air stream does not rise above the superior concha on inspiration and

is at an even lower level on expiration. Thus in ordinary quiet breathing the air does not impinge on the olfactory area directly but only as eddies coming off the main stream. When we attempt to smell a substance by sniffing, the air-stream reaches the olfactory area directly either because it is travelling more rapidly or because it is directed there by a change in the shape of the nostrils. If air cannot reach the olfactory areas, for example because of swelling of the mucosa during a cold in the head, the sense of smell is lost. Since the olfactory areas are covered with mucus it is reasonable to assume that to give rise to a sensation of smell an odorous substance must be dissolved in the secretions. Odorous substances must of course be volatile, and must be soluble in lipids.

Many substances lose their odour within a very few minutes of presentation. This may be a result of either a rapid adaptation of the end-organs or a washing away of the odoriferous molecules by mucus. Adaptation is at least part of the explanation since loss of odour tends to be specific, that is, any other odour, if quite different from the first, is easily perceived. The inability to perceive odours after prolonged exposure should perhaps be termed fatigue rather than adaptation. The term adaptation seems more appropriate in the case where the occupants of a stuffy room are quite unaware of their odour whereas a newcomer is at once unpleasantly conscious of it. Neutralization of one odour by another occurs to some extent but usually disagreeable odours, for example that of faeces, are disguised by overwhelming them with powerful, but more pleasant, odorous substances.

There has been, and still is, much speculation about the physical or chemical nature of the stimulus which gives rise to an olfactory sensation. The old classification of odours into four main groups, (1) fragrant, (2) acid or sour, (3) burnt and (4) caprylic (disagreeable or putrid) is not satisfactory. There is no simple relationship between chemical constitution and odour. Atomic bromine and chlorine have no odour but Br_2 and Cl_2 have. Similar odours may be given by substances with very different chemical constitution, for example hydrocyanic acid and nitrobenzene both smell of bitter almonds. Some sulphur containing compounds have a smell like bad eggs. If, however, the sulphur is replaced by oxygen the compound is odourless. In the case of optically active substances it frequently happens that one stereoisomer has a more intense smell than its mirror image.

Moncrieff has found that substances having similar odours are absorbed to similar extent on adsorbents such as active carbon even if they have quite different chemical constitutions; conversely substances having different odours behave differently towards the adsorbents used. He is tempted to suppose that some sort of selective adsorption goes on in the nose. In pursuing this idea Amoore has found that molecules with a similar odour have a similar shape and size although they might differ considerably in chemical formula. The odour of substances may depend on their fitting into specifically shaped receptor sites.

The minimum stimulus required to arouse a sensation of smell varies with different substances. Some substances, like mercaptan, chlorophenol and skatole, can be detected in the air at very high dilutions. The nose is in fact the most sensitive chemical detector known, its threshold of perception being more than one hundred times lower than the best of our gas chromatographs. It is possible to detect 0·01 mg of mercaptan in a room of 230 cubic metres (say 28 ft by 22 ft by 13 ft high). Since 0·01 mg of mercaptan contains 1×10^{17} molecules each sniff of 20 ml contains 1×10^{10} molecules, about enough to cover the receptive area with a layer one molecule thick. However, in the absence of some selective action by the olfactory epithelium it is difficult to understand how one molecule of mercaptan in fifty thousand molecules of air can produce an olfactory sensation or indeed how the adsorption of a small number of molecules can set up a nerve impulse.

Sympathetic efferent activity in the ethmoid nerve increases the olfactory response but whether this is a direct effect on the receptors or merely a vasomotor effect is not known. Patients suffering from adrenocortical insufficiency have increased olfactory sensitivity.

Adrian has made records of the electrical activity of the mitral cells in the olfactory bulb of the rabbit and has found that, if the spontaneous discharge of the cells has been suppressed by a general anaesthetic, discharges can be detected when air containing an odoriferous substance is drawn through the nose. The anterior part of the olfactory area is easily excited by esters (fruity smell) which are slightly soluble in water. Oil-soluble substances on the other hand excite the posterior area more easily.

Potentials have been recorded from microelectrodes placed in the olfactory epithelium of both man and animals. The spectrum of the frequency response in man ranges from 30 to 100 Hz and the frequency response is different for different odours but similar for similar odours. In animals experiments with microelectrodes have shown that an individual olfactory cell (primary receptor) responds to some odours and not to others. The possession of a relatively small number of primary receptors of this yes-or-no character could account for the ability to detect a very large number of odours. For example, if an animal possessed only one such receptor it would be able to detect two classes of substances, odorous and non-odorous; if it possessed two different types of receptors it could discriminate four classes; if it possessed twenty different receptors over a million different patterns are possible.

Many mammals react to odours (pheromones) emitted by members of the same or other species. For example the introduction of a strange male into a group of female rats may cause fetal reabsorption. The steroid hormones have an odour like that of musk and women but not men are said to be able to detect steroid odours. It is difficult to assess the influence of pheromones in man because of the complexity of his behaviour but these external chemical messengers may act at the subconscious level.

SENSE OF TASTE (GUSTATION)

The taste receptors or *taste buds*, about 9000 in the adult are placed mainly on the peripheral parts of the dorsum of the tongue, being most easily found in the groove surrounding the vallate papillae: some buds occur on the soft palate and a few on the epiglottis. In children they are much more numerous and much more widely distributed over the tongue and on the insides of the cheeks. The oval taste buds embedded in the stratified epithelium covering the tongue are usually described as containing thin fusiform taste cells surrounded by supporting cells like the staves of a barrel. Electron microscope pictures obtained by de Lorenzo suggest that the different appearances of the cells in the taste bud are due to their being at different stages of development. The taste cells have microvilli on their free surfaces. Experiments with tritiated thymidine (Chap. 20) have shown that taste cells are produced by mitotic division of cells at the edge of the bud which migrate to the centre of the bud where they disintegrate and disappear after a week. Since the afferent nerve fibres do not degenerate the nature of the receptor-nerve contact is rather difficult to imagine.

The *nerves of taste* (Fig. 36.2) are the lingual branch of the glossopharyngeal (IX) for the posterior third of the tongue, including the vallate papillae, while the anterior two-thirds are served by the chorda tympani branch of the facial (VII). The greater superficial petrosal nerve supplies gustatory fibres to the palate. Taste buds near the epiglottis are innervated by the vagus. All the nerve fibres pass centrally to the tractus solitarius, a long column of grey matter in the medulla oblongata. From here fibres cross the mid-line and pass to the posterior ventral nucleus of the thalamus and then on to the lower part of the postcentral gyrus and probably other areas nearby such as the lower surface of the frontal operculum and adjoining parts of the anterior insula. After destruction of the postero-medial ventral nucleus goats drink strong solutions of quinine, acid or salt which they had repeatedly refused before the operation. Stimulation of this area in intact unanaesthetized animals produces 'rejecting' movements of the jaw and tongue exactly as seen when a solution of quinine is squirted into the mouth. The lingual branch of the fifth nerve is the nerve of common sensibility for the anterior two-thirds of the tongue; the lingual branch of the glossopharyngeal carries both taste and common sensibility fibres for the posterior third.

Probably no true taste fibres pass into the brain stem by the fifth nerve but tactile, pain and thermal sensations conveyed by this nerve play an important part in the recognition of taste sensations. When the fifth nerve in man is cut or destroyed the sense of taste is immediately lost in the front of the tongue but it returns in some cases after a few hours, in others after years. There is no such separation

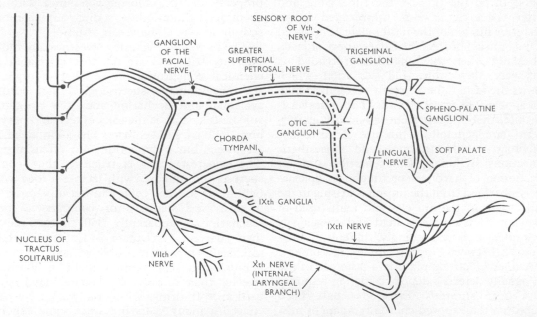

Fig. 36.2 Pathways of the taste fibres are given by bold lines. The interrupted line is an alternative route for taste from the anterior two-thirds of the tongue which appears to be present in some persons. Although these pathways are agreed to by most neurologists there is still some doubt as to details. Diagram modified and simplified from A. Brodal (1948). *Neurological Anatomy in relation to Clinical Medicine.* Oxford: Clarendon Press.

of routes for taste and common sensibility from the posterior part of the tongue since both enter the brain stem by the glossopharyngeal nerve. If the facial nerve is divided above its ganglion the sensation of taste is lost permanently on both the palate and the anterior part of the tongue; although the fifth nerve is intact, it is unable to provide a sensation of taste.

The sensibility of the tongue can be investigated by placing small drops of the test substances in solution in water on the protruded tongue while the subject holds his nose. Such experiments seem to show that there are four basic tastes, namely *sweet*, *sour*, *salt* and *bitter*. It is, however, difficult to get good evidence to support this simple classical description. For example many people confuse sourness and bitterness and most people say that the sweetness of lactose differs from that of glucose. Furthermore it has been shown that individual receptor cells are sensitive to many stimuli, that is they are not rigidly specific but an individual cell may be more sensitive to one kind of chemical stimulus than another. Sweet tastes are most easily perceived at the tip of the tongue, bitter at the back, sour at the edge and salt both on the tip and the edge. Sour and bitter tastes can also be appreciated at the posterior margin of the hard palate. The speed of recognition of the particular variety of taste, especially when the tongue is protruded, is relatively slow. The full sensation which allows the taste to be correctly recognized may not appear for several seconds. Tastes in everyday life, that is apart from laboratory experiments, are generally a mixture of gustatory, olfactory, tactile, thermal, and kinaesthetic sensations. By kinaesthetic sensation is meant the notion of hardness or toughness or elasticity obtained from the impulses arising in the end-organs of the muscles and joints used in chewing. If the nose is held and chewing is forbidden it is exceedingly difficult to distinguish between turnips, apples and onions.

A microelectrode place in a taste cell of the rat usually detects a depolarization in response to a taste stimulus. Electrodes have been placed on the exposed chorda tympani of man during operations on the middle ear. Action potentials were recorded when salt, sweet and bitter substances were placed on the tongue but the application of water reduced the spontaneous activity in the nerve. It thus appears that man, unlike the cat, dog and rhesus monkey, does not possess a specific water taste.

Sour-tasting materials all contain acids—vinegar contains acetic acid, citrous fruit citric acid, sour milk lactic acid. Whereas very weak acids do not have a sour taste, strong acids like hydrochloric acid, which dissociates almost completely in aqueous solution, have a sour taste at very high dilution. The sour taste of hydrochloric acid must be due to H^+ ions and not to Cl^- ions because 0.00125 N-NaCl is tasteless whereas 0.00125 N-HCl is sour. Organic acids like citric acid are, however, sourer than would be expected if hydrogen-ion concentration were the only factor determining sourness. It seems reasonable therefore to assume that both the anion and the undissociated molecule as well as the hydrogen ion play a part in determining the degree of sourness of an acid.

The typical salt substance is sodium chloride; calcium chloride and potassium chloride also have a salt taste but, since all chlorides do not have exactly the same taste, the cations must have a modifying influence. The salt taste is not, however, a specific property of the chloride ion since sodium bromide, sodium iodide, sodium sulphate and sodium nitrate all have salt tastes.

The factors determining sweetness are more puzzling still, and are not clearly related to chemical constitution. A number of organic substances, such as sucrose, glycerine, sodium cyclamate and saccharine, and a few inorganic substances, such as the salts of lead and beryllium, all have a sweet taste. The α-amino acids are sweet but polypeptides are often bitter. The sweetest sugar is fructose, then come sucrose and glucose, with lactose a long way behind. If a 2 per cent solution of sucrose is used as the standard of comparison then saccharine is apparently 350 times sweeter than sucrose. Cyclamate is less sweet than saccharine but more stable at cooking temperatures: in weak solutions it is 10 to 20 times sweeter than sucrose. Cyclamates have been used in 'soft' drinks without any apparent bad effect for many years but since some experiments have suggested that cyclamates may be carcinogenic their use as sweeteners has been abandoned.

A bitter taste is given by a variety of substances, for example caesium chloride, potassium iodide, magnesium sulphate, picric acid, and alkaloids such as quinine, strychnine and aloin.

Some tastes are not easily included in the above fourfold classification, for example alkaline and metallic tastes. It may be that in these cases the common chemical sensibility is stimulated. Freshly distilled water has an alkaline taste owing to the absence of carbon dioxide. Electrical stimulations of certain parts of the tongue give rise to an 'electric taste'. Some substances give rise to a persistent taste sensation, for example, saccharine and iodides. This is presumably due to the fact that they are excreted in the saliva.

Thus with the exception of the sour taste attempts to correlate chemical constitution with gustatory sensation have failed. To illustrate just how baffling the problem is, it may be mentioned that L-tryptophan is bitter, DL-tryptophan is sweet and D-tryptophan is nearly tasteless.

No instance has been reported in which one of the four taste sensations has been absent or impaired by disease. It has, however, been found that phenylthiocarbamide and allied substances are bitter to about 60 per cent of persons and tasteless (unless very high concentrations are used) to 40 per cent, and that the ability to taste these substances is inherited according to Mendelian laws. Most people cannot taste sodium benzoate, but a few describe it either as sweet or bitter.

The threshold concentration needed to arouse a sensation of taste, varies widely with different individuals and depends somewhat on the method of testing but the following are average values. Sour 0·0045 per cent HCl; salt, 0·055 per cent NaCl; sweet, 0·45 per cent fructose, 0·7 per cent sucrose, 0·01 per cent cyclamate, 0·001 per cent saccharine; bitter, 0·001 per cent brucine. The smallest difference in the strength of a test solution which can be distinguished is about 30 per cent.

McCance lowered the salt content of the blood of a number of subjects on a low-salt diet by making them sweat in a radiant heat bath. The subjects found that nearly all food stuffs had lost their flavour but the sense of taste returned within half an hour of taking salt by mouth.

Patients with adrenal cortical insufficiency have a very much lower threshold for the four modalities of taste. Treatment with deoxycorticosterone acetate does not alter the threshold but treatment with carbohydrate-active steroids brings it back to the normal level. D-Penicillamine used in the treatment of a number of conditions lowers the serum caeruloplasmin and serum copper and decreases the taste-sensitivity in all four modalities. Administration of copper by mouth restores taste-sensitivity. Diabetics have a slight hyposensitivity to glucose not correlated with the level of blood glucose.

The sense of taste shows adaptation. After cane sugar has been held in the mouth for 2 minutes the threshold for cane sugar may be raised ten times. At the same time the threshold for salt may be lowered a little. It is a matter of ordinary experience that successive draughts of a sweet drink become noticeably less sweet. A phenomenon like successive contrast (Chap. 37) is seen in taste; following on a sour stimulus distilled water may arouse a sweet sensation. Many examples of interaction between substances with easily recognized tastes could be given. Caffeine increases the sourness of citric acid and, as every coffee drinker knows, sucrose reduces the bitterness of caffeine. The temperature of the food affects gustation. Food has often more flavour when it is hot; iced food needs much more flavouring. Such a loss of flavour on cooling is not, however, universal; for example, the threshold for quinine is lower when the temperature of the solution is low. The flavour of food can be accentuated by adding flavour potentiators such as mono-sodium glutamate and certain 5′-nucleotides (disodium inosinate and disodium guanylate). These substances are effective in concentrations of the order of 100 parts per million but how they enhance flavour is quite unknown.

Symmetrical gustatory sweating is the name given to the sweating occurring on the head and face when irritants, such as chillies, are chewed; in a few persons it is aroused by eating chocolate. It is seldom seen in a cool climate or when the body is cooled but is experienced by nearly everyone in a tropical climate. This form of sweating is a reflex phenomenon initiated by excitation of the pain receptors in the mouth; the taste receptors are not involved.

REFERENCES

ADRIAN, E. D. (1956). The action of the mammalian olfactory organ. *Journal of Laryngology and Otology* **70**, 1–14.

AMOORE, J. E., JOHNSON, J. W. & RUBIN, M. (1964). The stereochemical theory of odor. *Scientific American* **210**, 42–49.

BAGSHAW, M. H. & PRIBRAM, K. H. (1953). Cortical organization in gustation (*Macaca mulatta*). *Journal of Neurophysiology* **16**, 499–508.

BEIDLER, C. M. (1961). Mechanism of gustatory and olfactory receptor stimulation. *Symposium of Sensory Communication* (Edited by W. A. Rosenbluth). M.I.T. Press and John Wiley.

VON BÉKÉSY, G. (1964). Sweetness produced electrically on the tongue and its relation to taste theories. *Journal of Applied Physiology* **19**, 1105–1113.

Bouquet and essence (1961). *Lancet* **ii**, 1391–1392.

GREEN, J. D. (1963). The function of the hippocampus. *Endeavour* **22**, 80–84.

HARRIS, W. (1952). The fifth and seventh cranial nerves in relation to the nervous mechanism of taste sensation. *British Medical Journal* **i**, 831–836.

KALMUS, H. & HUBBARD, S. J. (1960). *The Chemical Senses in Health and Disease.* Oxford: Blackwell.

KARE, M. R. & HALPERN, B. P. (Eds.) (1961). *Physiological and Behavioural Aspects of Taste.* University of Chicago Press.

KRUT, L. H., PERRIN, M. J. & BRONTE-STEWART, B. (1961). Taste perception in smokers and non-smokers. Smoking and food preferences. *British Medical Journal* **i**, 384–388.

KUEHNER, R. L. (Ed.) (1964). Recent advances in odor: theory, measurement and control. *Annals of the New York Academy of Sciences* **116**, 357–746.

MONCRIEFF, R. W. (1966). *Odour Preferences.* London: Leonard Hill.

MONCRIEFF, R. W. (1967). *The Chemical Senses,* 3rd edn. London: Leonard Hill.

MOULTON, D. G. & BEIDLER, L. M. (1967). Structure and function in the peripheral olfactory system. *Physiological Reviews* **47**, 1–52.

NEGUS, V. (1958). *The Comparative Anatomy and Physiology of the Nose and Paranasal Sinuses.* Edinburgh: Livingstone.

OAKLEY, B. & BENJAMIN, R. M. (1966). Neural mechanisms of taste. *Physiological Reviews* **46**, 173–211.

OHLOFF, G. & THOMAS, A. F. (1971). *Gustation and Olfaction.* New York: Academic Press.

SCHWARTZ, H. G. & WEDDELL, G. (1938). Observations on the pathways transmitting the sense of taste. *Brain* **61**, 99–115

STEVENS, S. S. (Ed.) (1954). *Handbook of Experimental Psychology,* Chap. 29. London: Chapman & Hall.

SUMNER, D. (1962). On testing the sense of smell. *Lancet* **ii**, 895–897.

WRIGHT, R. H. (1968). How animals distinguish odours. *Science Journal* **4**, 57–62.

ZOTTERMAN, Y. (Ed.) (1963). *Olfaction and Taste.* Oxford: Pergamon.

ZOTTERMAN, Y. (Ed.) (1967). Sensory mechanisms. *Progress in Brain Research* **23**, 1–154.

37 Structure of the eye

A horizontal cross section of the eye is shown in Figure 37.1. The eyeball, which is roughly

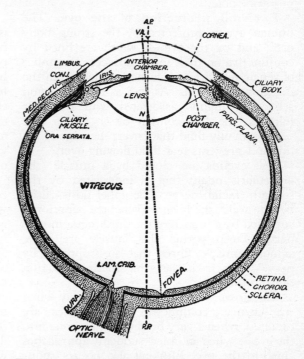

Fig. 37.1 Horizontal section of the eye. P.P., posterior pole; A.P., anterior pole; V.A., visual axis; N, nodal point; Lam. Crib., lamina cribrosa; Conj., conjunctiva. (Modified from Salzmann.) (Wolff's *Anatomy of the Eye and Orbit*.)

spherical and about 24 mm in diameter, has a tough fibrous coat whose opaque posterior part, the sclera, becomes continuous at the limbus with the strongly curved transparent cornea.

Light entering the eye through the cornea passes through the aqueous humour in the anterior chamber, the pupillary aperture of the iris, the lens and the vitreous body before striking the photosensitive retinal layer that lines the posterior two thirds of the eyeball. The retina is separated from the sclera by a vascular pigmented layer, the choroid, that is continued forward into the ciliary body and the iris.

The eye is normally directed so that the image of an object being fixated falls upon the fovea, a depression in the retina situated a short distance to the lateral side of the posterior pole of the eye. Nerve fibres arising in the retina run across the retinal surface to pass out of the eyeball through perforations in the sclera a short distance to the medial side of the posterior pole.

External protection of the eye. The human eye is protected by the strong bony orbit and by the eyelids which are lined by conjunctiva on their inner surfaces. A number of sensitive hairs (eye-lashes) project from the margins of the lids. As well as the sweat and sebaceous glands associated with the hair follicles, the lids contain tarsal glands that secrete an oily fluid that covers the edge of the lids and prevents tear fluid flowing over.

The eyelids are closed by a muscle (the orbicularis oculi) that is innervated by the seventh (facial) nerve. The upper lid is raised by a muscle (the levator palpebrae superioris) supplied by a branch of the third (oculomotor) nerve. Deep within this muscle are some smooth muscle fibres that are supplied by nerve fibres from the sympathetic plexus on the carotid artery.

Lid closure is brought about reflexly when the cornea, conjunctiva or eyelashes are touched, when a very bright light is shone into the eye or when an object suddenly approaches close to the eye. Blinking also occurs spontaneously about twenty times a minute and probably serves to renew the fluid film over the cornea. Since a blink lasts about 300 milliseconds we are normally blacked out for about one tenth of our waking time, although unaware of it. The eyelids can of course be closed voluntarily, although many people have difficulty in closing just one eye. The eyelids normally close during sneezing and sleep.

The fluid that moistens the conjunctiva and cornea is secreted partly by the lacrimal glands which lie in the upper and outer part of the orbit and partly by accessory lacrimal glands on the inner surfaces of the lids. Since tear fluid does not normally accumulate, its rate of secretion is presumably adjusted to compensate exactly for the rate of loss by evaporation.

When foreign bodies or other irritants get into the eye, the discharge of tear fluid from the lacrimal glands into the upper conjunctival sac is greatly accelerated by reflex activation of the parasympathetically innervated glands. If the rate of discharge is so great that drainage of the fluid into the naso-lacrimal duct through the small orifices (puncta lacrimalia) at the medial margin of each lid cannot remove the fluid fast enough, then tears spill over on to the cheek (lacrimation). Weeping may also occur in emotional circumstances.

The tear fluid produced during lacrimation is an isotonic solution of sodium chloride and bicarbonate having a pH of about 7·4 and a low protein content. The fluid, which contains a bactericidal enzyme, lysozyme, serves to wash foreign bodies or irritant materials out of the conjunctival sac. The fluid that normally fills the conjunctival sac and is spread by blinking to form a film over the cornea is rather more viscous, because of evaporation, and has an oily surface layer.

Cornea. The cornea is composed mainly of collagen fibres arranged in lamellae between which lie the fixed cells. Anteriorly the cornea is bounded by a condensed stromal layer that is covered by a stratified epithelium of remarkable regularity. The cornea obtains some of the oxygen required for its metabolism directly from the atmosphere through its anterior epithelium. The anterior surface of the cornea forms the major optical focusing component of the eye; its good optical properties rapidly disappear if it is not continuously moistened with tear fluid. At the edge of the cornea the epithelium continues as the conjunctiva; this covers the anterior part of the sclera and is reflected forwards to line the eyelids. Posteriorly the cornea is bounded by an elastic membrane and an endothelium. The cornea is richly supplied with free nerve endings; even slight damage to the corneal epithelium is intensely painful.

Anterior chamber and aqueous humour. The anterior chamber of the eye is filled with aqueous humour, a fluid whose composition is rather like that of plasma without the proteins. The aqueous humour is continuously being formed by the epithelial glands of the ciliary body (ciliary glands, Fig. 37.2). It flows forwards past the lens into the anterior chamber; from there it drains into the sinus

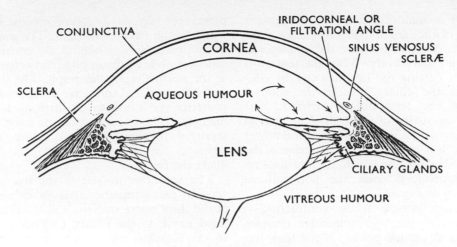

CONJUNCTIVA

CORNEA

IRIDOCORNEAL OR
FILTRATION ANGLE

SINUS VENOSUS
SCLERÆ

SCLERA

AQUEOUS HUMOUR

LENS

CILIARY GLANDS

VITREOUS HUMOUR

Fig. 37.2 To illustrate the probable source of aqueous humour in the ciliary glands and the routes of absorption into the circulation. The ciliary glands lie in the ciliary body which is an annular or ring-like structure. The suspensory ligaments of the lens are shown.

venosus sclerae (canal of Schlemm) and the venous plexus which lies within the sclera at the corneoscleral junction. By introducing inulin into the anterior chamber of the eye of an animal and measuring the rate at which its concentration subsequently falls, it has been estimated that about 1 per cent of the aqueous humour drains away each minute to be replaced by freshly formed fluid. If a dye is introduced into the anterior chamber of the intact eye it can be seen to appear in some of the small surface veins that connect with the intrascleral plexus and which are usually filled, not with blood, but with a clear fluid, presumably aqueous humour. The exact nature of the channels through which the aqueous humour gains access to the intrascleral plexus is not known, although they appear to be located near the iridocorneal angle.

The continuous production and flow of aqueous humour is associated with the existence of a greater pressure within the eyeball than without. Maintenance of this intraocular pressure (normally 10 to 20 mm Hg) is important because it maintains the rigidity of the eyeball. However, it is also important that the intraocular pressure should not rise unduly, as happens if the drainage channels become blocked, because a high intraocular pressure impedes the flow of blood in the retinal and choroidal circulations causing retinal damage (glaucoma).

Although the composition of aqueous humour is not unlike that of an ultrafiltrate of plasma it has been doubted whether the aqueous humour is in fact formed from plasma by simple filtration. The glucose and urea concentrations in the aqueous humour are substantially lower than those in the plasma; the low glucose concentration is probably due to its utilization by the lens and other avascular tissues whose metabolic requirements are provided by the aqueous humour.

The aqueous and vitreous humours, as well as the lens and cornea, all contain quite high concentrations of ascorbic acid (vitamin C). Although this is assumed to have a metabolic function no ocular changes are seen even in severe scurvy (p. 192).

The lens. The lens is composed of ribbon-like fibres arranged in concentric laminae. The cortex of the lens is rather softer than the central nucleus. The substance of the lens has a very high protein content (about 35 per cent) and is enclosed by a strong, elastic, membranous capsule which is attached to the ciliary body by suspensory ligaments (Fig. 37.2). The thickness of the lens capsule is not uniform; the part covering the posterior surface of the lens is uniformly thin, while the part covering the anterior surface is mostly thicker with a thin central portion. The anterior surface of the lens is much less curved than the posterior surface.

Tension is maintained in the suspensory ligaments of the lens by the intraocular pressure which tends to increase the diameter of the annular ciliary body. When the tension is released by cutting the ligaments, or by contraction of the ciliary muscle, the anterior surface of the lens bulges forward. The increased convexity of the central portion of the anterior surface increases the optical power of the lens. It is believed that restriction of the change in shape of the lens to its anterior surface is dependent upon the non-uniform thickness of its capsule.

The ciliary muscle whose contraction is responsible for the accommodative changes (alterations in optical power) of the lens, has been described as having different groups of fibres, though it is probably more realistic to consider the meshwork of interlaced fibres as a single muscle. The main nerve supply to the ciliary muscle is by parasympathetic fibres which run from a mid-brain nucleus together with the third nerve to the ciliary ganglion where they relay to pass in the short ciliary nerves to the eye. Parasympathetic stimulation causes the ciliary muscle to contract so that the eye is accommodated for near vision. There is probably also a certain degree of antagonistic sympathetic control of accommodation although the mechanism of this action is not clear (Fig. 48.13).

Although the lens has no blood supply it is a metabolically active tissue which continues to grow throughout life. Only about one-third of the energy required by the lens is obtained by oxidative metabolism.

As the lens ages it becomes harder, a probable factor in the loss of accommodation that occurs in middle age (presbyopia). Although the human lens is pale yellow from birth, having a pigment that absorbs strongly in the near ultraviolet, the senile lens is even more yellow and may be somewhat cloudy. The senile lens is very liable to become opaque (cataract). The biochemical basis of the senescent changes is not known.

The vitreous body is a transparent jelly-like substance that fills the posterior cavity of the eye. In fact the vitreous body is not simply a structureless protein gel but a tissue with an extensive, though delicate, skeleton of collagen-like fibres.

The iris. The iris is a more or less heavily pigmented screen containing muscle fibres which lies in front of the lens. The outer edge of the iris is hidden behind the corneoscleral junction while the inner edge forms the margin of the normally circular pupil. The iris has a well differentiated sphincter muscle which can constrict the pupil to as little as 1 mm in diameter. The iris also contains some radially arranged myoepithelial cells which lie posterior to the sphincter muscle and help to dilate the pupil.

The sphincter muscle of the iris is innervated by parasympathetic nerves with originate in the third nerve nucleus and travel with the third nerve to the ciliary ganglion (see Fig. 48.13); impulses are carried from the ganglion to the iris by postsynaptic fibres in the short ciliary nerves. Variations in the tonic level of activity in the parasympathetic system provide the major control of pupil diameter.

The dilator muscle (dilatator pupillae) has a sympathetic nerve supply. Preganglionic fibres leave the spinal cord mainly in the first and second thoracic roots and travel via the white rami communicantes into the sympathetic chain and thence up to the superior cervical ganglion. Postganglionic fibres arise in this ganglion and pass in the nerve plexus on the internal carotid artery to the ciliary ganglion. The fibres pass straight through the ciliary ganglion and run to the iris in the short ciliary nerves.

Although the sympathetically innervated dilator muscle normally plays only a small part in controlling pupil diameter, injuries of the cervical cord which interrupt the sympathetic pathway give rise to an obviously small pupil (miosis). Such a lesion which also paralyses the smooth muscle of the levator palpebrae superioris, causes the upper eyelid to droop (ptosis) and the eyeball to retract (enophthalmos). These signs constitute Horner's syndrome.

Effect of drugs on the intraocular muscles. Accommodation of the lens and constriction of the pupil can be produced by placing parasympathomimetic substances such as carbachol and eserine into the conjunctival sac from whence they reach the intraocular muscles by diffusion. Conversely parasympatholytic substances such as atropine cause accommodation to be relaxed and the pupil to dilate. Pupil dilatation can also be produced by activating the sympathetic system or simulating its action with drugs such as amphetamines or phenylephrine. Morphine causes intense pupillary constriction while intoxication by alcohol causes dilatation.

IMAGE FORMATION IN THE EYE

In a lens system light is refracted or bent out of its original direction when it passes from one medium into another with a different refractive index. The refractive indices of the cornea, aqueous humour and vitreous body are all very nearly the same as water (1·33) while the material of the lens has an effective refractive index of 1·42. Thus light is refracted when it passes into the eye through the air–cornea interface and again on entering and leaving the lens (Fig. 37.3).

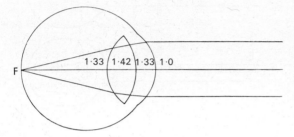

Fig. 37.3 Emmetropic (normal) eye. This diagram shows how the rays of light are refracted where there is a difference of refractive index between any two media. The figures give the approximate refractive indices of the media.

The power of the whole eye when unaccommodated for near vision is about 60 dioptres (the power of a lens in dioptres is the reciprocal of its focal length in metres). Some twenty dioptres of this power is contributed by the lens and forty dioptres by the anterior corneal surface. If refraction at the corneal surface is eliminated, as when swimming under water, the eye becomes very long-sighted and clear vision is impossible. Normal vision can, of course, be achieved underwater by wearing goggles which allow the refractive power of the air–cornea surface to remain intact. If the lens of the eye has to be removed because it becomes opaque a spectacle lens of approximately 10 dioptres must be supplied. The power of the spectacle lens that is needed is less than that of the lens that has been removed because it is placed further away from the retina.

ACCOMMODATION. When the gaze is suddenly transferred from a distant object to one nearer at hand the retinal images in the two eyes are at first both blurred and also disparate, that is they do not fall on corresponding points of the two retinae. After a latent period of about 160 milliseconds however, the axes of the eyes start to converge so as to move the retinal images of the nearer object on to corresponding retinal locations. After a longer latent period (about 360 milliseconds) the power of the lenses starts to increase so as to bring the images of the nearer object into focus. Although these corrections may take about a second to complete, the subject is usually unaware of their occurrence. The mechanism of accommodation is described on p. 732).

In a subject with emmetropia (that is normal vision) the range of accommodation is such that the images of objects at any distance from the eye greater than some minimum can be brought into focus on the retina. The amplitude of accommodation can be simply measured by finding the range of convex and concave (plus and minus) lenses that can be placed in front of the eye without causing a distant object to appear blurred. The amplitude of accommodation diminishes steadily throughout life (Fig. 37.4) until the condition of presbyopia is fully established at the age of fifty or sixty years.

An object that is as close to the eye as it can be without appearing blurred is said to be at the near point. In an emmetrope the near point is about 7 cm from the eye at 10 years of age. By the age of forty the near point has receded to about 20 cm and tasks such as threading a needle may become difficult. When presbyopia is fully established the near point may be 40 cm from the eye and it becomes difficult to read fine print without spectacles.

ERRORS OF REFRACTION. If a subject has normal vision then parallel rays of light entering the eye are brought to a focus on the retina when his accommodation is relaxed naturally or his ciliary muscle is paralysed with atropine. Such a subject is said to be an emmetrope. On the other hand if the subject is myopic, parallel rays are brought to a focus in front of the retina, while if he is hypermetropic (hyperopic, U.S.A.) parallel rays are focused behind the retina (Fig. 37.5).

The refractive error of a hypermetrope, or myope, can be corrected by placing, respec-

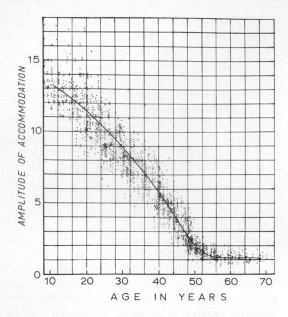

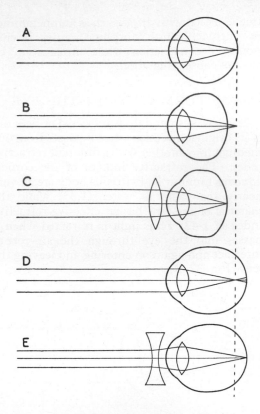

Fig. 37.4 This shows 1050 observations on the amplitude of accommodation to show how it varies with age. The accommodation was determined by rendering the patient emmetropic and measuring the distance of his near point from the anterior focus of the eye, that is from a point 13 mm in front of the cornea. (*After* A. Duane (1912). Studies of the accommodation. *Ophthalmoscope* **10**, 489.)

Fig. 37.5 Diagram illustrating the common refractive errors of the eye and the methods of correction. A, emmetropic eye. B and C, hypermetropic eye. D and E, myopic eye.

tively, converging or diverging lenses in front of the eyes (Fig. 37.5, C and E). The power (plus or minus) of the lens required to bring a distant object into sharp focus is a measure of the subject's refractive error.

Although a myope is unable to see distant objects clearly without spectacles, a hypermetrope can do so by exerting a certain amount of accommodative effort even when looking into the distance. In this latter case a convergent squint may develop because of the very close linkage that exists between convergence of the axes of the eyes and accommodation. In any case an uncorrected hypermetrope may suffer from fatigue and headaches from the continual effort of accommodation.

A refractive error arises when the power of the eye is not well matched to its axial length. Although the powers of the cornea and lens, as well as the length of the eye, show quite wide variations from person to person a suprisingly large proportion of the population have an insignificantly small overall refractive error (Fig. 37.6). This suggests that the three

variables are not randomly associated but must be partially correlated.

If often happens that the curvatures of the refracting surfaces of the eye are different along different meridians. This results in the focal length of the eye being different in different meridional planes (that is in different planes through the visual axis of the eye). If the shortest and longest focal lengths occur in meridional planes that are at right angles to each other, and if the focal length changes smoothly between these values, then the subject is said to have a regular astigmatism. When a subject with a regular astigmatism views a fan of radiating lines some appear sharply focused while those at right angles are blurred. The error can be corrected by providing a cylindrical spectacle lens of the correct power with its axis set at such an angle as to equalize the

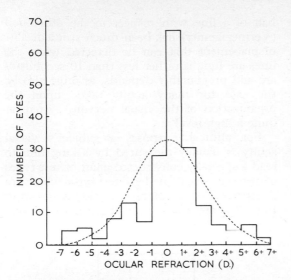

Fig. 37.6 The curve of distribution for ocular refraction in 194 eyes. The continuous lines indicate the actual observations while the dotted lines indicate the fitted normal curve. The fact that the former is more peaked than the latter suggests that, while the optical components vary over a wide range, in an individual eye these components are usually so correlated as to produce emmetropia. A. Sorsby, B. Benjamin, J. B. Davey, M. Sheridan & J. M. Tanner (1957). *Special Report Series. Medical Research Council. No. 293,* Fig. 8.) (*By kind permission of the Controller of H.M. Stationery Office.*)

focal lengths in the different meridional planes. If a spherical correction is also required it is usually formed on one side of the spectacle lens while the cylindrical surface is formed on the other.

If the corneal surface becomes irregularly distorted, due for example to healed ulceration, then an irregular astigmatism is present. In this case the only means of correction is a 'contact lens' which effectively replaces the irregular corneal surface by an optically regular one.

The magnitude of a subject's refractive error can be measured by placing lenses of different power in front of his eyes until the best visual acuity (p. 736) is obtained. Where difficulties of age or language prevent the co-operation of the patient, the refractive error can be estimated by retinoscopy. In this technique the examiner shines a light into the subject's eye and then checks whether the light reflected back from the fundus leaves the eye in a parallel beam.

THE PUPIL. If the image produced by an optical system is not correctly focused it is blurred. The extent of the blurring depends upon the magnitude of the focusing error and the diameter of the entrance pupil. As the diameter of the pupil is made smaller the effect of defocusing becomes less, the optical system being said to have a greater depth of focus. However, even when it is in best focus, the image formed by an optical system is usually somewhat degraded by spherical and chromatic aberrations and the effects of diffraction at the entrance pupil.

Spherical aberration is due to the central and peripheral parts of the lens system having different focal lengths while chromatic aberration arises because the focal length is different for different colours. The amount of blurring produced by these defects can be reduced by having a small diameter pupil so that only the central portion of the lens is used. Reducing the pupil size, however, reduces the amount of light reaching the retina and increases the blurring caused by diffraction. Since the resolving power of the nervous part of the visual system diminishes as the retinal light intensity is reduced there is, for every external light intensity, an optimum pupil size at which these various factors most nearly balance each other. It is found that over a wide range of light levels the natural diameter of the pupil of the eye corresponds closely to the optimum size.

The diameters of both pupils are normally the same and depend upon the amount of light falling on both retinae. Thus shining a light into one eye not only causes contraction of the pupil in that eye (direct light reflex) but also an equal contraction of the pupil of the other eye (consensual reflex). The pupil light reflex has a latent period of about 250 milliseconds and the contraction may not be complete for several seconds. Dilatation of the pupil in response to reduction of the light entering the eye may be even slower.

Contraction of the pupils also occurs when a subject looks at a near object (the near reflex). The increase in the depth of focus produced in this way probably helps to minimize the blurring resulting from the slight under-accommodation usual in near vision.

In certain diseases of the nervous system the light reflex is abolished although the pupil still contracts on accommodation. The lesion is probably in the pretectal area; in animals destruction of the superior colliculus alone does not abolish the light reflex although destruction of the pretectal area does.

The pupils contract during sleep, although they normally dilate if the eyes are closed. Dilatation of the pupils occurs in conditions such as fear, pain or asphyxia, presumably as part of the generalized sympathetic response.

VISUAL ACUITY

Visual acuity is the appreciation of fine spatial detail in a visual image. If a subject with normal vision looks at a well lit pattern of parallel light and dark bars of equal width he can distinguish the individual bars only if they subtend an angle at his eye of at least half a minute of arc. A visual acuity as high as this is only achieved over a small region of the subject's visual field around his fixation point and only when the level of illumination is high. Only a short way away from the fixation point the acuity is very much less, and even in the centre of the visual field visual acuity is reduced if the light is dimmed. The visual acuity of a subject is also much reduced if the contrast of the pattern is reduced, that is if the luminances of the light and dark regions of the pattern are made more nearly equal.

It is not surprising that the ultimate limit of 'grating' acuity should be reached with bars about half a minute of arc wide because this is approximately the angle subtended by the central foveal cones which are 2 μm in diameter. However, the effect of diffraction and aberrations in the lens system is to reduce the contrast of the detail of the retinal image below that of the object and it has been shown that this reduction in image contrast plays some part in determining the ultimate acuity. Optical factors (and these include the size of the elements of the receptor mosaic) are not alone in determining visual acuity. The reduction in acuity associated with lower contrast and lower levels of illumination, as well as that associated with peripheral vision, depend upon the nervous mechanisms involved.

Resolution of the separate elements of a pattern is not the only kind of task that taxes to the limit the ability of the visual system to detect spatial differences. The ability of a subject to detect the lateral displacement of one half of a line with respect to the other half (Vernier acuity) has been much studied. The displacement that can be detected when the lines are long is rather less than 10 seconds of arc and presumably depends, as must also be the case for many acuity tasks, upon the organization of the visual nervous system at quite a high level.

For clinical purposes a subject's visual acuity is usually measured by asking him to read a specially constructed chart of test types of different sizes. Snellens' test types which are constructed on a rectangular grid (as shown in Fig. 37.7) are widely used. Each line of letters

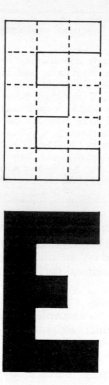

Fig. 37.7 The method of construction of Snellen's types.

of a different size is marked with the distance in metres at which the small squares subtend one minute of arc. A normal subject standing 6 metres from a well-lit chart should easily be able to distinguish the letters of the 6 metre line. If this is so his acuity is reported as 6/6. A subject with defective vision might only be able to read the 60 metre line; his acuity would be reported as 6/60. If the low acuity is due to a refractive error it will be raised by making the subject look through a small aperture which increases the depth of focus and reduces the amount of blurring produced by any error in focusing or by aberrations.

FIELD OF VISION

The visual field is that area of the inner surface of a sphere around the subject within which a luminous object can be seen. The extent of a subject's field of vision is measured with a perimeter. The subject covers one eye and looks with the other at a fixation point at the centre of the apparatus. A small white disc is brought from the periphery towards the fixation point until the subject indicates that he can see it. The angular distance of the disc from the visual axis is recorded and the measurement is repeated for another meridian. The shape of the field is dependent to some extent upon the shape of the face since it is restricted on the medial side by the nose and above by the supra-orbital margin (Fig. 37.8). Laterally there is no obstruction; with the eyes looking straight ahead objects can be seen up to about 100° away from the visual axis on each side. If the eyes are moved left and right the total field of vision around the head becomes quite considerable with a blind zone behind the head of as little as 130°.

Although an object may be detected when it is far from the centre of the visual field, especially if it is moving or flashing, visual acuity and colour recognition may be so poor that the object cannot be recognized. A subject's inability to recognize an object, although he is aware of its position, is in normal circumstances overcome by turning the eyes so that the image of the object falls upon the fovea. Flashing lights, or moving objects appearing in the periphery of the visual field, have a strong tendency to cause the appropriate involuntary changes of fixation.

If a small object is moved within the visual field until its image falls upon the optic nerve head, where there are no photo-receptors, it is no longer visible. Thus the normal visual field of each eye has a 'blind spot' or scotoma about 5 or 6° in diameter situated some 15° lateral to the fixation point. Subjects are usually quite unaware of the existence of this blind area as they may also be unaware of other scotomata resulting, for example, from local retinal damage.

THE RETINA

The retina is formed from the optic vesicle which grows out from the fore-brain of the embryo (Fig. 37.9) and is thus a part of the central rather than the peripheral nervous system. The outer wall of the optic vesicle

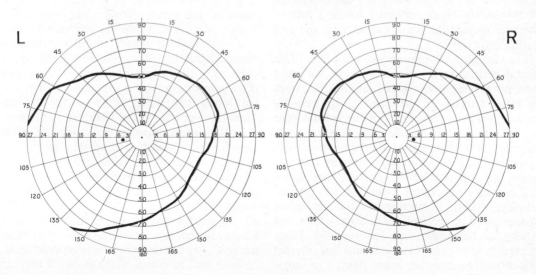

Fig. 37.8 The fields of vision. The black spot 15° lateral to the centre of the field is the blind spot.

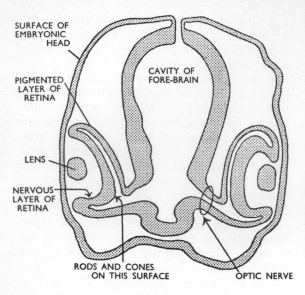

SURFACE OF
EMBRYONIC
HEAD

PIGMENTED
LAYER OF
RETINA

CAVITY OF
FORE-BRAIN

LENS

NERVOUS
LAYER OF
RETINA

RODS AND CONES
ON THIS SURFACE

OPTIC NERVE

Fig. 37.9 Schematic section of the embryonic head to show how the eye is developed in relation to the brain. (*After Gray's Anatomy.*)

forms the pigmented layer of the retina while the photoreceptors and nervous layers of the retina are formed from the inner wall. In the adult eye (Fig. 37.1) the retina lines the posterior part of the eyeball extending as far forward as the ora serrata, about 5 mm in front of the equator of the eyeball. Near the centre of the retina, a millimetre or so from the posterior pole of the eye, there is a depression in the retina, the fovea centralis. About 3 mm to the nasal side of the fovea the sclera is pierced by the optic nerve.

The retina has two blood supplies. The central retinal artery enters the eye with the optic nerve and spreads its branches out over the inner surface of the retina to nourish its inner layers; the tributaries of the retinal vein run alongside the arteries, the retinal vein passing out of the eye in the optic nerve. A separate choroidal system of vessels lies between the pigment epithelium and the sclera and supplies the outer layers of the retina. Damage to either vascular system can cause blindness. The retinal circulation does not appear to be under nervous control. High carbon dioxide concentrations and low oxygen concentrations produce some degree of vasodilatation. High arterial oxygen tension causes vasoconstriction.

The structures inside the eye (Plate 37.1) can be examined by the use of an ophthalmoscope. A beam of light is directed through the subject's pupil to illuminate the back of the eye (fundus oculi). The examiner looks along the beam of light into the subject's eye and if both his eye and the subject's eye are focused for infinity the retinal vessels and optic nerve head (optic disc) can be clearly seen but the retina itself, being nearly transparent, is not visible. The pigment layer of the retina hides the choroidal vessels so that the background of the fundus is a fairly uniform orange colour. The yellow colour of the macula lutea cannot usually be distinguished although its position, about two disc diameters to the temporal side of the optic disc, can be inferred from the absence of large blood vessels. The nerve fibres leaving the eye become myelinated, and therefore visible, only as they reach the optic disc.

FINE STRUCTURE OF THE RETINA. The actual photosensitive cells, the rods and cones (Figs. 37.10 and 37.11) lie in the outermost layer of the retina next to the pigment epithelium. Except in the foveal region, light reaching the photoreceptors must first pass through the other layers of the retina. The fovea is a small depression in the retina formed by a lateral displacement of the cells of the inner retinal layers. Over most of the retina, rods and cones are found side by side, although in the central fovea there are only very closely packed (rod-shaped) cones while in the most peripheral parts of the retina, where the receptors are widely separated, there are very few cones.

The rods and cones make synaptic connexion with bipolar cells and these in turn connect with the ganglion cells. The ganglion cells give rise to the optic nerve fibres which run towards the optic nerve head as the innermost layer of the retina. Ganglion cells, and to a lesser extent bipolar cells, can have quite extensively spreading dendrites so that each ganglion cell may be influenced by the activity of a large number of rods and cones. Further lateral interactions are mediated by the horizontal and amacrine cells (see Fig. 37.11). In all regions of the retina, except the central fovea, the ganglion cells can be seen to be connected, through intermediate bipolar cells, to rods as well as cones. Over most of the retina there are many more photoreceptors than ganglion cells although in the fovea the ratio of ganglion cells to cones may be more nearly one; however, this does not necessarily mean that even in the fovea each nerve fibre is connected to only a single cone.

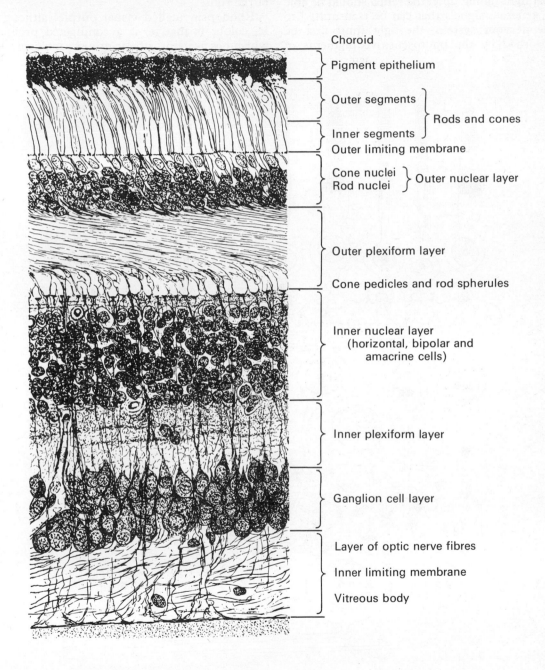

Fig. 37.10 Cross-section of adult human retina in the perifoveal region to show the layers. (S. Polyak (1941). *The Retina*, Fig. 37. University of Chicago Press.) Light traverses the retina to the rods and cones through all the other layers (upwards in the figure).

PHOTOPIGMENTS AND VISION

RHODOPSIN AND SCOTOPIC VISION. In order that light falling upon the retina should be able to generate signals that can be transmitted to the nervous system, the light must first be absorbed by the photosensitive pigments of the rods and cones. The first photopigment to be isolated from the retina was rhodopsin which was extracted from the outer segments of the rods.

Rhodopsin (called visual purple, although its colour is mauve) is a conjugated protein which has as its prosthetic group a molecule of

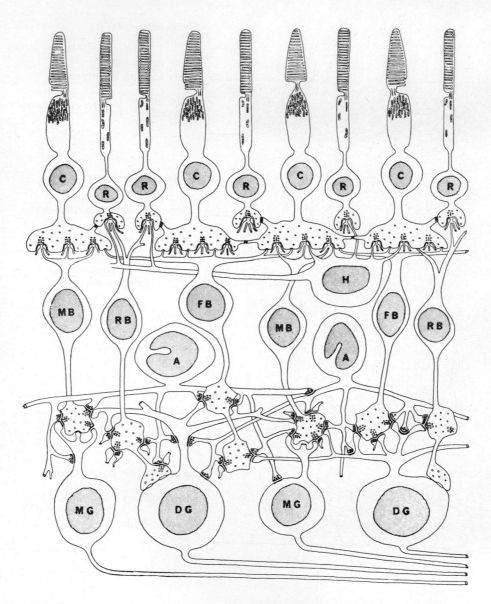

Fig. 37.11 A summary diagram of the contacts in the retina. R, rod; C, cone; MB, midget bipolar; RB, rod bipolar; FB, flat bipolar; H, horizontal cell; A, amacrine cell; MG, midget ganglion; DG, diffuse ganglion. (J. E. Dowling & B. B. Boycott (1966). *Proceedings of the Royal Society* B **166**, 80–111).

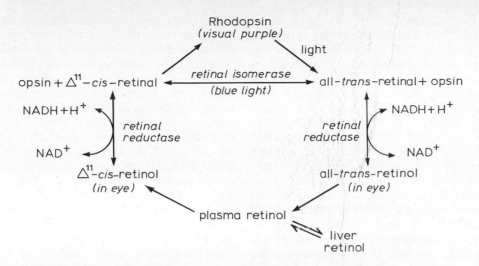

Fig. 37.12 The visual cycle. The *trans-* form of vitamin A (retinol), which predominates in the liver reserves and the blood, is the most active isomer in all physiological roles of the vitamin other than that of vision. For vitamin A to give rise to a visual pigment, it must be converted in the eye to a *cis-* isomer of retinal by the enzymes retinal reductase and retinal isomerase. The *cis*-retinal combines with the protein opsin to form rhodopsin. Light falling on the retina splits retinal from the protein opsin by isomerizing it to the *trans-* configuration. The liberated *trans*-retinal may then be either changed back to the *cis-* form by retinal isomerase to regenerate rhodopsin, completing an isomerization cycle in the eye or reduced to vitamin A by retinal reductase.

retinal (vitamin A aldehyde) present in the form of its 11-*cis* stereoisomer. It has been found that when a molecule of rhodopsin absorbs a quantum of light the retinal moiety undergoes a stereo-isomeric change to the all-*trans* form (Fig. 37.12). It is believed that it is this change that somehow results in the generation of a nervous signal (p. 811). Subsequently the retinal splits from the opsin; the free retinal and opsin no longer absorb visible light and the rhodopsin is said to be bleached. In the presence of the appropriate enzyme systems the retinal and opsin are reconjugated to regenerate rhodopsin. In cases of chronic vitamin A deficiency rhodopsin synthesis is impaired, dark-adaptation (p. 747) may be slow and incomplete and in extreme cases overt retinal degeneration may occur (p. 184).

The relative ease with which solutions of extracted rhodopsin can be bleached by lights of different wavelengths can easily be measured to provide the 'action spectrum' of rhodopsin. It is important to distinguish this 'action spectrum', which relates to the effect of light upon the pigment, from the 'absorption spectrum' which simply relates to the amount

of light that is absorbed without consideration of any change that it produces. Although the absorption and action spectra of rhodopsin come from quite different kinds of measurement, they are, in fact, very similar in shape because each quantum of light that is absorbed has the same chance of causing a pigment molecule to be bleached.

Rhodopsin in solution is bleached by lights with a range of wavelengths (λ) corresponding to the orange to violet part of the visible spectrum (Fig. 37.13) with an optimum in the

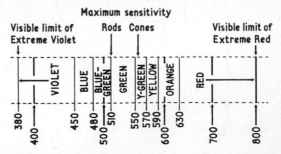

Fig. 37.13 The colour names given to the various parts of the spectrum. Wavelengths are in nm. Compare with the spectra shown in Plate 25.1.

green ($\lambda = 502$ nm). Rhodopsin also absorbs, and can be bleached by ultraviolet radiation.

The bleaching of photopigments in the living eye can be demonstrated by an ophthalmoscopic technique. A weak monochromatic light is shone on the retina through one half of the pupil. A fraction of this light, after passing through the retina, is reflected back from the pigment epithelium and choroid, passes back through the retina and leaves the eye through the other half of the pupil. The emerging light is allowed to fall on a sensitive photocell so that its intensity can be measured. If measurements of the intensity of the reflected light are made before and after the retinal photopigments are bleached by exposing them to a very bright light, then the fraction of the weak measuring light that returns through the pupil is found to be greater after bleaching, since less is absorbed. The increase in the fraction of light reflected, and the way in which it varies when the wavelength of the measuring or bleaching light is altered, can be used to investigate the absorption and action spectra of the photosensitive pigments of the retina. This technique has been used to show that the absorption spectrum of the photopigment in the peripheral retina is the same (due allowance being made for some absorption of light by the ocular media) as that of rhodopsin in solution.

When measurements are made of the relative abilities of low intensity lights of different wavelengths to produce visual sensation, it is found that the curve obtained (the equal energy scotopic luminosity curve, Fig. 37.14 (a)) is very similar to the action spectrum of rhodopsin. This, and the fact that those parts of the retina where the rhodopsin-filled rods are most closely packed function best at low light levels, has led to the belief that rhodopsin is the photosensitive pigment responsible for vision at low luminance levels (scotopic vision).

PHOTOPIC VISION. When a subject views a scene illuminated by white light of low intensity his visual acuity is low and he is unaware of any sensation of colour. However, when the illumination level is sufficiently raised objects can be seen in much greater detail and in their natural colours (Fig. 37.15). Vision at high luminance levels (photopic vision) appears to be mediated by retinal cones (high acuity being achieved by using the densely packed mosaic of cones in the fovea) while scotopic vision is mediated by peripherally sited rods.

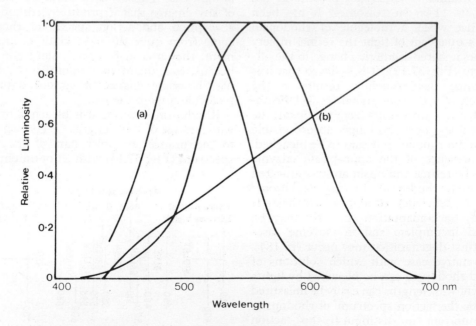

Fig. 37.14   The equal energy luminosity curve (a) for rod vision, as measured at low intensities in the extrafoveal area of the retina; (b) for cone vision, as measured at high intensities at the fovea (W. D. Wright (1944). *The Measurement of Colour*. London: Hilger).

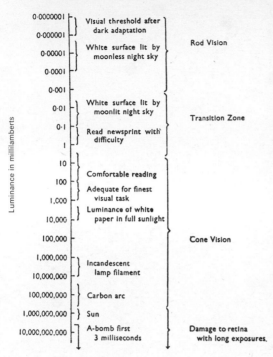

Luminance in millilamberts

0·0000001	} Visual threshold after dark adaptation } Rod Vision
0·000001	
0·00001	} White surface lit by moonless night sky
0·0001	
0·001	
0·01	} White surface lit by moonlit night sky } Transition Zone
0·1	
1	} Read newsprint with difficulty
10	}
100	Comfortable reading
1,000	Adequate for finest visual task
10,000	} Luminance of white paper in full sunlight
100,000	Cone Vision
1,000,000	} Incandescent lamp filament
10,000,000	
100,000,000	} Carbon arc
1,000,000,000	} Sun
10,000,000,000	} A-bomb first 3 milliseconds } Damage to retina with long exposures.

Fig. 37.15 Range of luminance to which the eye may be subjected with indications of the receptive mechanisms involved. (*By courtesy of F. W. Campbell.*)

Until recently, knowledge of the photosensitive pigments responsible for photopic vision had been indirectly obtained. However, measurements have now been made of the absorption spectra of the outer segments of a number of cones from primate and human retinae (Fig. 37.16). These measurements are technically extremely difficult to make and the results do not have high precision. However, it is quite clear that cones do not all contain the same pigment. The measured absorption spectra fall into three separate groups probably corresponding to the presence of three different photopigments. None of the cones has an absorption spectrum like that of the rod pigment rhodopsin (with a maximum absorption at about 500 nm).

The absorption spectra of two photopigments present in the human foveal cones have also been measured in the living eye by the ophthalmoscopic technique described earlier

(p. 742). The two foveal cone pigments detected in this way by Rushton have been called 'erythrolabe' (the more red sensitive pigment) and 'chlorolabe' (the more green sensitive one). The absorption spectra of these two pigments correspond to the two commoner types found in isolated cones (Fig. 37.16). Cones containing the blue-sensitive 'cyanolabe' seem to be relatively scarce and probably completely absent from the central fovea.

It has not proved possible to extract cone pigments from isolated mammalian retinae and the chemical nature of these pigments is therefore unknown. However, it is believed that, like rhodopsin, the cone pigments are conjugated proteins but that they differ from rhodopsin in the exact structure of the protein to which the retinal is bound.

When the relative energies of high intensity lights of different wavelengths needed to give a particular sensation of brightness are measured (Fig. 37.14 (b)), it is found that a yellow-green light of wavelength 560 nm is most effective. This wavelength lies between the absorption maxima of the cone pigments erythrolabe and chlorolabe (540 and 570 nm) which are probably largely responsible for determining the brightness of lights at the photopic level. If the equal energy luminosity curves of photopic and scotopic vision are compared (Fig. 37.14) it is evident that the whole scotopic sensitivity curve is displaced towards shorter wavelengths. The displacement of the peak scotopic sensitivity towards the blue is known as the 'Purkinje shift'.

ULTRAVIOLET. It is interesting to note that ultraviolet radiation (u.v.) is normally invisible not because the retinal photopigments are insensitive to u.v. but because the lens contains a yellow pigment which prevents the u.v. from reaching the retina. People who have had their lenses removed (because they have become opaque) can see u.v. quite well. Normal people may be aware of ultraviolet radiation as a vague haze because the lens pigment fluoresces strongly.

COLOUR VISION

TRICHROMACY. The quality of the sensation produced by a light of moderate or high intensity is dependent upon its spectral composition.

Structure of the Eye   743

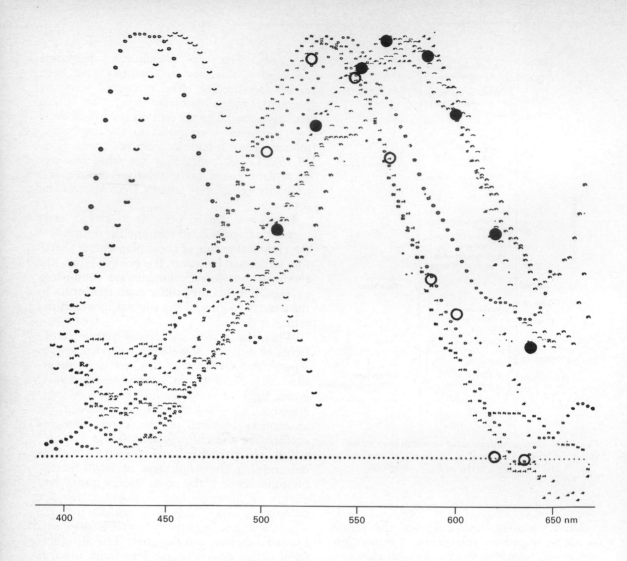

Fig. 37.16 Absorption spectra of outer segments of isolated human and primate cones (small symbols). (W. B. Marks, W. H. Dobelle & E. F. MacNichol (1964). *Science* **143**, 1181–1182.) Circles are spectra of erythrolabe (filled) and chlorolabe (open) in the living human eye (H. D. Baker & W. A. H. Rushton (1965). *Journal of Physiology*, **176**, 56–72). All curves are scaled to same maxima. Abscissa graduated in nm.

However, the spectral composition does not uniquely determine the sensation because different mixtures of monochromatic lights may appear identical. For two light stimuli to be indistinguishable it is only necessary for them to be alike in the values of three suitably chosen independent quantities. This threefold nature of colour vision is known as trichromacy.

The trichromacy of colour vision may be illustrated by the familiar fact that the stage lighting engineer, provided with red, green and blue lights, can, by mixing them in suitable proportions, reproduce almost any other colour. In fact, although any pale or unsaturated colour can be reproduced exactly, it is only possible to match approximately some pure spectral, that is 'saturated' colours. ('Satura-

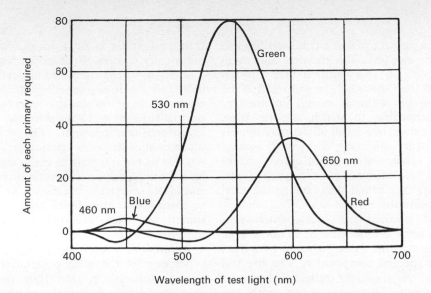

Fig. 37.17 The amounts of red (wavelength 650 nm), green (530 nm) and blue (460 nm) primaries required to match a constant quantity (measured in watts per unit solid angle) of a test light of variable wavelength. Where the ordinate of any curve is negative, this means that the corresponding primary must be added to the test light in order to match a mixture of the other two primaries. (*After* W. D. Wright (1928). *Transactions of the Optical Society* 30, **141**. *From* A. F. Rawdon-Smith (1938). *Theories of Sensation.* Cambridge University Press.)

tion' is the technical term for freedom of a colour from admixed white.) An exact match can always be obtained, however, if one of the three primary colours is added not to the other two, but to the test colour which it is required to match. For example, spectral yellow can nearly be matched with a mixture of red and green but, to perfect the match, a little blue must be added to the yellow to make it slightly paler. If the addition of a light to one of the two fields on which the matching is being carried out is by convention represented as equivalent to its subtraction from the other, then the amounts, positive or negative, of three specified primaries required to match a given light provide a set of three quantities which define uniquely the sensation which it produces. Figure 37.17 shows the result of measuring these quantities for monochromatic lights of all spectral wavelengths, when red, green and blue primaries are used. The primaries need not necessarily be red, green and blue, but these have the advantage as primaries that the quantities of them required

to match other colours are mostly positive, and when negative, are small.

BASIS OF TRICHROMATIC VISION. It is now generally accepted that the trichromatic nature of normal colour vision is determined by the existence in the retina of three kinds of photosensitive pigment molecules segregated in three different groups of cones (p. 743). If there are just three different pigments, and if the nature of the effect on a pigment molecule of absorbing a quantum of light is independent of its wavelength, which it is, then the trichromacy of normal colour vision is an inevitable consequence. Since the degrees of excitation of the three kinds of cone are the only independent variables, the sensation produced by a visual stimulus must be capable of description in terms of three quantities.

This is not to say that the perceived colour of a visual stimulus is entirely determined by its spectral composition. It is well known that the appearance of a coloured object is dependent upon the visual environment in which it is seen and the recent visual experience of the subject. If a subject places a coloured filter in front of one eye for a few minutes and then, after removing it, looks around first through one eye and then the other, he will see that the apparent colours of things around him are very different according to which eye he uses. Again, the colour of a small object may appear quite markedly different if viewed against different, large, strongly-coloured backgrounds. These effects do not disturb colour matches however much they change the subjective appearance.

It should be noted that objects which seem to have the same colour when lit by one source of light may well not match when lit by a light of different spectral composition. The spectral composition of the light reflected from an object depends, of course, upon both the illumination and the spectral variation in reflectivity of the object.

COLOUR BLINDNESS. Abnormalities of colour vision of genetic origin, generally known as colour blindness, are common in men (about 8 per cent) but much less common in women (0·4 per cent). The common abnormalities are inherited as sex-linked recessive characters.

Colour-blind subjects fall into a number of fairly sharply defined categories. Monochromats are quite unable to distinguish colours at all: a monochromat can match any two lights simply by adjusting their intensities. Monochromats are of two kinds. Rod monochromats have a luminosity curve like that of normal dark-adapted subjects; they see very poorly in bright surroundings and are presumed to lack functional cone mechanisms. Cone monochromats, on the other hand, appear to lack the rod mechanism: although their vision is more or less normal in bright surroundings, they see very badly when the illumination is reduced to scotopic levels. Monochromatism is rare, cone monochromatism exceptionally so.

Dichromats, who can match all colours with suitable mixtures of two primaries, are of three kinds. Protanopes and deuteranopes (each about 1 per cent of males) are often grouped together with the anomalous trichromats as 'red–green' blind. They have very little ability to discriminate colours at the red end of the spectrum, and thus confuse red, brown and green objects, though they can usually distinguish yellow objects by their higher reflecting power. Protanopes and deuteranopes differ from each other in the form of their photopic luminosity curves. Protanopes are relatively insensitive to red light and appear to lack the more red-sensitive pigment erythrolabe. Deuteranopes have luminosity curves similar to normal subjects and appear to have none of the green-sensitive pigment. The third kind of dichromatic vision is tritanopia. Tritanopes, who are as often female as male, are rare. They have normal colour discrimination at the red end of the spectrum, but they have little ability to distinguish blue from green. Tritanopes, sometimes called 'blue-blind', appear to lack the blue-sensitive pigment cyanolabe.

Anomalous trichromats, comprising nearly 6 per cent of the male population, resemble normal subjects in that they require three primaries to match all colours by colour mixture, but they require them in abnormal proportions. The cause of anomalous trichromacy is unknown: it has been suggested that anomalous trichromats may have abnormal pigments.

Colour-blind subjects, even when their colour-discrimination is very poor, are often unaware of their defect. In familiar situations they compensate for their defective colour-discrimination by an increased use of alternative clues, based in part upon prior knowledge of the usual colours of objects which they recognize by their shape. It may only be on rare occasions, when these clues are absent, that the defect becomes apparent.

The detection of colour-blindness may be important in selecting people for jobs in which it is necessary to be able to distinguish coloured markings or coloured light signals. A convenient quick test consists of a set of 'pseudo-isochromatic plates', of which Ishihara's are probably most widely used. Each plate is a card on which an array of multi-coloured dots is printed, so that a letter or figure is formed by dots of one colour, other colours forming the background. Some cards are designed to be read easily by the normal, but not the colour-blind, subject while others can be read only by the colour blind subject; some plates are interpreted differently by normal and colour-blind. This test is very efficient at separating the normal from the abnormal, but it does not distinguish well between different types of abnormality. To decide whether a subject with a mild abnormality can safely be employed in a particular occupation, a special test designed to imitate the task that has to be performed is often used.

Though hereditary colour-blindness is very much commoner, defects of colour vision can also be acquired as a result of disease of the retina, optic nerve or visual cortex. These acquired defects are usually accompanied by severe defects of visual acuity or of visual fields.

DARK-ADAPTATION

It is a familiar fact that immediately after going out from a brightly lit house on a dark night one sees badly but, after some time in darkness, one can see many objects which were at first invisible. The increase in visual sensitivity (fall in threshold) which occurs while the eyes are in darkness or near-darkness is called dark-adaptation, and the decrease in sensitivity caused by exposure to bright light is called light-adaptation. In subjects with chronic vitamin A deficiency dark-adaptation occurs more slowly and the ultimate sensitivity is less.

The following simple experiments indicate that light- and dark-adaptation are probably properties of the retina and not of the central nervous pathway: previous exposure of one eye to light or to darkness has no effect upon the threshold for stimuli presented to the other eye; if an eye is made temporarily blind by pressing it sufficiently firmly to stop its blood supply, light which falls on it while it is blind raises the threshold for stimuli which are presented after it has recovered from the pressure-blindness. In any case the phenomena of light- and dark-adaptation have been shown to be characteristic of retinal ganglion cells (p. 749).

The course of dark-adaptation after the eye has been adapted to a bright white light is shown in Fig. 37.18. Thresholds were tested with 1° circular fields of various colours placed 5° away from the fixation point. It is clear that the curve for orange test stimuli falls into two phases separated by a sudden change in gradient. With slightly different conditions of pre-adaptation a similar two-phase dark-adaptation curve can be obtained with white, yellow, green or blue test stimuli but, if the test stimulus is either deep red (to which the rods are very insensitive), or falls upon the fovea (where there are no rods), the second

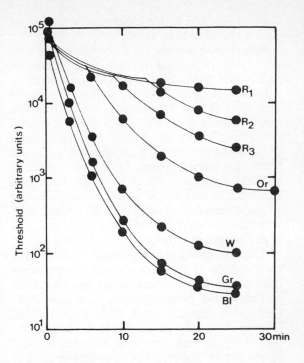

Fig. 37.18 Course of dark–adaptation for 1° circular fields of various colours, placed 5° from the fixation point. Wavebands were isolated by means of filters. Bl = blue, Gr = green, W = white, Or = orange; R_3, R_2 and R_1 are reds of successively longer wavelength. (From G. S. Brindley (1970). *Physiology of the Retina and Visual Pathway*. London: Arnold.)

phase is absent. Orange, yellow, green or blue test stimuli look coloured, however dim they seem, during the first phase but look white at threshold during the second phase. Deep red and foveal stimuli always appear coloured if they are visible at all. Vision during the first phase of dark-adaptation is attributed to cones and during the second phase to rods.

It appears, then, that both the rod and cone mechanisms, are rendered less sensitive by light-adaptation, but that the sensitivity of the rod mechanism is lowered by a much larger factor, so that for the first few minutes of dark-adaptation the threshold is determined by cones. At the end of this time the sensitivity of the cone mechanism is increasing only slowly but the sensitivity of the rod mechanism (which is inherently greater) is increasing rapidly. Thus, when the rods overtake the cones, there is a sudden increase in the rate at which the threshold falls.

When dark-adaptation is complete, the sensitivity of the retina is such that a flash of blue-green light from a field whose diameter subtends 10 seconds of arc at the eye may be seen if as few as 50 quanta of light enter the eye. Probably no more than 5 of these quanta are absorbed by rhodopsin molecules, and these absorptions must in general occur in different photoreceptors, since a 10 seconds of arc field illuminates about 350 rods. Thus each rod must be capable of being stimulated by a single quantum, though only if several such stimulations occur within a small time and area is any sensation produced.

There is strong circumstantial evidence that the decrease in sensitivity which accompanies light-adaptation depends upon the bleaching of the retinal photopigments and that dark-adaptation depends upon their regeneration. During dark-adaptation the logarithm of the threshold of both the cone and rod mechanisms can be shown to be inversely related to the concentration of the appropriate photopigment present. This relation rules out the simple idea that the sensitivity at any time is determined by the amount of photopigment available for catching quanta. The mechanism by which sensitivity and photopigment concentration are related must be considered as still unknown.

NEUROPHYSIOLOGY OF VISION

ELECTRICAL ACTIVITY IN THE RETINA. When light is absorbed by the photopigment in a retinal rod or cone it causes the cell membrane to become hyperpolarized. Although the mechanism underlying this effect is not yet known with certainty it seems likely that the action of light is to decrease the permeability of the cell membrane to sodium ions. It is possible that this change in permeability could be directly related to the light-induced change in the photopigment structure (p. 741) since it is now generally supposed that the photopigment is a constituent of the membrane of the outer segment of the photoreceptor cells (rods and cones).

The potential change generated in a photoreceptor outer segment is transmitted electrotonically to synapses on both bipolar and horizontal cells (Fig. 37.11). The bipolar cells in turn transmit a graded electrical signal to the ganglion and amacrine cells with which they are in contact, while the horizontal cells seem to be responsible for spreading an inhibitory effect to surrounding photoreceptors. In the ganglion cells are generated the action potentials which pass along the fibres of the optic nerve and tract to the lateral geniculate nucleus and the superior colliculus.

ELECTRORETINOGRAM. If the potential difference that exists between an electrode placed on the cornea and another on a remote part of the body is amplified and recorded it is found to change when the eye is illuminated. This response is the electroretinogram (E.R.G.). Recorded in this way the E.R.G. probably largely represents the summation over the whole eye of the electrical activity of bipolar cells. The human E.R.G. is of some diagnostic use; for example, in certain hereditary disorders of the retina (retinitis pigmentosa) which ultimately cause blindness, the E.R.G. disappears before there is much visual impairment.

RETINAL GANGLION CELLS. As a result of the extensive interconnexions that are formed between the cells of the retina, every ganglion cell responds to the pattern of illumination falling on a considerable area of the retina. Although in some mammals (for example rabbit and squirrel) some ganglion cells respond specifically to retinal images moving in a particular direction across the receptive field, the behaviour of all ganglion cells in the primate retina is less elaborate. In the cat and monkey (and probably in man) retinal ganglion cells have roughly circular receptive fields which are divided into functionally distinct concentric 'centre' and 'surround' regions. Light falling in one of these regions excites the ganglion cell, causing the frequency of its discharge to increase, while light falling in the other region has an inhibitory effect, causing the discharge frequency to decrease. As a consequence of this antagonism, light falling uniformly over the whole receptive field has little effect upon the activity of the ganglion cell. This is very clearly demonstrated in Fig. 37.19 which shows the responses of an 'off-centre' ganglion cell, that is a cell whose discharge is inhibited by light falling on the centre of its receptive field. It is also clear from

Fig. 37.19 that the response to a patterned stimulus can be very well maintained for many minutes, despite the initial adaptation which occurs. Some ganglion cells, particularly those

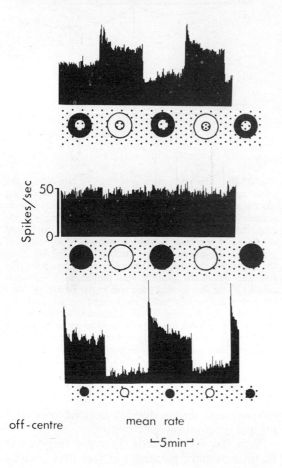

off-centre   mean rate

⊢5min⊣

Fig. 37.19 Responses of an off-centre ganglion cell in a cat retina to switching off and on either an annulus (top record), a large spot (middle record) or a small spot (bottom record). The small spot illuminated only the centre of the receptive field, the annulus only the surround, while the large spot stimulated both centre and surround. (P. O. Bishop & R. W. Rodieck (1965). *Symposium on Information Processing in Sight Sensory Systems.* California Institute of Technology.)

in the peripheral retina, have much lower discharge frequencies when they are uniformly illuminated, respond with a transient increase in discharge frequency whenever the illumination is changed and adapt much more quickly and completely to patterned stimuli. Such ganglion cells probably serve to send informa-

tion to the brain about the occurence of changes in the visual environment which require attention, rather than about the exact nature of the stimulating pattern.

When the retinal illumination is reduced to scotopic levels the behaviour of retinal ganglion cells usually changes, the surround mechanism becoming relatively less effective in antagonizing the action of the central region. This effect may be accompanied by an increase in the diameter of the central region of the receptive field and a change in the spectral sensitivity of the central mechanism. A ganglion cell which at high light levels has a spectral sensitivity typical of cones may, at low light levels, change to having one typical of rods. It is not surprising that ganglion cells should behave as though they were connected to both rods and cones since this is exactly what is shown by structural studies (Fig. 37.11). It is not clear, however, how the change in effectiveness of the connexions is brought about.

In some receptive fields the antagonistic regions have different spectral sensitivities. The significance of this organization in colour vision is not known.

As well as differing from each other in the ways already described, retinal ganglion cells have receptive fields of widely differing sizes. It is not simply that receptive fields in the central area of the retina are smaller than those in the peripheral retina, although this is in general true. Rather, in each area of the retina, the ganglion cells have overlapping receptive fields with a range of sizes. In different retinal areas the actual sizes may differ but the range stays much the same. Since the dimensions of the stimulus pattern to which a ganglion cell is most sensitive are related to the size of its receptive field, the existence of receptive fields of different sizes means that information about spatial details of different dimensions is transmitted by different ganglion cells.

LATERAL GENICULATE NUCLEUS. Retinal ganglion cells send their axons to both the lateral geniculate nucleus and the superior colliculus. Simultaneous records from single cells in the retina and lateral geniculate nucleus have shown that although the cells of the nucleus are influenced by the impulses in many optic tract fibres, they are often very strongly influenced by the discharges from one particular ganglion cell. It is thus not surprising that the behaviour of cells of the lateral geniculate nucleus is not very different from that of the retinal ganglion cells that drive them, though they usually adapt more rapidly and completely. Like ganglion cells they have concentrically organized receptive fields of various sizes many of which have colour-specific centre and surround regions. The antagonistic surround regions of the receptive fields of geniculate neurons are rather more diffuse than those of retinal ganglion cells though the significance of this difference is obscure. Cells in the primate lateral geniculate nucleus respond briskly to moving visual stimuli but they do not show any directional effects nor indeed any specific response to moving, as compared to flashing, stimuli.

The lateral geniculate nucleus receives afferent fibres from both eyes but individual cells can usually only be excited by stimulation of one eye. It has recently been found that most cells can be inhibited by stimuli falling on corresponding regions of the retina of the other eye. It has also been found that the activity of cells in the geniculate nucleus can be influenced by stimulation of the vestibular and other sensory systems. The functional significance of these influences is not yet known.

ACTIVITY OF CELLS IN THE VISUAL CORTEX. Axons from the cells of the lateral geniculate nucleus project in the optic radiations principally to the striate region of the occipital cortex. Here they form synapses mainly in the fourth layer of the cortex but there are many interconnexions to all cortical layers.

The responses of single neurons in the visual cortex were first extensively studied by Hubel and Wiesel. They found, initially in cats and later in monkeys, that the behaviour of cortical cells was very different from that of retinal ganglion cells or the cells of the lateral geniculate nucleus. Cortical cells were found to respond best, not to small spots or patches of light, but to straight lines or edges adjusted to have a particular optimum orientation. The orientation specificity of most cortical cells has been found to be rather high, the response typically falling to one half of the maximum when the stimulus is rotated about 15° to 20° in either direction from its optimum orientation (Fig. 37.20 (a)). It has recently been shown that kittens reared in a visually abnormal

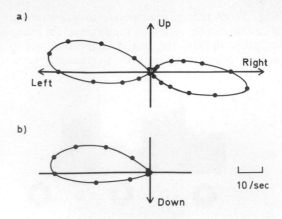

Fig. 37.20   The orientation specificity of two cells in the visual cortex of a cat. The cells (which had very low resting discharge frequencies in the absence of stimulation) were stimulated by a grating pattern moving continuously in one direction across their receptive fields. The frequency of discharge of the cells is plotted vectorially for different directions of movement. The distance of each point from the centre of the diagram shows the magnitude of the response when the grating was moved in a direction parallel to the line joining the point to the centre. (a) This cell responded equally well if the pattern moved to the left or right; it also responded well to patterns flashed on and off. (b) This cell did not respond at all to patterns moving to the right; it responded very much better to moving than to flashing stimuli. (*By courtesy of J. G. Robson.*)

environment do not develop cortical cells with normal orientation specificities.

Very few cortical cells show any response at all to stationary patterns; in fact they usually discharge at rather low frequencies unless they are being stimulated by patterns that are either being moved or flashed on and off. Some cortical cells are found to respond equally well to both flashing and moving stimulus patterns but others respond very much better if the stimulus is moved across the receptive field of the cell than if it is flashed in any position. The cells that seem to be specifically sensitive to moving patterns (Fig. 37.20 (b)) often respond only when the pattern moves in one direction, movement in the opposite direction producing little or no effect upon the cell.

If grating patterns made up of parallel light and dark bars are used as visual stimuli some cortical cells are found to respond rather selectively to patterns with a particular spacing between the bars. This behaviour recalls, even more strongly than that of ganglion cells, the frequency-selectivity of nerve cells in the auditory system. Indeed it has been suggested that a visual pattern may be analysed within the nervous system into its spatial frequency components in much the same way as a sound stimulus may be analysed into its temporal frequency components.

The ability of cortical cells to respond rather selectively to particular spatial stimulus patterns is presumably determined by the interplay of excitatory and inhibitory influences from different regions of their receptive fields. In this respect it is significant that the receptive fields of visual cortical cells are linearly, rather than circularly, organized, the excitatory and inhibitory influences coming from adjacent striplike regions of the field. It has been possible to show that some receptive fields are made up of several such antagonistic strips side by side. In general terms the striplike form of the receptive-field subdivisions can be presumed to account for the orientation specificity while the existence of multiple antagonistic subdivisions could provide the selectivity for patterns with particular periodicities.

Some cortical cells, called 'hypercomplex' by Hubel and Wiesel, do not respond well to lines or edges, even when they are set at their optimum orientation, unless they are of limited length. Such cells are probably driven by the outputs of the simpler cortical cells already described.

Most cells in the visual cortex can be excited by stimuli applied to either eye although one eye is usually dominant. If the stimulus pattern is presented simultaneously to both eyes then the response of the cell may be much greater than when either eye is stimulated alone. To obtain the maximum response, however, the stimulus pattern must be appropriately positioned on both retinae. With the eyes in their normal positions the optimal positioning of both retinal images of a real object can only be achieved if the object is at a certain distance from the eyes. It is probable that different cortical cells respond best to objects at different distances, thus providing a basis for stereoscopic vision (p. 754).

Perhaps rather surprisingly very few cortical cells respond in any discriminatory manner to coloured stimuli.

MOVEMENTS OF THE EYES

The eyeball is supported by the fatty tissues of the orbit which form a sort of socket for it. Within this socket the eyeball is free to rotate and also, since the socket is not rigid, to move a little from side to side and up and down. The eyeball is made to rotate by the action of three pairs of muscles. The lateral and medial recti rotate the eyeballs outwards or inwards about a vertical axis, but the other four muscles cause rotation around oblique axes so that they produce movements with both vertical and horizontal components. The extraocular muscles are supplied by the cranial nerves III, IV (superior oblique) and VI (lateral rectus) which originate from nuclei in the mid-brain. The actions of the extraocular and intraocular muscles (which are controlled from associated mid-brain nuclei) are very closely linked so that both eyes act together and so that conjugate movements, convergence and accommodation all occur together in a harmonious fashion. This prevents double vision (diplopia) by ensuring that the two retinal images always fall on corresponding points on the retinae in the two eyes. Corresponding points are distributed around the two foveae so that those in the right half of one retina correspond with those in the right half of the other; those in the upper half of one retina with those in the upper half of the other, and so on.

If the two eyes do not move sufficiently precisely together double vision may occur. However, when the two retinal images do not correspond with each other there may be a tendency for one of the retinal images to be suppressed so that it gives rise to no conscious perception. In children with refractive errors who squint, the image from the squinting eye is often suppressed to such an extent that vision in that eye progressively deteriorates unless use of the squinting eye is deliberately forced by frequent occlusion (that is covering up) of the other eye. When paralytic squint occurs diplopia is quite obvious to the patient. The false image seen by the squinting eye is projected as indicated in Figure 37.21.

FIXATION MOVEMENTS. When a subject looks around him his gaze is continually being shifted from one point to another in discrete jumps called 'saccades'. The nature of saccadic eye movements can readily be seen if the eyes

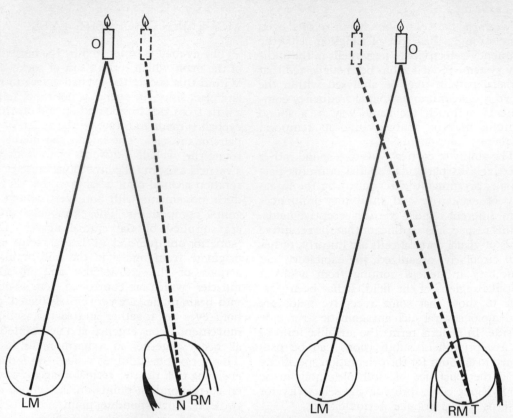

Fig. 37.21a Paralysis of the right lateral rectus. Rays reach both eyes from an object O the image of which falls on the left macula. The right eye is deviated inwards by the unopposed medial rectus, the image of O falling on an area N to the nasal side of the retina. This area possesses innate directional value (local sign) to the right. Since the eccentric area is less sensitive than the macula the image belonging to it is less clear (the false or 'ghost' image). The false image is to the same side as the paretic eye—homonymous diplopia. *(By courtesy of A. A. Douglas.)*

Fig. 37.21b Paralysis of the right medial rectus. Rays reach both eyes from an object O the image of which falls on the left macula. The right eye is deviated outwards by the unopposed lateral rectus the image of O falling on an area T to the temporal side of the retina. This area possesses innate directional value (local sign) to the left. Since the eccentric area is less sensitive than the macula the image belonging to it is less clear (the false or 'ghost' image). The false image is to the side opposite the paretic eye—crossed diplopia. *(By courtesy of A. A. Douglas.)*

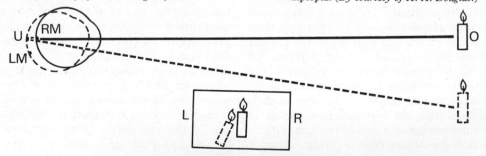

Fig. 37.21c Paralysis of the left superior oblique. The action of the superior oblique comprises depression, abduction and intorsion. In paralysis the eye is elevated and extorted by the inferior oblique and deviated inwards (adducted) by the superior and inferior recti. The image of O will fall on the macula of the right eye but, in the left eye, on the area U above and to the nasal side of the macula. The false image is therefore below and homonymous. It is the vertical element of the diplopia that is noticed by the patient. The inset shows the position of the images in detail. RM, right macula; LM, left macula. *(By courtesy of A. A. Douglas.)*

of a subject are observed while he is reading. A normal subject makes four or five separate saccadic fixation movements in scanning along each line of text (Fig. 37.22 (a)). Saccadic

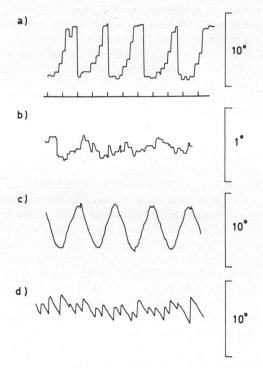

a)

10°

b)

1°

c)

10°

d)

10°

Fig. 37.22 Horizontal eye movements of a subject (a) reading, (b) steadily fixating a small object (note different vertical scale), (c) looking at a pendulum bob, (d) looking at a rotating drum. The time scale below record (a) shows seconds. *(By courtesy of J. G. Robson.)*

movements may involve rotation of the eyeballs through many degrees or through only a few minutes of arc. The saccadic rotation always occurs very quickly, its time course not being under voluntary control.

Even if a subject attempts to fixate a target as steadily as he can, his eyes do not stay perfectly still but continue to make small drifting and saccadic movements of which he is unaware (Fig. 37.22 (b)). In the absence of such small movements of the eyes, and hence of the image on the retina, vision of the outside world rapidly fades. A simple way of experiencing this fading can be achieved by looking at a large uniformly illuminated field (such as the blue sky) through a small pin-hole in a piece of card held very close to the eye. If

the pin-hole is moved from side to side about two or three times a second, then a fine network of lines will be seen superimposed upon the uniform background. These lines are the shadows of the small blood vessels which lie in front of the receptor layer of the retina. As soon as the movement of the pin-hole is stopped the shadows disappear. Experiments in which the retinal images of objects in the outside world are stabilized instrumentally have shown that the fading is most rapid and complete if the objects are small, but that it always occurs to an appreciable extent. It has been suggested that the fading is a result of sensory adaptation in the retina, but in view of the fact that many retinal ganglion cells have very well maintained discharges (p. 749), while adaptation is much more rapid and complete at higher levels in the visual pathway, it seems more likely that the fading is a result of adaptation at these higher levels.

If the object being fixated is moving then the subject's eyes make the appropriate smooth following movements (Fig. 37.22 (c)) upon which saccadic fixation movements may be superimposed. Smooth following movements of the eyes are produced reflexly whenever the gaze is transferred to a moving target; they cannot be produced voluntarily in the absence of such a target. The only kind of eye movement that can be made voluntarily is the saccade. If a subject in a moving vehicle looks out at the passing landscape then his eyes alternately make smooth following movements and saccadic jumps in the opposite direction (Fig. 37.22 (d)). The resulting oscillation is known as 'opto-kinetic nystagmus'. During the slow phase of the nystagmus the image of the passing scene remains more or less stationary on the retina, while in the fast phase the eyes are restored towards their undeviated position.

Smooth eye movements are also produced when a subject's head is rotated. If the head rotates through a small angle then an approximately compensating rotation of the eyes in the opposite direction is brought about by vestibular reflexes (Chap. 45). A larger or more prolonged rotation of the head results in a nystagmoid movement. These movements occur even if the eyes are closed and may persist for some time after cessation of a prolonged rotation.

At birth the fixation reflex is present but

weak, and eye movements are rather independent and unco-ordinated. By 5 to 6 weeks of age both eyes can simultaneously fixate an object, that is the conjugate fixation reflex is established, and a child can follow a moving target over a short range. By about 3 months objects of interest are voluntarily fixated and the beginning of co-ordinated eye and hand movements are seen. Fully co-ordinated movements of convergence and divergence do not usually appear before 6 months.

Although it has been found possible to elicit eye movements by electrical stimulation of many different nervous structures, the central mechanisms involved in the control of eye movements are not well understood.

PERCEPTION OF DEPTH AND DISTANCE. A subject with one eye covered can still estimate, with a fair degree of accuracy, the relative distances from him of objects in his field of vision, as well as the actual distance from him of objects of known size. The relative distances of objects can be judged from such clues as the relative sizes of their retinal images (if the objects are of similar actual size), the occlusion of part of a distant object by a nearer one, the greater amount of detail visible in nearer objects and, if there is relative movement between the objects in the field and the subject, the parallactic changes which result. If the subject can use both eyes a further clue to the absolute distance of an object from him is given by the degree of convergence of his eyes necessary to make the two retinal images of the object fall on corresponding retinal areas.

As well as these effects, a subject who examines a near object with both eyes can make use of the fact that he gets somewhat different (that is disparate) views of the object with each eye. If the disparity of the retinal images is not too great the subject is unaware of it although he perceives the object extending in depth in a way which never occurs if he uses one eye only. Binocular vision of this kind is called stereoscopic vision. The existence of stereoscopic vision is believed to depend upon the presence in the visual cortex of cells which respond specifically to retinal images with particular degrees of disparity (p. 733). It has recently been shown that if kittens are brought up in such a way that they are not allowed to use both eyes at the same time during a short critical period of their development, they do not establish the required neural connexions and do not achieve stereoscopic vision.

REFERENCES

BACY-Y-RITA, P., COLLINS, C. C. & HYDE, JANE E. (1971). *The Control of Eye Movements.* New York: Academic Press.
BRINDLEY, G. S. (1970). *Physiology of the Retina and Visual Pathway,* 2nd edn. London: Arnold.
CORNSWEET, T. N. (1970). *Visual Perception.* New York: Academic Press.
DAVSON, H. (Ed.) (1969). *The Eye,* 2nd edn, Vols. 1–4. New York: Academic Press.
DAVSON, H. (1972). *Physiology of the Eye,* 3rd edn. Edinburgh: Churchill Livingstone.
GRAHAM, C. H. (Ed.) (1965). *Vision and Visual Perception.* New York: Wiley.
VON HELMHOLTZ, H. (1909). *Handbuch der physiologischen Optik,* 3rd edn. Translated for the Optical Society of America (1924) as *Treatise on Physiological Optics.* English translation republished (1962). New York: Dover.
JULESZ, B. (1971). *Foundations of Cyclopean Perception.* University of Chicago Press.
KLING, J. W. & RIGGS, L. A. (Eds.) (1971). *Woodworth and Schlosberg's Experimental Psychology.* New York: Holt, Rinehart & Winston; London: Methuen.
LAST, R. J. (1968). *Wolff's Anatomy of the Eye and Orbit,* 6th edn. London: Lewis.
MOSES, R. A. (1970). *Adler's Physiology of the Eye: Clinical Applications,* 5th edn. St. Louis: Mosby.
MOTOKAWA, K. (1970). *Physiology of Color and Pattern Vision.* Berlin: Lange & Springer.
PERKINS, E. S. (Ed.) (1970). Recent research on the retina. *British Medical Bulletin* **26**, 99–184.
PIRENNE, M. H. (1967). *Vision and the Eye,* 2nd edn. London: Science Paperbacks.
PIRENNE, M. H. (1970). *Optics, Painting and Photography.* Cambridge University Press.
RUSHTON, W. A. H. (1971). Colourvision: an approach through the cone pigments. *Investigative Ophthalmology* **10**, 311–322.
WALLS, G. L. (1963). *The Vertebrate Eye and Its Adaptive Radiation.* New York: Hafner.
WEALE, R. A. (1963). *The Aging Eye.* London: Lewis.
WHITESIDE, T. C. D. (1957). *The Problems of Vision at High Altitudes.* London: Butterworths.
WYSZECKY, G. & STILES, W. S. (1967). *Color Science: Concepts and Methods, Quantitative Data and Formulae.* New York: Wiley.

38 Speech and hearing

Hearing is one of the perceptual processes by which animals are continually being informed about their environment. In man the general information from the ears, such as the proximity of a source of sound, is augmented by the more specific information contained in speech. Speech and hearing must be regarded as two complementary activities which subserve the function of communication. If complete deafness is congenital or occurs in infancy before vocalization has reached the stage of speech the child fails to develop the ability to speak, although he may be taught to speak intelligibly by special methods.

While there is great variation among children in the rate and manner of acquiring speech it is important to have some general idea of the way in which it is developed. The baby's first cries usually express hunger or discomfort, but if they always have the effect of attracting attention he may learn to cry for this reason alone. In three months he acquires several vowels and a few consonants and learns to smile. In the next six months he imitates spoken sounds but without any realization of their significance and at this stage his smiles may develop into laughter. The first meaningful use of words, perhaps connecting a sound with a toy or with himself, occurs when he is about one year old. The time at which this is accomplished depends on many factors, such as the state of health and level of intelligence, and it may be accelerated or delayed by learning to walk. At 18 months the vocabulary is about 12 words. In the second year the child may have a vocabulary of 80 words and he often uses one-word sentences, such as 'Up', meaning 'Lift me up'. At the end of the second year two-word sentences are uttered and by the age of five years sentences of five words are being used. Thereafter the vocabulary increases rapidly.

THE ORGANS OF SPEECH

So far as speaking is concerned, the lungs and chest wall act as bellows which drive air through the larynx between the two sharp folds, the *vocal folds* or vocal cords, and set them into vibration, much in the way that air blown through the lips makes them vibrate. Normally speech is produced on expiration although an abnormal type of speech can be produced on inspiration. The vocal folds stretch from the thyroid cartilage in front to the mobile arytenoid cartilages at the back of the larynx; the triangular space between them is called the *rima glottidis* or glottis. The muscles controlling the arytenoids (Fig. 38.1) open or close the rima glottidis, while the tension of the folds is regulated by the vocalis muscle lying in each fold and by the cricothyroid muscle which tilts the thyroid cartilage and so elongates the vocal folds (Fig. 38.2).

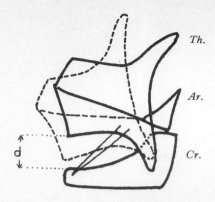

Fig. 38.2 Side view of the cartilages of the larynx. *Cr.*, cricoid cartilage; *Ar.*, left arytenoid cartilage; *Th.*, thyroid cartilage; *Th.*, thyroid cartilage. The dotted line shows the change in the position of the thyroid by the action of the crico-thyroid muscle, and the stretching of the vocal folds which results. The distance d can be seen by X-rays to diminish as the pitch of a note rises (D. Noel Paton (1920), *Essentials of Human Physiology*, 5th edn, p. 564. Edinburgh: Green & Son).

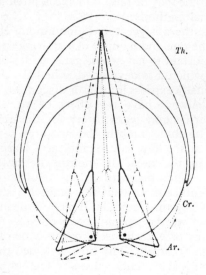

Fig. 38.1 A simplified representation of a cross-section of the larynx, to show the cricoid, *Cr.*; thyroid, *Th.*; arytenoid cartilages, *Ar.* The continuous line shows the parts at rest, the dotted line those under the action of the lateral crico-arytenoid muscle, and the dot-dash line those under the action of the posterior crico-arytenoid. The arrows show the direction of the pull of these muscles. The transverse and oblique arytenoid muscles (not shown) bridge the gap between the arytenoids; when they contract the arytenoid cartilages come together and the glottis is narrowed. The posterior crico-arytenoid muscles are abductors of the vocal folds; the other muscles can be classed as adductors (D. Noel Paton (1920), *Essentials of Human Physiology*, 5th edn, p. 564. Edinburgh: Green & Son).

When the interior of the larynx is inspected through a mirror placed obliquely against the soft palate, the posterior ends of the vocal folds are found to be about 8 mm apart in quiet respiration. In most people they remain in this position, half way between full adduction and full abduction, in quiet respiration; but in some (and in all cases on vigorous respiration) they open a little further on inspiration and come together a little on expiration. When the subject is asked to say 'ah' or 'ee' (that is, to phonate) the folds are seen to approximate. The frequency of a note can be regulated by the tension in the vocal folds. If the tension increases, the frequency rises and there may be considerable increase, up to 50 per cent, in the length of the folds. The vocal folds are lubricated by thin mucus secreted mainly in the ventricles which are spaces formed by folds of the laryngeal mucosa immediately above the vocal folds. In whispering the anterior two-thirds of the vocal folds are approximated, free escape of air occurring in the space between the two arytenoid cartilages posteriorly.

Small concentric or bipolar needle electrodes have been placed in the intrinsic laryngeal muscles in man to record action potentials. Even in quiet respiration there is some electrical activity which increases on inspiration in the adductors (vocalis and crico-thyroid mainly) whereas it is diminished in the abductor (the posterior crico-arytenoid). The process of phonation begins with an increase in the electrical activity in the adductor muscles which reaches a maximum just before the onset of the sound (Fig. 38.3) The time

The sound produced in the larynx, the laryngeal tone, is greatly modified by the acoustic properties of the mouth, throat and nasal cavities. The tongue, by alterations in its shape and position, has the main control over the resonant characteristics of the oral cavity (Fig. 38.4) but the positions of the lips and

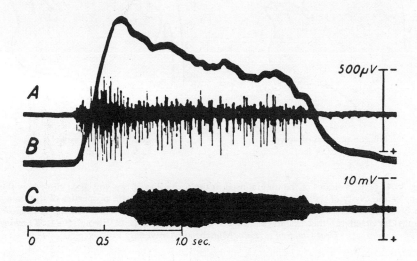

Fig. 38.3 Electrical activity in the left vocalis muscle during phonation. *A.* The action potential pattern. *B.* A trace electronically derived from *A* which gives the mean action potential amplitude. *C.* Microphone recording of sound ee, frequency 285 Hz. The subject was a 64-year-old woman with the right vocal cord immovable but with the left vocal cord moving normally. (K. Faaborg-Anderson (1957). *Acta physiologica scandinavica* **41,** Suppl. 140, 54.)

between the beginning of the electrical activity and the onset of the sound—0·35 to 0·55 sec— is necessary to allow the pressure to build up to the value needed to produce the sound. On the other hand in the posterior crico-arytenoid muscle (abductor) there is an inhibition of activity which begins just before sound is produced. An increase in the volume of the sound causes no change in electrical activity in the adductors but with a rise in pitch there is an increase in activity which indicates that the tension in the vocal folds is adjusted to a given pitch before the sound is actually produced. No change in the pattern of electrical activity is seen when different vowels are produced. When the subject is asked to think about production of a vowel without actually emitting any audible sound, the electrical activity of the adductor muscles increases.

jaws are also of importance in the production of speech sounds. The vowel sounds are produced by vibrations of the vocal folds (that is, are voiced), the air stream passing freely through the mouth. In the production of a consonant the air stream is either partially or completely obstructed so that it cannot issue freely from the mouth. Some of the consonants are very short, for example *t* and *p* and *k*, and may be regarded as particular ways of beginning or ending vowel sounds, but some consonants, such as *m*, can be spoken without a vowel. The larynx does not appear to be involved in the production of many consonants, that is the majority are unvoiced. A laryngeal component can, however, be easily added to some consonants to make them voiced; in this way *f* becomes *v*, and *s* becomes *z*.

The vagus nerve carries motor and sensory

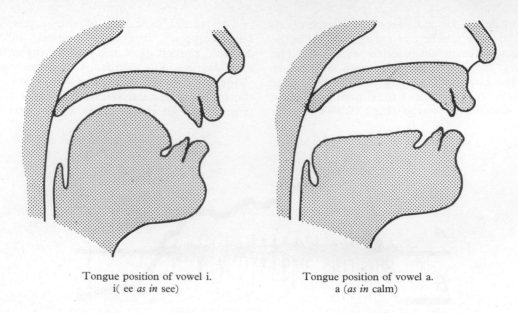

Tongue position of vowel i.
i(ee *as in* see)

Tongue position of vowel a.
a (*as in* calm)

Fig. 38.4 The positions of the tongue used in the formation of two vowels to show how the shape of the cavity of the mouth is altered. In this way the resonant properties of the cavity can be made to vary over a wide range. (*After* I. C. Ward (1948), *The Phonetics of English.* Cambridge: Heffer. Other examples are given by A. Ewing & E. C. Ewing (1964), *Teaching Deaf Children to Talk.* Manchester University Press.)

fibres to and from the larynx. The superior laryngeal branch after receiving a twig from the sympathetic divides into the internal laryngeal nerve which is sensory to the upper larynx, and the external laryngeal nerve which supplies the cricothyroid muscle. All the other intrinsic muscles are supplied by the recurrent laryngeal nerve which also gives sensory fibres to the larynx below the vocal folds After section of one recurrent nerve the fold on the denervated side lies in or close to the mid-line. The cricothyroid muscle (which is, of course, unaffected) stretches the paralysed fold. by tilting the thyroid cartilage. Full adduction or overadduction of the opposite vocal fold brings the folds together and in many cases phonation is normal. In bilateral recurrent nerve paralysis the folds lie motionless near the mid-line producing hoarseness and dyspnoea. If the superior laryngeal nerves are cut, the upper part of the larynx is anaesthetic and voice is hoarse because the vocal folds cannot be made tense.

There are many good reasons for regarding the upper resonators (mouth, throat and nose) and not the larynx as the main organs of speech, but perhaps the most dramatic evidence is furnished by the phenomenon of oesophageal speech. After complete removal of the larynx the trachea is made to open on the surface of the neck while the oesophagus retains its relation with the mouth. If the patient acquires the trick of swallowing air he can bring it up in a belch which sets the lower edge of the inferior constrictor of the pharynx (cricopharyngeus muscle) into irregular vibrations which are conveyed to the mouth cavity. The patient uses the same mouth positions as he did before his operation to modify this oesophageal sound and although the vibrations of this pseudolarynx are extremely irregular compared with those of the larynx, yet he can with practice produce oesophageal speech which, although rather hoarse, is readily intelligible. For natural and normal speech the larynx of course produces the sound and determines the fundamental pitch of the voice. But the variable resonances of the mouth and throat determine the quality of the sound and this is what conveys the information from speaker to listener, as is shown by the fact that the intelligibility of words is only slightly affected by singing them at different pitches. The mouth and throat are put into the proper position for the formation of speech by the help of the kinaesthetic sense which gives information to the brain of the position of the tongue, lips and jaws. In addition the ear, in both normal and oesophageal speech, is used to monitor the muscular adjustments so that the desired speech sounds are produced. We hear our own words by air conduction about 1 msec after they are spoken; if this auditory feed-back is delayed by electromechanical methods for say 0·2 sec the subject is apt to stutter and his speech is slowed. If a speaker is deaf the ear cannot function as a proprioceptor and his speech is defective; if a normal subject is asked to sing a simple musical scale when a strong masking noise smothers the sound of his own voice he sings badly out of tune. When a steady note is sung there is a periodic fluctuation in frequency of 5 or 6 Hz. These fluctuations are produced by the auditory feed-back mechanism because, if the sound of the voice is delayed electrically before reaching the subject's ears, the fluctuations are larger and occur more slowly.

ANALYSIS OF SPEECH SOUNDS

The sounds of speech and of musical instruments are very complex but are capable of analysis by physical and mathematical methods into a fundamental tone and a number of harmonics, or overtones, of higher frequency. In the middle of last century Helmholtz was able to make an analysis of such sounds using his own ear aided by resonators to obtain greater sensitivity. He could, however, pick out only the loudest harmonics and was unable to detect the weak components which are often important characteristics of speech sounds. The energy involved in speech is so small (10 to 25 microwatts in ordinary conversation) that without amplification no device is sufficiently sensitive to make an accurate record. The full complexities of speech and musical sounds were first revealed by using a microphone and valve amplifier connected to a galvanometer which recorded by a beam of light on a moving photographic film (Fig. 38.5). The vowels have a waveform which recurs regularly throughout.

their duration at the frequency of vibration of the vocal folds. The pattern is usually very complicated showing, in addition to the fundamental laryngeal tone, a considerable number of harmonics of much higher frequency, some of which may even be of greater amplitude than the fundamental. Consonants like *p*, *b*, *t*, *d* are so short—electrical engineers refer to them as 'transients'—that no sensation of pitch is given by them; sibilants, *s* and *z* may last as long as vowels and are characterized by small waves of very high frequency. Transitional consonants like *m*, *n*, *ng* also last for some time. Voiced consonants such as *z* and *v* have a waveform which recurs at the fundamental laryngeal frequency. A study of oscillograms of vowels suggests that each puff of air that comes up from the larynx sets the cavities of the mouth and throat into vibration. This vibration gradually diminishes until the next puff of air comes up from the larynx. This is well illustrated by the oscillogram of the vowel 'ah' as in 'father', when spoken by a man (Fig. 38.5A). When the same vowel is spoken by a woman (Fig. 38.5B) the time between the laryngeal puffs is reduced to about half with the result that the vibrations of the air in the mouth diminish very little before the next puff comes along. Although the two waveforms of 'ah' in Fig. 38.5, are quite different, the listener has no difficulty in recognizing that both speakers are pronouncing the same vowel. It is thus unlikely that the ear recognizes vowels by their

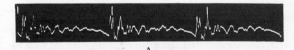

A

MALE VOICE. LARYNGEAL FREQUENCY 110 Hz

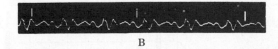

B

FEMALE VOICE. LARYNGEAL FREQUENCY 229 Hz

Fig. 38.5 Oscillograms of the vowel sound 'ah' as in 'father'. (A) Spoken by a man and (B) spoken by a woman. The vertical white lines are 0·01 sec apart. The ordinate gives a measure of air pressure at the speaker's mouth. (Harvey Fletcher (1929), *Speech and Hearing*, Fig. 37, p. 49. London: Macmillan.) (Bell Telephone Laboratories.)

waveforms and this is confirmed by the following considerations. When one listens to speech in a room, part of the sound arrives directly from the speaker to the ear and part indirectly after reflexion from the walls; the reflected waves are out of phase with the direct wave. In many gramophone or microphone amplifiers the higher frequencies are not emitted from the loudspeaker in their original phase relationship to the low frequencies. In both examples the original waveform is considerably changed, but the sound does not appear to be altered.

An acoustic spectrum is obtained by analysing a speech waveform to obtain the amplitude and frequency of the various components and some examples are given in Fig. 38.6. If spectra of the same vowel spoken at different pitches (or, to express it objectively, at different laryngeal frequencies) are compared it is found that, in spite of the alteration in the frequency of the fundamental as the pitch of the voice is raised, the overtones are relatively unchanged and a characteristic grouping of the higher frequency components can be recognized in all the spectra. Presumably it is this grouping or 'formant' which is recognized by the ear and clearly it is the higher frequencies produced in the mouth and throat which are important for differentiating and recognizing the vowels. Spectra of whole words have also been found to have recognizable similarities in spite of individual differences in pitch and quality. By the use of somewhat elaborate apparatus such spectra can be displayed on a cathode-ray oscillograph as the sounds are being spoken. Whether this so-called 'visible speech' would be useful to deaf persons remains to be seen.

The rate of vibration of the vocal folds, which can be easily found from oscillograms, gives a clue to the sex of the speaker. A deep-voiced man may have a laryngeal tone of 90 Hz, but

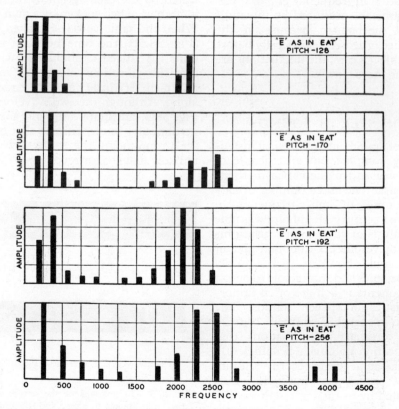

Fig. 38.6 Acoustic spectra of vowel sound 'EE' at different frequencies of the fundamental laryngeal tone (128, 170, 192 and 256 Hz) to show the general similarity of the spectra (Harvey Fletcher (1929), *Speech and Hearing*, Fig. 41, p. 54. London:Macmillan). (Bell Telephone Laboratories.)

in the average man it is usually between 125 and 145 Hz. The average woman's laryngeal tone is from 230 to 256 Hz, but in a shrill-voiced woman it may be as high as 300 Hz. The deepest bass note that can be sung is about 66 Hz, while the highest soprano note is about 1056 Hz. This is a range of four octaves.

The decibel notation. The energy involved in speech can be calculated from oscillograms or measured by suitable apparatus. The range is very great—from 0·001 microwatts in a very soft whisper up to 1000 microwatts in very loud talking. Since the ear, like the other sense organs, follows approximately the Weber-Fechner law (p. 707), which implies that equal steps on a logarithmic scale of intensity are equal steps on a loudness scale, it is more convenient to describe differences in power-level in the decibel notation:

Difference in power-level in decibels =
$\qquad$ 10 $\log_{10}$ (ratio of the two powers).

If the figures quoted above for speech are taken the difference in power-level between a soft whisper and loud talking is

$$10 \log_{10} \frac{1000}{0 \cdot 001} = 10 \log_{10} \frac{10^3}{10^{-3}} = 10 \log_{10} 10^6$$

$$= 10 \times 6 = 60 \text{ decibels, or } 60 \text{ dB.}$$

This formula shows that 1 dB is equivalent to a 26 per cent increase in power; 10 dB to a ten times increase in power; 20 dB to one hundred times increase in power.

The magnitudes of sounds may be stated in absolute values of intensity or sound pressure by the use of an agreed reference zero, namely 0·00002 Newton per square metre (N/m^2), which can be denoted by p ref. The 'Sound Pressure Level' (SPL) in decibels is given by 20 log p/p ref where p is sound pressure.

THE EXTERNAL EAR AND MIDDLE EAR

The mobile trumpet-like external ear (auricle) possessed by some of the lower animals may be helpful in localizing the source of a sound and in some animals the auricle twitches when loud sounds are heard. This can be used as a test of hearing. The auricle is not usually mobile in man. From the auricle, the external auditory meatus, about 2·5 cm long in the adult, passes down to the tympanic membrane. The skin of the outer half of the meatus possesses numerous ceruminous or wax-secreting glands which are modified sweat glands. Accumulation of wax in the external auditory meatus is a common cause of impaired hearing.

The tympanic membrane, about 1 cm in diameter, is composed of radial and circular collagenous fibres covered externally by stratified squamous epithelium continuous with the skin and internally by a mucous membrane of ciliated columnar epithelium. The outer surface of the membrane is concave; the inner convex surface, as seen from the middle ear or tympanic cavity, has the manubrium of the malleus (handle of the hammer) attached to it. The malleus articulates with the incus (anvil) and the incus with the stapes (stirrup), the base of which is attached by fibrous tissue to the margins of the fenestra vestibuli (oval window). These very light ossicles in the air-filled tympanic cavity link the ear drum to the inner ear (Fig. 38.9). The tensor tympani muscle, by its attachment to the root of the manubrium of the malleus, draws the tympanic membrane in, and at the same time, through the articulations between the three ossicles, pushes the stapes into the internal ear. The small stapedius muscle tends to have the opposite effect, namely to pull the stapes out of the internal ear. In response to sound, or to mechanical stimulation of the external auditory meatus, auricle or surrounding skin, these muscles contract reflexly, pulling in the tympanic membrane and tightening it and pushing the footplate of the stapes inwards, presumably because the tensor tympani is more powerful than the stapedius. The result is that sound transmission across the middle ear is impaired for tones of low frequency and the ear is partially protected from damage by loud sounds.

The ear drum and the ossicles act as a kind of transformer to convey vibrations of the light medium, air, to the denser, watery media in the internal ear. The area of the ear drum is about 90 mm^2, whereas the area of the base of the stapes is only 3·2 mm^2. The pressure on the fluid under the stapes is thus many times greater than the air pressure which makes the ear drum move. In the absence of this so-called impedance-matching function only a very small proportion of the energy of the sound would reach the inner ear as happens if the middle ear conducting function is abolished and the sound reaches the internal ear fluids directly. In the course of the condition called otosclerosis, a familial and progressive deafness beginning in young people, the base of the stapes become joined by bone to the margins of the fenestra vestibuli. This condition causes severe deafness (Fig. 38.17e).

The pharyngotympanic tube leads from the anterior part of the middle ear to the nasal part of the pharynx. The pharyngeal end is normally closed but during swallowing it opens and allows the air pressure in the middle ear to be equalized with the atmospheric. If the pressure is not equal on the two sides the drum is pushed in (as in the case of the descent of a diver) or drawn out (as on ascending in an aeroplane). Such abnormal displacement produces pain and impaired hearing and damage to the ear drum, but this discomfort can be avoided by frequent swallowing while the air pressure is altering. Reduced acuity of hearing may also result when the tube is blocked by the swelling of the mucous membrane which accompanies a 'cold'; in this case the oxygen in the middle ear is absorbed and the ear drum is pushed in by the external atmospheric pressure.

THE INTERNAL EAR

This chapter deals with cochlear function only. The semi-circular canals and otolith organs which are concerned with the maintenance of balance play no part in hearing; they are discussed in Chapter 45. A cross-section of the petrous temporal bone is difficult to understand until the basic structure of the cochlea, given in Figure 38.7 is appreciated. A diagram of a cross-section through the three canals is given in Figure 38.8. If the three canals are wrapped two and three-quarter times round the modiolus (the central pillar of the cochlea) the appearances seen in Figure 38.9, can be fully accounted for. The *basilar membrane*, about 35 mm long, is composed of fibres which run radially from the osseous spiral lamina of the modiolus to the spiral ligament, or crista basilaris, which binds the basilar membrane to the external bony wall of the cochlea. The fibres are short (0·04 mm) near the foramen vestibuli and increase gradually to the apex of the cochlea at the helico-trema, where they are about 0·5 mm long. The spiral organ of Corti is formed of about 4000 rods of Corti which make a spiral tunnel on top of the basilar membrane. On either side of this tunnel are hair cells whose bristle-like processes are attached to the underside of the tectorial membrane (Fig. 38.11). Filaments of the cochlear nerve lie on the exterior of the hair cells and pass to the spiral ganglion lying nearer the modiolus. The fluid in the scala vestibuli and scala tympani (the perilymph) is very like cerebrospinal fluid in chemical composition except for a somewhat higher protein content; the perilymph, however, may not be continuous with the cerebrospinal fluid (see Chap.

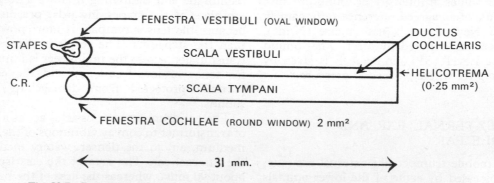

Fig. 38.7 Basic structure of the inner ear. If this system of three tubes (scala vestibuli, ductus cochlearis and scala tymphani) could be wrapped two and three-quarter times round the modiolus (central pillar of the inner ear) the actual anatomical relationships would be reached. C.R. = canalis reuniens which communicates with the saccule (see Fig. 38.9).

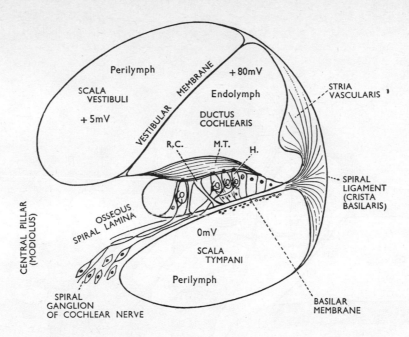

Fig. 38.8 A cross-section through the three canals of the cochlea to show the structures on the basilar membrane. H., hair-cells; M.T., membrana tectoria; R.C., rods of Corti. The reticular lamina is a net-like membrane holding in its apertures the hairs of the hair-cells (Fig. 38.11). The cochlear partition comprises the basilar membrane, the vestibular membrane and the structures contained in the ductus cochlearis. The endolymph in the ductus cochlearis is 80 mV positive with respect to the perilymph in the scala tympani. The perilymph in the scala vestibuli is 5 mV positive to the perilymph in the scala tympani. Note that the naked nerve fibres are about 0·15 μm in diameter as they arise from the hair cells. As they pass centrally they become myelinated (2 to 5 μm). The fine nerve fibres passing through the organ of Corti must be bathed in perilymph or fluid resembling it. All the sensory hair cells are richly innervated; the nerve endings are of two kinds, sparsely granular afferent and richly granular efferent.

45). The endolymph in the ductus cochlearis has a composition more like that of intracellular fluid with high potassium and low sodium concentrations; however, the main anion is chloride as in extracellular fluid. It is probably secreted by the cells of the stria vascularis (Fig. 38.8).

MECHANICAL ASPECTS OF COCHLEAR FUNCTION

If the air pressure in the external ear is increased, the tympanic membrane is pushed inwards, resulting in a movement of the chain of ossicles which moves the stapes so that it enters more deeply into the vestibular fenestra, thus displacing perilymph inwards into the scala vestibuli. If the movement is very slow, the fluid is displaced up to the helicotrema, then down the scala tympani, finally bulging the membrane of the fenestra cochleae outwards into the middle ear. The vestibular membrane, being flaccid, can be ignored in these movements. As a result of the movements of the stapes caused by sound the basilar membrane and its associated structures execute a complicated pattern of movements.

Of the many theories of cochlear function the resonance theory of Helmholtz seemed at first to be the most adequate. This regarded the basilar membrane as a series of tuned resonators, like the strings of the piano, in which the short transverse fibres at the base of the

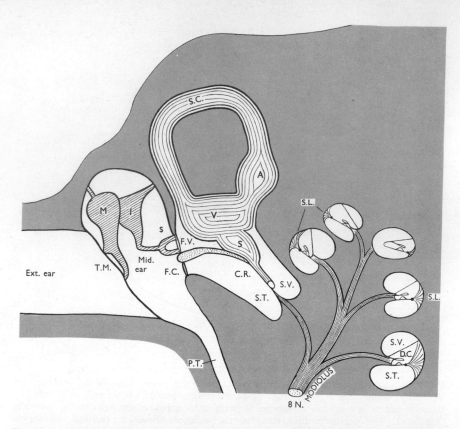

Fig. 38.9 Diagram of ear. T.M., tympanic membrane; M., malleus; I., incus; S., stapes; S.C., semicircular duct; A., ampulla; V., vestibule; S., saccule from which C.R., the canalis reuniens (cut across), connects with D.C., the duct of the cochlea; S.V., scala vestibuli; S.T., scala tympani; 8 N., eighth nerve; S.L., spiral ligament or crista basilaris; P.T., pharyngo-tympanic tube. In constructing the diagram the various structures have been shown in one plane but actually the cochlea lies anterior to the semicircular canals. (*After* McDowall and Keen). Further details of the labyrinths are given in Chap. 45.

cochlea reacted to high frequency tones, the long fibres at the apex to low frequency tones. Békésy has shown in several ways that these fibres are not in tension. For example, if the basilar membrane is cut the edges of the wound do not gape apart. Also pressure with a bristle gives a shallow circular depression. There are, however, differences in stiffness, this being greatest in the part of the basilar membrane nearest the stapes. In consequence the principle of tensioned resonators has had to be abandoned.

Békésy examined the movement of the human cochlear partition (see legend to Fig. 38.8) in fresh post-mortem material, in response to vibration applied to the fluid in the scala vestibuli by a small piston which took the

place of the stapes. Vibrations of known frequency and amplitude were imparted to the piston and the resulting movements of the cochlear partition were observed microscopically. The pattern of vibration of the basilar membrane and associated structures shows a travelling wave of transverse displacement which, starting from the base of the cochlea at high velocity, reaches a maximum and then dies away, with rapid reduction in amplitude and velocity (Fig. 38.10). The maximum amplitude is located towards the basal part of the cochlea for high frequencies and towards the apex for low frequencies. The higher the frequency the nearer to the artificial stapes is the maximum amplitude of vibration. Thus some degree of frequency analysis can be per-

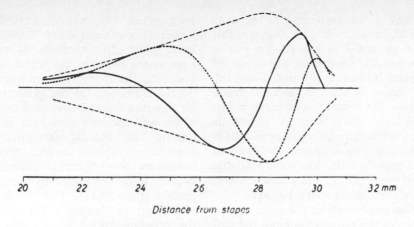

Fig. 38.10 The measured longitudinal bending of the cochlear partition for a tone of 200 Hz for two moments in time separated by a quarter period. (G. von Békésy (1947), *Journal of the Acoustic Society of America* **19**, 455.)

formed mechanically by the basilar membrane, although not in a sharply selective way, and it is clear that mechanical separation of frequencies cannot account for the known powers of pitch discrimination possessed by the ear. For further information on this matter we must look to electrophysiological investigations of cochlear function.

ELECTROPHYSIOLOGICAL ASPECTS OF COCHLEAR FUNCTION

The characteristics of the action of the cochlea have been greatly elucidated by the study of the electrical potentials generated in the inner ear. These potentials include

1. Steady or direct current (d.c.) potentials (endocochlear potential);

2. Alternating potentials, known as the cochlear microphonics;

3. Action potentials derived from activity in the auditory nerve.

Endocochlear potential. In the guinea pig cochlea, a potential was found between the endolymph and perilymph by Békésy. By inserting microelectrodes into the perilymphatic and endolymphatic spaces of the cochlea, he found that the endolymph in the ductus cochlearis, in the absence of any acoustic stimulation of the ear, was electrically positive

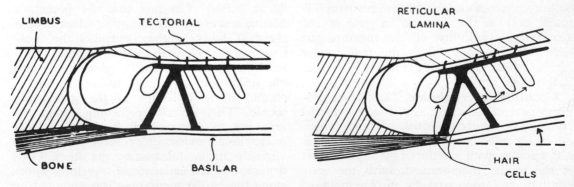

Fig. 38.11 The shearing action between the hair cells and the tectorial membrane, when the cochlear partition is deflected. This shearing action results in bending or deflexion of the hairs relative to the reticular lamina. (*From* H. Davis (1960), *Neural Mechanisms of the Auditory and Vestibular Systems*. Eds. G. L. Rasmussen & W. F. Windle. Springfield: Thomas.)

by some 80 mV to the perilymph in the scala tympani and scala vestibuli. The source of this endocochlear potential is not known but it disappears when the oxygen supply is cut off and is abolished if cyanide is injected into the endolymph space. When the microelectrode penetrated into the cellular structures bounding the cochlear duct, negative potentials were found: such internal negativity in cells is to be expected, and is, of course, well known in nerve and muscle cells. The total potential difference between the ductus cochlearis and the interior of the cells is about 140 mV.

The positive endocochlear potential almost certainly stops at the reticular lamina, the stiff membrane supporting the hair-bearing ends of the hair cells (Fig. 38.8 and 11) and at the hair-bearing surfaces of the cells themselves. Presumably everything on the side of this membrane towards the scala tympani is bathed in perilymph, not endolymph. The perilymph may be completely replaced by Locke's solution but if this solution is injected into the ductus cochlearis there is a fairly rapid failure of all electrical responses. Locke's solution with a raised potassium concentration (equal to that found by analysis of endolymph) can be injected into the cochlear duct without affecting the electrical potentials. Such a solution injected into the scala tympani causes a failure of action potential and cochlear microphonics. The bare nerve fibres in the organ of Corti are immersed in perilymph (see Fig. 38.8); they could not conduct impulses if they were bathed by a solution of high potassium content (see Chap. 39).

The large potential difference (that is, 80 mV) between endolymph and perilymph is found only in the cochlea. In spite of the anatomical continuity of the membranous labyrinth the potential in the semicircular canals, utricle and saccule is nearly zero (that is, less than 2 mV).

Cochlear microphonic. The cochlea produces potential changes as if it were a microphone. The steady potential difference which can be recorded between the ductus cochlearis and scala tympani is reduced by displacement of the basilar membrane towards the scala vestibuli and increased by displacement of the basilar membrane in the opposite direction, that is towards the scala tympani. These alterations in the potential are maintained so long as the deformation of the basilar membrane is continued. Such displacements have been made experimentally by applying pressure to the fluids in the scalae, and by gently touching the membrane itself with a needle. These alterations to the potential are caused by shearing forces between the tectorial membrane and the hair cells held in the reticular lamina. This shearing takes place in a direction transversely to the basilar membrane, and occurs as the membrane moves up and down on the 'hinge' provided by the osseous spiral lamina (Fig. 38.11). The mechanism can be visualized by flexing a book and observing the relative motion of the two covers. The alternating potential produced in this way when the basilar membrane is set into vibration by a sound is known as the cochlear microphonic. It can best be recorded between one electrode in the duct of the cochlea and another in the scala tympani, that is directly across the basilar membrane-hair cell complex. However, owing to electrical spread through the tissues, the microphonic can be most conveniently recorded between the round window and an 'indifferent' electrode on the neck.

The cochlear microphonic corresponds closely to the characteristics of the applied sound stimulus; it is virtually an electrical equivalent of the applied acoustic stimulus, and quite different from the action potentials in the auditory nerve. The potential changes follow the frequency of the applied sound stimulus at all audible frequencies; they resemble quite closely the wave-form of the stimulus; they increase in voltage with increase in sound energy up to a maximum of 5 mV; they have virtually no latent period. The fact that the potentials decline and eventually disappear when oxygen is cut off shows that they depend on the metabolism of the cells in this region. Tasaki, Davis and Legouix inserted small wire electrodes into the scalae and were able to record microphonics from a length of the guinea-pig cochlea. Their electrical recording system was arranged to provide simultaneous cochlear microphonic and auditory nerve traces, each virtually uncontaminated by the other. Confirmation of the existence of travelling waves along the basilar membrane was obtained by the pattern of cochlear microphonics produced at different parts of the cochlea by a sound stimulus; tones of low frequency caused waves

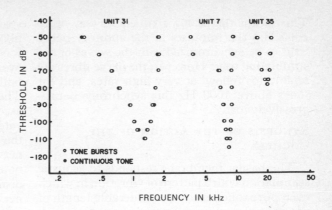

Fig. 38.12 Threshold response curves for three cat auditory nerve fibres. Each fibre responds best to some particular frequency, the 'characteristic frequency', but responds over a progressively wider band of frequencies as the sound intensity is raised. The response areas are not affected by whether the stimulus tone is short lasting (50 msec) or continues indefinitely. (*From* N. Kiang (1965), *Discharge Patterns of Single Fibers in the Cat's Auditory Nerve.* M.I.T. Press.)

which reached the apical parts of the basilar membrane, but the higher the frequency the nearer to the base was the wave extinguished. This work confirms in general that of Békésy in regard to the movement of the cochlear partition. In addition to the pure alternating component of the cochlear microphonic which reproduces the sound stimulus frequency, there is often observed, at high stimulus intensities, a steady displacement of the base line lasting for the duration of the stimulus. This is the so-called 'summating potential' and may be ascribed to the basilar membrane-hair cell complex vibrating non-linearly for large displacements.

Action potential of auditory nerve fibres. The estimate of the threshold sensitivity of the ear given on p. 771 shows that the amount of mechanical energy available for stimulating the nerve endings in the cochlea is, even for the every-day range of sounds, very small indeed. The tiny mechanical movements of the hair cells produced by such sounds must release a store of energy many times that of the original sound. Just how the nerve impulses are initiated is not understood. There has been considerable discussion about whether the cochlear microphonic corresponds to a receptor generator potential. It seems on the whole unlikely that the microphonic potential stimulates the nerve endings directly; electron microscope studies have revealed what appears to be a synaptic junction between the base of the hair cell and the nerve fibre, but so far no transmitter has been identified.

The responses of individual fibres in the auditory nerve have been investigated by a number of workers; a comprehensive study in the cat has been made by Kiang. All these studies show that there is for each fibre some characteristic sound frequency for which it is most sensitive (Fig. 38.12). As the stimulus intensity is raised, the fibre responds over a wider and wider frequency range which may reach as much as two or even three octaves at high intensities. It will be noted that the response curves are asymmetric triangles which qualitatively mirror the mechanical disturbance of the basilar membrane. The nerve fibres arise from the basilar membrane in an orderly manner, those with high characteristic frequencies arising from the basal end and those with lower characteristic frequencies arising progressively farther towards the apical end. Just as any single sound frequency sets in motion a considerable length of the basilar membrane, so too does the membrane in turn excite a large number of auditory nerve fibres. It can be inferred from curves like those of Fig. 38.12 that a single tone at a moderate intensity (say 40 dB SPL) may cause some 20 per cent of all the fibres in the auditory nerve to discharge. There is thus no question of discrimination being carried out in terms of excitation of specific auditory nerve fibres by specific frequencies. It can also be seen that increasing the stimulus intensity increases the number of excited fibres, as well as increasing the rate of discharge in individual fibres. Fibres arising from internal hair cells may have higher thresholds than those arising from external hair cells but clear evidence is lacking. There is a tendency for nerve impulses to be discharged preferentially for one direction of shear of the hair cells rather than the other, as the basilar membrane moves back and forth.

This means that at low frequencies (say up to 1000 Hz) the impulses in the auditory nerve tend to be synchronized with the waves of the stimulating tone. However, the nerve fibres are unable to 'follow' at very high rates, and for tones above 1000 Hz this synchrony is progressively lost.

SYNOPSIS OF THE ACTION OF THE COCHLEA

The basilar membrane responds to acoustic stimulation with a pattern of vibration in which even pure tones excite a considerable length of the membrane; travelling waves occur which die out at various distances from the base, the higher frequencies being eliminated nearer the base, the lowest frequencies reaching the apex. In this way are produced the cochlear microphonics which, in general, correspond to the pattern of mechanical excitation of the membrane; the microphonics in turn stimulate, possibly by the release of a chemical transmitter, the endings of the auditory nerve on the hair cells, setting up a pattern of activity in the fibres of the nerve. Higher intensities produce higher rates of discharge in the nerve fibres but a resting discharge is found and the highest frequency of discharge is quite modest; Kiang finds that fibres will not discharge continuously at much more than 150 impulses per second. The pattern of discharge is modified by inhibitory processes operating in the cochlear nucleus so that the second-order neurones show greater selectivity in their response to different frequencies of acoustic stimuli. It is also probable that at low frequencies (but not at high frequencies) the sense of pitch is conveyed by the frequency of the nerve impulses. The external hair cells are thought to provide greater sensitivity to sound than do the internal hair cells.

Central connexions of the cochlear nerve. The fibres of the auditory nerve leave the cochlea in an orderly manner so that they are arranged in order of their frequency sensiticity. As the nerve enters the brain stem, this orderliness is preserved and the individual fibres terminate in the cochlear nucleus in the same way. Second order fibres arising in the cochlear nucleus synapse in olivary and lemniscal nuclei and eventually reach the inferior colliculus. Throughout this pathway the same anatomical orderliness is preserved, so that the spatial relation of two given groups of active fibres remains the same in the colliculus as it was at the cochlea. However, owing to crossing over of some fibres in the medulla, inputs from both ears reach each colliculus (Fig. 38.14). There is no sharpening of the response as we ascend the system, so that a single tone activates just as high a proportion of neurones at the collicular level as it did at the auditory nerve level. The identification of a musical note appears to depend on the pattern of distribution of activity in the total array of nerve fibres rather than the activation of some specific neurone. Indeed the patterns of distribution of activity in the auditory pathway seem to mimic closely the corresponding acoustic spectra (Fig. 38.6) and this may indeed be how we 'recognize' a particular sound such as a vowel or a note produced on a musical instrument.

Bilateral removal of the auditory cortex does not seriously interfere with the ability to discriminate the pitch and intensity of sounds. However, it does interfere with the ability to recognize temporal sound patterns. A cat without its auditory cortex can, for example, still distinguish tone A from tone B, but cannot distinguish the sequence ABA from the sequence BAB. Neurones have been found in the auditory cortex which will respond to a rising frequency but not to a frequency falling through the same region (Fig. 38.13), while others respond only to falling and not to rising frequencies. Presumably such neurones play an important role in the discrimination of the patterns just referred to, and perhaps in responding to the complex frequency changes involved in speech.

The classical auditory pathway, as just described, from the ear to the cortex, is not the only one, since cats in which this pathway had been cut in the mid-brain could readily be aroused from sleep by loud sounds. The nerve impulses must be presumed to pass in a parallel pathway in the mid-brain reticular formation. There are also ascending pathways, as shown electrically, from the cochlea to the vermis of the cerebellum and from there to the cerebrum. In addition to these, the ascending pathway (Fig. 38.14) is paralleled throughout its length by a corresponding system of descending fibres terminating in the various nuclei. The complete significance of this

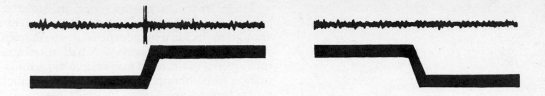

Fig. 38.13 Response of a neurone in the auditory cortex of the cat to a rising frequency. The signal bar represents a steady tone which rose in pitch at the point shown by about 10 per cent (1½ semi-tones) to a new steady value. The time taken to make the change was 50 msec. The neurone responded to the change, but not to either of the steady tones. Note that it did *not* respond when the frequency *fell* through exactly the same range. (*After* I. C. Whitfield & E. F. Evans (1965), *Journal of Neurophysiology* **28,** 655.)

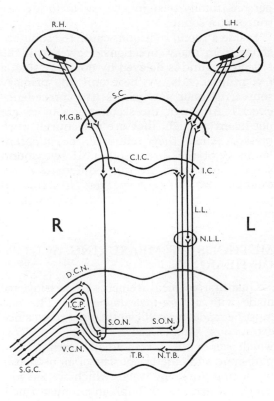

Fig. 38.14 Central connexions of the cochlear nerve. R.H. = Right hemisphere; L.H. = Left hemisphere; M.G.B. = Medial geniculate body; S.C. = Superior colliculus; I.C. = Inferior colliculus; C.I.C. = Commisure of the inferior colliculus; L.L. = Lateral lemniscus; N.L.L. = Nucleus of lateral lemniscus. D.C.N. = Dorsal cochlear nucleus. V.C.N. = Ventral cochlear nucleus. I.C.P. = Inferior cerebellar peduncle. S.G.C. = Spiral olivary nucleus. T.B. = Trapezoid body. N.T.B. = Nucleus of the trapezoid body. (*Courtesy of G. W. Pearce.*)

descending system remains to be determined, but it does appear, for example, to be capable of altering the acoustic threshold of groups of neurones in the brain stem nuclei.

LOCALIZATION OF THE APPARENT SOURCE OF A SOUND

The position of the source of a sound may be detected in either of two ways. If the head is kept still the observer can point to the source of the sound right or left of the saggital plane. Alternatively the head may be turned from side to side until the sound seems to be directly ahead. The second is the more accurate method, and can be made even more accurate if the ears are extended by trumpets, as in the apparatus used to locate aeroplanes before radar was available. In practice the eyes as well as the ears are used in localization. If the source cannot be seen, localization by hearing alone is sometimes quite difficult.

The ability to localize the source of a sound depends partly on the difference in loudness at the two ears and partly on the difference in time of arrival at the two ears. If the wavelength much exceeds the width of the head the sound waves readily pass round the head and reach the further ear with only a small and undetectable reduction in intensity; with such tones the difference in time of arrival (Fig. 38.15 A) is the clue to localization. Some subjects can detect time differences as small as 0·01 msec which corresponds to a path difference in air of 3·5 mm. When the frequency is such that the wavelength is nearly the same as the width of the head (about 1500 Hz) the

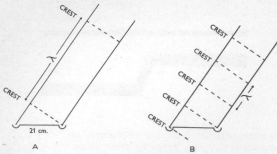

Fig. 38.15 (A) shows how a sound source is located by the crest of a wave (or some other distinctive part) reaching one ear before the other. The semicircles at the ends of the 21 cm horizontal line represent the ears. (B) shows how this method breaks down when the wavelength is short compared with the distance between the ears, that is 21 cm. In this case one crest arrives at the left ear as the following crest reaches the right ear. The velocity of sound in air is approximately 330 metres per second at 0°C and varies directly as the square root of the absolute temperature of the air.

phase difference gives ambiguous clues about localization as shown in Fig. 38.15 B. Above 1500 Hz the head casts a 'sound shadow'; these short waves cannot get round the head, just as the waves of the sea cannot get round a long breakwater. Consequently if the source of a high-frequency note is to one side of the median plane there is a considerable difference of intensity at the two ears which allows of localization. The superior olive is the first level at which binaural interaction occurs, and the lateral division of this nucleus is thought to play a role in localization. It has been shown that the probability of firing of a neurone in this nucleus depends on the relative time of arrival of a click stimulus at the two ears (Fig. 38.16).

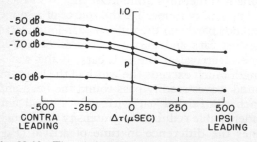

Fig. 38.16 The relative probability of firing (P) of cells in the lateral nucleus of the cat's superior olive as a function of the time difference ($\triangle\tau$) between the arrival of the click at the two ears. Curves are given for various intensity levels of the stimulus. (Note: sound travels about 6 inches in 500μsec). (*From* J. L. Hall (1965), *Journal of the Acoustical Society of America* **37**, 814–823.)

These two clues, difference in time of arrival and difference in intensity, are not enough to decide whether the source is in front of or behind the observer. Small rotational movements of the head may enable more information to be obtained. A familiar sound, such as the human voice, may be easily recognized as coming from behind the observer because the auricle shades the ear from high frequencies and so alters slightly the quality of the voice. This differential filtering effect of the auricle indeed varies continuously as the position of the sound source changes in azimuth and elevation. The resultant change of timbre is probably one factor which enables persons totally deaf in one ear to locate the source of a sound.

Inside a room, reflexions cause the sound to arrive from many directions. However, if the reflected sound is delayed by not more than a few milliseconds, we hear only the primary source, without any echoes (the 'precedence effect'). Although the secondary sources are not heard as such, they are not entirely suppressed, since they reinforce the apparent loudness of the direct sound. If two sound sources in a room are quite different, as for example two different speakers at a cocktail party, then it is usually possible by switching attention to suppress either source at will.

METHODS OF MEASURING ACUITY OF HEARING

Quantitative measurements of hearing are made with a *pure-tone audiometer*. The test must be made in a sufficiently quiet room since extraneous sounds cause masking (p. 773) which raises the threshold of hearing. The pure-tone audiometer consists of an oscillator producing pure sine waves which are fed to a telephone earpiece through an attenuator network to control the intensity of the sound. When the intensity of a given note is made very small it is inaudible but if the intensity is increased slowly then at a certain value the note is heard. This is the *threshold of audibility*. If the test is performed separately for each ear at a selected number of frequencies the threshold can be established and specified in terms of a standard set of values corresponding to the hearing of normal young persons. There are national and international standards for the normal threshold of hearing for pure tones. The current British Standard is BS 2479: Part 1: 1968; and Part 2: 1969.

The threshold of audibility is lowest, that is the auditory acuity is greatest, in the region 1000 to 4000 Hz; at lower and higher frequencies the sound intensity must be greater to produce an auditory sensation. The lowest and highest audible frequencies thus depend on the intensity of the tone used, but the average limits are 20 and 20 000 Hz in young persons. In old people the threshold in the higher frequencies is raised so that in practice their upper limit may be much lower, say 5000 Hz (see Fig. 38.17a).

By varying the intensity of the note produced by a pure-tone audiometer in a rhythmic fashion the minimum perceptible difference in intensity can be measured. Except near the threshold of audibility it is of the order of 1 dB. By varying the frequency of a note rhythmically the minimum fractional difference in frequency which is perceptible can be found; in the range 500 to 4000 Hz it is 0·3 per cent, but outside this range it is greater. A tone must persist for 10 to 15 msec before a pitch can be assigned to it but most people require the tone, especially if it is of low frequency, to last upwards of 100 msec before it supplies an experience of pitch.

The method of measuring intelligibility, first used by telephone engineers, is also useful for investigating deafness and the efficiency of deaf aids. A list of disconnected words containing a wide variety of speech sounds but without context is read out to a listener who writes down what he hears. The script is then checked with the original list and the percentage of sounds heard correctly—spelling errors are of course neglected—is usually referred to as the 'intelligibility'. Suitable word lists have been constructed by Fry (see References, p. 775).

When this test is made in the simplest possible conditions—that is, with the reader speaking directly to the listener in the same room—the intelligibility is high, but the vowel *e* as in 'ten' and the consonants *th*, *f* and *v* often give rise to errors. If now the listener hears the voice of the reader through a loudspeaker system which transmits all the speech frequencies up to 1000 Hz but cuts out all the higher frequencies the intelligibility may be only 40 per cent, although almost all the energy (85 per cent) of the original speech sounds is present. Sounds like *s*, *th* and *f* are particularly difficult to distinguish. This is an artificial high-frequency deafness. If, on the other hand, the loudspeaker transmits only the speech frequencies above 1000 Hz the intelligibility is now 86 per cent although only a small fraction of the original speech energy (17 per cent) is transmitted. In spite of this severe frequency distortion, ordinary speech with its context is easily understood although it does not, of course, sound natural.

The pure-tone audiometer can be used to investigate the hearing of persons with impaired hearing. Useful diagnostic information is obtained by measuring the amount by which the threshold is raised over the range of frequencies. Audiometers intended for clinical use are calibrated so that the zero mark on the intensity control at each frequency corresponds to the average threshold found in young people with normal hearing. This statistical concept of normal hearing is shown on the graph paper, used to record the results, as a straight line marked 0 dB hearing level (Fig. 38.17). If a person has impaired hearing to a particular tone its intensity must be increased *x* dB above the average threshold before it is heard; this is plotted as a point *below* the normal threshold in Fig. 38.17, and is known as a hearing loss of *x* dB at the frequency tested. If a number of tones is used the results fall into a curve called the *audiogram*. A few characteristic audiograms are shown in Fig. 38.17.

Important prognostic information can be gained from the audiogram, since it has been found that there is a definite relationship between the degree of deafness in the important speech frequencies, say 500 to 2500 Hz, and the quality of the speech. Deaf children learn to speak normally provided the loss is less than 30 dB; a greater degree of impairment prevents an infant from learning to speak naturally but may permit him to acquire speech with careful training. In the presence of a loss of 90 dB, however, even acquired speech is poor. Good hearing in the lower frequencies (up to 500 Hz, say) does not compensate for deafness in the high frequencies, since it is the latter which allows us to distinguish the different speech sounds.

Deafness due to middle-ear disease is so

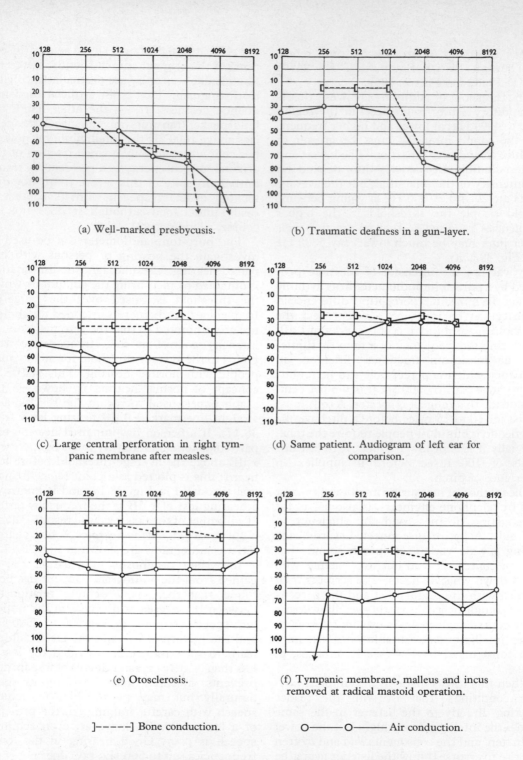

(a) Well-marked presbycusis.

(b) Traumatic deafness in a gun-layer.

(c) Large central perforation in right tympanic membrane after measles.

(d) Same patient. Audiogram of left ear for comparison.

(e) Otosclerosis.

(f) Tympanic membrane, malleus and incus removed at radical mastoid operation.

]------] Bone conduction.   O———O——— Air conduction.

Fig. 38.17 Audiograms from five different patients which illustrate conditiins mentioned in the text. Air conduction was tested by an ear-phone placed over the external auditory meatus. Bone conduction was tested by a vibrator on the mastoid process. The hearing level in dB is shown on the ordinate. o on the ordinate indicates the threshold of hearing in young healthy people. The abscissa is scaled (in octaves) in Hz. (*By courtesy of M. J. Gibson.*)

common that it is important to be able to test large numbers quickly. The pure-tone audiometer is not suitable because it is a rather uninteresting artificial test requiring a special room and considerable time; it can be applied only to one person at a time. The gramophone audiometer provides a somewhat empirical but yet practical test which can quickly pick out persons with defective hearing. Specially made records are played with an electrical pick-up which is connected to twenty or more single earphones to allow a whole class of children to be tested simultaneously. The children apply the earpiece to one ear at a time and hear a voice calling out numbers which they copy down. As the record proceeds the voice becomes fainter and fainter in steps of 3 dB. When the speech is too faint to be understood the children either make mistakes or stop writing. When the performance of normal children has been found with this apparatus, it is easy to detect children who have even minor defects of hearing.

Rinne's and Weber's tests. Sound vibrations can be conducted to the inner ear through the bones of the skull, but normally air conduction is much more effective. This can be shown by applying a vibrating tuning-fork (256 Hz) firmly to the mastoid process just behind the ear. When the sound dies away the prongs of the fork are brought near the external auditory meatus and the sound is heard once more, provided that the meatus is not blocked and that the ear drum and ossicles are intact. Rinne's test is then said to be positive.

Weber's test is carried out by placing a vibrating 256 Hz fork on the mid-sagittal line of the head (the vertex). If one ear is plugged by the finger to exclude room noise the sound produced by the fork appears to be louder in that ear. Thus in unilateral disease affecting the conducting system (middle-ear disease) Weber's test is referred to the diseased ear. When the organ of Corti or the cochlear nerve is damaged unilaterally Weber's test is referred to the normal ear.

Rinne's and Weber's tests are used in clinical practice to distinguish middle-ear (conduction) deafness from inner-ear (perceptive) deafness. In conduction deafness Rinne's test is usually negative and Weber's test indicates the diseased ear. In perceptive deafness Rinne's test is positive and Weber's test indicates the normal ear.

NOISE

Noise is not only to be defined in physical terms but also in psychological terms such as a sound which is disturbing or annoying. The loudness of sounds can be measured either subjectively or objectively. In the subjective method the loudness of a sound is compared with the loudness of a note of 1000 Hz which can be adjusted until the listener judges that it has the same loudness as the sound being investigated. When this equality has been achieved the intensity of the 1000 Hz note can be read from the instrument in dB above an arbitrary zero which has been internationally agreed as 2×10^{-5} N/m². If the reading is n dB the sound is said to have an intensity of n phons. It is more usual to employ objective sound level meters which have a microphone and a calibrated amplifier with a number of alternative frequency response characteristics based on those of the ear at different sound levels. Such instruments give a reading of the sound level in dB above the same reference level. This objective method does not necessarily give a true index of loudness. A range of phon values is given in Fig. 38.18.

Masking. Whenever there is a noise (as in a railway train) conversation becomes difficult and the speech power must be raised. This effect begins when the noise level reaches 20 phons and the loudness of speech is increased up to noise levels of 90 phons beyond which the intensity of the speech cannot compete with the noise. The increase in the loudness of the voice is often revealed if the noise suddenly ceases. The nature of this masking effect has been investigated with the pure-tone audiometer. First, in a quiet room the threshold of audibility is measured, then a masking tone is continuously sounded and the threshold is measured again. Masking tones of moderate intensity make it difficult to hear notes of neighbouring frequency but if the masking tone is very loud there is masking not only of neighbouring tones but also of all tones of higher frequency. For this reason loud tones of low frequency impair the intelligibility of speech.

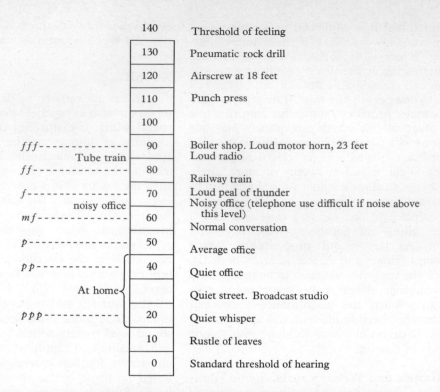

Fig. 38.18 A range of phon values. The phon is defined on p. 773. On the left are given the values of the musical *p* (piano) and *f* (forte) markings as used by the orchestral conductor Leopold Stokowski.

REFERENCES

BAST, T. H. & ANSON, B. J. (1949). *The Temporal Bone and the Ear*. Oxford: Blackwell.

BATTEAU, D. D. (1967). The role of the pinna in human localization. *Proceedings of the Royal Society* B **168**, 158–180.

BÉKÉSY, G. VON (1956). Current status of theories of hearing. *Science, New York* **123**, 779–783.

BÉKÉSY, G. VON (1960). *Experiments in Hearing*. New York: McGraw-Hill.

BÉKÉSY, G. VON & ROSENBLITH, W. A. (1951). *Handbook of Experimental Psychology* (ed. S. S. Stevens), pp. 1075–1115. London: Chapman & Hall.

BELL, A. (1966). *Noise*. Geneva: W.H.O.

BURNS, W. (1965). Noise as an environmental factor in industry. *Transactions of the Association of Industrial Medical Officers* **15**, 2–11.

BURNS, W. (1972). *Noise and Man*. London: John Murray.

BURNS, W. & ROBINSON, D. W. (1970). *Hearing and Noise in Industry*. London: H.M.S.O.

Committee on problems of Noise (1963). *Noise, Final Report*. London: H.M.S.O. Cmnd. 2056.

DAVIS, H. (1965). A model for transducer action in the cochlea. *Cold Spring Harbor Symposia on Quantitative Biology* **30**, 181–190.

DIX, M. R., HALLPIKE, C. S. & HOOD, J. D. (1949). Nerve deafness, its clinical criteria, old and new. *Journal of Laryngology and Otology* **63**, 685–698.

DRAPER, M. H., LADEFOGED, P. & WHITTERIDGE, D. (1960). Expiratory pressures and air flow during speech. *British Medival Journal* iv, 1837–1843.

ELDREDGE, D. H. & MILLER, J. D. (1971). Physiology of hearing. *Annual Review of Physiology* **33**, 281.

ENGSTROM, H., ADES, H. W. & ANDERSSON, A. (1966). *Structural Pattern of the Organ of Corti*. Stockholm: Almqvist & Wiksell.

EWING, A. & EWING, E. C. (1964). *Teaching Deaf Children to Talk*. Manchester University Press.

FLETCHER, H. (1929). *Speech and Hearing*. London: Macmillan. Also 2nd edn (1953). London: Macmillan.

FRY, D. B. (1961). Word and sentence tests for use in speech audiometry. *Lancet* ii, 197–199.

HALLPIKE, C. S. (1965). Clinical otoneurology and its contributions to theory and practice. *Proceedings of the Royal Society of Medicine* B **58**, 185–196.

IURATO, S. (1967). *Submicroscopic Structure of the Inner Ear*. Oxford: Pergamon Press.

KIANG, N. (1965). *Discharge Patterns of Single Fibers in the Cat's Auditory Nerve*. Cambridge, Massachusetts: M.I.T. Press.

LITTLER, T. S. (1965). *The Physics of the Ear*. Oxford: Pergamon.

NEGUS, V. E. (1962). *The Comparative Anatomy and Physiology of the Larynx*. London: Haffner.

PENFIELD, W. & ROBERTS, L. (1959). *Speech and Brain Mechanisms*. Oxford University Press.

POTTER, R. K., KOPP, G. A. & GREEN, H. C. (1947). *Visible Speech*. New York: Van Nostrand.

PRESSMAN, J. J. & KELEMEN, G. (1955). Physiology of the larynx. *Physiological Reviews* **35**, 506–554.

RASMUSSEN, G. L. & WINDLE, W. F. (Eds.) (1961). *Neural Mechanisms of the Auditory and Vestibular Systems*. Oxford: Blackwell.

RICE, C. E. (1967). Human echo perception. *Science*, New York **155**, 656–664.

STRANG, B. M. H. (1962). *Modern English Structure*. London: Arnold.

Symposium on Neuro-otology (1956). *British Medical Bulletin* **12**, 91–157.

Symposium on Speech (1961). *Nature, London* **192**, 601–603.

Symposium on hearing (1966). *Journal of the American Medical Association* **196**, 831–846.

Symposium on Hearing Mechanisms in Vertebrates (1968). Ciba Foundation.

TASAKI, I., DAVIS, H. & LEGOUIX, J. P. (1952). The space-time pattern of the cochlear microphonics (guinea-pig), as recorded by differential electrodes. *Journal of the Acoustical Society of America* **24**, 502–519.

TAYLOR, I. G. (1964. *Neurological Mechanism of Hearing and Speech in Children*. Manchester University Press.

WEVER, E. G. & LAURENCE, M. (1954). *Physiological Acoustics*. Princeton University Press.

WEVER, E. G. (1966). Electrical potentials of the cochlea. *Physiological Reviews* **46**, 102–127.

WHITFIELD, I. C. (1967). *The Auditory Pathway*. London: Arnold.

39 Neurone and synapse

THE STRUCTURE OF NERVE FIBRES

The nervous system is composed of an enormous number of nerve cells, or *neurones*, which are specially adapted for the handling of information in the form of *nerve impulses*. These impulses pass on from one neurone to another at junctions known as *synapses*. The processes involved in the passage of impulses along peripheral nerves or the fibre tracts of the central nervous system will be considered in the first part of this chapter, and the remainder of the chapter will be devoted to junctional transmission. Having described the characteristics of the two fundamental components of the nervous system, the nerve fibre and the synapse, we can then proceed to discuss its more complex aspects.

The first step towards understanding the mechanism of the nerve impulse is to obtain a clear picture of the structure of a nerve fibre. Each neurone comprises a cell body and a number of attached processes, one of which, usually much longer than the rest, is the *axon* or nerve fibre. No axon can continue to function for more than a short while after being severed from its cell body, but nevertheless the cell body does not appear to play a direct role in conduction of the nerve impulse, and it is therefore convenient to postpone until later a detailed account of its structure. Vertebrates have two main types of nerve fibre, the larger axons, 1 to 25 μm in diameter being *myelinated* and the smaller ones (under 1 μm) *non-myelinated*. Almost all invertebrates are equipped only with non-myelinated fibres, but some of these differ from vertebrate non-myelinated fibres in being very much larger. The great size of certain invertebrate axons has been extensively exploited by physiologists studying the propagation of nerve impulses, and isolated giant axons of the squid, up to 1000 μm in diameter, have proved to be especially valuable as experimental material. It may be noted at this point that vertebrate skeletal muscle fibres (10 to 120 μm) have much in common, as far as the spread of electrical activity along them is concerned, with non-myelinated invertebrate nerve fibres.

In essence, all nerve fibres consist of a long cylinder of cytoplasm, the *axoplasm*, surrounded by an electrically excitable *nerve membrane*. Since the electrical resistance of the axoplasm is fairly low, by virtue of the K^+ and other ions that are present in appreciable concentrations, while that of the membrane is relatively high, and since the salt-containing body fluids outside the membrane are again good conductors of electricity, nerve fibres have a structure analogous to that of a shielded electric cable, with a central conducting core surrounded by insulation, outside which is another conducting layer. As will be seen shortly, many features of the behaviour of nerve fibres depend intimately on their *cable structure*. The layer analogous with the insulation of the cable does not, however, consist solely of the high-resistance nerve membrane, owing to the presence of *Schwann cells*, which are wrapped around the *axis cylinder* in a manner which varies in the different types of nerve fibre. Details of the arrangement of the Schwann cells are beyond the resolving power of ordinary light microscopy, and have become clear only with the advent of the electron microscope. In osmic- or permanganate-stained sections, the membrane surrounding most types of cell is seen in the electron microscope as a structure about 7·5 nm thick, consisting of a pair of dark lines 2·5 nm across separated by a 2·5 nm gap; the membranes of both Schwann cells and nerve axons can readily be made out in electron micrographs, so that the relationship between the Schwann cell and the axon can be examined. In the small non-myelinated fibres of vertebrates the picture is somewhat variable. In some cases (Figs. 39.1 and 2), as was first described by Gasser for olfactory nerves, a single Schwann cell may serve as a multi-channel supporting structure partly enveloping a short stretch of a dozen or more tiny axons. Elsewhere, each axon may be more or less closely associated with a Schwann cell of its own, some being deeply embedded within the Schwann cell, and others almost uncovered. In the large invertebrate non-myelinated fibres the Schwann cells evidently

form a mosaic covering the whole surface of the axon, sometimes with only one or two layers, as in the giant axons of the cuttlefish (Fig. 39.3), and sometimes with a more elaborate arrangement of several inter-digitating layers of Schwann cell, as described for the squid axon by Geren and Schmitt. In all these non-myelinated fibres the 10 nm spaces between the Schwann cell membranes and the axon membranes communicate all along the length of the fibres with the extracellular space of the tissue, providing a relatively uniform pathway for the electric current which flows during the passage of an impulse. There is some evidence that this pathway can be sufficiently tortuous to prevent ions which move out through the axon membrane in the course of the impulse from mixing quickly with extracellular ions, and an effect of this kind may be important in mammalian C fibres (see p. 784 in the mechanism of production of after-potentials (see p. 807). However, for the immediate purpose of describing the way in which nerve impulses are propagated, non-myelinated fibres may be treated as if the external electrical resistance between different points on the outside of the membrane were uniformly low.

In the myelinated nerve fibres of vertebrates, each Schwann cell is responsible for laying down a section of the myelin sheath, by repeatedly enveloping the axis cylinder with many concentric layers of Schwann cell membrane, as may be seen in Figures 39.1 and 39.4. Various lines of evidence suggest that all cell membranes consist of a layer of lipid molecules, probably arranged with their long hydrophobic chains perpendicular to the plane of the membrane, sandwiched between tangentially orientated proteins and perhaps mucopolysaccharides (p. 76). In an adult myelinated fibre, neighbouring layers of Schwann cell membrane appear to be partly fused together, making a structure whose radial repeat distance is of the order of 17 nm as seen in the electron microscope or as determined by X-ray diffraction studies. Since myelin has a much higher lipid content than cytoplasm, it has a relatively high refractive index, and in unstained preparations has a characteristic glistening white appearance. This accounts for the different appearances of the grey matter of the spinal cord, consisting mainly of nerve cell bodies

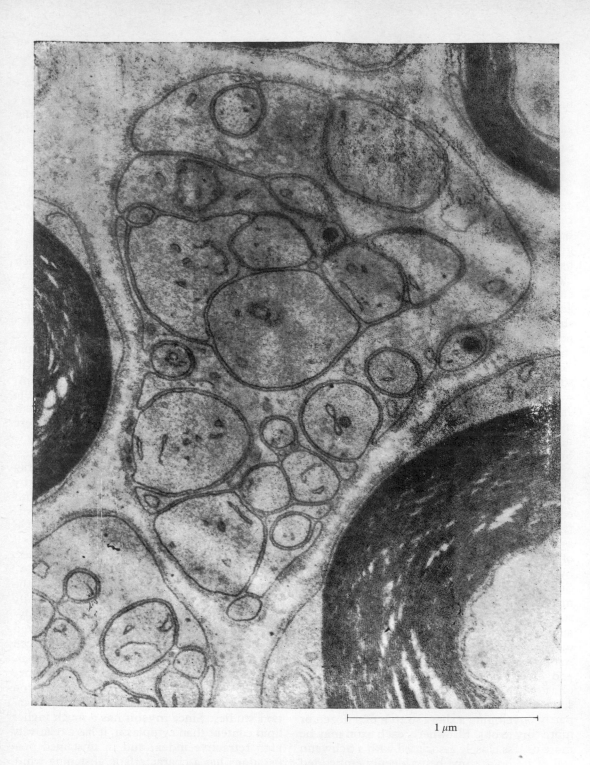

Fig. 39.1 Electron micrograph of a transverse section through fibres in the sciatic nerve of a 47-day-old mouse. In the centre is a Schwann cell with a number of non-myelinated fibres embedded in it. Parts of myelinated fibres may be seen at the sides. Magnification ×50 000.
(By courtesy of J. D. Robertson.)

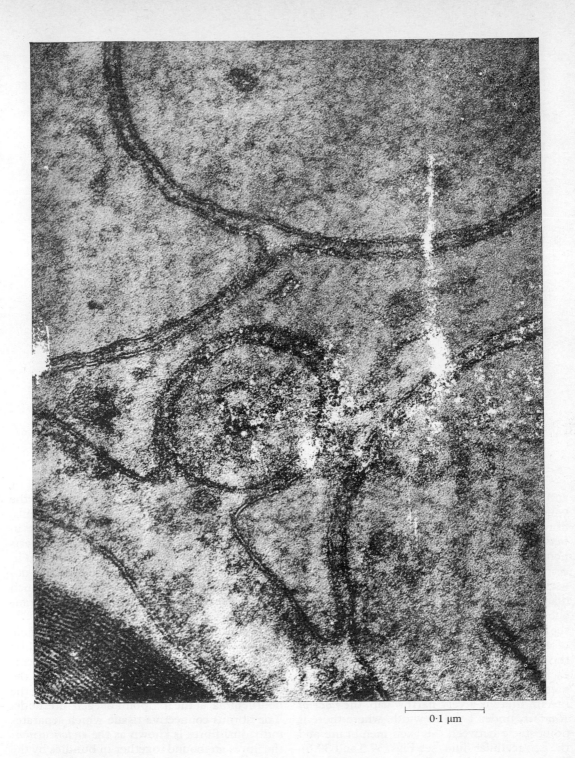

0·1 μm

Fig. 39.2 Part of Fig. 39.1 at higher magnification, showing the appearance of cell membranes as double dark lines with a gap separating the axons from the Schwann cell. Magnification × 210 000. (*By courtesy of J. D. Robertson.*)

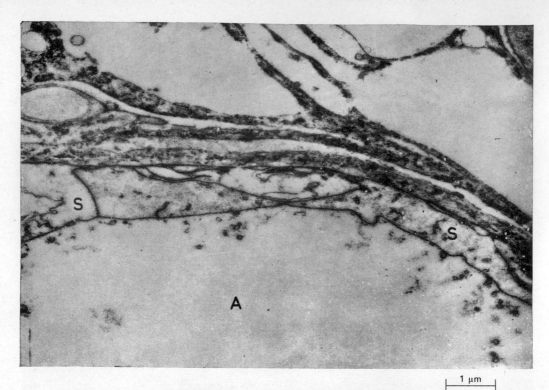

Fig. 39.3 Electron micrograph of a transverse section through a giant axon (diameter 200 μm) from a cuttlefish (*Sepia officinalis*). The axoplasm, A, is surrounded by one or two layers of Schwann cell, S, outside which is connective tissue. Magnification × 13 000. (*By courtesy of A. F. Huxley.*)

and supporting tissue, and of the white matter, which is mainly tracts of myelinated nerve fibres; it also accounts for the difference between the white and grey rami of the autonomic nervous system (see p. 861), containing respectively myelinated and non-myelinated nerve fibres. In a nerve fibre whose outside diameter is 10 μm, each stretch of myelin is about 1000 μm long and 1·3 μm thick, so that the myelin is built up of some 80 double layers of Schwann cell membrane. In larger fibres, the thickness of the myelin and the length of each segment of myelin, are proportionately greater. Between neighbouring segments of myelin there is a very narrow gap, the *node of Ranvier*, under 1 μm in width, where there is no obstacle between the axon membrane and the extracellular fluid (see Figs. 39.5 and 39.6). The external electrical resistance between neighbouring nodes of Ranvier is therefore relatively low, whereas the external resistance between any two points on the internodal

stretch of membrane is high because of the insulating effect of the myelin. The difference between the nodes and the internodes in accessibility to the external medium is the basis for the *saltatory* mechanism of conduction in myelinated nerve fibres, which will be described in detail later. Nerves may branch many times before terminating and the branches always arise at nodes.

In peripheral myelinated nerves the whole axon is usually described as being covered by a thin apparently structureless basement membrane, the *neurilemma*. The nuclei of the Schwann cells are to be found just beneath the neurilemma, at the midpoint of each internode. The fibrous connective tissue which separates individual fibres is known as the *endoneurium*: the fibres are bound together in bundles by the *perineurium*, and the several bundles which in turn form a whole nerve trunk are surrounded by the *epineurium*. The connective tissue sheaths in which the bundles of nerve fibres

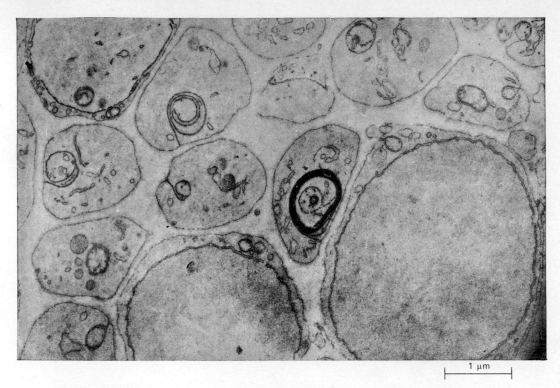

Fig. 39.4 Transverse section showing several stages in the development of myelin in a mouse sciatic nerve. Magnification × 17 520. (*By courtesy of J. D. Robertson.*)

are wrapped also contain continuous sheets of cells which prevent extracellular ions in the spaces between the fibres from mixing freely with those outside the nerve trunk. The barrier to free diffusion offered by the sheath is probably responsible for some of the experimental discrepancies between the behaviour of fibres in an intact nerve trunk and that of isolated single nerve fibres. The nerve fibres within the brain and spinal cord are packed together very closely, and are usually said to lack a neurilemma. Their myelin sheaths are laid down by oligodendroglial cells which seem to play a part similar to that of the Schwann cells in peripheral nerves. The myelin is also interrupted at nodes of Ranvier but, since the individual fibres are difficult to tease apart, the nodes are not easily demonstrated by histological techniques such as staining with silver nitrate.

CHARACTERISTICS OF THE NERVE IMPULSE

As observed with a pair of recording elec-trodes in contact with the outside of a nerve, the impulse consists of the passage along the nerve at a constant velocity of a short region which is electrically negative relative to the rest of the nerve. If both electrodes are on an intact part of the nerve, and if they are separated by a great enough distance, a *diphasic* recording of the type shown in Figure 39.7 (A), is obtained. Here the active region first appears under the left-hand electrode, then traverses the stretch between the electrodes, and finally reaches the right-hand electrode where it gives rise to a mirror-image deflexion on the recording instrument. If the nerve is crushed or cut under the right-hand electrode, the impulse never reaches this point, and a *monophasic* recording of the *action potential* is obtained (Fig. 39.7 (B)). Under many experimental conditions it is difficult to demonstrate the classical diphasic picture of Figure 39.7 (A) because of the impossibility of adequately separating the two electrodes. Thus in a frog nerve at room temperature the duration of the action potential is of the order of 1·5 msec, and

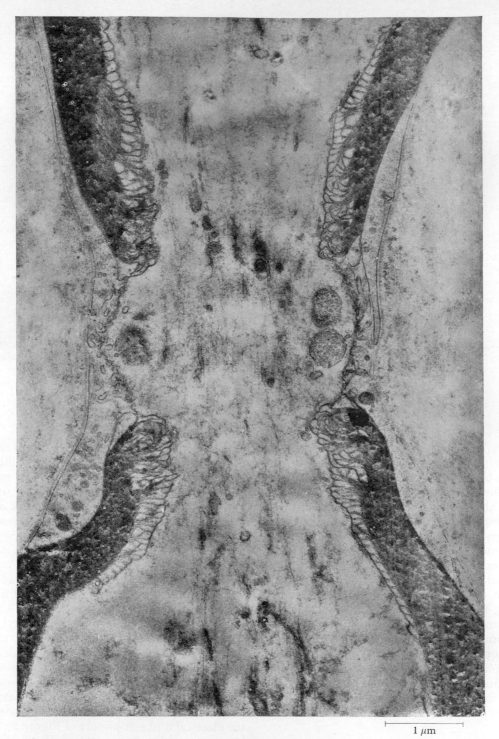

Fig. 39.5 Electron micrograph of a longitudinal section through a node of Ranvier in a frog sciatic nerve fibre. At the centre of the node the axon membrane bulges slightly, and there often appear round bodies containing numerous vesicles whose function has not yet been established. Magnification × 20 000. (*By courtesy of J. D. Robertson.*)

1 μm

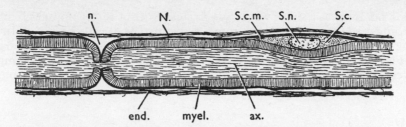

Fig. 39.6 Diagram of part of a normal vertebrate nerve fibre. The Schwann nucleus (S.n.) lies much nearer to the node of Ranvier (n.) than it would usually do, but otherwise the proportions are roughly correct. ax., axis cylinder: end., endoneurium; myel., myelin sheath, formed of multiple layers of Schwann cell membrane (S.c.m.); S.c., cytoplasm of Schwann cell; N., neurilemma. (*From* W. Holmes & J. Z. Young (1942). *Journal of Anatomy* **77,** 67.)

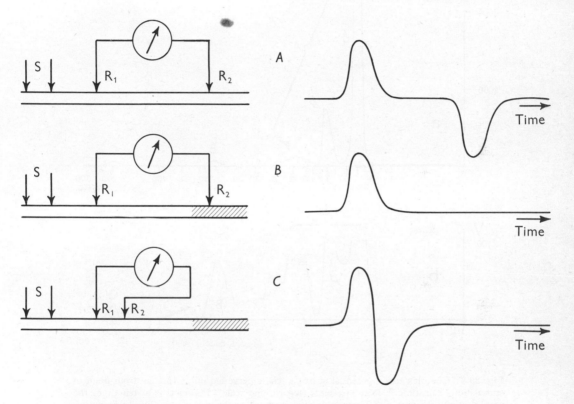

Fig. 39.7 Diagram illustrating the electrical changes accompanying the passage of a nerve impulse, as seen on an oscilloscope connected to external recording electrodes R$_1$ and R$_2$. S, stimulating electrodes. An upward deflexion is obtained when R$_1$ is negative relative to R$_2$. A, diphasic recording when R$_1$ and R$_2$ are both on intact part of nerve and separated by an appreciable distance. B, monophasic recording when nerve is cut or crushed under R$_2$. C, diphasic recording with R$_2$ moved on to intact nerve again, much closer to R$_1$.

the conduction velocity is about 20 m/sec; the length of the active region is therefore 30 mm, which is often as great as the maximum separation of the electrodes. The two phases of the recording then overlap one another to some extent, so that although the record is still diphasic there is no longer a flat portion between the phases (Fig. 39.7 (C)).

When a frog's sciatic nerve is very strongly stimulated (see below) all the nerve fibres in it are excited. A monophasic recording with an electrode placed near the point of stimulation shows the action potential as a single wave, but a similar recording made at some distance from the stimulating electrodes gives a complicated wave form (Fig. 39.8 (a)) in which three groups of waves (spikes) can be distinguished.

rates. Such a set of impulses, initiated at the same instant, is called a *volley*. In this particular experiment, for which a large American bull frog was used at room temperature, the distance from the stimulating to the recording

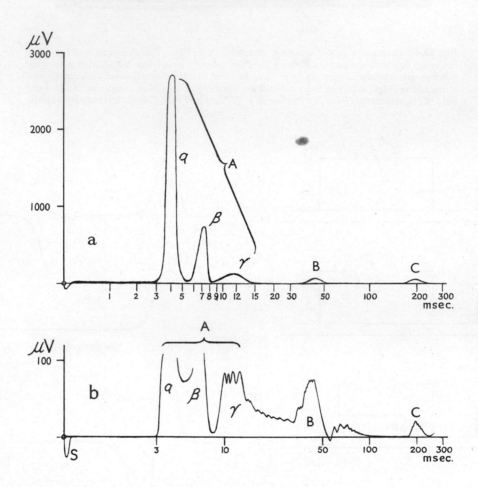

Fig. 39.8 Complex action potential of frog's sciatic nerve led off at 13·1 cm from point of stimulation. The time is in milliseconds (logarithmic scale). Figure b is at ten times the amplification of Figure a. S = stimulus artifact, marking the zero point. (Drawn from diagrams given by J. Erlanger & H. S. Gasser (1937). *Electrical Signs of Nervous Activity*. Philadelphia: University of Pennsylvania Press.)

These are usually labelled A, B and C; A is further subdivided into α, β and γ. Each part of the complex wave of Figure 39.8 (a) is contributed by a different group of nerve fibres; the action potentials all start from the point of stimulation at the same time, but travel toward the recording electrode at different

electrode was 13·1 cm. If the time for the foot of the wave to reach the recording electrode is read off the logarithmic abscissa of Figure 39.8 (a), it is easy to calculate that the rate of conduction was 41 metres/second for α, 22 for β, 14 for γ, 4 for B and 0·7 for C. The conduction velocities in mammalian nerves are somewhat greater (100 for α, 60 for β, 40 for γ, 10 for B and 2 for C), partly because body temperature is appreciably higher and partly because the fibres are larger.

The distribution of conduction velocities over such a wide range results from a variation in fibre diameter over an equally wide range. A large nerve fibre conducts impulses faster than a small one. Several other characteristics of a nerve fibre vary markedly with size. In particular, the smaller fibres require stronger shocks to excite them, so that the form of the volley in a mixed nerve is affected by the strength of the stimulus. If the stimulus is weak, only the α wave appears; if it is stronger then both α and β waves are seen, and so on. The size of the voltage change picked up by an external recording electrode also varies with fibre diameter; on theoretical grounds it would be expected to vary with $(diameter)^2$, but Gasser's reconstructions provide some support for the view that the relationship is a linear one. The consequence is that records of electrical activity in sensory nerves *in situ* show mainly the activity of the largest fibres and it is, indeed, difficult to see at all the action potentials in the smallest non-myelinated fibres.

Although there is a continuous range of fibre diameters in most nerves, it has proved hard to assign with confidence particular functions to particular sizes of fibres. The sensory root of the spinal cord contains fibres giving A (that is α, β and γ) and C waves; the motor root yields α, γ and B waves, the latter going into the white ramus. It is now generally believed that B fibres occur only in the preganglionic autonomic nerves so that the elevation labelled B in Figure 39.8 might be better reclassified as subdivision δ of Group A. The grey ramus, containing fibres passing from the sympathetic system, shows mainly C waves. The fastest fibres (α) are either motor fibres activating voluntary muscles or afferent fibres conveying impulses from sensory receptors in the muscles. The γ motor fibres are connected in mammals to intrafusal muscle fibres (see p. 828) and in amphibia to 'slow' (as opposed to 'twitch') muscle fibres. At least some of the fibres of the C group convey pain impulses. The C fibres are probably non-myelinated: they are found especially in autonomic post-ganglionic trunks. Myelinated sensory fibres in peripheral nerves have also been classified according to their diameters into Group I (20 to 12 μm), Group II (12 to 4 μm), and Group III (4 μm and under), with corresponding conduction velocities. The Group I fibres, which conduct impulses most quickly, are found only in nerves from muscles. They have been subdivided into Group I A, connected with annulo-spiral endings of muscle spindles, and the somewhat slower Group I B, carrying impulses from Golgi tendon organs. The slower fibres of Groups II and III transmit other modes of sensation in both muscle and skin nerves.

So far, the nerve impulse has been described only in terms of the picture seen when external electrodes are applied to the nerve. For many purposes this is the only practicable recording technique but, as a means for examining the mechanism of production of the action potential, it has severe limitations. It has been known for many years that in a resting nerve fibre there is a steady electric potential difference between the inside of the cell and the outside, the inside being relatively negative. This so-called *resting potential* gives rise to the *injury potential* recorded when one electrode is in contact with an uninjured part of the nerve and the other is in contact with a cut or crushed point (that is the condition for a normal monophasic recording with external electrodes). The existence of the resting potential was part of the basis for the idea put forward by Bernstein over seventy years ago that electrical activity in nerve fibres consisted in a temporary collapse of the membrane potential towards zero, this active region being propagated by a mechanism involving the spread of electric currents ahead of it. An important test of Bernstein's hypothesis was to compare the absolute sizes of the resting and action potentials, but with external electrodes it is very difficult to make the comparison with any confidence, because of uncertainties that arise in allowing for the short-circuiting effect of extracellular current pathways and for junction potentials between recording electrodes and tissue fluids. The test was not, in fact, made until 1939, when Hodgkin and Huxley in England and Curtis and Cole in the United States, first devised a technique for introducing an electrode inside a nerve fibre, and so measuring directly the potential difference across the resting and active nerve membrane.

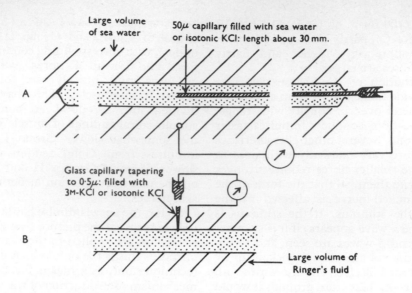

Large volume
of sea water

50μ capillary filled with sea water
or isotonic KCl: length about 30 mm.

A

Glass capillary tapering
to 0·5μ: filled with
3M-KCl or isotonic KCl

B

Large volume of
Ringer's fluid

Fig. 39.9 Methods for obtaining absolute values of resting potential and action potential. A, longitudinal insertion of internal electrode used for giant axon of the Squid, *Loligo*. B, transverse insertion of internal electrode used for muscle fibres and other cells (A. L. Hodgkin (1951), *Biological Reviews* **26,** 342).

Figure 39.9 (A) shows the type of electrode used to record the internal potential in a squid axon. A 50 or 100 μm electrode of this kind can be used only with an axon as large (500 to 1000 μm diameter) as that of the squid, but in 1949 Ling and Gerard devised another kind of micro-electrode applicable to a very wide variety of cells. They found that it was not difficult to make glass micro-pipettes with a tip diameter of 0·5 μm, and that such pipettes, filled with a KCl solution, could be thrust transversely into frog muscle fibres without tearing too large a hole at the point of entry, and so used for direct measurement of internal potential (see Fig. 39.9 (B)). Micro-electrodes of this type have since been used to observe resting and action potentials in every kind of excitable tissue, from cuttlefish axons to cardiac muscle fibres, and from motoneurones in the cat's spinal cord to the electroplates in the electric organs of fishes. Figure 39.10 shows some typical records, from which the following generalizations may be made:

1. There is a resting potential (inside negative) whose size ranges from 60 to 95 millivolts.

2. During the passage of an impulse the membrane potential is reversed for a short while; the size of this overshoot is of the order of 30 to 60 millivolts.

3. Although the sizes of the resting and action potentials thus vary remarkably little from tissue to tissue, the duration of the action potential (spike) shows considerable variation. In mammalian nerves the duration is under 1 millisecond, whereas in amphibian cardiac muscle fibres it is of the order of half a second.

4. There is also some variation in the shape of the spike. Thus the undershoot (positive phase) at the tail end of the spike is always present in isolated squid axons, though it is probably absent in the living animal. In frog muscle fibres the spike is characteristically followed by a relatively long-lasting negative phase, which may be related to the mechanism by which the contractile machinery is activated. In cardiac muscle fibres there is always a long plateau after the peak of the spike; the complicated shape of the spike recorded in the conducting (Purkinje) fibres of the sheep's heart (Fig. 39.10 (F)) is typical of cardiac muscle.

The first and most important conclusion drawn from micro-electrode studies was that Bernstein's hypothesis was only in part correct.

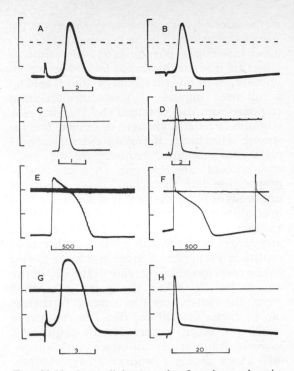

Fig. 39.10 Intracellular records of resting and action potentials. The horizontal lines (dashed in A and B) indicate zero potential level; negative downwards. Marks on voltage scales are 50 millivolts apart in every case. The number against each time scale gives the distance between the marks in milliseconds. In some cases the action potential is preceded by a stimulus artifact. The part of the action potential (usually about 40 mV) above the zero line is referred to as the overshoot. A, squid axon *in situ* at 8·5°C, using 0·5μm microelectrode. B, squid axon isolated by dissection, at 12·5°C, using 100 μm longitudinal electrode. C, myelinated fibre from dorsal root of cat. D, cell body of motoneurone in spinal cord of cat. E, fibre in frog's heart. F. Purkinje fibre in sheep's heart. G, electroplate in electric organ of *Electrophorus electricus*. H, isolated fibre from frog's sartorius muscle. (A and B recorded by A. L. Hodgkin and R. D. Keynes, *from* A. L. Hodgkin (1958). *Proceedings of the Royal Society* B **148**, 5; C recorded by K. Krnjevic; D *from* L. G. Brock, J. S. Coombs & J. C. Eccles (1952). *Journal of Physiology* **117**, 439; E recorded by B. F. Hoffman, F by S. Weidmann, *from* S. Weidmann (1956). *Elektrophysiologie der Herzmuskelfaser*, p. 20. Bern: Huber; G *from* R. D. Keynes & H. Martins-Ferreira (1953). *Journal of Physiology* **119**, 323; H *from* A. L. Hodgkin & P. Horowicz (1957). *Journal of Physiology* **136**, 17P.)

The membrane potential at the peak of the spike does not, as he supposed, merely collapse towards zero, but is reversed by a substantial number of millivolts. The current explanation for this fact, mainly due to Hodgkin and Huxley at Cambridge, will be discussed on pp. 791 to 794.

EXCITATION AND THE LOCAL-CIRCUIT THEORY

Before discussing the ionic basis of the mechanism of conduction of the nerve impulse, it is best to describe some facts concerned with the process of excitation, that is to say with the way in which the nerve impulse is set up. This order of treatment is, historically, that in which research on this subject proceeded, largely because it was difficult to make progress in studying the conduction mechanism before the introduction of internal micro-electrode techniques, whereas excitation could be investigated with comparatively simple methods. For the same reason, the student can more easily observe for himself some of the phenomena of excitation than perform experiments on the conduction mechanism.

Although a nerve can be stimulated by the local application of a number of agents—for example, electric current, pressure, heat, or chemical solutions such as KCl—it is most easily and conveniently stimulated by applying electric shocks. The most effective electric current is one which *depolarizes* the nerve membrane, that is to say which reduces the size of the electric potential across the membrane. The other agents listed above also produce their effect by causing a depolarization; pressure and heat do so by damaging the nerve membrane. It was suggested long ago that propagation of an impulse depends essentially on the flow of current in *local circuits* ahead of the active region which depolarizes the resting membrane, and causes it in turn to become active. This local circuit theory is illustrated in Figure 39.11; here it is the flow of current from region A to region B which results in movement of the active region towards the right. There are important differences between the current pathways in non-myelinated nerves (upper diagram) and in myelinated nerves (lower diagram), which will be dealt with later, but the principle is the same for both types of fibre in that the active region triggers the resting region ahead of it by causing an outward (depolarizing) flow of electric current. The local circuit theory has been universally accepted for some time, and is mentioned at this point in order to emphasize that in studying the effect of applied electric

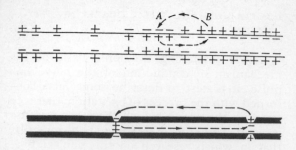

Fig. 39.11 A diagram illustrating the local circuit theory. The upper sketch represents a non-myelinated fibre and the lower sketch a myelinated fibre (A. L. Hodgkin (1957). *Proceedings of the Royal Society* B, **148**, 1).

currents we are not concerned with a non-physiological and purely artificial way of setting up a nerve impulse, but are examining a process which forms an integral part of the normal mechanism of propagation.

The first concept that must be established is that of a *threshold* stimulus. In the experimental arrangement of Figure 39.12, in which

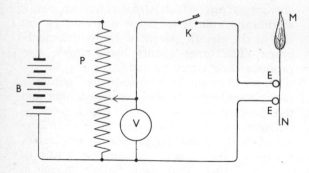

Fig. 39.12 B, battery; P, potentiometer; V, voltmeter; K, key; EE, non-polarizable electrodes; M, muscle; N, nerve.

an isolated nerve-muscle preparation is used, the key K can be closed with various settings of potentiometer P. The smallest voltage which gives a just perceptible muscle twitch is the minimal or threshold stimulus. It is the voltage which is just sufficiently large to stimulate one of the nerve fibres, and hence to cause contraction of the muscle fibres to which it is connected. If the nerve consisted only of a single fibre, it would be found that a further increase

in the applied voltage would not increase the size of the twitch. This is because conduction is an *all-or-none* phenomenon; the stimulus either (if it is subthreshold) fails to set up an impulse, or (if it is threshold or above) sets up a full-sized impulse. No intermediate-sized response can be obtained by varying the stimulus strength, though of course the response will change if certain external conditions (for example temperature or ionic environment) are changed. In a multi-fibre preparation like the sciatic nerve there are hundreds of fibres whose thresholds are spread over quite a wide range of voltages. Hence an increase in stimulus strength above that which just excites the fibre with the lowest threshold results in excitation of more and more fibres, with a corresponding increase in the size of the muscle twitch. When the point is reached where the twitch ceases to increase further, it can be taken that all the fibres in this nerve trunk are being triggered; this requires a *maximal* stimulus. A still larger (supra-maximal) shock does not produce a larger twitch.

A particularly clear example of the threshold behaviour of a nerve fibre is provided by the experiment shown in Figure 39.13. Here an isolated squid axon was being stimulated by applying brief shocks between a long wire inserted longitudinally into it and an external electrode; there was an internal recording electrode alongside the stimulating wire. Threshold for excitation was found to occur when a depolarizing shock of 12 nanocoulombs/cm^2 membrane was applied; at this shock strength the response was delayed for several milliseconds during which period the membrane was in a 'meta-stable' state, depolarized by about 10 mV, sometimes giving a spike and sometimes reverting to its resting state. The effect of increasing the shock strength was to reduce the waiting period but the size of the spike was not changed appreciably. The figure also shows the effect of applying inward (polarizing) currents; it will be seen that these had the effect of temporarily displacing the membrane potential, which then decayed exponentially back to its resting value. But no spikes were generated. Further explanations for this behaviour in terms of ionic permeability changes will be given in the next section.

An important variable in investigating the excitability of a nerve is the duration of the

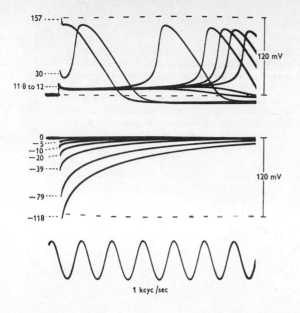

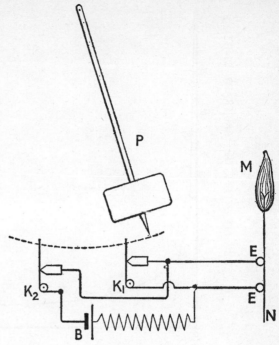

Fig. 39.13 Threshold behaviour of the membrane potential in a squid giant axon at 6°C. Shocks, whose strengths in nanocoulomb/cm² are shown against each curve, were applied to an internal wire electrode with a bare portion 15 mm long. Internal potential was recorded between a second wire 7 mm long opposite the centre of the stimulating wire and an electrode in the sea water outside. Depolarization shown upwards (A. L. Hodgkin, A. F. Huxley & B. Katz (1952). *Journal of Physiology* **116,** 424).

Fig. 39.14 Scheme of apparatus for sending in brief rectangular pulses of current to a tissue. P is a heavy pendulum which on being released knocks open two switches, K_1 and K_2, in quick succession. When K_1 and K_2 are both closed the nerve N is short-circuited by K_1. When K_1 is knocked open current passes through the nonpolarizable electrodes EE into the nerve. When K_2 is knocked open the battery circuit is broken. (Electronic devices for producing 'square waves' are now used.)

shock. The classical method of varying the duration is to use the type of apparatus shown in Figure 39.14, in which a pendulum operates two knock-down switches in quick succession. If the threshold is determined, it is found that for long shocks the applied current reaches an irreducible minimum; the threshold for a shock of infinite duration (obtained by using the apparatus of Figure 39.12 and simply closing the key) is known as the *rheobase*. When the duration is reduced, a stronger shock is needed, so that the curve relating shock strength to shock duration, which is known as the *strength-duration curve*, takes the form shown in Figure 39.15. It may be noted that the essential requirement to stimulate a nerve is that the membrane should be depolarized to a certain critical level whose existence is shown clearly by Figure 39.13. This critical level is, in effect, the rheobase. When the shock duration is reduced, a larger current needs to flow if the membrane potential is to reach the

critical level before the end of the shock. Rather roughly, this means that for short shocks a constant total quantity of electricity needs to be applied; in Figure 39.13 the shock strength was therefore expressed in nanocoulombs/cm² membrane.

One further definition may be helpful. The *chronaxie* of an excitable tissue was defined by Lapique as the duration of a threshold shock of strength equal to twice the rheobase. This quantity is a convenient way of expressing the excitability of a tissue since it is to some extent independent of experimental conditions such as size of electrodes and distribution of tissue resistances. However, the concept of chronaxie has not proved particularly valuable in recent research.

It has been known for a very long time that,

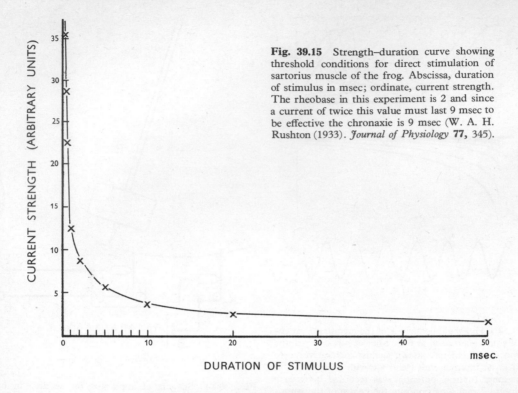

Fig. 39.15 Strength–duration curve showing threshold conditions for direct stimulation of sartorius muscle of the frog. Abscissa, duration of stimulus in msec; ordinate, current strength. The rheobase in this experiment is 2 and since a current of twice this value must last 9 msec to be effective the chronaxie is 9 msec (W. A. H. Rushton (1933). *Journal of Physiology* **77**, 345).

if a nerve is stimulated twice in quick succession, it may not respond to the second stimulus. In the period following a spike during which the excitability of the nerve is abolished or reduced, the nerve is said to be in a refractory state. The *absolute refractory period* is the brief interval after a successful stimulus when no second shock, *however large*, can elicit another spike. Its duration is roughly equal to the duration of the spike, which in mammalian A fibres at body temperature is of the order of 0·4 msec (see Fig. 39.16). In frog nerve at 15°C it lasts for about 2 msec (Fig. 39.17). It is succeeded by the *relative refractory period* during which a second response can be obtained if a large enough stimulus is used. After the relative refractory period a supernormal phase can often be detected, when the excitability is slightly greater than normal; this is correlated with the occurrence of an after-potential (see p. 807). The experiment on a mammalian nerve illustrated in Figure 39.16 shows the recovery after an impulse in a somewhat different way from the excitability curve plotted in Figure 39.17. In Figure 39.16 the second (testing) shock was supramaximal

throughout. When it fell within 0·4 msec of the conditioning shock there was no response (Fig. 39.16 (1)) but when the interval between the two shocks was gradually lengthened (Fig. 39.16 (2 to 5)) the second shock came within the relative refractory period of more and more of the fibres, and the second response progressively increased.

During the relative refractory period the spike size and conduction velocity are both subnormal as well as the excitability. Therefore, two full-sized impulses travelling down a long length of nerve must be separated by a minimum interval; if the interval is less the second one decreases. A mammalian A fibre can respond at a frequency of over 1000 impulses/second, but the spikes after the first one would be small, and would tend to decline during sustained stimulation. Recovery in such fibres is complete after about 3 msec, so that full-sized impulses can be carried at repetition frequencies of over 300/sec. Even this frequency is not often reached in the living animal, though a few sensory nerves may sometimes exceed it for short bursts of impulses.

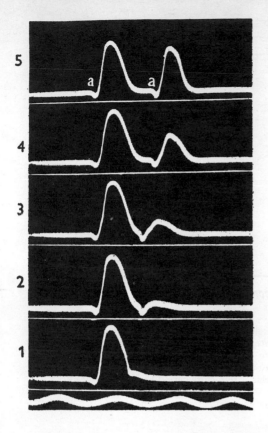

Fig. 39.16 Relative refractory period of A fibres of the phrenic nerve mapped in terms of the height of the response evoked by a supramaximal testing shock. The separation of the conditioning and testing shocks can be determined from the shock artifacts, a, a. At the interval of the lowermost record (about 0·5 msec) the nerve is completely refractory, but a small response is present in the next record. Conduction distance, 8 mm. Temperature, 38°C. The sine wave at the bottom of the figure gives time in msec (J. Erlanger & H. S. Gasser (1937). *Electrical Signs of Nervous Activity*, p. 172. Philadelphia: University of Pennsylvania Press).

As is shown clearly by Figure 39.13, in order to stimulate a nerve fibre it is necessary to depolarize the membrane by causing current to flow outwards. If the current is carried mainly by positively-charged cations, K^+ inside the fibre and Na^+ outside, then the direction of movement of these ions will be the same as the direction of the current (when current flows in a wire, the electrons which carry it are actually moving in the opposite direction to the current, since they are negatively charged). If the outward flow of current is to be brought about by the application of an external electrode, then the electrode must be a cathode. Note that *outward* current supplied from an *external* source reduces the membrane potential because it produces a voltage drop across the membrane resistance which opposes the resting potential. When the active process described in the next section has been initiated, further depolarization is caused by an *inward* current; the apparent contradiction arises because the source of the current is now the *internal* sodium battery (E_{Na} in Fig. 39.22). The simplest way of verifying that excitation takes place at a cathode is to set up the apparatus shown in Figure 39.12, with the lower stimulating electrode placed on the cut end of the nerve. It will be found that threshold for stimulation is lower when the electrode on the intact part of the nerve is a cathode than when it is an anode. Under some conditions, excitation will take place at an anode when the applied current is switched off; an explanation for such *anode break excitation* is given on p. 806. The experiment illustrated in Figure 39.18 shows another way of demonstrating that excitation normally occurs under a cathode. The shortest conduction time (record 1) occurred when electrode C was the cathode and B the anode, the longest (4) when A was the cathode and B still the anode. When B was the cathode there was an intermediate conduction time, which did not change much (records 2 and 3) when the anode was moved from A to C.

It should be clear from what has been said so far that nerve impulses can travel equally well in either direction. If a cathodal shock is applied at the midpoint of a nerve, impulses are propagated away from the stimulating electrode in both directions. In the living animal a given nerve fibre normally conducts impulses in one direction only, but this is because it is always stimulated at the same end—the peripheral end in sensory nerves and the central end in motor nerves. An impulse travelling in the abnormal direction is termed *antidromic*; such impulses occur in the *axon reflex* described on p. 562.

THE IONIC BASIS OF RESTING AND ACTION POTENTIALS

As has already been seen, the interior of all nerve and skeletal muscle fibres is 69 to 95 mV negative with respect to the outside. It was

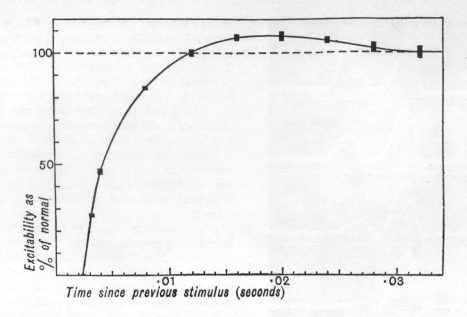

Fig. 39.17 Time course of the recovery of excitability (reciprocal of threshold) in a frog sciatic nerve after passage of an impulse. Temperature 15°C. The conditioning stimulus and the test stimulus were applied at electrodes 15 mm apart, so that about 0·5 msec should be subtracted from each reading to obtain the course of recovery under the test electrode. The absolute refractory period lasted 2 msec, and the relative refractory period 10 msec; they were succeeded by a supernormal period lasting for 20 msec (*from* E. D. Adrian & K. Lucas (1912). *Journal of Physiology* **44,** 114).

suggested by Bernstein that the resting potential in nerve fibres arose from the existence of a concentration gradient for potassium (high concentration inside, low outside), allied with a selective permeability of the cell membrane towards K^+ ions. This idea is now accepted, in a somewhat elaborated form, by most physiologists, and the principle that bioelectric potentials arise from a combination of ionic concentration gradients and specific membrane permeabilities has been extended to explain the generation of action potentials as well as resting potentials.

In a system of the type postulated by Bernstein, where the membrane separating two solutions is permeable to only one species of ion, a potential difference is set up across the membrane whose magnitude is given by the Nernst diffusion equation:

$$E_m = \frac{RT}{F} \log_e \frac{[\text{penetrating ion outside}]}{[\text{penetrating ion inside}]}$$

$$= 58 \log_{10} \frac{[\text{penetrating ion outside}]}{[\text{penetrating ion inside}]} \text{ mV}$$

E_m can be regarded as the equilibrium potential at which the force exerted by the concentration gradient tending to make the ions move into the weaker solution is exactly balanced by the force exerted by the voltage gradient tending to move them in the opposite direction. Strictly, the equation should be written with chemical activities instead of concentrations on the right-hand side; but for the present arguments it can be assumed that the activity coefficients are equal on the two sides. For a positively-charged cation, E_m is the inside potential with respect to the outside, so that if the concentration ratio is less than 1, E_m is negative. For K^+ ions, E_m will be written as E_K; for Na^+ as E_{Na}, and so on. It is now known that the resting membranes of nerve and muscle fibres are not in fact solely permeable to K^+ ions; they are also permeable to Cl^- ions, and are not totally impermeable to Na^+ ions. The potential developed across a membrane permeable to all these ions was calculated by Goldman on the assumption that the voltage gradient was uniform throughout the membrane. His 'constant field' equation can be written thus:

$$E_m = \frac{RT}{F} \log_e \frac{P_K[K]_o + P_{Na}[Na]_o + P_{Cl}[Cl]_i}{P_K[K]_i + P_{Na}[Na]_i + P_{Cl}[Cl]_o}$$

where the subscripts $_i$ and $_o$ refer to inside and outside respectively, and the Ps are permeability coefficients. It will be seen that if one of the permeability coefficients is overwhelmingly greater than the others, then this equation reduces to the Nernst equation. In other words, the Nernst equation shows the limiting value of the potential across a membrane which is exclusively permeable to one ionic species,

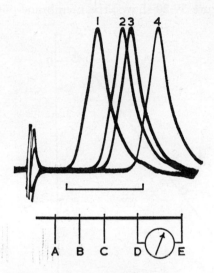

Fig. 39.18 Variation in conduction time with polarity and position of stimulus. Monophasic recordings of an earthworm giant fibre; time scale 1 msec. Shock strength just above threshold. Record 1, B+C−; 2, A+B−; 3, C+B−; 4, B+A−. Distances between electrodes were: AB, 10 mm; BC, 10 mm; CD, 14 mm. (*By courtesy of K. Krnjevic.*)

while the constant field equation can be used to calculate the compromise potential across a membrane permeable to several different ions.

The best evidence for the internal composition of excitable tissues has been obtained for frog muscle and for squid axons, whose axoplasm can be squeezed out like tooth-paste and

analysed with a minimum of contamination by extracellular fluid. Table 39.19 shows some values for the concentrations in the tissue water of K^+, Na^+ and Cl^- ions in freshly dissected squid axons and frog muscle, and the corresponding plasma concentrations. It will be seen that the concentration ratios for K^+ and Cl^- are of the same order (5 to 40 times) in the two

Table 39.19 Concentrations of ions in the cytoplasm and in the blood plasma of squid axons and frog muscle, and the equilibrium potentials calculated from the Nernst equation (inside relative to outside)

	Squid axon	Frog muscle	
$[K^+]$ inside	410	125	m-mole/kg H_2O
$[K^+]$ outside	22	2·6	m-mole/kg H_2O
E_K	−74	−98	millivolts
$[Na^+]$ inside	49	15	m-mole/kg H_2O
$[Na^+]$ outside	460	110	m-mole/kg H_2O
E_{Na}	+56	+50	millivolts
$[Cl^-]$ inside	123	1·2	m-mole/kg H_2O
$[Cl^-]$ outside	560	77	m-mole/kg H_2O
E_{Cl}	−38	−104	millivolts

From A. L. Hodgkin (1951). *Biological Reviews* **26**, 339–409; R. D. Keynes (1964). *Journal of Physiology* **169**, 690–705.

tissues, although the actual concentration of ions is substantially greater in the marine invertebrate (squid) than in the fresh water vertebrate (frog). The reader may wonder how a balance between anions and cations is achieved on each side of the membrane, since the total numbers of negative and positive charges must be almost equal; the numbers will not be *exactly* equal, because of the displacement of a few ions to set up the membrane potential, but in a very large cell like a squid axon it is only necessary to displace about 1 in 10^7 of the internal K^+ ions in order to give rise to the observed resting potential, so that for practical purposes equality can be assumed. In the plasma the predominant cation is Na^+, and there are also small concentrations of K^+, Ca^{2+} and Mg^{2+}. The principal anion is Cl^- but there is also a fair amount of HCO_3^- while negatively charged phosphate ions and plasma proteins make a small contribution to the balance. In the cytoplasm, there is more difficulty in drawing up a complete balance sheet. On the cation side, K^+ predominates, with low concentrations of Na^+ and Mg^{2+} and smaller amounts still of Ca^{2+}; there may also be some amino acids with a net positive charge. As far as anions are concerned, the concentration of Cl^- is low, at least in muscle, and the organic anions which make up the deficit in negative charge have not all been identified with certainty. In muscle the main contributions are probably made by phosphocreatine, ATP, hexosemonophosphate and proteins; in squid axoplasm the chief anions are isethionate ($CH_2OH.CH_2SO_3^-$) and the dicarboxylic amino acids, aspartic and glutamic. The important fact about all the intracellular anions except Cl^- is that they are relatively large molecules, to which the cell membrane can be regarded as being impermeable.

If it is correct to say that the resting potential arises from the potassium concentration grad-

ient, then there should be approximate agreement between the Nernst potential for K^+ and the measured resting potential. Exact agreement cannot be expected since there are technical difficulties in making some of the determinations; and the selectivity of the resting membrane towards K^+ is unlikely to be quite perfect. However, comparison of the resting potentials seen in Figure 39.10 with the theoretical potentials shown in Table 39.19 shows that the agreement is reasonably good. It will be seen that the Nernst potential for Cl^- is also of the same order as the resting potential, confirming that the membranes are permeable to this ion as well. Another test of the role of K^+ in determining the resting potential is to investigate the effect of varying external $[K^+]$. As Figure 39.20 shows, the membrane poten-

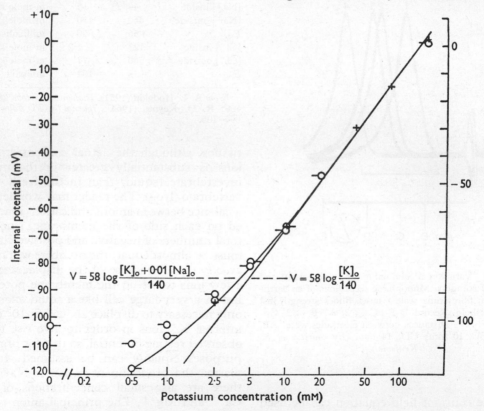

Fig. 39.20 Variation in resting potential of frog muscle fibres with external potassium concentration, $[K]_o$. The measurements were made in a chloride-free sulphate-Ringer's solution containing 8 mM-$CaSO_4$. Crosses are potentials measured after equilibrating for 10 to 60 min; circles are potentials measured 20 to 60 sec after a sudden change in concentration, —O after increase in $[K]_o$, O— after decrease in $[K]_o$. Note that for large $[K]_o$s the measured potentials agree well with the Nernst equation if $[K]_i$ is taken as 140mM, and that the deviations at low $[K]_i$s' can partly be explained by taking sodium permeability into account (*from* A. L. Hodgkin & P. Horowicz (1959). *Journal of Physiology* **148**, 135).

tial varies in the expected fashion with large [K]s, but when $[K]_o$ is lowered below normal the potential flattens off. This flattening has been used by some authors as a criticism of the viewpoint being advanced here, but is in fact exactly what would be predicted by the constant field equation, since when $[K]_o$ is very low the contributions to the resting potential of Na^+ and Cl^- permeabilities become relatively important. A further test of these ideas is to examine the rate of passage of radioactive potassium ions through the resting membrane since it should be appreciated that, although at equilibrium there is no net transfer of K^+ through the membrane, the equilibrium is a dynamic one, with equal and opposite fluxes of ions moving through the membrane all the time. Again there is tolerable agreement between theory and experiment, and there is no doubt that the resting membranes of nerve and muscle are permeable to labelled K^+ and Cl^- ions.

The basis for the theory of nervous conduction developed by Hodgkin, Huxley and Katz is that the concentration of Na^+ ions inside excitable cells is much lower than that in the body fluids outside. If one supposes that the permeability of the nerve membrane can change from its resting state, in which it is relatively permeable to K^+ and Cl^- but impermeable to Na^+, to an active state in which it becomes temporarily much more permeable to Na^+ than to any other ions, one can explain why the membrane potential should change from a value close to E_K (see Table 39.19) to a reversed potential which approaches E_{Na} (compare the values in the table with the overshoots seen in Fig. 39.10). The mechanism considered by Hodgkin and Huxley can be summarized as follows: when the membrane is depolarized by an outward flow of current, caused either by an applied cathode or by the proximity of an active region where the membrane potential is already reversed, its sodium permeability immediately rises, and there is a net inward movement of Na^+ ions, flowing down the sodium concentration gradient. If the initial depolarization is large enough, Na^+ enters faster than K^+ can leave, and this causes the membrane potential to drop still further; this extra depolarization increases the sodium permeability even more, accelerating the change of membrane potential

in a regenerative fashion. The linkage between sodium permeability and membrane potential forms, as shown in Figure 39.21, a 'positive

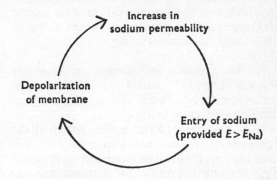

Fig. 39.21 Diagram illustrating the regenerative linkage between membrane potential and sodium permeability (*from* A. L. Hodgkin (1951). *Biological Reviews* **26**, 384).

feed-back' mechanism. The entry of Na^+ does not continue indefinitely, being halted partly because the membrane potential soon reaches a level close to E_{Na}, where the net inward driving force acting on Na^+ ions becomes zero, and partly because the rise in sodium permeability decays inexorably with time from the moment when it is first triggered, this process being termed 'inactivation'. After the peak of the spike has been reached, therefore, the sodium permeability is rapidly reduced (inactivated); at the same time, the potassium permeability of the membrane rises well above its resting value, and an outward movement of K^+ takes place, eventually restoring the membrane potential to its original level. It may be noted that the membrane potential would begin to return towards its normal level as soon as the Na^+ permeability was sufficiently inactivated, even if there were no delayed rise in K^+ permeability; but the rate of return would be slow, and the result of the increase in K^+ permeability is to ensure a more rapid restoration towards the resting potential than would otherwise be achieved. At the end of the spike, the membrane is left with the original potential level restored, but with its sodium permeability mechanism still inactivated. Lapse of further time allows the sodium permeability to be 'reactivated' to the quiescent state in which it is still very low, as is characteristic of the resting membrane, but is now ready once more to increase explosively if the membrane is retriggered.

Before describing the experimental evidence on which this concept rests, it may be helpful to recapitulate its most important features. These may be summarized:

1. The resting membrane is relatively permeable to K^+ (and Cl^-) ions; and impermeable to Na^+ ions. The resting potential therefore lies close to E_K.

2. At the peak of the action potential the permeability of the membrane to Na^+ has risen so much that the potential approaches E_{Na}, the limiting value for a membrane exclusively permeable to Na^+.

3. The sodium permeability of the membrane (P_{Na}) is controlled by the potential across the membrane in such a fashion that a relatively small degree of depolarization (reduction in potential) causes a large rise in permeability.

4. The potassium permeability of the membrane (P_K) is also controlled by the membrane potential, but whereas P_{Na} rises immediately when the potential is lowered and is subsequently inactivated, P_K only rises after a slight time delay, giving a maintained increase which is not inactivated (it is later reduced to normal by the repolarization of the membrane).

5. The separation in time of the permeability changes, P_{Na} rising quickly and then being cut off, P_K rising with a lag, is very important. Simultaneous increases in P_{Na} and P_K would merely result in an energetically wasteful interchange of Na^+ and K^+, and would not cause the membrane potential to vary in a useful manner.

6. The immediate source of energy for the electric currents which flow, carried by Na^+ and K^+ ions, during propagation of a nerve impulse, is the pre-existing ionic concentration gradient. Evidence as to how this is built up in the first place will be considered later.

7. After an impulse has passed along an axon, a little sodium has been gained and a little potassium lost. The quantities involved are small compared with the initial contents of the axon, so that in the course of a single impulse the internal ionic concentrations do not change to a measurable extent. It is the permeability of the nerve membrane which alters, not the concentration of ions within the cell.

Perhaps the most obvious prediction that can be made from this hypothesis is that conduction should be impossible in a medium which contains no sodium. In 1902 Overton published a paper about the effect on muscle excitability of substituting various ions for the normal constituents of frog Ringer's solution, and concluded that, although the Cl^- could be replaced by a variety of anions without abolishing excitability, Li^+ was the only cation able to substitute for Na^+; however, the significance of Overton's observations was overlooked for many years. Hodgkin and Katz made a preliminary test of the validity of the ionic hypothesis by showing that in squid axons the presence of Na^+ ions in the external medium was essential for conduction. In a choline chloride solution, or one in which sugar was substituted for all the NaCl, conduction was immediately blocked; again Li^+ seemed to be the only cation which could deputize for Na^+. Since this work was done, several examples have come to light of excitable tissues in which Na^+ and Li^+ are not the only cations which can sustain electrical activity. Thus barnacle muscle fibres can produce conducted spikes in solutions where the only cation is Ca^{2+}; and there is some evidence in the case of frog nerve and muscle that nitrogenous compounds such as hydrazinium, guanidinium or quaternary ammonium ions can be substituted for Na^+. These sodium-imitating substances are not normally present in body fluids, any more than Li^+ is, and it may well be an accident that their structure resembles Na^+ in the right way for them to penetrate the membrane through the Na-selective channels. There is, unfortunately, no good indication yet as to precisely how, at the molecular level, the cell membrane discriminates so markedly between different ions.

The next predictions that can be made from the ionic hypothesis concern the slope of the rising phase of the spike, and the extent to which the membrane potential is reversed at its peak. If the rising phase is due to an inrush of Na^+ ions, then one would expect its slope to be directly proportional to $[Na]_o$. This was verified by Hodgkin and Katz to be the case in squid axons, and it has since been demonstrated to hold good for various other types of nerve and muscle. If at the peak of the spike the membrane is, as suggested, far more permeable to Na^+ ions than to K^+ or Cl^-, then the potential should approach the Nernst potential for Na^+, E_{Na}, and should vary in a predictable way when either $[Na]_o$ or $[Na]_i$ is changed. This test has also been made successfully on a number of tissues, and the peak potential has generally been found to change by an amount close to the theoretical 17 millivolts calculated from the Nernst equation for a halving of the concentration ratio.

According to the ionic hypothesis, the nerve membrane can be represented electrically by the equivalent circuit shown in Figure 39.22.

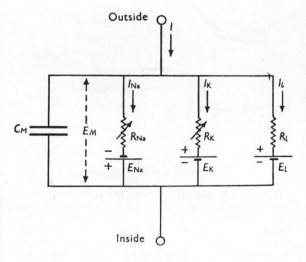

Fig. 39.22 Equivalent circuit for the nerve membrane described on this page R_{Na} and R_K vary with membrane potential and time; the other components are constant (*from* A. L. Hodgkin & A. F. Huxley (1952). *Journal of Physiology* **117**, 501).

The membrane has an electrical capacity because of its insulating properties and its ability to keep separate the charged particles on either side; in non-myelinated nerve fibres (and, as a matter of fact, in many other types of living cell) 1 cm^2 of membrane has a capacity of about 1 microfarad, while in muscle fibres the capacity is somewhat greater. Readers familiar with electrical units will appreciate that a capacity of this size implies that the insulating layer of the membrane must be exceedingly thin; if a reasonable value for the dielectric constant of lipid material is assumed,

a capacity of 1 μF/cm^2 is consistent with the thickness of less than 5 nm suggested by electron micrographs. The membrane must also be regarded as having an electrically conducting path across it, in parallel with its capacity, because ions are known to be able to move through it, albeit at a rate perhaps $1/10^7$ as great as that at which they would move across a water layer of equal thickness. This conducting path may be represented as three resistances in parallel, one being the channel through which Na^+ ions flow, another being the K^+ channel, and the third being the 'leak' channel for all other ions (probably mainly Cl^-). Each resistance has in series with it an appropriate battery; R_{Na} is pictured in series with E_{Na}, and so on. Re-stated in electrical terms, the mechanism described in the preceding paragraphs involves the following sequence of events:

1. In the resting membrane R_{Na} is much larger than R_K, that is the sodium conductance is much smaller than the potassium conductance. E_M is therefore close to E_K.

2. When a large enough depolarizing (outward) current is applied to the membrane, R_{Na} falls, because its value is controlled by the size of E_M at any instant. Current carried by Na^+ ions flows through the membrane, reversing the charge on the capacity C_M.

3. The flow of Na^+ current slows down when the peak of the spike is reached, partly because C_M is now recharged to a potential close to E_{Na}, and partly because the inactivation process begins to make R_{Na} rise once more.

4. A delayed fall in R_K now takes effect, and current carried by K^+ ions flows outwards through the membrane, recharging C_M to its original potential.

Knowing the electrical characteristics of the membrane, we can easily calculate the minimum amount of charge which has to be transferred through the membrane to alter the potential by the observed extent. Thus to change the potential across a 1 microfarad capacitance by 120 millivolts requires the transfer of 1.2×10^{-7} coulombs. One gram-molecule of a monovalent ion can carry 96 500 coulombs, whence the minimum inward movement of Na^+ ions to produce a 120 mV spike across a squid nerve membrane whose capacity is 1 μF/cm^2 is $1.2 \times 10^{-7}/96\,500 = 1.2 \times 10^{-12}$ mole/cm^2; a similar calculation for a frog muscle membrane whose capacity is 7·5 μF/cm² gives a minimum inward Na⁺ transfer of 9×10^{-12} mole/cm². A useful direct check of the validity of the ionic hypothesis is to measure the net entry of Na⁺ per unit area of membrane per impulse, to see whether or not it is greater than this figure. This was done by Keynes and Lewis for cephalopod (squid and cuttlefish) axons, and by Hodgkin and Horowicz for isolated frog muscle fibres, using radioactive isotope and radioactivation analysis techniques. It was found that in squid axons each impulse causes a net gain of about 4×10^{-12} mole Na⁺/cm², and a net loss of an equal quantity of potassium; in frog muscle fibres the net gain of Na⁺ was 16×10^{-12} mole/cm², and the net loss of K⁺ 10×10^{-12} mole/cm². It is not surprising that the experimental values should be somewhat larger than the calculated minima, this being the expected consequence of some degree of overlap between the changes in P_{Na} and P_K, but it would have raised serious difficulties for the theory had they been smaller. Measurements by Caldwell and Keynes of the net gain of labelled Cl⁻ in stimulated squid axons have given much smaller values (about $0·05 \times 10^{-12}$ mole/cm² impulse), but this finding is consistent with the tacit assumption that has been made in this account of the conduction mechanism that the chloride permeability of the membrane does not change during the passage of a nerve impulse.

The advantage of using radioactive isotopes to study the ionic movements during nervous activity is that the ions can be identified unambiguously; in electrical experiments on the other hand it is sometimes difficult to be certain what ion carries the current that is measured. However, the time resolution of tracer experiments is poor, since one can only observe the integrated effect of the passage of a fairly large number of impulses. In order to examine in detail the laws governing the flow of Na⁺ and K⁺ ions through the nerve membrane, that is to say to measure the variation in P_{Na} and P_K as functions of time and membrane potential, Hodgkin and Huxley used the 'voltage-clamp' technique for studying the membrane of the squid giant axon. A full description of their work is beyond the scope of this volume, but essentially the method involves changing the membrane potential uniformly over a fixed

area of membrane in a predetermined fashion, and measuring the resulting current flow through the membrane. The type of record obtained is illustrated in Figure 39.23, which

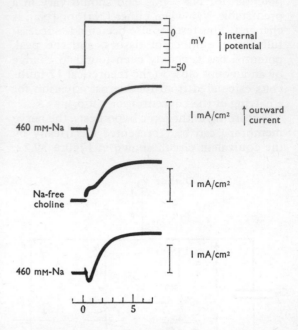

Fig. 39.23 An example of the kind of result obtained when applying the voltage-clamp technique to the squid giant axon. The top record shows the imposed change in membrane potential, a sudden depolarization of 65 mV. The lower three records show the resulting membrane currents in the presence and absence of external Na⁺ ions. Temperature 11°C (*from* A. L. Hodgkin & A. F. Huxley (1952). *Journal of Physiology* **116,** 449).

shows how depolarization of the membrane causes a brief phase of inward current flow, identified as carried by Na⁺ ions by the effect of substituting choline, followed by a maintained outward current flow carried by K⁺ ions. Analysis of a series of records with different imposed voltage steps gave information about the exact relationship between membrane potential and P_{Na} and P_K, and about the time course of the permeability changes, P_{Na} being increased immediately by depolarization but subsequently inactivated, and P_K being increased with a time lag but not inactivated. This analysis was used to construct the curves for the time courses of the separate permeability changes shown in Figure 39.24, which summarizes in pictorial form some of the events taking place during propagation of an impulse.

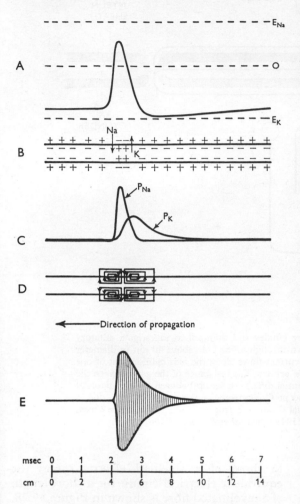

Fig. 39.24 Some of the events occurring during propagation of an impulse along a non-myelinated squid giant axon. The horizontal scale corresponds to a conduction velocity of 20 m/sec, and the diagrams can be considered to represent *either* the events at different points along the axon at the same moment in time, *or* the events at different times for one position. A, change in membrane potential; the dashed lines show zero potential and the Nernst potentials for Na^+ and K^+. B, polarity of potential difference across membrane, and approximate position of ion movements. C, variation in sodium and potassium permeabilities (*after* A. L. Hodgkin & A. F. Huxley (1952). *Journal of Physiology* **117**, 530). D, local circuit current flow. E, variation in total membrane conductance, as recorded by Cole and Curtis with an impedance bridge (*after* K. S. Cole & H. J. Curtis (1939). *Journal of General Physiology* **22**, 649).

SALTATORY CONDUCTION IN MYELINATED NERVES

In 1925 Lillie suggested that the function of the myelin sheath in vertebrate nerve fibres might be to restrict the inward and outward passage of local circuit current to the nodes of Ranvier, so causing the nerve impulse to be propagated along myelinated fibres from node to node in a series of discrete jumps. He coined the term 'saltatory' conduction for this kind of process, and supported the idea with some ingenious experiments on his iron wire model (an iron wire in nitric acid of the right strength acquires a surface film along which a disturbance can be propagated by local circuit action; the mechanism has several features analogous to those of conduction of a nerve impulse). Physiological experiments on saltatory conduction could not be performed until techniques for dissecting isolated myelinated fibres had been developed; Kato and his colleagues including Tasaki first did this in Japan about 1930. During the 1940's Tasaki produced strong support for the saltatory theory by showing that the threshold for electrical stimulation of an isolated fibre was much lower at each node than along the internodal stretches, and also that blocking by anodal polarization and by narcotics was more effective at the nodes than elsewhere. A particularly clear demonstration of the validity of the saltatory theory was provided by Huxley and Stämpfli, whose experiments were an extension of some earlier work by Tasaki and Takeuchi. This involved measurement of the longitudinal and transverse currents flowing at different points during the passage of an impulse along a single myelinated frog fibre. The principle of the method (Fig. 39.25) was to pull the fibre through a short glass capillary about 40 μm in diameter set in a partition between two troughs of Ringer's solution. The space around the nerve was narrow enough to have a total resistance of about 0·5 megohm, so that current flowing between neighbouring nodes outside the myelin sheath gave rise to a measurable potential difference between the two sides of the partition, which could be recorded on an oscilloscope. The records of longitudinal current showed (Fig. 39.26) that at all points along the outside of any one internode the current flow was roughly the same, both in magnitude and in timing; but the

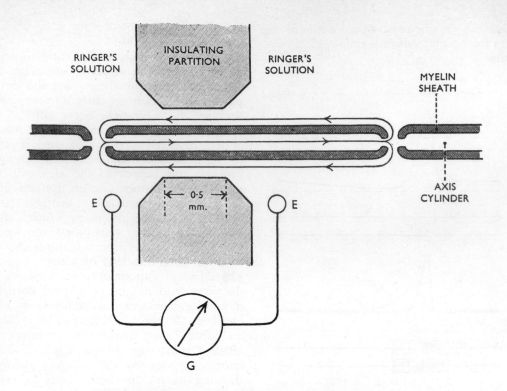

RINGER'S SOLUTION   INSULATING PARTITION   RINGER'S SOLUTION   MYELIN SHEATH

E   0·5 mm.   E   AXIS CYLINDER

G

Fig. 39.25 Diagram of the method used by Huxley and Stämpfli to investigate saltatory conduction in nerve. The nerve fibre is drawn through a fine hole about 40 μm in diameter in an insulator by a micromanipulator. The current flows along the axis cylinder out of one node and in the other node as indicated by the arrows. The resistance of the gap between the insulators being about half a megohm, a potential difference set up between the two pools of Ringer's solution is measured by the oscilloscope G connected to the two electrodes, one in each pool of Ringer's solution. The internodal distance in frog's myelinated fibres is 2 mm. (*After* A. F. Huxley & R. Stämpfli (1949). *Journal of Physiology* **108**, 315.)

peaks of current flow were displaced stepwise in time by about one-tenth of a millisecond as successive nodes were passed. To determine transverse current flow, neighbouring pairs of records were subtracted from one another, since the difference between the longitudinal currents at two positions must have arisen from current entering or leaving the axis cylinder transversely between these positions. This procedure gave the results illustrated in Figure 39.27, from which it is seen that over the internodes there was only a small leakage of outward current, but that at each node there was a brief pulse of outward current followed by a substantially larger pulse of inward current.

As was stressed at the beginning of this chapter, the structure of a nerve fibre is such that its passive electrical properties are similar

to those of a shielded electric cable; the equivalent electrical circuit for a short length of a myelinated fibre is shown in Figure 39.28. Owing to the cable structure of the nerve fibre, whenever a change in electric potential is developed across the membrane at a node, its effect spreads nearly instantaneously along the fibre, because of a flow of electric current through the network of resistances. This type of distribution of potential change, which depends only on the passive electrical behaviour of the system, is known as an *electrotonic* spread of potential. The dissimilarity between electric cables and nerve fibres arises because the materials composing the nerve are less suitable for obtaining electrotonic spread over long distances than those used by an electrical engineer; the central conductor, that

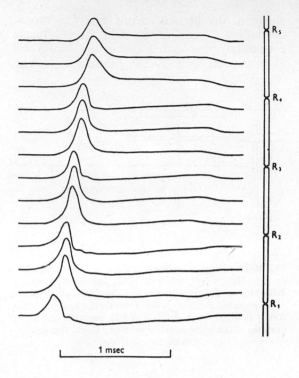

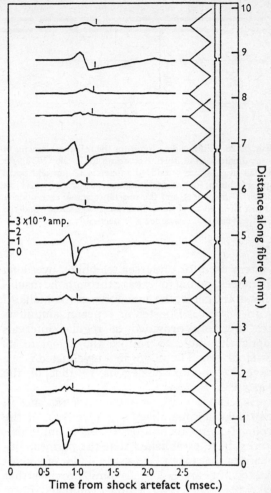

Fig. 39.26 Records obtained by the method shown in Fig. 39.25. Each cathode-ray oscillograph record is shown opposite the position on the fibre which the short glass capillary occupied when the record was obtained. All the records from any one internode are nearly the same and the peak of the disturbance occurs at the same time. R_1, R_2 to R_5 are nodes of Ranvier. (*After* A. F. Huxley & R. Stämpfli (1949). *Journal of Physiology* **108**, 321.)

Fig. 39.27 Membrane currents at different positions along an isolated frog fibre. Each curve shows the difference between the longitudinal currents, recorded as in Fig. 39.26, at the two points 0·75 mm apart indicated to the right. The vertical mark above each graph shows the time when the change in membrane potential reached its peak at that position on the fibre. Outward current is plotted upwards (*from* A. F. Huxley & R. Stampfli (1949). *Journal of Physiology* **108**, 327).

is to say the axis cylinder, has too high an electrical resistance, and the insulator, that is to say the nerve membrane and myelin sheath, too low a resistance. The values of the resistances shown in Figure 39.28 are approximately such that if a potential change, V, develops across one node, the change at the neighbouring node is 0·4 V, at the next node but one 0·16 V, and so on. In other words, the passive or electrotonic spread of potential decrements rather rapidly. However, the nerve membrane does *not* behave merely as a passive resistance, since the potential across it is capable of altering actively as a result of the sequence of permeability changes already described. Figure 39.27 shows that the myelin sheath exhibits only passive electrical behaviour, but that the membrane at the nodes

behaves in a more complicated way. The initial flow of outward current at each node arises from the electrotonic spread of activity from the preceding node; this triggers an active process which then allows current to flow inwards so amplifying the potential change handed on to the next node, and so on. The closest analogy with this mechanism is the

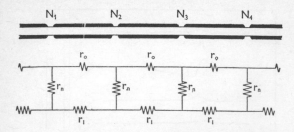

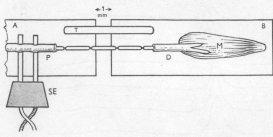

Fig. 39.28 Equivalent circuit for the resistive elements of a myelinated nerve fibre. According to Tasaki (1953), for a toad fibre whose outside diameter is 12 μm and nodal spacing 2 mm, the internal longitudinal resistance r_i is just under 20 megohms, and the resistance across each node r_n is just over 20 megohms. In a large volume of fluid the external resistance r_o is negligibly small.

Fig. 39.29 Diagram to illustrate the experiment described by A. F. Huxley & R. Stämpfli (1949). *Journal of Physiology* **108**, 315, to show the importance for conduction of the nerve impulse of current flowing outside the myelin sheath. A and B, microscope slides; P and D, proximal and distal portions of the nerve to the muscle M; T, wet thread; SE, stimulating electrodes.

system used for undersea telephony over long distances: owing to leakage through the insulation of the cable, the signal transmitted falls off with distance. Booster or repeater amplifiers are therefore provided at regular intervals along the cable to restore the signal to its original size. The nodes in a myelinated nerve have a function analogous to that of the repeaters.

Experiments by various workers, and in particular Frankenhaeuser's extension of the voltage-clamp technique to single amphibian fibres, have established that the permeability changes that take place at the nodes of a myelinated fibre are essentially similar to those that take place all along a non-myelinated fibre. Since the size of the fully-developed action potential is of the order of 120 mV, and since it is necessary to depolarize the membrane at the next node by only about 15 mV in order to initiate the active changes, the conduction mechanism embodies a considerable margin of safety. From these figures, and from the fact that the electrotonic potential decrements to about 0·4 from node to node, it can be predicted that the nerve impulse should be capable of being conducted past one or even two inactive nodes without blocking. As has been shown by Tasaki, this is indeed the case, and the impulse can successfully skip over two (but not three) nodes which have been treated with a local anaesthetic such as cocaine.

A simple experiment performed by Huxley and Stämpfli to demonstrate the importance of the external current pathway in propagation

along a myelinated nerve fibre is worth describing. The nerve of a nerve-muscle preparation was pared down until only one fibre was left (Fig. 39.29). Stimulation of the

nerve at P then caused a visible contraction of a motor unit in the muscle, M. The preparation was now laid in two pools of Ringer's solution on insulated microscope slides A and B in such a position that part of an internode, but not a node, lay across the 1 mm air gap separating the pools. At first, stimulation at P continued to cause a muscle twitch, but soon the layer of fluid outside the myelin sheath in the air gap was dried up by evaporation, and the muscle ceased to twitch. Conduction across the gap could, however, be restored by placing a wet thread, T, between the two pools. This proved that an action potential arriving at the node just to the left of the air gap could only trigger the node on the further side of the gap if there was a reasonably low-resistance electrical connexion between the pools. On reference to Figure 39.28 it will be seen that if r_o becomes at all large, the potential change at N_2 produced by a spike at N_1 will fall below threshold for excitation. As has been pointed out by Tasaki, a reservation needs to be made about this experiment. Unless special precautions are taken, the stray electrical capacity between the pools may provide, for a pulse of current, an alternative pathway whose impedance is low enough for excitation to occur on the far side of the gap in the absence of the wet thread if the threshold of the node N_2 is low. Even with such precautions, Tasaki finds that impulses are still able to jump the gap if the fibre has a really low threshold, probably because r_o cannot be raised high enough by simple evaporation. However, the fact that the experiment works in a clear-cut way only if the threshold is somewhat higher than it is *in vivo* does not prevent it from showing satisfactorily that there must be a low-impedance pathway between neighbouring nodes *outside* the myelin sheath if the nerve impulse is to be propagated along the fibre.

Since the electrotonic spread of potential along a nerve fibre is a virtually instantaneous process, it may be asked why the nerve impulse is not propagated more rapidly than it is. The explanation is that there is a definite delay at each node, of the order of 1/10 msec (see Fig. 39.26), which represents the time necessary for the ions to move through the capacitance of the membrane at the node and to reverse the potential across it. The main reason why myelinated fibres conduct impulses more rapidly than non-myelinated ones of equal diameter is that their total membrane capacity is smaller. In a non-myelinated fibre the delay in altering membrane polarization is in effect encountered all along the fibre and not just at the nodes. Conduction velocity is also affected by. the longitudinal electric resistance, and is decreased by a rise in either the internal or the external resistance. Thus a large non-myelinated axon conducts faster than a small one because of its lower internal resistance, while Hodgkin showed some years ago that the conduction velocity in a squid axon supported on a series of metal strips was greater if the strips were connected together electrically than if they were insulated from one another. It might be added that an increase of the distance between nodes in a myelinated fibre is not as advantageous as might be supposed, since the greater spacing is offset by an increase in the delay at each node. Huxley and Stämpfli pointed out that there is likely to be an optimum spacing of nodes for maximum conduction velocity, and that the spacing found in normal fibres is probably close to this optimum.

Myelinization not only increases conduction velocity but also, because of the lower total membrane capacity, reduces the exchange of Na^+ and K^+ ions that takes place during the passage of each impulse. It therefore results in considerable economy of energy expenditure. The ionic exchange has not yet been measured very accurately in myelinated fibres, but it is certainly much smaller than in non-myelinated fibres of comparable size.

Long before oscillograph records of action potentials could be obtained, the problem of measuring the speed of conduction of the nerve impulse was solved by Helmholtz (1852). His method is the basis of a well-known class experiment. The muscle of a frog's sciatic-gastrocnemius preparation is attached to a lever writing on a smoked drum. The nerve is first stimulated as far away from the muscle as possible; the latent period, that is the time from the moment of stimulation to the rise of the lever, is measured. This latent period includes the shock–response time of the nerve, the conduction time in the nerve, the delay at the myoneural junction, as well as mechanical delays in both muscle and recording lever. The latent period when the nerve is stimulated as near as possible to the muscle also includes all these delay times, but it is slightly shorter by the time taken for the impulse to pass between the two points of stimulation. The distance along the nerve between the points stimulated divided by the difference in the latent periods gives the speed of the nerve impulse. The speed found in this way is, of course, the average speed of conduction of the predominant size of nerve fibre.

Nerve Metabolism and Active Transport of Ions

The mechanisms described so far do not require a direct supply of energy from cell metabolism. Energy is certainly dissipated during the passage of a nerve impulse, since electric currents are generated which flow through the surrounding tissue and through the nerve itself; but the *immediate* sources of this energy appear to be the pre-existing concentration gradients for Na^+ and K^+ ions. Nerve metabolism is required in order to provide the energy to drive an active transport system which builds up the concentration gradients in the first place, and then reverses the loss of K^+ and gain of Na^+ which occur during nervous activity. The clearest evidence for this statement comes from experiments on the squid giant axon, where metabolic inhibitors can be shown to block the active transport mechanism without having any obvious effect on the ability of the axon to conduct impulses. Thus in the experiment illustrated in Figure 39.30, the effect of applying cyanide was to

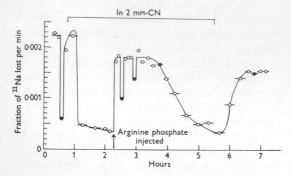

Fig. 39.30 The effect of the efflux of labelled sodium from a squid giant axon of first blocking metabolism with cyanide and then injecting a large quantity of arginine phosphate. Open circles show efflux with $[K]_o$ = 10mM, filled-in circles show efflux into a K-free solution. Immediately after the injection the mean internal concentration of arginine phosphate was 33 mM. Axon diameter 686 μm, temperature 18°C (*from* P. C. Caldwell, A. L. Hodgkin, R. D. Keynes & T. I. Shaw (1960). *Journal of Physiology* **152,** 561).

reduce the rate of extrusion of labelled sodium to less than a quarter of its initial value, but the size of the action potential was unchanged in the poisoned axon. This dissociation between the mechanism of the action potential, involving downhill ionic movements, and that of the recovery process, involving uphill movements against the concentration gradients, can be revealed in squid axons because of their exceptional size. The quantity of potassium lost by a squid axon in the course of a single impulse is less than a millionth part of its initial content, so that one would expect such an axon to be capable of conducting a large number of impulses even though its recovery mechanism were totally inoperative. The same argument probably applies to myelinated nerve fibres in which the economy in ionic exchange during the impulse, resulting from the presence of the myelin sheath, again means that many impulses can be conducted without large internal concentration changes; but for these fibres the

experimental evidence is less clear cut. The situation where conduction and recovery mechanisms cannot be clearly dissociated is that of excitable cells whose ratio of surface to volume is relatively large, that is to say in non-myelinated C fibres and in the cells of the central nervous system with their innumerable fine dendrites. Here a single impulse might lower the internal potassium concentration by over a thousandth part, so that, if recovery were blocked, the ionic reserves would suffice only for the conduction of a few hundred impulses. In such cells the recovery mechanism has to work continuously even to combat the resting leakage of K^+ and gain of Na^+, and this is an important reason for the dependence of the central nervous system on an adequate supply of both glucose and oxygen in the blood.

In the investigation of the active transport of ions in nerve and muscle radioactive isotopes have been very useful tools. The experiment of Figure 39.30 provides an example of the kind of approach that has been used. It will be seen that in an unpoisoned squid axon with an external $[K^+]$ of 10 mM the so-called 'sodium pump' operates at such a rate that about 2 parts in 1000 of the internal Na^+ are exchanged per minute. The absolute size of the sodium efflux depends on this 'rate constant' and on the internal $[Na^+]$, so that the system is self-regulating as far as the internal sodium level is concerned. Thus in an axon at rest, the internal $[Na^+]$ settles down at the level which makes the active efflux just balance the resting inward leakage (which is low, because P_{Na} is small, but is not zero). After a burst of impulses the internal $[Na^+]$ will be temporarily raised, so that for a while sodium efflux will be greater than influx, and the internal $[Na^+]$ will slowly fall back towards its steady-state value. Another important feature of the mechanism is that the outward movement of Na^+ is coupled in some way as yet not understood to an inward movement of K^+; the best evidence for the existence of some degree of coupling is that removal of external K^+ immediately reduces the Na^+ efflux, as Figure 39.30 shows. This behaviour is also exhibited by the sodium transport processes in many other tissues, such as muscle fibres and erythrocytes. The other question with which Figure 39.30 is concerned is that of identifying the metabolic intermediate which drives the sodium pump. In the cyanide-poisoned axon it has been found that the concentrations of ATP and arginine phosphate (the counterpart in many invertebrate tissues of the creatine phosphate in vertebrates, whose role, as described on p. 838, is to provide a readily available reserve supply of high-energy phosphate bonds) quickly fall to zero; at the same time the sodium efflux drops to a small fraction of its initial value, and the potassium influx is also cut down. As the figure shows, if a quantity of arginine phosphate is injected into the poisoned axon, restoring the concentration of ATP temporarily to its unpoisoned level, the sodium efflux is also restored for a while. Later, when the injected high-energy phosphate bonds have been used up, the efflux returns to the poisoned level. Finally, when the cyanide is washed away, and oxidative phosphorylation (p. 131) is allowed to start up again, ATP and arginine phosphate reappear in the axoplasm and the sodium efflux recovers. Experiments of this kind agree with studies on the Na- and K-activated ATPase systems first described by Skou for a preparation from crab nerve in showing that the energy for driving active transport mechanisms is derived, like that for muscular contraction (p. 838), from high-energy phosphate bonds. However, the details of operation of the sodium pump at the molecular level are still not completely understood.

Another important approach in considering the role of nerve metabolism is the study of thermal changes and oxygen consumption in resting and stimulated nerves. A. V. Hill and his colleagues, using specially designed thermocouples and galvanometers, first showed about 30 years ago that the conduction of nerve impulses was accompanied by the liberation of a minute amount of heat; it has also been shown that there is a concomitant increase in oxygen consumption and in the production of carbon dioxide. In the earlier work, it was possible to distinguish between the extra heat production occurring within one second of stimulation, which was termed the *initial heat*, and a slower heat production spread out over the following thirty minutes, termed the *recovery heat*. The initial heat formed a rather variable fraction (usually about one-twentieth) of the total extra heat, and in a *Maia* (spider crab) nerve at O°C amounted to about 8×10^{-6} joules/g $(2 \times 10^{-6}$ cal/g) nerve for a single impulse. In a myelinated frog nerve the extra heat production on stimulation was smaller than in crab nerve. The resting heat production of a frog nerve at 20°C was about 300×10^{-6} joules/g $(70 \times 10^{-6}$ cal/g) nerve per second, while the maximum additional heat production during continuous stimulation was about 170×10^{-6} joules/g $(40 \times 10^{-6}$ cal/g) nerve per second. The resting oxygen consumption of frog nerve is around 30 mm^3/g per hour, and during maximal stimulation the consumption may be doubled. For frog muscle the resting heat production and oxygen consumption are of the same order, but since the amount of energy liberated by 1 g of muscle in a single twitch may be many times that involved in the passage of a single nerve impulse, the rise in metabolic rate during activity may be much greater than it is in nerve.

The initial heat production in non-myelinated nerve fibres has now been reinvestigated with more rapid recording equipment, and it has become clear that it consists of an evolution of heat during the upstroke of the action potential followed by an absorption of heat during the downstroke. Thus in a single impulse at 0°C in *Maia* nerve, Abbott, Hill and Howarth found that the first thermal event was a temperature rise of about 9×10^{-6} °C; there was then a somewhat more gradual fall averaging 7×10^{-6} °C, leaving the net rise of around 2×10^{-6} °C that had been seen before with slower instruments. In the non-myelinated C fibres of the rabbit's cervical vagus nerve, Howarth, Keynes and Ritchie have recently estimated the initial positive heat for a single impulse at 5°C as 103×10^{-6} joules/g $(24 \cdot 5$ cal/g) nerve, and the initial negative heat as 93×10^{-6} joules/g $(22 \cdot 2 \times 10^{-6}$ cal/g) nerve. These figures are only larger than those for *Maia* because they take account of the inevitable overlap between the upstroke of the spike in the more slowly conducting fibres and the downstroke in the faster ones, which appreciably reduces the rise and fall in temperature that is actually recorded, but whose extent has not yet been accurately measured in *Maia* nerve. At least part of the positive initial heat must arise from the release during depolarization of electrical energy previously stored in the membrane capacity, while the negative heat results partly from the subsequent restoration of the charge at the expense of some of the thermal energy of the ions in the surrounding solutions; but this is probably not the whole story, and it seems likely that the process of charging and discharging the membrane is also accompanied by a change in its entropy, that is in the orderliness with which the phospholipid molecules are arranged. The time course of the positive recovery heat has not yet been studied in detail, but there is little doubt that it is associated with activity of the sodium pump.

Other phenomena observed in stimulated nerves which have not yet been fully explained are the production of small quantities of ammonia, changes in the amount of light scattered and in the optical retardation of the membrane, and a very slight swelling. The relatively slow changes in light scattering observed by D. K. Hill and Tobias seem to be connected with movements of water into and out of the fibres, producing the swelling that was measured directly in *Sepia* axons by D. K. Hill. In addition it has now been found that the action potential is accompanied by much faster optical effects whose origin is probably quite different. In polarized light experiments on squid giant axons, Cohen, Hille and Keynes have shown that the retardation of the membrane alters measurably with the potential across it, partly because of a small change in its thickness and partly because of a change in its birefringence analogous with the Kerr effect.

OTHER ASPECTS OF NERVE CONDUCTION

Threshold. As may be seen from Figure 39.13, excitation of a nerve fibre involves the rapid depolarization of the membrane to a critical level normally about 15 mV smaller than the resting potential. Depolarization to a lesser extent does not, of course, fail to increase the sodium permeability of the membrane; subthreshold depolarization does increase P_{Na}, but not to the point where Na^+ ions enter faster than K^+ ions can leave, so that at the end of the shock of the membrane is merely repolarized. The critical level for excitation is the membrane potential at which the net rate of entry of Na^+ becomes exactly equal to the net rate of exit of K^+ (plus a small contribution from entry of Cl^-). Greater depolarization than this tips the balance in favour of Na^+, and the regenerative process

already described takes over and causes a rapidly accelerating inrush of Na^+. After just subthreshold depolarization, when P_{Na} has been raised over an appreciable area of membrane, the repolarization process may be somewhat slow to begin with, and a *local response* may be observed.

Refractory period. At the end of the spike the membrane is left with its sodium permeability mechanism inactivated and its potassium permeability appreciably greater than normal. Both changes help to raise the threshold for re-excitation above normal. The partial inactivation of the sodium permeability means that even to raise inward Na^+ current to the normal critical level requires more depolarization than usual, and the raised potassium permeability means that the critical Na^+ current is actually greater than normal. Until the permeabilities for both ions have returned to their resting values and the sodium mechanism is fully reactivated, the shock necessary to trigger a second spike is above the normal threshold in size.

Accommodation. It has long been known that nerves are not readily stimulated by the application of slowly rising currents, because they tend to 'accommodate' to this type of stimulus. Accommodation arises from two causes:

1. Depolarization brings about a long lasting rise in potassium permeability.

2. Sustained depolarization semi-permanently inactivates the sodium permeability mechanism.

Both changes take place with an appreciable time lag after the membrane potential is lowered, so that they are not effective when a constant current is first applied, but do become important at long times. They also persist for some time after the end of a stimulus, and are thus responsible for the appearance of *post-cathodal depression*, that is to say for a lowering of excitability after application for some while of a weak cathodal current. As a result of accommodation, cathodal currents rising more slowly than a certain limiting value do not stimulate at all, since the rise in threshold keeps pace with the depolarization.

Anode break excitation. Another familiar phenomenon is the occurrence of excitation when an anodal current is switched off. This can readily be demonstrated in isolated squid axons or frog nerves, but is not seen in freshly dissected frog muscle or in nerves stimulated *in situ* in living animals. In general, the conditions which allow anode break excitation to be exhibited seem to be that the resting potential should be rather low, so that there is an appreciable resting leakage of potassium and gain of sodium. The nerve can then be considered to be in a state of mild cathodal depression, with its sodium permeability mechanism partially inactivated, and an abnormally high potassium permeability. The effect of anodal polarization is to reactivate the sodium permeability and to reduce potassium permeability, and this 'improved' state persists for a short while after the applied current is switched off. While it persists, the critical potential at which inward Na^+ current exceeds outward K^+ current may be temporarily above the membrane potential in the absence of external current. Hence when the current is removed, an action potential is initiated.

Stimulation by solutions of high potassium concentration. It is well known that application to excitable tissues of solutions containing high concentrations of potassium may have a stimulatory effect. This action is related to the depolarization caused by high external $[K^+]$, which is illustrated in Figure 39.20. However, if the high potassium is applied simultaneously over the whole surface of a fibre, its effect is usually to depress or even to abolish excitability since, as discussed in the two previous paragraphs, a maintained depolarization both inactivates P_{Na} and raises P_K. Thus, although at first sight the uniform depolarization produced by high $[K^+]$ might be expected to shift the membrane potential closer to the critical triggering level, and therefore to increase excitability, the critical level is itself altered by the changes in P_{Na} and P_K in such a way that larger currents have to be applied to the nerve in order to reach it. The statement that under these conditions excitability is lowered and not raised does not conflict with the common experimental finding that application of potassium solutions to many excitable cells makes them generate impulses. Stimulatory effects of this kind are apt to arise if the applied solutions do not penetrate uniformly into the tissue, because if one part of a nerve is depolarized

much more than another, this is analogous to the application of a local catelectrotonus, and a repetitive discharge may arise at the edge of the depolarized region.

The effect of calcium on excitable tissues. The actions on nerve and muscle fibres of solutions containing less than the normal amounts of Ca^{2+} and Mg^{2+} ions are sometimes dramatic. In squid axons, slight lowering of external $[Ca^{2+}]$ leads to an oscillatory behaviour of the membrane potential, while more drastic reduction of calcium causes a spontaneous discharge of impulses at a high repetition rate. Conversely, a rise in external $[Ca^{2+}]$ tends to stabilize the nerve membrane and to raise the threshold for excitation. Changes in external $[Mg^{2+}]$ have rather similar effects on peripheral nerves. Studies by Hodgkin and Frankenhaeuser with the voltage-clamp technique have shown that the curve relating P_{Na} to membrane potential is shifted in one direction along the voltage axis by lowering $[Ca^{2+}]$, and in the other direction by raising $[Ca^{2+}]$. However, the resting potential is rather insensitive to changes in $[Ca^{2+}]$. This readily explains the relationship between $[Ca^{2+}]$ and threshold, since a rise in $[Ca^{2+}]$ moves the critical triggering level away from the resting potential, while a fall in $[Ca^{2+}]$ moves the critical level towards it. If the reduction in external $[Ca^{2+}]$ is large enough, the critical level may even be shifted beyond the resting potential, causing a spontaneous discharge. Although calcium and magnesium have similar actions on the excitability of nerve and muscle fibres, they have antagonistic actions at neuromuscular junctions and possibly also at synaptic junctions between neurones. Thus calcium appears to increase the amount of acetylcholine released by a motor nerve ending, while magnesium decreases it (see p. 836). Although the tetany of hypoparathyroidism is, as described on p. 216, related to serum calcium levels, the relationship is a complicated one which probably involves actions on the central nervous system as well as those on peripheral nerves and muscles.

After-potentials. In many types of nerve and muscle fibre the membrane potential does not immediately return to the base-line at the end of the spike, but undergoes further small and relatively slow variations known as 'after-potentials'. The nomenclature of after-potentials dates from the period before the invention of intracellular recording techniques, so that a variation of membrane potential in the same direction as the spike itself is termed a *negative after-potential*, while a variation in the opposite direction (that is to say, a hyperpolarization relative to the normal resting potential) is termed a *positive after-potential*. As may be seen in Figure 39.10 (A) and (B), isolated squid axons display a characteristic positive phase, which is almost completely absent in the living animal, while frog muscle fibres have a prolonged negative after-potential (Fig. 39.10 (H)). In some mammalian nerves there is first a negative and then a positive after-potential; these are seen both in myelinated fibres (Fig. 39.31) and in non-myelinated fibres. A related

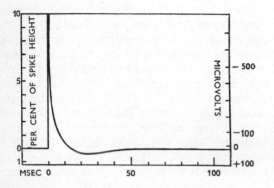

Fig. 39.31 Phrenic nerve at 37°C. The spike, lasting only 0·5 msec, leaves the base-line and goes above the figure; it then returns to the figure and is continued for about 10 msec as the negative after-potential, which is succeeded by a positive phase, the dip below the base line, lasting about 50 msec. (J. Erlanger & H. S. Gasser (1937). *Electrical Signs of Nervous Activity*, p. 148. Philadelphia: University of Pennsylvania Press.)

phenomenon, which is most marked in the smallest fibres, is the occurrence after a period of repetitive activity of a prolonged hyperpolarization of the membrane, known as the post-tetanic hyperpolarization.

It seems reasonable to suppose that after-potentials are always connected with changes in membrane permeability towards certain ions, but the precise sequence of changes probably varies with the type of fibre concerned. For example, in the isolated squid axon the positive phase arises because P_K is abnormally high at the end of the spike and P_{Na} very

low, so that the membrane potential temporarily comes close to E_K (Fig. 39.24). On the other hand, there is good evidence that production of the post-tetanic hyperpolarization which has been studied in frog nerves by Connelly and in mammalian C fibres by Rang, Ritchie and Straub, depends on an enhanced rate of electrogenic extrusion of Na^+ and of reabsorption of K^+ after a burst of impulses. There is also evidence that the presence of Schwann cells partially or wholly enveloping certain types of nerve fibre has important effects on the after-potential by slightly restricting the rate of diffusion of ions in the immediate neighbourhood of the nerve membrane.

The mechanism of the permeability changes. The most interesting problem raised by the recent work on the nerve impulse is to discover how, at the molecular level, the cell membrane manages not only to discriminate so successfully between Na^+, K^+ and other ions, but also to change its selectivity in such a subtle way. Very few clues are at present available. Nachmansohn has proposed that there is some sort of interaction in the membrane between acetycholine and a 'receptor protein'. This idea is based chiefly on the presence in all peripheral nerves of small, but very variable, quantities of acetylcholine, and on the blocking effect of high concentrations of certain anticholinesterases. Although acetylcholine may turn out to have a role in peripheral nerves, the evidence concerned with its 'nicotinic' action at autonomic synapses and at neuromuscular junctions is now so strong (pp. 831–835) that it seems safest to assume that its primary, if not its only function, is to act as a junctional transmitter. It is, unfortunately, impossible to propose any specific type of molecule as forming part of the Na^+ and K^+ transporting sites. It is clear that, since the permeability alters so steeply with membrane potential, either the movement at each site of a single multiply-charged particle must be involved, or else there must be a redistribution of four or five singly-charged particles, for example by the formation of a bridge across the membrane which permits the passage of ions only when all the particles are lined up. Since the Na^+ permeability changes can be completely blocked by very low concentrations of the Japanese puffer fish poison tetrodotoxin without any effect on K^+ permeability, while conversely tetraethyl ammonium ions can block the passage of K^+ without affecting Na^+, there seems no doubt that the ionic movements take place at two different types of site. Measurements by Keynes, Ritchie and Rojas of the amount of tetrodotoxin bound by nerve membranes have shown that in 1 μm^2 of membrane there are only some 50 Na^+ sites. Similar estimates for the number of K^+ sites have yet to be made.

The perfused axon preparation. An important advance in this field was the discovery by Baker, Hodgkin and Shaw in 1960 that a squid giant axon whose axoplasm has been squeezed out and wholly replaced with an artificial solution can still retain its ability to conduct impulses. Similar methods for perfusing giant axons have been introduced in the U.S.A. by Tasaki and his colleagues, and by Mullins and Brinley. The value of this new technique is, of course, that it enables the behaviour of the nerve membrane towards changes in the internal ionic concentrations to be studied in a way that was hitherto impossible. The results of applying it are entirely consistent with the explanation of the role of sodium and potassium ions in nervous conduction that has already been given and seem likely to lead to important new ideas about the mechanisms involved.

REGENERATION OF PERIPHERAL NERVES

When a mixed peripheral nerve is cut across there is paralysis and anaesthesia in the area supplied exclusively by that nerve; because of the wide overlap of sensory innervation, the completely insensitive area is much smaller than the total field supplied by the nerve. The interruption of the efferent fibres also causes sudomotor, pilomotor and vasomotor paralysis (see Chap. 44). The peripheral part of the nerve, that is the part containing nerve fibres no longer connected to their cell bodies, degenerates, and after a few days the affected muscles show a characteristic change in their strength-duration curve for electrical excitation (Fig. 39.15) and begin to atrophy. Whereas in a normally innervated muscle the strength-duration curve tends to be determined by that for the nerve endings, with greatest excitability at the point of entry of the motor nerve, the denervated muscle gives the curve typical of the muscle fibres themselves, with a much

greater chronaxie and an excitability that is almost the same all over the muscle belly. Another change in the denervated muscle fibres is that all parts of the muscle, and not just the region of the nerve endings, become sensitive to the application of acetycholine (p. 835).

If a nerve is crushed, for example by the end of a fractured bone, the axons may be divided but the nerve trunk may be held together by its fibrous tissue. The paralysis and anaesthesia are then exactly as before, and the peripheral part of the nerve degenerates, but no surgical operation is required to bring the severed ends of the nerve together. Pressure on a nerve by a tourniquet or by a crutch may cause a predominantly motor paralysis with little muscular wasting and with relatively little sensory loss, affecting chiefly the large myelinated motor nerve fibres. Complete recovery usually occurs within a few days or at most within a few weeks.

In spite of the relatively enormous length of the nerve fibre, its cell body exerts a great influence over even its most distant parts. If a nerve is cut and examined about a day later it is found that material has been extruded from the end of the central stump. Within a few days the myelin sheath in the peripheral stump breaks up into ovoid segments each somewhat shorter than the internodal distance. Phosphatase and esterase concentrations in the tissue are increased, suggesting that autolysis is being carried out by enzymes of the lysosomal system. A week after section the degenerating myelin stains black with osmic acid after treatment with bichromate (Marchi's method). This staining property, which has proved most useful in tracing the conducting tracts within the spinal cord, is then gradually lost. Shortly after nerve section the peripheral parts of the axons also fragment, and they and the myelin are subsequently removed by macrophages. This whole process is often called *Wallerian degeneration* because it was described first by Waller in 1862. In the central stump, where the fibres are still attached to their cell bodies, degeneration is usually confined to the centimetre or so next to the point of section, although a few of the central fibres and even their cell bodies may disappear, especially in very young animals. The cell body itself also shows changes which are known as *chromatolysis*; they are particularly well marked on cutting the nerve close to the cell body. The Nissl granules disappear from the dendrites and later from the main body of the cell, and the nucleus moves to one side of the cell. These changes, which are associated with an alteration in the excitability of the cell, begin on the first day and reach their maximum in the third week, after which the Nissl substance gradually reappears and the cell regains its normal appearance.

While Wallerian degeneration of the peripheral stump is proceeding, regeneration begins at the central end. Large numbers of little branches sprout from the cut fibres in all directions, forming an expanded mass or *neuroma*. Some of these branches enter the peripheral stump, where Schwann cells have proliferated and now form continuous tubes which guide the growing elements towards the periphery. When one branch has found its way satisfactorily into the peripheral stump, the other sprouting branches of that fibre disappear. Denervated organs such as muscle strongly attract new nerve fibres, probably by liberating a chemical factor which promotes their growth. Therefore neighbouring nerves also produce fresh branches. When new fibres reach the muscle, several of them may contact the same muscle fibre, but after the first functional end-plate has been formed, the other nerve branches degenerate.

The randomness in the healing process excludes any regular order in the return of function and makes complete recovery unlikely. However, there is sometimes a surprisingly good restoration of function after nerve section, which has been attributed to 're-education' of the controlling centres, perhaps through rearrangement of the relevant synaptic connexions in the spinal cord. The most complete recovery is obtained when the cut ends of the nerve are brought together very soon after the injury. The power of repair in nerves is not lost for many months, but the longer the delay in apposing the ends the poorer is the final result. This is partly due to the atrophy of the end-organs and partly to the atrophy and eventual disappearance of muscle cells. Improvement in muscle power and in sensation often continues for up to three years. Some of the properties of a muscle apparently depend on its particular nerve supply. Thus if

a 'slow' muscle such as soleus is re-innervated by fibres which previously supplied a 'fast' muscle like flexor digitorum longus, soleus becomes relatively fast.

Various methods are available clinically for determining the rate of recovery in a nerve. As regenerating fibres grow down the peripheral stump of a damaged nerve they re-innervate muscles in turn according to the distance to be travelled to each muscle. The time of recovery of motor function is taken as the time at which muscles show the first signs of voluntary contraction. The regenerated fibres are initially very much thinner than normal, and they may not regain their original thickness and conduction velocity for over a year; they probably never do if the nerve has been cut and not merely crushed. From anatomical dissections detailed information has been obtained about the lengths of the nerves to the muscles and, by dividing the distance from the point of section to the point of attachment of the nerve to the muscle cells by the time needed for recovery, it is found that a nerve fibre grows about 1·5 mm per day. Percussion over the non-myelinated tip of the advancing axon produces a tingling sensation which is known as Tinel's sign. This phenomenon advances at an average rate of about 1·7 mm per day, but it is not so satisfactory a method as the previous one for determining the rate of regeneration in a nerve. In calculating the time likely to be taken after nerve suture for the return of function it is well to add 6 or 8 weeks to allow for the possibility of an unusually slow rate of regeneration.

When the spinal cord is cut or the posterior roots are sectioned next to the cord, Wallerian degeneration occurs within the cord; although it may be followed by regeneration of a few new fibres within the central nervous system, restoration of function does not occur. The reason for this failure is not entirely clear, but two important factors seem to be scarring at the site of section, which prevents orderly penetration by regenerating fibres, and the absence of Schwann cell tubes. Unlike spinal neurones, cells in the brain itself usually degenerate irreversibly after section of their axons.

THE AFFERENT IMPULSE

So far we have considered the nerve impulse

produced artificially by electrical stimulation of a nerve. In the living body certain nerve fibres are connected to specialized endings and these, when suitably stimulated, generate continuous trains or phasic discharges of afferent impulses. As Adrian pointed out, the stimulus applied to an end-organ may be compared to the pressure on the trigger of a machine-gun; either it is strong enough to fire the bullets or it is too weak to do anything at all (all-or-nothing relationship). Because of its refractory period the nerve fibre cannot transmit a steady potential but only a a series of pulses, so that the message passing along the nerve may be likened to a stream of machine-gun bullets. If the abnormal conditions obtaining during the relatively refractory period are disregarded, the intensity of each impulse is constant or, to pursue the analogy, all the bullets are of the same size. The magnitude of the impulses cannot be altered by changing the strength of the stimulus. The only way, in fact, in which the message, carried by a given nerve fibre, can be graded to give the central nervous system information of the strength of the stimulus applied to the end-organ is by a variation in the total number of impulses and in the frequency with which they recur. Oscillograph records from even quite small nerves in the body may be too complex for analysis if many fibres are active and the action potentials from individual fibres are superimposed on one another. In the case of cutaneous sensory nerves the record may be simplified by stimulating only a very small area of skin, by cutting away unwanted receptors, or by cutting partly through the nerve. In this way the action potentials resulting from the activity of a single end-organ have been recorded. The form of these endings is very variable and, with the histological methods at present available, gives little clue to the particular stimulus to which the ending is sensitive. However, fibres have been isolated which carry signals from endings specifically sensitive to stretch, the chemical environment, temperature deviation, touch and pressure. The normal stimulus for the stretch sensitive endings is determined by the tissue in which they are deployed. Thus when arranged in the wall of a hollow organ, such as the carotid sinus, they signal fluid pressure (see p. 514 and p. 564), whereas arranged between the fibres of a tendon or linked mechanically to a con-tractile element composed of the intrafusal fibres of a muscle spindle (see p. 829) they signal longitudinal tension and act as muscle proprioceptors.

Adrian and Zotterman in 1926 first elucidated some of the properties of the muscle spindle by using the small sternocutaneous muscle of the frog. When the muscle was partly cut away records of the activity of one spindle were obtained from electrodes placed on the larger nerve trunk which is joined by the tiny nerve from the muscle. Matthews used the flexor digitorum brevis IV of the frog's foot, a muscle in which there is only one muscle spindle. Later in 1933 he recorded impulses from single spindle endings and tendon endings in the cat by cutting down a muscle nerve and leading from a single active nerve fibre. When the muscle is stretched by a weight the endings in a spindle discharge at a rate that depends on the weight used (Fig. 39.33) and continues for long periods with only a small decline in frequency (Fig. 39.32). This slight falling off is known as slow adaptation; it is a decline in excitability produced by the stimulus but, considering what we know already about the physiology of nerve, it is almost certainly not due to fatigue. Each end-organ has its own peculiar rate of adaptation; muscle spindles adapt slowly, pressure organs more quickly and touch and hair endings very rapidly. The nerve fibre itself, in contrast with the nerve ending, adapts most quickly of all, since it responds with only one impulse at the beginning of a constant stimulus. These observations are summarized in Figure 39.34.

By placing electrodes near the muscle spindle in a frog muscle Katz has found that, when the muscle is stretched, there is, in addition to the spikes (nerve impulses) shown in Figure 39.32, a negativity of the spindle. The amount of this electrical potential, the so-called *generator potential*, which gives rise to a train of impulses in the sensory nerve, depends on the amount and rate of stretching. The spikes but not the generator potential are abolished by a local anaesthetic. Katz has tried to account, at least partly, for these findings by supposing that the cell membrane becomes thinner on stretching so that its capacity increases and its potential falls, thus giving rise to the initial portion of the generator potential. In the muscle spindle the stretch stimulus may be produced by activity of the contractile elements within the spindle or stretching of the spindle as a whole if the muscle is stretched; in the living animal both these processes may interact. A fuller description of muscle spindles is given on p. 829.

A simpler arrangement is found at the joints where stretch receptors concerned with joint

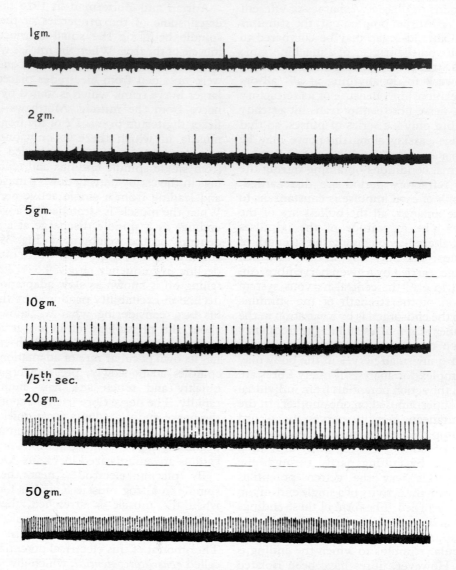

Fig. 39.32 Photographic recordings from a cathode-ray oscilloscope of action potentials in the sciatic nerve from a single-stretch receptor in the gastrocnemius muscle of a frog. Responses in other stretch-receptors had been eliminated by cutting down the sciatic nerve trunk until it contained only one active fibre. The load indicated on the diagram was applied to the tendo calcaneus one second before the beginning of each record. Close inspection of the records, especially those for 20 g and 50 g, shows that the space between individual spikes increases progressively from left to right. This decline in the frequency of discharge of the receptor is called adaptation. It is quite slow in the case of the stretch receptors. Some adaptation had, of course, occurred between the application of the weight and the beginning of the record in each case. Time marks represent $\frac{1}{5}$th sec, spike-height approximately 50 μV. (*By courtesy of B. L. Andrew.*)

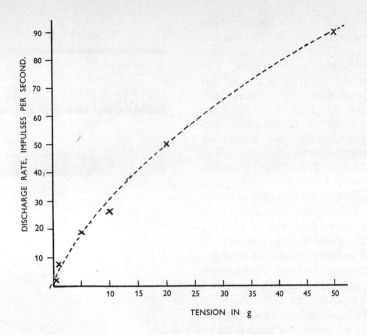

Fig. 39.33 Graph showing the relation between the tension applied to the tendon of a frog's gastrocnemius and the rate of discharge of nerve impulses from a single stretch-receptor within the muscle (calculated from Fig. 39.32).

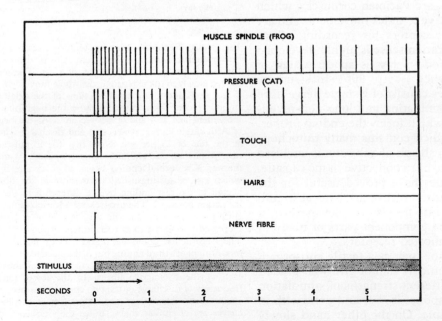

Fig. 39.34 Diagrams to illustrate the effect of applying a constant stimulus to various end-organs and to a nerve fibre. In the case of a nerve fibre a single impulse is aroused but in the end-organs a series of impulses of declining frequency is evoked. (*After* E. D. Adrian.)

position sense are found in the capsule; these are spray-like nerve terminals (Ruffini organs) several of which may share a single medium-sized (7 to 10 µm) sensory nerve fibre. The endings which, in any given position of the joint, lie on *stretched* connective tissue, discharge continuously. As parts of the joint are differentially stretched, or compressed, according to the position of the limbs, an individual joint proprioceptor may discharge impulses over a limited area of joint position (Fig. 39.35).

Larger Golgi endings with bunches of nerve sprays are found outside the capsules; each ending has its own nerve fibre (10 to 15 µm). These endings give a slower discharge than Ruffini organs, and may be associated with the joint ligaments. Numerous small Paciniform corpuscles give rapidly adapting responses and are very sensitive to quick movements. The fine free nerve endings probably subserve pain. Thus there is ample provision for recording all the movements of a joint and, since the discharges from the endings can be recorded in the cerebral cortex, it is probably largely by these means that we are aware of the position of our limbs.

Also near the joints, often under the tendons of the muscles, are Pacinian corpuscles which are again sensitive to quick movements and are probably the receptors for recording vibrations. The vibration sense is highly developed in mammals and its loss is an early sign of sensory deficiency (see p. 906). In structure these corpuscles consist of a single nerve fibre (4 to 7 µm) terminating in a long non-myelinated portion which forms the ending proper. The ending which contains many mitochondria but has no sheath of Schwann is enclosed in a thick onion-like connective tissue capsule. These corpuscles have proved useful for the study of the transference of mechanical stimuli into electrical impulses since they have in a simple form the component parts of many of the more complicated receptors.

An end-organ, such as a tactile end-organ, which adapts quickly cannot give well-graded responses to different strengths of stimulation because the response is over and done with in such a short time. On the other hand slowly adapting end-organs, like the muscle spindles, are able to give the central nervous system very accurate information about the intensity of

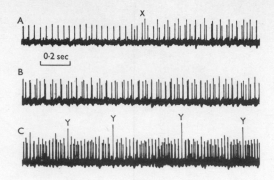

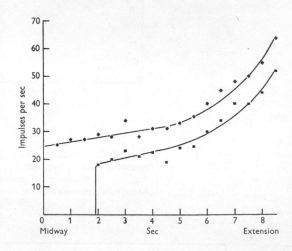

Fig. 39.35 Records of action potentials taken from a small sensory nerve filament arising from sensory endings in the thyroepiglottal joint of the larynx of a rat. The joint was moved slowly to full extension, the whole movement lasting 8 sec. A, B and C are small sections of the original continuous record taken respectively at the beginning of the movement and at 4 and at 6·5 sec later. At the beginning of A, a single ending was active and discharged impulses at the rate of 25 per sec. Note that the impulses are of nearly constant size and are regularly spaced. At the point marked X a second ending began to discharge impulses which can be distinguished from those in the fibre connected to the first ending by the difference in the size of the action potential. The progressive addition of new active endings as the stimulus (in this case joint extension) is increased is referred to as recruitment of sensory endings. The large action potentials marked Y are not due to the recruitment of a third ending but to the electrical summation of two impulses which have occurred simultaneously in each of the two active fibres. In the lower figure the frequency of discharge of the two endings has been plotted against time. Apart from the differences of threshold the curves are of similar shape. Since the frequency of both endings rises as the joint is extended it is easy to see that these endings signal the position of the joint to the central nervous system, that is they act as joint proprioceptors (B. L. Andrew (1954). *Journal of Physiology* **126,** 514).

the stimulus. Since a tactile stimulus is usually applied not to a single end-organ but to a fairly large area of skin a large number of end-organs is affected and so the information received at the central nervous system may be more complete than would be inferred from our knowledge of the properties of a single end-organ. Moreover when we touch an object we often run the fingers over it. This procedure causes stimulation of a large number of the quickly adapting end-organs and presents each with a rapidly changing stimulus so that each end-organ is continually re-excited. In this way a quickly adapting end-organ may be made to send a large number of impulses to the central nervous system.

THE SYNAPSE

The meeting-place of two neurones is called a synapse; at this point their surface membranes are very close together, but there is never any cytoplasmic continuity between them. The synapses between different types of excitable cell vary in their structure, but have some features in common. In general, the presynaptic fibre divides up into numerous fine branches which then end in greatly expanded terminals, *presynaptic knobs*, which make intimate contact with part of the membrane of the cell body or dendrites of the *postsynaptic* cell. A single anterior horn cell may be invested by as many as 1300 synaptic knobs (Fig. 39.36) derived from a large number of axons. In the part of the presynaptic knob closest to the postsynaptic cell, electron microscopy has shown that there is always a dense assembly of spherical vesicles about 50 nm in diameter (Fig. 39.37). The width of the extracellular space separating the pre- and postsynaptic membranes, the *synaptic cleft*, is another rather constant feature; it is generally about 25 nm.

The electrical resistance of the synaptic cleft is low enough to prevent any appreciable spread through the postsynaptic membrane of the electric current which flows around the presynaptic terminal when an impulse arrives there. Another factor which militates against electric excitation of the postsynaptic membrane by presynaptic action currents is the relatively greater area of the postsynaptic membrane. Except in a few special situations

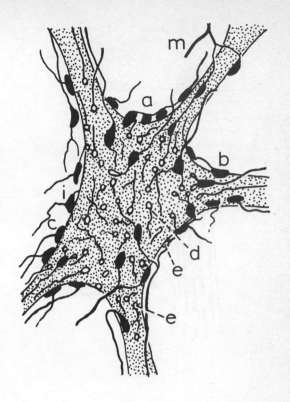

Fig. 39.36 This drawing of a motor cell of the spinal cord shows that it is invested by a large number of synaptic knobs from many neurones. There may be 1300 knobs on a single motor cell, occupying about one-third of its surface area. a, b, strong bulbs; c, i, terminal buttons; d, e, fine bulbs; m, terminal nerve fibre. S. R. Y. Cajal (1933). *Histology*, p. 290. London: Baillière, Tindall & Cox).

the coupling between the pre- and postsynaptic membranes is chemical rather than electrical. It depends on the release of a chemical transmitter substance from the presynaptic terminal, which then diffuses across the synaptic cleft and interacts with receptor sites on the postsynaptic membrane so as to cause a specific change in its ionic permeability. At some synapses the nature of this change is such as to produce a depolarization and hence an excitatory effect. At others the permeability change may tend to hold the membrane potential at its resting value, or may even cause a hyperpolarization; these synapses are inhibitory.

Some obvious distinctions between transmission of an impulse along a nerve fibre and transmission across a synapse follow directly from this description. In the first place,

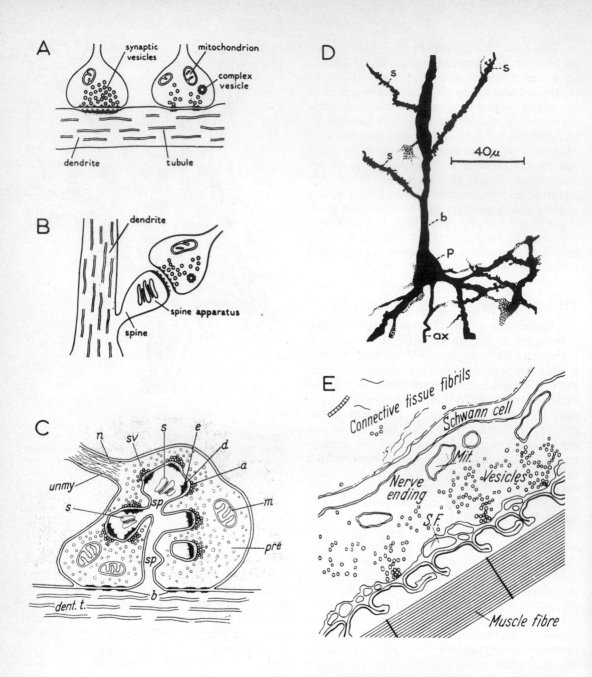

Fig. 39.37 The morphology of some synapses. A, two types of synaptic knob found on the soma and dendrites of cortical neurones. B, a dendritic spine of a neocortical pyramidal cell with its synapse. C, a more complicated synapse found on the spine (sp) of a dendrite of a hippocampal pyramidal cell. D, a Golgi preparation of a neurone from cat cerebral cortex showing spines (s) on its apical and basal dendrites, but not on the soma (p), axon (ax) or dendritic stumps (b) (*after* Whittaker & Gray). E, a frog motor end-plate, showing mitochondria (Mit), the characteristic folding of the synaptic cleft, and a few finger-like extensions of the Schwann cell (SF) beneath the nerve ending. (*After* Birks, Huxley & Katz.) (*From* J. C. Eccles (1964). *The Physiology of Synapses*. Berlin: Springer.)

although the impulse can travel equally well in either direction along a nerve fibre, its passage across a synapse is strictly unidirectional. No chemical transmitter is released when an antidromic impulse arrives at the postsynaptic membrane, so that such an impulse cannot travel backwards any further than the first synapse it reaches. Secondly, passage across a synapse involves a finite time delay, because the release of the transmitter, its diffusion across the synaptic cleft and its action on the postsynaptic membrane are relatively slow processes. In a spinal motoneurone, delay from this source, to which has to be added a smaller contribution from the slowness of conduction in the very fine presynaptic nerve branches, is about 0·4 msec. In autonomic ganglia it may be somewhat longer. As will be seen later (p. 909), a knowledge of the total delay in a complex reflex pathway is valuable in determining the number of synapses involved. Thirdly, the intervention of a chemical transmitter results in a susceptibility of synapses to blocking by a wide variety of drugs which have relatively little effect on conduction in nerve fibres. These may act either by interfering with the synthesis or release of the transmitter, or by changing the sensitivity of the postsynaptic membrane; some specific examples are discussed on p. 835. The last distinction, and by no means the least important, is that although some synapses—in particular the neuromuscular junction (p. 831)—operate in such a way that there is a one-to-one correspondence between presynaptic and postsynaptic impulses, the majority do not. In most neurones, a spatial and temporal summation of the several excitatory and inhibitory postsynaptic potentials takes place, and the postsynaptic membrane is triggered only by an appropriate combination of the impulses arriving along the presynaptic pathways. In contrast, therefore, to the all-or-none behaviour of peripheral nerves, synapses operate in a graded fashion, enabling the neurone to carry out an integration and sifting of the incoming information.

Many of our ideas about the mechanism of transmission across excitatory synapses have originated from micro-electrode studies of the neuromuscular junction. This work need not be described here, since it is discussed in detail on p. 831. It is important to note that there is no doubt as to the identity of the chemical transmitter at the motor end-plate, which is one of the places where acetylcholine exerts its 'nicotinic' effect, to use the term introduced by Dale, as opposed to its 'muscarinic' effect when it acts as a peripheral parasympathetic humoral transmitter. The characteristic action of nicotine in this context is to cause rapid excitation, followed by block; this action is antagonized by substances like curare. Muscarine, on the other hand, has a slower and longer-lasting action, and is antagonized by atropine. The action of nicotine on the motor end-plate occupies an honoured position in the history of the chemical transmission theory, since it was with frog muscle that Langley showed in 1907 that nicotine acted specifically on the junction rather than on the pre- or postjunctional fibres, its effect being strictly localized at the point of entry of the motor nerve.

Another relatively simple type of synapse where acetylcholine is certainly the transmitter is found in the ganglia of the autonomic system. Sympathetic ganglia have in fact provided particularly convenient material for studies on the chemistry of synaptic transmission, since it is not too difficult to perfuse their blood vessels and both to test the perfusate for the appearance of acetylcholine during electrical stimulation of the presynaptic nerve trunk and to see whether perfusion with acetylcholine sets up impulses in the postsynaptic fibres. These ganglia were also used in making some of the earlier electrical recordings of synaptic potentials, the superior cervical ganglion being conveniently accessible to electrodes without interrupting its blood supply. However, the potential changes observed with external electrodes were somewhat complex and as elsewhere internal microelectrodes were needed for a proper examination of the mechanism. Nishi and Koketsu were the first to succeed in penetrating the sympathetic ganglion cells of the frog, and their work has been extended by Ginsborg and his colleagues. Figure 39.38 shows that the spike recorded when the presynaptic nerve is stimulated orthodromically (O) differs from the antidromic spike (A) in being preceded by a small step, the *excitatory post-synaptic potential* (e.p.s.p.), which represents the depolarization of the postsynaptic membrane produced by the acetylcholine liberated from the presynaptic endings. The e.p.s.p. in a spinal motoneurone

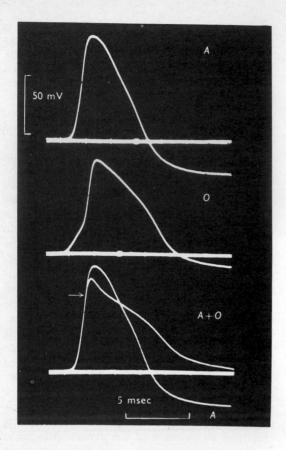

50 mV

A

O

A+O

5 msec

A

Fig. 39.38 The action potential recorded with an intra-cellular microelectrode in a frog sympathetic ganglion cell. A, antidromic response. O, orthodromic response. A+O, the effect on the antidromic response of initiating synaptic activity at the arrow. (*From* J. G. Blackman, B. L. Ginsborg & C. Ray (1963). *Journal of Physiology* **167**, 355.)

can be seen in Figure 39.39, and the similar junctional potential at the motor end-plate in Figure 40.7. The third record in Figure 39.38 shows the change in shape resulting from a superimposition of synaptic activity on the peak of an antidromic spike, and provides evidence about the equilibrium level towards which the membrane potential drops under the influence of acetylcholine. As Katz had pre-viously found for the frog motor end-plate, this equilibrium level is 10 to 20 mV below zero (that is internal potential still slightly negative); this suggests that the effect of acetylcholine is to cause a short-circuiting of the postsynaptic membrane by increasing its permeability simultaneously to Na^+ and K^+ ions. Another respect in which these syn-

apses behave similarly to the neuromuscular junction is in the occurrence of spontaneous miniature end-plate potentials (m.e.p.p.s) 1 to 3 mV in amplitude, which are apparently due to the release of small packets of acetylcholine (p. 834). The parallel with the motor end-plate extends to the effects on the m.e.p.p.s of various changes in the ionic concentrations of the external medium, but the number of quanta normally released by each presynaptic impulse has been estimated as only about twenty, as compared with several hundred for the motor end-plate.

The most important recent advances in our knowledge of the mechanisms involved in synaptic transmission have been made in Sir John Eccles's laboratory in Australia, where the techniques for using multibarrelled micro-pipettes have been further improved and applied extensively to motoneurones in the mammalian spinal cord. After inserting a double-barrelled micropipette into a moto-neurone, one barrel can be used for recording the junctional potentials set up by volleys in excitatory or inhibitory presynaptic pathways, while the other is used to alter the membrane potential in either direction by flowing electric current into or out of the cell. In this way (Fig. 39.39) the equilibrium potential level at which the e.p.s.p. disappears and beyond which it is reversed, was found to be slightly closer to 0 mV than it appears to be at the motor end-plate (see above). The equilibrium potential for inhibitory postsynaptic potentials (i.p.s.p.s) could be determined in a similar way; in the second experiment shown in Figure 39.39 it was about −80 mV. In order to investigate the ionic permeability changes underlying the e.p.s.p. and i.p.s.p., micropipettes can be filled with various salt solutions and the internal ionic concentrations of the motoneurones can then be altered by iontophoretic injection. The tentative conclusion from these studies is that during the e.p.s.p. the membrane becomes highly but unselectively permeable to cations (chiefly Na^+ and K^+), its anion permeability being little changed. During the i.p.s.p. the membrane becomes highly permeable to small anions like chloride, but remains relatively impermeable to anions whose hydrated dia-meter is more than 1·3 times that of the hydrated K^+ ion. There may also be a some-what smaller rise in K^+ permeability. The

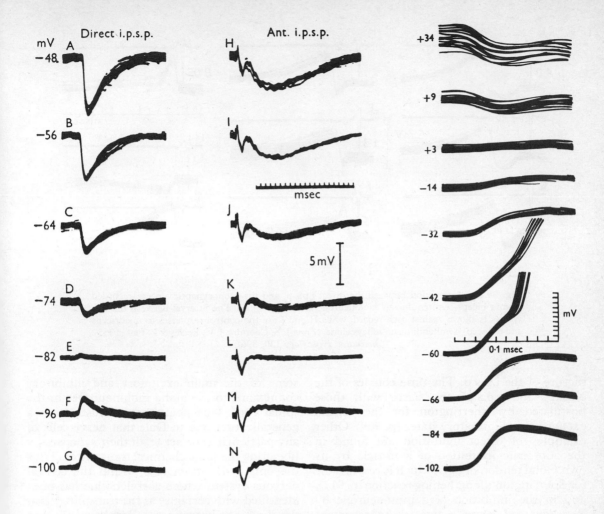

Fig. 39.39 Potentials recorded intracellularly from biceps-semitendinosus motoneurones by means of double-barrelled micropipettes. In each case 15 to 40 traces were superimposed, and the membrane potential was set at the level indicated by passing electric current from one of the barrels. Left-hand column shows the direct inhibitory action of a group Ia volley in the afferent fibres from the antagonist muscle quadriceps; central column shows the i.p.s.p. set up by an antidromic volley in the motor axon collaterals; right-hand column shows the e.p.s.p. elicited by a group Ia volley in the afferent fibres from the synergic muscle group. The reversal potential for the i.p.s.p. was about −80 mV, while for the e.p.s.p. it was close to 0 mV. (*From* J. S. Coombs, J. C. Eccles & P. Fatt (1955). *Journal of Physiology* **130,** 326 and 374.)

inhibitory mechanism in motoneurones would thus involve a combination of a rise in chloride permeability (which appears to be exclusively responsible for inhibition in crustacean muscle fibres) and of a rise in potassium permeability as takes place when acetylcholine acts on cardiac muscle fibres during vagal inhibition.

In the kind of inhibition so far considered there is a direct addition or subtraction of the e.p.s.p.s and i.p.s.p.s, as may be seen in Figure 39.40. A typical inhibitory connexion which operates in this fashion is that between a flexor muscle and its extensor antagonist, and vice versa. Reflex tests show that the latency of inhibition is short, as is consistent with the interpolation of a single interneurone (Fig. 39.41), and that its duration is of the order of 10 msec which fits well with the recorded

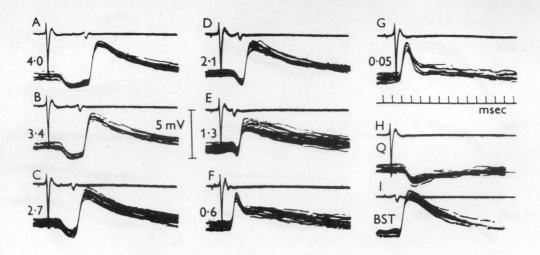

Fig. 39.40 Interaction between the direct i.p.s.p. and the monosynaptic e.p.s.p. recorded from a biceps-semitendinosus motoneurone as in Fig. 39.39. The interval between the test volleys is shown against each record, while H and I are the control responses to quadriceps and to biceps-semitendinosus volleys alone. (*From* J. S. Coombs, J. C. Eccles & P. Fatt (1955). *Journal of Physiology* **130,** 396.)

picture of the i.p.s.p. The time courses of the e.p.s.p. and i.p.s.p. also agree with those postulated by Sherrington for the central excitatory and inhibitory states (p. 909). Other examples of direct inhibition are found in the autogenic inhibition of a muscle by its own Golgi tendon organs, which is responsible for Sherrington's lengthening reaction (p. 917), and in the inhibition of a motoneurone by Renshaw cell axons activated by recurrent collaterals from neighbouring motoneurones. It should not be supposed, however, that this is the only type of inhibition. When, for example, cutaneous afferents are stimulated, reflex activity may be inhibited for several hundred milliseconds, particularly in flexor muscles, without any visible change in the postsynaptic membrane potential of the moto-neurones. It is now thought that this so-called presynaptic inhibition is brought about by inhibitory terminals situated on the excitatory presynaptic nerve endings (Fig. 39.41), which reduce the amount of transmitter released. Although presynaptic inhibition does not result in an intracellularly recorded i.p.s.p. it probably gives rise to the long-lasting dorsal root potentials which can be observed with external electrodes.

The diagram in Figure 39.41 summarizes

some of the main excitatory and inhibitory connexions to the alpha motoneurones in the spinal cord. One point that it illustrates is a generalization due to Dale that nerve cells of any particular type act at all their synapses by liberating the same chemical transmitter. Thus the only type of synapse within the central nervous system where acetylcholine has been identified with certainty as the transmitter is at the junctions between the Renshaw cells and the recurrent collaterals, which are of course branches from efferent axons which also release acetylcholine at the motor end-plates. On Dale's principle it is not surprising that it should now have been proved that inter-neurones intervene in the pathway between Ia afferents from muscle spindles (Fig. 39.41) and the inhibitory connexions on the antagonist motoneurones. Moreover, although the chem-ical identity of the transmitter involved in direct inhibition is not yet certain, it seems likely to be the same for all the direct inhibitory terminals, since all of them are blocked in a similar fashion by strychnine and tetanus toxin, whereas the most active blocking agent for the presynaptic inhibitory synapses appears to be picrotoxin.

As far as synaptic transmission within the brain itself is concerned, the picture is at present less clear. It is difficult to obtain intra-cellular records even from the largest cells, although Phillips and others have had some success in inserting micropipettes into the Betz cells of the motor cortex and the Purkinje cells of the cerebellum, and the properties of the giant Mauthner cells in fish brains have been examined in some detail. It has also proved hard to find well-defined interconnecting pathways which can be used for the evocation of e.p.s.p.s or i.p.s.p.s with any regularity. However, evidence is steadily accumulating that

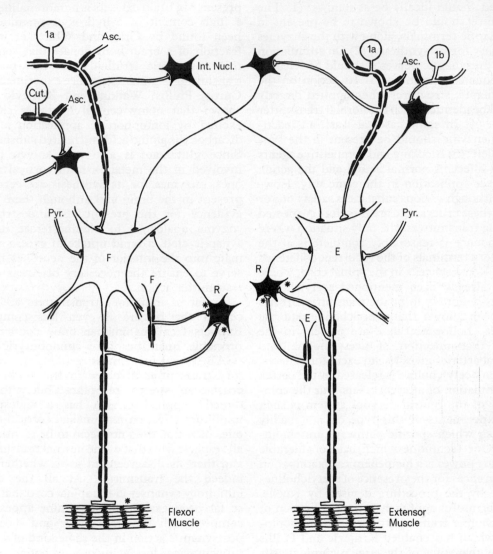

Fig. 39.41 A schematic diagram of some of the interconnexions between two flexor moto-neurones (F) and the motoneurones (E) of the antagonist extensor muscle. Inhibitory neurones and their synaptic terminals are shown in black. The recurrent collaterals make cholinergic connexions (*) with Renshaw cells (R). Ia, cell bodies of afferent fibres from annulo-spiral endings on muscle spindles within the corresponding muscles. Ib, cell body of an afferent fibre from a Golgi organ in the tendon of the extensor muscle. Cut., cell body of a cutaneous afferent which is connected to an inhibitory neurone having a presynaptic inhibitory terminal on one of the Ia excitatory endings. Int. Nucl., inhibitory neurones in the intermediate nucleus. Pyr., excitatory terminals of the descending pyramidal tract. Asc., branches of the afferents joining the ascending sensory fibre tracts.

the mechanisms which have been demonstrated in the spinal cord also operate in the brain.

We come now to the difficult question of the chemical identity of the transmitters in the central nervous system. The criteria to be satisfied should ideally be as follows: (1) The substance should be known to be present in presynaptic terminals, along with the enzymes necessary for its synthesis; (2) On stimulation of the presynaptic nerves it should be released in adequate quantities; (3) Its action on the postsynaptic structures when applied directly should be identical with the normal transmitter action; (4) In some cases at least, an inactivating enzyme should be present in the synaptic cleft; (5) Blocking and competitive agents should affect its normal action and the action on direct application in the same way. However, although various substances satisfy one or two of these criteria and are therefore suspected of being transmitters, the only situation where the evidence is reasonably complete is at the excitatory terminals of the recurrent collaterals on the Renshaw cells in the spinal cord, where as has already been mentioned acetylcholine appears definitely to be the transmitter. It has long been known that acetylcholine and the enzyme choline acetylase are present in the brain, that injection of acetylcholine into cerebral arteries gives rise to excitatory effects, and that acetylcholine is released in the cortex on stimulation of afferent tracts, but the complexity of the central nervous system is such that experiments of this type do not readily establish which specific pathways are cholinergic. One technique which has considerable resolving power is a histochemical examination of brain slices for the presence of acetylcholinesterase by the procedure devised by Koelle. Another useful approach is the application of acetylcholine from multi-barrelled micropipettes, which has enabled Krnjevic and Phillis to show that some of the large pyramidal cells in the deeper layers of the motor cortex can be activated by the local release of small quantities of acetylcholine. But although there is thus strong circumstantial evidence implicating acetylcholine as an important central transmitter, it is still not known for certain exactly which interconnexions within the brain are cholinergic.

A further candidate for consideration as a central transmitter on the grounds of its presence in brain extracts is noradrenaline, and a high content of 5-hydroxytryptamine has been found by Gray and Whittaker in the fraction of a brain homogenate that contains the presynaptic terminals. Another possible transmitter is glutamate, since experiments by Curtis, Phillis, Watkins and Krnjevic have shown that many central neurones can be excited by iontophoretic application of this dicarboxylic amino acid and related substances. Since glutamate is also well known to be involved in the metabolism of cortical neurones, enzymes for its synthesis are certainly present in the brain and although there is no evidence for the presence of a destructive enzyme analogous to cholinesterase in the synaptic cleft, a rapid uptake of excess glutamate into the interior of the neurones might serve to ensure the necessary briefness of its transmitter action.

As far as inhibitory transmitters are concerned the picture is even less complete. Extracts from mammalian brain contain appreciable quantities of γ-aminobutyric acid (GABA), which is almost certainly the inhibitory transmitter involved in the working of crustacean stretch receptors, but although directly applied GABA has a widespread inhibitory effect on mammalian cortical neurones its action does not seem to be identical in all respects with that of the normal transmitter, and there is disagreement as to whether it is indeed the transmitter. At all the direct inhibitory synapses in the spinal cord that have so far been examined, strychnine appears to compete with the transmitter and block its postsynaptic action in the same kind of way as tubocurarine does at nicotinic cholinoceptive receptor sites, while tetanus toxin works in a similar way to botulinum toxin at cholinergic endings by preventing the presynaptic release of transmitter.

REFERENCES

ADRIAN, E. D. (1928). *Basis of Sensation*. London: Christophers.

BAKER, P. F., HODGKIN, A. L. & SHAW, T. I. (1962). Replacement of the axoplasm of giant nerve fibres with artificial solutions. *Journal of Physiology* **164**, 330–354

BOWDEN, R. E. M. (1958). *Peripheral Nerve Injuries*. London: Lewis.

BUNGE, MARY B., BUNGE, R. P. & RIS, H. (1961). Ultrastructural study of remyelination in an experimental lesion in adult cat spinal cord. *Journal of Biophysical and Biochemical Cytology* **10**, 67–94.

CALDWELL, P. C. & KEYNES, R. D. (1960). The permeability of the squid giant axon to radioactive potassium and chloride ions. *Journal of Physiology* **154**, 177–189.

COHEN, L. B., KEYNES, R. D. & HILLE, B. (1968). Light scattering and birefringence changes during nerve activity. *Nature* **218**, 438–441.

CROSSLAND, J. (1960). Chemical transmission in the central nervous system. *Journal of Pharmacology, London* **12**, 1–36.

CURTIS, D. R., PHILLIS, J. W. & WATKINS, J. C. (1960). The chemical excitation of spinal neurones by certain acidic amino acids. *Journal of Physiology* **150**, 656–682.

DAVISON, A. N. (1969). Biochemistry and the myelin sheath. *Scientific Basis for Medicine Annual Review* 220–235.

DAVSON, H. (1970). *A Textbook of General Physiology*. 4th edn. London: Churchill.

ECCLES, J. C. (1957). *The Physiology of Nerve Cells*. London: Oxford University Press.

ECCLES, J. C. (1961). The mechanism of synaptic transmission. *Ergebnisse der Physiologie* **51**, 299–430.

ECCLES, J. C. (1964). *The Physiology of Synapses*. Berlin: Springer.

ERLANGER, J. & GASSER, H. S. (1937). *Electrical Signs of Nervous Activity*. Oxford University Press.

FRANKENHAEUSER, B. & HODGKIN, A. L. (1957). The action of calcium on the electrical properties of squid axons. *Journal of Physiology* **137**, 218–144.

GRANIT, R. (1955). *Receptors and Sensory Perception*. London: Oxford University Press.

GRAY, E. G. & WHITTAKER, V. P. (1962). The isolation of nerve endings from brain: an electron-microscopic study of cell fragments derived by homogenization and centrifugation. *Journal of Anatomy* **96**, 79–88.

GRAY, J. A. B. (1959). Mechanical into electrical energy in certain mechanoreceptors. *Progress in Biophysics* **9**, 285–324.

GUTH, L. (1956). Regeneration in mammalian peripheral nervous system. *Physiological Reviews* **36**, 441–479.

HILL, D. K. (1950). The effect of stimulation on the opacity of a crustacean nerve trunk and its relation to fibre diameter. *Journal of Physiology* **111**, 283–303.

HILL, D. K. (1950). The volume change resulting from stimulation of a giant nerve fibre. *Journal of Physiology* **111**, 304–327.

HILLE, B. (1970). Ionic channels in nerve membranes. *Progress in Biophysics and Molecular Biology* **21**, 1–32.

HODGKIN, A. L. (1951). The ionic basis of electrical activity in nerve and muscle. *Biological Reviews* **26**, 339–409.

HODGKIN, A. L. (1958). Ionic movements and electrical activity in giant nerve fibres. *Proceedings of the Royal Society* **148**, 1–37.

HODGKIN, A. L. (1963). *The Conduction of the Nervous Impulse*. Liverpool University Press.

HODGKIN, A. L. & HOROWICZ, P. (1959). Movements of Na and K in single muscle fibres. *Journal of Physiology* **145**, 405–432.

HOWARTH, J. V., KEYNES, R. D. & RITCHIE, J. M. (1968). The origin of the initial heat associated with a single impulse in mammalian non-myelinated nerve fibres. *Journal of Physiology* **194**, 745–793.

HUBBARD, J. I. (1970). Mechanisms of transmitter release. *Progress in Biophysics and Molecular Biology* **21**, 33–124.

KEYNES, R. D. (1970). Evidence for structural changes during nerve activity and their relation to the conduction mechanism. In *The Neurosciences: Second Study Program* (Edited by F. O. Schmitt), pp. 707–714. New York: Rockefeller University Press.

KEYNES, R. D. & LEWIS, P. R. (1951). The sodium and potassium content of cephalopod nerve fibres. *Journal of Physiology* **114**, 151–182.

KEYNES, R. D., RITCHIE, J. M. & ROJAS, E. (1971). The binding of tetrodotoxin to nerve membranes. *Journal of Physiology* **213**, 235–254.

KRNJEVIC, K. & PHILLIS, J. W. (1963). Acetylcholine-sensitive cells in the cerebral cortex. *Journal of Physiology* **166**, 196–327.

LING, G. & GERARD, R. W. (1949). The normal membrane potential of frog sartorius fibres. *Journal of Cellular and Comparative Biochemistry* **34**, 382–396.

MAGOUN, H. W. (Ed.) (1959). *Neurophysiology*. Vol. I, Section I of *Handbook of Physiology*. Washington: American Physiological Society.

McILWAIN, H. (1955). *Biochemistry and the Central Nervous System*. London: Churchill.

PHILLIPS, C. G. (1959). Actions of antidromic pyramidal volleys on single Betz cells in the cat. *Quarterly Journal of Experimental Physiology* **44**, 1–25.

RANG, H. P. & RITCHIE, J. M. (1968). On the electrogenic sodium pump in mammalian non-myelinated nerve fibres and its activation by various external cations. *Journal of Physiology* **196**, 163–221.

RITCHIE, J. M. & STRAUB, R. W. (1957). The hyperpolarization which follows activity in mammalian non-medullated fibres. *Journal of Physiology* **136,** 80–97.

ROBERTSON, J. D. (1960). The molecular structure and contact relationships of cell membranes. *Progress in Biophysics* **10**, 343–418.

SKOU, J. C. (1965). Enzymatic basis for active transport of Na$^+$ and K$^+$ across cell membrane. *Physiological Reviews* **45,** 596–617.

STAMPFLI, R. (1954. Saltatory conduction in nerve. *Physiological Reviews*. **34,** 101–112.

TASAKI, I. (1953). *Nervous Transmission.* Springfield: Thomas.

The neuron (1952). *Cold Spring Harbor Symposia on Quantitative Biology* **17**, 1–315.

WHIPPLE, H. E. (Ed.) (1964). Symposium on the nerve growth factor. *Annals of the New York Academy of Sciences* **118,** 147–232.

WHITTAKER, V. P., MICHAELSON, I. A. & KIRKLAND, R. J. A. (1964). The separation of synaptic vesicles from nerve-ending particles (synaptosomes). *Biochemical Journal* **90,** 293–303.

YOUNG, J. Z. (1949). Factors influencing the regeneration of nerves. *Advances in Surgery* **1,** 165–220.

40 Muscle

STRUCTURE OF SKELETAL MUSCLE

Even the smallest skeletal muscle seen in an anatomical preparation is made of of very large numbers of muscle cells, sometimes called muscle fibres. The individual cells, from 1 to 120 mm long or more in man, and from 10 to 100 μm in diameter have a tough outer membrane, the sarcolemma about 10 nm thick. Most of the space within each cell is occupied by numerous myofibrils about 1 μm in diameter, and the remainder by sarcoplasm (cytoplasm) containing several nuclei together with a network of tubules known as the sacroplasmic reticulum. The size of the cells and the amount of tissue lying between them increases with age; exercises which strengthen muscles and increase their bulk produce an increase in the cross-sectional area of the muscle cell. At the end of the muscle cells the myofibrils are in contact with the sarcolemma which at this position shows complex folds into which the collagen fibres of the tendon are fitted.

When examined by the ordinary light microscope muscle fibres appear to be transversely striated, and these striations are due to alternating zones of different refractive index within the myofibrils. The A (anisotropic) bands are birefringent, having a high refractive index, whereas the I (isotropic) bands have a lower refractive index. In most preparations of striated muscle tissue the A bands show up more deeply coloured. In the centre of each I band is a Z line (Fig. 40.1) which is, in fact, a disc of material running across the whole muscle fibre and joining the myofibrils to each other. The central region of the A band is paler and is known as the H (Hensen's) band (Fig. 40.1). The region between one Z line and the next is known as a sarcomere. The length of a sarcomere varies between 1·5 and 3·0 μm depending on the state of extension of the muscle fibres.

Electron microscope photographs (Fig. 40.2)

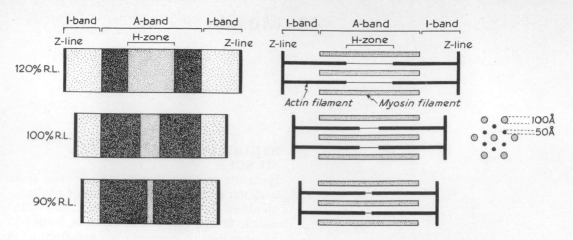

Fig. 40.1 Diagrammatic representation of the structure of skeletal muscle. The three diagrams on the left show the appearances as seen with the phase contrast or interference microscope (see p. 825) at various percentages of the resting length, R.L. Note that the A band remains constant in length, about 1·5 μm in the rabbit psoas. The diagrams on the right show the arrangement of the filaments in muscle in both longitudinal and transverse section; the latter should be compared with the middle portion of Fig. 40.2A. (*After* H. E. Huxley (1956). *Endeavour* **15**, 177–188.)

give more information about the structure of myofibrils. Each myofibril contains a system of longitudinal filaments arranged in a regular pattern (Fig. 40.1). In the I bands thin filaments consisting mainly of the protein actin are attached to the Z line (or Z disc as it may more appropriately be called). In a transverse section of the myofibril these thin filaments are seen to be arranged in a hexagonal pattern. The thin filaments extend into the A band where they interdigitate with a system of thicker filaments consisting mainly of the protein myosin. Here, too, there is a regular arrangement of filaments; the thick filaments are about 45 nm apart, and each thick filament is surrounded by six thin filaments (Fig. 40.2).

As the muscle changes in length, the thick and thin filaments slide over each other; during shortening the thin filaments move progressively further in between the thick filaments, and the I band becomes correspondingly narrower, though the A band remains the same width. At all but the shortest muscle lengths there is a region in the centre of the A band into which the thin filaments do not extend; this is the paler H zone, which becomes wider as the muscle fibre length increases.

Each thick (myosin) filament carries a series of side chains which project outward toward the six adjacent actin filaments; six such side chains occur in a helical arrangement over a distance of 40 nm. These side chains play an important part in the mechanism of muscle contraction which will be described later. The central part of the thick filament does not have them.

Amongst the myofibrils is a system of fine tubules, the sarcoplasmic reticulum, which is important in activating muscle contraction. A system of transverse tubules extend inward from the surface of the muscle fibre; these are, in fact, inward extensions of the extracellular space within the substance of the fibre. In some species these are situated at the Z line, but in mammals they are closer to the junctions of the A and I bands. A second system of tubules extends longitudinally through the sacroplasm amongst and around the myofibrils. This longitudinal component of the sarcoplasmic reticulum comes into close contact with the transverse tubules, but there is probably no actual continuity between them. (See also Fig. 26.9.)

SENSORY INNERVATION OF SKELETAL MUSCLE

Skeletal muscle contains many sense endings: some of these give rise to sensations of discomfort or pain when the muscle is fatigued and these can be regarded as protective,

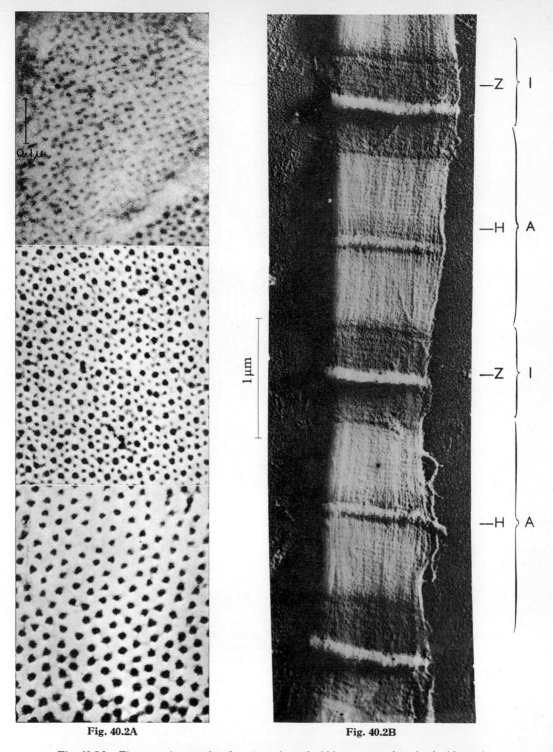

Fig. 40.2A   Electron micrographs of cross-sections of rabbit psoas muscle stained with osmic acid and phosphotungstic acid. Top: thin actin filaments in the I-band. Middle: double hexagonal array of filaments in the A-band. Bottom: simple hexagonal array of myosin filaments in the H-zone. (*By courtesy of H. E. Huxley.*)

Fig. 40.2B   Electron micrograph of toad muscle, platinum shadowed. I, isotropic band. A, anisotropic band. Z, Dobie's line. H, Hensen's band. (*By courtesy of M. H. Draper and A. J. Hodge.*)

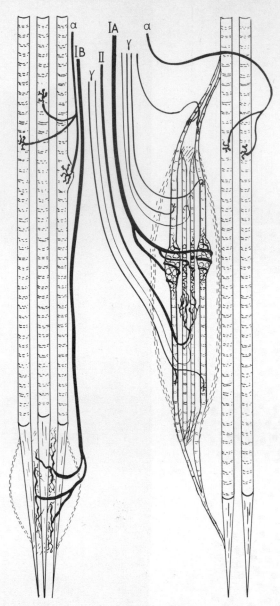

Fig. 40.3 Diagrammatic drawing of a muscle spindle and a tendon organ. The muscle spindle is seen on the right attached to extrafusal muscle fibres and tendon. It consists of small diameter intrafusal muscle fibres which are largely enclosed in a connective tissue capsule. Longitudinally the drawing is not to scale (the length of a spindle may be fifty times its width). Transversely in the drawing the width of the extrafusal muscle fibres represents a diameter of 40 μm; the intrafusal fibres are drawn to the same scale and represent diameters of about 20 μm for the two long fibres with nuclear bags at the equator of the spindle and about 10 μm for the two short fibres with nuclear chains at the equator. The group of nerve fibres shows the relative diameters of these fibres to each other. The largest fibre, marked IA, supplies the main primary afferent ending lying over the nuclear bags and chains. Fibre II goes to a secondary afferent ending on the nuclear chain fibres adjacent to the primary ending. Six small γ motor fibres of varying sizes supply motor endings on the intrafusal muscle fibres. The motor end-plates on the extrafusal muscle fibres are supplied by larger α nerve fibres. The remaining IB nerve fibre goes to the encapsulated tendon organ on the left; the branches of the afferent nerve ending lie between the tendons of a group of extrafusal muscle fibres.
(By courtesy of Sybil Cooper.)

preventing the individual from overworking his muscles. The remaining sense endings do not give rise to sensations but pass information to the central nervous system about the mechanical events in the muscle. This information is supplied to the motoneurones which control the motor units in the muscle and is used to make adjustments to the discharge of motor impulses. Branches of the afferent fibres carry duplicate information to higher levels of the nervous system, such as the cerebellum.

Two types of sensory nerve ending are present, tension receptors called Golgi tendon organs and length receptors called muscle spindles. The Golgi tendon organs (Fig. 40.3) are mounted on connective tissue lying in series with the muscle fibres. They issue signals in sensory nerves when the connective tissue is stretched by forces produced by the contraction of the muscle. Below a certain value of tension, called the threshold, the ending does not discharge nerve impulses, above this value the frequency of discharge rises as the tension rises. Inactive muscle can be extended, within physiological limits with very little rise of tension, so the extension of an inactive muscle by a joint movement has little effect on Golgi organs. However, many movements, for example, walking down a flight of stairs, involve elongation of an active muscle; in this example the activity of the thigh extensors acts as a brake opposing the descent. The discharge of a Golgi tendon organ is shown in Figure 40.4F. Note the ending discharges only during the phase of tension produced by the muscle contraction.

The spindle is much more complicated than the tendon organ; the basic element is a small bundle of skeletal muscle fibres a few mm long which lies within a fluid-filled capsule (Fig. 40.3). The fibres are unusual because they are much smaller in diameter than the other fibres; they are called intrafusal fibres to distinguish them from the large extrafusal fibres which make up the bulk of the muscle. The ends of the intrafusal fibres, which may protrude some distance beyond the poles of the spindle-shaped capsule, are attached to the connective tissue structure of the muscle. The bundle of intrafusal fibres has a motor innervation which is different from that of the extrafusal fibres. The motor fibres are small, and called either γ efferents or small motor fibres (Fig. 40.3).

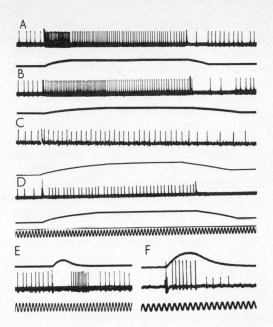

Fig. 40.4 Impulses originating in receptor organs in the soleus muscle of a decerebrate cat and recorded, by cathode ray oscilloscope, in single nerve fibres in the dorsal spinal roots. The ventral spinal roots were intact so that motor impulses may be travelling in both α and γ motor nerve fibres to the muscle to maintain tone in the extrafusal muscle fibres and to activate the intrafusal muscle fibres.

All the other muscles in the limb were denervated.

In A to D the second trace records the application and release of a 4 mm stretch to the muscle. The time signal of 50 Hz is shown below D.

A and B. Responses from the primary ending of a muscle spindle. In A the stretch is applied at 25 mm per sec, and in B at 10 mm per sec. In both cases the discharge rate rises rapidly during the dynamic part of each stretch, reaching 200 per sec in A and 150 per sec in B. The discharge is steady during maintained stretch and slows during release of stretch; after which it returns quickly to the resting rate.

C. Response from a secondary ending of a muscle spindle to a muscle stretch at 10 mm per sec. The increase in discharge rate during the dynamic period is much less marked than in B. The rate is steady during maintained stretch and shows some slowing on release of stretch.

D. Response from a tendon organ to a muscle stretch at 10 mm per sec. Owing to the high threshold of the organ the increase in discharge is slow to start. The discharge continues during the maintained stretch, but stops at once when the stretch is released and does not start again.

E and F. Responses of the three kinds of ending to a maximal shock to the muscle nerve. Tension is recorded in the upper traces and 50 Hz is shown below.

E. Primary ending. The discharge stops during the contraction but comes in at an increased rate during relaxation.

F. Secondary ending (small deflexions) and tendon ending (large deflexions). The secondary ending shows no response during contraction but give a discharge during relaxation. The tendon organ discharges during contraction but stops during relaxation. (*By courtesy of Sybil Cooper.*)

The motoneurones which give rise to these fibres lie scattered in the motoneurone pool supplying the extrafusal fibres but their activity is controlled differently. Each motoneurone controls a group of muscle fibres, called its motor unit; it is able to exercise this multiple control because the motor axon branches many times in the muscle. Extrafusal motor units range from several thousand in the large limb muscles to tens of fibres in small muscles of the hand. Intrafusal motor units are numerically small and endings from more than one intrafusal motoneurone make contact with an individual intrafusal muscle fibre, this is called polyneural innervation of the muscle fibre; polyneural innervation is not found in extrafusal muscle. Two types of intrafusal muscle fibres have been identified, their main anatomical difference lies in the arrangement of the cell nuclei, these are either closely packed in a group, these are called nuclear bag fibres, or arranged in a line, in which case they are called nuclear chain fibres; the contraction properties of the two types of fibre are different.

The main sensory nerve ending of the intrafusal muscle is found on the mid-capsular region called the equatorial zone. The contractile apparatus of the intrafusal muscle fibre is weakened at the equator, the cross-striation being very faint. The main sensory ending, the primary ending, consists of unmyelinated nerve terminals with many small swellings wrapped around the equatorial zone. It is sometimes called an annulo-spiral ending and is the terminal of a large myelinated afferent fibre.

In some spindles there are in addition sensory endings called secondary endings which lie a small distance away from the primary ending on the intrafusal muscle fibres. There is a great range of complexity in muscle spindles both in the number of intrafusal muscle fibres and the number of sensory endings. Ruffini, in the nineteenth century, suggested that spindles with a primary ending only should be called simple, those with a primary and one secondary should be called intermediate and those with more than one secondary should be called complex. This classification is still used, but the functional significance of the variation in sensory equipment is obscure.

The sensory nerves which form the primary endings are of large diameter, usually the largest in the muscle nerve, slightly larger than the fibres serving the Golgi tendon organs and about twice as large as the fibres serving the secondary endings. It is not possible to generalize about the sizes of these nerves since the spectrum of fibre sizes varies with different muscles and with different animals but the spindle primary afferents are described as group Ia, the Golgi tendon organ afferents as group Ib and the secondary ending afferents as group II (see Fig. 40.3).

If an isolated spindle is held in a slightly stretched position the sensory endings discharge nerve impulses steadily, if the length of the spindle is increased the frequency rises during the time of stretching and declines to a steady value which may only be slightly different from the initial frequency. The period during which the increase in length occurs is called the dynamic phase, and the increase in frequency during this phase is called the dynamic response. There is a marked difference in the behaviour of the primary and secondary endings during the dynamic phase. The primary endings have a noticeable dynamic response, the secondaries have little or none. This difference is well illustrated in Fig. 40.4A, B and C. Thus a nerve centre supplied with the output of both primaries and secondaries from a spindle could calculate the instantaneous length (from the secondary discharge) and the rate of change of length (from the dynamic response of the primary ending).

The spindle intrafusal muscle lies in parallel with the extrafusal muscle, that is changes in length imposed on the muscle when joints move are also equally imposed on the muscle spindles within the muscle. In a passively extended muscle the spindles signal stretch. The situation becomes more complicated when the muscle contracts. If the extrafusal muscle is made to contract, but the intrafusal muscle is inactive, the spindle shortens passively and the sensory discharge declines or stops. If the intrafusal muscle contracts in isolation it cannot, since it is very feeble, move the joints to which the muscle is attached; contraction of the muscular poles of the spindle stretches the equatorial zone of the spindle and leads to a vigorous discharge.

If the intrafusal and extrafusal muscles are

stimulated simultaneously the discharge from the spindle varies according to resistance to shortening experienced by the extrafusal fibres. If the muscle meets little resistance to shortening and so makes an isotonic contraction it brings together the points of attachment of the two ends of the spindle and thus the tension applied to the equatorial zone is slight and there is little afferent discharge. If the extrafusal contraction is isometric, or nearly so, the intrafusal contraction is concentrated on the equatorial zone and so a vigorous stimulation of the sensory endings occurs. Thus if the motor centres send simultaneous messages to the intra- and extrafusal muscle, the reply from the sense endings will indicate how much muscle shortening has occurred. It has been shown that the spindle discharge is relayed back to the motoneurones of the extrafusal muscle, where by excitatory synaptic action the discharge of the motoneurone tends to increase. Thus if the muscle meets resistance during shortening extra contractile force is called in by way of the spindles. However, this is a simplification because other influences act on the extrafusal motoneurone which can control it independently of its spindle feed-back. Movements are possible even after section of all sensory afferent feed-back from the muscle, but it seems likely that for certain types of closely controlled movement the spindles are necessary.

If a finger joint in man is anaesthetized position sense in the joint is greatly impaired although no muscle spindles have been disturbed. The impulses from the spindle play a part in the nervous control, at the subconscious level, of muscular activity both during movement and during sustained contractions.

When the skeletal muscles are examined histologically great variation is found in the density of muscle spindles and tendon organs. A few muscles, for example the cricothyroid and the extraocular muscles in many animals, have no spindles at all; some muscles have spindles and no tendon organs; others, like the diaphragm, have few spindles and tendon organs. The highest densities are found in the small muscles of the limbs such as the lumbricals and interosseous muscles. It seems reasonable to suggest that those muscles which take part in delicate movements need a more extensive sensory equipment.

MUSCLE ACTIVATION AND CONTRACTION

Muscles contract in response to activity in their motor nerves. Action potentials in motor nerve fibres reach the motor end-plates, where the process of neuromuscular transmission leads to a similar action potential in the corresponding muscle fibre; this then spreads in both directions along the muscle fibre initiating a mechanical contraction in the underlying myofibrils.

Nerve action potential
$\qquad\downarrow\qquad$ (Neuromuscular transmission)
Muscle action potential
$\qquad\downarrow\qquad$ (Contraction coupling)
Muscle contraction

The mechanism of generation and conduction of a muscle action potential is similar to that which occurs in a non-myelinated nerve fibre, and need not be described further here. It remains to describe neuromuscular transmission, the mechanism of contraction, and the process of contraction coupling.

NEUROMUSCULAR TRANSMISSION. As it approaches its termination, each motor nerve fibre divides into a number of smaller fibres, each of which supplies a single muscle fibre. Each action potential in the parent nerve fibre passes into all these divisions and excites each of the corresponding muscle fibres. This group of muscle fibres, along with the nerve fibre that supplies them, is known as a motor unit.

Each terminal branch of the motor nerve fibre ends on a muscle fibre in a motor end plate where the branching ending of the fibre comes into close apposition with the membrane of the muscle fibre (Fig. 40.5). The membranes of the nerve and muscle fibres are, in fact, separated by a cleft that is 50–100 nm wide, and the folding of the two membranes is such that this close apposition occurs over about 2000 μm^2 of membrane. This structure is analogous to that of a synapse in which two neurones come into close apposition. The motor end-plate, however, differs from synapses elsewhere in that each action potential in the nerve fibre gives rise to a single action potential in the muscle fibre; the motor end-plate behaves as a 'safe junction' which passes the signal on unchanged.

If a skeletal muscle is perfused with saline containing eserine (to prevent destruction of

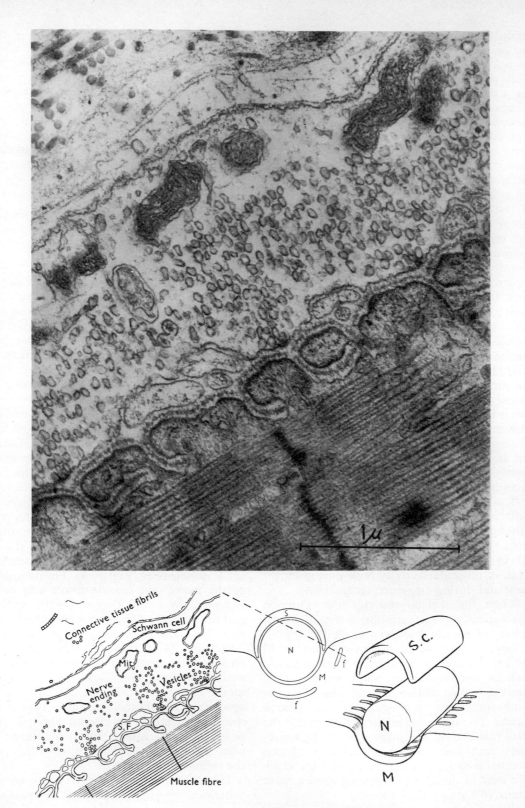

acetylcholine by cholinesterase), it can be shown that stimulation causes acetylcholine to appear in the perfusate. By injecting acetylcholine into the artery supplying a muscle a contraction is produced (Fig. 40.6); this

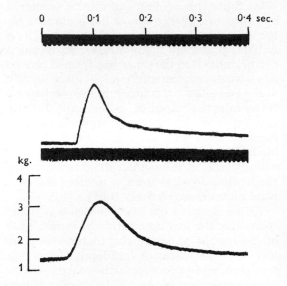

Fig. 40.6 Reactions of the normal mammalian muscle to acetylcholine. Upper record: maximal motor nerve twitch. Lower record: response (a very short tetanus) to *close* arterial injection, during temporary arrest of the circulation of the muscle, of 2·5 µg of acetylcholine in 0·25 ml of saline. For close arterial injection a cannula is placed in the main artery of the limb near the muscle and branches to all muscles but the one under investigation are tied. Time in 10 msec (G. L. Brown, H. H. Dale & W. Feldberg (1936). *Journal of Physiology* **87,** 402).

phenomenon differs from the action of acetylcholine as a peripheral autonomic humoral transmitter (p. 870) in not being abolished by atropine. The response is not a twitch but a short tetanus, because the injected acetyl-

choline cannot reach all the muscle fibres simultaneously. These observations were the original basis for the theory that when a nerve impulse arrives at the neuromuscular junction (motor end-plate) a small quantity of acetylcholine is liberated from the nerve ending and depolarizes receptor sites in the adjacent muscle cell membrane, setting up a conducted action potential which then causes contraction of the muscle fibre. The acetylcholine is quickly destroyed by the enzyme cholinesterase which is found in greatest concentration at the motor end-plate, so that normally a single nerve impulse only gives rise to a single twitch. However, in the presence of eserine the effect of acetylcholine is prolonged, so that a tetanus may occur instead of a twitch.

The theory of chemical transmission at the motor end-plate has been set on a firm experimental basis through the application of microelectrode techniques, chiefly by Katz and his colleagues. Micropipettes whose tip diameter is around 0·5 µm (Fig. 39.9) can not only be used to observe the local potential changes developed close to the motor end-plate, but can also be filled with acetylcholine chloride instead of KCl, so that by passing electric current outwards at the tip, acetylcholine ions can be delivered in small and well controlled quantities at selected places. Using micro-iontophoretic application of acetylcholine with frog muscle, del Castillo and Katz showed that whereas 10^{-15} to 10^{-16} mole of acetylcholine released just outside a motor end-plate could produce a local depolarization large enough to set up a muscle impulse, delivery of the same amount into the interior of the muscle fibre was ineffective. Working with rat diaphragm muscle, Krnjevic and Miledi were able to elicit large end-plate potentials with quantities of acetylcholine ten times smaller still. Krnjevic

---

Fig. 40.5 Electron micrograph of a longitudinal section through the neuromuscular junction of the frog. The small diagram on the left below is a reduced tracing of the electron micrograph. Mit., mitochondria (four are shown). S.F. Schwann finger, which extends a little way into the cleft between the nerve terminal and the muscle fibre. The 'exploded' diagram on the right shows the Schwann cell, S.C. lifted off the terminal branch of the motor nerve N which lies in a shallow gutter on the surface of the muscle fibre M. The gutter has semi-circular junctional folds. The small diagram in the centre is a transverse section through the nerve terminal N with the Schwann cell, S, covering it, except for the part in contact with the muscle fibre, M; the junctional folds are cut through at f and f. The plane of the electron micrograph is shown by the dashed line; note that the section passes through the Schwann cell at two places. (R. Birks, H. E. Huxley & B. Katz (1960). The fine structure of the neuromuscular junction of the frog. *Journal of Physiology* **150,** 134–144.)

and Mitchell have estimated the acetylcholine released from each nerve ending in an eserinized rat hemidiaphragm as 10^{-17} mole per impulse, so that there is now no great discrepancy between the amount demonstrably released and the amount that must be applied artificially to give a twitch. Since the micropipette can never be expected to approach the muscle membrane as closely as the nerve ending that it imitates, the discrepancy is unlikely to be abolished altogether; but it can no longer be used, as it was in the past, as an objection to the theory of chemical transmission.

It has sometimes been asserted that a mechanism involving the transfer of acetylcholine by diffusion from the nerve ending where it is released to the receptor sites on the post-junctional membrane where it acts would be too slow to account for the delay of under 1 msec in transmission across the motor end-plate. This criticism is invalid because the two membranes are separated by such a short distance that the time for diffusion would occupy only a fraction of a millisecond.

Observation of the potential changes at motor end-plates with intracellular micro-electrodes has enabled the mechanisms involved in the release of acetylcholine and its action on the post-junctional muscle membrane to be studied in some detail. One of the first points established by Fatt and Katz was that the amount of electric charge transferred across the post-junctional membrane as a result of the arrival of an impulse at the nerve ending corresponded to the passage of about 3×10^{-14} mole of a univalent ion. This figure is much larger than that for the acetylcholine released, or for the total amount of current that could be expected to flow out of the nerve ending, and it is clear that the acetylcholine acts as part of an amplifying mechanism which enables the relatively small nerve ending to produce sufficient depolarization of the much larger muscle structure to set up a propagated action potential in the muscle membrane. This depolarization is brought about by permeability changes in the post-junctional membrane whose nature has been examined by del Castillo and Katz, Takeuchi and others. By observing the effect of end-plate current flow at different times during a propagated muscle action potential, and at an end-plate whose

membrane potential was 'clamped' by current applied from an intracellular micropipette, it has become clear that during the action of acetylcholine the membrane potential is displaced towards a new equilibrium value which lies at about -14 mV (inside relative to outside). Since this equilibrium value is unaltered by substituting large anions like glutamate for the chloride in the external medium, it is unlikely that there is any increase in the anion permeability. On the other hand, small cations like ammonium can effectively replace sodium at the end-plate (in contrast to the mechanism for conducted impulses, where only lithium can deputize for sodium), while large ones cannot. The suggestion has therefore been made that acetylcholine enables all cations whose diameter is less than about twice that of the hydrated potassium ion to cross the end-plate membrane relatively freely.

An unexpected discovery which was made soon after the first application of intracellular micropipettes to study the motor end-plate was the existence of randomly occurring miniature end-plate potentials (m.e.p.p.s) 1 to 3 mV in amplitude. As can be seen in Figure

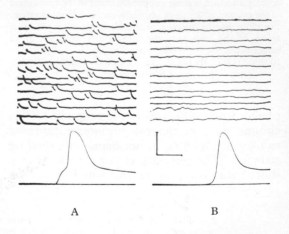

A   B

Fig. 40.7 The localization of spontaneous activity in a frog toe muscle. *A* was recorded at the end-plate, *B* at a point in the same fibre 2 mm away. The lower records, for which the voltage and time scales represent 50 mV and 2 msec, show the response to a nerve impulse. The upper records, taken with higher amplification and slower sweep speed (scales 3·6 mV and 47 msec), show the m.e.p.p.s at the end-plate but not further along the fibre. (P. Fatt & B. Katz (1952). *Journal of Physiology* **117**, 109–128.)

40.7, the m.e.p.p.s can be recorded only in the immediate vicinity of the end-plate and are not seen in the non-innervated part of the muscle fibre. Each m.e.p.p. represents the release of a packet or quantum of several thousand acetylcholine molecules, and although final proof has not yet been provided, there are good reasons for identifying these packets with the presynaptic vesicles seen in electron micrographs (Fig. 40.5). The normal end-plate potential (the inflexion preceding the spike in the lower part of Fig. 40.7A) is composed of two or three hundred m.e.p.p.s occurring simultaneously, the factor which suddenly increases the frequency of release of the acetylcholine packets from around 1 per sec to several hundred within 1 msec being the depolarization of the presynaptic nerve ending. In general, the frequency of the m.e.p.p.s is controlled by the condition of the presynaptic membrane, while their amplitude depends on the properties of the postsynaptic membrane. Thus substances like tubocurarine which compete with acetylcholine at the postsynaptic receptor sites reduce the size of the m.e.p.p.s and of the end-plate potential, while anticholinesterases prolong them both, but neither type of drug affects the numbers of quanta involved. In contrast, the antagonistic effects of calcium and magnesium on neuromuscular transmission depend mainly on a change in the number of packets of acetylcholine released by a single nerve impulse, calcium causing an increase and magnesium reducing the number; there are parallel alterations in the frequency of occurrence of the m.e.p.p.s, but no change in their size. The powerful muscular paralysing action of the toxin produced by *Clostridium botulinum* also results from an interference with acetylcholine release, both the frequency of the m.e.p.p.s and the number released per impulse being drastically cut down. However, none of these blocking agents has any effect on the number of acetylcholine molecules in each quantum, and hemicholinium-3 is the only drug which may act in this way, perhaps by competing with choline so as to disturb the resynthesis and storage of acetylcholine in the nerve terminal. Clinically, the most important neuromuscular blocking agents are tubocurarine and related compounds which have revolutionized the practice of anaesthesia by enabling the anaesthetist to supplement unconsciousness with controlled muscular relaxation.

In myasthenia gravis in man the quanta of acetylcholine are only about one fifth of the normal size and although the normal number of quanta may be discharged on the arrival of a nerve impulse the muscle fibres may fail to contract. In this condition there is severe muscular weakness going on to paralysis. It can usually be corrected by an anticholinesterase such as neostigmine. A hormone from the thymus may be responsible for reducing the size of the quanta. Excess of acetylcholine at a neuromuscular junction (cholinergic crisis) also causes muscular weakness. This condition may be mitigated by cholinesterase reactivators which are used as antidotes to 'nerve gases'.

It has long been known that in tissues innervated either with adrenergic or with cholinergic endings, the effect of denervation is to increase their sensitivity to the direct application of noradrenaline or of acetylcholine respectively, and this change can readily be demonstrated in muscle. A possible explanation for the denervation effect at adrenergic endings is given on p. 868, and in muscle it has now been shown that denervation causes the acetylcholine-sensitivity of the surface membrane, normally restricted to the end-plate region, to spread to all parts of the fibre. The phenomenon is reversible, and on regeneration of the nerve supply the acetylcholine-sensitive area is reduced once more to the part close to the new end-plates. Treatment of a limb with botulinum toxin, which blocks acetylcholine release while leaving the end-plates intact, has the same effect as denervation, and it may be that the restriction of the sensitive area in the normal muscle depends in some way on the occurrence of the m.e.p.p.s.

THE CONTRACTILE MECHANISM. Changes in the length of a muscle fibre are accompanied by movements of the thin (actin) and the thick (myosin) filaments in relation to each other with corresponding changes in the width of the I bands. A. F. Huxley has produced convincing evidence for the view that the forceful contraction of a muscle fibre is the result of a sequence of chemical reactions that actively draw the actin filament along the myosin filament.

A cyclical reaction is postulated in the course of which links form between the thin filaments and side chains of the thick filaments; the formation of such a link, however, is accompanied by a change in the mechanical properties of the thick filament, the side chains of which now tend to flex, taking the thin filaments along with them in a shortening movement. Having shortened, the cross-link now becomes within the orbit of an enzyme system which tends to break it down. The side chain then resumes its original position and the cycle can begin again with formation of a new cross link at another point on the thin filament. The sequence of events is illustrated in Figure 40.8.

Fortunately, the behaviour of contracting myofibrils can be examined after destruction

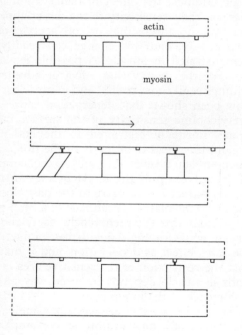

Fig. 40.8 Diagram showing, very schematically, the possible mode of action of cross-bridges. A cross-bridge attaches to a specific site on the actin filament, then undergoes some configurational change which causes the point of attachment to move closer to the centre of the A band, pulling the actin filament along in the required manner. At the end of its working stroke, the bridge detaches and returns to its starting configuration, in preparation for another cycle. During each cycle, probably one molecule of ATP is dephosphorylated. Asynchronous attachment of other bridges maintains steady force. (*From* H. E. Huxley (1971). *Proceedings of the Royal Society* B **178**, 139.)

of the membranes of the muscle fibre by immersion in glycerol at low temperatures. In such a preparation chemical agents can be applied directly to the myofibrils. Contraction occurs when ATP is added to glycerinated muscle fibres in the presence of free calcium ions and some magnesium ions; the course of this contraction ATP is converted to ADP and it is this conversion that provides the energy necessary for the mechanical work done during shortening. If, in such a preparation, the available supply of ATP becomes exhausted, the reaction ceases but the fibres remain stiff; a state that is analogous to the 'rigor mortis' that occurs when ATP disappears after death. This stiffness in the absence of ATP implies that the reaction has ceased with the filaments linked together, that is the ATP is used in the stage of uncoupling the thin filaments from the myosin side chains.

Calcium ions play a crucial part in the contractile process; in the absence of calcium ions the myofibrils remain inactive, and evidently no links are formed between the thick and thin filaments. But, on the other hand, an intracellular microinjection of calcium ions can initiate a local contraction of the myofibrils. Indeed, it is now becoming clear that a release of calcium within the muscle fibre is the physiological method of initiating a muscle contraction, and the removal of this free calcium brings the contraction to an end.

In addition to actin and myosin muscle contains the 'modulating' proteins *tropomyosin* and *troponin*, present in the thin filaments. They seem in some way to control the onset and end of contraction by providing a receptor mechanism for Ca^{2+}. In the absence of Ca^{2+} these proteins allow relaxation by inhibiting the primary actin–myosin interaction. It has been suggested that Ca^{2+} is bound to troponin, and tropomyosin exerts a predominantly stimulating effect on the Mg^{2+}-activated interaction between actin and myosin. In some such manner the presence of Ca^{2+} abolishes the inhibitory effect of these modulating proteins and permits the ATPase-stimulating and physicochemical interactions between actin and myosin to proceed.

The fact that energy is injected into the system by breakdown of ATP when cross links are broken down suggests that at the onset of activity cross links can be formed rapidly, and the contraction begun before any ATP is required. In this sense muscle is held charged and ready for immediate release.

CONTRACTION COUPLING. Contraction of myofibrils is initiated by a release of calcium within the fibre. Some of the slowly acting muscle fibres of invertebrates rely on the inward diffusion of calcium from the surface of the cell that follows each action potential. Such a process would be too slow for the large and faster acting fibres of most vertebrates, and a more elaborate system has developed. In the resting muscle calcium is stored in the sarcoplasmic reticulum that surrounds the myofibrils, so that although the overall calcium content of the fibre is high, the concentration of Ca^{2+} ions in the fluid that surrounds the actin and myosin filaments is very low indeed. When, during the passage of an action potential, the surface membrane of the muscle fibre becomes depolarized, this electrical change is conducted along the transverse tubules that extend into the fibre; the electrical change in the transverse tubule system then initiates the release of calcium ions from the longitudinal component of the sarcoplasmic reticulum. The calcium ions then diffuse the short distance into the myofibrils and contraction ensues. This release of calcium is then followed by an active process of pumping calcium back into the sarcoplasmic reticulum when the contraction is ended.

Since the transverse tubules are very fine (0.05 μm or less in diameter), the inward spread of depolarization is slow; Gonzales-Serratos has shown that there is a correspondingly slow inward spread of the contractile process in muscle fibres, the central myofibrils beginning to contract a number of milliseconds after those on the surface of the fibre.

CHEMISTRY OF SKELETAL MUSCLE

Skeletal muscle contains about 75 per cent of water, 20 per cent of protein, and 5 per cent of other materials. Of the structural proteins in muscle the most important are the contractile substances *myosin*, *actin* and *tropomyosin B*, which together comprise more than half the total protein content. The remaining proteins

include *albumin*, *globulin X* and stroma protein. All the enzymes which catalyse the various steps in the process of glycolysis are included in the protein fraction.

Muscle also contains, in the satcoplasm of the cell, the respiratory pigments, *myoglobin* and *cytochrome* which function in the transport of oxygen from the blood to the oxidizing systems. Since myoglobin does not show allosteric changes, as haemoglobin does, its oxygen dissociation curve is a rectangular hyperbola and its affinity for oxygen is higher than that of haemoglobin (Fig. 31.8).

After certain forms of injury such as automobile accidents, there may be prolonged crushing of muscle, which reduces its blood supply and finally brings about the death of the tissue. In such circumstances myoglobin diffuses into the general circulation and is excreted in the urine. In some people after very severe exercise the muscles become painful and weak and the dark brown urine contains myoglobin.

Only small amounts of lipid are found in muscle. Carbohydrate is present mostly as glycogen in amounts between 0.5 and 1.0 per cent.

Many of the simpler non-protein constituents of muscle are soluble in water and are usually termed *extractives*. They include creatine and creatine phosphate (phosphagen) (p. 386), adenosine monophosphate, lactic acid, inositol, carnosine and anserine.

One of the most important compounds in muscle is *adenosine triphosphate* (ATP) which is a 'high-energy' compound (see pp. 66 and 84). It is the energy released by the breakdown of ATP that is used in the performance of muscular work.

THE CHEMISTRY OF THE CONTRACTILE ELEMENT. *Myosin* is a fibrous protein of molecular weight 470 000. Its rod shaped molecule, about 160 nm in length consists of two identical very long polypeptide chains (mol. wt. 225 000) each containing about 1800 amino acids. Each chain is in the α-helical configuration and the two chains are wound round each other to form a double helical structure. At one end of the molecule both chains are folded into a globular 'head' occupying a very small proportion of the total length of the molecule. Exposure to trypsin splits myosin near the centre of the tail into two portions, a heavy fragment, *heavy meromyosin* containing the globular 'head', and a light fragment, *light meromyosin*.

The important discovery was made by Engelhardt and Ljubimowa in 1939 that myosin is itself the enzyme adenosine triphosphatase (ATPase) (or is inseparably associated with it). This enzyme catalyses the breakdown of ATP to ADP and its activity has now been located in the globular 'head' of the myosin molecule.

Actin, which was discovered in 1941 by Szent-Györgyi in Hungary, exists in two forms, G-actin (globular actin) and F-actin (fibrous actin). G-actin has a molecular weight of 46 000 and its molecule consists of a single polypeptide chain, globular in shape. Each molecule can bind one calcium ion very strongly and has a high affinity for ATP. Binding of ATP by G-actin results in polymerization to F-actin, one molecule of ATP being split to ADP and inorganic phosphate for each G-actin unit added to the F-actin chain. The ADP so formed remains attached to the G-actin units.

F-actin is made up of two long strands of G-actin units coiled round each other.

When solutions of myosin and actin are mixed, myosin binds actin at two specific sites, probably in the 'heads', to form *actomyosin*. The binding is associated with a large increase in viscosity and in birefringence of flow. In the presence of ATP and magnesium ions actomyosin dissociates into actin and myosin with a sharp drop in viscosity. At the same time ATP is·split by the myosin into ADP and inorganic phosphate and when all the ATP has been hydrolysed the actomyosin complex reforms. This splitting of the actomyosin complex by ATP is probably similar to the process of detaching side chains of the thick (myosin) filaments from the thin (actin) filaments in an intact muscle fibre or a preparation of glycerol extracted myofibrils.

creatine phosphate + ADP

creatine
$\rightleftharpoons$ creatine + ATP
kinase

THE SOURCE OF ENERGY FOR MUSCULAR CONTRACTION. While the immediate source of energy for muscular contraction appears to be ATP, the ultimate source is carbohydrate or fat.

It was first demonstrated by Fletcher and Hopkins in 1907 that fatigue and rigor in muscle are accompanied by the production of lactic acid, the source of which was shown by Parnas and Wagner in 1914 to be muscle glycogen. It was further pointed out that more lactic acid is produced in a muscle preparation stimulated in nitrogen than in a preparation stimulated in air, and that when oxygen is admitted to the former the lactic acid content diminishes, carbon dioxide is produced and some of the glycogen is restored. The mechanism of the formation of lactic acid from glycogen is, of course, the process of glycolysis originally described by Embden and by Meyerhof and discussed on p. 140.

In 1930 Lundsgaard showed that contraction could occur in muscles poisoned with iodoacetate which inhibits lactic acid formation (p. 315). In such preparations the tension production was proportional to the breakdown of the substance creatine phosphate (phosphocreatine, phosphagen) which had been discovered three years previously, simultaneously by Eggleton and Eggleton in Britain and Fiske and Subbarow in America. Lundsgaard originally suggested that in iodoacetate-treated muscles the energy for contraction came from creatine phosphate breakdown and that this might also be true for normal muscle, the role of glycolysis being to supply energy for the resynthesis of creatine phosphate.

We know now that this is not so. Although resting muscle contains five times as much creatine phosphate, an energy rich compound, as ATP, the creatine phosphate cannot be used as a direct source of energy for muscular contraction. It does, however, serve for the regeneration of ATP and ADP under the influence of the enzyme *creatine kinase*:

During recovery after exercise ATP is synthesized. Some of it reacts with creatine with the formation of creatine phosphate by the reversal of the above reaction.

The chemical processes which accompany muscular contraction (Fig. 40.9) may be summarized as follows:

1. The initial step is the breakdown of ATP with the sudden release of energy. Subsequent steps are concerned with the regeneration of ATP. An emergency mechanism capable of operating immediately is provided by creatine

phosphate which can transfer its high energy phosphate bond to ADP thus regenerating ATP; but this mechanism is of limited scope and serves only immediate needs.

and cytochromes, the resynthesis of ATP from ADP is largely the result of oxidation (respiration). In white muscle tissue on the other hand, such as the pectoral muscles of the domestic

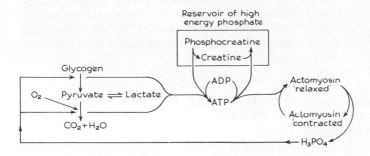

Fig. 40.9 The chemical basis of muscular contraction. Inorganic phosphate released from ATP by muscular contraction (right) is incorporated by the process of glycolysis (left) into compounds such as ATP and phosphocreatine containing high-energy phosphate bonds. The high bond energy of the ATP is then used in contraction. (*After* H. N. Munro.)

2. The second step is the breakdown of carbohydrate in muscle by the process of glycolysis to yield pyruvate (p. 144). The carbohydrate may be both stored glycogen and glucose from the blood; in either case it is converted into glucose 6-phosphate in the early stages of glycolysis (Fig. 18.4). The process results in the synthesis of ATP (p. 151) In the absence of adequate oxygen supplies, some of the pyruvate is reduced to lactate (p. 144).

3. The breakdown of ATP and glycolysis are both anaerobic processes. The final stage in muscle is the aerobic disposal of both pyruvate and lactate which are oxidized by molecular oxygen to yield carbon dioxide and water through the operation of the citric acid cycle (tricarboxylic acid cycle) which has been described in Chapter 11. In this process the mechanism of oxidative phosphorylation results in the formation of more than 30 molecules of ATP for each glucose unit oxidized (p. 152).

4. Much of the energy utilized in muscular work is also derived from fatty acids, especially in cardiac muscle (p. 473). During muscular exercise fat is mobilized directly as non-esterified fatty acids which are conveyed in the blood, loosely bound to serum albumin, to the muscles and there oxidized (p. 336). The final pathway of oxidation is of course by the citric acid cycle.

In red muscles, such as the leg muscles of active mammals, which are rich in myoglobin

fowl, which contain but little myoglobin and cytochromes, the resynthesis of ATP from ADP is largely the result of glycolysis. In both types of muscle carbohydrate is the major fuel during activity, while in the resting state mammalian skeletal muscle uses fatty acids and acetoacetate as fuel.

Since the initial process which yields the energy for contraction may be accomplished without oxygen, the body is in fact capable of sudden spurts of activity which require more oxygen than the lungs could take in or the blood could carry during the time of the exercise. The body is then said to be in *oxygen debt* (p. 852). The extra oxygen which is required to oxidize the lactic acid and pyruvic acid may be taken in over a long period after the exercise is ended.

Although, as is well known, lactic acid may be produced in muscle as the result of repeated contractions, pyruvic acid is to be regarded as a more fundamental product of muscle glycolysis. Normally, of course, it is oxidized at once to carbon dioxide and water though part may be converted back to glycogen by reversal of glycolysis during recovery after exercise. It may accumulate in the muscle under two sets of circumstances (1) when the rate of glycolysis is very rapid, or (2) if the oxygen available is insufficient for its immediate oxidation. Part of the pyruvate is then converted into lactate which, when the oxygen supply is again adequate, is oxidized back to pyruvate and so

Muscle 839

to carbon dioxide and water. During the time when the concentration of lactic acid is high in muscle a large proportion diffuses into the blood by which it is carried to the liver and converted into glycogen in the operation of the Cori cycle which is described in Chapter 18 (p. 314).

The oxidation of pyruvate by the citric acid cycle is an elaborate process (see p. 148) which requires among other things the presence of thiamine pyrophosphate (TPP) derived from thiamine. In the disorder called beri-beri (p. 185), which occurs when there is a severe dietary deficiency of vitamin B, even quite mild forms of exercise lead to an increase in the blood pyruvate because there is not sufficient cocarboxylase to bring about its rapid oxidation. The increase in blood pyruvate after exercise in a case of thiamine deficiency is shown in Fig. 40.10 compared with the results in a normal individual. This phenomenon has been used as a basis of a test for assessing deficiency of thiamine.

THE AEROBIC RECOVERY PHASE. It must not be imagined from what has already been said that muscle is an anaerobic tissue. Muscle tissue is essentially aerobic and, especially during active contraction, shows a high oxygen uptake and carbon dioxide output. Although the breakdown of muscle glycogen can take place under anaerobic conditions, the energy required for muscular activity is in the long run obtained as the result of oxidation by molecular oxygen carried to the muscle by the haemoglobin of the blood. The oxygen is transferred to the myoglobin (p. 620) (a form of haemoglobin occurring in muscle tissue) and is passed on in turn to the cytochrome (p. 129) of the oxidation enzyme systems in the interior of the muscle cells and is there utilized to bring about the oxidation of pyruvate or lactate or other intermediate products.

MECHANICAL PROPERTIES OF MUSCLES

When a muscle is activated its fibres contract or, if they are restrained from doing so, they exert tension on the tendon to which they are attached. The actual force that a muscle is able to exert is by no means constant and depends on the degree of activation, the muscle length, and the rate at which it is shortening (or being lengthened).

If a muscle is activated while its ends are rigidly fixed so that no shortening can occur, it is said to contract isometrically. The force during such an isometric contraction can be studied by a simple lever writing on a revolving smoked drum (Fig. 40.11). Electronic force transducers are now more often used.

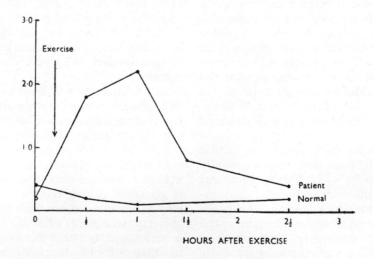

Fig. 40.10 Blood pyruvate (mg per 100 ml) after exercise (7900 ft. lb or 1090 kg m in 5 minutes) in a female aged 58 suffering from vitamin B deficiency, compared with the results in a normal subject.

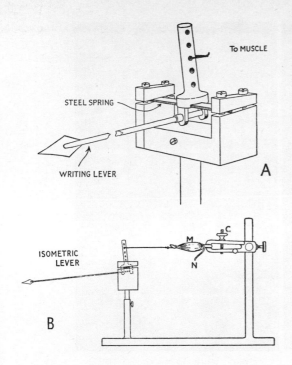

Fig. 40.11 Isometric lever. A shows the detail of construction and B the general arrangement of the apparatus when a frog muscle is used.

numbers of nerve fibres and therefore of muscle fibres with corresponding differences in twitch tension, but once the stimulating pulse is made large enough to excite all the muscle fibres a further increase does not lead to a further increase in tension.

If instead of single pulses, pairs of stimuli are delivered to the motor nerve and these are so timed that the second stimulus falls during the twitch contraction that followed the first, a second twitch grows out of the first one (Fig. 40.13), the tension rising to a higher level than

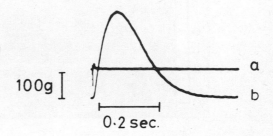

Fig. 40.12 Isometric twitch contraction of cat soleus muscle. (a) The action potential recorded from electrodes placed on the muscle surface; (b) tension measured at the muscle tendon. (*By courtesy of P. M. H. Rack.*)

THE MUSCLE TWITCH. When a muscle or its motor nerve is stimulated by a single electrical pulse of adequate amplitude, an action potential passes along the muscle fibre, after which there is a brief contraction, a muscle twitch. Figure 40.12 shows the tension record during a twitch in an isometric muscle. The muscle action potential is over in a few milliseconds, but the mechanical contraction that gives rise to the tension in Figure 40.12 lasted for very much longer.

The duration of the twitch is different in different muscles; in fast-acting muscles such as mammalian external eye muscles a twitch contraction may be effectively over in 40 msec, whereas the twitch of a slow muscle such as the cat soleus (Fig. 40.12) has a very much longer time course.

The twitch of a whole muscle is due to the summed activity of many muscle fibres, and its amplitude therefore varies with the number of fibres stimulated. By varying the strength of an electrical stimulus applied to the motor nerve it may be possible to excite different

in a single twitch. The closer together the two stimuli the higher is the tension so long as the second pulse is outside the refractory period of the nerve and muscle that follows the first impulse.

TETANUS. If a motor nerve is stimulated by a train of pulses the tension in the muscle rises progressively higher with each of the first few pulses until a plateau is reached (Fig. 40.14). If a high rate of stimulation is used the contraction is then continuous and a smooth tension record is obtained; when, however, the stimulus rate is low the tension is not smoothly maintained but fluctuates at the frequency of the stimulating pulses. The tension generated during the smooth 'fused' tetanus (or 'complete' tetanus) that accompanies stimulation at a high rate is many times larger than the twitch tension of that muscle.

In a living animal muscle fibres are seldom, if ever, activated at the high rates necessary for a smooth tetanic contraction, but since im-

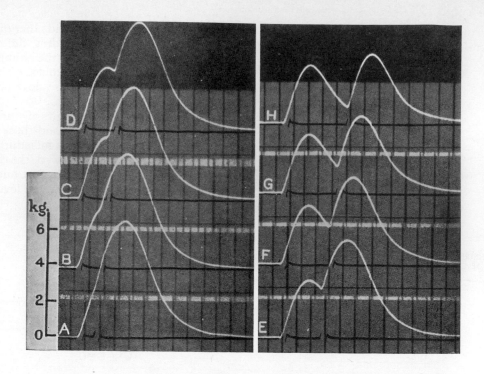

Fig. 40.13 Mechanical response of median head of gastrocnemius of a cat to double shocks to peroneal nerve. The mechanical response (white line) was recorded by an isometric frictionless mirror myograph. The electrical response (dark line) was recorded by a string galvanometer. A Lucas pendulum was used to give two maximal break shocks in quick succession to the nerve. Temperature, 34°C. Interval between stimuli: A, 24 msec; B, 32 msec; C, 40 msec; D, 48 msec; E, 57 msec; F, 69 msec; G, 77 msec; H, 88 msec. Time is shown by vertical lines 20 msec apart. (Sybil Cooper & J. C. Eccles (1930). *Journal of Physiology* **69,** 279).

pulses are normally delivered to different motor units at different times, the whole muscle contracts quite smoothly, although the rate of stimulation for each muscle fibre may be quite slow. This situation may be reproduced experimentally by subdividing the nerve supply to a muscle and stimulating different groups of fibres asynchronously. Figure 40.15 shows how the force increased with increasing stimulus rate in a slow and a fast muscle stimulated in this way. In the slow soleus muscle the tension was near maximal when the stimulus rate was about 30/sec, whereas in the faster gastrocnemius with its briefer response to a single impulse, stimuli had to be supplied at about 80/sec to achieve a near maximal force.

The cat soleus which is a slow postural muscle is in this respect well adjusted to the task of maintaining a steady tension with a minimum rate of stimulation.

THE EFFECT OF MUSCLE LENGTH ON TENSION. There is a characteristic relationship between muscle length and the tension that develops in an isometric tetanus (Fig. 40.16). The maximum tension develops when the muscle is at a length that usually approaches the maximum that it takes up in the intact animal; at lengths shorter than this the tension is smaller, and if the muscle is stretched beyond this optimum length the active tension also declines, though at extreme lengths the connective tissue in and around the muscle resists extension with a considerable force and dominates the situation (Fig. 40.16).

A. F. Huxley and his colleagues have shown that the form of the isometric length–tension curve can be explained in terms of the sliding filament theory of muscle contraction. The maximum active tension develops when the sarcomere lengths are such that there is sufficient overlap between thick and thin filaments in the myofibrils for all the side chains of each thick filament to have access to the thin filament so that a maximum number of cross links can form. When, however, the muscle is extended beyond this optimum length, the thin filaments are pulled out from among the thick ones, the overlap is less, and there are correspondingly fewer possible sites for cross link formation; the greater the extension the smaller is the overlap and the lower the tension.

When the muscle shortens from the optimal length each thin filament moves further in between the corresponding thick filaments, but this movement is impeded when (at a sarcomere length of 2·2 μm) each filament meets another one moving in the opposite direction. Some part of the contractile force is then taken up in compressing these filaments against or past each other so that the external force generated by the muscle is correspondingly less at the shorter length. If shortening proceeds further, a point will be reached (at a sarcomere length of 1·5 μm) at which ends of the thick filaments become compressed against the Z line and further impede shortening with a corresponding fall in tension at very short muscle lengths.

Although many features of the isometric length–tension curve can be explained in terms of the positions of filaments within the myofibrils, it seems likely that there are also other

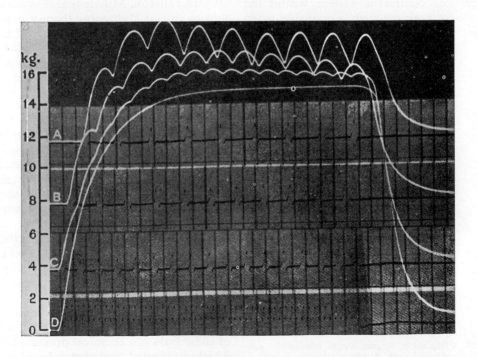

Fig. 40.14 Mechanical and electrical responses of median head of gastrocnemius of a cat to tetanic stimulation of popliteal nerve by a neon tube stimulating device. Temperature, 33°C. Rate of stimulation: record A, 19 shocks a second; record B, 23·5 shocks a second; record C, 35 shocks a second; record D, 315 shocks a second. Time is indicated by vertical lines 20 msec apart. (Sybil Cooper & J. C. Eccles (1930). Isometric responses of mammalian muscles. *Journal of Physiology* **69**, 383).

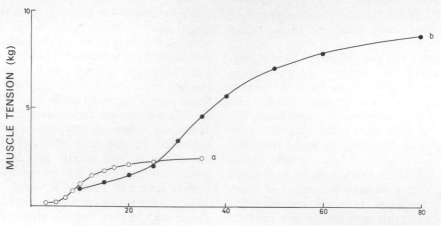

RATE OF STIMULATION (impulses per sec in each filament)

Fig. 40.15 The effect of stimulus rate on muscle tension. (a) Cat soleus; (b) medial head of cat gastrocnemius. In each case five different subdivisions of the motor nerve were stimulated sequentially to obtain a smooth contraction of the muscle at low rates of stimulation. (*By courtesy of P. M. H. Rack.*)

factors operating and, in particular, it is probable that the contractile system within the myofibrils is more completely activated when the muscle fibres are long than when they are short.

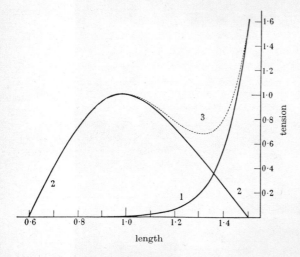

Fig. 40.16 Diagram of tension–length relations of sartorius muscle of frog or toad. (1) At rest passively stretched; (2) extra force developed during maximal tetanus; (3) dotted line, total force in maximal tetanus, sum of (1) and (2). The length is given as a fraction, or multiple, of the standard length in the body; the tension, of the maximum force developed. (*From A. V. Hill (1953). Proceedings of the Royal Society B 141, 113*).

THE EFFECT OF SPEED OF SHORTENING ON MUSCLE TENSION. Muscle is able to develop less tension during shortening than in an isometric contraction at the same length, and the more rapid the shortening the lower is the muscle tension; or conversely, the muscle cannot shorten so quickly against a large force as against a small one.

In this respect, the muscle behaves as though some internal friction or viscosity limits the speed at which it can shorten. The work of A. V. Hill, however, showed that simple friction or viscosity could not, in fact, account for this force–velocity relationship since the muscle does not during rapid shortening generate an appropriately larger amount of heat.

A. F. Huxley has pointed out that the sliding filament theory of muscle contraction offers an explanation for the force–velocity relationship in active muscle. Cross links form between thick and thin filaments in the myofibrils exerting a force that tends to draw the thin filaments along the thick ones. The formation of these links is a chemical process, the rate of which is limited and depends among other things on the availability of calcium ions; once formed, however, the links are broken down at a rate that depends to a large extent on how fast the filaments move bringing each link to a position at which enzymic breakdown becomes likely. When the filaments are stationary or movement is slow, the formation of cross links can proceed at a normal rate, and once formed each link may have a long life since it is relatively unlikely to be broken down. A large number of cross links are therefore present and tension is high. When, however, the filaments move over each other rapidly, the cross links come more quickly within reach of enzymic breakdown and therefore have a shorter life, so that at any time there are fewer of them and the tension is correspondingly smaller. A row of men pulling in a rope provides a useful analogy: if the rope is stationary or moving only slowly, all their hands are on it and the force is maximal. When, however, the rope is being pulled in more quickly there are always a number of hands off the rope preparing to take a new hold, and the force is correspondingly smaller.

THE SERIES ELASTIC ELEMENT. The mechanical properties of muscle are complicated by the fact that the contractile machinery of the muscle is linked to its insertion by an elastic coupling; a part of this elasticity is in the tendon and a part is in the muscle tissue.

This combination of a contractile system, which is limited in its velocity of shortening, in series with an elastic element explains the form of the isometric muscle twitch, and the way that a series of such twitches combine (Figs. 40.13 and 40.14). With each action potential the myofibrils are put into an 'active state' during which they shorten at the expense of the series elastic element. The tension that develops depends on the degree of extension of this series elastic element but the rate of this extension decreases when, with increasing tension, the contractile component of the muscle shortens more slowly. During a single twitch the 'active state' is over before the series elastic element has been extended very far, and the tension is correspondingly low. When, however, the myofibrils are activated a number of times in rapid succession, the process of stretching the series elastic element continues further since the contractile machinery does not become completely re-extended between impulses; a higher tension is then achieved. In the smooth tetanic contraction that occurs during stimulation at a high rate the series elastic element becomes extended to a length at which its tension is the maximum that the

muscle can achieve when internal movement has ceased.

VOLUNTARY MUSCLE CONTRACTION

THE CONTROL OF MUSCLE FORCE. The force exerted by a muscle may be modified either by altering the number of contracting motor units or by altering the rate at which each motor unit is activated. If two electrodes are placed close together on the skin over a muscle the action potentials in a few muscle fibres can often be recorded (Fig. 40.17); as voluntary contraction of the muscle becomes more powerful the rate of discharge of action potentials from each motor unit becomes more rapid, and more motor units are recruited to take part in the contraction (these appear as smaller spikes in Fig. 40.17 C and D, since they arise from motor units further from the recording electrodes). The mechanical contraction associated with such activity is quite smooth since the total muscle force is the sum of a larger number of asynchronous motor unit contractions.

OPTIMUM RATE OF WORKING. It is within ordinary experience that there is an optimum rate for carrying out any required muscular effort. If the rate is too high fatigue sets in; if the rate is too low energy is wasted. Many factors are involved, including the capabilities of the circulatory and respiratory systems. It is worth noting, however, that for each muscle there is, under experimental conditions, a speed of shortening at which its efficiency is greatest, and one might expect this to have an important bearing on the efficient performance of physical work.

Figure 40.18 shows how the efficiency of working of a subject at bicycle pedalling varied with the rate of pedalling. The efficiency was calculated from the extra oxygen consumption associated with a known work output.

FATIGUE. The loss of muscular power that occurs during fatigue may be due to failure at a number of different places including central synapses, the motor end plates and the contractile machinery. The work of Merton suggests that in a simple human limb movement it is probably the muscle fibres themselves that become fatigued. If a muscular movement of the hand is carried on to complete fatigue and at this moment the circulation in the arm is arrested by inflating a sphygmomanometer cuff there is no recovery of strength until the

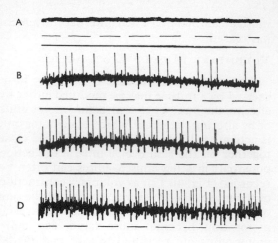

Fig. 40.17 Action potentials recorded on a cathode-ray oscillograph from a single motor unit of the human orbicularis oculi muscle by laying two pointed electrodes on the skin of the face, 0·25 mm apart. Such records are usually called electromyograms. A, lower lid relaxed; B, C, and D, slight, moderate, and submaximal voluntary contractions of the muscle. The smaller disturbances of the base line are due to the activity of adjacent motor units. Time signal, fifths of a second. (Tracings by courtesy of B. L. Andrew, and made according to the technique described by Gordon and Holbourn.)

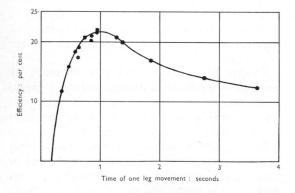

Fig. 40.18 The diagram shows the net efficiency of a subject pedalling a bicycle ergometer. The graph gives the relationship between the efficiency and the time of one leg movement. The curve is drawn through the observed points. The optimum time of one foot movement (half a pedal revolution) is 0·9 of a second. (S. Dickinson (1929). The efficiency of bicycle pedalling. *Journal of Physiology* **67**, 246).

cuff is released and the blood flow restored. Since occlusion of the circulation in the arm does not affect the central nervous system and yet delays recovery, the site of fatigue must be in the muscles themselves and not in the central nervous system. In the experiment illustrated in Figure 40.19A the contraction of the adductor muscle of the thumb was recorded together with the action potentials from the fibres of the muscle. The ulnar nerve was stimulated maximally every few seconds throughout the experiment. At the start each volley caused a twitch and an action potential. A maximum voluntary contraction (Fig. 40.19A) was then made by the subject; this gradually weakened over the next 2 minutes. When the adductor muscle was relaxed the twitches produced by stimulation of the nerve were quite small but they recovered rapidly. The action potentials, however, were about the same throughout the experiment. This shows that when fatigued the muscle fails to contract in response to impulses reaching it by its motor nerve. In Figure 40.19B, the circulation was occluded before the voluntary effort began; when the subject relaxed no twitches were seen until a little time after the cuff was released. The block cannot be at the neuromuscular junction since in this experiment also the muscle action potentials evoked by the stimulation of the nerve remained nearly constant throughout. The processes which follow the action potential and cause contraction are blocked. These observations suggest that fatigue is a peripheral phenomenon occurring in muscle itself.

The fatigue experienced in severe exercise is probably due to events at the synapses in the central nervous system. This may be regarded as a protective mechanism since central fatigue is manifest before there is any block at the neuromuscular junction and long before the muscle itself is incapable of contraction. It should be kept in mind that the word fatigue has also a psychological meaning; fatigue may, for example, arise through lack of interest in a routine and boring task. This is very different from the physiological meaning of fatigue.

Electromyography. Electromyography is of value in the diagnosis of various neuromuscular disorders in which the frequency or rhythm of the nerve impulses is affected. Such conditions include peripheral nerve injuries, irritative

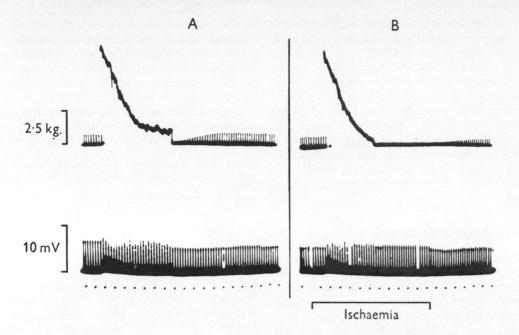

A   B

2·5 kg.

10 mV

Ischaemia

Fig. 40.19 Upper trace: a series of single twitches (evoked by maximal motor nerve shocks) preceding and following a maximal voluntary contraction. Lower trace: the corresponding action potentials. Time marker: dots at ½-min intervals. A: record obtained with circulation intact throughout. B: record obtained with circulation occluded for the period indicated. The records show that in the fatigued state the muscle does not respond to a motor volley, and that it does not recover from this condition so long as the blood supply is cut off; these findings lead to the conclusion that fatigue is peripheral, in the muscle itself. (P. A. Merton (1954). Voluntary strength and fatigue. *Journal of Physiology* **123**, 561.)

lesions of the nerve roots, and neuromuscular disorders such as the myotonias and myasthenia gravis. The electrical activity of a muscle is best investigated by inserting very fine needle electrodes into it and recording the potentials by an amplifier and cathode-ray oscillograph. In a normal muscle a voluntary contraction produces a series of responses (see Fig. 40.17) from motor units of the order of 1 millivolt, each of which lasts about 5 msec (Fig. 40.20a). Three weeks after denervation a muscle shows fibrillation due to the spontaneous activity of single muscle cells; these are rhythmically repeated potentials of the order of 0·05 mV which last about 1 msec or so. Fibrillation is increased by subcutaneous injection of prostigmine, an anticholinesterase (cf. p. 831); it is not abolished by anaesthesia. Fibrillation is not seen if the nerve is blocked by ischaemia or pressure or when the paralysis is due to interruption of the upper motor

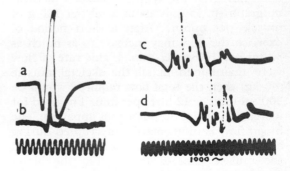

Fig. 40.20 Human electromyograms. (a) Motor unit action potential from normal muscle; amplitude 2 mV. (b) Single fibre potential from denervated muscle; amplitude 0·1 mV. (c) and (d). Polyphasic motor unit action potential during attempted voluntary contraction (no visible movement) of the extensor pollicis longus muscle during early re-innervation following complete interruption and suture of the radial nerve (H. Jasper & G. Ballem (1949). *Journal of Neurophysiology* **12**, 238–239).

neurone. When re-innervation occurs fibrillation disappears and low voltage motor unit potentials appear and gradually increase as the muscle recovers (Fig. 40.20 c and d).

Measurement of 'motor latency' has been found useful in the diagnosis of nerve compression. For example, the time elapsing between stimulation of the median nerve just above the wrist and the beginning of electrical activity in the thenar muscles, normally less than 5 msec, may be as much as 8 msec if the median nerve is compressed in the carpal tunnel.

MUSCULAR EXERCISE

Circulatory and respiratory adjustments. Even before exercise begins, that is before the circulatory and respiratory reflexes can be initiated, the heart quickens and breathing accelerates. This anticipatory response is presumably due to the activity of the cerebral cortex and has the obvious advantage that the body is made ready to meet the additional demands of the muscles as soon as the exercise begins (p. 561). It occurs before there are any changes in blood gases or pH. When the limbs begin to move the discharge from the carotid bodies is greatly increased and this causes an increase in respiration (p. 605). The blood flow through the muscle of the arm at rest as determined by the venous occlusion plethysmograph (p. 556) is about 2 ml per 100 g of muscle per minute. After a short period of exercise the flow may increase to as much as 30 ml per 100 g per min. If this rate of flow were maintained in all the skeletal muscle (40 kg) then the total flow required would be 300×40 ml, or 12 litres per min. The need for increased cardiac output is thus apparent. The blood flow is momentarily interrupted during each contraction because the pressure within the muscle exceeds that within the blood vessels lying around the muscle cells. While the active muscles show a vasodilatation the vessels in other parts of the body are constricted. An initial cutaneous vasoconstriction passes off when sweating occurs. The reduction in splanchnic blood flow in light exercise in man has been measured by bromsulphthalein; it fell from 1650 ml/min to 1250 ml/min. The actual volume of blood in the splanchnic area in these experiments fell from 17 to 10 per cent

of the total blood volume. The size of the vascular bed in the muscles is, however, so much increased that the total peripheral resistance is reduced and the blood pressure does not rise as much as might be expected from the raised cardiac output.

The cardiac output in exercise has been measured in various ways and representative results are given in Table 40.21 and Fig. 40.22.

Table 40.21 Approximate figures for the cardiac output and oxygen consumption during exercise of varying severity

Exercise	Cardiac output (l/min)	Total oxygen consumption of body (l/min)	Arterio-venous oxygen difference (ml/100 ml)
Rest	5	0·25	5
Walking 3·2 km/h (2 m.p.h.)	10	0·8	8
Walking 8 km/h (5 m.p.h.)	20	2·5	12
Running 12 km/h (7·5 m.p.h.)	25	3·0	12
Very severe exercise (max. athletic effort)	34	4·0	13

These values apply to calm conditions. Walking or running against a wind increases the oxygen consumption markedly (L. G. C. E. Pugh (1971). *Journal of Physiology*, **213**, 255–276). For example walking at 3·2 km/hr (2 m.p.h.) against a gale may cost 2 litres/min.

In persons in good training the pulse rate rises more slowly than in the untrained and this means that in the former a larger fraction of the increased cardiac output is obtained by increasing the stroke volume. Again, in athletes the blood pressure rises less, presumably by reason of a lowered vascular resistance in the muscles. It is, however, difficult to account satisfactorily for the changes in the circulation brought about by exercise. The return of the blood to the heart is, of course, aided by the muscular pump and the respiratory pump (p. 505) but there is no increase in the venous pressure in the chest or in the size of the heart (as seen by X-rays) until the exercise is severe. Even then the increase in stroke volume is much greater than the increase in diastolic volume. Thus we cannot readily invoke either Starling's Law (p. 481) or the effect of a rise in atrial pressure to account for the increase in output but in exercise, especially in athletes, the ventricles empty more completely, that is the increase in stroke output is accounted for by the lower residual volume (p. 484). There

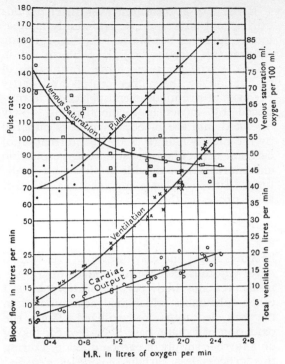

Fig. 40.22 Data for exercise of varying severity plotted as a function of metabolic rate (M.R.) on a stationary bicycle ergometer. Periods of work: 20 minutes or longer. Subject, D.B.D. (A. V. Bock, C. Vancaulaert, D. B. Dill, A. Fölling & L. M. Hurxthal (1928). *Journal of Physiology* **66**, 141).

is, however, no consistent increase in stroke volume, estimated by dye or Fick method, as the oxygen consumption increases. Even the rise of systolic blood pressure in exercise is difficult to understand; at rest a rise of blood pressure is reflexly prevented by the baroceptor mechanisms. If these are active in exercise a compensatory vasodilatation would be expected. But, in fact, the systolic pressure is raised with little alteration of the diastolic arterial pressure and vasoconstriction in areas other than the working muscles is maintained. Whether the slight hypoxia of exercise produces reflex changes through receptors in heart or lungs remains to be proved. It has become clear that the sensitivity of the baroreceptor reflex mechanism affecting the heart rate and blood pressure can alter with the changes that occur in exercise.

The complexity of the situation is well brought out by the recent experiments of Bristow on the human subject. He showed that the reflex sensitivity of the baroreceptor

mechanism to transient changes in blood pressure induced by sudden intravenous injections of phenylephrine could normally, but not always, be reduced by hypercapnia combined with hypo- or hyperoxia. Hypercapnia during breath holding with glottis open progressively lowered the sensitivity since the heart did not slow when there was a rise in blood pressure. He later showed that mild steady-state exercise also lowered the sensitivity to such rises but the effect disappeared quickly on ending the exercise. It would seem, therefore, that exercise induces special conditions which allow 'gearing up' of the subject's circulation and respiration to take the strain of the effort.

Respiration is quickened, as already mentioned, before exercise begins and during exercise minute volume increases to a steady level, unless the exercise is severe and exhausting. Although the exercising individual is conscious of his increased breathing, save at the start he is not distressed by it; his breathing cannot therefore be properly described as dyspnoeic. The rise in ventilation which occurs in the steady state of exercise is closely related to the increase in metabolic rate (measured for example by the rate of oxygen consumption) but the mechanism by which this occurs is still in dispute. Since in moderate exercise the arterial P_{CO_2}, P_{O_2} and pH are little altered from their resting values, either some other agent drives the chemoreceptors or their threshold is lowered so that they become more sensitive to the small oscillations arising from the alterations in alveolar P_{CO_2} and P_{O_2} occurring during a single respiratory cycle. Administration of oxygen during exercise always lowers the ventilation, so there must be a continuous hypoxic drive, even though the arterial P_{O_2} does not appear to fall below its resting value. As Bhattacharyya and his colleagues point out this has been taken to indicate a change in the sensitivity of the arterial chemoreceptors to the normal $H^+(CO_2)$ stimulus. The fact that in moderate exercise the arterial blood gas pressures do not alter appreciably from their resting values suggests that the hyperpnoea of exercise can be thought of as a compensating device tending to minimize the changes that the higher metabolic rate would otherwise cause.

The exercising muscle is not only more liberally supplied with blood but the low oxygen tension in the cells, the increased formation of carbon dioxide and acid metabolites all favour an increased uptake of oxygen from the blood. More oxygen is taken out of each ml of blood so that the arteriovenous oxygen difference is greater in exercise than at rest. Although by the simultaneous operation of all these factors large increases in oxygen requirements can be met, an oxygen debt (Figs. 40.26 and 27) is always incurred in exercise unless it is very mild.

The chemical changes in exercise. It is explained on p. 162 how the respiratory quotient is related to the kind of food which is being burnt. The R.Q. at rest is about 0·85 on the average and vigorous exercise causes a rise. This suggests that a slightly larger proportion of carbohydrate is being oxidized. It might be supposed that a clue to the type of fuel used could be derived from the R.Q. of the metabolism in excess of the resting value, that is, the ratio of excess carbon dioxide production over the resting value to the excess of oxygen consumption over the resting value. The R.Q. of the excess metabolism for a complete period of exercise and recovery usually lies between 0·85 and 1·0. It seems therefore that much the same fuel is burnt during exercise as at rest. However, Passmore and his colleagues exercised starving subjects on a treadmill and found a rise in the concentration of free fatty acids (that is NEFA, non-esterified fatty acids) in the plasma; the level of human growth hormone (HGH) in plasma also rose but the R.Q. fell. There is thus no doubt that fat is an important fuel in exercise in starving subjects and as exercise continues the contribution of fat increases. In the resting subject plasma NEFA is about 500 μ-equiv. per litre; moderate exercise doubles this value and this is apparently responsible for the increased utilization of NEFA by the muscles. If glucose is taken at half-hourly intervals during exercise these changes do not occur, that is there is no change in NEFA, HGH or R.Q. Introduction of fat into the duodenum of exercising subjects does not prevent the rise in NEFA and HGH. It seems therefore that ingested fat must first be deposited in the fatty tissues and released as NEFA before the muscles can use it. Thus if exogenous carbohydrate is not available the muscles use NEFA released from the fat stores by HGH. (See also Fig. 40.28.)

If the R.Q. is measured over short periods during and after exercise apparently anomalous results may be obtained. At the beginning of exercise carbon dioxide is blown off in excess, partly because of the stimulation of respiration by the higher centres and partly because, in strenuous exercise, lactic acid produced by muscular metabolism is slowly liberated into the blood, where it is buffered by the bicarbonate (p. 700) with formation of carbonic acid. The latter stimulates respiration and more carbon dioxide is blown off in the lungs. Since oxygen is involved in neither of these processes the ratio

$$\frac{CO_2 \text{ produced}}{O_2 \text{ used}}$$

becomes larger and may be over 1 (Fig. 40.23). After vigorous exercise has ended the oxygen intake remains high because lactic acid and other substances which had accumulated in the muscles must be oxidized. With the oxidation of lactate, base is set free for taking up a further supply of carbon dioxide. During this process less carbon dioxide is expelled from the blood via the lungs and so the R.Q. may fall below 0·7 (Fig. 40.24). The respiratory exchange ratio, CO_2 output/O_2 intake, in the steady state is the same as the metabolic respiratory quotient. The O_2 intake, since O_2 stores are negligible, is always the same as O_2 consumption but CO_2 output in an unsteady state may differ considerably from CO_2 production because (1) of reaction between bicarbonate and non-gaseous acids which are changing in concentration and (2) of changes in CO_2 in solution when the P_{CO_2} of the tissues changes. By measuring the mixed venous P_{CO_2} as an index of tissue P_{CO_2} it can be calculated that the body can store about 1 ml CO_2 per mm Hg rise in P_{CO_2} per kg body weight.

The increased energy output during exercise inevitably gives rise to an increased production of heat since only some 25 per cent of the energy liberated is converted into mechanical work. The raised blood temperature results in cutaneous vasodilatation (p. 557) and an increased heat loss. In strenuous exercise the augmented heat loss may be incapable of balancing the increased production of heat and the body temperature may rise several degrees centigrade during the exercise (Fig. 40.25).

Several writers in describing events during

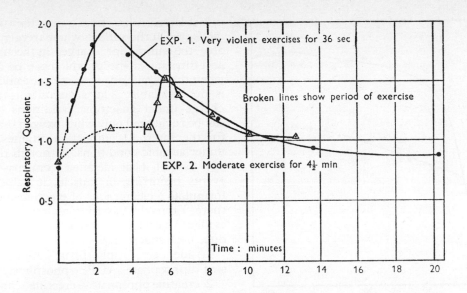

Fig. 40.23 Subject, C. N. H. L. The respiratory quotient during and after muscular exercise. These figures show the initial phase of recovery only. The final phase is shown in Fig. 40.24. (A. V. Hill, C. N. H. Long & H. Lupton (1924). *Proceedings of the Royal Society* B **96,** 468.)

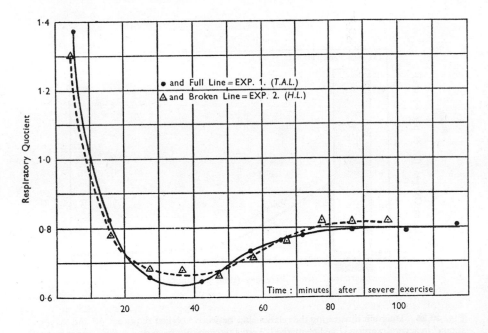

Fig. 40.24 The respiratory quotient after severe muscular exercise. Note that in the later phases, while carbon dioxide is being retained to compensate for that initially driven off, the respiratory quotient falls to a very low level, returning to its final value in about 80 minutes (A. V. Hill, C. N. H. Long & H. Lupton (1924). *Proceedings of the Royal Society* B **96,** 472).

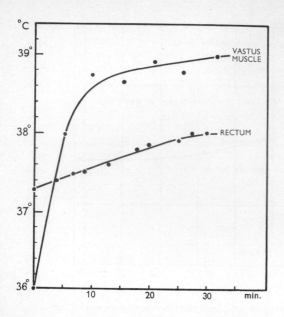

Fig. 40.25 Temperature measured in the lateral vastus muscle and in the rectum during work at 660 kg.m per min. (E. Asmussen and O. Boje (1950). *Journal of Physical Education* **42**, 48.)

muscular exercise have used the *submarine analogy*. On the surface where oxygen is available the submarine charges its accumulators and thus stores energy which can be called on when the submarine is submerged and oxygen is not available. The accumulators can discharge under anaerobic conditions but they can be recharged only when oxygen is available for the internal combustion engines, that is under aerobic conditions. Although this analogy cannot bear detailed comparison with events occurring in muscle, it helps one to formulate a clearer idea of reactions occurring there. From the description already given it is evident that three chemical reactions can produce heat anaerobically:

1. adenosine triphosphate (ATP)→adenosine diphosphate (ADP) + phosphoric acid;

2. creatine phosphate→creatine + phosphoric acid;

3. glycogen→pyruvic acid and lactic acid.

While the anaerobic reactions are occurring it can be supposed that the muscle accumulators are discharging. In a short burst of severe exercise, say a hundred yards' sprint lasting

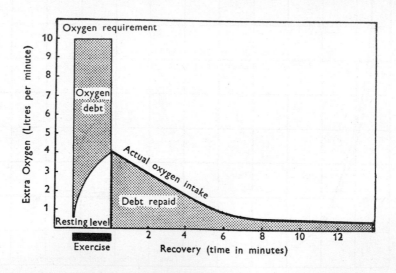

Fig. 40.26 Diagram illustrating the relationship between 'oxygen requirement' and 'oxygen intake' in severe exercise. The line marked 'Actual oxygen' has been drawn from experimental findings. The column marked 'Oxygen requirement' is constructed by making the shaded area marked 'Oxygen debt' equal to the shaded area marked 'Debt repaid.' (*After* M. G. Eggleton from K. Furusawa, A. V. Hill, C. N. H. Long & H. Lupton (1924). *Proceedings of the Royal Society* B **97**, 155.)

10 to 12 sec, the volume of oxygen that can be taken in is very small compared with the volume actually required and energy must, therefore, be derived from anaerobic processes. This results in an oxygen debt which is paid back through deep and frequent breathing that continues long after the exercise is over. The additional amount of oxygen consumed above the resting oxygen consumption reveals the amount of the oxygen debt (Fig. 40.26). During this period the accumulators are being charged, or, in other words, the three reactions listed above are proceeding in the reverse direction as the result of energy supplied by oxidation of lactic acid and other materials. In mild exercise the anaerobic mechanism may be called on only at the beginning (Fig. 40.27) because as exercise proceeds the circulatory and respiratory adjustments allow as much oxygen to be taken in as is required and a steady state is therefore reached although the oxygen debt is still unpaid. The initial oxygen debt is paid off at the end of the exercise. Even a small initial oxygen debt is apparently not paid off *during* the period of exercise. The body does not seem able to take in more oxygen than is required for current expenditure and it often fails to do even that. The lactic acid level in the blood during mild exercise such as walking does not rise above the resting value of 15 mg

per 100 ml. This is interpreted as meaning that lactic acid is disposed of as fast as it is formed. It does not, therefore, accumulate in the muscle tissue. In more vigorous exercise, however, lactic acid is formed so rapidly that the mechanisms for oxidizing it or reconverting it to glycogen are overloaded. It therefore appears in increased amount in the blood so that in severe exercise the blood lactate rises as high as 50 mg per 100 ml and sometimes up to 200 mg per 100 ml. The appearance of lactic acid in the blood is in some way related to the state of 'training' of the subject (Fig. 40.28). The difference between the reactions of the trained and the untrained male subject to running at a pace they could maintain for $1\frac{1}{2}$ hours is clearly shown by the findings of Johnson and his colleagues (1969) (Fig. 40.28). Samples of venous blood taken at the middle, end and during recovery showed that glucose does not seem to be an important source of energy. Its blood concentration altered little in either group. Lactate and pyruvate rose much higher in the untrained during the exercise. Plasma free fatty acids started lower in the athletes, rose a little during the run, then declined. In the untrained the rise during the exercise was much greater and on recovery was still above the athletes' level. Ketone bodies rose very little in the athletes during the exercise but

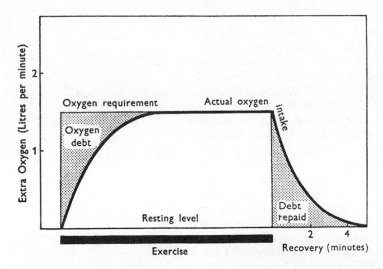

Fig. 40.27 Diagram illustrating the relationship between 'oxygen requirement' and 'oxygen intake' in mild exercise. The method of constructing this diagram is given in the legend to Fig. 40.29. (*After* M. G. Eggleton *from* K. Furusawa, A. V. Hill, C. N. H. Long & H. Lupton (1924). *Proceedings of the Royal Society* B, **97,** 155.)

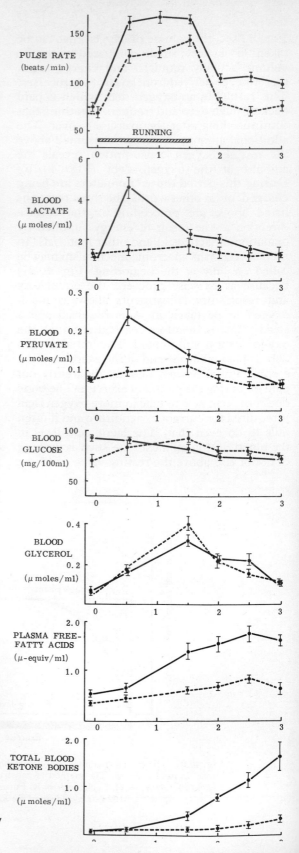

Fig. 40.28 Mean and standard deviation of the mean of observations in nine athletes (●----●) and eighteen untrained subjects (●———●) before and after running for $1\frac{1}{2}$ hours. The athletes ran much faster than the non-athletes. The athletes had a lower heart rate than the untrained subjects both during and after exercise. (R. H. Johnson, J. L. Walton, H. A. Krebs & D. H. Williamson (1969). *Lancet* **ii,** 452–455.)

rose markedly in the other group during the recovery period. In both the low levels during the run suggest that keto bodies were not an important fuel. The similar behaviour of glycerol in both groups during the exercise suggests that they both utilized fat to an equal extent but the athletes clearly oxidized fatty acids more efficiently than the untrained subjects. The glycogen–lactic acid mechanism is not quite so simple as may at first sight appear, since even in complete exhaustion a muscle still possesses considerable glycogen stores; thus the glycogen content of a frog muscle stimulated to exhaustion may fall only from 0·7 to 0·5 g per cent. Even relatively light exercise causes a lowering of muscle glycogen. The local glycogen store in the working muscles is a determining factor in the ability to perform long-term exercise, that is, the higher the muscle glycogen content the longer the performance time. A high glycogen concentration is necessary for the resynthesis of creatine phosphate and ATP in man. If glucose is infused the lowering of glycogen is of the same order whether or not glucose is administered during exercise, as if there was an obligatory breakdown of muscle glycogen. It is likely that the maximum oxygen debt which can be incurred is related to the maximum tolerable lactic acid production. But too close a correlation between oxygen debt and lactic acid level should not be looked for because the oxygen uptake after exercise declines rapidly while the blood lactate is slowly reduced; the blood lactate is taken up mostly by the liver and the heart. The main rise in blood lactate in exhausting exercise is usually after the exercise has stopped. Outward diffusion of lactate from muscle is relatively slow and exhaustion supervenes so rapidly that little lactate has had time to diffuse out.

854   Textbook of Physiology and Biochemistry

The effects of exercise of varying degrees of severity are given in Table 40.21 and Figure 40.22. It can be seen that the increase of oxygen consumption for any given exercise is much greater than the increase in cardiac output. The increased amounts of oxygen can be taken up in the lungs because the blood coming to them is less saturated than usual and because a greater quantity of blood is going through the lungs in unit time. The venous saturation has been calculated (see Fick principle, p. 502) and is plotted in Figure 40.22. This graph also shows that within the range of the observations the pulse rate increased from 70 to 165 per min whereas the cardiac output rose from 5 to 25 litres per min. A simple calculation shows that the stroke volume (output per beat) must have increased from 70 ml up to 150 ml. In exercise the sympathetic nerve impulses increase the heart rate and bring about more forceful atrial and ventricular contractions. The ejection time becomes slightly shorter. The increase in heart rate is achieved mainly by cutting down the duration of diastole with a consequent reduction of filling time. Thus above a certain heart rate (usually 130 to 140 beats/min) the increased minute volume of cardiac output is due largely to the increased heart rate rather than to increase in stroke volume. The amount put out at each beat, the stroke volume, remains stationary or may even decrease a little when the heart is nearing its maximum minute volume. In severe exercise the fit heart may increase its rate to 180/min or more; in young athletes, particularly in their early teens, the rate often exceeds 200 beats/min. The coronary blood flow is increased. One may speculate that in exercise the heart output is controlled through the aortic pressure acting directly on the heart while the heart rate is controlled by various reflexes.

The cardiovascular responses to static effort, or isometric exercise, have recently been receiving attention from Lind and others who have found that they differ considerably from the responses to rhythmic exercise. Common examples of static effort are lifting heavy objects, opening stiff windows and working with the arms overhead. The degree of effort in 'isometrics' is described in terms of the maximum voluntary contraction (MVC) which can be produced by a muscle group. A muscle group can maintain tension up to 10 or 15 per cent of MVC for a long time with a very small increase in heart rate and 5 to 15 mm Hg rise in blood pressure with some increase in blood flow during the time the tension is maintained. When the tension is increased beyond 20 per cent of MVC the blood flow through the muscle is not sufficient to supply the metabolic needs and fatigue sets in rapidly. At 20 per cent the contraction can be held for only 10 minutes, at 30 per cent for only 5 minutes and at 50 per cent MVC for 1 or 2 minutes only. At these values of MVC a steady state is not reached; the blood flow, heart rate and blood pressure rise continuously (Fig. 40.29) up to the point of fatigue. The rise in heart rate is not so much as in dynamic exercise but the rise in blood pressure is very much greater in isometric exercise due to a widespread vasoconstriction. A curious feature of isometric exercise is that the magnitude of the pressor response is nearly independent of the size of the muscle group involved. A 30 per cent MVC of forearm muscles (say 20 kg) produces the same pressor response as a 30 per cent MVC of leg muscles (70 kg). The pressor response to static effort is the same whether the subject is seated or standing or carrying out rhythmic exercise at the same time. Isometric exercise can probably be carried out safely by fit young people but it does very little to improve work performance or cardiovascular fitness. However, it may be dangerous for older people with weak cardiovascular systems. The pressor response in static exercise is probably brought about by release of some substance (perhaps potassium) from the active muscles which acts on as yet unidentified receptors in a muscle and produces reflexly a vasoconstriction.

Second wind. At the beginning of moderately severe exercise, say a long-distance run, the breathing becomes increasingly laboured; after some time the discomfort disappears and the subject says he has got his *second wind*. The phenomena of second wind are variable. In some people the transition from dyspnoea to hypernoea occurs suddenly, while in others the change is quite slow. The rate of transition has little to do with training as the phenomenon is often quite marked in the untrained person. During the dyspnoeic period the alveolar carbon dioxide tension is higher than normal, but when the subject gets his second wind the alveolar carbon dioxide tension falls and there

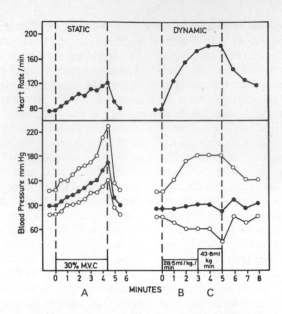

Fig. 40.29 Comparison of the heart rate and blood pressures (systolic, diastolic and estimated mean pressures in response to a fatiguing, sustained hand-grip contraction at 30 per cent MCV, and to an exhausting treadmill-walking test with stages of progressive severity shown as oxygen uptake/kg body weight/min. The results clearly show the great rise of blood pressure, and the small rise of heart rate during the sustained hand-grip, which contrast with the large rise in heart rate with little change in mean blood pressure during the fatiguing rhythmic exercise. A, 30 per cent maximum voluntary contraction (MCV); B, 28·5 ml/kg/min oxygen uptake; C, 43·8 ml/kg/min oxygen uptake. (A. R. Lind & G. W. McNicol (1967). *Canadian Medical Association Journal* **96,** 706–713.)

is a fall in the ventilation rate which is associated with the feeling of comfort. At the same time sweating may occur and the body temperature may rise. The respiratory quotient usually falls. A satisfactory explanation of this phenomenon is not yet forthcoming but it is likely that the early However, of heavy exercise is due partly to increase in metabolic CO_2 and partly to the liberation of lactic acid which displaces carbon dioxide from bicarbonate; this additional carbon dioxide causes stimulation of the respiratory centre. In time, when this additional carbon dioxide is eliminated by vigorous breathing and the circulatory adjustments to meet the demands of exercise are completed, less lactic acid reaches the blood and the body reaches a steady state in which respiration is less laboured.

Hormonal changes in exercise. The adrenal cortex is activated in exercise. A day's skiing produces a fall of up to 80 per cent in the circulating eosinophils; this could be described as beneficial and invigorating exercise. In long-distance swimmers the eosinophils may disappear almost entirely from the circulation while the output of 17-hydroxy-steroids in the urine may increase eightfold—an increase comparable to that occurring after major surgery. The serum growth hormone (HGH) rises in exercise. The raised HGH level falls quickly after the end of exercise in fit persons but in unfit persons HGH only slowly returns to the basal value. HGH probably initiates and maintains the mobilization of depot fat during exercise. The fall of plasma insulin during exercise allows the mobilization of fat and prevents the blood glucose falling too low.

SMOOTH MUSCLE

The muscle of the intestinal tract, the uterus, the spleen, and the blood vessels is called smooth, or visceral, muscle. Smooth muscle cells have a relatively simple structure; they are long (15 to 500 μm), narrow (2 to 20 μm) cells tapering towards the ends with a single long narrow nucleus about the middle of the cell. They may show a faint longitudinal striation but never a transverse. The electron-microscope shows myoactin filaments, 1 μm long and 5 to 15 nm across, to which myosin may be attached as small particles; cross links have not been seen between the filaments. These myofilaments are oriented obliquely at a few degrees to the long axis of the fibre so that they are inserted on one side of the fibre and pass over to the other side. They could be said to be in a parallel arrangement in contra-distinction to skeletal muscle where the contractile elements are in series. This parallel arrangement probably accounts for low velocity of shortening of smooth muscle and also for its ability to maintain a powerful contraction for long periods. Electron microscopy shows no protoplasmic continuity between cells but, where opposing membranes of adjacent cells are fused, the electrical resistance is likely to be low so that the tissue can behave as a syncytium. The sarcotubular system and the endoplasmic reticulum are poorly developed as compared with skeletal muscle. The sarcolemma, about 8 nm, is too fine to be seen by light microscopy. Smooth muscle cells may receive both sympathetic and parasympathetic nerve fibres which appear as a fine plexus of non-myelinated fibres 0·1 to 1·7 μm in diameter but not all are innervated. Synaptic junctions are extremely hard to find. Smooth muscle does not appear to be organized in motor units like skeletal muscle and its blood supply is relatively smaller.

The movements of smooth muscle are much slower than those of skeletal muscle but the former should not be thought of as being much weaker. The force exerted on the child by the smooth muscle of the uterus during parturition is very large indeed. Whereas skeletal muscle is paralysed when its nerves are cut, smooth muscle continues its rhythmic activity, though perhaps in a modified way, when its autonomic connexions have been severed. Whether this 'spontaneous' activity is myogenic, that is to say, an intrinsic property of the muscle fibres, is difficult to decide since most areas of smooth muscle contain nerve fibres and even ganglion cells. The sympathetic and parasympathetic nerves to a given group of visceral muscle fibres have usually opposite effects, one being motor and the other inhibitor, but the effect obtained on stimulation of, say, the sympathetic nerve is different in different organs and not necessarily the same in the same organ at different times. The tone of the muscle at the time of the experiment is one of the factors determining whether a motor or an inhibitory response is obtained.

The membrane potential (see p. 791) of smooth muscle cells of the colon has been measured by an intracellular electrode. It is about 60 mV on the average but in any one fibre the level fluctuates. Spike potentials, usually preceded by a slow depolarization, occur at about one per second, the average size being about 10 mV. Each spike is followed by a small production of tension. Acetylcholine causes an increase in the tension of the muscle, a fall in membrane potential and an increase in the frequency of the spikes; adrenaline produces exactly the opposite effects. Uterine cells have a low membrane potential, usually about 30 mV; it is doubled owing to increase in potassium permeability by treatment with oestrogen and progesterone. Large doses of oxytocin reverse the potential.

Although the strength-duration curves (p. 790) of smooth muscle have the same general shape as those of nerve or skeletal muscle the chronaxie is quite long, in the case of the frog's stomach about 50 msec and uterine muscle about 2 sec. Thus smooth muscle is relatively unresponsive to induced currents although it may readily be stimulated by the constant current from a battery (galvanic stimulation). When one part of a sheet of smooth muscle in the chick amnion which has no nerves is stimulated a wave of contraction passes over the whole sheet, but the fine nerve plexuses which usually accompany smooth muscle cells are presumably important for co-ordinated movements and reflex responses.

Smooth muscle is responsive to alterations

in temperature. Thus when the abdomen is opened and a loop of intestine is exposed to cool air it contracts but relaxes again if it is warmed. A fall in pH usually initiates a contraction, and an increase a relaxation. Because of its spontaneous activity and the effects just described it is difficult to say what the resting length of a piece of smooth muscle is.

The chemical processes involved in the contraction of smooth muscle are probably much the same as in skeletal muscle since it contains enzyme systems similar to those in skeletal muscle and calcium is also probably responsible for the activation of the contractile proteins. It uses mainly fatty acids and acetoacetate as sources of energy, and glucose to a lesser extent. The oxygen consumption of smooth muscle is only one-quarter of that of skeletal muscle at rest but smooth muscle when contracted consumes no more oxygen than it does when relaxed, and for this reason it is difficult to demonstrate fatigue in visceral muscle.

REFERENCES

ADAMS, R. D., EATON, L. M. & SHY, G. M. (Eds.) (1961). Neuromuscular disorders (the motor unit and its disorders). *Research Publications. Association for Research in Nervous and Mental Diseases* **38,** 1–813.

ANDREW, B. L. (Ed.) (1966). *Control and Innervation of Skeletal Muscle.* Edinburgh: Livingstone.

ASTRAND, P. O. (1956). Human physical fitness with special reference to sex and age. *Physiological Reviews* **36,** 307–335.

BERNSTEIN, N. (1967). *The Co-ordination and Regulation of Movements.* Oxford: Pergamon.

BEVEGÅRD, B. S. & SHEPHERD, J. T. (1967). Regulation of the circulation during exercise in man. *Physiological Reviews* **47,** 178–213.

BHATTACHARYYA, N. K., CUNNINGHAM, D. J. C., GOODE, R. C., GOODE, R. C., HOWSON, M. G., LLOYD, B. B. (1970). Hypoxia, ventilation, P_{CO_2} and exercise. *Respiration Physiology* **9,** 329–347.

BIANCHI, C. P. (1968). *Cell Calcium.* London: Butterworth.

BOURNE, G. H. (Ed.) (1960). *Structure and Function of Muscle,* 3 Vols. New York: Academic Press.

BOYD, I. A., EYZAGUIRRE, C., MATTHEWS, P. B. C. & RUSHWORTH, G. (1964). *The Role of the Gamma System in Movement and Posture.* New York: Association for the Aid of Crippled Children.

BRISTOW, J. D., BROWN, E. B. JR., CUNNINGHAM, D. J. C., GOODE, R. C., HOWSON, M. G. & SLEIGHT, P. (1968). The influence of ventilation, carbon dioxide and hypoxia on the baroreceptor reflex in man. *Journal of Physiology* **198,** 102–103P.

BRISTOW, J. D., BROWN, E. B. JR. CUNNINGHAM, D. J. C., GOODE, R. C., HOWSON, M. G., PICKERING, T. G. & SLEIGHT, P. (1969). Changes in the baro-receptor-cardiac reflex in exercise. *Journal of Physiology* **201,** 106–107P.

BÜLBRING, E., BRADING, A., JOKES, A. & TOMITA, T. (Eds.) (1970). *Smooth Muscle.* London: Arnold.

BULLER, A. J. (1965). Mammalian fast and slow skeletal muscle. *Scientific Basis of Medicine Annual Review* 186–201.

BURNSTOCK, G., HOMAN, M. E. & PROSSER, C. L. (1963). Electrophysiology of smooth muscle. *Physiological Reviews* **43,** 482–528.

CAIN, D. F., INFANTE, A. A. & DAVIES, R. E. (1962). Chemistry of muscle contraction. Adenosine triphosphate and phosphorylcreatine as energy supplies for single contractions of working muscle. *Nature, London* **196,** 214–216.

CAMPBELL, E. J. M. (1967). Exercise tolerance. *Scientific Basis of Medicine Annual Review* 128–144.

CARLSON, F. (1963). The mechanochemistry of muscular contraction. A critical revaluation of *in vivo* studies. *Progress in Biophysics* **13,** 262–314.

CATTON, W. T. (1970). Mechanoreceptor function. *Physiological Reviews* **50,** 297–318.

CHAPMAN, C. B. (Ed.) (1967). Physiology of muscular exercise. American Heart Association Monograph No. 15. *Circulation Research* **22,** Suppl. No. 1, I-1 to I-28.

COERS, C. & WOOLF, A. L. (1959). *The Innervation of Muscle. A Biopsy Study.* Oxford: Blackwell.

COOPER, SYBIL & DANIEL, P. M. (1963). Muscle spindles in man. *Brain* **86,** 563–586.

CUNNINGHAM, D. J. C. (1963). Some quantitative aspects of the regulation of human respiration in exercise. *British Medical Bulletin* **19,** 25–30.

ECCLES, J. C. (1964). *The Physiology of Synapses.* Berlin: Springer.

EDHOLM, O. G. (1967). *The Biology of Work.* London: Weidenfeld & Nicolson.

EICHNA, L. W. (Ed.) (1962). Vascular smooth muscle. *Physiological Reviews* **26,** Suppl. 5, 1–365.

FALLS, H. B. (Ed.) (1968). *Exercise Physiology.* New York: Academic Press.

FRICK, M. H. (1968). Coronary implications of hemodynamic changes caused by physical training. *American Journal of Cardiology* **22,** 417–425.

GERGELY, J. (Ed.) (1964). *The Biochemistry of Muscular Contraction.* London: Churchill.

GONZALES-SERRATOS, H. (1971). Inward spread of activation in vertebrate muscle fibres. *Journal of Physiology* **212,** 777–799.

GORDON, A. M., HUXLEY, A. F. & JULIAN, F. J. (1966). The variation in isometric tension with sarcomere length in vertebrate muscle fibres. *Journal of Physiology* **184,** 170–192.

GRANIT, R. (1966). *Muscular Afferents and Motor Control. Nobel Symposium—I.* London: Wiley.

HANSON, J. & LOWY, J. (1965). Molecular basis of contractility in muscle. *British Medical Bulletin* **21,** 264–271.

HILL, A. V. (1970). *First and Last Experiments in Muscle Mechanics.* Cambridge University Press.

HUXLEY, A. F. (1964). Muscle. *Annual Review of Physiology* **26,** 131–152.

HUXLEY, A. F. (1971). The activation of striated muscle and its mechanical response. *Proceedings of the Royal Society* B **178,** 1–27.

HUXLEY, A. F. & HUXLEY, H. E. (1964). A discussion on the physical and chemical basis of muscular contraction. *Proceedings of the Royal Society* B **160,** 434–536.

HUXLEY, H. E. (1965). The mechanism of muscular contraction. *Scientific American* **213,** 18–24.

HUXLEY, H. E. (1970). The mechanism of muscular contraction. *Science* **164,** 1356–1366.

HUXLEY, H. E. (1971). The structural basis of muscular contraction. *Proceedings of the Royal Society* B **178,** 131–149.

KATZ, B. (1966). *Nerve, Muscle and Synapse.* New York: McGraw-Hill.

KATZ, B. (1969). *The Release of Neural Transmitter Substances.* Liverpool University Press.

LIGHT, S. (Ed.) (1961). *Electrodiagnosis and Electromyography,* 2nd edn. Baltimore: Waverley Press.

LIND, A. R., McNICOL, G. W. & DONALD, K. W. (1966). Circulatory adjustments to sustained (static) muscular activity. In *Physical Activity in Health and Disease* (Edited by K. Evang & K. L. Andersen), pp. 38–63. Oslo: Universitetsforlager.

MARGARIA, R. (Ed.) (1967). *Exercise at Altitude.* Amsterdam: Exerpta Medica Foundation.

MARGARIA, R. (1972). The sources of muscular energy. *Scientific American* **226,** 84–91.

MATTHEWS, P. B. C. (1964). Muscle spindles and their motor control. *Physiological Reviews* **44,** 219–288.

MATTHEWS, P. B. C. (1972). *Mammalian Muscle Receptors and their Central Action.* London: Arnold.

MOMMAERTS, W. F. H. M. (1966). The rheology of muscle. *Laboratory Practice* **15,** 171–178.

MOREHOUSE, L. E. & MILLER, A. T. (1963). *Physiology of Exercise,* 4th edn. St. Louis: Mosby.

PAUL, W. M., DANIEL, E. E., KAY, C. M. & MONCKTON, G. (Eds.) (1965). *Muscle.* Oxford: Pergamon.

PETERS, R. A. (Ed.) (1956). Physiology of voluntary muscle. *British Medical Bulletin* **12,** 161–235.

PODOLSKY, R. J. (Ed.) (1965). Symposium. Excitation–contraction coupling in striated muscle. *Federation Proceedings* **24,** 112–1152.

RICHARDSON, A. T. (1962). Clinical electrodiagnosis. *Proceedings of the Royal Society of Medicine* **55,** 897–904.

RODAHL, K. & HORVATH, S. M. (Eds.) (1962). *Muscle as a Tissue.* London: McGraw-Hill.

SANDOW, A. (Ed.) (1959). Second conference on muscular activity. *Annals of the New York Academy of Sciences* **81,** 401–509.

SANDOW, A. (1965). Excitation–contraction coupling in skeletal muscle. *Pharmacological Reviews* **17,** 265–320.

Symposium (1967). The contractile process. *Journal of General Physiology* **50,** 1–292.

TIMMS, A. R. (1964). The coupling of electrical and mechanical activities in intestinal smooth muscle in relation to metabolism. *Transactions of the New York Academy of Sciences,* Ser. II **26,** 902–913.

DE VRIES, H. A. (1967). *Physiology of Exercise.* London: Staples Press.

WALTON, J. N. (Ed.) (1964). *Disorders of Voluntary Muscle.* London: Churchill.

WILLIAMS, D. J. M. (1966). Man powered aircraft. *Science Journal* **2,** 74–79.

ZACKS, S. I. (1964). *The Motor Endplate.* Philadelphia: Saunders.

41 Autonomic nervous system

In the discussion of the control of the heart, the blood vessels, and the movements of the alimentary canal, the sympathetic and parasympathetic nerve supply was considered in each case. Only the general features of the autonomic system, both afferent and efferent, will be described here.

The autonomic nervous system has central connexions which are still being investigated. The frontal lobe of the cerebral cortex has an influence on the cardiovascular system and the intestinal tract by virtue of its connexions with the hypothalamus but the main control of the autonomic system seems to be located in the hypothalamus itself. Important reflex centres, including the vagal centres, are found in the medulla oblongata. These subjects are considered in other chapters.

The autonomic fibres leave the cord in two great 'outflows', the thoracolumbar (sympathetic) and craniosacral (parasympathetic).

Autonomic nervous system

Parasympathetic outflow
- Cranial
  - Tectal (III)
  - Bulbar (VII, IX, X, XI)
- Sacral
  - Pelvic splanchnic nerves

Sympathetic outflow
- From all thoracic and upper two lumbar segments of cord

Unlike the somatic outflow which supplies the skeletal muscles the autonomic fibres to the viscera are interrupted by synapses. The fibre emerging from the central nervous system as a *preganglionic* fibre runs to synapse in a ganglion from which a second, *postganglionic*, fibre supplies gland or muscle cells. A preganglionic nerve impulse may, on reaching a synapse, set up impulses in as many as twenty postganglionic fibres. In the case of the sympathetic system the synapses occur in the ganglia of the sympathetic trunk or in more peripheral ganglia which lie along the anterior surface of the abdominal aorta. A few intermediate ganglia lie along the course of the white rami communicantes and also in relation to nerves other than those of the thoracolumbar outflow. Most of the sympathetic postganglionic fibres are distributed with the spinal nerves but some enter visceral branches. The parasympathetic ganglia, on the other hand, commonly lie on or in the organ supplied.

Ganglia can be located by the method used by Langley which depends on the fact that the alkaloid nicotine, after first stimulating ganglion cells, paralyses them by depolarization without having any effect on nerve fibres. Thus, if after painting a ganglion with nicotine the characteristic effect normally produced in a viscus by electrical stimulation of its autonomic nerve is abolished, this ganglion must contain synapses at which postganglionic fibres begin.

SYMPATHETIC OUTFLOW

The thoracolumbar outflow in man arises from all the thoracic segments of the spinal cord and from the upper two lumbar (Fig. 41.1); each fibre beginning in a cell of the lateral horn (the intermediolateral column) of the grey matter of the spinal cord emerges in an anterior root and leaves the spinal nerve in a *white ramus communicans* which passes to a sympathetic ganglion (Plate 41.1). Here the fibre may either end in a synapse or turn aside to run up or down in the sympathetic trunk to a neighbouring, or sometimes a distant, ganglion where there is a synapse. From the sympathetic trunk postganglionic fibres, chiefly non-myelinated, pass in the *grey rami communicantes* to every spinal nerve. They are distributed in the peripheral nerves to blood vessels and skin, sweat glands and pilomotor muscles, and in special autonomic nerves (splanchnic nerves) to the abdominal organs. Postganglionic fibres also reach the viscera in the nerve plexus surrounding the arteries which supply them. The adrenal medulla is exceptional in that it receives preganglionic sympathetic fibres directly from the spinal cord; the cells of the adrenal gland which produce adrenaline and noradrenaline are homologous with the cells of the peripheral sympathetic ganglia.

The sympathetic system is not essential to life and in human beings large parts of it have been removed without serious impairment of function. Cats which have undergone as complete a removal as possible of the sympathetic system appear to live normally if they are kept in a protected environment but they do not survive long if exposed to adverse conditions such as cold, loss of blood or general anaesthesia, the adjustment to which involves complex reactions within the body.

PARASYMPATHETIC (CRANIOSACRAL) OUTFLOW

The cranial part of the parasympathetic outflow is described in other chapters which deal with the organs it supplies. Myelinated preganglionic fibres are found in the third nerve going to the ciliary ganglion, in the seventh going to the submandibular and sphenopalatine ganglia, in the ninth to the otic ganglion, and in the tenth and eleventh cranial nerves to small ganglia and plexuses in the walls of the viscera. The postganglionic fibres are not myelinated and pass to the intra-ocular muscles, salivary and lacrimal glands, heart, lungs and abdominal viscera. The sacral outflow comes from cells situated laterally in the grey matter and leaves the cord by the anterior roots of the second and third sacral nerves as the pelvic splanchnic nerves (nervi erigentes). Their ganglia are usually in the walls of the viscera to which they run. The sacral parasympathetic supplies the colon, rectum, generative organs and urinary bladder. Parasympathetic nerves have not been proved to go to the limbs or to the surface of the body.

AFFERENT IMPULSES FROM THE VISCERA

The sympathetic and parasympathetic systems described above are efferent (effector) systems and it was in this sense that Langley used the expression 'autonomic nervous system' (1898). Alongside these efferent fibres, however, there is an approximately equal number of afferent (ingoing) fibres from the viscera which, like afferents from other parts of the body, have their cell bodies in the spinal ganglia of the posterior roots. It is convenient to describe them here as 'autonomic afferents' although many of these *viscerosensory* fibres form part of visceral reflex arcs and are not strictly concerned with autonomic activity. Some convey sensation of visceral pain to the brain; visceral pain can be relieved by sympathectomy. After section of the abdominal viscerosensory nerves the Pacinian corpuscles

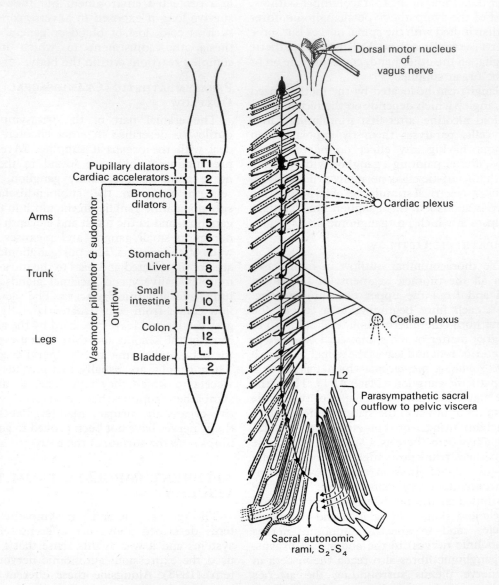

Fig. 41.1 The autonomic outflow. ———— Preganglionic axon. – – – Postganglionic axon. T, thoracic segment. L, lumbar segment. S, sacral segment. (J. C. White, R. H. Smithwick & F. A. Simeone (1952). *The Autonomic Nervous System*, 3rd edn, p. 24. London: Henry Kimpton.)

in the mesentery degenerate and there is apparently no essential difference except in size between viscerosensory autonomic fibres and afferent fibres from the skin. In the cat impulses in the splanchnic nerves have been traced into the cord, up the posterior columns of white matter on the same side to the nucleus gracilis on the same side and then across the mid-line to the opposite thalamus; more slowly running impulses go up the spino-thalamic tracts to the thalamus on both sides and also to the hypothalamus.

Records of activity in single fibres of the vagus show that very slowly adapting receptors are present in the stomach, intestine and urinary bladder. The rate of firing depends on the rate of distension of these viscera, being greater the more sudden the distension. The nerve endings are activated by passive disten-sion and by active contraction. Impulses pass from the receptors along non-myelinated fibres with a conduction velocity of less than 2·5 m/sec. Impulses from Pacinian corpuscles adapt very quickly; they are carried on large myelinated axons (conduction velocity 40 m/sec). The function of these corpuscles is unknown. Other rapidly adapting receptors exist, for example the receptors in the urethra responding to the flow of urine.

In a conscious, unanaesthetized person the abdominal viscera, provided that they are healthy, can be cut or burned without causing pain; this is clearly demonstrated when a knuckle of colon, brought through the ab-dominal wall by the surgeon, is opened pain-lessly some days later without any anaesthetic. Wolf and Wolff (p. 238) found that Tom's normal healthy gastric mucosa was not sensi-tive to pinching or electric stimulation but if the mucosa became hyperaemic and oedema-tous pain was readily evoked by these stimuli. In other words, the pain threshold of this viscus is normally high but it may be lowered by inflammatory changes. However, pain arises in the absence of inflammation if the appropriate kind of stimulation is applied. For example, pain is produced if the mesentery is pulled upon or injected with irritant material; since all the nerve fibres from the five metres of gut are crowded into the 15 cm long mesentery, traction on this structure must necessarily stimulate a much larger number of nerve fibres than does cutting or pinching of the gut

itself. Thus the production of pain by stimula-tion of visceral nerves depends on (a) the strength of the stimulus, (b) the number of nerve fibres stimulated and (c) the pain threshold at the moment of stimulation.

Pain impulses are conveyed almost entirely in sympathetic afferents. If a balloon passed into the small intestine is inflated the subject feels pain in the centre of the abdomen but after bilateral section of the splanchnic nerves inflation of the balloon does not cause pain. Vagotomy has no effect on pain produced by distension of the small intestine. Furthermore, patients with a complete transection of the spinal cord which renders the abdominal wall anaesthetic can not only experience intestinal colic but localize the pain in the centre of the abdomen. Learmonth reported that a patient who had been given a spinal anaesthetic to anaesthetize the lower part of the body experienced pain accurately localized to the bladder when the superior hypogastric plexus was crushed; the afferent fibres must of course have entered the cord above the level of the anaesthetized part of the cord.

It is well known that pain arising from a diseased organ may be projected to a definite position on the surface of the body (*referred pain*) but the mechanism of the phenomenon is somewhat obscure. A knowledge of the areas of localization of pain is, however, of great service in diagnosis and has been summarized in Table 41.2. The position at which a patient feels pain may coincide with the position of the tissue in which it is produced, or the pain may be felt at a distance from the place of its pro-duction. The pain due to pinching the skin on the surface of the body is felt by the subject at the point of stimulation but, when a transposed pedicle skin graft with intact innervation is pinched, the blindfolded patient experiences the pain in the position from which the graft was taken and not in its new site. Moreover, pain due to heart disease (angina pectoris) may be projected to the left arm even when this has been amputated. Pain is, therefore, projected to a position in the body image and not to any particular tissue. This phenomenon is further illustrated by the fact that pain, produced in organs such as the diaphragm and the testis which have migrated during development, is projected to the primary or original site of the organ. Rubbing the diaphragm at operation, or

Table 41.2 The segmental sensory innervation of the viscera

Organ	Superficial areas to which pain is referred	Sympathetic — Thoracic 1	2	3	4	5	6	7	8	9	10	11	12	Lumbar 1	2	Parasympathetic sacral 2	3	4	Afferent pathway from viscus
Heart	Precordium and inner arm	+	+	+	+	?													Middle and inferior cervical and thoracic cardiac nerves
Lung	No referred pain*		+	+	+	+	+	+											Inferior cervical and thoracic nerves (convey reflex impulses)
Liver and gall-bladder	Right upper quadrant and right scapula							+	+	+									Greater splanchnic nerve
Stomach	Epigastrium						+	+	+	+									Greater splanchnic nerve
Small intestine	Umbilicus									+	+	?							Greater splanchnic nerve
Colon {ascending	Suprapubic											+	+	+					Lumbar chains and pre-aortic plexus
{sigmoid and rectum	Deep pelvis and anus															+	+	+	Pelvic nerves and plexuses
Kidney	Loin and groin										+	+	+	+					Renal plexus via lowest splanchnic nerve and upper lumbar rami
Ureter	Loin and groin											+	+	+	+				Renal plexus and upper lumbar rami
Bladder {fundus	Suprapubic											+	+	+	+				Hypogastric plexuses
{bladder neck	Perineum and penis															+	+	+	Pelvic nerves and plexuses
Uterus {fundus and cervix	Suprapubic region and lower back, perineum											+	+	?		+	+	+	Hypogastric plexuses
Testes, vas deferens, seminal vesicles, prostate	Pelvis, perineum										+	+	+						Hypogastric plexuses

* Lung parenchyma is insensitive. Pain from larger bronchi is transmitted over somatic vagal axons. When disease spreads to parietal pleura pain is transmitted over intercostal nerves. (Derived from: J. C. White, R. H. Smithwick & F. A. Simeone (1952). *The Autonomic Nervous System*, 3rd edn, p. 136. London: Henry Kimpton; and J. J. Bonica (1968). Autonomic innervation of the viscera in relation to nerve block. *Anesthesiology* **29**, 793.)

electrical stimulation of the phrenic nerve, produces pain which is felt at the root of the neck, for the diaphragm develops in the neck, acquires its nerve supply there and later moves caudad. The testis develops near the kidney but low in the abdomen, and therefore testicular pain is felt in the lower abdomen and not in the testis itself. In other words the sensorium seems to make contact with, and to become aware of the position of, the organs in embryonic life and it continues to use the same reference map. The alimentary tract is a mid-line organ from the embryological point of view and pain arising in it or in any other mid-line organ should, according to this interpretation, be projected to various levels of the mid-line no matter whether the part concerned has migrated right or left. This is what is actually found. Thus in colic due to violent peristalsis of the intestine, pain is felt in the mid-line. In disorders of the stomach pain is felt in the mid-line high up in the abdomen; small-intestine pain is felt in the umbilical region; colon pain is situated in the mid-line in the lower part of the abdomen. The gall-bladder is embryologically a mid-line organ and true gall-bladder pain is felt in the epigastrium but, if the inflamed viscus irritates the diaphragm, pain is felt in the neck and right shoulder. It is unlikely that any nerve fibres from the gall-bladder or its peritoneal coverings reach the spinal cord by the phrenic nerve, since the gall-bladder can be painlessly removed under spinal anaesthesia in which the phrenic nerves and the diaphragm are functioning normally. Distension of the bile duct usually gives pain in the mid-line of the abdomen but sometimes in the back. Pain arising in the body and cervix of the uterus is felt in the mid-line about one and a half inches above the symphysis pubis.

When irritation of the parietal peritoneum results from visceral disease pain is sharply localized to the site of irritation and is associated with tenderness and often with spasm (rigidity) of the adjacent skeletal muscles. Thus the pain of early appendicitis is felt in the mid-line near the umbilicus and not until the parietal peritoneum, supplied by the somatic nerves, is irritated is the pain felt at the situation of the inflamed appendix, that is, generally, in the right iliac fossa. Similarly, although distension of the gall-bladder pro-duces pain in the centre of the epigastrium, inflammation of the fundus of the gall-bladder is associated with pain in the right hypochondrium in an area overlying the diseased organ.

Visceral pain can be modified by anaesthetization or irritation of the skin area in which the pain is felt. The pain of gastric ulcer is reduced by anaesthetization of the abdominal wall. The pain of angina pectoris can be relieved by anaesthetization of the painful area of the left arm and angina pectoris projected into an amputated limb is relieved by anaesthetization of the brachial plexus. Moreover, if, in a patient in whom angina pectoris is elicited by exercise, an area of skin is blistered by cantharidin the pain induced by the exercise may be felt in the blistered area. Even after complete cutaneous anaesthesia intended to relieve abdominal pain the patient may complain of a dull, aching, deep form of pain which is usually localized in or near to, the diseased organ itself. This pain is similar to that produced by stimulation of the coeliac ganglion and is relieved only by section of sympathetic nerves.

One possible 'explanation' of the results of experiments on diaphragmatic pain is as follows. Nerve impulses from the diaphragm and from the shoulder reach the same segment of the spinal cord and possibly the same area of the sensorium (Fig. 41.3), and by building up the central excitatory state determine the threshold for pain; this may be normally at the level represented at (a) in Fig. 41.3. Anaesthetization of the shoulder area reduces the number of impulses reaching the cord from the shoulder and so raises the threshold as shown at (b) and (c). Irritation of the diaphragm to a degree represented by the height of the triangle at (a) normally produces a shower of impulses which reach the threshold and pain is felt. After anaesthetization of the shoulder, however, a similar degree of irritation of the diaphragm represented by (b) now falls short of the threshold and there is no pain. A larger stimulus (large triangle at (c)) such as could be produced by a phrenic nerve crush can, however, reach the threshold and cause pain. The observations on angina pectoris already mentioned are also capable of being explained on this theory, if we postulate that blistering increases the number of afferent

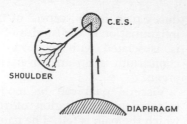

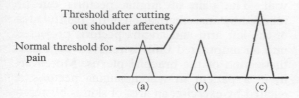

Fig. 41.3 Visceral pain. (a) When the diaphragm is rubbed pain is felt in the shoulder. (b) When the diaphragm is rubbed after anaesthetization of the shoulder no pain is felt. (c) Crushing the phrenic nerve after anaesthetization of the shoulder is painful. C.E.S. is central excitatory state. (F. R. Brown (1948). *Lancet* **i**, 386.)

impulses arriving at the central nervous system and so lowers the threshold for pain.

Autonomic reflexes. Just as in the somatic nervous system, there are, in the autonomic nervous system, reflexes both sympathetic and parasympathetic and these are described in various chapters of this book. For the most part the afferent impulses enter the cord from the viscera but some come from the muscles and skin. It is through such reflex arcs that diseases in the viscera can produce areas of hyperaesthesia and other effects on the skin and, on the other hand, that the application of heat to the skin for example can produce effects on the motility of the viscera. Most of these reflexes depend upon pathways involving the brainstem (for example baroreceptor reflexes) or hypothalamus (for example thermoregulatory reflexes) but some autonomic reflexes exist at a spinal level (p. 304). In addition to autonomic reflexes through the spinal cord there are others that operate through peripheral ganglia. Thus the reflex activity responsible for the controlled motility of the small intestine is a local affair the ganglion cells of which are in the myenteric and submucous plexuses in the wall of the gut. The inferior mesenteric ganglion likewise contains ganglion

cells through which local autonomic reflexes for the bladder and rectum operate. Axon reflexes (see p. 563) occur in the autonomic-nervous system, for example piloerection may occur in response to repetitive electrical stimulation of the skin but only if post-ganglionic sympathetic fibres are functioning.

HUMORAL THEORY OF TRANSMISSION OF THE EFFECT OF THE NERVE IMPULSE

Since the isolation of adrenaline in 1901 many similarities have been found between the effects of stimulation of the sympathetic nerves to a part and the effects on it of adrenaline injected into the circulation. This led to the idea, put forward in 1904 by T. R. Elliott, that when an efferent sympathetic nerve is stimulated adrenaline, liberated from some precursor at the nerve-endings, excites the effector organ. Dale in 1934 proposed the name *adrenergic* for those nerves which seem to operate by release of adrenaline. The nerves supplying arterioles are *adrenergic* whereas sympathetic fibres to sweat glands in man are *cholinergic*, that is they depend upon the release of acetylcholine. All parasympathetic fibres and all autonomic ganglia are also cholinergic (Fig. 41.4).

ADRENERGIC NERVES

Certain discrepancies between the effects of adrenaline and of sympathetic stimulation have been known for a long time, and it was postulated that two separate mediators, called sympathin E and sympathin I, were released at excitor and inhibitor sympathetic sites respectively. It is now clear, however, that all sympathetic adrenergic nerve terminals contain and release only noradrenaline, and that these nerve terminals contain no catechol O-methyl-transferase, the enzyme responsible for converting noradrenaline to adrenaline. However, during general sympathetic stimulation, the adrenal medulla releases both adrenaline (80 per cent) and noradrenaline (20 per cent) into the circulation. In the adrenal medulla, noradrenaline is a precursor of adrenaline, into which it is converted by methylation (Chap. 50). Noradrenaline is a powerful vasoconstrictor. If it is given by intravenous infusion in man, in doses of the order of 20 µg per min,

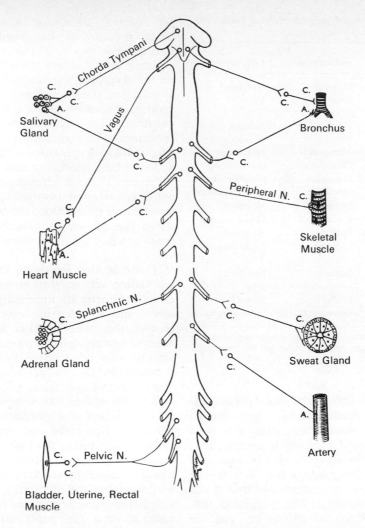

Fig. 41.4 Some examples of adrenergic and cholinergic nerve-endings. A, adrenergic nerve-ending. C, cholinergic nerve-ending. (From J. C. White, R. H. Smithwick & F. A. Simeone (1952). *The Autonomic Nervous System*, 3rd edn, p. 96. London: Henry Kimpton.)

both the systolic and the diastolic blood pressures are raised and in consequence cardiac slowing is reflexly produced so that the cardiac output is reduced (Fig. 28.15). Adrenaline, on the other hand, can in small doses (say 3 µg/min) augment the cardiac output without raising the blood pressure (Fig. 26.67). If larger doses of adrenaline of the order of 20 µg per min are infused, the heart accelerates and the systolic pressure increases, but the diastolic blood pressure is not increased. The explanation of these differences is that adrena-

line causes a dilatation of the vessels in the skeletal muscles and therefore a reduction of peripheral resistance. Noradrenaline by constricting the vessels in the muscles as well as those in the skin and viscera causes a rise in the total peripheral resistance. The actions of the two substances are compared in Table 41.5, which shows that in general they have similar effects.

α and β receptors. The action of adrenaline on smooth muscle has for long been known to be complex. For example, in the anaesthetized

Table 41.5 Table to compare the effects of L-adrenaline and L-noradrenaline in man

	Effect of	
	L-Adrenaline	L-Noradrenaline
Heart rate	Increase	Decrease*
Cardiac output	Increase	Variable
Total peripheral resistance	Decrease	Increase
Blood pressure	Rise	Greater rise
Respiration	Stimulates	Stimulates
Skin vessels	Constriction	Constriction
Muscle vessels	Dilatation	Constriction
Bronchus	Dilatation	Less dilatation
Eosinophil count	Increase	No effect
Metabolism	Increase	Slight increase
Oxygen consumption	Increase	No effect
Blood sugar	Increase	Slight increase
Central nervous system	Anxiety	No effect
Uterus in vivo in late pregnancy	Inhibits	Stimulates
Kidney	Vasoconstriction	Vasoconstriction

* On the *isolated* heart, where vagal reflexes are absent, noradrenaline increases heart-rate though less extremely than adrenaline.

cat when the vagi are cut the blood pressure is high; adrenaline produces a fall of pressure. In a spinal cat in which the brain has been destroyed the blood pressure is usually low; adrenaline then produces a rise of pressure. Again, for example, adrenaline causes the uterus of the rabbit to contract but that of the cat to relax. Ahlquist (1948) suggested that this double action of adrenaline can be accounted for by supposing that there are two kinds of tissue receptors with which it can combine. Thus when adrenaline produces constriction it is described as acting on α receptors and when it causes relaxation (or dilatation) it is said to combine with β receptors (Table 41.6). Noradrenaline acts almost exclusively on α receptors since it produces cutaneous and splanchnic vasoconstriction, a rise of blood pressure, dilatation of the pupil and relaxation of intestinal muscle. The drug isoprenaline acts almost entirely on β receptors since it increases the rate and force of the heart beat, causes dilatation of muscle vessels and relaxation of the bronchi and of the uterus. Adrenaline has both α and β actions. The α receptors can be blocked by ergotamine, piperoxan, phentolamine or phenoxybenzamine, so that,

for example, the action of adrenaline on the blood pressure of the spinal cat is reversed because the block of the α receptors (constricting effect) allows the action of adrenaline on the β receptors (relaxing effect) to be seen. Blocking of the α receptors, of course, greatly reduces the action of noradrenaline since this substance has seldom much effect on β receptors. The β receptors can be blocked by propranolol or practolol, which prevent the action of both adrenaline and noradrenaline on the heart. These β-adrenoreceptor antagonists are derivatives of isoprenaline. Patients suffering from raised blood pressure treated with adrenergic blocking agents are prone to hypotension when standing, especially after exercise.

There is now reason to believe that when adrenaline acts on β receptors it produces its effect indirectly by increasing metabolic processes and so releasing more energy; it increases, the concentration in the tissue of adenosinetriphosphate and creatine phosphate. On the other hand, adrenaline acts directly on α receptors to produce depolarization and contraction of muscle.

The adrenaline-like compounds, produced on stimulation of a sympathetic nerve, enter the circulation since they are only relatively slowly destroyed in the tissues by monoamine oxidase (Chap. 50). Their presence in the blood is shown by acceleration of the heart or withdrawal of the nictitating membrane, provided that these structures have been denervated about a week previously.

Denervation makes tissues which are supplied by sympathetic nerves more sensitive to adrenaline and particularly to noradrenaline. It is possible that the increase in sensitivity depends on the reduction in the amount of monoamine oxidase in the endings subsequent to degeneration of the nerve. Since this enzyme acts more rapidly on noradrenaline than on adrenaline the greater potentiation of the effect of noradrenaline after denervation is explained. Alternatively the increase in sensitivity may be explained by the loss of noradrenaline from the tissues after denervation. When the nerves are intact small amounts of noradrenaline are constantly discharged from the nerve endings to occupy some of the tissue receptors; after denervation the receptors are unoccupied so that a minute dose of

Table 41.6 α and β adrenergic receptor effects

	α Effects	β Effects
	Vasoconstriction	Vasodilatation (especially muscle)
	Intestinal muscle relaxation	Heart: increased force and rate; arrhythmias
	Pupillary dilatation	Bronchial relaxation
Noradrenaline	+ + +	+
Adrenaline	+	+ + +
Blocking drugs		
Propanolol ⎫ Practolol ⎭	− no blockade	+ + + blockade
Ergotamine ⎫ Pipcroxan ⎪ Phentolamine ⎬ Phenoxybenzamine ⎭	+ + + active blockade	− no blockade

(Slightly modified from R. H. Johnson & J. M. K. Spalding (1972). *Clinical Physiology of the Autonomic Nervous System.* Oxford: Blackwell.)

noradrenaline is now effective. Ephedrine, by inhibiting the action of monoamine oxidase, potentiates the action of adrenaline and noradrenaline and also increases the effects obtained by stimulation of adrenergic nerves. In animals, administration of thyroid hormone reduces the monoamine oxidase content of the liver and blood vessels and this may account for the exaggerated hyperglycaemia and rise of blood pressure after injection of adrenaline in hyperthyroid patients.

Organs supplied by postganglionic adrenergic nerves contain noradrenaline provided their nerve supply is intact; after denervation the noradrenaline almost entirely disappears but returns when the nerves regenerate. The nerve fibres must, therefore, be the only important source of noradrenaline. It is found that prolonged stimulation of an adrenergic nerve releases noradrenaline at a steady rate and does not reduce the noradrenaline content of the organ supplied; it must be assumed, therefore, that resynthesis is rapid enough to keep pace with the loss. The walls of blood vessels are rich in noradrenaline and there is a direct relationship between their content of noradrenaline and the number of nerve fibres they contain. Furthermore, the greater the supply of nerve fibres the more vigorously do they contract under the influence of noradrenaline. In conformity with this generalization the vessels of the skin and spleen, for example, react strongly to noradrenaline while those of the lung and brain respond only feebly.

Sympathectomy. Postganglionic fibres reach the blood vessels by two routes. Fibres from the sympathetic chain pass directly to plexuses around the aorta. These plexuses extend on to the large arteries of the abdomen, thorax and skull but reach only the proximal parts of the limbs. The sympathetic fibres to the vessels of the more distal parts of the limbs reach them through the peripheral nerves (Fig. 28.5). It follows that division of postganglionic nerve fibres (postganglionic sympathectomy) has little lasting effect on the circulation through the limbs. Preganglionic sympathectomy or removal of the sympathetic chain of ganglia (ganglionectomy) leads, however, to widespread and lasting dilatation of arteries and arterioles in the limbs which is greatest in the skin and least in the skeletal muscles. The reduction in vascular tone is not maximal after sympathectomy since further dilatation can be produced by local heating, by histamine, by exercise and by reactive hyperaemia.

The vasodilatation after sympathectomy of the legs is greatest on the day after the operation. By the fifth day the blood flow has fallen considerably to remain two or three times greater than that on the normal limb for many years. This persistent vasodilatation is confined to the skin; there is no lasting effect on muscle flow. It should be emphasized, however, that the influence of sympathectomy on vascular tone is far from uniform throughout the body.

CHOLINERGIC NERVES

In 1921 Loewi showed that, during stimulation of the vagus, Ringer's solution perfusing a frog's heart acquired a substance which could inhibit another frog's heart. Later work showed that the substance released from the heart is acetylcholine and nerves acting in this way were described by Dale as *cholinergic*.

Stimulation of certain sympathetic fibres, for example sympathetic vasodilator nerves to the muscles of the dog and also the cat, liberates acetylcholine. These nerves, then,

$$CH_3 \quad OH^-$$
$$CH_3\!-\!\overset{+}{N}$$
$$CH_3 \qquad CH_2.CH_2O.\overset{\displaystyle O}{\overset{\|}{C}}.CH_3$$

Acetylcholine

come into Dale's category of cholinergic nerves. Stimulation of the stellate ganglion causes an acetylcholine-like substance to appear in the coronary perfusate; presumably the sympathetic vasodilator fibres in the coronary arteries liberate acetylcholine. The sympathetic nerves to the adrenal medulla liberate acetylcholine which in turn evokes a secretion of adrenaline and noradrenaline. This is not an exception to the rule that the sympathetic nerves are adrenergic, since the adrenal medulla, being derived in the embryo from tissue which forms the sympathetic ganglia, is homologous with a sympathetic ganglion in which the transmitter is always acetylcholine. Although the fibres to the sweat glands are in fact postganglionic sympathetic fibres they act in most animals, including man, by releasing acetylcholine, that is they are cholinergic. The release of acetylcholine at the endings of pre-ganglionic nerve fibres in the sympathetic ganglia (p. 817) and at myoneural junctions (p. 833) is referred to elsewhere.

It has been assumed for a long time that the passage of a nerve impulse along a sympathetic postganglionic fibre releases noradrenaline directly. However it has been shown that stimulation of sympathetic fibres releases acetylcholine (ACh) as well as noradrenaline. Also it is well known that in the adrenal medulla ACh liberates adrenaline and noradrenaline. Burn has produced experimental evidence to show that in adrenergic nerves ACh is first released and in turn, by increasing permeability to Ca ions, it releases noradrenaline. In other words, one transmitter is first released which in turn releases the main transmitter. This mechanism would explain the finding that anticholinesterases, by increasing the effective concentration of ACh, tend to increase the release of noradrenaline. If the synthesis of acetylcholine is blocked by hemicholinium, stimulation of sympathetic nerves may evoke no response.

Cholinergic nerves were first demonstrated in the frog. In mammals, however, it was not possible to demonstrate that certain nerves acted by producing acetylcholine at their endings until it was discovered that both blood and tissues contain cholinesterases, the action of which is blocked by physostigmine (eserine) which combines with the enzyme. After its release from the nerve endings acetylcholine is quickly hydrolysed by the esterase to choline and acetic acid and its action is, therefore, local and transient. Only after an animal is eserinized does the acetylcholine released from parasympathetic nerve endings enter the circulation in sufficient quantity to produce effects at a distance. A number of compounds with a powerful anticholinesterase action have been synthesized, chiefly for use in chemical warfare and as insecticides. One of the best known of these is di-isopropylfluorophosphonate (p. 103).

Extremely sensitive and fairly specific pharmacological tests have been found to detect minute amounts of acetylcholine in the fluid or blood which has perfused an organ. The eserinized frog rectus abdominis muscle, for example, can detect 1 part in 10^9 of acetylcholine. Acetylcholine is produced in the nerve endings and not in the tissue cells, since after denervation the acetylcholine content of an organ declines as the nerves degenerate and finally disappears. Acetylcholine is stored in the nerve in an inactive form, probably bound to protein, which is not affected by cholinesterase, and it is only after its release from the nerve-ending as free acetylcholine that hydrolysis by cholinesterase occurs. A cholinergic neurone contains acetylcholine and choline acetylase along the whole course of the axon as well as at its endings. Acetylcholine acts on the effector tissue by depolarizing its cells. Denervation of tissues supplied by cholinergic nerves sensitizes them to acetylcholine.

We are accustomed to think of events in the nervous system in terms of milliseconds and it is difficult at first to fit in this rapidity of action with humoral transmission in which acetylcholine must diffuse from, say, the preganglionic nerve endings to the ganglion cell. The distance between endings and ganglion cell is, however, so small that there is no difficulty in explaining the synaptic delay of only a few milliseconds. Furthermore, acetylcholine is so quickly destroyed by cholinesterase that the acetylcholine released by one nerve impulse is reduced to ineffective amounts before the next impulse arrives at the synapse.

A diagram of the distribution of adrenergic and cholinergic nerves is given in Fig. 41.4. Where a tissue receives both sympathetic and parasympathetic fibres it is often found that the nerve impulses arriving by one system have the opposite effects to nerve impulses arriving by the other. For example, vagal impulses (parasympathetic) cause slowing of the heart and sympathetic impulses cause acceleration. When the vagal effect is blocked by atropine and the sympathetic effect by a β-adrenergic-blocking drug the heart beats at its 'intrinsic' rate, determined by the fastest pace-making cells, usually in the sinu-atrial node. Since such inhibitory and excitatory differences are almost universally found in the body it must be assumed that there is normally some kind of balance between the two systems. In other words, the autonomic system is a homeostatic mechanism. If the sympathetic system is eliminated by removal of the sympathetic trunks this balance is upset and the animal survives only if it is sheltered from situations which put too great a strain on its means of adjustment.

The importance of the autonomic nervous system for homeostasis in man is illustrated by observations, especially those of Guttman and Whitteridge, on men who have survived in good health after transverse lesions of the spinal cord. Consideration of Fig. 41.1 shows that a lesion above the first thoracic segment (T1) cuts off the whole of the thoracolumbar autonomic outflow from higher control. Tilting such a patient suddenly into the upright position causes a fall of blood pressure, a rise in pulse rate and loss of consciousness because there is no compensatory vasoconstriction. Neither is such a patient, since he has lost control of skin vessels, able to produce vasodilatation or sweating to keep the body temperature steady; if he becomes very cold the muscles below the lesion cannot shiver because they are paralysed. (The facial muscles and the muscles innervated by the cervical nerves can shiver.) These disabilities obviously become smaller the lower the lesion in the spinal cord. When the urinary bladder of a paraplegic patient is distended the vessels of the skin of the legs and of the muscles constrict and in a high transverse lesion the systolic blood pressure may rise as high as 260 mm Hg and the diastolic as high as 140 mm Hg. In a warm environment the body temperature may rise steadily whereas in a cold one hypothermia may develop. The rise of blood pressure is accompanied by flushing of the face with sweating and headache. In low lesions the flushing is presumably due to reduction of outflow of vasoconstrictor impulses in sympathetic nerves still in connexion with higher centres; but flushing of the face occurs even with lesions above the sympathetic outflow, perhaps as an axon reflex. In lesions below T6 bladder distension produces as marked a vasoconstriction of the legs but in spite of this there is relatively little effect on the blood pressure. This shows that with lower lesions the compensatory baroreceptor mechanisms are effective. The vagal reflexes which cause slowing of the heart, when the blood pressure rises, are intact and this may in part account for the maintenance of homeostasis in low transverse lesions of the cord. Effects similar to those just described for bladder distension may be produced by distension of the rectum.

REFERENCES

BURGER, A. (Ed.) (1967). *Drugs Affecting the Peripheral Nervous System*, Vol. I. London: Edward Arnold.

BURN, J. H. (1966). Adrenergic transmission. *Pharmacological Reviews* **18**, 459–470.

BURN, J. H. (1971). Release of noradrenaline from sympathetic endings. *Nature, London* **231**, 237–240.

BURN, J. H. (1971). *The Autonomic Nervous System*, 4th edn. Oxford: Blackwell.

DAVISON, A. N. (1958). Physiological role of monoamine oxidase. *Physiological Reviews* **38**, 729–747.

EMMELIN, N. (1965). Action of transmitters on the responsiveness of effector cells. *Experienta* **21**, 57–65.

FERRY, C. B. (1966). Cholinergic link hypothesis in adrenergic neuroeffector transmission. *Physiological Reviews* **46**, 420–456.

GILLESPIE, J. A. & DOUGLAS, D. M. (1961). *Some Aspects of Obliterative Vascular Disease of the Lower Limb*. Edinburgh: Livingstone.

GREENE, N. M. (1962). Physiology of sympathetic denervation. *Annual Review of Medicine* **13**, 87–104.

IGGO, A. (1966). Physiology of visceral afferent systems. *Acta neurovegetativa* **28**, 121–134.

JOHNSON, R. H. & SPALDING, J. M. K. (1972). *Clinical Physiology of the Autonomic Nervous System*. Oxford: Blackwell.

SHEPHERD, J. T. (1963). *Physiology of the Circulation in Human Limbs in Health and Disease*. Philadelphia & London: W. B. Saunders.

SYMPOSIUM (1969). The autonomic nervous system. *Anesthesiology* **29**, 621–862.

SYMPOSIUM on *Catecholamines* (1959). *American Society for Pharmacology and Experimental Therapeutics*. Baltimore: Williams & Wilkins.

TRIGGLE, D. J. (1965). *Chemical Aspects of the Autonomic Nervous System*. London: Academic Press.

VANE, J. R. (Ed.) (1960). *Adrenergic mechanisms*. Ciba symposium. London: Churchill.

WEYER, E. M. (Ed.) (1967). New adrenergic blocking drugs: their pharmacological, biochemical and clinical actions. *Annals of the New York Academy of Sciences* **139**, 541–1009.

WEYER, E. M. (Ed.) (1967). Cholinergic mechanisms. *Annals of the New York Academy of Sciences* **144**, 383, 936.

WHELAN, R. F. & DE LA LANDE, I. S. (1963). Action of adrenaline on limb blood vessels. *British Medical Bulletin* **19**, 125–131.

WHELAN, R. F. & SKINNER, S. L. (1963). Autonomic transmitter mechanisms. *British Medical Bulletin* **19**, 120–124.

WHITTERIDGE, D. (1956). The effects of distension of viscera. *Lectures on the Scientific Basis of Medicine* **4**, 305–310.

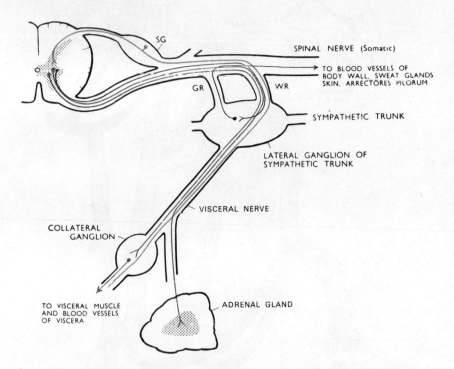

SG

SPINAL NERVE (Somatic)

TO BLOOD VESSELS OF
BODY WALL, SWEAT GLANDS
SKIN, ARRECTORES PILORUM

GR

WR

SYMPATHETIC TRUNK

LATERAL GANGLION OF
SYMPATHETIC TRUNK

VISCERAL NERVE

COLLATERAL
GANGLION

TO VISCERAL MUSCLE
AND BLOOD VESSELS
OF VISCERA

ADRENAL GLAND

Plate 41.1 Pathways of the sympathetic fibres from the lateral horn of the grey matter of the spinal cord. The fibres arising in the cells of the lateral horn, called preganglionic fibres, pass to a lateral ganglion or a collateral ganglion or to the adrenal medulla. These fibres may synapse with cells in these ganglia and these cells are continued as postganglionic fibres. The coeliac and hypogastric ganglia are examples of collateral ganglia. The fibre shown in blue represents afferent fibres from the viscera. The spinal nerve contains, in addition to postganglionic sympathetic fibres indicated in the diagram, afferent and efferent somatic fibres and is known as a mixed nerve. GR, grey ramus. WR, white ramus. SG, spinal ganglion.

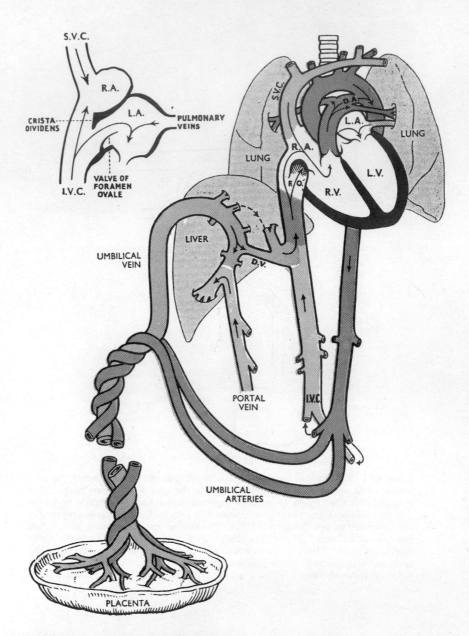

Plate 51.1 A plan of the fetal circulation. The upper end of the inferior vena cava opens directly into the left atrium through the foramen ovale (see inset) as well as into the right atrium. R.A. and R.V. = right atrium and ventricle. L.A. and L.V. = left atrium and ventricle. S.V.C. = superior vena cava. I.V.C. = inferior vena cava. D.A. = ductus arteriosus. D.V. = ductus venosus. F.O. = foramen ovale. (*By courtesy of G. S. Dawes.*)

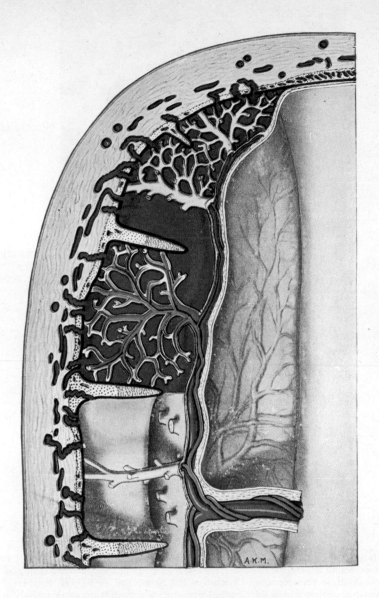

Plate 51.2 Scheme to show the essential features in placental structure which are found after the 60 mm stage. Three cotyledons, including a marginal one, are illustrated; the cotyledons are separated from each other on the maternal side by the septa. They each contain the group of villi which constitute the associated 'fetal' cotyledon. The villi branch freely and there are many adhesions between adjacent ones so giving a partially labyrinthine nature to the intervillous space. Such villi are shown in relief in the marginal cotyledon and in section in the adjacent one. The third cotyledon has been dissected to show the arrangement of the septa. The openings of the endometrial arteries and veins into the intervillous space through the basal plate are indicated and an attempt has been made to show the probable degree of oxygenation of the maternal blood in the intervillous space. A marginal sinus in the intervillous space has not been included in the scheme as our material does not show such a feature. About 70 uterine (spiral) arteries supply the mature placenta which has about 200 cotyledons of various sizes. (W. J. Hamilton & J. D. Boyd (1960). Development of the human placenta in the first three months of gestation. *Journal of Anatomy, London* **94**, 297–328.)

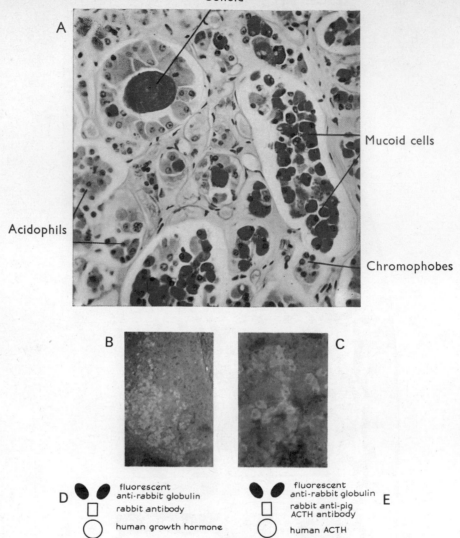

Colloid

A

Mucoid cells

Acidophils

Chromophobes

B   C

D
fluorescent
anti-rabbit globulin
rabbit antibody
human growth hormone

E
fluorescent
anti-rabbit globulin
rabbit anti-pig
ACTH antibody
human ACTH

Plate 52.1   A. Section of a 'normal' adenohypophysis from a woman of 74. It shows bright red mucoid (basophil) cells, orange-staining acidophils and faint pink or colourless chromophobes. Trichrome-P.A.S. method. × 500. B. Section of human adenohypophysis showing the localization of growth hormone exclusively in the acidophils (× 70). The hormone is stained bright yellow-green by the fluorescent antibody technique. The section was treated first with rabbit anti-human-growth-hormone antiserum and then, after washing, with fluorescein-conjugated sheep anti-rabbit-globulin antiserum. The antibody molecules become combined with the human hormone as shown in D. (The acidophil cells were identified by staining the next serial section by the method illustrated in A.) C. Section of human adenohypophysis showing the localization of ACTH in cells of the mucoid series (× 70). The hormone is stained bright yellow-green by the fluorescent antibody technique. These are the R-type cells of Adams & Pearse (1959). The section was treated with rabbit anti-pig-ACTH antiserum (which cross reacts with human ACTH) and then, after washing, with fluorescein-conjugated sheep anti-rabbit-globulin antiserum. The antibody molecules become combined with the human hormone as shown in E. (The mucoid cells were identified by staining the next serial section by the method illustrated in A.) (*By courtesy of A. G. Everson Pearse.*)

42 Temperature regulation

When calculated over a period of twenty-four hours the mechanical efficiency of the body (see p. 168) is relatively low. At least 95 per cent of the 13 MJ (3000 kcal) or so produced in the day is expended as heat, the remainder appearing as mechanical energy. This heat serves to maintain the temperature of the body but it must be got rid of as fast as it is generated: otherwise the temperature would rise. While the body temperature is remarkably constant in man, it is not so accurately regulated in other animals (Fig. 42.1). Those which are

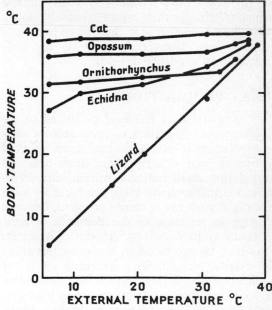

Fig. 42.1 Variation of body temperature of different species of animals after 2 hours in environmental temperatures of from 5°C to 35°C. This figure shows that there is a difference between the behaviour of homeothermic animals (like the cat), which can maintain a constant body temperature in spite of large changes in the external (ambient) temperature, and poikilothermic animals (like the lizard) where body temperature varies with that of the environment. Ornithorhynchus (platypus) and echidna are monotremes, phylogenetically primitive mammals. (C. J. Martin (1930). *Lancet* **ii,** 565.)

able to maintain a relatively constant body temperature in spite of great variations of external temperature are said to be *homeothermic*, whereas animals in which the body temperature varies with that of the environment are called *poikilothermic*, or cold-blooded. The former state obviously permits a more independent existence.

The relatively accurate control of body temperature is achieved by balancing heat loss against heat production or gain from outside. Since, however, heat is not produced uniformly throughout the body but more particularly in the muscles, in secreting glands and in the liver, whereas it is lost mainly through the skin and lungs, it is important that heat should be distributed rapidly. The body consists mainly of water which is a good conductor of heat and the continuous circulation and exchange of tissue fluids, blood and lymph serve to distribute heat evenly and rapidly throughout the body. Furthermore, the tissues of the body have a relatively high thermal capacity, that is they warm relatively slowly when heat is produced in them or on entry into a hot room. Conversely they do not cool instantly on exposure to cold.

HEAT PRODUCTION

A proportion of the heat produced in the body may be regarded as inevitable or *obligatory* since it arises as the result of vital activities such as respiration, heart-beat and circulation, maintenance of muscle tone, secretion and metabolism. Heat produced by such means, though not of course constant, does not change in response to the demands of temperature control. A large amount of heat can, however, be produced by the muscular effort of exercise voluntarily performed when the subject feels cold. Heat can also be produced by involuntary muscular activity when a subject is cold. This activity may be of two kinds. There is first an irregular but imperceptible activity of muscle units contracting out of phase with one another known as *thermal muscular tone*. Later the activity becomes regular and phasic and *shivering* occurs. Heat produced in these three ways may be regarded as *adjustable*. Shivering, the onset of which is not closely related to rectal temperature, can be initiated by afferent impulses from the surface

of the body, since, in a cold person, painful stimuli or a stream of cold air or water playing on the skin produces shivering within a few seconds. However, shivering has been produced in the intact fore-limbs of a dog, whose spinal cord had been transected, by the immersion of the paralysed and anaesthetic hind limbs in cold water. This experiment shows that, at any rate in the dog, there must be a central mechanism sensitive to a fall in temperature of the blood flowing through it. However, in the rabbit, whose spinal cord is cut across, either cooling of the isolated part of the cord or cooling of the limbs supplied by it causes shivering identical in rhythm with that seen when the central nervous system is intact. The rhythm of cold shivering seems to originate within the areas of the motoneurone pool. Heat cannot be produced by muscles paralysed by disease (Fig. 42.2), a relaxant

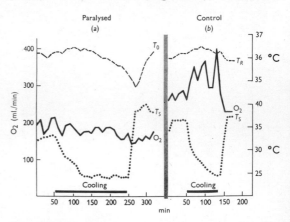

Fig. 42.2 Changes in total oxygen consumption (O_2) during cooling (ml/min, corrected to volume at $0°$); in oesophageal temperature (T_o); in rectal temperature (T_r); and skin temperature (T_s). (a) Paralysed patient. (b) Control subject. In the paralysed subject, virtually deprived of skeletal muscle, the oxygen consumption did not rise when he was cooled because he had no means of increasing his heat production. The skin temperature fell to a minimum of $17·6°C$ and the oesophageal temperature by as much as $1·8°C$. (R. H. Johnson & J. M. K. Spalding (1963). *Journal of Physiology* **166**, 24P.)

drug (like curare), or by an anaesthetic. The temperature of an anaesthetized patient thus tends to fall to that of the environment. For this reason surgical operating theatres are kept warm.

Heat of metabolism. The amount of heat produced by a man in a temperate climate depends on the kind and amount of food eaten and metabolized, variations in heat loss compensating for variations in the amount of heat produced at different levels of food intake. In very hot or very cold environments, however, there is a notable alteration in the appetite and in the amount of food consumed, the increased appetite during a cold day in winter being a familiar experience. Observations made during the 1939–45 war showed that the amount of food eaten by troops with access to as much food as they wanted varied with the local ambient temperature (Fig. 42.3) but the percentage of the total energy provided by protein, fat and carbohydrate varied very little with the climatic conditions; that derived from protein, about 12 per cent, was surprisingly constant.

may result in a decrease in the output of thyroid hormone presumably because of a reduction in the output of TSH (see Chap. 50). Low body temperature is often a feature of hypopituitarism (Chap. 52). However, in spite of these relationships between body temperature and the thyroid, adrenal and pituitary it is doubtful if the endocrine glands play an important role in the day to day control of body temperature. They may, however, play a part if the subject is moved quickly to a cold climate.

HEAT LOSS

Although the amount of heat produced in the body can be varied to some extent, it is

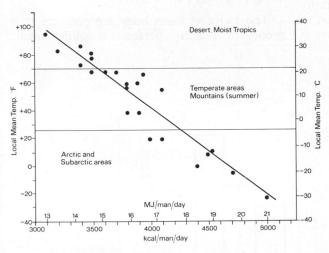

Fig. 42.3 Voluntary caloric intake, North American troops (averages for groups of 50 or more men). The diagram shows the relation between caloric intake and climatic conditions. The increased requirements in arctic conditions are partly caused by the impeding effect of heavy clothing (R. E. Johnston & R. M. Kark (1947). *Science* **105**, 378).

The metabolic rate can be increased considerably by an injection of adrenaline and it may be that sudden exposure to cold is followed by a discharge of adrenaline from the adrenal glands. This would account for the rise of blood pressure and constriction of the minute skin vessels which follows sudden exposure to cold. There is also evidence that exposure to cold increases the secretory activity of the adrenal cortex.

As might be expected increased activity of the thyroid gland, by increasing the metabolic rate, tends to raise the body temperature. Conversely diminished thyroid activity is associated with reduced metabolism and subnormal body temperatures. Exposure of animals to cold releases thyroid hormone into the circulation but sudden exposure to severe cold

mainly by variations in heat loss that the accurate control of body temperature is achieved. During ordinary life in a temperate climate our activities are such that it is more likely that the body becomes overheated than overcooled. With the onset of wintry conditions, however, the aged and infirm may suffer intensely from overcooling (p. 880), and may die of hypothermia.

Some 500 kJ are lost in warming the food and air taken into the body and another 1·5 MJ are consumed in evaporating water to saturate the air expired from the lungs. These 2 MJ represent, however, only about 16 per cent of the total of 12·5 MJ of heat which must be lost each day, leaving some 84 per cent to be lost through the skin.

The loss of heat by conduction, convection

and radiation is, of course, much influenced by environmental conditions such as ambient temperature, relative humidity and movement of air; it is greatest, for example, when the weather is cool and windy. It also depends on the state of the blood vessels in the subcutaneous tissues. Vasodilatation in these areas aids the loss of heat and vasoconstriction retards it. Under ordinary conditions, that is to say when sweating is not called into play, the greater part of the heat produced by the body is lost in this way. (Note that 1 kcal = 4·186 kJ.)

In the basal state there is a loss of body weight, due to evaporation from the lungs and to *insensible perspiration* of about 30 g per hour. This water passes through the skin by diffusion and osmosis, and not as the result of the activity of the sweat glands. Since the latent heat of vaporization of water at body temperature is

2·4 kJ/g (or 0·58 kcal/g) this water loss represents

$$30 \times 2\cdot4 \text{ kJ/h} = 72 \text{ kJ/h} = 1\cdot73 \text{ MJ/day}$$

or

$$30 \times 0\cdot58 \text{ kcal/h} = 17\cdot4 \text{ kcal/h} = 418 \text{ kcal/day}.$$

For an average man (see p. 169) the basal metabolism is 300 kJ/h or 72 kcal/h (7·2 MJ/day or 1728 kcal/day). Thus about one-quarter of the heat is lost by the skin and lungs even when no sweating occurs. It is usually reckoned that one-quarter of the heat loss occurs in this way whatever the level of activity.

BODY TEMPERATURE

The range of deep body temperature in a group of healthy persons is quite small.

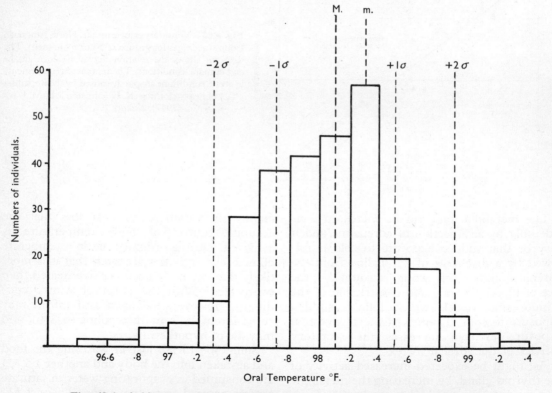

Fig. 42.4 A histogram showing the frequency distribution of oral temperature in 276 medical students. All temperatures were taken between 8 a.m. and 9 a.m. with the subjects seated in a warm class-room. M (arithmetic mean) = 36·7°C, 98·1°F; m (mode of the series) = 36·8°C, 98·3°F; Y (standard deviation) = 0·22°C, 0·4°F. The range was from 35·8°C, 96·4°F to 37·3°C, 99·2°F. Data from A. C. Ivy (1944). *Quarterly Bulletin of the Northwest University Medical School* **18**, 22; taken with modifications from Du Bois (1949). *Fever and the Regulation of Body Temperature*. Springfield, Ill.: Thomas.

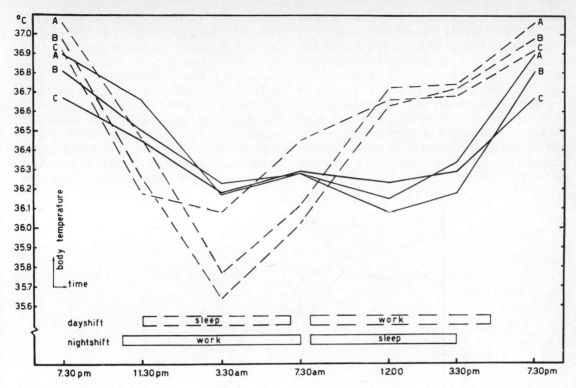

Fig. 42.5 Average body temperature curves of three subjects (A, B and C) in dayshift (– – – –) and nightshift (———). (From J. H. van Loon (1963). *Ergonomics* **6**, 267–273.)

Indeed the coefficient of variation of body temperature in man is one of the smallest for which quantitative data are available (Fig. 42.4).

If we take the temperature in the mouth of a healthy man at intervals during the day we find it varies from, for example, 35·8°C (96·5°F) to 37·3°C (99·2°F), being highest in the evening and lowest in the early morning; this pattern, largely unrelated to environmental conditions (Fig. 42.5), is an example of a circadian rhythm (see Fig. 32.13). Exercise, because it increases production of heat, may raise the oral temperature 1° or 2°C and violent exercise can produce temperatures of 39°C (102°F). In the female there are variations in body temperature during the menstrual cycle (Chap. 51); injection of certain β-steroids (pregnenolone and pregnanediol) causes a rise of body temperature. Temperatures in general tend to be higher in children and lower in the elderly and in neonates.

In temperate climates the body temperature is nearly always higher than the environmental temperature so that there is a continuous loss of heat through the skin, the temperature of which may fall as low as 17°C if the air is cold. There is thus a considerable temperature gradient in the superficial half-inch or so of tissue made up chiefly of subcutaneous fat and skin (Table 42.6). In other words the human body is insulated to a certain extent from its

Table 42.6 Temperatures of different tissues of the forearm determined electrically as rapidly as possible after removing clothes.

	Average temperature (°C)
Room	18·5
Skin	33·0
Subcutaneous tissue	33·6
Deep muscle	36·2
Mouth	36·9

H. Barcroft & O. Edholm (1946). *Journal of Physiology* **104**, 366.

environment. Many animals, especially aquatic animals living in cold regions, are insulated by a thick layer of fat in addition to a coat of fur. The thick layer of insulation provided by the superficial fat in obese persons may so limit heat loss that on a hot day they may suffer much more discomfort than thin persons.

The temperature of the skin of the body is by no means uniform. The temperature on the surface of a limb exposed to an environmental temperature of 18°C is higher in the proximal part of the limb than in the distal (Fig. 42.7).

then, varies to some extent according to its metabolic activity, to the rate of blood flow through it, and to the heat gradient between it and surrounding parts. The temperature of the surface of the liver in dogs can be as much as 1·0°C higher than the temperature of the arterial blood, and blood in the right side of the heart has been shown to be some 0·2°C warmer than blood in the left. In fact 'the conception that a man has an internal mass of tissue maintained at 37°C (98·6°F) surrounded by a thin layer with a steep temperature

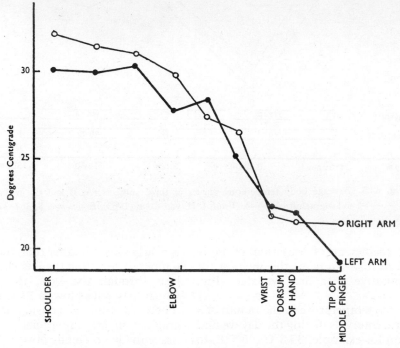

Fig. 42.7 Readings of skin temperature on the two arms of a healthy subject 30 minutes after the limbs had been exposed to an ambient temperature of 18°C.

However, the pattern of this *acral temperature gradient* depends very much on circumstances since the surface temperature of the distal tissue is much influenced by the temperature of the environment and by the degree of vasodilatation in the tissue.

Since most of the heat produced in the body is the result of oxidations, the main sources of heat are the most active tissues—the liver, secreting glands, and the muscles which together make up more than half the body weight. The temperature of a single organ,

gradient, such that the surface temperature is about 33°C (91·4°F) is common but inaccurate' (Bazett). Temperatures taken simultaneously in the mouth, axilla, and rectum usually differ by less than 1°C. Because the temperature recorded in the rectum is usually highest it has generally been assumed to be the best index of the central body temperature, say the temperature of the blood in the aorta. However, the rectal temperature may exceed the aortic by as much as 0·3°C; this suggests that heat may actually be produced in the rectum by bacterial action. On the other hand the temperature in the rectum may be lower than arterial temperature as the result of the arrival of cool blood from the legs. The temperature in the axilla is almost invariably somewhat lower (usually 1°F, 0·56°C) than that in the mouth but it is difficult to measure accurately with a clinical thermometer. It has been shown that the rectal temperature may vary only very little as the blood is rapidly warmed or cooled, and the same almost certainly applies to the temperature as recorded in the axilla. Mouth temperature, on the other hand, may show rapid alterations, the result, for example, of mouth breathing or of drinking warm or cold fluids (Fig. 42.8). Benzinger has shown that the temperature of the tympanic membrane gives the most meaningful correlations with heat losing activities such as vasodilatation and sweating. The ear drum and the hypothalamus share a common blood supply from the internal carotid artery.

Even a small rise of temperature produces an increase in heart rate, and often a feeling of discomfort and fatigue (malaise). In hot environments the efficiency of mental activity, as tested, for example, by accuracy of reception of Morse code, is reduced. This may be due to the difficulty of sleeping in hot climates. Body temperatures above 41·2°C (106°F) (*hyperpyrexia*) are accompanied by tachycardia, raised respiratory rate, weakness, headache, mental confusion, abnormal behaviour and finally loss of consciousness. A persistent temperature of over 43°C (109°F) is not compatible with life.

Spontaneous hypothermia. Normal persons exposed to cold without adequate protection begin to shiver as the body temperature falls; the heart rate increases, the skin vessels constrict and the blood pressure rises. These compensating reactions fail if the body temperature falls below 32°C (90°F) and shivering is replaced by muscular rigidity; consciousness

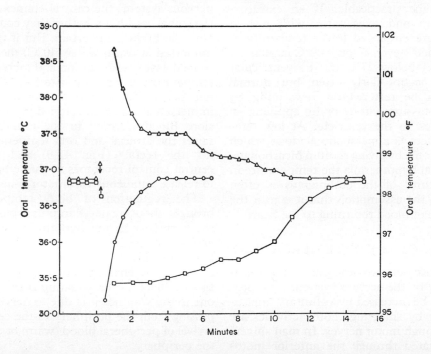

Fig. 42.8   Oral temperatures of the same subject taken with a standard 'half minute' mercury-in-glass clinical thermometer every 30 seconds in 'normal' conditions (O), showing that 3 to 4 minutes are required to obtain an accurate reading. Also shown are the effects on oral temperature of drinking 350 ml of hot tea (△) at 50°C (123°F) and 350 ml cool water (□) at 10°C (50°F); in both cases oral temperature took at least 10 minutes to approach the original level. (C. S. Leithead & A. R. Lind (1964). *Heat Stress and Heat Disorders*. London: Cassells.)

is impaired and reflexes are sluggish. Death usually occurs about 26°C (77°F). Immersion in cold water or the cooling effect of cold wind blowing through wet clothing accounts for many such deaths in young people in temperate climates. The temperature-regulating mechanism of old people is often defective so many elderly people die each winter from accidental hypothermia.

Artificial hypothermia. If shivering is abolished either by a general anaesthetic, or by a curare-like drug that paralyses the muscles, the body temperature can be reduced to 27°C (80°F) for example by packing the body about with ice. At this temperature the oxidation of glucose ceases and metabolism is reduced by some 30 to 40 per cent. The cardiac and respiratory rates are reduced, arterial pressure is low and the patient passes into a state of suspended animation from which he can completely recover if he is allowed to warm up slowly. In such a state the circulation can be interrupted for 10 to 15 minutes so that surgical operations on the heart and large blood vessels become practicable. If an extracorporeal pump and oxygenator with a heat exchanger are employed lower temperatures can be reached—even as low as 9°C in man. At temperatures below 27°C (80°F) ventricular fibrillation occurs fairly often but normal rhythm can be restored on rewarming by injecting potassium citrate or by applying an electric shock to the ventricle. At low temperatures there is a metabolic acidosis which can be overcome by giving sodium bicarbonate.

In artificial hypothermia the temperature in the lower third of the oesophagus is often recorded on the assumption that it is near the temperature of blood returning to the heart.

CONTROL OF BODY TEMPERATURE

The balance between heat gain and heat loss is controlled by the nervous system. Heat production can be increased by voluntary muscular effort or by shivering; both activities are effected through motor nerves. In man shivering is mediated through the anterior motoneurones, activated not by the pyramidal tract but by the lateral tectospinal or rubrospinal tracts. Heat loss can be altered by varying the amount of blood flowing through the vessels of the skin or it can be increased by sweating.

These activities are controlled through the sympathetic nervous system. Small adjustments of body temperature are made by altering the skin blood-flow; larger adjustments by shivering or sweating. In homeothermic animals other than man panting is a method of increasing heat loss and pilo-erection a method of conserving heat.

The usual class experiments on temperature regulation suggest that its control is complex and dependent partly on reflexes originating in the cutaneous receptors and partly on a central mechanism actuated by the arrival of blood from the periphery. If one hand is immersed in cold water the temperature of the other soon begins to drop, even if the circulation to the hand immersed had first been occluded by a cuff. The effect must be due to reflex vasoconstriction. Extensive vasoconstriction may cut down the total heat loss to such an extent that the oral temperature may actually rise a little. The afferent pathways for this reflex are in the cutaneous nerves and the efferent pathways are in the sympathetic nervous system; the central connexions are in the spinal cord but whether they communicate with the brain is uncertain. But if the hand is immersed in warm water (40°C) there is, after some delay, vasodilatation and increase in skin temperature in the opposite hand. This does not occur if the circulation to the hand immersed in water is occluded prior to immersion. Blood warmed in the immersed hand raises the arterial and oral temperatures (but not the rectal) (Fig. 42.9) and stimulates central control receptors in the hypothalamus to release cutaneous vasomotor tone.

The regulation of body temperature is brought about by nerve impulses passing from the periphery to the hypothalamus or to the arrival of peripheral blood at the hypothalamus. The observation of a response to a change in the environmental temperature such as vasodilatation or sweating does not enable one to say whether it is due to nerve impulses passing from the periphery to the centre or to arrival of peripheral blood (warm or cold) from the periphery.

The existence of heat regulating mechanisms in the hypothalamus has long been known. But even after transection of the spinal cord animals have some heat regulatory capacity. For example Thauer found that local heating

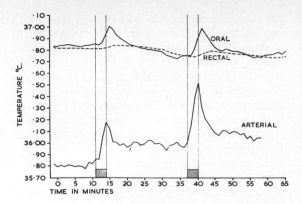

Fig. 42.9 Temperature measured in the mouth, rectum and contra-lateral subclavian artery in a normal subject during immersion of one forearm in a bath of warm water (shaded areas indicate duration of immersion). (From W. I. Cranston (1966). *British Medical Journal* **ii**, 69–75.)

of the cervical to sacral cord could raise the metabolic rate of dogs by as much as 50 per cent.

Since the body consists largely of water which has a great heat capacity, sudden exposure of even the nude body to heat or cold does not result in an immediate change of temperature or any apparent immediate response other than the subjective sensation of changed external temperature.

Attempts to discriminate between reflex control and central regulation operated by change in blood temperature have generally depended on stimulation of the skin of a particular region, usually a limb, before and after occlusion of the circulation. Brebner and Kerslake, using the human forearm as the test region, evoked sweating there by immersing the subject in a bath kept at mouth temperature; the subject's head was placed in a container with air saturated at this temperature. The body temperature gradually rose due to metabolism and the sweat rate rose proportionately. If the legs were exposed to the air and cooled the forearm sweat rate decreased, but not if the circulation to the legs had previously been arrested. Evidently the movement of cooled blood was an essential agent in inhibiting the action of the centre regulating vasodilatation and sweating but the precise role of the leg thermoreceptors could not be established beyond doubt. Benzinger tried to circumvent the objection that occlusion of the circulation is an abnormal circumstance. He applied graded radiant heat to the trunk and found that there was a graded sweat response in the forearm. Though the calibration of the recording apparatus was difficult it appeared that the sweat response lagged only 3·5 sec behind the change in heating. This is too short a time for movement of blood from trunk to forearm and suggests that the pathway was entirely neural but the method gave no evidence regarding the position of the central control.

Much work has been done on the central hypothalamic centres. Even in man a conflict between the central and peripheral receptors can be brought to light. If a subject is exposed to a rather warm environment his skin becomes warm and he starts to sweat. As time goes on the evaporation of sweat cools down his skin and must reduce or abolish afferent impulses from his cutaneous warmth receptors; yet he goes on sweating. This must be due to the continued stimulation of his central warmth receptors. Heating of even a restricted area of the body by radiant heat will soon raise the sweat rate but lower the skin temperature. On the other hand lowering the central temperature by eating ice-cream reduces the sweat rate and so raises the skin temperature. The central warmth receptors must now have become inactive despite the continued heating of the skin.

Sweating and central body temperature (as recorded near the tympanum) move concordantly. The central temperature is the determining factor, not the skin temperature, so far as sweating is concerned.

If a subject emerges from a room hot enough to have caused sweating and enters a very cold room, stimulation of his cutaneous cold receptors quickly stops the sweating, even if he shivers violently enough to keep his central temperature at levels which would otherwise have produced sweating. Rawson and Hardy whilst observing the reactions of paraplegic subjects noted that such an inhibitory effect from peripheral receptors could not be obtained if the subject's legs were suddenly cooled after prior heating had caused trunk sweating. The inhibitory effect observed in normal subjects is presumably due to reflex action initiated by the cold receptors of the skin rather than to a direct local effect on the sweat glands themselves.

Temperature Regulation 881

Benzinger maintains that the hypothalamic cold receptor centre in man is too weak to play a major role under ordinary circumstances. It must, however, be important in fur-covered animals in which the cutaneous receptors are well sheltered. He regards the anterior hypothalamic centre as the main thermostat control, the posterior cold receptors being merely relay stations where impulses from peripheral cold receptors are co-ordinated with inhibitory impulses from the central warmth receptor region.

But the central cold receptors can undoubtedly be brought into play in special circumstances. Downey, Chiodi and Darling studied seven patients with cervical cord section, keeping their sentient skin areas warm in a room at 29 to 35°C by covering them with warm clothing. In one set of experiments they cooled the insentient part of the body by immersion in a bath at 6°C. Shivering and increased oxygen uptake occurred when the ear drum temperature fell to about 35·6°C, even when the sentient part was kept warm; this would of course be a relatively small proportion of the total body surface. In another set of experiments they let the rest of the body also cool by exposing it to air at 19 to 24°C. The ear temperature fell about 0·7°C in an hour and thereafter at a slower rate. The skin temperature fell considerably, of course, and the oxygen uptake rose after the ear temperature had reached about 35·5°C, increasing thereafter as the temperature fell further. In these patients at any rate central cold receptors could bring about effects independently of impulses from cutaneous receptors.

If vasoconstriction and piloerection do not conserve enough heat to keep the temperature up, the metabolic rate increases (thermogenesis) and increases heat production. Any method of increasing heat production suffices but there is likely to be an orderly sequence: increased tone in skeletal muscles, shivering, hormonal activity, supplemented by voluntary action such as exercise, procuring extra clothing and shelter. Reckless movements of arms and legs generating large convection currents may lose rather than gain heat.

Hemingway showed that shivering induced by cooling the skin in dogs could not be abolished by warming the anterior hypothalamus with a thermode whereas warming the posterior hypothalamus did. In later work he recorded action potentials from fibres descending from the anterior to the posterior centre whilst stimulating the former electrically. He concluded that this showed the existence of inhibitory impulses passing from anterior to posterior hypothalamus when the former was warmed. He seems to ascribe a stronger role to the central cold receptors than Benzinger would allow.

In man shivering can raise the metabolic rate only 2 to 5 times. In a very cold environment unless he takes voluntary steps to insulate himself against further heat loss or to warm up his surroundings he may die from hypothermia.

The hypothalamus contains relatively high concentrations of 5-hydroxytryptamine (5HT) and noradrenaline. When injected into the anterior hypothalamus or into the ventral half of the third ventricle of the unanaesthetized cat 5HT causes a rise of body temperature and shivering, whereas adrenaline and noradrenaline cause a fall in body temperature and abolish shivering. It may therefore be that these substances act as transmitters. However, not all laboratory animals react in this way; perhaps they use different monoamines. The effect of injection of the catecholamines is observable within 2 minutes; injection of bacterial pyrogens into the anterior hypothalamus produces a rise of temperature only after quarter of an hour. Perfusion of the cat's cerebral ventricles with artificial cerebrospinal fluid containing excess calcium lowers body temperature and counteracts the hyperthermic effect of leucocyte pyrogen (the so called calcium brake). The latest candidate for the role of transmitter is prostaglandin E_1; it produces an obvious rise of temperature when injected in doses of 100 ng or less into the lateral ventricle of both cats and rabbits. When the temperature of the blood reaching the brain is raised certain neurones in the hypothalamus respond, usually with an increase in frequency of discharge. Vasodilatation occurs as a result of a reduction of efferent impulses in the sympathetic outflow.

Myers (1968) using the conscious rhesus monkey employed an ingenious crossed circulation technique to transfer C.S.F. or hypothalamic perfusate from a donor which could be heated or cooled to a recipient. On cooling the donor the recipient shivered; on heating the donor the recipient either showed little change or else a sharp drop in temperature. Clearly something had been transferred, though the transmitter need not have been 5-HT or noradrenaline; these substances on infusion might simply mimic its action.

Giarman and his colleagues induced fever in rabbits by intravenous injection of leucocyte pyrogen and then estimated the concentration of the amines in the appropriate regions of the hypothalamus. They also selectively reduced the brain content of 5-HT and noradrenaline before inducing fever. Though, as they admit, their results were not very striking it was obvious that there were changes in the amine stores of the hypothalamus in fever. When the store of 5-HT was low the rabbit had difficulty in cooling itself; when the store of noradrenaline was low the heat of the fever could not be maintained.

The role of 5-HT as a transmitter in the temperature regulating region has not been completely established but there is considerable evidence in favour of the idea that it is involved. The enzymes necessary for synthesis and decomposition are present and it can be stored in the neurones of the regulatory regions.

Removal of the cerebral cortex in the dog does not seriously impair temperature regulation and in man, too, control of body temperature is believed to be independent of the cortex. However, in both the dog and man emotional stress can cause a rise of body temperature. This may account for the finding that many patients when first admitted to hospital have a raised temperature.

In cases of haemorrhage into the mid-brain or pons the balance between heat production and heat loss is seriously disturbed and the body temperature may rise as high as 105°F, 40·6°C (*hyperpyrexia*). On the other hand it may become very low and the patient may behave like a poikilothermic animal. If he is covered with blankets and surrounded by hot-water bottles his body temperature quickly rises and may reach a dangerous level (Fig. 42.10). Such disturbances are probably the result of interruption of the pathway from the hypothalamus to the spinal cord.

SWEATING

In hot climates, when atmospheric temperatures exceed body temperature, heat loss must continue, although now radiation and convection actually supply heat to the body. The mere production of sweat, even in large quantities, does not get rid of heat. To accomplish this the sweat must be evaporated; every gram of water evaporated requires the expenditure of 2·42 kJ (0·58 kcal). The removal of 12·6 MJ (3000 kcal) per day, for example, requires the evaporation of $12\cdot6 \div 2\cdot42 = 5\cdot2$ kg of water per day. Evaporation of sweat occurs readily in a dry atmosphere but slowly or not at all in a humid one. Thus very hot humid environments are incompatible with life.

In very hot climates the body is heated by the solar radiation which in a tropical desert near noon may amount to 54 kJ/m^2/min or 13 kcal/m^2/min. After allowing for the area exposed and for reflexion, the heat absorbed by the body in these circumstances is about 17 kJ/min or 1 MJ/hr, which is more than twice the resting metabolism. A man marching in the desert at 4·8 km/h (3 m.p.h) produces, say 1·13 MJ/h. The total heat to be lost is therefore 2·13 MJ per hour which requires the evaporation of 880 g of water per hour. In such extreme circumstances a man may lose up to 10 litres per day of sweat containing as much as 30 g of salt. If water and salt are not taken at frequent intervals the blood becomes concentrated and the circulation may fail.

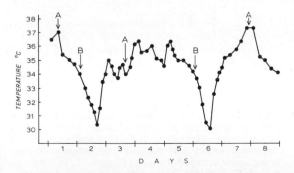

Fig. 42.10 Record of axillary temperatures from a female patient aged 68 unconscious for 3 months as the result of thrombosis of blood vessels at the base of the brain. At the arrows marked B extra blankets were applied; at A they were removed. The body temperature depended on external influences because the central thermo-regulatory mechanism had been destroyed.

Sweat is a slightly acid, watery fluid of variable composition but always hypotonic with respect to blood plasma. It contains only about 0·5 per cent of solids, mainly sodium chloride. The range of sodium chloride content is quite large; in young adults it is about 33 m-equiv./l (0·19 per cent NaCl); in persons over 50 years it averages 55 m-equiv./l (0·32 per cent NaCl). The dark cells (p. 713) are probably responsible for secretion of sodium chloride. The salt content is raised (perhaps two or three times) on a high salt diet; it is lowered by restriction of intake and by habitually working in a hot environment. Sweat usually contains about 0·03 per cent of urea, a concentration similar to that in blood, and also about 20 m-equiv./l of lactate produced by clear cells (p. 712). As the rate of secretion of sweat increases, the concentration of nitrogen in it falls, but the concentration of sodium may rise to 100 m-equiv./l because the sweat passes along the duct in too short a time to allow much reabsorption of sodium by the duct cells; in other words the resorptive capacity of the duct cells is limited. Micropuncture of the human sweat gland shows that sweat obtained from the coiled portion is nearly isotonic with body fluids; sweat obtained from the duct has a much smaller sodium content (see p. 712). The sodium content of sweat is also affected by aldosterone. This hormone acts relatively slowly on sweat glands; it increases the reabsorption of NaCl and therefore reduces the salt content of the sweat. The iron of sweat is mainly in the desquamated cells; cell-rich sweat contains 1·5 mg per litre while cell-free sweat contains 0·4 mg per litre. Sweat contains also 5 to 10 mg calcium per 100 ml and when sweating is profuse significant amounts may be lost. If the circulation to a limb is cut off by a tourniquet sweating is unaffected for several minutes before it decreases gradually to zero in half an hour or so. Therefore sweat must be produced by the activity of the cells and not by filtration.

The activity of the sweat glands may be studied by enclosing a limb in a plastic bag and collecting the sweat. The sweat from the forehead can be collected on filter paper enclosed in a suitably shaped chamber.

About 50 per cent of the total sweat produced on a hot day comes from the skin of the trunk, about 25 per cent from the lower limbs, and the remaining 25 per cent from the head and upper limbs. When sweating becomes profuse, so as to give rise to visible drops, it usually does so quite suddenly and at the same time all over the body. Thermal sweating is reduced if the person is short of fluid (dehydrated) and this may be the cause of 'dehydration fever'.

There are in fact two quite different types of sweating. After some time in a warm environment sweating begins almost simultaneously on all skin areas; it is, however, most noticeable on the forehead, upper lip, neck, chest and trunk, but to a much less extent of the palms of the hands or the soles of the feet. On the other hand, mental, emotional or sensory stimuli produce sweating mainly confined to the palms, soles and axillae; sweating in these areas should not be regarded as thermoregulatory. Both types of sweating are under nervous control, the temperature-regulating system from centres in the anterior part of the hypothalamus and the emotional system from centres probably in the cortex (Fig. 42.11).

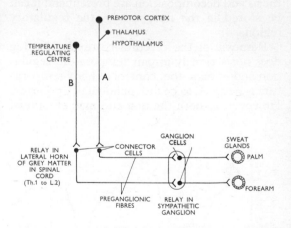

Fig. 42.11 Nervous pathways of control of sweating. A, pathway from cerebral cortex to sweat glands of palm. These are representative of the glands which take part in 'emotional' sweating. B, pathway from temperature regulating centre in hypothalamus to forearm (T. M. Chalmers & C. A. Keele (1952). The nervous and chemical control of sweating. *British Journal of Dermatology* **64**, 43–54).

Sweating usually begins when the skin temperature reaches 34·5°C and the rate of sweating is correlated with skin temperature rather than with deep body temperature.

Although the sweat glands are supplied by post-ganglionic sympathetic fibres the humoral effector substance is acetylcholine. After an injection of atropine sweating is much reduced. This can often be noticed in a hot operating theatre if the patient has had an injection of atropine before the operation to diminish bronchial secretion. The intravenous injection of adrenaline in man has no effect on sweating but the intra-arterial injection of adrenaline in relatively high dosage does produce sweating blocked by phentolamine but not by atropine.

Fox and Hilton have found a bradykinin-forming enzyme in sweat just as in saliva (p. 225). Presumably the enzyme released from the sweat glands acts on tissue-space proteins to produce bradykinin which is responsible for the accompanying vasodilatation.

Heat cramps. Cramping pains in the muscles, often those muscles in most active use, sometimes occur in men, for example stokers, who work in hot places. 'Stokers' cramp' occurs when the loss of large amounts of sodium-containing fluid in the sweat is made up by the drinking of quantities of water, or other fluid containing little or no sodium. The pains are probably due in some way to lowered sodium concentration; they are found in acute sodium depletion from other causes such as diabetic coma (p. 330) or Addison's disease (Chap. 50). They may be rapidly relieved by taking salt and can be prevented by drinking weak salt solution (0·25 per cent sodium chloride) instead of water. In Great Britain 8 to 12 g of salt are normally consumed each day and this is many times the minimum quantity needed to maintain metabolic balance; but in tropical heat it is wise to take more salt. Hard work in a hot climate may require the intake to be increased by an extra 7 to 21 g.

Heatstroke, heat hyperpyrexia, heat exhaustion. Sometimes after exposure to extreme temperatures as in the case of men working in hot, humid mines or in strong sunlight in the tropics, men may suddenly and quite unexpectedly collapse, become unconscious and die within a short time with body temperatures of over 41°C (106°F). Exposure to heat increases cardiac output and cutaneous circulation with often a rise of venous pressure. It seems likely, but by no means proved, that cessation of sweating and circulatory failure are the causes of the fatal hyperpyrexia. The symptoms are apparently due to a combination of salt loss, dehydration and hyperpyrexia with breakdown of the heat regulating mechanism and circulatory failure.

TEMPERATURE REGULATION IN THE NEWBORN

In newborn babies, especially if they are premature, the rectal temperature is normally about 1·1°C below the adult value which is usually not reached until some 10 to 14 days after birth. Thus, in the neonatal period, a rectal temperature of say 37·2 to 37·5°C is actually equivalent to a temperature of 38·3°C in an adult.

The body temperature of the newborn of all homeothermic animals is very unstable for at least several days; the premature human infant may have to be kept in an incubator for sometime (see however, p. 638). The two main reasons for this are that the minimum oxygen consumption of the newborn is relatively high, namely 4·6 ml/kg/min and that they are unable to regulate their heat loss efficiently. When the environmental temperature falls they are, however, able to increase their oxygen consumption several times without shivering (p. 334). Even so, the body temperature may fall (instead of rising) because of thin skin and poor insulation by subcutaneous fat.

The special site of heat production in the newborn is the *brown adipose tissue* (multilocular fat) lying deep to the (white) subcutaneous fat. The brown fat cells of rabbits can oxidize, for example, succinic acid twenty times faster than can white fat cells. The brown adipose tissue (Fig. 42.12) surrounding the shoulder-girdle of the human infant has a particularly rich venous drainage which passes round the spinal cord before entering the main veins to the heart. The metabolism of this specialized tissue seems to be governed by noradrenaline released from local nerve endings. A detailed description of brown fat is given on page 334.

The insulating property of white fat can be compared to that of a blanket whereas the brown fat can be regarded as thermogenic, like an electric blanket. Brown fat only slowly releases its fat in starvation.

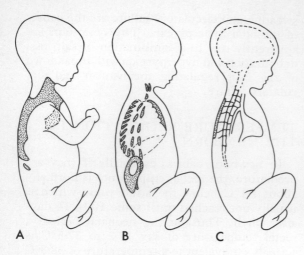

Fig. 42.12 A and B, distribution of brown adipose tissue in the baby. C, diagrammatic representation of the venous drainage from the interscapular pad. (From W. Aherne & D. Hull (1964). *Proceedings of the Royal Society of Medicine* **57**, 1172–1173.)

FEVER (PYREXIA)

Fever is produced if heat is applied to the body, or if heat is generated in the body, at greater rates than can be balanced by heat loss; hard exercise and hyperthyroidism (Chap. 50)

are examples of the latter. Certain steroids (especially aetiocholanolone) formed during the metabolism of adrenocortical and sex hormones are pyrogenic. Infection is a familiar cause of pyrexia but the thermo-regulatory system is readily upset by disease and many non-infective disorders run a febrile course.

Fever can be produced experimentally by the injection of a *pyrogen*, such as a vaccine of killed typhoid bacilli (bacterial pyrogen) (Fig. 42.13). About an hour after the injection the patient begins to feel uncomfortable; the skin becomes pale and cold due to constriction of the superficial blood vessels. There is 'goose-flesh' due to pilomotor activity (p. 712). Both effects are the result of sympathetic stimulation. The subject feels cold, shivers and may actually have severe clonic movements (rigor); the body temperature rises. About an hour later the subject begins to feel more comfortable, the skin is warm and dry and the subject a little flushed; heat production and heat loss are both increased so that the body temperature remains high. This stage is not well displayed in experimental pyrexia. When this phase is prolonged in disease circadian variations in body temperature are often similar to, but often larger than, those seen in

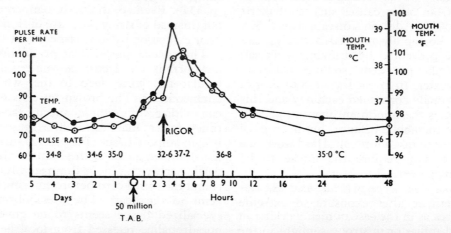

Fig. 42.13 Chart of mouth temperature and pulse rate in a male patient aged 37 who received an intravenous injection of typhoid vaccine (T.A.B.) containing 50 million organisms. An acute febrile reaction developed three hours later. The figures in the lower part of the chart indicate the temperature of the skin of the forearm measured electrically. At three hours after the injection the patient vomited and complained of pain in the back and legs and headache. He was pale and cold (skin temperature 32.6°C). Fifteen minutes later he developed a rigor. His mouth temperature increased rapidly and his skin became flushed and hot (37.2°C). Nine hours after the injection the patient felt much better. He was sweating profusely, his mouth temperature had fallen, but his skin temperature remained high.

healthy persons. During fever, in fact, temperature regulation is quite precise; apparently the 'thermostat' is working efficiently, but it is set at a higher level. In the third stage of pyrexia, the patient again feels discomfort. He is hot, flushed and sweating; his body temperature falls because the heat loss is greatly increased.

The evidence now seems strong that during the hour or so between the injection of bacterial pyrogen and the subsequent fever, a reaction takes place between a lipopolysaccharide from the bacteria and the leucocytes of the host as the result of which a polypeptide *endogenous (leucocyte) pyrogen*, is released. This does not dialyse, is precipitated by saturation with ammonium sulphate and destroyed by being heated at 70°C for 30 minutes. It produces fever rapidly, probably by a direct action on the anterior hypothalamus and preoptic areas; it causes fever when injected into the lateral cerebral ventricle of rabbits. Leucocyte pyrogen cannot be detected in the blood of persons with fever possibly because the concentration is too small (Fig. 42.14).

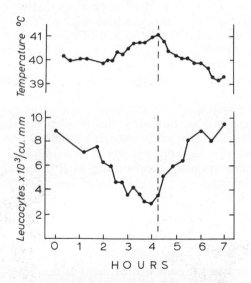

Fig. 42.14 Record of body temperature (in axilla) of a patient with relapsing fever. It is probable that as the infecting organisms (a spirochaete) and the leucocytes disappear from the blood, bacterial (endogenous) pyrogen is produced which raises the body temperature. The broken vertical line marks the disappearance of spirochaetes. (From T. P. C. Schofield, J. M. Talbot, A. D. M. Bryceson & E. H. O. Parry (1968). *Lancet* **i**, 58–62.)

CLOTHING

We are all accustomed to choosing our clothing according to the prevailing climatic and weather conditions. We can cope reasonably well with hot conditions by wearing light clothes, by remaining inactive and taking cooling drinks, but we are quite unable to produce enough heat to keep warm in severe cold. Certain combinations of central and cutaneous temperatures may also prevent an adequate rise in metabolic rate to maintain normal body temperature, as Benzinger has pointed out, since we cannot reflexly raise our metabolic rate by more than three or four times the basal value.

When the air is still, the layer of warm air round the body acts as an insulator because air is a poor conductor of heat. If wind plays on the naked body this layer of air is continuously being replaced and heat is lost by convection. In cold weather we diminish this loss by putting on clothes, the amount being adjusted by experience so that the regulation of body temperature is brought within the range of our physiological mechanisms. The thermal insulation provided by clothes depends on the air imprisoned in the meshes; the thermal insulation of an air space increases with its thickness up to a maximum of about half an inch, but if the space is loosely filled with textile material to prevent convection currents the insulation continues to be proportional to the thickness. For textiles of low bulk-density the insulation provided is the same whether the fabric is wool, cotton, kapok, or any other material. The essential insulator, then, is the air, the particular textile employed to imprison the air making practically no difference to its efficiency as an insulator. Clothes are warm when they entangle dry air in their fibres since dry air is a poor conductor of heat but they lose much of the insulating value if they become damp since damp air is a relatively good conductor. Clothing therefore should assist in maintaining a layer of warm still air close to the body but it must not hinder the escape of moisture. Mountaineers have also to contend with both cold and wind; the strength of the wind may be the dominating factor. It is essential to keep it out with leather or closely woven cloth but such garments hinder the escape of sweat.

Figures for thermal conductivity (in British units) show the necessity for keeping the above points in mind: water, 4·16; air at 22°, 0·17; woollen cloth, 0·28; cotton cloth, 0·49.

It is convenient to have a rating of the insulating value of clothing. The unit chosen is such that the business man's ordinary indoor suit has a value of 1 Clo. This allows maintenance of a comfortable skin temperature of about 33°C in a room at 21°C with air movement not over 30 cm/min and relative humidity not over 50 per cent when the subject's metabolic rate is 210 kJ/m^2/h (50 kcal/m^2/h), designated 1 Met. The best ordinary clothing has a value of about 4 Clo per inch of thickness.

Comfort is dependent largely on skin temperature, the optimum being about 33°C. This temperature can be maintained in a cold environment only if the clothing is properly chosen in relation to the activity of the subject.

People about to leave a sinking ship should of course wear approved life jackets but, almost as important, they should put on as much clothing as possible or wear approved survival suits. It is important to start with dry clothes and to try to keep them dry. Wet clothing may contain, say 3 kg of water, the evaporation of which at body temperature requires

$$3 \times 2·43 = 7·3 \text{ MJ } (3000 \times 0·58 = 1740 \text{ kcal})$$

which is the equivalent of a good emergency food ration for a day. The same principles apply to endurance walks on mountain and moor. Inner clothing must be kept dry by windproof outer clothing. Water-repellent exposure suits have been specially designed so that they can be continuously worn over the clothing. They prevent spray from soaking the clothing but at the same time allow water vapour derived by evaporation from the body to escape.

Shipwrecked men immersed in very cold water lose heat rapidly. Survival from immersion in sea water at 15·5°C is usually not more than 5 hours, at 0°C it is only 15 minutes, death being due to hypothermia. These times may be considerably reduced by drowning, caused by the reflex respiratory distress initiated by stimulation of cutaneous cold receptors. Vigorous swimming, even in water at 0°C, can produce nearly as much heat as is lost but this vigorous exercise cannot be main-

tained for very long except by people like channel swimmers. Experiments with subjects in cold water tanks have led to the conclusion that body temperature is best maintained by clinging to wreckage and that no attempt should be made to swim.

In hot conditions, especially under a broiling sun, lightweight cotton or linen clothing or some of the modern substitutes is usually worn. White garments reflect the heating rays of the sun. Black clothing may absorb over 90 per cent of the sun's infra-red radiation whereas white clothing absorbs only about 30 per cent. The radiant energy met indoors has a different wave length and white clothing is no protection. Low resistance to the transference of water vapour is a useful property of clothing used in hot climates. Cotton clothing is preferable to nylon since cotton threads act better as capillaries to carry water through the garment.

MEASUREMENT OF ENVIRONMENTAL CONDITIONS

The reactions of a human being to different climatic conditions depend on individual, subjective ideas of comfort and discomfort. It is not surprising therefore to find that no simple yet satisfactory method of estimating the suitability or otherwise of an environment either indoors or out-of-doors has been devised. Apart from the individual's preferences the physical factors involved are air temperature, air flow, humidity and incident radiant heat. Though air movement does increase markedly the cooling power of an atmosphere a certain amount is refreshing as it dissipates odours and removes the layer of damp air from the body surface.

Indoors the conditions should be such as to assist body temperature regulation and allow maintenance of a comfortable skin temperature without engaging too strongly the heat regulating mechanisms. Adjustment of the 'microclimate' where the individual actually sits or works may be difficult because of the incidence of radiant heat from the walls, the shape of the room, and the fact that air currents tend to circulate close to the walls. It is obviously important to find the factors which determine comfort so as to assist the heating and ventilating engineer to produce the optimum conditions.

The best test object is of course the individual himself and, as Bedford suggested, a number of individuals can be used to assess the comfort of a series of rooms on a suitably graded scale ranging from 'much too warm' through 'comfortable' to 'much too cool'. These judgments have been given numerical values in a seven point scale and various instruments have been used at the same time to measure the physical properties of the environment. Since the removal of the layer of moist air next the skin is essential for continuous heat loss and heat should be transferable from subject to surroundings rather than in the reverse direction the factors to be measured are fairly obvious. The instruments generally employed are:

(i) Wet and dry bulb thermometer to indicate air temperature and humidity.

(ii) The whirling psychrometer (sling hygrometer) or the Assmann psychrometer, where air is dragged over the thermometer bulbs at speeds in excess of 3 m/sec, are preferable to (i) since the effect of random air currents is eliminated.

(iii) A silvered kata thermometer or a vane anemometer to measure the rate of air movement.

(iv) A mat black globe thermometer to measure the radiation. Bedford has prepared tables and nomograms from which a single temperature indicative of the comfortableness of the atmosphere may be obtained, namely the 'effective temperature'. This is based on the temperature, humidity and rate of movement of the air. Where considerable radiant heat is present the 'corrected effective temperature' based on globe thermometer, humidity and air velocity is more suitable. A scale has been devised for both the fully clothed and the individual stripped to the waist.

The effective temperature gives in fact the temperature of a room with still, saturated air which would give the same sensation of warmth as the room being measured.

In Great Britain heating and ventilation engineers often use the 'equivalent temperature' which ignores humidity and considers only air temperature and velocity. This detracts from its value in hot conditions where sweating comes to be an important method of heat loss.

REFERENCES

ADOLPH, E. F. & ASSOCIATES (1947). *Physiology of Man in the Desert*. London: Interscience.

AHERNE, W. & HULL, D. (1966). Brown adipose tissue and heat production in the newborn infant. *Journal of Pathology and Bacteriology* **91**, 223–225.

ATKINS, E. (1960). Pathogenesis of fever. *Physiological Reviews* **40**, 580–646.

BEDFORD, T. (1946). Environmental warmth and its measurement. *M.R.C. War Memorandum* No. 17. London: H.M.S.O.

BEDFORD, T. (1948). *Basic Principles of Ventilation and Heating*. London: Lewis.

BENZINGER, T. H. (1969). Heat regulation: homeostasis of central temperature in man. *Physiological Reviews* **49**, 671–749.

DU BOIS, E. F. (1949). *Fever and the Regulation of Body Temperature*. Springfield, Illinois: Thomas.

BURTON, A. C. & EDHOLM, O. (1954). Man in a cold environment. *Physiological Society Monograph* No. 2. London: Arnold.

CONROY, R. T. W. L. & MILLS, J. N. (1970). *Human Circadian Rhythms*. London: Churchill.

COOPER, K. E. (1965). Quantitative observations in disordered temperature regulation. *Scientific Basis of Medicine Annual Review* 239–258.

COOPER, K. E. (1966). Temperature regulation and the hypothalamus. *British Medical Bulletin* **22**, 238–242.

COOPER, K. E. & ROSS, D. N. (1960). *Hypothermia in Surgical Practice*. London: Cassell.

COOPER, K. E., CRANSTON, W. I. & HONOUR, A. J. (1967). Observations on the site and mode of action of pyrogens in the rabbit brain. *Journal of Physiology* **191**, 325–337.

CRANSTON, W. I. (1966). Temperature regulation. *British Medical Journal* ii, 69–75.

CROWDY, J. P. (1968). Water requirements in hot countries. *Journal of the Royal Army Medical Corps* **114**, 116–122.

EDHOLM, O. G. & BACHARACH, A. L. (Eds.) (1965). *Exploration Medicine: being a Practical Guide for those going on Expeditions*. Bristol: Wright.

EDHOLM, O. G. & BACHARACH, A. L. (Eds.) (1965). *The Physiology of Human Survival*. London: Academic Press.

ELLIS, F. P. (1960). Ecological factors affecting efficiency and health in warships. *British Journal of Industrial Medicine* **17**, 318–326.

EULER, C. VON (1961). Physiology and pharmacology of temperature regulation. *Pharmacological Reviews* **13**, 361–398.

FERRER, M. I. (Ed.) (1958). *Cold Injury. Transactions of the Fifth Conference*. New York: Josiah Macy Foundation.

FOLK, G. E. (1966). *Introduction to Environmental Physiology*. London: Kimpton.

GAGGE, A. P., BURTON, A. C. & BAZETT, H. C. (1941). Practical system of units for description of heat exchange of man with his environment. *Science* **94**, 428–430.

GIARMAN, N. J., TANAKA, C., MONEY, J. & ATKINS, E. (1968). Serotonin, norepinephrine and fever. *Advances in Pharmacology* **6A**, 307–317.

GILLIES, J. A. (Ed.) (1965). *A Textbook of Aviation Physiology*. Oxford: Pergamon Press.

HARDY, J. D. (1961). Physiology of temperature regulation. *Physiological Reviews* **41**, 521–606.

HEMINGWAY, A. (1963). Shivering. *Physiological Reviews* **43**, 397–422.

HULL, D. (1966). Brown adipose tissue. *British Medical Bulletin* **22**, 92–96.

KEATINGE, W. R. (1969). *Survival in Cold Water: the Physiology and Treatment of Immersion Hypothermia and of Drowning*. Oxford: Blackwell.

KUNO, Y. (1956). *Human Perspiration*. Oxford: Blackwell.

LEITHEAD, C. S. & LIND, A. R. (1964). *Heat Stress and Heat Disorders*. London: Cassell.

LEWIS, H. E., FOSTER, A. R., MULLAN, B. J., COX, R. N. & CLARK, R. P. (1969). Aerodynamics of the human environment. *Lancet* **ii**, 1273–1277.

LINDBERG, O. (Ed.) (1970). *Brown Adipose Tissue*. Amsterdam: Elsevier.

VAN LOON, J. H. (1963). Diurnal body temperature curves in shift-workers. *Ergonomics* **6**, 267–273.

MACPHERSON, R. K. (Ed.) (1960). Physiological responses to hot environments. *Special Report Series*, Medical Research Council, London, 298.

MILLS, J. N. (1966). Human circadian rhythms. *Physiological Reviews* **46**, 128–171.

NEWBURGH, L. H. (1949). *Physiology of Heat Regulation and the Science of Clothing*. London: Saunders.

PARKES, A. S. (Ed.) (1961). Hypothermia and the effects of cold. *British Medical Bulletin* **17**, 1–73.

PICKERING, SIR GEORGE (1958). Regulation of body temperature in health and disease. *Lancet* **i**, 1–9, 59–64.

ROBINSON, S. & ROBINSON, A. H. (1954). Chemical composition of sweat. *Physiological Reviews* **34**, 202–220.

SELLS, S. B. & BERRY, C. A. (Eds.) (1961). *Human Factors in Jet and Space Travel*. New York: Ronald Press.

SMITH, A. V. (1961). *Biological Effects of Freezing and Supercooling*. London: Arnold.

SMITH, R. E. & HORWITZ, BARBARA A. (1969). Brown fat and thermogenesis. *Physiological Reviews* **49**, 330–425.

SNELL, E. S. & ATKINS, E. (1968). The mechanisms of fever. In *The Biological Basis of Medicine* (Edited by E. E. Bittar & N. Bittar), Vol. 2, pp. 397–419. London: Academic Press.

STOLL, ALICE M. (1964). Techniques and uses of skin temperature measurements. *Annals of the New York Academy of Science* **121**, 49–56.

Symposium on climate, environment and health. (1967). *Practitioner* **198**, 621–672.

WOLSTENHOLME, G. E. W. & BIRCH, J. (1971). *Pyrogens and Fever, Ciba Foundation Symposium*. Edinburgh: Churchill Livingstone.

43 Nervous system

The nervous system can be divided into *central* and *peripheral* portions. The central nervous system consists of the brain contained within the skull and the spinal cord lying in the vertebral canal. The peripheral part consists of forty-three pairs of nerves which leave the cerebrospinal axis and pass to the various organs of the body. The peripheral nerves contain fibres of two kinds, one group termed *afferent* or *sensory* fibres carries ingoing nerve impulses from the periphery to the central nervous system. The second group, the *efferent* or *motor* fibres, carries outgoing impulses from the central nervous system to muscles and glands and other organs. Each spinal nerve arises from the cord by two roots; the posterior (dorsal in animals) carries afferent fibres while the anterior (or ventral) is occupied by efferent fibres.

The nervous system can also be divided into two anatomical and functional components, the *somatic* (or cerebrospinal) and the *autonomic* (or visceral). The somatic component has a central part, the brain and spinal cord, and a peripheral part, the nerves of the body; it is concerned both with information arriving from the surface of the body, from the muscles and joints, and with the transmission of impulses to the muscles. The autonomic component may also be divided into a central part consisting of nerve cells and their processes which lie in the central nervous system and a peripheral part made up of fibres passing to and from the blood vessels and viscera.

The nervous system may be regarded as the main co-ordinator of the activities of the body. The afferent nerves bring to the central nervous system information about the external world from receptors which are sensitive to stimuli such as light, sound, temperature, or pressure, and also information about the internal state of the body (for example tension in muscles, distension in viscera). As a consequence of the arrival of these impulses the central nervous system sends impulses along the efferent nerves to produce appropriate movements of muscles or secretion of glands. The pathways taken by impulses after they arrive at the central nervous system are very complex, but the basic principle is that impulses are handed on from nerve cell to nerve cell until finally they emerge in an efferent nerve to produce a response in an effector organ such as a muscle or a gland. Sometimes this process rises to conscious levels but many of the activities of the body are regulated without our being aware of them. These activities controlled by the nervous system are called *reflex actions*, or simply *reflexes*, and the pathways followed in the nervous system along afferent, junctional and efferent neurones are called *reflex arcs*.

The cerebrospinal axis is developed out of a simple tube of ectoderm. The lumen of the cranial end of this neural tube becomes dilated and eventually produces the large ventricular system within the brain; the lumen of the caudal part of the tube remains very small and is recognizable in the adult as the tiny central canal of the spinal cord. In the human embryo about five weeks old three main dilatations of the cranial or cephalic end of the neural tube can be made out: 1, prosencephalon or forebrain vesicle; 2, mesencephalon or midbrain vesicle; and 3, rhombencephalon or hindbrain vesicle. The more important parts of the brain formed from these vesicles are shown in the schema on the next page, and a diagram of the main parts of the brain is given in Fig. 43.2.

In man the central nervous system contains 10 000 million neurones, all formed by the sixth month of intrauterine life; no mitotic figures can be found after birth. If tritiated thymidine, which is incorporated into the DNA of cells about to divide, is given intracerebrally to rats a few neurones are labelled. But this small production of neurones—if it does occur in man as well as the rat—is too small to be of importance in recovery from cerebral injuries. It is stated that about 10 000 neurones are lost per day in adult life. The neurones are very active in producing RNA (p. 364), the amount being proportional to the afferent stimulation. For example, the RNA content of certain neurones can be increased over 10 per cent by exercise; on the other hand retinal ganglion cells of animals kept in the dark have little RNA.

Each neurone of the central nervous system in man is surrounded by about ten glial cells, their delicately folded membranes covering the neurone except at the synaptic knobs and it is

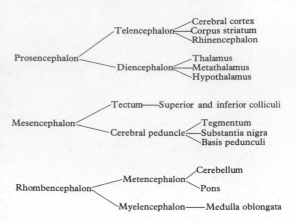

probable, but not yet directly demonstrated, that transfer of substances from blood to neurone occurs through the glial cells (Fig. 43.1). In tissue culture only those neurones

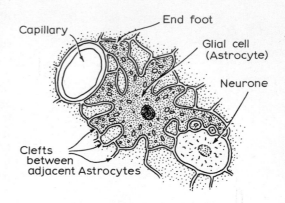

Fig. 43.1 Sketch of neurone-glial-capillary arrangement as seen in the electron microscope. An astrocyte is shown interposed between the endothelium of the capillary and the neurone. The cells, axons, dendrites and astrocytes are tightly packed with narrow clefts (dimension greatly exaggerated here) of about 15 nm between them. (From S. W. Kuffler (1967). *Proceedings of the Royal Society* B **168**, 8.)

surrounded by glia survive for any length of time. Glial cells contain only a tenth of the RNA of the neurones but have more lipid and very much more carbonic anhydrase (p. 623). The presence of this enzyme probably accounts for the passage of chloride ions into the cerebrospinal fluid against a chemical gradient.

Glial cells, unlike neurones, can multiply throughout life. The oligodendrocytes form myelin like the Schwann cells in the peripheral nerves; the astrocytes provide mechanical support and phagocytose degenerating neurones. The membrane potential of glial cells, about 90 mV, is due to the difference in potassium concentration between the interior of the cell

and that in the narrow clefts between them. Depolarizing currents passed through glial cells do not make them discharge.

In the descriptions of the anatomy of the nervous system which follow it is necessary to use certain technical terms. When referring to the spinal cord in the human subject the terms *anterior* and *posterior* have the usual anatomical meanings; in the case of animals *ventral* and *dorsal* are used. The *median plane* passes anteroposteriorly in man and divides the body into two symmetrical halves; any plane parallel

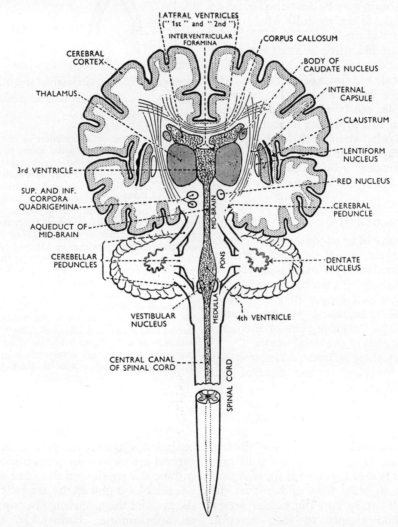

Fig. 43.2 Diagrammatic coronal section through the cerebral hemispheres, brain-stem and spinal cord to show the approximate relationships of the more important parts. Some of the structures are shown on one side only and the two halves of the cerebellum have been separated as they were in early development. The caudate and lentiform nuclei together form the corpus striatum. (*By courtesy of J. D. B. MacDougall.*)

to the median plane is called a *sagittal plane*. A vertical plane at right angles to this, for example a vertical plane passing through both external auditory meati, is called a *coronal plane*. It is convenient to refer to points in one vertical coronal plane as being above or dorsal to points in the same plane lying below or ventral. In the central nervous system the cord can be regarded as lower than the medulla, which is lower than the mid-brain, and the cerebral hemispheres are then referred to as the highest parts. To avoid any implication which may be contained in the words higher or lower the mid-brain may be described as being *rostral* or *cephalic* to the medulla which is the most *caudal* part of the brain.

The brains of primitive animals are much simpler than those of animals higher up the evolutionary scale but quite often certain parts of the primitive brain, a collection of nerve cells for example, can be recognized in higher animals because of a similarity in position or in relationship to other structures. Such areas are designated *palaeo*. Later developments in the evolutionary process with no counterparts in the primitive brain are designated with the prefix *neo*.

CEREBRAL CIRCULATION

The importance of an adequate blood supply to the brain scarcely needs stressing. Arrest of the cerebral circulation for more than about 5 seconds is followed by unconsciousness, and cerebral ischaemia of longer than 2 minutes causes irreparable damage to the grey matter of the cortex, the Purkinje cells of the cerebellum and the cells of the basal nuclei.

Since, in man, there is almost no anastomosis between the branches of the internal and external carotid arteries, the blood supply to the brain depends on the arterial circle of Willis at the base of the brain within the cranial cavity, into which blood is brought by the two internal carotid arteries and by the two vertebral arteries (Fig. 43.3). Studies in rabbits, in which the arrangement of these vessels resembles that in man, have shown that the internal carotid artery and the basilar artery share the blood supply to each cerebral hemisphere in such a way that there is normally, in a steady state, no interchange of blood between them. The arterial circle clearly provides for continued perfusion of all parts of

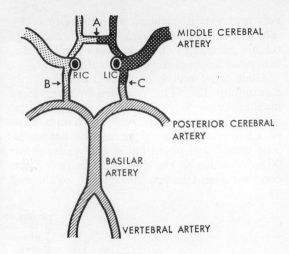

```
////////   supplied by basilar.
■■■■■■■■      „      „ left internal cartoid.
▨▨▨▨▨▨      „      „ right internal cartoid.
```

Fig. 43.3 Areas of distribution in man of the internal carotid arteries and the basilar artery within the circle of Willis (circulus arteriosus); the internal carotid arteries supply a larger volume of the brain than do the vertebral arteries. 'Dead points' occur at A (anterior communicating artery) and at B and C (posterior communicating arteries); obviously the dead points change if an artery is occluded. In normal persons the left vertebral artery is often larger than the right which may be very small; one or other posterior communicating artery may be narrow or absent. In fact in only 50 per cent of normal persons is the circle of Willis present as in the diagram. Movements of the head and neck cause considerable changes in blood flow in the vertebral arteries because they pass through the transverse processes of the atlas and because they are relatively fixed as they enter the foramen magnum. Since the lumen of the internal carotid artery can be constricted from 30 mm² to 5 mm² with only an insignificant reduction in the flow, this important artery could be said to be much larger than is haemodynamically necessary. (*After* R. Brain (1957). *Lancet* **ii**, 857–862.)

the brain even if a large vessel should become occluded. Digital compression of one common carotid artery in man normally causes a fall in intraocular tension on the same side. If it does not the blood flow in the artery is deficient and the eye and the brain must be supplied through an anastomotic channel. In the circle of Willis, 'the tides of the blood ebb and flow from moment to moment to achieve the sovereign purpose of making available a constant supply of blood to every part of the brain'

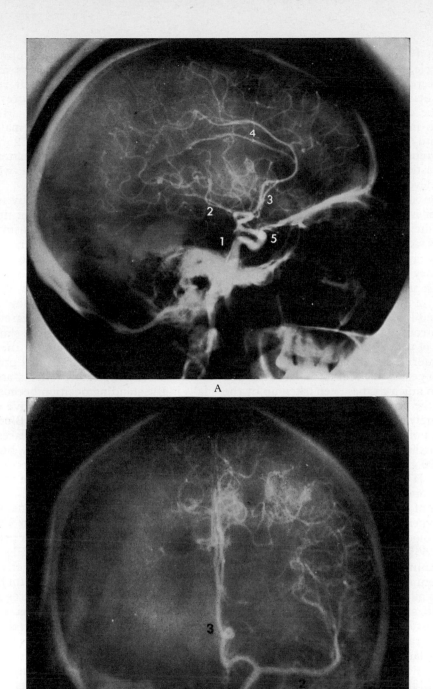

A

B

Fig. 43.4 Carotid angiogram to show the normal distribution of the ramifications of the internal carotid artery. This picture was obtained by injecting 10 ml of a radio-opaque solution rapidly into the left internal carotid artery through the skin of the neck. A, lateral view; B, antero-posterior view. 1. Internal carotid artery. 2. Middle cerebral artery. 3. Anterior cerebral artery. 4. Callosal arteries. 5. Ophthalmic artery. (*By courtesy of A. S. Bligh.*)

(R. Brain). Furthermore the branches of the arterial circle, that is the anterior, middle and posterior cerebral arteries, intercommunicate through their small ramifications over the cortex. The cerebral vascular system can be outlined in man by injecting a radio-opaque solution into the carotid artery (Fig. 43.4).

The cerebral vessels have a prominent internal elastic lamina but their muscle coat is less well-developed than that of arteries of comparable size elsewhere in the body. The grey matter is richly supplied with capillaries but the white matter, with less active metabolism, is less vascular. The venous sinuses of the dura carry away blood from the brain; in addition they also receive cerebrospinal fluid through the arachnoid villi (p. 685).

The rate of blood flow through the brain can be measured in man by causing the subject to breathe nitrous oxide, determining the concentration of the gas in samples of blood taken from the internal carotid artery and the jugular bulb over a period of 10 minutes. At the end of this time so much gas has been taken up by the brain that the brain, arteries and veins contain approximately equal concentrations of the gas. Application of Fick's principle (p. 502) shows that approximately 55 ml of blood flows through 100 g brain per minute under resting conditions. If the weight of the adult brain is 1400 g the total flow is about 770 ml per min or 15 per cent of the cardiac output. The carotid-venous oxygen difference is about 6 ml per 100 ml of blood. The oxygen consumption is therefore $(55/100) \times 6$ or $3 \cdot 3$ ml oxygen per 100 g per minute, which is not very different from the requirement of muscles during active work, or 46 ml oxygen for the whole brain. The energy utilization is thus about 20 W. The oxygen consumption varies relatively little over a wide range of P_{a,CO_2} and P_{a,O_2}. Infants with their large brains require a relatively large proportion of the total oxygen consumption. Cerebral blood flow is reduced in the elderly.

It is, of course, likely that even if the total blood flow remains constant there may be considerable regional differences in blood flow through different parts of the brain. The blood flow through the baboon cerebral cortex can be measured through the intact skull by injecting the γ-emitter ^{133}Xe in saline into the internal carotid artery and determining its clearance rate with scintillation counters placed over the skull. The decay curve shows a fast component and a slow component representing respectively the flow through grey and white matter. When the P_{a,O_2} and arterial blood pressure were within the normal range the mean cerebral blood flow was 52 ml per 100 g per min; the flow in the grey matter was 72 ml per 100 g per min and in the white matter 19 ml per 100 g per min.

The skull of the human baby is not completely ossified—at each beat of the heart the pulsations of the brain are transmitted to the anterior fontanelle. When he cries the venous pressure rises and the increase of intracerebral pressure bulges the anterior fontanelle. The fontanelle closes in the middle of the second year and the cranial contents now enclosed in the rigid cranium cannot increase in volume, but if an opening is made in the skull the brain tends to expand through it. In the adult, just as in the child, the intracranial pressure rises when violent expiratory efforts are made. It falls in the standing position because the pressure in the veins falls (to negative values) and venous drainage is facilitated. Blood entering the skull at each systole displaces a similar amount of blood from the veins, but if there is a prolonged increase of arterial pressure cerebrospinal fluid may also be forced out of the skull.

Cerebral blood flow remains remarkably constant over the range 60 to 130 mm Hg arterial pressure in man, in other primates such as the baboon and in the dog (Fig. 43.5). It is only when arterial (or more strictly, perfusion pressure) falls below about 60 mm Hg or when cardiac output falls below 3 litres per minute that the cerebral circulation becomes insufficient. The maintenance of a relatively high level of blood flow as pressure is reduced from physiological levels to about 60 mm Hg appears to depend upon two factors; one, an intrinsic factor which probably represents a response of vascular smooth muscle, which is comparatively slow being complete in 30 to 40 seconds. The other is neural and is mediated by the dilator pathway carried for part of its length by the VIIth cranial nerve. If this pathway is interrupted or if atropine is given, there is no longer vasodilatation and blood flow falls to low levels when the animal becomes hypotensive. The constancy of flow at high pressure

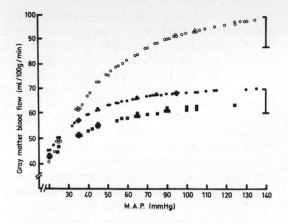

Fig. 43.5 The black circles show the relationship in the baboon between mean arterial pressure, MAP, and blood flow in the grey matter when the nerves are intact and when the P_{a,CO_2} and P_{a,O_2} are kept constant. The curve of flow is nearly flat between 60 and 130 mm Hg—this is the range of so-called autoregulation. The open circles show the effect of sympathectomy; there is no flat region of this curve. Stimulation of the sympathetic, black squares, brought the relationship back to normal. (I. M. James, R. A. Millar & M. J. Purves (1969). *Circulation Research* **25**, 80.)

(> 130 mm Hg) appears to depend largely upon an intact sympathetic pathway (Figs. 43.5 and 6).

The cerebral vascular responses to hypoxia and hypercapnia also depend in varying degree upon intrinsic and neural factors. If the constrictor fibres are cut, the flow response to both hypoxia and hypercapnia is enhanced. If the dilator pathway is cut, the response to hypercapnia is markedly reduced and the response to hypoxia is largely abolished. If both pathways are cut or if all the peripheral arterial chemoreceptors are denervated, blood flow may actually fall in response to hypoxia but there is clear evidence of a slow residual intrinsic response to CO_2.

From this we may deduce that in the intact animal the vascular response to changes in arterial pressure and CO_2 is determined by alterations in the balance of constrictor and dilator neural activity superimposed upon an intrinsic smooth muscle response. The function of the nerves under these conditions is (a) to cause an increase in the rate of response, (b)

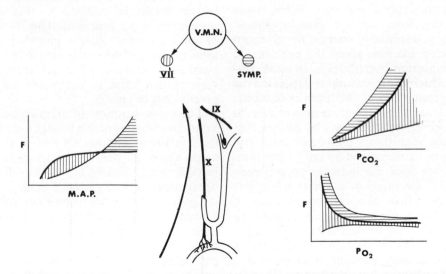

Fig. 43.6 A diagram to illustrate the neural pathways involved in the regulation of the cerebral vessels and the ways in which the responses to changes in arterial pressure (M.A.P.), P_{CO_2} and P_{O_2} are affected when the pathways are interrupted. Afferent fibres from the aortic and carotid groups of baroreceptors and chemoreceptors are shown in the Xth and IXth cranial nerves respectively together with an arrow representing afferent fibres from non-specific receptors. These fibres impinge upon vasomotor neurones (V.M.N.) which in turn affect activity in the VIIth cranial nerve (dilator) and sympathetic nerves (constrictor). In each of the relations between flow (F) and M.A.P., P_{CO_2} and P_{O_2}, the thick line represents the response with all nerves intact, the horizontal hatching the extent to which the response is altered when the sympathetic nerves are cut, vertical hatching that when the dilator nerves are cut. (*By courtesy of M. J. Purves.*)

to act as a fine control, and (c) probably to integrate the cardiovascular response with that which occurs in the body as a whole. With respect to hypoxia, the neural component appears to be of especial importance for if the chemoreceptor–vasodilator pathway is interrupted, none of the vascular responses which ensure adequate transfer of oxygen from capillary to tissue occur. Under normal conditions, the initiation of this balance of the neural components is from the peripheral arterial chemoreceptors and baroreceptors although it has been shown that stimulation of other peripheral receptors, for example pain, can cause transient constriction of cerebral vessels.

METABOLISM OF THE NERVOUS SYSTEM

By comparing the arterio-venous differences in oxygen, carbon dioxide and glucose it has been found that the R.Q. of brain is very nearly 1 and that the oxygen utilized is equivalent to the amount of glucose which disappears, that is 4·9 mg per 100 g per min. Brain tissue appears, therefore, to use glucose almost exclusively as a source of energy, the glycogen content being too low, about 0·1 per cent, to make an important contribution. In some conditions in which consciousness is depressed the oxygen consumption of the brain is reduced; for example, insulin hypoglycaemia reduces it by 20 per cent, diabetic coma by 50 per cent and surgical anaesthesia by 40 per cent. But sleep reduces consumption by only 3 per cent. Mental effort does not increase oxygen consumption of the brain as a whole but a few experiments indicate that the oxygen usage of the cortex near the central sulcus may be increased with a slight decrease in that of the white matter. These apparently contradictory findings show how difficult it is to find a relationship between function and metabolism in the brain.

In the cat the oxygen consumption of the cerebral cortex has been estimated by recording the fall of local P_{O_2} during local occlusion of the circulation. The resting value was 6 ml O_2 per 100 ml wet volume per min; when the cortex was stimulated the oxygen consumption went up by 80 per cent.

The synthesis of new proteins in the brain of the fetus and infant is quite rapid as might be expected from the rapid growth of the brain. Synthesis slows down gradually and in adult life it is much the same as in other tissues.

The water-soluble proteins of brain tissue are similar to those in other tissues and have a comparable turnover. However, brain contains two special types of protein. The lipoproteins of the myelin sheath contain protein with a turnover rate so low that they can for practical purposes be regarded as metabolically inert. The phosphoproteins of nerve cells contain phosphorus with a very rapid rate of turnover which is increased by electrical stimulation.

Nervous tissue is metabolically peculiar in being relatively isolated from the rest of the body owing to the difficulty with which many substances, including amino acids, pass into it from the blood, and in its very limited ability to store carbohydrate in the form of glycogen. The brain's dependence on glucose makes it very sensitive to hypoglycaemia. Direct utilization of glucose for energy production in the brain follows the tricarboxylic acid pathway but only about 25 per cent of the glucose used by the brain is oxidized in this way. The remaining 75 per cent is used for the formation of amino acids, chiefly glutamic acid and aspartic acid, which are used partly for protein synthesis but mainly for oxidation by pathways which exist in the nerve cell for the production of energy.

The concentration of free amino acids in brain tissue is very high, eight times higher than in blood plasma. Of this pool of amino acids about 15 per cent consists of γ-amino butyric acid (GABA) (p. 822) formed by decarboxylation of glutamic acid.

Cyclic AMP seems to be a key substance in brain function. Catecholamines and histamine increase the concentration of cyclic AMP in brain, the largest responses being found in cerebellar tissue. Brain tissue has the highest capability of any mammalian tissue of synthesizing cyclic AMP. Other parts of the C.N.S. are also highly active. There is probably a correlation between the distribution of adenylate cyclase and synaptosomes in the cerebral cortex. Cyclic AMP-dependent protein phosphokinases have been demonstrated in brain. Electroconvulsive shock treatment and anoxia cause elevated levels of cyclic

AMP. Combinations of depolarizing agents such as ouabain, batrachotoxin or high K^+ with adenosine or biogenic amines synergistically increase cyclic AMP formation. Hoffer has found that noradrenaline-containing axons make axo-dendritic and axo-somatic synapses upon Purkinje cells, the effect of noradrenaline being to reduce the mean discharge rate by increasing the time between single discharges. Cyclic AMP and drugs which inhibit cyclic AMP phosphodiesterase (for example methyl xanthines) mimic these effects, whilst prostaglandins antagonize them.

Injection of dibutyryl cyclic AMP into the mesencephalic reticular formation of cats induces catatonia, whereas injection into the cerebellum induces E.E.G. patterns similar to those of R.E.M. sleep and injection into the amygdala induces Jacksonian convulsions, convulsions and rage.

Cyclic AMP content of urine is lowered in cases of severe depression, and being elevated in the case of manic patients.

Cyclic AMP has also been implicated in the release of acetylcholine from nerve endings under the influence of adrenaline.

REFERENCES

BOWSHER, D. (1970). *Introduction to the Anatomy and Physiology of the Nervous System*, 2nd edn. Oxford: Blackwell.

BRADLEY, P. B. (Ed.) (1965). Pharmacology of the central nervous system. *British Medical Bulletin* 21, 1–90.

BRAZIER, M. A. B. (1968). *The Electrical Activity of the Nervous System*, 3rd edn. London: Pitman.

BROCK, M., FIESCHI, C., INGVAR, D. H., LASSEN, N. A. & SCHÜRMANN, K. (Eds.) (1969). *Cerebral Blood Flow*. Berlin: Springer.

CAMPBELL, H. J. (1965). *Correlative Physiology of the Nervous System*. London: Academic Press.

DUNN, J. S. & WYBURN, G. M. (1972). The anatomy of the blood–brain barrier: a review. *Scottish Medical Journal* 17, 21–36.

EIDUSON, S., GELLER, E. YUWILER, A. & EIDUSON, B. T. (1964). *Biochemistry and Behaviour*. Princeton, N.J.: Van Nostrand.

ELLIOTT, K. A. C. & JASPER, H. H. (1959). Gamma-amino-butyric acid. *Physiological Reviews* 39, 383–406.

ELLIOTT, K. A. C., PAGE, I. H. & QUASTEL, J. H. (Eds.) (1962). *Neurochemistry. The Chemistry of Brain and Nerve*, 2nd edn. Springfield: Thomas.

EVANS, C. R. & ROBERTSON, A. D. J. (1966). *Brain Physiology and Psychology*. London: Butterworths.

EYZAGUIRRE, C. (1969). *Physiology of the Nervous System. An Introductory Text.* Chicago: Year Book Medical Publishers.

GERARD, R. W. & DUYFF, J. W. (Eds.) (1964). *Information Processing in the Nervous System*. International Congress of the Union of Physiological Sciences. Amsterdam: Excerpta Medica.

GESSA, G. L., KRISHNA, G., FORN, J., TAGLIAMONTE, A. & BRODIE, B. B. (1970). *Role of Cyclic AMP in Cell Function. Advances in Biochemical Psychopharmacology* (Edited by P. Greengard & E. Costa), Vol. 3, p. 349. New York: Raven Press.

GLEES, P. & MELLER, K. (1968). Morphology of neuroglia. In *Structure and Function of Nervous Tissue* (Edited by G. H. Bourne), Vol. I, pp. 301–323. New York: Academic Press.

HARPER, A. M. (1967). Measurement of cerebral blood flow in man. *Scottish Medical Journal* 12, 349–360.

HARPER, A. M. (1969). Regulation of cerebral circulation. *Scientific Basis of Medicine Annual Review* 60–81.

HOFFER, B. J., SIGGINS, G. R. & BLOOM, F. E. (1970). *Role of Cyclic AMP in Cell Function. Advances in Biochemical Psychopharmacology* (Edited by P. Greengard & E. Costa), Vol. 3, p. 349. New York: Raven Press.

HYDEN, H. (1962). The neuron and its glia—a biochemical and functional unit. *Endeavour* 21, 144–155.

INGVAR, D. H. & LASSEN, N. A. (Eds.) (1965). *Regional Cerebral Blood Flow*. An international Symposium. Copenhagen: Munksgaard.

International Review of Neurobiology. Vols. 1–10 (1959–1967). London: Academic Press.

JAMES, I. M., MILLAR, K. & PURVES, M. J. (1969). Observations on the extrinsic neural control of cerebral blood flow in the baboon. *Circulation Research* 25, 77–93.

KUFFLER, S. W. (1967). Neuroglial cells: physiological properties and a potassium mediated effect of neuronal activity on the glial membrane potential. *Proceedings of the Royal Society* B 168, 1–47.

KUFFLER, S. W. & NICHOLLS, J. G. (1966). The physiology of neuroglial cells. *Ergebnisse der Physiologie* 57, 1–90.

LASSEN, N. A. (1959). Cerebral blood flow and oxygen consumption in man. *Physiological Reviews* **39**, 183–238.

MAGOUN, H. W. (1959). *Handbook of Physiology, Section I. Neurophysiology.* Washington: American Physiological Society.

MCILWAIN, H. & BACHELARD, H. S. (1972). *Biochemistry and the Nervous System,* 4th edn. London and Edinburgh: Churchill Livingstone.

NOBACK, C. R. (1967). *The Human Nervous System.* New York: McGraw Hill.

O'BRIEN, M. D. (1968). Physiological considerations in cerebrovascular disease. *American Heart Journal* **76**, 699–705.

PICKFORD, MARY (1969). *The Central Role of Hormones.* Edinburgh: Oliver & Boyd.

POYNTER, F. N. L. (1959). *The History and Philosophy of Knowledge of the Brain and its Functions.* Oxford: Blackwell.

Progress in Brain Research. Vols. 1–27 (1963–1967). Amsterdam: Elsevier.

QUASTEL, J. H. & QUASTEL, D. M. J. (1961). *The Chemistry of Brain Metabolism in Health and Disease.* Oxford: Blackwell.

RENFREW, S. (1967). *An Introduction to Diagnostic Neurology.* 3 vols. Edinburgh: Livingstone.

ROSENBLITH, W. A. (Ed.) (1961). *Sensory Communication. Contribution to the Symposium on Principles of Sensory Communication.* Cambridge, Mass.: M.I.T. Press.

RUCH, T. C. (1961). *Neurophysiology.* London: Saunders.

SHERRINGTON, C. S. (1947). *The Integrative Action of the Nervous System.* Cambridge University Press.

WALSH, E. G. (1964). *Physiology of the Nervous System,* 2nd edn. London: Longmans Green.

WALSHE, SIR FRANCIS (1965). *Further Critical Studies in Neurology and other Essays and Addresses.* Edinburgh: Livingstone.

WALTON, H. N. (1965). *Essentials of Neurology,* 2nd edn. London: Pitman.

WILLIAMS, D. (Ed.) (1957–1970). *Modern Trends in Neurology,* 2nd series, Vols. 1–5. London: Butterworth.

WYKE, B. (1965). *Brain Function and Metabolic Disorders.* London: Butterworth.

WYKE, B. (1969). *Principles of General Neurology.* Barking: Elsevier.

44 Spinal cord

THE PATHWAYS OF THE SPINAL CORD

The spinal cord extends in adult man from the upper border of the first cervical vertebra (the atlas) to the lower border of the first lumbar vertebra. It is about 1 cm in diameter on the average, being a little wider at the cervical and lumbar enlargements where the nerves to the arms and legs arise, because of the greater amount of grey matter there. There is no trace of segmentation on the surface of the spinal cord but it is convenient to describe it as if it were divided up into thirty-one segments from each of which arises a pair of *spinal nerves* (Fig. 44.1).

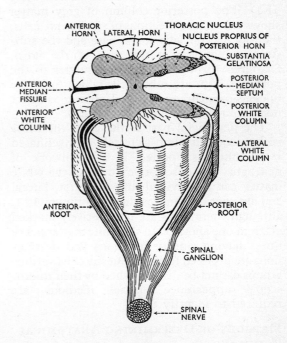

Fig. 44.1 A spinal segment from the thoracic region viewed from the left side. (After *Gray's Anatomy*, 30th edn., 1949, p. 924. London: Longman.)

The *posterior root* (dorsal root in animals) emerges from the postero-lateral aspect of the segment as a number of rootlets which fuse as they pass laterally to form a single trunk upon which is an enlargement, the *spinal* or *posterior root ganglion*. The *anterior root* (ventral root in animals) is formed by the fusion of a number of rootlets which arise in an irregular fashion from the antero-lateral aspect of the cord. The anterior root joins the posterior root beyond the ganglion to form the spinal nerve which emerges between the vertebrae.

The interior of the cord around the central canal is occupied by an H-shaped mass of *grey matter* consisting largely of nerve cells (Fig. 44.1). The shape of the grey matter varies from segment to segment but it shows three main parts, anterior, lateral and posterior columns or horns. From the large motor cells of the anterior horn axons pass in the anterior or motor root to the voluntary muscles. The small lateral horns, seen only in the thoracic and upper lumbar regions of the cord, contain cells of medium size, belonging to the sympathetic system, from which axons travel in the anterior roots to the white rami communicantes (Plate 41.1). The posterior column of grey matter shows two main collections of small nerve cells, the thoracic nucleus at the base and the substantia gelatinosa at the apex of the horn. Scattered throughout the grey matter of the posterior horn are other cells which form the nucleus proprius and give off fibres to connect neighbouring segments.

The peripheral part of the cord, the *white matter*, consists almost entirely of myelinated nerve fibres supported in a meshwork of neuroglia. For descriptive purposes the white matter can be divided into posterior, lateral and anterior columns as shown in Figure 44.1, although these areas are not precisely marked off from one another. In the white matter nerve fibres having similar functions and destinations are grouped together but since the various groups cannot be distinguished by their microscopic appearance special methods are required to identify them.

METHODS OF DETERMINING ANATOMICAL PATHWAYS

The degeneration which is a consequence of the cutting of a peripheral nerve has already been described (p. 809). The mapping out of

the degeneration after section or destruction of a part of the nervous system is a most important method for finding the routes taken by nerve fibres in the central nervous system. Experiments of this kind have been made very frequently in animals but, in view of the differences between the nervous system of man and that of the lower animals, degenerations produced in man by surgical operation or by war wounds give more valuable information.

Normal nerve fibres stain with osmic acid, OsO_4, which is reduced to a black product, probably a lower oxide or metallic osmium, by the reactive unsaturated fatty acids such as oleic acid found in the complex mixture of fats present in the myelin sheath. If normal fibres are treated with a mixture of osmic acid and potassium dichromate the unsaturated fatty acids are quickly oxidized by the dichromate so that they no longer react with the osmic acid which is, therefore, not reduced and no black colour is formed. In degenerating nerves, two weeks to two months after nerve section, the complex fatty material in myelin breaks down with the production of large amounts of unsaturated fatty acids which are not all oxidized by dichromate treatment so that the application of osmic acid simultaneously with potassium dichromate results in the formation of a black colour owing to the reduction of the osmic acid by such fatty material as has escaped oxidation by the dichromate. The degenerating nerve fibres are, therefore, stained black (Marchi's method). This technique does not show the non-myelinated components of nerves and thus can give only the position of the tracts and not the cells upon which the fibres end. The identification of the latter requires the staining of the terminal synaptic knobs (Fig. 39.36) or their immediate pre-terminal fibres. Synaptic knobs begin to swell and disintegrate a day or so after they are cut off from the nerve cell; the swelling and fragmentation reach a maximum in five to ten days and disappear slowly afterwards. The origins of nerve fibres can be inferred from the disorganization of the cell body and nucleus (chromatolysis, p. 815), which occurs when the axon is cut.

After three months, when degeneration is complete, Marchi's method is no longer applicable and the Weigert-Pal method is employed. The tissue is first stained black

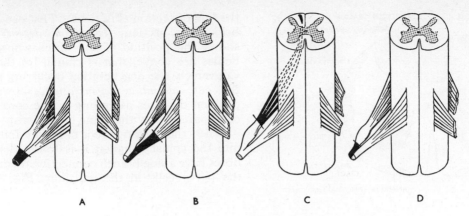

Fig. 44.2 Diagrams to show the site of degeneration (black) after division of the spinal nerve or its roots. A, division of spinal nerve below ganglion. B, division of anterior root. C, division of posterior root above ganglion. D, division of posterior root below ganglion. (*After* G. F. Yeo (1893). *A Manual of Physiology*, p. 478. London: Churchill.)

with haematoxylin and then treated with permanganate which takes the colour out of the degenerated areas but leaves the normal fibres a deep blue-black colour.

Further information about the connexions of nerve fibres is given by the study of transneuronal degeneration; a neurone may show chromatolysis and degenerative changes if the nerve fibres sending impulses to it have degenerated, but the only good example of this is the degeneration of certain cells in the lateral geniculate body after section of one optic nerve. (In fact after degeneration of the optic nerve fibres a number of non-optic synapses survive; but apparently not in sufficient amount to protect the lateral geniculate cells.) Another clue is furnished by observing the myelination of nerve fibres in the embryo, since different groups acquire their myelin sheaths at different times (p. 963). Histochemical tests for adrenaline and cholinesterase are now much used to show the position of adrenergic and cholinergic nerves. These classical methods described above which have been in use for many years still yield useful information. The use of micro-electrodes attached to sensitive amplifiers has made it possible to map out the pathways through the brain of nerve impulses started off by physiological stimuli such as sound or light touch or by artificial stimuli such as electrical pulses or local application of strychnine.

Study of the histological effects after section of the spinal nerves or their roots (Fig. 44.2) allows of a number of conclusions. Axons from motor cells situated in the anterior horn of the grey matter run in the anterior root. The fibres of the posterior root either pass a short distance into the cord to terminate around cells in the posterior horn, or pass up the cord in the posterior columns to the caudal part of the medulla oblongata where they end in the gracile and cuneate nuclei. The cell bodies of these posterior root fibres lie in the posterior root (spinal) ganglion.

When the spinal cord is cut across, degeneration occurs below (Fig. 44.3) and above

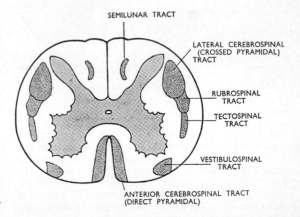

Fig. 44.3 Descending degeneration showing the position of the descending tracts in the human spinal cord. Intersegmental tracts lie between the grey matter and the descending fibres; the semilunar tract is also intersegmental. Magnification about × 5½.

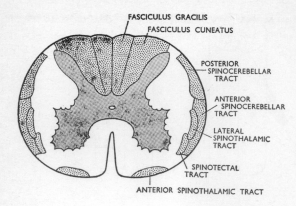

FASCICULUS GRACILIS
FASCICULUS CUNEATUS
POSTERIOR SPINOCEREBELLAR TRACT
ANTERIOR SPINOCEREBELLAR TRACT
LATERAL SPINOTHALAMIC TRACT
SPINOTECTAL TRACT
ANTERIOR SPINOTHALAMIC TRACT

Fig. 44.4 Ascending degeneration showing the position of the ascending tracts in the white matter of the human spinal cord. This diagram gives the usual view of the position of the spinothalamic tracts but their position in the anterolateral part of the cord may be more correctly given in Fig. 44.5. Magnification about × 5½

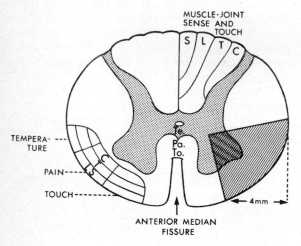

MUSCLE-JOINT SENSE AND TOUCH
S L T C
TEMPERA-TURE
Te.
Pa.
To.
PAIN
C
S
TOUCH
4mm
ANTERIOR MEDIAN FISSURE

Fig. 44.5 A diagrammatic representation of the approximate position of the main sensory paths in a cross-section of the spinal cord. In fact the separation of the various groups of fibres is by no means as sharp as shown. The letters indicate the relative positions of the fibres carrying impulses from C, the cervical; T, thoracic; L, lumbar; S, sacral root areas. Te, Pa, To, indicate place of decussation of the temperature, pain and touch fibres. Some of the touch fibres may decussate just posterior to the central canal. (*After* Foerster.) The cross hatched area shows the extent of the incision in the operation of anterolateral chordotomy performed usually in the upper part of the thoracic cord for relief of pain (see p. 909).

904 Textbook of Physiology and Biochemistry

the line of section (Fig. 44.4). That occurring below the section, *descending degeneration*, shows the position of the fibres whose cell bodies are above, that is cranial to, the cut. Conversely the degeneration occurring above the cut, *ascending degeneration*, shows the position of fibres ascending in the cord from cell bodies below the plane of section.

When the posterior root fibres are followed into the spinal ganglion it is found that the fibres have a side branch connecting them with the nerve cells in the ganglion (Fig. 44.6).

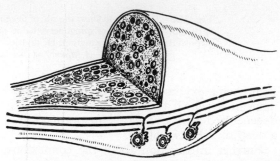

Fig. 44.6 Diagram of spinal root ganglion. (*After* H. S. D. Garven from Cajal and Carleton.)

Counts of microscopic preparations show that the number of cells is very nearly (97 per cent) the same as that of fibres. Presumably, therefore, nearly every fibre has a cell body in the ganglion. As each fibre enters the cord it divides into ascending and descending branches, each branch sending off numerous side branches at different levels in the cord (Fig. 44.7). The posterior root fibres have three main methods of termination: (1) a very few pass to the cells of the anterior horn of grey matter to form a synapse directly with the motor neurones of the same side, the mono-synaptic pathway, (2) a large number form synapses with cells at various positions in the grey matter of the posterior horn such as the cells of the nucleus proprius, the thoracic nucleus and the substantia gelatinosa, and from these secondary neurones impulses pass on to other parts of the central nervous system, (3) a large number of fibres pass up the same side in the posterior columns of the white matter of the cord to the cuneate and gracile nuclei in the medulla, sending collateral branches to the grey matter as they ascend.

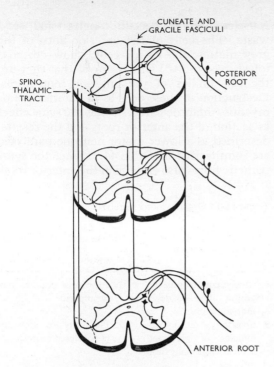

CUNEATE AND
GRACILE FASCICULI

POSTERIOR
ROOT

SPINO-
THALAMIC
TRACT

ANTERIOR ROOT

Fig. 44.7 Diagrammatic transparency of the spinal cord viewed from the anterior (or ventral) aspect to show the paths which conduct sensation through the cord (Sir Gordon Holmes (1960). *Introduction to Clinical Neurology*, 3rd edn reprint. Edinburgh: Livingstone.

This group includes fibres carrying impulses from the muscles (muscle–joint sense) and a proportion of the fibres subserving touch. The posterior columns, however, are not the sole route for this type of sensory information but may be particularly important in conveying information necessary for coordinated movement in space.

The secondary neurones of group (2) have various destinations. Many are simply short interconnecting or internuncial neurones (interneurones) carrying impulses to the motoneurones of the anterior horn of grey matter of the same or adjacent segments of the cord, but other secondary neurones have much longer paths. From the thoracic nucleus (Fig. 44.1) fibres pass to the posterior spinocerebellar tract of the same side; from other cells in the posterior horn fibres go up the anterior spinocerebellar tracts of both sides (Fig. 44.4) carrying impulses from the limb muscles to the cerebellar cortex (mainly ipsilateral). From the nucleus proprius and perhaps also from the

substantia gelatinosa the secondary neurones cross the mid-line, passing anteriorly or posteriorly to the spinal canal, and ascend in the contralateral cord as the spinothalamic tract (Fig. 44.7), the posterior part of which carries mainly the fibres of pain and temperature while the anterior part contains chiefly touch fibres (Fig. 44.5).

EFFECTS OF SECTION OF ROOTS OF SPINAL NERVES

For the localization of lesions of the spinal cord it is important to know the area of distribution of the posterior root fibres. The area of skin innervated by each root is called a *dermatome*. When a single posterior root is cut the area of anaesthesia is small because the skinfields of neighbouring posterior roots overlap considerably, any one patch of skin being supplied by two, or even three, posterior roots. To demonstrate the area of a dermatome the method of *remaining sensibility* was first used by Sherrington (1893) in the monkey and forty years later by Foerster in man. They cut three posterior roots above and three posterior roots below the root to be studied. This root, isolated as it were from its neighbours, is left intact. The skin area which remains sensitive to mechanical stimulation is then marked out. The results of numerous experiments of this kind can be plotted as a map of the dermatomes (Fig. 44.8).

Section of a series of posterior roots, or their destruction by disease, causes, in addition to the cutaneous anaesthesia just described, complete loss in the areas supplied of all forms of sensation whether cutaneous, proprioceptive or visceral. Since the ingoing part of the reflex arc is interrupted, spinal reflexes are abolished and muscle tone is reduced. Loss of the protective pain reflexes frequently results in damage to the skin which may be followed by ulceration. The interruption of the proprioceptive pathway to the cord prevents the patient from knowing, unless visually, the position of the muscles and joints in the affected area. As a result voluntary movements are carried out most inaccurately. Section of posterior roots also causes loss of vibration sense. When a heavy tuning fork is struck and laid with its base on the skin, especially over a bony prominence, the normal subject has a sensation of vibration which is conveyed by

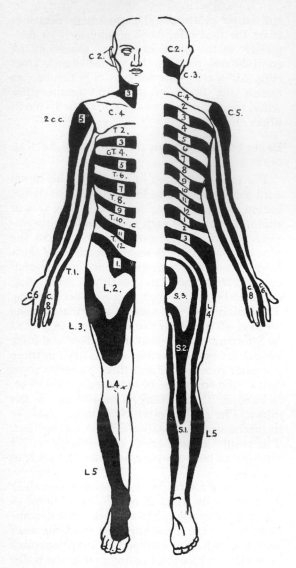

Fig. 44.8 The cutaneous areas supplied by the posterior nerve roots (dermatomes). W. N. Mann (1964). *Conybeare's Textbook of Medicine*, 14th edn., p. 751. Edinburgh: Livingstone.

muscles become flaccid, toneless and soon waste. The cells of the anterior horn of the grey matter, the motor cells, show chromatolysis (see p. 810) which proves that they are the cell bodies of the anterior root fibres. Destruction of these cells as in anterior poliomyelitis (infantile paralysis) has the same effect as section of the anterior roots and the results, described as a lower motor neurone paralysis, are summarized in Table 44.9. Since the sympathetic fibres also emerge in the anterior roots

Table 44.9 Organic motor paralysis

Upper motor (corticospinal) neurone (supranuclear paralysis)	Lower motor (nucleomuscular) neurone (nuclear and infranuclear paralysis)
1. Movements of muscle-groups affected, never individual muscles	1. Individual muscles may be affected
2. Spasticity or hypertonicity of paralysed muscles, especially limb muscles	2. Flaccidity or atonicity of paralysed muscles
3. May have superadded 'associated movements' (synkinesiae) on attempted voluntary movement	3. No 'associated movements'
4. No muscular atrophy, except from disuse	4. Atrophy of paralysed muscles
5. Electrical reactions normal	Responses diminished or reactions of denervation (degeneration)
6. Tendon reflexes in paralysed limbs present and usually increased	6. Tendon reflexes of paralysed muscles diminished and often absent
7. If foot affected, plantar reflex extensor in type	7. Plantar reflex, if present, is of normal flexor type
8. Slowness of residual voluntary movements	

(J. Purves-Stewart & C. Worster-Drought (1952). *The Diagnosis of Nervous Diseases*, 10th edn, p. 410. London: Arnold.

impulses passing along the afferent nerves and their continuations up the ascending columns of the cord. The spinal pathway for vibration sense includes fibres travelling in both the posterior and the lateral columns. Vibration sense is also impaired in thalamic lesions.

Section of the anterior roots causes paralysis of the muscles supplied, with complete loss of voluntary and reflex movements. The affected

(Plate 41.1) their section causes a loss of sweating in the area and a dilatation of the skin vessels (cf. p. 869).

TRANSECTION OF THE CORD AND SPINAL SHOCK

After spinal transection in animals the reflexes in the cord posterior (caudal) to the cut are depressed for a time, those in the cranial part of the nervous system remaining unaffected. In the frog spinal reflexes recover after two or three minutes and in the cat spinal reflexes may be depressed for an hour or more. In man the period of depression, that is total absence of reflexes below the transection varies from about two days to six weeks or longer. Slight movements of the toes or reflex penile contractions may sometimes be obtained a few hours after transection.

The state of depression accompanied by flaccidity of the muscles has been named *spinal shock*. It is not due to the trauma of section since it occurs if the continuity of the cord is interrupted by cooling or by the injection of local anaesthetic. It is probably related to the sudden interruption of impulses from higher, possibly cranial, parts of the nervous system. In the cat the state can be produced by partial section of the ventral part of the cord which includes division of the vestibulospinal pathways; to produce spinal shock in the primate it is necessary to section the corticospinal (pyramidal) tracts. This concept also explains the fact that, when recovery of reflex function has occurred, a second transection below the level of the original cut does not cause spinal shock in the remaining part of the cord. The reason for the recovery from spinal shock is not known.

Immediately after complete transection of the spinal cord in the thoracic region in man there is a total loss of voluntary movement in both legs (paraplegia) with loss of sensation (anaesthesia) over the lower part of the trunk and legs. At first the leg muscles are flaccid and no reflexes can be elicited. Reflex activity begins to appear first in the distal part of the limbs, later in the proximal parts, and the first response may be as early as two days or as late as six weeks. A number of cases of transection of the cord in man are followed by permanent flaccid paraplegia, possibly due to extensive trauma leading to damage to the anterior horns below the level of the section. The earliest reflexes to return are the flexor withdrawal movements following plantar stimulation. These include dorsiflexion (upturning) of the great toe (Babinski response). Later the flexor spasms become so pronounced that spread occurs to the autonomic outflow with emptying of the bladder and bowel and sweating. This mass reflex is particularly liable to develop in

the presence of infection of the bladder and bedsores. In the majority of cases the predominantly flexor activity is succeeded by a return of extensor reflexes so that the knee jerk and other reflexes are easily elicited and may be exaggerated. Generally the final outcome is one of predominantly extensor activity, although this is usually less marked than in incomplete sections of the cord. In complete sections of the cord paraplegia and anaesthesia below the line of section are permanent. Further details of the reflexes of spinal man are given on pp. 305, 670 and 871.

HEMISECTION OF SPINAL CORD

A hemisection produces paralysis of movement with loss of position sense on the same side below the level of the cut and on the opposite side loss of sensation of pain and temperature without loss of movement. The reason for this characteristic distribution of motor and sensory loss is clear from Figure 44.10. The band of hyperaesthesia on the same side is attributed to irritation of the neurones adjacent to the lesion. If the local lesion destroys at least three segments (see dermatomes, p. 905) there may be a band of impaired sensation in the dermatomes corresponding to the level of the lesion. When the anterior horn cells are involved a paralysis of the lower motor neurone type affects the muscles innervated by the corresponding segments. The muscles are represented in the cord by vertical groups of motoneurones extending over several segments; it is therefore difficult to locate the level of a lesion exactly by observing the extent of motor paralysis.

Disease causes incomplete transections of the cord more commonly than complete transections. In incomplete sections spasm of the extensor muscles is generally more pronounced than that of the flexor muscles. In animals extensor spasms occur when even a small portion of the corticospinal (pyramidal) tracts or the vestibulospinal tract is left intact, but when both are destroyed extensor reflexes disappear. In man the situation is more complex for flexor spasms may be a feature of both complete and incomplete lesions; in incomplete lesions extensor spasms however, generally predominate and in complete transections although extensor reflexes may be late in

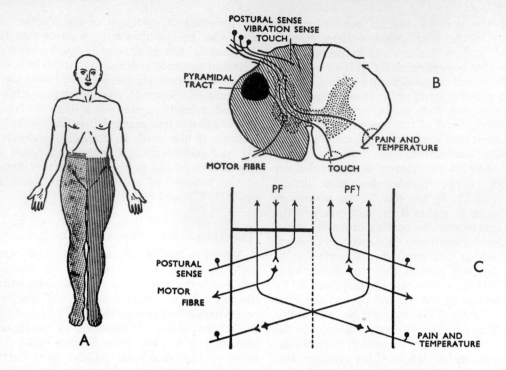

Fig. 44.10 This figure describes the Brown-Séquard syndrome arising from a lesion of the right half of the cord. In A the distribution of cutaneous sensory change is indicated by horizontal and vertical shading. The area of spastic paresis is indicated by stippling. In B, the shaded (right) half of the cord is the seat of the lesion and the diagram indicates the pathways interrupted. A diagrammatic longitudinal section is given in C. The black line indicates the site of the hemisection. The ascending and descending tracts which are interrupted are labelled. PF, pyramidal fibre. The resulting disorders are given in the table below.

ON THE SIDE OF THE LESION	ON THE OPPOSITE SIDE
Sensory	
At level of lesion. A band of cutaneous hyperaesthesia may be present. Alternatively, but only if three successive spinal segments are involved, there may be a band of impairment of cutaneous sensibility to touch, pain and temperature.	*At level of lesion.* No abnormality.
Below level of lesion. Tactile, painful and thermal sensibility normal. Postural and vibratory sensibility (p. 905) impaired or lost.	*Below level of lesion.* Impairment or loss of painful and thermal sensibility. Tactile, postural and vibratory sensibility normal.
Motor	
Below level of lesion. Spastic paresis of leg, increased tendon jerks, clonus, extensor plantar response.	*Below level of lesion.* No abnormality.

(Sir Francis Walshe (1963). *Diseases of the Nervous System*, 10th edn, p. 38. Edinburgh: Livingstone.)

appearing they may be dominant in the final state of the patient.

Unilateral chordotomy (see Fig. 44.5) does not make the other side of the body below the lesion permanently analgesic. Even for unilateral pain a bilateral chordotomy is successful only in one half of patients. In fact relatively few of the spinothalamic fibres reach the thalamus and it is likely that the tract consists of a few long fibres but mainly of multi-synaptic chains of short fibres of small diameter, forming a nerve net, ending at all levels of the neural axis which, obviously, is very difficult to interrupt effectively.

PROPERTIES OF SPINAL REFLEXES

FLEXION REFLEXES

If the foot of a decapitated frog is pinched the legs move quickly away from the harmful stimulus but, if the spinal cord of this brainless preparation is destroying by pithing, reflex withdrawal no longer occurs. This fundamental observation was made by Stephen Hales about 1730 but the word 'reflex' was not used until about forty years later, and serious investigation was delayed till the end of last century when Sherrington took up the subject.

Except for the monosynaptic reflex between the annulospiral discharge of a muscle spindle and certain motoneurones of the same muscle (p. 915), it is unusual for a reflex to be confined to one segment. In the withdrawal reflex in the frog just described there is a co-ordinated flexion of the ankle, knee and hip by synergistic groups of muscles innervated by a number of spinal segments. The idea of a spinal segment is convenient for anatomical description but a segment is not a functional unit since even simple reflexes involve many segments of the cord.

Since the activities of the spinal cord are greatly modified by the higher parts of the nervous system it is necessary for analysis of spinal reflexes to make a *spinal preparation*. The cat and dog have most frequently been used in such experiments. After the cord has been divided in the neck there is a period of spinal shock (p. 906), but if respiration is maintained by a pump the shock passes off and many reflexes can be elicited. The activity of the muscles to be studied is usually recorded by attaching them to isometric levers and the conditions of the experiment may be made as simple as possible by denervation of neighbouring structures or by removal of skin.

If an injurious stimulus, one which would be painful if applied to the intact animal, is applied to the foot of a spinal preparation the flexion reflex results in a withdrawal of the foot. This phenomenon is described by Sherrington as a *nociceptive response*. Withdrawal also occurs when any sensory nerve in the limb is stimulated electrically and this phenomenon is often made use of in experimental analysis of the properties of the spinal cord. For example, by varying the strength of the stimulus it is easy to demonstrate *irradiation*. A small stimulus causes a small response confined to a few muscles, whereas stronger stimulation involves many muscles in the limb and may cause movements in other limbs.

The time elapsing between the stimulus and the reflex response, the total latent period of a reflex, can easily be measured from a kymographic tracing. If the time taken for the nerve impulses to pass inward along sensory nerves and out again along the motor nerves is subtracted from the total reflex time, we have the *central reflex time* which in a spinal reflex must be due mainly to synaptic delay. Since the synaptic delay at the motoneurone is of the order of 0·5 msec the central reflex time can never be less than this. For a single volley set up by a single stimulus the central reflex time for a flexion reflex in the cat is about 4 msec. When two sublimal volleys reach the cord with only a short interval, say 7 msec, between them, the central reflex time for the flexion reflex evoked by the second volley is reduced, in some cases to 0·5 msec. The rise in *central excitatory state* (p. 911) produced by the first volley can be quickly enhanced by the second volley to a level at which the motor neurones discharge. A more complex explanation is that the first volley, although not sufficient to excite the motoneurones, sets the chains of internuncial neurones (Fig. 44.11) reverberating and in this way the motor neurones are continuously bombarded with impulses which facilitate the action of the next volley. Eccles has provided an electrophysiological explanation, namely that each excitatory impulse arriving at a motoneurone reduces the potential across the cell membrane (e.p.s.p. p. 817);

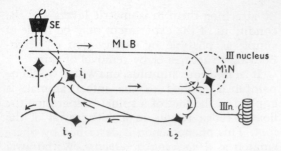

Fig. 44.11 Diagram after Lorente de Nó to show the principle of reverberating internuncial neurones. A stimulus applied through the electrodes, SE, can set up impulses along the medial longitudinal bundle, MLB, to the IIIrd nucleus, MN, which may not discharge and make the intraocular muscle contract until it receives other impulses passing along MLB or i_1. The side branch of MLB may excite i_1, and an impulse then passes, after a synaptic delay, to MN as indicated by the arrows. Alternatively impulses may be propagated along the closed chain of reverberating interneurones i_1, i_2, i_3, i_1. In this way a continuous bombardment of MN may persist for some time, keeping up the central excitatory state. (R. Lorente de Nó (1935). *American Journal of Physiology* **111**, 276 and **113**, 513; (1935–6). *Journal of Cellular and Comparative Physiology* **7**, 47; (1939). *Journal of Neurophysiology* **2**, 402.)

if the critical level for depolarization and discharge of a spike is not reached a rapid succession of such impulses can reduce the potential sufficiently to make the motoneurone discharge.

When a motor nerve to a muscle is stimulated by a single electrical impulse a single volley travels down to the muscle which responds by showing a sudden abrupt rise of tension followed by a quick relaxation, that is, a twitch. A single volley reaching the cord along an afferent nerve evokes a muscular response which is less abupt in onset and more sustained than a twitch; the latent period of the reflex contraction is, of course, longer than the latent period when the motor nerve is stimulated. The muscle tension may rise relatively slowly if the afferent volley does not bring all the accessible motoneurones into action at the one instant. The reflex contraction lasts somewhat longer and often shows a higher tension than a twitch produced by stimulation of the motor nerve. This is due to irregular repetitive firing of the motoneurones termed *after-discharge* which can be explained on the basis of the activity of reverberating

internuncial neurones (p. 909). The after-discharge also accounts for the relatively slow relaxation at the end of a reflex contraction.

If two stimuli are applied to an afferent nerve at an interval greater than the refractory period of the nerve two reflex muscle responses are obtained, but if the separation between the stimuli is less than the refractory period (say 0·5 msec) the second is ineffective. Another consequence of the properties of peripheral nerves is that two subliminal stimuli following one another at a very short interval (up to 0·2 msec) may elicit a reflex effect. This temporal summation is of no physiological importance since the smallest interval in which it occurs is less than the refractory period of the afferent nerve. A more important mechanism, *spatial summation* (simultaneous spinal induction), is shown by the following experiment (Fig. 44.12). The tibialis anterior muscle is attached

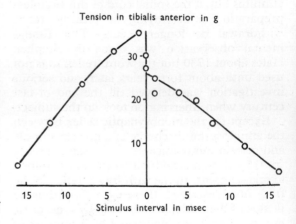

Fig. 44.12 To illustrate spatial summation. Reflex responses of the tibialis anterior muscle in a spinal cat to two stimuli (each one of which alone is just threshold) to two separate afferent nerves in the same limb at various intervals. To right of zero, the nerve of the lateral gastrocnemius muscle is stimulated first; to left of zero, the nerve of the medial gastrocnemius muscle first. Curve shows the relation between resulting tension and the stimulus interval (R. S. Creed, D. Denny-Brown, J. C. Eccles, E. G. T. Liddell & C. S. Sherrington (1932). *Reflex Activity of the Spinal Cord*, p. 32. Oxford: Clarendon Press).

to an isometric lever and the medial and lateral branches of the sciatic nerve are dissected out and laid on stimulating electrodes. The strength of stimulus (the threshold) required to be applied to each branch to produce a just perceptible reflex increase of tension in the muscle is found. If two such stimuli are sent in simultaneously to the spinal cord by both branches of the sciatic nerve quite a considerable tension is aroused reflexly—about 30 g in the experiment shown. If there is an interval between the stimuli the tension developed is less, but spatial summation is still detectable even when the interval between the stimuli is as long as 15 msec.

Sherrington supposed that the arrival of afferent impulses at the motoneurones increases the *central excitatory state* (c.e.s.) and that when this rises to a certain critical value the motoneurones discharge and a reflex contraction occurs (see also p. 909). If the experiment just described is considered in conjunction with Figure 44.13 (B), an explanation of spatial summation can be given. When a weak stimulus is sent in at *a*, one motoneurone discharges and a small contraction is obtained; at

the same time the excitatory state in the neighbouring motoneurones (within the dotted lines) is raised. This part of the motoneurone pool is described as the *subliminal fringe*. A similar state of affairs occurs when *b* is weakly stimulated but, if both *a* and *b* are stimulated at the same time, the subliminal fringes overlap and the c.e.s. in the area common to both may rise to a value at which a number of motoneurones discharge and produce a large muscular contraction.

If any afferent nerve of a limb is stimulated by a series of electrical impulses the flexor muscles of the limb show a reflex tetanus. This presumably corresponds, in an animal in a normal situation, to the continued withdrawal of a limb subjected to a painful stimulus. The tension developed depends on the strength of the afferent stimulation up to a maximum tension which is never as great as in a maximal tetanus obtained by direct stimulation of the motor nerve (Table 44.14). Sherrington ex-

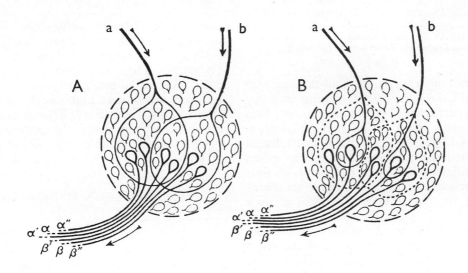

Fig. 44.13 A. Two excitatory afferents, *a* and *b*, with their fields of supraliminal effect in the motoneurone pool of a muscle. *a* activates by itself 4 units (α', α, α'' and β'); *b* by itself 4 (β', β, β'' and α'). Concurrently they activate not 8 but 6, that is, give a contraction deficit by occlusion of contraction in α' and β'. B. Weaker stimulation of *a* and *b* restricting their supraliminal fields of effect in the pool as shown by the continuous line limit: *a* by itself activates 1 unit; *b* similarly; concurrently they activate 4 units (α', α, β' and β) owing to summation of subliminal effect in the overlap of the subliminal fields outlined by dots. (Subliminal fields of effect are not indicated in diagram A). (C. S. Sherrington (1929). *Proceedings of the Royal Society* B, **105**, 338.)

plained this finding by suggesting that no single afferent nerve can reach all the motoneurones of a given muscle; some, however, command a larger fraction than others. This phenomenon he described as *fractionation*.

Table 44.14 To show the tension developed by the semitendinosus muscle of a cat made to contract directly and reflexly

	Isometric tension (g)	Percentage of maximum
Direct stimulation of the motor nerve to the semitendinosus muscle (maximal motor tetanus)	3000	100
Maximal stimulation of the afferent nerves named below yielded the tensions shown in the next column:		
Anterior tibial nerve	2900	97
A dorsal digital nerve	2400	80
Small sciatic nerve	1860	63
Nerve of sartorius	1500	50
External cutaneous (groin) nerve	830	28
Obturator (deep) nerve	630	21

(S. Cooper, D. E. Denny-Brown & C. S. Sherrington (1926). *Proceedings of the Royal Society* **100**, 460.)

The motoneurone is the *final common path* to the muscle but afferents arising in many different parts of the body converge on it. This convergence leads to *occlusion* as explained in Figure 44.13 (A). Strong stimulation of afferent *a* alone brings four motoneurones into action and stimulation of afferent *b* alone makes four motoneurones discharge; simultaneous stimulation of the two afferents, because of their overlap, brings only six, not eight, motoneurones into action. As would be expected, therefore, in experiments on the flexion reflex in a spinal animal it is found that the tension produced by the simultaneous maximal stimulation of two afferents is only a little greater than if either afferent is stimulated alone.

Although the motoneurone is the final common path it is now clear that motoneurones are not all of the same type. Thus it has been possible to distinguish fast and slow (phasic and tonic) motoneurones, the nerve cells concerned supplying a group of muscle fibres which is homogeneous histochemically.

In the flexion reflex the tension of the flexor muscle rises very suddenly (Fig. 44.15). This

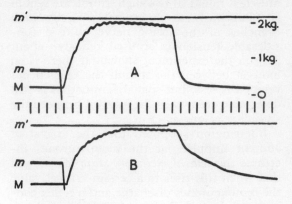

Fig. 44.15 Records of contraction of tibialis anterior of the cat. This muscle is an ankle-flexor. A, direct stimulation of the motor nerve to the muscle. B, stimulation of the ipsilateral popliteal nerve, an afferent nerve. Spinal preparation. M, myograph record. *m*, signals beginning of stimulation. *m'*, signals the end of stimulation. Stimulation frequency 40 sec. T, 0·04 sec. The tension calibration in kg applies to both traces. (E. G. T. Liddell & C. S. Sherrington (1923). *Proceedings of the Royal Society* B **95**, 301).

suggests that all the motor neurones involved discharge simultaneously (the reflex is said to be *d'emblée*) and there is no gradual *recruitment* of neurones. The latent period is very short—of the order of 10 msec. A smooth tetanus is produced only when a high rate of stimulation is used, say 40 per second. This description is very different from that of the extension reflex which is considered on p. 913. (See also Table 44.16.)

Table 44.16 Characteristics of flexion and crossed extension reflexes

	Flexion reflex	Crossed extension reflex
Latent period	Short (10 msec)	Long (40 to 100 msec)
Tension	Rises suddenly	Rises slowly (recruitment)
	Declines quickly	Declines slowly (after-discharge)
Stimulation for complete tetanus	40 per sec to ipsilateral afferent nerve	5 per sec to contralateral afferent nerve
Central excitatory state	Dies away quickly	Dies away slowly

If a reflex is repeatedly elicited the response of the muscle gradually declines and may eventually cease. Since direct stimulation of the motor nerve or the muscle itself is still able to produce a contraction it is evident that the *fatigue* must depend on changes in the central nervous system, probably as the result of changes at the synapses (for a contrary view see p. 846). The scratch reflex in the dog (p. 918), elicited by stimulation of the skin over the thorax, shows fatigue quite readily if the point of stimulation is kept constant but if this is changed slowly the rhythmic scratching movements can be greatly prolonged. In this example fatigue is likely to be due to changes in the afferent side of the reflex arc. Reflexes vary very much in their liability to fatigue; the scratch reflex is easily fatigued but the postural reflexes (p. 924) can be maintained almost indefinitely. Fatigue can be regarded as an increase of 'resistance' to passage of a nerve impulse across a synapse which may increase up to a complete block. It is obvious that quite apart from fatigue there must be a variation in the 'resistance' of the pathways throughout the central nervous system, for if this were not so, reflex responses would be unlikely to follow any definite pattern and an afferent impulse, because the interconnexions within the nervous system are so great, might send enormous numbers of motor neurones into action. This effect is seen in poisoning with strychnine or tetanus poisoning (lockjaw), where a slight touch on the skin may send the animal or man into generalized convulsions.

EXTENSION REFLEXES

If in a spinal preparation flexor withdrawal is evoked in a limb by the application of a noxious stimulus extension of the opposite limb takes place. Thus if a flexion reflex is induced in the left leg by a stimulus applied to the central end of the left lateral popliteal nerve the right quadriceps muscle (the extensor of the leg) contracts. This is a *crossed extension reflex*. The latent period, from 40 to 100 msec, is several times that of a flexor reflex. When this reflex is elicited a single afferent volley produces a succession of volleys from the motoneurones. Thus a slow rate of stimulation of the afferent nerve, say 5 per second, results in a smooth tetanus which could be produced by direct stimulation of the motor nerve only if

40 volleys per second arrived at the muscle. As stimulation of the afferent nerve continues the tension rises slowly as the number of motoneurones involved (or *recruited*) gradually increases. Because after-discharge of motoneurones (p. 910) is well marked in extension reflexes the muscle tension falls off more slowly than in a flexion reflex when the stimulation of the afferent nerve ceases.

The crossed extension reflex is elicited, as we have just seen, by stimulation of a *contralateral* afferent nerve. If, during the course of such a reflex, an *ipsilateral* afferent is stimulated there is, after a very short latent period, a large reduction of tension (Fig. 44.17). The

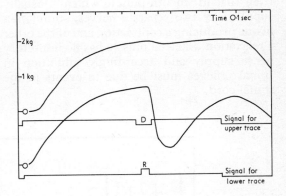

Fig. 44.17 Records of the contraction of the vastocrureus muscle in a decerebrate cat. Reflex tetanus from the crossed peronco-popliteal nerve stimulated electrically 48 times per second. The same intensity of stimuli was applied in both cases. In the upper trace 6 consecutive stimuli are omitted during a short circuit marked by a drop in the signal line D. In the lower trace there was no break of the stimulation of the excitatory afferent nerve but an intercurrent stimulation of the ipsilateral peroneal-popliteal nerve for the period marked by the rise in the signal line R, a period corresponding to the delivery of three consecutive stimuli to the contralateral afferent nerve. Time 0·1 sec. The same tension scale is applicable to both curves. (E. G. T. Liddell & C. S. Sherrington (1924). *Proceedings of the Royal Society* B, **95**, 142.)

amount of this inhibition depends on the strength of the stimulation of the ipsilateral afferent nerve but even a weak stimulus can cause some inhibition. Stimulation of the ipsilateral nerve alone, of course, elicits a flexion reflex; this is an artificial way of producing a flexion reflex which in the intact animal is usually elicited by applying a noxious stimulus to the skin. The flexion reflex, no matter how it is elicited, takes precedence over the extension reflex. During the inhibition of a crossed extension reflex, caused by stimulating an ipsilateral afferent nerve, the tension curve falls as quickly as at the end of a tetanus produced by direct stimulation of the motor nerve. Since an extension reflex dies away quite slowly it is likely that the inhibitory stimulus stops the discharge of the motor neurones. The short latent period of the inhibitory effect suggests that it is brought about by impulses which travel from the ipsilateral nerve to the ventral horn cells by a direct route with few synapses. This inhibition is quite different from that in smooth muscle which is usually supplied by two separate nerves, stimulation of one producing a contraction and of the other a relaxation. Skeletal muscle has no inhibitory nerve supply and accordingly inhibition in spinal reflexes must be due to events in the spinal cord.

Sherrington supposed that an afferent impulse which leads to inhibition of a reflex builds up a *central inhibitory state*, or c.i.s., and that this is capable of neutralizing or removing c.e.s. The tracings of experiments on inhibition (Figs. 44.17 and 18) show that c.i.s. can last for a considerable time—in some experiments up to 200 msec. The c.i.s. may be regarded as a process which makes polarized membranes of neurones more stable by increasing the potential difference across them, that is by hyperpolarization (p. 815).

STRETCH REFLEXES *Not with BLH around anymore!*

The knee jerk is an example of this reflex. By tapping the patellar tendon the muscle spindles within the quadriceps, the extensor muscle of the knee, are stretched and after a very short latent period the muscle contracts and the leg is jerked forward (Fig. 44.18). It is essential for the eliciting of the reflex that the quadriceps be initially partly stretched. The reduced (central) reflex time, about 2 msec, is so short that the reflex must be purely spinal

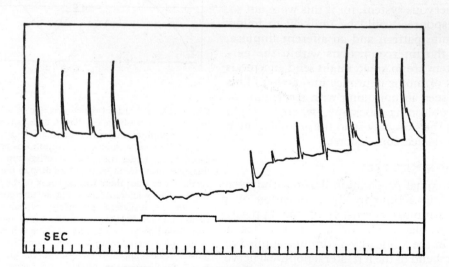

Fig. 44.18 Reflex inhibition of the knee-jerk. Records from above downwards: Tracing from preparation of the extensor muscles of the knee, recording a series of knee-jerks elicited at each alternate beat of a metronome. Signal line: weak repetitive stimulation of the central end of the hamstring nerve was applied during the time marked by the rise of the signal. Time in sec. At onset of stimulation the tonus of the extensor muscles at once fell and with it the knee-jerk was temporarily abolished. After cessation of the inhibiting stimulus the tonus and the knee-jerk quickly returned, and the latter became more brisk than before the inhibition. This increase in the reflex was described by Sherrington as a 'rebound' of the motoneurones. (C. S. Sherrington (1947). *The Integrative Action of the Nervous System*, p. 89. Cambridge University Press.)

and consist of a number of simultaneous mono-synaptic reflexes. It depends, like any other reflex, on the integrity of the reflex arc which consists of the afferent nerves, spinal cord and the efferent nerves. If the reflex is absent there must be an interruption somewhere in this arc. Since the character of this spinal reflex is greatly changed by impulses from higher parts of the brain its modifications are of con-siderable diagnostic value (p. 945) and also Fig. 46.10).

In animal experiments the knee jerk is inhibited if a flexion reflex is induced at the same time, as in Figure 44.18. Here the period of inhibition outlasts the inhibiting stimulus and may be followed by an increased excita-bility of the knee jerk as the result of the so-called *rebound* of the motor neurones.

Quantitative studies of stretch reflexes, sometimes called myotatic reflexes, are most easily made by increasing the length of a muscle attached to a myograph. The rise of tension which occurs on stretching is shown extremely well by the extensor muscles of a decerebrate preparation (M, Fig. 44.19). The

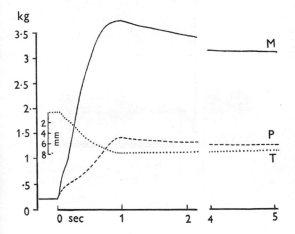

Fig. 44.19 Whole quadriceps of decerebrate cat. Mus-cular response to 8 mm stretch T before (M) and after (P) cutting the nerve to the muscle. Myograph multiplies tendon movement 62 times. Time in seconds. (E. G. T. Liddell & C. S. Sherrington (1924). *Proceedings of the Royal Society* B, **96**, 213.)

rise of tension is dependent on the integrity of the reflex arc but it is not abolished by anaesthetization of the tendon or by removal of the skin. After the nerve to the muscle is cut (P, Fig. 44.19) the rise of tension on stretching the muscle is relatively small and can be accounted for by the elastic properties of the muscle itself. Since the tension produced by stretching can be reduced by stimulation of an ipsilateral sensory nerve or by stretching an antagonistic flexor muscle there is no doubt about its reflex origin.

When a muscle is stretched the discharge in the large afferent fibres from its muscle spindles increases and produces by a pathway, which can be monosynaptic, a contraction in the same muscle which resists the applied stretch. The rate of discharge can be increased, even when the length of the muscle is fixed, by stimulating the γ efferents (see p. 831). The increased discharge from the annulospiral endings produces a reflex contraction of the muscle. The discharge in the large afferents, therefore, depends partly on the length of the muscle and partly on the length of the intra-fusal fibres (see Figs. 44.20 and 44.21).

When a muscle is contracting the discharge from its spindles declines and the reflex excitation of its motoneurones declines also. As the muscle relaxes when the contraction has ended the spindle discharge increases and the excitation of the α motoneurones increases. The muscular movement is thus prevented from being jerky by a reduction of tension during contraction and by a slowing of the fall of tension in relaxation.

If a quadriceps muscle is divided into four portions which can be separately stretched it is found that the rise of tension is confined almost entirely to the part extended. Pre-sumably the impulses from the muscle spindles excite motor neurones which innervate muscle cells in their immediate vicinity. In this way each part of the muscle contributes its share to resisting elongation.

These stretch reflexes are characteristic of the antigravity extensor muscles and they are especially noticeable in the decerebrate pre-paration (p. 926) in which the extensor muscles are firmly contracted. This state of contraction of the antigravity muscles is reflex in origin, since it is abolished by cutting the anterior or posterior roots or by section of the muscle tendon. This excessive tonus seen in decere-brate rigidity is similar in character to the spasticity which develops in patients with

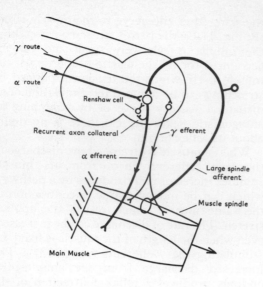

The γ efferents, by controlling the tension in the muscle spindles, control their sensitivity. The rate of discharge of the γ efferents can be altered by the 'higher centres' and may rise before the α motoneurones discharge. This latter finding has suggested that changes in tone could be brought about by an increase or decrease in the γ afferent discharge, the subsequent increase or decrease in the spindle afferent discharge causing through monosynaptic pathways an alteration in the reflex discharge of the α motoneurones. Evidence that this self-regulating mechanism is in action in posture is provided by the 'silent period'. While a steady voluntary contraction is maintained the muscle is made to twitch by an electric shock; during the twitch the electromyogram (Fig. 44.21) is reduced to a

Fig. 44.20 The muscle spindle lies in parallel with the main muscle and its fast-conducting afferent (large spindle afferent or Ia spindle afferent) is in synaptic connexion with the large α motoneurone supplying the main muscle fibres. A slow-conducting γ motor efferent (thin line) supplies the contractile poles of the spindle and thus can alter the bias on the spindle sensory ending. The muscle can be made to contract either by impulses from higher centres exciting the α motoneurone directly (the α route) or by impulses in the γ efferents (the γ route) which activate the muscle indirectly via the stretch-reflex arc (the 'follow-up' servo). A subsidiary feedback loop via the recurrent axon collateral and an inhibitory Renshaw interneurone may be concerned in stabilizing the response of the α motoneurone to its excitatory input. (P. H. Hammond, P. A. Merton & G. G. Sutton (1956). Nervous gradation of muscular contraction. *British Medical Bulletin* **12**, 214.)

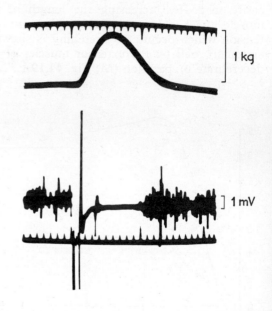

Fig. 44.21 The subject makes a steady voluntary contraction of 1 kg, and the associated action potentials (the electromyogram) are recorded. When the muscle is excited by stimulating the motor nerve with an electric shock, the electromyogram is 'silent' during the superimposed twitch. This is because impulses from the muscle spindles cease during the twitch and thus excitation is withdrawn from the α motoneurones. The record consists of ten traces superimposed, to demonstrate the great regularity of the phenomenon. Top and bottom traces: 10 msec and 100 msec. Second trace: twitch tension, roughly 1 kg. Third trace: action potentials recorded with a needle electrode. (P. A. Merton (1951). *Journal of Physiology* **114**, 187.)

hemiplegia. In the healthy subject skeletal muscle when fully relaxed is electrically silent but passive movement evokes a contraction to oppose the stretching force. Where there is a lesion affecting the upper motor neurone the stretch reflex is abnormally excitable and there is an abnormal degree of resistance to passive movement. A characteristic of muscle tonus and of stretch reflexes is that they can be maintained for long periods with only a small expenditure of energy and with, consequently, an absence of fatigue. However, recording from human spindle afferents during voluntary contraction, although it confirms that spindle discharge accompanies voluntary contraction, has failed so far to confirm that it initiates the α-motoneurone discharge.

straight line (that is, is 'silent'). The shortening of the muscle reduces the length of the muscle spindles with a consequent decline in the number of impulses in the large afferents, and it is this withdrawal of excitation from the α motoneurones which makes them silent. This pathway seems, however, hardly suited for carrying out rapid movements because of the slow conduction rate in the small fibres and because of the delay in the reflex circuit. Experiments on animals with deafferented limbs or deafferented respiratory muscles have indicated that voluntary and respiratory α motoneurones can be reached directly by descending pathways; the large calibre of the axons of the α motoneurones are eminently suitable for rapidly executed movements.

A decerebrate preparation (p. 926) with rigidity of the extensor muscles can be used to demonstrate other characteristics of the myotatic reflex. If the experimenter tries to flex the hind-limb at the knee there is first considerable resistance, since the myotatic reflex is evoked, but if the force is maintained the muscle slackens quite suddenly and then can easily be flexed. The sudden loss of resistance is similar to that felt when closing a pocket-knife; hence the name *clasp-knife reaction*. This reaction occurs only when great force is applied to the extensor muscle; it may therefore be regarded as a protective mechanism since the muscle gives way before the force is great enough to produce damage by tearing the fibres. The discharge from the Golgi tendon organs (p. 829), which have a high threshold, is believed to be the basis of this reaction; the impulses from these organs are inhibitory to the motoneurones of the synergic muscles and perhaps excitatory to the antagonists by a polysynaptic pathway. In this way the extension phase of stepping could be terminated and the flexion phase initiated.

After the limb of a decerebrate preparation is forcibly flexed in this way the extensor muscles are lengthened (*lengthening reaction*) and the new flexed position is maintained. If the flexed limb is now extended the tension in the extensor muscle is reduced because the ends are brought nearer, and presumably because the γ efferents take up the slack of the intrafusal fibres the afferent spindle discharge increases; the α motor neurone discharge is then increased reflexly so that the position of the limb is maintained. This is called the *shortening reaction*, since the fibres of the extensor muscle are shortened. These two modifications of the stretch reflex give the limbs plasticity, that is to say they allow the limbs to remain in the position in which they are placed.

RECIPROCAL INNERVATION

In addition to these 'static' reflexes, which are responsible for maintenance of posture, the spinal cord is concerned in the 'dynamic' reflexes involved in walking and running. These alternating contractions of the flexors and extensors of the limbs depend on production and inhibition of the appropriate reflexes. At any one instant the extensor muscle group of a limb is contracting and the antagonistic flexor group is relaxing, or vice versa. This involves *reciprocal innervation*, illustrated in Figure 44.22, which shows that in a reflex flexion movement there is an active contraction of the flexor muscles with a simultaneous relaxation of the extensor muscles. The converse occurs in a reflex extension movement. Reciprocal innervation is discussed on pp. 819–821 and the nerve pathways are described in Figure 39.41. As the limb flexes the extensor muscles are stretched and there is a reflex contraction of the extensor muscles with a reflex relaxation of the flexors. The resulting extension stretches the flexor muscles in turn and produces flexion of the limb with an inhibition of the extensor muscles. A rhythmic movement is thus set up in which the flexor and extensor reflexes come into action successively and the final common path is commanded alternately. This rhythmic response is readily evoked because after the occurrence of a reflex of one type it is easier to elicit the antagonistic reflex (Fig. 44.24). Figure 44.25 illustrates how the alternating movements of the limb which is the well-known pattern of walking may be brought about. At the instant represented in Figure 44.25b, a flexion reflex is elicited in the left fore-limb and an extension in the right fore-limb. This can be explained on the basis of the connexions shown in Figure 39.41. By longer connexions in the spinal cord the full walking pattern involving reciprocal activity in all four limbs is obtained.

The pattern of successive flexions and extensions brought about through reciprocal innervation can be overcome in several ways.

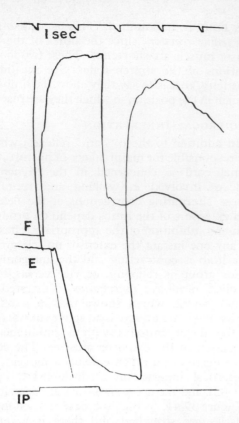

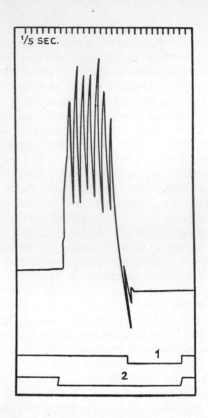

Fig. 44.22 Reciprocal reflex of antagonistic muscles of knee of a decerebrate cat. F, semitendinosus, a knee flexor; E, vastocrureus, a knee-extensor. IP, ipsilateral popliteal nerve stimulated during rise of signal line. The flexor muscle contracts and the extensor relaxes. The inhibition of the extensor is followed by marked rebound contraction, which is accompanied by a simultaneous relaxation of the flexor muscle. The myograph writer for extensor muscle is set a little to right of that for flexor muscle, in order that the two may clear each other: the ascent of F and the descent of E are therefore, in fact, practically synchronous. (C. S. Sherrington (1913). *Quarterly Journal of Experimental Physiology* **6**, 257.)

Fig. 44.23 The scratch reflex cut short by excitation of the skin of a digit of the opposite hind-foot. Records from above downwards: Time in ⅕ sec. Myograph curve: Scratch reflex. Signal 1: descent of signal marks the period of application of the stimulus to the opposite hindfoot. Signal 2: descent of signal marks the period of application of the stimulus exciting the scratch reflex. (C. S. Sherrington (1947). *The Integrative Action of the Nervous System*, p. 193. Cambridge University Press.)

For example, when the pad of the foot of an animal is pressed there is a *positive supporting reaction* in which the extensors and flexors contract simultaneously to convert the limb into a pillar to support the weight of the body.

A rhythmic reflex involving reciprocal innervation which has been much studied is the scratch reflex. This can be elicited in a dog which has been allowed to recover after section of the cervical spinal cord below the level of the phrenic nerves. Stimulation of the skin over the ribs results in rhythmic scratching movements (Fig. 44.23) by the hind-limb which appear to the observer to be purposeful. At any rate, the movements are co-ordinated accurately enough to remove the source of irritation. Thus this reflex may be said to possess 'localization'. If a noxious stimulus is applied to the foot during a scratch reflex the latter is completely inhibited (Fig. 44.23). As described already, flexion withdrawal reflexes which take the animal away from danger are prepotent, that is they have precedence over all other reflexes.

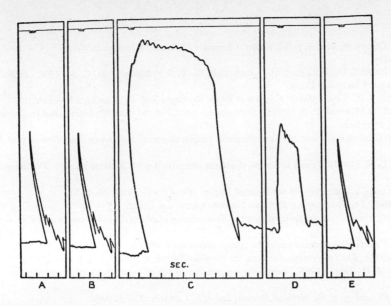

Fig. 44.24 Successive induction. Crossed extension reflex augmented by precurrent flexion reflex. Records from above downwards: Signal line: break-shocks of each stimulation evoking crossed extensor reflex. Myograph curve: crossed extensor reflex. This reflex was being elicited regularly by eleven break-shocks (unipolar faradization to skin of opposite foot) at 1 min intervals, stimulus and reflex being of low intensity. Time in seconds. Intervals of 1 min between A, B, C, D and E. In the interval between B and C a strong flexion reflex of the limb responding in the crossed extension reflex was provoked and maintained for 45 sec. The next following extension reflex C shows augmentation; this augmentation is also evident, though less, in the next crossed reflex D. In E, a minute later, the augmentation is seen to have passed off. (C. S. Sherrington (1947). *The Integrative Action of the Nervous System*, p. 210. Cambridge University Press.)

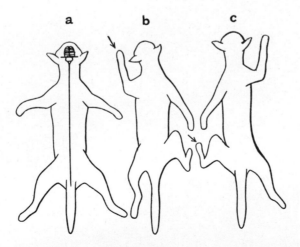

Fig. 44.25 a = Position under decerebrate rigidity (all four limbs extended). b = Change of attitude from *a* evoked by stimulation of left fore-foot. c = Change of attitude from *a* evoked by stimulation of left hind-foot. (C. S. Sherrington (1897–98). *Journal of Physiology* **22**, 330.)

REFERENCES

BULLER, A. J. (1963). The motor unit in reflex action. Chapter 4 in *Recent Advances in Physiology* (Edited by R. Creese). London: Churchill.

CREED, R. S., DENNY-BROWN, D., ECCLES, J. C., LIDDELL, E. G. T. & SHERRINGTON, C. S. (1932). *Reflex Activity of the Spinal Cord*. Oxford: Clarendon Press.

ECCLES, J. C. & SCHADÉ, J. P. (Eds.) (1964). *Progress in Brain Research*, Vol. 12. London: Elsevier.

ELDRED, E., GRANIT, R. & MERTON, P. A. (1953). Supraspinal control of the muscle spindle and its significance. *Journal of Physiology* **122**, 498–523.

GUTTMANN, L. (1952). Clinical significance of paraplegia in flexion or extension. *Journal of Nervous and Mental Diseases* **116**, 957–972.

HUNT, C. C. & PERL, E. R. (1960). Spinal reflex mechanisms concerned with skeletal muscle. *Physiological Reviews* **40**, 538–579.

KUHN, R. A. (1950). Functional capacity of the isolated human spinal cord. *Brain* **73**, 1–51.

LIDDELL, E. G. T. (1960). *The Discovery of Reflexes*. Oxford: Clarendon Press.

NOORDENBOS, W. (1959). *Pain. Problems pertaining to the Transmission of Nerve Impulses which give Rise to Pain*. London: Elsevier.

OSCARSSON, O. (1965). Functional organization of the spino- and cuneocerebellar tracts. *Physiological Reviews* **45**, 495–522.

SHERRINGTON, C. S. (1947). *The Integrative Action of the Nervous System*. Cambridge University Press.

SINGER, M. & SCHADÉ, J. P. (Eds.) (1965). Regeneration patterns in the nervous system. *Progress in Brain Research* **14**, 1–278.

WALSH, E. G. (1964). *Physiology of the Nervous System*, 2nd edn. London: Longmans.

WARTENBERG, R. (1953). *Diagnostic Tests in Neurology*. Chicago: Year Book Publishers.

WYKE, B. (1963). *Brain Functions and Metabolic Disorders*. London: Butterworths.

45 The brain stem and postural reflexes

The spinal cord is continuous within the skull with the brain stem, which includes the medulla oblongata, the pons and the mid-brain. Figure 45.1 shows that the medulla oblongata widens as it passes forward (rostrally) and the spinal canal opens out into a lozenge-shaped space, the fourth ventricle. On the ventral aspect of the medulla (Fig. 45.2) the pyramids and their decussation are prominent; a little rostral to this the pons forms a bridge of decussating transverse fibres gathered together into the two middle cerebellar peduncles which consist of fibres running into the cerebellum. On the dorsal aspect of the medulla (Fig. 45.1) the inferior cerebellar peduncles diverge and pass to the cerebellum. Rostral to the pons is the short mid-brain, the four colliculi lying on its dorsal surface. The mid-brain is joined to the fore-brain by the two cerebral peduncles.

A transverse section through the caudal, or lower, part of the medulla (Fig. 45.3) shows that the structure of the spinal cord is still recognizable at this level, but the grey matter is cut across by the fibres of the cerebrospinal tract passing from the pyramids across the mid-line to run down the cord in the lateral corticospinal or crossed pyramidal tract. These interlacing fibres passing obliquely across the cord form the decussation of the pyramids or the *motor decussation*. A section a little rostral to this (Fig. 45.4) shows the gracile and cuneate nuclei in which the fibres of the posterior columns of the white matter of the cord terminate. From these nuclei new fibres arise and, passing across the mid-line as the internal arcuate fibres, form the medial lemniscus (or medial fillet) which runs to the thalamus. This crossing can be called in physiological terms the *sensory decussation*. The spinothalamic fibres continue to occupy a ventrolateral position, as in the cord, but as they travel rostrally, first the touch fibres in the medulla, and then the pain and temperature fibres in the pons

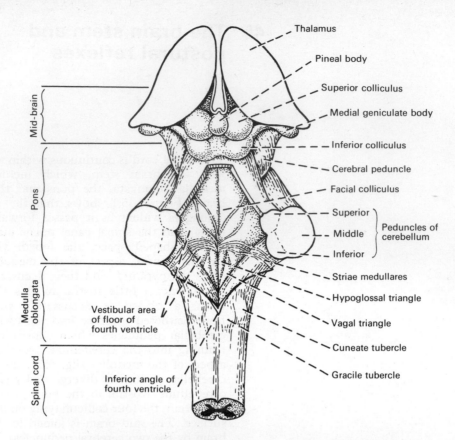

Fig. 45.1 Diagrammatic view of the human brain stem exposed from behind and above (approximately natural size). The cerebellum has been removed by section through the peduncles.

pass medially on each side and join the medial lemniscus. In the upper part of the medulla the central canal is displaced towards the dorsal surface of the medulla and opens out to give the appearances shown in Fig. 45.5.

In the pons (Fig. 45.6) the pyramidal fibres are separated into small bundles by the fibres of the middle peduncle of the cerebellum which, running at right angles to them, pass across the pons to the opposite cerebellar cortex. The nuclei pontis are small collections of grey matter scattered among the transverse fibres. The superior cerebellar peduncles (Fig. 45.1) are composed chiefly of fibres coming from the dentate nuclei of the cerebellum which cross in the mid-line (Fig. 45.7) before entering the thalamus or red nucleus.

The mid-brain (Figs. 45.7 and 8), only about 2 cm long, has a mass of grey matter centrally, around the aqueduct of the mid-

brain; in the ventral part of this grey matter lie the motoneurones of the third and fourth nerves. Dorsal to the aqueduct are the nuclei of the superior and inferior colliculi. A large part of the mid-brain is occupied by the cerebral peduncles which are cut across by the deeply pigmented cells of the substantia nigra into two parts, the base ventrally and the tegmentum dorsally. The base carries fibres from the cerebral cortex to the pons and spinal cord. In the tegmentum three decussations occur. The most caudal is the decussation of the superior peduncles of the cerebellum (Fig. 45.7); the rubrospinal tracts (Fig. 45.8) and the tectospinal tracts cross more rostrally at the level of the third nerve nucleus. The red nucleus, a prominent feature of the tegmentum in the upper part of the mid-brain, is part of the so-called extrapyramidal system. Fibres from this nucleus cross in the decussation of

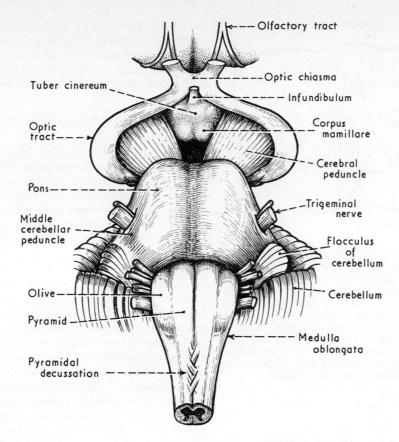

Fig. 45.2 View of ventral (lower) surface of medulla oblongata, pons and mid-brain of man (approximately natural size).

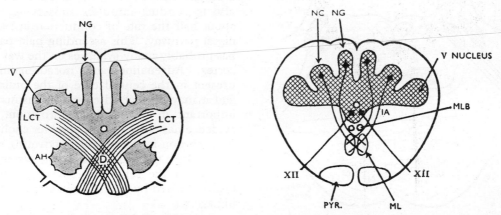

Fig. 45.3 Section through lowest part of the human medulla at the level of the *motor decussation*. AH = most rostral part of the anterior horn; D = decussation of cerebrospinal fibres (decussation of the pyramids); LCT = lateral cerebrospinal tract; V = spinal nucleus of fifth nerve, continuous caudad with the posterior horn of the spinal cord. (About 3 times actual size.)

Fig. 45.4 A section somewhat rostral to that in Fig. 45.3 to show the *sensory decussation*. NG = nucleus gracilis; NC = nucleus cuneatus; ML = medial lemniscus; XII = hypoglossal nerve; IA = internal arcuate fibres; PYR = pyramid; MLB = medial longitudinal bundle (fasciculus). (About 3 times actual size.)

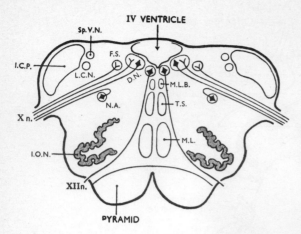

Fig. 45.5 Section through the upper part of the human medulla oblongata just cranial to the inferior angle of the fourth ventricle. The reticular formation occupies the space between the grey matter on the floor of the fourth ventricle and the pyramids. I.C.P. = inferior cerebellar peduncle; Sp.V.N. = spinal vestibular nucleus; L.C.N. = lateral cuneate nucleus; N.A. = nucleus ambiguus; D.N. = dorsal nucleus of vagus; F.S. = fasciculus solitarius; M.L.B. = medial longitudinal bundle; T.S. = tecto-spinal tract; M.L. = medial lemniscus; I.O.N. = inferior olivary nucleus. (About 4 times natural size.)

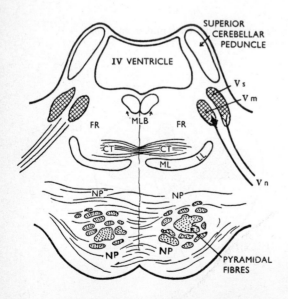

Fig. 45.6 Section through the upper part of the pons. FR = formatio reticularis; NP = nuclei pontis; ML, LL = medial lemniscus, lateral lemniscus; MLB = medial longitudinal bundle (fasciculus); V_s, V_m = sensory and motor nuclei of trigeminal nerve (V_n); CT = corpus trapezoideum. (About 4 times natural size.)

the rubrospinal tracts and pass down in the formatio reticularis (Fig. 45.7) of the pons and medulla oblongata to lie in front of the cerebrospinal tract in the cord. The tectospinal and tectobulbar tracts arise in the superior colliculi.

RETICULAR FORMATION OF THE BRAIN STEM

The work of Magoun and others from 1946 onwards has revealed the great importance of the reticular formation of the brain stem. It consists of scattered nerve cells lying in the central core of the brain stem which influences all parts of the central nervous system both rostrally and caudally. When the reticular formation is stimulated electrically motor activities either phasic or postural may be facilitated or inhibited according to the site of stimulation, but stimulation through implanted electrodes has failed to cause any inhibition in the intact conscious animal. The facilitating action of the reticular formation may be responsible for the exaggerated stretch reflexes which are expressed as spasticity.

The 'classical' afferent (lemniscal) pathways send lateral branches to the reticular core and this in turn influences the hypothalamus, which affects the anterior pituitary especially the output of ACTH. Impulses passing up from the reticular area to the cerebral cortex are concerned with modifying wakefulness and arousal. This extralemniscal pathway seems also to conduct impulses subserving pain at about half the rate of conduction in the lemniscal pathway. This ascending pain pathway has probably many synapses on the way to the cortex. Adrenaline and noradrenaline are present in the central core of the brain stem and it may be that adrenergic transmission is important in the reticular formation. Good evidence has also been obtained for cholinergic transmission in the reticular activating system especially in the rostral part of the system. The activity of the reticular activating system is much reduced by anaesthetics and tranquillizing agents.

DECEREBRATE RIGIDITY

The modifying and integrating influence of the brain stem on the spinal reflexes has been investigated by making sections through it at various levels to cut off the higher parts of the

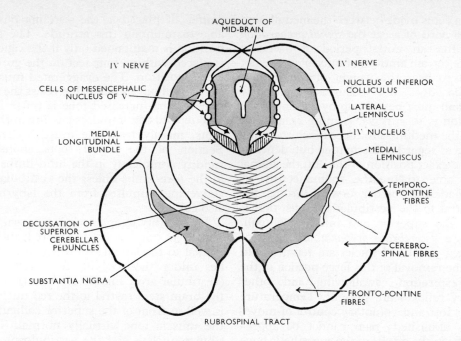

Fig. 45.7 Section through the mid-brain at the level of the inferior colliculi. The area between the medial lemnisci is occupied by the reticular formation. (About 5 times natural size.)

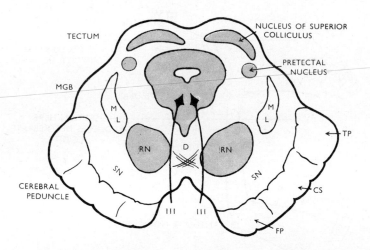

Fig. 45.8 Section through the mid-brain at the level of the superior colliculi. The pretectal nucleus is actually rostral to the plane of this section. The reticular formation occupies the space between the central grey matter and the substantia nigra. RN = red nucleus; MGB = medial geniculate body; D = decussation of rubrospinal tracts; ML = medial lemniscus; SN = substantia nigra; TP = temporopontine fibres; CS = cerebrospinal fibres; FP = frontopontine fibres. (About 5 times natural size.)

brain. If a cut is made between the medulla and the spinal cord of a cat the *spinal preparation* shows, after an initial period of flaccidity (p. 907), a certain amount of muscular tone not sufficient to bear the weight of the animal; respiration does not occur spontaneously (p. 603) and must be kept up by a pump if the preparation is to survive. A more cranial cut through the medulla produces much the same condition as regards muscle tone but, depending on the exact position of the cut, the animal may be able to breathe spontaneously. A section made still further forward through the pons rostral to the vestibular nucleus, that is between the superior and inferior colliculi, produces a *decerebrate preparation*; the cerebral cortex and basal nuclei are removed. In spite of the removal of this large portion of the brain, respiration, swallowing and other medullary reflexes are present. Temperature control is lost and voluntary control of movements is abolished; pain cannot be appreciated. When the effects of the anaesthetic have worn off, the cat shows decerebrate rigidity as described by Sherrington in 1898. The limbs of this preparation are extended, the tail is raised and the head elevated. The muscles acting against gravity are rigid so that the animal, if placed in the standing position, is able to maintain this attitude. The standing position is maintained only if the rigid animal is placed with its four feet on the ground in a stable position. The exaggerated muscle tone is seen only in muscles that resist the effect of gravity; the increased tone is reflex in origin and depends on impulses passing in the dorsal roots, mainly from the muscles themselves. Section of the dorsal roots abolishes the rigidity completely in the hind-limbs but not in the fore-limbs unless the vestibular nerves conveying impulses from the labyrinths are also cut.

For rigidity to appear it is necessary for the section to pass caudal to the red nucleus but cranial to the lateral vestibular nucleus which lies under the floor of the fourth ventricle (vestibular area, Fig. 45.1). A section through the brain stem rostral to the red nucleus, that is well in front of the superior colliculi, leaves the muscle tone virtually normal. The vestibular nucleus and the vestibulospinal fibres arising from it are essential for decerebrate rigidity, since section of the brain stem caudal to the nucleus or section of the anterolateral columns of the spinal cord, which carry the vestibulospinal fibres, abolishes it (Fig. 45.9).

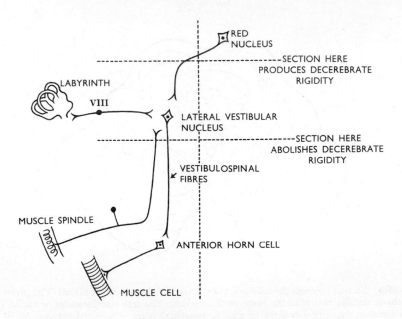

Fig. 45.9 Pathways involved in decerebrate rigidity in the cat. The section below the red nucleus also cuts through the reticulospinal inhibitory fibres.

Unilateral destruction of the vestibular nucleus abolishes rigidity on the same side (Fig. 45.10).

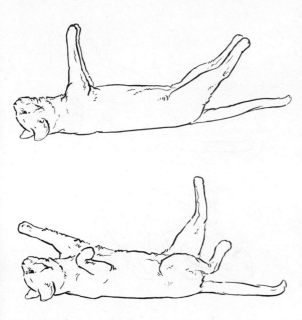

Fig. 45.10 A, decerebrate cat in the supine position to show rigidity of limbs. B, the same after destruction of the vestibular nucleus on the right side. (L. M. N. Bach & H. W. Magoun (1947). *Journal of Neurophysiology* **5**, 331.)

The cerebral cortex seems not to be deeply involved in the production of decerebrate rigidity since removal of the cerebral hemispheres alone does not produce rigidity unless the animal is lifted off the ground (p. 927). Furthermore, the cerebral cortex controls the muscles on the contralateral side of the body, whereas unilateral section between the red nucleus and the vestibular nucleus is followed by rigidity on the same side. The rigidity must depend, therefore, on pathways which go straight down the cord from the pons without any decussation, that is to say on tracts which have already decussated above the level of the section. We are thus led to conclude that decerebrate rigidity occurs in the cat when the vestibular nuclei no longer receive impulses from higher parts of the brain, which exercise an inhibitory influence on these nuclei. Since destruction of the red nuclei alone does not produce rigidity, the inhibitory impulses can-

not be derived from this source. The source of the inhibitory impulses is probably the suppressor reticular formation. When this is put out of action by decerebration the facilitatory reticular system is left unbalanced, the stretch reflexes are facilitated and decerebrate rigidity occurs. Although a cat deprived of its cerebral cortex (decorticate preparation) has a normal attitude and normal distribution of muscle tone when standing or walking, it shows rigidity when lifted off the ground. Thus the cortex may play a small part in inhibiting the vestibular nucleus although the impulses cannot be conveyed in the pyramidal tracts. Since the decorticate animal does not show rigidity when sitting or standing it is probable that the vestibular nuclei receive inhibitory impulses mainly from parts of the brain other than the cerebral cortex.

The γ efferents are specially active in decerebrate rigidity. The spindles discharge rapidly; the stretch reflexes are, therefore, very active and they are in fact responsible for the rigidity. If the dorsal roots of such a preparation are cut, the γ efferents still discharge at the same rate, but, since the discharge from the muscle spindles cannot reach the cord, the rigidity in the hind-limbs disappears (see p. 926). Pollock and Davis found that another type of decerebrate rigidity can be produced by tying off the arterial supply to the brain. In this case deafferentation (by cutting the dorsal roots) does not abolish the rigidity. This second type of rigidity, therefore, cannot be brought about by excitation of the γ efferents and must depend on direct activation of the α motoneurones; in fact the muscle spindles are discharging quite slowly.

Sherrington found that decerebrate rigidity persisted, or even increased, after cerebellectomy and furthermore a decerebrate preparation rendered flaccid by deafferentation is made rigid again by removal of the anterior cerebellum. It is now known that the cerebellum receives afferent influx from muscle spindles and Golgi tendon organs and that spindle activity declines after cerebellectomy. The rigidity must be caused by increased direct excitation of the α pathway, in other words, the operation has caused a switch-over from γ excitation to the α mode of excitation. India ink injection of the brains of cats decerebrated by Pollock and Davis's method shows that their procedure corresponds to a section between the forepart of the cerebellum and pons and thus corresponds to the classical decerebrate animal with the anterior cerebellum removed. The similarity of the reflexes in classical decerebrate animals and in ischaemically decerebrate animals suggests that the α and γ systems may often act together.

Decerebrate rigidity is occasionally seen in man. Since the function of the fore-limbs in man is very different from that in quadrupeds it is only to be expected that the attitude of the arms would also be different. Mid-brain lesions usually produce extension and pronation of the arms but higher lesions usually result in flexion (as in Fig. 45.11). The legs are always rigidly extended.

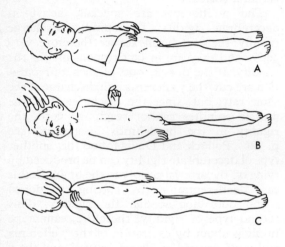

Fig. 45.11 Tonic neck reflexes shown in a decerebrate child. A, the characteristic position, flexion of the arms with extension of the legs. B and C, the position of the neck affects the tone of the muscles of the arms. (L. E. Davis 1925. *Archives of Neurology and Psychiatry (Chicago)* **13**, 572.)

TONIC NECK REFLEXES

The decerebrate cat has been used by de Kleijn, to study postural reflexes which arise in the neck muscles. To investigate the tonic neck reflexes it is necessary to exclude impulses from the labyrinths by performing a double labyrinthectomy.

When the head of this preparation is turned to the right the right limb is extended and the

left limb becomes relaxed. If the head is tilted so that the nose is raised the fore-limbs extend and the hind-limbs relax, as in Figure 45.12A.

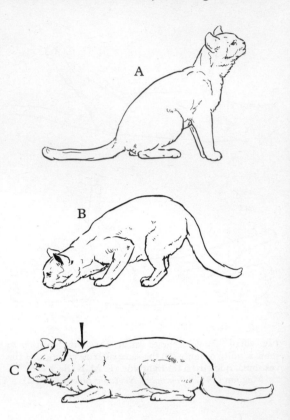

Fig. 45.12 Diagram to show the influence of neck reflexes on posture. Attitude of decerebrate animal when the head is pointed up, A, and pointed down as in B, and when pressure is applied to the neck as at C. Position A is seen in a normal animal when 'begging' at a table; position B when looking for a mouse under a dresser.

The opposite effect is seen when the nose is directed downwards (Fig. 45.12B). Pressure with the hand over the cervical vertebrae causes a relaxation of all four limbs (Fig. 45.12C). These reflex effects are rather slow, having a latency of several seconds. These alterations in the position of the head give rise to afferent impulses which enter the central nervous system by the first three cervical dorsal roots to initiate reflex alterations of tone in distant parts of the body and produce attitudes often observed in the normal dog and cat (Fig. 45.24). Similar reflexes may be seen in decerebrate or decorticate man (Fig. 45.11).

THE LABYRINTHS

The bony labyrinth (Fig. 45.14) consists of three communicating cavities, the vestibule, the semicircular canals and the cochlea, hollowed out of the petrous part of the temporal

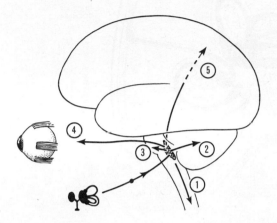

Fig. 45.13 Diagram to show the connexions of the vestibular fibres. 1, vestibulospinal tract; 2, to cerebellum via inferior cerebellar peduncle; 3, to reticular formation; 4, medial longitudinal bundle to nuclei of external eye muscles; 5, to cerebral cortex. Shaded area, vestibular nuclei.

bone, all containing clear fluid of high sodium content, the perilymph. The bony labyrinth communicates with the cerebrospinal fluid by way of the aqueduct of the cochlea. The membranous labyrinth (Fig. 45.15), consisting of the duct of the cochlea, the utricle and saccule and the three semicircular ducts, lies within the bony labyrinth and contains fluid called endolymph, which has a high potassium content. The membranous labyrinth is a closed system, the endolymphatic duct ending in a depression in the petrous bone surrounded by a venous plexus. The superior and posterior canals lie in vertical planes at right angles to one another while the lateral or horizontal canal is in a plane passing backwards and slightly down; the superior canal of one side is in a plane approximately parallel to the plane containing the posterior canal of the other side. The semicircular ducts, each about 15 mm long in man, are smaller in diameter than the bony semicircular canals which enclose them; each duct has a wider portion or ampulla con-

taining the ampullary crests (*cristae ampullares*), which possess hair cells covered with a mass of gelatinous material, the cupola. The cupola is composed of a network of fibres 10 nm in diameter (possibly sulphomucopolysaccharide) embedded in a protein matrix. Both the saccule and the utricle possess *maculae* with hair cells covered with a layer of jelly-like material containing particles of calcium carbonate (otoconia).

The fine nerve filaments from the hair cells in the cristae and maculae become myelinated as they pass centrally in the vestibular nerve, a division of the eighth cranial nerve, to the vestibular ganglion and then to the vestibular nuclei, as shown in Fig. 45.13. From these nuclei, fibres pass to at least five destinations. Some fibres run in the medial longitudinal bundle to the oculomotor nuclei through which changes in position of the eyes are effected. This is the pathway for nystagmus and counter-rolling. Fibres also pass in the medial lemniscus to the thalamus; others go in the inferior cerebellar peduncle to the cerebellar cortex, chiefly to the flocculonodular lobe. Some fibres pass into the reticular formation while others go in the vestibulospinal tract. In addition to these afferent impulses, efferent impulses from the lateral vestibular nuclei reach the hair cells. The function of these efferent fibres is still a matter for speculation

In the erect position in man the utricular maculae are approximately horizontal so that the hair cells project vertically upwards with the calcareous particles (otoconia) lying on them and the saccular maculae lie in a vertical plane (Fig. 45.15). The rate of discharge from the nerve endings varies according to the gravitational pull on the otoliths. The otolith organs could be thought of as 'out of level' indicators. Using very small electrodes Adrian picked up impulses from the vestibular nucleus in the floor of the fourth ventricle of a cat. The responses are of two different kinds, arising presumably in two different kinds of end-organ. One form of response depends on the position of the head in space and signals tilting and linear acceleration; it consists of slowly adapting trains of impulses which arise from gravitational pull on the otoliths. The second form of response, the rotation-controlled type, is produced by angular acceleration and

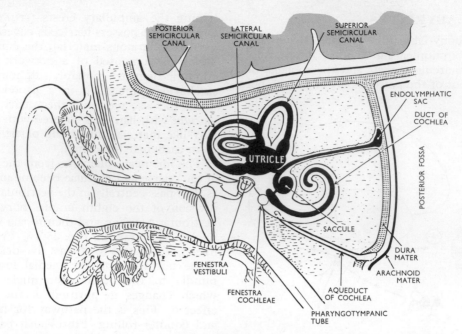

Fig. 45.14 The bony and membranous labyrinths. The tubes shown as solid black contain endolymph. (*By courtesy of J. D. B. MacDougall.*)

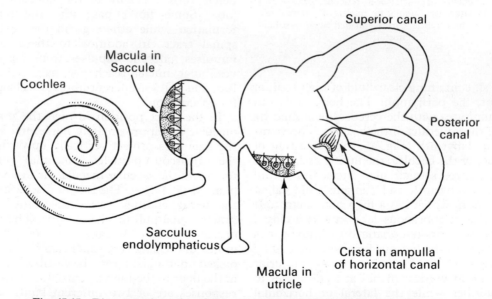

Fig. 45.15 Diagrammatic section of the membranous labyrinth. The macula of the saccule is vertical and the macula of the utricle is horizontal. The crista is situated at the dilated end of each semicircular canal (ampulla). The maculae contain bottle-shaped and cylindrical hair cells separated by supporting cells. Each hair cell possesses numerous cilia about 0·2μm in diameter and a single slightly thicker cilium (kinocilium) (Fig. 45.16). The sensory nerve endings attached to the hair cells send impulses into myelinated nerve fibres (diameter 2 to 9 μm). The hairs of the sensory cells pass into canals in a jelly-like material. In the case of the otolith organs this contains particles of calcium carbonate forming the otolith membrane. The macula in the ampulla is formed into a crest surmounted by a mass of jelly-like material, the cupola, which probably reaches right across the ampulla to the opposite wall.

is unaffected by the position of the head. These impulses must be derived from the cristae of the semicircular ducts since they are most

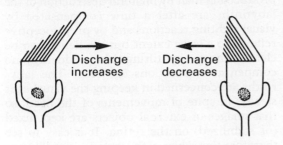

Fig. 45.16 Electron microscope studies have shown that vestibular sensory cells have one thick kinocilium and many thin stereocilia. When movement of lymph occurs towards a kinocilium the discharge increases, and when it occurs in the opposite direction the discharge decreases.

sensitive to rotation in one particular plane; rotation in planes at right angles produces little or no alteration of resting discharges. Rotation in the direction which displaces the cupola towards the utricle increases the frequency of discharge, rotation in the opposite direction diminishes the frequency. If a rotation which produced an increase in frequency is arrested the discharge rate declines and may fall to zero; it usually takes ten to thirty seconds to regain the normal resting discharge rate and this is presumably the time taken by the cupola to regain its resting position (Fig. 45.17). Each hair cell carries numerous fine cilia with a stronger cilium (kinocilium) nearer the utricle. (Fig. 45.16).

We can suppose that the cupolae of the semicircular canals operate as follows. When the

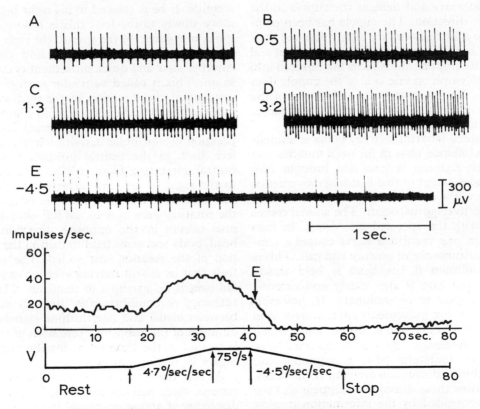

Fig. 45.17 Unit discharges of a tonic vestibular neurone of a cat. All records are from the same neurone. A is resting discharge. B, C, D, records taken shortly after beginning of rotation at 0·5, 1·3 and 3·2 degrees per sec per sec. E, record begins at point marked E on the graph, that is 2 sec after onset of 4·5° per sec per sec deceleration; it shows gradual reduction of discharge frequency to complete inhibition. V, angular velocity. (For further details see H. Shimazu & W. Precht (1965). *Journal of Neurophysiology* **28**, 991–1013.) (*By courtesy of W. Precht.*)

head is rotated the perilymph and endolymph in the semicircular canal lying in the plane at right angles to the axis of rotation tend to lag behind the movement of the head; the cupola is, therefore, deflected and the hair cells are stimulated. Movement of endolymph and of the cupola from the stereocilia to the kinocilium gives a drop in potential and increases the afferent discharge. Since all the hair cells are oriented in the same way, all respond similarly to a given displacement. In this way the semicircular canals give information about the angular acceleration of the head in space. If the rotational movement is continued to reach a steady angular velocity the lymph moves with the canal, the cupola returns to its normal position and the stimulation of the hair cells ceases. When the rotation of the head ceases the fluid, owing to its momentum, continues to move and deflects the cupola in the opposite direction. The cupola has been compared to a swing door or a highly damped pendulum, but it probably makes a fairly tight seal across the ampulla since India ink injected into the endolymph on one side of the cupola does not pass readily to the other. Rotation of the head may produce a sliding movement rather than a deflexion of the cupola.

Unilateral labyrinthectomy in lower animals leads to unequal tone in the neck muscles and the neck righting reflexes are brought into continuous action so that bending movements, rotation of the trunk and jerking movements of the eyes (nystagmus) occur. The animal comes to rest with the operated side down. In man section of one vestibular nerve causes a temporary disturbance of posture and gait. This is at a minimum if the head is held steady because the gaze is also steady and external objects appear to be stationary. If, however, side-to-side eye movements are required, as in crossing a street, vision is blurred and the gait becomes unsteady. This may be assumed to be caused by unilateral (that is asymmetrical) labyrinthine information arriving at the brain. After a time these disorders disappear and the gait is controlled by the information coming from visual and muscle-joint receptors.

Bilateral removal of the labyrinths in lower animals produces a very unsteady posture with great loss of muscular tone and ataxia, but in a higher animal this soon disappears as the animal learns to compensate for its disability by using its eyes and by proprioceptive reflexes. The unsteadiness returns if the animal is blindfolded. The disturbances of posture produced in man by bilateral destruction of the labyrinths are after a time compensated by visual righting reactions and by proprioceptive reflexes to a large extent but the deficit can be clearly shown by tilting him suddenly; no compensatory reactions are seen. The labyrinths are concerned in keeping the visual axes steady, in spite of movements of the head, so that images of external objects are kept fixed (or stabilized) on the retina. It is easy to see therefore that when such a patient is walking or is being wheeled across rough ground his vision is 'jumbled'.

The semicircular canals can be stimulated in man by rotating the subject in a special chair at a speed of say one revolution in two seconds. If he is rotated to his right both eyes move slowly to the left; this is followed by a quick movement of the eyes to the right which brings the eyes back to the mid position. Another slow and quick movement occurs and so on. This is called *vestibular nystagmus*. By convention the direction of the nystagmus is taken as the direction of the rapid component. The slow movement of nystagmus is an expression of vestibular activity while the rapid jerk back to the central position is due to central correcting factors. When the rotation to the right is suddenly arrested the sensory impulses from the labyrinths are altered as if the rotation were now to the left, and nystagmus occurs in the opposite direction. The head, body and arms tend to turn in the direction of the rotation just ended. The subject feels that he is still moving (vertigo) and may fall over if he attempts to stand up. This disturbance of equilibrium is due to confusion between visual and labyrinthine signals. The duration of the subjective sensation of turning is related to the time taken for the cupola to regain its resting position.

At rest the whole vestibular system is in a state of tonic activity manifested by a steady discharge of action potentials in its peripheral neurones. The vestibular tone of the two sides has equal and opposite actions on the eyes and skeletal muscles.

In the normal state the right vestibule tends to deviate the eyes to the left and the left vestibule deviates the eyes to the right. With normal tone the two effects cancel and the eyes remain in the straightahead position (Fig. 45.18). If

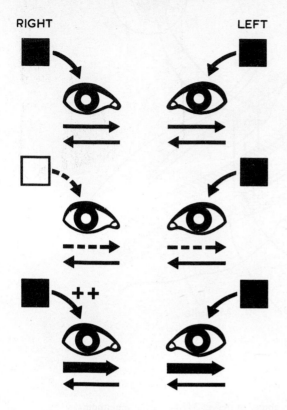

RIGHT   **LEFT**

Fig. 45.18 Illustrating the concept of vestibular tone. Upper diagram shows normal state. Middle diagram shows how a hypofunction of the right labyrinth causes nystagmus to the left. The lower diagram shows how hyperfunction of the right labyrinth causes nystagmus to the left. (A. G. D. Maran (1966). *Scottish Medical Journal* **11**, 382.)

the right vestibule is destroyed the normal tone exerted by the left side pushes the eyes to the right; this is corrected to the central position with a quick movement and so we get a nystagmus to the left. If the right side is hyperactive then the eyes are pushed to the left side, and corrected to the central position so that a nystagmus to the right occurs. The pathway followed by the nerve impulses involved in vestibular nystagmus is from the vestibular nerve to the vestibular nuclei in the brain stem; the slow component is produced by fibres passing in the medial longitudinal bundle (Fig. 45.19) and in the reticular formation which go to the nuclei of the eye muscles.

Slow component fibres send impulses to high threshold reticular neurones; when their threshold is reached they fire off to produce the quick component and at the same time they fire inhibitory neurones in the reticular formation which cut off a slow component discharge. The high threshold neurones then rest until their threshold is reached again and the cycle is repeated. If the inhibitor neurone is destroyed impulses in the median longitudinal bundle, which by-pass the reticular formation, are unopposed and a conjugate deviation of the eyes results.

The vestibular centres are not the sole sources of control of nystagmus. If a diencephalic dorso-lateral region, medial to the lateral geniculate body is stimulated, horizontal nystagmus with a quick component to the opposite side occurs. Direct stimulation of the vestibular nuclei results in a horizontal nystagmus to the same side. If both the vestibular and diencephalic nystagmogenic areas are stimulated at the same time the effects cancel each other out and if the stimulations are crossed, the effects are additive (Fig. 45.20). The diencephalic area moderates the nystagmogenic action of the ipsilateral vestibular nucleus and increases the action of the contralateral vestibular nucleus.

Excessive stimulation of the semicircular canals, especially if the rotational movement is about a horizontal axis, may produce the drowsiness, pallor, vertigo, salivation, nausea and vomiting characteristic of sea-sickness. The unpleasant sensations arise from a conflict of sensory information; the deck of the ship is accepted as horizontal and stationary because the eyes are fixed on it but the labyrinths in stormy weather signal rotational and unsteady movements. More than half of those subjected to these stimuli suffer from this debilitating condition. Some people suffer in a similar way if they travel on trains or buses or if they go on a swing. Although there may be a large psychological element in the production of their discomfort this is not necessarily the whole explanation; the removal of the nodulus (posterior lobe) of the cerebellum in dogs gives them complete protection from motion sickness. Healthy subjects in swings become sick most often if the vertical canals are subjected to acceleration, but people who suffer from swing-sickness are not necessarily subject to

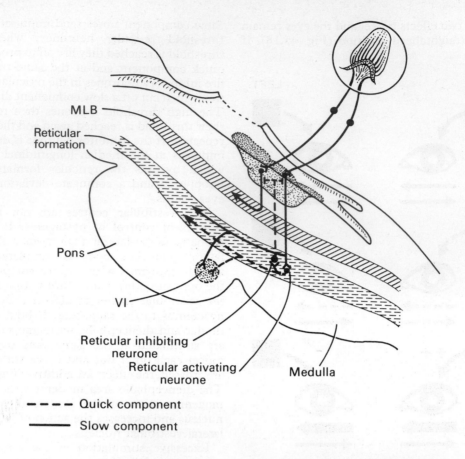

MLB

Reticular formation

Pons

VI

Reticular inhibiting neurone

Reticular activating neurone

Medulla

– – – – Quick component

———— Slow component

Fig. 45.19   The mechanism of nystagmus. VI, abducent nerve; MLB, medial longitudinal bundle. (A. G. D. Maran (1966). *Scottish Medical Journal* **11**, 379.)

sea-sickness. An intact vestibular system is necessary to produce motion sickness; deaf mutes and labyrinthectomized animals never experience it.

In certain circumstances, however, people have been sick without motion. For example, instructors on helicopter simulators are occasionally sick since they see the impression of motion but they know they are not moving; this leads to a conflict of information reaching the central nervous system. The central nervous system is used to a certain amount of 'normal acceleration and velocity'; for example, head movements, lying down quickly or standing up suddenly, motor-car or bus motion, are particular movements which the central nervous system accepts as a normal part of life. In other words, habituation has occurred and the

vestibular efferents prevent too much information leaving the vestibular system.

When one is sitting inside a plane that is in a turbulent air stream the vestibular impulses are increased due to the movement but the eyes see the relatively stable seats and bulkhead. Thus the information reaching the central nervous system is contradictory and results in sensory incongruity. It can be supposed that the central nervous system endeavours to resolve this conflict by asking the vestibular system for more information. It is probable that this is done via the vestibular efferent system—if inhibition lessens, then more information passes centrally.

The headache, pallor, perspiration, nausea and vomiting of motion sickness is due to vagal stimulation. The dorsal vagal nucleus is

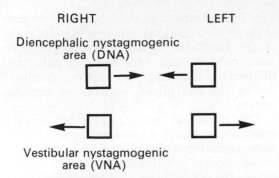

RIGHT   LEFT

Diencephalic nystagmogenic
area (DNA)

Vestibular nystagmogenic
area (VNA)

Fig. 45.20 Diagrammatic illustration of the two nystagmogenic areas. Stimulation of the DNA causes contralateral nystagmus and stimulation of the VNA causes ipsilateral nystagmus. (*By courtesy of A. G. D. Maran.*)

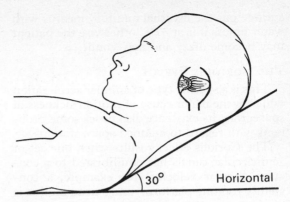

Fig. 45.21 When a caloric test is performed the patient lies supine with the head raised 30 degrees so that the crista is horizontal.

very near to the medial vestibular nucleus and there is a considerable degree of interconnexion between them. When the inhibitory influence is lifted from the vestibular system the increased neural activity overflows to some of the cells in the dorsal vagal nucleus and causes motion sickness.

It is well known that emotionally labile people are more prone to motion sickness than others; this is probably due to the fact that they cannot contain much sensory incongruity and lift the inhibitory influence on the vesticular system earlier than others.

With practice and training it is possible to habituate to vestibular stimuli produced by aerobatics and space flight so that motion sickness is overcome. However, the Coriolis effect is so enormous that it is more difficult to adapt to it. Thus sickness has become more common in space flights as the craft have become larger and the crew are able to move around more and produce more Coriolis effect (see below).

The canals may also be stimulated in man by pouring either warm (44°C) or cold (30°C) water through a tube into the external meatus of the ear (caloric tests) while the subject lies on his back with his head 30° forward. In this attitude the lateral canal is in the vertical plane with the ampulla uppermost (Fig. 45.21). The convection currents thus produced in the canal cause nystagmus, a twisting of the trunk and a feeling of giddiness. Syringing the ear with water at 44°C causes a flow of endolymph towards the ampulla of the lateral canal and a deflexion of the cupola with nystagmus to the

same side. The eye movements can be recorded from electrodes placed on the skin at the outer canthi of the eyes; this is called electronystagmography and a typical record is shown in Figure 45.22. Attempts, with the

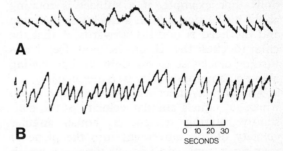

Fig. 45.22 Nystagmus as recorded by electronystagmography. When the eyes move to the right the pen moves upwards and when the eyes move to the left the pen moves downwards. Nystagmus to the right (named after the quick component) is represented as ⋀⋀ and nystagmus to the left is represented as ⁄⁄. Figure A shows a nystagmus to the right in response to a caloric stimulus with water at 30°C run into the left ear for 30 sec. Figure B represents a similar response with water at 30°C run into the right ear. If the patient fixes his eyes on a point during such a test nystagmus tends to be inhibited; with electronystagmography it is possible to record nystagmus with the eyes closed thus abolishing such inhibition. (*By courtesy of A. G. D. Maran.*)

eyes shut, to touch a given point may fail (pastpointing). Hallpike has shown that tests based on recording the duration of nystagmus after a standard caloric stimulus are valuable diagnostically in lesions of the vestibular apparatus and its nervous connexions. These physiological observations show that it is unwise to

Brain Stem and Postural Reflexes   935

syringe out the external auditory meatus with water unless it is at 37°C otherwise the patient may become dizzy and may vomit.

THE CORIOLIS EFFECT

This is a specific type of angular acceleration which is the major cause of motion sickness in spacecraft; its existence also poses some problems with regard to manned space stations.

The Coriolis effect results when one set of semicircular canals has equilibrated to a constant angular velocity (for example, a constantly rotating platform) and a head motion is made in a different plane (standing on a rotating platform and nodding). When a second set of canals is rotated into the plane of constant angular velocity an angular acceleration is imposed on them, and as the first set are rotated out of the plane of constant velocity an angular deceleration is applied to them.

The motion felt by the subject during this manoeuvre is in the plane of the remaining third set of canals (that is, the resultant direction). For example, if a subject is rotating about his vertical axis in a clockwise direction and the head is nodded forwards so that the chin touches the chest, he will feel he is turning clockwise to his right (that is, rolling about an anteroposterior horizontal axis).

The Coriolis effect is proportional to $\omega_1 \omega_2 t$, where ω_1 equals angular velocity of the constantly rotating system, ω_2 equals angular velocity of the movement into the plane of rotating system, and t equals time over which ω_2 acts. The Coriolis effect can take place even if the value of ω_1 is very small. If ω_1 is as high as 40 rev/min the slight head motions accompanying breathing can result in a Coriolis effect.

HABITUATION

Habituation is the process whereby normal individuals adapt to complex new stimuli or situations. With repeated exposure to vestibular stimulation many people can develop relative insensitivity. Consider, for example, the figure-skater ending her performance with a spin of about 7 rev/sec; she comes to a full stop in about 1 sec, which represents a deceleration of some 2000 degrees/sec^2—enough to pole-axe the untrained individual. Ballet dancers, pilots and seamen show a similar decreased response to stimuli which would be very uncomfortable to persons not habituated to the motion. Despite apparent hypofunction of the vestibular system these people have quite normal inner ears and can pass vestibular tests requiring very sensitive monitoring by the end organs and central nervous system.

The vestibular system, like other sensory systems can therefore perform some kind of auto-regulation which allows the central nervous system to perceive facts which are important to the organism and refuse admittance to information that is repetitive, irrelevant or inconsequential. As far as is known at the moment this auto-regulation is performed by the vestibular efferent system.

It is to develop habituation and a depression of sickness and giddiness that astronauts have such long periods of training on centrifuges and in weightless states.

THE OTOLITH SYSTEM

The utricle and saccule respond to linear acceleration, that is they discharge when linear motion (velocity) starts or stops. The acceleration causes the otoconia to slide and bend the sensory hairs which stimulate the nerve endings by a shearing force. Gravity constantly acts on the otolith system so that, if any other linear force acts, the effect on the otolith system is the resultant of the forces. This can be illustrated by the so-called 'oculogravic illusion' (Fig. 45.23). When a high-speed fighter plane takes off, the linear acceleration causes a posterior shearing force on the utricular maculae. This posterior force, when related to the constant downward gravity vector, causes a resultant so that the pilot feels he is too far 'nose-up'. If he does not believe his instruments and flies according to his sensation he will fly the plane into the ground.

In space flight under conditions of weightlessness the gravity vector ceases to act on the utricle and so this structure is partially 'deafferented'. The astronaut does not know which is 'up' and which is 'down' and is very liable to suffer from motion sickness.

Just as nystagmus is generated by stimulation of the semicircular ducts, compensatory eye movements are also generated by stimulation of the otolith organs. When the head is tilted to the left the eyes rotate slightly counterclockwise and remain there so long as the head

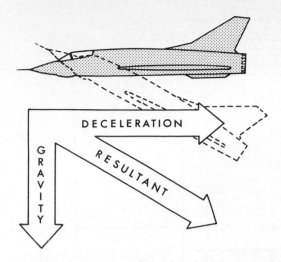

Fig. 45.23 The 'oculogravic illusion'. The resultant between the deceleration and gravity vectors gives the pilot the impression that the plane is 'nose up' when it is, in reality, flying straight. Any correction on the basis of sensation rather than instrument flying will result in a nose-down correction and a possible crash.

is tilted. The compensatory motion of the eyes about their visual (anteroposterior) axis secondary to a change in the relative direction of the gravity vector is called ocular counterrolling. This helps to keep the retinal image stabilized in the same way as nystagmus attempts to do.

Tonic Labyrinthine Reflexes

Magnus (1924) found that the reflex changes produced by changes in the position of the labyrinths affected the muscle tone in all four limbs equally. Reinvestigation by Roberts has not confirmed Magnus's findings. Roberts eliminated the neck reflexes of a cat by denervating the first three intervertebral joints in the neck and by holding the axis vertebra in a clamp. When the head was tilted towards the nose-down position the forelimbs extended and the hind limbs flexed; that is, the responses in the four limbs were not the same but in opposite directions. If the head were tilted towards the right-side-down position the right foreleg extended. Lateral rotations of the head were without effect because rotation of the head about a vertical axis did not affect the otolith organs.

Lateral movements of the head produce alteration in posture through the neck re-

flexes (p. 927) which may be in the opposite direction to those produced by the labyrinthine reflexes. Thus if the head is tilted nose-up the neck reflexes produce extension of the forelimbs and the labyrinthine reflexes a flexion of the forelimbs. In the intact animal the effects cancel one another and tilting the head nose-up extends all four limbs equally. The normal posture of an animal depends on the interaction of the labyrinthine and neck reflexes (Fig. 45.24).

Righting Reflexes

The decerebrate animal cannot right itself if it is pushed over. However, the animal with the thalamus and brain stem intact but with the cerebral cortex and cerebellum removed, the *thalamic animal*, possesses more adequate postural reflexes. The thalamic cat or dog shows normal posture and it can carry out walking movements. The muscle tone is normal when sitting or walking but if the animal is supported with the feet off the ground hyperextension of all four limbs occurs.

The thalamic animal shows positive supporting reactions which convert the limbs into pillars by simultaneous contraction of the flexors and extensors so that the body weight is taken on four columns. The actual contact of the feet with the ground produces an extensor response and at the same time the flexor muscles in the distal joints of the feet are stretched. This causes reflexly a contraction of all the muscles, extensors, flexors, abductors and adductors, which allows the limbs to carry the weight of the body. At the same time the vertebral column is made rigid by a contraction of the long muscles which run dorsally and ventrally along it. This supporting reaction is a static reflex and necessarily disappears when movements are carried out; the limb loosens (negative supporting reaction) when the extensor muscles of the distal joints are stretched. These supporting reactions, although shown quite well by the thalamic animal, are best seen in the intact animal.

In contradistinction to the animal with decerebrate rigidity, the thalamic animal possesses righting reflexes which ensure that the animal brings itself back to the upright position if it is made to deviate from it. The righting reflexes arise in the labyrinths and in the muscles of the neck and trunk. To study

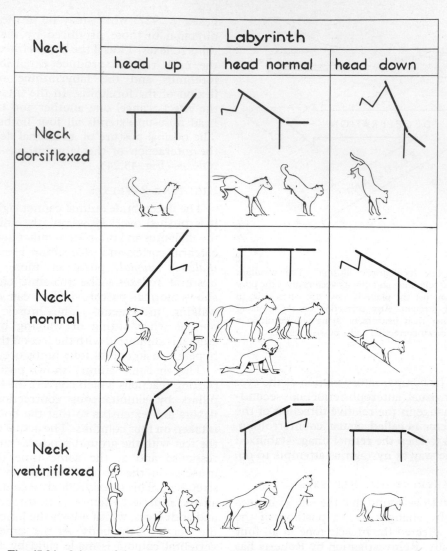

Fig. 45.24 Scheme of the interacting effects of labyrinthine and neck reflexes on the limbs of various animals. The middle column shows the neck reflexes acting alone. The middle row shows the labyrinthine positional reflexes acting alone. In the top left and bottom right corners, the neck and labyrinthine reflexes are in opposition; in the top right and bottom left corners thry reinforce one another. (T. M. Roberts (1968). Labyrinthine control of the postural muscles. *Third Symposium on the Role of the Vestibular Organs in Space Exploration*, p. 163. N.A.S.A.)

the components of the righting mechanism it necessary to employ one or more of the following procedures—labyrinthectomy, denervation of the neck, blindfolding. When the results of such experiments are put together it is possible to explain how an animal rights itself when put in an abnormal position. If a cat is held upside down above the ground and released, it turns over in mid air and lands on all four feet. In this manoeuvre probably the eye righting reflexes are the most important; they are dependent on the integrity of the visual area of the cortex in the occipital lobe. The animal, when it sees its surroundings, quickly brings its head to the normal position relative to the ground. The labyrinthine reflexes are also of some importance in correcting the position of the head since the otoliths inform the animal of the position of its head in space before it is released but a cat with congenital absence of the labyrinths (provided it was not blindfolded) could turn over without delay quite nimbly in mid air. A normal cat blindfolded and then released turns over after a delay of about 100 msec but makes a poor landing as it has no information about the position of the floor. McDonald investigated this old problem by high-speed (1500 frames per sec) cinephotography and showed that the rotation after release occurs in two principal stages: (1) the righting of the head and forepart of the body together; and (2) the rotation of the lumbar region, pelvic girdle and hindlimbs all in one, the whole turn taking less than 150 msec.

A further source of righting is displayed by a blindfolded labyrinthectomized animal laid on its side on a table. When the animal is released the head immediately rotates into the normal position relative to the ground and the position of the remainder of the body is corrected as before. If the animal is laid on its side again under a weighted board, so that the pressure on the body is the same on both sides, this righting reflex does not occur.

When the chin of an intact blindfolded cat is laid on a table, or when the fore-legs are brought into contact with the edge of the table, the feet are immediately placed on the table in a position ready to support the weight of the body. These are called *placing reactions*. When an intact animal is disturbed by being pushed sideways the normal position is regained by a series of hops (*hopping reaction*). The hopping and placing reactions are dependent on the integrity of the motor cortex.

The maintenance of the standing posture in man is necessarily a very complex affair. His centre of gravity is somewhere in the pelvis high above the very small base provided by the feet. It is therefore not difficult to understand why it takes the infant such a long time to acquire the ability to hold an erect stance. Little swaying movements occur continuously due to wandering away of the centre of gravity and its restoration to a stable position. The sensory information which governs these corrections comes from the eyes, the vestibular apparatus, muscles and the joints. If muscle-joint sense is diminished as in tabes dorsalis the patient becomes very unsteady if the eyes are closed. Electromyography (p. 847) shows that when the erect posture is being maintained easily and comfortably the antigravity muscles show surprisingly little activity. This economical maintenance must depend on early signalling of small deviations and prompt reflex correcting of them.

CONSCIOUSNESS

Consciousness can scarcely be defined but it involves a state of alertness in which the subject can react appropriately to the situation existing at the moment and which allows him to remember the events in the situation. Consciousness allows him to think and to be aware of his own personal identity. It was at one time assumed that, since the cerebral cortex carries out the intellectual functions, consciousness depends on that part of the brain. In anencephaly very little cerebral cortex is present and the mental status may be described as idiocy but the 'person' is not unconscious; he sleeps and wakes, sees and hears, utters crude sounds indicating pleasure or displeasure. It seems paradoxical that stimulation of an afferent nerve of an anaesthetized unconscious animal produces cortical activity; the cortex has received impulses from the periphery. However, anaesthetics have been shown to block the impulses passing to the cortex via the reticular system (p. 924). Accidental or surgical damage to the cortex alone does not produce unconsciousness but lesions of the upper brain stem and diencephalon, for example by haemorrhage, are often followed by unconsciousness. Injection into the cerebral ventricles of a variety of substances, such as adrenaline, noradrenaline, calcium chloride and also of some anaesthetics in doses which are quite ineffective intravenously produce a sleep-like state by acting on the nuclei lying in the wall of the third ventricle.

REFERENCES

ADRIAN, E. D. (1943). Discharges from the vestibular receptors in the cat. *Journal of Physiology* **101**, 389–407.

BOYD, I. A., EYZAGUIRRE, C., MATTHEWS, P. B. C. & RUSHWORTH, G. (1964). *The Role of the Gamma System in Movement and Posture.* New York: Association for the Aid of Crippled Children.

BRODAL, A. (1957). *The Reticular Formation of the Brain Stem: Anatomical Aspects and Functional Correlations.* Edinburgh: Oliver & Boyd.

BRODAL, A., POMPEIANO, O. & WALBERG, F. (1962). *The Vestibular Nuclei and their Connections, Anatomy and Functional Correlations.* Edinburgh: Oliver & Boyd.

DELAFRESNAYE, J. F. (Ed.) (1954). *Brain Mechanisms and Consciousness. A Symposium.* Oxford: Blackwell.

DENNY-BROWN, D. (1966). *The Cerebral Control of Movement.* Liverpool: University Press.

DE REUCK, A. V. S. & KNIGHT, J. (1968). *Myotatic, Kinesthetic and Vestibular Mechanisms.* Ciba Foundation Volume 1967. London: Churchill.

ECCLES, J. C. (1969). Inhibitory pathways of the central nervous system. Liverpool: University Press.

FELDBERG, H. (1959). A physiological approach to the problem of general anaesthesia and loss of consciousness. *British Medical Journal* **ii**, 771–782.

FISCHER, J. J. (1956). *The Labyrinth: Physiology and Functional Tests.* London & New York: Grune & Stratton.

GLASER, E. M. (1959). Prevention and treatment of motion sickness. *Proceedings of the Royal Society of Medicine* **52**, 965–972.

GRANIT, R. HOLMGREN, B. & MERTON, P. A. (1955). The two routes for excitation of muscle and their subservience to the cerebellum. *Journal of Physiology* **130**, 213–224.

GUEDRY, F. E. (1965). Psychophysiological studies of vestibular function. *Contributions to Sensory Physiology* **1**, 63–135.

HALLPIKE, C. S. (1967). Some types of ocular nystagmus and their neurological mechanisms. *Proceedings of the Royal Society of Medicine* **60**, 1043–1054.

IURATO, S. (1969). *Submicroscopic Structures of the Inner Ear.* Oxford: Pergamon.

JASPER, H. H., PROCTOR, L. D., KNIGHTON, R. S., NOSHAY, W. C. & COSTELLO, R. T. (Eds.) (1958). *Reticular Formation of the Brain.* Henry Ford Hospital International Symposium. London: Churchill.

JOSEPH, J. (1960). *Man's Posture: Electromyographic Studies.* Springfield: Thomas.

MAGOUN, H. W. (1958). *The Waking Brvin.* Springfield: Thomas.

MARTIN, J. P. (1967). *The Basal Ganglia and Posture.* London: Pitman Medical.

NASA (1968). *Third Symposium on the Role of the Vestibular Organs in Space Exploration.* Washington, D.C.: National Aeronautics and Space Administration.

O'LEARY, J. L. & COBEN, L. A. (1958). The reticular core. *Physiological Reviews* **38**, 243–276.

ROBERTS, T. D. M. (1967). *Neurophysiology of Postural Mechanisms.* London: Butterworth.

SCIENCE JOURNAL (1967). *The Human Brain* **3**, 42–117.

STABLE, J. (Ed.) (1971). *Vestibular Function on Earth and in Space.* Oxford: Pergamon.

TYLER, D. B. & BARD, P. (1949). Motion sickness. *Physiological Reviews* **29**, 311–369.

WERSÄLL, J. & FLOCK, A. (1965). Functional anatomy of the vestibular and lateral line organs. *Contributions to Sensory Physiology* **1**, 39–61.

WHITTERIDGE, D. (1956). Machinery of posture. *Advancement of Science* **13**, 104–110.

ZOTTERMAN, Y. (Ed.) (1967). Sensory mechanisms. *Progress in Brain Research* **23**, 155–168.

46 The cerebellum

The cerebellum, one-tenth of the mass of the brain, consists of a superficial layer of grey matter, the cortex, overlying a mass of white matter. It has a large number of fine parallel furrows on its surface (Figs. 46.2 and 3) which give it a surface area about one-third that of the cerebral hemispheres. A section through the cerebellum shows a tree-like arrangement of grey and white matter, the *arbor vitae*.

The cerebellum is developed from the cephalic portion of the hind-brain and in man can be divided on embryological and phylogenetic, as well as on functional grounds, into anterior, middle and posterior lobes. These lobes are demarcated by fissures which are evident at the third or fourth month of embryonic life (Fig. 46.1). The relative sizes of the three lobes in the adult are shown in Figures 46.2 and 3, and their main afferent connexions are given in Table 46.4. The anterior and posterior lobes are much older in a phylogenetic sense than the middle lobe and are, therefore, called *palaeocerebellar*. The great development of the middle lobe of the cerebellum (*neocerebellum*) in man has caused an alteration in the anatomical relations of the lobes as seen in the embryo, so that the posterior lobe eventually lies below and in front of the middle lobe.

The three cerebellar peduncles carry projection fibres connecting the cerebellum with other parts of the central nervous system. The majority of the fibres in the *superior cerebellar peduncles* pass from the cerebellar nuclei through the mid-brain where they decussate (Fig. 46.5) and then divide into descending branches to the upper brain stem and ascending branches which go to the red nuclei and to the thalami. The superior peduncles also convey the anterior spinocerebellar fibres which end mainly in the cortex of the anterior lobe. The *middle peduncles*, the largest afferent connexions of the cerebellum, carry impulses from the cerebral cortex. These travel to the

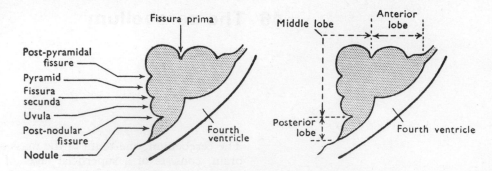

Fissura prima

Post-pyramidal fissure

Pyramid

Fissura secunda

Uvula

Post-nodular fissure

Nodule

Fourth ventricle

Middle lobe

Anterior lobe

Posterior lobe

Fourth ventricle

Fig. 46.1 A diagrammatic representation of a sagittal section through the cerebellum in a human four-months fetus. (*After* Gray.)

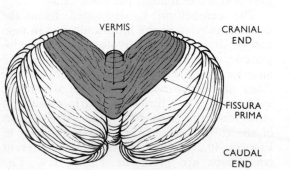

VERMIS

CRANIAL END

FISSURA PRIMA

CAUDAL END

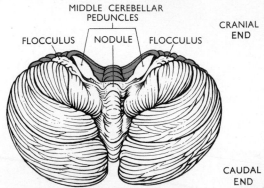

MIDDLE CEREBELLAR PEDUNCLES

FLOCCULUS   NODULE   FLOCCULUS

CRANIAL END

CAUDAL END

Fig. 46.2 The superior aspect of the cerebellum. The anterior lobe is stippled. The remainder belongs to the middle lobe. (*After Gray's Anatomy* (1954). 31st edn, p. 961. London: Longmans.)

Fig. 46.3 Inferior aspect of the human cerebellum. The flocculus and nodule (boldly outlined) form the posterior lobe. The anterior lobe is stippled. The remainder belongs to the middle lobe. (*After Gray's Anatomy* (1954). 31st edn, p. 964. London: Longmans.)

Table 46.4 Lobes of the cerebellum in Man

Lobe	Separating fissure	Phylogeny	Relative size	Main afferent connexions
Anterior		Palaeocerebellum	Small	Incoming spinocerebellar (also pontocerebellar)
	Fissura prima			
Middle		Neocerebellum	Large	Corticoponto-cerebellar
	Postnodular and floccular			
Posterior		Palaeocerebellum	Very small: nodule and flocculi only	Vestibular

The caudal parts of the middle lobe (uvula and pyramid), which receive the posterior spinocerebellar fibres, belong to the palaeocerebellum. It should be noted that there are several other schemes of classification of the parts of the cerebellum. In one such classification the posterior lobe comprises not only the nodule and the flocculus but also the uvula and pyramid.

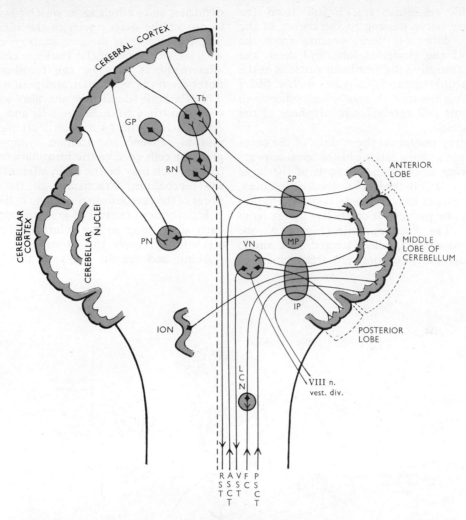

Fig. 46.5 A simplified scheme of the connexions of the cerebellum. Note that each side of the cerebellum is connected mainly with the same side of the spinal cord and with the opposite cerebral cortex. Three-quarters of the fibres in the peduncles carry impulses to the cerebellum; the remaining fibres, arising mainly in the dentate nucleus, pass in most cases to the opposite cerebral cortex via the middle peduncles. ASCT = anterior spinocerebellar tract; LCN = lateral cuneate nucleus; FC = fasciculus cuneatus; GP = globus pallidus; IP = inferior peduncle; MP = middle peduncle; ION = inferior olivary nucleus; PN = pontine nuclei; PSCT = posterior spinocerebellar tract; RN = red nucleus; RST = rubrospinal tract; SP = superior peduncle; Th = thalamus; VN = vestibular nuclei; VST = vestibulo-spinal tract; VIII n. = eighth cranial nerve. (See also Fig. 45.1, 5, 6 and 7.)

cells of the nuclei pontis from which new axons, crossing the mid-line in the pons, pass to the cortex of the middle lobe of the cerebellum of the opposite side. The *inferior peduncles* carry fibres from various brain stem nuclei and from the spinal cord as well as certain efferent fibres. The chief fibres are (1)

the posterior spinocerebellar tract which ascends without crossing and ends in the pyramid and uvula (the palaeocerebellar parts of the middle lobe); some fibres of this tract also go to the anterior lobe; (2) arcuate fibres from the lateral cuneate nuclei of both sides passing to an unknown destination; (3) fibres from the vestibular nuclei and from the vestibular nerve passing to the cortex of the posterior lobe; (4) fibres passing from the cortex of the posterior lobe and from the nucleus fastigii to the vestibular nucleus in the medulla oblongata; (5) olivocerebellar fibres passing via the inferior cerebellar peduncle to the vermis and cerebellar hemisphere of the opposite side.

The grey matter on the surface of the cerebellum has a characteristic histological appearance owing to the presence in it of the large Purkinje cells, but there are no local differences, such as exist in the cerebrum, sufficient to distinguish one part of the cerebellar cortex from another. The Purkinje cells (Fig. 46.6) are large flask-shaped cells arranged in a single layer; each has an arborescent arrangement of dendrites and a single axon which passes down to one of the deep nuclei of the cerebellum. The *climbing fibres* make synaptic connexions with the dendrites of the Purkinje cells; these afferent fibres may be the terminations of fibres from the vestibular and pontine nuclei. The Purkinje dendrites are also associated with dendrites from *basket cells* and *granular cells*. Each of the 14 million Purkinje cells is said to possess one million synapses. The granular cells receive the terminations of *moss* fibres which may be the main afferent fibres of the cerebellum. Practically all the afferent fibres of the cerebellum terminate in the cortex.

Independent centres of grey matter, four on each side, lying in the white matter of the cerebellum receive the axons of the cells of Purkinje and are the source of nearly all the

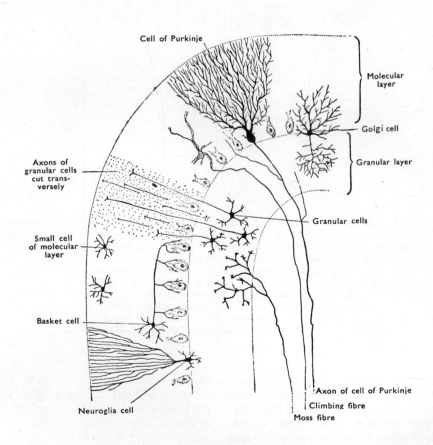

Fig. 46.6 A transverse section through a cerebellar folium (*Gray's Anatomy* (1954), 31st edn, p. 969. London: Longmans). (*After* Cajal & Kölliker.)

efferent fibres of the cerebellum. The nucleus dentatus, the largest and phylogenetically the youngest of the nuclei, is connected with the neocerebellum and through the superior cerebellar peduncles with the thalamus and cerebral cortex (areas 4 and 6) of the opposite side. The dentate nucleus also sends fibres across the mid-line to the red nucleus from which new fibres cross the mid-line and descend as the rubrospinal tract. The other, smaller, nuclei are the nucleus fastigii, the nucleus globosus and the nucleus emboliformis; the nucleus fastigii sends fibres to the vestibular nucleus and to the reticular formation of the medulla. The vestibular nucleus also receives fibres direct from the Purkinje cells. The globose and emboliform nuclei are most likely to be connected by the superior peduncles to the large cells of the red nucleus and the reticular formation of the mid-brain.

Much has been learned regarding the function of the fibre systems of the cerebellar cortex from experimental work with intracellular electrodes, much of it carried out by Eccles and his associates (Eccles, 1969—Fig. 46.7). It is clear that the output from the Purkinje cells to the cerebellar nuclei is exclusively inhibitory. The afferent fibres to the cerebellar cortex, namely the climbing fibres which connect directly with the Purkinje cells, and the mossy fibres which connect through a relay arising from the granule cells, all may exert an excitatory effect on the Purkinje cells. The Golgi cells provide an inhibitory feedback loop on mossy fibre input since their dendrites are excited by the parallel fibres, which are formed by branching of the axons of the granule cells, and their axons have an inhibitory action on the granule cells. The cerebellum thus contains an immensely complex and largely inhibitory system which receives its input from the higher centres and from the periphery. It delivers an inhibitory output to the cerebellar nuclei and the vestibular nucleus through which it exerts its action in the control and adjustment of movement.

Adrian inserted a fine wire about 1·5 mm deep into the cerebellum of a monkey under barbiturate anaesthesia and picked up afferent impulses when the limbs on the same side were pressed or moved (Fig. 46.8). A movement of the vibrissae on the face produced a discharge in the anterior part of the middle lobe. Dorsi-flexion of the wrist of the fore-limb or pressure on the hand produced discharges just craniad to the fissura prima, while pressure on the foot of the hind-limbs gave discharges well forward. Some afferent impulses to the cerebellum arose from the surface of the body but the majority came from the muscles, tendons and fasciae of hand and foot. These experiments show that spinocerebellar afferents end chiefly in the anterior lobe. It is now known that afferents from the skin and the viscera project to both sides of the anterior cerebellum. Recently Orcasson has found that the flow of impulses in this pathway to the cerebellum can be inhibited by painful stimulation; descending pathways from the cortex and brain (that is centrifugal pathways) have some control over the ascending pathway.

When the cerebral cortex is stimulated, either electrically or by application of a solution of strychnine, discharges are obtained in the cerebellum in the contralateral areas shown in Fig. 46.9. Thus the cerebropontine impulses reach cerebellar areas in which the foot is represented craniad and the hand caudad. This is the same order of distribution as in the case of the afferent spinocerebellar fibres but the cerebropontine afferents reach more caudad into the middle lobe. Snider and Stowell also studied the projection to the cerebellum and found visual and auditory receiving areas mainly in the middle lobe, the visual and auditory areas being almost coextensive; they also found a second proprioceptive receiving area on the posterior part of the cerebellum. Impulse traffic also passes in the opposite direction, namely from the cerebellum to the sensorimotor area, the visual area and the auditory area of the cerebral cortex. There is thus two-way traffic between corresponding areas of the cerebral cortex and the cerebellar cortex.

Extirpation (ablation) of parts of the central nervous system, although a useful method for investigating function, has given results which must be interpreted with care in the case of the cerebellum because it is difficult to remove this part of the brain without damaging the adjacent brain stem. Since the cerebellum possesses a modifying rather than an initiating function its removal does not destroy any reflexes although it usually alters their character (Fig. 46.10). It is important to note that there is never

Cerebellum

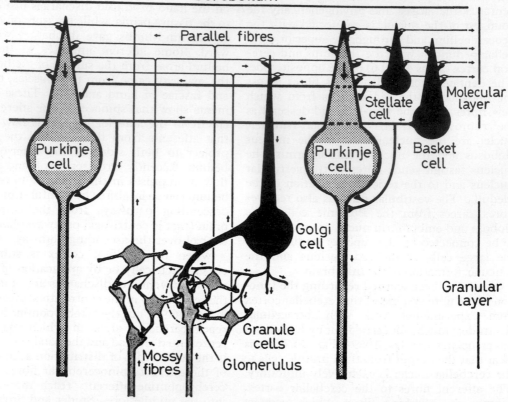

Fig. 46.7 Diagram showing the principal features that have been postulated for the mossy fibre input and the cerebellar glomerlus. The Golgi, stellate and basket cells are all inhibitory in action and are shown by convention in black. The broken line represents the glial lamella that ensheathes a glomerulus. The diagram is drawn as for a section along the folium and the main distribution of the basket and stellate cells would be perpendicular to the plane of the diagram, but they are also distributed as shown to a band of several Purkinje cells along the folium. The arrows indicate the direction of impulse propagation. (J. C. Eccles, (1969). *The Inhibitory Pathways of the Central Nervous System*, p. 89. Liverpool: University Press.

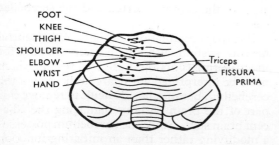

Fig. 46.8 Localization of discharges in the dorsal surface of the cerebellum from different parts of the fore- and hind-limb in a monkey. (E. D. Adrian (1943). *Brain* **66**, 298.)

any loss or disturbance of sensation after removal of the cerebellum; the gait is abnormal after decerebellation not because there is any disorder of the sensory pathways but because the muscles are not under normal control.

More than fifty years ago Luciani studied the effects of removal of the cerebellum. He found that in dogs the operation was followed by an increase in extensor tone and by clonic spasms. These signs passed off and in about ten days the decerebellate dog showed *asthenia* (weakness), *hypotonia* (reduction of the tone of the muscles) and irregular contractions of

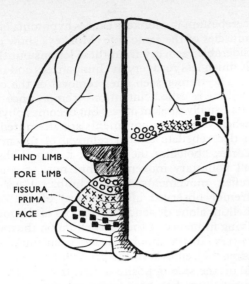

HIND LIMB
FORE LIMB
FISSURA PRIMA
FACE

Fig. 46.9 Cerebellar receiving area in the monkey, showing their connexion with different parts of the motor cortex. Stimulation of the area of the motor cortex marked O (leg area) produced potentials in the cerebellum in the area marked O; a similar relationship held for the areas marked X and ■. (E. D. Adrian (1943). *Brain* **66**, 301.)

muscles instead of the normal smooth, well-regulated muscle movements. This phase was followed by a certain amount of compensation in which standing and walking were much improved but recovery was never complete. The improvement is due largely to the use of the eyes, for if the eyes of an animal which has recovered from the initial effects of decerebellation are covered the gait at once deteriorates. Luciani also found, as have all subsequent investigators, that complete removal of one side of the cerebellum is followed by weakness and defects of muscular control on the same side of the body.

In healthy human beings a very small amount of tremor is always present. It consists of small adjustments of flexion and extension, or of abduction and adduction, which maintain body posture or the position of a limb. Analysis of records of tremor (Fig. 50.7) show that the main oscillations occur between 7 and 12 Hz. The peak around 9 Hz has been ascribed to the delay in the operation of the stretch reflex mechanism. However this cannot be the whole explanation since the tremor in a

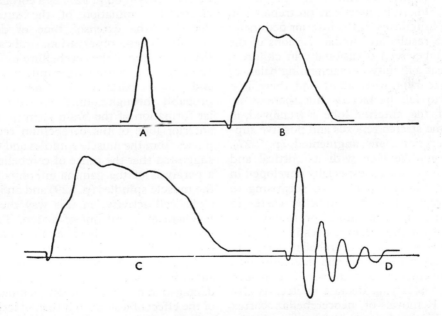

Fig. 46.10 A—tracing of a simple muscle twitch, B—of a normal knee jerk, C—of a knee jerk in a spastic limb, D—of a pendular knee jerk from a patient with an acute cerebellar lesion. (Gordon Holmes (1960). *Introduction to Clinical Neurology*, 2nd edn, reprint, p. 100. Edinburgh: Livingstone.)

The Cerebellum   947

completely deafferented human arm has also a peak frequency of 9 Hz. The synchronization of motor impulses at this frequency probably occurs in the spinal cord itself. Adrenaline increases the amplitude but not the frequency of finger tremor more in the normal than the deafferented hand; this suggests that the muscle spindles modulate the tremor. In cerebellar disease this tremor is greatly exaggerated and is especially obvious when movements are being carried out; the synergistic muscles do not act harmoniously. The tremor together with the hypotonia and asthenia gives the jerky staggering gait described as *cerebellar ataxia*.

The cerebellum possesses localization of function to a limited extent and partial removal shows that different parts vary in their regulatory functions. Each half (right or left) of the cerebellum modifies the tone and movements of the muscles on the same side of the body (see Fig. 46.5). If only the flocculonodular (posterior) lobe is removed in monkeys there is an oscillation of the head, a tendency to fall, and the animal walks with a staggering gait with the limbs abducted. This is due to overactivity of the vestibular apparatus since it does not occur if the labyrinths are previously destroyed. There is, however, no tremor and no change in reflexes. If the inferior peduncles are cut the results are similar. Lesions of the posterior lobe, seen occasionally in children, lead to great difficulty in maintaining balance; the feet are kept wide apart and there is a tendency to fall backwards but there is no tremor. If the anterior lobe is removed in monkeys the stretch reflexes and positive supporting reactions are augmented (p. 927). Since the middle lobe with its cortical and thalamic connexions is especially developed in the higher mammals, it is not surprising to find that removal of the cerebellar cortex in primates produces a more profound disturbance of function than it does in dogs and cats. Removal of the neocerebellar cortex in monkeys causes hypotonia and interferes with skilled movements, but tremor is not seen unless the underlying dentate nucleus is also removed. Removal of neocerebellar cortex together with the dentate nucleus disturbs the gait; the animal may leap and be unable to arrest its progress when confronted with an obstacle. Removal of one side of the neo-cerebellum in monkeys causes hypotonia of the muscles of the same side. Monkeys show considerable recovery from these lesions but there is much less recovery in chimpanzees and man.

The tremor seen after removal of the cerebellum is called 'intention' tremor since it is not usually present at rest but becomes obvious when voluntary movements are attempted. In a decerebellate animal removal of the motor cortex abolishes the tremor; it also abolishes all voluntary movements but not, of course, reflex movements of the limbs. Thus the tremor observed after removal of the cerebellum alone depends on the loss of the modifying influence of the cerebellum on the motor cortex. In the monkey it has been found that after the removal of the premotor cortex (area 6) on the side opposite to that from which the cerebellum has been removed the tremor is increased. Presumably, therefore, the premotor area plays an important part in the phase of compensation after a cerebellar lesion. The extensor spasms which follow decerebellation are presumably due to a release from cerebellar control. The increase of decerebrate rigidity which follows removal of the cerebellum or of the anterior lobe alone has been discussed on p. 927; as a corollary to this, electrical stimulation of the anterior lobe decreases the extensor tone of decerebrate rigidity. These observations indicate that in the intact animal the cerebellum must exercise an important co-ordinating influence on higher and lower parts of the nervous system, probably through connexions with the reticular formation of the brain stem (p. 924). The anterior lobe of the cerebellum receives impulses from the muscle spindles and Granit has suggested that the cause of cerebellar ataxia is a paralysis of the gamma efferents supplying the muscle spindles (p. 829) and an increase in alpha cell activity; in this way the feedback mechanism is put out of action. The patient misjudges weights because his (unconscious) sense of effort is much impaired.

The neocerebellum, because of its size and superficial position, is the part most often damaged in man and, therefore, more is known of the effect of lesions in it than of lesions in the palaeocerebellum which is more deeply placed. In gunshot wounds of the cerebellum, in which the vestibular nuclei do not appear to be involved, hypotonia with asthenia and dis-

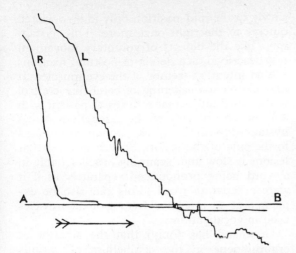

Fig. 46.11 The patient, who had a wound of the left side of his cerebellum, was asked to extend springs of equal strength with his two arms on a signal and to maintain the stretch at a level indicated by the line A, B. Note the slowness of the left arm in exerting power, the irregular tremulous character of the movement and the failure to maintain the final posture. The record was made on a slowly revolving kymograph. Read from left to right. (Gordon Holmes (1939). *Brain* **62**, 13.)

turbances of voluntary movement are the main effects (Fig. 46.11, 12). In unilateral lesions these effects are confined to the same side of the body as the lesion. Unilateral cerebellar injury in man always causes disturbances in both ipsilateral limbs and thus both the arm and the leg on the affected side show the same kind of abnormal behaviour. Hypotonia in man is shown by an alteration in the knee jerk (Fig. 46.10); since stretch reflexes depend on the tension in the muscles the knee jerk may be diminished in extent because of the hypotonia but the leg, if hanging free, tends to swing to and fro before coming to a stop (Fig. 46.10D). The stretch reflex remains since the alpha system is normal but is undamped because the gamma system is underactive. The diminished control by the gamma system explains many of the disorders of movement. The most prominent disturbance of movement is intention tremor—any voluntary effort is accompanied by tremor, which may be very severe. On attempting to touch the nose with the finger the face may be struck violently for not only is the force of the movement excessive but it is so badly directed that the nose may not be touched at the end of the movement (*dysmetria*, Fig. 46.12). On the other hand, willed movements may be carried out with an undue slowness and deliberation, a complex movement being *decomposed* into its several parts. Rapid movements of the limbs, such for example as rapid pronation and supination of the forearms, are faulty, the movements in the

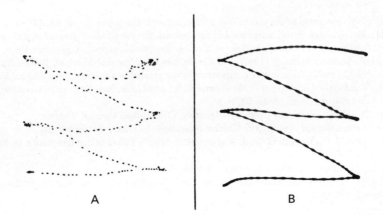

A   B

Fig. 46.12 Records obtained by photographing points of light attached to the tips of the forefingers as a patient with a lesion of the right side of the cerebellum attempted to move each finger slowly and accurately between series of luminous red points (not visible in the photographs) in a dark room. Each flash of light corresponds to 0·04 second. The range of each movement was about 75 cm. B is the record obtained from the left hand, A, that from the right. The irregularity in the rate and in the directions of movements of the affected hand, and the failure to arrest this finger accurately at the points, are well shown in A. (Gordon Holmes (1939). *Brain* **62**, 12.)

affected side being slower, irregular and jerky (*dysdiadochokinesis*, Fig. 46.13). If a patient

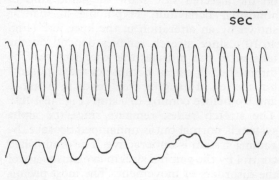

sec

Fig. 46.13 Tracings of rapid pronation-supination. The movements of the affected arm (below) were for a time regular though slower and of smaller amplitude than the normal, but later they became irregular and the arm became more or less fixed in supination. (Gordon Holmes (1939). *Brain* **62**, 22.)

suffering from loss of cerebellar function is asked to look to the right, the eyes are moved to the right quickly enough but they return slowly to the mid-position only to be deviated quickly to the right once more. This nystagmus, like the defects of voluntary movement just described, is a defect of postural fixation; it is an intention tremor of the eye muscles. It may denote a weakening of cerebellar control of the vestibulo-ocular reflexes. A patient with a lesion on one side of the cerebellum shows nystagmus with the quick component directed to the side of the lesion. Speech in cerebellar lesions is slow and 'scanning', each syllable in a word being pronounced separately as if it were a separate word. This can also be explained as an intention tremor of the muscles used in vocalization.

Crossland has found that the amount of ergothionene in the cerebellum of various animals is very much higher than in the cerebrum. When injected into the cerebral circulation of the decerebrate rabbit it, like his extracts of cerebellum, produces increased electrical activity in the cerebellum. The physiological meaning of these findings is still a matter for speculation.

REFERENCES

ADRIAN, E. D. (1943). Afferent areas in the cerebellum connected with the limbs. *Brain* **66**, 289–315.

ADRIAN, E. D. (1944). Localization in the cerebrum and cerebellum. *British Medical Journal* **ii**, 137–140.

ECCLES, J. C. (1969). *Inhibitory Pathways of the Central Nervous System*. Liverpool: University Press.

ECCLES, J. C., ITO, M. & SZENTÁGOTHAI, J. (1967). *The Cerebellum as a Neuronal Machine*. Berlin: Springer.

FADIGA, E. & PUPILLI, G. C. (1964). Teleceptive components of the cerebellar function. *Physiological Reviews* **44**, 432–486.

FULTON, J. F. (1949). *Functional Localization in the Frontal Lobes and Cerebellum*. Oxford: Clarendon Press.

HOLMES, G. (1939). Cerebellum of man. *Brain* **62**, 1–30.

JANSEN, J. & BRODAL, A. (1954). *Aspects of Cerebellar Anatomy*. Oslo: Johan Grundt Tanum.

LANCE, J. W. (1970). *A Physiological Approach to Clinical Neurology*. London: Butterworths.

MARSDEN, C. D., MEADOWS, J. C., LANGE, G. W. & WATSON, R. S. (1967). Effect of deafferentation on human physiological tremor. *Lancet* **ii**, 700–702.

MORUZZI, G. (1950). *Problems in Cerebellar Physiology*. Oxford: Blackwell.

SNIDER, R. S. & STOWELL, A. (1944). Receiving areas of the tactile, auditory and visual systems in the cerebellum. *Journal of Neurophysiology* **7**, 331–357.

47 The diencephalon

The forebrain comprises the *diencephalon* consisting of the structures bounding the third ventricle and the *telencephalon* which includes the cerebral hemispheres and the basal nuclei. In the embryo the lateral walls of the caudal end of the forebrain become thickened and develop two large nuclear masses, termed the *thalami*. Caudal to each thalamus two elevations, the medial and lateral geniculate bodies, appear; these constitute the *metathalamus*. The ventral and inferior parts of the lateral walls and the floor of the third ventricle form the *hypothalamus*.

THE HYPOTHALAMUS

The hypothalamus, estimated to weigh about 4 g, consists of the posterior perforated substance, the corpora mamillaria, the tuber cinereum, the infundibulum and the neurohypophysis; other nuclei are found in the lateral wall of the third ventricle below the hypothalamic sulcus and are listed in Fig. 47.1. For physiological purposes the nuclei of the hypothalamus can be divided into four groups: anterior (including the paraventricular and supra-optic nuclei), middle (including the tuberal, dorsomedial and ventromedial hypothalamic nuclei), posterior (including the posterior hypothalamic nuclei and the mamillary nuclei) and lateral. The supra-optic nucleus is noted for its extremely rich blood supply, there being 2500 capillaries per mm^3. The other nuclei are also well supplied with blood.

The hypothalamus receives impulses from many parts of the brain. The median forebrain bundle brings fibres from the cortex. Periventricular fibres come from the thalamus. The fornix brings fibres from the hippocampus to the mamillary and other nuclei; it was formerly believed that these fibres were concerned in olfaction but it is very doubtful if the hippocampus has any olfactory function in man. From the cerebral cortex fibres pass to the

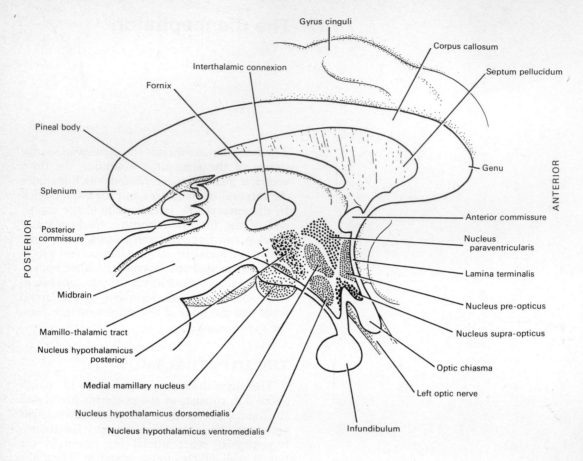

Fig. 47.1 A diagrammatic sagittal plane to show the hypothalamic area and adjacent structures. The hypothalamic nuclei have been projected on to the lateral wall of the third ventricle. × 2½. (*Partly after* W. E. Le Gros Clark (1936). *Journal of Anatomy* **70**, 204.)

medial nucleus of the thalamus and then to the hypothalamus. Fibres which ascend from the brain stem reticular formation to enter the hypothalamus seem to be concerned in maintaining alertness (see p. 924). The mamillary peduncles contain fibres from spinal nerves and probably from the ninth and tenth cranial nerves.

The hypothalamus sends fibres to many destinations. The mamillothalamic tract passes from the mamillary nucleus to the anterior thalamic nuclei from which impulses can be relayed to the gyrus cinguli. The dorsomedial nuclei of the thalamus are reached by periventricular fibres; this area of the thalamus projects in turn to the frontal cerebral cortex. Important descending fibres arise in the lateral

and posterior hypothalamic areas and pass to the reticular formation of the mid-brain, where they probably relay to the brain stem and spinal cord. Each side of the hypothalamus sends fibres to both right and left halves of the sympathetic system. The posterior lobe of the pituitary gland (neurohypophysis), derived from a pocket on the floor of the third ventricle, receives in man about 100 000 unmyelinated nerve fibres from the hypothalamus, chiefly in the supraoptico-hypophysial tract (Chap. 52). The anterior pituitary on the other hand is controlled by releasing factors passing from the hypothalamus into the capillaries supplying it (Fig. 52.2).

The hypothalamus has a great influence on many activities of the body. It affects all kinds of metabolic activities, it regulates body temperature, it influences the cardiovascular and alimentary systems, it controls the anterior and posterior lobes of the pituitary, it is concerned with appetite and thirst. These phenomena are considered in other parts of this book (see Index).

As mentioned earlier (p. 937), the decorticate animal with thalamus intact possesses normal muscular tone and shows righting reflexes. Its internal environment is well regulated by autonomic reflexes.

When the cerebral hemispheres of the cat are removed without damage to the hypothalamus the animal shows, especially if the operation is recent, disturbances of behaviour, generally called 'sham rage', which appear to occur spontaneously or in response to a very mild stimulus such as a pinch. The animal spits, bites, lashes its tail, growls, and increased activity of the sympathetic system is shown by erection of the hairs, dilatation of the pupils, salivation, increase of heart rate with extrasystoles and a rise of blood pressure. Animals deprived of the hypothalamic area but otherwise intact may show sham rage but never with the violence seen when this area is preserved.

The hypothalamus can be stimulated through electrodes introduced into it under general anaesthesia. When the animal recovers consciousness mild stimulation through these electrodes produces, after a latent period of a few seconds, an 'alerting reaction' or 'preparatory reflex' in which the ears are erected and the pupils dilated. Stronger stimulation produces the signs just described as sham rage and at the same time vasodilatation occurs in the skeletal muscles. This effect is mediated by atropine-sensitive sympathetic nerves which supply the arterioles of the muscles. These nerves do not take part in the ordinary homeostatic circulatory reflexes. The muscular vasodilatation is accompanied by vasoconstriction in the skin and intestine, increase in heart rate and central venous pressure because of venoconstriction; the increase of the blood pressure and heart rate indicates that the baroceptor (buffer) reflexes must be temporarily inhibited. Electrical stimulation of the hypothalamus and amygdala in man produces similar circulatory changes in muscles and skin, and similar changes occur in emotional situations such as cardiac catheterization. Vasodilatation through atropine-sensitive sympathetic nerves has been observed even when the stimulus appeared to be very mild, for example mental arithmetic.

Stimulation of the posterior hypothalamic areas in unanaesthetized animals produces sympathetic effects, such as the rise of blood pressure, and secretion of adrenaline from the adrenal medulla. On the other hand destructive lesions of the posterior hypothalamus cause a fall of blood pressure. The hypothalamus is not essential for sympathetic or parasympathetic effects since the cardiac reflexes controlling the heart rate function normally in an animal decerebrated through the pons; it is, however, called into play in conditions of stress.

Stimulation of a region in the anterior part of the hypothalamic area causes a contraction of the urinary bladder, a parasympathetic effect. Since the effect persists after degeneration of fibres from the cerebral cortex, this region seems to be a true parasympathetic centre so far as the bladder is concerned. Although stimulation of the anterior part has produced other parasympathetic effects, such as increased gastrointestinal activity and slowing of the heart, it may be that this is due to stimulation of fibres passing through the hypothalamic area from more cephalic parts of the brain.

The superficial nuclei of the hypothalamus are affected by drugs placed in the third ventricle (see also p. 882). Feldberg has found that in the unanaesthetized cat intraventricular injection of anticholinesterases produces a catatonic state in which the animal can maintain fixed postures for long periods. Tubocurarine injected intraventricularly produces a tremor resembling shivering by an action on the hypothalamus. This tremor is counteracted by adrenaline or noradrenaline. Tubocurarine in larger doses produces convulsions when it reaches the inferior horn of the lateral ventricle to act on the hippocampus.

The hypothalamus is essential for regulation of body temperature (p. 880). Animals deprived of this part of the brain are poikilothermic. The main reactions to a cold environment, vasoconstriction and shivering, are abolished if the caudal area of the hypothalamus is destroyed; as in the case of the sham rage this reaction to cold involves both sympathetic and somatic (muscle) activity.

The cranial area of the hypothalamus near the optic chiasma is sensitive to a rise of blood temperature and causes sweating and vasodilatation which lower body temperature. The descending pathways situated laterally in the caudal part of the hypothalamus carry impulses which cause sweating and cutaneous vasodilatation. Local heating of this area also causes these reactions and a local rise of temperature causes some of the neurones in this area to increase their rate of discharge. The hypothalamus is concerned with water, electrolyte and carbohydrate metabolism. Because the hypothalamus is depressed by general anaesthetics, especially the barbiturates, deeply anaesthetized patients have no power to regulate body temperature.

Pituitary tumours in man may be accompanied by obesity due to damage to the middle group of hypothalamic nuclei. If hypophysectomy in animals involves damage to the hypothalamus, there is genital atrophy and obesity due to increase in appetite (p. 171). The genital atrophy can be attributed to loss of the adenohypophysis. The relationship of the hypothalamus to ovulation and other reproductive functions is discussed in Chapter 51.

Injury to the hypothalamus, during surgical operations for example, may result in erosions or ulcers of the gastric mucosa, and is commonly followed by a persistent rise of body temperature; tumours of the posterior region may be associated with a fall of body temperature, a reduction in bodily activity (hypokinesia) and a tendency to fall asleep (hypersomnia). Disease of the mamillary bodies produces a remarkable state in which events of ordinary daily life are forgotten as quickly as they occur but events in childhood are well remembered.

THE THALAMUS

The thalami are two large masses of grey matter about 4 cm long placed on either side of the third ventricle and extending posteriorly, as the pulvinar, to overhang the superior colliculi. Each thalamus possesses three main nuclei; the anterior and medial form the *palaeothalamus*, while the lateral nucleus, which occupies the greater portion of the thalamus including the pulvinar, constitutes

the *neothalamus*. The geniculate bodies (the *metathalamus*) lie near the pulvinar, superolateral to the colliculi.

The thalamus can be regarded as a relay station on the ascending sensory pathways (Fig. 47.2). The anterior nucleus receives impulses from the mamillothalamic tract and transmits impulses to the gyrus cinguli. The medial nucleus receives some impulses via the hypothalamus from the viscera; most autonomic afferent fibres, however, relay in the lateral nucleus of the thalamus whence fibres pass to the sensory area of the cerebral cortex. Impulses can also pass from the frontal cortex to the medial nucleus and from there to the hypothalamus. The ventral portion of the lateral nucleus receives afferents from the medial, spinal and trigeminal lemnisci which carry impulses subserving proprioceptive, tactile, pain and temperature senses, and sends out efferents through the posterior limb of the internal capsule to the postcentral gyrus of the cerebral cortex. Thus the somatic sensory impulses originating in, say, the right side of the body cross over to the left thalamus which transmits them to the cerebral cortex on the left side only. Fibres from the surface of the body go to both thalami but the deep structures send fibres to the contralateral thalamus only. The lateral nucleus of the thalamus also receives fibres from the dentate nucleus of the cerebellum of the opposite side and relays impulses to the motor cortex (areas 4 and 6). Impulses from the cochlea pass through several synapses before reaching the medial geniculate body from which fresh impulses pass to the auditory cortex in the superior temporal gyrus (p. 974). The lateral geniculate body relays visual impulses from the retinae to the calcarine cortex (area 17) on the same side (p. 970). These *cortical relay nuclei* do not act in a random way but relay impulses from points on a limb, for example, to specific areas on the cortex; similarly every point on the retina has a corresponding area on the calcarine cortex in spite of the intervention of the relay in the lateral geniculate body.

In addition to these cortical relay nuclei there are *association nuclei* in the neothalamus which receive impulses from the relay nuclei and project to the association areas of the cerebral cortex, namely the prefrontal, parietal and occipital cortex. The dorsomedial nucleus

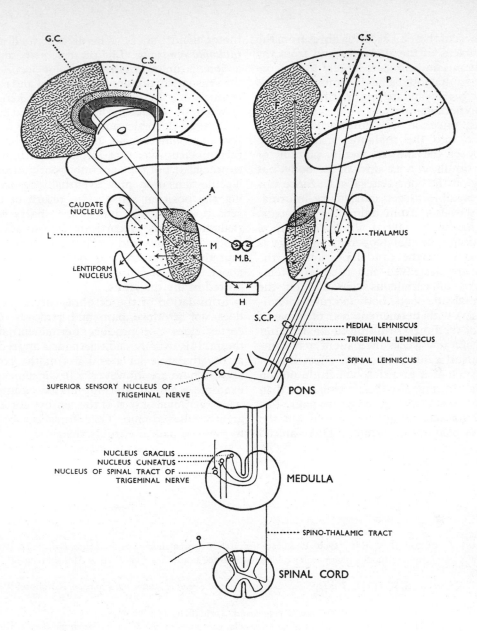

Fig. 47.2 Cortical and spinal connexions of the thalamus. G.C. = gyrus cinguli; C.S. = central sulcus with motor and somaesthetic areas on each side; F = frontal area; P = parietal area; A = anterior group of thalamic nuclei; M = medial group of thalamic nuclei; L = lateral group of thalamic nuclei; M.B. = mamillary bodies; H = hypothalamus; S.C.P. = superior cerebellar peduncle. On the upper right of the diagram the projections are to the lateral surface of the cerebral hemisphere: on the upper left the projections to the medial surface of the cerebral hemisphere are given. Internuclear connexions of the thalamus are shown only on the left. The lines joining the various parts of the thalamus and cortex have in most cases been given arrows to indicate that impulses may pass in either direction. The metathalamus has been omitted. The pulvinar is included in the lateral nucleus. (*By courtesy of G. W. Pearce.*)

is the largest of these; it receives fibres from the lateral nucleus of the thalamus and from the hypothalamus and projects largely to the cortex of the frontal lobe. The thalamus also sends impulses to the caudate and lentiform nuclei, and the hypothalamus.

In cats, injection of minute quantities of strychnine into the thalamus increases the sensitivity of the skin to tactile, painful, or thermal stimuli on both sides of the body, but especially on the contralateral side. Since it is scarcely possible to get satisfactory information on sensation from animal experiments, reliance has to be placed chiefly on clinical studies. When the thalamogeniculate artery is blocked in man, the caudal portions of the thalamus are destroyed and there is a temporary loss of cutaneous sensation on the contralateral side of the body (a crossed hemi-anaesthesia) with permanent loss of sense of position of the limbs resulting in ataxia. After the lapse of a few weeks the patient may complain of insufferable pain on the affected side. Stimuli, such as a pinprick, which do not on the normal side give rise to anything more than discomfort, may cause quite severe pain. The threshold for pain is actually raised but the reaction to pain is exaggerated. There are no motor disturbances. This is described as the *thalamic syndrome*. The reason for these over-actions is unknown; they do not occur when the sensory cortex alone is removed and cannot be due solely to a release from cortical control.

When the cerebral cortex is removed on one side (hemi-decortication) in man there is, for a few days, anaesthesia on the opposite side, then nociceptive, that is painful, stimuli are appreciated but they are only poorly localized. Tactile sensation may eventually return but not the position sense. The return of crude sensation is partly due to the ability of the thalamus on the decorticate side to subserve crude sensations and partly to bilateral representation of touch and pain as mentioned above. Position sense is not bilaterally represented in the thalamus.

Stimulation of the cerebral cortex (p. 970) does not produce pain and removal of the cortex does not produce permanent anaesthesia. It is usually said that pain is appreciated at the thalamic level and not at the cortical level. There are, however, a few instances, for example pain in an amputation stump, in which removal of part of the sensory cortex has greatly relieved pain. Thus it may not be right to say that pain is entirely thalamic.

REFERENCES

BROWN-GRANT, K. & CROSS, B. A. (Eds.) (1966). Recent studies on the hypothalamus. *British Medical Bulletin* **22**, 195–284.

FELDBERG, W. (1963). An unorthodox pharmacological approach to the brain. *Scientific Basis of Medicine Annual Review* 137–157.

FULTON, J. F. & RANSON, S. W. (1940). *The Hypothalamus and Central Levels of Autonomic Function.* Baltimore: Williams & Wilkins.

HAYMAKER, W. & ANDERSON, EVELYN (1969). *The Hypothalamus.* Springfield, Ill.: Thomas.

HILTON, S. M. (1965). Hypothalamic control of the cardiovascular responses in fear and rage. *Scientific Basis of Medicine Annual Review* 217–238.

GELLHORN, E. (1957). *Autonomic Imbalance and the Hypothalamus.* London: Oxford University Press.

McDONALD, D. A. (1951). W. R. Hess: the control of the autonomic nervous system by the hypothalamus. *Lancet* **i**, 627–629.

MARTIN, L. & MEITES, J. (Eds.) (1971). *Neurochemical Aspects of Hypothalamic Function.* New York: Academic Press.

MARTIN, L., MOTTA, M. & FRASCHINI, F. (Eds.) (1971). *The Hypothalamus.* New York: Academic Press.

OLDS, J. (1967). Emotional centres in the brain. *Science Journal* **3**, No. 5, 87–92.

REICHLIN, S. (1967). Functions of the hypothalamus. *American Journal of Medicine* **43**, 477–485.

48 The telencephalon

The telencephalon or end-brain is the most cranial (or rostral) part of the fore-brain and is represented in the adult by the cerebral hemispheres which consist of the olfactory system (collectively referred to as the rhinencephalon), the cortex or pallium and the corpus striatum (Table 48.1). The rhinencephalon at the base of the brain was so named because of its supposed connexion with the sense of smell. In fact this is only a minor function in man; the rhinencephalon is much larger in animals in which the sense of smell is well developed. The corpus striatum consists of large nuclei of nerve cells situated within the substance of the hemispheres. The pallium or cerebral cortex is divided into a small archipallium (part of the rhinencephalon) and the neopallium, which is the most highly developed part of the brain in man. The neopallium is composed histologically of iso-cortex. The allocortex with its three cell layers is found in the archipallium (see p. 961).

The superficial layer or cortex of the hemispheres consists of grey matter containing numerous nerve cells, while the deeper part or white matter is made up of fibres connecting the various parts of the hemispheres with one another and with lower parts of the central nervous system such as the brain stem and spinal cord.

The surface of the cerebral cortex is marked by furrows called sulci, the areas between sulci being called gyri or convolutions. The two main sulci are the lateral and the central sulcus (Fig. 48.2A). The area in front of (rostral to) the central sulcus is the frontal lobe; the area immediately behind the central sulcus is the parietal lobe. The most caudal (posterior) part of the cerebral hemisphere is the occipital lobe. The area below the posterior ramus of the lateral sulcus is the temporal lobe. Smaller sulci divide up each lobe into numerous gyri, the names of the main areas being given in Fig. 48.2. The term limbic lobe is often applied to

Table 48.1 Parts of the telencephalon

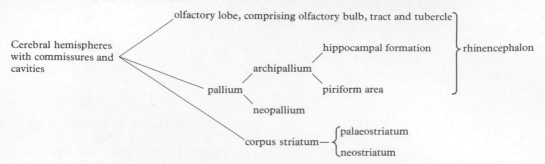

the region of the cerebral cortex which lies on the medial side of the hemisphere adjacent to the corpus callosum and the attachment of the

them gives rise to pain. For this reason removal of cerebrospinal fluid is apt to be followed by headache.

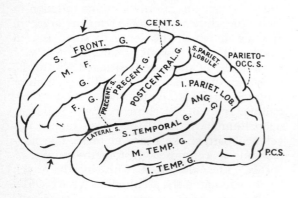

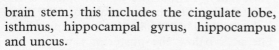

Fig. 48.2A The lateral aspect of the left cerebral hemisphere. S. FRONT G. = superior frontal gyrus. M. F. G. = middle frontal gyrus. I. F. G. = inferior frontal gyrus. PRECENT. S. = precentral sulcus. PRECENT. G. = precentral gyrus. POSTCENTRAL G. = postcentral gyrus. S. TEMPORAL G. = superior temporal gyrus. M. TEMP. G. = middle temporal gyrus. I. TEMP. G. = inferior temporal gyrus. S. PARIET. LOBULE = superior parietal lobule. I. PARIET. LOB. = inferior parietal lobule. CENT. S. = central sulcus. PARIETO-OCC. S. = parieto-occipital sulcus. P. C. S. = postcalcarine sulcus. LATERAL S. = lateral sulcus (posterior ramus). ANG. G. = angular gyrus. A line joining the arrows gives the approximate plane of section in the operation of prefrontal leucotomy.

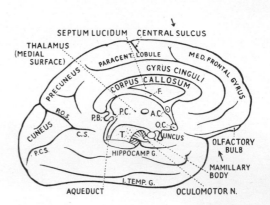

Fig. 48.2B The medial aspect of the left cerebral hemisphere. A. C. = anterior commissure. O. C. = optic chiasma. P.-O. S. = parieto-occipital sulcus. C. S. = calcarine sulcus. MED. FRONTAL GYRUS = medial frontal gyrus. P. C. S. = postcalcarine sulcus. F. = fornix. PARACENT. LOBULE = paracentral lobule. HIPPOCAMP. G. = hippocampal gyrus. P. B. = pineal body. P. C. = posterior commissure. T. = tegmentum. I. TEMP. G. = inferior temporal gyrus.

BASAL NUCLEI

brain stem; this includes the cingulate lobe, isthmus, hippocampal gyrus, hippocampus and uncus.

The brain and its coverings, the dura and pia mater may be cauterized or stimulated without the patient feeling pain but dilatation of the large arteries or veins or traction on

The basal nuclei (basal ganglia) include the corpus striatum (formed from palaeostriatum and neostriatum), the amygdaloid nucleus and the claustrum. The corpus striatum is divided by the anterior limb of the internal capsule into the caudate nucleus medially and the lentiform nucleus laterally (Figs. 48.3 and 48.6), the latter being further subdivided into putamen and globus pallidus. The caudate nucleus and putamen contain small cells and together

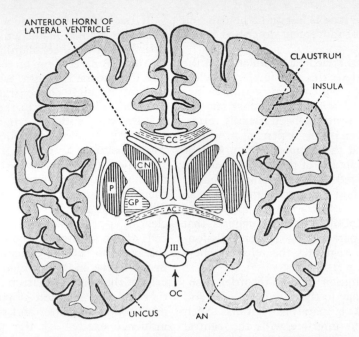

ANTERIOR HORN OF
LATERAL VENTRICLE

CLAUSTRUM

INSULA

CC

CN LV

P

GP

AC

III

OC

UNCUS

AN

Fig. 48.3 Coronal section through the adult human brain just behind the optic chiasma. Vertical shading, neostriatum (lentiform and caudate nuclei). Horizontal shading, palaeo-striatum (globus pallidus). LV = lateral ventricle. CN = caudate nucleus. P = putamen. GP = globus pallidus. CC = corpus callosum. OC = optic chiasma. AN = amygdaloid nucleus. F = anterior column of fornix. III = third ventricle. AC = anterior commissure.

represent the neostriatum which receives many afferent fibres from areas 4 and 6 of the cortex (Fig. 48.7) and also from the thalamus. The globus pallidus or palaeostriatum possesses large cells and may be regarded as the discharging centre of the basal nuclei. The efferent fibres leaving the palaeostriatum make up a large part of the descending extrapyramidal system; they go to the thalamus, hypothalamus, subthalamic nucleus, red nucleus, substantia nigra and reticular substance of the mid-brain. Relay fibres pass from these nuclei down the spinal cord to influence the motor cells of the anterior horn of grey matter. From the functional point of view we can group together with the basal nuclei proper, the red nucleus, the substantia nigra and the subthalamic body of Luys although the last three are in fact situated in the brain stem.

The basal nuclei, since they are quite prominent in primitive vertebrates, may be described as an old part of the fore-brain in a phylogenetic sense. They appear to have chiefly motor functions. Goltz in 1892 removed both cerebral hemispheres from a dog and

succeeded in keeping it alive for 18 months. During the day it walked up and down its cage and at night it curled up and slept. It did not appear to recognize its attendants or other dogs; it ate only when food was placed in its mouth. If the basal nuclei are removed from such decorticate animals their behaviour is not greatly altered. It seems probable, therefore, that the basal nuclei are activated by the cortex almost entirely and that when the cortex is removed they cease to function. Destructive lesions of the basal nuclei in monkeys do not produce any visible effects unless neighbouring areas such as the internal capsule are also damaged. Although stimulation of the basal nuclei by electrodes inserted into them seldom has any apparent effect in anaesthetized animals it may inhibit a movement of the muscles on the opposite side which is being carried out as the result of cortical stimulation, and it is known that impulses from areas 4 and 6 of the cortex go to the caudate nucleus, then to the globus pallidus and thalamus and back to area 4 of the cortex. If this circuit is interrupted the restraining influence of the basal nuclei on muscle tone is lost and athetosis and other disorders of movement described below are seen.

In an unanaesthetized cat stimulation of the amygdaloid nucleus through implanted electrodes evokes rage which seems genuine since it may, during such stimulation, attack another cat with which it is normally on friendly terms. In man stimulation of this region has produced rage and fear and the impulse to run. On the other hand stimulation of the caudate nucleus in unanaesthetized rhesus monkeys has a taming effect and during stimulation an aggressive animal can be readily approached and handled. Recent evidence suggests that the amygdala, part of the rhinencephalon-limbic complex, regulates the releasing hormone LRH (see p. 1082).

The most common clinical disorder associated with damage to the globus pallidus and substantia nigra is paralysis agitans. Such damage seems to interfere with the central connexions, concerned especially with visual and proprioceptive information, needed for normal postural reactions; postural fixation, righting, locomotion and phonation are all disturbed. The patient shows a tremor of which the principal component is present when the limb is at rest; in some subjects, however, the tremor is also present during voluntary movement. The affected muscles are rigid but, unlike spasticity, the rigidity generally affects flexor and extensor muscles equally. It is probably dependent on the gamma efferents since blockage of them by procaine results in its disappearance. The face has an expressionless or mask-like appearance. The muscles may show an intermittent (cog wheel) resistance when moved passively. The tremor disappears if such a patient develops a hemiplegia (paralysis of one side of the body) or if the motor cortex is destroyed. The tremor is not simply the result of release of the basal nuclei from cortical control since it is also abolished by destruction of the globus pallidus or of the ventrolateral nucleus of the thalamus. The site of origin of the tremor is unknown but it is clearly the result of disturbance of a complex neuronal circuit which includes the basal ganglia, thalamus and motor cortex. Paralysis agitans may be relieved surgically by stereotactic production of circumscribed lesions within the globus pallidus or the thalamus.

If the brain of a patient with Parkinson's disease is stained histochemically to show the distribution of catecholamines it is found that the dopamine content of the corpus striatum and the substantia nigra, where it is normally present in high concentration, is markedly reduced. This has led to the treatment of the disease with L-dopa, which penetrates the brain if it is taken by mouth to be converted into dopamine (p. 1017). This may be followed by clinical improvement particularly as regards the slowness of movement (bradykinesia) and there may also be relief of tremor and rigidity. The meaning of these observations is not yet clear. It appears that dopamine is formed in the cells of the substantia nigra and carried along the axons to become concentrated in granules in the nerve endings which connect with the cells of the corpus striatum. There is evidence that in this situation dopamine acts as an inhibitory transmitter and so may modify the influence exerted by the striatum on other basal nuclei such as the globus pallidus. The drug reserpine, which acts by depleting the brain of catechol amines including dopamine, may induce a clinical picture resembling paralysis agitans.

Disease of the corpus striatum in man, and injuries placed experimentally in this region in chimpanzees, may produce involuntary movements called *athetosis*. The lips, jaw and tongue may be affected but more frequently the hands show alternating movements of extension of the fingers with pronation at the wrist to full flexion with supination. Although these are associated with disease of the globus pallidus they may be due to compensatory extrapyramidal activity, since they are abolished by removal of the premotor area of the cortex (area 6). Damage to the caudate nucleus may result in quick, short, jerking movements, especially of the fingers, arm or face (*choreiform movements*). Small lesions destroying the subthalamic body of Luys cause *hemiballismus*, rhythmic, powerful, involuntary movements of the limbs on the opposite side; destruction of the neighbouring globus pallidus abolishes the movements.

CEREBRAL HEMISPHERES

Because the cerebral hemispheres are very much larger in man than in any of the primates, while the brain stem and spinal cord are not very different in size, we are led to conclude that human attributes are in some way associated with the greater development of the hemispheres. Such typically human activities as speaking and writing are extremely complex, involving the co-ordination of sensory information from the eyes, the ears and various proprioceptors to produce a very delicately adjusted motor response. This implies the participation of many parts of the cortex and suggests that the sensory and motor areas must be interconnected by very numerous pathways. When these activities are analysed further it is found that they involve memory, learning and thought. Damage to the human cortex, especially if widespread, interferes with memory and judgment and we are, therefore, accustomed to think of these as cortical functions although no specific cortical area associated with memory has been discovered; many years ago Lashley found that reduction of maze-learning ability in rats depended on the volume of cerebral cortex removed and not on its anatomical position. Since the memory of an event, such as a pleasant meal, involves the recollection of visual, olfactory, gustatory and visceral impression, it might be conjectured that memory is a function of the cortex as a whole rather than of any particular part of it.

There is a certain amount of circumstantial evidence that ribonucleic acid (RNA) synthesis is necessary for the storage of memory traces. The RNA concentration in neurones increases after learning procedures and RNA dietary supplements improve learning and retention in rats. If RNA synthesis in rats is blocked by large doses of inhibitor the learning process is sometimes depressed. However, we are still very far from being able to claim that RNA is the long-sought memory molecule.

HISTOLOGY OF THE CEREBRAL HEMISPHERES

In the fetus at six months three different cellular layers can be recognized in the grey matter of the cortex and this arrangement persists in the adult archipallium. In the neopallium further differentiation takes place, as shown in Table 48.4, and six layers can be made out in the adult (Fig. 48.5) but the pattern varies in different parts of the cerebral

Table 48.4 Structure of cerebral cortex

Six-month fetus	Adult	Number of layer in Fig. 48.5
Superficial or molecular layer of tangential fibres	Superficial or molecular layer of tangential fibres	1
Granular layer	Outer granular layer (small cells)	2
	Pyramidal layer (large	3
	Inner granular layer (small cells)	4
Infragranular layer	Ganglionic layer (large cells in motor cortex)	5
	Polymorphous layer	6

hemispheres. The study of such histological differences, called architectonics, has received much attention during the last eighty years, and the cortex has been mapped out into areas which are numbered according to the system introduced by Brodmann (Fig. 48.7). Area 4 of his classification, the *motor cortex*, is distinguished histologically by the large pyramidal cells and the giant cells of Betz which may be 30 to 120 µm in height (layer 5, Fig. 48.5). The *visuosensory cortex* (area striata or post-calcarine cortex) can be distinguished by the naked eye because the grey matter is divided by a white band due to an increase in the number of fibres in the outer band of Baillarger (Fig. 48.5). The thickness of the grey matter of the cerebral cortex in man varies from about 4 mm to 1·25 mm. It is thinner in the occipital region than in the anterior and posterior central gyri.

The white matter of the cerebral hemispheres consists of myelinated fibres which may run (1) as *association fibres* from one part to another part of the same hemisphere, or (2) as *commissural fibres* from one hemisphere to the other, or (3) as afferent or efferent *projection fibres* which carry impulses from or to the lower parts of the brain and the spinal cord. The neopallial parts of the cortex are joined by commissural fibres which pass across the midline in the corpus callosum (Fig. 48.3). The main efferent neopallial projection fibres pass

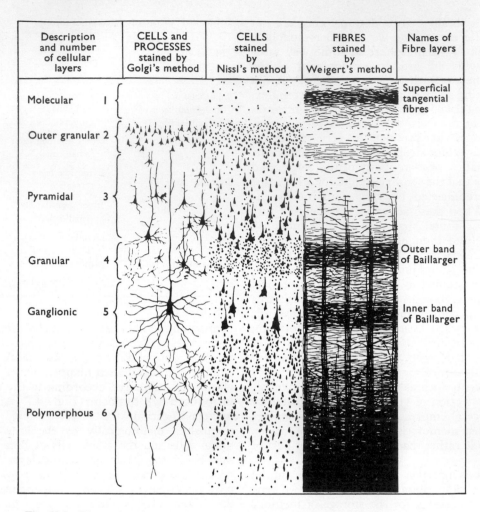

Description and number of cellular layers	CELLS and PROCESSES stained by Golgi's method	CELLS stained by Nissl's method	FIBRES stained by Weigert's method	Names of Fibre layers
Molecular 1				Superficial tangential fibres
Outer granular 2				
Pyramidal 3				
Granular 4				Outer band of Baillarger
Ganglionic 5				Inner band of Baillarger
Polymorphous 6				

Fig. 48.5 Diagram to show the layers of cells and fibres in the grey matter of the human cerebral cortex. (L. Luciani (1915). *Human Physiology*, Vol. 3, p. 539. London: Macmillan.) The main arrival area of fibres from the thalamus is said to be layer 4. The pyramidal cell layers are especially prominent in the frontal lobes. Layer 5 is characteristic of the motor cortex (area 4).

fan-wise as the corona radiata from all parts of the cortex, to run in the internal capsule which lies medial to the lentiform nucleus (Fig. 48.6).

Nerve impulses have been shown by electrical methods to pass across the corpus callosum from one hemisphere to the corresponding area of the other. If a conditional reflex is established in an animal in response to a stimulus applied to one side of the body it can generally be evoked by a contralateral stimulus. After section of the corpus callosum this is no longer possible but conditional reflexes can still be established independently for each side.

If the optic chiasma in the cat is divided to cut the decussating fibres, visual impulses from each eye pass only to the ipsilateral hemisphere. If an animal treated in this way learns a task involving pattern recognition through one eye it is able to perform the task equally well with the contralateral eye, but is unable to do so after section of the corpus callosum. Likewise monkeys can learn to perform tasks with one hand and this learning is immediately transferred to the contralateral hand. If the corpus callosum has been cut it is necessary for the animal to learn the task with each hand

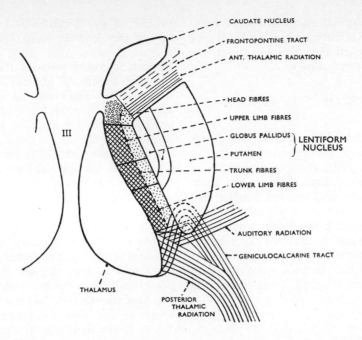

CAUDATE NUCLEUS
FRONTOPONTINE TRACT
ANT. THALAMIC RADIATION

HEAD FIBRES
UPPER LIMB FIBRES
GLOBUS PALLIDUS ⎫ LENTIFORM
PUTAMEN ⎬ NUCLEUS
TRUNK FIBRES
LOWER LIMB FIBRES

AUDITORY RADIATION
GENICULOCALCARINE TRACT

III

THALAMUS

POSTERIOR
THALAMIC
RADIATION

Fig. 48.6 Diagram of the internal capsule. The sensory fibres of any part of the body emerging from the thalamus (cross hatched section) are medial to the motor fibres (dotted section) of the same part. There is, however, a considerable admixture of motor and sensory fibres. (*After* S. W. Ranson & S. L. Clark (1951). *The Anatomy of the Nervous System*, 9th edn, p. 295. London: Saunders.)

separately. If the corpus callosum is sectioned after the task has been learned the untrained hand shows a slight drop in performance indicating that the trained hemisphere in the intact animal normally makes some contribution to the performance of the untrained side.

In man the effect of lesions of the corpus callosum has been studied in individuals born with agenesis of the corpus callosum and in patients who have had the corpus callosum divided in the treatment of epilepsy. After section of the corpus callosum learned and volitional activities persist independently in each hemisphere and learning takes place in each hemisphere separately and without the awareness of the other. This has surprisingly little effect on the behaviour and intelligence of the subject. It is of interest that while it is possible to communicate without difficulty with the dominant hemisphere the minor hemisphere has very little language capacity either for speech or writing. It is possible, however, to present the non-dominant hemisphere with test material and while it has little

verbal capacity it appears to be able to match objects presented to it visually or by touch. How far the minor hemisphere shows conscious awareness or capacity for ideation remains uncertain.

The fibres of the central nervous system acquire their myelin sheaths at different stages of development. The process of myelination begins at the sixth month of intrauterine life in the spinal cord and is seen at the eighth month in the sensory pathways going to the postcentral gyrus. Soon after this the afferent fibres of the visuosensory and auditosensory areas receive their myelin sheaths but the corticospinal (pyramidal) fibres are not completely myelinated until the child is three months old; the process continues in the so-called association areas until the eighteenth year. Since it has been shown that a nerve fibre can conduct impulses only relatively slowly until it is myelinated, we have a physical basis for the continuous process of mental development up to adolescence. Because they are not yet myelinated it is unlikely that in the new-born child the cerebrospinal motor fibres are functioning normally and it may be that his motor activities are carried out chiefly through the brain stem and spinal cord.

MOTOR AREAS OF THE CEREBRAL CORTEX

In 1870 Fritsch and Hitzig found that electrical stimulation of the frontal cortex of the dog caused movements of the opposite side of the body. Ferrier confirmed this a few years later in the monkey and showed further that removal of the area which on stimulation gave rise to movements of the hand resulted in paralysis of the hand. Numerous experiments in animals and man show that the area with the lowest threshold for electrical stimulation is area 4 of Brodmann (Fig. 48.7) and this area is

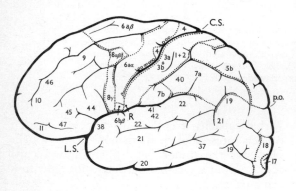

Fig. 48.7 Architectonic fields of the lateral aspect of the human cortex according to Brodmann's conventional numbering. C.S. = central sulcus; L.S. = lateral sulcus; p.o. = parieto-occipital fissure; R = respiration; 4 = true pyramidal cortex. Stimulation of areas 5, 6, 7, also 8, produces adversive movements if the stimulation is strong enough. Stimulation of 8, 18 and 19 produces adversive eye movements. Stimulation of 22 produces eye and ear adversive movements. Stimulation of 6b produces mastication. Some of these fields extend on to the medial aspect. The gyrus cinguli is numbered 24 rostrally and 23 caudad. The names of the main gyri are given in Fig. 48.2. (*After* W. Penfield & E. Boldrey (1937). *Brain* **60**, 393.)

called, therefore, the *motor area* although there are, in fact, several other cortical areas concerned with motor function. This most easily excited region of the cortex contains the giant Betz cells and lies in man in that part of the

precentral gyrus which is buried within the lips of the central sulcus. The following observations show that these cells are in fact important constituents of the motor area. In the first month of life before they are fully differentiated the electrical excitability of the area is low. By applying a hot plate for a few seconds to area 4 in the adult animal the superficial cells of the cortex can be destroyed while the Betz cells remain undamaged; it is then found that the excitability and spontaneous electrical activity of the area are practically unchanged. If the pyramids are sectioned in the medulla the Betz cells show chromatolysis.

The path of the axons of the Betz cells is shown by the degeneration which follows removal of area 4. The fibres pass down in the genu and in the anterior two-thirds of the posterior limb of the internal capsule in such a way that the fibres to the head are in front and those to the legs are posterior (Fig. 48.6). These fibres then pass through the centre of the basis pedunculi and through the pons to the pyramids, where about 80 per cent of the fibres cross to the opposite side in the decussation of the pyramids (Fig. 45.3) and pass down in the lateral column of the spinal cord as the lateral cerebrospinal tract (Fig. 44.3). There are about a million fibres in each pyramid but only about 34 000 Betz cells in each hemisphere, sufficient to account for only three per cent of all the pyramidal fibres; 30 per cent or so of the fibres arise from small or medium-sized pyramidal cells in the precentral gyrus. The origin of the remainder is unknown, but to produce total degeneration of the pyramidal tracts it is necessary to destroy the whole hemisphere. The majority of the fibres in the pyramidal tract, about one-third of which are non-myelinated, are quite small in diameter; only 1·7 per cent are in the range 10 to 22 μm and these may arise from the Betz cells. The vast majority of the fibres conduct relatively slowly. After removal of the motor area examination of the spinal cord shows degenerated synaptic knobs on the interneurones (p. 815) and not on the anterior horn cells. The corticospinal fibres, therefore, end at a position where the impulses carried by them may interact with impulses arriving by other descending tracts or through the posterior roots; the result of this interaction probably determines the nature of the signal transmitted by the interneurones to the motor cells of the anterior horns. Cortico-fugal fibres also pass to the reticular formation and to many of the nuclei in the brain, including the basal nuclei.

In all higher mammals, including man, the part of the motor area nearest the vertex of the skull is concerned with movements of the leg, the middle part with those of the arm, while the lowest part of area 4, near the lateral sulcus, is concerned with movements of the face. The area of the cortical representation of any part is in proportion to the complexity of movements and not to the size of the part. Thus cortical representation of trunk movements is quite small, but that of the thumb, fingers, lips and tongue which are involved in skilled movements is relatively very large. In Figure 48.8, an attempt is made to illustrate this dia-grammatically. Area 4 should not be regarded as a fixed mosaic of excitable cells with definite and circumscribed connexions with the periphery but as something with wide connexions, a certain degree of lability and a reserve of function as the following findings prove. Removal of a *small* part of the motor cortex does not result in permanent paralysis, the adjacent unharmed cortex eventually taking over the function of the ablated portion. The cortex is tolerant of destroying lesions and if such a lesion is restricted it may be symptomless. The connexions of any one area of the motor cortex are much wider than would be expected from the fixed mosaic idea. Thus after excision of a small part corresponding to the hand degenerated fibres can be traced to all parts of the spinal cord as far down as the lumbar region. Repeated stimulation of a single point on the cortex does not invariably lead to the same muscular movement. Finally, as described below, the actual map obtained varies with the method of stimulation.

In 1873 Hughlings Jackson described the initial signs of an epileptic fit due to a discharging lesion in the cortex in the following words. 'There are three parts where fits . . . mostly begin, (1) in the hand, (2) in the face or tongue or both, (3) in the foot. . . . When the fit begins in the hand the index finger and thumb are usually the digits first seized: when in the face the side of the cheek is first in spasm: when in the foot almost invariably the great toe.' It is difficult to believe that the cortical areas for the thumb, angle of the mouth and great toe (as shown on a classical map such as Fig. 48.8) have a specially high incidence of discharging lesions. In fact such lesions may occur at any part of the motor cortex. An explanation of this apparent anomaly is afforded by the experiments of Liddell and Phillips illustrated in Figure 48.9. They used single rectangular pulses of current of about 5 msec duration to stimulate the exposed motor cortex of the baboon. The area representing the movements of the opposite thumb and index varied according to the strength of the pulses (Fig. 48.9 (a)); the threshold was always lowest for the thumb complex. The areas for the thumb, hallux and mouth overlapped as shown in Fig. 48.9 (c) and when the strength of the pulse was sufficient movements of all three parts were

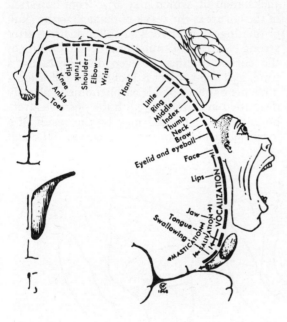

Fig. 48.8 Motor homunculus. The right side of the figurine is laid upon a cross-section of the hemisphere. This diagram gives the results of many experiments carried out at operation under local anaesthesia of the scalp. The exposed cortex was stimulated directly, usually by laying bipolar electrodes on it, the points of the electrodes being 3 mm apart. The stimulator provided 60 pulses of current per second. The potential required to elicit a movement varied between $\frac{1}{2}$ and 5 volts. Stimulation of the cortex over the areas marked in thick lines produced movements of the muscles in the parts labelled. (W. Penfield & T. Rasmussen (1950). *The Cerebral Cortex of Man*, p. 57. New York: Macmillan.)

obtained. If instead of single pulses a series of pulses at the rate of 50 per second was used, the simple large scale map (Fig. 48.9) of distal parts changed into the traditional map of finer grain in which movements of all parts of the body are represented (Fig. 48.8 was obtained in this way). The presence of these fields of low threshold explains adequately the common sites of onset of Jacksonian epilepsy.

If currents of suitable intensity and waveform are applied to area 4 (the motor area) in man, it is possible, according to some authors, to produce movements of single muscles on the opposite side of the body. Other investigators believe that area 4 deals with movements, not with individual muscles. The effect of cortical stimulation on limb muscles depends on the position of the limb and also on the position of the head. This is not to be wondered at since voluntary movements are always being modified by afferent sensory information. In the words of Penfield and Rasmussen, 'The movements produced by cortical stimulation are never skilled acquired movements but instead consist of either flexion or extension of one or more joints, movements which are not more complicated than those the new-born infant is able to perform.' Area 6, the premotor area, immediately rostral to area 4, contains no Betz cells and is not so easily

stimulated. Electrical stimulation of area 6 produces movements of small groups of muscles or more widespread movements involving perhaps a whole limb or a large part of the opposite side of the body (p. 968). If currents above threshold strength are applied to area 4, groups of muscles having the same action (synergic muscles) are activated but, if the current is made still stronger, movements begin in the same part of the body as before and spread to other parts of the body as adjacent motor cells are excited. The order of spread depends upon the anatomical arrangement of the excitable areas of the cortex; a movement starting in the thumb may spread to the fingers, forearm and shoulder, involving larger and larger groups of muscles, so that eventually the entire musculature on the opposite side of the body may be called into action. This is the sequence of events in a typical so-called Jacksonian fit, which may arise from irritation of the cortex as the result of damage to the skull or meninges. After such an epileptic fit, as after strong electrical stimulation, the excitability of the cortex declines or even disappears for a time, and the subject is inclined to fall asleep. This probably depends on phenomena already described in connexion with the positive after-potential (p. 807) and may be the result of a local fall in pH. Conversely a rise in pH, such

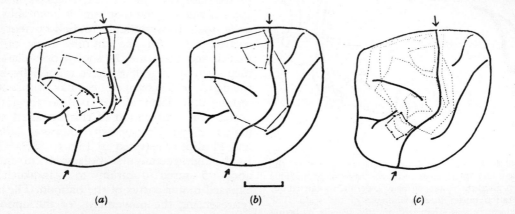

(a)   (b)   (c)

Fig. 48.9 Female baboon (*Papio papio*). 5 kg. Left Rolandic region, stimulated with single pulses 4·5 msec in duration. Scale 10 mm. Arrows mark the central fissure. (Traced from the photographic map.) (a) Extent of thumb-index-minimus area at strengths 1·2 mA, 1·6 mA and 2·7 mA. (b) Extent of hallux-middle toe area at strengths 2·05 mA and 4·7 mA. (c) Area for angle of mouth (solid line) at strength 2·35 mA. Thumb and toe areas indicated by dotted lines. (E. G. T. Liddell & C. G. Phillips (1950). *Brain* **73**, 130. See also E. G. T. Liddell & C. G. Phillips (1952). *Brain* **75**, 510–525.)

as is produced by hyperventilation, increases the excitability of the motor area.

Epileptic seizures occurring spontaneously or by electrical stimulation of area 4 may occasionally involve the limbs on the same side. These ipsilateral seizures presumably depend on fibres coming down from the cortex but their route is at present unknown. In the lightly etherized macacus monkey stimulation of certain parts of the facial area of the motor cortex produces movements of the face on the same side whereas stimulation of adjacent areas gives contralateral or bilateral facial movements. There are other indications that in man the motor cortex is able to influence muscles of the same side. For example the recovery of motor function after unilateral ablation of the motor cortex may be due in part to the activity of fibres from the ipsilateral motor cortex. Even a man whose motor cortex has been removed on one side can learn to walk, that is, to use muscles on both sides of the body. Another example is seen in severe infantile hemiplegia; the ipsilateral undamaged hemisphere assumes such complete control that removal in adult life of the damaged hemisphere causes no increase in disability.

Effects of removal of the motor cortex. Removal of the whole cerebral cortex in dogs and cats has very little effect on walking or running but the hopping and placing reactions (p. 939) are lost. In the monkey, ablation of area 4 produces a *hemiplegia*, that is a paralysis or loss of voluntary movements on the opposite side of the body; immediately after this operation the muscles are hypotonic and the condition is described as a *flaccid paralysis*. Voluntary movement begins to return in a few days; first at the proximal joints and somewhat later in the fingers and toes. Although a remarkable degree of recovery occurs eventually, it is usually possible to detect a slight residual clumsiness in the fingers, for example in picking up small objects. In man after destruction of the motor cortex recovery is much slower and dexterity of movements, especially the rapid and highly complex movements made with the fingers, is rarely regained. A mild degree of spasticity may eventually appear. The amount of the final disability is greater the greater the amount of motor cortex removed.

If numerous intersecting cuts are made into the grey matter of the sensorimotor cortex of the monkey very little upset of motor function ensues. Horizontal, that is tangential, intra-cortical transmission cannot, therefore, be of any great importance.

When both pyramidal tracts are divided at the level of the medulla (Fig. 45.2) in the monkey, the muscles become hypotonic so that the head sags and the abdomen protrudes. After some time, sitting and, later, walking become possible, but the animal is unwilling to move, and the ability to carry out accurate and delicate movements is lost. This experiment shows that a certain amount of voluntary movement can be carried out by extra-pyramidal fibres since the loss of movement on section of the pyramidal tracts is very much less than that seen when area 4 and area 6 of the cortex are both removed. There is thus no doubt about the importance in locomotion of the extrapyramidal pathways.

As already mentioned, removal of area 4 in the primate causes a flaccid paralysis. Some workers have claimed that if a *spastic paralysis* follows, as is seen in man after haemorrhage into the internal capsule, it is due to damage to area 6 lying in front of area 4. In man and other primates a spastic paralysis is characterized by an increase of flexor tone in the arms and of extensor tone in the legs. The hemiplegic patient, therefore, carries his paralysed arm flexed across his chest and his paralysed leg extended at the knee; there is loss of voluntary movement of the limb, but the limb is rigid and an attempt to flex it meets with resistance very like that observed in decerebrate rigidity; the tendon reflexes are exaggerated and clonus can readily be demonstrated. Apart from the paralysis all these abnormalities are due to over-activity of the spinal stretch reflexes (p. 913). In most patients with spasticity (due either to hemiplegia or Parkinsonism) injection of dilute procaine solution into the motor point of a muscle causes a reduction of spasticity before it causes loss of motor power. Since procaine blocks γ motor fibres long before it blocks α fibres it seems that spasticity depends on overactivity of γ fibres. Since lesions restricted to the pyramidal tract in primates produce a hypotonic paralysis, the spastic paralysis seen after extensive cortical ablation or cerebral haemorrhage in man must be presumed to be due to simultaneous interruption of pyramidal and extrapyramidal pathways.

Increase in tone and activity of reflexes has also been observed when the medial surface of the hemisphere just in front of the leg area has been damaged. Stimulation of this *supplementary motor area* in man has produced movements on the opposite side; removal does not cause paralysis but an increase in muscle tone.

When the corticospinal tract is interrupted by injury or disease the plantar reflex response is extensor instead of the normal flexor, that is to say firm stroking of the sole of the foot from behind forwards produces dorsiflexion of the great toe (Babinski's sign). In addition, if the stimulus is strong enough, there may be flexion of the ankle, knee and hip; this suggests that the reflex is actually part of a general flexor withdrawal reflex. Dorsiflexion of the great toe occurs also with destruction of cortical area 4 and if the cortical lesion involves the premotor area, stimulation of the sole of the foot produces not only dorsiflexion of the great toe but also a fanlike spreading of the other toes. An extensor plantar response is normally found in infancy; the change over to a flexion response occurs between the ninth and twenty-fourth month which is the time at which the myelination of the pyramidal tract becomes complete. Note, however, that an extensor plantar response may be seen temporarily after an epileptic fit and in biochemical disturbances, such as hypoglycaemia, affecting the nervous system in which there is no anatomical lesion of the pyramidal system.

Hines (1937) has found that in the monkey removal of a narrow strip of cortical tissue, 4s, lying immediately rostral to area 4 produces spasticity with very little paralysis, and it has been suggested that spasticity occurs by the release of the lower parts of the nervous system from the inhibitory control exerted by this narrow area of cortex. It was later found that stimulation of this so-called 'suppressor' area 4s inhibits muscular tone and the muscular movements which follow stimulation of area 4. This idea is not generally accepted.

Stimulation of the premotor area, for example area 6aα, usually causes movements of a small group of muscles—the same group that is activated by stimulation of the adjacent area 4. The effect depends on fibres running from area 6 to area 4 since after a cut has been made between the two areas, or after area 4 is removed, the character of the response is altered and stimulation of area 6aα yields only adversive movements consisting of rotation of the head and of the eyes to the opposite side with movements of the contralateral limbs. The impulses from area 6 must in these circumstances reach the spinal cord (probably by relaying in the reticular formation) by an extrapyramidal route. This is illustrated further by the finding that stimulation of area 6 produces movements of muscles even when the pyramids have been sectioned.

If a part of the premotor cortex, for example area 6a, is removed in the chimpanzee there is a slight weakness of the muscles of the opposite side but, although this soon disappears, skilled movements are permanently disturbed. This also applies to man after removal of the premotor cortex. A destructive lesion of the upper part of area 6 produces on the opposite side of the body a phenomenon called *forced grasping*. If the palmar surface of the hand or fingers is touched reflex flexion of the hand occurs to grasp the stimulating object and any attempt to withdraw the latter is followed by a firmer grasp. Since this reflex is normally present in infants its appearance in the adult is considered to be a regression to the infantile stage of the function of grasping. The grasp reflex is not seen after section of the pyramids alone, hence area 6 normally inhibits flexor activity through extrapyramidal pathways.

In addition to the main motor and premotor areas there are, as mentioned before, other areas of the cortex which are concerned in muscular movements. For example areas 44 and 45 together comprise the premotor area of the face (Broca's area); in right-handed persons lesions of this area on the left side produce motor aphasia, that is an inability to speak although the muscles of articulation are themselves not paralysed. This may be the basis for the well known association of right hemiplegia and aphasia; the left is usually the dominant hemisphere. Injection of barbiturate into one carotid artery produces a temporary loss of function of the parts of the brain supplied by the artery. If aphasia occurs it indicates that the barbiturated hemisphere is the dominant one for speech. Stimulation of the frontal eye fields in the posterior part of the second frontal convolution (area 8) in man produces conjugate deviation of the eyes to the opposite side without head movements and without visual hallucinations. Stimulation of the occipital eye fields (areas 18, 19) causes visual hallucinations and movements of the eyes to the opposite side. From area 17 visual hallucinations are obtained but not eye movements.

CORTICAL REPRESENTATION OF AUTONOMIC ACTIVITIES

The visceral or autonomic activities of the body seem to be controlled in the hypothalamus but there is a certain amount of cortical representation of autonomic activities, both sympathetic and parasympathetic. Stimulation of some points in areas 4 and 6 on the frontal lobe of the chimpanzee gives a rise of blood pressure and an increase of heart rate, but if the stimulating electrodes are moved only a short distance along the cortex the reverse effects may be obtained. Some of these effects may occur independently of the hypothalamus (p. 953). Injury to the cortex may produce vasomotor effects in man. Soon after a cerebral haemorrhage which gives rise to a hemiplegia there is cutaneous vasodilatation and an increase in sweating on the affected side; at the same time the surface temperature of the paralysed limbs is increased, although later it usually falls to subnormal levels. Autonomic effects are especially pronounced on stimulation of the orbital gyrus and cingulate gyrus. In dogs stimulation of the cingulate gyrus arrests the movements of the pyloric antrum. Since destruction of the hypothalamus or section of the vagus abolishes this effect, but section of the splanchnic nerves does not, the chain of connexions must be cerebral cortex, hypothalamus, medulla and vagus nerves to the stomach. Removal of the frontal lobes in monkeys increases the movements of the gastrointestinal tract and the animals become exceedingly hungry; stimulation of area 6 may inhibit peristaltic activity. The secretion of saliva in response to a conditioned stimulus depends on the integrity of the lower part of area 6a. Stimulation of area 8 in the monkey produces a dilatation of the pupil whereas stimulation of area 19 produces a constriction. The pathways for these autonomic effects are known with certainty in only a few instances.

FRONTAL ASSOCIATION AREAS

The prefrontal area consists of that part of the frontal lobe rostral to areas 6 and 8. Electrical stimulation of this area in the unanaesthetized patient may block thinking or cause fear; if the electrodes are in the infralateral surface of the frontal lobes the amount of talking increases. Removal of one prefrontal region produces very slight changes in behaviour in man. Removal of both prefrontal areas in monkeys results in restlessness and in defects in behaviour and learning which, however, only become obvious when quite complicated problems are presented to the animal. The normal chimpanzee given a complicated problem may be emotionally upset if it fails to solve it but after removal of both prefrontal areas this emotional response is not seen. In 1936 these findings were applied to man for the treatment of mental disorders by Moniz, a Portuguese neurologist, who devised the procedure of prefrontal leucotomy. The operation, performed through small holes made in the skull or in the roof of the orbit, divides the white matter of the prefrontal area (see Fig. 48.2) and interrupts the pathways between the cortex and the thalamus and hypothalamus as well as connexions with other cortical areas. The main pathway interrupted is that to the large dorsomedial nucleus of the thalamus. Bilateral leucotomy has been found to benefit patients with mental illness, especially when 'psychic tension' is a prominent feature. If the section is made too far back, certain undesirable sequelae such as labile emotional reactions, impaired moral sense and loss of initiative appear. Psychiatrists are searching for more restricted operations, for example section through the inferior frontal region, which can relieve depression without disturbing personality or intellect.

Injuries to the frontal lobes in children produce much more serious defects than they do in adults. The prefrontal lobes seem to be extremely important for the learning process in childhood. Wounds of the frontal lobe in adults may cause an increase in activity, restlessness, talkativeness, repetitiveness, impairment of character and behaviour but intelligence tests show in general little deterioration.

SENSORY AREAS OF THE CEREBRAL CORTEX

Since little information about sensation can be gained from animal experiments, except perhaps by the study of conditional reflexes, most of our knowledge must come from clinical investigation. For example, a conscious patient may describe his sensations when areas of the cortex exposed at operation are stimulated electrically. Data have also been obtained from the investigation of patients with traumatic or pathological lesions, although in such cases we are dependent on the co-operation of a patient who has a lesion the extent of which may be difficult to judge even at a later post-mortem examination. Information obtained in this way is not always entirely satisfactory.

The sensory areas of the cortex are the highest points in the brain reached by the impulses concerned in sensation. Although they are undoubtedly important as relay or junctional areas in the complicated pattern of the cerebral neurones, we are not justified in claiming that these areas are the seat of sensations and consciousness (p. 939). Although it is probably correct to say that there is no cortical representation of pain it can be argued that the cerebral cortex is necessary for the full appreciation of pain; after bilateral frontal leucotomy the pain is still felt but the emotional reaction is no longer present.

OCCIPITAL CORTEX

Although it has been known for more than a hundred years that removal of the cerebral cortex produces blindness without interfering with the reaction of the pupils to light, only comparatively recently has the precise area of the human occipital cortex involved been localized to the walls of the postcalcarine sulcus on the medial surface of the occipital lobe (Brodmann's area 17), occasionally extending a short distance on to its lateral surface.

The structure of the retina has already been described and the axons of the ganglion cells have been followed into the optic nerve. The path of the fibres after this has been found by studying the degeneration occurring after removal of different parts of the retina. The fibres from the nasal half of each retina in man cross over at the optic chiasma (Fig. 48.11 and 12) so that the fibres from the right halves of the two retinae pass into the right optic tract to the right lateral geniculate body from which new fibres go as the *optic radiations* to the calcarine cortex. The macular fibres concerned in central vision lie nearly centrally in the optic tracts and end on the posterior part of the lateral geniculate body, the macular relay fibres then going to the tip of the occipital lobe. Removal of the occipital cortex in man has often left some macular vision but this is likely to depend on incomplete removal of the area of cortex concerned rather than on bilateral cortical representation of the macula for which there is no anatomical basis. The results of correlation of the position of lesions with the shape of the visual field obtained by perimetry show that the centre of the field is represented in the tip of the occipital lobe at the posterior end of the postcalcarine sulcus, and that the upper half of each retina (lower half of the field of vision) is represented on the upper part of the visual area of the cortex, while the lower half of the retina is represented on the lower half. The area of the cortex connected with the macula and devoted to central vision is, as might be expected from the power of the macula to appreciate colour and fine detail, large compared with that subserving peripheral vision (Fig. 48.10). Removal of both occipital lobes in man results in permanent blindness, but in dogs and rats recovery is sufficient after a time to allow differences in light intensity to be distinguished.

The pathway from retina to cortex can be thought of in simple terms as six nerve cells, three in the retina (p. 739), one in the lateral geniculate body and two in the cortex. The visual pathway has been investigated in experiments in which an anaesthetized cat faced a wide screen at a distance of 1·5 m on which various patterns of white light were projected. The nervous system was penetrated at several places with micro-electrodes to record the responses of individual cells. The retinal area that on stimulation produces a response from a cell, no matter where the cell is situated in the visual pathway, is called the receptive field of the cell. The retinal ganglion cells, the cells of the lateral geniculate body and those of the occipital cortex receive many synapses, some of which are inhibitory and some of which are excitatory and at any one time a single cell may receive several excitatory or inhibitory impulses or a combination of both. The behaviour of the cells in the visual cortex was discussed on p. 750.

The areas adjacent to the occipital cortex, Brodmann's areas 18 and 19, are association areas in which co-ordination of eye reflexes

with other reflexes occurs. The interpretation of what is seen, for example a printed word, requires the presence of these association regions; their destruction causes visual agnosia.

Visual reflexes. When the superior colliculi are stimulated electrically dilatation of

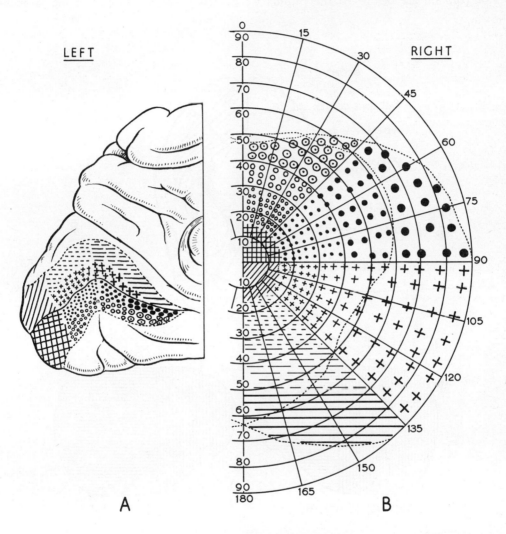

LEFT

RIGHT

A

B

Fig. 48.10 The cortical retina. A diagram of the probable representation of the different portions of the visual fields in the calcarine cortex. No great accuracy is claimed for this scheme. A. The postcalcarine fissure on the medial surface of the left occipital lobe is represented as widely opened. The macular area is relatively large, the peripheral area relatively small. B. The right halves of the field of vision are superimposed. Note that the centre of the field (maculae) is represented at the tip of the occipital lobe and that the lower half of the field of vision (upper half of the retinae) is represented in the cortex above the postcalcarine sulcus. The peripheral parts of the field of vision tend to be represented in the anterior portion of the calcarine area. (*After* diagrams by Sir Gordon Holmes (1918). *British Journal of Ophthalmology* **2**, 383; also (1945). *Proceedings of the Royal Society* B **132**, 348.) More recent information is given by Teuber *et al.* (1960) (see references).

the pupil is said to occur along with conjugate movements of the eyes towards the opposite side, that is lateral movements of both eyes simultaneously through equal angles. It is possible however that these results, reported many years ago, are due simply to spread of current to neighbouring structures. When the superior colliculi are destroyed the pupillary contraction on exposure to light remains but, if the pretectal area just rostral to them (Fig. 45.8) is destroyed at the same time, the light reflex can no longer be obtained. The explanation for this is clear from Fig. 48.12; some fibres from the retina travel in the optic nerve and optic tract but pass medial to the lateral geniculate bodies to end in the pretectal area

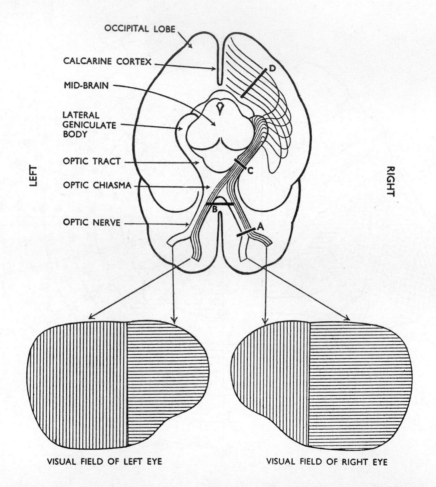

Fig. 48.11 The course of the fibres involved in vision from the right halves of the two retinae to the occipital cortex. These correspond to the left halves of the fields of vision. There is a relay in the lateral geniculate body. A lesion at A cutting through the right optic nerve results in complete blindness of the right eye. A lesion at B involving the rostral part of the optic chiasma interrupts the fibres from the nasal halves of the retinae and produces bitemporal hemianopia. A lesion at C divides the optic tract containing fibres from the right halves of both retinae and produces blindness in the left halves of the visual fields (left hemi-anopia); the pupil does not contract when a light is thrown on the right halves of the retinae. A lesion at D cutting through all the optic radiations also produces blindness in the right halves of the retinae, but the light reflex is normal. (Sir J. Conybeare (1961). *Textbook of Medicine*, 13th edn, p. 745. Edinburgh: Livingstone.)

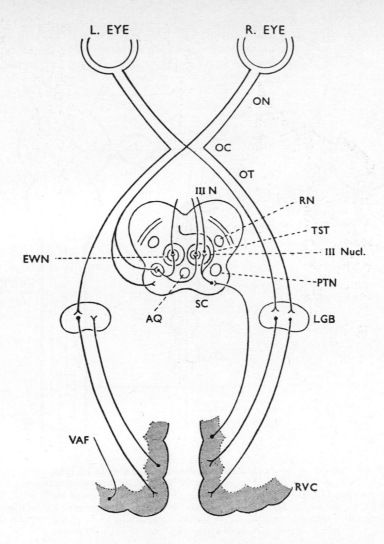

Fig. 48.12 Diagram of the visual pathways. L. Eye = left eye. R. Eye = right eye. ON = optic nerve. OC = optic chiasma. OT = optic tract. III N = third nerve. TST = tecto-spinal tract. III Nucl. = third nerve nucleus. AQ = aqueduct of mid-brain. RN = red nucleus. EWN = Edinger-Westphal nucleus. PTN = pretectal nucleus. SC = superior colliculus. LGB = lateral geniculate body. VAF = visual association body. RVC = right visual cortex. (*By courtesy of G. W. Pearce.*)

where the first synapse occurs. Second order relay neurones cross in the posterior commissure and reach the Edinger-Westphal part of the third nerve nucleus on both sides from which nerve fibres pass via the ciliary ganglion to the sphincter pupillae. The phenomenon of contraction of the pupil on accommodation but not on exposure to light, a common

finding in certain diseases of the nervous system, is probably due to interruption of the pathways in the pretectal region. The light reflexes remain after cortical ablation since they do not require a cortical pathway. The very much longer sympathetic pathway to the eye described on p. 732 is illustrated in Figure 48.13.

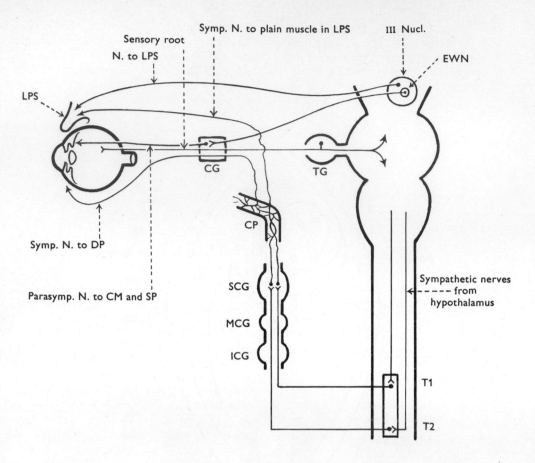

Fig. 48.13 Diagram of the course of the autonomic fibres to the eye. III Nucl. = oculomotor nucleus. EWN = Edinger-Westphal nucleus. LPS = levator palpebrae superioris. CG = ciliary ganglion. T1, T2 = first and second thoracic segments of spinal cord. SCG, MCG and ICG = superior, middle and inferior cervical sympathetic ganglia. TG = trigeminal ganglion. DP = dilatator pupillae. SP = sphincter pupillae. CM = ciliaris muscle. CP = plexus on internal carotid artery. (*By courtesy of G. W. Pearce.*)

THE TEMPORAL LOBES

The route taken by impulses from the cochlea to the medial geniculate body and to a relatively small area of auditory cortex in the superior temporal convolution has already been described (p. 768). The auditory cortex when in its normal state and free from anaesthesia, shows little of the tonotopic arrangement which is a feature of lower neural levels. In fact the cortex does not appear to be essential for pitch discrimination, but is concerned with the processing of more elaborate sound patterns and the way they change with time. It thus plays an essential role in the recognition of such things as speech and music.

In the cat destruction of one cochlea alone, or of one temporal lobe alone causes a hearing loss of 3 to 5 decibels. If in addition to removing one temporal lobe the opposite cochlea is destroyed, the hearing loss when the crossed fibres only are intact, increases to about 15 decibels, but the animal is by no means deaf. If after removal of one temporal lobe the cochlea on the same side is destroyed, the loss in the presence of the uncrossed fibres only is again about 15 decibels. The crossed and uncrossed fibres from the cochlea have, therefore, in the cat about equal acoustic significance. After ablation of both temporal lobes in the monkey the behaviour of the animal is essentially normal and words of command are apparently understood; although localization of the source of sound is poor, the animal is not deaf. Electrical stimulation of the superior temporal convolution in man gives subjective buzzing, clicking or booming sounds. Although ablation of one temporal cortex in man or even of one hemisphere has a negligible effect on hearing, location of the source of sounds may be faulty.

Vestibular impulses also end in the temporal lobes but the pathway is not yet known. Dizziness occurs when certain parts of the cortical area are stimulated in man but removal of the cortex does not produce any disturbance of equilibrium.

The uncus and amygdala are concerned with gustation and olfaction, the insula with sensory information from the alimentary tract. The hippocampus and amygdala play a part in the central control of autonomic activities. The hippocampus was formerly considered to be an olfactory centre but this idea has been abandoned. It probably has some relationship to emotional responses and to memory but the amnesia following hippocampal lesions may be due to disturbance of incoming and outgoing pathways. Loss of memory occurs with bilateral lesions of the hippocampus. The rhinencephalon includes these structures and comprises also the primitive allocortex, which is made up of three cell layers compared with the six layers of the isocortex. It forms part of the limbic system (p. 957) and has important connexions with the hypothalamus and there is evidence that this part of the brain plays an important part in emotional responses.

The effect of lesions in the temporal lobes in man is extremely variable. A circumscribed lesion of the superior temporal gyrus on the left side may give rise to pure word deafness in which the subject is unable to understand spoken words with no deficit in reading or verbal expression. A much commoner effect of damage to this part of the temporal lobe is central (jargon) aphasia, in which the patient can neither understand spoken words nor monitor his own speech so that jargon is spoken. A less serious disturbance is nominal aphasia in which the ability to recall names is affected. The temporal lobe has a function in reflation to memory and recall for although damage to one temporal lobe has little effect on the memory removal of both lobes may reduce the memory span to a few minutes. Penfield by stimulating points over a wide region of the temporal cortex in conscious patients has evoked recollections which were sometimes fragmentary but were often elaborate and vivid, such as the recollection of a piece of music. Lesions of the hippocampus are particularly liable to give rise to epileptic seizures and these may sometimes be associated with a dreamy state with vivid memories of the past.

THE PARIETAL LOBES

When the region of the postcentral gyrus in man (areas 3, 1 and 2) is stimulated electrically, sensations of touch and pressure are aroused. Occasionally feelings of warmth are elicited but never painful sensations, and no muscular movements occur if stimulation of the precentral gyrus is avoided. Sensation from the leg, trunk, arm and face on the opposite side of the body is represented in that order from above downwards and area representation in the sensory cortex (*somaesthetic area*) is, therefore, very similar to that of the motor cortex (area 4) lying immediately in front of it (Fig. 48.14). The postcentral convolution receives impulses via the thalamus from the medial lemnisci and spinothalamic tracts. Stimulation of the posterior lip of the precentral gyrus, which is usually regarded as part of the motor area, also gives rise to sensations similar to those arising on stimulation of the postcentral gyrus.

The sensory areas in animals have been mapped out by taking leads from the cortex to an amplifier and thence to an oscillograph or a loudspeaker. The experimenter, by touching various parts of the animal's body in turn, finds the area of skin giving maximum cortical activity as indicated by the height of the oscillograph deflexions or by the noise in the speaker. In this way the representation of the body surface—leg, arm and one side of the face from above downwards—on the contralateral postcentral gyrus has been confirmed. Stimulation of the skin activates neurones mainly in area 3 while the cells in area 2 just behind 3 are activated by joint movements. A small *secondary sensory area* has been found at the lower end of the postcentral gyrus.

Removal of the postcentral gyrus leads to impairment but not complete abolition of the

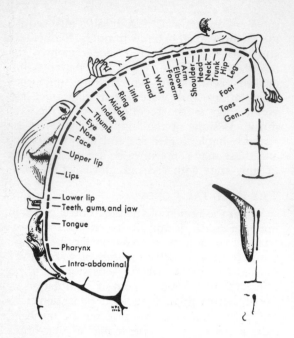

Fig. 48.14 Sensory homunculus. The right side of the figurine is laid upon a cross-section of the hemisphere, drawn somewhat in proportion to the extent of sensory cortex devoted to it. The length of the underlying black lines indicates more accurately the comparative extent of each representation. (W. Penfield & T. Rasmussen (1950). *The Cerebral Cortex of Man*, p. 44. New York: Macmillan.)

ability to detect differences in weight and texture and of stereognosis—the ability to recognize objects through touch. The latter implies more than superficial touch perception since temperature and deep pressure sensibilities as well as appreciation of weight (from joint receptors) are all involved. Tactile agnosia is the inability to recognize objects through touch when all relevant sensations are intact. It occurs in lesions of the posterior parietal lobe, possibly the supramarginal gyrus, and lesions of this part of the brain on the left side may produce bilateral tactile agnosia. This capacity to recognize, the failure of which gives rise to the different varieties of agnosia, is essentially a function of the cerebral cortex and depends on the ability of sensory perceptions to evoke the neuronal responses necessary to identify the object perceived.

A lesion of the left angular gyrus (Fig. 48.2A) in a right-handed person may result in inability to write (agraphia). This deficit is sometimes accompanied by inability to distinguish right from left, inability to recognize individual fingers and inability to calculate (Gerstmann's syndrome). Lesions in the parietal area in man may produce a loss of the ability to find the way even in the patient's own house, together with the loss of the ability to conceive space even in two dimensions. When we are aware of the position of our body relative to other objects around us and of the position of its various parts, head, trunk and limbs, to one another, we are also aware of changes in all these relationships. To this sort of awareness the term 'body image' has been assigned. Disturbances of the body image are particularly liable to occur when disease affects the parietal lobe of the right or non-dominant cerebral hemisphere. If this region is damaged there may be total neglect of one half of the body and the patient may have difficulty in carrying out such every-day acts as dressing and sitting down in a chair. Damage to the parietal lobe of the dominant hemisphere may sometimes make it impossible to carry out actions to command or even to perform complex actions at all. This inability to perform organized movements in the absence of paralysis is termed apraxia.

Taste has been localized in the lowest part of the postcentral gyrus. Lesions in this area cause disturbances of taste on the opposite side of the tongue as well as cutaneous sensory impairment on the face on the opposite side. Electrical stimulation gives rise to taste sensations. Impulses from each side of the tongue cross over almost completely to the opposite side in the medial lemniscus to the thalamus and so on to the cortex.

After removal or destruction of the postcentral gyrus there is a considerable recovery of sensation, a fact which may be explained by ipsilateral representation of sensation or by supposing that representation is not confined to the postcentral gyrus as is usually assumed. It is now thought that the sensory paths from the limbs and trunk are bilateral. A second somatic area containing ipsilateral representation is intermingled with the classical contralateral representation (Fig. 48.14). The lemniscal pathway goes almost exclusively to the contralateral side of the hemisphere; the anterolateral system is probably the main source of ipsilateral representation. Damage to the sensorimotor cortex on one side impairs the touch and pressure sense of the hand on the same side in both man and monkeys.

THE ELECTROENCEPHALOGRAM

Electrical potentials developed in the human brain can be detected on the surface of the head and records of these potentials, obtained by means of a string galvanometer, were first published by Berger in 1929. Nowadays electrodes are placed at various positions on the scalp and simultaneous recordings from different sites are made through a set of 8 or 16 amplifiers connected to ink-writing oscillo-graphs. Alterations in electrical potential occur rhythmically usually with a range of frequencies superimposed on one another, the dominant frequency being the alpha (α) rhythm at about 10 Hz and an amplitude of about 50 microvolts. This α rhythm is of highest amplitude postcentrally and probably arises from the occipital cortex around the visual association area. It is best seen when the subject is relaxed with eyes closed; when the eyes are opened and attention is paid to objects in the visual field, or when a mental task like mental arithmetic is undertaken, the rhythm disappears and is replaced by small rapid waves (Fig. 48.15). The α rhythm is, therefore,

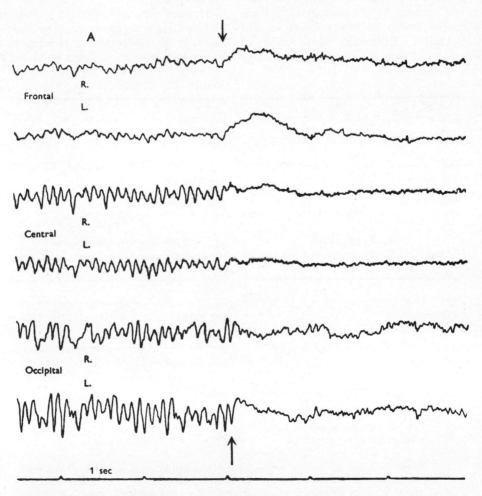

Fig. 48.15 The normal human electroencephalogram (e.e.g.) as seen in bipolar antero-posterior electrode placements. The dominant rhythm (11 Hz) spreads forward from the occipital regions. This α-rhythm is inhibited by eye opening which occurred at the arrow. All six traces were recorded simultaneously. (*By courtesy of D. A. Pond.*)

as Adrian puts it, a 'rhythm of inattention'. The α rhythm can scarcely be the product of nerve impulses since they pass in all directions through the complex cortical interconnecting pathways in the brain in a manner unlikely to give rise to a series of fairly regular occurring waves of electrical potential. Since such waves have been obtained in animals after section of the mid-brain, that is after removal of all afferents except the olfactory and optic nerves, they may represent the spontaneous activity of the cortical neurones but this matter is disputed. A recent speculation is that the α-waves result from the interaction between the corneoretinal potential and potentials produced by tremor of the extra-ocular muscles. However, some blind persons have a normal α-rhythm. The place of origin of the α-waves is still obscure. Delisle Burns has succeeded in isolating slabs of cerebral cortex in the cat by

undercutting them while still preserving their blood supply. Although the nerve cells in the slab retained their excitability in the unanaesthetized animal they showed no spontaneous electrical activity. These experiments suggest that rhythmical cortical discharges can be maintained only if the cells are driven from a source outside the cortex, that is only if the cortical cells receive afferent impulses from other parts of the brain.

More rapid oscillations are recorded further forward in the precentral region. These beta (β) waves which have a frequency of 16 to 25 Hz probably originate in the motor cortex; they may be blocked by movement of the part represented in the cortex underlying the electrodes. Theta (θ) rhythm with a frequency of 4 to 7 Hz is inconspicuous in healthy adults but may be the dominant rhythm of children in whom it may be blocked by visual attention

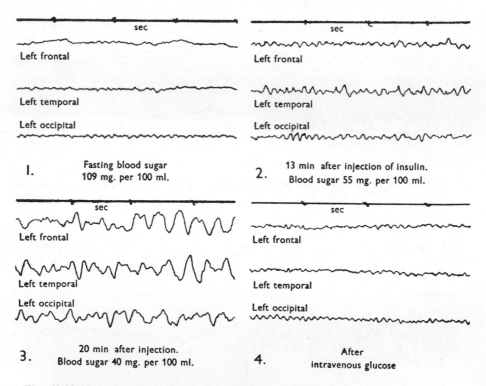

Fig. 48.16 Sequence of changes in human electroencephalogram during hypoglycaemia following intravenous insulin (0·5 unit per kg of body weight). Note that the appearance of slow activity in record 2 occurs first in the frontal regions and spreads to involve the whole cortex as the blood sugar falls (record 3). Complete and immediate recovery occurs following intravenous glucose. Time in minutes following insulin injection. All records are from the left frontal, temporal and occipital regions, the electrodes being in pairs along the head (bipolar recording). (*By courtesy of D. A. Pond.*)

and augmented by emotion. Delta (δ) waves of less than 4 Hz are not seen in the healthy waking adult, but occur in sleep (Fig. 48.18) and may be the dominant rhythm of infants. They also occur in hypoglycaemia (Fig. 48.16). Spindles, bursts of waves about 14 Hz, are characteristic of moderately deep sleep.

Young children below 18 months show electroencephalograms of low amplitude with little rhythmic activity. The α rhythm may not be fully established until the thirteenth year; before that slower rhythms constitute the dominant frequency.

When the eyes are exposed to a uniform field lit with a flickering light the potentials recorded from the occipital region occur at the same rate as the flicker up to 30 Hz as if the cells of the visual cortex were discharging synchronously in response to each flash of light (Fig. 48.17). This 'photic stimulation' can cause changes in the emotional state and may even induce an epileptic attack.

In disease of the brain normal rhythms are often absent from affected areas and may be replaced by slow activity. In epilepsy the synchronous discharge of large numbers of cortical neurones gives rise to high voltage waves which stand out against the background as spikes or slow waves. Paroxysmal discharges of this kind, however, occasionally occur in healthy individuals, particularly children; epileptic patients not uncommonly have a normal electroencephalogram.

The electroencephalogram has proved of value in the diagnosis of epilepsy and intracranial disease, particularly in defining the area affected by a pathological process.

SLEEP

The human nervous system cannot long continue to function without sleep. Sleep is distinguished from other forms of unconsciousness by the relative ease with which the waking conscious state is reached. Sleep is no doubt a cerebral phenomenon but the old notion that it is brought about by cerebral anaemia can no longer be maintained; the total blood flow through the brain of man during sleep is not altered although there may, of course, be local alterations of blood flow that cannot be measured. The heart rate, blood pressure (Fig. 27.10) and metabolic rate are reduced in sleep. In deep sleep the respiration rate is reduced, the alveolar P_{O_2} falls and the arterial oxygen saturation may fall from 96 to 87 per cent; at the same time the alveolar P_{CO_2} is raised to 46 mm Hg or more. Salivary and lacrimal secretions are reduced. The tendon reflexes are diminished. The volume of urine formed during sleep is small (p. 657 and Fig. 32.13); this permits a long period of unbroken sleep; plasma levels of hydrocortisone decrease quickly in the morning and more slowly towards midnight but increase sharply about 2 a.m.

Pavlov explained sleep as a process of internal inhibition (Ch. 49), but it cannot be entirely a cortical phenomenon since decorticate animals, as well as very young animals and human infants, show rhythmic waking and sleeping which, however, bear no relationship to day and night but are repeated several times during the twenty-four hours. The ability to sleep during the night and remain awake during the day is probably dependent on the development of conditional reflexes.

According to Bremer, a cat with its midbrain sectioned immediately behind the third nerve nucleus, the basilar artery remaining intact, exhibits sleep. Just as in deep sleep, the pupils become very narrow and the eyes look down and the electroencephalogram shows a rhythmic waxing and waning of the α-rhythm.

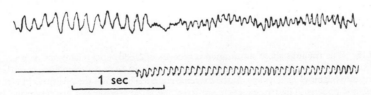

Fig. 48.17 Electroencephalogram from the occipital region in man, showing the effect of flickering light. Above, e.e.g.; below, record from photo-electric cell showing rate of flicker (18 per sec); the eyes were opened as the flicker began. (E. D. Adrian (1943). *Lancet* **ii**, 33.)

A cut made between the medulla oblongata and the spinal cord leaves the brain 'awake'. These findings led Bremer to the conclusion that afferent impulses reaching the fore-brain maintain wakefulness. Magoun believes that the cortex is kept awake, not by impulses travelling in the classical sensory paths of the lemnisci to the thalamus and cortex, but by impulses travelling up the central brain stem reticular formation to the posterior hypothalamus and the medial nuclei of the thalamus and then to the cortex. He showed that stimulation of the reticular formation and the diencephalic region just mentioned gave the e.e.g. picture of arousal with fast low amplitude waves but that lesions in the same area produced a state of sleep associated with a characteristic e.e.g. sleep record of high amplitude slow waves. However, this does not mean that the cortex does not receive impulses during sleep; e.e.g. records show that impulses reach the sensory areas of the cortex from the peripheral nerves. This is described as cerebral vigilance. This phrase does not mean consciousness. Even in deep sleep discrimination can be made between meaningful and meaningless auditory stimuli. Magoun showed further that lesions involving all the lemniscal tracts but sparing the central reticular formation did not produce sleep. Moreover, sleep was produced by injury to the reticular formation which spared the lemnisci. Sleep, according to these views, is a passive state from which we are aroused.

Prolonged sleep (hypersomnia) is associated with lesions of the posterior hypothalamus such as are encountered in encephalitis lethargica (sleepy sickness), and injection of fluid into this area, if it produces mechanical injury, also causes sleep. Hess found that stimulation of the posterior hypothalamus caused arousal, and from the same area he also obtained increased motor activity and excitability, dilatation of the pupils and a rise in blood pressure and heart rate. Such observations support the conception of a 'waking centre' in the posterior hypothalamus. Hess has produced 'sleep' by low frequency stimulation of the thalamus in the region of the interthalamic connexion.

The difficulty in all such experiments is that it is not possible to say in a direct way when an animal is truly asleep, and the experimenter is always unable to say whether an animal is exhibiting true sleep or whether it is merely showing a great decrease of activity.

Sleep occurs most readily when there is

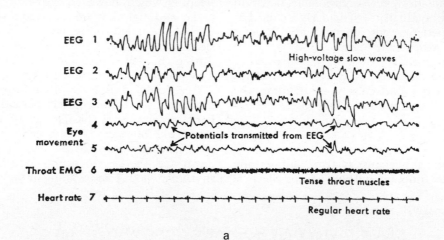

a

Fig. 48.18 The tracings show simultaneous records of electroencephalogram (e.e.g.) (channels 1–3), eye movements (channels 4–5), electromyogram (e.m.g.) from throat (channel 6) and heart rate (channel 7). The key on the extreme right in (b) shows the positions of electrodes and their connexions. The two excerpts are from the same tracing and are separated by two minutes. Excerpt (a), above, shows the orthodox or fore-brain phase of sleep and excerpt (b), opposite, shows the hind-brain phase of sleep in a normal adult. In excerpt (a) there are large slow waves and fast spindle activity in the e.e.g., the eyes are quiescent, the throat muscles

fatigue after muscular exertion; it has been suggested that a soporific substance produced during wakefulness is responsible for the onset of sleep, but this is rather doubtful since conjoined twins have each his own sleep rhythm. The best method of inducing sleep is to cut off afferent stimulation as much as possible by darkening the room and excluding noise. Warmth and a comfortable bed, by promoting muscular relaxation, reduce proprioceptive afferent impulses and facilitate sleep.

For over 50 years it has been suspected that CSF from sleep-deprived animals injected into the cisterna magna of other animals caused sleep in the recipient. This has been confirmed with newer techniques. But the nature of the substance and the physiological meaning of these experiments is by no means clear. It has recently been found that L-tryptophan in doses of 4 to 10 g in man has a sedative effect. It is just possible since 0·5 to 2 g of L-tryptophan are consumed in a normal diet that this amino acid is concerned in normal tiredness and sleep induction.

Sleep in man, as judged by e.e.g. records shows several cycles of alternation between the two categories illustrated in Fig. 48.18. The 'hind-brain' phase occupies about 20 per cent of sleeping time and the 'fore-brain' phase occupies 80 per cent. The hind-brain phase is especially related to dreaming and is characterized by rapid eye movements which can be described speculatively as scanning movements made in response to the visual imagery of the dream. Normally 'hind-brain' sleep does not develop until an hour or longer after sleep develops. In patients with narcolepsy who have an abnormal tendency to fall asleep 'hind-brain' sleep may develop almost immediately.

People vary greatly in the amount of sleep they 'need'. Young babies sleep most of the 24 hours but young adults may sleep any time between 5 and 9 hours. Old people have an even wider range. It is difficult to detect a loss of efficiency in carrying out mental tasks if sleeping time in young adults is reduced to 5 hours but efficiency falls sharply if it is reduced to 3 hours. It is not at all clear why sleep is necessary but a human being deprived of sleep for long periods, for example, 100 hours, becomes irritable and has hallucinations and delusions. After prolonged deprivation restoration of normal health may require a considerable time.

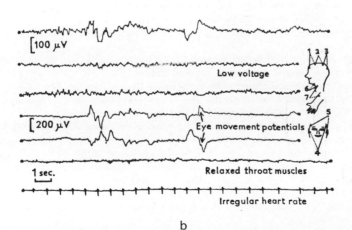

b

tense, though not as tense as in wakefulness, and the heart rate regular. In excerpt (b) the e.e.g. is of low voltage; but there is a short run of 'saw-toothed', 3-per-sec, waves in channels 1 and 3 just before the burst of eye-movement potentials in channels 4 and 5. The eye-movement potentials have also been picked up by the frontal electrode and appear in channel 1, and one or two small muscle-spike potentials from accompanying facial twitches are visible in this trace as well. The throat muscles have relaxed and there are occasional irregularities in the heart rate. (I. Oswald (1964). *British Medical Bulletin* **20**, 60–64.)

REFERENCES

ADEY, W. R. & TOKIZANE, T. (Eds.) (1967). Structure and function of the limbic system. *Progress in Brain Research* **27**, 1–489.

BATES, J. A. V. (1957). Observations on the excitable cortex in man. *Lectures on the Scientific Basis of Medicine* **5**, 333–347.

BRAIN, W. R. (1950). The cerebral basis of consciousness. *Brain* **73**, 465–479.

BRAIN, W. R. (1952). The contribution of medicine to our idea of the mind. *Rede Lectures*. Cambridge University Press.

BRAIN, LORD (1966). *Speech Disorders. Aphasia, Apraxia and Agnosia*, 2nd edn. London: Butterworths.

BRAZIER, MARY A. B. (Ed.) (1963). *First Conference on Brain Function, Los Angeles*, 1961. *UCLA Forum in Medical Sciences No. 1*. Berkeley: University of California Press.

BREMER, F. (1953). *Some Problems in Neurophysiology*. University of London: Athlone Press.

BURNS, B. DELISLE (1958). *The Mammalian Cerebral Cortex*. London: Arnold.

BRUTKOWSKI, S. (1965). Functions of prefrontal cortex in animals. *Physiological Reviews* **45**, 721–746.

CARMICHAEL, E. A. (1953–1954). Hemispherectomy and the localization of function. *Lectures on the Scientific Basis of Medicine* **3**, 92–104.

CRITCHLEY, M. (1953). *The Parietal Lobes*. London: Arnold.

DENNY-BROWN, D. (1962). *The Basal Ganglia and their Relation to Disorders of Movement*. Oxford University Press.

DENNY-BROWN, D. (1966). *The Cerebral Control of Movement*. Liverpool University Press.

DE REUCK, A. V. S. & PORTER, RUTH (Eds.) (1965). *Functions of the Corpus Callosum*. Ciba Foundation Study Group No. 20. London: Churchill.

Discussion on the logical analysis of cerebral functions (1968). *Proceedings of the Royal Society* B **171**, 276–386.

ECCLES, J. C. (Ed.) (1966). *Brain and Conscious Experience*. Springer-Verlag: Berlin.

FULTON, J. F., ARING, C. D. & WORTIS, S. B. (Eds.) (1948). The frontal lobes. *Research Publications. Association for Research in Nervous and Mental Diseases*. Baltimore: Williams & Wilkins.

GAZZANIGA, M. S. & SPERRY, R. W. (1967). Language after section of the cerebral commisures. *Brain* **90**, 131–148.

HARTMANN, E. (1967). *The Biology of Dreaming*. Springfield, Ill.: Thomas.

HORNYKIEWICZ, O. (1966). Dopamine (3-hydroxytyramine) and brain function. *Pharmacological Reviews* **18**, 925–964.

JOUVET, M. (1967). The states of sleep. *Scientific American* **216**, 62–75.

JOUVET, M. (1967). Neurophysiology of the states of sleep. *Physiological Reviews* **47**, 117–177.

KILOH, L. G. & OSSELTON, J. W. (1966). *Clinical Electroencephalography*, 2nd edn. London: Butterworths.

LASHLEY, K. S. (1964). *Brain Mechanisms and Intelligence*. New York: Hafner.

LASSEK, A. M. (1954). *The Pyramidal Tract. Its Status in Medicine*. Springfield, Ill.: Thomas.

McCALLUM, C. (1967). New waves in the brain. *New Scientist* **36**, 592–594.

MAGOUN, H. W. (1958). *The Waking Brain*. Springfield, Ill.: Thomas.

MAGOUN, H. W. & RHINES, RUTH (1947). *Spasticity. The Stretch Reflex and Extra-pyramidal Systems*. Springfield, Ill.: Thomas.

MARSHALL, W. H. (1959). Spreading cortical depression of Leâo. *Physiological Reviews* **39**, 239–279.

MARTIN, J. P. (1967). *The Basal Ganglia and Posture*. London: Pitman.

MORRELL, F. (1961). Electrophysiological contributions to the neural basis of learning. *Physiological Reviews* **41**, 443–494.

MYERS, R. E. (1965). *The Neocortical Commissures and Interhemispheric Transmission of Information in Functions of the Corpus Callosum*. Ciba Foundation Study Group No. 20. London: Churchill.

O'LEARY, J. L. & GOLDRING, S. (1964). D–C potentials of the brain. *Physiological Reviews* **44**, 91–125.

OSWALD, I. (1963). *Sleeping and Waking*. Amsterdam: Elsevier.

OSWALD, I. (1964). The experimental study of sleep. *British Medical Bulletin* **20**, 60–64.

PEELE, T. L. (1954). *The Neuroanatomical Basis for Clinical Neurology*. London: McGraw-Hill.

PENFIELD, W. (1958). *The Excitable Cortex in Conscious Man*. Liverpool University Press.

PENFIELD, W. & ROBERTS, L. (1959). *Speech and Brain Mechanisms*. Oxford University Press.

PFEIFFER, C. C. & SMYTHIES, J. R. (Eds.) (1959–). *International Review of Neurobiology*, Vol. 1. New York: Academic Press.

PHILLIPS, C. G. (1964). Experiments on single neurones within the central nervous system of vertebrates. *Scientific Basis of Medicine Annual Review* 81–101.

PHILLIPS, C. G. (1969). Motor apparatus of the baboon's hand. *Proceedings of the Royal Society* B **173**, 141–174.

REFSUM, S., LOSSIUS, H. M. & DIETRICHSON, P. (Eds.) (1962). *The So-called Extrapyramidal System*. Scandinavian University Books.

SCHULZ, D. P. (1965). *Sensory Restriction. Effects on Behaviour*. London: Academic Press.

SOLOMON, P., KUBZANSKY, P. E., LEIDERMAN, P. H., MENDELSON, J. H., TRUMBELL, R. & WEXLER, D. (Eds.) (1961). *Sensory Deprivation*. Harvard University Press.

STELLAR, E. & SPRAGUE, J. M. (Eds.) (1966). *Progress in Physiological Psychology*, Vol. 1. London: Academic Press.

Symposium on functional organization of the cerebral cortex (1947). *Federation Proceedings* **6**, 437–460.

TEUBER, H. L., BATTERSBY, W. S. & BENDER, M. B. (1960). *Visual Field Defects after Penetrating Missile Wounds of the Brain*. Harvard University Press.

VERNON, J. A. (1965). *Inside the Black Room*. London: Souvenir Press.

WALSHE, F. M. R. (1965). *Further Critical Studies in Neurology*. Edinburgh: Livingstone.

WOLSTENHOLME, G. E. W. & O'CONNOR, M. (Eds.) (1961). Ciba Foundation General Symposium on *The Nature of Sleep*. London: Churchill.

WYKE, B. (1963). *Brain Function and Metabolic Disorders*. London: Butterworths.

YOUNG, J. Z. (1966). *The Memory System of the Brain*. Oxford University Press.

49 Conditional reflexes

The methods for exploration of the function of the cerebral cortex given in the previous chapter nearly all involve more or less severe surgical procedures. The methods discussed here provide a fundamentally different approach to the problem, because the behaviour of the conscious intact animal is the subject of study.

The method most familiar to physiologists is that known as 'classical' or 'Type I' conditioning. At the end of the nineteenth century the Russian physiologist Pavlov was engaged in studying problems of digestion and in particular the problem of the control of salivation. It was soon found that this was not a simple physiological matter, since a large variety of factors, for example the sight, as well as the presentation, of food, would cause an animal to salivate and the experimental results were considerably influenced by the presence and behaviour of the observer. Pavlov, realizing the importance of these extraneous factors, gave up his work on the digestive glands and began a long investigation of 'conditional reflexes' which, as he said, opened the way to a study of the 'whole indivisible life'.

Pavlov used dogs in his experiments and, to avoid disturbing influences, carried out the work in quiet rooms specially built for the purpose. The animals were trained to stand still on a table, their movements restricted only by loops passing loosely round the legs to discourage struggling (Fig. 49.1). The experimenter observed the animal's behaviour unseen from another room from which by remotely controlled apparatus he could present food to the animal or apply various stimuli. A tube from the dog's exteriorized parotid duct was carried through into the observer's room so that the production of saliva could be measured. In these experiments food in the mouth is the so-called unconditional stimulus for salivation, but the response to be conditioned may be one of a number of glandular or muscular activities provided that they are,

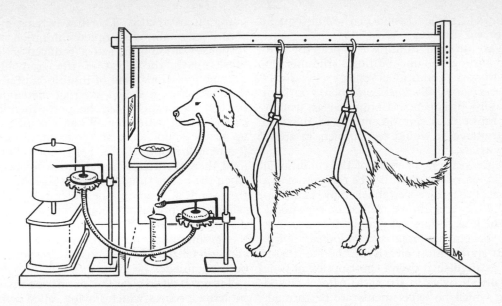

Fig. 49.1 A simplified diagram of the apparatus used in the investigation of conditional reflexes. The fluid from the parotid fistula passes down a tube and drops on a tambour connected to a second tambour which writes on a slowly moving drum. The observer can watch the dog through the window in the partition on the left.

at the outset of the experiment, reliably evoked by a specific stimulus (the *unconditional stimulus*) usually, though not necessarily, without special training. The experiment begins with the presentation of a new stimulus (the *conditional stimulus*), such as a sound, which is shown initially ·to have little effect upon salivary secretion; it normally evokes the 'orienting reflex', a complex response of alerting or arousal involving changes in heart rate, respiration, and other physiological functions, as well as overt signs of 'attention' to the stimulus.

The conditional and unconditional stimuli are now presented either simultaneously or with a short time interval between them on a number of occasions. This procedure is called *reinforcement*. As a result, the orienting reflex disappears and the conditional stimulus given alone now evokes a response resembling that originally produced only by the unconditional stimulus. Thus, in the particular experiment summarized in Table 49.2, after nine repetitions of the combined stimuli the conditional stimulus (the sound) was presented alone and the amount of salivation recorded; the strength of the conditional response was tested in the

Table 49.2 Development of a conditional reflex to a sound of 637·5 Hz

Number of times combination of sound with feeding had been performed	Extent of reflex in drops of saliva per 30 seconds	Latent period of conditional reflex in seconds
1	0	—
9	18	15
15	30	4
31	65	2
41	69	1
51	64	2

(From G. V. Anrep (1920). *Journal of Physiology* **53**, 367.)

same way after 15, 31, 41 and 51 repetitions of the combined stimuli. It can be seen that the response had become established both in regard to extent and to latent period when sound and food had been presented together 31 times. Conditioning is most readily achieved when the conditional stimulus precedes the unconditional; but conditioning can also be obtained when the unconditional stimulus comes first, provided that its characteristics are not such as to preoccupy the animal's attention to the exclusion of the subsequent conditional stimulus.

If a conditional stimulus is repeatedly presented without the unconditional stimulus, the strength of the conditional reflex progressively decreases (*experimental extinction*) until it ultimately disappears. Extinction is usually more rapid than reinforcement; in the case quoted above, it would be complete in about twenty trials.

The decrease in strength of the conditional reflex in extinction must not be thought of as a reversal of the process established by reinforcement. If a conditional reflex is repeatedly extinguished, it does not simply disappear but rather becomes a *negative conditional reflex*. This is shown most clearly by the fact that a reflex extinguished over a short period of time gradually recovers in strength. After a period during which no experiments are performed, presentation of the conditional stimulus again results in responses, though fewer than before extinction began. This *spontaneous recovery* shows that the effect of training has been suppressed rather than lost.

This conclusion is supported by the phenomenon of *disinhibition*. A recently established conditional reflex is unstable and may be inhibited by external stimuli which distract the animal's attention or which lead, as Pavlov put it, to an investigatory or orienting reflex. In the experiment described in Table 49.3, a conditional reflex was first set up to a visual stimulus and, after training, a relatively constant response, as judged by the flow of saliva, was obtained when the visual stimulus was presented. The strength of the reflex was then tested by presenting the conditional stimulus (the light) accompanied by the playing of a gramophone record. The first playing of the record led to almost complete inhibition of the response (amount of saliva reduced to one-tenth); later the effect of the gramophone became less. This kind of inhibition of a conditional reflex is called *external inhibition*. A parallel experiment can be performed after experimental extinction. When a response set up to the sound of a bell is extinguished by frequent ringing of the bell without giving food then eventually no salivation is obtained when the bell is rung. If at this stage the bell is rung and a gramophone record is played, salivation (a positive response) may occur. The distraction (external inhibition) provided by the gramophone has inhibited the negative conditional reflex and a positive response has been obtained.

Thus it is again clear that extinction involves an active process which Pavlov called *internal inhibition*. This process is also manifested in other situations. In the early stages of establishing a conditional reflex to a sound, for example the click of a metronome, any other sound may elicit salivation but after a time no sound differing greatly from that of the metronome will do so. Similarly, if a conditional reflex is set up to a tone of a particular frequency then it is found that a tone of a neighbouring frequency can also cause salivation but to a smaller extent than the original conditional stimulus. This effect is known as the *generalization* of the conditional response. The *specificity* of a conditional reflex can be greatly increased in the following way. Tone A is always accompanied by food whereas tone B of a different frequency never is; tone A when given alone always produces salivation. Tone B alone may at first produce salivation but when training

Table 49.3 External inhibition

						Percentage
Initial strength of conditional reflex						100
Strength of reflex at first application of gramophone record						10
,,	,,	second	,,	,,	,,	50
,,	,,	third	,,	,,	,,	65
,,	,,	fourth	,,	,,	,,	85
,,	,,	fifth	,,	,,	,,	90
,,	,,	sixth	,,	,,	,,	94
,,	,,	seventh	,,	,,	,,	100

(From C. Lovatt Evans (1928). *Recent Advances in Physiology*, p. 249. London: Churchill.)

has continued for some time tone B alone ceases to cause salivation. In other words the animal is able to distinguish between the two tones A and B, and the positive response to tone B has been turned into a negative one, as can be demonstrated by the inhibitory effect of tone B on other conditional reflexes in the same animal. This variety of internal inhibition is known as *differentiation*.

The limits of differentiation may be discovered by bringing the frequency of tone B nearer and nearer to that of tone A. In this way it has been found that the dog can distinguish very small intervals of pitch (an eighth of a tone) and that it can hear frequencies up to 60 000 Hz. Similar experiments with objects placed on the skin have shown that differences of temperature of 1°C can be recognized. The dog can readily distinguish between different shapes and varying degrees of luminosity, although he is unable to distinguish colours if very great care is taken to make their luminosity equal.

Conditional reflexes have been established with many different unconditional reflexes as a basis, ranging from the involuntary reactions of smooth muscle and glands to voluntary responses which would not usually be called 'reflexes'—indeed certain simple reflexes, such as the patellar and plantar reflexes, are extremely difficult to condition. The conditional stimulus may be any stimulus to which the organism is sensitive. Until lately, experimenters always used exteroceptive stimuli, but more recent research has shown the possibility of 'interoceptive' conditioning in which visceral stimulation serves as the conditional stimulus, the unconditional stimulus, or both. This suggests that in the normal course of life the activities of the viscera become extensively interlinked not only with one another, but also with the mechanisms involved in overt behaviour—a fact which may be of considerable significance in clinical medicine.

Conditional reflexes once established may themselves be used as a basis for further conditioning (*higher-order conditioning*). A striking illustration of this was provided by Hudgins, who trained human subjects to develop voluntary control over a normally involuntary response, the pupillary reflex. When a strong light is shone into the eye the pupil becomes smaller—this is the basic unconditional reflex.

If a bell is sounded whenever the light is shone into the eye then the bell becomes a conditional stimulus and the sound of the bell alone ultimately produces contraction of the pupil. The subject is now given a hand dynamometer (two bars kept apart by springs) with electrical switches so arranged that when he grasps it firmly and squeezes the bars together he closes the light and bell circuits; when he relaxes the circuits are broken. At first the subject squeezes at the command of the experimenter but later the bell and dynamometer are removed so that eventually the experimenter's commands to contract and relax are the conditional stimuli which produce contraction and relaxation of the pupil. In the next stage of training, the subject is asked to speak the words 'contract' and 'relax', then to whisper them, and finally to say them to himself. In this way the subject acquires voluntary control over his pupillary reflexes. Such an experiment, although fairly elaborate in its building up of one conditional reflex upon another, is not to be regarded as freakish since it has been carried out on numerous subjects; it suggests that there is less difference than is usually supposed between voluntary and involuntary reflex activity. A similar technique has been used to establish voluntary control of relaxation in patients suffering from chronic anxiety.

Indeed, when Pavlov's work first became widely known it was higher-order conditioning which attracted most attention since it seemed to indicate the possibility of regarding all the complex behaviour of men and animals as the progressive elaboration of a small repertoire of unconditional reflexes. For this reason the study of classical conditioning has come to be regarded mainly as a means of increasing our understanding of the process of learning.

It was thought at one time that classical conditioning was a form of 'association by contiguity' because reinforcement depends essentially on the conditional stimulus being presented contiguously in time with the unconditional stimulus. This idea led to the 'stimulus-substitution theory', namely that the conditional response is simply and completely transferred to a new stimulus. But this is an oversimplification. The choice of salivation, a component common to the conditional and unconditional responses, as the main index of conditioning gives the impression that the total responses are the same in the conditional and unconditional reflexes but, in fact, the chemical composition of the saliva is not identical in the two cases and in other experiments the difference is even clearer. For example, if the eye-blink normally evoked by a puff of air to the cornea is conditioned to a light, the blink changes in form, and becomes anticipatory in character, preventing the air-puff from reaching the eye. Salivation, in any case, is only a part of the unconditional response to food, and other components of the response may not be conditioned—the animal does not, after all, eat the bell. Moreover, in conditioning there are usually changes in heart rate and respiration, and movements are prevented by the restraint of the animal. The experimental procedure thus tends to obscure a situation which might be best described by saying that the animal learns to anticipate a certain sequence of events, and to respond appropriately.

Now, contiguity may explain learning when one stimulus is consistently and repeatedly followed by another provided the sequence is repeatedly experienced. A learned association between sensory events does not, however, account for the appropriateness of the response. As we have seen, even in the restricted classical conditioning situation, the animal does not respond to the bell in the same way as to food; and modification of response is the central feature of another conditioning method, *instrumental* or *Type II conditioning*.

This procedure is most simply demonstrated in an apparatus, such as the Skinner box, which is just big enough to permit free movement of a rat. From one wall projects a bar which can be arranged to release a pellet of food into a tray (Fig. 49.4). A hungry animal placed in this box at first demonstrates a great variety of movements. Eventually some movement results in pressing the bar. If no food is delivered, the rate of bar-pressing is low, and steady over a number of experimental sessions; but if pressing the bar is then made to produce a pellet of food, the rate of pressing soon rises sharply. The change of rate serves as a measure of the reinforcing effect of giving food.

As in classical conditioning, the procedure which strengthens a response is referred to as reinforcement; the procedure of omitting

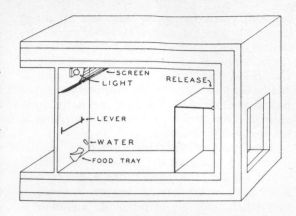

Fig. 49.4 A diagram of the apparatus used by Skinner in his studies of instrumental conditioning. One side has been cut away to show the chamber into which the animal is released through the door at the right. When the animal depresses the lever, apparatus behind the panel at the left automatically releases a pellet of food into the tray, or delivers a drop of water. (Skinner, B. F. (1938). *The Behaviour of Organisms*. New York: Appleton-Century-Crofts.)

reinforcement is likewise referred to as experimental extinction, and similarly leads to a decline in the strength of the conditional reflex. Spontaneous recovery, generalization and differentiation are also common to both procedures. The two types of conditioning do, however, differ in a number of ways. In classical conditioning, the response to be reinforced is already reliably elicited by a specifiable stimulus and is therefore under the experimenter's control; whereas in instrumental conditioning, the response to be reinforced appears spontaneously amongst a variety of reactions to the situation, and the experimenter must wait for it to occur. Skinner distinguishes *respondent* behaviour, consisting of responses directly elicited by known stimuli, from *operant* behaviour, consisting of responses which are emitted rather than elicited and for which there is no identifiable stimulus. He further suggests that this distinction is reflected in the two types of conditioning. In classical conditioning, the unconditional stimulus ensures that the response to be conditioned is already dominant in the situation, and its presentation invariably follows that of the conditional stimulus. The occurrence of the conditional response has no effect upon the reinforcement procedure, and indeed is not necessary to it. For example, the salivary response can be conditioned even when actual salivation is prevented by atropine. The essential feature of reinforcement in classical conditioning therefore appears to be the pairing of the conditional and unconditional stimuli. In instrumental conditioning, the response to be conditioned is not a dominant one, but is an essential part of the procedure because reinforcement is contingent upon its occurrence— the response is 'instrumental' in producing the reinforcement. An unconditional *stimulus* is usually not identifiable; it is the *response* which is correlated with reinforcement. The response appears to be selected from a number of reactions because of its advantageous effect on the individual (*principle of effect*).

Some features of learning, however, are not readily accounted for in this way. One of these is the effect of *partial reinforcement*. If the schedule of reinforcement in a Skinner box is modified so that only a small percentage of bar-pressings produce a pellet, extremely high response rates are recorded and resistance to extinction is greater than with one hundred per cent reinforcement. Similar results may be obtained with the classical conditioning procedure. Another anomaly is shown by the *sensory pre-conditioning* procedure devised by Brogden. If a tone and a light are repeatedly presented together, and a response is then conditioned to the light alone, subsequent presentation of the tone is now found to evoke the response, although the tone-light combination has never been reinforced.

The principle of effect offers the easiest explanation of instrumental conditioning but it has also been invoked to account for classical conditioning. The salivary response can be conditioned only if the animal is hungry, and food is therefore not only an unconditional stimulus for salivation, but also a reward; anticipatory salivation has the useful effect of preparing the mouth for food. Similarly, when a puff of air is used as an unconditional stimulus for the eye-blink, the conditional response may be said to have the advantageous effect of protecting the cornea. However, this response can be established just as readily if a blow on the cheek is used as unconditional stimulus; and in this case it is more difficult to see how the change in response is advantageous.

The complex phenomena of learning cannot yet be satisfactorily explained. The purpose of this account is to discourage a too simple interpretation of the classical conditioning situation, and the further assumption that this interpretation is an adequate explanation of all learning.

Pavlov himself was not in fact primarily concerned to formulate a theory of learning. His main interest in conditioning was as a technique for obtaining experimental control of higher nervous processes and so investigating their physiology. His procedure was to establish a number of positive and negative conditional reflexes, and then study the effect of one upon another at various intervals of time. For example, salivation was positively conditioned to the sound of a bell, and negatively to a light. Then if the bell was sounded immediately after the presentation of the light, the response was considerably greater than to the bell alone (*positive induction*). Conversely, presentation of the bell immediately followed by the light produced a state of exaggerated inhibition (*negative induction*), as shown by the fact that it was more than normally resistant to disinhibition. These are instances of a general tendency for excitatory and inhibitory processes to be reciprocally related.

Pavlov gave the name *analyser* to the peripheral end-organ together with its afferent nerves and the area of cortex receiving the afferent nerve impulses. For example, the auditory analyser consists of the organ of Corti (which as we already know makes an elementary analysis of sounds) the auditory nerve and the temporal lobes of the cerebral cortex. Afferent stimulation, he assumed, sets up at the corresponding cortical locus a state of excitation which then spreads with diminishing intensity over the entire sensory area or 'analyser' (*irradiation*). After a few seconds, the process reverses, and flows back to its starting point. The same phases of irradiation and concentration apply to the inhibitory processes initiated by negative stimuli. The irradiation of excitation over adjacent sensory areas is held to account for the generalization of a conditional reflex. When a reflex is differentiated, generalization is restricted by inhibition radiating from the cortical locus corresponding to the negative stimulus. A process of excitation or inhibition arising at any given point tends to be succeeded by the opposite process, and this after-effect accounts for the facts of induction. Irradiation, concentration, and induction are held to explain all the interactions between established reflexes.

Pavlov's account of cortical functioning is difficult to reconcile with neurone theory. He also committed himself to the view that conditioning is an exclusively cortical process. However, simple classical conditioning can be produced in the decorticate animal with little difficulty. On the other hand instrumental conditioning is not easily obtained in the absence of the cortex. The cortex is also important for classical conditioning where a differential response is required. Ablation of a primary sensory area before the beginning of training does not hinder the establishment of a discrimination involving that area. However, a similar operation carried out after training causes loss of the discrimination although it can be re-established unless the lesion is very large. Impairment tends to be proportional to the size of the lesion and the difficulty of the task. Ablation of association cortex adjacent to a primary sensory area has little effect, though removal of the whole primary and association areas of one modality may cause marked impairment of learning and retention. Among subcortical structures, the diffuse thalamic system, the dorsomedial nucleus and the posterior hypothalamus appear to be involved in avoidance conditioning.

The evidence from the effects of brain lesions might suggest that remarkably few areas of the brain are crucial for conditioning. However, although ablation of a cortical area may show that a given structure is not essential to conditioning, it cannot be concluded that the area does not normally participate in it. The best hope of investigating the normal physiology of conditioning appears to be by electroencephalography (e.e.g.) which offers a more direct approach to these central processes whose existence and characteristics Pavlov could only infer from behavioural evidence.

An unreinforced presentation of the stimulus to be conditioned evokes the orienting reflex, whose function is to increase the sensitivity of the analysers to an unexpected stimulus, so that the animal is more able to determine what is happening and to prepare an appropriate response. The orienting reflex is manifested in the e.e.g. as a disappearance of the alpha rhythm (p. 977) and its replacement by low-voltage fast activity (desynchronization). If the stimulus is repeated a number of times without reinforcement, the orienting reflex gradually disappears. Overt investigatory movements cease first; cortical desynchronization persists longer, but progressively increases in latency and decreases in duration until it is no longer discernible. This process is known as *habituation*. A habituated orienting reflex, however, shows spontaneous recovery, and is readily disinhibited by quite a slight change of stimulation. If, for example, a cat is habituated to a tone of 200 Hz, it will not respond to a tone of 220 Hz, but may respond to a tone of 230 Hz. In man, habituation to a given word is maintained to a word of similar meaning but different sound, while a word of different meaning but similar sound produces immediate dishabituation.

This marked selectivity of habituation argues against Pavlov's view that it reflects a process of cortical inhibition, since the fineness of the discrimination involved suggests that the cortical analysers must actively compare each stimulus with some representation of the habituated stimulus, and arouse an orienting response if it proves either novel or significant. This interpretation is supported by studies of the evoked potentials produced by stimulation, such as a flash of light. These are initially distributed widely over the cortex, but those from the non-specific association areas habituate rapidly, while those originating in the relevant projection area, for example the occipital cortex, do not. The persistence of the latter, even after the orienting response has habituated, shows that stimuli are still reaching the cortex for analysis. Control of the orienting reflex must therefore be subcortical, and is most plausibly attributed to the reticular formation, since electrical stimulation of this area produces effects identical with those evoked by sensory input. Other subcortical structures are likely to be involved, notably those of the limbic system, and especially the hippocampus. Electrical stimulation of this structure causes a marked arrest of ongoing movement, a pronounced inhibition of sensory input through the reticular formation, and desynchronization of the characteristic theta rhythm (p. 978). In conditioning, the appearance of this rhythm coincides with the onset of neocortical desynchronization; conversely, as the orienting response habituates, the hippocampal rhythm desynchronizes as synchronized activity appears in the neocortex. It appears to derive from strong reciprocal connexions with the reticular formation through the fornix and septum, and presumably represents hippocampal inactivity, since it is prominent during the early stages of learning and extinction of a variety of tasks known to be unaffected by hippocampal lesions. In tasks which are affected by such lesions, the characteristic deficit is difficulty in suppressing attention to stimuli which would normally evoke a response. Animals with hippocampal damage are unable to exclude irrelevant stimuli in external inhibition; to suppress stimuli associated with non-reinforcement, as in habituation of the orienting response; or to establish a negative conditional response. In man, the most striking effect of hippocampal damage is anterograde amnesia, which may well result from inability to protect memory traces from interference during the crucial phase of consolidation in long-term memory. The major function of the hippocampus thus appears to be the inhibition of the activating system of the lower brain upon repeated presentation of the same stimulus, so as to permit the organism to disengage its attention from a dominant stimulus in favour of novel or more significant events. This function is highly akin to Pavlov's concept of inhibition. A corresponding excitatory function of focusing attention on a stimulus associated with reinforcement may perhaps be ascribed to another limbic structure, the amygdala.

There is thus a coherent pattern of changes in the intrinsic rhythms of the hippocampus and neocortex, and in cortical evoked responses, during the initial phase of conditioning. It appears with the first introduction of the conditional stimulus, declines if this is not followed by reinforcement, and reappears, as might be expected, when reinforcement is introduced. It seems to reflect a state in which the animal is aware that something potentially significant has happened, but is uncertain what it is. As the conditional stimulus acquires more precise significance through repeated rein-

forcement, this pattern gradually habituates, and eventually disappears with the formation of a stable conditional response. It apparently

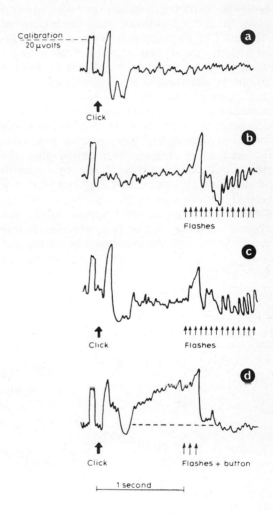

Fig. 49.5 The evoked responses and the CNV (contingent negative variation) recorded from the scalp vertex. In (a), the subject hears a click, which evokes the response shown. The stimulus in (b) is a series of flashes. If the click and flashes are provided one after another, and no connexion between the two phenomena has been suggested to the subject, the result, (c) is simply an addition of (a) and (b). However, if the subject is told that the click will warn him of the imminence of the flashes, when he will be required to make a response of some kind (for instance by pushing a button), then an 'expectancy wave' or CNV, as indicated by the dotted line, builds up between the two evoked responses. Each trace is an average of twelve trials. (*From* McCallum, W. C. (1967). *New Scientist* **36**, 592–594.)

Conditional Reflexes   991

represents those adjustments of attention necessary for conditioning to occur, rather than the process of acquisition itself.

The electrophysiological effect which most clearly parallels the establishment of the conditional response is a slow shift in cortical potential which develops in the interval between the conditional and unconditional stimuli. The phenomenon can also be demonstrated in the similar experimental arrangement used to measure the speed of human voluntary reaction to a stimulus, where it is well known that the provision of another stimulus as a warning signal substantially decreases the latency of reaction, presumably because preparatory adjustments are conditioned to it. Figure 49.5 shows that the simple temporal pairing of the two stimuli produces no change in the e.e.g.; but when the second stimulus is given significance by requiring the subject to respond to it, the evoked response to the first stimulus is followed by a slow surface-negative change of potential, recorded most prominently over the fronto-vertical region of the cortex. This potential change reaches a peak at the time of the evoked response to the second stimulus, and then returns abruptly to its original baseline when the subject responds. This effect, known as the 'contingent negative variation' (CNV), is essentially independent of the sensory modality and intensity of the stimuli, being contingent only upon their temporal association and upon the fact that some response is required to the second stimulus. As long as this requirement is maintained, the CNV shows no appreciable habituation; but if the requirement is removed, it extinguishes in a few trials. Its amplitude is depressed under conditions of partial reinforcement, and by external inhibition. It thus seems likely that the CNV represents a process of cortical 'priming' which reflects those preparatory adjustments which anticipate a response. Records from electrodes placed directly on the frontal cortex suggest that its electronegative

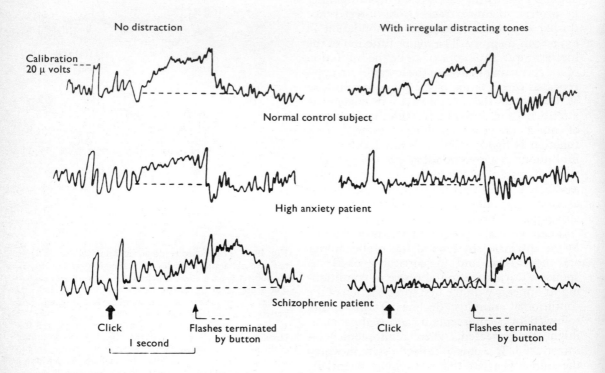

Fig. 49.6 A distracting noise does not have too great an effect on the CNV of a normal person, but in both high anxiety and schizophrenic psychiatric patients distracting tones all but eliminate the CNV (*From* McCallum, W. C. (1967). *New Scientist* **36**, 592–594.)

features arise in the most superficial layers, and probably represent depolarization of the apical dendrites.

The CNV is usually of rather small amplitude in schizophrenics and in patients with chronic anxiety. More remarkable, however, is the effect of distraction upon these patients. In normal subjects, the effect of external inhibition is transitory, and the CNV has largely recovered its amplitude after some twenty trials; whereas the psychiatric groups show little or no recovery (Fig. 49.6).

REFERENCES

BRAZIER, M. A. B. (Ed.) (1959). *The Central Nervous System and Behaviour. Transactions of the first Conference.* Josiah Macy, Jr. Foundation. (See also later Conferences.)

BROGDEN, W. J. (1939). Sensory pre-conditioning. *Journal of Experimental Psychology* **25**, 323–332.

BYKOV, K. (1959). *The Cerebral Cortex and the Internal Organs* (Translated by R. Hodes). Moscow: Foreign Languages Publishing House.

DOUGLAS, R. J. (1967). The hippocampus and behavior. *Psychological Bulletin* **67**, 416–442.

FRANKS, C. M. (Ed.) (1964). *Conditioning Techniques in Clinical Practice and Research.* New York: Springer.

GALAMBOS, R. & MORGAN, C. T. (1960). The neural basis of learning. Chap. 61 in *Handbook of Physiology, Section* 1. *Neurophysiology, III* (Edited by J. Field, H. W. Magoun & V. E. Hall). Baltimore: American Physiological Society.

GRAY, J. A. (Ed.) (1964). *Pavlov's Typology.* International Series of Monographs in Experimental Psychology, Vol. 1. Oxford: Pergamon.

HILGARD, E. R. & MARQUIS, D. G. (1961). *Conditioning and Learning,* 2nd edn (revised by G. A. Kimble). London: Appleton-Century.

HUDGINS, C. V. (1933). Conditioning and the voluntary control of the pupillary light reflex. *Journal of General Psychology* **8**, 3–51.

JOHN, E. R. (1961). High nervous functions: brain functions and learning. *Annual Review of Physiology* **23**, 451–484.

KIMBLE, D. P. (1968). Hippocampus and internal inhibition. *Psychological Bulletin* **70**, 285–295.

KLINE, N. S. (Ed.) (1961). Pavlovian conference on higher nervous activity. *Annals of the New York Academy of Sciences* **92**, 813–1198.

KONORSKI, J. (1948). *Conditioned Reflexes and Neuron Organization (A Criticism of Pavlov's Theories).* Cambridge University Press.

LYNN, R. A. (1966). *Attention, Arousal and the Orientation Reaction.* Oxford: Pergamon.

MCCALLUM, W. C. & WALTER, W. G. (1968). The effects of attention and distraction on the contingent negative variation in normal and neurotic subjects. *Electroencephalography and Clinical Neurophysiology* **25**, 319–329.

MARTIN, IRENE & LEVEY, A. B. (1969). *The Genesis of the Classical Conditioned Response.* Oxford: Pergamon.

MILLER, G. A. (1962). *Psychology: the Science of Mental Life.* New York: Harper & Row.

MORGAN, C. T. (1965). *Physiological Psychology,* 3rd edn. London: McGraw-Hill.

PAVLOV, I. P. (1927). *Conditioned Reflexes* (Translated by G. V. Anrep). Oxford University Press.

PAVLOV, I. P. (1955). *Selected Works* (Edited by Kh.S. Koshtoyants). Moscow: Foreign Languages Publishing House.

PAVLOV, I. P. (1958). *Experimental Psychology and other Essays.* London: Owen.

PAVLOV, I. P. (1964). *Lectures on Conditioned Reflexes* (Translated and edited by W. H. Gantt and G. Volborth), 2 vols. London; Lawrence & Wishart.

RAZRAN, G. (1961). The observable unconscious and the inferable conscious in current Soviet psychophysiology: interoceptive conditioning, semantic conditioning, and the orienting reflex. *Psychological Review* **68**, 81–147.

SKINNER, B. F. (1938). *The Behaviour of Organisms: an Experimental Analysis.* London: Appleton-Century.

SOKOLOV, Y. (1963). *Perception and the Conditioned Response.* Oxford: Pergamon.

WALTER, W. G., COOPER, R., ALDRIDGE, V. J., MCCALLUM, W. C. & WINTER, A. L. (1964). Contingent negative variation: an electric sign of sensorimotor association and expectancy in the human brain. *Nature* **203**, 380–384.

50 The endocrine glands

The term *hormone* is applied to a chemical substance which, produced in one part of the body, enters the circulation and is carried to distant organs and tissues to modify their structure and function.

A gland such as the thyroid is described as an *endocrine gland*, that is a ductless gland or gland of internal secretion, because the hormone is not conveyed along a duct but passes directly into the blood and into the lymphatics (high concentrations of hormones have been found in the lymph). The thyroid, testis, ovary and other glands are richly supplied with lymphatics which atrophy if the gland atrophies. The production of hormones is not, however, confined to the endocrine glands, for both secretin (p. 269) and cholecystokinin (p. 269) are hormones; hormones are produced by the kidney (Chap. 32) and noradrenaline and acetylcholine liberated at nerve-endings by nerve impulses (Chap. 41) may enter the circulation in sufficient amounts to produce effects in distant parts.

The biochemistry of hormone action is currently being extensively investigated but the picture is still very incomplete. Three sites of action for hormones can be postulated.

1. The cell membrane. (a) A hormone may act by altering the permeability or the active transport mechanisms for given molecules. Such effects may be seen in increases in glucose uptake by the target organs of the hormones ACTH, LH, TSH and others described in this and later chapters; in increases in amino acid transport into muscle cells by growth hormone; and in the effects of aldosterone on the movement of Na^+ through the kidney. Such effects could have a widespread effect on the level of metabolic activity in the cells, for example, when the transported substance had previously been present in small, rate-limiting quantities. It is difficult to distinguish these effects from those described in the following paragraphs; for example cyclic 3′:5′-AMP (p. 65) is known to affect some transport systems, and the synthesis of certain protein molecules which are component parts of membrane transport systems may be affected by hormonal effects on the nucleus. Present evidence is against the hormones themselves ever being components of transport systems.

(b) Hormones may act by activation of membrane bound enzyme systems which can then produce an intracellular metabolic controller. The only known example of this is adenylate cyclase, a membrane-bound enzyme found in all mammalian tissues so far studied, with the exception of the mature erythrocyte. Adenylate cyclase catalyses the conversion of ATP to cyclic 3′:5′-adenosine monophosphate (*cyclic AMP, cAMP*) and inorganic pyrophosphate (see p. 318). Cyclic AMP produced within the hormonally stimulated cell can alter the rates of cellular metabolism in a manner that is characteristic of the particular tissue, by activating certain protein phosphokinases (enzymes which can phosphorylate serine and threonine residues in protein molecules at the expense of ATP); the activated phosphokinase can then phosphorylate and change the activity of certain rate controlling enzymes. The various effects shown in different tissues will depend upon the specificity of the particular phosphokinase, and upon the range of phosphorylatable enzymes, that any tissue contains (see Chap. 18 for further information).

Among the hormones whose effects are mediated by AMP are: adrenaline (acting on liver and muscle glycogenolysis, heart action, fat cell lipolysis), neuromuscular transmission by increasing acetyl choline release; noradrenaline, in some actions on brain activity, and increase in melatonin synthesis in the pineal gland; glucagon, acting to increase fat cell lipolysis and liver glycogenolysis; ACTH, acting to stimulate steroidogenesis in the adrenal cortex; LH, acting to increase steroidogenesis in the ovary and testis; TSH, acting to promote thyroglobulin hydrolysis, iodination of tyrosine residues, and glucose oxidation in the thyroid gland; PTH, acting on the renal cortex to increase phosphate excretion, and upon bone to cause calcium resorption; vasopressin, acting on the renal medulla to increase water resorption.

Cyclic AMP is a nucleotide, and so does not penetrate membranes easily; thus its effects may be localized in the tissue in which its formation has been stimulated. It has been termed the 'Second Messenger' of endocrine systems, the 'First Messenger' being the hormone. Cyclic AMP is destroyed by a specific enzyme known as cyclic 3′:5′-adenosine monophosphate phosphodiesterase. Present evidence suggests that adenylate cyclase is a component of plasma membranes and that, depending on the tissue, it consists of one or more hormone receptor subunits linked through the membrane structure to the catalytic subunit in some way as yet unknown.

Insulin and prostaglandins seem to have some ability to decrease the formation of cyclic AMP, for example in adipose tissue.

2. Pre-existing intracellular enzyme systems. It might be expected that some hormones would be able to increase the activity of cytoplasmic mitochondrial systems directly, independently of any effects on protein synthesis or of any effects mediated by cyclic AMP.

It was suggested that oestrogens function as hydrogen carriers in transhydrogenase systems, that is that they functioned as co-factors, and that ACTH can directly activate glucose 6-phosphate dehydrogenase in the adrenal cortex. It has been claimed that insulin forms complexes with hexokinase, but the claims have received little support.

3. Sites of synthesis of new proteins. Jacob and Monod have shown that the transcription of structural genes for certain enzymes can be controlled by the presence or absence of certain metabolites (Chap. 21). It seems likely that certain hormones, especially those concerned with tissue growth and maintenance, may be able to alter rates of enzyme synthesis. That such is indeed the case can be shown by the inhibition of certain hormonal effects in the presence of inhibitors of protein or RNA synthesis, for example cycloheximide, puromycin and actinomycin D. It should be said that there is strong evidence that the inhibitory effects of the latter two compounds are not restricted to protein and RNA synthesis effects in higher organisms. There are many ways in which protein synthesis may be increased or decreased, not least by the availability of the necessary precursors. Some of the possible sites of hormonal action are indicated below:

(a) Effects on DNA synthesis—for example, formation of multiple copies of genes.

(b) Gene 'unmasking'—to produce increased synthesis of specific mRNA, or gene 'activation'—to increase the rate at which genes are transcribed.

(c) Increasing the amount of mRNA transported out of the nucleus (increasing nuclear mRNA synthesis does not necessarily mean more mRNA at the sites of protein synthesis in the cytoplasm unless such transport can also be increased).

(d) Stabilization of mRNA in the cytoplasm, that is increasing the useful lifetime of mRNA.

(e) Increasing the rate of translation of mRNA into protein by activation of the necessary ribosomal enzymes, or by changing the stability of polyribosomes or their ability to attach to the membranes of the endoplasmic reticulum.

(f) Increased production of new ribosomes, either to bring the ribosomal population up to operational strength or to create a new population of more active or more selective ribosomes which could preferentially transcribe mRNA for the particular enzyme whose synthesis is increased by the hormone.

There is good evidence for the involvement of protein synthesis effects in the actions of steroid hormones (corticoids, oestrogens, progestogens and androgens), growth hormone, thyroxine, ACTH and ecdysone (a hormone of insect development).

Steroid hormones show a familial pattern of mode of action; they accumulate in the nucleus of their target cells, which they enter after binding to specific proteins in the cytoplasm. The protein–steroid complex then enters the nucleus and can be shown to bind to nuclear chromatin. Experiments with progesterone and chick oviduct have shown that DNA, histone and three acidic proteins are necessary chromatin factors which facilitate the binding of the steroid–protein complex. The binding of chromatin can then, presumably, alter the extent of gene masking and subsequent transcription.

Growth-hormone-stimulated protein synthesis seems to be required for the increase in amino acid transport into muscle cells mentioned previously. Increases of RNA-synthesis in liver and fat cells seem to depend on prior protein synthesis. The work of Korner has shown that hypophysectomized animals have low ribosome populations in the liver, and the ribosomes which are present have poor translational ability. The nature of the defect, which can be reversed by growth hormone administration, is unknown but is located in the smaller ribosomal subunit (that is the 40S subunit) and it is possibly a deformed or absent protein factor involved in protein synthesis at the ribosome.

Thyroxine appears to increase basal metabolism, not as originally suggested by uncoupling oxidative phosphorylation in the mitochondria, but by causing specific synthesis of proteins involved in oxidative phosphorylation. This is achieved by increased production of ribosomes, together with possible effects on mRNA production and stability.

We have already indicated that ACTH increases corticosteroid synthesis by way of cyclic AMP; however, one of the effects of cyclic AMP seems to be to promote the translation of a stable mRNA which is present in the cytoplasm of adrenal cells. Cycloheximide, which inhibits such translation, blocks ACTH-induced steroid synthesis. Finally, it should be mentioned that cyclic AMP-dependent protein phosphokinases can phosphorylate histone and ribosomal proteins; this fact allows a possible explanation of its effect on protein synthesis.

The physiological actions of the various hormones have been explored in a number of ways; first by studying the effects of complete removal of an endocrine gland from the body and secondly by grafting experiments or injections of active extracts of the gland to try to restore the animal to normal; thirdly by observing the effects of the administration of large amounts of the hormone to normal animals. Conclusions based upon such experiments have in many cases been amplified by the study of disease in man since disorders due to overactivity, or to destruction, of the various endocrine glands are well recognized. The effects of over-secretion by an endocrine gland may be relieved or cured by the surgical removal of part of the overactive tissue. On the other hand the operative removal of too much tissue or the destruction of whole or part of an endocrine gland by disease gives rise to a 'deficiency syndrome' which is controlled, but never cured, by replacement therapy; that is to say by supplying extracts of the deficient gland obtained from animals or synthetic hormones or chemically related materials that have a similar effect.

This chapter deals with the thyroid, thymus and adrenal glands, together with a number of so-called 'local' hormones. The internal secretions of the parathyroids are dealt with in Chapter 14, of the pancreas, insulin and glucagon in Chapter 18, the gonads in Chapter 51 and the pituitary in Chapter 52.

THE THYROID GLAND

The thyroid gland develops in the floor of the embryonic mouth as a median diverticulum which grows downwards as a tubular duct with a bifurcated end that gives rise to the isthmus and part of the lateral lobes; the remainder of the lateral lobes is derived from the fourth pharyngeal pouch. The connexion with the mouth, the thyroglossal duct, disappears early in development. The thyroid gland in the adult lies at the root of the neck in front of the upper rings of the trachea; it is sometimes prolonged into the thorax behind the sternum. Accessory thyroid tissue is occasionally found along the line of the original thyroglossal duct.

The thyroid gland, weighing about 25 g in the healthy adult, consists of an enormous number of closed spherical vesicles, the walls of which in the resting phase of the gland are composed of a single layer of flattened epithelium. The vesicles, which vary in size from 0·05 to 0·1 mm in diameter, are filled by a structureless semifluid protein material, the 'colloid', containing the active principle of the gland. The thyroid gland has a remarkably rich blood supply; some 4 to 6 ml of blood per gram of tissue perfuse the normal gland per minute and the amount is greatly increased when the gland is overactive. The sympathetic nerves supplying the gland do not appear to have any direct effect on the secretory processes.

The thyroid gland is not essential to life but it contributes much to physical and mental well-being. In both man and animals it is necessary for growth and development and in amphibia for metamorphosis. The thyroid begins to trap iodine about the third month of fetal life but normal thyroid hormone synthesis is not established until the fourth or fifth months. At birth the gland contains relatively large amounts of colloid which gradually disappears during the first few weeks of extrauterine life. The secretion of thyroid hormones appears to be fairly consistent throughout life but compensatory hyperplasia tends to occur in response to the physiological changes of puberty and pregnancy.

The activity of the thyroid gland is regulated by variations in the level in the blood of thyroid-stimulating hormone (TSH) from the anterior pituitary (Chap. 52). The immediate action of this hormone is to release thyroid hormone from the gland. Subsequently it increases the uptake of iodine from the blood and promotes hyperplasia of the gland. The output of TSH from the anterior pituitary is itself governed by the blood concentration of thyroid hormone, a fall in the latter increasing the output and a rise diminishing it. This negative feedback mechanism is probably mediated through the production of thyrotropin releasing factor (TRF) by the hypothalamus which passes to the anterior pituitary in the portal circulation.

Desiccated thyroid gland is active when given by mouth. Even when it is standardized on the basis of its iodine content its physiological activity is variable and pure thyroxine is preferred for treatment of patients.

The thyroid gland contains a large amount of iodine (0·06 per cent), almost all of which is firmly bound to protein either in the cells lining the vesicles or in the colloid material within them. The characteristic protein of the colloid, thyroglobulin, though mainly found in the thyroid itself, has been found in the neighbouring lymphatic glands and in minute amounts in blood. On hydrolysis it yields several iodine-containing derivatives of tyrosine including mono- and diiodotyrosine and tri- and tetraiodothyronine (thyroxine).

Iodine in firm organic combination was shown to be an important constituent of thyroid tissue by Baumann in 1895, shortly after the therapeutic value of extracts of the thyroid glands of animals in the treatment of hypothyroidism had been shown by Murray.

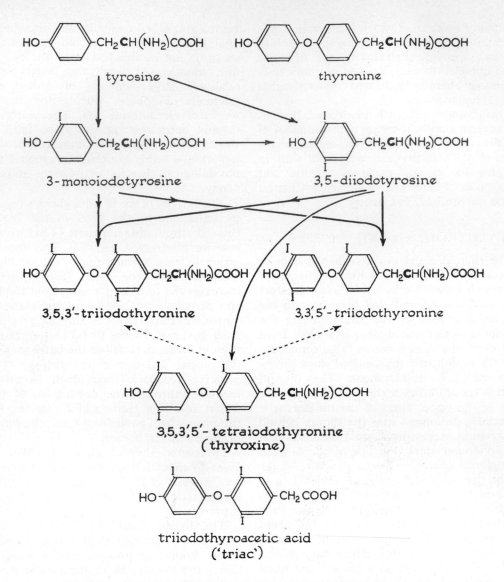

Fig. 50.1 Pathways leading to the biosynthesis of the thyroid hormones. Arrows in broken lines indicate processes which take place in other tissues as well. The synthetic analogue triac is biologically active, but much less so than the thyronine compounds.

This led to the preparation of more and more active extracts of thyroid tissue and culminated in the crystallization of thyroxine in 1919 by Kendall in the United States. The method of isolation was greatly improved by Harington in Britain and the chemical constitution of thyroxine was determined by him in 1927. In the following year thyroxine was synthesized by Harington and Barger. Its structure is shown in Fig. 50.1; the L-isomer is at least twenty times more active than the D form, as judged by calorigenic activity.

Thyroxine was at first thought to be the active principle of the thyroid gland although the time lag between its administration and its effects, about 24 hours, was difficult to explain. In 1951, however, Gross and Leblond, using the radioactive isotope ^{131}I, discovered a new iodine-containing compound in plasma which was later proved by Gross and Pitt-Rivers to be 3,5,3'-L-triiodothyronine (Fig. 50.1); it was synthesized in 1953. This compound is found both in the gland, though in much smaller amounts than thyroxine, and in the blood plasma of normal people and patients with overactive glands (hyperthyroidism). Triiodothyronine possesses all the biological properties of thyroxine and is more potent. For example, it is three to five times more active than thyroxine in raising the oxygen consumption of patients with hypothyroidism and this action occurs within four hours. When administration ceases, its effects rapidly disappear. Approximately one-third of the thyroxine secreted by the thyroid is metabolized to triiodothyronine which may be the active hormone at a cellular level although this has not been shown conclusively. While triiodothyronine is poorly bound to the plasma protein most of the thyroxine is transported in this way, mainly on the globulin fractions but a significant amount is bound to albumin. Only about one-third of the total protein-binding capacity (about 0·4 µg per ml) is taken up in a normal (euthyroid) person but this is increased to one half or more in hyperthyroidism.

In the first stage in the formation of the thyroid hormone iodine in the blood plasma as inorganic iodide is concentrated in the acinar cells of the thyroid. The gland's remarkable powers of accumulating iodide can readily be demonstrated with the aid of ^{131}I. This accumulation is accelerated by the thyrotrophic hormone (TSH) of the pituitary and is reduced by perchlorate or thiocyanate. Normally the iodide becomes organically bound as fast as it is taken up but, after administration of certain anti-thyroid agents (thiouracil and thiocarbamides, such as thiourea), the organic combination is blocked and a large amount of iodide remains in inorganic form in the gland (Fig. 50.2).

As iodide is absorbed by the acinar cells it is oxidized to iodine ($2I^- \rightarrow I_2 + 2e$) by an enzyme system requiring the participation of cytochrome. The iodine never appears as such in the free state but is immediately taken up by the tyrosine of the glandular protein to form monoiodotyrosine and diiodotyrosine (Fig. 50.2) as residues still incorporated in the protein molecule. These residues then undergo oxidative coupling with removal of one side chain so as to form derivatives of thyronine, thyroxine being formed from two diiodotyrosine units and triiodothyronine from one mono- and one diiodotyrosine unit (Fig. 50.1).

While triiodothyronine is produced by this coupling mechanism in the gland, it may also be formed by removal of one atom of iodine from thyroxine (Fig. 50.1). A *dehalogenating*

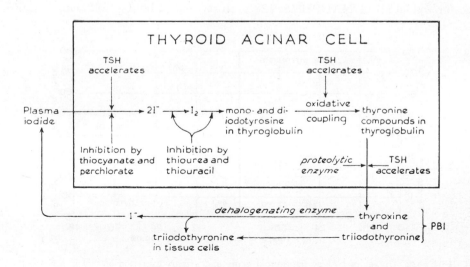

Fig. 50.2 Scheme of the metabolism of iodine in the thyroid cell.

enzyme which brings about this reaction is present in most body tissues as well as in the thyroid. For example, thyroidectomized mice or human subjects can convert thyroxine labelled with ^{131}I to triiodothyronine which can subsequently be recovered from the plasma.

The thyroxine and triiodothyronine produced in the gland itself are still in peptide linkage in the thyroglobulin molecule and are stored in the colloid. Under stimulation by TSH the thyroglobulin is broken down by a proteolytic enzyme to yield thyroxine (90 per cent) and triiodothyronine (10 per cent) which pass into the blood where, in combination with plasma globulin, they comprise the *protein-bound iodine* (PBI) of the plasma. Finally they enter the cells of the tissues where thyroxine is dehalogenated to triiodothyronine and inorganic iodide which is returned to the thyroid gland.

A considerable amount of thyroglobulin leaves the thyroid gland in the lymphatics and then reaches the blood; the amount is increased by administration of TSH. Thyroglobulin raises the metabolic rate, that is it has the properties of the thyroid 'hormone' (see below).

The wide range of effects of thyroid hormone suggests that it may well operate by regulating enzymes which control energy metabolism, for example by inhibition of transhydrogenases which are normally responsible for oxidation of NADPH (p. 128).

It has been shown that thyroxine causes mitochondria to swell *in vivo* as well as *in vitro*. This suggests that the primary action of the hormone is to increase the permeability of the mitochondrial membranes. Thyroxine is able *in vitro* to uncouple oxidation from phosphorylation. If such an uncoupling action were to occur in the intact organism much of the energy produced during the process of cellular oxidation would be lost as heat instead of being used in the synthesis of high energy phosphate compounds such as ATP. The calorigenic effect of the thyroid hormone could then be accounted for. Sokoloff and Kaufman disagree with this explanation: they are of the opinion that the thyroid hormone increases the rate of oxygen consumption because it increases the rate of incorporation of amino acids into protein (p. 378).

The isotopes ^{131}I and ^{132}I are of value in the diagnosis of thyroid disease in man. After a single dose by mouth (microcuries) in the form of sodium iodide about 2 per cent is eliminated in the faeces and sweat. Forty-eight hours after the dose has been given 80 per cent of the ^{131}I has either been taken up by the thyroid or excreted in the urine and the remainder is distributed throughout the body. A larger proportion is taken up by an overactive than by a normal gland, and this difference can be detected either by a suitable detector placed over the gland or by determining the amount of radioactivity in the urine passed during the forty-eight hours following administration

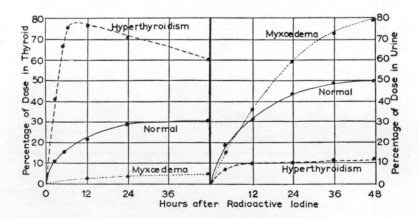

Fig. 50.3 Uptake of ^{131}I by the thyroid gland and urinary excretion of ^{131}I in a normal person and in typical cases of hyperthyroidism and myxoedema. (C. P. Stewart & D. M. Dunlop (1964). *Clinical Chemistry in Practical Medicine*, 6th edn. Edinburgh: Livingstone.)

(Fig. 50.3). ^{131}I in much larger doses (millicuries) may be used to suppress an overactive gland (hyperthyroidism); in this way the concentration of radioactive material is built up to a level at which the number of secreting cells is diminished by destruction or reduction of their reproductive capacity.

The use of ^{131}I in both animals and man has produced information about the control of the secretion of thyroid hormone. For example, the uptake of radioactive iodine is maximal some forty-eight hours after the injection of TSH in a normal person; in a patient with hypothyroidism due to diminished pituitary function, injections of thyrotrophic hormone (TSH) increase the uptake of radioactive iodine by the thyroid gland; the uptake is, however, unaffected when the hypothyroidism is the result of atrophy, or absence, of the thyroid. Thyroxine can also stop the release of ^{131}I from the thyroid gland for a considerable time. Exposure of rabbits to moderate cold increases the output of ^{131}I but intense cold, emotional stress, or physical trauma reduce the output of hormone. Injection of cortisone or ACTH reduces the output of ^{131}I from the thyroid, presumably mainly by a direct effect on thyroid hormone synthesis. Hypophysectomy causes a prompt and permanent decrease in the rate of release of ^{131}I from the thyroid. In normal persons the serum TSH activity is not detectable by the methods at present available.

The concentration of protein-bound iodine in the plasma is an index of the amount of thyroid hormone in the blood and of the secretory activity of the thyroid gland. The normal concentration is 3·5 to 7 µg per 100 ml.

Hypothyroidism. Thyroidectomy produces a marked fall in the metabolic rate, with subnormal temperature, slow pulse and respiration and mental apathy. Hypothyroidism in the human infant is described as *cretinism* (Fig. 50.5). Although the condition is not usually recognized until three or four months after birth, it is probable that the thyroid gland in these cases has never functioned. The child is small, mentally defective, with coarse, scanty hair and a thick, yellowish scaly skin. A rarer type of cretinism is known which tends to be familial and is associated with an enlarged thyroid gland. In such patients there is a

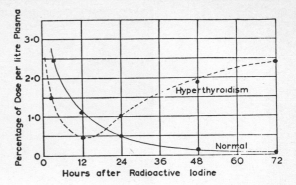

Fig. 50.4 The plasma ^{131}I after a test dose in a normal person and in a patient suffering from moderate hyperthyroidism. In the patient with hyperthyroidism the reappearance of radioactivity in the blood is the result of discharge of hormone containing ^{131}I from the active gland. At the end of 48 hr after the dose of isotope the hyperthyroid patient has therefore more protein-bound radioactivity than the normal. (C. P. Stewart & D. M. Dunlop (1964). *Clinical Chemistry in Practical Medicine*, 6th edn reprint. Edinburgh: Livingstone.)

defect in one of the enzymes responsible for the synthesis of thyroxine.

Severe hypothyroidism occurring in the adult is called *myxoedema* (Fig. 50.6) because of the puffiness of the hands and face. This is due to a thickening of the skin by the deposit in it of a semi-fluid material containing over 13 per cent of protein. Hypothyroidism occurs most commonly in middle-aged women. It is frequently associated with the changes of auto-immune thyroiditis in the gland manifested by lymphocytic infiltration and fibrosis. Antibodies to constituents of thyroid cells can be demonstrated in the blood and the condition is believed to have an auto-immune basis.

A patient with myxoedema feels cold and has a basal metabolic rate which may be 40 per cent below normal. Body temperature and pulse rate are reduced and the weight is increased. All bodily and mental processes are retarded. In primary thyroid failure the serum TSH is raised.

Hyperthyroidism. The administration of an excessive amount of thyroid hormone leads to an increase of the metabolic rate but the body temperature is not usually raised since the extra heat is got rid of by vasodilatation and sweating. Weight is lost as the result of in-

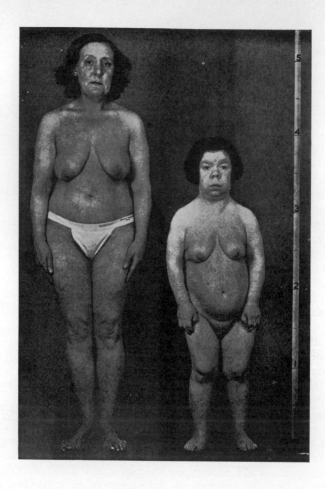

Fig. 50.5 On the right a cretin, aged 38 years, and on the left her normal sister, aged 48. The mental age of the cretin is about 4 years. She has a deep, croaking voice, coarse dry skin and hair. (*By courtesy of A. C. Crooke.*)

creased catabolism of tissue protein and oxidation of the stored fat. Similar effects are produced by an increase in the secretory activity of the thyroid gland often accompanied by an enlargement of the gland (a *goitre*). The patient has marked nervous excitability with an involuntary tremor (Fig. 50.7). He is breathless on exertion, has a rapid heart rate, a warm flushed skin and an anxious staring expression. Protrusion of the eyeballs (exophthalmos) may also be present and this may persist long after the other features have been abolished by treatment. In animals, but not in man, exophthalmos can be produced by TSH. In fact serum TSH levels have been shown to be subnormal in thyrotoxicosis. The increased concentration of circulating TSH in myxoedema is not accompanied by exophthalmos. Some evidence suggests that there may be an exophthalmos-producing factor (EPF) produced by the adenohypophysis. Some patients with hyperthyroidism have in their blood a long acting thyroid stimulator (LATS), a γ-globulin produced by the reticulo-endothelial system. Iodotyrosines are present in the blood of thyrotoxic patients but not in euthyroid persons.

Excess thyroid activity can be reduced by surgical removal of part of the thyroid gland (partial thyroidectomy). The administration

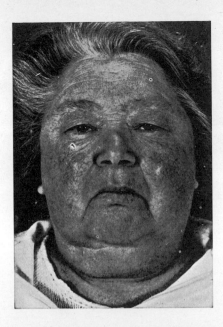

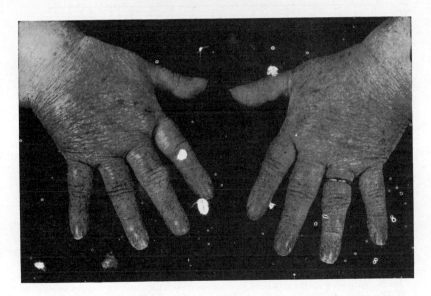

Fig. 50.6 Photographs of the face and hands of a woman, aged 69, suffering from myxoedema. Note the dull expression, scanty eyebrows and dilated vessels on the cheeks. The face and hands are puffy, and the skin of the hands is dry and wrinkled. (*By courtesy of W. T. Cooke.*)

of iodine greatly reduces the output of hormone and lowers the B.M.R. in a patient with hyperthyroidism (Fig. 50.8); it also brings back the histological appearance of the gland towards normal. The amount of iodine required to produce a maximal effect is of the order of 20 mg per day, and little is achieved by giving doses greatly in excess of this quantity. This dose should be compared with the daily intake adequate for the prophylaxis of endemic goitre (below), namely about 0·075 mg, and with the average daily intake of healthy people,

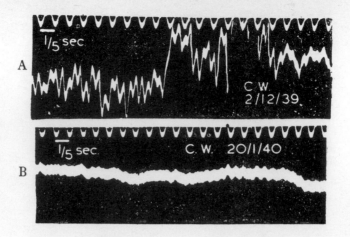

Fig. 50.7 Muscular tremor as shown by the magnified movements of the tip of the extended index finger of a patient, aged 20, with hyperthyroidism before operation (upper tracing), and 45 days after partial removal of the thyroid gland (lower tracing). (S. Lazarus & G. H. Bell (1943). *Glasgow Medical Journal* **140**, 77.)

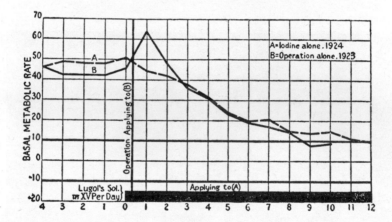

Fig. 50.8 The effect of iodine compared with that of partial removal of the thyroid gland in the untreated hyperthyroid patient. The solid line represents the average curve obtained by Segall & Means (1924) in uniodinized patients; the broken line the average curve for iodine response obtained by Starr, Segall & Means (1924). Lugol's solution containing 5 per cent iodine and 10 per cent potassium iodide was administered daily in a dose of 15 minims (=1 ml). Except for the temporary increase of B.M.R. after operation, the curves are remarkably similar.
(J. H. Means (1948). *The Thyroid and its Diseases*, 2nd edn. Philadelphia: Lippincott.)

about 0·1 to 0·2 mg. In spite of continued administration of iodine, however, the metabolic rate rises after two or three weeks and the signs and symptoms of hyperthyroidism reappear.

GOITROGENS OR ANTITHYROID SUBSTANCES. The action of a goitrogen is due to its ability to diminish the synthesis of thyroid hormone by interference with its production at a number of points. For example, thiocyanates and perchlorates interfere with the uptake of iodine by the gland, and the thiourea and imidazole compounds block the synthesis of organic iodine compounds in the gland. The latter group is used with much success in the treatment of hyperthyroidism (see Fig. 50.2).

Fig. 50.9 Endemic goitre in the Western Sudan. Note the massive enlargement of the thyroid gland.

Endemic goitre. Since iodine is an essential part of the active principle of the thyroid gland, it is to be expected that a diet deficient in iodine might lead to difficulties in the synthesis of thyroxine. These difficulties are shown by an enlargement of the thyroid gland ('goitre') (Fig. 50.9). In districts where the supply of iodine in the diet is deficient a large part of the population may be goitrous, hence the name endemic goitre. Exceptionally, in regions such as the Congo where the deficiency is very severe, hypothyroidism may occur. There is an inverse relationship between the intake of iodine and the incidence of goitre (Table 50.10), but factors other than an absolute deficiency of iodine are involved; thus goitre is much more common in females and in certain families.

The first large-scale trials of the effect of giving small amounts of iodine to a population in which goitre was endemic were made in Ohio in 1916. The results were quite remarkable as the following figures show; only 0·2 per cent of non-goitrous girls who took

Table 50.10 Prevalence of enlargement of the thyroid gland in children, aged 11 to 15 years, from four areas in which the drinking water varied greatly in iodine content

Area	Iodine content of water (μg per litre)	No. of children examined	Total no. of visible thyroid glands	Percentage incidence of visible thyroid glands
Okehampton, Devon	1·1 (very low)	298	78	26·2
North Oxfordshire	2·0 (low)	451	81	18·0
Windsor, Berks	10·1 (high)	461	31	6·7
Maldon, Essex	50·2 (very high)	527	13	2·5

From Table I of *M.R.C. Memorandum* (1948). No. 18. London: H.M.S.O.

iodine over the four years of the experiment showed an enlarged thyroid gland, whereas 27·6 per cent of those not taking iodine became goitrous. Since then, iodine has been administered to large populations as sodium or potassium iodide, as potassium iodate, or as iodized fat; the usual method is to add a small amount of iodide to the table salt. As a result of this measure the incidence of goitre in certain districts has fallen by about 90 per cent in the general population and the incidence of goitre in the new-born by the same amount. Therefore a subcommittee of the Medical Research Council recommended (1944) that all table salt sold in Britain should be iodized by the addition of one part of potassium iodide to 100 000 parts of sodium chloride, or of one part of potassium iodide to 40 000 parts of packeted table salt. These measures would, if put into effect, provide the daily iodine requirement of 200 µg. The goitre of pregnancy is likely to be a response to a low plasma inorganic iodine blood level due to the available iodine being dispersed in a greater space, for example, the expanded plasma and extracellular fluid spaces, the placenta and the fetal tissues.

THE ADRENAL (SUPRARENAL) GLANDS

The adrenal glands are situated one on each side of the vertebral column, closely applied to the upper poles of the kidneys. These glands contain two different kinds of tissue, *cortex* and *medulla* (Fig. 50.11), which have little physiological relationship and indeed have quite different embryological origins. Removal of both glands is fatal within a few days, but all the medullary tissue can be extirpated with

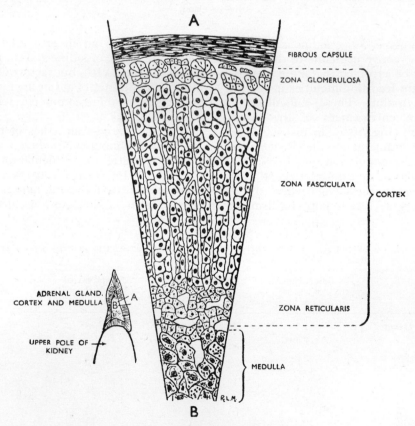

Fig. 50.11 Diagram of adrenal gland showing differentiation into medulla and cortex, the latter itself being divided into three zones not clearly demarcated from one another. The part of the gland from which the section was taken is shown alongside (AB).

little apparent effect; the adrenal cortex is therefore essential to life, while the medulla is not.

Structure of adrenal. The human adrenal gland can be divided into head, body and tail (most laterally). The head contains the largest proportion of medullary tissue; the proportion declines as we move laterally so that the tail contains hardly any medullary tissue. The gland is supplied by as many as 50 small arteries but is drained by only one main vein. The walls of the adrenal vein and its branches, both within and without the gland, are composed mainly of longitudinal muscle gathered into a few discrete bundles. Dobbie and Symington (1966) suppose that contraction of these muscle columns squeezes the thin-walled veins draining both cortex and medulla as they enter the muscular veins and so, by holding back blood in the various sinuses, increase temporarily the amount of blood in both medullary and cortical regions. A sudden relaxation of these muscular columns could, they suppose, produce an explosive release of hormones into the circulation.

THE ADRENAL CORTEX

The adrenal cortex is of mesodermal origin, being derived from the intermediate cell mass from which develop the primitive kidney and the gonads. Small amounts of tissue identical with the adrenal cortex have been found in many sites, mainly in the abdominal cavity. The adrenocortical cells appear to receive no nerve connexions.

The adrenal cortex is made up of three zones (Fig. 50.11) which in most animals are fairly distinct. In man (Plate 50.2) the outermost or gomerular zone is usually irregular and ill-defined, forming islands of cells lying under the capsule. The zona fasciculata in man is of variable width, and is filled with cells rich in lipid. During the treatment of tissue sections for histological examination, this lipid may be washed out leaving empty spaces in the cells which are accordingly described as 'clear cells'. The more 'compact' cells of the zona reticularis are relatively poor in lipid and form the innermost zone of the cortex.

After hypophysectomy the fascicular and reticular zones in man and other species atrophy and the secretion of cortisol and cor-

ticosterone ceases. By contrast, the glomerular zone and the secretion of aldosterone, known to occur in this zone, are unaffected.

Administration of adrenocorticotrophic hormone (corticotrophin, ACTH, Chap. 52), or exposure to stress which promotes ACTH secretion (Chap. 52), in addition to increasing production of cortisol and certain other steroids, decrease the amount of lipid and of ascorbic acid throughout the whole cortex. There is also an increase in weight and true growth of the gland. In man lipid depletion of the fascicular cells is most pronounced and these cells are transformed into 'compact' reticular zone cells, so that this zone appears to occupy more space in the cortex. The fascicular zone in man thus stores lipids (principally cholesterol esters) which may be used as precursors of steroid hormones under conditions of 'stress' in which the amount of circulating ACTH is increased. Normally the adrenal cortex may produce its hormones in both the fascicular and reticular cells from precursors carried in the blood.

ACTH has little, or no, effect on the glomerular cells or on aldosterone secretion, which is controlled by entirely different factors (p. 657).

Since 1930, more than forty steroids have been isolated from extracts of animal adrenal glands. Many of these compounds are precursors or metabolites of the active hormones. The basic structure of all adrenal steroids is the cyclopentenophenanthrene nucleus, the carbon atoms being numbered as indicated on page 57. The active hormones of the adrenal cortex are classified, according to their physiological activity, into three major groups —the *glucocorticoids*, the *mineralocorticoids* and the *androgens*. This classification is not rigid since glucocorticoids have some mineralocorticoid activity.

Additional carbon atoms are substituted at the positions shown on page 1008. C-21 steroids have predominantly glucocorticoid or mineralocorticoid activity with physiological effects on intermediary metabolism or salt and water metabolism respectively. C-19 steroids have androgenic activity.

The biosynthesis of adrenal steroids. Cholesterol within the adrenal gland is the precursor of all adrenal steroids. It may be synthesized there from acetate, and the adrenal

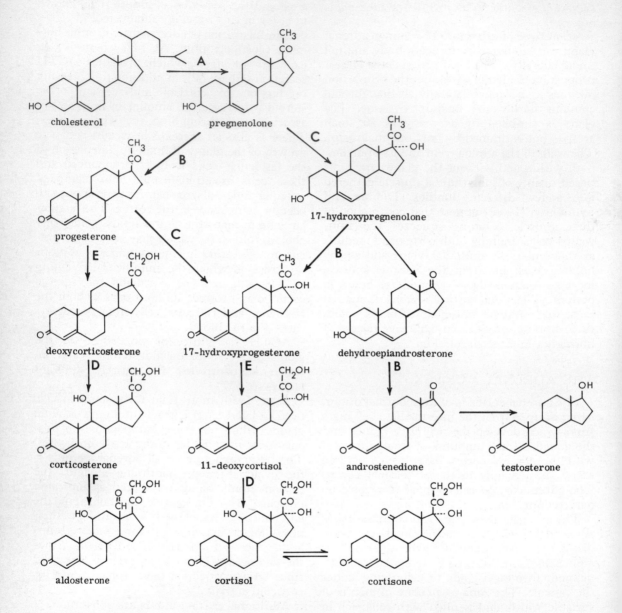

Fig. 50.12 Pathways of the biosynthesis of adrenal steroids. The enzymes involved are as
follows:
A, desmolase; B, Δ^5-isomerase 3β-hydroxydehydrogenase; C, 17α-hydroxylase; D, 11β-
hydroxylase; E, 21-hydroxylase; F, 18-oxidase. (*After V. H. T. James and J. Landon.*)

stores, under circumstances of high steroid output, may be replaced from circulating cholesterol. Cleavage of the cholesterol side-chain by the enzyme, desmolase, produces pregnenolone. This compound, in turn, under the action of the various enzyme systems present in the adrenal cortex, gives rise to the active hormones secreted by the adrenal. The pathways are shown in Figure 50.12.

On relatively rare occasions deficiency of one of the enzymes may occur, due probably to a genetic defect. The deficiency may be partial or total. As a result, there is an accumulation of steroid precursors with a deficiency in the formation of cortisol. The resulting low plasma-cortisol level activates the negative feedback mechanism (see below) with consequent increased output of ACTH which causes adrenal hyperplasia and, in some cases, depending on the site of the enzyme defect, increased production of adrenal androgens. The generic name for this condition is *congenital adrenal hyperplasia* and is usually detected within the first three years of life. However, it may escape notice until later life, presumably because the enzyme defect in such cases is incomplete and enough cortisol is produced to meet the demands of the patient even under circumstances of 'stress'. The alteration in secretion of adrenal androgens is reflected by varying degrees of abnormality of phenotypic sex in the patient. The commonest defect is 21-hydroxylase deficiency, accounting for about 90 per cent of all cases of congenital adrenal hyperplasia. The metabolites of the cortisol precursors can be detected in the urine and identification of these metabolites may indicate the site of enzyme defect. Table 50.13 illustrates the commoner defects, the major urinary abnormalities and clinical consequences of the enzyme defects. The condition is treated by administration of replacement doses of glucocorticoids which suppress the excessive output of ACTH, and subsequently reduce androgen production where this occurs. Biosynthetic pathways in the gonads have some enzymes in common with the adrenal cortex. Deficiencies of these enzymes can therefore occur in both tissues.

Glucocorticoids. The most important of these substances is cortisol (hydrocortisone, Compound F). In normal adult subjects it is secreted at a rate of between 10 and 30 mg per day, but the cortisol secretion rate may be considerably higher during times of 'stress'. When measured in plasma, there is a definite and marked circadian variation in the circulating level of cortisol, being highest before awakening in the morning (4 a.m. to 8 a.m.) and lowest during the early hours of sleep. The peak level lies between 10 and 25 µg/100 ml falling to a level of less than half the morning value or less than 10 µg/100 ml between midnight and 3 a.m. The variation in cortisol secretion is secondary to a circadian variation in ACTH secretion. The variation is uninfluenced by light since it persists in completely blind subjects, but may be altered by changes in the pattern of sleeping and waking as long as these changes are maintained for several days. It may be absent during times of stress and certain abnormalities of the hypo-

Table 50.13 The defects in congenital adrenal hyperplasia

Enzyme defect	Hormone deficiency	Urine excess	Clinical presentation
3β-Hydroxydehydrogenase	Cortisol, aldosterone	Δ5,3β-ol steroids particularly DHA	Salt loss; male—hypospadias; female—mild virilism
17-Hydroxylase	Cortisol, aldosterone, androgens	Tetrahydrocorticosterone Tetrahydrodeoxycorticosterone	Immature female, hypertension, hypokalaemic alkalosis
21-Hydroxylase	Cortisol, aldosterone	Pregnanetriol	Masculinization, ± salt loss
11-Hydroxylase	Cortisol, aldosterone	Tetrahydro-11-deoxycortisol	Masculinization, hypertension
18-Hydroxylase	Aldosterone	Tetrahydrocorticosterone	Salt loss

thalamo-pituitary-adrenocortical system such as Cushing's syndrome (see below).

Most of the cortisol secreted by the adrenal cortex is reversibly bound to protein in the blood and is biologically inert, since only free cortisol is physiologically active. At concentrations up to 20 μg/100 ml, about 95 per cent of cortisol is bound to an α-globulin (transcortin). Transcortin provides highly specific binding but the capacity is low. When plasma levels of cortisol exceed 20 μg/100 ml, the steroid is bound to albumin; the binding is less specific but the capacity is much greater. Protein binding provides a buffer mechanism which mops up cortisol when the plasma free levels are high and releases the hormone when the plasma free level is low. Corticosterone with qualitatively similar activity is secreted in about one tenth the amount of cortisol.

By definition, the glucocorticoids are important in the regulation of carbohydrate metabolism. They promote the synthesis of glycogen in the liver and encourage gluconeogenesis from protein by accelerating protein catabolism. These actions are accompanied by a rise in the concentration of glucose in the blood. Cortisol also reduces the number of circulating eosinophils and has anti-inflammatory and anti-allergic properties, although these effects are more obvious in situations where excess cortisol is produced (Cushing's syndrome) or when synthetic steroids are administered therapeutically. The glucocorticoids also possess limited intrinsic mineralocorticoid properties; they act on the distal tubule of the kidney to promote sodium and water retention. When the adrenals are the seat of a primary disease process (Addison's disease), as a result of tuberculosis or autoimmune adrenalitis, the metabolic effects of the steroid deficiency are shown by hypoglycaemia and excessive sodium loss in urine. The administration of replacement doses of synthetic glucocorticoids and mineralocorticoids corrects the metabolic abnormalities. The excessive production of cortisol (Cushing's syndrome) or the administration of large amounts of synthetic glucocorticoids may result in hyperglycaemia (or even diabetes mellitus) and excessive sodium and water retention.

In addition to a circadian rhythm of ACTH secretion and subsequent cortisol release, two other mechanisms are important in the control of glucocorticoid secretion. The first of these is the circulating level of cortisol itself. The median eminence of the hypothalamus contains receptors which are sensitive to the level of plasma cortisol (or synthetic glucocorticoids). A negative feedback mechanism operates by which a plasma glucocorticoid level lower than normal for any particular time of day induces the release of *corticotrophin releasing factor* (CRF), a short chain polypeptide (Chap. 52) which is carried to the anterior lobe of the pituitary gland by the pituitary portal circulation. It induces the release of ACTH which acts on the adrenal cortex to increase the output (and biosynthesis) of glucocorticoids and consequently of adrenal androgens. The resulting increase in plasma glucocorticoid levels is in turn monitored by the hypothalamic receptors with subsequent reduction in CRF, ACTH and cortisol secretion. The feedback mechanism ensures that the level of cortisol is maintained within relatively narrow limits.

Stress and the hypothalamo-pituitary-adrenal axis. The hypothalamo-pituitary-adrenal axis is intimately involved in maintaining homeostasis during times of 'stress'. This is a non-specific term and physiological adaptation to 'stress' is a complex and ill-understood phenomenon. The normal subject under stress is capable of generating an increased output of steroid which, in an ill-defined manner, allows rapid physiological adjustment to the situation which might be caused by infection, emotion or trauma (including surgery). This response to stress overrides the normal feedback mechanism and also the circadian variation in steroid secretion. Adrenalectomized animals are unable to maintain their body temperature in a cold environment. Patients with untreated adrenocortical insufficiency are unduly sensitive to the effects of stressful situations and often have gross functional gastrointestinal abnormalities and low blood-pressure. Unless such persons are treated with glucocorticoids and mineralocorticoids, the outcome may be serious or even fatal. The stress response is therefore the second of the additional mechanisms controlling steroid release and is considered in many instances to originate at the level of the cerebral cortex. Neurohumoral mechanisms stimulate CRF release from the hypothalamus with an eventual increase in the plasma steroid level as a result of the increased ACTH output.

The metabolism of cortisol. The major site for glucorticoid breakdown is the liver. There are three main steps in cortisol metabolism: saturation of the A ring, hydroxylation at the C-20 position and cleavage of the side chain at C-17. Saturation of the double bond at the C-4 to C-5 position in the A ring and reduction of the ketone group at C-3 results in *tetrahydrocortisol* (THF). Cortisol and cortisone are interchangeable in the liver and the metabolism of cortisone produces *tetrahydrocortisone* (THE). Both compounds are made water-soluble by conjugation with glucuronic acid at the C-3 position to facilitate excretion by the kidney. Hydroxylation at the C-20 position of THF and THE before conjugation allows further metabolism of a proportion of these compounds to *cortol* and *cortolone* respectively. A small amount of these 20-hydroxylated compounds is further metabolized by side-chain cleavage to form 11-hydroxy-17-oxosteroids. Less than 1 per cent of cortisol is excreted free and unchanged and a similar amount is excreted as 6β-hydroxy-cortisol. A schematic representation of the breakdown is shown in Figure 50.14. The

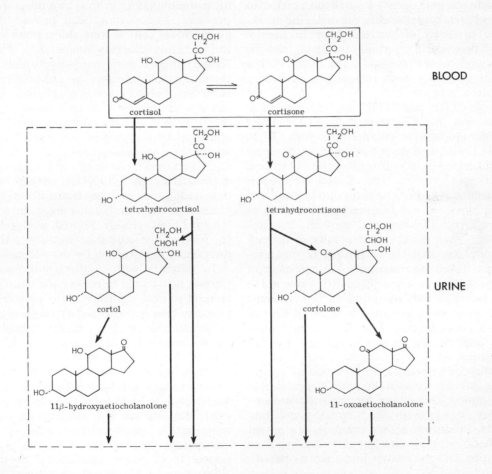

Fig. 50.14 The metabolic breakdown products of cortisol.

relative proportions of metabolites may be considerably altered by the administration of drugs such as phenobarbitone, phenytoin, phenylbutazone and progestogens which are known to induce enhanced enzymatic activity at selected sites.

Rough estimates of adrenocortical activity may be made by the estimation of urinary metabolites (Fig. 50.14). Apart from a few specialist centres where individual metabolites may be measured, most laboratories rely for routine purposes on group analysis. The nomenclature of urinary steroids is confusing but the following account may serve to distinguish between the two main groups: 17-hydroxycorticosteroids and 17-oxosteroids (ketosteroids). 17-Hydroxycorticosteroids are principally the urinary metabolites of cortisol and 17-oxosteroids consist of metabolites of the adrenal androgens, a small proportion of cortisol metabolites and, in the male, the breakdown products of the testicular hormones. 17-Hydroxycorticosteroids in urine can be measured by two methods, the Porter-Silber technique which measures compounds with a dihydroxyacetone side chain at C-17 for example THF and THE, or after chemical conversion to 17-oxosteroids. This latter method measures compounds with 17,20 ketol (17α-hydroxyprogesterone), 17,20 glycol (pregnanetriol), dihydroxyacetone (THE and THF) and glycerol (cortols and cortolones) side chains at C-17. The compounds measured by this method are frequently referred to as 17-oxogenic (17-ketogenic) steroids. It is important when interpreting results of urinary steroid assay that the technique of measurement is taken into account. Measurement of 17-oxosteroids is normally of little value in the assessment of adrenal function. It is of some value when there is a gross increase in adrenal androgen production. The most potent androgen, testosterone, is *not* measured by this technique.

Synthetic steroids, with modified structures, are much used in clinical practice in replacement therapy or for their anti-inflammatory or non-specific 'stress'-controlling effects. Prednisolone, for example, is a potent synthetic steroid that causes less sodium retention and potassium loss than its parent, cortisol.

Mineralocorticoids. Hypophysectomized

Prednisolone

animals do not require sodium chloride in order to survive and there is little or no atrophy of the glomerulosa cells in the adrenal cortex. Thus it was long suspected that a sodium-retaining hormone, independent of pituitary control, might be secreted by these cells. This hormone is the powerful salt-retaining steroid of the adrenal cortex, aldosterone (Fig. 50.15), which produces a measureable effect in man at a dose of 100 µg per day. It has now been proved that the glomerulosa cells secrete aldosterone and that this secretion is largely independent of ACTH. Thus patients with panhypopituitarism excrete normal amounts of aldosterone. The secretion of aldosterone is governed by changes in extracellular fluid volume, by alterations in the volume of the vascular compartment; it is increased by decrease in plasma sodium or by an increase in plasma potassium. A peptide *renin*, secreted by cells of the juxtaglomerular apparatus in the kidney has enzymic properties and catalyses the transformation of a plasma protein angiotensinogen into angiotensin I and closely related angiotensin II (p. 659). This substance stimulates the production of aldosterone by the adrenal cortex.

In patients with Addison's disease aldosterone is not able to restore the ability of the kidneys to deal efficiently with a water load, a property that is possessed by cortisol.

Aldosterone is present in normal blood plasma to the extent of some 2 to 15 ng per 100 ml and is eliminated in the urine in amounts of about 5 to 25 µg per day. It should be noted that methods for the measurement of aldosterone reduction products in urine do not exist, and that the free aldosterone measured represents a small fraction only of an administered dose of the hormone. For this reason it is more satisfactory to measure aldosterone secretion rates. A tracer dose of tritium labelled aldosterone is injected into a

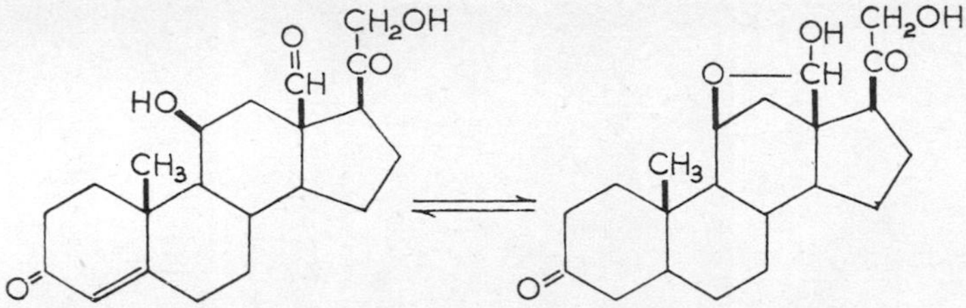

Aldosterone

Fig. 50.15 In solution the aldehyde form (left) of aldosterone appears to be in equilibrium with the hemiacetal form (right).

vein and urine is collected for the subsequent 24 hours. The urine is hydrolysed and the free aldosterone released is extracted and purified until its specific activity (S.A.) is constant. This S.A. is less than that of the injected steroid on account of dilution by the hormone secreted in the body and the amount secreted is readily calculated. The secretion rate is usually observed with the subject on a high and on a low sodium chloride intake in order to suppress and to stimulate aldosterone production respectively. Normal results are 300 to 1000 μg/24 h with salt restriction and 0 to 150 μg/24 h for a high salt diet. The urinary excretion of aldosterone is increased in many oedematous states such as chronic cardiac,

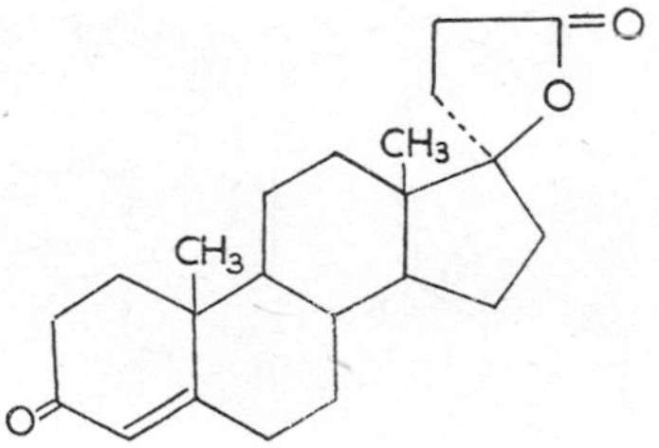

Spironolactone

hepatic or renal disease (secondary aldosteronism), in patients with certain tumours of the adrenal cortex (primary aldosteronism or Conn's syndrome) and when the daily intake of sodium falls below about 10 milli-equivalents per day.

There is an obvious need for aldosterone 'inhibitors' for therapeutic use as diuretics in oedema. Such a substance is the synthetic steroid spironolactone. This apparently competes with aldosterone at the level of the kidney tubule blocking sodium retention and potassium excretion.

Androgens. The masculinizing properties of the adrenal androgens are weak. Although small amounts of testosterone are elaborated by the adrenal, these are relatively insignificant compared with the amount produced by the testes. However, raised circulating testosterone and androstenedione levels may be found in virilized women (Fig. 50.16). The physiological significance of dehydroepiandrosterone sulphate is uncertain although quantitatively it is the most important steroid secreted by the adrenal. In general, therefore, the importance of the androgens is unknown but they have anabolic properties and may be involved in nitrogen retention.

Assessment of cerebro-hypothalamo-pituitary-adrenocortical function. Measurements of urinary 17-oxosteroids and 17-hydroxycorticosteroids give only a rough indication of adrenocortical activity. The measurement of the cortisol secretion rate by an isotopic dilution technique gives the best measure of basal adrenocortical function. The development of rapid simple techniques for the measurement of plasma 11-hydroxycorticosteroids (11-OHCS), of which in the normal subject the principal component is cortisol, has been of great value in assessing the response of the adrenals during dynamic function tests. The adrenal cortex itself may be directly stimulated by an intramuscular injection of ACTH or a synthetic polypeptide (tetracosactrin) which contains the amino acids 1 to 24 that occur in natural ACTH. The rise in plasma 11-OCHS over the baseline value is an indication of adrenocortical re-

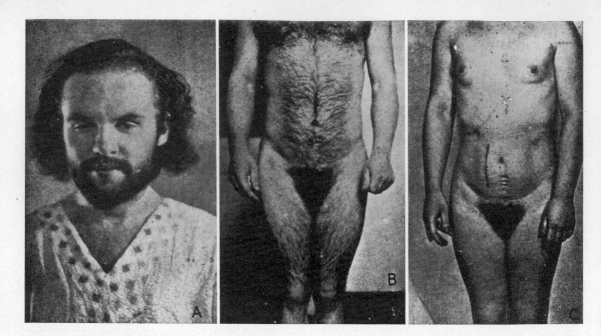

Fig. 50.16 Adrenogenital syndrome in a female patient, aged 17, with an adrenal cortical tumour. A and B were taken before operation; C, two years and ten months after removal of the tumour. (F. D. W. Lukens & H. D. Palmer (1940). *Endocrinology* **26**, 941.)

sponsiveness. Before embarking on provocative tests of the cerebro-hypothalamo-pituitary-adrenocortical (CHPA) axis, normal adrenocortical responsiveness to ACTH or a synthetic analogue must be ensured since all the other tests of CHPA function depend on an intact end-organ (adrenal cortex). These tests are at present indirect, relying on the response of the adrenal cortex to reflect changes in the output of trophic hormones at the higher levels. It is hoped that in the near future, the introduction of reliable methods of ACTH measurement will allow more direct assessment of the function of the higher centres of the CHPA axis.

The 'stress' response may be simulated by a number of procedures of which the most commonly used is the induction of hypoglycaemia by intravenous administration of insulin. Hypoglycaemia stimulates the entire cerebro-hypothalamo-pituitary-adrenal axis with a consequent rise in the plasma 11-OHCS levels. Patients with a lesion of the hypothalamus or pituitary gland fail to respond to insulin hypoglycaemia although the response to ACTH is normal.

The feedback mechanism may be tested by administering to the patient small amounts of a highly potent synthetic glucocorticoid such as dexamethasone. This steroid is used in such

Dexamethasone

small dosage that it is not detected in plasma 11-OHCS or urinary 17-OHCS measurements. In the normal subject, the plasma 11-OHCS level falls after dexamethasone indicating normal feedback responsiveness. While this test measures the ability of the hypothalamus to respond to high plasma steroid activity, the response to low levels can be investigated by the administration of metyrapone. Metyrapone inhibits the action of

Metyrapone

Fluorocortisol

11β-hydroxylase and when administered in doses of 4·5 g per day, inhibits the formation of cortisol. The immediate precursor, 11-deoxy-cortisol, is secreted by the adrenal cortex. This compound fails to inhibit CRF and ACTH secretion which in normal subjects are then secreted in excess, stimulating the adrenal cortex to produce even more 11-deoxycortisol, the urinary metabolites of which are similar to those of cortisol and are also measured with other 17-hydroxycorticosteroids (17-OHCS). A rise of 10 mg per 24 hours of 17-OHCS output over the baseline value is considered to indicate normal hypothalamo-pituitary function provided the baseline value is within normal limits. Patients with hypothalamic or pituitary disease fail to respond to the decreased level of cortisol and so the output of 11-deoxycortisol is not increased.

Hypo- and hyper-function of the adrenal cortex. Destruction of the adrenal glands by disease in man (Addison's disease) is the chronic counterpart of the acute disorder produced in animals by the removal of both adrenals. The chief symptoms described by Addison in 1855 are 'anaemia, general languor and debility, remarkable feebleness of the heart's action, irritability of the stomach and a peculiar change in the colour of the skin'. These symptoms are accompanied by a low blood pressure, loss of weight, hypoglycaemia, extreme sensitivity to insulin, excessive loss of sodium and chloride in the urine and an impaired elimination of water by the kidneys. Patients with Addison's disease can usually be kept in good health provided they take some 12·5 to 25 mg of cortisone daily by mouth. In many patients the powerful synthetic salt-retaining compound 9α-fluorocortisol must also be given by mouth in doses varying between 0·1 mg on alternate days to 0·2 mg daily

It is remarkable and convenient for patients that cortisol, cortisone and their analogues used in treatment are effective orally. Many other steroid hormones, including aldosterone,

have to be administered by injection to avoid inactivation by the liver.

Hyperfunction of the adrenal cortex may affect any or all of its hormones and may be due either to functional overactivity of the secreting tissue, adrenal hyperplasia, or to a tumour and a number of different syndromes can be recognized. For example, Cushing's syndrome is the result of over-production of cortisol and many of its characteristics can, in fact, be produced by administration of large doses of cortisone or cortisol; obesity, glycosuria, hirsutism and osteoporosis are its main features.

THE ADRENAL MEDULLA

The adrenal medulla, which has a common origin with the ganglion cells of the sympathetic nervous system from the primitive neuro-ectoderm, consists of masses of polyhedral cells separated by large blood sinuses.

The cells contain granules stained blue by ferric chloride or brown by salts of chromic acid (*chromaffin tissue*). Chromaffin tissue is also present, especially in infants, along the aorta near the origin of the inferior mesenteric artery (organ of Zuckerkandl) and chromaffin cells are present in human skin, scattered around nerves and blood vessels, and throughout the alimentary canal (enterochromaffin cells). The adrenal medulla is richly supplied by a plexus of sympathetic nerve filaments derived mainly from the greater splanchnic nerve. The efferent fibres are myelinated pre-ganglionic fibres, the cells of the medulla representing the sympathetic ganglion cells and their post-ganglionic fibres. Afferent impulses have been detected by Niijima and Winter in small filaments of the adrenal nerve. The spontaneous rate of firing decreased after intravenous injection of adrenaline, noradrenaline or acetylcholine or stimulation of the splanchnic nerve but a rise of general blood pressure had no effect on the rate of firing. The location and function of these chemosensitive cells are not known.

The active principles of the medulla are easily extracted and when injected into animals they produce a sharp rise of blood pressure. It is not surprising, therefore, that this was one of the first hormonal effects to be discovered. In 1894 Schäfer and Oliver injected a watery extract of the adrenal into a dog and obtained an immediate large rise of blood pressure which was, however, transient (Fig. 50.17). It was later found that this property belonged to the medulla only and in 1900 one active principle was isolated and named *adrenaline* (or epinephrine). Much later it was found that normal adrenal tissue from man and many other animals contains *noradrenaline* (nor-epinephrine or arterenol) as well as adrenaline

and they are often referred to together as *catechol amines*. The ratio of adrenaline to noradrenaline in fresh human adrenals is 4:1.

adrenaline

noradrenaline

Adrenaline has an asymmetric carbon atom (**C**) and therefore exists in two optically active

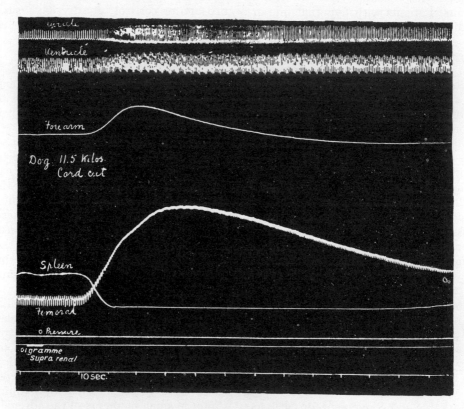

Fig. 50.17 Reproduction of an original figure of G. Oliver & E. A. Schäfer (1895). *Journal of Physiology* **18**, 246, to show the effect of an extract of the adrenal in increasing the rate of the heart, raising the arterial pressure, causing decrease of splenic volume and increase in the volume of the forelimb. Time marks every 10 sec.

forms. That prepared from the gland is L-adrenaline which is some fifteen times more active in raising the blood pressure of the spinal cat than is the D-isomer. Adrenaline in solution is very readily oxidized and inactivated but blood and tissue fluids contain reducing agents, such as glutathione and ascorbic acid, which can protect it from rapid oxidation. Adrenaline and noradrenaline are metabolized in the tissues in two stages (1)

methylation by catechol-O-methyltransferase to the corresponding 3-methoxy derivatives, which are pharmacologically inactive and (ii) oxidative deamination by monoamine oxidase of these derivatives under the influence of monoamine oxidase (p. 116) to yield 3-methoxy-4-hydroxymandelic acid which is excreted in the urine (Fig. 50.18). This second step is inhibited by the monoamine oxidase inhibitor iproniazid (1-isonicotinyl-2-

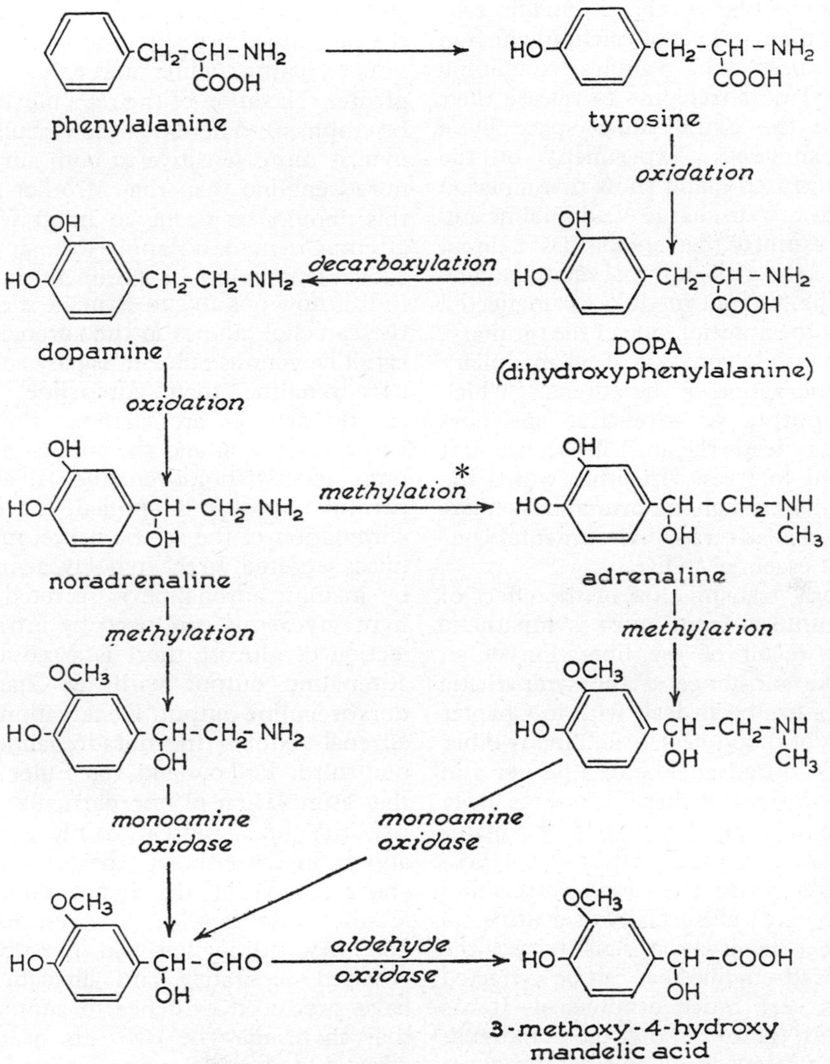

Fig. 50.18 Biosynthesis and breakdown of adrenaline and noradrenaline.
* The methyl group comes from S-adenosyl methionine.

isopropyl hydrazine), a drug which is used in the treatment of certain mental diseases (p. 116).

The steps by which adrenaline is synthesized by the chromaffin tissue are outlined in Fig. 50.18. Experiments on rats with the isotope ^{13}C have shown that phenylalanine, which can be converted to tyrosine (p. 385), is the precursor of both noradrenaline and adrenaline in the adrenal glands.

The medullary secretion is under the control of the splanchnic nerves which are cholinergic, that is, they act by liberating acetylcholine (ACh) from their endings in the medulla. ACh depolarizes the chromaffin cells and increases their uptake of calcium which in some way causes the granules containing adrenaline and noradrenaline to release their contents into the extracellular space by a process of exocytosis. Experiments on the denervated perfused gland show that a period of anoxia leads to a discharge of adrenaline and it may be presumed that anoxia has a direct effect on the cells of the adrenal medulla. The function of the adrenal medulla is unaffected by removal of the anterior lobe of the pituitary.

Neither complete removal of all medullary tissue nor denervation of the adrenals, which reduces the output of adrenaline and noradrenaline, handicaps the animal seriously if it is not exposed to stress. In other words the adrenal medulla, though important in an emergency and under adverse environmental conditions, is not essential to life.

The humoral transmission of the effect of the nerve impulse from most sympathetic fibres as the result of the liberation of an adrenaline-like substance at the sympathetic nerve-endings has been dealt with in Chapter 41. The walls of blood vessels and many other tissues contain noradrenaline (0·5 µg per g in the rabbit's aorta) and it therefore seems likely that noradrenaline plays a part in the maintenance of vascular tone. After sympathectomy, at a time when the blood vessels and other tissues are abnormally sensitive to noradrenaline ('denervation sensitivity'), the amount of noradrenaline that can be extracted from them is very much diminished. It has been suggested, therefore, that the noradrenaline normally present in many tissues reduces their sensitivity to this material by occupying receptors; after sympathectomy the noradrenaline store diminishes and many receptors are left free so that a relatively small dose of noradrenaline can now make the tissue react.

The effects of injection of adrenaline or noradrenaline are dealt with in more detail on pp. 866–868 and summarized in Table 41.5.

Since pure adrenaline and noradrenaline are readily available it is unnecessary to have a biological method for their standardization. In physiological research, however, it may be desirable to determine the amounts of noradrenaline and adrenaline in an extract or perfusate. This can be done by using organs which respond differently to the two substances ('differential bioassay'). For example, the cat's blood pressure is more easily raised by noradrenaline, while adrenaline produces a greater relaxation of the rat's uterus. It should be emphasized here that the vascular system of man is more sensitive to both adrenaline and noradrenaline than that of other species and this should be borne in mind whenever an attempt is made to apply to man conclusions drawn from animal experiments.

It is now possible to estimate the amounts of the catechol amines in the adrenal vein of the cat. The venous effluent usually contains more noradrenaline than adrenaline. When the carotid arteries are clamped there is reflex vasoconstriction and the adrenal effluent contains mostly noradrenaline. If, however, a painful stimulus is applied, for example by stimulation of the sciatic nerve, more adrenaline is secreted. In the hypoglycaemia produced by insulin, adrenaline is secreted but in the hyperglycaemia produced by intravenous injection of glucose there is a reduction of the adrenaline output with no change in the noradrenaline output. Denervation of the cat's adrenal reduces the noradrenaline output to one-third. Folkow and von Euler have found that stimulation of one particular area of the hypothalamus causes mainly adrenaline to appear in the effluent whereas stimulation of another area of the hypothalamus releases mainly noradrenaline. All these findings suggest that adrenaline and noradrenaline are released separately and histochemical tests have produced evidence to support the idea that there may be two sets of cells in the adrenal medulla, one responsible for secreting adrenaline and the other for secreting noradrenaline.

The excretion of the catechol amines (Table

50.19) provides further information about the activity of the adrenal medulla since 2 to 3 per cent of injected catechol amines are excreted in the urine. The absence of adrenaline after

Table 50.19 Urinary excretion of catechol amines in man (U.S. von Euler (1955), *Lancet* **ii**, 151–154; J. P. Shillingford (1967). *Lancet* **i**, 127.)

Condition	Urinary excretion in ng/min	
	Adrenaline	Noradrenaline
Rest	2–5	16–27
Mild exercise	2 times	2 times
Strenuous exercise	10 times	10 times
Mental stress	Increased	No effect
Insulin hypoglycaemia	10 times	No effect
Tilt, recumbent to head up	Little change	Increased 8–15
After adrenalectomy	0	8–15
Chromaffin cell tumours	Slightly increased	100 times or more
Myocardial infarction	Up to 35 times	Up to 8 times

adrenalectomy suggests that the urinary adrenaline comes from the medulla; the urinary noradrenaline output continues and must therefore, come from extramedullary sources such as the adrenergic nerves. In strenuous exercise the increase in output is correlated with fatigue rather than with the amount of work.

On the basis of these findings von Euler makes a contrast between the two hormones as follows. Adrenaline, by producing vasoconstriction in the skin and vasodilatation in the muscles, moves blood from the skin to the muscles; it stimulates metabolism and mobilizes glycogen as glucose. These preparations are valuable for 'fight or flight'—that is the emergency function as suggested by Cannon. Noradrenaline on the other hand produces a general vasoconstriction (except in the coronary arteries) but its effects on metabolism are inconspicuous. Noradrenaline may be regarded as a pressor hormone required for maintenance of blood pressure.

The amount of catechol amines in the blood can be assessed by fluorimetry. Human blood plasma contains about 0·3 µg of noradrenaline per litre and 0·06 µg of adrenaline per litre. Healthy man excretes not more than 50 µg/ 24 h of catechol amines and adrenaline accounts for less than 15 per cent of this.

Hypofunction of the adrenal medulla has not been recognized as producing any clear-cut clinical syndrome. Actively secreting tumours (*phaeochromocytomata*) of the chromaffin cells of the adrenal medulla or of the organs of Zuckerkandl are known to produce large quantities of noradrenaline with a small amount of adrenaline, and their discharge at intervals into the blood may produce elevations of arterial pressure. As much as 2 mg of catechol amines may then appear in the urine per day.

THE THYMUS

The thymus consists of a peripheral cortex densely packed with lymphocytes and a central medulla in which the lymphocytes are less numerous. The medulla also contains reticular cells probably belonging to the reticuloendothelial system and the unique 'Hassall's corpuscles', spherical structures composed of concentric layers of spindle-shaped cells whose function is unknown.

The thymus in mammals arises from the endoderm of the third and fourth branchial clefts and usually consists of two or more lobes on each side of the midline. In the evolutionary development of the immune response, the thymus appears at the same time as immunological activity and it is the tissue in which during intrauterine life lymphocytes can first be recognized.

At birth the thymus is the major lymphoid organ, the spleen and lymph nodes being undeveloped; in a young rat the thymus produces about two million lymphocytes per hour, many more than are needed for the growth of the gland. The thymus therefore contributes considerably to the lymphocytes found in the peripheral blood.

Two types of lymphocytes have been described recently. One type, thymus-dependent (T-lymphocyte), is involved in immunological reactions of the cellular type such as rejection of a foreign-tissue graft (allograft reaction). The T-lymphocyte is derived probably from a stem cell in the bone marrow, migrates to the thymus and is 'processed' there to become 'immunologically competent'. When thymectomy is performed in neonatal mice the

T-lymphocytes disappear and the mice develop a wasting disease, fail to reject allografts and show other signs of loss of immunological competence. Thymectomy in adult mice does not have the same effects. There is evidence that the thymus produces a diffusible factor which is important in the immunological competence of the lymphocytes.

The other type of lymphocyte is not thymus-dependent (B-lymphocyte) and is derived from cells in the bone marrow. The B-lymphocyte is involved in antibody production, becoming a plasma cell when stimulated by an antigen. B-lymphocytes disappear in the chicken following excision of the bursa of Fabricius and antibody production falls after bursectomy. In mammals the equivalent of the bursa may be the tonsils, lymphoid tissue of the intestines and appendix.

An abnormal thymus is often found in adults with the disease *myasthenia gravis* in which the skeletal muscles are weak and there is a derangement of neuromuscular transmission. Surgical removal of the large thymus may produce striking improvement; the reason for this is not fully understood but is probably

tamine-N-methyl transferase which inactivates it.

Injection of melatonin into female rats decreases the weight of their ovaries: pineal tumours in man are associated with delayed sex development. Removal of the pineal in rats increases ovarian activity. If a rat is exposed to light melatonin synthesis is reduced and ovarian activity is increased. The afferent pathway of this effect begins in the eyes—the efferent is likely to be in the sympathetic supply of the pineal which comes from the superior cervical ganglion.

'LOCAL HORMONES'

Great interest is being shown at the present time by physiologists and pharmacologists in substances which are released locally and produce local effects, such as vasodilatation or vasoconstriction, in the surrounding tissue. The so-called local hormones include histamine, 5-hydroxytryptamine, bradykinin, adrenaline, noradrenaline and acetylcholine. The amines in this list are examples of the *biogenic amines*.

$$ \underset{\text{histidine}}{\underset{\displaystyle HN\diagdown_{\underset{H}{C}}\diagup N}{CH=C-CH_2-CH-COOH \atop NH_2}} \quad \xrightarrow{\;-CO_2\;} \quad \underset{\text{histamine}}{\underset{\displaystyle HN\diagdown_{\underset{H}{C}}\diagup N}{CH=C-CH_2-CH \atop NH_2}} $$

related to the hormone thymin which depresses neuromuscular transmission. It has been suggested that there is an excess production of thymin by the abnormal thymus in myasthenia gravis (see also p. 835).

THE PINEAL BODY

The pineal body, often visible in X-rays of the skull because of its tendency to calcify, is a small structure about 8 mm long and 4 mm wide lying on the dorsal aspect of the superior colliculi. The gland contains much serotonin (5-HT) which is converted by the enzyme hydroxyindole-O-methyl transferase to melatonin (N-acetyl-5-methoxytryptamine). Melatonin is the most potent substance yet found to constrict the melanocytes of the frog's skin. The gland also contains histamine and his-

Histamine. Large amounts of histamine are present in the lumen of the intestine due apparently to decarboxylation of histidine by bacteria. It is present in many tissues in a combined inactive form. There is histamine in the skin, in skeletal muscle, in nerve fibres, in the wall of the intestine, and in the liver and lung. The submucosa and muscular coats of the small intestine contain relatively large amounts of histamine but it is difficult to account for this by absorption since the intestinal mucosa contains histaminase. In several animals there is a good correlation between the histamine content of a tissue and the number of mast cells (basophil leucocytes) it contains but these cells are not the only source of histamine. The mast cells also contain heparin which is released along with histamine when the animal is injected with a histamine liberator. The effects of histamine on the skin (triple response p. 569 and itching p. 719) and on gastric secretion (p. 260) have already been described. Urticaria, or nettle-rash, may be presumed to be due at least partly to a release of histamine and there is no doubt that histamine increases the passage of particulate matter across capillary walls and causes the membrane to become permeable to protein molecules and water.

The physiological role of histamine has long been mysterious but some light has been shed on it by Kahlson and his colleagues who are more concerned with the histamine-forming capacity of the tissues than with their histamine content. The body acquires histamine from the metabolism of its cells and by absorption from the intestine. Kahlson has found that germ-free rats fed a histamine-free diet excreted as much histamine as ordinary rats on the same diet and that the histamine content of the tissues of the two groups was the same. The males excreted histamine mainly as biologically inactive methyl histamine, the females in the free, biologically active, form. The excretion was greatly increased in pregnancy and promptly fell if the fetuses were removed; the fetal liver is the main source of histamine. Similar findings have been obtained in human pregnancies. The placenta contains histaminase which may protect the mother from histamine produced by the fetus. Arrest of the formation of histamine by the embryo leads to arrest of growth and death of the embryo. Granulation and wound tissue in the rat form large amounts of histamine but contain very little. Artificial increase of extracellular histamine does not increase wound healing. In general, histamine formation has not been proved to be characteristic of growing tissues. While the physiological role, if any, of histamine is still an enigma, its participation in the anaphylactic response and related syndromes is undeniable.

Hydroxytryptamine. The serum obtained after blood clots has vasoconstrictor properties due to 5-hydroxytryptamine (5-HT, serotonin). 5-HT is released from the blood platelets when they disintegrate during clotting and may assist haemostasis. Human blood contains 0·1 to 8·0 µg of 5-HT per ml, nearly all of which is in the platelets (50 µg 5-HT per g of platelets). When blood platelets are in-

tryptophan

5-hydroxytryptophan

5-hydroxytryptamine (serotonin), 5-HT

5-hydroxyindoleacetic acid, 5-HIAA

cubated with 5-HT they take up appreciable amounts. Much of the 5-HT in blood seems to be derived from the argentaffin cells of the gastrointestinal tract since the concentration is much lower after large parts of the intestine have been removed at operation. The intestinal mucosa contains about 6 µg per g of 5-HT. The brain contains about 0·2 µg per g. Stimulation of some peripheral nerves, including the vagus, leads to release of 5-HT from the brain.

5-HT is derived from dietary tryptophan by preliminary hydroxylation to 5-hydroxytryptophan which is then decarboxylated. It is oxidized in the tissues to 5-hydroxyindoleacetic acid (5-HIAA) under the influence of monoamine oxidase. Some 5 to 10 mg of 5-HIAA are normally excreted in the urine every 24 hours but when there is a tumour of the argentaffin cells of the small intestine (carcinoid tumour) very much larger amounts of 5-HIAA are excreted in the urine and the concentration of 5-HT in the blood may also be raised.

A large number of antimetabolites of 5-HT are known, one of the most important being lysergic acid diethylamide (LSD). If 30 µg of LSD are taken by mouth there are mental disturbances and hallucinations which may be due to antagonism in the brain of the action of 5-HT. Monoamine oxidase inhibitors increase the amount of 5-HT in the brain; they are used in the treatment of depression.

Prostaglandins. The prostaglandins (PGs) are unsaturated fatty acid derivatives containing 20 carbon atoms. Their nomenclature is based on the parent saturated fatty acid prostanoic acid (Fig. 50.20) in which a five membered ring is formed by ring closure between carbon atoms 8 and 12. For example PGE is designated 9-keto-11α,15α-dihydroxy-prost-13-enoic acid. All the prostaglandins have unsaturated carbon chains and the different series are characterized by variations in the 5-membered ring. For example prostaglandins of the E series have a keto group on position 9 while those of the F series have a hydroxyl group at this position (Fig. 50.20). All the prostaglandins are formed in the animal body by ring closure and oxygenation of essential polyunsaturated fatty acids such as arachidonic acid (p. 62).

Although originally isolated from human seminal plasma and sheep seminal vesicles PGs occur naturally in the tissues of many organs such as brain, heart, lung, kidney, stomach, gut and iris, as well as uterus, placenta and testis.

The fatty-acid precursors seem to be stored as part of the phospholipids of all membranes. PGs are rapidly synthesized from these precursors and are released at their site of action. They are rapidly inactivated in tissue by 15-hydroxy-prostaglandin dehydrogenase (PGDH) and Δ^{13} prostaglandin reductase and also by oxidation in the liver. The concentrations of plasma PG are therefore low.

Although PGs have been shown to exert many striking pharmacological actions on all the main systems of the body their physiological roles are still uncertain. In the repro-

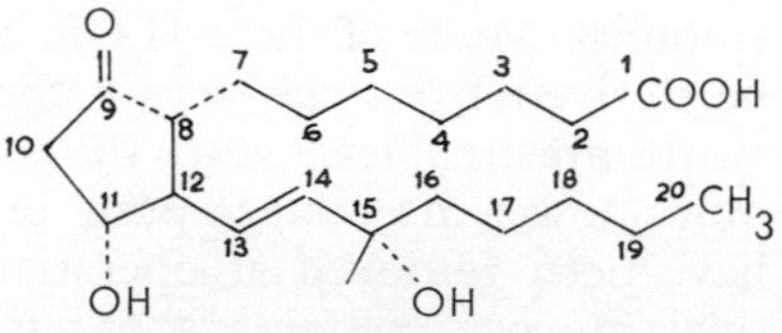

PGE$_1$

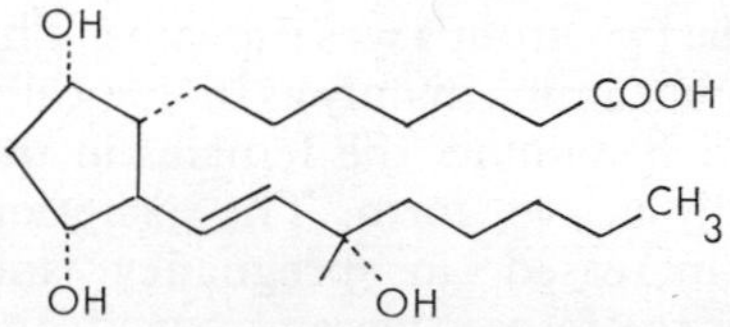

PGF$_{1\alpha}$

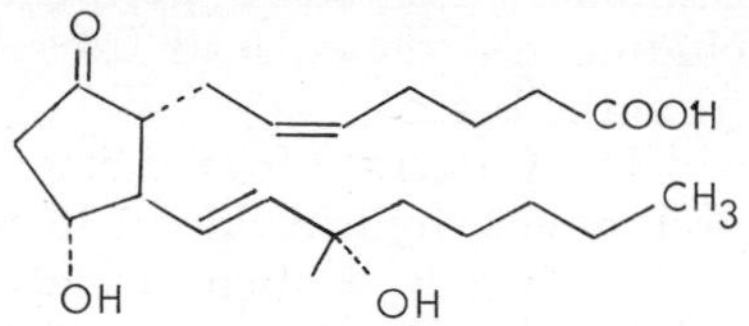

PGE$_2$

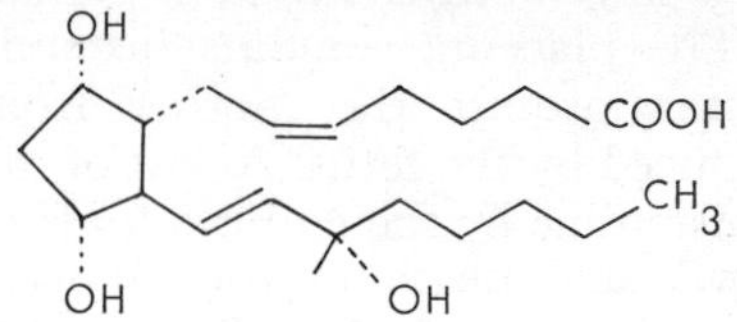

PGF$_{2\alpha}$

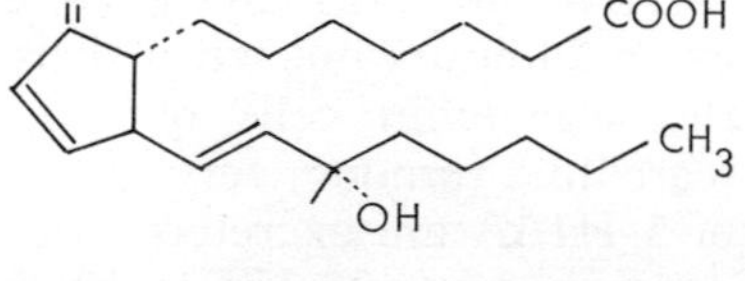

PGA$_1$

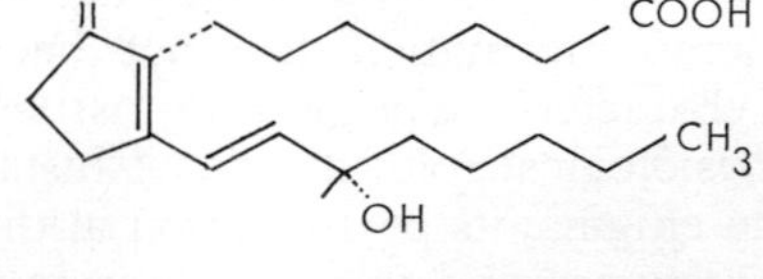

PGB$_1$

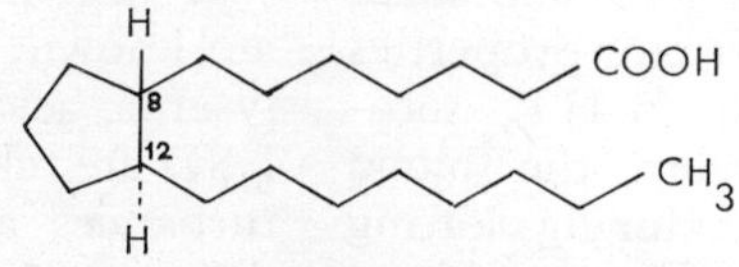

Prostanoic acid

Fig. 50.20 Structure of the prostaglandins.

ductive system PGE_1 and PGE_2 increase the contraction of the pregnant uterus, but inhibit the motility and tone of the non-pregnant uterus. PGE_1 and $PGF_{2\alpha}$ are given intravenously and intravaginally to induce therapeutic abortion. PGF_2 increases the motility of both sperm and uterine tubes.

In the gastrointestinal system PGs increase motility and diminish gastric acidity. They increase renal blood flow, inducing a sodium and water diuresis, but also redistribute blood from medulla to cortex, thus antagonizing renin production. They also antagonize the action of vasopressin and increase the motility of bladder and ureters.

PGs of different groups exert opposing effects on the cardiovascular, respiratory and haematological systems. Thus, while they exert positive inotropic and chronotropic effects (Chap. 26) on the heart and increase capillary permeability PGEs and PGAs reduce the blood pressure in systemic arteries and veins while PGFs increase it. PGEs and PGAs increase regional arterial blood flows, including the coronary flow, but PGFs diminish regional arterial flow. Bronchial dilatation is produced by PGE, and $PGE_{2\alpha}$ but bronchial constriction by $PGF_{2\alpha}$. PGE_1 inhibits platelet aggregation, while PGE_2 stimulates it.

PGs even inhibit spinal reflexes and the release of noradrenaline after stimulation of sympathetic nerves. They can cause contraction of the iris (irins) and cutaneous hyperaemia. They seem to act as local regulators of hormonal activity *in vitro*; they increase the levels of cyclic AMP and the synthesis of hormones in many endocrine organs, and in the target tissues they can modulate the activities of many hormones or drugs mediated by cyclic AMP.

Evidence is accumulating that PGs are mediators of inflammation, and some anti-inflammatory drugs, such as aspirin, have been shown to inhibit the biosynthesis of PG *in vitro*. It has recently been suggested that prostaglandins play a part in the hypothalamic control of body temperature.

Kinins. Like the angiotensins kinins are vasoactive polypeptides liberated from plasma proteins by endogenous enzymes. The term 'kinin' is a general one indicating a polypeptide that causes hypotension, contracts most isolated smooth muscle but relaxes rat duodenum.

Kinins increase vascular permeability, cause pain when applied to the exposed base of a blister on human skin, and produce bronchoconstriction in the guinea-pig. The plasma substrate is termed a 'kininogen', the proteolytic enzymes that release the kinins are termed 'kininogenases' and those that inactivate them are known as 'kininases'. According to this recently recommended terminology trypsin, pepsin and some snake venoms are all kininogenases but endogenous enzymes that rapidly and specifically liberate a kinin from kininogens in the plasma are termed 'kallikreins' because they were found in extracts of pancreas.

The structures of two naturally-occurring kinins, *bradykinin* and *kallidin* were elucidated about ten years ago and subsequently they, and over 170 analogues, have been synthesized. Bradykinin is

Arg–Pro–Pro–Gly–Phe–Ser–Pro–Phe–Arg

and was so named because guinea-pig ileum responded more slowly to it than to histamine or acetylcholine.

The kinins cause hypotension on systemic administration and an efflux of water from small blood vessels on local administration. The fall in systemic arterial pressure is associated with a rise in cardiac output. The decrease in peripheral resistance and increased blood flow are due to a direct relaxing action of kinins on vascular smooth muscle. The efflux of water from systemic capillaries is probably the result both of an increase in transmural hydrostatic pressure and a decrease in transmural colloid-osmotic-pressure.

Hilton has suggested that kinins act as mediators of functional vasodilatation in various glands. For example when the trunk of a subject is heated enough to cause sweating of the exposed forearm the kinin concentration of subcutaneous fluid in the arm is increased five-fold—enough to account for the active dilatation of skin vessels.

Large amounts of kininogen are present in man so that there are potentially large amounts of plasma kinins. Plasma kinins are, however, only part of a complex system, being rapidly inactivated after their release and broken down to products indistinguishable from other constituents of the body. They are of great pharmacological interest but their physiological role is still uncertain.

REFERENCES

ADAMS, D. D. (1965). Pathogenesis of the hyperthyroidism of Grave's disease. *British Medical Journal* i, 1015–1020.

AURBACH, G. D. & POTTS, J. T. (1967). Parathyroid hormone. *American Journal of Medicine* **42**, 1–8.

BELL, E. T. & LORAINE, J. A. (Eds.) (1967). *Recent Research on Gonadotrophic Hormones.* Edinburgh: Livingstone.

BERGSTRÖM, S. (1967). Prostaglandins: members of a new hormonal system. *Science* **157**, 382–391.

BERGSTROM, S., CARLSON, L. A. & WEEKS, J. R. (1968). The prostaglandins: a family of biologically active lipids. *Pharmacological Reviews* **20**, 1–48.

Biosynthesis and Secretion of the Adrenocortical Steroids (1960). Biochemical Society Symposia, No. 18. Cambridge University Press.

BUFFONI, F. (1966). Histamine and related amine oxidases. *Pharmacological Reviews* **18**, 1163–1199.

CAMERON, M. P. & O'CONNOR, M. (Eds.) (1964). *Brain–Thyroid Relationships.* Ciba Foundation Study Group No. 18. London: Churchill.

CATT, K. J. (1971). *An ABC of Endocrinology.* London: Lancet.

CHATTEN, J. (1964). The thymus in systemic disease. *American Journal of Medical Science* **248**, 715–727.

Ciba Foundation Colloquia on Endocrinology, Vols. 1–12. London: Churchill.

COLLIER, H. O. J. (1962). Kinins. *Scientific American* **207**, 111–118.

COPE, C. L. (1965). *Adrenal Steroids and Disease.* London: Pitman.

COUPLAND, R. E. (1965). *The Natural History of the Chromaffin Cell.* London: Longmans.

CURRIE, A. T., SYMINGTON, T. & GRANT, J. K. (Eds.) (1962). *The Human Adrenal Cortex.* Edinburgh: Livingstone.

DE GROOT, L. J. (1965). Current views on formation of thyroid hormones. *New England Journal of Medicine* **272**, 243–250.

DENTON, D. A. (1965). Evolutionary aspects of the emergence of aldosterone secretion and salt appetite. *Physiological Reviews* **45**, 245–295.

EICHLER, O., FARAH, A., HERKEN, H. & WELCH, A. D. (1970). Bradykinin, kallidin and kallikrein. *Handbook of Experimental Pharmacology,* Vol. 25, pp. 1–768.

FRYE, B. E. (1967). *Hormonal Control in Vertebrates.* London: Collier & Macmillan.

GANN, D. S. & EGDAHL, R. H. (1965). *Serotonin.* London: Elsevier.

GARATTINI, S. & VALZELLI, L. (1965). *Serotonin.* London: Elsevier.

GARDINER-HILL, H. (Ed.) (1958–67). *Modern Trends in Endocrinology.* Vols. 1–3. London: Butterworth.

GILLESPIE, J. S. (1966). *The Mode of Action of Catecholamines on Smooth Muscle* in Pickles, V. R. & Fitzpatrick, R. J. (Eds.), *Memoirs of the Society of Endocrinology,* Vol. 14. Cambridge University Press.

GOLDSTEIN, G. & MACKAY, I. R. (1969). *The Human Thymus.* London: Heinemann.

GRANT, J. K. (1968). The biosynthesis of adrenocortical steroids. *Journal of Endocrinology* **41**, 111–135.

GRANT, J. K. & HALL, P. E. (1971). Laboratory investigation of the human hypothalamic-pituitary-adrenocortical system. *Scottish Medical Journal* **16**, 157–167.

GRAY, C. H. & BACHARACH, A. L. (1967). *Hormones in Blood,* 2nd edn, Vols. 1 and 2. London: Academic Press.

HARRISON, T. S. (1964). Adrenal medullary and thyroid relationships. *Physiological Reviews* **44**, 161–185.

HOCH, F. L. (1968). Biochemistry of hyperthyroidism and hypothyroidism. *Postgraduate Medical Journal* **44**, 347–362.

HORTON, E. W. (1969). Hypotheses on physiological roles of prostaglandins. *Physiological Reviews* **49**, 122–161.

KAHLSON, G. & ROSENGREN, ELSA (1968). New approaches to the physiology of histamine. *Physiological Reviews* **48**, 155–196.

KARIM, S. M. M. (1971). Prostaglandins. *British Journal of Hospital Medicine* **5**, 555–563.

KARLSON, P. (1965). *The Mechanism of Hormone Action.* New York: Academic Press.

KARLSON, P. & SEKERIS, C. E. (1966). Biochemical mechanisms of hormone action. *Acta endocrinologica* **53**, 505–518.

LARAGH, J. H. & KELLY, W. G. (1964). Aldosterone; its biochemistry and physiology. *Advances in Metabolic Disorders* **1**, 218–262.

LEWIS, G. P. (1962). Bradykinin, biochemistry, pharmacology and its physiological role in controlling local blood flow. *Scientific Basis of Medicine Annual Review* 242–258.

LITWACK, G. & KRITCHEVSKY, D. (1964). *Actions of Hormones on Molecular Processes.* New York: Academic Press.

LORAINE, J. A. & BELL, E. T. (1966). *Hormone Assays and their Clinical Application,* 2nd edn. Edinburgh: Livingstone.

McGOWAN, G. K. & SANDLER, M. (Eds.) (1967). *Symposium on the Thyroid Gland.* London: published by *Journal of Clinical Pathology.*

MALMEJAC, J. (1964). Activity of the adrenal medulla and its regulation. *Physiological Reviews* **44**, 186–218.

MARTINI, L. & GANONG, W. F. (1966–67). *Neuroendocrinology,* Vols. 1, 2. London: Academic Press.

MILLER, J. F. A. P. (1964). Functions of the thymus. *Scientific Basis of Medicine Annual Review* 218–234.

MILLER, J. F. A. P. & OSOBA, D. (1967). Current concepts of the immunological function of the thymus. *Physiological Reviews* **47**, 437–520.

MILLS, I. H. (1964). *Clinical Aspects of Adrenal Function.* Oxford: Blackwell.

NAKANO, J. (1971). Prostaglandins and the circulation. *Modern Concepts Cardiovascular Disease* **40**, 49–54.

PAASONEN, M. K. (1965). Release of 5-hydroxytryptamine from blood platelets. *Journal of Pharmacy and Pharmacology* **17**, 681–697.

PAGE, I. H. (1969). *Serotonin*. Chichester: Wiley.

PARSONS, V. & RAMSAY, I. (1968). Thyroid and adrenal relationships. *Postgraduate Medical Journal* **44**, 377–384.

PICKFORD, MARY (1969). *The Central Role of Hormones*. Edinburgh: Oliver & Boyd.

PICKLES, V. R. & FITZPATRICK, R. J. (Eds.) (1966). Endogenous substances affecting the myometrium. *Memoirs of the Society of Endocrinology*. Cambridge University Press.

PINCUS, G. (Ed.) (1947–67). *Recent Progress in Hormone Research*, Vols. 1–23. New York: Academic Press.

PITT-RIVERS, R. & TROTTER, W. R. (1964). *The Thyroid Gland*. London: Butterworth.

PRUNTY, F. T. G. (Ed.) (1962). The adrenal cortex. *British Medical Bulletin* **5**, 89–173.

QUAY, W. B. (1965). Indole derivatives of pineal and related neural and retinal tissues. *Pharmacological Reviews* **17**, 321–345.

ROSS, E. J. (1960). *Aldosterone in Clinical and Experimental Medicine*. Oxford: Blackwell.

SAVIN, C. T. (1969). *The Hormones. Endocrine Physiology*. Boston: Little, Brown & Co.

SHARP, G. W. G. & LEAF, A. (1966). Mechanism of action of aldosterone. *Physiological Reviews* **46**, 593–633.

SPICKETT, S. G. (Ed.) (1967). *Endocrine Genetics. Memoirs of the Society of Endocrinology* No. 15. Cambridge University Press.

SUTHERLAND, E. N., OYE, I. & BUTCHER, R. W. (1965). The action of epinephrine and the role of the adenyl cyclase system in hormone action. *Recent Progress in Hormone Research* **21**, 623–646.

SYMINGTON, T. (1964). The pathological physiology of the human adrenal cortex and its relationship to hypercorticalism. *Scientific Basis of Medicine Annual Review* 15–31.

SYMINGTON, T. (1969). *Functional Pathology of the Human Adrenal Gland*. Edinburgh: Livingstone.

TATA, J. R. (1964). Basal metabolic rate and thyroid hormones. *Advances in Metabolic Disorders* **1**, 153–189.

TEPPERMAN, J. (1968). *Metabolic and Endocrine Physiology*, 2nd edn. London: Year Book Medical Publishers.

The thyroid gland (1960). *British Medical Bulletin* **16**, 92–169.

TROTTER, W. R. (1962). *Diseases of the Thyroid*. Oxford: Blackwell.

UNGAR, G. (Ed.) (1965). Physiological functions of histamine. *Symposium Federation Proceedings* **24**, 1293–1352.

VILLEE, C. A. & ENGEL, L. L. (Eds.) (1961). *Mechanism of Action of Steroid Hormones*. London: Pergamon.

Vitamins and Hormones (1959–67). Vols. 17–24. New York: Academic Press.

VON EULER, U. S. & ELIASSON, RUNE (1968). *Prostaglandins*. London: Academic Press.

WAYNE, E. J., KOUTRAS, D. A. & ALEXANDER, W. D. (1964). *Clinical Aspects of Iodine Metabolism*. Oxford: Blackwell.

WILLIAMS, R. H. (1963). *Textbook of Endocrinology*, 3rd edn. London: Saunders.

WOLSTENHOLME, G. E. W. & PORTER, RUTH (Eds.) (1966). *The Thymus: Experimental and Clinical Studies*. London: Churchill.

51 Reproduction

THE TESTES

The testes develop during embryonic life in the abdominal cavity. The gubernaculum testis (a mesenchymal structure not containing muscle) connects each to the area of the skin which later forms the scrotum. As development proceeds the testes travel to the scrotum which they reach about the eighth month of intrauterine life. Sometimes this process is delayed; if the testes are retained in the abdominal cavity in extrauterine life (*cryptorchidism*) spermatozoa are not formed (Fig. 51.1D) because the temperature of the abdominal cavity is too high. The incidence of undescended testis at one year is 0·7 per cent and in young adults 0·5 per cent; in other words 2 out of 7 descend spontaneously usually in the first year of life. The undescended testis is said to begin to degenerate at 6 years. In cases of partial descent it may be possible at 9 or 10 years (not later) to correct the position of the testes surgically and spermatogenesis may eventually occur normally. Descent of the testes may sometimes be aided by injections of chorionic gonadotrophin (p. 1049). The cutaneous muscle of the scrotum, the *dartos muscle*, by contracting when the environment is cold and relaxing when it is warm so regulates the distance between the testes and the warm groin that the scrotal contents are kept 2 to 4°C below the temperature of the abdominal cavity. At this temperature spermatogenesis proceeds normally.

Each testis is an oval body weighing about 25 g, about 5 cm long and 3 cm across, covered by the tunica albuginea, a bluish-white membrane of fibrous tissue, to which are attached fibrous septa dividing up the testis into some 250 compartments each containing about three convoluted *seminiferous tubules*. On cross-section these tubules, each of which is about 70 cm long, show a basement membrane on which three irregular layers of epithelial cells can be distinguished (Fig. 51.1A). The outer-

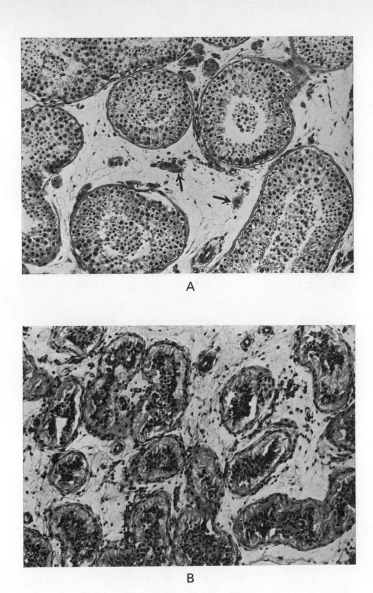

Fig. 51.1, A and B A. Normal adult male testis. Note especially the intensive proliferative activity in every tubule, the thin basement membranes, and the relatively scanty interstitial (Leydig) cells (examples are arrowed) in the interstitial tissue. This is a surgical biopsy and most of the loose cells in the centre of each tubule are the result of pressure during removal: the minute dots of spermatozoon heads can be recognized among them however. ×105. B. Testis in long-standing pituitary deficiency. The tubules are shrunken and their walls greatly thickened. The nature of the small dark-staining inactive cells within is not obvious at this magnification, but they consist of Sertoli cells and spermatogonia, without any spermatogenic activity. No Leydig cells can be seen. Masson's trichrome ×105.

most layer of cubical cells or *spermatogonia* give rise by cell division (p. 1101) to the second layer, the *spermatocytes*, which are large cells with large nuclei. Two types of spermatocyte are found in this layer, namely the primary spermatocytes containing the somatic or diploid number of chromosomes (p. 1101) and the secondary spermatocytes formed from them by

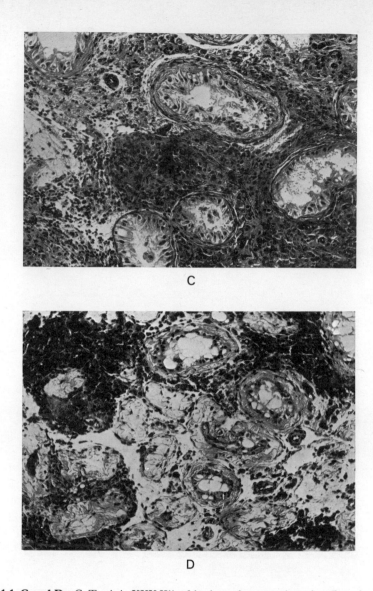

C

D

Fig. 51.1, C and D C. Testis in XXY Klinefelter's syndrome to show the effect of a normal pituitary on a defective testis. In this condition, as a result of an abnormal chromosome constitution (XXY), most of the tubules show total atrophy, and of the remainder most, such as those seen here, are lined by plump Sertoli cells only, with no germ cells. A rise in ICSH occurs and leads to a proliferation of Leydig cells: the dark masses between the tubules here consist almost entirely of Leydig cells. Masson's trichrome × 105. D. Testis in cryptorchidism. There is some oedema of the stroma and hyalinization of the tubules. The germinal epithelium is wholly atrophic; the few cells that remain in the lining of the tubules are Sertoli cells. The darkly-staining masses are sheets of Leydig cells. This undue prominence of Leydig cells is often found in the cryptorchid testis: it represents, however, a relative and not absolute increase in number. Haematoxylin and eosin × 105. (*A, B and C by courtesy of B. Lennox; D by courtesy of W. W. Park.*)

reduction division. The latter, containing only the haploid number of chromosomes, give rise by a further division to the spermatids, small polyhedral cells found in the third layer. The nucleus of the spermatid forms the head of the spermatozoon; the Golgi complex forms the cap or acrosome which is probably a lysosome. During this transformation the spermatids cluster around the *supporting cells of Sertoli* which are large phagocytic cells rich in glycogen and lipids projecting from the basement membrane. It is likely that these cells produce a steroid hormone necessary for spermatogenesis. The spermatozoa leave the Sertoli cells and lie around the inner margin of the germinal epithelium where they can easily be recognized by their densely staining heads and long curved tails pointing into the lumen of the tubule. At this stage they are immature, being not yet capable of fertilizing an ovum. The high refractive index of ripe spermatozoa shows that they must contain nearly 50 per cent of solid matter. About 10^7 spermatozoa are produced per day per g of testis. By the use of radioactive tracers which after injection are incorporated into spermatozoa or by studying the cells damaged by X-radiation it has been found that in the mouse the time required for spermatogenesis and maturation is 34 days. In man it is probably 75 days. Spermatozoa are first produced at puberty and spermatogenic

activity is maintained into old age so that there is no definite end to reproductive life in the male. The germinal cells are easily damaged by ischaemia. In the rat occlusion of the testicular artery for one hour leads to irrevocable destruction of all spermatogenic tissue; only the cells of Sertoli survive and act as macrophages.

From the seminiferous tubules (Fig. 51.2) the spermatozoa pass into the *rete testis*, a series of channels in the fibrous stroma of the posterior part of the testis. About a dozen efferent ductules pass from the upper part of the rete testis into a single canal which by its convolutions makes up the *epididymis* where the spermatozoa mature and are stored until ejaculation takes place. Production continues whether ejaculation occurs or not and spermatozoa not ejaculated are reabsorbed in the *vas deferens*. The canal of the epididymis, about six metres long, is continued into the vas deferens which rises up over the brim of the pelvis and passes down to the *prostate gland* at the base of the urinary bladder. Here it is joined by the duct of the *seminal vesicle* to form the short ejaculatory duct which passes through

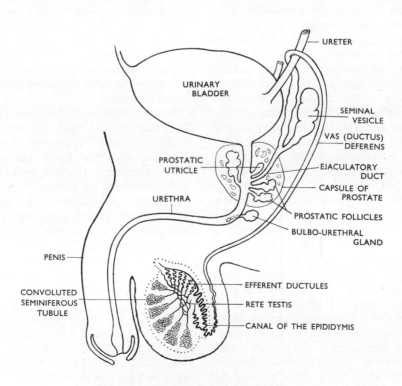

Fig. 51.2 Diagram of the male sex organs.

the prostate and enters the prostatic urethra. The two ejaculatory ducts enter the prostatic urethra separately, one on each side of the opening of the prostatic utricle. The seminal vesicles do not normally store spermatozoa. The prostate, surrounding the commencement of the urethra, is a glandular organ containing much muscle tissue and numerous follicles which communicate with the prostatic urethra by about a dozen small ducts. About 2·5 cm distally in the penile part of the urethra are the openings of the ducts of the two bulbo-urethral glands.

At coitus there is ejaculated into the vagina 2 to 5 ml of semen containing 40 to 100 million spermatozoa per ml; the amount varies from man to man and also with frequency of emission. The lower limit for fertility is 20 million per ml. The first portion ejaculated contains most of the sperms. From 60 to 80 per cent of the sperms are of normal shape and, after incubation for one hour at 37°, 50 per cent are motile. The sperms from the testis collect in the dilated end (ampulla) of the vas deferens until they are swept out by fluid from the seminal vesicles. Less than ten per cent of the ejaculate is spermatozoa, the remainder being seminal plasma. The volume of the semen is unaffected by double vasectomy. The semen clots soon after ejaculation but it is liquefied shortly after by fibrinolysis; then the spermatozoa become fully motile. The composition of seminal plasma is rather variable from one individual to another and consecutive samples from the same individual may differ widely. Its pH is about 7·4 and it contains some 0·1 to 1·0 per cent citric acid, 0·1 to 0·5 per cent fructose, 0·01 per cent ascorbic acid together with glyceryl phosphorylcholine, inositol, sorbitol and various enzymes mentioned below. Seminal plasma contains fibrinogen, thromboplastin but not prothrombin or thrombin. The seminal plasma acts as an activator and as a diluent for the spermatozoa which are tightly packed in the epididymis. The hyaluronidase of semen is carried by the spermatozoa but a certain amount leaks out into the plasma; it acts on hyaluronic acid found in mucus and so allows the sperms to pass more readily through the cervix to the uterus and tubes. After castration, fructose, citric acid and phosphatase are not found in the seminal plasma but they reappear when testosterone is given.

Seminal fluid has an extremely high creatine-phosphokinase activity.

The thin prostatic secretion forms about 20 per cent of the volume of semen; it gives seminal fluid its peculiar odour. It is colourless and slightly acid (pH 6·5) due to the presence of citric acid; it contains the proteolytic enzymes, fibrinolysin and fibrinogenase, and also acid phosphatase. The amount of acid phosphatase in the prostate is very small until puberty but in adult life the concentration is several hundred times greater than in other organs. The acid phosphatase of the serum is not derived from the prostate since it is the same in women and in men before and after prostatectomy. The acid phosphatase of male urine is due to admixture of prostatic secretion. In newly ejaculated semen this enzyme breaks down the phosphorylcholine to free choline and orthophosphate. The output of the enzyme in monkeys is increased by injection of androgens. The prostatic secretion contains a heat-stable compound of low molecular weight with antibacterial properties.

The seminal vesicles provide about 60 per cent of the volume of the semen. Their secretion is yellow owing to the presence of flavins and is sometimes deeply pigmented. It contains a large amount of potassium and also fructose, ascorbic acid, ergothioneine and phosphorylcholine. The seminal vesicles are the main source of the prostaglandin (p. 1022) found in semen.

Spermatozoa after ejaculation obtain energy from the seminal fluid and vaginal secretions by the anaerobic breakdown of the fructose to pyruvate with the formation of ATP from ADP. If oxygen is present, pyruvate is oxidized with liberation of energy. The spermatozoa within the epididymis are, because of the lack of oxygen and glycolysable sugar, metabolically inactive; they can survive in this situation for perhaps a month. Outside the body, spermatozoa survive only a short time at body temperature, a few days at 4°C and a few years at −79°C.

INTERNAL SECRETIONS OF THE TESTES

The testis grows a little during the first two years of life and remains about 1 g in weight till about 11 years of age when there is a rapid spurt to 15 g reached at 17 years. The prostate and the other accessory glands increase in size at the same time. The first external sign of pubescence, the beginning of rapid growth of the penis, is seen at 11 years on the average. The pubic hair soon becomes darker and extends laterally; the hair in the axillae and on the upper lip darkens and finally the pubic hair extends to the umbilicus and the hair on chin and face becomes stronger and darker. About the middle of pubescence the larynx enlarges and the pitch of the voice falls about an octave. The vocal folds lengthen by 10 mm, that is by almost half their original length. At this time the muscles develop rapidly and the pattern of behaviour alters. As Figure 51.3 shows there is considerable variation in the onset of puberty.

These bodily and mental changes depend on the testes since they do not occur if these organs are removed before puberty. Castration, that is removal of the testes, has been performed on domestic animals for thousands of years with the object of making the flesh more pleasant to eat or of making the animals more docile. Castration has also been performed on man for many centuries for various reasons. If the operation is carried out before puberty the

eunuch usually becomes a tall man because of the prolonged period of growth of the long bones. Usually rather fat, he retains the treble voice of the boy, the penis remains small and hair does not grow over the face and abdomen in the typical masculine fashion; the common masculine form of baldness does not develop and sebaceous secretion is diminished. Temporal recession of the hair-line in early middle age does not occur in eunuchs or in women. The skin of eunuchs does not tan on exposure to the sun unless male hormone has been administered. The prostate, seminal vesicles, and scrotum remain infantile and the thymus persists. Such an individual has no sexual desire and is both impotent and sterile. Castration after puberty produces muscular weakness, atrophy of the prostate and seminal vesicles. The histological changes in the pituitary are described on p. 1081. Although psychological changes are common, sexual desire may be retained, but the individual is sterile although he is potent, that is able to copulate. Bilateral vasectomy (removal of a short portion of each vas deferens with ligature of the open ends) is frequently used in man

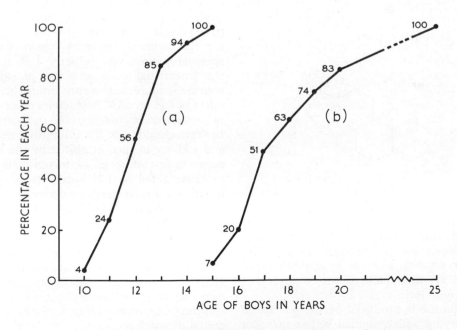

Fig. 51.3 Cumulative frequency curves of (a) earliest detectable onset of pubescence (see text) and (b) full sexual maturity in 1475 normal boys. The graphs were constructed from the data of W. A. Schonfeld (1943). Primary and secondary sexual characteristics. *American Journal of Diseases of Children* **65**, 535–549.

as a means of sterilization. It causes no loss of weight of the testis nor change in its histology; spermatogenesis continues and spermatozoa collect in the epididymis where they are taken up by phagocytosis. The prostate and seminal vesicles are unaffected. If, even many years later, the vasa are reunited, in many cases sperms reappear in the semen and fertility is restored.

Testosterone, the most potent androgen, was first obtained in 1935 from the testes of bulls. In the form of an ester with propionic acid, it is used in clinical medicine. It has to be given intramuscularly because, on absorption from the gut into the portal circulation it is transformed by the liver into substances of lower physiological activity. Methyl testosterone, a synthetic substance, may however, be given under the tongue, since it is absorbed

Androsterone

Testosterone

from the buccal mucosa. Fluorohydroxymethyltestosterone is five times more active both as an androgen and as an anabolic agent (see p. 1034) than methylestosterone and is effective by mouth.

Testosterone is produced by the *interstitial cells* of Leydig which contain yellow pigment granules and possess a rich supply of nerves. They lie scattered between the seminiferous tubules (Fig. 51.1A). Exposure to X-rays

destroys the seminiferous tubules without affecting the interstitial cells which, by producing testosterone, maintain the secondary sex characteristics. The interstitial cells, which occupy about 1 ml in each testis, contain dehydrogenases necessary to metabolize various precursors to testosterone (p. 1046). The Sertoli cells also produce some testosterone. Testosterone has been found in the spermatic vein blood, in amounts ranging from 40 to 160 µg per 100 ml. The total production is of the order of 4 to 12 mg per day. These amounts are increased by administration of human chorionic gonadotrophin but decreased by administration of androgens or oestrogens. Smaller amounts are also secreted by the adrenal cortex and the ovaries, and may be increased in certain pathological virilizing conditions; the peripheral tissues are able to convert less potent androgens to testosterone. The mean concentrations of testosterone in peripheral venous blood plasma in adults under 35 years of age are, for men 0·75, and for women 0·05 µg per 100 ml. The synthetic pathway for testosterone starts from acetate to cholesterol to pregnenolone and then either by way of progesterone to testosterone or by way of dehydroepiandrosterone to testosterone. Both follicle stimulating hormone (FSH, p. 1049) and luteinizing hormone (LH, p. 1049) together with testosterone are necessary for spermatogenesis. At puberty LH stimulates the interstitial cells and the plasma testosterone is increased twenty times through the effect of LH on cAMP which increases the rate of conversion of cholesterol to pregnenolone in the mitochondria of the interstitial cells. FSH and LH are in turn inhibited by the feedback action of testosterone. Testosterone is thought to cause a fall in LH secretion; large doses impair spermatogenesis probably by inhibiting the secretion of FSH. Thus the relationship between the gonadal hormones and the hypothalamus is similar to that in the female but cyclical variations in secretion do not occur in the male. When the testes are removed, that is after castration, this inhibitory influence is removed and the urinary output of gonadotrophins increases. Gonadotrophins are secreted in small amounts by boys; the amount increases slowly as they grow older until puberty when there is a further increase.

Androsterone which possesses much less

androgenic activity is a urinary excretory product derived in part from the metabolism of testosterone; however, since the greater part of injected testosterone cannot be accounted for as urinary metabolic products and, since androsterone may also be an excretory product of a substance such as androstenedione produced in the adrenal cortex, neither androsterone secretion nor 17-oxosteroid excretion

Androstenedione

generally is a satisfactory index of testosterone production in the body. Aetiocholanolone which differs from androsterone only in the configuration of the H atom at C-5 is another important metabolite of testosterone. These metabolites along with dehydroepiandro-

sterone are the principal 17-oxosteroids present in urine in which they occur as water-soluble conjugates with sulphuric acid. Separation and determination of individual 17-oxosteroids may best be achieved by gas-liquid chromatography.

The amounts of total 17-oxosteroids excreted in the urine are given in Table 51.4. The range of variation among healthy individuals is considerable. Since the adrenal cortex is responsible for the production of some of the urinary 17-oxosteroids it is not surprising to find that overactivity of the cortex, due to hyperplasia or a tumour or to stimulation of the cortex by ACTH, is characterized by a great increase in 17-oxosteroid excretion. After stress, for example a major surgical operation, the output of 17-oxosteroids is increased for several days. The excretion of more than 35 mg of 17-oxosteroids per day is suggestive of an adrenocortical tumour; values as high as 2000 mg per day have been obtained in proved cases of this condition and the steroid which predominates is dehydroepiandrosterone. Certain tumours of non-endocrine origin, notably carcinoma of the bronchus, cause an increased

Table 51.4   Urinary excretion of steroids

	17-oxosteroids mg/24 hours	Oestrogen µg/24 hours	Pregnanediol mg/24 hours	17-hydroxy-corticosteroids mg/24 hours†	Testosterone µg/24 hours
Boys age 5	1·7	<1	0·7	2–4	—
Girls age 5	1·7	<1	0·7	2–4	—
Boys before puberty	5	1	0·7	6	<5
Girls before puberty	5	2	0·7	6	<5
Adult man, aged 25	14 (8–23)	10	0·9	7–20	50–60
Adult woman, aged 25	9 (5–17)	13–56‡	0·8–4·2‡	4–14	5–15
Senile man	5	10	—	3–12	20
Senile woman	3	5	0·6	2–9	—
Pregnant woman	11	100 to 40 000	10 to 50	6–24	—
Castrate man	7	1	—	—	5
Oophorectomized woman	2	1	0·7	4–14	—
Female with Addison's disease (failure of adrenal cortex)	0	—	0·2★	<6	—
Male with Addison's disease	<5	—	0·4★	<6	40

These values are approximate only and wide variations from them are met with in individual cases. A dash in the table means that information could not be found in the literature. The chemistry of the corticosteroids is discussed on pp. 1007 to 1013.

★ After bilateral adrenalectomy.
† Method of Norymberski et al.
‡ Varying during the menstrual cycle.

excretion of 17-oxosteroids, due to an unexplained hyperplasia of the adrenal cortex. In adrenocortical insufficiency (Addison's disease) the 17-oxosteroid excretion is reduced (see p. 1015).

The testes produce, probably in the seminiferous tubules, a certain amount of oestrogenic steroid hormone. The richest natural source of oestrone is in fact the testis of the stallion. Testosterone given intramuscularly increases not only the 17-oxosteroid excretion but also oestrogen excretion, since a small amount of testosterone is converted in the body to oestrogen. Feminization has been seen in association with tumours of the interstitial cells and of the cells of Sertoli; these are therefore possible sites of oestrogen production in the testes. The adrenal cortex also contributes to the oestrogen found in the urine of men. The functional significance of the oestrogen produced in males is not known. It is perhaps better to regard testosterone and oestradiol as growth hormones secreted in different proportions in males and females, rather than to continue with the old ideas of male and female hormones. Since changes in the anterior pituitary occurring after castration cannot be reversed by injection of testosterone alone, other hormones (possibly oestrogens) produced in the testes may regulate the output of gonadotrophins (p. 1049) from the anterior pituitary. The androgens and oestrogens found in the urine after castration are derived from the adrenal cortex.

Injection of androgens into castrated mammals prevents the atrophy of penis, scrotum, prostate, seminal vesicles, epididymis and bulbourethral glands which normally follows removal of the testes, and the increase in weight of the prostate of the castrated rat after androgen treatment is the basis of a method of bioassay. Injection of androgens can also restore to normal the sexual behaviour of castrated mammals. Androgens can maintain normal spermatogenesis and prevent the testicular atrophy which follows hypophysectomy, provided that they are administered before there is atrophy of the germinal epithelium. Either gonadectomy or hypophysectomy produces equally prompt atrophy of the accessory glands. Thus the testis, in common with other steroid-secreting endocrine glands, does not appear to store hormone.

In cases of delayed puberty in boys administration of testosterone has produced hair in the typical male distribution and also growth of the larynx. Testosterone is usually regarded as the hormone responsible for transforming the boy into the man. But it can only be one of the factors since in testicular failure testosterone may produce only slight growth of hair or external genitals. After castration, however, testosterone propionate can produce erections of the penis and allow coitus but is of no value in the treatment of psychogenic impotence. Males castrated before puberty do not become bald, but administration of androgen produces baldness if the subject is of the genotype destined to become bald. Baldness of some degree appears in about half of all adult males; those possessing homozygous recessive genes do not become bald even if given androgens. Repeated injections of testosterone in men have been reported to depress spermatogenesis, presumably by depressing the output of pituitary gonadotrophin, but in a few cases, after the injections cease, spermatogenic activity is for a time greater than before.

In undernourished and emaciated persons injection of testosterone propionate has an anabolic effect, nitrogen and other tissue-forming materials such as potassium, calcium and phosphorus being retained. This useful therapeutic effect is complicated by the masculinization effect. Various derivatives of testosterone, such as methandienone (1-dehydro-17α-methyl testosterone), have a much higher ratio of anabolic to androgenic potency than testosterone propionate and these so-called 'non-virilizing androgens' are used clinically.

If the diet of experimental animals is inadequate in amount or deficient in vitamins, especially A, B or E, spermatogenesis is impaired and the accessory sex organs regress and become insensitive to testosterone. In male rats such changes can be counteracted by administration of testosterone or anterior pituitary extracts. Inadequate feeding probably depresses the anterior pituitary so that the gonadotrophic stimulus to the testes declines and consequently their output of male hormone also diminishes. Small amounts of cadmium damage the testes but zinc and selenium seem to be necessary for their function.

THE FEMALE GENITAL ORGANS

The internal genital organs in the female consist of the ovaries, the uterine tubes, the uterus and the vagina (Fig. 51.5). The *uterus* is a pear-shaped muscular organ about 7·5 cm long with a flattened triangular lumen, the upper part of which communicates through the *uterine tubes* with the peritoneal cavity, while the lower part is continuous through the narrow canal of the *cervix* with the *vagina* (7 to 11 cm long) which passes to the exterior. The peritoneum covering the uterus is continuous with that lining the rest of the abdominal cavity. The walls, about 1 cm thick, are composed of smooth muscle fibres (the *myo-*

metrium), among which is a considerable amount of fibrous tissue. The muscle tissue of the uterus is prolonged into the round ligaments which respond to hormones much as the muscle of the uterus. There is very little elastic tissue in the body of the uterus but it is abundant in the cervical region. The mucous membrane or *endometrium* lining the interior, and continuous with the mucous membrane lining the uterine tubes and cervix, is covered by a columnar epithelium which dips down into the stroma of the endometrium to form simple tubular glands. The epithelium covering the interior of the tubes and uterus, with the exception of the lower part of the cervix, is provided with cilia which tend to move ova or

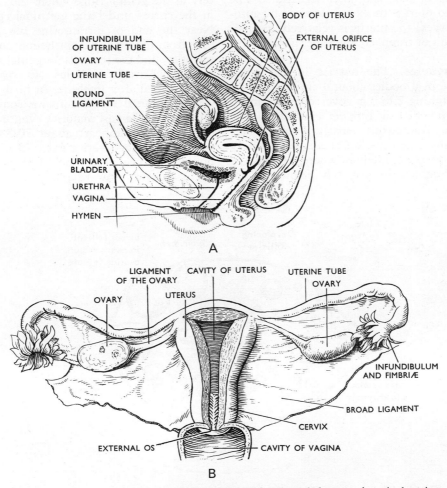

Fig. 51.5 Diagrams to show the main features of the female genital organs. A, sagittal section through the female pelvis. The uterus is shown in the so-called normal, anteflexed position; it is however retroflexed in 20 per cent of normal women. B, the genital organs removed from the body (posterior aspect).

secreted material towards the exterior. The columnar cells of the cervix give place abruptly, close to its external orifice, to the stratified squamous epithelium covering the vaginal portion of the cervix. Human cervical secretion contains β-glucuronidase. The vagina is lined by a stratified squamous epithelium 0.2 mm thick without any glands. During reproductive life the vagina is invaded by *Lactobacillus acidophilus* (Döderleins's bacillus), a rod-shaped organism which keeps the vaginal pH between 4.9 and 3.5 by producing lactic acid from glycogen. The acid inhibits the growth of pathogenic bacteria which otherwise might invade the vagina. The perineum and vulva are supplied by the pudendal nerves which convey pain and other sensations. The uterus and uterine tubes are supplied by autonomic nerves; the uterus is relatively insensitive but minor trauma of the tubes gives rise to pain.

The ovaries. The ovaries in the adult female are oval bodies about 4 cm by 2 cm by 1 cm, weighing 2 to 8 g, covered by a cubical epithelium called the *germinal epithelium*. The stroma or framework consisting of fibrous tissue with some muscle fibres is particularly dense below the epithelium where it forms the tunica albuginea. Immediately below this are numerous *vesicular ovarian (Graafian) follicles* in various stages of development or degeneration (Fig. 51.6). There is considerable difference of opinion as to whether interstitial cells analogous with those of the testis are present in the adult human ovary; they can, however, be recognized in the ovary of the fetus.

Very early during development of the fertilized ovum the cells which give rise to the future sex cells are segregated from those forming all the other tissues of the body. The primordial sex cells in man arise either from the endoderm or the overlying mesoderm of a small area of the yolk sac and migrate to the wall of the hind gut and then along its mesentery to the genital ridge where they are found in the cortex under the germinal epithelium. Near the end of intrauterine life groups of cells from the germinal epithelium form themselves around the sex cells (oogonia) to give the primordial ovarian follicles. No oogonia seem to be formed after fetal life. At birth the ovary weighs about 0.3 g; it contains some 750 000 follicles but the vast majority eventually disintegrate and disappear; about 70 000 follicles are found in the ovaries from 25 to 40 years, but after 40 years less than 10 000 are present. From puberty onwards, however, a limited

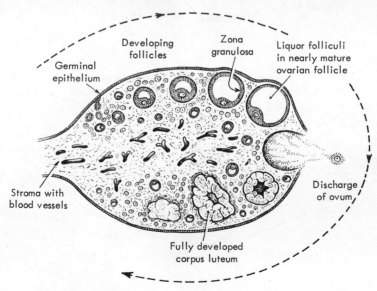

Fig. 51.6 A diagrammatic representation of a section made obliquely through an adult human ovary to show the stages in the development of the vesicular ovarian or Graafian follicle and the process of ovulation and finally the formation of a corpus luteum. The processes occur in the sequence indicated by the arrows.

number mature in succession so that various stages of their development may be seen in one adult ovary. In each follicle the oogonium or primitive ovum or oocyte can be recognized by its larger size. As the follicle begins to mature (Fig. 51.6) the cells investing the ovum multiply rapidly and a cavity filled with *liquor folliculi* appears among them giving the follicle the naked eye appearance of a clear vesicle on the surface of the ovary; the outer layer of investing cells is called the *stratum granulosum* and the layer around the ovum is the *cumulus ovaricus*. The maturing ovum is readily dis-

tinguished by its central position and relatively large size (0·13 mm); it is bounded by the elastic *zona pellucida*. An extension from each cumulus cell through the zona allows a granular liquid, perhaps food material, to be emptied into the perivitelline space around the ovum. The ripe follicle is surrounded by a *tunica interna* consisting of spindle cells with numerous capillaries and a more compact *tunica externa*. The follicle begins to enlarge about the ninth day of the menstrual cycle and in four days grows from 2 to 15 mm in diameter. A digestive enzyme appears in the

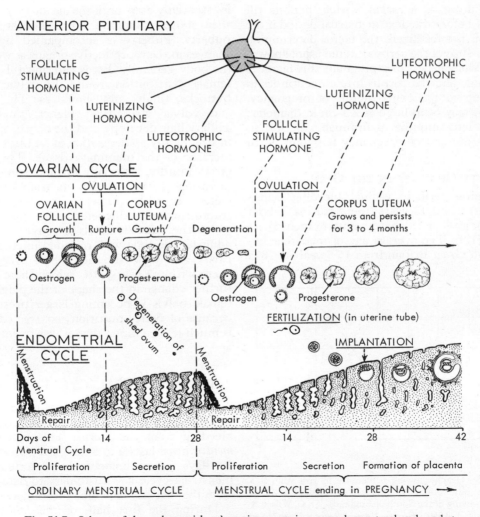

Fig. 51.7 Scheme of the endometrial and ovarian events in a normal menstrual cycle and at the commencement of pregnancy. Whether or not luteotrophic hormone acts during the human menstrual cycle is a matter of controversy. All the other hormonal effects illustrated in this diagram are well substantiated. (*After* McNaught & Callander.)

liquor folliculi and weakens both tunicas. At *ovulation* the follicle ruptures and the yellow coloured ovum, surrounded by 3000 granular cumulus cells (*corona radiata*), is set free into the abdominal cavity, the first polar body having been already formed. The ovum then passes into a uterine tube where it becomes free of cumulus cells. The empty space in the ruptured follicle is filled with serous exudate and a varying amount of blood that is gradually replaced by large cells containing yellow pigment (*lutein cells*) which are mainly hypertrophied cells of the stratum granulosum although a few (*paralutein cells*) may be derived from the tunica interna. The resulting structure, known as a *corpus luteum*, persists till shortly before the next menstrual period if the ovum is not fertilized; the factor determining the life-span of the corpus luteum is not known. If the woman becomes pregnant this factor is overcome and the corpus luteum continues to grow up to the third month of pregnancy, when it may be as large as 2·5 cm in diameter. In the later months it diminishes in size so that at the end of pregnancy it is only 1 cm across.

PUBERTY CHANGES IN THE GIRL

In girls periodic vaginal bleeding begins between 9 and 18 years (the *menarche*); about half the girls of 13½ have begun to menstruate (Fig. 51.8). The average age of menarche has fallen in Great Britain from 15·5 years in 1855 to 13·1 years in 1957. Many factors affect the age of menarche. Since the mean difference in monozygotic twins is only 2 or 3 months whereas in unrelated women it is 19 months, genetic factors must be involved but living conditions and the standard of nutrition are significant factors in determining the age of onset of menstruation. There is no basis for the popular idea that the menarche is earlier in tropical than in temperate climates. Comparison of different races in U.S.A. shows that in the same environment there is very little difference in the time of onset of menstruation. Breast development may be seen more than 2 years before the menarche. The growth curve is essentially even until the age of 10 years and then shows a definite spurt associated with puberty. Puberty is accompanied by many changes in the reproductive organs as well as in the body generally. Fat is laid down in typical female distribution, round the pelvic girdle, buttocks, thighs and shoulders. The uterus and vulvar structures enlarge, pubic and axillary hair develops, and the breasts increase in size mainly by deposition of fat but by some increase of the mammary tissue. The pelvis grows rapidly, especially in the region of the sacro-iliac joints, so that its transverse diameter (breadth) increases relative to the antero-posterior diameter. The changes by which the infantile pelvis is converted to the adult female type with rounded inlet, shallow cavity and wide subpubic angle are said to occur within 2 years, after which there is little further change in the shape of the pelvis. The female pelvis is becoming larger presumably because of better nutrition; contracted pelvis is much less common than it was twenty years ago.

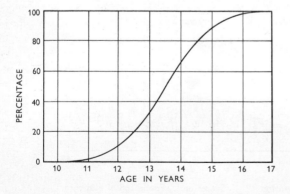

Fig. 51.8 Graph showing percentage of girls who, at a given age, have attained the menarche (standard based on 2590 girls living in the South of England). (D. C. Wilson & I. Sutherland (1950). *British Medical Journal* **ii**, 862. See also D. C. Wilson & I. Sutherland (1960). *Journal of Obstetrics and Gynaecology of the British Empire* **67**, 320–322.)

THE MENSTRUAL CYCLE

From the time of puberty the endometrium undergoes cyclical changes which result in bleeding from the uterus. This bleeding or *menstruation* lasts 4 to 6 days and recurs every 28 days approximately. In describing the menstrual cycle it is customary to date events from the first day on which bleeding occurs.

During the first half, or the *follicular* or *proliferative phase* of the cycle, an ovarian follicle enlarges and comes to the surface of the ovary, ovulation occurring about the fourteenth day. The subsequent fourteen days in which the corpus luteum is active are referred to as the *luteal* or *secretory phase*. The corpus luteum degenerates just before the next menstruation (Fig. 51.6 and Fig. 51.7).

On the first three days or so of the menstrual cycle (Fig. 51.9D) the more superficial layers of the endometrium degenerate and are cast off with a certain amount of haemorrhage, leaving the endometrium about 1 mm thick. Markee has studied the bleeding process by transplanting small fragments of endometrium into the anterior chamber of the eye of a monkey where they can easily be observed microscopically through the transparent cornea. In the follicular and luteal phases of the cycle the circulation through the spiral arteries supplying the endometrium is continuous. When the time of menstruation approaches the circulation in the spiral arteries supplying the superficial parts of the endometrium slows down and then stops for some hours so that the endometrium blanches and then degenerates. After a time these arteries open again and blood escapes through the walls of capillaries, venules and arterioles with the formation of subepithelial haematomata. The cause of the vasoconstriction is not clear.

The menstrual flow consists of blood mixed with mucus and at first contains numerous leucocytes. The blood clots and is simultaneously broken down by plasmin formed by the activation of the fibrinolytic system. The fibrinolytic activity of the endometrium and the concentration of fibrin degradation products in the plasma are greatest at the time of menstruation. The volume of the blood lost varies greatly in different women but it is usually 6 to 50 ml, a loss of 3 to 24 mg of iron and 1 to 10 g of protein; greater losses are often associated with hypochromic anaemia (p. 426). Menstrual fluid contains prostaglandins probably derived from the endometrium in the premenstrual phase.

At the end of menstruation the endometrium is regenerated (third to fourth day) from the deeper parts of the glands next the myometrium, which are not affected by the vasoconstriction and are not cast off. During the *phase of proliferation*, occupying from the fifth to the fourteenth day, new capillaries grow out from the spiral arterioles, the endometrial glands progressively increase in length and tortuosity and the cells of the stroma hypertrophy so that at the end of this phase the endometrium is some 3 to 4 mm thick (Fig. 51.9A and B). During the next fourteen days the glands continue to grow, becoming more irregular in outline and much convoluted and showing secretory activity. About the end of this *secretory phase* the glands are dilated and contain mucus and glycogen; their walls are folded and the free edges of the cells are indistinct. The superficial cells of the stroma enlarge and since they possess more cytoplasm they give the stroma as a whole a more tightly packed appearance. In the deep layers the cells are also enlarged but the stroma has a more open and oedematous appearance (Fig. 51.9C). The basal part of the endometrium next to the muscle does not show oedema or hypertrophy. These premenstrual changes may be regarded as preparation for a fertilized ovum and when they are complete the endometrium is about 6 or 7 mm thick. If the ovum is not fertilized small collections of blood (haematomata) appear in the endometrium and the degenerative changes which result in menstruation begin again.

These histological details are derived from examination of small samples of endometrium obtained by inserting a curette through the cervix. The experienced histologist can tell from the microscopic appearances whether or not ovulation is likely to have occurred and can give with some degree of certainty the time of ovulation relative to the day of curettage. The precise day in any individual cycle cannot, however, be given since ovulation may occur any time between the sixth and twentieth day of the menstrual cycle. At the time of ovulation the cervix secretes watery mucus rich in glucose which may facilitate penetration of spermatozoa.

Bleeding normally takes place, as just described, from endometrium in the secretory phase but periodic bleeding can occur even if the secretory stage is not reached and the endometrium, just before bleeding, is histologically still in the proliferative phase. Since the secretory phase of the endometrium is dependent on an active corpus luteum (p. 1038) it can be assumed that in these cases ovulation has not occurred and that the corpus luteum has not developed; bleeding of this kind is described as *anovular menstruation*.

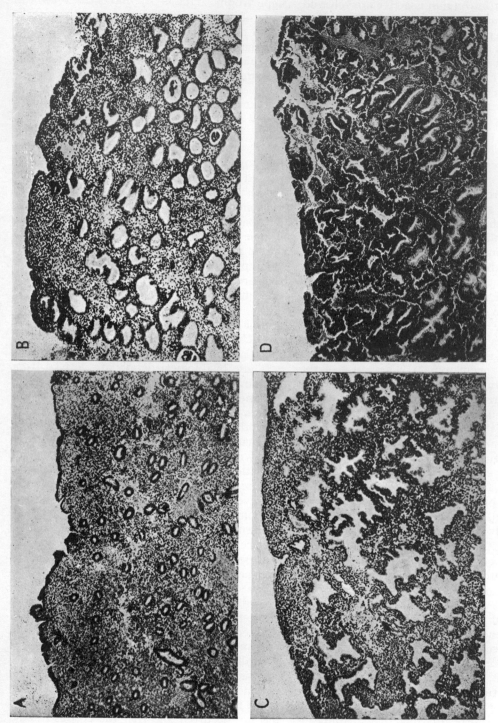

Fig. 51.9 Human endometrium at different stages of the menstrual cycle (×45). Stained with haematoxylin and eosin. A. Reparative or early proliferative phase: 6th to 7th day of cycle. B. Late proliferative phase: 12th to 13th day of cycle. C. Secretory phase: 23rd to 24th day of cycle. D. Commencing menstruation. (*By courtesy of W. W. Park.*) See also R. W. Noyes, A. T. Hertig & J. Rock (1950). Dating the endometrial biopsy. *Fertility and Sterility* **1**, 3–25.

The activity of the muscle of the non-pregnant human uterus can be recorded by placing in its cavity a rubber bag, filled with fluid under pressure, which is connected to a recording manometer (Fig. 51.10).

In the follicular phase the uterus shows small irregular waves, each lasting about 30 sec, increasing in frequency at ovulation. In the luteal phase much larger waves lasting a minute or so are seen (Fig. 51.10); in the late luteal phase the human uterus is relaxed by adrenaline and contracted by noradrenaline.

Although the uterus in other mammals in the non-pregnant state reacts to the oxytocic principle from the neurohypophysis (p. 1093) by contraction, the human myometrium in similar circumstances is stimulated by vasopressin (ADH) and not by oxytocin.

It is usually said that in healthy women menstruation occurs at intervals of 28 days but there is a considerable range of variation among individuals. In a recent American survey the mean length was 29·1 days with a standard deviation of 7·46 days. In any one woman successive cycles may vary in length by one to two days. At the onset of menstruation, and also when the menopause is nearly reached, the cycles may be irregular and anovulatory. The process is absent during pregnancy, may be suppressed during lactation and ceases finally at the *menopause*, usually between 45 and 55 (Fig. 51.11) the mean being about 50 years. In 1850 the mean age was about 46 years. When cyclical uterine and ovarian changes cease the uterus, breasts and external genitalia slowly atrophy but sexual desire

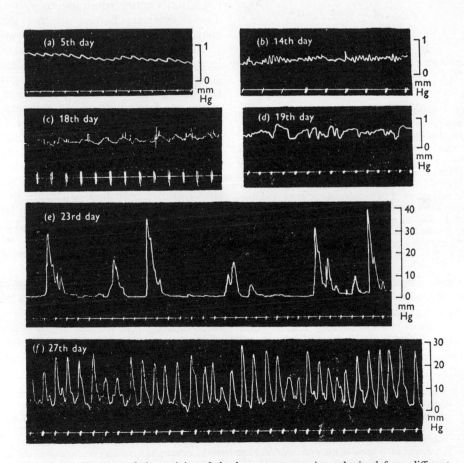

Fig. 51.10 Tracings of the activity of the human myometrium obtained from different patients at various stages of the menstrual cycle. All time markers in min. (W. J. Garrett (1956). Some observations on the human myometrial cycle. *Journal of Physiology* **132**, 553–558.)

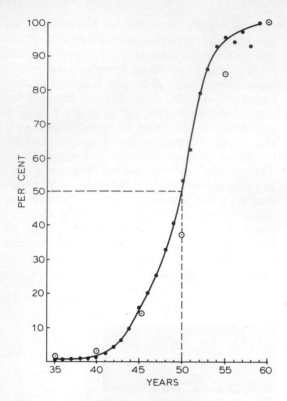

Fig. 51.11 Cumulative frequency for incidence of the natural menopause. The continuous curve ●——● was derived from data obtained from women in the London and Midland regions of the United Kingdom. The figures were obtained by a study group in the D.P.H. course, London School of Hygiene and Tropical Medicine under Margot Jefferys. The values indicated by ⊙ show the cumulative age-specific prevalence of post-menopausal women at five-year intervals. These were calculated by P. R. J. Burch from the data of F. Benjamin (1960). *South African Medical Journal* **34**, 316. At 25 and also at 30 the incidence was 0·22 per cent; at 55 98·6 per cent and at 65 100 per cent.

This period of involutionary change, the *climacteric,* is often attended by vasomotor phenomena such as hot flushes and by giddiness, excessive sweating and temporary emotional disturbances. Similar uterine and vasomotor changes occur after an artificial menopause induced for therapeutic purposes by surgical removal or destructive irradiation of the ovaries. The intensity of the associated symptoms varies greatly from woman to woman: in most women they are slight.

OESTROUS CYCLE

The simplest oestrous cycle occurs in rodents such as the rat and mouse. The mature female of these species accepts the male only at the period of heat or *oestrus,* which lasts about a day and recurs every four or five days; ovulation occurs spontaneously at this time. These cyclical changes do not occur after spaying, that is removal of both ovaries.

The discovery by Stockard and Papanicolaou in 1917 that the vagina of rodents undergoes characteristic changes during the cycle can be regarded as the starting-point of much of the modern work on sex hormones. These changes can be followed by microscopic examination of the cells in 'vaginal smears' made by passing a wire loop into the vagina and spreading the material so obtained on a glass slide. In the rat at the height of oestrus the smear shows cornified squamous cells almost entirely, while at the end of oestrus large numbers of leucocytes appear (Fig. 51.12). During dioestrus only a small amount of cellular material, chiefly leucocytes and epithelial cells, can be obtained.

The vaginal changes in woman are by no means so clear-cut as in the rodents but cyclical alterations can be made out. During reproductive life there is next to the muscle coat a single layer of small basophilic basal cells. On top of this are several rows of intermediate

often persists for some years. Because ovulation ceases corpora lutea cannot be formed and progesterone production fails. The endometrium does not show secretory changes and is thus either proliferative or atrophic. Pregnancy is exceedingly rare after the age of 50.

Fig. 51.12 Cellular changes in the rat vagina. A is a vaginal smear from a rat spayed fourteen days previously. It contains degenerating leucocytes and a few cells from the vaginal epithelium. C is a transverse section of the vagina of the same animal. The epithelium is quite thin. B is a vaginal smear from a spayed rat given a large dose of oestradiol dipropionate a week before the smear was made. The cells are large non-nucleated squames. D is a transverse section of the vagina of the same animal. Note how the oestrogen treatment has increased the thickness of the vaginal epithelium. There is considerable cornification of the superficial layers in D. When these cells are cast off they appear as squames (B). (*By courtesy of W. W. Park.*)

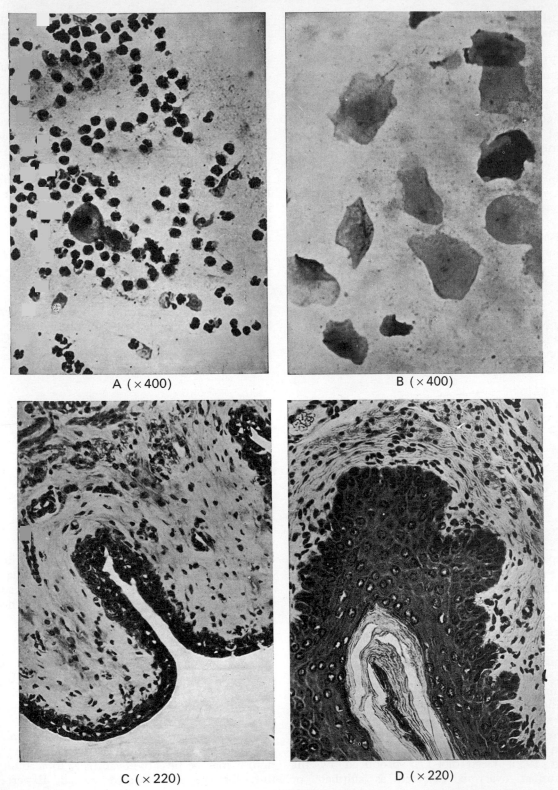

A (×400)

B (×400)

C (×220)

D (×220)

cells which are large and clear with well-marked cell boundaries. A layer of acidophilic, cornified cells lies superficial to these layers. These cells are constantly being shed from the surface of the vaginal epithelium and contribute to the slight white vaginal discharge which is present in all women of reproductive age. Vaginal smears obtained in the first half of the human menstrual cycle show more and more cornified (superficial zone) cells up to the fourteenth day, when leucocytes suddenly become abundant; in the second half of the cycle the proportion of cornified cells declines and the intermediate zone cells show clumping which is a characteristic progesterone effect. At menstruation the vaginal smear consists largely of red blood cells.

OVARIAN HORMONES

OESTROGENS. After oophorectomy (spaying), the uterus of the mouse and rat shrinks and the vaginal epithelium is reduced to a few layers of cells (Fig. 51.12C). By showing that the injection of extracts of liquor folliculi into glucuronide, or as oestrone, oestriol or other closely related substances. The theca interna seems to be the main site of oestrogen production. Steroid oestrogens are only slightly active when taken by mouth and are therefore given parenterally. Ethinyl oestradiol; a synthetic substance, is effective by oral administration. The action of injected oestradiol is short but more prolonged effects can be obtained by administering oestradiol benzoate or dipropionate, or by intramuscular implantation of a pellet of oestrone which is slowly absorbed. Prolonged oestrogenic effects are, however, more simply and cheaply achieved by the oral administration of a synthetic oestrogen, such as stilboestrol.

Information is accumulating about the nature of the biochemical actions of oestrogens at the cellular level. They may alter the structure and activities of enzymically active proteins but the primary site of action has not yet been located. The activity of a transhydrogenase is increased in the presence of oestrogens and this enzyme is present in tissues which are target organs for oestrogens, for

Oestradiol-17β Oestrone   Oestriol

spayed mice produced vaginal cornification (Fig. 51.12D) Allen and Doisy in 1923 found a method for testing oestrogenic activity. (It has been discovered recently that the concentration of steroid hormones in the liquor folliculi may be 1000 times that in the serum.) All substances producing oestrus and vaginal cornification in spayed mice or rats are said to be oestrogenic substances or *oestrogens* although they need not be steroids. As a result of the application of the vaginal cornification test *oestrone* was isolated from urine in 1929 and *oestriol* by Marrian in 1930, and the presence in ovarian tissue of oestradiol-17β was proved in 1935. Oestradiol-17β produced by the human ovary is excreted as sulphate or example the genital tract and the acini of the breast. Oestrogens also alter the polymerization of acid mucopolysaccharides and so modify the physiochemical properties of the ground substance of connective tissues.

Oestrogenic steroids, in common with most other steroid hormones, undergo metabolic changes, mainly in the liver. Since only a fraction of the administered hormones can be accounted for as known metabolites these substances have often been said to be 'inactivated' or 'destroyed' by the liver. With the discovery of increasing numbers of new metabolites and new methods for their quantitative determination, it may be possible eventually to account fully for the administered steroid. Experiments with ^{14}C-labelled steroids provide no evidence for breakdown of the basic ring structure.

Oestrogens can produce not only the vaginal changes of oestrus in rodents but also the enlargement and distension of the uterus which occur naturally at oestrus. If spayed mice are given large doses of oestrogens they will mate but in higher mammals, certainly in man, mating behaviour is mainly under the control of the nervous system; it can occur in the absence of testes or ovaries. Androgens are probably the principal erotogenic hormones in both men and women.

After intravenous administration of a colloidal suspension of oestradiol-17β the spontaneous activity of the human uterus is increased and its response to oxytocin is enhanced.

After removal of the ovaries from the rabbit the concentration of actomyosin (AM) in the uterus falls to one-fifth of that in the uterus at oestrus; the concentration of AM is restored by giving oestrogens and so is the content of energy-rich phosphate. As pregnancy in woman advances the myometrium increases in weight up to the 32nd week and the maximum AM concentration is also reached at this time.

HORMONE OF THE CORPUS LUTEUM. It has long been suspected that the corpus luteum produces a hormone since, when actively secreting corpora lutea are present in the ovary of mammals, oestrus is absent and the endometrium shows characteristic changes. In the cow, oestrus and ovulation are suppressed when there is a persistent corpus luteum but if this is expressed from the ovary by manipulation per rectum the animal comes into heat again and can be mated. Furthermore in the presence of corpora lutea the endometrium of the guinea-pig, rabbit and monkey reacts to trauma, produced experimentally by inserting a piece of silk through the uterus, by the *decidual reaction* in which the connective tissue cells at the site of the injury enlarge. One method of testing ovarian extracts for luteal activity depends on their action on the endometrium of the rabbit. The animal is first spayed and treated with oestrone to stimulate growth of the endometrium. If the animal is then injected with active extracts of corpus luteum, the endometrial picture changes to the fronded appearance, known as *progestational*

proliferation. With the help of this test crude extracts were prepared in 1929, and in 1934 the pure hormone *progesterone* was isolated (Fig. 51.13). In the liver this compound is reduced to *pregnanediol* and conjugated with glucuronic acid to be excreted in the urine as a biologically inactive glucuronide. About 10 to 40 per cent of injected progesterone can be recovered in the urine and about 5 per cent in the bile as conjugated pregnanediol. The human ovarian follicle produces oestrone and oestradiol-17β whereas the main product of the corpus luteum is progesterone. The ovarian stroma produces, at least *in vitro*, mainly androgens, namely dehydroepiandrosterone, androstenedione and testosterone. (See Fig. 51.13.) The granulosa and thecal cells of the ovary can convert precursor pregnenolone to oestrogens but the granulosa cells show a marked preference for converting it to progesterone.

A small amount of the urinary pregnanediol seems to be derived from the adrenals; bilateral adrenalectomy markedly reduces the output in both men and women.

Unless very large doses are used progesterone has no action on a uterus which has not been prepared ('primed') by oestrogen and, since the actions of progesterone and oestrogens are in some respects antagonistic, a delicately adjusted quantitative relationship between them is necessary to produce the maximum effect. Thus the changes produced in the endometrium by injections of progesterone in the rabbit or monkey are reduced or even completely prevented by large doses of oestrogen; the maximal progestational effect cannot, however, be obtained below a certain level of oestrogenic stimulation. Although progesterone seems to inhibit uterine activity in some animals it does not do so in women.

Pregnanediol

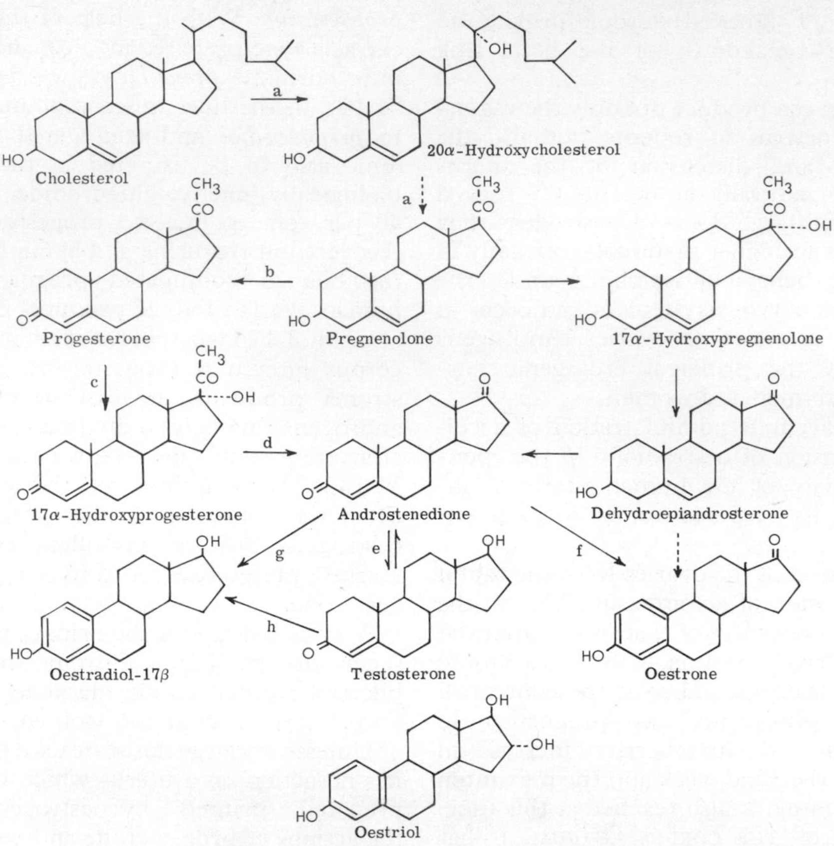

Fig. 51.13 Steroid biosynthesis in the mammalian ovary.

It seems that the principal function of progesterone, in conjunction with oestrogenic hormones, is to build up the endometrium for the reception of a fertilized ovum and to maintain the conditions necessary for its further growth.

BIOSYNTHESIS OF OESTROGEN AND PROGESTERONE

All steroids are produced initially from acetate which is a product of carbohydrate, fat and protein metabolism. It is known as 'active acetate' when combined with coenzyme A and the compound is known as acetyl-coenzyme A. When this is available almost all tissues appear to be capable of producing cholesterol (p. 343) which is the precursor steroid for all steroidogenesis in the human ovaries.

Oestrogens are formed from cholesterol by various enzyme systems (hydroxylases and dehydrogenases) present in the ovarian tissues. These enzymes act at specific sites on the carbon chains which form the steroid molecule. Cholesterol is synthesized to pregnenolone by cleavage of the side-chain at C-17 (Fig. 15.13a) and pregnenolone to progesterone by transfer of the C-5 and C-6 double bond (Δ^5 position) to C-4 and C-5 (Δ^4 position) (Fig. 15.13b) and dehydrogenation of the hydroxyl group at C-3 by 3β-ol dehydrogenase which is present in ovarian tissue. Progesterone is converted to 17α-hydroxy progesterone (Fig. 15.13c). This is an enabling step to allow the side-chain at C-17 to be reduced still further in size. When the side-chain at C-17 is removed it is replaced by an α-oxo(keto) group and androstendione (Fig. 15.13d) is formed. This androgen is a precursor of testosterone (Fig. 15.13e) in the testes, but in the ovaries it is a precursor of oestrone (Fig. 15.13f). In this conversion a hydroxyl group is introduced at C-3, with unsaturation of ring A and removal of the angular methyl group at C-10. Oestradiol-17β is manufactured in the ovary by changing the oxo(keto) group at C-17 to a hydroxyl group (Fig. 51.13g). However, this steroid can also be produced from testosterone by aromatization of ring A, replacement of the oxo(keto) group at C-3 by an hydroxyl group and removal of the angular methyl group at C-10 (Fig. 51.13h).

The steroidogenesis of progesterone in the liver principally involves reduction of the Δ^4-3-oxo(keto) group to form pregnenolone. This reduction gives rise to several asymmetric carbon atoms and as a result there is produced a large range of stereoisomeric metabolites. Further reduction at C-20 produces pregnanediol, the principal metabolite of progesterone. These metabolites are conjugated almost exclusively with glucuronic acid and they do not have any progestational activity.

SYNTHETIC 'HORMONES'. Large numbers of oestrogenic compounds having little or no chemical resemblance to the naturally occurring sex hormones have been synthesized. Stilboestrol was one of the first nonsteroidal oestrogens to be synthesized and the discovery of this substance by Dodds in 1938 was a landmark in endocrinology. Ethinyl oestradiol is closely related to the natural oestrogen oestradiol-17β. It is about twenty-five times as potent as stilboestrol and its 3-methyl ether derivative is widely used in oral contraceptive preparations.

For use in treatment these compounds have the great advantage that they are easily and cheaply prepared in bulk and are active when given by mouth. Oestrogens, both synthetic and natural, have proved of value in the treatment of cancer of the prostate and of some cases of cancer of the breast.

For a long time progesterone was the only available compound with progestational activity but we have now a number of compounds which are not only more active than progesterone itself but are active when taken by mouth. Some are related to progesterone itself; a second group includes compounds such as 17α-hydroxyprogesterone caproate; a third group includes compounds related to testosterone or 19-nortestosterone. Ethisterone, related to testosterone, was the first orally active compound to be used but it is now superseded by 19-nortestosterone derivatives such as norethisterone and norethynodrel (see also p. 1053).

THE HORMONAL CONTROL OF THE MENSTRUAL CYCLE

After oophorectomy in women the menstrual cycle is abolished, the vagina atrophies and the vaginal smear consists almost entirely of leucocytes; at the same time the uterus atrophies and the endometrium becomes very thin. Treatment with oestrogens causes the walls of the vagina to become cornified so that the smear contains squamous cells and the vaginal secretion returns to its normal acid pH (p. 1036). The uterine muscle hypertrophies and the endometrium proliferates to reach eventually the stage characteristic of the middle of a normal menstrual cycle. At the same time the glands of the cervix uteri begin to secrete again. If daily injections are stopped or if the dosage is suddenly reduced the uterine mucosa becomes oedematous and about 5 days later there is bleeding from the endometrium. This is called *withdrawal bleeding*. If, however, treatment is continued when the endometrium is showing proliferative changes, and progesterone as well as oestrogen is injected in suitable dosage a typical premenstrual endometrium can be produced. This treatment increases the vascularity of the endometrium and the spiral arteries have been seen growing

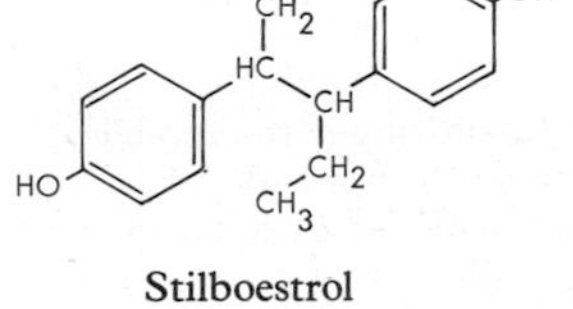

Stilboestrol
(Diethylstilboestrol)

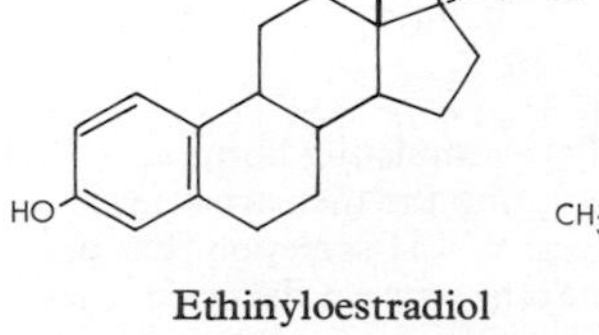

Ethinyloestradiol

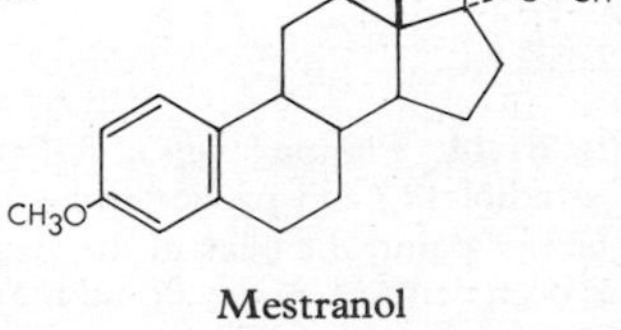

Mestranol

towards the surface of the endometrium in an endometrial transplant in the anterior chamber of the eye of a monkey injected with these hormones. Cessation of hormonal treatment is followed as before by bleeding but in this case it takes place in a premenstrual endometrium just as in the normal menstrual cycle.

Since the uterine changes of the menstrual cycle can be imitated by injection of oestradiol and progesterone, it is a fair assumption that the ovarian hormones are responsible for these changes in the normal menstrual cycle; oestradiol produced in the first half of the cycle would bring about the changes of the follicular phase while oestradiol and progesterone together would be responsible for the changes occurring in the luteal phase. This view is supported by the study of the urinary excretion of the end-products of the metabolism of the ovarian hormones. Oestrogens are excreted throughout the cycle rising from values of 20 µg/24 hours at the beginning of the cycle to 65 µg/24 hours at the time of ovulation (ovulatory peak). In the luteal phase of the cycle there is a secondary rise to about 40 µg/24 hours on about the 23rd day of the cycle. The daily excretion of pregnanediol in the proliferative phase is about 1 to 2 mg/24 hours and in the luteal phase about 4 mg/24 hours. The actual production of progesterone can be assumed to be about 15 mg/day.

The blood levels of oestradiol and progesterone vary significantly during the menstrual cycle (Fig. 51.14). The peak level of oestradiol, 100 ng per 100 ml blood, is found in the day before ovulation. The highest values of progesterone are found in the mid-luteal phase of the cycle—15 ng/100 ml blood—and

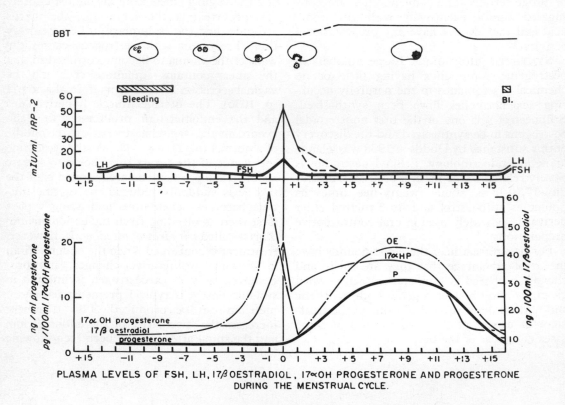

PLASMA LEVELS OF FSH, LH, 17β OESTRADIOL, 17αOH PROGESTERONE AND PROGESTERONE DURING THE MENSTRUAL CYCLE.

Fig. 51.14 Plasma levels of follicle-stimulating hormone (FSH), luteinizing hormone (LH), oestradiol-17β and progesterone during the menstrual cycle in human beings and monkeys. The key point, the peak of the surge of LH secretion that precedes ovulation by a day or so, has been denoted by day 0, and the other days in the cycle as minus or plus days from this time. BBT, basal body temperature. (*By courtesy of F. Naftolin*)

at the same time the blood oestradiol level rises. The first rise of oestrogen 'production' parallels the maturation of the ovarian follicle and the second rise after the transient fall corresponds with the growth and regression of the corpus luteum.

Plasma renin concentration and also plasma angiotensin rise during the luteal phase of the cycle and these changes are probably responsible for the increased secretion of aldosterone observed at this time.

GONADOTROPHIC HORMONES

Our knowledge of the controlling influence of the adenohypophysis on the ovary and testis began in 1927 with the pioneering work of Smith and Engle and also Zondek and Aschheim. The hormones involved are termed *gonadotrophins.* If the pituitary gland is removed from an immature animal the gonads remain infantile and, since the development of the secondary sex organs, the clitoris, uterus and vagina in the female and the penis, prostate and seminal vesicles in the male, is dependent on the hormones of the gonads, these organs also remain immature. We have seen already that the secondary sex organs atrophy after gonadectomy (castration) in spite of the presence of the pituitary gland, and we can make the important generalization here that the adenohypophysis has no direct influence on the secondary sex organs but only an indirect one through the gonads.

In the human male and female two gonadotrophins are produced by the pituitary gland and another gonadotrophin is produced by the placenta. In rats and mice (but probably not in man) another pituitary hormone, prolactin, maintains the activity of the corpus luteum, that is it has a luteotrophic effect. It was therefore called luteotrophin or LTH.

The two pituitary gonadotrophic hormones are glycoproteins: follicular stimulating hormone (FSH) with a mol. wt. about 35 000 and luteinizing hormone (LH) which seems to exist as two loosely attached units with a mol. wt. of about 20 000 each. FSH is so named because it stimulates primordial follicles in the ovary to grow and eventually one follicle enlarges greatly and becomes a vesicular (Graafian) follicle. Why only one follicle (and very rarely more) out of the many apparently available should enlarge is not clear. Under the influence of a peak of LH which occurs in the middle of the menstrual cycle, possibly in response to a small mid-cycle peak of oestrogen this ovarian follicle ruptures and the ovum is released. This is the phenomenon called *ovulation.* Changes in the granulosa cells left behind in the ovary lead to the formation of a corpus luteum in the second half of the cycle (p. 1038). Thus the gonadotrophins play a fundamental part in the menstrual cycle in the human female.

The control of these hormones is complex and delicate. The delicacy has been exemplified by the difficulty in using them in the treatment of infertility in women of reproductive age who fail to ovulate. Many workers have found that in those cases in which ovulation has been successfully induced many multiple births have occurred. The cause of the multiple ovulations is now known to be brought about by excessive stimulation of the ovary with FSH.

Patients undergoing this treatment for infertility are given parenterally a mixture containing mainly FSH with some LH over a few days in divided doses. The response of the ovary is indicated by the urinary output of oestrogen. If this is satisfactory, but not excessive, human chorionic gonadotrophin (HCG) is given for its LH-like effect in the hope that ovulation will follow. Excessive stimulation of the ovary can cause not only multiple ovulation and in consequence multiple births but in extreme cases ovarian haemorrhage.

Because animal FSH has undesirable antigenic effects the FSH must be prepared from human material, either from the pituitary glands of cadavers or from the urine of post menopausal women. The LH effect at midcycle is achieved with HCG from pregnant women's urine since it resembles closely LH and is much more easily obtained than LH.

The stimuli for the release of gonadotrophins (as indeed for other pituitary trophic hormones (Chap. 52)) are *releasing hormones.* These hormones are small molecular weight polypeptides. The investigation of the nature of these compounds is hampered not by their complexity, which is the problem with gonadotrophins, but by the great difficulty in obtaining sufficient material for study; the quantity in each hypothalamus is minute.

The hypothalamus appears to control the menstrual cycle. If the woman has a period every 28 days then this periodicity exists in the hypothalamus or its associated nuclei. The cycle of release of these factors can be modified by signals coming back from the gonads, and also by signals coming from the higher centres of the brain, so that stresses of various kinds can affect the menstrual cycle by way of the higher centres and the hypothalamus. It is not uncommon for girls leaving home, sitting examinations, or losing a relative to have a disturbance of menstruation even to the point of amenorrhoea. During the Second World War such events were not uncommon in service women and also in civilian women particularly during the London blitz. Amenorrhoea was the rule in concentration camps.

The releasing hormone enters a portal system in the hypothalamus, and travels in the blood stream to the pituitary to exert its effect there. This arrangement ensures that releasing factor is present in a concentrated state in the pituitary but very much diluted in the general circulation. The releasing hormone that we are concerned with here is luteinizing hormone releasing factor (LRH). Partially purified LH releasing hormone (LRH) from pig pituitaries has recently become available. Subcutaneous injection of porcine LRH produced a four or five times increase of serum LH. The preparation also had some FSH releasing activity since it doubled the serum FSH. However ovulation did not occur in any of the subjects: this may simply mean that the dosage was insufficient. LRH when available in pure form will certainly have many important clinical applications (see also p. 1082).

The pituitary gonadotrophins are also affected by the gonadal hormones themselves in a negative feedback mechanism and in general it can be said that the production of FSH and LH is reduced by a rise in the secretion of oestrogens and progesterone by the ovary (Fig. 51.15). The main effect of the synthetic oestrogens and progestins contained in the contraceptive pill is to interfere with the pituitary gonadotrophin secretion and thus with ovulation. The interrelations of gonadotrophins with steroid hormones are complex.

In young children the secretion of gonadotrophic hormones is very low; it rises slowly throughout childhood until with approaching

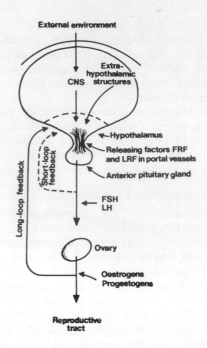

Fig. 51.15 The hypothalamus and control of ovulation. The central nervous–pituitary–ovarian axis. Ovarian function is regulated by the blood concentration of anterior pituitary gonadotrophic hormones (follicle-stimulating hormone, FSH; luteinizing hormone, LH). The secretion of FSH and LH is in turn dependent on the transportation of releasing factors (follicle-stimulating hormone, releasing factor, FRF; luteinizing hormone releasing factor, LRF) from hypothalamic nerve terminals to the anterior pituitary gland by the hypophysial portal vessels. The hypothalamus itself appears to act as a major integrative centre. It appears to have some autonomous function in maintaining anterior pituitary activity. This in turn is modulated by (i) neural inputs from extrahypothalamic brain structures (such as the amygdaloid nuclei), some of which mediate environmental influences, and (ii) hormonal feedback through the long-loop (solid line), and possibly (and therefore denoted by an interrupted line) short-loop systems. (G. W. Harris & F. Naftolin (1970). *British Medical Bulletin* **26**, 3–9.

puberty the secretion of both hormones increases rapidly to the adult levels. These levels are maintained in the male until the onset of senility; in the female there are cyclic changes during reproductive life and then a further increase after the menopause when the inhibitory influence of the ovarian steroids is much reduced.

In human pregnancy the urine contains very large amounts of gonadotrophic material which is formed in the placenta and is therefore called human chorionic gonadotrophin (HCG). HCG is biologically and immunologically virtually identical with LH. At term, that is, at the very end of pregnancy, the human placenta contains 5 i.u. of HCG per g wet weight. The active principle is a glycoprotein which is thought to exist as two loosely bound units, each with a molecular weight of about 30 000. The mechanism controlling HCG is not known. It is only in woman that large quantities of gonadotrophin are excreted throughout pregnancy; in the chimpanzee, monkey and giraffe CG is excreted in the urine in early pregnancy only. The serum of pregnant mares from the 45th to the 150th day of pregnancy is a rich source of gonadotrophin (PMS) a glycoprotein which is produced by the endometrium; PMS is not excreted by the kidney and, therefore, remains in the blood for a considerable time.

HORMONAL BASIS OF PUBERTY AND THE CLIMACTERIC

An immature female rat can be brought to sexual maturity by implantation of tissue from the adenohypophysis of mature animals. After this procedure the animal goes into heat and allows mating. This experiment suggests that puberty is brought about by the discharge of gonadotrophic material from the adenohypophysis. In the immature male rat pituitary transplants can produce development of the seminiferous tubules but not spermatogenesis. The explanation of the delay in the occurrence of puberty cannot be that the pituitary does not produce gonadotrophins until that time because it has been found that the adenohypophysis contains a small amount of gonadotrophin even at birth, and implantation of pituitaries of immature animals into immature female rats can also bring about the changes of puberty. Furthermore, it is now known that if pituitary glands taken from rats a few days after birth are implanted below the pituitary stalk of hypophysectomized adult female rats, they become vascularized by the hypophysial portal vessels and full adult reproductive function is maintained. The most likely explanation for the delay in the appearance of sexual maturity is that, although the immature adenohypophysis contains gonadotrophins, these are not released from it in sufficient quantity until puberty to bring about maturation of the gonads. Even so, the stimulus responsible for the increased secretory activity of the anterior lobe of the pituitary at the time of puberty has still to be found; a clue may be provided by the observation that tumours responsible for precocious puberty most often involve the mamillary bodies. Poor nutrition delays the onset of puberty probably by depressing pituitary function (cf. p. 1034).

The gonads themselves play a part in determining the onset of puberty since it is only when the animal has arrived at this stage of its growth that the gonads begin to respond readily to gonadotrophic hormones. That there is a parallel increase in sensitivity of the secondary sex organs as puberty approaches is indicated by the fact that in immature animals large doses of androgens have little effect on the seminal vesicles, while at puberty quite small doses can cause an increase in size; similarly very large doses of androgens are necessary to induce spermatogenesis in immature animals but, once the process has begun, it can be maintained in mature hypophysectomized animals with a lower dosage. All these observations suggest that an increased production of pituitary hormones together with an enhanced responsiveness of the gonads and secondary sex organs results in the changes of puberty. Girls before puberty have sometimes a cyclical pattern of oestrogen excretion.

The phenomena of the climacteric in women are not due simply to decline of oestrogen production (Table 51.4). The excretion of oestrogens rises temporarily and then declines but even older women still excrete some oestrogenic material; the origin of this is obscure since the urinary oestrogen does not necessarily fall to zero after combined oophorectomy and adrenalectomy. Hot flushes (p. 1042) are associated with low oestrogen levels; they are not associated with increased production of gonadotrophins and may occur in hypophysectomized women. The menopause does not develop suddenly; the biochemical and other changes are spread over several years. Women in their late thirties, long before the menopause, may show levels of LH ten times the normal. After the menopause the gonadotrophin content of the pituitary is increased and the urinary output of gonadotrophins rises (p. 1050); this increased output is presumably related to the falling off of ovarian activity, since it can be reduced to the premenopausal level by administration of natural or artificial oestrogens in adequate dosage. After the menopause there is usually a fall in basal metabolic rate and a rise in body weight. Both changes could be attributed to a fall in thyroid activity. In many women sexual desire and capacity continue for a number of years after either a natural or an artificial menopause.

Thus the changes at the end of reproductive life seem to depend partly on a decreased power of the gonads to produce their hormones and partly on a diminished ability of the gonads to respond to the pituitary hormones.

OVULATION AND FERTILITY

The determination of the time of ovulation in woman is a very interesting physiological problem and one which has important implications in the study and treatment of infertility. An ovum can be fertilized at a mating which precedes ovulation only if the spermatozoa placed in the genital tract survive until ovulation occurs; fertilization can occur after ovulation only so long as the ovum remains capable of being fertilized. If, as is almost certainly true, the period of viability of both spermatozoa and ova is short the time of ovulation determines to a large extent the time during which mating is fertile.

VIABILITY OF SPERMATOZOA AND OVA. Microscopic examination of spermatozoa gives little information as to their fertilizing power; at most it can be said that semen containing a large number of abnormally shaped spermatozoa is unlikely to bring about a pregnancy. Normal spermatozoa retain their motility long after they have lost the power to fertilize; motile sperms have been found in human cervical mucus three days after coitus and in the uterine tubes seven days after coitus. Direct experiments are required in order to solve the problem of viability. Animal experiments, because of enormous species differences, can give no clue to the viability of human spermatozoa or ova. The results of artificial insemination by donors (A.I.D.) indicate that the period during which human sperms are able to fertilize an ovum is not more than 24 hours. One factor limiting the survival

of spermatozoa after insemination is the relatively high temperature of the vagina and uterus as compared with that within the scrotum.

The ovum has an even shorter life than the spermatozoon. Histological examination shows that the unfertilized ovum quickly degenerates in the uterine tubes. It is not possible to obtain satisfying evidence in man but Farris thinks that the fertilizable life of the human ovum is not more than eight hours.

TIME OF OVULATION. In a number of mammals, including the rabbit and the cat, coitus produces reflexly a release of LH from the adenohypophysis and so brings about ovulation but it is almost certain that ovulation in woman occurs spontaneously, that is, it is not induced by coitus. The finding of an actual ovum is, of course, the best proof that ovulation has recently occurred and in a few cases human ova have been obtained by washing out the uterine tubes and collecting the fluid which emerged from the fimbriated end; the ova were found only at the middle of the cycle (about the fifteenth day) and not at other times. Inspection of human ovaries during abdominal operations has shown that ovulation usually occurs between the twelfth and fifteenth day of the menstrual cycle. The change from the proliferative to the secretory phase of the endometrium, which depends on the presence of a corpus luteum, also occurs about the middle of the cycle. Although it is difficult to obtain information about mating in the human subject analysis of cases in which conception resulted from a single insemination on a known day of the cycle shows, on the assumption that the ovum is short-lived, that ovulation can occur at any time between the sixth and twentieth day of the cycle. In agreement with this is the fact that artificial insemination in women is usually most successful about the middle of the cycle, although some women become pregnant only when inseminated during the first week. It has been claimed that the slight fall of body temperature followed by a slight rise, usually less than 0·5°C, which occurs about the middle of the menstrual cycle (Fig. 51.14) indicates the time of ovulation but direct inspection of the ovaries shows that the rise of temperature may occur before ovulation or may be delayed as much as three days after ovulation. The slight elevation of body temperature in the second half of the cycle has been attributed to the presence of progesterone in the circulation; at any rate it can be imitated by giving progesterone after a preparatory course of oestrogens in women with amenorrhoea (absence of menstruation). The claim that the cervical mucus contains increased amounts of a reducing substance at the time of ovulation has been abandoned. A maximum peak of oestrogen excretion occurs about the middle of the cycle and has been termed 'ovulation peak' and has been taken to indicate ovulation. The time of occurrence in the cycle may be as early as 4 days before or as late as 6 days after mid-cycle.

Administration of progesterone to a rabbit inhibits ovulation by reducing the output of pituitary gonadotrophin since she can be made to ovulate by injection of gonadotrophin. Presumably this mechanism prevents super-fetation in pregnancy. The capacity to ovulate in these animals in response to coitus returns shortly after cessation of the injections of progesterone. On the basis of this information Pincus in 1956 gave a group of normal women 300 mg of progesterone by mouth from the 5th to the 25th day of the menstrual cycle. Examination of endometrial biopsies and vaginal smears, as well as records of basal temperature, suggested that ovulation had not occurred and indeed there were no pregnancies in the group. A number of synthetic steroid compounds have progestogen effects when given orally to women (p. 1047); they act on the pituitary suppressing the formation of FSH and LH and they reduce the amount of glandular epithelium in the endometrium which causes a relative increase in stromal tissue. These 'pills' also alter the chemical and physical composition of cervical mucus so that spermatozoa may fail to penetrate into the uterine cavity. It has been suggested that these compounds prevent ovulation by a direct action on the ovaries and not indirectly through the pituitary.

In postmenopausal women dithiocarba-moylhydrazine derivatives inhibit the pituitary and reduce the urinary output of gonado-trophin. In normally menstruating women these non-steroid compounds, like the progestogens, suppress ovarian activity. Clomiphene, an analogue of a non-steroidal oestrogen, chlorotrianisene, stimulates the human ovary as judged by the increase in output of steroid hormones. It is used therefore to stimulate ovulation in infertile women with ovarian failure.

PREGNANCY

Conception occurs after a fertile mating in which the spermatozoa deposited in the vagina reach an ovum shortly after ovulation. Before coitus the skin of the male flushes. The heart rate increases and breathing becomes deeper. The penis becomes erect, increasing in length from 9·5 to 17 cm by engorgement of the erectile tissue of the corpora cavernosa and spongiosa. Their arterioles dilate and the veins in some way, not satisfactorily explained, obstruct the outflow. The dartos and cremaster muscles contract so as to elevate the testes. The testes themselves become congested. The passage of the penis into the vagina is facilitated by the mucus which exudes from its walls.

The nipples of the woman become erect by contraction of their smooth muscle, the areolae and breasts become engorged. The skin is flushed. Within 10 to 30 seconds of sexual stimulation the vagina, although it has no glands, exudes mucus. Later the clitoral glans becomes tumescent and the clitoris is retracted against the anterior border of the symphysis pubis. The vagina becomes lengthened and distended, especially the inner two-thirds. The outer third shows great vaso-congestion and is described as the orgasmic platform. The uterus is raised.

Coitus culminates in the orgasm in which semen is ejaculated by rhythmic contractions of bulbocavernosus and ischiocavernosus muscles, and of the smooth muscle of the vasa deferentia, epidydimis, prostate and seminal vesicles. Orgasm in the woman consists of rhythmical contractions of vaginal and uterine muscle and dilatation of the cervix which should occur simultaneously with the male orgasm.

After ejaculation of the highly buffered semen the vaginal pH increases to 7 and may not return to pH 4 for 10 hours or so. Sperm motility is favoured by a neutral pH and inhibited by a low pH. X-ray photographs of a cap filled with radio-opaque fluid of the consistency of semen and fitted over the cervix provide no evidence that the uterus sucks up vaginal fluid during coitus. After the orgasm the cervix remains dilated for 20 to 30 minutes.

Coitus is accompanied by psychic excitement in both partners and must be regarded primarily as a cerebral event although the reflex changes in the genitalia are governed by lumbar and sacral spinal centres. Stimulation of the sacral (parasympathetic) outflow in the pelvic splanchnic nerves (p. 861) causes erection of the penis or clitoris by dilatation of their blood vessels and these nerves have, therefore, been called the *nervi erigentes*. Somatic nerves arising in the sacral region of the cord pass out in the pudendal nerves to the ischiocavernosus and bulbocavernosus muscles in both sexes and, therefore, both somatic and autonomic nerves are concerned in producing erection.

The internal organs of reproduction in both sexes are supplied by sympathetic fibres from the lumbar outflow travelling in the hypogastric nerves. Stimulation of these nerves in nonpregnant women produces a contraction, followed by an inhibition, of the body and cervix of the uterus. After section of the hypogastric nerves in man the power of ejaculation but not of erection, is lost. Failure of ejaculation is common in patients with high blood pressure who are being treated with adrenergic-blocking drugs.

Although only one spermatozoon is necessary for the fertilization of an ovum, pregnancy does not occur unless many millions are deposited in the vagina. The male ejaculate (p. 1030) usually contains 100×10^6 to 200×10^6 spermatozoa per ml; fertile males have seldom less than 20×10^6 per ml. Only a tiny fraction of this number reaches the site of fertilization. Shettles has incubated human ova covered with adhering cumulus cells in the presence of tubal mucosa. The cumulus cells were dispersed; neither semen nor hyaluronidase (p. 48) had this effect. But the cumulus cells may not be such a barrier as has been supposed for the phase contrast microscope has shown male and female pronuclei and sperm tails in newly fertilized rat ova which were still covered with cumulus cells. Shettles found no evidence that the ova possessed a trophic influence on spermatozoa. The latter seemed to move about quite randomly and sometimes passed by quite close to the ovum. If however,

a spermatozoon touched the ovum it remained in contact. When the fertilizing spermatozoon passed through the zona pellucida into the ovum the zona became impenetrable to further spermatozoa. Freshly ejaculated spermatozoa do not appear to be able to enter the ovum; they acquire this capacity after a few hours stay in the reproductive tract of the female. During capacitation the membranes surrounding the acrosome (p. 1029) disintegrate and disappear. Progesterone, which is known to disrupt lysosomes, is secreted by the cumulus cells and it is tempting to suppose that this hormone acts on the sperm when it is in contact with the cumulus cells; the resultant disintegration of the external membranes probably liberates the enzymes necessary for penetration of the cumulus cells and the zona pellucida. Fertilization occurs usually in the uterine tubes, probably within 12 hours of coitus in man, and the ovum then passes along the tube into the uterine cavity, undergoing cleavage all the way, so that, as far as can be estimated, it is implanted in the endometrium five days after fertilization at the 16-cell stage. The transport may be facilitated by peristaltic or ciliary activity or by the secretion of the uterine tube which, presumably under oestrogenic stimulation, is greatly increased at ovulation. The secretions in the tube probably form a stream, which helps to convey the ovum to the uterus. Recently it has been suggested that these secretions may coat the ovum with a protein membrane called the oölemma which is a necessary prerequisite for the embedding of the ovum in the endometrium.

In the cow spermatozoa reach the uterine tubes $2\frac{1}{2}$ minutes after insemination—during this time they must travel about 50 cm. Since this far exceeds their unaided rate of progress (100 μm per sec) the spermatozoa must be transported by ciliary action or by uterine contractions which may be excited by oxytocin released from the neurohypophysis. The observations that coitus in man and animals causes ejection of milk from the mammary glands and that in cows mating causes strong uterine contractions afford circumstantial evidence for this release. It is known, however, that conception may occur after removal or denervation of the neurohypophysis. The biological advantage of rapid transport is not clear since, as mentioned above, newly deposited sperms do not have the capacity to fertilize.

When an ovum is fertilized the corpus luteum persists instead of degenerating at the end of the menstrual cycle and menstruation does not occur. The hormonal or nervous mechanism which is responsible for this is not known with certainty but it is likely that chorionic gonadotrophin (p. 1050) is produced very early by the embedding blastocyst in sufficient quantity to cause the pituitary to continue the production of LH which is necessary for the maintenance of the corpus luteum. This ensures a continuous production of progesterone to maintain the secretory changes in the endometrium and prevent it from breaking down in menstruation. Failure to menstruate at the expected date is thus one of the first presumptive signs of pregnancy in woman. The developing ovum reaches the uterus when the endometrium is in the premenstrual phase and sinks into it about the seventh day by digesting away the superficial layers; cells and cell debris taken up by the trophoblast presumably provide food material for the growth of the blastocyst. In the rat progesterone with oestradiol (but not progesterone alone) can successfully bring about implantation so long as the blastocysts survive; LH may also be involved. The endometrium continues to grow (Fig. 51.7) during pregnancy, since there is no menstrual degeneration, and it may eventually reach 10 mm or more in thickness. The three layers described in the premenstrual endometrium become more obvious (Fig. 51.16). The superficial layer of the stroma becomes compact and the greatly enlarged stromal cells are known as *decidual cells*. The value of this decidual reaction is not clear since it occurs a considerable time after implantation. It may protect the uterus against invasion by trophoblast. In the much thicker spongy layer below this the glands are widely dilated so that there is little stromal tissue. In the limiting or boundary layer next the uterine muscle the glands are tortuous but are not dilated. Since a large part of the endometrium is cast off with the placenta at parturition it is often referred to as the *decidua*; after parturition the endometrium is reformed from the limiting layer. In the decidua basalis under the developing embryo there is from an early stage a great dilatation of the maternal blood vessels and small finger-like outgrowths of the outer layer of the blastocyst, the *chorionic villi*, grow into them by erosion of the decidua. This penetration by the villi is aided by obliteration of small arteries of the decidua causing necrosis and the formation of large spaces in the decidua which fill with maternal blood. As the villi are soon invaded by mesoderm carrying fetal blood vessels, the fetal and maternal circulations are brought very close to one another and in this way the placenta is formed. The human placenta is described as haemochorial, since the chorionic villi dip directly into maternal blood spaces. There is, however, no direct connexion between the two vascular systems and blood does not normally pass from the mother to the fetus or vice versa. Intervening between the fetal and maternal circulations are only three thin layers (Fig. 51.17), namely the fetal vascular endothelium, the connective tissue of the villi and the trophoblast, across which the transfer of nutrient materials readily occurs. The fetus, because it is partly derived from its father, is a homograft and, since the gestation period is longer than the homograft rejection period, by all the 'laws' it should be rejected. This may happen occasionally. The embryo possesses transplantation antigens from a very early stage and cannot be immunologically immature. Fetuses removed from the uterus of pregnant rats and rabbits and transplanted in the abdominal muscles are rejected like any other homografts. Grafts placed in the pregnant uterus of animals survive for a considerable time if surrounded by decidual cells but there is as yet no explanation of the remarkable immunological privilege of the fetus. There may be an immunological buffer zone between mother and fetus—perhaps a layer of fibrinoid material.

In view of the increased blood flow to the uterus and other tissues it is not surprising to find that the cardiac output increases in pregnancy up to a maximum of 40 per cent. In early pregnancy most of the increase goes to the kidneys—the GFR is markedly increased—later the uterus receives a larger share. Perhaps as a consequence of the increased GFR there is a tendency to depletion of sodium and this in turn may be responsible for the raised levels of renin, angiotensin and aldosterone so frequently found in the maternal plasma.

Hyperventilation is a feature even of early pregnancy and causes a fall of alveolar P_{CO_2} but because of compensatory renal excretion of bicarbonate the blood pH remains unchanged.

As pregnancy advances the products of conception increase in size and eventually occupy the whole of the uterine cavity. The growing fetus attached by its umbilical cord to the placenta is bathed by the amniotic fluid which is contained within the 'bag of membranes' consisting of the amnion and chorion. The volume of the amniotic fluid can be estimated (see blood volume, p. 408) by the dye-dilution technique. The maximum volume, about 1100 ml, is reached at the 38th week of gestation, it decreases by 145 ml per week thereafter. Amniotic fluid is fairly freely exchanged with the maternal fluids and its protein and amino acid content approaches that of blood plasma; it is mainly an excretory product of the amniotic epithelium but it also contains urea and bile pigments (0·8 µg per ml) derived from the fetus. The fetus swallows amniotic fluid and absorbs water and solid constituents in its intestine; the water absorbed is excreted by the fetal kidneys into the amniotic cavity and is again swallowed by the fetus which in this way regulates the volume of the amniotic fluid. Renin is found in higher concentration in the amniotic fluid than in the maternal blood plasma; it may be excreted by the fetal kidney. Prostaglandins, E_2, $F_{1\alpha}$ and $F_{2\alpha}$, are found in the amniotic fluid especially at the end of pregnancy. Monoamine oxidase produced in the placenta passes into the amniotic fluid in large amounts. Amniotic fluid always contains a number of fetal cells and the sex of the fetus can be discovered by their sex chromatin content (p. 1102). The function of the amniotic fluid is to provide space for fetal growth and movements and to distribute the pressure due to uterine contractions evenly over the fetus.

THE HORMONAL CONTROL OF PREGNANCY

In all mammals the ovary is necessary in the early stages of pregnancy since the hormone from the corpus luteum (progesterone) is responsible for preparing the endometrium for the reception of a fertilized ovum but after this preliminary stage the ovaries can be removed in several species, including man, without disturbing the pregnancy. The placenta apparently provides all the hormonal material necessary for the continuation of pregnancy. Since both oestrogen and pregnanediol are excreted by pregnant oophorectomized women it can be assumed that oestrogen and progesterone are produced in the placenta by both cytotrophoblast and syncytiotrophoblast (Fig. 51.17).

The endocrine function of the conceptus begins in early pregnancy about the time of implantation when the primitive trophoblast secretes HCG which increases the life span of the corpus luteum that supplies oestrogen and progesterone necessary for the growth and development of the uterus and fetus. As pregnancy advances there are marked changes in the blood levels of certain hormones. It is now known that steroid metabolism during pregnancy involves a complex interrelationship between the mother, the placenta and the fetus. Almost all steroid hormones are derived from acetate in the form of acetyl-coenzyme A (p. 343). The placenta does not appear to be capable of synthesizing hormones from this simple precursor and therefore it must utilize maternal cholesterol, present in large amounts in the maternal blood, for steroidogenesis. Perfusion and incubation studies suggest that the fetus is capable of synthesizing steroids from acetate but the main precursor of steroids in the fetus is pregnenolone (Fig. 51.13) which is synthesized in the placenta from maternal cholesterol. The fetal adrenals appear to produce a substance that is either a precursor metabolized by the placenta to oestrogens or is a stimulus to placental production of oestrogens. Since the maternal adrenal glands contain both oestrogen and progesterone they may play a part in the maintenance of pregnancy. The main oestro-

Fig. 51.16 Decidua (the modified endometrium) at the 10th week of pregnancy. Stained with haematoxylin and eosin. A. An area at a low magnification ($\times 13$) showing the thickened uterine lining divided into a deep *stratum spongiosum* and a superficial *stratum compactum*. B. An area in the *stratum compactum* showing the greatly enlarged stromal or *decidual* cells ($\times 70$). C. An area in the *stratum spongiosum* showing the large *pregnancy glands* ($\times 70$). (*By courtesy of W. W. Park.*)

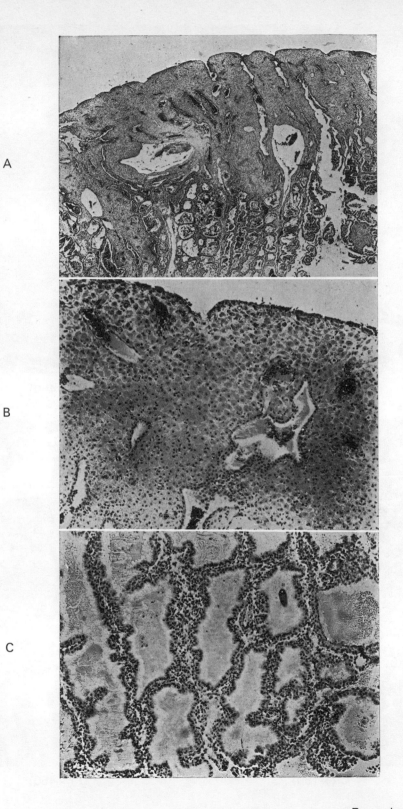

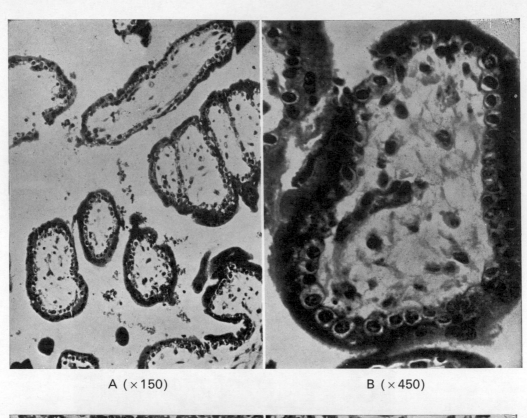

A (×150)　　　　　　　　　　　　　B (×450)

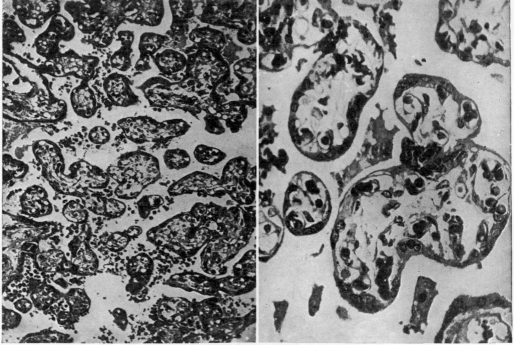

C (×150)　　　　　　　　　　　　　D (×450)

gen in placental tissue is oestriol, oestrone and oestradiol being present in much smaller amounts. Progesterone can be extracted from the human placenta; the total amount increases as pregnancy advances. At term the placenta produces about 250 mg progesterone per 24 hours. The human placenta has a high gonadotrophin content, especially early in pregnancy, but the function of the gonadotrophic hormones is not clear; since a high excretion of these hormones occurs early in pregnancy in women (Fig. 51.24) at a time when the corpus luteum is active, they probably maintain the secretion of progesterone by the corpus luteum until this function is assumed by the placenta. In agreement with this idea is the finding that injection of chorionic gonadotrophin into non-pregnant women increases the pregnanediol excretion and delays the onset of the next menstrual period. The corpus luteum may therefore be maintained either by gonadotrophins or by oestrogens.

PRODUCTION AND EXCRETION OF HORMONES IN PREGNANCY

Oestrogens. The oestrogens are excreted in the urine chiefly as glucuronides, but some may be excreted as sulphates. The daily output rises during pregnancy. Figure 51.18 shows the excretion of oestriol, the main oestrogen metabolite, throughout pregnancy. The high output (about 1000 times that in the non-pregnant state) reached at the end of gestation is maintained up to parturition after which the excretion rapidly declines to the non-pregnant level. However, the amount of oestrogen eliminated in the urine is not a reliable guide to the amount produced in the body, since the proportion metabolized in the body, mainly in the liver, is quite unknown (see p. 1044).

The plasma levels of oestrogens in the maternal peripheral veins rise as pregnancy advances (Table 51.19). The fetus is probably actively engaged in the metabolism of placentally produced oestrogens, because it has been shown that, when a pre-viable fetus is injected with diluted blood containing oestriol or oestradiol-17 β, increased amounts of conjugated oestrogens are found in fetal tissues. The fetal adrenals can convert circulating pregnenolone sulphate into dehydroepiandrosterone which is a very important oestrogen precursor readily transformed by the placenta into oestrone and oestradiol-17β but not oestriol. The placenta cannot synthesize oestrogens from simple precursors: it depends on steroids mainly of fetal origin. The fetus by its facility for conjugation appears to thrive in an environment free from highly active oestrogens. However, the highly individual oestrogen environment may help to promote fetal growth. The oestriol levels in the umbilical vessels at term are much higher than those in the maternal peripheral vessels (130 μg/100 ml plasma in the umbilical vessels and 18 μg/100 ml in the maternal peripheral blood). The production of oestrogens depends on there being a live fetus with functional adrenal glands and an intact feto-placental circulation. After fetal death the urinary excretion of oestriol falls; in the presence of an anencephalic fetus with poorly developed adrenals the oestriol excretion is low. The placenta possesses enzymes lacking in the fetus but the fetus has enzymes lacking in the placenta; the two together, the feto-placental unit, is a complete system for steroid biosynthesis.

Progesterone and pregnanediol. A very small amount of progesterone, only 5 μg per litre in late pregnancy, is excreted in the urine, but large quantities of pregnanediol are excreted.

Fig. 51.17 Sections of human placental tissue. A and B show chorionic villi at the 10th week of pregnancy. C and D show chorionic villi at term. The villi have more numerous and finer terminal branchings than in the earlier placental tissue. Haematoxylin and eosin. The early villus consists of a core of mesodermal tissue containing an occasional blood vessel surrounded by an inner and an outer layer of trophoblastic epithelium. The cells of the inner layer are relatively large and clear and form the Langhans' cell layer or *cytotrophoblast*. The outer layer is composed of smaller, darker nuclei lying irregularly in a background of cytoplasm which shows no cell boundaries: this is the syncytium or syncytial layer or *syncytiotrophoblast*. As the villus ages the cells of the cytotrophoblast become fewer and usually disappear entirely, leaving only the irregular syncytiotrophoblast as a covering for the villus: at the same time the blood vessels within the core of the villus gradually increase in number until they may occupy 50 per cent of the cross-sectional area of a villus during the last weeks of pregnancy. (*By courtesy of W. W. Park.*)

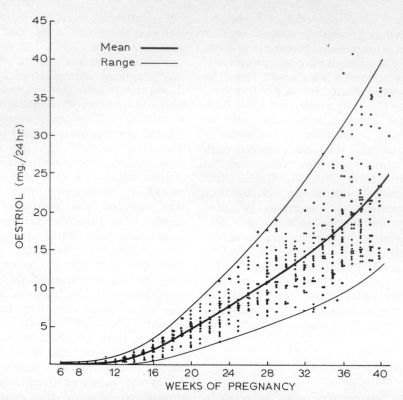

Fig. 51.18 Urinary oestriol excretion during 36 normal pregnancies (baby weight over 3 kg) showing fitted mean, maximum and minimum ($P = 0·05$) values. (Coyle, M. G. & Brown, J. B. (1963). *Journal of Obstetrics and Gynaecology of the British Commonwealth* **70**, 225.)

Table 51.19 Oestrogens in maternal plasma (mean readings in µg/100 ml

Oestrogens	Weeks of pregnancy							
	12	16	20	24	28	32	36	40
Oestriol	0·20	0·75	1·63	2·43	2·79	3·26	5·99	9·71
Oestrone	0·39	0·66	1·63	2·80	.3·73	3·92	4·53	5·44
Oestradiol	0·05	0·10	0·27	0·53	0·65	0·90	0·89	1·19

From Evelyn J. Roy & Rachel Mackay (1962). *Journal of Obstetrics and Gynaecology of the British Commonwealth* **69**, 13–17.

Figure 51.20 shows that the excretion increases during pregnancy, rises to its highest value at the 36th to 38th week and falls a little in the last weeks of pregnancy but the excretion is still high at the time of labour; delivery is followed by a rapid drop in excretion (Fig. 51.23).

The mother's plasma contains from 1 to 5 µg of progesterone/100 ml in early pregnancy, and as pregnancy advances the level rises to an average of 15 µg/100 ml (Fig. 51.22). Radioactive studies have shown that about 75 mg of progesterone are produced per day at mid pregnancy and about 250 mg per day at the end. A metabolite of progesterone, 20α-hydroxyprogesterone, also found in maternal peripheral plasma, increases as pregnancy advances to a maximum of 10 µg/100 ml.

The plasma progesterone level remains high during labour but a few minutes after parturition progesterone cannot be detected. The rapid disappearance is no doubt related to the fact that the half-life of progesterone in the plasma is about 5 minutes. The levels of plasma progesterone are higher in the maternal uterine vein, which is nearer the site of production, than in the maternal arm vein (Fig. 51.21) and are even higher in the fetal umbilical vessels. The level is, however, higher in the umbilical vein 102 µg/100 ml than in the umbilical artery 57 µg/100 ml. The fetus uses

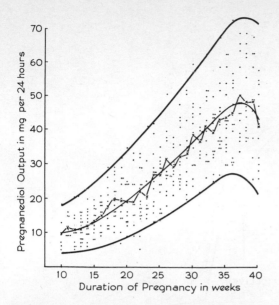

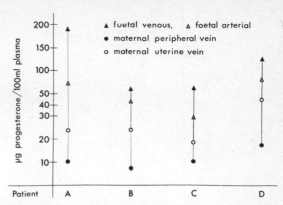

Fig. 51.21 Maternal and fetal blood plasma progesterone levels in 4 patients A, B, C and D. (M. Greig, M. G. Coyle, W. Cooper & J. Walker (1962). *Journal of Obstetrics and Gynaecology of the British Commonwealth* **69**, 772.)

Fig. 51.20 Scatter diagram of pregnanediol excretion in normal pregnancy (10th to 40th week). Smooth curves have been calculated and drawn to represent the average excretion (fine line) together with the limits of 'normal' excretion. (C. S. Russell, C. G. Paine, M. G. Coyle & C. J. Dewhurst (1957). *Journal of Obstetrics and Gynaecology of the British Empire* **64**, 650.)

the progesterone produced in the maternal placenta for the production of corticosteroids.

Gonadotrophic substances. The presence of gonadotrophins in the urine of pregnant women was first demonstrated by Aschheim and Zondek in 1928, and the detection of these

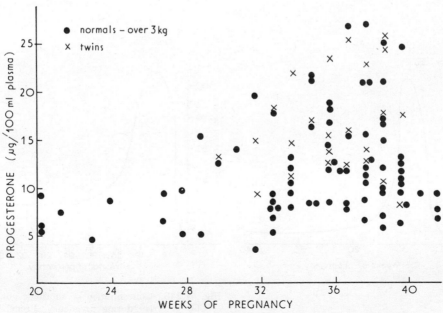

Fig. 51.22 Concentration of progesterone in the plasma of normal single and twin pregnancies. (M. Greig, M. G. Coyle, W. Cooper & J. Walker (1962). *Journal of Obstetrics and Gynaecology of the British Commonwealth* **69**, 772.)

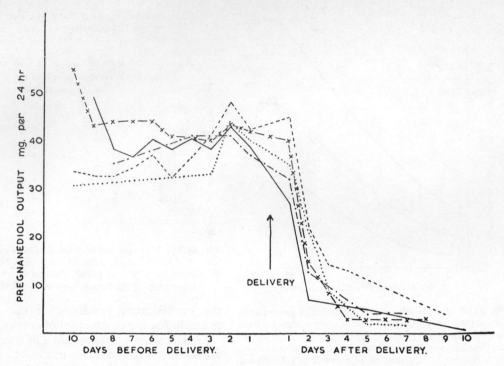

Fig. 51.23 Urinary excretion of pregnanediol in women before and after labour. (M. G. Coyle, F. L. Mitchell & C. S. Russell (1956). *Journal of Obstetrics and Gynaecology of the British Empire* **63**, 564.)

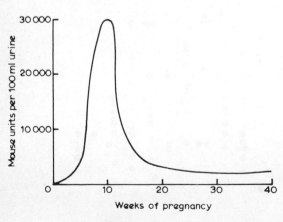

Fig. 51.24 Concentration of HCG in *urine* during pregnancy.

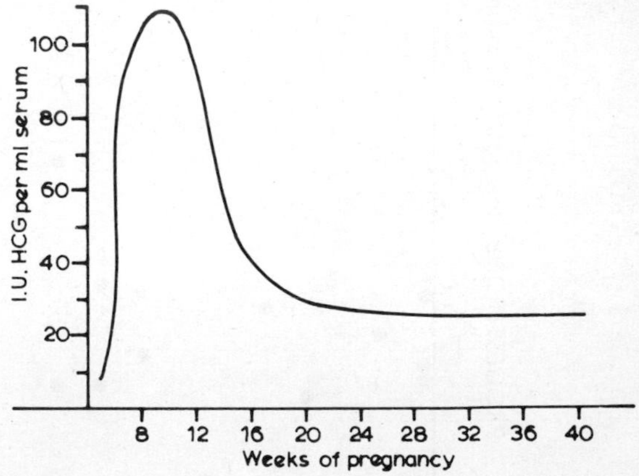

Fig. 51.24b Concentration of chorionic gonadotrophin (HCG) in *serum* during pregnancy. From the data of Mishell, Wide & Gemzell (1963). Figures are from F. E. Hytten & I. Leitch (1964). *The Physiology of Human Pregnancy*. Oxford: Blackwell.

substances forms the basis of all the reliable human pregnancy diagnosis tests. Quantitative measurement (Fig. 51.24) shows that gonadotrophin (CG) excretion begins about the time of the expected, but missed, menstrual period, rises to a peak about ten weeks later and then falls to a much lower but still easily detectable value, around which it fluctuates until the end of pregnancy (Fig. 51.24). The amount of gonadotrophin in the serum parallels this curve. The amount of gonadotrophin seems to depend on the number of cells of the cytotrophoblast in the placenta. It is known that by standards of light microscopy the number of these cells increases rapidly in early pregnancy and then the number decreases until they can scarcely be seen at term. Within a few days after parturition the excretion falls to non-pregnancy levels (the test in fact is now measuring LH). After it is injected gonadotrophic material from the pituitary is partly destroyed in the body and partly excreted and, therefore, the amount eliminated in the urine may be much less than that actually produced in the pituitary and placenta.

Growth hormone. The amount of growth hormone in the maternal circulation during pregnancy and lactation is much increased. High levels of growth hormone are also found in the fetus (see p. 1085).

PREGNANCY DIAGNOSIS TESTS

In pregnancy the placenta produces very large quantities of HCG, which is for all practical purposes identical with LH. However, the quantities of HCG in pregnancy are of an entirely different order of magnitude from those of LH in the menstrual cycle. The level of HCG increases very rapidly early in pregnancy (see Fig. 51.24) and this is the basis of the laboratory method of confirming pregnancy. Until the mid 1960s biological tests were used to detect the presence of HCG based on the original work of Aschheim and Zondek in 1927 but recently immunological tests have superseded biological tests. The principle of the immunological tests is as follows. HCG is prepared from the urine of pregnant women and injected into a rabbit; because HCG is a foreign protein it acts as an antigen and the rabbit produces anti-HCG antibodies in its serum. When a suspension of latex particles (or of red cells) is coated with HCG and mixed with anti-HCG serum agglutination of the particles (or red cells) occurs due to the reaction between the HCG and the anti-HCG. If urine containing HCG (that is urine from a pregnant woman) is added to anti-HCG serum the reaction between the HCG and anti-HCG leaves no free anti-HCG. On subsequent addition of latex particles coated with HCG no agglutination occurs. The absence of agglutination indicates that the urine contained HCG and constitutes a positive pregnancy test.

HCG is produced by the cytotrophoblastic cells of the placenta even in the entire absence of a fetus. Indeed the urinary excretion in most hydatidiform moles where no fetus is present is even higher than in pregnancy; in fact the finding of such high levels may influence the differential diagnosis. In the rare cancer of pregnancy, choriocarcinoma, which can follow a normal pregnancy but more often follows hydatidiform mole, HCG is also excreted in considerable quantities and this is of great diagnostic value. The state of the tumour can be accurately monitored in these patients and recrudence of growth can be quickly detected. This unique possibility of sensitive monitoring of the tumour growth has contributed greatly to the extremely successful treatment now possible.

WATER RETENTION IN PREGNANCY

The total body water in pregnant women has been measured by the deuterium oxide method (p. 675) and the results are summarized in Table 51.25. There is good agreement between the estimates and measured increases in water for women with little or no oedema at 20 and 30 weeks. The discrepancy at term of 1 to 2 litres must be due to additional water storage.

In pregnancy the blood volume increases so that at term it may be 20 to 100 per cent (average 30 per cent) over the non-pregnant level. Since the average increase in total red cells is about 20 per cent the haemoglobin concentration in the peripheral blood falls but this is not a true anaemia. After delivery the blood volume returns to the non-pregnant level in a few hours.

The diuresis after drinking water is smaller in a pregnant than a non-pregnant woman. During the greater part of pregnancy sodium

Table 51.25 The water component of weight gain in pregnancy compared to the measured increase in body water

Component	Water content (g)		
	20 weeks	30 weeks	40 weeks
Fetus	264	1185	2343
Placenta	153	366	540
Liquor amnii	247	594	792
Added uterine muscle	483	668	743
Added mammary gland	135	270	304
Plasma	506	1058	920
Red cells	32	98	163
Total	1820	4239	5805
Measured increase			
No oedema	1740	4300	7500*
Leg oedema	1810	4290	7880*
Generalized oedema	2230	5740	10 830*

* Extrapolated from 30–38 week gain.
From F. E. Hytten, A. M. Thomson & Nan Taggart (1966). *Journal of Obstetrics and Gynaecology of the British Commonwealth* **73**, 553–561.

is retained to the extent of 3 g per week, and this retention would lead to a rise in the osmotic pressure of the extracellular fluid unless more than 300 ml of water are also retained. The plasma volume increases gradually and reaches 125 per cent of the non-pregnant value at term, but, since the colloid osmotic pressure, because of the fall in the serum albumin from 3·8 g to 3 g per 100 ml, is actually some 20 per cent lower than in the non-pregnant woman, it is difficult to see the reason for the increase. The retention of water and sodium which results in a deuterium space of 38 litres at term may be related to the increased production of oestrogens in pregnancy, since oestrogen therapy in women has produced an increase of up to 20 per cent in blood volume with a lowered haematocrit. The increase in plasma progesterone may be responsible for the increase in aldosterone excretion up to 100 µg per day; its sodium retaining effect could account partly for the increase of water in the body. After parturition the excretion falls to the non-pregnant value of 10 µg per day.

Newton found that mice increase in weight during pregnancy and regain their initial weight after parturition. If the fetuses are destroyed about the fourteenth day of pregnancy by applying pressure through the abdominal wall, the body weight is maintained at the high pregnancy level so long as the placentae are retained. When the placentae are expelled at full term (twentieth day) there is a rapid loss of weight and a return to the level before pregnancy; the loss in weight is probably due to a loss of body water and is about eight times greater than the weight of the expelled placentae. Similar changes in body weight are observed in monkeys in which the fetus has been removed without disturbing the placenta. The hormonal mechanism of water retention is not known, but progesterone and aldosterone may be involved. The plasma renin activity doubles during pregnancy. The plasma volume in woman returns to the non-pregnant value about 2 months after delivery.

The figures quoted above are for water only. The total gain in weight during pregnancy is about 12·5 kg.

The gain in weight is greater than would be expected from the gain in water. The difference can only be accounted for by an increase in depot fat. A gain of 12·5 kg is an addition of 25 per cent to the body weight of an 8-stone (50 kg) woman. An even greater average increase in weight has been reported by other investigators but many obstetricians now advise that the gain in weight during the whole of pregnancy should not exceed 12·5 kg. A sudden gain of weight, say 0·5 kg or more in a week in the later stages of pregnancy, is often a sign of impending toxaemia.

Thomson and Hytten calculate that about 0·8 kg protein and about 3·5 kg fat are added during pregnancy. The protein is mainly in the uterus and its contents while the additional fat is laid down in the maternal (non-reproductive) tissues. The total value of the food material stored during pregnancy is about 170 MJ or 40 000 kcal. The metabolism of the fetus, uterus and additional maternal tissue gradually increases as pregnancy proceeds; the cumulative requirement for these purposes is around 150 MJ or 35 000 kcal. The total additional requirement is therefore about 320 MJ or 75 000 kcal. It is difficult to make allowance for the daily activities of the mother—this depends much on her domestic circumstances. A primigravida increases her intake (over the pre-pregnancy level of about 8·4 MJ or 2000 kcal per day) quite early in pregnancy and on the average consumes during the last two-thirds of pregnancy about 1·5 MJ or 350 kcal extra per day. Often, however, the mother tends to reduce her activity and may consume only half of this extra amount. The diet recommended in pregnancy is described on p. 203. The increase in maternal stores (about 3 kg) which the mother possesses after parturition may be used up in breast feeding. In pregnancy in women the β cells of the pancreas increase in number and the serum insulin rises. This suggests that pregnancy puts a strain on carbohydrate metabolism and indeed pregnancy is occasionally followed by diabetes mellitus.

FUNCTIONS OF THE PLACENTA

The early stages in the formation of the placenta have already been described (p. 1055). The organ increases in weight steadily throughout pregnancy and at term its weight is about 500 g, that is one-sixth of that of the child. Patches of degeneration commonly seen in the placenta at the end of pregnancy are apparently due to local coagulation of the maternal blood in the area affected. The circulation through the placenta is illustrated in Plate 51.2.

A 3 kg human fetus requires 15 to 25 ml oxygen per minute. If the arterio-venous oxygen difference of the blood supplying the uterus is 7 ml per 100 ml the fetal requirement could be met with a blood flow of 200 to 350 ml per min. Since the blood flow, estimated by the Fick principle, is from 500 to 750 ml per min it seems likely that the fetus could obtain all the oxygen it requires. The uterine blood flow has, however, to supply oxygen to the uterus and placenta as well as to the fetus.

The human placenta can be localized by exploring the abdominal wall over the uterus with a counter to find the area of highest radioactivity after an intravenous injection of ^{24}NaCl or after labelling the mother's red cells with ^{11}Co. If 10 microcuries of ^{24}NaCl is injected into the placental site it is found that the half-period of clearance is about 20 sec. The volume of the placental pool is difficult to estimate but if it is assumed to be 250 ml the blood flow can be calculated to be about 600 ml per min. The blood flow is reduced by uterine contractions and by exercise.

The umbilical vein taking fetal blood from the placenta has a P_{O_2} up to 40 mm Hg and the umbilical artery has a P_{O_2} of 8 to 16 mm Hg immediately after delivery but probably 20 to 30 mm Hg *in utero*. The P_{O_2} gradient across the placenta, that is between the intervillous space (containing maternal arterial blood) and the umbilical vein, is therefore considerable; since the placental exchange membrane at full term is about 14 m^2 and the amount of tissue between the fetal vessels is quite small (see Fig. 51.17) oxygen can easily cross the placental membrane by diffusion. The P_{CO_2} difference between the umbilical artery and the intervillous space is said to be about 7 mm Hg but this is likely to be too high a value.

The P_{O_2} of the blood going to the fetus is remarkably low when compared with the adult value of 100 mm Hg for arterial blood at sea level. The fetus, unlike people living at a high altitude, is hypoxic and hypercapnic. The fetus might be said to be better equipped for oxygen transport than its mother. It has a higher red cell count, a faster heart rate and a larger heart (relatively) than has the adult. The oxygen carrying capacity of the baby's blood at term is 20 to 25 ml per 100 ml and its haemoglobin can be saturated at a lower P_{O_2} than the maternal haemoglobin if both are at the same pH; it appears that the pH gradient *in vivo* is about 0·02 so that *in vivo* the fetal blood has a somewhat greater affinity for oxygen. It is difficult to account for this 'shift to the left' (see p. 620). It may be caused by the Bohr effect, that is by loss of CO_2 which passes through the placenta to the mother's blood but it cannot be entirely explained by the fact that three-quarters of the fetal haemoglobin is in the fetal form. If the fetus were usually hypoxic, reduction of atmospheric pressure should be very dangerous but this is not so. The fetuses of sheep living at 15 000 feet (4570 metres) above sea level showed normal P_{O_2} levels and there was no rise in blood lactate such as would indicate anaerobic metabolism. The placental oxygen gradient was however half the normal. The adaptation to high altitude can be attributed partly to an increase in the amount of haemoglobin in the fetal circulation. These findings show that at sea level the fetus may not in fact be hypoxic or at least that it is not called on to use up all its powers of adaptation to hypoxia.

Because of the inaccessibility of the fetal and uterine vessels quantitative information is so meagre that it is impossible to describe gas transfer across the placenta in anything more than a tentative way. Samples of blood for analysis can be obtained only in special circumstances, for example after the fetus is delivered from the uterus—but the P_{O_2} and P_{CO_2} in the various parts of the fetal circulation after delivery are unlikely to be the same as they were when the child was still *in utero*. Again a blood sample can be obtained from the intervillous space but we have no way of locating the needle relative to the inflow or outflow. The direction of the maternal blood flow in the intervillous space is quite unknown.

The maternal blood sugar is higher than the fetal and a rise in blood sugar in the maternal circulation is followed by an increase in the fetal blood sugar. Since the latter returns quickly to the resting value, the sugar must be utilized or stored in the fetal tissues. Glucose passes through the placenta from a zone of high concentration to a zone of lower concentration but the amount transferred is not proportional to the difference in concentration as would be expected in simple diffusion. A carrier system has been postulated but there is no clue as to its nature. The fetal liver glycogen varies more rapidly than the glycogen stored in the decidual tissue of the placenta; the former fluctuates with the fetal blood sugar while the latter is relatively constant in the absence of excessive tissue demands. Insulin does not appear to pass across the human placenta.

Fat and also a lipase have been found in the human placenta but there is no evidence that any hydrolysis of fat occurs there. By injecting phospholipids marked with ^{32}P into the circulation of pregnant rabbits it has been found that although the placenta takes up large amounts of phospholipids it does not transmit them to the fetus. Fat seems to be able to pass through from the maternal to the fetal circulation but the mechanism is not known. Some of the fetal fat is derived from carbohydrate. When labelled inorganic phosphate ($Na_2H^{32}PO_4$) is injected into a pregnant rabbit, its plasma phospholipids are after a time labelled with ^{32}P. The radioactivity of the phospholipids extracted from the fetal liver and placenta is much higher, 4 hours after injection, than that of the fetal or maternal blood phospholipids. These two fetal organs must therefore be sites of active synthesis of phospholipid.

The free amino acid concentration in maternal plasma is about 25 per cent lower than that in a non-pregnant woman. The concentration in the fetal blood (umbilical vein) is higher than that in the maternal blood; the fetal/maternal concentration ratio varies for the different amino acids between 1·2 and 4. Injection of amino acids into the mother increases their concentration in the fetal blood. As might be expected from the rate of increase in fetal weight, protein synthesis in the fetus is at its maximum at the end of pregnancy and must depend on active transfer of amino acids across the placenta. Although there is a proteolytic enzyme in the placenta its function, if any, in protein transport is unknown. The passage of the Rh-antigen from the fetus to the mother and the return passage of the anti-Rh material into the fetus (p. 412) show that unchanged protein must be able, although in nutritionally insignificant amounts, to pass between mother and child.

Even if the mother is anaemic the child is usually born with a normal haemoglobin concentration. It may, however, quickly develop anaemia since even in the most favourable circumstances the fetal stores of iron are relatively low. Studies with ^{51}Fe as a ferrous salt have shown that a pregnant woman absorbs a much larger proportion of ingested iron than a non-pregnant woman and that the iron may pass very quickly via the placenta to the fetus. A pregnant woman must absorb 400 to 500 mg Fe to supply the fetus and placenta and the increased red cell mass; because of amenorrhoea some iron is saved but the usual loss of 1 mg Fe per day continues (p. 424). Anaemia is, therefore, fairly common in pregnant women. Although folic acid is often deficient in the last few weeks of pregnancy only a few women show a megaloblastic anaemia at that time (cf. p. 423); 300 µg of folic acid per day is sufficient to prevent deficiency. The fetal blood calcium and inorganic phosphorus are always above the maternal level and the considerable demands made on the mother for calcium and phosphorus may exceed her intake of these materials; calcium and phosphorus are then drawn from the mother's bones and teeth. For these reasons the daily intake of calcium and phosphorus in the last three months of pregnancy should be at least 1·5 g and 2 g respectively.

Vitamin A passes through the placenta to the fetal liver, especially if the diet contains fat. In deficiency of this vitamin in man and animals fertilization and implantation may fail and in some animals, if the deficiency is gross, fetal malformation may occur. Vitamin C passes from mother to fetus as dehydroascorbic acid and gets quickly into the fetal red cells. Both vitamin D and vitamin E also pass through the placenta. The former is stored in the fetal liver; the latter is essential for the maintenance of pregnancy in some animals, though not necessarily in man. The importance of vitamin K in the prevention and treatment of haemorrhagic disease of the newborn is referred to on p. 197. All anaesthetics used to relieve pain at parturition pass through the placenta and if given in sufficiently high dosage to the mother they produce fetal anaesthesia.

In general substances of small molecular weight such as oxygen, carbon dioxide and perhaps glucose (mol. wt. 180) pass through the placental barrier in either direction by diffusion. Substances of higher molecular weight either do not pass at all (for example, sucrose, mol. wt. 342) or are transported in some special way in which numerous enzymes, notably phosphatase, are likely to be concerned.

The placenta acts as a heat exchanger between fetus and mother; the fetal temperature is about 0·5°C above the deep body temperature of the mother.

The placenta functions adequately for about 40 weeks only. If parturition is delayed the gaseous and other needs of the fetus may not be adequately met. The child born after a prolonged pregnancy has a reduced chance of survival.

THE FETUS

The human fetus is very small up to the middle of pregnancy and consequently the nutritional demands from the mother's point of view are negligible, but from this time onward the protein intake should be increased by about 6 g per day to provide for the formation of the fetus, placenta and mammary glands. More than half the weight is acquired in the final six to eight weeks (Table 51.26) and during this time, if the mother does not receive sufficient food, the fetus draws on the maternal

Table 51.26 Composition of the human fetus in grams

	Body	Water	Protein	Fat	Ash
End of 5th lunar month	300	260	22	3·5	1·5
End of 7th lunar month	1000	800	100	25·0	30
End of 10th lunar month (birth)	3200	2420	400	350·0	90
Average of daily deposition:					
1. Throughout fetal life	11·4	8·6	1·4	1·25	0·32
2. In last three lunar months	26·2	19·3	3·57	3·87	0·71
3. In last month	35·7	23·6	6·4	5·0	2·0

A. St G. Huggett (1941). *Physiological Reviews* **21**, 442.

tissues. The birth weight of the infant is not, therefore, governed by the mother's diet unless there is severe undernutrition. Individual birth weights of the children of a multiple pregnancy are on the average lower than those of children of single birth. The mean birth weights of singletons, twins, triplets and quadruplets born in Britain are respectively 3·38, 2·40, 1·82 and 1·40 kg. The rate of fetal growth in multiple pregnancies may be retarded in the last few weeks, perhaps because the smaller size of the placenta restricts fetal growth or because the maternal circulation cannot supply sufficient nutrients.

During the last months of pregnancy the glycogen content of the liver, skeletal muscles and heart rises to several times the adult level. These easily available carbohydrate reserves may help the child to survive the changes in food and oxygen supply as well as the drop in environmental temperature occurring after birth. Brown adipose tissue can be recognized in 28-week human fetuses. It occurs chiefly between the muscles of the neck and back and the full-term infant has a considerable store of brown fat in the subcutaneous white adipose tissue. Exposure to cold after birth increases his oxygen consumption and therefore his heat production but he shivers very little. It is likely that cold causes release of adrenaline which increases the metabolism of the brown adipose tissue. (See p. 334 and Fig. 42.12.)

By means of artificial insemination Walton and Hammond were able to cross a Shire mare with a Shetland stallion and a Shetland mare with a Shire stallion and to show that the birth weight of the progeny was determined by the size of the mare; even after birth the growth of the foal showed only a small paternal influence. There is evidence of a similar maternal factor determining the size of the offspring in human beings.

At least two other factors influencing birth weight are social class and prematurity. On the average the infant's birth weight is lower, the lower the social group to which the mother belongs; the reason for this is undoubtedly complex and the differences are not likely to be attributable to dietary causes alone. Premature birth, or birth at any age of an infant which is small for its gestational age is associated with a higher death rate during the first month of extrauterine life (neonatal mortality). Furthermore a child whose birth weight is much below the average may for a long time be unable to reach the weight appropriate for his age, though this does not mean that he is weakly or mentally retarded.

Fetal circulation. The sheep is a very convenient animal for investigating the fetal circulation because the uterus does not contract as soon as the lamb is delivered. Also catheters may be implanted chronically into the fetal vessels and fluids *in utero*. When a lamb is delivered and the umbilical cord is tied, respiration begins and a number of circulatory adjustments takes place. The pathway taken by the circulating blood can be studied before and after respiration has begun by taking a number of X-ray photographs of the fetus during the injection into the circulation of a radio-opaque material. The information afforded by this angiographic method has been supplemented by Dawes and his colleagues at Oxford who have measured the percentage oxygen saturation of the blood at various points in the fetal circulation of the sheep (Fig. 51.28).

The fetal circulation (Fig. 51.27A and Plate 51.1) is quite different from that in the adult, chiefly because the heart has to pump the blood through the placenta as well as through the fetal body. Of the blood passing down the descending aorta only a small part goes to the lower limbs, the major part passing by the two *umbilical arteries* to the fetal side of the placenta. After taking up oxygen there the blood is returned in the *umbilical vein* and flows either through the liver or by means of a by-pass, the *ductus venosus*, directly to the inferior vena cava which receives also the blood from the alimentary canal and from the lower extremities. At the cardiac end of the inferior vena cava the blood is divided by the edge of the interatrial septum, the *crista dividens*, into two streams, a large stream which goes directly through the foramen ovale into the left atrium and a somewhat smaller stream which reaches the right atrium. In the left atrium the blood coming from the inferior vena cava is mixed with a small pulmonary bloodstream and then passes to the left ventricle which pumps it to the arch of the aorta. This blood, the most highly oxygenated blood available in the systemic arteries of the fetus, first supplies the heart by the coronary arteries and then the head and upper part of the body, the remainder reaching the descending aorta. The reduced blood returning from the head drains into the superior vena cava and right atrium where it is mixed with the reduced blood coming from the coronary sinus and with the right stream from the inferior vena cava which has a high oxygen content. The blood then passes to the right ventricle, the output of which is delivered partly into the lungs via the pulmonary artery but mainly to the aorta through the *ductus arteriosus*, a wide vessel about the same diameter as the aorta at the point where they meet. In this way the main stream of blood from the right ventricle by-passes the lungs and is distributed to the thorax and extremities as well as to the placenta through the umbilical arteries.

Figure 51.28 shows the percentage oxygen saturation of blood in eight of the principal vessels of the mature lamb before respiration is allowed to take place. From these figures the distribution of the blood flow can be calculated. The right atrium receives blood of 67 per cent saturation from the inferior vena cava and of 25 per cent from the superior vena cava and these streams are mixed and ejected into the pulmonary artery at a saturation of 52 per cent. This means that nearly two parts of inferior caval blood are mixed with one part of superior caval blood. On the other hand the left atrium receives inferior caval blood and pulmonary vein blood in the ratio of 4·5 to 1. The greater proportion of blood flow from the inferior vena cava into the left side of the heart explains the observation that the oxygen content of the

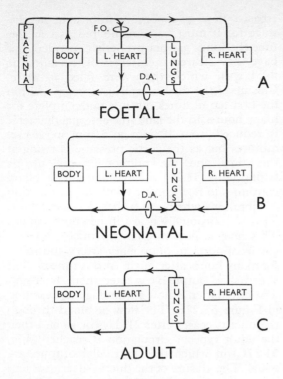

FOETAL   A

NEONATAL   B

ADULT   C

Fig. 51.27 Much simplified diagrams to illustrate the changes in the circulation at birth as described in the text. In A the two ventricles are working in parallel to drive blood from the great veins to the arteries. B is the condition reached a few minutes after birth when the cord is tied and the foramen ovale F.O. closes. When the ductus arteriosus D.A. finally closes the adult circulation C is established with the two ventricles working in series. (G. V. Born, G. S. Dawes, J. C. Mott & J. G. Widdicombe (1954). *Cold Spring Harbor Symposia on Quantitative Biology* **19**, 106.)

brachiocephalic artery supplying the head is larger than that of the umbilical artery. The fetal tissues receive oxygen at only one-third of the pressure obtaining in the adult. Inspection of Figure 51.27A shows that, since the two ventricles are not in series (as they are in the adult, Fig. 51.27C), there is no need for their outputs to be equal. There is little difference in the thicknesses of the walls of the two ventricles in the fetus and presumably little difference in the work done by them. The flow through the various tissues as a percentage of the combined output is, lungs, 10; other fetal tissues, 35; placenta, 55; foramen ovale, ductus arteriosus and aortic isthmus, 40. These figures have been confirmed by other methods retaining the fetus *in utero*.

When the mother is made to breathe air containing a low percentage of oxygen the fetal oxygen saturation falls; fetal heart rate and blood pressure increase and umbilical blood flow also increases. Since the increased umbilical flow and rise of blood pressure can be produced by injecting adrenaline or noradrenaline into the lamb, it is possible that the mechanism compensating for maternal asphyxia is a reflex release of sympathetic amines originated by stimulation of the chemoreceptors of the aortic bodies. When the fetus is very severely asphyxiated and the arterial oxygen saturation falls below 15 to 20 per cent the heart begins to slow; this is a useful indication to the obstetrician of fetal distress.

At a normal birth the umbilical cord is torn as the animal is delivered and the umbilical arteries constrict. The umbilical vessels are extremely sensitive to stretch and also contract firmly if, for example, an attempt is made to insert a needle into them to obtain blood. This is a local response of the smooth muscle and not a reflex adjustment since these vessels, which are outside the fetal body, have no nerves.

At birth the cord is tied (or in animals is ruptured) and breathing begins. The sudden cessation of umbilical flow increases the peripheral resistance suddenly and causes a rise of fetal blood pressure; at the same time the arterial oxygen content falls. When breathing begins the expansion of the lungs is associated with a great decrease in pulmonary vascular resistance; hence the rate of blood flow through the lungs increases five- to ten-fold within a minute or so. Consequently the pressure in the left atrium rises; the pressure in the inferior vena cava falls as the blood flow from the placenta is cut off and the valve of the foramen ovale (Plate 51.1) closes. Thus from within a minute of breathing all the venous blood enters the right atrium and passes through the lungs (Fig. 51.27B). About a week after birth the valve of the foramen ovale becomes fused to the atrial wall. The ductus venosus closes by collapse or contraction of its junction with the umbilical vein probably as a result of the decrease in transmural pressure when the umbilical flow is arrested.

Before the animal breathes the pulmonary arterial pressure exceeds the pressure in the descending aorta. The decrease in pulmonary

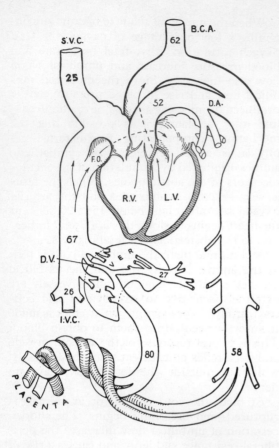

Fig. 51.28 Diagram of the fetal circulation in the lamb. The figures indicate the percentage oxygen saturation of blood withdrawn simultaneously from various vessels and averaged from determinations on six lambs. I.V.C., inferior vena cava; S.V.C., superior vena cava; D.V., ductus venosus; F.O., foramen ovale; D.A., ductus arteriosus; B.C.A., brachio-cephalic artery. (G. V. R. Born, G. S. Dawes, J. C. Mott & J. G. Widdicombe (1954). *Cold Spring Harbor Symposia on Quantitative Biology* **19**, 103.)

vascular resistance results in a fall in pulmonary pressure and in a reversal of the direction of blood flow through the ductus arteriosus (Fig. 51.27B); that is to say within a few minutes of birth blood flows from the aorta to the pulmonary artery and may contribute half or more of the total pulmonary blood flow. This stream of blood flowing through the left heart, the ductus arteriosus and the lungs increases the pulmonary blood flow and leads to a substantial increase of carotid oxygen saturation; temporary occlusion of the ductus at this stage causes a fall of systemic arterial oxygen saturation. This in-

creased flow is especially important at this stage for it must be remembered that at this time the lungs are not properly expanded, nor have they cleared themselves of the amniotic fluid with which they were filled *in utero*. Closure of the ductus arteriosus begins within the first ten minutes but it is not complete for many hours. In the lamb the external diameter is reduced from 10 mm to 5 mm in several minutes but as the aortic pressure is rising at this time and the pulmonary pressure is falling (Fig. 51.29), a large amount of blood continues to flow through with the production of vibrations which can be heard as a murmur. This is occasionally heard in newborn babies. This reverse flow through the ductus arteriosus from aorta to pulmonary artery continues for many hours after birth. At this time the left ventricular output is greater than the right. The circulation is now at the stage represented in Figure 51.27B. The flow of blood through the ductus ceases after 21 days or so and then the adult type of circulation is reached (Fig. 51.27C) in which the ventricular outputs are equal. The closure of the ductus arteriosus and ductus venosus is finally made permanent by a proliferation of the lining endothelium which takes several months to complete and eventually the obliterated vessels are represented only by fibrous cords.

There are good reasons for believing that the circulatory changes at birth in the human fetus are similar. Although the umbilical cord continues to pulsate for some time after the child is delivered it is customary to tie the cord as soon as the child is born. This may not, however, be the best practice because, if drainage of blood from the placenta to the child is allowed by holding it below the level of the vulva until pulsation in the cord has ceased, it may gain up to 100 ml of blood and so increase its stores of iron. The blood volume of the newborn infant is about 275 ml. Before respiration begins the placenta contains 85 ml of blood but only 40 ml if the cord is clamped after the child has breathed; the difference, 45 ml, is presumably accommodated in the baby's pulmonary circulation. The human umbilical cord varies in length from 30 to 129 cm (mean 61 cm). The blood flow in the human cord is 8·5 ml per min in a 12 week fetus and 80 ml per min in a 28 week fetus. The systolic blood pressure of the child at birth is about

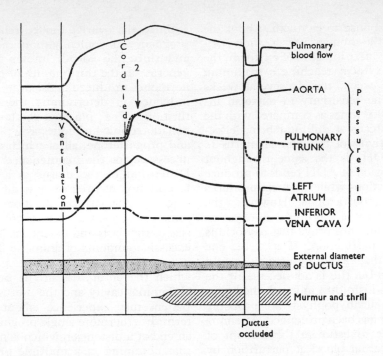

Fig. 51.29 Diagram to illustrate the changes which occur in the circulation of a lamb when ventilation is begun before the umbilical cord is tied. At 1 the mean pressure in the left atrium rises above that in the inferior vena cava and thereafter the valve of the foramen ovale is closed. At 2 the pressure in the pulmonary trunk falls below that in the descending aorta and the direction of blood flow through the ductus arteriosus is reversed. Temporary occlusion of the ductus arteriosus now reduces pulmonary blood flow and abolishes the murmur and thrill which have appeared in the pulmonary trunk. (G. V. Born, G. S. Dawes, J. C. Mott & J. G. Widdicombe (1954). *Cold Spring Harbor Symposia on Quantitative Biology* **19**, 104.)

70 mm Hg and this rises to 90 mm Hg at 6 months.

PARTURITION

Labour occurs in some 55 per cent of all women at approximately 280 ± 7 days from the first day of the last menstrual period; in 25 per cent it occurs before the 273rd day and in 20 per cent after the 287th day. There is no evidence that the length of the menstrual cycle has an influence on the duration of pregnancy. Since the non-pregnant uterus shows spontaneous activity (Fig. 51.10) it is not surprising to find that contractions occur early in pregnancy; they are small and usually separated by periods of an hour or so in which the uterus is quiescent. From the twentieth week the contractions are greater in amplitude and much more frequent, although they remain irregular until the last few weeks of pregnancy when rhythmic contractions begin. The interval between the contraction waves in late pregnancy is usually from five to ten minutes. The contractions which deliver the child at full term may be regarded as an exaggeration of this spontaneous uterine activity.

It is not easy to account for the onset of parturition in hormonal terms. The oestrogen and progesterone blood levels continue to rise (see Table 51.19 and Fig. 51.22) until parturition has actually occurred and the final decline in the production of these hormones is the result rather than the cause of parturition. Even very large doses of progesterone do not delay labour or affect the response to oxytocin. About the fourth month the human uterus shows increasing activity after infusion of oxytocin, the sensitivity to oxytocin reaching a maximum about the 34th week, that is several weeks before term. The sensitivity to oxytocin increases 200 to 1000 times as compared with the non-pregnant uterus, 1 milliunit being effective. The sensitivity of the human uterus to vasopressin (ADH) is the same throughout pregnancy. Infusion of ADH tends to produce irregular contractions while oxytocin produces regular uterine contractions. However, the increasing reactivity of the uterus to oxytocin up to full term does not prove that oxytocin is responsible for parturition. If it were, one would expect that hypophysectomy would delay parturition but this is not necessarily the case. In diabetes insipidus, in which the hypothalamus and neurohypophysis are probably damaged, labour has been prolonged. Attempts have been made to estimate the amount of oxytocin in maternal blood at parturition by injecting maternal plasma into the arterial supply of the mammary gland of a lactating rabbit, but even in the second stage the amount is scarcely above the threshold of the method. Immunoassay methods for oxytocin are being developed. The plasma of pregnant but not of non-pregnant women contains oxytocinase, produced by the syncytiotrophoblast, which destroys oxytocin. The half-life of injected oxytocin in pregnancy is 2 to 4 minutes only. The oxytocinase activity increases with gestational age up to a maximum at term but there is no decline before or during labour; it disappears after parturition. It seems likely, though not yet proved, that parturition is brought about by oxytocin.

It was shown by Sir James Simpson in 1856 that the nervous system is not responsible for the uterine contractions at parturition. In his experiments on pigs, destruction of the thoracic and lumbar cord did not interfere greatly with delivery although the contractions of the abdominal muscles were abolished. These muscles, which normally aid the uterus to expel the fetus, are, of course, under the control of the spinal cord and higher centres. Near full term the onset of labour may be hastened by emotional factors but it is not affected by section of the sympathetic and parasympathetic nerves.

Since the ovaries can be removed early in pregnancy in women without causing abortion and since the corpus luteum begins to degenerate in the third month of pregnancy, the hormones produced in the ovary can have little influence in determining the onset of parturition. The placenta certainly produces gonadotrophic hormones as well as oestrogens and progesterone, all of which are apparently necessary for the maintenance of pregnancy. Distension of the uterus as a factor in the termination of pregnancy is difficult to assess since in women the duration of pregnancy remains constant in spite of great variations in size of the fetus and in spite of the presence of excessive amounts of amniotic fluid (hydramnios); even in extrauterine pregnancies in which the embryo is in the uterine tube or abdominal cavity and the uterus is empty the patient may experience labour pains at full term. Furthermore single pregnancies come to an end at a distension which is much less than that obtaining in a multiple pregnancy long before parturition. If distension of the uterus were the only exciting factor in parturition it would be expected that it would occur when a certain total litter weight was reached. This is not so since the total litter weights in singletons, twins, triplets and quadruplets at birth are 3·38, 4·79, 5·45 and 5·58 kg respectively. It is likely however that the uterus becomes less tolerant of distension as pregnancy advances since the mean period of gestation decreases in the four categories just given in the order 280·5, 261·6, 246·8 and 236·8 days.

Two twin pregnancies in women have been reported in which one twin was present in each horn of a bicornuate uterus. There was an interval of several weeks between the births of the children. It is difficult to see how conditions at the same time in one woman can be suitable for the maintenance, as well as the termination, of pregnancy unless there is some local factor which triggers off the mechanism of parturition. It has long been suspected that the human fetus has some influence on the length of gestation. If it has a poorly developed pituitary gland, as in anencephaly with hypoplasia of the adrenal cortex, gestation is often prolonged. Although the evidence, as just given, for a fetal triggering of parturition in women is rather slender there is no doubt of fetal influence in the sheep. Electrocoagulation of pituitary glands of fetal lambs prolongs gestation for many days. This of course causes adrenal hypoplasia and in conformity with this, fetal adrenalectomy results in prolongation of pregnancy. Conversely infusion of ACTH into fetal lambs is followed by parturition in a few days. Cortisol given to the ewe has no effect but given to the fetuses causes parturition. One can speculate that unknown factors acting on the fetal hypothalamus lead to production of ACTH and in turn of corticosteroid which produces myometrial activity ending in parturition. It is not yet possible to say whether these findings are applicable to human parturition.

The muscular activity of the human uterus can be studied by passing a catheter through the cervix, or through the abdominal wall and through the uterine wall, into the amniotic cavity and by connecting it to a manometer which records the variations in pressure. The slight contractions of the uterus which occur throughout pregnancy increase the pressure in the uterine veins and assist the movement of blood through the intervillous placental spaces and so aid the oxygenation of the fetal blood. Towards the end of pregnancy, but long before labour begins, the basal amniotic pressure is about 5 mm Hg with slow oscillations of 1 to 5 mm Hg above this value. At the onset of labour these rhythmic contractions increase and the basal intrauterine pressure rises to 8 or 12 mm Hg. In the *first stage* of normal labour the intrauterine pressure rises slowly with each contraction to a peak of 30 to 50 mm Hg above the basal value and then more slowly falls to the basal level. These contractions, which occur at intervals of two to five minutes, can be felt by the hand placed on the abdominal wall. When the pressure is above a certain value, which varies with different patients, but is in the region 25 to 35 mm Hg, the mother is conscious of a labour pain, that is a pain commencing in the lumbosacral region and radiating to the front and down the thighs. This pain is relieved or abolished by injecting a local anaesthetic alongside the cervix. By pressing on the amniotic fluid the uterine contractions dilate first the upper part of the cervical canal and later the os uteri. The membranes finally rupture and a small quantity of amniotic fluid escapes. In the *second stage* of labour the intrauterine pressure at the height of a contraction

may rise to 110 mm Hg and the pressure between contractions is usually about twice that recorded in the first stage. The head of the child is slowly forced through the pelvis and is born, the remainder of the body following almost at once. During this stage voluntary contraction of the abdominal muscles can double the intrauterine pressure, so that the combined pressure caused by the contractions of the uterine and of the abdominal muscles may be as much as 260 mm Hg. It has been calculated that the force on the fetal head produced by the uterine contractions alone is between 4 and 8 kg, but in well-developed women the abdominal muscles can increase the expulsive force up to 18 kg or more.

Alvarez and Caldeyro-Barcia of Montevideo have recorded uterine activity by placing small balloons (0·02 ml) into the myometrium at various positions. When the muscle contracts the pressure rises in the balloons and is recorded. They find that the contraction wave in labour starts at a pace-maker at the uterine end of one uterine tube, either right or left; the wave is propagated mainly downwards to the cervix at a speed of one or two cm per sec and reaches its peak in 50 sec or thereby. The contraction lasts longer in the parts near the pace-maker and diminishes in intensity as it passes down the uterus.

The pressure within the uterus at the height of a contraction may be sufficient to squeeze out the maternal blood from the uterine vessels into the vena cava and even occasionally into the aorta if the intrauterine pressure rises above the maternal arterial pressure. Thus at the height of each contraction the venous pressure is considerably raised, and this may be dangerous in patients with cardiac disease. The arterial pressure increases slightly during a labour contraction but quite considerably if the abdominal muscles contract at the same time. The fetal heart rate falls during a uterine contraction, the amount of the fall depending on the rise of intrauterine pressure.

After the birth of the child uterine contractions cease for a time, usually five to fifteen minutes, and then in the *third stage* of labour they begin again and the placenta and membranes are expelled. With the detachment and delivery of the placenta a raw bleeding surface is left but the average loss of blood is only 300 ml. The mechanisms which limit blood loss are not clearly understood. The contraction of the uterus no doubt plays a part; when the child is allowed to suck oxytocin is released and the uterus contracts firmly but does not go into a prolonged spasm. However, administration of oxytocin makes no apparent difference to the blood loss. During the *puerperium*, that is the period immediately following parturition, the uterus shrinks until, when involution is complete, it is only a little larger than in the virgin state.

The average maternal oxygen uptake in the last hour of labour ranges from 285 to 700 ml/min and this indicates that labour may be described as mild to moderately heavy work (see p. 169). The blood lactate rises from a resting value of 13 mg/100 ml up to 24 or more; this is higher than would be expected from exercise like cycling which involves many muscles. In labour a small mass of muscle is vigorously exercised and in these circumstances lactate cannot all be oxidized as quickly as it is produced.

THE MAMMARY GLAND

The mammary gland consists of numerous lobules made up of clusters of rounded alveoli embedded in fatty tissue. The secretion of the cells lining the alveoli passes into small ducts, and these unite with ducts from neighbouring alveoli to form the lactiferous tubules which radiate towards the mammary papilla (usually called nipple or teat) and open on its summit. Beneath the areola, just before it reaches the nipple, each tubule has a dilatation, the ampulla, which acts as a milk reservoir. Smooth muscle fibres are found in the nipple and around the ducts. The alveoli are surrounded by myoepithelial cells. In the non-pregnant woman the alveoli are small but during pregnancy they enlarge and the cells lining them increase in number.

Both sexes possess mammary glands but they remain rudimentary in the male. In the girl enlargement of the breasts is one of the signs of puberty; there may be a transient slight enlargement during the luteal phase of the menstrual cycle but full development is achieved only as a result of pregnancy. In a non-pregnant woman an increase in the size of the mammary gland can be produced by injection of oestrogen but a greater effect can

be obtained by treatment with oestrogen plus progesterone. Experiments on rats show that oestrogen seems to be responsible chiefly for growth of the ducts, the combination of oestrogen and progesterone causing, in addition, alveolar growth. The full development of the lactating breast in these animals requires prolactin, somatotrophin, oestrogen, progesterone, insulin, thyroid hormone and adrenal corticoids. In other words successful lactation requires a normal hormonal environment. Although during pregnancy in woman it is often possible by pressure to expel a few drops of milk from the breasts, a free flow of milk does not occur until the second or third day after the birth of the child. It seems that during pregnancy the mammary glands are kept in an unresponsive state by the direct action of the large amount of oestrogens and progesterone in the circulation but the lactogenic hormones cannot bring about milk secretion unless the mammary gland has been prepared by the ovarian hormones. The decline in the amount of these hormones in the blood after parturition allows the lactogenic hormone, *prolactin*, from the anterior pituitary to act directly on the alveolar epithelium. The amount of prolactin in the blood and pituitary is not altered by the rise in oestrogens; it may be liberated from the pituitary by oxytocin. Prolactin may not be the only lactogenic hormone secreted by the adenohypophysis and it is likely that adrenocorticotrophin supports lactation in an indirect way by keeping up the supply of milk precursors. Since pituitary grafts in sites remote from the pituitary fossa can maintain to some extent the secretion of milk, lactation seems largely independent of hypothalamic control. Unfortunately prolactin does not increase output in women with inadequate milk production. During the puerperium administration of large amounts of orally-active progestogen and oestrogen can stop the flow of milk probably by a direct action on the breast tissue and perhaps by inhibiting the pituitary.

Although the building up of secreting breast tissue and the initiation of lactation is mainly a hormonal affair there is no doubt that regular removal of milk from the breast is essential to maintain the flow. A woman with a strong, healthy, hungry child who sucks vigorously and empties the breasts, secretes more milk than one with a small weak child. In dairy cattle more frequent milking produces more milk. Suckling may act by releasing simultaneously corticotrophin, prolactin, oxytocin and ADH.

In 1962 a polypeptide with considerable lactogenic activity on certain animals was found in large amounts in human placental tissue. The structure of this human placental lactogen (HPL) is related to that of human growth hormone (HGH). HPL is synthesized by the syncytiotrophoblast and is just detectable in maternal serum by radioimmunoassay from the fifth week, increasing up to 1 to 5 μg per ml at term; it disappears shortly after delivery of the placenta. Near term the daily production is 1 to 2 g. The physiological role of HPL is yet uncertain but it is likely to play a part in the enlargement of the breast and in the initiation of lactation. Many metabolic effects have been tentatively attributed to HPL but its main value at present may be as an indicator of placental function, a fall in plasma HPL being often followed by abortion.

It is well known that lactation is promoted by tranquillity and inhibited by emotional upsets (perhaps through the hypothalamus) but it has not been possible to demonstrate experimentally the existence of a nervous control. Sympathectomy does not affect milk production and no one has demonstrated secretory nerve fibres in the mammary gland. Lactation cannot be fully successful unless the nursing mother is having a generous diet and an indication of the daily milk production in such circumstances is given in Figure 51.30.

In the rat suckling causes, after a latent period of $\frac{1}{2}$ to $1\frac{1}{2}$ min, by contraction of the myoepithelial cells a rapid rise of intramammary pressure which transfers the milk rapidly to the offspring sucking at the teats. If this mechanism is abolished by anaesthesia or by emotional disturbances the amount of milk obtained by the young is greatly reduced. This 'let-down' phenomenon has nothing to do with production of milk but merely with the ejection of milk which is already formed in the gland. Injection of small doses of oxytocic

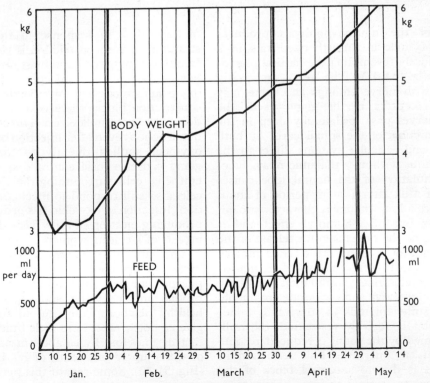

Fig. 51.30 Weight and test-feeding chart of a healthy breast-fed baby. The amount of milk obtained at each feed is measured by weighing the baby before and after suckling. (*After* P. Evans (1950). *British Medical Journal* **i**, 1128.)

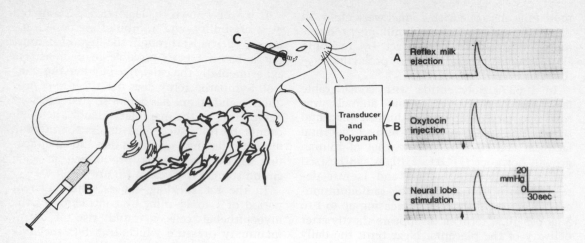

Fig. 51.31 Reflex oxytocin release from the neural lobe of the pituitary gland during suckling. The anaesthetized rat readily exhibits the milk-ejection reflex and provides an excellent preparation (shown on the left of the figure) in which to demonstrate that milk-ejection is brought about by oxytocin release from the neurohypophysis. Milk-ejection, as indicated by intramammary pressure recordings (shown on the right of the figure), occurred after: A, the litter of ten pups had been placed on the nipples; B, a rapid intravenous injection of 2·5 mu of oxytocin; C, stimulation of the neural lobe (1·6 mA, 50 pulses/sec) for 3 seconds. In each case the rise in intramammary pressure was similar both in amplitude and duration. (*By courtesy of B. A. Cross, J. B. Wakerley and D. W. Lincoln.*)

extracts of the neurohypophysis produces ejection of milk in anaesthetized rats in a manner closely resembling the natural phenomenon (Fig. 51.31). The teats are richly supplied with nerve endings and there may even be receptors in the gland itself; the afferent nerves pass ipsilaterally deep in the lateral funiculus of the spinal cord, by both lemniscal and reticular systems direct to the hypothalamus without a thalamic relay. Electrical stimulation of the neural lobe causes ejection of milk after a latent period of about 30 sec; such a long latency is characteristic not of a reflex but of a humoral effect and is assumed to be the time taken for release of the hormone and for its transport to the mammary gland (Fig. 51.31). After removal of the neurohypophysis electrical stimulation of these regions is ineffective and the young are unable to obtain sufficient milk to survive. If a lactating rabbit is forcibly restrained, the milk yield obtained by the young is reduced in some experiments to one-fifth. This reduction in most cases is due to a central block of the release of oxytocin since an intravenous injection of oxytocin restores milk ejection. In a few cases the inhibition may be due to the release

of adrenaline in response to the emotional stimulus. If vasoconstriction is produced in a lactating mammary gland by stimulation of the nerves to it, or by injection of adrenaline, oxytocin cannot cause milk ejection. The vasoconstriction blocks the access of oxytocin but does not put the milk ejection mechanism itself out of action; even after injection of adrenaline, oxytocin applied directly to the mammary gland causes a local contraction of the myoepithelial cells surrounding the alveoli—these cells also still respond to mechanical stimulation. Injection of oxytocin has proved useful in producing ejection of milk from the engorged breasts of lactating women.

The fluid (colostrum) produced in the breasts during the first two or three days of the puerperium is small in amount and has a high protein and a low fat and sugar content. A healthy newborn baby put to the breast or given a bottle takes in so little fluid or nutrient that he may lose by catabolism and excretion of water up to one tenth of his birth weight (Fig. 51.30). Some regard this period of semi-starvation as physiological. The intake of breast milk and the growth curve of a healthy baby are shown in Figure 51.30. When milk begins to flow freely on the fourth or the fifth day the central cells of the mammary alveoli are discharged into the fluid, where they appear as the so-called colostrum corpuscles.

Mammary metabolism. The R.Q. of actively secreting mammary tissue is usually over 1·0 because of the synthesis of fat from carbohydrate (see p. 163). The mammary gland is an efficient milk producer since, provided the maternal energy requirements have already been fully met, about 60 per cent of the energy value of additional food is recovered in the milk. To avoid negative nitrogen balance during lactation about 2 g of protein per kg of body weight per day must be supplied to the mother. The protein content of the milk of malnourished mothers is not significantly different from that of well-nourished mothers; in the former case the protein must be derived from the mother's tissues. A daily production of 570 ml of milk (see Fig. 51.30) is equal to an energy loss of 1·6 MJ (400 kcal). The required increase in food intake may take place during pregnancy or during lactation or both. Lactating mothers do, in fact, eat more than non-lactating mothers and usually draw on their stores of fat so that they lose weight. About 150 mg iron are required for lactation. About 300 mg calcium is lost in the milk per day but the serum calcium does not fall even if the calcium intake is low; calcium must obviously be taken from the bones.

Venous blood from the udder of a lactating cow contains about 10 mg less glucose per 100 ml than the arterial blood. By applying the Fick principle (p. 502) the total blood flow through the gland can be calculated from the arteriovenous glucose difference and the output of sugar in the milk if it is assumed that the lactose of the milk is derived entirely from the glucose of the blood. Calculations based on this assumption give very high values for the blood flow and it appears that the volume of blood which passes through the mammary gland is 400 to 500 times the volume of milk secreted. Experiments with isotopically labelled precursors such as glucose, acetate and amino acids labelled with ^{14}C and fatty acids labelled with ^{3}H, have conclusively proved that the lactose of milk is formed entirely from blood glucose as also is the glycerol of milk fat; that the long chain fatty acids are derived from blood fats; that the short chain fatty acids which are peculiarly abundant in milk fat are synthesized from acetate; and that the milk proteins are formed from the free amino acids of the blood.

A wide range of drugs, including purgatives and antibiotics, likely to be taken by a nursing mother can enter the breast milk.

REFERENCES

ANDERSON, J. M. (1971). Transplantation—nature's success. *Lancet* **ii**, 1077–1082.

ASSALI, N. S. (Ed.) (1968). *Biology of Gestation*, 2 vols. New York: Academic Press.

AVERILL, R. L. W. (1966). The hypothalamus and lactation. *British Medical Bulletin* **22**, 261–265.

BARCROFT, J. (1946). *Researches on Pre-natal Life*. Oxford: Blackwell.

BARRON, D. H., METCALFE, J., MESCHIA, G., HUCKABEE, W., HELLEGERS, A. & PRYSTOWSKY, H. (1963). Adaptations of pregnant ewes and their fetuses to high altitude. *Symposium on the Physiological Effects of High Altitude*, pp. 115–129. Oxford: Pergamon.

Biology of the prostate and related tissues (1963). *National Cancer Institute. Monograph No.* 12. Bethesda: U.S. Dept. of Health, Education and Welfare.

BISHOP, D. W. (1962). Sperm motility. *Physiological Reviews* **42**, 1–59.

BOYD, J. D. & HAMILTON, W. J. (1955). The cellular components of the human ovary. Chapter 3, pp. 50–78 in *Modern Trends in Obstetric and Gynaecology* (Edited by K. Bowes). London: Butterowrth.

BOYD, J. D. & HAMILTON, W. J. (1970). *The Human Placenta*. Cambridge: Heffer.

CALDEYRO-BARCIA, R. & HELLER, H. (Eds.) (1961). *Oxytocin*. Oxford: Pergamon.

CALDEYRO-BARCIA, R. & POSEIRO, J. J. (1960). Physiology of the uterine contraction. *Clinical Obstetrics and Gynaecology* **3**, 386–408.

CAREY, H. M. (Ed.) (1963). *Modern Trends in Human Reproductive Physiology*, Vol. 1. London: Butterworth.

CATT, K. J. (1970). ABC of endocrinology, IV—reproductive endocrinology. *Lancet* **i**, 1097–1104.

CROSS, B. A. (1966). Neural control of oxytocin secretion. In *Neuroendocrinology*, Vol. 1 (Edited by L. Martini & W. F. Ganong), pp. 217–259. London: Academic Press.

CROSS, K. W. & DAWES, G. S. (Eds.) (1966). The foetus and the new-born. *British Medical Bulletin* **22**, 1–96.

DAWES, G. S. (1961). Changes in the circulation at birth. *British Medical Bulletin* **17**, 148–153.

DAWES, G. S. (1967). New views on O_2 transfer across the placenta. *Scientific Basis of Medicine Annual Review* 74–89.

DAWES, G. S. (1968). *Foetal and Neonatal Physiology*. Chicago: Year Book Medical Publishers.

DICZFALUSY, E. (1967). *Steroid Metabolism in Pregnancy and in the Foetal-Placental Unit in Hormonal Steroids* (Edited by L. Martini, F. Franschini & M. Motta). London: Excerpta Medica.

DONOVAN, B. T. & VAN DER WERFF TEN BOSCH, J. J. (1965). *The Physiology of Puberty*. London: Arnold.

ECKSTEIN, P. (Ed.) (1959). *Implantation of Ova*. Cambridge University Press.

EIK-NES, K. B. (1964). Effects of gonadotrophins on secretion of steroids by the testis and ovary. *Physiological Reviews* **44**, 609–630.

FRYE, B. E. (1967). *Hormonal Control in Vertebrates*. London: Collier-Macmillan.

GARDINER-HILL, H. (Ed.) (1961). *Modern Trends in Endocrinology*. London: Butterworth.

GRADY, H. G. & SMITH, D. E. (Eds.) (1963). *The Ovary*. London: Baillière, Tindal & Cox.

GRAY, C. H. & BACHARACH, A. L. (Eds.) (1961). *Hormones in Blood*. London: Academic Press.

HYTTEN, F. E. & LEITCH, I. (1971). *The Physiology of Human Pregnancy*, 2nd edn. Oxford: Blackwell.

IRVINE, W. J. (1969). *Reproductive Endocrinology*. Edinburgh: Livingstone.

JACKSON, H. (1966). *Antifertility Compounds in the Male and Female*. Springfield, Ill.: Thomas.

KLOPPER, A. & DICZFALUSY, E. (Eds.) (1969). *Foetus and Placenta*. Oxford: Blackwell.

KON, S. K. & COWIE, A. T. (Eds.) (1961). *Milk: the Mammary Gland and its Secretion*, 2 vols. New York: Academic Press.

LANG, W. R. (Ed.) (1959). The vagina. *Annals of the New York Academy of Sciences* **83**, 77–358.

LINZELL, J. L. (1959). Physiology of the mammary glands. *Physiological Reviews* **39**, 534–576.

LORAINE, J. A. (1970). *Sex and the Population Crisis*. London: Heinemann.

LORAINE, J. A. & BELL, E. T. (1971). *Hormone Assays and Their Clinical Application*, 3rd edn. Edinburgh: Livingstone.

LORAINE, J. A. & BELL, E. T. (1967). *Fertility and Contraception in the Human Female*. Edinburgh: Livingstone.

McLAREN, A. (1966). *Advances in Reproductive Physiology*, Vol. I. London: Academic Press.

MARTINI, L. & GANONG, W. F. (1966–67). *Neuroendocrinology*, Vols. I, II. London: Academic Press.

MASTERS, W. H. & JOHNSON, VIRGINIA E. (1966). *Human Sexual Response*. London: Churchill.

METCALFE, J., BARTELS, H. & MOLL, W. (1967). Gas exchange in the pregnant uterus. *Physiological Reviews* **47**, 782–838.

MANN, T. (1964). *The Biochemistry of Semen and of the Male Reproductive Tract*. London: Methuen.

PARK, W. W. (Ed.) (1966). *The Early Conceptus, Normal and Abnormal*. Edinburgh: Livingstone.

PARKES, A. S. (1966). *Sex, Science and Society*. London: Oriel Press.

PETROW, V. (1966). Steroidal oral contraceptive agents. In *Essays in Biochemistry*, Vol. 2, pp. 117–145. London: Academic Press.

PICKLES, V. R. & FITZPATRICK, R. J. (Eds.) (1966). Endogenous substances affecting the myometrium. *Memoirs of the Society for Endocrinology* **14**.

PINCUS, G. (1965). *The Control of Fertility*. London: Academic Press.

PRUNTY, F. T. G. (1966). Androgen metabolism in man—some current concepts. *British Medical Journal* **ii**, 605–613.

RHODES, P. (1969). *Reproductive Physiology for Medical Students*. London: Churchill.

RICE, B. F. & SAVARD, K. (1966). Steroid hormone formation in the human ovary—IV. Ovarian stromal compartments; formation of radioactive steroids from acetate-1-^{14}C and action of gonadotrophins. *Journal of Clinical Endocrinology* **26**, 593–609.

ROTHSCHILD (Lord) (1962). Spermatozoa. *British Medical Journal* **ii**, 743–749 and 812–817.

SAUNDERS, F. J. (1968). Effects of sex steroids and related compounds on pregnancy and on development of the young. *Physiological Reviews* **48**, 601–643.

SHEARMAN, R. P. (1972). *Human Reproductive Physiology*. Oxford: Blackwell.

SHETTLES, L. B. (1960). *Ovum Humanum*. London: Hafner.

SWYER, G. I. M. (Ed.) (1970). Control of human fertility. *British Medical Bulletin* **26**, 1–91.

Symposium on the placenta (1962). *American Journal of Obstetrics and Gynecology* **84**, 1541–1798.

THOMSON, A. M. (1971). Nutrition in pregnancy. *British Journal of Hospital Medicine* **5**, 600–612.

THOMSON, A. M. & HYTTEN, F. E. (1961). Calorie requirements in human pregnancy. *Proceedings of the Nutrition Society* **20**, 76–83.

TILLINGER, K.-G. (1957). Testicular morphology. *Acta endocrinologia (Copenhagen)* **24**, Suppl. 30.

VILLEE, C. A. (Ed.) (1961). *Control of Ovulation*. Oxford: Pergamon.

VLIEGENTHART, J. F. G. & VERSTEEG, D. H. G. (1967). The evolution of the vertebrate neurohypophysial hormones in relation to the genetic code. *Journal of Endocrinology* **38**, 3–12.

VON EULER, U. S. & ELIASSON, RUNE (1968). *Prostaglandins*. London: Academic Press.

WALKER, J. & TURNBULL, A. C. (Eds.) (1959). *Oxygen Supply of the Human Foetus*. Oxford: Blackwell.

WHITE, J. C. & BEAVER, G. H. (1959). Foetal haemoglobin. *British Medical Bulletin* **15**, 33–39.

WOLSTENHOLME, G. E. W. & CAMERON, M. P. (Eds.) (1961). *Progesterone and the Defence Mechanism of Pregnancy.* London: Churchill.

WOLSTENHOLME, G. E. W. & KNIGHT, JULIE (Eds.) (1972). *Lactogenic Hormones.* A Ciba Foundation Symposium. Edinburgh: Churchill Livingstone.

WOLSTENHOLME, G. E. W. & O'CONNOR, MAEVE (Eds.) (1967). *Endocrinology of the Testis.* London: Churchill.

WOLSTENHOLME, G. E. W. & O'CONNOR, MAEVE (Eds.) (1969). *Foetal Autonomy.* London: Churchill.

WOODRUFF, M. F. A. (1958). Transplantation immunity and the immunological problem of pregnancy. *Proceedings of the Royal Society* B **148**, 68–75.

ZUCKERMANN, S. (1962). *The Ovary.* 2 vols. London: Academic Press.

52 The pituitary body

The pituitary body or *hypophysis cerebri* is an ovoid structure, about 0·6 g in the adult, attached by a short stalk, the infundibulum, to the base of the brain between the optic chiasma and the mamillary bodies. It occupies the hypophysial fossa of the sphenoid bone which is 11 mm long, 13 mm wide and 8 mm deep. The size of the human hypophysis varies with age, sex, weight and height; in pregnancy the gland becomes much enlarged. Since the hypophysis occupies the whole of the fossa it is clear that a pituitary tumour can grow only by compressing structures within the fossa and later only by pressing on the base of the brain itself. Because the optic chiasma lies immediately over the diaphragma sellae (which roofs in the gland) an enlarging pituitary body, by exerting pressure on the optic nerves, chiasma, or tracts, can produce defects in the field of vision.

Table 52.1 gives a classification of the various parts of the hypophysis, based upon its embryology, which is useful in terms both of morphology and of function (see also Fig. 52.3).

The tissues of the *adenohypophysis* are derived from the somatic ectoderm of the buccopharyngeal pouch, Rathke's pouch, which grows upward towards the third ventricle of the developing brain. The *neurohypophysis*, on the other hand, has its origin in a prolongation from the floor of the diencephalon (Fig. 47.1), which grows downwards towards the base of the skull so that it eventually lies immediately behind the buccopharyngeal pouch. The line of demarcation between the two parts is represented in the human adult by a few isolated cystic spaces. A small amount of tissue histologically resembling the adenohypophysis, but only one thousandth of its volume, lies in the mucoperiosteum of the roof of the human nasopharynx. It contains prolactin and growth hormone but its function, if any, is unknown.

Table 52.1 Subdivisions of the pituitary body (terminology recommended by the International Commission on Anatomical Nomenclature)

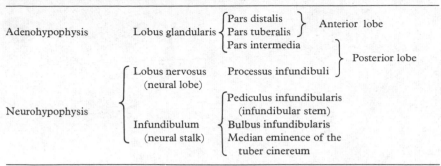

Adenohypophysis	Lobus glandularis	Pars distalis	Anterior lobe
		Pars tuberalis	
		Pars intermedia	Posterior lobe
Neurohypophysis	Lobus nervosus (neural lobe)	Processus infundibuli	
	Infundibulum (neural stalk)	Pediculus infundibularis (infundibular stem)	
		Bulbus infundibularis	
		Median eminence of the tuber cinereum	

THE ADENOHYPOPHYSIS

The pars distalis of the adenohypophysis accounts for the greater part of the pituitary body. Its dark red colour depends on the presence of a very large number of thin-walled blood vessels ramifying throughout the tissue.

Several kinds of cells can be distinguished in the pars distalis by special staining techniques. The periodic-acid-Schiff (PAS) method, which stains carbohydrates, should pick out glycoprotein hormones by staining cell granules containing FSH, ICSH, TSH and leave unstained the polypeptide hormones such as ACTH. The PAS-positive cells, now called mucoid (formerly basophil), can be divided into at least two (and probably five) varieties according to their staining properties. Other cells, staining with orange G, and therefore described as acidophil, account for half the cells in the adult human subject. Certain cells called chromophobes do not take up stains easily (see Plate 52.1). The staining reactions give one indication of the hormones produced by individual cells. The mucoid cells alter in appearance after removal of gonads or thyroid; the red-staining mucoids have been named gonadotrophs, and the blue-staining mucoids thyrotrophs. ACTH (and also MSH) is almost certainly produced by the red-staining mucoid cells (Plate 52.1A). In animals, however, the cells producing MSH are situated in the pars intermedia. The fluorescent antibody technique (Plate 52.1B and C) has located GH in human acidophils; these cells probably produce prolactin (p. 1088). (Luteotrophic action does not occur in man; LTH and prolactin are the same.) Histological examination of pituitary tissue from patients who have died, for example, from hyperthyroidism or Cushing's disease (p. 1010) has shown predominance of certain types of cell which are presumably responsible for the production of specific hormones (TSH and ACTH).

The pars distalis does not have an arterial blood supply; it is supplied by portal vessels. The superior hypophysial arteries supply capillary loops in the stalk and the adjacent median eminence (P in Fig. 52.2a) and these drain into the long portal vessels, LPV, which then supply about 90 per cent of the pars distalis. The remainder of the pars distalis is supplied by a similar portal system drained by the short portal vessels (see legend to Fig. 52.2b). If the stalk is cut through, the gonads and the thyroid gland atrophy but they recover when the long portal vessels regenerate. If the pars distalis of an animal is removed from the sella turcica (D in Fig. 52.3) and placed under the capsule of the kidney or in the temporal lobe of the brain the cells of the transplant survive but the target organs atrophy. This shows that a transplanted anterior lobe has little endocrine influence. If the transplant is returned to its original situation the target organs recover provided that the portal vessels revascularize it. These experiments show clearly the importance of the hypothalamic connexions. The activities of the cells of the pars distalis are controlled by the nerve cells of the hypothalamus which send axons to the capillary beds (P in Fig. 52.2a). The nerve endings liberate chemical substances, *releasing hormones*, into these capillaries which carry them to the cells of the pars distalis to control the liberation of their hormones into

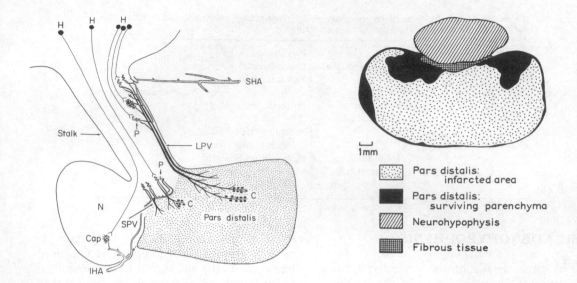

⬚	Pars distalis: infarcted area
◼	Pars distalis: surviving parenchyma
▨	Neurohypophysis
▦	Fibrous tissue

Fig. 52.2a Schematic diagram of the arrangements for the blood supply of the pituitary. The larger adeno-hypophysis is on the right; the smaller neurohypophysis is on the left. H, nerve cells of the hypothalamus; LPV, long portal vessels; SPV, short portal vessels; N, posterior lobe; P, primary capillary bed; Cap, secondary capillary bed; SHA, superior hypophysial artery; IHA, inferior hypophysial artery; C, anterior lobe cells. (J. H. Adams, P. M. Daniel & M. M. L. Prichard (1964). *Endocrinology* **75**, 120–126.)

Fig. 52.2b This is a tracing of a horizontal section through a human pituitary gland 30 hours after stalk section; that is at a time before much shrinkage could occur. The volume of the infarct (dead cells) was 90 per cent of the whole pars distalis. Since a complete transection of the stalk severs all the long portal vessels (LPV) the cells of the infarct must have been supplied by the LPV. The remaining 10 per cent, namely the small part near the neural lobe, must be supplied by the short portal vessels (SPV). The epithelial cells of the pars distalis are not supplied with nerves: chemical transmitters (releasing factors) must be carried from the hypothalamus to the cells in the portal vessels. (J. H. Adams, P. M. Daniel & M. M. L. Prichard (1964). *British Medical Journal* **ii**, 1619–1625.)

the circulation. The release of TSH, ACTH, GH, FSH and LH is increased by their appropriate releasing hormones whereas the release of MSH and prolactin is reduced by theirs. The structure of four of these compounds is known. Thyrotrophin-releasing hormone (TRH) is a tripeptide ((pyro)Glu–His–Pro(NH$_2$)) which, given intravenously, produces a rapid rise in serum TSH provided that the subject's anterior pituitary gland is healthy. A significant rise of TSH is produced by 50 μg given intravenously. In some circumstances TRH increases the TSH hormone even if the pituitary is diseased. TRH also increases secretion of prolactin. Luteinizing hormone-releasing factor (LRH) is a decapeptide ((pyro)Glu–His–Trp–Ser–Tyr–Gly–Leu–Arg–Pro–Gly(NH$_2$)); it brings about release of LH and, to a smaller extent, of FSH. Growth hormone-releasing hormone (GRH) has 10 amino acids in the sequence Val–His–Leu–Ser–Ala–Glu–Glu–Lys–Glu–Ala. Melanocyte-stimulating hormone release-inhibiting

hormone (MRIH) reduces the release of MSH; it is a tripeptide (Pro–Leu–Gly(NH$_2$)) containing the three amino acids of the C-terminal end of oxytocin.

The releasing hormones are formed at the ends of axons in various regions of the hypothalamus revealed by localized stimulation or localized lesions. TRH seems to be produced anteriorly in the hypothalamus, LRH in the anterior and middle regions, CRF in the posterior region and GRH in a wide area in the ventral hypothalamus.

The hormonal activity of the pars distalis is also influenced by the hormones of the target organs, that is the thyroid gland, adrenal cortex and gonads. When, for example, the thyroid hormone reaches the cells of the pars distalis the output of thyroid stimulating hormone (TSH) is reduced and as a result the output of thyroid hormone from the thyroid

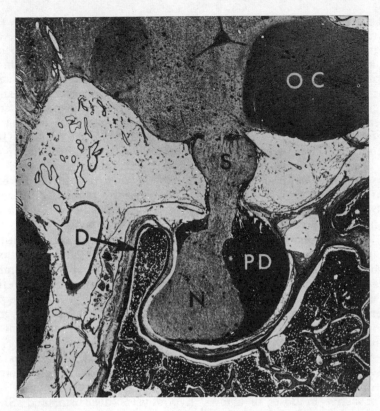

Fig. 52.3 Sagittal section through the hypothalamus and pituitary fossa of a normal monkey. OC, optic chiasma; S, pituitary stalk; PD, pars distalis (anterior lobe of the pituitary); N, neural lobe (infundibular process); D, dorsum sellae. Haematoxylin and eosin. × 10. (P. M. Daniel, L. W. Duchen & M. M. L. Prichard (1964). *Journal of Pathology and Bacteriology* **87**, 385–393.)

gland is reduced. This mechanism, by analogy with electronic circuits, is known as 'feedback'. Since small implants of oestrogen placed in the hypothalamus inhibit FSH and LH secretion it seems that feedback effects are due to the action of oestrogen on a receptor in the hypothalamus. Cortisol in the floor of the hypothalamus (median eminence) suppresses ACTH secretion more effectively than cortisol placed in the anterior pituitary. The major site of thyroxine feedback is in the anterior pituitary; thyroxine also acts, but much less effectively, on hypothalamic cells. The feedback mechanism works relatively slowly as if it controlled the rate of synthesis of the hormones whereas the 'releasing factor' mechanism acts quickly as if it caused release of stored hormone. The secretory activity of the adenohypophysis is also influenced through the nervous system. Many kinds of mental and physical activity, coitus and suckling for example, influence pituitary activity. It seems likely that nervous and emotional stimuli, as well as the feedback mechanisms, act via the hypothalamus.

THE HORMONES OF THE ADENOHYPOPHYSIS

Methods used to determine the function of the adenohypophysis have included removal or implantation of pituitary tissue and the administration of more or less pure preparations derived from them. The study of pituitary syndromes in man has yielded important information.

Since certain of these hormones produce their main effects indirectly through their actions on other endocrine glands ('target organs') they are called *trophic hormones*, and have been given names ending in 'trophic' or 'trophin'. This group includes the pituitary gonadotrophins and the thyrotrophic and adrenocorticotrophic hormones. On the other hand, some anterior pituitary hormones, the growth hormone and the lactogenic hormone, prolactin, act directly upon tissues.

The thyrotrophic hormone (thyroid stimulating hormone, TSH thyrotrophin). Injections of acid or alkaline extracts of the adenohypophysis increase the functional activity of the normal thyroid; conversely ablation of the adenohypophysis results in atrophy of the thyroid gland and a pronounced fall in metabolic rate which can be restored to normal by injections of pituitary extracts. In man hypopituitarism is accompanied by a low basal metabolic rate and even by myxoedema, and conversely in acromegaly where the adenohypophysis is overactive the basal metabolism is frequently above the normal. After thyroidectomy in animals hypophysectomy is without effect on metabolic rate. These effects are due to thyroid-stimulating hormone (TSH). This mixture of glycoproteins (average mol. wt. 30000) is active in doses of a few micrograms and has a direct action on thyroid tissue since it produces hyperplasia in slices of thyroid in tissue culture and in thyroid transplants. Work with radioactive iodine shows that TSH controls the rate of iodine uptake by thyroid tissue and that it influences the synthesis of thyroxine from diiodotyrosine (see p. 1000).

The concentration of TSH in the blood, approximately 25 milli-units per ml, is raised in myxoedema, sometimes as much as a hundredfold. In hyperthyroidism, too, the blood contains an increased concentration of a TSH-like substance. The concentration in the blood is increased by cold conditions and reduced by a hot environment. TSH is normally secreted at a rate exceeding 2 international units per day by the mucoid cells under the control of cells in the anterior part of the hypothalamus, since secretion ceases if bilateral lesions are placed experimentally in that area. TSH disappears rapidly from blood, less than 5 per cent of an injected dose remaining at 1 hour.

It is now generally agreed that the exophthalmos of hyperthyroidism is not due to overactivity of the thyroid for, even when the thyroid is grossly overactive, exophthalmos may be absent and it may appear for the first time after the hyperthyroidism has been relieved by thyroidectomy. Moreover, exophthalmos is not produced in animals by administration of thyroid but it may follow injections of adenohypophysial extracts even in thyroidectomized animals. It is believed, therefore, that exophthalmos is produced by another substance from the adenohypophysis to which the name EPS or exophthalmos-producing substance has been given. The protrusion of the eyeballs is caused by oedema localized to the retro-orbital tissues.

The adrenocorticotrophic hormone, ACTH (corticotrophin). Hypophysectomy is invariably followed by considerable atrophy of the zona fasciculata and zona reticularis of the adrenal cortex (p. 1007) but the zona glomerulosa and the adrenal medulla are unaffected. Daily administration to animals of extracts of adenohypophysis is regularly followed by marked hyperplasia of the zona fasciculata and the zona reticularis. In conditions known to be associated with hypofunction of the adenohypophysis (dwarfism and Sheehan's syndrome) there is marked atrophy of the adrenal cortex and evidence of cortical hypofunction. Contrariwise a secreting tumour in the pituitary (basophilic adenoma) is accompanied by hypertrophy of the adrenal cortex and clinical signs of overactivity.

The physiological properties of ACTH are attributable to stimulation of the adrenal cortex, a subject already referred to on page 1007. Briefly, the main effects produced by injection of ACTH (corticotrophin) in man are as follows: (1) increased blood flow to the adrenal glands and hypertrophy of the cortex, (2) increased secretion of cortisol and corticosterone and DHA by the adrenal cortex, and (3) the depletion of lipids (mainly cholesterol esters) and ascorbic acid from the adrenal cortex. The secretion of ACTH is regulated by cortisol—a negative feedback mechanism.

Intravenous injection of insulin or pyrogen causes rapid liberation of ACTH and has been used in tests of the normality of hypothalamic–pituitary–adrenocortical function (see also the use of Metyrapone, p. 1015). Lysine-vasopressin is also used to test pituitary–adrenal function; it acts directly on the anterior pituitary to release ACTH and indirectly by acting on the hypothalamus but this is not the releasing factor in man. The insulin hypoglycaemia test is regarded as the most useful. ACTH is destroyed by the digestive enzymes and therefore has to be given by injection. The hormone is rapidly inactivated by liver and kidney and its half-life in the blood is less than two minutes. Stress or the reaction of the organism to potentially injurious changes in its environment causes secretion of corticotrophin releasing factor from the hypothalamus.

Corticotrophic activity in plasma may be measured by biological assay in which the unknown preparation is compared with a stable standard preparation. The assay method commonly used is based on the estimation of the depletion of ascorbic acid in the adrenal glands which corticotrophin causes when it is injected in hypophysectomized rats (Sayer's assay). The activity (potency in units) of clinical preparations is tested against the current (Third) International Standard for corticotrophin. ACTH may also be assayed by measuring the corticosterone released by it from rat adrenals or the cortisol released from sheep adrenals. Radioimmunoassay methods are now available. Morning plasma ACTH values in man vary from 12 to 35 pg/ml. When hypoglycaemia is induced by insulin the plasma ACTH may rise to 400 pg/ml.

Maximal stimulation of the adrenal glands in man is achieved by a dose of 20 to 30 units given by intramuscular injection every six hours. The response of the adrenal cortex to an injection of ACTH is shown most easily by an increase in the output of 17-hydroxycorticosteroids in the urine above the normal value (p. 1011) or by a rise in the plasma cortisol (or 11-hydroxycorticosteroid) concentration. In Addison's disease (p. 1015) in which the cortex is atrophied or destroyed this test produces little or no rise in steroid output.

ACTH was isolated in 1943 from pituitaries of both pig and sheep and was synthesized in 1963. Hog ACTH yields eight active polypeptides. One of these, the β-ACTH with a molecular weight of 4500, has the following sequence of 39 amino acids (for contractions see pp. 24–26): Ser-Tyr-Ser-Met-Glu-His-Phe : Arg-Trp-Gly-Lys-Pro-Val-Gly-Lys-Lys-Arg-Arg-Pro-Val-Lys-Val-Tyr-Pro-

Asp-Gly-Ala-Glu-Asp-Gln-Leu-Ala- Glu-Ala-Phe-Pro-Leu-Glu-Phe. Corticotrophins from various species differ in the part of the chain shown in heavy type. For example, in human corticotrophin the sequence is Asp-Ala-Gly-Glu-Asp-Gln-Ser-Ala. The *active* portion of the molecule, the first 24 amino acids, is the same in all species but the sequence 25 to 39 is responsible for its immunological activity. A synthetic preparation (Synacthen) containing these 24 amino acids is used in tests and therapy.

Gonadotrophic hormones. Removal of the adenohypophysis from immature animals reduces the rate of growth (see Fig. 52.4a), prevents the development of the gonads and of the secondary sex organs. Since the two pituitary hormones responsible for the development of the gonads have no direct action on the secondary sex organs, but only an indirect one through the ovaries and testes, they have been named gonadotrophic hormones. Human follicle stimulating hormone (FSH) is a glycoprotein with a molecular weight of 41 000 containing 254 amino acid residues. The pituitary contains about 2 μg of FSH. Human luteinizing hormone (LH), also a glycoprotein of molecular weight 15 000 to 25 000, is so-called because of its action in females. Interstitial cell stimulating hormone (ICSH) is the same substance as LH; the name describes its action in males. The actions of these hormones are described in pp. 1049 to 1051.

The growth hormone (somatotrophin). Experimental removal of the pituitary gland or failure of the gland in young animals causes growth to cease. Daily injections of potent extracts not only increase the rate of growth of an injected rat as compared with its littermate but also extend the period of growth, with the result that the injected animal attains a size more than twice its untreated control (Fig. 52.4b). The sequence of the 188 amino acids in human growth hormone (HGH), mol.-wt. 21 500, has been worked out. The human pituitary contains about 5 mg of HGH and about 500 μg are produced daily. HGH is secreted throughout life, the levels being highest in the newborn. Injections of HGH in man cause retention of nitrogen for synthesis of protein and also of intracellular potassium and phosphorus. At the same time sodium and

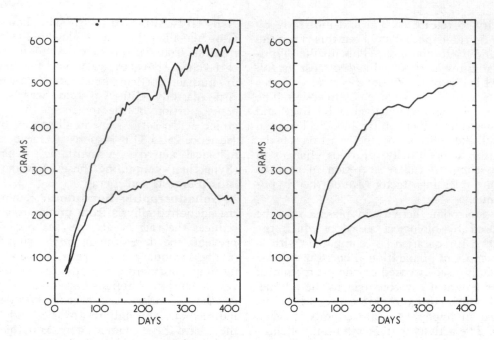

Fig. 52.4a The upper curve shows the growth of a normal rat. The lower curve shows the body weight of a litter mate which was hypophysectomized at 40 days when it weighed 150 g. After the initial post-operative loss of weight the rate of growth was much slower. (P. M. Daniel, L. W. Duchen & M. M. L. Prichard (1964). *Quarterly Journal of Experimental Physiology* **49**, 245.)

Fig. 52.4b The effect of an extract of the adenohypophysis on growth. The upper line is the growth curve of a female rat treated for 400 days, the lower line that of an untreated litter-mate. (Redrawn from H. M. Evans (1923–24). *Harvey Lectures* **19**, 212.)

chloride are retained with a resultant increase in extracellular fluid. Growth hormone increases absorption of calcium and magnesium from the gut and increases the excretion of both in the urine. There is a raised level of free fatty acids in the blood and a lowering of the R.Q. and an increase of urinary ketones—all indications of increasing fat metabolism. The glucose tolerance is impaired. In general the effects on metabolism are the reverse of the effects of insulin. At the cellular level GH may exert its effect on protein synthesis and growth by controlling the production of messenger RNA.

In 1937 F. G. Young reported that dogs given daily injections of crude pituitary extracts developed a permanent diabetic state, with complete degeneration of pancreatic islet tissue, which persisted after the injections were stopped (p. 324). The degeneration is apparently due to prolonged stimulation and subsequent exhaustion of the islets by the

continuous increased production of insulin in response to the pituitary extract. A more important effect of GH in dogs is an inhibition of glucose uptake by the tissues; since plasma-insulin levels remain high, exhaustion is not likely. In man the diabetogenic effect of HGH is much less marked than in cats and dogs and is seen only when islet function is reduced.

Since diabetes mellitus can be produced in dogs and cats by injection of growth hormone it is not surprising to find that acromegalic patients are frequently diabetic.

Assays of HGH by radioimmune techniques (Hunter and Greenwood, 1964) have shown that levels in the plasma of adults fluctuate widely throughout the 24 hours. After the ingestion of glucose or any normal meal containing carbohydrate, the hormone is not detectable in the plasma (<2 ng/ml). Peaks of secretion may occur before the next meal, during the night and always if the subject takes moderate exercise in the post-absorptive state (compare pp. 854 to 856). The size of the rise in concentration is very variable (to levels from 5 to 60 ng/ml) and takes place rapidly, the maximum level being reached usually between 30 and 60 minutes after the start of the rise; the level then falls abruptly. This fall occurs even if no food is taken; but if the subject continues to starve, another peak appears later. Figure 52.5 shows three peaks occurring in the plasma of a medical student whilst walking 28 miles in 7 hours without food. The size of the peaks and their frequency throughout the day and night appear to be greater in growing children than in adults but, owing to

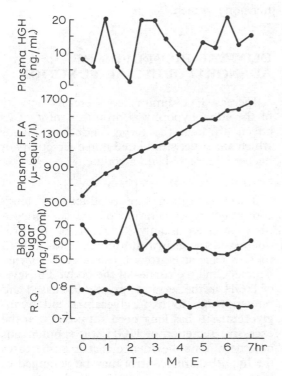

Fig. 52.5 Plasma growth hormone and FFA, blood sugar and R.Q. during a walk at 6·5 km per hour (4 m.p.h.) in a fasting state, of 45 km (28 miles) in a 20-year-old male. (W. M. Hunter, C. C. Fonseka & R. Passmore (1965). *Quarterly Journal of Experimental Physiology* **50**, 406.) See also Fig. 40.28.

their abrupt onset and their irregular occurrence, it is impossible to demonstrate the pattern of secretion throughout the 24 hours without numerous venepunctures. Consequently the effects of age, sex, physique and manner of life upon secretion have not been thoroughly investigated. Pituitary dwarfs produce no detectable hormone, but patients with acromegaly have reached levels up to 250 ng/ml. The most effective way of stimulating HGH secretion is to lower the blood sugar by insulin. A small lowering of blood sugar may be ineffective but a reduction to 50 per cent (which is obviously unphysiological) consistently raises plasma HGH. Ingestion of glucose immediately abolishes HGH secretion in normal persons but 3 or 4 hours later there is a rebound rise of plasma HGH. HGH secretion is also increased by infusion of large doses of arginine and other amino acids, by glucagon and by stressful stimuli such as anxiety or trauma. Administration of corticotrophin causes, about an hour later, a rise of circulating HGH.

Growth hormone is found in the plasma only when lipid is providing a greater proportion of the metabolic fuel than carbohydrate; the hormone appears to be a lipid-mobilizing factor. It has been shown to promote lipolysis in isolated adipose tissue cells though the effect is uncertain and more difficult to demonstrate than the lipolytic action of adrenaline and noradrenaline.

It is possible to look upon insulin and growth hormone as complementary agents, each regulating the supply of energy to the tissues. Insulin is active after a meal, when glucose is probably the main fuel; growth hormone is active in the starving state.

The immediate stimulus to growth hormone secretion is not known and the occurrence of peaks is not closely related to the levels of either glucose or free fatty acids in the blood or to changes in these levels. Furthermore we do not know the factors responsible for the rises and falls in the plasma during starvation, when the subject is apparently in a steady state.

In the fetal and neonatal period GH does not seem necessary for normal growth. Indeed, a normal birth weight is characteristic of true pituitary dwarfs and of the anencephalic human fetus which may have no pituitary gland. It is unlikely that the maternal HGH is responsible for fetal growth since HGH does not seem to be able to pass from mother to fetus. There is no doubt of the necessity of GH for growth in later life but the correlation between plasma GH and growth rate is so variable that it is suspected that the immunoactive HGH may not always be a measure of the circulating growth principle.

The lactogenic hormone (prolactin, also called mammotrophin or luteotrophin). Lactation depends on the pituitary for it ceases after hypophysectomy; it may be maintained after the operation by pituitary extracts. Prolactin may be assayed by injecting it into the mammary ducts of pseudopregnant rabbits and observing the local lactogenic response, or by observing the lactogenic response of mammary tissue in organ culture.

Prolactin is now known to be a separate hormone in man but it has similar properties to growth hormone and placental lactogen. Its secretion (and that of MSH) from the anterior pituitary is inhibited by a hypothalamic factor termed prolactin-release-inhibiting hormone. Immunofluorescent stains show that prolactin and growth hormone come from separate types of acidophil cells. The actions of prolactin are described on page 1074.

The melanophore-stimulating hormone (MSH). Hogben and Winton first showed that extracts of the pituitary cause darkening of the skin of frogs and that after hypophysectomy frogs became light in colour. These changes are due to the fact that the skin of the frog, like that of many cold-blooded animals, contains large cells with branching processes containing granules of melanin (melanophores). When the melanin is aggregated in the body of the cell the animal appears pale in colour, but when the pigment is dispersed throughout the cell body and its processes the skin is dark. The distribution of pigment in the skin, and therefore the blending of the animal with its environment, is under hormonal control, adrenaline producing blanching and the pituitary hormone darkening. These experiments are important since they provide clear evidence that a hormone can govern the distribution of materials within a cell. Though it is found in extracts of the neurohypophysis this hormone is elaborated in the pars intermedia of the adenohypophysis of animals.

MSH activity is now ascribed to two peptides, α- and β-MSH, both of which resemble ACTH chemically. The heptapeptide shown in heavy type (below) in the formula for human β-MSH is essential for activity. It has been found in α-MSH, β-MSH and corticotrophin in all animal species studied so far. β-MSH, isolated from human pituitary glands, is a polypeptide containing 22 amino acids in the following sequence: Ala-Glu-Lys-Lys-Asp-Glu-Gly - Pro - Tyr - Arg - **Met - Glu - His - Phe - Arg-Try-Gly-**Ser-Pro-Pro-Lys-Asp.

Many preparations of ACTH possess much melanophore-expanding activity and even pure ACTH has a little. The secretion of this material is said to be increased in Addison's disease (p. 1010), in which the skin is pigmented. It is not raised in the dark-skinned races. Injections of MSH have produced menstrual bleeding in amenorrhoeic women but no function has been found for this hormone in men.

CLINICAL DISORDERS OF ADENOHYPOPHYSIAL FUNCTION

Increased or diminished secretory activity of the adenohypophysis produces most of its effects through the target endocrine glands which are either stimulated into overactivity or become depressed and atrophic.

GIGANTISM AND ACROMEGALY

These conditions are produced by long-continued hypersecretion of the adenohypophysis and are mainly the result of overproduction of the growth hormone. Thus in both there is excessive growth, though to a varying degree, of all the tissues of the body. The level of HGH in the serum is greatly increased and remains high in hyperglycaemia and hypoglycaemia. It has long been supposed that the excessive secretion of HGH was spontaneous but there is now evidence that in some cases the hypothalamic control may be deranged as shown by the variable results of alteration of blood glucose. The progestational agent medroxyprogesterone acetate (MPA), which acts on the hypothalamus and suppresses gonadal and adrenal function, also reduces HGH secretion provoked by arginine infusion or by insulin-induced hypoglycaemia. MPA also reduces HGH level in acromegalic patients. The tranquillizer chlorpromazine also reduces serum HGH probably by an action on the hypothalamus; this presumably interferes with the metabolism of catecholamines in the hypothalamus.

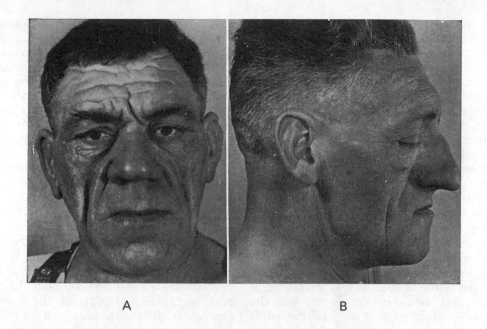

A   B

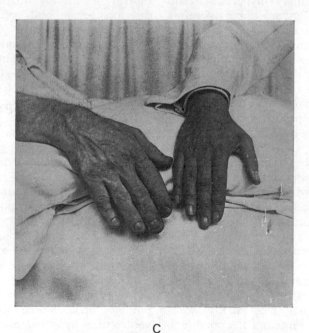

C

Fig. 52.6 A. Front view of the face of a patient with acromegaly to illustrate the gross thickening of the soft tissues. (*By courtesy of R. Greene and The Practitioner*.) B. Profile of the face of another patient to show early changes in the bones of the face. Note the prominent supra-orbital ridges and the wide angle of the jaw. C. The hand of an acromegalic patient contrasted with the hand of a normal man.

Gigantism is produced by excessive growth of bone before the epiphyses have closed, with the result that the patient may attain a height of 2·1 to 2·6 metres (7 to 8½ feet). The growth of the limbs is relatively greater than that of the trunk and the tallness is largely the result of excessive growth of the legs. The internal organs are also hypertrophied and the muscles, at least in the early stages, may be very well developed. Such individuals are usually mentally subnormal and die before the age of twenty.

Acromegaly (Fig. 52.6) occurs in adults after the epiphyses have closed. The bones, therefore, cannot grow in length but they become thickened and deformed; the internal organs are enlarged as in gigantism. The most striking signs of acromegaly are the generalized coarsening of the features due to thickening of the skin and subcutaneous tissues, and the progressive enlargement of the head, the hands and the feet. Occasionally sexual function is increased with mammary hyperplasia and lactation. Other findings are hyperglycaemia and glycosuria, raised B.M.R. and increased concentration of growth hormone in the blood (p. 1085). Since the excessive secretion from the adenophypophysis is often associated with a tumour consisting of acidophil cells the acromegalic patient may have symptoms referable to pressure upon the optic chiasma.

Cushing's syndrome. This syndrome (p. 1010) is due to excessive secretion of adrenal corticosteroids. It may be due to a tumour of the adrenal cortex or to excessive secretion of ACTH which produces adrenal hyperplasia. The detection of ACTH in the blood by radioimmunoassay suggests that pituitary overactivity is the prime cause.

Pituitary dwarfism. Insufficiency of the adenohypophysial secretions in childhood results in delayed growth and a form of dwarfism, with permanently retarded sexual development. Pituitary dwarfs are usually well-proportioned, alert and intelligent, and able to earn their own living on the stage or in the circus, but occasionally other symptoms of hypopituitarism may be found such as those described below. It is not easy to diagnose HGH deficiency since TSH and ACTH may be deficient at the same time. The plasma GH rises relatively little in insulin-induced hypoglycaemia in true hypopituitary dwarfs. How-

ever, the nitrogen retention in response to HGH treatment (at least 2·5 mg per week) is relatively high. The growth response of true pituitary dwarfs is very good. It is remarkable that a hormone with a half-life in the blood of less than 30 minutes should have such long-lasting effects. Diabetogenic effects have not been seen in children with deficiency of growth hormone treated over a lengthy period with HGH.

Panhypopituitarism (Sheehan's syndrome). Adenohypophysial insufficiency in the adult may be due to destruction of the adenohypophysis following an interruption of its circulation at parturition (*post partum* necrosis) or to compression of the adenohypophysis by a tumour. There is genital atrophy, gradual regression of secondary sex characteristics, and loss of body hair. The basal metabolism is reduced, the body temperature is often subnormal, and the heart rate and arterial pressure low. Carbohydrate metabolism is grossly disturbed and patients develop severe and sometimes fatal hypoglycaemia. Treatment with cortisone and small doses of testosterone and thyroxine may relieve the condition but rarely succeeds in restoring the patient to normal health.

THE NEUROHYPOPHYSIS

The neurohypophysis or posterior lobe of the pituitary is a prolongation of the floor of the third ventricle (see Figs. 52.2a, 52.3 and 52.7). It receives a small supply of sympathetic fibres, probably vasomotor, derived from the carotid plexus but they do not seem to play any part in regulating the secretion of the hormones. The neurohypophysis is, however, connected with the hypothalamic nuclei by a large number of fibres, estimated at 100 000 in man, collectively known as the hypothalamo-hypophysial tract, of which the main components arise in the supraoptic and paraventricular hypothalamic nuclei (Fig. 52.7). The neurohypophysis is supplied, as shown in Fig. 52.2a, by the inferior hypophysial arteries.

HORMONES OF THE NEUROHYPOPHYSIS

The neurohypophysial hormones are formed by nerve cells in the paraventricular and supraoptic nuclei of the hypothalamus (Figs. 47.1 and 52.7). The hormones pass slowly

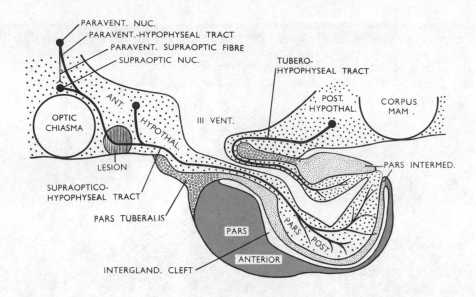

Fig. 52.7 The nervous connexions of the pituitary gland of the cat. Diabetes insipidus is produced by destruction of the posterior part of the pituitary gland or by a lesion of the supra-optico-hypophyseal tract. (C. Fisher, W. R. Ingram, W. K. Hare & S. W. Ranson (1935). *Anatomical Record* **63**, 31.)

down the axons of the nerve cells; radioactive tracers show that in the rat the hormones arrive in an hour or so at the dilated nerve terminals in the posterior lobe. Here the hormones are stored until they pass into the blood. Gomori's stain picks out material (neurosecretory material, NSM) in the cells and axons which may be the carrier of the hormones. If the stalk is cut NSM accumulates in the hypothalamus and disappears from the posterior lobe. At the same time the hypothalamo-hypophyseal nerve tracts distal to the section degenerate. After a few weeks new nerve fibres accompanied by NSM begin to sprout out from the hypothalamus and the posterior lobe is fully re-innervated in a year. This capacity for vigorous regeneration is, according to Daniel, unique among the tracts of the central nervous system. This may be related to the fact that they are unmyelinated fibres or that they are neurosecretory in nature.

The polypeptide hormones are attached in the hypothalamus to cysteine-rich proteins called neurophysins (mol. wt 25 000), which may be regarded as carrier proteins. There are at least two neurophysins: neurophysin I is linked to oxytocin, and neurophysin II to vasopressin (ADH). It is likely but not quite certain that vasopressin is synthesized in the supraoptic nuclei and oxytocin in the paraventricular nuclei. The hypothalamus can synthesize vasopressin *in vitro*.

The neural lobe is shown by the electron microscope to consist of a tangle of unmyelinated nerve fibres surrounding the pituicyte-glial cells. The nerve fibres end in dilatations, from 0·5 to 5 μm in diameter, often placed near a capillary. The dilatations contain granules or vesicles about 0·1 μm in diameter (Fig. 52.8). The granules correspond to the Gomori-positive material seen with the light microscope. Large aggregations probably correspond to the bodies described by Herring in 1908. Electron microscopy cannot distinguish two types of particles but oxytocin may be carried in the slightly more dense ones and vasopressin in less dense ones. Douglas has shown that the release of the posterior lobe hormones from the nerve endings of the hypothalamo-hypophysial tract is calcium dependent. Release may follow a depolarization of the endings which allows influx of calcium.

The injection of an aqueous extract of the pituitary gland into animals (Fig. 52.9) was shown by Oliver and Schäfer in 1895 to pro-

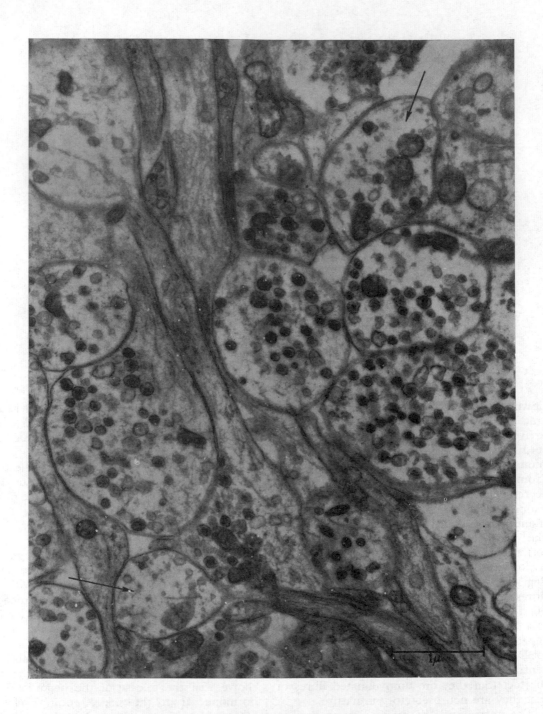

Fig. 52.8 General medium-power view of rabbit neurohypophysis showing unmyelinated nerve fibres in longitudinal section and a number of neural swellings. Some of the nerve fibres are continuous with these swellings. Most of the swellings contain small granules about 130 nm in diameter with an electron-dense centre. These are referred to as neurosecretory granules. In some swellings (arrows) much smaller, apparently empty, vesicular structures, 20 to 30 nm in diameter, can be seen. Note the spaces between some of the neural swellings. Fixed OsO_4, stained lead hydroxide. $\times 24\,000$. (R. Barer, H. Heller & K. Lederis (1963). *Proceedings of the Royal Society* B **158**, 338–416.)

duce a marked increase of arterial pressure which lasted about ten times longer than that following an injection of adrenaline giving a similar rise of pressure. Simple aqueous extracts of the posterior pituitary also bring about contraction of the uterus (oxytocic effect, Fig. 52.9); they increase arterial pressure (vasopressor effect); they reduce the output of urine (antidiuretic effect).

The physiological effects mentioned above are produced by two hormones *oxytocin* and *antidiuretic hormone (vasopressin)* both of which have been synthesized (1953). Both are cyclic octapeptides, of molecular weight 1025 and 1102 respectively, differing in only two amino acids (Fig. 52.10). The hormones seem to be produced at different sites and may even be released separately. Whatever the stimulus

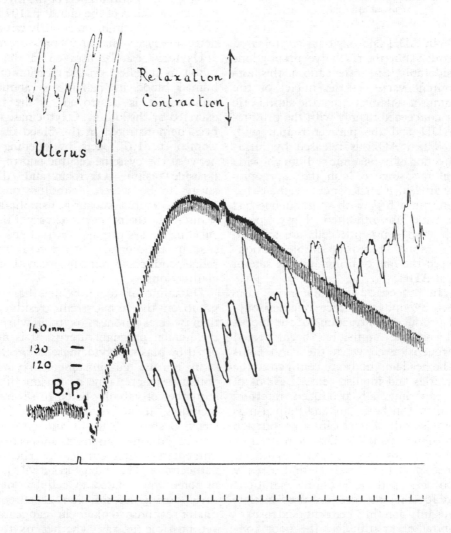

Fig. 52.9 The original figure which showed the oxytocic action of pituitary extract. The simple extract used had, in addition to its oxytocic effect, vasopressor (that is ADH) activity. The sympathetic motor effects had been abolished by a previous dose of ergot. At the signal, extract from 0·4 g dried ox-pituitary was given intravenously, producing the rise of blood pressure and contraction of the uterus. Pregnant cat. Time in 10 sec. (H. H. Dale (1906). *Journal of Physiology* **34**, 195.)

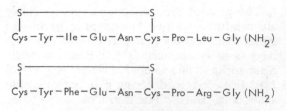

S――――――――――――――――S
| |
Cys – Tyr – Ile – Glu – Asn – Cys – Pro – Leu – Gly (NH$_2$)

S――――――――――――――――S
| |
Cys – Tyr – Phe – Glu – Asn – Cys – Pro – Arg – Gly (NH$_2$)

Fig. 52.10 The structures of human oxytocin (above) and human antidiuretic hormone, arginine-vasopressin (below). Vertebrates all have the same oxytocic factor (upper formula). The antidiuretic factor, however, shows species differences. For example in the pig the arginine residue is replaced by lysine.

applied, both ADH and oxytocin are released simultaneously but their relative proportions vary considerably. Since the ratio of the hormonal content varies in the nuclei of the hypothalamus it is deduced that the supraoptic nucleus is concerned mainly with the production of ADH and the paraventricular with oxytocin. When ADH is released by intracarotid injection of hypertonic sodium chloride solution in rats some cells in the supraoptic and paraventricular nuclei increase their discharge rate but a few (with spontaneous fast discharge rate) are inhibited. Fewer paraventricular than supraoptic cells are affected by this procedure. There is no simple relationship between the rate of firing and the rate of secretion of ADH.

Oxytocin. Electrical stimulation of the supraoptico-hypophyseal tract releases oxytocin and produces contractions of the uterus (provided that the animal has been treated with oestrogens) even when the connexions between the head and body are confined to the carotid arteries and jugular veins. Lesions of the tract, experimentally produced, interfere with parturition in both cats and guinea-pigs but the precise role of oxytocin in parturition is not yet known (p. 1072). Oxytocin itself or synthetic oxytocins (for example syntocin) with only a small pressor effect find a ready application in obstetrics. Dilute solutions (from 2 to 10 units per litre) are used in slow intravenous infusion (in 5 per cent dextrose) to induce labour or to stimulate the poor contractions of inert labour. Oxytocin alone is little used to contract the uterine muscle to control obstetric haemorrhage since ergometrine is somewhat more active. However, recently an injection combining syntocin and ergometrine has found favour in the prevention of post partum haemorrhage. The release of oxytocin is prevented by high blood alcohol levels. Intravenous administration of alcohol giving blood levels of 200 to 400 mg/100 ml is used to prevent premature onset of labour.

The oxytocic hormone, then, reinforces contractions of the uterus during parturition. It also increases the motility of the uterus during coitus (p. 1053) and gives rise to the rapid ejection of milk from a secreting mammary gland by stimulation of the myoepithelial cells in the ducts of the gland (p. 1075). A dose of 4 μg of oxytocin intravenously can give a just detectable rise in mammary pressure in women.

Oxytocin can be assayed in the lactating rabbit by its effect on the ejection of milk. In human blood its half-life is about 1 to 4 minutes. It is destroyed in the tissues or excreted by the kidney. Oxytocinase which has been demonstrated in the blood of pregnant women (p. 1072) acts by opening the link between the cysteine residue and the adjacent tyrosine residue. Oxytocin and ADH seem always to be secreted simultaneously in response to coitus, suckling, osmotic and other stimuli, but the proportions vary. For example suckling causes uterine contractions with only a small antidiuresis; intravenous hypertonic saline produces an antidiuresis without uterine contractions.

Desamino-oxytocin, which has no amino group on the first cysteine residue, is more than twice as potent as oxytocin when tested in the human pregnant uterus. It is not inactivated by plasma oxytocinase *in vitro*.

In normal circumstances oxytocin is as potent as isoprenaline in causing dilatation in a number of vascular beds. In several species, including man, oxytocin induces vasoconstriction some 30 to 40 min after the intravenous administration of an oestrogen, and immediately after surgical or chemical sympathectomy. The change in activity is probably in some way related to cellular metabolism, since during an infusion of adrenaline the dilator response to oxytocin reappears. It is not yet possible to say whether oxytocin is of physiological importance to the vascular system.

Antidiuretic hormone. ADH (vasopressin) acts directly on the collecting tubules of the kidney. It may increase the permeability to water of this part of the nephron by causing a local release of hyaluronidase which alters the permeability characteristics of the intercellular cement substance (p. 661). The state of dehydration, in which fluid must be conserved in the body, leads to an appreciable increase in the amount of antidiuretic activity in the infundibular process. As little as 1 m-u of ADH given intravenously to a normal person in water diuresis causes a reduction in urinary output. This dose could at most raise the blood level by only 0·0002 m-u per ml. The normal blood level must therefore be very low indeed and present methods are not sensitive enough to measure it even in dehydrated subjects. Less than 10 m-u of ADH are excreted per day in a normally hydrated person and even in severe dehydration only about 10 m-u per hour appear in the urine. Night urine usually contains more ADH than day urine. ADH has a mild milk ejection action on the mammary gland and a slight oxytocic activity. The half-time for disappearance from the plasma is about 5 minutes.

If ADH is injected into an animal from which the liver and kidneys have previously been removed the hormone disappears from the blood much more slowly than it does in the intact animal. Furthermore, large doses produce a greater rise of arterial pressure when injected into the femoral vein of an animal than when given into the splenic vein. These experiments suggest that ADH is destroyed, or removed from the circulation, by the liver. Relatively rapid removal of ADH from the circulation is clearly necessary for effective homeostatic control of fluid balance.

As already mentioned the vasopressor effect of posterior pituitary extracts was the first to be discovered. Pressor fractions constrict arteries and capillaries by a direct action on their walls and so raise the blood pressure and make the skin pale. The increase of arterial pressure is followed by reflex slowing of the heart and passive dilatation of the cerebral and renal vessels, which do not apparently participate in the generalized vasoconstriction. These actions are regarded as pharmacological rather than physiological because removal of the hypophysis does not cause a fall of blood pressure and because amounts of hormone, too small to produce a rise of blood pressure,

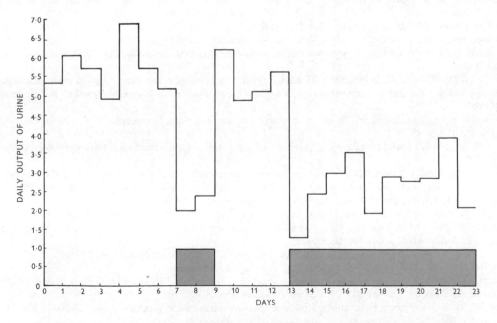

Fig. 52.11   The effect of pituitary extract on the daily output of urine of a male patient, aged 24, with diabetes insipidus. During the first shaded area three subcutaneous injections a day were given on each of two successive days. During the second shaded area the patient was treated with daily injections of the same extract.

produce a marked diminution in the output of urine.

Recently it has been found that desamino-antidiuretic hormone when tested on the rat has about three times the antidiuretic potency of the naturally occurring, arginine containing, hormone.

Experimental removal of the neurohypophysis, or division of the supraoptico-hypophyseal tract, is followed by increased elimination of water (polyuria). Lesions of the hypothalamic area supplying the posterior lobe (Fig. 52.7) produce *diabetes insipidus* in which very large volumes of urine may be produced (Fig. 52.11). Daily injections of ADH or pituitary extracts can reduce the urinary output to normal unless the renal tubules are unable to respond to ADH (nephrogenic diabetes insipidus). Verney has established that the output of ADH from the pituitary is governed mainly by the osmotic pressure of the blood (p. 661). Superimposed on this, however, is the influence of the central nervous system through which the output of hormones is affected by emotional stress and by changes in the external environment (pp. 657 and 664). Stimulation of the central end of the vagus or ulnar nerve and of certain areas in the brain stem causes (as does any painful stimulus) liberation of ADH and antidiuresis. The impulses pass in a reticular pathway to the thalamus and then to the hypothalamus. But this is not the only pathway to the supraoptic nuclei.

REFERENCES

ACHER, R. (1960). Biochemistry of the protein hormones. *Annual Review of Biochemistry* **60**, 547–576.

BARER, R., HELLER, H. & LEDERIS, K. (1963). The isolation, identification and properties of the hormonal granules of the neurohypophysis. *Proceedings of the Royal Society* B **158**, 388–416.

BARGMANN, W. (1966). Neurosecretion. *International Review of Cytology* **19**, 183–201.

BERDE, B. (Ed.) (1968). Neurohypophysial hormones and similar polypeptides. *Handbook of Experimental Pharmacology* **23**, 1–967.

BROWN-GRANT, K. & CROSS, B. A. (Eds.) (1966). Recent studies on the hypothalamus. *British Medical Bulletin* **22**, 195–277.

CATT, K. J. (1970). *ABC of Endocrinology*. London: *Lancet*.

Ciba Foundation Colloquia on Endocrinology, Vols. 1 to 14. London: Churchill.

Comparative aspects of neurohypophyseal morphology and function (1963). *Symposia of the Zoological Society of London* **9**, 1–183.

DANIEL, P. M., DUCHEN, L. W. & PRICHARD, M. M. L. (1964). The cytology of the pituitary gland of the Rhesus monkey: changes in the gland and its target organs after section of the pituitary stalk. *Journal of Pathology and Bacteriology* **87**, 385–393.

Discussion on polypeptide hormones. Posterior pituitary hormones and neurophysiology (1968). *Proceedings of the Royal Society* B **170**, 3–47.

EVERETT, J. W. (1964). Central neural control of reproductive functions of the adenohypophysis. *Physiological Reviews* **44**, 373–431.

EZRIN, C. & MURRAY, S. (1963). The cells of the human adenohypophysis in pregnancy, thyroid disease and adrenal cortical disorders. *Colloques Internationaux du Centre National de la Recherche Scientifique* **128**, 183–199.

FRASER, R. (1963). Human growth hormone. *Scientific Basis of Medicine*, 36–52.

HARRIS, G. W. & DONOVAN, B. T. (Eds.) (1966). *The Pituitary Gland*, Vols. 1–3. London: Butterworth.

HARRIS, G. W., REED, M. & FAWCETT, C. P. (1966). Hypothalamic releasing factors and the control of anterior pituitary function. *British Medical Bulletin* **22**, 266–272.

HARRIS, I. (1960). Chemistry of pituitary polypeptide hormones. *British Medical Bulletin* **16**, 189–195.

HELLER, H. & CLARK, R. B. (1962). *Neurosecretion*. London: Academic Press.

HOFER, H. O. (1968). The phenomenon of neurosecretion. In *Structure and Function of Nervous Tissue*, Vol. 1 (G. H. Bourne, ed.), Ch. 11, pp. 461–517. New York: Academic Press.

HUNTER, W. M. & GREENWOOD, F. C. (1964). Studies on the secretion of human pituitary growth hormone. *British Medical Journal* i, 804–807.

JAMES, V. H. T. & LANDON, J. (Eds.) (1968). *The Investigation of Hypothalamic–Pituitary–Adrenal Function*. London: Cambridge University Press.

KNUBIL, E. & HOTCHKISS, J. (1964). Growth hormone. *Annual Review of Physiology* **26**, 47–74.

KOIZUMI, K., ISHIKAWA, T. & BROOKS, C. M. (1964). Control of activity of neurons in the supraoptic nucleus. *Journal of Neurophysiology* **27**, 878–892.

McCANN, S. M. & PORTER, J. C. (1969). Hypothalamic pituitary stimulating and inhibiting hormones. *Physiological Reviews* **49**, 240–284.

McSHAN, W. H. & HARTLEY, M. H. (1965). Production, storage and release of anterior pituitary hormones. *Ergebnisse der Physiologie* **56**, 264–296.

MILLS, E. & WANG, S. C. (1964). Liberation of antidiuretic hormone: location of ascending pathways. *American Journal of Physiology* **207**, 1399–1404.

PEARSE, A. G. E. & VAN NOORDEN, S. (1963). The functional cytology of the human adenohypophysis. *Canadian Medical Association Journal* **88**, 462–471.

PINCUS, G. (Ed.) (1948–67). *Recent Progress in Hormone Research*. New York: Academic Press.

PRADER, A. (1967). Dwarfism, hypopituitarism and growth hormone. *Archives of Disease in Childhood* **42**, 225–257.

PURVES, H. D. & ADAMS, D. D. (1960). Thyroid stimulating hormone. *British Medical Bulletin* **16**, 128–132.

ROOS, P. (1968). Human follicle-stimulating hormone. *Acta endocrinologica* **131** Suppl., 1–93.

SAYERS, G., REDGATE, E. S. & ROYCE, P. C. (1958). Hypothalamus, adenohypophysis and adrenal cortex. *Annual Review of Physiology* **20**, 243–274.

SCHINDLEP, W. J. (1962). Hypothalamic neurohumoral control of pituitary function. *Proceedings of the Royal Society of Medicine* **55**, 125–130.

SHEEHAN, H. L. & WHITEHEAD, R. (1963). The neurohypophysis in postpartum hypopituitarism. *Journal of Pathology and Bacteriology* **85**, 145–169.

SONENBERG, M. (Ed.) (1968). Growth hormone. *Annals of the New York Academy of Science* **148**, 289–571.

STUTINSKY, F. (Ed.) (1968). *Neurosecretion*. Berlin: Springer-Verlag.

SWANSON, H. E. & EZRIN, C. (1960). Natural history of the delta cell of the human adenohypophysis in childhood, parenthood and pregnancy. *Journal of Clinical Endocrinology* **20**, 952–966.

WOLSTENHOLME, G. E. W. & BURCH, JOAN (Eds.) (1971). *Neurohypophysial Hormones*. Ciba Foundation Study Group No. 39. London: Churchill Livingstone.

53 Chromosomes and heredity

We have seen that cells are adapted to the specific function which they have to perform in the different tissues and that they are therefore often highly differentiated morphologically. This differentiation begins during the embryonic period and continues during growth until the adult state is reached. With increasing specialization, however, there is progressive loss of the ability to multiply. In some tissues, such as epidermis and bone marrow, certain cells remain undifferentiated and these can multiply readily. In glandular tissues, such as liver and thyroid, all the cells normally are differentiated but they still have a limited power of regeneration should the organ suffer any damage. The ability to replace any worn out or damaged cell has disappeared in cells of the central nervous system which in the adult have lost the capacity to divide. During differentiation the morphological changes which occur in the cell are less marked in the nucleus than in the cytoplasm. Since the nucleus contains DNA (deoxyribonucleic acid) which carries heritable information in the sequence of its four bases along its very great length (p. 68), it is important that any new cells should have the same DNA complement as the parent cell, and hence the same chromosome content, for the DNA of the cell is shared out among the individual chromosomes and is tightly coiled up within them. This orderly division of nuclear material is known as *mitosis*. In preparation for mitosis the DNA doubles itself during interphase by replication as described on page 352. At the outset of a division the nuclear material becomes condensed into discrete individual chromosomes which can be seen to have formed two identical halves called *chromatids* which remain joined at the centromere. The nuclear membrane then disappears, a spindle of hyaline fibres is formed between the two centrosomes and the 'split' chromosomes line up on the equator. The chromatids then separate and travel towards

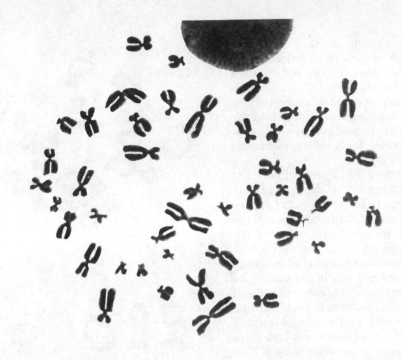

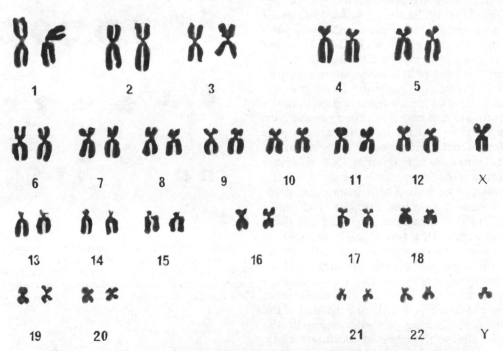

Fig. 53.1 Above, mitotic metaphase spread of a lymphocyte obtained from a culture of blood from a normal human male. Below, chromosomes from the same cell arranged as a karyotype. This cell has been stained with acetic orcein which does not allow each chromosome pair to be distinguished with certainty. For convenience they are designated by groups. Group A, chromosomes 1–3; group B, chromosomes 4–5; group C, chromosomes 6–12; group D, chromosomes 13–15; group E, chromosomes 16–18; group F, chromosomes 19–20; group G, chromosomes 21–22. (*By courtesy of M. J. W. Faed.*)

the poles of the spindle so that each new nucleus has identical genetic information to that of the parent nucleus. The cytoplasm then divides and two new daughter cells are formed, each containing the same DNA as the parent. In unicellular organisms mitotic division of the cell results in the formation of two new individuals, a special form of reproduction known as binary fission.

In man, and most of the higher animals, the chromosome complement is derived equally from the two parents. Their cells contain two numerically equal sets of chromosomes and the individual is said to be *diploid*. Each of the chromosomes of one set can be paired up with those of the other and these 'like' chromosomes of maternal and paternal origin are said to be *homologous*. In man the *diploid number* is 46 and this number of chromosomes is present in all somatic cells. These may be seen at mitosis (Fig. 53.1). There are 44 *autosomes* and 2 *sex chromosomes*. The latter are designated XX in the female and are the same shape and size but in the male they are designated X and Y and are clearly distinguishable. The *haploid number* 23 is best recognized when homologous pairs come together at meiosis (Fig. 53.2) at the onset of gamete formation although, of course, it is not until the pairs have separated and the cell divided that true haploid cells are present.

The process of sexual reproduction involves the conjugation of two cells called *gametes* with union of their nuclei to form a *zygote* or fertilized ovum. The gametes arise within the gonads from cells which remain in a relatively undifferentiated state until the onset of the reproductive period. These primordial germ cells contain the diploid number of chromosomes and if the gametes derived from them were to retain this full somatic number of chromosomes the number in the zygote would be doubled as would also the amount of DNA. Such an increase is avoided by the occurrence of *meiosis* in which there are two consecutive divisions of the cell but only one division of the chromosomes during the maturation of the gametes. Before the onset of the first meiotic (reduction) division the nuclear material resolves itself into chromosomes and the homologous pairs come together to form *bivalents*. During this period of close apposition of the homologues one of the chromatids of one chromosome may exchange material with one

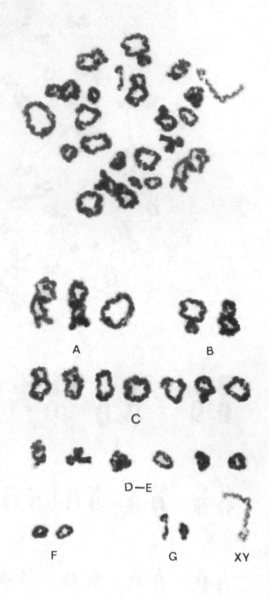

Fig. 53.2 Above, meiotic plate from a spermatocyte from a normal male. The homologous chromosomes are paired. Below, karyotype of the same cell. The XY bivalent is easily recognized but the autosomes are less readily identified and are shown here in groups. (*By courtesy of Ann Chandley.*)

of the chromatids of the other. The cross-over points or *chiasmata*, which can be seen at the appropriate stage of meiosis, have an important role to play in the redistribution of the genetic material. Separation of the pairs of chromosomes accompanied by cell division then occurs resulting in two cells, each with a haploid chromosome constitution. A further, second division then follows in which the chromosomes divide in a manner similar to that seen in mitosis but this time, because of the crossing over which took place earlier, the chromatids are not identical and therefore the resulting daughter cells carry different genetic information. Thus each female gamete or ovum and each male gamete or spermatozoon contains only the haploid number of chromosomes (in man, 23) and only half the usual amount of DNA. It should be noted that in ordinary mitosis the diploid chromosome complement is *retained by* the daughter cells but in meiosis the diploid number is *shared between* the daughter cells. The random distribution of the homologous chromosomes results in a shuffling of the chromosomes so that different combinations of maternal and paternal genetic material (genes) come to reside in the germ cells produced.

In the seminiferous tubules (p. 1027) each primary spermatocyte undergoes meiosis, the first reduction division giving two secondary spermatocytes which on their second division give four spermatids all of the same size (Fig. (Fig. 53.3). In the transformation of a spermatid into a spermatozoon the nucleus is retained in the head and the centrosomes in the middle piece, and the much reduced cytoplasm forms a thin investing layer from which the axial filament of the tail projects.

During maturation of the ovum analogous changes occur, but the cytoplasm is not divided equally between the daughter cells (Fig. 53.3). The ovum (or primary oocyte) undergoes a reduction division so that each of the daughter cells has the haploid number of chromosomes; one of the two daughter cells, the secondary oocyte, retains most of the cytoplasm; the other smaller cell is cast off as the *first polar body*. The secondary oocyte now divides again and throws off another polar body; the first polar body may or may not divide again.

In most mammals, including the human species, fertilization occurs in the uterine tube by the entry of a spermatozoon into a mature

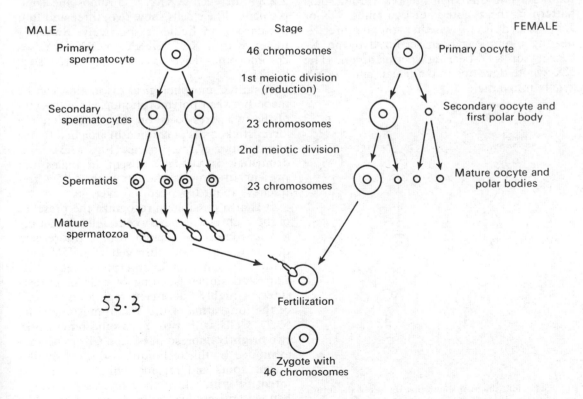

Fig. 53.3 The development of the cells involved in the reproductive process.

ovum, the zygote formed by this union having the diploid number of chromosomes. Thus a new individual begins life with a chromosome number and a DNA complement derived equally from its parents and this number is usually retained in all its somatic cells since all subsequent divisions are mitotic.

SEX DETERMINATION

It has now been demonstrated that the Y chromosome is male-determining in man and that the X chromosome carries some, if not all, of the female-determining genes. As mentioned on p. 1100 the sex chromosomes are XX in the female and XY in the male. Since at meiosis the homologous pairs of chromosomes separate all the ova contain a single X chromosome. In males the X and Y chromosomes behave as a pair and at the reduction division X passes into one secondary spermatocyte and Y into the other. The subsequent meiotic division results in two X-bearing and two Y-bearing spermatids which ultimately develop into two X-bearing and two Y-bearing spermatozoa. It follows from this that the sex chromosome pattern in the zygote is of two kinds, XX or XY, according to whether the mature X-bearing ovum has been fertilized by an X-bearing or a Y-bearing spermatozoon. The XX zygote develops into a female and the XY zygote into a male (Fig. 53.4).

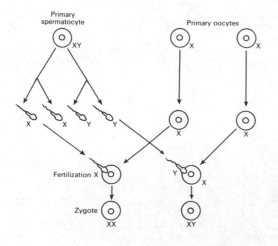

Fig. 53.4 This diagram shows that the sex of the future individual is determined by the kind of spermatozoon which fertilizes the ovum.

It is possible to diagnose the chromosomal sex of an individual by examination of cells from somatic tissues such as skin, buccal mucosa, nerve and many others. A small particle staining deeply with dyes like cresyl violet, the so-called 'Barr body' or 'sex chromatin body' may be seen in contact with the nuclear membrane in the nuclei of 40 to 90 per cent of somatic cells in the normal female (the incidence varies in different tissues) but rarely, if ever, in the normal male. It is now established that the sex-chromatin body is derived from a single X chromosome. This X chromosome is metabolically inactive; this presumably prevents an overdose of X-borne genes (the male cell, it will be remembered, has only one X chromosome). The number of sex chromatin bodies seen within a nucleus is therefore one less than the number of X chromosomes in that nucleus. Thus a male with one X chromosome (XY) shows no sex-chromatin body; a female with two X chromosomes (XX) shows one sex-chromatin particle. This pattern is maintained in abnormal individuals who have an XXX, XXXX or even XXXXX chromosome complement. Their cells show two, three and four sex-chromatin bodies respectively. Similarly males with XXY, XXXY or XXXXY sex chromosomes have one, two or three such bodies.

A similar morphological distinction can be made between polymorphonuclear leucocytes of the two sexes based on the recognition of a 'drumstick' mass of chromatin attached to one of the lobes of the nucleus (Fig. 53.5). This drumstick is virtually absent in males but present in females though with a low frequency, about 1 per cent on average.

It is also possible to recognize the presence of the Y chromosome in a cell. Cells containing a Y chromosome stained with a fluorescent quinacrine dye and observed with UV light show a bright spot in interphase cells (Fig. 53.6) and similar staining of cells in mitosis shows a highly fluorescent region on the end of the long arms of the Y chromosome (Fig. 53.7). Cells with two Y chromosomes show two brightly fluorescent bodies. Chromosomes examined by this technique also show smaller bright spots and regions which vary in the intensity with which they fluoresce giving a banded appearance to the chromosomes. The resulting pattern is consistent for each chromo-

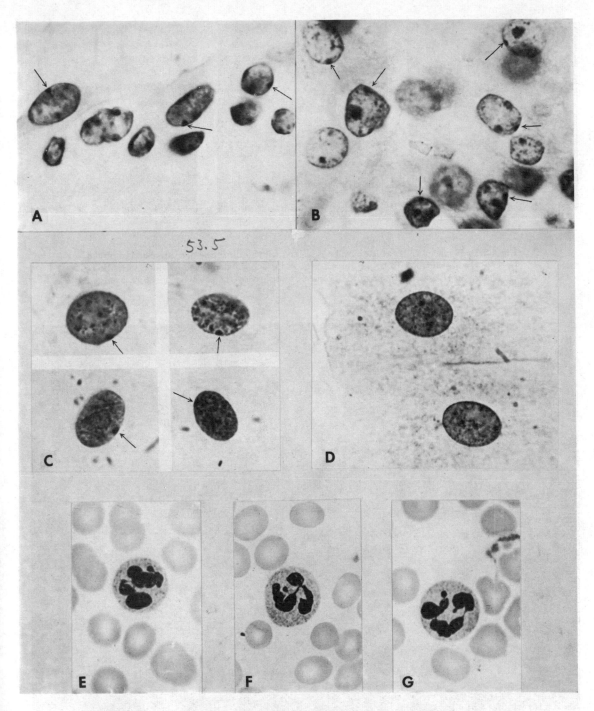

53.5

Fig. 53.5 Photomicrographs showing the sex chromatin particle in A, vascular endothelium; B, stratified squamous epithelium; C, epithelial squames from buccal mucosa. Photograph D also shows buccal epithelial squames but from a male; the sex chromatin particle is absent. Neutrophil polymorphonuclear leucocytes in E, F and G are from a female; they show the characteristic 'drumstick'. The magnification in all cases is of the order of 1400 times. (*Figures A and B are by courtesy of W. W. Park, C and D by courtesy of B. Lennox and M. Ferguson-Smith, E, F and G, by courtesy of W. M. Davidson.*)

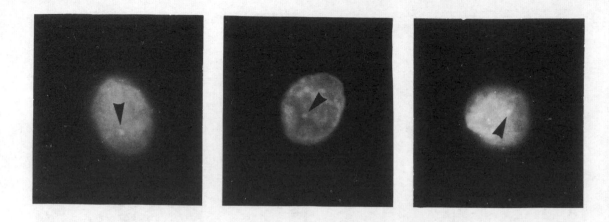

Fig. 53.6 Nuclei of cells, which have been stained with a quinacrine derivative and observed by U.V. light, obtained from the buccal mucosa of a normal male. The fluorescent body characteristic of a cell with a Y chromosome is indicated by the arrowheads. (*By courtesy of M. J. W. Faed.*)

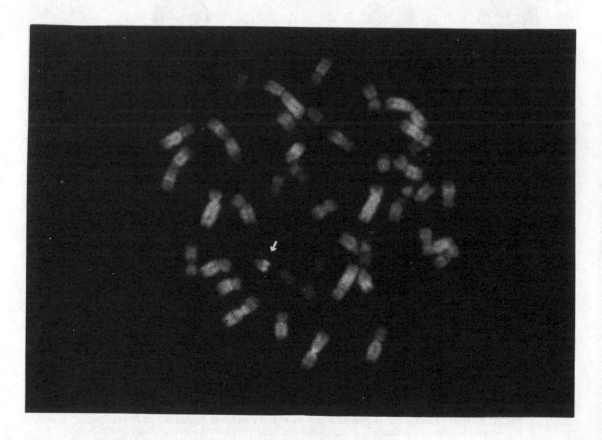

Fig. 53.7 Mitotic metaphase spread derived from the same male as the cell in Fig. 53.1. This cell has been stained with a fluorescent quinacrine derivative. The chromosomes show characteristic fluorescent patterns which permit each pair to be identified. The end of the Y chromosome is particularly bright (arrowed). (*By courtesy of M. J. W. Faed.*)

some pair so that it is now possible to identify them all. A similar banded effect can be achieved by more conventional chromosome stains after appropriate pretreatment of the preparations.

SEX RATIO

Since in the course of spermatogenesis X and Y spermatozoa are produced in equal numbers, a 1:1 sex ratio in the fertilized eggs would be expected. In other words, in a large group of offspring there should be an equal number of males and females. It is probable that the *primary sex ratio*, that is the ratio between males and females at conception (fertilization), considerably exceeds unity but there is no satisfactory explanation for this. It might be because the ova have a slight ability to select the Y spermatozoa in preference to the X spermatozoa but this is unlikely. It might be that conditions in the genital tract, perhaps pH or temperature, favour the Y spermatozoa, but various attempts to control the sex of the offspring, for example by altering the pH of the vagina, have met with no success. Control of sex by means other than selective abortion is unlikely to be achieved until some practical method is found of separating the X spermatozoa from the Y spermatozoa. Sperm carrying a Y chromosome can now be identified by means of the fluorescent techniques mentioned above but only fixed material can be used.

In England and Wales in 1962 for every 106 boys born alive there were 100 girls born alive; this gives a *secondary sex ratio* of 106 to 100. Approximately 130 male fetuses perish during the last three months of intrauterine life for every 100 female fetal deaths and the proportion of male-conceptus deaths is even higher in the third and fourth months of pregnancy. The primary sex ratio may, therefore, greatly exceed 106 to 100 to allow for the wastage of males; one estimate based on the nuclear 'sexing' of aborted placental tissue puts it as high as 111 to 100: however, since it is now known that about 7 per cent of spontaneous abortions have a single X and no Y chromosome and these are chromatin negative this figure must be regarded as still tentative. Since the death rate in males is higher we may conclude that they are more 'fragile' than the females. If the conditions which attend pregnancy are unfavourable to the fetus, because of

disease or malnutrition for example, there may be more fetal deaths than normal and it is to be expected that the higher fetal mortality would affect the males particularly; thus the secondary sex ratio might be relatively low. The expectation that the secondary sex ratio would be a biological yardstick for comparing the social conditions of different countries has not been fulfilled. For example, the secondary sex ratio in England and Wales has been 106 since the year 1946 while the infant mortality (which might be used as an indication of social conditions) has fallen from 43 per 1000 live births in 1946 to 18 in 1969. Inspection of a list of sex ratios and corresponding infant mortality rates for 1960 for fifty countries revealed no consistent relationship between the two.

The selective elimination of the male occurs after birth as well as before and has nothing to do with the greater occupational hazards to which he is exposed; his shorter life is probably related in some way to his higher metabolic rate. The composition of the various age groups in the population of England and Wales

Table 53.8 England and Wales home population 1969 (thousands)

Age	M	F
0–4	2105·8	2002·4
5–9	2037·2	1934·7
10–14	1752·9	1663·0
15–19	1701·8	1652·1
20–24	1892·0	1868·6
25–29	1566·6	1532·8
30–34	1513·2	1443·1
35–39	1496·1	1424·5
40–44	1518·6	1492·0
45–49	1667·4	1680·8
50–54	1360·4	1425·3
55–59	1457·2	1587·5
60–64	1305·4	1492·4
65–69	1023·0	1301·2
70–74	642·4	1050·2
75–79	402·1	765·7
80–84	208·0	475·3
85 and over	101·9	283·2

Table 53.9 Sex ratio (M : F) England and Wales, at stated ages, on the assumption that the 1969 cohort of births would be subject to 1969 death rates throughout life

Exact age (years)	Sex ratio (males per 100 females)
0	105·8
5	105·2
10	105·1
15	105·1
20	104·8
25	104·5
30	104·3
35	104·1
40	103·9
45	103·5
50	102·5
55	100·5
60	96·6
65	89·7
70	79·7
75	67·1
80	54·2
85	41·2

in 1969 is given in Table 53.8. This shows an excess of males in each group up to 40–44 years; the group with equal numbers is somewhere between 40–44 and 45–49, the latter being the first, that is youngest, to have an excess of females. In old age (85 years and over) there was more than twice the number of women as compared with men. In 1946 parity occurred earlier than in 1969 with a male excess up to 15–19 years, and female excess at 20–24 and in all subsequent age groups. Comparisons of existing populations, as given in Table 53.8, may not demonstrate purely biological influences because of selective migration and war casualties. Table 53.9 has been constructed by applying life table mortality only to the 1969 births (410 052 males and 387 486 females). This calculation is somewhat arbitrary but it is at least free from the effects of migration and war casualties. The Table shows that equality in number is expected a little after 55 years of age and thereafter there is a female excess increasing to over two and a half times the male population at 85 years and over.

To be born is a more dangerous adventure for the male than the female and, since there is a higher selective mortality after birth as well as before, the expectation of life is greater at all ages in the female.

PHENOMENA OF HEREDITY

The most important constituent of the chromosomes is of course DNA (p. 68). Its double stranded helix is coiled tightly in each chromosome thread. The unit of genetic information, the *gene* or *cistron*, is in fact a segment of the DNA molecule containing on the average about 600 base pairs, and the genetic message is carried in the sequence of bases along the DNA strand, written in the four letter language of the four bases.

In the process of *transcription* this message is transferred to messenger RNA (p. 355) which carries it to the ribosomes on which it is *translated* into the 20 letter language of the amino acids in the proteins (p. 364). A cistron is in fact that section of the DNA strand which carries the genetic information for the correct sequence of amino acids in the polypeptide chain in one particular protein (or enzyme)—this is the principle summarized in the phrase 'one gene, one polypeptide'.

The genes, being segments of the DNA strand, are arranged in orderly manner along the length of the DNA molecule in the chromosome. Each gene for any particular characteristic has its counterpart in the corresponding locus on the homologous chromosome and these two genes form an *allelic pair*. When both loci of a pair carry genes with the same characteristic, say tallness, then the individual is said to be *homozygous* with respect to that character; when one of a pair carries tallness and the other shortness the individual is *heterozygous*. The matter may be made plain by considering an actual example originally described by Mendel, who crossed tall (TT) pea-plants with dwarf (tt) ones (Fig. 53.10). The first generation were all tall plants. Since they all possessed the Tt pair of allelic genes T must be *dominant* over t and t must be *recessive*. By 'selfing', that is by self-fertilizing, these Tt plants he obtained one-quarter dwarfs and three-quarters tall plants. If the dwarfs were selfed they bred true; their gene construction

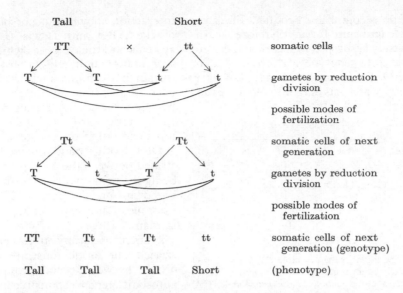

Tall		Short		
TT	×	tt		somatic cells
T	T	t	t	gametes by reduction division
				possible modes of fertilization
Tt		Tt		somatic cells of next generation
T	t	T	t	gametes by reduction division
				possible modes of fertilization
TT	Tt	Tt	tt	somatic cells of next generation (genotype)
Tall	Tall	Tall	Short	(phenotype)

Fig. 53.10 To illustrate the inheritance of dominant and recessive characters.

must, therefore, have been tt. Only one-third of the tall plants of this generation bred true; they were TT. The remainder were tall hybrids Tt (heterozygous). In man the character of height is controlled not by one but by many genes, so that the single symbol T employed above is inapplicable. The symbol T could, however, be employed to represent the ability to taste phenylthiourea in low concentrations. 'Tasters' have one or two dominant 'T' genes and 'non-tasters' have two recessive 't' genes.

The reassortment of genetic material in meiosis has already been mentioned (p. 1100). Since the members of each pair of chromosomes separate and pass into the daughter cells, allelic genes must also segregate quite independently from genes on other pairs of chromosomes. Genes on the same chromosome may stay together as a gene string, that is they exhibit *linkage*; on the other hand they may be separated by *crossing-over*, that is the interchanging of corresponding segments of homologous chromosomes. The result of this reassortment of genetic material is that members of a succeeding generation differ from their parents and differ among themselves. Some of the offspring may have a combination of characteristics with an enhanced survival value while others may possess characteristics with a reduced value; the latter offspring are eliminated by the pressure of natural selection.

In this way a species adapts to changes in its external environment.

Since each individual plant or animal possesses in its chromosomes a large number of genes which are assorted during development of the germ cells, it follows that the zygotes produced by their union must vary enormously in constitution. Any two children in a family are not absolutely alike unless, as occasionally happens, they arise from one fertilized ovum. Such individuals are referred to as monozygotic or identical twins.

INHERITANCE OF BLOOD GROUPS

Blood groups provide an example of inheritance in man. As long ago as 1908 it was suggested that the ABO blood groups were inherited and in 1924 Bernstein published his theory that the four groups, A, B, AB and O, were inherited as Mendelian characters through the action of three allelic genes A, B and O. An individual receives from each parent one of these three genes, and can therefore belong to one of the following *genotypes*:

$$OO \quad AO \quad AA \quad BO \quad BB \quad AB$$

The presence on the red cell of the antigens A and B, the products of the genes A and B, can be recognized by their reaction with the equivalent anti-bodies anti-A and anti-B. The gene O, however, appears to be an *amorph*, producing no recognizable product. The antisera, anti-A and anti-B can therefore distinguish only the *phenotypes*, A and B, and cannot differentiate the genotype *AA* from *AO*, or *BB* from *BO*. Such genotypes may, however, be disclosed by analysis of family blood groups (Fig. 53.11).

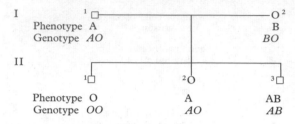

Fig. 53.11 The fact that the first child (II [1]) is Group O shows that each parent must possess an *O* gene. This in turn establishes, that the genotype of the second child (II [2]) is *AO*.

Bernstein's triple allele theory can be extended to include the existence of the subgroups A_1 and A_2, and also some rarer variants, for example, A_3, B_w. Other ABO variants, however, such as O_h and some examples of A_x and A_m cannot be explained by postulating 'new' alleles at the ABO locus. It has been suggested that these variants may be due to modifying or suppressor genes interfering with the action of the *ABO* genes.

The spate of observations carried out on the blood group antigens and antibodies, particularly in the last thirty years, has added much to our knowledge of the genetics of the blood group systems but at the same time has left many questions still to be answered. For example, the blood group antigens M and N discovered in 1927 appeared to be controlled by two allelic genes *M* and *N* giving the straightforward genotypes *MM*, *MN* and *NN*. Since then numerous other antigens have been detected, some of which can be attributed to alleles of *M* and *N* (for example M_2, M^g, N_2) but others can not, although studies of their inheritance in many families have shown them to be closely associated with *M* and *N*, for example, S, s, U, Mi^a, V_w and others.

Similarly, the continuing discoveries of further antigens within other blood-group systems, such as Rh and Kell, have made their genetic interpretation increasingly complex.

Of great genetical interest has been the interaction shown to occur between the genes of the ABO and Lewis (Le) blood group systems and the *Se* genes which determine the secretion of soluble blood group substances in certain body fluids such as the saliva. It has been shown that although this relationship between the *ABO*, *Le* and *Se* genes exists, these genes are not linked but segregate independently from one generation to the next.

On the other hand linkage has been demonstrated between the *Se* and *Lu* genes, the latter being the genes which determine the Lutheran blood group antigens. This was the first example of autosomal linkage to be discovered in man.

Numerous family investigations have been carried out to demonstrate whether linkage occurs between blood group genes and the autosomal genes responsible for other inherited characteristics both normal, for example, red hair and abnormal, for example, polydactyly, elliptocytosis, phenylketonuria. The rarity of most of the hereditary anomalies in man reduces the chance of such investigations being fruitful but two such searches have been successful. Linkage has been demonstrated between the genes of the Rh blood group system and the gene causing at least one type of hereditary elliptocytosis, and between the *ABO* genes and the gene responsible for the condition known as 'nail-patella syndrome'.

In 1962 it was recognized that the gene responsible for a newly discovered blood group antigen (Xg^a) must be carried on the X chromosome. With this exception all of the numerous blood group antigens detected have been shown to be determined by autosomal genes.

The Xg groups of families, members of which suffer from sex-linked constitutional defects (see below) are being investigated to help to plot the relative positions of some of the genes on the X chromosome.

SEX-LINKED INHERITANCE

Some genetic features, although transmitted by a gene located on an autosome, are only expressed in one or other sex. Such sex-limited genes should be distinguished from those which are located on the X chromosome such as those responsible for certain constitutional defects, such as colour-blindness which are commoner in men than in women. The explanation lies in the fact that these characters are recessive. A male (X'Y) whose single X chromosome has the defect-causing gene is affected with colour-blindness but a heterozygous female (X'X) has normal colour vision because her normal X chromosome is sufficient for this. In other words the recessive character does not appear in the presence of the dominant normal gene. A female carrier (XX') could have normal sons (XY) and colour-blind sons (X'Y); her daughters would all be apparently normal but half of them (X'X) would be carriers like their mother. Homozygous females (X'X') show the defect; in the absence of mutation they can only be the offspring of a colour-blind father and a carrier or colour-blind mother. Red-green colour blindness is common (8 per cent) in males but is uncommon (0·4 per cent) in females and there are many pedigrees which illustrate the typical sex-linked recessive mode of inheritance. This inherited defect is quite common because it has only trivial disadvantages. It is quite otherwise with haemophilia (p. 396) which is inherited in exactly the same way. The defect is severe and is likely to cause death before the age of reproduction. Haemophilia is therefore rare in men and almost unknown in women.

MUTATION

On the whole the gene is an extremely stable structure, and the enzymatic process for which it is responsible is carried out with great exactitude. When a cell divides each gene first produces an exact replica of itself and one of these genes goes into each new cell. Occasionally, however, mistakes occur in the replication of the DNA so that a 'wrong' base finds a place in the new DNA strand. Alternatively, bases may be inserted or deleted. In this way mutations arise. Thus haemophilia may appear in a member of a family which has no previous history of such an abnormality. The rate of mutation may be speeded up experimentally by treatment with certain mutagenic substances, for example mustard gas, or by exposure of the individual to ionizing radiation. At present we can only guess at the number of mutations which a given dose of X-rays will produce in human beings. It is believed that if the doses received by the gonads and red bone marrow of radiation workers do not exceed 5 rems* per year, direct radiation effects are extremely unlikely. This dose limit takes into account the possibility of production by radiation of undesirable dominant mutations likely to affect the children and grandchildren of the individual. A further restriction of 5 rems averaged over a whole population and a generation is imposed to minimize the possibility of undesirable hereditary changes due to 'recessive' mutations occurring many generations hence.

---

*The rem = absorbed dose in radiation × modifying factors, to allow for the varying biological potency of different types of radiation.

REFERENCES

BOYER, S. H. (Ed.) (1963). *Papers on Human Genetics.* New Jersey: Prentice-Hall.

CARTER, C. O. (1969). *An ABC of Medical Genetics.* London: Lancet.

CAVALLI-SFORZA, L. L. & BODMER, W. F. (1971). *The Genetics of Human Populations.* London and San Francisco: Freeman.

CLARKE, C. A. (1964). *Genetics for the Clinician,* 3rd edn. Oxford: Blackwell.

COURT-BROWN, W. M. (1967). *Human Population Cytogenetics.* Amsterdam: North-Holland.

COURT-BROWN, W. M., HARNDEN, D. G., JACOBS, PATRICIA A., MACLEAN, N. & MANTLE, D. J. (1964). *Abnormalities of the Sex Chromosome Complement in Man.* Special Report, Medical Research Council No. 305. London: H.M.S.O.

DARLINGTON, C. D. (1964). *Genetics and Man.* London: Allen & Unwin.

DAVIDSON, J. N. (1972). *The Biochemistry of the Nucleic Acids,* 7th edn. London: Chapman & Hall.

EMERY, A. E. H. (1971). *Elements of Medical Genetics,* 2nd edn. Edinburgh: Churchill Livingstone.

FALCONER, D. S. (1960). *Introduction to Quantitative Genetics.* Edinburgh: Oliver & Boyd.

FORD, C. E. & HARRIS, H. (Eds.) (1969). New aspects of human genetics. *British Medical Bulletin* **25**, 1–114.

GIBLETT, E. R. (1969). *Genetic Markers in Human Blood.* Oxford: Blackwell.

HARRIS, H. (1970). *The Principles of Human Biochemical Genetics*. Amsterdam: North-Holland.
JACOBS, P. A., PRICE, W. M. & LAW, PAMELA (1970). *Human Population Cytogenetics*. Edinburgh University Press.
MCKUSICK, V. A. (1969). *Human Genetics*, 2nd edn. New Jersey: Prentice-Hall.
MOORE, K. L. (Ed.) (1966). *The Sex Chromatin*. Philadelphia: Saunders.
RACE, R. R. & SANGER, R. (1968). *Blood Groups in Man*, 5th edn. Oxford University Press.
STERN C. (1960). *Principles of Human Genetics*. San Francisco and London: Freeman.
ROBERTS, J. A. F. (1970). *An Introduction to Medical Genetics*, 5th edn. Oxford University Press.
STEVENSON, A. C. (Ed.) (1961). Human genetics. *British Medical Bulletin* **17**, 177–259.
VALENTINE, G. H. (1969). *The Chromosome Disorders*. London: Heinemann.
WATSON, J. D. (1970). *Molecular Biology of the Gene*. 2nd edn. New York and Amsterdam: Benjamin.

54 Growth and senescence

We have already described in various parts of this book some of the conditions necessary for normal growth. An adequate supply of food materials is required including protein of high biological value, certain essential amino acids and vitamins; but all the nutritional factors concerned in growth are not yet known. For example the difference in biological value of animal and vegetable proteins fed to animals cannot be entirely ascribed to the difference in their amino acid content. Liver and yeast both contain as yet unidentified substances (not vitamin B_{12}) which can promote growth in children as well as in rats. An important factor in the internal environment of the growing organism is a properly balanced supply of hormones because the first effect of failure of almost any of the glands of internal secretion in young animals is a diminished growth rate.

A survey in 1959 in England and Wales gave the average birth weight of boys as 7·53 lb (3·42 kg) and of girls as 7·21 lb (3·28 kg). Infants born at full term whose weight is under average are probably normal individuals who develop into small adults. However, most underweight babies are premature, thrive less well and have a smaller chance of survival than children born at full term; it is also widely believed that infants born prematurely remain small at all ages. But if they escape serious defects like cerebral palsy their intellectual development is almost normal. Improvement of the diet of pregnant women decreases neonatal mortality which is due at least partly to prematurity (defined by international agreement as less than 2·5 kg at birth). The influence of maternal nutrition on the birth weight of the child is, however, quite small for, when the mother's diet is deficient in a minor degree, the child grows *in utero* at the expense of the maternal tissues and so reaches a normal weight at birth; it is only when the mother's diet is grossly deficient that the birth weight is subnormal. On the other hand, the birth weight of the child cannot be increased by supplementing the maternal diet. A factor of importance in determining the size of the offspring is the parity of the mother, for the later children in a family are on average 0·2 kg heavier than the first-born. The size of the mother has already been mentioned on p. 1068 as influencing the size of the offspring so that a small mother, fortunately for herself and her baby, usually has a small child. As a rule the fetuses of diabetic mothers, no matter the size of the mother, are larger than usual, presumably because they have pancreatic hyperplasia and hyperinsulinism.

The rate of growth in infancy and adolescence has accelerated during the last sixty years or more and this change may be attributed, at least partly, to improved nutrition. The final adult height has been increasing slowly during the past hundred years and it is reached at an earlier age. Experiments show that poor feeding of young animals retards development and that the full potential of growth is not attained by later improvement of the diet. The growth of the very young baby is not influenced by its pituitary. Thus both the anencephalic fetus and the baby without a pituitary show normal growth *in utero* and the latter sometimes for a year or more after birth. The cretin (p. 1001), likewise, may be normal in weight at birth and grow normally for a time after that.

At birth the *rate* of growth is high; it falls off slowly until the adolescent growth spurt is reached. In Britain this occurs at 14 years in boys and 12 years in girls. The rate then falls off to zero when the constant adult level is reached.

In man the growth in length of the legs occurs late in adolescence and contributes considerably to the final stature; it has been said that the best-fed parts of the population are relatively long limbed. Maximum height is reached in boys between 18 and 19 years and in girls between 14 and 18 years; about the fifth or sixth decades some reduction in height often occurs because degenerative changes in the vertebral column lead to slight curvature (kyphosis) especially in the thoracic region. At the beginning of this century in Great Britain the weight of both men and women increased regularly with age between 20 and 60 but at the present time men on the average keep at the same weight from age 25 onwards. On the other hand women, whether single or married, with or without children, still show a gradual increase in weight from age 25 onwards.

Man reaches the zenith of his physical powers at the age of twenty years or so and sooner or later his bodily tissues, even if they do not become affected by disease, gradually begin to deteriorate. It is not at all clear why this should be so.

It has been assumed for many years that cells are potentially immortal but that ageing is a characteristic of the whole organism. This assumption is based on the claim of Alexis Carrel to have kept chicken fibroblasts (derived from chick heart cells) alive in tissue culture for seventeen years, that is indefinitely. But this work has not been confirmed and in later tissue culture experiments the cells have eventually degenerated. Cells from human embryos grown in tissue culture divide about 50 times and then disintegrate. In view of these new findings it seems likely that the somatic cells have only a limited existence.

The *average life span* of man is determined by infection, disease or accident. In England and Wales during the last 70 years the *expectation of life* at birth, that is to say the number of years a new-born baby may be expected to live, has increased enormously (Fig. 54.1) In 1965 it was 68·5 years for males and 74·7 years for females. The comparative longevity of women is not due to their being less exposed to environmental (for example industrial) hazards because it is found also in closed communities. The increased life expectancy of the child (Fig. 54.1) is due to the elimination of many causes of death; but the expectation of life of the middle-aged has increased very little.

Under present conditions the *maximum life span* of man is about 110 years but it is not easy to identify circumstances which favour *longevity* or, to put it another way, factors that occasion *senescence*. Genetic factors certainly play a part; people who live to be 90 or more have a higher percentage of long-lived parents than control subjects. In determining the life-span of the offspring the longevity of the mother is more important than that of the father. However, the greater length of life of sons of long-lived parents as compared with those of relatively short-lived parents is a matter of only some 2 to 4 years.

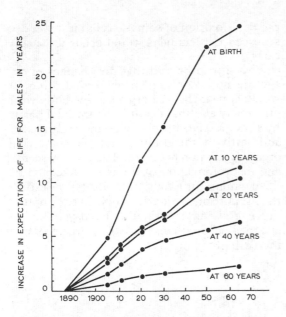

Fig. 54.1 Increase in expectation of life for males at different ages from 1881–1965. While more people are living to a greater age, the expectation of life at older ages is not increasing much. (Figures for England and Wales published by the Registrar General.)

Although every centenarian asked how he has reached his venerable age will usually offer advice on the kind of food to eat, it is difficult to obtain reliable information on the effect of diet on longevity in man. There is, for example, no evidence of any adverse influence of additional dietary protein on the life-span or the ageing process in man and, therefore, there is no reason for restricting the protein intake of old people. As much as 1·4 g of protein per kg body weight per day may be necessary to avoid a negative nitrogen balance in old people. Carbohydrate metabolism may be impaired in old age, since during a glucose tolerance test (p. 327) the blood sugar often reaches a higher level and returns to the initial value more slowly than in young adults. An adequate vitamin intake is essential for a healthy old age. The basal oxygen requirement and therefore the calorie requirement is reduced only a few per cent in old age. An excessive calorie intake must, of course, be avoided by the elderly just as by the middle-aged since there is no doubt that obesity reduces the expectation of life.

The process of ageing (*senescence*) does not necessarily affect the organs and tissues equal-ly. In some people one of the earliest signs is presbyopia (p. 733); in others deafness. Some are disabled by degenerative arthritis, others by the effects of arterial disease in the brain or heart. Often the examination of the bodies of persons who have died of old age does not reveal gross abnormality in any vital organ. Nearly all the organs show a decrease in weight. Atrophy of one organ may affect other organs which depend on it and so accelerate senescence. Damage to an organ by infection or by any pathological process may similarly speed up the ageing process.

We judge the age of a person mainly by the appearance of his skin which in old age shows wrinkling, thinning and loss of elasticity. The electron microscope shows that the collagen network in the skin of exposed areas in old people is defective, the fibres being short, bent and frayed. The collagen of unexposed sites is normal. The percentage of ash in bones tends to rise slightly in old age and the marrow cavity may become enlarged; X-ray photographs of elderly persons show that there is, especially in women, a reduction in the amount of bone substance (*osteoporosis*). The healing of fractures in the aged is sometimes retarded. Old people are often in negative calcium balance. The changes in the elasticity of the arteries are illustrated on p. 529. The calcium content of arteries and cartilage increases in old age but there is no alteration in the blood calcium level. The cholesterol content of arterial walls, of the cornea and the lens increases often without any obvious change in blood cholesterol. Renal function becomes progressively less efficient in old age; after the age of forty there is a progressive reduction in renal blood flow, creatinine clearance and GFR; the blood urea tends to rise.

Tests of muscle strength show that the maximum is reached somewhere about twenty years of age and is retained to at least forty years; thereafter there is a slight decline, which becomes marked only after seventy years, when the maximum work output may be only 50 per cent of that of the young adult. These findings may be related to the histological changes in the muscles of old people described on p. 825. Increasing difficulty in carrying out physical exercise is a well-known characteristic of old age but there may be no change in the apparent efficiency (p. 168) with which muscular work of moderate severity is performed. The output of creatine, which may be taken as an index of muscle mass, declines very little in old women but quite markedly in old men.

In young women about 30 per cent of the body weight is fat; in old women fat makes up 45 per cent. Since the body weight increases with age the total fat content may increase by 50 per cent. In healthy men fat forms about 30 cent of the body weight at all ages. The body water expressed as a percentage of the body weight decreases with age (Table 33.4).

Observation of rat colonies shows that there is no correlation between the state of the ovaries or testes and senility. The deterioration of the gonads could hardly be responsible for the deterioration of somatic cells; if it were so senescence in eunuchs would occur unusually early.

Thus it can be said that the changes of old age are not very sudden or very striking. It would appear that the organs of the body wear out slowly and that death is determined partly by a progressive atrophy of all the body tissues and partly by the gradual failure of some vital organ. A solution to the problems of senescence and of prolongation of life is not likely to be forthcoming for a long time. In the meantime the proportion of elderly people in the population is growing; and their social, psychological and medical problems (geriatrics) are becoming increasingly important.

REFERENCES

ANDERSON, W. F. & ISAACS, B. (Eds.) (1964). *Current Achievements in Geriatrics.* London: Cassell.

ANDREW, W. (1952). *Cellular Changes with Age.* Springfield, Ill.: Thomas.

ANDERSON, W. F. & COWAN, N. R. (1965). Body weight in older people. *Clinical Science* **29**, 33–39.

BOURNE, G. H. (Ed.) (1962). *Structural Aspects of Ageing.* London: Pitman.

BOYNE, A. W., AITKEN, F. C. & LEITCH, I. (1957). Secular change in height and weight of British children including an analysis of measurements of English children in primary schools: 1911–1953. *Nutrition Abstracts and Reviews* **27**, 1–18.

CAMERON, R. (1965). The twilight land of change and decay. *British Medical Journal* **1**, 1085–1088.

COHEN, C. (1965). Demographic aspects of the ageing population. *Scottish Medical Journal* **10**, 420–424.

COMFORT, A. (1964). *The Biology of Senescence,* 2nd edn. London: Routledge & Kegan Paul.

COMFORT, A. (1965). *The Process of Ageing.* London: Weidenfeld & Nicolson.

COMFORT, A. (1966). The prevention of ageing in cells. *Lancet* **ii**, 1325–1329.

DOBBING, J. (1967). Growth of the brain. *Science Journal* **3**, 81–86.

ELLIS, R. W. B. (1951). Phases of postnatal growth. *British Journal of Nutrition* **5**, 151–157.

ENGLE, E. T. & PINCUS, G. (Eds.) (1956). *Hormones and the Ageing Process.* London: Baillière, Tindall & Cox.

HARRISON, G. A., WEINER, J. S., TANNER, J. M. & BARNICOT, N. A. (1964). *Human Biology.* Oxford: Clarendon Press.

ILLINGWORTH, R. S. (1960). *The Development of the Infant and Young Child.* Edinburgh: Livingstone.

JOSEPH, N. R. (1971). Physical chemistry of aging. *Interdisciplinary Topics in Gerontology* **8**. Basel: Karger.

LANSING, A. I. (1951). Some physiological aspects of ageing. *Physiological Reviews* **31**, 274–284.

LANSING, A. I. (Ed.) (1952). *Cowdry's Problems of Ageing,* 3rd edn. Baltimore: Williams & Wilkins.

LEBLOND, C. P. & WALKER, B. E. (1956). Renewal of cell populations. *Physiological Reviews* **36**, 255–276.

LONSDALE, KATHLEEN (1967). Physics and ageing. *Advancement of Science London*, Sept. 1967, 11–30.

LOWE, C. R. & GIBSON, J. R. (1955). Changes in body weight associated with age and marital status. *British Medical Journal* **ii**, 1006–1008.

McCANCE, R. A. (1962). Food, growth and time. *Lancet* **ii**, 621–626, 671–676.

McKEOWN, T. & RECORD, R. G. (1952). Observations on foetal growth in multiple pregnancy in man. *Journal of Endocrinology* **1**, 386–401.

MEDAWAR, P. B. (1957). *The Uniqueness of the Individual.* London: Methuen.

MONTAGNA, W. (Ed.) (1965). *Advances in Biology of Skin,* Vol. 6, *Ageing.* Oxford: Pergamon.

NAGORNYI, A. V., NIKITIN, V. N. & BULANKIN, I. N. (1963). *The Problem of Ageing and of Longevity.* Moscow: Medgiz.

NORRIS, A. H., LUNDY, T. & SCHOK, N. W. (1963). Trends in selected indices of body composition in men between the ages of 30 and 80 years. *Annals of the New York Academy of Sciences* **110**, 623–639.

PARNELL, R. W. (1958). *Behaviour and Physique. An Introduction to Practical and Applied Somatometry.* London: Arnold.

SHELDON, J. H. (1960). Problems of an ageing population. *British Medical Journal* **i**, 1223–1236.

SHOCK, N. W. (Ed.) (1962). *Biological Aspects of Ageing.* New York: Columbia University Press.

SILBERBERG, M. & SILBERBERG, R. (1955). Diet and life span. *Physiological Reviews* **35**, 347–362.

SINCLAIR, D. (1969). *Human Growth after Birth*. London: Oxford University Press.

SMITH, C. A. (1959). *The Physiology of the New Born Infant*. Oxford: Blackwell.

STREHLER, B. L. (1962). *Time, Cells and Ageing*. London: Academic Press.

STREHLER, B. L. (Ed.) (1964). *Advances in Gerontological Research*. London: Academic Press.

TANNER, J. M. (Ed.) (1960). *Human Growth. Symposia of the Society for the Study of Human Biology*, Vol. 3. London: Pergamon.

TANNER, J. M. (1962). *Growth at Adolescence*, 2nd edn. Oxford: Blackwell.

TANNER, J. M. (1965). The trend towards earlier physical maturation. *Biological Aspects of Social Problems* (Edited by J. E. Meade & A. S. Parkes), pp. 40–65. Edinburgh: Oliver & Boyd.

THOMSON, A. M. (1951). Human foetal growth. *British Journal of Nutrition* **5**, 158–166.

VERZAR, F. (1963). *Lectures on Experimental Gerontology*. Springfield, Ill.: Thomas.

WEIR, J. B. DE V. (1952). The assessment of the growth of school children with special reference to secular changes. *British Journal of Nutrition* **6**, 19–33.

WELFORD, A. T. (1958). *Ageing and Human Skill*. Oxford University Press, for the Nuffield Foundation.

WELFORD, A. T. (1962). On changes of performance with age. *Lancet* **i**, 335–339.

WIDDOWSON, ELSIE M. (1970). Harmony of growth. *Lancet* **i**, 901–905.

YOUNG, C. M., BLONDIN, J., TENSUAN, R. & FRYER, J. N. (1963). Body composition studies of 'older' women, thirty to seventy years of age. *Annals of the New York Academy of Sciences* **110**, 589–607.

UNITS AND MEASURES USED IN PHYSIOLOGY AND BIOCHEMISTRY WITH ACCEPTED ABBREVIATIONS AND WITH CONVERSION FACTORS

PREFIXES APPLIED TO SYMBOLS FOR VARIOUS UNITS: THESE ARE NEVER USED ALONE

T = tera = 10^{12} $\quad$ G = giga = 10^9 $\quad$ M = mega = 10^6 $\quad$ k = kilo = 10^3 $\quad$ c = 10^{-2}
m = milli = 10^{-3} $\quad$ μ = micro = 10^{-6} $\quad$ n = nano = 10^{-9} $\quad$ p = pico = 10^{-12}
$\qquad\qquad\qquad$ f = femto = 10^{-15} $\qquad$ a = atto = 10^{-18}

MASS

gram $\qquad$ g $\qquad\qquad\qquad\qquad$ milligram $\qquad$ mg
kilogram $\quad$ kg $\qquad\qquad\qquad\qquad$ microgram $\qquad$ µg
1 kg $\quad$ = 1000 g $\qquad\qquad\qquad$ millimicrogram $\quad$ ng
1 g $\qquad$ = 1000 mg = 1 000 000 µg $\quad$ micromicrogram pg
1 kg $\quad$ = 2·20 lb
1 tonne = t = 1000 kg = 0·984 tons
1 g $\qquad$ = 15·4 grains $\qquad\qquad\qquad$ 1 lb $\quad$ = 453·6 g = 0·4536 kg
1 g $\qquad$ = 0·035 oz $\qquad\qquad\qquad$ 1 ton $\quad$ = 1016 kg
1 slug $\quad$ = 14·59 kg $\qquad\qquad\qquad$ 1 oz $\quad$ = 28·4 g
$\qquad\qquad\qquad\qquad\qquad\qquad\qquad$ 1 grain = 65 mg

LENGTH

metre $\qquad$ m $\qquad\qquad$ micron $\qquad$ µm $\qquad\qquad$ yard $\quad$ yd
centimetre cm $\qquad\qquad$ millimicron $\quad$ nm $\qquad\qquad$ foot (feet) ft
millimetre mm $\qquad\qquad$ Angstrom unit Å $\qquad\qquad$ inch $\qquad$ in
1 m $\quad$ = 100 cm = 1000 mm = 10^6 µm = 10^9 mµ = 10^{10} Å $\quad$ 1 Å = 10^{-10} m = 0·1 nm
1 µm = 10^{-6}m = 1000 nm =10000Å
1 nm = 10^{-9} m = 10 Å $\qquad\qquad$ 1 m $\qquad$ = 39·37 in = 1·094 yd
1 cm = 0·394 in $\qquad\qquad\qquad$ 1 in $\qquad$ = 2·54 cm = 0·0254 m
1 km = 3281 ft = 0·6214 mile $\qquad$ 1 mile $\quad$ = 1760 yd = 5280 ft = 1·609 km
1 yd $\quad$ = 0·9144 m $\qquad\qquad\qquad$ 1000 ft = 0·3048 km
$\qquad\qquad\qquad\qquad\qquad\qquad\qquad$ 1 nautical mile = 1·852 km

AREA

The SI unit of area is the square metre, m^2

1 sq in $\qquad$ = 645·2 mm^2
1 sq ft $\qquad$ = 0·093 m^2
1 m^2 $\qquad$ = 1·196 sq yd = 10·764 sq ft
1 sq yd $\quad$ = 0·836 m^2
1 hectare = 10^4 m^2 = 1·076 × 10^5 sq ft = 1·196 × 10 $\,$ sq yd = 2·471 acres
1 acre $\qquad$ = 4840 sq yd = 4047 m^2 = 0·4047 hectare
1 sq mile = 640 acres = 2·59 km^2 = 259 hectares

VOLUME

The SI unit of volume is the cubic metre, m^3

1 cubic millimetre $\quad$ 1 mm^3 $\qquad\qquad\qquad\qquad$ millilitre $\quad$ ml
litre $\qquad\qquad\qquad$ l $\qquad\qquad\qquad\qquad\qquad$ microlitre $\quad$ µl
cubic centimetre $\qquad$ c.c.
The volume of 1 kg of water at the temperature of maximum density (4°C) is
1 l. The millilitre (ml) is one thousandth part of a litre. The cubic centimetre
(c.c.) is based on the standard of length, that is the metre, and is very nearly
but not quite the same as the ml (see below).

1 l $\qquad$ = 1000 ml $\qquad\qquad\qquad$ 1 ml $\qquad$ = 1000 µl
1 l $\qquad$ = 35·196 fluid ounces (oz) $\quad$ 1 µl $\qquad$ = 1·000028 mm^3
1 l $\qquad$ = 1·76 pint $\qquad\qquad\qquad$ 1 oz $\qquad$ = 28·4 ml
1 pint = 568 ml $\qquad\qquad\qquad$ 1 gallon = 4·55 l
1 pint = 20 fluid ounces $\qquad\qquad$ 1 gallon = 0·00455 m^3
1 c.c. = 0·061 in^3 $\qquad\qquad\qquad$ 1 cu in $\quad$ = 1·639 × 10^{-5} m^3
1 ml $\quad$ = $\frac{1}{1000}$ l = 1·000028 c.c. $\qquad$ 1 cu ft $\quad$ = 0·028 m^3 = 28·32 l
To convert from g per 100 ml to grains per fluid ounce multiply by 4·375.
1 cu ft water = 62·39 lb = 28·32 kg.

Density

The SI unit of density is the kilogram per cubic metre, kg m^{-3}
$$1 \text{ lb per in}^3 = 2\cdot768 \times 10^4 \text{ kg m}^{-3}$$
$$1 \text{ lb per ft}^3 = 16\cdot02 \text{ kg m}^{-3}$$

Time

hour	hr	h		second	sec	s
minute	min			millisecond msec (occasionally written σ)		

1 hr = 60 min = 3600 sec
1 sec = 1000 msec = 1 000 000 μsec

Pressure

1 normal atmosphere = 760 mm Hg = 760 torr
= 29·92 in Hg
= 1033 g wt. per cm^2
= 14·70 lb/in^2
= 1013 millibars
= 1 013 250 dyn/cm^2
= 101·3 Nm^{-2}

1 mm H$_2$O = 0·1 cm × 0·999972 g/cm^3 × 980·67 cm/sec^2
= 98·06 dyn/cm^2 = 0·0736 mm Hg

1 mm Hg = 1333 dyn/cm^2 = 1 atmosphere ÷ 760 = 1 torr = 133·3 Nm^{-2}
1 Nm^{-2} = 1·019 × 10^{-2} g/cm^2

To convert cm H$_2$O to mm Hg multiply by 0·736

1 lb/in^2 = 7·031 × 10^2 kiloponds/m^2
= 6894 Nm^{-2}

1000 lb/in^2 = 0·703 kg/mm^2
1 millibar or 1 mb = 1000 dyn/cm^2

Heat

The calorie (cal), also known as the small calorie or gram calorie, is the amount of heat required to raise the temperature of 1 g of water from 14·5°C to 15·5°C. (This unit is deprecated.) Thermodynamic temperature is customarily expressed in °C, degrees Celsius (degrees Centigrade) denoted by t. Degree Kelvin (K) are reckoned from absolute zero so that $t = T - 273\cdot15$ where T is the absolute temperature. The name Kelvin (symbol K) is to be used instead of degree Kelvin (°K).

The large calorie (kcal), also known as the kilogram calorie or kilocalorie, is the amount of heat required to raise the temperature of 1 kg of water from 14·5°C to 15·5°C.

1 kcal = 1000 cal = 4·1868 J   1 B.T.U. = 0·252 kcal   1000 mcal = 1 cal

Conversion from Fahrenheit to Centigrade scale of temperature or vice versa may be made by using the formula—

$$9/5 \text{ C} = \text{F} - 32$$

where F is the temperature in degrees Fahrenheit and C is the temperature in degrees Centigrade.

°C	−40	−10	0	10	20	30	35	37	40	45	100
°F	−40	14	32	50	68	86	95	98·6	104	113	212

Electrical Units

ampere, A	volt, V
milliampere, mA	millivolt, mV
watt, W	ohm, Ω
farad, F	megohm, MΩ
microfarad, μF	joule, J
coulomb, C	

One ampere (A) is a flow of 6 × 10^{18} electrons per second past a given point. If I amperes are flowing in a wire the quantity (Q) of electricity passing a point in the wire in t sec is given by $Q = It$ coulombs.

Ohm's Law. The ratio of the potential difference (V) between the ends of conductor to the current (I) flowing in it is a constant, called the resistance (R). $R = V/I$.

The heat in watts (W) produced in such a resistance is given by $W = I^2R$ or $W = IV$.

PHOTOMETRIC UNITS

Candela (cd). The unit of luminous intensity. A full radiator (black body) at the melting point of platinum emits 60 candelas per square centimetre. The candela is approximately equal to the obsolete 'standard candle'.

Lumen (lm). The unit of luminous flux. It is the flux emitted into unit solid angle by a uniform point source of one candela.

Lux or Metre-candle (lx). The unit of illumination, one lumen per square metre. It is approximately the illumination of a surface placed one metre from a standard candle.

Units of Luminance (*Objective Brightness*). Two surfaces with the same *luminance* in a given direction look equally bright when viewed under similar conditions from this direction by a human observer. The internationally recognized unit of luminance is the candela per square metre ($cd\ m^{-2}$). The luminance of a surface in a given direction expressed in $cd\ m^{-2}$ is the number of lumens per steradian emitted in this direction by any small element of it, divided by the area in square metres of the orthogonal projection of this element on a plane perpendicular to the given direction. Extensively used but no longer official units are the millilambert, equal to $10/\pi$ or $3 \cdot 183\ cd\ m^{-2}$, and the foot-lambert, equal to $3 \cdot 426\ cd\ m^{-2}$.

MISCELLANEOUS

cycles per second, c/s, Hz
normal temperature and pressure, s.t.p.
acceleration due to gravity $= g$

One dyne is the force which would produce an acceleration of 1 cm/sec/sec if acting on 1 g. $1\ dyn = 10^{-5}\ N$.
SI unit of force is the newton, N, that force which produces an acceleration of 1 m per sec per sec acting on 1 kg.

1 N	$= 10^5$ dyn
1 pound-force	$= 4 \cdot 448$ N
1 poundal	$= 0 \cdot 138$ N
1 kilogram-force	$= 9 \cdot 806$ N (deprecated)

One erg is the work done when a force of one dyne·acts through one cm in the direction of the force. $1\ erg = 10^{-7}\ J$.
SI unit of work is the joule, the work done when a force of 1 newton acts through 1 m in the direction of the force. $1\ joule\ (J) = 10^7\ ergs = 0 \cdot 239\ cal = 1$ newton-metre.

1 kilogram force, 1 kilopond = wt. of 1 kg at standard g
$= 9 \cdot 806$ N.
$1\ kJ = 10^3\ Nm = 737 \cdot 6$ ft lb.
$1\ ft\ lb = 0 \cdot 1383\ kg\ m = 1 \cdot 356\ J$.
Standard $g = 980 \cdot 7$ cm per sec per sec.
$\quad\quad\quad = 32 \cdot 174$ ft per sec per sec.
Power is the rate of doing work. The unit of power is the watt.
1 watt = 1 joule per sec.
1 horse power (h.p.) = 33,000 ft lb min = 746 W.
1 kcal = 427 kg m.
Osmolarity is expressed in osmoles per litre of solution; osmolality in osmoles per kilogram of water.

1 mile per hour = 88 ft per min
$\quad\quad\quad\quad = 1 \cdot 467$ ft per sec
$\quad\quad\quad\quad = 0 \cdot 447\ m\ s^{-1}$
$\quad\quad\quad\quad = 26 \cdot 82\ m\ min^{-1}$.
1 mile per gallon = 2·83 litres per km.

Table of Approximate Atomic Weights

C	Ca	Cl	F	Fe	I	H	K	Mg	N	Na	O	P	S
12	40	35·5	19	56	127	1	39	24	14	23	16	31	32

The set of metric rules known as *système internationale* or SI was proposed first in 1954 and was officially adopted in 1960 by nearly thirty countries. The SI has six basic quantities: metre, m; kilogramme, kg; second, s; ampere, A; degrees Kelvin, K; and candela, cd. In this edition of our textbook we have used SI units but recognizing that for the older reader at least this is a transition period we have often given the equivalent in the superseded units. In other words we recognize that for a long time some units, for example, calories, have got into the language of the layman and like the names of the old coins will probably remain for many years. There are indeed a few units like parsec and electronvolt which do not fit happily into SI.

A great deal of statistical and medical information together with conversion tables is to be found in the following publications:

DIEM, K. (1970). *Documenta Geigy Scientific Tables*, 7th edn. Manchester: Geigy Pharmaceutical Co. Ltd.
HALL, L. J. (1965). *Biological Laboratory Data*. London: Methuen.
Metrication in Scientific Journals. (1968). The Royal Society Conference of Editors. London: The Royal Society.
SUNDERMAN, F. W. & BOERNER, F. (1949). *Normal Values in Clinical Medicine*. Philadelphia: W. B. Saunders.
S.I. the International System of Units (1970). Translation approved by the International Bureau of Weights and Measures of its publication 'Le système International d'Unités. London: H.M.S.O.
Symbols, Signs and Abbreviations recommended for British Scientific Publications (1969). London: The Royal Society.
The change to metric units (1967) *Nature (London)* **216**, 1272–1274.
Units, Symbols and Abbreviations—A Guide for Biological and Medical Editors and Authors (1971). London: The Royal Society of Medicine.

Table A Average weights for men and women according to height and age

Height (in shoes)*	Weight in pounds (in indoor clothing)					
	Ages 20–24	Ages 25–29	Ages 30–39	Ages 40–49	Ages 50–59	Ages 60–69
Men						
5 ft 2 in	128	134	137	140	142	139
5 ft 3 in	132	138	141	144	145	142
5 ft 4 in	136	141	145	148	149	146
5 ft 5 in	139	144	149	152·	153	150
5 ft 6 in	142	148	153	156	157·	154
5 ft 7 in	145	151	157	161	162	159
5 ft 8 in	149	155	161	165	166	163
5 ft 9 in	153	159	165	169	170	168
5 ft 10 in	157	163	170	174	175	173
5 ft 11 in	161	167	174	178	180	178
6 ft 0 in	166	172	179	183	185	183
6 ft 1 in	170	177	183	187	189	188
6 ft 2 in	174	182	188	192	194	193
6 ft 3 in	178	186	193	197	199	198
6 ft 4 in	181	190	199	203	205	204
Women						
4 ft 10 in	102	107	115	122	125	127
4 ft 11 in	105	110	117	124	127	129
5 ft 0 in	108	113	120	127	130	131
5 ft 1 in	112	116	123	130	133	134
5 ft 2 in	115	119	126	133	136	137
5 ft 3 in	118	122	129	136	140	141
5 ft 4 in	121	125	132	140	144	145
5 ft 5 in	125	129	135	143	148	149
5 ft 6 in	129	133	139	147	152	153
5 ft 7 in	132	136	142	151	156	157
5 ft 8 in	136	140	146	155	160	161
5 ft 9 in	140	144	150	159	164	165
5 ft 10 in	144	148	154	164	169	†
5 ft 11 in	149	153	159	169	174	†
6 ft 0 in	154	158	164	174	180	†

* 1 in heels for men and 2 in heels for women.
† Average weights not determined because of insufficient data.
Source: Metropolitan Life Insurance Company.

Table A gives figures for *average* height and weight at various ages for adults. The figures were obtained as a result of an investigation called 'The Build and Blood Pressure Study, 1959'. This study combines the experiences of 26 large life insurance companies in the United States and Canada between 1935 and 1954. It involved the observation of nearly five million insured persons over periods up to 20 years. Measurements were made with the subjects wearing *indoor clothing and shoes*. The clothing was estimated to weigh 7 to 9 lb for men and 4 to 6 lb for women. Heels added about 1 in to the height of men and 2 in to the height of women.

A more detailed investigation of heights and weights of the population of the U.S.A. will be found in: *Weight by Height and Age of Adults* (1966). U.S. Department of Health Education and Welfare. National Center for Health Statistics, Ser. 11, No. 14: Washington.

Table B The following units and conventions for reporting the normal concentration ranges for constituents of human body fluids are recommended (1972) by the Association of Clinical Biochemists. They are based on the SI system. Note that the SI unit of pressure is the pascal (Pa) which will ultimately replace mmHg for blood gas measurements.

BLOOD/PLASMA/SERUM

	Approximate normal ranges		Multiplication factor for converting from present units to new units
	Convention commonly used at present	Recommended convention	
Amino acid nitrogen	4 to 8 mg/100 ml	3 to 6 mmol/l	0·714
Ammonium	40 to 80 μg/100 ml	22 to 44 μmol/l	0·554
Ascorbate	0·2 to 0·7 mg/100 ml	11 to 40 μmol/l	57·0
Bicarbonate	24 to 32 m-equiv./l	24 to 32 mmol/l	1·0
Bilirubin (total)	0·3 to 1·0 mg/100 ml	5 to 17 μmol/l	17·1
Caeruloplasmin	30 to 60 mg/100 ml	2·0 to 4·0 μmol/l	0·0667
Calcium (total)	8·5 to 10·5 mg/100 ml	2·12 to 2·62 mmol/l	0·249
Carbon dioxide (P_{CO_2})	34 to 46 mmHg	4·5 to 6·0 kPa	0·133
Carotene	50 to 300 μg/100 ml	1·5 to 5·5 μmol/l	0·0186
Chloride	95 to 105 m-equiv./l	95 to 105 mmol/l	1·0
Cholesterol	140 to 260 mg/100 ml	3·6 to 6·7 mmol/l	0·0258
Copper	80 to 150 μg/100 ml	13 to 24 μmol/l	0·157
Cortisol	10 to 25 μg/100 ml	276 to 690 nmol/l	27·6
Creatinine	0·7 to 1·4 mg/100 ml	60 to 120 μmol/l	88·4
Glucose	45 to 85 mg/100 ml	2·5 to 4·7 mmol/l	0·0555
Haemoglobin (Fe)	12·0 to 17·5 g/100 ml	7·5 to 11·0 mmol/l	0·621
Haptoglobin (as haemoglobin (Fe) binding capacity)	30 to 180 mg/100 ml	19 to 112 mmol/l	0·621
Iron	80 to 160 μg/100 ml	14 to 28 μmol/l	0·179
Total iron binding capacity (as iron)	250 to 400 μg/100 ml	45 to 70 μmol/l	0·179
Ketones (as acetoacetate)	0·8 to 1·4 mg/100 ml	0·08 to 0·14 mmol/l	0·099
Lactate	3·6 to 13·0 mg/100 ml	0·4 to 1·5 mmol/l	0·112
Lead	10 to 40 μg/100 ml	0·5 to 1·9 μmol/l	0·0483
Lipids (total)	400 to 1000 mg/100 ml	4·0 to 10·0 g/l	0·01
Magnesium	1·8 to 2·4 mg/100 ml	0·7 to 1·0 mmol/l	0·411
Non-protein nitrogen	20 to 30 mg/100 ml	14 to 21 mmol/l	0·714
Oxygen (P_{O_2})	90 to 110 mmHg	12 to 15 kPa	0·133
Phenylalanine	0·7 to 1·2 mg/100 ml	0·05 to 0·07 mmol/l	0·0605
Phosphorus (inorganic)	2·5 to 4·5 mg/100 ml	0·8 to 1·4 mmol/l	0·323
Phospholipid-phosphorus	5 to 10 mg/100 ml	1·6 to 3·2 mg/100 ml	0·323
Potassium	3·8 to 5·0 m-equiv./l	3·8 to 5·0 mmol/l	1·0
Proteins—total	6·2 to 8·2 g/100 ml	62 to 82 g/l	10
—albumin	3·6 to 5·2 g/100 ml	36 to 52 g/l	10
—globulins	2·4 to 3·7 g/100 ml	24 to 37 g/l	10
—fibrinogen	200 to 400 mg/100 ml	2·0 to 4·0 g/l	0·01
—IgG	500 to 1600 mg/100 ml	5 to 16 g/l	0·01
—IgA	125 to 425 mg/100 ml	1·25 to 4.25 g/l	0·01
—IgM	47 to 170 mg/100 ml	0·47 to 1·70 g/l	0·01
—IgD	1 to 14 mg/100 ml	0·01 to 0·14 g/l	0·01
Protein bound iodine	4·0 to 7·5 μg/100 ml	0·31 to 0·59 μmol/l	0·0788
Pyruvate	0·4 to 0·7 mg/100 ml	0·05 to 0·08 mmol/l	0·115
Sodium	136 to 148 m-equiv./l	136 to 148 mmol/l	1·0
Transferrin	120 to 200 mg/100 ml	16·2 to 27·0 μmol/l	0·135
Triglyceride	25 to 150 mg/100 ml	2·8 to 16·9 mmol/l	0·113
Urate	2 to 7 mg/100 ml	0·12 to 0·45 mmol/l	0·0595
Urea	15 to 40 mg/100 ml	2·5 to 6·0 mmol/l	0·166

Table B *continued*

URINE

	Approximate normal ranges		Multiplication factor for converting from present units to new units
	Convention commonly used at present	Recommended convention	
Amino acid nitrogen	50 to 300 mg/24 h	4 to 20 mmol/24 h	0·0714
Ascorbate	20 to 50 mg/24 h	114 to 285 µmol/24 h	5·7
Calcium	100 to 300 mg/24 h	2·5 to 7·5 mmol/24 h	0·0249
Catecholamines (as adrenalin)	10 to 100 µg/24 h	0·05 to 0·55 µmol/24 h	0·00546
Copper	10 to 50 µg/24 h	0·2 to 0·8 µmol/24 h	0·0157
Coproporphyrins (I and III)	100 to 200 µg/24 h	0·15 to 0·3 µmol/24 h	0·00153
Creatine	0 to 50 mg/24 h	0 to 0·4 mmol/24 h	0·00763
Creatinine	1·0 to 2·0 g/24 h	9 to 17 mmol/24 h	8·84
Glucose	0 to 0·2 g/100 ml	0 to 11 mmol/l	55·5
5-Hydroxyindole acetate (5 HIAA)	3 to 14 mg/24 h	16 to 73 µmol/24 h	5·23
Lead	30 to 80 µg/24 h	0·14 to 0·39 µmol/24 h	0·00483
Magnesium	80 to 120 mg/24 h	3·3 to 4·9 mmol/24 h	0·0411
Nitrogen	10 to 20 g/24 h	0·7 to 1·5 mol/24 h	0·0714
Oxalate	20 to 40 mg/24 h	0·23 to 0·46 mmol/24 h	0·0114
17-Oxogenic steriods	10 to 20 mg/24 h	35 to 69 µmol/24 h	3·47
17-Oxosteroids	10 to 25 mg/24 h	35 to 85 µmol/24 h	3·47
Phosphorus	0·5 to 1·5 g/24 h	16 to 48 mmol/24 h	32·3
Porphobilinogen	0·2 to 2 mg/24 h	0·9 to 8·8 µmol/24 h	4·42
Potassium	40 to 120 m-equiv./l	40 to 120 mmol/l	1·0
Pregnanediol	0 to 1 mg/24 h	0 to 3·1 µmol/24 h	3·12
Protein	0 to 20 mg/100 ml	0 to 0·2 g/l	0·01
Sodium	100 to 250 m-equiv./l	100 to 250 mmol/l	1·0
Urate	0·5 to 2·0 g/24 h	30 to 120 mmol/24 h	59·5
Urea	1·5 to 3·0 g/100 ml	250 to 500 mmol/l	166
Urobilinogen	0·3 to 3·0 mg/24 h	0·5 to 5 µmol/24 h	1·69
Uroporphyrins (I and III)	0 to 25 µg/24 h	0 to 0·03 µmol/24 h	0·0012

Index

A-band, 825
  'a wave', atrial, 491
  'a wave', jugular phlebogram, 491
  'c wave', 491
  'v wave', 493
Abdominal muscles, 593
  pain, 863
  pressure, 595
  segment, 228
ABO system, 411
Absorption, disorders of, 300
  in colon, 295
    intestine, 293, 306
    mouth, 295
    stomach, 259
  of amino acids, 298
    carbohydrates, 296
    fat, 299
    glycine, 299
    iron, 425
    methionine, 299
    polypeptides, 299
    protein and amino acids, 298
    spectrum, 741, 744
    vitamin A, 184
    water, 7, 295
Absorption spectra of hæmoglobin
    derivatives, plate 25.1
Acalculia, 976
Acceleration, angular, and semi-
    circular ducts, 929
Accessory food factors, 182
  muscles of respiration, 593
Acclimatization, 633
Accommodation, 732
  amplitude, 733, (fig.) 734
  atropine, 732
  eserine, 732
  in nerve, 806
  in the eye, 733
  pathway, 972
  range of, 733
  reflex, 733
Acetaldehyde, 145
Acetate, 331
Acetazolamide, 625
Acetoacetyl coenzyme A, 339, 344
Acetone, 338
  bodies, 338
Acetoacetic acid, 338
Acetylcholine, 103, 512, 581, (fig.)
    582, 808, 818, 822, 833,
    870, 885, 1018
  on muscle, 833
Acetylcholinesterase, 822
Acetyl-coenzyme A, (fig.) 112, 148,
    337
Achalasia, 231
Acid secretion, 245
Acidaemia, 14, 330

due to loss of base, 702
  carbon dioxide retention, 700
  excessive production of acids,
    700
  of renal origin, 702
  respiratory, 700
Acid-base balance, 665
  clinical disturbances, 699
  disturbances, non-respiratory, 699
    respiratory, 699
  metabolism, 673
  status, measurement of, 693
  tubular regulation of, 665
Acidity constant, 16
Acidophil cells, 1081
Acidosis, 484, 692, 696
  metabolic, 700
  respiratory, 699
Acid phosphatase, 1030
  production in stomach, 245
Acids, definition of, 15
Acinus, 223
Aconitase, 150
Aconitate hydratase, 150
Acoustic spectrum, (fig.) 760
ACP, 341
Acridine, 372
Acromegaly, 321, 1086, 1087, 1088,
    (fig.) 1089
Acrosome, 1029, 1054
ACTH (adrenocorticotrophic
    hormone), 378, 1007, 1084
  circadian variation in, 1009
  chemical formula, 1085
Actin, 464, 826, 837, 838
  chemistry, 838
  filament, (fig.) 826
  myosin interactions, (fig.) 476,
    477
  mechanism, (fig.) 476
Actinomycin, 131, 995
  D, 372
Action potential, 781, (fig.) 786,
    791, (fig.) 846
  of frog's sciatic nerve, (fig.) 784
  ionic basis of, 791
  on the myocardium, 467
  spectrum, 741
Activation energy, 88
  concept, 88
  of enzymes, 92
'Active acetate', 1046
Active site of enzymes, 100
Active transport, 293
  of glucose, (fig.) 297
  mechanism, 247, 295
Actomyosin, 838
Acuity of hearing, methods of
    measuring, 770
Acyl carrier protein (ACP), 341

Acyl CoA dehydrogenase, 337
Adaptation, 2, 717, 811, (fig.) 812
  in olfaction, 723
    end organs, 718
    pain sense, 718
    tactile sense, 717
    taste sense, 727
    temperature sense, 715
Addisonian anaemia, 423
Addison's disease, 664, 1010, 1015
Adenine, 64
  deaminase, 360
  phosphoribosyltransferase, 350
Adenohypophysis, 1080
  blood supply, 1081
  cells of, 1081
  clinical disorders of, 1088
  control of, 1081
  growth hormone, 1085
  hormones of, 1083
  inhibition by ovarian hormones,
    1083
  melanophore-expanding hormone,
    1088
Adenosine, 64, 519
Adenosine-3', 5'-cyclic monophos-
    phate (cyclic AMP), 321,
    898, 994, 995
Adenosine diphosphate (ADP), 66,
    84, 111, 317, 836, 837
Adenosine monophosphate (AMP),
    65
  3, 5'-adenosine monophosphate
    (3, 5'-AMP), 158
  triphosphate (ATP), 66, 76, 84,
    111, 317, 349, 473, 836,
    837
    in muscular contraction, 837
Adenosinetriphosphatase, 838
S-adenosyl methionine, 343, 359
Adenyl cyclase, 158, 995
Adenylosuccinic acid, 349
ADH, 524, 651, 661
  half-life, 663
  urine output, 1095
  vasopressin activity, (fig.) 1095
Adhesiveness of platelets, 401
Adiadochokinesis, 950
Adipose tissue, 51, 334
  brown, 1067
  water content, 674
  white, 334
Adolescent growth spurt, 1112
ADP, 66, 84, 111, 317, 836
Adrenal cortex, 58, 61, 324, 1007,
    1034
  and water loss, 663
  ascorbic acid in, 192
  control of secretion of, 1010
  function of, 1007

Adrenal cortex, hormones
 of, 58, 1056
 hyperfunction, 1015
 hyperplasia, (table) 1009
 hypofunction, 1015
 in carbohydrate metabolism,
  1010
  exercise, 856
  stress, 1010
  structure of, 1006
 cortical insufficiency, 1015
 glands, 1006
 zones, (fig.) 1006
 medulla, 861, 866, 1015
 steroids, biosynthesis of, (fig.)
  1008
Adrenalectomy, 322, 664, 1010, 1045
Adrenaline, 116, 319, 321, 323, 343,
  (fig.) 510, 511, (fig.) 513,
  566, (figs.) 567–568, (fig.)
  572, 631, 866, 868, 875,
  1016
 action of, (fig.) 868
 and blood pressure, 566, 567
  bronchial muscle, 598
  effects on blood vessels, 1018
  fat, 343
  glycogenolysis, 321
  heart output, (fig.) 510, 513,
   567
  intestine, 285
  kidney, 656
  metabolic rate, 875
  muscles, 866
  nerve endings, 866
  noradrenaline, biosynthesis and
   breakdown, 1017
  effects compared, (table) 868
  on respiration, 631
  salivary secretion, 225
  spleen, 585
  veins, 572
 apnœa, 631
 asthma and, 598
 biosynthesis and breakdown, 1017
 blood-sugar level and, 323
 denervation and, 868
 destruction of, 1017
 emergency function, 1019
 estimation of, 1019
 formula, 1017
 glycosuria, 325
 in blood, 1019
  exercise, 319
  urine, (table) 1019
 receptors, 867
 urinary output, 664
Adrenals and body temperature, 875
 ascorbic acid in, 1097
 fetal, 1056, 1059
 in pregnancy, 1056
 water balance, 663
Adrenal steroids, properties of,
 1007
Adrenergic blocking agents, 869
 fibres, 866
β-Adrenergic receptors, myocardial,
 484
Adrenocortical hormone, (ACTH)
 378, 1007
 biosynthesis of, 1008
 insufficiency, 1010, 1015

tumour, 1010, 1015
Adrenocortical virilism, (fig.) 1014
Adrenocorticotrophic hormone
 (ACTH), 378, 1084
 assay, 1085
Adrenogenital syndrome, (fig.) 1014
Adult height, 1112
Aerobic recovery in muscle, 840
Aesthesiometer, 717
Aetioporphyrin, 450
Afferent arteriole of glomerulus,
 642, (fig.) 644, (fig.) 646,
 648
 fibres, 810
 impulses, 810, 811
  from viscera, 861
 nerves, 5, 891
 of heart, 512
After-discharge, 910
After-load, (fig.) 476, 479
After-potentials in nerve, 807, 808
Agammaglobulinæmia, 405
Age and basal metabolic rate, (fig.)
 167
 urinary steroids, 1033
 oxygen requirement, (table) 595
 blood pressure, (fig.) 536
 pulse rate, (fig.) 517
 respiratory rate, (table) 595
Ageing, process of, 1113
Agglutination of red cells, 410
Aggregated lymph follicles, 287, 308
Aglomerular kidney, 653
Agraphia, 976
A.I.D., 1052
Air, alveolar, 602
 atmospheric, 602
 breathed, volume of, 600
 comparison of expired, alveolar
  and inspired, (table) 602
 composition of respired air, 602
 conduction, (fig.) 772
 dead space, 600, 613
 expired, 602
 flow, rate of, 887
 functional residual, 595, 598
 hunger, 330
 inspired, 602
 passages, 589
 tidal, 598
 velocity, 888
Aircraft at high altitude, 634
 passenger, 634
Airway resistance, 596
Ala (alanine), 23
ALA (δ-amino lævulic acid), 189,
 451
 synthetase, 451
Alanine (Ala), 23
β-Alanine, 23, 360
Albumin, 401, 405
 in osmotic equilibrium, 405
 plasma, (table) 403
Albuminuria, orthostatic, 648
Alcohol, 181, 679, 1094
 and pupil, 733
 calorific value of, 181
 dehydrogenase, 145
 thirst and, 679
Alcoholic fermentation, 145
Aldehyde oxidase, (fig.) 1017
Aldolase, 107, 142
Aldosterone, 61, 524, 650, 659, 661,

663, 680, 1012, 1049
 and sodium excretion, 659
 excretion in pregnancy, 1055
 formula, 1013
 inhibitors, 1013
 in blood plasma, 1012
 œdematous states, 1013
Aldosteronism, 1013
Alerting reaction, 953
Alimentary canal, sites of absorp-
 tion, (fig.) 294
 turnover of water and electrolytes,
  (table) 678
Alkalæmia, 14, 699
Alkaline phosphatase, 104, 116, 212
 tide, 249
Alkalosis, 692, 696, 700
 metabolic, 702
 renal, 700
Alkaptonuria, 120, 385
Alkylating agents, 354
Allantoin, 360, (fig.) 361
Allantoinase, 360
Alleles, 1108
Allelic genes, 1106, 1108
 pair, 1106
Allocortex, 957
Allograft reaction, 1019
Allopurinol, 116, 360
'All-or-none' phenomenon, 788, 811
Allosteric activation, 114
 changes, 447
 effect, 620
 inhibitor, 113
Alloxan, 64, 326, 327
Alpha discharge, 828
 efferents, 784, 821, 828, 927
 helix, 34
 motoneurones, 915
 receptors, 867
 rhythm, 977
Altitude and alveolar gases, 633
 effect of high, (fig.) 633
 oxygen saturation, 633
Alveo-capillary gas transfer, 615
Alveolar air, 602, 613, 614
 apparatus for collecting, (fig.) 602
 at high altitudes, 633
 capillary difference, 615
 carbon dioxide, (table) 602
 ducts, (fig.) 589
 gas at high altitude, 633
 gas transfer, 615
 membrane, 590
  area of, 590
  electronmicrographs, (fig.) 616
 oxygen tension, (table) 617
 P_{CO_2}, (table) 617
 ventilation, (fig.) 628
Alveoli, 590
 fluid lining of, 613
Alveolo-capillary surface, 458
Alveolus, 223
 lung, (fig.) 616
Amenorrhœa, 1053
Amethopterin, 190, 349
Amidopyrine, 259
Amine precursor uptake and decarboxylation (APUD), 251
Amino acid, 23
 absorption, 298
 activation, 364
 ammonia formation, 379, 382

Amino acid, antiketogenic or glucogenic, 381
catabolism, 378
classified, 24, 386
decarboxylase, 383
essential, or indispensable, 199, 386
excretion, 390
fate after absorption, 298
glucogenic, 381
intravenous infusion, 375
ketogenic and antiketogenic, 381
nitrogen of blood, 375
pool, 408
requirements, 199
sulphur containing, 388
transamination, 377
transferase, 107
urea from, 381
D-Amino-acid oxidase, 137
L-Amino-acid oxidase, 77, 378
Amino-aciduria, 390
renal, 390
Aminoacyl-tRNA synthetase, 364
γ-Aminobutyric acid (GABA), 822,
898
p-Aminobenzene sulphonamide, 115
p-Aminobenzoic acid, 115, 190
ε-Aminocaproic acid (EACA), 26,
398
β-Aminoisobutyric acid, 360
δ-Amino lævulic acid (ALA), 189,
451
synthetase, 451
δ-Aminolaevulinate dehydratase, 451
Aminopterin, 190, 349
Amino-sugar, 44
Aminotransferase, 116, 379
Ammonia, 379, 389, 665
formation of, 382
Ammonium ion, 697
mechanism, 666
Amnesia, 991
Amniotic fluid, 1056
Amniotic pressure, 1073
Amorph, 1107
AMP, adenosine monophosphate,
65, 66, 349
cyclic, 65
3′, 5′-AMP, 65
Amphetamine, 116
Ampholytes, 27
Amphoteric electrolytes, 27
Ampulla of semicircular canal, (fig.)
764
of bile duct, (fig.) 277
of Vater, (fig.) 277
Ampullary crests, 929
Amygdala, 975
Amygdaloid nucleus, 958
Amylase, 226, 268
pancreatic, 268
salivary, 226
serum, 269
Amylobarbitone, 131
Amylopectin, 46
Amylose, 46
Anabolic action, 378
effect, 1034
hormones, 378
steroids, 378, 1034
Anabolism, 2
Anæmia, 245, 258, 423, 1039
and copper, 424

hæmolytic, 421
hypochromic, 426
in pregnancy, 1066
iron deficiency, 424, 426
megaloblastic, 423
pernicious, 423
Anæmic hypoxia, 635
Anærobic delayed heat, muscle, 852
Anæsthesia, 907
Anal canal, 302
sphincter, 304, 305
Analyser, 989
Anaphylactic response, 1021
shock, 598
Anastomosis, arteriovenous, 285
Androgens, 58, 60, 1007, 1013, 1032,
1034
amount excreted in urine, (table)
1033
non-virilizing, 1034
Androstenedione, 1033, 1045
Androsterone, 60, 1032
Anencephalic fetus, 1087
Anencephaly, 1072
Aneurysm, 539
Angina pectoris, 519, 863
Angiocardiography, (fig.) 498–499
Angiogram, carotid, (fig.) 895
Angiostomy, 119
Angiotensin, 568, 659, 1012, 1049,
1055
constitution of, 657
infusion, (fig.) 660
Angiotensinogen, 657
Angular gyrus, 976
notch, (fig.) 234
stomatitis, 187
Aniline, 259
Anions, 10
Annulus fibrosus, 461
Anodal polarization, 807
Anode break excitation, 791, 806
Anorexia, 330
Anovular menstruation, 1039
Ansa subclavia, (fig.) 510, 511
Anterior cerebrospinal tract, (fig.)
903
chamber of the eye, 730
fontanelle, 896
poliomyelitis, 906
root of spinal cord, 902
section of, (fig.) 903
spinocerebellar tract, (fig.) 904
spinothalamic tract, (fig.) 904
Anterolateral chordotomy, (fig.)
904
Antibiotics, 24, 372
and biosynthesis, 372
Antibodies, 37, 299, 409
immune, 410
naturally occurring, 411
production of, 1020
in organs, (fig.) 405
Anticholinergic drugs, 249
Anticholinesterases, 870
Anticoagulants, 198, 396, 397, (fig.)
397
Anti-codon, 364
Antidepressive drugs, 116
Antidiuresis, 649
Antidiuretic effects, (fig.) 662
of pituitary extract, 1093

hormone (ADH), 524, 651, 661,
680, 1094
desamino, 1096
structure of, (fig.) 1094
Antidromic impulses, 562, 791, 817
spike, 818
Antigen–antibody reaction, 37
Antigens, 37, 409
Antihæmophilic globulin (AHG),
394
Antihistamine, 260
Antiketogenic amino-acids, 381
Antimetabolites, 1021
Anti-oxidant, 53, 184, 196
Anti-parallel, 71
Antiplasmin, 398
Antirachitic substance, 774
Antithrombin, 397
Antithyroid agents, 1004
Antitoxins, 410
Antral mucosa, diagram (fig.) 252
Antrum of stomach, 253
pyloric, 253
Anuria, 666
Anus, sphincters, 305
Anvil, 761
Aorta, 458, 460, 461, (fig.) 493, (fig.)
499
valve, (fig.) 462
volume-elasticity curves, (fig.) 529
Aortic and ventricular pressure, (fig.)
495
blood flow, (fig.) 492
bodies, 630
chemoreceptors, 515
impedance, 480
pressure, (fig.) 480
sinuses, 493, (fig.) 499
sound, 500
valve, (fig.) 462
Apatite, 209, 212
Apex beat, 502
cardiogram, (fig.) 492, 502
Aphasia, 968
jargon, 975
motor, 968
nominal, 975
Apnœa, 629
adrenaline, 631
Apocrine cells, 712
glands, 712
Apoferritin, 426
Apomorphine, 261, 293
Appendicitis, pain of, 865
Appetite, 171, 250
juice, 250
APUD, 251
Aqueduct of mid-brain, 922, (fig.)
925
Aqueous humour, 730
escape from eye, 731
production, 730, (fig.) 731
Arachidonic acid, 51
Arachnoid mater, 685
villi, (fig.) 685, 896
Arbor vitæ, 941
Archipallium, (table) 958
Architectonic fields of human cortex,
(fig.) 964
Architectonics, 961
Arcuate arteries, 643
Area 4, 964

Area 6, 964
Area, intestine, 285
Area, surface, 166
Arg (arginine), 25, 380
Argentaffin cells, 234, 287, 1021
Arginase, 380
Arginine (Arg), 25, 380
Argininosuccinate, 380
Ariboflavinosis, 187
Aristotle's experiment, 717
Arousal effect of reticular formation, 924, 980
Arrector pili muscle, (fig.) 709
Arrhythmia, respiratory, 532
  sinus, 484, 532
Arterial and mixed venous blood compared, (table) 617
Arterial blood and cerebrospinal fluid compared, (table) 689
  blood gases, 613
  circle of Willis, 894
  disease, 344
  oxygen tension, (fig.) 622
  Pco_2, (table) 617
  pH, (table) 617
  plasma, (fig.) 690
  pressure, 532
    age variations, (fig.) 536
    and breathing, (fig.) 536
    mean, 574
    variations of, (table) 534
  pulse, 456, 520, 530
    wave, origin of, 493
  resistance, 481
  stretch receptors, 537
Arterioles, resistance to flow in, 537
  wall to lumen ratio, 540
Arteriovenous anastomoses, 285, 644
Artery, 4, 458
  and vein compared, (fig.) 528
  elasticity, (fig.) 529
  histology, (fig.) 528
  structure, 527
  volume-elasticity curve, (fig.) 529
Artificial hypothermia, 880
  insemination by donor (AID), 1052
  respiration, 608
  ventilation, 523, 608
Arytenoid cartilages, 756
Ascending degeneration in cord, (fig.) 904
Aschheim-Zondek test, 1063
Ascorbic acid (vitamin C), 132, 192, 731, 1007
  deficiency and wound healing, (plate) 193
  estimation of, 192
  in adrenal glands, 1007
    aqueous humour, 731
    blood, 193
    bones, 192
    foods, (table) 193
    leucocytes, 193
    platelets, 193
    scurvy, 192
  oxidase 136, 219
  synthesis, 192
Asp (aspartic acid), 25
Asparagine, 25
Aspartate aminotransferase, 117, 379, 473

carbamoyl transferase, 36, 115, 372
Aspartic acid, (Asp), 25
Asphyxia, 598, 608, 626
  and cortical cells, 608
Aspirin, 306, 401, 425
Association fibres, 961
Asthenia, 946
Asthma, 598
  bronchial, 596
Astigmatism, 734, 735
Astrocytes, 892
Astronaut, 936
Astrup method, 693, 695
Ataxia, 932, 948
  cerebellar, 948
Athetosis, 960
Athletes, 511, 517
Atmospheric air, 602
Atomic weights, 1118
ATP, 66, 76, 84, 111, 158, 317, 349, 473, 836
  hydrolysis of, 84
  –ase, 390, 489
Atria, 458
Atrial 'a wave', 491
  'c wave', 491
  diastole (fig.) 491
  fibrillation, 491
  flutter, 521
  pressure, (fig.) 492, (fig.) 496
  receptors, 516
  reflexes, 516
  repolarization, 489
  sound, 502, (fig.) 523
  standstill, 523
  systole, 490, (fig.) 491, (fig.) 492, 502, 521
  transport function, 484, 491
Atrioventricular bundle (bundle of His), 461
  conduction, 470
  junctional system, 470
  node (fig.) 469
  ring, 470
  valves, 491, 493
Atrium, 461
Atropine, 224, 249, 271, 511, (fig.) 514, 562, 565, 598, 679, 885, 732
  and accommodation, 732
  –sensitive sympathetic nerves, 953
Audible frequencies, range of, 770
Audibility, threshold, 770
Audiogram, 771
Audiometer, 770
  gramophone, 773
  pure tone, 770
Auditory acuity, 770
  cortex, 768
  nerve fibres, action potential of, 767
  pathway, 768
Auerbach's plexus, 234
Auricle, 761
Auscultatory method, blood pressure, 533
Autoimmune reaction, 1001
Autoimmunizing thyroiditis, 1001
Autonomic activities, cortical representation, 969
  fibres to eye, 974

ganglia, 817
  nervous system, 860
  outflow, (fig.) 862
  reflexes, 866
Autoradiography, 122
Autoregulation, 558, 897
Auxotonic contraction, (fig.) 479, 493
A. V. bridge, 543
A.V. bundle, 461
  node, 470, (fig.) 497
  valves, closure of, 491
    opening of, 502
Avidin, 190
Avogadro's number, 72
Axerophthol (vitamin A), retinol, 183
Axilla, temperature, 879, 883
Axillary apocrine glands, 712
Axis cylinder, 777
Axon, 780
  giant, (fig.) 780
  ions in, (table) 793
  perfused, 809
  reflex, 562, 791
Axoplasm, 776
Azathioprine, 351

Babinski response, 907
Babinski's sign, 968
Baby, body weight, 1075
  brown fat, 885, 1067
  GFR, 667
  growth, curve of, (fig.) 1075
  iron content of, 426
  metabolic rate, 168
  milk taken, (fig.) 1075
  oxygen consumption, 885
  premature, 1111
  temperature, 885
Bacteria, 24, 49, 76, 78
Bacteria-free animals, 307
Bacteria of intestine, 307
Bacterial pyrogen, 886
Bacteriophages, (fig.) 70, 78
Bacterium, 72
'Bag of membranes', 1056
Baillarger, band of, 961
Baldness, 1031, 1034
Band of Baillarger, 961
Barbiturate, 629
Barium enema, (fig.) 301, 302
  meal, (fig.) 234, (fig.) 235, 285, 303
    transit time, 285
  sulphate, 234
Baroceptor activity, 514
Baroceptors, 514, (fig.) 515, 631
  arterial, 537
'Barr body', 1102
Basal blood pressure, 535
Basal ganglia, 958
Basal metabolic rate, B.M.R., 165, 166, (fig.) 167, 1003
  metabolism, 166
  nuclei, 958
  temperature, 1052
Base deficit, 693
  definition of, 15
  excess, 260, 694
  pairs, 68
Basic electrical rhythm (BER), 283
Basilar membrane, 762, 764
Basket cells of cerebellum, 944

Basilophil leucocytes, 430, 1020
Basophilic adenoma, 1084
Beer, 181
Belching, 758
Bell cardiometer, (fig.) 493
'Bends', 638
Benedict-Roth apparatus, 164
Beri-beri, 185, 840
Bernoulli's theorem, 539
Bernstein's hypothesis, 785
Beta receptors, 867
Betz cells, 964
Bicarbonate, standard, 693
Bicycle ergometer, 168
  pedalling, (fig.) 846
Bile, 273, 452
  acids, 58, 60, 273
  canaliculus, (fig.) 276
  composition of, 273
  discharge of, 278
  duct, 273
  pigments, 438, 452
  salts, 273, 295, 299
    and absorption, 295
      circulation, (fig.) 277
    enterohepatic circulation, 277
    function of, 279
    half-life, (fig.) 281
    in fæces, 277
    pool, 280
    properties of, 279
  volume, (table) 678
Biliary canuliculi, 273, (fig.) 276
  diversion, effect of, (fig.) 278
  function, tests of, 280
  system, 273
Bilirubin, 278, 421, 438, 452
  diglucuronide, 278
  index, (fig.) 282
  output, (fig.) 282
  serum, 453
Bilirubinglucuronide, 453
Biliverdin, 452
Binary fission, 1100
Binaural interaction, 770
Binocular vision, 754
Bioassay, differential, 1018
Biochemical lesion, 182
Biogenic amines, 1020
Biological emulsifiers, 60
  oxidation, 125
  values of proteins, 200, 1111
Biosynthesis of fatty acids, 341
  lipids, 341
  nucleic acids, 347
  proteins, 365
Biotin, 189
Bipolar ECG lead, (fig.) 487, 489
Birth, 5
  change in circulation, 1069
  weight, 1067, 1072, 1111
    and parity, 1112
Bitter taste, 726
Bivalents, 1100
Biventricular volume, 492
'Black eye', 441
Bladder, urinary, 669
  filling, 669
  innervation, (fig.) 668
  in paraplegia, 670
  nervous control, 670
  pain, 669

pressure in, 669
reflexes, 871
sensation, 669
sphincters, 669
structure, 669
volume/pressure curve, (fig.) 669
Blastocyst, 1055
Bleeding of newborn, 197
Bleeding time, 401
Blind loop, 307
'Blind spot', 737
Blindness, 970
  colour, 746, 1108
Blinking, 730
  and blackout, 730
Blister base and pain, 718
Blood, 391
  acid base status, 691
  agglutinogens, 226
  alcohol, 181
  amino acid nitrogen, 375
  ammonia, 380
  arterial and venous compared,
    (table) 617
  −brain barrier, 436
  buffers, 692
  calcium in, 213
  carbon dioxide of, 622
  cells, red, (fig.) 416, 419
    altitude and number, (fig.)
      422
    life of, 420
    mean corpuscular volume,
      416
    origin of, 418
    white, 429
    life of, 429
  cholesterol, 344
  circulation of, 456
  coagulation, 391
    factors, synonyms, (table),
      394
  clotting, 391
    effect of temperature, (fig.) 392
    factors, (table) 394
    time, 401
    vitamin K deficiency and, 197
  coagulation, 391
  constituents, (table) 676
  defibrinated, 397
  flow, 494
    calculation of, 494
    coronary, 518, 519
    in liver, 581
    in various vascular beds, (fig.)
      459, (table) 504, 518, 519,
      524
    measurement in limb, by
      plethysmograph, (fig.) 540
    resting, (table) 541
      in different organs, (table)
        541
  flow through kidneys, 644, 655
    muscle, 848
    brain, 896
  formation, 418
  gas apparatus, (fig.) 618
  gas exchange, 458
  gases, in, 613, 617
  glucose, (fig.) 274, 323
    effect of meals, (fig.) 326
  groups, 410
    family, 1108

group antibodies, 410
  inheritance of, 1107
  substances, 410
    in stomach, 245
hydrogen-ion concentration, 695
in infancy and childhood, 441
  pulmonary circuit, 581
  splanchnic area, 848
  vein, 584
lactate, 316, 317
lactic acid, 314
  in exercise, (fig.) 314
laminar flow of, 538
loss, 409
  at menstruation, 1039
measurement of, 540
mixed venous, 502
muscle pump, 505, (fig.) 506
oxygen dissociation curves, (fig.)
  620
oxygen in, 620
pH, 691
phosphorus, 219
pigments, 444
  relationship, (fig.) 449
plasma, 9
platelets, 193, 399, 434, 719
potassium, 632
pressure and age, (fig.) 536
  and kidneys, 644
    splanchnic nerve, 561
  apparatus for recording, 532
  at various points in vascular
    system, (fig.) 537
  auscultatory method, 533
  basal, 536
  casual, 534
  during 24h, (fig.) 535
  in exercise, 848
    man, (fig.) 540
  normal, 545
protein, 401
redistribution, 524
regeneration, 422
respiratory functions, 612
respiratory pump, 505
return to the heart, 481
Rh factor, 412
shift reflexes, 564, 572
spectra, (fig.) 741, 744
-sugar, 318, 319, 327
  and adrenaline, 323
    exercise, (fig.) 1087
    hunger, 172
    insulin, 320
  curves, (figs.) 326, 327, 328, 329
    hormonal control of, 324
  in diabetes, (fig.) 329
    mother and fetus, 1066
thromboplastin, intrinsic, 395
transfusion, 413
turbulent motion of, 539
uric acid, 360
velocity, 539
vessels, chemical control, 555
  critical closure, 558
  cross sectional area, 528
  distension of, 529
  laminar flow, 538
  nervous regulation of, 559
  structure of, 528
viscosity, 539
volume, 408, 524, 633

Blood, volume, and age, (table) 409
  formulæ, 409
    in children, (table) 409
      pregnancy, 1063
        various organs, (table) 541
          splanchnic, 561
Blue-blind, 746
Blushing, 560
B.M.R. (basal metabolic rate), 165,
  166, (fig.) 167, 1003
Body, composition of human, 7
  image, 976
  temperature, 876, 1052
    nervous control of, 880
  water, 674
    and fatness, (table) 674
    at various ages, (table) 675
    distribution of, (table) 674
    in man, (table) 674
    in pregnancy, 1063
Bohr effect, 621
Bomb calorimeter, 51, 161
Bonds covalent, 10
Bone, 193
  and parathyroid hormone, 213
    vitamin D, 210
  composition, 209
  conduction, 772
  crystals, 209
  growth, 210
  marrow, 418, 430, 437
  modelling, 210
Bony labyrinth, 929, (fig.) 930
Botulinus toxin, 224, 822
Bradykinesia, 960
Bradykinin, 563, 885, 1023
Bradykinin-forming enzyme in
  sweat, 563, 885
Brain, 5
  and glucose, 316, 898
  angiogram, (fig.) 895
  arterial circle, (fig.) 894
  autoregulation, 896
  blood flow, 896
  circulation through, 894
  coronal section, (fig.) 959
  diagram, (fig.) 893
  hypoglycaemia, 898
  main parts, 892
  metabolism, 898
  mid-, 921
  oxygen consumption, 896
  reticular formation, 924
  R.Q., 898
  stem, 921
  weight, 896
Bran, 179
Branching enzyme, 155, 156
Bread, 179
  brown, 179
  fortification of, 179
Breaking-point, 630
Breast development, 1074
Breathalyser test, 181
Breath-holding, 629
  time, 629
Breath sounds, 598
Breathing, 588
  Cheyne-Stokes, 634
  effect of loading, (fig.) 607
  periodic, 634
  rate of, 594
  reflex alterations of, 605

reserve, 600
Breathlessness, 542
Brightness, range of, (fig.) 743
Brilliant cresyl blue, 419
'Brisket disease', 635
Broca's area, 968
Brodmann's areas of cortex, 961
Bromsulphthalein, 848
Bronchi, 589
Bronchial arteries, 577
  asthma, 596
  circulation, (fig.) 459
  epithelium, 589
  murmur, 598
    and histamine, 598
  muscle, 597
  spasm, 598
  veins, 577
Bronchioles, 589
Bronchitis, 637
Bronchoconstriction, 598
Bronchodilatation, 598
Bronchopulmonary anastomoses, 577
Brønsted-Lowry defition, 16
Brown adipose tissue, 334, 885, 1067
  fat in baby, 885, 1067
Brown-Séquard syndrome, (fig.)
  908
Bruise, 452
Brunner's glands, 242, 290, (fig.)
  292, 293
Brush border, (fig.) 288, 290
Buccopharyngeal pouch, 1080
Bucket-handle movement, 593
Buffer, 694
Buffering action of hæmoglobin, 624
  plasma, 691
  systems, plasma, 691
Buffers, 18
Buffers, of blood, 692
  plasma, 622, 691
Buffy coat, 415
Bulbogastrone, 256
Bulbourethral gland, 1029
Bundle, atrio-ventricular (A.V.),
  461
Bundle of His, 461, 468, (fig.) 469
Bundle branches of cardiac special-
  ized conducting system,
  469
Butanol, 12
Butter, 177
Buttermilk, 177
Butyric acid, 50

'c wave' (atrial), 491
  in jugular phlebogram, 491
Cabin 'altitude', 634
Cable structure of nerve fibre, 777,
  800
Cadmium, 1034
Cæcum, 285, 302
Caeruloplasmin, 219, 405
Caffeine, 64, 181
Caisson workers, 638
Calcarine cortex, (fig.), 971
Calciferol, 193
  in foods, (table) 195
  toxic effects, 195
Calcification, mechanism of, 211
Calcitonin, 214
Calcium, 835, 936, 882

absorption, 195, 213
and myocardial cell, 464, 470, 471
balance, 212
  in old age, 1113
content of body, 212
deficiency, 213
in blood, 213
  diet, 212
  plasma, 216
  skeleton
  sweat, 213
  urine, 213
intake of, 212
ionized, 213, 217
ions, 836
  level in plasma, 213, 216
losses, 213, 1077
metabolism, 212
on excitable tissue, 807
reabsorption, 213
requirements, 212
role in myocardial contraction, 471
Calculus, dental, 226
Caloric stimulation of semicircular
  canals, 935
Calorie, 161
  requirement of man, 169
    in old age, 1113
  value of expired air, 165
  values, 161
Calorific value of oxygen, 163
Calorimeter, bomb, 162
Calorimetric records of heat loss,
  (fig.) 567
Calorimetry, direct, 162
  indirect, 162
cAMP, 66, 318, 319, 321, 323
cAMP cascade, (fig.) 318
Canal of Schlemm, 731
Canalis reuniens, (fig.) 764
Cancer of breast, 1047
  of prostate, 1047
Cane sugar, 45
Cantharidin, 718
Capacitation, 1054
Capacity vessels, 551, 553
  control of, 570
Capillaries, number of, 570
Capillary, 4, 458, 543
  and interstitial fluid, (fig.) 682
    water exchange, (fig.) 682
  bed, (fig.) 544
  blood flow in, 545
  chemical control, 569
  electronmicrograph, (fig.) 542,
    (fig.) 549
  endothelium, 683
    pathways through, 548
  filtration, 406, 550
  flow in, 545
  fragility, 550
    in lung, 615
    in nail bed, 547
    in skin, 545
  nervous control, 570
  of muscle, 570
  permeability, 547, 550, 683
  per unit area, 617
  pores, 683
  pressure, 545
  resistance, 550
  rouget cells, 543

Capillary structure, (fig.) 543, (fig.) 549, (fig.) 616
  wall, passage of fluid through, (fig.) 682
Capacitance, arterial, 458
  venous, 458
Capsid, (fig.) 70, 78
Capsomere, (fig.) 70, 78
Carbamino compounds, 623
  hæmoglobin, 624
  reaction, (fig.) 625
Carbamoyl-aspartate, 347
  -phosphate, 347, 380
Carbohydrate, 40
  absorption, 296
  calorie value, 161
  digestion of, 292
  metabolism, 311, 317, 335
    control of, 317
    hormonal control of, 319
    in adipose tissue, 335
    in pituitary gland, 324
    in the ruminant, 331
    organization and control, 311
  protein-sparing action, 201
  requirements, 204
  storage, 154
Carbon dioxide and respiration, 627
  as anæsthetic, 609
  carriage of in blood, 622
  dissociation curve, (fig.) 623
  effect on oxygen dissociation curve, 621
  effects of increasing, (fig.) 628
  electrode, (fig.) 619
  excessive loss of, 700
  excretion of by lungs, 701
  in alveolar air, 628
    blood, (table) 617
    plasma, (table) 617
    red cells, (table) 617
  narcosis, 636
  retention, 696
  solubility coefficient, 691
  tension in alveoli, 627
  transport of, 625
  vasomotor centre, 512, 559
Carbon isotopes, 121
Carbonium ion, 102
Carbon monoxide, 448, 617, 635
  poisoning, 608, 609, 636
  tetrachloride, 336
Carbonic anhydrase, 219, 623, 698, 892
Carboxyhæmoglobin (COHb), 448
  absorption spectrum, plate 25.1
Carboxypeptidases, 291
Carcinoid tumour, 1021
Carcinoma of bronchus, 1033
Cardiac arrest, 523
  arrhythmia, 519
  biventricular volume, (fig.) 493
  catheter, (fig.) 494
  catheterization, 493
  cells, action potentials of, 467
  cycle, 490, (fig.) 491, 492, (fig.) 488, (fig.) 489
  electrical activity, total unbalanced, 471, 487
  excitation–contraction coupling, 471
  failure, 523
  impulse, 502

index, 503
  isopotential field map, (fig.) 472
  massage, 523, 609
  metabolism, 464
  murmur, 502
  muscle, 336, 461
  nerves, (fig.) 508
  notch, (fig.) 234
  output, 502, 506, (fig.) 507, 521
    and adrenaline, (fig.) 510, 513, 567
    at rest, (fig.) 504
    by Fick principle, 494, (fig.) 513
    dye dilution method, 503, (fig.) 504
    factors modifying, 502
    formula, 504
    in anæmia, 503
      exercise, 503, 504, 517
      pregnancy, 1055
    indicator dilution, 503, 504
    techniques for measurement, 503
    ventricular, 461
  pacemaker, 467
  pain, 519, 863
  reflexes, 512
  response to exercise, 517
  rhythms, disordered, 519
  sarcomere, 464, (fig.) 474
  shadow in chest, X-ray, (fig.) 463
  tamponade, 505
  valve cusps, 462
  ventricular excitation, 471, 487, (fig.) 488
    and the ECG, diagram of, (fig.) 487
Cardio-acceleration centre, 512
  -aortic nerve, 512
  -inhibitory centre, 512
Cardiometer, (fig.) 493, (fig.) 510
Cardio-pulmonary bypass, (fig.) 458, 472
Caries, 220
Carnitine, 337
β-Carotene, 183
Carotenoids, 159, 183
Carotid angiogram, (fig.) 895
  artery compression, (fig.) 564
  body, 515, 630, 632
    blood flow, 518
    chemoreceptors, 628
  sinus, (fig.) 514, (fig.) 515, (fig.) 560, (fig.) 571, 572, 631
    nerves, 514
    reflex, 514
Carpopedal spasm, 217
Carriage of carbon dioxide, 622
Carrier, 1108
Carr-Price reaction, 183
Cartilage, 208, 209
  epiphysial, (fig.) 211
  hyaline, 209
Casein, 176
Castration, 1031, 1032, 1034, 1049
Catabolic hormones, 378
  response to injury, 204
Catabolism, 2
  of nucleotides, 359
  of purines, 360, (fig.) 361
Catalase, 137, 426
Catalyst, 89, 91
Catalytic subunit, 115

Cataract, 717
Catatonic state, 953
Catechol, 136
Catecholamines, 882, 1016, 1018
  in blood, 1019
  in urine, 1019
Cathodal shock, 791
Cathode excitation, 791
Cations, 10
  and anions in intracellular fluid, (table) 675
Cauda equina, destruction of, 670
Caudate nucleus, (fig.) 893, 958, (fig.) 959
Casual blood pressure, 534
C-cells, micrograph, 214
CCK-PZ (cholecystokinin-pancreo-zymin), 270, 271
Cell, 2
  diagram of, (fig.) 75
  division, 1098
  envelope, 75
  membrane, 56, (fig.) 69, 76, 466, 994
    action potential, 467
    depolarization, 468
    myocardial, potentials, 467, 478
    pacemaker, 467
    polarization, 468
    resting potential, 467
    threshold, 467
  organelle, 75
  sap, 77
  Sertoli, 1029, 1032, 1034
α-Cell, 319
β-Cell, 319
Cells, immortal, 1112
  myocardial, specialized conducting, 46
    working, 464
  of adenohypophysis, 1081
Cellulose, 47, 303, 308, 334, 343
  digestion of, 292
Central chemoreceptors, 628
  cyanosis, 636
  dogma, 354
  excitatory state, 909, 911
  inhibitory state, 914
  nervous system transmitters, 822
  pillar, 763
  reflex time, 909
  sulcus, 957
  venous pressure, 506, 551
Centrosomes, 1098
Cephalic phase of gastric secretion, 250
Cephalin, 54
Cereals, 179
Cereal products, composition of, (table) 180
Cerebellar ataxia, 948
  lesions, 950
  peduncle, 921, 941
  receiving area, (fig.) 947
Cerebellectomy, 927
Cerebellum, 941
  and decerebrate rigidity, 915, 919, 924, 948
  gait, 947
  knee jerk, 947
  motion sickness, 933
  muscle spindles, 948
  ataxia, 948

Cerebellum, connexions, (fig.) 943
  cortical connexions, 948
  damage to, 949
  flocculonodular lobe, 948
  hypotonia, 949
  inferior peduncles, 948
    aspect, (fig.) 942
  intention tremor, 948
  lobes, 942
  peduncles, 941
  removal of, 946
  superior aspect, (fig.) 942
Cerebral blood flow, 896
  circulation, 894
  coronal section, (fig.) 959
  cortex, 961
    cells of, (fig.) 962
    motor areas of, 964
    oxygen consumption, 898
    sensory areas, 970
    structure, (table) 961
  frontal areas, 969
  giant Betz cells, 964
  grey matter, 959
  hemisphere, 960
    histology of, 961
  hypoxia, 633
  ischæmia, 894
  metabolism, 898
  motor areas, 964
    homunculus, (fig.) 965
  occipital cortex, 961
  parietal lobes, 961, 964
  peduncle, 948
  temporal lobes, 974
  ventricles, shape, 689
  vigilance, 980
Cerebro-hypothalamo-pituitary-
    adrenocortical   (CHPA)
    axis, 1014
Cerebrospinal cisternal fluid, (table)
    689
  fluid, 628, 685, 687, 896
    composition, (table) 688
    pressure, 688
    production, 686
    volume, 688
  tracts, 964
Cerebrum, motor areas, 964
Ceruminous glands, 761
Cervical mucus, 1036, 1053
Cervix, 1035
  uteri, 1047
C.e.s. (central excitatory state), 911
CG, 1026, 1049, 1051
  chimpanzee, 1051
  giraffe, 1051
Chain termination signal, 367
Chalone, 256
Cheese, 116, 177
Cheilosis, 187
Chemical control of blood vessels
    555
  equilibrium, 80
  score of proteins, 201
  senses, 722
  transmission, 815
Chemoreceptors, 515, (fig.) 516, 628,
    629
Chemoreflexes, 515, 565
Chenodeoxycholic acid, 273

Chest, circumference, 593
  expansion, 599
  X-rays, (fig.) 463
Cheyne-Stokes respiration, 634
Chiasma, 1100
Chitin, 48
Chloramphenicol, 373
Chloride ions in blood, (table) 617
  shift, 624
Chlorolabe, 743
Chlorophyll, 159, 184
Chloroplasts, 72, 159
Chlorpromazine, 1088
Chocolate, 181
Cholanic acid, 60
Cholecalciferol, 193
Cholecystectomy, (fig.) 282
Cholecystokinesis, 278, (fig.) 282
Cholecystokinin-pancreozymin, 242,
    269
Choledocho-duodenal sphincter,
    control of, 279
Choleglobin, 452
Cholic acid, 273
Choleresis, 278
Cholestanol, 59
Cholesterol, 57, 273, 277, (table)
    333, 343, 1009, 1046
  biosynthesis, 344
  in blood, 344
  metabolism, 343
Cholic acid, 60, 273, 277
Choline, 54, 191
  acetylase, 822
Cholinergic fibres, 866
  nerves, 870
Cholinesterase, 103, 870
Chondrocytes, 209
Chondroitin sulphate, 48
Chorda tympani, 224, (fig.) 227,
    726
Chordæ tendineæ, 461, 472, 491
Chordotomy, 909
Choreiform movements, 960
Choriocarcinoma, 1063
Chorionic gonadotrophin (CG), 1026
  human, 1049, 1051
  villi, 1055, (fig.) 1059
Choroid plexuses, 686
Christmas disease, 396
  factor, (table) 394
Chromaffin cells, 1019
  cell tumours, 1019
  tissue, 1015
Chromatic aberration, 735
Chromatid, 1098
Chromatin, 75, 357
Chromatography, 28
Chromatolysis, 810, 964
Chromophobe cells, 1081
Chromosome, 72, 75, 1098
  homologous, 1100
  number, 1100
Chronaxie, 789
Chvostek's sign, 217
Chylomicrons 299, 333, 407
Chyme, 244, 291
Chymotrypsin, 35, (fig.) 92, 104,
    (fig.) 105, 106
Chymotrypsinogen, 268
Cilia, 590

Ciliary activity, 590
  body, 729
  glands, 730
  muscle, 729, 732
Ciliated columnar epithelium, 590
Cingulate gyrus, 969
Circadian rhythm, (fig.) 658, (fig.)
    877
  variation in ACTH, 1009
Circle of Willis, (fig.) 894
Circulation, 456, (fig.) 459, 526
  birth changes, 1068
  cerebral, 894
  general plan of, (fig.) 458
  in fetus, (fig.) 1069, (fig.) 1070,
    (plate)
  of blood, (fig.) 458
  myocardial, 462
  peripheral, resistance of, 458
  pulmonary, 458
  renal, (fig.) 459
  systemic, 458
  time, 496
  through liver, 581
    lungs, 577
    spleen, 585
Circulatory failure, 575
Circulus arteriosus, (fig.) 894
'Circus movement', 522
C.i.s. (central inhibitory state), 914
Cisternae, myocardial, 472
Cisternal puncture, (fig.) 686
Cisterna magna, 686
Cistron, 355, 369, 1106
Citrate, and blood coagulation, 394
Citrate synthase, 149
Citrated plasma, 394
Citric acid cycle, 139, 140, 148
Citrulline, 26, 380
Clasp-knife reaction, 917
Classical conditioning, 984
Claustrum, (fig.) 893, (fig.) 959
Claws, 34
Clear cells of sweat glands, 712
Clearance, renal, 653, (fig.) 654
Climacteric, 1042
  hormonal basis of, 1051
Climate and food intake, 875
Climbing fibres of cerebellum, 944
Clinical thermometer, 879
Clitoris, 1053
Clomiphene, 1053
Clonus, 908
Clostridium botulinum, 835
Clot retraction, 391
Clothing, 887
Clotting, blood, 391
  factors, 393
  of milk, 177
  time, 401
CMP, 65, 66
CNU (contingent negative variation),
    (fig.) 991
CO_2 narcosis, 636
Coagulation and temperature, (fig.)
    392
  and platelets, 392
Coagulation 'cascade', (fig.) 395
  of blood, factors in, (table) 394
  time, 401
Coal dust in lung, (fig.) 591

Cobalamin (vitamin B_{12}), 190, 191, 422, 423
  deficiency, 423
  demyelinization, 423
  half-life, 423
  in fæces, 423
  liver, 191
Cobalt, 220
Cobamide, 191
  coenzyme, 423
Co-carboxylase, 148, 185, 341
Cochlea, 762
  cross section, (fig.) 763
  destruction of, 974
  electrophysiological aspects of, 765
  endolymph, 766
  mechanical aspects of, 763
  perilymph, 766
  synopsis of action, 768
Cochlear duct, (fig.) 762, 763
  hair cells, 765, 931
  microphonic, 765
  nucleus, 768
  nerve, 768
    central connexions, 768
  partition, (fig.) 763, 764, (fig.) 765
Cocoa, 181
Coding strand, (fig.) 356
Cod-liver oil, 195
Codon, 354, 363
Coefficient of friction in joints, 212
Cœliac disease, 287, 301
Coenzyme, 109, 111
  A, 67, 111, (fig.) 112, 189, 335
  Q, 130
Coffee, 181
Cog-wheel resistance, 960
Coitus, 1030, 1052, 1053, 1094
  and ovulation, 1052
Colchicine, 287
Cold, 335
  and body temperature, 873, 874, 875
  effects on skin, 557
  spots, 714
Colic, 306, 865
Collagen, 31, 34, 193, 208
  fibres, 193
Collagenase, 211
Collecting tubule, 651
Colliculus, inferior and superior, (fig.) 922, (fig.) 925
  facial, 922
Colloid, 20
  of thyroid, (fig.) 123
  osmotic pressure, 21, 682
Colon, 302, (fig.) 302
  absorption in, 295
  bacteria of, 308
  gas in, 308
  haustration of, 304
  movements, 303
  nerve supply, 304
  X-ray photograph, (fig.) 301, (fig.) 302
Colostomy, 304
Colostrum, 299, 1076
  corpuscles, 1077
Colour-blindness, 746, 1108
  names, 741
  vision, 743

Colours, primary, 745
Columnar cells, 289
Coma, diabetic, 330
  hypoglycaemic, 321
Combustion of foodstuffs, 4
Comfort, 888
Commissural fibres, 961
Common chemical sensibility, 722
Compensatory pause, 521
Competitive inhibition, (fig.) 98
Complementarity, 68
Complete heart block, 491
  tetanus, 841
Compliance, 524, 597
Composition of cereal products (table) 180
  human body, 7
  fruit and vegetables, (table) 180
  living tissues, 5
  man, 7
Concentric corpuscles of Hassall 1019
Condensed milk, 177
Condensing enzyme, 149
Conditional reflex, 984
  analyser, 989
  differentiation, 987
  external inhibition, 986
  extinction, 986
  internal inhibition, 986
  negative, 986
  reinforcement, 986, 988
  specificity, 986
  stimulus, 985
Conditioning, 984
  classical, 984
  higher order, 987
  instrumental, 988
  'Type I', 984
  'Type II', 988
Conducting system, specialized, of the heart, 469
Conduction rate, 803
  saltatory, 799
  speed of, in nerve, 803
  through heart, 470
  velocity, 783, 785
  myocardial, (table) 470
Cones of retina, 738
  absorption spectra, 744
Congenital adrenal hyperplasia, 1009
Conjugate base, 16
  movement, 751
Conjugating agents, 384
Conjunctiva, 729, 730, 731
Connective tissue, 208
Conn's syndrome, 1013
Consciousness, 939
Consensual light reflex, 735
Conservation of energy, 3
  matter, 3
Consonants, 757, 759
  voiced, 759
Constant element, 335
  field equation, 793
Constipation, 304
Contingent negative variation (CNV), 991
Contraceptive preparations, 1047, 1050
Contact lens, 735
Contractile mechanism, 835

state, 518
Contractility, 509
Contraction, auxotonic, 480
  coupling, 837
  isometric, (fig.) 476
  isotonic, (fig.) 476
  of sarcomeres, 470
  ventricular, 480
Convergence, 751
Conversion factors, 1116
Convoluted tubules, 642
Cooking, 198
Copper, 136, 219, 261, 424
  and taste, 727
Coprophagy, 308
Coproporphyrin, 451
Coprostanol, 59
Coprosterol, 56
Corepressor, 372
Cori cycle, (fig.) 314
Coriolis effect, 935, 936
Corium, 709, 711
Cornea, 730
  refracting power, 733
Corona radiata, 1038
Coronal plane, 894
Coronary arterial anastomosis, (fig.) 497
  circulation, 496
  system (human), (fig.) 497
  arteriography, 496
  artery, (fig.) 461, 462
    nerves to, 519
  blood flow, 518
  sinus, 461, 468, 494, 518
  thrombosis, 117, 473
Corpora cavernosa, 1053
  mamillaria, 923, 951, 954
  spongiosa, 1053
Corpus callosum, 961
  section of, 962
  luteum, 1038
    hormone of, 1045
  mamillare, (fig.) 923
  striatum, 957, 958
  trapezoideum, (fig.) 924
Corresponding points, 751
Corrin nucleus, 190
Corrinoids, 190
Cortex, cerebral, 957
  architectonic fields, (fig.) 964
  evoked potentials, 990
  frontal, 969
  motor, 961
  occipital, 970
  parietal, 961, 964
  retina, (fig.) 971
  sensory areas, 970
  temporal, 974
  visual, 750
  visuosensory, 961
Corti, organ of, 762, (fig.) 763
Cortical cells and oxygen lack, 608
  representation of autonomic activities, 969
  retina, (fig.) 971
Corticosterone, 1008, 1084
Corticotrophin, 1007, 1084
  and carcinoma of bronchus, 1033
  constitution, 1085
  releasing factor (CRF), 1010

Index 1131

Cortisol, 61, 322, 324, 1009, 1084
and eosinophils, 1010
metabolism, 1011
formula, 1011
Cortisone, 664
and water loss, 664
Cortol, 1011
Cortolone, 1011
Costo-diaphragmatic recess, 594
Cough, 607, 688
syncope, 505
Coumarin, 197
Counter-current theory, 650
Counter-rolling, 929, 937
Covalency, 10
Covalent bonds, 10
Craniosacral outflow, 860
Cream, 177
Creatine, 386
formation, (fig.) 388
in children's urine, 387
kinase, 838
output in old age, 1114
phosphate, 464, 838
phosphokinase, 111, 116
Creatinine, 111, 386, 654
clearance, 654
Creatinuria, 388
Crenation of red cells, 21
Cretinism, 1001, (fig. 1002), 1112
CRF, 1010
Cricopharyngeus muscle, 228
Cricothyroid muscle, 756
Crista basilaris, (fig.) 764
dividens, 1068
Cristæ ampullares, 929
Cristæ mitochondriales, 75
Critical closing pressure, 558
Cross bridges, muscle, 836
Crossed extension reflex, 912, 913
pyramidal tract, (fig.) 903
Cross link formation, 843
Cross-matching of blood, 413
Crossing-over of chromosomes, 1107
Cryoprecipitate, 396
Cryoscopy, 680
Cryptorchidism, 1026
Crypts of Lieberkühn, 287, 290
C.s.f. (cerebrospinal fluid), 628, 685
687
CTP, 342, 348
Cumulus ovaricus, 1037, 1054
Cuneate nucleus, 921
Cupola, 929, 931
Cuproproteins, 109
Curare, 874
Curettage, 1039
Cushing's syndrome, 1010, 1090
Cutaneous pain, 718
sensations, anatomical basis, 719
modalities of, 719
senses, 714
Cyanide, 131, 636
Cyanide poisoning, 636
Cyanocobalamin, 191, 220, 423
Cyanolabe, 743
Cyanosis, 632, 636
central, 636
peripheral, 636
Cyclamate, 726
Cycle, Krebs, citric acid, tricar-
boxylic acid, 140
Cyclic AMP, 321, 898, 994, 995

Cyclic 3′,5′-adenosine monophos-
phate (3′,5′-AMP), 65,
158, 214
Cycloheximide, 373, 995, 996
Cysteine, 25, 388
Cystic duct, 273, (fig.) 277
Cystine, 25, 388
Cystitis, 669
Cystometrogram, (fig.) 669
Cytidine, 65
monophosphate (CMP), 65
triphosphate (CTP), 342, 348,
372
Cytochemistry, 75
Cytochrome, 129, 426, 636, 837
oxidase, 129
reaction in gastric mucosa,
(fig.) 247
Cytoplasm, 75
Cytosine, 64
Cytotrophoblast, 1056, (fig.) 1058,
1063

Dairy produce, composition of,
(table) 178
Dale's principle, 820
Damping vessels, (fig.) 527
vessels, control of, 569
Dark adaptation, 747, 748
Dark cells of sweat glands, 712
Dartos muscle, 1026
dATP, 350
D.B.P., 533
dCTP, 349
Dead space, 600, 613
Deafness, 771
Deamidation, 82
Death, 1114
Debranching enzyme, (fig.) 157
Decarboxylation, 383
Decerebellation, 946
Decerebrate preparation, 928
rigidity, 915, 919, 924, 948
and cerebellum, 948
Decibel notation, 761
Decidua, 1055, (fig.) 1056
Decidual cells, 1055
reaction, 1045
Decompression sickness, 638
Decorticate animal, 953, 959
child, (fig.) 928
Decussations in tegmentum, 922
motor, 921
pyramidal, 921, 923, 964
sensory, 921, 923
Deep cardiac plexus, 508
Defæcation, 305
time of, 306
Defibrination, 397
Degeneration, 5
ascending, (fig.) 904
descending, (fig.) 903
in peripheral nerve, 809
in spinal cord, 903
Wallerian, 809
Deglutition, 228
Dehalogenating enzyme, 999
Dehydration, 679, 680, 1095
fever, 884
L-dehydroascorbic acid, 192
7-dehydrocholesterol, 57, 193
Dehydroepiandrosterone, 1045

Dehydrogenases, 128
Dehydroretinol, 183
Delta waves, 979
Demyelinization, 423
Denaturation, 71, 93
of protein, 36
Dendrites, (fig.) 816
Denervated muscle, 847
Denervation, 847, 868
sensitivity, 1018
Density gradients, 74
of body, 168
Dental caries, 195
Dentate nucleus, (fig.) 943
Dentine, 212
Deoxycholic acid, 60
11-Deoxycorticosterone (DOC),
1008
11-Deoxycortisol, 1008
Deoxyribonuclease, 357
Deoxyribonucleic acid (DNA), 67,
68
Deoxyribonucleoprotein, 68
Deoxyribonucleoside, 65
2-Deoxyribose, 63
Deoxythymidine 5′-monophosphate
(dTMP), 349
Depolarization, 468, 787
myocardium, 467
Depressor nerves, 508, 513
Depth, perception of, 754
Dermatitis in pellagra, (fig.) 188
Dermatome, 905, 906
Dermis, 709, 711
Desamino-antidiuretic hormone,
1097
-oxytocin, 1094
Descending degeneration in cord,
(fig.) 903
Descent of base of heart, 493, 502
Desmosine, 26
Desmosomes, 710
Detoxication, 384
Detrusor muscle, 668
Deuteranopes, 746
Deuterium, 122
oxide, 7
Dexamethasone, 324, 1015
Dextran, 647
Dextrins, 47
DFP, 103, (fig.) 105
dGTP, 350
DHA, 1084
DHCC, 195
Diabetes insipidus, (fig.) 1091, 1096
nephrogenic, 1096
mellitus, 120, 318, 328, 329, 336,
339, 344, 345
and growth hormone, 1086
pituitary, 1086
Diabetic coma, 330
ketosis, 330, 700
mothers and size of baby, 1112
Diads of the sarcoplasmic reticular
systems of the heart, 466
Diagnostic use of enzymes, 116
Dialysis, 21, 666
Diaphragm, 593
pain, 865
Diarrhœa, 298, 304, 306, 681
Diastasis, ventricular, 491, (fig.) 492
Diastole, 460, 474, 490, 519

Diastolic pressure, 533
Dichromats, 746
Dicoumarin, 396
Dicoumarol, 197
Dicrotic wave, 531
Diencephalon, 951
Diet and growth, 1111
  composition of, 311
  'normal', 199
Diethylstilboestrol, 1047
Differentiation, 987
Diffusion, 20
Digestion in intestine, 291
Digitalis, 484, 491
Diglyceride, 51
Dihydro-orotic acid, 347
Dihydroxy-acetone, 40, 142
1:25 Dihydroxycholecalciferol, 195,
  213, 216
3,4-Dihydroxyphenylalanine
  (DOPA), 192, 385
Di-isopropyl phosphorofluoridate,
  (DFP), 103, 430, 434, 870
Dilatator pupillæ, 732
Dilution technique, (table), 675,
  (table) 677
Dimethylisoalloxazine, 67
Dinitrophenol, 247, 297
2,4-Dinitrophenol, 132
Diodone clearance, 655
Diodotyrosine, 997
Diodrast, 655
Dioptre, definition, 733
Dipeptidases, 291
Dipeptide, absorption, 298
Diphasic recording, 781
2,3-Diphosphoglycerate (DPG),
  621
1,3-Diphosphoglyceric acid, 142
5-Diphosphomevalonic acid, 344
Diploid number of chromosomes,
  1100
Diplopia, 751, (fig.) 752
Direct calorimetry, 162
  light reflex, 735
  pyramidal tract, (fig.) p. 903
Disaccharidase, (fig.) 291, 296
Disaccharide, 40, 45
α-Discharge, 828
γ-Discharge, 830, 915, 927, 948, 960
Discrimination, 705
  of touch, 717
Disinhibition, 986
Dissociation constant, 13
  curve of blood, 620
Distal convoluted tubule, 651
Dithiocarbamoylhydrazine, 1053
Diuresis, 661
  in pregnancy, 1063
Diuretics, 664
Diver's palsy, 638
Diverticulitis, 304
Diving, 638
Division of cell, 1098
Dizziness, 933
DNA, 67, 68, 287
  bases in, 67
  complement, 1102
  damaged, repair of, 358
  dependent RNA polymerase, 357
  helix, 68
  molecule, 1106

nucleotidyltransferase, 353
  polymerase, 349, 353
  replication, 352, (fig.) 353
  synthesis, (fig.) 352
D/N ratio, 327
D.O.C., 1008
Döderleins's bacillus, 1036
Dominant genes, 1106
DOPA, 192, 385, 1017
Dopamine, 383, 960, 1017
Dorsal nucleus of vagus, (fig.) 924
Dorsomedial nucleus of thalamus,
  955, 969
Double helix, 68, 353
  vision, 751
Dough, 179
Douglas bag, 165
  technique, 502
DPG, 621, 635
Dried milk, 177
Drinking of saline and water, (fig.)
  662
Dropsy, 683
Drowning, 608, 610, 630
'Drumstick' mass of chromatin, 1102
Dry eye xerophthalmia, 184
dTMP, 349
dTTP, 349
Ductless gland, 994
Ductus arteriosus, 1068
  cochlearis, (fig.) 762
  deferens, 1029
  venosus, 1068
Duocrinin, 293
Duodenal aspirate, (fig.) 282
  bulb, 235
  distension on gastric emptying,
    242
  glands, 290, (fig.) 292, 293
    histology, (fig.) 292
  juice, 266
  receptors, 239, 240
Duodenum, 290
  acid in, 239, 240
  pH, 241
Dura mater, 685
Dwarfism, 1084, 1087, 1090
Dye-dilution method, 408
Dynamic equilibrium of body proteins, 376
Dysdiadochokinesis, 950
Dysmetria, 949
Dyspnœa, 627, 632, 856

EACA, 398
Ear, diagram of, (fig.) 764
  drum, 879
  external, 761
  internal, 762
  middle, 761
  ossicles, 761
  sensitivity of, 761
ECG, leads, 487
Eccrine sweat glands, 712
Eck fistula, 120
Ectopic foci, 520
Edinger-Westphal nucleus, (fig.) 973
Edman reagent, 32
EDTA, 213, 394
E.e.g. (electroencephalogram), 977,
  991
Effective filtration pressure, 648

Efferent nerves, 4, 891
α-Efferents, 828
γ-Efferents, 830, 915, 927, 948, 960
Efficiency of human body, 168
Eggs, 178
Egg white, 190
Ejaculate, 1030
Ejaculation, 1053, 1054
Ejaculatory duct, 1030
Ejection phase of ventricular con-
    traction, 493
Elastic fibres, 596
  resistance, 596
  tissue, arterial, 530
Elastin, 208
Electrocardiogram (ECG), 471, (fig.)
    472, 484
  direct, of cardiac conducting
    system, (fig.) 496
  excitation, electrical, of myo-
    cardium, (fig.) 488
  'exploring' electrode, 489
  'indifferent' electrode, 489
  in hyperkalæmia, (fig.) 489
  in hypokalæmia, 489
  intracardiac, 496
  lead systems and terminology, 489
  leads, bipolar, 487, 489, 490
    standard limb, 490
    unipolar, 487, 489
    V, 489
  PR interval, 490
  QRS complex, 490
  ST segment, 490
  T wave, 490
  U wave, 490
Electrocardiograph, 485, 519
Electroencephalogram, 977, 991
  in hypoglycæmia, 978
Electrolyte balance, 673
  composition of body fluids, (table)
    675
  homeostasis, 679
  loss, (table) 678
    and adrenal cortex, 663
  metabolism, 673
  turnover, (table) 678
Electrolytes daily intake, (table) 678
  in body water, 675
  osmotic pressure of, 675
Electromyography, 846, 939
Electron transfer mechanism, 132
Electronystagmography, 935
Electrophoresis, 28, (fig) 403
Electroretinogram, 748
Electrostatic forces, 12
Electrotonic spread of potential, 800
Elements in body, (table) 7
Embden-Meyerhof pathway, 140
Emboli, 607
Emboliformis, nucleus, 945
Emetic centre, 261
Emetics, 261
Emmetropia, 733
Emotional stress reflexes, 565, 573
Emotion, sweating, 884
Emphysema, (fig.) 597
Emulsification, fat, 293
Enamel, 212
Encephalitis lethargica, 980
End-diastolic volume, 504
    pressure, 483

Endemic goitre, 1005
Endergonic reaction, 86
Endocardium, 527
Endocochlear potential, 765
Endocrine glands, 994
Endogenous (leucocyte) pyrogen, 886, 887
  nitrogen metabolism, 200
  protein metabolism, 376
Endolymph, 766, 932
Endometrial cycle, 1038
Endometrium, 1035, 1039, 1055
  in menstrual cycle, (fig.) 1037
    pregnancy, 1055
Endoneurium, 780
Endonuclease, 357, 359
Endopeptidase, 291
Endoplasmic reticulum, 67, (fig.) 69
Endosperm, 179
Endothelial cells, 527
End-plate, 831, 833
'End-tidal sampling', 602
Enema, Soap and water, 306
Energy, 4
  changes, 83
  exchange, 161
  expenditure by different species, (table) 166
  output estimation, 161
    of a clerk, (table) 170
    coal miner, (table) 170
  -rich compound, 87
  requirement of man, 169
  source of, 4
  storage, 50
  value of foodstuffs, 163
  value of oxygen, 163
Enolase (phosphopyruvate hydratase), 143
Enophthalmos, 732
Enoyl CoA hydratase, 337
Enterogastrone, 242, 256
Enteroglucagon, 323
Enterohepatic circulation, 273, 277, 279, 299
Entero-insular axis, 273
Enterokinase, 268, 280, 290
Enteropeptidase, 268, 280, 290
Environment, internal, 2
  external, 3
Environmental conditions, measurement of, 888
Enzyme, 90
  action, 99
  active site, 99
  competitive inhibitors, 97
  deficiencies, 298
  effect of pH, 93
    temperature, 92
  forming system, (fig.) 119
  induction, 369
  inhibition, 97
    by drugs, 115
  kinetics, 95
  location on villus, (fig.) 290
  regulatory, 112
  secretion in stomach, 249
  specificity, 91
  -substrate complex, 95
Enzymic, modification of nucleic acids, 359
Eosinophilia, 432

Eosinophil leucocyte, 430
E-oxidation reduction potential, 137
Ependyma, 686
Ephedrine, 116
Epicholesterol, 59
Epidermis, 709 (fig.) 710
Epididymal fat pad, 334
Epididymis, 1029
Epileptic fit, 965, 967
Epineurium, 780
Epiphyseal cartilage, 210, (fig.) 211
EPF, 1002
EPS (exophthalmus-producing substance) 1084
E.p.s.p., 817, (fig.) 819, 821
Equal energy luminosity curve, (fig.) 745
Equilibrium constant, 81, 82
Equivalent, 9
Erection, 1054
E.R.G. (electroretinogram), 748
Ergocalciferol, 193
Ergometer, bicycle, 168, (fig.) 846
Ergometrine, 1094
Ergosterol, 57
Ergotamine, (table) 869
Ergothionene, 950
Erotogenic hormones, 1045
Errors of metabolism, 120, 384
Erythema, 711
Erythritol, 12
Erythroblast, 418
Erythrocyte, 313, 415, 418
  destruction, 439
  diameter, 417
  glucose in, 417
  glucose-6-phosphate dehydrogenase, 417
  high altitudes, 422
  life of, 420
  metabolism, 417
  sedimentation rate, E.S.R.), 416
  shape, (fig.) 416
  stroma, 417
Erythrolabe, 743
Erythron, 421
Erythropoiesis, 418, 421
  control of, 422
  sites of, 421
Erythropoietin, 418, 422, 635
Escherichia coli, 72, (table) 77, 112, 115, 369
Eserine, 293, 831, 870
  and accommodation, 732
ESF, 422
ESR, 416
Essential amino acids, 199, 386
  fatty acids, 51, 204
  nutrients, 120
Ethanol, 12
Ethanolamine, 54
Ethidium, 372
Ethyl alcohol, 145
Ethinyl oestradiol, 1044, 1047
Ethylene diaminetetra-acetate (EDTA), 213, 394
  glycol, 98
    poisoning, 99
Eunuch, 1031, 1114
Eustachian tube, 762, (fig.) 764
E values, 135
Evans blue, 408

Evaporated milk, 177
Exchange vessels, control of, 569
Exchangeable sodium and potassium, 676
Excitation effect of calcium, 807
  of nerve, 788, 806
Excitatory post-synaptic potential (e.p.s.p.), 817
Excretion of H^+ ion, (fig.) 665
Exercise, 316, 431, 484, (table) 504, 519, (fig.) 565, 572, 578, 616, 630, 838, 846
  anaerobic, 852
  and appetite, 173
    HGH, 850, 1086
    F.F.A., 1087
    protein requirement, 202
    respiration, 849
  arterial oxygen tension, 849
  arterio-venous difference, (table) 848
  blood, lactic acid, (fig.) 850, 854
    pressure, 848
    pyruvate, (fig.) 854
  cardiac output, (table) 848, 855
    response to, 504
  chemical changes in, 850
    blood, 854
  growth hormone, 856
  hyperæmia, 556
  isometric, 855
  lactic acid, 832
  muscle temperature, 852
  muscular, 848
  N.E.F.A., 850
  oxygen consumption, (table) 848
    debt, 853
  pulse rate, 849
  reflexes, 565
  residual volume, 848
  respiration, 849
  RQ, 850
  stroke volume, 848
  vasodilatation, 556, 562
  venous saturation, 855
  ventilation, (fig.) 849
Excitation, electrical of myocardium, 467
  and E.C.G., (fig.) 488
Exergonic reaction, 86
Exocrine gland, 223
Exogenous protein metabolism, 376
Exonuclease, 357
Exopeptidases, 291
Exophthalmos, 1002, 1084
Exophthalmos-producing factor (EPF), 1002, 1084
Expansion of lungs, 594
Expectation of life, 1112
Expiratory centre, (fig.) 605
  reserve volume, 598
Expired, air 602
  energy value, 165
  resuscitation, 609
'Exploring' electrode (ECG), 489
Extension reflexes, 913
Extensor plantar response, 968
External auditory meatus, 761
  ear, 761
  cardiac massage, 523, 609
  inhibition, (table) 986
  sphincter of bladder, 668

1134 Index

Extinction of conditional reflex, 986
Extracellular fluid, 674, (table) 675
  hydrogen-ion concentration of, 689
  volume, (table) 677
Extraction of flour, 179
Extractives, 178, 837
Extrafusal fibres, 829
Extrapyramidal system, 959
Extrasystole, 519
Exudative enteropathy, 299
Extrinsic coagulation system, 394
Eye, accommodation, 733
  autonomic fibres to, (fig.) 974
  focal length, 733
  fundus of, 738
  horizontal section, (fig.) 729
  image formation, 733
  -lashes, 730
  -lids, 730
  movements of, 751, 935
  optics of, 733
  power, 733
  protective mechanisms, 729
  refraction, 733
  structure of, 729

Facial colliculus, (fig.) 922
FAD, flavine-adenine dinucleotide, 67, 128, 187, 337
Fæces, 299, 304, 307, 308, 453
  composition of, 309
  fat, 309
  ions of, 309, (table) 678
  odour, 309
  water in, 309
Fainting, 505
  post-haemorrhage, (fig.) 566
Fallout, 220
Faraday, definition of, 138
Farnesyl pyrophosphate, 344
Farnoquinone, 197
Fasciculus cuneatus, (fig.) 904
  gracilis, (fig.) 904
  solitarius, 261, (fig.) 924
Fat, 6
  absorption, 299
    disorders of, 301
    electronmicrograph, (fig.) 300
  brown, 334, 885
  calorie value of, 161
  depots, 336
  digestion of, 292
  droplets (chylomicrons), 300, 333
  formed from glucose, 334
    from carbohydrate, 334
  in fæces, 309
    liver, 336
    old age, 1114
  melting point, 52
  metabolism, 336
    hormonal control of, 343
    in adipose tissue, (fig.) 335
  oxidation, 336
  rancid, 53
  requirements, 204
  stores, 334
Fatigue, 838, 855, 913
  of muscle, 845
Fatness and body water, (table) 674
Fatty acid, 50, 464
  oxidation, scheme of, (fig.) 337

synthetase, 341
  complex, 341
FDP, 399
Feathers, 34
Fechner's law, 707
Feedback, 3, 1083
  inhibition, 113, 119, 369
  mechanisms, 1050
  positive, 795
Feeding centres, 171
Feline leukaemia, 78
Female genital organs, (fig.) 1035
Feminizing tumours, 1034
Fertilization, 1052
Fenestra cochleæ, (fig.) 762
Fenestra vestibuli, 761
Fermentation, 90
Ferredoxin, 159
Ferrihæmochrome, 450
Ferrihæmoglobin, 447
Ferritin, 425, 438
Ferrochelatase, 451
Ferrohæmochrome, 449, 450
Fertility, 1052
Fertilization, 1054, 1101
Fetal adrenals, 1059
  asphyxia, 1069
  circulation in the lamb, (fig.) 1070
  death, 1059
  distress, 1069
  growth, 1087
  haemoglobin, 442
Fetus, 585, 1067
  amino acids in blood, 1066
  and oestrogens, 1059
  anencephalic, 1112
  asphyxia, 1069
  blood sugar, 1066
  breathing, 1069
  circulation, 1068, (fig.) 1069
  composition, 1067
  fat, 1066
  food requirements, 1064
  glycogen, 1066, 1067
  haemoglobin (HbF), 1065
  heart rate, 1069
  high altitude, 1065
  initiation of breathing, 1069
  iron, 1065
  of diabetic mother, 1112
  oxygen saturation of blood, 1065
  Po_2 of blood, 1065
  respiration, 601
  sugar, 1066
FEV$_1$, 600
Fever, 887
  dehydration, 884
FFA, 333
Fibrillation, atrial, (fig.) 522, 552
  muscle, 847
  ventricular, (fig.) 521
Fibrin, 391
  degradation products (FDP), 399
  electronmicrograph, (fig.) 393
Fibrinase, 393
Fibrinogen, 392 (table) 391, 401, 405
  molecule (fig.) 392
  stabilizing factor, 393 (table) 394
Fibrinoligase, 393
Fibrinolysis, 398
  dilute whole blood clot lysis time technique, 399

euglobulin lysis time technique, 399
  fibrin plate method, 399
Fibrinolytic system of plasma, (fig.) 398
Fibrotendinous ring, 461
Fibrous proteins, 34
Fick principle, 494, (fig.) 503
Field of vision, 737
'Fight or flight', 1019
FIGLU, 190
Fillet, medial, 921
Figure-skater, 936
Filtration angle, 731
  capillary, 406, 550
  fraction, 655
  in kidney, 647
  pressure, 648
Final common path, 912
'Finger-printing', 29, 444
First messenger, 995
Fish, 179
Fixation, 751
  reflex, 753
Flaccid paralysis, 967
Flatus, 308
Flavin-adenine dinucleotide (FAD), 67, 128, 187, 337,
  mononucleotide (FMN), 67, 128
  nucleotides, 67
Flavoproteins, 109, 128, 187
Flavour potentiators, 727
Flexion and extension reflexes compared, (table) 912
  reflexes, 909
  withdrawal reflex, 909
Flexor spasms, 907
Flicker, 979
Flocculonodular lobe, 948
Flocculus of cerebellum, (fig.) 942
Flour, 179
Flow, in blood vessels, 538
  capillaries, 545
Fluid, 4
  compartments, 673
  exchange in tissues, 681
Fluorescent antibody technique, 1081
Fluorine, 220
9α-Fluorocortisol, 1015
1-fluoro-2, 4-dinitrobenzene, 32
Fluoro-hydroxymethyltestosterone, 1032
Flutter, atrial, (fig.) 521
FMN, 67, 128
Folacin, 190
Folic acid, 115, 190, 422
  antagonists, 190
  deficiency, 423
Follicle-stimulating hormone (FSH), 1032, (fig.) 1048, 1049
Follicular hyperkeratosis, 184
  phase, 1038, 1048
'Follow-up' servo, 916
Fontanelle, anterior, 896
Food, 4
  composition of, 176
  deprivation of, 673
  recommended daily allowances, (table) 203
Foot, venous pressure, 552
Foramen ovale, 1068, (fig.) 1069

Force-velocity relation in myocardial contraction, 476
  in tachycardia, 518
Forced expiratory volume (FEV), (fig) 601
  grasping, 968
Forebrain, 892
'Foreign body' giant cells, (fig.) 440
Formaldehyde, 27
Formant, 760
Formatio reticularis, (fig.) 924
Formed elements of blood, 415
Formiminoglutamic acid (FIGLU), 190
Formol titration, 27
N-Formyltetrahydrofolic acid, 349
Fornix, 951
Four letter language, 1106
Fourth ventricle, 924
Fovea, 738
Fractionation, 912
F.R.C., 595
Free energy, 83
  change, 134
  fatty acids (FFA), 333, 850
    and exercise, (fig.) 1087
Frequency analysis, 765
  distortion, 771
  minimal difference, 771
Friction in joints, 212
Frog's heart, 512
  sinus venosus, 512, (fig.) 513, (fig.) 514
Frontal association areas, 969
  lobes, 969
Fructose, 42, 44, 311
  conversion to glucose, 311
  diphosphatase, 159
Fructose 1,6-diphosphatase, 146, 153
Fructose 1, 6-diphosphate, 142
Fructose-1-phosphate, 311
Fruits, composition of, (table) 180
FSH (follicle-stimulating hormone), 1032, 1048, 1049
Fumarase, 151
Fumarate hydratase (fumarase), 107, 150, 151
Functional residual capacity (FRC), 595, 598
Fundus of eye, 738
Furan, 42
Furanose structure, 42

G cell, 251
ΔG, free energy change, 85
$\Delta G°$, standard free energy change, 85
GABA, 822, 898
Gait and cerebellum, 947
Galactokinase, 312
Galactosaemia, 120, 312
Galactose, 44, 311
Galactose 1-phosphate, 312
β-Galactosidase, 370
Galactosides, 43
Galactosuria, 325
Gall-bladder, 273, (fig.) 277, 865
  absorptive capacity, 280
  contraction, 279
  effect of secretin and CCK-PZ, 280
  function, tests of, 280

movements of, 279
  pressure, 279
  vagotomy, 279
  X-ray photograph, (figs) 283, 284
Gall-stone, 57, 278
Gamma discharge, 830, 915, 927, 948, 960
  efferents, 830, 915, 927, 948, 960
Gamete, 1100
Ganglion, autonomic, 817
  cells in retina, 748
  spinal, 904
GAR, 349
Gas, partial pressure, 612
  solubility coefficient, 613
  tension, 612
  transfer, alveolo-capillary, 615
Gases in blood, 613
  in water, solution of, 612
Gastrectomy, 325
Gastric absorption, 293, 306
  acid, 241
    production, 245
    secretion, diagram of mechanisms, (fig.) 255
  blood group substances, 245
  contents, osmolarity of, 241
  distension, 252
  emptying, 238, 241
    duodenum, 241
    emotion, 242
    gravity, 241
    products of digestion, 242
    rate, 241
  enzymes, 249
  fistula, 238, 256
  function, 233
    tests of, 259
  gland, 243, (fig.) 245
  inhibitors, 240, 256
  intrinsic factor, 245, 423
  juice, 246
    chemical excitation, 249
    enzymes, 249
    histamine and, 256
    ions of, 247
    pH of, 249
    secretion of, 249
    volume, (table) 678
  motility, 233, 235, 238
  mucosa, 242, 243
    blood flow through, 259
    cytochrome oxidase, (fig.) 247
    electronmicrograph, (fig.) 247
    histology, (fig.) 244
  mucus, 243
    visible, 244
  muscle, 233, 237
  nerves, 233
  outline, (fig.) 234
  peristalsis, 235, 238
  phase of gastric secretion, 251
  pits, 242
  potential, 247
  pouch, (fig.) 248
  pressure, 235
  pyloric sphincter, 238
  secretion, 234, 242, 244
    and histamine, 256
    control of, 249
    gastric phase, 251
    inhibition of, 256
    insulin test of, 260

intestinal phase, 252
  stress, 255
  nervous excitation, 250, 254
Gastrin, 250, 270, 285, 293, 303, 319
  actions of, 253
  formula, 252, 270
  secreting D cell, (fig.) 265
Gastrins I and II, 252
Gastrocolic reflex, 305
Gastroenterostomy, 239
Gastro-oesophageal junction, 238
  sphincter, 231
Gastro-intestinal mucosa renewal, 242, 287, 302
  tract transit time, 309
Geiger counter, 122
Gelatin, 178
Generator potential, 811
Gene, 72, 120, 355, 369, 1106
  activation, 996
  unmasking, 996
Genetic code, 354, 363
  defect, 120
  information, 71
Geniculate bodies, 954
  cells, 970
Genital organs, female, (fig.) 1035
  male (fig.) 1029
Genotype, (fig.) 1107
Germ cell, 1100
Germ-free animals, 307, 1021
Germinal cells, 1029
  epithelium, 1029, 1036
Gerstmann's syndrome, 976
Gestation period in singletons and twins, 1072
GFR, 646, 649, 652, 653, 664
Giddiness, 935
Gigantism, 1088
Girl puberty changes in, 1038
GIT, 320
Glands of Brunner, 290, 293
Glands, endocrine, 994
Glaucoma, 731
Gliadins, 180
Glial cells, 688, 892
Globin, 444
Globose nucleus, 945
Globulin, 401, 405
α_1-Globulin, 405
α_2-Globulin, 405
β-Blobulin, 405
γ-Globulin, 405
Globus pallidus, 960
Glomerular arterioles, 656
  capillaries, 649
  capsule, 641
  filtrate, 648
  filtration rate, GFR, 646, 649, 652, 653, 664
    rates in new-born, 667
  fluid, method of obtaining (fig.) 647
  function, 647
Glomerular arterioles, membrane, 648
  pressure, 648
Glomerulus, 641
  afferent arteriole, 642, (fig.) 644, (fig.) 646, 648
  electron micrograph, (fig.) 645
Glomus cells, 630

Glossopharyngeal nerve, 515, 725
Glottis, 756
Glu (glutamic acid), 25
Glucagon, (fig.) 270, 279, 293, 319, 323, 343
  -secreting cell, (fig.) 265
Glucocorticoids, 1007, 1009
Glucogenic amino acids, 381
Glucokinase, 141, 154
Gluconeogenesis, 315, 323, 327, 339
  pathways of, (fig.) 315
Gluconic acid, 43, 159
Glucosamine, 44
Glucose, 41, 42, 464, 473
  absorption, 295, 296
  active transport, 297
  and brain, 316, 898
  blood, in pancreatic insufficiency, (fig.) 274
  -fatty acid cycle, 340
  in blood, in man 327
    in cow, 331
  oxidase, 137, 159, 327
  oxidation, 159
  6-phosphatase, 146, 315
  1-phosphate, 43, 312
  6-phosphate, 43
  6-phosphate dehydrogenase, 117, 152, 417
  phosphate isomerase, 141
  synthesis, 146
  T_m, 652
  tolerance test, 318, 327
Glucosides, 43
Glucostatic theory of hunger, 172
Glucuronic acid, 43
Glucuronides, 44
Glutamate as transmitter, 822
Glutamate dehydrogenase, 378, 665
Glutamate-oxaloacetate transaminase (GOT), 117, 379, 473
Glutamic acid (Glu), 25, 665
Glutaminase, 382, 665
Glutamine, 25, 382, 665
Glutathione, 28, 389, 417
Glutathione-insulin transhydrogenase (GIT), 320
Glutelins, 180
Gluten, 180, 301
Gly (glycine), 23
Glyceraldehyde, 40
Glyceraldehyde 3-phosphate, 142, 336, 342
Glyceraldehyde 3-phosphate dehydrogenase, 142
Glycerol, 50, 342
α-Glycerol phosphate, 335, 342
Glycerose, 40
Glycinamide ribonucleotide (GAR), 349
Glycine (Gly), 23, 208, 273, 384
Glycocalyx, 290, 297
Glycocholic acid, 277
Glycocyamine, 387
Glycogen, 47, 315, 467, 473
  breakdown, 157
  in liver, 312
    muscle, 312
  molecule, 155
  storage disease, 315
  synthesis, 154, 156, 318
  to blood glucose, (fig.) 314
  UDP glucosyl transferase, 155

Glycolipids, 56
Glycolysis, 139, 140, 144, 839
Glycolytic pathway, 315
Glycoproteins, 47, 209
Glycoside, 43
Glycosuria, 325
  emotional, 325
GMP, 65, 66, 349
Gnotobiotics, 307
G/N ratio, 327
Goblet cell, 289
Goitre, 1002, 1005
  endemic, 1005
  exophthalmic, 1002
  in children, (table) 1005
Goitrogens, 1004
Golgi body, (fig.) 75
  cells, 945
  endings, 814
  region, 266
  tendon organ, 829, 917, 927
Gomori stain, 1091
Gonadectomy, 1049
Gonadotrophic hormones, 1049, 1085
  action on pituitary, 1049
  in urine, 1061
Gonadotrophin, 1049
  of placenta, 1059
Gooseflesh, 886
GOT, 117, 379
Gout, 116, 351, 360
Graafian follicle, (fig.) 1036
Gracile nucleus, 921
Gradient calorimeter, 162
  centrifugation, (fig.) 74
Gramophone audiometer, 773
Granular cells of cerebellum, 944
  leucocytes, 430
Granulocyte, 430
Granulosa cells, 1049
Grasp reflex, 968
Grey matter of spinal cord, 902
Grey ramus communicans, 861
Group O, 411
Growth, 2, 1087, 1111
  and diet, 1111
  hormone, 173, 322, 324, 343, 378, 1063, 1088
    and diabetes, 324
    exercise, (fig.) 1087
    insulin, 324
  in pregnancy, 1063
  plasma, (fig.) 1087
  rate of, 1111
GTP, 65, 66, 349
Guanethidine, 238
Guanido-acetic acid (glycocyamine), 387
Guanine, 64
  deaminase, 360
Guanosine, 65
  monophosphate (GMP), 65
  5'-triphosphate (GTP), 65
Gubernaculum testis, 1026
Gustation, 724, 975
Gustatory sweating, 727
Gyrus, cerebral, 957
  cinguli, 954

Habituation, 934, 936, 990
Hæm, 444, 449
  biosynthesis of, 450

Hæmatin, 449
Hæmatocrit, 408, 415, 416
  value of body, 408
Hæmatoidin, 439
'Hæmic' murmurs, 539
Hæmin, 449
Hæmochorial placenta, 1055
Hæmochromogen, 450
Hæmodialysis, 666
Hæmoglobin, 37, 113, 369, 377, 415, 444
  A, 444
  A_2, 444
  absorption spectrum, (plate) 25.1
  and oxygen content, 620
    dissociation curve, 620
  at birth, 441
  at high altitude, (fig.) 422
  as buffer, 624
  breakdown 452
  chains, 444, (fig.) 446
  dissociation curve, 620
    effect of CO_2 on, 620
    effect of temperature on, (fig.) 621
  estimation, 448
  fetal, 442
  imidazole groups, 625
  in lungs and tissues, (fig.) 625
  level, infancy, 441
  mean corpuscular concentration (MCHC), 448
  molecular weight, 620
  molecule, (fig.) 445
  quaternary structure, 447
  F, 444
  S, 33
  sickle cell, 447
  sigmoid relationship, 620
  to oxyhæmoglobin, speed of, 622
  variants, 447
Hæmoglobinopathy, 427
Hæmolysis, 410, 418
Hæmolytic anæmia, 421
  disease, 421
    of newborn, 412, 455
  jaundice, 454
Hæmophilia, 396, 1109
Hæmopoiesis, 418
Hæmorrhage, 484
  effects of, 409
  into pons, 883
Hæmosiderin, 425, 438
Hæmostasis, 401
Hæmostatic interactions, (fig.) 402
Hageman factor, (table) 394
Hair, 31, 713, 716
  cells, (fig.) 765, 931
  growth of, 713
Haldane effect, 635
  gas analysis apparatus, (fig.) 602
Haldane's blood gas apparatus, (fig.) 618
Half-life, of isotopes, 121
Hamburger phenomenon, 624
Hammer, 761
Haploid number, 1100
Haptoglobins, 406
Harvey, William, 456, (fig.) 457, 525
Hashimoto's disease (auto-immunizing thyroiditis), 1001

Hassall's corpuscles, 1019
Haustration of colon, 304
Haversian systems, 210
H band, 825
HbF, 1065
Hb phenotypes, 406
HCG, 1049, 1051, (fig.) 1062
Hearing, 755
  acuity, 770
  and cortex, 768
  loss, 771
  masking, 770, 773
  measurements, 770
  method of measuring intelligibil-
      ity, 771
  Rinne's and Weber's tests, 773
  threshold, 770
Heart, 4, 456
  action potentials, (fig.) 467,
      (fig.) 469
  afferent nerves, 512
  anatomy, 458
  annulus fibrosus, 461
  apex beat, 502
  block, 469
    complete, 522, (fig.) 523
    incomplete, 522, (fig.) 523
  bruits, 503
  capillaries, 519
  cardiac impulse, 502
    output in man, 503
    reflexes, 508
  cardio-aortic nerve, 508
  carotic sinus nerve, 508
  carotid body, 515, 630, 632
  catheterization, 493
  cells, action potential, 469
  chambers of, 458, 460, (fig.) 498–
      499
  conduction through, 467
  coronary arterial tree, (fig.) 497
    circulation, 497, 519
    vessels, 497
  descent of base, 493, 502
  dimensions of orifices, (table) 460
  effect of adrenaline, (fig.) 510, 511,
      (fig.) 513
    hypoxia, 519
    potassium, (fig.) 489, 512
    vagus on, 508, (fig.) 509
  electrical excitation, 467
  end-systolic and end-diastolic
      volume, 504
  failure, 523
  fatty acids, 464
  fibrotendinous ring, 461
  filling, 484
  flutter and fibrillation, 521, 522
  histology of, 465, (fig.) 466
  human, isolated, perfused, 486
  intercalated discs, 471
  law of, 474, 475, 481
  -lung preparation, (fig.) 481
  massage, 523, 609
  mechanical efficiency, 481
  metabolism, 317, 464, 473
  movements, 503
  murmur, 503
  muscle, 336, 461
    cells, 466
  nervous regulation of, 508
  non-contractile components in,
      482

output, 502, 506, (fig.) 507, 521
pace-maker, 467
pain, 519, 863
  nerves, 512
papillary muscle, (fig.) 461, 471,
    472, 473, (fig.) 474, (fig.)
    478, 479
position in thorax, 464
power, 481
rate, 516, 517
  and emotion, 517
    respiration, (fig.) 849
    temperature, 517
  at various ages, (fig.) 517
  depressor nerve, 513
  in exercise, 517, (figs.) 854, 849,
      854, 856
  vagus on, 508
reflexes, 508
refractory period, 469
residual volume, 504
return of blood to, 504
sequence of events in the cardiac
    cycle, 490
shadow of normal, (fig.) 463
shape of, 464
size of, 458, 462
  in chest X-ray, 464
sounds, (fig.) 492, 496, 500, 501
speed of conduction, (fig.) 470
structure of, 458
sympathetic stimulation, (fig.) 510
thickness of, (table) 460
vagus nerve, 508
valves, 461
venous return, 481, 504
volume of, 461
weight of, (table) 460
work of, 481
Heat cramps, 885
  exhaustion, 885
  hyperpyrexia, 885
Heat loss, 875
  of metabolism, 317, 875
    shortening, muscle, 931
  production, 166, 874
    of adult mammals, 166
    in muscle, 852, 874
    in nerve, 805
  regulating mechanisms, 880
Heatstroke, 885
  units, 1117
Heavy hydrogen, 122
  nitrogen, 122
  water, 122
Heidenhain pouch, (fig.) 248
Height adult, 1116
  maximum, 1116
Helicotrema, (fig.) 762
α-Helix, 34
Helmholtz, theory of action of
    cochlea, 763
Hemianæsthesia, 956
Hemianopia, (fig.) 972
Hemiballismus, 960
Hemicholinium-3, 835
Hemi-decortication, 956
Hemiplegia, 916, 960, 967, 969
Hemisection of spinal cord, 907
Hemispheres, cerebral, (fig.) 958
Henderson-Hasselbalch equation,
    17, 691

Henle's loop, 642, 649
Heparin, 48, 394, 397, 1020
Hepatectomy, 120
Hepatic artery, 582, 584
  ligature of, 583
  duct, 273, (fig.) 277
  jaundice, 454
  veins, 581
Heredity, 1098, 1108
Hering-Breuer reflex, 601, 606
Herpes virus, 369
  zoster, 563
Herring bodies, 1091
Heterozygous genes, 1106
Hexadimethrene bromide, 397
Hexamethonium, 250
Hexonic acid, 43
Hexosamine, 44
Hexosediphosphatase, 146
Hexose, 40
Hexose monophosphate shunt, 140
Hexose to pentose conversion, 152
Hexuronic acid, 43
HGH (human growth hormone),
    1085, 1088
  and exercise, 856
5-HIAA, 1021
Hiatus hernia, 231
Hibernating animal, 7
Hiccup, 608
High altitude, 580, 621, 628, 630,
    634, 635, 638
  and red cells, (fig.) 422
  effect of, (fig.) 633
  oxygen pressures at, (fig.) 633, 634
High-energy compounds, 76, 87
Hind brain, 892
  sleep, 981
Hippocampal formation, (table) 958
  gyrus, (fig.) 958
Hippocampus, 975, 991, 1042
Hippuric acid, 384
Hirudin, 397
His, bundle of, 461
His (histidine), 26, 190, 1020
Histalog, 260
Histaminase, 1021
Histamine, 256, 259, 383, 569, 598,
    1020
  and gastric secretion, 256
Histamine forming capacity (HFC),
    (fig.) 256, 1021
    of gastric mucosa, (fig.) 256
  and bronchial muscle, 598
  pregnancy, 1021
  gastric juice and, 256
Histidine (His), 26, 190, 1020
  operon, 372
Histiocytes, 435
Histone, 31, 68, 357, 996
Histotoxic hypoxia, 636
Homeostasis, 2, 871
Homeothermic animals, (fig.) 873,
    874
Homocysteine, 388
Homogenate, 119
Homogentisate oxygenase, 385
Homogentisic acid, 385
Homograft reaction, 433
Homologous chromosomes, 1100
Homozygous genes, 1106
Honey, 181

1138   Index

Hopping reaction, 939
Hormonal control of menstrual cycle, 1047
  parturition, 1071
  pregnancy, 1056
Hormone, 5
  definition, 994
  lactogenic, 1088
  local, 1020
  method of action, 994
  of neurohyphophysis, 1090
  production and excretion in pregnancy, 1059
  sites of action, 994
  synthetic, 1047
Horner's syndrome, 732
Hot flushes, 1042
Houssay dog, 324
Howell-Jolly bodies, 439
HPG (human pituitary gonadotrophin), 1049
HPL (human placental lactogen) 1075
H-region, 825
H-substance, 411
5-HT, serotonin, 116, 1021
Human body, composition of, 7
  brain, coronal section, (fig.) 959
  chorionic gonadotrophin (HCG), 1051
    growth hormone HGH, 1063, 1085
  milk, 176
  pituitary gonadotrophin (HPG), 1049
  placental lactogen (HPL), 1075
Humidity, 889
Humoral system, 5
  theory, 866
Hunger, 171, 969
Hunting reaction, 557
Hyaline cartilage, 209
Hyaluronic acid, 48, 212
Hyaluronidase, 48, 1030, 1095
Hydatidiform mole, 1063
Hydramnios, 1072
Hydrocephalus, 687
Hydrochloric acid and parietal cell, (fig.) 248
Hydrocortisone, 664, 1009
Hydrogen bond, 11, 68
  electrode, 496
Hydrogen-ion concentration, 14
  of blood, 695
    extracellular fluid, 689
    plasma, 695
  pump, 247
  renal regulation of, 696
Hydrogen-ions, secretion of, 245, (fig.) 248
Hydrogen peroxide, 418
Hydrogenation of fats, 178
Hydrolases, 106
Hydroperoxidases, 137
Hydrotropic power of bile salts, 279
β-Hydroxy acyl dehydrogenase, 337
β-Hydroxybutyric acid, 338
25-Hydroxycholecalciferol, 195
Hydroxycobalamin, 191
11-Hydroxycorticosteroid, 1013
17-Hydroxycortiscosteroids, 1012
5-Hydroxyindoleacetic acid, (5-HIAA), 1021

Hydroxymandelic acid, 1017
β - Hydroxy - β - methylglutaryl coenzyme A, 338, 344
Hydroxymethyl tetrahydrofolic acid, 384
17α-Hydroxyprogesterone, 1047
Hydroxyproline (Hyp), 26, 34, 208
5-Hydroxytryptamine (5HT), 116, 284, 287, 383, 719, 832, 882, 1021
5-Hydroxytryptophan, 1021
Hyp (hydroxyproline), 26, 34, 208
Hyperæmia after exercise, 556
  reactive, 556
Hyperæsthesia, 907
Hyperbaric oxygen, 609, 638
Hypercalcæmia, 216
Hypercapnia, 484, 630, 632, 636, 637
Hyperchlorhydria, 259
Hyperglycæmia, 1018
Hyperglycæmic-glycogenolytic factor, 323
Hyperinsulinism, 322
Hyperkalæmia and electrocardiogram, (fig.) 489
Hyperkeratosis, follicular, 184
Hyperkinetic circulation, 518
Hyperlipæmia, 333
Hypermetropia, 733
Hypernatraemia, 680
Hyperparathyroidism, 216
  and bones, (fig.) 215
Hyperphagia, 172
Hyperpnœa, 630
Hyperpolarization, 808, 914
Hyperpyrexia, 879, 883
Hypersomnia, 954, 980
Hypertension, 661
  pulmonary, 633
Hypertensive crisis, 631
Hyperthyroidism, 325, 1001, 1084
  effect of iodine, 1003
  goitrogens and, 1004
  muscular tremor before and after thyroidectomy, (fig.) 1004
Hypertonic solution, 21
Hypertonicity, 915, 967
Hyperventilation, 518
  in pregnancy, 1056
Hypervolæmia, 664
Hypocalcæmia, 216
Hypocapnia, 630
Hypogammaglobulinæmia, 405
Hypogastric nerves, 304, (fig.) 668
  plexus, 304
Hypoglossal triangle, (fig.) 922
Hypoglycæmia, 260, 321, 1018
  and electroencephalogram, 978
  hunger, 171
Hypokalæmia, 681
  and electrocardiogram, (fig.) 489
Hypokinesia, 954
Hyponatraemia, 681
Hypoparathyroidism, 216
Hypophysectomy, 1001, 1034, 1072, 1084, (fig.) 1086, 1088
  and carbohydrate metabolism, 324
Hypophysial arteries, (fig.) 1082
  portal vessels, 1081, (fig.) 1082
Hypophysis cerebri, 1080
Hypopituitarism, 1090

Hypoproteinæmia, 299, 406
Hypothalamo
  – pituitary – adrenocortical function, 1084
Hypothalamo-hypophysial tract, 1091
Hypothalamus, 171, 560, 561, 563, 566, 661, 880, 951, 1049, 1081, 1084, 1090
  and diabetes insipidus, 1091
    gastric secretion, 255
    milk ejection, (fig.) 1076
    obesity, 171
    ovulation, 1050
    temperature regulation, 882
    thirst, 679
  body temperature and, 880
  effects of drugs, 953
  nuclei, 951, (fig.) 952
  sham rage and, 953
  stimulation of, 953
Hypothermia, 93, 875, 879
  artificial, 880
Hypothyroidism, 1001
Hypotonia, 946
Hypotonic solution, 21
Hypovolæmia, 663
Hypoxanthine, 64, 116, 360
  -guanine phosphoribosyl-transferase, 350
Hypoxia, 484, 515, (fig.) 516, 519, (fig.) 582, 629, 632, 634
  and lung vessels, 559
    pulmonary artery pressure, 580
  cerebral, 633
  types of, 634
Hypoxic hypoxia, 635

I band, 825
Icterus, 197, 454, 719
ICSH, 1085
'Ideal' alveolar air, 613
Idioventricular rhythm, (fig.) 523
IgA, 308
IgG, 37
Ile (isoleucine), 24
Ileectomy, (fig.) 281
Ileocolic junction, 285
  sphincter, 303
Ileostomy, 306
Ileum, 277, 285, (fig.) 289, (fig.) 294
Image, false, (fig.) 752
Imbecillitas (oligophrenia) phenylpyruvica, 120, 385, 1108
Imidazole compounds, 1004
  groups in blood, 625
Immune antibodies, 410
Immunity and gastrointestinal tract, 308
  passive, 299, 405
Immunoassay of insulin, 320
Immunoelectrophoresis, (fig.) 404
Immunoglobulin (IgG), 413
Immunoglobulins, 37, 308
Immunological competence, 1019
  reactions, 1019
Immunological pregnancy test, 1063
Immunosuppressive agents, 351
IMP (inosine 5-monophosphate), 66, 349
IMP (integrating motor pneumotachograph), 165

Impotence, 1031
Inborn errors of metabolism, 120, 384
Incisura angularis, (fig.) 235
Incontinence 670
Incus, 761
Indican, 386
Indicator dilution, 503, 504
  dyes, 138, 503
Indifferent electrode (ECG), 489
Indirect calorimetry, 162
Indispensable amino acids, 199, 386
Indole, 308, 309, 386
Indoxyl, 386
Induction of enzymes, 369
Infancy, basal metabolism in, 168
  changes in blood during, (fig.) 442
  oxygen poisoning, 638
  renal function, 667
  respiratory rate, (table) 595
Infantile paralysis, (table) 77, 906
Inferior colliculus, (fig.) 925
  peduncles of cerebellum, 948
Infertility, 1052
Inflow tract of cardiac ventricles, 458
Influenza (table), 77
Infra-red rays, 589
Infundibulum, 1042
Inheritance, 1098, 1108
  of blood groups, 1107
  sex-linked, 1108
Inhibition, 97, 113, 819
  competitive, 97
  external, 986
  internal, 986
  of knee jerk, (fig.) 914
  presynaptic, 820
Initial heat, 805
Initiation codon, 365
Inhibition of spinal reflex, 913
Inhibitory post-synaptic potentials, (i.p.s.p.s.), 818
Injury, 837
  catabolic response to, 204
  potential of nerve, 785
Inner ear, structure of, (fig.) 762
Innervation of bladder, (fig.) 668
  reciprocal, 821, 917
Inorganic phosphate, 84
Inosine, 5′-monophosphate (IMP), 349
Inosonic acid, 349
Inositol, 45, 54
  hexaphosphate, 45, 179
Inotropic drugs, 484
  response, positive, 511
    negative, 509
Inotropy, 484
Insemination, artificial, 1052
Insensible perspiration, 876
Inspiratory capacity, 598
  centre, (fig.) 605
  reserve volume, 598
Inspired air, 602
Instrumental or type II conditioning, 988
Insula, (fig.) 959
Insulin, 33, 260, 319, 975, 1066
  and gastric secretion, 260
    zinc, 320
  antagonists, 322, 330
  assay, 320

blood-sugar and, 321, (fig.) 322
hyperinsulinism, 322
hypoglycæmia, 321
in plasma, (fig.) 318, 320
mode of action, 322
molecule, species differences, 320
plasma, 320
protamine zinc, 320
radio-immunoassay of, 320
secreting cell, (fig.) 265
secretion, 321, 325
structure, (fig.) 319
test of gastric secretion, 260
Insulinase, 321
Integrating motor pneumotacho-graph (IMP), 165
Intelligibility of speech, 771
Intention tremor, 948
Interalveolar cell-islets, 263
Interatrial septum, 510
Intercalated discs, 466, 471
Intercostal muscles, 593
Interlobular arteries, 643
Intermediate ganglia, 861
Internal arcuate fibres, 921
  capsule, 962, (fig.) 963, 964
  cardiac massage, 523
  ear, 762
Internal environment, 2
  hydrocephalus, 687
  inhibition, 986
Interneurones, 905
Interstitial cells of Leydig, (fig.) 1027, 1032
  of ovary, (fig.) 1027, 1032
  -cell-stimulating hormone (ICSH), 1085
  fluid, 674
    composition, (table) 675
Interventricular septum, 469, 471, (fig.) 497
Intestinal flora, 307
  lymphocyte, 308
Intestinal mucosa, renewal of, 287
  obstruction, 308
  secretion, control of, 292
Intestinal tract, (fig.) 234
  villi, 285, (fig.) 286
Intestine, absorption, 294, 295
  area, 285
  bacteria, 307
  blind loop, 307
  brush border, (fig.) 288
  cells, replacement, 287
  cul-de-sac, 307
  digestion, 292
  electrical rhythm, 283
  enzymes, 290
  epithelium, (fig.) 287
  flora, 307
  gas, 308
  glands, 287, 290
  histamine, 1020
  histology, (fig.) 287
  juice, 290
    composition of, 292
  lacteal, (fig.) 287
  large, 302
    absorption in, 300
    barium enema, (fig.) 301, (fig.) 302
  capacity, 303

histology, (fig.) 303
  movements of, 303
  rate of passage of meal, 285, 303
  length, 283
  lymph follicle, (fig.) 289
  microvilli, 285
  motility, 233
  movements, neural control, 284
  mucosa and 5HT, 1021
  orientation, 285
  pain, 863
  phase of gastric secretion, 252
  plexuses, 234
  protein loss, 299
  resection, 295
  secretion control of, 293
  sites of absorption, (fig.) 294
  tract plan, (fig.) 234
  transit time through, 309
  vagotomy, 296
  villi, 285, (fig.) 286
  water absorption, 7
    secreted into, (table) 678
Intracardiac electrocardiogram, (fig.) 495, 496, 519
  phonocardiogram, 500
  pressures in man, (table) 494, (fig.) 495, (fig.) 496
Intracellular fluid, 674, 675
  potentials, (fig.) 787
Intracranial pressure, 687
Intrafusal muscle fibres, 829
Intragastric pressure, 236
  volume, 236
Intramyocardial pressure, 518
  tension, 483
Intra-ocular pressure, 731, 894
Intraœsophageal pressure, (fig.) 597
Intrapleural pressure, 594, 595, (fig.) 596
Intrarenal pressure, 648
Intrasplenic pressure, 585
Intrathoracic pressure, 484, 493, 501, 505
Intrauterine pressure, 1073
Intravascular fluid, 675
Intravenous pyelogram, (fig.) 642
Intraventricular pressure, 483
Intravesical pressure, (fig.) 669
Intrinsic blood thromboplastin, (table) 394
  coagulation system, 394
  factor, 245, 423
Inulin, 47, 647, 653
Invertase, 45
Inward current, 791
Iodide, 227
Iodine, 997
  and endemic goitre, 1005
    output of thyroxine, (fig.) 1004
  compounds, radio-opaque, (fig.) 283, (fig.) 284
  daily intake, 1063
  excretion after dose of ^{131}I, (fig.) 1000
  in hyperthyroidism, (fig.) 1064
    plasma after test dose of ^{131}I, (fig.) 1000
  saliva, 227
  thyroid gland, 997
  isotope, (fig.) 1000
  metabolism in thyroid, (fig.) 999

Iodine, protein bound, 1000
  radioactive and thyroid, (fig.) 1000, 1001
Iodized salt, 1006
Iodoacetate and muscle, 838
Iodotyrosine, 997, (fig.) 998
Ion fluxes and cardiac transmembrane potentials, 468
Ionic movements in nerve, 804
Ionization, 13
Ionizing radiation, 354
Ions in axon, (fig.) 793, (table) 793
    cytoplasm of nerve, (table) 793
    muscle cell, (table) 793
I.p.s.p., 818
Iproniazid, 116, 1017
Irins, 1023
Iris, 732
  functions of, 735
  innervation of, 732
  sphincter, 732
Iron, 444, 1039
  absorption, (table), 425, (fig.) 426, (fig.) 427
  catalase, 426
  daily allowance, (table), 203
  deficiency anæmia, 424, 426
  in body, 424
    foods, (fig.) 427
    lactation, 424
    new-born, 426
    plasma, 424
    pregnancy, 424
    sweat, 884
  intake, 425
  loss, 424
  menstrual loss, 424, 1039
  metabolism, 424
  requirements, 424
  reserves, 424
  serum, 424
  transport, 424
Iron-binding capacity of plasma, 425
Irradiation, 909, 989
  in reflexes, 909
Irreversible shock, 575
Ischaemia, myocardial, 519
Ishihara's plates, 746
Islands of Langerhans, 263, 273, 319
Islet cells of pancreas, 263, 273, 319
Islets, interalveolar cell, 263, 273, 319
Isocitrate dehydrogenase, 150
Isocitric acid, 150
Isocortex, 957
Isodesmosine, 26
Isoelectric point, 36
Isoenzymes, 116
Isoleucine (Ile), 24
Isomerase reactions, (fig.) 110
Isomerases, 109
Isometric contraction, (fig.) 474, (fig.) 476, 840
  exercise, 855
  lever, (fig.) 841
  phase, 491
Isoniazid (isonicotinic hydrazide), 189
Isopentenyl pyrophosphate, 344
Isopotential maps, 471, 486

Isoprenaline, 868
Isosthenuria, 657
Isotonic contraction, (fig.) 474, (fig.) 476, 493
Isotonic lever, (fig.) 474
  solutions, 21
  twitch, (fig.) 841
Isotopes, 120
  half-life of, 121
  radioactive, 121
Isovolumetric ventricular relaxation (fig.) 479
  contraction, (fig.) 479, 491
Isozymes, 116
Isthmus, 958
Itching, 719
  and jaundice, 719

J (joule), 161
Jacksonian epilepsy, 966
Jargon aphasia, 975
Jaundice, 197, 454, 719
  hæmolytic, 454
  hepatic, 454
  itching in, 719
  obstructive, 197, 454
  pre-hepatic, 454
Jejunum, 287
Joint proprioceptors, 814
  sense, 831
Joints, 212
Joule (J), 161
Jugular phlebogram, 519
  pulse, 519, 552
  vein, 551
  venous pressure, (fig.) 492
Junctional tissue, cardiac, 461
Juxta-glomerular cells, 422, 643, (fig.) 646, 659, (fig.) 660
Juxta-medullary glomeruli, 643

K, significance of, 81
K_a, 16
Kallidin I, 1023
Kallikrein, 225
Karyotype, (fig.) 1099
Kcal (definition), 161
Keratin, 31, 34, 710
Keratinocytes, 710
Keratomalacia, 184
Kernicterus, 455
Keto-acidosis, 330
Keto acids, 381
β-Ketoacyl-ACP-reductase, 341
β-Ketobutyric acid, 338
Ketogenic amino acids, 382
Ketone bodies, 330, 338
Ketonuria, 338
Ketosis, 171, 330, 339
  diabetic, 330, 700
  in lactation, 340
Kidney, 641
  adrenaline, 656
  aglomerular, 653
  ammonia formation and, 665, 666
  anatomy, 641
  and acid-base balance, 665
  blood pressure, 644
  glutamine, 666
  œdema, 664
  pH of blood, 665
  water sodium loss, 663

blood flow, 644, 655
calyces, 641
clearance, 653
corpuscle, 642
cortex, 641
counter-current theory, 650
emotion and, 664
failure, 666
filtration, 646
glomerular function, 646
gluconeogenesis, 316
H^+ excretion, 665
hilum, 641
in newborn, 667
pregnancy, 1055
juxtaglomerular cells, 643
juxtamedullary glomeruli, 643
maximum reabsorptive capacity, 652
    secretory capacity, 653
medulla, 641
nerves, 646
œdema, 664
plasma flow, 655
renal function in the newborn, 667
response to ingestion of water and salt, 662
structure, 641
tubular function, 649
vessels, 643
water diuresis, 661
weight, 641
Kinase, 348
Kinase kinase, 318
Kinetic energy, 89
Kinins, 568, 1023
Kilocalorie, (kcal) definition, 161
Kinocilium, 931
'Kiss of life', 609
Knee jerk, (fig.) 914, 947
Korotkov sounds, 533
Krebs tricarboxylic acid cycle, 140, 148
Kulchistsky cells, (fig.) 287
Kupffer cells, 436, 437, 582, 584
Kwashiorkor, 204
Kynurenine, 386
Kyphosis, 1112

Labelling, isotopic, 120
Labile factor, 394
Labour, 1071
  oxygen requirement, 1074
  pace-maker, 1073
  pain, 1073
  stages of, 1073
Labyrinthectomy, 932
Labyrinthine reflexes, 937
Labyrinths, 929, (fig.) 930
  bony and membranous, (fig.) 930
  caloric tests, 935
  connexions of vestibular fibres, (fig.) 929
  removal of, 932
Lacrimal glands, 730
  sac, 730
Lacrimation, 730
a-Lactalbumin, 176
Lactase, 297
Lactate, 82, 316, 464, 473
  dehydrogenase, 116, 117, 144
  in blood, 314

Lactation, 5, 1074, 1077
  and emotion, 1075
  dietary requirements, 1075, 1077
  output of milk, (fig.) 1075
Lacteal, 285, 301, 684
  of intestine, 683
Lactic acid, 314
  and fatigue, 838
  in blood, 314
    muscles, 838
Lactic dehydrogenase, 37
Lactobacillus acidophilus (Döder-
    leins's bacillus), 1036
Lactogenic hormone, 1074, 1088
β-Lactoglobulin, 176
Lactose, 45, 369
Lævulose, 44
'Laguno', hair, 713
Laminar flow of blood, 538
Langhans' cell layer, (fig.) 1058
Laplace, law of, 235, 551, 590
Laplace's equation, 235, 551, 590
  absorption in, 306
  bacteria in, 307
  capacity, 303
  movements, 303
  nerve supply, 304
  secretion in, 306
  section, (fig.) 303
  size, 303
Large lymphocyte, 432
Laryngeal muscle, 756
  tone, 757
    male, 760
    female, 761
Laryngectomy, 758
Laryngismus stridulus, 217
Laryngoscope, 756
Larynx, 756, 1031
  cross-section, (fig.) 756
  nerves, 758
  side view, (fig.) 756
Latent heat of evaporation, 7
  solidification, 7
  vaporization of water, 876
Lateral cerebrospinal tract, (fig.) 903
  geniculate body, 970
  geniculate nucleus, 749
  horn of spinal cord, 901
  lemniscus, (fig.) 924, (fig.) 925
  spinothalamic tract, (fig.) 904
  sulcus, 954
  vestibular nucleus, 926
LATS, 1002
Law, Fechner's, 707
  Laplace's, 235, 483, 551
  of conservation of energy, 3, 174
    conservation of matter, 3
    specific nerve energy, 706
    the heart, 474, 475, 481
  Poiseuille's 539
  Starling's, 474, 475, 481
  Weber's, 706
Laxatives, 304
LBM (lean body mass), 168, 674
L-Dopa, 960
Lean body mass, (LBM), 168, 674
  tissue solids (LTS), (table) 674
Learning, 987
Lecithin, 53, 277, 343
Lecithinase A, 54
Left atrial pressure 496

coronary artery flow, (fig.) 492
Lemniscus, lateral, (fig.) 921, 924
  medial, 921, 924
Length of intestine, 283
Length-tension relation of myocar-
    dial contraction, 473
Lengthening reaction, 917
Lens, 731
  refractive power, 733
Lentiform nucleus, (fig.) 893
Leptocytosis, 439
Leptomeninges, (fig.) 685
Lesch-Nyhan syndrome, 351
'Let-down' phenomenon, 1075
Leu (leucine), 24
Leucine (Leu), 24
Leucocyte, 193, 313, 429
  and fever, (fig.) 432
  basophil, 430
  destruction, 432
  differential count, (table) 431
  drumstick, 1102
  eosinophil, 430
  formation, 429
  granular, fate of, 432
  myeloid (granular), 429
  neutrophil, 429
  number of, 431
  polymorphonuclear, 429
  promoting factor, (LPF), 431
  pyrogen, 432, 886
Leucocytosis, 431
Leucopenia, 430, 431, 432
Leucotomy, prefrontal, 969
Levator ani, 305
  palpebrae superioris, 730
Leydig, cell of, (fig.) 1027
LH (luteinizing hormone), 1032,
    1049, 1051
Lifeboat rations, 681
Life expectation, 112, (fig.) 1113
  span, average, 1112
  of red cells, 420
Ligases, 108
Light reflex, 735
  units, 1117
Limbic lobe, 957
Lineweaver and Burke equation, 97
Lingual nerve, 562
Linkage, 1107, 1108
Linoleic acid, 51, 204
Linolenic acid, 51, 204
Lipæmia, 330
Lipase, 268, 279, 299, 473
Lipid content of blood plasma,
    (table) 333
  metabolism, 333
  stores, 334
Lipids, 50
  biosynthesis, 341
  in plasma, 333
Lipoate, 341
Lipoic acid, 185, 186
β-Lipoprotein, 405
Lipoproteins, 56, 184, 334, 406
Liquor amnii, 1056
  folliculi, 1037, 1044
Lithocholic acid, 60
Litter weight, 1072
Liver, 9, 436
  and carbohydrate metabolism, 314
    fat metabolism, 336

biliary system, 273
blood channels, 584
  flow, 584
cells, 273
circulation through, 581
fat, 336
gluconeogenesis 315, 323, 327, 339
glycogen, 312, 315
in carbohydrate metabolism, 311
lymph, 584
microscopic structure, (fig.) 276,
    582, (fig.) 584
parenchyma, 581
protein metabolism, 380
sinusoids, 582
temperature, 878
urea formation, 380
vitamin A in, 184
vitamin B_{12} in, 191
vitamin D in, 195
Lobe, temporal, 974
Local circuit theory, 787
  hormones, 320, 1020
  response, 806
Localization of pain, 863
  reflex, 918
  sound, 769
  touch, 717
Lockjaw, 913
Long-acting   thyroid   stimulator
    (LATS), 1002
Longevity, 1112
Long sight, 733
Loop of Henle, 642, 649
Lower motor neurone paralysis,
    (table) 906
LPF, 431
LSD (lysergic acid diethylamide),
    1021
LTH, 1049, 1074
Lumbar puncture, 686
Luminosity curve, 742
Lung inflation reflex, 566
Lungs, 4
  air passages of, 589
  alveolus, (fig.) 616
  blood flow, 578
  capacity total, 600
  capillary, 578, 615
    pressure in, 579
  changes at birth, 1069
  ciliary activity, 590
  circulation in, 577
  coal dust in, (fig.) 591
  compliance, 524, 597
  dead space, 600, 613
  excretion of carbon dioxide, 701
  elasticity, 595
  expansion of, 594
  oxygen consumption, 591
  peripheral resistance, 579
  pleura, 592
  pressure volume curve, (fig.) 572
  primary lobule, (fig.) 592
  secondary lobule, 591
  specific gravity, 601
  stretch receptors, (fig.) 607
  structure of, 589
  subdivisions of volume, (fig.) 599
  surface tension, 590
  surfactant, 591
  tests of ventilatory efficiency, 600
  time in capillary, 616

Lungs, vessels, 559
  viscosity, 659
  volume, (fig.) 599
  weight, 590
Luteal phase, 1039, 1048
Lutein cells, 1038
Luteinizing hormone (LH), 1032, (fig.) 1049
Luteotrophin, 1049, 1088
Lyases, 106, 107
Lymph, 683
  composition, (table), 683
  flow, 685
  liver, 584
  nodes, 437, 684
  nodules, 302
  pressure, 682
Lymphatics, 683
Lymphoblast, (fig.) 419
Lymphocytes, 432, 685, 1019, 1020
  activated, 432
  and immunity, 433
  antibodies, 405, 433
  circulation, 433
  fate of, 433
  γ-globulin, 405
  irritation, 432
  life span, 432
  stress, 432
Lymphoid tissue, 437
Lysergic acid diethylamide (LSD), 1021
Lysine, 25
  vasopressin, 1084
Lysolecithin, 54
Lysosomes, 77, 287, 1029
Lysozyme, 35, 49, (fig.) 99, 100, (fig.) 103, 730

Macrocytes, 423
Macrophages, and antigens, 441
  bronchial, (fig.) 439
Macrophage system, 434, (fig.) 438
Macula densa, 642, (fig.) 646
  lutea, 738
  of saccule, 929
  of utricle, 929
Magnesium, 219
  and the myocardial cell, 472
Main d'accoucheur, 217, (fig.) 218
Maize, 189
Malabsorption, 297
Malaria, 428
Malate dehydrogenase, 128, 151
Malic acid, 151
Malleus, 761
Malonylcoenzyme A, 341
Malpighian bodies, 585
  corpuscle, 642
Maltase, 268, 298
Maltose, 45
Mamillary bodies, 954
Mamillothalamic tract, 954
Mammary gland, 1074
  blood flow, 1077
  effect of hormones, 1074
  metabolism, 1077
  myoepithelial cells, 1074

Mammotrophin, 1088
Man, composition of, 6
Mandelic acid, 1017
Manganese, 219
Mannose, 44
Manometer, 533
MAO (maximal acid output), 260
MAO (monoamine oxidase), 116
Maple syrup urine, 390
Marchi's method, 809, 902
Margarine, 52, 178
Marrow, 210
Masking, 770, 773
Mass reflexes, 992
  spectrometer, 122
Mast cells, 1020
Mastication, 222
Maternal nutrition, 1111
Mating, 1045, 1053
  behaviour, 1045
Maximal acid output (MAO), 260
Maximum breathing capacity, (MBC), 600
  rate of tubular reabsorption, 652
  reabsorptive capacity, 652
  secretory capacity, 653
  tubular capacity (T_m), 652
  voluntary contraction (MVC) 855
Max-Planck respirometer, (fig.) 165
Mayer waves, (fig.) 560
MBC, 600
MCHC, 448
MCV, 416
Meal, rate of passage through intestine, 285, 303
Meals, (fig.) 205
Mean arterial pressure, 533
  cell hæmoglobin concentration (MCHC), 448
  corpuscular volume (MCV), 416
Meat, 178
  extracts, 252, 256, (fig.) 257
Mechanical efficiency of human body, 168
Mechanism of calcification, 211
Meconium, 307
Medial fillet, 921
  geniculate body, (fig.) 922, (fig.) 925
  lemniscus, 921
  longitudinal bundle, (fig.) 924
Median eminence, 1081
  forebrain bundle, 951
  plane, 893
Medroxyprogesterone acetate (MPA), 1088
Medulla, adrenal, 861, 866, 1015
  and respiration, (fig.) 606
  oblongata, 921
Megacolon, 306
Megakaryocytes, 434
Megaloblast, 423
Megaloblastic anaemia, 190, 423
Meiosis, 1100
Meiotic plate, (fig.) 1100
Meissner's corpuscle, (fig.) 709
Meissner's plexus, 234
Melanin, 109, 136, 385, 711
Melanocytes, 711
Melanophore expanding activity, 1088
  stimulating hormone (MSH), 1088

Melanophores, 1088
Melanuria, 385
Melatonin (N-acetyl-5-methoxy-tryptamine), 1020
Membrane, 53
  currents, (fig.) 801
  of nerve, 797
  permeability, 792
  potential, 792
    cardiac, 512
    and sodium permeability, (fig.) 795
  semi-permeable, 20
Membranous labyrinth, 929 (fig.) 930
Memory, 961, 975
Menaphthone, 197
Menarche, 1038
Mendelian inheritance, 1107
Meninges, 685
Menopause, 1041, (fig.) 1042, 1050, 1051
Menstrual blood loss, 1039
  cycle, 1038, 1050
    basal temperature, 877, 1052
    diagram of events in, (fig.) 1037
    hormonal control of, 1047
    length, 1041
  flow, 1039
  loss of iron, 424, 1039
  myometrial movements, (fig.) 1041
  prostaglandins, 1039
Menstruation, 1038
  anovular, 1039
Mental work, 169
M.e.p.p.s. 818, 834
Mercaptan, 724
2-Mercaptoethylamine, 67
6-Mercaptopurine, 351
Meromyosin, 838
Mesencephalon, 892
Mesobilirubinogen, 453
Messenger RNA (mRNA), 72, 364
Mestranol, 1047
Met (methionine), 25, 388
Metabolic acidosis, 692, 695, 696, 700
  alkalosis, 692, 695, 696, 702
  pool, 120, 376
  rate and external temperature, 875
  rates of males and females, (fig.) 167
  water, 7, 681
Metabolism, 2
  basal, (fig.) 167
  intermediary, 118
  of brain, 898
    glucose in brain, 898
    heart, 464
    nervous system, 898
Metaphase, mitotic, (fig.) 1099
Metarterioles, 569
Meta-stable state, 788
Metathalamus, 954
Methæmalbumin, 449
Methæmoglobin, 447, 448
Methæmoglobinæmia, 448
Methandienone, 1034
Methanol, 12, 99
Methionine (Met), 25, 388
  absorption of, 299

Methyl alcohol, 98
  poisoning, 99
  malonic acid (MMA), 191
  testosterone, 1032
Methylation of DNA and RNA, 359
Methylmalonyl-CoA, 343
  mutase, 191, 343
N-5, N-10 Methylene tetrahydro-
  folic acid, 190
Methyltransferase, 359
Metyrapone, 1014
Mevalonic acid, 344
Micellar particles, 279
Micelles, 279, (fig.) 293, 299
Michaelis constant, 96
Michaelis-Menten equation, 96, 113
Microcirculation (fig.) 546
Micro-electrode, 786
Microglia, 436
Micropuncture methods, (fig.) 660
Microsomes, 76
Microvilli, 285
  dimensions of, (table) 289
  gastric, (fig.) 244, (fig.) 246
Micturition, 667, 669
  reflexes, 671
Mid brain, (fig.) 925
  and temperature, 883
  aqueduct, 922, (fig.) 925
Middle ear, 761
Middle peduncles of cerebellum, 941
Milieu interieur, 2
Milk, 5, 176, 195
  chemical composition of (table) 176
  clotting of, 177
  comparison of human with cow's,
    (table) 176
  composition of, 176
  condensed, 177
  dried, 177
  ejection of, 1054, 1075, 1076 (fig.)
    1076, 1094
  evaporated, 177
  flow, 1074
  minerals in, (table) 176
  output, (fig.) 1075
  pasteurized, 176
  production, daily, 1075, 1077
  secretion of, 1074
  skim, 177
  sterilized, 177
Milliequivalent, definition of, 9
Millimole, definition of, 9
  of gas, table 10
Milling, 179
Milliosmole, definition of, 21
Mineralocorticoids, 1007, 1012
Miniature end-plate potentials,
  (m.e.p.p.s), 818, 834
Minute volume, 627
Miosis, 733
Missed period, 1055
Mitochondria, (fig.) 69, 75, 133, 148,
  337
Mitomycin C, 372
Mitosis, 1098
  mitotic, division, 1098
  metaphase, (fig.) 1099
Mitral incompetence, 502
  stenosis, 502, (table) 504, 581
  valve, 461, 502
Mixed venous blood, 617
  gases in, table 617

MMA (methyl malonic acid), 191
Modalities of cutaneous sensation,
  719
Modiolus, (fig.) 763
Molar solution, 10
Molarity, definition of, 10
Mole, definition of, 9
Molecular biology, 1
  weights of proteins, 31
Monoamine oxidase (MAO), 116,
  219, 868, 1017
  inhibitors, 1021
Monochromats, 746
Monocistronic messenger, 355
Monocytes, 433, 435
Monoglyceride, 51
Monoiodotyrosine, (fig.) 989
Monomeric form, 37
Mononucleotide biosynthesis, 347
Monophasic recording, 781
Monopteroylglutamic acid, 190
Monosaccharides, 40, 41
Monosynaptic pathway, 915
  reflex, 909
Morphine, 293, 629
Moss fibre, (fig.) 944
Motion sickness, 933
Motoneurone pool, (fig.) 911
Motoneurones, 817, 829
α-Motoneurones, 915
Motor aphasia, 968
  areas of cerebral cortex, 964
    supplementary, 968
  cell of spinal cord, (fig.) 815
  cortex, 961, 965
    effect of removal of, 967
    histology, 961
    ipsilateral projection, 967
  decussation, 921, (fig.) 923
  end-plate, 831, 833
  homunculus, (fig.) 965
  latency, 848
  nerves, 4
  unit, 830
Mountain sickness, 629, 632
Mouth, 222
  absorption in, 295
  temperature, 876
Mouth-to-mouth ventilation, 609
Mouth-to-nose ventilation, 609
Movements of small intestine, 281
MPA, 1088
mRNA, 72, 355, 364, 996
MSH, 1088
Mucin, 226
Mucoid cells, 1081
Mucopolysaccharides, 47, 209
Mucoproteins, 47
Mucosa, (fig.) 234
Mucous cells, (fig.) 244
  secretion, 243
Mucus, 243
  dissolved, 244
  visible, 244
Müller's law, 706
Multimeric form, 37
Murmur, bronchial, 598
Murmurs, hæmic, 539
  organic, 539
Muscarinic effect, 817
Muscle, 387, 825
  action potentials, (fig.) 846

activation, 831
active state, 844
adenosine triphosphate in, 837
aerobic recovery phase, 840
afferent nerves, 810
anærobic heat, 852
bands, 825
blood flow through, 848
blood supply in exercise, 848
capillaries of, 570
cell, (fig.) 465, (fig). 466, 825
  ions in, (table) 793
chemical transmission, 833
chemistry of skeletal, 837
contraction, 826, 831
  source of energy for, 838
creatine, 387
cross-bridges, (fig.) 836
crushing, 837
depolarization, 834
denervation, 809
effect of acetylcholine, 833
efficiency, (fig.) 846
elasticity, 844
electrical response, (fig.) 929
electromyography, 846
electronmicrographs, (fig.) 847
end-plate, 831
exercise, 315, 838
  cardiac output and oxygen
    consumption, 848
  fatigue, 845
  oxygen debt, 853
  respiratory quotient during and
    after, (figs.) 851
  second wind, 855
extractives, 837
fatigue, 838, 845
fatty acids, 850
fibres, 825
fibrils, 825
fibrillation, 847
glycogen, 312, 314, 316, 854
glycolysis, 839
heat production in, 850, 852
innervation, 831
iodoacetate-treated, 838
isometric contraction, (fig.) 474,
  (fig.) 476, 840
isotonic contraction, (fig.) 474,
  (fig.) 476, 493
joint sense, 831
lactic acid, 838
length and tension, 842
mechanical properties, 840
mechanical response, (fig.) 829
metabolism, 314, 315
miniature end-plate potentials,
  818, 834
motor end-plate, 831
  unit, 830, 831
nerves of, 831
optimum speed of working, 845
oxygen supply in exercise, 853
phosphorylase, 158
pump, 505
pyruvic acid, 840
rate of contraction, 845
refractory period, 910
relaxation period, 910
resting potential, (fig.) 795
rigor mortis, 836
sensory innervation, 826

Muscle, silent period, 916
  skeletal, 4, 313, 825
  slow, 841, 842
  smooth, 857
  source of energy, 838
  spindle, 829, 915
    drawing of, (fig.) 828, (fig.) 916
    number of, 831
    rate of discharge, (fig.) 829
    responses, (fig.) 829
  strength in old age, 1113
  structure of skeletal, 825
  temperature in exercise, 852
  tension and speed of shortening, 844
  tension/length relationship, (fig.) 844
  tetanus, 841
  tone, 926
  tremor, 947
  twitch, 841, 947
  uterine, 1071, 1072, 1073
  vasodilatation, 953
  visceral, 857
  viscosity, 844
  voluntary contraction, 845
Muscular atrophy, (table) 906
  contraction, characteristics of, 826, 831
    chemical basis of, (fig.) 839
    source of energy for, 838
  dystrophy, 196
  exercise, 848
  tone, 932
    thermal, 874
  tremor, (fig.) 1004
  vasodilatation, 853
Muscularis mucosæ, 234
Mutarotation, 41
Mutation, 354, 370, 1109
Myasthenia gravis, 835, 1020
Myelin, (fig.) 781, 902
  sheath, 777
  sheaths in central nervous system, 903
Myelinated fibres, 903
  equivalent circuit, (fig.) 802
  nerves, 776
Myelination in central nervous system, 963
Myeloid leucocytes, 429
Myenteric plexus, 234
  reflex, 237
Myocardial actin-myosin interaction, 477
  β-adrenergic receptors, 484
  cells, 464, (fig.) 465, (fig.) 466
    action potentials, various types, (fig.) 469
    and calcium, 464
      magnesium, 464
      potassium ions, 512
    conducting, 467
    pacemaking, 467
    specialized, 464, 467
    structure of, 464
    supranormal phase, 468
    working, 464
  contractility, 484, (fig.) 485
  depolarization, (fig.) 486
  excitation, wavefront, 486
  fibrillation, 521
  hypoxia, (fig.) 489

infarction, 116, 473
intercalated disc, (fig.) 466
ischaemia, 519
membrane currents (fig.) 486
metabolism, 473
  anaerobic, 519
repolarization, (fig.) 486
Myocardium, 461
  contractile state, 479
  depolarization, (fig.) 467, (fig.) 486
  electrical excitation, 467
  initial length of, 479
  non-contractile components in, 482
  order of contraction, 472
    relaxation, 472
  oxygen consumption of, 473
  repolarization, (fig.) 467, (fig.) 486
Myoepithelial cells, 712
Myofibrils, 825
Myoglobin, 35, 113, 837
  oxygen dissociation curve, (fig.) 620
Myograph, 473, (fig.) 474
Myometrium, 1035, 1041, 1045
  movements in menstrual cycle, (fig.) 1041
Myoneural junction, 831
Myopia, 733
Myosin, 34, 464, 826, 837
  ATPase, 464, 472, (fig.) 476, 477
  chemistry, 837
  filament, (fig.) 826
Myotatic reflex, 915, 917
Myxœdema, 1001, (fig.) 1003, 1084

N-acetyl-5-methoxytryptamine (melatonin), 1020
NAD, 67, 109, 187
NADP, 67, 109, 187
NADPH, 341
Nail-bed capillaries, 545
Nails, 31, 712
Narcotics, 636
Nasal cavities, 589
Nasolacrimal duct, 730
'Natriuretic' hormone, 664
Near point, 733
  reflex, 735
Neck reflexes, 928, 937
NEFA, 333, 336, 340
Negative after-potential, 808
  conditional reflex, 986
  feedback, 1010
  induction, 989
  supporting reaction, 937
Neocerebellum, 941
Neonatal circulation, 1069
Neonatal mortality, 1111
  respiration, 601
Neopallium, table 958
Neothalamus, 954
Nephrectomy, 667
Nephrogenic diabetes insipidus, 1096
Nephron, 641, (fig.) 644
Nernst diffusion equation, 792
Nerve, accommodation, 807
  action potentials, 784, 786, 791, 831, 846
  adrenergic, 866
  afferent, 5, 810
  afferent impulse, 810

after-potentials, 807
  and calcium, 807
    potassium, 807
  anode break excitation, 806
  autonomic, 860
  cholinergic, 866
  compression, 848
  conduction, 806
    velocity, 785, 847
  crushed, 809
  degeneration, 809
  depolarization, 787
  efferent, 4
  equivalent circuit, (fig.) 797, (fig.) 802
  excitation, 788
    threshold, 806
  fibre, rate of growth, 810
  fibres, A, B, and C, 784
    classification, 785
    size, 776, 785
    structure of, 776, 800
  heat production, 805
  impulse, 781
    characteristics of, 781
    diagram of events, (fig.) 799
    propagation, (fig.) 799
    summary, 796
  ionic movements, 791, 804
  membrane, 777, 797
    capacity, 797
    equivalent circuit for, (fig.) 797, (fig.) 802
  metabolism, 803, 805
  motor, 4
  myelinated, 776
  non-myelinated, 776
  oxygen consumption of, 905
  permeability changes, 795, 797
  propagation of impulse, (fig.) 799
  rate of conduction, (fig.) 847
  recovery, 810
  refractory period, 790
  regeneration of, 809, 810
  repair, 809, 810
  saltatory conduction, 780, 799
  section, 809, 901
  sensory, 5
  somatic, 891
  spikes, 784
  stimulation, 788
  supply to blood vessels, 561
    heart, 508
  thermal changes, 805
  threshold, 806
  transmitter, 833, 866
  volley, 784
Nervi erigentes, 1054
Nervous system, 891
  metabolism, 898
Nettle-rash, 1021
Neuraminic acid, 49
Neurilemma, 780
Neurohumoral transmitters, action of on the myocardium, 467
Neurohypophysis, 952, 1080, 1090
  electronmicrograph, (fig.) 1092
  hormones of, 1090
Neuroma, 810
Neuromuscular block, 846
  junction, (fig.) 832
    electronmicrograph, (fig.) 832
  transmission, 831

Neurone, 776
Neurone-glial-capillary arrangement, (fig.) 892
loss, 892
Neurones, internuncial, 905
number of, 892
of central nervous system, 892
Neuropathy, 423
Neurophysin, 1091
Neurosecretion, 1091
Neurosecretory material (NSM), 1091
Neutral red, 245
Neutrality, 14
Neutrophil polymorph, 429
Newborn infants, 168, 601
adipose tissue, 885
hæmolytic disease, 421
hæmorrhage, 197
renal function in, 667
respiration, 601
temperature, 885
Nexus, myocardial, 467
N-formyl methionine, 365
N-formyltetrahydrofolic acid, 349
Niacin, 187
Nicotinamide, 66, 187
-adenine dinucleotide (NAD), 67, 187
phosphate (NADP), 67, 187
and pellagra, 187
nucleotides, 66
Nicotine, 861
Nicotinic acid, 187
and tryptophan, 187
bound form in maize, 188
daily requirements, 188
effect, 817
in foods, table 188
Night blindness, 741
Ninhydrin, 27
Nipple, 1074
Nissl granules, 810
Nitrogen balance, 199, 200, 204
in old age, 1113
endogenous output, 200
excretion, 389
premortal rise, 205
in fæces, 309
urine, 200
losses, in skin and sweat, 713
washout test, 613
Nociceptive response, 909
Nocturia, 657
Node, atrioventricular, 468
of Ranvier, 780, (fig.) 782, (fig.) 783, 799
sinu-atrial, 468
Nodulus of cerebellum, 933, (fig.) 942
Noise, 773
Nominal aphasia, 975
Nomogram, (fig.) 163
Non-enzymic hydrolysis, 100
Non-esterified fatty acids (NEFA), 333, 336,
Non-myelinated fibres, 848
Non-polar solvents, 11
Non-secretors, 245
Non-virilizing androgens, 1130
Noradrenaline, 116, 238, 484, 511, 566, 568, 631, 822, 866, 868, 1016

and adrenaline, comparison of effects, (table) 868
muscles, 868
and denervation, 868
biosynthesis and breakdown, (fig.) 1017
in blood, 1019
intravenous infusion, (fig.) 567, 866
receptors, 866
respiration, 631
synthesis of, 1017
urinary excretion, table 1019
Norethisterone, 1047
Norethynodrel, 1047
'Normal' diet, 199
solution, 9
Normoblast, 418
Nortestosterone, 1047
Norymberski method, 1033
Nose, 589
and sense of smell, 722
in respiration, 589
stream of air through, 723
NSM (neurosecretory material), 1091
Nuclease, 268, 357
Nucleic acid, 63
biosynthesis, 351
enzymic modification, 359
metabolism, 347
Nuclei pontis, 922
Nucleolus, 75
Nucleoside, 64
Nucleotide pyrophosphorylases, 350
sequence, 68
Nucleotides, 63, 65
catabolism at, 359
Nucleus, 75
cuneate, 921
dentatus, 944
fastigii, 44
gracile, 921
pretectal, (fig.) 925
red, 926
vestibular, 927
Nutrition, 1111
Nychthemeral rhythm, (figs.) 658, 877
Nystagmogenic areas, (fig.) 935
Nystagmographic record, 935
Nystagmus, 929
mechanism of, (table) 934
optokinetic, 753
vestibular, 932

Oatmeal, (fig.) 180
Obesity, 171, 173
Obstructive jaundice, 197, 454
Occipital cortex, 970
lobe, 954, 970
Occlusion, 912
Ocular counterrolling, 937
Oculogravic illusion, (fig.) 937
Oddi, sphincter of, 273
Odours, classification of, 723
Oedema, 406, 524, 683
and aldosterone, 1013
kidney, 664
in beri-beri, 185
pulmonary, 579
Oesophageal pressure, (fig.) 230
speech, 758

sphincter, 231
Oesophagostomy, 250
Oesophagus, abdominal segment, 228
lower sphincter, 231
peristaltic wave, 231
pressure changes, (fig.) 229
structure of, 228
Oestradiol, 61, 1044
Oestriol, 1045
excretion in pregnancy, (fig.) 1060
Oestrogens, 58, 61, 1044
action of, 1044
biosynthesis, 1046
excretion, 1045
in menstrual cycle, 1047
pregnancy, 1059
from testis, 1034
inactivation by liver, 1044
in adrenal cortex, 1056
maternal plasma, 1060
plasma, 1060
mammary gland and, 1074
synthetic, 1047
Oestrone, 61, 1034, 1044
Oestrous cycle, 1042
Oestrus, 1042
Old age, 1113
Oleic acid, 51
Olfaction, 722, 975
adaptation, 723
Olfactory area, 724
bulb, 724
cells, 723
epithelium, (fig.) 723
lobe, table 958
tract, (fig.) 923
Oligodendroglial cells, 780
Oliguria, 666
Olive, (fig.) 923
Olive oil, 52
One-stage prothrombin time, 394
Oocyte, 1037, (fig.) 1101
Oogonia, 1036
Oolemma, 1054
Oophorectomy, 1044, 1047, 1056
and pregnancy, 1056
Opening snap, 502
Operant behaviour, 988
Operator site, 370
Operon, 370, (fig.) 371
Ophthalmoscope, 738
Ophthalmoscopic view of eye, plate 37.19
Optic chiasma, (fig.) 923, 962, 970
disc, 970
nerve, 970
pathways, 970
radiations, 970
tract, (fig.) 923
Optokinetic nystagmus, 753
Ora serrata, 738
Orbicularis oculi, 730
Order of contraction, myocardium, 472
relaxation, myocardium, 472
Organ of Corti, (fig.) 763
Organophosphorus inhibitors, 103
Orgasm, 1053
Orgasmic platform, 1053
Orienting reflex, 985
Ornithine, 26, 380
-arginine cycle, 380
carbamoyl transferase, 372

Orotic acid, 347
Orotidine 5′-monophosphate, 348
Orthopnœa, 600
Orthostatic albuminuria, 648
Osmic acid, 902
Osmolality, 680
Osmole, definition of, 21
Osmometry, 680
Osmoreceptors, 239, 661
  in duodenum, 239
Osmosis, 20
Osmotic pressure, 20
  and colloid, 21, 682
  diuresis, 657
  of extracellular fluid, 679
    plasma, 406
Ossicles, 761
Ossification, 210
Osteoblasts, 210
Osteoclasts, 210
Osteocytes, 210
Osteoid matrix, 210
Osteomalacia, 195, 216
Osteoporosis, 212, 1113
Otoconia, 929
Otolith organs, 929
  system, 936
Otosclerosis, 762, (fig.) 772
Outflow tract of cardiac ventricles, 458
'Out-of-level' indicators, 929
Output, cardiac ventricular, 503
Outward current, 791
Oval window, 761
Ovarian follicles, maturation, 1036
  hormones in urine, 1048
  primordial, 1036
Ovarian hormones, 1044
Ovariectomy, effects of, 1044, 1047
  in pregnancy, 1056
Ovary, 1036
  diagram of changes in, (fig.) 1036
  follicles (Graafian), (fig.) 1036
  interstitial cells, 1036
  hormones, 1044
  number of follicles, 1036
  steroid biosynthesis, 1046
Overflow incontinence, 670
Overshoot, 795
Ovulation, 1038, 1039, 1049, 1052
  and hypothalamus, 1050
  and temperature, 1052
  FSH and LH in, 1053
  inhibition by progesterone, 1053
  multiple, 1049
  peak, 1053
  time of, 1052
Ovum, 5, 1101
  viability, 1052
Oxalate and blood clotting, 394
Oxalated plasma, 397
Oxaloacetate, 339
Oxaloacetic acid, 147, 152
Oxidation biological, 125
  definition of, 126
  of fats, 336
  -reduction potential, 134, 137
β-Oxidation, 336
Oxidation-reduction potential, 134
Oxidative phosphorylation, 76, 131, 133, 152, 464
Oxidoreductase reactions, 109

Oxidoreductases, 108
Oximeter, 619
Oximetry, 504
17-Oxogenic steroids, 1012
α-Oxoglutaric acid, 379
17-Oxosteroids, 1033
  excretion, table 1033
Oxycalorimeter, 162
Oxygen administration, 637
  and cortical cells, 608
  atmospheric, partial pressure of, (fig.) 633
  calorific value of, 163
  capacity of blood, 618
  consumption, (fig.) 164, 517
    of brain, 896
    nerve, 905
    resting, 168
  content of blood, 618
  debt, 853
  dissociation curve, 447
    of haemoglobin, 620
  effect of increased, 637, 638
  effect of lack, 631
  electrode, (fig.) 619
  hyperbaric, 626, 638
  in arterial blood, (fig.) 617
    blood, estimation, 618
  inhalation at high altitudes, 633
  lack, 631
  poisoning, 638
  pressure at high altitudes, (fig.) 633
  saturation, (fig.) 633
    at high altitudes, (fig.) 633
    of arterial blood in exercise, (fig.) 849
    fœtal blood, 1065
  supply, effect of increased, 637
  tension, arterial, 622
    in alveolar capillary, 615, 632
  tent, 637
  therapy, 638
  toxicity of, 638
  transfer in placenta, 1065
  transport in blood, 620
Oxyhæmoglobin, 447
  dissociation curve of, (fig.) 621
  rate of formation, 622
Oxyntic cells, 234
Oxytocic action of pituitary extract, (fig.) 1093
Oxytocin, 1054, 1072, 1093
  and alcohol, 1094
    coitus, 1094
  desamino, 1094
  half-life, 1072
  in pregnancy, 1072
  milk ejection, 1076, 1094
  on uterus, (fig.) 1093
  structure of, (fig.) 1094
Oxytocinase, 1072, 1094
Ozone, 638

Pace-maker, myocardial, 467
  fibre, action potentials, (fig.) 467
  potential, 470, 512, (fig.) 513, (fig.) 514
'Pacing', cardiac, artificial, 511, 518
Pacinian corpuscles, 814, 862
Packed cell volume (PCV), 415
PAH, 652, 653, 655

Pain, 956
  abdominal, 863, 865
  adaption, 718
  and asphyxia, 718
    blister base, 718
  angina pectoris and, 863
  diaphragm, 863
  double sensation, 718
  end-organs, 718
  fibres in sympathetic nerves, 863
  gall-bladder, 865
  gastric, 865
  in heart disease, 863
    urinary bladder, 864
  intestinal, 863
  localization of, 863
  minimum stimulus, table 718
  pathway, (fig.) 905
  -producing substance (P.P.S.), 718
  projected, 863
  referred, 863
  sense, 718
  skin, 718
  spots, 718
  testicular, 864
  threshold, 718, 865, 866
  uterine, 864
  visceral, 863
Palæocerebellum, 941
Palæostriatum, 959
Palæothalamus, 954
Pallium, table 958
Palmitic acid, 51
Pancreas, 263
  acinar cells, 266
  anatomy of, 263, (fig.) 277
    phenylalanine, 272
  α-cells, (fig.) 266
  control of secretion, 270
  diagram, (fig.) 277
  endocrine cells, (fig.) 265
    effects, 273
  evocative test, 271
  exocrine cells, (fig.) 265
  gastrointestinal hormones on, 273
  glucagon, 266
  histology, (fig.) 264
  islets, (fig.) 204, 273
    histology, (fig.) 266
  secretion, 266, 272
    of electrolytes, 266
Pancreatectomy, 329
Pancreatic amylase, 268
  enzymes, 266, 268
  function, tests of, 271
  insufficiency, (fig.) 274
  juice, 266
  amylase, ions of, 268
  RNase, 357
  secretion, 266
    bicarbonate, 271
    cephalic phase, 270
    control of, 270
    gastric phase, 270
    intestinal phase, 271
Pancreatitis, 272
Pancreozymin, 242, 269
Paneth cells, (fig.) 287, 289
Panhypopituitarism, 1090
Panting, 880
Pantothenic acid, 67, 189, 341

Paper chromatography, 122
Papillary muscles, (fig.) 461, 471, 472, 473, (fig.) 474, (fig.) 478, 479, 491
  length-tension curve, 473
Papilloedema, 687
Para-amino-hippuric acid (PAH), 652, 653, 655
Parabiotic rats, 174
Parahæmatin, 450
Paralutein cells, 1038
Paralysis, 906
  agitans, 960
  and temperature, 874, 881
  flaccid, 967
  spastic, 967
Paraplegia, 670, 907
Parasympathetic outflow, 860
Parasympathomimetic drugs, 224
Parathyroid extract, effect on serum calcium, (fig.) 216
  glands, 213
  gland structure, 213
    action of, 214
    control, 213
    removal of, and tetany, 216
    tumour and bones, (fig.) 215
    tumours of, 214
  hormone (PTH), 213, 214
    half life, 214
Parathyroidectomy, 214
Parietal cell, 242, (fig.) 245, 246
  and maximal acid output, (fig.) 259
  electron micrograph, (fig.) 246
  lobes, 975
  mass (PCM), 259
  peritoneum, 865
  secretion, 247
  lobe, 957
  pleura, 592
Parity and birth weight, 1112
Parkinsonism, 960
Parotid gland, 222
Pars distalis, 1081
  intermedia, 1088
Partial pressure, 612
  of carbon dioxide in arterial blood, 615, 617
    carbon dioxide in plasma, 615, 622
Partition chromatography, 29
Parturition, 1060, 1071, 1094
  and adrenals, 1073
    hormonal control, 1071
    intrauterine pressure, 1073
    nervous system, 1072
    oxygen requirements, 1074
    pregnanediol, (fig.) 1062
    progesterone, 1060
  blood loss, 1073
PAS (periodic acid Schiff), 1081
Passenger aircraft, 634
Passive immunity, 299, 405
Past-pointing, 935
Pasteur effect, 90
Pasteurized milk, 176, 177
α-Pathway, 927
Pavlov pouch, (fig.) 248
PBI (protein-bound iodine), 1000
PCM, 259
PCV, 415

Pedal spasm, (fig.) 218
Pedicels, 643
Peduncles of cerebellum, 921, 941
Pellagra, 187, (fig.) 188
Pelvic plexus, 304
  splanchnic nerves, 304, 668
Pelvis, female, 1038
Pendular movement, 284
Penicillin, 49, 373
Penis, 1053
Pentagastrin, 253, 260
Pentose phosphate pathway, (fig.) 140, 152, 341
  to hexose conversion, 153
Pentoses, 40, 44, 63
Pentosuria, 325
Pepsin, 249
  secretion, 257
Pepsinogen, 249
Peptic cells, 242
Peptide, 27, 28
  bond, 28
Percentage saturation of blood, 618
Perchlorate, 1004
Perhydrocyclopentenophenanthrene ring, 57
Pericardium, 460, 462, 505
Perichondrium, 209
Pericytes, 436, 543
Perilymph, 766, 932
Perimeter, 737
Perimetry, 970
Perineurium, 780
Periodic acid Schiff (PAS) technique, 1081
  breathing, 634
Peripheral circulatory failure, 575
  cyanosis, 636
  neuropathy, 423
    regulation of 539
  resistance, 458, (fig.) 458, 537, 579
  vision, 737
Peristalis in colon, 303, 305
  intestine, 283, 303
  œsophagus, 231
  stomach, 235
Peristaltic rush, 284
Peritoneal dialysis, 667
Peritoneum, pain, 865
Perl's reaction, 426
Permeability changes, mechanism of, 808
  coefficients, 793
  of capillaries, 547, 550, 683
Permeases, 134
Pernicious anæmia, 245, 258, 260, (fig.) 420, 423
Peroxidase, 137
Peroxisomes, 77
Perspectival occlusion, 754
Perspective, 754
Perspiration, insensible, 876
Petechiæ, 551
Peyer's patches, 287, 308
PGA, (pteroylglutamic acid), 190
pH (hydrogen-ion concentration) 14
  of blood, 691
  gastrointestinal contents, 258
Phæochromocytoma, 1019
Phagocytosis, 438
Pharyngotympanic tube, 762

Phe (phenylalanine), 26
Phenoltetrabrompthalein sodium sulphonate, 848
Phenothiazines, 261
Phenotype, (fig.) 1107, 1108
Phenoxybenzamine, 868
Phenylalanine (Phe), 26, 272, 384, 1018
Phenylisothiocyanate, 32
Phenylketonuria, 120, 385, 1108
Phenylthiocarbamide and taste, 727
Phenylthiourea, 1107
Pheromones, 724
Phlebogram, 491, (fig.) 553
Phlorizin, 120, 297, 327, 653
Phlorizinization, 120
Phlorizinized dog, 381
Phon, definition, 773
  values, (fig.) 774
Phonation, 756, 757
Phonocardiogram, 496, 500
Phosphagen, 838
Phosphatase, acid, 1030
  alkaline, 104, 116, 212
  prostatic, 1030
Phosphate absorption, 219
  excretion, 665
  in blood, 212, 219
    bones and teeth, 212
    foods, 219
    plasma, 219
  of urine, 665
  requirement, 219
Phosphatidic acid, 342
Phosphatidylcholine, 53
Phosphatidylethanolamine, 54
Phosphatidylinositol, 54
Phosphatidylserine, 54, 342
Phosphocreatine, 473
Phosphodiesterase, 358
Phosphofructokinase, 141, 158, 311, 317
Phosphogalactose uridyl transferase, 312
Phosphoglucomutase, 154, 317
6-Phosphogluconate dehydrogenase, 153
6-Phosphogluconic acid, 153
Phosphoglycerate kinase, 142
2-Phosphoglyceric acid, 143
3-Phosphoglyceric acid, 143
Phosphoglyceromutase, 143
Phospholipids, 53, 277
Phosphopyruvate hydratase (enolase), 143
Phosphoribosyl pyrophosphate (PRPP), 347
5-Phosphoribosylamine, (5-PRA), 349, 360
Phosphors, 122
Phosphorus, 219
  compounds in blood, 219
  in plasma, 219
Phosphoryl group transfer, 107
Phosphorylase, 155, 317, 323
  kinase, 317
  phosphatase, 158
Phosphorylation, 108, 131
  oxidative, 131
Photic stimulation, 979
Photometric units, 1118
Photopic vision, 742

Photopigments, 740, 743
  bleaching of, 742
  in retina, 742
Photosensitive pigments, 742
Photosynthesis, 139, 159, (fig.) 160
Phrenic nerves, 593
Phylloquinone, 197
Physiological jaundice, 442
Physostigmine, 870
Phytic acid, 45, 179
P_i, definition of, 84
Pia mater, 685, 958
Picrotoxin, 820
Pigmentation of skin, 710
Pill, contraceptive, 1050, 1053
Pilocarpine, 293
Pilomotor activity, 886
Pineal body, 922, 1020
Pinocytotic vesicles, (fig.) 542
Pin-point pupils, 732
Piriform area, table 958
Pitch, 771
  of voice, 757
Pituicyte-glial cells, 1091
Pituitary, anterior lobe, 1082
  acromegaly, 1086, 1087
  adenohypophysis, table 1081
  adrenocorticotrophic hormone
    (ACTH), 378, 1084
  and releasing factors, 1082
  antidiuretic hormone, 1094
  blood supply, (fig.) 1082
  body, 1080
  carbohydrate metabolism, 322,
    323, 324
  cells of, 1081
  deficiency, testis in, (fig.) 1027
  dwarfism, 1087, 1090
  gland, 952
  gonadotrophic hormones and,
    1049
  growth hormone, 1085
  kidney and, 661, 1094
  lactogenic hormone, 1088
  nervous connexions of, (fig.) 1091
  neurohypophysis, (table) 1081
  oxytocic hormone, 1093
  portal vessels, (fig.) 1082
  posterior lobe, 1082
  section through, (fig.) 1083
  subdivisions of, table 1081
  thyrotrophic hormone (TSH),
    1084
  tumours, 954
  undernutrition, 1051
pₓa, 17
Placenta, 1051, 1055, 1056
  amino-acid transfer, 1066
  and hormones, 1059
  blood flow, 1065
  circulation in, plate 51.x
  cotyledons, plate 51.x
  enzymes in, 1059, 1067
  functions of, 1065
  gas transfer, 1065
  hæmochorial, 1055
  iron transfer, 1066
  localization, 1065
  oxygen transfer, 1065
  sugar transfer, 1066
  vessels, P_{O_2}, 1065
  villi, 1055
Placental tissue, (fig.) 1058

Placing reactions, 939
Planes in central nervous system,
    893
Plantar reflex, (fig.) 906, 968
Plasma, 415
  albumin, 401
  and cerebrospinal fluid, chemical
    composition, table 688
    lymph, chemical composi-
      tion of, table 683
    urine composition compared,
      table 647
  buffers, 691
  calcium, 213, 216
  cells, 1020
  chemical composition, table 617
  cholesterol, 344
  citrated, 394
  clearance, 653, (fig.) 654
  composition of, table 647
  flow through kidneys, 644
  gases in, (table) 617
  globulin, 401
  growth hormone, (fig.) 1087
  hydrogen ion concentration, 695
  immunoelectrophoresis, (fig.) 404
  insulin, (fig.) 318, 320, (fig.) 329
  ionic composition, (fig.) 690
  iron, 424
  lipids, 333
  lipoproteins, 406
  oxalated, 397
  proteins, 376, 401, table 403, 683
    deficiency of, 683
    diagram of sizes (fig.), 407
    escape from capillaries, 683
    formation of, 407
    functions of, table 403
    hypoproteinæmia, 299
    molecules of, (fig.) 407
    osmotic pressure, 406
    shape of, (fig.) 407
  renin, 1049
  thromboplastin antecedent (PTA),
    factor XI, 393
  ultrafiltrate, 682
  venous, 11
  volume, 408, (table) 677
    in pregnancy, 1064
Plasmalogens, 56
Plasmapheresis, 407
Plasmin, 398
Plasminogen, 398
  activators, 399
  inhibitors, 398
Plasticity, 917
Platelets, 193, 399, 434, 719
  adhesiveness, 401
  aggregation, 400
  and clotting, 392
  and 5-HT, 1021
  clot retraction, 400
  in coagulation, 392
  life of, 434
  number in blood, 434
  structure, (fig.) 435
  viscous metamorphosis, 400
β-Pleated sheet, 34
Plethysmograph, 532, (fig.) 540, 579
Pleura, 592
Pleural space, 592
Plexus, myenteric, (fig.) 234
  submucous, (fig.) 234

PMS (pregnant mare serum), 1051
Pneumotaxic centre, 603, 605
Pneumothorax, 595, 607
P/O ratio, 131, 132
Podocytes, 643
Poikilothermic animals, (fig.) 873,
    874
Poiseuille's Law, 539
Polar body, (fig.) 1101
  molecules, 11
  solvents, 11
Polarization, 468
Poliomyelitis, (table) 77, 906
Polybrene, 397
Polycistronic messenger, 357, 369
Polycythæmia, 539, 635, 636
Polydipsia, 330
Polygraph, 519
Polyhexosamines, 48
Polymorphonuclear leucocytes, 429
  leucocytosis, 430, 431
Polyneural innervation, 830
Polyneuritis, 185
Polynucleotide ligase, 358
Polypeptide synthesis, (fig.) 366
Polypeptides, absorption of, 299
  definition of, 28
Polyphenol oxidase, 136
Polyribosome, 367, 996
Polysaccharide, 40, 45
Polysome, 77, 367, (fig.) 368
Polyunsaturated fatty acids, 51, 62
Polyuria, 330, 657, 1096
Polyuronides, 48
Pons, (fig.) 922, (fig.) 923
Pontine hæmorrhage, 883
Population, England and Wales,
    1969, table 1105
Porcine cholecystokinin-pancreozy-
    min, structure, (fig.) 270
Porphin, 450
Porphobilinogen, 451
Porphyria, 452
Porphyrin, 450
  excretion, 452
Porphyrinuria, 452
Portal blood flow, 582
  hypertension, 584
  system of pituitary, 1081
  tracts, 581
  vein, 285, (fig.) 459, 582
  venous pressure, 583
  vessels of pituitary, 1081 (fig.) 1082
Position sense, 831
Positive after-potential, 807, 808
  feedback, 795
  induction, 989
  supporting reaction, 918, 937
Postcalcarine cortex, 970
  fissure, (fig.) 958, (fig.) 971
  sulcus, 970
Postcentral gyrus, (fig.) 958
Posterior columns of spinal cord, 902
  root fibres, 903
    ganglion, 903, 904
    of spinal cord, 903
      section of, (fig.) 903
  spinocerebellar tract, (fig.) 904
Postganglionic fibres, 860
Post-ischæmic hyperæmia, (fig.) 556
Post-partum necrosis of pituitary,
    (fig.) 1082
Postural reflexes, 921

Potassium, 306, 489, 610
  absorption and secretion, 652
  and electrocardiogram, (fig.) 489
  membrane potential, 512
  myocardial cell, 470, 471, 512
  depletion, 681
  exchangeable, 676
  flux, 793, 796
  in cells, (table) 675, (fig.) 793
    extracellular fluid, (table) 677
    human body, (table) 676, (table) 677
  intake, 632
  loss by kidney, 681
  permeability, 793, 796
  secretion, 649
  stimulation by, 793
Potatoes, 180
Potential, action, myocardial, 468
    nerve, 786
  energy, 161, 168
  injury, 785
  intracellular, (fig.) 787
  membrane, 792
  myocardial, 468
  resting, 785
  reversed, 787
  synaptic, 817
  transmembrane, resting and action, myocardial, 468
Pouch, Heidenhain, (fig.) 248
  Pavlov, (fig.) 248
Power, ventricular, (fig.) 480
PPS, 718
5-PRA, 349
PR interval, 490
Precapillary sphincter, 569
Precardial chest lead (ECG), 490
Precedence effect, 770
Precentral gyrus, (fig.) 958
Prednisolone, 324, 1012
Prefrontal cortex, 969
  leucotomy, 969
Preganglionic fibres, 860
Pregnancy, 670, 1050, 1053
  aldosterone, 1055
  anæmia in, 1066
  and HGH, 1063, 1085
  and gonadotrophin excretion, 1063
    histamine, 1021
    hormonal control, 1056
    insulin, 1065
    œstrogen excretion, 1061
    pancreas, 1065
    pregnanediol excretion, 1061
    protein requirements, 1064
    sebaceous activity, 713
    water retention, 1063
  blood volume, 1064
  calcium and phosphorus requirements in, 1066
  conception, 1052
  diagnosis tests, 1063
  duration, 1071
  endometrium, 1055, (fig.) 1057
  excretion of hormones, 1059
  histamine in, 1021
  hormone control of, 1056
  iron in, 1066
  maintenance of, 1056
  multiple, 1067, 1072
  nutritional requirements, 1064
  plasma volume, 1064

plasma renin, 1064
pregnanediol, 1059
production of hormones, 1059
progesterone, 1059
protein requirements, 1066
renin, 1064
salt retention, 1064
serum insulin, 1065
signs of, 1055
sodium retention, 1064
tests, 1063
water diuresis, 1063
water retention, 1063
weight gain, 1064
Pregnanediol, 61, 1045
  excretion, (fig.) 1061
  in pregnancy, 1060, (fig.) 1061
  urinary excretion of, (fig.) 1062
Pregnant mare serum (PMS), 1051
Pregnenolone, 1056
Pre-load, (fig.) 476
Premotor area, 968
Premature babies, 1111
  contraction, 519
Prematurity, 1111
Premenstrual endometrium, (fig.) 1040
Premotor area, 966
Presacral nerve, (fig.) 688
Presbycusis, (fig.) 772
Presbyopia, 732, 733
Pressor reflexes, 559
Pressure abdominal, 595
  arterial, 532, 533
    and age, (fig.) 536
  atrial, (fig.) 492, (fig.) 496
  capillary, 545
  colloid osmotic, 21, 682
  critical closing, 558
  diastolic, 533
  factors determining arterial, 536
    regulating arterial, 537
  filtration, 648
  hepatic veins, 583
  in arteries, capillaries, veins, (fig.) 538
    brachial and tibial artery, (table) 534
    brachial artery, (fig.) 578
    glomerular capillaries, 648
    left atrium, (fig.) 579
    œsophagus, (fig.) 229
  pulmonary artery in relation to altitude, (table) 580
    right ventricle, (fig.) 578
    salivary gland, (fig.) 227
  intracardiac, (fig.) 492
  intracranial, 687
  intrapleural, 594, 595
  intrasplenic, 583
  intrathoracic, 484, 493, 501, 505
  intrarenal, 648
  jugular, (fig.) 492
  load, ventricular, 473
  mean arterial, 533
  measurement of arterial, 533
  negative intrapleural, 484, 493, 501, 505
  osmotic, 20, 682
  pulmonary artery, 577
  pulse, 532
  systolic, 533
  transmural, 557

venous, 551
Pressure/volume curve, (fig.) 572
Pressurization of aircraft, 634
Presynaptic inhibition, 819
  knobs, 815
  vesicles, (fig.) 816
Pretectal nucleus, (fig.) 736, 935, 973
Price-Jones' curve, (fig.) 417
Prickle cells, (fig.) 710
Primary colours, 745
  lobule of lung, 639, (fig.) 592
  sex ratio, 1105
  structure of proteins, 32
Primordial ovarian follicle, 1036
Principle of effect, 989
PR interval, 490
Pro (proline), 26, 34, 208
Procaine, 967
Procarboxypeptidases, 268
Pro-erythroblast, 418
Progestational proliferation, 1045
Progesterone, 58, 61, 1045, 1053, 1056, 1059
  and ovulation, 1053
    parturition, 1060
    temperature, (fig.) 1048
  biosynthesis of, 1046
  excretion in pregnancy, 1059
  half-life of, 1060
  in blood, (fig.) 1061
    human placenta, 1059
    maternal and fetal plasma, (fig.) 1061
    plasma, 1048
  synthetic, 1047
Proinsulin, 319
  molecule, (fig.) 319
Projection fibres, 961
  of pain sense, 863
    sense of smell, 722
    touch sense, 717
Prolactin, LTH, 1074
Proliferative phase, 1038
Proline (Pro), 26, 34, 208
Promotor site, 370
Pronation-supination, (fig.) 950
Propanol, 12
Propranolol, 511
Proprioceptive pathway, (fig.) 904
Proprioceptors, 814
Proptosis, 1002
Prosencephalon, 892
Prostaglandin E, 882
Prostaglandins, 51, 61, 343, 899, 1022, 1056
Prostate gland, 1029, 1030
  secretion, 1030
Prostatic phosphatase, 1030
Prosthetic groups, 109
Prostigmine, 847
Protamine, 68, 320, 397
  zinc insulin, 320
Protanopes, 746
Proteases, 103, 268
Protein, 23, 30
  absorption of, 298
  allosteric, 113, 114, 447, 620
  anabolism and androgens, 1034
  biological value, 201
  biosynthesis, 365
  blood, 401
  -bound iodine, 1000
  calorie value of, 161

Protein, chemical score, 201
  daily allowance, (table) 203
  deficiency, 202
  denaturation, 93
  digestion, 291
  dimensions of various, (fig.) 407
  dynamic equilibrium, 376
  gastrointestinal secretion, 299
  half-life, 377
  hydrolysates intravenous adminis-
    tration of, 378
  immunological reactions of, (fig.)
    404
  intake and nitrogen equilibrium,
    (fig.) 201
    in hard work, 202
  insufficiency, 202
  in urine, 648
  isoelectric point, 36
  life span, 377
  loss from gut, 299
  metabolism, 375, (fig.) 383
    in the ruminant, 390
    regulation, 378
  molecular weight, 31
  osmotic pressure of plasma, 406
  plasma, 376, 401, (table) 403, 683
  primary structure, 32
  quaternary structure, 36
  requirements, 199
    in exercise, 202
      old age, 1113
      pregnancy, 1066
  secondary structure, 34
  sensitivity, 299
  sparing action of carbohydrate and
    fat, 201
  specific dynamic action, 376
  storage, 375
  structure, 31
  synthesis, 72, 363, (fig.) 365
    control of, 369
  tertiary structure, 35
Proteinuria, 648
Prothrombin, 197, (table) 394
  time, 197
    one-stage test, 394
    two-stage test, 395
Protoporphyrin, 450
Provitamins, 184
Proximal convoluted tubule, 642
  (fig.) 644
PRPP, 347, 349
Pruritus, 719
  and jaundice, 719
Pseudo-isochromatic plates, 746
Pseudolarynx, 758
Psychic juice, 250
Psychrometer, whirling, 889
PTA, 394
Pteroyl-glutamic acid (PGA), 190
Pteroyl heptaglutamate, 190
  triglutamate, 190
PTH, 213
Ptosis, 732
Ptyalin, 226
Puberty, 5, 1034, 1050
  hormonal basis, 1051
  in boy, 1034
    girl, 1038
  onset of, 1031
Pubescence in female, 1038
  in male, 1031, (fig.) 1031

Pubic hair, 1031
Pudendal nerves, 670
Puerperium, 1074
Pulmonary alveoli, (fig.) 616
  arterial hypertension, 633
  artery, 458
    pressure, 577
    hypoxia, 580
    trunk of, (table) 460, (fig.) 461
  blood volume, 581
  capillaries, 578
  capillary bed, (fig.) 578
    pressure, 579
  circulation, 577
    time, 579
  depressor reflex, 581
  hypertension, (fig.) 580
  œdema, 524, 579, 581
  resistance, 579
  veins, (fig.) 499
  venous pressure, 524
Pulse, arterial, 530
  collapsing, 532
  deficit, 532
  jugular, (fig.) 553
  pressure, 532
  rate, 532
    and emotion, (table) 535, (fig.)
      565
    at various ages, (fig.) 517
  venous, 552
  wave, 520, 530
    shape of, 531
    velocity, 531
Pulvinar, 954
'Pump-handle' movement, 593
Puncta lacrimalia, 730
Pupil, 735, 972
  in accommodation, 735
  reflexes, 735
  size of, 735
Pure-tone audiometer, 770
Purgatives, 306
Purine bases, 64
  biosynthesis, 349
  catabolism, 360, (fig.) 361
  mononucleotides, 349
Purines, 64
Purkinje cells, 944, 945, (fig.) 946
  network, 468, (fig.) 469
Purkinje-Sanson images, 753
'Purkinje shift', 743
Puromycin, 321, 373, 995
Purpura, 401, 551
Putamen, 958
P wave, 490
PV (portal vein), (fig.) 459
Pyelogram, intravenous, (fig.) 642
Pyloric antrum, 234
  stenosis, 702
Pyloroplasty, 239
Pylorus, 239
Pyramid, (fig.) 924
Pyramidal decussation, 922
  fibres, (fig.) 924
  tract, 964
Pyran, 42
Pyranose sugars, 42
Pyrexia, 886
  artificial, 886
Pyridoxal, 189
  phosphate, 189, 379, 451
Pyridoxamine, 189

Pyridoxine, 188, 424, 427
Pyridoxol, 189
Pyrimidine, 63
  bases, 63
  biosynthesis, 347
  catabolism, 360
Pyramids, 921
  decussation of, 921
Pyrogen, 886
  endogenous, 887
Pyrophosphatase, 212
Pyrrole ring, (fig.) 449
Pyruvate, 82, 148, 464
  blood after exercise, (fig.) 840
  kinase, 143
Pyruvic acid, 143

QRS complex, 490
Quaternary structure of haemo-
    globin, 447
  proteins, 36
Quinacrine, (fig.) 1104

Racemose gland, 223
Radioactive isotopes, 121
Radioautography, 122
Radio-immunoassay of insulin, 320
Radio pill, (fig.) 285
Rage, 953
Rancidity, 53
Random coil, 71
Ranvier, node of, (fig.) 780, (fig.)
    782, (fig.) 783, 799
Rapid eye movements, 981
Rate of breathing, 594
  recovery in nerve, 810
Rathke's pouch, 1080
Reabsorption, maximal tubular, 652
Reactive hyperæmia, 519, 556
Reading, 753
Rebound of motor neurones, 915
Receptive relaxation, 235
  field of cell, 970
Receptor generator potential, 767
Receptors α and β, 561, 867
β-Receptors, 598
  cardiac blocking drugs, 511
Recessive genes, 1106
Reciprocal innervation, 821, 917
Recovery heat, 805
Recruitment of neurones, 912
Rectal temperature, 874
Rectum, 302
  effects of distension, 306
  pressure in, 305
  nerve supply, 304
Recurrent laryngeal nerve, 758
Red blood cells, 21, 53, 415
  agglutination of, 410
  at high altitudes, (fig.) 422
  carbon dioxide in, 622
  counts from birth–12 years, (fig.)
    422
  destruction of, 439
  diameter, 417
  formation, 418
  glucose-6-phosphate dehydrogen-
    ase, 417
  in capillary, (fig.) 547, 616
  life of, 420
  lysis of, 410
  metabolism of, 47
  micrograph, (fig.) 416

Red blood cells, number, 416
  old, 421
  origin, 418
  oxygen and carbon dioxide, (table) 617
  sedimentation, 417
  shape, (fig.) 416
  size, (fig.) 416
  time in pulmonary capillary, 616
  volume, 408
Red–green blind, 746
Red nucleus, (fig.) 893, (fig.) 925, 926
Redox indicators, 138
Redox-potential E, 137
'Reducing' diet, 325
Reduction, definition of, 126
  division, 1100
Referred pain, 863
Reflex or reflexes, 892
  actions, 892
  after-discharge, 910
  after spinal transaction, 907
  arc, 892
  autonomic, 866
  cardiac, 512
  characteristics of flexion and extension, (table) 912
  chemoreceptor, 565
  conditional, 986
  crossed extension, (fig.) 912, 913
  definition, 892
  d'emblee, 912
  emotional stress, 565
  exercise, 565
  extension, 913
  fatigue of, 913
  flexion, 909, (fig.) 912
    and extension compared, (table) 912
  gastrocolic, 305
  grasp, 968
  Hering-Breuer, 601, 606
  inhibition of, (fig.) 914
  irradiation, 909
  labyrinthine, 937
  light, 735
  lung inflation, 566
  monosynaptic, 909
  myenteric, 237
  myotatic, 915, 917
  neck, 937
  plantar, 968
  postural, 921
  reciprocal, 821, 917
  righting, 937, 938
  scratch, (fig.) 918
  spinal, 909
  stretch, 914, 927
  time, 909, 914
  tonic labyrinthine, 937
  vasoconstriction, 559
  vasomotor, 563
  visual, 971
  withdrawal, 969
Refractive errors, 734
  index of media, (fig.) 733
Refractory period, 791, 910
  of heart, 469
    nerve, 790, (fig.) 791
    skeletal muscle, 910
Regeneration of nerves, 809, 810
Regulator gene, 370

Regulatory enzymes, 112
  kinetics of, 114
  subunit, 115
Reinforcement, 985, 986
Relative refractory period, (fig.) 791
Relaxation of sarcomere, 472
Releasing factors, 1049, 1050, 1091
Releasing hormones, 1049, 1050, 1081
Remaining sensibility, 905
Renal acidæaim, 702
  acute circulatory renal failure, 646
  blood flow, 644
    supply, (fig.) 646
    vessels, 643
  capsule, 641
  clearance, 653, (fig.) 654
  compensation, 698, 699
    of acidosis, 699
    of alkalosis, 699
  concentrating system, 650
  control of water and sodium loss, 663
  corpuscle, 642
  correction of acidosis, 701
    alkalosis, 702
  dialysis, 666
  disease, 666
  excretion of acids and bases, 665, 696
  failure, 666
  function in new-born, 667
  glycosuria, 325, (fig.) 326
  nerves, 646
  plasma flow, 655
  regulation of acid-base balance, 665, 696
  secretion of K^+, H^+, 652
  threshold, 652
  tubules, 649
Renin, 646, 657, 659, 1012, 1055
  and arterial pressure, (fig.) 660
  –angiotensin–aldosterone system, 657
  in amniotic fluid, 1056
    juxtaglomerular cells, (fig.), 660
    plasma, 1049
  source of, 659
Rennet, 177
Rennin, 177
Renshaw cells, 820, (fig.) 916
Repair in nerves, 809, 810
  of damaged DNA, 358
Replication of DNA, 352, 354
Repression of enzyme synthesis, 369, 372
Reproduction, 2, 5, 1026
Resection of intestine, 295
Reserpine, 960
Residual volume, 598, 848
Resistance, elastic, 596
  in pulmonary circuit, 579
  of lungs, 595
  peripheral, 458, 537
  vessels, control of, 555
  viscous, 596
Resonance theory of Helmholtz, 763
Respiration, (fig.) 140
  and blood pressure, (fig.) 536
    heart rate, (fig.) 849
    vagus nerve, 606
  artificial, 523, 608
  at birth, 601

cell, 125
  central control, 603
  centre, 603
  chamber, 163
  chemical regulation, 627
  Cheyne-Stokes, 634
  deglutition, 229
  effect of adrenaline, 631
  effect of acidaencia carbon dioxide, 627
    high altitude, 580, 621, 628, 630, 634, 635, 638
    loading, (fig.) 607
    oxygen lack, 631
  exercise, 849
  fetal and neonatal, 1069
  Hering-Breuer reflex, 601, 606
  hydrogen ion, 696
  in acidæmia and alkalæmia, 699, 700
  in plants, (fig.) 140
  mechanics of, 592
  movements of, 593
  muscles of, 593
  nervous regulation, 605
  periodic, 634
Respiratory acidosis, 692, 695, 699
  alkalosis, 692, 695, 700
  arrhythmia, 532
  bronchioles, 589
  centre, 603
    and stimulation of cells of, 628
  compensation of acidosis, 701
  data at different ages, (table) 595
  functions of blood, 612
  minute volume, 627
  movements, recording of, 593
  processes, (fig.) 588
  pump, 505
  quotient, R.Q., 162, 850
    during and after muscular exercise, (fig.) 851
  rate, 594
    at birth, 601
      different ages, (table) 595
  resistance, 596
  sounds, 598
  tests, 600
Respirometer, Max-Planck, (fig.) 165
Respondent behaviour, 988
Resting potential, 785, (fig.) 786, 794
  of cardiac cells, 468
    nerve 786, 791
      ionic basis of, 791
  oxygen consumption, 168
  tidal volume, 598
Retching, 231, 261
Retention incontinence, 670
Rete testis, 1029
Reticular formation, 261, 924, 980
  lamina, (fig.) 763
  system, 924
Reticulocytes, 418
Reticulocytosis, 422
Reticulo-endothelial cells, 420
  system, 420
Reticulum cell, 437
Retina, 737
  and cortex, 971
  blood supply, 738
  contacts in, 740

Retina, corresponding points, 754
 electrical activity of, 748
 ganglion cells, (fig.) 739, (fig.)
  740, 748, 970
 in vitamin A deficiency, (fig.) 184
 number of rods and cones, 738
 structure of, 738
Retinal (vitamin A aldehyde), 183,
  (table) 741
Retinal artery, 738
 blockage of, 738
Retinitis pigmentosa, 748
Retinol (vitamin A), 183, (table) 741
Retrolental fibroplasia, 638
Reverberating internuncial neurones,
  909, (fig.) 910
Reynold's number, 539
Rheobase, 789
Rhesus system, 411
Rh antigen, 410
 factor, 411
 immunization, 410
  prevention of, 412
 negative blood, 412
 positive blood, 412
Rhinencephalon, 957, (table) 958
Rhodopsin, 184, 740
 absorption spectrum of, (fig.) 741,
  742
 regeneration of, (fig.) 741, 742
Rhombencephalon, 892
α-Rhythm, 977
Rhythm of inattention, 977
Rhythmic segmentation, 284
Rhythmicity, innate, myocardial, 467
Ribitol, 67
Rib movements, 593
Riboflavin, 186
 deficiency of, 187
 in foods, (table) 187
Ribonuclease, 35, 36, 357
Ribonucleic acid (RNA), 67, 961
Ribonucleoside, 65
Ribose, 44, 63
Ribosomal RNA (rRNA), 72
Ribosome, (fig.) 69, 72, 76, 364, 996
Ribulose 5-phosphate, 153
Rice, 186
Rickets, 193, (fig.) 194, 216
 radiological appearance, (fig.) 196
'Rickety rosary', 195
Right atrial pressure, 506
 to left shunt, 615, 635
 ventricular pressure, (fig.) 492
Righting reflexes, 937, 938
Rigidity, decerebrate, 924, 948
 in man, 928, 948
 decorticate, (fig.) 928
Rigor, 838
 mortis, 836
Rima glottidis, 756
Rinne's test, 773
RNA, (ribonucleic acid), 67, 892,
  961
 biosynthesis of, (fig.) 356
 -dependent RNA polymerase, 357
 memory, 961
 messenger, 72, 364
 polymerase, 349, 355
 synthesis, 355
 transfer, 72, 364
Rod, density, 738
 luminosity curve, 742

monochromats, 746
Rods of Corti, 762, (fig.) 763, (fig.)
  765
 retina, 738
Rostral, definition, 894
Romanowsky stain, 429
Rouget cells, 543
Roughage, 179, 181
Rouleaux, 416
Round ligaments, 1035
 window, 761
R.Q., respiratory quotient, 162,
  850
 and ventilation/perfusion ratio,
  (fig.) 614
rRNA, 355
Rubrospinal tract, 922
Ruffini organs, 814
Rumen, 47, 292
Ruminants, 308, 343
 carbohydrate metabolism in, 331
 protein metabolism in, 390

Saccadic movement, 751, 753
Saccharine, 727
Saccule, 936
 maculæ of, (fig.) 930
S-adenosylmethionine, 343, 359
Saggital plane, 894
Saline, depletion, 680
 hypertonic, 679
 overload, 680
Saliva, 225
 composition of, 226
 functions of, 226
 histological basis of secretion, 228
 iodide, 228
 mechanism of secretion, 227
 volume per day, 226
Salivary digestion, 227
 glands, 222
  microscopic appearance, (fig.)
   223
  nerve supply, 223
Saltatory conduction, 780, 799
Salt depletion, 883
 free diet, 664
 in sweat, 883
 intake, 663
 taste, 726
SAMI (socially acceptable monitor-
  ing instrument), 165
Sanger reagent, 32
S.A. node, (fig.) 469, (fig.) 497
Saponification, 52
Sarcolemma, myocardial, 466, 825
Sarcomere, myocardial contraction
  of, 470
 relaxation of, 472
Sarcoplasm, 825
Sarcoplasmic reticulum, (fig.) 465,
  466, 826, 837
Satiety centre, 171
Saturated colour, 744
 fatty acids, 50
Sayer's assay, 1085
S.B.P., 533
Scala tympani, (fig.) 762
 vestibuli, (fig.) 762
Scanning speech, 950
Schizophrenia, 993
Schlemm, canal of, 731
Schilling test, 423

Schwann cells, 777
 sheath of, 777, (fig.) 778, (fig.)
  779, (fig.) 783
Scintillation counting, 122
Sclera, (fig.) 729
Scotoma, 737
Scotopic vision, 740, 749
Scratch reflex, 918
Scrotum, 1026
Scurvy, 192, 424
 skin wound, plate 13.x
S.D.A. (specific dynamic action),
  168, 202, 376
Sea-sickness, 261, 933
Sebaceous gland, 713
Sebum, 713
Secondary motor area, 968
 sensory area, 975
 sex ratio, 1105
 structure of proteins, 34
Second messenger, 995
 wind, 855
Secretagogues, 251, 256
Secretin, 242, 269, (fig.) 270, 272,
  279, 285, 293
Secretin-pancreozymin test, (fig.)
  271, 275
Secretion of H-ion, 652
Secretors, 245
 of blood group substances, 410
Secretory phase, 1039
 Tm, 652
Section of anterior roots of spinal
  cord, 902
Sedimentation rate, 416
 coefficient, 75, (fig.) 77
Segmentation, 281
Selective angiocardiography, 496,
  (fig.) 497
Selenium, 1034
Sella turcica, 1080
Semen, 1030
Semi-circular canals, 929
 and angular acceleration, 929
 and rotation, (fig.) 931
 caloric stimulation, 935
 ducts, 929
Semi-conservative replication, 352
Semilunar valves, 456, 491
 closure of, 493
 opening of, 493
Seminal plasma, 1030
 vesicles, 1029
Seminiferous tubules, 1027, 1101
Semipermeable membrane, 20
Senescence, 1112
Sense, chemical, 722
 pain, 718
 skin, 714
 smell, 722
 taste, 724
 temperature, 715
 touch, 716
Sensibility, remaining, 905
Sensory areas of cerebral cortex, 970
 decussation, 921, (fig.) 923
 homunculus, (fig.) 976
 innervation of skeletal muscle, 826
 of viscera, 864
 nerves, 5
 paths, in spinal cord, (fig.) 904
 secondary, 975
Separated plasma, 622

Septum, cardiac interventricular, (fig.) 461
  interatrial, 458
  interventricular, 458, (fig.) 460
    membranous, (fig.) 461
Sequestrene, 394
Serine, (Ser), 24, 190, 268,
  esterases, 103
Serosa, 234
Serotonin (5-HT), 1020, 1021
Serous cells, 223
Sertoli, supporting cells of, 1029, 1032, 1034
Serum, 391
  amylase, 269
  electrophoresis of, (fig.) 403
  iron, 425
  transaminase (SGOT), 379, 473
Servo, follow up, 916
Sex cells, 1037
  chromatin, 1102, (fig.) 1103
    body, 1102
  chromosomes, 1102
  determination, 1102
  hormones, 60
  -linked inheritance, 1108
  of fetus, 1036
  ratio, 1105
    at stated age, (table) 1106
Sexual behaviour, 1034
SGOT, 379, 473
Sham feeding, (fig.) 250, 256
  rage, 953
Sheehan's syndrome, 1084, 1090
Shingles, 563
Shipwreck survivors, 681, 888
Shivering, 880, 882, 886
Shock, 409, 575, 666
  irreversible, 575
  traumatic, 575
Short sight, 733
Shortening reaction, 917
Shunts, 285, 644
Sialic acid, 49
Sibilants, 759
Sickle cell disease, 427
  haemoglobin, 447
Sickness, decompression, 38
Silent period, 916
Silk, 34
Sinu-atrial (S.A.) node, (fig.) 469
  (fig.) 497
Sinus arrhythmia, 484, 532
Sinus venosus sclerae, 731
  venosus, 512
Sinuses, aortic, 462
Sinusoids, 436
Skatole, 308, 309, 386, 724
Skeletal muscle, 4, 825
  chemistry of, 837
  sensory innervation, 826
  structure of, 825
Skim milk, 177
Skin, 709
  arteriovenous anastomoses, 285, 644
  bacteria of, 712
  blood flow, 570
  capillary, 570
  circulation in, 570
  colour, 710
  effect of cold, 882
    light, 711

endings, 719
in old age, 1113
insulation, 877
of eunuchs, 1031
pain, 718
pigmentation, 710
reflexes, 880
routes of penetration, (fig.) 714
sensations, 714
  anatomical basis, 719
sense of touch, 716
senses, 714
structure of, (fig.) 709
temperature, 877, 878
  sense, 714
Skinfold thickness, 168
Skinner box, 988
Sleep, 255, 534, 605, 730, 736, 768, 898, 979, 1009
  deprivation, 981
  electroencephalogram in, (fig.) 980
  rapid eye movements, 981
  requirements, 981
Sleepy sickness, 980
Sliding filaments, 835
  theory, 843
Small intestine, 281
  movements of, 281
  pain, (fig.) 864
Smell, sense of, 722
Smokers, 617
Smooth muscle, 857
  arterial, 530
Snake venom, 397
Sneezing, 607, 730
Snellen's test types, 736
Soap, 52
Socially acceptable monitoring instrument (SAMI), 165
Sodium and myocardial cell, 470
  potassium in human body, table 676
  benzoate and taste, 727
  chloride, absorption, of 296
  depletion, 681
  efflux in nerve, (fig.) 799
  exchangeable, 676
  intake, normal, (table) 678, 883
  in digestive juices, (table) 292
    excretions, (fig.) 678
    plasma, (fig.) 676
    the human body, (table) 677
  loss by kidney, 663
  permeability and membrane potential, (fig.) 795
  -potassium pump, 489
  pump, 804
  space, (fig.) 677
Solar radiation, 883
Solubility coefficient, 613
Solvents, 11
Somæsthetic area, 975
Somatotrophin, 1085
Sound intensity, 705, 761
  localization of, 769
  pressure level (SPL), 761
  shadow, 770
  velocity, 598
Sour taste, 726
Space flight, 936
Spasticity, 915, 967
Spastic paralysis, 967
Spatial summation, (fig.) 910

Spaying, 1044
  effects of, 1047
Specialized conducting tissue, myocardial, 467
Special senses, 705
Specific activity, 122
  dynamic action, 168, 202, 376
  nerve energy, 706
Specificity of conditional reflex, 986
  enzymes, 91
Spectrum, (fig.) 741
  acoustic, (fig.) 760
Speech, 755
  acquiring of, 755
  energy, 761
  in cerebellar lesions, 950
  intelligibility, 771
  laryngectomy and œsophageal, 758
  œsophageal, 758
  organs of, 756
  power, 759
  rate of vibration of the vocal cords, 760
  sounds, analysis of, 759
Spermatids, (fig.) 1101
Spermatocytes, 1027, 1101
  primary, (fig.) 1101
  secondary, (fig.) 1101
Spermatogenesis, 1026, 1029
  temperature, 1026
Spermatogonia, 1027
Spermatozoa, 5, 1026, 1030, (fig.) 1101
  life of, 1030, 1052
  metabolism, 1030
  motility, 1052
  rate of movement, 1054
  transport, 1052
Spherical aberration, 735
Sphincter ani, 304, 305
  ileocolic, 303
  œsophageal, 231
  of Oddi, 273
  precapillary, 544
  urethræ, 668
  vesicæ, 668
Sphingomyelins, 54
Sphingosine, 54
Sphygmomanometer, 533
Spike, 786
Spinal anæsthesia, 670, 863
  cord, 901
    and blood pressure, 559
      defæcation, 305
      micturition, 670
    anterior root, (fig.) 901
    ascending paths, (fig.) 904
    Brown-Sequard syndrome, (fig.) 908
    cross-section, (fig.) 901
    degeneration after section of nerves, (fig.) 903
    dermatomes, 905, 906
    descending paths, (fig.) 903
    hemisection of, 907
    methods of determining pathways, 902
    nerves, 901
      effects of section, 905
    paraplegia, 670, 863, 907
    pathways, 901, 902
    posterior root, (fig.) 901
    preparation, 926

Spinal anæsthesia, reflexes, 909
  repair, 809, 810
    root ganglion, (fig.) 904
    section of, 871
    segment, (fig.) 901
    sensory paths, (fig.) 904
    transection, 304, 559, 906
    transverse lesions, 670
  ganglion, 904
  grey matter, 902
  nerves, effects of section, 905
  preparation, 909
  reflexes, 909, 913
  reflex inhibition, 913
  root ganglion, (fig.) 904
  roots, 902
  shock, 305, 907
Spindle, 829, 915
Spindles, number of, 831
Spinocerebellar tract, (fig.) 904
Spinotectal tract, (fig.) 904
Spinothalamic tract, (fig.) 905
Spiral arteries, 1039, 1047
  ligament, (fig.) 764
  organ of Corti, (fig.) 764
Spirits, 181
Spirometer, (fig.) 163, 502, 594, (fig.) 599
  Benedict-Roth, 164
  recording, (fig.) 164
Spironolactone, 1013
Splanchnic blood flow in exercise, 848
  nerve, 561
Splanchnicectomy, 561
Spleen, 436, (fig.) 583
  and blood destruction, 436
    platelets, 434
  angiogram (fig.) 583
  arterioles, 585
  blood vessels, (fig.) 586
  cells, 585
  circulation through, 585
  contraction of, 585
  nerves, 585
  pressure in, 583
  reservoir function, 585
  structure of, 585
  volume, (fig.) 1016
Splenectomy, 439, 585
'Spreading factor,' 48
Squint, 751 (fig.) 752
ST segment, 490
Stable factor, (table) 394
  isotopes, 120
Stagnant hypoxia, 636
Standard bicarbonate, 694
Standard free energy change, 85
Standing posture, 484
Stapedius muscle, 761
Stapes, 761
Starch, 46, 163
Starling's experiments on isolated heart, 481
  hypothesis, 682
  law, (fig.) 474, 475, 481
Starvation, 171, 204, 309, 316, 323, 339, 376
  ketosis, 205, 700
Steady state, 95
Stearic acid, 51
Steatorrhœa, 180, 301, 307
Stercobilin, 308, 453

Stercobilinogen, 453
Stereognosis, 976
Stereochemistry of the steroids, 59
Stereocilia, 931
Stereoisomeric forms, 24
Stereoscopic vision, 751, 754
Steroid hormones, 996
Steroids, 57
  anabolic, 378, 1034
  biosynthesis of in ovary, 1046
  excretion of, (table) 1033
  nomenclature of, (fig.) 58
  stereochemistry of, 59
Sterols, 55
  absorption of, 306
  in faeces, 309
  synthesis, 278
Stethograph, 594
Stethoscope, 500
Stilbœstrol, 1047
Stimulus, 2
  substitution theory, 987
  threshold, 706
Stirrup, 761
Stokers' cramp, 885
Stomach, 233
  absorption from, 259, 295
  and emotions, 255
    guanethidine, 238
    noradrenaline, 238
  atropine, 237
  barium meal, (figs.) 234, 235, 236
  cardiac notch, 234
  cephalic phase of digestion, 280
  distension, 239
  emptying, 239
    and gravity, 241
  enzymes, 249
  fistula, 238, 256
  fundus, 234
  gastric phase, 251
  gastrin, 250, 270, 285, 293, 303, 319
  glands, 243
  histamine, 256
  Heidenhain pouch, 248
  hydrochloric acid, 245
  inhibitors, 240, 256
  innervation, 233, 260
  insulin, 260
  intestinal phase, 252
  intrinsic factor, 245, 423
  juice, 246
  local mechanisms, 254
    nerves, 254
  maximum output of acid, 260
  motility, 233
  movements, 234, 238
  mucosa, 243
  mucosal pattern, (fig.) 236
  mucous cells, (fig.) 244
  mucus, 243
  muscle, 233
  nerves, 237
  osmolarity and, (fig.) 240
  parietal cell, (fig.) 245
  Pavlov pouch, 248
  peristalsis, 235, 238
  pH, (fig.) 241, (fig.) 260
  pressure in, 236
  protective function of, 261
  reservoir, 234
  secretion, 244

  secretion control, 249
  shape, (fig.) 234
  tests of function, 259
  vagus, 234
  vomiting, 231, 261
  water absorption, 7
Strabismus, 751 (fig.) 752
Stratum compactum, (fig.) 1057
  corneum, 709, 710
  germinativum, 709
  granulosum, 710, 1037
  lucidum, (fig.) 710
  spongiosum, (fig.) 1057
Stream tube, 539
Strength-duration curve, 789, 857
Streptokinase, 399
Streptomycin, 373,
Streptozotocin, 326, 327
Stress, 1009, 1010
  and gastric secretion, 255
Stretch receptors, 537, 607
  reflexes, 914, 927
Striated duct cells, 223
Stria vascularis, (fig.) 763
Striæ medullares, (fig.) 922
Stroke output, 503
  volume, 481, 483, 493, 503, 518, 848
    in exercise, 848
Strontium, 220
Structural gene, 370
Strychnine, 822, 913, 945, 956
Stuart-Power factor, (table) 394
Stuffiness, 589
Subacute combined degeneration, 423
Subarachnoid space, 685, (fig.) 686
Subcutaneous fatty layer, 712
Subdural space, (fig.) 685
Subliminal fringe, 911
Sublingual gland, 222
Submandibular gland, 222
Submarine analogy, 850
Submucosa, (fig.) 234
Submucous nerve plexus, 234
Substantia gelatinosa, (fig.) 901
  nigra, 960
Substrate, 94
Subthalamic body of Luys, 959
Successive induction, (fig.) 919
Succinate dehydrogenase, 98, 131, 151
Succinic acid, 151
Succinyl-coenzyme A, 191, 343
Succus entericus, 290
Suckling, 1075, 1094
Sucrase, 297
Sucrose, 45, 726, 727
Sugars, 40
Sulphæmoglobin, 449
Sulphanilamide, 115
Sulphonamide, 115
Sulphonyl ureas, 321
Sulphur-containing amino acids, 389
  excretion, 389
  metabolism, 388 (fig.) 389
Sulphydril group, 67
Summating potential, 767
Sunburn, 711
Sun, exposure to, 711
  tan, 711, 1031

Sunlight and rickets, 193
  and skin, 711
Superfetation, 1053
Superior cerebellar peduncles, 941
  colliculus, 749
  olive, 770
  temporal convolution, 974
Supporting cells of Sertoli, 1029
  reaction, 937
Suppressor genes, 1108
Supraoptic nucleus, 661, 951
Supraoptico-hypophysial tract, (fig.)
  1091
Supramarginal gyrus, 976
Suprarenal glands, 1006
Supraventricular ectopic beat, 520
  extrasystoles, 520
Surface area of body, 166
  formula, 166
  tension, 590
Surfactant, 591, 615
Suspensory ligaments (fig.) 731
S values, 75, (fig.) 77
Svedberg units, 75
Swallowing, 228
Sweat, 883
  amount, 883
  bradykinin, 885
  calcium in, 213
  composition of, 488
  electrolytes, 678
  glands, 712, 884
    nerves to, 885
  ions of, (table) 678
  salt in, 883
  water loss, 883
Sweating, 883
  effect of atropine, 885
    heat loss, 883
  'emotional', (fig.) 884
  gustatory, 727
  insensible, 876
  two types of, 884
Sweet clover disease, 396
Sweet taste, 726
Swing-sickness, 933
Sympathectomy, 869
Sympathetic ganglion, (fig.) 561
  nerves, effect on heart, 511
    to blood vessels, (fig.) 561
  outflow, 860
  pathway, (plate) 41.1
  tone, 511
Synacthen, 1085
Synapse, 776, 815
  chemical transmitter, 817, (fig.)
    867
  cleft, 815
  excitatory post-synaptic potential
    (e.p.s.p.) 817
  inhibitory post-synaptic potential
    (i.p.s.p.) 818
  morphology of, (fig.) 816
  potentials, 817
  transmitters, 817, (fig.) 867
Synaptic cleft, 815
  delay, 909
  knobs, 815, (fig.) 816
Syncope, 505, 573
Syncytiotrophoblast, 1056, (fig.)
  1058, 1075
Synovial fluid, 212
  membrane, 212

Synthesis of glucose, 146
Synthetic hormones, 1047
Syntocin, 1094
Systemic venous pressure, 824
Systole, 460, 474, 490, (fig.) 499, 518
Systolic pressure, 533
  in brachial and tibial artery (fig.)
    534

T_1 (transfer factor), 167
T_{co_2} and T_{o_2}, 617
Tachycardia, 518, 521
Tachypnœa, 605
Tactile agnosia, 976
  end-organ, 814
  sensation, 716
Tæniæ coli, 302
Tanning of skin, 710, 1031
Tarsal glands, 730
Tartar, 226
Taste, 724, 976, 1107
  adaptation, 727
  and adrenal cortical insufficiency,
    727
    copper, 727
    phenylthiourea, 1107
    temperature, 727
  basic, 726
  buds, 724
  classification, 726
  cortical representation, (fig.) 976
  fibres, pathways of (fig.) 725
  nerves of, 725
  threshold, 727
  sour, 726
  water, 726
'Tasters', 1107
Tastes, four, 726
TBW (total body water), 6
Taurine, 273, 389
Taurocholic acid, 277, 389
Tear fluid, 730
Teat, 1074
Tectorial membrane, (fig.) 765
Tectospinal tract, 922
Tectum, 892
Teeth, 212
  ascorbic acid deficiency, 193
  calculus, 226
  fluorine, 212
  tartar, 226
  vitamin D deficiency, 195
Tegmentum, 922
Telemetering capsule, 258
Telencephalon, 957, (table) 958
Telogen, 713
  defluvium, 713
Temperature, acral gradient, 878
  and blood clotting, (fig.) 392
    heart rate, 517, 879
    hypothalamus, 880, 882
    ovulation, 1052
    sense, 714
    thermometers, 879
    thyroid, 1001
  body, 876
    control of, 871, 880
  circadian variation, 877
  conversion factors, 1117
  effect on blood vessels, 557
    chemical reactions, 92
  extremes of body, 880

frequency distribution of, (fig.)
  876
exercise, (fig.) 852
gradient in skin, 877
  acral, 878
in axilla, 879
  dayshift and nightshift, (fig.)
    877
  menstrual cycle, 877, 1052
  newborn, 885
  paralysis, 874, 881
  tympanic membrane, 879
limits, 879
mouth, 876
muscle, 852
nervous control of body, 880
of air in trachea, 589
  anæsthetized patient, 874
  different tissues of forearm,
    (table) 877
oral, histogram, (fig.) 876, (fig.)
  879
pathway, (fig.) 904
receptors, (fig.) 715
rectal, 878
regulation, 873
  in newborn, 885
rise of basal, 1052
sense, 714
shivering, 874, 880, 882, 886
skin, 878
ventricular fibrillation, 880
Temporal lobe, 974
  and hearing, 974
  vestibular impulses, 975
Tendon organs, 917
  drawing of, (fig.) 828
  responses, (fig.) 829
Tendons, 208
Tensor tympani muscle, 761
Terminal cisternæ, (fig.) 465, 466
  hair, 713
Tertiary structure of proteins, 35
Testes, 1026
  and œstrogens, 1034
    vitamin E, 196
  cells of Leydig, (fig.) 1027
    Sertoli, 1029, 1034
  descent of, 1026
  in cryptorchidism, (fig.) 1028
    pituitary deficiency, (fig.) 1027
    XXY syndrome, (fig.) 1028
  internal secretions of, 1030
  interstitial cells, (fig.) 1027, 1032
  normal, micrograph, (fig.) 1027
  temperature, 1026
  undescended, 1026
Testicular artery, 1029
  pain, 864
Testosterone, 60, 1030, 1032, 1045
  biosynthesis, 1033
  in plasmaa, 1032
  propionate, 1034
Tests of gastric function, 259
  hypothalamic pituitary–
    adrenocortical function,
    1084
Tetanus, 841, 913
  poisoning, 913
  reflex, 911
  toxin, 822
Tetany, 216, 630
Tetracosactrin, 1013

1156   Index

Tetracycline, 373
Tetradotoxin, 808
Tetrahydrocortisol (THE), 1011
Tetrahydrocortisone (THE), 1011
Tetrahydrofolic acid, 190
Tetraiodothyronine, 997
Tetrose, 40
Thalamic animal, 937
  syndrome, 956
Thalamogeniculate artery, 956
Thalamus, 954
  connexions of, (fig.) 955
  dorsomedial nucleus, 954, 969
  pain, 956
  posterior ventral nucleus, 725
  ventrolateral nucleus, 960
Thalassæmia, 428
THE, 1011
Thebesian drainage, 616
Theca interna, 1044
Theobromine, 64
Thermal muscular tone, 874
Thermogenesis, 882
Thermometer, clinical, 879
  globe, 889
  kata, 889
  wet and dry bulb, 889
Thermoreceptors, 881
Thermoregulatory reflexes, 563
'Thermostat', 887
THF, 1011
Thiamin, 186, 840
  and beri-beri, 185
  daily requirement, 186
  in blood and tissues, 186
    foods, (table) 186
  pyrophosphate (TPP), 148, 185,
    341
Thiocarbamide, 999
Thiochrome, 185
Thioctic acid, 186
Thiocyanate, 1004
Thiokinase, 337
Thiolase, 338, 339
Thiouracil, 64, 999
Thiourea, 999, 1004
Third nerve nucleus (Edinger-
    Westphal nucleus), (fig.)
    973
  ventricle, 953
Thirst, 679
  and alcohol, 679
    diabetes mellitus, 330
  central receptors, 679
Thiry-Vella fistula, (fig.) 293
Thoracic duct, 285, 684
  flow, 285
  nucleus, 902
Thoracolumbar outflow, 304
Thoroughfare channel, 543
Threonine (Thr), 24
  deaminase, 113
Threshold, 706
  of audibility, 770
    hearing, 770
  nerve, 806
  pain, 718, 865, 866
  two-point, 717
Thrombasthenia, 401
Thrombin, 393
Thrombocytes, 193, 399, 434, 719
Thrombocytopenia, 401

Thromboplastin, (table) 394
  antecedent (factor XI), 394
  extrinsic, 394
  intrinsic, 394
  generation test, 395
Thrombopoietin, 434
Thrombosis, 401
Thrombosthenin, 400
Thrombus, 401
Thymectomy, 1019
Thymidine, 287, 430
  tritiated, 287, 892
Thymine, 64, 190, 1020
  dimers, 358
Thymus, 835, 1019
  -dependent lymphocyte (T-lym-
    phocyte), 1019
  gland, 432
  immunogenesis, 1019
Thyrocalcitonin, 214
Thyroglobulin, (fig.) 123, 997
Thyroglossal duct, 997
Thyroid colloid, 997
  dessicated, 997
  development, 997
  disease, diagnosis of, 1000
  endemic goitre, (fig.) 1005
  enlargement in children, (table)
    1005
  gland, (fig.) 123, 997
    actions, 1000
    activity, 875
    and body temperature, 975
      carbohydrate metabolism, 325
      metabolism, 999
      uncoupling of oxidation, 1000
  hormone, 378, 997, 998, 999
    biosynthesis of, (fig.) 999
  long acting thyroid stimulator
    (LATS), 1002
  micrograph, (fig.) 123
  hyperthyroidism, 1001
  hypothyroidism, 1001
  -stimulating hormone (TSH), 997
  uptake of ^{131}I, (fig.) 1000
Thyroidectomy, 214
Thyroiditis, autoimmune, 1001
Thyronine, 999
Thyrotoxicosis, 1002
Thyrotrophic hormone (TSH), 999,
    1084
Thyrotrophs, 1081
Thyroxine, 997
Tickle, 719
Tidal volume, (fig.) 597, 598
  at different ages, (table) 595
'Tight junctions' (nexuses), myo-
    cardial, (fig.) 466, 467,
    471
Tilting, 564
Tinel's sign, 810
Tissue culture, 119
  fluid, composition, 735
    water content, (table) 6
Tissues, composition of, (table) 7,
    (table) 675, (table) 677
  fluid exchange in, (table) 676
Titratable acidity, 666
T-lymphocytes, 685
T_m, 652
Tocopherol, 196
Tolazoline, 868

Tolbutamide, 321, 330
Toluidine blue, 397
Tonic labyrinthine reflexes, 937
  neck reflexes, 928
Tonometer, 620
Tonotopic arrangement, 974
Tonus, 915
Total body water (TBW), 6, 674
  lung capacity, 600
Touch, 716
  localization, 717
  pathway, (fig.) 904
  postcentral gyrus and, 975
  projection, 717
  spots, 716
    histology, 716, 719
  threshold pressure, (table) 716
  two-point threshold, 717
Toxins, 410
TPP (thiaminpyrophosphate), 148,
    185
Tracers, isotopic, 121
Trachea, 589
Tracheostomy, 610
Tractus solitarius, 725
Training, 848
Tranexamic acid, 398
Transaldolation, 153
Transamidination, 387
Transaminase, 116, 189, 379
  in blood serum (SGOT), 379
Transamination, (fig.) 110, 377, 379
Transcortin, 1010
Transcription, 355, (fig.) 356, 995,
    1106
Transection of spinal cord, 304,
    559, 906
Transfer factor (T_1), 617
Transfer RNA (tRNA), 72, 355, 364
Transferases, 107
Transferrin, 405, 425
Transfusion, blood, 413
Transhydrogenase, 995
Transients, 759
Transitional epithelium, 668
Transketolation, 153
Translation, 364, 996
Transmethylation, 387, 388
Transmitter, central nervous system,
    822
  chemical, 815
  inhibitory, 822
Transneuronal degeneration, 903
Transverse tubule system, 837
Tranylcypromine, 116
Trasysol, 398
Traumatic shock, 575
Travelling wave, 764, 767
Tremor, in cerebellar disease, 960
  in hyperthyroidism, 1002, (fig.)
    1004
  intention, 948
  muscle, 947, 1004
  paralysis agitans, 960
Triad of the sarcoplasmic reticular
    systems of the heart, 466,
    471
Tricarboxylic acid cycle, 140, 148
Trichromacy, 743, 746
Trichromatic vision, basis of, 745
Tricuspid valve, (fig.) 461
Trigger zone, 261
Triglyceride, 51

Triiodothyroacetic acid (triac), (fig.) 998
Triiodothyronine, (fig.) 998, 999
Trinitrin, 293
Triolein, 51
Triose, 40
Triosephosphate isomerase, 142
Tripalmitin, 51
Tripeptide, definition of, 28
Triple-helix, 34, 208
Trisaccharide, 40
Tristearin, 51
Tritanopes, 746
Tritiated thymidine, 287, 892
Tritium, 122
tRNA (transfer RNA), 72, 355, 364
Trophoblast, 1055
Tropocollagen, 34
Tropomyosin, 464, 472, 836, 837
Troponin, 464, 472, 477, 836
Trousseau's test, 217
Trp (tryptophan), 26, 187, 386, 1021
Trypsin, (fig.) 92, 104, (fig.) 105, 106, 268, 290
Trypsinogen, 268, 290
Tryptophan (Trp), 26, 187, 386, 1021
  pyrrolase, 372
  synthetase, 372
TSH, 997, 999, 1001, 1084
T-tubes, 837
Tuber cinereum, 1042
Tubocurarine, 835
Tubular absorption, 650
  capacity, 652
  control of water and sodium, 661
  fluid osmolarity, (fig.) 650
    pH, 665
  function, 649
  hydrogen ion secretion, 698
  maximum, 652
  reabsorption, 649
  regulation of acid-base balance, 665
  secretion, 652
Tumour viruses, 78
Tunica adventitia, 527
  albuginea, 1027, 1036
  externa, and interna, 1037
  intima, 527
  media, 527
Turbulence, 539
T wave, 490
Twitch, 947
Two-point threshold, 717
Tympanic membrane, 761, 879
  perforation, (fig.) 772
'Type I' conditioning, 984
'Type II' conditioning, 988
Typhoid vaccine (TAB), (fig.) 886
Tyr (tyrosine), 26
Tyramine, 116, 178
Tyrosinase, 109, 136, 219
Tyrosine (Tyr), 26, 384, 997, 1018
Tyrosinosis, 385

Ubiquinones, 130
UDP-galactose, 312
UDP glucose, 67, 154
UDP glucuronate, 453
UDP glucuronyl transferase, 442, 453

UDPG pyrophosphorylase, 154
UFA, 333
Ultraviolet light, 711
  and vitamin A, 183
    vitamin D, 193
  on skin, 711
  radiation, 354, 358, 711, 743
Umbilical arteries, 1068
  cord, tying of, 1069
  vein, 1068
UMP, 65, 66, 348
Unconditional stimulus, 985
Unconsciousness, 939
Uncoupling agents, 133
Uncus, 975
Undernutrition and pituitary, 1051
Undescended testis, 1026
Unesterified fatty acids (UFA), 333
Unipolar ECG lead, (fig.) 487, 489
Units and measures, 1116
Unsaponifiable fraction, 57
Upper motor neurone paralysis, table 906
Uracil, 64
Uræmia, 666
Urea, 161, 380
  and ornithine-arginine cycle, 380
  clearance, 655
  excretion, 118
  formation, 379, 381
  in blood, (table) 647
    colon, 307
    renal tubules, (table) 659
    urine, 118, (table) 659
  synthesis, diagram, (fig.) 381
Urease, 91
  pressure, 667
Ureter, 667, (fig.) 1029
  transplantation of, 306
Ureteric muscle, 667
Urethra, (fig.) 1029
Uric acid, 64, 116, 360
  in blood, 360
Uricase, 360
Uridine diphosphoglucose (UDPG) 67, 154
  5'-monophosphate (UMP), 65, 348
  triphosphate (UTP), 154, 348
Urinary bladder, 667, (fig.) 1029
  nitrogen output, (fig.) 201, 659
  output and emotion, 664
Urine, 118, 657
  ammonia, 665
  and glomerular filtrate compared, (table) 659
    plasma compared, (table) 647
  calcium in, 213
  colour of, 657
  composition, 657, (table) 659
  creatinine, (table) 647
  diabetes insipidus, 1091, 1096
  diurnal variations in volume, (fig.) 658
  flow after drinking water, (fig.) 662
    and ureteric pressure, 667
  ions of, (table) 678
  leucocytes of, 657
  minimum amount, 677
  obligatory volume, 677
  odour, 657
  osmolarity, 657
  output, (fig.) 1095

  and ADH, (fig.) 1095
  ovarian hormones, 1048
  pH, 665
  specific gravity, 657, 680
  steroid excreted in, 1033
  titratable acidity, 666
  volume of, 657, (fig.) 658, 664, 667
    circadian rhythm, (fig.) 658
Urobilin, 453
Urobilinogen, 421
Urocanic acid, 711
Urochrome, 666
Urokinase, 399
Uronic acid, 43
Uropepsinogen, 249
Uroporphyrinogen, 451
Urticaria, 1021
Uterine tubes, 1035
Uterus, 1035, 1041
  activity, 1045
  actomyosin, 1045
  and oxytocin, 1073
  at parurition, 1073
  bicornuate, 1072
  contraction, 1073
  distension, 1072
  endometrial reaction to ovarian hormones, 1048
  involution, 1074
  pain, (table) 865
  pressure in, 1073
  tubes, (fig.) 1035
UTP, 154, 348
Utricle, 936
  maculæ, 929
U wave, 490

'V leads' (ECG), 489
'V wave' (atrial), 493
Vaccinia, (table) 77
Vagal tone, 511
Vagina, 1035, 1053
  cellular changes in, (fig.) 1043
  cornification, 1044
  pH, 1036, 1053
  smear, 1042, 1047
Vaginal bleeding, 1039
  mucus, 1036
  pH, 1053
  smear, 1042, 1047
Vagotomy, 238, 255, 271, 296, (fig.) 560, 604, 863
  and gall-bladder, 299
    gastric motility, 238
    hunger, 171
  bilateral, 238
Vagus nerve, 249, 508
  and stomach, 234
  block, 606
  depressor fibres in, 467, 508, 513
  effect on heart, 508
  escape, 511
  gastrin, 250
  to heart, 808, (fig.) 509
Valine (Val), 24
Valsalva manœuvre, 483, 505, 553, 564, 575
Valve, aortic, 460, (fig.) 462, 573
  atrio-ventricular, (fig.) 461
  dimensions, table, 460
  mitral, (fig.) 461, (fig.) 462
  pulmonary, (fig.) 462

Valve, semilunar, 456
  venous, (fig.) 457, 458
Van den Bergh reaction, 454
Van der Waals forces, 11
Van Slyke apparatus, 618
Vaporization, latent heat, 876
Variable element, 335
Varicose veins, 552
Vasa recta, 651
Vas deferens, 1029
  ligature of, 1031
Vasectomy, 1031
Vasoconstrictor nerves, 559
  reflexes, 559
Vasodilatation, 869
  after sympathectomy, 869
  in skeletal muscles, 953
Vasodilator nerves, 561, 565, (fig.) 566
  reflexes, 561
Vasomotion, 569
Vasomotor centre, 512, 559
  control, 555
  reflexes, 563
Vasopressin, 1072, 1093
Vasopressor effect of pituitary extract, 1093
Vaso-vagal syncope, 574
Vegetables, 181
  composition of, table 180
Vein and artery compared, (fig.) 528
Veins, 4, (fig.) 457, 458, 551, 572
  as capacity vessels, 553
  of arm, (fig.) 457
  structure, (fig.) 528
  valves in, (fig.) 457
  volume-elasticity curve, (fig.) 529
'Vellus' hair, 713
Velocity constant, 81
  of blood, 539
  of pulse wave, 531
Vena cava, inferior, (table) 460, (fig.) 461
  valve of, (fig.) 461
Venoconstriction, 553, 572
Venom, snake, 397, 1023
Venomotor responses, 571
Venous admixture, 615
Venous blood gases, 613, 617
  plasma, composition of, 10
  pressure, 551, 552
    and emotion, 573
    in foot, 552
  pulse, 552
  return of blood, 481, 506, (fig.) 507
  tone, 571
  valves, (fig.) 547
Ventilation, 587, 592
  artificial, 484, 608
  dynamic tests of, 600
  positive-pressure, 484
Ventilation-perfusion ratio, 613, (fig.) 614
  relationship, 614
Ventricle, 458, 461
  left, contraction of, 479
  extrasystoles, (fig.) 520
  fibrillation, (fig.) 521
  lateral, (fig.) 687, (fig.) 959
  pressure, 473
  tachycardia, 518, 521
Ventricular activation, (fig.) 472

asystole, (fig.) 523
automatism, (fig.) 523
auxotonic contraction, (fig.) 479, 493
contractility, (fig.) 484
  increased, (fig.) 480
contraction, 491, (fig.) 492
  isotonic phase, 493
  rapid ejection phase, (fig.) 492
  slow ejection phase, (fig.) 492
  velocity of, (fig.) 480
depolarization, 489
diastasis, 491, (fig.) 492, 493
diastole, (fig.) 491
diastolic filling pressure, 483
  volume, 483, 484
end-diastolic volume, 479, 484, (fig.) 485
ectopic beat, 520
ejection, isotonic phase, 493
end-diastolic volume, 479, 484, (fig.) 485
filling, rapid phase, (fig.) 492, 493, 501
  slow phase, (fig.) 492, 493
isometric contraction, (fig.) 479, 491, (fig.) 492
  relaxation, (fig.) 479, (fig.) 492, 493
load, 473
muscle, end-diastolic length, 481
  volume, 484
output, 481, 503
performance, 483
power, 480
pressure, 473
relaxation, isovolumetric, (fig.) 479
repolarization, (fig.) 489
standstill, (fig.) 523
systole, (fig.) 491, 502
tachycardia, 520
volume, 473, (fig.) 480, 493
  end-diastolic, 479, 484
  increased, (fig.) 480
volumetric relaxation, (fig.) 479
work, 480
Ventriculogram, 689
Venules, 550
Verdohæmoglobin, 452
Vermiform appendix, 302
Vernier acuity, 736
Vertigo, 933
Vesicular murmur, 598
  ovarian follicle, 1036
Vessels, exchange, 543
Vestibular area, (fig.) 922
  connexions, (fig.) 929
  membrane, (fig.) 763
  nerve impulses, 931
  neurone, (fig.) 931
  nucleus, 927
  nystagmus, 932
  sensory cells, 931
Vestibule, (fig.) 764
Vestibulospinal tract, (fig.) 903
Vibration sense, 814, 905
Vicia faba, 418
Villus, intestinal, 285
  location of enzymes on, (fig.) 290
  movements of, (fig.) 286
  chorionic, (fig.) 1055, (fig.) 1059

Virilism, 1013, (fig.) 1014
Viruses, (fig.) 70, 77
Viscera, afferent impulses from, 861
  segmental sensory innervation of, table 864
Visceral muscle, 857
  pain, 865
  sensation, 865
Viscerosensory fibres, 862
Viscosity of blood, 539
Viscous metamorphosis, 400
  resistance, 596
Visible speech, 760
Vision, 729, 733
  accommodation, 733
  acuity, 736
  binocular, 754
  blackout, 730
  colour, 743
  dark adaptation, 747
  double, 751
  errors of refraction, 733
  field of, 737
  iris, 732
  peripheral, 737
  reflexes, 735, 971
  retina, 737
  stereoscopic, 751, 754
  under water, 733
Visual acuity, 736
  cortex, 750
  cycle, (fig.) 741
  field, (fig.) 737
  pathway, 972
  purple, 740
  reflexes, 971
Visuosensory cortex, 961
Vital capacity, 593, 599
Vitamin A, 183, 210, (fig.) 741
  absorption of, 184
  aldehyde, 183, (table) 741
  deficiency and retina, (fig.) 184, 741, 747
    and spermatogenises, 1034
  in blood, 184
    foods, 184
    plasma, (fig.) 104
Vitamin B complex, 66, 185
  biotin, 189
  cobalamin, 190
  folic acid, 190
  lipoic acid, 185
  nicotinamide, 187
  pantothenic acid, 189
  pyridoxine, 188
  riboflavine, 186
  thiamin, 186
Vitamin B_2 (thiamin), 185
Vitamin B_6 (pyridoxine), 188
Vitamin B_{12}, 190, 191, 245, 295, 307, (fig.) 420, 422
Vitamin C, 192, 208
  and wounds, plate 13.3
  in foods, table 193
    leucocytes, 193
    scurvy, plate 13.2
Vitamin D, 193, 210, 213
  in foods, (table) 195
  rickets, 193
Vitamin D_2 (ergocalciferol), 193
Vitamin D_3 (cholecalciferol), 193
Vitamin E, 196
Vitamin K, 197, 307, 396, 442

Vitamins, 181
  and cooking, 198
  list of, table 182
Vitreous body, 732
Vocal cords, 756, 1031
  folds, 756
    rate of vibration, 760
Vocalis muscle, (fig.) 757
Volley, nerve, 784
Voltage-clamp technique, 798, 801
Volume conductor (fig.) 488
  – elasticity curves, (fig.) 529
  load (ventricular), 473
Voluntary apnœa, 629
  hypernœa, 630
  muscle contraction, 845
Vomiting, 231, 261
  prolonged, 262
von Frey's bristles, 716
von Gierke's disease, 315
Vowel sounds, 757
Vowels, 759
  oscillograms of, 759
VQ ratio, 613, (fig.) 614

Wakefulness, 924
Waking centre, 980
Walking, 829, 853
  pattern of, 917
Wallerian degeneration, 809
Warfarin, 198
Warm spots, 714
Wart virus, (fig.) 70
Waste, 198
Water, 6, 9
  absorption, 7, 295
  adrenal cortex and, 663
  and electrolytes, daily turnover,
      676
    hæmostasis, 679
  as solvent, 11
  balance, 7
  body, 674, 675
  compartments, 674
  content of human tissues, 6, 674
  daily intake, (table) 678
  depletion, 680
  deprivation of, 673
  dissociation, 13
  diuresis, (fig.) 662
    inhibition of, (fig.) 662

flow through capillary wall, (fig.)
    682
  hard, 52
  homeostasis, 679
  in fæces, (fig.) 678
  intake, 676, 677
    and urine flow, (fig.) 662
  internal daily turnover, 678
  in tissues, 673
  intoxication, 680
  latent heat of evaporation, 7
  loss, 663, 677
    and adrenal cortex, 663
  lost in lungs, 677
  metabolism, 673
  molecule, 10
  needed for metabolism of foods,
      (table) 681
  of human tissues, (table) 6
  pituitary and, 649, 661
  produced by oxidative meta-
      bolism, 7, (table) 681
  properties of, 9
  retention in pregnancy, 1063
  total body, (table) 677
  turnover, (table) 678
Waves, Mayer, (fig.) 560
Waxes, 53
Wax-secreting glands, 761
Weber-Fechner law, 761
Weber's law, 706
Weber's test, 773
Wedge pressure, 579
Weigert-Pal method, 902
Weight gain in pregnancy, 1064
  of lung, 590
Weir's formula, 165
Wet and dry bulb thermometer, 889
Wheat germ oil, 179, 196
  gluten, 180, 301
  grain, composition of, 179
Whey, 177
Whispering, 756
White adipose tissue, 334
  cells, 193, 313, 429
  matter of spinal cord, 902
  ramus communicans, 861
Wholemeal, 179
Willis, circle of, 894
Wilson's disease, 219
Windkessel, 526, 530

Wines, 181
Wirz countercurrent theory, 650
Withdrawal bleeding, 1047
  reflex, 909
Word deafness, 975
Work, 140
  energy requirements, 169
  gradings, 169
  optimum rate of, 845
  ventricular, (fig.) 480
Wound healing, 202
  in ascorbic acid deficiency, plate
      13.3, 193

Xanthine, 64, 360
  oxidase, 116, 137, 360
X chromosome, 1102
Xeroderma pigmentosum, 359
Xerophthalmia, 184
Xerosis conjunctivæ, (plate) 13.1a.
Xerostomia, 679
Zinc in insulin, 320
Zona pellucida, 1037
X-ray crystallography, 35
XX in female, 1102
  zygote, 1102
XY in male, 1102
  zygote, 1102

Yawn, 608
Y chromosome, 1102
  fluorescent body, (fig.) 1104
Yeast, (fig.) 73, 90
Yellow enzyme, 186

Zinc, 199
Zein, 129
  and insulin, 320
Z line, (fig.) 465, (fig.) 466, 825
Zona fasciculata, 1084
  glomerulosa, 1084
  pellucida, 1054
  reticularis, 1084
Zone centrifugation, 74
Zuckerkandl, organ of, 1015
Zwitterion, 27
Zygote, 1100, (fig.) 1101
Zymogen granules, 268
Zymogens, 268